International Conference on Advances in Energy Systems and Environmental Engineering (ASEE17)

E3S Web of Conferences Volume 22 (2017)

Wroclaw, Poland
2 - 5 July 2017

Part 1 of 2

Editors:

Bartosz Kazmierczak
Malgorzata Kutylowska
Katarzyna Piekarska
Hussam Jouhara
Jan Danielewicz

ISBN: 978-1-5108-5194-8

Printed from e-media with permission by:

Curran Associates, Inc.
57 Morehouse Lane
Red Hook, NY 12571

Some format issues inherent in the e-media version may also appear in this print version.

Printed by Curran Associates, Inc. (2017)

For additional information, please contact EDP Sciences – Web of Conferences at the address below.

EDP Sciences – Web of Conferences
17, Avenue du Hoggar
Parc d'Activité de Courtabœuf
BP 112
F-91944 Les Ulis Cedex A
France

Phone: +33 (0) 1 69 18 75 75
Fax: +33 (0) 1 69 28 84 91

contact-edps@webofconferences.org

Additional copies of this publication are available from:

Curran Associates, Inc.
57 Morehouse Lane
Red Hook, NY 12571 USA
Phone: 845-758-0400
Fax: 845-758-2633
Email: curran@proceedings.com
Web: www.proceedings.com

TABLE OF CONTENTS

PART 1

USING A COUPLED INDUCTOR CONTROLLED BY FUZZY LOGIC TO IMPROVE THE EFFICIENCY OF A BUCK CONVERTER IN A PV SYSTEM 1
Abouchabana Nabil, Haddadi Mourad, Rabhi Abdelhamid, El Hajjaji Ahmed

EXPERIMENTAL INVESTIGATION AND CFD SIMULATION OF MULTI-PIPE EARTH-TO-AIR HEAT EXCHANGERS (EAHES) FLOW PERFORMANCE 10
Amanowicz Lukasz, Wojtkowiak Janusz

URBAN STORMWATER – GREYWATER MANAGEMENT SYSTEM FOR SUSTAINABLE URBAN WATER MANAGEMENT AT SUB-WATERSHED LEVEL 18
Arora Amarpreet Singh

ESTIMATION OF THE TEMPERATURE SPATIAL VARIABILITY IN CONFINED SPACES BASED ON THERMAL IMAGING 25
Augustyn Grzegorz, Jurasz Jakub, Jurczyk Krzysztof, Korbiel Tomasz, Mikulik Jerzy, Pawlik Marcin, Rumin Rafal

ASSESSING HIGH SHARES OF RENEWABLE ENERGIES IN DISTRICT HEATING SYSTEMS – A CASE STUDY FOR THE CITY OF HERTEN 33
Aydemir Ali, Popovski Eftim, Bellstädt Daniel, Fleiter Tobias, Büchele Richard

A NEW TEST PROCEDURE TO EVALUATE THE PERFORMANCE OF SUBSTATIONS FOR COLLECTIVE HEATING SYSTEMS 41
Baetens Robin, Verhaert Ivan

INFLUENCE OF SILANE ON THE STRUCTURE OF POLYSTYRENE PREPARED BY SOL-GEL COATINGS VIA UV CURING 49
Balbay Senay, Acikgoz Caglayan

COMPARISON OF METHODS FOR SOLIDS RETENTION TIME DETERMINATION AND CONTROL 57
Balbierz Piotr, Knap Marta

USE OF CEMENT-FLY ASH-BASED STABILIZATION TECHNIQUES FOR THE TREATMENT OF WASTE CONTAINING AROMATIC CONTAMINANTS 65
Banaszkiewicz Kamil, Marcinkowski Tadeusz

THE INFLUENCE OF MULTIFAMILY APARTMENT BUILDING OCCUPANTS ON ENERGY AND WATER CONSUMPTION – THE PRELIMINARY RESULTS OF MONITORING AND SURVEY CAMPAIGN 73
Bandurski Karol, Hamerla Milosz, Szulc Jowita, Koczyk Halina

ENVIRONMENTAL IMPACT ASSESSMENT OF COAL POWER PLANTS IN OPERATION 81
Bartan Ayfer, Kucukali Serhat, Ar Irfan

VARIABILITY OF ROAD TRAFFIC NOISE RECORDED BY STATIONARY MONITORING STATIONS 89
Bakowski Andrzej, Radziszewski Leszek

THE EFFECT OF THE THERMAL INERTIA ON THE THERMAL TRANSFER IN BUILDING WALL 96
Bellahcene Lahcene, Cheknane Ali, Bekkouche SMA., Sahel Djemal

CYTOTOXICITY AND GENOTOXICITY PROPERTIES OF PARTICULATE MATTER FRACTION 2.5 µM 103
Belcik Maciej K., Trusz-Zdybek Agnieszka, Zaczynska Ewa, Czarny Anna, Piekarska Katarzyna

THE USE OF MOVING BED BIO-REACTOR TO LAUNDRY WASTEWATER TREATMENT 111
Bering Slawomira, Mazur Jacek, Tarnowski Krzysztof, Janus Magdalena, Mozia Sylwia, Morawski Antoni Waldemar

ENERGY SAVING BY USING NATURAL ENERGY FROM THE SHALLOW GROUND DEPTHS – MANY YEARS OPERATING RESULTS 120
Besler Maciej, Skrzycki Maciej, Cepinski Wojciech

EFFECTS OF CLIMATE FACTORS AND VEGETATION ON THE CO_2 FLUXES AND $\sigma^{13}C$ FROM RE-ESTABLISHED GRASSLAND 128
Bezyk Yaroslav, Dorodnikov Maxim, Sówka Izabela

APPLICATION OF STOCHASTIC APPROACH BASED ON MONTE CARLO (MC) SIMULATION FOR LIFE CYCLE INVENTORY (LCI) OF THE RARE EARTH ELEMENTS (REES) IN BENEFICIATION RARE EARTH WASTE FROM THE GOLD PROCESSING: CASE STUDY 137
Bieda Boguslaw, Grzesik Katarzyna

FOOD WASTE IN CENTRAL EUROPE – CHALLENGES AND SOLUTIONS 143
den Boer Jan, Kobel Przemyslaw, Dyjakon Arkadiusz, Urbanska Klaudia, Obersteiner Gudrun, Hrad Marlies, Schmied Elisabeth, den Boer Emilia

PERFORMANCE OF MECHANICAL BIOLOGICAL TREATMENT OF RESIDUAL MUNICIPAL WASTE IN POLAND 151
den Boer Emilia, Jedrczak Andrzej

ORGANIC MATTER AND HEAVY METALS CONTENT MODELING IN SEWAGE SLUDGE TREATED WITH REED BED SYSTEM 159
Boruszko Dariusz, Dabrowski Wojciech, Malinowski Pawel

ENERGY AND ENVIRONMENTAL EVALUATION OF COMBINED COOLING HEATING AND POWER SYSTEM 166
Bugaj Andrzej

MODELING THE DEFROST PROCESS IN COMPLEX GEOMETRIES – PART 1: DEVELOPMENT OF A ONE-DIMENSIONAL DEFROST MODEL 174
van Buren Simon, Hertle Ellen, Figueiredo Patric, Kneer Reinhold, Rohlfs Wilko

COMPARISON OF POWER CURVE MONITORING METHODS 183
Cambron Philippe, Masson Christian, Tahan Antoine, Torres David, Pelletier Francis

AN INTEGRATED LOW CARBON ENERGY SOLUTION TO COOKING FUEL, TAILORED TO NIGER STATE'S RURAL POPULATION 191
Carvell Aaron, Price-Allison Andrew, Birch Calum, Green Toby, Harijan Khanji, Maihankuri Sheidi, Raji Abdulganiy, Uqaili Mohammed, Dupont Valerie

ACID EMISSIONS MONITORING NEEDS IN CERAMIC TILE INDUSTRY: CHALLENGES DERIVED FROM NEW POLICY TRENDS 200
Celades Irina, Gomar Salvador, Romero Fernando, Chauhan Amisha, Delpech Bertrand, Jouhara Hussam

INDUSTRIAL APPLICATIONS OF THE AIR DIRECT-CONTACT, GRAVEL, GROUND HEAT EXCHANGER 208
Cepinski Wojciech, Besler Maciej

THE USE OF MULTI CRITERIA ANALYSIS TO COMPARE THE OPERATING SCENARIOS OF THE HYBRID GENERATION SYSTEM OF WIND TURBINES, PHOTOVOLTAIC MODULES AND A FUEL CELL 216
Ceran Bartosz

THE EFFECT OF BIOLOGICALLY AND CHEMICALLY SYNTHESIZED SILVER NANOPARTICLES (AGNPS) ON BIOFILM FORMATION 224
Chojniak Joanna, Biedron Izabela, Mendrek Barbara, Plaza Grazyna

USE OF CFD MODELLING FOR ANALYSING AIR PARAMETERS IN AUDITORIUM HALLS 232
Cichowicz Robert

POWER PLANT WASTES CAPITALIZATION AS GEOPOLYMERIC BUILDING MATERIALS 240
Ciobanu Gabriela, Litu Loredana, Harja Maria

ASSESSMENT OF THE POSSIBILITY OF USING DATA MINING METHODS TO PREDICT SORPTION ISOTHERMS OF SELECTED ORGANIC COMPOUNDS ON ACTIVATED CARBON 249
Dabek Lidia, Szelag Bartosz, Picheta-Oles Anna

ENERGY CONSUMPTION MODELING DURING DAIRY SEWAGE PRETREATMENT 257
Dabrowski Wojciech, Zylka Radoslaw, Malinowski Pawel, Boruszko Dariusz

A REVIEW ON WASTE HEAT RECOVERY FROM EXHAUST IN THE CERAMICS INDUSTRY 265
Delpech Bertrand, Axcell Brian, Jouhara Hussam

EU ENERGY POLICIES ACHIEVEMENT BY INDUSTRIES IN DECENTRALIZED AREAS 273
Destro Nicola, Stoppato Anna, Benato Alberto, Schiro Fabio

MIXED CONVECTION IN A THREE-DIMENSIONAL VENTILATED CAVITY WITH TWO ISOTHERMAL HEATING SECTIONS 282
Doghmi Hicham, Abourida Btissam, Belarche Lahoucine, Sannad Mohammed, Ouzaouit Meriem

THE USE OF MATHEMATICAL MODELS FOR DIAGNOSIS OF ACTIVATED SLUDGE SYSTEMS IN WWTP 291
Drewnowski Jakub, Zmarzly Marcin

ANALYSIS OF A COMBINED HEATING AND COOLING SYSTEM MODEL UNDER DIFFERENT OPERATING STRATEGIES 299
Dzierzgowski Mieczyslaw, Zwierzchowski Ryszard

INVESTIGATING THE WATER CONSUMPTION FOR ELECTRICITY GENERATION AT TURKISH POWER PLANTS 308
El-Khozondar Balkess, Koksal Merih Aydinalp
EXPERIMENTAL STUDY OF AIR EVAPORATIVE COOLING PROCESS USING MICROPOROUS MEMBRANES 316
Englart Sebastian
CO_2 CONCENTRATION AND OCCUPANCY DENSITY IN THE CRITICAL ZONES SERVED BY THE VAV SYSTEM 324
Etoua Evina Ghislaine, Kajl Stanislaw, Lamarche Louis, Beltran-Galindo Javier
THE INFLUENCE OF MECHANICAL GEAR ON THE EFFICIENCY OF SMALL HYDROPOWER 332
Ferenc Zbigniew, Sambor Aleksandra
TEMPERATURE INFLUENCE ON THE FAST PYROLYSIS OF MANURE SAMPLES: CHAR, BIO-OIL AND GASES PRODUCTION 339
Fernandez-Lopez Maria, Anastasakis Kostas, De Jong Wiebren, Valverde Jose Luis, Sanchez-Silva Luz
THE IMPACT OF THE HOT TAP WATER LOAD PATTERN IN THE INDUSTRIAL HALL ON THE ENERGY YIELD FROM SOLAR COLLECTORS 346
Fidorów-Kaprawyl Natalia, Dudkiewicz Edyta
THE ANALYSIS OF THE DIFFERENCES BETWEEN THE RESULTS OF THE THERMAL RESPONSE TEST AND THE DATA FROM THE OPERATION OF THE BRINE-TO-WATER HEAT PUMP'S VERTICAL EXCHANGER 354
Fidorów-Kaprawy Natalia, Stefanowicz Ewelina, Mazurek Wojciech, Szulgowska-Zgrzywa Malgorzata, Bryszewska-Mazurek Anna
SOLAR RADIATION – TO – POWER GENERATION MODELS FOR ONE-AXIS TRACKING PV SYSTEM WITH ON-SITE MEASUREMENTS FROM ESKISEHIR, TURKEY 363
Filik Tansu, Basaran Filik Ümmühan, Gerek Ömer Nezih
MICROBIOLOGICAL STIMULATION OF PHYTOREMEDIATION PROCESS USING SALVINIA NATANS TO MERCURY CONTAMINED WATER 370
Filyarovskaya Viktoriya, Sitarska Magdalena, Traczewska Teodora, Wolf Mirela
DYNAMIC ENVIRONMENTAL CONTROL MECHANISMS FOR PNEUMATIC FOIL CONSTRUCTIONS 379
Flor Jan-Frederik, Wu Yupeng, Beccarelli Paolo, Chilton John
CLOSING OF WATER CIRCUITS – A GLOBAL BENCHMARK ON SUSTAINABLE WATER MANAGEMENT 388
Fröhlich Siegmund
SEASONAL COEFFICIENT OF PERFORMANCE FOR GROUND SOURCE HEAT PUMP AND GROUNDWATER ONE IN BIALYSTOK 396
Gajewski Andrzej
CARBON AND ENERGY FOOTPRINT OF ELECTROCHEMICAL VINEGAR WASTEWATER TREATMENT 403
Gerek Emine Esra, Yilmaz Seval, Koparal A. Savas, Gerek Ömer Nezih
SIZING PROCEDURES FOR SUN-TRACKING PV SYSTEM WITH BATTERIES 411
Gerek Ömer Nezih, Filik Ümmühan Basaran, Filik Tansu
NATURAL VENTILATION SYSTEMS TO ENHANCE SUSTAINABILITY IN BUILDINGS: A REVIEW TOWARDS ZERO ENERGY BUILDINGS IN SCHOOLS 419
Gil-Baez Maite, Barrios-Padura Ángela, Molina-Huelva Marta, Chacartegui Ricardo
THE SLAG ORIGINAL FROM THE PROCESS OF SEWAGE SLUDGE INCINERATION SELECTED PROPERTIES CHARACTERISTIC 428
Glowacka Anna, Rucinska Teresa, Kiper Justyna
ENERGY CROPS CULTIVATED ON THE SLAG FROM INCINERATION OF THE SEWAGE SLUDGE ENERGY VALUE ASSESSMENT 436
Glowacka Anna, Tarnowski Krzysztof, Bering Slawomira, Mazur Jacek, Kiper Justyna, Woloszyk Czeslaw
APPLICATION OF POROUS MATERIALS IN OIL SUBSTANCES SEPARATION FROM WATER 442
Golub Adam, Piekutin Janina
INFLUENCE OF PCMS ON THERMAL BEHAVIOR OF BUILDING WALLS: EXPERIMENTAL STUDY USING THE WALLS OF A REDUCED SCALE ROOM 449
Gounni Ayoub, Mabrouk Mohamed Tahar, Kheiri Abdelhamid, El alami Mustapha
PRETREATMENT METHODS AS A MEANS OF BOOSTING METHANE PRODUCTION FROM SEWAGE SLUDGE AND ITS MIXTURES WITH GREASE TRAP SLUDGE 456
Grosser Anna, Neczaj Ewa

ELECTRODIALYSIS POTENTIAL FOR FRACTIONATION OF MULTICOMPONENT AQUEOUS SOLUTIONS 465
Grzegorzek Martyna, Majewska-Nowak Katarzyna
INCINERATION, PYROLYSIS AND GASIFICATION OF ELECTRONIC WASTE 473
Gurgul Agnieszka, Szczepaniak Wlodzimierz, Zablocka-Malicka Monika
MONITORING OF WATER SUPPLY CONNECTIONS AS AN ELEMENT TO REDUCE APPARENT LOSSES OF WATER? 480
Gwozdziej-Mazur Joanna
REMOVAL OF OXYTETRACYCLINE FROM AQUEOUS SOLUTIONS BY HYDROXYAPATITE AS A LOW-COST ADSORBENT 487
Harja Maria, Ciobanu Gabriela
MONITORING OF DRUG RESISTANCE AMPLIFICATION AND ATTENUATION WITH THE USE OF TETRACYCLINE-RESISTANT BACTERIA DURING WASTEWATER TREATMENT 495
Harnisz Monika, Korzeniewska Ewa, Niestepski Sebastian, Osinska Adriana, Nalepa Beata
MODELLING THE DEFROST PROCESS IN COMPLEX GEOMETRIES – PART 2: WALL-FUNCTION BASED COUPLING TO A MULTI-REGION CFD SOLVER 504
Hertle Ellen, van Buren Simon, Figueiredo Patric, Kneer Reinhold, Rohlfs Wilko
MULTICRITERIA OPTIMIZATION APPROACH TO DESIGN AND OPERATION OF DISTRICT HEATING SUPPLY SYSTEM OVER ITS LIFE CYCLE 513
Hirsch Piotr, Duzinkiewicz Kazimierz, Grochowski Michal
POLLUTION OF SOILS (PB, CD, CR, ZN, CU, NI) ALONG THE RING ROAD OF WROCLAW (POLAND) 522
Holtra Anna, Zamorska-Wojdyla Dorota
A NOVEL DESIGN OF REACTIVE DISTILLATION CONFIGURATION FOR 2-METHOXY-2-METHYLHEPTANE PROCESS 531
Hussain Arif, Abdul Qyyum Muhammad, Minh Le Quang, Jimin Hong, Lee Moonyong
COMPREHENSIVE STUDY FOR ANAMMOX PROCESS VIA MULTISTAGE ANAEROBIC BAFFLED REACTORS 540
Ismail Sherif, Tawfik Ahmed
CONCEPTUAL DESIGN OF MULTI-SOURCE CCS PIPELINE TRANSPORTATION NETWORK FOR POLISH ENERGY SECTOR 548
Isoli Niccolo, Chaczykowski Maciej
TOWARDS ENERGY NEUTRALITY OF WASTEWATER TREATMENT PLANTS VIA DEAMMONIFICATION PROCESS 556
Janiak Kamil, Lojek Andrzej, Muszynski-Huhajlo Mateusz
STUDY OF HEAT EXCHANGE PROCESSES IN A SCREW ASH COOLER INSTALLATION 564
Janowska Barbara, Regucki Pawel, Andruszkiewicz Artur, Wedrychowicz Wieslaw
INDIRECT CONTACT FREEZE WATER DESALINATION FOR AN ICE MAKER MACHINE – CFD SIMULATION 572
Jayakody Harith, Al-Dadah Raya, Mahmoud Saad
SIMULATION OF THERMAL ENVIRONMENT IN A THREE-LAYER VINYL GREENHOUSE BY NATURAL VENTILATION CONTROL 580
Jin Tea-Hwan, Shin Ki-Yeol, Yoon Si-Won, Im Yong-Hoon, Chang Ki-Chang
TEMPORAL AND SPATIAL COMPLEMENTARITY OF WIND AND SOLAR RESOURCES IN LOWER SILESIA (POLAND) 590
Jurasz Jakub, Wdowikowski Marcin, Kazmierczak Bartosz, Dabek Pawel
ANAEROBIC CO-DIGESTION OF SEWAGE SLUDGE AND MOLASSES 599
Kalemba Katarzyna, Barbusinski Krzysztof
SINGLE AND COMPETITIVE ADSORPTION OF OMPS BY CARBON NANOTUBES – MECHANISM AND FITTING MODELS 606
Kaminska Gabriela, Dudziak Mariusz, Bohdziewicz Jolanta, Kudlek Edyta
STABILITY ASSESSMENT OF THE CHEMICAL COMPOSITION OF THE TREATED MINING WATER USED TO REPLENISH THE COOLING CIRCUIT IN JAWORZNO III POWER PLANT – POWER PLANT II 617
Karpinski Marcin, Kmiecik Ewa
AIR TEMPERATURE GRADIENT IN LARGE INDUSTRIAL HALL 625
Karpuk Michal, Pelech Aleksander, Przydrózny Edward, Walaszczyk Juliusz, Szczesniak Sylwia
THE ANALYSIS OF THE POSSIBILITY OF USING 10-MINUTE RAINFALL SERIES TO DETERMINE THE MAXIMUM RAINFALL AMOUNT WITH 5 MINUTES DURATION 634
Kazmierczak Bartosz, Wartalska Katarzyna, Wdowikowski Marcin, Kotowski Andrzej

COMPARING ENERGY PAYBACK AND SIMPLE PAYBACK PERIOD FOR SOLAR PHOTOVOLTAIC SYSTEMS 643
Kessler Will
ANALYSIS OF THE IMPACT OF SIMULATION MODEL SIMPLIFICATIONS ON THE QUALITY OF LOW-ENERGY BUILDINGS SIMULATION RESULTS 649
Klimczak Marcin, Bojarski Jacek, Ziembicki Piotr, Keskiewicz Piotr
THE QUATERNARY GROUNDWATER AS THE LOW TEMPERATURE ENERGY SOURCE FOR HEAT PUMPS IN MALOPOLSKA PROVINCE 657
Kmiecik Ewa, Tomaszewska Barbara, Mazurkiewicz Justyna
MODIFICATION OF RHODAMINE WT TRACER TESTS PROCEDURE IN ACTIVATED SLUDGE REACTORS 664
Knap Marta, Balbierz Piotr
DETERMINATION OF THE SOLAR TRANSMITTANCE FOR THE TRANSLUCENT SHUTTER WITH PCM IN LIQUID AND SOLID STATE 672
Komerska Anna, Ksionek Dariusz, Rosinski Marian
ANALYSIS OF PROFITABILITY OF USING A HEAT RECOVERY SYSTEM FROM GREY WATER DISCHARGED FROM THE SHOWER (CASE STUDY OF POLAND) 678
Kordana Sabina, Slys Daniel
CHANGE-OVER NATURAL AND MECHANICAL VENTILATION SYSTEM ENERGY CONSUMPTION IN SINGLE-FAMILY BUILDINGS 686
Kostka Maria, Szulgowska–Zgrzywa Malgorzata
REMOVAL OF COPPER IONS FROM AQUEOUS SOLUTIONS BY MEANS OF MICELLAR-ENHANCED ULTRAFILTRATION 695
Kowalska Izabela, Klimonda Aleksandra
INFLUENCE OF THE SINGLE EGR VALVE USABILITY ON DEVELOPMENT OF THE CHARGE DIRECTED TO INDIVIDUAL CYLINDERS OF AN INTERNAL COMBUSTION ENGINE 701
Krakowian Konrad, Kazmierczak Andrzej, Górniak Aleksander, Wróbel Radoslaw
GEOSTATISTICAL METHODS IN THE ASSESSMENT OF THE SPATIAL VARIABILITY OF THE QUALITY OF RIVER WATER 709
Krasowska Malgorzata, Banaszuk Piotr
INFLUENCE OF SPUTTERING DEPOSITION PARAMETERS ON ELECTRICAL AND OPTICAL PROPERTIES OF ALUMINIUM-DOPED ZINC OXIDE THIN FILMS FOR PHOTOVOLTAIC APPLICATIONS 717
Krawczak Ewelina, Agata Zdyb, Gulkowski Slawomir, Fave Alain, Fourmond Erwann
BIOARCHITECTURE – A NEW VISION OF ENERGY SUSTAINABLE CITIES 724
Krzeminska Alicja, Zareba Anna, Dzikowska Anna
INFLUENCE OF ENVIRONMENTAL POLLUTION WITH CREOSOTE OIL OR ITS VAPORS ON BIOMASS AND SELECTED PHYSIOLOGICAL GROUPS OF MICROORGANISMS 731
Krzysko-Lupicka Teresa, Cybulska Krystyna, Kolosowski Pawel, Telesinski Arkadiusz, Sudol Adam
THE OCCURRENCE OF HEAVY METALS AND METAL-RESISTANT BACTERIA IN WATER AND BOTTOM SEDIMENTS OF THE STRASZYN RESERVOIR (POLAND) 739
Kulbat Eliza, Sokolowska Aleksandra
DETERMINING RESIDENTIAL ENERGY CONSUMPTION-BASED CO_2 EMISSIONS AND EXAMINING THE FACTORS AFFECTING THE VARIATION IN ANKARA, TURKEY 748
Kus Melike, Akan Perihan, Koksal Merih Aydinalp, Gullu Gulen
EXPERIMENTAL DETERMINATION WITH EPR-METHODS OF DOSE LOADS ON LOCAL POPULATION IN INHABITED LOCALITIES ADJACENT TO THE TAILING POND "KOSHKAR-ATA" (KAZAKHSTAN) 757
Kuterbekov Kairat, Sadykov Nurllah, Zhumadilov Kassym, Nurgaliyeva Rano, Kabyshev Aset, Abseitov Erbolat, Kurmanzhanov Askar
ELECTRON THERMAL EMF FOR $NA_XCU_{2-X}S$ 762
Kuterbekov Kairat, Balapanov Malik, Ishembetov Rais, Kubenova Marzhan, Baitasov Talgat, Kabyshev Asset, Azhibekov Aidos, Kenzhebatyr Bekmyrza, Alibay Temirulan
APPLICATION OF REGRESSION TRESS FOR PREDICTION OF WATER CONDUITS FAILURE RATE 767
Kutylowska Malgorzata
THE MEASURES FOR ACHIEVING NZEB STANDARD OF RETROFITTED EDUCATIONAL BUILDING FOR SPECIFIC POLISH LOCATION – CASE STUDY 775
Kwiatkowski Jerzy, Mijakowski Maciej, Trzaski Adrian
RESEARCH OF CO_2 CONCENTRATION IN NATURALLY VENTILATED LECTURE ROOM 782
Laska Marta, Dudkiewicz Edyta

THE IMPORTANCE OF MONITORING RENEWABLE ENERGY PLANTS: THREE CASE HISTORIES 790
Lazzarin Renato
BATCH OROGRAPHIC INTERPOLATION OF MONTHLY PRECIPITATION BASED ON FREE-OF-CHARGE GEOSTATISTICAL TOOLS 798
Ledvinka Ondrej
THERMODYNAMIC ANALYSIS OF A COMBINED-CYCLE SOLAR THERMAL POWER PLANT WITH MANGANESE OXIDE-BASED THERMOCHEMICAL ENERGY STORAGE 806
Lei Qi, Bader Roman, Kreider Peter, Lovegrove Keith, Lipinski Wojciech
THE EFFECTIVENESS OF THE BIODEGRADATION OF RAW AND PROCESSED POLYSTYRENE BY MEALWORMS 815
Leluk Karol, Hanus-Lorenz Beata, Rybak Justyna, Bozek Magdalena
MODERN PROPOSAL OF METHODOLOGY FOR RETRIEVAL OF CHARACTERISTIC SYNTHETIC RAINFALL HYETOGRAPHS 823
Licznar Pawel, Burszta-Adamiak Ewa, Lomotowski Janusz, Stanczyk Justyna

PART 2

APPLICATION OF MULTI-ENZYMATIC HYDROLYSIS FOR IMPROVING THE EFFICIENCY OF THE BIOGAS PRODUCTION IN SOLID WASTE FERMENTATION PROCESS IN OSTRÓDA WWTP 830
Lipinski Kamil, Umiejewska Katarzyna
SYNTHESIS OF MESOPOROUS CERIUM COMPOUND FOR CO_2 CAPTURE 839
Liu Guiqing, Tatsuda Kou, Yoneyama Yoshiharu, Tsubaki Noritatsu
MODELING AN IMPACT OF ROAD GEOMETRIC DESIGN ON VEHICLE ENERGY CONSUMPTION 847
Luin Blaž, Petelin Stojan, Al-Mansour Fouad
THE TRANSMISSION CHARACTERISTICS OF INDOOR PARTICLES UNDER DIFFERENT VENTILATION CONDITIONS 854
Lv Yang, Wang Haifeng, Wei Shanshan
SCIENTIFIC APPROACH AND PRACTICAL EXPERIENCE FOR RECONSTRUCTION OF WASTE WATER TREATMENT PLANTS IN RUSSIA 861
Makisha Nikolay, Gogina Elena
TRANSFER RESERVOIR AS A NEW SOLUTION FOR TRANSFER OF STORMWATER TO WATER RECEIVERS 867
Malmur Robert
VARIOUS METHODS OF HEAT SUPPLY FOR A BUILDING WHICH IS OPERATED PERIODICALLY DURING THE YEAR 876
Maletka Marek, Laska Marta
APPROACH TO THE IMPLEMENTATION OF LINEAR-ASSISTED-BASED BATTERY CHARGERS 884
Martínez-García Herminio, García-Vílchez Encarna
ON THE MODELLING OF LINEAR-ASSISTED DC-DC VOLTAGE REGULATORS FOR PHOTOVOLTAIC SOLAR ENERGY SYSTEMS 891
Martínez-García Herminio, García-Vílchez Encarna
THE INFLUENCE OF THE SMALL PLATINUM CLUSTERS ON HYDROGEN SORPTION PROPERTIES 899
Martyla Agnieszka, Przekop R. E., Osinska-Broniarz Monika, Kopczyk M., Rybka J. D., Kirszensztejn P.
ANALYSIS OF ENERGY CONSUMPTION AT THE RZESZÓW WASTEWATER TREATMENT PLANT 907
Maslon Adam
ANALYSIS OF THE ENERGY POTENTIAL OF MUNICIPAL SOLID WASTE FOR THE THERMAL TREATMENT TECHNOLOGY DEVELOPMENT IN POLAND 915
Midor Katarzyna, Jaderko Karolina
CFD ANALYSIS OF A FULL-SCALE CERAMIC KILN MODULE UNDER ACTUAL OPERATING CONDITIONS 922
Milani Massimo, Montorsi Luca, Stefani Matteo, Venturelli Matteo
CO-DIGESTION OF SEWAGE SLUDGE FROM EXTERNAL SMALL WWTP'S IN A LARGE PLANT 930
Miodonski Stanislaw

NANOMATERIALS IN THE ENVIRONMENT 938
Mrowiec Bozena
CYCLE ANALYSIS OF MCFC/GAS TURBINE SYSTEM 945
Musa Abdullatif, Alaktiwi Abdulsalam, Talbi Mosbah
SHORTCUT NITRIFICATION/PARTIAL NITRITATION START-UP FOR REJECT WATER TREATMENT IN A SBR 955
Muszynski-Huhajlo Mateusz, Miodonski Stanislaw
PRODUCTION OF BIO-FERTILIZER FROM MICROWAVE VACUUM PYROLYSIS OF WASTE PALM SHELL FOR CULTIVATION OF OYSTER MUSHROOM (PLEUROTUS OSTREATUS) 962
Nam Wai Lun, Su Man Huan, Phang Xue Yee, Chong Min Yee, Liew Rock Keey, Ma Nyuk Ling, Lam Su Shiung
INFLUENCE OF NANOFLUIDS ON THE EFFICIENCY OF FLAT-PLATE SOLAR COLLECTORS (FPSC) 970
Nejad Marjan B., Mohammed H.A., Sadeghi O., Zubeer Swar A.
THE INFLUENCE OF USING HEAT STORAGE WITH PCM ON INLET AND OUTLET TEMPERATURES IN SUBSTATION IN DHS 980
Nogaj Kinga, Turski Michal, Sekret Robert
ATMOSPHERIC CO_2 CAPTURE FOR THE ARTIFICIAL PHOTOSYNTHETIC SYSTEM 987
Nogalska Adrianna, Zukowska Adrianna, Garcia-Valls Ricard
THE MATHEMATICAL MODEL OF RAINWATER CATCHMENT IN WROCLAW 996
Nowakowska Monika
IMPACT ASSESSMENT OF TREATED WASTEWATER ON WATER QUALITY OF THE RECEIVER USING THE WILCOXON TEST 1005
Ofman Piotr, Puchlik Monika, Simson Grzegorz, Krasowska Malgorzata, Struk-Sokolowska Joanna
CALIBRATION OF THE COMPUTER MODEL DESCRIBING FLOWS IN THE WATER SUPPLY SYSTEM; EXAMPLE OF THE APPLICATION OF A GENETIC ALGORITHM 1013
Orlowska-Szostak Maria, Orlowski Ryszard
ANALYSIS OF FREEZE PROTECTION METHODS FOR RECUPERATORS USED IN ENERGY RECOVERY FROM EXHAUST AIR 1021
Pacak Anna, Jedlikowski Andrzej, Pandelidis Demis, Anisimov Sergey
ENERGY DEMAND AND THERMAL COMFORT OF HVAC SYSTEMS WITH THERMALLY ACTIVATED BUILDING SYSTEMS AS A FUNCTION OF USER PROFILE 1029
Palaszynska Katarzyna, Bandurski Karol, Porowski Mieczyslaw
PROPOSITIONS OF IMPROVEMENT OF THE CROSS-FLOW M-CYCLE HEAT EXCHANGERS IN DIFFERENT AIR-CONDITIONING APPLICATIONS 1036
Pandelidis Demis, Anisimov Sergey, Rajski Krzysztof, Brychcy Ewa
ORGANIC RANKINE CYCLE – REVIEW AND RESEARCH DIRECTIONS IN ENGINE APPLICATIONS 1043
Panesar Angad
CHEMICAL TREATMENT OF WASTEWATER FROM FLUE GAS DESULPHURISATION 1051
Pasiecznik Iwona, Szczepaniak Wlodzimierz
THE INFLUENCE OF HEAT SINK TEMPERATURE ON THE SEASONAL EFFICIENCY OF SHALLOW GEOTHERMAL HEAT PUMPS 1059
Pelka Grzegorz, Lubon Wojciech, Sowizdzal Anna, Malik Daniel
CONCENTRATION OF POLYCYCLIC AROMATIC HYDROCARBONS IN WATER SAMPLES FROM DIFFERENT STAGES OF TREATMENT 1065
Pogorzelec Marta, Piekarska Katarzyna
IMPACT OF THE CIRCULATION SYSTEM ON THE ENERGY BALANCE OF THE BUILDING 1073
Polarczyk Iwona, Fijewski Michal
IMPACT OF THE CIRCULATION SYSTEM ON DOMESTIC HOT WATER CONSUMPTION 1079
Polarczyk Iwona, Fijewski Michal
INVESTIGATION OF THE BIOFUEL FLUE AND PRODUCER GASES CLEANING EFFICIENCY USING ESP 1086
Poškas Robertas, Sirvydas Arunas, Poškas Povilas, Striugas Nerijus, Pedišius Nerijus, Valincius Vitas
SEASONAL CHANGES IN QUALITY OF WASTEWATER FROM FRUIT AND VEGETABLE INDUSTRY 1093
Puchlik Monika, Ignatowicz Katarzyna
A NOVEL VORTEX TUBE-BASED N_2-EXPANDER LIQUEFACTION PROCESS FOR ENHANCING THE ENERGY EFFICIENCY OF NATURAL GAS LIQUEFACTION 1099
Qyyum Muhammad Abdul, Wei Feng, Hussain Arif, Ali Wahid, Sehee Oh, Lee Moonyong

THE CHOICE OF PRIMARY ENERGY SOURCE INCLUDING PV INSTALLATION FOR PROVIDING ELECTRIC ENERGY TO A PUBLIC UTILITY BUILDING – A CASE STUDY 1107
Radomski Bartosz, Cwiek Barbara, Mróz Tomasz M.

NUMERICAL MODELLING OF MULTI-VANE EXPANDER OPERATING CONDITIONS IN ORC SYSTEM 1115
Rak Józef, Blasiak Przemyslaw, Kolasinski Piotr

RECOVERY OF USEFUL CHEMICALS FROM PALM OIL MILL WASTEWATER 1123
Ratanaporn Yuangsawad, Duangkamol Na-Ranong, Teruoki Tago, Takao Masuda

NUMERICAL MODELLING OF SERIES-PARALLEL COOLING SYSTEMS IN POWER PLANT 1131
Regucki Pawel, Lewkowicz Marek, Kucieba Malgorzata

A SIMPLIFIED MODEL OF A MECHANICAL COOLING TOWER WITH BOTH A FILL PACK AND A COIL 1139
Van Riet Freek, Steenackers Gunther, Verhaert Ivan

EFFECT OF OPERATING CONDITIONS ON PERFORMANCE OF SILICA GEL-WATER AIR-FLUIDISED DESICCANT COOLER 1146
Rogala Zbigniew, Kolasinski Piotr, Gnutek Zbigniew

ORGANIC AND ELEMENTAL CARBON BOUND TO PARTICULATE MATTER IN THE AIR OF PRINTING OFFICE AND BEAUTY SALON 1154
Rogula-Kopiec Patrycja, Pastuszka Józef S., Rogula-Kozlowska Wioletta, Mucha Walter

THE USE OF LICHENS IN POST-SMELTING DUMPS RECLAMATION – PRELIMINARY RESULTS OF EXPERIMENTAL CULTIVATION OF SELECTED SPECIES ON SLAG SUBSTRATE 1162
Rola Kaja, Osyczka Piotr

ASSESSMENT OF THE AERMOD DISPERSION MODEL OVER COMPLEX TERRAIN WITH DIFFERENT TYPES OF METEOROLOGICAL DATA: TRACY POWER PLANT EXPERIMENT 1169
Rzeszutek Mateusz, Szulecka Adriana, Oleniacz Robert, Bogacki Marek

STUDY ON WATER EVAPORATION RATE FROM INDOOR SWIMMING POOLS 1178
Rzeznik Ilona

PROCESS SIMULATION AND COMPARISON OF BIOLOGICAL CONVERSION OF SYNGAS AND HYDROGEN IN BIOGAS PLANTS 1186
Salman Chaudhary Awais, Schwede Sebastian, Thorin Eva, Yan Jinyue

SEDIMENTS FROM STORMWATER DRAINAGE SYSTEM AS SORBENTS OF ORGANIC AND INORGANIC POLLUTANTS 1194
Salata Aleksandra, Dabek Lidia

THE INFLUENCE OF HYDRAULIC CONDITIONS ON COAGULATION PROCESS EFFECTIVENESS 1202
Sambor Aleksandra, Ferenc Zbigniew

GAS FIRED BOILERS: PERSPECTIVE FOR NEAR FUTURE FUEL COMPOSITION AND IMPACT ON BURNER DESIGN PROCESS 1209
Schiro Fabio, Stoppato Anna, Benato Alberto

EXPERIMENTAL STUDY OF OPTICAL AND ELECTRICAL PROPERTIES OF ZNO NANO COMPOSITES ELECTRODEPOSITED ON N-POROUS SILICON SUBSTRATE FOR PHOTOVOLTAIC APPLICATIONS 1217
Selmane Naceur, Cheknane Ali, Gabouze Nourddine, Maloufi Nabila, Aillerie Michel

OPTIMIZATION FOR ENERGY EFFICIENCY OF UNDERGROUND BUILDING ENVELOPE THERMAL PERFORMANCE IN DIFFERENT CLIMATE ZONES OF CHINA 1226
Shi Luyang, Liu Jing, Zhang Huibo

HYBRID CAPACITIVE DEIONIZATION WITH ANION-EXCHANGE MEMBRANES FOR LITHIUM EXTRACTION 1233
Siekierka Anna, Bryjak Marek

EVALUATION OF THE CORRELATION BETWEEN CONCENTRATION OF VOLATILE ORGANIC COMPOUNDS AND TEMPERATURE OF THE EXHAUST GASES IN MOTOR VEHICLES 1241
Skretowicz Maria, Wróbel Radoslaw, Andrych-Zalewska Monika

ANALYSIS OF APPLICATION OF ALTERNATIVE DRIVE SYSTEMS FOR INTERNATIONAL HEAVY-DUTY TRANSPORT ON WROCLAW-DRESDEN-PRAGUE ROUTES 1248
Skretowicz Maria, Sroka Zbigniew

ENVIRONMENTAL ASPECTS AND RENEWABLE ENERGY SOURCES IN THE PRODUCTION OF CONSTRUCTION AGGREGATE 1256
Skrzypczak Izabela, Kokoszka Wanda, Buda-Ozóg Lidia, Kogut Janusz, Slowik Marta

THE INFLUENCE OF GLYCERIN ON NITROGEN REMOVAL IN WASTEWATER TREATMENT WITH ACTIVATED SLUDGE 1264
Smyk Joanna, Ignatowicz Katarzyna

ADAPTIVE HEAT PUMP AND BATTERY STORAGE DEMAND SIDE ENERGY MANAGEMENT 1271
Sobieczky Florian, Lettner Christian, Natschläger Thomas, Traxler Patrick

BIOMASS UNIVERSAL DISTRICT HEATING SYSTEMS 1279
Soltero Victor Manuel, Rodríguez-Artacho Salvador, Velázquez Ramón, Chacartegui Ricardo

ENVIRONMENTAL ASPECTS OF THE GEOTHERMAL ENERGY UTILISATION IN POLAND 1288
Sowizdzal Anna, Tomaszewska Barbara, Drabik Anna

CHARACTERISTICS OF SELECTED ELEMENTS OF THE AIR QUALITY MANAGEMENT SYSTEM IN URBAN AREAS IN POLAND 1297
Sówka Izabela, Kobus Dominik, Chlebowska Stys Anna, Zathey Maciej

HYDRAULIC OPTIMIZATION AND MODELING OF HYDRO-CYCLONE-SYSTEMS FOR TREATMENT AND PURIFICATION OF ANY KIND OF WATERS 1305
Spangemacher Lars, Fröhlich Siegmund, Buse Hauke

CHANGES IN MICROBIOLOGICAL COMPOSITION OF SOILS AND SOIL CONTAMINATION WITH DRUG-RESISTANT BACTERIA CAUSED BY THE USE OF SEWAGE SLUDGE IN NATURE 1313
Stanczyk-Mazanek Ewa, Pasonl Lukasz, Kepa Urszula

EVALUATING THE FINANCIAL EFFICIENCY OF ENERGY AND WATER SAVING INSTALLATIONS IN PASSIVE HOUSE 1321
Stec Agnieszka, Mazur Aleksandra, Slys Daniel

COD FRACTIONS CHANGES IN THE SBR-TYPE REACTOR TREATING MUNICIPAL WASTEWATER WITH CONTROLLED PERCENTAGE OF DAIRY SEWAGE 1329
Struk-Sokolowska Joanna, Rodziewicz Joanna

LABORATORY TESTS OF OVERPRESSURE DIFFERENTIAL SYSTEMS FOR SMOKE PROTECTION OF LOBBIES 1336
Szalanski Pawel, Misinski Jacek

PRELIMINARY SANITARY ANALYSIS OF SUPPLY AND EXHAUST AIR OF VENTILATION UNITS WORKING AT SPECIAL ROOMS 1344
Szczesniak Sylwia, Trusz-Zdybek Agnieszka, Piekarska Katarzyna

DETECTION OF COMFORTABLE TEMPERATURE BASED ON THERMAL EVENTS DETECTION INDOORS 1352
Szczurek Andrzej, Maciejewska Monika, Uchronski Mariusz

MEASUREMENT OF ACETATES IN AIR USING DIFFERENTIAL ION MOBILITY SPECTROMETER 1360
Szczurek Andrzej, Maciejewska Monika, Zajiczek Zaneta, Maziejuk Miroslaw

PREDICTION OF WASTEWATER QUALITY INDICATORS AT THE INFLOW TO THE WASTEWATER TREATMENT PLANT USING DATA MINING METHODS 1368
Szelag Bartosz, Barbusinski Krzysztof, Studzinski Jan, Bartkiewicz Lidia

IMPLEMENTATION OF FERRIC HYDROXIDE-BASED MEDIA FOR REMOVAL OF TOXIC METALLOIDS 1376
Szlachta Malgorzata, Wójtowicz Patryk

INTEGRATED ASSESSMENT OF SOIL QUALITY AFTER APPLICATION OF THE BIOGAS FERMENTATION RESIDUES – A LABORATORY EXPERIMENT 1384
Telesinski Arkadiusz, Cybulska Krystyna, Platkowski Maciej, Strek Michal, Jarnuszewski Grzegorz, Wronska Ilona, Mularewicz Piotr, Kajdan Tomasz, Biczak Robert, Kolosowski Pawel

A NUMERICAL INVESTIGATION OF LAMINAR FORCED CONVECTION IN A SOLAR COLLECTOR WITH NON-CIRCULAR DUCT 1392
Teleszewski Tomasz Janusz

COMPARISON OF THE AVAILABILITY OF GROUNDWATER INFORMATION SOURCES IN POLAND WITH OTHER EUROPEAN COUNTRIES. KNOWLEDGE INVENTORY FOR HYDROGEOLOGY RESEARCH – PROJECT KINDRA 1400
Tomaszewska Barbara, Dendys Marta, Tyszer Magdalena

EXPERIMENTAL PROCESS INVESTIGATION OF A LATENT HEAT ENERGY STORAGE SYSTEM WITH A STAGGERED HEAT EXCHANGER WITH DIFFERENT PHASE CHANGE MATERIALS FOR SOLAR THERMAL ENERGY STORAGE APPLICATIONS 1408
Tsolakoglou Nikolas P., Koukou Maria K., Vrachopoulos Michalis Gr., Tachos Nikolaos, Lymberis Kostas, Stathopoulos Vassilis

A METHOD OF DETERMINING THE THERMAL POWER DEMAND OF BUILDINGS CONNECTED TO THE DISTRICT HEATING SYSTEM WITH USAGE OF HEAT ACCUMULATION 1417
Turski Michal, Sekret Robert

THE EFFICIENCY OF MACROPOROUS POLYSTYRENE ION-EXCHANGE RESINS IN NATURAL ORGANIC MATTER REMOVAL FROM SURFACE WATER 1423
Urbanowska Agnieszka, Kabsch-Korbutowicz Malgorzata

INTAKE POWER MEASUREMENT AS A CRITERION FOR CONTROL OF HVAC SYSTEMS 1431
Walaszczyk Juliusz Jerzy, Przydrózna Aleksandra, Przydrózny Edward

THE EFFECT OF INFLATION RATE ON THE COST OF MEDICAL WASTE MANAGEMENT SYSTEM 1441
Walery Maria Jolanta

SELECTED HEAVY METALS SPECIATION IN CHEMICALLY STABILISED SEWAGE SLUDGE 1449
Wisniowska Ewa, Wlodarczyk-Makula Marla

ABSORPTION COOLING SOURCES ATMOSPHERIC EMISSIONS DECREASE BY IMPLEMENTATION OF SIMPLE ALGORITHM FOR LIMITING TEMPERATURE OF COOLING WATER 1457
Wojdyga Krzysztof, Malicki Marcin

INVESTIGATIONS OF HOT WATER TEMPERATURE CHANGES AT THE PIPE OUTFLOW 1463
Wojtkowiak Janusz, Oleskowicz-Popiel Czeslaw

THE APPLICATION OF IMPEDANCE MEASUREMENT TO ASSESS BIOFILM DEVELOPMENT ON TECHNICAL MATERIALS USED FOR WATER SUPPLY SYSTEM CONSTRUCTION 1470
Wolf Mirela, Traczewska Teodora, Grzebyk Tomasz

THE EFFECTIVENESS OF REMOVING PRECURSORS OF CHLORINATED ORGANIC SUBSTANCES IN PILOT WATER TREATMENT PLANT 1478
Wolska Malgorzata, Szerzyna Slawomir, Machi Justyna, Molczan Marek, Adamski Wojciech, Wisniewski Jacek

RHEOLOGICAL PROPERTIES OF DISINTEGRATED SEWAGE SLUDGE 1486
Wolski Pawel

HOW EFFECTIVE IS AERATION WITH VORTEX FLOW REGULATORS? PILOT SCALE EXPERIMENTS 1493
Wójtowicz Patryk, Szlachta Malgorzata

PROSPECTS FOR THE USE OF SMR AND IGCC TECHNOLOGIES FOR POWER GENERATION IN POLAND 1500
Wyrwa Artur, Suwala Wojciech

AN APPLICATION OF DATA MINING IN DISTRICT HEATING SUBSTATIONS FOR IMPROVING ENERGY PERFORMANCE 1509
Xue Puning, Zhou Zhigang, Chen Xin, Liu Jing

MECHANICALLY ENHANCED ELECTROSPUN NANOFIBERS FOR WASTEWATER TREATMENT 1517
Yalcinkaya Fatma

INFLUENCE OF CROSS-SECTIONAL RATIO OF DOWN COMER TO RISER ON THE EFFICIENCY OF LIQUID CIRCULATION IN LOOP AIR LIFT BUBBLE COLUMN 1524
Yamamoto Tatsumi, Kawasaki Hiroyuki, Mori Hidetoshi

MICROWAVE PYROLYSIS USING SELF-GENERATED PYROLYSIS GAS AS ACTIVATING AGENT: AN INNOVATIVE SINGLE-STEP APPROACH TO CONVERT WASTE PALM SHELL INTO ACTIVATED CARBON 1530
Yek Peter Nai Yuh, Liew Rock Keey, Osman Mohammad Shahril, Wong Chee Chung, Lam Su Shiung

ENERGY PRODUCTION FOR ENVIRONMENTAL ISSUES IN TURKEY 1539
Yuksel Ibrahim, Arman Hasan, Demirel Ibrahim Halil

THE ROLE OF ENERGY SYSTEMS ON HYDROPOWER IN TURKEY 1547
Yuksel Ibrahim, Arman Hasan, Demirel Ibrahim Halil

ZERO ENERGY BUILDINGS IN THE LOGISTICS WAREHOUSE SYSTEMS 1555
Zajac Pawel, Kwasniowski Stanislaw

ENERGY SUSTAINABLE CITIES. FROM ECO VILLAGES, ECO DISTRICTS TOWARDS ZERO CARBON CITIES 1563
Zareba Anna, Krzeminska Alicja, Lach Janusz

ENERGY POTENTIAL OF THE MODIFIED EXCESS SLUDGE 1572
Zawieja Iwona

EVALUATION OF EXTERNAL AND INTERNAL IRRADIATION ON URANIUM MINING ENTERPRISE STAFF BY TOOTH ENAMEL EPR SPECTROSCOPY 1580
Zhumadilov Kassym, Ivannikov Alexander, Khailov Artem, Orlenko Sergei, Skvortsov Valeriy, Stepanenko Valeriy, Kuterbekov Kairat, Toyoda Shin, Kazymbet Polat, Hoshi Masaharu

DOSIMETRY STUDY OF EAST KAZAKHSTAN RESIDENTS BY TOOTH ENAMEL EPR SPECTROSCOPY 1587
Zhumadilov Kassym, Ivannikov Alexander, Skvortsov Valeriy, Stepanenko Valeriy, Rakhypbekov Tolebay, Hoshi Masaharu

ENHANCEMENT OF SEDIMENTATION AND COAGULATION WITH STATIC MAGNETIC FIELD 1594
Zielinski Marcin, Debowski Marcin, Hajduk Anna, Rusanowska Paulina

OPTIMIZATION OF ENERGY COST IN WATER SUPPLY SYSTEM 1602
Zimoch Izabela, Bartkiewicz Ewelina

A REVIEW OF PHOTOVOLTAIC CELLS COOLING TECHNIQUES 1610
Zubeer Swar A., Mohammed H.A., Ilkan Mustafa

CHARACTERISTICS OF LARGE THERMAL ENERGY STORAGE SYSTEMS IN POLAND 1621
Zwierzchowski Ryszard

IMPACT OF EXTERNAL CONDITIONS ON ENERGY CONSUMPTION IN INDUSTRIAL HALLS 1631
Zabnienska-Góra Alina

INTENSIFICATION OF ANAEROBIC DIGESTION EFFICIENCY WITH USE OF MECHANICAL EXCESS SLUDGE DISINTEGRATION IN THE CONTEXT OF INCREASED ENERGY PRODUCTION IN WASTEWATER TREATMENT PLANTS 1639
Zubrowska-Sudol Monika, Podedworna Jolanta, Bisak Agnieszka, Sytek-Szmeichel Katarzyna, Krawczyk Piotr, Garlicka Agnieszka

ENERGY EFFICIENCY OF A SOLAR DOMESTIC HOT WATER SYSTEM 1648
Zukowski Miroslaw

WATER PIPE NETWORK AS A HEAT SOURCE FOR HEAT PUMP INTEGRATED INTO A DISTRICT HEATING 1656
Jadwiszczak Piotr, Niemierka Elzbieta

Author Index

edp sciences

E3S Web of Conferences

Statement of Peer review

In submitting conference proceedings to E3S Web of Conferences, I certify to the Publisher that I adhere to the **Policy on Publishing Integrity** of the journal in order to safeguard good scientific practice in publishing.

1. All articles have been subjected to peer review administered by the proceedings editors.
2. Reviews have been conducted by expert referees, who have been requested to provide unbiased and constructive comments aimed, whenever possible, at improving the work.
3. Proceedings editors have taken all reasonable steps to ensure the quality of the materials they publish and their decision to accept or reject a paper for publication has been based only on the merits of the work and the relevance to the journal.

Title, date and place of the conference

International Conference on Advances in Energy Systems and Environmental Engineering (ASEE17)
2-5 July 2017
Wrocław, Poland

Proceedings editor(s):

Bartosz Kaźmierczak
Małgorzata Kutyłowska
Katarzyna Piekarska
Hussam Jouhara
Jan Danielewicz

Date and editor's signature 11-10-2017

E3S Web of Conferences **22**, 00001 (2017) DOI: 10.1051/e3sconf/20172200001
ASEE17

Using a coupled inductor controlled by fuzzy logic to improve the efficiency of a Buck converter in a PV system

Nabil Abouchabana[1,*], *Mourad* Haddadi[2], *Abdelhamid* Rabhi[3], and *Ahmed* El Hajjaji[3]

[1]LSMF laboratory, University of Laghouat, BP 37G, Ghardaia Road, Laghouat 03000, Algeria
[2]LDCCP Laboratory, E N P, 10 avenue H. Badi BP 182 El Harrach Algiers, Algeria
[3]MIS Laboratory. University of Picardie Jul. Verne, 33 rue Saint Leu.80039 Amiens Cedex1, France

Abstract. Photovoltaic generators (PVG) produce a variable power according to the solar radiation (G) and temperature (T). This variation affects the sizing of the components of DC / DC converters, powered by such PVG, and make it difficult. The effects may differ from one component to another. The main and critical one is presented by the inductor, the element that stores the energy during sampled periods. We propose in this work an auto-adaptation of these inductor values to maintain optimal performance of the power yield of these converters. Our idea is to replace the inductor by a coupled inductor where this adjustment is made by the addition of an adjustable electric field in the magnetic core. Low current intensities come from the PVG supply the second inductor of the coupled inductor through a circuit controlled by a fuzzy controller (FC). The whole system is modeled and simulated under MATLAB/SIMULINK for the control part of the system and under PSPICE for the power part of the system. The obtained results show good performances of the proposed converter over the standard one.

1 Introduction

The purpose of the present paper is to find a topology designed to photovoltaic applications to ensure minimum losses whatever climate changes. A Photovoltaic generators (PVG) are characterized by non-linear I (V) characteristics with different points of maximum power (MPP) [1, 2]. These characteristics depend on the level of radiation (G) and the temperature of PV cells that compose it [1, 2, 3]. To address the latter influence, specific control laws were designed and developed since 1968 to now [3, 4], to allow these PVG to produce their maximum electrical power. This type of control is called Maximum Power Point Tracking (MPPT) [4, 5].

The main function performed by these controls is to ensure, at all times, a perfect adaptation between the PVG and load by making the operating point in the maximum power point (MPP), this adaptation is assured by a DC/DC converters (figure 1) [5].

* Corresponding author: n.abouchabana@lagh-univ.dz

 DOI: 10.1051/e3sconf/20172200001

The problem now is that the power at the input of the converter changes depending to diverse parameters such as solar radiation and temperature [1–5], which causes a deterioration of the converter power yield because this yield is maximum only for the pre-sizing of the DC/DC converter [6].

The aim of the paper is to propose a method allows us to maintain the converter power yield as maximum as possible when any change comes in the power input of the DC/DC converters used in PV applications.

The paper organizes as follows: as a starting point, we focus on a standard PV system with standard DC/DC converter commanded by an MPPT controller as presented in figure (1) to calculate the power yield of the DC/DC converter used in this application for various radiation levels (figure 3). We can improve its operating in his PPM using simulation tests. We have used the 'Simulink' environment of MATLAB to simulate different parts (solar panel, P&O-MPPT), the DC-DC converter is simulated using the PSpice, SLPS bloc is used (see figure 2). After a discussion of the power yield result in section 2, Section 3 presents our proposed topology to improve the power yield. Simulation results for DC/DC converter using a coupled inductor on PV application and conclusions are provided in sections 4 and 5 respectively.

2 PV System with standard Buck converter

Photovoltaic systems have become globally accepted as a practical and feasible tool to generated power. One of the major problems of PV systems is that the output power of PV panels is highly dependent on solar radiation and ambient temperature [1–3]. Hence, loads cannot be directly connected to the output of PV panels. To force the PV generator to work as possible in its optimum regime, the commonly adopted solution is to introduce a static converter [5], which acts like source-load adapter as shown in figure 1.

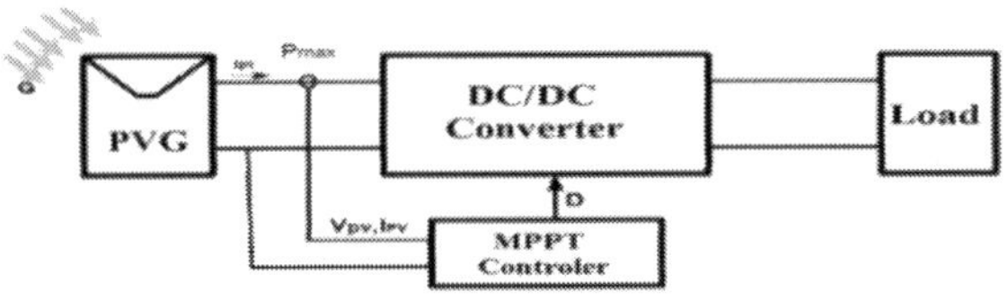

Fig.1. Elementary chain of a PV conversion used a DC/DC converter controlled by MPPT controller.

The MPPT controller delivers the action of appropriate control to track the maximum power point at any time. The choice of the conversion structure is performed depending on the load [2]. The P&O method is an MPPT algorithm used in the field of MPPT technics. P&O method requires measures on the PVG output voltage Vpv and its output current Ipv. It detects the MPP and gives us the output voltage where the power is maximum [2, 3, 5].

In our application, we use a buck topology for the DC/DC converter, which is controlled by a P&O-MPPT controller, as an adaptation between the PVG and the load; the Buck electrical circuit is presented in figure 2.

The pre-sizing of the components in this Buck depends on: input current i_{in}, and input voltage V_{in}. Components to size are: inductor L, capacitors C_1 and C_2, selection of the diode D, and the switching transistor K. When components C, D and K are well chosen the energy loss in them will be neglected [6]. The element will influence on the Buck losses is the inductor L because its sizing depends directly on the input Voltage and current (V_{in}, i_{in}). The Buck converter proposed in this paper is designed to operate in the continuous conduction modes (CCM) which means that the inductor current is always higher than zero. The inductance value is presented by equation (1) [6].

$$L_{min} = \alpha^2(1-\alpha)\frac{Vi}{0{,}1.f.Ii} \quad (1)$$

Where L_{min} is the minimum inductance, α is the duty cycle, V_i is the input voltage, I_i is the input current, and f is the frequency of the PWM signal.

2.1 DC/DC converter power Yield

The Power yield « η » of a system is defined by the ratio of output energy to the input energy. It is always between 0 and 1. A system is more effective if its Power yield tends to 1 [7]. The Power yield of the DC/DC converter is defined and calculated by:

$$\eta = \frac{P_{Load}}{P_{PV}} \quad (2)$$

2.2 Simulation tests

In simulations tests, we have used the Orcad environment of PSpice to simulate the buck converter as figure 2 shows. We use ORCAD because it is based on the SPICE industry standard and provides access to model libraries developed by manufacturers. Also by using PSpice we can use the *PSpice Model Editor* to model a magnetic core from a datasheet.

The whole system is simulated by using the co-simulation between Orcad and Simulink, SLPS bloc is introduced. SIMULINK is used to implement control part of the system, see figure 2.

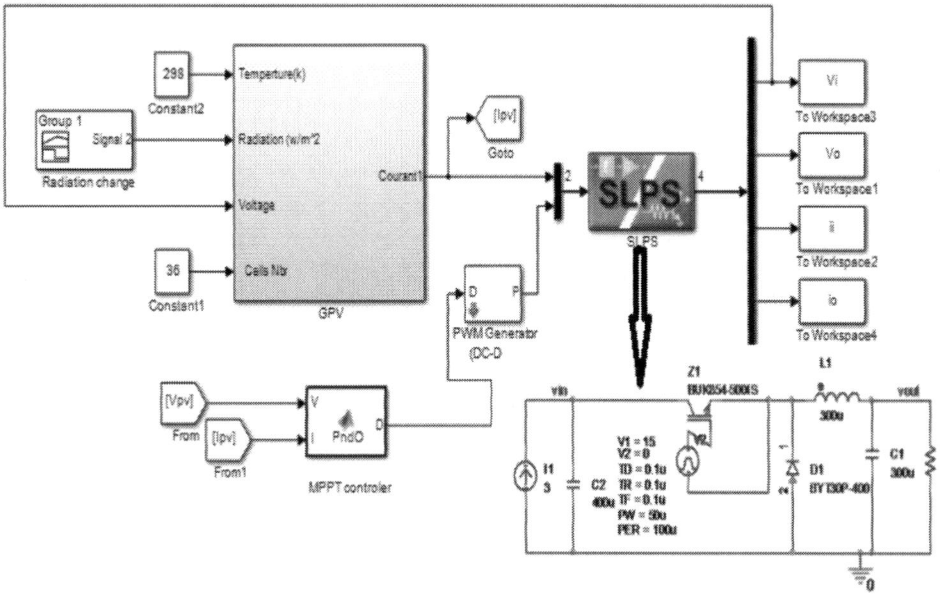

Fig. 2. Used Simulation Blocs.

The Simulation Blocs components are: PV panels as a power source (made by mathematical model [8, 9]), DC-DC buck converter implemented using PSPICE that is considered as an adaptation bloc between PVG and the load, in figure 2 implemented circuit in PSpice for the SLPS Bloc is presented. The system is controlled using a P&O algorithm which is modeled in Simulink .

2.3 Improvement of the MPPT controller and Power yield calculation

Figure 3 presents the curve of radiation change used in simulation and efficiency of the P&O controller used in our application for different solar radiation changes. After assembling the whole system, we calculate the Power yield of the DC/DC converter used for different level of radiation, the trace of power yield is presented also in figure 3.

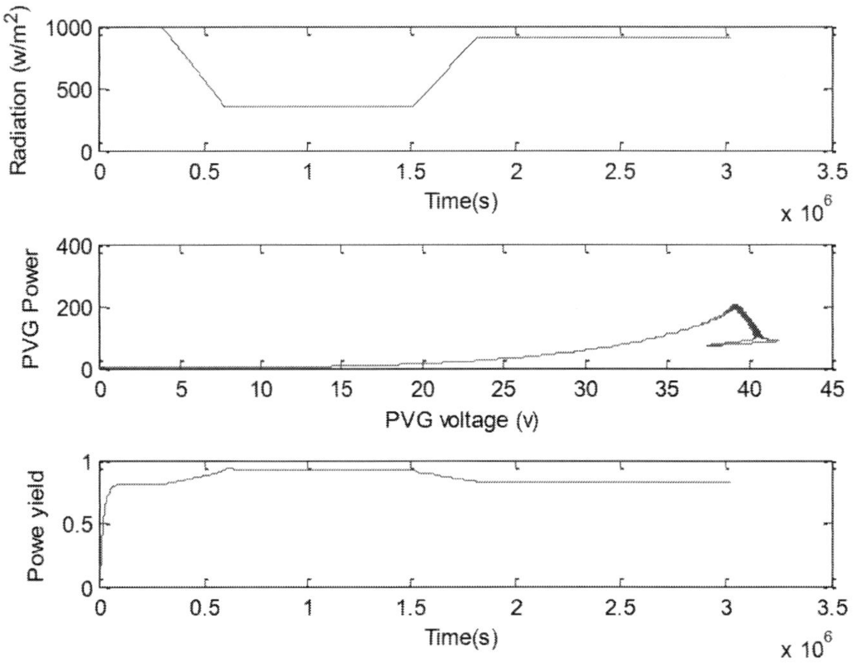

Fig. 3. Radiations curve, P=f(V) characteristics, Power yield.

We remark that there is similarity, in shape, between the power curve and the radiation curve. The system tracks the maximum power the GPV can produce under the measured radiation. It is known that the converter's power yield is maximum only around the nominal current that is used to size the convert's elements. When the current varies as in figure 3 (the current is the image of the radiation) the converter's power yield decreases and can't be maximum all the time and therefore the topology of the converter have to be improved. According to the above, we have to find a topology where the power yield of a converter is almost maximum and constant regardless to the input power.

3 Proposed DC/DC converter

The sizing of conventional dc-dc converter elements is heavily based on the nominal input value. In table 1, a pre-sizing of inductor values for different current values is given using equation (1).

Table 1. Sized inductor value for different current values.

I(A)	0.5	1	1.5	2	2.5	3	3.5	4	4.5
L(μH)	1100	530	353	265	212	177	167	133	118

In case of PV applications where the current varies steadily, the sizing of the inductor becomes difficult and using only one value of the inductor results in more power loss. In the following section we propose a solution to fix the mentioned problem

3.2 Proposed solution

According to Table 1, we remark that the values of inductor in a DC/DC converter using in PV application must change depends on current input changes.

To adjust the inductor values automatically, in order to maximize the power yield of these converters, an external circuit has been added. The new topology is presented in figure 4. In the proposed topology, a coupled inductor is used instead of a simple inductor where (L1) value is adaptive to the input current (I_{PVG}). The current I_2 that flows through in (L2) is controlled to generate a variable electromagnetic field and therefore changing the L1 value. The reference current I_{2ref}, that I2 must follow, is calculated basing on the measured current I_{PVG}. To realize the coupled inductor we add another coil (2nd coil) to the inductor magnetic core used for inductor of the DC/DC converter (1st coil). The two coils operate in mutuality between them and are connected as in figure 4. The tow electromagnetic fields created by the tow coils constitute the total electromagnetic filed in the core.

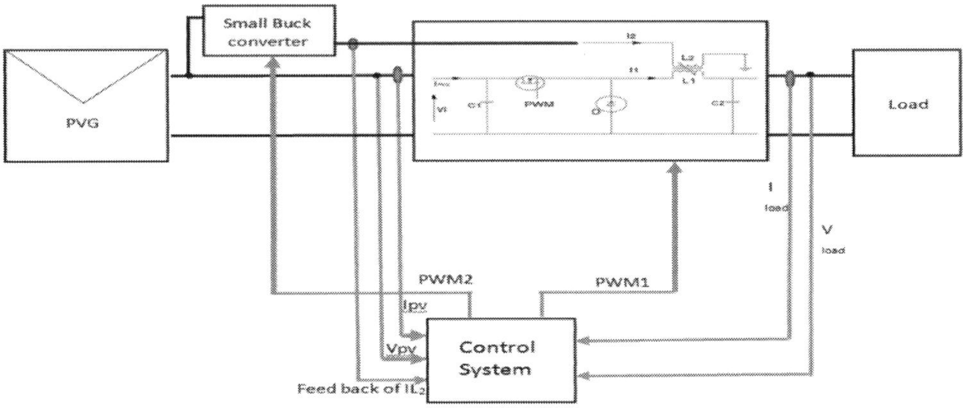

Fig. 4. PV System using Buck converter with a coupled inductor.

After sizing base on the reference [11], the coupled inductor in our application is composed of the magnetic core ETD 49 and a copper winding.

It is known that in B=f(H) characteristic, the linear region can be long or short depending on the inductor core size [12]. When the core increases in size, the maximum current, which can be reached in the linear region, increases in value [12].

In the conventional power converters, the inductor operation point needs to meet the linear region of the B-H curve under all operating conditions [6, 12].

3.2.1 Inductor and coupled inductor

The following equations present the relationship between electromagnetic field and flux and inductance [10].

$$\varphi = B.S \tag{3}$$

Where B is the magnetic flux, and S is the cross section area [m^2].

$$B= H.\mu \tag{4}$$

Where H is the electromagnetic field and μ is the permeability [H/m]

$$\varphi_T = \frac{N^2.i}{R} \tag{5}$$

Where: R is the magnetic reluctance, N is turns numbers and I is the flows current
Any change in the electromagnetic fields results in a change in the inductors values. Inductance L is defined as the ratio between total flux φ_T and current i associated to a winding N [10].

$$L = \frac{\varphi_T}{i} \tag{6}$$

For the coupled inductor: L1 and L2 are the self-inductance for the two coils and M is the mutual inductance between L1 and L2. If the fluxes are additive M is positive, and if the fluxes are subtractive M is negative [10]. Sign (M) =a

$$\begin{cases} \varphi_{1Tot} = \varphi_{11Tot} + a.\varphi_{12Tot} = L_1 i_1 + M i_2 \\ \varphi_{2Tot} = a.\varphi_{21Tot} + \varphi_{22Tot} = M i_1 + L_2 i_2 \end{cases} \tag{7}$$

Where φ_{1Tot} and φ_{2Tot} are the flux generated by L1 and L2 respectively.

The flux is proportional to the current that flows into the inductor [12, 13]. In our proposed converter structure, the inductor core is sized to operate in the linear region where the maximum current, that flows into the inductor, do not exceed one third (1/2) of the nominal PVG current. This means that the inductor will operate in the saturation zone when the nominal PV current flows into it, and therefore another solution is required.

Our proposed solution is to add another coil in order to generate a second flux in the magnetic core. Depending on the direction of the second flux, the later can be added or subtracted from the first one [6, 13]. The total flux is the result of the addition or subtraction in the magnetic core using the second coil and to avoid saturation figure 5.

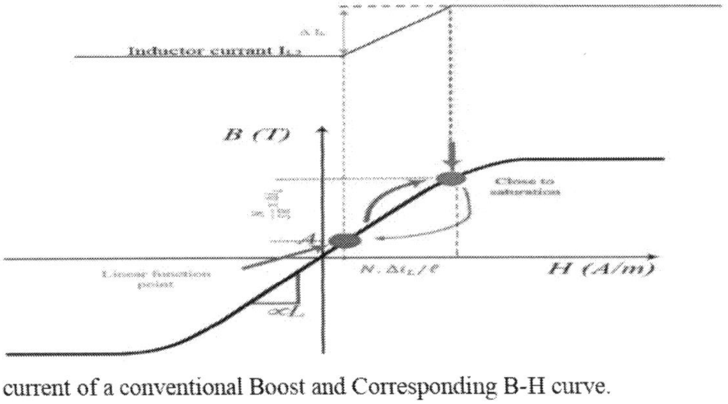

Fig. 5. Inductor current of a conventional Boost and Corresponding B-H curve.

Where L is inductance, l is core length, N is number of turns in the winding and Δi_L is the current change.

3.2.2 SPICE modeling of ETD49 N67 core from datasheet

The PSpice model of a core can be obtained from the B-H curve of ferrite from the datasheet and from the geometrical data of the core. We use *Model Editor* of PSpice to model the core used for the coupled inductor in simulation of the circuit presented by figure 4. A calculation data of its parameters will be done, and a Model of Core ETD49 will be created and able to use.

4 Simulations tests

To validate our idea simulations tests have been carried out. Two systems are simulated. One uses a standard BUCK as presented in figure 2. The second uses a coupled inductor as figure 4 shows. We used two similar PV panels model. Two model PSpice circuit for the SLPS bloc: one for the standard Buck circuit and the other for the Buck with the coupled inductor. The two inductors are made by the same magnetic core (ETD49), and the same resistive load is used.

A fuzzy controller (FC) controls the small buck converter presented in figure 4. FC is used to control the inductor currant i_2, this currant must have the same value as currant reference comes from theory calculation basing on PV currant measurement i_{PVG}. We have chosen the buck topology because the output current of a buck is always greater than its input current, so we can ensure that our second windings consume low current intensities even when i_2 has intensities greater than the powered current of buck input.

4.1 Design of the Fuzzy controller

The average output current of the PVG changes according to irradiation. When the latter is increased, the operating point of the inductor in the B-H curve changes and getting close to saturation zone and vice versa. The value of the inductor in this case must follow currant changes. From this observation, the idea of fuzzy control came up. Considering the error between the current (i_2) and its reference (i_{ref}) comes from the calculation basing on i_1 sensing value, we can decide whether incrementing, decrementing or keeping unchanged the value of the duty cycle. One of the advantages of this strategy is that only one measurement is needed; which is of the inductor currant i_2. The fuzzy algorithm has been chosen with two linguistic input variables (error (E) and error change (DE)). Each variable has five membership functions. The output of the algorithm represents the value that we should add to the duty cycle. The fuzzy rules are summarized in Table 2.

Table 2. Fuzzy controller rules.

DE\E	NB	NS	Z	PS	PB
NB	PB	PS	PS	NS	NB
NS	PB	PS	Z	NS	NB
Z	PB	PS	Z	NS	NB
PS	PB	PS	Z	NS	NB
PB	PB	PS	PS	NS	NB

NB: Negative Big, NS: Negative Small, PB: Positive Big, PS: Positive Small, Z: Zero

The figure 6 shows the result of the FC: reference current (I_{ref}) in red color and measurement current (I_{meas}) in blue. We remark that The FC shows a good tracking performance.

DOI: 10.1051/e3sconf/20172200001

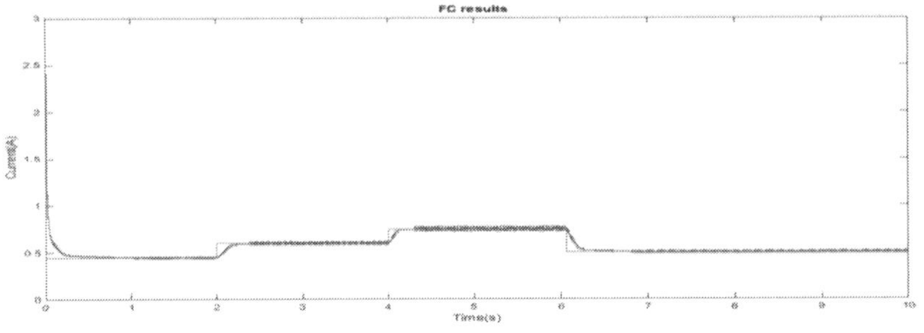

Fig. 6. Current through 2nd coil and its reference.

4.2 Simulation results

To verify the advantages of the introduced concept, the powers yield of the system with and without coupled inductor are compared in figure 7 under same operating conditions.

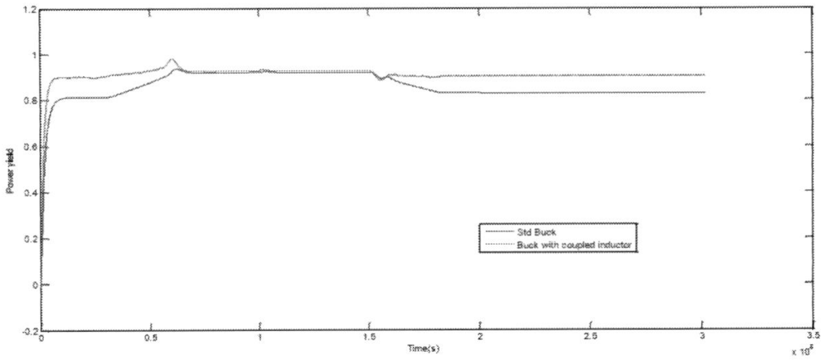

Fig. 7. Comparing power yield.

The results show that the power yield with coupled inductor is improved. We note that the curves of the power yield with coupled inductor (red curve) and without coupled inductor (blue curve) figure 7 meet at a certain stage. It is the stage where current in inductor have the value of pre-sizing.

5 Conclusions

The Buck converter with coupled inductor was analyzed and evaluated for PV application in this paper. Coupled inductor provides some advantages. Firstly, the lower inductor characteristic can be obtained (core size) by adding a second winding. Secondly, coupled inductor improves power yield of the DC/DC converter in PV application. A fuzzy controller is used to control the current in the second coil of the coupled inductor. The main advantage of the fuzzy controller is that no prior knowledge is needed about the system parameters and that only one measurement is required. The simulation of the proposed structure allowed us to improve Power yield of the DC/DC converter used in PV system.

References

1. J.A.R. Hernanz, J.J.C. Martn, I.Z. Belver, J.L. Lesaka, E.Z. Guerrero, E.P. Prez, (ICREPQ'10), Granada, 1–5, (2010)

2. M. Ramli, S. Twaha, K. Ishaque, Y. Al-Turki, RSE Reviews **67**, 144 (2017)
3. D. Verma, S. Nema, A.M. Shandilya, S. K. Dash, RSE R **54**, 1018–1034 (2016)
4. A.F. Boehinger, IEEE AES-4, **1**, 102–111 (1968)
5. J. Santos, F. Antunes, A. Chehab, C. Cruz, S. E. **80**, 772–778 (2006)
6. M.H. Rashid, *Power Electronics Handbook*, (California, Academic press, 2001)
7. G. Chateigner, M. Boës, D. Bouix, J. Vaillant, D. Verkindère, *Manuel De Génie Electrique*, DUNOD, ISBN 978-2-10-048499-7
8. C.S. Chiu, IEEE Transactions on EC **25**, 4, 1123–1132 (2010)
9. M. Park, I.K. Yu, Proc. IECON, 2040–2045 (2004)
10. Techniques de l'ingénieur, «Circuits magnétiques – Principes» (10August 2010)
11. http://schmidt-walter-schaltnetzteile.de/smps_e/abw_smps_e.html
12. P.L. Wong, Q. Wu, P. Xu, B. Yang , F.C. Lee, IEEE, 0-7803-5864-3 (2000)
13. M. Hirakawa, Y. Watanabe, M. Nagano, K. Andoh, S. Nakatomi, S. Hashino, T. Shimizu, IEEE, 978-1-4244-2893-9 (2009)

Experimental investigation and CFD simulation of multi-pipe earth-to-air heat exchangers (EAHEs) flow performance

Łukasz Amanowicz[1,*], and *Janusz* Wojtkowiak[1]

[1]Poznań University of Technology, Faculty of Civil and Environmental Engineering, ul. Berdychowo 4, 60-965 Poznan, Poland

Abstract. In this paper the experimentally obtained flow characteristics of multi-pipe earth-to-air heat exchangers (EAHEs) were used to validate the EAHE flow performance numerical model prepared by means of CFD software Ansys Fluent. The cut-cell meshing and the k-ε realizable turbulence model with default coefficients values and enhanced wall treatment was used. The total pressure losses and airflow in each pipe of multi-pipe exchangers was investigated both experimentally and numerically. The results show that airflow in each pipe of multi-pipe EAHE structures is not equal. The validated numerical model can be used for a proper designing of multi-pipe EAHEs from the flow characteristics point of view. The influence of EAHEs geometrical parameters on the total pressure losses and airflow division between the exchanger pipes can be also analysed. Usage of CFD for designing the EAHEs can be helpful for HVAC engineers (Heating Ventilation and Air Conditioning) for optimizing the geometrical structure of multi-pipe EAHEs in order to save the energy and decrease operational costs of low-energy buildings.

1 Introduction

1.1 Application of earth-to-air heat exchangers

Earth-to-air pipe-type heat exchangers (EAHEs) are used as a part of the energy efficient ventilation systems with heat recovery, enabling additional energy recovery thanks to the accumulative properties of a ground and quite stable ground temperature at a depth of about 2 m [1]. Fresh air flows through the pipes and heats up in the winter or cools down in the summer. Its additional advantage is preventing the air-to-air heat exchanger in air-handling unit against freezing in cold and moderate climates. The simple payback time of EAHE in Italian climates can be 5–9 years [2]. Performance evaluation and life cycle cost analysis of EAHE were also done in the New Delhi (India) [3]. EAHE coupled to a photovoltaic system was investigated in Greece [4]. EAHE performance for greenhouses was investigated in Turkey [5]. The potential of EAHE for low energy cooling of buildings in Algeria was investigated in [6]. The EAHEs were also investigated in France [7].

* Corresponding author: lukasz.amanowicz@put.poznan.pl

1.2 Multi-pipe earth-to-air heat exchangers

In buildings such as factories, markets [8], green houses [5], offices or swimming pools, for high demand of the fresh air, multi-pipe (registry type) heat exchangers are used to decrease total pressure losses and to decrease the area necessary for system installation. A proper designing of such exchangers is more complicated than single pipe structures, because much more parameters have to be taken into account. For the estimated calculations, the equal airflow division between all of parallel pipes is assumed but experimental investigations revealed the weakness of this assumption [9].

There are many papers focused on the issue of mathematical modelling of EAHEs operation [10–20]. Majority of mentioned take into account one-pipe structures of exchangers and focus attention on EAHEs thermal performance. Measuring results of pressure losses and of airflow division between parallel pipes are not presented.

In this work the attention is drawn only to the EAHEs flow characteristics. Thermal performance of the EAHEs is not a field of this work. Numerical flow performance model built with CFD code (Ansys Fluent software) is validated for typical 5-pipe heat exchangers of various pipes length and various main pipes diameters.

2 Scope, experimental investigation and CFD model

2.1 Scope of investigation

Both experimental and numerical investigations of earth-to-air multi-pipe heat exchanger flow performance were conducted for 5 branch-pipe exchangers having different main pipe diameters to the parallel pipe diameter ratio and different lengths of the parallel pipes: 1) $d_{main} = d, L = 76d$, 2) $d_{main} = 2.3\ d, L = 76d$, 3) $d_{main} = d, L = 271d$. Those values were chosen from the typical range of use to show the influence of geometrical parameters d_{main} and L on the flow performance.

2.2 Experimental investigation

The experimental investigations of EAHEs models in a scale 1:4 were conducted to measure the total pressure losses of EAHE and to measure the airflow in each parallel pipe. The most important data of the experimental set-up shown in Fig. 1 are summarized in Table 1. Commercial polypropylene smooth pipes were used to build the heat exchangers models.

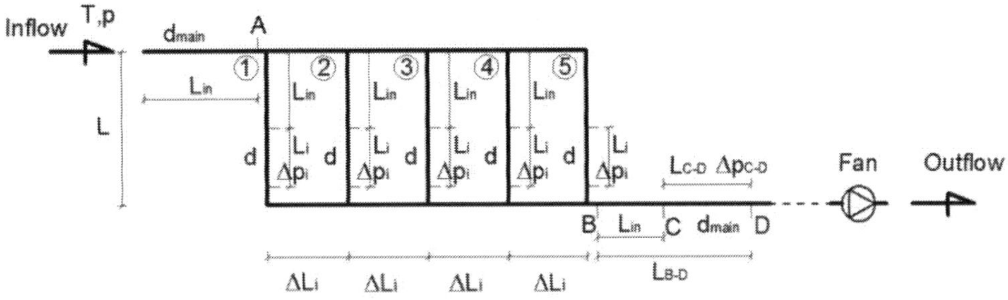

Fig. 1. Schema of the experimental set-up.

A simple and non-invasive method for airflow measurement was implemented to avoid interaction of measuring tools with airflow in pipes. Airflow in each pipe of the exchanger

 DOI: 10.1051/e3sconf/20172200002

was calculated based on pressure losses Δp_i at the measuring sector of each branch L_i. The same method was used for total airflow V_{tot} calculations. Each measuring sector was located in a fully developed region produced by at least 30 diameter entrance length. Only turbulent (typical) flows were taken into account. Formula for calculation of friction factor λ used in this method was selected based on literature review and preliminary investigations with laboratory rotor gas-meter. Several formulas collected in [21] such as: Blasius, Walden, Haaland, Altszul, Colebrook-White, Chen, Moody, Pham, to mention only a few, were taken into account. Finally the simplest Blasius equation: $\lambda = 0.3164/Re^{0.25}$ was selected. For this equation the accuracy of branch V_i (Eq. 1) and total V_{tot} (Eq. 2) airflow, calculated from rearranged Darcy-Weisbach formula, was estimated to be ±5% with the 95% confidence [22].

Table 1. Geometry of the experimental set-up.

Symbol	Description	Value	Dimensionless
d_i	internal diameter of parallel (branch) pipes	0.0461 m	1 *d*
d_{main}	internal diameter of main pipes (inflow and outflow manifolds)	0.0461 m (DN50) 0.1058 m (DN110)	1 *d* (DN50) 2.3 *d* (DN110)
L [m]	Length of branch pipe	3.5 m 12.5 m	76 *d* 271 *d*
L_{in} [m]	Length of the developing flow sector	1.5 m	32.5 *d*
L_i [m]	Length of measuring sector of branch pipes	1.85 m	40.1 *d*
$L_{C\text{-}D}$ [m]	Length of measuring sector of main pipe	1.35 m	29.3 *d*
$L_{B\text{-}D}$ [m]	Length of sector BD in case of: d_{main} = DN50 d_{main} = DN110	2.85 m 3.36 m	61.8 *d* 31.8 *d*
ΔL	distance between parallel pipes	0.28 m	6 *d*

$$\dot{V}_i = 3600 \cdot \left(\frac{2\Delta p_i \cdot d_i^{1.25}}{0.3164 \cdot \rho \cdot L_i \cdot \upsilon^{0.25}} \right)^{\frac{1}{1.75}} \cdot \frac{\pi \cdot d_i^2}{4} \tag{1}$$

$$\dot{V}_{tot} = 3600 \cdot \left(\frac{2\Delta p_{C\text{-}D} \cdot d_{main}^{1.25}}{0.3164 \cdot \rho \cdot L_{C\text{-}D} \cdot \upsilon^{0.25}} \right)^{\frac{1}{1.75}} \cdot \frac{\pi \cdot d_{main}^2}{4} \tag{2}$$

ρ – air density, kg/m^3,
v – air kinematic viscosity, m^2/s.

The following quantities were measured during experiments (see Fig. 1):

- pressure losses at measuring sector for each i-branch: Δp_i [Pa]
- pressure losses at measuring sector of outlet pipe: $\Delta p_{C\text{-}D}$ [Pa]
- pressure losses between A point and D point: $\Delta p_{A\text{-}D}$ [Pa]
- temperature of air: T [°C]
- barometric pressure: p [Pa].

Total pressure losses were calculated with Eq. (3) as pressure losses between points A and D diminished by pressure losses at sector BD (Fig. 1), where a fully developed flow was assumed. Basic information about measuring equipment is presented in Table 2.

$$\Delta p_{tot} = \Delta p_{A\text{-}D} - \frac{L_{B\text{-}D}}{L_{C\text{-}D}} \cdot \Delta p_{C\text{-}D} \tag{3}$$

 DOI: 10.1051/e3sconf/20172200002

Table 2. Experimental apparatus and its precision.

Measured value	Apparatus	Precision
T [°C]	Laboratory thermometer	± 0.5 °C
p [Pa]	Laboratory barometer	± 100 Pa
	Micromanometer with range 0 ÷ 50 Pa	± 0.05 Pa
Δp_i, $\Delta p_{C\text{-}D}$, $\Delta p_{A\text{-}D}$ [Pa]	Micromanometer with range 50 ÷ 500 Pa	± 0.5 Pa
	Micromanometer with range 500 ÷ 1990 Pa	± 3,0 Pa
L [m], L_i [m], ΔL [m]	Measuring tape	± 1.0 mm

2.3 Earth-to air multi-pipe heat exchanger CFD model

The geometry of earth-to-air multi-pipe heat exchanger model was created in the ANSYS Design Modeller. Parametric run was established to generate different types of exchanger structures. The list of the parameters and the range of its variability used in numerical calculations were the same as in the experimental investigations and are given in the Table 1. The example of investigated geometry of exchanger is shown in the Fig. 2.

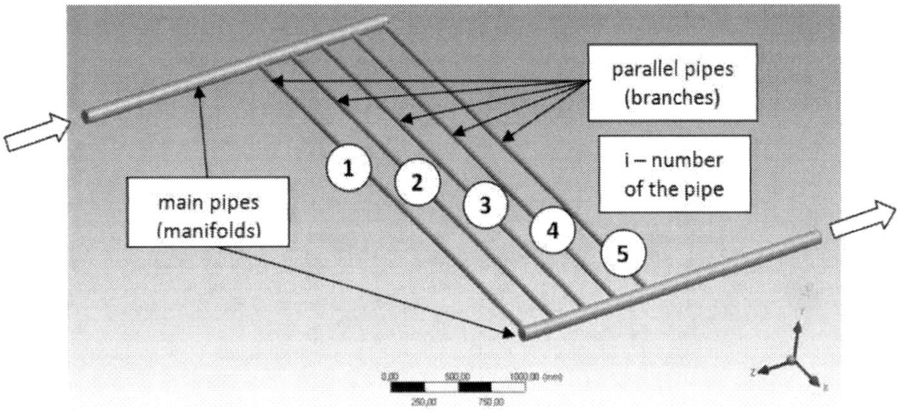

Fig. 2. 5-pipe heat exchanger: $d_{main}/d = 2.3$, $L = 76d$, $\Delta L = 6d$.

Geometry discretization was done with Ansys Meshing. To diminish the number of elements the "Cut Cell" method was used resulting in the most of hexahedral volumes. Tetrahedral volumes were created at the geometry intersections and in the boundary layer.

The density of mesh and inflation parameters were chosen with the method of grid independent solution [23], where the mesh was improving in every next step. Automatically (program controlled) method of inflation was selected. Finally the mesh with 8 layers of inflation with fine relevance center of 50 was accepted. Further increasing of mesh elements and additional inflation layers did not change the results significantly.

The k-ε realizable turbulence model with default coefficients values and enhanced wall treatment was used.

2.4 Effect of air heating on EAHE flow characteristics

To determine the effect of air heating on the airflow division uniformity and the total pressure losses a few simulations were done. Calculations were performed for pipes with an external diameter D = 50 mm, wall thickness 1.95 mm (internal diameter d = 46.1 mm), length of 76d, d_{main}/d = 1 and 3. Air heating was simulated assuming constant temperature of the outer surface of the pipe wall 8°C and various external air temperature. Thermal conductivity of the pipe material: k = 0.22 W/(mK). The percentage differences between adiabatic (constant temperature of +20°C assumed) and non-adiabatic (air heated from −20°C to about 5°C) CFD simulation are not higher than 7% and confirm that the effect of air heating on the airflow division and total pressure losses is negligible. Thanks to that it is possible to investigate the flow characteristics of the EAHEs without taking into consideration changes of air temperature in the exchanger.

3 CFD model validation and discussion

3.1 Pressure losses

In Figs. 4 and 5, a comparison of experimental and numerical results is presented in the form of Δp = f(V_{tot}). As it is seen the results are in good agreement and in general, the discrepancies are smaller than 10%.

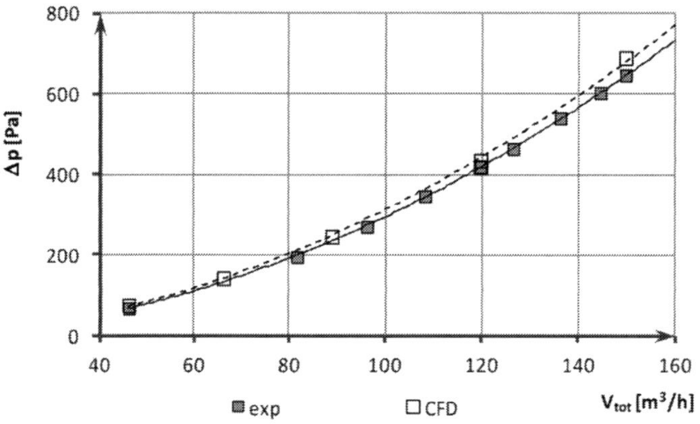

Fig. 4. Comparison of experimental and numerical investigation of total pressure losses for EAHE: 5 pipes, d = 0,0461 m, $d_{main} = d$, $L = 76d$, $\Delta L = 6d$.

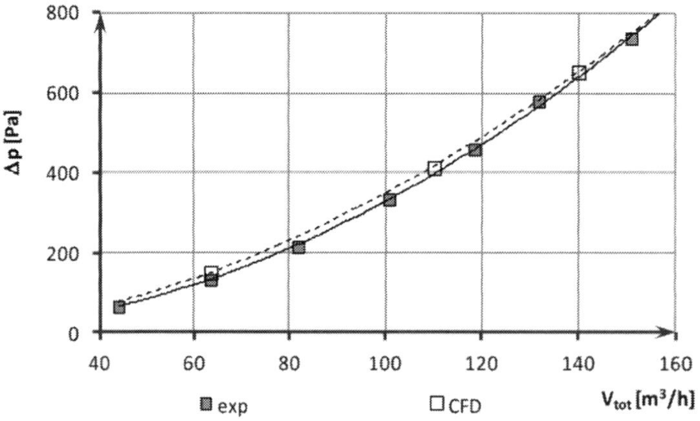

Fig. 5. Comparison of experimental and numerical investigation of total pressure losses for EAHE: 5 pipes, d = 0,0461 m, $d_{main} = d$, $L = 271d$, $\Delta L = 6d$.

3.2 Airflow division

A comparison of the numerically and experimentally obtained values of airflow in the exchanger parallel pipes is shown in the Figs. 6 to 9. The results are presented in form of graphs: $V_i = f(V_{tot})$. The average difference between results of airflow calculations and airflow measurements for small flow rates is not higher than ±15%. For larger flow rates the agreement of experimental and numerical results is much better.

The analysis of airflow division in multi-pipe EAHEs leads to the conclusion, that the percentage share of airflow in the given branch pipe in the total airflow is approximately constant [9] within a wide range of flow rates. It should be noticed that in some structures the maximum airflow can be almost 5 times higher than the minimum one. It is a significant airflow division non-uniformity that should be taken into account in the thermal performance analysis.

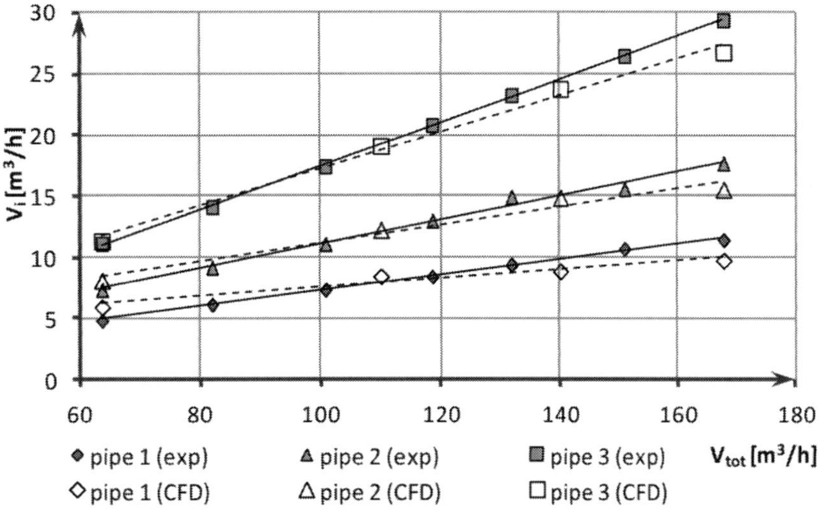

Fig. 6. Comparison of experimental and numerical investigation of airflow division for EAHE: 5 pipes, $d = 0{,}0461$ m, $d_{main} = d$, $\Delta L = 6d$, $L = 271d$, pipes 1 to 3.

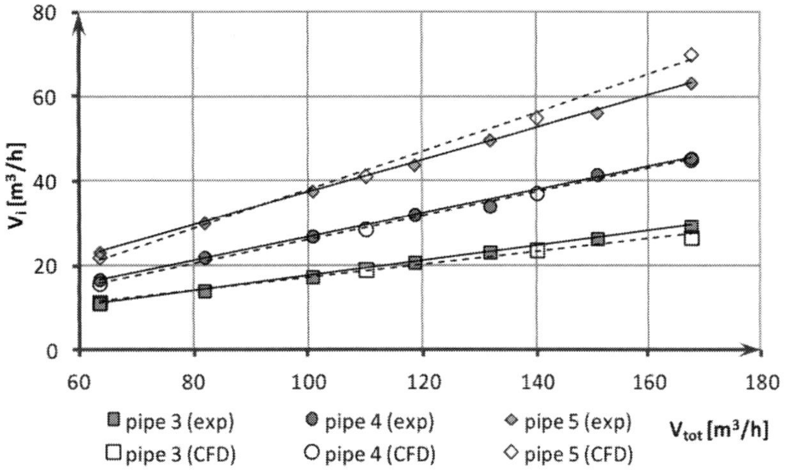

Fig. 7. Comparison of experimental and numerical investigation of airflow division for EAHE: 5 pipes, $d = 0{,}0461$ m, $d_{main} = d$, $\Delta L = 6d$, $L = 271d$, pipes 3 to 5.

 DOI: 10.1051/e3sconf/20172200002

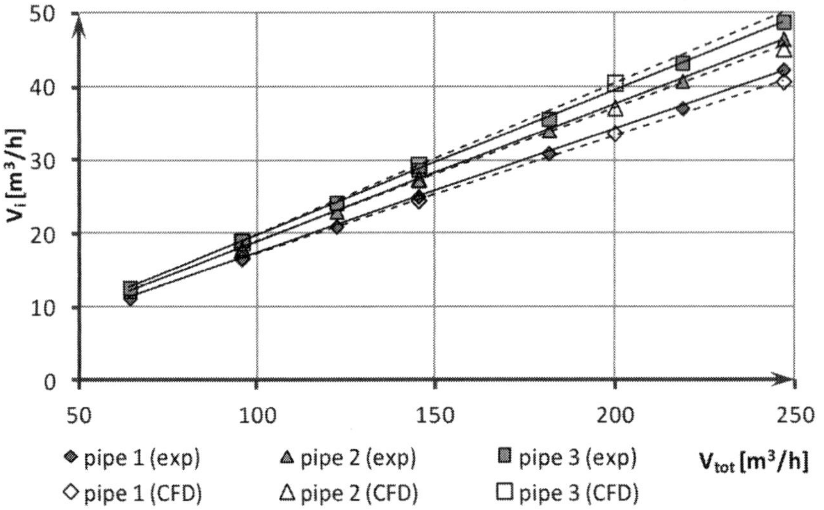

Fig. 8. Comparison of experimental and numerical investigation of airflow division for EAHE: 5 pipes, d = 0,0461 m, d_{main} = 2.3d, ΔL = 6d, L = 76d, pipes 1 to 3.

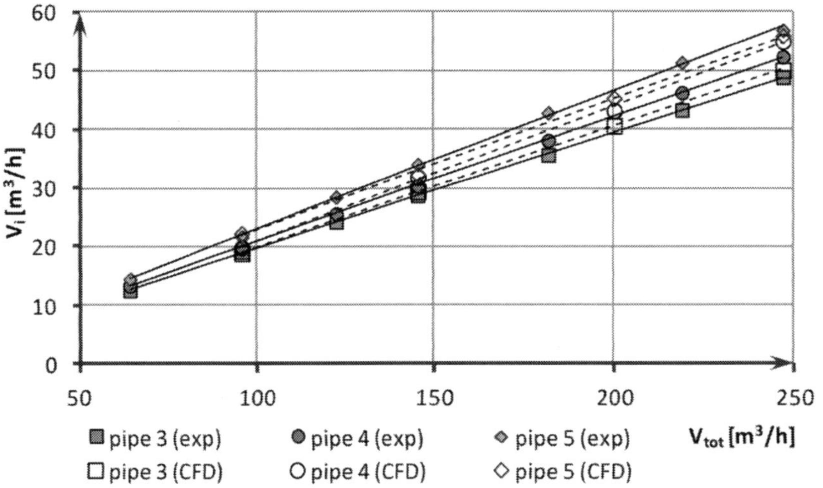

Fig. 9. Comparison of experimental and numerical investigation of airflow division for EAHE: 5 pipes, d = 0,0461 m, d_{main} = 2.3d, ΔL = 6d, L = 76d, pipes 3 to 5.

4 Engineering application and conclusion

In the designing process of EAHEs it is not only important to know the heating or cooling capacity of the device but also to predict the cost of its operation which depends on the total pressure losses. There are many geometrical parameters of EAHEs that influence both the total pressure losses and airflow division between parallel pipes. In practice the airflow division between parallel pipes is ignored in the design process although it can influence the total pressure losses, thermal performance and acoustics.

The total pressure losses of different structures can differ considerably and the airflows in pipes of multi-pipe EAHE are not equal. It was shown that for the EAHEs having the same diameter of the main and parallel pipes the airflow in the parallel pipe located nearest to the EAHE outflow can be more than 5 times larger than airflow in the pipe nearest to the EAHE inflow. The reason of that is the unfavorable static pressure distribution over the length of the exchanger with manifolds of constant and relatively small diameter. It means that the assumption of uniform airflow distribution among the parallel pipes is not always justified and has to be make very carefully.

Validated in this paper CFD model of multi-pipe EAHE can be used for designing and optimization of the EHAE structures. The model should be especially useful in untypical cases for which simplified methods are useless.

References

1. C.O. Popiel, J. Wojtkowiak, B. Biernacka, Exp. Therm. Fluid Sci. **25**, 301–309 (2001)
2. F. Ascione, L. Bellia, F. Minichiello, Renew. Energ. **36**, 2177–2188 (2011)
3. A. Chel A, G.N. Tiwari, Energ. Buildings **41**, 56–66 (2009)
4. A. Argiriou, S. Lykoudis, C. Balaras, D. Asimakopoulos, J. Sol. Energ.-T. ASME, **126**, 620–625 (2004)
5. O. Ozenger, L. Ozenger, Energ. Policy **38**, 5141–5150 (2010)
6. A. Sehil A, A. Hasni, M. Tamali, Energy Procedia **18**, 496–506 (2012)
7. S. Thiers, P. Peuportier, Sol. Energy **82**, 820–831 (2008)
8. A. Gorka, M. Szymanski, T. Mroz, *Proceedings of the 34th AIVC Conference*, Athens (2013)
9. Ł. Amanowicz, J. Wojtkowiak, Central Heating, Heating, Ventilation **41**, 208–212, 220 (2010) (in Polish)
10. V. Badescu, Renew. Energ. **32**, 845–855 (2007)
11. V. Badescu, D. Isvoranu, Appl. Energ. **88**, 1266–1280 (2011)
12. V. Bansal, R. Misra, A. D. Ghansyan, J. Mathur, Energ. Buildings **42**, 645–648 (2010)
13. M. Bojić, N. Trifunovic, G. Papadakis, S. Kyritsis, Energy **22**, 1151–1158 (1997)
14. M. Cucumo, S. Cucumo, L. Montoro, A. Vulcano, Heat Mass Transfer **51**, 506–516 (2008)
15. M. De Paepe, A. Janssens, Energ. Buildings **35**, 389–397 (2003)
16. G. Florides, S. Kalogirou, Renew. Energ. **32**, 2461–2478 (2007)
17. P. Hollmuller, B. Lachal, Energ. Buildings **33**, 509–518 (2001)
18. J. Pfafferott, Energ. Buildings **35**, 971–983 (2003)
19. A. Trząski, B. Zawada, Build. Environ. **46**, 1436–1444 (2011)
20. T. Tzaferis, D. Liparakis, M. Santamouris, A. Argirion, Energ. Buildings **18**, 35–43 (1992)
21. D.C. Rennels, H.M. Hudson HM, *Pipe Flow. A practical and comprehensive Guide*, (John Wiley & Sons, Hoboken, New Jersey, 2012)
22. H.W. Coleman, W.G. Steele, *Experimentation and Uncertainty Analysis for Engineers (2nd Ed.)*, New York, John Wiley & Sons, Inc. (1999)
23. J.H. Ferziger, M. Perić, *Computational Methods for Fluid Dynamics*, (Springer-Verlag second edition, 1999)

Urban stormwater – greywater management system for sustainable urban water management at sub-watershed level

Amarpreet Singh Arora[1,*]

[1]Marwadi University, Faculty of Engineering, Department of Environmental Science and Engineering, Rajkot-360003, Gujarat, India

Abstract. Urban water management involves urban water supply (import, treatment and distribution of water), urban wastewater management (collection, treatment and disposal of urban sewage) and urban storm water management. Declining groundwater tables, polluted and declining sources of water, water scarcity in urban areas, unsatisfactory urban water supply and sanitation situation, pollution of receiving water bodies (including the ground water), and urban floods have become the concerns and issues of sustainable urban water management. This paper proposes a model for urban stormwater and sewage management which addresses these concerns and issues of sustainable urban water management. This model proposes segregation of the sewage into black water and greywater, and urban sub-watershed level stormwater-greywater management systems. During dry weather this system will be handling only the greywater and making the latter available as reclaimed water for reuse in place of the fresh water supply. During wet weather, the system will be taking care of (collection and treatment) both the storm water and the greywater, and the excess of the treated water will be disposed off through groundwater recharging. Application of this model in the Patiala city, Punjab, INDIA for selected urban sub-watersheds has been tried. Information and background data required for the conceptualization and design of the sub-watershed level urban stormwater-greywater management system was collected and the system has been designed for one of the sub-watersheds in the Patiala city. In this paper, the model for sustainable urban water management and the design of the Sub-watershed level Urban Stormwater-Greywater Management System are described.

1 Introduction

Urbanization directly impacts water quantity and quality owing to construction of urban infrastructure and changes in landscape and runoff conveyance networks [1, 2]. The increased percentage of impervious surfaces lead to increased runoff volumes, whilst a variety of anthropogenic activities produce a range of pollutants such as nutrients, solids and organic matter, which accumulate on catchment surfaces [3–8]. As a result, both hydrologic and water quality characteristics are significantly affected [9–12]. The

* Corresponding author: amarpreetsingh.arora@marwadieducation.edu.in

stormwater management should in fact be an integral part of sustainable urban water management with focus on localized or decentralized collection, treatment and disposal of the stormwater may be through, reuse and groundwater recharging. But this should be done without causing urban floods and pollution of urban groundwater resources. To address these concerns, a research study was conducted at a model city Patiala (29°49' and 30°47' north latitude, 75°58' and 76°54' east longitude), an erstwhile princely city of the Punjab state of India.

This paper proposes an integrated model for the recycling of both wastewater (predominantly greywater) and urban stormwater as both stormwater and greywater are usually of better quality than untreated sewage or industrial discharge [13] and as such has better public acceptance for utilisation. An experimental treatment system was developed for one of the sub-watershed of the Patiala city. This system during its first phase was run on stormwater only and then greywater was introduced into the system during non-rainy period. The outcomes of this research can aid the urban water managers in better water management of the urban areas.

2 Material and methods

2.1 Selection of the urban catchments

Five urban sub-watersheds were identified within Patiala to serve as catchments for the study. The location of the study areas is shown in Figure 1.

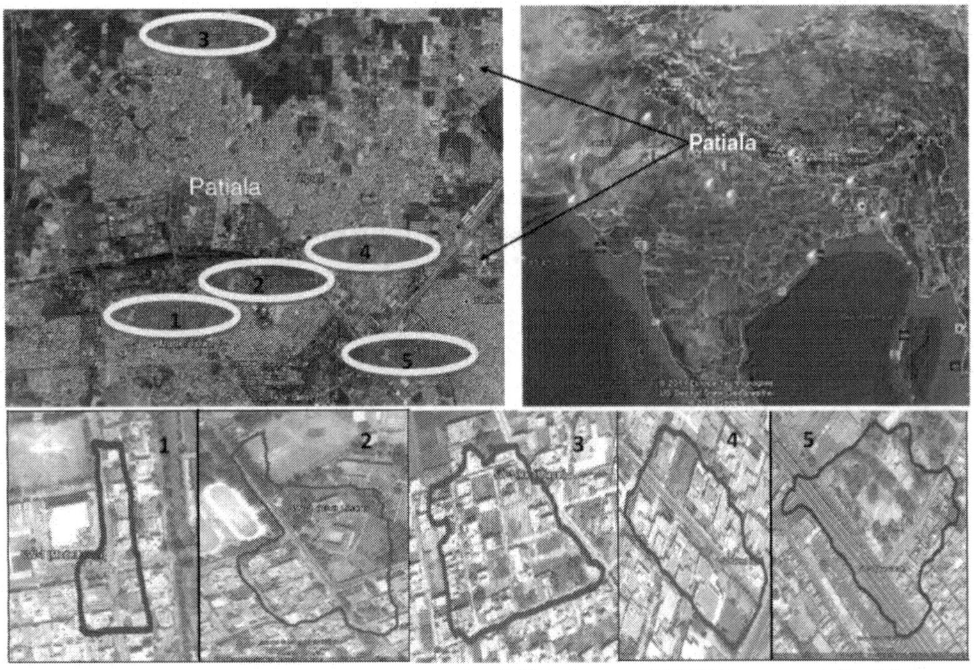

Fig. 1. Map showing the locations of the sub-watersheds studied.

The five catchments encompass a wide spectrum of land uses, housing densities and other land development characteristics. Civil Lines area is a predominantly residential catchment. Manjit Nagar, located near Seona village, is a mix of rural and residential acreage. Preet Nagar contains mixed urban development. Model Town is a predominately commercial area near Tagore theatre. Lastly, Bus Stand area is a heavily travelled and polluted catchment. The study catchments were chosen within a five kilometer radius as to have uniformity in the geological, topographical and climatic variables. Despite differing land use forms, all five catchments were prone to flooding even during modest of rainfall. Information on the land uses was obtained through the Department of Town and Country Planning Punjab, Patiala Municipal Corporation, local zoning maps and physical survey visits.

2.2 Development of sustainable urban water management model for the city

Sustainable and environmentally sound management of urban water should incorporate the following:

- Avoiding stormwater flooding in urban areas
- Maximizing the local groundwater recharging by the storm water
- Avoiding the groundwater contamination from the polluted stormwater and greywater
- Minimizing and/or eliminating stormwater runoff beyond the urban limits
- Re-using the greywater as reclaimed water

The urban area can in fact be divided into a multitude of sub-watersheds and decentralized urban water management systems at the urban sub-watershed levels may be used. The management system will include the following three components:

- Sub-watershed level combined stormwater and greywater collection system
- Water treatment system
- Groundwater recharge system

Development of an integrated urban water management system comprising of these three components and serving the above mentioned objectives has been attempted for the Patiala city, Punjab, India. Five sub-watersheds, all suffering from storm water flooding even with small rainfall events (as low as 4 mm rainfall), were chosen from within the city for the development of water management systems.

The accumulated storm water of each of the five sub-watersheds during 12 rainfall events was sampled and characterized for knowing and assessing the pollution levels of the stormwater. Hydrological methods have been used for estimating the stormwater generation rates of the sub-watersheds. The assessed/estimated stormwater characteristics and quantities have been used to conceptualize and design the storm water treatment and disposal system. An experimental water treatment system has been developed and studied for its performance in removing the pollutants from the stormwater.

The management system conceptualized would consistently satisfy the treated effluent quality requirements for disposal through the groundwater recharging. Operation and control of the conceptualized system will not require any human intervention especially during the rainfall events. Human intervention has been limited to the annual prescheduled maintenance (may be prior to the monsoon).

Storm water characterization studies have indicated removal of suspended solids and treatment for the BOD/COD strength reduction as the essential features of the conceptualized treatment system. Viability of the treatment through the highly varying and random intervening dry period (ADD) was identified as the essential feature [14].

For ensuring this, collection, conveyance and treatment of even the greywater along with the storm water of the sub-watershed has been envisioned. Such a strategy is believed to make available of the treated greywater as reclaimed water for reuse within the sub-watershed.

 DOI: 10.1051/e3sconf/20172200003

3 Results and discussion

3.1 Pilot scale experimental treatment system

The experimental treatment plant was designed and fabricated. Line diagram of the experimental treatment plant is shown in Figure 2.

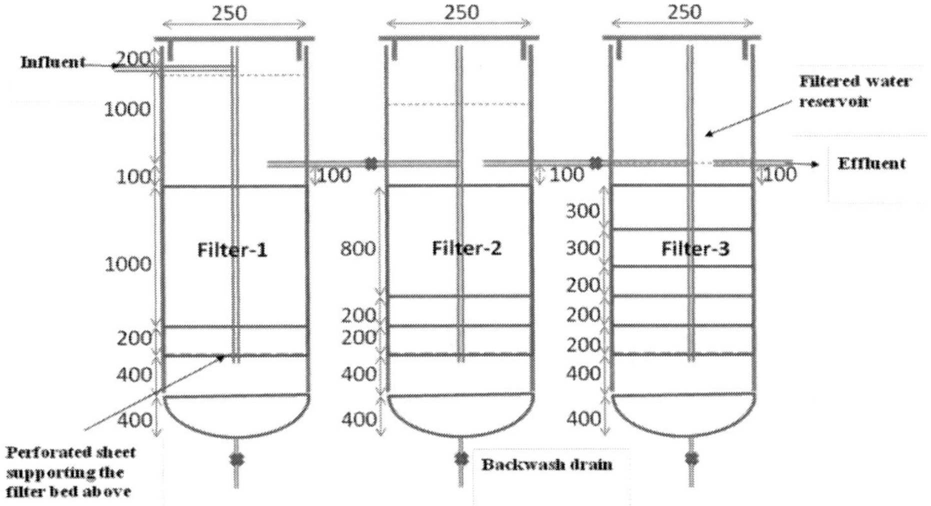

Fig. 2. Three-stage Multi-grade up-flow roughing Filter.

Stormwater treatment unit was installed near the western gate of the Thapar University, Patiala for evaluating its performance in terms of removal efficiency. This site was chosen because being a low lying area with inadequate stormwater conveyer facilities the area outside the gate was prone to flooding during rains. Stormwater samples were collected from the inlet and outlet of the unit and were analyzed for BOD_5 at 20°C, COD, TSS, TDS, Turbidity, Total and Fecal Coliform count. Results obtained have been listed in the Table 1.

Table 1. Results of the removal efficiency of Three-stage Multi-grade up-flow roughing Filter.

Parameter	Method/Reference	Inlet	Outlet	Removal efficiency
BOD (mg/L)	APHA (2005) "Manual standard method" 21st edition (5210: B)	9.2–22.5	0.7–1.2	92–95%
COD (mg/L)	Closed Reflux Method, APHA (2005): (5220: B)	112–240	28–46	69–81%
TSS (mg/L)	Total suspended solids dried at 103-105 °C, APHA (2005): (2540: D)	345–434	23–30	89–94%
TDS (mg/L)	Total dissolved dried at 180°C. APHA (2005): (2540: C)	215–332	196–240	30–37%
Turbidity (NTU)	Nephelometric Method	1180–1572	18.6–40	96–98.8%
Total Coliform (MPN/100 mL)	Serial dilution method, APHA (2005): (9221:B,C)	430000–730000	940–2400	99.4–99.8%
Fecal Coliform (MPN/100 mL)	Serial dilution method, APHA (2005): (9221:B,C)	230000–310000	70–750	99.6–99.9%

 DOI: 10.1051/e3sconf/20172200003

The treatment plant was first run with tap water so that the dissolved and suspended impurities of the filter media were washed out. The treatment unit was run on-site on storm water sample immediately after the rainfall event. The plant was run during three rainfall events. Each time after running at constant flow rate for the duration, much longer than the HRT of the treatment plant, water samples were collected from both the inlet and the outlet. The collected samples were then analysed for BOD, COD, TSS, TDS, Turbidity and Coliform count for evaluating the performance (removal efficiency) of the treatment plant.

3.2 Conceptualized system for sub-watershed level stormwater-greywater management

The urban stormwater management system conceptualized is supposed to perform three core functions: stormwater collection, treatment and disposal of the treated stormwater through reuse as reclaimed water and through ground water recharging. The system in addition to the stormwater will also be handling the greywater of the sub-watershed.

The stormwater treatment system conceptualized for the urban sub-watersheds is schematically shown in Figure 3. The circular design has been formulated taking into account the less availability of land in developing countries.

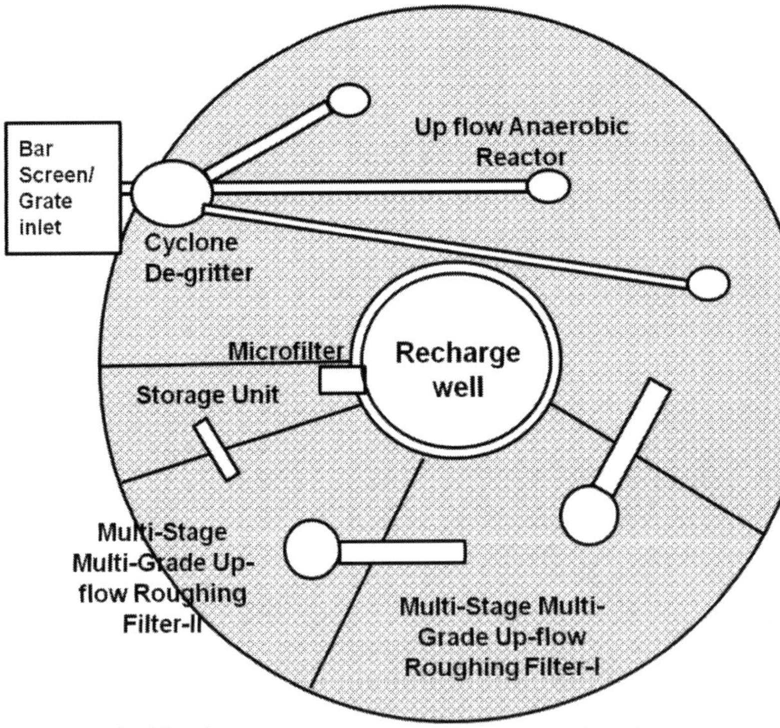

Fig. 3. Sub-watershed level stormwater-greywater management system.

Stormwater generated in the sub-watershed will flow as sheet flow for the first few meters distance then as shallow concentrated flow and enter the storm sewer inlet. If gutters are provided then the stormwater flow will be through gutters prior to entry into the storm sewer inlets. The storm sewer inlets are connected to the storm sewer through catch basins. The sewer inlet with bar screen or grating will remove larger particles like leaves, rags, etc from the stormwater. Catch basins will help in the removal of grit and floating material from the stormwater. Greywater will also be allowed to enter the storm sewers, designed to carry the greywater during dry weather at above the self cleansing velocity and stormwater during wet weather at below the storm sewer scour velocity.

Under normal conditions (dry weather conditions) treated stormwater will be held in the storage and will be pumped and supplied as reclaimed water to the nearby areas for reuse and no treated effluent will flow into the recharge well.

Under wet weather conditions the treated stormwater will be overflowing the storage provided and will get into the recharge well through a micro-filter or slow sand filter. Stormwater diversion would be provided for the case when combined stormwater and greywater flow is beyond the capacity of the system.

Different units of the treatment system can be sized to handle the expected stormwater at 90 percentile rainfall event. Provisions may be made at the grate inlet for diverting and transporting out excess stormwater flows. Filtered water storage facility can be sized to store the maximum greywater flows expected. Hydraulic design of the treatment system may be such that all units will remain submerged in water and no pumping is required for the stormwater and/or greywater to flow through the treatment system.

The treated water collected into the storage facility can be pumped and reused as reclaimed water. Treated water overflowing the storage facility can be disposed off through ground water recharging. The overflowing water may be finally filtered through a micro-filter or slow sand filter and then allowed into a recharge well for ground water recharging. All the units of the stormwater treatment system may be arranged peripheral to the recharge well. The micro-filter/slow sand filter may be provided over the recharge well. Hydraulic design of the system can be such that the treatment system will remain submerged in water throughout.

4 Conclusions

Recycling of stormwater as well as greywater is an important strategy for improving urban water cycle management, given the current and increasing stresses on water resources throughout urban centres of India, and much of the world. In rapidly growing cities, the absence of infrastructure creates a demand for small-scale wastewater treatment and improved sanitation facilities. Decentralised water management has a potential for alleviating the water-related problems. Since there is insufficient water available, wastewater and stormwater should be recognised as a significant, growing and reliable water resource. The reclaimed water can fulfil the demand for toilet flushing and could certainly contribute to relieving the pressure on the ground water sources. Also the system can help in avoiding floods during the rainy season. Integrating greywater recycling could prove to be a very useful strategy not only in keeping treatment unit in live condition but also providing a fixed amount of water volume as it is independent of seasonal variations. Lastly if sustainable water management is to be achieved, the three urban water streams – potable supply, wastewater and stormwater—should be integrated into the management of the total urban water cycle.

Acknowledgement

Author would like to extend thanks to Marwadi University, Rajkot for providing the platform to share my research on an International platform and School of Energy and Environment of Thapar University, Patiala for providing the infrastructural and laboratory facilities for under taking this research work.

 DOI: 10.1051/e3sconf/20172200003

References

1. A. Goonetilleke, E. Thomas, S. Ginn, Understanding the role of land use in urban stormwater quality management, J. Environ. Mgmt. **74**, 31–42 (2005)
2. J. Marsalek, W.E. Watt, B.C. Anderson, Water Sc. Tech. **53**, 175–183 (2006)
3. US EPA, *Results of the Nationwide Urban Runoff Program, Vol. 1-Final Report, NTIS access number PB84-18552,* (Washington, DC, 1983)
4. H. Ha, M.K. Stenstrom, Water Research **37**, 4222–4230 (2003)
5. B.E. Hatt, A. Deltic, T.D. Fletcher, J. Environ. Mgmt. **79**, 102–113 (2006)
6. P. Mahbub, A. Goonetilleke, G.A. Ayoko, P. Egodawatta, T. Yigitcanlar, Water Sc. Tech. **63**, 2077–2085 (2011)
7. G.F. Birch, C. Matthai, M.S. Fazeli, Urban Water J. **3**, 69–77 (2006)
8. M.F. Chow, Z. Yusop, S.M. Shirazi, Environ. M. Assess. **185**, 8321–8331 (2013)
9. G.P. Codner, E.M. Laurenson, R.G. Mein, *Hydrologic effects of urbanization: a case study, Proceedings of the Hydrology and Water Resources Symposium*, IE, Australia, 201–205 (1988)
10. ASCE, J. Hyd. Div. ASCE **101**, 449–468 (1975)
11. R.G. Mein, A.G. Goyen, *Urban Runoff, Civil Engg. Trans.* IE, Australia CE30, 225–238 (1988).
12. J.J. Sansalone, J.P. Hird, F.K. Cartledge, M.E. Tittlebaum, Water Environ. Research **77**, 348–65 (2005)
13. V.G. Mitchell, R.G. Mein, T.A. McMahon, Aust. J. of Water Resources **6**, 31–43 (2002)
14. A.S. Arora, A.S. Reddy, Bul. of Environ. Cont. and Tox. **92**, 36–43 (2014)

E3S Web of Conferences **22**, 00004 (2017) DOI: 10.1051/e3sconf/20172200004
ASEE17

Estimation of the temperature spatial variability in confined spaces based on thermal imaging

Grzegorz Augustyn[1], *Jakub* Jurasz[1,*], *Krzysztof* Jurczyk[1], *Tomasz* Korbiel[2], *Jerzy* Mikulik[1], *Marcin* Pawlik[1], and *Rafał* Rumin[1]

[1]AGH University, Faculty of Management, Department of Engineering Management, ul. Mickiewicza 30, 30-059 Kraków, Poland
[2]AGH University, Faculty of Mechanical Engineering and Robotics, Department of Mechanics and Vibroacoustics, ul. Mickiewicza 30, 30-059 Kraków, Poland

Abstract. In developed countries the salaries of office workers are several times higher than the total cost of maintaining and operating the building. Therefore even a small improvement in human work productivity and performance as a result of enhancing the quality of their work environment may lead to a meaningful economic benefits. The air temperature is the most commonly used indicator in assessing the indoor environment quality. What is more, it is well known that thermal comfort has the biggest impact on employees performance and their ability to work efficiently. In majority of office buildings, indoor temperature is managed by heating, ventilation and air conditioning (HVAC) appliances. However the way how they are currently managed and controlled leads to the nonhomogeneous distribution of temperature in certain space. An approach to determining the spatial variability of temperature in confined spaces was introduced based on thermal imaging temperature measurements. The conducted research and obtained results enabled positive verification of the method and creation of surface plot illustrating the temperature variability.

1 Introduction

The human perception of comfort can be interpreted as a reaction to the observed/noticeable conditions present in given environment. This state results from a multitude of factors such as thermal comfort, air quality, lighting, sounds (noise). More and more often the state of comfort is being represented by a matrix of indices which combine specified parameters describing given surrounding and its conditions. Both, sounds and noise are quantities which can be measured directly and described by a single parameter. Whereas, assessing the quality of climate forces us to consider and combine several factors and parameters.

Accordingly to the norm [1] the thermal comfort can be defined as a state of mind which express satisfaction from the thermal environment. This state is a consequence of an equilibrium between heat generated by human body as a result of metabolism and its loses of heat to the surrounding environment [2]. Tenants who are occupying given building are performing various activities which are associated with varying levels of performed physical activity – which directly translates into generating some quantity of heat. To operate in

* Corresponding author: jjurasz@zarz.agh.edu.pl

comfortable state, tenants choose their clothing based on performed occupation, age or gender. Those factors may have various impact on the perception of the observed climate in given room. In consequence, it is an arduous task to ensure that all users/tenants will have optimal thermal conditions in their working space. However, appropriately designed heating, ventilating and air-conditioning systems may be a solution which will lead to a situation in which a majority of tenants will feel comfortable [3].

Thermal comfort is influenced by a multitude of factors which in general can be divided into two major categories: environmental ones (air temperature, wind speed, air humidity, surface radiation temperature and the asymmetry of a vertical temperature distribution); personal/individual ones like metabolic rate, acclimatization or clothing insulation. So far a significant amount of papers which has been reviewed [4, 5] has been dedicated to the idea of measuring thermal comfort by placing sensors or minimizing the energy consumption. However, they seemed to avoid the problem of an uneven distribution of temperature in various states of given space (heating, cooling etc.) and its impact on the heating, ventilation and air-conditioning systems (HVAC).

This study is a part of a larger project which aim is to establish the certain number and locations of temperature measuring sensors in order to maximize the efficiency of the HVAC system and minimize the energy consumption. Therefore, the aim of the research presented in this study was to assess the validity of thermal imaging in measuring the temperature distribution in confined spaces.

2 Thermal imaging as a measuring method

To assess the thermal comfort in given confined space one should use a dedicated tool in form of Microclimate Measure (for example MM-01) [6]. Accordingly to the PN-EN 15251 regulations it is admissible to measure the temperature only when measurements are long-lasting and the temperature of radiation can be estimated [7]. It is possible to measure temperature by means of wireless sensors connected to the computer or/and by measuring charts (black matte sheet of paper) which temperature is estimated based on thermal imaging. We have assumed that the radiation coefficient of the considered sheet of paper is equal to 0.9 [8, 9].

2.1 Thermal imaging

Temperature can be measured based on thermal imaging, which is a convenient and reasonable solution in research concerning buildings and their interior [10]. For the purpose of this study we have used thermal imaging Flir i7 which measuring range expands from -20 to +250°C and measurements accuracy is ±0.1°C and spatial resolution is 140 x 140 pixels. The parameters of this comply with PN-ISO 7726:2001 and ASHRAE 55:1992. The potential problem may be the measuring accuracy but it is important to underline that the minimal temperature difference which can be distinguished by a human being is estimated to be ± 0.5°C. According to the [7] and the purpose of our study we set up the measuring range at 0–40°C and the measuring accuracy at ± 0.4°C. Measuring black cards have been placed in the system 12/10 (rows/columns) and the distance between cards was evenlyplaced. The total number of measuring points (cards) was 120.

2.2 Air flow temperature impact on measurements

Prior to the measurements in the test room, two independent tests – hot and cold air flow - were run in order to confirm the possibilities of using thermal imaging for the purposes of thermal comfort calculation. The procedure is based on measurement of matte black sheet of paper. The tests were performed at a height with range from 1 to 2 meters in three different points in the room: 1 – on black matte sample sheet of paper, 2 – on the cooling/heating source, 3 – on the adjacent surface. The objective of these tests was to verify the possibility of performing measurements in changing surroundings conditions (flow of cold or hot air) as shown on Fig. 1. For further tests a procedure based on matt black sheet of paper was used. Which is sensitive to the temperature changes in the measured environment.

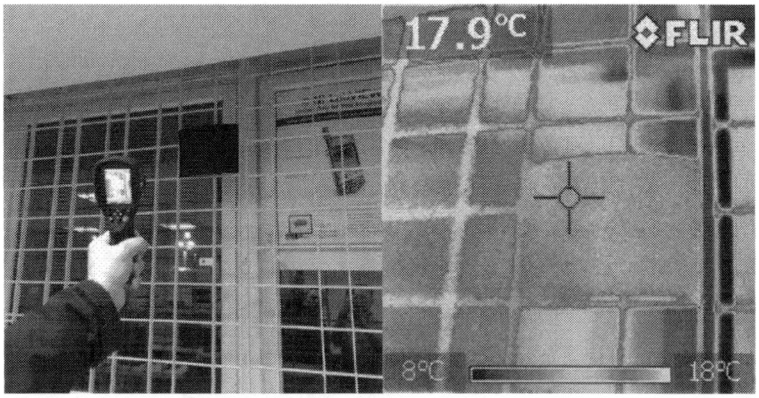

Fig. 1. Temperature measurements near cold source (window).

2.3 Measuring procedure impact on readouts

In order to verify the accuracy of results obtained by measuring of the temperature using thermal imaging and check if they are dependent on the measuring angle, distance or influence of hot and cold sources in the imaging field the following experiments have been proposed. Each of them has been repeated five times by two independent operators and realized for a black matte sheet of paper. The aim of this procedure was to check the impact of different measurement positions (angle, distance, etc.) on the results accuracy. An example of performed measurements is presented on the Fig. 2, where Table 1 contains the obtained results based on the below 15 experiments:

1. lens was perpendicular to matte black sheet of paper (referred below to as “it”) and they were 0.5 meter apart;
2. lens was perpendicular to it and they were 2.0 meter apart;
3. lens was at 45° angle to it and they were 0.5 meter apart, irradiation was perpendicular to the camera lens;
4. lens was at 45° angle to it and they were 0.5 meter apart, irradiation was parallel to the camera lens;
5. lens was at 45° angle to it and they were 0.5 meter apart;
6. lens perpendicular to it and they were 0.25 meter apart, heat source (radiator) is not visible in thermal imaging;
7. lens perpendicular to it and they were 0.5 meter apart, heat source (radiator) covers 50% of the taken image;
8. lens perpendicular to it and they were 2.0 meters apart, heat source (radiator) covers is a background of the taken photo;
9. lens perpendicular to it and they were 0.25 meter apart, heat (cold) source (window) is not visible in thermal imaging;
10. lens perpendicular to it and they were 0.5 meter apart, heat (cold) source (window) covers 50% of the taken image;

11. lens perpendicular to it and they were 2.0 meters apart, heat (cold) source (window) covers is a background of the taken photo;
12. lens perpendicular to it and they were 2.0 meters apart, heat and cold sources) cover one side of the photo;
13. lens perpendicular to it and they were 0.25 meter apart, heat and cold sources are not visible in thermal imaging;
14. lens perpendicular to it and they were 0.5 meter apart, heat and cold sources are visible in thermal imaging, black matte sheet of paper covers 50% of image;
15. lens perpendicular to it and they were 0.5 meter apart. There is the same distance from cold and heat source, and both are visible on the photo.

Table 1. Results of the control of measurements accuracy test.

Operator/ Measurement serie		**Experiments**														
		1	**2**	**3**	**4**	**5**	**6**	**7**	**8**	**9**	**10**	**11**	**12**	**13**	**14**	**15**
A	1	21.8	22.1	22.0	22.2	22.3	27.1	26.8	27.2	22.5	23.5	23.8	26.8	25.4	25.0	25,1
	2	21.9	22.1	22.0	22.0	22.3	26.9	26.7	27.3	22.7	23.4	23.7	26.8	25.3	25.0	25,1
	3	21.8	22.2	21.9	21.9	22.3	27.0	26.8	27.4	22.5	23.4	23.8	26.9	25.2	24.9	25,3
	4	21.6	22.1	22.0	22.0	22.3	27.1	26.8	27.2	22.4	23.3	23.7	26.8	25.2	24.9	25,2
	5	21.7	22.1	22.0	22.0	22.3	26.6	27.0	27.2	22.4	23.4	24.0	26.7	25.3	25.0	25,3
	Mean	21.7	22.1	21.9	22.0	22.3	26.9	26.8	27.2	22.5	23.4	23.8	26.8	25.2	24.9	25,2
	Range	0.30	0.10	0.10	0.30	0.00	0.50	0.30	0.20	0.30	0.20	0.30	0.20	0.20	0.10	0,20
B	1	21.8	21.7	22.1	22.1	22.2	26.9	27.2	27.4	23.1	23.2	23.8	26.8	25.3	25.1	25,1
	2	21.8	21.6	21.9	21.9	22.3	26.9	27.1	27.4	23.0	23.2	23.9	26.6	25.5	25.0	25,1
	3	21.9	21.7	22.0	22.0	22.5	27.0	27.2	27.4	23.0	23.3	23.9	26.7	25.3	25.2	25,0
	4	21.9	21.9	22.1	22.1	22.3	27.0	26.7	27.2	23.1	23.4	23.9	26.8	25.4	25.2	25,0
	5	21.6	21.6	22.0	22.0	22.5	27.2	27.1	27.1	23.1	23.2	23.8	26.8	25.4	25.3	25,1
	Mean	21.8	21.7	22.0	22.0	22.3	27.0	27.0	27.3	23.0	23.2	23.8	26.7	25.3	25.1	25,0
	Range	0.30	0.30	0.20	0.20	0.30	0.30	0.50	0.30	0.10	0.20	0.10	0.20	0.20	0.30	0,10
Mean (X_{avg})		21,7	21.9	22.0	22.0	22.3	26.9	26.9	27.2	22.7	23.3	23.8	26.7	25.3	25.0	25.1

In case of the operator A the mean measured temperature in case of all 15 experiments was equal to 24.2°C and their range was 0.22°C. In case of the operator B the results were respectively 24.25°C and 0.24°C. For the sake of thermal comfort measurements the obtained results are acceptable because a human being is not capable of distinguishing absolute temperature changes smaller than 0.5°C.

 DOI: 10.1051/e3sconf/20172200004

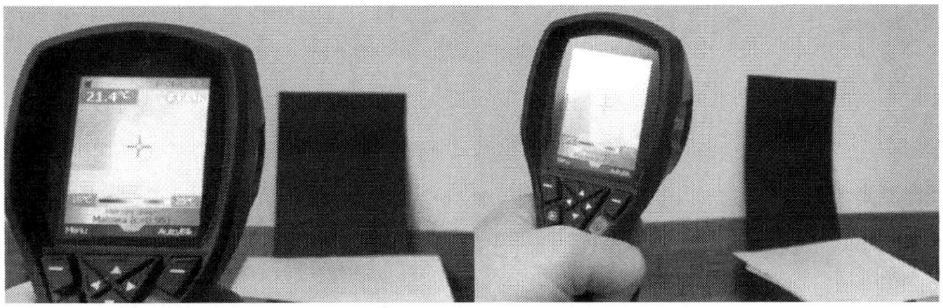

Fig. 2. Experiment no. 1 (left) and 4 (right), thermal imaging position and matte black sheet of paper.

The second test was performed in order to estimate the temperature differences observed on a single measuring black matte sheet of paper, depending on which one out of five points was the lens aimed. Fig. 3 depict the distribution of considered points whereas Table 2 summarizes the results.

Fig. 3. Location of measuring point during the second part of the experiment.

Table 2. Results of the control of measurements accuracy test – point's location.

Operator/ Measurement serie		**Measurement points on the black matte sheet of paper**				
		1	**2**	**3**	**4**	**5**
A	1	21.9	21.9	22.0	22.2	22.3
	2	21.9	21.9	22.1	22.0	22.0
	3	21.7	21.9	21.8	22.0	21.9
	Mean	21.83	21.90	21.97	22.07	22.07
	Range	0.20	0.00	0.30	0.20	0.40
B	1	22.4	22.5	22.2	22.6	22.7
	2	22.3	22.3	22.4	22.6	22.7
	3	22.4	22.2	22.5	22.2	22.3
	Mean	22.37	22.33	22.37	22.47	22.57
	Range	0.10	0.30	0.30	0.40	0.40
Mean (X_{avg})		22.10	22.08	22.22	22.27	22.32

The performed tests revealed that in case of the operator A the mean measured temperature was equal to 21.96°C and the set of performed measurements had range of 0.22°C. The second operator had slightly higher mean and respectively larger range which was 0.3°C. The results of the conducted tests suggested that the black matte sheet of paper is characterized by a uniform distribution of temperature and this does not impact the measurements accuracy and in consequence they might be used for estimating the temperature.

 DOI: 10.1051/e3sconf/20172200004

3 Materials – area of study

All measurements were performed in a lecture hall being a part of the Faculty of Management at the AGH University in Kraków. During measurement the room was not occupied by students and therefore it was possible to simulate the impact of various thermal conditions on the temperature distribution (severe draughts during winter, overheating etc.). The room is controlled by the Building Managements System (BMS). BMS controls a set of convection radiators with electric valves and fan coil unit connected to the building air handling unit installation. In the considered lecturing hall all inlets were located along the northern wall of the room, while outlets are located along the southern one. During measurements the ventilation was operating at its 20% of capacity (about 200 m^3/h) whereas the volume of the lecturing hall is close to 300 m^3. The lecturing hall is presented on the Fig.4 in panoramic image.

Fig. 4. Lecturing hall where measurements by means of thermal imaging and matte black sheet of paper were performed.

3.1 Measuring scenarios

Measurements were taken in an unoccupied lecturing hall (at a stable/fixed state). Over this period the parameters of the HVAC system were as follow: the intake and exhaust of air was 200 m^3/h which equals to 20% of the ventilation capacity; the temperature of the supplied air was 20°C; the radiators was set up in the BMS system to achieve the desired temperature 22°C.

4 Results and discussion

Final measurements were taken in the lecturing hall described in the previous section. Temperature was estimated for an evenly distributed grid of reference points which were located at 1.2 meters above the floor (which is the height at which temperature is usually measured and then used in calculating the thermal comfort). Photos presented on Fig. 5 depict the real object (lecturing hall) and corresponding view from thermal imaging below. Please note that black matte sheet of paper has not have the temperature of the object to which they are attached to, but the surrounding air temperature.

Based on the performed measurements we were able to reconstruct the spatial variability of temperature. For purpose of demonstration we present only results obtained for one of the scenarios explained in the section 3.1. In this scenario measuring sheet of papers were attached to the selected points in the lecturing hall based on previously determined grid. Majority of reference points have been attached to backs of the chairs, where some of them were pinned to the wall surface (as shown in the middle photo on Fig. 5). By attaching them to the wall we were able to determine the temperature of the parietal layer not the wall itself.

Fig. 5. Demonstrative photo of the lecturing hall with placed measuring sheets of paper in various locations: normal photo and thermal image.

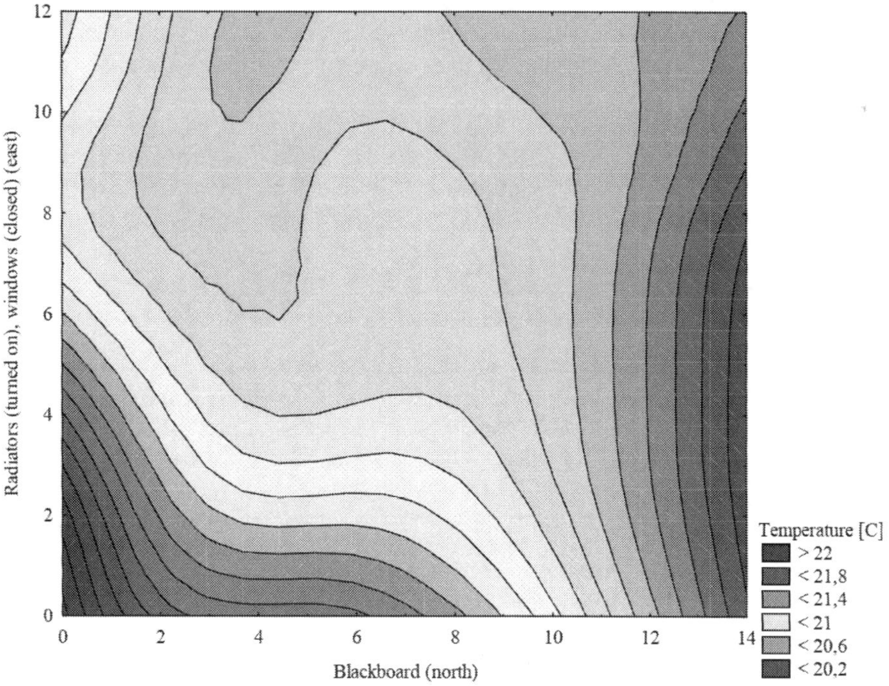

Fig. 6. Temperature distribution in lecturing hall accordingly to the scenario 1.

Additionally, it is important to underline that the sensor which controls the air handling unit was located in the main intake channel of the whole building and as the consequence its

 DOI: 10.1051/e3sconf/20172200004

operation was independent from the temperature observed in the lecturing hall. Sensor used for controlling radiators was located on the east/south corner. Please note, that the temperature observed in this location (Fig. 6. – which was created based on over 120 measuring point and distance-weighted least squares method available in Statistica software) is significantly lower than in the north/east corner were radiators and server (additional heat source) were located. This situation leads to a non-uniform temperature distribution.

The results show the uneven temperature distribution over the room. As the temperature has the major influence on the thermal comfort parameter we can see that measuring the temperature parameter at one single location and supply the feed from the single sensor is inadequate in terms of proper room temperature representation. Other scenarios will very likely confirm that this uneven temperature distribution will increase its impact on the general thermal comfort in the room. The uneven distribution is caused by heating and cooling devices placement but this also affects the thermal comfort for the space around them. In general the common approach assumes the evenly spaced temperature and does not includes the location of heating and cooling sources.

5 Conclusions

Obtained results indicate that the presented method is sufficient for performing measurements in lecturing hall. Created chart (Fig. 6) visualized the distribution of temperature in the considered confined space, which is not uniform. Such differences point to the need of using more sensors to correctly operate and control the Building Management Systems (BMS) and ensure that the thermal comfort will be at sufficient level for all occupants. The next step is the verification of the obtained results by comparing them with various measuring tools like InviNets sensors, mercury-in-steel thermometer. Further works will concentrate on investigating various scenarios of room occupancy and their impact on the temperature distribution. Additionally, all measurements will be performed by means of several different measuring systems.

This work was financed by the NCBiR research and scientific project development no. POIR.01.01-00-2008/15 and research grant no. 11/11.200.322

References

1. W. Olesen, Bjarne Ashrae J. 21–28 (2004)
2. P.O. Fanger, *Thermal Comfort* (Arkady, 1974) (in Polish)
3. I. Sudoł-Szopińska, A. Bogdan, Bezpieczeństwo pracy, **5** (2007) (in Polish)
4. E. Halawa, J. Van Hoof, Energy and Build. **51**, 101–110 (2012)
5. A.K. Mishra, M. Ramgopal, Build. Environ. **64**, 94–106 (2013)
6. P. Kiełbasa, P. Budyn, Inżynieria Rolnicza **11**, 109, 105–113 (2008) (in Polish)
7. https://www.omega.com/manuals/manualpdf/M5150.pdf *User's manual Flir i7* retrived 15.02.2017
8. T.S. Wiśniewski, Energia i budynek, **2**, (2007) (in Polish)
9. K. Krystian, Instalreporter **10**, 59 (2015) (in Polish)
10. M. Pešek, M. Pavelek, J. Eng. Mech.-ASCE **20**, 3/4, 327–335 (2013)

Assessing high shares of renewable energies in district heating systems – a case study for the city of Herten

Ali Aydemir[1,*], *Eftim* Popovski[1], *Daniel* Bellstädt[1], *Tobias* Fleiter[1], and *Richard* Büchele[2]

[1] Fraunhofer ISI, Breslauer Straße 48, 76139 Karlsruhe, Germany
[2] TU Wien, Energy Economics Group (EEG), Gusshausstrasse 25-29, A-1040 Vienna, Austria

Abstract. Many earlier studies have assessed the DH generation mix without taking explicitly into account future changes in the building stock and heat demand. The approach of this study consists of three steps that combine stock modeling, energy demand forecasting, and simulation of different energy technologies. First, a detailed residential building stock model for Herten is constructed by using remote sensing together with a typology for the German building stock. Second, a bottom-up simulation model is used which calculates the thermal energy demand based on energy-related investments in buildings in order to forecast the thermal demand up to 2050. Third, solar thermal fields in combination with large-scale heat pumps are sized as an alternative to the current coal-fired CHPs. We finally assess cost of heat and CO2 reduction for these units for two scenarios which differ with regard to the DH expansion. It can be concluded that up to 2030 and 2050 a substantial reduction in buildings heat demand due to the improved building insulation is expected. The falling heat demand in the DH substantially reduces the economic feasibility of new RES generation capacity. This reduction might be compensated by continuously connecting apartment buildings to the DH network until 2050.

1 Introduction

Germany intends to cut greenhouse gas emissions by 80 up to 95 percent in 2050 compared to 1990. To achieve the reduction goals several states in Germany implemented climate protection laws. In North Rhine-Westphalia the "Klimaschutzgesetz NRW" was enforced in 2013. As a consequence among others local heat and energy planning is required by the communities. Such planning requires information about the future heat demand and resulting potentials for renewable energy sources (RES) heat supply, especially with regard to district heating (DH) networks. This paper is on the DH network in Herten, located in North Rhine-Westphalia.

The district heating network in Herten consists majorly of two large sub-networks. The first system provides heat for residential and public buildings located in the northern part of

* Corresponding author: ali.aydemir@isi.fraunhofer.de

the city, whereas the second one provides heat for buildings located in the southern part. Both larger sub-networks are connected to each other; however capacities of this interconnection are limited creating a bottleneck. Furthermore, several island networks exist such as the island network located in the southern part providing heat for commercial buildings and plants where the heat is supplied by the local waste incineration plant. According plans from the local utility the demand in the southern part will be fully covered by heat from the waste incineration plant in future. For the northern part this will probably not the case, as the operator of the plant is negotiating also with other interested parties from neighbour cities. Furthermore huge investments are necessary to overcome the bottleneck between northern and southern part of the DH network. The northern part of the DH network is currently provided with heat from coal-fired CHPs. Thus, the focus of this local case study is to show pathways on how to substitute the current heat supply in the northern DH network with centralized renewable technologies.

2 Methodological approaches

Several earlier studies have assessed the DH generation mix without taking explicitly into account future changes in the building stock and the heat demand (cf. [2, 3). Thus our approach is constructed to fill this gap in order to show the relevance of future demand projections with regard to system design. The approach consists of three steps, combining stock modelling, energy demand forecasting and the simulation of energy technologies.

1. First, we construct a detailed residential building stock model for Herten using remote sensing together with a typology for the German building stock.
2. Second, we us a bottom-up simulation model that calculates the annual energy demand for heating and cooling based on energy-related investments in buildings to forecast the heating demand up to 2050.
3. Third, we design a potential new mix of new renewable generation units for the DH taking hourly demand profiles into account. The mix consists of solar thermal fields in combination with large scale heat pumps. Together they are capable to substitute the current coal-fired CHPs nearly fully.
4. We finally assess the new generation mix with regard to cost of heat and CO_2 reduction for two scenarios.

2.1 Residential building stock model

The building stock model constructed distinguishes between regional location and type of a building. To construct the building stock five main sources of information are used as follows:

- A printed map from the climate concept 2020+ fractioning the city in so-called energetic city space types 1.
- A digital map from the land registry office containing basic (plot) areas of buildings in Herten.
- The online available GIS-system TIM-online provided by the state of North Rhine-Westphalia 5.
- The building typology for Germany prepared by the Institut Wohnen und Umwelt GmbH (IWU) in Darmstadt 6.
- The census database for Germany. In the database, the number of flats differentiated by size categories is contained differentiated by cities 7.

The building stock is constructed in two steps. The first step was to rework the map of energetic city space types. This was based on the reason that the plot areas of buildings contained in the map from the land registry office do not appear to be homogenous within all areas differentiated in the map of energetic city space types. The second step was to assign to each area from the new map a building type from the building typology for Germany prepared by IWU. Therefore the new map has been compared with the map of energetic city space types and orthoimages from the TIM-online system visual. Based on this visual comparison a building type had been assigned to each area.

2.2 Heat demand forecasting

The present heat demand for the building stock differentiated by building type has been calculated based on data from the local utility. The local utility evaluated the energy demand of the buildings in the city by analyzing the demand for natural gas and district heating at the points of delivery to the buildings. As a result the present heat demand for the building types had been defined. Based on this heat demand and building stock projections are calculated up until 2030 and 2050 using the simulation model INVERT/EE-Lab 4. The model allows simulating the impact of current building standards and renovation on the evolution of the future building stock and the resulting energy demand. The projection is based on a current policy scenario for Germany taking the renovation of buildings, the demolition and the construction of new buildings into account. Energy-related investments are simulated based on a combination of approaches from decision theory as well as diffusion theory.

Averages of resulting rates are given in the following as indication: average rate of modernization 1.5%, new construction rate 0.4%, demolition rate 0.5%. A substantial decrease in heat demand from about 460 GWh in 2014 to little above 200 GWh in 2050 can be observed. The corresponding specific thermal energy demand for space heating and hot water per year and square meter is given in Table 1.

Table 1. Estimated final energy demand for space heating and hot water [kWh/m^2·a].

Building type	2014	2030	2050
Detached house	185	115	89
Terraced house	150	97	81
Apartment building	150	85	64
Large apartment building	83	64	54

2.3 Design of RES heat generation units

The technical detailed modelling of potential new generation units has two goals. First, technical feasible assumptions on the heat delivered to the DH by renewable energies shall be derived. This assumption shall serve for the scenario definition in chapter 3. Second, current cost for a modelled potential new mix of RES supply units in the DH shall be indicated. This is to indicate challenges for such systems in the results section (chapter 5).

For the proper design of new generation units an hourly demand profile for the district heating network is needed. This is calculated based on the given annual consumption data for the community by using energyPRO 8. In principle the profile is generated by fractioning the annual demand of the district heating network into hourly demand values. To do so 80% of the district heating demand is modelled to be linear dependent on ambient temperatures accounting for space heating. A threshold value of 15°C ambient temperature is used. Above this threshold, no space heat is demanded. The residual 20% - share of the district heating demand is modelled independent from ambient temperatures, assuming that it is mainly due to residential hot water demand.

Technology and cost assumptions are used to model the heat generation of the potential new thermal supply units in the DH and to calculate the resulting levelised cost of heat (LCOH). For the calculation of the LCOH the annuity method is applied and it is defined based on equation (1) and (2). CRF represents the capital recovery factor which is derived by taking the interest rate (*i*) and economic lifetime of the unit (*n*) into account. C_I represents the investment necessary to buy the unit and $C_{O\&M}$ the corresponding annual cost for operation and maintenance. Finally HS stands for the heat supplied by the unit to the DH within a year.

$$LCOH = (CRF \cdot C_I + C_{O\&M} + C_{fuel\&tax\&levies}) \div HS \quad (1)$$

$$CRF = (i \cdot (1 + i)^n) \div ((1 + i)^n - 1), \text{ with} \quad (2)$$

For the potential new mix of renewable generation units a solar district heating system together with a large scale heat pump is chosen. For the solar district heating system this refers to solar thermal collector fields in combination with pit storages. Capital costs are taken from a guideline 10. In the guideline capital cost for various system sizes are given. Based on this we modelled cost functions setting system sizes into relation to capital cost. Both cost functions reflect economies of scale, i.e. larger fields or storages have lower specific cost than smaller ones. Costs of land are derived from a Geographical Information System from the state North Rhine-Westphalia 11.

Herten and is neighbour cities are former coal mining cities. As a consequence a local mining company annually pumps over 100 million cubic meters of water to the surface in the region to guarantee the stability of the ground and the safety of the drinking water. The temperature of the pit water is around 20°C and can serve as a heat source for large scale heat pumps. To define technical parameters for the heat pump (such as the COP) two existing case studies are used as reference 12. The first case study is located in Bochum. In the study a heat pump using pit water as heat source has already been erected and tested. The second case study is from Helsinki. In the study sewage water is used as heat source for a large scale heat pump in the DH network 13. Capital cost for heat pumps are based on the "Technology Data for Energy Plants" database published by the Danish Energy Agency 14. For the heat pumps we have in contradiction to the solar fields no cost function, so that no economies of scale are taken into account. Electricity cost for 2011 are based on Eurostat electricity prices for industry (consumption band: ID, Excluding VAT and other recoverable taxes and levies, 113.9 EUR/MWh) and price projections are based on own calculations.

The system design is carried out by first defining the size of the solar thermal collector field including storage volume and afterwards defining the thermal capacity of the heat pump. Please note that in the project system designs has been carried out for several sub-networks. However, as the behaviour is equivalent for all sub-networks the description in the following is only on one sub-network. For the design two criteria are applied, the LCOH and the heat delivered to the DH. The heat delivered to the DH by the solar field and the heat pump is set in reference to the overall demand and called heat fraction. The design approach is stepwise as described in the following.

- First, the solar field and storage size is set, so that the LCOH of the solar thermal collector field including storage is minimal (including storage).
- Second, the thermal capacity of the heat pump is set, so that the present heat fraction is maximized for the combined system as much as possible.

The share of the heat provided by the solar field compared to the overall demand (solar fraction) for varying field and storage sizes is given in Fig. **1**. Referred to the heat produced, the LCOH for the solar size finally chosen is around 22 EUR/MWh.

 DOI: 10.1051/e3sconf/20172200005

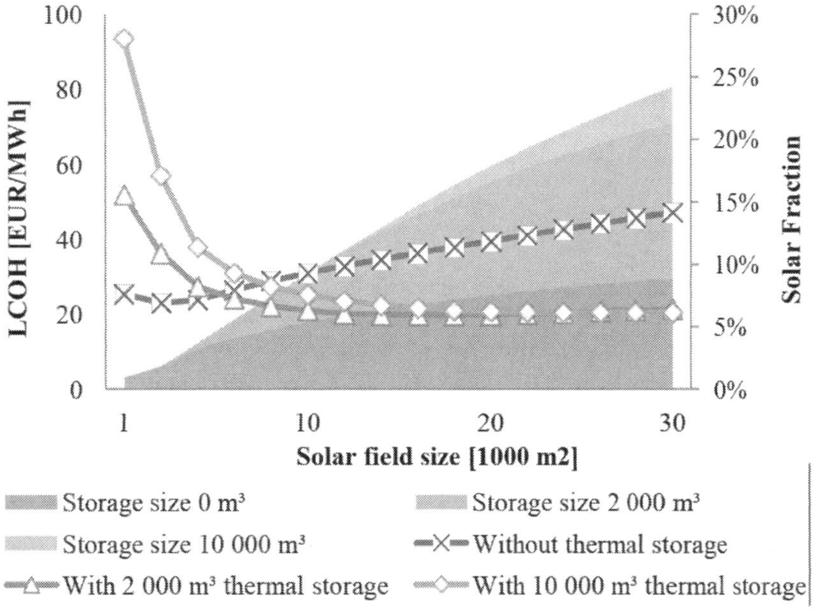

Fig. 1. Exemplary solar field analysis for one sub-network.

For the heat pump the maximum share of heat delivered to the DH network compared to the overall demand reaches a maximum at approximately 85%, which is based on two reasons. The first reason is that the solar field has a lower LCOH than the heat pump. As the final objective set is to minimize cost whenever it is possible, the solar field and storage are preferred in the operation strategy. The second reason is that the demand during very cold hours requires high inflow temperature in the district heating network, which can't be supplied by the heat pump. However, this is the minor reason. Modelling the heat pump without solar fields and heat storages a maximum heat pump fraction of approximately 99% can be achieved based on ambient temperatures in the city in 2011. Referred to the heat produced, the LCOH for the heat pump designed is around 70 EUR/MWh. Including thermal losses in the distribution system the overall LCOH for the combined system (solar + heat pump) is around 75 EUR/MWh. Both based on the demand in 2011.

It is obvious that the design approach chosen tends on the one hand towards relatively small solar thermal collector fields and on the other and towards high capacities for the heat pump. The argumentation for this approach was to fully substitute the heat delivered by the coal fired CHPs by the new units also for present conditions. However, the heat pump has higher LCOH than the solar thermal collector field. Thus lower capacities for the heat pump result in lower overall LCOH for the combined system. This is even more relevant behind the background that the heat demand forecast predicts a substantially falling demand in future. Consequently it is debatable if a smaller heat pump capacity than chosen might make more sense. A better way to size the heat pump with regard to LCOH might be to define the thermal capacity, so that the capacity factor is maximised with regard to the future demand projected (i.e. with regard to the demand in 2030 or 2050).

3 Scenario definitions

Four scenarios described in the following are used to calculate future CO_2 reduction and LCOH for the new units designed (solar + heat pump). The LCOH are calculated for a so

 DOI: 10.1051/e3sconf/20172200005

called socio-economic perspective (1.5% discount rate, no taxes and profit margins, no external costs). Thus, this perspective only reflects cost without taking possible earnings required by operators into account. To paint a whole picture the residual heat demand not covered by the DH network is also taken into account. The assumed share of heat generation technologies supplying the residual (de-central) heat demand is based on projections from 15 for Germany. In general the scenarios differentiate between assumptions on the DH network expansion and mix of supply units. A summary is given in Table .

- **Scenario 1: No additional district heating expansion**. The relative share of connected/supplied houses remains constant as in the present. 95% of the demand in the northern DH network is supplied by the heat pump and solar thermal fields. The southern DH network is supplied by the waste incineration plant. Furthermore the heat demand for residential housing develops as forecasted in 2.2.
- **Scenario 2: Moderate district heating expansion.** The number of connected/supplied houses is increased, so that the final energy demand for district heating from the residential housing in 2050 equals more or less to the demand in 2011. The share of the district heating network increases from 28% in 2011 to 36 and 68% in 2030 and 2050 respectively. Heat generation units and assumed heat demand forecast are the same as in Scenario 1.
- **Forecasted:** The district heating network is not expanded and still supplied by the existing coal-fired CHP plants. The residential heat demand forecast is the same as in Scenario 1.
- **No renovation:** The district heating network is not expanded and supplied by the existing coal-fired CHP plants. The heat demand forecast is not the same as in Scenario 1 as the heat demand stays constant as in 2011. This reflects a scenario where households do not renovate.

Table 2. Scenario overview.

	Heat Sources DH (central)	**Heat Sources de-central**	**Heat demand forecast**	**DH expansion**
Scenario 1	North: 95% HP + solar South: waste incineration	As projected in 15for Germany	Reduction as modelled in 2.2.	no
Scenario 2				yes
Forecasted	As today			no
No renovation			No reduction	no

To calculate the capital cost for the expansion of the district heating network a formula based on Persson et al. 16 is applied, which combines both heat and building density. In Persson et al. the formula is used to analyze potential capital cost for district heating expansion for areas with different heat building densities within a city (plot ratios) under varying market shares for district heating.

4 Results

In Fig. 2 the LCOH for the new system designed and the CO_2 reductions for residential buildings compared to 2011 are presented for the scenarios. It can be seen that the LCOH for the modelled new generation mix (heat pump + solar) strongly depends on the future district heating demand. In Scenario 2, where the density is increased costs are 10 to 35% lower than in the corresponding Scenario 1 with no DH expansion.

 DOI: 10.1051/e3sconf/20172200005

CO_2 reductions range around 60% for 2030 compared to 2011 for scenarios, which take the forecasted demand reduction in residential housing into account (Scenarios 1&2 and Forecasted). From this reduction 33%pts. are due to renovation. Further 26% pts. are mainly due to exchanging de-central heat devises as projected in 15 for Germany. This share is quite large, because in Herten still large shares of residential heating are based on heating oil and coal (21% in 2011). Finally, in Scenario 1&2 further 4%pts. reduction are achieved by the new modelled generation units in 2030.

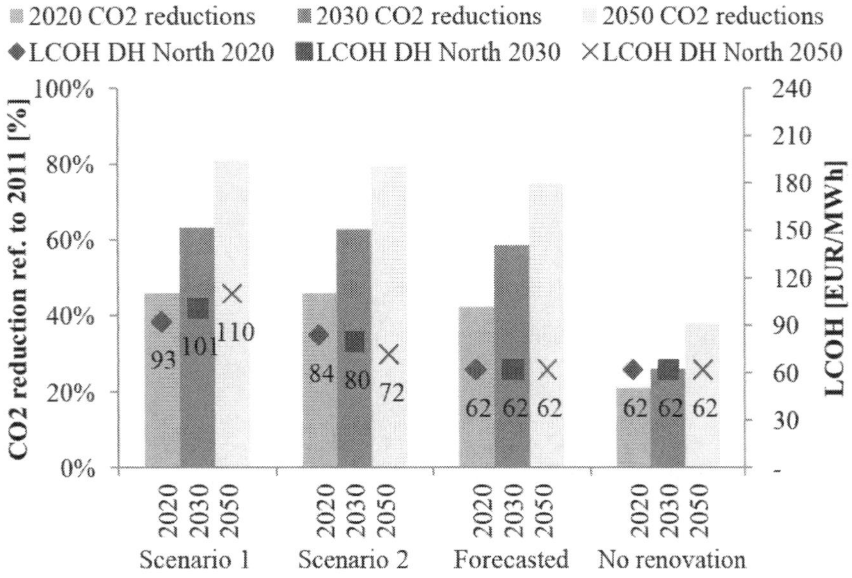

Fig. 2. LCOH for new units and CO_2 emissions reduction per scenario.

5 Conclusions

Up to 2030 and 2050 a substantial reduction in buildings heat demand due to improved building insulation is expected. The falling heat demand in the system can substantially reduce the economic feasibility of new RES generation capacity, particularly for systems with high upfront cost such as heat pumps. Nevertheless, the reduction in heat demand can be compensated by continuously connecting apartment buildings to the DH network as there is still enough heat demand in the whole community. This is a lever to compensate the increasing price. Including solar thermal fields reduces the LCOH substantially compared to using heat pumps only. Anyhow, the LCOH for the combined system is by far higher compared to the reference system. The major reason for that is the relatively high LCOH for the large scale heat pump. Thus, further evaluations should focus on how to reduce LCOH for heat pumps. Such evaluations might cover technical and political means. For example it could be examined how to further improve the overall system efficiency by improving components and system design. However, as in our case study round about two-thirds of the LCOH for the large scale heat pump is determined by electricity cost the relevance of political means should not be underestimated. In the electricity cost assumed several levies are included implicitly by using a statistical value from Eurostat for a group based on consumption ranges. These levies might have large influences on the LCOH calculated. Finally, the gas infrastructure is a major competitor to DH expansion, especially with regard to the erection of new buildings and renovation. As a higher density of the DH

 DOI: 10.1051/e3sconf/20172200005

network decreases the cost of RES in DH by tendency, the competition is from major relevance and should be analyzed more in detail in future for Germany.

Acknowledgements

This paper is based on work packages carried out within the project progRESsHEAT. The project is funded by the the European Union's Horizon 2020 research and innovation programme. Furthermore we thank the anonymous reviewer for the helpful comments increasing the quality of this paper.

References

1. W. Rüdiger, B. Tenberg, *Hertener Klimakonzept 2020+ Ein Masterplan für 100% Klimaschutz in Herten*, Jung Stadtkonzepte (2013)
2. A. Dalla Rosa, R. Boulter, K. Church, S. Svendsen, Energy **45**, 960–974 (2016)
3. M. Köfinger, D. Basciotti, R.R. Schmidt, E. Meissner, C. Doczekal, A. Giovannini, Energy **110**, 95–104 (2016)
4. Homepage of Invert/EE-Lab model, Internet http://www.invert.at/ (accessed on 23.02.2017)
5. TIM-online, Geographical Information System, Internet: http://www.tim-online.nrw.de/tim-online/initParams.do?role=default (accessed on: 23.02.2017)
6. Institut für Wohnen und Umwelt, Deutsche Gebäudetypologie – Systematik und Datensätze (2005)
7. Statistisches Bundesamt: Zensus 2011: Gebäude und Wohnungen (2014). Internet: https://www.destatis.de/DE/Methoden/Zensus_/Zensus.html(accessed on: 23.02.2017)
8. Homepage of energyPRO model. Available at: http://www.emd.dk/energypro/ (accessed on 23.02.2017)
9. J. Conrad, T. Schmid, C. Pellinger, Erzeugung von Fernwärmelastgängen aus Temperaturprofilen, Energiewirtschaftliche Tagesfragen **9**, 52–54 (2015)
10. J.E. Nielsen, R. Battisti, Solar district heating guidelines, Feasibility study, (2012), Internet: www.solar-district-heating.eu (accessed on: 23.02.2017)
11. Aktuelle Informationen zum Immobilienmarkt in NRW: http://www.boris.nrw.de/ (accessed on: 23.02.2017)
12. C. Geimer, Thermal heat from pit water, FIZ Karlsruhe (2013)
13. S. Foster, J. Love, I. Walker, M. Crane, Heat Pumps in District Heating – Case Studies, DEEC (2016)
14. The Danish Energy Agency, Technology data catalogue, downloaded in March (2016) Internet: https://ens.dk/en/our-services/projections-and-models/technology-data
15. Öko-Institut; Fraunhofer ISI: Klimaschutzszenario 2050. 1. Modellierungsrunde. Hg. v. Bundesministerium für Umwelt, Naturschutz, Bau und Reaktorsicherheit. Öko-Institut e. V.; Fraunhofer-Institut für System- und Innovationsforschung ISI (2014)
16. U. Persson, S. Werner, Appl. Energy **88**, 3, 568–576 (2011)

A new test procedure to evaluate the performance of substations for collective heating systems

Robin Baetens[1,*], and *Ivan* Verhaert[1]

[1]University of Antwerp, Faculty of Applied Engineering, Energy and Material in Infrastructure and Buildings, Campus Groenenborger, Groenenborgerlaan 171, 2020 Antwerp, building Z

Abstract. The overall heat demand of a single dwelling, existing out of space heating and domestic hot water production, decreases due to higher insulation rates. Because of this, investing in efficient and renewable heat generation becomes less interesting. Therefore, to incorporate renewables or residual heat on a larger scale, district heating or collective heating systems grow in importance. Within this set-up, the substation is responsible for the interaction between local demand for comfort and overall energy performance of the collective heating system. Many different configurations of substations exist, which influence both local comfort and central system performance. Next to that, also hybrids exist with additional local energy input. To evaluate performance of such substations, a new experimental-based test procedure is developed in order to evaluate these different aspects, characterized by the two roles a substation has, namely as heat generator and as heat consumer. The advantage of this approach is that an objective comparison between individual and central systems regarding performance on delivering local comfort can be executed experimentally. The lab set-up consists out of three different subsystems, namely the central system, the domestic hot water consumption and the local space heating. The central system can work with different temperature regimes and control strategies, as these aspects have proven to have the largest influence on actual performance. The domestic hot water system is able to generate similar tap profiles according to eco-design regulation for domestic hot water generation. The space heating system is able to demand a modular heat load.

* Corresponding author: robin.baetens@uantwerpen.com

1 Introduction

Today, we face great challenges concerning energy. The consequences of climate change and the depletion of fossil fuels like oil and natural gas, have led to following commitments of the EU, compared to 1990 [1].

- increasing the share of renewable energy sources by 20% in 2020 and by 27% in 2030
- increasing energy efficiency by 20% in 2020 and by 27% in 2030
- reducing emissions by 20% in 2020, by 40% in 2030 and by 80–95% in 2050

As heat is responsible for about 40-50% of our energy usage, increasing the overall efficiency of this heat utilization is crucial in order to achieve these objectives. A first step should always be reducing demand, e.g. by improving the insulation rate of our building stock. Secondly, the heat generation has to become more sustainable by increasing the share of renewables and/or its efficiency. However, with a decreasing heat demand, investments in efficient local heat generators become less beneficial. From this point of view a, rather 'old', heating design strategy becomes an option again, namely collective heating systems or district heating.

It has been shown that if the potential of district heating will be fully exploited, the target of 2050 of the EU to reduce energy consumption can be realized, even with a cost reduction of 15% for the heating and cooling demands of buildings [2]. If the combustion systems at the end users can be eliminated, this will also mean a safer heat production (no more gas in city and buildings) and less space occupation. On top of that, local pollutants will decrease, which will have a positive effect on air quality in city areas and hence health of the inhabitants [3]. Next, district heating enables the use of excess heat from e.g. industrial processes. With the city of Antwerp as the example par excellence, it is not surprising, Persson et al. (2014) found a very high spatial correlation of 46% between heat excess and heat demand in the EU27. Also, in terms of storage, thermal grids have an advantage with the inexpensive storage methods for heat (compared to e.g. batteries in an electric grid).

With respect to the acceptance of collective heating systems, worries arise whether collective heating systems are effective in reaching local comfort demand. The substation is the link between this local comfort demand and energy efficient collective heating system.

As can be expected, the type of substation will have a large influence on both local comfort (meeting heat demand with little time delay) and on energy efficiency of the district heating network (ensuring and/or allowing low distribution temperatures and efficient control strategies).

A diversity of products are available on the market. Therefore, it is necessary to develop a proper test procedure, which addresses the different KPIs a substation should meet, allowing a proper comparison between different types and if possible to compare them with local heat generation.

In this paper a lab set-up is built which allows to execute different test procedures for substations. Special attention is paid to evaluate dynamic behaviour in order to gain the necessary parameters to model different kinds of substations within a heating network. Next to that, to allow comparison with local heat generators the set-up is built in such a way, that they can be tested under the same circumstances.

After the design of a lab set-up, different test procedures are evaluated.

 DOI: 10.1051/e3sconf/20172200006

2 Description of the test facility

Originally a test facility was built to test heat generators according to NBN EN 13203-1 [5] and 2 [6]. Acknowledging the lack of in-depth test results for substations, the test facility was rebuilt and expanded. The present lab set-up consists out of three different subsystems, namely the central system, the domestic hot water consumption and the local space heating. The central system (CS) should be able to simulate behaviour of different kinds of collective heating networks. Therefore, it is crucial that it can work with different temperature regimes and control strategies, as these aspects have been proven to have the largest influence on actual performance. To evaluate end-use comfort, the domestic hot water system (DHW) is able to generate similar tap profiles according to eco-design regulation for domestic hot water generation. The space heating system (SH) is able to demand a modular heat load.

2.1 Central system

A small household combi boiler with an approximated output of 30 kW provides energy to a 150 l buffer. A 0…10V regulated circulation pump provides different flow rates. Through smart switching of ball valves different pipe lengths varying from 10 m to 50 m can be added. These variations simulate the effects of long distant stations and how the supply/return temperature varies over time with a fluctuating thermal load. Based on NBN EN 13203, sensors are placed as close as possible to the station eliminating the need for calculating losses in the central or DHW system. The sensors of each subdivision consist of a rapid response thermometer in each line, a differential pressure transmitter between supply and return, a flow transmitter in the hot line and finally for backup a heat meter installed in the cold line. The combination of high precision temperature and flow sensors allows the calculation of thermal energy flow, backed up by the heat meter, which is more accurate, but has a lower measuring frequency, approximately 16 s between refreshes compared to 1s. The measurement of the paired differential pressure offers insight in the pumping losses due to the substation.

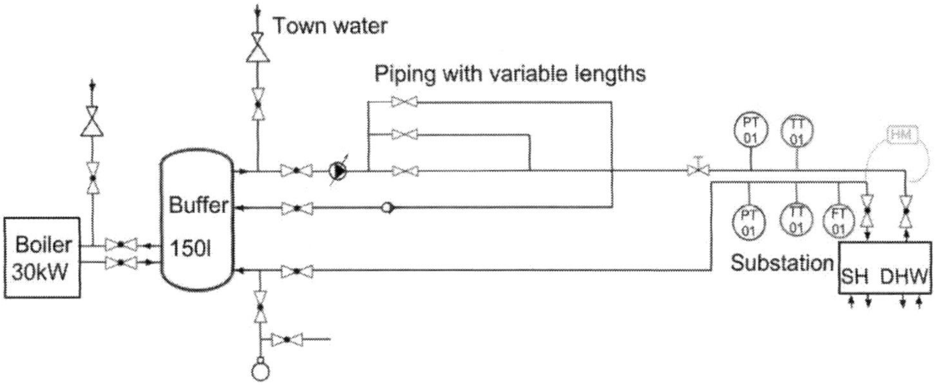

Fig. 1. Primary circuit, heat supply central system.

 DOI: 10.1051/e3sconf/20172200006

2.2 Domestic hot water consumption

Using a PLC in the configuration, it is possible to simultaneously read sensor values and drive actuators. Through electromagnetic valves and mechanically adjustable dynamic flow restrictors, a tapping cycle, as described in NBN EN 13203-2 is applied. Besides the energy performance at the selected tapping cycle, also other aspects can be examined referring to the behaviour of the unit at comfort level. For instance when the unit gets temperature shocked, albeit from a cold start or with a constant withdrawal of SH.

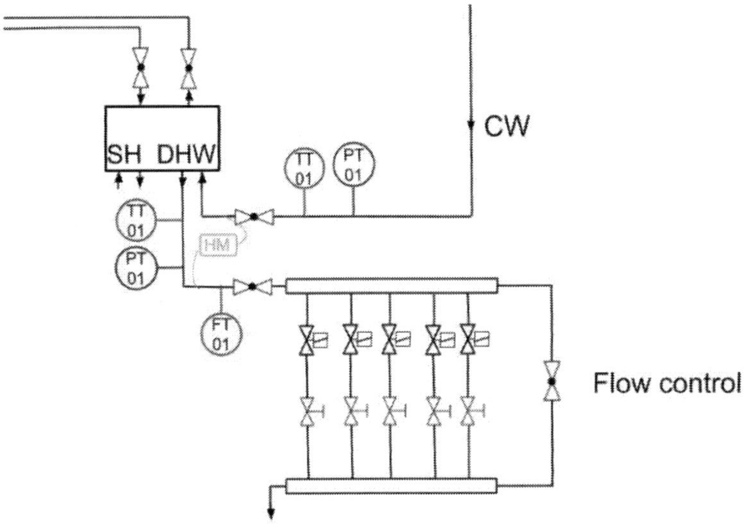

Fig. 2. Secondary circuit, Domestic hot water consumption.

Cold water (CW) needs to be provided at a constant temperature of 10 ± 2°C. A 300 l buffer is kept at a temperature close to 10°C. To maintain a uniform temperature without stratification, a circulation pump is powered when no withdrawal is made.

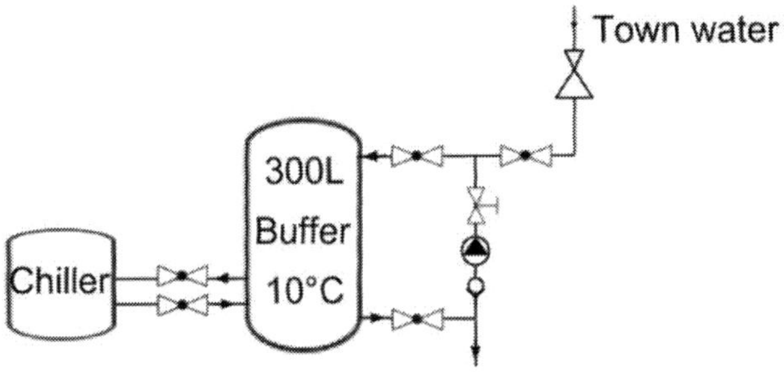

Fig. 3. Secondary circuit, Controlled cold water supply.

 DOI: 10.1051/e3sconf/20172200006

2.3 Local space heating

Fan coils placed in a cold storage, kept at a fixed temperature, simulate the local space heating demand. Two fan coils at maximum fan speed can each subtract approximately 3kW, a third fan coil at maximum speed can subtract an additional 5 kW. Each fan coil has three different fan speeds, each demanding a different flow rate of hot water (HW). Prior to the start of the test a fan speed of minimum, medium or maximum is selected. In initial test phases, fan speed is adjusted manually. Hydronicaly, the same dynamic valves are used as for the tapping cycles, with the minor adjustment of an additional electronic actuator. Using additional flow transmitters as feedback for the PLC, automated adjustments can be made to control the dissipation of heat. Future adjustments shall, if proven helpful, entail a stepless adjustment of fan speeds. This upgrade will allow us to continuously adjust the heat dissipation between 2 kW and 11 kW, instead of selecting one of 27 combinations.

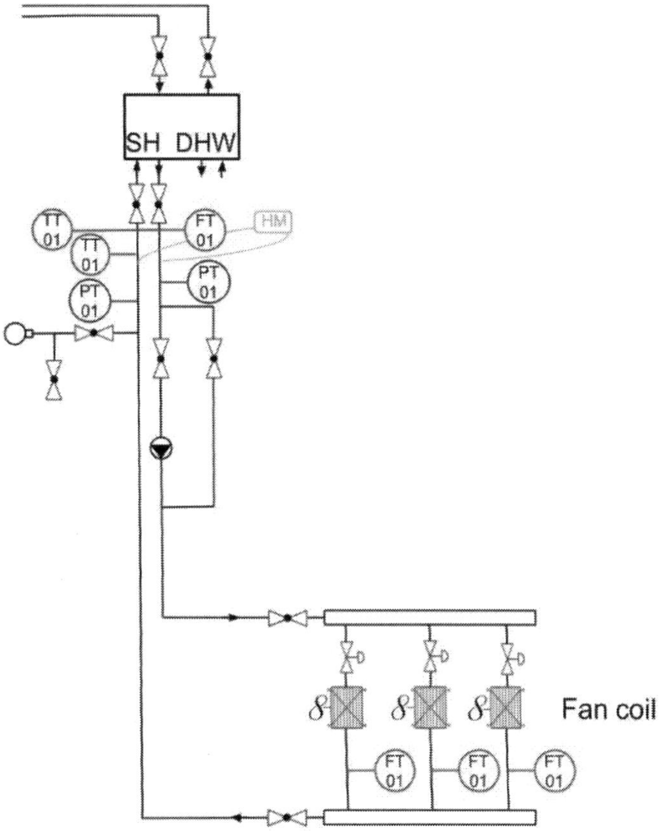

Fig. 4. Secondary circuit, Local space heating.

3 Test procedure

As stated previously a test method has been standardized in NBN EN 13203 for testing gas fired appliances. For space heating (SH) the existing standard (test at 30% and 100% nominal load) provides little insight in the dynamic behaviour of the heat generator.

Also how a heat generator addresses a DHW demand combined with a heat demand for SH, is not yet described. For this reason, a new method was developed. In general, the test procedures can be partitioned in two segments. One being assessments of performance of hot water deliveries and the other being assessment of energy consumption. Basically still referring to NBN EN 13203-1 [5] and 2 [6], respectively, but with the addition of a withdrawal of energy for space heating. An additional and highly influential parameter is the temperature regime at which the central supply works. The required electrical energy for the controller of the substation is also logged using a Beckhoff component.

3.1 Assessments of performance of hot water deliveries

An assessment of performance of hot water delivery can be described as how the unit reacts temperature wise and timewise to different variations in boundary conditions. Two main directions can be investigated. First maintaining a constant withdrawal of SH and allowing the DHW to vary, secondly, maintaining DHW and adjusting the heat dissipation. Investigating the units behaviour i.e., temperature fluctuations in SH, required heat-up time for DHW and its ability to uphold this temperature is a straight forward indication of the units comfort capacity. In order to make a correct assumption, different temperature regimes of the primary circuit also need to be included as a boundary condition. Influenced by the manufacturer's control strategy and individually determined for each test case, flexibility towards lower temperatures in the central supply is investigated.

3.1.1 Fixed SH, varying DHW

Based on a test prescribed towards independent test laboratories it is of interest that the return temperature of the central system remains above a certain threshold. In this case a central supply of 80°C and a return of 30°C with a maximum of 40°C needs to be maintained. The required power to dissipate is, prior to test, adjusted accordingly. A test sequence exists of doubling the flow rate at the DHW supply every minute until reaching the maximum applicable flowrate provided by the manufacturer, followed by halving the flow rate every minute until zero. This for no longer than seven minutes. The test sequence to impose is repeated for four times over the course of 24 h.

A revision to this procedure is to use a fixed value for heat dissipation instead of a 'fixed' return temperature. To determine this value a typical apartment dwelling has been chosen in which the living room, bath room and bedroom are being heated. With an extreme outside temperature of -8°C, a required power of 2,5 kW; 1 kW and 0,5 kW results in a total decrease of 4kW to be withdrawn from the fan coils [7]. Test procedures described in NBN EN 13203-1 are applied and all data is logged for 24 h.

3.1.2 Fixed DHW, varying SH

The same principle stands as with previous methods, but instead of an oft-recurring condition, being tapping of DHW, a slower les frequent occurring of added temperature shock by extra heat dissipation is investigated. Essentially adding a room to be heated or raising the requested room temperature. Though not the most acclaimed results, they can teach us more about the behaviour of the applied priority regulation of the unit and thereby its comfort level.

3.2 Assessment of energy consumption

As the assessment of performance teaches us how the unit reacts comfort wise, the assessment of energy consumption teaches us how efficient the unit is. Similar to NBN EN 13203-2, a tapping profile of different energy withdrawals is applied to the substation's DHW supply, as stated in 1.2. These tapping profiles are designed to match behaviour of certain sized families and dwellings and are divided starting from small (S) to four times extra-large (4XL). In a certifying environment it is the choice of the manufacturer to determine which size pattern to apply. In a research environment different variations are put to test.

By measuring input and output heat, efficiency and heat losses of the unit can be determined. By either choosing to apply SH, a prediction of summer and winter efficiency can be determined.

3.3 Cold start

A final assessment to be made is how the unit reacts from a cold start, albeit through comfort or efficiency. Imagine coming home from a long holiday and immediately wanting to heat up the house and take a shower at the same time. What will receive priority, how will return temperatures react, how fast can the unit heat up and what is its efficiency during warmup?

4 Conclusion

The construction and provision of considered test facility proves to fill up a highly anticipated gap in the research fields as well as with manufacturers.

A lab set-up is developed, which is able to simulate a high number of variations in control strategies and district heating lay-outs. Besides this, the set-up allows comparison with local heat generators.

A number of test procedures are evaluated which can be implemented to gather realistic data on dynamic behaviour of substations, allowing use in simulations.

Comparable studies will not only result in more energy efficient units, more appropriate dimensioning and selecting will result in more energy efficient buildings together with more accurate building simulation programs. Combined with a well dimensioned and fine-tuned Thermal Energy Storage tank (TES) these improvements will allow for a far more energy efficient implementation of hybrid configurations with renewable energy sources, especially for individual district heating systems.

References

1. EC, "Horizon 2020 Work Programme: Secure, clean and efficient energy" (2015)
2. D. Connolly, H. Lund, B.V. Mathiesen, S. Werner, B. Möller, U. Persson, T. Boermans, D. Trier, P.A. Ostergaard, S. Nielsen, Ener. Pol. **65** (2014)
3. M. A. Ancona, M. Bianchi, L. Branchini, F. Melino, Ener. Proc. **45** (2014).
4. U. Persson, B. Möller, S. Werner, Ener. Pol. **74** (2014)
5. nbn, "NBN EN 13203-1: Gas fired domestic appliances producing hot water - Part 1: Assessment of performance of hot water deliveries," **1** (2015)
6. nbn, "NBN EN 13203-2: Gas-fired domestic appliances producing hot water - Part 2: Assessment of energy consumption," **2** (2015)
7. F. Van Riet, H. El Khaoui, F. Hulsbosch, G. Steenackers, and I. Verhaert,. "Exploring the novel software Hysopt : a comparison of hydronic heat distribution systems of an apartment building", REHVA Aalborg (2016)

 DOI: 10.1051/e3sconf/20172200007

Influence of silane on the structure of polystyrene prepared by sol-gel coatings via UV curing

Senay Balbay[1,*], *and Caglayan* Acıkgoz[2]

[1]Bilecik Seyh Edebali University, Vocational School. Department of Chemical Technologies, 11030 BİLECİK, Turkey
[2]Bilecik Seyh Edebali University, Faculty of Engineering, Department of Chemistry and Process Engineering, 11030 BİLECİK, Turkey

Abstract. Light, heat, oxygen, moisture, ozone, atmospheric pollution and biological effects are the most important effectives wreak to chemical degradation in the polymer structure. In result of chemical degradation on the polymer consist of problems such as discoloration, brittleness, surface cracks, perspiration, crumbling, smell, surface acidity. In this work, it is aimed to improve the problem of the polystyrene (PS) material against chemical degradation. For this reason, PS is coated with silica sol-gel hybrid coating. Silica sol-gel was synthesized by using vinyltrimethoxysilane (VTMS) as a cross-linker and tetraethylorthosilicate (TEOS) as a silica source. Firstly, four different pre-treatment technique (oven, vacuum oven, lyophilizer and freezing) was studied to determine the most suitable pre-treatment technique for coating on PS substrate of sol-gel prepared with initial formulation (S1). A freezing technique gave the best results for coating sample. The change of surface colour of coated PS was measured by CIE L*a*b* methods. Secondly, the most suitable curing agent (Irgacure 184, Irgacure 819, Darocur 1173 and TiO_2 as crystalline anatase phase) was determined to coat the sol-gel on PS. It was determined to the lowest yellowing of PS surface hybrid coated as UV curing of TEOS sol modified by VTMS and TiO_2 as photo-initiators. Finally, the chemical and morphological structure of the coated PS samples was determined by FT-IR and SEM instruments,respectively.

1 Introduction

Today plastic materials are widely used similarly by the automotive, building, white goods and household appliance industry, with a use that has constantly grown over the last few years. The use of plastics in building applications is popular in the developing world because of the low cost and the ease of use of plastic components compared to other materials. One of the disadvantages of using polymers is that they degrade when they are

*corresponding author: senay.balbay@bilecik.edu.tr

used in high temperature conditions or in outdoor applications. When polymers are used in outdoor applications, the environment negatively influences the servicelife. Polymer degradation can be caused by heat (thermal degradation), light (photodegradation), ionizing radiation (radio degradation), mechanical action or by fungi, bacteria, yeasts, algae and their enzymes (biodegradation)[1].Plastics sensitive to high temperatures (e.g. PMMA, PET) have to be coated by UV-curing systems since the thermally cured systems described show only poor results when cured below 100°C [2].Today's hybrid coatings are among other methods also prepared by the sol–gel process used of silica aerogels (silica nanoparticles, siliconalkoxides) and used for many different applications in material science and engineering [3]. Hybrid materials for coating applications can exhibit properties such as optical transparency, specific electrical and mechanical behavior, thermal and weathering resistance, abrasion and impact resistance [4].

Feng et all were studied an immunoassay for bisphenol a based on direct hapten conjugation to the polystyrene surface of micro titer plates. Recoveries were between 70% and 142% [5]. Deng et all were presented A facile method of coating polystyrene (PS) particles with organo-silica and their functionalization. By adding the organo-silane precursor into PS aqueous solution in presence of ammonia, an organo-silica shell could be coated on PS particles directly [6].

The object of this study was to improve the ability to yellowing of PS against chemical degradation by coating on PS substrate of the silica sol-jellies synthesized using tetraethylorthosilicate(TEOS) as silica source and vinyltrimethoxysilane (VTMS) as coupling agent.

2 Experimental

Materials: Titanium (IV) isopropoxide $Ti[OCH(CH_3)_2]_4$ as crystalline anatase phase was supplied by Fisher Scientific. TEOS (tetraethylorthosilicate) as silica source and vinyltrimethoxysilane (VTMS) as coupling agent were provided by abcr - Gute Chemie. Darocur 1173 (2-hydroxy-2-methyl-1-phenylpropan-1-one), Irgacure 184 (1 hydroxycyclohexyl phenyl ketone) and Irgacure 819 (phosphine oxide, phenyl bis(2,4,6-trimethyl benzoyl) as photoinitiator were provided by Ciba Specialty Chemicals.

Table 1. Formulation of UV-curable hybrid coatings.

Components	S1	S2	S3	S4	S5
TEOS (ml)	1.85	3.70	3.70	3.70	3.70
VTMS (ml)	1.85	-	-	-	-
C_2H_5OH (ml)	9.8	9.8	9.8	9.8	9.8
1 M HCl (ml)	1.4	1.4	1.4	1.4	1.4
H_2O (ml)	0.8	0.8	0.8	0.8	0.8
Irgacure 184 (g)	1	1	-	-	-
Irgacure 819 (g)	-	-	1	-	-
Darocur 1173 (g)	-	-	-	1	-
TiO_2 (g)	-	-	-	-	1
1 M NH_4OH (ml)	1.6	1.6	1.6	1.6	1.6

Hybrid sol synthesis: The sol-gel precursor was prepared by using TEOS, VTMS, ethanol as solvent, distilled water for hydrolysis, 1 M HCl as a catalyst, 1 M NH_4OH for gelling and photo-initiators for UV curing. The compositions of all formulations are also given in Table 1. In the first step of the sol synthesis, partial hydrolysis and condensation of the organoalkoxysilane was carried out under an acid catalysis of 1 M HCl, distilled water, TEOS or VTMS and ethanol with vigorous stirring. The mixing for sol formation was hold at 65°C for 3 hours in oven. Secondly, a photo initiator for UV curing put to the mixture. The mixture was magnetically stirred at room temperature. Then 1 M NH_4OH was added to the mixture and stirred at room temperature until a gel solution was obtained. Thirdly, the gel solution for aging was washed with 5 ml distilled water and put on 4 ml ethanol to solution. Then, the solution was aged at 40°C for 3 days in oven [7].

Coating procedure: Polystyrene(PS)test substrates (75 mm × 100 mm × 4 mm) were used as substrates in all coating applications. Surface of polystyrene test substrates were cleaned with methanol prior to coating. Five formulations represented in Table1 were applied on the PS substrates using an applicator. The pretreatment technique was carried out to the coating substrates and the substrates were irradiated by using a high pressure UV-lamp [(Osram, 300 W), radiation intensity of 1.55 $W/m^2/nm$. λ max = 340 nm] at 40°C for 1 hour.

Method: Experimental studies were carried out in two series studies. Firstly, four different pre-treatment technique (oven, vacuum oven, lyophilizer and freezing) was studied to determine the most suitable pre-treatment technique for coating on PS substrate of sol-gel prepared with initial formulation (S1).The change of surface colour of coated PS was measured by CIE L*a*b* methods. The oven pre-treatment was carried out at 60°C for 24 hours. The vacuum oven pre-treatment was carried out at 40°C for 24 hours. The lyophilizer pre-treatment was applied for 24 hours. The freezing pre-treatment was performed at – 80°C for 24 hours. Secondly, the most suitable curing agent (Irgacure 184, Irgacure 819, Darocur 1173 and TiO_2 as crystalline anatase phase) was determined to coat the sol-gel on PS. The uncoated PS substrate was determined which yellowing for 26 hours by using ATLAS UV test. For this reason, the yellowing analysis for 26 hours by using ATLAS UV test was applied to the all coating substrates. Finally, surface chemical modification, surface morphology and change of surface colour of coated PS samples were determined by FT-IR, SEM and CIE L*a*b* methods spectrophotometer, respectively.

Characterization: The colorimetric valuesof yellowing color changes of virgin PS (PS), yellowing virgin PS(YPS) and after yellowing analysis and curing of the hybrid material coating test substrates (PS-$S1_{oven}$, PS-$S1_{vacuum\ oven}$, PS-$S1_{lyophilizer}$, PS-$S1_{freezing}$, PS-S2, PS-S3, PS-S4, PS-S5)in theCIE Lab color space were measured by on a Color-Eye automatic differential colorimeter (CM-2600d. Konica Minolta Optics Inc. Japan).The color difference between the test substrates was calculated by the formula $\Delta E = [(\Delta L^*)^2 + (\Delta a^*)^2 + (\Delta b^*)^2]^{1/2}$. $\Delta L^* = L_{ref} - L_{coated}$. , $\Delta a^* = a_{ref} - a_{coated}$, $\Delta b^* = b_{ref} - b_{coated}$ (ref: reference sample and coated: coated PS) [8]. In this color measurement system, parameter L is the color lightness (L = 0 for black and L=100 for white), a is the green(-)/red(+) axis, and b is the blue(-)/yellow (+) axis[9].Changes of chemical structure of PS, YPS andPS-$S1_{oven}$, PS-$S1_{vacuum\ oven}$, PS-$S1_{lyophilizer}$, PS-$S1_{freezing}$, PS-S2, PS-S3, PS-S4, PS-S5were determined by ATR-FTIR spectroscopy. Spectra were measured in the range of 400 to 4000 cm^{-1} with a Perkin Elmer Spectrum 100 instrument. The morphological structure of PS, YPS and PS-$S1_{oven}$, PS-$S1_{vacuum\ oven}$, PS-$S1_{lyophilizer}$, PS-$S1_{freezing}$, PS-S2, PS-S3, PS-S4, PS-S5 were studied using the following electron microscopes: SEM-ZEISS Supra 40VP.

 DOI: 10.1051/e3sconf/20172200007

3 Results and Discussion

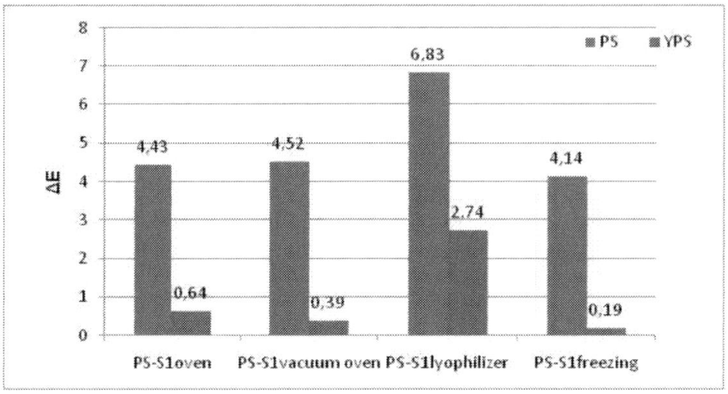

Fig. 1. Colour differences of hybrid coatings of the pre-treatment techniques (to PS and YPS).

Color differences on hybrid coatings of the pre-treatment techniques are showed in Fig. 1.The colorimetric values of yellowing color changes of virgin PS (PS), yellowing virgin PS(YPS) and coated on PS substrate of sol-gel prepared with initial formulation (PS-S1) were measured. ΔE value according to PS of YPS was determined as 4.24. The ΔE value of PS-S1$_{freezing}$ according to PS is lower than the ΔE value of YPS according to PS. In addition, the ΔE value of PS-S1$_{freezing}$ according to YPS is the lowest according to ΔE value of other pre-treatment applications.

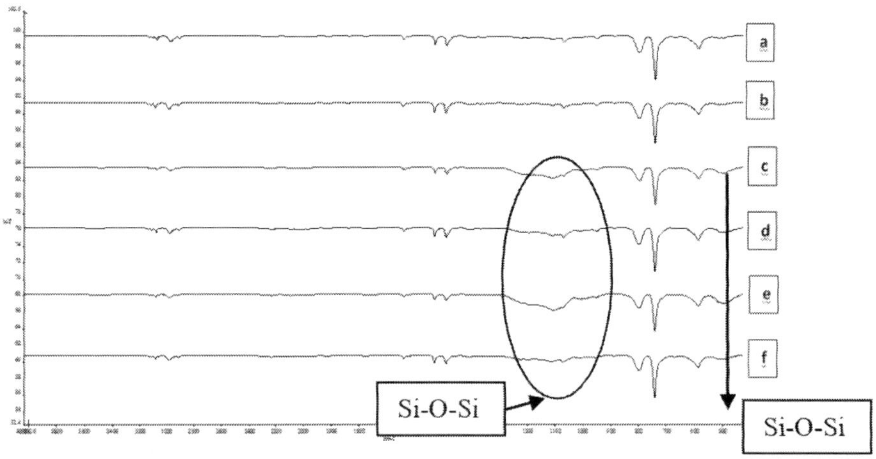

Fig. 2. FT-IR spectra of hybrid coatings of the pre-treatment techniques (a: PS, b:YPS, c: PS-S1$_{oven}$, d: PS-S1$_{vacuum\ oven}$, e:PS-S1$_{lyophilizer}$, f: PS-S1$_{freezing}$)

FT-IR spectra of hybrid coatings of the pre-treatment techniques are given in Fig. 2. The characteristic stretching vibration and bending vibration peaks of Si–O–Si occurred at 1077 cm^{-1} and 465 cm^{-1}, respectively [15]. These peaks were not observed in the YPS and PS spectra. However, vibration peaks of Si–O–Si were observed in PS-S1$_{oven}$, PS-S1$_{vacuum\ oven}$, PS-S1$_{lyophilizer}$, PS-S1$_{freezing}$ surfaces applied of the all pre-treatment.

 DOI: 10.1051/e3sconf/20172200007

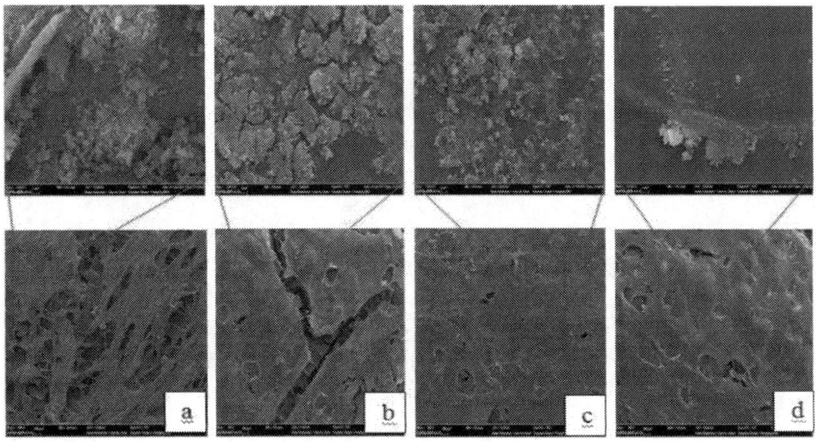

Fig. 3. SEM images of hybrid coatings of the pre-treatment techniques (a: PS-S1$_{oven}$, b: PS-S1$_{vacuum\ oven}$, c:PS-S1$_{lyophilizer}$, d: PS-S1$_{freezing}$ and top: 5000 mag., bottom: 100 mag.).

SEM images of hybrid coatings of the pre-treatment techniques are given in Fig. 3. Accordingly, it was observed to more homogenous of PS-S1$_{freezing}$ surface applied of the freezing pre-treatment.

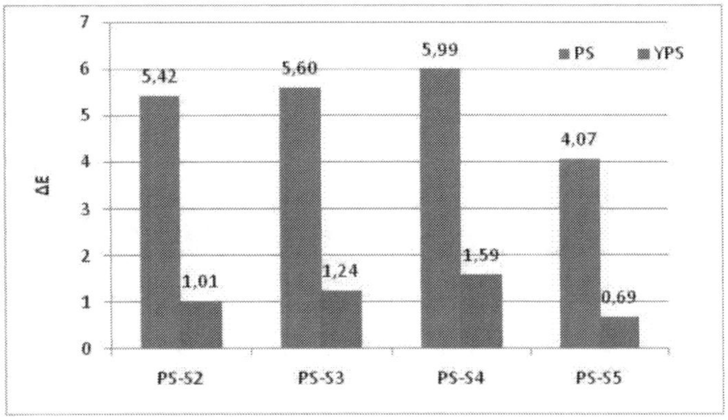

Fig. 4. Colour difference of the photo-initiator for UV curing on hybrid coatings (to PS and YPS).

Color differences of the photo-initiator for UV curing on hybrid coatings are showed in Fig. 4. ΔE value of PS-S5 used of TiO_2 as photo-initiator according to PS and also YPS has the lowest value.

FT-IR spectra of hybrid coatings of the different photo-initiators for UV curing are given in Fig. 5. The vibration peaks of Si–O–Si were also observed in PS-S5 surfaces prepared with different photo-initiators as well as PS-S1$_{oven}$, PS-S1$_{vacuum\ oven}$, PS-S1$_{lyophilizer}$, PS-S1$_{freezing}$ surfaces applied of the all pre-treatment.

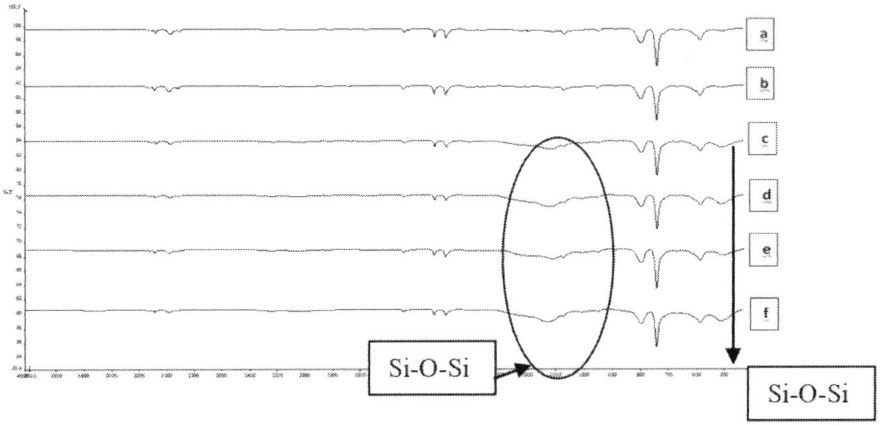

Fig. 5. FT-IR spectra of hybrid coatings of the different photo-initiators for UV curing (a: PS, b: YPS, c: PS-S2, d: PS-S3, e:PS-S4, f: PS-S5).

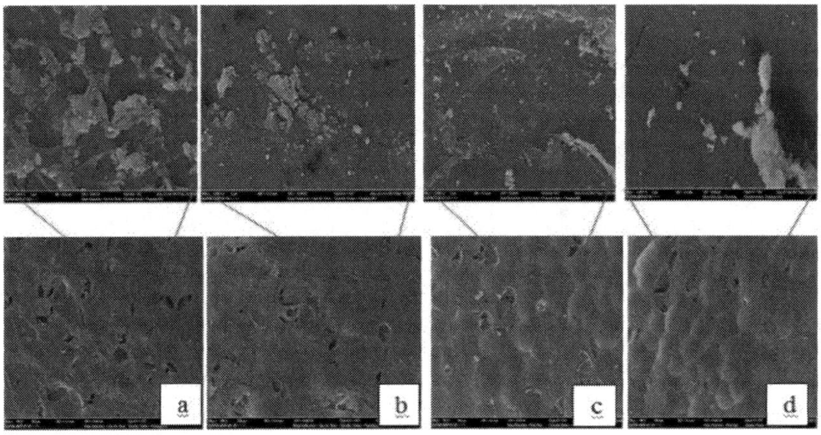

Fig. 6. SEM images of hybrid coatings of the different photo-initiators for UV curing (a: PS-S2, b: PS-S3, c:PS-S4, d: PS-S5 and top: 5000 mag., bottom: 100 mag.).

SEM images of hybrid coatings of the different photo-initiators for UV curing are given in Fig. 6. Accordingly, it has been observed more homogeneous surface in PS-S5 used of TiO_2 as a photo-initiators.

Colour differences in PS surfaces of yellowing analysis and coated from only TEOS sol and TEOS sol modified by VTMS are given Fig 7.ΔE value of PS-S5 by TEOS sol as a silica source according to PS is lower than ΔE value of PS-S1 by TEOS sol modified VTMS as coupling agent according to PS. However, when ΔE values of the hybrid coating are compared with ΔE value of YPS, ΔE value of PS-S1 by TEOS sol modified VTMS is lower. In study, Aydemir was determined that surface roughness had a significant effect on color change [10]. ΔE value of PS-S1 by TEOS sol according to PS is low due to the roughness of the cracks on the surface.

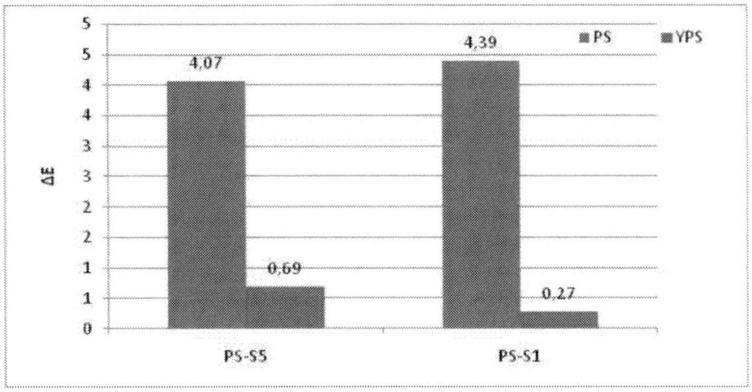

Fig. 7. Colour differences of the TEOS sol (PS-S5) and TEOS sol modified by VTMS(PS-S1) (to PS and YPS).

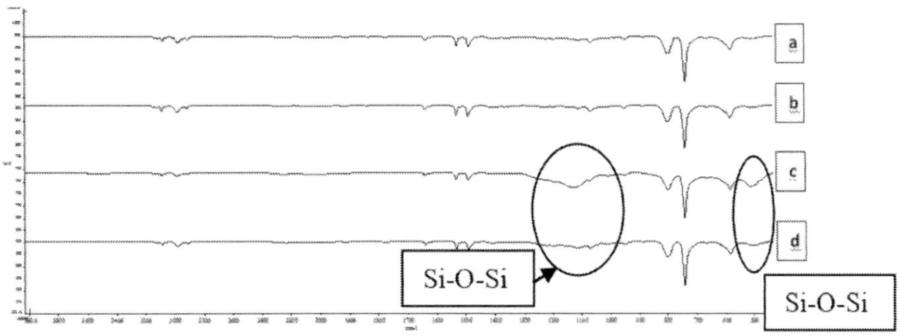

Fig. 8. FT-IR spectra of hybrid coatings of TEOS sol and TEOS sol modified by VTMS (a: PS, b: YPS, c: TEOS(PS-S5), d: TEOS+VTMS(PS-S1)).

FT-IR spectra of hybrid coatings of TEOS sol and TEOS sol modified by VTMS are showed in Fig 8. It was observed that the vibration peaks Si - O - Si were more significant in PS-S5surfaces prepared with TEOS sol in Fig 8(c). Xiao and Hao determined that the vibration peaks Si - O - Si were more specifically to not occur of three dimensional network in nano silica sol modified by 3-Trimethoxysilyl-propyl methacrylate (TMSPM) as silicon coupling agent [11]. In Fig. 8d was observed that the vibration peaks Si - O - Si were less obvious in PS-S1 by TEOS sol modified by VTMS. This shows to occur of three dimensional network in PS-S1 by TEOS sol modified by VTMS.

 DOI: 10.1051/e3sconf/20172200007

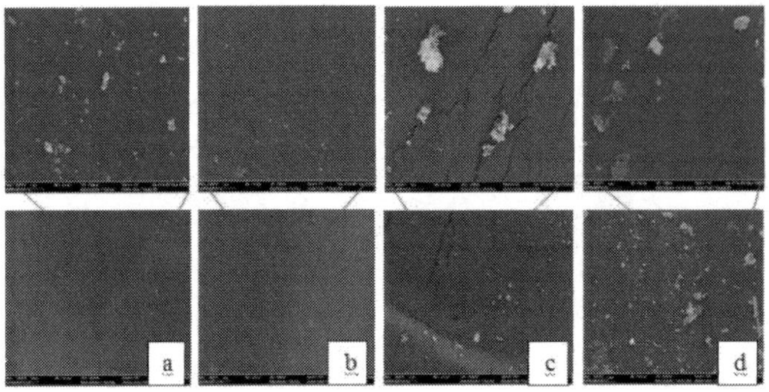

Fig. 9. SEM images of hybrid coatings of TEOS sol (PS-S5) and TEOS sol modified by VTMS (PS-S1)(a: PS, b: YPS, c: TEOS, d: TEOS+VTMS and top: 20000 mag., bottom: 5000 mag.).

SEM images of hybrid coatings of TEOS sol and TEOS sol modified by VTMS are given in Fig 9. Accordingly, it was determined that cracks were formed on surface of PS-S5 by TEOS sol. It has been observed that a crackless and homogeneous surface formed on surface of PS-S1 by TEOS sol modified by VTMS as coupling agent.

4 Conclusion

This study was to improve the ability to yellowing of PS against chemical degradation by coating on PS substrate of the silica sol-jellies synthesized using tetraethylorthosilicate (TEOS) as silica source and vinyltrimethoxysilane (VTMS) as coupling agent. It was determined to the lowest yellowing of PS surface hybrid coated as UV curing of TEOS sol modified by VTMS and TiO_2 as photo-initiators. In result of yellowing analysis, this hybrid coating is crack-free and homogenous due to effect cross-linking of VTMS. Plastic materials such as PS have low surface energy, surface energies are increased by corona application before bonding or coating process. In this study, the hybrid coatings were successfully coated onto the PS surface under UV light without the corona application on the PS surface. The hybrid coated PS of TEOS sol modified by VTMS as applied by freeze pre-treatment and UV curing can be improve that the various degradation problems for different areas used of PS such as automotive, building, white good.

References

1. E. Yousif, R. Haddad, Yousif and Haddad Springer Plus, **2** (2013)
2. S. Sepeur, N. Kunze, B. Werner, H. Schmidt, UV curable hard coatings on plastics
3. K. Khezri, H. Mahdavi, Microp. and Mesop. Mat. **228** (2016)
4. I. Šlamborová, V. Zajícová, J. Karpíšková, Exnar. P. Mat. Sci. and Eng. C . **33** (2013)
5. Y. Feng, B. Ning, P. Su, H. Wang, C. Wang, F. Chen, Z. Gao, Talanta, **80** (2009)
6. T.S. Deng, H.J Bongard, F. Marlow, Mat. Ch. and Phy. **162** (2015)
7. D. Güler, Gazi University, MSc Thesis (2012)
8. A. Özcan, İstanbul Business Univ. J. of Sci. **14** (2008)
9. C. Liu, M. Ye, A. Han, J. Li, Ceramics Inter. **41** (2015)
10. C. Aydemir, Marmara J. of Sci. **3** (2014)
11. X. Xiao, C. Hao, Colloids and Surfaces A: Phys-ch and Eng. Asp. **359** (2010)

Comparison of methods for solids retention time determination and control

Piotr Balbierz[1,*], and *Marta* Knap[1]

[1]Wroclaw University of Science and Technology, Faculty of Environmental Engineering, Wybrzeze Wyspianskiego 27, 50-370 Wroclaw, Poland

Abstract. Solids retention time (SRT) is the basic and most important process parameter in design, operation and control of activated sludge (AS) systems, especially with nitrification and biological nutrient removal. However, determination of SRT at the full-scale wastewater treatment plant (WWTP) may not be straightforward and the significance of proper SRT control is often underestimated by plant operators, which may lead to severe operational problems, including complete nitrification failure. This paper presents different methods for SRT determination and control, including simple hydraulic method and phosphorous mass balance method. The comparison is based on an extended data set from a large WWTP in south-west Poland, both in stable and transient operational conditions. The results of the comparison show that during stable operational conditions, all the SRT determination methods give similar results, thus the simple hydraulic method can be applied during standard operation, even at small WWTPs, which lack laboratory facilities. Hydraulic method is based on continuous on-line flowrate measurements, therefore it gives opportunity to implement an algorithm to automatically control SRT and sludge wastage from the system.

1 Introduction

Solids retention time (SRT) is the basic and most important process parameter in design, operation and control of activated sludge (AS) systems, especially with nitrification and biological nutrient removal. SRT influences not only the required bioreactors volume, thus the investment costs, but also oxygen demand, thus the operational costs, as well as microbiological sludge composition, thus settling properties and achieved treatment effects. Furthermore, SRT impacts waste activated sludge (WAS) production and chemical composition, hence the potential biogas production in case of anaerobic sludge digestion [1–3].

This paper aims to presents the theoretical and practical aspect of SRT determination and control at full-scale WWTPs, as well as different methods for SRT determination, including simple hydraulic method and phosphorous mass balance method. The comparison is based on an extended data set from a large WWTP in south-west Poland, both in stable and transient operational conditions.

* Corresponding author: piotr.balbierz@pwr.edu.pl

1.1 The theoretical and practical aspects of SRT determination

SRT (*SRT*, d) is the mean residence time of AS flocs in the system, defined as the ratio of the AS mass in the system (M_X, g) and the daily net sludge production (ΔX, g/d) (total sludge load entering the system and generated in the system, less the sludge load disappearing in the system), according to the formula:

$$SRT = \frac{M_X}{\Delta X} \tag{1}$$

In case of a steady-state system, the daily net sludge production is equal to the daily load of sludge discharged from the system so that the SRT can be determined in a relatively simple manner. The sludge mass in the system includes the mass of sludge accumulated in the bioreactor (M_R, g) and the secondary settling tanks (M_{SST}, g). On the other hand, the daily sludge load discharged from the system includes the sludge load discharged with wasted activated sludge (ΔX_{WAS}, g/d) and the sludge load discharged in the treated effluent (ΔX_E, g/d). Thus Eq. 1. assumes the form:

$$SRT = \frac{M_R + M_{SST}}{\Delta X_{WAS} + \Delta X_E} \tag{2}$$

However, full-scale plants rarely work in steady-state, due to loading variations, temperature fluctuations, and changes in process parameters, thus making the correctness of the SRT determination using the classical method questionable. Generally, it is assumed that in case of real systems operating under stable operating conditions, the SRT determined by Eq. 2. allows the correct estimation of the actual SRT, especially if the average, rather than instantaneous, values for a certain period are taken into account.

However, in case of plants far from steady state, *e.g.* due to significant changes in the incoming load or the WAS load, the use of the classical method can lead to serious errors, since the determined SRT is only a numerical value not reflecting the actual SRT in the system. Therefore, it cannot be the basis for predicting the performance of the system. As an example, one can imagine a steady-state system working at a 5-day SRT, where the WAS load is rapidly decreased twice – as a consequence, the classical SRT increases immediately from 5 to 10 days, which is physically impossible because the sludge cannot "age" by more than 1 d/d. In fact, the time required for the system to reach the new steady state at 10-day SRT will be several days (usually a period of 3–5 time constants, in this case SRT is the time constant).

The need to determine the actual SRT in technical systems working in conditions far from steady-state led Vaccari et al. [4] to suggest the idea of dynamic SRT (SRT_D, d), which was developed and generalized by Takacs [5], who expressed it in the form of the following differential equation, where F_P (g/d) is the daily gross sludge production:

$$\frac{dSRT_D}{dt} = 1 - \frac{SRT_D \cdot F_P}{M_X} \tag{3}$$

The considerations and simulations by Takacs [5] showed that the dynamic SRT determined according to Eq. 3 in steady-state simplifies to the classical equation (Eq. 2), while in transient conditions it better reflects the dynamics of the AS process and correctly determines the actual SRT, and thus allows to predict the performance of the system.

It should be noted, however, that the use of dynamic SRT in practical applications is severely limited due to difficulties in reliable determination of gross sludge production due to the lack of necessary data for real systems. For this reason, the classical SRT determination method is usually used at WWTPs, which is not a mistake if the underlying assumption of the steady-state system and the resulting constraints is remembered.

In case of the AS system shown in Figure 1, for which further discussion will be conducted, the SRT according to the classical method (Eq. 2) can be determined as:

$$SRT = \frac{V_R \cdot X_R + M_{SST}}{Q_{WAS} \cdot X_{WAS} + Q_E \cdot X_E} \tag{4}$$

where V_R (m^3) is the bioreactor volume, X_R (g/m^3) is the bioreactor sludge concentration, Q_{WAS} (m^3/d) and X_{WAS} (g/m^3) are the flowrate and concentration of WAS, Q_E (m^3/d) and X_E (g/m^3) are the flowrate and suspended solids concentration of treated effluent.

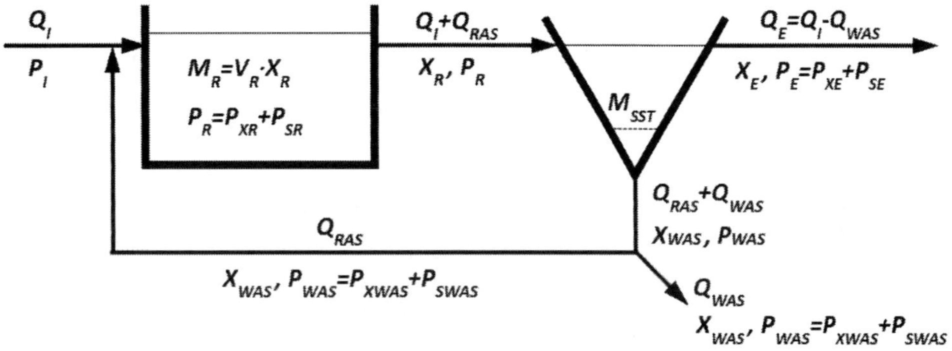

Fig. 1. Schematic diagram of AS system for the SRT considerations (description in the text).

In practical applications, simplifying assumptions are often made regarding the mass of sludge in the secondary settler and the sludge load in the effluent. The mass of sludge in settlers is often neglected, because in systems with biological nutrient removal, settlers should be operated at a low sludge layer, so the mass of sludge accumulated therein is usually relatively small, moreover, difficult to estimate without the sludge concentration profile at the depth of the settler. It is also acceptable to omit the load of sludge discharged in treated wastewater, if the suspended solids concentration is low. With these two assumptions, equation 4. simplifies to:

$$SRT = \frac{V_R \cdot X_R}{Q_{WAS} \cdot X_{WAS}} \tag{5}$$

Therefore, to determine the SRT, it is necessary to know the values of at least four parameters. However, bioreactor volume (V_R) is generally known and constant for a given system, and the WAS flowrate (Q_{WAS}) is usually measured on-line with satisfactory resolution and accuracy. On the other hand, determination of sludge concentration usually requires laboratory analyses, results of which are delayed due to the time required to perform them, ranging from a few hours to several days, depending on whether the plant has its own laboratory. Sludge concentration in the bioreactor (X_R) is relatively constant, so even grab samples taken once a day give reliable results, whereas WAS concentration (X_{WAS}) varies significantly during the day (cf. Fig. 2), which means that successive grab samples or a daily composite sample would be needed for a meaningful result.

From the point of view of WWTP operator, the basis for SRT control, to maintain proper operating conditions in the system, is to discharge the proper load of excess sludge during the day, by applying proper WAS flowrate for the current WAS concentration.

However, due to the variability of WAS concentration, determination and control of SRT at the full-scale WWTP is not as straightforward. Moreover, the importance of proper

SRT control is often underestimated or not fully understood by plant operators, which may lead to severe operational problems, including complete nitrification failure.

Practical problems with SRT control, urge plant operators to use other methods to determine the required excess sludge disposal. Commonly used methods are based on maintaining: a constant sludge concentration in the bioreactor, less often a constant food to microorganism ratio (due to the need for laborious and long-lasting BOD_5 measurements), but also unreliable constant sludge settled volume after 30 minutes, which is strongly dependent on variable settling properties. These methods may be adequate for the operation of small organic removal systems, but do not ensure the control of the pre-set SRT, and are therefore inadequate for nitrifying and nutrient removal systems.

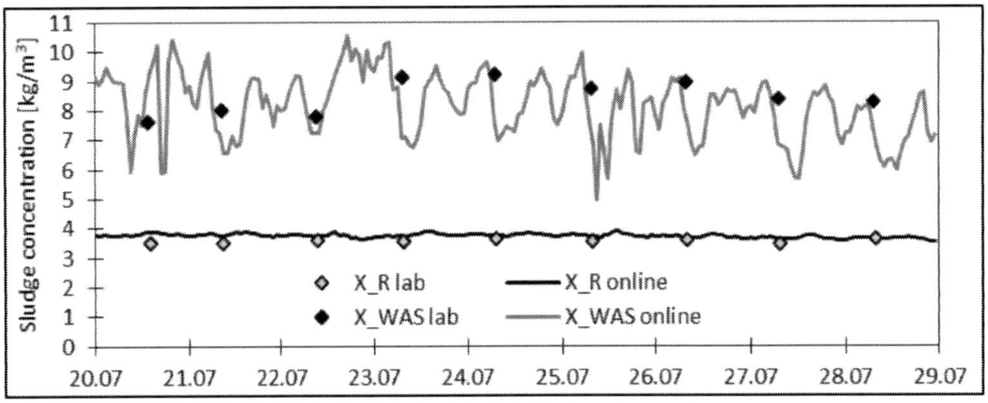

Fig. 2. Variation of sludge concentrations in bioreactor (X_R) and WAS stream (X_WAS) based on laboratory analyses (lab) and on-line measurements (online) [data from Wroclaw WWTP].

There is a very simple yet reliable method of hydraulic SRT control proposed by Garrett [6], which assumes the discharge of excess sludge directly from the bioreactor, so that the WAS concentration is equal to the bioreactor sludge concentration ($X_{WAS} = X_R$), which simplifies Eq. 5 to:

$$SRT = V_R / Q_{WAS} \tag{6}$$

Using this method, maintaining the SRT of n days requires a daily WAS discharge equal to $1/n$ of the bioreactor volume, and does not require determination of sludge concentration. This method, however, is rarely used in practice, because its greatest advantage is also its biggest disadvantage, since the WAS stream is not thickened.

1.2 Alternative SRT determination methods

The importance of SRT as a process parameter of AS systems and identified problems with its determination and control, make this issue still a valid research problem. In recent years, several papers have been published that offer an original approach to this problem.

In response to problems with reliable determination of WAS concentration, Greenwood et al. [7] proposed a method for hydraulic SRT control, which is based on the sludge mass balance for bioreactors and secondary settlers. This method eliminates the need for sludge concentration analyses, because to determine the SRT it requires only the flowrates of the influent wastewater Q_I (m^3/d), recirculated Q_{RAS} (m^3/d) and wasted Q_{WAS} (m^3/d) AS, according to the equation:

$$SRT_Q = \frac{V_R \cdot (Q_{RAS} + Q_{WAS})}{Q_{WAS} \cdot (Q_I + Q_{RAS})} \tag{7}$$

The use of this method is particularly attractive in practical applications because it is based on on-line measurements, which allows to implement an automated algorithm to control excess sludge discharge, depending on the operating conditions of the system, to maintain the pre-set SRT. Derivation of this method, however, presupposes some simplifications as it neglects the sludge load discharged with treated effluent and assumes that there is no significant change in the mass of sludge in secondary settling tanks. Therefore, the obtained results should be periodically verified using the classical SRT determination method.

Puig et al. [8] have focused on improving the reliability of the SRT determination, particularly for the mathematical modelling of AS systems. It is recognized that for a dependable simulation study the SRT values should be known with 95% accuracy, so it is necessary to check the quality of sludge production data [9, 10].

The tool that allows to verify and correct the quality and coherence of operational data is the mass balance determined for a given facility. The mass balance for the biological treatment unit can be calculated for e.g. COD, total nitrogen, total phosphorus or flow. However, only in the last two cases, if the measurements are conducted in all wastewater and sludge streams, we will be dealing with a closed balance, which will allow for its verification. Whereas for COD and nitrogen the mass balance will be open due to the component present in the gas phase - respectively CO_2 or N_2, which is usually not measured. Unfortunately, total phosphorus (as well as COD and total nitrogen) is usually only determined in wastewater streams, which makes it impossible to check the mass balance.

On the basis of the total phosphorus mass balance for the biological treatment unit, Puig et al. [8] proposed an alternative method to determine the SRT based on the particulate phosphorous concentration in sludge. Derivation of this method requires the assumption that there is no significant release of phosphorus from the sludge in the secondary settler. It is also necessary to know the particulate phosphorus concentration in sludge, which is calculated as a difference between the analytically determined total and dissolved phosphorus concentrations.

The first variant of this method allows to determine the SRT based on the particulate phosphorous mass in the system and the particulate phosphorus load discharged from the system according to the formula:

$$SRT_{PE} = \frac{V_R \cdot P_{XR}}{Q_{WAS} \cdot P_{XWAS} + Q_E \cdot P_{XE}} \tag{8}$$

where P_{XR} (g P/m^3), P_{XWAS} (g P/m^3) and P_{XE} (g P/m^3) are the particulate phosphorous concentration in the bioreactor, WAS and effluent, respectively.

The second variant allows to determine the SRT based on the particulate phosphorus mass in the system and the total phosphorus load entering the system (not necessary to determine particulate phosphorus in the WAS) according to the equation:

$$SRT_{PI} = \frac{V_R \cdot P_{XR}}{Q_I \cdot (P_I - P_{SE})} \tag{9}$$

where P_I (g P/m^3) is the total phosphorous concentration in the influent, and P_{SE} (g P/m^3) is the soluble phosphorous concentration in the effluent.

It is also possible to derive a third variant, which does not require the particulate phosphorous determination at all, but because it is based on the difference between the total and soluble phosphorus concentrations in the effluent, it will be characterized by a high scatter of results:

$$SRT_{PN} = \frac{V_R}{Q_I} \cdot \frac{X_R}{X_E} \cdot \frac{P_E - P_{SE}}{P_I - P_{SE}} \quad (10)$$

where P_E (g P/m^3) is the total phosphorous concentration in the effluent.

Taking into account the importance of the SRT as a process parameter of AS systems and the practical difficulties encountered in reliable determination and control of SRT, the authors find it worthwhile to compare the results obtained using various methods for the operational data of the full-scale WWTP under transient and stable operating conditions.

2 Results and discussion

The practical verification and comparison of above-mentioned methods were carried out for operational data from Wroclaw WWTP, gathered during an 8-month period between January and August, including an intensive sampling campaign 20–28.07.

The comparison included SRT values determined by the classical method based on sludge concentrations measured in the laboratory (SRT_{XL}) and online (SRT_{XS}), hydraulic method (SRT_Q), and phosphorus mass balance based methods: based on the total phosphorus load entering the system (SRT_{PI}) and the particulate phosphorus load discharged from the system (SRT_{PE} – only for the intensive sampling period when the phosphorus was analysed in the WAS stream).

The first half of the analysed 8-month period was characterized by very unstable and transient operational conditions. Due to problems with the operation of the sludge drying plant and limited discharge of excess sludge for over four months, in the first half of January the SRT of sludge in bioreactors of Wroclaw WWTP significantly exceeded 100 d, and the AS concentration in bioreactor was almost 10 kg TSS m^{-3}, despite only a 50% loading of WWTP, due to the renovation of the main sewer.

The nominal WAS discharge capacity was restored around mid-January, and intensive WAS discharge was begun to restore reasonable values of SRT and AS concentration as quickly as possible. However, the stable operation of the plant with effective SRT control (based on the implemented SRT_Q hydraulic method) was reached only around July 1st.

The daily values of SRT determined by different methods in the 8-month period are shown in Figure 3.

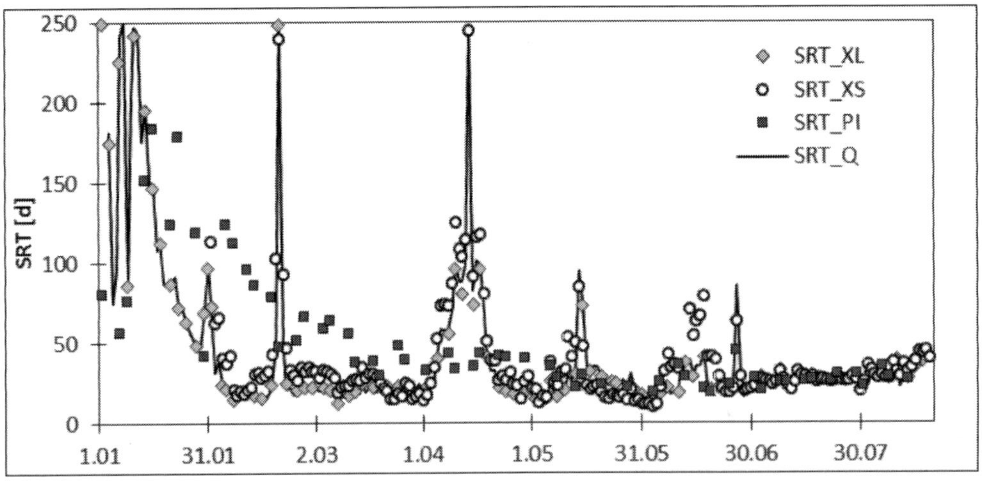

Fig. 3. Comparison of SRT determination methods in transient conditions.

The visual evaluation of the obtained results shows that the values of SRT_{XL}, SRT_{XS} and SRT_Q in the whole analysed period are consistent. In contrast, values determined by the SRT_{PI} method in the initial period are significantly higher and coincide with the results of other methods only from mid-May. The differences observed in the initial period are the result of the intensive WAS discharge after a period of operation at a very long SRT – as a result the instantaneous SRT values determined on the basis of discharged WAS load are significantly lower than the actual SRT. The SRT_{PI} method, unlike the other methods, does not include the WAS load in the calculation and thus is not susceptible to rapid changes in the discharged WAS load. Thus, in such a situation the SRT_{PI} method seems to be a better estimator of the actual value of SRT. However, it is sensitive to changes in the total phosphorus load in the influent and soluble phosphorus load in the treated effluent (changes in phosphorus removal efficiency).

A more accurate comparison of mean SRT values determined by different methods was carried out for the period 1.07.–15.08., characterized by stable operation of WWTP with effective SRT control, with special attention given to the intensive sampling campaign 20–28.07.2013 (Figure 4). Relatively short (as for the averaging of the SRT), 9-day intensive sampling period was taken into account, since only for this period laboratory measured phosphorous concentrations in the WAS stream were available.

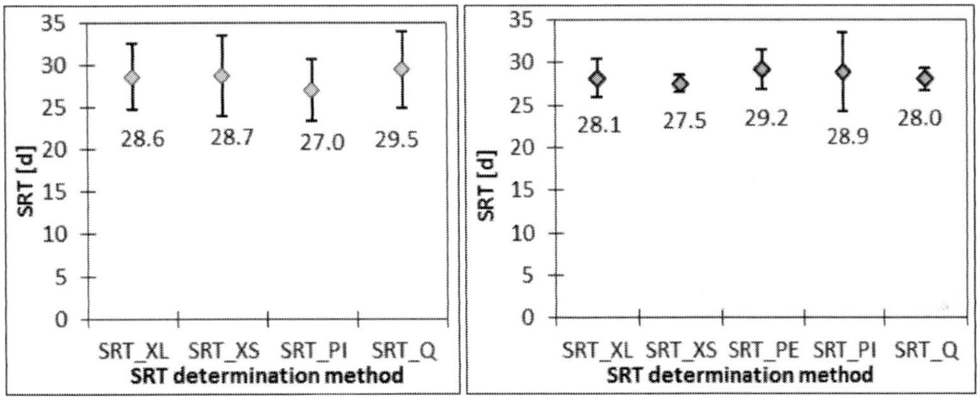

Fig. 4. Comparison of SRT determination methods during stable conditions 1.07–15.08 (left) and intensive sampling period 20–28.07 (right).

Based on the obtained results it can be stated that in the period of stable operation of WWTP, the SRT values determined by all compared methods are similar – the differences between individual methods and the classical SRT_{XL} method (as well as the mean of all methods) generally do not exceed 1 day (< 5% of the reference value) – such accuracy is completely sufficient for the operation of the WWTP.

This means that, under typical stable operating conditions, the quality of the simple hydraulic SRT control method, which is based on on-line measurements of wastewater and sludge flowrates, is comparable to the conventional SRT method. Therefore, the SRT_Q method can be successfully used for the purpose of WWTP operation (with periodic verification by the classical SRT method). It is therefore possible to implement a simple and effective SRT control even in treatment plants where, due to too low frequency of sludge concentration analysis, it is not possible to control the SRT using the classical method.

 DOI: 10.1051/e3sconf/20172200008

3 Conclusions

The SRT is the most important process parameter of AS systems; however, in unstable operational conditions typical for full-scale WWTPs, the SRT values calculated with the classical method may be unreliable and not representative. Therefore, several new SRT determination methods were proposed in the recent years to overcome those limitations.
The results of the comparison show that during stable operational conditions, all the SRT determination methods give similar results, thus the simple hydraulic method SRT_Q can be applied during standard operation, even at small WWTPs, which lack laboratory facilities. Hydraulic method is based on continuous on-line flowrate measurements, therefore it gives opportunity to implement an algorithm to automatically control SRT and sludge wastage from the system.

However, during periods of rapid changes in the discharged WAS load, the SRT_{PI} method, based on the phosphorus balance (using the total phosphorus load entering the system) provides better estimation the actual SRT value than the other methods.

Acknowledgements

The authors gratefully acknowledge co-funding from the National Centre for Research and Development (grant no. PBS2/B9/25/2014) and the Municipal Water and Sewage Company in Wroclaw.

References

1. Metcalf & Eddy, *Wastewater Engineering: Treatment and Resource Recovery* (G. Tchobanoglous, F.L. Burton, H.D. Stensel, ed., McGraw-Hill Education, 2014)
2. J. Makinia, *Mathematical modelling and computer simulation of activated sludge systems* (London, New York, IWA Publishing, 2010)
3. G.A. Ekama, Water Sci. Technol. **61**, 7 (2010)
4. D.A. Vaccari, T. Fagedes, J. Longtin, Biotechnol Bioeng. **27**, 5 (1985)
5. I. Takacs, *Experiments in Activated Sludge Modelling*. PhD thesis, Ghent University, Ghent (2008)
6. M.T. Garrett, Sewage Ind. Wastes **30**, 3 (1958)
7. S. Greenwood, C. Anderson, M. Rieth, T. Stein, *Constant SRT calculated from liquid flows improves BNR activated sludge performance*, Central States Water Environment Association. (2002) [online] http://www.cswea.org/papers/
8. S. Puig, M.C.M. van Loosdrecht, J. Colprim, S.C.F. Meijer, Water Res. **42**, 18 (2008)
9. D. Brdjanovic, M.C.M. van Loosdrecht, P. Versteeg, C.M. Hooijmans, G.J. Alaerts, J.J. Heijnen, Water Res. **34**, 3 (2000)
10. S.C.F. Meijer, M.C.M. van Loosdrecht, J.J Heijnen, Water Res. **35**, 11 (2001)

DOI: 10.1051/e3sconf/20172200009

Use of cement-fly ash-based stabilization techniques for the treatment of waste containing aromatic contaminants

Kamil Banaszkiewicz[1,*], and *Tadeusz* Marcinkowski[1]

[1] Wroclaw University of Science and Technology, Faculty of Environmental Engineering, Wybrzeze Wyspianskiego 27, 30-370 Wroclaw, Poland

Abstract. Research on evaluation of evaporation rate of volatile organic compounds from soil beds during processing is presented. For the experiment, soil samples were prepared with the same amounts of benzene and stabilized using a mixture of CEMI 42.5R cement and fly ash from pit-coal combustion. Solidification of soils contaminated with BTEX hydrocarbons using hydraulic binders involves a risk of releasing vapours of these compounds during homogenization of waste with stabilizing mixture introduced and its dilution with water. The primary purposes of the research were: analysis of benzene volume emitted from soil during stabilization/solidification process and characterization of factors that may negatively affect the quality of measurements/the course of stabilization process. Analysis of benzene emission intensity during the process was based on concentration (C_6H_6) values, recorded with flame-ionization detector above the surface of reacting mixture. At the same time, gaseous contaminants emitted during waste stabilization were passed through pipes filled with activated carbon (SCK, Anasorb CSC). Benzene vapours adsorbed on activated carbon were subjected to analysis using gas chromatograph Varian 450-GC. Evaporation characteristics of benzene during processing contaminated soils revealed the stages creating the highest danger to workers' health, as well as a need for actions connected with modification of technological line.

1 Introduction

The main cause of soil degradation are chemical contaminants emerging as a result of human activity – their composition and volume are beyond possibilities of their treatment by the natural environment. Organic compounds introduced to soil, depending on the kind of ground, may occur in the following forms: gaseous - dispersed in pores or adsorbed on soil particles; liquid – immiscible with water or dissolved in water; or solid. As a result of migration they can cause pollution of significant areas of pedosphere, as well as surface and ground water.

* Corresponding author: kamil.banaszkiewicz@pwr.edu.pl

Aromatic hydrocarbons belong to main pollutants of soil and ground water occurring in the vicinity of oil deposits, petroleum gas producing plants and storage areas for gasolines and other petroleum-derived substances. Due to toxic, mutagenic or cancerogenic properties of organic solvents even their small amounts in water or soil creates a serious risk to human health and requires instantaneous actions to reduce their propagation and remove such substances from the natural environment.

Benzene is used in numerous processes as solvent or base material for further synthesis. It has common applications in such branches as rubber industry, pharmaceutical industry and paints, lacquers and plastics industry [1]. Overall benzene production in EU is estimated at about 7 million tonnes annually [2] Due to its high volatility, ability to create high concentrations in air, inflammability and carcinogenicity, all processes with large evaporation area and little chance of hermetization are highly dangerous.

Remediation of contaminated area, i.e. cleaning and removing contaminants emerged as a result of routine industrial activities or unexpected failure may be performed by elimination, decomposition, immobilization or chemical transformation into less toxic, less mobile and less reactive forms. The choice of method of cleaning/neutralizating contaminated soil depends on volume and physico-chemical properties of introduced pollutants, as well as on geohydrologic conditions [3].

One of the technologies widely used for immobilizing inorganic and (not so often) organic constituents of hazardous waste is stabilization/solidification (S/S) process using hydraulic binders. S/S of contaminated soil may be performed *in situ* – in the place where pollution occurred, or *ex situ*, e.g. in a waste processing plant [4]. Reduction of organic and inorganic pollutants mobility is a result of the reactions of precipitation, hydrolysis, oxidation/reduction, complexation or adsorption [5, 6]. Significant impact of organic pollutants on cement hydration process restricts possibilities of S/S technology [6], therefore the efficiency of immobilization of organic contaminants in cement depends strongly on possibilities of their physical lock-up [7]. Additionally, in order to reduce the impact of organic pollutants on the course of S/S process, various cement additives are applied, such as: fly ash, powdery clays, quick lime, hydrated lime, gypsum and natural resins.

The effect of organic substances on slower fixation and hardening of cement was observed e.g. during mixing cement with water, methanol and 2-monochloroaniline, in volumetric proportions of 100/76/4 [8]. Basic cement reactions remained the same, but they were running slower. For example, Gussoni et al. [9], on the basis of their research using magnetic resonance techniques, report formation of big pockets containing liquids during the first hours and days of hydration in presence of methanol and dichloroaniline. Using X-ray diffractometry, the same authors showed that a solution containing the two compounds mentioned retains its ability to slow-down hydration even after 6 months of maturation of examined material [9]. Vipulanandan [10] reports that the presence of phenol in amount of 10% causes about sevenfold elongation of total time of cement solidification.

Despite significant limitations in application of cement for stabilization of wastes containing hydrocarbons, such applications are in use and we have relations [11], according to which it is possible to solidify properly wastes containing some polycyclic aromatic hydrocarbons. Indeed, particles of polycyclic aromatic hydrocarbons cannot be built into the cement matrix, but still their physical encapsulation is possible to some extent [12]. S/S with cement was also proposed as a method of processing waste (e.g. sludges from petrochemical industry), containing non-polar organic compounds, such as aromatic hydrocarbons, halogenated hydrocarbons and polychlorinated biphenyls, that is feasible on specific conditions [13].

Experimental data on parameters of leaching hydrocarbons from petroleum refinery sludges stabilized/solidified with various amounts of cement are scarce and concern only the Trzebinia Refinery and the Czechowice-Dziedzice Refinery. S/S with cement and similar materials should be recognized as a technology that is economically justified and can be implemented in a relatively short time – it was confirmed in some Polish petroleum refineries.

Solidification of waste with hydraulic binders requires its mechanical mixing in order to disperse in the entire volume of waste and to dilute with water. This is a process during which (in case of processing waste contaminated with aromatic hydrocarbons) poses a risk of releasing vapours of volatile organic compounds. This concerns mainly the stage of material homogenization and initiation of hydration process with dilute water. The primary purposes of the research were: (a) analysis of benzene volume emitted from soil during S/S process (based on continuous measurements of concentrations of gaseous contaminants above the surface of reacting mixture), (b) verification of continuous measurements on the basis of chromatographic analysis of benzene vapours adsorbed on activated carbon, (c) characterization of factors that may negatively affect the quality of measurements/the course of stabilization process.

2 Materials and methodology

2.1 Soil samples and binders

Soil bed used in the research was poorly coherent and was characterized by hydration at the level of W = 3.41%, and low content of organic substances ($1.97\%_{d.m.}$). As a result of low cohesion, the material had a high permeability. The structure of used soil allowed easy dispersion of binding materials in entire volume, but at the same time created a risk of significant emission of gaseous pollutants from intergranular spaces.

In the experiment a sample of 200 g soil was weighted, put into a container and contaminated with benzene (p.a.grade, Chempur) in amount of 0.5 mL. Then containers with samples were shut close, to reduce evaporation of gaseous contaminants. In order to distribute solvent in entire volume of material, tightly closed samples were placed in somersault shaker (Heidolph Reax 20/8). Time of mixing was 120 s. Prepared samples were stored for 24 h in the temperature of about 7°C.

Soil contaminated with benzene was processed with a mixture of cement and fly ash from pit-coal combustion. The content of each component in binding mixture amounted to 70% and 30%, respectively. Cement used in the experiment was up to PN-EN 197-1 standard, had a durability class of 42.5 and a high early strength: CEM I 42.5 R. This material does not contain any mineral additives, 95-100% of its mass is clinker, 0–5% secondary components (inorganic materials from clinker manufacturing). The main minerals of Portland clinker are di- and tricalcium silicates (alite and belite) that together constitute 74% of its mass [14, 15] Portland cement CEM I 42.5 R is a binder characterized by large amounts of heat released during hardening, as well as high rate of its releasing. In case of solidifying soils contaminated with organic solvents, higher temperature of hydration (apart from the risk of microfractures when the limit value is exceeded) [16], may cause increase in volatile contaminants emissions [17].

Additional component used in research was fly ash from pit-coal combustion. Introduction of this material leads to extended time of mixture fixation, as well as decrease in its specific density. As a result, application of cement and fly ash mixture in the same amount as Portland cement CEM I alone leads to introduction of greater volume of binding material and therefore to a final product characterized by lower permeability [16]. Addition of fly ash, parallel to reduction of cement content, causes decreased contraction shrinkage of mixture.

When replacing a part of cement with fly ash also the content of non-burnt coal (roasting loss) should be concerned. Coal increases water absorbance of ash and as a result may have adverse impact on hardiness of a product containing cement-fly-ash mixture. Due to spherical shape of grains, siliceous fly ash considerably improves workability, tightness and strength of a concrete [18–20]. Ashes used in reported research were characterized by roasting loss at the level below 5%.

2.3 Measuring equipment for volatile organic compound

Continuous measurements of concentrations of benzene emitted from reacting mixture during the solidification process were performed using the MicroFID flame-ionization detector (Photovac). The concentration meter was calibrated for methane, whereas concentrations measured were converted into volume of benzene in the air. The volume of sample taken by the instrument 0.6 L/min, benzene concentration was recorded in 5 seconds intervals.

At the same time, gaseous contaminants emitted during waste stabilization were passed through pipes filled with activated carbon (SCK, Anasorb CSC), using two-channel aspirators LAT ASP – 3 II. Rate of flow was ca. 60 L/h, corresponding to the volume of 4.7 to 5.1 L of process gases, aspired to activated carbon bed in particular phases of process.

Vapours of volatile organic compounds adsorbed on activated carbon were subjected to concentrations analysis using gas chromatograph Varian 450-GC and columns Varian VF-WAXms (30 m x 0.25 mm ID DF: 0.25 μm). In order to do that, activated carbon from absorbing pipe was put into 5 cm^3 flasks, then 2 mL of carbon disulphide (CS_2) was added. Flasks were shut close and left for 20 min., with their content shaken from time to time. Then the solution from above the carbon layer was taken into 2 mL vial and placed in autosampler of chromatograph. Desorption coefficient of the method was 80%.

Tests were performed in the following conditions: the temperature of furnace (column) – 110°C for 10 min.; temperature of batchmeter – 225°C; split – 1/20, temperature of detector – 250°C; carrier gas – helium; flow rate of carrier gas – 1mL/min; volume of injection – 5 μL.

2.4 Methodology of the stabilization/solidification process

S/S of soil contaminated with benzene was performed in a reactor for cement mortars Tecnotest B205/X5. Air samples from above the reacting mixture were taken through teflon pipes introduced into the reactor bowl. All channels ended on the same depth, about 2 cm below the rim of the bowl. Pipes with activated carbon were mounted on two channels, the third one was connected directly with the flame-ionization detector (VOC measuring equipment). S/S process was performed in two stages: (1) waste homogenization phase – homogenizing the mixture of waste and the binding blend. (2) hydration phase – initiation of fixation process in binders through introducing diluent water. The amount of water was determined on the basis of research on workability of mixtures containing stabilizing agents and contaminated soil.

The whole process lasted for 10 minutes, 300 seconds each of phases. Detailed material composition of solidified waste samples is presented in Table 1. Every time before the start of test series the ambient temperature and humidity were measured using Hanna Instruments HI 9564 thermo-hygrometer. After each gas sampling all ducts were washed with clean air in order to remove residues of gases and dusts accumulated in ducts. Measuring series started with background measurement (about 60 s) and then concentrations in teflon ducts with the aim of verifying purity of air and possible additional cleansing of ducts. In order to reduce influence of external factors on gaseous contaminants emission, all materials (binders and waste) were mixed with a constant speed (planetary speed 62 rpm (clockwise); beater speed 140 rpm (anticlockwise)).

Table 1. Material composition of composites produced from soil contaminated with benzene, as well as ambient conditions (humidity and temperature).

Sample	Soil	Benzene	Portland cement	Fly ash	Temperature	Humidity
Sample 0	200 g	0.5 mL	0 g	0 g	20.1±0.5^{0}C	25.4±3%$_{RH}$
Sample 80	200 g	0.5 mL	56 g	24 g	20.1±0.5^{0}C	25.4±3%$_{RH}$
Sample 120	200 g	0.5 mL	84 g	36 g	20.1±0.5^{0}C	25.4±3%$_{RH}$
Sample 160	200 g	0.5 mL	112 g	48 g	20.1±0.5^{0}C	25.4±3%$_{RH}$

3 Discussion and results

Every process of neutralization of waste contaminated with volatile organic compounds is characterized by technical parameters that may determine their emissions during processing. In case of S/S processes, as actions causing release of volatile or semi-volatile compounds we can count: (1) mixing waste with binding components, (2) diluting mixture with water that can displace volatile organic compounds from adsorption places, (3) increase in temperature of reacting mixture during hydration of binding factors (e.g. mixture of cement and lime) [5]. Knowledge of characteristics of gaseous contaminants emission enables evaluation of the process dynamics, as well as effectiveness of actions that stabilize releasing vapours of volatile organic compounds.

The total volume of benzene emitted in the first phase was a few dozen times greater than in the phase of hydration (Table 2). Results of continuous measurements confirmed that the course of emission of the solvent analyzed was very dynamic. The highest momentary concentration of benzene in the reactor (in all tests, irrespective of binding mixture dose) was recorded in the first 40 s of the process. Intensity of releasing benzene vapours might me closely related to mixing of contaminated soil. Applied rotary speed – 140 rpm (suggested by the manufacturer of universal planetary mixer for asphalt and cement) might be too high for material contaminated with volatile organic compounds. Application of lower speed might have a positive effect and reduce a sudden emission.

Table 2. Amount of benzene adsorbed on activated carbon (GC analysis).

Phase of process	Sample 0	Sample 80	Sample 120	Sample 160
Homogenization phase	65 mg	49 mg	41 mg	37 mg
Hydration phase	not measured	1.02 mg	0.71 mg	0.78 mg

Besides, a significant emission of dusts from binders was observed during homogenization of soil with binding components (the 1st phase of the process). Emitted solid pollutants were sucked in together with air from the reactor bowl. A considerable part of dusts settled on walls of ducts and was captured on a filter before combustion chamber of the MicroFID detector. In order to reduce effect of pollination on the quality of measurements, each time the ducts were blown through with clean air. Also in this case reduction of mixer speed might prove positive.

 DOI: 10.1051/e3sconf/20172200009

The research confirmed positive impact of hermetization of waste using cement-ash mixture. Along with increased dose of binding components, a decrease in benzene concentration in reactor was observed. In case of the sample 160 during the phase of homogenization, the amount of solvent adsorbed on activated carbon was lower by 43% (Fig. 1). For the sample solidified with the lowest amount of mixture (80 g) the emission was lower by 24%. A reduction of benzene vapours mobility in the first phase of process did not cause its increase in the phase of diluting mixture with water. The average concentrations of benzene in the second phase of process were 190 to 270 mgC_6H_6/m^3. For the samples 120 and 160 the amount of volatile organic compounds above the surface of mixture were comparable and amounted to 190 and 196 mgC_6H_6/m^3, respectively.

For continuous measurements a similar characteristics of changes was observed. A decrease in total amount of released benzene, along with an increase in stabilizing mixture mass was 19, 25 and 35%, respectively (Fig. 1).

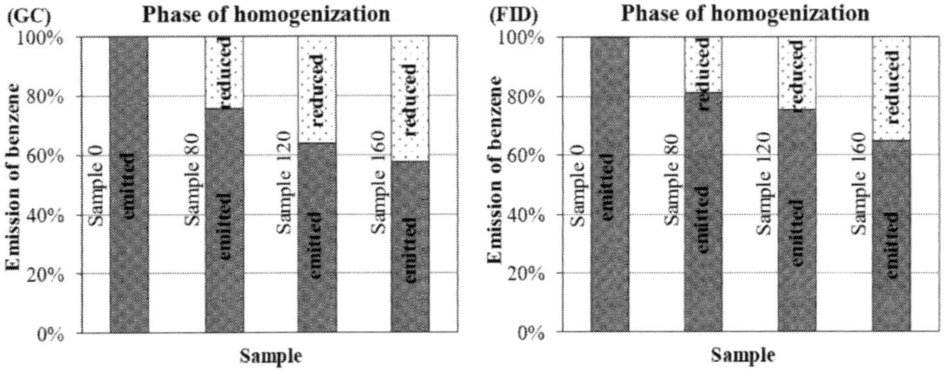

Fig. 1. Reduction of the total benzene emission during the process of S/S - GC/FID.

Evaporation characteristics of benzene in processing contaminated soils did not show a significant effect of binding mixture dose on dynamics of benzene vapors releasing during the phase of homogenization (Fig. 2). As a result of introduction of hydraulic binders a decrease was observed in maximum momentary benzene concentration by: 19, 24 and 43%. Suppression of sudden evaporation was a result of (among others) sealing the soil with fine (powdery) binding materials. One of the components of binding mixture was fly ash that is characterized with greater specific area, compared with cement.

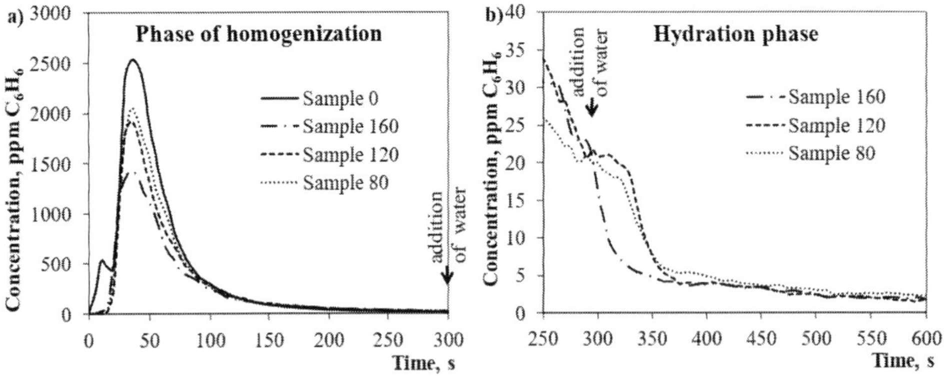

Fig. 2. Amount of benzene emission during the S/S process: a) homogenization phase, b) phase of hydration.

Another neuralgic phase of the S/S process is the stage of hydration. In order to initialize concretion of hydraulic binders, introduction of diluent water is necessary. The dose of water is closely connected with the amount of cement used, as well as with water absorption capability of mineral additives (e.g. fly ash) and soil being neutralized. In case of neutralizing soils contaminated with volatile organic compounds, suppression of their emission in the phase of homogenization through sealing soil with binding mixture may cause releasing of greater amounts during hydration.
According to the process methodology, water was introduced in 300 s. In case of sampled being solidified with cement-ash mixture in amount of 80 and 120 g a phenomenon of displacing benzene vapours adsorbed in micro/macropores by water was observed (on-line measurements) (Fig. 2).

Total amount of released benzene calculated on the basis of momentary concentration measurements in hydration phase did not correspond to the mass adsorbed on activated carbon (Fig. 3). Inconsistencies concerned a sample solidified with a dose of 120 g binding mixture. In case of continuous measurements benzene emissions in hydration phase for the samples 120 and 80 were very similar (Fig. 3).

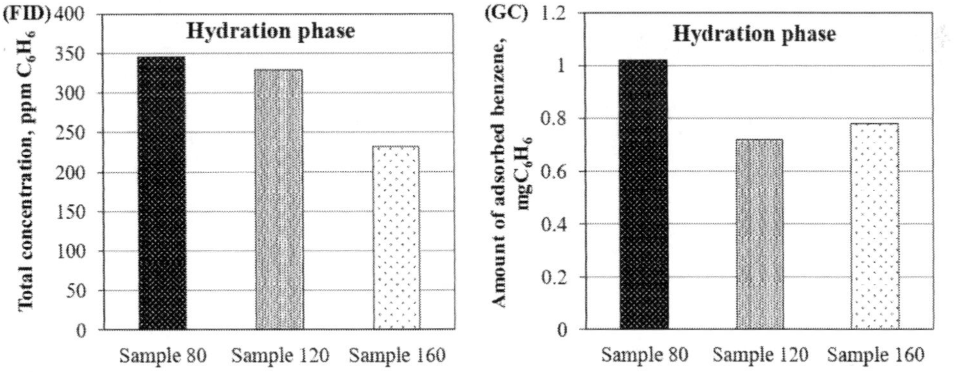

Fig. 3. Total amount of gaseous contaminants: released in the course of S/S process (FID), adsorbed on activated carbon in the phase of hydration (GC).

The amount of benzene fixed on the surface of activated carbon in the second phase of process for the samples 120 and 160 was 0.71 and 0.78 mg, respectively. Recorded variations may be a result of measurement error caused by clogging the sucking ducts by solid particles (dusts). Besides, differences in flow rates of process gases sucked in by measurement equipment also should be concerned. The volume of air sample taken into flame-ionization detector was 0.6 L/min. A flow rate of air passing through pipes with activated carbon (SCK, Anasorb CSC) was about 1 L/min. Total volumes of analyzed air inside the reactor were: ca. 4.7–5.1 L (GC), ca. 3 L (FID).

4 Conclusions

Choice of the remediation method is determined mostly by a kind of contaminants and their concentration. Reported method may be effective in case of application for soils contaminated at the same time with mineral and organic compounds, and as a technique supporting another process of remediation. In reported research the greatest extent of reduction of benzene emission was achieved in the sample 160, in which a binding mixture made 80% of the mass of soil being neutralized. Large dose of stabilizing components causes increase of mass/volume of waste. The content of cement if obtained composites may be lowered by application of additional components sorbing organic contaminants, e.g.

activated carbon, powdery clay or rubber. Also micro-aggregates should be taken into account in choice of composition of stabilizing mixtures.

The essential factor affecting the quality of measurements are dusts from fine-grained binders. Intensity of dusting may be controlled through lowered rotary speed of a mixer, which in turn will prolong the time of homogenizing and diluting mixture. Large amounts of solid contaminants cause occlusion of dust-filters.

References

1. A. Kalina, Principles and methods of assessing the working environment 2000, **16**, 3, 149–158 (2000) (in Polish)
2. P. Stepnowski, E. Synak, B. Szafranek, Z. Kaczyński, *Monitoring and analytics of pollutants in the environment* (The University of Gdańsk Press, 2010) (in Polish)
3. K. Radwan, Z. Ślosorz, J. Rakowska, Safety & Fire Technique **3**, 107–114 (2012) (in Polish)
4. U.S. EPA: *Treatment Technologies for Site Cleanup: Annual Status Report* (Twelfth Edition), EPA-542-R-07-012 (2007)
5. U.S. EPA/Bone B.D. *Review of scientific literature on the use of stabilisation/solidification for the treatment of contaminated soil, solid waste and sludges*, Science Report SC980003/SR2 (2004)
6. U.S. EPA, Innovative Site Remediation Design and Application **4**, *Stabilization/ Solidification*, EPA 542/B-97/007. Office of Solid Waste and Emergency Response (1997)
7. S. Trussell, R.D. Spence, Waste Manage. **14**, 6, 507–519 (1994)
8. N.I. Sora, R. Pelosato, D. Botta, G. Dotelli, J. Eur. Ceram. Soc. **22**, 1463–1473 (2002)
9. M. Gussoni, F. Greco, F. Bonazzi, A. Vezzoli, D. Botta, G. Dotelli, I.N. Sora, R. Pelosato, L. Zetta, Magn. Reson. Imag. **22**, 877–889 (2004)
10. C. Vipulanandan, S. Krishnan, Cement Concrete Res. **23**, 4, 792–802 (1993)
11. V.M. Hebatpuria, H.A. Arafat, H.S. Rho, P.L. Bishop, N.G. Pinto, R.C. Buchanan, J. Hazard. Mater. **B 70**, 117–138 (1999)
12. E. Mulder, J.P. Brouwer, J. Blaakmeer, J.W. Frenay, Waste Manage. **21**, 247–253 (2001)
13. M.D. LaGrega, P.L. Buckingham, J.C. Evans, *Hazardous Waste Management* (McGraw-Hill, New York, NY, 1994)
14. W. Kurdowski, *Cement chemistry* (Polish Scientific Publishers PWN, Warszawa, 1991) (in Polish)
15. W. Kurdowski, *Chemistry of building materials* (AGH University of Science and Technology Press, Kraków, 2003) (in Polish)
16. Z. Giergiczny, *Cement, aggregates, concrete in offer of Górażdże Group. Types, properties, application* (Chorula, 2011) (in Polish)
17. U.S. EPA, *In Situ Treatment Technologies for Contaminated Soil*, EPA 542/F-06/013, (2006)
18. J. Jasiczak, P. Mikołajczak, *Technology of concrete modified with admixtures and additives* (Poznan University of Technology, Alma Mater, 2003) (in Polish)
19. J. Newman, B.S. Choo, *Advanced Concrete Technology* (Elsevier Ltd., 2003)
20. H.F.W. Taylor, *Cement chemistry* (ACADEMIC PRESS, London, 1990)

The influence of multifamily apartment building occupants on energy and water consumption – the preliminary results of monitoring and survey campaign

Karol Bandurski[1,*], *Miłosz* Hamerla[2], *Jowita* Szulc[3], and *Halina* Koczyk[1]

[1]Poznan University of Technology, Faculty of Civil and Environmental Engineering, Institute of Environmental Engineering, ul. Berdychowo 4, 60-965 Poznań, Poland
[2]HM-PROJEKT Miłosz Hamerla, 77-400 Złotów, Poland
[3]Freelance engineer, Poland

Abstract. Occupants' attitudes and behavior have a significant influence on energy and water consumption in buildings. To provide more robust solutions, energy efficient applications should consider occupant-building interaction. However, there is a question to be answered: which aspects of lodging and occupant behavior cause the most substantial increase in consumption of these mediums. Thus, the aim of this study is to investigate the influence of household characteristics and occupants' behavior on level and variability in utilities consumption. The study uses the results of a measuring campaign and the questionnaire. The measuring campaign was carried out to monitor the consumption of energy used for space heating and domestic hot water, as well as electricity, gas and water. The questionnaire specifically focused on household characteristics and occupants' behavior. The research was carried out in four apartment buildings, all consisting of more than 100 apartments. Data from approximately 100 households was gathered and analyzed; the survey's respond rate was almost 50%. A quantitative analysis of the results confirms the assumption that both household characteristics and occupants' behavior (e.g. the use of heating control) are important factors for utilities consumption. Further work with the obtained data is planned in terms of including of greater number of apartments, assessment of ventilation effectiveness, as well as analysis of heat transfer between the apartments.

1 Introduction

Approach to operation of residential buildings has a significant influence on use of energy, water and air pollution in large cities in developing countries. With its households responsible for 25% of direct energy consumption, Poland could serve as a good example [1]. At the same time, the share of heating and cooling households in emissions of CO_2 is 10%, while their share in emission of particle matter (PM_{10} and $PM_{2.5}$) is nearly 48% [2].

* Corresponding author: karol.bandurski@put.poznan.pl

This is the area with a great potential in terms of reducing overall energy consumption and air pollution in cities. It is recognized, that development of reliable energy-saving and low-emission technological solutions in civil engineering requires certain consideration of the human factor in analysis [3], as it is one of the main factors that can have impact on buildings operation [4]. The same applies to water consumption, especially in the context of *smart cities* concept [5]. However, the answer to the question of what behavior is responsible for it requires some further research. Similar studies have been already carried out in different regions of the world, e.g. [6–11]

This article contains the analysis of readings and survey data, obtained in Poland (a developing country) and tries to find answers to the following questions:

1. What level of variability is observed in the use of water, gas, electricity and heat energy in the analyzed apartments?
2. Is this variability in consumption of energy observed mainly between households or does it also fluctuate over time?
3. What causes these differences between the households?
4. Is there any interdependence between the consumption of particular mediums by particular apartments?

2 Methodology

The analysis was conducted using the results of measurements taken in multifamily communal buildings (blocks of flats) in a large city in Poland. The data was obtained in 2015–2016 for the following mediums: heat energy for space heating (SH), domestic hot water (DHW), electric energy (EE), gas (G) and water (W). The data covers about one hundred apartments from four buildings of similar design that differ only by numbers of stairways and positioning. The sources of data were:

- questionnaires (respond rate 50%);
- weekly readings of EE, W and G consumption by households (8 weeks);
- daily readings of EE, W and G consumption by households (9 days);
- daily readings of heat energy used by households for SH and DHW (during one year).

This article is the first, brief analysis of the obtained data, so for SH and DWH the measurements only apply for internal apartments of buildings which differ only by their positioning. All of them comprise two rooms, a kitchen and a bathroom, with the total area of 46 m^2 and are characterized by heat transfer coefficient of about 31 [W/K]. In total there are 32 such apartments in the monitored buildings, but it was not possible to gain complete sets of data for all of them. For separate analysis of all other mediums (EE, W, G), all available measurements were taken into account (not only those coming from internal apartments).

Major part of the analysis was carried out using box plots where: the x sign is the mean of the data set, the horizontal band inside the box is the median, and bottom and top of the box show the first (Q1) and the third (Q3) quartiles. Meanwhile, the vertical lines (whiskers) represent the lowest and the highest values of data set within the range from Q1-1.5*(Q3-Q1) to Q3+1.5*(Q3-Q1).

3 Results

The results of the study are presented below as answers to questions placed in the introduction, and are grounded on the analyzed data.

3.1 Differentiation by the mediums use

Fig. 1 shows differentiation in consumption of energy (with differentiation by objectives and mediums) and water (by analyzed apartments). It turns out, even without data from apartments with extreme values of obtained measurements, the differences in energy use between apartments can reach factor five mark and six for water.

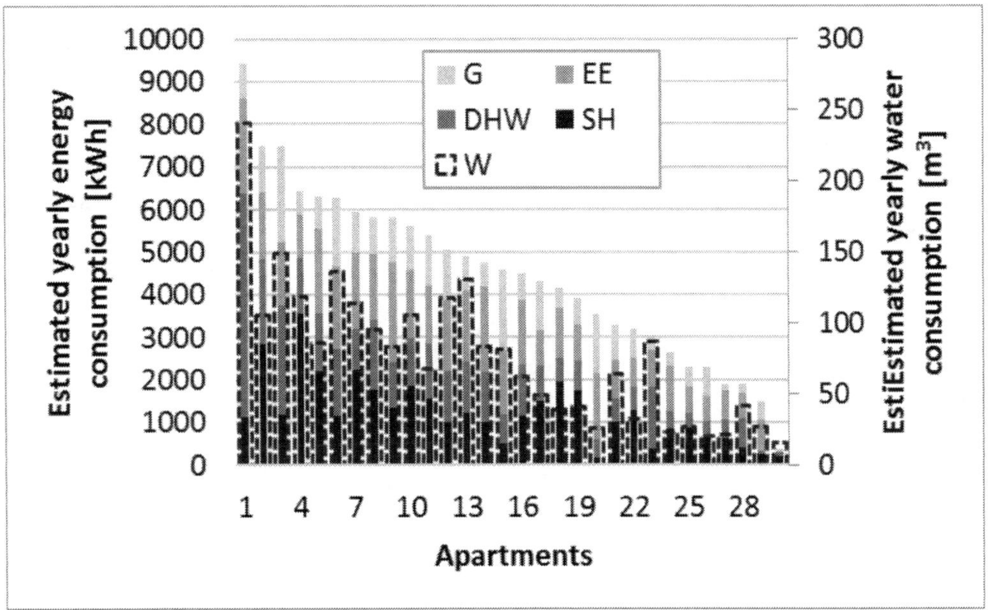

Fig. 1. Estimated yearly energy and water consumption for analysed apartments.

3.2 Weekly vs daily variations in consumption as compared to differences between particular households

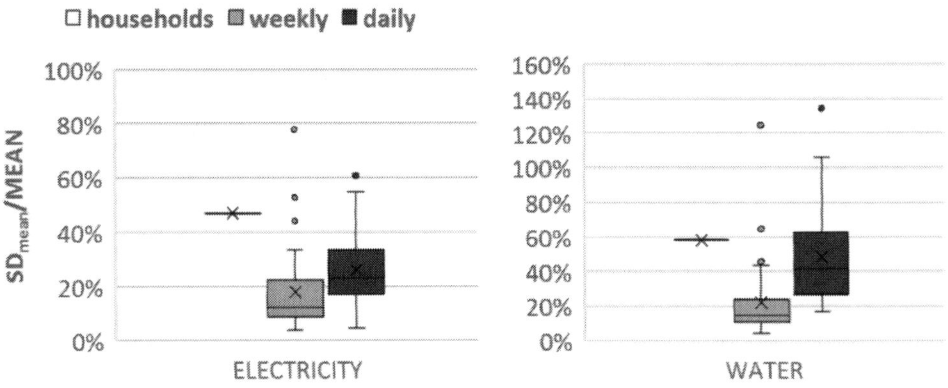

Fig. 2. Weekly, daily and households diversity in energy and water consumption.

For every household with relevant data available, mean daily and weekly values of electric energy and water consumption were defined on the grounds of the measurements taken, as well as the mean standard deviation of these values. Then the mean and mean standard deviation of the use of EE, W and G over the measurement period for all households was

DOI: 10.1051/e3sconf/20172200010

determined. The ratio of SD_{mean} to the mean was adopted as the measure of variation. Fig. 2 shows the obtained values; one can notice that independently of the medium, the largest differentiation can be found between households, and the most common recurrence can be observed in monitored weeks for a given household. The differentiation between monitored daily data is more distinct than between weeks. At the same time the value of SD_{mean}/MEAN for 24 h use differ greatly for particular households. Similar results were obtained by analysis of variations in G use, as well as by using a different measure of variation: the ratio of the difference between the highest and the lowest value to the median.

3.3 What does cause households consumption diversity?

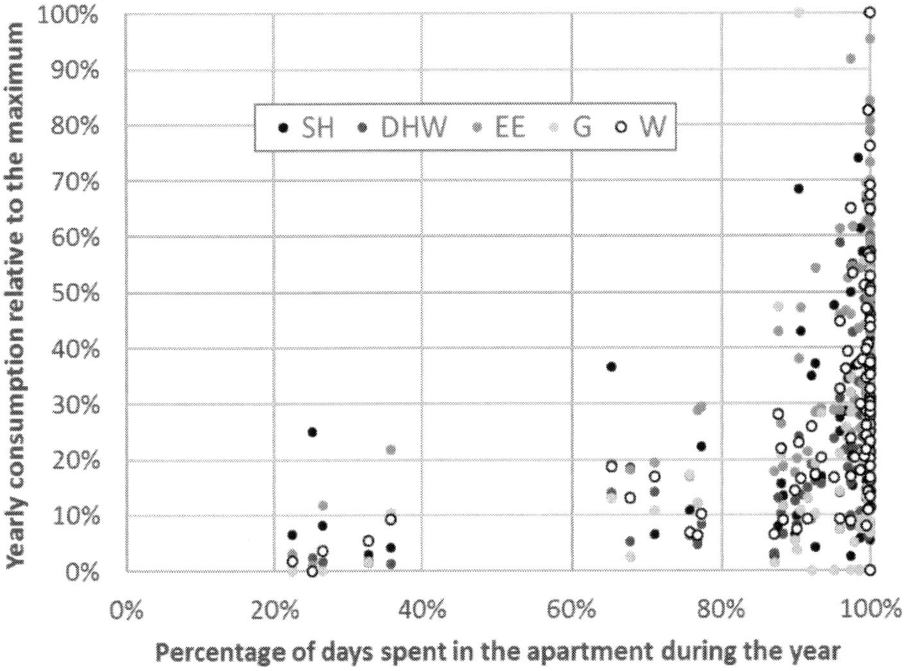

Fig. 3. The use of mediums as the function of number of days in the year when the apartment was occupied (lived in).

The primary reason for variations in the mediums use for particular apartments is the fact that some of them were occupied only for certain periods of time during the year. It can be stated on the grounds of data obtained from 24 hour readings of heat meters in the apartments, as shown in Fig. 3.

Another very important factor is the number of people in particular households. The use of most mediums grew with the increase of this number – the fastest rise was in EE and the slowest was in G. A very interesting correlation was observed in case of SH use, as shown in Fig. 4 – left. The rising trend was observed only for single- or two-person household, meanwhile decrease was observed in three- and more person households. This phenomenon might have appeared because of two reasons. First, three or four-personal households are usually families who prefer an active lifestyle and rarely stay at home, and thus can use the heating less (Fig. 4 - right). Second, the greater the household, the higher internal heat gains are obtained. These two reasons may be interlaced. The results of measurement campaign

 DOI: 10.1051/e3sconf/20172200010

also imply inappropriate ventilation of the rooms. For hygienic reasons multi-personal households require more intensive ventilation. Yet, as SH use in the most populated apartments is the lowest, it can be assumed that the fresh air inflow was over-reduced by the residents.

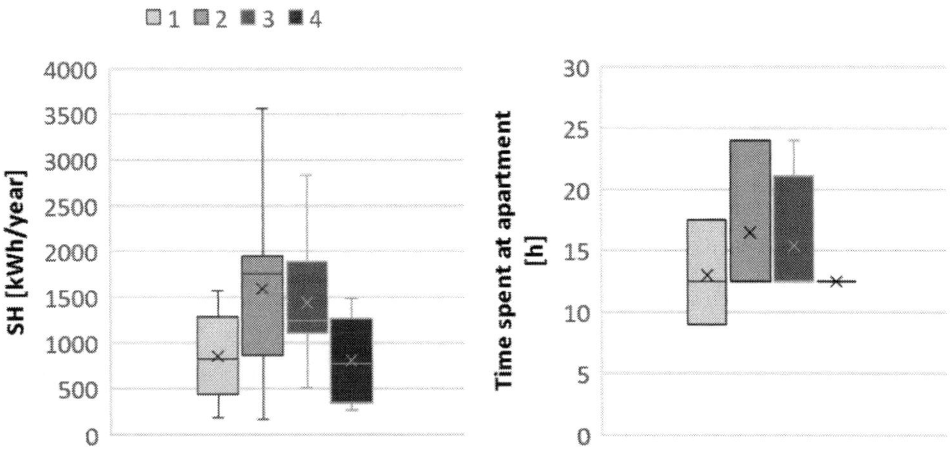

Fig. 4. The use of SH by households and time (during 24 hours) the residents spent in the apartment as the function of number of persons in households.

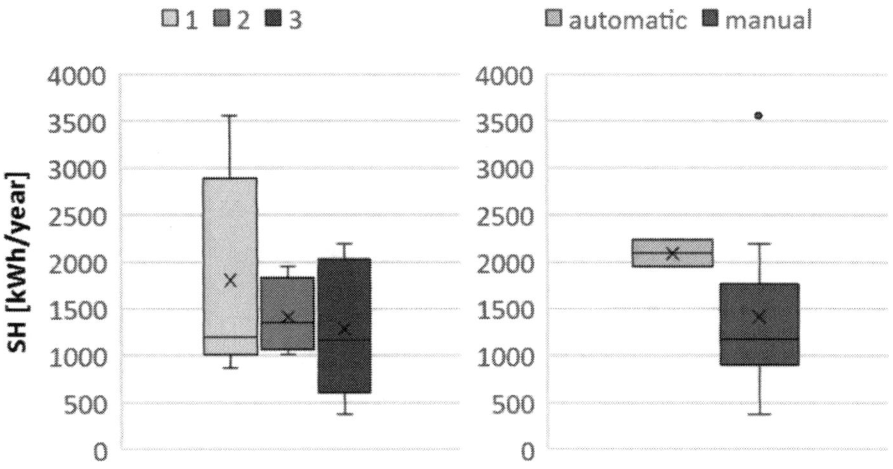

Fig. 5. Use of SH by households as the function of number of rooms without heating and the method of heating adjustment.

The use of SH is also influenced by method of adjustment of heating and ventilation devices. The following three aspects were chosen to be discussed.

- The number of heated rooms - the apartments under analysis comprise four rooms (kitchen, bathroom and two other rooms). Fig. 5 shows that the fewer rooms are heated the lower the average use of SH is. The effect is reinforced by the fact that apartments with lower temperature are heated by adjacent apartments with higher temperature.

 DOI: 10.1051/e3sconf/20172200010

- The method of heating adjustment - the residents can use in their apartments both programmable setting units and thermostatic valves. Fig.5 shows significant variations in SH use in apartments where heating is adjusted manually, as compared to those where programmable setting units are used. The probable reason for such differences could be the assumption that use of programmable unit alone leads to heating of all rooms, while manual adjustment of valves allows for meeting the real needs only.
- Adjustments by opening and closing windows (Fig. 6). The results of the measurement campaign show how excessive limiting of ventilation can decrease SH use, and lack of coordination in terms of ventilating and heating rooms can increase SH use.

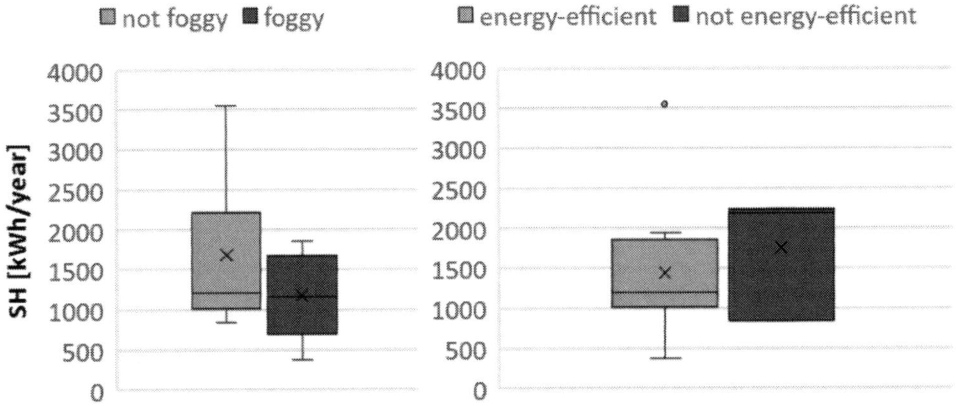

Fig. 6. SH use and ventilation. The diagram on the left presents SH use with the division into apartments where windows are steamed up in winter and those where they are not. The diagram on the right presents SH use with the division into apartments whose residents avoid simultaneous airing and heating of rooms (energy efficient use) and those whose residents do not care.

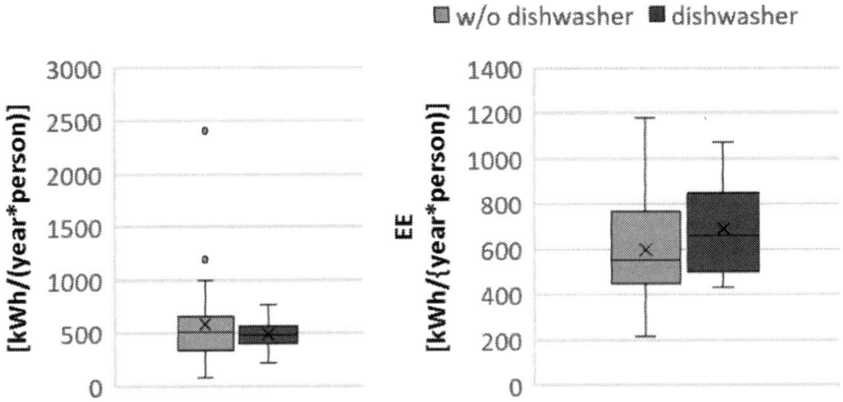

Fig. 7. The impact of dishwasher on the use of DHW and EE.

The use of dishwasher can serve as a good example of interdependence of different mediums use. Fig. 7 shows the use of *DHW* and EE per person by households with and without a dishwasher. One can see significant differences in use of EE, however, it is not so

DOI: 10.1051/e3sconf/20172200010

definite for *DHW* (*W* use has similar profile). The cause for such variations to appear can be ineffectiveness of dishwashers as compared to manual washing up. However, it has to be mentioned that households with dishwashers are characterized by narrower range of variations in DHW and EE use. The obtained readings show that higher effectiveness of manual washing up is strongly dependent on particular households, while the use of dishwashers gives more "homogeneous" results.

3.4 Correlation between utilities consumption

If the use of mediums depends on lifestyle, it probably would be possible to notice correlation in particular mediums use by households. As presented in Table 1, such correlations do exist. The most evident one is between DHW and W, as well as the one between DHW and EE or W and EE.

Table 1. Coefficient of determination (R^2) for correlation between the use of analyzed mediums by households.

	SH	DHW	EE	G	W
SH	1	0.18	0.16	0.10	0.23
DHW	x	1	0.40	0.16	0.68
EE	x	x	1	0.14	0.57
G	x	x	x	1	0.13
W	x	x	x	x	1

4 Conclusion

The article discusses in brief some of the results of mediums' use measurement campaign in multifamily buildings: cold water (W), electric energy (EE), gas (G), energy for space heating (SH) and domestic hot water (DHW). The presented data was analyzed in terms of differentiation in the use of mediums and possible reasons of such variability. It was proved that occupants can have significant influence on consumption of the mediums: characteristics of households as well as methods of heating and ventilation adjustment. On the basis of the conducted analysis it can be also stated that the problem of ineffective natural ventilation or its ineffective use is widespread and common. This is why development of modern solutions for energy-efficient multifamily constructions, which is still in high demand, requires an appropriate design not only in terms of the building's shape and its technical equipment, but also methods of control and adjustment of microclimate inside in the context of needs and behavior of its occupants.

This article's follow-up will be comprehensive analysis of data obtained from measurements and questionnaires. It will include normalized data concerning SH use, as this type of data for apartments with various heat loss coefficients cannot be compared directly. The data will also allow for estimation of ventilation-related heat losses for particular apartments and assessment of its effectiveness as well as analysis of the influence of adjacent apartments on each other in terms of heat losses/gains.

DOI: 10.1051/e3sconf/20172200010

Acknowledgment

We would like to thank the following people for supporting our measuring campaign: Małgorzata Fengler and Patrycja Mazurkiewicz – students; Mr. Jacek - keeper of the buildings; Sylwia Hemmerling from ZZN Sp. z o. o.; Wojciech Michalak, Agnieszka Staniewska, Małgorzata Piwińska from ZKZL Sp. z o. o and Łukasz Filip from Kamstrup.

References

1. W. Tkaczyk, Eds., *Gospodarka paliwowo-energetyczna w latach 2014 i 2015* (GUS, Warszawa 2016)
2. *Rachunki ekonomiczne środowiska*, 2016. [Online]. Available: http://stat.gov.pl/obszary-tematyczne/srodowisko-energia/srodowisko/rachunki-ekonomiczne-srodowiska,7,1.html
3. P. Hoes, J.L.M. Hensen, M. G.L.C. Loomans, B. de Vries, D. Bourgeois, Energy Build. **41**, 3 (2009)
4. H. Yoshino, S. Chen, Eds., *Total Energy Use in Buildings: Analysis and Evaluation Methods (Annex 53)*, (IEA EBC An. AECOM 2016)
5. F. Mizuki, K. Mikawa, H. Kurisu, Hitachi Rev. **61**, 3 (2012)
6. O. Guerra-Santin, L. C. M. Itard, Build. Res. Inf. **38**, 3 (2010)
7. R. K. Andersen, *10th International Conference on Healthy Buildings* (2012)
8. R. Haas, H. Auer, P. Biermayr, Energy Build. **27**, 2 (1998)
9. A. Al-Mumin, O. Khattab, G. Sridhar, Energy Build. **35**, 6 (2003)
10. J. Ouyang, K. Hokao, Energy Build. **41**, 7 (2009)
11. G. M. Huebner, J. Cooper, K. Jones, Energy Build. **66** (2013)

Environmental impact assessment of coal power plants in operation

Ayfer Bartan[1], *Serhat Kucukali*[2,*], and *Irfan* Ar[1]

[1]Gazi University, Chemical Engineering Department, Maltepe, Ankara, Turkey
[2]Cankaya University, Civil Engineering Department, 06790, Ankara, Turkey

Abstract. Coal power plants constitute an important component of the energy mix in many countries. However, coal power plants can cause several environmental risks such as: climate change and biodiversity loss. In this study, a tool has been proposed to calculate the environmental impact of a coal-fired thermal power plant in operation by using multi-criteria scoring and fuzzy logic method. We take into account the following environmental parameters in our tool: CO, SO_2, NO_x, particulate matter, fly ash, bottom ash, the cooling water intake impact on aquatic biota, and the thermal pollution. In the proposed tool, the boundaries of the fuzzy logic membership functions were established taking into account the threshold values of the environmental parameters which were defined in the environmental legislation. Scoring of these environmental parameters were done with the statistical analysis of the environmental monitoring data of the power plant and by using the documented evidences that were obtained during the site visits. The proposed method estimates each environmental impact factor level separately and then aggregates them by calculating the Environmental Impact Score (EIS). The proposed method uses environmental monitoring data and documented evidence instead of using simulation models. The proposed method has been applied to the 4 coal-fired power plants that have been operation in Turkey. The Environmental Impact Score was obtained for each power plant and their environmental performances were compared. It is expected that those environmental impact assessments will contribute to the decision-making process for environmental investments to those plants. The main advantage of the proposed method is its flexibility and ease of use.

1 Introduction

Measures for environmental control in modern coal-fired thermal power plants may be useful in reducing SO_2, NOx, and particulate matter emissions. It has become important to calculate the environmental impacts of such power plants with the cumulative impact assessment, which has also taken place at the legal level. Coal-fired thermal power plants have been subject to many studies because of their high environmental impact. When studies in the literature are evaluated, it is concluded that the following gaps are identified:

* Corresponding author: kucukali@cankaya.edu.tr

existence of deeper studies mostly on more specific issues, (ii) lack of studies to evaluate environmental parameters as a whole, (iii) lack of studies on the development of cumulative environmental impact analysis [1]. Moreover, previous studies aimed to calculate the environmental impact index of coal-fired thermal power plant at the planning stage [1]. However, this study focused on calculating the environmental impact index of a thermal power plant in operation. A new tool has been proposed to calculate the environmental impact index, which is used in combination with fuzzy logic membership functions and multi-criteria scoring technique. This study evaluates the environmentally monitored parameters of a coal-fired thermal power plant in operation.

2 Environmental impact index parameters membership functions

Environmental parameters imposed by the relevant Regulations for monitoring in thermal power plants in Turkey are given in Table 1.

Table 1. Environmental parameters related to coal-fired thermal power plants in operation.

Name of Monitored Parameters	**Monitored Parameter Unit**		**Regulation**
Emissions	SO_X	mg/Nm^3	Regulation on Control of Industrial Air Pollution
	NO_X	mg/Nm^3	Regulation on Control of Industrial Air Pollution
	CO	mg/Nm^3	Regulation on Control of Industrial Air Pollution Implementing Regulation on Greenhouse Gas Emissions
	Particulate Matter	mg/Nm^3	Regulation on Control of Industrial Air Pollution
Water Temperature	Temperature	^0C	Water Pollution Control Regulation
Waste Management	Fly Ash and Bottom Ash	Kg / year	Regulation on the Regular Storage of Wastes Environmental Impact Assessment Regulation Waste Management Regulation Notice on recycling of Some Non-dangerous wastes

In this study, environmental parameters were selected which were continuously monitored in thermal power which are required by the legislation. However, some parameters have been left out of the assessment because they are not actively monitored by the responsible government agency. In summary; In Turkey, eight important environmental parameters that can be assessed consistently in thermal power plants or objectively evaluated by documented evidence have been evaluated. These are: SO_2, NOx, CO, particulate matter (dust), water temperature (T_w), fly ash, bottom ash, water intake for cooling water (Fish Friendly Pump). In the environmental legislation CO emission allowable value is defined as 200 mg/Nm^3 (Fig. 1); while the limit value for NO_X emissions is 600 mg/Nm^3 for coal-fired power plants (Fig. 2).

 DOI: 10.1051/e3sconf/20172200011

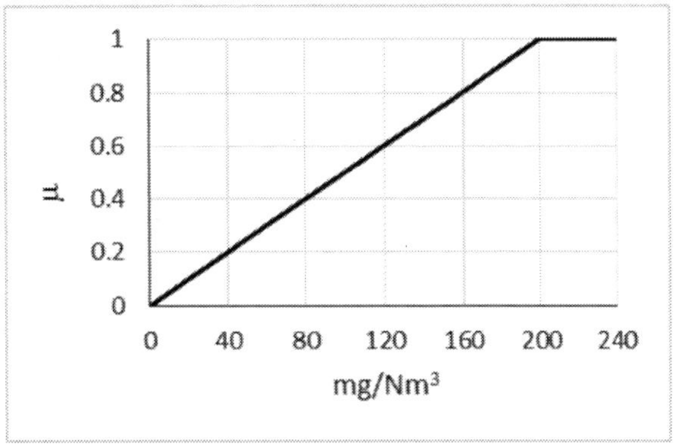

Fig. 1. Environmental impact score membership function of CO emissions.

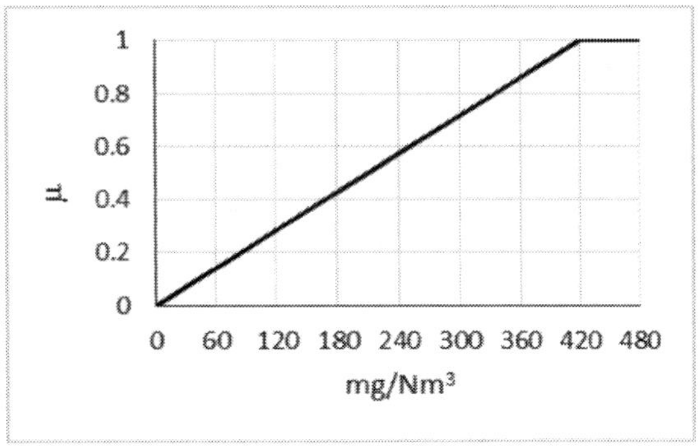

Fig. 2. Environmental impact score membership function of NOx emission.

The SO_2 emissions are set at a limit value based on the fuel thermal power and the installed power of the Cates thermal power plant is stated as 314 MW. According to this, in the legislation the limit value is defined as linear decrease as 2000–400 mg/Nm^3. The environmental impact value when exceeding the specified threshold value shall be considered as 1 (one) if it does not exceed 0 (zero).

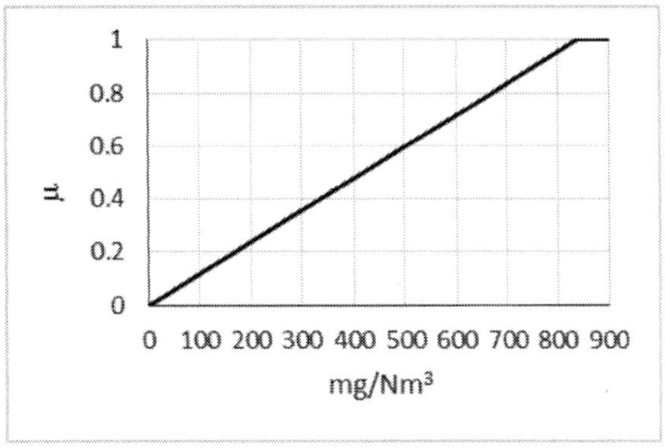

Fig. 3. SO_2 emissions environmental impact score membership function.

 DOI: 10.1051/e3sconf/20172200011

Dust emissions are not dependent on fuel heat power, but for all plants the threshold value is defined as 100 mg/Nm^3 in the legislation (Fig. 4).

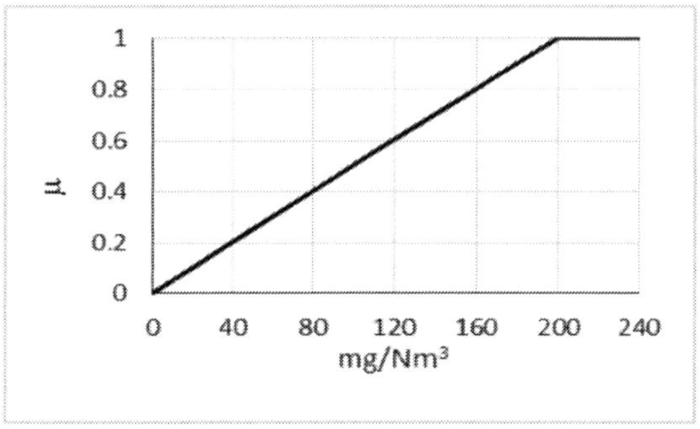

Fig. 4. Environmental impact score membership function of particulate matter.

It was stated that, regardless of the dilution capacity of the marine environment, the temperature of the waters to be discharged to the sea would not exceed the limit value of 35°C in the legislation (Fig. 5). Accordingly, the limit value was chosen as 35°C by taking environmental legislation as expert judgment.

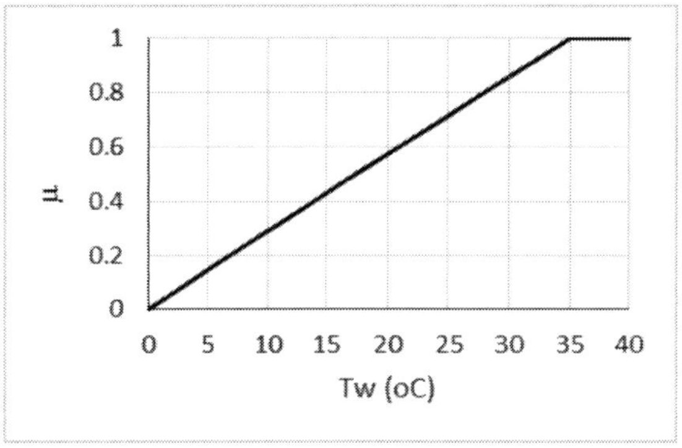

Fig. 5. Environmental impact score membership function of discharged water temperature.

For bottom ash, on evidence basis, membership functions are defined in Fig. 6.

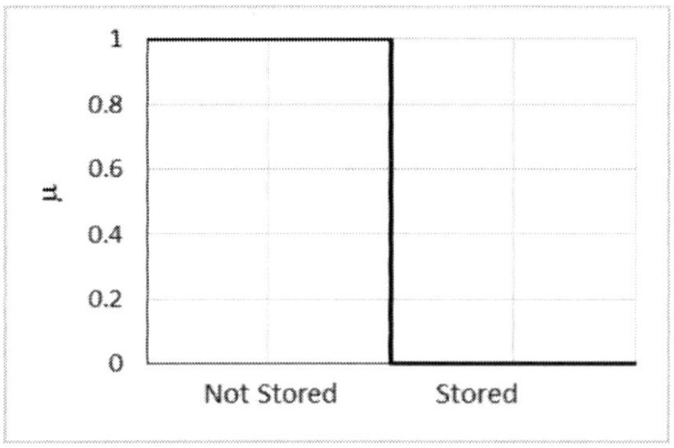

Fig. 6. Bottom ash environmental impact scoring membership function.

The thermal power plant is being sold for use in fly ash concrete industry. Fly ash is considered to be environmentally beneficial because it is not left as waste and forms the input part of different industries as recycling (GD). For this reason, as the membership function is sold, the impact score is evaluated as zero (0). Otherwise it will be taken as one (Fig. 7).

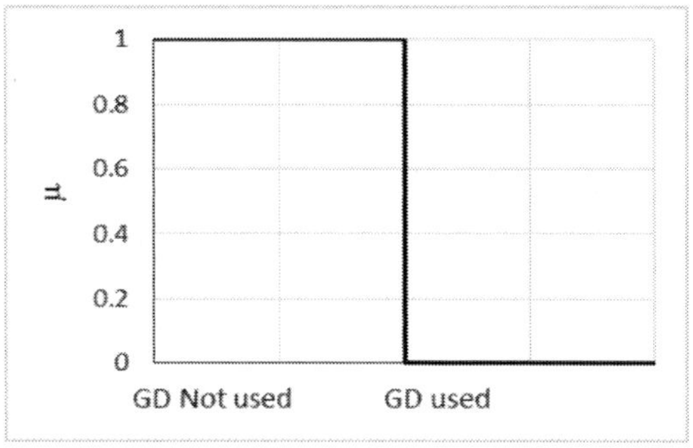

Fig. 7. Fly ash environmental impact scoring membership function.

When the water intake structure is evaluated, it is emphasized that fish friendly water intake should be used. Since the use of fish-friendly pump was not considered at this point, the impact score was evaluated as 1 because the water intake structure had a negative effect on aquatic organisms (Fig. 8).

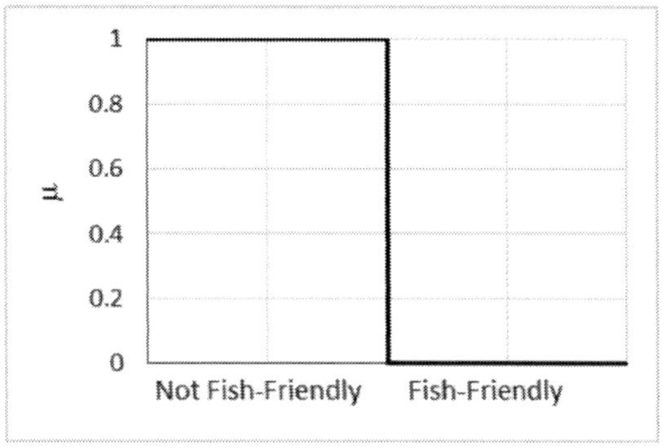

Fig. 8. Environmental impact score membership function of fish-friendly water intake.

3 Calculation of the environmental impact index

Previous studies in the literature aimed to calculate the environmental impact index of the coal-fired thermal power plant at the planning stage. In addition, studies on the environmental impact of a coal-fired thermal power plant operating in the literature have focused on a single environmental parameter, neglecting the cumulative effect. In this study, it was focused on calculating the environmental impact index of a thermal power plant in operation. In the calculation of the Environmental Impact Index; measured data, documented evidence (i.e. drawings, contracts), and site visits were taken as basis. A similar method has been previously developed by Kucukali [2] for the calculation of the Environmental Risk Index of river type hydroelectric power plants in operation, adapted to Coal-fired thermal plants. However, Kucukali [2] did not use the fuzzy logic membership

 DOI: 10.1051/e3sconf/20172200011

functions in his work. In the establishment of fuzzy logic membership functions, Ak et al. [3] method has been employed. The threshold values determined by the legislation in the establishment of the limit values of the fuzzy logic membership functions. Multi-criteria scoring technique was used in the proposed method. Here, a Plant Environmental Impact Index (EII) value is calculated from the following formula:

$$CSI = w_1 \times S_1 + w_2 \times S_2 + \ldots w_n \times S_n \tag{1}$$

in which, w_1 is the weight ratio of the related parameter, S_1 is the score of the corresponding parameter calculated from the fuzzy logic function, and n is the number of environmental parameters. In this study, 8 environmental parameters were used and all parameters were given equal weight. The EII varies from 0 to 1 and the high value of the EII indicates that the plant has a high environmental impact. The value of the EII of a plant is calculated on a monthly basis, depending on the change in operating and working conditions. One of the most important parts of the proposed method is that the measured environmental parameters are transformed into a dimensionless common measure at 0–1 interval.

4 Application of the proposed tool to the thermal power plants in turkey

The proposed method is based on 4 thermal power plants in Turkey. The locations of these thermal power plants in Turkey is shown on map in Fig. 9 and the technical characteristics of those power plants are presented in Table 2.

Table 2. Characteristics of the coal-fired thermal power plants for which the Environmental Impact Index is calculated in the scope of the study.

Power Plant Name	In Operation since	Installed Power (MW)	Cooling Water
Tunçbilek, Kütahya	1956	365	Cooling tower
Yeniköy, Muğla	1986	420	Cooling tower
Yeniköy, Muğla	1993	630	Aegean Sea
Soma(B), Manisa	1981	990	Cooling tower

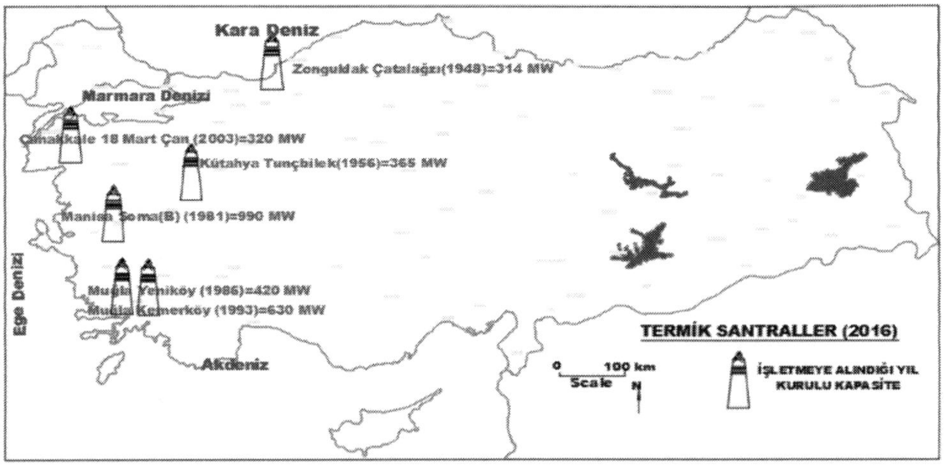

Fig. 9. Locations of the Thermal Power Plants in Turkey, for which Environmental Impact Indexes are calculated within the scope of the study.

The calculation of the Environmental Impact Index takes into account the gases emitted (CO, SO_2, NO_X), the effect of the water intake on aquatic life, the discharge cooling water temperature, the fly ash, and the bottom ash parameters. The hours the plant does not work are not taken into account. The values between 01.10.2014 and 31.10.2014 are taken into account for all power plants. Thus, each of the 8 identified environmental impact parameters were examined separately. Percentage of observed values per hour in the environmental parameters scored at 5% of the time (Fig. 10) and other parameters were scored according to documented evidence. In addition, it is necessary to make a site visit to the plant where the environmental performance is measured and to observe the examined environmental parameters in the proposed method. Then, the Environmental Impact Index of each plant was obtained by adding up the cumulative effect of these parameters. In the evaluation 0-1 interval, environmental impact index was calculated by fuzzy logic application according to membership status. If the Environmental Impact Index is close to 0, it will be considered as a fully environmentally friendly plant. As far as the result goes from 0 to 1, the environmental impact will be considered as a high facility. The Environmental Impact Index for four more coal-fired thermal power plants was found. The results are given in Table 3.

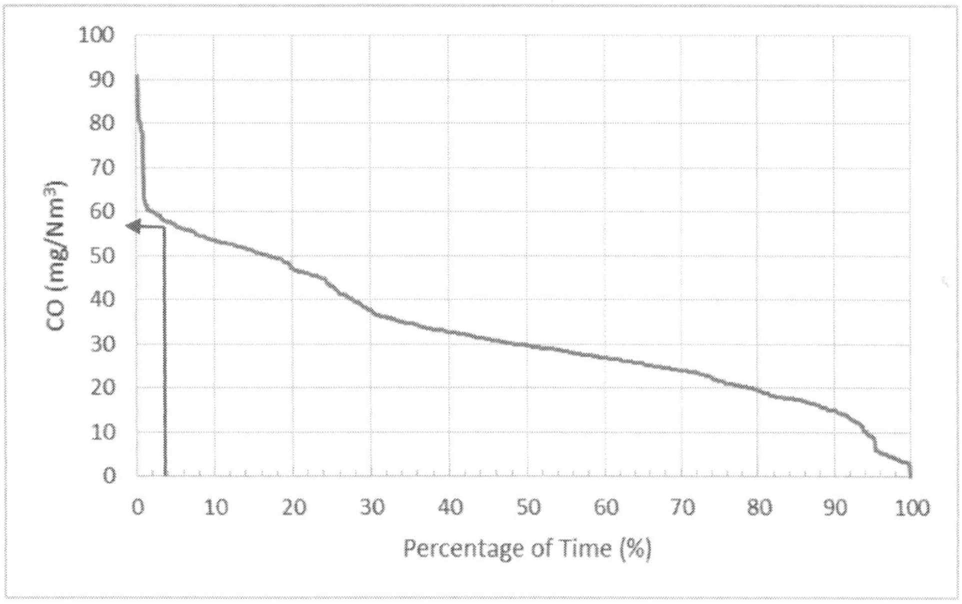

Fig. 10. CO emission duration curve for Kemerköy coal fired thermal power plant and the 5% value. The corresponding fuzzy membership value is read from Fig. 1 as 0.28. (Sampling period: 01/10/2014-31/10/2014 for hourly-averaged values).

Table 3. Environmental impact scores of the studied thermal power plants in operation (monthly evaluation between 01/10/2014-31/10/2014).

Coal Fired Thermal Power Plant	CO	NOx	SO_2	Particulate matter	T_w	Fish Friendly Pump	Fly Ash	Bottom Ash	Environmental Impact Index
Kemerköy	0.28	0.91	1	0.89	0.82	0	1	0	0.61
Yeniköy	0.05	0.91	1	0.94	0	0	1	0	0.49
Tunçbilek	0.07	1	1	1	0	1	0	0	0.52
Soma	0.15	0.78	1	1	0	0	1	0	0.48

5 Conclusions

For the environmental impact assessment of coal-fired power plant a composite model, which is a combination of statistical analysis and fuzzy logic method, is developed. Thus, proposed tool considers the environmental legislation as an expert in order to determine the boundaries of fuzzy membership functions. The advantage of this method is flexibility and ease of use. But, the proposed method have some limitations. It can be used only during the operation phase and it requires environmental monitoring data of power plant. Therefore, the data availability and collection could represent a limitation. The proposed method has been applied to several coal-fired power plants in Turkey and Environmental Impact Index has been calculated for these power plants. It is expected that the proposed tool will contribute to determine the environmental performance of the thermal power plants in operation. Also the tool can be used in the decision-making process related to environmentally related investments. The results ıf the case studies demonstrated that the proposed tool can easily be used by the professionals to estimate the environmental impact of a coal power plant in the operation phase.

References

1. A. Bartan, PhD Thesis, Gazi University (2017)
2. S. Kucukali, Energy Sustain. Dev. **19** (2014)
3. M. Ak, E. Kentel, S. Kucukali, Renew. Sust Energ. Rev. **68** (2017)

Variability of road traffic noise recorded by stationary monitoring stations

Andrzej Bąkowski[1,*], and *Leszek* Radziszewski[1]

[1]Kielce University of Technology, Aleja Tysiąclecia Państwa Polskiego 7, 25-314 Kielce, Poland

Abstract. The paper presents the analysis results of equivalent sound level recorded by two road traffic noise monitoring stations. The stations were located in Kielce (an example of a medium-size town in Poland) at the roads out of the town in the direction of Kraków and Warszawa. The measurements were carried out through stationary stations monitoring the noise and traffic of motor vehicles. The RMS values based on A-weighted sound level were recorded every 1 s in the buffer and the results were registered every 1 min over the period of investigations. The registered data were the basis for calculating the equivalent sound level for three time intervals: from 6:00 to 18:00, from 18:00 to 22:00 and from 22:00 to 6:00. Analysis included the values of the equivalent sound level recorded for different days of the week split into 24h periods, nights, days and evenings. The data analysed included recordings from 2013. The coefficient of variation and positional variation were proposed for performing comparative analysis of the obtained data scattering. The investigations indicated that the recorded data varied depending on the traffic routes. The differences concerned the values of coefficients of variation of the equivalent sound levels.

1 Introduction

The Resolution of the European Committee on the requirement of developing, making accessible and updating noise maps has again drawn the attention of communities to environmental noise defined as a factor greatly affecting comfort of life [1]. Noise assessment relies on various measurands, including short- or long-term noise indicators [2]. Acoustic measurements are made at selected sites of the assessment area and last from a few hours to several days. In the next step, models and simulations are used to determine the values of these measurands and their uncertainties. Determining long-term indicators, which should be representative of the whole calendar year, is especially difficult. To this end, traffic noise and vehicle monitoring systems using permanent monitoring terminals were installed in some cities to record the values of the measurands throughout the year. Such systems were constructed in Bydgoszcz, Gdańsk, Kielce, Szczecin, and Wrocław. The outcome of these stations, available on the Internet in real time, describes the changes in traffic structure and volume and indicates hazards arising from the negative effects of noise on people present in a particular part of the city. Kielce has more than ten such stations,

* Corresponding author: abakowski@tu.kielce.pl

both in the centre and on the outskirts. This study analyses measurement results, the equivalent sound level, recorded by two of the stations. Kielce was chosen as an example of a medium-size town (a population of approximately 200,000) located in the southern part of central Poland. Temperatures within a year range approximately from -5°C in January to +17°C in July. Average monthly precipitation is from 34 mm in October to 96 mm in July. The wind, predominantly from the south and west, reaches an average speed of about 10 km/h over a year. Kielce gets on average 70 days a year of snow on the ground.

2 Traffic noise and volume monitoring stations

The data under analysis were recorded by permanent, automatic sound and traffic volume monitoring stations located in Krakowska Road and in Warszawska Road [3]. Krakowska Road, with four lanes of traffic separated by a 3-m grass median, is the major part of the Kraków-bound thoroughfare functioning as both the transit route and the city street. The structural condition of the bituminous pavement is good. The noise monitoring station is located at the site with the following coordinates 50°51'37.8"N and 20°35'41.2"E. It is situated between two busy intersections. Opposite the road, there is a large handball hall and the bus depot of the Municipal Public Transport Service. The microphone for measuring acoustic pressure is situated at a distance of 4 m from the edge of the roadway at the height of 4 m.

Data under analysis comes from records for 2013. The database of the recorded data, created in the form of calculated values of the equivalent sound level, comprises 905 records. Due to technical difficulties, the database is not complete and covers only the period between 8 January and 6 December 2013 split into three sub-intervals of a 24-hour interval: day time, evening time and night time. Technical problems in maintaining continuous measurement of data are frequent at 24h stations and lead to increased uncertainty as for the measurement results reported.

Warszawska Road is a two-lane road out of Kielce. The function of this road has changed. It was originally designed for heavy vehicle loads. Currently, after transformation, the road is being used mainly for municipal transport and, to a slight extent, as a transit route to Warszawa. The structural condition of the bituminous pavement is good. The noise monitoring station is located at the site with the following coordinates 50°54'15"N and 20°40'32"E. The measured equivalent sound level data from this station cover the 2012 to 2013 period, with the sample size of 1022 records.

The monitoring terminals mounted in Kielce include a sound level meter, a safety camera and a weather station. The acoustic measurements were carried out using SVAN 958A device, which is a four-channel digital vibration analyser and a class 1 sound level meter, operating within the measuring frequency range 0.5 Hz to 20 kHz, depending on a microphone used. The microphone for the acoustic pressure measurements is mounted at a distance of 4 m from the edge of the road at a height of 4m. The measuring microphone used in this study was a Microtech Gefell MK250 free-field, prepolarised 1/2" condenser microphone with a sensitivity of 50 mV/Pa, SV 12L preamplifier, a frequency range 3,5 Hz to 20 kHz and a dynamic range from 15 dB to 146 dB. The temperature range within which the device is operable is from -50°C to 100°C. The resolution of the signal RMS detector is 0.1 dB. The measurements were carried out 24 hours a day. The RMS values of the A sound level were registered in the buffer every 1 s and the results were recorded every 1 minute. The equivalent sound level was calculated on this basis for three time intervals, i.e., from 6:00 to 18:00, from 18:00 to 22:00 and from 22:00 to 6:00. Traffic volume was measured with a digital radar 245 MHz by WAVETRONIX. Weather data were recorded by the VAISALA WTX 510 automatic meteorological station.

 DOI: 10.1051/e3sconf/20172200012

2 Measurement results

Examples of the equivalent sound level data recorded by the monitoring stations in Krakowska and Warszawska roads are shown in Fig. 1.

a) b)

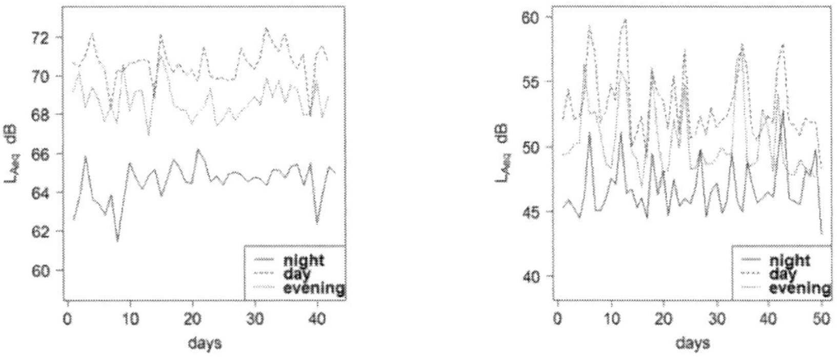

Fig. 1. Equivalent sound level L_{Aeq}, split into time sub-intervals for Mondays in 2013, recorded in Kielce a) Krakowska Rd, b) Warszawska Rd.

Analysis of the L_{Aeq} curves in Fig. 1 and of the box plots in Fig. 2 shows that in most cases, the distribution of the analyzed variable values is not normal. Also, compared with Warszawska Road, the L_{Aeq} parameter in Krakowska Road has noticeably higher values, by about 16 dB. In all cases, the data show certain values that can be regarded as atypical (outliers). This phenomenon is thus characterized by high randomness, which is consistent with the findings reported in the literature [4].

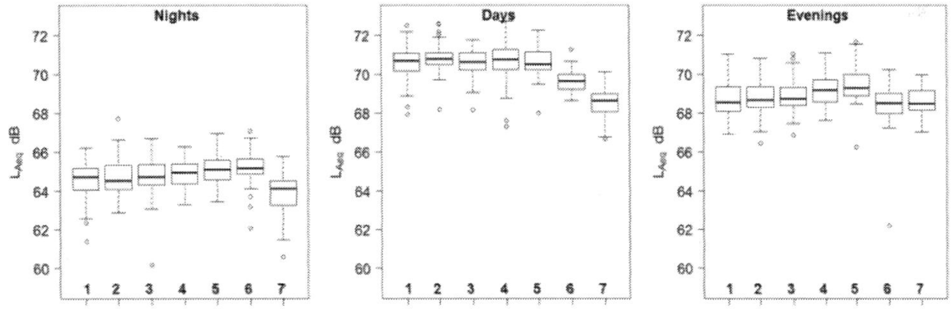

Fig. 2a. Equivalent sound level, L_{Aeq}, split into time sub-intervals for all week days in 2013, recorded in Kielce - Krakowska Rd., (denotation: 1 – Monday, 2 – Tuesday, 3 – Wednesday, 4 – Thursday, 5 – Friday, 6 – Saturday, 7 – Sunday).

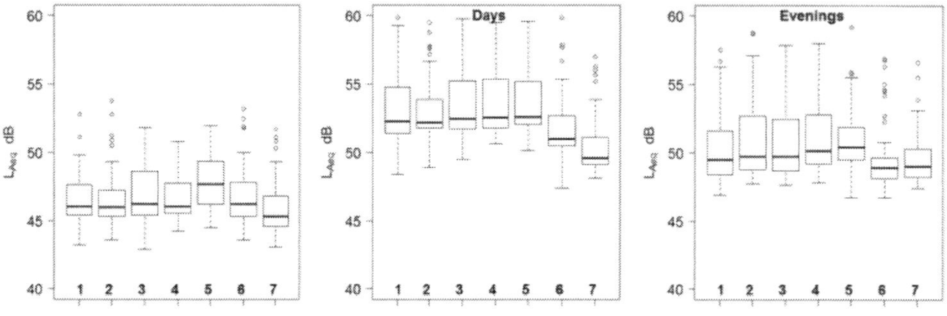

Fig. 2b. Equivalent sound level, L_{Aeq}, split into time sub-intervals for all week days in 2013, recorded in Kielce - Warszawska Rd.

The most commonly used measurand for the assessment of $L_{Aeq,T}$ expressed in terms of decibels, is defined as follows:

$$L_{Aeq,T} = 10 \cdot \log\left[\frac{1}{T}\int_0^T \left(\frac{p_A(t)}{p_0}\right)^2 dt\right] = \ 10 \cdot \log\left[\left(\frac{p_{A_{RMS}}}{p_0}\right)^2\right] \quad (1)$$

where:

T – is the total time of measurement, s

$p_A(t)$ - is the A-weighted acoustic pressure,

p_0 – is the reference acoustic pressure, standardized as 20 μPa.

The non-linear character of logarithmic function changes is a limitation that impedes the determination of, for example, standard deviation or measurement uncertainty. Therefore, the authors of this paper decided to determine the RMS sound pressure in the T period from equation 1 and use this parameter in further analysis.

$$p_{A_{RMS}} = p_A = \sqrt{10^{(0.1 \cdot L_{Aeq,T})} \cdot p_0^2} \quad (2)$$

In this study, the authors analysed the measurement data expressed in terms of Pa to be able to compare the fixed components (the mean and the median) and variable components (deviation from the mean) of the signals recorded. The logarithmic scale of the equivalent sound level ($L_{Aeq,T}$) can hinder a direct comparison of measurement data.

Figure 3 and Table 1 summarize the calculated RMS values, $p_{A_{RMS}}$, for Krakowska and Warszawska roads (on Mondays).

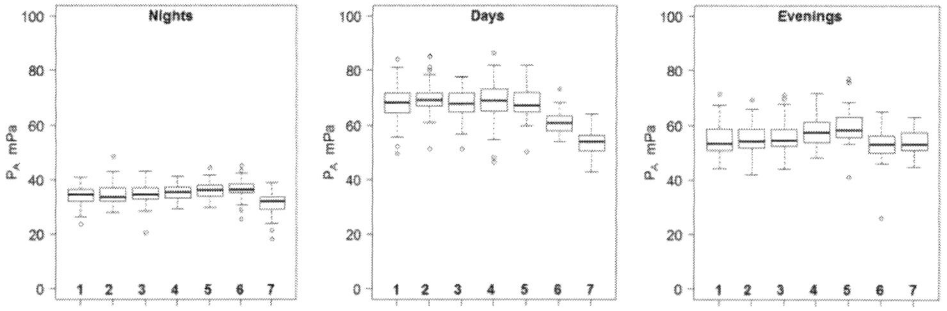

Fig. 3a. Root means square (RMS) acoustic pressure for days of the week split into time sub-intervals, recorded in 2013 in Krakowska Rd.

 DOI: 10.1051/e3sconf/20172200012

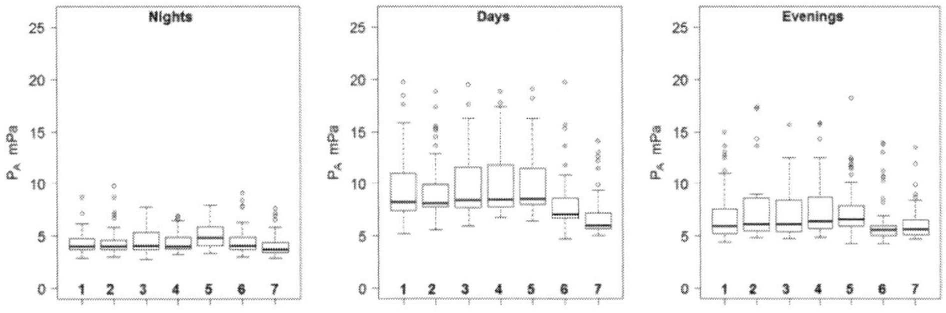

Fig. 3b. Root means square (RMS) acoustic pressure for days of the week split into time sub-intervals, recorded in 2013 in Warszawska Rd.

The median, the mean value, standard deviation, the coefficient of variation, the quartile deviation and the positional variation used further in the analysis were determined from the following dependencies. The mean value of $p_{A_{RMS}}$ can be determined as the expected value:

$$\overline{p_{A_{RMS}}} = \frac{1}{n}\sum_{i=1}^{i=n} p_{A_{RMS}} \tag{3}$$

Standard deviation of the measurement data can be determined from:

$$\sigma_{p_{A_{RMS}}} = \sqrt{\frac{1}{n-1}\sum_{i=1}^{i=n}\left(p_{A_{RMS}} - \overline{p_{A_{RMS}}}\right)^2} \tag{4}$$

The acoustic pressure variation was determined using the coefficient of variation, quartile deviation [5] and positional variation. The coefficient of variation can be calculated from:

$$COV_{p_{A_{RMS}}} = COV = \frac{\sigma_{p_{A_{RMS}}}}{\overline{p_{A_{RMS}}}} \cdot 100\% \tag{5}$$

This coefficient is a relative, dimensionless measure of acoustic pressure dispersion. The quartile deviation was determined from:

$$Q_{31} = 0.5 \cdot \left(Q_3\left(p_{A_{RMS}}\right) - Q_1\left(p_{A_{RMS}}\right)\right) \tag{6}$$

Quartile deviation is an absolute but dimensional measure of acoustic pressure dispersion. When referred to the median, quartile deviation gives positional variation, which can be determined from:

$$V_Q = \frac{Q_{31}}{Med} \tag{7}$$

Table 1. Root means square (RMS) acoustic pressure $p_{A_{RMS}}$ recorded on Mondays at Krakowska and Warszawska sites.

Time interval	Med mPa	$\overline{p_{A_{RMS}}}$ mPa	$\sigma_{p_{A_{RMS}}}$ mPa	COV %	Q_{31} mPa	V_Q %	SW test	JB test
Krakowska Rd.								
24h interval	52.67	52.18	15.32	29.36	14.53	27.59	0.00	0.03
Night time	34.48	33.89	3.59	10.60	2.15	6.23	0.20	0.11
Day time	68.36	68.19	7.19	10.54	3.46	5.06	0.24	0.70
Evening	53.34	54.89	5.96	10.86	3.89	7.29	0.19	0.21
Warszawska Rd.								
24h interval	6.18	7.06	3.34	47.28	1.75	28.32	0.00	0.00
Night time	4.01	4.47	1.15	25.82	0.52	12.96	0.00	0.00
Day time	8.24	9.65	3.42	35.39	1.71	20.75	0.00	0.00
Evening	5.97	7.04	2.66	37.76	1.17	19.60	0.00	0.00

Table 1 compiles the values of the medians, the expected values, standard deviation, the coefficient of variation, quartile deviation, positional variation and the results of Shapiro-Wilk (SW) and Jarque-Bera (JB) tests. Note that the COV is calculated using every recorded observation, including those that may be outliers. Thus, the outliers are not eliminated unless the positional measure of variation, V_Q, is used. Its numerator is calculated by subtracting the first quartile from the third quartile (Q_{31}), covering the middle 50% of the recorded data.

The Shapiro-Wilk test and the Jarque-Bera test provided sufficient evidence to reject the null hypothesis about normal distribution of the $p_{A_{RMS}}$ parameter for Warszawska Road, whereas no such evidence was found for Krakowska Road.

The values of L_{Aeq} recorded on subsequent Mondays in 2013 at the Krakowska Road site are compiled in Fig. 1a. The diagram indicates that on some days the L_{Aeq} parameter increases or decreases noticeably, especially at the time interval denoted as "day time". Analysis of the equivalent sound level for the 24 h period at the Krakowska Road site in 2013 indicated that the mean value of $L_{Aeq,T}$ was about 67.9 dB. The $L_{Aeq,T}$ values change with week days and time sub-intervals, as shown in Fig. 2a. Figure 2a summarizes the box plots for the year 2013 data but split into week days and 24 h sub-intervals: night time, day time and evenings. The diagrams show that the median of equivalent sound level:- for night times increases slightly from Monday to Saturday up to 65 dB and decreases on Sunday by about 1 dB, for day times is about 71 dB from Monday to Friday and decreases to about 69 dB on Saturday, for evening time is about 68 dB on Monday, Tuesday, Wednesday, Saturday and Sunday and increases on Thursday and Friday by 1 dB.

Analysis of the L_{Aeq} values recorded on subsequent Mondays of 2013 shown in Fig. 1b indicated that at the Warszawska site irregular variability over time is close to the impulse variability. This phenomenon is evident for the data recorded for each time sub-interval – several points can be indicated at which L_{Aeq} is higher by at least 5 dB that the neighbouring values. Analysis of the equivalent sound level for the 24 h period at the Warszawska Road site in 2013 indicated that the mean value of $L_{Aeq,T}$ was about 50 dB. These values vary with week days and time sub-intervals. Figure 2b summarizes the box plots for the year 2013 data but split into week days and 24h sub-intervals: night time, day time and evenings. The diagrams show that the median of equivalent sound level: for night times on Monday, Tuesday, Wednesday, Thursday, Saturday and Sunday is 46 dB and increases by 2 dB on Friday, for day times is about 52 dB from Monday to Friday and decreases to about 49 dB on Saturday and Sunday, for evenings is about 49 dB on Monday,

Tuesday, Wednesday, then increases by 2 dB on Thursday and Friday to decrease up to 49 dB on Saturday and Sunday.

On Saturday and Sunday evenings, the monitoring station recorded a lot of high-value atypical data. Analysis of the median values for all the data from 2013 and for the sub-intervals within the 24h interval, Fig. 2, indicates the following differences:

1. Warszawska Road – between day-night time sub-intervals: 5.9 dB, between day-evening time sub-intervals: 2.5 dB, and between evening-night time sub-intervals: 3.4 dB
2. Krakowska Road - between day-night time subintervals:5.6 dB, between day-evening time sub-intervals: 1.6 dB, and between evening-night time sub-intervals: 4 dB.

Analysis of changes in the V_Q value indicates that it was the lowest for the night time subintervals in both roads. Its highest value was recorded in Krakowska Road for the evening sub-intervals and in Warszawska Road for day time sub-intervals.

3 Conclusions

Comparison of the data obtained from both stations shows clear differences between the COVs of $p_{A_{RMS}}$. These differences are about 29% for Krakowska Road and about 47% for Warszawska Road. A considerable difference, of about 18 dB, was also observed between the medians of the equivalent sound level. Analysis of the medians, with measurement uncertainty taken into account, for the sub-intervals indicates that regardless of the monitoring station site, the differences between the values of the equivalent sound level are as follows: about 6 dB between day-night sub-intervals, 2 dB between day-evening sub-intervals, and about 4 dB between evening-night subintervals.

The coefficient of variation calculated for the $p_{A_{RMS}}$ parameter at the Krakowska Road site was about 29%, and for the Warszawska Road site about 47%, as shown in Table 1. The values of this parameter for particular sub-intervals vary but in Warszawska Road they are always about three times as high as in Krakowska Road. It has to be noted that higher interquartile ranges Q_{31} were observed for Krakowska Road. However, the relative values of the interquartile ranges, related to the median (positional variation V_Q), were higher in the case of Warszawska Road (comparable values were obtained only for 24h intervals)

References

1. Directive 2002/49/Ec of the European Parliament and of the Council, Official Journal of the European Communities – 18.07.2002
2. A. Jagniatinskis, B. Fiks, O. Zaporozhets, Procedia Engineering **134**, 301–308 (2016)
3. A. Bąkowski, L. Radziszewski, Z. Skrobacki, Procedia Engineering **177**, 281–288 (2017)
4. B. Przysucha, W. Batko, A. Szeląg, Arch. Acoust. **40**, 183–189 (2015)
5. A. Bąkowski, L. Radziszewski, M. Žmindak, Procedia Engineering **177**, 297–302 (2017)

The effect of the thermal inertia on the thermal transfer in building wall

Lahcene Bellahcene[1,*], *Ali* Cheknane[2], *SMA.* Bekkouche[3], and *Djemal* Sahel[4]

[1]Mechanical laboratory, Department of Mechanical Engineering, Amar Thelidji University of Laghouat BP 37G road of Ghardaïa, Algeria
[2]Laboratory for the Study and Development of Dielectric Materials and Semiconductor, Amar Telidji University of Laghouat BP 37G road of Ghardaïa, Algeria
[3]Applied Research Unit on Renewable Energies, URAER, B.P. 88, ZI, Gart. Taam, Ghardaïa, Algeria
[4]Gas Fuels and Environment Laboratory, Mechanical Engineering Faculty, University of Sciences and the Technology -Oran, USTO-MB, BP 1505, El-M'Naouer, Oran 31000, Algeria

Abstract. In a hot and dry climate, the design and construction of buildings involve the adoption of combination between shape of building envelope and construction materials. The objective of this work is to study the thermal behavior of a multilayer wall submitted to varying climatic conditions. We have proposed four configurations of an element of an outer wall. A numerical simulation was used to understand the phenomenon of thermal inertia, especially its influence on the resulting temperatures. The study is based on the modeling of heat transfer in a 2D unsteady-state using a computational fluid dynamics (CFD) code. The comparison of numerical results was affected with an available experimental data and shows a satisfactory agreement. In addition, this work highlights the importance of the study of the thermal inertia of the wall in order to ensure a comfortable indoor climate of building located in hot and dry climate.

1 Introduction

In desert regions, unsuitable adaptation of the building design and its indoor environment can result in a greater need for active climatisation and, thus, increased the using energy. The design and construction of a building involve the adoption of three elements: typology, shape and technology. The shape of a building is a crucial factor to achieving the relationship between cost and comfort [1]. Some properties that are related to building shape and which influence heating and cooling requirements are: compactness index, the height of walls, climate, and the characteristics of the building envelope. The previous characteristics are important variables of the energy requirements for maintaining the building at a comfortable temperature [2]. The thermal performance of the building envelope can make a significant contribution to decrease the needed energy. So, it is important to insulating the envelope of the building [3]. The determination of the temperature of the walls that constitute the building envelope may consider as an essential

* Corresponding author: belahcene.ste@gmail.com

DOI: 10.1051/e3sconf/20172200013

part of an efficient building design. The thermal performance of an opaque wall depends on how well its materials transfer the heat flux through it, and the weather conditions [4]. The building envelopes propose an essential element between occupied building spaces and weather conditions. Thermal insulation in the walls is often designed for static performance based on its thermal resistance [5]. Karlsson et al. [6] show the energy used for maintaining of comfortable indoor temperatures depends on the thermal storage capacity of walls in the contact with the indoor air. They studied the effects of increasing of the thermal storage capacity of building materials. Thermal inertia of buildings can help to reduce such consumption, improve comfort, and even replace HVAC (Heating, Ventilation and Air-Conditioning) systems. This thermal inertia is usually associated to heavy wall construction, but in fact, there is other parameters that can affect a significantly this property [7]. Many researchers reported the heat transfer in multilayer walls. Tamene [8] studied numerically heat transfer through an exterior multilayer wall of a house subjected to real conditions. Moreover, Ibrahim et al. [9] examined the energy behaviour of the buildings' multilayer exterior wall structures. The aim of the present work was to find the best wall structure and the number and position of insulation layers within exterior walls for continuous heating, intermittent heating, and no heating operation modes for some commonly used construction materials.

Generally, the thermal inertia is a character used for modeling heat transfers and it is a property of mass material related to thermal conductivity and capacity volumetric heat. In the envelope of building thermal transfer aspect, thermal inertia is responsible of the reduction of inside air temperature peaks and for the delay between the accumulation of energy and its respective release. Also, it depends on two factors namely time lag and decrement factor, they are influenced by thermo-physical properties of the material, thickness and wall position's. In the heat storage capacities the essential characteristics are time lag and decrement factor [10]. Numerous works have been performed in this field. Kontoleon et al. [11] analyzed in detail the impact of the concrete density and conductivity variations on decrement factor (f) and time lag (φ) on six insulated wall assemblies. Mavromatidis et al. [12] developed a numerical model of a multilayer wall. They used the time log and decrement factor for a building wall including different configurations of multilayer thermal insulation.

In this work, the study was devoted to the numerical simulation by using a computational fluid dynamics (CFD) code; this code is used to simulate the temperature in a 2D unsteady-state of a multilayer wall. We have proposed four configurations of an element of an outer wall. The main purpose of this study is to characterize the thermal inertia of building wall in hot and dry climate.

2 Mathematical models

In this section, a mathematical model is proposed to assess the energy performance of a multilayer wall. This model applies only to heat exchanges. A wall of four layers; two layers of cement mortar, a heavy stone layer and a layer of plaster. The wall is subjected to varying climatic conditions, where (L) and (H) represent the length and height, respectively. The direction of the heat transfer is in the x direction. So, x = 0 is for outer surface and x = L for inner surface. The values of inner and outer convection heat transfer coefficients are 9 (W/m^2K) and 17 (W/m^2K) respectively in period of summer season in Ghardaïa region. The proposed mathematical model contains certain assumptions that are important considerations that have already been tested and do not have an influence on the creation and use of this model. It is assumed as: Heat transfer through the walls is assumed unidirectional and perpendicular to these walls.

 DOI: 10.1051/e3sconf/20172200013

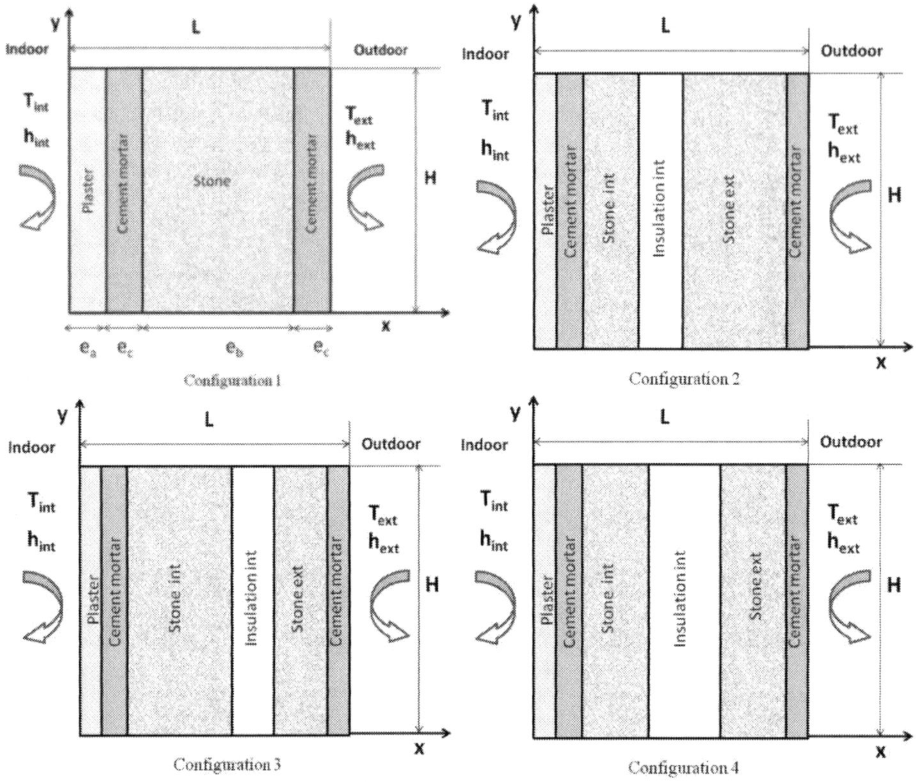

Fig. 1. Multilayer wall configurations studied with boundary conditions.

The distribution of the temperature on the outer and inner surfaces of the wall is uniform. Consider a multilayer wall thickness (L) without internal source and in thermal equilibrium at the initial time at temperature T_0. This wall subject to the boundary conditions of Dirichlet: T(0, t) and T(L, t) is shown schematically in fig.1. It is assumed that the thermophysical properties of the wall are constants. The materials considered are those generally used in Algeria in the construction of buildings, their thermal physical properties are shown in Table 1.

Table 1. The thermal physical properties using in the simulation.

	λ [W/m.K]	**ρ [kg/m³]**	**Cp [J/kg.K]**
Mortar	1.15	1900	864
stone	2.8	2600	702
plaster	0.4	1440	840

2.1 Mathematical formulation

All papers should be concise regardless of length. Long quotations should be avoided by referring to sources. Illustrations and tables, where they help clarify the meaning or are necessary to demonstrate results properly, are desirable, but they should be kept to a practicable minimum, and included on individual pages at the end of the manuscript.

Given these assumptions, the heat balance equations are unsteady two-dimensional conduction equation in solid material:

$$\rho c \frac{\partial T}{\partial t} = \frac{\partial}{\partial x}\left(\lambda \frac{\partial T}{\partial x}\right) + \frac{\partial}{\partial y}\left(\lambda \frac{\partial T}{\partial y}\right) \quad (1)$$

Boundary conditions:

$$-\lambda_a \frac{\partial T}{\partial x} = h_{ext}\left(T_{fe} - T_{ew}\right), \quad x = 0 \quad (2)$$

$$-\lambda_c \frac{\partial T}{\partial x} = h_{\text{int}}\left(T_{iw} - T_{fi}\right), \quad x = L \quad (3)$$

$$\frac{\partial T}{\partial y} = 0, \text{ at } y = 0, \quad \frac{\partial T}{\partial y} = 0, \quad y = H \quad (4)$$

Initial conditions:

$$T_w = T_0 \text{ , } t = 0, \text{for } 0 \leq x \leq L, \quad (5)$$

$$T_{ext}(t) = T_1 + \Delta T_1 \sin(wt) \quad (6)$$

$$T_{\text{int}}(t) = T_2 + \Delta T_2 \sin(wt) \quad (7)$$

Where: $w = 2\pi/\tau$, τ = *1440min*, T_1 = 38.5°C, T_2 = 38°C, ΔT_1 = 8°C, ΔT_2 = 8°C.

3 Results and discussion

Analysis by the experimental study of the thermal inertia in an exterior wall of a building on measured the temperatures each tin minutes (Δt=10min). It measures the temperatures on the left and right side of the wall. Knowledge of the evolution of temperatures makes it easier Knowledge of the evolution of temperatures allows bettering understanding the phenomena of thermal inertia, storage and retrieval of heat in the wall. This section discusses thermal inertia effect on measured indoor air temperatures in a building in an arid region of Algeria, the measurements make at the Unit for Applied Renewable Energy Ghardaïa.

Fig. 2a shows the validation of the simulation model by comparing it with the measured data during a very hot period. The results of numerical simulation effected with two days in the period studied by considering the climate of the city of Ghardaïa (20–21 July 2008). By analyzing the variation of temperature with time, the results shows in Fig. 2b that the fluctuation of the temperature wave in configuration (1) is smaller than the temperature wave in the others configurations environ 1°C.

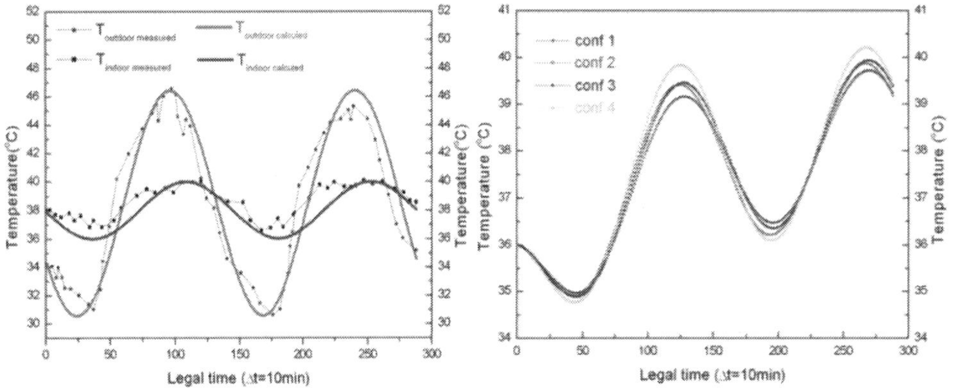

Fig. 2. (a): Validation of numerical model with experimental results and (b): temperatures resultant of wall configurations.

For the day of July 20, the calculated minimum temperature of the outer surface of the wall is 31.10°C; the minimum measured value is 31.07°C. The calculated and measured maximum values are respectively 45.70°C and 39.90°C. Furthermore, the inner surface temperature of the wall including the minimum value is calculated to 36.08°C, while its measured minimum value is 36.90°C. The calculated maximum value is reached 39.90°C and the maximum value measured is 40.10°C. All measured data was found in good agreement with the simulated data, we can see a very good consistency between the results obtained by the two methods.

3.1 Time lag (φ) and decrement factor (*f*)

The time lag and decrement factor are assumed as important parameter to determine the thermal inertia of different building materials, when the thermal behavior of a wall is studied of unsteady state boundary conditions. The time it takes for the temperatures wave imposed on the left side to propagate from the right side named as "time lag" (φ) and the decreasing ratio of its amplitude during is named as "decrement factor" (f) [12].
The time lag (φ) and decrement factor (f) are defined by the following equations [14].

$$\varphi_{min} = t_{Ti,min} - t_{To,min} \tag{8}$$

$$\varphi_{max} = t_{Ti,max} - t_{To,max} \tag{9}$$

$$f = \frac{T_{i,max} - T_{i,min}}{T_{o,max} - T_{o,min}} \tag{10}$$

$t_{To,min}$, $t_{Ti,min}$, $t_{To,max}$ and $t_{Ti,max}$ are the times when the temperatures are imposed on its left and right side of the wall surfaces reach the minimum and maximum values.
$T_{o,min}$, $T_{i,min}$, $T_{o,max}$ and $T_{i,max}$ are respectively the minimum and maximum temperatures imposed on left and right side of the wall.

Fig. 3 shows the time lag (φ) and decrement factor (f) calculate of walls with thermal insulation. By analyzing the change in temperature as a function of time, we notice a thick heavy stone wall is not isolated in these conditions. But, it will slow down by its inertia, the input of heat flux. In principle, the indoor ambient must remain cool in summer thanks to their Stone walls with high heat storage capacity. These thick walls have a strong time log to bring the freshness the night in the wall and the return day. For this, the stone will become the ideal material: dense, capable of storing and conductive to promote the storage provided there is fresh night. Indeed, that's the problem in these arid regions: in summer, the outdoor ambient temperatures remain high almost all the time even during the night. So, the thermal inertia consists of an accumulation of heat in the envelope for an interior return radiation. The complexity of this phenomenon is that the flow of heat through the shell successively increases the temperature of materials which allows increasing at the same time the air temperature habitat. The results support the conclusion that the mathematical models are good enough to predict the effective thermal behavior of the building under real weather conditions.

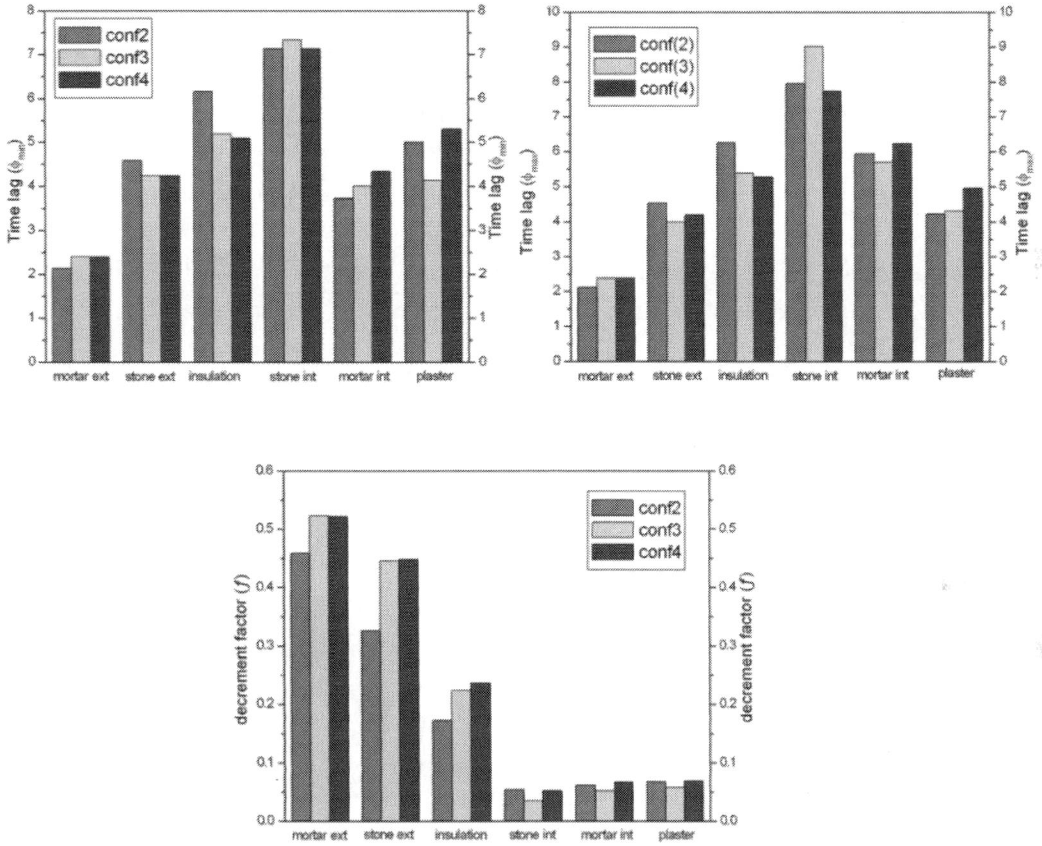

Fig. 3. (a): The time lag (φ_{min}), (b): The time lag (φ_{max}) and (c): decrement factor (f).

4 Conclusions

A numerical simulation is used to analyze and descript the phenomenon of thermal inertia in the walls. We wanted to show through this study that the high inertia walls are the main passive means to improve the thermal comfort in the hot and dry climate zone.

The results obtained are experimentally validated, and mainly showed that the wall deadens and delays the variations in the temperature between inside and outside of the building. The choice of construction materials plays an essential role in the comfort level of the buildings and reducing the energy consumption. Indeed, the inertia of the wall decreases and delays the effect of external conditions and is particularly suited to the hot and dry climate. This is clear from the results.

References

1. S.M.A. Bekkouche, T. Benouazb, M. Hamdania, M.K. Cheriera, M.R. Yaichec, N. Benamranea, J. Buil. Eng. **1**, 42 (2015)
2. S.M.A. Bekkouche, T. Benouazb, M.K. Cheriera, M.R. Hamdani, M.R. Yaichec, N. Benamranea, J. Ene. Buil. **66**, 678 (2013)
3. S.M.A. Bekkouche, T. Benouazb, M.K. Cheriera, M. Hamdani, M.R. Yaichec, R. Khanniche, J. Therm. Sci. **17**, 349 (2013)
4. K.J. Kontoleon, Th.G. Theodosiou, K.G. Tsikaloudaki, J. App. Ene. **112**, 325 (2013)
5. D.E.M. Bond, W.W. Clark, M. Kimber, J. App. Ene. **112**, 235 (2013)
6. J. Karlsson, L. Wadsö, M. Öberg, J. Ene. Buil. **60**, 146 (2013)
7. J.A. Orosa, A.C. Oliveira, J. Renew. Ene. **37**, 89 (2012)
8. Y. Tamene, S. Abboudi, C. Bougriou, Revue des éne. Ren. **12**, 117 (2009)
9. M. Ibrahim, P.H. Biwole, E. Wurtz, P. Achard, Buil. Env. **81**,112 (2014)
10. S.C. Ng, K.S. Low, N.H. Tioh, Ene. Buil. **43**, 2956 (2011)
11. K.J. Kontoleon, T.G. Theodosiou, K.G Tsikaloudaki, App. Ene. **112**, 325 (2013)
12. L.E. Mavromatidis, M. EL Mankibi, P. Michel, M. Santamouris, App. Ene. **92**, 480 (2012)
13. C. Tonelli, M. Grimaudo, Buil. **83**, 89 (2014)
14. H. Asan, Y.S. Sancaktar, Ene. Buil. **28**, 159 (1998)

E3S Web of Conferences **22**, 00014 (2017) DOI: 10.1051/e3sconf/20172200014
ASEE17

Cytotoxicity and genotoxicity properties of particulate matter fraction 2.5 μm

Maciej K. Bełcik[1,*], *Agnieszka* Trusz-Zdybek[1] , *Ewa* Zaczyńska[2], *Anna* Czarny[2] and *Katarzyna* Piekarska[1]

[1]Faculty of Environmental Engineering, Wrocław University of Science and Technology, Wybrzeże S. Wyspiańskiego 27, 50-370 Wrocław, Poland
[2]Institute of Immunology and Experimental Therapy, Polish Academy of Science, ul. Rudolfa Weigla 12, 53-114 Wrocław, Poland

Abstract. In the ambient is more than 2,000 chemical substances, some of them are absorbed on the surface of the particulate matter and may causes many health problems. Air pollution is responsible for more than 3.2 million premature deaths which classifies it as a second place environmental risk factor. Especially dangerous for health are polycyclic aromatic hydrocarbons and their nitro- and amino derivatives which shows mutagenic and carcinogenic properties. Air pollutions were also classified by International Agency for Research on Cancer to group which carcinogenic properties on human were proved by available knowledge. Air pollutions, including particulate matter are one of the biggest problem in Polish cities. World Health Organization in report published in May 2016 set many of Polish cities on the top of the list most polluted in European Union. The article presents results of mutagenicity, genotoxicity and cytotoxicity researches conducted on a particulate matter fraction 2.5 μm collected during all year long in Wroclaw agglomeration. The material were collected on filters using high-flow air aspirator and extracted using dichloromethane. Additionally it was fractionated into 2 parts containing: all pollutants and only polycyclic aromatic hydrocarbons. Dry residue of this fractions were dissolving in DMSO and tested using biological methods. Biological methods include mutagenicity properties which are investigated by *Salmonella* assay (Ames assay). Other biological method was comet assay and 4 parameter cytotoxicity test PAN-I assay. Results of the conducted experiments shows differences in mutagenic, genotoxic and cytotoxic properties between seasons of collection and between volume of dust pollutions fractions. The worst properties shows particles collected in autumn and winter season and this containing only polycyclic aromatics hydrocarbons. Results showed also some correlations in results obtained during different methods and properties.

[*] Corresponding author: maciej.belcik@pwr.edu.pl

DOI: 10.1051/e3sconf/20172200014

1 Introduction

There is more than 2,000 chemical substances in the ambient air. Some of them are in a gas form, the other are absorbed on the surface of the particulate matter. Among this ones which are absorbed on the surface of particulate matter there are inter alia polycyclic aromatic hydrocarbons, phenols, organic compounds and many more [1, 2].

Particulate matter with all absorbed substances can penetrate the human respiratory system may cause health problems among which it is worth mentioning asthma, pneumonia, decreased lung function and cardiovascular system. The smaller particulate matter can penetrate deeper parts of airways and reach directly into alveoli. All these factors cause health problems induced by dust pollutions are observed not only in groups exposed to environmental risk (children and elderly people) but also in health part of population [3, 4].

Recently in May 2016 the World Health Organization present the report with the average concentrations of particulate matter fraction 10 μm and 2.5 μm obtained in cities among the world. The data presented by WHO comes from local air quality monitoring centres. In a report appears many of Polish cites which are on the top the polluted regions in European Union. One of the mentioned cities is Wrocław – the urban and industrial agglomeration, where air pollutions comes from three main sources: low-emissions, industrial emissions and urban traffic. Although the air quality in our city was improved in recent years and it has been observed reduction in annual average concentrations of particulate matter, there are still happened smog situations especially in winter season. The biggest problem in particulate pollution in Wrocław agglomeration is low emission caused by low quality coal combusted for heating houses not connected to the central heating network [5].

This article presents results of biological tests carried out on dust particles collected in Wrocław agglomeration during all year long. Tests conducted during the research included mutagenicity *Salmonella* test, genotoxicity comet assay and cytotoxicity PAN-I assay.

2 Materials and methods

2.1 Particulate matter collection and extraction

Material for research was particulate matter air pollutions fraction 2.5 μm collected Wrocław agglomeration (Poland). In Wrocław is a capital city of Lower Silesia region in which lives about 630 thousand residents with average density of population 2.1 thousand people per square kilometre. Samples were collected using high-flow air aspirator DHA-80 manufactured by Digitel Enviro-sense with cooperation with Regional Inspectorate for Environmental Protection. Samples were collected on filters made of glass fiber, which was replaced every 24 hours. This operation allows to obtain one sample which representatives one day of collection. The collected particulate matter consisted fraction with a diameter between 2.5 μm and 0.01 μm. Samples were taken during all year long and was divided into 4 seasons: spring from April to June, Summer from July to September, Autumn from October to December and Winter from January to March.

The test area was located on the corner of Wiśniowa street and Powstańców Śląskich street in the city center, directly near one of the major routes in the southern part of Wrocław. In close neighborhood place of sampling are located single- and multi-family residential buildings, the majority of which are not connected to the municipal heating network. The heating of this buildings is carried out by individual source of heating.

Dust extraction was carried out in 16-hour cycle with a 15-minutes reflux in conditions of reduced light. As the extraction solvent was used dichloromethane in an amount of 350 cm^3 for each sample. After extraction samples were placed in a vacuum evaporator for concentrating and vaporization solvent to dryness [6, 7].

In addition, the dry extracts were fractionated to obtain clear fractions containing polycyclic aromatic hydrocarbons (PAH). This fraction with the starting extract gives material subjected to biological tests [7, 8].

2.2 A549 cell line

Cytotoxicity tests were performed on the A549 cell line which are adenocarcinomic human alveolar basal epithelial cells (American Type Culture Collection, Cell Culture Line 185). The choice of adherent A549 cells allows to stimulate conditions as close as in real ingress finest fractions of particulate matter pollutions to the alveoli. A549 cell line is widely used as an in vitro model for various studies, including cytotoxicity studies of air pollutions [9, 10].

Breeding of A549 cell line was carried out in culture medium Dulbeco containing 10% addiction of calf serum, 2mM L-glutamine, 100 units/cm^3 penicillin and 100 μg/cm^3 streptomycin. Cell culture was incubated at 37°C and 5% of carbon dioxide atmosphere.

2.3 Biological tests

2.3.1 Salmonella assay

In the *Salmonella* assay was used two Salmonella typhimurium TA98 strain. This was a gift from K. Sugiyama and M. Yamada from Division of Genetics and Mutagenesis, National Institute of Health Sciences, Tokyo, Japan. Characteristic of TA98 strain allows him to check mutagens caused phase change read, which transform strain to form nonhistidinedependent.

Experiment was conducted in the absence and in the presence of microsomal fraction S9, obtained from liver of Wistar rats and activated with Aroclor 1254. The microsomal S9 fraction was used with the purpose of metabolic activation of promutagens. Protein content in the fraction, as determined by Lowry's method, was 64.44 mg/cm^3. S9-mix was used in experiments, and the S9 content in the S9-mix was 4% (v/v).

Salmonella assay was conducted in the KADO procedure. This procedure allows to save the volume of samples tested and volume of rat leaver microsomal fraction. Reducing the amount of input sample, while provided optimal results in this procedure allows preincubation in temperature of 37°C and 90 min.

2.3.2 Comet assay

Comet assay was performed using OxiSelect™ Comet Assay Kit from Cell Biolabs Inc. according to the methodology described in the instruction of use on A549 cell line. This test can be divided into three stages: exposure the cells for the particulate matter, cells immobilization in agar gel and electrophoresis with result read.

The study was performed using method of directly contact samples on cell line A549. On the 24-well plate was prepared cell culture in density 1 x 10^6 cells/cm^3 incubated by 24 h. After this time on the plate was applied particulate matter sample with appropriate dilutions and leaved to incubation for 48 hours at temperature 37°C and 5% carbon dioxide atmosphere. Assay was carried out in two replications for each concentration of dust.

Moreover on the plates was prepared positive and negative controls which were DMSO and phenol.

In second step after the incubation the cells were scraped from the vessel and resuspended with PBS without Mg^{2+} and Ca^{2+} solution to obtain a cell density level 1 x 10^5 cells/cm^3. Cells were transferred to the slides and compared with agarose in 1:10 ratio.

Electrophoresis was performer in an alkaline solution at voltage of 23 V and 300 mA for a period of 30 min. Reading in the presence of Vista Green DNA Dye was made using epifluorescence microscope Nikon Eclipse 90i. For analysis of the results was used CASPLab program. It was observed not less than 100 comets in each sample.

2.3.3 PAN-I cytotoxicity test

Cytotoxicity tests were performed using PAN-I assay containing four single tests to evaluate different parameters defining toxic effects on cells. Parameters measured by PAN-I assay was: integrity of cell membrane (LDH test), mitochondrial activity (XTT test), lysosomal activity (NR test) and total protein content (SRB test). All tests were performed on the same breeding of A549 cells in the 96-well plates. On the monolayer cell culture it was applied particulate matter fraction 2.5 μm extracts with appropriate dilutions. The exposure time was 48 hours. After that it was performed all toxicity tests starting from LDH according to the instruction of use. As a positive controls was used solvent (DMSO) and as a negative control Triton X-100.

Only LDH test was performed in the supernatant of the culture medium with using kinetic results reading by the time 1 hour at 22°C and 2 min interval. Other tests were used on the original cells washed after each one with PBS solution. Results of the PAN-I assay was reading by the spectrophotometer μQuant with a specific wavelengths. The used wavelength was as follow: 340 nm for LDH test, 480 nm for XTT test and 540 nm for NR and SRB tests. In XTT, NR and SRB test it was also used reference wavelength of 690 nm.

3 Results and discussion

3.1 *Salmonella* assay

Mutagenicity of the air polluting particulates increases in heating season and decreases in summer [11]. As it has been proved, there is an inverse relationship between the mutagenicity and ambient temperature, and the maximum mutagenicity values are reached under winter conditions [12].

Results obtained in *Salmonella* assay are presented below. In case of TA98 strain no matter with S9 or not the potential mutagenicity samples were raw extract of pollutants labeled as "all" in the highest tested concentration of 25 m^3/cm^3. For this concentration the highest value of MR parameter (Mutagenicity Ratio) was 5.6 in case of autumn sample with S9 fraction, this sample reached also the highest value of MR with all tested samples with TA98 strain. The lowest potential mutagenicity for initial concentration was observed in summer samples tested with and without metabolic activation. The MR value in this cases reached 2.1. The value MR $\geq$ 2,0 was also obtained by samples of all pollutants in concentration 12.5 m^3/cm^3, but only examined with the absorbance of S9 fraction in spring – MR = 2.6 and winter MR = 3.0.

Samples collected during autumn season shows values of MR on the level 3.3 with metabolic activation and 2.0 without. In two instances the concentrations 6.25 m^3/cm^3 also have demonstrated the potential mutagenicity. This samples was for autumn and winter with metabolic activation, which shows MR value 2.2. Other samples tested by TA98 strain did not exceed the rate of mutagenicity above 2.0, which can proves their mutagenicity character. Higher values of MR obtained samples collected during autumn and winter period compared to this collected in spring and summer.

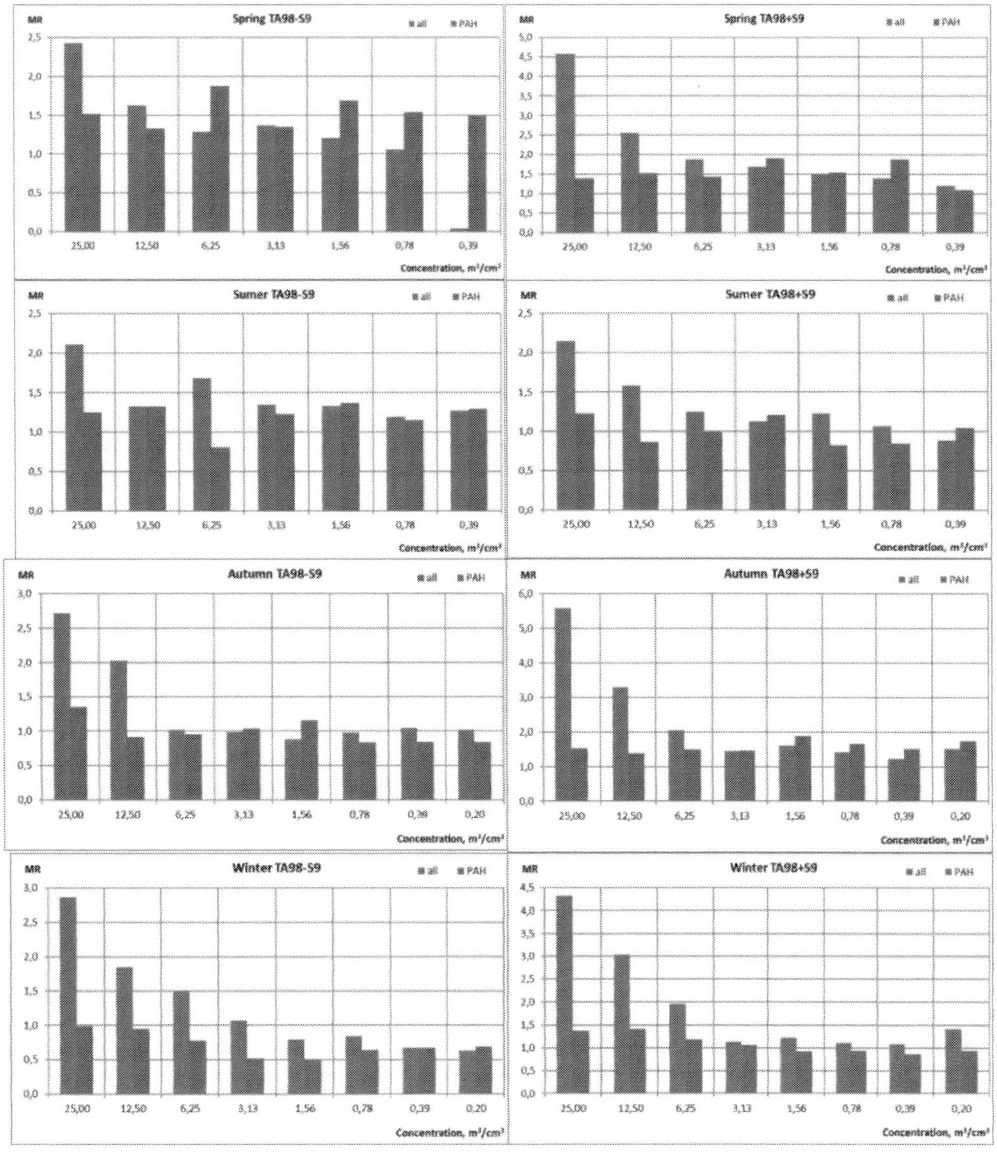

Fig. 1. Mutagenic Ratio of particulate matter fraction 2.5 μm extracts without S9 fraction (on the left) and with S9 fraction (on the right).

When analyzing samples containing only a PAH fraction it can be observed that none of them values higher than 2.0 of mutagenicity ratio. The highest value 1.9 was obtained for spring sample without metabolic in concentration 6.25 m^3/cm^3.

Claxton et al. (2004) reviewed results of examinations conducted on different parameter then mutagenic ratio. Comparing to this examinations performed in Europe showed in Czech Republic from 2 to 107 revertants/m^3 in winter, and from 2 to 23 revertants/m^3 in summer, whereas in Germany, in densely populated, with heavy traffic, highly industrialized regions, noted were more than 100 revertants/m^3 of air [13].

3.2 Comet assay

The results of comet assay are presented as Tail Moment Length, which presents average distance from the center of the comet head to the center of its tail. It was measured at least 100 cells at each dilution of the test sample.

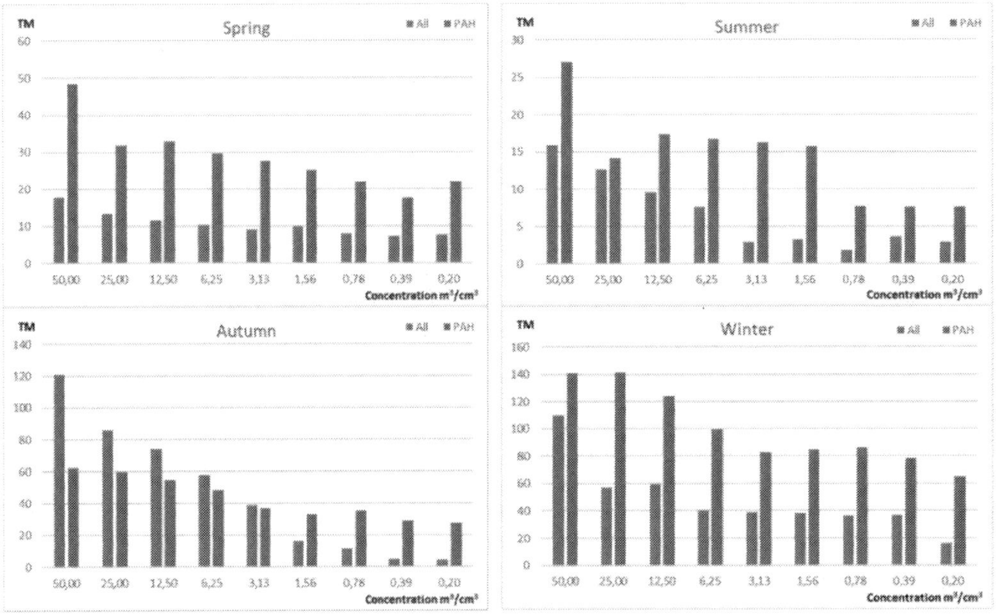

Fig. 2. Tail Moment parameter of particulate matter fraction 2.5 μm extracts.

Tail Moment Length for control samples (culture medium with 1% DMSO) was 20.972. The highest value of the Tail Moment Length parameter showed PAH fraction for winter season from 141.254 to 64.944, exceeding the level of control sample even in the lowest concentration tested. Genotoxic properties of particulate matter were also obtained for autumn and spring season. During summer months only the largest tested concentrations showed Tail Moment parameter above 20.972. In extracts containing all absorbed pollutants the highest genotoxic properties showed samples collected in autumn. Not much lower values were obtained for winter samples. Spring and summer values did not exceed control sample.

Research performed in Brazil on a V79 cell line shows various tail intensity. Tested concentration of 10 μg/cm^3 of organic extracts shows genotoxic effect of cells from 18.4 to 44.09 in winter-autumn season and form 7.20 to 36.13 in spring-summer season [14].

3.3 PAN-I cytotoxicity test

The results of cytotoxicity assays are shown as EC_{50}, which presents the concentration causing 50% inhibition of observed effect. The results are shown in the table below.

The results presented in the table show much higher values in the case of LDH test in comparison with the other three tests. This fact is result in disorders related to reading the samples containing color, which are undoubtedly particulate matter extracts, in this type of test. In the other three toxicity testing obtained results were similar to each other, wherein it should be noted that the XTT assay result in most cases shows slightly higher than those obtained in the NR and SRB tests.

Lowest EC_{50} values, which mean higher toxicity was showed for all pollutants sample collected during amount season. It was 2.810 for SRB test, 2.769 for NR test and 5.120 for XTT test.

Table 1. EC_{50} concentrations of particulate matter extracts.

Sample		LDH	XTT	NR	SRB
Spring	All	23.000	11.682	5.687	4.674
	PAH	24.000	11.926	5.270	6.340
Summer	All	> 50.000	30.757	18.773	15.570
	PAH	25.000	34.642	22.906	18.668
Autumn	All	26.000	5.120	2.769	2.810
	PAH	> 50.000	25.648	11.855	8.486
Winter	All	43.000	17.300	6.045	5.634
	PAH	> 50.000	9.143	7.917	5.622

Low EC50 values are also obtained for winter and spring samples (both all and PAH fraction). The highest values which means lowest toxic properties showed samples collected during summer months from 15.570 in SRB test to 30.757 for XTT.

Impact of the particulate matter and organic extracts were also investigated by other authors. Among many results of researches it was measured apoptosis an viability of A549 cell line and cytotoxic effects using different parameters including LDH test, MTT test, which is alternative to the XTT, and MDH test [10, 15–17].

4 Conclusions

The presented results lead to the following conclusions:

1. It has been observed difference between on obtained results in various seasons of collection.
2. It has been proven worse properties of particulate matter collected during autumn-winter period in comparison to spring-summer ones.
3. *Salmonella* assay shows higher mutagenicity effect in all absorbed pollutions than in PAH fraction.
4. In comet assay only in autumn season all absorbed pollutants showed higher genotoxicity then PAH fraction. In other seasons higher tail moment parameter was observed in relation to PAH fraction.
5. It has been shown inaccuracy of LDH test in conducted formula. It is proposed to change the method of carrying out, to not read the results in supernatant.
6. It has been shown similarity the results of EC_{50} obtained in the NR test and SRB test and lower sensitivity of the XTT test.
7. It has been confirmed effectiveness of using A549 cell line in researches of particulate matter fraction 2.5 μm.

The investigations were co-financed within the framework of the order No. 0401/0068/16 with the specific subsidy granted for the Faculty of Environmental Engineering Wroclaw University of Science and Technology (W-7) by the Minister of Science and Higher Education.

References

1. D.K.M Alves, F. Kummrow, A.A. Cardoso, D.A. Morales, G.A. Umbuzeiro, Environ. Mol. Mutagen. **57**, 41 (2015)
2. L.D. Claxton, P.P. Matthews, S.H. Warren, Mutat. Res. **567**, 347 (2004)
3. J. Burns, H. Boogaard, R. Turley, L.M. Pfadenhauer, /a./m. van Erp, A.C. Rohwer, E. Rehfuess, Cochrane Database of Systematic Reviews, *Interventions to reduce particulate matter air pollution and their effect on health*, (Cochrane Library, 2014)
4. S.S. Lim, Lancet **380**, 2224 (2014)
5. Global Urban Ambient Air Pollution, WHO Report, May 2016
6. B. Binková, M. Černá, A. Pastorková, R. Jelínek, I. Beneš, J. Novák, R.J. Šrám, Mutat. Res-Fund. Mol. M. **525**, 43 (2003)
7. J. Leníček, M. Sekyra, K. Bednárková, I. Beneš, F. Šípek, Int. J. Environ. An Ch. **77**, 269 (2000)
8. L. Østby, S. Engen, A. Melbye, I. Eide, Mutagenicity testing of organic extracts of diesel exhaust particles after fractionation and recombination. Arch. Toxicol. **71**, 314 (1997)
9. K. Piekarska, M. Zaciera, A. Czarny, E. Zaczyńska, Environ. Prot. Eng. **35**, 37 (2009)
10. M.E. Salcido-Neyoy, Y. Sánchez-Pérez, A.R. Osornio-Vargas, M.E. Gonsebatt, J. Meléndez-Zajgla, R. Morales-Bárcenas, P. Petrosyan, E.D. Molina-Servin, E. Vega, N. Anzano-León, C.M. Garcia-Cuellar, Environ. Pollut. **203**, 175 (2015)
11. L.D. Claxton, P.P. Matthews, S.H. Warren, Mutat. Res. **567**, 347 (2004)
12. R. Crebbelli, S. Fuselli, L.T. Baldassarri, G. Ziemacki, A. Carere, R. Banigni, Int. J. Environ. Health Res. **5**, 19 (1995)
13. M. Černá, D. Pochmanová, A. Pastorková, I. Beneš, J. Leníček, J. Topinka, B. Binková, Mutat. Res. **469**, 71 (2000)
14. A.T. Lemos, C.T. de Lemos, A.N. Flores, E.O. Pantoja, J.A.V. Rocha, V.M.F. Vargas, Chemosphere **159**, 610 (2016)
15. M. Dergham, C. Lepers, A. Verdin, F. Cazier, S. Billet, D. Courcot, P. Shirali, G. Garçon, Environ. Res. **137**, 256 (2015)
16. U.S. Akhtar, N. Rastogi, R.D. McWhinney, B. Urch, C.W. Chow, G.J. Evans, J.A. Scott, Toxicology Reports **1**, 145, (2014)
17. G. Mazzarella, A. Lucariello, A. Bianco, C. Calabrese, T. Thanassoulas, L. Savarese, A. Fiumarella, V. Esposito, A. de Luca, In Vivo **28**, 557 (2014)

The use of moving bed bio-reactor to laundry wastewater treatment

Sławomira Bering[1,*], *Jacek* Mazur[1], *Krzysztof* Tarnowski[1], *Magdalena* Janus[1], *Sylwia* Mozia [2], and *Antoni Waldemar* Morawski[2]

[1]West Pomeranian University of Technology, Szczecin, Faculty of Civil Engineering and Architecture, al. Piastów 50, 70-311 Szczecin, Poland
[2] West Pomeranian University of Technology, Szczecin, Institute of Chemical and Environment Engineering, ul. Pułaskiego 10, 70-322 Szczecin, Poland

Abstract. Large laboratory scale biological treatment test of industrial real wastewater, generated in industrial big laundry, has been conducted in the period of May 2016–August 2016. The research aimed at selection of laundry wastewater treatment technology included tests of two-stage Moving Bed Bio Reactor (MBBR), with two reactors filled with carriers Kaldnes K5 (specific area – 800 m^2/m^3), have been realized in aerobic condition. Operating on site, in the laundry, reactors have been fed real wastewater from laundry retention tank. To the laundry wastewater, contained mainly surfactants and impurities originating from washed fabrics, a solution of urea to supplement nitrogen content and a solution of acid to correct pH have been added. Daily flow of raw wastewater Qd was equal to 0.6–0.8 m^3/d. The values of determined wastewater quality indicators showed that substantial decrease of pollutants content have been reached: BOD_5 by 94.7–98.1%, COD by 86.9–93.5%, the sum of anionic and nonionic surfactants by 98.7–99.8%. The quality of the purified wastewater, after star-up period, meets the legal requirements regarding the standards for wastewater discharged to the environment.

1 Introduction

Rising prices of water taking and wastewater discharging force industry to improve their water and wastewater management. Studies aimed at selection of laundry wastewater biological treatment technology for the large industrial laundry Albatros (Fliegel Textilservice group), localized nearby Szczecin city, have been undertaken. Project was conducted onsite and real laundry wastewater has been used. Currently Albatros laundry generates ca. 600 m^3/d of wastewater average daily however maximum daily flow may reach the level of 800 m^3/d.

The quality of laundry wastewater depends on the types of washed materials. Generally wastewater contains dissolved and solid impurities washed from processed fabric as well as washing agents [1, 2].

* Corresponding author: slawomira.bering@zut.edu.pl

Among, proposed in the subject literature, methods of treatment of wastewater containing surfactants, physicochemical, chemical (mainly oxidation), biological aerobic and anaerobic [3] as well as combined methods are indicated as effective for laundry wastewater treatment. A number of authors tested also membrane purification of laundry wastewater. Positive effects has been reached by coagulation followed by membrane ultrafiltration [4, 5]. Pre-precipitation gives effective charge neutralization, decreased fouling and increased critical flux [5]. Coagulation has also been used as stand-alone process of laundry wastewater treatment [6]. Electrocoagulation and electroflotation processes were effective in COD removal with efficiency on the level of 62%. For Fenton process the reductions of COD and TOC were 85% and 69% [8]. The removal of the organic compounds was mainly carried out by Fenton-based Fe^{3+} coagulation rather than Fenton oxidation [8].

The total content of surfactants on the level below 2 mg/L, in the outflow wastewater, has been reached by method based on preliminary coagulation then by flotation (DAF) followed by sand filtration, ozonation and eventually Granulated Activated Carbon (GAC) filtration [9].

Biological methods are also used for laundry wastewater treatment. These methods, similar to the coagulation, generates significant amounts of sludge however biological excess sludge could be used for the land reclamation which is an advantage of biological methods [10]. In the processes of anaerobic wastewater treatment by expanded granular sludge bed (EGSB) reactor [11, 12] and up-flow anaerobic sludge blanket reactor [13] the efficiency of removal of anionic surfactants was on the level of 76% [11] and 78% [13]. Membrane bioreactors (MBR), moving bed bioreactors (MBBR) as well as sequencing batch reactors (SBR) have been used for aerobic laundry wastewater treatment. Pilot MBR plant with microfiltration membrane reduce the COD of raw laundry wastewater by around 90%. [14]. Part of the MBR permeate was subsequently treated by reverse osmosis (RO) filtration [14]. Results of MBBR test showed the possibility of reducing BOD to 12.7 mg/L and TOC to 6.8 mg /L [15]. In laboratory test of SBR COD at the outlet was not higher than 184 mg O_2/L and COD removal efficiency was equal to 82–91% [1].

Laundry wastewater treated in MBBR were further successfully purified by membrane and advanced oxidation water renovation processes [16].

The aim of undertaken research was testing of the efficiency of moving bed bioreactor (MBBR) technology with Kaldnes K5 carriers in Albatros laundry wastewater treatment. Research has been conducted onsite in real conditions with usage of real raw wastewater discharged from the laundry.

2 Materials and methods

Large laboratory-scale moving bed bio reactor (MBBR) has been installed onsite in the large industrial laundry Albatros (Fliegel Textilservice group), localized nearby Szczecin city. The real, mechanically pretreated, wastewater was pumped continuously to MBBR. Studies have been done during the period from May to August 2016.

2.1 Materials and equipment

The laundry wastewater, contained mainly surfactants and impurities originating from washed fabrics. The Albatros laundry mostly washing linens from high standard hotels and wastewater contains mainly chemicals used for wet washing and impurities removed from linens. Composition of washing agents used in the laundry was analyzed. Used surfactants fulfill biodegradation criteria described in the regulation (EC) No 648/2004 of the European

Parliament and the Council of 31 March 2004 on detergents. Dozing of detergents is automated. The raw wastewater used for research has been taken from the 350 m^3 capacity equalization tank in which wastewater from wet washing processes and from regeneration of ion exchange columns, used for water softening, is collected. In the equalization tank heat is recovered and wastewater is cooled down to the temperature of ca. 40°C.

To improve conditions of biological treatment the water solution of urea, as nitrogen source, has been dozed (20 mg per 1 L of wastewater) to the raw wastewater. Phosphorus has not been added. Based on pH measurements in the first MBBR tank sulphuric acid water solution (H_2SO_4 ca. 10%) has been used for automated pH control (above pH 8).

The research included tests of two-stage moving bed bio reactor (MBBR), with two reactors filled with carriers Kaldnes K5 (specific area – 800 m^2/m^3) and total capacity equal to 400 dm^3, have been realized in aerobic condition with coarse bubble aeration system. The scheme of the tested wastewater treatment system is presented in Figure 1. Microorganisms, typical for active sludge and biological bed, grew on the Kaldnes K5 carriers (Figure 1) floating in the wastewater.

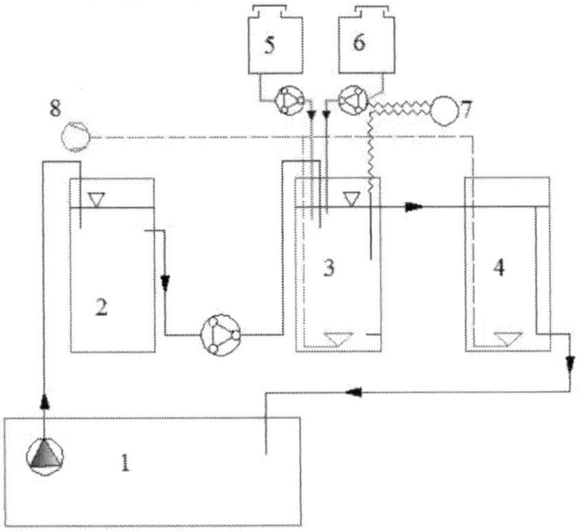

1 - 350 m3 capacity equalization tank
2 - buffer tank
3 - reactor *MBBR - 1 stage*
4 - reactor *MBBR - 2 stage*
5 - urea pump
6 - acid pump
7 - pH measurment
8 - blower

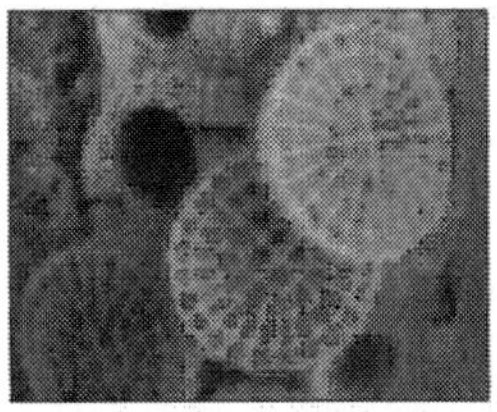

Fig. 1. The technological scheme of MBBR. The photograph of Kaldnes K5 carriers.

Start-up of MBBR has been conducted with wastewater daily flow Qd equal to 0.6 m^3/d (mean hourly flow Qh = 25 L/h) and, in the time of the tests, the daily flow equal to 0.8 m^3/d (mean hourly flow Qh = 33 L/h) has been reached. The source of activated sludge for inoculate of MBBR was communal wastewater treatment plant in Gryfino town. WWTP in Gryfino is recipient of the Albatros laundry wastewater which means that activated sludge is already adopted to that type of wastewater. Mixed liquor suspended solids (MLSS) in the taken, from a thickener, activated sludge was equal to ca. 15 g/L. Concentration of dissolved oxygen in MBBR tanks was maintained on the level of 2–4 mg O_2/ L.

2.2 Analytical methods

Wastewater quality has been determined by:
- chemical oxygen demand (COD) according to the PN-ISO 6060:2006,
- 5-day biochemical oxygen demand (BOD_5) according to the ISO 15705:2002,
- total nitrogen according to the PN-EN 25663 and LCK 238 cuvette tests (Hach-Lange),
- nitrate nitrogen by Spectroquant 14773 test (Merck),
- total phosphorus according to the EN ISO 6878 and LCK 350 cuvette tests (Hach-Lange),

 DOI: 10.1051/e3sconf/20172200015

- anionic and non-ionic surfactants according to the PN-EN 903 and PN-ISO 7875-2 as well as by LCK 332 and LCK 333 cuvette tests (Hach-Lange),
- total suspended solids TSS/mixed liquor suspended solids MLSS by gravimetric method,
- chlorides according to the PN-ISO 9297:1994,
- pH and conductivity (pH/EC/T CPC-505 laboratory pH/conductivity meter).

For checking the stable operation of the wastewater treatment system, in the time of MBBR working parameters changes (e.g. start-up, change in wastewater flow), only COD or BOD and COD has been determined (11 follow-up and 10 monitoring samples – 21 samples in total).

3 Discussion of results

Tests of laundry wastewater treatment efficiency using MBBR technology were included in the study. The quality of the laundry raw wastewater was determined and wastewater treatment test was conducted.

3.1. Raw wastewater quality

The values of determined raw wastewater quality indicators are presented in the Table 1.

Table 1. Quality of mechanically treated Albatros laundry raw wastewater (sampling 05–08.2016).

	Parameter/Indicator	Unit	Allowable limit	Value	
				Max.	Min.
1	pH		6.5–9.5	8.5	7.7
2	COD	mg O_2/L	125	1087	479
3	BOD_5	mg O_2/L	25	538	271
4	Total Suspended Solids TSS	mg/L	35	191	104
5	conductivity	µS/cm		3085	1800
6	Chlorides	mg/L	1000	642	225
7	Total phosphorus	mg P/L	2	6.73	2.45
8	Total Nitrogen	mg N/L	30	13.11	9.64
9	Anionic Surfactants (MBAS)	mg/L	5	9.2** 26.1***	5.56** 17.4***
10	Nonionic Surfactants (BiAS)	mg/L	10	69.9** 74.2***	38,2 31.6***

**result according to the standards methods of surfactants determination (anionic surfactants PN-EN 903 - determination of anionic surfactants by measurement of the methylene blue index MBAS; PN-ISO 7875-2 – determination of non-ionic surfactants using Dragendorff reagent) [17]; ***results according to the cuvette tests (LCK 332 for anionic surfactants and LCK 333 for nonionic) [17]

Characteristic for the investigated laundry raw wastewater, in comparison to communal sewage, are high content of surfactants (anionic and nonionic), low content of total nitrogen and slightly elevated level of phosphorus. Periodically high concentration of chlorides have been also observed.

Differences in the results of anionic and nonionic determination between standard and cuvette tests methods are also confirmed by previous research [17]. Variability in the total phosphorus concentration is most likely caused by a occasional use of washing agents with increased content of phosphates

DOI: 10.1051/e3sconf/20172200015

The quality of raw wastewater has been determined. Ratio of chemical oxygen demand (COD) to five day biochemical oxygen demand (BOD_5), in the period of conducted research, varied in the range of 1.6 to 2.09. The amounts of, necessary for proper wastewater biological treatment, nutrients have been also determined. The ratio of five day biochemical oxygen demand to total nitrogen and total phosphorus BOD_5/N/P has been calculated. Given in the literature value of that ratio, for proper wastewater biological treatment, is equal to 100/5/1 [18]. The median of the BOD_5/N/P ratio, calculated based on the results of quality determination of ten wastewater samples, was equal to 100/2.97/1.25 which indicated that there is deficiency of nitrogen and periodical deficiency of phosphorus in the laundry wastewater. The amount of phosphorus in raw wastewater related to the amount of 100 units of BOD_5 varied in the range of 2.15 –0.68. Value below one have been recorded in three samples.

3. 2. Laundry wastewawater treatment efficiency tests

After five weeks MBBR start-up the permanent reduction of COD and BOD_5 values have been reached. Obtained COD and BOD_5 values were below the allowable limits given in the appendix no 4 of Regulation of the Minister of Environment of 18 November 2014 establishing conditions to be met for wastewater discharged into water or ground (…) [19]. The allowable limits are equal to 125 mg O_2/L for COD and 25 mg O_2/L for BOD_5. Obtained results of COD and BOD_5 treated wastewater determination, in relation to the allowable limits, are presented in figures no 2 (COD) and 3 (BOD_5).

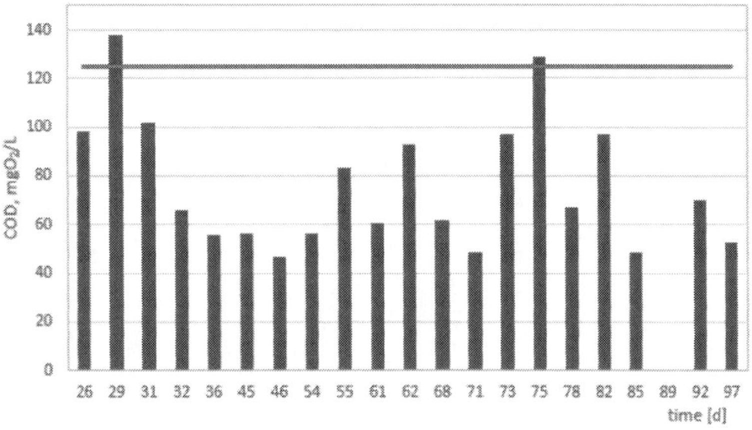

Fig. 2. COD in treated wastewater. ▮ results of COD, — COD allowable limit.

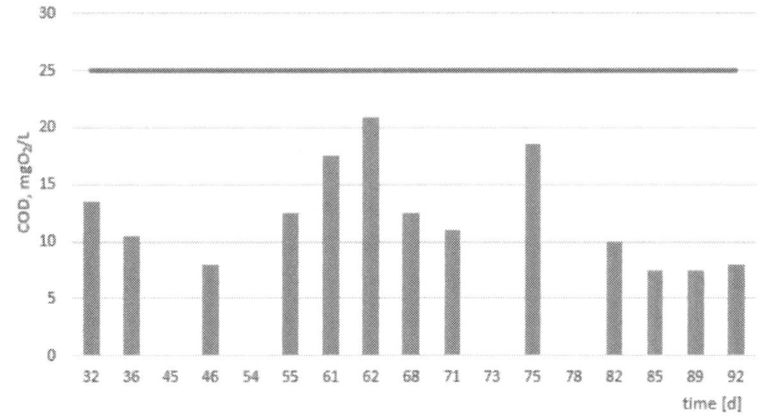

Fig. 3. BOD_5 in treated wastewater. ▮ results of BOD, — BOD allowable limit.

Determined values of BOD_5 indicator for treated wastewater, for MBBR stable operation periods, reached the values of 7.5–17.5 mg O_2/L while COD values varied in the range from 46.7 to 102.0 mg O_2/L. Exceeding of allowable COD value has been observed in the start-up period and in emergency situation caused by raw wastewater feeding pump failure (75th day). Short increase of COD value has been noticed after wastewater flow increase (62th day) however allowable value has not been exceeded. In the period of stable operation the MBBR treatment efficiency, defined as ratio of removed pollution load to inflowing pollution load, was equal to 94.7–98.1%for BOD_5 and 86.9–93.5% for COD. Efficiency of BOD_5 and COD removal is presented in figure 4.

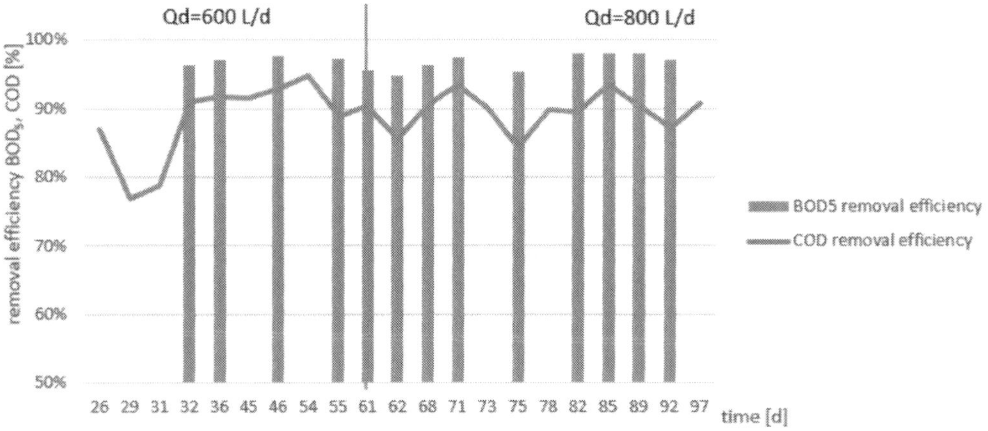

Fig. 4. COD removal efficiency and BOD_5 removal efficiency of laundry wastewater.

Removal efficiency on the level of 90% noticed in three successive measurements has been assumed as indicator of MBBR stable operation. Initial, lower COD removal efficiency shows the start-up period. Removal efficiency on the level of 90% has been reached in 32rd day of experiment. That result has been confirmed in the two next samples. After increasing of raw wastewater flow (in 62rd day) and consequently higher biological bed loading, the short decreasing of treatment efficiency has been observed. For the flow equal to 0.8 m^3/d stabilization in the treatment efficiency, on the level of 90%, has been reached after 10 days. Breakdown of the process, observed in day 75, can be explained by raw wastewater feeding pump failure. For 1 day there was no wastewater inflow, but aeration system and dosing systems have been working. Three days after emergency

DOI: 10.1051/e3sconf/20172200015

efficiency return to the level under 90% which shows relative flexibility of biological system for accommodation to rapid changes in real wastewater flow.

The variability in efficiency of BOD_5 removal in relation to the changes in biological bed active surface loading has been calculated (figure 5).

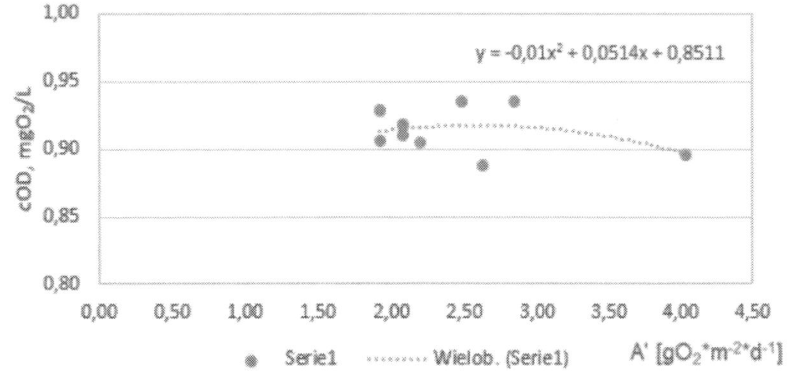

Fig. 5. BOD_5 removal efficiency versus biological bed active surface loading (A').

Loading A' [$gO_2 \cdot m^{-2} \cdot d^{-1}$)] has been calculated based on the formula given below [20]:

$$A' = L_{BOD} / F_a \tag{1}$$

where: L_{BOD} – BOD_5 load in the raw wastewater inflowing to the MBBR [g/d]; F_a – active surface of biological bed.

$$F_a = F_s \cdot V_{BB} \cdot V_{\%} \tag{2}$$

where: F_s – specific area of carriers, for Kaldnes K5 F_s = 800 m^2/m^3; V_{BB} – capacity of MBBR tanks, V_{BB} = 0.400 m^3; $V_{\%}$ – filling degree of tanks by Kaldnes carriers, n = 0.33. Calculated active surface F_a is equal to 107 m^2.

The quadratic polynomial trendline (correlation coefficient R^2 = 0.153) for BOD removal efficiency versus biological bed active surface loading is shown in the figure 5. The decrease of removal efficiency with increase of biological bed active surface loading is visible. It can be assumed that in the range of tested loading (1.9–4 g $O_2 \cdot m^{-2} \cdot d^{-1}$) biological bed operated with expected organic pollutants removal efficiency. Because of low levels of ammonia, nitrate and total nitrogen nitrification process has not been used.

Concentration of total phosphorus in raw and treated wastewater is shown in figure 6.

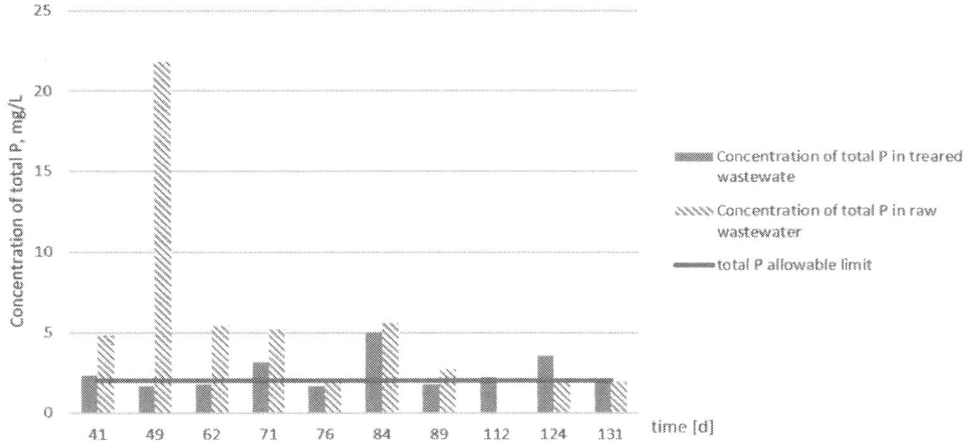

Fig. 6. Concentration of total phosphorus P in raw and treated wastewater.

Content of the total phosphorus in the raw laundry wastewater varied however most frequently it concentration was only slightly above the allowable limit. Treatment processes, conducted in aerobic conditions, not always allowed to reduce total phosphorus content to the level below allowable limit. In raw wastewater, in 3 samples (in 46th, 54th, 89th days), ratio BZT_5/P was lower than 1 which means that significant deficit of phosphorus took place [18] however, as it is shown in figure 4, substantial decrease in removal efficiency of BOD_5 and COD has not been observed despite of flow raising and emergency brake in power delivery. Biological bed's microorganisms had assured, constant delivery of raw wastewater with stable composition and optimal for them temperature (25–35°C) which can reduce demand for a phosphorus compounds. At the same time weak excess sludge sedimentation ability has been observed [21].

The surfactants removal efficiency was equal to 98.7–99.7% for anionic, 98.6–99.8% for nonionic ones and 98.7–99.8% for the sum of anionic and nonionic surfactants.

4 Conclusions

1) Application of Moving Bed Bio Reactor technology for Albatros laundry wastewater treatment gives the good results. The quality of treated wastewater is in accordance with the regulation establishing conditions to be met for wastewater discharged into water or ground [19]. Removal of nitrogen is not required.
2) Removal efficiency of organic pollutants, determined as BOD_5 and COD, is equal to 94.7–98.1 % for BOD_5 and 86.9–943.5 for COD.
3) Deficit of nitrogen in raw wastewater means necessity of nitrogen nutrient dozing.
4) In the range of tested biological bed active surface loading (1.9–4 $g \cdot m^{-2} \cdot d^{-1}$) biological bed operated with BOD_5 removal efficiency on the level of 94.7–98.1%.
5) Content of the total phosphorus in the raw laundry wastewater varies and periodically a deficit of phosphorus, in relation to removed BOD5 load, is observed. In aerobic biological treatment of investigated laundry wastewater, elevated up to ca. 5 mg/L, content of phosphorus cannot be reduced down to the allowable limit according to the regulation establishing conditions to be met by wastewater discharged into water or ground.
6) Observed deficit of phosphorus does not affect the efficiency of BOD_5 removal.

The research was financed by The National Centre for Research and Development in Poland within the Applied Research Programme (project No. PBS2/B9/23/2013).

References

1. J. Podedworna, M. Żebrowska-Sudoł, Gaz, Woda i Technika Sanitarna **4**, 21 (2007)
2. S. Bering, J. Mazur, K. Tarnowski, Civil and Environ. Eng. Rep. **6**, 33 (2011)
3. K. Jardak, P. Drogui, R. Daghrir, Environ. Sci. Pollut. Res. **23**, 3 (2016)
4. H-C. Kim, X. Shang, J.H. Huang, B.A. Dempsey, J. Membrane. Sci. **456**, 167 (2014)
5. X. Shang, H-C Kim, J.H. Huang, B.A. Dempsey, Sep. Purif. Technol. **147**, 44 (2015)
6. E.A.M. Ribeiro, F.R.B. De Souza, F.A., G.R. Do Amaral, G.R. Filho, R.M.F De Sousa, J.G. Vieira, T.F. De Rezende Costa, J.P. Thompson, R.M.N. De Assunção, S.C. Canobre, Quim. Nova **40**, 1 (2017)
7. C-T. Wang, W-L. Chou, Y-M. Kuo, J. Hazard. Mater. **164**, 81 (2009)

8. M. Vilve, A. Hirvonen, M. Sillanpää, J. Hazard. Mater. **164**, 1468 (2009)
9. I. Ciabattia, F. Cesarob, L. Farallia, E. Fatarellaa, F. Tognottia, Desalination **245**, 451 (2009)
10. K. Tarnowski, S. Bering, A. Iżewska, Mazur J. Przem. Chem. **95**, 8 (2016)
11. T.P. Delforno, A.G.L. Moura, D.Y. Okada, M.B.A. Varesche, Bioresource Technol. **154**, 114 (2014)
12. T.P. Delforno, A.G.L. Moura, D.Y. Okada, I.K. Sakamoto, M.B.A. Varesche, Bioresource Technol. **192**, 37–45 (2015)
13. D.Y. Okadaa, T.P. Delfornoa, A.S. Estevesa, I.K. Sakamotoa, I.C.S. Duarteb, M.B.A. Vareschea, Bioresource Technol. **128**, 1 (2013)
14. J. Hoinkis, V. Panten, Chem. Eng. Process **47**, 1159 (2008)
15. B. Altenbaher, M. Levstek, B. Neral, S.Š. Turk, Tekstil **59**, 340 (2010)
16. S. Mozia, M. Janus, P. Brożek, S. Bering, K. Tarnowski, J. Mazur, A.W. Morawski, Environ. Sci. Pollut. R. **23**, 19 (2016)
17. J. Mazur, S. Bering, K. Tarnowski, Przem. Chem. **95**, 8 (2016)
18. Z. Sadecka, *Podstawy biologicznego oczyszczania ścieków* (Seidel – Przywecki, 2010)
19. Regulation of the Minister of Environment of 18 November 2014 establishing conditions to be met for wastewater discharged into water or ground and on substances particularly harmful to the aquatic environment. (Journal of Laws of 2006, no. 1800)
20. ATV A 135 guideline
21. Y. Azimi, D.G. Allen, P. Seto, R. Farnood, Ind. Eng. Chem. Res. **53**, 31 (2014)

Energy saving by using natural energy from the shallow ground depths – many years operating results

Maciej Besler[1], *Maciej* Skrzycki[1], and *Wojciech* Cepiński[1]

[1]Wroclaw University of Science and Technology, The Faculty of Environmental Engineering, ul. Norwida 4/6, 50-373 Wroclaw, Poland

Abstract. We pay back more and more larger attention on solutions which saving energy produced from conventional fuels. This is possible to obtainment in significant quantities in fields in which use up the large quantities of energy. The formation the microclimate of interiors is an example of such situation. Especially in the case air conditioning, heating and mechanical ventilation. There is, however, a possibility of energy saving as well as considerable reducing the pollution coming from combustion of raw materials by utilising the natural renewable energy from the shallow ground. In the paper the results gained during several year of continuous measurement on the exchanger were presented. In summer periods an air cooling occurs 10–12 K, e. g. from +30 °C to +20 °C. In winter on the other hand, a preparatory preheating of the air is possible, e.g. from -18°C to about ± 0°C. It is then possible to obtain for the air conditioning system the total energy needed for cooling purposes at the summer periods, or up to 50% of the ventilation heat energy in winter picks.

1 Introduction

There is a widespread aspiration that the energy needs of the world are increasingly met by renewable resources. Maintaining the current way of producing energy mainly from fossil fuels is not sustainable in a long period of time. At least for two reasons: one is the depletion of these resources, and the second is the need to protect our natural environment. More and more often we notice that the environment is no longer able to absorb the residue of combustion processes, especially CO_2, SO_2, NO_x, which the most damaging effects are acid rain and global warming. In the Central European climate for the purpose of shaping the microclimate is 40% of total energy production and so much pollution is getting into the environment. Environmental cleanliness requirements are still growing. For the health and well-being of residents and building users a good microclimate is essential. Optimal microclimate promotes full and rapid regeneration of the body. It is also indispensable for proper reception of artistic experiences (e.g. At concerts, opera, theatre). With the increase in quality also increase the cost of manufacturing and maintaining the appropriate microclimate of the premises. Technology is changing in new buildings. Buildings are increasingly tight. Where there was sufficient heating and gravity ventilation in the past, today, well-ventilated mechanical ventilation or air conditioning is required. Mechanical ventilation systems allow for a better microclimate and, at the same time, reduce operating costs, favoring especially energy savings. An example of saving energy for heating and cooling in small and medium-sized ventilation systems is a diaphragm ground heat exchanger.

2 The solution idea

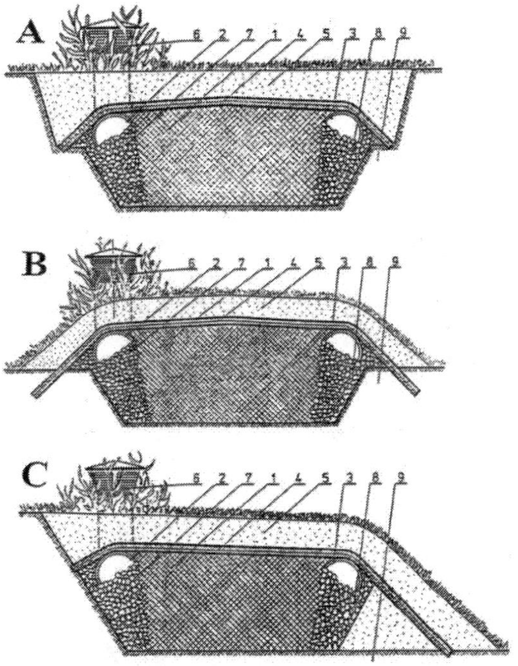

Fig. 1. Various proposal of heat exchanger design: A – recessed, B – partially recessed C – elevated above the level of the existing terrain and located on the slope. 1-accumulating (actual) bed; 2 conducting channel; 3-collecting channel; 4-heat-moisture insulation; 5-covering blanket; 6-drawing air inlet; 7-air distributing bed; 8-air collecting bed; 9-native ground.

The idea of the solution consists in the thermal energy from a depth of 4–5 m below the ground. At this depth in the Mid-European climate prevails almost a constant temperature of +10°C (± 1.5°C). The thing is to utilize a part of the practically unlimited stock of the heat energy contained at this depth. However, getting the outside air through a gravel bed placed 4–5 m below the ground surface level would be very troublesome and expensive, and often simply impossible, because of relatively high level of the underground water or rocky ground. It was thus decided to replace the thermal resistance of the ground layer by that of an insulating layer of a value corresponding to that of the 5 meter thick layer of wet ground. This way the isotherms corresponding to a deep ground layer are brought up to a higher level, just below the insulating layer. Such a solution enables to locate the heat exchanging gravel bed at a smaller depth (half recess), or even at a ground surface level, e.g. at an scarp. In the figure 1 is shown the intersection through the typical medium size ground exchanger. The effects are very close to those of 4–5 m depth. The heat exchanger can be located next to the building, under the lawn or even beneath the building between

 DOI: 10.1051/e3sconf/20172200016

the structural elements. The plants can be planted in a suitable coarse soil layer above the exchanger.

3 Experimental results

Obtained in the extruded and operating heat exchanger for many years, the results confirm the validity of the above assumptions. In the warm season, on gravel storage beds located below the surface, air is cooled by up to 10÷12K, for example from +30°C to +20°C. In addition, drying of the flowing air stream is also often added. Rarely, when the relative humidity of external air is less, drying is also rare. Conversely, in cold weather, the air stream passing through the bed is pre-heated for example from -18°C to 0°C or a temperature of several degrees above zero while moisturizing the flowing air.

In the case of obtaining air temperatures exiting the heat exchanger at +16 – +20°C for many days in the summer, it is possible to obtain even the total cooling demand for the air conditioning [4]. In winter the share of energy saved by the preheating of the air on the bed can reach up to 50% (for temperature conditions of -18°C). Figure 2 shows an exemplary design of a single ground heat exchanger. Where a larger airflow is required, a two-bed double exchanger may be used. Double heat exchangers are also applicable to the projected alternating bed operation. It is advantageous in the case of continuous operation of the ventilation system and allows for the self-regeneration of the beds. Switching mode can occur automatically or in a simple manual version.

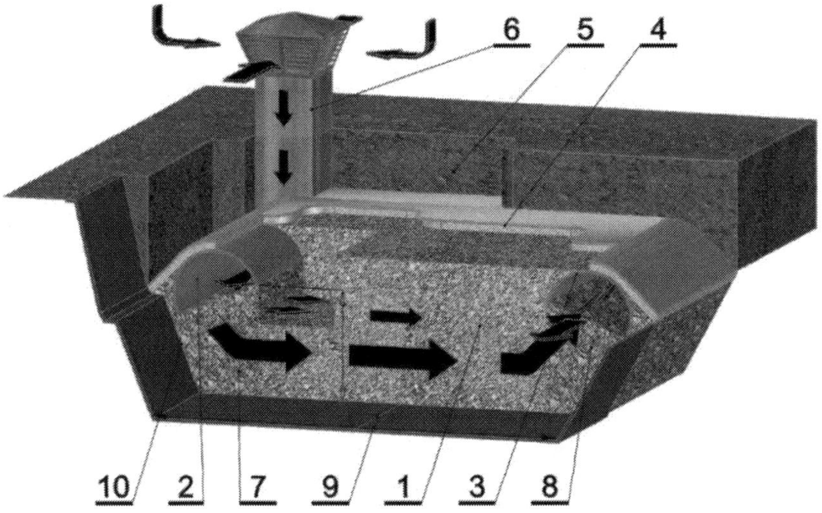

Fig. 2. Exemplary performance of ground heat exchanger: description of elements just like at fig. 1 (1-accumulating (actual) bed; 2-conducting channel; 3-collecting channel; 4-heat-moisture insulation; 5-covering blanket; 6-drawing air inlet; 7-air distributing bed; 8-air collecting bed; 9-native ground, 10 – geotextile [3].

The heat exchanger bed has the characteristics of the storage bin, both in short periods of work – several days as well as seasonal work. This means that the parameters of leaving air stream are very slow to follow the rapidly changing parameters of external air stream in the intake of the heat exchanger.

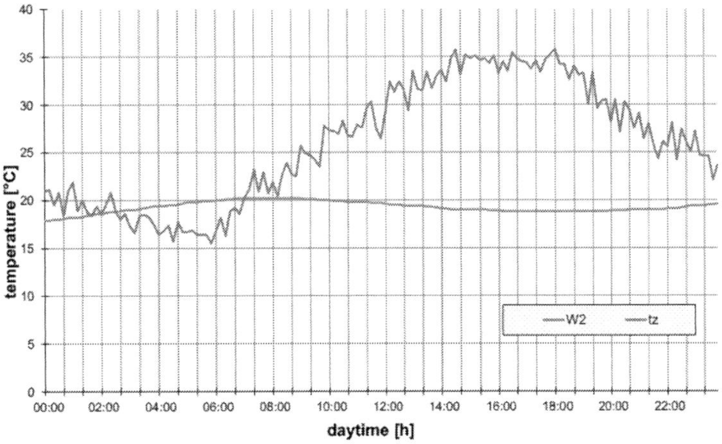

Fig. 3. Experimental results in the example summer period.

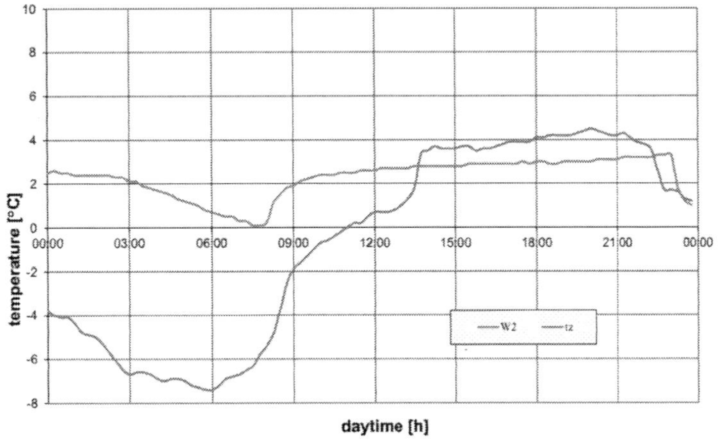

Fig. 4. Experimental results in the example winter period.

Seasonal changes of the air flow parameters from ground heat exchangers affect the cooling of the bed at the end of the winter season and heating at the end of the bed operation in the summer season. Consequently, different bed temperature is obtained in autumn and spring at similar external air temperatures. The variations in gravel temperatures in the ground heat exchangers in the whole year show figures 5–8 [3]. Based on the measurements, the expected air temperatures exiting the heat exchangers according to the different outdoor air parameters were graphed, dividing the year into the so-called the "climate quarters" as follows: December, January, February – I quarter, March, April, May – II quarter, June, July, August – III quarter, September, October, November – IV quarter.

 DOI: 10.1051/e3sconf/20172200016

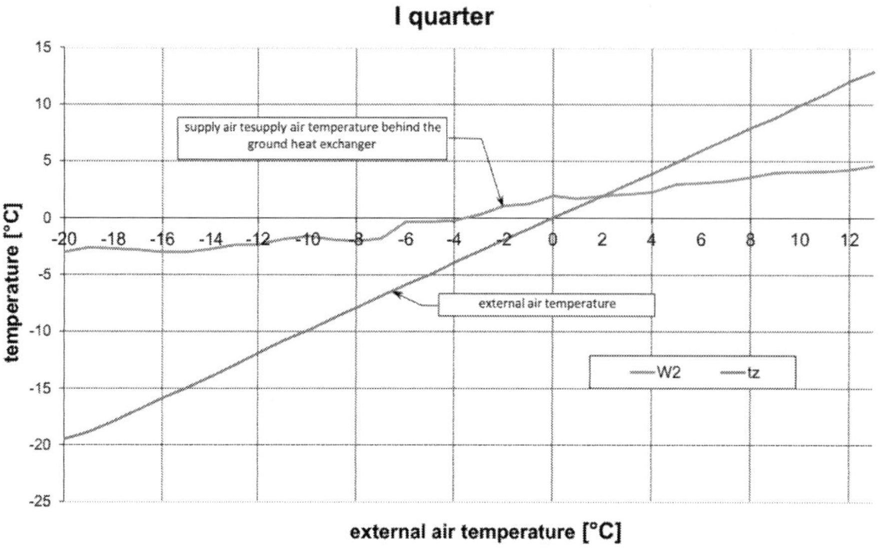

Fig. 5. Air temperatures exiting the exchanger as a function of outside air temperature for the first quarter (climate).

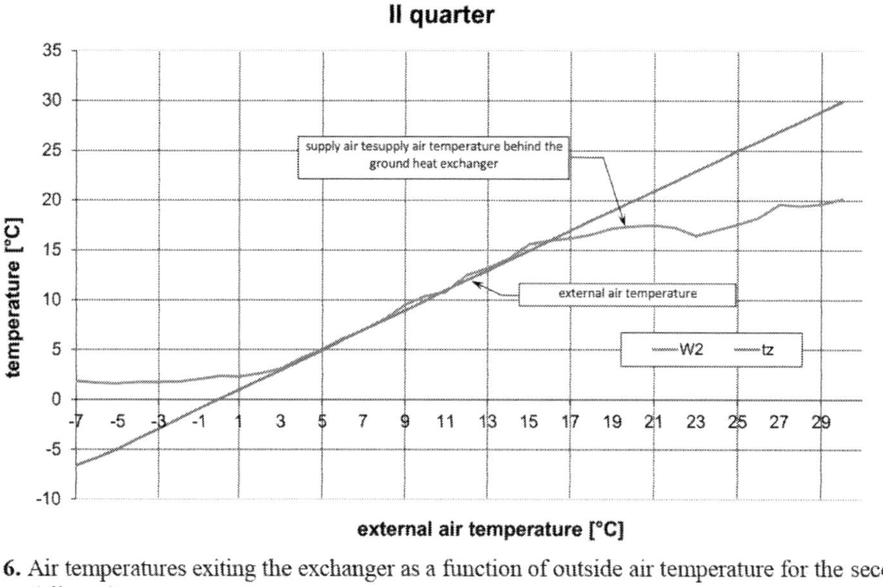

Fig. 6. Air temperatures exiting the exchanger as a function of outside air temperature for the second quarter (climate).

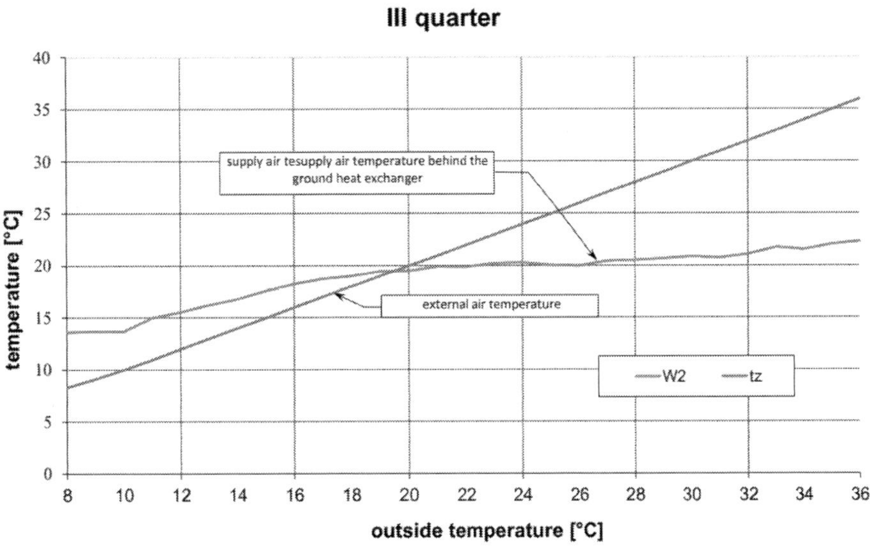

Fig. 7. Air temperatures exiting the exchanger as a function of outside air temperature for the third quarter (climate).

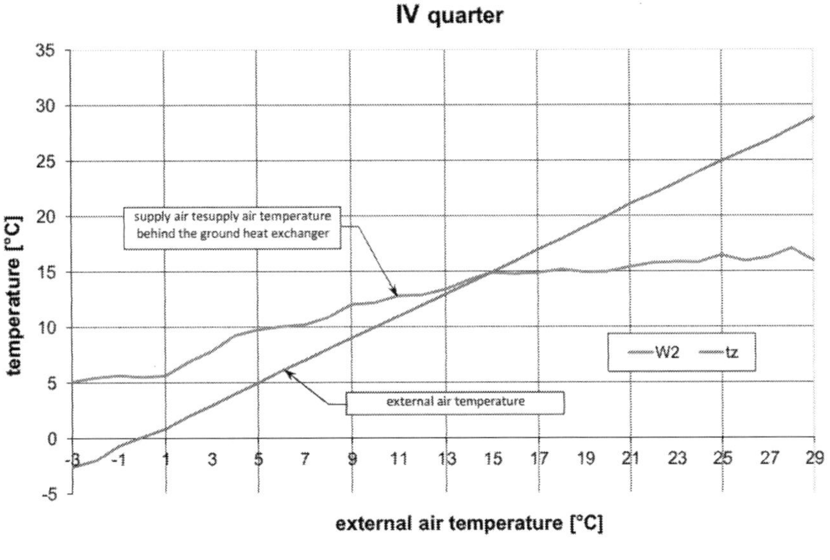

Fig. 8. Air temperatures exiting the exchanger as a function of outside air temperature for the fourth quarter (climate).

The change of conditioned air obtained in operating heat exchangers in the winter seasons shows the phenomenon of taking up moisture by the air stream. Conversely, during the summer seasons there is noticeable drying of air flow. The relative humidity of the air stream exiting the ground exchanger is in the range of 75–85%, which is a limitation for air conditioning systems (in case of the need for significant drying of air) when the air temperature in bed is lowered to +16 – +18°C during the summer season. Sample air changes for the winter and summer period on i-x graph are shown in Fig. 9.

 DOI: 10.1051/e3sconf/20172200016

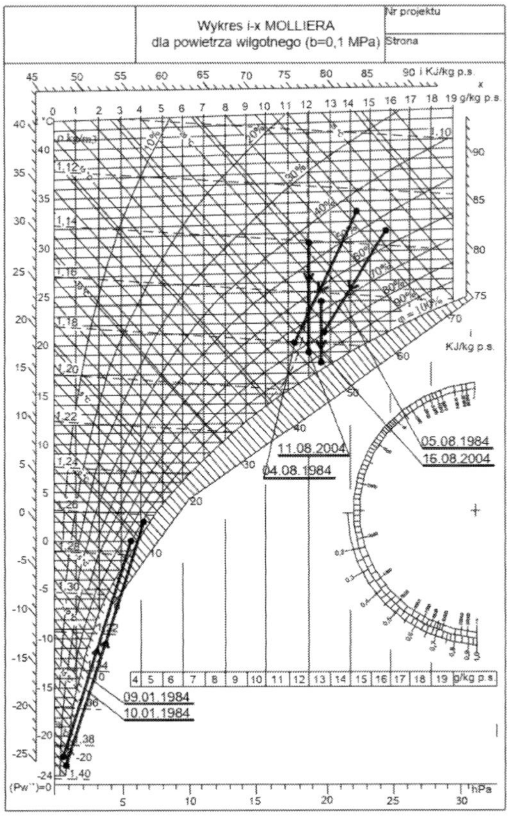

Fig. 9. The i-x graph: examples of characteristic of air flow in the heat exchanger.

3.1 The problem of air purity

After 14-years of non-stop exploitation an evaluation of the purity of air leaving the membraneless heat and mass exchangers was made. The investigation concerned the exchangers working in the Lab. of the Chair of Air Conditioning and District Heating at the Wrocław University of Technology since 1980. The investigations were carried out by the Provincial Epidemiological Station in Wrocław. The results are shown below in the tab. 1.

Table 1. Results of the investigations of air purity in ground exchangers in Wrocław.

No of. meas	Meas. point	Total No. of bacteria	Number					Total No of fungi
			Actinomycetales	Pseudomonas fluorescens	Staphylococus Type of hemolysis		
					α	β	
417	inlet	10615	3319	288	no incr.	no incr.	3485
418	outlet	680	157	no incr.	no incr.	no incr.	1231

The final conclusion is as follows: "In effect of the investigation carried out it has to be stated that the air – after passing through the heat exchanger – contains much less cells of micro-organisms than that contained at the inlet". At this opportunity it should be explained that – on the average – the pure air contains 3000–5000 fungi in 1 m^3; the air containing 5000–10000 fungi in 1 m^3 is concerned as one possibly influencing negatively upon the environment, it is usually assumed that the air which may endanger the environment contains more than 10000 fungi per 1 m^3.

3.2 Recapitulation

Basing on the investigation results obtained during the many years exploitation one may conclude that:

- The introduction of BGWCiM allows extracting from the ground in winter periods, at the picks, up to 50% of ventilation heat energy, and in summer periods - the whole cooling energy for air conditioning.
- The energy inputs for overcoming the flow resistance are negligeable, because the ratio of energy input to the energy gained reaches a value of 1:40, whereas for example in the heat pump installations the average ratio is about 1:3.5.
- There are no contraindications as to locating the BGWCiM. They may be placed under a lawn, under a building or even on the ground surface with effects not much deviating from those obtained at 4–5 m depth.
- They may be built on to the existing ventilating systems.
- For the construction of the BGWCiM only low-cost materials are used, mainly gravel and natural stones.
- Refund of money spent for building the BGWCiM occurs after about 2–4 years, which speaks well about profitableness of such enterprises.

References

1. G.J. Besler and others: *Membraneless heat and mass ground heat exchanger,* Wroclaw University of Technology Patent PRL No 128261; Patentschrift DDR 159206 Membranloser Boden- Wärme- und Stoffaustauscher; Schweiz Nr. 28/74; BRD Nr. P-311117.953.3.13
2. M. Besler, W. Cepiński, Pol. Inst. **5** (2012) (in Polish)
3. M. Besler, *Research of effectiveness of ground energy in environmental engineering* PhD Study, Faculty of Environmental Engineering, Wroclaw University of Technology, Wroclaw 1997 (in Polish)
4. W. Cepiński, *Efficiency of natural energy acquisition*, PhD Study, Faculty of Environmental Engineering, Wroclaw University of Technology, Wroclaw 2010 (in Polish)
5. A. Schwitalla, *The shaping of internal microclimate using gravel beds*, PhD Study, Faculty of Environmental Engineering, Wroclaw University of Technology, Wroclaw 2007 (in Polish)

DOI: 10.1051/e3sconf/20172200017

Effects of climate factors and vegetation on the CO_2 fluxes and $\delta^{13}C$ from re-established grassland

Yaroslav Bezyk[1,*], *Maxim* Dorodnikov[2], and *Izabela* Sówka[1]

[1]Unit of Ecologistics and Environmental Risk Management, Faculty of Environmental Engineering, Wroclaw University of Science and Technology, Poland
[2]Dept. of Soil Science of Temperate Ecosystems, Georg August University of Gottingen, Gottingen, Germany

Abstract. The relationship between stable carbon isotope composition ($\delta^{13}C$-CO_2) of soil CO_2 flux, vegetation cover and weather conditions was investigated in a short-term campaign at a temperate re-established grassland in Germany. During August-September 2016, we measured surface CO_2 flux with a closed-chamber method at high and low soil moisture content ('wet', 'dry'), with and without above ground vegetation ('planted', 'clear-cut') and estimated the effects of treatments on respective $\delta^{13}C$-CO_2 values. The concentration and stable carbon isotope composition of CO_2 were determined using the gas chromatography and mass spectrometry analyses. The $\delta^{13}C$-CO_2 of the soil fluxes decreased over sampling time for the 'dry-warm' conditions and canopy manipulation. The ecosystem-derived $\delta^{13}C$-CO_2 values (corrected for the atmospheric $\delta^{13}C$-CO_2) which included predominately soil- and rhizosphere respiration were −26.2 ± 0.8‰ for the 'dry-warm' conditions and decreased down to −28.1 ± 1.4‰ over a period of 28 days from late August to the end of September. The decrease coincided with the lowering of CO_2 flux and could be attributed to changes in plant physiological processes at the end of the vegetation season. Though the removal of shoots did not significantly affect the $\delta^{13}C$-CO_2 values as compared with the control, the pattern of further $\delta^{13}C$-CO_2 decrease (down to −28.8 ± 0.8‰) supported the role of living vegetation in a contribution of ^{13}C-enriched CO_2 to the ecosystem respiration.

1 Introduction

Soils as source and sink of the carbon dioxide (CO_2) are an important component in the global carbon (C) balance. The gas exchange between the atmosphere, vegetation and soil is controlled by the complex mechanisms related to various physical (temperature, moisture) and biochemical soil properties (microbial communities and activity, organic matter content), regional geographical features and meteorological conditions (duration of vegetation period, photosynthetically active radiation, precipitation, etc.) [1]. The environmental and biological mechanisms can strongly influence the carbon isotopic

* Corresponding author: jaroslaw.bezyk@pwr.edu.pl

composition of ecosystem-respired CO_2 and play a major role in controlling processes of respiratory isotopic fractionation [2]. In order to predict the response of C balance to environmental changes, it is necessary to determine effects of different climate factors including diurnal and season temperature changes, precipitation level as well as the role of live vegetation (aboveground biomass, plant ground cover) in an ecosystem.

Many studies have examined the main driving factors affecting the soil CO_2 respiration rates, but most of these studies provide contradictory results in relations of soil temperature and soil moisture content. Analysis of field observations in large part demonstrates that the soil temperature is the primary factor determining the rates of soil respiration [3, 4], while other numerous data considers that the effect of temperature is constrained by soil moisture availability [5, 6]. The impact of moisture on soil respiration and microbial activity is more complex than temperature because moisture availability also depends on physical characteristics of soil (texture, porosity, and organic matter content) [7]. Generally, the separating the effect of temperature on respiration from the effect of moisture or vice versa in the field face difficulties, since these parameters tend to vary continuously and inversely.

Grasslands are one of the highly important terrestrial biome types since grasslands hold a large portion of the soil C and serve as C sink. They cover approximately one-quarter of the global land surface, but many of these areas have been turned into managed lands (e.g. farming and agriculture) [8]. Although C sequestration is generally associated with native grassland areas, it was found that a large portion of the CO_2 fixation and C sequestration occurs in re-established grasslands [9, 10]. In such restored grasslands, especially in perennial systems, the microbial community composition could be similar to natural grasslands [11] but the response of such ecosystems to environmental changes, i.e. the sustainability of C balance in them, still remains largely unknown.

Therefore, it is critically important to deepen our knowledge on the mechanisms of C turnover in restored grasslands and the useful technique for this – isotope composition method. The carbon isotopic composition of CO_2 has significant potential as a tool to understand the influence of the environmental changes on carbon transformation mechanisms and predict of the future carbon balance of terrestrial ecosystems. So, the goal of this study was to follow the changes in soil CO_2 fluxes and their $\delta^{13}C$ values in response to the varying climate conditions (temperature and precipitation), and link those changes with the activity of vegetation (via canopy manipulation) in a re-established grassland. We hypothesized that (i) the surface CO_2 flux will be higher under warm weather conditions ('dry-warm') as compared to cool rainy weather ('wet-cool') due to higher temperature and accelerated soil respiration; (ii) a contribution of plant-derived respiration to the total ecosystem respiration should decrease towards the effects of the treatment on vegetation cover compared to a non-treated plot ('planted', 'clear-cut' conditions), as well as due to end of a vegetation season irrespectively of temperature and moisture regimes.

2 Materials and methods

2.1 Study site

The study site was a re-established grassland on anthropogenic soil located in the Forest Botanical Garden ('Forstbotanischer Garten') of Georg-August-University Göttingen in the north-eastern part of the city of Göttingen, Germany (9°57'48.4"E, 51°33'25.2"N) [12]. The dominating plant communities were herbaceous species. The main soil characteristics (0-10 cm) are shown in table 1.

Table 1. The main characteristics of the soil before the beginning of the experiment.

Parameters (units)	Value
Soil bulk density (g/cm^3)	1.12
pH in H_2O	7.58
Organic C (% of dry mass)	4.15
Total N (% of dry mass)	0.28
Soil C:N ratio	15.33
Microbial biomass C (mg C kg^{-1})	980.66
Microbial biomass N (mg N kg^{-1})	114.48

The mean annual temperature at the study site is 8.4°C, the mean annual precipitation is 628 mm, with mild winter and moderately humid summer, and the annual average wind speed is 12 km/h from South-West (SW) [13]. The weather conditions data on the day of the sampling: air temperature, photosynthetic active radiation (PAR) relative humidity, precipitation (all the data are from a meteorological station of the University of Gottingen) [14] are summarised in table 2.

Table 2. Microclimate conditions (temperature and PAR, precipitation) during the treatments, August-September 2016.

Parameters (units)	August 12, 2016 ('wet' conditions)			August 26, 2016 ('dry' conditions)			September 28, 2016 ('planted', 'clear-cut')		
	Min	Max	Aver	Min	Max	Aver	Min	Max	Aver
Air Temp. (°C)	16.2	18.9	17.8	16.2	34.8	27.3	14.3	20.5	18.1
PAR (μmol m^{-2} s^{-1})	-0.02	296.7	113.1	1.5	1482	887.7	-0.01	934	169.3
Precipitation (mm)	0.11	0.11	0.11	0	0	0	0	0	0
Real humidity (%)	88.1	95.3	92.0	47.2	83.2	50.8	65.0	72.0	81.1

The short-term campaign took place in August-September 2016. Despite the descending vegetation season, there were a number of weather events characterized as "hot and dry" (with air temperatures above 28°C and ca. 8 consecutive days of '0' precipitation) and 'wet and cool' (t = +17° C; 0.11 mm precipitation) [14].

2.1 Experimental set-up

For ecosystem CO_2 flux measurements (soil + vegetation), a static chamber sampling system was installed. Chambers (n = 6) consisted of two parts: a 'cap' - polypropylene plain tubing of inner diameter 24 cm and 20 cm in height - used to collect the emitted gases; and PVC collar (the lower section) of 24 cm in diameter and 10 cm in height, inserted ~3–4 cm into the soil. The chambers were tightly placed on the PVC collars, to omit leakage from chambers to the atmosphere. The chamber volume was ca. 8 L and covered 0.014 m^2 area. During each measurement, the CO_2 was collected with plastic gas-tight 60-ml syringes equipped with 3-way stopcocks at intervals of 15 min over 45 minutes of chamber exposition time. Such time intervals were chosen to achieve qualitative $\delta^{13}C$ signal in CO_2. Part of gas samples was immediately transferred into pre-evacuated 12-ml glass vials and the rest were kept in 60-ml plastic syringes. All the measurements occurred between 2 and 4 PM time.

The experimental plot had 6 randomly established replicate chambers. Chambers are placed on the soil surface including roots and small vegetation e.g. grass in the chamber area during 1-st and 2-nd treatment groups. The incubation experiments were maintained at the treatment-dependent temperature and moisture content: defined as high precipitation treatment and high soil moisture content (concluded on August 12, 2016) and high ambient temperature and low moisture content (August 26, 2016) called in this study 'wet', 'dry' conditions. During 3-rd treatment (September 28, 2016), few days before measuring, the soil respiration, all aboveground plant materials inside the chamber collars were cut to the ground to eliminate aboveground plant respiration ('planted', 'clear-cut' conditions). The above-mentioned treatment was further divided into two parts connected to environmental and vegetation factors affecting a total respired CO_2 from a soil surface: 'planted' related to changes in weather (under 'warm') conditions according to the previous treatment periods; and 'clear-cut' as control manipulation for studying of both plant physiology (shoots and root biomass distribution) and surface soil features contributing a variation of soil respiration rates in re-established ecosystems.

The soil temperature and soil water content were also determined at 10 cm in soil depth. The soil chemical properties: total nitrogen and carbon contents, soil acidity (pH), soil microbial biomass carbon (MBC), were determined by collecting soil samples at relevant depth (7–10 cm) adjacent to each PVC collar. The soil moisture content was measured the oven-drying method. The soil pH was measured using the electrode method in volumetric proportions of 1:3 soil:water-extract. The total C/N ratio of soil was determined using the dry combustion method [15]. Microbial biomass C and N were measured by the chloroform fumigation-extraction (CFE) technique [16].

The CO_2 concentrations were analysed on a gas chromatograph (GC 6000 VEGASERIES 2, Carlo Erba Instruments) equipped with Electron Capture Detector (ECD). The stable carbon isotope composition ($\delta^{13}C$) of CO_2 were analysed by the CRDS (Cavity Ring-Down Spectroscope, G2131-i Picarro Inc., Santa Clara, CA). All the analyses were conducted at the Department of Soil Science of Temperate Ecosystems, Georg August University of Gottingen, Germany.

2.2 Calculations and statistical analysis

The CO_2 flux was calculated as a change of concentrations over time using a linear function of the slope derived from a geometric mean regression. Gas concentrations were converted from volumetric into mass units using the Ideal Gas Law considering atmospheric pressure and temperature. For each measurement, the results are presented as means of 3–5 replicates ± standard error. Pearson correlation was employed to examine relationships between the CO_2 and $\delta^{13}C$-CO_2 values, and environmental variables. A one-way ANOVA followed by a Tukey test at $P < 0.05$ was used to identify the significance of differences between the variables [17]. Effects of the ambient temperature, related humidity and their interaction on the total ecosystem respiration were carried out using repeated measures ANOVAs ($P > 0.05$).

A two-end-component mixing model, the 'Keeling plot' method [18] was used to identify the stable carbon isotope composition of the soil CO_2 respiration out of the mixture with atmospheric $\delta^{13}C$-CO_2. As the atmospheric $\delta^{13}C$-CO_2, the value of 8‰ from Atmospheric Concentrations of CO_2 from Mauna Loa, (Hawaii, 2015) was used [19].

3 Results

Compared to the average C:N ratios for the typical grasslands in Germany (value in range 10 to 11) [20], the measured C:N ratio in top soil 10 cm at the experimental plot was substantially higher (15.3, Table 1). Significant increase in C:N ratio in re-established grasslands is associated with an effect of land conversion and illustrates the large capability of soil C storage as compared to N [21].

In general, the total CO_2 flux from a soil surface saturated after about 45 min. The significant positive linear relationships were marked between the CO_2 concentration at the

 DOI: 10.1051/e3sconf/20172200017

soil surface and the chamber exposure time (approximately 45 min.): the 'wet-cool' treatment (R^2 = 0.73, $p < 0.05$), the 'dry-warm' treatment (R^2 = 0.99, $p < 0.05$), the 'planted' treatment (R^2 = 0.98, $p < 0.05$), the 'clear-cut' treatment (R^2 = 0.98, $p < 0.05$) (Fig. 1).

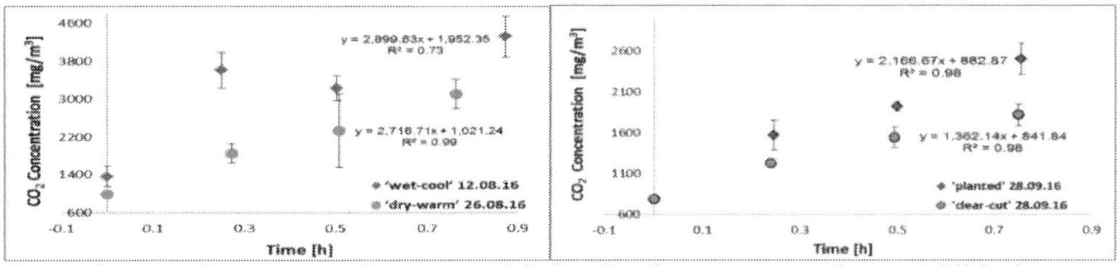

Fig. 1. The mean CO_2 concentration increment during chamber exposure at the re-established grassland. Three dates of measurements and two plant treatments are shown: 12.08.2016 and 26.08.2016 ('wet-cool' and 'dry-warm' weather conditions, respectively); 28.09.2016 (canopy manipulation: 'planted', 'clear-cut'). Error bars show the standard deviation across sub-plots ('wet' 'dry' scenario, n=6; 'planted', 'clear-cut' scenario, n=3). The slope of CO_2 concentration increment over time of a chamber exposition was linearly approximated to estimate the flux.

The ecosystem CO_2 flux did not differ significantly between weather conditions and vegetation manipulations (Fig. 2). Furthermore, the environmental factors tested by ANOWAs (temperature (R^2 = 0.58; F = 2.71, df = 3, p>0.2412) and air humidity (R^2 = 0.79; F = 7.49, df = 3, p > 0.1114)) in a short-term perspective showed positive correlations with total respiration rate, but had no significant effect ($P < 0.05$) on CO_2 flux from ecosystems. Carbon dioxide fluxes during weather condition treatments varied from mean 1640 mg CO_2 m^{-2} h^{-1} ('wet-cool' treatment) to 1555 mg CO_2 m^{-2} h^{-1} ('dry-warm' treatment) and to a minimum 1222 mg CO_2 m^{-2} h^{-1} ('planted' treatment). However, comparing these data at the level of significance ($P < 0.1$) showed a little, but still significant differences between 'wet-cool' and 'warm'/'planted' treatments indicating that flux from ecosystems is depending on changes in ambient temperature. The treatments on vegetation cover 'planted'(soil + vegetation CO_2 respiration), 'clear-cut' '(soil CO_2 respiration) represented the CO_2 flux decreased by 36.9% indicating a significant ($p > 0.0111$) role of vegetation on ecosystem-derived CO_2 flux. There was no obvious effect of temperature changes, tested by ANOWAs (R^2 = 0.74; F = 5.64, df = 3, p > 0.1407) on a CO_2 respiration from soil without plant shoots ('clear-cut' treatment).

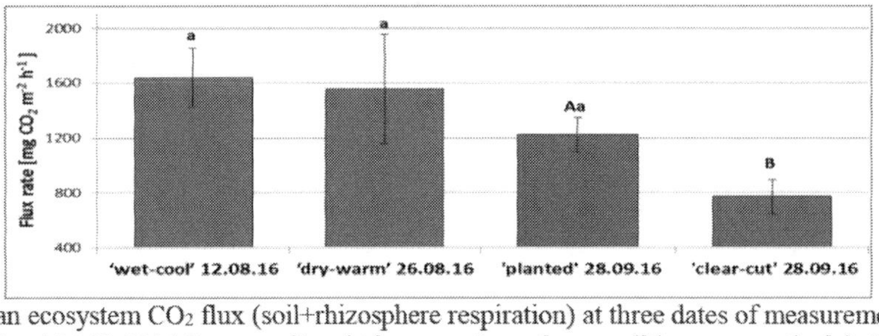

Fig. 2. Mean ecosystem CO_2 flux (soil+rhizosphere respiration) at three dates of measurements: 12.08.2016 and 26.08.2016 ('wet-cool' and 'dry-warm' weather conditions, respectively); 28.09.2016 (canopy manipulation: 'planted', 'clear-cut'). Error bars show the standard deviation of soil CO_2 flux. Significance assessed by one-way ANOVA and Bonferroni test for pairwise comparisons ($P < 0.05$) and denoted in the graph by different case letters. Capital letters indicate which measurements were significantly different and low-case letters are considered non-significant.

There was a positive correlation between CO_2 concentrations and the related $\delta^{13}C$-CO_2 values in different treatments (R^2 = 0.88–0.97, $p < 0.05$) (Fig. 3). The $\delta^{13}C$-CO_2 of the ecosystem and soil ('clear-cut' conditions) respired CO_2 varied in a range from

−11.3 to −23.7‰ and from −10.4 to −23‰ for weather conditions and canopy manipulation, respectively.

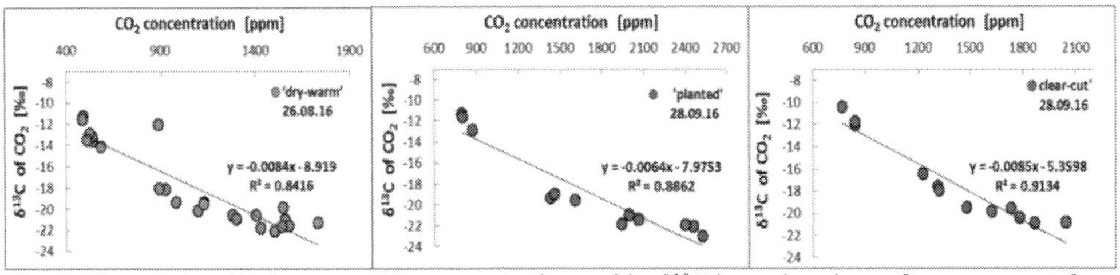

Fig. 3. The relationship between CO_2 concentration and its $\delta^{13}C$ isotopic values of measurement data on 26.08.2016 and 28.09.2016 ('dry-warm' and 'planted') as well as of canopy manipulations ("planted" and 'clear-cut'). There were no data on $\delta^{13}C$-CO_2 available for 'wet-cool' weather conditions.

The mean $\delta^{13}C$ value of the ecosystem (soil + vegetation) CO_2 estimated by Keeling plot method was -26.2‰±0.8 under 'dry-warm' conditions and −28.0‰±1.4 in 'planted' treatment (Fig. 4).

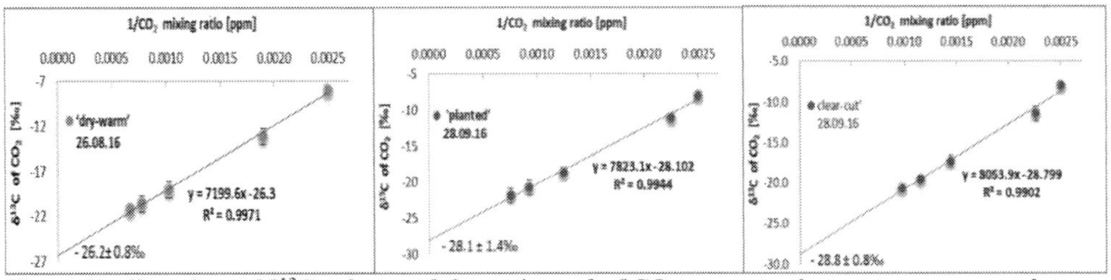

Fig. 4. Keeling plots of $\delta^{13}C$ values and the reciprocal of CO_2 concentrations at measurement data on 26.08.2016 ('dry-warm'); and 28.09.2016, and canopy manipulation: 'planted', 'clear-cut'. Error bars show the standard deviation of the ratio. Data for atmospheric $\delta^{13}C$ (CO_2) value (-8‰) and CO_2 concentration (400.71 ppm) were taken from Atmospheric Concentrations of CO_2 from Mauna Loa, Hawaii, 2015 [19].

The estimated $\delta^{13}C$-CO_2 values of ecosystem respiration in re-established demonstrate the sensitivity of changes in CO_2 ecosystem respiration and its isotopic composition relative to the variation in weather conditions and the role of living vegetation (Fig. 5). It is consistent with the fact that after the treatment on excised shoots ('clear-cut'), the contribution of root respiration was about half (60–65%) of total respiration derived from soils.

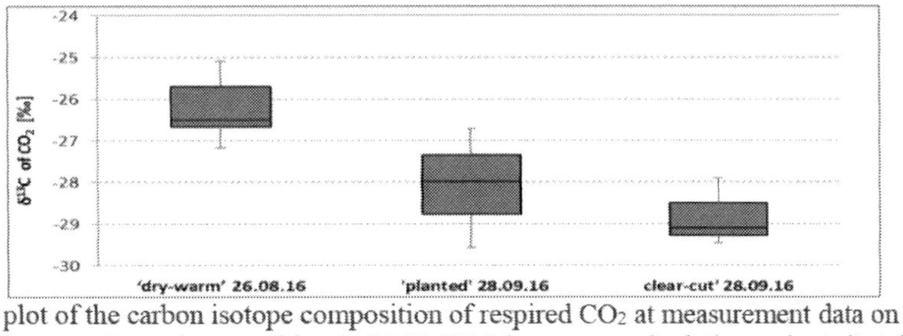

Fig. 5. A box plot of the carbon isotope composition of respired CO_2 at measurement data on 26.08.2016 ('dry-warm' weather conditions); 28.09.2016 (canopy manipulation: 'planted', 'clear-cut'). The box plots represent the spread of 50% (25–75%) of the data range. The horizontal lines indicate the median value. The error bars show upper/lower quartile range (minimum-maximum) of the data values.

 DOI: 10.1051/e3sconf/20172200017

4 Discussion

The relationships between ecosystem CO_2 fluxes and environmental factors (weather and vegetation) differ substantially in terms of little, but still, significant differences in temperature and precipitation conditions and high significance of vegetation manipulation conditions. The CO_2 flux from ecosystem was positively correlated with ambient temperature and precipitation and has an average level of significance ($p < 0.1$) between treatments ('wet-cool' and 'warm'/'planted' measurements). The significant interactive effect ($p < 0.05$) between the CO_2 flux and vegetation cover was detected. Cumulative CO_2 flux from ecosystems was highly affected (decreased more than three times) by separating, manipulation on aboveground (removing shoots) versus belowground (roots and microorganisms) respiratory CO_2 sources.

We hypothesized that the surface CO_2 flux will be higher under warm weather conditions ('dry-warm') as compared to cool weather ('wet-cool') due to a higher temperature accelerated soil respiration. Despite the difference was not significant: the CO_2 flux under 'wet-cool' conditions was only by 5% higher than under 'dry-warm' weather. It could be a result of larger soil moisture when water replacing the soil pore space and may increase the CO_2 efflux (physical phenomenon). Increasing was also possible since vegetation respiration was more pronounced under wet conditions. The removal of vegetation under 'warm' conditions decreased the surface CO_2 flux approximately 1.6 times compared with control (with canopy) under the same conditions. The decrease was mainly attributed to the exclusion of shoot respiration and partly due to suppression of root- and rhizosphere respiration.

The difference between the treatments in the $\delta^{13}C$-CO_2 value was mainly due to factors that affect the isotopic fractionation (biotic and abiotic variables) during of aboveground (biomass, plant ground cover) versus belowground (root + rhizosphere) respiratory CO_2 sources. Indeed, the $\delta^{13}C$-CO_2 decrease after treatment on vegetation cover ('clear-cut' manipulation), may reflect a decline in the contribution of CO_2 from vegetation respiration to the total ecosystem respiration.

The weather conditions had an impact on carbon isotopic composition of ecosystem-derived CO_2 mainly due to effects of plants physiology in dry and wet periods [3]. The $\delta^{13}C$-CO_2 values of ecosystem respiration for the 'warm' conditions decreased insignificantly by 2‰ over a period of 28 days (from late August to the end of September). This may be explained by the lowering of photosynthetic activity (partly reflected in CO_2 flux decrease, Fig. 2) at the end of the vegetation season and probably, microbial activity.

Despite there was no significant difference (the decline from 28.1 to 28.8‰) between the $\delta^{13}C$-CO_2 values in 'planted' and 'clear-cut' treatments, the pattern of further decrease of $\delta^{13}C$-CO_2 as compared with earlier dates highlighted the role of living vegetation in a contribution of relatively enriched ^{13}C to the entire ecosystem flux. Additional variation in ecosystem respiration and $\delta^{13}C$-CO_2 flux can be explained by the sum of other environmental conditions such as soil moisture and temperature, as well as by pH and the soil organic matter quality (C:N ratio) in this grassland. Therefore, detailed studies on linking soil organic matter quality, microbial activity and vegetation properties with ecosystem CO_2 fluxes and environmental conditions are necessary for the estimation of current and future C balance.

5 Conclusion

Contrasting weather conditions and vegetation cover, a primary temperature were the main factors controlling ecosystem CO_2 flux in re-established grasslands. However, the differences between treatments in the significance of the relationships between flux rates

 DOI: 10.1051/e3sconf/20172200017

and changes in environmental variables were identified. In particular, measured respiration rates were strongly correlated with temperatures and changing water content, but the most significant effects on ecosystem CO_2 fluxes was determined by manipulating vegetation cover. The relationships between carbon isotopic composition of ecosystem-derived $\delta^{13}C$-CO_2 and environmental conditions showed a minor difference between 'wet' and 'dry' weather treatments as well as slight responsibility for the changing in a contribution of living vegetation. As the processes controlling ecosystem isotope discrimination are more complex and highly unpredictable, the application of continuous monitoring of the $\delta^{13}C$-CO_2 signal in respiration to predict the response of ecosystems to environmental changes supposed to have the highest potential.

Acknowledgements

The investigations were co-financed within the framework of the project No. 0402/0077/16 with the specific subsidy granted for the Faculty of Environmental Engineering Wroclaw University of Science and Technology (W-7) by the Minister of Science and Higher Education to conduct research and development work and related tasks contributing to the development of young scientists and doctoral students in 2016/2017.

References

1. N. Edwards, R. Norby, Plant & Soil **206**, 85–97 (1999)
2. Y. Salmon, N. Buchmann, R.L. Barnard, PLoS ONE **11**, 3: e0151583 (2016)
3. E. Davidson, I. Janssens, Nature **440**, 165–173 (2006)
4. D. Martin, J. Beringer, L. Hutley, I. McHugh, Agri. & Forest Meteor. **147**, 58–70 (2007)
5. D. Asensio, J. Penuelas, J. Llusia, R. Ogaya, I. Filella, Soil Biol. & Biochem. **39**, 10 (2007)
6. D. Kim, R. Vargas, B. Bond-Lamberty, M.R. Turetsky, Biogeosci. **9**, 2459–2483 (2012)
7. C. Gabriel, L. Kellman, Biogeosci. Discuss. **8**, 1369–1409 (2011)
8. Q. Ashton Acton, Iss. in Industrial., Applied, and Env. Chemistry (2013)
9. R. Phillips, M. Eken, M. West. Environ. Manage. **55**, 1191–1199 (2015)
10. A. Iwaasa, P.M. Schellenberg, B. McConkey, Prairie Soils & Crops **5**, 85–95 (2012)
11. M. Potthoff, K. Steenwerth, L.E. Jackson, R.E. Drenovsky, K.M. Scow, Biol. & Biochem. **38**, 1851–1860 (2006)
12. Forstbotanischer Garten. Forest Botanical Garden and Arboretum of the University of Göttingen. Available at http://www.uni-goettingen.de/en/forest-botanical-garden
13. Climate Göttingen. Available at https://www.meteoblue.com/en/weather
14. Weather station Experimental Botanical Garden University of Göttingen. Available at http://www.uni-goettingen.de/en/weather-station-experimental-bot-garden/83165.html
15. I. Matejovic, Commun. Soil. Sci. Plant Anal. **24** (17–18), 2213–2222 (1993)
16. R. Joergensen, Soil Biol. & Biochem. **28**, 25–31 (1996)
17. J.R. Taylor, Oxford Univ. Press (1982)
18. C.D. Keeling, Geochimica et Cosmochimica Acta, **13** (1958)

19. Atmospheric Concentrations of CO_2 from Mauna Loa, Hawaii. Available at http://cdiac.ornl.gov/trends/co2/recent_mauna_loa_co2.html
20. E. Solly, I. Schöning, S. Boch, E. Kandeler, S. Marhan, B. Michalzik, J. Müller, J. Zscheischler, S. E. Trumbore, M. Schrumpf, Plant & Soil **382**, 203 (2014)
21. J. Reeder, G. Schuman, R. Bowman, Soil & Tillage Res. **47**, 339–349 (1998)

E3S Web of Conferences **22**, 00018 (2017)
ASEE17
DOI: 10.1051/e3sconf/20172200018

Application of stochastic approach based on Monte Carlo (MC) simulation for life cycle inventory (LCI) of the rare earth elements (REEs) in beneficiation rare earth waste from the gold processing: case study

Bogusław Bieda[1,*], and *Katarzyna* Grzesik[2]

[1]AGH University of Science and Technology, Faculty of Management, al. Mickiewicza 30, 30-067 Kraków, Poland
[2]AGH University of Science and Technology, Faculty of Mining Surveying and Environmental Engineering, al. Mickiewicza 30, 30-059 Kraków, Poland

Abstract. The study proposes an stochastic approach based on Monte Carlo (MC) simulation for life cycle assessment (LCA) method limited to life cycle inventory (LCI) study for rare earth elements (REEs) recovery from the secondary materials processes production applied to the New Krankberg Mine in Sweden. The MC method is recognizes as an important tool in science and can be considered the most effective quantification approach for uncertainties. The use of stochastic approach helps to characterize the uncertainties better than deterministic method. Uncertainty of data can be expressed through a definition of probability distribution of that data (e.g. through standard deviation or variance). The data used in this study are obtained from: (*i*) site-specific measured or calculated data, (*ii*) values based on literature, (*iii*) the ecoinvent process *„rare earth concentrate, 70% REO, from bastnäsite, at beneficiation*". Environmental emissions (e.g, particulates, uranium-238, thorium-232), energy and REE (La, Ce, Nd, Pr, Sm, Dy, Eu, Tb, Y, Sc, Yb, Lu, Tm, Y, Gd) have been inventoried. The study is based on a reference case for the year 2016. The combination of MC analysis with sensitivity analysis is the best solution for quantified the uncertainty in the LCI/LCA. The reliability of LCA results may be uncertain, to a certain degree, but this uncertainty can be noticed with the help of MC method.

1 Introduction

All of the REEs, were finally identified in the 20th century. Promethium (Pm), the rarest, was not identified until 1945, and pure lutetium (Lu) was not refined until 1953 ([1–2]). Detailed history of REE production is presented in Castor and Hedrick [3].

* Corresponding author: bbieda@zarz.agh.edu.pl

 DOI: 10.1051/e3sconf/20172200018

Rare-earth elements (REEs) are a group of 17 elements with similar chemical properties, including 15 in the lanthanide group, yttrium, and scandium because of their similar physical and chemical properties [2]. For more than two decades, at least 95% of annual global supply of the rare-earth elements (REEs) has been provided by Chinese rare-earth producers. Mining companies are now actively seeking for new exploitable rare-earth deposits, white old mines ae being reopened [4]. Life cycle assessment (LCA) takes a holitic approach and provides a complete view of the environmental impacts over the entire life cycle of a proces or product, from raw material extraction and aquisition, manufacturing, transportation and didtribution, use and maintenance, reuse and recycle, and all the way to disposal and waste management [2, 5, 7].

2 LCI data quality analysis

2.1 Data preparation

The LCI is a critical component as it is the data foundation of the LCA study [2]. The approach adopted by ISO 14040 is to compile the inventory based on the inputs and outputs fromeach of the processes (referred to as unit processes) involved in a product's life cycle [2, 8, 9].

An LCI analysis usually requires a large amount of data. The uncertainty of these parameters directly affects the outcome of any environmental impact method [10]. The overall uncertainty of an LCI is usually dominated by a few major uncertainties [11]. The use of stochastic model helps to characterize the uncertainties better than a purely analytical mathematical approach.

The aim of this study were to develop a stochastic approach for LCA technique limited to LCI uncertainty assessment for study for rare earth elements (REEs) recovery from the secondary materials processes production applied to the New Krankberg Mine in Sweden.

2.2 Uncertainty analysis of LCI

At the LCI level, Monte Carlo (MC) simulation was used. Stochastic nature of the MC method is based on random numbers [12–13]. The uncertainty, at LCI level, introduced into inventory due to the cumulative effects of input uncertainty and variability of inventory data was qualified by using expert judgment-based approach [14]. Uncertainty carried out in this study lead to a transparent increase in confidence in the life cycle impact assessment (LCIA), and finally LCA findings [14]. The complexity of many practical situations often requires simulation [15]. The MC method is recognized as an important tool in science [16]. MC simulation uses distributions to generate realistic random values. The benefits of simulation modeling approach are: (*1*) understanding of the probability of specific outcomes, (*2*) ability to pinpoint and test the driving variables within a model, (*3*) a far more flexible model; and (*4*) clear summary charts and reports [10]. In this work the MC sampling was carried out using an Excel® spreadsheet and Crystal Ball®, a software package which generates random numbers for a probability distribution. A large number of trials is required to obtain accurate results. The MC analysis-simulation is the only acceptable approach for U.S. Environmental Protection Agency (EPA) risk assessments [17]. Simulation models are generally easier to understand than many analytical approaches [15]. Bieda [13] quoted definition of the uncertainty presented in the Commission Decision of 18 July 2007 guidelines for the monitoring and reporting of greenhouse gas emissions pursuant to Directive 2003/87/EC of the European parliament and of the Council. Definition is defined as follow: "*a parameter associated with the result of the determination of a quantity that characterizes the dispersion of the values that could reasonably be attributed to the particular quantity, including the effects of systematic as well as of random factors and expressed in per cent and describes a confidence interval around the mean value comprising 95 % of inferred values taking into account any asymmetry of the distribution of values*" [13].

2.3 Results

Data used in the study has been obtained from the following sources:
• site-specific measured or calculated data - the secondary materials processes production applied to the New Krankberg Mine in Sweden
• value based on information in the literature
• the ecoinvent process *„rare earth concentrate, 70% REO, from bastnäsite, at beneficiation.*

In the present study we discuss and modeled our LCI based on the proposed process for the beneficiation of REE in the flotation tailings from new Kankberg mine [18].

After the flotation stage, the concentrate that contains a mix of phosphates (apatite and monazite) can be further enriched through magnetic separation thanks to the paramagnetic property of monazite (apatite is non-magnetic) [18]. Magnetic separation leads to the production of a concentrate containing phosphate (P_2O_5) content (monazite mainly), cerium(Ce), lanthanum (La) and neodymium (Nd) [18]. The simulation results after 10,000 trials have been presented in the form of frequency forecast charts (Figs. 1, 2, 3 and 4), and and statistic reports (Tab. 1). The confidence interval is 95%. The total forecast value ranges of Ce, La, Ne and P_2O_5 forecast value amounted to the mean values of 170.02 ppm, 89.78 ppm, 70.00 ppm and 0.70%, respectively, ranging from 136.98 ppm to 203.37 ppm (see Fig. 1), from 72.06 ppm to 107.19 ppm (see Fig. 2), from 56.58 ppm to 83.76 ppm (see Fig. 3), and from 0.14% to 0.20% (see Fig. 4), respectively. The frequency chart is a histogram of the outcome variable that includes all values within 2.6 standard deviations from the mean, which represents approximately 99% of the data [15]. Just below the horizontal axis at the extremes of distribution there are two small triangles, called endpoint grabbers. The confidence limits, presented in the frequency charts, are fixed using above-mentioned grabbers (the area of the frequency charts covered by them is darker) [15].

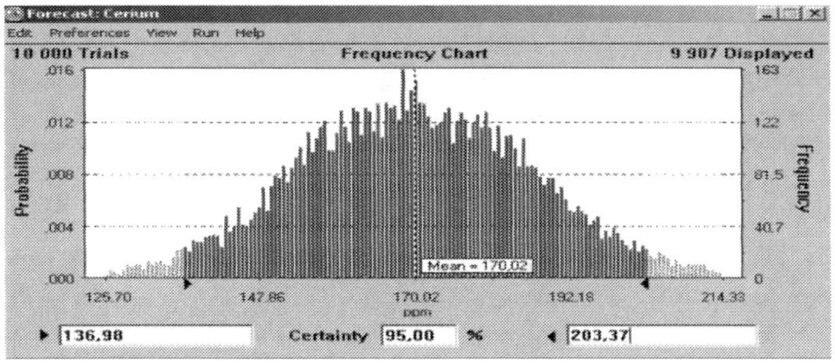

Fig. 1. Forecast frequency chart: Cerium (95% confidence level).

DOI: 10.1051/e3sconf/20172200018

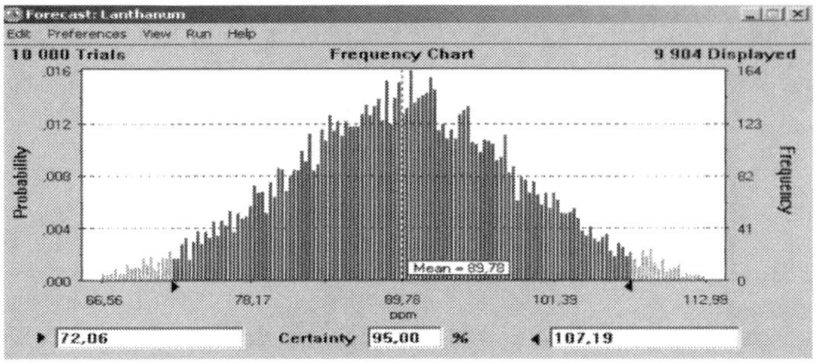

Fig. 2. Forecast frequency chart: Lanthanum (95% confidence level).

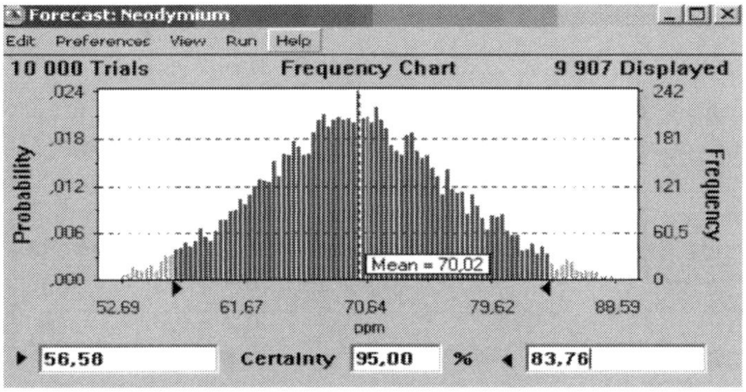

Fig. 3. Forecast frequency chart: Neodymium (95% confidence level).

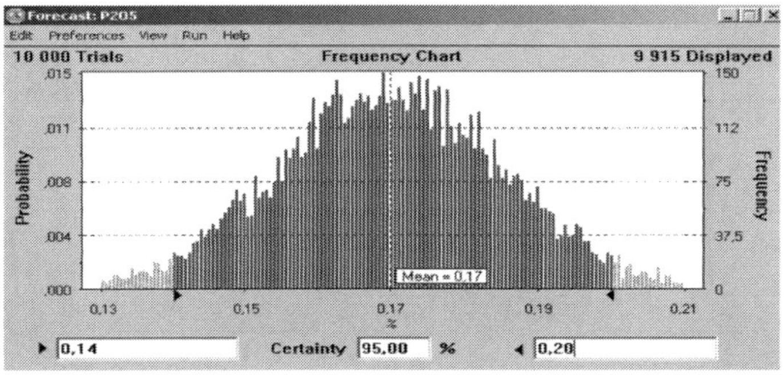

Fig. 4. Forecast frequency chart: P_2O_5 (95% confidence level).

Table 1. Statistics of outcomes from the simulation (Ce, Ne, La, P_2O_5)

Statistic	Forecast: Cerium (Cell A1) Value	Forecast: Lanthanum (Cell A2) Value
Trials	10 000	10 000
Mean	170,02	89,80
Median	169,89	89,82
Mode	---	---
Standard Deviation	17,04	9,09
Variance	290,52	82,56
Skewness	0,01	0,01
Kurtosis	2,99	3,03
Coeff. of Variability	0,10	0,10
Range Minimum	95,07	56,93
Range Maximum	232,13	123,71
Range Width	137,06	66,78
Mean Std. Error	0,17	0,09

DOI: 10.1051/e3sconf/20172200018

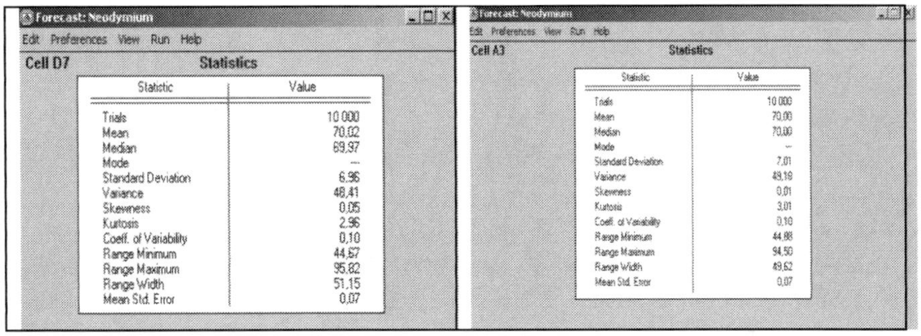

Statistic	Value
Trials	10 000
Mean	70,02
Median	69,97
Mode	---
Standard Deviation	6,96
Variance	48,41
Skewness	0,05
Kurtosis	2,96
Coeff. of Variability	0,10
Range Minimum	44,67
Range Maximum	95,82
Range Width	51,15
Mean Std. Error	0,07

Statistic	Value
Trials	10 000
Mean	70,00
Median	70,00
Mode	---
Standard Deviation	7,01
Variance	49,19
Skewness	0,01
Kurtosis	3,01
Coeff. of Variability	0,10
Range Minimum	44,88
Range Maximum	94,50
Range Width	49,62
Mean Std. Error	0,07

3 Conclusions

The aim of the study was to using a stochastic modelling based on the MC simulation for LCI to the rare earth elements (REEs) in beneficiation rare earth waste from the gold processing: case study, as well as to promote the use of uncertainty approach in environmental science. Uncertainty analysis in the LCA methodology has received increasing attention over the last years.

The use of MC simulation allows for saving in time and resources. Application of the MC simulation into the LCA of the production of rare earths, may give support in the interpretation of LCA results and permit to better understanding of many analytical approaches. Generally, in a deterministic model, all data is known, while in a probabilistic model, data is presented and described by probabilistic distributions.

LCA/LCI data is full of uncertain numbers, and the MC analysis is a useful approach of quantifying parameter uncertainty in LCA studies.

Thanks to uncertainty analysis, a final result is obtained in the form of value range. The results obtained from this work can help practitioners and decision makers in the environmental engineering. Lack of uncertainty analysis in LCI has influence on the LCIA results, and finally on the LCA outcomes.

The results of the study were the base for decision making process. Moreover, these results move the LCI study on the rare earth elements (REEs) recovery from the secondary materials processes production one step forward.

Recommendations and outlook

The research described in this paper can also serve as the basis for future scientific work. The potential direction for future research is to integrate LCA and risk assessment for industrial processes. A complementary paper about LCIA will be produced.

Acknowledgments

This publication and research was completed within ENVIREE project (ENVIronmentally friendly and efficient methods for extraction of Rare Earth Elements from secondary sources) funded by NCBR, within the 2nd ERA-NET ERA-MIN Joint Call Sustainable Supply of Raw Materials in Europe 2014.

DOI: 10.1051/e3sconf/20172200018

References

1. J. Emsley, *Nature's Building Blocks: An A-Z Guide to the Elements* (Oxford, England: Oxford University Press 2001)
2. J. Navarro, F. Zhao, **2** 1–17. http://home.3frontiersin.org/ (2014)
3. S.B. Castor, J.B. Hedrick, *Rare Earth Elements*, *Industrial Minerals and Rocks* (Edited by Jessica Elzea Kogel, Nikhil C. Trivedi and James M. Barker. Society for Mining, Metallurgy and Exploration. http://www.rareelementresources.com/i/pdf/RareEarths-CastorHedrickIMAR7.pdf , 769–792, 2006)
4. K. Binnemans, P.T. Jones, B. Blanpain, T. Van Gerven, J. Clean Prod. **99**, 1–22 (2015)
5. M.A. Curran, *Cincinnati, OH: National Risk Management Research Laboratory*, Office of Research and Development, US Environmental Protection Agency (2006)
6. H. Lehtinen, A. Saarentaus, J. Rouhiainen, M. Pitts, A. Azapgic, Manchester, UK: Europe Innova. (2011)
7. S. Lizin, S. Van Passel, E. De Schepper, W. Maes, L. Lutsen, J. Manca, Energy Environ. Sci. **6**, 3136–3149 (2013)
8. S. Suh, G. Huppes, J. Clean Prod. **13**, 687–697 (2005)
9. EU Commission, *Critical Raw Materials for the EU. Report of the Ad-Hoc Working Group on Defining Critical Raw Materials.* (Technical Report. Brussels: European Commission 2010)
10. G. Sonnemann, F. Castells, M. Schumacher, *Integrated Life-Cycle and Risk Assessment for Industrial Processes.* LEWIS PUBLISHERS. A CRC Press Company (2004)
11. B. Bieda, R. Tadeusiewicz, Intl. Trans. in Op. Res. **15**, 103–119 (2008)
12. B. Bieda, Sci. Total Environ. **442**, 489–496 (2013)
13. B. Bieda, Sci. Total Environ. **481**, 649–655 (2014)
14. M. Guo, R.J. Murphy, Sci. Total Environ. **435–436**, 230–243 (2012)
15. J.R. Evans, D.L. Olson, *Inroduction to simulation and risk analysis* (Prentice Hall. Inc. A Simon & Schuster Company Upper Saddle River, New Jersey 07458, 1998)
16. K. Binder, *Introduction* (In: Binder K, editor. The Monte Carlo Method in Condensed Matter Physics. Topics in Applied Physics, **71**, Springer, 1–22 (1995)
17. B. Finley B, D. Proctor, P. Scott, N. Harrington, D. Pasutenbach, P. Price, Risk Anal. **1**, 533–553 (1994)
18. Y. Menard, M. Alastair, *ENVIREE – Deliverable 2.1 – Report on the most suitable combined pre-treatment, leaching and purification processes*, 6–27 (2017)

Food waste in Central Europe – challenges and solutions

Jan den Boer[1,*], *Przemysław* Kobel[1], *Arkadiusz* Dyjakon[1], *Klaudia* Urbańska[1], *Gudrun* Obersteiner[2], *Marlies* Hrad[2], *Elisabeth* Schmied[2], and *Emilia* den Boer[3]

[1]Wroclaw University of Environmental and Life Sciences, Institute of Agricultural Engineering, ul. Chelmonskiego 37a, 51-630 Wroclaw, Poland
[2]University of Natural Resources and Life Sciences, Department of Water, Atmosphere and Environment, Institute of Waste Management, Muthgasse 107, 1190 Vienna, Austria
[3]Wrocław University of Science and Technology, Faculty of Environmental Engineering, Wybrzeże Wyspańskiego 27, 50-370 Wrocław, Poland

Abstract. Food waste is an important issue in the global economy. In the EU many activities aimed at this topic are carried out, however in Central Europe is still quite pristine. There is lack of reliable data on food waste quantities in this region, and not many preventive actions are taken. To improve this situation the STREFOWA (Strategies to Reduce and Manage Food Waste in Central Europe) was initiated. It is an international project (Austria, Czech Republic, Hungary, Italy, Poland), founded by the Interreg Central Europe programme, running from July 2016 to June 2019. Its main purpose is to provide solutions to prevent and manage food waste throughout the entire food supply chain. The results of STREFOWA will have positive economical, social and environmental impacts.

1 Introduction

Food is one of the largest existing product flows. It is often suggested, that up to half of all food grown is lost or wasted and thus not eaten [1]. Estimates on the global level of food wastage by the FAO state that 'roughly one-third of the edible parts of food produced for human consumption, gets lost or wasted globally, which is about 1.3 billion ton per year [2]. Parfitt presented results of an overview on food waste quantity studies, coming up with estimates for the post consumer food waste for the United Kingdom of 25%, for the United States up to 15%, for The Netherlands 8–11% and for Turkey 9% [3].

In Central Europe however, the available data on food waste generation is limited. In the frame of a recent 7th Framework European project, Fusions, an attempt was done to obtain data from EU Member States. The results for the Central European Member states are shown in table 1 below.

* Corresponding author: jan.denboer@upwr.edu.pl

 DOI: 10.1051/e3sconf/20172200019

Table 1. Assessment of food waste data provided by Central European Member States within the Fusions project [4].

Country	**Production**	**Processing**	**Wholesale and logistics**	**Retail and markets**	**Redistribution**	**Food service**	**Household**
Bulgaria	no data	no data	no data	no data	no data	no data	no data
Croatia	low quality	low quality	low quality	low quality	no data	low quality	low quality
Czech republic	low quality	low quality	low quality	low quality	no data	low quality	low quality
Estonia	no data	low quality	sufficient quality	sufficient quality	no data	low quality	sufficient quality
Hungary	no data	no data	no data	no data	no data	no data	no data
Latvia	no data	no data	no data	no data	no data	no data	no data
Lithuania	low quality	sufficient quality	low quality	low quality	no data	low quality	low quality
Poland	no data	no data	no data	no data	no data	no data	no data
Romania	no data	no data	no data	no data	no data	no data	no data
Slovakia	low quality	low quality	no data	no data	no data	no data	no data
Slovenia	low quality	low quality	sufficient quality	sufficient quality	no data	low quality	low quality

In general the data on food waste generation in Central Europe is of low quality or is completely lacking. Exceptions are smaller countries, that have an over-average GDP, like Estonia and Slovenia. These countries provided with data that was partly of sufficient quality. However the larger countries, like Poland and Romania did not provide with any data. In figure 1 an overview is given on the food waste generation data availability in the European Union.

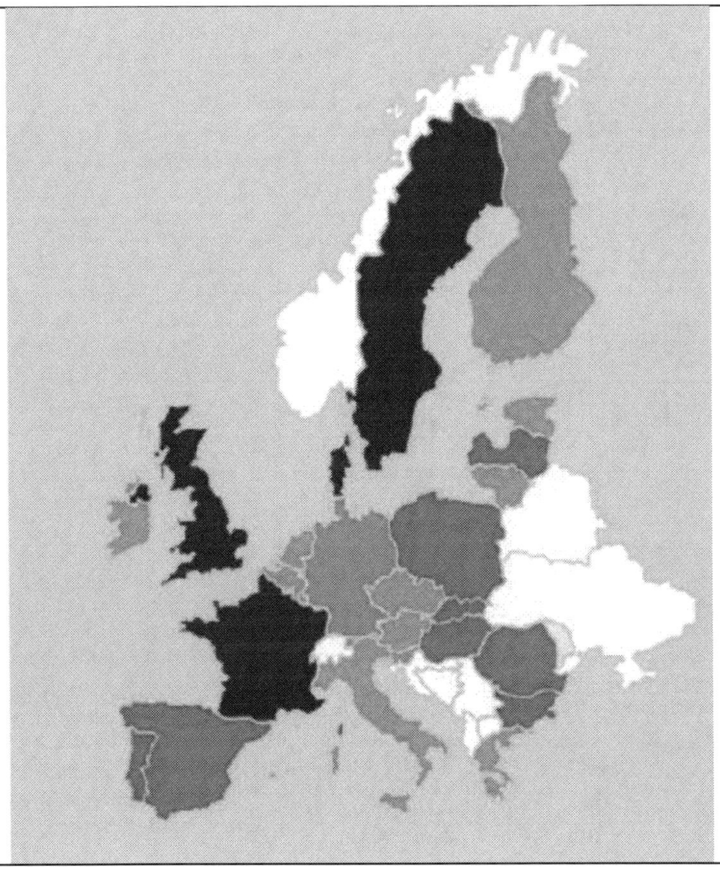

insufficient data quality
1 red
2. yellow
↓ 3. light blue
4. dark blue
5. green
high quality data

Fig. 1. Food waste data provided by Member States within the Fusions project [4].

The overall assessment of the food waste data quality as provided in figure 1 shows that the situation is far from positive for most Central European countries. However, based on generic indicators for food waste generation in specific sectors and the size of those sectors in the considered countries (e.g. number of inhabitants, number of hotel stays, number of meals prepared in restaurants and catering) estimated of the amounts of generated food waste can be made. BIO Intelligence Service came up with an European wide overview for several sectors in the year 2010, partly based on older (2006) data [5]. In table 2 the figures for the Central European Member States are provided.

Table 2. Estimates of food waste generation in Central European Member States (kg/inh.yr). Adjusted from [5].

Country	**Manufacturing**	**Households**	**Other sectors**	**Total**
Bulgaria	49	39	4	91
Czech republic	35	25	11	71

 DOI: 10.1051/e3sconf/20172200019

Country	Manufacturing	Households	Other sectors	Total
Estonia	179	62	27	268
Hungary	116	40	31	186
Latvia	55	35	5	95
Lithuania	62	31	69	162
Poland	170	53	9	233
Romania	22	31	49	102
Slovakia	64	25	19	108
Slovenia	21	36	32	89

Apart from the lack of precise data in Central Europe, the striving to solutions is as well focused mainly on Western and Northern Europe. This is both true for research and demonstration project funded by the European Commission (e.g. FUSIONS, Food Use for Social Innovation by Optimising Waste Prevention Strategies, FP7; REFRESH, Resource Efficient Food and dRink for Entire Supply cHain, Horizon2020; FORWARD, Food Recovery and Waste Reduction, Life Long Learning Programme), national research and demonstration programmes (e.g. in Austria, Germany, The Netherlands, Denmark, Sweden, the UK) and national and transnational activities of NGO's.

For the reasons mentioned above, the project STREFOWA was initiated. STREFOWA stands for *'Strategies to Reduce and Manage Food Waste in Central Europe'*. It is funded within the Interreg Central Europe programme. Although in Central Europe the exact amounts and composition of generated food waste is rather unknown, the main focus of the project is not on retrieving reliable data. The acquiring of such data is a costly and time consuming process. Apart from the fact, that it is harder to find funding for food waste data research than for implementation of solutions, it is also obvious that the food waste problem also exists in Central Europe. Therefore the main focus of the STREFOWA project is to demonstrate new as well as existing solution to prevent and manage food waste in a Central European context.

2 STREFOWA project: Background

The project *'Strategies to Reduce and Manage Food Waste in Central Europe'*, STREFOWA, runs from July 2016 to June 2019. It is implemented in Austria, the Czech Republic, Hungary, Italy and Poland.

2.1 Goal and objectives

The main goal of the STREFOWA project is to improve food waste management in selected Central European urban areas by fostering food waste prevention as well as proper treatment and therefore reduce greenhouse gas emissions.
This main goal can be divided into three objectives:

- To improve **prevention** of food waste in urban areas. By establishment of strategies to avoid food waste along whole value chain (agriculture – commerce/gastronomy – households) through training programmes and awareness raising, stakeholder participation processes and cooperation strategies. These will be tested via different pilot actions;
- To improve **treatment** of food waste in urban areas. Definition of options to collect and treat food waste in sustainable way and introduction of new collection schemes and test facilities to separate organic from residual waste as well as introduction of new innovative treatment options for food waste;
- To improve **knowledge** on food waste management. All findings will end up in the interactive, tailored to different users (private persons, local authorities, restaurateur, retailers, teachers, waste management authorities...) web based decision support tool which will provide best practice guidelines, training material, guidelines etc. adjusted to all relevant public and private actors.

2.2 Consortium

Within the STREFOWA project a number of project partners with various geographical, cultural and expertise backgrounds is working together under the lead of the University of Natural Resources and Life Sciences, Vienna. The consortium consists of the following partners:

1. University of Natural Resources and Life Sciences, Vienna (ABF-BOKU, Austria)
2. Glopolis, p.s.c (Glopolis, Czech Republic)
3. Waste Management Association Mid-Tyrol (ATM, Austria)
4. Province of Rimini (Province of Rimini, Italy)
5. Langhe Monferrato Roero Development Agency (LAMORO, Italy)
6. Wrocław University of Environmental and Life Sciences (WUELS, Poland)
7. Bay Zoltán Nonprofit Ltd. for Applied Research (BZN, Hungary)
8. Federation of Polish Food Banks (FPFB, Poland)
9. Square Bracket eU (Square Bracket eU, Austria)
10. SPAR Hungary Ltd. (SPAR, Hungary)

Apart from the project partners, three institutions are associated to the project, not being a formal project partner: the Municipal Department 22 of the city of Vienna, the Municipality of Asti and Waste Treatment Ahrental Ltd.

3 STREFOWA project: Approach

3.1 Intended project outputs

The STREFOWA project aims to reach overall outputs in the area of transnational cooperation, tools, training & education as well as demonstration. The intended project outputs that will be reached by the end of the project duration are:

- **Food Waste Prevention Tool.** All relevant project outcomes will be included in an interactive web based tool providing information on food waste prevention and management tailored to different users (consumers, restaurateurs, retailers, authorities…) in English and localised language to partner countries. The Tool will include best practice guidelines along the food supply chain, training and education material, outcomes of the pilot activities and information on environmental benefits. The results of the Food Waste Hackathons that will be organised in the frame of the project also serve as input for the Tool.
- **Pilot Actions on the reduction and management of food waste**. Demonstration projects on food waste prevention will be performed. According to indentified deficits and existing best practice models proper approaches for food waste prevention will be implemented in various functional urban areas. The optimisation of biowaste (esp. food waste) collection will be tested for hotels. Separate collection biowaste collection will be introduced and tested in at least one functional urban area. The content of food waste in residual waste will be reduced via specific sorting aggregates.
- **Guidelines and training on food waste prevention and management.** For all sectors along the food chain guidelines and training programmes aiming at the prevention and management of food waste will be developed. Pilot trainings will be implemented and all materials will be compiled in a Handbook on food prevention and treatment approaches
- **Action plan on long-term transnational umbrella organisation on food waste.** An association will be formed existing of a conglomerate of food waste prevention and management experts that will cooperate beyond the time frame of the project on newly developing themes in the field.

3.2 Pilot Actions

An important mean to reach the project objectives is the implementation of pilot activities. In functional urban areas located in the project partner's countries it will be demonstrated, that approaches and methods for the prevention, collection and treatment of food waste will lead to positive results in the designated cities. These measures have been used in other geographic regions already and were identified as 'best practices'.

In table 3 an overview of the intended Pilot Actions is given.

Table 3. Planned Pilot Actions in the STREFOWA project.

Pilot Action	Description	Urban functional area
Food waste prevention		
agriculture and food industry	redistribution	Vienna
retail	reuse, donation to foodbanks	Warsaw, Prague, Asti, Hungary
gastronomy	hotels and restaurants: donation and prevention	Vienna, Rimini
schools	education of pupils, influencing households	Vienna, Prague, Asti
consumers	raising awareness in households	Vienna, Rimini
Food waste collection		

Pilot Action	Description	Urban functional area
gastronomy	optimisation of separate collection of food waste in hotels	Rimini
households	introduction of separate collection of biowaste from consumers	Wrocław
Food waste treatment		
sorting out of food waste from residual waste	additional sorting aggregate in an existing mechanical-biological pretreatment plant. The organic fraction will be digested anaerobically.	Innsbruck

The outcomes of the pilot actions will be assessed both on their economic effectiveness (how much efforts it costs to come to the results) and their environmental impacts. For the environmental impacts a Life Cycle Assessment approach will be used, resulting in the amounts of avoided Greenhouse Gas emissions for each Pilot Action.

4 Conclusions

In today's world of declining natural resources and increasing environmental degradation, reducing the amount of lost or wasted food is one of the most important priorities.

In the EU a variety of programs and activities on this important topic is carried out at the international, national and local level. STREFOWA is in line with this pan-European trend but focuses its efforts on the relatively pristine area of Central Europe.

Its main purpose is to provide solutions to prevent and manage food waste. Project activities focus on all aspects of the food supply chain. The project results are intended to be used by professionals within the food supply chain (e.g. farmers, retailers, restaurateurs), authorities (regional, municipal, etc.), NGOs and the general public.

Implementation of the project will cause following positive effects:

- the improvement of resource efficiency (to avoid food loss);
- the improvement of the social situation (donation of excess food to feed the needy);
- the improvement of energy efficiency (use of food waste for energy production);
- the prevention of climate change (reduction of greenhouse gases emissions).

Acknowledgements

This paper emerged in the frame of the project STREFOWA 'Strategies to Reduce and Manage Food Waste in Central Europe', which is co-financed by the Interreg CENTRAL EUROPE Programme that encourages cooperation on shared challenges in central Europe. With 246 million Euro of funding from the European Regional Development Fund, the programme supports institutions to work together beyond borders to improve cities and regions in Austria, Croatia, Czech Republic, Germany, Hungary, Italy, Poland, Slovakia and Slovenia.

References

1. J. Lundqvist, C. de Fraiture, D. Molden, *Saving water: from field to fork—curbing losses and wastage in the food chain.* In *SIWI Policy Brief.* Stockholm, Sweden: SIWI (2008)
2. FAO. *Global food losses and food waste – Extent, causes and prevention*, Rome (2011)
3. J. Parfitt, M. Barthel, S. Macnaughton, Phil. Trans. R. Soc. **365** (2010)
4. Å. Stenmarck, C. Jensen, T. Quested, G. Moates, *Estimates of European food waste levels.* ISBN 978-91-88319-01-2. Stockholm (2016)
5. BIO Intelligence Service. *Preparatory Study on Food Waste Across EU27*, final report-2010-54 for the European Commission (DG ENV) Directorate C - Industry, ISBN: 978-92-79-22138-5. BIO Intelligence Service in association with Umweltbundesamt and AEA (2010)
6. BIO Intelligence Service. *Assessment of resource efficiency in the food cycle,* Final report prepared for European Commission (DG ENV) in collaboration with AEA, Dr Donal Murphy-Bokern, Institute of Social Ecology Vienna and Institute for Environmental Studies (2012)

Performance of mechanical biological treatment of residual municipal waste in Poland

Emilia den Boer[1,*], and *Andrzej* Jędrczak[2]

[1] Wrocław University of Science and Technology, Faculty of Environmental Engineering, Wybrzeże Wyspiańskiego 27, 50-370 Wrocław, Poland
[2] University of Zielona Góra, Faculty of Civil Engineering, Architecture and Environmental Engineering, ul. Prof. Z. Szafrana 1, 65-516 Zielona Góra, Poland

Abstract. The number and capacity of mechanical-biological treatment (MBT) plants in Europe increased significantly in the past two decades as a response to the legal obligation to limit the landfilling of biodegradable waste in landfills and to increase recycling and energy recovery from waste. The aim of these plants is to prepare residual municipal waste for recovery and disposal operations, including especially separation and stabilization of the easily biodegradable fraction (the biofraction). The final products of MBP technology are recyclables, stabilate, high calorific fraction which is used for the production of refuse derived fuel (RDF) and the remaining residual fraction. The shares of the output fractions, especially of the recyclables and RDF determine the overall efficiency of MBT technology in diverting waste from landfills. In this paper results of an assessment of one exemplary MBT plant are provided. The analysis was performed within a comparative study in which 20 selected MBT plants in Poland were subject to a detailed analysis, focusing, both at the design parameters as well as operational ones. The selected plant showed relatively higher overall materials recovery efficiency. With the view to circular economy targets, increased automation of the mechanical waste treatment will be required to support achieving high level diversion from landfills. The study reviled that stabilisation of biofraction should be improved by a better control of process conditions, especially moisture content.

1 Introduction

Majority of the municipal residual waste in Poland is treated in regional mechanical-biological treatment (MBT) plants. There are 127 regional MBP plants with combined mechanical processing capacity of approximately 9.4 million Mg/year, and the biological about 4.1 million Mg/year. The total amount of municipal waste collected in Poland in 2015 was 10863 thous. Mg, of which 8326 thous. Mg was collected as residual waste [2]. Approximately 8250 thous. Mg of residual waste was treated in MBT plants. In the period 2015–2016 six new incineration plants have been commenced with the total capacity of

* Corresponding author: emilia.denboer@pwr.edu.pl

974 thous. Mg/year. As the summary capacity of MBT plants and waste incinerators exceeds the amount of residual waste, which is decreasing due to rising levels of separate collection, the competition on the waste market becomes severe. The main purpose of the MBT technology is to recover materials for recycling and energy recovery and to stabilise the biodegradable waste before landfilling. In Poland the MBT plants accept both the residual waste as well as separately collected waste to sort and prepare it for recycling.

1.1 Requirements of MBT technology

1.1.1 Mechanical treatment

Mechanical processing of mixed municipal waste is aimed at separating fractions for material and/or energy use, and the fraction requiring further biological treatment. Waste generated in the process of mechanical treatment of residual municipal waste are to be processed in accordance with the waste hierarchy. The fraction of at least 0-80 mm needs to be separated and stabilised in the biological part of a MBT plant, which can be conducted under aerobic or anaerobic conditions. Three types of biological treatment are defined: aerobic stabilization, anaerobic stabilisation and biodrying.

1.1.2 Biological treatment

The biostabilisation under aerobic conditions is carried out in accordance with the following requirements: waste is processed, including turning of the waste, for a period of 8 to 12 weeks in total; for at least two first weeks the process takes place in a closed reactor or in the hall, with active aeration and prevention of the entry of untreated process air to the atmosphere, until reaching the respiration activity (AT_4) value of less than 20 mg O_2/g dry mass (dm). The process is continued in open windrows (maturation) until AT4 reaches < 10 mg O_2/g dm.

The biostabilisation under anaerobic conditions is carried out in two stages:

- in the first stage mesophilic fermentation for at least 20 days or thermophilic fermentation for at least 12 days needs to be carried out
- in the second stage aerobic stabilization in a closed reactor or in the hall, with active aeration for at least two weeks should be carried out. It is allowed to carry out the second stage in the open windrows, while waste should aerated by turning at least once a week, for at least 3 weeks.

The biodrying is carried out under aerobic conditions, with active aeration in the closed reactor or hall for at least 7 days. The aim of biodrying is to reduce the moisture content of waste in order to improve the calorific value of the combustibles, which are used for the refuse derived fuel (RDF) production.

1.2 The challenges of the European circular economy

The circular economy package, adopted by the Commission on 2 December 2015, includes legislative proposals on waste [3], with long term targets to reduce landfilling and increase recycling and reuse. The targets set for municipal waste are very ambitious, including:

- a common EU target for recycling 65% of municipal waste by 2030;
- a common EU target for recycling 75% of packaging waste by 2030;
- a binding landfill target to reduce landfill to maximum of 10% of municipal waste by 2030.

Those targets have already been incorporated to the newly adopted National Waste Management Plant (KPGO2022) [1]. In Poland in 2015 only 23.3% of municipal waste collected separately [2]. The recycling level of the total municipal waste in 2015 amounted to 26.4%. From the past data it is clear that increase of the separate collection levels progresses very slowly in Poland. It is certain that accomplishment of the required recycling levels will require both actions – improving the separate collection rates and enhancing recovery of recyclables from residual waste. This can be done with help of modern sorting technologies which are applied in the Polish MBP plants. The national waste management plan stipulates that energy recovery should not exceed 30% of the total mass of municipal waste. Moreover the regional plans assume that landfilling rate of MBT residues should be limited to 30% of the plant input.

In this paper technology used in one of more advanced MBT plants in Poland is presented. Efficiency of the plant is assessed in the context of the future recycling targets and the obligation to divert waste from landfills. The results presented here are based on a large scale study of 20 MBT installations in Poland commissioned by the General Directorate for Environmental Protection [4].

2 Selected MBT plant

The MBT plant „Eko-MAZURY" Sp. z o.o in Siedliska has been selected for a detailed presentation and analysis in this paper, due to its relatively high technological advancement, and high score in the overall performance assessment. The plant is located in the north-eastern part of Poland. It receives municipal waste from the area of 12 communes, with the total population of 159 thousand people. The capacity of the mechanical part of the plant is 59 000 Mg and of the biological part – 19 000 Mg per year.

The MBT process takes place in two basic technological facilities: sorting hall and in the aerobic stabilization part. The aerobic stabilisation is conducted in two stages. The intensive stabilisation takes place in a concrete hall. The second stage - maturation takes place in open windrows. The technological scheme of the installation is shown in Fig. 1. The plant accepts both residual municipal waste and separately collected waste.

2.1 Technology description

2.1.1 Mechanical treatment of waste

Residual waste passes through a beg opener (BRT, type MSW2-13-15) or is loaded directly to the conveyor loading channel. Optionally waste can be shredded (KOMPTECH shredder). Waste is transported on a conveyor to the preselection cabin, where 6 people crew sorts our items, which are likely to hinder or disrupt the sorting process, i.e. construction waste, glass, cardboard boxes, large bags and waste foil, waste large metal and hazardous waste. From the preselection cabin waste is transported on the conveyor belt to the rotating sieve drum (length 12 m, diameter 3.0 m, SUTCO Polska), in which the splitting into two fractions takes place: the biofraction (< 80 mm) and the coarse fraction (> 80 mm). The biofraction passes through a magnetic separator (Magnetix, type SNK120-130) for the separation of ferrous metals, and the final sorting cabin to separate the aluminum cans, PET bottles, films and other non-biodegradable waste before unloading it into the aerobic stabilization hall.

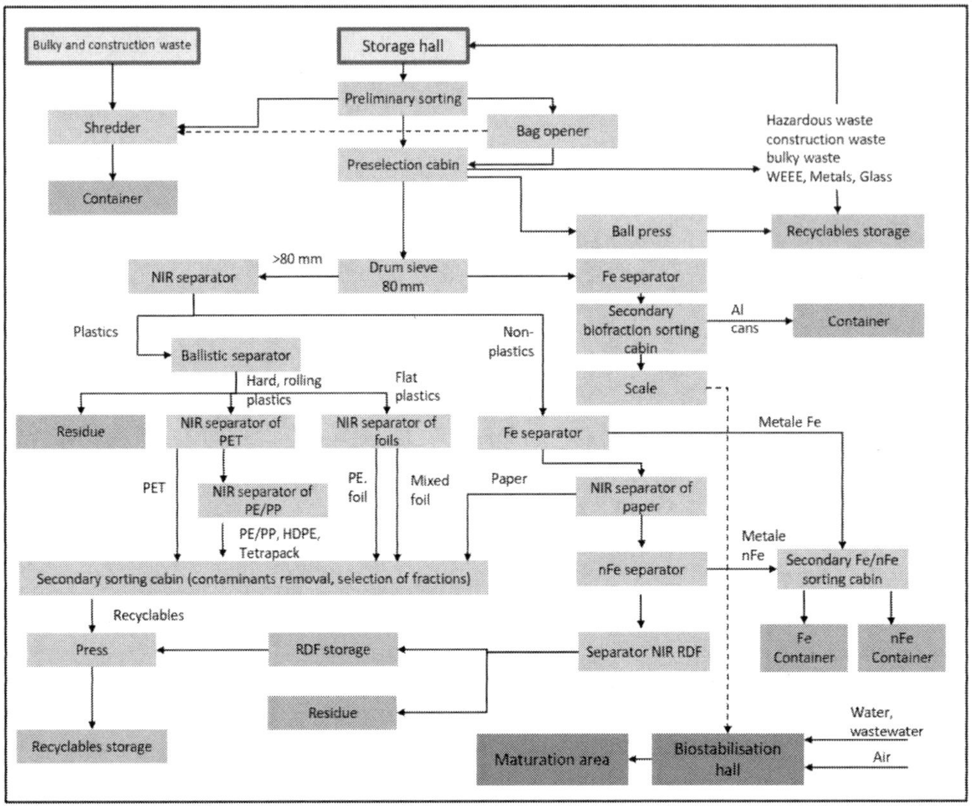

Fig. 1. Scheme of the MBT technology in Siedliska.

The coarse fraction (> 80 mm) if fed to the further sorting process in order to recover recyclables and the high calorific fractions for waste derived fuel (RDF) production.the sorting line consists of the following devices: a near infra-red (NIR) optical separator (TITECH/TOMRA) to separate plastics; a ballistic separator to separate rotating and flate plastics (SUTCO Polska); five further NIR optical separators to sort out PET, foils, paper, PP/PE and refuse derived fuel (pre-RDF); a magnetic separator for ferrous metals (Magnetix typ SNK120-130, and an eddy current separator for non-ferrous (mostly aluminium) metals (Magnetix typ SWP150). The automated separation is followed by the secondary (manual) sorting cabins (24 persons controlling the recyclables streams and 3 – the biofraction). The recyclables and pre-RDF fraction are transported to the channel press baler and transferred to external recycling/recovery companies. The remaining waste stream of coarse fraction (>80 mm) is gathered in the container and disposed of in a landfill. The whole sorting technology involves 69 conveyor belts with the total length of 800 m.

2.1.2 Biological treatment of waste

The biofraction separated on a rotating screen (< 80 mm) is subject to two stages biological processing: the intensive aerobic stabilization and maturation. The intensive aerobic stabilization is performed in a BIOFIX system comprising a closed hall with 4 tunnels (total capacity of 2920 m^3) with an automatic loading system; an automatic waste turning machine (TOPTURNX 53); moistening with process water circulation system; forced aeration system and an automatic unloading mechanism for the stabilised waste. The process of intensive aerobic stabilization takes place within four weeks. The temperature

 DOI: 10.1051/e3sconf/20172200020

range is 25–70°C. The resulting stabililate (referred to as stabilate I) is discharged automatically on the outdoor maturation area. The maturation process takes from a few to several weeks until the required stability is reached (AT4 < 10 mg O_2/g s.m.). The stabilate is screened on the mobile rotating screen. The fraction < 20 mm can be used for constructing the biological landfill cover (which is a recovery operation). The remaining stabilate is landfilled.

3 Assessment of the overall technology performance

Table 1 provides data on the volume of emissions and energy consumption by the plant operation, based on yearly average values. The wastewater is purified by a reverse osmosis technology. Waste gases pass through a water scrubber and a biofilter to remove part of the organic pollutants and mitigate odour problem. The biofilter surface is - 400 m^2 and the yearly amount of air from stabilisation hall let though a biofilter is approx. 54820 thous. m^3.

Table 1. Data on total emissions and energy consumption of the MBT plant.

Media consumption		**Emissions volume**	
Electric energy, [kWh /Mg]	Water [m^3/Mg]	Wastewater [m^3/ Mg]	Waste gases [m^3/ (Mg · h)]
35	0.01	0.14	14.3

The average composition of input waste (based on manual sorting analyses) is presented in Fig. 2.

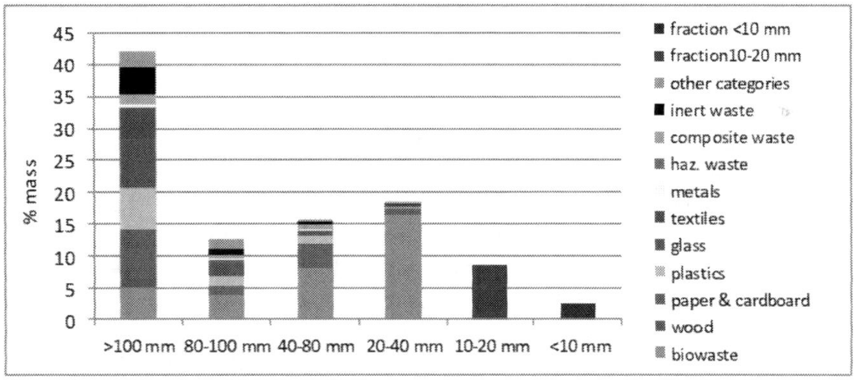

Fig. 2. Composition of input (residual) waste, based on [4].

From Fig. 2 it can be seen that biowaste is contained mainly in fraction 20-40 mm and to a lesser extent in fraction 40–80 mm. Recyclables occur mainly in fractions >100 mm and 80–100. Paper is also present in fraction 40-80 mm, however this is rather a low quality hygienic paper. Further characteristics of the input (residual) waste is provided in Table 2.

Composition of output waste fractions separated in the mechanical part of the plant – the biofraction (< 80 mm) and the residue of the coarse fraction (> 80 mm), after passing through the sorting process, are presented in Fig. 3. The coarse fraction after sorting still contains a significant amount of plastics. The other materials include textiles and other categories of waste (mainly diaper and other hygiene items).

Table 2. Characteristics of the Siedliska MBT plant input waste.

	Size fraction [mm]						Overall residual waste
	< 10	10–20	20–40	40–80	80–100	> 100	
Recyclables content [%]	0.0	0.0	7.9	36.5	44.6	52.6	35.0
Share of biodegradable fractions [%][1)]	47.9 (15.7)	75.3 (30.2)	95.4 (0.4)	77.7 (4.1)	43.8 (5.6)	39.1 (2.0)	60.7 (1.6)
Moisture content [%][1)]	35.4 (1.2)	29.5 (14)	61.6 (3.3)	51.2 (3.1)	44.7 (2.0)	43.4 (1.4)	46.1 (2.9)
Loss of ignition [% dry mass][1)]	36.9 (2.4)	30.7 (14.4)	64.2 (4.2)	70.8 (2.8)	80.0 (2.1)	84.4 (2.5)	70.2 (5.0)

[1)]standard deviation values in brackets

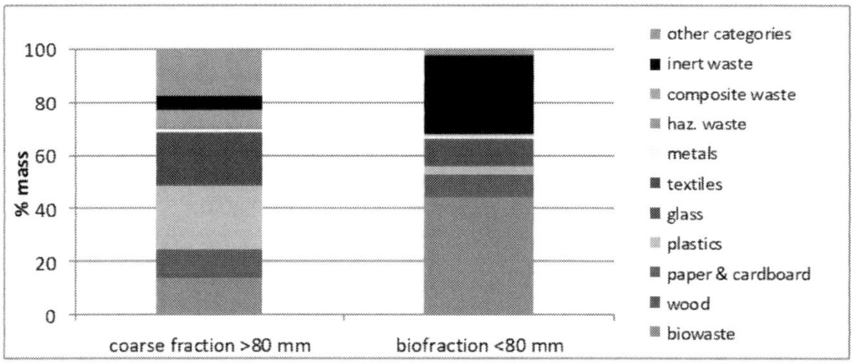

Fig. 3. Composition of the residual coarse fraction and the biofraction.

The destination of the residual coarse fraction is "other recovery" at external companies (e.g. energy recovery) or landfilling. Landfill acceptance criteria for this fraction include [5]: dissolved organic carbon (DOC) < 800 mg/kg dm; total dissolved solids (TDS) < 60 g/kg dm and the additional parameters (from January 2016): loss of ignition (LOI) < 8% dm; total organic carbon (TOC) < 5% dm; higher heating value (HHV) < 6 MJ/kg dm. Properties of the coarse residues and the biofraction are presented in Table 3. The coarse fraction does not meet the current landfill acceptance criteria.

The biofraction is subject to two stage aerobic stabilisation. The stabilate after intensive stabilisation (stabilate I) and after maturation (stabilate II) has been analysed. Results are presented in Table 3. The acceptance criteria for stabilate at landfills are: *1)* AT4 < 10 mgO_2/g dm or *2)* LOI < 35% dm and TOC < 20% dm or *3)* the loss of organic matter during stabilization measured by LOI or TOC must be greater than 40%. Both stabilate I and stabilate II showed very low moisture content. Stabilate I fulfilled already the landfill acceptance criteria with respect to AT4, TOC and LOI. The organic matter loss during intensive stabilisation measured by LOI amounted to app. 24%. The sample of stabilate II did not show any further progress in the decomposition rate. The reason for it was probably the very low moisture content, which hindered the microbiological activity. In some of the examined plants [4] the same problem was observed. The suboptimal process parameters hinder the stabilisation of the biofraction. On the other hand results of the analyses of the biofraction show rather low LOI and biodegradables content, which results in stabilate I meeting the landfill criteria.

Table 3. Characteristics of the coarse fraction and biofraction separated in the MBT plant.

	Coarse fraction > 80 mm	Biofraction < 80 mm	Stabilate I	Stabilate II
Recyclables content [%]	36.1	22.6		
Share of biodegradable waste [%]	36.2	53.2		
Moisture content [%]	44.6	30.8	24.0	25.7
LOI [% dry mass]	74.5	26.6	20.6	26.8
TOC [% dry mass]	42.2	16.3	12.6	14.3
HHV [MJ/kg dry mass]	17.5			7.3
DOC in eluate [mg/kg dry mass]	5460			3727
TDS in eluate [mg/kg dry mass]	45990			38690
AT_4 [mg O_2/g dm]			4.0	3.8
GB_{21} [Ndm^3/kg dm]			10.3	11.5
CH_4 [% v/v]			54.9	54.7
CO_2 [% v/v]			43	43.2

Fig. 4 shows the overall mass balance of the presented Siedliska MBT plant. The recyclables and the stabilate for land recultivation are waste fractions which are recovered from waste stream. The share of recyclables amounted to 11,0%. Part of > 80 mm fraction is transferred to "other recovery" at external company (most likely for RDF production). The remaining stabilate, non-composted fractions and residual fraction > 80 mm were landfilled.

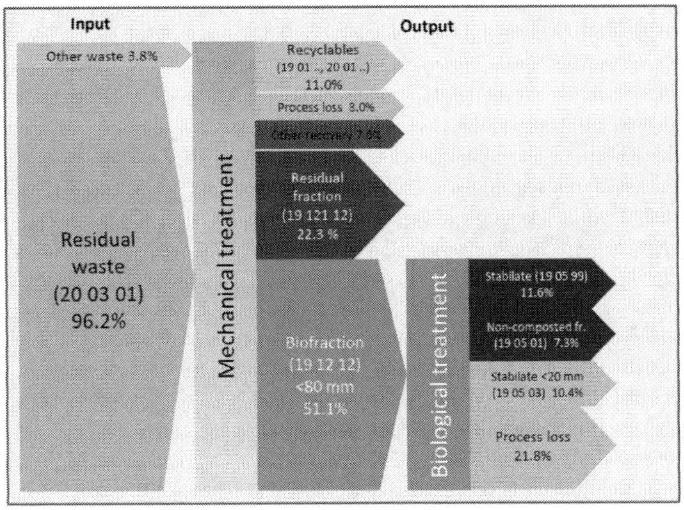

Fig. 4. Mass balance of residual waste treatment in the presented MBT plant.

3.1 Performance of the Siedliska MBT plant against other MBP plants

Mass balance of the Siedliska MBT plant against the other plants included in the assessment [4] is presented in Fig 5. Assessment of the overall efficiency of waste recovery in individual installations takes into account the shares of recycling, RDF recovery and landfilled waste in relation to the plant input. The average recycling rate of all plants was 6.1% (range: 0.04% -15.0%). The plant in Siedliska showed second highest recycling rate

(11.0%). Analyses show that there is a clear correlation between the share of recycling and the number of mechanical devices per 10 000 Mg of the capacity of the plant [4].

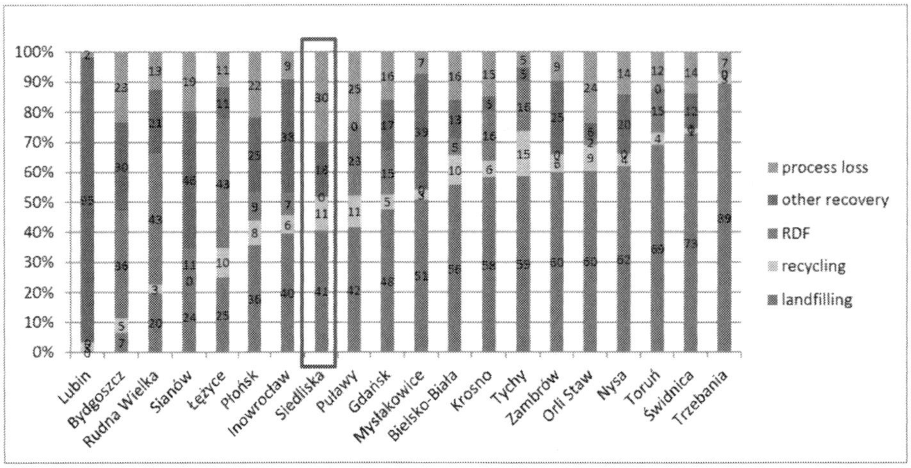

Fig. 5. Mass balance of the MBT plants included in the MBT comparative study [4].

The average share of RDF production is 12% (0 – 43%). The coarse fraction after separation of recyclables is used for RDF, however not in all plants. In some cases there is high share of "other recovery" (as in Siedliska plant). It includes the use of stabilate for land recultivation (normally as the biological landfill cover) and the transferring of plant outputs for recovery at other companies (RDF production). The average landfilling rate of waste after MBT is 45.9% (6.5%–69.2%).

4 CONCLUSIONS

In this paper an analysis of the selected MBT technology was presented. MBT technology can support waste management systems in its development towards circular economy targets. However, the existing plants need to be upgraded to reach higher recyclables and RDF recovery levels. The coarse fraction >80 mm remaining after separation of recyclables and RDF does not meet the current landfill acceptance criteria. It should be recovered as low quality RDF fraction.

References

1. KPGO2022. *National waste management plan 2022*, Warszawa, 2016. Uchwała nr 88 Rady Ministrów z dnia 1 lipca 2016 r. (poz. 784)
2. GUS 2016, Environment protection, Warszawa
3. COM (2015) 595. *Proposal for a Directive of the European Parliament and of the Council amending Directive 2008/98/EC on waste*, Brussels, 2.12.2015
4. A. Jędrczak, E. den Boer, *Final report of the 3rd stage of the study to carry out waste tests in 20 plants for mechanical and biological waste treatment*, a study commissioned by The General Directorate for Environmental Protection, Zielona Góra (2015)
5. Dz.U. 2015 poz. 1277 regulation of the Minister of Economy of 16 July 2015 r. on the acceptance of waste to landfills

Organic matter and heavy metals content modeling in sewage sludge treated with reed bed system

Dariusz Boruszko[1,*], *Wojciech* Dąbrowski, and *Paweł* Malinowski

[1]Białystok University of Technology, Faculty of Civil and Environmental Engineering, Department of Technology in Environmental Protection and Engineering, ul. Wiejska 45A, Białystok, Poland

Abstract. The long process of sludge stabilization (7–15 years) remarkably reduces the organic matter content and causes the process of sludge humifaction. This paper presents the results of using low-cost methods of sludge treatment in the wastewater treatment plant located in Zambrow, Podlaskie Province. The results of studies on the organic matter and heavy metals content in sewage sludge after treatment in a reed bed system are presented. The aim of the research was to evaluate and model organic matter and heavy metals concentrations during sewage stabilization in reed bed lagoons. The lowest concentration, below 1.3 mg/kg DM of the examined seven heavy metals was mercury (Hg). The highest concentration, exceeding 1300 mg/kg DM was zinc (Zn). The obtained results for the heavy metals in sewage sludge from the reed bed lagoons in Zambrow show that the average content of the analyzed heavy metals is approximately 1620 mg/kg DM. The results of the study demonstrate a high efficiency of low-cost methods used in Zambrów WWTP in terms of the quality of the processed sludge. Sewage sludge from the lowest layer of the reed lagoon (12-14 years of dewatering and transformation) is characterized by the lowest organic matter and heavy metals content. The higher a sediment layer lies, i.e. the shorter the time of processing, the higher is the heavy metals content. This indicates a great role of reeds in the accumulation of these compounds.

1 Introduction

Low-input methods of processing sewage sludge are characterized, first of all, by their simple design, technology and ease of operation. These methods allow using only a limited number of process equipment due to which the consumption of electrical energy is low. Utilization of natural processes occurring in the environment allows maintaining low operation costs [1].

Reed lagoons are successfully used for long-term stabilization and dehydration of sludge from wastewater treatment. They are easy to construct and operate and consume small amounts of energy [2, 3]. And what is perhaps the most important factor of STRB: any type of sewage sludge of organic kind can be stabilized [4]. For more than thirty years this technology, among others in Denmark, has been applied with success. The initial content of organic matter in the sediment, which is between 50% and 65%, and low load of the lagoon surface allow for the drainage of sewage sludge to a degree not lower than obtained through using filter presses, that is up to the value of 20%–30% DM [3, 5, 6]. The content of heavy

* Corresponding author: d.boruszko@pb.edu.pl

DOI: 10.1051/e3sconf/20172200021

metals in stabilized sewage sludge is often crucial for the possibility of its use in agriculture. Their excessive content in sewage sludge may be toxic and dangerous to the environment.

Organic matter content in a particular layer can be modeled using statistical analysis. Some heavy metal concentrations also depend on the content of the organic matter and can be modeled afterwards. The obtained models can be used to predict a specified content in a layer given its initial condition and time from its sedimentation.

2 Material and methods

2.1 Sampling

The samples were collected from 2012 to 2014 at 4 points (from one bed) evenly spaced along the symmetry axes in the 3-meter profile of the sediment fill. Sludge samples from the same layers were mixed and their quality was determined.

Sampling was performed using a special probe to enable the extraction of sewage sludge from the full depth of the lagoon. Samples were averaged and sludge samples from each 0.5 m section were tested. In 2013 the service of the oldest lagoon was completed after 14 years. The studies reflect the following STBR cycles:

- end of 2012 – after the last 12th year of STBR supply with sewage sludge at full load,
- end of 2013 – after transitional year at STBR supply with sewage sludge at minimum load,
- end of 2014 – year after STBR closing.

2.2 Analytical procedure

The analysis of the obtained material included examining heavy metal content, dry mass and organic matter, among others. The examinations were carried out in the Department of Technology in Engineering and Environmental Protection laboratory according to the valid norms. Sludge samples were treated with mineralization in HACH mineralizator with the use of sulphuric acid and hydrogen peroxide mixed with nitric and hydrochloric acid in ratio 1:3. For further analysis mineralizators were filtered through MN 616 G paper filter. Determination of cadmium, nickel and total chromium content was done in samples of mineralizators with the use of atomic absorption spectrometer Perkin-Elmer 4100 ZL with transversely heated graphite cuvette and Zeeman-effect background correction. Determination of mercury content was done in samples of mineralizators by means of cold steam technique with the use of atomic absorption spectrometer Perkin-Elmer 4100 ZL equipped in add – on device FIAS-200. Determination of zinc, lead and copper content was done in samples of mineralizators with the use of atomic absorption spectrometer Varian SpectrAA 20 Plus by means of flame atomization. Dry matter and organic matter were determined by using standard methods [7].

2.3 Statistical model

Organic matter decay was modeled using first order [8] reaction:

$$y(t,l) = I(t,l)e^{-kt} \tag{1}$$

where: $y(t,l)$ – organic matter content in layer l after time t; $I(t,l)$ – initial organic matter content at the moment of sedimentation in layer, that after time t becomes layer l; k – decay rate

By taking logarithm of (1), a linear model is obtained, which is suitable for regression:

$$\ln\left(y(t,l)\right)=\left[\ln\left(I(t,l)\right)\right]+\left[-k\right]t \tag{2}$$

By using sedimentation process specification and year of sampling, it is possible to calculate the year of sediment forming (further called starting year) in a given layer, and the time which has passed since its forming. Due to possible differences in organic matter content during layer formation, additional 0–1 variable will be included, corresponding to different starting years in a full model:

$$\ln\left(y(t,l_i)\right)=\left[a_0\right]+\sum_{k\in start\ years}\left[a_k\right]id_{i,k}+\left[-k\right]t$$

$$id_{i,k}=\begin{cases}0 & \Leftrightarrow i\neq k\\ 1 & \Leftrightarrow i=k\end{cases} \tag{3}$$

$$\ln\left(I(t,l_i)\right)=a_0+a_{start\ year(i)} \qquad I(t,l_i)=e^{a_0+a_{start\ year(i)}}$$

where: i – index over all samples; k – index over start years; a_k – coefficients of additional 0–1 variable.

After initial estimation, variables related to statistically insignificant coefficients (at 5% significance level) were taken out from model iteratively, until none were left. The final model will be presented in the original form, along with directly calculated coefficients.

Due to the nonlinearity of connection between organic matter and heavy metal content, general additive model (GAM) [9] was used. A dependent variable (heavy metal content) was linearly modeled by smooth function of an independent variable (organic matter content):

$$z=b_0+b_1f\left(y\right) \tag{4}$$

where: z – heavy metal concentration; y –organic matter content; $f(y)$ – a smooth function; b_0, b_1 – regression terms

Smooth function of an independent variable was estimated using thin-plate regression splines [10] over existing data points. Due to a complicated form of smooth functions, only model coefficients along with a graphical model representation with confidence intervals will be presented. Statistically insignificant coefficients (at 5% significance level) will be marked out, along with an insufficient linear fit (coefficient of determination less than 0.650).

3 Results and discussion

The sludge is stabilized and dehydrated. The degree of sludge dehydration is low and ranges from 2.9% in 2012 to 3.1% in 2014. The lowest dry matter content was recorded in the top layer of the vertical profile at 12.6%. The highest content of dry matter, which was 14.5%, was characteristic of sludge that was deposited on the reed lagoon for the longest period of time (the lowest layer in the vertical profile). In the middle section of the vertical profile of sludge at the depth of 1.5–2.0 m, a remarkable reduction in dry matter content was observed: to 10.4% in 2012 and 13.2% in 2014. The dry matter content in STRB in different depths of the vertical profile is an indicator of dehydration. The degree of dehydration is on one hand dependent on the degree of water evaporation from the lagoon surface and, on the other hand, on gravitational concentration and uptake by reed roots [11]. Definitely higher dry matter contents, which were above 29% in sludge stabilized in STRB, were achieved in studies by Matamoros et al. [12]. The content of organic matter oscillated within a wider range. The lowest organic matter content of 49.0% DM was recorded in the deepest layer of sludge in

 DOI: 10.1051/e3sconf/20172200021

2014, i.e. one year after closing the reed lagoon. The highest organic matter values were observed in the highest sludge layer (sludge that was drained and stabilized for the shortest time); these values ranged from 58.1% DM to 62.2% DM. Organic matter accumulated in the sludge over the years undergoes biodegradation and stabilization, leading to a reduction in dry matter content. Similar results of the organic matter reduction by 10%–12% were obtained by Nielsen [6] and in studies conducted in Poland in Darżlubie [2].

The average sum of heavy metals in sludge ranged from 1607.2 mg/kg DM in 2014 to 1635.2 mg/kg DM in 2012. The highest content, among all the analyzed heavy metals, was recorded for zinc with its mean concentration of 1301.1 mg/kg DM. Zinc is considered the most essential microelement for plant growth and its very high concentration (above 2500 mg/kg DM) becomes toxic [13]. The lowest content among all of the tested heavy metals was shown by mercury at an average of 1.3 mg/kg DM. The analysis of heavy metals contents in particular layers (depths) of sewage sludge reveals an increase in zinc and copper levels along with their processing time in the reed lagoon. The highest contents of these metals were recorded at the depth of 1.5 m–2.0 m, which amounted to 1412.3 mg Zn/kg DM and 247.0 mg Cu/kg DM. The opposite trend was observed in the case of lead and chromium, the overall content of which decreased with the sludge layer depth in the reed lagoon. The highest average loss of the total content was observed in relation to chromium, the concentration of which fell by 26.4%.

3.1 Model of organic matter and heavy metal content

Statistical analysis was performed in R environment, version 3.3.0 [14]. For modeling heavy metal content, mgcv package [15] GAM implementation was used.

The final model for organic matter contained only one significant interaction term for the layer from 2012 year. All other samples were modeled using one initial organic matter content. Details of the model are presented in Table 1 and Fig 1.

Table 1. Model for organic matter content.

	estimate	**p-value**
a_0	4.148	< 0.0001
a_{2012}	0.048	0.0443
k	0.019	< 0.0001
R^2	0.922	

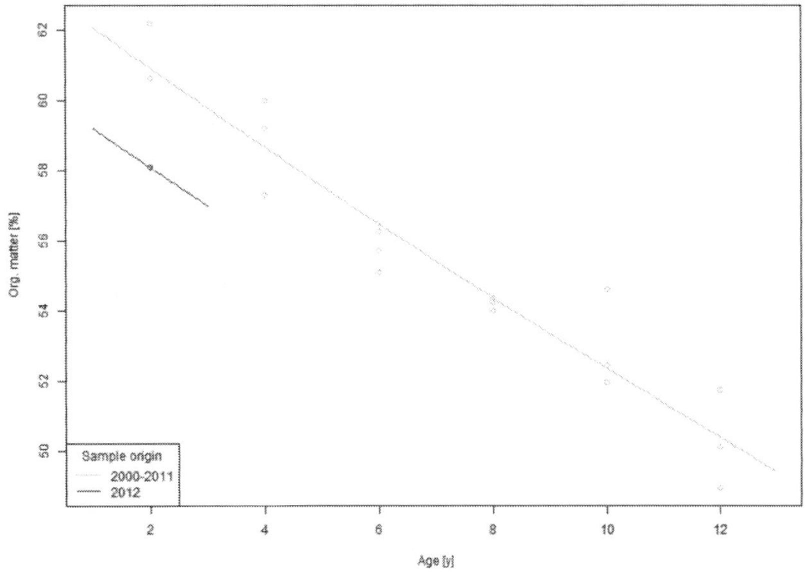

Fig 1. Model for organic matter content.

The obtained model in an explicit form is given by the following equation:

$$y(t,l)=\begin{cases}63.293\times e^{-0.019t} & \Leftrightarrow start\ year(l)\neq 2012\\ 60.348\times e^{-0.019t} & \Leftrightarrow start\ year(l)=2012\end{cases} \quad (5)$$

Details of the models for heavy metal concentration are presented in Table 2, where especially low fit or statistically insignificant coefficients are marked red. Fitted models were also presented in Fig 2 with confidence intervals.

Table 2. Heavy metal content model details.

	Pb		Cr		Cu		Cd	
	estimate	p-value	estimate	p-value	estimate	p-value	estimate	p-value
b_0	30.185	<0.0001	53.387	<0.0001	210.611	<0.0001	2.971	<0.0001
b_1	4.203	<0.0001	4.816	<0.0001	2.969	<0.0001	1.000	<0.0001
R^2	0.941		0.927		0.854		0.674	
	Zn		**Ni**		**Hg**			
	estimate	p-value	estimate	p-value	estimate	p-value		
b_0	1303.500	<0.0001	19.272	<0.0001	0.045	<0.0001		
b_1	3.899	0.0061	1.000	0.0799	1.000	0.1960		
R^2	0.688		0.179		0.102			

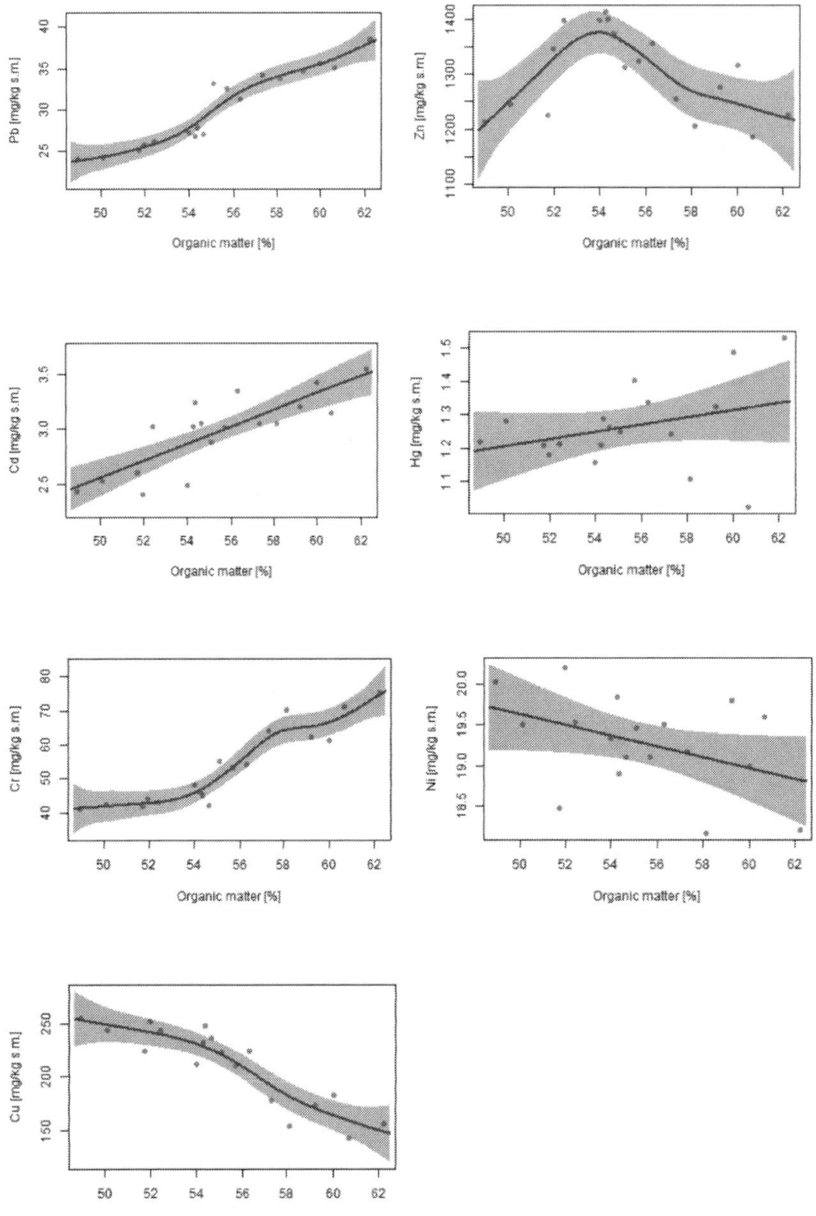

Fig 2. Heavy metal content models.

4 Conclusions

From the study results obtained from the long-term stabilization and dehydration of sewage sludge in the reed lagoon, the following conclusions can be drawn:

1. An increase in dry matter and decline of organic matter occurred in the stabilized sewage sludge.
2. Statistical analysis showed very strong correlations with the organic matter of the sludge layer.

Organic matter content can be modelled with very high approximation using a first-order decay model. The content of many (but not all) heavy metals depends on organic matter concentration. Using obtained models, the content of Pb, Cr, Cu can be estimated with high confidence. The content of other metals like Cd and Zn can still be approximated with satisfactory precision using organic matter concentration. In the overall perspective these conclusions mean that the obtained models can be used for estimating heavy metals concentration depending on time which has passed since layer formation.

Acknowledgments

The study was conducted as a research project S/WBiIŚ/3/2014 in Faculty of Building and Environmental Engineering of BUT and financed by Ministry of Science and Higher Education.

References

1. D. Boruszko, Eng. Environ. Prot. **13**, 1, 29–42 (2010) (in Polish)
2. W. Zwara, H. Obarska-Pempkowiak, Water Sci. Technol. **41**, 1, 65–68 (2000)
3. S. Nielsen, E. Peruzzi, C. Macci, S. Doni G. Masciandaro, *Stabilisation and mineralization of sludge in reed bed system after 10-20 years of operation*. (13th International IWA Specialist Group Conference on Wetland Systems for Water Pollution Control, Perth, Australia, 2012)
4. E. Uggetti, I.Ferrer, E. Llorens E, J. Garcia. Biresour. Technol. **101**, 9, 2905–2912 (2010)
5. S. Nielsen, Water Sci. Technol. **48**, 5, 101–109 (2003)
6. S. Nielsen, Water Sci. Technol. **63**, 5, 941–947 (2011)
7. Standard Methods for the Examination of Water and Wastewater (21th edn, American Public Health Association/American Water Works Association/Water Environment Federation, Washington DC, USA, 2005)
8. K. Kołecka, *Long-term changes of the speciation of heavy metals in sewage sludge treatment in reed lagoons* (Gdańsk University of Technology, 2007) (in Polish)
9. T. Hastie, R. Tibshirani, Statistical Science **1**, 3, 297–318 (1986)
10. S.N. Wood, Journal of the Royal Statistical Society (B) **65**, 1, 95–114 (2003)
11. K. Kołecka, H. Obarska-Pempkowiak. Environ. Prot. Eng. **34**, 3, 13–20 (2008)
12. V. Matamoros, L.X. Nguyen, C.A. Arias, S. Nielsen, M.M. Laugen, H. Brix, Water Res. **46**, 12, 3889–3896 (2012)
13. A. Pathak, M.G. Dastidar, T.R. Sreekrishnan, J. Environ. Manag. **90**, 8, 2343– 2353 (2009)
14. R Core Team, *A language and environment for statistical computing* (R Foundation for Statistical Computing, Vienna, Austria 2016). https://www.R-project.org/
15. S.N. Wood, Journal of the Royal Statistical Society (B) **73**, 1, 3–36 (2011)

E3S Web of Conferences 22, 00022 (2017) DOI: 10.1051/e3sconf/20172200022
ASEE17

Energy and environmental evaluation of combined cooling heating and power system

Andrzej Bugaj[1,*]

[1]Wroclaw University of Science and Technology, Faculty of Environmental Engineering, ul. Norwida 4/6, 50-373 Wroclaw, Poland

Abstract. The paper addresses issues involving problems of implementing combined cooling, heating and power (CCHP) system to industrial facility with well-defined demand profiles of cooling, heating and electricity. The application of CCHP system in this particular industrial facility is being evaluated by comparison with the reference system that consists of three conventional methods of energy supply: (a) electricity from external grid, (b) heat from gas-fired boilers and (c) cooling from vapour compression chillers run by electricity from the grid. The CCHP system scenario is based on the combined heat and power (CHP) plant with gas turbine–compressor arrangement and water/lithium bromide absorption chiller of a single-effect type. Those two scenarios are analysed in terms of annual primary energy usage as well as emissions of CO_2. The results of the analysis show an extent of primary energy savings of the CCHP system in comparison with the reference system. Furthermore, the environmental impact of the CCHP usage, in the form of greenhouse gases emission reductions, compares quite favourably with the reference conventional option.

1 Introduction

Conventional thermoelectric power plants convert only about 35% of primary energy into electricity. The rest of primary energy in the form of heat is usually dispersed and can be considered as an energy loss. One of the methods of increasing the efficiency of electricity generation is combined heat and power (CHP) production known also as cogeneration. CHP plant transforms over 80% of primary energy to usable energy in the form of heat and power. That high conversion efficiency translates also into improved environmental impact giving considerable reduction in emissions of greenhouse gases. The ideal situation is to run a CHP plant throughout the whole year with full utilization of produced heat and power. In Polish climate while the heat from CHP unit is being used in heating, ventilating and air conditioning installations during winter time, afterwards there is always an excess of available heat that is a by-product of electricity generation in the rest of the year. A method of using this excess heat is based on expanding the CHP plant to combined cooling, heating and power generation process also known as tri-generation. The tri-generation or the CCHP system is in fact CHP plant connected to absorption chiller fired

* Corresponding author: andrzej.bugaj@pwr.edu.pl

with the heat produced in CHP unit. In this way the heat would not be wasted in summer season due to the lack of heat demand but instead it can be used effectively to make cooling energy, e.g. for air conditioning purposes.

CCHP system considered for the particular application in the presented study can use gas reciprocating engines or gas turbines with absorption cooling units. LiBr-water absorption cycle is chosen because the cooling effect is needed mostly in air conditioning installations with cooling water temperatures always above 5°C. The choice of the most appropriate CCHP system arrangement depends on such factors as heat/power ratio, the temperature level of the required heat output and variations of heating, cooling and power demand.

Numerous literature positions illustrate the specific benefits of using CCHP systems in comparison with conventional alternatives. Apart from a better conversion of primary energy and consequently reduction of greenhouse gases emissions also economic benefits are evident for CCHP options [1]. The same total energy produced in tri-generation process is costs about 30% less than by conventional ways. The energy supply from the CCHP plant is more reliable than the electricity from the grid. Additionally, the tri-generation units ensure some increase in the electricity grid stability. During hot summer time there would be a significant relief in the grid since cooling process changes from compression to absorption cycles. That further improves efficiency because summer demand peaks are often served by utilities through inefficient stand-by units and/or overloaded transmission lines.

The CCHP system consists of two parts, CHP unit and absorption chiller. There is some variety of these two plants and thus one should choose the appropriate type of the plant for the particular application. Since it is assumed that a continuous power demand prevails on the site and thermal energy can be utilized throughout the year a gas turbine type of CHP is chosen. As for the absorption chiller it is decided to have the LiBr-water unit of two different effect types The first one is a single-effect appliance with small COP value of 0.65 and correspondingly low temperature of medium running the unit. The second one is a two-effect unit with higher COP value of 1.2 and likewise a need of higher temperature to run the chiller.

2 Case study description

The case study presented in this paper addresses issues regarding a problem of introducing CCHP system to an industrial facility with well-defined demand of cooling, heating and electricity. The industrial facility is a pharmaceutical factory producing adhesive dressings and plaster materials. The factory is currently being expanded by installing additional production lines. All new production lines should be housed in air conditioned spaces. The production processes are assumed to be run of 16 to 20 hours per day with 5 to 6 days schedules depending on the product demand. Until now air conditioned spaces use heating energy provided by gas fired boilers and cooling energy is supplied from vapour compression type chillers. All electricity needs are covered by external electrical grid. The projected process of production expansion poses a question whether it is worthwhile attitude to increase the existing ways of energy supply that mean additional boilers, larger compression chillers and bigger electricity demand from the grid. Or perhaps one should consider energy production on the site with CCHP system. The most economic operation option is to run the CCHP plant through the year with almost full utilization of produced on site heat and power.

2.1 Conventional and CCHP systems

The application of CCHP system in this particular industrial facility should be assessed in comparison with a conventional option that is a baseline scenario called Case 1. This reference scenario encompasses three standard ways of providing to the site: (a) electricity from the external grid, (b) heat from the gas-fired boilers and (c) cooling from vapour compression chillers run by electricity from the grid, as it is illustrated in fig.1. Electricity taken from the grid is used to cover a power load of production lines and to generate cooling energy in compression chillers. Conversely, natural gas is supplied to the boilers ensuring all heating needs.

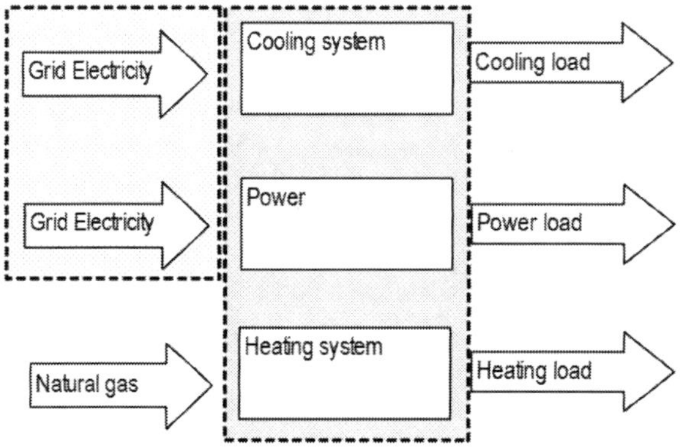

Fig. 1. Conventional system – Case 1.

The CCHP scenario called Case 2 is based on the operation of the CHP plant producing all needed electricity while recovered heat is being used to generate cooling energy in an absorption chiller and to cover a heating load, as it is illustrated in fig.2.

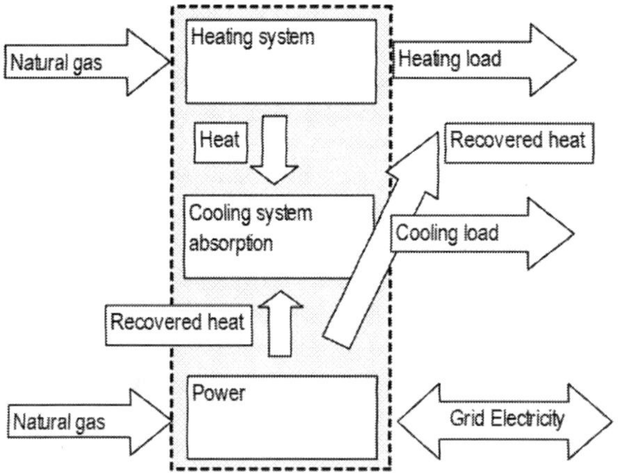

Fig. 2. CCHP system – Case 2.

In the general practice of using CHP units, occasionally a certain amount of cooling energy could be also generated in vapour compression chillers supplied with electricity from the CHP unit. But in the case discussed it is not a viable option because there would be an excess of heat from the CHP unit in summer season. That excess heat could be otherwise wasted and an overall efficiency of the system would decrease considerably. Additionally, the heating system is equipped with gas boilers which could be used as backup.

3 System performance analyses

Most of the studies on CCHP system performance relate to energy, environmental and economic evaluations [2, 3]. The authors use primary energy savings ratio and CO_2 emission reduction ratio to assess the energy usage and environmental impact of the system, respectively. In general, it is well known that CCHP systems are rather difficult to design and operate in facilities with distinctively varying power, cooling and heating demands, as it is a case in residential and commercial buildings. However, some investigations were carried out for these types of buildings and reported in technical literature [4–7]. In contrast, very few researches can be found on the CCHP performance in cases with more uniform power and thermal loads, e.g. in data centres building. A comprehensive research on this type of site is described in [8, 9]. This paper appraises the CCHP performance on the industrial site where power, heating and cooling loads fluctuate in different mode than in the separate buildings due to thermal needs of the production lines.

Three elements of the presented study: data patterns, the simulation model and performance evaluation constitute comprehensive procedure for the system operation and efficiency assessment. In addition, a sensitivity parameter analysis can be used to optimize design of the system according to prevailing demand profiles. Data patterns describing power and thermal demands serve as inputs to the simulation model. The model is based on existing and validated procedures of TRANSYS program taking into account equipment specifications, system configurations and operational schedule. This report focuses mostly on using the simulation model results in the process of energy and environmental evaluation of the CCHP system in the similar way like in the approach described in [9].

The primary energy consumption of the system is obtained by multiplying the entire amount of energy consumed on site by its primary energy factor that takes into account all losses occurring throughout conversion, transmission, storage and distribution. Hence, the primary energy consumption is used as the basis for energy performance evaluation. This energy consumption of the conventional system in Case 1, PE_{conv}, is calculated as follows:

$$PE_{conv} = (E_{el} + E_{cool}/COP_c)F_{el} + (E_{hb}/\eta_t)F_g \tag{1}$$

where E_{el} is the power load integrated over time, E_{cool} is the cooling energy generated by compression chiller, COP_c is coefficient of performance of the chiller, E_{hb} is the heating load taken as heating energy produced by gas boilers with total efficiency of η_t. E_{hb} can also be interpreted as the amount of energy covering all central heating, ventilation and technological needs. F_g and F_{el} are primary energy factors for gas heating and electricity, respectively. The following values of these factors were taken for the study: $F_g = 1.1$ and $F_{el} = 3.0$. Consequently, the primary energy consumption of the CCHP system in Case 2 is determined in the following way:

$$PE_{CCHP} = (E_{el} + E_{cool}/COP_{abs} + E_h)\,F_g + (E_{hbb}/\eta_t)F_g \tag{2}$$

where E_{cool} in this case is the cooling energy generated by absorption chiller characterized by COP_{abs}, E_h is heating energy produced by the CCHP system and E_{hbb} denotes energy taken from the gas backup boilers.

In addition to the primary energy rating, a carbon dioxide apprising was also performed as stated in the standard EN 15603:2008. The combined environmental impact of all greenhouse gas compounds is commonly normalized to the specific effect of CO_2 and all emissions are expressed in CO_2 equivalents. For the purpose of this study the emissions are just expressed in the mass of CO_2. Emission factors for gas usage, EF_g and electricity production, EF_e representative in the local energy market were introduced in order to calculate actual emissions. The following values are currently used in Poland: $EF_g = 0.202$ kgCO$_2$/kWh and $EF_e = 0.812$ kg CO$_2$/kWh. The annual CO_2 emissions for the conventional system of the Case 1 (AE_{conv}) are computed in the following way:

$$AE_{conv} = (E_{el} + E_{cool} / COP_c)EF_e + (E_{hb}/\eta_t)EF_g \quad (3)$$

Consequently, for the CCHP system of the Case 2 the annual CO_2 emissions are found from the following formula:

$$AE_{CCHP} = (E_{el} + E_{cool}/COP_{abs} + E_h)EF_g + (E_{hbb}/\eta_t)\, EF_g \quad (4)$$

4 Results and discussion

The foremost task in the process of optimizing the CCHP system performance is to match its operation with existing heating, cooling and power load profiles. Fig.3 shows the monthly power demands as well as heating and cooling loads occurring on site.

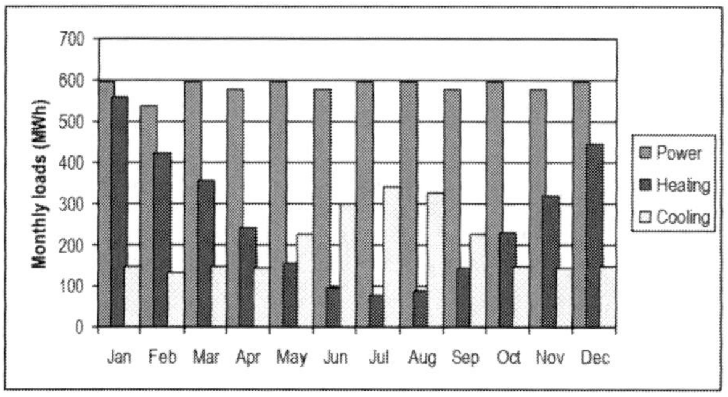

Fig. 3. Power, cooling and heating monthly loads.

Monthly load profiles of all three forms of energy shown in Fig.4 illustrate a relation between the profiles throughout a year.

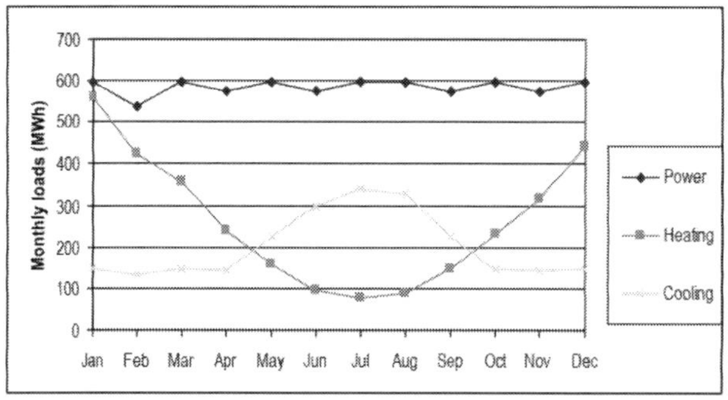

Fig. 4. Annual power, cooling and heating profiles.

 DOI: 10.1051/e3sconf/20172200022

Maximum heating needs occur during winter months but heating is also required during summer mostly due to production lines demand. The cooling energy is not only required in summer for air conditioning installations but there is also quite a considerable demand of cooling for manufacturing purposes. On the whole, all energy load profiles on site look rather suitable for the usage of the CCHP system.

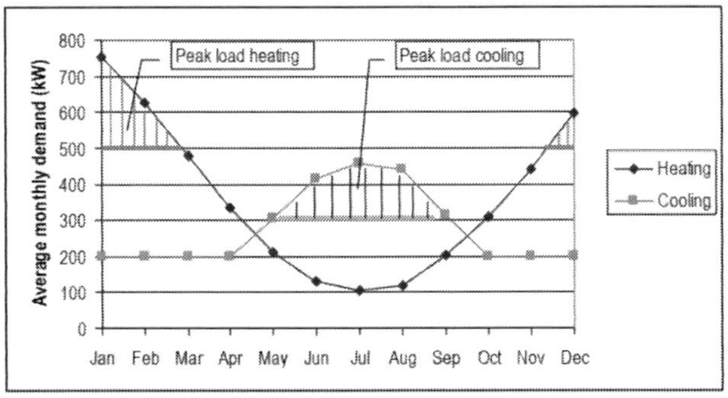

Fig. 5. Average monthly demand of heating and cooling.

However, the problem with the instantaneous capacity demands can take place when production and HVAC needs (expressed in kW) attain their maximum values at the same time resulting in either peak load heating or peak load cooling as shown in Fig.5.

Fig.6 shows how the CCHP system can adapt to the monthly power, heating and cooling loads of the facility. The total heating load in the form of heat recovered from the CCHP system is increased at that moment to the value covering the heating and cooling needs. The cooling energy is then generated by heat in the absorption chiller.

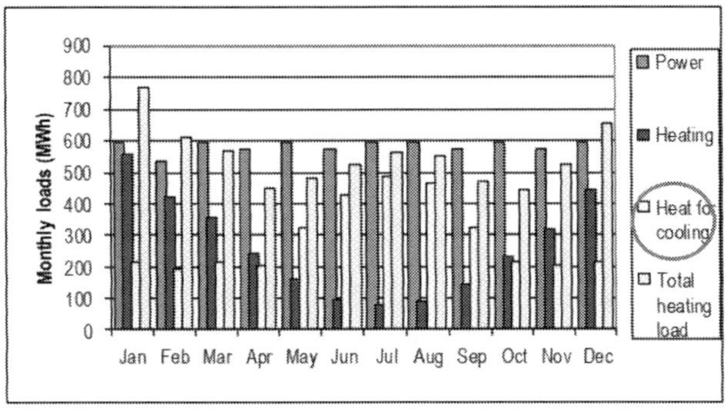

Fig. 6. Monthly loads of power, heating and cooling – Case 2.

The relation between all four load profiles in a yearly framework is presented in Fig. 7. Fig. 8 depicts entire annual primary energy consumption for the conventional system of Case 1 and for the CCHP system of Case 2. This total primary energy is additionally allocated to the specific installations such as power, heating and cooling systems. The usage of total primary in the CCHP system is 24% smaller than in the conventional system. In power utilization the reduction in Case 2 versus Case 1 ranges up to 32%. Alternatively, in heating production Case 2 gives an increase of 55% in comparison with Case 1. In contrast, cooling generation in Case 2 offers primary energy reduction of 67% as opposed to Case 1.

 DOI: 10.1051/e3sconf/20172200022

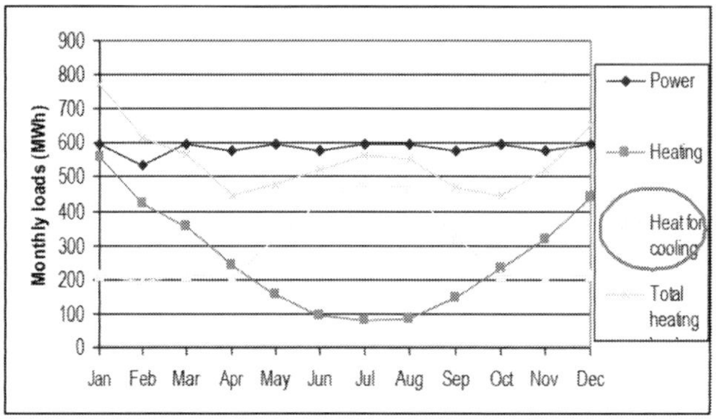

Fig. 7. Annual power, cooling and heating profiles – Case 2.

Fig. 9 illustrates annual CO_2 emissions for Case 1 of the conventional system and Case 2 of the CCHP system. The total CO_2 emissions are similarly assigned to power, heating and cooling installations. The total annual emissions in the CCHP system is around 36% lower than in the conventional system. The emission reduction in power utilization in Case 2 is just 7% lower than in Case 1.

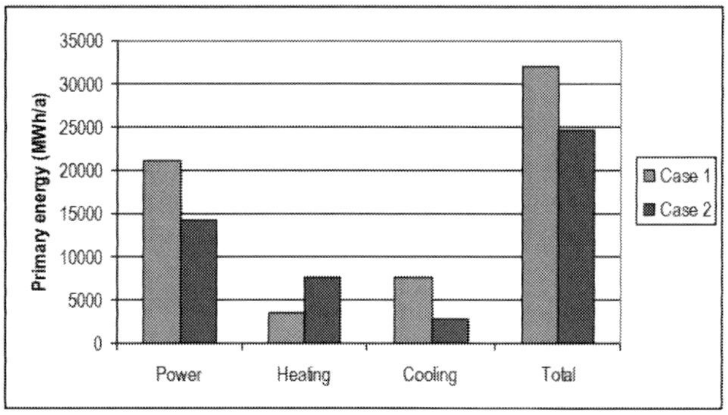

Fig. 8. Primary energy consumption – Case 1 and Case 2.

Alternatively, in heating energy production the emission reduction in Case 2 reaches 90% in contrast to Case 1. Furthermore, cooling generation in Case 2 offers no CO_2 emission.

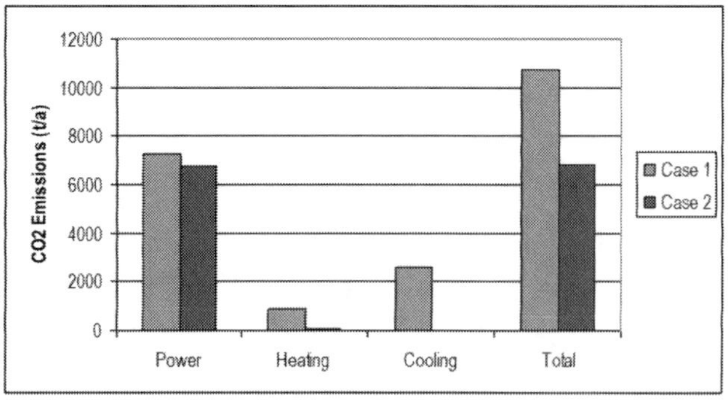

Fig. 9. Annual CO_2 emissions – Case 1 and Case 2.

5 Conclusions

This paper presents an energy and environmental evaluation of the CCHP system in the context of a case study with the conventional system in the industrial facility. According to some evaluation results the following conclusions can be presumed. (1) A relatively stable power demand and technological cooling and heating needs associated with the HVAC loads can match in quite favourable way with the CCHP system operation. Therefore, the industrial facility could be rather suitable site for CCHP applications. (2) The energy performance assessment based on the comparison of primary energy consumption in two systems (conventional and CCHP) indicates that the CCHP system attains better performance than the conventional one. (3) Likewise, the environmental performance appraisal based on the total annual CO_2 emissions proves that the CCHP system is more environmentally benign than the conventional one. (4) Additionally, there also exist several other opportunities to optimize the operation of the proposed CCHP system in this particular industrial facility. One of these measures could be a replacement of a single – effect absorption unit with a two-effect unit with increasing effectiveness of cooling generation almost twofold. Another operation optimizing measure would an improvement of demand side management on site, especially with grid electricity and backup boilers.

References

1. M. Ebrahimi, A. Keshavarz, *Combined cooling, heating and power* (Elsevier, 2015)
2. H. Li, L. Fu, K. Geng, Y. Jiang, Energy and Buildings **38**, 253–257 (2006)
3. F. Fang, L. Wei, J. Liu, J. Zhang, Energy **46**, 211–220 (2012)
4. P,J, Mago, L.M. Chamra, Energy and Buildings **41**, 1099–1106 (2009)
5. M. Carvalho, L.M. Serra, M.A. Lozano, Energy **36**, 3779–3790 (2011)
6. Q. Gu, H. Ren, W. Gao, J. Ren, Energy and Buildings **51**, 143–152 (2012)
7. G. Pagliarini, C. Corradi, S. Rainieri, Appl. Therm. Eng. **44**, 150–158 (2012)
8. G.L. Guizzi, M. Manno, M. Kuosa, Energy **41**, 56–64 (2012)
9. D. Xu, M. Qu, Energy and Buildings **67**, 176–186 (2013)

Modeling the defrost process in complex geometries – Part 1: Development of a one-dimensional defrost model

Simon van Buren[1] , *Ellen* Hertle[1], *Patric* Figueiredo[1], *Reinhold* Kneer[1], and *Wilko* Rohlfs[1,*]

[1]RWTH Aachen University, Institute of Heat and Mass Transfer, Augustinerbach 6, 52062 Aachen, Germany

Abstract. Frost formation is a common, often undesired phenomenon in heat exchanges such as air coolers. Thus, air coolers have to be defrosted periodically, causing significant energy consumption. For the design and optimization, prediction of defrosting by a CFD tool is desired. This paper presents a one-dimensional transient model approach suitable to be used as a zero-dimensional wall-function in CFD for modeling the defrost process at the fin and tube interfaces. In accordance to previous work a multi stage defrost model is introduced (e.g. [1, 2]). In the first instance the multi stage model is implemented and validated using MATLAB. The defrost process of a one-dimensional frost segment is investigated. Fixed boundary conditions are provided at the frost interfaces. The simulation results verify the plausibility of the designed model. The evaluation of the simulated defrost process shows the expected convergent behavior of the three-stage sequence.

1 Introduction

The phenomenon of frost formation occurs in various technical applications including the cooling of humid air such as air fin coolers. Frost is the deposition of water onto cold surfaces as porous structures of ice crystals [1]. The requirements for frost formation are that the substrate surface temperature is below the freezing temperature of water and that the temperature of the moist ambient air is below the dew point. According to Hayashi [3], the process of frost formation can be divided into three characteristic periods: crystal growth, frost layer growth, and full frost layer growth. Further, a preceding phase of nucleation is named as the initial period of frost formation. The characteristics of frost are closely related to the physical processes during the formation periods. The complex ramified structures of the frost layer lead to local inhomogeneities in properties (e.g. porosity, density, specific heat capacity) through the frost layer. The process of frost formation is experimentally and numerically widely studied (e.g. [4, 5]).

In technical applications, isolating frost layers reduce the cooling capacity of heat exchangers due to the low heat conductivity of frost. To maintain the heat transfer performance, the temperature gradient in the heat exchanger must be increased, reducing the energy efficiency of the device. Simultaneously, a frost layer reduces the air flow rate

* Corresponding author: rohlfs@wsa.rwth-aachen.de

through the heat exchanger. At a certain point, the flow rate becomes too low to ensure stable operation conditions and only a strong increase in energy consumption allows for the continuation of the cooling mode. Consequently, the frost layer on cooled surfaces cannot grow indefinitely but must be removed regularly. Electric defrosting and defrosting by hot gaseous refrigerants are most commonly used in industrial applications [6, 7].

2 Existing models for defrosting

One of the first analytical approximations of non-stationary and stochastic [8] defrost processes was conducted by Sanders [1]. He discussed defrosting of a thin frost layer and formulated two different melting models based on a one-dimensional heat balance within the layer. In the first stage, the temperature of the frost layer increases due to wall-side heating. This process continues until the frost temperature at the wall-frost interface reaches the melting temperature. Further heating results in a phase change with a decreasing frost layer thickness. For this process, Sanders [1] presented two different modeling approaches (Fig. 1).

The first model approach assumes a constant temperature throughout the frost layer, a constant heat flux from the wall to the frost and a constant heat flux from the frost to the ambient air. Sanders [1] assumes a frost layer continuously attached to the wall. Consequently, the frost/air interface moves towards the wall. The second model approach accounts for a developing air gap δ_g between the wall and the frost layer, causing additional thermal resistance and thus slowing down the defrost process. Furthermore, meltwater is assumed to diffuse into the remaining frost layer, where it refreezes at the opposite interface due to heat loss to the ambient. Note that the two model approaches presented by Sanders describe two limiting cases. The defrosting does not proceed homogeneously and the detachment of the frost layer will only occur locally and not on the entire substrate surface [9].

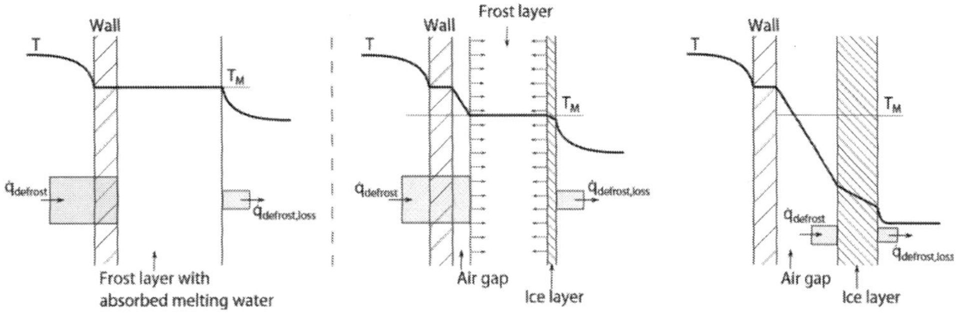

Fig. 1. Illustration of the defrost model by Sanders [1]. Alternative A: Absorption model (left), Alternative B: Air gap model (middle, right).

Machielsen [10] introduces a method to design an optimally balanced refrigeration system based on the model of Sanders [1]. They only consider the second alternative with an air gap between substrate surface and frost layer and divide the defrosting process into three phases.

Hoffenbecker [11] developed a transient numerical model to simulate the hot gas defrost cycle of an industrial air-cooling evaporator, using a simplified circular geometry. The model estimates the time and energy input for defrosting.

Dopazo [2] propose a transient model for defrosting of a round-tube plate fin heat exchanger. The defrost process is divided into six different stages, as shown in Fig. 2. Different governing equations are formulated for a representative tube and fin element for each stage. The model is validated using experimental data.

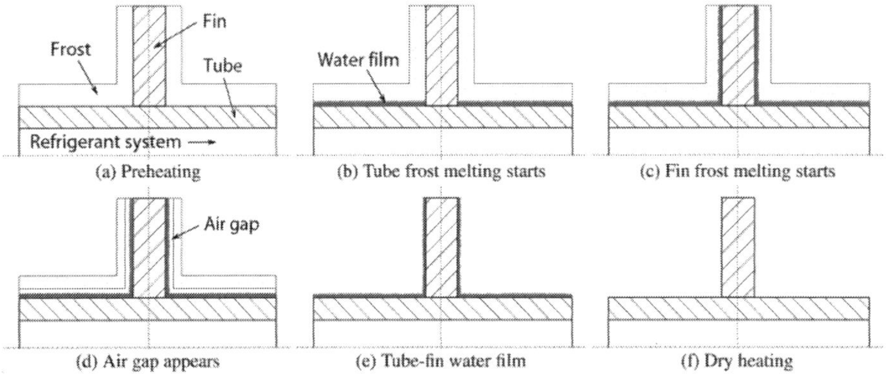

Fig. 2. Six stages of defrosting according to Dopazo [2].

Wellig [12] developed a model to calculate the simultaneous heat and mass transfer during the melting of ice and frost deposits. To consider the changing air temperature and humidity along the fin, a stepwise calculation for several small elements along the fin depth was employed.

The defrost process on an electrically heated cylindrical tube has been analyzed by Sherif [13]. In their quasi-steady state approach, the frost melts at the heated surface and is immediately drained. The frost remains attached to the tube, while the frost surface moves towards the heated tube wall.

Alebrahim [6] presents a two-dimensional electric defrost model for a tube fitted with circular disk fins. The analysis is divided into two stages. First the frost formation stage is analyzed to calculate the temperature distribution within the tube and the frost. Secondly, the defrosting is and further divided into two sub-phases, a pre-melting phase and a melting phase. The author uses the enthalpy method to calculate the defrost time and the temperature distribution within the frost layer during the defrost process. In contrast to prior approaches the model allows to determine the temperature distribution in a two-dimensional space.

Experimental data on the defrost process and detailed research on various defrost methods has been conducted in Switzerland within the framework "Improvement of the Defrosting Process using Air-Cooled Evaporators" [14–16].

Mohs [9] developed a multi-stage defrost model which predicts the transient heat and mass transfer rates for a vertical, plain fin. The defrost process is divided into three separate stages, with each stage solved separately. The final state of each stage is used as an initial condition for the subsequent stage.

Nasr [17] experimentally studies the performance of air-to-air heat exchangers for ventilation purpose throughout the frosting and defrosting process. For the evaluation of the energy consumption, the crucial defrost time ratio is introduced. The influence of the fin-and-tube heat exchanger surface characteristics on the frosting and defrosting (e.g. required energy for the frost melting) is experimentally investigated by Wang [18].

3 Design and model for defrosting

We assume a frost layer of thickness δ normal to the substrate surface that is exposed to an ambient air flow, heating or cooling the frost layer from above. To reduce the computational effort simulating the complex defrosting process, some simplifications are necessary. Lateral heat conduction in the frost layer is assumed to be negligible compared to heat transport in normal direction (valid for thin frost layers or frost layers with homogenous boundary conditions). Secondly, mass transfer by diffusion within the frost layer and at the interface is the neglected. For implementation as a wall-function in a CFD solver (OpenFOAM) the frost layer is modeled with no physical expansion into the air region.

The above-mentioned simplifications allow for a wall-function formulation of the entire defrost process introducing an enhanced thermal coupling of the air and wall region. The thermal coupling condition of the wall and the frost region is the wall-frost interface temperature T_{WF} as well as the wall-side heat flux. In analogy, the temperature T_{AF} and the corresponding heat flux couple the frost and ambient air at the frost surface.

3.1 Frost model

The wall function model accounts for different scenarios of defrosting, including heating or cooling of the frost layer at the wall/frost boundary or the air/frost boundary. Like the models of Dopazo [2] and Hoffenbecker [7], the defrost process is subdivided into different characteristic stages. These stages include *preheating, development of a water film*, and the *development of an air gap*:

During the initial *preheating phase* the temperature in the entire frost layer is below the melting point T_{Melt}. Heat induction from the wall side causes melting and the *development of a thin water film* between the wall and the decreasing frost layer. When the water film exceeds a critical thickness, additional melt water is assumed to drain out, causing the *development of an air gap* between the water film (with the critical thickness) and the frost.

If the temperature of the frost layer reaches the melting point at the frost-air interface, air-side melting occurs. In this case, melt water is assumed to drain off, such that no additional resistance arises. A minimum frost layer thickness for the static stability is assumed. If this minimum thickness is surpassed, the remaining frost is assumed to drop off and the defrost process ends.

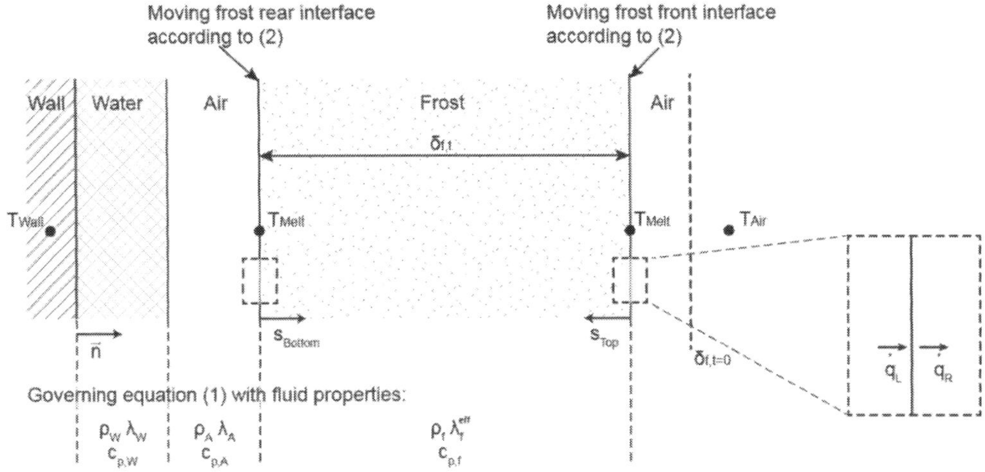

Fig. 3. Sketch of a frost layer with its governing equations and properties according to the region.

A schematic of the different regions in the frost layer model is presented in Fig. 3. The temperature across the regions changes according to the one-dimensional energy equation

$$\partial T_f/\partial t = \lambda_f^{eff}/(c_{p,f}\, \rho_f)\, \partial^2 T_f/\partial x^2 \qquad (1)$$

Material properties, e.g. thermal conductivity λ, specific heat capacity c_p, and density ρ depend on the specific region (frost, water and air).

The frost layer is bounded by the two interfaces S_{Bottom} and S_{Top}. The two interface positions move in the melting phase according to

$$\partial s_f/\partial t = -\,(q_t|_{x=sL} + q_t|_{x=sR})/(h_{Melt}\, \rho_f) \qquad (2)$$

with $q_t = -\lambda_f^{eff}\, \partial T_f/\partial x$. Consequently, a localized heat sink with a discontinuity in the heat flux $q_t|_{x=sL}$ and $q_t|_{x=sR}$ drives the phase change.

Required initial conditions are the frost layer thickness δ, the local frost temperature $T(x)$ and the local frost porosity $\varepsilon(x)$ which is assumed to be constant over the defrost process (no densification). The porosity ε is defined as the ratio of the frost pore volume V_p to the to the total volume V_{tot}. Initially, the frost interfaces S_{Bottom} and S_{Top} are located at the wall and the position δ, respectively. For the boundary conditions, a time-dependent wall temperature $T_{WF}(t)$ and ambient air temperature (at the frost-air interface), $T_{AA}(t)$ are required. Due to the moving frost boundary edges described by eq. 2, the system has two further degrees of freedom. Thus, the temperature at the melting interface is set to the melting point of the frost.

3.2 Frost properties

Frost layers are usually characterized by a strong inhomogeneity. Porous air and ice compositions change along the frost height leading to a local gradient in porosity. Frost properties are modeled in agreement to previous studies [1, 19, 20]. Density, ρ_f, effective thermal conductivity, λ_f^{eff}, and specific heat capacity, c_f, of the frost layer are derived by the local frost porosity ε and the physical properties of the components ice, air and water [20]:

$$\rho_f = (1-\varepsilon)\, \rho_{ice} + \varepsilon\, (\rho_{vapor,f} + \rho_{air,f}) \approx (1-\varepsilon)\, \rho_{ice} + \varepsilon\, \rho_{air}. \qquad (3)$$

$$\lambda_f^{eff} = 0.001202\, \rho_f^{\,0.963} \qquad (4)$$

$$\rho_f\, c_f = \varepsilon\, \rho_a\, c_a + (1-\varepsilon)\, \rho_{ice}\, c_{ice} \approx (1-\varepsilon)\, \rho_{ice}\, c_{ice} \qquad (5)$$

4 Numerical implementation

The model has been implemented in MATLAB prior to an implementation as wall function in the CFD solver. The locations of the frost boundaries S_{Top} and S_{Bottom} are discretized in accordance to the underlying mesh of the frost section, avoiding complex mesh adjustments of moving grids. The individual cells are assigned the thermophysical properties of the containing material (water, frost, air) and changed accordingly throughout the course of the defrost process.

The computation of the presented set of differential equations is carried out by the *ode23s* solver provided by MATLAB. Discrete movement of the frost boundaries and the connected reallocation of cells is realized by event functions that are activated upon a complete shift of the frost interface to the neighboring mesh interface.

Investigating the mesh dependency, a converging behavior is observed, showing variations of 5% at 25 cells and 1% at 100 cells. Thus, a mesh resolution of 100 cells is used for the presented results.

 DOI: 10.1051/e3sconf/20172200023

5 Results

The model approach outlined and described in the previous sections is first applied to study the defrost process of a one-dimensional homogeneous frost layer (ε = *0.5)* of 1 mm of thickness with a linear temperature profile from 258 to 270K a time invariant boundary conditions (T_{Wall} = 281K, T_{Air} = 276K, α = 25W/(m^{2}K)). The minimal frost- and maximal water-layer thickness are both 0.2 mm.

5.1 Dissection of the defrost process

Figure 4 shows the temperature profiles of the melting frost layer at four instants of time. In the initial phase of the defrost process (at t = 0.1 s, see Fig. 4a) the temperature profile in the frost layer changes rapidly. Owing to the high temperature at the frost-wall interface, rapid melting occurs and a thin water layer develops within the first 0.1 s. Consequently, the *preheating phase* ends with the beginning of melting and the *development of a water film* in less than 0.1s. The temperature minimum in the center of the frost layer indicates a strongly transient heat transfer process in the initial phase.

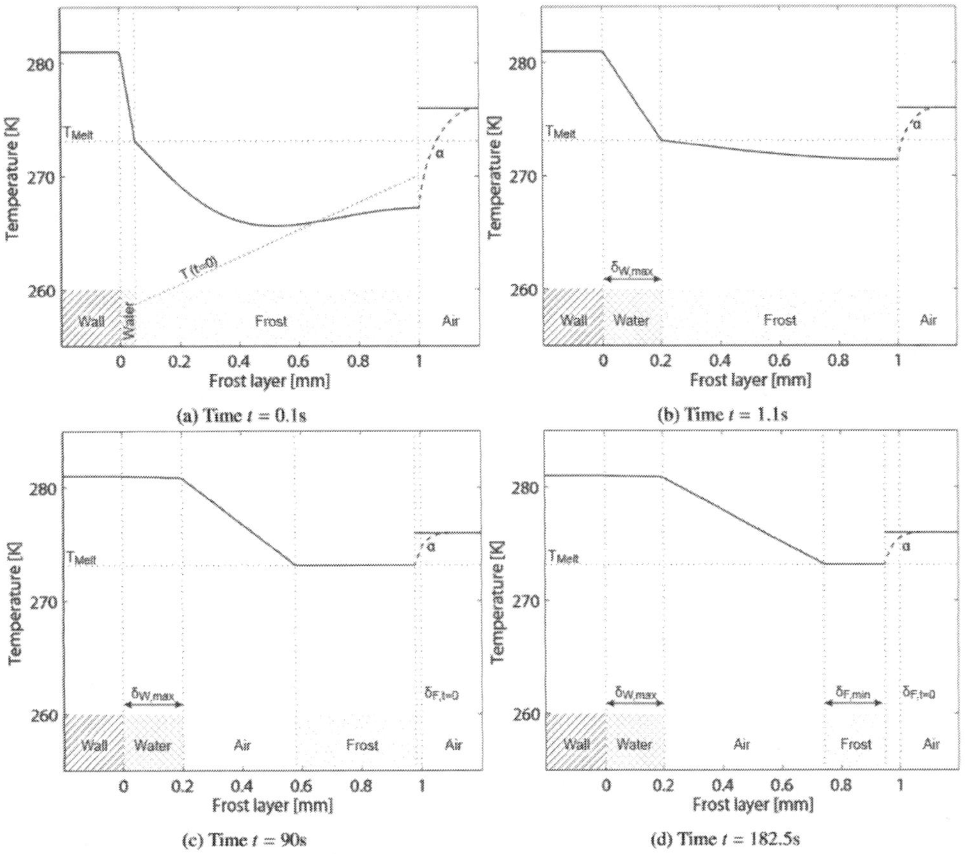

(a) Time t = 0.1s

(b) Time t = 1.1s

(c) Time t = 90s

(d) Time t = 182.5s

Fig. 4. Temperature profiles of a defrost process at different times. Convection at the ambient air-frost interface is indicated by the dashed line labeled α. Vertical dashed lines refer to the interfaces between the regions.

At t = 1.1 s (Fig. 4b), the water film has increased in size to 0.2 mm (critical water film thickness). The model assumes now that further melt water drains out, not remaining in the frost-wall gap – end of phase "*development of a water film*". The flattened temperature

DOI: 10.1051/e3sconf/20172200023

profile inside the frost layer and the straight profile in the water layer indicate a quasi-stationary defrost process.

At t = 90 s (Fig. 4c) an air gap has formed (phase "*development of an air gap*"). The size of the water film remains at the critical thickness of 0.2 mm. Owing to the lower thermal conductivity of air compared to water and frost, the temperature drops across the air gap, indicating that the air gap presents the most significant resistance in the defrost process. In addition to the wall-sided melting process, air-sided effects are observed at 90 s. The temperature of the remaining frost section is marginally below the melting point.

At t = 182.5 s (Fig. 4d) the frost layer has decreased down to the critical frost layer thickness of 0.2 mm. Now, the remaining frost layer is assumed to drop off, completing the defrost frost process

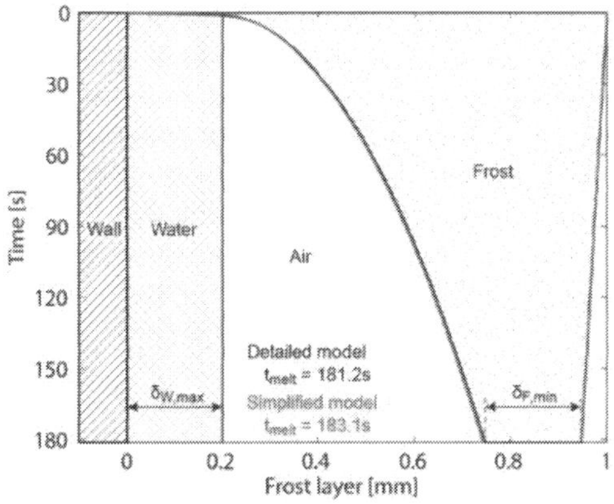

Fig. 5. Remaining frost layer and water and air regions over the defrost time. Black: Presented, detailed model including discretized temperature information of the frost section. Red: Simplified model only considering the latent phase change.

In Fig. 5 the progress of the melting process is plotted over the defrost time (black line) by illustrating the individual regions (melting frost, water film and air gap). The beginning of the defrost process is characterized by wall sided melting in which the water film of critical length develops rapidly. The following increasing air gap acts as an additional thermal baffle, slowing down the process. Thus, the wall sided boundary shows a concave behavior over time. On the other, air-sided frost interface, a linear reduction of the frost layer is observed. The linear defrost behavior at the frost surface indicates an increasing importance of air-sided melting in thicker frost layers, as it is opposed the decreasing wall-sided melting process.

In addition to the black lines showing the results of the detailed model (Eq. (1) and (2)), the red line illustrates the results of a further simplified model approach. The approach neglects the local change of temperature, only considering the dominating heat of fusion and the melting process of the frost interface defined by Eq. (2). The simplification is valid due to the high value of the heat of fusion h_{fus}=333.5 kJ/kg compared to the specific heat capacity of the frost c_p=1.8 kJ/(kg K) and the low increase in temperature in the preheating phase.

Comparing the results of the detailed model (black line) with the simplified approach (red line) a good accordance of the defrost process is observed. Likewise, the calculated defrost times of the frost layer match. In conclusion, the simplified model is likewise able to describe the defrost process, strongly reducing the numerical effort.

5.2 Parameter variation

For further investigation of the defrost model, a parameter variation of the air temperature is performed. The range of 275 K to 300 K is chosen for both temperatures, always ensuring a two-sided defrost process. All remaining parameters are adopted from the set-up previously described (compare Tab. 2).

An increase in ambient air temperature accelerates the defrost process. Fig. 6a shows a defrost time of approximately 85 s at an air temperature of 300 K, whereas, an air temperature slightly above the melting temperature, defrosting takes about 190 s. The convex behavior of the defrost time over the variation of fixed air temperatures as boundary conditions indicates the importance of the air temperature especially close to the melting point. In Fig. 6b, the location of the remaining frost layer before its thickness undercuts the critical stability criteria is displayed. With an increasing air temperature, the remaining frost layer moves from the initial air interface towards the wall.

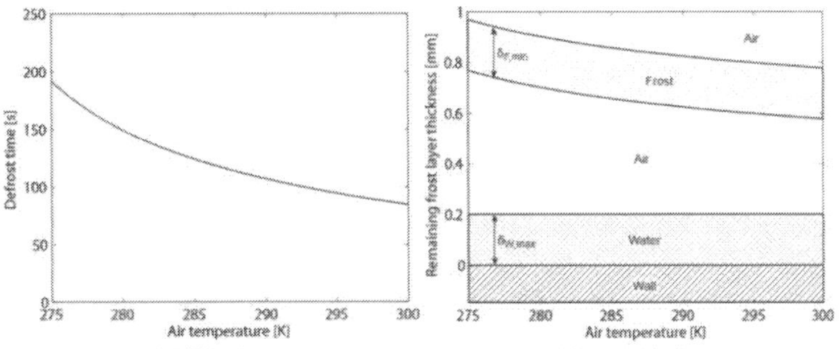

(a) Defrost time (b) Remaining frost layer

Fig. 6. Variation of the air temperature T_a.

6 Conclusion

A numerical approach for a one-dimensional transient defrost model accounting for heat conduction is presented. Fixed boundary conditions are applied to the frost interfaces. This simplified approach allows for an integration of the defrost model in a CFD solver (OpenFOAM) using a zero-dimensional wall function. In accordance to previous work, a multi stage model is applied (e.g. [1, 2]).

For validation and verification, the multi-stage model is implemented in MATLAB, evaluating the defrost process of a one-dimensional frost segment. The simulation results verify the plausibility of the designed model. The evaluation of the simulated defrost process shows the expected behavior of the three-stage sequence. Likewise, the variation of boundary conditions results in plausible reactions from the defrost model. A convergent behavior of the numerical implementation is also conducted.

W.R. is associated postdoctoral researcher in the graduate school GRK 1856 "mobilEM" funded by the German Research Foundation (DFG).

References

1. C. Sanders, Ph.D. thesis, University of Delft (1974)
2. J.A. Dopazo, J. Fernandez-Seara, F.J. Uhia, R. Diz, Int. J. of Refrigeration **33**, 829–839 (2010)
3. Y. Hayashi, A. Aoki, S. Adachi, K. Hori, J. Heat Transfer **99**, 239–245 (1977)

4. S. Anisimov, A. Jedlikowski, D. Pandelidis, Int. J. Heat Mass Transfer **90**, 201–217 (2015)
5. X. Wu, S. Hu, F. Chu, Applied Thermal Engineering **95**, 95–105 (2016)
6. A. Alebrahim, S. Sherif, J. of Mechanical Engineering Science **216**, 655–673 (2002)
7. N. Hoffenbecker, Master's thesis, University of Wisconsin (2004)
8. N. F. Aljuwayhel, Ph.D. thesis, University of Wisconsin (2006)
9. W. F. Mohs, Ph.D. thesis, University of Minnesota (2013)
10. C. Machielsen, H. Kerschbaumer, Int. J. of Refrigeration **12**, 283–290 (1989)
11. N. Hoffenbecker, S. Klein, D. Reindl, Int. J. of Refrigeration **28**, 605–615 (2005)
12. B. Wellig, M. Imholz, M. Albert, K. Hilfiker, 9th IEA Heat Pump Conference, Zurich CH (2008)
13. S. Sherif, M. Hertz, Int. J. of Energy Research **22**, 85–92 (1998)
14. S. Bertsch, P. Hubacher, M. Ehrbar, SFOE (2002)
15. B. Hubacher, M. Ehrbar, SFOE (2000)
16. P. Hubacher, C. Bernal, M. Ehrbar, S. Schwendener, SFOE (2004)
17. X. Wu, S. Hu, F. Chu, Applied Thermal Engineering **95**, 95–105 (2016)
18. F. Wang, C. Liang, M. Yang, C. Fan, X. Zhang, Applied Thermal Engineering **75**, 1126–1132 (2015)
19. M. Kandula, Tech. rep., ASRCAerospace, John F. Kennedy Space Center, FL, USA (2010)
20. K. Prölss, G. Schmitz, The Modelica Association **1**, 509–516 (2006)

E3S Web of Conferences **22**, 00024 (2017) DOI: 10.1051/e3sconf/20172200024
ASEE17

Comparison of power curve monitoring methods

Philippe Cambron[1,*], *Christian* Masson[1], *Antoine* Tahan[1], *David* Torres[2], and *Francis* Pelletier[2]

[1]École de Technologie Supérieure, 1100 Notre-Dame O, Montréal, QC, Canada
[2]Arista Renewables Energies, 2648 Desjardins, Montréal, QC, Canada

Abstract. Performance monitoring is an important aspect of operating wind farms. This can be done through the power curve monitoring (PCM) of wind turbines (WT). In the past years, important work has been conducted on PCM. Various methodologies have been proposed, each one with interesting results. However, it is difficult to compare these methods because they have been developed using their respective data sets. The objective of this actual work is to compare some of the proposed PCM methods using common data sets. The metric used to compare the PCM methods is the time needed to detect a change in the power curve. Two power curve models will be covered to establish the effect the model type has on the monitoring outcomes. Each model was tested with two control charts. Other methodologies and metrics proposed in the literature for power curve monitoring such as areas under the power curve and the use of statistical copulas have also been covered. Results demonstrate that model-based PCM methods are more reliable at the detecting a performance change than other methodologies and that the effectiveness of the control chart depends on the types of shift observed.

1 Introduction

The reliability and availability of Wind Turbines (WT) are two key aspects used to improve and insure the development and integration of wind energy in current grids. The development of robust monitoring is a way of contributing to this objective. Interesting works have been published lately on various monitoring condition techniques for WT [1, 2]. Among the monitoring methods, some authors proposed approaches to monitor the production of a WT using Power Curve Monitoring (PCM) [4–9].

The power curve of a WT is the relationship between the wind speed and the power output. It is also viewed as the performance of the WT. Therefore, any performance degradation will be reflected in the power curve: less power will be produced for a given wind speed. In the past years, various methods for PCM have been developed, and each one of them reported interesting results. Nowadays, these methods are receiving more attention from wind farm operators as their WT are starting to age. It is now common to encounter

* Corresponding author: philippe.cambron.1@ens.etsmtl.ca

DOI: 10.1051/e3sconf/20172200024

wind farms that have been in operation for more than ten years. As a wind farm ages the performance of its WT degrades with time as reported in [3].

However, these PCM methods were developed using data from different industrial wind farms. Therefore, it is not possible to clearly establish the potential of each PCM method. The objective of the current work is to compare some PCM with a common dataset. Two case studies from industrial WTs will serve as the basis for comparison.

Four different PCM methodologies will be compared. The first one from [4] uses data mining methods for the power curve model together with a Shewhart control chart for the monitoring. This approach will be in direct comparison with the one proposed by [5] which is based on the binning method defined in the IEC 64100-12-1 standard. For this second PCM method, an Exponentially Weighted Moving Average (EWMA) control chart is use to establish if the power curve sustained a change. These firsts two PCM can be classified as model-based monitoring methods since they rely on difference between the observed and the predicted behavior to determine the presence of any anomaly. Two other PCM approaches discussed are the use of the Performance Index (PI) proposed by [6] and the monitoring of power curve copulas developed by [7].

2 Data Source

2.1 Power Curve of a Wind Turbine

The power curve of a WT is the relationship between the wind speed and the output power. It is also a function of the WT operating modes. Figure 1 from previous work presents the power curve of a WT and illustrates the four zones corresponding to the operating modes [5]. Zone I is where the wind speed is lower than the cut-in speed. In this region, the power of the WT is null, since there is not enough power in the wind to move the rotor. Between the cut-in speed and the rated speed, the power of the WT is a function of the cubic wind speed. The rated speed is where the power reaches its nominal value. In Zone III, which is between the rated speed and the cut-out speed, the power of the WT is maintained to a constant value corresponding to the nominal power by varying the blade pitch angle. For a wind speed greater than the cut-out speed (Zone IV), the WT is shut down to protect its structure and components from extreme loads.

Since in Zone III the WT power is imposed, performance degradation will not be noticeable. So as recommended by [8] and done in [5], values with wind speed greater than the rated wind speed will be removed from the dataset.

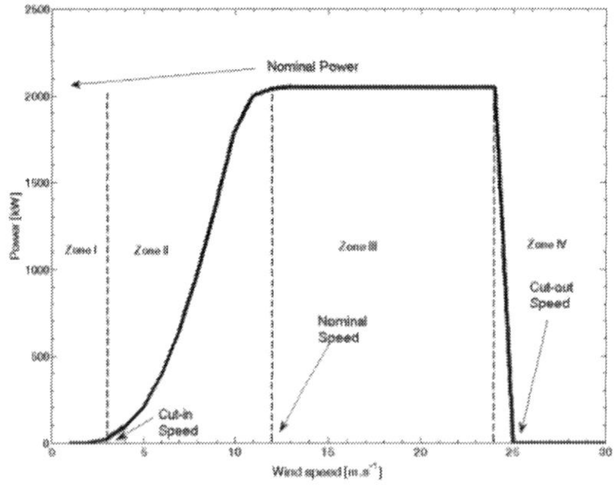

Fig. 1. Power Curve and operating modes of a WT. From [5].

2.2 Case Studies

The PCM methodologies will be compared using data from an industrial wind farm located in Canada. The WTs are MW class and pitch regulated. This wind farm has been in service since 2005 and data is archived at frequency of 1 Hz. However, as suggested by the IEC standard, 10-minute averages are use rather than the high frequency data to reduce the noise and the effect of autocorrelation. Data have been filtered for outliers caused by events such as sensor malfunctions, database unavailability or malfunction and icing with the method proposed in [9]. The timestamp of the date is the beginning of the time period.

Two cases of WT performance degradation will provide the common reference for the comparison of the PCM methods. The first case is an example of an erroneous activation of a curtailment algorithm of a WT resulting in a sudden underperformance of the WT. The second case presents a small and progressive performance degradation caused by blade leading edge erosion.

The nacelle wind speed will be use because met masts are not available for these two WTs. This wind speed will also be corrected for the effect of air density has suggested by the IEC standard (all further mention of wind speed take account of the air density correction).

3 Power Curve Monitoring Methodologies

Each PCM method will be briefly introduced. References to the original work are included for more details on each one of them. All PCM methods will be compared on their ability to detect a small change in the power performance and on the time needed for the detection of the behaviour change presented by the two different case studies. The first case is a sudden shift corresponding caused by a control algorithm change. For the second case, which is about a progressive degradation, it is not possible to clearly define when the issue started. Therefore, the date of detection will be use as a PCM performance metric.

3.1 Model-Based Monitoring

In model-based monitoring, the energy output coming from the WT is predicted with a model. Then, the residuals (differences between the prediction and the observation) are analysed. For a valid model, and without any on-going under-performance, the residuals should be ergodic (same statistical behaviour in time). However, a change of energy output coming from the WT will result in a significant change or trend in the residuals. Thus, the power residuals will be monitored for any occurrence of under-performance. Two different models will be compared, the bins method and the random forest.

3.1.1 Bins Model

The bins model consists of splitting wind speed data by an interval of 0.5 $m.s^{-1}$. The power curve is obtained by taking the average of the power in each bin. This model is widespread in the industry since it is recommended by the IEC standard. This model is used in [7].

The power residuals are normalized using the average and the standard deviation of the power in the respective bin, allowing treating all the bins simultaneously rather than bin by bin. For more on this transformation, the reader should consult [7].

3.1.2 Random Forest

In [4], a model using a random forest is use to predict the WT power. Random Forest is a non-parametric data mining methods constructed by an ensemble of randomly generated regression trees that can be used for prediction. See [10] for more on random forest. The random forest used for the prediction of the power will have the wind speed and the air density as inputs factors. The power residuals $R(t)$ in this case are transformed to normalized residuals:

$$R(t) = \frac{P(t)-\hat{P}(t)}{P(t)} = \frac{\Delta P(t)}{P(t)} \quad (1)$$

where $P(t)$ is the observed power and $\hat{P}(t)$ the expected power.

3.1.3 Control Charts

Control chart are use in Statistical Process Control (SPC). They act in a way like a continuous statistical testing. A statistical property of the process is plotted trough time (individuals, sample average, range, sample standard deviation, etc.) If the data of the control chart is outside of the control limits, the process is reported as statistically different. A $\bar{X}$ Shewhart chart is use in [4] and a EWMA chart is use in [5]. The process monitored here is the power residuals. The residuals are averaged on a three-day period to reduce the effect of noise as discussed in [5]. The EWMA chart smoothes the residuals, acting as a high frequency filter. The smoothing parameter (λ) is adjustable and takes values between 0 and 1. If $\lambda = 0$ the smoothing is at its highest possible point and if $\lambda = 1$, there is no smoothing. A complete definition of control charts is included in [11]. Here, as recommended in [5], $\lambda = 0.2$ will be use.

3.2 Copulas Monitoring

Copulas are coming from probability theory and statistics. They are a multivariate probability distribution for which the marginal probability distribution of each variable is uniform. According to Sklar's theorem, any joint distribution can be transformed to a copula with uniform marginal distribution. The copula is an illustration of the dependence structure between the variables [7].

Copulas have been use for PCM by [7]. The power curve can be considered as a multivariate probability distribution function of the wind speed (v) and the power (P). It is therefore possible to transform this multivariate distribution to a copula by transforming the random variables v and P into two new random variables with uniform marginal distribution $u \sim U[0,1]$ and $w \sim U[0,1]$. Figure 2 presents the multivariate distribution of the wind speed and power and its transformation to a copula. Since the copula represents the dependence structure between the wind speed and the power, any change in the WT performance will be reflected in the copula. As shown in Figure 2, the copula of a power curve has a linear aspect. A reference copula is first constructed, and this transformation from the original probability distribution to the new probability distribution is use afterwards with date from the test periods. Therefore, a change on the energy output coming will be easily detected in the copula. The advantage of the use of copula in PCM is that no model is required, hence limiting the uncertainty inherent in the use of a model.

The residuals between the baseline period copula and the copula of the period under evaluation ($\tilde{R}$) is used as the metric to determine if there is a change in the WT behaviour. The suggested threshold for the determination of a behaviour change of are $|\tilde{R}| > 0.02$.

 DOI: 10.1051/e3sconf/20172200024

Authors suggested a two-month baseline period and a month of data for each evaluation period.

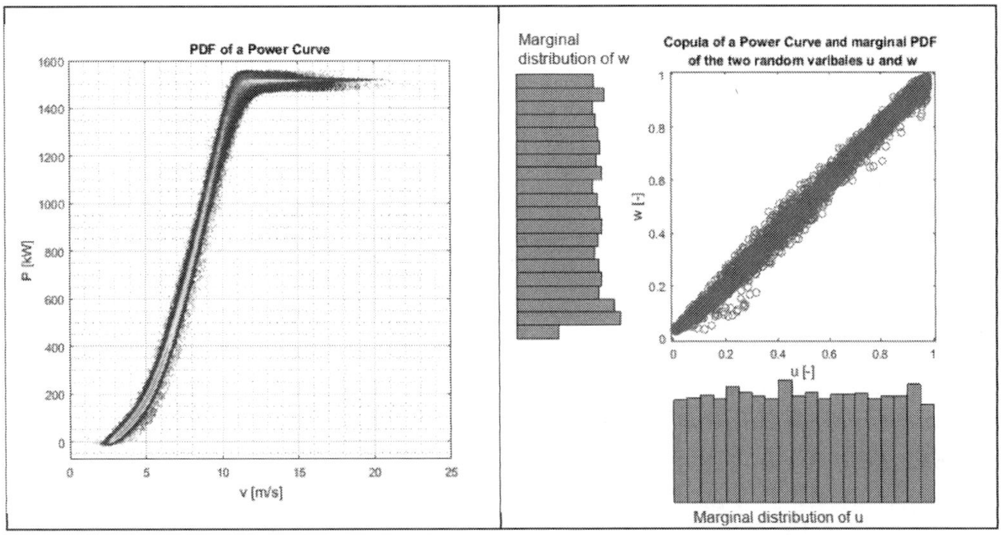

Fig. 2. On the left: Kernel density estimate of the joint probability of wind speed and power output with the colour in function of the probability (blue for 0, red for 1). On the right: Copula of a Power Curve along with the uniform marginal distribution of the random variable u and w.

3.3 Performance Index

The Performance Index (PI) for the performance assessment of a WT was proposed for the performance assessment of a WT in [6]. The power curve is separated in n different wind speed regions. In their original work, the authors proposed two different regions (below and above the nominal wind speed). Its calculation is based on the area under the power curve is the following way:

$$PI_i = \frac{S_i^T - S_i^R}{S_i^R} \quad (2)$$

where PI_i is the performance index in the region i, S_i^R is the area under the power curve in the region i during the reference period and S_i^T is the area under the curve for the period under evaluation. The final PI is obtained with a weighted sum of all the PI_i with the weight corresponding to the wind distribution. The calculation is the following:

$$PI = \sum_{i=1}^{n} WD_i \cdot PI_i \quad (3)$$

where WD_i is the percentage of wind distribution in the region i for the period under evaluation. The use of the PI for PCM implies the use of a power curve model. In the original work, a linear hinge model was used. However, the effect of a different model on the PCM will be covered by the comparison between the bin model and the random forest. Also, to isolate the results of the PI as a metric for PCM, the PI will be used with the bin model. As in [6], the PI will be calculated monthly.

4 Results

4.1 Model-Based Monitoring

4.1.2 Shewhart Control Chart

Figure 3 reports the results of the two model-based PCM when used with a $\bar{X}$ Shewhart chart. The normalized residuals for the bins method are expressed by $\|R\|$. For case study A, both power curve model-based PCM could easily detect the shift in the power curve; by 2012-05-08 for the random forest and by 2012-04-15 for the bin model. However, neither of the two model-based PCM methods can detect the performance degradation in Case B, which is therefore not presented. Also, both methods return a false positive around 2010-02-14.

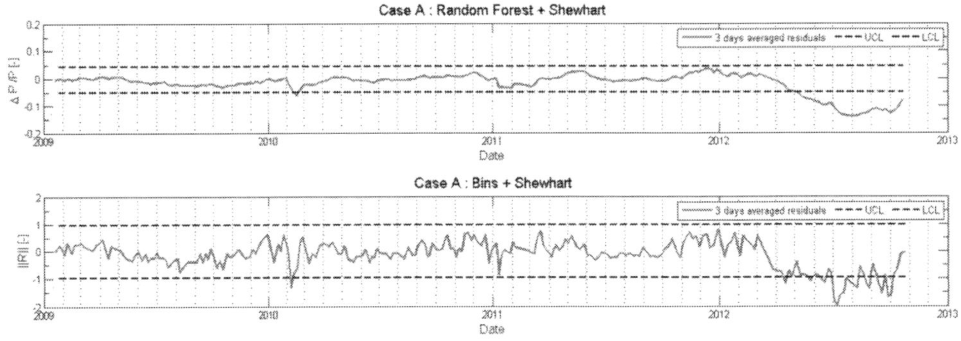

Fig. 3. Power Curve Monitoring with the $\bar{X}$ Shewhart chart for both models.

4.1.2 EWMA Control Chart

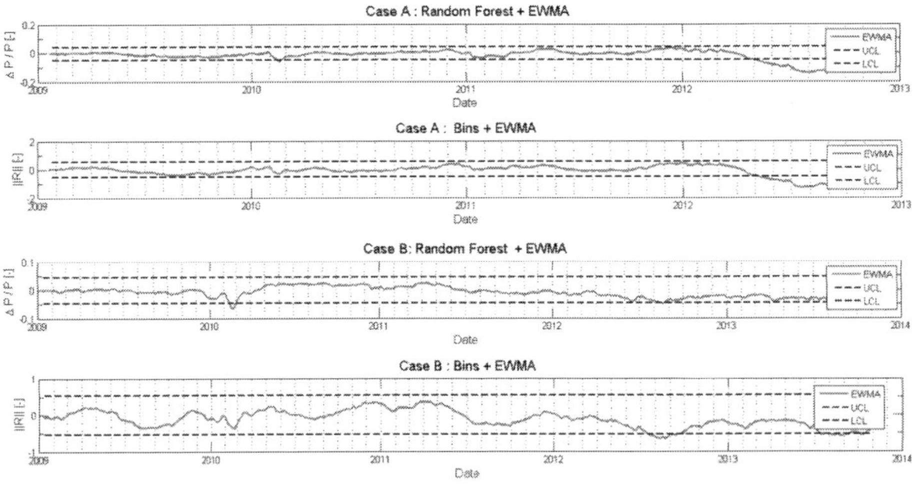

Fig. 4. Power Curve Monitoring with the EWMA chart for both models.

Figure 4 presents the results of the two model-based PCM when use this time with an EWMA chart. For Case A, both power curve models could detect easily the shift in the power curve; by 2012-05-02 for the random forest and by 2012-05-13 for the bins model. However, for the case study B, the PCM method based on the bins model performed better than the one based on the random forest. The bins + EWMA method found the change of behaviour on 2012-07-21 while the random forest + EWMA reported the change a month later, on 2012-08-22.

The false positive in Case A is no longer observed, due to the smoothing effect of the EWMA. It was expected in Case A that the Shewhart chart performed better than the EWMA chart. The smoothing of the EWMA is known to enhance the detection of small shifts while reducing the reactiveness to sudden variations [11].

4.2 Copula Monitoring

The results of PCM with copula are presented in Figure5. This method could detect the shift in power curve in the dataset observed in Case A. The detection date is 2012-05-01. This result is like the one obtained with the use of the random forest with the EWMA and the bins method with the EWMA. The need for a full month of data to calculate the copula, since it requires a large sample, is a drawback for copula monitoring. However, the use of a rolling window of a month for the evaluation period with a daily calculation of the copula might be able to overcome this issue.

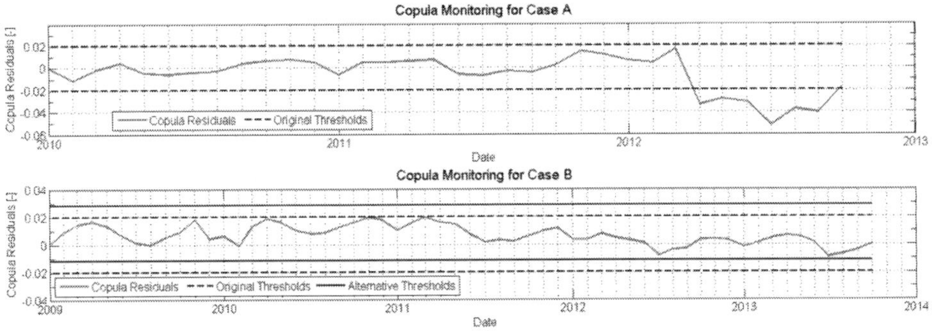

Fig. 5. Power Curve Monitoring with Copula Monitoring.

However, the copula method was not able to detect the change in the power curve for the case study B using the threshold provided in the original work ($|\tilde{R}| > 0.02$). The choice of this threshold is based under the assumption that, in absence of any issue, the value of $\tilde{R}$ will remain close to 0. However, it can be seen on Figure 5 that the value of the copulas residuals $\tilde{R}$ is above 0 at the beginning of the reference period, pointing toward a systematic bias. We suggested that the thresholds should be defined with regard of the average copula residuals from a reference period of a year. Using these alternative thresholds, the copula PCM method would almost have could detect the under-performance by 2012-07-01.

4.3 Performance Index

The result of PCM with the PI is reported on Figure 6. It is difficult to identify if the PI detected the behaviour changes in the two case studies as it is unclear what threshold to use. For Case A, a change in the PI can be observed by 2012-04-01. However, it is not possible to tell if the PI has detected the change of behaviour in Case B.

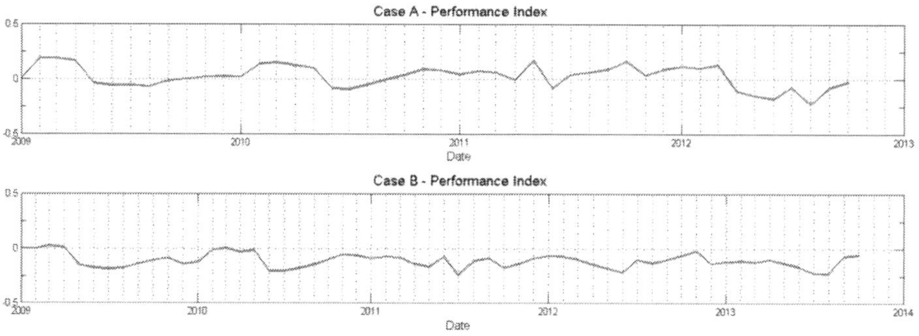

Fig. 6. Power Curve Monitoring with the Performance Index.

5 Conclusion

Four different PCM methods were compared using common set of data. The dataset consisted of two case studies from WT in industrial wind farms that suffered a performance change. The model-based PCM used with a EWMA control chart was the only approach able to detect both behaviour changes. However, model-based PCM along with the Shewhart chart and the copula PCM provided better results for case A. The smoothing effect of the EWMA chart is enhancing the ability to detect small shifts while reducing the reactiveness to sudden important shifts. Here, both models studied presented similar results. A combination of both control charts could be an interesting SCADA monitoring strategy.

Not all PCM methodologies present in the literature have been covered in this work and new PCM approaches are still interesting as results may be improved. Further work on the benchmarking of other PCM methods would be a logical follow-up to the current work.

This research was made possible with the help of the National Science and Engineering Research Council of Canada (NSERC) and Quebec's Nature and Technology Research Fund (FRQNT) joint Industrial Innovation Scholarships Program. Special thanks go to the participating industrial partner.

References

1. Z. Hameed, Y.S. Hong, Y.M. Cho, S.H. Ahn, C.K. Song, JRESER **13**,1–39 (2009)
2. F.P. Garcia Marquez, A.M.Tobias, J.M. Pinar Perez, M.Papaelias, JRENE **46**, 94–101 (2012)
3. I. Stafell, R. Green, JRENE **66**, 775–786 (2014)
4. A. Kusiak, H.Zheng, Z.Song, JRENE **34**, 1487–1793 (2009)
5. P. Cambron, R. Lepvrier, C. Masson, A.Tahan, F. Pelletier JRENE, **94**, 126–135 (2016)
6. R. de Andrade Viera, M. Sanz-Bobi, S. Kato ICRERA, 31–36 (2013)
7. S. Gill, S. Bruce, S. Galloway, JSE **3**, 94–101 (2012)
8. S. Butler, J. Ringwood, F. O'Connor, Sys. Tol. 389–394 (2013)
9. J. Park, J. Lee, K. h, J. See, IEEE Trans. Energy Convers **29**, 119–128 (2014)
10. L. Breiman, JMLR **45**, 5–32 (2001)
11. D. Montgomerry, Introduction to statistical quality control (2007)

E3S Web of Conferences **22**, 00025 (2017) DOI: 10.1051/e3sconf/20172200025
ASEE17

An integrated low carbon energy solution to cooking fuel, tailored to Niger state's rural population

Aaron Carvell[1], *Andrew* Price-Allison[1], *Calum* Birch[1], *Toby* Green[1,*], *Khanji* Harijan[2], *Sheidi* Maihankuri[3], *Abdulganiy* Raji[4], *Mohammed* Uqaili[5], and *Valerie* Dupont[1]

[1]School of Chemical and Process Engineering, University of Leeds, Leeds, LS2 9JT, UK
[2]Department of Mechanical Engineering, Mehran University of Engineering & Technology, Jamshoro, Pakistan
[3]Mechanical Engineering, Petroleum Training Institute, Effurun, Delta State, Nigeria
[4]Department of Agricultural and Environmental Engineering, University of Ibadan, Ibadan, Nigeria
[5]Department of Electrical Engineering, Mehran University of Engineering & Technology, Jamshoro, Pakistan

Abstract. Niger State (Nigeria) was selected as a case study of renewable, affordable and user friendly clean energy provision in remote areas of developing countries. Niger state has 80% of its 4.5 million population living in rural agrarian areas with low literacy rates, there is a lack of wind thus eliminating wind as widely available potential power source. Based on the assessment of the local large insolation, the type of agricultural, biomass and husbandry resources, this study selected the design of anaerobic digestion units processing mostly animal and human waste, and whose heating and power requirement would be entirely provided by solar photovoltaic/thermal to maintain optimum efficiency of the biogas production. The designs was carried out at the scale of up to 15 household demand (community scale). Volume and therefore the production of biogas maybe increased or decreased in the design considered, and local, low cost resilient material were proposed. The proposed system was costed for a community of 24 people, demonstrating the potential for clean and renewable gas production economically.

1 Introduction

A range of influencing factors must be considered when choosing alternative low carbon energy solutions including; landscape, resource availability, resource and technology reliability and cost. Understanding the problems faced by these populations and of the nature and limits of their energy resources is required.

Nigeria is the most populous nation in Africa, the population is supported by a fertile landscape, a long rainy season with high annual rainfall and high solar irradiance throughout most of the nation. This large population is maintained despite an extremely low life expectancy, estimated at around 53 years [1]. Most of Nigeria's population live without

* Corresponding author: pmtag@leeds.ac.uk

DOI: 10.1051/e3sconf/20172200025

access to electricity, with electrification rates of around 40%, and of those that do have access, the needs of only around 30% are adequately met, with frequent black-outs of varying length and timing. The power crisis that has been affecting Nigeria for many years now has been exacerbated by some key failings; poorly maintained/aging equipment and infrastructure, disruptions in supply of oil/gas and high level corruption [2]. In this aspect it is very similar to the current situation of Pakistan.

The energy usage figures for Nigeria provided by the International Energy Agency show the huge proportion of energy production coming from biomass, at 41.5% [3], however contrasted with the final energy consumption, for which biomass accounts for 86.2% [3], the existing importance of renewables in Nigeria can be seen.

Niger State, located in Western Nigeria, is the largest of Nigeria's 36 states, it accounts for almost 10% of the land area of the country, although its population is only 3.9 Million, 80% of which are rural dwellers [4] with literacy levels lower than the national average (37.5%) [4], thus its economy is mostly based on agriculture, where "archaic" farming practices are still in use[5]. Eighty five percent of its land area is arable helped by long rainy season. The large rivers and their tributaries running through the state also give the opportunity for irrigation, not yet fully exploited [6]. It is rare for net electricity generation to exceed 50% of its capacity, the record high for 2016 being 5075 MW [7]. Niger State's large agricultural industry producing a lot of crop residue (from sugarcane, yam, cassava, onion, sweet potatoes) and animal wastes (poultry, cattle, goats, sheep), it benefits from good solar irradiance, adequate hydro potential but no significant wind power potential [8-10].

It is shown that consistent levels of solar radiation and sunny hours are received year long, with a slight decrease in August and increase in November for the heights of the wet and dry seasons respectively. This, along with this other data and literature has led the German Society for International Cooperation (GIZ) in partnership with the Nigerian government to highlight solar as a rapidly dispersible power generation source [8]. Numerous studies of varying depth and detail have looked at the hydro potential of Nigeria as a whole, an assessment of available data and literature showed that of a total 277 potential SHP sites 61 were located within Niger State. Throughout Nigeria, biomass is being burnt as a cooking fuel [9]. As detailed above, both solar and hydro are viable options for water heating (solar thermal) and power generation. Niger State was chosen as location for a case-study of integrated low carbon energy solutions for remote rural areas of the developing world.

The primary objectives of the study were (a) to propose a low cost, low emissions, renewable alternative cooking fuel to charcoal for the remote rural population of Niger State, (b) to embed low carbon technologies to the proposed process's associated heat and power demand, (c) to provide process design characteristics variable scale for community use, and (d) to determine associated costs by making use of local materials and low maintenance requirements adjusted to local population constraints.

2 Experimental method

2.1 Selection of technology according to landscape and scale

The viability of a given technology in a given area is dependent upon certain factors; suitability of scale, availability of feedstock/input, geography and geology of the area, average sunlight/winds speed/rainfall. All of these factors were considered approaching the design in order to provide a flexible and reliable energy solution for as broader range of situations as possible. A simplified yes/no method has been created to determine the capability of a technique suitable to Niger State's energy needs. Furthermore the scalability of each technology, from domestic to regional use, determines its potential in rural and remote areas. For both tables a legend has been created for ease of understanding and uniformity throughout. A tick merely defines the likely probability of this technology working for the suggested situation, whilst a cross is an unlikely or impractical use. A question mark defines a situation information is lacking so a definite yes/no decision cannot be made; or the use of this technology is situational, such as seasonal changes, economics or no examples currently exist.

Table 1. Applicability of each representative technology in a landscape.

	Hydro	Wind	AD	Solar	Tidal	Geother
Flood Plains	✓	✓	✓	✓	✗	?
Plains	✗	✓	✓	✓	✗	?
Hills	✓	✓	✓	✓	✗	?
Mountains	✓	✓	✓	✓	✗	?
Tundra	✓	✓	✗	✓	✗	?
Dessert (HOT)	✗	?	?	✓	✗	?
Dessert (COLD)	✗	✓	✗	✓	✗	?
Savanah	✓	?	✓	✓	✗	?
Jungle	✓	✗	✓	✗	✗	?
Forest	✓	✗	✓	✗	✗	?
Marsh/wetlands	?	?	✓	✓	✗	?
Delta/River	✓	✓	✓	✓	✓	?
Coastline	✓	✓	✓	✓	✓	?
Atoll	✗	✓	✓	✓	✓	?
Volcanic	✗	✓	?	✓	✗	✓

Table 2. Applicability and scalability of each representative technology.

Applicable use of each technology based upon home powerage		Scale					
		Household	Multiple House	Community	Multiple community	Urban centre	Regional
Tech	Pico Hydro	✓	✗	✗	✗	✗	✗
	Micro Hydro	✓	✓	✓	✗	✗	✗
	Small Hydro	✗	✗	✓	✓	✗	✗
	Hydro (+10MW)	✗	✗	✗	✗	✓	✓
	Tidal	?	✓	✓	✓	✓	✓
	Solar	✓	✓	✓	✓	✓	✓
	Solar Tower*	✗	✗	✗	✗	✓	✓
	Small wind	✓	✓	✗	✗	✗	✗
	Wind	✓	✓	✓	✓	✓	✓
	AD	✓	✓	✓	✗	✓	✓
	Biofuels	✗	✗	✗	✗	✗	✓
	MSW	✗	✗	✗	✗	✓	✓

2.2 Quality function deployment

Quality function deployment (QFD) is a tool used to relate customer requirements to the technical aspects of the design. There are four main aspects to the QFD. The customer requirements are shown on the left hand side of the house of quality (HOQ). This is a list of all that the customer requires from the design e.g. clean biogas. The customer requirements often are produced through interviews, however the customer requirements in this study have been discovered through literature. The technical requirements are listed along the top of the HOQ. The technical requirements should be measureable and should be solutions to the customer's requirements e.g. size and volume. An additional row is added to show if the technical requirement needs to be maximized or minimized. The main block of the house contains the inter-relationships between the customer requirements and technical requirements. The relationships are described as either direct, strong, weak or indirect.

In out QFD diagram (available as supplementary data), these relationships have been colour coded to make for ease of viewing. The roof of the HOQ allows links to be formed between the technical aspects of the design. The question is asked 'how does one requirement affect the performance of another?' The relationship maybe positive such as the relationship between corrosion resistance and material lifetimes. Negative such as the relationship between simplicity and the size and volume of the system, or no relationship such as the relationship between materials used and the plant location.

2.3 Objectives tree

The objectives tree is the third technique used in the present study to produce the best low carbon energy system for Niger State's rural population's alternative cooking fuel. Starting with the top level requirement of 'Integrated rural energy solution: community/domestic scale anaerobic digestion – solar PVT/thermal system', the tree is split into functional requirements and operational improvements. The functional requirements are highlighted in red and the operational requirements are blue, again to make for ease of viewing (see supplementary data). The functional requirements are then spilt into providing a range of solutions to meet the requirement e.g. mesophilic or thermophilic for a microorganism population. The operational improvements are however more specific and are more of an aim e.g. consistent temperature. Within the objectives tree the shape of the design begins to take place, allowing ideas to flow about what needs to be included in the design to meet customer needs. The importance of a simple design became incredibly apparent during the objectives tree. A simple design leads to simple maintenance, repair and installation.

3 Results and Discussion

Using the methodologies described in 2 to the geographical area of Niger State (Savanah and flood-planes, Table 2), and to small community application representative of remote agricultural populations (Table 3), two energy resources stand out as suitable low carbon energy solutions: anaerobic digestion and solar. Added too these the outcomes of the QFD and Objective tree, the proposed solution to the low carbon energy challenge within Niger State's provision of alternative cooking fuel converged to proposing biogas generated by an integrated solar Photovoltaic-Thermal (PVT) and Anaerobic Digestion (AD) unit, using mesophilic technology for the community scale. The AD system was characterised by a fixed volume corresponding to communities of up to 15 households, an agitator powered by the electricity from the solar PVT, a heating system provided by the solar PVT, an inlet and outlet, mesh in order to keep floating biomass submerged, thermometer, entrance hatch and a H_2S scrubber. Community scale AD plant (ADP) cost is fundamentally dependent upon integration of solar PVT and volumetric capacity. Total biogas production is predominantly affected by feedstock type and availability.

3.1. Design of a community anaerobic digester

The AD tank is 20 m^3 in volume, it is designed to have flat walls yet a rounded bottom with no sharp corners to minimize sludge accumulation in corners. The AD tank is to be made from bricks and cement with a polyethylene (PE) lining. The materials are widely available as well as being economically affordable. The materials provide the best solution regarding simplicity, economically affordability and longevity of the system.

The digestion medium will be agitated in order to increase the efficiency of the AD system. The medium is also held down by a mesh to prevent floating waste from not being digested. The agitator will be powered by the electricity produced by solar PVT and should therefore not require any external electricity to run.

The solar PVT is also providing hot water to the anaerobic digester. Through a series of copper pipes hot water will be distributed throughout the system. The hot copper pipes will aim to keep the medium at 37°C, and thus running at optimum efficiency. The temperature of the digester will be monitored by a thermometer displayed at the top of the system.

3.1.1. Material requirements

Typically AD reactors, in an attempt to improve deployability and reduce cost, utilize poured concrete, brick and stone materials often resulting in enhanced porosity and gas-loss over time [11]. In addition, fixed-dome designs may incorporate reinforced-concrete, fiberglass-reinforced-polyester and steel with additional internal mortar layers applied so as to maintain anaerobic-conditions within the ADP [11]. Concrete is applied due to ease of application, capacity for air-retention and water-retention, and resistance to thermal inertia [12].

Traditionally, alkaline concrete is directly exposed to the solid-liquid phase of the digestion medium resulting in corrosive deterioration through biogenic sulphuric-acid generation [13]. Corrosive decline of the structure may be maintained at 1mm/year [13], dependent upon internal and operational conditions. In addition, volatile fatty acids and ammonia accumulation within the solid/liquid phase may contribute in "concrete leeching" [13] with dissolved-CO_2 generation [11] may be considered a source of concrete carbonization [12]. Such processes result in reduced efficiency and costly repair-works with risk of gas liberation and solid/liquid discharge within the adjacent environment [11].

3.1.2. Biogas scrubbing

Of all the impurities commonly present in biogas H_2S is the most pressing when used as a energy solution in this fashion, used as an alternative cooking fuel it would be damaging to health and co-ignited in a diesel engine it would cause rapid engine wear.

H_2S remediation may be achieved through conversion to sulphurous-acid via passage between a solid-filter-pack of iron-wool, the transfer conversion of iron-wool to Ferric-Sulphide (FeS) negates the presence of detrimental H_2S within the biogas-compound [14]. In addition, limitations in iron-wool surface area inhibits conversion potential, as such low-cost wood-chippings with iron-oxide (Fe_2O_3) may be applied so as to enhance surface-area ratios [14]. Approximately 20 g of H_2S may be remediated per 100g of wood-chipping iron-oxide filter-pack [15].

3.1.3. Agitation

The application of mixing is established as a method of improved operational performance within ADP's [16]. Advantageous performance is achieved through augmented nutrient-distribution and digestion medium inoculation; uniformity of heating; negation of scum and foam formation; biogas bubble release and disturbance of feedstock sedimentation [17]. Rates of mixing tend to be low, usually within the range of 15–50 *rpm,* a mechanical propeller type stirrer is identified for universal application [17].

3.1.4. Shape and size

ADP operational size is determined as a function of feedstock volume availability (kg), biogas potential (m^3/kg) and consumption demand (m^3/kg) specific to the location of deployment [18,19]. Attainment of ADP size adequacy should be conceived with consideration for capital and labour expense associated with excessive construct volume and material investment [18]. In addition, variability in the presumed feedstock capacity for biogas generation [20] should be considered when assessing the volumetric dimensions of presumed systems [19]. A linear relationship may be assumed between feedstock mass, supportable population and integrated ADP volume (m^3) [18]. In addition, ADP size

 DOI: 10.1051/e3sconf/20172200025

maintains a direct impact upon land-use patterns, fresh-water investment and heating intensity.

A numerical estimation of consumer population (Þ), per ADP device, is determined a prerequisite when considering total biogas demand and operational plant size [19]. The total daily-biogas requirement (G_r) per consumer population (Eq.1) is initially considered in volumetric ADP design.

Notable variation in average household size is also presented with the greatest proportion of people living within 8 person households (*Þ*) across Niger State [21]. Cooking requirements per inhabitant (*C*) are estimated to be 0.227m^3 [22,23] with lighting requirement per household (*L*) within the region of 0.125m^3 per gas-lamp per hour ($L_1 \cdot L_2$) [22]. Assuming lighting requirements of 2 lighting hours per day the total estimated biogas demand is 2.09m^3. The estimation of 2m^3 biogas demand is equivalent to 1805.11 kg/year and 722.05 kg/year of firewood and charcoal respectively [22].

A total biogas potential (A) is dependent upon both number of livestock (z), mass of feedstock production per day (kg) (z_1), and biogas potential per unit of substrate (m^3/kg) (x) [22, 23]. A total of 7 head of cattle per household, a dung production rate of 10kg per day [22] and biogas potential of 0.03m^3/kg is (equation 1). As such a total biogas potential of 2.10m^3 is estimated for this case study.

$$G_r = [(Þ \times C) + (L \times L_1 \times L_2)] \tag{1}$$

An estimation of digester liquid phase volume (V_d) (equations 2 & 3) is calculated for dry matter feedstock proportion (*M*) of *10%* [24], a dry feedstock density (Ψ) of 50kg/m^3 [22] and a hydraulic retention time (t_r) of 30 days [22]. As such a total digester volume of 4.18m^3 would be required for the liquid phase proportion of the ADP structure. An additional 10% volume size is determined for gas generation disengagement over time [22] suggesting a minimum ADP volumetric capacity of 4.598m^3.

$$A = (z \times z_1 \times x) \tag{2}$$

$$V_d = \frac{\left[(Þ \times C) + \frac{(L \times L_1 \times L_2)}{x}\right]}{\Psi} \times t_r \tag{3}$$

The proposed set up of the integrated solution is shown above in Figure 1, the biogas scrubbing unit, which on the diagram is on the reactor but would be a small and mobile unit. The submerged piping serves to distribute heat from the solar PVT around the reactor.

3.1.5 PVT specifications

The specifications of the solar PVT system associated with the chosen design are defined here to determine the required coverage according to reactor size. Reactor size represents the rough estimate of the water content of the reactor; Coverage required refers to the area to be covered with the PVT panels and is determined using Equation 4. The required temperature is lower than ideal because a 1°C rise leads to a 50% increase in required coverage.

$$CR = (RT - AT) * Erq * (RS*1000) \tag{4}$$

Parameters in Eq. 4 are defined in Table 3, along with their values evaluated for the chosen design.

3.2 Estimating Initial Capital Investment for Community Scale Anaerobic Digester

Significant capital cost (Cc), associated with large-scale systems, is identified as a characteristic inhibiting ADP deployment. However, in terms of energy production capacity, ADP systems present the most cost effective renewable energy resource, an aspect of paramount importance in developing nations. Enhancement in operational volume is synonymous with increase in capital-cost. As previously suggested, volume requirement is the result of biogas potential, feedstock supply, density and hydraulic retention time. The development of community-scale ADP, and associated capital-cost, is therefore dependent upon feedstock typology and availability, biogas demand and hydraulic retention time.

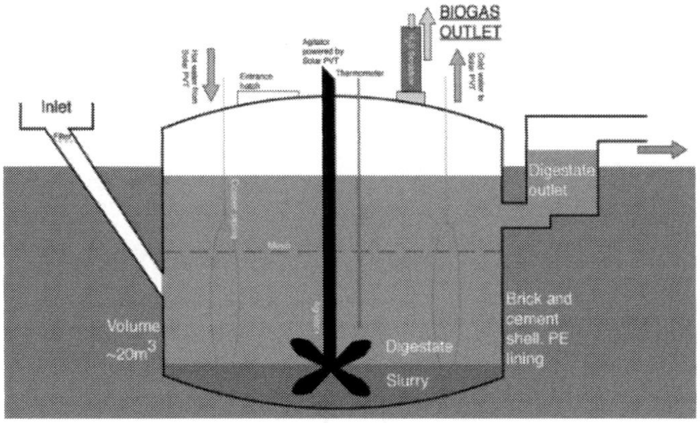

Fig. 1. Community scale design for an AD plant using solar thermal.

Table 3. Specifications of the solar PVT system.

Energy to raise 1 litre of water by 1°C	**0.001161**	**kWh**	**Erq**
Reactor Size	15	m	RS
Annual average temp (2015)	27.4	°C	AT
Required temp	29	°C	RT
Energy required	27.9	kWh	
Nigeria average solar radiation	7	kWh/m^2	
Expected eff. of PVT	0.45		
Daily generation per m^2	3.15	kWh	
Coverage required	8.8	m^2	CR

An estimation of the volumetric ADP requirement was established for varying numbers of community consumers (Equations 2 and 3). An estimated cost formula was developed integrating fundamental requirement and material volume for an ADP system of varying size. Results indicate variation in initial capital investment depending upon integration of solar PVT and volume of ADP digester. Estimations suggest a community of 24 people would require a reactor size of 12.7 m^3 at a cost of 4613.77 US$ when PVT is integrated. Removing the PTV system significantly reduces the cost to 1338.79 US$ for a system of equal volume. In addition, the total initial capital investment is shown to reduce, in proportion to ADP volume, with an increase in size.

4 Conclusion

The proposed solution to provision of clean alternative cooking fuel to solid biomass for remote communities in Niger State is biogas generated by anaerobic digestion mesophilic technology with digester temperature control as well as agitator power provided by hybrid solar PV-T panel. The system also features a design allowing flexible range of digester volume, an inlet and outlet, an internal mesh keeping floating biomass submerged, a thermometer, an entrance hatch and a wood chipping iron oxide based H_2S scrubber, and is made of local construction materials. Total biogas production is predominantly affected by feedstock type and availability. The solution has potential to save lives and increase life expectancy by reducing the amount of harmful gases inhaled by cooking with poor quality fuels in inefficient cook stoves.

Acknowledgments

The EPSRC is gratefully acknowledged for GCRF grant EP/P51097X/1 (Integrated Low Carbon Energy solutions for Remote Rural Area project) and EP/L014912/1 (CDT bioenergy) programme at The University of Leeds. Supplementary data available at http://eprints.whiterose.ac.uk/ under title name of paper.

References

1. United Nations, Department of Economic and Social Affairs. Population division. (2015)
2. A.C. Ohajianya, O.E. Abumere, I.O. Owate, E. Osarolube, Int. J. Eng. Sci. Inv. **3**, 7 (2014)
3. IEA. Nigeria: Balances for 2014 (2014)
4. The World Bank. Population Total (2016)
5. I.U. Abubakar, M.A. Yamusa, Int. J. Agri. Food. Sci. Tech. **4**, 3 (2013)
6. O.A. Olayo, J. Food. Sec. **2**, 1 (2014)
7. V.O. Oladokun, O.C. Asemota, Renew. Sust. Energy. Rev. **52** (2015)
8. K. Ley, J. Gaines, A. Ghatikar, GmbH (GIZ) (2014)
9. IEA, Statistics (2016)
10. IEA, National Energy Policy (2016)
11. C. Gunnerson, D. Stuckey, World Bank Technical Paper **49** (1986)
12. C. Voegel, A. Bertron, B. Erable, Sci. Total. Environ. **571** (2016)
13. A. Koenig, F. Dehn, Biosys. Eng. **147** (2016)
14. A. Agarwal, M. Shukla, Int. J. Oil. Gas. Coal. T. **2**, 4 (2009)
15. K. Krich, D. Augenstein, J. Batmale, J. Benemann, B. Rutledge, D. Salour, Clean Concepts (2005)
16. D. Stafford, Biomass **2**, 1 (1982)
17. A. Lemmer, H. Naegele, J. Songermann, Energies **6**, (2013)
18. K. Singh, S. Sooch, Energ. Convers. Manage. **45**, 9–10 (2004)
19. S. Sakar, K. Yetilmezsoy, E. Kocak, Waste. Manage. Res. **27** (2009)
20. L. Makai, M. Molinas, GHTC (2013)
21. National Population Commission (Nigeria) (2006)

22. G. Rai, *Non-Conventional Energy Resources* (New Delhi, 2002)
23. E. Alokore, B. Ebangu, Agri. Eng. Int. **15**, 1 (2013)
24. T. Pullen, *Anaerobic Digestion, Making Biogas, Making Energy* (The Earthscan Expert Guide, Routledge, 2015)

Acid emissions monitoring needs in ceramic tile industry: challenges derived from new policy trends

Irina Celades[1,*], *Salvador* Gomar[1], *Fernando* Romero[1], *Amisha* Chauhan[2], *Bertrand* Delpech[2], and *Hussam* Jouhara[2]

[1]Instituto de Tecnología Cerámica - Asociación de Investigación de las Industrias Cerámicas - Universitat Jaume I, Av. Vicent Sos Baynat, s/n, E12006, Castellón, Spain
[2]RCUK Centre for Sustainable Energy Use in Food Chains, Institute of Energy Futures, College of Engineering, Design and Physical Sciences, Brunel University, London, Uxbridge, Middlesex, UB8 3PH, UK

Abstract. The emission of acid compounds during the manufacture of ceramic tiles is strongly related to the presence of precursors in the raw materials and/or fuels used, with some exceptions such as the production of thermal NO_X. The stages with the potential to produce significant emissions of these compounds have been identified as the suspension spray drying and tile firing stages. The monitoring of emission levels of acid pollutants in these stages has turned in a great importance issue from a regulatory and industrial aspect. The DREAM project (https://www.spire2030.eu/dream) will tackle the regulation of acidic emissions focusing in the firing stage. The initial stages of the project have made it possible to identify the design requirements for the monitoring system. This will allow the control of acid pollutants emissions and other key parameters such as pressure, flow, temperature and humidity. One of the tasks developed has been the review and compilation of current emissions monitoring systems detailing technical specifications such as: position (in situ or extractive), measurement principle and frequency. The future policy trends in air pollution are encouraging the continuous monitoring across the European industry. The present document assesses the advantages regarding environmental impact control, highlighting the main challenges for the ceramic tile industry.

* Corresponding author: irina.celades@itc.uji.es

DOI: 10.1051/e3sconf/20172200026

1 Background

Emission of acid compounds during the manufacture of ceramic tiles is strongly dependant to the presence of precursors in the raw materials and/or fuels used, with some exceptions such as the production of thermal NO_X. Therefore, these compounds can be generated in the stages where combustion processes take place and/or where the processed materials reach high temperatures. Consequently, the stages that could have significant emissions of these compounds are the suspension spray drying and, mainly, tile firing stages [1].

In the specific case of the ceramic kilns, the emission of acid compounds entails the necessity to treat and monitor the exhaust gases emitted to ensure the accomplishment of the Emission Limit Values (ELVs) applied.

As the ceramic industry is affected by the Industrial Emissions Directive (known as IED), the Best Available Techniques (BATs) and ELVs associated, are reviewed periodically [2]. Next review will start during 2018 and could imply a tightening of the current ELVs associated to BATs and/or including new pollutants or new requirements regarding industrial emissions monitoring.

This is in part the rationale for DREAM, a European project funded by the EC through the SPIRE-H2020 Program, which aims to design, develop and demonstrate a radically improved architecture for ceramic industrial furnaces which include reduced emissions obtained by substantially enhancing emissions abatement system, beyond pollutants and ELV applied nowadays.

1.1 European Regulatory Framework BREF

In the EU, the application of IED obliges companies of the Directive to obtain an Integrated Environmental Authorisation (IEA), which includes ELV for different atmospheric pollutants and per manufacturing stages as shown in annex 1. Public authorities set permit conditions and fix the ELV applicable for each installation with regards to the Best Available Techniques (BAT). The proceedings have been implemented by European Commission as a EU Decision, a document containing the parts of a BAT reference document, their description, information to assess their applicability, the emission levels associated with the best available techniques (ELV-BAT), associated monitoring, associated consumption levels and, where appropriate, relevant site remediation measures.

The BAT findings from the Ceramic manufacture process have not yet been published or categorised under a directive, so 2007 BREF should be used. In near future, it is expected a revised and updated version of ELV will be produced with regards to the existing IED directive.

Table 1 highlights the emission values defined for firing stage in ceramic tile manufacture in the EU as ELV-BAT, the values are considered as a maximum reference for fixing ELV in national IEA.

The IED dictates the required monitoring methods and requires sampling and analysis conducted on all polluting substances as per CEN-standards. Alternative standards apply if CEN standards are not available such as ISO, national or other international standards with an equal quality and scientific integrity.

Table 1. ELV-BAT in firing stage in the BREF for ceramic tile manufacture, in mg/Nm^3: concentration in standard conditions: 273 K and 101,3 kPa, as a daily average value.

Parameter	ELV-BAT [1)]
Fluoride stated as HF	1–5[2)]
Chloride stated as HCl	1–30[3)]
NO_X, stated as NO_2 for kiln gas temperatures < 1300°C	< 250
NO_X, stated as NO_2 for kiln gas temperatures > 1300°C	< 500
SO_X stated as SO_2. Sulphur content in raw material < 0.25%	< 500
SO_X stated as SO_2. Sulphur content in raw material > 0.25%	500–2000[4)]

[1)] The ranges depend on the content of the pollutant (precursor) in the raw materials, i.e. for firing processes of ceramic products with a low content of the pollutant (precursor) in the raw materials, lower levels within the range are BAT and for firing processes of ceramic products with a high content of the pollutant (precursor) in the raw materials, higher levels within the range are ELV BAT.
[2)] The higher BAT level can be lower depending on the characteristics of the raw material.
[3)] The higher BAT level can be lower depending on the characteristics of the raw material. Also, the higher ELV BAT should not prevent the re-use of waste water.
[4)] The higher BAT level only applies to raw material with an extremely high sulphur content.

1.2 Acid compound emissions in the ceramic industry

The main acidic compounds are released during the firing of ceramic tiles and are identified as: HF, HCl, SO_X and NO_X. Among them, fluorine compounds are present within exhaust streams and are characterised as ceramic industry pollutants. The concentration of fluorine based emission stems from the presence and level of fluorine ions in clays used as raw materials in ceramic tile manufacture.

Fluorine ions replace OH- groups in the crystalline mica structures, and other present clay minerals such as montmorillonite, illite, kaolinite, etc. [3, 4]. Fluorine compound emissions usually start with the dehydroxylation of these minerals at temperatures of the order of 500–700°C [5–7]. The major compounds that form are hydrofluoric acid, silicon tetrafluoride and, to a lesser extent, alkaline fluorides in particle form, the presence of these alkaline fluorides being practically negligible [8-10]. In the presence of water vapour, a typical situation in industrial combustion kilns, fluorine is mainly emitted as hydrofluoric acid [11].

The formation of chlorine based emissions mainly stem from the presence of chlorine ions in the water used during the manufacturing process. Many clays and mixtures contain trace levels of chlorine. The production of chlorine based emissions occur during the firing process at temperatures above 850°C, from the decomposition of chlorine-containing mineral salts. In addition, the decomposition of organic compounds that contain chlorine leads to HCl emissions in the 450–550°C range [12].

On the other hand, sulphur based emissions stem from the sulphur content in the raw materials and the type of fuel used. The clays used in ceramic tile manufacture can contain sulphites in the form of pyrite or calcium and magnesium sulphates such as gypsum, and sulphur in organic compounds. Fossil fuels can also generate sulphur emissions. Natural gas is the most widely used fuel, and contains practically no sulphur in its composition; however, if fuel oil, coal or coke is used, sulphur emissions can be higher [12].

Nitrogen compounds are emitted in the form of nitrogen oxides, and are related to the formation of thermal NO_X in high-temperature processes when a reaction takes place between

 DOI: 10.1051/e3sconf/20172200026

nitrogen and oxygen in the combustion air. This reaction is encouraged in processes that unfold at high temperature (particularly at temperatures above 1400°C).

However, NO_X formation can even be significant at process temperatures below 1200°C, when the burners run at high flame temperatures [12]. The decomposition and combustion of nitrogen compounds present in the raw materials, admixtures, or fuels can also be a source of NO_X at lower temperatures.

1.3 Current acid pollutant emissions in ceramic tile manufacture

Previous studies conducted in the ceramic sector showed that main fluoride and chloride compounds mainly emitted as HF and HCl, respectively, are the most significant due to the presence of H_2O in industrial kiln. In many cases emissions exceeds the currently proposed ELV–BAT, Figure 1 [1]. Therefore, according to these results and considering ELVs from BREF, many companies need to introduce preventive and/or corrective measures to reduce HF emissions in the kilns. In contrast, in all cases studied, the SO_2 and NO_X emissions lie below the ELV–BAT in firing.

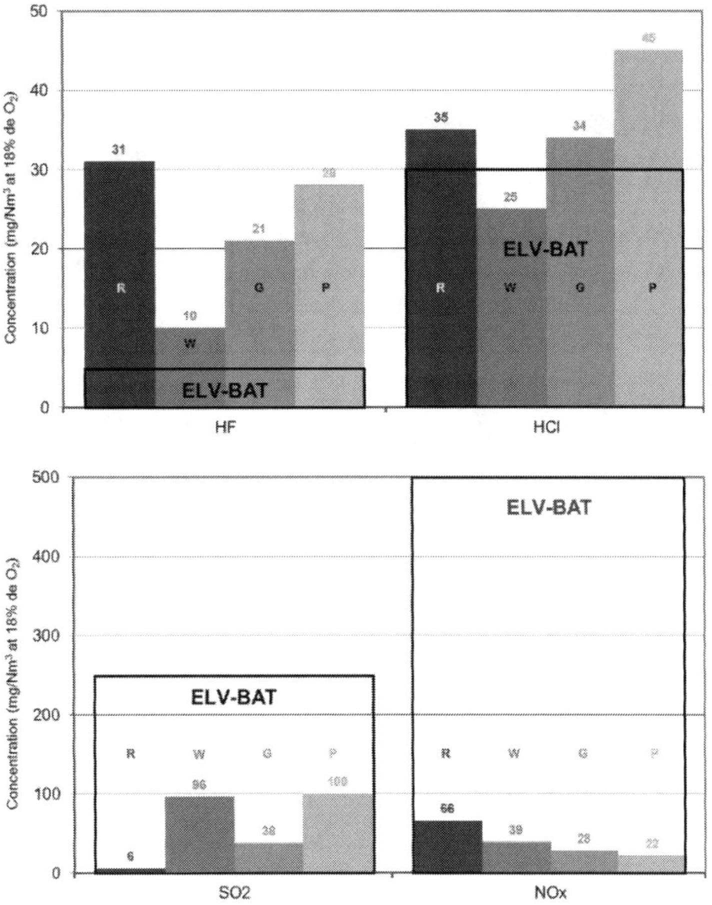

Fig. 1. Median acid pollutant concentrations for earthenware (red (R) and white (W)-firing body), glazed stoneware tile (G) and porcelain stoneware tile (P) in the firing stage.

Acid emissions vary considerably but not due to the variation in acid compound contents in the raw material, but also because of differences in the manufacturing technology used. During firing stage, factors such as: larger shapes, sizes and thickness, technological

DOI: 10.1051/e3sconf/20172200026

advancements, increasing personalization/customization trends, etc., are all major influencers in the variation of emissions. All these new product developments make it necessary to update the acid emission levels and documentation in ceramic tile firing.

2 Acid emissions measurement methods in channelled emissions from ceramic tile industry

Under this legislative framework and the current emissions regulations, monitoring is a key issue to consider regarding legislation accomplishment.

Stack emissions monitoring approaches can be classified into two types: periodic monitoring and continuous emissions monitoring systems (CEMS). One approach is not inherently superior to the other; both have their own strengths and weaknesses depending upon the application. In general, it is recognised that CEMS provide increased confidence for both regulatory purposes and process control [13].

2.1 Periodic measurements

Periodic measurements can be done through manual systems or automatic systems. The manual systems are single measurement methods in which a representative stack sample is extracted and analysed by logical sequences operations, next to the stack or at laboratory. By the other hand, automatic systems analyse pollutants through a measurement device that obtains gas stream concentration.

Even if the manual method looks to be quite simple, several factors can strongly influence the final results. The competencies and knowledge of the accredited laboratory operators performing the SRM measurements have to be carefully checked.

Moreover, the currently required periodic measurement, does not represent a realistic value of the environmental impact generated in the firing stage, due to the high temporal variability of the processed material, as it has been mentioned above.

The measurement methodology used to determine the acid gaseous pollutants emitted in the ceramic industry is described in various specific test standards, outlined in Table 2. As described below, these methodologies include manual and automatic measurement methods, ISO or EN standard. The EN standards produced in Table 2 are considered as Standard Reference Method (SRM) by CEN (the European Committee for Standardization).

Table 2. Standards on air pollutant measurements.

Pollutant	Reference standard	Period	SRM (CEN)
Chlorides stated as HCl	EN 1911	Manual	Yes
Fluorides stated as HF	ISO 15713	Manual	No
SO_2	EN 14791	Manual	Yes
NO_X stated as NO_2	EN 14792	Automatic	Yes

The European SRM for HCl measurement describes three alternative analytical techniques for determining gaseous chlorides content emitting to atmosphere from ducts and stacks. All compounds which are volatile at the temperature of filtration and produce chloride ions upon dissolution during sampling are measured by this method, which gives therefore the volatile inorganic chlorides content of the waste gas. This European Standard can be used as a SRM; provided the overall uncertainty of the method is less than 30% relative at the daily Emission Limit Value (ELV) for incineration and large combustion plants or at the ELV prescribed by the specific regulations for other plants.

On the other hand, HF can be measured in a similar way to HCl as described in an ISO standard, because CEN has not published a specific manual method for HF. This method does not determine the overall uncertainty of the method but it is considered the same uncertainty as HCl (30%) due to similar reactive nature and measurement technique [13].

Both methods may not reflect the chemical nature of the compounds, typically these gaseous chlorides and fluorides are measured manually at the filtration system. The drawback to such a measuring method is the lack of identification of hydrogen chloride (HCl) or hydrogen fluoride (HF), and are passed through the filtration system.

For SO_2, SRM is a manual method, whereas that for NO_X the SRM is an automatic analyser using chemiluminescence absorption principle. Both methods, manual for SO_2 and automatic for NO_X, are specifics for SO_2, NO and NO_2, without complications with other species. In fact, it is possible the use of automatic devices for SO_2 when the equivalence with the SRM has been demonstrated. The uncertainty regarding to SO_2 and NO_X are less than 20% and 10%, respectively.

Compliance assessment monitoring for SO_2 and NO_X are not problematic due to their emissions level values in ceramic kiln, contrary to what happens with chlorides and fluorides. For this reason, the rest of the article focuses in the selection of the automated measurement system best value for money for HF and HCl.

2.2 Continuous emission measurement systems (CEMS)

The on-line measurement of a pollutant through a CEMS enables systematic monitoring of fluctuations within an exhaust stream. On-line measurements are increasingly becoming of great interest, since this allows the observation of fluctuations in the emission with across a specific time frame. The study of fluctuating emissions allows a direct relationship to be made about changes occurring in the manufacturing process, or the addition of emissions treatment. The addition of online systems provides information for appropriate decision-taking and correct eco-management, for instance by optimizing emission treatment systems.

There is no legal requirement to implement a CEMS in the ceramic industry but the installation of CEMS is in the best interest of the ceramic industry. For example, a better knowledge of HF emissions behaviour has been obtained using continuous monitoring device adapted to the characteristics of the ceramic industry [1, 5, 6, 10], showing that:

- HF emissions are continuous and constant in time, their value depending mainly on the mass flow and the type of product being processed. This always happen if the ceramic kiln is running in a steady state.
- Modifications in realistic industrial manufacturing conditions, such as in heat treatment and in the dry bulk density of the tiles produced, does not significantly reduce HF emissions.
- Tile glazing, performed to provide certain surface properties, effectively reduces HF emissions.

The measurement of HCl and HF by CEMS is technically difficult due to the amount of specialised systems. The systems have been created to identify specific chlorides but this does not include other inorganic chlorides or fluorides. As the manual method measures total inorganic salts of chloride and fluoride, they are measuring the pollutants specified in the BAT conclusions.

The measuring technology for the HF and HCl in CEMS are based on optical technology and mainly on infrared absorption method. The attenuation of infrared light passing through a sample cell is a measure of the concentration of gas of interest in the cell, according to the Lambert- Beer law. Not only HCl or HF but also most hetero-atomic molecules absorb infrared light, in particular, water and CO_2 have broad bands that can interfere with the measurement of HCl or HF. Different technical solutions have been developed to suppress cross-sensitivity, instability and drift in order to design automatic monitoring systems with acceptable properties. For instance: Gas Filter Correlation (GFC), Tunable Diode Laser (TDL), Fourier Transform Infrared Spectroscopy (FTIR) and newer developments like: Optical Feedback Cavity Enhanced Absorption Spectroscopy (OFCEAS) and Differential Optical Absorption Spectroscopy (DOAS).

All CEMS have to be tested prior to their installation on site in order to prove that they are suitable for the intended monitoring purposes. The tests are performed by an accredited laboratory in accordance with the EN 15267, part 3 and part 4, and the tests results are summarized in a QAL1 certificate. Individual contributors to the total measuring uncertainty are identified and quantified.

The Certified Measuring Range (CMR) can be no higher than 2.5 times the daily ELV for large combustion plants (1.5 times the daily ELV for waste incinerators). The global uncertainty, expressed as a % of the CMR, must be less than 75% of the maximum uncertainty allowed by the IED. The remaining 25% is left for other uncertainty sources such as the variation of the uncertainties between two CEMS of the same type, the representativeness of the sampling point, the peripheral measurements like temperature, O_2, etc.

For HCl and HF, the maximum uncertainty allowed by the IED is 40% of the daily ELV. Consequently, the global uncertainty, calculated on the basis of the QAL1 certificate for HF and HCl CEMS is 30% of the ELV given in mg/Nm^3.

In conclusion, compared with limits value applied in EU, HF and HCl emissions need to be corrected by appropriate cleaning systems before such emissions are released into the air. According to the BREF ceramic manufacture industry, for firing ceramic tiles fluorides and chlorides have to be tested using periodic methods. The fact that on-line measurement has not been considered as a legal requirement in ceramic industry, it is recognised that continuous monitoring provide increased confidence for both regulatory purposes and process control, in general.

From the process control point of view, the use of CEMS is of great interest because supply information which could relate emissions at stacks with operating parameters of the kiln, raw materials or of cleaning system.

From the regulatory purposes point of view, the use of CEMS strengthens the control on the cleaning system used and offers better levels of confidence and transparency of the environmental behaviour to the public authority and the general society. It is necessary to consider aspects such as operation activities related to quality assurance schemes (via EN 14181), as it is technically challenging to generate certified reference material for checking monitoring system's integrity for acid compounds.

In general, the economic impact derived from the use of CEMS is highly relevant for the industry, considering the investment required, the maintenance and calibration costs. In the ceramic tile industry, a medium size facility may have between 2 to 4 kilns. So, this additional cost should be considered by public authorities and companies during regulations development adjusting control necessities to the real environmental impact.

This project has received funding from the European Union's Horizon 2020 research and innovation programme under grant agreement No 723641.

References

1. E. Monfort, I. Celades, S. Gomar, F. Rueda, J. Martinez, Bol. La Soc. Esp. Ceram. Y Vidr. **50**, 179–184 (2011)
2. Directive 2010/75/EU, (2010).

3. B. FABBRI, Cerám. Inf. 33–36 (1992)
4. J. De Reymer, A. Jong, Castellón Faenza Ed. Ibérica, 1071–1076 (1993)
5. E. Monfort, J. García-Ten, I. Celades, M.F. Gazulla, S. Gomar, Appl. Clay Sci. **38**, 250–258 (2008)
6. J. García-Ten, E. Monfort, P. Gomez, S. Gomar, J. Ceram. Process. Res. **7**, 75–82 (2006)
7. S.J. Chipera, D.L. Bish, Clays Clay Miner. **50**, 38–46 (2002).
8. E. Monfort, I. Celades, S. Gomar, M.F. Gazulla, H. Adams, J. Tulipe, Ind. Ceram. Verriere. 30–36 (2005)
9. E. Monfort, I. Celades, S. Gomar, M.F. Gazulla, H. Adams, J. Tulip, Ind. Ceram. Verriere. 50–56 (2004)
10. E. Monfort, J. García-Ten, I. Celades, S. Gomar, J. Fluor. Chem. **131**, 6–12 (2010)
11. E. Monfort, M.F. Gazulla, I. Celades, P. Gomez, M. Bigi, M. Tonelli, Am. Ceram. Soc. Bull. **82**, 31–35 (2003)
12. E.I. Bureau, *Reference document on best available techniques in the ceramic manufacturing industry*, (2007)
13. E. Agency, *Technical Guidance Note M2*, www.mcerts.net

Industrial applications of the air direct-contact, gravel, ground heat exchanger

Wojciech Cepiński[1,*], *Maciej* Besler[1,†]

[1]Wrocław University of Science and Technology, Faculty of Environmental Engineering, Department of Air Conditioning, Heating, Gas Engineering and Air Protection, ul. Norwida 4/6, 50-373 Wrocław, Poland

Abstract. The paper describes the analysis of possibility of using the air direct-contact, gravel, ground heat exchanger (Polish acronym BGWCiM), patented at the Wroclaw University of Science and Technology to prepare air for conditioning rooms in the industry. Indicated the industry sectors where the application may be the most beneficial.

1 Introduction

1.1 Reasons for raising the issue

The urgent need to save energy obliges to more widespread use of solutions enhancing generation of natural energy which exists in nature and is easily accessible. One of methods to use it is the air direct-contact, gravel, ground heat ex-changer (Polish acronym BGWCiM), tested thoroughly at the Wroclaw University of Science and Technology and the advantages of which will be presented. Industry is one of the most important sectors of a national economy. It is worth mentioning that together with agriculture and construction, industry consumes nearly half of all energy produced. Heavy industry has got an important share. In the processing industry, most of the energy is consumed by the food, chemical, mineral, steel and paper industries [7]. Greater efficiency in the use of energy in the economy, and especially in the industrial sector, has got a significant impact on production costs, profits of enterprises, as well as competitiveness of their products on the world market. It also contributes to faster development of enterprises.

1.2 The air direct-contact, gravel, ground heat exchanger

For many years, research has been carried out at the Institute of Air Conditioning and District Heating of the Wroclaw University of Science and Technology on efficient capture of natural heat (and also cooling) from a small depth of ground, in the air direct-contact, gravel, ground heat exchanger for ventilation and air conditioning purposes [1–3].

* Corresponding author: wojciech.cepinski@pwr.edu.pl

DOI: 10.1051/e3sconf/20172200027

In the exchanger, the outside air is led horizontally through a 3–5 m long accumulation bed (fig. 1) [1]. Following the contact of the air flowing between the filling of the bed, its temperature is brought closer to the filling's temperature, which during peak load results in heating of the outside air in winter from -18°C to about 0°C, and cooling it in summer from +30°C to about +20°C.

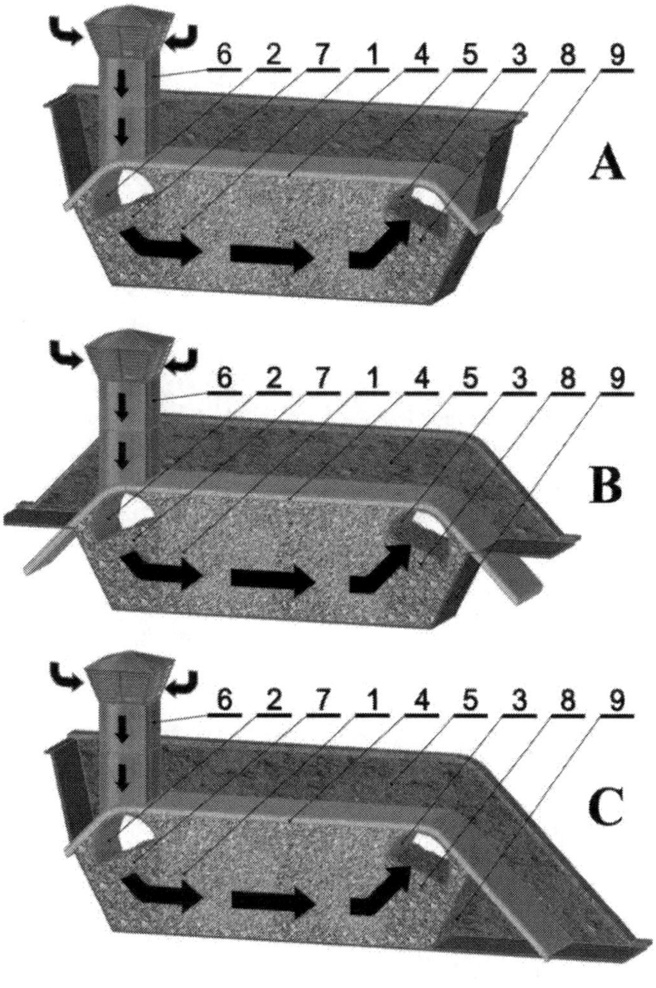

Fig. 1. Various proposals for the exchanger construction: A – recessed, B – partially recessed, C – elevated above the level of the existing terrain and situated on the slope, 1 – accumulation bed, 2 – distribution channel, 3 – collecting channel, 4 – thermal/humidity insulation, 5 – exchanger cover, 6 – air intake, 7 – distribution bed, 8 – collecting bed, 9 – natural ground.

Experience has shown that the accumulation bed should be made of washed gravel, gravel or rock grit having a hydraulic diameter of 15–40 mm but without fractions smaller than 3 mm. Stones having a hydraulic diameter of 80–100 mm constitutes a distribution layer. Thermal/humidity insulation should be made of 10 cm polystyrene protected with two layers of double foil. The entire area should be covered with about a half-meter layer of soil in order to enable planting of shrubs or lawn cultivation.

The heat exchanger (distribution and accumulation beds) should not exceed 5 m length and 2.5 m height. Studies have shown that continuous operation of a 5 m long bed during a summer day results in an optimal shift of a heat wave in the filling. As a consequence, lowest temperatures of the air leaving the exchanger are achieved at the highest outdoor temperatures [1, 2].

The nominal air velocity, depending on operating mode, may range from 0.05 to 0.20 m/s. Such values result from the heat and mass exchange conditions be-tween the air and the filling, as well as from resistance of the airflow across the exchanger.

A sprinkler system can be performed under the layer of thermal/humidity insulation. Based on a study [1], it was found that sprinkler pipes having a nominal diameter of ø 25 mm, perforated with ø 2 mm holes made at an angle of 90° at intervals of 10 cm are the cheapest and best for this purpose. The optimal distance between pipes is 0.5 m. The sprinkler system enables possible periodic disinfection and washing of the bed. It also facilitates intensification of heat and mass exchange processes in the accumulation filling, especially in the context of increasing the relative humidity of the air which leaves it. The research conducted by the Sanitary-Epidemiology Station in Wroclaw on an exchanger which has been working for many years without sprinkling did not show any microorganisms harmful to human health in the air discharged from the exchanger. On contrary – it was concluded that the number of particles from air pollution following the exchanger got reduced to half of the value contained in atmospheric air [2]. Thus, the exchanger is additionally a specific kind of filter. These results are also confirmed by publications concerning other exchangers of this type operating in Poland. The experiments [2] of hydrating of the bed during its operation show that it does not cause a significant decrease in the air temperature. It only, due to moisturising, significantly increases relative humidity to values near to saturation, i.e. $\varphi = 100\%$. Additionally, during the course of the study, it was noted that similar parameters are obtained by humidifying only the inlet section of the filling [2]. This allows for possible savings when constructing systems where it is important for some operating periods to achieve higher relative air humidity at the outlet. In addition, what is important, hydraulics tests of the filling performed during sprinkling of the inlet section only, as well as of the entire filling, did not show increase in resistance of the airflow across the bed.

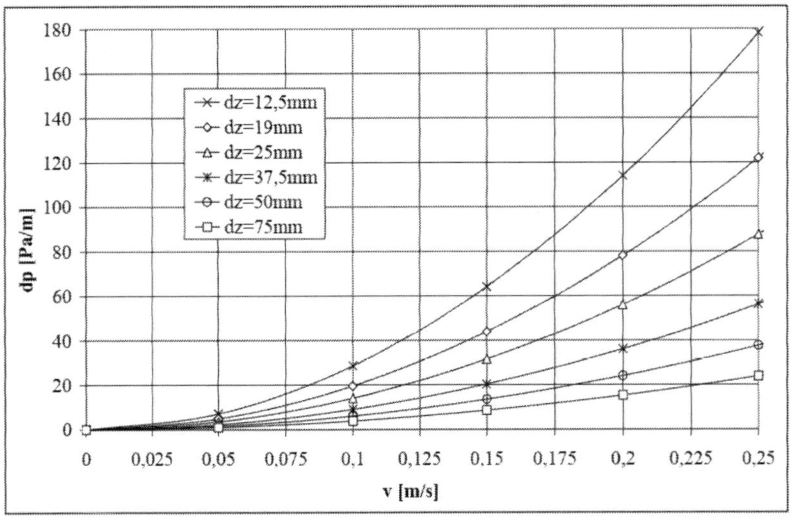

Fig. 2. Practical chart to determine the pressure loss in the bed according to [3].

Fig. 2 shows a practical, usable during designing, experimentally determined dependency between air flow velocity (v) and pressure loss (dp) of the accumulation bed for different hydraulic diameters used in BGWCiM.

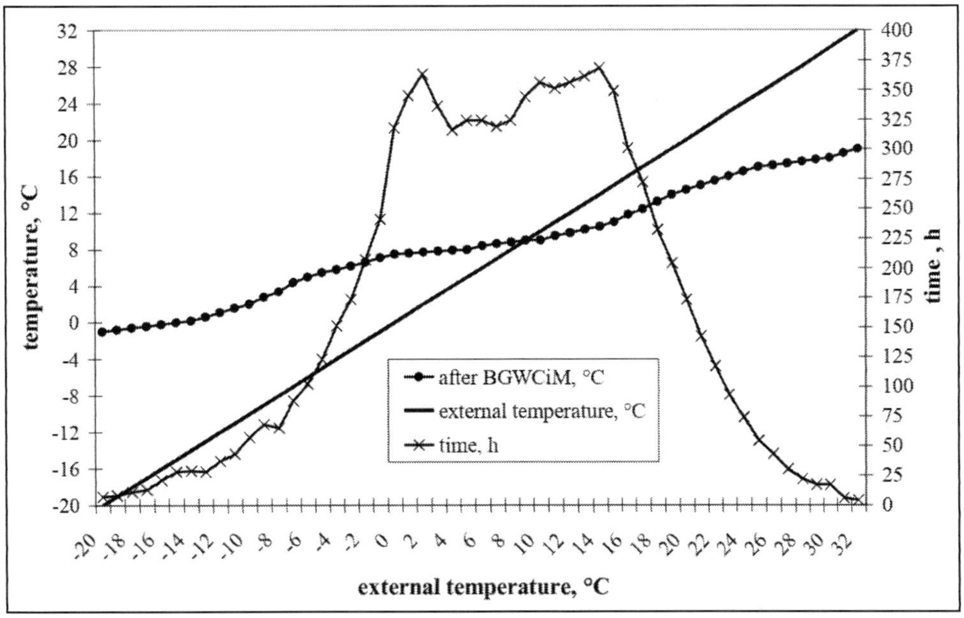

Fig. 3. Average air temperature values following the ground exchanger and a number of hours of occurrence of given outdoor temperatures during a year for A type of heat exchanger, 5 m length, 0.14 m/s air velocity, operated year-round continuously [2, 3].

In fig. 3 average values of air temperature after passing through BGWCiM and average times of occurrence of given outdoor temperature data for Wroclaw [2] were presented. They were indicated and statistically calculated on the basis of many years of measurements. This chart can be used in designing and provide the basis for performance of any analyses connected with the full-year operation of the device.

As a result of air contact with the ground in BGWCiM, the soil gets drained during periods of hot summer, moisturised in winter and cleansed. The above is achieved only by negligible energy inputs necessary to overcome the airflow resistance through the bed.

2 Conditioning rooms in the industry

2.1 Required air parameters

Each type of production, depending on the technology used, requires very specific microclimate conditions (table 1) to produce a given product of possibly the highest quality. What must not be overlooked is also the human aspect. In order to achieve high productivity, the best possible conditions must also be created in the work area. It often happens that the production requirements and the comfort conditions of the people differ. It is when the technology is more important. Ventilation in an industrial plant constitutes a service component rather than a core business. However, production would be impossible without the optimal conditions of the microclimate. An important issue for the ventilation system to be economically viable. Therefore, it must provide appropriate parameters for both the technological process and people, ensuring minimal investment and operating costs [4].

Table 1. Recommended temperature and humidity ranges for various industrial processes [6].

No.	Industry	Plant type	Temperature	Relative humidity
-	-	-	°C	%
1	*2*	*3*	*4*	*5*
1	Libraries	Books warehouse	21–25	40–50
		Reading rooms	21–25	35–55
2	Breweries	Fermentation rooms	4–8	60–70
		Barrels of malt	10–15	80–85
3	Printing works	Paper warehousing	20–26	50–60
		Printing	22–26	45–60
		Multicolour printing	24–28	46–50
		Photographic printing	21–23	60
		All other works	21–23	50–60
4	Furs	Warehousing	5–10	50–60
		Drying	43	-
5	Museums	Paintings	18–24	40–55
6	Bakeries	Flour warehouse	15–25	50–80
		Yeast warehouse	0–5	60–75
		Dough making	23–25	50–60
		Sugar warehouse	25	35
7	Mushroom cultivation	Growth period	10–18	-
		Warehousing	0–2	80–85
8	Linoleum production	Oxidation of linseed oil	32–38	20–28
		Printing	26–28	30–50
9	Confectionery	Warehousing (dried fruit)	10–13	50
	industry	Soft sweets	21–24	45
		Production of hard sweets	24–26	30–40
		Packing of hard sweets	24–28	40–45
		Chocolate production	15–18	50–55
		Chocolate wrapping	24–27	55–60
		Chocolate packing	18	55
		Chocolate warehousing	18–21	60–65
		Fruitcake and wafers production	18–20	50
10	Electrical engineering	General production	21	50–55
	industry	Production of thermo- and, hygrostats	24	56–55
		Production with small tolerances	22	40–45
		Production of isolations	24	65–70
11	Pharmaceutical	Semi-products warehousing	21–27	30–40
	industry	Manufacture of tablets	21–27	35–50
12	Photographic	Production of normal films	20–24	40–65
	industry	Production of non-flammable films	15–20	45–50
		Treatment of films	20–24	40–60
		Darkroom	21–22	45–50
		Photographic studio	22–23	40–50
		Films warehousing	18–22	40–60
13	Rubber industry	Warehousing	16,24	40–50
		Manufacturing	31–33	-
		Vulcanisation	26,28	25–30
		Surgical materials	24–33	25–30
14	Paper	Machines	22–30	50–60
	industry	Paper warehouses	20–24	50–60
15	Tobacco industry	Raw tobacco warehousing	21–23	50,65
		Tobacco preparation	22–26	75–85
		Tobacco warehousing	24–25	70–75
		Tobacco steeping	32	85–88
		Manufacture of cigarettes and cigars	21–24	55–65
		Packing	23	65

Table 1. continued…

No.	Industry	Plant type	Temperature	Relative humidity
-	-	-	°C	%
1	*2*	*3*	*4*	*5*
16	Textile industry	Bath	22–25	40–50
	- cotton	Carding machine	22,25	50–55
		Comber	22–25	45–55
		Textile stretching machine	22–25	50–55
		Roving frame	22–25	50–55
		Ring frame	22–25	40–45
		Spooling, spinning, trimming, pulling warp	22–25	60–70
	Textile industry	Preparation	18–20	80
	- flax fibres	Carding mill	20–25	50–65
		Spinning	24–27	65–75
		Weaving	27	65–70
	Textile industry	Preparation	27–29	60
	- wool	Carding mill	27–29	65–70
		Spinning	27–29	50–60
		Weaving	27–29	60–70
		Equipping	24	50–60
	Textile industry	Preparation	27	60–65
	- silk	Spinning	24–27	65–70
		Weaving	24–27	60–75
	Textile industry	Carding, spinning	21–25	65–75
	- rayon	Weaving	24–25	60–65
17	Match	Production	18–22	50
	industry	Warehousing	15	50
18	Mechanical	Offices, warehousing, general assembly	20–24	35–55
	plants	Precision assembly	22–24	40–50
19	Matches	Production	22–23	50
		Drying	21–24	60
		Warehousing	16–17	50
20	Leather	Drying	20–52	75
		Warehousing	10–18	40–60
21	Plastics	Thermosetting injection-moulded plastics	27	25–30
		Cellophane coated	24–27	45–65
22	Optical industry	Application	24	45
		Grinding	27	80
23	Ceramics	Treatment of refractory materials	43–66	50–90
		Storage of clay	16–27	35–63
		Decoration of ceramics	24–27	48
24	Electronics	Winding of reels	22	15
		Semiconductor assemblies	20	40–60
25	Electrical devices	Production and metrological tests	21	50–55
		Assembly and calibration of thermostats	24	50–55
		Assembly and calibration of hygrostats	24	50–55
26	Precision mechanics	Assemblies having strict dimension tolerances	22	40–45
		Assembly and tests of measuring devices	24	60–63
27	Distributor apparatus	Fuses and circuit breakers	23	50
	and switchers	Winding of power capacitors	23	50
		Paper insulation warehousing	23	50

2.2 Use of BGWCIM

In fig. 4 there are areas presented which show the nature of the variability of the outdoor air and the air following BGWCiM, recorded in the climatic zone of Poland.

 DOI: 10.1051/e3sconf/20172200027

Moreover, in this drawing the so-called climatic curve is indicated, which shows average outdoor air parameters of Poland's geographical area and typical parameters of thermal comfort.

When referring to the air parameters following BGWCiM to the air parameters required in production processes (room temperature and air conditioning temperature), it can be seen that the use of the exchanger for air preheating will not always be cost effective. During the cold season, except for processes requiring very low humidity in the room (dry rooms), preheating and humidification will be almost always beneficial. This differs in the case of the transition and warm periods, when too much humidification of the air in the ground heat exchanger, usually close to the saturation parameters, will cause the need to dry it, which will be associated with additional costs which can exceed the savings resulting from the temperature drop.

Beneficial effects of using BGWCiM should be expected in air treatment systems for air conditioning of rooms with relatively high temperature and high relative humidity (e.g. textile industry, tobacco industry, etc.).

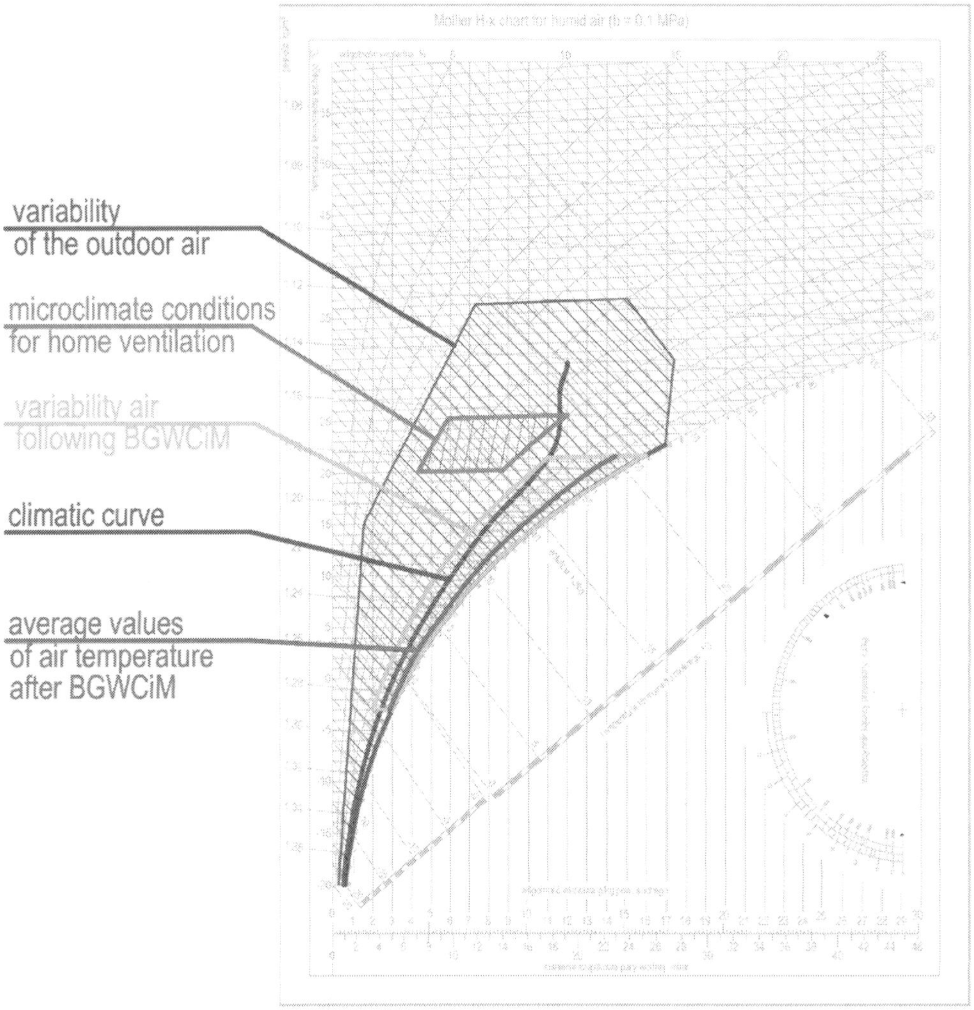

Fig. 4. Ranges of changes in air parameters.

3 Conclusion

The decision to use BGWCiM to pre-prepare the air in the process of air conditioning of production rooms should be each time preceded by an in-depth analysis of the full-year operation of the system with a full assessment of possible effects. Restrictive requirements for microclimate parameters in many industries will not always be achievable with BGWCiM alone or with BGWCiM serving to pre-prepare the air. Wherever the analysis confirms the suitability of its use, savings of up to 30% per year of energy consumed can to be expected, compared to BGWCiM-free system while reducing power and size of other air treatment devices (e.g. heaters, radiators, heat sources, chillers, humidifiers, etc.) [2].

References

1. G.J. Besler and others, The air direct-contact, gravel, ground heat exchanger, Patent No. 128261 PRL, Wroclaw University of Science and Technology, Wrocław 1980 (in Polish).
2. M. Besler, Research on energy efficiency of soil in the environmental engineering, PhD thesis, Wrocław 1997 (in Polish)
3. W. Cepiński, Efficiency of natural energy acquisition, PhD thesis, Wrocław 2010 (in Polish)
4. A. Pełech, Ventilation and air conditioning. Fundamentals, Publishing House of Wroclaw University of Science and Technology, Wrocław 2008 (in Polish)
5. S. Przydrożny, Ventilation, Publishing House of Wroclaw University of Science and Technology, Wrocław 1991 (in Polish)
6. A. Słomka, Ventilation and air conditioning for work rooms Library of the PIP Training Center, 2007 (in Polish)
7. Consumption of fuels and energy carriers in 2015, Central Statistical Office (GUS), Warsaw 2016 (in Polish)

 DOI: 10.1051/e3sconf/20172200028

The use of multi criteria analysis to compare the operating scenarios of the hybrid generation system of wind turbines, photovoltaic modules and a fuel cell

Bartosz Ceran[1,*]

[1]Poznań University of Technology, Faculty of Electrical Engineering, Institute of Electrical Power Engineering, Poland

Abstract. The paper presents the results of the use of multi-criteria analysis to compare hybrid power generation system collaboration scenarios (HSW) consisting of wind turbines, solar panels and energy storage electrolyzer - PEM type fuel cell with electricity system. The following scenarios were examined: the base S-I- hybrid system powers the off-grid mode receiver, S-II, S-III, S-IV scenarios -electricity system covers 25%, 50%, 75% of energy demand by the recipient. The effect of weights of the above-mentioned criteria on the final result of the multi-criteria analysis was examined.

1 Introduction

Fully distributed generation is playing an increasingly important role in the power system, especially wind and photovoltaic plants. While observing an increasing saturation of a power system by this type of sources, it is extremely important to gain knowledge about planning of the burden between unconventional and conventional sources based on carbon, which will continue to be the primary energy source in the coming years.

While planning collaboration between renewable sources with conventional ones, energy aspects of examined electricity generation technologies should be considered (efficiency process of processing of primary energy into electricity), economic aspects (cost of generating electricity) and environmental aspects (exhaust fumes emissions) [1].

The paper proposed the use of method of "programming compromise" to conduct a multi-criteria analysis [2, 3], aimed at selecting the best variant of the load distribution between the hybrid power generation system consisting of wind turbines, photovoltaic modules and energy storage electrolyzer – fuel cell and the electricity power system including the above mentioned aspects.

* Corresponding author: bartosz.ceran@put.poznan.pl

2 The analyzed hybrid power generation system

Scheme of the analyzed receiver supply system is shown in Figure 1 [4, 5]. The excess energy produced from renewable sources is used for electrolysis process (decomposition of water into hydrogen and oxygen). Hydrogen is compressed and stored, and then used in a fuel cell to produce electricity, at a time when it is needed by customer.

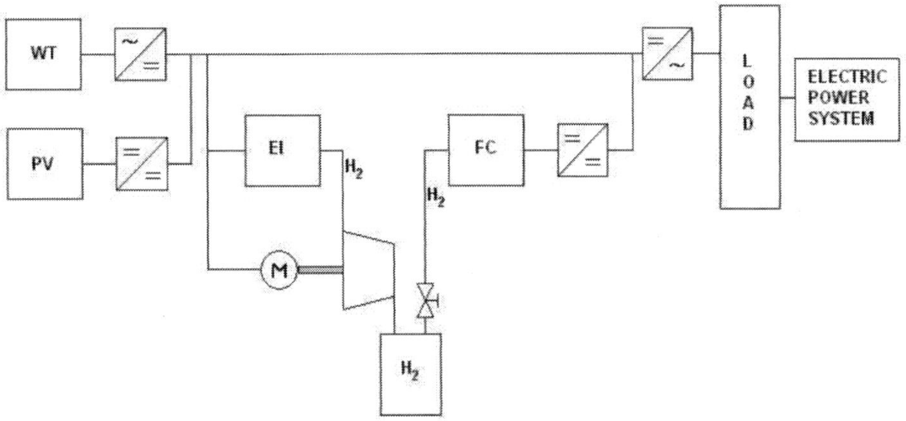

Fig. 1. Flowchart-hybrid power generation system.

The balance of power in the system, depends on the ratio of power generated by RES sources to the demand of the recipient. In the case where the demand is less than the power generated balance sheet equation takes the form [6].

$$P_{load} = P_{PV} + P_{WT} - P_{elek} - P_{comp} + P_{sys} \tag{1}$$

where: P_{load} – power consumed by the recipient, P_{PV} – power generated by the photovoltaic installation, P_{WT} – power generated by the wind plants, P_{elek} – power consumed by electrolyzer, P_{comp} – power consumed to drive the compressor, P_{sys} – power drawn from the power system.

If the demand exceeds the power produced by sources the receiver power describes the formula:

$$P_{load} = P_{PV} + P_{WT} + P_{FC} + P_{sys} \tag{2}$$

where: P_{FC} – power generated by the fuel cell stack.

3 Scenarios of cooperation and decision-making criteria

For the analysis were adopted two customer profiles shown in figures 2 and 3 with the marked scenarios of load distribution between the hybrid power generation system and electric power system. The base scenario S-I is a power of the receiver by hybrid power generation system in off-grid mode, without the electric power system. Another scenarios S-II, S-III, S-IV include covering parts of the load by the system in the following order: 25%, 50% and 75% of energy demand is covered by the electric power system.

DOI: 10.1051/e3sconf/20172200028

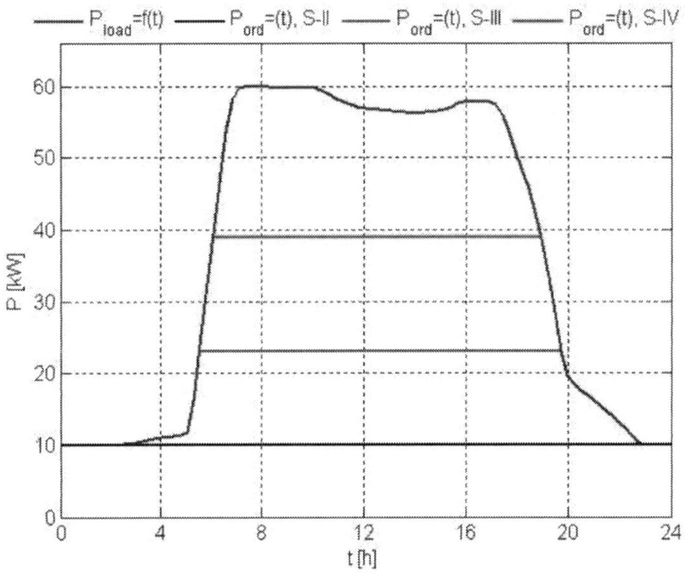

Fig. 2. Energy profile of the recipient 1 - work scenarios.

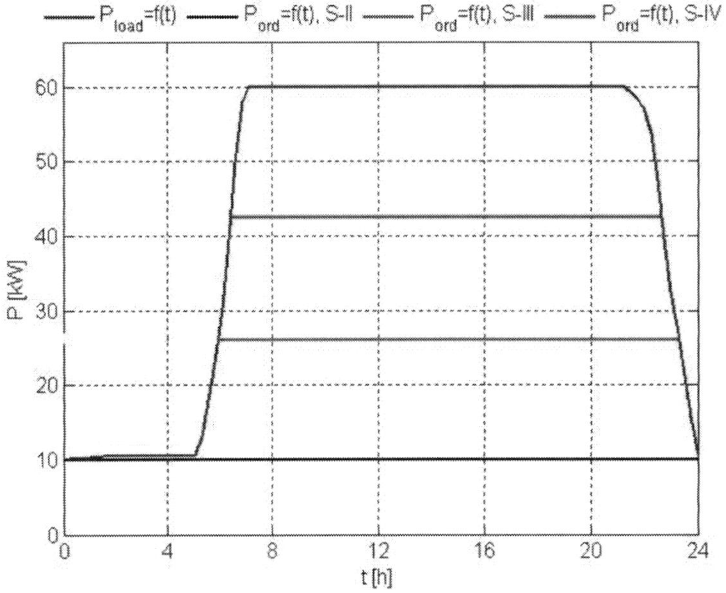

Fig. 3. Energy profile of the recipient 2 - work scenarios.

For each cooperation scenario a set of devices has been chosen (table 1) of the hybrid power system based on the energy analysis described in the paper [7].

Table 1. Equipment forming a hybrid power system for scenarios S-I…S-IV.

	Scenarios of power system loads							
	Profile 1				**Profile 2**			
Devices in hybrid system	S-I	S-II	S-III	S-IV	S-I	S-II	S-III	S-IV
Wind turbines [kW]	3x50	2x50	2x50	50	4x50	3x50	2x50	50
Panels PV-250 [Wp]	380x250	325x250	52x250	31x250	373x250	318x250	197x250	97x250
Fuel cell [kW]	50	50	40	20	50	50	40	20
Electrolyzer [kW]	150	110	100	50	200	150	100	50

As the decision criteria for the multi-criteria analysis the following indicators has been adopted:
k_1 – unit usage of additional hydrogen by the hybrid system,
k_2 – unit cost of energy production by the hybrid system,
k_3 – the indicator of carbon dioxide emission by hybrid power system and electric power system,
k_4 – the capacity of utilization of power ordered from the electric power system.

$$k_1 = \frac{M_{H_2}}{E_{load}} [\frac{kg_{H_2}}{kWh}] \quad (3)$$

where: M_{H2} – the amount of utilized additional hydrogen for backup purpose, E_{load} – the demand for energy by the recipient.

$$k_2 = \frac{K_e}{E_{HPGS}} [\frac{PLN}{kWh}] \quad (4)$$

where: K_e – operating costs of hybrid system, E_{HPGS} – the amount of energy produced by hybrid power generation system

In the case of S1 scenario, the cost of 1 kWh generation in a hybrid system is almost twice bigger than an average energy cost charged by public consumers, which is rated at 0.56 PLN/kWh. This high price is associated with high depreciation costs, which are assumed at the level of 10% of investment expenditure. As far as this instance is concerned, taking into consideration current prices of fuel cells, an investment expenditure of hybrid system, depending on particular scenario, is rated between 1.5 million and 3 million PLN.

$$k_3 = \frac{E_{sys} \cdot WE_{CO_2} + M_{H_2} \cdot CO_2 / H_2}{E_{load}} [\frac{kg_{CO_2}}{kWh}] \quad (5)$$

where: E_{sys} –energy input from the electricity system, WE_{CO2} – indicator of carbon dioxide emission in electricity system, CO_2/H_2– the amount of carbon dioxide formed in the production of 1 kg of hydrogen in the steam reforming process of hydrocarbons

$$k_4 = \frac{E_{sys}}{P_{ord} \cdot t} [\frac{kWh}{kWh}] \quad (6)$$

where: P_{ord} – power ordered from the electric power system

4 Comparing the scenarios of work using multi-criteria analysis

Multi-criteria analysis is a mathematical method which allows to select the most favorable solution, the so-called. scenario, in the light of adopted criteria. One of the methods of multi-criteria analysis is a method of "compromise programming" which ranks considered scenarios by distance from the so-called target point X (ideal point).

A mathematical record of distance measure of considered variant from the ideal point describes the formula:

$$L_\alpha(S_n) = \sum_{m=1}^{M} w_m^\alpha (x_m^{'} - x_{nm}^{'})^\alpha \tag{7}$$

where: L_α – a measure of divergence of considered scenario S_n from the ideal point, w_m – weight factor m of this criterion, α – exponent measuring the deviation strategy from the ideal point, x_m '- *mth* coordinate of the ideal point, x_{nm}' - normalized value of the evaluation criterion.

The most favorable scenario in the light of the adopted criteria complies with the relation:

$$S_j = S^* \Leftrightarrow L_\alpha(S_j) = min\, L_\alpha(S_n), \text{ dla } n = 1, 2.....N; \alpha = 1,2...\infty \tag{8}$$

where: S^* - selected scenario.

Comparison of the best scenario in the light of the adopted criteria with the other scenarios is accomplished through the threshold of acceptability PA defined as:

$$PA = \frac{p}{100} \cdot min\, L_\alpha(S_n) \tag{9}$$

where: p – a percentage value of the distance of the best scenario from the ideal point.

Acceptable scenarios meet the inequality:

$$L_\alpha(S_n) - min\, L_\alpha(S_n) < PA \tag{10}$$

Mathematical notation of multi-criteria decision-making problem consisting of four scenarios S (I-IV), examined using four criteria k_1-k_4 represents a decision matrix called the matrix of payments (Table 2):

Table 2. Payment matrix of decision problem for scenarios S-I...S-IV.

	Load scenarios of the power system							
	Profile 1				**Profile 2**			
Decision criteria	S-I	S-II	S-III	S-IV	S-I	S-II	S-III	S-IV
k_1 [kgH_2/kWh]	0.233	0.163	0.104	0	0.259	0.197	0.089	0
k_2 [PLN/kWh]	1.15	1.32	1.33	1.51	1.08	1.18	1.22	1.36
k_3 [$kgCO_2$/kWh]	0.28	0.43	0.55	0.62	0.32	0.43	0.52	0.62
k_4 [kWh/kWh]	0	1	0.8	0.7	0	1	0.85	0.79

Table 3. The normalized payment matrix for the scenarios S-I...S-IV.

	Load scenarios of the power system							
	Profile 1				**Profile 2**			
Normalized decision criteria	S-I	S-II	S-III	S-IV	S-I	S-II	S-III	S-IV
k_1(min)	0.00	0.30	0.55	1.00	0.00	0.30	0.55	1.00
k_2(min)	1.00	0.53	0.50	0.00	1.00	0.53	0.50	0.00
k_3(min)	1.00	0.58	0.22	0.00	1.00	0.58	0.22	0.00
k_4(max)	0.00	1.00	0.80	0.70	0.00	1.00	0.80	0.70

5 The results of multi-criteria analysis

Studying the impact of weights of decision criteria on the outcome of multi-criteria analysis was started with the variant in which all the criteria were assigned the same value of the weight wm. Next the prevailing effect of one of the criteria was tested in relation to the others and its effect on the results of multi-criteria analysis using the following assumptions: prevalence of one criterion took place till the moment when as a result of multi-criteria analysis was indicated a favoured scenario by this criterion.

The degree of acceptability p adopted at a low level equal to 5%. Due to this another scenarios were taken into account which were in the distance no more than 0.05*minLα(Sn) from the best solution. The analysis was conducted for two criterion of aggregating value α equal to 1 and 2. The scenarios are in order from most to least favorable. Acceptable variations are indicated with *. Multi-criteria analysis results are shown in Tables 4, 5 and Figures 4, 5.

Table 4. Multi – criteria analysis results – profile 1.

Ratio of weight of criteria	**Recipient profile 1**	
w_1:w_2:w_3:w_4	α =1	α =2
1:1:1:1	**S-II*** → S-III → S-I → S-IV	**S-II*** → S-III → S-I → S-IV
2:1:1:1	**S-II*** → **S-III*** → **S-IV*** → S-I	**S-III*** → S-IV → S-II → S-I
3:1:1:1	**S-IV*** → S-III → S-II → S-I	**S-IV*** → S-III → S-II → S-I
1:2:1:1	**S-I*** → **S-II*** → S-III → S-IV	**S-II*** → S-III → S-I → S-IV
1:1:2:1	**S-I*** → **S-II*** → S-III → S-IV	**S-II*** → S-I → S-III → S-IV
1:1:1:2	**S-II*** → S-III → S-IV → S-I	**S-II*** → S-III → S-IV → S-I

Table 5. Multi – criteria analysis results – profile 2.

Ratio of weight of criteria	**Recipient profile 2**	
$w_1:w_2:w_3:w_4$	$\alpha = 1$	$\alpha = 2$
1:1:1:1	**S-II*** → S-III → S-I → S-IV	**S-III*** → **S-II*** → S-I → S-IV
2:1:1:1	**S-III*** → S-IV → S-II → S-I	**S-III*** → S-IV → S-II → S-I
3:1:1:1	**S-IV*** → **S-III*** → S-II → S-I	**S-III*** → S-IV → S-II → S-I
1:2:1:1	**S-II*** → S-I → S-III → S-IV	**S-II*** → S-I → S-III → S-IV
1:3:1:1	**S-I*** → S-II → S-III → S-IV	**S-II*** → **S-I*** → S-III → S-IV
1:1:2:1	**S-II*** → S-IV → S-III → S-IV	**S-II*** → S-I → S-III → S-IV
1:1:3:1	**S-I*** → S-II → S-III → S-IV	**S-II*** → **S-I*** → S-III → S-IV
1:1:1:2	**S-II*** → S-III → S-IV → S-I	**S-II*** → **SIII*** → S-IV → S-I

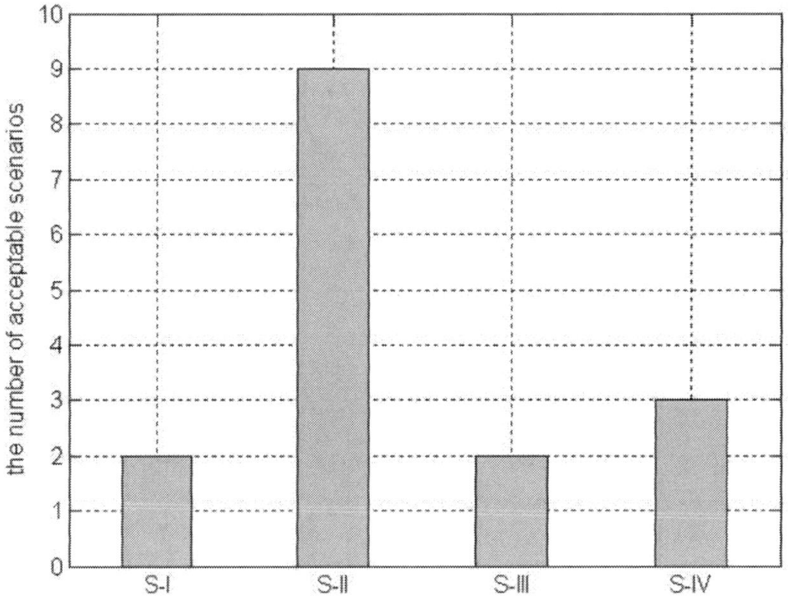

Fig. 4. Multi – criteria analysis results – profile 1.

DOI: 10.1051/e3sconf/20172200028

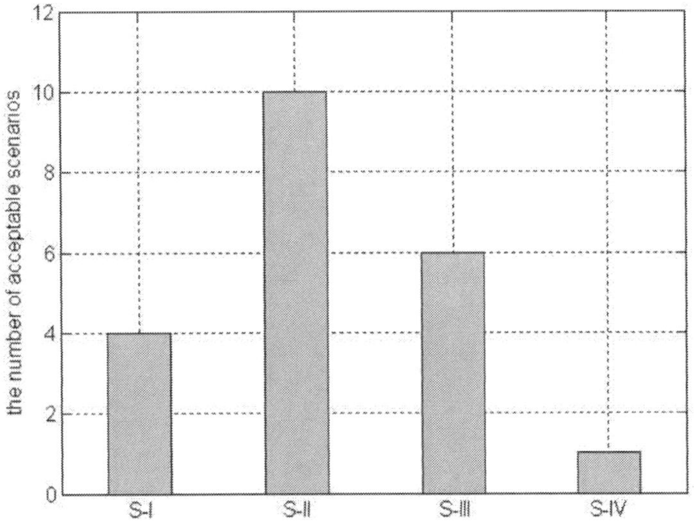

Fig. 5. Multi – criteria analysis results – profile 2.

6 Conclusions

The research allows to formulate the following conclusions:

- It is not possible to completely replace the classic power system with renew-able energy sources combined in hybrid power generation systems. The most frequently chosen scenario is S-II, where the electric power system covers the base load of the recipient.
- Planning the load distribution between the hybrid power generation system, consisting of wind turbines, solar panels and energy storage cell electrolyzer - fuel cell and the electric power system can be seen as the decision-making task that must be resolved according to various criteria (energy, economic, environmental) using multi-criteria, decision-making methods.
- The proposed analysis model can be a starting point for optimizing the operation of hybrid power systems based on sources of stochastic nature of work in the electric power system. It is possible to expand this model, e.g. by expanding by new decision criteria.

Reference

1. A. Nadimi, F. Adabi, IJRER **6**, 2 (2016)
2. Z. Mucha, J. Mikosz, A. Generowicz, Environment Engineering **4**, 145–155 (2012)
3. W. Wójcik, Z. Mucha, A. Generowicz, Acta Sci. Pol. **13**, 3, 101–114 (2014)
4. B. Ceran, Q. Hassan., M. Jaszczur, K. Sroka, E3S Web Conf. **14**, 01020 (2017)
5. R. Szczerbowski, B. Ceran, E3S Web Conf. **10**, 00090, (2016)
6. C. Wang., M. H. Nehrir, IEEE Transactions 957–964, (2008)
7. B. Ceran, K. Sroka, Acta Energetica **23**, 2, 36–42 (2015)

The effect of biologically and chemically synthesized silver nanoparticles (AgNPs) on biofilm formation

Joanna Chojniak[1,*], *Izabela* Biedroń[1], *Barbara* Mendrek[2], and *Grażyna* Płaza[1]

[1]Institute for Ecology of Industrial Areas, Environmental Microbiology Unit, ul. Kossutha 6, 40-844 Katowice, Poland
[2]Centre of Polymer and Carbon Materials, Polish Academy of Sciences, ul. M. Curie-Sklodowskiej 34, 41-819 Zabrze, Poland

Abstract. Bionanotechnology has emerged up as integration between biotechnology and nanotechnology for developing biosynthetic and environmental-friendly technology for synthesis of nanomaterials. Different types of nanomaterials like copper, zinc, titanium, magnesium, gold, and silver have applied in the various industries but silver nanoparticles have proved to be most effective against bacteria, viruses and eukaryotic microorganisms. The antimicrobial property of silver nanoparticles are widely known. Due to strong antibacterial property silver nanoparticles are used, e.g. in clothing, food industry, sunscreens, cosmetics and many household and environmental appliances. The aim of the study was to compare the effect of silver nanoparticles (AgNPs) synthesized biologically and chemically on the biofilm formation. The biofilm was formed by the bacteria isolated from the water supply network. The commonly used crystal violet assay (CV) was applied for biofilm analysis. In this study effect of biologically synthesized Ag-NPs on the biofilm formation was evaluated.

1 Introduction

Nanoparticles of metals are currently an area of intense research due to a wide variety of their potential applications in many areas, for example in biomedical, agricultural, environmental fields. Nanoparticles have specific properties and chemical reactivity [6, 17, 19]. One of them is high surface area to volume. The antimicrobial activity of AgNPs seems to be a function of the surface area to effectively interact with microorganisms. Large surface area of nanoparticles enhances the interaction with microbes and results in a wide spectrum of antimicrobial activities. Silver nanoparticles (AgNPs) are the most popular material to using as the antimicrobial agent.

In the AgNPs production the physical, chemical and biological approaches are applied. Biological synthesis involves biochemical conversion of a substrate, whereas the chemical

[*] Corresponding author: j.chojniak@ietu.pl

one is based on the reduction of chemical compounds. The formation of NPs by physical and chemical routes requires toxic chemicals which results in toxic side effects [6]. Green synthesis also provides the cost effective, non-toxic, large-scale, and high-output nanoparticle products [13]. In biologically mediated nanoparticle synthesis, the microbial enzymes or plant phytochemicals having reducing or antioxidant properties are usually responsible for the reduction of metal ions into respective metal nanoparticles [5].

Several reports have successfully demonstrated that AgNPs have antimicrobial activity against a broad range of Gram-negative and Gram-positive including the pathogenic bacteria [9]. To determine the antibacterial effectiveness of AgNPs and consequences of their release into environment, studying the effect of AgNPs on biofilms is important because bacteria are often present in biofilm communities. To survive in harsh environment, bacteria attach to surface, and produce extracellular polymeric substrates (EPS) to create more complex structures, called biofilms [11].

Most of the bacteria in drinking water distribution system is attached to surfaces of pipes as a biofilm. Biofilms in water distribution system can pose financial and functional problems (impact on taste, odour and colour of water) and can be harbour of pathogenic microorganisms. Moreover old mature biofilms can reduce effective pipe diameters. All these negative effects of the presence of biofilm cause associated with maintaining the water supply network. Bacteria from drinking water biofilms are more resistant to disinfectants than planktonic microbes. Also biofilms are hard to remove from the pipes surfaces without dismantling them which is an expensive and practically impossible to do in the drinking water distribution system [1, 23].

In this context, the effect of silver nanoparticles obtained chemically (chem-AgNPs) and biologically (bio-AgNPs) on antimicrobial and antibiofilm activities against bacterial strains isolated from the water supply network biofilms was investigated.

2 Materials and Methods

2.1 Bacterial strains used in the experiment

The high biofilm-forming strains isolated from biofilm samples collected from drinking water distribution system in Wroclaw were used in the study. Taxonomic identification and characterization of these strains have been described previously by Biedron et al. [12]. The strains were used to test the antibacterial and antibiofilm properties of AgNPs synthesized biologically and chemically. *Bacillus subtilis* strain was isolated form the petroleum contaminated soil in Czechowice- Dziedzice in Poland was used in the biological synthesis of Ag-NPs [3]. The strain was identified and characterized in previous studies [7].

2.2 Synthesis of silver nanoparticles

2.2.1. Growth of Bacillus subtilis

The bacterial cultures were grown aerobically in Luria-Bertani (LB) at 30°C for 96 h with constant shaking (110 rpm). Then, the freshly grown bacterial culture was centrifuged and the supernatant was collected and filtered through sterile 0.22-μm filter to sterile flasks. Cell free supernatant was subjected to synthesis of AgNPs [9].

2.2.2.Biological and chemical synthesis

In the extracellular production of AgNPs, a silver nitrate solution was added to 50 mL of the *Bacillus subtilis* culture supernatant to final concentration of 1 mM, and the mixture was kept for stirring at 200 rpm at 30°C for 48 h under dark conditions. The bioreduction of silver ions was monitored at regular intervals on a UV-Vis spectrophotometer (300–700 nm) for the formation of peaks to confirm the synthesis of silver nanoparticles. During the synthesis, the color change was observed from yellow to dark brown.

The chemical synthesis of AgNPs was performed as the control for biological synthesis. The chemical synthesis of AgNPs was carried out using ascorbic acid as reductant. 5 mL of ascorbic acid [2]. The reaction solutions changed from colorless to light yellow after 1 h of reaction. After 96 h of the reaction the precipitate appeared. The concentration of biological and chemical AgNPs measured by atomic absorption spectroscopy (AAS) was around 165 mg/L in solutions.

2.3 Evaluation of Ag-NPs effect on the biofilm formation

2.3.1 Growth of the bacteria

The strains isolated from the biofilm samples were grown on LB agar plates at 22°C for 48 hours. The single colony of bacteria was introduced to 150 mL of LB broth, and incubated at 22°C at 150 rpm for 48 hours. After incubation, the cultures were centrifuged and the pellet of bacteria was re-suspended in phosphate buffer saline (PBS). The each bacterial suspension was adjusted to 0.5 of McFarland's standard with PBS (approximately 10^7 CFU/ml) prior to use as an inoculum in experiments.

2.3.2. Biofilm inhibition assay

A volume of 120 µl of bacterial suspension was added to each well of 96-well plate. The biologically and chemically synthesized silver nanoparticles were added to the bacterial suspensions in the appropriate volume 15 µl of AgNPs to 105 µl of bacterial culture and incubated 72h at 22°C. The biofilm amount in a 96-well plate was quantified by crystal violet (CV) staining according to the method by Christensen et al. [8] with some modifications. The concentration of CV was determined by measuring the optical density of solution at 620 nm using a microtiter plate reader. The bacterial suspensions were used as controls. The experiment was performed in triplicates and average values were obtained.

2.4 Evaluation of antibacterial properties of chem- and bioAgNPs towards selected bacteria

2.4.1 Evaluation of antibacterial activity of chem- and bio-AgNPs by MT microplates method

The experiment was carried out in the MT2 MicroPlates™ (Biolog). They contain only mineral salts and redox dye to colorimetrically indicate the activity of bacteria. The bacterial cultures were prepared as described in 2.3.1. 105 µl of the strains suspensions were added to each microplate well, and 15 µl volumes of bio- or chem-AgNPs solutions. The microplates were incubated in the OmniLog Reader (Biolog) for 84 hours at 22°C. During the incubation, the colour development was recorded by camera every 15 minutes. We calculated from raw data the average value of color development and plotted a kinetic

DOI: 10.1051/e3sconf/20172200029

curve for each tested variant. The bacterial suspensions without AgNPs were used as the control. The experiment was done in three replicates.

2.4.2 Determination of antibacterial activity of chem- and bio-AgNPs by disc diffusion method

Bacterial cultures were obtained as described in section 2.3.1. 100 µl of each bacterial suspension (0.4–0.5 McFarland's standard, approximately 10^7–10^8 CFU/ml) was spread onto Mueller-Hinton agar plates. The sterilized disks were put on these freshly prepared lawns, and then 10 µl chem- and bio-AgNPs solutions were added on the discs. The plates were incubated at 22°C for 72 h. The diameter of inhibition zone around each disc was measured. Experiments were conducted in triplicates and average values of inhibitory zone were determined.

3 Results and Discussion

The physical properties of AgNPs formed in bio- and chemical synthesis are discussed by Mendrek et al. [3]. The antibiofilm effect of bio-AgNPs and chem-Ag-NPs is presented on Figure 1. The antibiofilm activity was observed after 72 hours of incubation. The reduction in biofilm formation was observed in 10 tested strains as a result of biologically synthesized AgNPs.

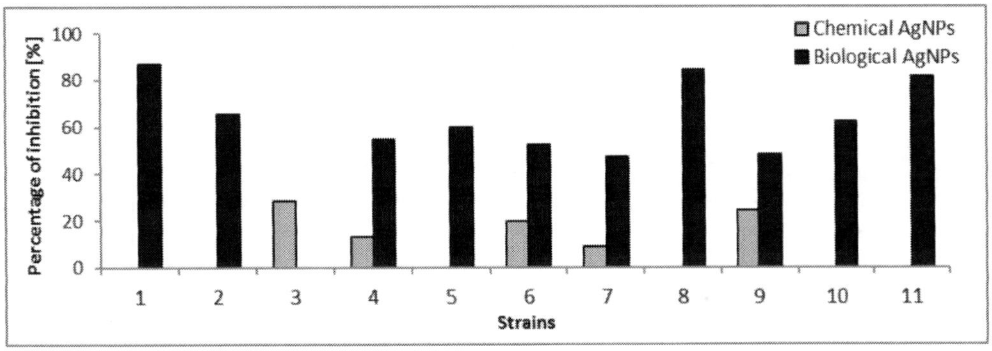

Fig. 1. The inhibition in biofilm formation by selected strains. Strains: 1. *Advenella incenata*, 2. *Moraxella osloensis* or *Enhydrobacter aerosaccus*, 3. *Kocuria rhizophila* or *Psychrobacter* sp., 4. *Microbacterium oxydans*, 5. *Bacillus thuringiensis*, 6. *Advenella incenata*, 7. genus *Advenella*, 8. *Micrococcus luteus*, 9. genus *Micrococcus*, 10. genus *Micrococcus*, 11. genus *Pseudomonas*.

However, the potency of bio-AgNPs biofilm inhibition was different. The maximum antibiofilm activity of biological AgNPs were observed for *A. incenata* (87% reduction), *M. luteus* (84% reduction) and strain belong to the genus *Pseudomonas* (81% reduction). No effect of biosilver nanoparticles was observed only for bacteria identified as *K. rhizophila* or *Psychrobacter sp.* The reduction of biofilm formation as an impact of chemically synthesized AgNPs were observed in five strains. In the case of these strains the reduction in biofilm formation were relatively low and ranged from 9% to 29%. The highest precentage of biofilm inhibition by chem-AgNPs were observed for *K. rhizophila* or *Psychrobacter* sp. and *Micrococcus* sp., 29% and 24%, respectively.

3.1 Antibacterial activity of chem- and bio-AgNPs

In Table 1 the results from the agar diffusion method are presented. The inhibition zone was observed for bio-AgNPs against almost all selected strains. However, there was no bio-AgNPs activity against *M. oxydans* (Fig. 3B). The highest antibacterial effect was observed against *A. incenata* and strain belong to the genus *Microccocus* with inhibition zone of 16 mm for both strains (Fig. 3A). Biological Ag-NPs showed also high antibacterial effect against *A. incenata, B. thuringiensis, Pseudomonas* sp., and *K. rhizophila (or Psychrobacter* sp.), with inhibition zone of 15, 15, 14, 13 mm, respectively.

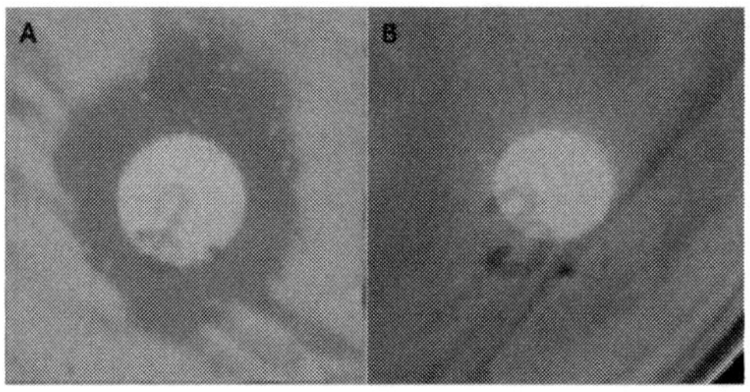

Fig. 3. Antibacterial activity: A. *Advenella incenata* and bio-AgNPs; B. *Microbacterium oxydans* and bio-AgNPs.

Table 1. Antibacterial activity spectrum of silver nanoparticles synthesized biologically and chemically obtained from disc diffusion method.

Strains	**Diameter of inhibition zones (average values., mm)**		
	Chem Ag- NPs	**Bio- AgNPs**	**Control**
1. *Advenella incenata*	0	15	0
2. *Moraxella osloensis* or *Enhydrobacter aerosaccus*	0	10	0
3. *Kocuria rhizophila* or *Psychrobacter* sp.,	0	13	0
4.*Microbacterium oxydans*	0	0	0
5. *Bacillus thuringiensis*	0	15	0
6. *Advenella incenata*	0	16	0
7. genus *Advenella*	0	9	0
8. *Micrococcus luteus*	0	9	0
9. genus *Micrococcus*	0	16	0
10. genus *Micrococcus*	0	10	0
11. genus *Pseudomonas*	0	14	0

The results obtained from the MT assay are presented on Figure 2. Based on the results obtained from the analysis of the kinetic curve was found that all strains in control samples have reached the plateau phase between 5 and 35 hour of experiment (Fig. 1A–K). Based on kinetic curves there were no differences between the control and samples incubated with bio- AgNPs for strain having the sequence nearest *M. osloensis* or *E. aerosaccus* (Fig. 1 B), *K. rhizophila* or *Psychrobacter* sp.(Fig. 1C), *A. incenata* (Fig. 1F), strains belong to the genus *Advenella* (Fig. 1G) and genus *Micrococcus* (Fig. 1I). A clear effect of the chem-AgNPs was observed for strain belongs to the genus *Micrococcus*. Based on the kinetic curve analysis it can be concluded that the microbial growth is inhibited (Fig. 1J). In other cases, although the initial growth inhibition were observed, inhibitory effect of nanoparticles has been overcome for *M. oxydans* after 42 hours of incubation with chem-AgNPs (Fig.1D), for *A. incenata* after 55 h (Fig.1A), in case of *M. luteus* and strain belongs

 DOI: 10.1051/e3sconf/20172200029

to the genus *Pseudomonas* after 70 h (Fig. 1.H, K) and for *B. thuringiensis* after 75 hour of experiment (Fig.1E).

The kinetic curves obtained after incubation with a bio-AgNPs show very weak effects of bio-AgNPs on the tested strains. Based on kinetic curves of eight strains there were no differences between the control and samples incubated with bio- AgNPs (Fig.1 A, B, D, E, G, H, J and K). In the case of the strain identified as *K. rhizophila* or *Psychrobacter sp* in the first 27 hours it was observed inhibition effect, after this time there has been growth of the activity of the bacteria and the inhibitory effect of nanoparticles has been overcome (Fig. 1 C). A similar trend was observed in the case of *A. incenata* (Fig. 1 F), and strain belongs to genus *Micrococcus* (Fig. 1 I) where the increase in activity was observed, respectively 55 and 60 hours of incubation with bio-Ag-NPs.

In this study, biogenic silver nanoparticles had potent antibiofilm effects. Antibacterial effects of AgNPs have been previously studied [14, 16], but there are a few studies on effects of AgNPs against bacterial biofilm. In most of the research the pathogenic bactera was applied [18].

While, the study reported by Sheng et al. [22] describes the effect of chemically AgNPs on intact wastewater biofilms from a local wastewater treatment plant. The results indicate that AgNP treatment decreased microbial community diversity but did not significantly affect the microbial community function. This provides direct evidence for the functional redundancy of microbial community in engineered ecosystems such as wastewater biofilms.

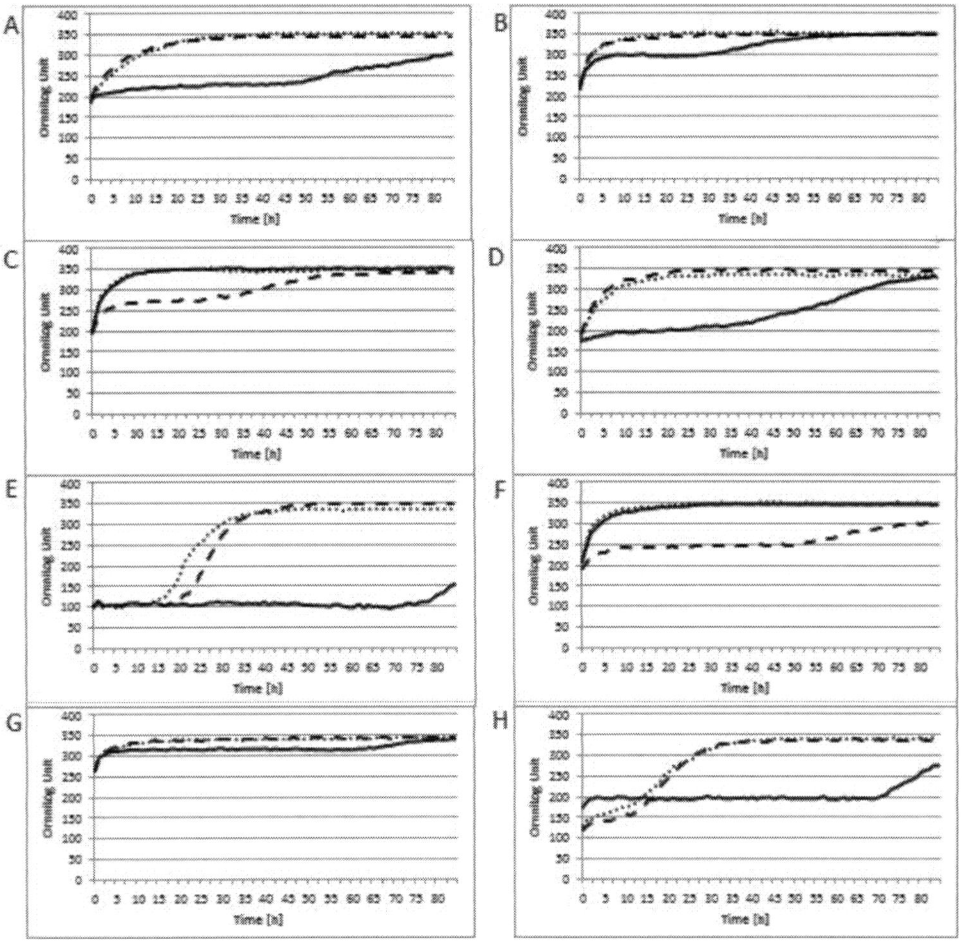

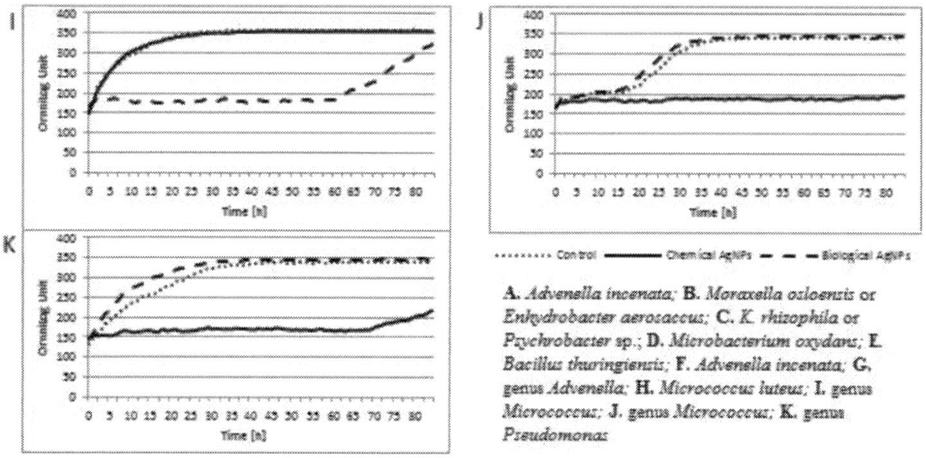

Fig. 2. The scheme of bacteria sensitivity profile to chem- and bio-AgNPs obtained by using the MT microplates. Bacteria sensitivity profile displayed in the form of kinetic graphs of the their growth.

Also, inhibition of nitrification byAgNPs has been relatively well-characterized in the previous research. It was presented that nitrification can be inhibited by AgNPs even at concentrations lower than1mg Ag/l [23]. Ammonia oxidizing bacteria (AOB) are more sensitive to AgNPs than nitrite oxidizing bacteria (NOB) [26]. The effects of AgNPs on other microbial functional groups in biological wastewater treatmentsystems are far less well studied. Biofilms are difficult to break due to their extracellular matrix. Grun et al. reported that highly reactive Ag+ ions bind to biological molecules, like proteins and polysaccharides within the EPS. Probably, AgNPs or released from them free Ag+ ions may be involved in blocking the formation of some compounds of EPS [1]. Kalishwaralal et al. said that exopolysachharide synthesis is arrested by AgNPs, and microorganism cannot form biofilm [15]. Which may explain the impact of nanoparticles to inhibit the formation of the biofilms. Low antimicrobial effects on the selected strains observed in our study might by associated with low concentration of AgNPs in our experiment. The penetration rate of the biofilm may also differ between strains. The ability of silver nanoparticles to agglomerate may also have meaning to their the activity [17]. Therefore, Nookola et al. [19] demonstrated antibacterial and antifungal properties of chemical silver nanoparticles on environmental strains. The nanoparticles were coated by citrate, which would prevent aggregation [19]. Rai et al. shown that silver nanoparticles can be used as effective antimicrobial agents for Gram-negative and Gram-positive bacteria, including antibiotic-resistant bacteria [16]. However, in our study AgNPs are not so effective in killing bacterial strains (the results from MT microplates), but we observed promising activity biol-AgNPs on the inhibition of biofilm formation. From the results from our study the biological silver nanoparticles can be involved in biofilm prevention. Silver nanoparticles, as opposed to chemical disinfectants, are not strong antioxidants and these are chemicals that behave indifferently in water [20].

4 Conclusions

1. No relations of results obtained from two applied antibacterial methods was noted.
2. Biologically synthesized AgNPs was more active against tested bacteria than chemically synthesized AgNPs in traditional disc diffusion method.
3. Higher antibacterial activity and reduction of biofilm formation was observed for biogenic AgNPs.

4. The further research and development are necessary to complete the knowledge on bio-AgNPs effects towards environmental bacteria and biofilm formation process.

The study was supported by the National Science Centre (Poland) (Project No. 2013/09/B/NZ9/01759 and No. 2015/19/N/NZ9/01621).

References

1. A. Grün, J. Meier, G. Metreveli, G. Schaumann, W. Manz, Environ. Sci. Pollut. Res. **24277** (2016)
2. B. Mendrek, J. Chojniak, M. Libera, B. Trzebicka, P. Bernat, K. Paraszkiewicz, G. Płaza, J. Disp. Sc. Technol. (2016)
3. C. Berry, S. Story, D. Altman, R. Upchurch, W. Whitman, D. Singleton, G. Płaza, R. Brigmon, *Biological Treatment of Petroleum and Radiological Contaminated Soil* (Oxford University Press, 2006)
4. C.L. Keat, A. Aziz, A. Eid, N.A. Elmarzugi, Biores. Bioprocess **47** (2015)
5. C. Malarkodi, C. Rajeshkumar, S. Paulkumar, M. Vanaja, G. Gnanajobitha, G. Annadurai, Bioinor. Chem. Appl. **347** (2014)
6. D. Nath, P. Banerjee, Environ. Toxicol. Pharmacol. **997** (2013)
7. G.A. Płaza, M. Pacwa-Płóciniczak, Z. Piotrowska-Seget, R. Brigmon, E. Król, *Characterization of Bacillus strains producing biosurfactants* (Springer Science Business Media, 2015)
8. G.D. Christensen, W.A. Simpson, J.J. Younger, L.M. Baddour, F.F. Barrett, D.M. Melton, and E. H. Beachey, J. Clin. Microbiol. **996** (1985)
9. G. Franci, A. Falanga, S. Galdiero, M. Rai, G. Morelli, M. Galdiero, Molecules, **8856** (2015)
10. G. Płaza, J. Chojniak, B. Mendrek, B. Trzebicka, L. Kvitek, A. Panacek, R. Prucek, R. Zboril, K. Paraszkiewicz, P. Bernat, IET Nanobiotechnol. **62** (2016)
11. H.C. Flemming, J. Wingender, Nat. Rev. Microbiol. **623** (2010)
12. I. Biedroń, T. Traczewska, G. Płaza, Rocz Ochr Sr. **86** (2016)
13. I. Roy, A. Gaur, A. Jain, S. Bhattachaya, V. Rani, Environ. Toxicol. Pharmacol. **807**, (2013)
14. K. Chaloupka, Y. Malam, A. M. Seifalian, Trends Biotechnol. **580** (2010)
15. K. Kalishwaralal, S. BarathManiKanth, S. R. K. Pandian, V. Deepak, S. Gurunathan, Colloids Surfaces B Biointerfaces **340** (2010)
16. M. Rai, A. Yadav, A. Gade, Biotechnol. Adv. **27**, 76–83 (2010)
17. N.K. Palanisamy, N. Ferina, A.N. Amirulhusni, Z. Mohd-Zain, J. Hussaini, L.J. Ping, R. Durairaj, J. Nanobiotechnology **2** (2014)
18. P. Chaudhari, S. Masurkar, V. Shidore, S. Kamble, Nano-Micro Lett. **34** (2012)
19. S. Nookala, N. Tollamadugu, G. Thimmavajjula, Adv. Nano Res. **97** (2015)
20. W. Li, F. Wang, J. Zhang, Sci. Total Environ. **499** (2016)
21. Y. Yang, M. Li, C. Michels, H. Moreira-Soares, P.J. Alvarez, Environ. Toxicol. Chem. **2234** (2014)
22. Z. Sheng, Y. Liu, Water Research, **6039** (2011)
23. Z. Yuan, J. Li, L. Cui, B. Xu, H. Zhang, C. Yu, Chemosphere, **1404** (2013)

DOI: 10.1051/e3sconf/20172200030

Use of CFD modelling for analysing air parameters in auditorium halls

Robert Cichowicz[1,2,*]

[1]Lodz University of Technology, Faculty of Architecture, Civil and Environmental Engineering, Al. Politechniki 6, 90-924 Łódź, Poland
[2]Venture Industries Sp. z o. o., Mokra 27, 05-092 Łomianki – Kiełpin, Warszawa, Poland

Abstract. Modelling with the use of numerical methods is currently the most popular method of solving scientific as well as engineering problems. Thanks to the use of computer methods it is possible for example to comprehensively describe the conditions in a given room and to determine thermal comfort, which is a complex issue including subjective sensations of the persons in a given room. The article presents the results of measurements and numerical computing that enabled carrying out the assessment of environment parameters, taking into consideration microclimate, temperature comfort, speeds in the zone of human presence and dustiness in auditory halls. For this purpose measurements of temperature, relative humidity and dustiness were made with the use of a digital microclimate meter and a laser dust particles counter. Thanks to the above by using the application DesignBuilder numerical computing was performed and the obtained results enabled determining PMV comfort indicator in selected rooms.

1 Introduction

The assessment of the interior environment should be carried out in relation to numerous physical-chemical factors, because their changes can influence the quality of the interior air [1]. The above is the consequence of the fact that in rooms at the same time there can stay different groups of people, and accordingly there can be noted a variable level of air pollution and different amount of heat and humidity [2–3].

One of the most important parameters of air quality is CO_2 concentration in the air, because it influences the effectiveness of work of people in a given room (concentration above 1000 ppm causes the feeling of discomfort and drowsiness [4–5]). Whereas there exists yet another equally significant factor having a negative influence on wellbeing. The factor is dustiness in a room, which even in a very low concentration can be a threat for health [6–7]. Due to this reason each system of mechanical ventilation installed in a building occupied by people should be designed to provide for the necessity to remove the appearing pollution and gains of heat and humidity [8]. It should also be remembered that when carrying out renovations of ventilation systems and controlling the internal

* Corresponding author: robert.cichowicz@p.lodz.pl

environment it is important to concentrate also on thermal comfort of the persons staying in a given room [9].

When analysing the sensation of comfort while staying in a room one of the basic needs of human organism should be taken into account, which is maintaining a stable body temperature. In order to fulfil this condition it is necessary to maintain the balance in a number of parameters. The elements of heat exchange between a person and the environment are the following: energy expenditure, air temperature, average temperature of radiation, partial pressure of vapour, air speed and resistance for heat conductivity through clothing (the equation describing the phenomena taking place between a human organism and the environment has been determined by Fanger). In order to determine the distribution of comfort indicator it is necessary to determine the value of metabolism of the people staying in the room and the insulation properties of their clothing. A person's metabolism is defined as the energy returned to the environment obtained by means of chemical processing of eaten food (its unit is 1 met, which means 58.2 W/m^2 of the returned energy and characterises a person at rest) [10].

The assurance of thermal comfort means the necessity to determine whether specific parameters have been fulfilled. The parameters include temperature proper for a given room, relative humidity and, in case of mechanical ventilation being installed in the room, appropriate speeds of air flow in the human presence zone (the data is usually needed to carry out various types of computer analyses) [6, 11].

Also other parameters that significantly influence the conditions sensed by the persons in a room, such as temperature of partitions, type of heating system and distribution of temperatures in a room, have to be specified [11].

All these parameters generate a great amount of data, difficult to analyse each separately in a transparent and fast way. Therefore a PMV indicator is used. The indicator provides for 6 variable parameters, such as: relative humidity, temperature, air speed, average temperature of radiation, the level of activity of persons in the room and insulation properties of clothing. The range of the indicator includes values from -3 (sensation of being cold) to +3 (sensation of being hot). Whereas the level of comfort in a given zone of the room is assured when the indicator remains within the range from -0.5 to +0.5 [6, 12]. In order to correctly interpret the obtained results cross-sections of the room with distribution of the indicator at the height of 1.1. above the floor has to be analysed in case of auditory halls, because this height is analogous to the location of heads of people in a seated position.

Determining proper conditions for a given room demands defining its purpose. An auditory hall is a place that can accommodate as many as a few dozen or even a few hundred people. In rooms of such dimensions it is critical to maintain comfort in the zone of human presence. It is a common case that the temperature or air speed outside of this zone can be far from accepted standards. The reason for that can be the lack of possibility to supply the room with an appropriate amount of air and at the same time to maintain in the whole cubature of the room of the air speeds that are not sensed by the rooms' users (the problems usually occur in high rooms) [3].

In accordance with the standard ISO 7730:2005 [13] for rooms having the properties of auditory halls the air temperature in the heating season should be maintained within the range from 20°C to 24°C, and in the summer season between 23 and 26°C. Requirements for relative humidity are 40% to 60% during winter and 30–60% in summer. Yet it has to be remembered that when analysing the comfort of seated persons, the speed of air should be controlled in the zone of human presence so as not to exceed 0.19 m/s in summer and 0.16 m/s in winter [13].

2 Numerical modelling

Computational Fluid Dynamics (CFD) enables analysing fluids flow together with heat exchange between them. CFD is more and more popular in modelling the phenomena taking place in interiors. Thanks to the use of this method already at the stage of architectural and building structure design it is possible to analyse the best solution for the whole building as well as for its ventilation or the air-conditioning system, and thus it is possible to reduce the costs of prospective possible failures or modernisation [3].

CFD modelling enables providing for proper ventilation openings, speeds of the intake air and the range of the streams of an assumed speed. It is also possible to optimise a given design from the perspective of assuring the best possible comfort conditions in a room, taking into account convection movements caused by the presence of people in the analysed space [14].

The existing computer software enables modelling by discretizing variable equations into algebraic equations, which is associated with assuming certain approximations, i.e. limiting the model with the use of one of the following methods: differences, capacity or finite elements. It is also always important to introduce specific boundary conditions into a mathematical model [15]. Calculations in computer applications are usually performed with the use of a three-dimensional net, which describes the analysed space with a selected accuracy [14, 16].

In order to start the work on a selected building or a room it is necessary to specify its shape in the application. This stage is very important and time consuming. The analyses presented in the article have been performed in DesignBuilder, an application characterised by intuitive use and by adjustment to modelling of any type of building. The application enables analysing models form the perspective energetic effectiveness, building costs and optimisation of interior systems. DesignBuilder also has a module for performing CFD calculations for a building as well as for its surroundings [17]. Also by introducing required input parameters it is possible to carry out numerical computing, thanks to which distribution of temperature in a room and range of air stream of a characteristic speed are obtained [3, 14, 18].

3 Methodology of measurements

Experimental research was carried out in auditory halls. The first of the halls, marked as A has the total area of 118.68 m^2 (the highest point in the hall is located at the height of 6.20 m) and is designed for 110 persons. The hall is equipped with a system of mechanical intake-exhaust ventilation with a constant capacity of 5550 m^3/h. Another hall, marked B, has no windows, and is the largest of the analysed rooms (its area is 199.21 m^2, and the maximum height is 5.90 m). The last of the analysed halls, marked as C, is a mirror reflection of the auditory hall A.

In all the three auditory halls a microclimate research was carried on, and thus information necessary to perform the CFD analysis in DesignBuilder application was obtained. A digital meter of microclimate BABUC/A was applied. With the use of this meter the air parameters, among others the temperature of the surroundings and relative humidity were determined. Also the temperature of surfaces of partitions in the room was measured. Additionally such parameters as metabolism, the average relative humidity and insulation properties of clothing were determined (obtained decomposition indicator PMV). The measurements were supplemented with the research on dustiness of the outdoor air. They were performed with the use of laser particle counter Icleen PCM. In the halls measurement points were determined (Fig. 1). In these points two series of measurements

 DOI: 10.1051/e3sconf/20172200030

were performed (in the halls A and C they were made with windows closed and open), and each measurement had a duration of 10 minutes ($\sum = 2 \times 3 \times 10$ minutes).

4 Results of measurement and numerical calculations

The following parameters were measured for all the auditory halls: the temperature of the surface of the external or inner walls, if applicable – windows temperature, the temperature of the supplied air and the temperature of radiators. This information was introduced into the computer application as boundary conditions for further analyses (table 1).

Table 1. Parameters necessary for numerical analyses performing (boundary conditions).

Temperature of external walls	19.0°C
Windows temperature	17.0°C
Temperature of inner walls	21.0°C
Temperature of supplied air	21.5°C
Temperature of the surface of central heating elements	30.0°C

In order to perform calculations and simulations of air flows in particular auditory halls the values of inflowing and removed stream were provided in post-completion documentation of the mechanical intake-exhaust ventilation system.

In the auditory hall A the model was assumed in which the room was filled with students in order to present the most unfavourable conditions. The effect of such an assumption is Fig. 1, which presents distribution of temperatures with the vectors of air speed in the analysed room, with special stress put on the human presence zone.

When analysing the presented results of calculations (Fig. 1) a great influence of persons in the room on distribution of temperature and air speed in the room can be noted. Considerable heat gains from people cause additional convection movements, and the niche, visible to the right on the cross-section, has a negative influence on the comfort of being in the room due to warm air that gathers there. A solution to this problem could be accepting the concept proposed by Koper P. and Lipska B. M. [3], who suggest that the best way to divide the air is by application of an upper nozzle air supply from the back of the room together with a low exhaust along the stairs. Such an arrangement could also improve distribution of temperature in a room, because the application of ceiling anemostates would heighten temperature in the human presence zone.

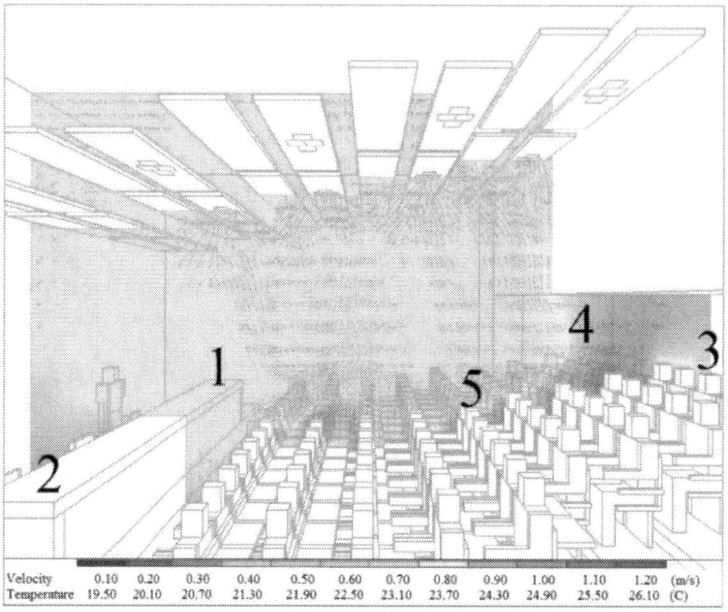

Fig. 1. Distribution of temperature and the vectors of speed on the cross-section of the auditory hall.

Fig. 2 and 3 present distribution of PMV indicator, which was shown from two different levels, in order to assess thermal comfort of the persons staying in the zone of the room designed for human presence. The distribution was achieved by introducing the metabolism of persons staying in the room at the level of 1.2 met [13], insulation properties of clothing as 1 clo and the average relative humidity calculated on the basis of measurements as 34%.

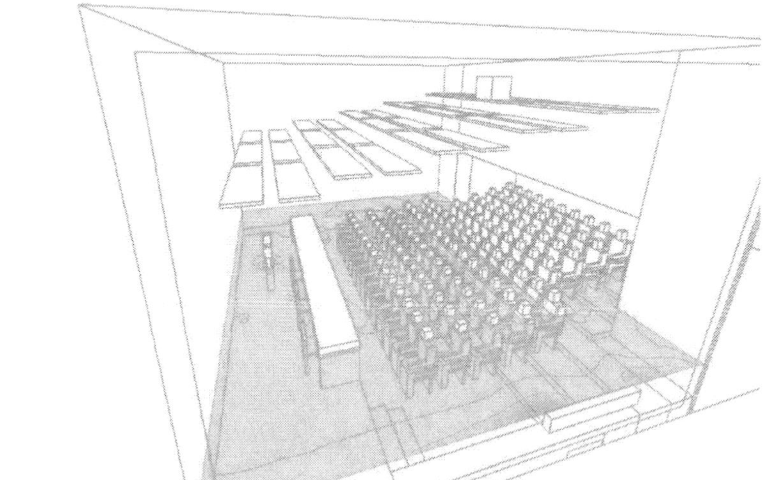

Fig. 2. Distribution of PMV indicator at the height of 1.1 m.

In the zone of human presence the obtained values of PMV indicator are included within the recommended range of values from -0.5 to 0.5. Such a distribution of indicator suggests fulfilment of all conditions of comfort. Whereas low temperature of windows' surface causes the situation in which in their proximity the indicator drops to the value below -0.5. Yet due to lack of sitting places in this area, this situation has no direct influence on the sensation of thermal comfort in the room.

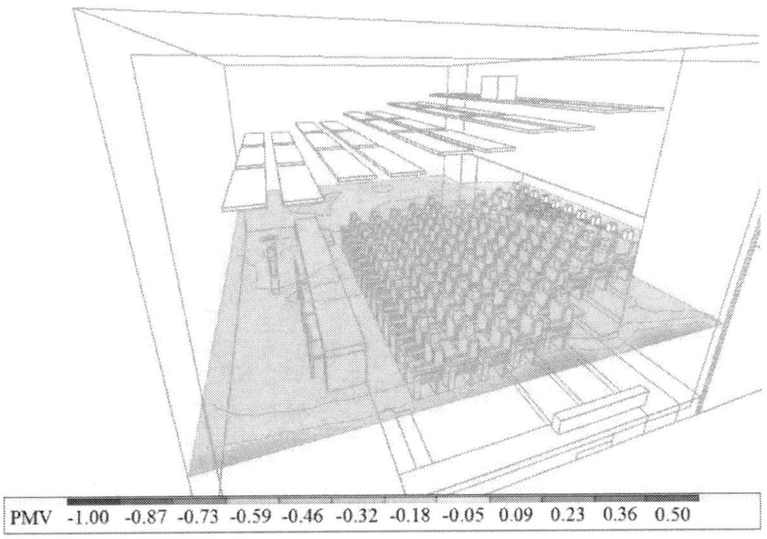

Fig. 3. Distribution of PMV indicator at the height of 2.1 m.

Fig. 4 presents the change of value of dustiness referred to the period of performing the measurements in all the three auditory halls.

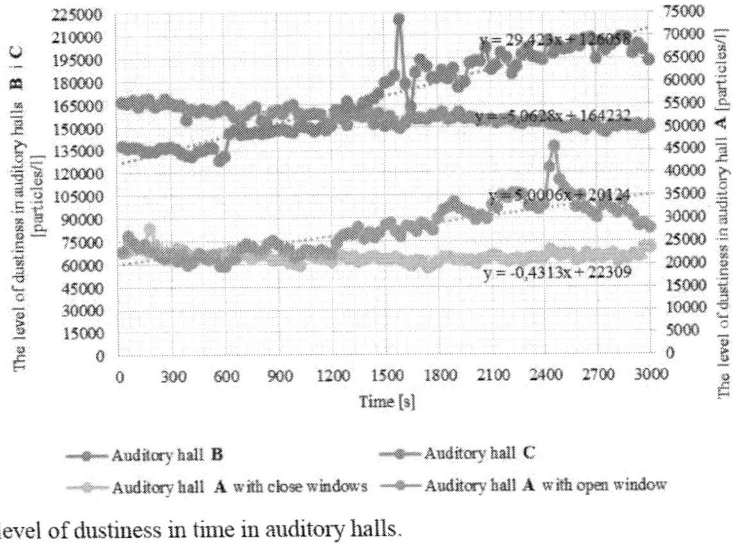

Fig. 4. The level of dustiness in time in auditory halls.

During the measurements in auditory hall B the ventilation system was switched on, and measurements in each following point indicated the increase of dustiness values (blue line on Fig. 4). This visible relation between the mechanical ventilation system work and the increased level of dustiness has also been observed by other scientists, According to Parker J. L., Larson R. R., Eskelson E., Wood E. M. and Veranth J. M. [19] a mechanical ventilation system is the main factor causing dustiness in rooms. While on the basis of measurements made in the auditory hall C, a constant level of dustiness was observed when the ventilation system was switched off (orange line in Fig. 4). This relation was confirmed by the results of measurements performed in hall A (yellow line in Fig. 4) because with the ventilation system switched off the level of particles concentration in the whole room remained between 21000 and 22000 particles/l, while opening of the windows caused the

 DOI: 10.1051/e3sconf/20172200030

increase of dustiness in time (grey line in Fig. 4). It can also be seen while analysing the levels of dustiness depending on the location of a given measurement point (Fig. 5).

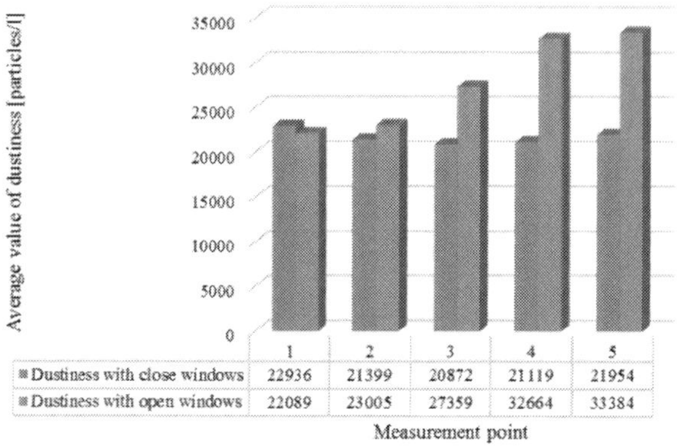

Fig. 5. The average value of dustiness in measurement points in room A with closed and open windows.

The results of measurements presented in Fig. 6 show a constant temperature of around 20°C. It can be noticed that opening of the windows caused a drop of relative humidity from 31.5% to 27.5%. The average value of dustiness from all measurements is 27700 particles/l. The lowest dustiness was measured in point 1, and it was lower from the average dustiness in the hall by 20.3%. While in the 5th measurement point the dustiness was the highest, and it exceeded the average dustiness by 20.5%.

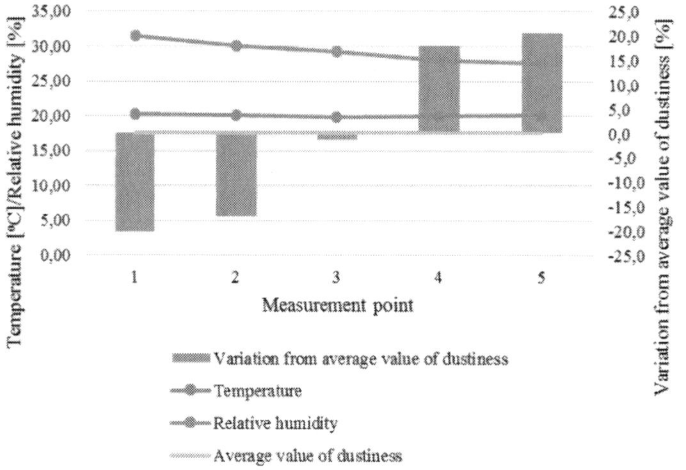

Fig. 6. Average dustiness, temperature and relative humidity in measurement points in room A.

5 Summary

Numerical calculations performed with the use of CFD enabled obtaining precise distributions of temperature and air speed in selected auditory halls. This in turn enabled determining the PMV indicator for all rooms, thanks to which it was possible to assess the conditions of thermal comfort in the zone of human presence. By analysing the obtained results it was found out that in all the rooms conditions for parameters of the environment temperature and air speed are compliant with the standard ISO 7730:2005 [13], and the PMV indicator ranges within the recommended -0.5 to 0.5 (of course with the assumed values for metabolism, insulation properties of clothing and relative humidity).

On the basis of measurements of air dustiness the influence of mechanical ventilation system on the increase of dustiness in a given room was stated. The dustiness in the analysed auditory halls increased significantly during ventilation through airing. Whereas when analysing the dependence of the number of dust particles in the function of relative humidity in a room, a drop of dustiness together with the increase of relative humidity was observed.

Acknowledgements

The author thank Mr. Jakub Florek MSc. for its contribution to the measurement of air samples and for his assistance technical support in graphic design work.

References

1. M. Kavgic, D. Mumovic, Z. Stevanovic, A. Young, Energy and Buildings **40**, 1334–1343 (2008)
2. A. Leavey, Y. Fu, Building and Environment **85**, 287–297 (2015)
3. P. Koper, B.M. Lipska, Ciepłownictwo, Ogrzewnictwo, Wentylacja **41**, 7–8 (2010)
4. R. Cichowicz, H. Sabiniak, G. Wielgosiński, Ecol. Chem. Eng. S. **22**, 1, 61–71 (2015)
5. I. Salma, Atmospheric Environment **64**, 219–228 (2013)
6. ISIAQ-CIB Task Group TG 42: Performance criteria of buildings for health and comfort, CIB **292** (2004)
7. Z.B. Klaić, Geofizika **32** (2015)
8. R. Cichowicz, G. Wielgosiński, A. Targaszewska, Ecol. Chem. Eng. S. **23**, 1, 49–60 (2016)
9. Y. Fu, M. Sha, Proceedings – International Conference on Distributed Computing Systems **8**, 73–82 (2014)
10. S. Gilani, M. Khan, W. Pao, Energy Procedia **75**, 1373–1379 (2015)
11. P.O. Fanger, *Thermal comfort* (Arkady, Warszawa, 1974)
12. A. Garnier, J. Eynard, M. Caussanel, S. Grieu, Journal of Process Control **24**, 6, 750–762 (2014)
13. ISO 7730:2005: Ergonomics of the thermal environment. Analytical determination and interpretation of thermal comfort using calculation of the PMV and PPD indices and local thermal comfort criteria. (2005), with amendments
14. B. Lipska, Ciepłownictwo, Ogrzewnictwo, Wentylacja, **10** (1999)
15. J.K. Calautit, B.R. Hughes, Building and Environment **80**, 71–83 (2014)
16. T. Mikeska, J. Fan, Energy and Buildings **107**, 59–67 (2015)
17. http://www.designbuilder.co.uk/
18. B. Lipska, A. Palmowska, P. Ciuman, P. Koper, Instal, **3** (2015)
19. J.L. Parker, R.R. Larson, E. Eskelson, E.M. Wood, J.M. Veranth, Indoor Air **18**, 5, 386–393 (2008)

Power plant wastes capitalization as geopolymeric building materials

Gabriela Ciobanu[1], *Loredana* Litu[1], and *Maria* Harja[1,*]

[1]"Gheorghe Asachi" Technical University of Iasi, Faculty of Chemical Engineering and Environmental Protection, Prof. dr. doc. D. Mangeron Street no. 73, 700050, Iasi

Abstract. In this innovative study, we are present an investigation over the properties of geopolymeric materials prepared using ash supplied by power plant Iasi, Romania and sodium hydroxide solutions/pellets. Having as objective a minimum consumption of energy and materials was developed a class of advanced eco-materials. New synthesized materials can be used as a binder for cement replacement or for the removal/immobilization of pollutants from waste waters or soils. It offers an advanced and low cost-effective solution too many problems, where waste must be capitalized. The geopolymer formation, by hydrothermal method, is influenced by: temperature (20–600°C), alkali concentration (2M-6M), solid /liquid ratio (1–2), ash composition, time of heating (2–48 h), etc. The behaviour of the FTIR peak of 6M sample indicated upper quantity of geopolymer formation at the first stage of the reaction. XRD spectra indicated phases like sodalite, faujasite, Na-Y, which are known phases of geopolymer/zeolite. Advanced destroyed of ash particles due to geopolymerisation reaction were observed when the temperature was higher. At the constant temperature the percentage of geopolymer increases with increasing of curing time, from 4–48 h. Geopolymer materials are environmentally friendly, for its obtaining energy consumption, and CO_2 emission is reduced compared to cement binder.

1 Introduction

Geopolymers are a relatively new amorphous alumino-silicate with best properties such as mechanical resistance, thermal stability, durability, hardness, chemical stability, corrosion stability, etc. [1, 2]. Geopolymers show characteristics of low shrinkage and excellent heat resistance. Geopolymer binder can be used as a partial or complete replacement of conventional cement to produce concrete [3-5]. Geopolymer concretes can be used in building, repairing infrastructures, in livestock, construction etc.

The geopolymers have empirical formula: $Me_x[-(SiO_2)_y-AlO_2]_x\ zH_2O$, where y is 1, 2 or 3; Me is an alkali cation, such as potassium or sodium, and x is the degree of polymerisation [4–9].

Geopolymer binders can be produced from materials containing Si and Al ions. These can be raw materials or wastes (by-products) such as fly ash, ground granulated blast,

* Corresponding author: maria_harja06@yahoo.com

furnace slag, metakaolinit or any other waste sources of Si and Al [5, 6, 10–13]. The material must be thermal activated in presence of sodium or potassium hydroxide, sodium metasilicate and water. Most frequently geopolymer, a novel eco-material, are prepared from fly ash and alkaline solution by different activation methods: sol–gel or diffusion methods at different temperatures. In the first stage raw materials are dissolved in high pH alkaline solution, process facilitated by heating, after this the geopolymers obtained are precipitated. In the process of the geo-polymerization reactions, polysialates, polysialate siloxo, and polysialate disiloxo are formed [14–16].

By using fly ash, first environmental and economic benefits of obtaining geopolymer eco-materials are due to capitalization a wastes from energy manufacture, 75% aren't used yet [7]. The second benefit is determined by the geopolymer binder is a low-CO_2 cementious material; this technology can save up to 80% of CO_2 emissions caused by the cement industry [9, 17]. On the other hand the geopolymerisation of fly ash offers environmental benefits, such as lower consumption of natural resources (limestone, clay etc.) and lower consumption of energy.

Previous publications on geopolymer materials investigated the effect of curing procedures, composition of the alkaline solution on strength development and hydration products, influence of temperature [18–23]. Other researchers demonstrated that geopolymer concrete is expensive comparatively with ordinary concrete [24].

In this paper, we present an investigation over the properties of geopolymeric materials prepared using ash supplied by power plant Iasi, Romania and sodium hydroxide solutions/pellets. Having as objective a minimum consumption of energy and materials was developed a class of advanced eco-materials. New synthesized materials, having pozzolanic properties, can be used as a binder for cement replacement or for the immobilization of pollutants from wastewaters or soils. It offers an advanced and low cost-effective solution to many problems, where waste must be capitalized. By minimizing the temperature of activation, and recycling the hydroxide solution, the cost of geopolymer is lower compared with the ordinary Portland cement.

2 Experimental

The fly ash (FA) sample supplied from power plant Iasi, was dehydrated in an oven at 105°C, characterized and used throughout all experiments. The conditions used for the synthesis of geopolymeric eco-materials are presented in Table 1.

Table 1. Conditions of geopolymer synthesis.

Sample	Method	Synthesis conditions	Activation time
G1, G2	Direct activation	1 /2 ratio, 6 M, 20°C	7 days, 21 days
G3, G4		1 /2 ratio, 2 M, 50°C	4 h, 8 h
G5, G6		1 /2 ratio, 4 M, 50°C	4 h, 8 h
G7, G8		1 /2 ratio, 2 M, 80°C	4 h, 8 h
G9, G10		1 /2 ratio, 4 M, 80°C	4 h, 8 h
G11/ G12		1 /2 ratio, 6 M, 80°C	4 h, 48 h
G13, G14	Fussion	1/1 ratio fly ash/NaOH pellets, 600°C	1 h, 2 h
G15, G16		1/2 ratio fly ash/NaOH pellets, 600°C	1 h, 2 h

It can be observed from Table 1 that the synthesis of new eco-materials was carried out at ambient temperature, 50°C and 80°C, respectively, in an alkaline medium with NaOH

 DOI: 10.1051/e3sconf/20172200031

concentrations of 2 M, 4M and 6M, respectively, corresponding to a mixture with solid/liquid ratios of 1 : 2. In the case of G13–G16 samples synthesis was carried at 600°C, with solid/solid ratios of 1:1 and 1: 2 using solid pellets, for 1 and 2 h, by fusion method. After treatment, the geopolymer samples were crystallized for 18 h; subsequently, they were filtered, washed at 8-8.5 value of pH, and dried at oven at constant mass. In this study NaOH was produced by Chimcomplex Company.

SEM and EDAX were carried out with a Quanta 3D instrument AL99/D8229 (FEI Company), XRD analysis was conducted with an X'Pert PRO MRD X-ray diffractometer (PANalytical). The IR analysis was performed using a DIGILAB FTS 2000 spectrometer.

3 Result and discussion

For the studied fly ash the following chemical composition was obtained: SiO_2 – 49.82%; Al_2O_3 – 14.70%; Fe_2O_3 -4.44%; CaO-6.21%; MgO-0.96 %; SO_3 -1.18%; et all., in accord with literature [24-26]. The sum of SiO_2 + Al_2O_3 + Fe_2O_3 is more than 70%, the fly ash from Iasi power plant is class F [25]. The ratio SiO_2/Al_2O_3 recommended the fly ash as raw materials for synthesis of geopolymeric and zeolite, fact demonstrated in previously research [27, 28]. The chemical composition, obtained by energy dispersive spectroscopic (EDAX) and elemental analysis of the synthesized materials, for four selected geopolymeric materials is presented in table 2. Its constituent particles have: Si, O, Al, K, Na, Mg, Ca Fe, Ti and C.

Table 2. Chemical composition of geopolymeric materials.

Sample	O	Na	Mg	Al	Si	K	Ca	Ti	Fe
Fly ash	43.32	0.79	0.6	19.09	30.81	1.75	1.05	0.54	2.05
G2	42.85	2.44	0.93	17.33	28.89	1.66	1.44	0.91	2.07
G6	42.93	4.42	0.78	17.37	23.38	2.09	1.19	0.79	2.51
G9	42.77	7.75	1.18	16.38	23.22	0.48	1.40	0.93	2.28
G16	40.37	10.87	1.28	16.31	24.10	0.61	1.69	0.82	2.07

The FTIR spectra obtained for four geopolymeric material are presented in Fig. 1.

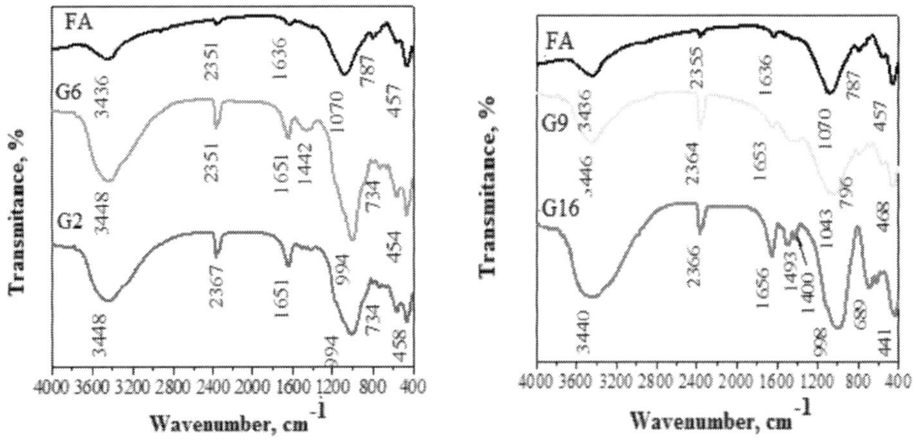

Fig. 1. FTIR spectra for fly ash and synthetized materials.

DOI: 10.1051/e3sconf/20172200031

From Fig. 1 it can be observed that comparing with fly ash the most relevant band is moving in the direction of wavenumber decreasing from 1075–1080 cm^{-1} to 995–1001 cm^{-1}, that means appearance of new bending vibrations in the structure of synthesized material. Same representative picks were recorded between 993 cm^{-1} and 1001 cm^{-1} corresponding to asymmetric stretching Si–O–Si and Al–O–Si. The metakaolinite (Si–O–1050 cm^{-1}) band moving to 990 cm^{-1}. Bands corresponding to alumino-silicates compounds are around 995 cm^{-1}. Other characteristics bands for inorganic polymer (geopolymer) are 695 and 445–460 cm^{-1}, when there is occur substitution of Si by Al [13, 29, 30].The bands that occur within 500–650 cm^{-1} indicating the silicate and alumino-silicate vitreous phase [2, 4, 5,18, 29].

Geopolymerisation rates are determined from the intensity variation of the bands associated to the geopolymer gel network/unreacted fly ash. In the first step have an initial period (induction), followed by gel evolution by network formation. The period of induction is diminished by increasing the alkaline solution concentration. The increasing of NaOH concentration determines increase rate of network formation [30].

Influence of temperature over materials morphology is presented in Fig. 2. Experimental studies had shown that a proportion of spherical particles of initial fly ash are found in the materials obtained at temperatures up to 100°C, while increase temperature to 600 ° C determines an advanced destroy of particles [31, 32].Temperature is an important parameter in obtaining geopolymeric materials. Increasing of the temperature determines accelerated of chemical reactions with changing of crystalline network, Fig. 2.

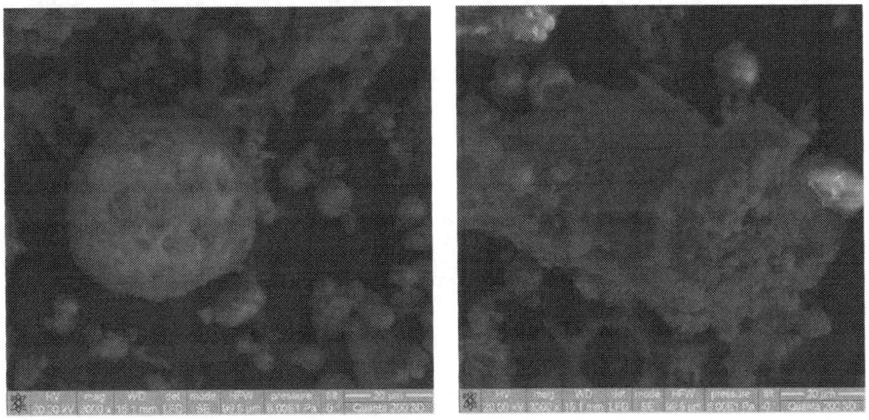

Fig. 2. Influence of temperature, activation method (G5 and G9).

The temperature increase, for fusion method, causes the most significant changes, with appearances of rhombohedra crystals, which mainly growing on the old solid, Fig.3.

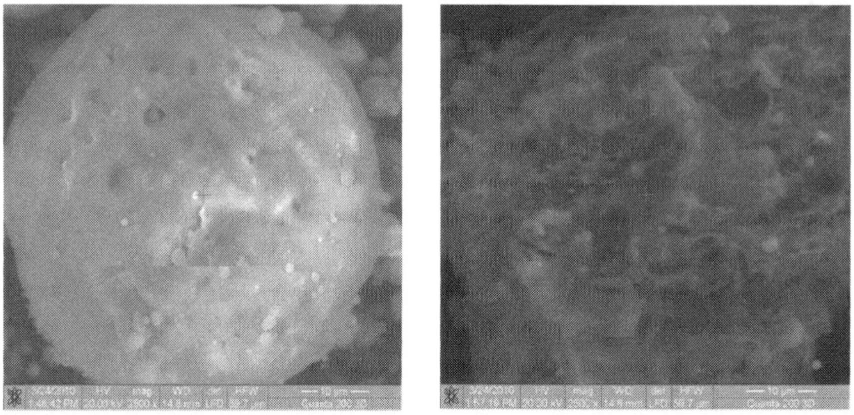

Fig. 3. Comparison between activation method and fusion method (G6 and G16).

At temperatures before 50°C is observed spherical unmodified ash particles, on these increases the new geopolymeric particles. Geopolymerisation degree is too significant at higher temperature, Fig. 4.

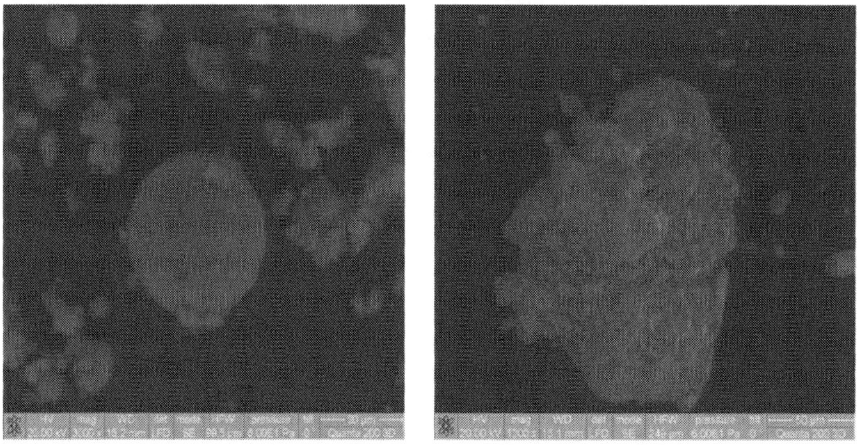

Fig. 4. Comparison between activation method and fusion method (G1 and G15).

Contact time influences the degree of geopolymerisation, Fig. 5. It were studied the influence of contact time at different temperature of curing.

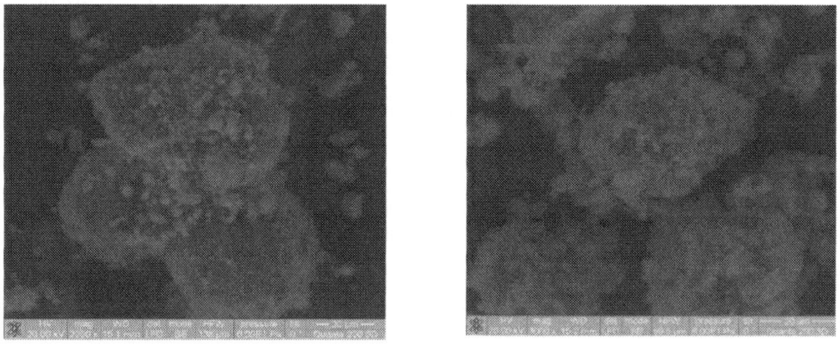

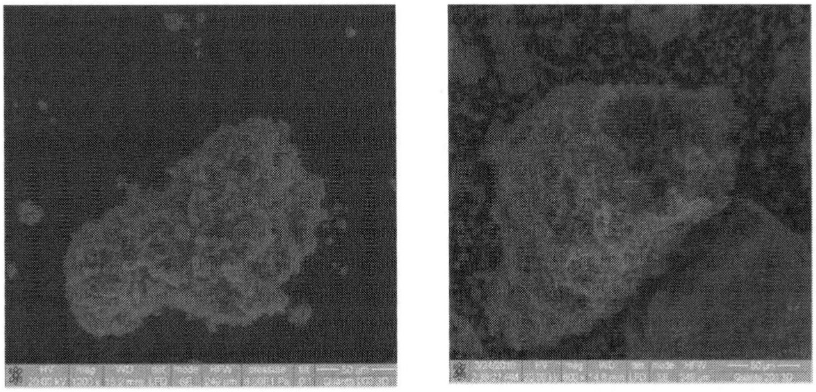

Fig. 5. Influence of the contact time (G5, G6, G13 and G14).

The synthetized geopolymeric materials contain multiple diffraction peaks (Fig. 6), that demonstrated appearance of new compounds compared with initial ash [26]. Identification of the peaks was performed according to the program PCPDF Win (1999). XRD analysis shows that the analyzed samples contain: quartz (Q), mullit (M), sodalite (S), feldspar (F), chabasite (Cha), clinotobermorit (CT), gismodine (Gis), Linde L (L) herchelite (Her), faujasit (Fau), Na-Y, tobermorite (T) [32]. From Fig. 6 it can be seen that sodalite was formed in all samples, it has a high stability at different temperatures [27]. It can be seen also that at materials synthesized at low temperatures peaks corresponding to quartz are intense, which demonstrated a relatively low degree of geopolymerization. Na-Y compound appear only at relatively high temperatures. G12 and G16 also shows, peaks corresponding Fau and Linde L.

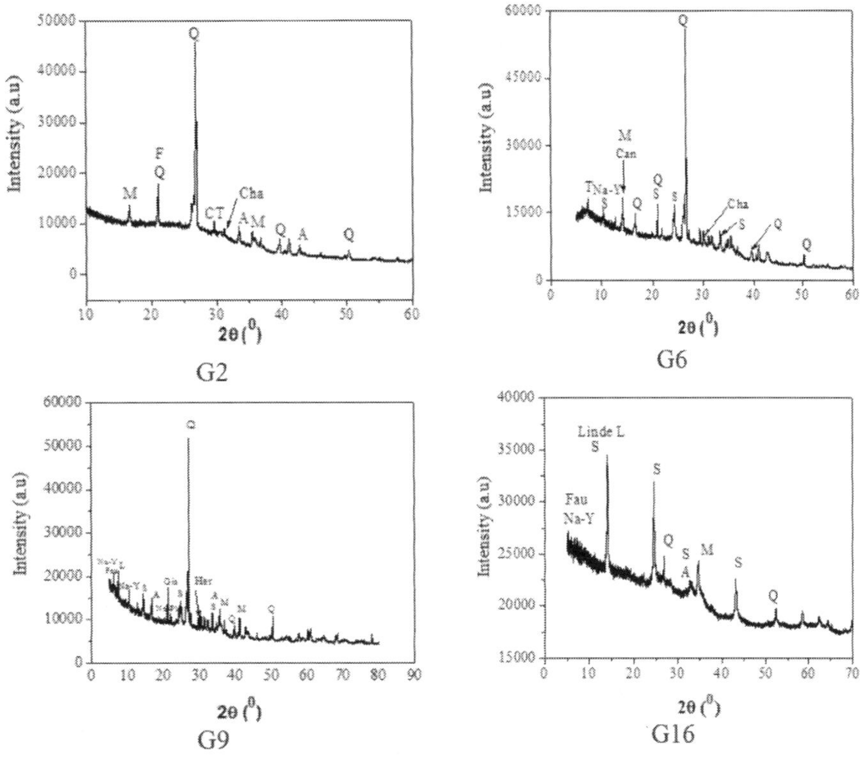

Fig. 6. XRD analysis for geopolymeric materials.

The trend of using of by-product materials, in the area of materials production, is likely when economic effectiveness increasing during fabrication [33]. In this context is necessary an economic evaluation of replacement of cement with geopolymer from fly ash as a possible alternative material [34]. The basic input is the cost of materials for the manufacture of geopolymer from fly ash as a substitute for cement. Evaluation is carried out based on analysis of the possibilities to reduce the total price of material costs with possible using in building material.

The method used for economic evaluation is the classical cost calculation. On the other hand, the cost of the geopolymer obtaining depends on the conditions of synthesis.
Our economic evaluation start from activation method at 80°C, solid/liquid ratio 1/2, 4 M NaOH solution, 4 h of treatment. Experimental results demonstrated that geopolymer namely G9 have chemical and technological characteristics, comparable with geopolymer obtained by diffusion methods, needed for used as cement replacement.

In actual context (capitalization of waste is priority), the price of one ton of fly ash is very cheap. The price of alkaline liquids needed to the make the geopolymer powder is 185–200 \$/t, water 1.05 \$/m^3 and caloric energy 56.28 \$/Gcal. The price needed for obtaining 1 t of geopolymer powder is 77.07 \$, in our country the price of Portland cement is 104.65 \$/t. In this condition the price of fly ash-geopolymer is with 26.3% cheaper than Portland cement, in accordance with literature estimated to10 to 30 percent [35, 36].

The calculation does not take into consideration that the NaOH solution can be recycled, after concentration correction to 4 M. Then, if geopolymerisation is realized into the power plants would save the heating agent, flue gas at 120–140°C after the economizer, before eviction to desulphurization or discharged. With these improvements geopolymers becomes cheaper than Portand cement, ensuring significant savings in price and energy. On the other hand reduction of CO_2 emissions caused by the cement industry, determine price reduction by carbon-credit that currently has a redemption value of about 5–6\$/eCO_2 [37].

4 Conclusion

Taking into account the concept of sustainable development, the by-product of one industry can be a raw material for other industries. The capitalization of waste has an important economic impact on the environment, by saving energy, raw materials and reducing pollution caused by storage of waste.

Having as objective a minimum consumption of energy and materials was developed a class of advanced eco-materials. New synthesized materials, having pozzolanic properties, can be used as a binder for cement replacement. It offers an advanced and low cost-effective solution to many problems, where waste must be capitalized.

The geopolymer formation, by hydrothermal route, is influenced by: temperature, alkali concentration, solid /liquid ratio, ash composition, time, pH etc.

Geopolymer samples were characterized by SEM, EDAX, FTIR, XRD, BET and pozzolanic activity. Regarding the characterization of synthesized materials we conclude:

- FTIR analysis demonstrated that the higher peak intensity was recorded for 6 M alkali solution. The behavior of the FTIR peak of 6M sample indicated upper quantity of geopolymer formation at the first stage of the reaction.
- XRD spectra indicated that the intensity of crystalline phases was increased in alkali concentrated solutions.
- Phases like sodalite, faujasite, Na-Y were identified by XRD which are known phases of geopolymer/zeolite.
- Advanced destroyed of ash particles due to geopolymerisation reaction were observed from SEM study when pellets of NaOH were used and the temperature was higher.

• At the constant temperature the percentage of geopolymer increases with increasing of curing time, from 4 to 48 h.

• The price of fly ash-geopolymer is with 26.3% cheaper than Portland cement, it serves as a better alternative for replacement of Portland cement in the building industry.

Geopolymer materials are environmentally friendly, for its obtaining moderate energy consumption, and CO_2 emission is reduced compared to ordinary cement binder.

This work was performed with the support of Doctoral school – Faculty of Chemical Engineering and Environmental Protection.

References

1. A. M.Aguirre-Guerrero, R. A. Robayo-Salazar, R. M. de Gutiérrez, Appl. Clay Sci. **135**, 437 (2017)
2. T. Bakharev, Cem. Concr. Res. **35**, 658 (2005)
3. Z. Abdollahnejad, F. Pacheco-Torgal, J.B. Aguiar, C. Jesus, Key Eng. Mater. **634**, 113 (2014)
4. M. Harja, M. Bărbuţă, M. Gavrilescu, Env. Eng. Manag. J. **8**, 513 (2009)
5. M. Harja, M. Bărbuţă, M. Gavrilescu, Env. Eng. Manag. J. **8**, 1021 (2009)
6. E.A. Azimi, M.M. Al Bakri Abdullah, L.Y. Ming, H.C. Yong, K. Hussin, I.H. Aziz, Rev. Adv. Mater. Sci. **44**, 273 (2016)
7. T. Bakharev, Cem. Concr. Res. **35**, 1233 (2005)
8. J.C. Swanepoel, C.A. Strydom, Appl. Geochem. **17**, 1143 (2002)
9. J. Davidovits, J. Davidovits (Ed.), Proceedings of the World Congress Geopolymer, Saint Quentin, France (2005)
10. A. Hajimohammadi, J. van Deventer, Waste Biomass Valor. **8**, 225 (2017)
11. M. Barbuta, N. Taranu, M. Harja, Env. Eng. Manag. J. **8**, 1145 (2009)
12. M. Harja, M. Barbuta, Adv. Sci. Letter **19**, 455 (2013)
13. M. Criado, A. Fernández-Jiménez, A. Palomo, Micr. Mesop. Mater. **106**, 180 (2007)
14. J. Davidovits, US Patent 4472199 (1984)
15. J. Davidovits, USPatent 4349386 (1982)
16. H.A. Abdel-Gawwad, S.A. Abo-El-Enein. HBRC J. **12**, 13 (2016)
17. B.C. McLellan, R.P. Williams, J. Lay, A.Van Riessen, G.D. Corder, J. Cleaner Prod. **19**, 1080 (2011)
18. A. Palomo, M.W. Grutzeck, M.T. Blanco, Cem. Concr. Res. **29**, 1323 (1999)
19. G. Habert, J.B. D'Espinose de Lacaillerie, E. Lanta, N. Roussel, 2nd International Conference on Sustainable Construction Materials and Technologies (2010)
20. J. Davidovits, Concretes, **1**, SRIBM, Kiev, Ukraine (1994)
21. A. Narayanan, P. Shanmugasundaram, Energ. Build. **138**, 539 (2017)
22. I. Nikolić, M. Tadić, D. Đurović, R. Zejak, B. Mugoša, Env. Prot. Eng. **41**, 2 (2015)
23. J. Giesekam, J.Barrett, P. Taylor, A. Owen, Energ. Build. **78**, 202 (2014)
24. Z. Sarbak, M. Kramer-Wachowiak, Powder Technol. **123**, 53 (2002)
25. M. Harja, M. Barbuta, L. Rusu, N. Apostolescu, Env. Eng. Manag. J. **7**, 289 (2008)

26. R. D. Bucur, C. Cimpeanu, M. Barbuta, G. Ciobanu, G.Paraschiv, M. Harja, J. Food Agric. Env. **12**, 943 (2014)
27. F. Noli, G. Buema, P. Misaelides, M. Harja, J. Radioanal. Nuclear Chem. **303**, 2303 (2015)
28. G. Buema, F. Noli, P. Misaelides, D. M. Sutiman, I. Cretescu, M. Harja, J. Radioanal. Nuclear Chem. **299**, 381 (2014)
29. W.K.W. Lee, J.S.J. van Deventer, Langmuir **19**, 8726 (2003)
30. C.A. Rees, J.L.Provis, G.C. Lukey, J.S.J. van Deventer Langmuir **23**, 9076 (2007)
31. R.C. Ciocinta, M. Harja, D. Bucur, G. Buema, J. Food Agric. Env. **11**, 1108 (2013)
32. S. Curteanu, G. Buema, C.G. Piuleac, D. M. Sutiman, M. Harja, J. Ind. Eng. Chem. **20**, 597 (2014)
33. G. Mucsi, J. Lakatos, Z. Molnár, R. Szabó, The 9th International Conference "Environmental engineering (2014)
34. Š. Vilamová, M. Piecha, Acta Montan. Slov. **21**, 139 (2016)
35. N.A. Lloyd, B.V. Rangan, 35th Conf. Our world in concrete & Structures (2010)
36. D. Hardjito, B.V. Rangan, Research Report GC 1 (2005)
37. https://www.investing.com/commodities/carbon-emissions-historical-data

 DOI: 10.1051/e3sconf/20172200032

Assessment of the possibility of using data mining methods to predict sorption isotherms of selected organic compounds on activated carbon

Lidia Dąbek[1,*], *Bartosz* Szeląg[1], and *Anna* Picheta-Oleś[2]

[1]Kielce University of Technology, Faculty of Environmental, Geomatic and Energy Engineering, al. Tysiąclecia Państwa Polskiego 7, Kielce, Poland
[2]Marshal's Office of the Świętokrzyskie Voivodeship, al. IX Wieków 4, Kielce, Poland

Abstract. The paper analyses the use of four data mining methods (Support Vector Machines, Cascade Neural Networks, Random Forests and Boosted Trees) to predict sorption on activated carbons. The input data for statistical models included the activated carbon parameters, organic substances and equilibrium concentrations in the solution. The assessment of the predictive abilities of the developed models was made with the use of mean absolute error (MAE), mean absolute percentage error (MAPE), and root mean squared error (RMSE). The computations proved that methods of data mining considered in the study can be applied to predict sorption of selected organic compounds on activated carbon. The lowest values of sorption prediction errors were obtained with the Cascade Neural Networks method (MAE = 1.23 g/g; MAPE = 7.90% and RMSE = 1.81 g/g), while the highest error values were produced by the Boosted Trees method (MAE=14.31 g/g; MAPE = 39.43% and RMSE = 27.76 g/g).

1 Introduction

Ever increasing industrialization is associated with the generation of wastewater that has complex chemical composition, and contains hazardous, carcinogenic and mutagenic substances. Due to enormous quantitative and qualitative diversity of industrial wastewater, its treatment involves the application and combination of many different methods. They include biodegradation, coagulation, sorption, oxidation and filtration [1–3]. Literature data [3–6] indicate that sorption on activated carbons is one of the most effective means of the removal of organic compounds. That is due to carbons extended porous structure and specific chemical properties of their surface. To efficiently remove given pollutants from aqueous solution by means of sorption, it is necessary to choose appropriate process conditions. That concerns the physicochemical properties of both the adsorbate and activated carbons

* Corresponding author: lidiadabek@tu.kielce.pl

(magnitude of specific surface area, pore distribution, size and volume, surface chemical properties). Sorption capacity is determined on the basis of experimental investigations and the fitting of the theoretical curves (Langmuir, Freundlich isotherms and others) to the results obtained in measurements. It is a complex task that requires substantial labour input and generates high costs while the results obtained describe only one particular adsorbent – adsorbate system. However, both the literature on the subject [1–12] and technological practice provide databases of sorption results for activated carbon-adsorbate systems that have already been investigated. Therefore, it seems reasonable to make use of those data to predict sorption in the systems that show comparable properties. Literature data [5, 11–15] indicate that mathematical models using multiple regression equations, artificial neural networks and their numerous modifications could be employed to this end. The models are developed on the basis of the dataset, which is partitioned into the training set, and the testing and validating set. That is intended to ensure satisfactory predictive abilities of the model [16–18]. At the training stage, parameters in the model structure are estimated, the next two stages involve the assessment of the predictive powers of the developed mathematical models. The condition of model usability is met if the model has satisfactory predictive ability with respect to all three sets.

The application of mathematical models to sorption predictions was discussed in many studies, which concerned both metal sorption on active carbons and that of organic compounds [5, 12–14]. However, it should be noted that in a majority of cases, the models presented by researchers described the sorption of single pollutants on selected activated carbons. In addition, more attention was given to the sorption process conditions than to physicochemical properties of activated carbons. As a result, the practical applicability of such models is limited as they cannot be used for predicting sorption amount of another adsorbate group on different activated carbons. Consequently, it is reasonable to develop statistical models for the prediction of sorption on activated carbons, which account for the diversity of carbon physicochemical properties, and also different adsorbates.

On the basis of previous results [10] on the sorption of dyes (crystal violet and phenol red) and para – chlorophenol on four different activated carbons that have brand names of WDex, WG12, F200S and F200R, in this study an attempt was made to develop a mathematical model for the prediction of sorption of the same adsorbates, but on different activated carbons that have parameters similar to those used while developing the model.

2 Methodology

To develop a mathematical model for predicting sorption (output), the data (discrete variables) that constitute the characteristics of WDex, WG12, F200S and F200R activated carbons were used [3,8-10]. They include the following: dechlorination ability (Ab. De.), iron content (Cont. Fe.), methylene number (L. Met.), iodine number (L. J.), detergent number (L. Det.), ash content (Cont. ash), soluble substance content (Cont. s. sol.), specific surface area (S), and values of initial and equilibrium concentrations for the following adsorbates: crystal violet, phenol red and p-chlorophenol. The variables mentioned above were inputs for the mathematical model. The methodology followed in sorption experiments and in the determination of physico- chemical properties of activated carbons was discussed in details in the study by Dąbek et al. [3, 10]. The characteristics of activated carbons selected for investigations are presented in Table 1. Equilibrium concentrations of the examined adsorbates ranged 1.32–64.41 mg/dm^3 for phenol red, 0.42–15.85 mg/dm^3 for crystal violet and 4.52–84.52 mg/dm^3 for p-chlorophenol, respectively. Additionally, because of their diversified composition and chemical properties, the adsorbates selected for the investigations were modelled as linguistic variables.

Table 1. Physicochemical characteristics of activated carbons selected for investigations [3, 8–10].

Carbon	Ab.De	Cont.Fe.	L.Met.	L.J.	L.Det.	Cont. ash	Cont.s. sol.	S	V
	cm	mg/g	cm^3	mg/g		%	%	m^2/g	cm^3/g
WDex	5	0.50	36	990	20.6	20.64	1.07	1050	0.95
WG12	3	0.44	32	1230	32.0	8.85	2.48	980	0.89
F200S	7	0.43	24	710	12.7	7.28	1.16	720	0.72
F200R	6	0.50	16	770	15.7	7.84	1.74	800	0.86

In this study, Support Vector Regression (SVR), Boosted Trees (BT), Random Forests (RF), and Cascade Neural Networks (CNN) methods were applied for predicting sorption on selected activated carbons. The methods selected for the study were divided into two model groups. In complex models, complex structures (regression support vectors, neural networks) are generated at the training stage, and also it is necessary to compute many parameters. The other group comprises much simpler models, for the implementation of which a much smaller number of parameters is needed. Therefore, one of the objectives of this study was to compare the results of sorption simulations obtained with complex models (CNN, SVM) with those produced by models that have a simpler structure (BT, RF).

Prior to the development of the models, input data and discrete output data were normalised using the min-max transformation. Additionally, to ensure the correctness of the training process, the data (N = 50) were partitioned into three sets, namely the training (75%), testing (25%) ones. In the study, the measurement results of crystal violet sorption on F 200R activated carbon were employed to validate the statistical model.

Support Vector Machines (SVM) comprise a group of methods developed by Vapnik [21] originally for entirely classification purposes, and later on also for regression problems (SVR). Due to the fact that dependence holding between the model output and input variables could be non-linear, in the method, N – dimensional space is non-linearly transformed into K – dimensional space of features, which has a greater size. In this study, to predict sorption, Support Vector Regression method was applied with the radial kernel function intended to minimise the functional having the following form:

$$\sum_{i=1}^{m} \frac{C}{m} \cdot |y_i - f(x_i)|_\varepsilon + \frac{1}{2} \cdot \|f\|_k^2 \quad (1)$$

where: $|y_i - f(x_i)|_\varepsilon = \max\{0, |y_i - f(x_i) - \varepsilon\}$, ε – permissible error value, $\|f\|$ - rule f in the Hilbert space, c – constant selected by the user depending on value ε [19], m – size of the training set, $f(x_i)$ – value of function $f(x)$ at point x_i, described by the equation:

$$f(x) = \sum_{i=1}^{N_{sv}} (\alpha_i - \alpha_i') \cdot K(x, x_i) + w_0 \quad (2)$$

in which: w_0 – deviation, N_{sv} – number of support vectors, which depends on C and ε, α_i, α_i' – Lagrangian multipliers, $K(x,x_i)$ – kernel function with radial basis functions [19]

Boosted Trees (BT) are an implementation of the stochastic gradient boosting method applied to classification and regression problems [17]. The main concept behind the method is the creation of a series of decision trees, with each successive tree used to identify the remainders generated by the previous one. The computations demonstrated that for some estimation and prediction problems, boosted trees produced predictions closer to actual values than it was the case with solutions obtained while using single regression trees.

The random forest algorithm was proposed by Leo Breiman [16] and it originated from the bootstrap method. At the first stage, k – fold sampling for the training dataset of size n is performed, in which repetitions are allowed. Next, based on the sets obtained, regression trees are formed. The process of their construction was modified compared with the original algorithm, so that every node of the tree could be split in the best manner. That is done not on the basis of all attributes, but those randomly selected ones (independent variables). In this way, k-number of regression trees are obtained, which form a forest. The later provides a basis to determine prediction, which comes as the model output. The overall prediction is computed as an arithmetic mean of individual predictions for individual trees.

Neural networks could be used for optimisation, classification, the simulation of linear and non-linear processes and also for control [18, 20, 21]. As regards the modelling of the processes taking place in the natural environment, the network termed as multilayer perceptron (MLP) is used. One of the modifications of ANN of the multilayer perceptron-type is a cascade neural network (CNN). The latter has additional weighted connections from each layer input, and from each layer to a successive one. Cascade network having a higher number of layers can be used to model complex non-linear dependences. The optimal structure of the neural network for sorption amount predictions was sought computing the measures of fit of simulation results to measurements (MAE, MAPE). That was done for a number of neurons in the hidden layer (3–20) adopted in the study and for linear, exponential (exp), sine (sin), sigmoidal (sigm) and hyperbolic tangent activation functions. In the output layer, linear activation function was assumed. Based on studies [22] on cascade neural networks, three connections linking inputs to successive CNN layers were adopted.

In the analyses, to assess the predictive powers of the models for sorption value prediction, mean absolute error (MAE), mean absolute percentage error (MAPE), coefficient of correlation (R), and root mean squared error (RMSE) described in studies by Szeląg and Gawdzik [19, 20] were employed.

3 Results

On the basis of the determined parameters that describe physicochemical properties of activated carbons and the values of equilibrium concentrations for individual adsorbates, statistical models were developed to predict sorption using Support Vector Machines, Boosted Trees, Random Forests, and Cascade Neural Networks methods. Table 2 shows the computed parameters of fit (MAE, MAPE, RMSE, R) of computational results to sorption value measurements. Table 3 lists the parameters that describe the structure of the five selected statistical models, developed using the CNN method, which predict sorption in the best way. Additionally, to visualise the results, a comparative analysis of experimentally determined sorption isotherms and values computed using statistical models is presented in Figs. 1–3.

The analyses indicate that the value of parameter C in eq. (1) in the model obtained with the SVM method was 7. As regards both BT and RF methods, the number of additive trees, on the basis of which the regression models were devised, was not higher than 150, which means the models were not over-trained. The data presented in Tables 2 and 3 show that the lowest errors in the fit of computational results to sorption measurements were obtained for the CNN method, whereas the highest error values were found when the BT method was used. With respect to CNN-based models (Table 3), it was observed that the number of neurons in the individual hidden layers ranged from 6 to 10. Also, the lowest error values (MAE, MAPE, RMSE) were produced using the statistical model, in which the neuron number in individual hidden layers was 7, and the activation function was a hyperbolic tangent dependence. The analyses performed for the study show the values of mean absolute and percentage errors are higher for the models based on BT and RF methods than it is the case for SVM and CNN methods.

Table 2. Comparison of predictive abilities of sorption prediction models developed using BT, RF, and SVM methods.

Method	training				test			
	MAE	MAPE	RMSE	R	MAE	MAPE	RMSE	R
BT	12.29	38.46	22.73	0.812	14.32	39.43	27.76	0.808
RF	5.32	24.28	8.26	0.991	5.7	25.49	8.93	0.989
SVM	2.78	14.94	4.06	0.995	2.91	15.45	4.26	0.994

Table 3. Comparison of predictive abilities of sorption prediction models developed using the CNN method.

Neurons	Activation function	training				test			
		MAE	MAPE	RMSE	R	MAE	MAPE	RMSE	R
9	tanh	1.77	8.74	2.61	0.997	1.88	9.3	2.72	0.996
6	sigm	2.02	9.9	2.95	0.995	2.15	10.53	3.07	0.994
10	tanh	2.05	9.92	2.95	0.995	2.18	10.55	3.07	0.994
7	sin	1.84	9.06	2.75	0.996	1.96	9.64	2.87	0.995
8	exp	1.85	9.07	2.76	0.996	1.97	9.65	2.88	0.995

It should be noted that values of errors in the fit of computational results to measurements, for the RF method are lower than those determined using the BT method. Additionally, it was concluded that for the prediction of the sorption of selected organic compounds, the models that have more complex structure produce lower values of MAE, MAPE, RMSE errors (Tables 2 and 3) than those obtained using the models based on the modified method of regression trees.

The predictive powers of the statistical models developed for sorption prediction were confirmed in the investigations of Meena et al [13]. The researchers applied artificial neural networks to model the phenomenon of crystal violet dye removal and obtained the value of R = 0.965. The results reported by Yang et al. [11], who received a good fit of computational results to measurements (R = 0.997), support the previous findings.

a)

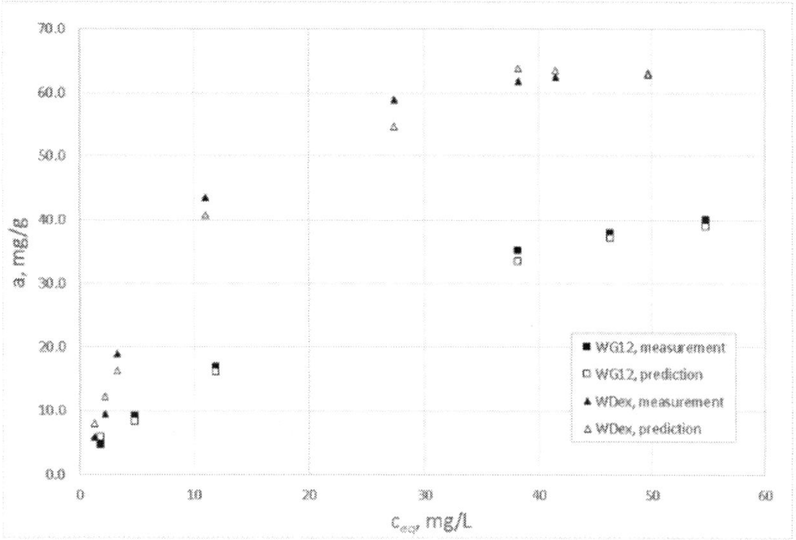

b)

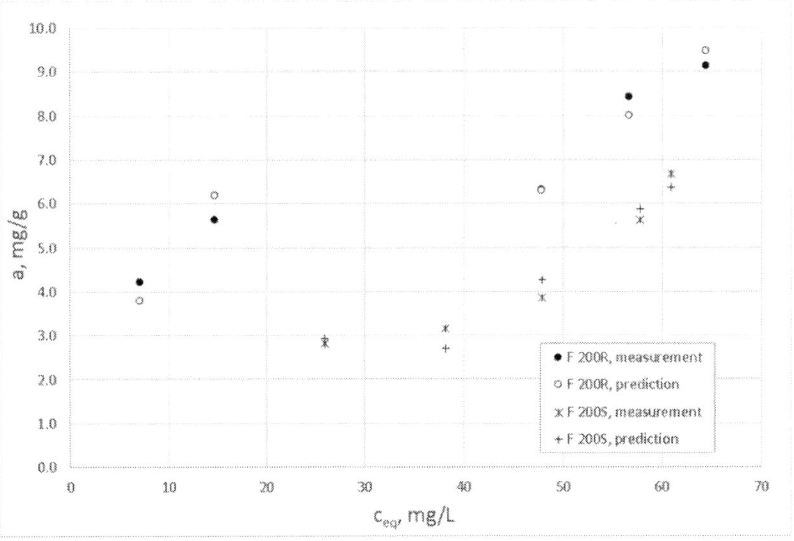

Fig. 1. Comparison of the results of measurements and computations of phenol red sorption: a) on WG12 and WDex activated carbons ; b) on F 200R and F 200S activated carbons.

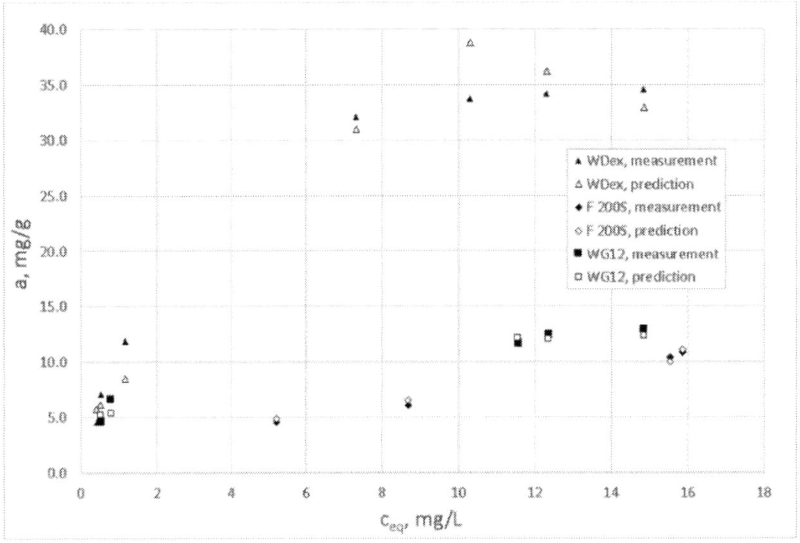

Fig. 2. Comparison of the results of measurements and computations of crystal violet sorption on WDex, F 200S, and WG12 activated carbons.

The values of errors (Tables 2 and 3) are comparable with the results (0.777–0.999) obtained with classic methods (Langmuir, Freundlich isotherms and others) reported by Dąbek et al. [10]. The researchers used Langmuir model to predict sorption on WDex, F200S, F200R and WG12 activated carbons. The analyses conducted by Depci et al. [12] produced similar results.

The results of sorption computations with the use of the CNN method for the validating set, presented in Fig. 4, confirm that the developed statistical model shows satisfactory predictive abilities. In the case under consideration, error values are MAE = 0.85 mg/g;

DOI: 10.1051/e3sconf/20172200032

MAPE = 12.68%; RMSE = 0.86 mg/g. The results of computations presented above indicate it is possible, using data mining methods, to predict sorption on selected activated carbons in the presence of different organic substances.

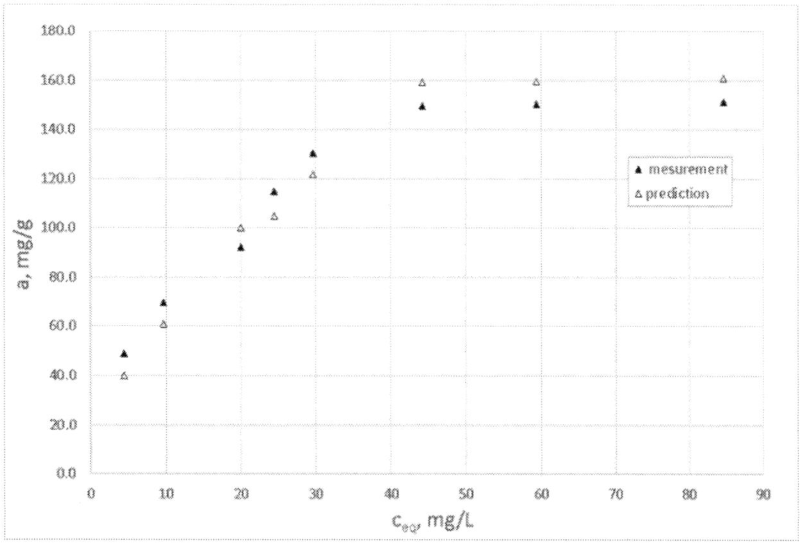

Fig. 3. Comparison of the results of measurements and computations of para-chlorophenol sorption on WDex activated carbon.

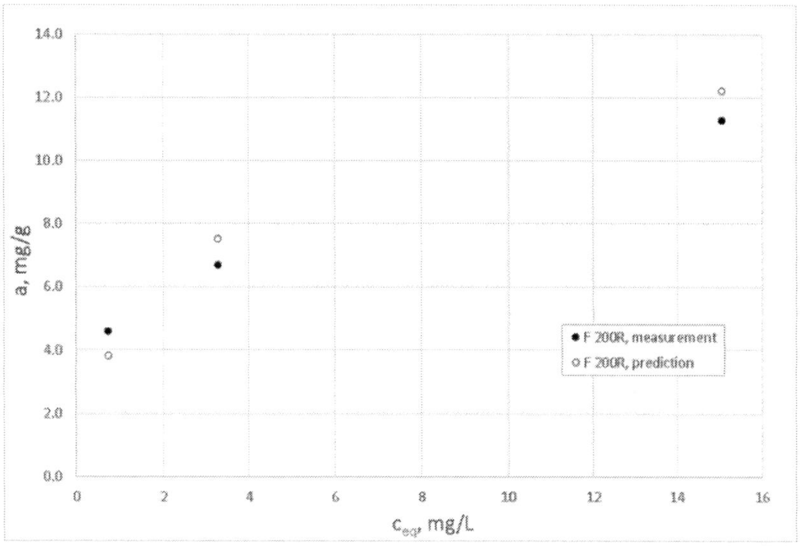

Fig. 4. Comparison of the results of measurements and computations of crystal violet sorption on F 200R activated carbon.

4 Conclusions

The computations performed for the study showed that the course of sorption on selected activated carbons with diversified properties, in the presence of various organic substances can be modelled using Boosted Trees, Random Forests, Support Vector Machines, and

Cascade Neural Networks. In the models under consideration, the lowest error values in sorption prediction were obtained using the CNN method, whereas the BT method produced the highest error values. The computations also indicated that lower error values were found for complex models (Neural Networks, Support Vector Machines) than for the models that required estimation of a smaller number of parameters in their structure.

The results of introductory investigations and satisfactory predictive abilities of the mathematical model developed for predicting sorption of various adsorbates on carbons of diversified properties seem promising. However, it is necessary to verify the model, and to evaluate its usability for different active carbons. Due to a limited amount of data available for analysis, it is also necessary to work out methodology that would account for random selection of data for the training and testing sets, respectively.

References

1. V.K Gupta, V.Suhas, J. Environ. Manage. **90**, 8 (2009)
2. V.P. Santos, M.F.R. Pereira, P.C.C. Faria, J.J.M. Órfão, J. Hazard. Mater. **162**, 2–3 (2009)
3. L. Dąbek, E. Ozimina, A. Picheta-Oleś, Ecol. Chem. Eng. A. **19**, 3 (2012)
4. R.Ch. Bansal, M. Goyal, *Activated Carbon Adsorption* (Taylor & Francis Group, Baca Raton-London-New York-Singapore, 2005)
5. S. Azizian, M. Haerifar, H Bashiri, Chem. Eng. J. **146**, 36 (2009)
6. A.P. Terzyk, G. Rychlicki, P.A. Gauden, P. Kowalczyk, Water Encyclopedia, Volume Oceanography; Methodology; Physics and Chemistry; Water Law; and Water History Art and Culture, 404–408, Willey (2005)
7. N. Daneshvar, S. Aber, F. Hosseinzadeh, Global Nest J. **10**, 1 (2008)
8. L. Dąbek, E. Ozimina, A. Picheta-Oleś, Environ. Prot. Eng. **38**, 1 (2012)
9. L. Dąbek, E. Ozimina, A. Picheta-Oleś, Ecol. Chem. Eng. A. **17**, 11 (2010)
10. L. Dąbek, E. Ozimina, A. Picheta-Oleś, Rocznik Ochrona Środowiska **13** (2011)
11. Y. Yang, X. Lin, B. Wei, Y. Zhao, J. Wang, Int. J. Environ. Sci. Te. **11**, 4 (2013)
12. T. Depci, A. R. Kul, Y. Onal, E. Disli , S. Alkan, Z. F. Turkmenoglu, Physicochemical Problems of Mineral Processing **48**, 1 (2011)
13. M. Meena, K.Srinivasan, Pollution **50**, 10414 (2012)
14. A. Khataee, A.Khani, Inter. J. Chem. React. Engin. **7**, A5 (2005)
15. N. Prakash, S.A.Monikandan, J. Hazard. Mater. **152**, 1268 (2008)
16. L. Breiman, Journal Machine Learning **45**, 1 (2000)
17. J. Friedman, Comput. Stat. Data An. **38**, 4 (2002)
18. S. Dellana, D. West, Environ Modell Softw. **24**, 1 (2009)
19. B. Szeląg, J Gawdzik, Pol. J. Environ. Stud. **25**, 4 (2016)
20. B. Szeląg, J Gawdzik, Pol. J. Environ. Stud. **26**, 1 (2017)
21. V. Vapnik, *Statistical Learning Theory* (John Wiley and Sons, New York 1998)
22. S. B. Al-Batah, M.S Alkhasawneh, L.T Tay, U.K. Ngah, H. H. Lateh, M.T.A Isa, Math. Probl. Eng. **2015**, 1 (2015)

E3S Web of Conferences **22**, 00033 (2017) DOI: 10.1051/e3sconf/20172200033
ASEE17

Energy consumption modeling during dairy sewage pretreatment

Wojciech Dąbrowski[1,*], *Radosław* Żyłka, *Paweł* Malinowski, and *Dariusz* Boruszko

[1]Białystok University of Technology, Faculty of Civil and Environmental Engineering, Department of Technology in Environmental Protection and Engineering, ul. Wiejska 45A, Białystok, Poland

Abstract. The research was conducted in a dairy WWTP located in north-eastern Poland with the average flow of 546 m^3d^{-1} and PE 11500 in 2016. Energy consumption was measured with the help of Lumel 3-phase network parameter transducers installed within the plant. The modeling was conducted based on the quantity and quality of raw sewage, after its screening, averaging and dissolved air flotation. The following parameters were determined: BOD_5, COD, N-total and P-total. During the research period, 15 measurement series were carried out. Pollution loads removed in primary treatment varied from 167.0 to 803.5 kgO_2d^{-1} and 1205.9 to 10032 kgO_2d^{-1} for BOD_5 and COD respectively. The energy consumption share during dairy pretreatment in relation to the total energy consumption was in the range from 13.8 to 28.5% with the mean value of 18.7% during the research period. Energy consumption indicators relating to removed pollution loads for primary treatment were established with the mean values of 0.74 and 0.83 $kWhkg^{-1}d^{-1}$ for BOD_5 and COD respectively. An attempt was made to determine the influence of raw sewage characteristics and pretreatment efficiency on energy consumption of the object. A model of energy consumption during pretreatment was estimated according to the experimental data obtained in the research period. It was modeled using the linear regression model and principal component analysis.

1 Introduction

High demand for Polish dairy products on the global market means significant increase of milk production. The amount of milk produced in Poland raised from 11575 million liters in 2005 to 12859 million liters in 2015. Podlaskie Voivodeship is an absolute leader of milk production in reference to 1 ha of agricultural land. The milk produced in 2015 amounted to 2424 lha^{-1} whereas the average value of this indicator within the whole country was 884 $l{\cdot}ha^{-1}$ [1]. A continuous increase of the amount of produced milk results in larger amounts of post-production sewage which requires treatment. Dairy sewage, as a product of cleaning technological lines after the production of milk powders, fat products and cheese, is characterized by high content of organic pollution, fat and grease. Such sewage content determines the usage of intensive mechanical pretreatment prior to biological treatment, which is most often conducted based on sewage averaging and dissolved air flotation (DAF).

* Corresponding author: dabrow@pb.edu.pl

 DOI: 10.1051/e3sconf/20172200033

The research previously conducted in that field has shown that electric energy consumption during pretreatment process constitutes 18% of the total energy consumption in the facility [2]. It was also proven by the analysis of 601 wastewater treatment plants conducted by Longo et al. [3] in which the achieved percentage share of pretreatment in the total energy consumption ranged between 5 and 18%. The authors of the research conducted in Irish municipal wastewater treatment plants report this value at 1.8–9.1% [4]. It constitutes a substantial part of the costs of energy consumption of a facility. Additionally, the environmental issue should be taken into consideration. Parravicini et al. [5] state that in two analyzed models of municipal wastewater treatment plants, electric energy consumption was responsible for even 59.9% of the facility's total carbon footprint. It has encouraged the authors to make an attempt at creating a mathematical model of electric energy consumption during dairy sewage pretreatment. The purpose of devising such a model is to determine the influence of electric energy consumption of pretreatment devices, in reference to removed pollutants load, on the total energy consumption in the facility. The obtained model might serve to optimize the costs of industrial wastewater treatment.

In the course of the research, electric energy consumption by particular subsystems of the pretreatment unit and the removed pollutant loads were determined. It allowed to establish energy consumption indicators in reference to 1 kg of removed load. A mathematical model was designed on the basis of the research data from 12 measurement series conducted from January to June 2016.

2 Material and methods

2.1 Dairy WWTP characteristics and evaluation of energy consumption

Bielmlek sewage treatment plant applies sequential flow. The average sewage flow in 2016 was 546 m^3d^{-1}, with the personal equivalent (PE) of 11500. The sewage treatment process has two phases: mechanical and biological. Pretreatment is completed through filtering through a sieve with the diameter of 6 mm, sedimentation with the use of a grit chamber, averaging in an averaging tank, and pressure flotation supported by adding coagulants and aqueous solution of polyelectrolytes. Additionally, the mechanical treatment unit consists of pumps, mixers and primary aeration system in the averaging tank. Biological treatment is carried out in two sequence batch reactors (SBR). Treated sewage is discharged directly to a reception tank – the river Biała. Sewage sludge generated in the process is subjected to mechanical dewatering and used as fertilizer (Fig. 1).

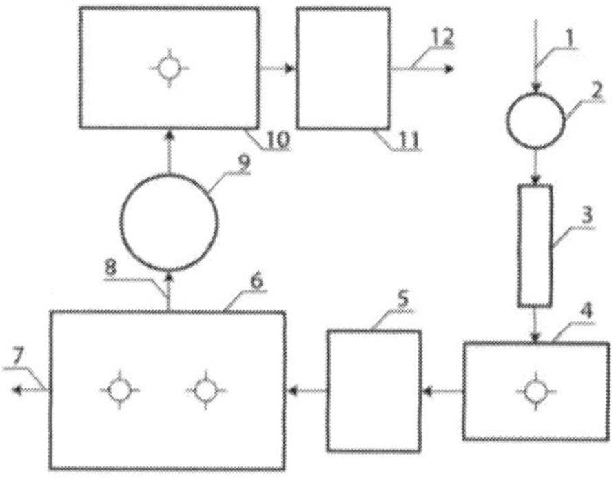

Fig. 1. View and flow diagram of Bielmlek WWTP; 1 – raw wastewater, 2 – screen, 3 – sand trap, 4 – averaging tank, 5 – DAF flotator, 6 – SBR reactors, 7 – treated wastewater, 8 – excessive sludge, 9 – gravity thickener, 10 – aerobic stabilization tank, 11 – screw press, 12 – dewatered sludge.

The installed power of the treatment plant is 234.7 kW, 27.1% of which is constituted by pretreatment devices (Table 1). It should be mentioned that the grit chamber is a device with gravitational flow and manual sand discharge and for that reason it was not included in the study.

Sewage quality tests were conducted in S.M. Bielmlek laboratory. The sewage samples were collected as average values from a work shift in the production plant. Parameters such as BOD_5, COD, N-total and P-total were determined in the samples. The research was completed in 12 measurement series, during which electric energy consumption was monitored for both the whole facility and its individual parts.

Table 1. Power installed of Bielmlek WWTP.

Item	Power installed [kW]	Percentage share of total [%]
Pretreatment:	63.6	27.1
screen	3.5	1.5
DAF	19.3	8.2
pumps	22.4	9.5
mixers	7.4	3.2
blower	11	4.7
Biological treatment	104.5	44.5
Sludge management	56.6	24.1
Air deodorization	10	4.3
In total	234.7	100

2.2 Energy consumption measurement system

Installed within the plant, Lumel 3-phase network parameter transducers were used to measure energy consumption. The current transformers send signals, which are then received by the transducers. Precise measurement is possible thanks to current transformers connected to every single electric device within a plant. Supervisor control data system (SCADA) collects the information from the current transformers and makes it possible to view the results online, archive and process them, as well as to control the wastewater treatment process.

2.3 Statistical model

Statistical analysis was performed using Principal Component Regression (PCR) [6]. It is a two-step method. In the first step, 5 original input variables (loads of: BOD_5, COD, N-total, P-total and flow) are normalized to zero mean and unit variance. After normalization, the variables are transformed linearly(or equivalently – rotated) by performing principal component analysis [7], to set of principal components (PC) – new variables (1).

$$x_i' = \frac{x_i - \mathbf{1}\mu_i}{\sigma_i} \qquad PC_j = \sum_{i=1}^{5} a_{ji} x_i' \tag{1}$$

where: i, j – indexes (from 1 to n); x_i – original independent variable along with its normalization factors - mean μ_i, and standard deviation σ_i; $\mathbf{1}$ – vector of ones; x_i' - normalized independent variable; PC_j – j-th principal component; a_{ij} – linear coefficients of PCA transformation

 DOI: 10.1051/e3sconf/20172200033

Due to the aforementioned normalization, the total variance (sum of variable variances) of the transformed dataset is equal to its dimensionality (5). Each additional PC, given the previous ones, is a direction that is orthogonal to them and maximizes the remaining variability of the transformed dataset. New variables are also not correlated – they are independent of each other. The second step is the standard linear regression with energy consumption E as a dependent variable. For the analysis, only selected subset of PCs that cumulatively reflects more than 90% of dataset variance are taken as independent variables. PCs whose linear regression coefficients were statistically insignificant (at 5% significance level) were iteratively removed from the calculated model until no such components were left.

$$E = \alpha_0' + \sum_{k \in sel(j)} \alpha_k' PC_k \quad E = \alpha_0 + \sum_{i=1}^{5} \alpha_i x_i \tag{2}$$

$$\alpha_0 = \alpha_0' - \sum_{i=1}^{5} \frac{\mu_i}{\sigma_i} \left(\sum_{k \in sel(j)} \alpha_k' a_{ki} \right) \quad \alpha_i = \frac{1}{\sigma_i} \sum_{k \in sel(j)} \alpha_k' a_{ki} \tag{3}$$

where: k – index (over selected $sel(j)$ principal components); α_k', α_i – linear coefficients of the regression for: principal components (estimated) and original variables (calculated from estimations using).

The final model (2) will be presented in 2 forms: using selected PCs and in the original variables space. While the coefficients for the selected PC are directly estimated by regression, the coefficients for the original variables (3) can be calculated from estimations and transformation (1).

PCR offers a number of benefits over usual linear regression in input variable space[6]. Due to normalization, it is insensitive to the scale of the original variables. Due to decorrelation, it is more numerically stable. It offers dimensionality reduction – only a given subset of PCs, reflecting most of the dataset variability, could be used in regression model, making a starting model simpler. Effectively, PCR can explore the original problem space (the dataset) in its more natural dimension, that is usually lower than the number of original input variables.

3 Results and discussion

3.1 Energy consumption ratio

The research results were presented in Table 2. Pretreatment devices took on average 217.5 $kWhd^{-1}$ of electric energy, which was almost equal to the energy intake of sewage treatment devices, which consumed on average 230.3 $kWhd^{-1}$. Among the pretreatment devices, pumps and mixers were responsible for the highest electric energy intake which was 76.2 and 73.7 $kWhd^{-1}$ respectively. Pretreatment energy consumption indicators, in relation to 1kg of the removed substance, reached the highest average value of 132.32 $kWhkg^{-1}$ for total phosphorous. Removing 1 kg of organic substance expressed as BOD_5 and COD required respectively 0.74 and 0.83 kWh of electric energy. In comparison, in 607 Slovakian municipal wastewater treatment plants analyzed by Bodík and Kubaská [8], 2.27 kWh was the average amount necessary to remove 1 kg of BOD_5. During the research period the whole facility consumed on average 1132.7 kWd^{-1} with the biggest share belonging to biological treatment, which was on average at 565.3 $kWhd^{-1}$. Energy indicator referred to 1 m^3 of treated sewage fluctuated between 1.53 and 3.3 $kWhm^{-3}$with the average value of 2.31 $kWhm^{-3}$. In comparison, in municipal wastewater treatment plants located in the USA, the indicator varies between 0.086–1.119 $kWhm^{-3}$ [9], whereas in the Chinese city of Shenzhen it is at 0.20 ± 0.06 $kWht^{-1}$ [10].

Table 2. Values of energy consumption (E) and energy consumption indicators (EI).

Item	Mean value (range)
Pretreatment	
E [$kWhd^{-1}$]	217.5 (119.8–435.8)
screen	1.2 (0.2–3.5)
DAF flotator	29.6 (8.7–58.6)
pumps	76.2 (29.6–103.6)
mixers	73.7 (44.4–95.7)
blower*	36.9
EI_{BOD5}[$kWhkg^{-1}$]	0.74 (0.21–1.89)
EI_{COD}[$kWhkg^{-1}$]	0.83 (0.12–3.03)
EI_N[$kWhkg^{-1}$]	13.70 (4.38–32.03)
EI_P[$kWhkg^{-1}$]	132.32 (23.47–494.52)
EI_Q [$kWhm^{-3}$]	0.43 (0.26–0.77)
Biological treatment	
E [$kWhd^{-1}$]	565.3 (371.5–838.0)
Sludge treatment	
E [$kWhd^{-1}$]	230.3 (118.4–431.9)
*Air deodorization**	
E [$kWhd^{-1}$]	28.2
Other†	
E [$kWhd^{-1}$]	91.5 (70.2–127.1)
In total WWTP	
E [$kWhd^{-1}$]	1132.7 (784.3–1537.1)
EI_Q [$kWhm^{-3}$]	2.31 (1.51–3.13)

*due to periodic and irregular operation of device during research period range was not determined
†cf. Fig. 2

In the course of calculations, the percentage share of the pretreatment devices in the total electric energy consumption and electric energy consumption by pretreatment itself were established (Fig. 2). In the hierarchy of the most energy consuming devices of the technological line, pretreatment comes third with the share of 19%. It correlates with the authors' results of the research conducted in dairy and meat wastewater treatment plants, where the share of pretreatment devices in the total energy intake was 17% [11]. Among pretreatment devices, DAF flotation is worth noticing for its low share of electric energy consumption at 14%. It should be mentioned that DAF flotation during the research period was responsible for respectively 47.6%, 37.6%, 47.6% and 33.0% of the total BOD_5, COD, $N_{tot.}$ and $P_{tot.}$ removed loads.

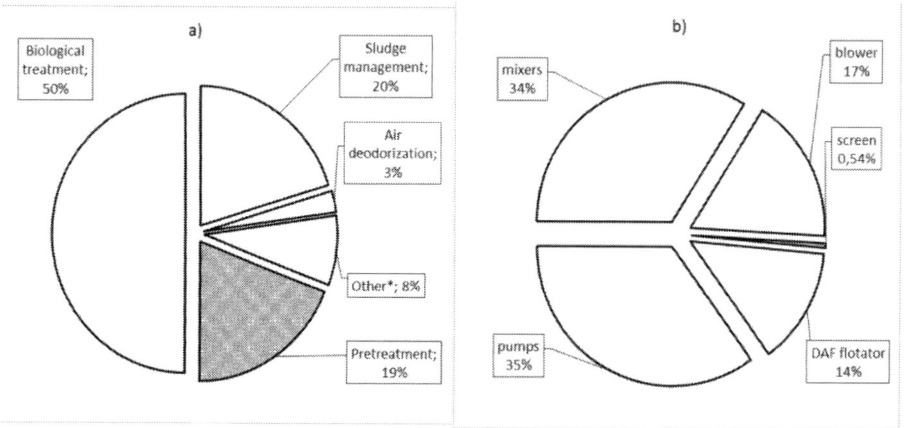

Fig. 2. Percentage share of energy usage during research period; a) whole WWTP, b) pretreatment; *staff's welfare facilities, lighting and measuring equipment.

3.2 Model of energy consumption due to PCR

Statistical analysis was performed in R environment, version 3.3.0 [12]. Energy consumption was modeled using the PCR method. Table 3 presents normalization factors.

Table 3. Normalization factors.

	BOD_5	COD	N-total	P-total	Q
μ	386.01	4170.51	18.83	2.85	502.58
σ	219.71	2665.34	6.45	1.93	95.53

In the next step the PCs were calculated. (Table 4) presents variance, percent of total variance (explained variability) for each PC, along with cumulative explained variability.

Table 4. Variances of PCs.

Parameter	PC_1	PC_2	PC_3	PC_4	PC_5
Variance	2.21	1.47	0.85	0.41	0.06
Explained variability [%]	44.2	29.4	17.0	8.2	1.2
Cumulative explained variability [%]	44.2	73.6	90.6	98.8	100

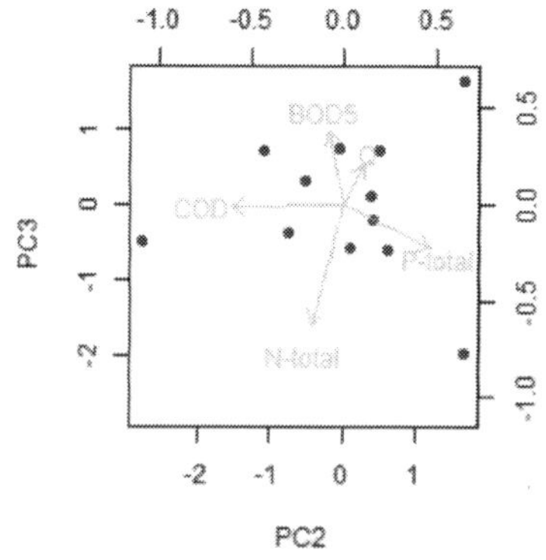

Fig. 3. View on original dataset from perspective of final model PCs.

Only 3 first PCs were sufficient to explain 90% of the dataset variance. Those components were retained for the second step – linear regression. Additionally, the linear coefficient for PC_1 was insignificant, so it was removed from the model. The final model contained PC_2 and PC_3. Figure 3 presents a biplot – a view on 5-dimensional space of normalized original input variables, from the perspective of PC_2 and PC_3 [13].The bottom and left axes represent PC_2 and PC_3 coordinates of observations (blue dots, fig. 3). The remaining axes represent PC_2 and PC_3 coordinates of unit vectors of the normalized original variables (green arrows, fig. 3). The final model on PC scale is presented in table 5, and graphically in Fig. 4.

Table 5. Final model parameters – PC scale.

Parameter	Estimate	p-value
Intercept	217.492	<0,0001
PC_2	36.473	0.0007
PC_3	66.403	<0.0001
R^2	0.891	

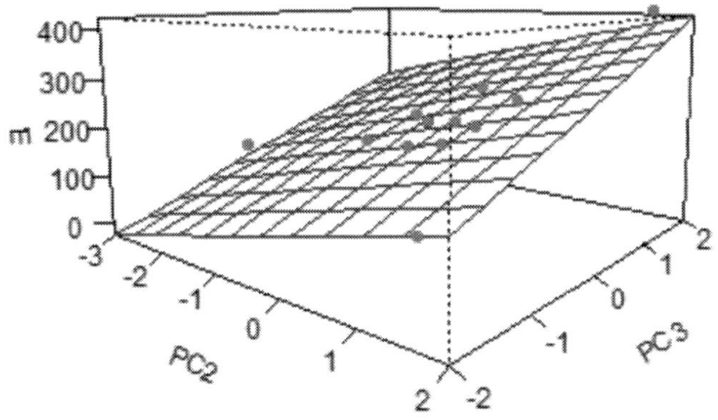

Fig. 4. Final model of energy consumption during dairy sewage pretreatment on PCs scale.

Equation (4) presents the calculated model in PCs and original input variable scale. While the second equation gives an impression that 5 variables can be set independently, from the regression point of view the whole model has only 2 degrees of freedom, not 5.

$$E = 217.492 + 36.473 \times PC_2 + 66.403 \times PC_3$$
$$E = 0.127 \times BOD_5 - 0.011 \times COD - 9.296 \times N_{tot} + 1.413 \times P_{tot} + 0.238 \times Q + 264.920 \qquad (4)$$

4 Conclusions

Dairy WWTPs usually require additional pretreatment in comparison to the municipal ones. Operations such as screening, averaging and air flotation are commonly applied and have a substantial share in the electric energy consumption in the functioning of dairy WWTPs. The analyzed data set obtained as a result of research conducted in a dairy sewage treatment plant allowed to determine the dependencies between five main technological parameters (the loads of removed pollutants and hydraulic flow) and energy consumption during individual pretreatment processes of dairy sewage treatment plants. The obtained model in the form of an equation allows to estimate the expected energy consumption depending on the given initial conditions. The chosen analysis model presents at the same time the mutual dependencies between these variables, exposing the actual, hidden number of parameters which control the aforementioned model.

Optimizing the process of dairy sewage pretreatment is vital not only from the energy consumption standpoint, but also has an important influence on biological processes, which are responsible for the final degree of sewage treatment and water quality in a receiver.

Acknowledgements

The research was funded by Research Project conducted in the department of Technology in Engineering and Environmental Protection No.S/WBiIŚ/3/2014, as well as with the technical and scientific cooperation between the BUT and Bielmlek Dairy Cooperative in Bielsk Podlaski, Poland.

References

1. Central Statistics Office, *Statistical Yearbook of Agriculture* (Statistical Publishing Establishment, Warsaw, 2017)
2. W. Dąbrowski, R. Żyłka, P. Malinowski, Envir. Res. **153** (2017)
3. S. Longo, B.M. d'Antoni, M. Bongards, A. Chaparro, A. Cronrath, F. Fatone, J.M. Lema, M. Mauricio-Iglesias, A. Soares, A. Hospido, Appl. Eng. **179** (2016)
4. L. Fitzsimons, M. Horrigan, G. McNamara, E. Doherty, T. Phelan, B. Corcoran, Y. Delaure, E. Clifford, J. Clean. Prod. **131** (2016)
5. V. Parravicini, K. Svardal, J. Krampe, Energy Procedia **97** (2016)
6. I. T. Jolliffe, ,J. R. Stat. Soc. Series C **31**, 3 (1982)
7. K. Pearson, Philos. Mag. **2**, 11 (1901)
8. I. Bodík, M. Kubaská, Environ. Prot. Eng. **39**, 2 (2013)
9. H. Wang, Y Yang, A. A. Keller, X. Li, S. Feng, Y. Dong, F. Li, Appl. Energy, **184** (2016)
10. W. Li, L. Li, G. Qiu, J. Clen. Prod. (to be published),
11. W. Dąbrowski, R. Żyłka, M. Rynkiewicz, J. Environ. Eng. **17**, 3 (2016)
12. R Core Team, *R: A language and environment for statistical computing. R Foundation for Statistical Computing*, (Vienna, Austria. URL,2016) (https://www.R-project.org/)
13. K. R. Gabriel, Biometrika, **58** (1971)

E3S Web of Conferences 22, 00034 (2017) DOI: 10.1051/e3sconf/20172200034
ASEE17

A review on waste heat recovery from exhaust in the ceramics industry

Bertrand Delpech[1,*], *Brian* Axcell[1], and *Hussam* Jouhara[1]

[1]Institute of Energy Futures, RCUK Centre for Sustainable Energy Use in Food Chains (CSEF), Brunel University London, Uxbridge, UB8 3PH, UK

Abstract. Following the energy crisis in 1980, many saving technologies have been investigated with attempts to implement them into various industries, one of them is the field of ceramic production. In order to comply with energy saving trends and environmental issues, the European ceramic industry sector has developed energy efficient systems which reduced significantly production time and costs and reduced total energy consumption. The last achievement is of great importance as the energy consumption of the ceramic process accounts for a significant percentage of the total production costs. More precisely, the firing stage consumes the highest amount of energy during the whole ceramic production process. The use of roller kilns, fired by natural gas, involves a loss of 50% of the input energy via the flue gas and the cooling gas exhausts. This review paper briefly describes the production process of the different ceramic products, with a focus on the ceramic sector in Europe. Due to the limited on waste heat recovery in the ceramic industry, other high temperature waste heat recovery applications are considered in the paper, such as in concrete and steel production, which could have a potential use in the ceramic industry. The state of the art technologies used in the ceramics industry are reviewed with a special interest in waste heat recovery from the ceramic process exhaust stacks and energy saving technologies.

1 Introduction

Ceramic components are defined as non-organic, non-metallic materials that are consolidated using heat. The solidification of the ceramic based products takes place in a high temperature kiln, usually for a prolonged duration. However, the application of high temperature heat is by definition an energy intensive process [1, 2]. The energy consumption is reflected in the associated cost contribution, and there is a clear indication that a significant percentage of the total production costs is for energy consumption [3]. Studies on energy saving and the quality of ceramic products have highlighted that the implementation of energy saving technologies is crucial in response of the worldwide energy crisis and environmental issues but also for product quality and cost reduction [4–6].

* Corresponding autor: bertrand.delpech@brunel.ac.uk

The European ceramics sector is divided in two different sectors: "traditional ceramics" (wall and floor tiles, tableware and sanitary ware) and "advanced ceramics" (electronical ceramics, technical ceramics, and bio ceramics) [7]. The tile ceramic industry is the largest sector within the traditional ceramics. Tile manufacture represents 75% of the total energy consumption; sanitary ware 15% and the table ware 10% (Fig. 1).

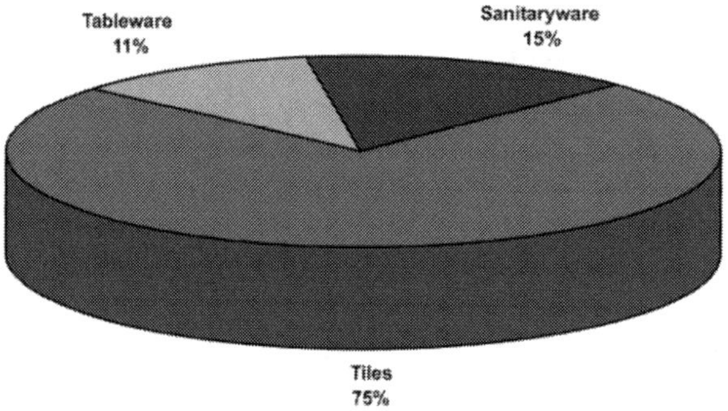

Fig. 1. European ceramics industry by energy consumption [7].

2 The ceramic tile manufacturing process

The mass-production manufacturing process of ceramic prismatic goods (shown in Fig. 2) is composed of four main stages: preparation of raw material, shaping, drying and firing [8, 9].

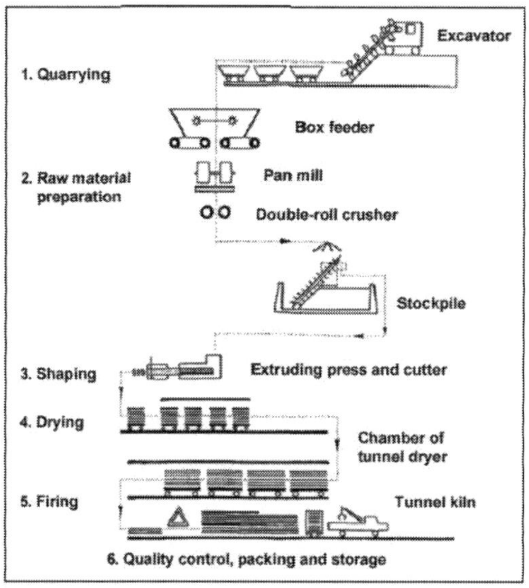

Fig. 2. Schematic view of preparation of extruded bricks in masonry brick manufacture [10].

2.1 Preparation of raw materials and shaping/forming of ware

A vast range of raw materials can be used by the ceramic industry to implement and expand the products range depending on the chosen product functionality. A combination of natural and synthetic materials can be used. Products manufactured for such as wall and floor tiles, household ceramics and sanitary ware products consist of aluminium silicates and clay based materials. The prepared raw materials are mixed in controlled proportions and homogenised to achieve chemical uniformity. Then the raw material is passed through a mechanical/hydraulic press. The shaping/forming stage is conducted when the material reaches the desired viscosity; typically, a softer (wetter) mix will require a higher heat energy during the drying stages which will eventually lead to higher drying shrinkage. The shaping process is called the soft-mud process, the machine allow the mud to flow in a pre-sanded mould to obtain the shape. An external vibration can be applied to the mould to ensure a complete fitting and to consolidate the mix. When the raw material has achieved the desired the shape of the mould, the part is send to the drying stage [10].

2.2 Process of the drying stage

Traditionally, clay products were dried naturally by standing in the ambient air; usually during summertime, and left until dried. The modern ceramic industry needs a drying process that is efficient in both drying speed and low wastage. Heat for the drying air is mainly supplied by a combination of gas heaters and hot air recovered from the cooling zone of the firing kiln. The heat for drying air can be supplied by cogeneration or other fuels.

The response of different materials depends on the drying process and this subsequently affects the drying times. We can identify two different drying stages during this process, the first stage is an initial warming-up period in high humidity conditions followed by the main drying stage using hotter and drier air. The most difficult drying aspect is the removal of any residual water content left within the product

The most common type of dryer is the tunnel dryer. These dryers consist of long tunnel structures with a line of carts loaded with the ceramic clay products. This continuous drying is achieved by high temperature air flowing within the tunnel. As heat is applied to the product, the humidity content within the tunnel will gradually increase. The length of the tunnel depends on the material and the moisture content in the product. Due to the high intensity heat requirements, approximatively 26% of the total CO_2 emissions are produced during the drying stage [11]. This drying stage is crucial, in order to prepare the product for the firing stage [10].

2.3 The firing stage

The purpose of the firing stage is to increase the structural integrity of the product, for example the mechanical strength, abrasion resistance, dimensional stability, resistance to water and chemicals, and fire resistance. The product is fired at a temperature which depends on the chemical composition of the ceramic clays, the typical firing range is 750°C–1800°C. The firing requires a significant amount of energy, and this is reflected in the contribution to emissions. Recent data suggest that the firing stage contributes 57% of the total CO_2 emissions of the tiles ceramics process [11].

Tunnel kilns are commonly utilise in the ceramics industry for the firing stage [10]. As for the drying stage, the products are placed on carts. The carts are moved through the kiln at set intervals.

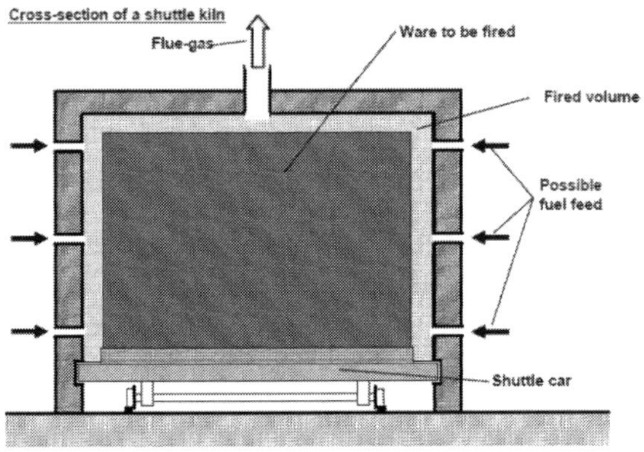

Fig. 3. Cross-section of a shuttle kiln [10].

After the firing, the products are cooled, machined and polished.

3 Optimisation of the drying stage

Several optimisations have been tested and applied in the drying stage such as [7]:

- Optimisation of the recirculation of drying air: using more sophisticated ventilation techniques to control fundamental parameters such as relative humidity, temperature and flow rate improved the efficiency of the hot-air dryer.
- Waste Heat recovery: Clean hot air from the cooling exhaust of the firing kilns is available and can significantly improve the efficiency of the process. Waste heat recovery can provide up to 100% of potential energy saving for the drying process.
- Pulsed hot air: A periodically interrupted flow allows the use of higher drying air temperature; this solution allows enough time for the moisture to migrate from the centroid to the surface. These higher drying air temperatures allows a reduction of 40 min when compared with the process with a classic roller dryer.
- Microwave drying: Microwave assisted drying has two obvious advantages. First of all, only the object is heated, whilst the chamber remains cool and so the energy to heat the drying chamber is saved. The microwave also heats the centre of the body and not only the surface; this promotes moisture migration to the surface. Water absorbs the microwave better than the raw materials, which considerably accelerate the drying. The use of microwave technology can significantly reduce the drying time (from 7 to 30 minutes, depending on the object/material), with a higher energy efficiency [12].

4 Waste heat recovery

A significant number of investigations have been conducted regarding the waste heat recovery from the kiln cooling stage. The best available technique for waste heat recovery in ceramic industry have been summarised by [10].

4.1 Recovery of excess heat from roller kilns

Hot air from the cooling zones of tunnel kilns is usually used in the drying stage, and added to the hot air from gas burners. This method of recovering hot air can be managed only if the

DOI: 10.1051/e3sconf/20172200034

length of the pipes (distance between the cooling and the drying zone) is limited. A significant amount of insulation is needed over the pipe section to limit thermal losses. A large amount of energy is saved using this technique of heat recovery. Some processes also use a heat exchanger to recover the heat from the cooling zone to preheat the combustion air and the air for the drying stage. The application is limited due to the production of acidic combustion gases and other implications such as foaming. Some applications use a thermal oil as a working fluid to transfer the recovered heat to the drying and firing stages [10, 13].

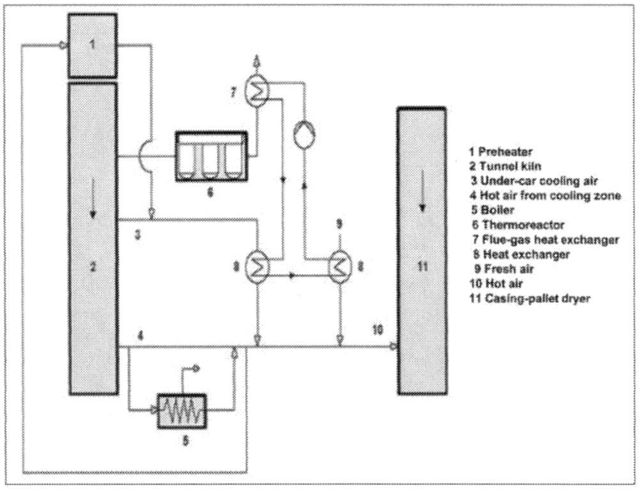

Fig. 4. Schematic view of a combined heat recycling system.

4.2 Cogeneration/combined heat and power plant

Cogeneration systems can be applied in the ceramic sector due to the simultaneous need of heat and electricity. The cogeneration technology has been applied in the ceramic industry to recover the waste heat from the cooling stage [10].

Figure 5 highlights the possible configuration used for a cogeneration system. The hot air from the drying stage is used in the dryer and in the cogeneration system via a heat exchanger placed in the kiln cooling zone.

Fresh air from the factory is injected into the system (high radiation from the kiln and the drying can be recovered). The fresh air is mixed with the hot air coming from the cooling stage and the hot air coming from the cogeneration system and other gas engine emissions from the factory. Then, the hot air is sent in the dryer, a gas burner maintains the desired heat in the dryer. If the hot air temperature is too high, the uncooled gases from engines are withdrawn immediately. With this process, 10% to 50% of the heat input can be saved [9, 14].

 DOI: 10.1051/e3sconf/20172200034

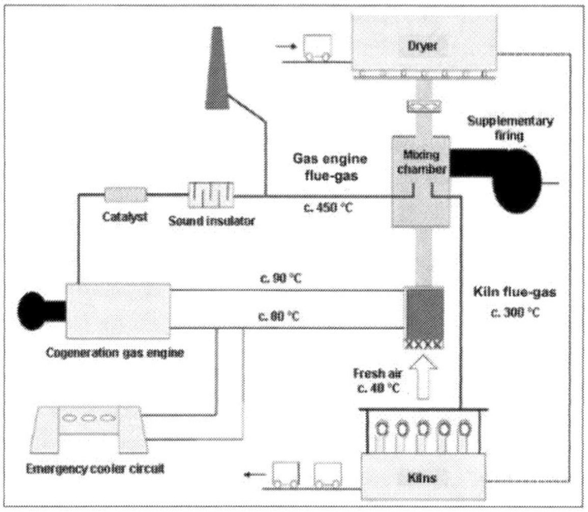

Fig. 5. Schematic view of hot air generation using a cogeneration system [9].

4.3 Organic Rankine cycle [15, 16]

The application of the ORC for the ceramic industry has been investigate by B. Peris et al. [15]. The ORC has proven its efficiency for waste heat recovery from low grade heat sources. Similar to the Rankine cycle, the operative principle consist of recovering the waste thermal energy from a heat source via a working fluid to an expander in order to produce mechanical work which is subsequently converted to electricity.

The heat is taken from exhaust gases from a ceramic furnace via a collector heat exchanger to the ORC (Fig. 6).

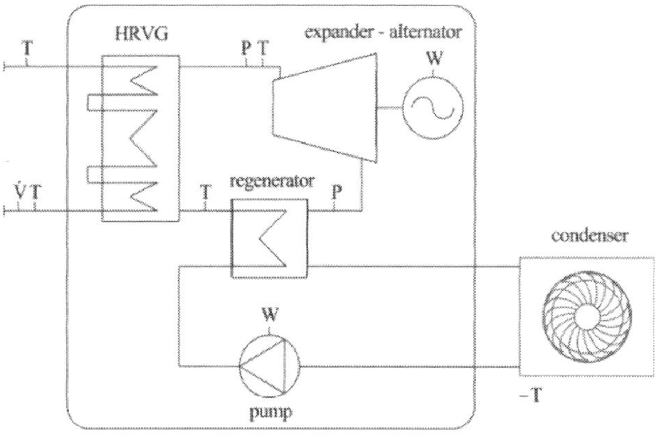

Fig. 6. Schematic of an Organic Rankine Cycle [10].

A simulation of the ORC system during a typical year of production has been conducted. The final results show an energy production for the whole year in excess of 115 MWh. This ORC could save around 237 MWh of primary energy and avoid about 31 tonnes per year of atmospheric CO_2 emissions [15]. The payback of the ORC is around 4-5 years [15].

DOI: 10.1051/e3sconf/20172200034

4.4 Heat pipe systems

Waste heat recovery using heat pipe based heat exchangers has been conducted in many industrial sectors but not the ceramic tile sector. W. Srimuang and P. Amatachay have compiled a review of the applications of heat pipe heat exchangers for heat recovery in air-to-air waste heat recovery [17]. They used 3 types of heat pipe, a conventional heat pipe, two-phase thermosyphon, and oscillating heat pipes (Fig. 7). The use of heat pipe in waste heat recovery present many advantages. The design of the heat pipe ensures no cross contamination between the exhaust air and the air supply. The effectiveness of a heat pipe based heat exchanger is higher than for a conventional heat exchanger. The heat pipe system tend to be more compact, with fewer mechanical parts and minimal pressure drops. Such a system ensures a complete separation between the hot and cold flows and a high reliability, with minimal need for maintenance [18].

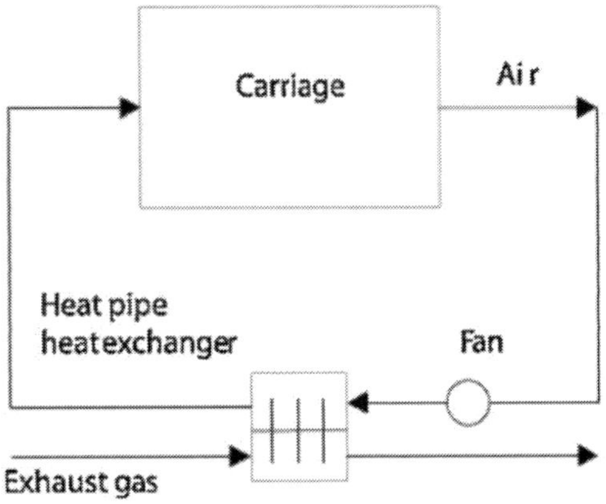

Fig. 7. Schematic of a heat pipe based waste heat recovery system [17].

5 Conclusion

The energy crisis of the 1980s initiated the development of many energy saving technologies. Over the last 20 years, the technologies developed and implemented allow a reduction of greenhouse gas emissions and fossil fuel consumption.

Technological innovations have been applied in large scale ceramic plants and are primarily focused on energy saving systems, for example, by control of air parameters during the drying process, pulsed air drying, and microwave drying and especially by waste heat recovery. Technologies have also been developed to improve the roller kiln technology with combined heat and power generation and again waste heat recovery. Current research in the ceramics sector is more focused on product quality, process optimisation and reducing the environmental impact of the large scale processes. Further improvements could be implemented with systems such as heat pipe heat exchangers and Organic Rankine Cycles.

References

1. E.F.S. Ciacco, J.R. Rocha, A.R. Coutinho, Appl. Therm. Eng. **113**, 1283 (2017)
2. S. Kuhtz, C. Zhou, V. Albino, D.M. Yazan, Energy, **35**, 364 (2010)
3. H. Reh, Tableware and Sanitaryware Plants Follow Developments in the Tile Industry (Ceram. Engng. Sci. Proc, 1995)
4. J. de Beer, E. Worrell, K. Blok, Stud. Environ. Sci. **65**, 1081 (1995)

5. E.A. Abdelaziz, R. Saidur, S. Mekhilef, Renew. Sustain. Energy Rev. **15**, 150 (2011)
6. M. Croucher, Energy Policy **45**, 304 (2012)
7. C. Agrafiotis, T. Tsoutsos, Appl. Therm. Eng. **21**, 1231 (2001)
8. A. Mezquita, J. Boix, E. Monfort, G. Mallol, Appl. Therm. Eng. **65**, 102 (2014)
9. O. Rentz, A. Schmittinger, R. Jochum, F. Schultmann, **44** (2001)
10. European Comission, **210** (2007)
11. J. Peng, Y. Zhao, L. Jiao, W. Zheng, L. Zeng, Energy Procedia **16**, 467 (2012)
12. Anonymous **5**, 230 (1995)
13. A. Mezquita, J. Boix, E. Monfort, G. Mallol, Appl. Therm. Eng. **65**, 102 (2014)
14. J. Beltran, Cogeneration Systems in the Ceramics Tile Sector, Proceedings of the Workshop on New Technologies for the Rational Use of Energy in the Ceramics Tiles Industry, DG for Ene (EC, Castellon de La Plana, Spain, 1994)
15. B. Peris, J. Navarro-Esbrí, F. Molés, A. Mota-Babiloni, Energy **85**, 534 (2015)
16. H. Tian, G.Q. Shu, in *Org. Rank. Cycle Power Syst.* 613–636 (2017)
17. W. Srimuang, P. Amatachaya, Renew. Sustain. Energy Rev. **16**, 4303 (2012)
18. H. Jouhara, H. Merchant, Energy **39**, 82 (2012)

EU energy policies achievement by industries in decentralized areas

Nicola Destro[1,*], *Anna* Stoppato[1], *Alberto* Benato[1], and *Fabio* Schiro[1]

[1]Department of Industrial Engineering (DII), University of Padova, Via Venezia 1, 35131 Padova, Italy

Abstract. Energy Roadmap outlined by the European Commission sets out several routes for a more sustainable, competitive and secure energy system in 2050. All the outlined scenarios consider energy efficiency, renewable energy, nuclear energy and carbon capture and storage. In this paper, more attention has been devoted to the energy efficiency issue, by the identification of new micro and small networks opportunity fed by hybrid plants in the North-East of Italy. National energy balance and national transmission system operator data allowed to collect industrial energy consumptions data on the investigated area. Applying industrial statistics to the local energy needs allows to collect a dataset including consumption information by factory and by company structure (size and employees) for each industrial sector highlighting the factory density in the area. Preliminary outcomes from the model address to the exploitation of local by-product for energy purposes.

1 Introduction

The European Commission policy is offering significant opportunities for public and private investment to support growth, employment and structural transformation. The new generation of programmes under the European Structural and Investment Funds (ESI Funds) 2014–2020 is focused on 11 thematic objectives. Some of them are related to strengthening research, technological development and innovation, competitiveness of Small and Medium Enterprises (SMEs), supporting the transition towards a low-carbon economy [1]. The industry energy efficiency levels are expected to be improved up to 2050 in several industrial sectors. Iron, steel, chemicals and pharmaceuticals industries are the most involved in this process because it is expected an increment of their energy consumption due to the production growth despite the energy intensity improvement. In pulp and paper industries an energy intensity improvement and an energy consumption reduction is expected despite a gradual increase in production rates. In the non-metallic minerals sector energy intensity is expected to remain flat for ceramics, cement and glass sectors as well as for non-ferrous metals industries. Food and drink sector is subject to a continuous improvement leading to the energy consumption reduction despite the growth of production. Energy consumption of the petroleum refineries sector is assumed to decrease the energy consumption due to the past energy efficiency improvement and due to the demand reduction [2].

2 Italian Company Structure

The Italian data warehouse of the Industry and Services Census 2011 contains a wealth of detailed information on the main features of businesses, non-profit and public institutions, disaggregated at the regional level. The census reference date is 31st December 2011. From census returns, it is possible to derive information related to the main structural characteristics of businesses, non-profit and public institutions: legal form, economic activity, human resources and geographic location. The Italian National Institute of Statistics arranges economic activity using the ATECO 2007 classification which is the national version of the European nomenclature, Nace Rev. 2 [3].

The area investigated in this paper is composed by four Italian Regions, often regarded as North-East of Italy. The considered regions are: Trentino Alto Adige, Veneto, Friuli-Venezia Giulia and Emilia Romagna. Using the Census 2011 a first description of the overall area under investigation is summarized in Table 1.

Table 1. Comparison between Italian and North-East enterprises structure.

	Number of enterprises				**Number of employees**			
Size class of employees	**3–19**	**20–99**	**over 100**	**Total**	**3–19**	**20–99**	**over 100**	**Total**
Italy	191030	29098	4615	224743	1325280	1125403	1158713	3609396
North-East	50494	9595	1592	61681	371281	371126	359312	1101719
North-East [%]	26.4%	33.0%	34.5%	27.4%	28.0%	33.0%	31.0%	30.5%

In Table 1 enterprises with less than three employees are not considered due to their little energy consumption. Note that these neglected enterprises account for the 85% of the whole industrial employees in the North-East of Italy. The regions account for the 27.4% of the Italian enterprises and the 30.5% of the industrial labour force. In the North-East the 81.9% of enterprises has a workforce between 3 and 19 employees, the 97.4% of enterprises has a workforce between 3-99 employees while just the 2.6% of enterprises has more than 100 employees.

Table 2. Comparison between Italian and North-East enterprises structure.

	Enterprises		**Local units of enterprises**	
	number of active units	**number of employees**	**number of active units**	**number of employees**
Extraction of crude petroleum and gas	0.00%	0.00%	16.55%	10.44%
Coke and refined petroleum products	9.88%	4.18%	12.36%	5.77%
Electricity, gas and steam supply	25.13%	14.68%	21.45%	22.67%
Water supply and management	16.33%	15.76%	16.86%	16.78%
Iron and steel, non-ferrous metals	22.02%	21.75%	23.03%	24.04%
Chemical and pharmaceutical	23.05%	17.20%	23.54%	20.01%
Non-metallic minerals	25.19%	37.77%	26.41%	37.28%
Transport equipment	27.00%	17.05%	25.97%	16.45%
Machinery	31.03%	35.31%	31.18%	35.78%
Mining and quarrying	17.50%	18.76%	19.05%	19.40%
Food products, beverages and Tobacco	21.79%	29.45%	22.59%	28.66%
Paper products and print	25.62%	30.45%	25.68%	29.61%
Wood products	31.83%	36.80%	32.00%	37.58%
Constructions	21.19%	22.71%	21.32%	22.89%
Textiles, apparel, leather	23.42%	25.58%	23.50%	25.63%
Other manufacturing	31.93%	36.64%	32.12%	36.75%

DOI: 10.1051/e3sconf/20172200035

From the employees classification it is possible to claim that the 33.7% of them are working in enterprises with 3-19 employees, the 67.4% are working in enterprises with 3-99 employees while the 32.6% are employed in enterprises with more than 100 employees. The percentage of industrial enterprises and local unit of enterprises on the region are related to the corresponding Italian amount.

3 Energy consumption and cogeneration potential

National energy balance published by Eurostat [4] provides information about industrial energy supplies. In this investigation the year 2013 has been considered as reference year. For the industrial sector and the energy industries the energy balance states 31.6 Mtoe of energy consumption. Assuming the whole consumption of fossil fuels for heating purpose and a conversion efficiency of 90% the heat request is estimated equal to 20.9 Mtoe. Heat is fulfilled for 19% by derived heat that includes self-production from combined heat and power plants (CHP). Electricity request is 10.7 Mtoe and fulfilled for 15% by self-produced electricity [5].

Table 3. Energy consumption of Italian enterprises.

	Total products	**Solid fuels**	**Oil (total)**	**Gas**	**Total Renewables**	**Wastes (non ren.)**	**Derived heat**	**Electricity**
	ktoe	ktoe	ktoe	ktoe	ktoe	ktoe	ktoe	ktoe
Extraction of crude petroleum and gas	835	0	0	786	0	0	18	31
Coke and refined petroleum products	4894	35	3033	278	0	0	1082	467
Extraction of coal	5	0	0	0	0	0	2	4
Electricity, gas and steam supply	721	0	0	27	0	0	345	348
Water supply and management	521	0	0	0	0	0	0	521
Iron and steel, non-ferrous metals	5663	1923	92	1779	0	0	88	1781
Chemical and pharmaceutical	4121	1	469	1053	7	71	1243	1276
Non-metallic minerals	4984	241	1520	2004	96	202	108	814
Transport equipment	365	0	0	0	0	0	79	286
Machinery	3356	0	252	1387	1	0	15	1701
Mining and quarrying	116	0	23	34	0	0	0	59
Food products and Tobacco	2657	0	160	1174	30	0	266	1028
Paper products and print	2020	0	63	613	0	0	564	781
Wood products	404	0	0	31	89	0	27	256
Constructions	361	0	33	217	0	0	0	111
Textiles, apparel, leather	1167	0	81	586	0	0	40	459
Other manufacturing	1067	3	42	19	52	8	129	813

Eurostat energy balance provides detailed information of the produced energy by self-producers, in particular combined heat and power plants fed by fossil fuels: 1264 ktoe of electricity and 2000 ktoe of heat.

The detailed industrial energy consumption has been determined and is listed in Table 3.

From a literature review the estimation of the efficiency and the characteristics of combined heat and power plants has been obtained [5]. The wide range of industrial enterprises leads to analyse the system with average values for each industrial sector highlighting just the mean characteristics of each one. One of the most important tasks in CHP exploitation is the local use of heat. This topic can be summarized for each type of industry considering the amount of heat and its temperature. These information can be used to analyse the industrial cogenerative potential.

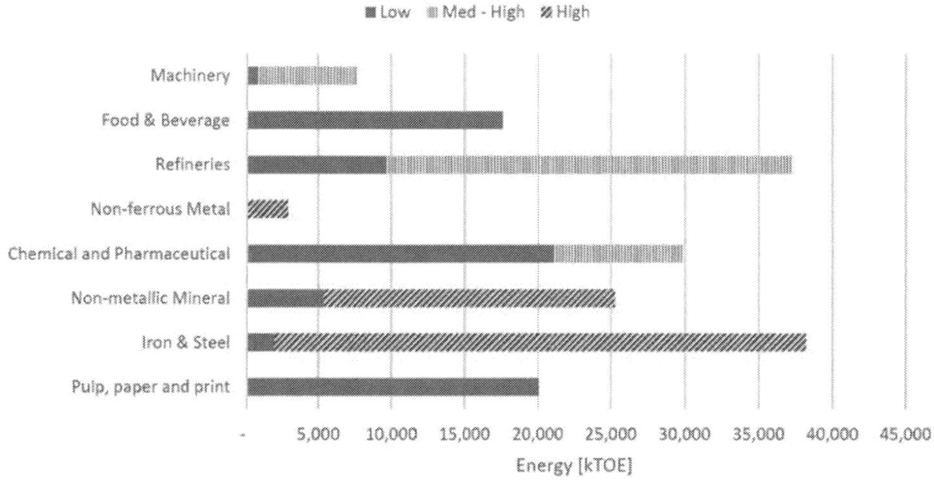

Fig. 1. Estimation of process heat temperature range [6].

For the main industrial sectors, the use of heat is categorized by temperature and by amount: it is possible highlighting the high percentage of industrial heat used at low temperature and the low amount of heat used at high temperature. Table 4 can lie out to identify the industry with high potential of cogeneration.

Table 4. Estimation of process heat temperature percentage.

	low	**med-high**	**High**
	< 250°C	**250–600°C**	**> 600°C**
Machinery	11%	89%	0%
Food & Beverage	100%	0%	0%
Refineries	26%	74%	0%
Non-ferrous Metal	0%	0%	100%
Chemical and Pharmaceutical	70%	30%	0%
Non-metallic Mineral	22%	78%	0%
Iron & Steel	4%	96%	0%
Pulp, paper and print	100%	0%	0%

The analysis of the industrial energy consumption allows to evaluate the energy efficiency improvement potential of each industrial sector through the installation of new combined heat and power plants.

Based on the data reported in Table 5, the actual CHP plants productions are 1497 ktoe of heat and 1300 ktoe of electricity. Estimated further productions from new CHP plants are

1845 ktoe of heat and 1746 ktoe of electricity without considering economic and financial factors. Heat from boiler is determined as difference of fossil fuel consumption of each industrial sector and actual CHP consumption, assuming a boiler conversion efficiency of 90%. Electricity from grid is determined as difference between electricity use and CHP actual production. The percentage of further heat from CHP, CHP efficiency, and Heat/Electricity ratios, are estimated considering statistical data for each application.

Table 5. Estimation of new CHP production in Italy.

	Actual Heat from CHP	Actual Electricity from CHP	Heat from boilers	Electricity from grid	**Further heat from CHP**	Further heat from CHP	Heat/Electricity CHP	Further electricity from CHP
	ktoe	Ktoe	ktoe	ktoe	**%**	ktoe		ktoe
Chemical and Petrochemical	406	393	472	883	**0.42**	200	0.9	222
Coke Ovens and Refineries	805	603	1427	-135	**0.11**	151	0.9	168
Food and Tobacco	162	192	874	899	**0.24**	209	1.1	190
Iron and Steel	53	49	2913	1525	**0.06**	176	0.7	251
Machinery	13	11	1447	1691	**0.50**	724	1.5	482
Non-Ferrous Metals	2	3	383	205	**0.06**	23	0.7	33
Non-Metallic Minerals	28	23	3331	791	**0.06**	203	0.8	254
Textile and Leather	19	15	562	445	**0.28**	156	1.1	142
Wood and W. Products	9	11	6	245	**0.55**	3	0.9	4

Outlined heat and electricity production improvements are function of the industrial sectors characteristics. Again, based on the data reported in Table 5, there are sectors with high percentage of possible CHP improvement like the Chemical, the Food and the Textile due to the low temperature heat requests. For other sectors, like the Machinery and the Wood product, CHP production improvements are of less interest due to the high temperature requested that could still affect the electrical efficiency.

4 Local Energy Balance and CHP potential

The knowledge of the local energy balance can help regulators to assess the effectiveness of current regulations and to identify new strategies suitable for efficiency improvements.

To get the local industrial energy balance starting from the national industrial energy balance it is possible to follow several pathways. The first step is to through improve the knowledge of the local area. The knowledge of industrial structure can help to identify industrial energy requests; also, the knowledge of the average number of employees for each industrial sector can give information about the industrial consumption in the investigated area. From the Italian TSO (Transmission System Operator) it is possible to get information about local energy consumption on a specific area. All those indicators can help to evaluate the local energy balance. This approach has been applied to the North-East regions of Italy and the three outlined balances present an absolute error of 3.2% on the total energy consumption of the area. This result permits to verify the relation between national and local consumptions that are linked to the number of industries and employees and to the local energy consumption. In this paper, it is studied and computed the balance obtained considering as reference information, the local energy consumption in four regions of the North-East of Italy (Veneto, Trentino Alto Adige, Friuli Venezia Giulia and Emilia Romagna); the result is summarized in Table 6.

The effective outlined method can be applied also to smaller areas, for example rural areas to gather local information. To prove its potential, the method has been applied to a small region including three towns in the countryside. The outlined energy balance for the towns can help to investigate local opportunity. The local energy balance is summarized in Table 7.

Table 6. Energy consumption of North-East Italian enterprises.

	Total products	**Solid fuels**	**Oil (total)**	**Gas**	**Total Renewables**	**Wastes (non ren.)**	**Derived heat**	**Electricity**
	ktoe	ktoe	ktoe	ktoe	ktoe	ktoe	ktoe	ktoe
Extraction of crude petroleum and gas	26	0	0	25	0	0	1	1
Coke and refined petroleum products	123	1	76	7	0	0	27	12
Extraction of coal	0	0	0	0	0	0	0	0
Electricity, gas and steam supply	249	0	0	9	0	0	119	120
Water supply and management	85	0	0	0	0	0	0	85
Iron and steel, non-ferrous metals	1156	392	19	363	0	0	18	363
Chemical and pharmaceutical	843	0	96	215	1	15	254	261
Non-metallic minerals	1948	94	594	784	37	79	42	318
Transport equipment	52	0	0	0	0	0	11	40
Machinery	1068	0	80	441	0	0	5	541
Mining and quarrying	22	0	4	6	0	0	0	11
Food products, beverages and Tobacco	952	0	57	420	11	0	95	368
Paper products and print	676	0	21	205	0	0	189	261
Wood products	207	0	0	16	46	0	14	132
Constructions	120	0	11	72	0	0	0	37
Textiles, apparel, leather	214	0	15	108	0	0	7	84
Other manufacturing	312	1	12	6	15	2	38	238

Table 7. Energy consumption of enterprises in the small selected rural area.

	Total products	**Solid fuels**	**Oil (total)**	**Gas**	**Total Renewables**	**Wastes (non Ren.)**	**Derived heat**	**Electricity**
	ktoe	ktoe	ktoe	ktoe	ktoe	ktoe	ktoe	ktoe
Chemical and Petrochemical	0.201	0	0.032	0.046	0	0.004	0.056	0.064
Food and Tobacco	2.746	0	0.167	1.234	0.009	0	0.243	1.093
Iron and Steel	0.364	0.153	0.006	0.092	0	0	0.002	0.111
Machinery	1.605	0	0.134	0.650	0	0	0.011	0.810
Non-specified (Industry)	0.683	0.001	0.012	0.010	0.034	0.001	0.091	0.534
Textile and Leather	1.562	0	0.059	0.760	0	0	0.059	0.684
Wood and W. Products	0.895	0	0	0.063	0.189	0	0.060	0.583

Following the previously described approach it is possible to determine the local CHP potential as possibility for new local energy production.

Table 8 shows how it is possible to improve the local energy production and in which sectors the improvement could be more effective. The results achieved showed that high potential improvement could be obtained in the food sector but also in the machinery and textile sectors.

Table 8. Estimation of new CHP production of three towns.

	Heat from boilers	Electricity from grid	**Further heat from CHP**	Further heat from CHP	TH/EL CHP	Further electricity from CHP
	ktoe	ktoe	%	ktoe		ktoe
Chemical and Petrochemical	0.028	0.046	0.423	0.012	0.9	0.013
Food and Tobacco	0.874	0.929	0.239	**0.209**	1.1	0.190
Iron and Steel	0.174	0.088	0.060	0.010	0.7	0.015
Machinery	0.675	0.797	0.500	**0.338**	1.5	0.225
Non-specified (Industry)	-0.067	0.496	0.487	0.000	0.9	0.000
Textile and Leather	0.656	0.649	0.278	**0.182**	1.1	0.166
Wood and W. Products	-0.104	0.515	0.549	0.000	0.9	0.000
Total				**0.751**		**0.609**

DOI: 10.1051/e3sconf/20172200035

5 Local by-product exploitation in food and beverage sector

Local investigation carried out with Energy Service Companies helps to verify the real industrial structure of the considered towns. The reference area highlights a rural structure with high concentration of enterprises linked to the agriculture and food processing. High vinery concentrations and intensive wine cultivation suggests the exploitation of agriculture by-products for energy use improving the use of local renewable energy sources.

The area under investigation is composed by three towns and the total surface is 64 square kilometres. The area dedicated to grape growing is 18 square kilometres. The grape vine cultivation is subjected to the plants management, and during the cold months grapevines are trimmed getting wood by-product (the so called "grape canes") from the cultivation. The wood by-product can be collected and dried in order to be used as fuel in a biomass furnace and producing heat for industrial purposes.

Preliminary analysis provides the specific wood by-product production. Trimmed wood has a relative humidity of 45%. The wood yield is 270 t/km^2. After a natural drying process, the relative humidity becomes 10% with a wood yield of 150 t/km^2. The lower heating value of the dry biomass can be estimated as 4.6 kWh/kg while the collection cost is 100 €/t of dry wood. The cost consider the use of energy for collection and wood processing [7].

The use of local wood by-product for energy purposes allows a CO_2 reduction due to a cutback avoiding of fossil fuels utilization. Switching from natural gas to grape canes for heat production can provide a CO_2 reduction of 2500 tCO_2 / year

Literature investigations provided information about energy consumption in wineries. For a typical facility with a production of 3 million of litres of wine per year the electrical energy consumption is 11 kWh/100 litres of wine, the thermal energy consumption is 1 kWh/100 litres of wine. Energy consumption can vary from 3 kWh/100litres for facilities with more than 5 million of litres of product to 25 kWh/100litres for facilities with less than 25 kWh/100 litres [8].

The use of pruning wood in an Italian winery has been investigated by the use of a pilot plant [9]. Winery has about 2.5 km^2 of vineyards with a yearly production of 150 tons of pruning wood. Biomass is burned in a boiler which generates 720 MWh of heat. Energy is used for space and water heating and for space cooling by means of an absorption chiller. The use of pruning residues can provide the 42% of the overall energy need of the winery.

The use of local wood by-product for energy purpose can be carried out in local industries not necessary linked to vineyards. The exploitation can be made in the local industrial area with the cooperation of the consortium for industrial development. The reduction of fossil fuel consumption in local enterprises can be reached using biomass collected from local agricultural enterprises. The vineyards are the main providers of wood for energy use but is not excluded the utilization of different sources like woodlands maintenance. However, additional investigations are needed to verify the technical and economic feasibility of wood by-products exploitation for energy purposes in the selected area.

6 Conclusions

The developed method can support the energy investigation on Italian areas, providing contribution to the identification of energy efficiency opportunity. Using national energy balance is possible to determine the local energy balance through local information like number of employees, number of enterprises and electrical energy utilized in the selected area. One possible outcome of the model is the identification of CHP potential improvements in the investigated region. The method can be applied also to local community and small areas. Cooperation on research and detailed investigations on site,

 DOI: 10.1051/e3sconf/20172200035

showed local opportunity and potential exploitation of local resources. Renewable energy sources available on-site can improve the by-products exploitation following local development. The exploitation of renewable sources can support local industrial sectors. In-depth investigations carried out in cooperation with local enterprises are necessary to better identify the technical and economic potential of the by-products exploitation.

References

1. "Priorities for 2014–2020 - Regional Policy - European Commission." [Online]. Available: http://ec.europa.eu/regional_policy/en/policy/how/priorities. [Accessed: 28-Apr -2017]
2. "Energy efficiency in industry will rise, study finds - European Commission." [Online]. Available: https://ec.europa.eu/energy/en/news/energy-efficiency-industry-will-rise-study-finds. [Accessed: 28-Apr -2017]
3. "ATECO (Classification of Economic Activity) 2007." [Online]. Available: https://www.istat.it/en/archive/17959. [Accessed: 28-Apr -2017]
4. Eurostat, "Energy balances." [Online]. Available: http://ec.europa.eu/eurostat/web/energy/data/energy-balances. [Accessed: 28-Apr - 2017]
5. GSE-Gestore Servizi Energetici, "Assessment of the national potential for high-efficiency cogeneration and efficient district heating," 2015
6. "ICF: Strategic Consulting & Communications for a Digital World." [Online]. Available: https://www.icf.com/. [Accessed: 28-Apr -2017]
7. Saorin, "Studio di fattibilità di due filiere per la valorizzazione energetica dei sarmenti di vite," 2013
8. T.U. of M. (UPM), "Tesla, Transfering Energy Save Laid on Agroindustry."
9. "Progetto Biomasse: Energia Rinnovabile per le Aziende Agricole derivante da scarti di Potature dei Vigneti - ERAASPV - GELSO GEstione Locale per la SOstenibilità ambientale." [Online]. Available: http://www.sinanet.isprambiente.it/. [Accessed: 28-Apr-2017]

DOI: 10.1051/e3sconf/20172200036

Mixed convection in a three-dimensional ventilated cavity with two isothermal heating Sections

Hicham Doghmi[1,*], *Btissam* Abourida[1], *Lahoucine* Belarche[1], *Mohammed* Sannad[1], and *Meriem* Ouzaouit[1]

[1]Ecole Nationale des Sciences Appliquées d'Agadir, BP 1136 Agadir, Maroc.

Abstract. Mixed convection in a three-dimensional ventilated cavity discretely heated has been analyzed numerically by the finite volume method. The results are presented in terms of streamlines, temperature distribution and average Nusselt number; for different combinations of thermal controlling parameters namely, Reynolds and Richardson numbers. The computational results indicate that the heat transfer's rate is strongly affected by the mentioned parameters.

1 Introduction

The study of the heat surplus evacuation problems has been the subject of extensive research for many years, due to their importance in electronic equipments, air-cooling process and design. The obtained results showed that mixed convection is the most simple and low-cost mode of cooling operation. In addition, this phenomenon is also involved in many engineering applications such as the thermal design of buildings, furnace design, air conditioning, nuclear reactors and others... This practical interest explains the existence of various studies such as the numerical approach of Papanicolaou & Jaluria [1, 2] who studied the mixed convection in a ventilated rectangular cavity provided with a localized heat source. Their principal results showed that the average Nusselt number increases by increasing the solid wall thermal conductivity. The cooling rate was also found to be higher when the outlet flow opening was located near the bottom of the vertical wall. Hsu et al. [3] conducted a numerical study of the mixed convective flow inside an enclosure with a partially dividing partition. They concluded that the maximum heat transfer rate dissipated from the source is obtained when the outflow opening is located at the lower part of the vertical wall, and a better cooling system was designed by placing the heat source as close as possible from the cold stream inlet opening. Rahman et al. [4] investigated numerically the effects of Reynolds and Prandtl numbers on mixed convective flow and heat transfer characteristics in a ventilated cavity in presence of a heat-generating solid circular obstacle placed at the center. They indicated that the average Nusselt number at the heated surface for different values of Ri can be increased with the increasing of the Reynold and Prandtl

* Corresponding author: hicham.gte@gmail.com

numbers. On the other hand, Moraga & López [5] compared numerically the mixed convection in a two- and three-dimensional air-cooled cavity They found that a 3-D model must be used to capture the fluid mechanics for Ri = 10 when $10 \leq Re \leq 250$, and to calculate the global Nusselt number when Re = 500 for Ri < 1.

Hence, the literature review shows that the majority of the previous works consider the case of two-dimensional model, while very few studies have been conducted for 3D numerical simulation. Therefore the three-dimensional approach gives a better modelization of the fluid flow and heat transfer within the cavity. This explains the choice of our present physical model which intended to investigate numerically the effects of conjugated Richardson and Reynolds numbers on the thermal transport and fluid flow phenomena in the considered cubical enclosure.

2 Problem formulation

The considered geometry is presented in Figure 1. It consists of a three-dimensional ventilated cavity with two heating square portions, similar to the integrated electronic components, are located on the vertical wall, while the rest of the considered wall is adiabatic. The opposite vertical wall is maintained at a cold uniform temperature, while the other four walls are adiabatic. An external air -flow enters the enclosure through rectangular cross section at the top of the left wall and exits from a channel, with the same cross section, placed at the bottom of the right vertical wall to extend the cavity and make possible the use of developed boundary conditions at the outlet for the fluid flow and heat transfer. In fact, many tests has been done in order to check the influence of the channel length on the outlet boundary conditions. Finally, a length of L has been adopted as it gives a good compromise between the results accuracy and the computational time.

The cooling flow is air having a Prandtl number Pr of 0.71 and is assumed to be laminar and incompressible. Viscous dissipation is negligibly small and all other fluid properties are assumed constant except the fluid density giving rise to the buoyancy forces (Boussinesq approximation).Under the above assumptions, the Governing equations of heat transfer and fluid flow can be expressed in three-dimensional dimensionless form as follows:

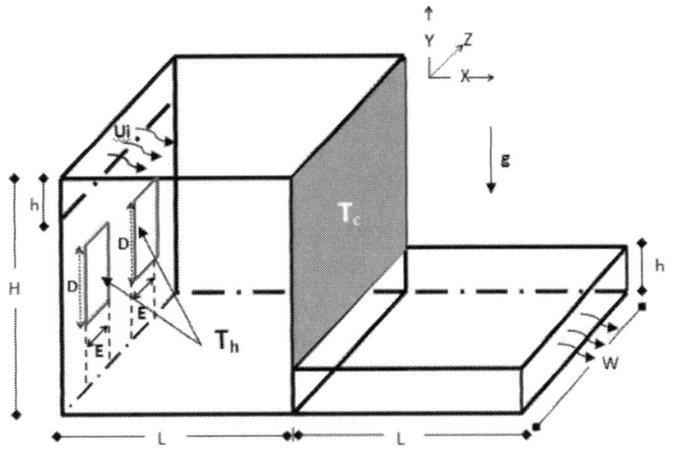

Fig. 1. Physical model and coordinate system.

$$\frac{\partial U}{\partial X}+\frac{\partial V}{\partial Y}+\frac{\partial W}{\partial Z}=0 \quad (1)$$

$$U.\frac{\partial U}{\partial X}+V.\frac{\partial U}{\partial Y}+W.\frac{\partial U}{\partial Z}=-\frac{\partial P}{\partial X}+\frac{1}{Re}\left(\frac{\partial^2 U}{\partial X^2}+\frac{\partial^2 U}{\partial Y^2}+\frac{\partial^2 U}{\partial Z^2}\right) \quad (2)$$

$$U.\frac{\partial V}{\partial X}+V.\frac{\partial V}{\partial Y}+W.\frac{\partial V}{\partial Z}=-\frac{\partial P}{\partial Y}+\frac{1}{Re}\left(\frac{\partial^2 V}{\partial X^2}+\frac{\partial^2 V}{\partial Y^2}+\frac{\partial^2 V}{\partial Z^2}\right)+\frac{Gr}{Re^2}q \quad (3)$$

$$U.\frac{\partial W}{\partial X}+V.\frac{\partial W}{\partial Y}+W.\frac{\partial W}{\partial Z}=-\frac{\partial P}{\partial Z}+\frac{1}{Re}\left(\frac{\partial^2 W}{\partial X^2}+\frac{\partial^2 W}{\partial Y^2}+\frac{\partial^2 W}{\partial Z^2}\right) \quad (4)$$

$$U.\frac{\partial \theta}{\partial X}+V.\frac{\partial \theta}{\partial Y}+W.\frac{\partial \theta}{\partial Z}=\frac{1}{RePr}\left(\frac{\partial^2 \theta}{\partial X^2}+\frac{\partial^2 \theta}{\partial Y^2}+\frac{\partial^2 \theta}{\partial Z^2}\right) \quad (5)$$

The non-dimensional variables are defined as follows:

$$X=\frac{x}{L}, Y=\frac{y}{L}, Z=\frac{z}{L} \quad (6)$$

$$U=\frac{u}{u_i}, V=\frac{v}{u_i}, W=\frac{w}{u_i}, P=\frac{p-p_0}{\rho_0.u_i^2}, \theta=\frac{T-T_c}{T_h-T_c} \quad (7)$$

Where p_0 and ρ_0 are respectively the reference pressure and density. T_h is the temperature at the heated surface, T_c the temperature at the cooled surface, p is the pressure and (U, V, W) are the velocity components.

In the above equations, the parameters Pr, Gr and Re denote the Prandtl number, Grashof number and the Reynold number, respectively. These parameters are defined as:

$$Re=\frac{L.u_i}{\nu}, Gr=\frac{gbL^3(T_h-T_c)}{\nu^2}, Ri=\frac{Gr}{Re^2}, Pr=\frac{\nu}{\alpha} \quad (8)$$

Where β, ν and α are the thermal expansion coefficient, the kinematic viscosity and the thermal diffusivity, respectively.

The boundary conditions, associated to the problem are:

- $U=V=W=0$ on the rigid walls of the enclosure;
- $U=1, V=W=0, \theta_c=0$ at the inlet;
- $\theta_h=1$ on the left vertical heated sections; $\frac{\partial\theta}{\partial n}=0$ elsewhere on the wall
- $\theta_c=0$ on the right vertical coold wall;
- $\frac{\partial\theta}{\partial n}=0$ on other vertical and horizontal walls and the right and left sections near the opening ("n" is the normal direction to the considered wall);
- $\frac{\partial U}{\partial X}=0, \frac{\partial\theta}{\partial X}=0, V=W=0$ at the outlet.

The local Nusselt number and the average Nusselt characterizing the heat transfer at the heated walls are respectively defined by:

$$Nu_{local}(Y,Z)=\frac{\partial\theta}{\partial X} \quad \text{at} \quad X=0 \quad (9)$$

DOI: 10.1051/e3sconf/20172200036

$$Nu_h = \frac{2}{\varepsilon^2}\int_{\frac{L-\varepsilon}{2}}^{\frac{L+\varepsilon}{2}}\int_{\frac{L-\varepsilon}{4}}^{\frac{L+\varepsilon}{4}} Nu(Y,Z)dYdZ + \frac{2}{\varepsilon^2}\int_{\frac{L-\varepsilon}{2}}^{\frac{L+\varepsilon}{2}}\int_{\frac{3L-\varepsilon}{4}}^{\frac{3L+\varepsilon}{4}} Nu(Y,Z)dYdZ \tag{10}$$

3 Numerical method

The incompressible Navier-Stokes and energy equations are discretized by the finite volume method developed by Patankar [6] adopting the power law scheme for the convective terms. The SIMPLEC algorithm is used to couple momentum and continuity equations. The discretized equations are iteratively solved using an Alternating Direction Implicit (ADI) scheme. The system of algebraic equations is solved iteratively by means of the Thomas algorithm. Convergence of the numerical code is established according to the following criterion:

$$\sum_{i,j,k=1}^{imax,jmax,kmax} \frac{\left|\phi_{i,j,k}^{n+1} - \phi_{i,j,k}^{n}\right|}{\left|\phi_{i,j,k}^{n}\right|} \leq 10^{-4} \tag{11}$$

Where ϕ represents a dependent variable U, V, W, T, and P, the indices i, j, and k indicate the grid positions, n represents the iteration number.

The present computational code, used to solve the governing equations, was validated after a comparison between our results and those presented by Moraga & López [5] in terms of temperature distribution and velocity profiles presented in Fig 2a and Fig 2b respectively. Hence, a good agreement was found with the Three-dimensional results of Moraga & López [5]. Uniform grid sensitivity tests were also conducted for different sets of the governing parameters and showed that the grid size 91 × 71 × 71 was the most appropriate for the present study.

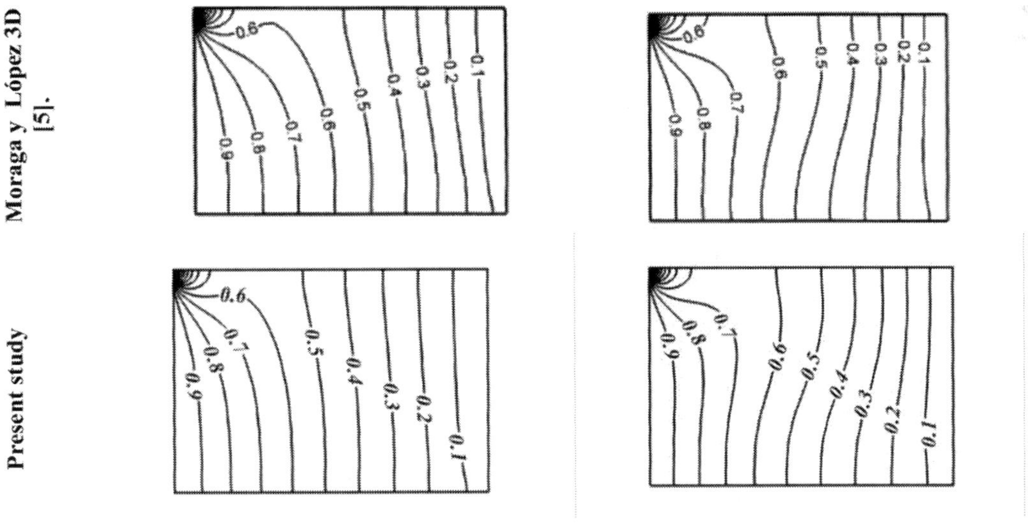

Fig. 2a. Comparison between the isotherm obtained at the middle plane Z = 0.5 and those of Moraga and López [5].

DOI: 10.1051/e3sconf/20172200036

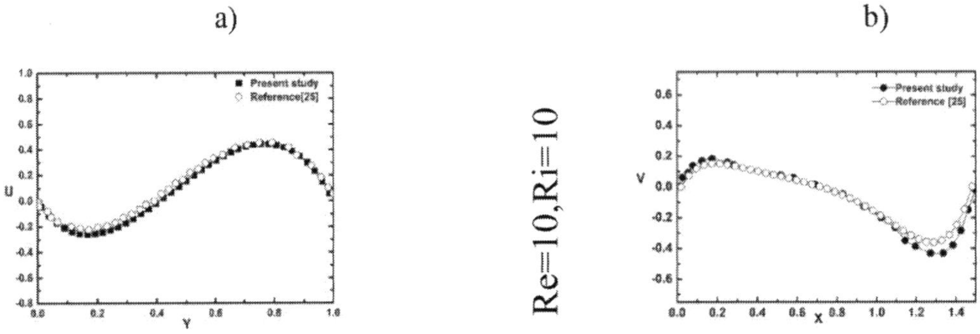

Fig. 2b. U velocity component at X = 0.75, Z = 0.5 and b).V velocity component at Y = 0.5, Z = 0.5

4 Results and discussion

The aim of this paper is to study the effects of the Richardson and Reynolds numbers on the thermal transport and fluid flow phenomena in the cavity. The numerical computations are conducted for a wide range of Richardson number $0 \leq Ri \leq 10$, Reynolds number $50 \leq Re \leq 100$, which correspond to a laminar regime. The relative height of the opening $B = h/L$, the Prandtl number, the relatives height of the heated portions $\varepsilon = D/L$ and their width $\varepsilon/2$ are maintained respectively at constant values: 1/8, 0.71, 0.3 and 0.15.

4.1 Characteristics of the flow and the thermal fields

In order to give a global overview of the cavity, 3D streamlines and isotherms are respectively shown in figures 3a and 3b, for Re = 50 and Ri = 1. It is seen that the fluid flow consists of a clockwise rotating cell which appear inside the cavity under the open lines of the forced flow. These last ones are directed to the outlet. The interaction between the incoming jet flow and the heated portions leads to important changes in the temperature distribution through the cavity toward the cooled wall as seen in the corresponding isotherm.

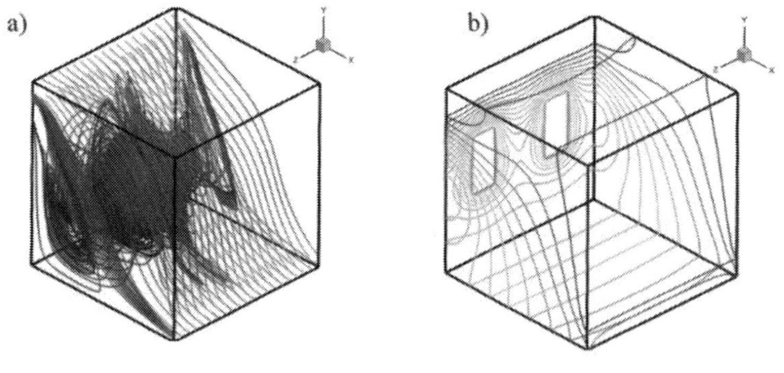

Fig. 3. 3-D streamlines (a) and isotherm (b) for Re = 50 and Ri = 1.

The effects of the Reynolds and Richardson numbers on the flow structure at the plane Z = 0.5 and temperature distribution are shown in Fig. 4a–b. In fact, a previous analysis of the isotherms and streamlines in different plans ($0 \leq Z \leq 1$) shows a perfect symmetry with respect to the plane Z = 0.5. This is due to the geometrical symmetry and adopted thermal

DOI: 10.1051/e3sconf/20172200036

boundary conditions. Hence, the planes Z = 0.5 , X=0.025 and Y=0.5 were chosen for the following presentations. Indeed, these plans are characterized by higher activity and presented adequately the dynamical and thermal fields within the cavity.

The results show the existence of three kinds of heat transfer regimes, which depends on the buoyancy parameter Ri values. The first structure is shown in Fig. 4 for Ri = 0 and low Reynolds numbers (Re = 50). In this case, the inflow is not strong enough, the open lines pattern conserves the same form and occupy almost the whole cavity until the outlet. A small vortex is located beneath the open lines of the forced flow. Furthermore, the size of the vortex increases by increasing Re to 100. Due to this increased inertia force, the corresponding isotherms are stratified over the two heating sections due to the imposed boundary conditions while the lower isotherms are uniform and parallel to each other in the remaining parts of the cavity which signify a dominant conduction heat transfer.

At higher Re values, Re = 200, the flow structure become intense and the intensity of the forced flow the open lines and the vortex size increases. Moreover, a secondary counter clockwise vortex appears in the upper corner of the cavity. The examination of the isotherm lines shows an improvement in the exchange at the cold wall of the cavity.

The higher temperature regions are concentrated near the heated portions while the thermal gradient increases and are submitted to a slight deflection. Indeed, pointing out that vigorous actions of inertia forces are the main role of the flow transport.

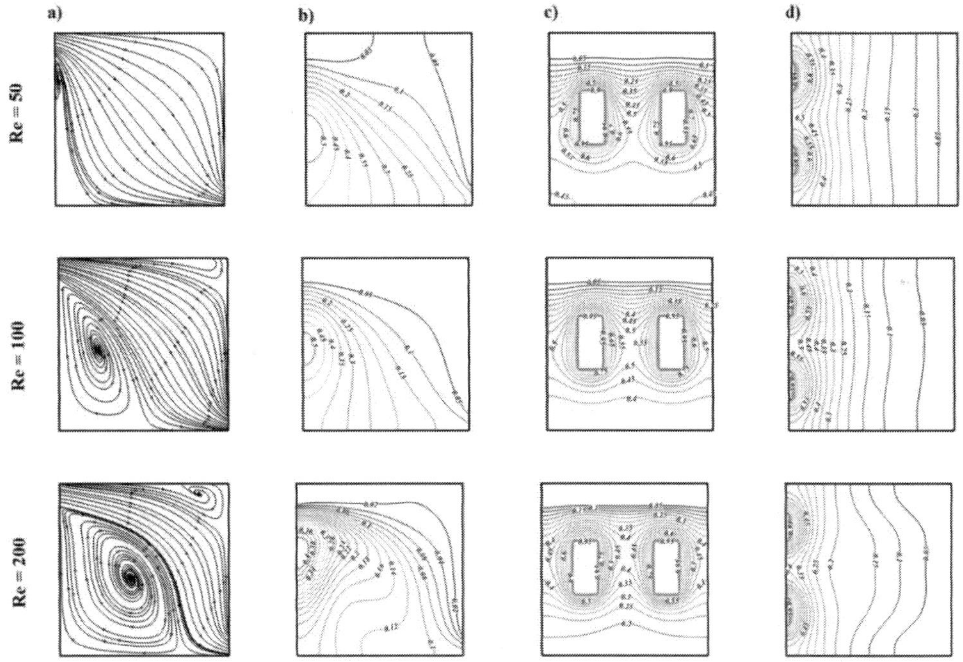

Fig. 4. Streamlines (a) and isotherms (b) at the mid-plane XY (Z = 0.5), isotherms (c) at the plane YZ (X = 0.025) and isotherms at the mid-plane XZ (Y = 0.5) for Ri = 0 and different values of Re.

The second structure is shown in Fig. 5 in terms of dynamical and thermal fields for Ri = 1 and different Re. The behaviour of convective heat transfer is the same as discussed previously in Fig 3 on 3D-model, in the case of low Reynolds number. However, the circulation in the flow becomes larger and the size of the vortex increases by increasing Reynolds number. The corresponding isotherms are more tightened at the vicinity of the heated portions and the gradient temperature becomes larger due to thermal interaction between the heated portions and the incoming cold flow which ensure well the installation of mixed convection.

 DOI: 10.1051/e3sconf/20172200036

At higher value of Richardson number (Ri = 10), the role of the free convection in the cavity becomes more significant (Fig. 6) and consequently the increase of all considered Reynolds numbers is accompanied by a reduction of the convective cell. This can be attributed to the dominance of thermal buoyancy effect where the boundary layers become

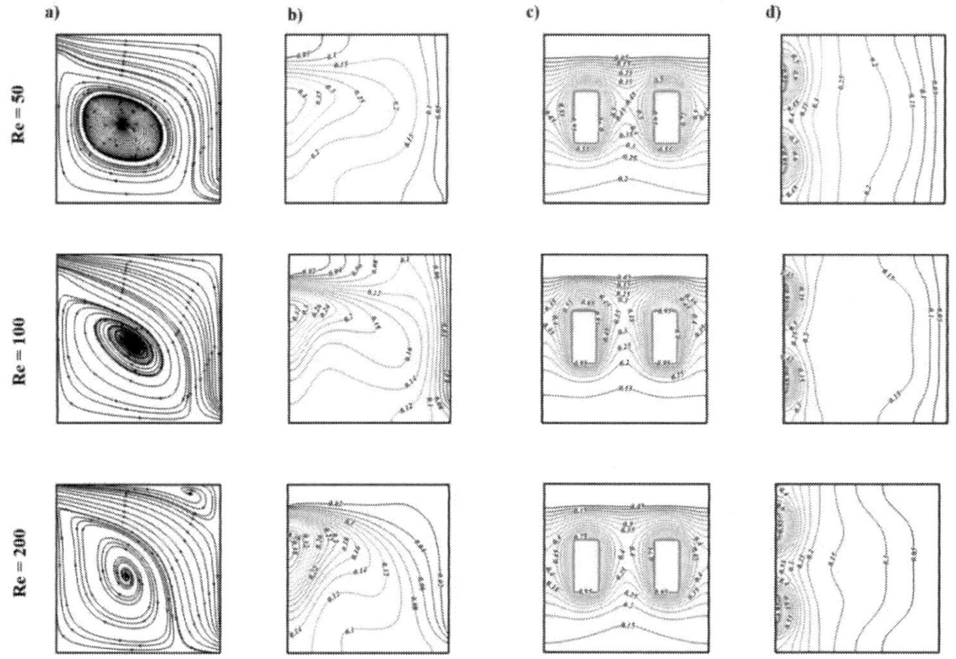

Fig. 5. Streamlines (a) and isotherms (b) at the mid-plane XY (Z = 0.5), isotherms (c) at the plane YZ (X = 0.025) and isotherms at the mid-plane XZ (Y = 0.5) for Ri = 1 and different values of Re.

DOI: 10.1051/e3sconf/20172200036

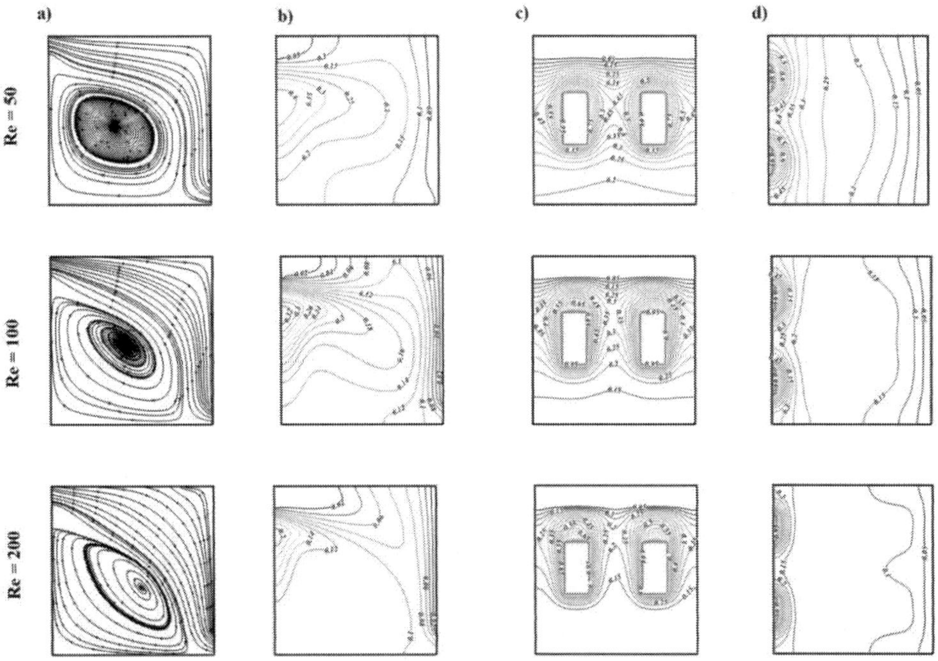

Fig. 6. Streamlines (a) and isotherms (b) at the mid-plane XY (Z = 0.5), isotherms (c) at the plane YZ (X = 0.025) and isotherms at the mid-plane XZ (Y = 0.5) for Ri = 10 and different values of Re.

4.2 Heat transfer enhancement

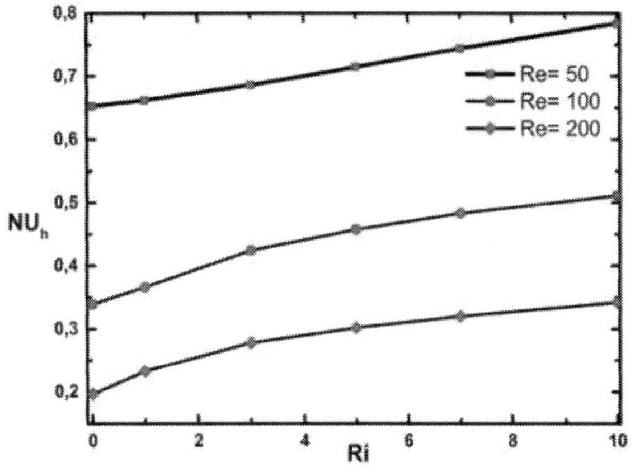

Fig. 7. Variation of the total average Nusselt number with Ri for different Re.

In order to analyze the thermal performance in our configuration, we illustrate in Fig. 7, the total average Nusselt number evaluated on the two heating sections versus Ri, for different value of Re. As expected, the total average Nusselt number increases with the Richardson number. This can be explained by the fact that the buoyancy effects are the dominant mode of mixed convection transport. In contrast, for a given number Ri, the

DOI: 10.1051/e3sconf/20172200036

Nusselt number decreases by increasing Re. This is due to the acceleration of the cold forced flow through the cavity along the cold wall which does not mix well with the internal heated air.

5 Conclusion

A numerical study has been carried out in the case of laminar three-dimensional mixed convection discretely heated from the side. The results show that the controlling parameters Re and Ri leads to important modifications of dynamical and thermal structures. The increase of the Richardson numbers enhances significantly the heat transfer. however, the forced flow support the heat transfer for low Reynolds numbers and plays an opposing role for the high values of Re.

References

1. E. Papanicolaou, Y. Jaluria, Numer, Heat. Tr. **18**, 4, 427–461 (1991)
2. E. Papanicolaou, Y. Jaluria, Numer, Heat. Tr. **23**, 4, 463–484. (1993)
3. T.H. Hsu, P.T. Hsu, S.P. How, Numer, Heat. Tr. **31**, 6, 655–683 (1997)
4. M.M. Rahman, S. Parvin, N.A. Rahim, M.R. Islam, R. Saidur, M. Hasanuzzaman, Appl. Math. Model **36**, 5, 2056–2066 (2012)
5. N.O. Moraga, S.E. Lopez, Numer, Heat. Tr. **45**, 8, 811–824 (2004)
6. S. Patankar, CRC press (1980)

The use of mathematical models for diagnosis of activated sludge systems in WWTP

Jakub Drewnowski[1,*], and *Marcin* Zmarzły[2]

[1]Gdansk University of Technology, Faculty of Civil and Environmental Engineering, ul. Narutowicza 11/12, 80-233 Gdansk, Poland
[2]ZAPSOFT Ltd., Al. Kasztanowa 3A, 51-125 Wrocław, Poland

Abstract. In this study diagnosis of activated sludge systems in wastewater treatment plant (WWTP) was investigated. Diagnosis of technical objects can be realized in many ways. One of the divisions of the diagnostic methods include modelling with or without a model of the object. The first of these is the analysis of the symptoms for which, based on the parameter values, the abnormality in the diagnosed objects are sought. Another way is to use models of objects undergoing diagnosis. In this case, the diagnosis comes down to a comparison of information from the response object model or the estimated parameters of the model with data from the real object. The aim of this study was to evaluate an innovative concept of the possible use the mathematical model and computer simulation in the diagnosis and control of activated sludge systems in WWTP.

1 Introduction

Research in diagnosis systems has been very active since the 1970s. Numerous fault diagnosis methods are reported in the industrial and academic literature. Common know was fault detection and diagnosis (FDD) methods, which was a family of approaches finalized to detection and location of faults on processing systems. The term "fault" is generally defined as "an unpermitted deviation of at least one characteristic property of the system from the acceptable, usual, standard condition" [1]. The main aim of FDD methods consists in determining the type, size and location of the most possible fault, as well as its time of detection in order to improve product quality, plant efficiency, as well as the ability to meet environmental and safety constraints.

The task of diagnosis systems in this study was the earliest and most accurate detection of irregularities in the operation of diagnosed process, object etc. based on on-line data. Such a system allows the operator of WWTP to unload the object and monitor operating parameters of the object in order to diagnose abnormal process. Diagnostics of technical objects is based on two main methods: symptomatic diagnostics and modelled support. In the first of these cases, object are performed on the basis of known diagnostic parameters, while in the second variant the diagnosed model is used. Diagnosis supported WWTP control is comparing the results obtained on the real object with the results obtained

* Corresponding author: jdrewnow@pg.gda.pl

from the model and then determining, so-called "Residues occurrence of a fixed value may indicate an emergency" or "Closely linked with data acquisition is the detection of faulty behaviour from process data" [2, 3].

The purpose of this study is diagnosis of activated sludge systems in wastewater treatment plant (WWTP). Diagnosis of technical objects can be realized in many ways. One of the divisions of the diagnostic methods include modelling with or without a model of the object. The first of these is the analysis of the symptoms for which, based on the parameter values, the abnormality in the diagnosed objects are sought. Another way is to use models of objects undergoing diagnosis. In this case, the diagnosis comes down to a comparison of information from the response object model or the estimated parameters of the model with data from the real object. The aim of this study was to evaluate an innovative concept of the possible use the mathematical model and computer simulation in the diagnosis and control of activated sludge systems in "Klimzowiec" WWTP. located in Chorzów (Poland).

2 Material and methods

Figure 1 presents a general diagram of the model diagnosis of studied plant under full-scale condition. The model is powered by data from the real object, based on which the object is simulated. The next step is to compare modelling results with the actual results of measurements from the real object. On this basis, residual generation is performed. Then the evaluation of the residue values is made on the basis of which the detection of possible malfunctions is performed (Figure 1).

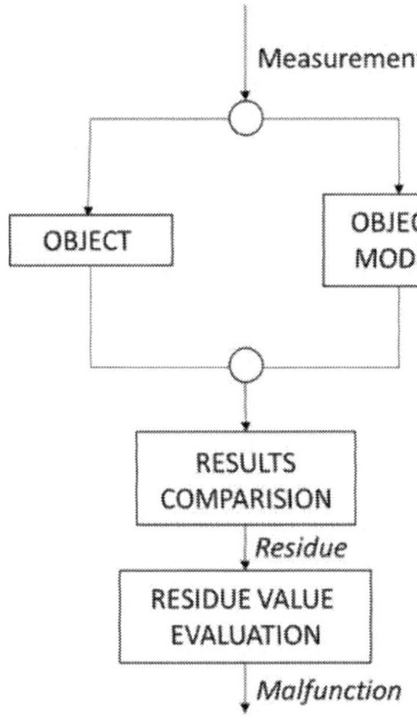

Fig. 1. The diagram of the model diagnosis of studied plant under full-scale condition.

Object modelling can take place in several different ways. The most popular modelling methods include:

 DOI: 10.1051/e3sconf/20172200037

- Mathematical model - approximate object description was created using a mathematical software
- Real model - most often, the system is similar to that of a researcher, but simplified and thus simpler to study its behaviour:
- Theoretical model - hypothetical design describing a given object in a simplified way, the parameters of the object are reduced to the relations between the most important ones.
- Nominal model - is a deterministic model whose uncertainty is defined by a particular scenario.

Regardless of the modelling methodology, the main problem is the correct representation of the modelled object. This may be due to a number of factors, such as too much measurement data, what causes that all data in the model cannot be taken into account. The reason of this situation is the increasing complexity of parameters that simulation time is too long and the model will no longer be justified. The solution to this problem is to simplify the model. However, it may be that excessive model simplification may reduce the accuracy of the model so that simulation results may prove useless.

2.1 Study side

The "Klimzowiec" WWTP is a mechanical and biological treatment plant based on the use of activated sludge in the Bardenpho system. The WWTP is managed by the Chorzów-Świętochłowice Water and Wastewater Utility. Klimzowiec WWTP consists of pre-denitrification zones and defatting zones as well as three denitrification bioreactors and five nitrification. Each of the nitrification bioreactors has three independent oxygenation zones. The average effluent flow is approximately 25,900 m^3/d which is equivalent around 200,000 RLM. The Klimzowiec WWTP also has six secondary settlers of which only four are used at the same time.

2.2 Model description

The Klimzowiec WWTP model, created in software WEST 2014 SP3, was used to diagnose measuring devices installed in the objects. The mathematical model was developed in WEST (Wastewater Treatment Plant Engine for Simulation and Training) software and used for diagnostics (Figure 2). The ASM2d (Activated Sludge Model No.2d) was used to create the simulation.

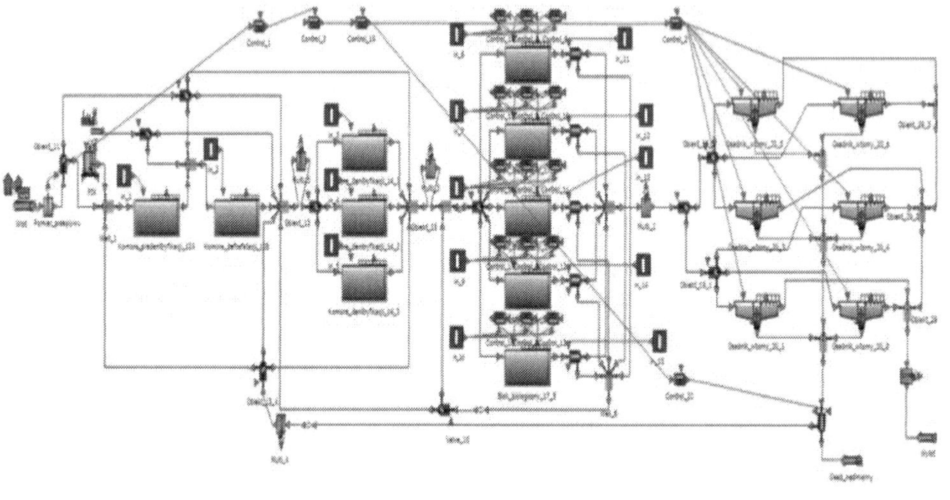

Fig. 2. The existing model of "Klimzowiec" WWTP created in the WEST software.

The model is enhance, among on-line data such as:
- wastewater flow Q with sewer parameters (COD, TSS, TN, TP),
- the amount of wastewater brought along with the wastewater parameters (COD, TSS, TN, TP),
- temperature in pre-denitrification, anaerobic (P removal), denitrification, nitrification tanks etc.,
- the concentration of oxygen in each of the three sections of biological reactors,
- amount of external and internal recirculation.

The standard library of WEST software blocks lacked the means to reproduce the control of sludge recirculation. The WEST software gives the ability to add and modify existing models. This allowed to create a control element for the concentration of suspensions in the effluent, respectively. Control of this parameter is achieved by adjusting the amount of excess sludge discharged from the system.

2.3 Previsys and diagnostics system

PreviSys was used in order to start the diagnostics of the sensors, as it was necessary to create a system that allows two-way communication with the SCADA system. It enabled to inform the dispatcher about the problem in easy and readable way and is closely linked with data acquisition is the detection of faulty behaviour from process data. This system can be divided into three parts: PreviBox, PreviSoft and PreviSCADA.

PreviBox is a device that, when used in conjunction with or replacing wastewater automation systems, implements optimization algorithms. The complete device consists of a PLC with modules suitable for communication with devices at a given WWTP such as multitouch panel, industrial PC, router, and a switch.

PreviSoft is based on advanced algorithms used in industrial automation. For control, the ASM (Activated Sludge Model) is used. To run the modeling process, PreviSys can use both general purpose software (MATLAB®) and specialized modelling program for wastewater treatment simulation (WEST®, GPS-X®).

PreviSCADA is software installed on an industrial PC built into PreviBox, which can additionally serve as a supervisory system (SCADA) for smaller treatment plants. The

 DOI: 10.1051/e3sconf/20172200037

software includes, among others, specific modules: communication, databases, alarm, reporting, visualization, web server. The PreviSys PreviSCADA software acts as an object data logger, so that the user can see locally how PreviSys works.

In summary, PreviSys is a smart, complete solution that can be used to control the work of a WWTP, which, in oposition to SCADA software, uses advanced control and diagnostics algorithms. It also allows to create a mathematical model of a controlled object. The model can be used both in the control algorithms of the WWTP as well as in the diagnostics of the sensors.

2.4 Downloading data to online model

In order to have the current data in the diagnostic system, it is necessary to automate the simulation execution. For this purpose, the operating system task schedule was used. It initiates a batch script that manages all the steps involved in retrieving current data and activating a mathematical model. This script runs every 10 minutes. The first step in the script is to retrieve the current SCADA data, then convert the data into a format suitable for the mathematical model. Then the simulation of the object's work is started. The last step is to save the resulting data in the appropriate files.

3 Results and Discussion

The task of diagnostic systems is the earliest and most accurate detection of irregularities in the operation of diagnosed object. Such a system allows the operator to unload the object because it does not need to monitor operating parameters of the object in order to diagnose abnormal process. As the operation of the model has been automated in order to simulation data, that were available from WWTP. The next step was the comparison of model outputs with actual on-line measurements (e.g. N-NH_4) at WWTP. Based on this, MATLAB® software was used to designate a residue. Subsequently, a classification of resides will be made, depending on which further actions are taken. If residuum exceeds the qualifier value, it was considered to be informative of the likely occurrence of a sensor failure and the information will be sent to the dispatcher. Information is sent using the Modbus tk python library. This library is for communication using the Modbus protocol. The information is sent to the PLC located in the PreviSys and then read by the SCADA system. This way, when a sensor fails to diagnose a sensor, SCADA will display the message. According that, the dispatcher will decide on further action.

DOI: 10.1051/e3sconf/20172200037

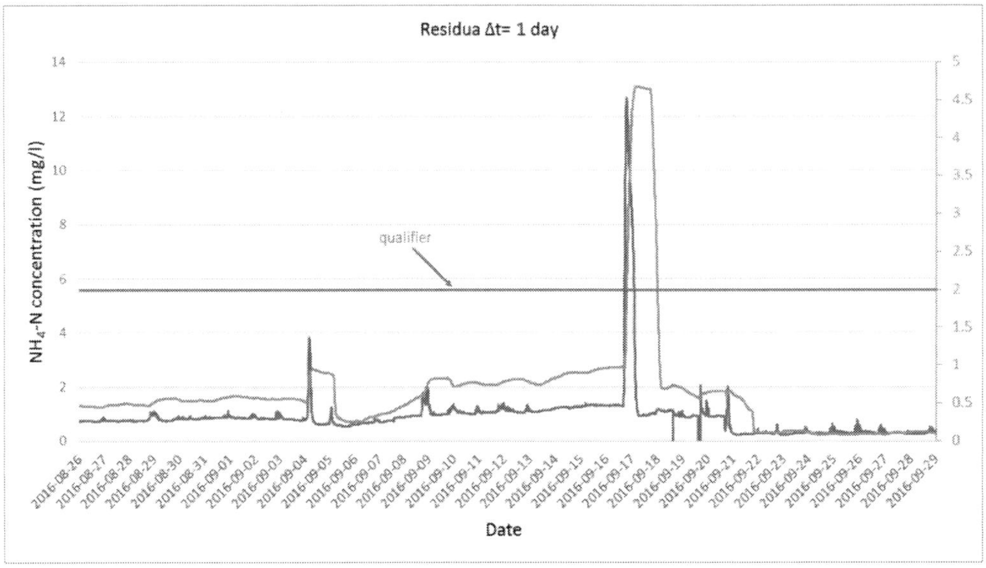

Fig. 3. The residue value and ammonia nitrogen concentration changes in the biological reactor of the studied WWTP.

The Figure 3 shows the residue value for $\Delta t = 1$ day and the value of ammonia nitrogen concentration changes in the biological reactor of the studied WWTP. Simulation and calculation of residual values takes place every 10 minutes. The chart contains data from approximately 34 days between August 26, 2016 and September 29, 2016. On the left axis the concentration of the measured ammonium nitrogen is shown and the right value of the residue is shown on the left. A horizontal line at the value of 2 is residua, which is the qualifier value. In this case, the qualifier value was set to operator and residue is greater than 2, so this value was exceeded, the dispatcher was informed of the possible calibration of the probe each time.

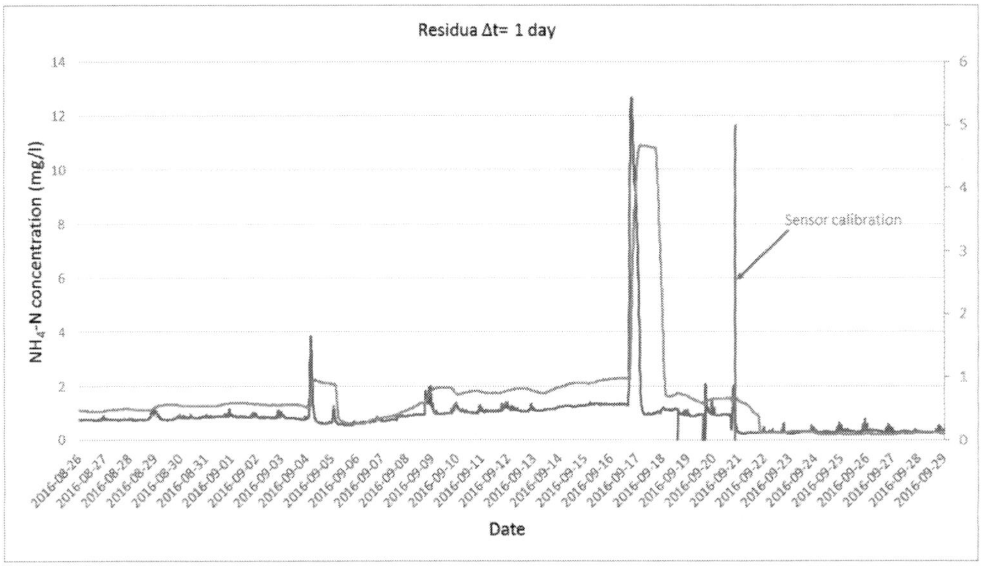

Fig. 4. The measurement results from the N-NH_4 probe after calibration process.

After informing the dispatcher about possible problem with the N-NH_4 probe, further steps have been taken as it was presented in the Figure 4. The dispatcher was informed about the situation on 17 September. Then, the N-NH_4 probe was cleaned on September 19th and on September 21st (marked on the vertical line). The probe was calibrated based on data from the laboratory. After calibration process, the measurement results from the N-NH_4 probe have dropped by about 0.5 mg/l, which was a good information for the WWTP.

Interest in this field, monitoring systems for fault detection and diagnosis (FDD) in biological processes and on-line measurements, specifically, WWTPs has started around 30 years ago with an early application of model identification and diagnosis by Beck [4] based on a fuzzy control. Since then, several alternative approaches were presented, mainly based on statistical considerations. Fuente [5] compare a number of established techniques for estimating the process parameters when a fault occurs in a simplified nonlinear model for a real plant. Yoo [6] proposed the identification of different operating conditions and discrimination between sustained and simple faults by using databased methods for dimensionality reduction and clustering. In a similar way, Lee [7] focused on the identification of faulty sensors and the reconstruction of the measurements by estimating the process dynamics; here, a dynamic principal component analysis (PCA) was successful in effectively detecting the faults, especially when the abnormal condition was caused by a single sensor failure. In this study the step further in this field was made. First of all, it was created a mathematical model of the studied plant (calibrate and automate model launching), which allowed access to current data for comparison with on-line measurements from the object. Based on this comparison, the residue was determined. The last step in the operation of the sensor diagnostics system is to evaluate the residues and then, depending on the results sent to the operator, information about possible sensor damage. The system allows for defining parameters of residual determination and qualifier functions. This makes it possible to customize the diagnostic system to any type of variable, regardless of the magnitude of its variation and the speed with which changes in the value of the measurement occur. Based on the calculated residue it is possible to determine the level of the qualifier after which the information for the dispatcher with possible error in the operation of the sensor is reported. Moreover, this study presents innovative approach in this field in comparison with the more recent work. For example, Ciappelloni [8] proposed a combination of model and databased techniques to describe a pilot scale aerobic sequence batch reactor (SBR); PCA is again used to describe different process cycles and a decision algorithm was developed to classify different process faults. The SBR was also studied by Wimberger and Verde [9] with the goal of evaluating the detectability and is possibility for analytical and signal based fault detection and isolation. A different approach was proposed by Nejjari [10], where an extended Luenberger observer is adopted for the residual generation of sensor faults. Zumoffen and Basualdo [11] focused on the need for an efficient fault tolerant control on WWTP. The term “fault” is generally defined as “an unpermitted deviation of at least one characteristic property of the system from the acceptable, usual, standard condition” [1]. Nevertheless, the databased approach to fault detection methods can achieve only limited validity, mainly because vast amounts of data are required but not necessarily available in a WWTP. The FDD technology is still developing and should continue to improve as a part of future "Smart Control System" at WWTP. The important point is that wastewater utilities can be heroes instead of burdens by working towards a goal of biochemical process improvement and control for energy neutrality at WWTP. Essentially, for these reasons the causal fault diagnosis approach is believed to have a promising Nonlinear modelling and adaptive monitoring with fuzzy and multivariate statistical methods in biologic wastewater treatment plants.

4 Conclusions

Research in the field of activated sludge systems diagnosis in wastewater treatment plants has been very active since the 1970s and numerous fault diagnosis methods are reported in the industrial and academic literature. For example FDD methods are a family of approaches finalized to detection and location of faults on processing systems. The main challenge of FDD methods consists in determining the type, size and location of the most possible fault, as well as its time of detection in order to improve product quality, plant efficiency, as well as ability to meet environmental and safety constraints [3].

Based on the presented data, it can be concluded that the diagnostics of the sensors using a mathematical model is an innovative method allowing to simplify and accelerate the diagnosis of measurement sensor problems. A commercial software WEST was used to model the WWTP configuration and for the dynamic simulation of the biochemical process, but the reason of greatest interest is the implementation of a special system and an appropriate communication software with SCADA as a development of mathematical model for the diagnosis of the plant. The causal fault diagnosis approach is believed to have a promising potential and should be developed regarding to online measurement equipment is still needed because of challenging conditions for reliable measurements at WWTPs. Closely linked with data acquisition is the detection of faulty behaviour from process data.

References

1. R. Isermann, (Heidelberg, Springer, 2006)
2. C.F. Lindberg. PhD thesis, System and Control Group, Uppsala University, Sweden (1997)
3. H. Haimi, M. Mulas, K. Sahlstedt, R. Vahala, Helsinki University of Technology Water and Wastewater Engineering, Espoo (2009)
4. M.B. Beck, A. Latten, R.M. Tong, International Institute for Applied Systems Analysis (1978)
5. M.J. Fuente, P. Vega, M.B. Zarrop, M. Poch, CEP **4**, 8, 1089–1098 (1996)
6. C. Yoo, P.A. Vanrolleghem, I.B. Lee, JofB **105**, 12, 135–163 (2003)
7. C. Lee, S. W. Choi, I.B. Lee, ChILS **70**, 2, 165–178 (2004)
8. F. Ciappelloni, D. Mazouni, J. Harmand, L. Lardon, WST **53**, 1, 169–177 (2006)
9. D. Wimberger, C. Verde, CEP **16**, 11, 1344–1353 (2008)
10. F. Nejjari, V. Puig, L. Giancristofaro, S. Koehler, 17th IFAC World Congress, Seoul, Korea (2008)
11. D. Zumoffen, M. Basualdo, IEChR **47**, 15, 5464–5481 (2008)

Analysis of a combined heating and cooling system model under different operating strategies

Mieczysław Dzierzgowski[1], and *Ryszard* Zwierzchowski[1,*]

[1]Department of Heating and Gas Engineering, Faculty of Building Services, Hydro and Environmental Engineering, Warsaw University of Technology, ul. Nowowiejska 20, 00-653 Warsaw, Poland

Abstract. The paper presents an analysis of a combined heating and cooling system model under different operating strategies. Cooling demand for air conditioning purposes has grown steadily in Poland since the early 1990s. The main clients are large office buildings and shopping malls in downtown locations. Increased demand for heat in the summer would mitigate a number of problems regarding District Heating System (DHS) operation at minimum power, affecting the average annual price of heat (in summertime the share of costs related to transport losses is a strong cost factor). In the paper, computer simulations were performed for different supply network water temperature, assuming as input, real changes in the parameters of the DHS (heat demand, flow rates, etc.). On the basis of calculations and taking into account investment costs of the Absorption Refrigeration System (ARS) and the Thermal Energy Storage (TES) system, an optimal capacity of the TES system was proposed to ensure smooth and efficient operation of the District Heating Plant (DHP). Application of ARS with the TES system in the DHS in question increases net profit by 19.4%, reducing the cooling price for consumers by 40%.

1 Introduction

Cooling demand for Air Conditioning (AC) purposes has grown steadily in Poland since the early 1990s. The main clients are large office buildings and shopping malls located in city centers. In widespread conventional cooling systems, demand for air conditioning is met by centrally installed compressor chillers in Compressor Refrigeration System (CRS) producing chilled water streamed to receivers – fancoils or coolers in air handling units, using a dedicated hydraulic installation inside the building. Electricity is the driving medium for producing chilled water in CRS. There is an increasing scarcity and cost associated with the medium, especially in light of recent summertime blackouts in Poland. There is a steady increase in the demand for electricity in the summer and in the price for cooling capacity for the end user [1. Both factors have an adverse impact on large office buildings, shopping malls, sports and culture centers in cities as main customers for

* Corresponding author: ryszard.zwierzchowski@is.pw.edu.pl

cooling, and on the attractiveness of cooling supply generated from heat using absorption chillers in the Absorption Refrigeration System (ARS) [2–5].

2 System description

The District Heating System (DHS) subject to analysis supplies heat to a town of 20,000 inhabitants in central Poland. The DHS consists of a District Heating Plant (DHP) with total power of around 7 MW, District Heating Network (DHN) with total length of 3.1 km and 39 District Heating Substations (DHSu). The layout of the DHS is shown in Fig. 1.

The DHP consists of 5 heat only boilers, 3 of them with heating power of 1750 kW each are connected in a cascade scheme and operate during the heating season, while the other 2 with heating power of 970 kW each are designated to operate during the summer season. All boilers are Condensing type Boilers (CB). The DHN supply heat to consumers connected to the system through 39 indirect, 2-funtion DHSu, i.e., supply heat for Central Heating (CH) and Domestic Hot Water (DHW) purposes. The main piping of the DHN features nominal diameters 200 and 150 mm, others are 25–100 mm, all of them pre-insulating piping.

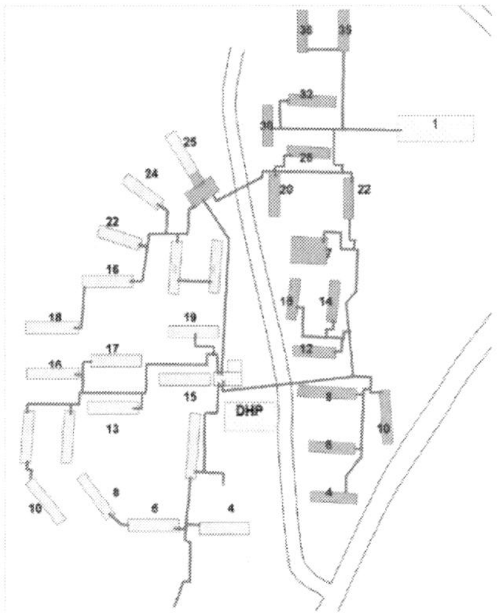

Fig. 1. Layout of present state of analyzed DHS.

The current total heat demand of consumers is 5.0 MW and by category: 4.0 MW for CH and 1.0 MW for DHW purposes, but the real total heat demand is 3.4 MW (2.6 MW for CH and 0.8 MW for DHW). During the last three years, the annual amount of heat sold to consumers is 31 000 GJ and heat transportation losses are 10–11.5%, but during the summer season losses reach 25%. The quantity of degree-days in that period is in the range 3684.4–3981.0. Pressure disposal in the DHP varies in the range 180–220 kPa (average is 188 kPa), while water supply and return temperatures are in the range 70–94.1°C (average is 71.6°C), and 40–64.1°C (average is 46.5°C) respectively. The real value of total flow coefficient K_v of the DHN varies from 19.6 to 127.4 m^3/h (average is 47.4 m^3/h) and the drop in water temperature in the DHN is 11.8–31.5 K (average is 23°C).

Analyses of operational data have indicated that currently there is insufficient adaptation of operation conditions of the DHP to real consumer demand for heat for CH purposes. The biggest value of this discrepancy, which is characterized through the

DOI: 10.1051/e3sconf/20172200038

Coefficient of Heat Load (CHL), is below 0.43 and appears in the transition period with the outside temperature of +3°C, which is most of the time of the heating season.

Figure 2 shows in schematic form the DHS with the Thermal Energy Storage (TES), which produces and distributes heat for both heating and cooling purposes. In the analyzed case, heat is produced in in the DHP furnished with 5 CB for CH and DHW purposes during the heating season, and chilled water for Air Conditioning (AC) and DHW during the summer season.

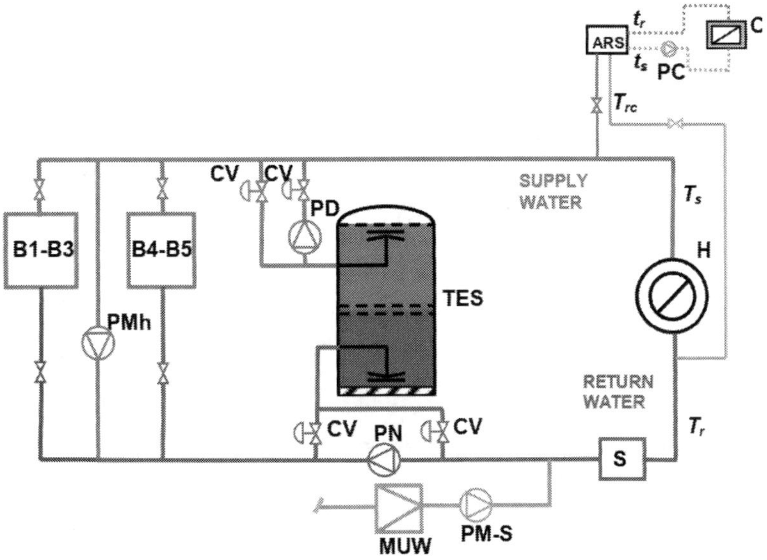

Fig. 2. Schematic layout of the DHS with TES which supplies heat for CH, DHW and AC purposes. Descriptions of symbols: B - boiler, C – cold consumers, CV – control valve, H – heat consumers, MUW – make-up water station, PC – chilled water pumps, PD – TES pump, PMh – hot water mixing pumps, PM-S - make-up water and pressure stabilizing pumps, PN – network pumps, S – sludger.

3 Absorption Chillers

3.1 Application of ARS for DHS

In Poland, most DHS supply only heat to their clients, but during the summer season heat demand falls significantly in most urban areas which, taking into consideration the significantly lower price growth rate of heat in comparison e.g. to the electricity price growth rate, defines the potential use of DHS during the summer to power ARS [6].

Also benefiting from increasing heating demand in summer would be CHP plants, as they could boost electricity production due to the increased heat demand.

3.1 Calculation of hot water flow rate for supplying ARS

To give an idea of increased heat demand, an example is given below of the calculation of hot water flow increase for the purposes of supplying ARS with various cooling capacities from 500 kW to 1000 kW. For the purposes of the calculation some assumptions regarding the absorption chillers Coefficient of Performance (COP) were made [7, 8].

$$\dot{m}_{ARS} = \dot{Q}_{ARS}/(COP \cdot \Delta T \cdot C_w) \quad (1)$$

 DOI: 10.1051/e3sconf/20172200038

where: $\Delta T = T_s - T_{rc}$

The above formula makes it possible to calculate the supply water mass flow needed to produce 100% cooling capacity. For example, for 500 kW of cooling capacity we need 38.99 Mg/h of hot water at 85°C/70°C, 750 kW needs 58.48 Mg/h and 1000 kW needs 77.97 Mg/h, taking into consideration an average COP of 0.735 for the ARS.
To give greater insight, example calculations have been made using the following formula:

$$CF = \dot{m}_{ARS(r)}/\dot{Q}_{ARS(r)d} \tag{2}$$

The *CF* is applied to gain the hot water mass flow.

$$\dot{m}_{ARS}^{CF} = \dot{m}_{ARS} \cdot CF \tag{3}$$

4 Operational analyses of DHS due to introduction of cooling generation

A detailed analysis of this tri-generation system was performed. The analysis covers the effect of increase in supply temperature, both on efficiencies of heat generation in CB and chilled water production in ARS. Also, changes in water flowrate, return temperature and pressure disposal were analyzed for water supply temperatures in the range 70°C to 85°C.

Efficiency of heat generation in condensing type gas boilers was approximated on the basis of real experimental data through polynomial in the following form:

$$\eta_b = a_6 \cdot T_{av}^6 + a_5 \cdot T_{av}^5 + a_4 \cdot T_{av}^4 + a_3 \cdot T_{av}^3 + a_2 \cdot T_{av}^2 + a_1 T_{av} + a_0 \tag{4}$$

where: $T_{av} = (T_s - T_r)/2$

Useable power of DHP was calculated from the equation:

$$\dot{Q}_{DHP} = \dot{V}_g \cdot W_u \cdot \eta_b \tag{5}$$

Heating power balance for the system could be expressed in the following way:

$$\dot{Q}_{DHP} = \dot{Q}_L^{sp} + \dot{Q}_{DHW} + \dot{Q}_{ARS} + \dot{Q}_L^{rp} \tag{6}$$

Required power to generate DHW could be calculated as:

$$\dot{Q}_{DHW} = \dot{m}_{DHW} \cdot (T_s - T_r) \cdot C_w \tag{7}$$

Losses of power from supply DHN were calculated from the equation:

$$\dot{Q}_L^{sp} = (T_s - t_e)/R_{rp-te} \tag{8}$$

Losses of power from return DHN are the following:

$$\dot{Q}_L^{rp} = (T_r - t_e) \cdot R_{rp-te} \tag{9}$$

Required heating power of ARS was estimated by heat effects in cooling spaces:

$$\dot{Q}_{ARS} = \dot{Q}_{ee} + \dot{Q}_{el} + \dot{Q}_{em} + \dot{Q}_{ent} + \dot{Q}_{ep} + \dot{Q}_{et} \tag{10}$$

4.1 Analysis of increase in supply water temperature

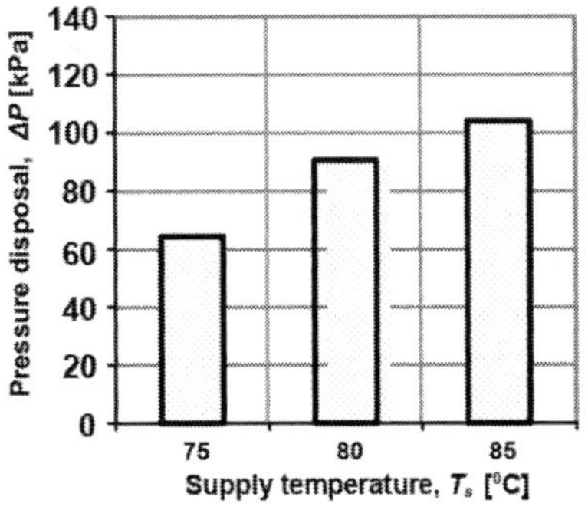

Fig. 3. Influence of supply temperature T_s on required pressure disposal ΔP for DHS.

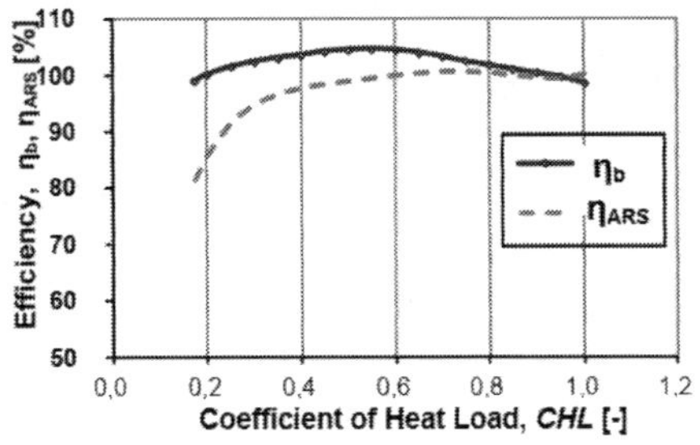

Fig. 4. Influence of CHL on efficiency of heat and cool generation.

As a result of increase in supply water temperature T_s for DHS, increase in heat losses from supply piping is evident. Simultaneously, the following can be observed: decrease in required pressure disposal for DHS and in flow rate of water supplying DHSu (Fig. 3). Simultaneously, the working conditions and effectiveness of chilled water production by absorption chillers are improved [9–11]. Fig. 4 presents changes in efficiency of heat generation in the CB and cool generation in absorption chillers vs. Coefficient of Heat Load (CHL).

4.2 Analysis of heat supply for chilled water production purposes

Seasonal, simulation calculations of heat demand for chilled water production by absorption chillers were carried out for habitable-office building.

Fig. 5 shows changes in flowrate of water and heat demand for DHW and chilled water for the analyzed building on 8 selected days in May.

4.3 Operation of TES and its influence on DHS

Operational analyses of the DHS were performed for the purpose of selecting the appropriate size of the TES in order to increase the operational efficiency of the existing

DOI: 10.1051/e3sconf/20172200038

CB especially for DHW and ARS. Investment costs for the ARS and the TES system application in the DHS were also assessed [12].

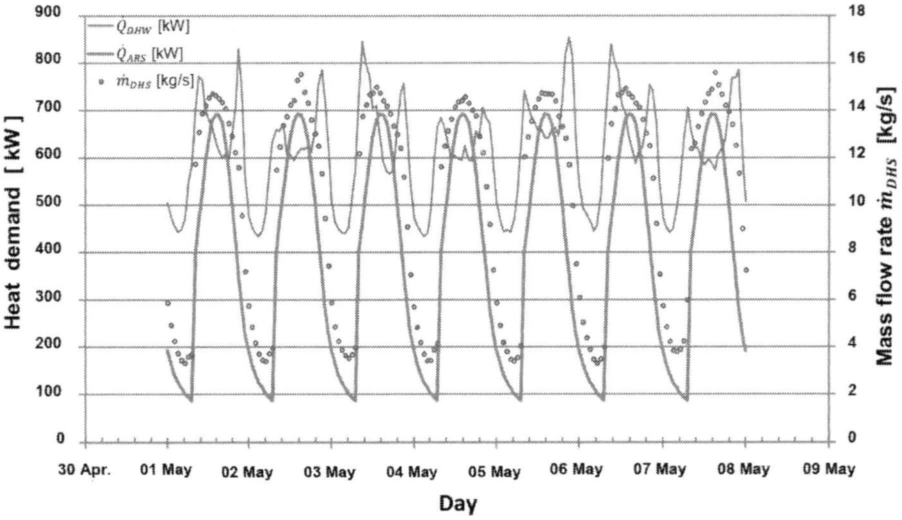

Fig. 5. Hourly changes in heat demand for chilled water production, cool demand and total flowrate of network water for the analyzed building and eight selected days in May.

Operational analysis of the DHS covered supply and demand side, i.e., heat generation by the CB and heat consumption by existing consumers for DHW purposes and new consumers for ARS purposes. Heat consumption analysis for the summer season was carried out in order to increase operational efficiency of the ARS installation and to investigate the application of TES systems in the DHS. Heat storage in the DHS and its effect on power and operation of the CB was investigated i.e., heat accumulation by (i) the DHN, and (ii) the non-pressure TES system was analyzed [13].

Finally, the calculation results were shown for the required capacity of the TES systems assuring continuous and efficient operation of the CB installation in the DHS, especially during the summer season. Also presented were some advantages of implementing the TES system in the DHS in the case where the CB and ARS are applied. Simulation of the DHS with the TES system using real, operational data for the system situated in central-southern Poland (real total heat demand is 3.4 MW i.e. 2.6 MW for CH and 0.8 MW for DHW) was performed to determine the optimum capacity of the TES tank from an operational point of view. The optimum capacity of the TES tank from an operational point of view means that the boilers can operate for a long time practically at constant capacity. The DHS heat demand fluctuation can be covered by the TES system. For the optimum operating conditions of the CB the volume of the TES system is approx. V_{TES} = 55 m^3. Changes in heating capacity of the CB and heat demand by the DHS during 2 weeks' operation in the summer season for boilers with total capacity of 970 kW are shown in Fig. 6.

Changes in the charging and discharging processes of the TES system for optimum operational heating capacity of the CB were analyzed. Fig. 6 shows also changes in the charging and discharging processes of the TES taking into consideration the capacity of the DHN, during 2 weeks' operation in the summer season for optimum boiler power.

DOI: 10.1051/e3sconf/20172200038

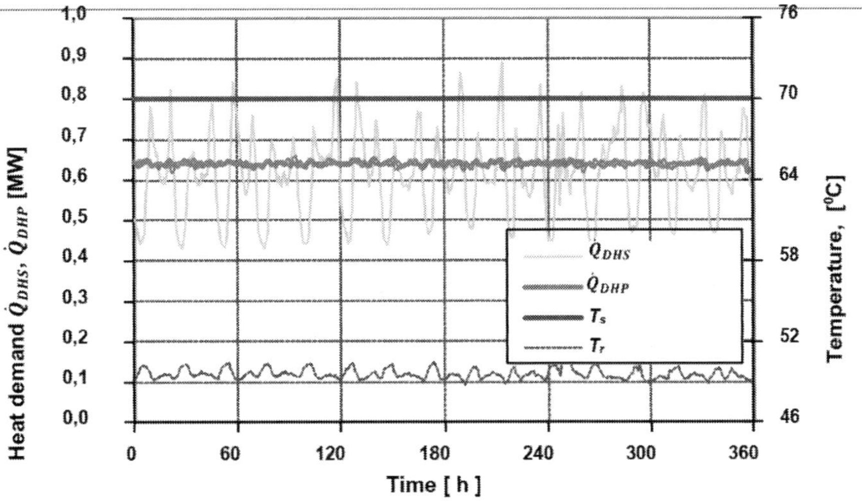

Fig. 6. Changes in boiler power, DHS heat demand, T_s and T_r with the TES application during 2 weeks in the summer.

5 Influence of cooling generation on conditions of DHS

Results of detailed analysis show that additional heat production for chilled water purposes has a major effect on technological and economical operation of the existing system, especially in the summer season. the required condition to increase water supply temperature Ts, causes increased total heat losses of the DHN, slightly decreased efficiency of heat production by the CB and increased fuel consumption. Influence of cooling generation on operational and economic conditions of DHS is shown in Fig. 7. Moreover, existing infrastructure and flow capacity of DHN will be much better utilized when additional heat production for chilled water starts, e.g., total heat losses of DHN will drop from the current 25% to 19%. Taking into consideration the present condition of heat generation for DHW purposes in the summer season, when the supply temperature of network water T_s is 70°C, and the new condition when T_s is raised to 85°C due to chilled water production, the following items shown in Table 1 will change.

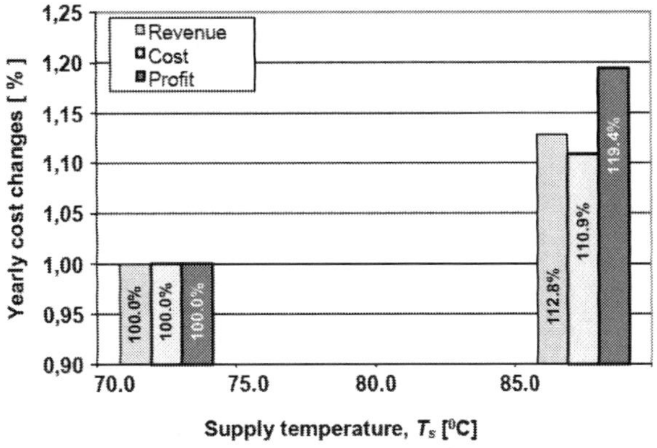

Fig. 7. Changes in costs, revenue and profit after ARS application.

Table 1. Influence of cooling generation on conditions of DHS.

Increase in	**fuel costs**	**pumping costs**	**heat losses costs**	**operating costs**	**total costs**	**total revenue**	**profit**
[%]	3.5	46.6	35.3	10.9	27.7	12.8	19.4

6 Conclusions

In this paper, computer simulations were performed for different supply temperature T_s in DHN, assuming as input, real changes in the basic parameters of the DHS (heat demand for DHW and ARS, flow rates, pressure disposal). On the basis of the above-mentioned calculations and taking into account investment costs of the ARS and the TES system, an optimal size (capacity) of the TES system was proposed for the considered DHS.

Finally, the capacity of the TES system was 55 m^3 ensuring smooth and efficient operation of the CB. Application of the ARS with the TES system in the DHS in question increases net profit by 19.4%, reducing the cooling price for consumers by 40% – from 0.60 PLN/kWh to 0.35 PLN/kWh. The existing CB can operate with high constant efficiency in the wide range of changes in DHS heat demand.

Expected advantages of applying the ARS and TES system in the DHS are as follows:

- increase in overall, operational efficiency of heat generation, by operation of the CB close to optimal conditions of the units,
- decrease in fuel consumption due to continuous operation of the CB with capacity within the limits of high efficiency,
- decrease in the emission of pollutants to the atmospheric air,
- smooth and continuous operation of the CB, which leads to higher availability of the unit and increased failure-free operation time,
- decrease in the percentage of heat losses from DHS, especially during the summer season (from 19% to 25%),
- decrease in the percentage of water losses in the DHS related to water expansion with temperatures changes (the TES can additionally play the role of an expansion tank).

Nomenclature

Symbol	Description	Symbol	Description
a_0–a_6	constants (-)	$\dot{Q}_L$	power of heat losses of DHN (kW)
C_w	specific heat (kJ/(kg K))	$\dot{Q}_L^{rp}$	power of heat losses of ret. pipe (kW)
K_v	Flow Coefficient (m³/h)	$\dot{Q}_L^{sp}$	power of heat losses of sup. pipe (kW)
$\dot{m}_{ARS}$	water mass flow rate for ARS (kg/s)	$R_{rp\text{-}te}$	thermal resistant of return piping to external air(K/kW)
$\dot{m}_{ARS}^{CF}$	water mass flow rate considering *CF* (kg/s)	$R_{sp\text{-}te}$	thermal resistant of supply piping to external air(K/kW)
$\dot{m}_{ARS(r)}$	relative mass flow rate for ARS (%)	t_e	external air (ambient) temperature (°C)
$\dot{m}_{DHS}$	DHS mass flow rate (kg/s)	t_r	return chilled water temperature (°C)
$\dot{m}_{DHW}$	DHW mass flow rate (kg/s)	t_s	supply chilled water temperature (°C)
$\dot{Q}_{ARS}$	cooling power (kW)	T_{av}	average water temp. in DHN (°C)
$\dot{Q}_{ARS(r)d}$	relative cooling power demand (%)	T_r	return water temperature in DHN (°C)
$\dot{Q}_{DHP}$	heating power of DHP (kW)	T_{rc}	return water temp. from ARS (°C)
$\dot{Q}_{DHS}$	heating power of DHS (kW)	T_s	supply water temperature in DHN (°C)
$\dot{Q}_{DHW}$	heating power required for DHW production (kW)	$\dot{V}_g$	volumetric natural gas flow rate (m³/s)
$\dot{Q}_{ee}$	power of heat effects from electrical equipment (kW)	V_{TES}	volume of the TES (m³)
$\dot{Q}_{el}$	power of heat effects from lighting (kW)	W_u	fuel net calorific value (kJ/nm³)
$\dot{Q}_{em}$	power of heat effects from moisture (kW)	ΔP	pressure disposal (difference) (kPa)
$\dot{Q}_{ent}$	power of heat effects by none-transparent walls (kW)	ΔT	temperature difference (°C or K)
$\dot{Q}_{ep}$	power of heat effects by people (kW)	η_{ARS}	operational efficiency of ARS (%)
$\dot{Q}_{et}$	power of heat effects by transparent walls (kW)	η_b	operational efficiency of CB (%)

References

1. R. Zwierzchowski, M. Dzierzgowski, Proceedings of the ASME **40**, 365–371 (2000)
2. E. Cardona, A. Piacentino, F. Cardona, Appl. Therm. Eng. **26**, 1427 (2006)
3. P. Chicco, P. Mancarella, Renew. Sustain. Energy Rev. **13**, 535 (2009)
4. M.A. Lozano, M. Carvalho, L.M. Serra, Energy **34**, 2001 (2009)
5. M.A. Lozano, J.C. Ramos, L.M. Serra, Energy **35**, 794 (2010)
6. K. Kavvadias, A.P. Tosios, Z. Maroulis, Energy Convers. Manag. **51**, 833 (2010)
7. P.W. Gerland, R.W. Gerland, Absorption Chillers: Technology for the Future, Energy Engineering **94**, 6 (1997)
8. O. Kaynakli, M. Kilic, Energy Convers. Manag. **48**, 599 (2007)
9. A. Smith, R. Luck, P.J. Mago, Energy Build. **42**, 2231 (2010)
10. P.J. Mago, L.M. Chamra, Energy Build. **41**, 1099 (2009)
11. S.M. Mohammadi, M. Ameri, Energy Build. **67**, 453 (2013)
12. I. Dincer, M.A. Rosen, *Thermal Energy Storage. Systems and Applications, Second Edition* (John Wiley & Sons, Ltd., Chichester, England, 2011)
13. R. Zwierzchowski, *The International Conference on Energy Efficiency and Control of Air Pollutants from Utilization of Fossil Fuels* (Wroclaw, Poland, 181, 2009)

Investigating the water consumption for electricity generation at Turkish power plants

Balkess El-Khozondar[1,*], *and Merih Aydınalp* Koksal

[1]Department of Environmental Engineering, Hacettepe University, 06800, Beytepe, Ankara, Turkey

Abstract. The water-energy intertwined relationship has recently gained more importance due to the high water consumption in the energy sector and to the limited availability of the water resources. The energy and electricity demand of Turkey is increasing rapidly in the last two decades. More thermal power plants are expected to be built in the near future to supply the rapidly increasing demand in Turkey which will put pressure on water availability. In this study, the water consumption for electricity generation at Turkish power plants is investigated. The main objectives of this study are to identify the amount of water consumed to generate 1 kWh of electricity for each generation technology currently used in Turkey and to investigate ways to reduce the water consumption at power plants expected to be built in the near future to supply the increasing demand. The various electricity generation technology mixture scenarios are analyzed to determine the future total and per generation water consumption, and water savings based on changes of cooling systems used for each technology. The Long-range Energy Alternatives Planning (LEAP) program is used to determine the minimum water consuming electricity generation technology mixtures using optimization approaches between 2017 and 2035.

1 Introduction

Energy production is a water–intensive process in which water is used for fuel extraction, fuel processing, and electricity generation. Processes such as water treatment, desalination, water transportation, pumping require energy to provide water services. The energy and water related processes are closely interacted, thus this interaction is called water-energy nexus.

The global demand for electricity, and hence electricity generation, is continuously increasing and this results in an increase in the amount of water consumed in these processes. The amount of water consumption depends on the technology used for electricity generation. Majority of the global electricity is generated at thermoelectric power plants which use significant amount of water especially at the cooling systems. Once through, wet closed loop, and dry cooling are the most common cooling systems used at thermoelectric power plants [1]. The once through cooling system withdraws a large amount of water from a nearby water body, such as ocean, river, lake, pond or canal. After this water passes through the condenser, it is sent back to the water body where it is withdrawn. This type of cooling system has low capital and operating costs, and has low water consumption which is indirectly consumed at

* Corresponding author: balkess@hacettepe.edu.tr

higher temperatures [1–3]. In the wet cooling towers, the water is pumped back to the cooling tower after condensing the steam. While this system does not withdraw large amounts of water, it consumes more water than the once through system during the evaporating process at the cooling tower [1, 4]. Another method used at thermoelectric power plants is the dry cooling system in which air flows across the steam in tube arrays to condense it [5].

In some cases, the water deficiency can be a problem restricting the electricity generation at location where there is water scarcity. Hence, it is important to investigate the total amount of water used per generated electricity and also to specify the water consumption of the various processes during electricity generation at power plants. Since the most water consumption at the power plants occur at the cooling systems, it is important to focus on analyzing the water consumption at various types of cooling systems.

In the open literature, there are many studies on determining the amount of water consumption per electricity generation (L/MWh) at the power plant. Macknick et al. analyzed the water consumption at various US power plants in 2011. They took into account the water consumption in the operational phase only. The results of their study showed that thermal power plants with wet cooling tower systems consume more water per unit of generated electricity than the once through cooling systems. Furthermore, the coal and lignite power plants consume the highest amount of water among all types of plants. Whereas; photovoltaic (PV) and wind as renewable power plants, and concentrated solar power (CSP), Stirling Engine technology, and natural gas combined cycle power plants with dry cooling systems have the lowest water consumption factors (WCF) [6].

Delgado and her colleagues did not stop in just examining the water consumption data obtained from power plants, they also developed a simple model to determine the water consumption per generated electricity for a whole range of thermal power plants. The model is based on analyzing the water and heat flows in the power plants. In another study, Dale and his colleagues developed a model using two popular simulation tools (LEAP and WEAP) to define the impact of the climate change on the regional water and energy systems. In this study, data from some power plants at California is used to check the vulnerability of energy and water planning systems [7].

In US, many studies are conducted to compare the water consumption of alternative cooling systems of the electricity generation power plants. The systems are compared from several points, such as cost, the amount of water consumption, and environmental impacts. As a result, the amount of water consumption reduces nearly by 95% when wet cooling systems are replaced by dry cooling systems [8]. However, the dry cooling systems have lower efficiencies due to the heat capacity difference of water and air, and higher capital costs than those of the wet ones.

This study covers two main objectives; one is to identify the amount of water consumed (m^3) to generate 1 GWh of electricity based on each generation technology currently used in Turkey, and the other one is to investigate ways to reduce the water consumption per electricity generation at power plants expected to be built in the near future to supply the increasing demand. This paper contains four sections. In the first section, general background information about the study is presented. The second section includes the methodology that is used in water consumption calculations and the application of scenarios to the model. In the third section, the results and discussions are provided. The conclusions are given in the fourth section of the paper.

2 Methodology

The methodology of this study consists of multiple steps which can be divided into three main stages; data gathering and analysis, model development, and scenario applications. The following sections describe these three stages in detail.

2.1 Data gathering and analysis

The installed capacity, electricity production, cooling system type, fuel type, and technology data of all power plants operating in 2016 at Turkey (approximately 1581power plants of which 361 are thermoelectric power plants) are gathered from various sources [9–13]. The water consumption data of Turkish power plants are not available for public access. Thus, WCF (m^3/GWh) for all power generation technologies currently operating in Turkey are gathered from several studies [6, 14, 15]. These factors are then used to calculate the water consumption of 361 thermoelectric power plants using the below equation.

$$PPWC = \sum_i^M \sum_j^N \left(GEN_{i,j} \times WCF_{i,j}\right) \tag{1}$$

where, PPWC is the water consumption of all thermoelectric power plants (m^3), GEN is the amount of electricity generated (GWh) at each thermoelectric power plant, WCF is the water consumption factor based on generation technology and cooling system type (m^3/GWh), j is the power plant with specific cooling system in each generation technology, j is the power plant with specific generation technology, M is the number of power plant with specific cooling system in each generation technology, and N is the number of power plant with specific generation technology.

2.2 Model development

The electricity generation model based on various generation technologies is developed using Long Range Energy Alternatives Planning (LEAP) until 2035 taking 2016 as the base year. LEAP supports various electricity generation technologies and facilitates the tracking of the energy generation based on the available resources and demand estimates. LEAP operates at two levels; where built-in equations are used for energy, emissions, cost-benefit, etc. calculations, or these equations can be entered as spreadsheet-like expressions by the user [16].

In the demand analysis part, the annual electricity demand estimates between 2017 and 2035 obtained from [17] are entered to examine how the total electricity generation based on various technologies will be dispatched over the study period by applying various scenarios. In this study, the demand is specified as a single branch from the electricity generation over the analysis time.

Electricity generation, distributing, and dispatching losses are placed in the transformation analysis section in LEAP. Power plant lifetime, efficiency, maximum availability, and operation and investment costs are the technical parameters in LEAP. The analysis process starts with calculating the electricity requirements for each year according to the dispatch rule, followed by the cost calculations. These steps are repeated for each year, for each module such as natural gas, coal, solar, etc. In this study, the transformation section just includes one branch which is electricity generation. The electricity generation includes the resources modules which are used to supply the electricity demand. As the distribution losses are not included in this study, the dispatch rule is used as the full capacity and the technical factors are entered for each module. The optimization model is based on water consumption for this study, however there is no factor that takes into account the water consumption in LEAP. Thus, the variable operation and maintenance cost parameter of the model is replaced by the WCF and the other costs are assumed to have a value of one unit for all modules. In addition to this, the renewable resources are limited in the maximum capacity section in LEAP using the potential of Turkey for each type [17]. The "Interpolation (Interp.)" function of LEAP is used in this study to calculate the straight-line change between specified pairs of data years and values [16]. After developing the model at LEAP, the generation technology mixtures between 2017 and 2035 are optimized based on the scenarios given in Table 1.

Table 1. Description of the scenarios applied to the model.

Scenario Name	Scenario Description
Business as usual (BAU)	The distribution of each generation technology will not change over the period of the study and average WCF of the cooling technologies will be used for each generation technology.
Optimization (OPT)	Generation technology distribution will be optimized based on minimum water consumption.
Dry cooling system (DCS_BAU)	The distribution of each generation technology will not change over the period of the study and the cooling systems for all technologies, except nuclear, will be assumed as dry cooling system starting in 2016.
DCS Optimization (DCS_OPT)	Optimization will be conducted based on dry cooling system for all technologies, except nuclear, starting in 2016.

Below are the assumptions applied to the scenarios listed Table 1.

- All current power plants are assumed to operate between 2017 and 2035.
- The renewable energy potential is assumed to be used gradually during the study period.
- The share percentage of natural gas power plants is assumed to be between 30–40% over the years.
- The first unit of the nuclear power plants will start to operate in 2022, and the remaining units will gradually start operating until 2029.

3 Results and discussions

In the first phase of the study, the detailed power plant data is analyzed to determine the total water consumption based on cooling systems for each technology. As presented in Table 2, majority of natural gas and coal/lignite, all biogas, half of geothermal power plants use wet cooling towers which consume more water than once through or dry cooling systems.

Table 2. WCF and 2016 sharing percentages of the cooling systems based on installed capacities.

Technology/Fuel Type	Wet Cooling Towers	Once Through System	Dry Cooling System
Natural gas	749 m^3/GWh – 48%	378 m^3/GWh – 16%	7.5 m^3/GWh – 37%
Geothermal	5299 m^3/GWh – 55%	Not Available (NA)	511 m^3/GWh – 45%
Coal and lignite	2600 m^3/GWh – 37%	946 m^3/GWh – 53%	106 m^3/GWh – 10%
Biogas	889 m^3/GWh –100%	Not Available	132 m^3/GWh – NA
Nuclear	2543 m^3/GWh – NA	1018 m^3/GWh – NA	Not Available

By using the WCF determined from open literature for each cooling type used in each technology, total water consumption of each technology is calculated using Equation 1. The WCF for solar power plants is taken as 96 m^3/GWh. These values are then used to determine the water intensity (WI) of each technology. As it can be seen in Table 3, the maximum amount of water is consumed at the coal/lignite power plants in 2016, since it has high electricity generation sharing percentage in parallel with the relatively high WCF as given in Table 2. While the geothermal power plants have generation share less than 2%, they resulted in high water consumption and the highest WI among all technology types.

Table 3. Water consumption and water consumption intensity (WI) for 2016.

Technology Type	Generation (GWh)	Generation Share, %	Water Consumption (Thousand m^3)	Water Intensity (WI) (m^3/GWh)
Solar	612	0.2%	59	96
Wind	15,370	5.7%	0	0
Geothermal	4,214	1.6%	13,333	3,164
Biogas	2,008	0.7%	1,790	891
Hydropower	67,067	24.8%	0	0
Natural gas	90,993	33.6%	38,292	420
Coal/Lignite	90,262	33.4%	134,114	1,485
Total	270,526	100%	187,588	693

As presented LEAP program is used to determine the share distributions of the technologies between 2017 and 2035. For BAU and DCS scenarios, the shares of technologies in 2016 are assumed to be constant until 2022 as presents in

Fig. 1. After the nuclear power plants start to operate in Turkey in 2022, the remaining generation is then divided based on the shares of 2016 for both scenarios.

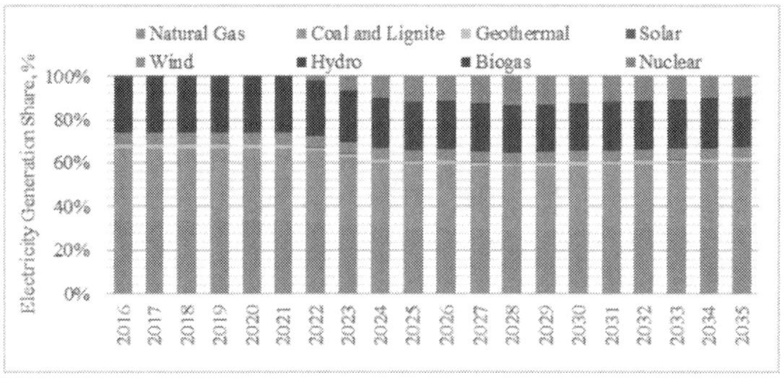

Fig. 1. Annual electricity generation shares for the BAU and DCS_BAU scenarios.

The shares of technology types for the OPT scenario determined based on the optimization of minimum water consumption at LEAP program is presented in Fig. 2. As it can be seen here, the shares of technologies with high WCF, such as geothermal and coal/lignite power plants, are replaced by the ones with lower WFC. Even though, the coal/lignite power plants have high WCF, their shares increase slightly after 2029 to supply the forecasted demand since other resources have reached their potential of dispatching.

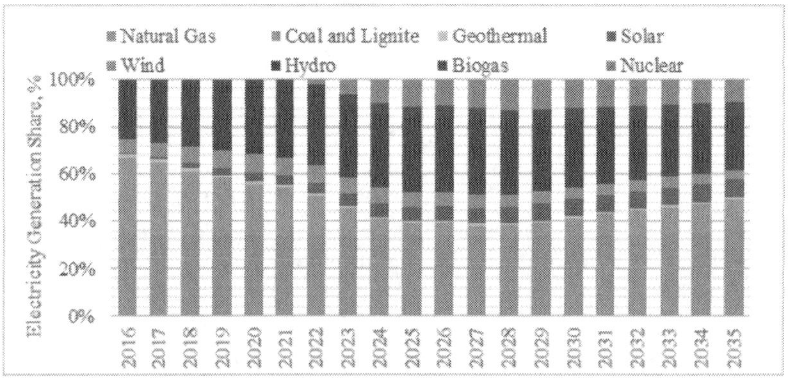

Fig. 2. Annual electricity generation shares for the OPT scenario.

 DOI: 10.1051/e3sconf/20172200039

The last scenario applied to the LEAP model was the DCS_OPT which is based on optimizing the water consumption using dry cooling systems for all technologies. As can be seen in Fig. 3, unlike OPT scenario results, the shares of biogas and geothermal power plants decrease over the years due to their high WCF for dry cooling systems. Similar to the results of OPT scenario, the share of coal/lignite power plants start to increase after 2029.

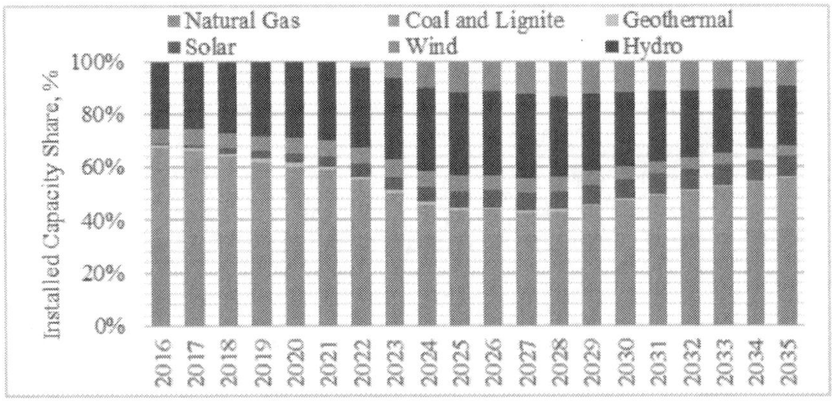

Fig. 3. Annual electricity generation share for the DCS_OPT scenario.

Using the estimated generation of each technology from LEAP model, the water consumption of each technology is calculated to determine the total water consumption between 2017 and 2035 as presented in Fig. **4**.

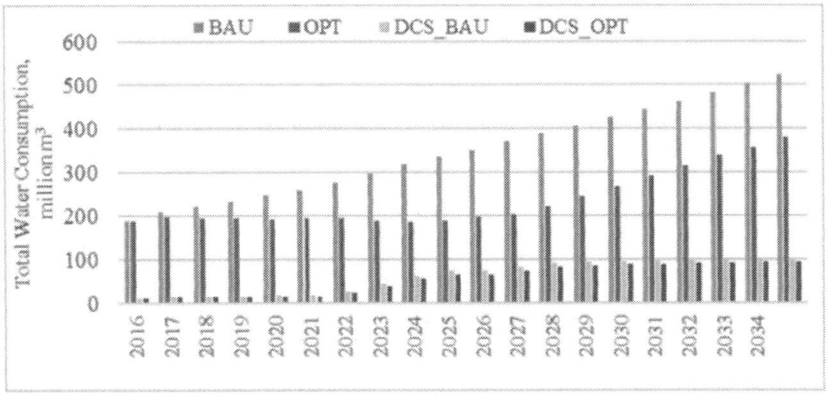

Fig. 4. Total water consumption over the study period based on four scenarios.

As it can be seen clearly from this figure, BAU scenario results in the highest water consumption in all years throughout the estimation period. As the electricity demand in 2035 is expected to be the double the amount in 2016, the total water demand is estimated to increase almost three folds by 2035 based on BAU scenario. Furthermore, the nuclear power plants will roughly increase the water consumption once they start to operate gradually between 2022 and 2028. Total water consumption reduces significantly when the technology shares are optimized based on WCF as done at OPT and DCS_OPT scenarios. The use of dry cooling systems for all technologies reduces the water consumption about 80%. Even the dry cooling systems result in high decreases in the water consumption, they have lower heat transfer efficiencies compared to wet cooling systems, due to the heat capacity difference of water and air. In addition, these systems have higher capital costs than those of the wet ones. The annual WI determined based on each scenario is presented in Table 4. As it can be seen here, WI based on BAU scenario increases by 5% in 2035, whereas WI based OPT scenario decreases by 25%.

Table 4. Water intensity, WI, (m^3/GWh) based on four scenarios.

Scenario	2016	2020	2025	2030	2035
BAU	693	693	731	732	726
OPT	693	539	414	460	525
DCS_BAU	47	47	159	163	141
DCS_OPT	47	41	142	150	132

In a similar project conducted in California, US, the annual WI was determined for the projected electricity generation [18]. The authors estimated that the WI would decrease from 1083 m^3/GWh in 2010 to 965 m^3/GWh in 2035. In another study conducted in Texas, US, the WI for the electricity generation in 2010 was quantified as 1287 m^3/GWh [19]. The WI determined in these studies are relatively higher the one estimated in this study. This is due to high share of nuclear power plants in the generation mixture and the high usage of wet cooling systems.

4 Conclusion

In this study, power plant data are gathered from various resources to determine the water consumption of all current power plants. The data analysis showed that the most common cooling system in Turkish power plants is the wet cooling system, which consumes the highest amount of water among all other cooling systems. Total water consumption in 2016 is determined as 187 million m^3, and the WI for 2016 is calculated as 693 m^3/GWh. To optimize the technology mixture between 2017 and 2035 based on four scenarios, LEAP simulation program was used. The first scenario is BAU where 2016 technology shares are assumed to stay constant. Based on BAU technology mixture, the total water consumption is expected to increase by almost three folds parallel to the estimated increase in demand. The scenario in which water consumption is optimized (OPT) resulted in about 25% reduction in water consumption in 2035. In the third and fourth scenarios where it was assumed dry cooling systems are used for all thermoelectric power plants, expect nuclear, the water consumption and WI were reduced by almost 80%. Even though, the dry cooling systems achieve high decrease in the water consumption, they have low efficiencies due to the heat capacity difference between water and air. Also, these systems have higher capital costs than those of the wet ones. So, the use of hybrid cooling systems can be recommended in which the system will use the wet component in the hot days and the dry component in the normal days to decrease the water consumption.

References

1. P. A. Torcellini, N. Long, J. Ron, Consumptive water use for US power production, NREL conference paper (2003)
2. B. Lamya, G. Boardman, J. Bigger, J. Energ. Eng.-ASCE **138**, 246–257 (2012)
3. E. Mielke, L. D. Anadon, V. Narayanamurti, Water consumption of energy resource extraction, processing, and conversion, Belfer Center for Science and Int. Affairs (2010)
4. M. J. Rutberg, Modeling water use at thermoelectric power plants, PhD diss., Massachusetts Institute of Technology (2012)

5. T. Havey, California's Coastal Power Plants: Alternative Cooling System Analysis, Golden Co., California (2008)
6. J. Macknick, R. Newmark, G. Heath, K. Hallet, A review of operational water consumption and withdrawal factors for electricity generating technologies, NREL, U.S. (2011)
7. L.L. Dale, N. Karali, D. Millstein, et al., Climatic Change **132**, 223 (2015)
8. Comparison of Alternate Cooling Technologies for California Power Plants, EPRI, Palo Alto, CA, and California Energy Commission, Sacramento, CA (2002)
9. Turkey electricity energy production capacity projection of 5 years (2015–2019), [Online]. Available: www.teias.gov.tr [Accessed December 2016]
10. Enerji Atlasi, [Online]. Available: http://www.enerjiatlasi.com/. [Accessed December 2016]
11. Enka engineering for a better future, (2016). [Online]. Available: http://www.enka.com/ [Accessed 1 January 2017]
12. Green Power Through Innovation, (2016). [Online]. Available: http://exergy-orc.com [Accessed 1 January 2017]
13. A. Ashwood, D. Bharathan, Hybrid cooling systems for low-temperature geothermal power production, 1–62 (2011). Available at: http://digitalscholarship.unlv.edu
14. Energy demands on water resources: Report to Congress on the interdependency of energy and water, U.S. Department of Energy, U.S. (2006)
15. T. J. Feeley, T. J. Skone, G. J. Stiegel, A. McNemar, M. Nemeth, B. Schimmoller, J. T. Murphy, L. Manfredo, Energy **33**, 1, (2008)
16. User guide for version LEAP, (2015). [Online]. Available: www.energycommunity.org [Accessed 20 December 2016]
17. Blue Book (Mavi kitap), MENR, (2015), Available: www.enerji.gov.tr [Accessed 20 December 2016]
18. H. Cooley, J. Fulton, P. H. Gleick, N. Ross, P. Luu, Water for energy, Report from Pacific Institute, USA (2011)
19. B. Scanlon, R. C. Reed, I. Duncan, W. F. Mullican, M. Young, Int. J. Environ. Sci. Technol. **47**, 19 (2013)

Experimental study of air evaporative cooling process using microporous membranes

Sebastian Englart[1,*]

[1]Wrocław University of Science and Technology, Faculty of Environmental Engineering, Department of Air Conditioning, Heating, Gas Engineering and Air Protection, ul. Norwida 4/6, 50-373 Wrocław,

Abstract. This article describes the potential use of microporous membranes in evaporative cooling applications for air conditioning. The structure of membrane contractor and the measuring device are described. On the basis of the results of the measurements air cooling effectiveness coefficient has been determined.

1 Introduction

Evaporative cooling is an air cooling method which uses the heat of vaporization of water. Numerous studies analyzed the possibilities of using the evaporative cooling in various options, to maintain comfort conditions in living quarters and to lower air temperature in industrial facilities [1]. This type of air cooling can be characterized by less energy consumption compared to conventional solutions used in air conditioning [2–10]. The disadvantages of evaporative cooling include dependence on the outside air parameters and the lack of possibility of air drying. In the case of direct cooling (DEC) an additional disadvantage is also a direct contact with water. To avoid direct contact of air with water, Johnson et al. [11] proposed the use of membranes for this process. Such a solution is intended to prevent the passage of bacteria, including Legionella, from water to air. The use of a synthetic membrane, which is a porous, semi-permeable barrier between the liquid and the air, allows for a kind of semi-direct evaporative cooling (SDEC).

Membranes are currently used in various industries for separation processes. They may also be used in heating, ventilation and air conditioning (HVAC) systems for various processes such as air treatment, for example: dehumidification or humidification or the evaporative cooling in question [12–19].

The results of experimental research on air evaporative cooling using the membrane module will be presented later in the article.

2 Performance indices

While evaluating and selecting the method of evaporative cooling a few indicators should be taken into account. One of the indicators that can be used to compare devices and

* Corresponding author: sebastian.englart@pwr.edu.pl

processes of air evaporative cooling is their effectiveness. Besides the effectiveness criterion, also the energetic aspect of the evaporative cooling is important.

2.1 Effectiveness

The evaporative cooling has its limitations resulting from the ability of the air to absorb moisture. Those maximum capabilities are clearly determined by the temperature of the air that is being humidified, the amount of moisture in it and the atmospheric pressure.

The equipment efficiency is determined by the real air cooling effectiveness. The effectiveness of the heat exchange can be defined as a relation of the temperature difference in one of the fluids to the difference of temperatures on the heat exchanger inlet. Thus for the water and air membrane heat exchanger the effectiveness can be represented by the following equation:

$$\varepsilon_{tw} = \frac{t_{d1} - t_{d2}}{t_{d1} - t_{w1}} \tag{1}$$

In the studies [18, 19], where the membrane module was used, the above equation was proposed to determine the effectiveness of the sensible cooling.

In order to compare commercial evaporative coolers the following saturation effectiveness indicator is used [20]:

$$\varepsilon_{twb} = \frac{t_{d1} - t_{d2}}{t_{d1} - t_{wb1}} \tag{2}$$

For some structural solutions for the evaporative coolers the dew point effectiveness is determined [8]:

$$\varepsilon_{tdp} = \frac{t_{d1} - t_{d2}}{t_{d1} - t_{dp1}} \tag{3}$$

2.2 Energy efficiency ratio

The energy effectiveness of cooling equipment can also be determined using the energy effectiveness ratio (*EER*). For the complete thermal load the *EER* indicator is determined as a ratio of the cooling power to the electrical power consumption:

$$EER = \frac{Q_c}{P} \tag{4}$$

This indicator is used to compare the efficiency of air conditioning equipment. Taking into account only a drop in the outside air temperature, the cooling power of the direct evaporative cooler can be calculated using the following formula [21]:

$$Q_c = m_a c_{pa} (t_{d1} - t_{d2}) \tag{5}$$

The main energy expenditures associated with the membrane semi-direct evaporative cooling (SDEC) operation result from the air and water transport. Hence, the demand for electrical power for the air transport can be represented by the following formula:

$$P_a = \frac{m_a \Delta p_m}{\rho_a \eta} \tag{6}$$

For the water transport pumping system:

$$P_w = \frac{m_w \Delta p_p}{\rho_w \eta} \tag{7}$$

The total demand for electrical power for the evaporative cooling operation using the membrane module, will amount to:

$$P = P_a + P_w \tag{8}$$

For the direct evaporative coolers Watt [21] proposed the introduction of the following parameter:

$$UC = \frac{Q_c}{t_{d1} - t_{wb1}} \tag{9}$$

The unit cooling *UC* determines the relation of the sensible heat removed from air to the temperature difference (entering air wet bulb depression). Consequently it becomes possible to eliminate the influence of external weather conditions and to bring the comparison to a common base. Watt [21] also proposed the introduction of evaporative *EER* indicator to the direct evaporative cooling:

$$eEER = \frac{UC}{P} \tag{10}$$

2.3 Required parameters of the indoor air

It is necessary to evaluate the influence of an evaporating cooling device on the quality of the indoor microclimate of the premises. The air parameters (temperature, humidity) obtained by means of the evaporative cooler and the conditions required in the room will determine the applicability of the solution. In the areas intended for permanent human stay, specific temperatures, relative humidity and air speed are required, depending on the season and physical activity of the people. The appropriate air parameters are to ensure thermal comfort of the people using the room. The ASHRAE comfort zone for the summer period [1] is indicated on the psychometric chart (Fig.1).

Skrzyniowska [22] indicates that the optimum values of the indoor air for the summer period are as follows: temperature 18–26°C, relative humidity 40–60%, maximum air speed of 0.3–0.6 m/s. The closer range of those values should be determined after taking into account physical activity of the people inside. The limits of air in the ventilated facilities without the possibility of cooling can amount up to 70% of the relative humidity and 3 to 5°C of the temperature above the ambient temperature. In industrial spaces, the air parameters in a room (temperature, relative humidity) will often be subordinated to the technological processes conducted there.

One of the methods for determining the cooling capacity of the environment can be the effective temperature. This is a fictitious temperature to which combinations of temperature, humidity and air speed correspond. Such combinations should give the same feeling of comfort. A fragment of the nomogram for determining effective temperatures, with the marked zone of comfort for the summer period is included in Figure 2. It has been developed on the basis of [22].

 DOI: 10.1051/e3sconf/20172200040

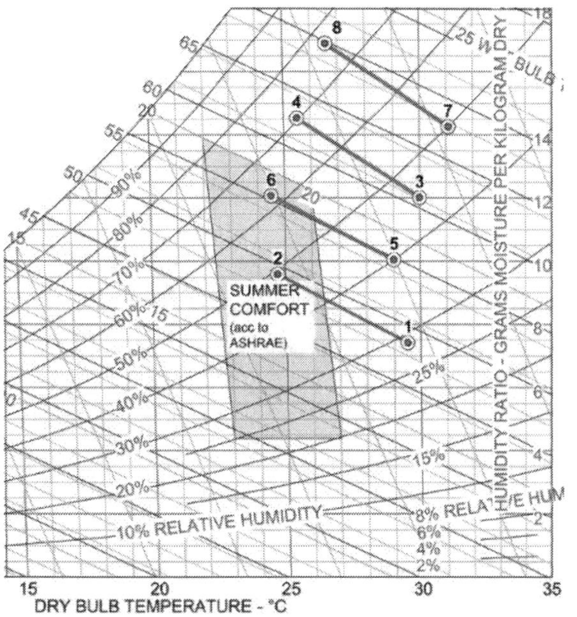

Fig. 1. ASHRAE comfort zone [1] with examples of changes of air in a pilot plant for the semi-direct evaporative cooling.

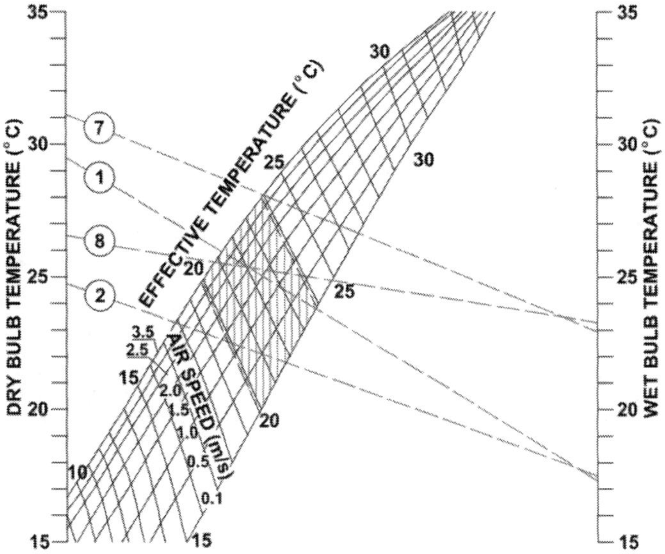

Fig. 2. The cooling capacity of air in the monogram of effective temperatures (developed on the basis of [22]).

3 Experimental setup

For the process of semi-direct evaporative cooling (SDEC), a membrane module with a capillary membrane of polypropylene (PP) was used. The parameters of the membrane were as follows: the inner diameter of the membrane of 1.67 mm, the outer diameter of the

 DOI: 10.1051/e3sconf/20172200040

membrane of 2.71 mm, the average pore diameter of 0.43 μm, the maximum pore size of 0.55 μm. In the module, 744 membranes of a total surface area of 3.04 m^2 and a packing density of 667 m^2/m^3 were used. The membranes were situated perpendicularly to the air flow direction (fig. 3).

Fig. 3. View of membranes in the module.

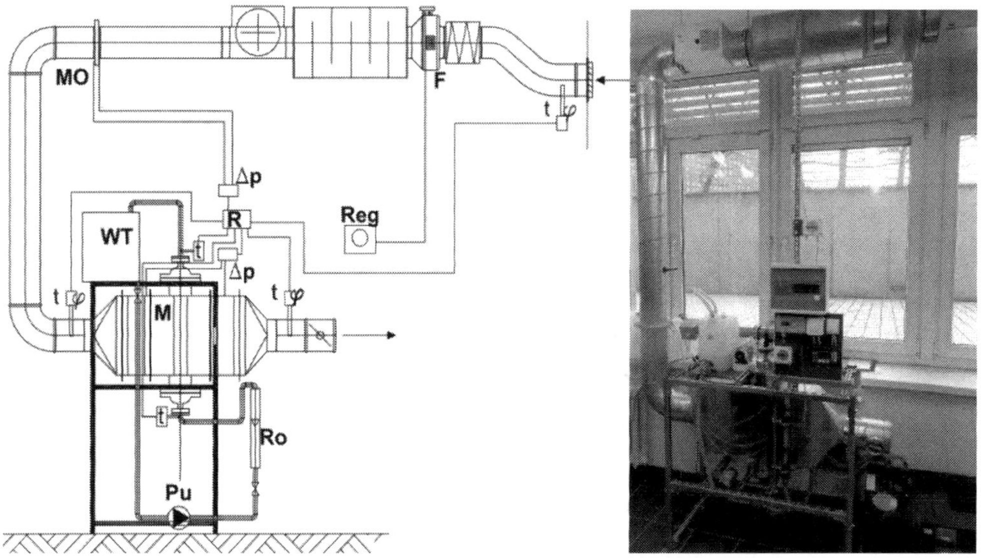

Fig. 4. Schematic diagram and view of the test bench. F-fan, MO-measuring orifice, Ro-rotameter, R-recorder, WT-water reservoir, M-membrane module, Pu-pump, Reg-thyristor controller, sensors of: t – temperature, φ – humidity, Δp – differential pressure.

To evaluate the effectiveness of the evaporative cooling , where the membrane module had been applied the measurements were taken on the test bench which is shown in Figure 4. The measuring bench was equipped with devices for measuring the following parameters:

- outdoor air temperature and humidity,
- air temperature and humidity at the inlet and outlet of the membrane module,
- the air flow rate in the duct,
- temperature of water upstream and downstream the module,

- the water flow rate.

The measurements were performed at constant flow of air and a variable flow of water. The air flow was 95 m^3/h. The water flow through the internal part of capillaries, forced with a pump, ranged from 50 kg/h to 200 kg/h. The sample air changes obtained in the pilot plant during the tests are shown on the psychometric diagram (Fig. 1). Points 1, 3, 5, 7 determine the air parameters before the module (outside air), points 2, 4, 6, 8 represent the state of the air after cooling in the membrane module.

4 Results and discussion

In the process of semi-direct evaporating cooling which was tested in the module the temperature decreased and the humidity increase simultaneously.

The air changes indicated in Fig. 1 show that the evaporative cooling in the tested unit was insufficient to maintain comfort conditions. After taking into account the potential heat gains in the room, further cooling of the air to the required inlet air temperatures will be necessary. In the case of 1–2, and 5–6 changes, after the application of further cooling, it is possible to obtain the air inlet temperature which ensures thermal comfort without dehumidification. The solution with the initial evaporative cooling can be operated as supportive for conventional air coolers.

In the case of 7–8 changes, it is also possible to obtain a relatively favorable cooling effect related to the effective temperature. Fig. 2 includes the lines presented as numbers 1, 2, 7, 8, which link the dry and wet bulbs scale. The line numbers correspond to the air parameters before and after the module, according to the numbering of the points in Fig. 1. The intersection of each line between the air speed curves determines the effective temperature for a particular speed. In the case 1 and 2 both the outside air and the cooled air fall within the limits prescribed for the summer period. In the case 7 and 8 the correct effective temperature, on condition that the minimum speed of the air is 1 m/s, can be obtained only after cooling the air in the membrane module. In the absence of classic air cooling, the use of the semi-direct evaporative cooling may, in some cases, improve the parameters of the air in the room in comparison to other rooms ventilated only with the outside air. Such a solution can be applied to partially improve the microclimate of industrial premises where there is no need to ensure thermal comfort conditions.

The effectiveness ε were determined based on the equations (1–3). Fig. 5 shows the results obtained for an exemplary measurements series, with respect to the pilot plant operation time. The values obtained are somewhat lower than in the case of classical direct evaporative cooling, but comparable to the processes where the membranes are used [18, 23]. The changes in the rate of water flow through the module, within the range of the stream variation, caused no significant changes in the temperature effectiveness. A similar effect was observed at study [23]. Instead, Johnson et al. [11] proposed resignation from the circulating pump to replace it with the gravitational flow of water into the membrane module.

EER and *eEER* indicators set by equations (4) and (10) are presented in Fig. 6. The obtained *EER* for the presented measurement series averaged from 13.6 to 18.5 and the *eEER* from 1.10 to 1.41. With the water flow decreasing, the demand for electric energy for the water pumping decreased, and thus the increase of *EER* and *eEER* was possible.

In the case of conventional air conditioners the most effective units are characterized by *EER* of approximately 3.20. According to ASHRAE [1], *EER* for the two-stage indirect-direct evaporative cooling range between 3.7–11.3. For the analyzed solution Bruno [8] obtained an average value of 11.5 ratio with top temporary values reaching 13.

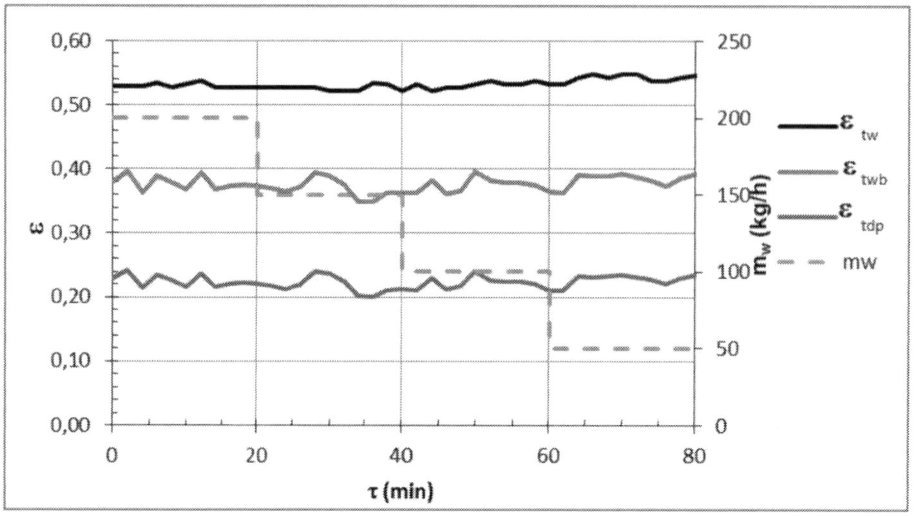

Fig. 5. Changes in the effectiveness and the water flow during the system operation.

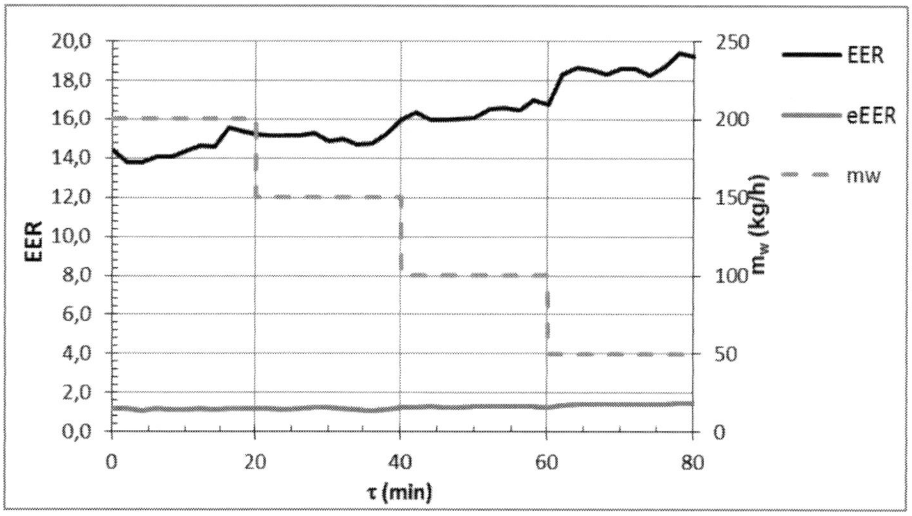

Fig. 6. Changes of *EER*, *eEER* indicators and the water flow during the system operation.

5 Conclusions

In the air conditioning technology drying the air is not always necessary to maintain its appropriate parameters. The semi-direct evaporative cooling results in the increase of air humidity which may limit applicability of the device. The direct evaporative cooling where the membrane module is used can be applied in hot and dry climate conditions. It can also be used as a supportive solution, cooperating with classic solutions, in temperate climates in periods of low relative humidity (approximately 40%). Moreover the use of the membranes makes it possible to achieve satisfactory temperature coefficient and satisfactory energy effectiveness factors.

DOI: 10.1051/e3sconf/20172200040

Nomenclature

c_p – specific heat (J/(kgK))
eEER – evaporative EER ((W/W)/K)
P – total electric power demand (W)
Q_c – cooling capacity (W)
Δp_m –pressure drop in membrane module (Pa)
ε – effectiveness
η – device efficiency
EER – energy efficiency ratio (W/W)
m – mass flow rate (kg/s)
t – temperature (K)
UC – unit cooling (W/K)
Δp_p – pump delivery head (Pa)
ρ – density (kg/m^3)

Subscripts:
1 – inlet, *2* – outlet, *a* – air, *d* – dry bulb, *dp* – dew point, *wb* – wet bulb, *w* – water

References

1. ASHRAE, *Handbook of HVAC Applications* (ASHRAE, Atlanta, GA, USA, 2011)
2. G. Heidarinejad, M. Bozorgmehr, S. Delfani, J. Esmaeelian, Build. Environ. **44**, 2073–2079 (2009)
3. G. Heidarinejad, M. Heidarinejad, S. Delfani, J. Esmaeelian, Energ. Buildings, **40**, 1946–1953 (2008)
4. G. Chiesa, M. Grosso, Build. Environ. **94**, 263–272 (2015)
5. J. R. Camargo, C. D. Ebinuma, J. L. Silveira, Int. J. Refrig. **28**, 1124–1132 (2005)
6. C. Lertsatitthanakorn, S. Rerngwongwitaya, S. Soponronnarit, Biosyst. Eng. **93**, 213–219 (2006)
7. H. Caliskan, I. Dincer, A. Hepbasli, Energ. Buildings **43**, 1461–1472 (2011)
8. F. Bruno, Energ. Buildings **43**, 3475–3483 (2011)
9. S. Anisimov, D. Pandelidis, J. Danielewicz, Energy **80**, 452–464 (2015)
10. H.T. El-Dessouky, H.M. Ettouney, W. Bouhamra, Trans. Inst. Chem. Eng. A **78**, 999–1009 (2000)
11. D. W. Johnson, C. Yavuzturk, J. Pruis, J. Membr. Sci. **227**, 159–171 (2003)
12. J. Woods, Renew. Sust. Energ. Rev. **33**, 290–304 (2014)
13. S. Bergero, A. Chiari, Appl. Therm. Eng. **21**, 1119–1135 (2001)
14. S. Bergero, A. Chiari, Energ. Buildings, **42**, 1976–1986 (2010)
15. S. Bergero, A. Chiari, Energy **36**, 5261–5273 (2011)
16. C. Isetti, E. Nannei, A. Magrini, Energ. Buildings **25**, 185–193 (1997)
17. N. T. Charles, D. W. Johnson, J. Membr. Sci. **319**, 44–53 (2008)
18. L.Z. Zhang, Z.X. Li, T.S. Zhong, L.X. Pei, J. Membr. Sci. **427**, 1–9 (2013)
19. L.Z. Zhang, Int. J. Heat. Mass. Tran. **55**, 5861–5869 (2012)
20. L.Z. Zhang, Energy **31**, 1228–1242 (2006)
21. J. R. Watt, *Evaporative Air Conditioning Handbook* (Chapman and Hall, New York, 1986)
22. D. Skrzyniowska, Cz. T. Environment Engineering **28**, 3, 15–35 (2012)
23. L.Z. Zhang, S.M. Huang, Int. J. Heat. Mass. Tran. **54**, 1055–1063 (2011)

CO_2 concentration and occupancy density in the critical zones served by the VAV system

Ghislaine Etoua Evina[1], *Stanislaw* Kajl[1,*], *Louis* Lamarche[1], and *Javier* Beltran-Galindo[1]

[1]École de technologie supérieure, 1100, Notre Dame Street West, Montreal, H4L 4N5, Canada

Abstract. This article presents the results obtained from monitoring a VAV system with highly diversified zone occupancy density are presented in the article. The investigated VAV system meets the load for 72 zones (68 perimeters and 4 interiors) consisting of classrooms, offices, conference rooms, etc. with highly diversified occupancy densities from 1.875 to 2.5 m^2/person for the classrooms and from 10 to 15 m^2/person for the offices. The monitoring shows that the CO_2 concentration can exceed the set point in the critical rooms. Simulation results are also presented in the article to show that it is often impossible to adjust the operation of such VAV systems because the adjusted System Outdoor Air Fractions, % OA, can reach 100% even where the zone CO_2 concentration is not respected. The presented monitoring and simulation results were obtained in the winter, with the VAV system operating at partial load and with the minimum outdoor air flowrate required by the economizer system. As shown in the article, to respect the zone set point CO_2 concentration in such period, the VAV system must operate mostly at a %OA equal to 100% instead of its minimum value. To circumvent this, the supply zone air flow rate may have to be designed taking into account the CO_2 concentration resulting from the critical zones occupancy density.

1 Introduction

A VAV (variable air volume) system, recognized for its energy efficiency, is a system which controls the fan air flow rate according to the thermal comfort of occupants. The supply air temperature is held relatively constant, but could be moderately reset depending on the season. It must always be low enough to meet the cooling load in the most demanding zone. VAV terminal units measure supply airflow rate and control that flow in response to room temperature. The static pressure regulator controls motor-fan speed to maintain the inlet pressure in the VAV boxes. The minimum throttling ratio in the zones is 20%, while the system ratio is 50%. The supply zone air flow rate required is maintained by a specific control loop and depends on the zone temperature. In the perimeter zones, the supply zone air flow rate is at its minimum before the electric baseboard starts to heat. The main disadvantage of a VAV system is that when it is operating at partial load, the outdoor air flow rate supplied to the zones served by the system could be too low, and could result in poor air quality [2, 3]. This paper presents a study of CO_2 concentration in the zones of

* Corresponding author: stanislaw.kajl@etsmtl.ca

 DOI: 10.1051/e3sconf/20172200041

the existing VAV system, operated with an air-side economizer cycle to control the outdoor air flow rate.

2 VAV system description and monitoring

The investigated VAV system meets the load for 72 zones (68 perimeters zones with electric baseboards and 4 interiors zones), consisting of classrooms, offices, conference rooms, etc. The design system air flowrate $V_{ps,design}$ is 17.314 m^3/s and the minimum system outdoor air fraction is 15% of this $V_{ps,design}$ flowrate. The system outdoor air flowrate V_{oa} is controlled by economizer cycle, but the measurement of CO_2 concentration sensor is also used to control the V_{oa} flowrate if the 1000 ppm is exceeded. The system supply air temperature set point changes linearly within the 14 to 18°C range as a function of the outdoor temperature. It is corrected by adding a value of between -2 and +2°C when the fan airflow rate varies from 90 to 50%. The zone occupancy density is hifhly diversified, and varies from 1.875 to 2.5 m^2/person for the classrooms and from about 10 to 15 m^2/person for the offices. The system operates by applying RACO2-DAV (Return Air CO_2 Control – Demand Controlled Ventilation) [3, 6] with only one CO_2 concentration sensor to maintain the return CO_2 concentration set point.

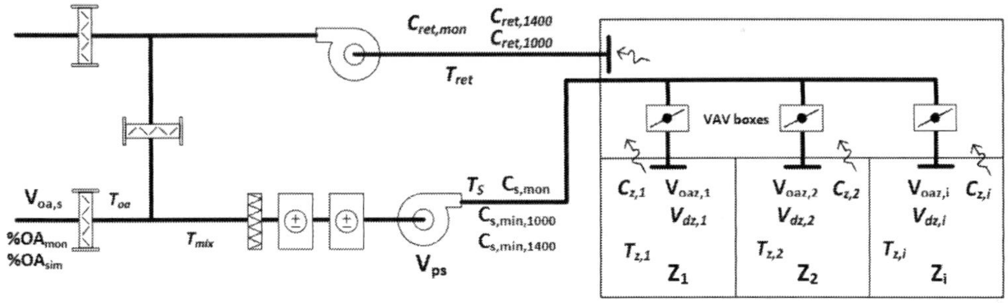

Fig. 1. Diagram of the VAV system with variables used.

Since, in the VAV systems, the system outdoor air fraction %OA is the same for all zones served, and since the CO_2 is only generated by occupants of these zones, the CO_2 concentration could respect the set point in the return duct by exceeding it in the critical zones with high occupancy density. To demonstrate the problems encountered in this system, we monitored of this CVCA system with the following considerations and steps:

- The winter is chosen for this monitoring because the VAV system operates during this period at the minimum $V_{oa,s}$ flowrate required by the economizer system;
- The classrooms Z_1 to Z_7 are identified as the critical zones in terms of occupancy density, and are presented in Table 1;
- The following data are measured by the VAV control system :
  - o The zone supply airflow rate $V_{dz,i}$
  - o Supply air T_s, return air T_{ret}, outdoor air T_{oa}, mixing air T_{mix} and zone air T_{zi} temperatures;
  - o Applying the fan laws, fan speed in rpm used to determine the fan supply air flowrate V_{ps};
  - o CO_2 concentration of return air $C_{ret,mon}$, where $_{mon}$ means monitoring.
- Missing data, such as the CO_2 concentration in the critical zone $C_{z,I}$, are measured by temporarily installing new CO_2 sensors.

The system outdoor air fraction $\%OA_{mon}$ ($V_{oa,s}/V_{ps}$) is calculated by applying the mass conservation and the energy balance using T_{mix}, T_{ret}, T_{oa} and V_{ps} data. It should be noted that the minimum %OA value of the investigated VAV system was 15%, but sometimes T_{ret} and T_{mix} were close or very close, and gave $\%OA_{mon}$ values lower than 15%. In those cases the minimum 15% was used in the monitoring data.

The zone outdoor airflow rate $V_{oaz,i}$ is calculated as $V_{oaz,i}=\%OA_{mon}*V_{dz,i}$ and is used to determine the $V_{oaz,i,pers}=V_{oaz,i}/P_{z,i}$ when $P_{z,i}$ is the zone population. The critical zones characteristics are as follows:

Table 1. Critical zone characteristics.

Zone	$P_{z,i}$	Area m²	m²/pers design	$V_{dz,design}$ l/s	$V_{dz,design}$ l/s/pers	$V_{dz,min}$ l/s
Z_1	30	62,1	2.07	272	9.1	81.7
Z_2	30	56,3	1.875	272	9.1	81.7
Z_3	45	86,3	1.912	500	11.1	113
Z_4	55	101,6	1.838	750	13.2	194.6
Z_5	48	92,7	1.931	600	12.5	180
Z_6	48	101,6	2.123	600	12.5	180
Z_7	50	106,5	2.132	600	12.0	180

The data acquisition was realized during the following periods: 2015: February 12, 13, and 25, and March 10 and 11; 2016: January 28, 29 and 30, February 1, 2 and 3, and March 9 and 10. Each day of monitoring covers three courses periods: 8:30 AM to 12:00 PM, 1:30 PM to 5:00 PM and 6:00 PM to 9:30 PM.

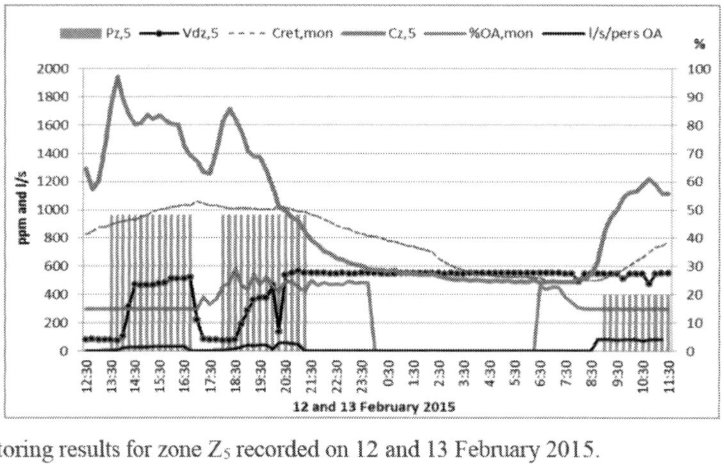

Fig. 2. Monitoring results for zone Z_5 recorded on 12 and 13 February 2015.

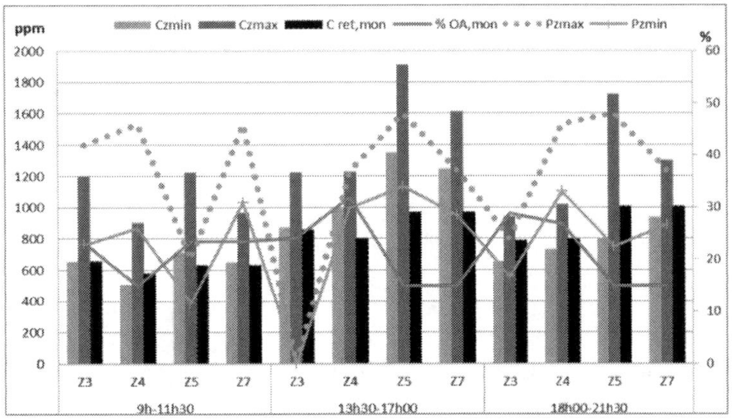

Fig. 3. Summary of recorded data for zones Z_3, Z_4, Z_5 and Z_7.

Figure 2 shows the results obtained in zone Z_5 from noon on February 12 to noon on February 13, 2015, the day when the worst results for $C_{z,i}$ are recorded. Figure 3 shows the min and max values of CO_2 concentration $C_{z,i}$ only the zones Z_3, Z_4, Z_5 and Z_7, obtained in 2015. The mean values of population, $\%OA_{mon}$, and of return air CO_2 concentration, $C_{ret,mon}$, are also presented in this figure. The following observations may be drawn from an analysis of the results in Figure 2:

- The CO_2 concentration $C_{z,5}$ is very high and widely exceeds the 1000 ppm limit. It varies with the population and with the $V_{oaz,i}$ using $\%OA_{mon}$ and $V_{dz,i}$ presented in the Figure 2. Finally, it depends on the $V_{oaz,i,pers}$ (l/s*person);
- The maximum CO_2 concentration value $C_{ret,mon}$ is 1033 ppm, but the $C_{ret,mon}$ mostly respects the 1000 ppm;
- The population $P_{z,i}$ is determined as the number of students registered in the courses;
- As mentioned above, when the $\%OA_{mon}$ presented in the figure is equal to 15%, it means that it could be lower according to the energy balance used in the calculation of $\%OA_{mon}$;
- The results show that V_{dz} is really a function of the zone temperature because it is not at its maximum (600 l/s) even if the CO_2 concentration exceeds the 1000 ppm limit.
- The fan is in operation during the night because of low outdoor air temperatures.

Concerning Figure 3, it can be noted that:

- The maximum values of the zone CO_2 concentration $C_{z,i}$ often exceed 1000 ppm;
- Zone Z_4 is the largest and most ventilated zone, and the $C_{z,i}$ of this zone exceeds the limit only slightly;
- The best factor explaining the CO_2 concentration overrun may be the l/s/person ratio, but it is calculated by taking into account the zone $V_{oaz,i}$ and the $P_{z,i}$. As already mentioned, $\%OA_{mon}$ (used to determine $V_{oaz,i}$) is calculated using the energy balance, and is sometimes corrected to 15%. The $P_{z,i}$ is based on the number registered students in the courses and it could be different from the real $P_{z,i}$. The l/s/person ratio can therefore sometimes be inaccurate.
- The CO_2 concentration of zone Z_3 is not normal during the PM period when the population is zero. This zone could very well have been occupied even if according to the course schedule, it should not have been.

The question now is how the operation of this VAV system should be adjusted in order to avoid exceeding the concentration in critical zones.

3 Simulation of investigated system

3.1 $C_{z,i}$ model and calculation

The CO_2 zone concentration $C_{z,i}$ is determined through the model proposed by [1]:

$$C_{z,i} = \frac{1}{\frac{V}{\Delta\tau} + V_{dz,i}} \cdot \left[\frac{V}{\Delta\tau} \cdot C_{z,i-1} + V_{dz,i} \cdot C_{s,mon} + 4900 \cdot P_{z,i}\right] \quad (1)$$

where the zone air supply air flowrate $V_{dz,i}$, and the zone population $P_{z,i}$ are recorded by monitoring, V is the zone volume, $\Delta\tau$ is the calculation step, and 0.0049 is the CO_2 generated by a person in l/s (4900 ppm). The CO_2 concentration of supply air $C_{s,mon}$ is calculated by equation 2 with $V_{oa,s}$ according to $\%OA_{mon}$ and $C_{ret,mon}$ from monitoring, and the CO_2 concentration of outdoor air $C_{oa,s}$ is equal to 450 ppm.

$$C_{s,mon} = (1 - \%OA_{mon}) \cdot C_{ret,mon} + \%OA_{mon} \cdot C_{oa,s} \quad (2)$$

Figure 4 shows the comparison between the calculated and measured $C_{z,i}$ for zone Z_5 on February 12, 2015.

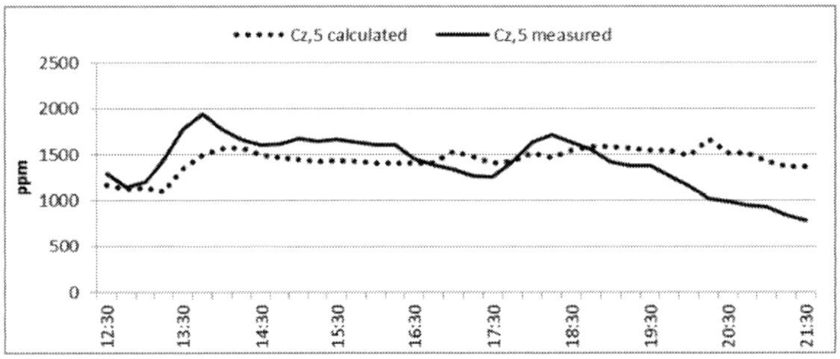

Fig. 4. Comparison between the calculated and measured CO_2 concentration in zone Z_5.

3.2 VAV-F system and application of $SACO_2$-DCV control system

The VAV-F system was created to simulate the corrected operation by applying the new SACO2-DCV (Supply Air CO_2 Control–Demand Controlled Ventilation) [4, 5] control strategy in order to respect the zone CO_2 concentration set point. This system includes only the critical zones presented in Table 1.

Since the CO_2 concentration is widely exceeded in the monitoring, the first question is: was the investigated VAV system well designed to respect the 1000 ppm set point? To answer this question, the $C_{z,i,design}$ was calculated for each critical zone, i.e., for all zones of the VAV-F system operating in design conditions. The model applied was that of the ASHRAE 62.01 Standard (Ventilation Rate Procedure). The following assumptions were used: (*i*) Zone Air Distribution Effectiveness (Ez) and Occupation Diversity (D) are equal to 1.0, and (*ii*) the zone air supply air flow rate $V_{dz,i}$, and zone population $P_{z,i}$ are the design values. This model determines factors such as the Zone Ventilation Efficiency, the System Ventilation Efficiency, the Supply Outdoor Air Fraction for each zone and finally the System Outdoor Air Fraction (Y) required in the air supply by system V_{ps}. If the CO_2 outdoor air concentration $C_{oa,s}$ is considered to be 450 ppm, the $C_{z,i,design}$ could be determined by the following equation:

$$C_{z,i,design} = C_{oa,s} + \frac{4900 \cdot P_{z,i,design}}{V_{dz,i,design} \cdot Y} \tag{3}$$

Table 2. Data used to determine the CO_2 concentration in the design condition.

Zone	$V_{dz,design}$ l/s	$P_{z,max}$	Area m²	R_a l/s/per s	Ra l/s/m²	Y	V_{oz} l/s	Z_p	$C_{z,i,design}$ ppm
Z_1	272	30	62.1	5	0.6	0.61	187.2	0.69	1400
Z_2	272	30	56.3	5	0.6	0.61	183.8	0.68	1400
Z_3	500	45	86.3	5	0.6	0.61	276.8	0.55	1180
Z_4	750	55	101.6	5	0.6	0.61	336.0	0.45	1100
Z_5	600	48	92.7	5	0.6	0.61	295.6	0.49	1100
Z_6	600	48	101.9	5	0.6	0.61	301.1	0.50	1100
Z_7	600	50	106.5	5	0.6	0.61	313.9	0.52	1150

The results in Table 2 effectively show that the CO_2 concentration $C_{z,i,design}$ in the zones is not respected, particularly in zones Z_1, Z_2 and Z_3. It therefore follows that these design parameters are not well selected for the investigated system.

3.3 Simulation results

The monitoring results show that the $C_{z,i}$ is widely exceeded in the critical zones. The principal question is then whether it is possible to adjust the CO_2 concentration of supply air Cs to meet the $C_{ret,mon}$ 1000 ppm set point? To answer this question, the following methodology for each monitoring instant is applied:

- Calculation of $C_{S,mon}$ for the operation during monitoring by equation 2 with V_{ps}, $C_{ret,mon}$, and $\%OA_{mon}$ from monitoring;
- Calculation of $C_{z,i,sim}$ of each zone by equation 1 using $C_{s,mon}$ determined before and $V_{dz,i}$ and $P_{z,i}$ from monitoring. This is presented in the Figure 5;
- Determination of the set point of $C_{z,i}$ for the zones. This could be the maximal value of $C_{z,i,sim}$ obtained in the preceding step but, as indicated in Figure 6, it mostly exceeds 1000 ppm, and the $C_{z,i}$ set point is taken as 1000 ppm and as 1400 ppm, which is the maximal value of $C_{z,i,design}$ in Table 2;
- Calculation of $C_{s,min,1000}$ and $C_{s,min,1400}$ i.e., the minimum values required to meet the $C_{z,i}$ set points for the zones (1000 and 1400 ppm) determined in the previous step;
- Calculation of the new adjusted System Outdoor Air Fractions $\%OA_{sim}$ in each instant (t) by equation 2 using $C_{s,min,1000}$ or $C_{s,min,1400}$ and $C_{ret,1000}$ and $C_{ret,1400}$ equal to 1000 and 1400 ppm, respectively. The $C_{ret,mon}$ from monitoring is also used for this calculation, and three curves of $\%OA_{sim}$ ($\%OA_{sim,1000}$, $\%OA_{sim,1400}$, $\%OA_{sim,mon}$) are presented in Figure 6.

$$C_{S,min}(t) = C_{z,i,setpoint}(t) - \frac{4900 \cdot P_{z,i}(t)}{V_{dz,i}(t)} \tag{4}$$

The results obtained by applying this methodology are presented in the Figures 5 and 6 for zone Z_5 and the VAV-F system respectively. The following remarks may be made:

- $C_{z,5}$ exceeds 1000 or 1400 ppm even if $\%OA_{sim}$ is 100%, but we see that $\%OA_{sim}$ is also 100% even if the $C_{z,5}$ is lower than 1000 or 1400 ppm. This happens for two reasons:

  - o The $C_{oa,s}$ is considered and limited to 450 ppm;
  - o The $\%OA_{sim}$ equal to 100% is required by other zones that Z_5 presented in Figure 5;
- Taking into account the $C_{ret,mon}$, as shown in Figure 6, the $\%OA_{sim}$ equals to 100% is also required by $C_{s,sim}$ determined using $C_{ret,mon}$.

The question now is why the application of SACO2-DCV is not sufficient to respect the CO_2 concentration $C_{dz,i}$ in the critical zones. One reasons for this is that the $V_{dz,i}$, which is controlled by the control loop, depends only on the zone temperature $T_{z,i}$ and, on the other hand, the design (maximum) supply zone air flowrate $V_{dz,i,design}$ is determined only as a function of the zone load. To meet the $C_{z,i}$ set point with the existing cooling charge in the critical zones when the %OA is already at 100%, the $V_{dz,i}$ should be increased. To that end the T_s (the system supply air temperature) must also be increased, but, depending on the cooling or heating demand, this can lead to overheating or overcooling of other zones. In the case of VAV systems with diversified zone occupancy density, the design and the operation control of these systems must be optimized.

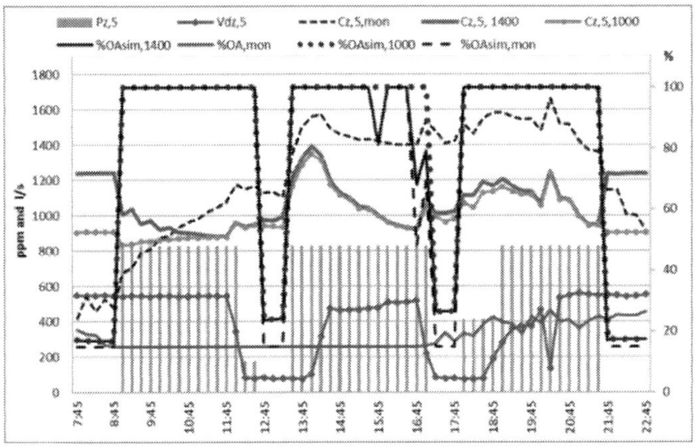

Fig. 5. Zone Z5 simulation and monitoring results recorded on February 12, 2015.

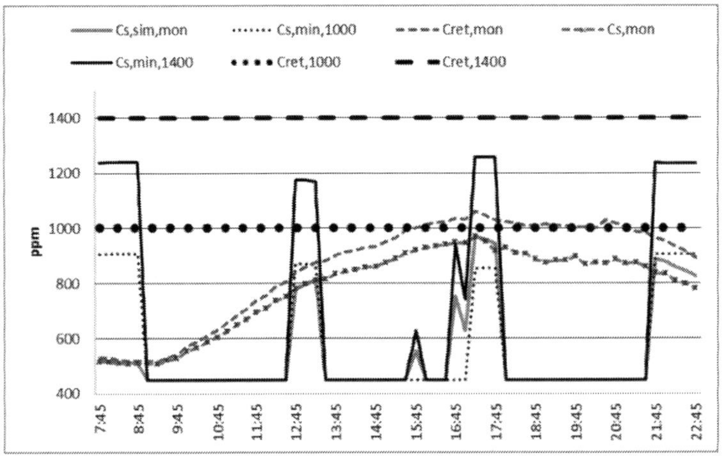

Fig. 6. System supply CO_2 (C_s) simulation and monitoring results recorded on February 12, 2015.

3 Conclusion

When the VAV systems are operating at partial load, the outdoor air flow rate supplied to the zones could be too low and result in poor air quality.

The results obtained when monitoring a VAV system with a highly diversified zone occupancy density show that the CO_2 concentration can exceed the set point in critical rooms. One reasons for this is that the $V_{dz,i}$, which is controlled by the control loop, depends only on the zone temperature $T_{z,i}$ and, on the other hand, the design (maximum) supply zone air flowrate $V_{dz,design}$ is determined only as a function of the zone load.

It is sometimes impossible to adjust the operation of such a VAV system because the adjusted System Outdoor Air Fractions % OA could reach 100% even if the zone CO_2 is not respected, as can be seen in this article.

The monitoring and simulation results presented were obtained in the winter during the operation of the VAV system at partial load, and with the minimum %OA outdoor air flowrate required by the economizer system. As shown in the article to respect the zone set point CO_2 concentration in such a period, this VAV system must operate mostly at an %OA equal to 100% instead of minimum. It would seem that to avoid such operations when the VAV system must meet the loads of zones with diversified occupancy densities, the design supply zone air flow rate should be determined while taking into account the CO_2 concentration resulting from the critical zones occupancy density.

References

1. S.A. Mumma, ASHRAE IAQ Applications, 21–23 (2004)
2. S.J. Emmerich, A.K. Persily, ASHRAE Transactions **103**, 229–243 (1997)
3. M. Hydeman, S. Taylor, J. Stein, E. Kolderup, T. Hong. California Energy Commission (2003)
4. N. Nassif, ASHRAE Transaction **118**, 300–307 (2012)
5. N. Nassif, S. Kajl, R. Sabourin, HVAC&R Research **3**, 459–486 (2005)
6. D. Warden, ASHRAE Journal **46**, 10, 26–30 (2004)

The influence of mechanical gear on the efficiency of small hydropower

Zbigniew Ferenc[1,*], and *Aleksandra* Sambor[1]

[1]Wroclaw University of Science and Technology, Faculty of Environmental Engineering, Wybrzeze Wyspianskiego 27, 50-370 Wroclaw, Poland

Abstract. Pursuant to the "Strategy of development of renewable energy", an increase in the share of renewable energy sources in the national fuel-energy balance up to 14% by 2020 is planned in the structure of usage of primary energy carriers. The change in the participation of the clean energy in the energy balance may be done not only by the erection of new and renovation of the already existing plants, but also through an improvement of their energetic efficiency. The study presents the influence of the mechanical gear used on the quantity of energy produced by a small hydropower on the basis of SHP Rzepcze in Opole province in 2005–2010. The primary kinematic system was composed of a Francis turbine of a vertical axis, a toothed intersecting axis gear of 1:1 ratio, a belt gear of a double ratio. After a modernization the system was simplified by means of reducing the intersecting axis gear and the double ratio of the belt gear. The new kinematic system utilized a single-ratio belt gear of a vertical axis. After the kinematic system was rearranged, a significant improvement of efficiency of the small hydropower was concluded, which translates into an increase of the amount of energy produced.

1 Introduction

Pursuant to the "Strategy of development of renewable energy" [1], approved by the Council of Ministers of 5 September 2000, an increase in the share of renewable energy sources in the national fuel-energy balance up to 14% by 2020 is planned in the structure of usage of primary energy carriers. Currently the usage of renewable energy in Poland is estimated to be 11.4% [2, 3]. Renewable energy sources can have a significant contribution in the energy balance of individual municipalities, guaranteeing regional independence and energy security, as well as adding to an increase of the supply of energy in underdeveloped power-transmission areas.

The following shall be included in renewable energy: hydro-power, wind power, solar power, geothermal power and biomass power. In Poland the share of water energy in total electric energy production is currently on the level of 2.3% [2]. Small hydropower including the objects of installed capacity up to 5 MW play a major role. Currently in Poland there are approximately 700 small hydropower. These objects are environmentally-friendly (do not cause fundamental changes in landscapes and natural habitats, as compared with big

* Corresponding author: zbigniew.ferenc@pwr.edu.pl

hydroelectric plants) and positively influence the hydrological balance of adjoining areas, increasing the surface-based and underground retention of water, which influences the levels of groundwater and its melioration. Small hydropower allow to monitor the quality and level of water on a regular basis. Not only do they impose a need to maintain constant technical and operational efficiency of hydrotechnical facilities (stages of fall, weirs, canals, etc.), but also the maintenance and current usage of flumes [5]. Moreover, small hydropower improve the parameters of power grids in the so-called power output terminals, as well as reduce the costs of their modernization. What is more, they affect the reduction of loss related to power transmission.

The change in the participation of the clean energy in the energy balance may be done not only by the erection of new and renovation of the already existing plants, but also through an improvement of their energetic efficiency. The study presents the influence of the mechanical gear used on the quantity of energy produced by a small hydropower on the basis of SHP (SHP: Small HydroPower) Rzepcze in Opole province in 2005–2010.

2 Original power transmission system

Originally, the kinematic system was composed of the 110/170 cm Francis turbine of a vertical axis, manufactured by J.M. Voith in 1936, of the following rated capacity: Q = 1.6 m^3/s, N = 72 kM, n = 120 RPM, H = 4.1 m (fig. 1), the toothed intersecting axis gear of a 1:1 ratio, as well as the belt gear of a double ratio.

The technical condition of the turbine was assessed sufficient. Numerous corrosion pits were spotted on the surface of rotor blades. Damaged pins and liners were found in the guide apparatus of the turbine [6].

Fig. 1. The 110/170 cm Francis turbine from 1936.

Power take-off from the shaft to the drive shaft was done by means of the intersecting axis toothed gear. Its technical state was unsatisfactory. Its teeth were broken off and the inter-tooth clearances were too big, giving evidence of too big a backlash in the bearing of the turbine shaft. The intersecting axis gear transmitted power to the belt gear of a horizontal axis (fig. 2) my means of a disengaging jaw coupling. The diameter of the pulley was

 DOI: 10.1051/e3sconf/20172200042

1200/800 mm. This pulley was primarily used to drive a bucket elevator of a corn elevator. The same overhead shafting was equipped with a 1700mm-diameter pulley to power a generator [6].

Fig. 2. The belt gear.

3 Potential spots of reducing power loss in the power transmission system

The possible spots where power loss in the power transmission system can be achieved:

- inlet channel,
- turbine bearing,
- intersecting axis gear,
- belt gear,
- connection with power generator (coupling),
- properly selected power generator.

The study concerns the problem of energy loss solved my means of modernizing the gear in the power transmission system.

In the first run, the current power transmission system, i.e., the turbine, the toothed intersecting axis gear (fig. 3) and the belt gear underwent a renovation. These works were finished in the second half of 2004. Commissioning was done on 4 December 2004, whereas a 72-hour test run commenced on 8 January 2005. Successively, the plant was connected to the power grid permanently.

 DOI: 10.1051/e3sconf/20172200042

Fig. 3. The toothed intersecting axis gear after the renovation.

Figure 4 presents the factual energy production in SHP Rzepcze in 2005 as dependent on the amount of water flowing through the turbine pit.

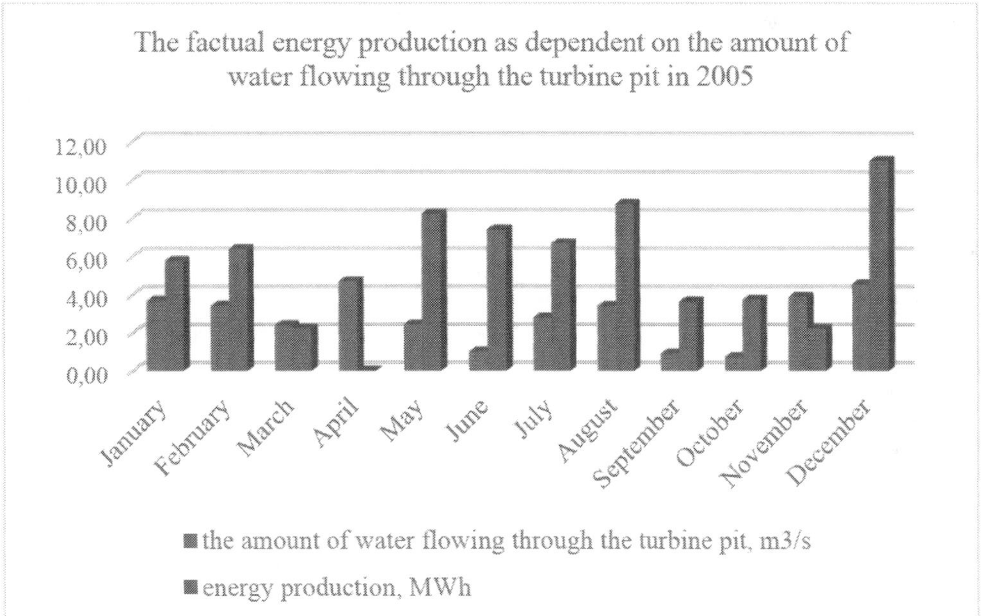

Fig. 4. The factual energy production as dependent on the amount of water flowing through the turbine pit in 2005.

On the basis of figure 4, it may be concluded that since April 2005 the amount of power generated was on the average level of 4.8 MWh with the rate of turbine pit water flow of 3.2 m^3/s. The degree of energy output as the quotient of energy production to turbine flow rate was only 1.5. Therefore, it was decided that a modernization of the existing power transmission system should be done in order to achieve the maximum efficiency of energy production and, as a consequence, optimal utilization of all elements of the power transmission system. The works commenced in April 2005. Figure 4 clearly exhibits that

 DOI: 10.1051/e3sconf/20172200042

after the modernization had been done, the degree of power output from the turbine pit water flow increased to 2.6 on average.

4 Modernized power transmission system

The modernization of the power transmission gear was done in phases, with gradual elimination of power loss points. In the new kinematic system a single-stage belt gear of a horizontal axis was used, the task of which was to adjust the rotation of the asynchronous generator to the power grid. The gear was composed of a 1840 mm pulley made of iron cast, a 105 mm intermediate shaft supported with three bearings, as well as a small 880 mm pulley (fig. 5). The ratio was 8.41.

Fig. 5. The modernized belt gear.

After the kinematic system was rearranged, a significant improvement of efficiency of the small hydropower was concluded, which translates into an increase of the amount of energy produced (fig. 6).

Figure 6 clearly exhibits that the energy production increased nearly three times, from approx. 66 MWh/year to 180 MWh/year. The rate of energy output for first quarter in 2005–2008 increased by 6.5 on average, within 6.2 to 6.7. Figure 6 presents distinguished changes in the amount of energy produced depending on season, which directly translates in the state of water level in the river and the water flow. The best period in relation to energy production was Autumn to Spring. The months of outages the reason of which was too small water flow were months of minor energy production.

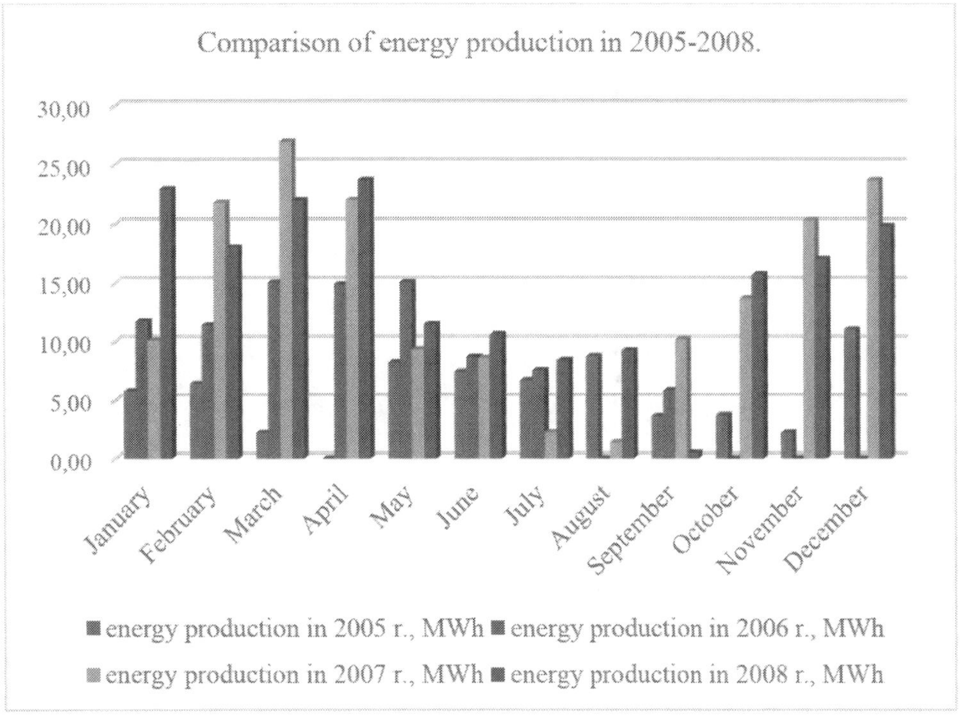

Fig. 6. Comparison of energy production in 2005–2008.

In order to eliminate the two-stage ratio of the gear and to increase the efficiency and reliability of the plant the belt gear was reconstructed to a single-stage belt gear of a vertical axis (fig. 7), which allowed to eliminate the toothed gear from the power transmission system. This was done in October 2008. During the reconstruction the turbine shaft was extended, bearings, including the main axial bearing (slide to rolling bearing) were replaced. The reconstruction allowed to increase the distance between the turbine axis and the generator axis from 3 to 6 meters, which allowed to increase the wrapping angle of the small pulley by 15%.

Fig. 7. Belt gear of a vertical axis.

The total scope of works allowed to eliminate lost motion and to reduce belt tension, as well as to increase the level of reliability of the plant. The power output increased by approximately 10% to 7.2 (fig. 8).

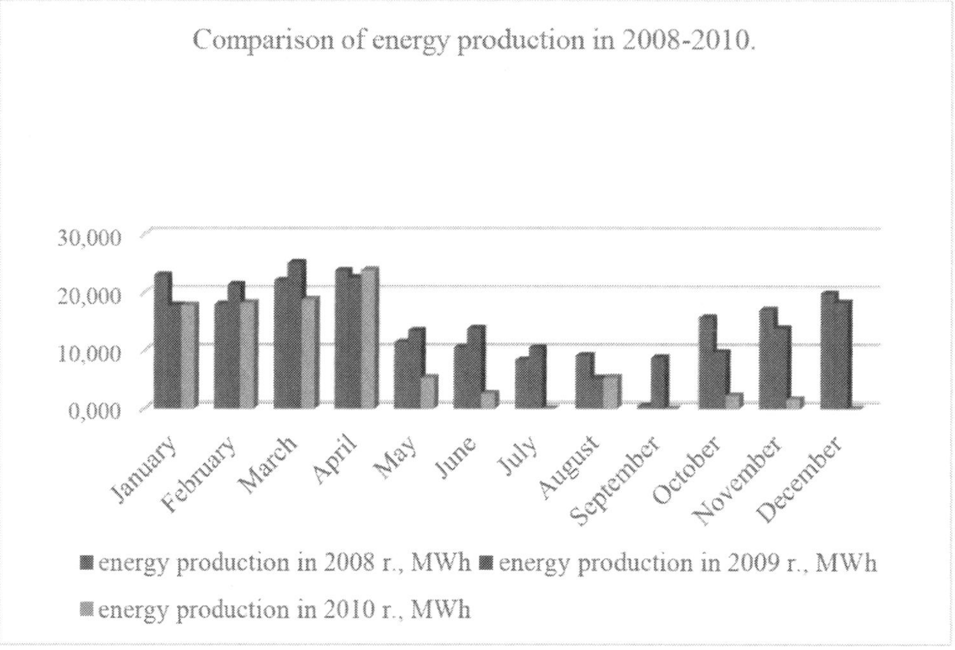

Fig. 8. Comparison of energy production in 2008–2010.

5 Summary

On the basis of the data presented, it was concluded that the efficiency of small hydroelectric plants can be increased by improving power transmission systems. The example described hereinabove clearly shows the increase of power production after the modernization of friction nodes and intermediate gears. The power output increased nearly five times (1.5 to 7.2).

Acknowledgments

This publication was made possible by the statutory activity of the Faculty of Environmental Engineering Wroclaw University of Science and Technology (W7/K2 0401/0069/16), financed by the Ministry of Science and Higher Education in 2017.

References

1. Ministerstwo Środowiska, *Strategia Rozwoju Energetyki Odnawialnej* (2000)
2. A. Hildebrandt, Pomeranian Economic Review (in polish), **10** (2014) (in Polish)
3. Statistical Pocketbook, EC, Renewable energy statistics (2016)
4. Eurostat, Energy from renewable sources (2016)
5. Z. Piec, A.Sambor, Energetics, **10** (2010) (in Polish)
6. Cieśla E., Paczkiewicz A., *Inventory of a dam facility and a hydroelectric plant,* "Promel" Water Projects and Land Reclamation Office in Opole (1983) (in Polish)

Temperature influence on the fast pyrolysis of manure samples: char, bio-oil and gases production

Maria Fernandez-Lopez[1,*], *Kostas* Anastasakis[2], *Wiebren* De Jong[2], *Jose Luis* Valverde[1], and *Luz* Sanchez-Silva[1]

[1]Department of Chemical Engineering, University of Castilla-La Mancha. Avda. Camilo José Cela 12, Ciudad Real, 13071
[2]EDP Sciences, Production Department, 91944 Les Ulis Cedex AFaculty of Mechanical, Maritime and Materials Engineering, Process and Energy Department, Delft University of Technology, Leeghwaterstraat 39, Delft, 2628 CB, The Netherlands

Abstract. Fast pyrolysis characterization of three dry manure samples was studied using a pyrolyzer. A heating rate of 600°C/s and a holding time of 10 s were selected to reproduce industrial conditions. The effect of the peak pyrolysis temperature (600, 800 and 1000°C) on the pyrolysis product yield and composition was evaluated. Char and bio-oil were gravimetrically quantified. Scanning electron microscopy (SEM) was used to analyse the char structure. H_2, CH_4, CO and CO_2 were measured by means of gas chromatography (GC). A decrease in the char yield and an increase of the gas yield were observed when temperature increased. From 800°C on, it was observed that the char yield of samples Dig R and SW were constant, which indicated that the primary devolatilization reactions stopped. This fact was also corroborated by GC analysis. The bio-oil yield slightly increased with temperature, showing a maximum of 20.7 and 27.8 wt.% for samples Pre and SW, respectively, whereas sample Dig R showed a maximum yield of 16.5 wt.% at 800°C. CO_2 and CO were the main released gases whereas H_2 and CH_4 production increased with temperature. Finally, an increase of char porosity was observed with temperature.

1 Introduction

Bioenergy production from renewable energies has been attracted an increasing attention due to the depletion of fossil fuel reserves and the environmental problems derived from their use. Nowadays, the oil consumption rate is about 95 million barrels per day whereas that of natural gas is 9.5 billion cubic meters per day. Therefore, reserves will satisfy 49 years of oil and 70 years of natural gas supply [1]. In this regard, biomass is considered to be one of the most viable energy source [2]. Currently, its contribution to the world annual energy production is about the 10% [3]. Among the different type of biomass,

* Corresponding author: maria.fernandez@uclm.es

livestock manure (LSM) is generated from animal fats, carcasses and excrements. The accumulation of LSM has associated some environmental problems such as subterranean and surface water contamination as well as ground and air pollution, odours and greenhouse gases (GHG) and ammonia emissions. In this sense, LSM production accounts for an estimated 14% and 64% of the world GHG and NH_3 emissions [4]. Therefore, this waste has been traditionally used as fertiliser and landfill. However, these traditional uses have to be changed due to land limitations and the strict regulations [5]. The number of studies about the use of LSM as biomass to produce energy or other added-value products by anaerobic digestion of thermochemical processes has increased in recent years [6–9]. There are many different technologies which have been applied for biomass utilization. Among them, fast pyrolysis is a promising process from which biomass can be converted into liquid, gas and solid products in absence of oxygen and in a short reaction time [10]. The gases mainly consist of small and simple molecules such as carbon dioxide and carbon monoxide, hydrogen and methane whereas light and heavy hydrocarbons are the predominant liquid products. Ash and fixed carbon are the solid products. The distribution of these pyrolytic products depends on the biomass composition and the operating conditions [11]. Solid and gas products are favoured with operating conditions of slow pyrolysis (heating rates up to 100°C/min and high gas residence times) while the liquid yield can be maximized with fast pyrolysis conditions (heating rates from 10 to 200°C/s and gas residence time from 0.5 to 10 s) [12–14]. In this sense, Xing et al. (2016) [15] used the analytical pyrolysis-gas chromatography coupled to mass spectrometry (Py-GC/MS) to identify the main compounds generated during the pyrolysis of wood dust and determine the influence of copper and potassium used as catalysts on the pyrolytic products obtained. Ding et al. (2016) [16] studied the production of valuable single ring aromatic compounds from the catalytic pyrolysis of waste tires by Py-GC/MS. They also analysed the catalytic thermal degradation of D-limonene and polybutadiene rubber (BR) to produce these valuable aromatics and proposed a catalytic pyrolysis mechanism based on the product distribution obtained. On the other hand, Gao et al. (2017) [17] investigated the influence of the pyrolytic reaction temperature of rapeseed stalk on the composition and yield of three pyrolytic products to determine the optimum operating temperature for the pyrolysis process. They also performed an economic analysis based on the use of a moving bed reactor for the pyrolytic process.

In this work, the fast pyrolysis of three manure samples was carried out in a pyroprobe pyrolyzer and the three pyrolytic products, namely pyrolytic gas, pyrolytic bio-oil and char, were analysed by gas chromatography (GC) and scanning electron microscope (SEM). The effect of the pyrolysis peak temperature on the product yields, the composition of the gas and the char structure was evaluated. Subsequently, the compositional differences of the products obtained by the three manure samples at different temperatures were presented.

2 Materials and methods

2.1 Biomass samples

In this work, three dry animal solid waste from the province of Québec (Canada) were selected as biomass feedstock: a swine (SW) manure sample pretreated by a bio-drying process and two dairy manure samples before (Pre) and after (Dig R) an anaerobic digestion process. Pre and Dig R samples were selected to assess the influence of the chemical changes during the biological process on the pyrolytic products obtained from the fast pyrolysis process here studied. These samples were dried in an oven for 5 h, milled and sieved to an average particle size between 150 and 200 μm.

On the other hand, the three manure samples were characterized by elemental analysis, thermogravimetric analysis (TGA), bomb calorimetry and atomic emission spectroscopy inductively coupled plasma (ICP-AES). Carbon (C), hydrogen (H), nitrogen (N), oxygen (O) and sulphur (S) were measured in an elemental analyser following the standard UNE-EN 15104:2011 and was expressed as mass percentage of each element in dry basis. Moisture, ash, fixed carbon and volatile matter content were obtained by proximate analysis by TGA and according to ASTM D 3172-73(84) standard. The high heating value (HHV) of the samples was determined using a Parr 1356 bomb calorimeter according to the UNE standard 164001:2005 EX by their combustion under an oxygen atmosphere. Finally, the mineral content was obtained by ICP-AES analysis. Table 1 shows the results of the manure biomass characterization obtained by all the techniques previously commented.

Table 1. Physico-chemical characterization of manure samples (Pre, Dig R and SW).

	Ultimate Analysis (wt. %)*					Proximate Analysis (wt. %)				Bomb calorimeter
Sample	**C**	**H**	**N**	**S**	**O****	**Moisture**	**Ash**	**Volatile matter**	**Fixed carbon****	**HHV (MJ/kg)**
Pre	51.43	6.72	1.74	0.52	39.59	3.45	20.78	64.09	11.68	18.4
Dig R	51.50	6.48	1.27	0.55	40.20	2.90	28.09	56.68	12.33	16.9
SW	45.00	6.40	1.90	0.71	45.98	4.41	9.72	66.30	19.57	16.1

Mineral content and Cl^-, CN^- content (ppm)										
	Al	**Ca**	**Cr**	**Cu**	**Fe**	**Mg**	**Na**	**P**	**Si**	**Cl^-**
Pre	516	16997	73	-	3800	6426	37445	16207	537	42866
Dig R	238	22214	127	-	3589	8720	6392	21093	525	20325
SW	-	10070	-	142	4626	4465	2336	17568	278	18779

*daf: dry ash free; **obtained by difference

2.2 Pyrolysis experiments

Fast pyrolysis experiments of manure samples were carried out in a commercial pyroprobe pyrolyzer (CDS5200) using nitrogen as the carrier gas with a constant flow rate of 17 mL/min. 30 mg of sample was introduced in a quartz tube in each experiment. Raw material was placed in the middle with quartz wool on both sides. Then, the quartz tube with the sample was located into the pyroprobe and heated up to the pyrolysis peak temperature at 600°C/s. Finally, the temperature was kept for 10 s to reproduce industrial conditions [18]. Three pyrolysis peak temperatures for each manure biomass sample were studied: 600°C, 800°C and 1000°C. Each experiment was repeated at least twice to ensure data reproducibility.

The gas products evolved passed through a trap where the condensable gases were captured at 50°C. Subsequently, permanent or non-condensable gases were collected in a syringe at room temperature and then analysed by GC using a micro-gas chromatograph (Varian CP-4900). CO_2, CO, CH_4 and H_2 were the main gases produced and analysed. Finally, biomass manure samples and pyrolytic chars obtained after fast pyrolysis were analysed by SEM.

Char and bio-oil were gravimetrically quantified and the pyrolytic product yields were calculated as follows:

*Char yield (%) = (mg of residue after pyrolysis/mg of initial sample weight)*100* (1)
*Bio-oil yield (%) = (mg of bio-oil/mg of initial sample weight)*100* (2)

On the other hand, the gas yield of each gas (CO_2, CO, CH_4 and H_2) was calculated using the results from GC and the ideal gas law (considering 293 K and 1 bar).

*Individual Gas yield (%) = (mg of gas/mg of initial sample weight)*100* (3)

Therefore, the gas yield was calculated as:

*Gas yield (%) = (Σ(CO_2, CO, CH_4 and H_2 mg)/mg of initial sample weight)*100* (4)

3 Results

The influence of three different peak pyrolysis temperatures (600, 800 and 1000°C) on the product yields and compositions during fast pyrolysis (600°C/s) in a pyroprobe pyrolyzer was studied for the three manure samples considered in this study. Devolatilization is the main process which takes place during fast pyrolysis, leading to three pyrolytic products, namely char (solid residue after pyrolysis), bio-oil (condensable gases) and gaseous compounds. These products were individually measured and analysed. Fig. 1 shows the product yield results with the standard deviation for samples Pre (A), Dig R (B) and SW (C) in a dry basis (d.b.).

Generally, the devolatilization product yields followed the same trend for the three samples studied. A decrease of char yield and an increase of gas and bio-oil yields were obtained with the increase of the peak pyrolysis temperature. As it can be observed in Fig. 1, a decrease in the char yield from 52 to 31, 60.5 to 45.2 and 35 to 28 wt.% and an increase of gas yield from 7.7 to 16.2, 7.8 to 14.4 and 10.3 to 17.3 wt.% were observed when the peak pyrolysis temperature increased from 600 to 1000°C for samples Pre, Dig R and SW, respectively. Char yield of samples Dig R and SW were constant (46 and 28 wt.%, respectively) from 800°C which indicated that the primary devolatilization reactions stopped. This fact was also corroborated for both samples with the evolved gas analysis, where a constant CO_2 production was also observed. In agreement with the proximate analysis (Table 1), sample Dig R at 1000°C appeared to be the highest value of final char residue and ash and FC content.

On the other hand, the bio-oil yield slightly increased with temperature except for sample Dig R. The maximum bio-oil yields were 20.7 and 27.8 wt.% at 1000°C for samples Pre and SW, respectively and 16.5 wt.% at 800°C for sample Dig R.

 DOI: 10.1051/e3sconf/20172200043

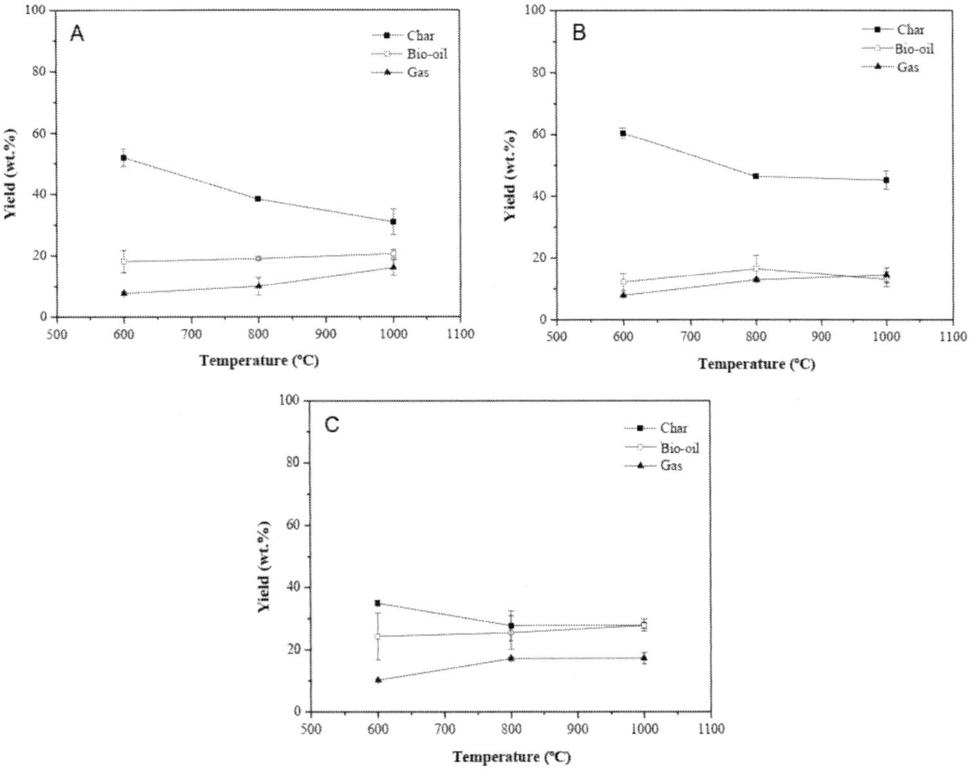

Fig. 1. Char, bio-oil and gas yield during fast pyrolysis of samples Pre (A), Dig R (B) and SW (C).

Concerning the gas fraction, a micro-gas chromatographer was used to analyse and quantify the main gaseous species evolved during fast pyrolysis. Fig. 2 shows the mass yield (d.b.) of the main gases evolved during fast pyrolysis of samples Pre (A), Dig R (B) and SW (C) as a function of the peak pyrolysis temperature.

As it can be seen in Fig. 2, there was an increase in the yield of all species with the increase of the temperature, which indicated that the formation reactions of these gaseous species were thermally favoured. Gases evolved can be either the products of primary decomposition reactions by the bond breaking of cellulose, hemicellulose or lignin, or those coming from secondary cracking ones [18].

CO_2 and CO were the main released gases for all the samples in the range of temperatures studied [19]. H_2 and CH_4 production increased when the temperature also did. In Fig. 2, the increase of CO_2 production in all the range of temperatures for sample Pre could indicated that primary devolatilization reactions did not stop at high temperatures whereas for samples Dig R and SW the production of CO_2 was stable from 800°C on as it was previously mentioned.

Fig. 3 shows the SEM images of raw sample Pre (A) and the char remained after fast pyrolysis at 600°C (B), 800°C (C) and 1000°C (D). According to these SEM images, the char morphology changed with the pyrolysis peak temperature [19]. Overall, it was observed that the porosity increased as the pyrolysis temperature increased due to the release of volatiles compounds during thermochemical degradation [20, 21]. The higher the pyrolysis temperature, the lower the pore sizes of the char [19]. The pore network observed for chars suggest their potential use as soil amendment. The porosity can play an important role for helping as a storage of water in soil, serving as habitat for a variety of

DOI: 10.1051/e3sconf/20172200043

microorganism, and help in the root movement in soils [20]. On the other hand, these chars could be also be used depending on the pore size as gas adsorbent.

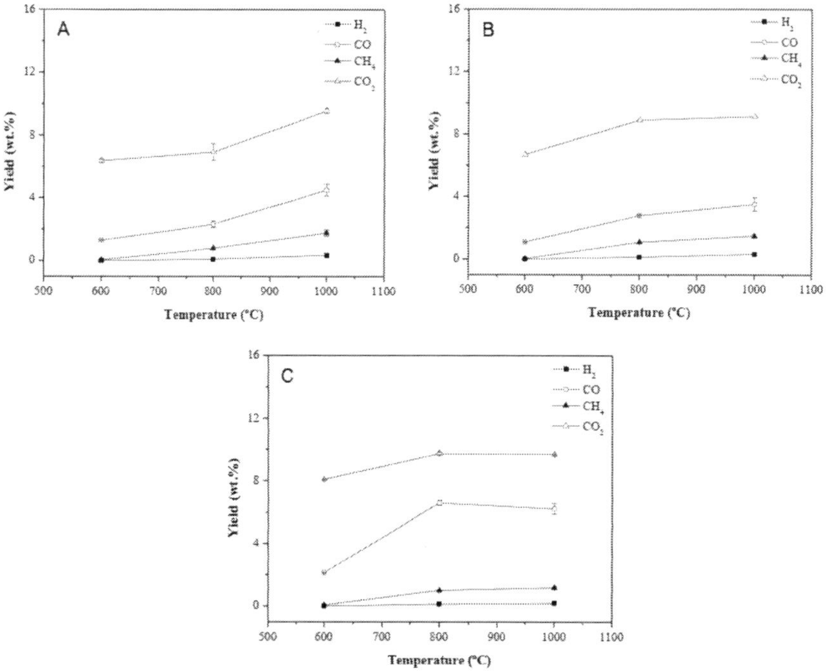

Fig. 2. H_2, CH_4, CO and CO_2 mass yield obtained during fast pyrolysis of sample Pre.

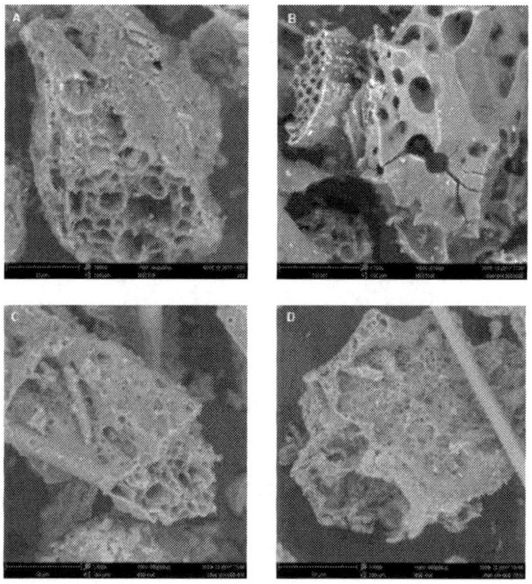

Fig. 3. SEM images of sample Pre (A) and chars obtained from this sample during fast pyrolysis at 600°C (B), 800°C (C) and 1000°C (D).

Authors acknowledge the financial support from the Regional Government of Castilla-La Mancha (Project PEII-2014-007-P) and also from the University of Castilla-La Mancha of Spain (UCLM grant). The authors also thank to the Centre National en Électrochimie et Technologies Environnementales (CNETE) of Canada for its in-kind contribution.

References

1. BP (2016)
2. F. Pierobon, M. Zanetti, S. Grigolato, A. Sgarbossa, T. Anfodillo, R. Cavalli, Appl. Energ. **150**, 185–195 (2015)
3. WEC
4. L. Loyon, C.H. Burton, T. Misselbrook, J. Webb, F.X. Philippe, M. Aguilar, M. Doreau, M. Hassouna, T. Veldkamp, J.Y. Dourmad, A. Bonmati, E. Grimm, S.G. Sommer, J. Environ. Manage. **166**, 1–11 (2016)
5. J.-P. Cao, X. Huang, X.-Y. Zhao, X.-Y. Wei, T. Takarada, Fuel **140**, 477–483 (2015)
6. U. Ekpo, A.B. Ross, M.A. Camargo-Valero, P.T. Williams, Bioresour. Technol. **200**, 951–960 (2016)
7. N. Ruiz-Gómez, V. Quispe, J. Ábrego, M. Atienza-Martínez, M.B. Murillo, G. Gea, Waste Manage. **59**, 211–221 (2017)
8. M.S. Hussein, K.G. Burra, R.S. Amano, A.K. Gupta, Fuel **189**, 428–435 (2017)
9. K.G. Burra, M.S. Hussein, R.S. Amano, A.K. Gupta, Appl. Energ. **181**, 408–415 (2016)
10. A.V. Bridgwater, Biomass Bioenergy **38**, 68–94 (2012)
11. J. Akhtar, N. Saidina Amin, Renew. Sustain. Energy. Rev. **16**, 5101–5109 (2012)
12. L. Chen, X. Zhang, L. Sun, H. Xu, H. Si, N. Mei, Energy Fuels **30**, 10222–10227 (2016)
13. W.N.R.W. Isahak, M.W.M. Hisham, M.A. Yarmo, T.Y. Yun Hin, Renew. Sustain. Energy. Rev. **16**, 5910–5923 (2012)
14. M.I. Jahirul, M.G. Rasul, A.A. Chowdhury, N. Ashwath, Energies **5**, 4952–5001 (2012)
15. S. Xing, H. Yuan, Huhetaoli, Y. Qi, P. Lv, Z. Yuan, Y. Chen, Energy **114**, 634–646 (2016)
16. K. Ding, Z. Zhong, B. Zhang, J. Wang, A. Min, R. Ruan, J. Anal. Appl. Pyrolysis **122**, 55–63 (2016)
17. Y. Gao, X. Wang, Y. Chen, P. Li, H. Liu, H. Chen, Energy **122**, 482–491 (2017)
18. K. Anastasakis, I. Kitsiou, W. de Jong, Fuel Process. Technol. **142**, 157–166 (2016)
19. A.M. Widiyannita, R.B. Cahyono, A. Budiman, Sutijan, T. Akiyama, AIP Conference Proceedings **1755**, (2016)
20. D. Rehrah, R.R. Bansode, O. Hassan, M. Ahmedna, J. Anal. Appl. Pyrolysis **118**, 42–53 (2016)
21. N. Claoston, A. Samsuri, M.A. Husni, M.M. Amran, Waste Manage. Res. **32**, 331–339 (2014)

The impact of the hot tap water load pattern in the industrial hall on the energy yield from solar collectors

Natalia Fidorów-Kaprawy[1,*], and *Edyta* Dudkiewicz[1]

[1]Department of Air Conditioning, Heating, Gas Engineering and Air Protection, Faculty of Environmental Engineering, Wrocław University of Science and Technology, ul. Norwida 4/6, 50-373 Wrocław, Poland

Abstract. The systems using solar energy, popular in Poland, can be used to supply hot water for the installation used by employees of industrial halls. In manufacturing plants, employing a large number of people, the demand for hot water is practically constant throughout the year and is characterized by periodic use at the end of each work shift. Dynamics of the hot water consumption depends on the number of shifts as well as working days and holidays. Additionally the maximum hot tap water demand occurs in the whole period of installation operation. In polish climatic conditions the solar collectors' systems have the largest capacity in the summer, while in winter they need to be assisted. Beside that the supply of renewable energy is uneven and depends on weather conditions. In the paper the one-hour step analysis concerning the dependence of the load pattern of the hot tap water preparation system on the energy yield from solar collectors had been performed.

1 Introduction

Systems using renewable energy sources are the most promising solution to cope with increasing cost of energy production, global warming and pollutant emissions. In last years, wind and solar power generation technologies achieved high reliability and their market share increased [1]. Currently in Poland solar collectors' manufacturing sector is the best developed one among RES (renewable energy sources), producing both for export and internal market. Polish market of solar collectors for several years is at the forefront of the markets not only in Europe but in the world, both in terms of number and the surface area of annually installed solar thermal systems. According to the International Energy Agency, in 2013 Poland was in 8th place in the world ranking of 57 leading countries in installing solar collectors and 3rd in Europe in terms of solar installations sales with the share of 9%. In year 2014 Polish market was in 4th place in Europe in solar collector sales [2, 3]. This has pushed investors, especially those investing in large cube halls, where the demand for energy is exceptionally high (figure 1 [4]), towards using this kind of RES. Indicators of the primary energy use indicators given in figure 1 [4] concern the level of the non-renewable

* Corresponding author: natalia.fidorow@pwr.edu.pl

 DOI: 10.1051/e3sconf/20172200044

energy (contained in sources), which is necessary for covering the heat demand for heating, ventilation and hot water production.

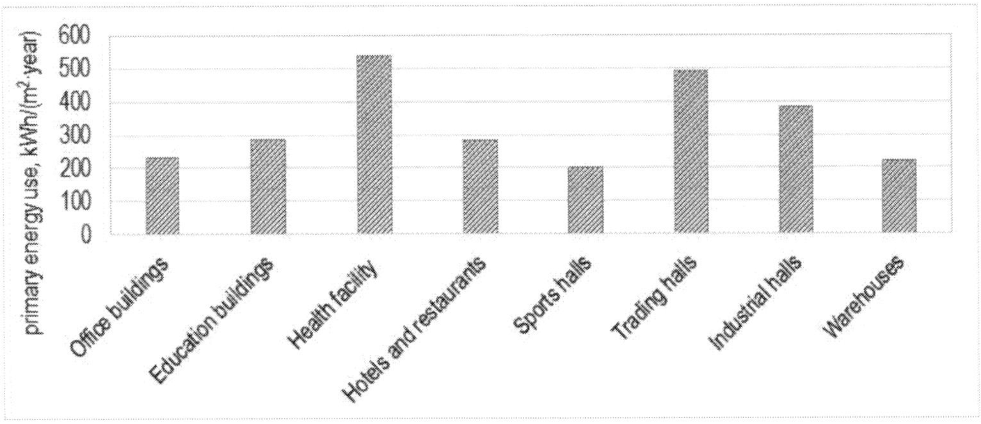

Fig. 1. The weighted average primary energy use indicator in various kinds of non-residential buildings [4].

In industrial and production halls it is possible to use solar collectors for the purposes of the hot tap water preparation, the office spaces heating and technological heat installations. Because of the high level of the technological processes' complication, the use of the heat from solar collectors for the purposes of technological installation requires the individual approach and precise analysis of the industrial process characteristics in order to ensure the integration of the solar and industrial installation. In installations of the special purpose often some new technical developments of solar collectors are required [5]. However the most frequent and popular use of solar collectors in the industrial halls is for the purposes of the hot tap water preparation for the employees. It is mostly because of the favorable ratio of the system investment costs to the obtained energy gains [3].

The supply of renewable energy is uneven and depends on weather conditions but it has been assessed, that in Polish weather conditions the hot tap water preparation installation using solar collectors may cover about 60% of the annual hot tap water demand. The amount of gained energy is higher in period from May to September and reaches up to 90% of the demand while in the remaining period it falls to about 30% [6]. Installation of the solar thermal collectors in the industrial halls unused during the summer because of the holiday season would be a mistake.

In production halls hot utility water demand has periodical peaks when showers and washbasins are used at the end of each work shift. The maximum demand for hot water occurs in whole installation working period and in practice is equal during the year. The periods of installation usage are dependent on the number of working shifts and working/holiday days during a year. Usually, the water draw-off after the first shift is larger than those after the second and third shifts owing to the larger manning during the first shift and due to some office employers working in this time. Duration of draw-off periods is generally determined and lasts for about 30 minutes. Between the main draw-off periods, there is generally low or no demand of hot water caused by sporadic use by employees from the hall and office. The daily load pattern of hot tap water in production halls is not convergent to the supply of renewable energy from solar collectors [7]. How the research [7, 8] shows the useful energy, and therefore the profitability, will decrease if the demand is not distributed throughout the day or focused on the end of the day. Therefore the assessment of the impact of the hot tap water load pattern in the industrial hall on the energy yield from solar collectors is important issue in terms of the energy efficiency.

In this paper authors proposed the use of solar collectors for the purposes of the hot tap water preparation in the industrial halls with 210 employees working in one, two or three shifts. The energy yield from solar collectors was calculated basing on the one-hour step analysis. Additionally authors provided the energy gains which are the results of the calculations by f-chart method for comparison. The aim of the present work is to assess the impact of load pattern of the hot tap water for industrial hall's employees on the energy yield from solar collectors.

2 Description of the system

A hot tap water preparation system for 210 workers of typical production halls located in Wroclaw, Poland was assumed. This system is based on solar collectors being the basic heat source in summer (renewable heat source). For the weather conditions prevailing in Poland, the system with solar collectors has the highest capacity in summer, but in winter season it needs to be assisted by the auxiliary thermal energy source independent on weather conditions [6]. In case of analyzed system a gas boiler shall be used to cover the energy shortage from solar collectors.

A work organization in production halls from Monday to Friday (or to Saturday morning for III shift) is shown in table 1. Water consumption for bath after work completion was assumed as 50 liters per person. The temperature of cold water of 5°C and hot water to bath under shower of 40°C was assumed [9]. Hot tap water is expected to be consumed within 1 hour. The daily energy demand for the hot tap water preparation purposes is the same in each production halls, regardless on the number of work shifts.

Table 1. The work organization in the production halls.

Hall no 1	**Hall no 2**		**Hall no 3**		
I shift	I shift	II shift	I shift	II shift	III shift
6:00–14:00	6:00–14:00	14:00–22.00	6:00–14:00	14:00–22.00	22:00–6:00
210 workers	105 workers	105 workers	70 workers	70 workers	70 workers

The flat plate solar collectors were chosen for the system, each of surface area equal to 1.873 m^2, with optical efficiency of 84.5% and heat loss coefficient equal to 4.1 $W/(m^2 \cdot K)$. The number of solar collectors required in the system is equal to 66 and have been selected basing on the widely used f-chart method [10]. Such number of collectors allow for their maximum use in summer months. The hot water storage unit volumetric capacity of 10 000 liters was assumed and is consistent with general guidelines (close to daily water consumption and close to 75 liters per square meter of the solar collector aperture area [9]).

A gas boiler was assumed as the conventional heat source. Appropriate power, determined during analysis, allows for maintaining the minimum usable water's temperature of 45°C for bath [9] in the storage tank. The operation of gas boiler was assumed on the basis of time schedule as there is no need to keep the high temperature of hot tap water in storage tank continuously. During working days preset temperature of hot tap water in the storage tank is 35°C at two hours before the end of shift and 45°C at one hour before the end of shift. At all other times and during holidays, the boiler would not be turned on irrespective of the water temperature in the storage tank.

3 Method

The analysis focused on the energy shares calculations depending on the hot tap water load pattern. The one-hour step analysis allows taking into account the daily and weekly hot tap water load pattern. As the f-chart method had been used for the calculation of the needed solar collectors' aperture area, authors decided to show the energy shares calculated by this method for comparison.

While using f-chart method data on average monthly temperatures and total monthly radiation on the southern surface inclined at 45° necessary to run calculations were taken from the website of the Ministry of Infrastructure and Construction [11]. Heat demand for production of hot tap water was calculated from the formula:

$$Q_{htw} = 5/7 \cdot l_d \cdot m_d \cdot c_w \cdot (t_{htw} - t_{cw}) \cdot 10^{-3}/3600 \tag{1}$$

where: l_d – number of days in a month (-), m_d – daily demand for the hot tap water (kg), c_w – specific heat of water (kJ/(kg·K)), t_{htw} – temperature in the hot water tank (°C), t_{cw} – temperature of the cold water (°C).

The 5/7 coefficient in formula (1) was introduced to take into account the water draw-off over 5 days per week. The f-chart method does not consider periods of the maximum hot water consumption or the number of consumption periods per day.

In one-hour step analysis the data for each of the 8,760 hours of one year were analyzed. To calculate solar collectors' gains the Hottel-Whillier-Bliss equation was applied [10]:

$$Q_{sol} = q \cdot A_a = F_R(\tau\alpha)_e \cdot G \cdot A_a - F_R U_L \cdot (t_{in} - t_a) \cdot A_a \tag{2}$$

where: A_a – aperture surface of the solar collector or battery of collectors (m^2), $F_R(\tau\alpha)_e$ – maximal efficiency of the solar collector ($t_{in} = t_a$) (-), $F_R U_L$ – heat loss coefficient of the solar collector (W/(m^2 K)), G – intensity of solar radiation (W/m^2), t_{in} – temperature of the working fluid in the solar collector inlet (°C), t_a – ambient temperature (°C).

The temperature of the liquid in a tank in each hour had been determined from energy balance equation using simple Euler integration. The equation was solved at the end of time increment:

$$t'_{htw} = t_{htw} + \Delta\tau \cdot Q/(m \cdot c_w) \tag{3}$$

where: t'_{htw} – temperature in the hot water tank before the heat delivery/withdrawal (°C), t_{httw} – temperature in the hot water tank after the heat delivery/withdrawal (°C), $\Delta\tau$ – method time step (3600 s), Q – source thermal power or the average power transferred during one hour (W), m – mass of water gathered in the hot water tank (kg).

It was assumed that for standard operation, a circulation pump of collectors shall be turned on only when energy gains from solar collectors are greater than heat loss. The system must operate in such a way to protect the storage tank from water overheating by dissipating excess energy or the system turning off by stopping the circulation pump (emergency turning off).

The results of the one-hour step calculations showed the sources activity and enabled the determination of the sources impact (renewable and conventional one) on the water temperature in the tank. The analysis of the outcomes enabled the system elements sizing improvement and revealed the necessity of the circulation pomp's control strategy implementation to protect water in storage tank from overheating and to minimize the necessity of energy dissipation.

4 Results and discussion

The results of the analyses are presented below. The f-chart analysis results (not taking into account the dynamics of hot tap water load pattern) are compared to the one-hour step analysis results for three various hot tap water load patterns depending on the number of work shifts (table 1). In figure 2 the shares of the energy coming from the solar collectors and the gas boiler in subsequent months are shown. The 4 columns for each month from left to right concern: the energy shares calculated using f-chart method and the energy shares calculated using the one-hour step calculation method for one, two and three working shifts. The outcomes of the analysis made with one-hour step show that even in summer the gas boiler is needed for the maintenance of the demanded temperature of the water before the end of the work shift. Those situations are more frequent in case of two- and three-shift work, hence the greater share of the conventional energy from gas boiler in those cases.

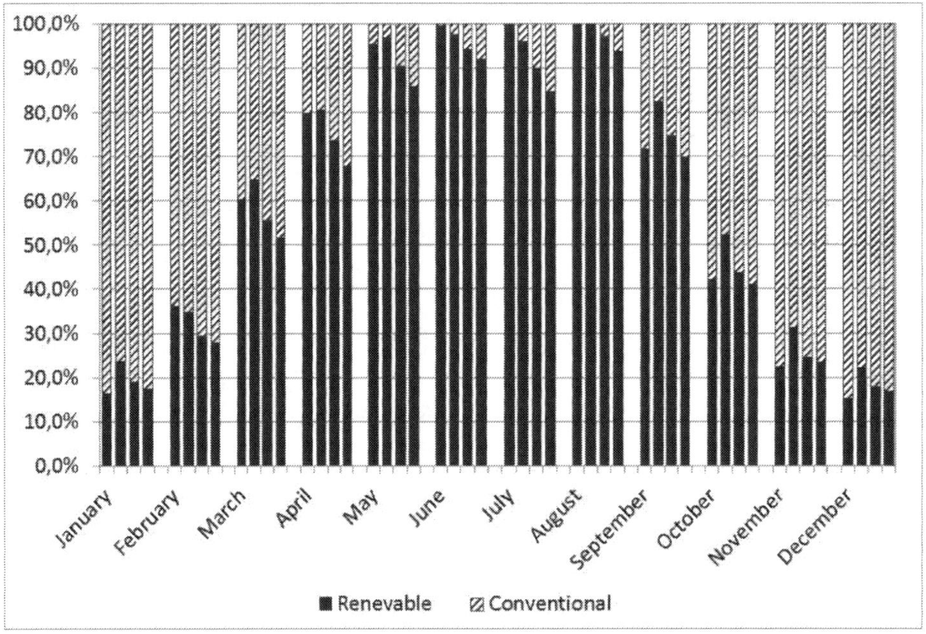

Fig. 2. The monthly energy shares from renewable and conventional sources for all analysed cases.

Table 2 summarizes average annual contributions of energy gained from the solar collectors and from the gas boiler, used to prepare hot tap water, as well as the amounts and shares of the energy that couldn't have been used and have to be dissipated. The amount of energy that couldn't have been used result from the necessity of the circulation pump turn off while the temperature in hot water tank is too high (above 90°C). Those amounts of energy are not included in energy balance. Energy gained from the solar collectors is the whole amount of energy transferred from the solar collectors to the hot water tank, some of which is later dissipated because of the water in the hot water tank overheating risk. The sum of the renewable energy amount (after dissipation) with the energy gained from the gas boiler covers the 100% of the energy demand for the hot tap water preparation purposes. Additionally table 2 contains the characteristic parameters describing the installation elements (the number of solar collectors, the gas boiler required power rate, and the hot tap water tank volume) and the yearly average temperature of the water in the tank.

The number of solar collectors and resulting total active surface of the absorbers, determined using f-chart method and the volume of the hot water tank had been assumed the same for all analyzed cases. The yearly average temperatures of the water in the tank for three cases analyzed in one-hour step are the calculated values and for f-chart method is the value assumed for the calculations.

Different values of the yearly energy demand for the purposes of the hot tap water preparation for f-chart method and one-hour step method result from the different assumptions. In one-hour step method all non-working days (weekends and holidays) have been adequately considered and in f-chart method only the share of the working days, equal to 5/7 have been assumed (formula (1)).

Table 2. The average annual shares and yields of renewable and conventional energy.

		f-chart	I shift	II shifts	III shifts
Collectors number		66	66	66	66
Gas boiler power rate [kW]		220	220	120	120
Hot water tank volumetric capacity [l]		10000	10000	10000	10000
Average temperature in hot water tank [°C]		45.0	32.1	43.2	47.7
Energy demand for hot tap water purposes [kWh]		111515	108643	108643	108643
Energy gained from the solar collectors	Amount [kWh]	68938	75473	69616	67790
	Share [%]	61.8%	69.5%	64.1%	62.3%
Energy gained from the gas boiler	Amount [kWh]	42741	37498	43980	47560
	Share [%]	38%	34.5%	40.5%	43.7%
Energy that couldn't have been used	Amount [kWh]	-	4064	3950	3451
	Share [%]	-	3.7%	3.6%	3.2%
Energy dissipated via solar collectors	Amount [kWh]	164	4327	4953	6564
	Share [%]	0.2%	4.0%	4.6%	6.0%
Renewable energy (after dissipation)	Amount [kWh]	68774	71146	64663	61226
	Share [%]	**61.7%**	**65.5%**	**59.5%**	**56.3%**

As it can be seen in table 2 the shares of the particular energy balance elements differ between f-chart and one-hour step analysis as well as in one-hour step among different hot tap water loads (resulting from the number of shifts). The biggest share of the utilized solar energy equal to 65.5% occurs when one shift work is assumed. When the hot tap water load pattern characteristic for two shift work occurs, the share of the solar energy is lower and equal to 59.5%; at the same time it is similar to the solar energy share obtained using f-chart method (61.7%). The lowest renewable energy share equal to 56.3% had been obtained while analyzing the system with the tot tap water load pattern characteristic for three work shifts. After each hot tap water consumption in cases of two and three shifts work the medium temperature in the hot water tank is higher than in case of one shift work (especially during the transition period and the winter), so less energy may be transferred from solar collectors to the hot water tank.

While analyzing the system in one-hour step it can be seen that during some days the collectors deliver excess energy (especially on the weekend when there is no energy demand) so it needs to be dissipated, hence the higher value of the energy dissipated from solar collectors in this method. In general, it can be said that the f-chart method based on monthly average values do not take into account the inconsistency between the solar energy supply and hot tap water demand, which influences the energy gains form solar collectors. This may be an issue while considering the industrial objects with not untypical hot tap water load patterns.

The share of the energy that couldn't have been used is the biggest in case of one shift work (3.7%), slightly lower for two shifts work (3.6%) and the smallest for three shifts work (3.2%). The results from one-hour step analysis show that the situations in which the solar energy cannot be used occur most often from June to August, during the weekends and on Mondays till 14:00, when the water accumulated in the tank has temperature above

 DOI: 10.1051/e3sconf/20172200044

90°C. In case of three shift work the last water consumption before the weekend occurs on Saturday at 6:00 a.m., hence the temperature in the hot water tank is slightly lower at the weekend beginning in comparison to the one and two shifts work. As a result the greater amount of the solar energy may be gathered on Saturday and hence the lower share of the energy that couldn't have been used in this case.

The share of the energy dissipated via solar collectors is lowest in case of one shift work (4.0%) and bigger for cases of work in two or three shifts (4.6% and 6.0%). This is the result of the higher average temperature in the water tank in case of two and three shifts (43.2°C and 47.7°C); the temperature equal to 65°C, when the energy dissipation begins (if possible), is reached earlier.

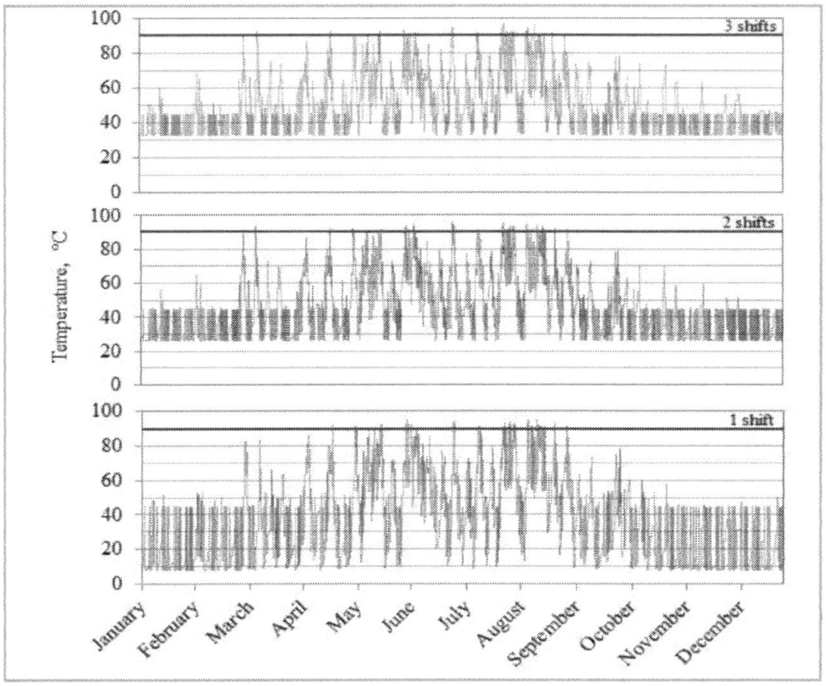

Fig. 3. Temperature in hot water tank along the year for one-, two- and three- shifts work.

In figure 3 the average temperature in the hot water tank during the year is shown. For one shift work the lowest average water temperature in the tank is 8.3°C, for two shifts work 26.6°C and for three shifts work 32.8°C. The water temperature reaching values above 90°C result from the step of the method. The high temperature of the water in the tank in the halls working for two and three shifts causes that, according to the system operation assumptions, the boiler have to heat up the water by 12.2 to 18.4 K during two hours in order to obtain the required temperature of 45°C before the end of the shift. This allows for selection of almost twice smaller power rate of boiler, than in the case of hall working for one shift, where the water has to be heated up by 36.7 K during two hours. The power rate of the gas boiler assumed while analyzing the system with f-chart method is 220 kW (the same as for one shift work). It should ensure the water heat up during two hours.

5 Conclusion

The analysis of the influence of the hot tap water load pattern on the energy yield from solar collectors carried out in the paper concerns the systems applied in industrial halls. In such halls the specificity of the hot tap water load pattern lies in the lack of necessity of the high level of water temperature maintenance during whole day, but only at the end of work shift. In the systems with other hot tap water consumption dynamics the influence of the hot tap water load pattern on the energy yield from solar collectors, the share of gas boiler and whole system performance may be different.

Both: higher average temperature of the water in the tank after the water consumption as well as higher average temperature during whole year cause the lower share of energy gained from solar collectors and necessity of dissipation of greater amount of energy. The highest yearly yield from the solar collectors equal to 65.5% of the annual energy demand had been reached in case of the hall with one shift work, where there is one maximal water consumption during whole day after the period of the largest operation of the sun (at 14:00). The need for the hot tap water preparation for the 210 workers in short time requires the high power rate of the gas boiler (220 kW), greater than in case of other hot tap water load patterns, still the share of the energy delivered by the gas boiler during whole year is in this case the lowest one (34.5%). In the hall with one shift work the solar collectors cover 97.5% of the energy demand for hot tap water purposes in period from May to August.

The analysis of the hot tap water load pattern influence on the energy yield from solar collectors in the industrial hall showed that the magnitude of the individual water consumption incident and the fall of average water temperature in the tank after the water consumption have the great influence on the amount of energy gained from the sun. Increased number of shifts and hence lower amount of water consumed at one causes less temperature drop in the tank and the maintenance of higher yearly average temperature in the tank, which results in lower energy gain from solar collectors.

References

1. R. Cozzolino, L. Tribioli, G. Bella, Energy **106**, 774–789 (2016)
2. G. Wiśniewski, A. Więcka, J. Bolesta, P. Czajka, *Polish manufacturing industry for renewable energy equipment* (Warszawa, 2016) (in Polish)
3. E. Dudkiewicz, Building halls (2017)
4. *Financing of buildings' energy efficiency improvement in Poland* BPIE (Warszawa, 2016) (in Polish)
5. J. Chodura, Rynek Instalacyjny, **12**, (2011)
6. K. Pytel, A. Kłos, Scientific Papers of Rzeszów University of Technology. Civil and Environmental Engineering (Zeszyty Naukowe Politechniki Rzeszowskiej. Budownictwo i Inżynieria Środowiska), **59**, 2, 645–652 (2012) (in Polish)
7. E. Dudkiewicz, N. Fidorow-Kaprawy, Energy (2017) (accepted for publication) http://dx.doi.org/10.1016/j.energy.2017.03.061
8. F.R. Mazarrón, C.J. Porras-Prieto, J.L. García, R.M. Benavente, Energy Conversion and Management **113**, 16–26 (2016)
9. K. Çomaklı, U. Çakır, M. Kaya, K.Bakirci, Energy Conversion and Management **63**, 112–117 (2012)
10. J.A. Duffie, W.A. Beckman, *Solar Engineering of Thermal Processes* (4th ed. John Wiley & Sons, Inc., Hoboken, United States, 2013)
11. The website of the Ministry of Infrastructure and Construction: http://mib.gov.pl/2-Wskazniki_emisji_wartosci_opalowe_paliwa.htm; 2016 [accessed 15.01.2015]

The analysis of the differences between the results of the thermal response test and the data from the operation of the brine-to-water heat pump's vertical exchanger

Natalia Fidorów-Kaprawy[1,*], *Ewelina* Stefanowicz[1], *Wojciech* Mazurek[1], *Małgorzata* Szulgowska-Zgrzywa[1], and *Anna* Bryszewska-Mazurek[1]

[1]Wroclaw University of Science and Technology, Faculty of Environmental Engineering, Wybrzeże Wyspiańskiego 27, 50-370 Wrocław, Poland

Abstract. The article discusses the principles and the problems of obtaining an accurate data input for the design of brine-to-water heat pump's vertical exchangers. Currently, the most accurate method is the thermal response test (TRT). Unfortunately, the test procedure has its limitations and the quality of the results depends on many factors that cannot be fully controlled during the test. As an illustration of the problems, the results of the TRT were presented. The test was executed on the vertical boreholes (one actively regenerated and one not actively regenerated during the summer) which are parts of the operating heat pump system. The test results were compared to the data from the device's operation, in particular with the measurements of the undisturbed ground temperature profiles and the actual unit energy gains from the boreholes. The level of difference between the results of the test and the data from the operation of the boreholes under the real load and the threats concerning the boreholes overload were shown. Additionally the performance differences between the actively regenerated and not actively regenerated boreholes have been emphasised.

1 Introduction

The thermal response test (TRT) is one of the basic tools that allow for the ground parameters data acquisition in the process of the heat pumps' low heat sources design. The procedure of the tests conducted nowadays does not differ much from one developed by Mogensen in 1983 [1]. The general application of TRT as the examination of the ground thermal conductivity found use in 90's and very quickly become the important diagnostic tool. The borehole heat transfer ability may vary in a large extent. It may be in range from 10 to 100 W/m depending on the local geological conditions, the probe type, the quality and the application precision of the filling between the pipes and the ground [2]. The TRT involves making the test borehole or boreholes (in case of the bigger installations) and the

* Corresponding author: natalia.fidorow@pwr.edu.pl

measurement of the effective ground thermal conductivity – it covers the thermal conductivity of the ground as well as the pipe and the filling. The heat transferring agent, transporting known amount of heat is pumped into the borehole. The flow and return parameters are registered. The measurement is carried out until the achievement of the stabilization of the heat flow in the ground; usually 50 to 70 hours (2–3 days) [1, 3]. The test results, according to the information found in literature, while maintaining the required test conditions assure the precision of the ground thermal conductivity estimation on the level of 10% [4, 5], what significantly raises the accuracy of the brine-to-water heat pump's vertical boreholes length design. The main problems and restrictions regarding the TRT procedure concern: the measurement of the undisturbed ground temperature profile, the heating power stabilization during the test, the measurement duration, the influence of the advection and the ground water flow on the measurement, the influence of the instantaneous weather conditions, the influence of the transition heat loss and the quality of the borehole and the filling application. In this paper the issues concerning the measurement of the undisturbed ground temperature profile and the ground water flow will be discussed.

There are a few methods for the measurement of the undisturbed ground temperature for the purpose of TRT. It should be noticed that the borehole heat resistance determined by TRT is very sensitive to the undisturbed ground temperature value. Usually this measurement is carried out by the slow pumping of the heat transferring agent through the borehole and the fluid temperature measurement. To increase the accuracy of the measurement it is recommend in [6] immersing the temperature sensor in the pipe and measuring the temperature on various depths. In the literature, more methods are described: for example using the immersed sensors or the optical fibres. However, they are not popular because of the procedure cost increase.

The mobile TRT is mainly used for the determination of the effective thermal conductivity of the ground. It should be kept in mind that the ground is a heterogeneous form. The thermal conductivity might vary with the depth. Moreover, the heat transfer in the ground is not only in the form of conduction and may be influenced by the ground water flow or advection. As a result, the thermal conductivity measured during TRT may contain some information concerning those factors influence and it is better to call it the effective thermal conduction. According to [7] the ground water flow influences the TRT results. The results of the simulations [8] proved, a that relatively low flow rate may significantly increase the heat transfer. In the article [9] the TRT simulation had been performed using two-dimensional models of the borehole and the ground. The tests had been simulated using various values of the U coefficient (characterizing the flow rate of the ground water, expressed in m/s). Generally, the higher flow coefficient, the more the TRT result (the effective conductivity) differed upwards from the ground thermal conductivity. At the same time, the lower was the thermal conductivity of the ground the greater was the influence of the ground water flow on the TRT results.

Basing on that research it should be noticed that the results of the test conducted under the occurrence of ground water flow may significantly alter the assumptions made for the design purposes and cause the operational problems. In this paper authors decided to show the extent of the problem using the exemplary installation. Data serving as the illustration of the problem have been obtained during the TRT tests conducted on the boreholes which are parts of operating heat pump system. The outcomes of the test procedure have been presented and compared with the system operational data, particularly with the measurements of the real temperature profiles of the boreholes and the real specific energy extraction rates. The level of divergence among the test results and the operational data of the system working under the real load has been presented. As the examined system, apart from the measurement devices, has the possibility of boreholes' regeneration, the TRTs were performed on two boreholes: the one which is actively regenerated during summer (Borehole 1) and the one which is not actively regenerated (Borehole 3). The results of TRT and the operational data analysis have been compared for those two boreholes.

2 Installation and measuring devices

The system on which the study and the analysis have been performed acts as the heat source for the CTE (Energy Technology Centre) office building in Świdnica since 2012. The underfloor heating system works for the 533 m^2 building ground floor area and is designed to deliver around 24.6 MWh of energy during the heating season. The energy is delivered by the BW 117 type brine-to-water heat pump with the nominal thermal power of 17.6 kW, the cooling capacity of 13.8 kW and the COP of 4.4 (B0/W35). The low heat source consists of five boreholes (the single U-pipe (PE 40 x 4.0 mm) surrounded by bentonite filler), each 78 m deep, 6 m apart. The heat transferring agent is the propylene glycol solution in water (1032 kg/m^3, 3900 kJ/(kg·K), concentration 34%). Three boreholes are equipped in measuring devices. There are Pt 1000 type, class B (permissible deviation ± 0.3 K, standard deviation for a B type measurement is 0.17 K) temperature sensors located every 5 meters about 10 cm from the polyethylene pipe in the bentonite (16 pieces in each borehole). In the boreholes' measurement wells there are the Endress&Hausner Proline PROMAG electromagnetic flow meters (1% accuracy) and Endress&Hausner Omnigrad M TST90 coupled temperature sensors (accuracy of a temperature difference measurement of ± 0.05 K and an individual sensor accuracy of ± (0.15 + 0.002·|t|)°C). The summary volumetric flow in the system is around 2.8 m^3/h (the minimum flow rate for the heat pump is 2541.4 l/h). The assumed flow rate for each borehole in case of 5 boreholes performance is around 0.55 m^3/h (the real operational data are 0.51 m^3/h for the tested boreholes number 1 and 3). In case of 3 boreholes performance (for the purpose of the research) the flows are: in the borehole number 1 – 1.02 m^3/h and for the borehole number 3 – 0.98 m^3/h.

3 Thermal response test

The thermal response tests presented in this article were conducted in October 2015. Tests were carried for two boreholes: actively regenerated during the summer Borehole 1 and not actively regenerated Borehole 3. During each test the fluid temperature changes and the flow rate was recorded as a function of time (t). The test consisted of three parts. The objective of the first part of the test was to determine the undisturbed ground temperature. The measurement was carried out using sensors mounted in the measurement well and by slow fluid pumping (0.15 m^3/h). The next stage concerned the parameters' stabilization. It was carried out with the flow matching the operational values (0.55 m^3/h) and lasted for about 12 hours. The stabilization step had been followed by the proper measurement. The borehole model which enables the determination of the thermal conductivity of the ground is created basing on this last measurement step. The calculations were made based on the borehole linear model described by the equation 1 [10].

$$T(t) = (Q/4 \cdot \pi \cdot H \cdot \lambda) \cdot [\ln(4 \cdot a \cdot t/r^2) - 0.5772] + R_b \cdot (Q/H) + T_0 \quad (1)$$

where: T – mean fluid temperature (°C), t – time (h), Q – specific heat load (W), H – borehole depth (m), λ – effective ground thermal conductivity (W/(m·K)), a – thermal diffusivity (m^2/h), R_b – effective borehole thermal resistance (m·K)/W), T_0 – undisturbed ground temperature (°C), r – radial distance from line source (m).

Figure 1 shows the results of the first measurement step, which was the slow brine pumping through the borehole (temperature measured in the well), compared with the measurements made by the temperature sensors in borehole (temperature measured in the borehole). In the figure, the considerable differences in both measurement results are visible. The ground water impact on the temperature profile has been significant. The measurement results enable the calculation of approximate average temperature of the borehole. In addition it has been observed that, the mean temperature of the actively regenerated Borehole 1 is about 1 K higher than the not regenerated Borehole 3.

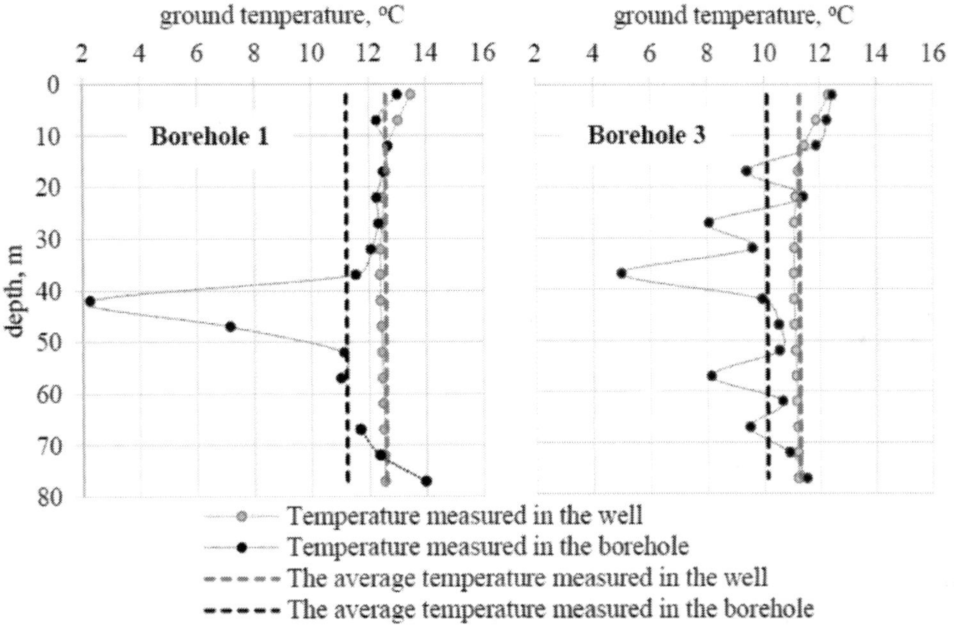

Fig. 1. The undisturbed ground temperature profile in Boreholes 1 and 3.

The figure 2 shows the TRT results for two heat pump's boreholes (actively regenerated Borehole 1 and not actively regenerated Borehole 3). Basing on TRT measurement data a graph in semi logarithmic coordinate system was created. The graph enables to designate a straight trend line of the measurement results' last part (after 12 hours of measurement). The equation describing this line is [10]:

$$T(t) = s \cdot ln\ t + b \tag{2}$$

The slope of the trend line determines the value of the average effective thermal conductivity of the ground, which was calculated using equation (3) [10]:

$$\lambda = q_0 / (4 \cdot \pi \cdot s) = m \cdot C_f \cdot (T_1 - T_2) / (4 \cdot \pi \cdot H \cdot s) \tag{3}$$

where: s – slope of the trend line, b – constant, m – mass flow rate of the brine (kg/s), q_0 – specific heat load (W/m); C_f – specific heat (kJ/(kg·K)), T_1 – inlet brine temperature (°C), T_2 – outlet brine temperature (°C)

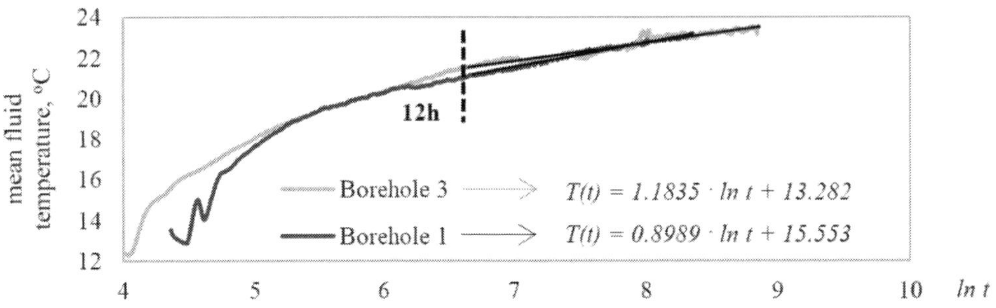

Fig. 2. The brine temperature recorded during TRT in boreholes 1 and 3.

The estimated values of thermal conductivity are presented below. The difference in the values of thermal conductivity of ground may be caused by the intense flow of ground water (observed in both boreholes) but it might be also a result of the active regeneration of the Borehole 1. The tests lasted for 115 h for the first borehole and for 70 h for the third borehole (measurement time is much longer than suggested in the literature), despite that the stabilisation of the temperature had not been achieved. It should therefore be noted that the disorders resulting from the groundwater flow in the area under the study are of the great importance from the point of view of system design, which has a direct impact on its size and subsequent its costs both investment and operational.

Table 1. TRT measurements and calculations results.

	Borehole 1	**Borehole 3**
Effective ground thermal conductivity, (W/(m·K))	4.18	3.09
Specific heat load, (W/m)	47.2	45.86
Mass flow rate, (kg/s)	0.15	0.14
Trend line slope	0.8989	1.1835

Calculated ground conduction coefficient values allow for assumption of the specific power extraction rates basing on which the length of ground heat exchanger is designed. In the literature there are a few references assigning the specific power extraction rates per meter of the borehole to the ground thermal conductivity coefficients. Basing on the literature references, the unit yield for the tested boreholes should be about 55 W/m [11] (according to the recommendations for 2000 h). In order to verify the theoretical parameters of the ground, in the following part of this article the results of increasing the low heat source extraction rate to the value resulting from TRT have been presented.

4 The measurements of the heat pumps' low heat source performance

As mentioned before the analysed brine-to-water heat pump installation had been operated with five 78 meters long boreholes serving as the low heat source. The source size results in maximum borehole energy extraction rate q_{max}= 35 W/m which is the effect of the heat pump's cooling power. For the purposes of the TRT results verification the change in the system operation had been implemented and two out of five boreholes had been cut off. As a result the maximum borehole extraction rate should have risen to q_{max} = 60W/m. Boreholes number 1(actively regenerated) and 3 (not actively regenerated) underwent the detailed analysis.

In figures 3 and 4 the system performance in subsequent labour cycles have been presented (the heat pump labour cycle is defined as the period from the compressor switching on to its switching off). The calculations have been performed in one-minute step and afterwards averaged in subsequent labour cycles. The ambient air temperature (T_a), the brine temperature at the borehole number 1 and 3 outlet ($T_{brine,1}$ and $T_{brine,3}$) and the average power specific extraction rate of the boreholes number 1 and 3($q_{(B1)avg}$ and $q_{(B3)avg}$) have been calculated. The maximum power specific extraction rate taken from Borehole 1 and 3 ($q_{(B1)max}$ and $q_{(B3)max}$) have been also determined for each labour cycle.

4.1. Operation with five boreholes

The performance of the low heat source of the brine-to-water heat pump is presented in the Figure 3. The ambient air temperature (T_a) during two weeks presented in the Figure 3 (at the turn of December 2014 and January 2015) ranged between +10°C and -10°C. The Borehole1 maximum power specific extraction rate was around $q_{(B1)max}$ = 39W/m and Borehole 3 – around $q_{(B3)max}$ = 33 W/m. The average power extraction rate for boreholes also differed one from another and were around 4–5 W/m greater for Borehole 1. When the ambient air temperature was low, the average power extraction rates of boreholes have been getting close to the maximum values. Despite the greater load of the Borehole 1, the brine temperature on this borehole outlet ($T_{brine,1}$) was higher than for the Borehole 3 ($T_{brine,3}$).

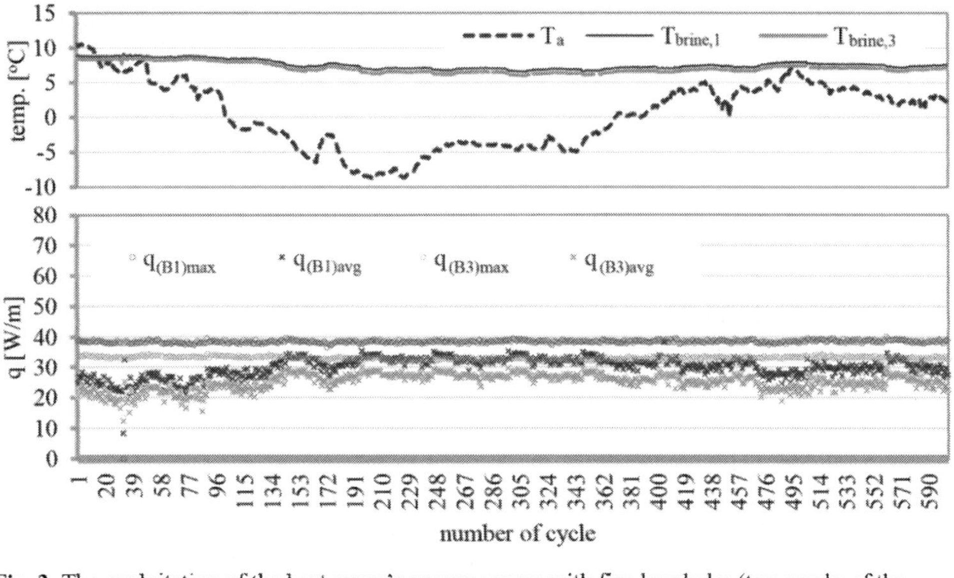

Fig. 3. The exploitation of the heat pump's energy source with five boreholes (two weeks of the operation in December 2014 and January 2015).

DOI: 10.1051/e3sconf/20172200045

4.2. Operation with three boreholes

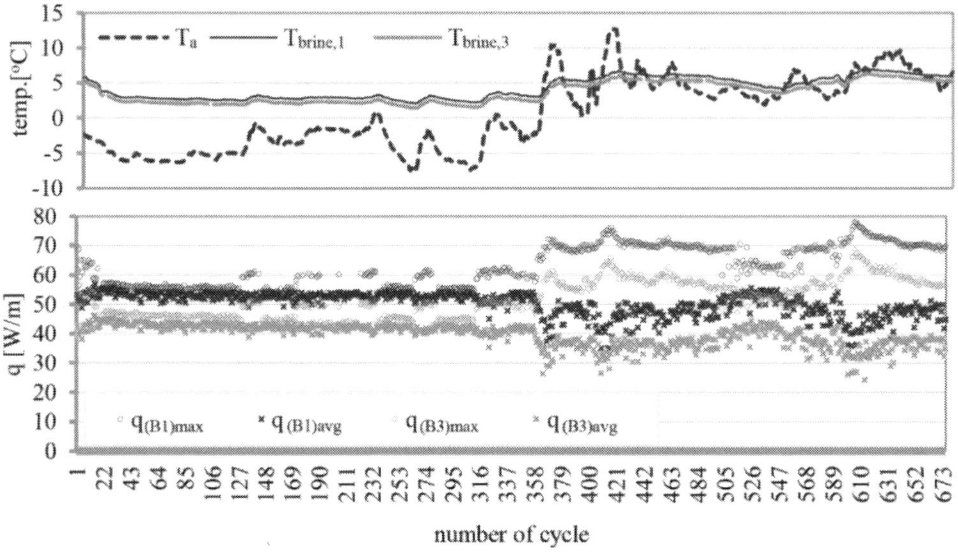

Fig. 4. The exploitation of the heat pump's energy source with five boreholes (two weeks of the operation in February 2017).

The performance of the brine-to-water heat pump's low heat source (operation with three boreholes) is presented in the Figure 4. The ambient air temperature (T_a) during two weeks presented in the Figure 4 (February 2017) also ranged between +10°C and –10°C. The Borehole 1 maximum power specific extraction rate was around $q_{(B1)max}$ = 70W/m and Borehole 3 – around $q_{(B3)max}$ = 60W/m. The average power extraction rate for boreholes also differed one from another and were around 10 W/m higher for Borehole 1. When the ambient air temperature was low, the average power extraction rates of boreholes also had been getting close to the maximum values. At the same time the increase of the load made the maintenance of the long-term power extraction rate on the level of 60–70 W/m impossible. Such operation results in the brine temperature decrease at the boreholes outlets form 7–8°C (typical for the examined installation) to 2°C. Despite the greater load of the borehole number 1, the brine temperature on this borehole outlet ($T_{brine,1}$) is higher than for the borehole number 3 ($T_{brine,3}$).

4.3. Discussion

The measurement outcomes summary of the heat pump's low heat source working with five and three boreholes are set together in Table 3. The outcomes concern the borehole number 1 (actively regenerated) and 3 (not actively regenerated). The summary applies to the periods illustrated in Figures 3 and 4. The average ambient air temperature and hence the average heat load have been approximately equal in both cases. When the system is operated with five boreholes, the maximum specific power extraction rates reach 38.5 W/m. In case of the actively regenerated borehole this value is about 20% higher than for not actively regenerated one. Switch to the operation with three boreholes alters the boreholes specific power extraction rates significantly. They increase by more than 50% for the Borehole 3 and more than 60% for the Borehole 1. In neither of the cases the parameters that allow for the brine temperature maintenance on the level characteristic for the operation with five boreholes are not achieved. The temperature falls by more than 50%. At the same time it may be noticed that the influence of the borehole regeneration is greater for the system operated with three boreholes. The heat output of the regenerated borehole is 27% higher in this operation type, than the not actively regenerated one.

Table 2. The summary of the system with three and five boreholes performance data.

	Borehole 1			Borehole 3			
	q_{max}	q_{cycle}	$T_{B,avg}$	q_{max}	q_{cycle}	$T_{B,avg}$	T_a
type of exploitation	W/m	W/m	°C	W/m	W/m	°C	°C
five boreholes	38.5	30.1	7.6	33.1	25.1	7.1	0.4
borehole 1 / borehole 3	*116%*	*120%*	*107%*	-	-	-	-
three boreholes	62.5	50.2	4.3	51.7	39.5	3.6	0.4
borehole 1 / borehole 3	*121%*	*127%*	*119%*	-	-	-	-
three / five boreholes	*162%*	*167%*	*57%*	*156%*	*157%*	*51%*	*100%*

Due to the necessity of the heating maintenance with the high heat pump's efficiency it is extremely important to achieve the suitable specific power extraction rates of the borehole. The ground overload and failure to maintain the appropriate specific power extraction rate will quickly result in the excessive fall of the temperature of the brine (glycol solution in water) leaving the boreholes. In Figure 5 the relation between the brine temperature and the ambient air temperature in analysed periods and examined boreholes for both operation configurations have been presented. Two effects can be observed while analysing the charts: the brine temperature is higher at the outlet of borehole with regeneration (regardless of the operation mode) and the change in the trend line slope when changing the operation mode from 5 to 3 boreholes.

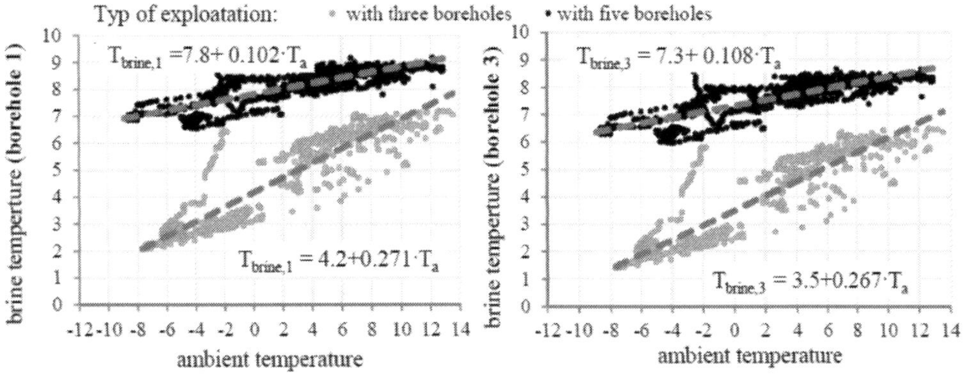

Fig. 5. The relation between the ambient air temperature and brine temperature in heat pump system.

5 Conclusions

The achievement of the specific power extraction rates received basing on the TRT results was not achieved during operation in examined installation. The low heat source load increase results in the severe brine temperature drop both in actively regenerated and not regenerated boreholes' outlets. In the case of exploitation with three boreholes, the analysis results show the strong relation between the external temperature (defining the building thermal load) and the brine temperature at the boreholes' outlets, which will influence the system operational conditions in following years. The ground water flow is the important factor that influences the accuracy and disturbs measurement results of TRT performed on the examined system. Both the TRT results as well as the outcomes of performed analyses show the differences of the regenerated and not actively regenerated boreholes, proving the benefits resulting from the energy storage in summer.

References

1. J.D. Spitler, S.E.A. Gehlin, Renew. Sust. Energ. Rev. **50**, 1125–1137 (2015)
2. J. Acuna, P. Mogensen, B. Palm, Distributed thermal response test on a u-pipe borehole heat exchanger, 11th International Conference on Thermal Energy Storage, Stockholm, Effstock, June 14–17 (2009)
3. C. Zhang, Z, Guo, Y. Liu, X. Cong, D. Peng, Renew. Sust. Energ. Rev. **40**, 851–867 (2014)
4. W.A. Austin, Development of an in situ system for measuring ground thermal properties, Master thesis (Oklahoma State University, 1998)
5. S. Javed, J.D. Spitler, P. Fahlén, ASHRAE Trans. **117**, 1, 13–21 (2011)
6. W. Austin, C. Yavuzturk, J.D. Spitler, ASHRAE Trans. **106**, 365–379 (2000)
7. B. Sanner, M. Reuss, E. Mands, J. Müller, Thermal response test – experiences in Germany, Proceedings to Terrastock, Campus Universität Stuttgard, 177–182 (2000)
8. S. Gehlin, G. Hellström, Renew. Energy **28**, 2221–2235 (2003)
9. A. Liuzzo-Scorpo, B. Nordell, S. Gehlin, Geothermics **56**, 119–127 (2015)
10. T. Śliwa, A. Gonet, **28**, 3 (2011)
11. PORT PC, Wytyczne projektowania, wykonania i odbioru instalacji z pompami ciepła, część 1 – Dolne źródła do pomp ciepła, 01/2013

Solar radiation – to – power generation models for one-axis tracking PV system with on-site measurements from Eskisehir, Turkey

Tansu Filik[1,*], *Ümmühan* Başaran Filik[1], and *Ömer Nezih* Gerek[1]

[1]Anadolu University, Faculty of Engineering, Eskisehir, 26555, Turkey

Abstract. In this study, new analytic models are proposed for mapping on-site global solar radiation values to electrical power output values in solar photovoltaic (PV) panels. The model extraction is achieved by simultaneously recording solar radiation and generated power from fixed and tracking panels, each with capacity of 3 kW, in Eskisehir (Turkey) region. It is shown that the relation between the solar radiation and the corresponding electric power is not only nonlinear, but it also exhibits an interesting time-varying characteristic in the form of a hysteresis function. This observed radiation-to-power relation is, then, analytically modelled with three piece-wise function parts (corresponding to morning, noon and evening times), which is another novel contribution of this work. The model is determined for both fixed panels and panels with a tracking system. Especially the panel system with a dynamic tracker produces a harmonically richer (with higher values in general) characteristic, so higher order polynomial models are necessary for the construction of analytical solar radiation models. The presented models, characteristics of the hysteresis functions, and differences in the fixed versus solar-tracking panels are expected to provide valuable insight for further model based researches.

1 Introduction

The share of solar power generation inside the overall electricity generation is rapidly increasing, thanks to the developments in solar photovoltaic (PV) technology [1]. Solar PV systems convert the energy acquired from sun light (photons) into electricity (voltage), using the, so called, PV effect. Obviously, the total amount of average solar radiation value at a certain point is the most important indicator to determine the total electrical energy generation from the sun. This value naturally changes according to the geo-location of the panel, the time of the day (which determines the arrival angle of the sun lights), and the climatic conditions.

Modern electrical grid operations require an accurate estimation of power generation (including renewables, such as sunlight and wind). Non fuel-based power generation systems, such as solar PV panels, impose a challenging energy estimation problem because of the high uncertainty and statistical variation of the data [2, 3]. In the literature, various

* Corresponding author: tansufilik@anadolu.edu.tr

short-term forecasting models are proposed for wind and solar radiation. The mean hourly global solar radiation can usually be modelled analytically using geo-astronomical calculations and/or time series analysis of a well-structured solar radiation data [4, 5]. These models can give the hourly global solar radiation values of an arbitrary region in the world. However, in order to accurately calculate the electricity output of the PV panels, other models are necessary. There are some models which depend on solar radiation on PV panels and ambient temperature [6]. These models discover the nonlinear relation between the solar radiation and the produced electrical energy.

In this study, we approach the modelling problem through simultaneous observation of solar radiation values, temperatures, solar power output of active sun tracking and fixed PV systems in a controlled environment in Eskisehir. A comparison of fixed and active solar tracking system on the same geo-location in Central Anatolia (Eskişehir) is also a novel contribution of this particular work. All these region specific parameters are recorded with very high resolution (below one minute intervals).

A critical observation is that the conversion of real averaged solar radiation values into real electrical power exhibits a hysteresis function relation for both the fixed and active tracking PV panels. The hysteresis is in the form that the nonlinear relation between a solar radiation value and its corresponding electrical power changes in time. This time variance was not reported or modelled in the literature before. As a next novel contribution, the nonlinear and time-varying relation between solar radiation and power generation is modelled as three-parts hysteresis function, corresponding to the morning, noon, and afternoon time segments. Finally, the third novel contribution of this work is achieved by performing the above analysis on both fixed and (single axis) active sun tracking systems. Therefore, efficient system models are expected to be achieved for two widely popular PV panel types.

2 Experimental Setup

Figure 1 shows an actual photo of the controlled experimental setup, where all the measurements are recorded. The experimental setup is inside the Anadolu University 2 Eylul Campus in Eskisehir region (39.78 latitude, 30.52 longitude, and 794 m altitude with respect to sea level). The installed solar energy generation system consists of 39 solar PV panels with 260 W capacity. The overall energy generation system is established as a hybrid renewable energy system which consists of both 6 kW on-grid (3 kW fixed on the ground, 3 kW solar tracker) and 4 kW off-grid (roof mounted) solar panels combined with a 1 kW wind turbine. The active sun tracking system ads up to 3 kW through 12 panels with 260 W capacity, each. (as can be seen on the right hand side of Figure 1). Similarly, the fixed panel system also adds up to 3 kW with 12 panels of 260 W capacity (as can be seen on the left hand side of Figure 1). The recording station at the back is equipped with the necessary inverters, acquisition cards, computers and data storage units. Detailed technical specifications of the system can be reached from [7].

DOI: 10.1051/e3sconf/20172200046

Fig 1. The Installed fixed (left side) and sun tracking solar PV panels (right side) with same capacity (3 kW) in Iki Eylul Campus of Anadolu University, Eskisehir, Turkey.

Figure 2 (a) shows the outdoor sensors which measure the solar radiation values, temperature values, wind speed / direction, humidity, pressure, etc., and Figure 2 (b) shows a photo of the real-time data recording system.

(a) (b)

Fig 2. (a) Outdoor sensors, and (b) multichannel real time data recording system.

The generated electricity of active tracking and fixed PV panels are separately recorded on inverter with solar logging system which is shown in Figure 3. Also the global radiation values are instantaneously (with 10 seconds resolutions) recorded in data recording system for year 2016.

Fig 3. On-grid Inverter (Fronius)

3 Solar Radiation to Power Generation

Classical models for the estimated conversion of solar radiation value into generated electricity power typically use average solar radiation, static values of PV parameters, and a fixed ambient temperature [6]. Practically, in order to roughly calculate the electrical potential of annual solar energy output from a PV system can be calculated as:

$$E = A \cdot r \cdot G \cdot \rho \quad (1)$$

where A is the total solar panel area in m^2, r is the panel efficiency (yield) parameter, G is the annual average solar radiation on tilted panels, and ρ is the performance ratio of the overall system. This global formula naturally gives a rough annual electricity generation value, and it does not take into account instantaneous conversion values.

Solar radiation, however, is a dynamic process that depends on the geographic location and atmospheric conditions. The black-labelled curve ($g(t)$) in Figure 4 shows averaged global radiation values with the blue ($E_{fix}(t)$) and red ($E_{trac}(t)$) curves corresponding to the instantaneous electrical powers generated from fixed and solar tracking PV panels, respectively. The plotted functions are averages of the daily measurements as a function of time over a 24-hour duration (starting from 00:00 at night and ending at 23:59). Using the same approach of Eq. 1, it can be shown that the instantaneous electricity output value for a PV panel can be calculated as:

$$E(t) = A\, r(t)\, g(t)\, \rho(t), \quad (2)$$

with the same parameters (only converted into functions of time) as in Eq. 1.

Unfortunately, the performance ratio and efficiency of the overall system also change with time (and temperature), as expected. Therefore, the need for an accurate model of $r(t)$ and $\rho(t)$ is necessary. In order to see the time varying behaviour of these parameters, a plot of real averaged electricity output values versus real averaged global radiation values of the active sun tracking and fixed panels are shown in Figure 5. Two very important observations can be made from the curves in Figure 5:

- The sun tracking system (in red) provides a higher conversion ratio than the fixed panel (this can be seen due to the red curves being above the blue curves).
- Sun rising (morning) conversion plots are *different* from the conversion curves of afternoon hours (exhibiting a hysteresis shape) for both the fixed and the sun tracking PV systems. In other words, a certain solar radiation value (say, g) causes the production of *different* electric power values depending on whether the value appears in the morning or in the afternoon.

It is, therefore, argued that there is a hysteresis phenomenon in the conversion function, probably due to the ambient temperature difference between morning and afternoon hours. The noon time is observed to constitute a threshold region. This hysteresis function means a time varying behaviour, therefore nonlinear models are required. In this study we propose a piecewise model (consisting of three-part functions) that considers morning, noon and afternoon hours separately. The proposed approach is used for the modelling of both fixed and solar tracking PV systems.

DOI: 10.1051/e3sconf/20172200046

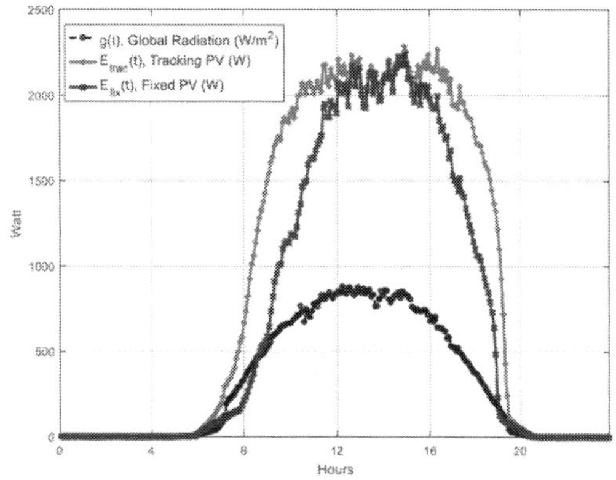

Fig. 4. Real averaged electricity generations of the installed tracking ($E_{trac}(t)$) and fixed ($E_{fix}(t)$) PV systems with the measured global radiation value ($g(t)$) between 1st July and 31st Dec., 2016.

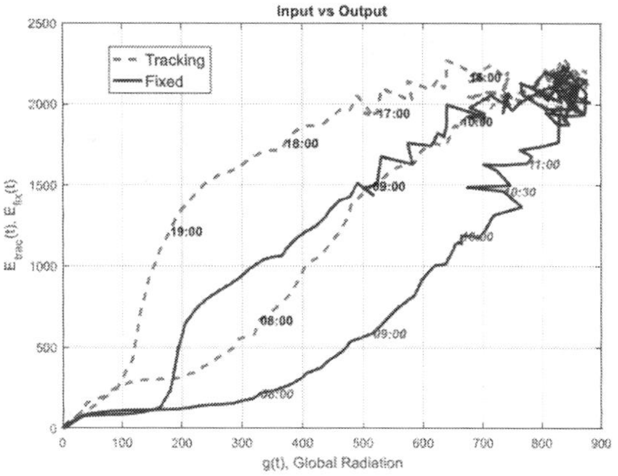

Fig. 5. Real averaged electricity output values versus real averaged global radiation values of the active sun tracking and fixed panels (blue for fixed panel, red for sun tracking system). Certain hour points are shown on the plots.

The proposed three-part functions for fixed and active sun tracking systems are as follows, respectively;

$$E_{fix}^{m}(t) = \begin{cases} a_1 + b_1 g(t) + c_1 g^2(t), & 0 < t \leq 11.5 (morning) \\ a_2 + b_2 g(t), \quad , & 11.5 < t < 15.4\ (noon/evening) \\ a_3 + b_3 g(t) + c_3 g^2(t), & 15.4 < t \leq 0\ (evening/night) \end{cases} \tag{3}$$

$$E_{trac}^{m}(t) = \begin{cases} \alpha_1 + \beta_1 g(t) + \gamma_1 g^2(t), & 0 < t \leq 11 (morning) \\ \alpha_2 + \beta_2 g(t), \quad , & 11 < t < 17\ (noon/evening) \\ \alpha_3 + \beta_3 g(t) + \gamma_3 g^2(t), & 17 < t \leq 0\ (evening/night) \end{cases} \tag{4}$$

The coefficient parameters in these equations are estimated using minimum least square optimization from the recorded values between 1st July and 31st December, 2016. The least square optimized function coefficients for the records are presented in Table 1 for both the active sun tracking and fixed systems. The model generated energy generation values (over actual power values) are shown in Figure 6 (a). The fitted model conversion curves and real conversion curves are presented in Figure 6 (b).

Table 1. Model (mapping) function coefficients.

E(t)	Fixed PV			Tracking PV		
	a	b	c	α	β	γ
Morning	12.515	-0.433	0.003	1.233	1.844	0.0014
Noon/Evening	1907.6	0.0002	-	2010	0.0002	-
Evening/Night	-18.439	3.093	-0.0003	-19.87	7.5741	-0.0072

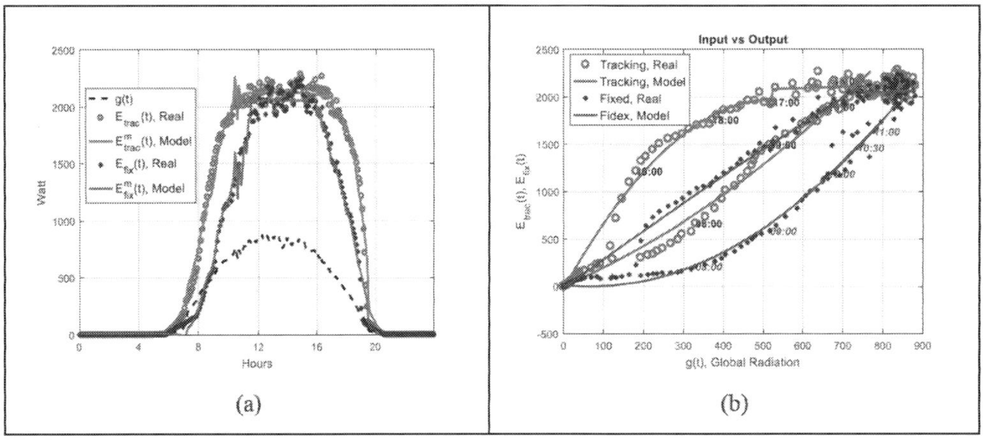

Fig. 6. Model vs. real values of (a) generated electric power, and (b) conversion functions.

Finally, real electricity generation and the estimated model electricity generation values are given in Table 2 for the remaining data set. It is shown that proposed model gives accurate results.

Table 2. Real electricity generation versus estimated model generation values.

kWh	Real Generation	Model Generation
Tracking PV	26.362	26.516
Fixed	20.672	20.834

4 Conclusion and discussion

In this work, solar radiation and energy generation data obtained from our experimental set-up are used to illustrate various important aspects of the radiation-to-energy conversion phenomenon. First, it was observed that the conversion is not only non-linear, but it also has a strong time-varying characteristic. Particularly, the conversion rate is consistently higher in the afternoon hours as opposed to the morning hours, although the radiation value plot is necessarily symmetric around the noon time. This observation induced a novel hysteresis

 DOI: 10.1051/e3sconf/20172200046

function modelling approach through separate piece-wise quadratic functions for the morning, noon, and afternoon. The resulting hysteresis model is found to achieve a small squared error when compared to the actual recordings.

The second important observation is that the efficiency gain of the sun tracking PV panels is clearly visible from the conversion curves. The curve for the sun tracking system is consistently tracing above the conversion values of the fixed panel.

The time varying hysteresis effect of the conversion curve is expected to be due to the ambient temperature and the operation characteristics of the PV materials. However, this correspondence needs to be examined and reported as a possible follow-up of this research.

This work was supported by Anadolu University Research Fund under contract no: 1505F512

References

1. B. Parida, S. Iniyan, R. Goic, Renew. and Sust. Energy Reviews **15**, 3, 1625–1636 (2011)
2. H. Farhangi, IEEE Power and Energy Mag. **8**, 1 (2010)
3. P. Varaiya, F.W. Felix, J.W. Bialek, Proc. of the IEEE **99**, 1, 4–57 (2011)
4. S.N. Kaplanis, Renew. Energy **31**, 6, 781–790 (2006)
5. C. Gueymard, Solar Energy **68**, 3, 285–303 (2000)
6. H. Yang, L. Lu, W. Zhou, Solar Energy **81**, 1, 76–84 (2007)
7. T. Filik, U.B. Filik, Anadolu Univ. J. of Sci. and Tech. A–Appl. Sci. and Eng. **18**, 1 (2017)

Microbiological stimulation of phytoremediation process using *Salvinia natans* to mercury contamined water

Viktoriya Filyarovskaya[1,*], *Magdalena* Sitarska[2], *Teodora* Traczewska[2], and *Mirela* Wolf[2]

[1]US Pharmacia Sp. z o.o., u. Ziębicka 40, 50-507 Wrocław, Poland
[2]Wroclaw University of Science and Technology, Faculty of Environmental Engineering, Wybrzeze Wyspiańskiego 27, 50-370 Wroclaw, Poland

Abstract. An alternative to traditional cleaning methods of heavy metals in the water environment is phytoremediation. They efficiency depends on used technological process conditions as well as plant species. One of the most dangerous metallic elements mercury plays a particular role, which is a trace element and a physiologically foreign in living organisms. Mercury has a high degree of toxicity with strong affinity to thiol groups. This may cause an adverse effect on the enzymatic processes and consequently inhibiting the physiological functions. Because of high risk for human health, water environment treatment from mercury is essential proecological action. Mercury removal studies were conducted using *Salvinia natans* pleustofit, sampled from its natural water environment. In the first step, epiphytic bacteria, which was resistant to high concentrations of mercury (0,6 mgHg/l), was isolated from the plant and than selected by the tiles gradient mthod. In the next step, the identification using molecular biology methods was made. In the following step plant *Salvinia natans* was exposure to high levels of mercury in the presence of the three isolated *Pseudomonas* strains with exceptional resistance characteristics to environmental factors. Has been found a positive bacteria effect on the plant condition because the selected strains belong to *Pseudomonas* species producing materials supporting plant growth. The use of microbial stimulation to phytoremediation by hyperaccumulator *Salvinia natans* can multiply the effectiveness of the process.

1 Introduction

Environmental pollution by heavy metals, i.e., Pb, Cu, Fe, R, Ni, Zn, Cr, Cd, Mo and metalloids (Se, As) is a serious global problem resulting from the intensive progress of civilization [1–3]. One of the most dangerous elements on Earth is mercury, since it occurs in liquid, gas or a solid form and is an element of a high volatility, easily spreading in the atmosphere [4]. It is necessary to control the presence of mercury in various parts of the environment, due to its emissions from a variety of industries, including chloralkaline,

* Corresponding author: W.Filarowska@usp.pl

electrical, pharmaceutical, as well as from the process of combustion of coal and oil [5, 6]. Aquatic ecosystems are worthy of attention. They are particularly vulnerable to contamination with heavy metals, including mercury, as a result of surface runoff, discharges and dry or wet deposition from the atmosphere, because in this environment, as a result of microbiological changes, mercury can be present in a particularly toxic form [7, 8]. Due to the toxicity of mercury, it is necessary to control and effectively remove it. Therefore, the ions of the element should be removed from the water before the water treatment. The used methods could be of biological nature, such as accumulation in tissues of plants, for example in a known hyperaccumulator *S. natans*. In addition, the literature review reports the removal of mercury from the water by bacteria of the genus *Psuedomonas* and *Kleibsiella* [9].

Therefore, the aim of this study was to clarify the role of bacteria accompanying *S. natans* in the process of phytoremediation of water contaminated with mercury. It was assumed that bacteria *S. natans* can support the process of phytoremediation. The objective was to explain the process.

2 Materials and methods

The research material was the floating fern *Salvinia natans*, collected at the estuary of the Oława river to the Odra river (Wrocław, Polska). The endophytic bacteria from the plants were collected according to scheme no. 1.

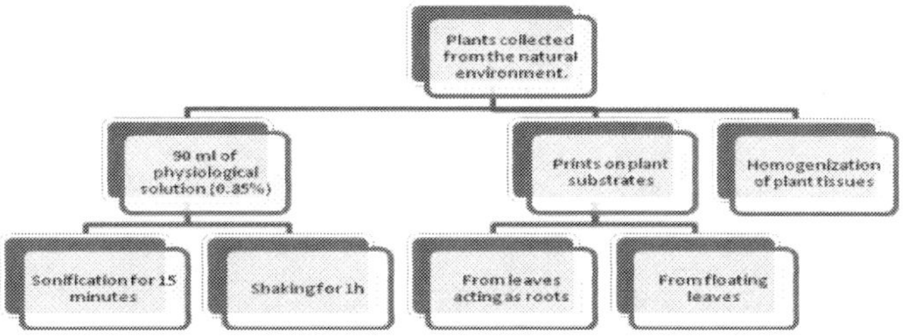

Fig.1. The scheme of bacteria isolation from *S. natans*.

The preliminary identification of Gram-negative strains resistant to mercury was performed with biochemical methods in accordance to the Bergey scheme [10].

The identification was carried out with the help of methods determining the physiological and biochemical traits of microorganisms and methods of molecular biology. The evaluation scheme for taxonomical identity is shown in fig. 2.

Selected Gram-negative bacteria resistant to high concentrations of mercury have been subjected to preliminary identification using the API 20E test (bioMerieux). The test strip consists of 20 microtubes containing dehydrated substrates and integrated oxidase assay. The test has been conducted according to manufacturer's instructions. The chamber with the microtubes has been incubated for 18h in temperature of 26 degrees Celsium, as those were the best growth conditions for the isolated bacteria. The results were interpreted with the help of apiweb software.

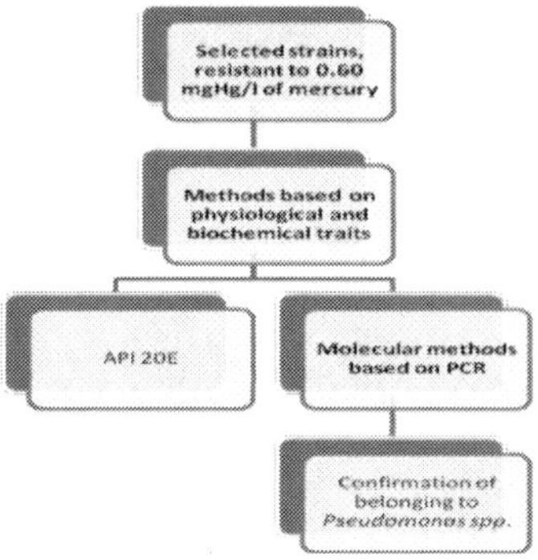

Fig. 2. Scheme of identification of strains resistant to 0.60 mgHg/l of mercury.

Isolation of the nucleic acids strains was performed using a set of regents Genomic Mini (A&A Biotechnology) according to the manufacturer's instructions. Amplification was performed using specific primers PA-GS-F of the sequenceGACGGGTGAGTAATGCCTA and PA-GS-R of the sequence CACTGGTGTTCCCTTCCTATA, yielding products of approximately 618bp [11]. The reaction mixture (25 µl) consisted of 1U BIO-X-ACT Short Polymerase (Bioline), 250 µM a mixture of DTP, 0,4 µM of each primer, 5 µl OptiBuffer, 10ng isolated DNA and 2 µM $MgCl_2$. Conditions of the reaction have been shown in table no.1.

Table 1. Amplification conditions [12].

Reaction phase	Temperature [°C]	Time [s]	Number of cycles
Initial denaturation	95	120	1
Denaturation	94	20	25
Hybridization	54	20	
Elongation	72	40	
Final extension	72	60	1

Control of the obtained PCR products has been carried out using electrophoretic separation (40min, 100V) on agarose gel 1%) in presence of the intercalating agent (ethidium bromide 0.5 µl/ml).

Phytoremediation has been conducted in reactors containing 500 ml of modified Hoagland medium with a mercury concentration of 0.3 mgHg/l with 5 g of plant biomass of *S.natans*.

To maintain a instable conditions, the experiment was carried out in FD 147 Inox phytotron of the Biosell company, equipped with fluorescent bulbs 18W/965 Biolux of the OSRAM company, with the day/night cycle (12h/12h), at the temperature of 22°C/15°C and air humidity of 40% (Fig. 14).The analyses were carried out on the 7th, 14th, and the 21st day.

Before the evaluation of the effect of epiphytic bacteria on the process of phytoremediation, the plants were washed with 70% ethanol and then placed for 3 hours in the bacterial solution of the selected strain (5 in the McFarland scale).After that time, the plants were transferred to the prepared culture media inoculated with 5 ml of bacterial solution. The control were plants incubated in the medium without mercury, and the positive control were plants exposed to mercury, but without the tested microorganisms.

To evaluate the toxicity of mercury (II) to the bacteria, the Microtox test was used, utilizing concentrations: 0.15; 0.2; 0.3; 0.4; 0.5; 0.6 mgHg/l. During an initial assessment of

 DOI: 10.1051/e3sconf/20172200047

the impact of mercury on the condition of *S. natans*, the plants were placed in the Hoagland medium containing 0.035; 0.075; 0.150; 0.200; 0.300; 0.350; 0.400; 0.500 mgHg/l.

In the culture vessels, 5 specimens with similar morphology were place. The specimens were incubated for 14 days in optimal conditions. After that time, a photographic documentation of chlorotic and necrotic changes of the leaves was taken.

The total protein content in the tested plants was measured using the Lowry method modified by Eggsteinand Kreutz [13] in homogenates of plant tissues obtained from 0.1 g of fresh tissues suspended in 7 ml 1M NaOH (homogenizer of the Ultra-TurraxTube Driver type). The amount of protein was read from the calibration curve at a wavelength of 750 nm on a spectrophotometer T80+ UV/VIS (PG Instruments Ltd). The amount of assimilation pigments was determined in extracts obtained through homogenization of 0.1 g of fresh plant tissues in 7 ml 90% acetone with the use of Ultra-TurraxTube Driver homogenizer. The amount of *a* chlorophyll has been read from respective calibration curve at a wavelength of 663 nm, and 645 nm, and for the chlorophyll *b* 645 nm on a spectrophotometer T80+ UV/VIS (PG Instruments Ltd).

In order to control the growth of the biomass, the dry weight of the plants was measured after the experiment. The plants were dried at room temperature to the air-dry state. Because the determination of the dry mass of plants before the experiment is not possible, a mean dry mass has been experimentally determined. For this purpose, 10 samples with 5 g of fresh plants with different physiological condition have been prepared, which were weighted, then dried at room temperature to a solid weight. Obtained results were averaged, thus obtaining a mathematical converter. The mercury content was analyzed by atomic absorption spectrometer AMA 254 Altem company [14]. This method does not require sample preparation. Analysis was carried out in aqueous solution and on air-dry weight of the plant.

3 Results

Although *S. natans*is considered to be a hyperaccumulator of heavy metals, under the influence of high concentrations of mercury chlorosis and necrosis were observed, especially after 14 days of exposing the plant to the tested element in concentrations above 0.3 mgHg/l (fig. 3.). It should also be noted that the fist tested concentration is a value close to the one recorded by [15] in lakes Miedwie and Reko. Therefore, an assessment was made of the influence of that concentration on physiological parameters of *S. natans* (after 7, 14 and 21 days of exposition). There was a significant decrease of protein content in plants relative to control samples such that after 21 days for the concentration 0.15 mgHg/l it was 75%, for concentration of 0.2 mgHg/l 60%, and for concentration of 0.3 mg/l 53%. The toxic effects manifested also in the decrease of the amount of assimilation pigments with simultaneous disturbance of the proportions of chlorophyll *a/b* (figures 4 and 5), as a higher concentration of chlorophyll b was observed than the concentration of chlorophyll a, what can be clearly observed for most of the tested samples after 7 and 21 days of experiment. At the same time, for the concentration of 0.3 mgHg/l a very significant decrease of both chlorophyll *a* and *b* has been observed, what shows an advanced process of chlorosis.

Since the objective was also to stimulate the process of phytoremediation by the bacteria characterized by a high (from 0.3 to 0.6 mgHg/l) resistance to mercury, firstly the assessment of the toxicity of mercury was performed using Microtox test with bacteria *Vibrio fischeri* (fig. 6). As can be seen, the concentration of 0.3 mgHg/l has induced an effect of fluorescence decay, exceeding the limit for the III hazard class (50% < PR < 100%), because after 5 minutes of the process, the inhibition was 68.68, and after 15 min, 96.26.

 DOI: 10.1051/e3sconf/20172200047

Accompanying microorganisms obtained by cavitation in physiological solution of prolonged (1 h) shaking in physiological solution, direct print of floating leaves and those acting as roots and as a result of homogenization of tissues, were inoculated on a nutrient medium containing mercury at the concentration of 0.15 mgHg/l.

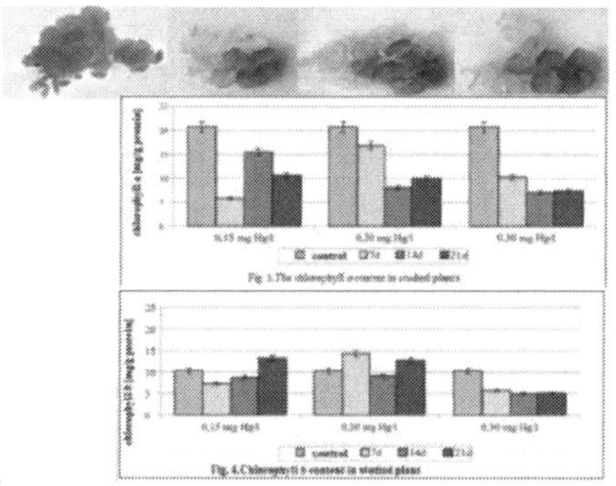

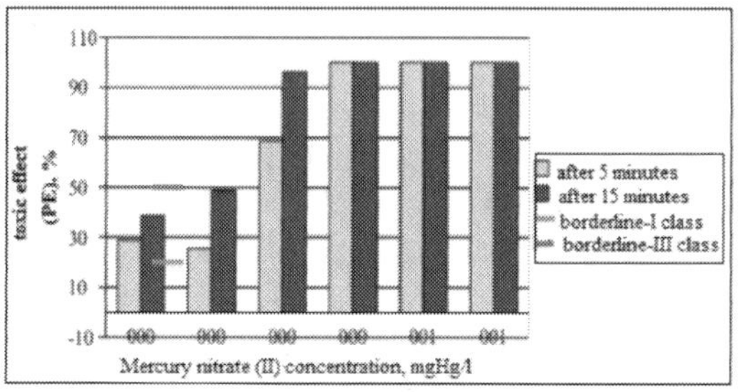

Fig. 5. Reduction of *Vibrio fischeri* bacteria bioluminescence.

Accompanying microorganisms obtained by cavitation in physiological solution of prolonged (1 h) shaking in physiological solution, direct print of floating leaves and those acting as roots and as a result of homogenization of tissues, were inoculated on a nutrient medium containing mercury at the concentration of 0.15 mgHg/l. As a result, 51 strains of bacteria were obtained (fig. 6). During selection of strains on gradient plates 0.00–0.3 mgHg/l 14 strains were obtained. Resistance to mercury in the range of 0.3–0.6 mgHg/l of 14 strains was evaluated with the gradient plates and disc-diffusion method. It was found, that in the pool of tested strains, 5 showed a remarkable tolerance to mercury and those were selected for the stimulation of the phytoremediation process. It needs to be emphasized that those strains have been obtained as a result of one-hour-long shaking of *S. natans* in a physiological solution. For the purpose of preliminary identification of the bacteria, first the API E20 test was used, on the basis of the result of the Gram staining method, which has shown that those are G(-)bacilli. The results of growth on selective-differentiating and chromogenic media shown their belonging to *Pseudomonadaecae*.

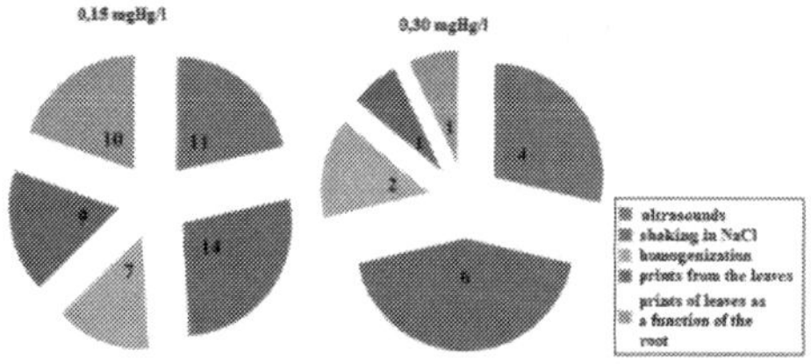

Fig. 6. Effectiveness of methods used to isolate mercury resistant bacteria in different concentrations.

For epiphytic strains: 1-26-27, 4-8-22, and st.pr. the highest degree of similarity to the reference species was obtained for *Pseudomonas aeruginosa* and they were one 77.5%, 67.1% and 43.6%, respectively. Because the bacterium belongs to the clinical pathogens, if these results were confirmed by genetic studies, the applicability of this species in water treatment would not be possible due to the epidemiological threat. Strain 3-22-22 was identified as *Pseudomonas fluorescens* or *Pseudomonas putida* in equal similarity (44.5%), and only 27.9% of traits pointed to *Pseudomonas aeruginosa*.

Based on the results obtained in the biochemical tests, the identification of the two strains (3-22-22, 1-26-27) has been recognized by the program as "a good for a genus", while the other three (2-26, st.pr., 4-8 -22) for "questionable for a genus" in connection with the discrepancy of the results obtained in biochemical tests. The API 20E method is based on the basic biochemical features designed for the detection of most enterobacteria, its application to environmental microbes may not take into account the high volatility of the so-called wild strains for which in most biochemical assays there is no reference data. Moreover, the test is designed for low-maintenance bacillus from the family *Pseudomonaceae*, what could also influence the accuracy of the measurement. The results of API 20E can be considered illustrative, and helpful in choosing the conditions for further analyses, as well as in choosing primers for PCR reactions. From five strains resistant to the highest concentration of mercury, DNA was isolated, which then was amplified by PCR reaction with the use of appropriate primers. For all tested DNA samples, the result of the reaction was positive, yielding products of the size 618bp (fig. 7). It confirms the belonging of the tested strains to the genus *Pseudomonas*.

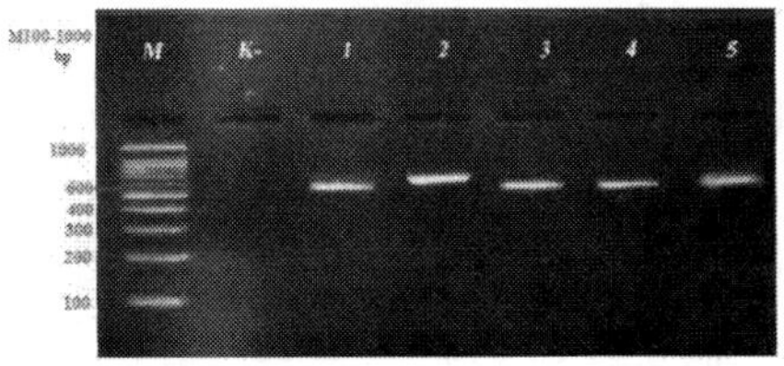

Fig. 7. Photo of agarose gel with PCR products: M - size marker (100-1000 bp); "K-" - negative control; 1-st.pr. strain, 2-strain 2-26; 3 -strain 3-22-22; 4 - strain 1-26-27; 5 - strain 4-8-22.

This genus has 245 species, and a detailed analysis of each strain requires further studies with the use of techniques of sequencing the order of DNA nucleotide base pairs.

Visual evaluation of the plants carrying the process of phytoremediation, stimulated microbiologically by bioaugmentation, in the presence of the strains 4-8-22, and 3-22-22 showed significant deterioration in their condition compared to controls (fig. 8). Many necrotic changes have been observed already in the 7th day of the experiment. This result prompted us to reject these two strains in further studies on phytoremediation process because, despite the high resistance to mercury concentration (0.6mgHg / l), there was no positive influence on *S. natans*.

Different results were obtained for the process of stimulating the process of phytoremediation involving strains 2-26, st.pr., 1-26-27.Control plants showed very high growth of dry mass. In the last day of experiment, its value increased to 1.97 g, which is more than seven-fold increase compared to the initial weight (0.26 g), what confirms the optimum conditions for the experiment.

In the case of the control plants exposed to mercury already at the 7th day of the experiment (Fig. 72) progressive chlorotic and necrotic changes in leaf could be observed. However, in *Salvinia natans* in the presence of selected microorganisms (2-26, 1-26-27, st.pr.) the necrosis of tissues was minor and concerned individual leaves.

The increase in weight of plants stimulated by the presence of bacteria 2-26 was similar as in the control, which suggests the reduction of the toxic effect presumably as a result of microbiological transformation of mercury. Also, for strain st.pr after 21 days of culture, the increase in biomass was the same as in the control against the strain 2-26 which in this case suggests the adaptation of bacteria to such a high concentration of mercury, and then their participation in the transformation of the element. The least favorable effect on plant growth was observed for strain 1-26-27.

DOI: 10.1051/e3sconf/20172200047

S. natans against the tested bacterial strains maintained a correct ratio of the amount of chlorophyll a and b, whereas the amount of total chlorophyll gradually decreased, but only slightly differed from the amounts in controlled organisms. By contrast, without microbial stimulation, after 21 days this amount was lower by 50% compared to control.

From the graphs it can be seen that the correct proportions of chlorophyll a to chlorophyll b were preserved (approx. 1:2). In addition, in both cases similar trends can be observed. On the 14th day of the experiment, the pigment content significantly decreased in almost all the studied plants.

The exceptions being tissues of the control plants and with the addition of strain st.pr. For *S. natans* control, the content of chlorophyll a decreased proportionally from 74.81 mg/g of protein (day 7) to 57.74 mg/g of protein (day 21), and chlorophyll *b* from 40.88 mg/g of protein (day 7) to 32.73 mg/g protein (day 21).

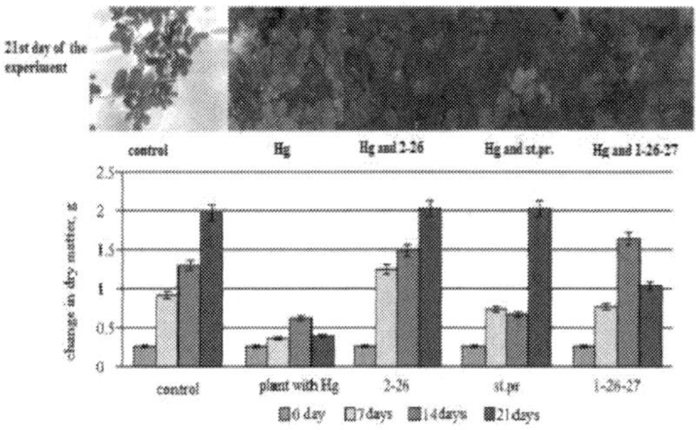

Fig. 8. Increase *S.natans* dry matter in the presence of mercury and strains.

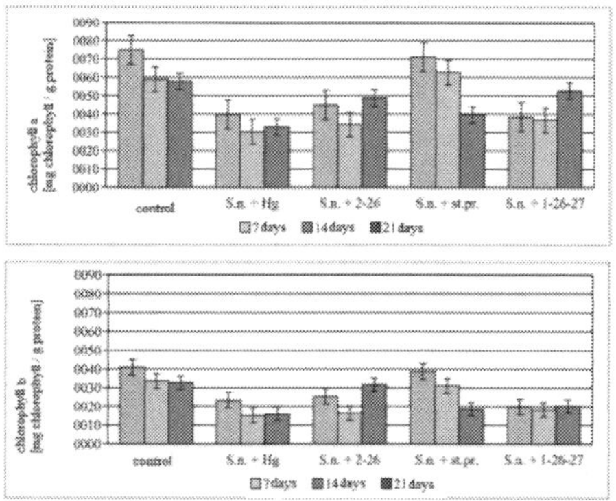

Fig. 9. Amount of chlorophyll in tested plant.

In contrast, in *S. natans* tissues exposed to mercury with the strain st.pr., the amount of chlorophyll content decreased from 71.03 mg/g of protein to 39.58 mg/g of protein forchlorophyll *a*, and from 38.71 mg/g of protein to 18.58 mg/g of protein for chlorophyll b.

As is apparent from the obtained results, the protein synthesis is not translated into a number of photosynthetically active pigments. This indicates a decrease in the living activity of plants, because the inhibition of photosynthesis process may result in a shift in the energy balance of the cell.

The highest content of chlorophyll a on the last day of the experiment characterized the control plants (57.74 mg/g of protein) and *Salvinia natans* in the presence of microorganism 1-26-27 (52.67 mg/g of protein).In the case of chlorophyll b, it was the highest in control plants (40.88 mg/g of protein), and plants in the presence of strain 2-26 (31.68mg/g of protein).The lowest value of the pigmentin both cases characterized the plants without the bacterial help grown in the presence of mercury.For chlorophyll a it amounted to 32.89 mg/g of protein, and for chlorophyll *b* – 15.77 mg/g of protein. This indicates a possible impairment of photosynthesis under the influence of mercury ions.

In the case of plants containing bacterial strains, an increase of the content of the element in tissues along with prolonged exposure time as compared to plants without the microorganisms can be seen (Fig. 10). The greatest accumulation of mercury was found for *Salvinia natans* in the presence of the strain on the 2-26 in the 21st day – 693.2 mg/kg DM, which is approx. 22% more than in the case of plants without the help of bacteria, which fully emphasizes the nature of the stimulus effect of the tested bacterial strains on process of mercury accumulation by *S. natans*. It should be emphasized that although the accumulated element impairs such physiological processes as protein and chlorophyll synthesis, it did not have a significant impact on the growth of biomass. In the case of the other variants, after 21 days, the cumulative amount of mercury amounted in the presence of 1-26-27 to581.9 mg/kg DM and 632.4 mg/kg DM in st.pr. (Fig. 11).

4 Summary

The study has shown that in the case of used strains of bacteria resistant to high concentrations of mercury, not only measurable parameters as biomass growth and assimilation pigment content per gram of protein show its stimulating effect on plants. Visual assessment and condition of plants can allow to determine only minor in necrotic and chlorotic changes caused by mercury. Therefore, the significant reduction in toxic effects in *S. natans* by strains of *Pseudomonas,* with simultaneous increase by more than 20% of its accumulation capacity should be emphasized.

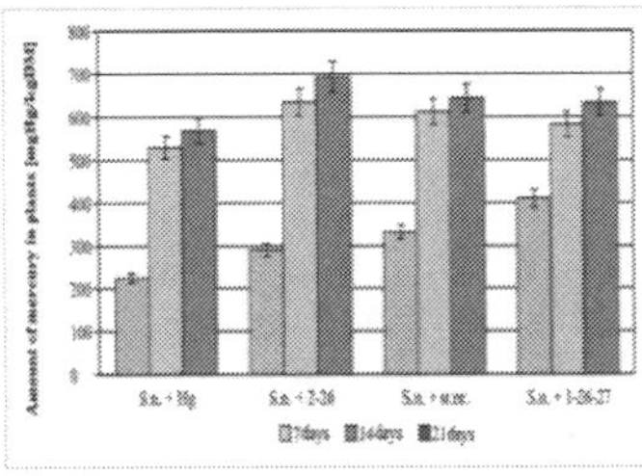

Fig.10. Amount of mercury in plants.

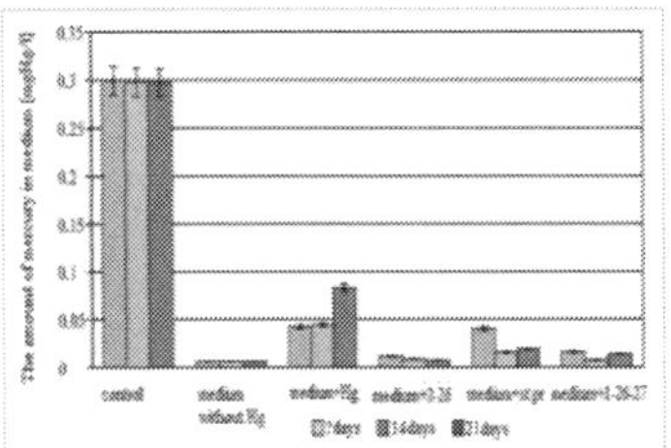

Fig. 11. The amount of mercury in Hoagland's medium.

In conclusion, it should be emphasized that, in the environment, especially among epiphytic microflora, bacterial strains with very specific characteristics can be found, the presence of which is difficult to explain with the influence of processes such as adaptation to high concentrations of anthropogenic pollution. Used *S. natans* was taken from the Oława river. A river, whose waters are used for the purposes of water supply for Wroclaw and in which the permissible concentrations of heavy metals are not exceeded. Nevertheless, among its epiphytic bacteria were *Pseudomonas* strains showing resistance to 0.6mgHg/l. The results indicate that in the use of stimulated phytoremediation processes of waters contaminated with mercury, it may be possible to find technology to remove that biologically very toxic element from water used for the purposes of water supply. Moreover, it seems that the relatively safe disposal of plant biomass contaminated by mercury can be achieved through drying and thermal utilization.

References

1. J. Koc-Jurczyk, Environmental Engineering **34**, 166–172 (2013) (in Polish)
2. S.Manahan, *Toxicology of the environment* (PWN, 2010) (in Polish)
3. M. Piontek, Z. Fedyczak, K. Łuszczyńska, H. Lechów, Environmental Engineering, Science Notebooks **155**, 35 (2014) (in Polish)
4. J. Sonke, L. Heimbürge, A. Dommergueb, Cr. Chim. Geo. **345**, 5–6 (2013)
5. G. Liu, Y.Cai, N. O'Driscoll, *Environmental chemistry and toxicology of mercury* (Wiley, John Wiley and sons, Inc., New Jersey, 2012)
6. R. Bernhof, J. Environ. Public Health **2012**, (2011)
7. S.Clemens, Biochimie **88**, 11 (2006)
8. V. Mishra, B.Tripathi, H.Kim, J. Hard. Mater. **172**, 2–3, 749–754 (2009)

9. H. Dash, S. Das, Int. Biodeter. Biodegr. **75**, 207–213 (2012)
10. S. Silver, L. Phung, J . Ind. Micr. Bio. **32**, 11–12, 587–605 (2005)
11. G. Garrity, G. Brenner, J. Don, R. Noel, T. James, Bergey's Manual of Systematic Bacteriology, **2** (2005)
12. T. Spilker, T. Coenye, P.Vandamme, J. Microbiol. **42**, 5, 2074–2079 (2004)
13. A. Grobelak, M. Kacprzak, M. Fijałkowski, J. Ecol. 14, 276–280 (2010)
14. B. Nowak, M. Grzegorczyk, M. Czaplicka, U. Zielonka, Environ. Prot. Eng. **39**, 1, 75–85 (2013)
15. H.Lowry, N. Rosebrough, A. Farr, R. Randall, JBC. **193**, 1, 265–275 (1951)
16. B. Nowak, M. Grzegorczyk, M. Czaplicka, U. Zielonka, Environ. Prot. Eng. **39**,1, 75–85 (2013)
17. P. Daniszewski, R. Konieczny, International Letters of Chemistry, Physics and Astronomy **8**, 279–287 (2013) (in Polish)

DOI: 10.1051/e3sconf/20172200048

Dynamic environmental control mechanisms for pneumatic foil constructions

Jan-Frederik Flor[1,*], *Yupeng* Wu[1], *Paolo* Beccarelli[1], and *John* Chilton[1]

[1]University of Nottingham, Department of Architecture and Built Environment, United Kingdom

Abstract. Membrane and foil structures have become over the last decades an attractive alternative to conventional materials and building systems with increasing implementation in different typologies and scale. The development of transparent, light, flexible and resistant materials like Ethylene Tetrafluoroethylene (ETFE) has triggered a rethinking of the building envelope in the building industry towards lightweight systems. ETFE foil cushions have proven to fulfil the design requirements in terms of structural efficiency and aesthetic values. But the strategies to satisfy increasing demands of energy efficiency and comfort conditions are still under development. The prediction and manipulation of the thermo-optical behaviour of ETFE foil cushion structures currently remain as one of the main challenges for designers and manufacturers. This paper reviews ongoing research regarding the control of the thermo-optical performance of ETFE cushion structures and highlights challenges and possible improvements. An overview of different dynamic and responsive environmental control mechanisms for multilayer foil constructions is provided and the state of the art in building application outlined by the discussion of case studies.

1 Introduction

The energy performance of the building envelope is becoming increasingly important since the building sector is held responsible for more than one-third of the worldwide energy consumption, according to the International Energy Agency. Dynamic environmental control mechanisms in building envelopes are a common strategy to balance between the reduction of solar heat gains and achieving good natural illumination while reducing energy consumption for cooling and artificial lighting at the same time. The integration and automatization of these mechanisms into complex fenestration systems and high-tech building facades has become a reality and opens possibilities for further optimisation of the energy performance and improvement of comfort conditions in the built environment. Responsive building envelopes can adapt to changing environmental conditions, or user demanded requirements and enhance the performance of the building. The developments in mechatronics, sensors and computer technology during the last decades, have made possible miniaturisation, mass production and successive cost reduction, which have

* Corresponding author: jan.flor@nottingham.ac.uk, janflor@gmx.de

allowed designers to integrate sensors and actuators into building elements making them truly responsive to the environment and occupant needs.

Pressurised multi-layered cushions of thin Ethylene Tetrafluoroethylene (ETFE) foil are increasingly applied in building envelopes and are now considered state of the art since the contributions of Frei Otto initiated the evolution of lightweight building over sixty years ago [1]. Lightweight and transparent materials play an important role in this development and ETFE is probably one of the most outstanding. ETFE has become a true alternative to glass and other transparent materials mainly because of its excellent material characteristics like low weight, flexibility, high translucency, tensile strength and good weathering performance. However, the thermal and optical behaviour of ETFE multilayer cushions is somehow still a topic of ongoing research, and the difficulty of predicting their energy performance has led in the past to reported problems of overheating of indoor spaces during summertime. Dynamic environmental control systems, including responsive shading mechanisms, could be a possible solution for the negative effects of overheating and glare, but despite their broad application in complex fenestration systems, they have not yet arrived at the same level of development in membrane and foil constructions. Advances in controlling the optical and thermal properties of the cushion system through, mainly, static adaptation measures are now common practice in the industry. Currently, the most widely used techniques are frit prints of reflective inks and combinations of different layer type compositions. However, these techniques are not sufficient during peak climatic conditions, and the integration of dynamic environmental control mechanisms into multilayer ETFE cushions remains a design objective alongside multiple approaches from the past. In search for technical solutions to respond to pressing environmental issues, new building techniques and materials are under constant development for more efficient and better performing building envelopes [2]. The following literature review is an attempt to provide an overview of the latest developments and state of the art in the building sector of membrane structures regarding the environmental control of multilayer foil constructions.

2 Building envelopes of ETFE

ETFE foil has been increasingly employed as a building material for building envelopes over the past decades [3]. Although used initially because of its high transparency only for temporary agricultural coverings and greenhouses, ETFE is now a recognised alternative for conventional transparent building materials such as glass [4]. The acceptance of the material among designers and the building industry can be attributed mainly to its excellent properties like transparency, flexibility, tensile strength and robustness towards weathering [5]. Since the first application of ETFE foil in a permanent building for the Mangrove Hall project in Burgers' Zoo, Arnheim, Netherlands in 1982 [6], important technological advances have been made. New form-finding methods and simulation software, as well as automated production and manufacturing processes together with a highly specialised design industry, have contributed to its widespread use in building envelopes. The absence of homogenised design codes and standardised testing methods remain as an obstacle, but ETFE coverings are now considered to be a reliable and safe solution for performance orientated building envelopes [7]. Recent iconic building projects like the Eden Project in St. Austell, UK (2000) (Fig. 1.), Allianz Arena in Munich Germany (2005) and the Water Cube swimming pool in Beijing China (2008) exemplify the outstanding performance and architectural attractiveness of ETFE as a building material [8, 9].

Fig. 1. Eden Project, St. Austell, United Kingdom (2000).

ETFE as a material for building envelopes is normally employed as an extruded thin foil in multilayer cushions which are pneumatically stabilised structural systems. The cushions can be composed of various layers of foil with air chambers in between. These foils are cut and welded together according to pre-calculated optimal shapes of equilibrium state. The foil cushion is clamped into a rigid frame system, typically of extruded aluminium profiles, inflated with air and maintained by means of a blower system at a predetermined pressure which enables the system to withstand external loads and maintain geometric stiffness. The number of layers of the cushion can vary in order to improve the structural or thermal resistance of the system. The foil thicknesses may also vary, according to the expected loads. Different tints or printed patterns can be applied, in response to the thermal and optical requirements [10].

Different building systems for ETFE foil are currently in use, but according to a case study from Moritz in 2007 [11], ETFE is most commonly employed in pressurised multilayer cushions systems (87%) and significantly less in mechanically pre-stressed single layer systems (10%), which however are getting used more often in recent years. According to the study ETFE is generally used in buildings related to occupancy for offices and housing (28%), swimming pools (18%), exhibition spaces and pavilions (15%). The building elements usually covered by ETFE are atria (35%), hall roofs (33%), canopies (14%) and building facades (8%) among others, ranging in a covered area between 100 and 10000 m^2. The predominant cushion geometry found in the analysed projects was reported as rectangular (60%) within a range of other hexagonal, triangular, circular or rhombic geometries. A comprehensive study by Schiemann [12] about the structural behaviour of ETFE foils proposed a classification of cushion systems according to the layer number and configuration which also is of interest to the understanding of the building physics regarding optical and thermal performance. Several experimental designs with different layer build ups and combinations of materials have been carried out so far [13, 14], including novel double layer ETFE panels which are pretensioned by heat shrinking and don't require an internal air pressure [15]. Nevertheless, few of these novel systems have been tested in any commercial construction so far.

Even though the majority of studies on ETFE have focused in the past on the structural efficiency of the building system, recently the interest in current research has been shifted increasingly towards the energy performance of ETFE envelopes.

3 Responsiveness in textile building envelopes

Active or responsive building envelopes are commonly referred to as shelter systems which can adapt autonomously, using sensors and actuators, to changing environmental conditions, or user and activity demanded requirements [16]. While the concept is not entirely new, as can be learned from several historical examples, like the Roman velarium or Venetian blinds, the idea of autonomously self-adapting buildings has emerged only recently [17], fulfilling the technological innovations envisioned by modernist scholars like Sigfried Ebeling almost a century ago [18]. The development of environment responsive membrane structures with automated mechanisms is indeed very recent and although several projects are reported in literature most of them are prototypes or of an experimental nature [19]. A schematic diagram of the functional relations of a responsive building envelope incorporating a dynamic shading mechanism within a multilayer ETFE cushion is shown below (Fig. 2).

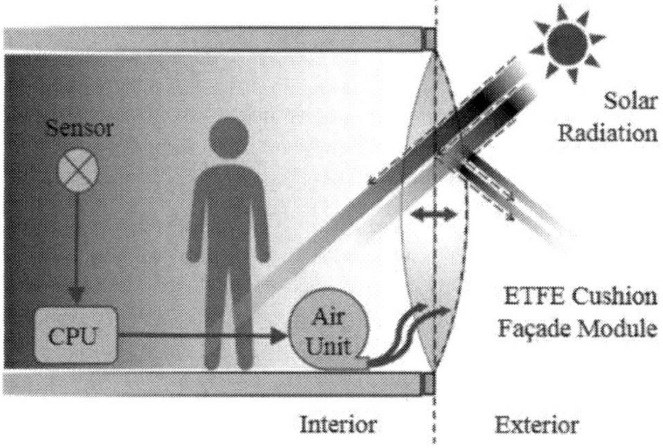

Fig. 2. Schematic diagram of a responsive building envelope incorporating a dynamic shading mechanism within a multilayer ETFE cushion.

4 Dynamic environmental control mechanisms for multilayer foil construction

True adaptability to indoor requirements and outdoor climate conditions is in this sense a relatively new approach as well for ETFE cushion systems considering the advances which have been made with other materials or building systems like, for example, retractable textile membrane structures, glass and complex fenestration systems. This could be owing to the fact that ETFE has been in use for only a relatively short period as a construction material but it is also related to its very distinctive physical characteristics which are still in the process of being investigated. A variety of different environmental control strategies have been developed for multilayer foil constructions (Fig. 3). But very few systems have been tested so far in commercial buildings.

The currently most widely used active mechanism for ETFE cushions is a triple layer system which employs positive and negative, reflective printing patterns on the middle layer and the inner or outer layer. These printed layers fit exactly when overlapped, covering a 100% of the cushion surface. The middle layer position can be moved by the control of the air pressure within the cushion chambers, making it geometrically overlap, either with the inner or outer layer. This allows adjustment to achieve a partial percentage of light translucence or total coverage, depending on the layer position (Fig. 4). This mechanism has been reported in several studies [3, 9, 10, 20–22] and was first employed in a building for the World Exposition pavilion "Duales System Deutschland (DSD) Cycle Bowl" in Hannover, Germany in 2000 [8, 9, 23]. The project comprised an 1800 m^2 active shading façade composed of a triple layer system with positive/negative frit patterns. In the same year another innovative building complex, the "Festo Technology Center", an office building with a 2655 m^2 roof structure of triple layer ETFE cushions was constructed in

 DOI: 10.1051/e3sconf/20172200048

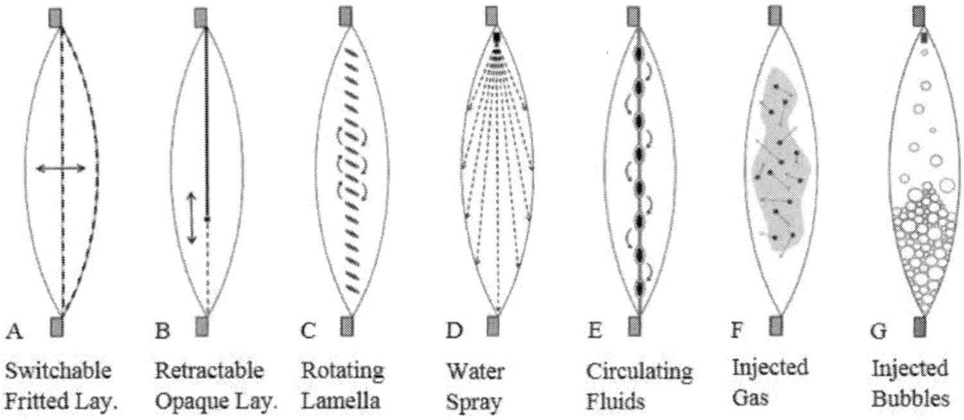

Fig. 3. Environmental control mechanisms for multilayer foil constructions.

Germany, and later in a scaled version for a branch project in France. A similar system as in the DSD project was employed, with the same mechanism of a switchable middle layer, but with a different frit pattern. In this case, the frit pattern was laid out as a squared checkerboard pattern. This project was followed by the "Kingsdale School" project in the UK in 2004, where a courtyard was covered with a dynamic ETFE roof which employed a rectangular frit pattern (Fig. 5). Several other projects incorporated the same or a similar mechanism in recent years and are listed below (Table 1). In addition to the mechanism mentioned above, very few alternative systems have been applied in commercial buildings. Retractable opaque middle layers have been employed in a 56 m² ETFE skylight of a gas-station in Munich, Germany in 2004 [24]. Venetian blinds have been incorporated into membrane envelopes only in a few built projects. One case is the Association for Medical Information (GFI) in Munich-Riem, Germany, built in 2003, which incorporates solar shading lamellas inside the air chamber of a triple layer cushion covering an area of 120 m² [25]. The Media-Tic building in Barcelona, Spain built in 2009 is an experimental building which integrated the above discussed dynamic triple layer shading mechanisms into the façade. In the same building, a novel shading mechanism based on gas injection was used for the first time, where the solar transmittance of the cushion was modified by pumping nitrogen into the air chambers of the ETFE cushions [9, 22].

Recently, there has been a spate of interest in the more unconventional mechanism, incorporating spray, liquids and bubbles as a strategy for dynamic environmental control mechanisms at an experimental stage. Mainini et al. investigated the influence of water spray on the thermal performance of a mechanically pre-stressed double layer ETFE panel as a strategy for reducing solar heat gains on building facades during summer climate conditions [26].

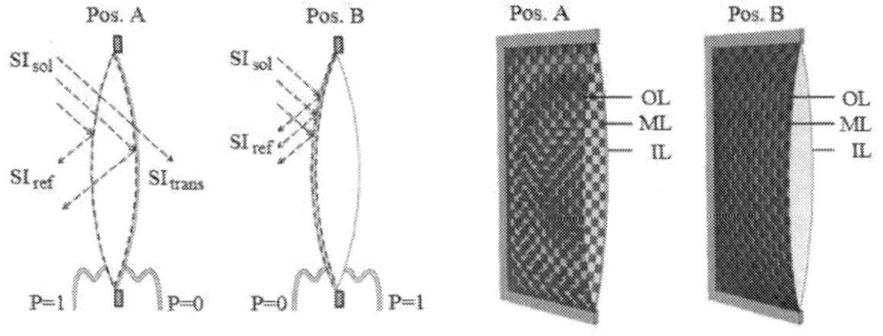

Fig. 4. Dynamic triple layer shading mechanism for pneumatic foil constructions, functional diagram and rendering of section in perspective view.

Fig. 5. Project Images: Duales System World Exposition Pavilion, Hannover Germany; Festo, Paris, France ; Kingsdale School, London, United Kingdom (left to right), source: © Vector Foiltec, http://www.vector-foiltec.com (accessed 30th April 2017).

Table 1. List of projects incorporating multilayer foil constructions with integrated dynamic environmental control mechanisms.

Project Name	Building	Application	Mechanism	Area m²	Year
Duales System World Exposition	Pavilion	Façade	A	1800	2000
Festo Technology Center	Office	Roof	A	2655	2000
Association for Medical Information (GFI)	Office	Roof	C	120	2003
Kingsdale School	School	Roof	A	5000	2004
Pasadena Art Center	Cultural	Skylight	A	690	2004
Allguth GmbH Service Center	Gasstation	Skylight	B	56	2004
The Mall Athens	Mall	Roof	A	-	2005
Archibald Vivian Hill Building	Research	Roof	A	547	2007
ZEP Leisure Park	Multiuse	Skylight	A	-	2009
JinSo Pavilion	Restaurant	Roof	A	600	2009
Lancaster University, ISS Building	University	Skylight	A	-	2009
Media-TIC Building	Office	Façade	F	2500	2010
St. Bartholomew's Hospital	Hospital	Roof	A	405	2013
RCS Pavilion EXPO2015	Pavilion	Skylight	A	72	2015
Urban Algae Folly EXPO2015	Canopy	Roof	A	-	2015
Mechanism Type (referring to Fig. 3): A – Switchable fritted Layer, B – Retractable Opaque Layer, C – Rotating Lamella, F – Injected Gas					

The effectiveness of different spray nozzles in relation to the thermal performance of the ETFE panel was tested. However, it was found that the system performance improvement was rather limited, as the water spray, while effectively reducing the surface temperature of the foil, had only a small influence on the mean radiant temperature. This was mainly attributed to the fact that water is a weak absorber of solar radiation. The effectiveness of the system was therefore strongly reliant on the volume of water sprayed on to the foil. Previously a study with a similar approach had been carried out [21]. The evaporative cooling effect of water on the middle layer of a triple layer foil construction was investigated. This study found great potential in improving the thermal behaviour of foil constructions with water. Numerical calculations and experiments under laboratory conditions showed a reduction of the inner layer surface temperature of up to 5°C. Other research projects have explored the possibility of achieving environmental control of multilayer foil constructions with tinted fluids [27]. The fluids are circulated in pre-formed channels in-between the foil layers of the cushion. The system is switchable from a transparent to a translucent state. This is achieved by changing the tint concentration of the circulating fluid. The system is able to modify the light transmission but also has the capacity to absorb and remove heat from the cushion. A general U-value of 0.64 W/(m²K) was calculated for the system. The optical performance was tested according to DIN EN 410 and showed a absorption of 95.7%, a reflection of 4.3% and a solar transmission of nearly 0% using a 7.5% black tinted fluid. A similar system was used for a showcase pavilion construction at the Milan EXPO 2015 [28]. But instead of tinted fluids, an algae substrate was pumped through the ETFE panels. The algae substrate would filter eventually the solar radiation and absorb part of it and provoke photosynthesis of the algae. The growth of algae enhances a biological self-regulating effect on the shading performance of the canopy structure. Another highly experimental mechanism was tested for double layer foil constructions in greenhouses where for thermal and solar control soap bubbles where injected into the air chamber [29]. This allowed the system to be switched from the initial transparent condition to a highly insulated and light scattering state. Prototypes have been built and the thermal performance of the system was investigated. The results of CFD simulations showed that liquid bubbles can improve the performance of greenhouses. The temporary increase of the thermal insulation with liquid bubbles reduces the energy demand for heating in winter and the risk of overheating during the summer season [30].

5 Conclusions

A variety of environmental control mechanisms for multilayer foil constructions has been outlined and their function and application in building envelopes has been discussed through a range of case studies. It can be concluded that besides the perceived fascination in the public opinion about adaptive building envelopes and energy saving façade systems the topic of the dynamic shading mechanisms for multilayer cushions appears to be neglected by systematic research studies. Even though mentioned in several studies, the building physic aspects of dynamic foil constructions have not been addressed thoroughly. Nevertheless, they are continuously applied in new construction beside the vagueness of understanding of their influence on the energy performance of buildings. There have been very few published reports directly addressing the problem of the thermal and optical behaviour of these systems and their effect on the energy performance of buildings remains as a research objective. This represents a risk and opportunity at the same time, which needs to be addressed and which bears great potential for improving the system performance of both, building and envelope. It is believed that a thorough understanding of the systems' optical and thermal behaviour is indispensable to predict and assess their effectiveness in controlling the climatic factors, enhancing occupant comfort conditions and improving the energy performance of the building. Literature provides only limited information on the effectiveness of the reviewed systems. An explicit description of their optical and thermal properties and the complex physical mechanisms of light and heat transfer are available only to a very limited extent.

This initial review can only be the first step towards a broader understanding of dynamic environmental control mechanisms for multilayer ETFE cushions. More data from monitoring of operating buildings is required as well as new approaches towards the simulation of their dynamic behaviour with numerical calculation methods. Great potential is seen in these tasks for further research.

 DOI: 10.1051/e3sconf/20172200048

Acknowledgements

This work was supported by the Faculty of Engineering at the University of Nottingham through a PhD studentship awarded to Jan-Frederik Flor.

6 References

1. F. Otto, *Das hängende Dach: Gestalt und Struktur* (1954)
2. G. Hausladen, *Climate skin: Concepts for building skins that can do more with less energy* (2007)
3. A. Gómez-González, J. Neila, J. Monjo-Carrio, Procedia Eng. **21**, 125 (2011)
4. T. Hanke, Bautechnik **90**, 239 (2013)
5. S. Tanno, *ICBEST '97 – Proceedings* (1997)
6. S. Robinson-Gayle, M. Kolokotroni, A. Cripps, S. Tanno, Constr. Build. Mater. **15**, 323 (2001)
7. J. Hu, W. Chen, B. Zhao, D. Yang, Constr. Build. Mater. **131**, 411 (2017)
8. A. LeCuyer, *ETFE. Technology and Design* (2008)
9. J. Chilton, Proceedings of the Institution of Civil Engineers – Construction Materials, Construct. Mater. **166**, 343 (2013)
10. J. Knippers, J. Cremers, M. Gabler, J. Lienhard, *Construction Manual for Polymers + Membranes* (2011)
11. K. Moritz, *ETFE-Folie als Tragelement* (2007)
12. L. Schiemann, *Tragverhalten von ETFE-Folien unter biaxialer Beanspruchung* (2009)
13. J.F.J. Max, G. Reisinger, T. Hofmann, J. Hinken , H.-J. Tantau, A. Ulbrich, S. Lambrecht, B. v. Elsner, U. Schurr, Energ. Build. **50**, 298 (2012)
14. M. Karwath, *Textile Composites and Inflatable Structures IV* (2011)
15. B.A.J. Martin, B. Lau, P. Beccarelli, J. Chilton, Y. Wu, *Proceedings of the International Association for Shell and Spatial Structures (IASS) Symposium - Future Visions* (2015)
16. A. Wyckmans, *Intelligent building envelopes* (2005)
17. M. Barozzi, J. Lienhard, A. Zanelli, C. Monticelli, Procedia Eng. **155**, 275 (2016)
18. S. Ebeling, *Der Raum als Membran* (2016)
19. D. Cardoso, D. Michaud, L. Sass, *Predicting the future: 25th eCAADe Conference Proceedings* (2007)
20. H. Poirazis, *Proceedings of Building Simulation* (2009)
21. F. Xie, *A novel clear foil cushion construction incorporating an additional water layer* (2011)
22. M. Juaristi, A. Monge-Barrio, *Advanced Building Skins* (2016)
23. B. Forster, M. Mollaert, *European design guide for tensile surface structures* (2004)
24. K. Moritz, T. Brengelmann, L. Schiemann, Detail **7/8**, 798 (2005)
25. H. Köster, *Entwicklung v. retroflekt. Oberflächenkonturen f. Licht lenkende Fassadenbauteile u. Entwicklung zugeordneter Herstellungsverfahren* (2003)
26. A.G. Mainini, A. Speroni, A. Zani, T. Poli, Procedia Eng. **155**, 352 (2016)

27. T. Hanke, H. Michael, M. Köhler, *ETFE-Solarthermie: schaltbare Verschattungen für Membrankissen aus Ethylen-Tetrafluorethylen-Hochleistungs-Copolymerfolie mit integrierter Solarthermienutzung* (2016)
28. http://www.ecologicstudio.com
29. http://www.solarbubblebuild.com
30. G. Gan, Build. Environ. **44**, 2486 (2009)

Closing of water circuits – a global benchmark on sustainable water management

Siegmund Fröhlich[1,*]

[1]Process Engineering @ Hochschule Emden/Leer, Constantiaplatz 4, D26723 Emden

Abstract. Access to clean water resources has always been a crucial factor in the history of mankind. Now, in the 21st century, water, as an increasingly scarce resource, will take a strategic role for the future development of global populations. As the former UN Secretary General Dr. Dr. Boutrous Boutrous Ghali predicts: "The wars of the 21st century will be fought not over oil, they will be fought over water." [1]. In nine global examples will be demonstrated the different ways of dealing with water resources. That are: Mexico City, Egypt, Libya, DOW Terneuzen, Los Angeles, Israel, China and Singapore and also global trends, such as, scarcity & rural exodus and salinization of soil. Thereby, he explains the different kinds of water management to be observed. The most relevant prognosis of the WHO is, that to the end of 21st century Africa's population will grow over proportionally from 1 billion now up to nearly 4 billion [9]. That is why all efforts need to be concentrated on helping Africa create a sustainable economic development. The first and by far most important strategic step is to assure access to clean water resources in the rural and mostly arid regions of the continent. The lecturer shows several technological proposals on how to overcame problems like: water scarcity, rural exodus, salinization of soil and others. Such technologies could be successfully implemented in sustainable development programs in African countries.

1 Global benchmark of sustainable water management

In nine global examples inclusind four typical phenomena it will be shown how different – depending on local and economic situation – the water management worldwide is.

1.1 Egypt as river oasis

Water management in Egypt has been exceptionally from the beginning. Only 4% of Egyptian soil are suitable for agricultural use. This includes the riverside and, especially, the Nile delta, which is highly cultivated and densely populated. Due to constantly increasing population growth many new channels and irrigated land areas are currently under construction.

* Corresponding author: procesconsult@t-online.de

 DOI: 10.1051/e3sconf/20172200049

Over centuries, the White Nile River has been transporting million tons of intensely fertile mud to nourish the adjacent cultivation areas. Also, volcano ash from highlands of Ethiopia, transported by the river, served as the best natural fertilizer for the Egyptian soil, and has assured the country's prosperity over thousands of years. Another positive side effect concerns the salt, which has been transported away from the fields by the yearly floodings [6].

This changed with the construction of the Aswan High Dam in 1971. The water capacity of the dam is: $\approx$ 130.000–170.000 km^3 water. It is worldwide the 7th biggest water reservoir which delivers $\approx$ 2.1 GW-electrical power – to provide the Egyptian industry with power. However, after the commissioning of the dam, only a small amount of natural fertilizer reaches the agricultural areas of Egypt.

That means the dam has both positive and negative effects on the industry and agricultural landscape. On the one hand, it prevents devastation by flooding and brings power for industrialization. On the other hand, the salt load of the conventional oasis cannot be washed out of the soil and consequently compromises the fertility of the fields. As such, Salinization of the soil currently poses the greatest threat to Egyptian agriculture producing million tons of non-fertile salty soil every year [6].

To overcome that, the lecturer is proposing the establishment of mobile soil washing plants in which the salt content could be extracted to reuse the purified soil. With a throughput of 125 m^3/h the plant can wash and purify up to 100 tons of desalinated soil per day, which can then be reintroduced to the agricultural cycle.

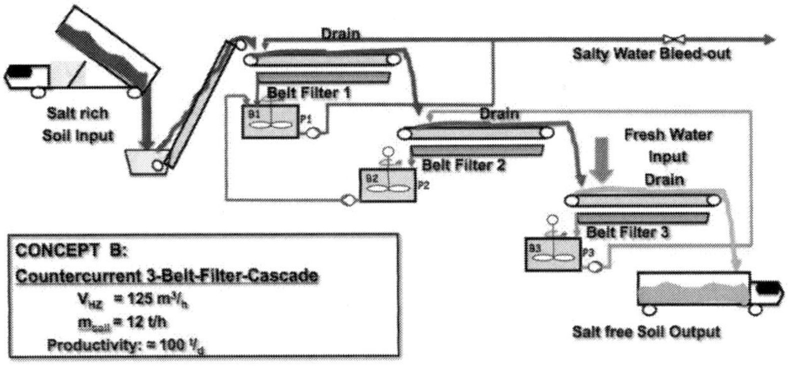

Fig. 1. Mobile Soil Washing Units for Egypt Salty Soil Remediation [7].

1.2 Ground water overexploitation in Mexico megacity

Until the conquista, the Spanish conquest of Mexico in 1519, the Aztec Empire was ruled by its capital city Tenotchtitlan. The population, approx. 100 000 inhabitants, was relatively big for a city of that time. The city was situated 2200 m above sea level within the holy Texcoco Lake – accessible only via five dams. Access to the city was strictly controlled and regulated. However, very unfortunate for the Aztecs, their ancient beliefs in myths and "white goddesses that will revisit the earth" misled them to giving the Spanish conquerors easy access to their holy city, who exploited it thereupon.

Tenochtitlan was devastated, the holy lake drained and the Hispanic colony town of Mexico City was erected in the same place. Today, Mexico is one of the worlds' so-called Mega Cities counting approx. 22 million inhabitants. Only 20% of drinking water can be acquired via rainfall from surrounding mountains. The remaining portion is acquired from contaminated ground water. Caused by the tremendous thirst of Mexico City more than 3000 wells were drilled to acquire ground water from a depth of up to 400 m. But the

DOI: 10.1051/e3sconf/20172200049

ground water is highly polluted, because the lion share of municipal and industrial wastewaters is not purified before being discharged into the ground water stock.

In addition, the ground water level constantly keeps declining, increasing the costs and difficulty to acquire water. However, during spring season heavy rainfall and flooding cause temporary collapses of the city's drinking water grid. It is outdated and extremely morbid which is why up to 1/3 of drinking water disappears in leakages. All in all, this means that each citizen "consummates" round about 300 litres of water daily. Only the small share of rich people living in Mexico City can afford to buy good drinking water [8].

Only 10% of Mexico City's wastewaters are treated in municipal wastewater treatment plants. Industrial wastewater is generally not treated at all. There is even no basic separation between "grey" and "black water" in order to reuse the lower polluted share of water. After just one single usage the tremendous portion of 170 m^3/sec (620.000 m^3 per hour) of Mexico City's wastewaters, the so called "aqua negra" – is directly discharged into ocean.

Several actions need to be started in order to overcome this situation:

1. It starts initially by raising awareness to change the situation through education in schools and industry.
2. The next step is to repair the old water grid, replace it in parts and extend it which is an essential step to lower the overall water consumption.
3. Then, new water retention and buffer reservoirs need to be established.
4. Furthermore, modern municipal wastewater plants need to be installed.
5. Finally, water recycling and multi usage systems need to be introduced to industry and municipal management authorities.

Although there are plenty of good proposals, the government has remained unwilling to change or even acknowledge the urgency of the situation. Mr. Lopez Meija, Biology Expert of the Environmental Dpt. of Mexico City is quite sceptic. To him, as long as living in the city is more attractive for the people than living on the countryside, no politician can counteract rural depopulation and tremendous growth of the Megacity Mexico [8].

1.3 Worldwide rural depopulation and megacities growth in Africa

The biggest challenge of the 21st Century lies in finding sustainable ways of how to manage water as the rare resource it is while considering the estimated global population growth by the year 2100.

Fig. 2. Global population shifts in the 21st century [9].

On the basis of data provided by [9]. The African continent faces the most dramatic demographic development of this century. According to this prognosis, the number of Africans now, will have quadrupled up to the year 2100.

As a consequence, migration from Africa to Europe will intensify dramatically. To plan ahead and avoid massive future migration flows, the European governments have to

DOI: 10.1051/e3sconf/20172200049

launche plenty of sustainable development aid programs in African countries to counteract the African emigration and rural depopulation.

In light of this prognosis, the lecturer proposes, that in costal regions with stabile wind conditions it would be very helpful to establish wind converters to provide the energy for water desalination plants. Those plants could be installed in movable containers for a quick installation, especially, in case of natural disasters and rising local demand [10].

At a daily production rate of approx. 1.000 m^3 of such a plant, a water demand for up to 20.000 people could be covered. In countries with good maintenance service e.g. reverse osmosis plants could be established to desalinate sea water. In countries with less maintenance level more robust vapour compression plants could be installed. For those robust technology maintenance service could be assured by local mechanical workshops. Such modular systems are not only suitable for villages and small cities but also for hotels und industrial purposes as well.

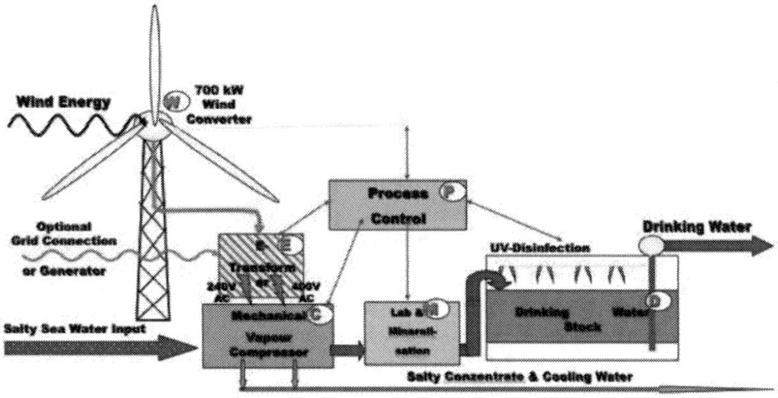

Fig. 3. Mobile Desalination Plants for Sustainable Development [10].

1.4 The Great Man Made River – GMMR-Project in Libya

Currently, almost every Libyan river faces water scarcity. This comes as no surprise, considering the fact that 95% of Libya's land is dominated by a barren desert landscape – allowing agriculture only in the coastal regions. The ground water stocks by the coast are, however, heavily contaminated with salt. Plenty solutions to solve Libya's problem with water scarcity were explored to satisfy the rising water demand of Libya.

Quite fascinating, that under the Sahara Desert there is an approx. 373.000 billion km^3 underground fossil water reservoir from the Ice Age. The so called Nubian Aquifer is the largest sweet water basin of the world. Following experts' prognosis, approximately 9.000 billion km^3 of that water deposit could be used. As UN-Centre for Environmental Development of the Arabic Region & Europe – CEDARE in Cairo calculate – Those waters are good for maximum of 4.860 years. [11]

In the aftermath of the discovery of the underground lake, the Great Man-Made River (GMMR)- Project was established. Mr. Abdelmagid El Gaood from the Libyan Ministry of Agriculture, is the general project chief who advised to build 467 water wells in the centre of Sahara which are currently accessing the reservoir and supplying 6 million m^3 of water per day. Some of CO_2-, Mn- und Fe-containing waters had to be purified in special water treatment plants. The concrete water pipes have to be resistant at 60 bar pressure.

To date, this is the most monumental water project on earth. The total project shall be finished by the year 2030 with estimated costs of approx. 27 billion €. According to Libyan

 DOI: 10.1051/e3sconf/20172200049

experts, the water amount is sufficient for the next 50 up to 250 years. By 25 €-Cent per m^3 is such mammoth project the cheapest one in drinking water production [11].

1.5 Water overexploitation by the city of Los Angeles

The City of Los Angeles has approx. 3,8 million inhabitants and it is the 2nd biggest in the USA. The wider LA region counts 17,8 million inhabitants. The LA metropolis is ranked on place 17th of all metropolis regions of the world by the number of inhabitants. LA consummates more than 50 % of the water resources of the total California State. Only a great irrigation grid network can supply the demand of such an immensely thirsty region.

Therefore, a gigantic wastewater treatment plant was built. The American Public Works Association (APWA) has chosen the so called "Hyperion Project" as one of the Top Ten Projects of the 20th century, which had the same ranking as Panama Chennel, Hoover Dam & the Golden Gate Bridge. [13]

The Biomass of the plant was utilized initially in a so called Biosolids-to-Fertilizer-Program as fertilizer for agriculture. The Environmental Protection Agency (EPA) stopped this procedure in 1980 because of the accumulation and increase of heavy metals in the soil. In 1998, the treatment plant was reconstructed and its capacity was adjusted by 800.000 m^3/d. Nowadays, the biosludge of the plant is converted into biogas. By means of such treatment the biosludge discharging could be minimized by 95%.

An exceptional method of water saving and reusing was developed here to satisfy the huge water demand Los Angeles. By infiltration of purified waste water into the ground, this water re-joins the ground water stock which is additionally filtered by the natural Volcanic porous rock. This special kind of rock contains Zeolites and Bentonite minerals, that additionally purify the water when it seeps through the porous rocks. The water quality is constantly monitored and controlled in monitoring wells. Later, in production wells, the purified water can be again retrieved and serve as drinking water.

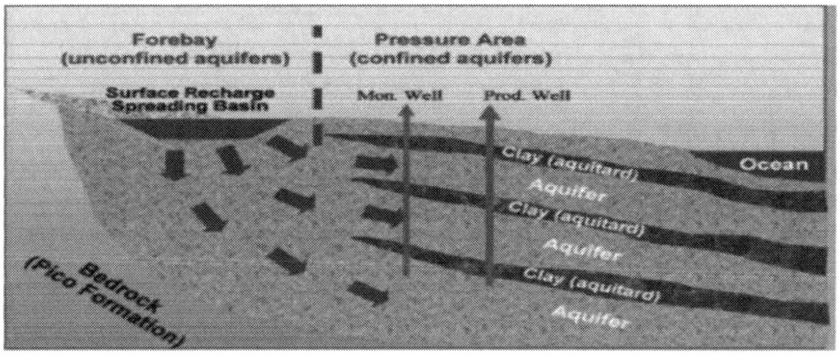

Fig. 4. Recovery & Reuse of Treated Wastewater in a Final Purification by Penetration of Volcanic Rocks [12]

The careful dealing with the water resource in California led to new markets and helped to increase the economic development of the region thanks to a number of activities:

1. The awareness and acceptance of the public to start such non-conventional ways of saving and reusing purified water.
2. Awareness raising in schools and public places.
3. Giving new incentives and impulses to municipal waste water treatment.
4. Installing efficient ground water pumps that help close water circuits to minimize water loss.
5. Finally, installing new desalination plants that further increase water safety [5].

 DOI: 10.1051/e3sconf/20172200049

1.6 Xiaolangdi dam as a pattern for new infrastructure projects in China

The rising energy demand of China can be covered only by a mid-term fossil energy sources policy. Gigantic infrastructure programs of hydro- and renewable energy power stations have a longer term for realisation. According to Yves Rannou, Alstom Chef in CHN, China's hydropower is already exceeding by 15 GW per year (2015).

The new extraordinary dam project shall help to solve several infrastructural problems:

1. The Yellow River sediments rises the river level by 10 cm a year. The Xiaolangdi Dam shall store the clay river sediments for the next 20 years.
2. 30 million tons of clay mud are discharged a year by the dam base outlets.
3. In the last 13 years, over 390 million tons of clay mud discharged.
4. The dam also prevents flooding

Fig. 5. Clay Mud Discharge at the Xiaolangdi Dam Base in China [15].

1.7 Sustainable water management of DOW Terneuzen & its region

DOW Chemical and the Biesbosch Waterboard Region have realised the most successful integrated Municipal & Industrial Wastewater Management in Europe. For the DOW Company it was essential to grant safe access to 22 million m^3/year to assure the prosperity of the chemical company.

DOW as a polymer and membrane producing company has made a strategic decision to utilize its own Reverse Osmosis membranes widely to cover its own great water demand.

In 2010, 35% of DOW's fresh water demand was covered by the Biesbosch Region. 65% had to be acquired by internal & external Reverse Osmosis water recycling action. By the year of 2015 – Thanks RO & Energy Recovery – a recycling ratio for fresh water of nearly 80% could be realized. The target of the year 2020 is – Thanks RO & Energy Recovery – a 100% recycling ratio for fresh water. This means that the company is on its way to become the best practice example in Sustainable Water Management in Europe.

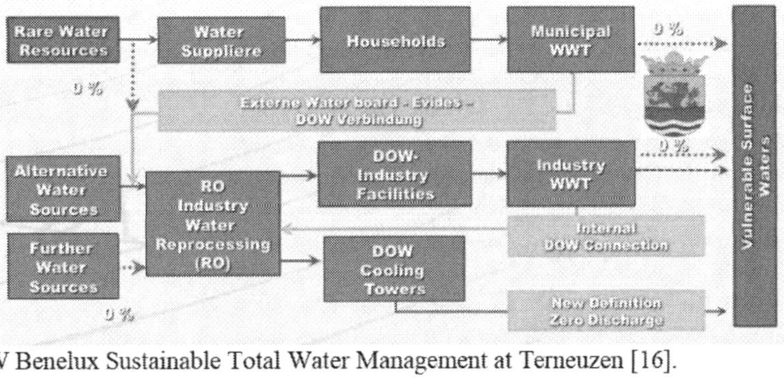

Fig. 6. DOW Benelux Sustainable Total Water Management at Terneuzen [16].

1.8 The Jordan River water resources in permanent dispute in Near East

Sharing such a very rare resource as Jordan water in that part of the earth is quite difficult. The demographic growth of the neighbour nations has led to a complete disappearing of the Jordan River at entering the Dead Sea since more than 20 years. Intensive agriculture have brought the Jordan River to dray up - The water level of the Dead Sea is permanently descending. The south part of the Dead Sea is drying out. There are even plans to construct a pipeline from the Red Sea - including a power station - to save the Dead Sea.
Therefore unconventional solutions in water usage had to be chosen:

Israel had to choose and to define new ways and politics to get reliable water resources:

1. Israel engineers were the technological pacemakers in terms of droplet irrigation.
2. Rain buffering and storage of rainwater resources in ground water reservoirs was essential to survive in that region.
3. Aquaponic – an inventory fish farming and vegetable production was born.
4. The recycling of purified municipal wastewater was successfully utilized similar as in California.
5. Finally three great sea water desalination Plants – utilizing energy recycling modules – have assured very safely the rising water demand of the country [14].

By those step by steps measures Israel is nowadays able to export „virtual water" in kind of agricultural products into the global market [17]. And the droplet irrigation, invented in Israel, is actually implemented in any successful agricultural farming around the Mediterranean Sea.

1.9 Global benchmark No. 1: total water management of the megacity Singapore

Only 40 years ago, today's agile metropolis city Singapore had totally insufficient water supply. Located on an island, without any lakes or water retention basins (to buffer the storm rainfall) nearby, almost all water had to be imported by the neighbouring country Malaysia. Nowadays, Singapore is the leading world nation, which could change his great disadvantages completely by an inventory water policy and technology [17].

That situation could be reached only:

1. By wide programs for awareness changes on water policy.
2. By extended education programs at schools and in the industry.
3. By setting of appropriate prices for water consumption.
4. By erection of new water retention- and buffer basins.
5. By exceptional unique total recovery & recycling of waste water.
6. By implementation of modern technologies for seawater desalination.

So the megacity Singapore is nowadays – thanks its excellent economic prosperity – the shiny and Best Practice Example of Sustainable Total Water Management in the world.

DOI: 10.1051/e3sconf/20172200049

2 Conclusions

Presented nine global examples demonstrate how different – depending on local and economic situation – the water management worldwide is. In rich countries there are all technological and economical solutions available to support the inhabitants with excellent quality and adequate portion of drinking water. In wide regions of Africa, Asia and South America the population suffers much of extremely intensifying scarcity.

But the biggest challenge of the 21st Century lies in finding sustainable ways of how to manage water as the rare resource it is while considering the estimated global population growth by the year 2050 and 2100. To plan ahead and avoid massive future migration flows – especially from Africa - the European governments has to launch plenty of sustainable development aid programs in African countries to counteract the African emigration and rural depopulation.

In the light of this prognosis, the lecturer proposes two possible technological solutions helping to minimize the global scarcity problems:

- Wind power driven water desalination plants and
- Mobile soil desalination plants,

which could be implemented in those development aid programs.

References

1. www.sourcewatch.org/index.php?title=Water_wars
2. K. Hancke, Wasseraufbereitung, Springer (2003), ISBN 978-3540068488
3. W. Lorch, *Handbook of Water Purification* (Ellis Horwood Books, 1998)
4. P. Gleick, Basic water requirements for human activities: Meeting basic needs Pacific Institute for studies in development, environment and security in Water International **21**, 83–92 (1996)
5. M. Elimelech, W.A. Phillip, Energy, Technology, and the Environment, Science, **333** (2011)
6. 3Sat-Mediathek: Dirk Steffens "Die Macht der Elemente #2 Wasser"
7. Siegmund R. Fröhlich: Own Proposal "Mobile Soil Washing Units"
8. www.daserste.de/information/politik-weltgeschehen/weltspiegel/videos/mexiko-wasser-fuer-reiche-jauche-fuer-arme-100.html
9. www.welt.de/wissenschaft/umwelt/article117275819/Menschen-haben-bald-nicht-mehr-genug-Nahrung.html
10. S.R. Fröhlich: Own Proposal "Smart Wind Powered Desalination Plants"
11. www.zeit.de/wissen/umwelt/2010-12/libyen-wasser
12. www.wrd.org/water_quality/recycled-water-groundwater-recharge.php
13. www.water-technology.net/projects/hyperion
14. Stover, R.L., Low Energy Consumption SWRO Energy Recovery, Inc., San Leandro, CA USA.
15. www.water-technology.net/projects/shanxi
16. /www.dow.com/benelux/
17. www.scientificamerican.com/article/israel-proves-the-desalination-era-is-here/
18. www.cscollege.gov.sg/Knowledge/ethos/Issue%202%20Apr%202007/Pages/Water-Management-in-Singapore.aspx

Seasonal coefficient of performance for ground source heat pump and groundwater one in Białystok

Andrzej Gajewski[1,*]

[1]Bialystok University of Technology, Faculty of Civil Engineering and Environmental Engineering, Department of HVAC Engineering, ul. Wiejska 45a, 15-351 Białystok, Poland

Abstract. European Economic Area (EEA) states declare to contain greenhouse gases emissions at 20% by 2020, whereas European Union (EU) does 40% before 2030, which result in encouragement to apply low-carbon technologies. Coefficient of Performance (*COP*) and Seasonal Coefficient of Performance ($SCOP_{net}$) are obtained using temperature measurement done by The Institute of Meteorology and Water Management – National Research Institute (IMGW-PIB) at the weather station in Bialystok for ten-year period. The first variant is ground source heat pump (GSHP) and the second one is groundwater source heat pump (WSHP) which can be equipped with separating heat exchanger (SHE) optionally. In both cases heat is generated for heating system only. Ground temperature is determined from Baggs (1983) formula using Oleśkowicz-Popiel et. al. (2002) adaptation to Polish climate and substituting the local constants achieved by Biernacka (2010). Water temperature in a groundwater basin is obtained from Kowalski (2007) equation. Estimation is done in each hour of heating season. All COP values are higher than 3.5 required by EU (2013). $SCOP_{net}$ are as follows: 6.12, 5.86, 5.03 for WSHP, WSHP+SHE, GSHP respectively. Insomuch as WSHP needs only two boreholes it is recommended to the areas beneath ones a groundwater basin is located.

1 Introduction

Continuous temperature measurement since half of 19th century led into conclusion about climate changes on the Earth. It is commonly assumed the reason of temperature increase is in greenhouse gases (methane, carbon dioxide, water vapour, ozone, nitrous oxide, halogenated hydrocarbons, etc.) emissions. There are either natural or anthropogenic sources of these gases. It is supposed that carbon dioxide contribution in temperature rise is between 9% and 26%. Moreover Keeling Curve indicates higher carbon dioxide concentration from 316 ppm in 1960 to 385 ppm in 2010 at Mauna Loa Observatory (Hawaii) [1]. These observations lead 27 countries to sign a protocol to limit greenhouse effect (the protocol was ratified by Poland on 2 December 2002). Inasmuch as the agreement was concluded in Kyoto, on 11 December 1997, it is known as Kyoto Protocol. After expiry date of Kyoto Protocol

* Corresponding author: a.gajewski@pb.edu.pl

which was on 31st December 2012, the states joined in European Economic Area agreed to limit greenhouse gases emissions by at least 20% of the 1990 levels before the year 2020. European Commission submitted a conception of 2030 climate and energy framework which includes targets for reducing greenhouse gases emissions by at least 40% before 2030 and encourages to apply low-carbon technologies [2].

The aim of the paper is estimation of seasonal coefficient of performance for ground source heat pump and groundwater source heat pump. Assessment is based on meteorological measurements conducted by The Institute of Meteorology and Water Management – National Research Institute (IMGW-PIB) at the weather station in Bialystok in the years 2003– 2012.

2 Computation

Computing is used to the indication of more efficient lower heat source for heat pump which supplies heat for commercial building with total design heat load Q_d = 92.73 kW which is located close to the northern border of Białystok. There are following input: external design temperature t_e = -22°C, internal one t_i = 20°C, design inflow temperature in heating system is at 55°C. Since, instant electric water heaters warm tap water to domestic heat water a heat pump unit supplies heat only to the radiators. Insomuch as the building is supposed to be constructed on a location which is situated about 100 m above major groundwater basin 218 [3] water to water heat pump is considered as the second heat generator for the building.

Estimation of seasonal coefficient of performance ($SCOP_{net}$) is based on external temperature measurements done by IMGW-PIB at the weather station in Białystok. The temperature data are taken from 1st January 2003 to 31st December 2012. Applying these data mean temperature in each hour in Julian year (that lasts 365,25 days i.e. 8766 hours) is calculated, the results are plotted in Fig. 1 as "external temperature". These mean values rounded to one decimal place are used as bin temperature t_j in further calculations.

The evaluation is done for every hour in heating season which lasts from 21st September to 10th May. If amid investigated hour h_j heat pump should supply heat current heat load $P_h(t_j)$ is determined:

$$P_h(t_j) = Q_d \frac{t_i - t_j}{t_i - t_e} \quad [kW] \tag{1}$$

Heating system stops whenever external temperature t_j is 15°C at least and then $P_h(t_j)$ = 0. Since flow temperature is adjusted by a weather compensator along a curve with exponent 0.6 the curve is approximated by quadratic equation. Temperature in the heating system changes along the curve but in cannot be less than 35°C to provide at least 10°C temperature difference between the radiator and the room. As a result we obtain heating system temperature dependence on external temperature:

$$\begin{cases} t_f = -0.0049t_j^2 - 0.5794t_j + 33.602 \quad [°C] & for\ t_j \le -2.5°C \\ t_f = 35°C & for\ t_j > -2.5°C \end{cases} \tag{2}$$

According to EN 14825:2016 $SCOP_{net}$ is calculated as follows:

$$SCOP_{net} = \frac{\sum_{j=1}^{n} h_j \left[P_h(t_j) - \sup(t_j) \right]}{\sum_{j=1}^{n} h_j \left[\frac{P_h(t_j) - \sup(t_j)}{COP_{bin}(t_j)} \right]} \quad [-] \tag{3}$$

where:
sup(t_j) is the required capacity of an fossil fuel backup heater for corresponding temperature t_j [kW],
$COP_{bin}(t_j)$ means coefficient of performance at bin temperature t_j [-].

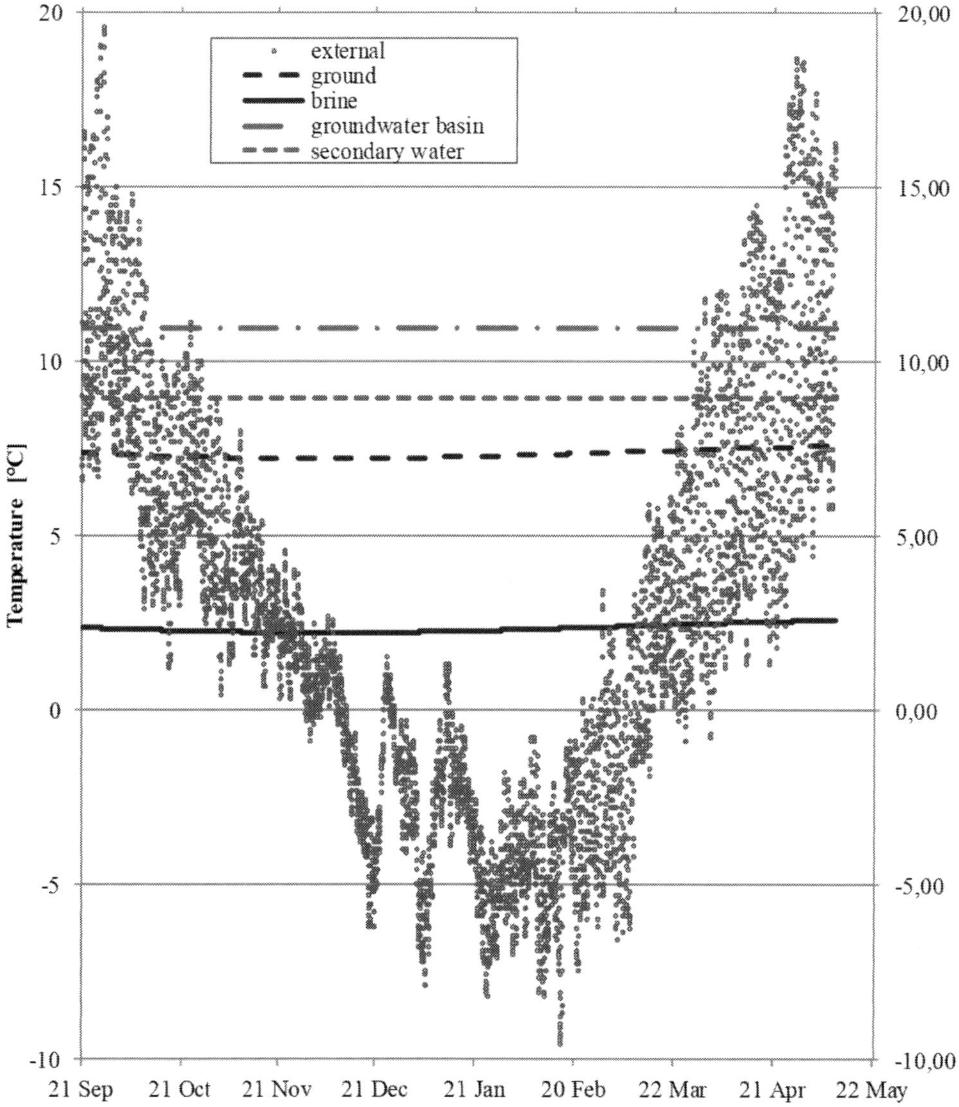

Fig. 1. Hourly temperature changes along average heating season: mean external temperature, ground and brine temperature for GSHP calculations, groundwater basin temperature for WSHP calculations and secondary water temperature for WSHP+SHE calculations.

Inasmuch as heat pump unit is greater than 50 kW it must be equipped with buffer tanks with capacity $V = 2000$ dm^3, thus heat pump unit warms water in buffer tanks to temperature obtained from heating curve (2). Heat pump works when temperature in buffer tanks t_t is lower at least 4°C than t_f, which prevents to turn on heat pump for short time period. Hence, the procedure of buffer tanks performance and ones' efficiency of energy storage $\eta_t = 98\%$ is included into analysis. To proper estimate buffer tank charging and discharging processes amid heating season the calculations are done as follows. Thus ones differ from EN 14825:2016.

Heat losses of buffer tanks Q_j are obtained from formula

$$Q_j = \left[i_{j-1} - i(20°C) \right] \rho_{j-1} V (1 - \eta_t) \quad [kJ] \tag{4}$$

where:
i_{j-1} – specific enthalpy of water in buffer tanks in previous hour [kJ/kg],
$i(20°C)$ – specific enthalpy of water in buffer tanks at room temperature [kJ/kg],
ρ_{j-1} – water density at temperature in previous hour [kg/m^3].

Then, we can establish temperature in buffer tanks in given hour t_{tj}

$$t_{t\,j} = t_{t\,j-1} - \frac{P_h(t_j) \cdot 3600 + Q_j}{\rho_{j-1} V c_p} \quad [°C] \tag{5}$$

where:
t_{tj-1} – water temperature in buffer tanks in previous hour [°C],
c_p – water specific heat [kJ/(kg·deg)].

Since 29th February occurs once per four years length of h_j is equal to one quarter hour on this particular day, in other days duration of h_j equals 1 hour. Forasmuch as there is not any backup heater formula (3) reduces to

$$SCOP_{net} = \frac{\sum_{j=1}^{5592} h_j P_h(t_j)}{\sum_{j=1}^{5592} \frac{h_j P_h(t_j)}{COP_{bin}(t_j)}} \quad [-] \tag{6}$$

where 5592 is number of hours in a heating season containing a leap year, whereas $COP_{bin}(t_j)$ is linear interpolated from manufacturer data [4].

2.1 The first variant

Brine to water heat pump Vitocal 300-G Pro type BW 302.B120 was selected as heat generator in the first option. Its the most important technical data are written down in Table 1.

To determine ground temperature Baggs [5] formula achieved for Australia continent was used. Therefore, the formula must have been adopted to Polish climatic conditions, which was done by Oleśkowicz-Popiel et. al [6]. Eventually, temperature profile in a ground is a function of depth z and time τ:

$$t_g(z,\tau) = (t_m + \Delta t_m) - 1.07 k_v A_s \exp\left(-0.00031552 z a^{-0.5}\right) \cdot \\ \cdot \cos\left[\frac{2\pi}{365}\left(\tau - \tau_o - 0.018335 z a^{-0.5}\right)\right] \tag{7}$$

The local constants in the formula (7) were achieved by Biernacka [7] and they are presented in Table 2.

Temperature difference between ground and brine is based on a use experience in the location and it is equal to 5 deg. Since, ground temperature changes along borehole and it is dependent on time its mean value from ground level to depth h=100 m at each hour in heating season is obtained from formula:

 DOI: 10.1051/e3sconf/20172200050

$$\bar{t}_g(\tau)=\frac{\int_0^h t_g(z,\tau)dz}{h} \qquad (8)$$

where $t_g(z,\tau)$ results from formula (7). The results of Eq. (8) are shown in Fig. 1 as "ground temperature". Hence, brine temperature, which is temperature of lower heat source, is 5 deg less than $\bar{t}_g(\tau)$ and it is presented in Fig. 1 as "brine temperature".

Table 1. Technical data of sized ground and water heat pump units, an extract from technical guide [4].

		Vitocal 300-G Pro type BW 302.B120	**Vitocal 300-W Pro type WW 302.B125**
Output data to EN 14511		B0/W35, 5 K spread	W10/W35, 5 K spread
Rated heating output	kW	117.2	112.1
Power consumption	kW	24.4	18.3
Coefficient of performance (COP)	-	4.8	6.1
Fluid in primary circuit		brine	water

Table 2. The values of the constants in Baggs formula (7) at Białystok, cf. Biernacka [7].

difference between ground temperature below shallow zone and average annual air temperature	Δt_m	[deg]	2.2
vegetation coefficient	k_v	[-]	0.85
the amplitude of the annual air temperature	A_s	[deg]	12.1
the soil thermal diffusivity	a	[m²/s]	6.00E-07
the phase shift of the air temperature wave	τ_0	[d]	22
average annual air temperature	t_m	[°C]	7.407659

2.2 The second proposition

One unit of Vitocal 300-W Pro WW 302.B125 is sized as alternative choice. Its the most important information is in Table 1. Just as previous one it must supply heat through a buffer tank and Kowalski [8] formula is applied to determine groundwater basin temperature

$$t(z)=t_m+A+g_g(z-h)\quad [^\circ C] \qquad (9)$$

where all the quantities are specified in Table 3. Since temperature below shallow zone is supposed to be constant all the year, achieved water temperature in groundwater basin is equal to 10.98°C and it is marked as "groundwater basin temperature" in Fig. 1.

Table 3. Data to calculation groundwater basin temperature.

depth of shallow zone	h	[m]	5
depth of the groundwater basin	z	[m]	100
correction factor due to true altitude	A	[-]	0.856
geothermal gradient	g_g	[deg/m]	1/35

Forasmuch, we do not know in advance whether groundwater ingredients satisfy manufacturer's chemical requirements we consider a variant with separating heat exchanger (SHE) which would be built in the primary circuit in negative case. In that case water temperature of heat source in evaporator will be lower at 2°C according to manufacturer data [4] (cf. "secondary water temperature" in Fig. 1).

3 Results

Coefficient of performance values in each hour during heating season are plotted in Fig. 2. $SCOP_{net}$ values resulting from Eq. (6) are written down in Fig. 2 as the labels.

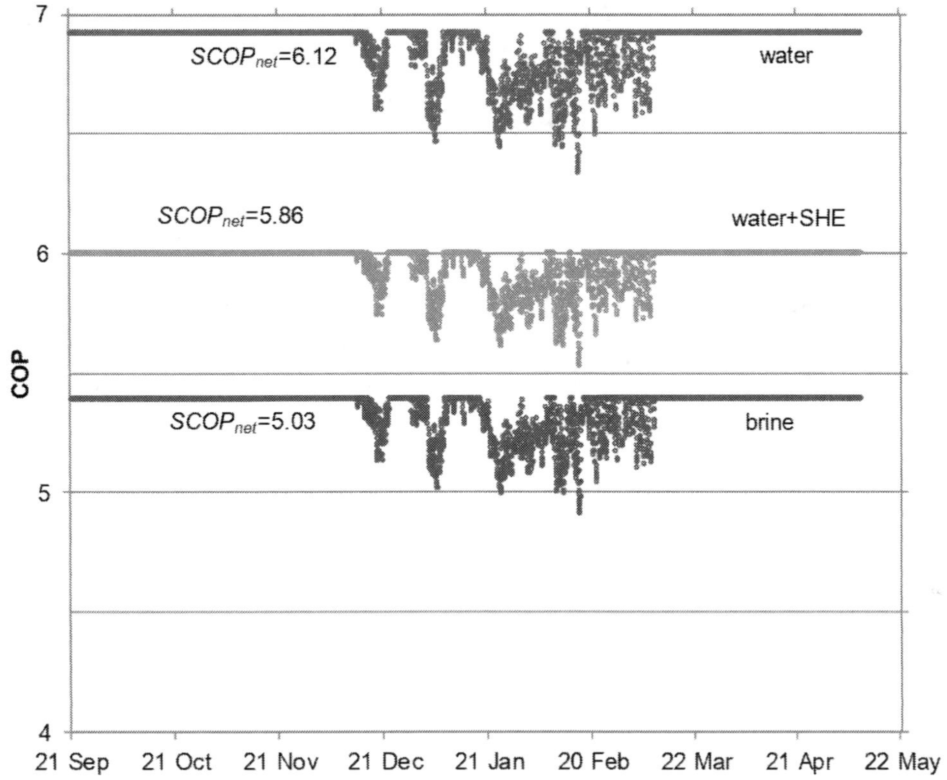

Fig. 2. COP_{bin} interpolated from manufacturer data [4] and $SCOP_{net}$ obtained from Eq. (6) for three considered variants.

During heating season heat losses are function of external temperature because other factors are exactly or more less unchangeable, thus lower outside temperature causes bigger losses. Moreover lower outdoor temperature results in higher temperature in heating system. All these factors eventuate in lower values of bit *COP* and as a consequence bring about less value of $SCOP_{net}$. Heat pump is profitable when primary energy used during electricity production is less than heat transferred to a heating system. Taking into account product of the efficiencies of both power generation and electrical grid in Poland heat pump must achieve COP = 3.5 at least (cf. Rubik [9], European Commission [10], and paper [11]). In all the investigated cases the requirement is fulfilled.

Higher *COP* values are in the case of water to water heat pump, regardless of possessing SHE.

4 Concluding remarks

The major share in heat pump investment expenditures belongs to borehole drilling. If we assume specific heat extraction rate at 50 W/m [4] we will need 19 U-tube ground heat exchangers at 100 m depth each. In the case of water heat pump despite of their rated heating output only two boreholes are needed the first bore for delivery well and the second one for return well. Hence, less interference with the environment is necessary to achieve the goal. Moreover, the fixed costs can be much lower when the groundwater source heat pump is applied. Furthermore, groundwater source heat pump is more effective as it results from the above computations. In conclusion, if there is a groundwater basin under assumed location water source heat pump should be considered as primary variant of heat generator for heating system in the buildings with design heat losses at about 100 kW.

Meteorological data have been available thanks to a kindness of The Institute of Meteorology and Water Management - National Research Institute (IMGW-PIB). The scientific research were financed by Bialystok University of Technology as Dean's project at Department of HVAC Engineering S/WBiIŚ/4/2017.

References

1. *Atmospheric CO2 from Continuous Air Samples at Mauna Loa Observatory, Hawaii, U.S.A.* http://cdiac.ornl.gov/trends/co2/graphics/SIOMLOINSITUTHRU2008.JPG [12.05.2016]
2. European Council of the European Union, http://www.consilium.europa.eu/en/policies/climate-change/reform-eu-ets/ [23.05.2017]
3. *The groundwater of the voivodeship capitals in Poland* (Państwowy Instytut Geologiczny, Warszawa, 2007) (in Polish)
4. *Viessmann, VITOCAL brine/water and water/water heat pumps, single and two-stage, 89 to 290 kW, Technical guide*, May 2013
5. S.A. Baggs, Sol. Energy **30**, 4, 351–366 (1983)
6. C. Oleśkowicz-Popiel, J. Wojtkowiak, I. Prętka, FCEE **1**, 2, 151–164 (2002)
7. B. Biernacka, Civil and Environmental Engineering **1**, 5–9 (2010) (in Polish)
8. J. Kowalski, *Hydrology with the basics of geology* (UWP, Wrocław, 2007) (in Polish)
9. M. Rubik, COW **6**, 3–6 (2008) (in Polish).
10. 2013/114/EU: Commission Decision of 1 March 2013 establishing the guidelines for Member States on calculating renewable energy from heat pumps from different heat pump technologies pursuant to Article 5 of Directive 2009/28/EC of the European Parliament and of the Council
11. A. Gajewski, J. Siergiejuk, K. Szulborski, Energy Build. **65**, 197–204 (2013)

Carbon and energy footprint of electrochemical vinegar wastewater treatment

Emine Esra Gerek[1,*], *Seval* Yılmaz[1], *A. Savaş* Koparal[1], and *Ömer Nezih* Gerek[1]

[1]Anadolu University, Faculty of Engineering, 26555 Eskişehir, Turkey

Abstract. Electrochemical treatment of wastewaters that are rich in organic compounds is a popular method, due to its acidic nature that avoids biological treatment. In many cases, the pollution hazard is considered as the chemical oxygen demand (COD) from active carbon, and the success of the treatment is measured in terms of how much this specific parameter is reduced. However, if electricity is used during the treatment process, the treatment "itself" has manufacturing and operational energy costs. Many of the studies consider energy utilization as a monetary cost, and try to reduce its amount. However, the energy cost of the treatment also causes emission of carbon at the energy producing side of the closed loop. This carbon emission can be converted into oxygen demand, too. Therefore, it can be argued that one must look for the total optimal carbon efficiency (or oxygen demand), while reducing the COD. We chose a highly acidic wastewater case of vinegar production, which is a popular food product in Turkey, to demonstrate the high energy consumption and carbon emission problem of the electrochemical treatment approach. A novel strategy is presented to monitor total oxygen demand simultaneously at the treatment and energy production sides. Necessity of renewable energy utilization and conditions on process termination points are discussed.

1 Introduction

Treatment of wastewater is applied for eliminating biological or chemical hazardous materials. Toxicity or poisonous behaviors of contaminants obviously need careful treatment. There is a wide literature regarding various types of treatment methods that target various contaminants inside different media. A thorough classification of treatment strategies for different targets is presented in [11]. From this categorization and due to various other reports, it can be seen that electrochemical treatment is a popular method that is used particularly for food production wastewaters [1–4]. The pollution sometimes requires careful treatment of a particularly harmful material. However, in many cases, the hazard is considered as the chemical oxygen demand (COD) from active carbon inside the raw wastewater. Electrochemical process, therefore, targets the oxygen demanding active molecules inside the aqueous environment. Elimination of the oxygen demand from acidic environments, such as the vinegar wastewater, prohibits a successful utilization of

* Corresponding author: eegerek@anadolu.edu.tr

biological treatment tools. Similarly, various membrane structures sometimes quickly spoil in the acidic environment prior to a successful treatment. These observations leave the electrochemical approach as a reasonable alternative [1].

Unfortunately, the treatment process "itself" has manufacturing and operational energy costs. Electrical energy utilization during COD removal (via electrochemical treatment methods) causes emission of carbon at the energy producing side of the closed loop. Therefore, it can be argued that one must look for the total and optimal carbon efficiency, which reduces the COD of the wastewater up to such an extent that the emitted carbon (or consumed oxygen) due to the spent energy remains "worth" the treatment.

The energy efficiency regarding electrochemical treatment of food industry wastewaters is a relatively known issue [2–4]. Besides, several researchers have attacked the problem of energy efficiency for different wastewater types [5–7]. A distinct novelty of the proposed research here (as opposed to the parametric optimization approaches of the described literature survey) is the assessment of the overall electrochemical treatment process with a common denominator of oxygen demand (or carbon emission, thereof). For example, in [2], the researchers have investigated energy consumption in electrochemical treatment of sugar, fruit juice and whey production wastewater. They have concluded various electrical energy consumption levels at the end of experiments with several settings, including applied voltage, electrolyte concentration, wastewater concentration, etc., in terms of kWh energy required per COD in mass. Unfortunately, the oxygen demand equivalent of the utilized kWh electric energy was not evaluated and an overall COD was not proposed. In [3], an energy consumption estimation was performed and in [4], both the consumed energy and the reduction in COD was monitored as a function of time. Again, the consumed energy was not set at an equivalent oxygen demand figure to assess the overall efficiency. It was observed that the oxygen demand equivalent of the consumed electrical energy was not stated as an issue in other types of electrochemical wastewater treatment applications, either [5–7]. In [6, 7], very detailed analyses regarding the monetary costs of the plant installations and treatment runs were achieved for various cost parameters, including the electrical energy. But the oxygen demand cost of the generated electricity was not mentioned. In a recent study by the authors [8], the oxygen equivalent of the utilized electrical energy was briefly examined for the case of tannery wastewaters, which also contain a poisonous contaminant of chromium.

In this particular study, the proposed "oxygen equivalent" assessment strategy is researched through the electrochemical treatment of vinegar production wastewater, which is known to have a highly acidic nature. Vinegar is a popular product in Turkey, and due to the acidity of its waste, electrochemical processes are the most preferred and successful treatment methods [9, 10]. In this work, parallel plate iron and iridium electrodes were selected as two alternatives for comparison. The treatment experiments were repeated several times at different electrical current densities (mA/cm^2) and at various initial pH settings for the batch processes. Both the reduction in COD and the consumed electrical energy were monitored and recorded (as explained in the Materials and Methods section).

Unlike conventional papers which report the energy consumption in terms of the utilized kWh of electrical energy [2–7], this work converts the energy to the production level oxygen demand. The conversion is performed via averaging electric production plants throughout the country in terms of their CO_2 emissions, and then by calculating the oxygen part of the required emission from the atomic mass ratios. For this part, the average CO_2 emissions of electric generation plants are researched and it was observed that, as of 2012, the average CO_2 emission is 0.534 kg per kWh in Turkey, and 0.600 kg per kWh in the USA [12]. These values correspond to only averaged grid values and in case the plant uses local and renewable energy sources, the following analysis gets invalid.

Once the energy is converted to the oxygen demand, the COD reduction (at the treatment plant side) and oxygen demand (at the energy generation side) may be combined as an "overall oxygen demand" on the same time axis, which would be considered as a novel contribution of this article. Since one component of the curve decreases (due to treatment) and the other component linearly increases (due to power generation), their combined value is expected to reach to a "minimum" oxygen case at a certain point. It can be, then, argued that the time point of the described minimum value essentially corresponds to the real efficient point of the closed loop system. Even though the treatment process may reach to lower COD levels for longer treatment times, the marginal COD gain would become not worth the additional oxygen demand requirement due to used energy.

The above described situation is exemplified on a high energy-consuming electrochemical treatment case of vinegar production wastewater treatment. Effects of varying the electrode types and initial pH values are also investigated and reported.

2 Materials and Methods

2.1 Sample Acquisition and Experimental Set-Ups

Electrochemical treatment experiments were conducted over real Vinegar Production Plant wastewater samples that were obtained from a large factory in Eskişehir. The samples were collected in polypropylene bottles and stored in a refrigerator at 277K until experiments (on par with ISO 5667-1, 2007; EN ISO 5667-1/AC, 2007; ISO 5667-3:2012; ISO 5667-10: 2002 standards). The wastewater samples were measured to have a COD value of about 11000 mg/L, a conductivity of close to 1.80 mS/cm and an initial pH value near to 4. Due to its particular restrictions, the initial COD values were diluted to around 2000 mg/L for the iridium coated electrode utilization.

The COD determination was achieved using prepared potassium di-chromate ($K_2Cr_2O_7$), acidic silver-sulfate solution, sulfuric acid solution (H_2SO_4), and iron aluminum sulfate solution [$(NH_4)_2Fe(SO_4)_2.6H_2O$] as a titrant. In order to improve conductivity, the support electrolyte of sodium sulfate (Na_2SO_4, Merck) was also used. The pH value was adjusted using sodium hydroxide (NaOH, Merck).

Apart from the constructed reactor cabin, the experiments also required the utilization of an electrical power supply (STATRON, type 3262), a multimeter (Fluke 26-III model), stirrer (IDL, Hotplate Stirrer), peristaltic pump (Materflex L/S, model 7518-00), analytic weight scale (OHAUS, Explorer Pro), pH- and conductance-meter (Thermo Scientific Orion STAR A215), centrifugator (Nüve NF 800 R) and thermos-reactor (Merck Spectroquant TR 420.

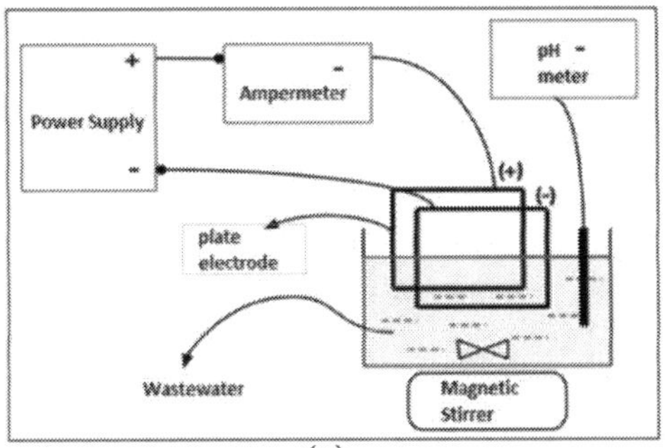

(a)

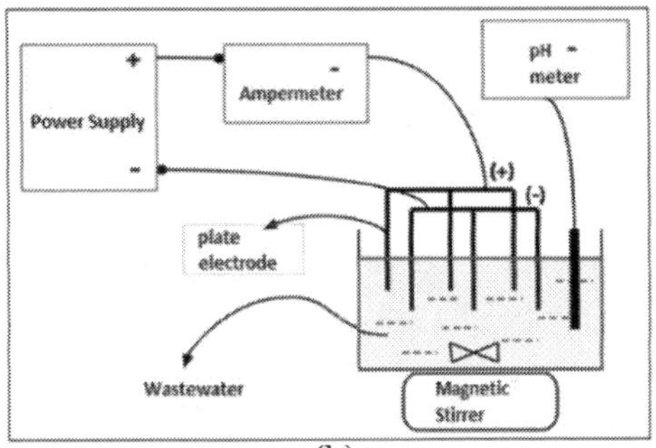

(b)

Fig. 1. Parallel plate electrochemical treatment set up illustrations for (a) iridium coated and (b) iron plate electrodes.

Two types of electrodes were used: iron and iridium, both in the parallel-plate form. The iron electrode structure consisted of 6 plates that provides a surface area of 100 cm^2. Iridium coated plate version had two plates, making a smaller surface area. In order to reduce plate worn-out and oxidation on the surfaces, occasional polarity switching was applied. The resulting reactor for iridium electrodes and iron electrodes are briefly illustrated in Figures 1 and 2, respectively.

2.2 Electrochemistry

Corresponding to the type of the electrode, two different electrochemical reactions occur in the treatment tank. Electro-coagulation happens when iron electrodes are used according to the following reactions.

At the anode side:

$$4Fe \rightarrow 4Fe^{+2} + 8e^- \quad (1)$$

$$4Fe^{+2} + 10H_2O + O_2 \rightarrow 4Fe(OH)_3 + 8H^+ \quad (2)$$

At the cathode side:

$$8H^+ + 8e^- \rightarrow 4H_2 \quad (3)$$

making an overall eqation of :

$$4Fe + 10H_2O + O_2 \rightarrow 4Fe(OH)_3 + 4H_2 \quad (4)$$

Detailed explanations of the electro-coagulation process can be found in [13, 14].

When the iridium coated electrodes are used, an oxidation process occurs. The equations for this process is same for various metals including Ti, Pt, Ru, Pb, Sn, etc. Using a common metal oxide notation: MO_x, the equations start by forming oxidization of organic matter and producing hydroxyl radicals:

$$MO_x + H_2O \rightarrow MO_x(OH^-) + H^+ + e^- \quad (5)$$

The hydroxyl radical is absorbed on the electrode surface and organic matter decays:

$$R + MO_x(OH)_z \rightarrow (z/2)O_2 + zH^+ + 2e^- + MO_x \quad (6)$$

Again, details of the process can be reached from [14–17].

Performance differences of these two reactions will be illustrated and discussed in the experimental results section.

2.3 Electrical energy production equivalent oxygen demand

In order to convert the consumed energy (kWh) into the demanded oxygen amount for the energy generation process, energy generation profile of the country must be investigated. In general, a combination of various electric power sources contributes to the overall electric supply. Each electric plant has its own CO_2 emission per-kWh. The overall CO_2 emission can be evaluated using the available ratios of each generator plant type [18]. Even for non-fossil fuel based electric generation systems (such as hydro-electric plants, nuclear plants and renewable energy plants), an equivalent emission rate has been evaluated [19, 20]. The most current CO_2 emission averages per-kWh can be obtained from public reports of EIA, WEA or IEA [21–23]. A recent electrical energy source distribution study regarding the situation in Turkey can be found in [12]. According to these most recent studies (2012), the shares of each generator type and their contribution amounts are given in Table 1. These rates are not expected to wildly change within the near future. Taking all these energy contributors into account and from the efficiencies of each source, the equivalent CO_2 emission per kWh can be obtained as 0.534 kg. As a comparison, the same CO_2 emission in USA is around 0.600 kg/kWh (according to EIA reports).

Table 1. Distribution of Turkey's electric energy generation in 2012.

Source	Amount (MW)	Ratio
Hydroelectric	19620	34
Natural Gas	17164	30
Coal	12395	22
Wind	2261	4
Other	5902	10

Using the molecular mass formula, it can be easily seen that 32/44 of the CO_2 mass corresponds to oxygen. By calculation:

$$N = 0.534 \times 32/44 = 0.388\ kg/kWh = 388\ g/kWh \text{ of oxygen} \quad (10)$$

in mass per consumed energy for the treatment (in Turkey). This rate can be compared with the values of energy consumption that we have obtained in our experiments.

3 Experimental results

The first set of experiments were conducted using an iron electrode as described above. The first comparison was for the effect of electrical current densities at the original (raw) pH value of 4.11. A detailed plot is provided in Figure 2, where the COD values at the treatment side is drawn blue and the equivalent oxygen demand of the electrical energy production is drawn red. It can be seen that their sum (shown in green) does not necessarily decrease in time.

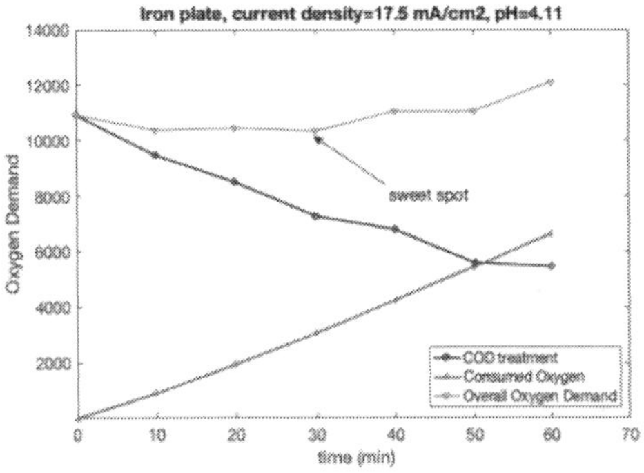

Fig. 2. Oxygen demands on the treatment (blue) and energy generation (red) sides with a total oxygen demand (green) plot using iron plate electrodes, 17.5 mA/cm^2 current density and pH = 4.11.

The overall increase starts at 30 min and any further process causes an overall oxygen demand of higher amounts. For a compact comparison of the eventual situation, the electrode configuration is tested further at 20 mA/cm^2 and 22.5 mA/cm^2 at the same pH of 4.11 in Figure 3(a). It can be seen that the better efficiency is achieved at 22.5 mA/cm^2. Therefore, at that electrical current density the effect of pH level was observed in Figure 3(b) by altering the initial pH value from 4.11 to 7 and 9.

From Figure 3(b), it can be seen that alkalinity positively effects the treatment efficiency, where lower overall COD values could be achieved without contributing much oxygen demand for its energy consumption. Numerically, with the iron electrode, the lowest COD was achieved as 700 mg/L at the end of 60 minutes with a current density of 22.5 mA/cm^2 and a starting pH of 9. The contribution of oxygen demand due to the utilized electricity was 5930 mg. Yet, the overall oxygen demand goes down from 11200 to 6630.

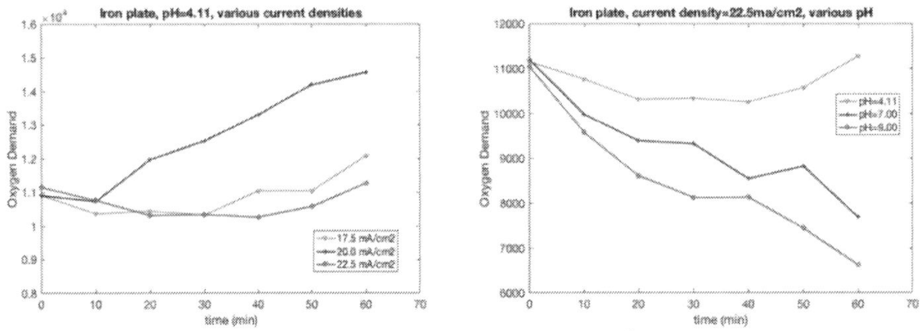

(a) (b)

Fig. 3. Total oxygen demands of the operation with (a) pH = 4.11 and varying current densities, and (b) current density = 22.5 mA/cm^2 and varying pH values.

The experimental results for iridium coated electrodes show that these settings were clearly more inefficient in terms of energy, hence the oxygen cost for the electricity generation immediately invalidates the marginal COD reduction. After conducting experiments at various current densities and various pH values, the best results were achieved at a current density of 50 mA/cm^2 at the original pH value of 4.03 (as shown in Figure 4).

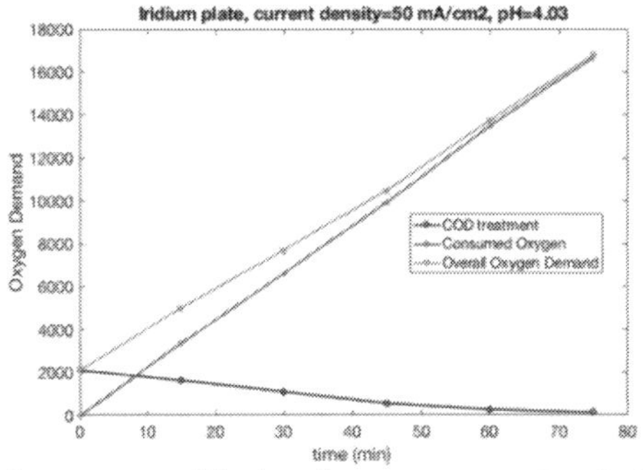

Fig. 4. Oxygen demands on the treatment (blue) and energy generation (red) sides with a total oxygen demand (green) plot using iridium plate electrodes, 50 mA/cm^2 current density and pH = 4.03.

It can be seen that the overall oxygen cost curve (green) starts to immediately increase. Numerically, the initial COD was 2100 mg/L, and at the end of 75 minutes, the COD reduces to 137 mg/L whilst the required energy was 43 kWh/m3, corresponding to an oxygen consumption of 16672 mg!

4 Conclusions and discussions

Despite the popularity of electrochemical treatment of wastewaters, the focus on the energy cost was limited to the economical aspects of the overall treatment [4–7]. Unfortunately, the economic cost of energy production (and consumption, thereof) is not directly compatible with the environmental impact of the direct and indirect wastes. The direct waste for this particular work is the wastewater of an actual vinegar production factory in Eskişehir. The impact of the wastewater is measured in terms of the COD. Using a novel approach to compare the reduction of COD in the wastewater to the average oxygen consumption during electric energy generation, a combined oxygen effect is obtained. The contributors can be plotted on the same time axis, and can be added to assess the overall oxygen demand (as in Figures 2–4). It is concluded that the treatment process should normally continue until the minimum cumulant oxygen demand is achieved (a time point, where the oxygen demand starts to increase). Some electrode types (e.g., iridium coated) are too inefficient even to *start* the electrochemical treatment process. Sadly, many of the energy consumption values reported in [1–7, 9, 10] as compared to the reduced COD values are even higher than the energy consumption values reported herein. In order to constitute a valid and reasonable environmental effectiveness, the following are proposed:

- The regulation COD limits of certain wastewaters are arbitrarily set, and they do not take into account the treatment effect. The new limits must be achieved as a result of such studies.
- The best electrode, pH and current densities must be studied prior to applying electrochemical treatment in industrial plants.
- If electrochemical processes are a necessity, a real treatment plant MUST use local renewable electric generation systems, and the treatment process must not use grid electricity. In that case, the oxygen cost of electricity generation would become technically zero.
- The proposed novel "oxygen denominator" approach must be tested over various other wastewater types as further studies.

This research is funded by Anadolu University Research Fund under contract no: 1601F026.

References

1. C. Barrera-Diaz, G. Roa-Morales, L. Avila-Cordoba, T. Pavon-Silva, B. Bilyeu, Ind. & Eng. Chem. Res. **45**, 1, 34–38 (2006)
2. G. Guven, N.A. Perendeci, K. Ozdemir, A. Tanyolac, J. Chem. Tech. **87**, 4 (2012)
3. M. Panizza, P.A. Michaud, G. Cerisola, C. Comninellis, Electrochem. Comm. **3**, 7, 336–229 (2001)
4. M. Kobya, E. Demirbas, J. Water Process Eng. **8**, 64–74 (2015)
5. B.K. Korbahti, K.M. Turan, J. Turkish Chem. Soc. **3**, 3, 75–92 (2016)
6. E. Demirbas, M. Kobya, Proc. Safety & Env. Prot. **105**, 79–90 (2017)
7. S.T. Lokendra, P. Mondal, J. Env. Mgmt. **190**, 102–112 (2017)
8. E.E. Gerek, S. Yılmaz, A.S. Koparal, O.N. Gerek, Int. Conf. Sust. Water Proc., Sitges, Spain (2016)
9. M.Z. Justin, D. Vrhovsek, A. Stuhlabacher, T.G. Bulc, Desalination **1**, 6, 100–109 (2009)
10. P.J. Strong, J.E. Burgess, Bioremediation J. **12**, 2, 70–87 (2008)
11. Woodard & Curran, Inc., Ch. 7 – *Industrial Waste Treatment Handbook, Second Ed.* 149–334 (2006)
12. Ü.B. Filik, T. Filik, O.N. Gerek, *Handbook of Clean Energy Systems*, **4**, 9, 1981–1995 (2015)
13. E. Butler, Y-T. Hung, T. Yeh, M.S. Al Ahmad, Water **3**, 495–525 (2011)
14. C. Comninellis, G. Chen, G. (Ed.'s), *Electrochemistry for the Environment* (Springer, 2010)
15. D. Rajkumar, K. Palanivelu, N. Mohan, J. Env. Sci. & Health, Part A: Tox/Hazard. Subst. Environ. Eng. **36**, 19, 1997–2010 (2001)
16. C. Chen, A.J. Bloomfield, S.W. Sheehan, Ind. Eng. Chem. Res. (to appear, 2017)

17. C.A. Martinez-Huitle, D.C. de Moura, D.R. da Silva, Chem. Eng. Trans. **41**, 373–378 (2014)
18. E.H.S. Ralph, H-H. Rogner, K. Gregory, Energy Policy **31**, 13, 1315–1326 (2003)
19. B.K. Sovacool, Energy Policy **36**, 8, 2950–2963 (2008)
20. X. Varun, I.K. Bhat, R. Prakash, Energy Policy **13**, 5, 1067–1073 (2009)
21. US Energy Information Administration: How much carbon dioxide is produced per kilowatthour when generating electricity with fossil fuels?, Online at: https://www.eia.gov/tools/faqs/faq.cfm?id=74&t=11
22. World Energy Assessment, United Nations Development Programme (UNDP), UN Department of Economic and Social Affairs (UNDESA) and World Energy Council (WEC) Joint Report, NewYork, USA (2000)
23. International Energy Agency (IEA) report, Online at https://www.iea.org

Sizing procedures for sun-tracking PV system with batteries

Ömer Nezih Gerek[1,*], *Ümmühan Başaran* Filik[1], and *Tansu* Filik[1]

[1]Anadolu University, Faculty of Engineering, Eskisehir, 26555, Turkey

Abstract. Deciding optimum number of PV panels, wind turbines and batteries (i.e. a complete renewable energy system) for minimum cost and complete energy balance is a challenging and interesting problem. In the literature, some rough data models or limited recorded data together with low resolution hourly averaged meteorological values are used to test the sizing strategies. In this study, active sun tracking and fixed PV solar power generation values of ready-to-serve commercial products are recorded throughout 2015–2016. Simultaneously several outdoor parameters (solar radiation, temperature, humidity, wind speed/direction, pressure) are recorded with high resolution. The hourly energy consumption values of a standard 4-person household, which is constructed in our campus in Eskisehir, Turkey, are also recorded for the same period. During sizing, novel parametric random process models for wind speed, temperature, solar radiation, energy demand and electricity generation curves are achieved and it is observed that these models provide sizing results with lower LLP through Monte Carlo experiments that consider average and minimum performance cases. Furthermore, another novel cost optimization strategy is adopted to show that solar tracking PV panels provide lower costs by enabling reduced number of installed batteries. Results are verified over real recorded data.

1 Introduction

The share of renewable energy generations from solar and wind inside the overall electricity generation is rapidly increasing. The most critical drawbacks of using these renewable energy sources can be listed as their unpredictable nature, dependence on seasonal changes, and the variations of solar and wind energy, which may not match with the time distribution of load demand. For a stand-alone renewable energy system (wind-photovoltaic-battery), in order to continuously satisfy energy demand with lowest investment costs, a thorough sizing methodology is necessary. Several mathematical approaches are proposed in the literature [1–3, 5, 6], most of them being obtained under relatively low resolution (i.e. hourly) wind / solar radiation readings) data recording. For a valid and accurate renewable energy system sizing result, it is typically required to have at least one-year high resolution (dense) wind speed, temperature, solar radiation, and energy demand values. These values depend on several atmospheric events, as well as the physio-chemical characteristics of the energy harvesting machinery (i.e., PV panels and turbine engines). Nevertheless, the sizing methods normally stop the process after a fixed period of data recording and training. Then, the algorithms run a time series analysis to come up with an optimized sizing that would minimize loss of load probability (LLP) if the same days would have appeared in the future [5, 6]. Unfortunately, the performance of a determined size may not show the same performance in the future conditions, even if the overall sun light and wind conditions do not statistically change. For a more reliable sizing result, methods that involve wide-sense models and forecasting of wind and solar radiation could be considered [2]. An analysis for a wind energy integrated systems was developed in [8] to form a production simulation tool. In general, linear regression and ARMA based methods are preferred for wind speed modelling [9], whereas solar energy models range from geographical calculations, to linear models, to artificial neural networks [10]. In a recent study, the solar model was even constructed as a surface in two dimensions, with axes being hours and days [11]. In that particular work, stochastic data generation functions were fed with transition models for wind speed and electricity demand, and the 2D modelling approach was adopted for solar radiation modelling.

In our experiments, active sun tracking and fixed PV solar power generation values of ready-to-serve commercial products are recorded throughout years 2015–2016 in Eskişehir (an inland place around 30–31° north latitude). At the same period, several outdoor parameters (solar radiation, temperature, humidity, wind speed/direction, pressure) are recorded with high resolution. Finally, in order to perform the sizing analysis using the demand side, hourly energy consumption values of a standard 4-person household are recorded from a constructed household emulator in our university campus in Eskisehir, Turkey. Using the above-mentioned set-up, parametrized random process models are extracted for wind speed, temperature, solar radiation, energy demand and electricity generation. Such modelling enables us to simulate the stand-alone renewable energy system's performance over several years. The Monte Carlo experimentation approach [7] that is used for modelling is explained in Sec. 2.2. By using these random process models to generate renewable data, it is possible to test the sizing results both in the average and in the worst-case sense. Finally, using a new sizing and cost analysis approach [3], a comparison between the size obtained from a fixed solar panel and an active solar tracking panel (under the same solar conditions) is made. The new model approaches the sizing problem with a combination of source cost and its corresponding LLP cost [3]. It is observed that the overall battery requirements for an active solar tracking system is sufficiently lower than that of a fixed panel system, making the cost of the system smaller, despite the initial higher cost of the tracker system.

2 Data acquisition and modelling

In order to come up with exact simultaneously time stamped data of solar radiation, wind speed, energy from PV, energy from wind, temperature, and actual energy demand, rigorous data acquisitions are achieved from a controlled experimental setup as shown in Figure 1. This experimental setup is inside the Anadolu University 2 Eylul Campus in Eskisehir region. The described data are logged in hard drives and in cloud based systems. The solar part of the produced energy was from (i) a single-axis active sun tracking PV panels, (ii) fixed PV panels, and iii) roof-mounted panels. These are accompanied by wind turbines. All of the utilized machinery were ready-to-serve commercial products, including the acquisition and recording systems. The energy demand was emulated using an actual construction (inside the same campus) which hosts a standard 4-person household with all average electrical consuming electronics and lighting products. The electrical machineries (such as refrigerator, lights, etc.) were run according to reported average usage values in Turkey. Detailed technical specifications of the system can be reached from [4].

 DOI: 10.1051/e3sconf/20172200052

2.1 Data sets

Portions of the collected 20 seconds interval wind speed, temperature and solar radiation and pressure values are shown in Figure 2(a), (b), (c), and (d), respectively. Neither the presented wind speed, nor the solar radiation values correspond to the generated electrical energy. The energy conversion result is illustrated in Figure 3.

Fig. 1. (a) 12 panels (260 W) with one-axis (right) and without (left) solar tracker platform (b) averaged total daily power generation values of tracking and fixed panels.

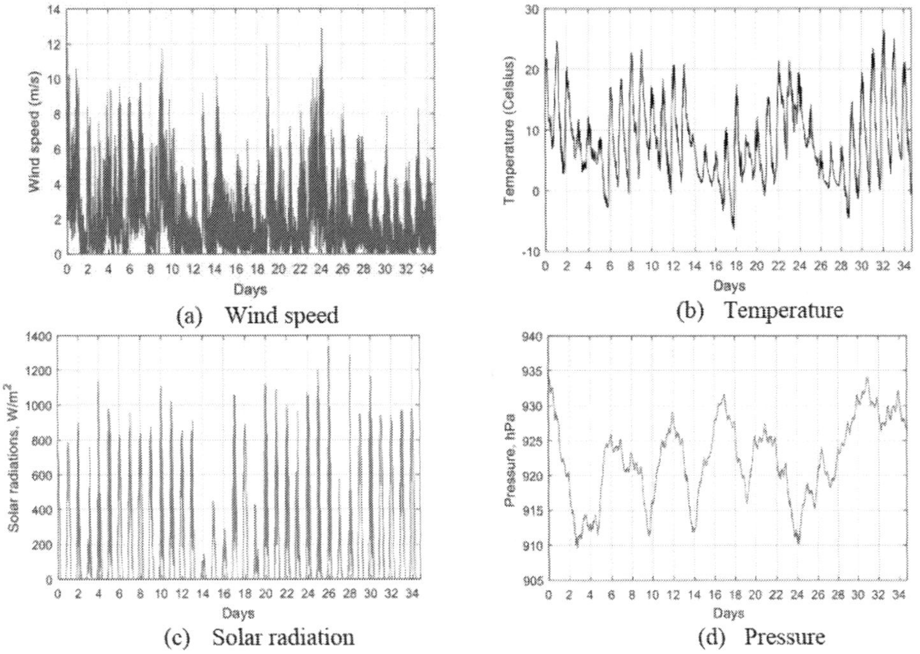

Fig. 2. Sample 20 seconds interval recoded (a) wind speed (b) temperature (c) solar radiation and (d) pressure values.

The electricity generation output of the active sun tracking and fixed PV panels are also recorded with 5 minutes' intervals as in Figure 3 (a) and (b), respectively. These figures use a compact 2D representation, where the produced electrical power is illustrated as a z-axis value, and the x-y axes consist of day-hour values. It can be seen that the power values have similar values along the same hour of consecutive days. Solar-wise, a highly varying season of March is selected here for the illustration of a relatively tough example. The geo location of the solar panels is (39.81 E, 30.54 N), corresponding to an inland region in Anatolia. Similar observations are expected to be observed in around inland geographical regions with latitude around 28–33.

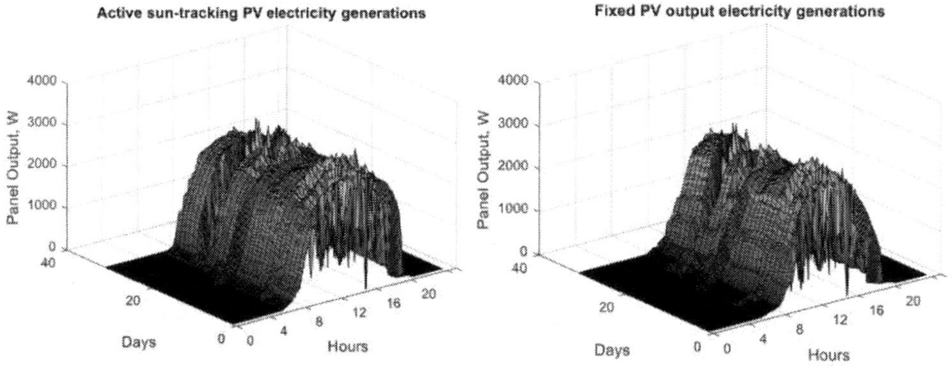

Fig. 3. The recorded real electricity generation output of the active sun tracking (left) and fixed PV panels (right) during March 2016 in Eskişehir region @ (39.81 E, 30.54 N).

The final data of interest for sizing applications are the actual prices and installation costs of the units. A list of unit costs for the experimental set-up items is given in Table 1. These values are per-unit scaled versions of real system components. Since the total cost of the inverter and battery charging regulator is directly proportional to the component costs they are not included in the calculations.

Table 1. Unit costs of ready-to-serve commercial products within Figure 1.

Component	Cost ($)
Fixed PV panel (260 W)	160
Active-sun track PV panel (260 W)	312
Wind Tribune (260 W)-scaled	500
12 V, 200 Ah Battery	350

2.2 Data modelling

In this part, we propose a method to extend the recorded data to various scenarios in order to better assess the sizing result. The proposed extension method helps us to observe more accurate average performance of the sizing, as well as situations corresponding to worst case scenarios. It is claimed that the described data generation models can be extended for

 DOI: 10.1051/e3sconf/20172200052

different geographical locations using by simply changing the defined parameters (such as mean/variance functions). Then, a sizing procedure can be simulated and compared for more stable results. In general, stochastic parameters are estimated from data, and applied to a random process generator that exhibits data for the extension purposes [7].

2.2.1 Wind data

A sample portion of the collected 20-seconds-interval wind speed values (corresponding to March) is plotted in Figure 4 (a). Each day (during March) is represented with a different color, and 24-hour data are plotted on top of each other. The average of the days along the hour axis is shown as solid black. Since the acquisition resolution is very high, time-windowed mean and variance values also exhibit a continuous variation (Figure 4 (b)).

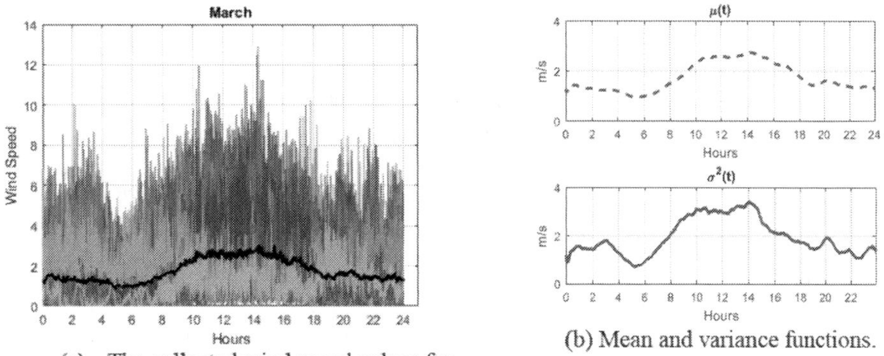

(a) The collected wind speed values for March.

(b) Mean and variance functions.

Fig. 4. Sample wind speed plots.

Again, due to the high resolution the effective randomness of the wind speed data the time averages fit into locally normal or chi-squared distributions, while the distribution parameters (mean and variance) slowly change in time. This time varying distribution can be expressed according to the formula in Eq. 1.

$$w_i(t) = \mathrm{N}\,(\mu_i(t), {\sigma^2}_i(t)) \tag{1}$$

where $\mu_i(t)$ *and* ${\sigma^2}_i(t)$ are timely varying mean and variance functions of the ith month. Therefore, the model slowly varies with the month index, *i*, and also varies in time, *t*.

For the modelling of solar radiation and the corresponding electrical energy, atmospheric conditions are also considered. However, a rougher set of values are used for modelling the atmospheric conditions: open, partly or fully cloudy, as explained in the next subsection.

2.2.2 Solar radiation and power output models for active sun tracking PV

Both the solar radiation and the corresponding electrical power depend on the extra-terrestrial radiation model as well as the atmospheric conditions in the PV location at the given date. Figure 5 shows three example recorded electricity generation value plots of single-axis active sun tracking (blue curves) and fixed PV panels (red curves) for sample sunny, partly cloudy and cloudy days. They all have different characteristics. Since all the

 DOI: 10.1051/e3sconf/20172200052

efficiencies and electricity generation characteristics are changing according to these three situations we propose a state varying procedure. An expected-but-important observation is that the blue curves corresponding to the sun tracking PV is consistently outperforming the analogous fixed panel energy production values.

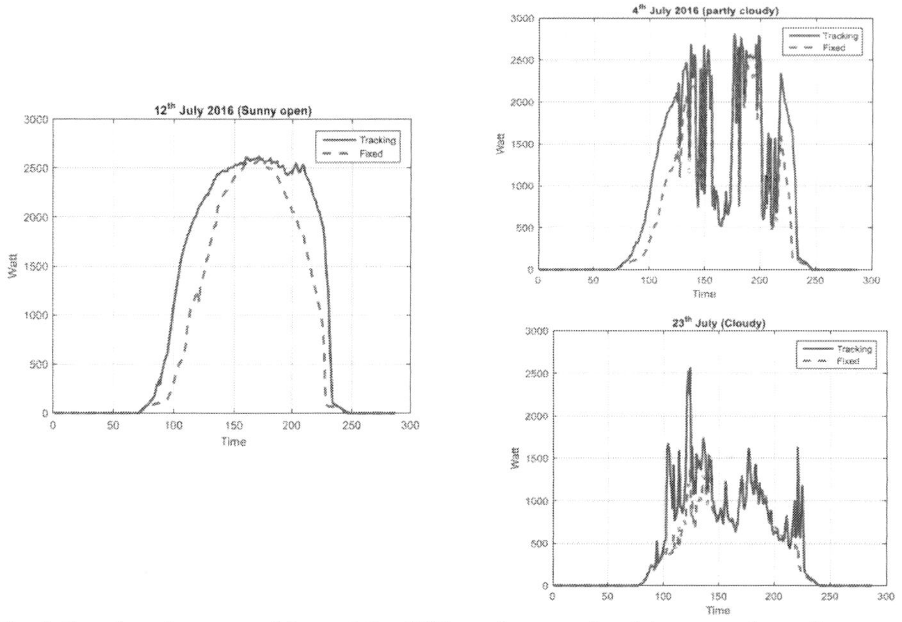

Fig. 5. Sample active sun tracking and fixed PV panel output electricity generations of sunny, partly cloudy, cloudy days.

The proposed state-varying atmospheric model can be illustrated with a 3-state chain as illustrated in Figure 6. First, for each month all the state probabilities are calculated according to the total number of sunny, partly cloudy and cloudy days. Then state transitions are counted to obtain transition probabilities.

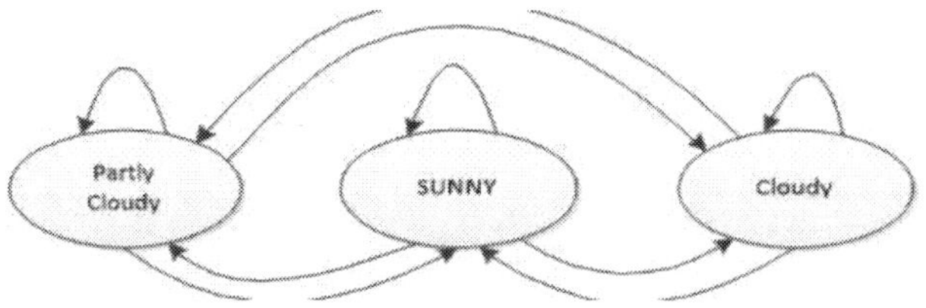

Fig. 6. Three atmospheric condition states with transition routes illustrated as arrows.

For a sunny day solar radiation values are modelled as truncated normal random process with calculated mean and variance functions as:

$$g_{i,s}(t) = \mathrm{N}\left(\mu_{i,sunny}(t), \sigma^2{}_i(t)\right) \tag{2}$$

For a partly cloudy day the cloud arrival probability is calculated with Poisson point process which is notated as $\mathrm{P}\left(\lambda_i(t)\right)$. With this model, solar radiation value can be generated by switching between the mean function of sunny and cloudy days' mean

 DOI: 10.1051/e3sconf/20172200052

functions. Poisson process will decide the switch instants of the mean functions according to the distributions:

$$g_{i,pc}(t) = \mathrm{N}\left((\mu_{i,cloudy}(t), (\mu_{i,sunny}(t)\ ), \sigma^2{}_i(t) \right) \tag{3}$$

$$g_{i,cloudy}(t) = \mathrm{N}\left(\mu_{i,cloudy}(t)\ , \sigma^2{}_i(t) \right) \tag{4}$$

Once an extended set of data is generated from the above model, The sizing result that is obtained from the actual data can be tested for average and worst case scenario analysis.

2.2.3 Electricity demand model

Depending on the season, the electricity demand varies in the household. The yearly average electricity demand is measured as 6.38 kWh. Figure 7 depicts three sample daily demand profiles.

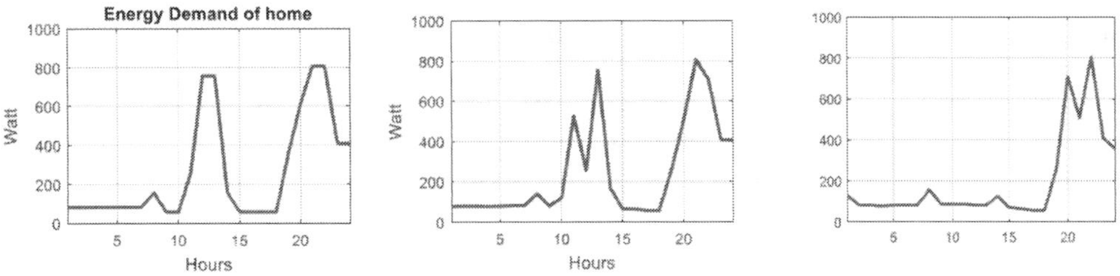

Fig. 7. Sample electric power demand profiles for three separate days in different seasons.

3 Sizing procedure

The actual recorded data are used for sizing the system to a minimal cost situation. Then, the achieved size is tested over the model simulations for improved assessment. During the sizing process, a new and efficient procedure is adopted [3]. The procedure uses all available renewable energy sources and starts by assuming maximum battery size (yielding a capacity to store all excess electric power and dispatches the storage whenever necessary). Then, gradually, the battery size is reduced while monitoring the LLP costs. On an axis of battery size, the LLP cost curve forms a decreasing plot. However, the batteries themselves contribute to the cost, producing a linearly increasing curve on the same axis. The combination of these two costs form the total cost, which exhibits a minimum value as illustrated in Figure 8. This optimal point is, then, tested against model generated data.

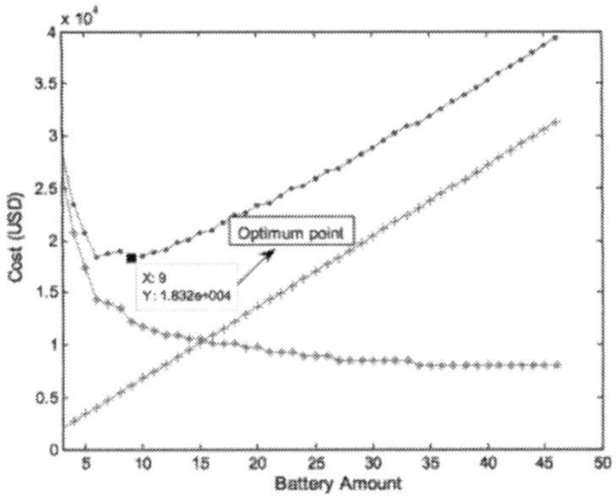

Fig. 8. Costs for LLP (red) and battery (blue cross) adding up to the total cost (blue dotted).

4 Results and discussions

Using the recorded data, the system which contains fixed PV panels (260 W, \$160) and the wind turbines (260 W, \$500) provided a minimum cost point at the battery size of 4, corresponding to an LLP level of below 0.0100 and a total cost of 160+500+4 x 350 = \$2060. When this system was tested for the model generated data, the LLP cost was observed to have an average value of 0.0092 with a worst case value of 0.0120. Then, the system is considered to have the sun tracking PV panels (260 W, \$312) together with the same wind turbines. As expected, the sizing produced a minimum cost at a point with less batteries (3 of them) at the same LLP levels. Although the initial cost of this version is higher than the one with fixed panels, the total cost is still lower (312+500+3x350 = \$1862). Besides, the larger values achieved by the tracking system (blue curves in Figure 5) pays off well to perform better on the worst case scenarios with an LLP of 0.0092 on the average and 0.0109 at the worst case. If the building performs critical electricity-dependent operations, or if the life cycle assessment of the electrical products is considered, the worst case improvement of 0.0109 versus 0.0120 might be even more critical than the total investment cost optimization of \$1862 versus \$2060, which is rather marginal. It is concluded that utilization of a sun tracker PV panels is more efficient as it demands less battery storage at the same or better LLP levels. The situation is expected to be valid in similar geographical locations (i.e. inland regions with latitude 28–33), but such a conclusion requires more experimental works and verifications throughout the world.

References

1. W. Zhou, C. Lou, Z. Li, L. Lu, H. Yang, Applied Energy **87**, 2, 380–389 (2010)
2. H. Yang, L. Lu, W. Zhou, Solar Energy **81**, 1, 76–84 (2007)
3. F.O. Hocaoğlu, O.N. Gerek, M. Kurban, Solar Energy **83**, 11, 2019–2028 (2009)
4. T. Filik, U.B. Filik, Anadolu U. J. of Sci. & Tech.-A: App. Sci. & Eng. (to appear, 2017)
5. S. Diaf, D. Diaf, M. Belhamel, M.A. Haddadi, A. Louche, Energy Policy **35**, 5708–5718 (2017)
6. A.R. Prasad, E. Natarajan, Energy **31**, 1943–1954 (2006)
7. R.Y. Rubinstein, D.P. Kroese, *Simulation and the Monte Carlo Method (3.E.)* (Wiley, 2017)
8. N. Maisonneuve, G. Gross, IEEE Trans. Power Sys. **26**, 4 (2011)
9. A. Naimo, Procedia – Soc. & Beh. Sci. **108**, 187–196 (2014)
10. T. Kahlib, A. Mohamed, K. Sopiah, Ren. & Sust. Energy Rev. **16**, 5 (2012)
11. F.O. Hocaoğlu, O.N. Gerek, M. Kurban, Solar Energy **82**, 8 (2008)

Natural ventilation systems to enhance sustainability in buildings: a review towards zero energy buildings in schools

Maite Gil-Baez[1,*], *Ángela* Barrios-Padura[1], *Marta* Molina-Huelva[1], and *Ricardo* Chacartegui[1]

[1]University of Seville, Avda. Reina Mercedes n°2, 41012 Seville, Spain

Abstract. European regulations set the condition of Zero Energy Buildings for new buildings since 2020, with an intermediate milestone in 2018 for public buildings, in order to control greenhouse gases emissions control and climate change mitigation. Given that main fraction of energy consumption in buildings operation is due to HVAC systems, advances in its design and operation conditions are required. One key element for energy demand control is passive design of buildings. On this purpose, different recent studies and publications analyse natural ventilation systems potential to provide indoor air quality and comfort conditions minimizing electric power consumption. In these passive systems are of special relevance their capacities as passive cooling systems as well as air renovation systems, especially in high-density occupied spaces. With adequate designs, in warm/mild climates natural ventilation systems can be used along the whole year, maintaining indoor air quality and comfort conditions with small support of other heating/cooling systems. In this paper is analysed the state of the art of natural ventilation systems applied to high density occupied spaces with special focus on school buildings. The paper shows the potential and applicability of these systems for energy savings and discusses main criteria for their adequate integration in school building designs.

1 Introduction

Energy consumption in EU due to buildings operation is 40% of the global energy consumption, and its main fraction is due to HVAC systems operation with 60%. To accomplish with EU regulations [1] and the Zero Energy Buildings [2] is required a drastic reduction of this energy consumption for indoor conditioning. Different recent studies and publications analyse natural ventilation systems potential to provide indoor air quality and comfort conditions with minimum electric power consumption. As an example of the interest of this subject, in last ten years more than ninety studies of natural ventilation systems are recorded in Scopus.

* Corresponding author: gmaitegil@gmail

 DOI: 10.1051/e3sconf/20172200053

In this paper is analysed the state of the art of natural ventilation systems applied to high density occupied spaces with special focus on the educational sector, showing the potential and applicability of these systems for energy savings under different classification criteria. It is structured in four sections: firstly design and demand profiles in schools; then indoor environmental quality (IEQ) is analysed; in third section actual situation is exposed and in the last section main conclusions are presented.

2 School design and demand profiles in schools

One of the most important groups of civil buildings is composed by schools. In the Mediterranean region there are more than 80000 schools [3], being a big energy consumer. This sector consumes 4% of the energy in the commercial sector in Spain [4].

R&D activity in school buildings is intense with relevant international R&D projects., [3, 5]. The school building segment is unique in the building domain since it has specific typologies, users and usage patterns, energy infrastructures, energy uses and functions. Specific characteristics of schools are the next:

a) Schools design is based in a rational organization of the spaces. Classrooms are usually grouped and there is a corridor to access them. This situation is presented in figure 1. In addition, classrooms have wide dimensioned windows in order to allow natural illumination of the space, as well as there are other high windows to corridor to be used in both cross ventilation and illumination.

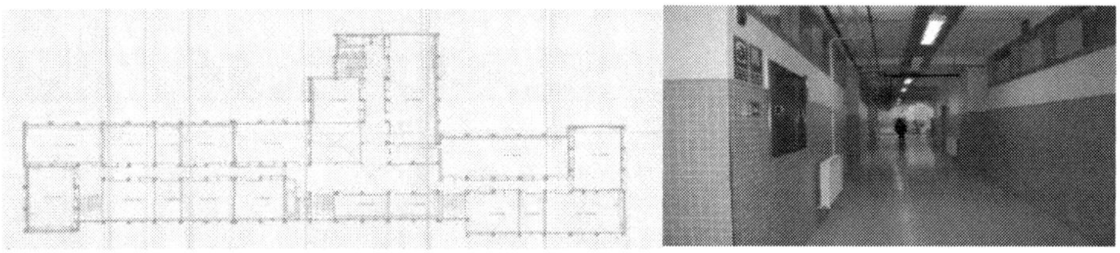

Fig. 1. Typical school building plant distribution. a) left plant, b) right corridor windows.

b) Schools, in special in mild climates, have reduced heating requirements, due to the mild weather and internal gains produced by high density occupation. Nevertheless, ventilation requirements are intense and they are present along the whole school year. Typical heating and ventilation patterns for a Mediterranean school are shown in figure 2.

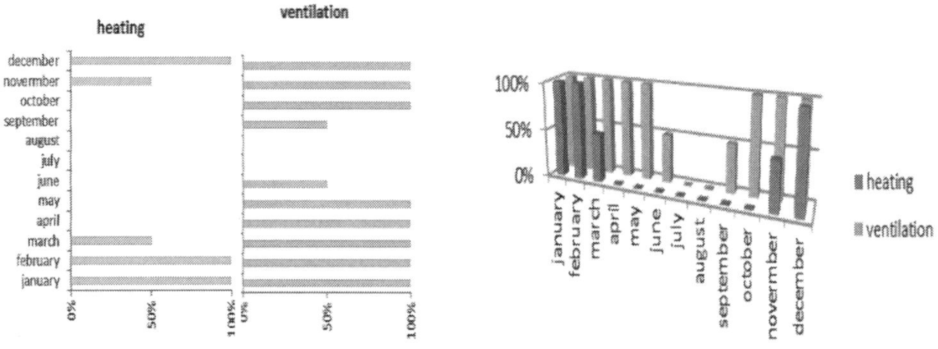

Fig. 2. Mild climate school heating and ventilation patterns. a) Left. Heating profile estimation in terms of heating days percentage by month and ventilation profile estimation in terms of percentage of ventilation days per month. b) Right. Comparison between percent of days per month in which heating and/or ventilation energy requirements for a school in mild climate.

c) Schools operate half of the year days, around 175 days, with a summer break of 2 months and a half in Spain, and Christmas and Easter holidays. In other countries, where the summer break is shorter, the number of school days is similar. A typical occupation pattern of school occupation in a school of Mediterranean region is presented in fig. 3 (left).
d) Classrooms present non-permanent occupation, due to breaks between lessons, and internal displacements and changes of class into the school building. In figure 3 (right) is shown the variation of occupation profile in classrooms in a Mediterranean school.

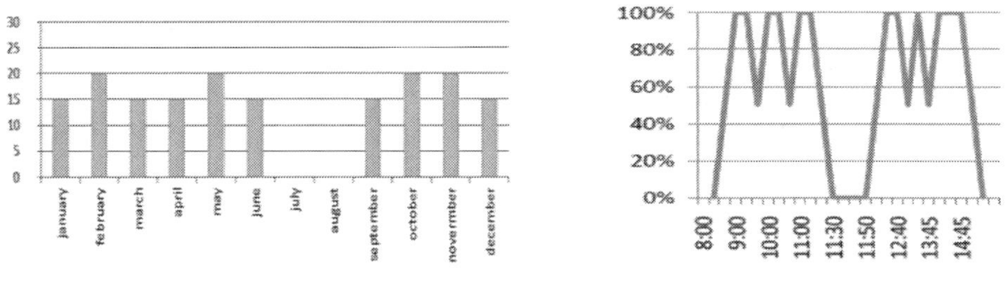

Fig. 3. A) Left. Typical pattern of school days distribution in Mediterranean region as function of the number of days per month B) Right. Occupation rate in a classroom from Monday to Friday.

So, given that all these characteristics define the educational use, determining the way of use and operation of their buildings, so they must be taken into account as basic aspects in the design of buildings and their facilities. In this pathway is interesting CIBSE's School design group publication TM57:2015 Integrated school design [6], with specific recommendations for school buildings.

2 Indoor environmental quality (IEQ) in schools

Indoor environmental quality (IEQ) is the result of the thermal, visual, acoustic comfort and indoor air quality (*7*). Most used Key performance Indicators (KPIs) used to characterize IEQ evolution in the classroom are CO_2 level, temperature and humidity.

Indoor Air Quality control in schools is required to ensure the best performance in students and teachers due to time expend indoor. This situation has been studied in different countries, Greece [8–10]. In Serbia, [11] in USA, [12] Denmark and in [14] Portugal. The relationship between ventilation rates and learning is studied in [14] by Clements-Crome et al. considering the occupancy density in classrooms (1.8–2.4 m^2/person) while Toftum et al. analyses in [12] the relationship between ventilation mode and learning outcome.

Pollutants levels were studied in Portuguese case in [13, 15] as well as in [10] and its relationship with outdoor air quality. Gennaro et al in [16] analyse different contaminants present indoor and give design recommendations as adequate ventilation systems. Nevertheless, most of these research studies identify IAQ with CO_2 levels as an indicator of the bio-effluents in high density occupied spaces [17, 18]. According to technical regulations, CO_2 levels must be below a ppm limit [19, 20]. In [21], Chenari et al resume values in different countries, being these between 400–1500 ppm, considering EN13779 and SIA 382/1 and DIN 1946-2 standards respectively.

Thermal comfort has been studied specifically in school buildings due to children comfort perception is different to the adults. Romana et al. express thermal comfort definitions [22] in terms of subjective (expresses satisfaction with the thermal environment) and objective condition (looking at the human body as a thermodynamic system that exchanges heat with the surrounding physical environment). In this way, Teli et al. realised in [23] that children perceiving overall comfort is not always related to their thermal state and they suggest that current adult-based comfort standards may not apply to school children.

Almeida et al. in [24] in Portugal and. Trebilcock et al. in Chile follow the same line [25]. Furthermore Romana et al. in [7], validated Fanger's approach in naturally ventilated Italian classrooms and Dear et al. [26] in Australia studied adaptive thermal comfort in school classrooms showing an acceptable summertime range of temperatures from 19.5 to 26.68°C.

Therefore, it is necessary to look for a balance between the energy usage to achieve/maintain thermal comfort and the ventilation requirements to guarantee indoor air quality [27]. Sometimes, both conditions are in conflict, with situations as is described in different publications and works, as the ones studied by Wargocki and Wyon [28], Almeida and Freitas [19] or Pereira [29]. For the case of a school building in mild climate, the analysis of ventilation and heating requirements in terms of percent of days per month, figure 2, right, evince that ventilation requirements along the full year for ventilation is around three times global heating requirements.

3 Thermal and air quality in schools. Actual situation

Several reviews expose the evidence of inadequate IAQ in schools, as in [9, 11, 13, 14, 31–35], and its negative effects on health producing allergic diseases and asthma [15]. Under Wargocki and Wyon research [28], this is due on one hand to the reduced financial resources for the maintenance of systems and inadequate upgrading of school buildings, but in other hand it is also due in warm climates to inadequate air supply and ventilation designs, oriented to preserve energy. Considering that in south Europe and Mediterranean climate, weather conditions are similar and different to north Europe, is necessary to develop specific solutions and standards in order to improve IEQ in schools guaranteeing thermal and quality indoor conditions. Therefore, the application of standards and regulations based on experiences in cold climate regions cannot be applied to new school buildings designs in Mediterranean areas, and the pathway for reaching low-energy or near-zero-energy buildings must be different to avoid unsuitably high temperatures in airtightness highly insulated classrooms. This issue as has been observed in low energy housing in Denmark as showed by Oropeza and Østergaard in [35]. They showed that a dwelling considered as a passive house focused on heating savings presented clear overheating problems during warmer session.

3.1 Mechanical systems in schools

The construction standards resulted in less airtightness in buildings constructed in twentieth century. Ventilation was controlled opening doors and windows [7, 24], as way to control contaminant levels in classrooms [37]. In last two decades, the use of mechanical ventilation systems has been extended to achieve IAQ saving heating energy [38] promoted by favourable technical regulations. They integrate heat recovery solutions in order to reduce energy consumptions, as well as night cooling and [38]. Nevertheless, these mechanical systems don't offer in schools the indoor quality expected [39]. In addition, in classrooms because of the high internal heat gains, the heating period is shorter than for other utilizations like residential buildings. Therefore heat recovery will be cost efficient only in very cold regions [5].

With the focus on health, MVS are related with Sick Building Syndrome (SBS). Comparison between natural ventilation and mechanical ventilation in schools were studied in [12, 40]. Ben-David and Waring studied [41] the impacts of natural versus mechanical ventilation in offices. Pereira et al. in [42] compared risk assessment for airborne infectious diseases between natural ventilation and a split-system air conditioner, concluding that the risk of airborne contagion were much lower opening windows and doors.

From the school buildings point of view in a mild climate as the Mediterranean climate, mechanical systems cannot be the optimal solution. In Portugal Almeida et al describes in

[30] how even installed these mechanical systems are not being used regularly. The use of MVS in school buildings do not give the expected results in their lifetime due to lack of resources for operation and maintenance and their minor contribution to IAQ.

However, current HVAC regulations in Spain [19] develops a detailed technical framework for MVS to ensure indoor air quality in school buildings. Regulation allows alternative HVAC systems to the mechanical ones but technical framework is not developed yet. It indirectly favours MVS system installation and design against other options that have to give additional steps for certify regulation accomplishment. This situation is similar in other countries as Swiss, as is exposed in [43] or in Portugal, where there are research works in order to improve ventilation energy consumption as Pereira in [29], but natural systems are not considered.

In figure 4 are shown pictures of Mechanical Ventilation Systems in an Andalusian school installed under this regulation. It supposes a relevant capital cost, additional oversized electricity installations for it, including transformers, ducts integration in buildings and maintenance of the whole set of direct and indirect installations.

Fig. 4. Mechanical Ventilation system in an Andalusian school. Roof tops.

Therefore it is necessary to seek alternative solutions to ensure Indoor Air Quality within the ambient conditions and specific reality of mild climate schools [28, 39]. Taking into account the set of equipment, the associated capital, operational and maintenance costs and the relative low benefit derived from these systems under real operation of the schools, it is clear the interest of promoting Natural Ventilation Systems in schools applications.

3.2 NVS in schools

Several regulations of different countries prescribe natural ventilation systems or at least consider them in designing buildings and schools.

Most important documented experience is shown in UK where the Education Department has published a specific regulation about ventilation in school buildings (BB101) [20]. This document presents natural ventilation as the first solution that must be considered in the design of scholar buildings, while mechanical ventilation systems are considered as the last and non-desirable solution due to investment, operational and maintenance costs and requirements, that can derive in problems related with SBS and other health implications. Besides it has supported the growing of commercial companies with natural ventilation devices specially designed to scholar environment [44–47]. In other countries, NVS are favourite or at least considered to ventilate schools, as in German [48], Australia [49, 50], Ireland [51], and New Zealand [52].

Recent R&D works in several countries are oriented to assess the effect of using these systems in school buildings design, to guarantee IAQ reducing energy consumption. For example in Italy, Stabile et al. in [38] studied manual airing and Stazi et al. in [54] automatic windows. In Saviese-Switzerland, Flourentzou et al. [55] studied the use of a high stack in comparison with a complex mechanical ventilation system with heat recovery and in

Thailand, Nitatwichit et al. in [56] studied natural ventilation as well as a cooling passive system.

Different natural ventilation strategies are studied, as Kotani et al present in [56] a review of Cross-Ventilation research papers and stack effect is also studied in [47] with the design of windcatchers for schools. Furthermore in Germany, Hellwig et al. [57] studied the possibility of reactivating historic ventilation stacks. Similar devices were usual in 20th century schools in Andalusia, as is shown in figures 4. However technical design evolution in buildings, evolution of regulatory framework, end user changes of habits, and the lack of experimentation and an adequate knowledge framework supporting design (for this application and climate) have displaced them, with no new building in the last 30 years until 2017 where new ones are under construction.

Fig. 5. Left) Cross ventilation in an Andalusian school. Right) Natural Ventilation system in an Andalusian school constructed in 1985.

Considering that, as is seen in [58] air flow depends on window size and position, typical windows design in schools, provides an adequate solution to propose the use of natural ventilation, combining different strategies as cross ventilation and stack effect, as propose Hellwig et al. in [57] reactivating historic ventilation stacks.

Andalusian Government, with more than 4500 schools, is developing a design program in 2017, in which an experimental natural ventilation system in classrooms is under installation in a set of new school buildings in order to prove and validate its operation, to assess their performance, generate new knowledge about the application to schools and give support to the future development of regional technical regulations.

ClimAct project, in the framework of the program Interreg-Sudoe, is in development since June of 2016 to December of 2019. Within this project different strategies to accomplish a transition to a low carbon economy in schools will be analysed. Under this project focus, natural ventilation systems are going to be analysed in school buildings design in Andalusia, Spain, to achieve this goal.

4 Conclusions

In this paper natural ventilation systems have been analysed as feasible option to improve the sustainability of buildings in mild climate with special focus on schools. Main derived conclusions applied to schools are:

1. Indoor Air Quality, required to assure the best performance in students and teachers, is controlled with ventilation. Mechanical systems are widely extended using heat recovery systems in order to achieve the efficiency of the HVAC system. In mild climates, requirement of heating is usually lower than for ventilation, so HVAC with heat recovery system could not be the optimal solution to enhance the efficiency of the building. These systems consume electric power in the operation and the internal heat gains can make unnecessary to recover heat.

2. Natural Ventilation Systems are highlighted as a feasible and sustainable strategy to reduce energy use in buildings. In addition, they can be used for passive cooling strategies with significant energy savings.
3. A Natural Ventilation System present particularities to consider in their design:
   - It must be present in first steps of design. They are affected by building location, type of building and dominant winds.
   -Window type and its position in the façade is key in the design. In schools, wide windows and windows to corridors are an opportunity to be used in natural ventilation strategies. In addition, cross ventilation and use of stacks can increase ventilation rates.
   -There is a clear different thermal subjective perception of children compared with adults standards, so different values should be considered in NVS for schools.
4. Although the interest of Natural Ventilation Systems in school buildings and in Mediterranean and mild climate is collected in an increasing number of publications and R&D works, further research in Mediterranean climate is required, in order to:
   - Prove NVS as a strategy to achieve the goal of Zero Energy Buildings, both in schools and other types of buildings.
   - It is required further research and testing experiences, differentiated from the developed in Northern countries. It would support the knowledge framework for design criteria and regulatory framework for high density occupation spaces, as schools. Once generated this basis, diffusion of the principles of natural ventilation operation, design constrains and applicability limits and advantages among technical actors, promotors as well as designers, engineers, etc., is a key activity.
   - Commercial offer development is a challenge for a greater implantation, to look for, as the example of UK where these systems are highly developed.

Acknowledgments

This study has been performed as part of the ClimACT project (SOE1/P3/P0429EU) within the Interreg Sudoe Programme, funded by European Regional Development Funds. We acknowledge Andalusian Government and APAE secondary schools by the support and collaboration.

References

1. E.L.P. Europeo, E.L. Consejo, D.E.L. a Uni, Directiva **2002**, 65–71 (2003)
2. Comisión Europea, Bol. Of. **L135**, 13–35 (2010)
3. ZEMeds, ZEMeds- Promoting renovation of schools in a Mediterranean climate up to nearly Zero-Energy Buildings. *www.zemeds.eu*, (available at 21/09/2016%5Cnhttp://www.zemeds.eu/)
4. D.A. Krawczyk, Energy Build. **81**, 30–37 (2014)
5. H. Erhorn-kluttig, H. Erhorn, Energy Procedia **48**, 1468–1473 (2014)
6. CIBSETM57, *Integrated school design* (2015)
7. F. Romana, E. Ianniello, B. Igor, Build. Environ. **67**, 129–137 (2013)
8. P. Barrett, F. Davies, Y. Zhang, L. Barrett, Build. Environ. **89**, 118–133 (2015)
9. K.K. Kalimeri et al., Atmos. Pollut. Res. **7**, 1128–1142 (2016)
10. M. Jovanovi, B. Vu, V. Turanjanin, M. Zivkovi, **77**, 42–48 (2014)
11. I. Air, T.A. Journal, B. Munksgaard, I. Air, 109–121 (2007)
12. J. Toftum et al., Build. Environ. **92**, 494–503 (2015)

13. P.N. Pegas, et al., 455–468 (2011)
14. A.I.N. Press **43**, 362–367 (2008)
15. M. Severo, E. Ramos, J. Madureira, H. Barros, E.D.O. Fernandes, **96**, 198–205 (2016)
16. G. De Gennaro, P. Rosario, D. Annamaria, D. Loiotile, 467–482 (2014)
17. L. Chatzidiakou, D. Mumovic, A. Summerfield, (2015), doi:10.1177/0143624414566244
18. V. Turanjanin, B. Vu, M. Jovanovi, N. Mirkov, I. Lazovi, **77**, 290–296 (2014)
19. E.Y.T. Ministerio de Industria, 137 (2013).
20. Bb101, *Building Bulletin 101 Ventilation of School Buildings* (2006)
21. B. Chenari, J. Dias Carrilho, M. Gameiro da Silva, Renew. Sustain. Energy Rev. **59**, 1426–1447 (2016)
22. F. Romana, B.W. Olesen, B. Igor, G. Riccio, Energy Build. **81**, 326–336 (2014)
23. D. Teli, et al., **3218** (2017), doi:10.1080/09613218.2013.773493
24. R.M.S.F. Almeida, N.M.M. Ramos, V.P. De Freitas, Energy Build. **111**, 64–75 (2016)
25. M. Ya, M. Trebilcock, J. Soto-mu, R. F. Martin, **114** (2017), doi:10.1016/j.buildenv.2016.12.036
26. R. De Dear et al., Build. Res. Inf. **0**, 1–16 (2017)
27. L. Chatzidiakou, D. Mumovic, A. J. Summerfield, **8975** (2017), doi:10.1080/17508975.2012.725530
28. P. Wargocki, D.P. Wyon, Build. Environ. **59**, 581–589 (2013)
29. M.D. Pereira, (2016)
30. R.M.S.F. Almeida, V.P. De Freitas, Energy Build. **81**, 127–140 (2014)
31. U. Heudorf, Gesundheitswesen **69**, 408–14 (2007)
32. J.T.M. Rosbach et al., **1–10** (2013)
33. J.C. Rufo et al., Porto Biomed. J. **1**, 142–146 (2016)
34. J.M. Daisey, W. J. Angell, M. G. Apte
35. M. Turunen, et al., Int. J. Hyg. Environ. Health **217**, 733–739 (2014)
36. I. Oropeza-Perez, P.A. Stergaard, Appl. Energy **114**, 520–530 (2014)
37. L. Stabile, M. Dell, A. Frattolillo, A. Massimo, A. Russi, Build. Environ. **98**, 180–189 (2016)
38. Y. Wang, F.Y. Zhao, J. Kuckelkorn, X.H. Li, H.Q. Wang, Energy Build. **81**, 9–17 (2014)
39. R.M.S.F. Almeida, V.P. de Freitas, Energy Procedia **78**, 3132–3137 (2015)
40. A.K. Mishra, M. Ramgopal, Build. Environ. **84**, 181–188 (2015)
41. T. Ben-David, M.S. Waring, T. Ben-david, M.S. Waring, Build. Environ. **104**, 1–17 (2016)
42. M. Luiz, M. Pereira, R. Vilain, A. Tribess, L. Morawska, **1–5** (2015)
43. F. Flourentzou, S. Pantet, (2014)
44. B. Richard, J. Kaiser, S. Abdul, Appl. Energy **92**, 606–627 (2012)
45. M. Dehnavi, M. H. Ghadiri, H. Mohammadi, H. Ghadiri, **2**, 559–564 (2012)
46. B.M. Jones, R. Kirby, Build. Environ. **44**, 1925–1934 (2009)
47. B.M. Jones, R. Kirby, B.M. Jones, R. Kirby, **3315** (2017), doi:10.1080/14733315.2010.11683886

 DOI: 10.1051/e3sconf/20172200053

48. H.-J. Moriske, R. Szewzyk, (available at http://www.umweltbundesamt.de/index-e.htm)
49. *Building schools in the 21st century and Current thinking about learning for a lifetime* (2009)
50. I. Division, E. C. Development (2011)
51. G. Document (2009)
52. D. Quality, L. Spaces
53. F. Stazi, F. Naspi, G. Ulpiani, C. Di, Energy Build. **139**, 732–746 (2017)
54. F. Flourentzou, S. Pantet, K. Ritz, Int. J. Vent. **0**, 1–12 (2017)
55. C. Nitatwichit, Y. Khunatorn, N. Tippayawong, **3839** (2017), doi:10.1080/02533839.2008.9671413
56. H. Kotani, T. Goto, M. Ohba, T. Kurabuchi, **3315** (2017), doi:10.1080/14733315.2009.11683848
57. R. Hellwig, M. Sedlmeier, C. Tanzer, **3315** (2017), doi:10.1080/14733315.2015.11684076
58. P. Heiselberg, E. Bjørn, P. V Nielsen, **3315**, 90–100 (2017)

 DOI: 10.1051/e3sconf/20172200054

The slag original from the process of sewage sludge incineration selected properties characteristic

Anna Głowacka[1,*], *Teresa* Rucińska[2], and *Justyna* Kiper[1]

[1]EDP Department of Sanitary Engineering West Pomeranian University of Technology in Szczecin, al. Piastów 50, 70-311 Szczecin, Poland
[2]EDP Department of Building Physics and Building Materials West Pomeranian University of Technology in Szczecin, al. Piastów 50, 70-311 Szczecin, Poland

Abstract. This work characterizes the physical and chemical properties of slag from combustion of municipal sewage sludge in "Pomorzany" waste treatment plant in Szczecin. The technology of sludge management is based on drying the sludge in low-temperature belt driers, to a content level of at least 90%, dry mass., and then burning in a grate boiler with mobile grate. The research of the slag resulting from combustion of municipal sewage sludge was conducted using reference methods, presenting images from a scanning electron microscope. The tested waste contained from 16.300 to 23.150% P_2O_5 completely soluble in strong acids, pH 8.03, mineral substance 98.73% dry mass. The content of heavy metals did not exceed the permissible amount specified in the Regulation of the Minister of Agriculture and Rural Development of 18 June 2008 on the implementation of certain provisions of the Act on fertilizers and fertilization (Journal of Laws of 2008 No. 119, item. 765). The screening trials showed that 48.4% are fractions of 630 μm–1.25 mm. The results show that the waste code 19 01 12 may be used as: alternative source of phosphorus for direct application to soil treatment, for production of organic – mineral fertilizers and as construction aggregate for production of concrete mortars.

1 Introduction

In highly industrialized countries, management and disposal of sewage sludge is a logistical, economic and above all an environmental problem. Sewage sludge is characterized by a high organic load, chemical pollutions. It also constitutes a health risk, due to the presence of pathogenic bacteria, viruses and other pathogenic organisms [1–4]. The increase in the degree of sewage treatment increases the amount of sludge. Therefore, an appropriate manner of its disposal must be ensured, allowing for safe use that does not endanger people and the environment. Management and disposal of sewage sludge is achieved through a variety of processes, such as composting, drying, incineration,

* Corresponding author: anna.izewska@zut.edu.pl

pyrolysis and methane fermentation [1, 2, 5, 6]. The main areas of sewage sludge utilization are agriculture and land reclamation. Its disposal is most commonly performed by composting, incineration and storage.

In 2013 in Poland, 540.3 thousand Mg of dry mass of municipal sewage sludge was produced [7] of which approximately 41% was disposed of by storage and accumulation in treatment plants (in landfills, lagoons and sedimentation ponds). Further development trends were, among others, usage in agriculture (20%) and thermal conversion (14%). The basic technological processes of disposal of sewage sludge currently used in Polish waste treatment plants include: composting, drying, incineration, pyrolysis and methane fermentation [1, 8, 9].

In accordance with Annex 4 to the Regulation of the Minister of Economy of 16 July 2015 on admission of waste for storage in landfills, non-stabilized sludge cannot be stored as of 1st January 2016.

Therefore, one of the best methods of disposal of sewage sludge is the utilization of thermal methods [2, 5, 10,11]. Products resulting from the combustion of sewage sludge are not neutral to the environment. The quantity and quality of pollutants depend on the chemical composition of the deposits, the combustion technology and the exhaust gas treatment. In many countries, incineration of sewage sludge is common. Systems designed for this purpose are equipped with systems for control and removal of pollutants in exhaust gases [12].

The aim of the study was to characterize the physical and chemical properties of slag resulting from combustion of municipal sewage sludge in "Pomorzany" waste treatment plant in Szczecin and to determine the possibility of using it as: an alternative source of phosphorus for direct application to soil treatment, production of organic – mineral fertilizers and as construction aggregate for manufacturing concrete mortars .

2 Sludge management in the „Pomorzany" waste treatment plant in Szczecin

The "Pomorzany" Waste Treatment Plant in Szczecin ensures the required parameters of treated sewage. However, the greater the effectiveness of treatment, the more waste remains after the purification process.

Waste removed at the start of the mechanical treatment – the screenings – the residue of sifting via the gratings and sand - suspended mineral solids removed at the sand traps, are subjected only to rinsing to reduce the organic parts content, and their content of water is reduced – by compression in case of screenings or separation on a screw conveyor for sand. The main role of sludge processing is to minimize their amount and volume through draining part of the water and neutralization of the sludge – so that it does not pose a threat to the environment. The first phase of handling sludge is to drain part of the water in the process of compaction. Due to different properties of the initial and excessive sludge, compaction is carried out in different ways [13]. The initial sludge is periodically received from the sedimentary hoppers of primary settling tanks and then is gravitationally concentrated. The concentrated sludge from the bottom is periodically received and transported to the mixed sludge tank. The thickened excess sludge is also fed to this tank. The excess sludge is periodically received from the active sludge chambers. The biological structure of sludge means that most of the water contained in it is water bound in the cells. Before feeding the sludge into the compaction device, it is conditioned using polyelectrolytes. The general flow diagram of sludge management is shown in Figure 1.

The initial sludge is gravitationally pre-compacted up to about 5% of dry weight, the excessive sludge concentrated in mechanical compactors contains about 6% of dry

DOI: 10.1051/e3sconf/20172200054

weight. The initial and excessive sludge, after concentration, is discharged to the concentrated sludge tank and from there, by means of pumps, it is forced into two closed septic tanks. Inside them, mesophilic fermentation is performed at a temperature of approx. 37°C. The biogas produced inside the chamber is stored in two double-shell tanks with a capacity of 1500 m^3 each, and after desulfurization using the biosulfex method (the process results in elemental sulfur), is used as fuel in cogeneration units.

The Department of Water Supply and Sewage Ltd. in Szczecin has chosen a technology of thermal disposal of sewage sludge. This technology is based on dehydration of sludge by means of four conveyor presses, up to a dry weight content of about 20%, drying in contact dryers up to 90–92% dry weight and burning in two boilers of two steam boilers with power of 2460 kW.

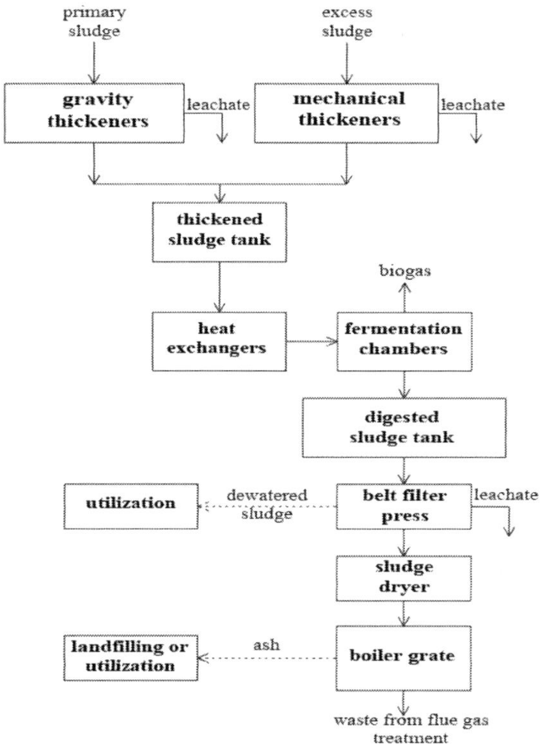

Fig 1. Block diagram of the sludge management technologies.

3 Research methods

The study used slag after combustion of municipal sewage sludge (waste code 19 08 05) in the "Pomorzany" Waste Treatment Plant, code 19 01 12.

The analytical physicochemical properties of the tested slag were performed using reference methods, in accordance with applicable standards and instructions pH - PN-EN 12176: 2004, dry weight - BS EN 12880: 2004, mineral and organic substances - PN-EN 12879:2004, phosphorus - PN-EN ISO 6878:2006, Ca, Co, Cu, Fe, K, Cd, Mg, Mn, Na, Ni, Pb, Zn - PN-EN 13657:2006.

4 Discussion of results

In 2016, 16485.3 Mg of sewage sludge was transferred for incineration in the "Pomorzany" waste treatment plant, with dry weight content of 20%, with a monthly average of 1498.7 Mg (table 1).

Table 1. Mass of municipal sewage sludge [in Mg] of average content 20% d.m. received in Pomorzany Sewage Treatment Plant in Szczecin in 2016

Specification	January	February	March	April	May	June	July	August	September	October	November	December
Mass [MG]	744. 1	925. 0	962. 7	210. 9	232. 8	611. 3	431. 9	98. 9	339. 2	n.d. *	377. 7	751. 0
Sum	16485.3											
Average	1498.7											
Maksimu m	2032.8											
Minimu m	377.7											

n.d.* - no data

Slag is formed as a result of incineration of dehydrated sludge in a mechanical grid boiler.

This sludge was sifted on a vibrating shaker and fractions were separated. Table 2 lists the particle size of the slag fractions. with content of grains with individual diameters. Studies have shown that 29.2% was formed by the fraction of < 500 µm. and 0.3% was formed by fractions of 5 and 4 mm.

Figures 2 and 3 show the appearance and structure of the slag samples in crude form.

Fig. 2. Slag from the combustion of sewage sludge. fraction 800 µm.

Fig. 3. Slag from the combustion of municipal sewage sludge. fraction < 500 µm.

Figures 2 and 3 show the appearance and structure of the slag samples in crude form.

The chemical and physical properties of the test material are presented in Table 3 and Table 4. The total content of individual heavy metals in the material collected in 2015 forms a series of decreasing values of Zn > Cu > Mn > Fe > Ni > Pb > Co> Cd. Analyzing the amount of elements identified in previous years. we can see a large increase in the concentration of zinc and copper. respectively amounting to 1925.07 mg Zn kg^{-1}·dry weight and 1591.56 mg Cu kg^{-1}·dry weight. The content of lead. cadmium and manganese has decreased. In 2015. the ashes from incineration of sewage sludge contained five times more potassium than in 2011-2013. Analyzing the standard deviations. the smallest variation of the chemical composition of ash may be observed in case of the amounts of soluble forms of phosphorus and total content of calcium. magnesium and cadmium.

Table 2. Granulation of slag from the combustion of municipal sewage sludge.

The diameter of the mesh sieve	Mass fraction of residua the sieve [g]	Percentage a fraction
< 6.3 mm	533.40	7.9
5.0 mm	17.75	0.3
4.0 mm	18.03	0.3
3.15 mm	66.77	1.0
2.5 mm	188.14	2.8
2.0 mm	280.16	4.1
1.25 mm	1210.56	17.8
800 μm	1362.58	20.1
630 μm	702.47	10.4
500 μm	412.87	6.1
< 500 μm	1977.19	29.2
Summ	6769.92	100.0

The chemical and physical properties of the test material are presented in Table 3 and Table 4. The total content of individual heavy metals in the material collected in 2015 forms a series of decreasing values of Zn > Cu > Mn > Fe> Ni > Pb > Co > Cd. Analyzing the amount of elements identified in previous years. we can see a large increase in the concentration of zinc and copper. respectively amounting to 1925.07 mg Zn kg^{-1}·dry weight and 1591.56 mg Cu kg^{-1} · dry weight. The content of lead. cadmium and manganese has decreased. In 2015. the ashes from incineration of sewage sludge contained five times more potassium than in 2011–2013. Analyzing the standard deviations. the smallest variation of the chemical composition of ash may be observed in case of the amounts of soluble forms of phosphorus and total content of calcium. magnesium and cadmium.

Slag samples obtained for research was characterized with alkaline reaction. It had dry weight content of 95.50% and 1.27% of organic matter (Table 4).

Table 4. The physical characteristics of slag.

Specification	pH	Conductivity [mS/cm]	Humidity [% d.m]	Dry mass [% d.m]	Mineral substance [% d.m]	Organic matter [% d.m]
Slag	8.03	2.835	6.50	93.50	98.73	1.27

While searching for ways of utilizing sewage sludge in the form of slag. an attempt was made to evaluate slag's suitability for being used as fine-grained aggregate for cementitious mortars. Hence. it was necessary to determine density of the waste and. for this purpose. pycnometer method was used in accordance with norm PN-88/B-04481. Due to conducted research. density result of 2.86 g/cm^3 was obtained. This is a value higher compared to typical natural aggregate – quartz sand (2.65 g/cm^3). Furthermore, a microscope evaluation was also conducted with the use of both stereo (Figure 4a) and electron (Figure 4b) microscopes.

Images on Figure 4 clearly indicate the porous structure of slag. From one side, this a positive trait due to the fact that it results in low volumetric density and provides an

DOI: 10.1051/e3sconf/20172200054

advantage of convenient thermal insulation. On the other side however. it results in lower level of mechanical durability and increased water absorption.

Table 3. The chemical composition of ash from the incineration of municipal sewage sludge.

Year of sampling	The content of soluble forms P_2O_5 [%]			The total content [%]			The total content [mg·kg^{-1} d.m.]							
	in strong mineral acids	2% citric acid	in the water	K	Ca	Mg	Cd	Cu	Mn	Ni	Pb	Zn	Co	Fe
2011	22.050	12.080	7.640	4.260	6.220	3.870	4.160	476.200	411.400	74.700	93.000	745.100	-	-
2013	16.300	9.980	5.140	4.960	6.970	4.670	2.390	310.300	476.900	18.000	84.300	820.190	-	-
2015	23.150	-		26.417		4.795	0.300	1591.560	367.960	66.760	17.960	4209.920	13.970	179.880
Średnia	20.500	11.030	6.390	11.879	6.595	4.445	2.283	792.687	418.753	53.153	65.087	1925.070	-	-
Maksimun	23.150	2.080	7.640	26.417	6.970	4.795	4.160	1591.560	476.900	74.700	93.000	4209.920	-	-
Minimun	16.300	9.980	5.140	4.260	6.220	3.870	0.300	310.300	367.960	18.000	17.960	745.100	-	-
Standard deviation	3.679	1.485	1.768	12.595	0.530	0.502	1.932	696.800	54.841	30.701	41.044	1979.094	-	-

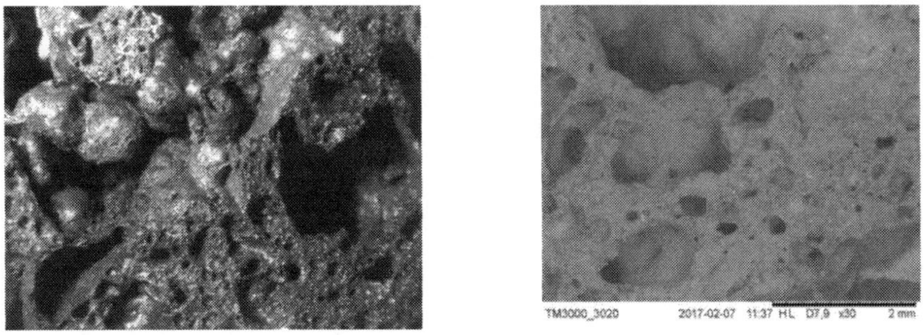

Fig. 4. Image of microstructure of slag enlarged: a) 10 x, b) 30 x.

In order to utilize this waste as a fine-grained aggregate for construction mortars. it is necessary to pass it through a set of sieves corresponding with mesh size for screening natural sand – Figure 5. which will allow slag to be combined into the mortar mixture.

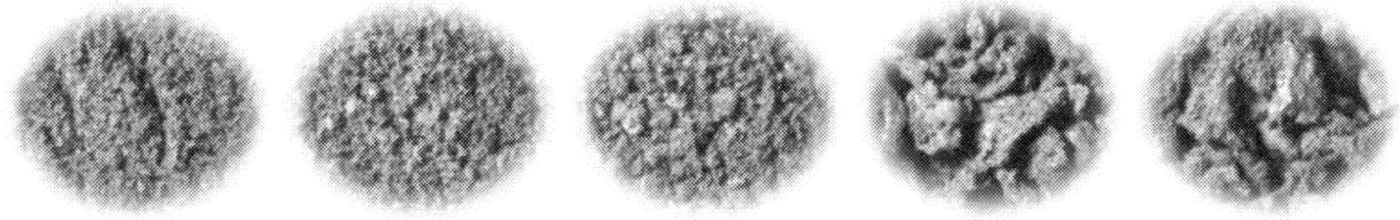

Fig. 5. Fine-grained recycling aggregate with fraction division. From left 0/0.063; 0.125/0.25; 0.25/0.5; 0.5/1; ½.

Nonetheless, it is worth remembering that slag's porous structure would require a special procedure of regulating the amount of water dosed into the mortar mixture. This will comprise the sum of batched water and water added. which results from the aggregate's absorption while the components are being mixed. This has both advantages and disadvantages. What is a positive aspect is the fact that water absorbed by the slag's grains during maturing period encourages self-care in the grain-cementitious matrix contact area. As a consequence. it enhances the adhesion of both elements. Furthermore. one can also expect increase of tensile strength of the composite due to the added waste [14]. The disadvantage is the fact that an excessive amount of water. unused during the hydration reaction process. leaves microporous material while evaporating during the maturing period. which undermines the structure of the obtained composite. Nonetheless. bearing in mind different needs regarding the properties of cementitious mortars. the usage of slag allows for obtaining lower volumetric density compared to traditional mortars. more convenient thermal insulation properties as well as an interesting coloring. Furthermore. this also complies with expectations set out in Regulation no 305/2011 of European Parliament and Council (EU) of 9th March 2011. regarding the introduction of construction products manufactured with the use of secondary recycled and raw materials to the EU market.

Scanning electron microscope **Hitachi TM 3000** equipped with EDS unit was used in order to determine the elemental composition of the analyzed slag – Figure 6. This allowed to observe the elemental concentration spectrum (Figure 7).

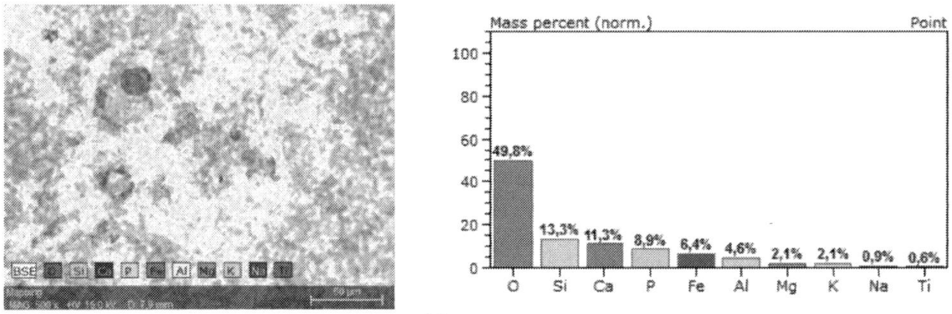

a) b)

Fig. 6. a) Map of identified elements on the slag's observed surface; (b) along with their percentage share.

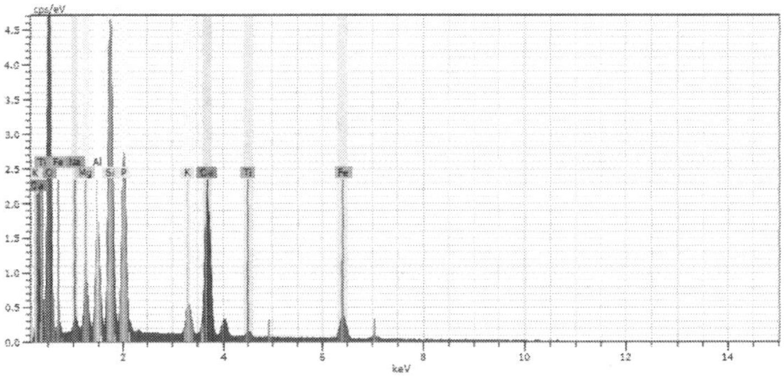

Fig. 7. Energy spectrum of elements identified on the slag's observed surface.

As indicated by the presented summarizing of elements – Figure 6b), the analyzed slag either lacks heavy metals or their concentration is so low that it cannot be detected by EDS. This provides the guarantee of safe use of this waste in construction mortars as well as floor and plaster coatings. Initial research in this matter has already been launched by the authors.

5 Conclusions

1. The slags resulting from the process of thermal treatment of sewage sludge in the "Pomorzany" waste treatment plant have a neutral reaction and contain 98.73% mineral matter.
2. The chemical composition of slag does not exceed the permissible norms contained in the Regulation of the Minister of Agriculture and Rural Development of 18 June 2008 on the implementation of certain provisions of the Act on fertilizers and fertilization. Journal of Laws 2008 No. 119. item. 765.
3. The content of phosphorus in the slag ranged from 16.300 to 23.150% P_2O_5. This waste may be used as: alternative source of phosphorus for direct application to soil fertilization and for organic – mineral fertilizer production.
4. The obtaining of slag granulation corresponding to the grain size of the sand. without the necessity of further treatment. and the marked density and microscopic analysis. provide the basis for further research into the use of this recycled material in construction mortars. As already mentioned. such research is already under way. and the test results obtained so far for essential technical characteristics. including mechanical strength. confirm the usefulness of slag in cement composites technology.

References

1. I. Białobrzewski, M. Mikš-Krajnik, J. Dach, M. Markowski, W. Czekała, K. Głuchowska, Waste Management, **43** (2015)
2. J.B. Bień, *Sewage sludge. Theory and practice* (Częstochowa, 2002) (in Polish)
3. E. Iacovidou, D.G. Ohandja, N. Voulvoulis, J. Env. Management, **112** (2012)
4. J. Mata-Alvarez, J. Dosta, M.S. Romero-Güiza, X. Fonoll, M. Peces, S. Astals, Renewable and Sustainable Energy Reviews, **36** (2014)
5. J.B. Bień, Environmental Protection Engineering **15**, 4 (2012) (in Polish)
6. P. Cornel, A. Meda, S. Bieker, *Wastewater as a source of energy. nutrients and service water. Treatise in Water Science* (Elsevier Verlag, 2011)
7. Environmental Protection 2014. Ochrona Środowiska 2014. Statistical yearbooks GUS (Warszawa, 2015) (in Polish)
8. J.B. Bień, E. Neczaj, M. Worwąg, A. Grosser, D. Nowak, M. Milczarek, M. Janik, Inżynieria i Ochrona Środowiska **14**, 4 (2011)
9. J.B. Bień, B. Bień, Inżynieria Ekologiczna, **45** (2015) DOI: 10.12912/23920629/60592
10. B. Chefetz, Y. Hadar, Y. Chen, Acta Hydrochimca et hydrobiologica **26**, 3 (1998)
11. J. Perez, J. Munoz-Dorado, T. de la Rubia, International Microbiology, **5** (2002)
12. S.J. Balogh, L. Liang, Water Air Soil Pollution, **80** (1995)
13. Operating data wastewater treatment plant "Pomorzany". Received from Zakład Wodociągów i Kanalizacji Sp. z o.o. in Szczecin (2014)
14. A.M Neville, *Properties of Concrete*, (Polish Cement Association, 758–762, Kraków 2012)

Energy crops cultivated on the slag from incineration of the sewage sludge energy value assessment

Anna Głowacka[1,*], *Krzysztof* Tarnowski[1], *Sławomira* Bering[1], *Jacek* Mazur[1], *Justyna* Kiper[1], and *Czesław* Wołoszyk[1]

[1]Department of Sanitary Engineering, West Pomeranian University of Technology in Szczecin, al. Piastów 50, 70-311 Szczecin, Poland
[2]Department of Soil Science, Grassland and Environmental Chemistry, West Pomeranian University of Technology in Szczecin, al. Piastów 50, 70-311 Szczecin, Poland

Abstract. In 2011–2013, research on the fertilizer value of slag from the incineration of municipal sewage sludge as an alternative source of phosphorus was carried out. The research scheme included 5 variants (in 4 repetitions) fertilization cultivated for grain with mineral fertilizers and ash. (P1, P2 and P3 – consecutive doses of phosphorus from ash) from municipal sewage sludge combustion: NK, NPK, NK+P1, NK+P2 and NK+P3. The obtained results indicate that the average of the three years of research, the value for the straw spring rape heat of combustion was 15.99 MJ/kg d.m., corn straw 16.20 MJ/kg d.m., triticale straw 17.06 MJ/kg d.m. and Miscanthus 17.34 MJ/kg d.m. The highest value of combustion heat for spring rape straw and miscanthus performed for objects fertilized with NK + P3 – 16.08 MJ/kg d.m. (Spring rape) and 17.57 MJ/kg d.m. (Miscanthus); For corn straw objects fertilized with nitrogen and potassium – 16.35 MJ/kg d.m. and triticale straw objects fertilized with NPK and NK + P2 – 17.10 MJ/kg d.m. Straw calorific value of tested plants was lower than the combustion heat by an average of 6.97% (triticale) to 7.38% (spring rape).

1 Introduction

In accordance with the National Waste Management Plan 2014 [1], by 2022, as part of the municipal sewage sludge management, the mass of sludge processed thermally will be increased and the use of nutrients from the sludge will be maximized. Strategy for development of municipal sewage sludge by means of thermal processing according to [2] is fully valid; in some countries of the European Union (Switzerland) it covers almost 100% of weight, while in others (Germany) it exceeds 50%. Currently, 11 incineration

* Corresponding author: anna.izewska@zut.edu.pl

plants of municipal sewage sludge operate in Poland and the need to increase thermal processing in the sludge management strategy will force the construction of subsequent incineration plants, especially in large urban areas [3].

Combustion of municipal sewage sludge generates a new kind of waste e.g. ashes and slag and produces gases polluted by dust, oxides of carbon, nitrogen, sulphur, heavy metals and hydrocarbons [4].

Wzorek [5] stated that in accordance with "the principles of sustainable development", the search for alternative sources of phosphorus, and in particular, the possibility of its recovery and recycling of waste should be classified as priority issues of the phosphoric industry. She points out the most important actions, such as recovery of phosphorus from urban and industrial waste water, sewage sludge, poultry manure and from meat industry waste. In order to avoid phosphorus deficit, some researchers propose to release soil phosphorus reserves unavailable to plants or from waste with phosphorus content e.g. by means of using appropriate microorganisms causing solubilisation of phosphorus compounds [6–8].

By analysing chemical composition, attention was drawn to the high content of phosphorus in ashes from the incineration of municipal sewage sludge, which can exceed even 9.61 $gP \cdot kg^{-1}$ d.m. [9].

The purpose of the research was to determine the influence of slag from the combustion of municipal sewage sludge on the contents of macronutrients in the straw, the calorific value and the heat of the combustion of plants.

2 Research material and methods

In the years 2011–2013, in the Agricultural Experimental Station in Lipnik near Stargard, tests have been carried out to fertilize plants using slag (fraction of less than 2 mm) from the combustion of municipal sewage sludge.

A field experiment was conducted on brown and red, incomplete soil produced from light, dusty clay sand, with average deep sediments of light clay. The scheme of the research covered five variants of fertilization with mineral fertilizers and slag from the combustion of municipal sewage sludge: NK, NPK, NK + P1 (P1 – 1 dose of ash – 50 kg $P_2O_5 \cdot ha^{-1}$), NK+P2 (P2 – 2 dose of ash – 100 kg $P_2O_5 \cdot ha^{-1}$), NK+P3 (P3 – 3 dose of ash – 150 kg $P_2O_5 \cdot ha^{-1}$). The slag from the combustion of municipal sewage sludge came from the Sewage Treatment Plant "Pomorzany" in Szczecin. The waste code – 19 01 12. Direct soil application of slag as phosphorous fertilizer was conditioned by the fulfilment of the requirements defined in the Regulation of the Minister of Agriculture and Rural Development [10]. The slag contained from 16.30 to 22.05% of total P_2O_5 soluble in strong mineral acids and from 9.98 to 12.08% of P_2O_5 soluble in 2% citric acid (Table 1). This waste was applied under: grain maize (variety DKC-FAO 240), spring rape (variety Larissa – 2011, variety Markus – 2012, variety Clipper – 2013), spring triticale (variety Milkaro), and giant miscanthus. Plants were also fertilized with nitrogen and potassium, in accordance with the recommended doses.

The research scheme included 5 variants (in 4 repetitions) fertilization cultivated for grain with mineral fertilizers and ash. (P1, P2 and P3 – consecutive doses of phosphorus from ash) from municipal sewage sludge combustion: NK, NPK, NK+P1, NK+P2 and NK+P3. Nitrogen pre-sowingly was used in the form of ammonium sulphate (20% N and 24%S), and ammonium nitrate for top dressing (34% N). On the other hand, phosphorus in the dose of 21.80 kg $P \cdot ha^{-1}$ (enriched superphosphate 17.44% P) and potassium in the dose of 91.30 kg $K \cdot ha^{-1}$ (potassium salt 49.80% K) in mineral fertilizers were used only pre-sowingly annually. What is more, ash was also used pre-sowingly in the following doses:

 DOI: 10.1051/e3sconf/20172200055

P1 – 21.80, P2 – 43.60 and P3 – 65.40 kg·ha^{-1} of phosphorus dissolvable in strong mineral acids.

The slag from the combustion of municipal sewage sludge, which provided a substitute for phosphorous fertilizer, came from Sewage Treatment Plant "Pomorzany" in Szczecin. The chemical composition is given in Table 1.

Table 1. Chemical composition of ash from municipal sewage sludge combustion [11].

Year of study	Content of forms soluble P [%]			Total content of [%]		
	in strong mineral acids	in 2 % citric acid	in water	K	Ca	Mg
2011–2012	9.613	5.266	3.331	4.260	6.220	3.870
2013	7.106	4.351	2.241	4.960	6.970	4.670

Total content of [mg·kg^{-1} d.m.]						
Year of study	Cd	Cu	Mn	Ni	Pb	Zn
2011–2012	4.160	476.2	411.4	74.70	93.00	745.1
2013	2.390	310.3	476.9	18.00	84.30	820.2

Total phosphorus content (P) soluble in strong mineral acids in the ash used in 2011–2012 was 9.61% and in 2015 – 7.11%. Moreover, ashes contained 4.26 and 4.96% of potassium, 6.22 and 6.97% of calcium and 3.87% of magnesium in 2011–2012 and 4.67% in 2013.

Heavy metals content did not exceed quantities established under the Regulation of the Minister of Environment (2010) on Municipal Sewage Sludge.

Test plants included: spring rape, maize, spring triticale and miscanthus. The mass of individual parts of the yield was defined, samples were taken and dried at 60°C. In such dry straw, after creating a medium-sized object samples, it was determined: the dry matter using the dry oven test (105°C); total N and S – using the CNS analyser manufactured by the Costeh company; P – using the Barton method; K, Ca and Mg using the ASA method, after mineralization of samples in the mixture (3:1) of nitric acid (V) and chloric acid (VII). Heat of combustion with the determined calorific value (according to PN-EN-ISO 9931:2005) was measured in the IKA 2004 calorimeter with uncertainty of measurement of the calorific value of 44 kJ/kg d.m. Multiple comparison of average values was conducted using the Tukey procedure, at p = 0.05. The standard deviation (SD) for the content of the macronutrients in a test plant was calculated according to STATISTICA 10 [12].

3 Results and discussion

Total macronutrients content in the straw (spring rape, maize, triticale and miscanthus) was presented as the average from the years of research (Table 2 and 3). The average nitrogen content (15.950 ± 1.161 gN·kg^{-1} d.m.) in the spring rape straw from individual objects was not diverse and ranged from 14.110 to 17.330 gN·kg^{-1} d.m. The average N content in the maize straw was 8.370 ± 0.559 gN·kg^{-1}d.m., and ranged from 7.641 to 9.002 gN·kg^{-1} d.m. In the spring triticale straw: 7.136±0.160 gN·kg^{-1} d.m., and in miscanthus: 8.450 ± 0.273 gN·kg^{-1} d.m.

Small diversity of the nitrogen content in the plant straw is probably associated with the same dose of this component in all fertilizing variants. The research conducted by [13], shows that even a big difference in the dose of nitrogen or the type of the phosphorous fertilizer does not cause significant changes in the content of this nutrient in maize.

Greater differences were expected in the case of the phosphorus content in plant straw, as doses in individual objects were varied. In the plant straw from objects NK, NPK, NK+P1 and NK+P2, similar phosphorus content was found, which varied: in spring rape 4.240–4.580 gP·kg^{-1} d.m., in maize 4.155–4.501 gP·kg^{-1} d.m., in spring triticale 1.292–1.336 gP·kg^{-1} d.m. and in miscanthus 3.740–4.420 gP·kg^{-1} d.m. The same dose of potassium (91.30 kgK·ha^{-1}) was used in all objects in the form of potassium salt. However, in versions with the ashes from the combustion of sewage sludge, additional potassium was used (from 6.74 to 20.22 kgK·ha^{-1} within a period of three years), which had no effect on its contents in the maize straw (12.211 ± 0.220 gK·kg^{-1} d.m.) (Tab. 2). In conclusion, it was found that the test plants were well nourished with potassium [14].

The biggest average calcium content (Tab. 3) was found in the spring rape straw and amounted to 18.370 ± 1.192 gCa·kg^{-1} d.m., and the smallest in the spring triticale straw 3.073 ± 10.416 gCa·kg^{-1} d.m.

The content of macronutrients in plants depends on several factors, including climate, soil and variety. By analysing the chemical composition of four varieties of maize, [15] obtained smaller phosphorus content (2.010 g P·kg^{-1} d.m.), potassium (1.980 gK·kg^{-1} d.m.), calcium (0.300 gCa·kg^{-1} d.m.), and magnesium (1.170 gMg·kg^{-1} d.m.), more in the maize and miscanthus straw than in the presented studies.

The average sulphur content in the straw of the test plants (Tab. 3) was not significantly varied and amounted to: in spring rape 3.200–5.510 gS·kg^{-1} d.m., in maize 0.820–1.050 gS·kg^{-1} d.m., in spring triticale 1.003–1.376 gS·kg^{-1} d.m. and in miscanthus 0.709–0.958 gS·kg^{-1} d.m.

Table 2. The average of the three-year content of macronutrients in the straw of test plants.

Objects	N				P				K			
	Spring rape	Maize	Spring triticale	Miscanthus	Spring rape	Maize	Spring triticale	Miscanthus	Spring rape	Maize	Spring triticale	Miscanthus
NK	14.110	8.401	7.172	8.100	4.090	3.821	1.349	3.670	12.000	12.155	9.272	10.570
NPK	15.150	7.855	7.421	8.850	4.240	4.286	1.363	3.900	14.780	11.866	9.360	10.060
NK + P1	16.600	8.946	6.946	8.240	4.580	4.155	1.336	4.420	14.210	12.441	8.776	9.680
NK + P2	16.570	7.629	7.057	8.660	4.120	4.287	1.363	3.740	13.450	12.135	8.801	8.870
NK + P3	17.330	8.963	7.083	8.400	4.330	4.501	1.292	3.970	13.450	12.457	9.885	9.510
Average	15.950	8.359	7.136	8.450	4.270	4.210	1.340	3.940	13.580	12.211	9.219	9.740
Standard deviation	1.161	0.547	0.160	0.273	0.176	0.224	0.026	0.263	0.935	0.220	0.409	0.567

Table 3. The average of the three-year content of macronutrients in the straw of test plants.

Objects	Ca				Mg				S			
	Spring rape	**Maize**	**Spring triticale**	**Miscanthus**	**Spring rape**	**Maize**	**Spring triticale**	**Miscanthus**	**Spring rape**	**Maize**	**Spring triticale**	**Miscanthus**
NK	18.280	9.040	3.187	5.536	0.640	10.870	0.207	1.265	3.200	0.880	1.376	0.832
NPK	16.750	11.220	3.373	5.719	0.680	10.040	0.207	1.370	4.760	0.820	1.150	0.928
NK + P1	19.350	12.210	3.500	6.535	0.660	11.080	0.207	1.429	4.690	0.870	1.176	0.958
NK + P2	17.480	11.100	2.988	5.321	0.770	10.970	0.182	1.382	5.480	0.850	1.094	0.885
NK + P3	20.020	9.940	2.317	5.552	0.810	10.160	0.168	1.345	5.510	1.050	1.003	0.709
Average	18.370	10.700	3.073	5.733	0.710	10.620	0.194	1.358	4.730	0.890	1.160	0.862
Standard deviation	1.192	1.099	0.416	0.421	0.066	0.435	0.016	0.054	0.838	0.081	0.123	0.088

The obtained results indicate that the average value from three years of research for heat of combustion from spring rape straw was 15.99 MJ/kg d.m., from maize straw 16.20 MJ/kg d.m., from spring triticale straw 17.06 MJ/kg d.m., and from miscanthus 17.34 MJ/kg d.m. The biggest value of the heat of combustion from spring rape and miscanthus straw occurred for objects fertilized with NK+P3 – 16.08 MJ/kg d.m. (spring rape) and 17.57 MJ/kg d.m. (miscanthus). For maize straw objects fertilized with nitrogen and potassium – 16.35 MJ/kg d.m., and spring triticale straw for objects fertilized with NPK and NK+P2 – 17.10 MJ/kg d.m. The calorific value of straw from test plants was smaller than the heat of combustion by 6.97% (spring triticale) –7.38% (spring rape) (Table 4).

Table 4. The average of three years of burning heat and calorific value of straw test plants.

Objects	Heat of combustion				Calorific value			
	Spring rape	**Maize**	**Spring triticale**	**Miscanthus**	**Spring rape**	**Maize**	**Spring triticale**	**Miscanthus**
NK	15.93	16.35	17.06	17.21	14.82	15.17	15.86	15.98
NPK	15.96	16.27	17.10	17.31	14.60	15.09	15.91	16.08
NK + P1	15.98	16.25	17.09	17.19	14.59	15.06	15.89	15.97
NK + P2	16.02	16.24	17.11	17.41	14.62	15.06	15.91	16.19
NK + P3	16.08	15.89	16.97	17.57	14.90	14.71	15.78	16.35
Average	15.99	16.20	17.07	17.34	14.70	15.02	15.87	16.11
Standard deviation	0.052	0.160	0.051	0.140	0.129	0.159	0.049	0.142

4 Conclusions

1. The average yield of grain and straw in particular years and for from three years of research was not significantly different between the objects fertilized with mineral fertilizers (NK and NPK) and with mineral fertilizers including the ash from the combustion of municipal sewage sludge.
2. The average content of N, K, Mg, and S in the straw of the test plants was not diverse. Only the highest dose of the ash from the combustion of sludge increased the total phosphorus content in the straw by 6.18%, compared with NPK.
3. The average value from three years of research for heat of combustion from spring rape straw was 15.99 MJ/kg d.m., from maize straw 16.20 MJ/kg d.m., from spring triticale straw 17.06 MJ/kg d.m., and from miscanthus 17.34 MJ/kg d.m. The biggest value of the heat of combustion from spring rape and miscanthus straw occurred in objects fertilized with NK+P3 – 16.08 MJ/kg d.m. (spring rape) and 17.57 MJ/kg d.m. (miscanthus).
4. The use of slag resulting from thermal processing of sewage sludge as phosphorous fertilizer does not affect the contents of macronutrients in the test plants. In most cases, fertilization with raw slag improved energy performance of the plant straw.

The research was financed from the Ministry of Science and Higher Education under the research Project No: NN 305 1723 40

References

1. The Monitor Polish 2010 National Waste Management Plan 2014. M.P. No 101, item.1183 (in Polish)
2. T. Pająk, Eng. and Prot. of Env. **13**, 1 (2010) (in Polish)
3. J. Bień, Eng. and Prot. of Env. **15**, 4 (2012) (in Polish)
4. Ł.Fukas-Płonka, http://www.osady.pl/artykuly/termiczne_metody_przerobki_osadow.pdf (access 28.05.2013.) (2013) (in Polish)
5. Z. Wzorek, *Monograph/Cracow University of Technology. Chemical Engineering and Technology* **356**, (2008) (in Polish)
6. H. Rodriguez, R. Fraga, Biotech. Adv. **17**, (1999)
7. B. Sundara, V. Natarajan, K. Hari, F.C. Res. **77**, (2002)
8. A. Saeid, M. Labuda, K. Chojnacka, H. Górecki, Chem. Ind. **91**, 5 (2012) (in Polish)
9. A. Iżewska, Cz. Wołoszyk, Eco. Chem. and Eng. **20**, 9 (2013) (in Polish)
10. Regulation of the Minister of Agriculture and Rural Development of 18 June 2008 on exercising some provisions of the Act on Fertilisers and Fertilisation, Journal of Laws of 2008 No 119, item 765, (2008) (in Polish)
11. A. Iżewska, Cz. Wołoszyk, J. Elementology **20**, 2 (2015)
12. Statistical Yearbook Of The Republic Of Polish 2013, Central Statistical Office, (Warszawa, 2014) (in Polish)
13. J. Potarzycki, Agricultura **8**, 3 (2009) (in Polish)
14. W. Szczepaniak, W. Grzebisz, J. Potarzycki, J. Elementology **2**, 533–548 (2014) (in Polish)
15. B. Gąsiorowska, A. Makarewicz, A. Nowosielska, F. Agronomica, **28**, 3 (2011) (in Polish)

 DOI: 10.1051/e3sconf/20172200056

Application of porous materials in oil substances separation from water

Adam Gołub[1,*], and *Janina* Piekutin[1]

[1]Bialystok University of Technology, Department of Technology in Engineering and Environmental Protection, ul. Wiejska 45A, 15-351 Białystok, Poland

Abstract. The aim of the study was to determine the ability of the four porous materials: birch bark, cork, glass wool, and polyurethane foam to reduce the mineral oil index and the concentration of n-alkanes C_7H_{16}-$C_{38}H_{78}$ as well as to select the most efficient materials. Model solutions of gasoline, diesel oil, and distilled water with the following values of mineral oil index were prepared to tests: 52 μg/dm^3, 68 μg/dm^3 and 73 μg/dm^3. Then, studies were carried out using a dynamic method, wherein the columns were filled with adsorbents tested, and in each of three testing series, 500 mL of the model solution at constant bed load of 1,0551 m^3/m^{2}h was filtered through the column. After filtration, the collected sample had volume of 250 mL. The collected samples were subject to determination of mineral oil index and concentrations of n-alkanes from C_7H_{16} to $C_{38}H_{78}$. Studies have shown that the most effective materials to lower the mineral oil index and the concentrations of n-alkanes in water are birch bark and glass wool.

1 Introduction

The composition of oil pollution getting into the water varies, but it is mostly a crude oil and its derivatives (fuels, lubricants) [1]. There are many sources of such substances in water. The most important are: natural exudation of oil from the seabed and the ocean, drilling in the sea bottom, leaking of installations for extraction and transportation of oil, river and sea transport, emergency splashes (e.g. as a result of a tankers collision or crash), road and air transport, petroleum waste from industrial installations, storm water from urban areas [2–6]. Petroleum products spilled on the surface of water immediately spread in all directions, seeking to obtain the thinnest possible layer on the surface. The process of spreading is dependent on the viscosity of petroleum products, wind power, flow velocity, size of water reservoir, water waving, and type of medium [7].

Removal of oil contamination from water is a very complex process that is influenced by many factors: type of the oil, properties of water-oil emulsion (chemical or mechanical), degree of dispersion, presence of other compounds in water (e.g. emulsifiers, stabilizers, surfactants). For this reason, there is no universal method for oil removal usable in any situation [8].

* Corresponding author: adamgolub92@gmail.com

There are many different methods for removal of toxic oils from the water. One such method is based on sorption, i.e. the physical method of de-oiling, during which two processes occur simultaneously: absorption of oil within the pores and voids of sorbent as well as oil adsorption on its surface [6, 15]. Among sorbents, materials of a porous nature are the most commonly used. They are solids having a high specific surface area resulting from the presence of a system of open and closed spaces, where adsorption of the substance occurs. Sorbents are readily available materials, easy to use, inexpensive, and environment friendly [7].

The aim of the study is to determine the ability of the four porous materials: birch bark, cork, glass wool, and polyurethane foam to reduce the mineral oil index and the concentrations of n-alkanes C_7H_{16} - $C_{38}H_{78}$ as well as to select the most efficient materials in this view.

2 Materials and methods

2.1 Subject of research: porous materials

In order to conduct the study upon removal of the oil substances from model water, 4 porous materials were used: birch bark, cork, glass wool, and polyurethane foam. Birch bark and cork were mechanically ground to obtain a particle size of 0.1–1 mm. Glass wool was cut into small squares of 2 × 2 cm. Pieces of polyurethane foam had dimensions of 1 × 1 cm.

Basic model water consisted of distilled water enriched with a mixture of fuels: gasoline and diesel oil. The study included petroleum products expressed as a mineral oil index (IOM) and aliphatic hydrocarbons (C_7H_{16} to $C_{38}H_{78}$). The testing included model water at three difference IOM and n-alkanes concentrations describing them with adequate series. The first series was IOM: 52 µg/dm^3 and ∑n-alkanes 8.3940 µg/dm^3, the second series consisted of IOM 68 µg/dm^3, ∑n-alkanes 9.4276 µg/dm^3, while the third one IOM 73 µg/dm^3, ∑n-alkanes 10.4276 µg/dm^3.

Table 1. The results of determination of mineral oil index and concentrations of n-alkanes C_7H_{16}-$C_{38}H_{78}$ in model solutions used for the tests.

Additive	Model solution [µg/dm^3]			Additive	Model solution [µg/dm^3]		
	52	68	73		52	68	73
IOM	51.6	67.8	73.1	$C_{23}H_{48}$	0.0039	0.0069	0.006
C_7H_{16}	0.0011	0.0031	0.0044	$C_{24}H_{50}$	0.0287	0.0329	0.0383
C_8H_{18}	0.0014	0.0024	0.0023	$C_{25}H_{52}$	0.3259	0.4064	0.4638
C_9H_{20}	0.0078	0.0073	0.008	$C_{26}H_{54}$	0.4299	0.5359	0.4811
$C_{10}H_{22}$	0.1053	0.1128	0.1485	$C_{27}H_{56}$	0.0109	0.0148	0.0125
$C_{11}H_{24}$	0.2163	0.2466	0.2449	$C_{28}H_{58}$	0.0892	0.0957	0.0787
$C_{12}H_{26}$	0.157	0.1913	0.1867	$C_{29}H_{60}$	0.6992	0.7094	0.7346
$C_{13}H_{28}$	0.0298	0.0379	0.0364	$C_{30}H_{62}$	0.845	0.9087	0.7509
$C_{14}H_{30}$	0.0058	0.0064	0.0058	$C_{31}H_{64}$	0.1084	0.1173	0.1988
$C_{15}H_{32}$	0.0046	0.0065	0.0084	$C_{32}H_{66}$	1.7973	1.8624	2.1507
$C_{16}H_{34}$	0.0042	0.0067	0.0077	$C_{33}H_{68}$	1.3796	1.4497	2.1261
$C_{17}H_{36}$	0.0125	0.0893	0.1185	$C_{34}H_{70}$	0.006	0.0151	0.0361
$C_{18}H_{38}$	0.1125	0.1461	0.1552	$C_{35}H_{72}$	0.1363	0.2286	0.5075
$C_{19}H_{40}$	0.1147	0.1497	0.2206	$C_{36}H_{74}$	1.4998	1.7223	1.3425
$C_{20}H_{42}$	0.0069	0.0075	0.0096	$C_{37}H_{76}$	0.0069	0.004	0.0067
$C_{21}H_{44}$	0.0135	0.0159	0.0227	$C_{38}H_{78}$	0.0748	0.081	0.0823
$C_{22}H_{46}$	0.1588	0.207	0.2313	∑n-alkanes	8.3940	9.4276	10.4276

2.2 Methodology of technological research

The removal process was conducted using a dynamic method applying model water with different concentrations of petroleum substances. Volume of 500 mL of model solution at constant load of the filter bed 1.0551 m^3/m^2h was filtered through the column filled with adsorbent (porous material); after filtration, the collected sample volume was 250 mL. All materials for the experiment were washed with distilled water.

2.3 Methodology of analytical testing

The collected samples were subject to determination of the mineral oil index (IOM) and the concentrations of n-alkanes from C_7H_{16} to $C_{38}H_{78}$. Determinations of individual samples were performed using gas chromatograph coupled with mass spectrometer VARIAN 4000. Separation of analytes was conducted on a column VF-5MS of dimensions 30 m × 0.25 mm × 0.2 mm. The stationary phase for the column was composed of polydimethylsiloxane with 5% share of phenyl groups.

Procedure for preparing the sample for analysis was divided into several stages: fixing, extraction, purification, and concentration. The sample fixing process consisted of acidification using inorganic acid up to a pH of about 2.

For the isolation of compounds included in the n-alkanes mixture, the liquid-liquid extraction system was applied. The isolation process was carried out at room temperature equal to ± 20°C on a magnetic stirrer at a speed of about 800 rpm, with 50 mL of dichloromethane and hexane mixture at a volume ratio of 2:1. After separation of aqueous and organic layers, the eluate was transferred to a volumetric flask and then subjected to florisil treatment and drying over anhydrous sodium sulfate p.a. The extract was separated from the water sample and concentrated to a volume of 1.5 mL. Concentrated extract was subject to separation and detection on the GC/MS system applying the following parameters:

- volume of injected sample: 1 μL,
- no splitting the carrier gas stream (splittless mode),
- injector temperature: 250 °C,
- carrier gas flow rate: 1 mL/min,
- initial oven temperature: 40 °C – isothermal 5 minutes,
- final oven temperature: 300 °C – isothermal 20 minutes,
- temperature increment: 10 °C/min,
- transfer line temperature: 230 °C,
- ion source temperature: 180 °C,
- arrange of scanned masses: 40-400 m/z.

The carrier gas: helium of 6.0 purity.
The results of the test substances contents are the average of three replicates.

3 Results and discussion

The effectiveness of lowering the mineral oil index and the sum of the concentrations of n-alkanes C_7H_{16}-$C_{38}H_{78}$ on materials used for the tests are shown in Figure 1. The maximum degree of IOM removal for particular sorbents is: 82% in the case of birch bark, 78.8% for glass wool, 74.8% for cork, and 73.2% for the polyurethane foam. The highest degree of n-alkanes sum removal, however, is as follows: birch bark 97.1%, glass wool 96.8%, cork 95.7%, polyurethane foam 95.5%. Efficiency of removal of both the IOM and n-alkanes can be ranked as follows: birch bark, glass wool, cork, polyurethane foam.

 DOI: 10.1051/e3sconf/20172200056

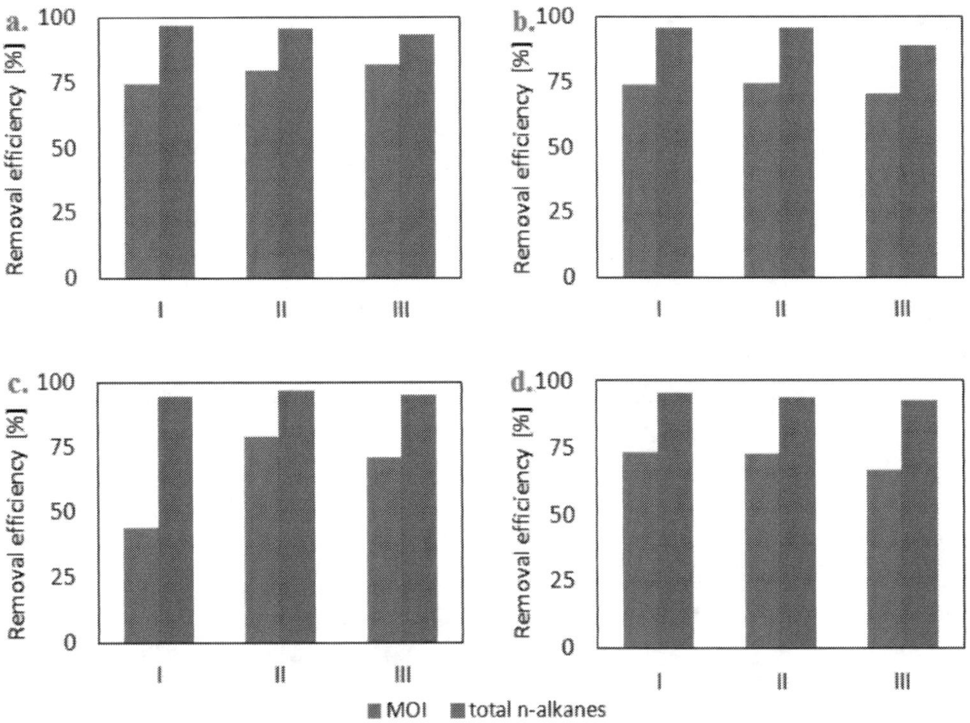

Fig. 1. Comparing the degree of removal of mineral oil index and the sum of n-alkanes C_7H_{16}-$C_{38}H_{78}$ by individual porous materials from model water after sorption process: a. birch bark, b. cork, c. glass wool, d. polyurethane foam.

In Figure 2, a bar graph indicates the degree of removal of particular n-alkanes, and a line graph indicates the concentration of each n-alkanes after the sorption process. It was found that tested sorbents removed below 70% of n-alkanes as follows: C_7H_{16}, C_8H_{18}, $C_{14}H_{30}$-$C_{18}H_{38}$, $C_{20}H_{42}$, $C_{21}H_{44}$, $C_{23}H_{48}$, $C_{37}H_{76}$ and $C_{38}H_{78}$. The best removal effect – above 90% - was achieved for the following n-alkanes removal: C_9H_{20}, $C_{11}H_{24}$, $C_{12}H_{26}$,$C_{22}H_{46}$, $C_{25}H_{52}$, $C_{26}H_{54}$, $C_{29}H_{60}$-$C_{36}H_{74}$.

It has been found (Figure 2) that in the first series of tests, over 95% removal was obtained for 17 n-alkanes in the case of birch bark, 16 for the cork, 14 for polyurethane foam, and 13 for glass wool. In the second series of studies, above 95% removal for 17 n-alkanes using birch bark and cork was recorded, for 16 n-alkanes using glass wool, and for 14 using polyurethane foam. In the third series of tests, more than 95% of 16 n-alkanes was removed in the case of birch bark, and glass wool, 10 in the case of polyurethane foam and 6 in the case of cork. For all series of tests, over 95% removal was obtained for 50 n-alkanes using birch bark, 45 using glass wool, 39 applying cork, and 38 using polyurethane foam.

DOI: 10.1051/e3sconf/20172200056

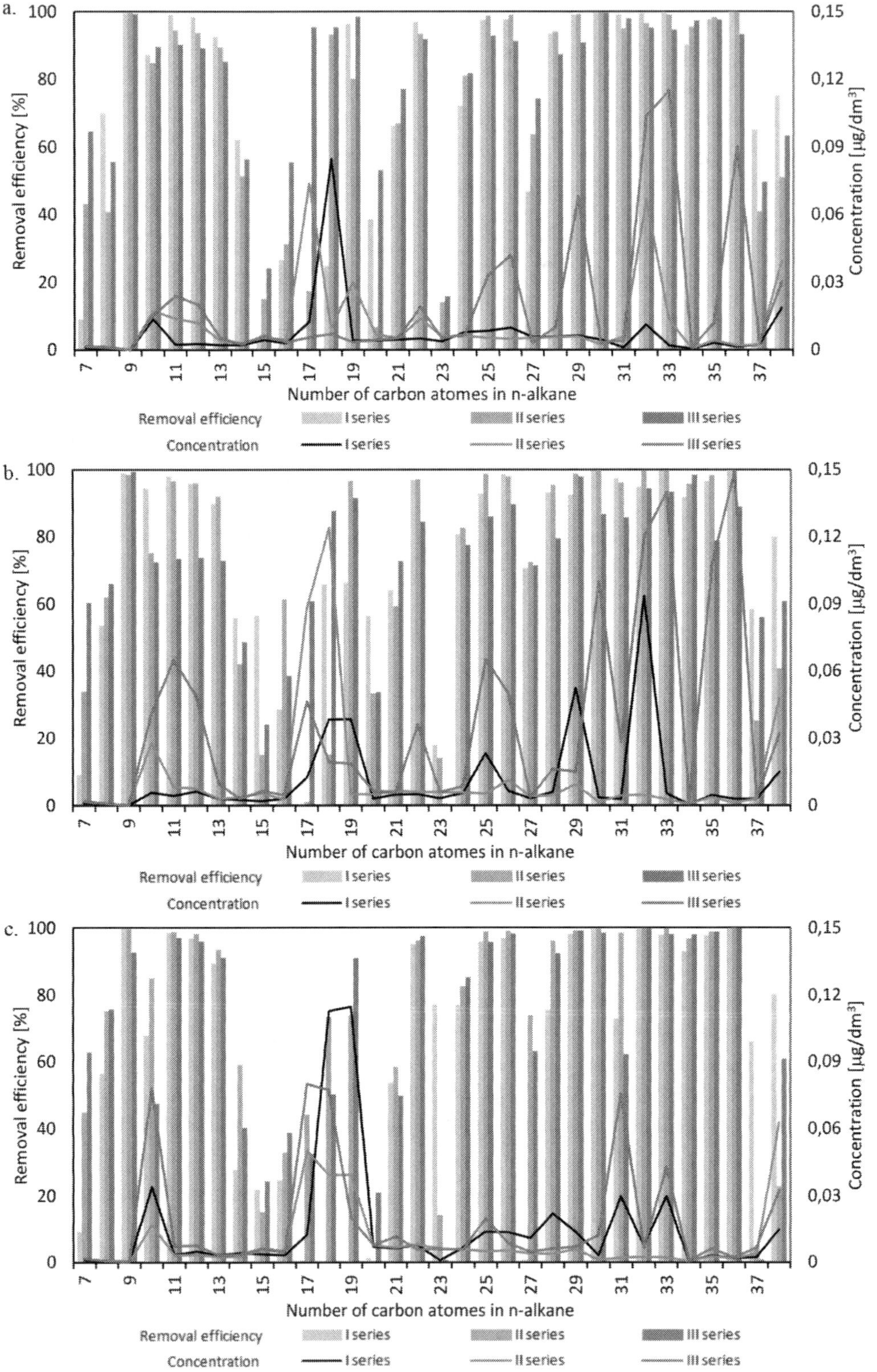

DOI: 10.1051/e3sconf/20172200056

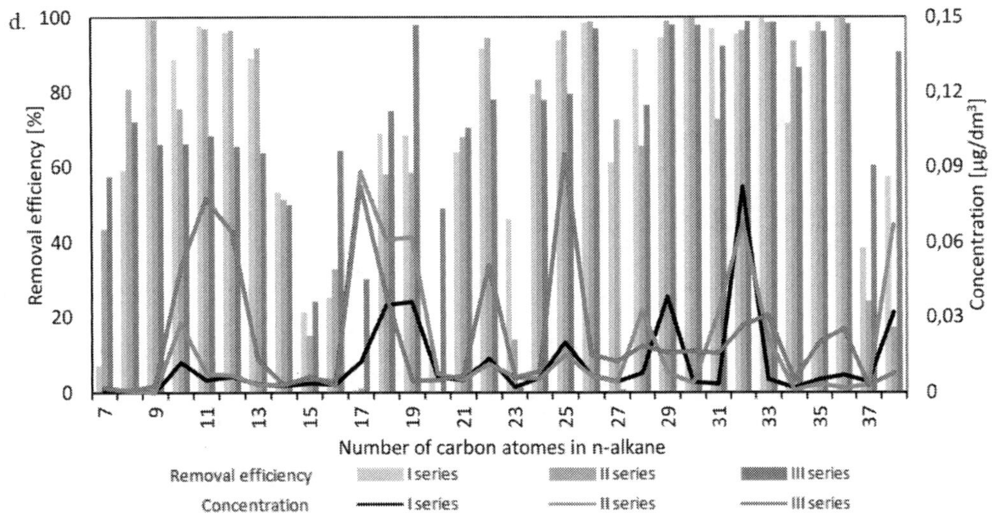

Fig. 2. Comparison of removal efficiency of n-alkanes C_7H_{16}-$C_{38}H_{78}$ and their concentrations in solutions after sorption process on individual porous materials: a) birch bark, b) cork, c) glass wool, d) polyurethane foam.

Sorbents are materials with high affinity to oils and a little to water, which due to two simultaneously occurring processes: adsorption (involving the oil adhesion to the surface of the sorbent, the extent of which depends on the viscosity of the oil) and absorption (consisting in filling the pores of the material by oil, which can be taken out of the water along with the sorbent due to the capillary force) allow for the removal of toxic oil substances from water in a non-invasive manner [9]. Among the most important material properties affecting the efficiency of oil removal from water, a high absorption capacity is mentioned in relation to the most of the oily substances [10–17]. The experiment demonstrated that the most effective materials in lowering the n-alkanes concentrations within the range of C_7H_{16}-$C_{38}H_{78}$ (these alkanes compose the oil contamination [2]) include birch bark and glass wool.

Studies describing the separation of oil contamination from water by sorbents tend to focus on the removal degree of the total content of oily substances without considering of lowering the concentration of the individual components. Some of these studies [18] report that loose natural wool fibers are able to absorb 74% of oil pollution from water, the recycled wool-based nonwoven material – 78%, while inorganic sorbent sepiolite absorbs up to 86% of oil pollution. Other studies [19] indicate that the silkworm coco on waste reduces the amount of oil by more than 50% after 20 minutes contact with the model solution. Maximum removal efficiencies obtained in the above experiment are higher than those given in the literature, however, it should be noted that in the cited literature data, a total mass of removed oily substances was the basis for the sorbent efficiency determination, while in the present experiment, there were petroleum products expressed as IOM and hydrocarbons of n-alkanes group.

4 Summary and Conclusions

Based on the studies, it was found that the applied porous materials removed the oil substances with varying efficiencies. It was also shown that not all of them remove at the same time the IOM and tested aliphatic hydrocarbons. Achieved results make it possible for a broader analysis of their application in the sorption process of oil products from water. The experiment compared the degree of mineral oil index removal and concentrations of n-alkanes C_7H_{16}-$C_{38}H_{78}$ during sorption on 4 porous materials: birch bark, cork, glass wool,

 DOI: 10.1051/e3sconf/20172200056

and polyurethane foam. After analyzing the test results, the following conclusions can be drawn:

- birch bark is characterized by the maximum effectiveness in reducing both IOM and the concentration of n-alkanes,
- glass wool is characterized by high removal efficiency of n-alkanes at various concentrations of oily substances in water,
- polyurethane foam is the least efficient in terms both of the IOM reduction and concentration of n-alkanes.

References

1. W. Meissner, Wiadomości ekologiczne **51**, 1, 17–34 (2005)
2. T. Graczyk, Ł. Piskorski, *Protecting marine environment from petroleum pollution* (Wydawnictwo Uczelniane Politechniki Szczecińskiej, Szczecin, 1996) (in Polish)
3. J.M. Gutteter-Grudziński, *Study on the efficiency of deoiling bilge water using hydrocyclone sections and coalescing porous baffles* (Wydawnictwo Naukowe Akademii Morskiej w Szczecinie, Szczecin, 2012) (in Polish)
4. Ż. Polkowska, M. Błaś, *Presence of selected groups of compounds in water from the airport plate. State and anthropogenic changes of water quality in Poland* (Wydawnictwo Uniwersytetu Łódzkiego, Łódź, 2010) (in Polish)
5. Z. Pruszak, *Sea water. An outline of physical processes and environmental engineering* (Wydawnictwo IBW PAN, Gdańsk, 2003) (in Polish)
6. W. Wawrzyniak, *Seas and oceans pollution. Sources and substances* (Wydawnictwo Uczelniane Politechniki Szczecińskiej, Szczecin, 2004) (in Polish)
7. W. J. Tic, P. M. Pijarowski, Przem. Chem. **94**, 3 301–306 (2015)
8. E. Gomółka, Ochr. Sr. **463/1**, 23, 39–43 (1985)
9. A. Norizan, E. Mohanad, A. Ayman, Y. Chun-Yang, Environ. Technol. **33**, 4, 481–486 (2012)
10. M.O. Adebajo, R.L. Frost, J.T. Kloprogge, O. Carmody, J. Porous Mat. **10**, 3, 159–170 (2003)
11. A. Bayat, G. Vakili-Nezhaad, Chem. Eng. Technol. **28**, 12, 1525–1528 (2005)
12. O.A. Galblaub, I.G. Shaykhiev, S.V. Stepanova, G. Timirbaeva, Process Saf. Environ. **101**, 88–92 (2016)
13. M. Husseien, A.A. Amer, A. El-Maghraby, N.A. Taha, Int. J. Environ. Sci. Te. **6**, 1, 123–130 (2009)
14. C. Hyungmin, M.C. Rinn, Environ. Sci. Technol. **26**, 772–776 (1992)
15. Z. Jiang, J. Ge, Y. Zhou, Z. Wang, D. Chen, S. Yu, H. Jiang, NPG Asia Materials, **8** (2016)
16. M. Radetic, V. Ilic, D. Radojevic, R. Miladinovic, D. Jocic, P. Jovancic, Chemosphere **70**, 525–530 (2008)
17. O.V. Rotar, D.V. Iskrizhitskaya, A.A. Iskrizhitsky, A.A. XV International Scientific Conference Chemistry and Chemical Engineering in XXI century, 145–150 (2014)
18. V. Rajakovic, G. Aleksic, M. Radetic, Lj. Rajakovic, J. Hazard Mater. **143**, 494–499 (2007)
19. H. Moriwaki, S. Kitajima, M. Kurashima, A. Hagiwara, K. Haraguchi, K. Shirai, R. Kanekatsu, K. Kiguchi, J. Hazard Mater. **165**, 266–270 (2009)

Influence of PCMs on thermal behavior of building walls: experimental study using the walls of a reduced scale room

Ayoub Gounni[1,*], *Mohamed Tahar* Mabrouk[2], *Abdelhamid* Kheiri[2], and *Mustapha* El alami[1]

[1]LPMMAT, Faculté des sciences Ain Chock, Université Hassan II, Km8 Route d'el Jadida B.P 5366 Maarif Casablanca 20100 Maroc
[2]Université de Lorraine, Lemta, UMR CNRS 7563, 2 avenue de la Forêt de Haye, Vandœuvre-lès-Nancy, F-54500 France

Abstract. Using Phase-Change Materials (PCM) for lightweight building applications can increase equivalent thermal mass and provide energy savings. In the present experimental work, the heat transfer performance testing of some building walls, with or without PCM, is carried out using a reduced-scale cubic room (the test-cell). The cubic cell is heated by an incandescent bulb placed on its centre, and it is housed in an air-conditioned large-scale room that allows to control the ambient air temperature. The effect of the double PCM layer and of its location relatively to the outside surface of the wall is tested and discussed in terms of overall transmitted heat flux and in terms of reduction of the inside and outside surface temperatures. Findings shows that the additional inertia introduced by the PCM leads to a reduced overall heat flux transmission by the wall and to a lesser daily temperature amplitude on the surface of the wall that enhances the thermal comfort inside the building. In the next step of this work, the case of sandwich walls with air gap, and with wood and PCM layers will be considered.

1 Introduction

Building envelopes significantly impact the human thermal comfort. Improving the thermal performance of these envelopes can help achieve the desired thermal comfort. This can be obtained by enhancing the thermal resistance and inertia of the walls [1]. However, design limitation of the thickness and the weight of the wall implies that the enhancement of the thermal resistance and inertia cannot be easily set on the desired values. An alternative solution to improve the thermal performance of these envelopes is to enhance its thermal inertia and storage capacity by integrating Phase Change Materials (PCMs).

Recently, the integration of the PCM in building envelopes have been successfully validated for reducing heating and cooling energy demand and increasing the thermal comfort [2, 3]. The most previous research papers showed a reduction in peak heat flux across the PCM enhanced walls, a time delay related to the peak cooling-heating load and a reduction of the fluctuation of the indoor air temperature [4].

* Corresponding author: gounni.ayoub@gmail.com

However, there are some design challenge still pending: the most critical one is the integration method of the PCM into building envelop i.e. the optimal thickness and location in the building wall. In a recent review, Konuklu et al. [5] highlighted the technique of the PCM integration into building. There is a class of technics that are easy to implement, among which direct incorporation, immersion and macro-encapsulation. But those techniques faces the serious problem of the leakage of the PCM. The second class of PCM integration technics is the micro-encapsulation of the PCM. This last method, proposed as a solution to overcome the problem of the leakage of the PCM seems to be the most suitable for the PCMs integration in the layers of the building sandwich wall.

In this paper, the used PCM layer is laminated within two aluminium sheets and constitute a part of the sandwich wall. This integration method is referred as the thermal shield [6, 7]. The aim of this research paper is to experimentally investigate the effectiveness of the PCM layer in reducing the surface temperature and the heat flux across the wall. The used test-cell simulates in reduced scale and under laboratory conditions a room of a building with sandwich walls. The effect of the PCM layer and of its location relatively to the outside surface of the wall is presented and discussed in terms of overall transmitted heat flux and in terms of reduction of the inside and outside surface temperatures.

2 Experimental device

2.1 The used PCM

The used PCM in the present study consists of a panel, with two aluminium sheets 130 microns thick that contains a solid copolymer (ethylene, 40%) and paraffin (60%) with properties shown in the Table 1. The panel edges are sealed with an aluminium tape 75 microns thick. The final form of the composite PCM is a flexible panel, 0.5 cm thick, as shown in the Fig. 1. It can be integrated directly into the structure during the construction of the building or added during the building renovation.

Table 1. Thermal properties of PCM.

Fusion temperature	21.7°C–31 °C
Latent heat of fusion (0°C–30°C)	> 70 kJ/kg
Total heat storage capacity (Temperature range 0°C to 30°C)	~ 140 kJ/kg
Thermal conductivity (solid phase)	0.18 W/(m·K)
Thermal conductivity (liquid phase)	0.14 W/(m·K)

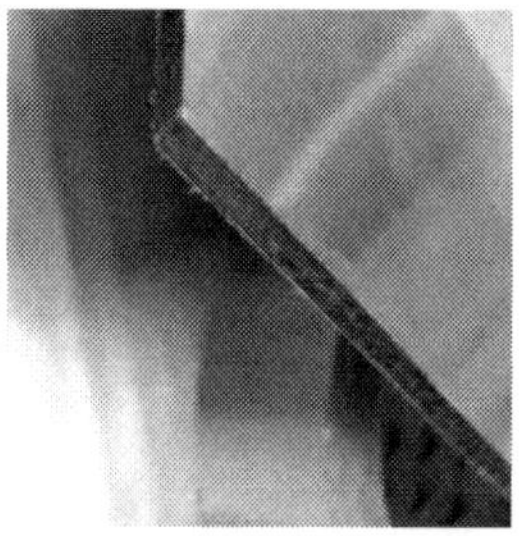

Fig. 1. PCM used in this study.

 DOI: 10.1051/e3sconf/20172200057

2.2 Test-cell

The heat transfer performance testing of the walls is carried out using a cubic room built with reduced scale, here termed as test-cell. The vertical walls of this test cell are removable and exchangeable in order to discuss the thermal insulation and inertia with or without PCM of several configurations. The test cell dimensions are 0.4 m x 0.4 m x 0.4 m. The heating of the test cell is provided by an incandescent bulb placed at the centre of the cubic cell. The test-cell is housed in an air conditioned large-scale room. The used air conditioning system permits the control of the temperature of the ambient air, Fig. 2. The four vertical wall of the test-cell are: a glazed wall (wall 1), a wooden wall (Wall 4) and two PCM walls (Wall 2 and Wall 3). The successive layers with their respective thickness are given from indoor toward outdoor as shown in the Fig. 3. At real scale, the thicknesses would be 15 cm and 5 cm, respectively, for the wood (or equivalent) layer and the PCM layer.

Notice that the inside surfaces of the three walls Wall 1, Wall 2 and Wall 3 are the same (wood surfaces) and hence they have the same radiation properties, i.e. they have the same radiative absorptivity and emissivity.

Fig. 2. Test cell placed in a room with controlled temperature.

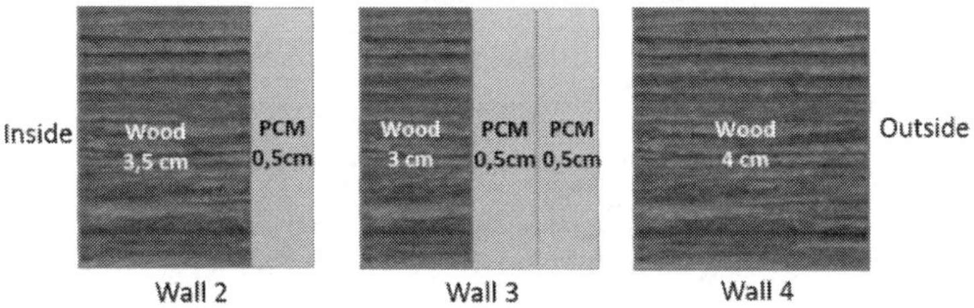

Fig. 3. Composition of test cell walls.

2.3 Measurement devices

Using k-type thermocouples (2/10 mm), with 0.1°C as error, the temperature measurements at the indoor and at the outdoor of the test-cell and the parietal temperatures of the walls (i.e. interior and exterior surfaces) are performed. The heat flux density through each wall is measured using a 2.54 cm diameter heat flux meters, with an accuracy of ±3% and a 0.3 s response time. The flux meters are placed on the inside of the surface of the walls. All the temperature and flux measurements was carried out with time intervals of 5 minutes.

 DOI: 10.1051/e3sconf/20172200057

3 Results and discussion

3.1 Operatives conditions

The experimental test-cell is sets up in such a way that its interior and its exterior sides can simulates the interior or the exterior condition of a real room. The experimental data analysed concern two cycles of heating with different time period. In the first cycle, the heat source inside the test cell is switched "on" for a successive time duration of 11.6 h, and then switched "off" for 3.75 h. In the second cycle, the heat source is switched "on" for 4 h and then switched "off" for 2 h. The test-cell indoor maximum temperature set-point is 37°C. Furthermore, in order to maintain a sufficient temperature difference between the outdoor and indoor of the test-cell, the outdoor air temperature of the test-cell, i.e. the laboratory room temperature is kept constant (Fig. 4) at 19°C thanks to the aforementioned air conditioning system. Under these conditions, the interior of the test cell simulates the exterior of a building at real scale in the summer season.

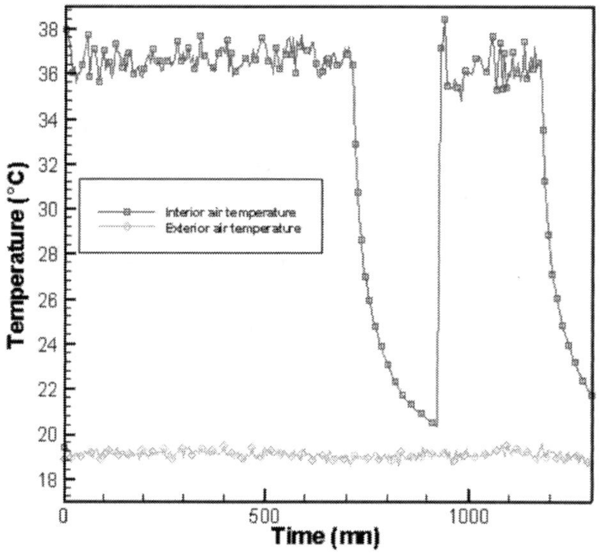

Fig. 4. The exterior and interior air temperature.

In the following parts, a qualitative study based on the parietal temperature profiles will be performed in the first section. A quantitative study will be performed in the second part of the paper in which the heat flux densities across the four vertical walls are quantified.

3.2 Thermal profiles

Fig. 5 and 6 show the inside surfaces temperatures of the four walls during the two heating-cooling cycles. As shown in those figures, the inside surface temperature of the Wall 1 (glass) is the lowest. This behavior indicates a less thermal resistance and also a less thermal inertia. Because of those two facts, it releases a huge heat flux to the other side. In the three other walls, the massive and resistive layers of wood and PCM decrease the heat flux that crosses the wall and hence, as shown in Fig. 5, reduce the outside surface temperature of those walls, i.e. Wall 2, Wall 3 and Wall 4.

During the heating phase, it is noticed from Fig. 5 that the outside surface temperature of the glazed wall (Wall 1) is the higher. This confirms the fact that the glass had the higher thermal conductivity.

On the other hand, it can be clearly seen that PCM walls (Wall 2 and Wall 3) can reduce significantly the outside surface temperature compared to the wooden wall (i.e. Wall 4). This occurs because PCM layer reaches its interval melting temperature and then would melt as a function of temperature, which translate to a reduction on the surface temperature. Recalling that in these operative conditions, the outside surface of the walls simulates the inside surface temperature of a real room in summer conditions, the lower is this surface, the lower is the heat flux rate transmitted and the best is the thermal comfort in the room. This is the reason why the comparison between the walls is made only on the wall surface temperature.

The outside surface temperature of the walls with PCM layer (i.e. Wall 2 and Wall 3) is the same with few more reduction for the wall with double PCM layer (i.e. Wall 3). This is because the second PCM layer of Wall 3 don't occur a fully melting which translate to a very few reduction on the outside surface temperature compared to Wall 2. Actually, the first PCM layer of the Wall 3 stores the majority of heat coming from heat source, which prevent the second PCM layer to occur a fully melting.

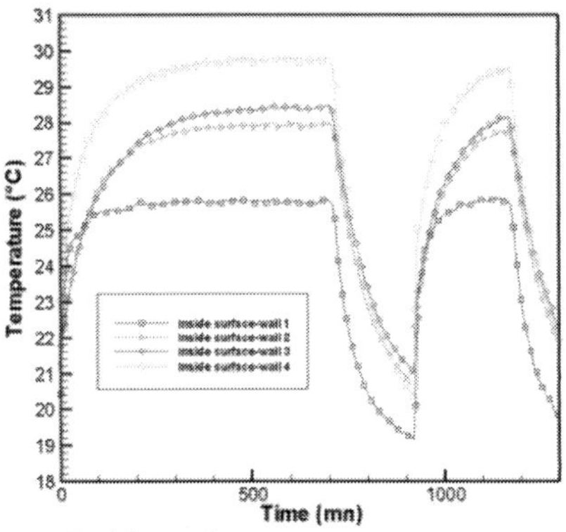

Fig. 5. Inside surface temperature.

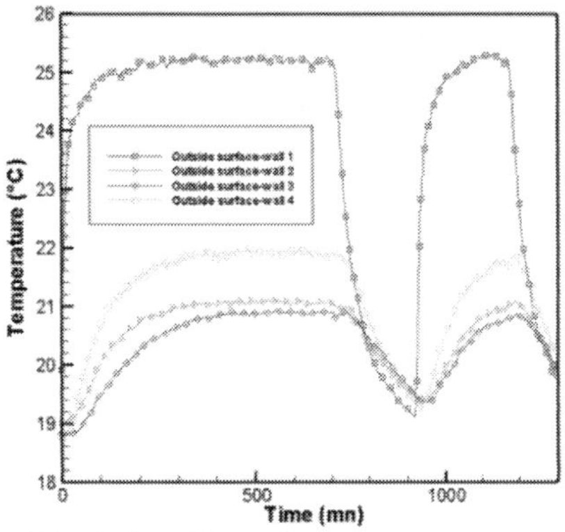

Fig. 6. Outside surface temperature.

Based on this finding, it is concluded that, the PCM layer reduces significantly the inner surface temperature of a real room that experiences summer conditions. The double PCM layer (Wall 3) seems to achieve a less reduction of the surface temperature (Fig. 6) than the single PCM layer (Wall 4) but, in the same time its surface temperature (Fig. 5) is higher than the one of Wall 4. To have a comprehensive understanding about the effect of PCM layer on the thermal Inertia of the wall, a quantitative study is done on the following section.

3.3 Heat flux densities through the walls

The fluxmeters sensors are placed at the inside walls surfaces in order to compare the absorbed flux for the four walls. Fig. 7 shows the heat flux across the four walls at this location. Firstly, it can be seen that the flux density related to the glass wall (Wall 1) is greater than the ones of the other walls. However, the walls that contains the massive layers (i.e. wood and PCM) has the lowest absorbed heat flux. This measure results confirm the results about heat flux that one can deduct from the thermal profiles of Fig. 5.

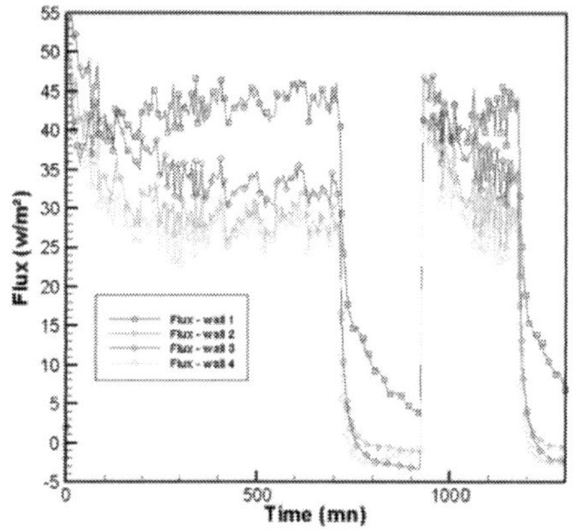

Fig. 7. heat flux across the four walls at the inside surface.

3.4 Stored heat flux

In this section, the stored amount of heat in the PCM layer contained in the two PCM walls (i.e. wall 2 and wall 3) is presented and discussed. To achieve this objective, the conductive heat flux, which crosses the wood layer of PCM walls, is calculated based on the following equations:

$$q_{cd} = \frac{\lambda}{e} \Delta T \tag{1}$$

with
λ: Thermal conductivity of the wood layer which is 0.14 W/(m.K);
e: Thickness of the wood layer which is 3.5cm and 3cm, respectively, related to the Wall 2 and Wall 3.
ΔT: is the measured temperature difference between the inside and outside wood surface (°C).
Then, the heat flux stored into PCM layer is quantified as following:

$$q_{st} = q_{in,meas} - q_{cd} \tag{2}$$

where
$q_{in,meas}$: Measured inlet heat flux related to the Wall 2 and Wall 3 (W/m²).

First of all, it should be emphasized that the contact between the wall layers is assumed to be perfect. This means that the thermal resistance between the wall layers is zero. The stored heat flux related to the Wall 3 is quantified between the two PCM layers. This means that the temperature difference ΔT is calculated between the two PCM layers. Fig. 8 shows the stored and released heat flux in the PCM layer of the PCM walls. Based on the comparison between the figures 7 and 8, during the heating period, the PCM layer stores the major part of the inlet heat flux. We notice that the double PCM layer stores more heat flux. This can be obviously explained by the thickness of the double PCM compared to the single PCM. Therefore, we can conclude that, under these conditions, the walls outfitted with single and double PCM behave in the same way. During the heating off period, the PCM layer releases (negative flux rate in Fig. 8) the absorbed heat flux, its value reaches 12 and 16 W/m², respectively, for the single and double PCM layer and then decreases until 0 W/m². The time period of the heating off was not sufficient to estimate the time needed to release the entire heat, i.e. to have a value of 0 W/m².

DOI: 10.1051/e3sconf/20172200057

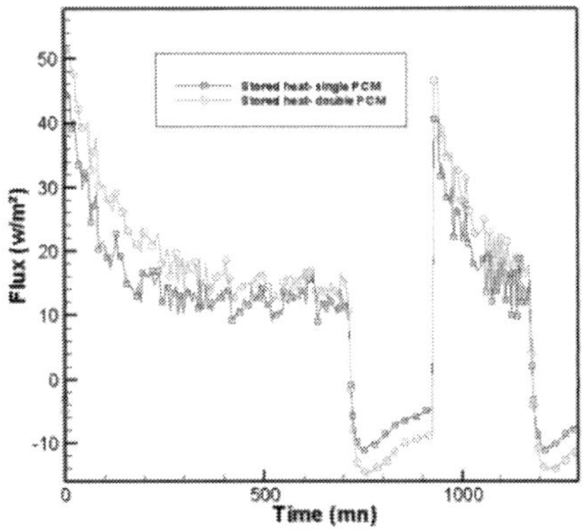

Fig. 8. Stored heat flux in the single and double PCM layer.

4 Conclusion

In this experimental work, the thermal performance of a building wall that contains a single or two PCM layers is evaluated, in term of the surface temperature and of the stored heat flux. A test-cell cavity at reduced scale placed in a thermally controlled room was used. The results shows that: The PCM layer reduces the surface temperature by storing a large amount of absorbed heat flux and then increases the thermal Inertia of the wall because the inside surface temperature of the walls outfitted with PCM layers is reduced due to its important latent heat of fusion. The thermal comfort is thus enhanced by walls with PCM layers.

- The time needed to release the stored heat flux is over 4 hours.
- Under the experimental condition, the wall with double PCM layer behaves like the wall with single PCM. This happens because in the second PCM layer of the wall with double PCM layer occur only a partial melting.
- Further tests are recommended with large heating off period in order to estimate more accurately exactly the time needed to release the entire stored heat. Indeed, a higher heat source set-point temperature is recommended to reach more quickly the interval melting temperature of the second PCM layer of the wall with double PCM layer.

References

1. X. Jin, A. Medina, X. Zhang, Appl. Energ. **106**, 72–78 (2013)
2. A. de Gracia, L. Cabeza, Energy Build. **103**, 414–419 (2015)
3. H. Jamil, M. Alam, J. Sanjayan, J. Wilson, Energy Build. **133**, 217–229 (2016)
4. X. Jin, A. Medina, X. Zhang, Appl. Energ. **106**, 72–78 (2013)
5. K. Pielichowska, K. Pielichowski, Progress in Materials Science **65**, 67–123 (2014)
6. K. Lee, A. Medina, E. Raith, X. Sun, Appl. Energ. **137**, 699–706 (2015)
7. X. Jin, A. Medina, X. Zhang, Energy **73**, 780–786 (2014)

Pretreatment methods as a means of boosting methane production from sewage sludge and its mixtures with grease trap sludge

Anna Grosser[1,*], and *Ewa* Neczaj[1]

[1]Czestochowa University of Technology, Institute of Environmental Engineering, ul. Brzeznicka 60 A, Czestochowa, Poland

Abstract. The main objective of this study was to determine the applicability of the selected pretreatment methods as a means of intensification of methane production from sewage sludge as well as its mixtures with grease trap sludge. The addition of the fat rich material to the digester treating sewage sludge resulted in an increased methane yield as well as volatile solids (VS) removal of up to 36% (from 134.75 mL/g VS to 182.84 mL/g VS). Furthermore, thermochemical pretreatment of the co-digestion mixture resulted in an approximately 76% higher methane yield as compared to the untreated sewage sludge. The energy balance showed that, for both materials ultrasonic pretreatment and thermochemical pretreatment has an energy self-sufficiency. All of the tested models fit the experimental data with coefficients of determination higher than 0.96.

1 Introduction

Energy recovery from sewage sludge provides an opportunity not only for sustainable management of this waste, but it is also an interesting option from the energy production standpoint [1]. Biogas which is one of the renewable energy carriers can be used as a substitute for fossil fuels in electricity production or as a fuel for combustion vehicles [2]. Rising fossil fuel prices and the growing interest in renewable energy sources, justifies the attempts of intensifying biogas production from sewage sludge [3, 4]. Disadvantages of anaerobic digestion of sewage sludge such as long retention time of sludge in the digester or a low degree of volatile solids (VS) removal are an additional reason for the seeking of new solutions for enhancing the efficiency of this process [3, 5]. From the available intensification options for anaerobic digestion of sewage sludge, two seem particularly interesting, namely co-digestion with other organic wastes with high energy potential (e.g. fatty wastes) and pretreatment of the feedstock prior to its introduction to the digester, methods combining both of the mentioned should also be taken into consideration [6, 7]. Benefits as well as drawbacks of the techniques mentioned above are well documented in literature [3, 4, 6, 8, 9]. However, the overwhelming part of these publications relates to sewage sludge. Nevertheless, in the context of the co-digestion process, pretreatment of

* Corresponding author: agrosser@is.pcz.czest.pl

feedstock also seems to be an interesting approach. However, the available data is still limited due to the fact that pretreatment methods employed for co-digestion have become more widely described in literature only starting from the year 2011. Therefore, more research needs to be carried out, for example in order to assess how pretreatment affects the process performance. For these reasons, the objective of this study was to determine the applicability of the selected pretreatment methods as a means of intensification of methane production from sewage sludge as well as its mixtures with grease trap sludge. Fat rich material was selected due to its high energy potential. As literature data show, theoretically, about 1.425 L of biogas can be produced from 1 g of VS lipids, while only 0.830 L, 0.921 L can be produced respectively from 1 g of VS carbohydrates and protein [10]. In the context of the legal requirements related to them, such as veterinary regulations i.e. the European Commission Regulation (EC) 1069/2009 and its corresponding Implementing Regulation (EC) 142/201, it seemed reasonable to test the thermochemical pretreatment (Tch) as one of the methods. Especially that the conditions of the proposed method (similar to the biological pretreatment (BP) defined as temperature phased anaerobic digestion (TPAD) with hyperthermophilic conditions for hydrolysis) are consistent with one of the types of thermal pretreatment which is recommended for sanitising animal by-products from the meat-processing industry (namely processing the material at 70°C for at least 60 min.) [2]. The suggested addition of sodium hydroxide also may intensify the process due to fact that sodium at low concentrations is necessary for the methanogenic bacteria, because it plays a significant role during the formation of adenosine triphosphate (ATP) as well as the oxidation of the NADH coenzyme [11, 12]. Moreover, saponification (i.e. reaction between a hydroxide and fats or more precisely triglycerides) results in the formation of long chain fatty acid salts, which may lead to the decrease in the toxicity of these compounds as well as improve the contact between microorganisms and digested materials, thereby enhancing their biodegradability [13, 14]. Furthermore, thermochemical as well as ultrasonic pretreatments (UD) are indicated as the most effective and economically viable methods [15]. Additionally, in the study kinetic parameters were estimated using the following methods: the transference function, the modified Gompertz model and the logistic function. Additionally, an energy balance evaluation, which is often neglected in literature, was also carried out [15].

2 Materials and methods

2.1 Materials

The biochemical methane potential (BMP) assays were carried out for two kinds of materials. Sewage sludge (Ss) (a mixture of primary sludge and waste activated sludge) which was collected from a wastewater treatment plant (WWTP) in the Silesia region of Poland. The second material was the co-digestion mixture (CdM) containing (on volatile solids basis): 30% grease trap sludge and 70% sewage sludge. Grease trap sludge was obtained from a meat processing plant (the Silesia region of Poland). Characteristics of both substrates as well as the inoculum (from the WWTP) is shown in Table 1.

2.2 Experimental procedure

During the experiment the following pretreatment methods were tested: ultrasonic (UD), thermochemical (Tch) and biological pretreatment (BP). The highest dose of sodium was below the inhibition threshold reported by Appels et al. [11], namely 3,500 mg/L. Regardless of the method, the experiment was divided into two stages. In the first phase the

 DOI: 10.1051/e3sconf/20172200058

selection of the most favourable pretreatment conditions took place. Based on the obtained results in the second phase methane potential assays were conducted for the optimal conditions for each type of pretreatment. Selection was made based on the changes of volatile fatty acids (VFAs), total organic carbon (TOC) and chemical oxygen demand (COD) in the supernatant. The BMP assays were carried out using customised equipment consisting of Dreschel bottles, fermentation bottles, an inductive stirring system and biogas collection unit. The batch medium for the BMP assays was prepared according to the procedure described by Owen et al. [16]. The inoculum before introduction to the fermentation bottles was "degassed" by preincubation in an incubator (37°C) for 5 days. The pH values of the feedstock in the fermentation bottles were adjusted to 7.0 using 1 M HCl or 1 M NaOH. Additionally, before closing the fermentation bottles, their headspaces were flushed with nitrogen gas. The bottles were mixed using magnetic stirrers only during gas measurements. Details of the experiment are shown in Fig.1.

Table 1. Characteristics of the batch mixtures used in the study.

Indicator	Unit	Material		
		Inoculum	Sewage sludge	Co-digestion mixture
TS	g/L	15.31–19.51	23.7–24.98	30.32–31.83
VS	g/L	9.92–12.4	18.12–18.72	24.78–25.51
pH	-	7.43–7.91	5.73–6.43	5.63–5.83
VFAs	$mgCH_3COOH/L$	674–846	2606–3314	2714–3371

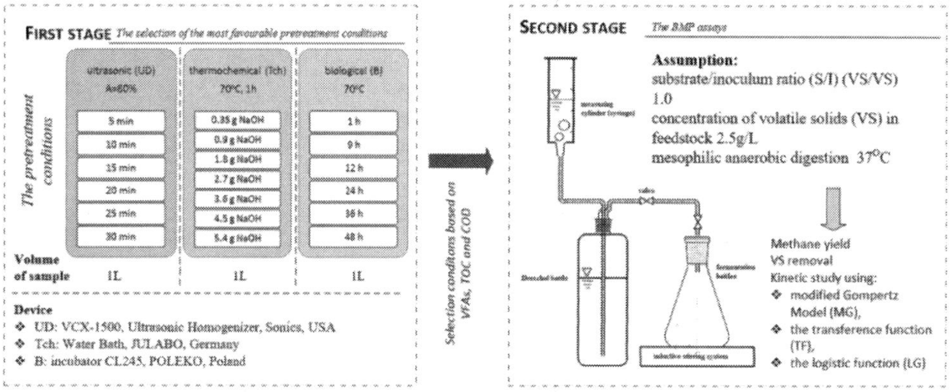

Fig. 1. Scheme of the experiment.

2.3 Analyses and calculations

Measurement of pH values, VFAs, COD and TOC in the supernatant (obtained by centrifugation of the samples at 12100 rcf for 15 minutes using the Eppendorf centrifuge type 5804, followed by filtering through filter papers (3w)) as well as total solids (TS) and volatile solids (VS) were determined according to APHA [17]. In turn biogas production was measured using the water displacement method, while the content of methane in the biogas was analysed by a portable gas analyser (NANOSENS DP-27 BIO+, Poland). All obtained gas measurement results were calculated at standard temperature and pressure (STP). Additionally, based on the obtained results, the kinetic parameters were estimated using the three following models: modified Gompertz model, the logistic function and the transference function. The equations used in these estimations are presented in study by Li et al. [18]. The estimations of kinetic parameters were made in STATISTICA software using non-linear regression. The above-mentioned software was also used to determine how the tested pretreatment methods affect VFAs, COD and TOC as well as methane yield (one way-ANOVA). In the case of finding statistically significant differences between data a post hoc Tukey honest significance difference (HSD) was carried out. Additionally, the hydrolysis constant rate was estimated using the first-order kinetic model in accordance with the equation presented in [19]. Moreover, for all of the tested pretreatment methods the energy balance was made. Equations and some theoretical assumptions applied for the energy balance came from the work of Dhar et al. [20] and Cho et al. [21].

3 Results and discussion

3.1 Selection of the most favourable pretreatment conditions

The one-way ANOVA showed significant differences between tested pretreatment methods (see results of the Tukey test in Table 2). Values of COD and TOC for both materials increased gradually with the increase in the sonication time. This is in agreement with the results of many studies reported in the literature [3, 9]. The highest values of COD (TOC) were observed for the 30 min. exposure, for which, in comparison to the control sample (without conditioning) an increase of up to 157% (83%) and 183% (116%) for Ss and CdM, respectively were recorded. While, for the thermochemical pretreatment, the observed increase of COD (TOC) content was in the range from 111% to 337% (from 71% to 264) and 111% to 601% (from 53% to 413%) for Ss and CdM, respectively. In turn for BP, the improvement in COD and TOC content were proportional to the incubation time, reaching values of up to 153% and 93% for Ss and 163% and 84% for CdM. The obtained results (Table 2) also showed that the tested pretreatment methods had the slightest impact on the VFAs concentrations. The longest sonification pretreatment time resulted in a 3% VFAs increase for both materials, while saponification allowed for an increase of this indicator by 15% and 8% for Ss and CdM, respectively. An opposite trend was observed for BP. In the case of this method VFAs concentrations decreased along with the extension of the incubation time.

As shown in Fig. 2 the values of specific energy input (the amount of energy supplied per the initial concentration of total solids in the substrate) for sewage sludge and co-digestion mixture ranged from 4,232 kJ/kgTS to 18,386 kJ/kgTS and from 3,263 kJ/kgTS to 15,134 kJ/TS, respectively. Thus, for almost all of the tested conditions the values were in the typical range reported in the literature, namely from 1,000 to 16,000 kJ/kg TS and depends on the content of total solids in the feedstock [3]. For both materials the total solids content was in the optimal range for sonication (i.e. between from 2.3% to 3.2% TS) [22].

 DOI: 10.1051/e3sconf/20172200058

Table 2. The changes of volatile fatty acids (VFAs), total organic carbon (TOC) and chemical oxygen demand (COD) in the supernatant during the experiment (cases marked with the same letter are not significantly different, p > 0.05).

Methods	**Ss**			**CdM**		
	COD (mg/L)	TOC (mg/L)	VFAs (mg/L)	COD (mg/L)	TOC (mg/L)	VFAs (mg/L)
UD0	5918±207a	1870±9a	3314±20a	6458±153a	1950±2a	3371±20ab
UD5	10050±348b	2270±2b	3343±75ab	11369±692b	2510±9b	3326±91a
UD10	11575±246c	2700±15c	3343±17ab	12863±585c	3100±8c	3371±20ab
UD15	12763±390cd	3010±1d	3349±49ab	14031±303c	3360±2d	3371±52ab
UD20	13175±74d	3100±5e	3371±71ab	15644±277d	3890±1e	3463±69ab
UD25	14538±231e	3440±1f	3417±20ab	16813±255d	4440±2g	3486±52b
UD30	15238±1060e	3420±16f	3463±34b	18319±660e	4220±5f	3463±59ab
Tch0	4818±219a	1500±23a	2606±69a	5115±62a	1670±20a	2714±77a
Tch0.35	10213±370b	2570±93b	2709±91a	10800±336b	2550±79b	2806±10b
Tch0.9	11525±98b	2750±23b	2857±20b	13542±208c	3280±50c	2834±20bc
Tch1.8	13475±337c	3700±93c	2846±34b	19781±353d	5000±89d	2869±20bcd
Tch2.7	15788±232d	4230±62d	2869±20b	28981±865e	6880±105e	2874±10bcd
Tch3.6	17538±509e	4540±80e	2949±34bc	33213±477f	7660±80f	2909±10cde
Tch4.5	21056±879f	5500±90f	3006±20c	34100±152f	8610±38g	2943±10de
Tch5.4	21063±641f	5460±92f	3006±20c	35900±1126g	8580±99g	2989±10e
BP0	4818±219a	1500±2a	3314±20e	5115±62a	1670±2a	3371±20c
BP1	7406±198b	1470±8a	2720±20a	8538±280b	1750±2b	2974±12b
BP9h	10233±611c	2400±1b	2769±12ab	11244±321c	2410±5c	2949±0b
BP12h	10958±644cd	2510±2c	2811±0b	11650±402cd	2730±2d	2897±24a
BP24h	11394±311de	2780±1d	2903±52c	12400±213cde	2840±0e	2949±0b
BP36h	12169±307ef	2890±30e	2957±12c	13050±1096de	3040±2f	2880±0a
BP48h	13033±263f	2990±2f	3051±0d	13488±734e	3080±7g	2949±0b

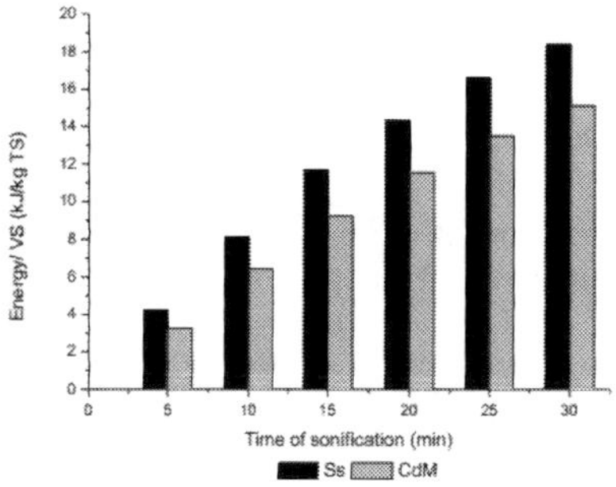

Fig. 2. The specific energy input during sonication of the samples.

The pretreament conditions for the BMP assays were selected based on the results shown in Table 2 as well as the analyses of specific energy inputs. They were as follows: UD 5, 10, 15 and 20 min. of sonification time; Tch 3.6, 4.5 and 5.4 g/L dose of sodium, and for BP 24, 36 and 48 h incubation time at 70°C. It should be noted that the chosen retention time for BP is acknowledged in literature as an optimum for the TPAD pretreatment of waste activated sludge [23].

3.2 BMP assays results

The observed methane yields for Ss as well as for CdM are mostly lower in comparison with the data found in literature. This discrepancy may be explained by the impact of the following factors: the study scale (batch assays, semi-continuous or continuous systems), the origin/source of waste, the content of VS, the characteristics of the inoculum (the origin, activity, concentration), composition of the basic medium, the substrate-inoculum ratio as well as the experiment conditions or lack of results given for standard conditions (STP) [24, 25]. Despite this, the addition of fat rich material into the feedstock resulted in an increased methane yield of approximately 36% (from 134.75 mL/g VS to 182.84 mL/g VS) (Fig. 3.A), which is a result higher than the one reported by Davidsson et al. [26], who noted an increase of the specific methane yield by 9–14 %. The one-way ANOVA showed significant differences between the tested pretreatment methods (Fig. 3A). However, the kind of the used pretreatment method had a significantly bigger effect on the methane yields noted for Ss ($F = 16.7$, $p = 0.00$) than for the CdM (F 3.5, $p = 0.07$). The highest increase of efficiency of the anaerobic digestion for both materials were obtained for the samples prepared using the thermochemical pretreatment. This action resulted in an increased methane yield of up to 184.28 mL/g VS (the sewage sludge pretreated with 4.5 g NaOH) compared to 134.75 mL/g VS noted for the control sample. While, thermochemical pretreatment of CdM resulted in an approximately 30% higher methane yield as compared to the untreated mixture, what is more, this value was 76% higher than that recorded for anaerobic digestion of sewage sludge without conditioning. Thus, the enhancement of methane production via saponification of the CdM was significantly higher than the results reported in the literature. For instance Carrere et al. [13] reported a very slight impact of this method on the methane potential for batch co-digestion of waste-activated sludge and fatty wastewater (an enhancement ranging from 4 to 7% – similar to the results obtained in this study for saponification with a NaOH dose higher than 3.6g/L). Also, Battimelli et al. [14] noted a very little increase in the volume of biogas (from 1222 to 1284 and from 1126 to 1166 for aeroflotation fat and carcass fat, respectively). In turn, Li et al. [27] noted an improved ultimate methane production by $9.9 \pm 1.5\%$for a mixture of waste activated sludge and fat, oil grease treated at $pH = 10$ and 55°C. In turn for UD pretreatment the noted methane yield for the treated materials generally increased with the increase of specific energy input (Fig. 3A). The methane yield was enhanced by 6–21% for the digester fed with sonified sewage sludge, while, in the literature biogas enhancement ranges from 24% to 140% for batch systems, these differences in the efficiency of the UD pretreatment may be linked with sewage sludge characteristics [9]. In comparison to the control sample, methane yield increased by approximately 22% (from 182.85 mL/g VS to 222.41 mL/g VS). Thus, this study showed an opposite tendency that the ones reported by Li et al. [27] or Luste et al. [28], who noted that ultrasonic pretreatment does not always lead to any significant improvement in the ultimate methane production. Furthermore, sometimes the results obtained for the treated materials are lower in comparison to the digestion without ultrasonic pretreatment. While for BP of Ss, the maximum methane yield was reached for an untreated sample. The opposite trend was observed for CdM. This method allowed for obtaining an approximately 18% higher methane yield compared to the control sample (untreated CdM). Generally, the use of the selected methods resulted in an increase of volatile solids removal (Fig. 3B). The only exception being the case of the biologically pretreated samples, for which regardless of incubation time, lower volatile solids removal compared to the control sample was observed.

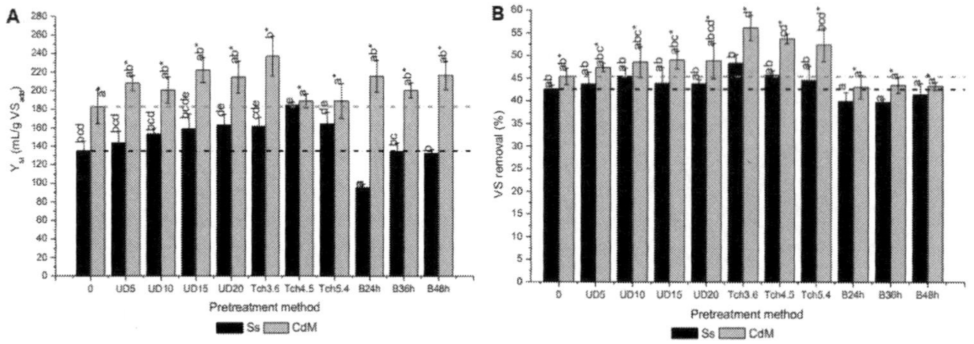

Fig. 3. The methane yield (A) and VS removal (B) of untreated and treated samples (bars marked with the same letter are not significantly different, $p > 0.05$; *- Tukey test for CdM).

3.3 Kinetic study

The Tch as well as UD considerably enhanced the hydrolysis reaction rate constant. While for BP an opposite trend was observed. In the case of Ss this kinetic parameter increased from 0.17 to 0.38/day and from 0.17 to 0.29/day for UD and Tch, respectively. It should be noted that the value for untreated sewage sludge was in the range reported by Zhen et al. [19], Zhang et al. [29] or Vavilin et al. [30]. In turn, for the untreated CdM as compared to the sewage sludge significantly lower hydrolysis reaction rate constant was noted (0.11/day). Tch resulted in approximately 173% higher value of this parameter as compared to the untreated mixture (from 0.11/day to 0.30/day), while for UD this kinetic parameter increased up to 209% (from 0.11/day to 0.34/day). The values of the coefficient of determination were higher than 0.96 indicating that all of the applied kinetic models showed a high degree of fit to the measured data. Moreover, the kinetic analyses showed that generally Tch and UD accelerated the maximum rate of methane production.

3.4 Energy aspect

The publications regarding pretreatment methods presented in scientific journals focus primarily on the aspect of anaerobic digestion efficiency enhancement resulting in the increase of methane yield as well as volatile solids removal. While, energy assessments are usually overlooked. This is shortsighted when taking into account that not all of the proposed pretreatment methods in scientific reports have an energy self-sufficiency, meaning that the potential gains do not offset the energy requirements of these methods.

In this study, the net energy production (defined as the difference between the output and input energy) was evaluated. As shown in Fig.4 the calculated net energy production showed that for Tch as well as UD the energy gain not only covers the requirements of the method itself but also allows for a significant improvement of the WWTPs profitability.

While, in the case of BP, the energy analyses showed that it is not economically viable due to energy consumption exceeding the energy obtained from the methane production (as shown in Fig. 4). These observations are similar to results noted by Dhar et al. [20], who reported that for anaerobic digestion of municipal waste activated sludge, ultrasound (1000 kJ/kg total suspended solids) and thermal pretreatments (50–90°C) are economically justified, since they allow to reduce operating costs by $44–66/ton dry solids in comparison to conventional anaerobic digestion without pretreatments. Also, Cano et al. [15] noted that thermal pretreatments have an energy self-sufficiency to be implemented in a wastewater treatment plant.

DOI: 10.1051/e3sconf/20172200058

The values for UD as well as Tch shown in Fig. 4. are much lower than the ones reported in literature. For instance, Serrano et al. [31] obtained a net energy production of 1963 J/gVS for co-digestion of pre-treated sewage sludge and strawberry extrudate. In turn, Cho et al. [21] reported a positive net energy production of 6300 J/g VS for anaerobic digestion of mixed-microalgal biomass after alkali pretreatment.

Moreover, the energy consumption during UD as well as Tch pretreatments were lower than the limit suggested by Cano et al. [15] (energy requirements should be below 0.2 kWh/m^3).

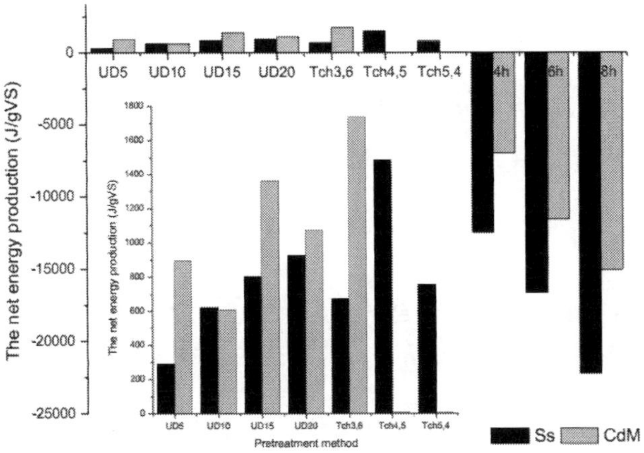

Fig. 4. The net energy production for pretreatment methods tested in this study.

4 Conclusion

The experimental results showed that both UD and TCh can improve the solubilisation of organic matter measured as the change of COD and TOC content positively affecting the methane yield. While BP only enhanced the methane production for CdM in comparison to the untreated sample. In terms of methane yield the experimental results showed the most promising approach seems to be the co-digestion of a Tch pretreated feed. The energy balance showed that only UD and Tch are energy self-sufficient.

Acknowledgements

The study was supported by the BS/PB-401-301/13.

References

1. V.K. Tyagi, S.L Lo, Renew. Sust. Energ. Rev. **25** 708–728 (2013)
2. J. Grim, P. Malmros, A. Schnürer, Å. Nordberg, Energy **79** 419–427 (2015)
3. H. Carrere, G. Antonopoulou, R. Affes, F. Passos, A. Battimelli, G. Lyberatos, G.I. Ferrer, Bioresour. Technol. **199** 386–397(2016)
4. P. Neumann, S. Pesante, M. Venegas, G. Vidal, Rev. Environ. Sci. Biotechnol. **15**, 2, 173–211 (2016)
5. A. Grosser, E. Neczaj, Energ. Convers Manage. **125**, 301–308 (2016)
6. J. Mata-Alvarez, J. Dosta, M.S. Romero-Güiza, X. Fonoll, M. Peces, S. Astals, Renew. Sust. Energ. Rev. **36**, 412–427 (2014)

7. F.A. Shah, Q. Mahmood, N. Rashid, A. Pervez, I.A. Raja, M.M. Shah, Renew. Sust. Energ. Rev. **42**, 627–642 (2015)
8. J. Ariunbaatar, A. Panico, G. Esposito, F. Pirozzi, P.N. Lens, Appl. Energy **123**, 143–156 (2014)
9. S. Pilli, P. Bhunia, S. Yan, R.J. LeBlanc, R.D. Tyagi, R. Y. Surampalli, Ultrason. Sonochem. **18**, 1, 1–18. (2011)
10. M.M. Alves, M.A. Pereira, D.Z. Sousa, A.J. Cavaleiro, M. Picavet, H. Smidt, A.J. Stams, Microbial. Biotechnology **2**, 5, 538–550 (2009)
11. Y. Chen, J.J. Cheng, K.S. Creamer, Bioresour. Technol. **99**, 4044–64 (2008)
12. L. Appels, J. Baeyens, J. Degrève, R. Dewil, Prog. Energ. Combust **34**, 6, 755–781 (2008)
13. H. Carrere, Y. Rafrafi, A. Battimelli, M. Torrijos, J.P. Delgenes, C. Motte, Chem. Eng. J. **210**, 404–409 (2012)
14. A. Battimelli, H. Carrère, J.P. Delgenès, J. P. Bioresour. Technol. **100**, 15, 3695–3700 (2009)
15. R. Cano, S.I. Pérez-Elvira, F. Fdz-Polanco, Appl. Energy **149**, 176–185 (2015)
16. W.F. Owen, D.C. Stuckey, J.B. Healy, L.Y. Young, P.L. McCarty, Water Res. **13**, 6, 485–492 (1979)
17. APHA. *Standard methods for the examination of water and wastewater* (American Public Health Association, Washington. DC, 1999)
18. L. Li, X. Kong, F. Yang, D. Li, Z. Yuan, Y. Sun, Appl. Biochem. Biotechnol. **166**, 5, 1183–1191 (2012)
19. G. Zhen, X. Lu, T. Kobayashi, Y.Y. Li, K. Xu, Y. Zhao, Appl. Energy **148**, 78–86 (2015)
20. B.R. Dhar, G. Nakhla, M.B. Ray, Waste Manage. **32**, 3, 542–549 (2012)
21. S. Cho, S. Park, J. Seon, J. Yu, T. Lee, Bioresour. Technol. **143**, 330–336 (2013)
22. K.-Y. Show, T. Mao, D.-J. Lee, Water Res. **41**, 20, 4741–4747 (2007)
23. H. Ge, P.D. Jensen, D.J. Batstone, Water Res. **45**, 1597–1606 (2011)
24. F. Raposo, M.A. De la Rubia, V. Fernández-Cegrí, R. Borja, Renew. Sust. Energ. Rev. **16**, 1, 861–877 (2012)
25. L. Angelidaki, M. Alves, D. Bolzonella, L. Borzacconi, J.L. Campos, A.J. Guwy, J.B. Van Lier, Water Sci. Technol. **59**, 5, 927–934 (2009)
26. A. Davidsson, C. Lövstedt, J. La Cour Janse, C. Gruvberger, H. Asperge, Waste Manage. **28**, 6, 986–992 (2008)
27. C. Li, P. Champagne, B.C. Anderson, Bioresour. Technol. **130**, 187–197 (2013)
28. S. Luste, S. Luostarinen, M. Sillanpää, J. Hazard Mater. **164**, 1, 247–255 (2009)
29. P. Zhang, C.J. Lin, J. Liu, P. Pongprueksa, S.A. Evers, P. Hart, Renewable Energy **68**, 304–313 (2014)
30. V.A. Vavilin, B. Fernandez, J. Palatsi, X. Flotats, Waste Manage. **28**, 6, 939–951 (2008)
31. A. Serrano, J.A. Siles, M. C. Gutiérrez, M.Á. Martín, **90**, 25–33 (2015)

Electrodialysis potential for fractionation of multicomponent aqueous solutions

Martyna Grzegorzek[1,*], and *Katarzyna* Majewska-Nowak[1]

[1]Wrocław University of Science and Technology, Department of Environmental Engineering, Wybrzeże Stanisława Wyspiańskiego 27, 50-370 Wrocław, Poland

Abstract. The paper aimed at the evaluation of the batch electrodialysis (ED) run in the course of treatment and desalination of various aqueous mixtures containing both mineral (sodium fluoride, sodium chloride) and organic substances (dyes or humic acids). The commercial ED stack (PCCell Bed) equipped with standard anion-exchange and cation-exchange membranes or monovalent selective anion-exchange membranes was used. The ED experiments were performed at a constant current density (1.56 or 1.72 mA/cm^2). The mechanism of ion migration as well as membrane deposition for variable solution composition and various membrane types was analyzed The calculated mass balance and electrical energy demand for each ED run were helpful in evaluating the membrane fouling intensity. It was found that the presence of organic substances in the treated solution had a minor impact on energy consumption, but rather strongly affected chloride flux. The extent of organics deposition was significantly lower for monovalent selective anion-exchange membranes than for classic anion-exchange membranes.

1 Introduction

Membrane technology is being increasingly used in the treatment of natural water as well as industry wastewater. Most of the currently conducted studies are focussed on water reuse and reducing pollution impact on the environment. Among membrane processes, the electrodialysis (ED) seems to be very viable technology for simple treatment and desalination of polluted natural and process water. However, electromembrane processes offer also a great opportunity to the recovery of valuable substances, separate charged components from neutral substances, and fractionation of mono- and multi-valence ions of the same sign. These unique advantages make the electrodialysis very attractive in view of environmental and economic benefits. On the other hand, the mechanism of mass transport through the specially designed ion-exchange membranes becomes more complex than through the classic ion-exchange membranes, especially when multicomponent solutions are treated by ED.

Membrane fouling is one of the key obstacles for ensuring cost-efficient operation of membrane systems. In the course of electromembrane processes, the anion-exchange

* Corresponding author: martyna.grzegorzek@pwr.edu.pl

membranes are especially prone to fouling by negative charged organics. As a result of this phenomena electrical resistance of the ED stack and power consumption increase.

Among many organic contaminants present in natural and industrial waters, the most studied as potential foulants are: methylamine [1], natural organic matter (NOM) [2–4], surfactants [5, 6], pestides [7], and dyes [8]. The mechanism of organic ion transport trough the anion-exchange membranes is as yet uncertain. Zhang at el. [1] studied the ED process of salt solutions containing various organics of ionic type, which varied in molecular weights. They found that the greater the molecular weight of organic particles the better their retention in diluate cells. Application of monovalent selective anion-exchange membranes resulted in improving of organics separation from mineral components. Similar results were found by Kabsch-Korbutowicz et al. [9] – the ED process with monoselective anion-exchange membranes enabled better regeneration of spent brines contaminated by NOM, than standard ED. Detailed investigation of electrodialytic fractionation of small organic ions from mineral salts revealed that the process efficiency was influenced by current density, organic ion size and charge, as well as by the type of functional groups fixed to organic particles [10]. It is interesting to note that the big organic ions (of molecular weight above 130 kDa) can be deposited at the membrane surface, thus creating an additional bipolar layer and alternating the membrane selectivity.

Contrary to popular opinion, surfactants can also caused a serious fouling of the anion-exchange membranes. Lee et al. [5] found that sodium dodecyl benzene sulfonate (SDBS) generated intensive surface fouling when its concentration was higher than the critical micelle concentration. However, single SDBS particles were able to migrate into membrane structure, thus increasing the electrical resistance and hydrophobicity of the anion-exchange membranes. Analyzing the fouling phenomena in electromembrane processes, the interactions between solutes should also be taken into account. Banasiak et al. [7] proved that the presence of humic acids (HA) in the solution treated by ED reduced deposition of endosulfan (pesticide) at the ion-exchange membranes due to competitive sorption.

The paper aimed at the evaluation of the ED process run during treatment and desalination of various aqueous mixtures containing both mineral (sodium fluoride, sodium chloride) and organic substances (dyes or humic acids).

2 Methods and materials

2.1 Membranes and electrodialysis equipment

In the experiments on the desalination of organics-salt mixtures the laboratory electrodialytic installation PC Cell Bed-1-System was used. The ED stack (PCCell 64002 model) consisted of 10 cell pairs, thus 10 anion-exchange membranes (standard PC-SA, PCA GmbH, Germany or monovalent selective ACS Neosepta, Tokuyama Soda, Japan) and 11 cation-exchange membranes (standard PC-SK, PCA GmbH, Germany or standard CMV Selemion Asahi Glass, Japan) were applied. Each membrane had an effective surface area of 64 cm^2.

The ED system operated in a batch mode, i.e. both diluate and concentrate circulated in the stack till the end of desalination process. The electrode solution tank had a volume of 9 dm^3 and was filled with NaCl solution (0.05 mol/dm^3), whereas volumes of the external concentrate and diluate tanks amounted to 2 dm^3 (each). Flow rate of diluate and concentrate was maintained at 90 dm^3/h, giving liquid flow velocity of 6.25 cm/s in the electrodialyzer. The maximum value of current intensity which can be obtained with the use of this installation was equal to 5 A.

 DOI: 10.1051/e3sconf/20172200059

2.2 Reagents

Model solutions containing organic compounds and mineral salts were used during the ED experiments. They were prepared based on distilled water. Anionic organic dyes varying in molecular weight were chosen as example of anthropogenic contaminants (Merck, Zachem) (Table 1). Dye concentration amounted to 10 mg/dm^3. Synthetic solutions containing natural organic matter (NOM) were prepared by dissolving humic acids (HA, Aldrich) at dosages of 5, 10 and 15 mg/dm^3.

All experimental solutions contained mineral salt - sodium chloride (NaCl) at concentration of 1 g NaCl/dm^3. Besides, NOM solutions were supplemented with sodium fluoride (NaF). Concentration of fluoride ions in the tested solutions amounted to 5, 10, 100, 200 mg F^-/dm^3.

Table 2. Characteristics of the experimental dyes.

Parameter	Methyl Orange	Indigo Carmine	Amido Black	Titan Yellow	Direct Green	Direct Blue	Direct Black
Symbol	MO	IC	AB	TY	DG	DB	DB
λ_{max}, nm	465	610	618	399	370	577	585
Molecular weight, Da	327	466	615	696	878	1029	1084
$LogK_{ow}$[1]	3.29	1.53	3.37	4.75	-	-	5.04

[1]calculated with the use of VCCLAB (http://www.vcclab.org)

2.3 Methodology

The electrodialysis desalination process was carried out for water salt solution and model salt solution containing organic contaminants. Generally, the electrodialysis was focused on salt separation from organics, thus at the process beginning the diluate and concentrate chambers contained solutions varied in composition – concentrate tank contained only NaCl solution (1 g/dm^3), whereas diluate tank contained various solutions: (1) sodium chloride (NaCl, 1 g/dm^3); (2) mixture of mineral salts (NaF, 5–200 mg F^-/dm^3 and NaCl, 1 g/dm^3); (3) mixture of dye (10 mg/dm^3) and NaCl (1 g/dm^3); (4) mixture of HA (5–15 mg/dm^3) and salts (NaF, 5–200 mg F^-/dm^3 and NaCl, 1 g/dm^3).

The ED experiments were performed a constant current density (1.56 mA/cm^2 for dye solutions and 1.76 mA/cm^2 for HA solutions). The process was terminated when the maximum voltage value was reached (i.e. 24.4 V). The determined limiting current density (for initial salt concentration of 1 g/dm^3), assuming 90 and 95% desalination efficiency was equal to 3.27 and 1.64 mA/cm^2, respectively.

In the course of ED process both diluate and concentrate samples were collected every 5–10 minutes and analyzed for organic compound concentration, as well as chloride and fluoride content. For quick evaluation of desalination efficiency the electrolytic conductivity was measured.

The dye concentration was determined spectrophotometrically at the wavelengths (λ_{max}) listed in Table 1, which corresponded to the maximum absorbance of the sample. The humic acids (HA) content was also detected spectrophotometrically at the UV wavelength of 254 nm. The spectrophotometer Hitachi U-1900 was used for absorbance measurements. The fluoride content was determined by a colorimetric analysis with SPADNS reagent (at the wavelength of 580 nm). The spectrophotometer DR 2000 (HACH) was applied for fluoride ion analysis. Concentration of chloride ions in salt solutions was determined by

Mohr method (PN-ISO 9297:1994). The electrical conductivity was monitored with the use of conductometer Elmetron CC-41.

The anion flux through the membranes was calculated according to the following equation:

$$J = \frac{V_t C_t - V_i C_i}{A_m t} \quad (1)$$

where: J – ion flux (mol/m^{2}h), A_m – total surface area of the anion-exchange membranes (m^2), V_i, V_t – volume of the concentrate at the beginning of the process and at a time t, respectively (m^3), C_i, C_t – anion concentration at the beginning of the process and at a time t in the concentrate cells, respectively (mol/m^3), t – process duration (h).

The amount of organics deposited in the ED/membrane system was calculated by using a mass balance according to the following equation:

$$M = (V_{id}C_{id} + V_{ic}C_{ic}) - (V_{fd}C_{fd} + V_{fc}C_{fc}) \quad (2)$$

where: M – mass of dyes or HA accumulated in the ion-exchange membranes (mg), V_{id},V_{fd} – initial and final volume of diluate (m^3), V_{ic},V_{fc} – initial and final volume of concentrate (m^3), C_{id}, C_{fd} – initial and final concentration of dyes or HA in the diluate (mg/dm^3), C_{ic}, C_{fc} – initial and final concentration of dyes or HA in the concentrate (mg/dm^3).

Changes of voltage were monitored for each ED run and the electrical energy demand was calculated according to the method given in [11].

3 Results

3.1 ED run under the presence of humic acids

Generally, the ED of fluoride solutions (containing HA) resulted in 88–91% of F^- ions removal and in 90-97% desalination efficiency (detailed data not shown). The variation of the mean chloride flux in relation to various composition of treated solution is given in Fig.1. The results obtained show significant influence of humic acids as well as increasing concentration of fluoride ions on chloride transport through the standard ion-exchange membranes. The chloride flux decreased from 780 mmol/m^{2}h for solution containing only Cl^- and low amount of F^- ions (5 mg F^-/dm^3) to 330 mmol/m^{2}h for solution supplemented with humic acids (5 mg HA/dm^3) and high concentration of F^- ions (200 mg F^-/dm^3).

It seems that the effect of increasing content of fluoride on chloride migration through the anion exchange membranes is much greater than the presence of organic matter. This relation becomes clear by keeping in mind that all experiments were performed at the constant current density and constant chloride concentration. Thus, increasing fluoride concentration simply results in the increase of total dissolve solids and decreasing desalination rate.

The addition of HA into desalted solution brought about considerable worsening of Cl^- ion flux. This drop in ion flux was approximately constant, irrespectively of the HA concentration. This observation seems to be in a some contradiction with the results shown in Fig. 3, where masses of accumulated HA are given. The amount of deposited NOM was proportional to the HA initial concentration, however this phenomena was not reflected in a significant deterioration of Cl^- ion flux. On the other hand, a slight worsening of the mean fluoride ion flux with increasing HA content can be detected (Fig. 2). Generally, the mean F^- ion flux was proportional to the initial F^- content and varied from 5.7 to 215 mmol/m^{2}h.

 DOI: 10.1051/e3sconf/20172200059

It was higher for solutions containing mineral salts only in comparison to solutions with HA content.

It appears that the observed relationships can be explained by considering the mobility and radius of ions present in model solutions. The ionic mobility of Cl^- and F^- ions is equal to 7.92 and 5.70 m^2/Vs, respectively [3], whereas their hydrated radiuses amount to 0.332 and 0.352 nm. Thus, Cl^- ions are more preferably transported than F^- ions under the presence of organic membrane foulant.

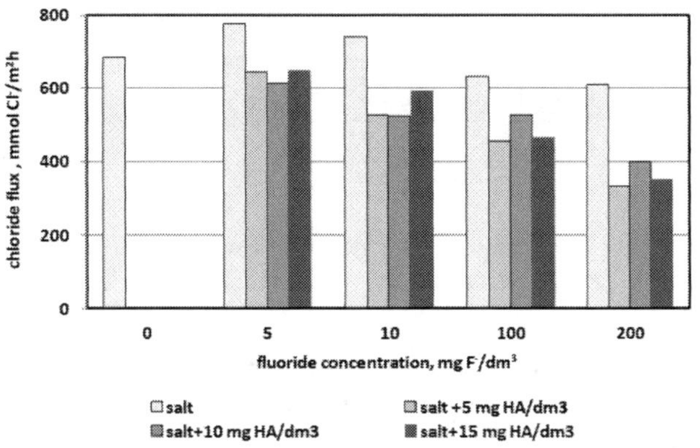

Fig. 1. Chloride flux versus fluoride and humic acids (HA) concentration (salt – 1 g NaCl/dm^3, current density – 1.72 mA/cm^2).

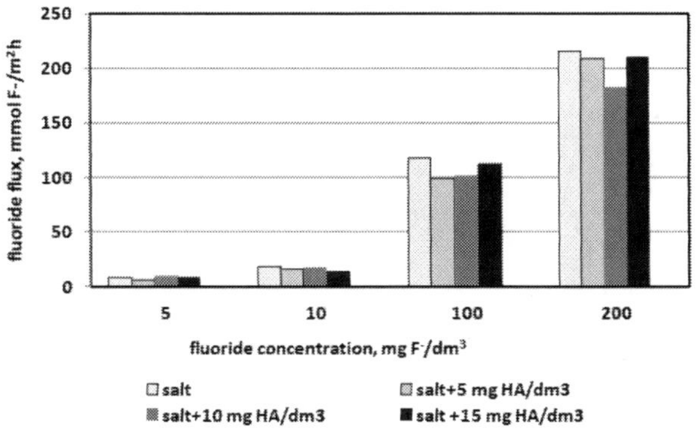

Fig. 2. Fluoride flux versus humic acids (HA) concentration (salt – 1 g NaCl/dm^3, current density – 1.72 mA/cm^2).

3.2 ED run under the presence of organic dyes

The ED experiments of organic dye-salt mixtures revealed excellent desalination efficiency, however the concentrate streams were less contaminated by dyes when monovalent selective anion-exchange membranes were used (detailed data not shown).

The variation of mean chloride flux during the ED desalination of dye solutions is shown in Fig. 4. In this stage of experiments the ED process was performed with the use of standard as well as monovalent selective anion-exchange membranes. Basically, the Cl^- ion flux varied in a small extent – from 480 to 590 $mmol/m^2h$ (with the exception of MO solution). The differences between chloride fluxes for standard and monoselective anion-exchange membranes were not remarkable, however some relations can be inferred.

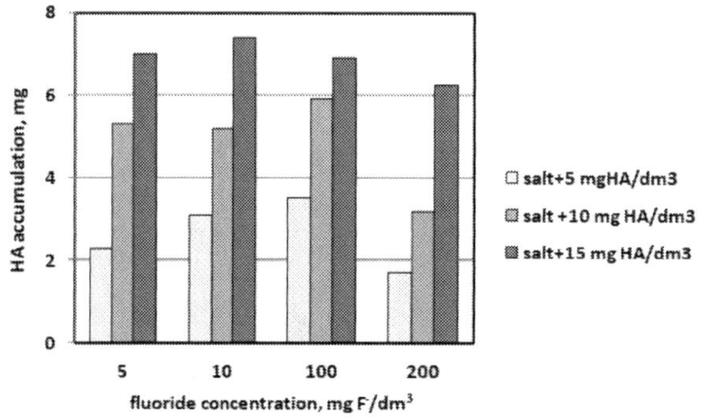

Fig. 3. Amount of humic acids (HA) accumulated in membranes versus fluoride and HA concentration (salt – 1 g $NaCl/dm^3$, current density – 1.72 mA/cm^2).

In principle, the monovalent selective anion-exchange membranes were characterized by a slight lower chloride flux than standard membranes when chloride solution was treated. This seems quite obvious due to additional negative-charged layer at the membrane surface, which increases membrane electrical resistance.

Comparing the membrane performance in view of chloride removal from dye solutions, it can be concluded that in most cases the Cl^- flux was higher for monovalent selective anion-exchange membranes than for standard membranes. However, for DG and TY solutions this observation was not so evident.

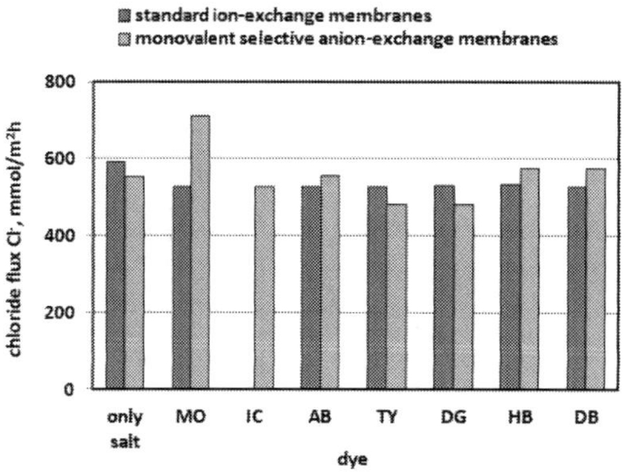

Fig. 4. Chloride flux in the course of ED desalination of model solutions containing various organic dyes (dye concentration – 10 mg/dm^3; current density – 1.56 mA/cm^2).

Dyes are known as organic substances of hydrophobic nature (Table 1) and great fouling potential. This is reflected in Fig. 5, which shows the extend of dye accumulation in the anion-exchange membranes. The membrane fouling caused by dyes was very intensive in relation to standard membranes. It is interesting to note that the extent of dye adsorption was influenced mainly by their molecular weights – the lower the molecular weight, the greater the dye mass deposited on membrane. This observation suggests facilitated penetration of small dye particles into membrane matrix. The application of mono-selective anion-exchange membranes resulted in almost complete elimination of dye adsorption and improvement of chloride flux.

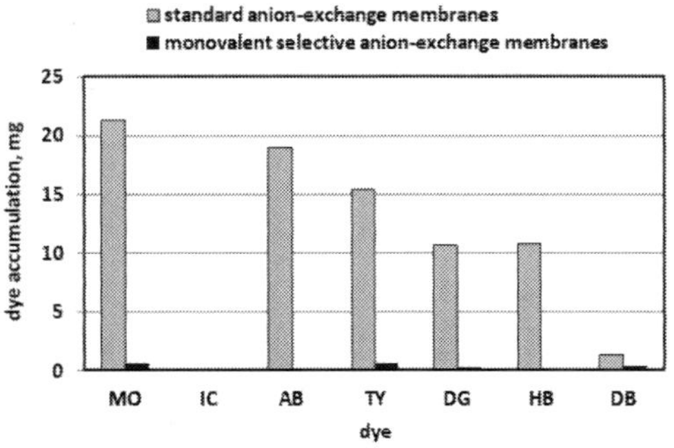

Fig. 5. Amount of dyes accumulated in membranes in the course of ED desalination of dye solutions (dye concentration – 10 mg/dm^3; current density – 1.56 mA/cm^2).

3.4 Electrical energy demand

Generally, the energy consumption is an essential factor in practical applications. On the other hand, it can be a measure of the correctness of the ED process run. The calculated energy consumption for dye and fluoride solutions is given in Table 3 and Fig. 6, respectively.

Table 3. Energy consumption (EC) for ED desalination of dye-salt mixtures (salt – 1 g NaCl/dm^3, current density – 1.56 mA/cm^2).

Dye symbol	MO	IC	AB	TY	DG	HB	DB	H_2O
Standard anion-exchange membrane PC-SA								
EC, kWh/m^3	0.25	-	0.37	0.46	0.39	0.40	0.41	0.27
Monovalent selective anion-exchange membrane (ACS)								
EC, kWh/m^3	0.38	0.40	0.49	0.55	0.48	0.48	0.49	0.47

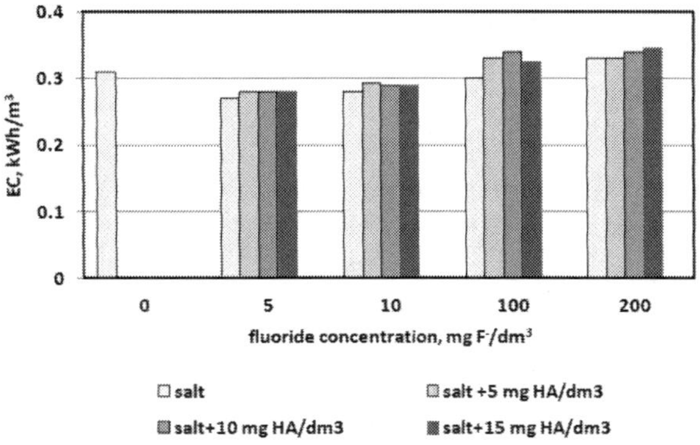

Fig. 6. Energy consumption (EC) versus fluoride and humic acids (HA) concentration (salt – 1g NaCl/dm^3, current density – 1.72 mA/cm^2).

From the results obtained (Table 3) in can be inferred that the power consumption is almost of the same value, irrespectively of the type of organic dye. However, a slight shift towards higher power demand was observed with the increasing of dye molecular weight. This observation is valid for both anion-exchange membranes used. It is interesting to note that the ED process with monovalent selective anion-exchange membranes exhibited higher energy consumption than ED with standard anion-exchange membranes. This can be attributed to the higher electrical resistance and denser structure of ACS membranes than PC-SA membranes.

Unexpectedly, it can be concluded that humic acids slightly increased the energy demand in the course of ED process with fluoride solutions (Fig. 6). Although the HA deposition at/in membranes was confirmed (Fig. 2), the variation in energy used with increasing HA content was less than 10%. Fluoride concentration was the main factor which influenced the energy demand in the case of fluoride solutions. This finding seems to be clear when Faraday law is taken into account.

4 Conclusions

1. The presence of humic acids (HA) in the salt solutions treated by electrodialysis caused a considerable worsening of chloride flux, irrespectively of HA concentration. The fluoride flux decreased with the increasing HA content. The drop of F^- ion flux was especially significant when HA concentration amounted to 10 mg/dm^3.
2. Although the HA deposition in the ion-exchange membranes was confirmed, the increase in the electrical energy demand with increasing HA content was less than 10%.
3. The use of monovalent selective anion-exchange membranes resulted in almost complete elimination of dye deposition and improvement of chloride flux.

The financial support of the Faculty of Environmental Engineering (Wroclaw University of Science and Technology, grant no. 0402/0080/16) is greatly appreciated.

References

1. Y. Zhang, B. Van der Bruggen, L. Pinoy, B. Meesschaert, J. Membrane Sci. **332**, 104 (2009)
2. H.J. Lee, D.H. Kim, J. Cho, S.H. Moon, Desalination **151**, 43 (2002)
3. L.J. Banasiak, A.I. Schäfer, J. Membrane Sci. **334**, 101 (2009)
4. H.J. Lee, S.H. Moon, S.P. Tsai, Sep. Purif. Technol. **27**, 89 (2002)
5. H.J. Lee, M.K. Hong, S.D. Han, J. Shim, S.H. Moon, J. Membrane Sci. **325**, 719 (2008)
6. V. Lindstrand, G. Sundström, A. Jönsson, Desalination **128**, 91 (2000)
7. L.J Banasiak, B. Van der Bruggen, A.I. Schäfer, Chem. Eng. J. **166**, 233 (2011)
8. S. Caprarescu, A.R.Miron, V. Purcar, A.L. Radu, A. Sarbu, D. Ion-Ebrasu, L.I. Atanase, M. Ghiurea, Water Sci. Technol. **74**, 2462 (2016)
9. K. Kabsch-Korbutowicz, J. Wisniewski, S. Kliber, A. Urbanowska, Desalination **280**, 428 (2011)
10. Y. Zhang, L. Pinoy, B. Meesschaert, B. Van der Bruggen, AIChE J. **57**, 2070 (2011)
11. R. Rautenbach, *Membrane processes* (Wiley, 1989)

Incineration, pyrolysis and gasification of electronic waste

Agnieszka Gurgul[1,*], *Włodzimierz* Szczepaniak[1], and *Monika* Zabłocka-Malicka[2]

[1]Wrocław University of Science and Technology, Faculty of Environmental Engineering, Wybrzeże Wyspiańskiego 27, 50-370 Wrocław, Poland
[2]Wrocław University of Science and Technology, Faculty of Chemistry, Wybrzeże Wyspiańskiego 27, 50-370 Wrocław, Poland

Abstract. Three high temperature processes of the electronic waste processing: smelting/incineration, pyrolysis and gasification were shortly discussed. The most distinctive feature of electronic waste is complexity of components and their integration. This type of waste consists of polymeric materials and has high content of valuable metals that could be recovered. The purpose of thermal treatment of electronic waste is elimination of plastic components (especially epoxy resins) while leaving non-volatile mineral and metallic phases in more or less original forms. Additionally, the gaseous product of the process after cleaning may be used for energy recovery or as syngas.

1 Electronic waste

Rapid economic growth has resulted in an increase of production of electrical and electronic equipment (EEE) which is one of the fastest growing global manufacturing and it is still developing. Waste electric and electronic equipment (WEEE) is now considered as one of the fastest growing waste streams in EU, approximately 3–5% per year [1]. The amount of electric and electronic equipment put on market in 2013 in EU reached 8.1 million tonnes. In 2013, 3.6 million tonnes of WEEE were treated, of which 2.7 million tonnes were recovered.

The electric and electronic equipment includes a broad range of goods. Directive 2012/19/EU [2] defines 'electrical and electronic equipment' or 'EEE' as equipment which is dependent on electric current or electromagnetic fields in order to work properly. This type of waste includes large and small household appliances and IT equipments as computers, mobile phones, video and audio equipment. Furthermore, more appliances that were previously considered as electrical equipment as refrigerators are becoming 'electronic' through the addition of programmable microprocessors [3]. The development in production of electric and electronic waste is a cause of various environmental and economical problems. This type of waste contains diverse toxic substances that could be released to environment during landfilling or inappropriate processing. Therefore, this type of waste can pose considerable environmental and health risk if treated inadequately. On

* Corresponding author: agnieszka.gurgul@pwr.edu.pl

the other hand, recycling of electronic waste is an important subject not only from the point of waste treatment but also from the recovery aspect of valuable materials. The recycling of WEEE offers opportunities of making secondary raw materials available on the market, especially valuable metals. In Table 1 it is presented concentration of metals in different electric and electronic equipment waste samples.

Table 1. Concentration of metals in different electronic scrap samples [4].

Electronic waste	wt.%					ppm (wt.)		
	Fe	Cu	Al	Pb	Ni	Ag	Au	Pd
TV board	28	10	10	1.0	0.3	280	20	10
PC board	7	20	5	1.5	1	1000	250	110
Mobile phone	5	13	1	0.3	0.1	1380	350	210
Portable audio	23	21	1	0.14	0.03	150	10	4
DVD player	62	5	2	0.3	0.05	115	15	4
Calculator	4	3	5	0.1	0.5	260	50	5
Printed circuit boards	12	10	7	1.2	0.85	280	110	-

Essential part of almost all of the electronic products are printed circuit boards (PCBs). Waste printed circuit boards (WPCBs) contain lots of valuable resources together with hazardous materials. This type of waste contains heavy metals (i.e., lead, cadmium) and hazardous substances that may pollute environment if they are not properly treated. However, a significant amount of valuable materials contained in WPCBs make them worth being recycled. Example value of some metals from PCB is presented in Table 2. The precious metals (gold, palladium, silver) make up more than 70% of the value of PCB that could be recovered.

Table 2. Composition and the intrinsic values of PCB metals [5].

Component	Wt.%	Value (£/kg)	Intrinsic value of PCB	
			(£/kg of PCB)	(%)
Gold	0.025	14200.00	3.55	65.4
Palladium	0.010	6200.00	0.62	11.4
Silver	0.100	250.00	0.25	4.6
Copper	16.0	3.30	0.53	9.7
Aluminium	5.0	1.20	0.06	1.1
Iron	5.0	0.10	0.01	0.1
Tin	3.0	8.10	0.24	4.5
Lead	2.0	1.30	0.03	0.5
Nickel	1.0	13.20	0.13	2.4
Zinc	1.0	1.20	0.01	0.2

Due to high content of valuable metals in WPCBs, many researchers were using various methods to separate metals and nonmetals from this type of waste. Those methods include e.g. electrostatic separation, gravity separation, magnetic separation, hydrometallurgical and pyrometallurgical methods. Electronic waste represent a challenging problem mainly due to complexity of the material that consist of low value plastics and valuable metals that are integrated. Developing new clean technologies for recycling of WPCBs is a significant challenge. Processing of the waste should optimize recovery of materials and minimize final waste volume and process emissions. Thermal processes (incineration, pyrolysis and gasification) may be alternative solution for processing complex electronic waste. The purpose of thermal treatment of electronic waste is elimination of organic components (e.g. plastics) while leaving non-volatile mineral and metallic phases in more or less original forms that could be recovered. Additionally, the gaseous, liquid or solid products after cleaning may be used for energy recovery.

2 Incineration of electronic waste

Pyrometallurgical processes are currently the most important method of electronic waste processing. They are the final stage of the recovery chain, comprising in Europe approximately ten thousand collectors, one thousand dismantling installations and one hundred pre-processing units [6]. This stream of waste may be processed in two ways: smelting and incineration.

2.1 Smelting

There are only 3 metallurgical plants in Europe smelting and refining metal-bearing waste streams: UMICORE N.V., AURUBIS A.G. and BOLIDEN. The first two companies use ISASMELT™ technology [7, 8], whereas Boliden uses Kaldo rotary furnace. In the Isasmelt™ technology, a vertical furnace was applied for smelting electronic waste and copper scrap mixture (plus necessary fluxes and recycled fly ashes). The furnace is fed from the top. The air, enriched in oxygen up to 90 vol.%, is blown into the furnace also from the top. In the process, the melt is intensively mixed by the process gases – this mixing is essential for fast equilibration of the melt. Smelting is carried out at temperature of 1200°C (approximately.) Molten metal and slag are tapped from the furnace bottom. Precious metals are dissolved in molten copper. Further processing is typical for copper metallurgy: the copper is refined, electrolyzed and precious metals are recovered from anode slimes. According to Umicore [8], it is possible to smelt up to 1000 tons of material per day in the Isasmelt™ furnace.

2.2 Incineration

Contrary to smelting, formation of melts (in the first place oxide ones, as slags) is avoided during incineration. Such process is useful in the small scale processing of the electronic waste with high content of precious metals for further processing by hydrometallurgical routes. Typical rotary furnaces seem most useful in such processing, however stationary fluidized bed was also tested for electronic waste incineration in the laboratory scale[9]. Samples of the printed circuit board were heated for 3 minutes at 840–850°C and at 6.2 vol.% of oxygen. Weight losses of the incinerated material were equal to approximately 21 wt.%. Product of incineration was a brittle substance, suitable for chemical and mechanical processing. It was pulverized below 0.2 mm and was leached with sulfuric acid with concentrations from 30 to 95%. With 95% concentration of sulfuric acid is possible to obtain almost complete leaching of copper (99.95%) and precious metals (e.g. 97% for Ag) [10].

3 Pyrolysis of electronic waste

Pyrolysis is a thermochemical decomposition of organic material at high temperature in the absence of oxygen. It involves changes of chemical composition and physical phase. In the pyrolysis process organic material, that is a part of electronic waste, is decomposed to molecular products that can be use e.g. as fuel. Pyrolysis of organic materials produces gas and liquid products and leaves a solid residue rich in carbon (char). Proportions of solid, liquid and gas yields depend on material that is pyrolyzed and process conditions. Zhou [11] conducted experiments of vacuum pyrolysis of waste printed circuit boards at 240°C. Results of the pyrolysis experiments are presented in Table 3.

Table 3. Pyrolysis yields (wt.%) of the two types of WPCBs [11].

	Type -A				Type-B			
	Test 1	**Test 2**	**Test 3**	**Average**	**Test 1**	**Test 2**	**Test 3**	**Average**
Solid yield	68.7	70.1	69.7	69.5	76.1	75.6	75.4	75.7
Liquid yield	28.2	27.3	27.9	27.8	19.5	20.3	20.1	20.0
Gas yield	3.1	2.6	2.4	2.7	4.4	4.1	4.5	4.3

The solid residue consisted of the glass fiber, electronic components, metals and other inorganic materials that could be recycled for further processing. Gaseous product of the process was rich in CO, CO_2, CH_4 , H_2, etc. which makes this gas valuable for combustion. Energy recovered from combustion of the gas may be used in the process and therefore make pyrolysis a self-sustain (Fig.1).

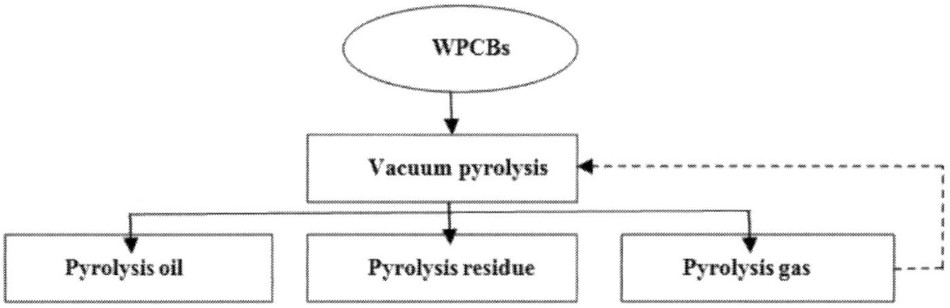

Fig. 1. Pyrolysis of WPCBs with recovery of syngas [10].

The residue of pyrolysis contained various metals (e.g., Cu, Fe, Al, Au, Ag, Pt, Pd etc.) that could be recovered in further processing. Content of those metals may vary depending on type of waste that is processed. Concentration of metals that were identified in the solid residue of various electronic waste pyrolysis is presented in Table 4.

Table 4. Concentration of metals in pyrolysis solids (wt.%) [12].

	Mg	**Al**	**Ca**	**Mn**	**Fe**	**Ni**	**Cu**	**Zn**	**Sn**	**Pb**
PE wires	0.6	72.4	< 0.5	< 0.5	0.9	< 0.5	16.4	< 0.5	< 0.5	< 0.5
Table phones	< 0.5	4.7	1.0	< 0.5	4.7	0.6	18.0	1.6	0.6	1.0
Mobile phones	< 0.5	0.6	< 0.5	< 0.5	< 0.5	0.8	12.4	2.8	0.8	0.6
Circuit boards	< 0.5	8.7	1.8	0.6	3.9	0.5	36.4	3.5	4.6	3.8

The metals may be separated and recovered in further processing. Each metallic element may be recovered in different physical and chemical processes, according to its characteristics. Pyrolysis can be considered as an alternative method of recycling electronic waste, because during the process organic content of waste is decomposed to product that can be used for energy recovery – the gas and to the solid combination of char and nonorganic materials, including metals, that can be recovered. In addition, pyrolysis significantly reduces the volume of the waste which is beneficial for further treatment. The combination of mechanical methods (separation, shredding), pyrolysis and chemical processing could increase efficiency of electronic waste recovery.

4 Gasification of electronic waste

Gasification seems to be an ambiguous concept in the literature and often is identified with pyrolysis (i.e. transformation of organic materials into gaseous species). However, we would like to define gasification as a process that requires an additional reagent, in contrast to pyrolysis for which only temperature determines generation of gaseous species. Therefore gasification, not pyrolysis, is the process that enables complete elimination of organic fractions to the gaseous phase at low oxygen potential of the system. Typical gasifying reagents are: steam, oxygen (air, oxygen enriched air) or carbon dioxide (much less). Considering steam gasification, there is an important difference between steam gasification and steam/oxygen/air processes – the former is endothermic (allothermal), whereas the later one is usually balanced as autothermic.

Gasification or pyrolysis are not cited in the literature as developed methods for metals recovery from electronic waste [13–15]. On the other hand, gasification was mentioned as accepted technology of municipal solid waste processing [16]. Recently, there were presented laboratory experiments of printed circuit boards gasification in molten carbonates by steam [17–19]. There is also a paper reporting an experiment of processing of plastics from printed circuit board to hydrogen by pyrolysis and steam conversion [20].

Taking into account total volume and composition of electronic wastes it seems, particularly from the Circular Economy point of view, that not gaseous product (syngas) but inorganic residue, especially metals, should be in the first place recovered. We believe, that direct steam gasification is a promising method of this recovery. It was demonstrated first time for the RAM waste, the data storage device comprising of FR-4 board and set of integrated circuits [21]. The original item, the crude product of gasification and some of materials which were the result of simple physical separation of the product are presented in Figure 2. Finally, 6 fractions were gathered: (from the integrated circuit) 1.1 – dies, 1.2 – ferromagnetic elements of leadframes, 1.3 – residue of encapsulation and (from PCB) 2.1 – copper foil, 2.2 – coarse metallic elements, 2.3 – fine fraction. Concentrations of some, typical for electronic waste metals in these fractions are presented in Figure 3. The differences between fractions and high level of precious metals are evident. There was also almost no char in the residue after gasification of organic fraction – epoxy resins.

Fig. 2. Direct steam gasification of the RAM waste [15] (general illustration).

 DOI: 10.1051/e3sconf/20172200060

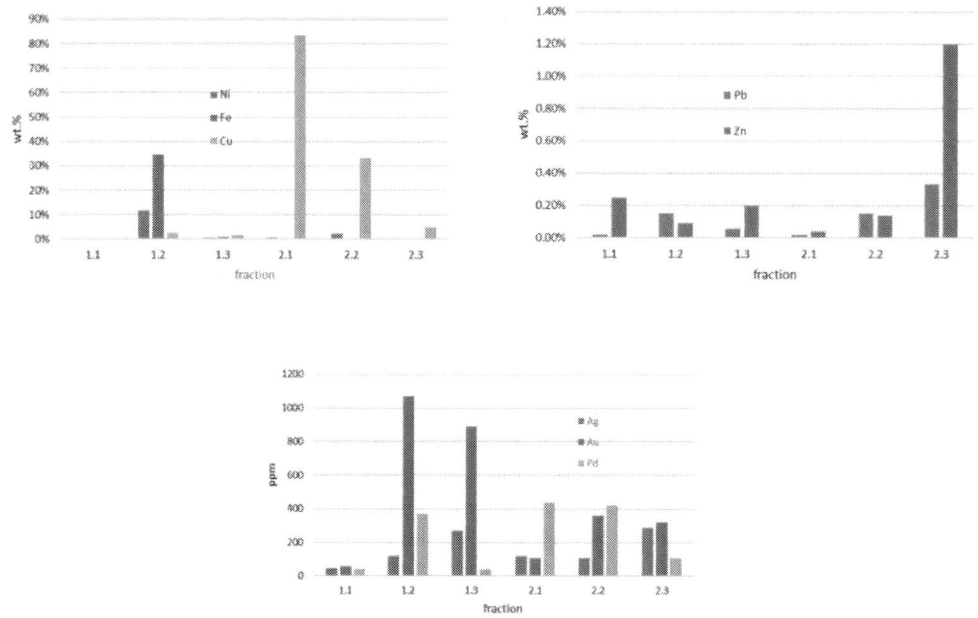

Fig. 3. Concentration of metals in the final product (fractions – description in the text) of steam gasification of the RAM waste [15].

Figure 4 is a good illustration of effectiveness of the steam gasification process for electronic waste. Gasification experiment was performed in a quartz reactor with samples of the raw FR-4 board at 860°C for few hours [A. Zielińska, Laboratory of HT Processes, W7/PWr]. It may be noticed that the structure of the board was fully opened and the epoxy resin, which bonded fiber glass and copper layers, was completely removed.

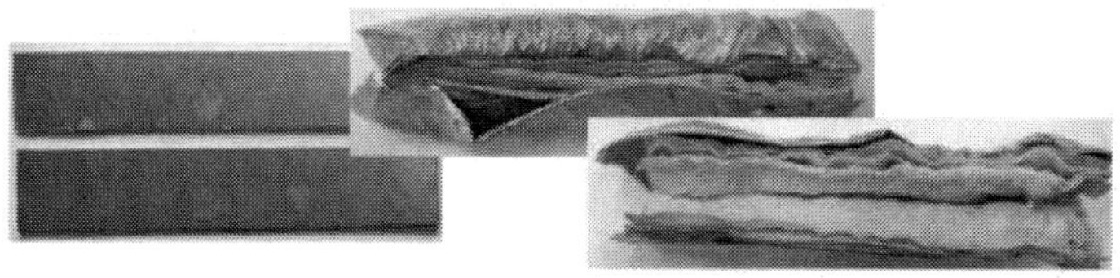

Fig. 4. Steam gasification of a raw FR-4 board [A. Zielińska, Laboratory of HT Processes, W7/PWr].

It should be kept in mind that allothermal steam gasification is useful not only in electronic waste processing – examples of gasification of the carton packaging, abrasive disc, bones, PCV-Cu wire and another electronic waste: PC ISA card, have been described in the accompanying paper [W. Szczepaniak, M. Zabłocka-Malicka, Steam gasification for waste valorization with energy recovery].

5 Conclusion

Increasing stream of electronic wastes is a valuable resource of metals, in the first place precious and copper, but also many others, ferrous and non-ferrous ones. Classical pyrometallurgical smelting processes are nowadays the major method of this waste processing. Generally, these processes are focused on precious metals and copper recovery. Organic materials (e.g. epoxy resins) are additional source of the heat; other components are collected in the slag or fly ash fractions. From this point of view, high concentration of chlorine or bromine is disadvantageous. However, there are undertaken efforts to develop alternative technologies, as incineration without smelting, pyrolysis or gasification. It seems, that the method of allothermal steam gasification is particularly interesting, because enables complete elimination of organic fractions without high oxygen potential of the gaseous phase, collecting volatile contaminants (including halides) in the aqueous condensate and production of H_2-rich syngas. It should be noted that the air is not used in this process.

Acknowledgments

The work was financed by statutory activity subsidies from the Polish Ministry of Science and Higher Education for the Faculty of Environmental Engineering and Faculty of Chemistry of Wroclaw University of Science and Technology.

References

1. Eurostat (http://ec.europa.eu/eurostat/web/waste/key-waste-streams/weee)
2. Directive 2012/19/EU of The European Parliament and of the Council on waste electrical and electronic equipment (WEEE) (2012)
3. B.R. Babu, A.K. Parande, C.A. Basha, Waste Manage. Res. **27**, 307–318 (2007)
4. J. Cui, L. Zhang, J. Hazard. Mater. **158**, 228–256 (2008)
5. Y.J. Paryk, D.J. Fray, J. Hazard. Mater. **164**, 1152–1158 (2009)
6. C. Hageluken, *Exploring cross-border collaboration in R&D,* EIT-KIC-RM Seminar, September 14th, KU Leuven (2012)
7. G.R.F.A. Flores, S. Nikolic, P.J. Mackey, JOM **66**, 823–832 (2014)
8. M. Caffarey, *Recycling Electronic End of Life Materials*, SERDC Recycling Summit (2012)
9. A. Woynarowska, J. Baron, S. Kandefer, W. Żukowski, Przem. Chem. **92**, 997–1005 (2013)
10. A. Woynarowska, *Termiczna utylizacja odpadów elektronicznych w reaktorze fluidyzacyjnym* (Politechnika Krakowska, Kraków, 2014)
11. Y. Zhou, K. Qiu, J. Hazard. Mater. **175**, 823–828 (2010)
12. I. de Marco, B.M. Caballero, M.J. Chomon, M.F. Laresgoiti, A. Torres, G. Fernandez, S. Arnaiz, J. Anal. Appl. Pyrolysis **82**, 2, 179–183 (2008)
13. L. Zhang, Z. Xu, J. Clean. Prod. **127**, 19–36 (2016)
14. Y. Lu, Z. Xu, Resour. Conserv. Recy. **113**, 28–39 (2016)
15. M. Kaya, Waste Manage. **57**, 64–90 (2016)
16. U. Arena, Waste Manage. **32**, 625–639 (2012)
17. S. Shang, K. Yoshikawa, H. Nakagome, T.Kamo, Appl. Energy **101**, 815–821 (2013)
18. J.A. Salbidegoitia, E.G. Fuentes-Ordonez, M.P. Gonzalez-Marcos, J.R. Gonzalez-Velasco, T. Bhaskar, T. Kamo, Fuel Processing Technol. **133**, 69–74 (2015)
19. S. Zhang, Y. Yu, Procedia Environ. Sci. **31**, 903–910 (2016)
20. J.C. Acomb, M. Anas Nahil, P.T. Wiliams, J. Anal. Appl. Pyrolysis **103**, 320–327 (2013)
21. P. Mońka, W. Szczepaniak, M. Zabłocka-Malicka, Czas. Tech. 1-Ch **108**, 119–126 (2011)

Monitoring of water supply connections as an element to reduce apparent losses of water?

Joanna Gwoździej-Mazur[1,*]

[1]Department of Environmental of Engineering Systems, Bialystok University of Technology, ul.Wiejska 45E, Bialystok 15-351, Poland

Abstract. Measuring instruments are designed to measure a given physical value, to process the obtained information and forward it to the observer. They are designed to perform specific tasks in specific working conditions and meeting the envisaged requirements. The most important requirement to be met by measuring instruments, is to preserve the established metrological characteristics. The basic and most common instrument for measuring the volume of flowing water is the water meter. Selecting the right water meter in the operating conditions is not an easy issue. The problem has been further intensified by decrease of water consumption which began in the 90s of the twentieth century and continuing to the present day. As a result, there has changed the structure of water consumption in both the residential and industrial applications. In this situation, a right selection of the optimal water meter it is an important case.The article presents the results of research in the field of characteristic flows in the water supply connections in multi-family housing using modern monitoring systems. It has been presented the calculated inequality ratio of water consumption, which can be helpful when designing a plumbing systems. In addition, the structure of water consumption due to the typical flow ranges was determined.

1 Introduction

The technical progress seen today involve great measuring capabilities that have not been used on such a wide scale until now. There are currently many modern meters for registering water flow on the market. Devices like digital loggers, ultrasound, static, or electromagnetic, together with computer software, make it possible to quickly display detailed monitoring data. The benefits of such methods are significant, including determination of correct water meter selection in a given building or determining the efficiency of internal water supply systems [1–8].

The goal of this paper is to investigate actual water flow curves on the water service pipe of a multi-family building under operating conditions as well as to determine non-uniform distribution factors and determine the structure of water consumption according to characteristic intervals describing the accuracy of measuring devices.

* Corresponding author: j.mazur@pb.edu.pl

 DOI: 10.1051/e3sconf/20172200061

2 Monitoring of water service pipes

The subject of monitoring water service pipes has been present ever since water consumption was first metered in buildings. Since then, recipients have had doubts as to the metered water consumption [2, 5, 9–11]. As technology developed, so too did monitoring. The development of meter monitoring was researched as the world progressed technologically. George Theodore Paraskevakos was a pioneer in this field, and it was he who invented the sensor-based monitoring system while working for an aircraft manufacturer from Alabama in 1972. At first, this system used digital data transmission. It was used to monitor alarm systems in hospitals, fire safety systems. In 1977, George Theodore Paraskevakos started an enterprise called Metretek Instruments, which is manufacturing remote reading systems and different types of meters under the name Honeywell to this day. Automatic meter reading (AMR) – a telemetric technology involving automatic downloading of data from water, gas, heat and electricity meters and its transmission to a central database for analysis and billing. This is a way to save the human labor associated with visiting residents and reading meters. Both building administrators and residents can gain greater control of water consumption, among other things, through access to current data combined with analysis. AMR technology consists of solutions (including mobile, portable and stationary networks) based on telephone platforms (both wired and wireless), radio transmission (RF) or PLC transmission [12]. The priority underlying the introduction of automatic meter reading was not to minimize costs linked to reading, but to gain the capability of accessing a meter's data, which could not be acquired otherwise. Many water meters are installed in places that are difficult or inconvenient to access, like wells, and it is also necessary to meet with the real estate owner to conduct inspections - so many water companies began to invest in remote reading systems. The development of remote readings introduced monitoring on water service pipes, not limited to just data on water consumption. As technology advanced, it was possible to register additional data from the water meter and archive it. The development of an advanced remote reading measuring structure introduced a new type of water meter, namely ultrasound and electromagnetic water meters. The combination of a traditional water meter and a data reading module led to the creation of a single, modern measuring device, which became a tool for remote control and monitoring of water consumption [9, 10].

The continuous development of the measuring structure caused the definition of automatic meter reading (AMR) to evolve into advanced metering infrastructure (AMI). Advanced metering infrastructure is an integrated set of components: intelligent meters, modules and communication systems, concentrators and loggers, enabling two-way communication via various media and technologies between the central system and selected meters. Such a network makes it possible to gather data on the consumption of individual recipients, transmit control signals to devices and configure them remotely [12].

Water meters in advanced metering infrastructure are called smart meters because they process gathered data. Such metering on water service pipes forms an entire metrological network that facilitates water management. The operation of such a monitoring network provides time savings thanks to the speed of reading a water meter and limits apparent losses, which include unauthorized consumption [1, 2, 13–17].

Apparent losses are mainly the result of low accuracy of measurement of the volume of flowing water and non-simultaneity of readings of devices registering the amount of water supplied to the system (main water meter) and consumed by recipients (sub-meters on premises) [1, 2, 9, 10, 16].

Actual losses are mostly the result of a poor technical condition of the water supply system. Failures of a system's pipes, joints and fittings have the greatest impact on the size of these losses.

Benefits of an advanced water service pipe monitoring system:
- precision of flow readings,
- many parameters that can be read,
- leak detection in the system.

Despite such great technological expansion in the field of water service pipe monitoring systems in recent years, there are countries that ban the development of this technology. In India, there are many legal barriers relating to operation on radio frequencies, regardless of signal strength [8].

3 Research methodology

Monitoring of water consumption was conducted in multi-family buildings using static water meters with iPERL remnant field technology from Sensus (fig. 1).

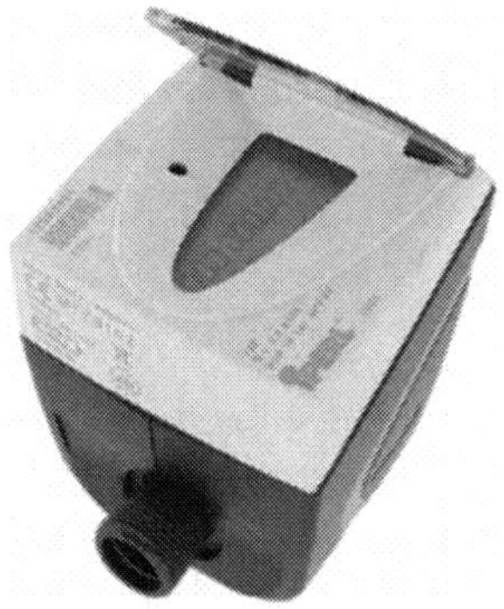

Fig. 1. iPERL water meter from the Sensus company used in tests of flow registration on the water service pipe [18].

The development of the patented remnant field technology has made the next generation of mag meter technology possible. Remanence is the magnetization left behind in a material after the external magnetic field is removed. Remanent (or residual) magnetic field technology effectively helps solve the power demands of traditional mag meter technology to serve as a viable option for the residential market. The new metering technology uses a pulse of energy to magnetize small strips of remanent material. The magnetic field is reversed each time a pulse of energy is emitted to magnetize the material, thereby inducing voltages of opposite polarity on each field flip and eliminating the effects of electrode offset which can interfere with measurement accuracy. The design of the front-end amplifier and sigma-delta analog to digital conversion contribute to improved low flow measurement accuracy over traditional battery-powered mag meters. This is critical for the detection of leaks, which is important for both the consumer and the utility company. The remanent material holds its magnetic field strength without requiring continual current consumption. In fact, remanent technology meters also conserve energy by employing low frequency field flipping to switch the magnetic field. As a result, the energy required to power the meter over the course of its lifetime is significantly reduced. This allows the meter to be powered by a traditional battery cell for the 20-year life span required for the residential market. Another key advantage of sigma-delta analog to digital conversion is the ability to continuously record flow with no gaps in measurement. This is a significant advantage for a residential water meter where the flow is frequently intermittent [18].

8 multi-family buildings located in the Northeast of Poland were studied. Flowrate measurements on the water service pipe covered a period of 4 months.

An exemplary diagram of one month digital recording in one of the tested buildings, shown on figure 2.

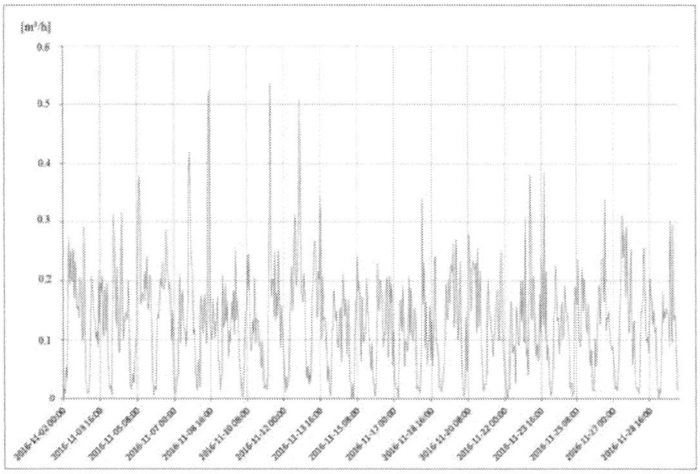

Fig. 2. Registered, average hourly water flow on the water supply connection in the tested multi-family building.

4 Analysis of test results

Analysis of hourly water consumption as a percentage in the studied buildings shows that flowrates up to 0.25 dm^3/h had the greatest share. Together, they make up approx. 82% of all minimum flows on the water service pipe. The detailed distribution of minimum flowrates depending on intervals describing the class of measuring devices installed on the water service pipe is presented in figure 3.

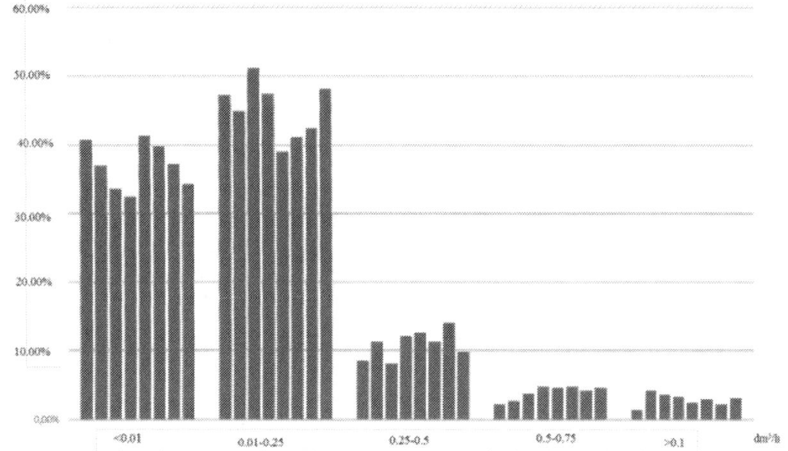

Fig. 3. Structure of hourly water consumption in selected intervals in multi-family buildings.

Based on the analysis, the amount of apparent losses can be determined depending on the minimum volume stream. Apparent losses consist of unauthorized consumption – e.g. theft as well as losses occurring as a result of incorrect measurements and readings. The actual distribution of the minimum volume stream presented above may serve as the basis for selecting water meter's for water service pipes to multi-family buildings, and thus minimize apparent losses.

The demand for and consumption of water by its recipients is not uniform throughout the year, and it is subject to seasonal, monthly and daily fluctuations. A series of factors affect these fluctuations, including geographical location, time of the year, weather conditions and time of day. Just as there are annual, monthly and weekly fluctuations, daily fluctuations of water distributions also occur.

Daily (N_{dp}) and hourly (N_{hp}) non-uniformity were determined over the course of monitoring of water flowrate in service pipes. Factor values were calculated from the following formulas:

- daily non-uniformity under measuring conditions:

$$N_{dp} = Q_{dmax}/Q_{dsr} \quad (1)$$

- hourly non-uniformity under measuring conditions:

$$N_{hp} = Q_{hmax}/Q_{hsr} \quad (2)$$

where:

Q_{dmax} – maximum daily water consumption, m^3/d,
$Q_{dśr}$ – mean daily water consumption, m^3/d
Q_{hmax} – maximum hourly water consumption, m^3/h,
$Q_{hśr}$ – mean hourly water consumption, m^3/h.

Daily and hourly non-uniformity factors are presented in table 1 and table 2.

Table 1. Value of daily non-uniformity factor in multi-family buildings.

N_{dp} for 25%–75% fit	Mean value	Median
1.44 – 1.51	1.48	1.49

Table 2. Value of hourly non-uniformity factor in the studied multi-family buildings.

N_{hp} for 25% - 75% fit	Mean value	Median
3.20–3.42	3.31	3.32

From the perspective of water meter selection, hourly flowrates are more meaningful, as they reflect actual maximum flows to be expected on a given water service pipe rather accurately. Thus, the hourly non-uniformity factor, determined under actual conditions, can be used to determine maximum flowrate and may serve as the basis for determining maximum flowrate on the service pipe and for water meter selection.

The next stage of analysis of the test results was to determine actual cold water consumption values per resident. Water consumption may differ in every household depending on appliances and awareness of how to save on utilities. Monthly and daily values of water demand per resident are presented in figure 4 and figure 5.

 DOI: 10.1051/e3sconf/20172200061

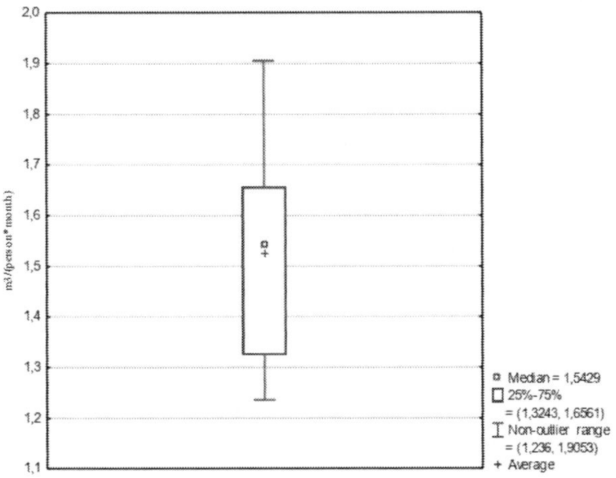

Fig. 4. Mean monthly cold water consumption per resident in the studied multi-family buildings.

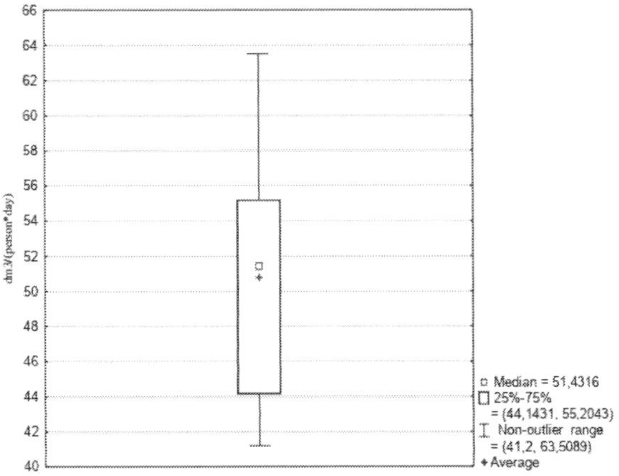

Fig. 5. Mean daily cold water consumption per resident in the studied multi-family buildings.

Unit water demand factors make it possible to calculate total water demand in the entire building. The monitored buildings had a mean monthly consumption equal to 1.52 m^3/ resident·month. To determine the actual water flowrate on the water service pipe of a building that is under construction in a manner that may serve for selecting a water meter with the proper measuring accuracy and thus contribute to limiting apparent losses, measured values should be multiplied by the number of residents, in a manner accounting for the time horizon.

5 Summary

Thanks to monitoring of quantities like minimum, mean and maximum volume stream, as well as water consumption, it can be assessed whether a given system is working correctly and measuring devices have been selected properly. Monitoring is the chief diagnostic tool for control of water distribution, stabilizing water meter management and allowing for rapid reaction in the event of emergencies and over the course of planned operations.

Monitoring a water service pipe in a multi-family building will make it possible to learn about the structure of water consumption, in detail, according to the intervals defining the accuracy of measuring devices. The charts that are obtained may serve as the basis for selecting water meters installed on water service pipes, thus limiting apparent losses.

Analysis of test results confirms that the studied buildings are characterized by variability of water demand within daily and hourly periods, as substantiated by monitoring of water flowrates in water service pipes. The determined actual daily and actual hourly non-uniformity factors can be used in water demand calculations for a given multi-family building.

Analysis showed a depressed structure of mean daily water consumption per person, amounting to 50.8 dm^3/resident · day, on average.

After installation of a water meter, with readings based on residual magnetic field technology ensuring a linear measuring range, on the water service pipe in the studied buildings, apparent losses in the form of unregistered flows on the water service pipe reached up to 30% of daily consumption, even at the lowest flow values.

Studies have been carried out in the framework of working No. S/WBiIŚ/02/2014 and financed from the funds for science MNiSW.

References

1. M. Rimeika, R. Albrektien, *Analysis of Apparent Water Losses, Case Study*, The 9th International Conference ENVIRONMENTAL ENGINEERING, Vilnius, Lithuania (2014)
2. G.L. Richards, M.C. Jonson, S.L. Barfuss, J. Am. Water Works Ass. **105**, 5 (2010)
3. W. Koral, Instal **5** (2005)
4. T. Cichoń, J. Królikowska, A. Królikowski, Gaz Woda Tech. **6** (2013)
5. F. Arregui, R. Cobacho, E. Cabrera, V. Espert, J. Water Res. Pl.-Asce **137**, 1 (2011)
6. E.H. Johnson, Water SA **27**, 4 (2001)
7. F. Arregui, E. Cabrera, R. Cobacho, *Integrated Water Meter Management* (IWA Publishing, 2006)
8. K. Billewicz, *Smart metering – Intelligent measuring system* (Wydawnictwo Naukowe PWN, Warszawa, 2011) (in Polish)
9. A. Criminisi, C. Fontanazza, G. Freni, G. Loggia, Water Sci. Technol. **60**, 9 (2009)
10. P. Mukheibir, R. Stewart, D. Giurco, K. O'Halloran, Water J. **39**, 8 (2012)
11. A. Lambert, R. McKenzie, *Practical Experience in using the Infrastructure Leakage Index,* IWA Conference Leakage Management – A Practical Approach, Cyprus, (2002)
12. A. Bochena, *Treasures of the Earth - Water. Where does the water come from in tap?* (Wydawnictwo Oulxl Media Sp. z o.o., Bydgoszcz, 2015) (in Polish)
13. T. Cichoń, J. Królikowska, Ecol. Eng. **48** (2016)
14. Water Audits and Loss Control Programs. Manual of Water Supply Practices M36. 3rd ed. AWWA. 283 (2009)
15. F. Arregui, E. Cabrera, R. Cobacho, J. Gracia-Serra, Water Pract. Tech. **1**, 4 (2006)
16. P. Tuz, J. Gwoździej-Mazur, Rynek Instalacyjny **9** (2012)
17. P. Tuz, Rynek Instalacyjny **12** (2012)
18. Materials company Sensus

 DOI: 10.1051/e3sconf/20172200062

Removal of oxytetracycline from aqueous solutions by hydroxyapatite as a low-cost adsorbent

Maria Harja[1] and *Gabriela* Ciobanu[1,*]

[1]"Gheorghe Asachi" Technical University of Iasi, Faculty of Chemical Engineering and Environmental Protection, Prof. dr. docent Dimitrie Mangeron Rd., no. 73, 700050, Iasi, Romania

Abstract. The present paper involved a study of the adsorption process of the oxytetracycline drug from aqueous medium by using the hydroxyapatite nanopowders as adsorbent materials. The batch adsorption experiments were performed by monitoring the solution pH, contact time, adsorbent dosage and drug solution concentration. At pH 8 and ambient temperature, high oxytetracycline removal rates of about 97.58% and 89.95% for the uncalcined and calcined nanohydroxyapatites, respectively, were obtained. The kinetic studies indicate that the oxytetracycline adsorption onto nanohydroxyapatite samples follows a pseudo-second order kinetic model. The maximum adsorption capacities of 291.32 mg/g and 278.27 mg/g for uncalcined and calcined nanohydroxyapatite samples, respectively, have been found. So, the conclusion can be drawn that the hydroxyapatite shows good adsorption ability towards oxytetracycline.

1 Introduction

In recent years, soil and water pollution by the antibiotic residues is a matter of growing preoccupation [1]. Among different groups of antibiotics, tetracyclines are of special concern because large quantities are applied in the therapy of human and animal infections and also at sub-therapeutic levels in animal feed as growth promoters. But, only a small amount of tetracyclines can be metabolized or absorbed by humans and animals. Residues of these antibiotics are accumulated in environmental waters and soil and they may lead to occurrence of resistant species [2]. The antibiotics occurrence in water environment exposes human beings and animals to constant low concentrations of antibiotics through drinking water contamination. A soil sorption study indicated that oxytetracycline, a major member of tetracyclines, is strongly adsorbed in soil regardless of soil type [3]. Broad scale monitoring studies have found oxytetracycline at concentrations as high as 300 μg/kg in some soils and up to 15 μg/L in some ground and surface waters [4]. The World Health Organization (WHO) recommends a guideline value of less than 1 ppb of antibiotic residues in the aquatic environment and less than 100 ppb in soil respectively [5].

Therefore, efficient and inexpensive treatment methods for the removal of such compounds from the environment were developed. A low-cost strategy to remove

* Corresponding author: gciobanu03@yahoo.co.uk

antibiotics from effluents at sewage treatment plants, after a traditional treatment, is to include a solid phase adsorption step. Adsorption of tetracyclines, oxytetracycline in the present case, on isolated clays, organic clays, soils, organic matters and marine sediment has been previously investigated [6–8].

Apatites are the most commonly occurring phosphate minerals on the Earth's crust, and consequently the primary natural sources of phosphorous in the biosphere [9]. Apatites are important in a great variety of natural and industrial processes. They are used as raw material in fertiliser production and in the remediation of contaminated soils [10]. Hydroxyapatite (HA), $Ca_{10}(PO_4)_6(OH)_2$, is similar to the biological apatite that is the main constituent of mammalian bone and teeth, and it is therefore a promising material for bone and tooth implants [11–13]. HA has been investigated as adsorbent for removing many toxic metal ions, dyes and other contaminants from aqueous solutions [14, 15].

The present work involved a study of the adsorption process of the oxytetracycline (OTC) from aqueous solutions by HA as adsorbent material. The reason for choosing HA as adsorbent is his non-toxicity, adsorption properties and low-cost material. Besides, the adsorption behaviour of OTC on HA has not yet been studied in the literature till now. Kinetic adsorption experiments were carried out to establish the effect of time on the adsorption process and to determine the adsorption rate for OTC removal.

2 Experimental

2.1 Materials

Calcium hydroxide $Ca(OH)_2$, orthophosphoric acid H_3PO_4 (85 wt.%), sodium hydroxide NaOH, hydrochloric acid HCl, ethanol $CH_3–CH_2–OH$, and oxytetracycline hydrochloride $C_{22}H_{24}N_2O_9{\cdot}HCl$ were purchased from Sigma-Aldrich (Germany), all being of analytical grade and used without further purification.

2.2 Preparation and characterization of nanohydroxyapatite adsorbents

The uncalcined and calcined HA samples labeled *HA-uc* and *HA-c*, respectively, were prepared by using $Ca(OH)_2$ and H_3PO_4 as raw materials, as presented elsewhere [16, 17]. The HA powder was synthesised by the drop-wise addition of a 250 mL solution of $Ca(OH)_2$ (0.1M) to an appropriate amount of H_3PO_4 (0.1M) aqueous solution to achieve predetermined Ca/P atomic ratio of 1.67, under magnetic stirring for 1 h at 60±1°C. The pH was continuously monitored and adjusted to 11±0.5 by adding NaOH (1M) aqueous solution. The suspension was aged for 3 h at 60±1°C and then filtered and washed with ethanol and triply distilled water. The obtained powder was calcined at 800°C in an electrically heated furnace in order to increase its crystallinity.

The phase composition and average crystallite size of the as obtained HA powders were obtained by X-ray diffraction (XRD) with X'PERT PRO MRD diffractometer (PANalytical, Netherlands) using monochromatic CuKα radiation (λ = 0.15418 nm). The morphology and chemical composition of the samples were studied by scanning electron microscopy (SEM) coupled with energy dispersive X-ray spectroscopy (EDX) with QUANTA 200 3D microscope (FEI, Netherlands). The specific surface areas were evaluated by fitting the Brunauer-Emmett-Teller (BET) equation to the N_2 adsorption isotherms recorded by a Quantachrome Nova 2200e Win2 apparatus. The pH at the point of zero charge (pH_{pzc}) of the HA samples was determined by the pH drift method [18].

2.3 Batch adsorption experiments

All adsorption experiments were carried out using the batch equilibrium technique. Various OTC solutions with different initial concentrations are prepared by dilute of the OTC (1000 mg/L) stock solution. The influence of various parameters such as initial solution pH, contact time, HA dose and OTC initial concentration on removal of OTC by HA samples (*HA-uc* and *HA-c*) was investigated by varying one parameter at a time and keeping the remaining other parameters as constant. HCl and NaOH solutions were used to adjust the solution pH. All pH measurements were realized with a Multi-Parameter Consort C831. The concentration of OTC in solutions was analysed by UV-Vis Jasco V-550 spectrophotometer at wavelength of maximum adsorption value of OTC (λ_{max} = 355 nm). Each experiment was repeated three times and the given results were the average values.

The adsorption capacity (q_t, mg/g) at time t and the removal rate (R,%) were determined using following equations:

$$q_t = \frac{C_0 - C_t}{m} \cdot V \quad (1)$$

$$R = \frac{C_0 - C_t}{C_0} \cdot 100 \quad (2)$$

where C_0 and C_t are the initial and t time's concentrations of OTC in the solution (mg/L), V is the volume of solution (L), and m is the mass of the adsorbent (g).

2.4 Adsorption kinetics models

The adsorption process order and the rate constant were estimated by applying the pseudo-first order kinetic [19], pseudo-second order kinetic [20] and intraparticle diffusion [21] models (Table 1), and the model with higher correlation coefficient (R^2) was chosen.

Table 1. The adsorption kinetics models.

Model	Linear equation	Model parameters	Comments
Pseudo-first order kinetic model	$\log(q_e - q_t) = \log q_e - \frac{k_1}{2.303} \cdot t$ (3)	k_1 = pseudo first-order rate constant (min^{-1})	k_1 and q_e values were calculated from the plot of log(q_e - q_t) versus t; q_e = adsorption capacity at equilibrium (mg/g)
Pseudo-second order kinetic model	$\frac{t}{q_t} = \frac{1}{k_2 \cdot q_e^2} + \frac{1}{q_e} \cdot t$ (4)	k_2 = pseudo-second order rate constant (g/mg·min)	k_2 and q_e values were calculated from the plot of t/q_t versus t
Intraparticle diffusion model	$q_t = k_{id} \cdot t^{1/2} + c$ (5)	k_{id} = intraparticle diffusion rate constant (mg/g·min$^{0.5}$); c = intercept (mg/g)	k_{id} and c values were calculated from the plot of q_t versus $t^{1/2}$

3 Results and discussion

3.1 Characterization of nanohydroxyapatite adsorbents

The HA nanopowders were prepared and characterized by XRD, SEM-EDX and BET methods, as reported in previous papers [16, 17]. The uncalcined and calcined HA samples have hexagonal crystal structure, crystal sizes smaller than 100 nm, and a specific surface area of 316 m^2/g and 139 m^2/g, respectively. The morphology of *HA-uc* and *HA-c* samples is shown in SEM micrographs (Fig. 1). The *HA-uc* sample contains needle-like shape crystallites. Instead, in *HA-c* sample the particles are agglomerated in spherical-like shape aggregates with an average diameter smaller than 200 nm. Also, the micrographs show irregular and porous surfaces with intergranular porosity.

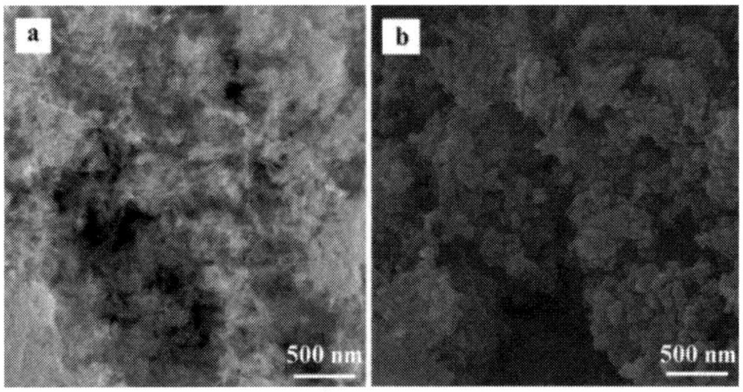

Fig. 1. SEM images of *HA-uc* (a) and *HA-c* (b) hydroxyapatite nanopowders.

3.2 Adsorption of oxytetracycline from solutions

3.2.1 Effect of initial pH

Figure 2a shows the effect of pH (from 2 to 10) on OTC adsorption onto HA samples (at initial OTC concentration 50 mg/L, adsorbent dose 2 g/L, contact time 1 h, temperature 20±1°C). The optimum adsorption was found to be at pH = 8, with OTC removal rate of about 97.58% and 89.95% for *HA-uc* and *HA-c* samples, respectively. Lower adsorption of calcined sample can be explained by the calcination process which resulted in a significant decrease of the specific surface area due to particles coalescence and densification. So, *HA-uc* sample is able to hold higher amounts of dyes than *HA-c* sample.

The removal efficiency of OTC on both HA samples showed a increase when pH increased from 2 to 8, and then decreased significantly when pH was above 8. This trend can be understood considering the interactions between OTC species and HA surface.

The existence of acid/base-active phenolic hydroxyl and amine groups in OTC (an amphoteric antibiotic) determines that OTC may exist in different ionic species at different solution pHs. The OTC molecule has three acid dissociation constants depending on the pH (pK_a = 3.57, 7.49 and 9.44) and existed as cationic form in strong acid solutions at pH < 3.57, zwitterion at 3.57 < pH < 7.49 and anionic form at pH > 7.49 [22].

Likewise, HA acts as an amphoteric material depending on the solution pH [23]. The pH at the point of zero charge (pH_{pzc}) of *HA-uc* and *HA-c* samples was found to be 6.5 and 6.65, respectively. Hence, for pH < pH_{pzc} the surface of *HA-uc* and *HA-c* samples gets positively charged (due to $\equiv CaOH_2^+$ sites) and for pH > pH_{pzc} is negatively (due to

≡OPO_3H^- sites), indicating that the HA surfaces exhibited zwitterionic properties, in good agreement with literature data [24].

The observed pH effect on the adsorption of OTC onto *HA-uc* and *HA-c* samples is related to the molecular structure of OTC and the functional groups present on the surface of HA. The sorption mechanism of OTC on HA surface at pH = 8 can be established via surface complexation. Physical mechanisms such as van der Waals forces attraction and hydrogen bonding between OTC molecules and –OH groups on the surface of HA samples may also contribute to surface complexation mechanism in OTC adsorption.

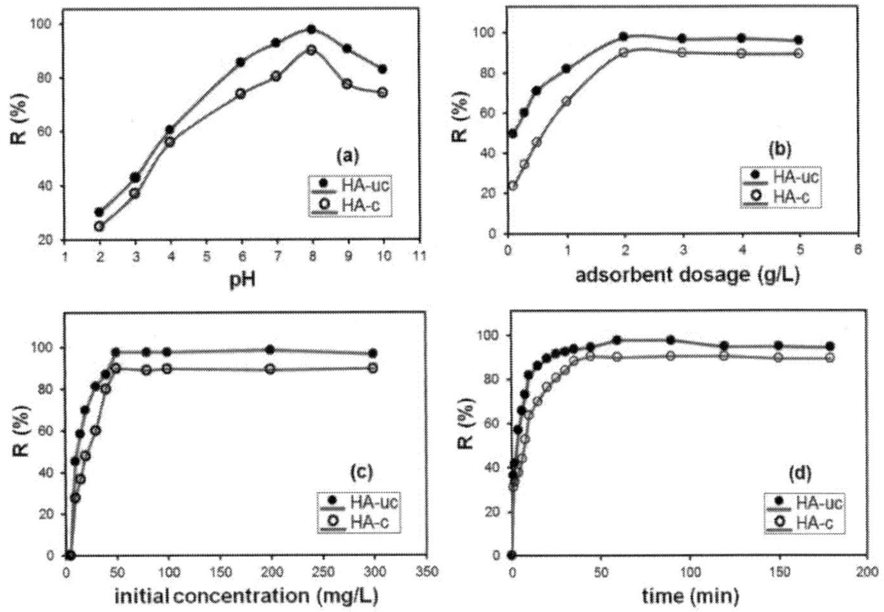

Fig. 2. Effects of initial solution pH (a), HA adsorbent dosage (b) OTC initial concentration (c) and contact time (d) on OTC adsorption on *HA-uc* and *HA-c* samples.

3.2.2 Effect of adsorbent dosage

The effect of adsorbent dosage on the OTC removal by HA samples was evaluated in the range of 0.1–5 g/L adsorbent, keeping all other parameters constant (pH = 8, initial OTC concentration 50 mg/L, contact time 1 h, temperature 20±1°C). As seen in Fig. 2b, for both HA samples, the OTC removal rate increased with increasing adsorbent dosage and the maximum was attained at 2 g/L adsorbent dose. The increase of OTC removal was due to the increase in surface area and active surface sites available of HA samples. Therefore, in the following investigations the adsorbent dosage of 2 g/L was used.

3.2.3 Effect of initial OTC concentration

The effect of initial solution concentration was studied by varying the initial OTC concentration from 10–300 mg/L in the test solution, keeping all other parameters constant (pH = 8, adsorbent dose 2 g/L, contact time 1 h, temperature 20±1°C). The amount of OTC adsorbed increased with increase in initial drug concentration tending to saturation at higher drug concentrations (Fig. 2c). At lower concentrations, because the adsorption sites of adsorbent are more, an increase in OTC removal was achieved. At higher concentrations, lower adsorption yield is due to the saturation of adsorption sites. On the other hand, the removal degree remained constant for OTC concentration exceeding 50 mg/L. This value was used in the following investigations.

3.2.4 Effect of contact time

Adsorption of OTC onto HA samples at different contact time was studied, keeping all other parameters constant (pH = 8, initial concentration of OTC 50 mg/L, adsorbent dose 2 g/L, temperature 20±1°C). The effects of contact time on the removal of OTC by HA samples are shown in Fig. 2d. Results showed that the adsorption was rapid at the first 15 min and gradually increased with increasing contact time until equilibrium was attained, which was considered at 1 hour. At beginning of adsorption, there existed enough sorption sites on the HA surface which was beneficial to adsorption of OTC. With increasing contact time, sorption sites on the surface of HA samples were close to saturation which made adsorption of OTC transfer from the surface into the void of HA samples.

3.2.5 Adsorption kinetics models

In order to analyse the adsorption kinetics for the adsorption of OTC onto *HA-uc* and *HA-c* samples, the pseudo-first order, pseudo-second order and intraparticle diffusion models were applied. Figure 3 shows the fitted plots of adsorption kinetic models for OTC by HA samples. The validity of each model is checked by the corresponding correlation coefficients (R^2) as well as the experimental and calculated data summarized in Table 2.

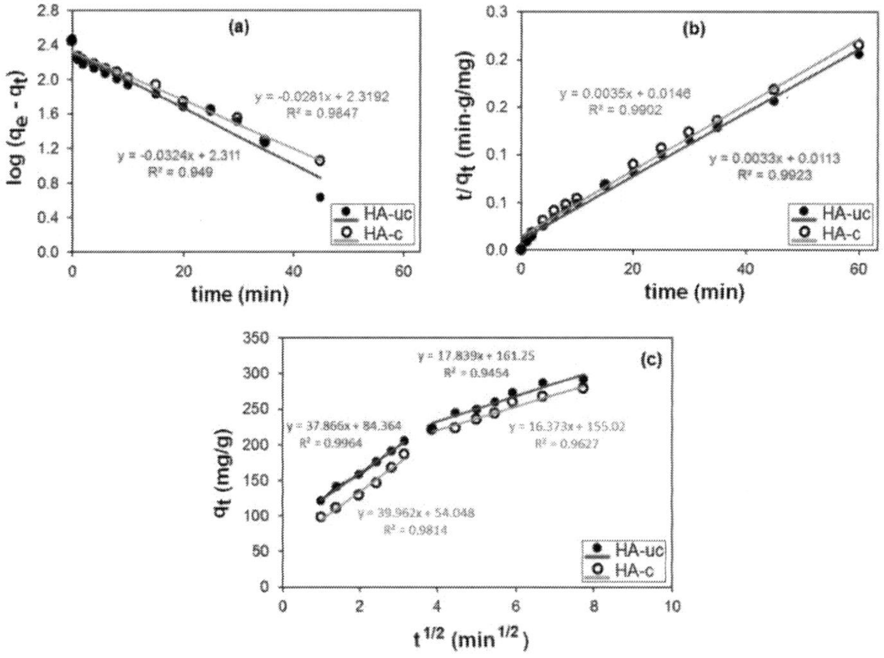

Fig. 3. The pseudo-first order (a), pseudo-second order (b) and intraparticle diffusion (c) kinetic models for OTC adsorption on *HA-uc* and *HA-c* samples.

The R^2 values obtained by the pseudo-second order kinetic model (Fig. 3b) were very high, and the calculated adsorption capacity values ($q_{e,calc}$) in Table 2 were in good agreement with the experimental results ($q_{e,exp}$), suggesting the applicability of the pseudo-second order kinetic model to describe the adsorption kinetics data of OTC onto HA samples. In this case, the adsorption rate limiting step may be chemisorption and the adsorption of OTC occurs probably via surface complexation reactions at specific adsorption sites on HA surface.

To better understand the mechanism of adsorption, the intraparticle diffusion model was applied. The plot of this model (Fig. 3c) could be evaluated in two linear parts. In the first part (phase I) is occurred OTC rapid uptake and the second part (phase II) is associated with the intraparticle diffusion consisting after the accomplishment of external surface coverage by the previous process. The first period is the external surface adsorption or instantaneous diffusion stage, during which a large amount of OTC are rapidly adsorbed by the exterior surface of the adsorbent. When the adsorption of exterior surface reaches saturation, the diffusion resistance increases, leading to the decreased diffusion rate, it is the second period in the adsorption process.

Consistent with the results shown in Fig. 2d, the OTC adsorption by *HA-uc* and *HA-c* samples was achieved at two stages: the rapid surface adsorption (external surface adsorption) in the first stage (first 15 min) and the slow intraparticle diffusion (internal surface adsorption) in the second stage (15–60 min). After 60 min, the adsorption reached equilibrium.

Table 2. Kinetic parameters for OTC adsorption on *HA-uc* and *HA-c* samples (initial pH = 8, adsorbent dose 2 g/L, initial OTC concentration 50 mg/L, temperature 20±1°C).

Kinetic model		**Parameter**	**Sample**	
			HA-uc	***HA-c***
		$q_{e,exp}$ (mg/g)	291.32	278.27
Pseudo-first order kinetic model		$q_{e,calc}$ (mg/g)	204.64	208.55
		k_1 (min^{-1})	0.0746	0.0647
		R^2	0.9490	0.9847
Pseudo-second order kinetic model		$q_{e,calc}$ (mg/g)	303.03	285.71
		k_2 (g/mg·min)	0.000963	0.000839
		R^2	0.9923	0.9902
Intraparticle diffusion model	I	$k_{id(I)}$ (mg/g·min$^{1/2}$)	37.866	39.962
		$c_{(I)}$	84.364	54.048
		$R^2_{(I)}$	0.9964	0.9814
	II	$k_{id(II)}$ (mg/g·min$^{1/2}$)	17.839	16.373
		$c_{(II)}$	161.25	155.02
		$R^2_{(II)}$	0.9454	0.9627

Finally, because of the relatively low-cost of the starting materials and the route used, the resultant hydroxyapatite nanopowders have the potential to be cost-effective adsorbents for the removal of OTC from aqueous solutions.

4 Conclusions

In the present study the adsorption of the oxytetracycline (OTC) from aqueous solution on hydroxyapatite (HA) nanopowders was investigated. The batch adsorption experiments were performed and the effects of the initial solution pH, contact time, initial OTC concentration and adsorbent dosage on the OTC adsorption by HA nanopowders were determined. Adsorption process is dependent on pH, and the optimum pH value was 8. The adsorption capacity of OTC on uncalcined and calcined samples (*HA-uc* and *HA-c*) is up to 291.32 mg/g and 278.27 mg/g at experimental condition, respectively. Removal rates of about 97.58% and 89.95% for the *HA-uc* and *HA-c* samples, respectively, were obtained. The kinetic study made evident a fast adsorption process reaching the equilibrium at a contact time of 60 min and pH = 8, for both samples. It was found that the pseudo-second order equation describes in the best way the adsorption kinetics of OTC on calcined and uncalcined HA nanopowders. The adsorption of OTC onto HA samples follows the intraparticle (pore) diffusion model. This means that the adsorption process is not only a surface phenomenon, but also a rate-limiting process in the micro- and nano- porous structure of HA. The obtained results suggest that the OTC adsorption process on HA samples occurs into two main stages. The first stage involves bulk diffusion of OTC to the surface of the adsorbent, while the second one involves intraparticle or pore diffusion processes. The adsorption of OTC proceeds probably via surface complexation reactions at specific adsorption sites on HA surface. The results of present study are indicative of hydroxyapatite as a new alternative low-cost adsorbent for removing the oxytetracycline from wastewaters.

References

1. J.P. Bound, N. Voulvoulis, Chemosphere **56**, 1143–1155 (2004)
2. P. Gao, M. Munir, I. Xagoraraki, Sci. Total Environ. **421–422**, 173–183 (2012)
3. M. Rabolle, N.H. Spliid, Chemosphere **40**, 715–722 (2000)
4. D. Fatta-Kassinos, S. Meric, A. Nikolaou, Anal. Bioanal. Chem. **399**, 251–275 (2011)
5. WHO, *Pharmaceuticals in drinking water* (WHO Press, Geneva, 2012)
6. A.A. MacKay, B. Canterbury, J. Environ. Qual. **34**, 1964–1971 (2005)
7. M. Rabolle, N.H. Spliid, Chemosphere **40**, 715–722 (2000)
8. P. Kulshrestha, R.F. Giese, D.S. Aga, Environ. Sci. Technol. **38**, 4097–4105 (2004)
9. E.H. Oelkers, E. Valsami-Jones, Elements **4**, 83–87 (2008)
10. D.A.C. Manning, Elements **4**, 105–108 (2008)
11. S.V. Dorozhkin, J. Funct. Biomater. **6**, 708–832 (2015)
12. G. Ciobanu, G. Carja, O. Ciobanu, Surf. Coat. Tech. **202**, 2467–2470 (2008)
13. G. Ciobanu, S. Ilisei, C. Luca, G. Carja, O. Ciobanu, Prog. Org. Coat. **74**, 648–653 (2012)
14. Y. Nishiyama, T. Hanafusa, J. Yamashita, Y. Yamamoto, T. Ono, J. Radioanal. Nucl. Ch. **307**, 1279–1285 (2016)
15. S. Wang, H. Liu, W. Liu, Q. Zuo, J. Biomater. Tiss. Eng. **6**, 433–439 (2016)
16. G. Ciobanu, A.M. Bargan, C. Luca, Ceram. Int. **41**, 12192–12201 (2015)
17. G. Ciobanu, A.M. Bargan, C. Luca, JOM **67**, 2534–2542 (2015)
18. M.N. Khan, A. Sarwar, Surf. Rev. Lett. **14**, 461–469 (2007)
19. S. Lagergren, K. Sven. Vetenskapsakad. Handl. **24**, 1–39 (1898)
20. Y.S. Ho, G. McKay, Chem. Eng. J. **70**, 115–124 (1998)
21. W.J. Weber jr., J.C. Morris, J. Sanit. Eng. Div. **89**, 31–60 (1963)
22. S.A. Sassman, L.S. Lee, Environ. Sci. Technol. **39**, 7452–7459 (2005)
23. K. Kadlec, K. Adamska, A. Voelkel, Talanta **147**, 44–49 (2016)
24. Å. Bengtsson, S. Sjöberg, Pure Appl. Chem. **81**, 1569–1584 (2009)

DOI: 10.1051/e3sconf/20172200063

Monitoring of drug resistance amplification and attenuation with the use of tetracycline-resistant bacteria during wastewater treatment

Monika Harnisz[1,*], *Ewa* Korzeniewska[1], *Sebastian* Niestępski[1], *Adriana* Osińska[1], and *Beata* Nalepa[2]

[1]Department of Environmental Microbiology, University of Warmia and Mazury in Olsztyn, ul. Prawocheńskiego 1, 10-957 Olsztyn, Poland
[2]Department of Food and Industrial Microbiology, University of Warmia and Mazury, Plac Cieszyński 1, 10-726 Olsztyn, Poland

Abstract. The objective of this study was to monitor changes (amplification or attenuation) in antibiotic resistance during wastewater treatment based on the ecology of tetracycline-resistant bacteria. The untreated and treated wastewater were collected in four seasons. Number of tetracycline- (TET^R) and oxytetracycline-resistant (OTC^R) bacteria, their qualitative composition, minimum inhibitory concentrations (MICs), sensitivity to other antibiotics, and the presence of *tet*(A, B, C, D, E) resistance genes were determined. TET^R and OTC^R counts in untreated wastewater were 100 to 1000 higher than in treated effluent. OTC^R bacterial counts were higher than TET^R populations in both untreated and treated wastewater. TET^R isolates were not dominated by a single bacterial genus or species, whereas *Aeromonas hydrophila* and *Aeromonas sobria* were the most common in OTC^R isolates. The treatment process attenuated the drug resistance of TET^R bacteria and amplified the resistance of OTC^R bacteria. In both microbial groups, the frequency of *tet*(A) gene increased in effluent in comparison with untreated wastewater. Our results also indicate that treated wastewater is a reservoir of multiple drug-resistant bacteria as well as resistance determinants which may pose a health hazard for humans and animals when released to the natural environment.

1 Introduction

Microbial resistance to drugs poses a growing threat for human health and life. The wide-spread use of antibiotics in the treatment of bacterial diseases has contributed to the development of strains that are resistant to this group of drugs. The problem has been escalating dramatically by the increased use of antimicrobial agents in medicine, detergent production, agriculture, breeding and cosmetic industry [1–3]. Multiple drug resistance has developed and spread rapidly among various bacterial groups within a short timeline of

* Corresponding author: monika.harnisz@uwm.edu.pl

evolution. The problem is deepened by the existence of large bacterial populations, short generation times and effective mechanisms of genetic information exchange [4].

Wastewater treatment plants (WWTPs) are flagship projects aiming to prevent environmental pollution. Wastewater purified of suspended matter, excess organic substances and biogenic elements can be reused to irrigate fields or evacuated to surface waters [6, 7]. Municipal treatment plants are supplied with wastewater generated by households, hospitals, public utility buildings, industrial plants as well as rain water. This combination of organic substances, biogenic elements and trace amounts of antibiotics supports the development and survival of drug-resistant bacteria [8]. The use of activated sludge and biological membrane systems in the treatment process contributes to the exchange of genetic information between bacteria of the same or different species, including microorganisms from unrelated families [9]. The above facilitates the exchange of genes located on mobile plasmids, including those encoding drug resistance [10].

WWTPs are important reservoirs of antibiotic resistance of both commensal and pathogenic organisms [11, 12]. Drug-resistant bacteria and resistance genes can be released from the plant into the natural environment. To mitigate this threat, the presence of antibiotic-resistant bacteria has to be determined in untreated and treated wastewater. The resulting data support an evaluation of the effect of the treatment process on changes in bacterial resistance to antibiotics.

The microbial indicator has been defined by Nielsen et al. [13] as "microbial parameter that represents properties of the environment or impacts to the environment, which can be interpreted beyond the information that the measured or observed parameter represents itself". Microbial bioindicators could be based on functional or structural diversity of the community [14]. In the case of changes in the antibiotic resistance of bacteria, structural bioindicators are specific bacterial species and genera which are cultured on selective media and functional bioindicators comprise groups of bacteria resistant to a specific drug or a class of drugs. They are grown on nutrient agars, such as TSA, with the addition of antibiotics, such as tetracyclines which, despite a rapid increase in bacterial resistance, continue to be widely used in human and veterinary medicine.

The objective of this study was to monitor changes (amplification or attenuation) in antibiotic resistance during wastewater treatment based on the ecology of tetracycline- and oxyteracycline-resistant bacteria in untreated and treated wastewater. In Poland, tetracycline is one of the most popular drug of choice in human medicine, whereas oxytetracycline is widely used in the treatment of both humans and animals. Akinbowale et al. [15] noticed, that minimum inhibitory concentrations of these two drugs can be different, so we thought the populations of tetracycline- and oxytetracycline-resistant bacteria can be different, also. Therefore, we decided to study both microbial groups.

2 Materials and Methods

2.1. Wastewater treatment plant

The investigated site was the Łyna Municipal Waste Treatment Plant in Olsztyn. The plant's process line comprises mechanical, biological and chemical treatment sections, as well as sludge processing units. The plant has the following technical specification: treatment system - activated sludge, average processing capacity – 60,000 m^3/d, wastewater type - municipal wastewater, mechanical treatment devices – screenings, grit chamber and pre-sedimentation tank, biological treatment devices – separation chambers, aeration chambers and secondary sedimentation tanks, sedimentation devices – closed and open digestion chambers, belt filter press, incinerator.

2.2. Sample collection

The experimental material comprised samples of untreated (from grid chamber) and treated (effluent of secondary sedimentation tank) wastewater collected on four occasions: in April, July and October 2012 and in February 2013. Wastewater samples were collected into sterile bottles, transported to the laboratory at a temperature of 4°C and processed on the day of collection.

2.3. Heterotrophic plate counts and counts of tetracycline- and oxytetracycline-resistant bacteria

Heterotrophic plate counts (HPC), counts of tetracycline-resistant bacteria (TET^R) and counts of oxytetracycline-resistant bacteria were determined in plates containing the TSA medium (Oxoid) without/with tetracycline and oxytetracycline supplementation, respectively. The plates were cultured at a temperature of 30°C for 24h. According to CLSI guidelines [16], pathogens with MIC values of the tested antibiotics at ≥ 16 µg/mL are regarded as resistant. In this study, tetracycline and oxytetracycline were applied at a concentration of 32 µg/mL to isolate highly resistant bacteria. MIC value of these antibiotics were determined by the agar dilution method with final antibiotic concentrations in the range of 32 to 512 µg/mL. Cultured colonies of HPC, TET^R and OTC^R were counted, and the results were stated in terms of colony forming units (CFU) per mL of wastewater. Colonies with various phenotypes were isolated from the TSA medium supplemented with tetracycline and oxytetracycline. A total of 40 isolates were obtained from each medium (in summary 80 isolates). The strains were stored on the LB medium (Merck) with 10% glycerol at -70°C for further analyses.

2.4. Antibiotic susceptibility testing

TET^R and OTC^R strains isolated in the study were subjected to sensitivity tests against nine antimicrobials from five classes: (1) beta-lactams: mezlocillin (MEZ 75 µg), ampicillin (AMP 10 µg), piperacillin (PRL 75 µg); amoxicillin/clavulanic acid (AMC 20/10 µg), ceftazidime (CAZ 30 µg), cefotaxime (CTX 30 µg); (2) aminoglycosides: gentamicin (CN 10 µg), tobramycin (TOB 10 µg); (3) macrolides: erythromycin (TOB 10 µg); (2) tetracyclines: tigecycline (TGC 15 µg); (4) trimethoprim/sulfamethoxazole (SXT 1.25/23.75 µg); (6) phenicols: chloramphenicol (C 30 µg); and (6) fluoroquinolones: ciprofloxacin (CIP 30 µg). All disks were supplied by Oxoid. Resistance was estimated by measuring the inhibition zone according to the guidelines of CLSI [16].

2.5. Identification of isolates and multiplex PCR of tetracycline-resistant genes

Genomic DNA was isolated by the use of thermal lysis [14]. TET^R and OTC^R isolates were identified by 16S rRNA gene sequencing in accordance with a previously described method [14].

The mechanism of molecular resistance to tetracyclines was analyzed with the use of five out of around 40 known determinants [17]. They were selected based on source data indicating that *tet*(A), (B), (C), (D) and (E) genes are the most popular determinants of resistance in *Enterobacteriaceae* and *Aeromonas* sp. bacteria [18, 19]. Multiplex PCR was conducted in line with the methodology proposed by Nawaz et al. [18]. For all reactions, standard PCR mixtures without DNA template were used as negative controls. Plasmids carrying *tet* genes or the 16S rRNA gene verified by sequencing were used as positive controls.

2.6. Data analyses

One way analysis of variance (ANOVA, StatSoft Inc., 1984–2017) was used to check differences in the abundance of the studied bacterial groups between the analyzed sites.

3 Results

HPC analyses revealed differences in the size of TET^R and OTC^R bacterial populations, subject to the site and time of sampling. TET^R and OTC^R concentrations in untreated wastewater were 100 to 1000-fold higher than in discharged effluent ($p = 0.006$ and $p = 0.003$ for TET^R and OTC^R, respectively) (Table 1).

Table 1. The share of tetracycline- and oxytetracycline-resistant strains incubated at a temperature of 30°C (TET^R30°C and OTC^R30°C) in total heterotrophic bacterial counts incubated under identical conditions (HPC30°C).

Bacterial group		**Date**	**Sampling site: Untreated wastewater**	**Sampling site: Treated wastewater**
TET^R 30°C [1)]	CFU/cm^3	04.2012	6.18×10^4	8.40×10^2
		07.2012	1.08×10^5	1.26×10^3
		10.2012	1.15×10^5	3.80×10^2
		02.2013	3.36×10^4	7.02×10^2
OTC^R 30°C [2)]		04.2012	3.51×10^5	6.61×10^2
		07.2012	1.23×10^5	1.52×10^3
		10.2012	1.99×10^5	1.86×10^3
		02.2013	2.74×10^5	6.50×10^3
HPC 30°C [3)]		04.2012	5.01×10^6	3.67×10^4
		07.2012	9.70×10^6	7.40×10^4
		10.2012	9.60×10^6	2.05×10^4
		02.2013	4.88×10^6	1.21×10^5
Share of TET^R 30°C in HPC 30°C	%	04.2012	1.23	2.3
		07.2012	1.1	1.7
		10.2012	1.2	1.85
		02.2013	0.7	0.6
Share of OTC^R 30°C in HPC 30°C		04.2012	7	1.8
		07.2012	1.3	2.1
		10.2012	2.1	9.1
		02.2013	5.6	5.4

[1)] – tetracycline-resistant bacteria cultured at 30°C after 2 days of incubation;
[2)] – oxytetracycline-resistant bacteria cultured at 30°C after 2 days of incubation;
[3)] – total heterotrophic bacterial counts determined at 30°C after 2 days of incubation;

Maximum TET^R levels were noted in October and July 2012 in untreated and treated wastewater, respectively. The highest OTCR counts in inflows and outflows of WWTP were reported in April 2012 and February 2013. OTCR bacteria were more abundant than TETR bacteria in both untreated and treated wastewater samples ($p = 0.012$). In both incoming and outgoing wastewater, TETR had the lowest share of HPC in February 2013, and the highest – in April 2012. OTC^R was characterized by a different share of HPC. The percentage share of OTC^R bacteria in untreated sewage was the lowest in July 2012 and the highest in April 2012. As regards treated wastewater, the lowest and the highest percentages of OTC^R bacteria were noted in April 2012 and October 2013, respectively.

In TET^R bacteria, the predominance of a single genus or species was not observed. In the group of 40 OTC^R isolates, eight *Aeromonas hydrophila*, ten *Aeromonas sobria*, one *Aeromonas jandei* and two *Aeromonas allosacharphila* strains were identified and found to predominate (Table 2).

In the group of TET^R bacteria isolated from untreated wastewater, *Acinetobacter baumanii* was characterized by very low sensitivity to drugs (MIC of TET – 256 μg/mL, MIC of OTC – 512 μg/mL, MDR to seven groups of drugs). The least drug-sensitive microorganism isolated from the OTC^R group in untreated wastewater was *Aeromonas sobria* (MIC of TET – 32 μg/mL, MIC of OTC – 256 μg/mL, resistant to five classes of UW - untreated wastewater, TW - treated wastewater; MDR to: B – β-lactams, A-Aminoglycosides, M-Macrolides, T-Tetracyclines, S-Sufonamides/Trimethoprim, P-Phenicols, F-Fluoroquinolones; nd - not detected; - resistance to less than three classes of drugs antibiotics). The most frequently observed pattern of resistance was insensitive to β-lactams, aminoglycosides and macrolides (Table 2).

Table 2. Diversity of tetracycline- and oxytetracycline-resistant bacteria, minimum inhibitory concentrations of tetracycline (TET) and oxytetracycline (OTC), the resistance profile, multi-drug resistance (MDR) and determinants of resistance of strains isolated from UW and WW.

Origin		Identification results (number of strains)	MIC of TET/OTC	Resistance profile (number of strains)	MDR	Genes
UW	TET^R	*Acinetobacter baumanii(2)*	256/512	MEZ, CAZ, CTX, CN, TOB, E,SXT,C,CIP(1) AMP,AMC,PRL,CAZ,CTX,E,TGC,C(1)	BAMTSPF BMTP	*tet*(B) *tet*(B)
		Acinetobacter junii(2)	96/512	MEZ,AMP,AMC,PRL,CAZ,CTX,E,TGC,C(1) AMP,PRL,CAZ,CTX,E,TGC,C(1)	BMTP BMTP	*tet*(A),*tet*(B) *tet*(B)
		Burkholderia cepacia(3)	96/128	AMP(1) AMP,PRL,CAZ,CTX,E,TGC,C(1) AMP,AMC,PRL,CAZ,CTX,E,TGC,C(1)	- BMTP BMTP	*tet*(A) *tet*(A) *tet*(A)
		Klebsiella pneumoniae(1)	96/512	AMP,E,TGC(1)	BTP	*tet*(E)
		Pasteurella pneumotropica(1)	32/128	MEZ,AMP,AMC,PRL,CAZ,CTX,E,TGC,C(1)	BMTP	nd
		Plesiomonas shigelloides(3)	96/128	MEZ,AMP,AMC,CAZ,CTX(1) MEZ,AMP,AMC,PRL,CAZ,CTX,E,TGC,C(2)	- BMTP	*tet*(E) *tet*(E)
		Pseudomonas putida(1)	64/128	MEZ,AMP,AMC,PRL,CAZ,CTX,E,TGC,C	BMTP	*tet*(D)
		Salmonella arizonae(3)	128/512	AMP,CTX(1) MEZ,AMP,AMC,PRL,CAZ,CTX,E,TGC,C(1) AMP,AMC,PRL,CAZ,CTX,E,TGC,C(1)	- BMTP BMTP	*tet*(D) *tet*(D) *tet*(D)
		Serratia marcescens(3)	128/512	MEZ,AMP,AMC,CAZ,CTX,CIP(1) MEZ,AMP,AMC,PRL,CAZ,CTX,E,TGC(1) AMP,AMC,PRL,CAZ,CTX,E,TGC,C(1)	- BMT BMTP	*tet*(D) *tet*(D) *tet*(D)
		Sphingomonas multivorum(1)	64/512	AMP,PRL,CAZ,CTX,E,TGC,C(1)	BMTP	nd
	OTC^R	*Aeromonas hydrophila(4)*	128/256	AMP,E(2) AMP,AMC,E(2)	- -	*tet(A)* *tet(B)*
		Aeromonas sobria(6)	32/128 32/256 32/128 32/128	AMP,E(1) AMP,SXT,C,CIP(2) AMP,CAZ,CTX,CN,TOB,E,SXT,C,CIP(2) AMP,MEZ,CAZ,CTX,CN,TOB,E(1)	- - BAMSP BAM	*nd* *tet(A),tet(B)* *tet(A),tet(B)* *nd*
		Citrobacter	128/128	MEZ,AMP,AMC,CAZ,CTX(1)	-	*tet(A)*

		freundii(2)	96/128	MEZ,CAZ,CTX,CN,TOB,E(1)	BAM	*tet(A)*
		Escherichia coli(3)	96/256	MEZ,AMP,AMC,CAZ,CTX(1) MEZ,CAZ,CTX,CN,TOB,E(2)	- BAM	*tet(A)* *tet(A)*
		Pseudomonas putida(1)	96/128	SXT,C,CIP(1)	SPF	*tet(A)*
		Sphingomonas multivorum(2)	32/128	AMP,CTX(1) MEZ,CAZ,CTX,CN,TOB,E(1)	- BAM	*tet(A)* *tet(A)*
		Yersinia pestis(2)	32/128	MEZ,AMP,AMC,CAZ,CTX(1) MEZ,CAZ,CTX,CN,TOB,E(1)	- BAM	*tet(E)* *tet(E)*
TW	TET^R	*Acinetobacter baumanni(2)*	32/256	MEZ,AMP,PRL,E,CIP(2)	BMF	*tet*(A)
		Acinetobacter junii(2)	32/512	MEZ,AMP,PRL,E,CIP(1) MEZ,AMP,PRL,E,SXT,CIP(1)	BMF BMF	*tet*(A) *tet*(A)
		Klebsiella pneumoniae(3)	32/256 128/256	AMP,AMC(2) MEZ,AMP,AMC,CAZ,CTX(1)	- -	*tet*(E) *tet*(E)
		Pasteurella pneumotropica(3)	32/256 128/256	MEZ,AMP,PRL,E,SXT,CIP(2) MEZ,AMP,AMC,CAZ,CTX(1)	BMSF -	*tet*(A) *tet*(A)
		Pseudomonas putida(3)	32/512 32/512 128/512	AMP,E(1) CTX,E(1) MEZ,AMP,PRL,E(1)	- - -	*tet*(A) *tet*(A) *tet*(A)
		Salmonella arizonae(2)	32/512 128/512	MEZ,AMP,CTX,E(1) MEZ,AMP,PRL,E(1)	- -	*tet*(E) *tet*(E)
		Serratia marcescens(3)	32/128 256/512	MEZ,AMP,AMC,PRL,E,SXT,CIP(1) E,CIP(2)	BMSF -	*tet*(A),*tet*(B) nd
		Vibrio parahaemolyticus(2)	32/128	MEZ,AMP,PRL,CAZ,CTX(1) MEZ,AMP,AMC,PRL,E,SXT,CIP(1)	- BMSPF	*tet*(E) *tet*(E)
	OTC^R	*Aeromonas allosaccharophila(2)*	32/512	MEZ,AMP,AMC,PRL,CTX,E,TGC,SXT,C(1) MEZ,AMP,PRL,E,TGC,SXT(1)	BMTSP BMTS	*tet*(A) *tet*(A)
		Aeromonas hydrophila(4)	128/512	AMP,E(1) AMP,AMC,E(1) AMP,AMC,E(1) AMP,AMC,E(1)	- - - -	*tet*(A) *tet*(A) *tet*(B) *tet*(E)
		Aeromonas jandei(1)	32/512	MEZ,AMP,AMC,PRL,CTX,E,TGC,C(1)	BMTP	*tet*(A)
		Aeromonas sobria(4)	128/512(4)	MEZ,CAZ,CTX,CN,TOB,E,SXT,C,CIP(4)	BAMSPF(4)	*tet*(A)(2) *tet*(B) -
		Citrobacter freundii(1)	128/256	MEZ,AMP,AMC,CAZ,CTX,E,SXT(1)	BMS	*tet*(A)
		Flavimonas oryzihabitans(1)	128/128	PRL,CAZ,TOB,E(1)	BAM	*tet*(A)
		Pantoea spp.(1)	96/128	MEZ,AMP,AMC,CAZ,CTX(1)	-	*tet*(D)
		Pseudomonas putida(1)	128/512	AMP,AMC,CAZ,TOB,E,TGC,C(1)	BAMTP	*tet*(A)
		Salmonella spp.(2)	32/96	MEZ,AMP,PRL,E,CIP(2)	BMF	*tet*(E)
		Shewanella putrefaciens(2)	32/96	AMP,AMC,CAZ,E(2)	-	*tet*(A),*tet*(B)
		Vibrio metschnikovii(1)	32/96	MEZ,AMP,PRL,CAZ,CTX(1)	-	*tet*(A)

In treated wastewater samples, TET^R group isolates which were the least sensitive to the tested drugs were *Vibrio parahaemoliticus*, *Klebsiella pneumoniae* and *Pasteurella pneumotropica* (MIC of TET – 32 μg/mL, MIC of OTC – 128 or 256 μg/mL, resistant to four or five classes of antibiotics). In the OTC^R group, the highest drug resistance was reported for *Aeromonas sobria* (MIC of TET – 128 μg/mL, MIC of OTC – 512 μg/mL,

DOI: 10.1051/e3sconf/20172200063

MDR to six classes of drugs). The isolates from treated wastewater were characterized by multiple drug resistance. In the TET^R group 45% of isolates were multidrug resistant to three or four groups of antibiotics, in the OTC^R group 60% isolates were multidrug bacteria (Table 2). Regardless of the sampling site, the MIC values of tetracycline were always lower than oxytetracycline MICs (Table 2). In both types of analyzed wastewater, tetracycline MICs were determined in the range of 32 to 256 µg/mL. The MIC values of OTC ranged from 96 to 512 µg/mL in untreated sewage, and from 128 to 512 µg/mL in treated wastewater. In treated wastewater samples, the majority of TET^R and OTC^R isolates were characterized by the presence of the *tet*(A) gene which was noted in 60% strains. In the group of TET^R strains showing the lowest levels of drug resistance, the presence of the *tet*(A) and *tet*(B)gene was noted (*Vibrio parahaemoliticus* and *Klebsiella pneumoniae*). In treated wastewater, OTC^R strains of the genus *Aeromonas* featured *tet*(A), *tet*(B) and *tet*(E) genes (Table 2).

4 Discussion

The treatment process decreased TET^R and OTC^R bacterial counts by two orders of magnitude. The abundance of OTC^R bacteria was somewhat higher in both untreated and treated wastewater samples. There is a general scarcity of data regarding the populations of wastewater-borne heterotrophic bacteria resistant to tetracycline and oxytetracycline. In studies of treated wastewater, Kim et al. [20] and Huang et al. [11] determined TET^R counts in the range of 10^2 - 10^3 CFU/mL, and their findings are consistent with our results. Vast similarities are also observed in TET^R bacteria's share of HPC which was determined at 0 to 2% by Kim et al. [20], Huang et al. [11] and Munir et al. [21] and at 1.61% in this study. It should be noted, however, that tetracycline concentrations reached 16 µg/ml in the experiment described by Huang et al. [11], compared with 32 µg/mL in a study by Kim et al. [20] and in the present study. Then, bacteria resistant to tetracyclines constitute a small part of HPC in treated wastewater - it may be due to of lack natural resiatance to tetracyclines among bacteria. Tet^R isolates were not marked by a predominance of a single bacterial species or genus, whereas bacteria of the genus *Aeromonas* were the most abundant microorganisms in the group of OTC^R isolates. *Aeromonas* bacteria are popularly observed in both untreated wastewater [22] and treated effluent from fish farms [23, 24]. Huys et al. [25] and Rhodes et al. [26] found more than 85% bacteria of the genus *Aeromonas* to be oxytetracycline-resistant. In this study, relatively large populations of *Enterobacteriaceae* microorganisms were reported in both, TET^R and OTC^R isolates, and similar observations were made by Guillaume et al. [27] and Moura et al. [28]. A comparison of MIC values and multiple drug resistance of the least resistant TET^R isolates in samples of untreated and treated wastewater indicates a drop in resistance to tetracycline, oxytetracycline and a decrease in multiple drug resistance as a result of sewage treatment. The reverse was observed in the OTC^R group which resistance to tetracycline and oxytetracycline decreased with an increase in the number of multiple drug-resistant strains. In the TET^R group, the majority of isolates obtained from untreated wastewater were resistant to four classes of antibiotics, whereas bacterial isolates from treated wastewater samples were resistant to one or two groups of antibiotics. The majority of OTC^R bacteria identified in untreated wastewater were not multidrug resistant, however four isolates found in treated effluent samples were resistant to of all analyzed antibiotics. Our findings are consistent with the results of published studies [29]. The results of our previous study [30] and other authors' findings [31, 32] testify to the high drug resistance of TET^R and OTC^R bacteria. According to Hall et al. [33], resistance to a single antibiotic could lead to selective resistance to other drugs and, ultimately, multiple-drug resistance. In untreated wastewater isolates, there was not the predominant gene in the TET^R group, *tet*(A) was most frequently determined in the OTC^R groups, whereas the predominance of *tet*(A) gene was noted in both bacterial groups from treated effluent. Sandalli et al. [34] and Tao et al. [19] have demonstrated the predominance of the *tet*(A) gene in *Enterobacteriaceae* bacteria, and Harnisz et al. [14] - in bacteria of *Aeromonas* genus. The *tet*(A) gene, which encodes the removal of tetracyclines from the cell with the involvement of transport proteins, is often found on plasmids, including conjugating plasmids [35]. In this study, the high frequency of *tet*(A) noted in untreated and treated wastewater, could be attributed to the gene's presence on mobile plasmids. In the work of Guillaume et al. [27], *tet*(A) was found to be highly stable in *Salmonella* sp. isolated from treated wastewater. The presence of *tet*(B), *tet*(E) and, rarely, *tet*(D) determinants has also been noted in TET^R and OTC^R isolates, whereas the *tet*(C) gene was not found. Similar results were reported by Börjesson et al. [36], Sandalli et al. [34] and Tao et al. [19]. Auerbach et al. [37] observed a wide variety of tetracycline resistance genes in wastewater, including *tet*(C) which was not identified in this study. The above could attest to the "regional" occurrence of tetracycline resistance determinants, subject to the climate, the intestinal microbiota of the local population and various antibiotic treatments. In this experiment, the simultaneous presence of two resistance determinants, *tet*(A) and *tet*(B), was noted only in two strains of the family *Enterobacteriaceae*: *Serratia marcescens* and *Shewanella putrefaciens* and three isolates of genus *Acinteobacter* and *Aeromonas*. According to Tao et al. [19], around 30% bacteria have at least two genes encoding tetracycline resistance. Similar findings were reported in studies of other bacterial species by Akinbowale et al. [15], Henriques et al. [31] and Nikolakopoulou et al. [32].

5 Conslusions

Based on the results of this study, it can be concluded that the drug resistance of TET^R bacteria was attenuated and the drug resistance of OTC^R bacteria was amplified in the course of the treatment processes. In both bacterial groups, the frequency of the *tet*(A) gene was high in untreated and treated wastewater. Our results also indicate that treated wastewater is a reservoir of multiple drug-resistant bacteria as well as resistance determinants which may pose a health hazard for humans and animals when released to the natural environment.

References

1. F. Baquero, J.L. Martinez, R. Canton, Curr. Opin. Biotech. **19**, 260–265 (2008)
2. J. Williams-Nguyen, J.B. Sallach, S. Bartelt-Hunt, A.B. Boxall, L.M. Durso, J.E. McLain, R.S. Singer, D.D. Snow, J.L. Zilles, J. Environ. Qual. **45**, 394–406 (2016)
3. K.S. Le Corre, C. Ort, D. Kateley, B. Allen, B.I. Escher, J. Keller, Environ. Int. **45**, 99–111 (2012)
4. L. Boto, Proc. R. Soc. B. **277**, 819–827 (2010)
5. Clinical and Laboratory Standards Institute, Performance standards for susceptibility testing, M100–S15. CLSI, Wayne, Pennsylvania, USA (2005)
6. E. Korzeniewska, M. Harnisz, J. Environ. Manage. **128**, 904–911 (2013)
7. M.L. Luprano, M. De Sanctis, G. Del Moro, C. Di Iaconi, A. Lopez, C. Levantesi, Sci. Total Environ. **571**, 809–818 (2016)
8. A. Schlüter, R. Szczepanowski, A. Pühler, FEMS Microbiol. Rev. **31**, 449–477 (2007)
9. R. Szczepanowski, B. Linke, I. Krahn, Microbiol. **155**, 7, 2306–2319 (2009)

10. A. Osińska, M. Harnisz, E. Korzeniewska, Environ. Sci. Pollut. Res. **23**, 10818–10831 (2016)
11. J.J. Huang, H.Y. Hu, S.Q. Lu, Y. Li, F. Tang, Y. Lu, B. Wei, Environ. Int. **42**, 31–36 (2011)
12. A. Osińska, E. Korzeniewska, M. Harnisz, S. Niestępski, Sci. Total Environ. **577**, 367–375 (2017)
13. M.N. Nielsen, A. Winding, Technical Report 388, National Environmental Research Institute, Denmark (2002)
14. M. Harnisz, E. Korzeniewska, S. Ciesielski, I. Gołaś, Sci. Total Environ. **505**, 704–711 (2015)
15. A.L. Akinbowale, H. Peng, M.D. Barton, J. Appl. Microbiol. **100**, 1103–1113 (2006)
16. Clinical and Laboratory Standards Institute (CLSI). Performance standards for susceptibility testing, M100–S15. Wayne, Pennsylvania, USA (2014)
17. Chopra, M. Roberts, Microbiol. Mol. Biol. R. **65**, 232–260 (2001)
18. M. Nawaz, K. Sung, S.A. Khan, A.A. Khan, R. Steele, Appl. Environ. Microbiol. **72**, 6461–6466 (2006)
19. R. Tao, G.G. Ying, H.C. Su, H.W. Zhou, J.P.S Sidhu, Environ. Pollut. **158**, 2101–2109 (2010)
20. S. Kim, H. Park, K. Chandran, Chemosphere **78**, 1071–1077 (2010)
21. M. Munir, K. Wong, I. Xagoraraki, Water Res. **45**, 681–693 (2011)
22. V. Figueira, I. Vaz-Moreira, M. Silva, C.M. Manaia, Water Res. **45**, 5599–5611 (2011)
23. M. Harnisz, S. Tucholski, Ecol. Eng. **36**, 4, 466–470 (2010)
24. M. Harnisz, E. Korzeniewska, I. Gołaś, Chemosphere **128**, 134–141 (2015)
25. G. Huys, D. Gevers, R. Temmerman, M. Cnockaert, R. Denys, G. Rhodes, R. Pickup, P. McGann , M. Hiney, P. Smith, J. Swings, J. Syst. Appl. Microbiol. **24**, 1, 122–130 (2001)
26. G. Rhodes, G. Huys, J. Swings, P. McGann, M. Hiney, P. Smith, J. Swings, Appl. Environ. Microbiol. **66**, 3883–3890 (2000)
27. G. Guillaume, D. Verbrugge, M.L. Chasseur-Libotte, W. Moens, J.M. Collard, FEMS Microbiol. Ecol. **32**, 77–85 (2000)
28. A. Moura, I. Henriques, R. Ribeiro, A. Correia, J. Antimicrob. Chemoth. **60**, 1243–1250 (2007)
29. Y. Zhang, C.F. Marrs, C. Simon, C. Xi, Sci. Total Environ. **407**, 3702–3706 (2009)
30. M. Harnisz, I. Gołaś, M. Pietruk, Ecol. Indic. **11**, 663–668 (2011)
31. I.S. Henriques, F. Fonseca, A. Alves, M.J. Saavedra, A. Correia, Lett. Appl. Microbiol. **47**, 526–533 (2008)
32. T.L. Nikolakopoulou, E.P. Giannoutsou, A.A. Karabatsou, A.D. Karagouni, J. Microbiol. **46**, 633–640 (2008)
33. A.R. Hall, D.C. Angst, K.T. Schiessl, M. Ackermann, Evol. Appl. **8**, 261–272 (2015)
34. C. Sandalli, O.B. Ozgumus, A. Sevim, World J. Microbiol. Biot. **26**, 2099–2103 (2010)
35. A.S. Schmidt, M.S. Bruun, I. Dalsgaard, J.L. Larsen, Appl. Environ. Microb. **67**, 5675–5682 (2001)
36. S. Börjesson, A. Mattsson, P.E. Lindgren, J. Water Health **8**, 2, 247–256 (2010)
37. E.A. Auerbach, E.E. Seyfried, K.D. McMahon, Water Res. **41**, 5, 1143–1151 (2007)

Modelling the defrost process in complex geometries – part 2: wall-function based coupling to a multi-region CFD solver

Ellen Hertle[1], *Simon* van Buren[1], *Patric* Figueiredo[1], *Reinhold* Kneer[1], and *Wilko* Rohlfs[1,*]

[1]Institute of Heat and Mass Transfer, RWTH Aachen University, Augustinerbach 6, 52056 Aachen, Germany

Abstract. Presence of frost in air coolers reduces the performance and thus has to be removed frequently. A method commonly applied for frost removal is hot-gas defrosting. For the efficient design of this periodically performed operating sequence, a precise prediction of the process is essential. A CFD simulation is a capable tool for this evaluation. This paper presents an integration of a defrost model into a CFD solver of the software OpenFOAM. The multi-region properties implemented in the existing code are adapted for the metal fin and tube structure and the air flow channel. The extension of the solver for evaluating the frost layer is accomplished by an interfacial wall function which models the frost characteristics and performs the thermal coupling. The extended solver is tested in a simplified environment, focusing on a fin-tube section. The simulated course of the defrost process and the total defrost time are discussed.

1 Introduction

Frost formation occurs in different technical applications involving cooling of moist air and other gases. In general, frost grows on surfaces for which the temperature is both below the dew point of the surrounding moist air and below the freezing point of water. In most cases frost formation is an undesired phenomenon, creating an isolating layer of ice which negatively affects the performance of heat exchangers. Avoiding frost formation is not always possible, such that defrosting strategies are important for air cooling systems [1]. Relevant aspects of the applied methods are defrosting time, pressure drop across the evaporator, energy consumption, overall efficiency, and defrost cost [2].

The first theoretical model describing the defrost process has been developed by Sanders [3], who divided the entire process into different stages including preheating and melting. Machielsen [1] used the model of Sanders [3] to determine the optimal cooling-defrost cycle. Using a simplified circular geometry, Hoffenbecker [4] investigated the defrost time and required defrost energy in industrial air-cooling evaporators.

A more detailed model was developed by Dopazo [5], dividing the defrost process into six different stages, including preheating, different resistances in water and air layers, and

* Corresponding author: rohlfs@wsa.rwth-aachen.de

evaporation. The above mentioned models use a strongly simplified geometry, neither accounting for natural convection nor varying air temperature in the evaporator. Further, only the melting process at the wall-frost interface is considered, and not at the opposed air-frost interface. A modelling approach for frost formation in complex geometries was for instance developed by Ellgas [6], coupling a commercial CFD-code with a frost model. The model was applied to study the frosting of an injector nozzle and blocking effects of the fluid region due to frost growth was accounted for. Kim [7] presented a numerical model for predicting frost behaviour on a heat exchanger fin under frosting conditions, taking into account fin heat conduction and three-dimensional air flow. Lenic [8] proposed a transient two-dimensional model of frost formation that includes air and frost subdomains as well as a boundary condition on the air-frost interface. A subsequent study determined the optimal duration of the cooling cycle and the optimal starting point for the defrost process [8].

In our approach, a three-dimensional CFD code is coupled with a multi-stage defrost model. The CFD code considers different solid and fluid regions which are coupled by a wall-function including the defrost model. Thus, spatially and temporally varying temperatures at the frost-wall and the frost-air interface can be analysed throughout the defrost process. With this model, defrosting of air-coolers with complex shapes can be investigated.

2 Modelling approach

Fig. 1 shows a wall, partially covered by a frost layer of thickness δ_f, which is situated between the wall and a gaseous atmosphere (air). In the frost region, T_{WF} and T_{AF} denote the temperatures at the wall-frost and the air-frost interfaces, respectively. Due to the thermal capacity of the frost and the latent heat required for the phase change, heat fluxes at both interfaces are not identical.

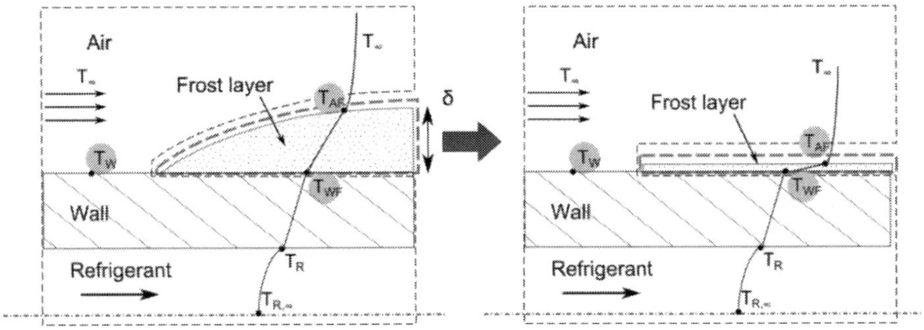

Fig. 1. Modelling approach with an infinitesimal thin frost layer. One-dimensional frost model (left) and zero-dimensional modelling approach (right).

The expansion of the frost layer into the fluid region can influence the flow, due to a partial narrowing of the flow channel. To fully resolve the effects resulting from the frost layer's expansion on the air flow, a computational method including an adaptive grid [6] or an interface tracking method is necessary [9], which require an high computational and modelling effort. Alternatively the expansion of the frost layer in the three-dimensional domain can be reduced to an infinitesimally thin layer between wall and air region (see Fig. 1). This allows a simple description of the frost layer as a wall function in the CFD model, coupling the thermal boundary conditions of the solid-fluid interfaces. Note that this approach neglects the expansion of the frost layer and thus does not account for flow channel narrowing or blocking. Nevertheless, the virtual and time-dependent frost layer thickness is integrated in the defrost model and remains a crucial parameter significantly influencing the energy requirement for defrosting.

2.1 Conservation equations in the solid and fluid regions

The problem of defrosting in heat exchangers involves solid and fluid regions, for which a multi-region approach is required. In the solid region, the three-dimensional transient equation for energy conservation in terms of enthalpy reads

$$\frac{\partial(\rho_{Wall} h)}{\partial t} = \nabla \cdot (\alpha_{Wall} \nabla h) \tag{1}$$

In the fluid region, the three-dimensional energy conservation equation with the additional terms of convection yields

$$\frac{\partial(\rho_{Air} h)}{\partial t} + \nabla \cdot (\rho_{Air} U h) - \nabla \cdot (\alpha \nabla h) = \frac{\partial p}{\partial t} \tag{2}$$

where α is the thermal diffusivity of the fluid. The energy equations of the respective regions are coupled at the interface. In the case of a frost covered wall, a frost-specific wall function imposes the temperature and the heat flux based on the interaction with the frost model. Additionally the conservation equations for mass and momentum in the fluid region are taken into account. Turbulent models in the fluid region can be considered by an effective thermal diffusivity and an effective viscosity.

2.2 Mathematical description of the frost layer

The proposed model describes the transient defrost process of a frost layer. It is integrated into a wall function and a one-dimensional approximation of the frost layer is applied, which considers heat conduction and melting in normal direction to the wall. The reduction to a one-dimensional model is justified by the small thickness of the frost layer compared to its lateral expansion. Both wall-side and air-side defrosting can be taken into account.

The frost layer is assumed to be a porous structure consisting of areas of ice and air gaps [10]. Depending on the separate stages of the defrost process (preheating, incipient melting and appearance of an air gap) the one-dimensional transient heat conduction equation is solved for each discretized element i

$$\frac{\partial T_i}{\partial t} = \frac{\lambda_i}{c_{p,i}\rho_i}\left(\frac{\partial^2 T_i}{\partial x^2}\right) \tag{3}$$

The thermal properties correspond to the three different regions of water, frost and air. During the defrost process, the initial frost thickness $\delta_{f,t=0}$ decreases until it falls below the stability limit, whereupon the remaining frost is assumed to drop down. The lateral displacement of the frost layer boundaries s_f due to the melting process is defined by

$$\frac{\partial s_f}{\partial t} = -\frac{1}{h_{melt}\rho_f}\left(\dot{q}|_{x=s_L} + \dot{q}|_{x=s_R}\right) \tag{4}$$

The quantities $\dot{q}|_{x=s_L}$ and $\dot{q}|_{x=s_R}$ denote the heat flux across the melting interface. The thermal properties of the porous frost layer are determined based on the porosity ε by averaging the properties of air and ice according to Le Gall [11]. With the assumption of a negligible contribution of the ambient air [12] the frost density yields $\rho_f \approx (1-\varepsilon)\rho_{ice}$. The thermal conductivity λ_f^{eff} in the frost is based on the empirical correlation suggested by O'Neal [13]:

$$\lambda_f^{eff} = 1.202 \cdot 10^{-3} \rho_f^{0.963} \quad (5)$$

A more detailed description of the model is provided in part I [14]. The boundaries of the frost model and the neighbouring regions (solid and fluid) are coupled by temperature and heat flux (temperature gradient). As an input parameter in the frost model either the temperature or the heat flux can be imposed. The quantity that is not handed from the CFD to the defrost model is returned to the CFD model. At the solid-frost interface, the temperature of the solid is imposed in the frost model, while the heat flux results. Contrary, the heat flux is enforced at the air-frost interface in order to avoid oscillations, which arise due to a high difference in the thermal diffusivity.

3 Numerical calculations

For the numerical simulations the finite volume code package OpenFOAM is employed. Its multi-region solver *chtMultiRegionFoam* uses a partitioned approach, which is applied to solve the governing equations in the solid and fluid region. The energy equations of the respective regions are coupled at the boundary interface through the present wall function. A first order implicit (bounded) Euler method has been used for the time-stepping treatment. For the velocity-pressure coupling the PIMPLE algorithm has been applied.

The calculation of the defrost process is conducted for each time step and each boundary cell. This yields an updated value for the heat flux at the solid interface and an updated value for the temperature at the gas interface. For the numerical calculations the one-dimensional frost layer of thickness δ_f is spatially discretized into N elements. The governing equations for the different stages during defrosting, together with the thermophysical properties, define a system of ordinary differential equations. This system with the respective boundary conditions at the two interfaces is solved using a solver based on the Rosenbrock method (rodas23 as described in [15]). The mesh and the fluid properties in the melting zone are adapted based on the displacement of the frost layer.

4 Test case definition

The developed defrost model is simulated for a three-dimensional defrost process on a simplified fin geometry. The geometry is shown in Fig. 2 and consists of a plain fin with a height of H and width of 2B.

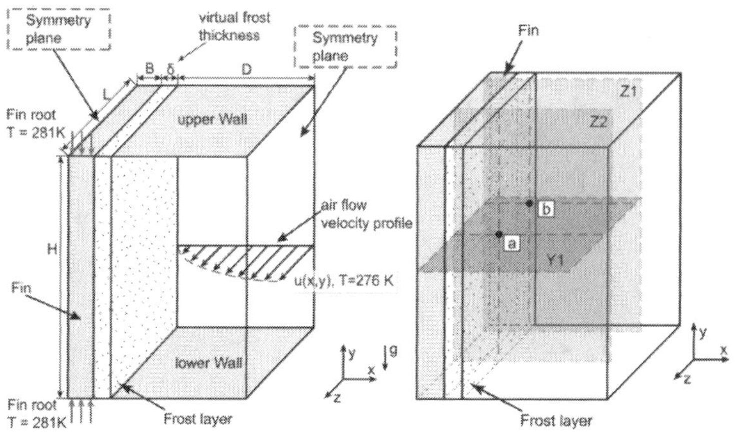

Fig. 3. Geometry of the test case (left) and position of the evaluation planes and points (right).

The fin, together with the upper and lower wall confine a flow channel of width 2D. Due to symmetry reasons, the fin and half of the channel define the numerical domain. The no-slip boundary condition is imposed at the channel walls and at the frost surface. The fin has an initial temperature of 270K and the other two channel walls are adiabatic.

Laminar air flow enters the channel in z-direction Pr = 0.7 and a parabolic velocity profile (superpositioned in x- and y- direction) is imposed at the inlet. Based on the hydraulic diameter of the rectangular channel, the Reynolds number of the flow is Re $\approx$ 60.

The fin is covered by a frost layer with a thickness of δ_F and a porosity of ε. The porosity is assumed to be constant over the defrost process (no densification). The model assumes a maximum water film layer of $\delta_{W,max}$ and a minimal frost layer thickness (stability limit) of $\delta_{F,min}$. The values are based on prior works of [5] and presented in Tab. 1.

Table 1. Input parameters for the three-dimensional simulation.

Parameter		Unit	Value
Fin height	H	mm	14
Fin length	L	mm	8
Fin width	B	mm	1.5
Channel width	D	mm	10
Average air inlet velocity	U	m/s	0.05
Fin root temperature	T_{Wall}	K	281
Air inlet temperature	T_{Air}	K	276
Initial frost temperature	T_F	K	270
Initial frost thickness	δ_F	mm	1
Minimal frost layer thickness	$\delta_{F,min}$	mm	0.2
Maximal water layer thickness	$\delta_{W,max}$	mm	0.2
Frost porosity	ε	-	0.5
Mesh size (3D-model)	Δx, Δy, Δz	mm/cell	0.125 (0.25)
Mesh size frost	Δx_f	mm/cell	0.01 (0.02)
Time step size	Δt	s	0.001

5 Results

In the following section, numerical simulation results of the test case are presented for a set of cross-sections and positions at different instances of time. Figure 4 illustrates the location of the sampling planes (Z1, Z2) and points (a-b). The end time of the simulation is 200s. After 150 s, the frost layer thickness falls below the stability threshold at any position of the wall. Thus, according to the defrost model, the wall is covered by a remaining water layer.

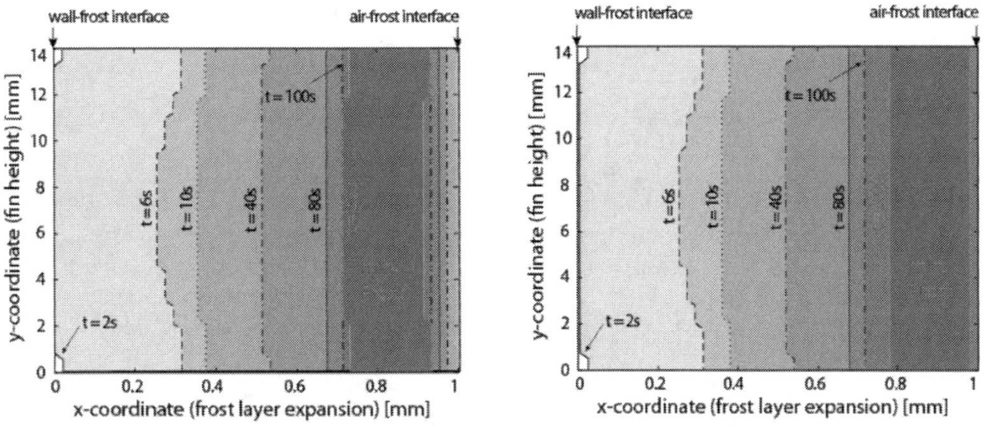

(a) Cross section Z1 according to Fig. 4 (b) Cross section Z2 according to Fig. 4

Fig. 5. Melting process of the frost borders for two different streamwise locations. For both cross sections wall side heating is dominant, while air-side melting mainly occurs at the beginning of the channel. Cross sections Z1 (left) and Z2 (right) according Fig. 6.

Figure 3 presents the defrost process for two planes perpendicular to the air flow (Z1 and Z2). Due to the higher heat transfer rate from fin to frost, the wall-side melting process is significantly stronger than the air-side one. As a consequence, a large gap (filled with water and air) between the fin and the remaining frost layer develops, acting as an additional thermal resistance, slowing down the defrost process. Five seconds after the defrost process has started, the frost layer thickness has decreased by at least 0.2 mm, which corresponds to the maximal thickness of the water layer. The heat input on the top and the bottom side of the fin causes an inhomogeneous melting rate, with a minimum in the fin centre (y = 7mm). During the defrost process, the melting rate becomes more homogeneous over the fin height, since the thermal resistance of the air gap dominates the process (the thermal resistance in the fin becomes negligible). Differences in the air-side melting process are shown by a comparison of the cross sections Z1 and Z2. While no significant air-side melting appears at cross section Z2, the frost layer decreases by approximately 0.1 mm close to the air inlet (Z1). The temporal development of the frost layer at the two sampling positions (a-b) is presented in Fig. 7.

DOI: 10.1051/e3sconf/20172200064

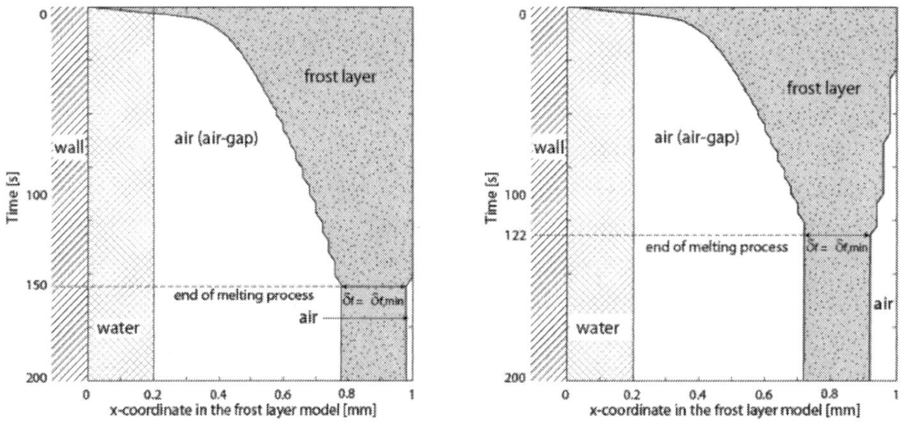

(a) Probe location **a** according to Fig. 2 (a) Probe location **b** according to Fig. 2

Fig. 7. Temporal development of the frost layer at two distinct locations a (left) and b (right).

The three different regions (water, air, and frost) are indicated by different patterns and the time increases from top to bottom in each plot. For both sampling positions, the wall-side melting process is very similar up to the point in time when $\delta = \delta_{F,min}$. In the beginning of the process, the water layer develops, reaching its maximum thickness at t = 3.5s. With the increasing thickness of the air gap and the increasing thermal resistance, the fin-side melting process slows down, whereas the air-side melting process (apparent at position a) reveals a linear behaviour due to the constant flow rate. The temperature distributions for fin, frost, and air regions are presented in Fig. 8 at two different locations (Z1 and Z2) and two different instances of time t = 6s and t = 100s.

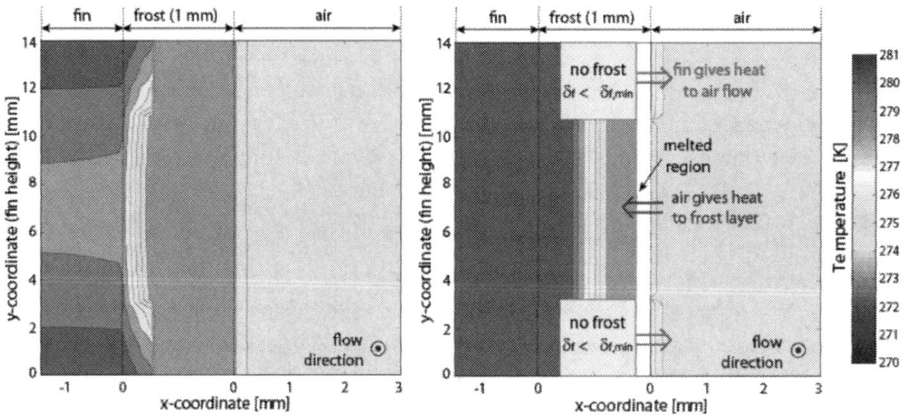

(a) Section close to the air inlet, t=6s. (b) Section close to the air inlet, t=100s.

Fig. 8. Temperature field for a section close to the air inlet at t = 6 s and t = 100 s.

The temperature field close to the air inlet at t = 6 s shows a strong temperature gradient over the fin height. This gradient is caused by the heating of the fin root and the transient boundary condition within the first five seconds. In addition, a smaller temperature gradient is apparent over the fin width, resulting from the heat transfer to the frost layer. The temperature distribution in the frost is characterized by two different regions. In the center region (3 < y < 11 mm), the water layer between the frost and the wall exhibits a linear temperature gradient. Close to the fin roots, the maximum water layer thickness is exceeded such that an additional air gap develops. Due to the different thermal conductivity of air and water, a strong temperature gradient in the air gap establishes. The remaining frost layer is characterized by an almost homogeneous temperature, just below melting temperature. Note that there are no significant variations of the temperature distribution in the fin and the frost with respect to the flow direction (z-coordinate) at t = 6 s.

For t = 100 s, the temperature in the entire fin and the covering water layer is similar to the temperature of the fin root in both sections. The dominating thermal resistance of the air gap limits the heat transfer from the water layer to the frost. In Fig. 8, two different regions exist. Close to the fin roots, the frost layer thickness falls below the stability limit. Thus, the remaining frost layer drops down and a water layer remains. As a consequence, the air temperature increases due to a wall-side heating above the air inlet temperature. In the center region, the defrost process is ongoing with a heat flux from the ambient air to the frost. The white area illustrates the amount of frost melted by air-side defrosting. In the center section of the channel (not shown), the increased temperature boundary layer reduces the air-side defrosting. Over the entire height of the fin, the frost layer thickness has not reached the stability limit yet.

6 Conclusion

In this study, a one-dimensional defrost model developed in [14] has been included in a multi-region three-dimensional finite volume code through a wall-function. The wall-function couples the temperatures and the heat transfer rates (temperature gradients) of the different regions. The defrost model itself considers the frost layer, and additional resistances due to a water layer and an air gap. Wall- and air-side heating can be applied resulting in a two-sided melting process.

Despite the fact that the presented approach does not account for the expansion of the frost layer into the air region, the wall-function based model can be used to simulate the defrost process in large and complex geometries.

A simple three-dimensional test case of a frost-covered fin is designed to identify the capabilities of the model. Results of first numerical simulations show local variations of the defrost process due to the temperature distribution along the height of the fin and a developing boundary layer in the air flow.

References

1. C. Machielsen, H. Kerschbaumer, Int. J. Refrig. **12**, 5 (1989)
2. A. Alebrahim, S. Sherif, Proceedings of the Institution of Mechanical Engineers, Part C: Journal of Mechanical Engineering Science **216**, 6 (2002)
3. C. T. Sanders, *Ph.D. thesis*, University of Delft (1974)
4. N. Hoffenbecker, S. Klein, D. Reindl, Int. J. Refrig. **28**, 4 (2005)
5. J.A. Dopazo, J. Fernandez-Seara, F.J. Uhia, R. Diz, Int. J. Refrig **33**, 4 (2010)
6. S. Ellgas, M. Pfitzner, M. Numer, Heat Tr. A-Appl. **53**, 5 (2007)
7. J.-S. Kim, K.-S. Lee, S.-J. Yook, Int. J. Heat Mass Tran. **52**, 11–12 (2009)
8. K. Lenic, A. Trp, B. Frankovic, Int. J. Heat Mass Tran. **52**, (2008)
9. C. Hirt, B. Nichols, J. Comput. Phys. **39**, 1 (1981)
10. Y.-X. Tao, R.W. Besant, K.S. Rezkallah, Int. J. Heat Mass Tran. **36**, (1993)
11. R. Le Gall, J. Grillot, Int. J. Heat Mass Tran **40**, (1997)
12. K. Prölss, G. Schmitz, The Modelica Association **1**, (2006)

13. D. O'Neal, D. Tree, ASHRAE Trans. **91** (1985)
14. S.H. van Buren, E. Hertle, P. Figueiredo, R. Kneer, W. Rohlfs (2017)
15. A. Sandu, J. Verwer, J. Blom, E. Spee, G. Carmichael, Atmos. Environ. **31** (1997)

 DOI: 10.1051/e3sconf/20172200065

Multicriteria optimization approach to design and operation of district heating supply system over its life cycle

Piotr Hirsch[1,*], *Kazimierz* Duzinkiewicz[1], and *Michał* Grochowski[1]

[1]Gdansk University of Technology, Faculty of Electrical and Control Engineering, ul. Narutowicza 11/12, 80-233 Gdansk, Poland

Abstract. District Heating (DH) systems are commonly supplied using local heat sources. Nowadays, modern insulation materials allow for effective and economically viable heat transportation over long distances (over 20 km). In the paper a method for optimized selection of design and operating parameters of long distance Heat Transportation System (HTS) is proposed. The method allows for evaluation of feasibility and effectivity of heat transportation from the considered heat sources. The optimized selection is formulated as multicriteria decision-making problem. The constraints for this problem include a static HTS model, allowing considerations of system life cycle, time variability and spatial topology. Thereby, variation of heat demand and ground temperature within the DH area, insulation and pipe aging and/or terrain elevation profile are taken into account in the decision-making process. The HTS construction costs, pumping power, and heat losses are considered as objective functions. Inner pipe diameter, insulation thickness, temperatures and pumping stations locations are optimized during the decision-making process. Moreover, the variants of pipe-laying e.g. one pipeline with the larger diameter or two with the smaller might be considered during the optimization. The analyzed optimization problem is multicriteria, hybrid and nonlinear. Because of such problem properties, the genetic solver was applied.

1 Introduction

Space heating is a major component in overall heat consumption in Europe. At the same time heat uses amount to 80% of total energy consumed in residential houses. District Heating (DH) is an attractive solution for space heating, especially when it is supplied by Combined Heat and Power (CHP) plant [1, 2]. Recovering the waste heat from any kind of thermal power plant allows to greatly increase the plant energy efficiency and is an effective method for cutting carbon emissions. Several studies have shown that conventional CHP plants are highly attractive from economic and environmental perspective [3], however the DH area have to be close to the plant site. Another important Heat Source (HS) for DH systems can be CHP Nuclear Power Plant (NPP). Although the CHP NPP can be a source of a carbon-

* Corresponding author: piotr.hirsch@pg.gda.pl

free heat for residential and industrial use, it is seldom considered as a viable HS for DH. The main obstacle in a nuclear heat utilization is a necessity of a long distance heat transportation, as NPPs are most often located far from dense urban areas. For these reasons, the NPPs are considered solely for electricity production under global zero emission scenarios [4].

Despite the potential investment costs on pipelines and heat losses on transportation, the CHP NPP is still being investigated. In [5] a 77 km long 1000MW HTS has been analyzed for possible CHP production at the Loviisa 3 NPP project in Finland. The simulations presented by the authors showed, that only 11 MW of the heat was lost on transportation. In [1] economic estimations were performed for the French Nogent-sur-Seine NPP located 110 km east of Paris, showing that heat transportation of 1500 MW over long distance can be cost effective, with the payback time less than 10 years.

DH networks are often subject to detailed analyze in order to find optimal pipes sizing [6] or minimize the heat losses [7,8]. Similar approaches can be utilized for DH supply system, that would transport heat over long distances, from the HS to the DH area. In [2, 9] desired characteristics of the HTS efficiency analysis method are discussed. The method should utilize optimization techniques, to find the design minimizing construction costs, thermal losses and pumping power. The design variables should be: inner pipe diameter, insulation thickness, supply and return temperatures and number of pumping stations. The performance of the HTS changes in time due to the energy market shifts as well as pipe and insulation aging processes. Therefore, it was decided to analyze the HTS over presumed system life cycle. Other important conditions that affect the HTS efficiency are: the heat demand and its annual variability, terrain profile and annual ground temperature distribution. The method presented in the paper is extended with the pipe-lying strategy evaluation, allowing consideration of multiple, smaller pipelines working in parallel. Moreover, the revised mathematical model is presented, resulting in much better computing performance, allowing effective genetic solver application.

2 Decision model for optimization based efficiency analysis

The decision model for the HTS optimization based efficiency analysis consists of the mathematical model of objectives and constraints, and optimization problem formulation. Upon the decision model formulation, the following assumptions on the HTS operation are considered: the heat power delivered to the DH is equal to the heat demanded; the outlet and inlet temperatures of the HS heat exchanger are the supply and return temperatures, respectively; the pumping stations pressure gain is constant and independent to heating water flow rate. Further assumptions on the decision model computation are: only steady states are considered, the minor pressure losses are neglected.

Evaluation of the HTS design and operation over its life cycle must take into account the HTS operation conditions and parameters variability in time. The time-variant processes affecting the HTS are occurring on significantly different time scales. The pipeline and insulation aging are examples of processes with time constants of several years. It is convenient to approximate the impact of these processes as fixed over some intervals. This approach leads to Long Horizon Time Discretization (LHTD) of the HTS life cycle L_C, with $\Delta\tau_L$ time step, into $\sum_{\tau_L}$ intervals indexed by t. As a result, for selected parameters such as: pipeline roughness, thermal conductivity, and the price of heat and electricity, it is possible to acquire different constant values for each life cycle interval. Depending on the decision-maker needs and knowledge, the LHTD time step can be either fixed or variable. The HTS is also subjected to significant annual variability due to annual heat demand and ground temperature changes. To deal with these conditions, the Short Horizon Time Discretization (SHTD) is proposed. The SHTD time base is one year, which is divided with the variable time step $\Delta\tau_Y$ into $\sum_{\tau_Y}$ intervals indexed by y, in accordance to discretized, structured graph of annual heat demand (see Fig. 3b.). As a result, $\sum_{TY}$ sub-models are created for each of the LHTD intervals. Each sub-model is corresponding with a different heat demand level (different part of the year), and is characterized by different flows and temperatures. Although the duration of each LHTD interval may be decades, its parameters are assumed constant over that period and, consequently, it may be covered by only one set of $\sum_{TY}$ sub-models, the results of which are multiplied according to the LHTD time step.

According to the HTS topology scheme shown in Fig. 1, the HTS is a network composed of links and nodes. The links are pipeline sections, pipeline sections with the pumping stations, or heat exchangers. The nodes are allocated according to the spatial discretization, which is introduced in order to deal with the HTS spatiality and the resulting terrain elevation profile (see Fig. 3a.) impact on the nodal pressure and on the pumping stations localization. The terrain elevation profile, given with the base resolution of Δl [m], is chosen as basis of the discretization, with the nodes being allocated at the points of the terrain profile change. Therefore, the HTS of length L [km] is discretized with the variable step Δl_i [m] that is a multiple of Δl, resulting in N nodes in each direction, indexed by i. The two directions are denoted with the second subscript u.

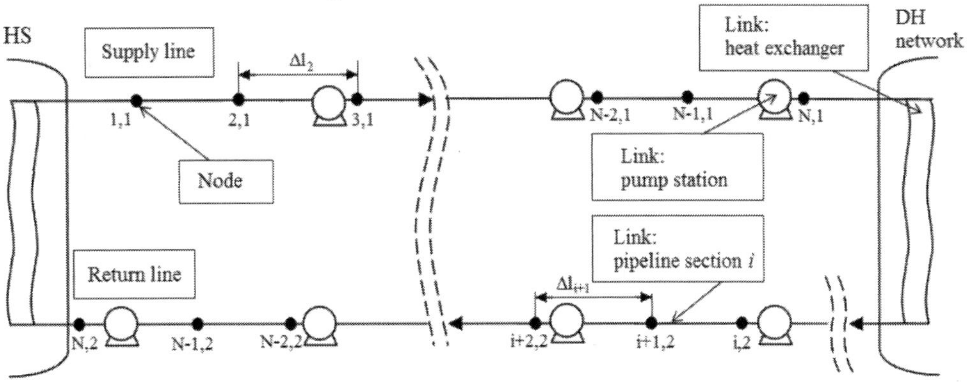

Fig. 1. The HTS topology scheme.

2.1 Thermal and hydraulic mathematical model

The mathematical model formulation utilizes reduced notation for conciseness. The nodal variables (describing a quantity in each node) have the subscripts i,u,y,t, meaning respectively: the node number, the line, the SHTD interval, the LHTD interval. In reduced notation, unless necessary, only the i subscript is listed.

The main purpose of the thermal model is to calculate the heat losses on transportation. The losses must be taken into account to ensure assumed equilibrium of the heat delivered and demanded. Considering the pipeline section of length of Δl_i [m] with diameter Φ [m], external insulation of thickness δ [m] and thermal conductivity λ W/(m·K)] in which superheated water at temperature T_i [K] is flowing (Fig. 2.), the linear heat losses can be calculated as:

$$\Delta Q_i = Q_{i+1} - Q_i = \frac{-2\pi \cdot \Delta l_i \cdot \lambda_t \left(T_i - T_{g_y}\right)}{\log(1 + 2\delta/\Phi)} \tag{1}$$

where Q is the heat power at given node [W] and T_g is the ground temperature [K]. Note that the insulation thermal conductivity and the ground temperature in (1) are discretized according to LHTD and SHTD respectively. Another important feature of the presented

 DOI: 10.1051/e3sconf/20172200065

linear heat loss model is approximation of superheated water temperature over the section with T_i, which is the section inlet temperature. Different approximation methods, such as inlet-outlet average or logarithmic mean can also be utilized, however if the section length is small enough, the outcomes differences are negligible. The exact value of 'small enough' section length is dependent on the case study and can vary within range from 100 meters up to several kilometers. The loss of the heat translates in direct proportional manner into the temperature loss:

$$\Delta T_i = T_{i+1} - T_i = \frac{\Delta Q_i}{\dot{m}_{y,t} C_p} \tag{2}$$

where C_p is the specific water heat capacity [J/(kg·K)] at 100°C, and $\dot{m}_{y,t}$ is the mass flow of the heating water [kg/s] at a given time step. Assuming the absence of the heating water phase change in the heat exchanger and the heat exchanger efficiency η_e, the heating water mass flow can be calculated from the amount of the heat exchanged and the temperature drop on the heat exchanger. This applies for both DH and HS side heat exchangers:

$$\dot{m}_{y,t} = \frac{Q_{DH\,y,t}}{\eta_e \Delta T_{DH\,y,t} C_p} = \frac{Q_{HS\,y,t}}{\eta_e \Delta T_{HS} C_p} \tag{3}$$

In Eq. (3) $Q_{DH\,y,t}$ is the heat delivered to the DH network [W], with the temperature drop of $\Delta T_{DH\,y,t}$ [K]. Similarly, $Q_{HS\,y,t}$ is the heat absorbed from the HS [W] at temperature drop ΔT_{HS} [K]. The heat delivered to the DH network is equal to the heat demand and is known at any time, while the amount of heat needed from HS has to be calculated as a sum of total heat lost on transportation in both directions and the heat delivered. The same applies to temperatures - ΔT_{HS} is equal to the difference between the supply and return temperatures, which are inputs to the model, while $\Delta T_{DH\,y,t}$ has to be calculated using Eqs. (1–3).

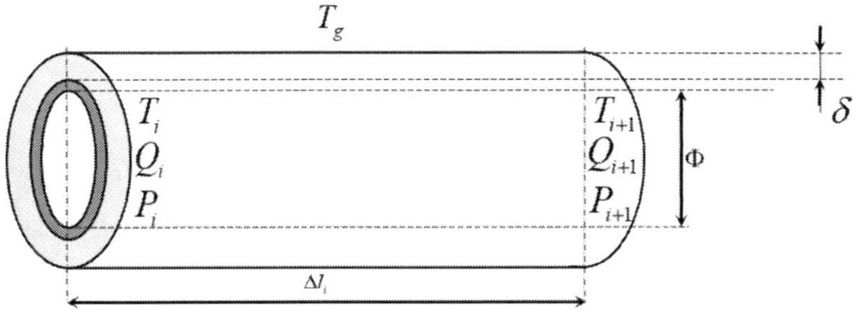

Fig. 2. Pipeline section scheme with associated variables and parameters.

The proposed thermal model is characterized by a dense computational nodalization that is required because of the temperature approximation accuracy. At the same time, the spatial discretization based on terrain elevation profile is much sparser. It can be shown, that the thermal model can be calculated only for the selected nodes, without the need of computing all the base nodes in between.

Let us consider the ratio of the heat losses in two successive pipeline sections, as shown in Eq. (1):

DOI: 10.1051/e3sconf/20172200065

$$\frac{\Delta Q_{i+1}}{\Delta Q_i} = \left.\frac{T_{i+1} - T_{g_y}}{T_i - T_{g_y}}\right|_{T_{i+1}=T_i+\frac{\Delta Q_i}{\dot{m}_{y,t}C_p}} = 1 + \frac{\Delta Q_i}{\dot{m}_{y,t}C_p \cdot \left(T_i - T_{g_y}\right)}$$
$$= 1 - \frac{2\pi \cdot \Delta l_i \cdot \lambda_t}{\dot{m}_{y,t}C_p \cdot \log(1 + 2\delta/\Phi)} = k_{y,t} \quad (4)$$

It can be observed, that the successive heat losses ratio is constant and lower than 1 at any given time step. The same applies to the temperature losses ratio, as the two quantities are proportional. It can be shown, that $\Delta T_{i+1}/\Delta T_i$ also equals to $k_{y,t}$. Therefore, the heat or the temperature loss at any node can be found from the geometric sequences: $\Delta Q_i = k^{i-1}\Delta Q_1, \Delta T_i = k^{i-1}\Delta T_1$. By applying geometric series, the heat power and the temperature at any node can be calculated:

$$Q_i = \frac{\Delta Q_1(1 - k^{i-1})}{1 - k} + Q_1, T_i = \frac{\Delta T_1(1 - k^{i-1})}{1 - k} + T_1 \quad (5,6)$$

Using the extended thermal model (Eqs. (1-6)), the heating water temperature and the transported heat power can be found at any node that is allocated at a multiple of Δl. Proposed thermal model can be solved either as the system of equations, or using sequential procedure as described in [2].

The hydraulic model is a base for calculation of the number of pumping stations required and their location, as well as the electrical power used on pumping. Since considered HTS supply and return lines are buried in a single trench, the terrain elevation has no impact on the number of pumping stations, as its impact on the pressure profile cancels out. However, the terrain has still strong impact on the pumping stations locations.

The pressure drop between consecutive computational nodes ΔP_i [Pa] is affected by: the linear pressure losses $P_{l_{y,t}}$ [Pa] (minor pressure losses are neglected), the elevation difference pressure P_{z_i} [Pa] and the possible pressure gain P_g [Pa] due to pump station allocation at the inlet node. As a consequence to assumptions of fixed pipeline diameter and closed HTS structure (without branching nodes), the flow velocity is the same for any node at any given time step. Thus, the linear pressure loss over the pipeline section is constant for any given time step and can be described by the Darcy-Weisbach equation:

$$P_{l_{y,t}} = f\frac{\rho \cdot \Delta l_i \cdot v_{y,t}^2}{2\Phi} \quad (7)$$

where ρ is the density of water [kg/m^3], $v_{y,t}$ is the heating water flow velocity [m/s] calculated from the mass flow and f is the Darcy-Weisbach friction coefficient, which for fully turbulent flow (Reynolds number>4000) and the inner pipe roughness of ξ [mm] can be calculated as shown in [1]. In order to determine the pressure at any node allocated at multiply of Δl, an arithmetic series can be utilized to find the linear pressure losses, while the elevation pressure is equal to the difference between the initial and the nodal elevation pressures, resulting in Eq. (8):

$$P_i = P_1 - (i - 1)P_{l_{y,t}} - \rho g \cdot (z_i - z_1) + n_i \cdot P_g \quad (8)$$

where z is the HTS elevation head [m], g is the Earth gravity constant [m/s^2] and n_i is the number of pumping stations allocated between the nodes *1* and *i*. The pumping stations can be added at the node, when the nodal pressure decreases below the safety value or it can be subject to further analysis.

 DOI: 10.1051/e3sconf/20172200065

2.2 Optimization problem formulation

As a result of the time discretization, a number of sub-models are established. Each sub-model is described by a set of variables. Some of these variables are the same for each sub-model (i.e. inner pipe diameter, insulation thickness, supply and return temperatures), while the others, like nodal pressure or temperature, are independent. While all variables together constitute decision space $\boldsymbol{x_d}$, the former are visibly more important for the decision making process. Therefore, categorization of the decision space is proposed: the linking variables shall be called design vector, denoted $\boldsymbol{x_{d_1}}$, while the rest compose simulation vector $\boldsymbol{x_{d_2}}$. According to this categorization, the decision space is represented as $\boldsymbol{x_d} = [x_{d1} \quad x_{d2}]^T$.

The decision variables are subjected to a set of nonlinear equality constraints $h(\boldsymbol{x_d})$ that constitute the mathematical model. Furthermore the upper and lower bounds, as well as inequality constraints $g(\boldsymbol{x_d})$, can be considered on the decision space due to technological and operational limitations. The range of considered supply temperatures and maximal heating water flow velocity can be examples of such constraints.

The HTS design and operation is assessed on the aspects of its technical and economic value. Technical value can be represented by terms of thermal and transport efficiency. One way to indicate the HTS thermal efficiency is to estimate its life cycle heat losses on transportation. It can be done by calculating firstly the heat losses of each sub-model $\Delta Q_{L_{y,t}}$ [W] (Eq. (9)), followed by the annual heat losses ΔQ_{A_t} [GJ/year] (Eq. (10)) and the total heat losses ΔQ_{TL} [GJ] on transportation (Eq. (11)).

$$\Delta Q_{L_{y,t}} = Q_{1,1,y,t} - Q_{k,1,y,t} + Q_{1,2,y,t} - Q_{k,2,y,t} \tag{9}$$

$$\Delta Q_{A_t} = \sum_{y=1}^{\Sigma\tau_Y} \left[k_Q \cdot \Delta\tau_{Y_y} \cdot \Delta Q_{L_{y,t}}\right], \Delta Q_{TL} = \sum_{t=1}^{\Sigma\tau_L} \left[\Delta\tau_{T_t} \cdot \Delta Q_{A_t}\right] \tag{10,11}$$

where: k_Q=8.64·10^{-5} is the unit conversion coefficient, $\Delta\tau_{T_t}$ is the time step of interval t [years], $\Delta\tau_{Y_y}$ is the time step of sub-model y [days].

The transport efficiency can be represented by means of energy used on heating water pumping. At first annual pumping energy W_{A_t} [MWh/year] is calculated using Eq. (12), then the total pumping energy W_{TL} [MWh] (Eq. (13)):

$$W_{A_t} = \sum_{y=1}^{\Sigma\tau_Y} \left[\frac{k_w}{\eta_p} \cdot \Delta\tau_{Y_y} \left(n_{y,t} \cdot p_g + p_{mg_{y,t}}\right) \cdot \dot{V}_{y,t}\right], W_{TL} = \sum_{t=1}^{\Sigma\tau_L} \left[\Delta\tau_{T_t} \cdot W_{A_t}\right] \tag{12,13}$$

where k_w=24 is the unit conversion coefficient, η_p is the pump efficiency, $n_{y,t}$ is the number of pumping stations set for the given sub-model, p_g is the pressure added by the pump station [MPa], $p_{mg_{y,t}}$ is the pressure added by the main pump station [Pa] and $\dot{V}_{y,t}$ is the volumetric flow rate [m^3/s].

The economic value of the HTS is estimated using empirical construction cost function:

$$C_C = 2L \cdot f_d(\Phi) + c_{ps} \cdot n_{1,\Sigma\tau_L} \tag{14}$$

where C_c is the overall HTS construction cost [m.u.], $f_d(\Phi)$ is the pipeline material and labor costs, as a function of pipeline diameter, and c_{ps} is the cost of pumping station construction [m.u.]. The objective functions Eqs. (9–14) can be represented as the objective vector

DOI: 10.1051/e3sconf/20172200065

$\boldsymbol{J}(x_d) = [C_C(\boldsymbol{x_d}) \quad \Delta Q_{TL}(\boldsymbol{x_d}) \quad W_{TL}(\boldsymbol{x_d})]^T$, leading to general optimization problem formulation:

$$\begin{gathered} \min_{x_d} \boldsymbol{J}(x_d) \\ \text{s.t. } h(\boldsymbol{x_d}) = 0 \\ g(\boldsymbol{x_d}) \leq 0 \\ \boldsymbol{x_{d_{lb}}} \leq \boldsymbol{x_d} \leq \boldsymbol{x_{d_{ub}}} \end{gathered} \tag{15}$$

3 Case study and solver

The optimization problem is multicriteria, hybrid and nonlinear. The combination of these properties makes it difficult to solve. In an attempt of simplifying the optimization problem, the weighted-sum scalarization method is proposed. Eq. (11) is multiplied by the heat and (13) by the electricity prices, allowing the objective functions summation. Resulting optimization problem, with pipe diameters and insulation thickness being integers and other decision variables being real, is solved using genetic algorithm solver.

The optimization problem formulation can be extended with the possibility of different pipe-lying strategies and variants evaluation. Depending on the maximal heat demand and its annual variation, it may be cost-effective to divide the pipeline into several smaller ones, working alternately or simultaneously. This functionality is introduced by allowing a number of parallel pipelines, each having the same parameters except for inner pipeline diameter, and additional decision variable, that limits the maximal heat provided by the single pipeline. If the heat demand exceed the heat provided by the current working pipelines, additional pipeline is considered, increasing the construction costs accordingly to its diameter.

The presented methodology has been verified and analysed based on the case study of a potential 40 km HTS. The heat source would be intended NPP, located by Lake Żarnowiec, Northern Poland, while the main heat customer would be Gdynia and its surroundings. Knowledge of the NPP and DH location allows specification of the terrain elevation profile (Fig. 3a.) as well as the annual ground temperature and the DH heat demand (Fig. 3b.). The number and length of the SHTD intervals can be read from Fig. 3b. Three 20 years LHTD intervals are considered, allowing variability of certain parameters: pipe roughness $\xi = [0.1, 0.15, 0.2]$ [mm], insulation thermal conductivity $\lambda = [0.02, 0.03, 0.05]$ [W/(m·K)], heat price $p_h = [12, 18, 24]$ [€/GJ], and electricity price $p_e = [60, 64, 68]$ [€/MWh]. Other case study parameters can be found in [2], with the exception of the heat exchanger efficiency, that in this study is selected as 1. A maximum of two possible pipelines was assumed, with the same bounds for each: $[0 \quad 373.15 \quad 323.15] \leq [\Phi \quad T_S \quad T_R] \leq [2 \quad 393.15 \quad 353.15]$. The insulation thickness of 0.2 and 0.3 [m] is considered, with the latter increasing the pipeline construction costs estimation function (Eq. (16)) by 15%.

$$f_d(\Phi) = 1.5 \cdot 10^7 \cdot \Phi^2 + 2 \cdot 10^6 \cdot \Phi + 5 \cdot 10^5 \tag{16}$$

DOI: 10.1051/e3sconf/20172200065

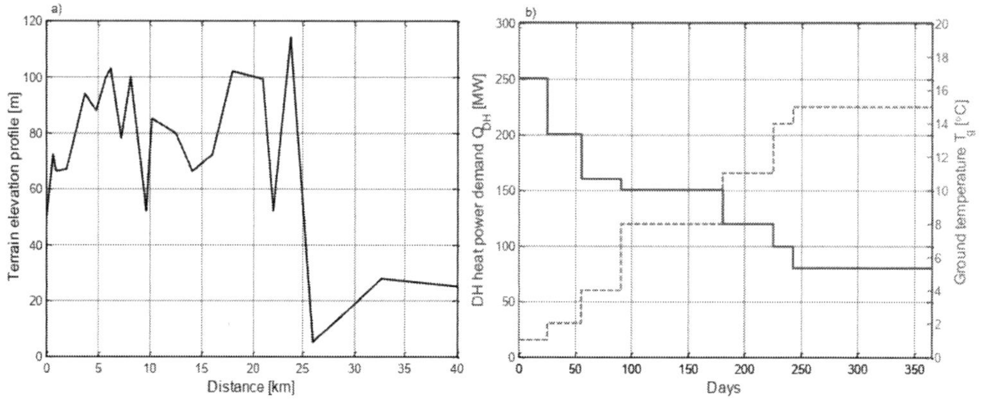

Fig. 3. a) The case study terrain elevation profile, b) structured graph of annual ground temperature and heat demand.

The results of the optimization procedure for the analysed case, together with the decreased and increased heat demand, are presented in Table 1. Each time the maximal heat provided by single pipeline is greater than heat demand, therefore only one pipeline is selected and one pipe diameter shown. It would require greater heat demand for the two smaller pipes case to be viable. It can be observed, that benefits of thicker insulation outweigh the 15% construction costs increase for each of the heat demand cases.

Table 1. Optimization results.

Heat demand	Decision variables				Objective functions			
	Inner pipe diameter [m]	Supply temperature [K]	Return temperature [K]	Insulation thickness [m]	Construction costs [M€]	Heat loss costs [M€]	Pumping power costs [M€]	Total costs [M€]
80%	0.606	393.15	323.15	0.3	171.97	68.56	38.99	247.74
100%	0.659	393.15	323.15	0.3	195.22	72.91	48.89	317.03
120%	0.707	393.15	323.15	0.3	217.79	76.84	58.33	352.95

4 Conclusions

The proposed method for optimized efficiency analysis of the HTS, can be utilized as a part of the decision support system for evaluation of projects associated with the heat transportation. An example of such project is a Combined Heat and Power NPP, where part of the technical and economic feasibility assessment task concerns high power, long distance HTS. With the appliance of the proposed method, the optimal design and operation strategy of the HTS can be found, concerning its construction costs, thermal and transport efficiency, while subjected to a number of operation and technological constraints. The constraints are a result of mathematical model and governing equations, additional can be imposed by decision-maker to ensure specific HTS operation, i.e. heating water flow velocity or heat loss over the pipeline section.

Important feature of the proposed method is an ability to evaluate projects in long term or in particular, over the project life cycle. Long Horizon Time Discretization allows consideration of the pipe and insulation aging phenomena, as well as the heat and electricity prices changes. At the same time, conditions of the time scale of the month or even days are still taken into account due to Short Horizon Time Discretization. In addition, the proposed method takes into account the spatial topology of the HTS to evaluate terrain elevation impact on the nodal pressure and localization of the pumping stations.

The decision model formulation allows investigation of different pipe-lying strategies. Depending on the case study and decision-maker preferences, the solver determines whether one bigger or several smaller pipelines are more cost effective. As a result a set of integer design variables are found by solving the optimization problem: each pipeline inner diameter, insulation thickness, number of pumping stations, as well as two real variables: supply and return temperatures. Moreover, all of the simulation variables, such as nodal pressure or temperature are calculated, allowing further HTS safety analyzes.

References

1. H. Safa, Int. J. Electr. Power Energy Syst. **42**, 553–559 (2012)
2. P. Hirsch, K. Duzinkiewicz, M. Grochowski, R. Piotrowski, Appl. Energy **182**, 164–176 (2016)
3. A. Colmenar-Santos, E. Rosales-Asensio, D. Borge-Diez, E. Collado-Fernández, Appl. Energy **162**, 892–907 (2016)
4. K. Tokimatsu, S. Konishi, K. Ishihara, T. Tezuka, R. Yasuoka, M. Nishio, Appl. Energy **162**, 1483–1493 (2016)
5. M. Paananen, T. Henttonen, 20th Int. Conf. on Structural Mechanics in Reactor Technology, Espoo, Finland, 2056 (2009)
6. T. Asim, R. Mishra, L. E. Kollar, S.R. Pradhan, Int. J. Press. Vessels Pip. **113**, 40–48 (2014)
7. A. Dalla Rosa, H. Li, S. Svendsen, Energy **36**, 2407–2418 (2011)
8. Y. Başoğul, A. Keçebaş, Energy **36**, 6156–6164 (2011)
9. P. Hirsch, M. Grochowski, K. Duzinkiewicz, 20th Int. Conf. on Methods and Models in Automation and Robotics, 1044–1049 (2015)

Pollution of soils (Pb, Cd, Cr, Zn, Cu, Ni) along the ring road of Wrocław (Poland)

Anna Hołtra[1,*], and *Dorota* Zamorska-Wojdyła[1]

[1]Wrocław University of Science and Technology, Faculty of Environmental Engineering, Wybrzeże Wyspiańskiego 27, 50-370 Wrocław, Poland

Abstract. The concentrations of metallic pollution in soils and plants along the ring road of Wrocław, Poland, have been determined. Environmental samples were collected from the surface layer of the profile within 2–3 m from the edge of the road. The analysis of metals (Pb, Cd, Cr, Zn, Cu and Ni) has been carried out through FAAS and GFAAS methods. The mineralizates of soils and plants were prepared in HNO_3, 65% supra pure, using the Microwave Digestion System. The pH and conductivity of the soil solutions were measured to evaluate their active and exchangeable acidity and the salinity of the soils. The index of the enrichment of soils in metals (*Wn*) and the bioaccumulation coefficient (*WB*) have been determined. Also, histograms of the frequency of the occurrence of metals in the environmental samples and the Pearson's correlation coefficients were presented. The results of metal concentrations in soils were compared to the geochemical background in uncontaminated soils of Poland. The assessment of the results in the soils was also made relative to the standard, according to the Polish Ministry of Environment Regulation from September 1st, 2016. During the assessment of the bioaccumulation coefficients of metals in plants a reference was made to the content of undesirable substances in feed in agreement with the Polish Ministry of Agriculture and Rural Development Regulation from January 23rd, 2007.

1 Introduction

Wrocław is one of the most developed cities in Poland in terms of transport infrastructure. Pollution along the roads is caused by exhaust emissions from motor vehicles, dust emissions, leakage of fluids from installations. Another environmental threat is the surface runoff from roads and contaminated sites, e.g. due to the road maintenance in winter. Degradation of soils in Wrocław is caused by the increase of heavy metals content and fluorine, petroleum substances, polycyclic aromatic hydrocarbons, as well as the salinity of the soil solution. The content of heavy metals, polycyclic aromatic hydrocarbons and petroleum substances in soils depends primarily on traffic, distance from the road, landform (the road runoff) and the chemical composition of fuels and lubricants, and supplies used in the motor vehicles (the abrasion of tires and brake pads). Organization of traffic – frequent

* Corresponding author: anna.holtra@pwr.edu.pl

 DOI: 10.1051/e3sconf/20172200066

braking, accelerating and impediments in traffic are affected to the urban soil pollution. According to data presented in Athanasopoulou & Kollaros [1], the emission of metals from engine oil into roadside soils was as follows: Cd (83%), Cr (33%), Zn (28%) and Ni (25%). The emission from the abrasion of pavement was at the level of 12% of Cr and 11% of Ni, while tyre wear caused the emission of Se (58%), Zn (33%) and Ni (15%). The amount of the released zinc is 2.4–3.2 µg/km with the car tires consumption of about 0.2 mg/km [2]. Therefore, this explains the presence of zinc oxide in the drain water of ca. 1.5–2% of the tire weight. The most important source of metal emission is dust from the brake wear. Grigoratos & Martini [3] cited the concentrations of the following heavy metals (mg/kg): Cu (70–210 000), Pb (4–1 290), As (< 2–110 in PM_{10}), Ni (80–730), Zn (120–27 300), Cr (135–12 000).

Intensive traffic is conducive to the accumulation of trace metals in the upper soil layers [1, 4–6]. Metallic pollutions depending on the physico-chemical properties of soil, the presence of humus and colloids of soil are affected by the biocenosis in soil. Plants in a contaminated environment can develop adaptive or defensive mechanisms. For that reason, the plants' metal content increases proportionately to their growing period in the contaminated soil. But plants often lose their properties of selectively accumulating the contaminants in tissues. Consequently, the development of plants is disturbed and the species' diversity declines. Contaminations of dust and gas contribute to the degradation of soils and vegetation in the zone of up to 500 m from the road edge [1, 4, 7, 8].

2 Materials and methods

Environmental samples were collected along the downtown ring road in Wrocław, encircling the central districts of the city at a distance of 3 to 5.5 km from the city's main square. Places of sampling are located at the busiest road junctions, starting with the crossing of the Armii Krajowej Street with the following streets: Krakowska Street (sample no. 1), Bardzka Street (no. 2), Borowska Street (no. 3) and Ślężna Street (no. 4). Then, the ring road of Wrocław runs west to the crossing of Wiśniowa Avenue with Powstańców Śląskich Street (no. 5). Another point of sampling is at the crossroads of Hallera Street and Grabiszyńska Street (no. 6). Two locations were on the crossings of Klecińska Street with Szwajcarska Street (no. 7) and Klecińska Street with Strzegomska Street (no. 8). Three samples were collected at the crossings of Na Ostatnim Groszu Street with: Bystrzycka Street (no. 9), Legnicka Street (no. 10) and Popowicka Street (no. 11). In the northern part of the ring road of Wrocław the samples were taken from the crossing of the Millennium Bridge Street and Osobowicka Street (no. 12), also along the Nowak-Jeziorański Street at the crossings with: Obornicka Street (no. 13) and Żmigrodzka Street (no. 14). Samples of plants – grasses and dandelion (*Taraxacum officinale* F.H. Wigg.) – were collected from the same locations as soils samples.

Soil and plant samples were collected approximately 3 m from the edge of the downtown ring road of Wrocław. A total of 84 environmental samples were collected in Autumn (September) 2016. Soil samples were taken from the surface layer to a depth of 25 cm. Plant samples were harvested with root system. Each research area (location) consisted of 6 measurement positions. Samples were collected within the distance of ca. 2 meters from each other. All soil samples were sieved after drying. The process of microwave mineralization was prepared according to the PN-ISO 11465:1999. Into a sample of 0.2 g of dry matter (DM) 8 ml of nitric acid 65%, Supra pure, was added. The mineralization process was made with the use of Start D device (Milestone). Additionally, measurements of pH in water and 1 M KCl and the electrical conductivity (EC) in the soil solutions were made according to the PN-ISO 10390:1997. The soil solutions were prepared in a stoichiometric soil-to-solution ratio of 1:2.5 (m:v). The zinc, copper, nickel concentrations have been determined through the Flame Atomic Absorption Spectroscopy (FAAS) method. The lead, chromium and cadmium concentrations have been determined through the Graphite Furnace Atomic Absorption Spectroscopy (GFAAS) method. In the spectroscopy analysis iCE 3500 Solaar Thermo device (Thermo Scientific) [PN-ISO 11047:2001] was used. The arithmetic mean and the standard deviation were calculated with the six individual samples collected in measurement location.

Statistical calculations and visualisations have been carried out using the Microsoft Excel software. The pollution index of soil was calculated as the ratio of the average content of chemical element in the soil from one location to the geochemical background. The bioaccumulation coefficient of metal in a plant was calculated as the ratio of the average content of chemical element in a plant to the average content of metal in the soil from one location.

Chemical analyses have been carried out in the certified Laboratory of Toxycology and Environmental Research in the Faculty of Environmental Engineering at the Wrocław University of Science and Technology.

3 Results and discussion

The pH and salinity of soil solutions affect the behavior of metals in the soil environment. The values of pH show mostly alkaline character of soils (pH above 7.2 in 1 M KCl and 7.4 in H_2O). Lee et al. [9] confirmed a higher pH of soil solutions nearest to the road. Three samples are neutral. Only sample no. 12 has a pH of 5.26 in 1 M KCl and 6.25 in H_2O, showing a slightly acidic character of the soil. Heavy metals in soils with acidic character could be a risk to the plants [10]. All soil samples are of poor salinity. Generally, the soil solutions have conductivity of less than 25 mS/m. The highest value of conductivity (25.44 mS/m) was measured in sample no. 3.

Parameters characterizing the content of trace elements in the surface layer of soil and plant samples collected along a downtown ring road in Wrocław are presented in Table 1.

The biggest differences in the content of heavy metals in soils between locations were identified for zinc and lead, coefficient of variation (V) being 81% for Zn and 69% for Pb. The content of other metals in soils also varied, as evidenced by the 50% higher values of coefficients of variation for copper, chromium and cadmium. The lowest value of coefficient of variation was noted for nickel (44%).The maximum concentrations of Zn, Pb, Cr were observed in soil sample no. 11. In this location the average zinc content of 502.20 mg/kg DM exceeds the limit value for soils of group I (500 mg Zn/kg DM), according to the Polish Ministry of Environment Regulation from September 1st, 2016, regarding the manner of conducting the assessment of the contamination of ground [11]. High content of several metals was also observed in locations no. 7, 5 and 1. In the soil samples there were no exceedances of the content of remaining trace metals. Lower concentrations of cadmium are as compared to other metals correspond with the results of other researchers [4].

Table 1. The content of metals in soils and plants present along the downtown ring road in Wrocław.

Parameter	Zn	Cu	Cr	Pb	Ni	Cd
	Soils					
	[mg/kg of DM]					
Minimum	56.25(no.14)	11.27(no.14)	0.91(no.2)	10.77 (no.14)	5.33(no.13)	0.10(no.2)
Maximum	507.01(no.11)	51.81(no.12)	19.04(no.11)	104.12 (no.11)	26.31 (no.8)	0.59(no.1)
Arithmetic mean, x	146.08	27.25	9.49	32.82	12.76	0.28
Standard deviation, s	118.10	13.51	5.46	22.81	5.61	0.16
Mediana	123.00	22.94	9.27	31.40	11.77	0.20
V[%], V= (s/x)·100	81	50	58	69	44	57
	Grasses					
Minimum	36.18(no.8)	11.24(no.8)	1.86(no.10)	0.94(no.12)	0.82(no.10)	0.02(no.8)
Maximum	162.10(no.9)	30.23(no.9)	7.49(no.3)	13.59(no.7)	11.69(no.3)	0.41(no.1)
Arithmetic mean, x	97.36	16.45	4.42	5.94	5.48	0.13
Standard deviation, s	34.97	6.00	1.54	3.36	2.74	0.09
Mediana	95.34	13.04	4.16	5.33	4.78	0.12
V[%], V= (s/x)·100	36	36	35	56	50	67
	Taraxacum officinale					
Minimum	26.61(no.6)	2.01(no.1)	0.41(no.9)	0.25(no.9)	0.58(no.12)	0.01(no.9)
Maximum	74.15(no.12)	19.47(no.12)	4.08(no.13)	1.54(no.13)	4.63(no.8)	0.23(no.2)
Arithmetic mean, x	51.28	13.14	1.91	0.67	2.39	0.09
Standard deviation, s	16.16	3.72	1.02	0.33	1.24	0.06
Mediana	49.64	12.97	1.92	0.63	2.01	0.09
V[%], V= (s/x)·100	32	28	53	50	52	64

Analysis of the issue in Poland shows the highest content of heavy metals in soils in the Upper Silesia area (1st Maja Street in Ruda Śląska). Metal concentration ranges were as follows: 46.9–3404.2 mg Zn/kg, 9.0–378.4 mg Cu/kg, 16.5–2257.0 mg Pb/kg, 0.3–28.2 mg Cd/kg [12]. Mean content of trace elements in upper surface of soils along the heavily trafficked exit road in Warsaw amounted to 137.0 mg Zn/kg, 16.7 mg Cu/kg, 41.5 mg Pb/kg, 0.67 mg Cd/kg, 19.4 mg Cr/kg, 12.6 mg Ni/kg [13]. In soils along the South-North highway which goes through the heart of the city of Łódź concentrations of 17.8 mg Cu/kg, 50.6 mg Pb/kg, 0.42 mg Cd/kg, 25.5 mg Cr/kg, 9.92 mg Ni/kg were observed [6]. Contents of Zn and Cu along communication routes in the city centre of Wrocław amounted to 156.6 and 46.9 mg/kg respectively [5]. Mean amounts of metals in urban soil from the roadside in Torino (Italy) were as follows: 183 mg Zn/kg, 90 mg Cu/kg, 149 mg Pb/kg, 191 mg Cr/kg, 209 mg Ni/kg [14]. Soils along the highway route E 75 from Belgrade (Serbia) to Preševo (Macedonia) were contaminated with 25.3 mg Cu/kg, 40.9 mg Pb/kg, 47.8 mg Ni/kg [7]. On other continents, the ranges of metal concentrations along the main highways of Dubai, Toronto (no. 400, 401, 404), Istanbul (E-5) were, respectively: for Zn (1.23–46.6, 39.3–1367.0, 81.3–366.8, 56.8–290.9, 190.9–1852.0 mg/kg), Cu (0.94–5.81, 53.9–286.1, 113.6–392.1, 76.2–230.3, 47.3–1358.5 mg/kg), Pb (6.9–113.3, 118.3–205.0, 32.5–378.7, 124.7–212.2, 61.4–555.4 mg/kg), Cd (0.39–0.93, 0.47–0.50, 0.46–0.95, 0.48–0.56, 0.80–6.70 mg/kg), Ni (3.3–73.8, 34.0–80.0, 32.0–327.0, 16.0–62.0, 10.4–65.7 mg/kg) [1]. In roadside soils from D100 highway in Turkey the average content of lead was 120.1 mg Pb/kg [15].

Metal concentrations in grasses and dandelion (*Taraxacum officinale*) were much lower than in soils (table 1). Locations of grass samples with the highest concentrations of metals do not coincide with the locations of dandelion samples with the highest concentrations of metals. High content of several metals was observed in locations no. 9, 7, 6 for grasses and sampling sites no. 12, 13, 8 for dandelion. The content of metals in plants is characterized with lower volatility: Cd (67% for grasses/64% for dandelion), Pb (56%/50%) and Ni (50%/52%), and for Cr of 53% in dandelion (*Taraxacum officinale*).

Arithmetic means of heavy metal contents in *Taraxacum officinale* from Warsaw (Poland) were as follows: 75.6 mg Zn/kg, 14.3 mg Cu/kg, 3.1 mg Pb/kg, 0.41 Cd/kg, 4.7 mg Cr/kg and 2.2 mg Ni/kg [13]. Leaves and roots of dandelion in Ruda Śląska (Poland) showed the following ranges of chemical elements: 92.3–625.9 mg Zn/kg (leaf), 49.5–802.6 mg Zn/kg (root), 3.3–626.0 mg Cu/kg (leaf), 4.0–52.4 mg Cu/kg (root), 2.14–165.0 mg Pb/kg (leaf), 0.44–699.7 mg Pb/kg (root), 0.68–4.4 mg Cd/kg (leaf), 0.27–9.7 mg Cd/kg (root) [12]. Content of Zn and Cu in grasses along communications roads in the city centre of Wrocław (Poland) was 101.5 mg Zn/kg and 16.3 mg Cu/kg [5]. In three species of grasses (*Dactylis glomerata* L., *Arrhenatherum elatius*, *Alopecurus pratensis* L.) growing along S2 expressway areas near Siedlce (Poland) the following content of metals and coefficient of variation (V) were found: for lead 2.7 mg/kg (25%)/2.5 mg/kg (28%)/3.3 mg/kg (33%) and for cadmium 0.23 mg/kg (45%)/0.19 mg/kg (34%)/0.18 mg/kg (33%) [16]. Traffic pollutions of Zn, Cu, Pb, Cd, Cr and Ni in grass (*Lolium perenne* L.) from city centre areas exposed to heavy vehicular traffic in Karatay (Turkey) amounted to: 41.9–58.2 mg Zn/kg, 6.1–9.4 mg Cu/kg, 1.9–3.4 mg Pb/kg, 0.14–0.16 mg Cd/kg, 14.5–25.7 mg Cr/kg, 6.6–13.8 mg Ni/kg [17].

In addition, an evaluation of the degree of the contamination of soil with heavy metals was presented in relation to the geochemical background in Poland. Values of trace metals in geochemical background are as follows: 50 mg Zn/kg DM, 0.3 mg Cd/kg DM, 30 mg Pb/kg DM, 15 mg Cu/kg DM, 10 mg Ni/kg DM, 20 mg Cr/kg DM [18]. For every metal at each location the pollution index of soil was calculated [4]. The average value of all pollution indexes for the metal is the following: zinc (Wn 2.92), copper (1.82), nickel (1.28), lead (1.09). Values of Wn above unity indicate medium level of the contamination of soils with heavy metals (fig. 1a). Wn indexes for chromium and cadmium are 0.47 and 0.95, respectively. It means a lack of soils enrichment in relation to the geochemical background. The average values of all pollution indexes (Wn) of soils by trace elements were arranged into the following order: Zn > Cu > Ni > Pb > Cd > Cr. The median of Wn for soils enrichment with: zinc is 2.32, copper is 4,57, lead is 1.75, nickel is 1.05, cadmium is 0.68 and chromium is 0.24.

Pollution indexes (arithmetic mean) of top soil layer along expressway E30 in Warsaw (Poland) presented in descending order are as follows: Pb (3.91) > Zn (3.52) > Cd (3.25) > Ni (2.59) > Cu (1.90) [4]. Wn indexes for Zn and Cu in urban soils of the city centre of Wrocław (Poland) amounted to 4.80 and 8.12, respectively [5]. In roadside soils of Torino, (Italy) the Wn indexes formed the following order: Pb (7.5) > Cu (3.3) > Zn (2.9) > Ni (2.8) > Cr (2.0) [14].

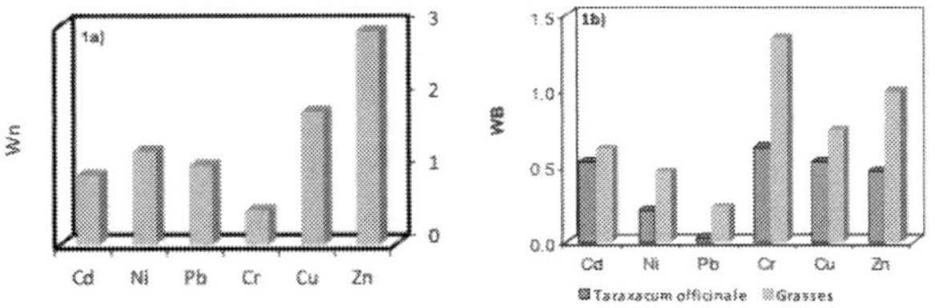

Fig. 1. *Wn* indexes for soils and *WB* coefficients for grasses and *Taraxacum officinale*.

The bioaccumulation coefficients were calculated to assess the degree and direction of mobility of heavy metals. They reflect the ability of the plant to accumulate metals from the soil and inform about their movement from the soil solution to the aerial parts of the plant. The average value of all bioaccumulation coefficients (*WB*) was higher in grasses than in dandelion (*Taraxacum officinale*) for all trace elements (Fig. 1b). *WB* coefficients of grasses formed the following order: Cr > Zn > Cu > Cd > Ni > Pb. The avarage values of *WB* in *Taraxacum officinale* can be presented in the following order: Cr > Cu > Cd > Zn > Ni > Pb. During the evaluation of metal bioaccumulation in plants, the medium level accumulation of zinc, copper, nickel, lead and cadmium (*WB* 0.1–1) was shown [18]. The accumulation of chromium in grasses amounting to *WB* 1.34 indicates the ease accumulates in plants. The medians of *WB* for grasses and *Taraxacum officinale,* respectively, are as follows: Zn 0.69 / 0.45; Cu 0.71 / 0.50; Pb 0.23 / 0.02; Ni 0.48 / 0.17; Cd 0.34/0.22 and Cr 0.40/0.32.

Mean bioaccumulation coefficients for Zn and Cu in grasses collected from the roadsides of the city centre of Wrocław (Poland) were 0.71 and 0.42, respectively [5]. Mean WB indexes of Zn, Cu, Pb, and Cd content in dandelion in relation to soil in Upper Silesia were below of unit, due to very strong contamination of soils in Ruda Śląska area. WB indexes for relationship of dandelion root/soil were as follows: Cd (0.57) > Cu (0.37)>Pb (0.22)>Zn (0.19), and for relationship of dandelion leaves/soil could bepresented in the following order: Cu (0.94) > Cd (0.33) > Zn (0.22) > Pb (0.07) [12]. Similarly, low values of WB for *Taraxacum officinale* samples from Warsaw were observed, in descending order: Cu (0.86) > Cd (0,61) > Zn (0.55) > Cr (0.24) > Ni (0.17) > Pb (0.07) [13].

The concentrations of heavy metals in grasses and dandelion (*Taraxacum officinale*) were compared to the permissible concentrations of undesirable substances in feeds according to the Polish Ministry of Agriculture and Rural Development Regulation from January 23rd, 2007 [19]. In grasses exceedances of the content of the following metals were recorded: copper in 99% of samples (acceptable limit is 10 mg Cu/kg DM), zinc in 91% of samples (100 mg Zn/kg DM) and lead in 12% of samples (10 mg Pb/kg DM). In *Taraxacum officinale* the only exceedance was that of copper content which was observed in 58% of samples. Due to the exceeding of the limit of the concentrations of heavy metals in plants, the biomass is not suitable for the animal feeds. It can be used for industrial purposes.

Linear correlation of the content of heavy metals in the soil may be the result of geochemical relationship of chemical elements in the parent rock or may inform about the sources of pollutions of the soil. Pearson's correlation coefficients (*r*) for relationships of metals are the following: Zn-Pb of 0.94 (significantly correlated – fig. 2), Pb-Cu and Zn-Cu of 0.66 and 0.60 (moderately correlated), respectively. It may point to the same source of metals' origin in soils (pollution from motor transport). Positive correlation coefficients of Pb-Cr (0.36), Zn-Cr (0.34) and Cd-Ni (0.22) show weak relationship between heavy metals in the soil.

The Pearson's correlation coefficients between different metals in soil, within the shortest distance (0.5 m) from the E30 expressway (Warsaw, Poland) shown significantly or moderately correlation of pairs of Ni-Cu (0.90), Zn-Cd (0.73), Zn-Cu (0.66), Cu-Cd (0.65) and Pb-Cu (0.57) [4]. The strongest positive correlations in Łódź (Poland) were identified for Cr-Cu (0.95) and between Cd and other trace element (Cu, Pb, Cr, Ni) averaging 0.80–0.83 [6]. Pearson's correlation coefficients for urban soils in the city centre of Wrocław (Poland) amounts to 0.51 for the pair Zn-Cu [5]. In the case of urban soils taken from Torino (Italy) positive and significant correlation was observed for Zn-Cu (0.89), Zn-Pb (0.71) and Ni-Cr (0.70). The following pairs of metals were moderately correlated: Pb-Cu (0.64), Cu-Cr (0.54) and Zn-Cr (0.51) [14].

The Pearson's correlation coefficients for relationships of Zn-Cu (*r* 0.81 – fig. 2) and Pb-Cu (*r* 0.56) in grasses are as high as in soils. Also noticed in grasses was a strong correlation of Pb-Cr (*r* 0,82 – fig. 2) and moderate correlation in pairs of Zn-Ni (*r* 0,52) and Cu-Cd (*r* 0,42). Weak relationships of metal pairs were observed for Pb-Zn (*r* 0,31), Zn-Cd and Cu-Ni (*r* 0,29), Ni-Cr (*r* 0,37), Cd-Ni (*r* 0,32). In dandelion (*Taraxacum officinale*) a very strong linear relationship was noted for Pb-Cr (r 0.95 –fig. 2), similarly as in grasses. Positive values of *r* coefficients were also observed in the following pairs of metals: Pb-Cd (0.46), Zn-Cu (0.40), Cd-Cr (0.36), Ni-Cr (0.33), Pb-Ni (0.27), Zn-Ni (0.26). For other pairs of metals in soils and plants no linear correlation was observed.

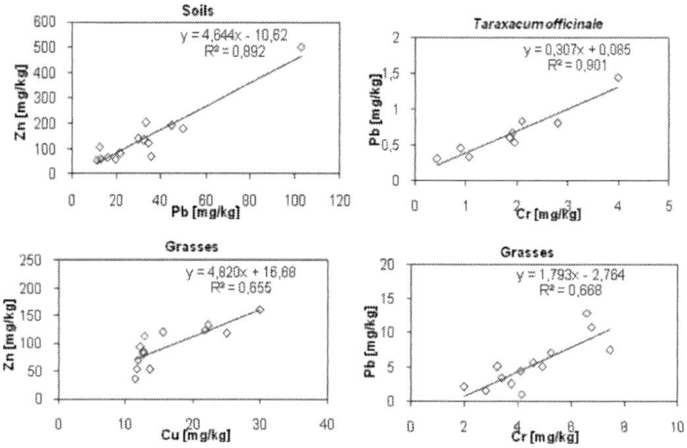

Fig. 2. Significantly correlated relationships of metal concentrations in soils and plants.

Statistically significant relationship between pH and the content of heavy metals in soils was not established. In case of copper the reported Pearson's correlation coefficient was only – 0.50 (moderate dependence). Radziemska et al. [4] reported a 0.46 linear correlation between pH and Ni content in soil. Additionally, Cd, Pb and Cu were very weakly correlated with pH.

Weak linear correlations were observed in relationships between metal content in soil and metal content in grass. The highest values of Pearson's coefficient were observed for copper (*r* 0.47) and zinc (*r* 0.38). No linear relationship between Cr (in grasses)/Cr (in soil) was found. Correlation between urban soil and grass from the city centre of Wrocław (Poland) showed higher value of r factor for copper: 0.63 (r 0.26 for Zn) [5]. Metals in samples from Karatay (Turkey) were moderately correlated: Cu (0.70), Ni (0.69) and Pb (0.44). Metals of Zn, Cr and Cd in soil and grass (*Lolium perenne* L.) were also weakly correlated [17].

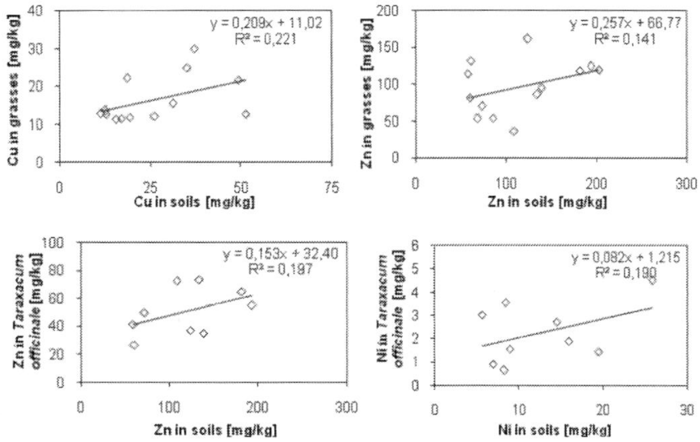

Fig. 3. Linear relationships of metal concentrations in soils and plants.

Relationship between metal content in soil and dandelion (*Taraxacum officinale*) is weakly correlated in the case of zinc and nickel (*r* 0.44). For other trace elements no linear relationships in *Taraxacum officinale* was observed (fig. 3). For samples from Warsaw (Poland) strong correlations were observed for Zn, Pb and Cd in soil/dandelion (*Taraxacum officinale*): 0.73, 0.70 and 0.57, respectively [13]. Pearson's correlation coefficients in urban soils / root of *Taraxacum officinale* for Upper Silesia (Poland) were as follows: Cu (0.85), Zn (0.60), Cd (0.49), for Pb no coefficient values were reported. In the case of dandelion leaves the coefficients were as follows: Cu (0.88), Cd (0.77), Pb (0.51), for Zn no coefficient values were reported [12].

Comparison of the frequency histograms for the occurrence of metals in soils and plants reveals deviations from the bell curve caused by substantial diversification of the content of metals in the population of investigated samples. It is due to different locations of sampling along the downtown ring road of Wrocław and also due to the diversified intensity of motor traffic. Kurtosis is mostly below zero in soils and plant samples (a Flatten-Gaussian distribution). Histograms are right-skewed, therefore most results are lower than the average value, and the arithmetic mean is higher than median. Distribution close to Gaussian curve was observed for Ni and Cr in all environmental matrices and also in the case of Zn in the grasses.

4 Conclusions

1) In location no. 11 (the crossings of Na Ostatnim Groszu Street with Popowicka Street – table 1.) the limit value of 500 mg Zn/kg DM for soils of group I, as established by the Polish Ministry of Environment Regulation from September 1st, 2016, was exceeded.
2) The coefficients of variation in urban soils are very high for Zn (81%) and Pb (69%), and close to 50% for Cr (58%), Cd (57%), Cu (50%). It may stem from the non-uniform distribution of contaminants between locations due to diverse intensity of traffic motor transport.
3) Average values of pollution indexes indicate medium level of the contamination of urban soils in Poland: Zn (2.92) > Cu (1.82) > Ni (1.28) > Pb (1.09) > Cd (0.94) > Cr (0.47).
4) Average contents of heavy metals in plants are much lower than in soils about 1.5–2.5 times in grasses, and 2–5.5 times in dandelion.
5) Due to the exceedance of the content of Cu in grasses, the biomass is not acceptable for use in animal feeds, according to the Polish Ministry of Agriculture and Rural Development

DOI: 10.1051/e3sconf/20172200066

Regulation from January 23rd, 2007. Also in the samples of *Taraxacum officinale* exceedances of the content of Cu were observed (except for samples no. 1 and 6).
6) The coefficients of variation of Cd, Pb, Ni in biomass and also for Cr in dandelion are above 50%.
7) The bioaccumulation coefficients in grasses and dandelion (*Taraxacum officinale*) are arranged as follows: Cr > Zn > Cu > Cd > Ni > Pb and Cr > Cu > Cd > Zn > Ni > Pb, respectively.
8) The content for chromium and lead in grasses and dandelion are strongly correlated with each other (r of 0.82/0.95, respectively). Similarly strong correlation in grasses was observed for zinc and copper (r 0,81). In the case of soils the correlation coefficient for zinc and lead equals 0.94.

The investigations were co-financed within the framework of the order No. 0401/0068/16 with the specific subsidy granted for the Faculty of Environmental Engineering Wroclaw University of Science and Technology (W-7) by the Minister of Science and Higher Education.

References

1. A. Athanasopoulou, G. Kollaros, AJER **5**, 12 (2016)
2. Å. Lindgren, Sci. Total Environ. **189/190** (1996)
3. T. Grigoratos, G. Martini, Environ Sci. Pollut. Res. **22**, 4 (2015)
4. M. Radziemska, J. Fronczyk, Int. J. Environ. Res. Public Health **12**, 10 (2015)
5. A. Hołtra, D. Zamorska-Wojdyła, Environ. Prot. Eng. **42**, 3 (2016)
6. M.I. Szynkowska, A. Pawlaczyk, E. Leśniewska, T. Paryjczak, Pol. J. Environ. Stud. **18**, 6 (2009)
7. R.N. Pivić, A.B. Stanojković Sebić, D.L. Jošić, Pol. J. Environ. Stud. **22**, 5 (2013)
8. K. Dzierżanowski, S.W. Gawroński, Environ. **50** (2011)
9. M.A. Lee, L. Davies, S.A. Power, Environ. Pollut. **163** (2012)
10. S. Guala, F.A. Vega, E.F. Covelo, Environ. Monit. Assess. **185**, 1 (2013)
11. Journal of Laws from September 1st, 2016 pos. 1395 (Poland)
12. E. Królak, Pol. J. Environ. Stud. **12**, 6 (2003)
13. B. Gworek, A. Dećkowska, M. Pierścieniak, Pol. J. Environ. Stud. **20**, 1 (2011)
14. M. Biasioli, R. Barberis, F. Ajmone-Marsan, Sci. Total Environ. **356**, 1–3 (2006)
15. F. Çelenk, F.T. Kiziloğlu, Int. J. Res. Agric. and For. **2**, 5 (2015)
16. K. Jankowski, et. all, J. Elem. **19**, 1 (2014)
17. Onder S., Dursun S., Gezgin S., Demirbas A., Pol. J. Environ. Stud. **16**, 1 (2007)
18. A. Kabata-Pendias, M. Piotrowska. T. Witek, Evaluation of contamination degree of soils and plants by heavy metals and sulfur. (IUNG, Puławy, 1993)
19. Journal of Laws from January 23rd, 2007 no. 20 pos. 119 (Poland)

A novel design of reactive distillation configuration for 2-methoxy-2-methylheptane process

Arif Hussain[1], *Muhammad Abdul* Qyyum[1], *Le Quang* Minh[1], *Hong* Jimin[1], and *Moonyong* Lee[1,*]

[1]Process System Design & Control Laboratory, School of Chemical Engineering, Yeungnam University, Dae-Dong 214-1, Gyeongsan, 712-749, South Korea

Abstract. The study aims to reveal the possibility of reactive distillation (RD) in the 2-methoxy-2-methylheptane (MMH) production process. MMH is getting more industrial and academic interests as a gasoline additive to replace methyl tert-butyl ether. Traditionally, MMH is obtained by carrying out the reaction in the reactor followed by three distillation columns. The high yield of MMH could be achieved by keeping the large reactor size or by using the large excess of 2-methyl-1-heptene (MH). Both former and latter strategies are associated with the high capital and operating costs. To solve these problems, this study proposed an innovative RD configuration to take synergistic benefits of reaction and separation involved. This innovative RD configuration allows the production of MMH with significantly lower capital, operating and total annual costs. For desired MMH yield, the result demonstrates that the proposed RD configuration can reduce energy, capital, and total annual costs up to 7.7, 31.3, and 17.1%, respectively, compared to a conventional process. Furthermore, the influence of some important design parameters on the RD column performance was also explored to overcome the temperature limitation of acid resin catalyst inside the reactive zone of the RD column.

1 Introduction

The invention of the lead-containing compounds to be added to gasoline for reducing the knocking tendency has gained tremendous importance [1]. Methyl tert-butyl ether (MTBE) was considered as an important "oxygenate" to raise the oxygen content of gasoline. However, the restrictive energy and environmental policies have banned the use of MTBE as a fuel additive in gasoline due to its solubility in water and its widespread detection in groundwater [2]. This led to raising interests in higher molecular weight ethers that will exhibit significantly decreased solubility. One possible alternative is the chemical MMH produced through etherification reaction between MH and MeOH. However, the main barrier of this approach is the high investment and operating costs associated with the conventional process of making MMH that uses the reactor followed by three sequenced distillation columns. The high yield of the desired product could be achieved by keeping the large reactor size or by using the large excess of MH. Both former and latter strategies are associated with the high capital and operating costs. For the specified yield of MMH, Luyben [3] studied the optimum

* Corresponding author: mynlee@ynu.ac.kr

values of reactor vessel sizes and the recycle flow rates to minimize the total annual cost. Furthermore, Griffin et al. [4] studied the effect of high and low values of an equilibrium constant on the optimum operating conditions for MMH plant. All these researchers used the conventional MMH process in their studies. The reaction mechanism of MMH process that includes both the equilibrium limited and irreversible reaction is a possible candidate for a more complicated multifunctional reactor technology like RD that offers significant advantages over the conventional reactor-column sequence. However, the RD column design for the MMH process is challenging due to the maximum allowable temperature (by 400 K) of its acid resin catalyst. By designing the innovative equipment and implementing more energy-efficient production processes, the possible solutions to overcome the constraints imposed on various raw materials and products can be found. Process intensification is the novel design strategy aiming to improve existing processes by reducing the equipment sizes, energy consumption, and overall plant footprint. RD falls into the category of process intensification because of its potential to integrate a conventional reactor column sequence into a single shell to improve the performance of reactive processes. Presently, more than 150 RD columns with annual capacities of 100-3000 ktonnes are being operated worldwide [5]. A huge scholarly data deals with the modeling, simulation, and industrial applications on RD can be found elsewhere [6–9]. The industrial RD columns for etherification reactions especially for MTBE process has grown tremendously and globally over the last decades with improved capital and operating savings. The MTBE process is, in fact, the largest application of RD in terms of a number of columns and production capacity [10], but the future is uncertain due to the legislation that prohibits its use. The growing interest in the chemical MMH to replace MTBE needs to increase capacity in a cost-effective way to meet the accelerated growing market demand. This study proposed an innovative RD configuration that allows the production of MMH with significantly lower capital and operating costs. The temperature limitation while designing RD configuration was overcome by selecting an appropriate reactive zone, feed locations, reflux ratio, catalyst holdup, and suitable column pressure. This study emphasized the potential benefits of the proposed RD column configuration to replace the conventional energy-intensive MMH production process.

2 Process studied

The proposed study is based on the conventional flowsheet (Fig. 1) for making MMH, taken from the work of Luyben [3] and Griffin et al. [4] to verify the feasibility of RD configuration. The reaction mechanism involves the desired reversible reaction (Eq. 1) between methanol (MeOH, CH_3OH) and 2-methyl-1-heptene (MH, C_8H_{16}) to form 2-methoxy-2-methylheptane (MMH, $C_9H_{20}O$). The same reactants producing dimethyl ether (DME, C_2H_6O) and 2-methyl-2-heptanol (MHOH, $C_8H_{18}O$) as by-products in an undesirable irreversible reaction (Eq. 2).

$$CH_3OH + C_8H_{16} \leftrightarrow C_9H_{20}O \quad (1)$$

$$2CH_3OH + C_8H_{16} \rightarrow C_2H_6O + C_8H_{18}O \quad (2)$$

The corresponding rate equations in terms of the mole fraction for the desired and undesired reactions are represented by (Eq. 3) and (Eq. 4), respectively:

$$R_D = k_{Df}X_{MeOH}X_{MH} - k_{Dr}X_{MMH} \quad (3)$$

$$R_U = K_U(X_{MeOH})^2 \quad (4)$$

where k_{Df} and k_{Dr} are the forward and backward rate constants for the desired reaction. K_U is the rate constant for the undesired reaction. Table 1 lists the values of reaction rate constants and the activation energies for both reactions [4].

Table 1. Kinetic parameters [4].

		kmol $s^{-1}kgcat^{-1}$			kJ/kmol
k_{Df}	=	6.7 x 10^7	E_{Df}	=	90000
k_{Dr}	=	2.1 x 10^{-6}	E_{Dr}	=	900
k_U	=	1.3 x 10^9	E_U	=	105900

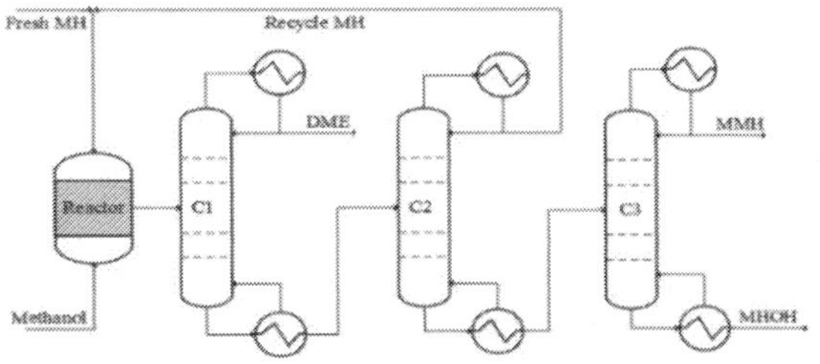

Fig. 1. Conventional flowsheet of making MMH [3–4].

3 Phase equilibrium

All the unit operations were simulated using Aspen Plus® built in "UNIQUAC" physical property method. The components MMH and MHOH were generated using the molecular structure (Fig. 2) because they were not available in the Aspen databank. The result of the base case was retrieved to confirm the phase equilibrium diagrams for both reaction and separation. Figs. 3a and 3b show the T-xy diagrams for separating DME/MeOH and MMH/MHOH used in this simulation, respectively.

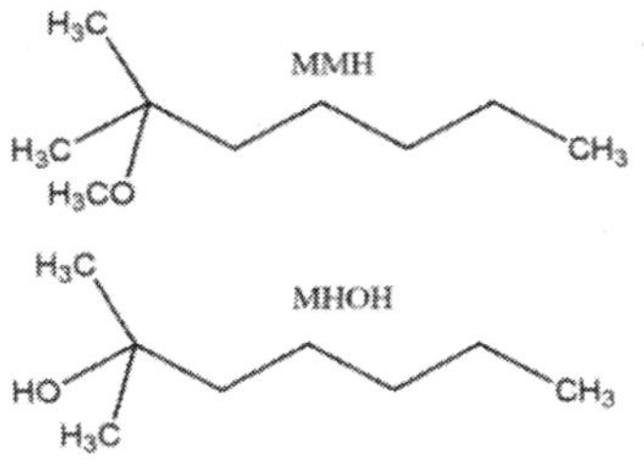

Fig. 2. Molecular structure for MMH and MHOH.

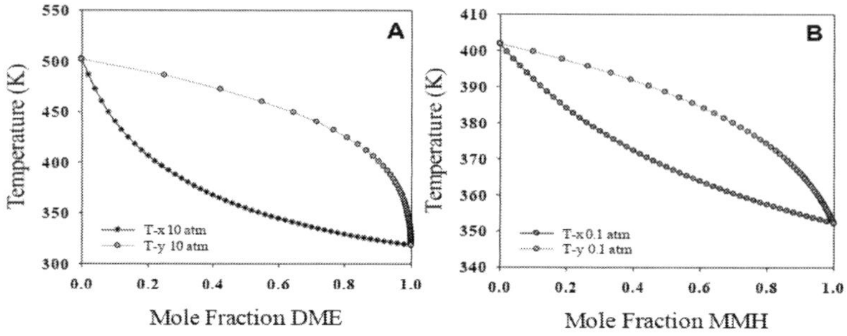

Fig. 3. VLE diagram (A) DME/MeOH at 10 atm, and (B) MMH/MHOH at 0.1 atm.

4 Proposed configuration

4.1 RD–Column

The newly proposed configuration based on RD column is shown in Fig. 4. The RD column has total 35 stages. There is no rectifying section, the reactive zone is between stage 2–12 and the stripping section at the bottom. The stages are numbered according to Aspen Plus stage notation with stage 1 and 35 being the condenser and reboiler, respectively. The fresh MeOH stream as a low boiling point reactant is fed on stage 13, while the fresh MH stream as a heavier reactant combined with a recycle stream that gives the total of 128.98 kmol/h is introduced on stage 1. The column is operated at 1 atm. The high MH/MeOH ratio makes this process be a suitable candidate for the RD implementation. A total estimated load of 291.6 kg of acid resin catalyst per stage was used within the reactive zone.

4.2 Column–C2

The column C2 serves the purpose of delivering high purity DME to be utilized as a green energy source as a distillate product, while the unreacted MeOH and MH to be recycled from the bottom. The distillate from the RD column is fed to stage 5 of a 12-stage distillation column. The normal boiling point of DME is 248.2 K. The operating pressure of C2 column is set at 10 atm, that permits the use of cooling water in the condenser. The small distillate stream of 0.513 kmol/h with 99.9 mole % DME gives the condenser duty of 0.0044 MW. The bottom stream contains the unreacted MeOH and MH, which is recycled to RD column. The reboiler duty was 0.86 MW with the base temperature of 501.9 K that uses high–pressure steam. The column diameter from Aspen tray sizing was 0.9 m.

 DOI: 10.1051/e3sconf/20172200067

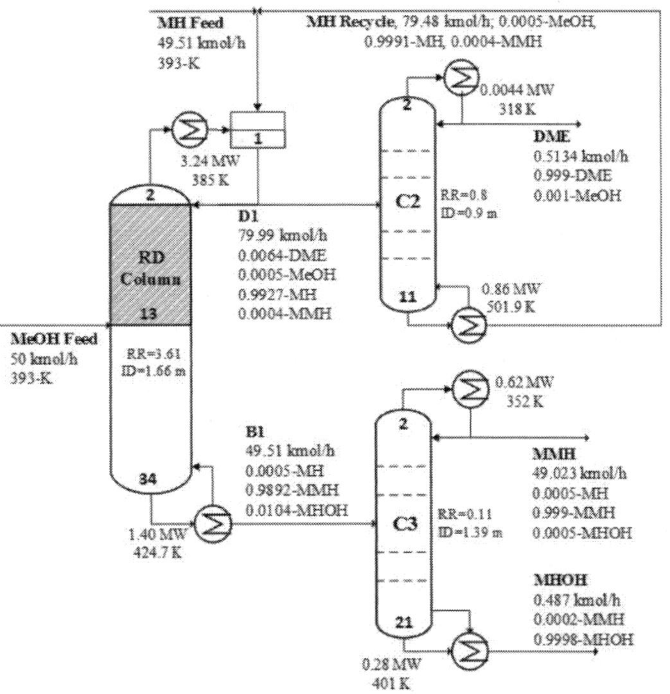

Fig. 4. Proposed RD configuration for MMH process.

4.3 Column–C3

The column C3 receives a feed stream from the bottom of RD column. This column has total 22 stages with stage 12 being the feed location. The distillate product was 49.02 kmol/h with 99.9 mole % MMH. The bottom was 0.487 kmol/h with 99.9 % of undesired MHOH. This column operates at 0.1 atm and gives 0.62 and 0.28 MW condenser and reboiler duties, respectively. The reflux drum temperature was 352 K, and the bottom temperature was 401 K that uses the medium pressure steam. The column diameter was estimated at 1.39 m.

5 Design economics

For an economic evaluation, the capital costs (column shell, heat exchangers, catalyst) and the energy costs (steam in the reboilers) were taken into consideration. The economic assessment of the proposed configuration for the desired yield of MMH was carried out in terms of total annual cost (TAC), expressed by Eq. 5:

$$\text{TAC} = \frac{\text{Capital cost}}{\text{Payback period}} + \text{Energy Cost} \tag{5}$$

The basis for economics and equipment sizing are summarized in Table 2 taken from Douglas [11] and Turton et al. [12]. The payback period of three years and energy costs for LP, MP, and HP steams were taken similarly as in the base case for a fair comparison.

DOI: 10.1051/e3sconf/20172200067

Table 2. Basis for economic and equipment sizing.

Column Shell Cost	$17640 * (D)^{1.066} * (L)^{0.65}$
Column Diameter (m)	Aspen Tray Sizing
Column Length (m)	NT trays with 0.61m spacing plus 20% extra length.
Heat Exchanger Capital costs	
Condenser	
Capital Cost ($)	$7296 * (A_{con})^{0.65}, A_{con} = \frac{Q_{con}}{U_{con} + \Delta T_{con}}$
Heat Transfer Coefficient	$U_{con} = 0.852\ kW/(K.m^2)$
Differential Temperature (K)	ΔT_{con} = Reflux Drum Temp − 318 K
Reboiler	
Capital Cost ($)	$7296 * (A_{reb})^{0.65}, A_{reb} = \frac{Q_{reb}}{U_{reb} + \Delta T_{reb}}$
Heat Transfer Coefficient	$U_{reb} = 0.568\ kW/(K.m^2)$
Differential Temperature (K)	$\Delta T_{reb} = 34.8$ K
Energy Cost ($/year)	LP steam = $7.78/GJ MP steam = $8.22/GJ HP steam = $9.88/GJ
Catalyst Cost ($)	$10/kg

6 Results and discussion

6.1 RD column

The bottom product at stage 35 produces the desired MMH and small quantity of undesired MHOH product. The distillate product is a small quantity of DME which is produced as a by–product together with unreacted MeOH and excess MH to be recycled. MeOH and MH conversions are 99% and 38.5%, respectively. The last stage temperature of the reactive stage was 399 K, which satisfied the temperature constraint in the reactive zone. The reboiler duty was 1.40 MW with the bottom temperature of 424.7 K, which uses the medium pressure steam. The column diameter from Aspen tray sizing was 1.66 m. Figs. 5a and 5b show the temperature and composition profile of RD column.

 DOI: 10.1051/e3sconf/20172200067

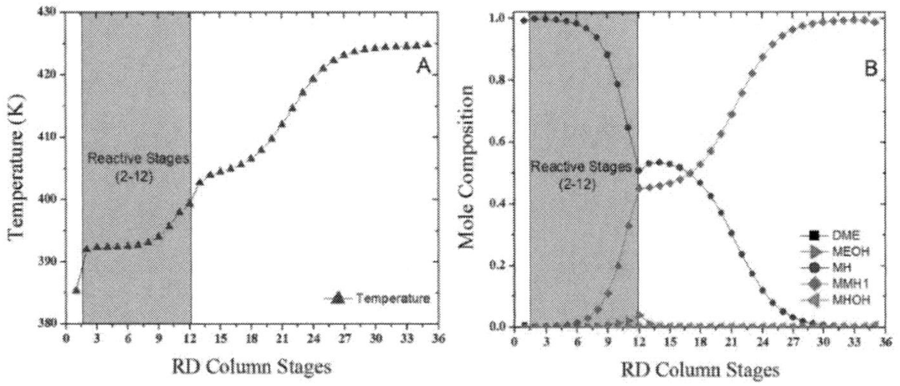

Fig. 5. (A) Column temperature profile and (B) Mole composition profile.

6.2 Column C2

The distillate contains the small flow rate of 0.5149 kmol/h with 99.9% DME purity. The condenser duty was 0.0044 MW with reflux drum temperature of 318 K. The bottom stream is mostly excess MH with the small amount of unreacted MeOH that is recycled back to the RD column. The reboiler duty was 0.859 MW with the base temperature of 501 K. The column diameter from Aspen tray sizing was 0.9 m. Figs. 6a and 6b show the temperature and composition profiles of C2 column, respectively.

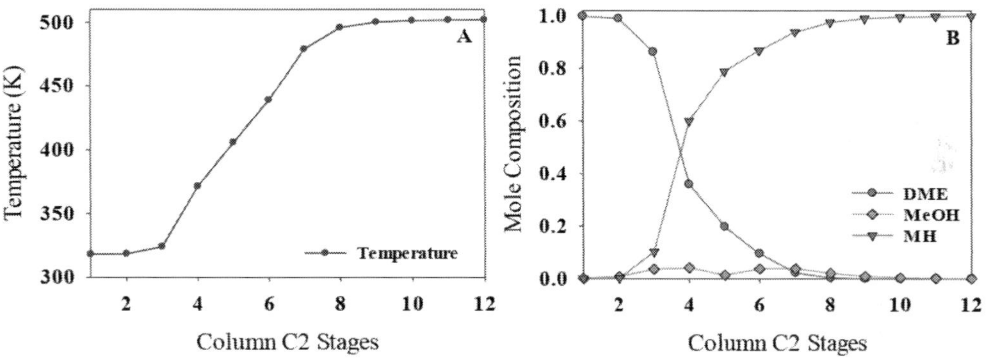

Fig. 6. Column C2 (A) Temperature profile, and (B) Mole composition profile.

6.3 Column C3

The column separated the bottom stream from the RD column into MMH and MHOH. The distillate is 49.02 kmol/h with 99.9% MMH. The condenser duty was 0.62 MW with reflux drum temperature of 352 K. The bottom stream is 0.4870 kmol/h with 99.99 % MHOH. The reboiler duty was 0.28 MW with the base temperature of 401 K that uses the medium pressure steam. The column diameter was 1.39 m. The temperature and composition profiles of C3 column are shown in Figs. 7a and 7b, respectively.

 DOI: 10.1051/e3sconf/20172200067

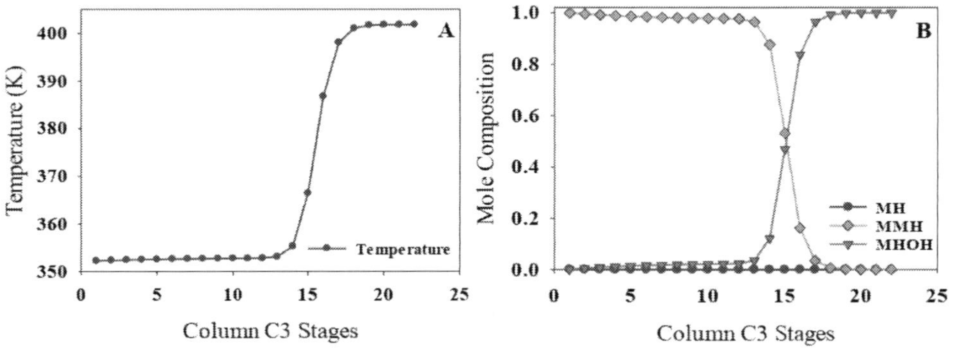

Fig. 7. Column C3 (A) Temperature Profile, and (B) Mole composition profile.

7 Economic analysis

The economic analysis was carried out for the proposed RD configuration in terms of TAC. For the desired yield of MMH, the proposed RD configuration can reduce energy, capital, and TAC up to 7.7, 31.3, and 17.1%, respectively. Table 3 summarized the economic analysis of the proposed RD configuration.

Table 3. Economic comparison of RD configuration with the base case.

Economic Analysis	Base Case	Proposed Case
Reactor Cost		
Reactor Capital (106$)	0.1091	-
Catalyst (106$)	0.12	-
C1/ RD Column		
H(m)	8.05	24.15
ID (m)	1.084	1.66
Catalyst Cost (106$)	0	0.035
Column Vessel Cost (106$)	0.095	0.389
Exchanger Costs (106$)	0.128	0.201
Energy Costs (106$/year)	0.4424	0.398
C1 Column		
H(m)	30.01	7.32
ID (m)	2.044	0.9
Column Vessel Cost (106$)	0.5668	0.077
Exchanger Costs (106$)	0.213	0.09
Energy Costs (106$/year)	0.2244	0.267
C2 Column		
H(m)	15.37	14.64
ID (m)	1.589	1.39
Column Vessel Cost (106$)	0.249	0.215
Exchanger Costs (106$)	0.1284	0.087
Energy Costs (106$/year)	0.141	0.079
Total Capital Cost (106$)	1.6082	1.105
Total Energy Cost (106$/year)	0.808	0.745
TAC (106$/year)	1.344	1.114

DOI: 10.1051/e3sconf/20172200067

8 Conclusions

An energy efficient configuration based on RD column was proposed for MMH production. A desired yield of MMH in the conventional process is usually obtained by keeping the large reactor size or by using the large excess of MH. Both strategies are expensive in terms of capital and operating costs. The proposed configuration uses the RD technology to obtain the desired MMH yield at significantly lower capital, and operating costs. The result demonstrates that the RD configuration can significantly reduce energy, capital, and total annual cost by 7.7, 31.3, and 17.1%, respectively compared to the conventional process.

This work was supported by Basic Science Research Program through the National Research Foundation of Korea (NRF) funded by the Ministry of Education (2015R1D1A3A01015621) and by Priority Research Centers Program through the National Research Foundation of Korea (NRF) funded by the Ministry of Education (2014R1A6A1031189).

References

1. S.P. Srivastava, J. Hancsok, *Fuels and Fuel-Additives* (Wiley, 2014)
2. A. Soren, A. Elzinga, J. Environ. Econ. Manage. **67**, 258 (2014)
3. W.L. Luyben, J. Ind. Eng. Chem. Res. **49**, 11564 (2010)
4. D.W. Griffin, D.A. Mellichamp, M.F. Doherty, J. Ind. Eng. Chem. Res. **48**, 8037 (2009)
5. G.J. Harmsen, J. Chem. Eng. Process **46**, 774 (2007)
6. A.A. Kiss, D.J.P.C. Suszwalak, J. Comput. Chem. Eng. **38**, 74 (2012)
7. A. Kumar, P. Daoutidis, J. AlChE **45**, 51 (1999)
8. M.A. Arfaj, W.L. Luyben, J. AlChE **50**, 1462 (2004)
9. R. Jacobs. R. Krishna, J. Ind. Eng. Chem. Res. **32**, 1706 (1993)
10. W.L. Luyben. Y.C. Cheng, *Reactive Distillation Design and Control* (Wiley, 2008)
11. J.M. Douglas, *Conceptual Design of Chemical Processes* (McGraw-Hill, New York, 1988)
12. R. Turton, R.C. Whiting, W.B. Shaelwitz, *Synthesis and Design of Chemical Processes* (Prentice Hall, New York, 2009)

 DOI: 10.1051/e3sconf/20172200068

Comprehensive study for Anammox process via multistage anaerobic baffled reactors

Sherif Ismail[1,*], and *Ahmed* Tawfik

[1]Egypt-Japan University of Science and Technology (E-JUST); Environmental Engineering Department, P.O. Box 179 – New Borg El Arab City – Postal Code 21934, Alexandria, Egypt

Abstract. Continuous anaerobic ammonia oxidation (Anammox) process in multistage anaerobic baffled (MABR) reactor was investigated. The reactor was operated for approximately 150 days at constant hydraulic retention time (HRT) of 48 h and was fed with synthetic wastewater containing nitrite and ammonium as main substrates. The MABR was inoculated with mixed culture bacteria collected from activated sludge plant (41.6 g MLSS/L and 19.1 g MLVSS/L). The MABR reactor exhibited excellent performance for the start-up of Anammox process within a period of 35 days. The start-up period was divided into four successive phases; cell lysis, lag, activity elevation and steady state. Total inorganic nitrogen (TIN) removal efficiency of 96.8± 0.9% was achieved at steady state conditions, corresponding to nitrogen removal rate (NRR) of 50.2±1.7 mg N/L·d. Moreover, the effect of HRT on the Anammox process was assessed with applying five different HRTs of (48, 38.4, 28.8, 19.2 and 9.6 h). Decreasing HRT from 48 to 9.6 h reduced the removal efficiencies of NH_4-N, NO_2-N and TIN from 97.7±2.2 to 49.0±9.8%, from 95.7±1.9 to 71.0±8.5% and from 96.8±0.9 to 57.9±9.1%, respectively, that corresponding to reduction in NRR from 50.8±1.2 mg N/L·d at HRT of 48 h to 32.5±5.0 mg N/L·d at HRT of 9.6 h.

1 Introduction

Several types of wastewater such as wastewater from fertilizer industry, dewatering of digested sludge and landfill leachate have high concentrations of ammonium and low concentrations of biodegradable organic compounds (low C/N ratio). Nitrogen compounds presented in such a wastewater create a significant adverse environmental impact on the receiving water bodies because of their role in eutrophication, consumption of oxygen content of receiving waters, and their toxicity to aquatic species, including human beings [1]. Conventionally, the removal of nitrogen (ammonium) is accomplished by the combination of nitrification and denitrification processes [2]. Both of these processes are energy consuming and are associated with high costs. Moreover, these processes have an additional environmental impact due to high biomass production and greenhouse gas (CO_2, N_2O, etc.) emission, which promote global warming [3, 4].

* Corresponding author: sherif.ismail@ejust.edu.eg, ahmed.tawfik@ejust.edu.eg

Anaerobic ammonium oxidation which was initially discovered in a denitrifying fluidized bed reactor in 1995 [5], has been reported as a novel and promising biotechnology for the treatment of ammonium-rich wastewater especially with low C/N ratio. Under anaerobic or anoxic conditions, Anammox bacteria, which belong to the order *Planctomycetales*, accomplish autotrophic ammonium (NH_4^+) oxidation to produce dinitrogen (N_2) gas while using nitrite (NO_2^-) as an electron acceptor and CO_2 as carbon source, producing scanty amounts of nitrate (NO_3^-) (Eq. (1)) [6,7]. This process offers several advantages over conventional nitrification-denitrification systems including higher nitrogen removal rate, lower operational cost and less space requirement [4]. The application of the Anammox process in wastewater treatment results in significant reduction in energy (60%) and greenhouse gas emission (90%) compared to those of traditional biological nitrogen removal processes [8, 9].

$$NH_4^+ + 1.32\ NO_2^- + 0.066\ HCO_3^- + 0.13\ H^+ \rightarrow 1.02\ N_2 + 0.256\ NO_3^- + 0.066\ CH_2O_{0.5}N_{0.15} + 2.03\ H_2O \quad (1)$$

Many studies have been carried out to investigate the effectiveness of Anammox process in various configurations, and these studies have achieved distinct reactor performances such as up-flow anaerobic sludge blanket (UASB) [10, 11], sequencing batch reactor (SBR) [12, 13], membrane bioreactor [14, 15] and anaerobic baffled reactor (ABR) [16, 17]. However, the ABR reactor is known as a highly efficient anaerobic reactor configuration compared with other anaerobic systems. The ABR system has several advantages such as longer retention times, less vulnerable to check loadings, the ability to separate the anaerobic metabolic phases in addition to its simple design since no special gas separation, no packing material, no moving parts and no mechanical mixing are required [18]. The anaerobic baffled reactor (ABR) has been described as a series of up-flow anaerobic sludge blanket (UASBs), which does not need granulation for its operation. Therefore, it has lower start-up period than the other high rate reactors [19]. A typical ABR consists of a series of vertical baffles that the liquid flows alternately upward and downward between the partitions, and on its upward passage the waste flows through an anaerobic sludge blanket.

The Anammox process treating ammonia rich wastewater was investigated insufficiently using anaerobic baffled reactor. Therefore, the main objective of this study is to investigate the effectiveness of multistage anaerobic baffled reactor (MABR) configuration on the start-up of Anammox bacterial community, assess the potential of using MABR for continuous Anammox process, and study the effect of hydraulic retention time (HRT) on nitrogen removal efficiencies.

2 Materials and methods

2.1 Experimental set-up of multistage anaerobic baffled reactor (MABR) and operating condition

A multistage anaerobic baffled (MABR) reactor fabricated from Perspex was designed and used for the experimental study. Fig. 1 shows a schematic diagram of the lab-scale multistage anaerobic baffled reactor (MABR) used in this investigation. The working volume of the Anammox MABR reactor was 20 L, and the total dimensions of the reactor were 77.5 cm in length, 12.5 cm in width, and 33 cm in height. The MABR system consists of five sequential compartments ended with settler for excess sludge removal. The reactor compartments were separated by vertical standing where inside each compartment, downcomer and riser regions were constructed using a slanted edge (45°) vertical baffle to direct the flow evenly through the riser and to increase the contact time between the substrate and bacterial community. The

 DOI: 10.1051/e3sconf/20172200068

downcomer to riser volume ratio amounted to 1:3. Each compartment was equipped with three sampling ports to withdraw the wastewater and sludge samples. The working dimensions for each compartments were 12.5 × 12.5 × 25 cm. Black cloth was used to cover the reactor preventing light penetration in order to inhibit the growth of photosynthetic bacteria [20]. The produced gas was collected via porthole on the top of the reactor as shown in Fig. 1.

Synthetic wastewater was continuously fed from the tank into the reactor using a peristaltic pump (Masterflex - USA, Cole-Parmer Instrument Company) to control the flow rate of influent feeding. The reactor was operated at constant flow rate of 10 L/d and HRT of 48 h at ambient temperature (24±6°C). The influent concentrations of ammonia and nitrite were maintained both at 50 mg N/L in order to keep the nitrite concentration at nontoxic level with an average nitrogen loading rate (NLR) of 52.2±3.74 mg N/L·d. The pH of the influent was adjusted to be 7.5±0.3 with 1 M HCL. Anaerobic condition in the feeding tank was maintained by flushing the influent synthetic wastewater with argon gas before feeding to the reactor [11]. Throughout the operation, no sludge was intentionally removed from the reactor.

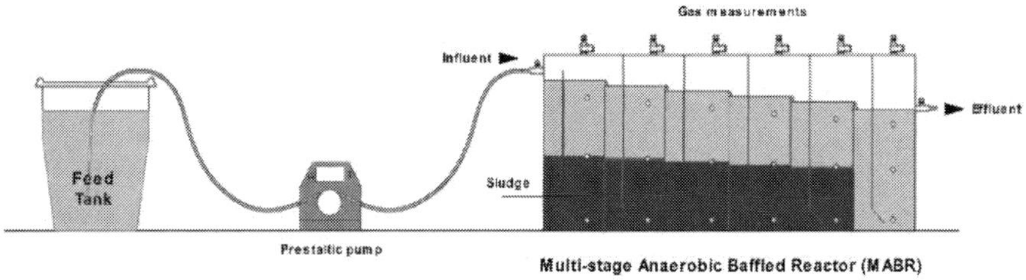

Fig. 1. Schematic diagram of the experimental MABR reactor.

2.2 Synthetic wastewater and inoculation

A synthetic ammonia rich wastewater was used for the continuous experiments. Ammonium and nitrite was supplemented to mineral medium in the form of $(NH_4)_2SO_4$ and $NaNO_2$, respectively and was added as required. The composition of the mineral medium was KH_2PO_4 (27.2 mg/L), $CaCl_2{\cdot}2H_2O$ (180 mg/L), $MgSO_4{\cdot}7H_2O$ (300 mg/L) and $KHCO_3$ (500 mg/L). The trace elements were supplied at a dosage of 1 mL per L of wastewater of trace elements solution 1 and 2. Trace elements solution 1 contained (per 1 L distilled water) $FeSO_4$ (5 g), EDTA (5 g); and trace elements solution 2 contained (per 1 L distilled water) EDTA (15 g), $MnCl_2{\cdot}4H2O$ (0.99 g), $CuSO_4{\cdot}5H_2O$ (250 mg), $ZnSO_4{\cdot}7H_2O$ (430 mg), $NiCl_2{\cdot}6H_2O$ (190 mg), $CoCl_2{\cdot}6H_2O$ (240 mg), H_3BO_4 (14 mg), $NaSeO_4{\cdot}10H_2O$ (210 mg), $NaMoO_4{\cdot}2H_2O$ (220 mg) and $NaWO_4{\cdot}2H_2O$ (50 mg) [21].

The ABR reactor was seeded with mixed culture bacteria obtained from the thickener of activated sludge treatment plant located in Alexandria, Egypt. The collected seeding sludge was further concentrated in gravitational settling tank for 24 h. Afterwards, the concentrated sludge was filtered through sieve No. 10 to remove coarse suspended solids, and then this concentrated sludge was used for inoculation. The concentration of mixed liquor suspended solids (MLSS) and mixed liquor volatile suspended solids (MLVSS) of the seed sludge were 41.6 and 19.1 g/L, respectively. The reactor was initially inoculated with 10 L sludge (50% of working volume) resulting in inoculum to substrate ratio of 5.2 mg N/g VS.

2.3 Analytical methods

Samples from the influent and the treated effluent were collected three times a week for analysis of ammonium nitrogen (NH_4-N), nitrite nitrogen (NO_2-N), nitrate nitrogen (NO_3-N), pH and oxidation reduction potential (ORP). All analyses were carried out according to the standard methods for examination of water and wastewater [22]. In addition, physical and chemical parameters were also analyzed for each compartment of the reactor at each operational condition's steady state. All measurements were performed in triplicates and average values were presented. Analysis of liquid samples was recorded (mean ± standard deviation) during steady state operation. Nitrogen removal rate was calculated as the sum of ammonium and nitrite consumption.

3 Results and discussion

3.1 The start-up of Anammox bacteria in MABR reactor (days 1–150)

The lab scale multistage anaerobic baffled reactor MABR was used to study the treatment of ammonia rich wastewater by Anammox process. The reactor was operated at constant hydraulic retention time HRT of 2 days. The influent ammonia and nitrite was adjusted to be 50 mg/L each, corresponding to total nitrogen loading rate of 50 mg N/L·d. The occurrence of the Anammox activity was demonstrated through a simultaneous depletion of nitrite NO_2-N and ammonia NH_4-N coupled to a small production of nitrate NO_3-N. The data presented in Fig. 2, 3 and 4 revealed that the start-up period was divided into four distinct phases based on the amount of ammonia removed: cell lysis phase (1–7 days), lag phase (8–23 days), propagation or activity elevation phase (24–35 days) and stationary phase (after 35 days).

At the beginning of the Anammox MABR start-up, the concentration of NH_4–N was increased. Accordingly, the ammonium removal efficiency was below zero. The results in Fig. 2 shows that the ammonia concentration was increased from 59.6 mg/L to 99.3 mg/L with ammonium removal of -66.6% in the beginning as shown in Fig 3. This phase named cell lysis phase as NH_4-N and organic matter were released from denitrifying bacteria autolysis due to the initial adverse conditions of the reactor [23]. The NO_2-N was completely consumed in this phase (Figs. 2 and 3). This removal could be attributed to the prevalence of denitrification process. The organic released from cell lysis can be used as an electron donor by heterotrophic denitrifying bacterial communities for denitrification process [24]. The sludge lysis phase lasted for only one week.

Lag phase lasted for 15 days (days 8–23). In this phase, the average effluent ammonium concentration and the average ammonium removal efficiency were 38.6±3.4 mg/L and 21.9±5.9%, respectively. The average ammonium removal rate was 5.4±1.5 mg/L·d, suggesting a juvenile Anammox activity in MABR reactor. Similarly, the high nitrite removal efficiency and small amount of nitrate production confirmed denitrification as well as cell lysis still continued in this phase (Fig. 2). Moreover, the effluent NO_3-N concentration of increased progressively. The effluent NO_3-N concentration of 4.3 mg/L was achieved in day 23, which revealed itself to start a slight Anammox activity (Fig. 2).

In activity elevation phase, continuous increase in ammonia and nitrite removal was observed. Figs. 2 and 3 shows that, during activity elevation phase, a significant increase in ammonia removal efficiency was monitored. The ammonia removal efficiency increased from 19.5% on day 23 to 94.9% on day 35. The ammonium removal rate (ARR) and total nitrogen removal rate (NRR) were increased from 4.74 to 21.28 mg N/L·d and from 29.85 to 45.55 mg N/L·d, respectively, which suggested that Anammox performance progressively enhanced in this phase. Nitrite concentration in the effluent were still the same as mentioned

DOI: 10.1051/e3sconf/20172200068

in the cell lysis and lag phase (Fig. 2). NO_3-N concentration started to increase during activity elevation phase. The average nitrate concentration of 8.4 ± 1.2 mg/L was achieved in the effluent of the reactor. At the end of this stage, the nitrite removal to ammonium conversion molar ratio was 1.14. Whereas, nitrate production to ammonium removal ratio was 0.17, which were nearly to the theoretical stoichiometric ratio of 1.32 and 0.26 through the Anammox process [25]. The MABR was successfully started-up on day 35.

During the stationary phase (steady state), ammonium removal was continuously monitored (days 36–150). The influent concentrations of NH_4-N and NO_2-N were maintained 50 mg/L. Nitrogen removal efficiency of the system was relatively stable. Ammonium, nitrite and total nitrogen removal efficiencies were 97.7±2.9%, 94.6±3.3% and 96.2±2.2% at HRT of 2 days and NLR of 52.2±3.7 mg/L·d, respectively corresponding to NRR of 50.2±3.9 mg N/L·d. The average effluent NH_4-N and NO_2-N were 1.2±1.6 and 2.8±1.6 mg/L, respectively. The production of NO_3-N increased stepwise, and the maximum NO_3-N concentration in the effluent was 21.4 mg/L on day 120. The NH_4-N, NO_2-N and NO_3-N stoichiometric ratio was 1: 1: 0.21, i.e. nearly the same as the theoretical ratio of 1:1.32:0.26. These results indicated that the Anammox reactor had reached a steady state.

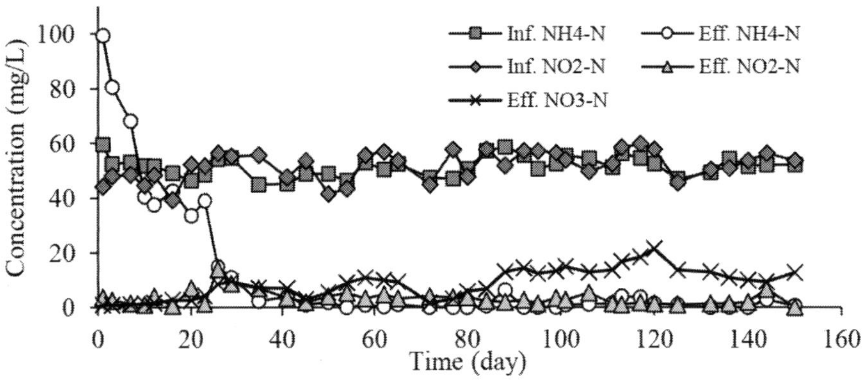

Fig. 2. Time course of nitrogen compounds in influent and effluent in the MABR reactor.

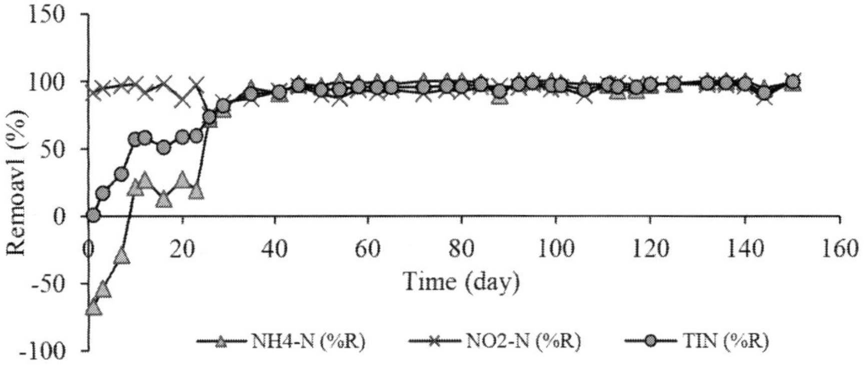

Fig. 3. Time course of nitrogen compounds removal efficiencies in the MABR reactor.

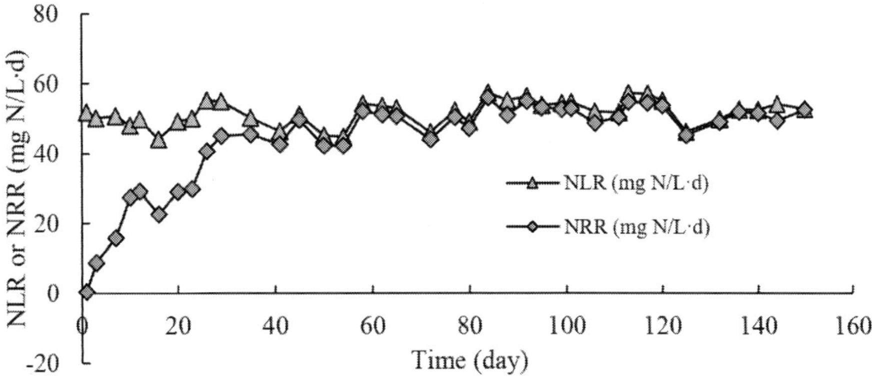

Fig. 4. Time course of NLR and NRR in the MABR reactor.

Based on the above results, the MABR reactor exhibited excellent performance for the start-up of Anammox process within a period of 35 days using activated sludge (mixed culture bacteria, 41.6 g MLSS/L and 19.1 g MLVSS/L) as inoculum. Tang, Zheng, et al. [24] demonstrated that an Anammox reactor was successfully started up on day 73. Tsushima et al. observed Anammox reactions after a 55-day the start-up process [26].

3.2 Behavior of Anammox process in the MABR reactor at different HRTs

Throughout the reactor's operation, the influent ammonium and nitrite concentrations were maintained constant whereby the nitrogen loading rate (NLR) was progressively increased by shortening HRT. The nitrogen removal efficiencies in MABR reactor is depicted in Table 1. Five different HRTs (48, 38.4, 28.8, 19.2 and 9.6) were applied in order to investigate the effect of HRT on Anammox process performance, which accompanied by changing NLR. The results obtained revealed that decreasing the HRT from 48 to 9.6 h reduced the removal efficiencies of NH_4-N, NO_2-N and TIN from 97.7±2.2 to 49.0±9.8%, from 95.7±1.9 to 71.0±8.5% and from 96.8±0.9 to 57.9±9.1%, respectively, that corresponding to decrease in NRR from 50.8±1.2 mg N/L·d at HRT of 48 h to 32.5±5.0 mg N/L·d at HRT of 9.6 h. Short HRT probably reduced the substrate utilization efficiency as the contact time between the microorganisms and substrate was insufficient resulting in high residual values of Nitrogen ammonia. The residual values of NH_4-N and NO_2-N were increased from 1.2±1.1 and 2.1±1.1 mg/L to 26.0±7.1 and 13.8±8.4 mg/L while decreasing the HRT from 48 to 9.6 h. these observation is in agreement with Tang et al. [6] who mentioned the same behaviour in up-flow anaerobic sludge bed (UASB) reactor were initially inoculated with anaerobic granular sludge taken from a paper mill wastewater treatment plant.

Table 1. Performance of Anammox process in MABR at different HRTs.

Parameters	**HRTs (h)**				
	48	**38.4**	**28.8**	**19.2**	**9.6**
Eff. NH_4-N (mg/L)	1.2±1.1	5.2±3.9	11.1±5.1	18.1±6.4	26.0±7.1
Eff. NO_2-N (mg/L)	2.1±1.1	3.4±1.4	4.0±1.4	5.9±1.5	13.8±8.4
DO (mg/L)	0.02±0.03	0.03±0.04	0.05±0.03	0.1±0.2	0.26±0.04
NH_4-N removal (%)	97.7±2.2	89.5±8.4	78.0±10.5	64.4±12.8	49.0±9.8
NO_2-N removal (%)	95.7±1.9	93.0±2.2	91.6±2.0	82.2±2.4	71.0±8.5
TN removal (%)	96.8±0.9	91.0±4.1	84.2±4.3	75.0±5.3	57.9±9.1
NRR (mg N/L·d)	50.8±1.2	48.1±1.1	44.8±1.0	40.4±3.4	32.5±5.0

4 Conclusion

In this study, Anammox process can be readily started up from conventional activated sludge in MABR reactor within 35 days. The cultivation process can be divided into four phases; cell lysis, lag, activity elevation and steady state. The ammonia concentration was increased from 54.9±3.9 mg/L (influent) to 82.6±15.6 mg/L in the effluent during cell lysis phase. Anammox activity in lag phase, amounted to 21.9±5.9%. The major portion of Anammox performance was occurred in activity elevation phase, where the nitrogen removal was increased from 19.5% to 94.9%. Anammox reaction process was nearly completed in the stationary phase which strongly indicated that the reactors were attained steady state conditions where the reactors were operated at nitrogen loading rates (NLRs) ranging from 44.1 to 57.5 mg N/L·d. The performance of Anammox process was relatively stable. Ammonium, nitrite and total nitrogen removal efficiencies were 97.7±2.9%, 94.6±3.3% and 96.2±2.2% at HRT of 2 days and NLR of 52.2±3.7 mg/L·d, respectively corresponding to NRR of 50.2±3.9 mg N/L·d. The average effluent NH_4-N and NO_2-N were 1.2±1.6 and 2.8±1.6 mg/L, respectively. Furthermore, the results obtained indicated the negative effect of decreasing the HRT in the performance of MABR reactor. The NH_4-N, NO_2-N removal efficiencies were reduced to 49.0±9.8% and 71.0±8.5% respectively at HRT of 9.6 h.

The first author is grateful for the Egyptian ministry of higher education which granted him a full PhD scholarship and for Egypt Japan University (EJUST) and Japan International Cooperation Agency (JICA) for providing all the facilities to accomplish in this work.

References

1. D. Paredes, P. Kuschk, T.S.A. Mbwette, F. Stange, R.A. Muller, H. Koser, Eng. Life Sci. **7**, 13–25 (2007)
2. S. Tomar, S.K. Gupta, Appl. Microbiol. Biotechnol. **99**, 9245–9254 (2015)
3. A. Daverey, S.H. Su, Y.T. Huang, S.S. Chen, S. Sung, J.G. Lin, Water Res. **47**, 2929–2937 (2013)
4. M.S.M. Jetten, M. Wagner, J. Fuerst, M. Van Loosdrecht, G. Kuenen, M. Strous, Curr. Opin. Biotechnol. **12**, 283–288 (2001)
5. A. Mulder, A.A. van de Graaf, L.A. Robertson, J.G. Kuenen, FEMS Microbiol. Ecol. **16**, 177–183 (1995)
6. C.J. Tang, P. Zheng, C.H. Wang, Q. Mahmood, J.Q. Zhang, X.G. Chen, L. Zhang, J.W. Chen, Water Res. **45**, 135–144 (2011)
7. L.F. Ren, S.Q. Ni, C. Liu, S. Liang, B. Zhang, Q. Kong, N. Guo, Environ. Sci. Pollut. Res. **22**, 2925–2934 (2015)
8. Y. Ren, N. Yan, Q. Wen, Z. Fan, T. Wei, M. Zhang, J. Ma, Chem. Eng. J. **175**, 1–7 (2011)
9. S.Q. Ni, J.Y. Ni, D.L. Hu, S. Sung, Bioresour. Technol. **110**, 701–705 (2012)
10. B. Ma, Y. Peng, S. Zhang, J. Wang, Y. Gan, J. Chang, S. Wang, S. Wang, G. Zhu, Bioresour. Technol. **129**, 606–611 (2013)
11. C.J. Tang, P. Zheng, L.Y. Chai, X.B. Min, Int. Biodeterior. Biodegrad. **82**, 141–148 (2013)
12. Y. Tao, D.W. Gao, Y. Fu, W.M. Wu, N.Q. Ren, Bioresour. Technol. **104**, 73–80 (2012)
13. N. Chamchoi, S. Nitisoravut, Chemosphere **66**, 2225–2232 (2007)
14. D. Akgul, C.K. Aktan, K. Yapsakli, B. Mertoglu, Biodegradation **24**, 399–412 (2013)

15. T. Wang, H. Zhang, D. Gao, F. Yang, G. Zhang, Bioresour. Technol. **122**, 78–82 (2012)
16. J.J. Yu, H. Chen, Q. Guo, Z.Z. Zhang, L.X. Guo, R.C. Jin, Ecol. Eng. **83**, 19–27 (2015)
17. C. Chen, F. Sun, H. Zhang, J. Wang, Y. Shen, X. Liang, Bioresour. Technol. **216**, 571–578 (2016)
18. H. Chen, Q.Q. Chen, Z.J. Shi, J.J. Xu, A.N. Liu, L.L. He, Y.H. Wu, M.L. Shi, R.C. Jin, Ecol. Eng. **92**, 229–235 (2016)
19. A. Elreedy, A. Tawfik, Energy Procedia **74**, 1071–1078 (2015)
20. H. Li, S. Zhou, W. Ma, G. Huang, B. Xu, Desalination **286**, 436–441 (2012)
21. A.A. van de Graaf, P. de Bruijn, L.A. Robertson, M.S.M. Jetten, J.G. Kuenen, Microbiology **142**, 2187–2196 (1996)
22. APHA, 25th ed., Am. Public Heal. Assoc. Water Environ. Fed, Washington DC, USA, 1995
23. J. Guo, S. Wang, J. Lian, H. Hao, W. Guo, Y. Liu, Y. Song, Bioresour. Technol. **220**, 641–646 (2016)
24. C.J. Tang, P. Zheng, Q. Mahmood, J.W. Chen, J. Ind. Microbiol. Biotechnol. **36**, 1093–1100 (2009)
25. M. Strous, M. Jetten, Appl. Environ. Microbiol. **63**, 2446–2448 (1997)
26. I. Tsushima, Y. Ogasawara, T. Kindaichi, H. Satoh, S. Okabe, **41**, 1623–1634 (2007)

Conceptual design of multi-source CCS pipeline transportation network for Polish energy sector

Niccolo Isoli[1], *Maciej* Chaczykowski[1,*]

[1]Warsaw University of Technology, Department of Building Facilities, Hydro and Environmental Engineering, ul. Nowowiejska 20, 00-653 Warsaw, Poland

Abstract. The aim of this study was to identify an optimal CCS transport infrastructure for Polish energy sector in regards of selected European Commission Energy Roadmap 2050 scenario. The work covers identification of the offshore storage site location, CO_2 pipeline network design and sizing for deployment at a national scale along with CAPEX analysis. It was conducted for the worst-case scenario, wherein the power plants operate under full-load conditions. The input data for the evaluation of CO_2 flow rates (flue gas composition) were taken from the selected cogeneration plant with the maximum electric capacity of 620 MW and the results were extrapolated from these data given the power outputs of the remaining units. A graph search algorithm was employed to estimate pipeline infrastructure costs to transport 95 MT of CO_2 annually, which amount to about 612.6 M€. Additional pipeline infrastructure costs will have to be incurred after 9 years of operation of the system due to limited storage site capacity. The results show that CAPEX estimates for CO_2 pipeline infrastructure cannot be relied on natural gas infrastructure data, since both systems exhibit differences in pipe wall thickness that affects material cost.

1 Introduction

The long-term warming of the climate system is mainly driven by CO_2 emissions, therefore the internationally agreed objective of limiting the increase in global average temperatures to well below 2°C above pre-industrial levels requires significant reductions in CO_2 emissions. In the context of necessary reductions by developed countries as a group, the European Union member states are committed to reducing greenhouse gas emissions to 80–95% below 1990 levels by 2050 [1]. The reductions of greenhouse gas emissions will put particular pressure on energy systems, which need to be almost emission-free despite higher demand. The European Commission proposed several scenarios of energy system transformation for achieving its 2050 emissions target. In a low nuclear energy scenario, assuming that no new nuclear reactors are being built (besides reactors currently under construction), around 32% penetration of CCS in power generation has been predicted. In the case of industrial applications carbon capture and storage is expected to account even for half of the global emissions cuts required by 2050. Indeed, CCS is the only technology

* Corresponding author: maciej.chaczykowski@is.pw.edu.pl

that can significantly reduce CO_2 emissions from coal- and gas-fired power plants as well as from energy intensive-industries.

Until recently research objectives on CCS transport were mostly focused on predicting the cost of pipeline transport [2], while route planning received relatively less attention. Chandel et al. [3] describe the economies of scale that can be achieved for CO_2 transport by collecting CO_2 emissions from multiple power plants into a trunkline that pipes the emissions to a single storage site. As the CO_2 flow rate that can be handled by a trunkline is increased, the levelized cost of transporting it declines exponentially. They conclude that low transportation costs over long distances open up the possibility for developing CCS systems that connect CO_2 sources to far away storage sites where storage costs are relatively low. Weiths and Wiley [4] employ genetic algorithm for designing the spanning tree type CCS network. Optimal CCS network transporting CO_2 from each source to at least one sink should not have loops, since recirculation would increase the pumping power required by the network, unless there is the advantage due to redundancy in parallel configuration of the pipelines. Different transport and storage options resulting in different network topology were discussed in the study of offshore CO_2 transportation cases for South Korea by Zahid et al. [5] and Jung et al. [6]. CO_2 transport options from coal-fired power plant to offshore storage sites near Japan‘s coastline, including liquefied CO_2 by ship, CO_2 hydrate by ship and pipeline, were studied by Suzuki et al. [7]. In the study by Jain et al. [8] greedy algorithm was used to calculate the optimal distance between the source and the sink for CCS infrastructure in eastern India. It was assumed that future pipelines for transportation of CO_2 would be built along the railway network. Least cost optimisation model of an integrated CO_2 capture, transportation and storage infrastructure for the UK over four time periods up to year 2050 was presented in the study by Elahi et al. [9]. The physical characteristics of a potential pipeline network for CO_2 transportation in the Humber Region in the UK area have been discussed by Luo et al. [10]. Hetland et al. [11] present the status of the large European CCS demonstration projects with particular emphasis on transport systems development. The costs of 15 transport scenarios involving the use of 3 pipelines and 5 offshore storage sites in Guangdong province, China were evaluated by Bai et al. [12]. It has been concluded that cost assessment methodology during the design-price evaluation should integrate local prices. Some key design issues that must be considered for the development of large scale CO_2 transportation network are reviewed in the study by Han et al. [13]. They conclude that to ensure the safe and cost-effective transportation, the concentrations of impurities in CO_2 stream should be restricted in an appropriate range. Similar conclusions were drawn in the studies by Chaczykowski and Osiadacz [14] and Wetenhall et al. [15], where the effect of impurities on pumping power and transport costs was observed, regardless of the assumed network structure and geometry.

The main objective of this study is to identify an optimal group selection of power plants, which contribute 32% to national power generation capacity for achieving European Commission 2050 emissions target. The optimal selection of power plants has been made based on a tree branched pipeline network design and a capital expenditures analysis for each unit. The design-price evaluation integrates local prices. Finally, the comparison to the current natural gas infrastructure capital expenditures is made.

2 Basic assumptions and input data

CO_2 streams for power plants have been calculated based on the analysis carried out for "Siekierki" power plant, located in Warsaw, whose flue gas composition was obtained from plant operator. A basic post combustion installation based on MEA absorption process was assumed and with the use of process simulator, taking the overall efficiency of

 DOI: 10.1051/e3sconf/20172200069

90%, the flow rate of 951 m^3/h of captured CO_2 was predicted. The volume stream from the above plant was considered as a reference case, by looking at we have extrapolated the numbers for the other power plants. The results are presented in table 2.1

The operating pressure range of the pipeline infrastructure was set to (8–15) MPa in order to maintain the post combustion CO_2-rich mixture in supercritical phase. As a result the pipeline wall thickness for the whole network was chosen equal to 25 mm. X70 steel has been selected for pipe material with the density of 7850 kg/m^3.

Material costs are calculated based on pipeline mass [16], with the assumption of unit price of 1000 € per 1 tonne of X70 steel. These costs were assumed to represent 30% of total costs including assembly works.

2.1 Power plants in Polish energy sector

The basic parameters of major fossil fuel based power plants in Polish energy sector are presented in Table 2.1. Predicted CO_2 streams are extrapolated values based upon the reference case calculations for the Siekierki power plant input data.

Table 2.1. Technical data of power plants.

No.	Power plant	Fuel	Power MW	CO_2 stream m^3/h	Latitude	Longitude
1	Rogowiec	Lignite	5 298	8 127	51°16'36"N	19°18'11"E
2	Kozienice	Black coal	2 820	4 326	51°35'08"N	21°33'04"E
3	N. Czarnowo	Black coal	1 984	3 044	53°11'42"N	14°29'06"E
4	Połaniec	Black coal, Biomass	1 811	2 778	50°25'57"N	21°16'50"E
5	Rybnik	Black coal	1 775	2 723	50°05'55"N	18°32'42"E
6	Jaworzno	Black coal, Biomass	1 535	2 355	50°12'16"N	19°16'12"E
7	Bogatynia	Lignite	1 499	2 300	50°54'27"N	14°57'14"E
8	Brzezie	Black coal	1 492	2 289	50°45'39"N	17°52'30"E
9	Pątnów	Lignite	1 200	1 841	52°18'27"N	18°15'29"E
10	Łaziska Górne	Black coal	1 155	1 772	50°09'17"N	18°50'37"E
11	Będzin	Black coal	820	1 258	50°19'26"N	19°07'45"E
12	Trzebinia	Black coal	666	1 022	50°09'35"N	19°28'14"E
13	Ostrołęka	Black coal, Biomass	647	993	53°04'58"N	21°34'21"E
14	Warszawa	Black coal, Biomass	620	951	52°13'56"N	21°00'30"E
15	Pątnów	Lignite	474	727	52°18'27"N	18°15'29"E
16	Skawina	Black coal, Biomass	440	675	49°58'30"N	19°49'42"E
17	Stalowa Wola	Black coal	330	506	50°34'34"N	22°03'40"E
18	Konin	Lignite	198	304	52°13'39"N	18°15'41"E

2.2 Identification of storage site

The selected geological structure is located in northwest area of the Baltic Sea. Based on the Polish Geological Institute report [17], there are four wells in the sea that satisfy the CO_2 storage criteria with total capacity of 0.9 Gt. Storage depth of the structure is 800–2800 m and its porosity is higher than 9%.

3 CO_2 transmission system design

It has been assumed that pipeline system would be used to transport dense-phase, CO_2-rich mixtures from capture plants to the offshore storage site. Pipeline transportation has the advantages of being cost-effective and readily expandable. The CO_2 pipe network, except from pipeline interconnector to the storage site, consists of tree branches running parallel to the existing natural gas transmission system infrastructure. The advantages arising from such arrangement are threefold:

1. The existing natural gas pipeline safe separation distances, i.e. the distances at which natural gas pipelines can be safely sited near a community, could be considered as beneficial in terms of CO_2 pipeline risk analysis. These distances are defined in terms of the distances needed to protect against a specified heat flux from the fire. There have been limited studies to date concerning the determination of safety separation distances for CO_2 pipelines. Mahgerefteh et al. [18] performed the modelling of product loss during a pipeline rupture, which is an important factor related to the establishment of a safe separation distance therefrom. Comparison of CO_2 outflow data with those for the rupture of the same pipeline containing natural gas indicates a significantly greater amount of CO_2 released.
2. The pumps installed in booster stations along the pipelines to compensate for the pressure losses and elevation changes and to ensure a required flow of CO_2 would be able to be driven by gas turbines fuelled with natural gas, which would not require significant additional investments in gas pipeline or electricity infrastructure.
3. Linear infrastructure investment projects, such as roads, railway lines, telecommunication lines, electric power lines and pipelines for transmission of gas and petroleum products have a negative ecological effects in terrestrial and aquatic ecosystems. It is clear that the impacts associated with the use of existing infrastructure corridor would be quite different compared to those posed by newly built infrastructure.

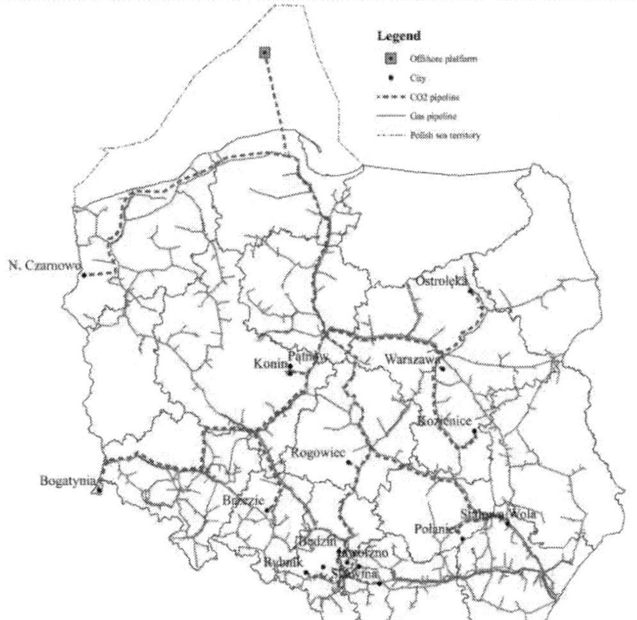

Fig. 3.1. Optimal infrastructure design including all power plants.

The capital cost of a pipeline project is largely a function of its diameter and length, although other factors, such as operating pressure, and various risk factors, are also significant. One of the challenging problems in pipeline infrastructure design calculations is the shortest path problem minimizing transport costs. Shortest path programming is a well-established subject in mathematics and computer science. Dijkstra's algorithm solves the single-source shortest-path problem on directed graph with a nonnegative edge weights. This algorithm was used as a subroutine in this work. However, Dijkstra's algorithm fails to form a tree that includes every specific vertex. Furthermore, optimal CO_2 pipeline infrastructure parallel to the natural gas transmission system do not necessarily should have shortest paths to the storage site due to the overlapped paths that could be used as trunk lines with reduced pipe material coverage. In fact, the priority level of this requirement can be very high as in the case of data transmission applications [19]. In this study we propose the following algorithmic approach to the preliminary design of CO_2 transmission system infrastructure running parallel to an existing natural gas pipeline infrastructure:

1. Determine shortest paths along existing natural gas pipelines from power plants to storage site and mark all power plants as unexamined.
2. Set the unexamined plant with the longest path as current plant, mark it as examined and save its value in a matrix of distances.
3. Find the first/next unexamined plant to the current plant by shortest path routing.
4. Shortest path from the unexamined plant to the storage site forms a subtree of the currently registered tree in a matrix of distances? Yes: Set it as current plant, mark examined and fulfil its value in a distance matrix; No: Mark the plant as examined and go to pt. 3
5. Are all the plants examined? Yes: Go to pt. 6., No: Go to pt. 3.
6. Matrix of distances completed? Yes: End; No: Mark all power plants with missing routes as unexamined and go to pt. 2. to continue with the new start point.

The results of optimal infrastructure design calculations for all power plants based on the above procedure are presented in Fig. 3.1. Given the topology of the network, appropriate diameter sizing calculations were performed using the steady state pipeline flow model from [20] with maximum flow velocity criterion of 20 m/s. The results are shown in Table 3.1 in the form of the corresponding flow weighted unit costs of the CO_2 transport infrastructure for each power plant. The total transportation distance and total CAPEX for this case scenario were 2544 km and 2 167.8 M€, respectively.

Table 3.1. Transport infrastructure CAPEX for individual power plants.

Power Plant	Total cost (M€)	Flow weighted unit cost EUR/(m^3/h)
Będzin	59.3	45 615
Bogatynia	230.4	100 193
Brzezie	149.9	65 514
Jaworzno	116.2	45 615
Konin	9.6	31 701
Kozienice	274.4	63 434
Łaziska Górne	103.2	45 615
N. Czarnowo	232.9	76 508
Ostrołęka	82.6	83 247
Pątnów I Pątnów II	78.6	30 578
Połaniec	207.5	74 679
Rogowiec	277.0	34 091
Rybnik	158.9	45 615
Skawina	45.6	45 615
Stalowa Wola	40.0	79 195
Trzebinia	59.1	45 615
Warszawa	42.3	44 489

DOI: 10.1051/e3sconf/20172200069

4 Optimal CO_2 transportation network for selected 2050 EC roadmap scenario

Mechleri et al. [21] presented an approach for the optimal design of CO_2 transport pipeline. They demonstrate that where CCS plants are deployed in an energy system characterised by high share of intermittent renewable energy, the resulting displacement of CCS power plant generation by renewable energy generation leads to a decade-on-decade declining flow of CO_2 through the transport infrastructure. This means that the right-sizing of CO_2 transport infrastructure requires the assumption of a reduced penetration of CCS in power generation without compromising the ability to accommodate future capacity. Therefore, the realistic scenarios should assume reduced penetration of CCS in power generation sector. In this study 32% penetration of CCS in power generation as predicted in [1] is considered.

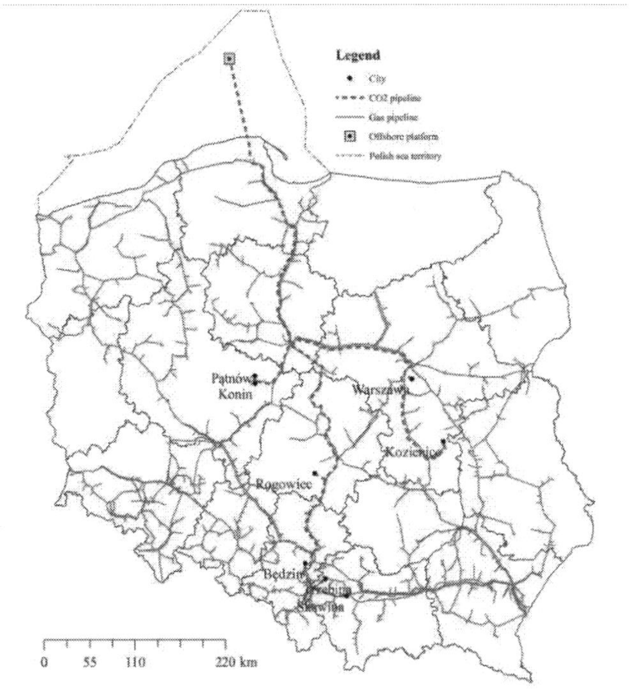

Fig. 4.1. Optimal CO_2 transport route regarding infrastructure CAPEX.

The selection of CO_2 sources was made by taking into account the flow weighted unit cost of transport infrastructure (per m^3/h) required for each power plant. Power plants with high costs were discarded and the choice was made among the ones with low values, having in mind that total power output of the selected units should contribute towards 32% in total national power generation capacity. Infrastructure design procedure described above was again employed for this scenario followed by diameter sizing calculations resulting in technical specification of the infrastructure as presented in Tables 4.1 and 4.2.

Annual amount of transported CO_2 is 168 460 305 m^3 which is 95 MT. Storage site capacity is ~900 MT [17] which gives for this case scenario 9 years of operational time under full load conditions of power plants. Total CAPEX is 612.6 M€.

 DOI: 10.1051/e3sconf/20172200069

Table 4.1. Infrastructure costs for specific branches.

Power Plant No.	CO_2 stream (m^3/h)	Length (km)	DN	Pipeline cost (M€)	Cost incl. assembly works (M€)
Main branch					
(16+12+11+1)+(9+15+18)+(2+14)	19 231	252	650	98.7	328.9
Skawina - Rogowiec branch					
16	675	41	150	3.6	12.1
12	1 022	21	200	2.5	8.4
12+16	1 697	2	250	0.3	1.0
11	1 258	5	200	0.6	2.0
16+12+11	2 955	130	300	24.0	79.9
1	8 127	15	450	4.0	13.3
16+12+11+1	11 082	31	500	9.2	30.8
Pątnów - Konin branch					
9+15	2 568	9	250	1.4	4.6
18	304	3	150	0.3	0.9
9+15+18	2 872	19	300	3.5	11.7
Kozienice - Ostrołęka branch					
2	4 326	150	350	30.6	101.9
14	951	11	200	1.3	4.4
2+14	5 277	19	350	3.9	12.9

Table 4.2. Power plants individual CAPEX.

Power Plant	Total cost (M€)	Flow weighted unit cost EUR/(m^3/h)
Rogowiec	174.9	4 415
Kozienice	186.5	26 005
Pątnów	58.9	5 850
Warszawa	23.0	7 060
Konin	7.3	6 973
Będzin	61.0	31 390
Trzebinia	56.9	38 605
Skawina	44.1	48 295

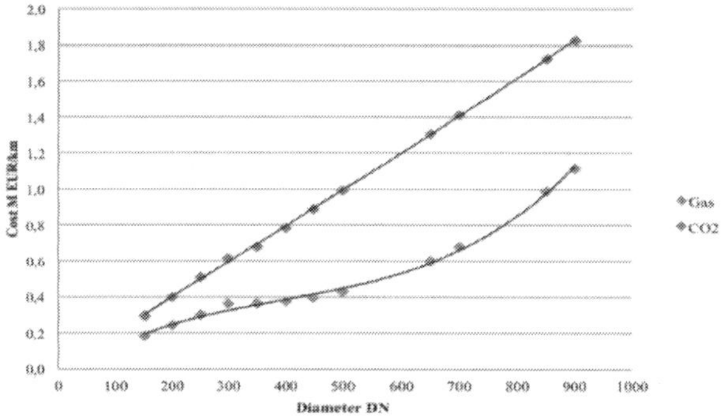

Fig. 4.1. Comparison between natural gas and CO_2 pipeline costs incl. assembly works.

The estimation of CO_2 pipeline infrastructure cost cannot be achieved based on natural gas infrastructure data. The difference lays in maximum operating pressure (MOP) of the pipelines, which in turn determine the pipe wall thickness. The CO_2 networks will be operated under much higher MOP compared to onshore natural gas networks currently operated by Polish transmission system operator, which leads to an increased material cost due to greater pipe wall thickness. These differences in costs are presented in Figure 4.1

DOI: 10.1051/e3sconf/20172200069

5 Summary

The operational time for selected geological structure is too short to consider it as a long-term delivery storage site despite fulfilling physical criteria for CO_2 storage. In order to accommodate future capacity additional onshore storage locations, or preferably supra-national scale offshore transport network as proposed in [22], should be considered. Future, detailed calculations of the pipeline options should be on a GIS platform with the help of 3D spatial analysis and consideration of terrain factors, and should also include the costs of compression facilities.

References

1. European Commission. *Energy Roadmap 2050* (2011)
2. M.M.J. Knoope, A. Ramírez, A.P.C. Faaij, Int. J. Green. Gas Cont. **16**, 241–70 (2013)
3. M.K. Chandel, L.F. Pratson, E. Williams, Energy Convers. Manage. **51**, 12, 2825–34 (2010)
4. G.F. Weihs, D.E. Wiley, Int. J. Green. Gas Cont. **8**, 150–168 (2012)
5. U. Zahid, U. Lee, J. An, Y. Lim, C. Han, Environ. Prog. Sust. Energy **33**, 978–92 (2014)
6. J.Y. Jung, C. Huh, S.G. Kang, Y. Seo, D. Chang, Appl. Energy **111**, 1054–60 (2013)
7. T. Suzuki, M. Toriumi, T. Sakemi, N. Masui, S. Yano, H. Fujita, H. Furukawa, Energy Proced. **37**, 2989–2996 (2013)
8. P. Jain, K. Pathak, S. Tripathy, Energy Proced. **37**, 3233–3241 (2013)
9. N. Elahi, N. Shah, A. Korre, S. Durucan, Energy Proced. **63**, 2655–62 (2014)
10. X. Luo, M. Wang, E. Oko, C. Okezue, Appl. Energy **132**, 610–20 (2014)
11. J. Hetland, J. Barnett, A. Read, J. Zapatero, J. Veltin, Energy Proced. **63**, 2458–66 (2014)
12. B. Bai, X. Li, Y. Yuan, D. Zhou, P. Li, Acta Geotech. **9**, 1, 115–26 (2014)
13. C. Han, U. Zahid, J. An, K. Kim, C. Kim, Curr. Opin. Chem. Eng. **10**, 42–8 (2015)
14. M. Chaczykowski, A.J. Osiadacz, Int. J. Green. Gas Cont. **9**, 446–56 (2012)
15. B. Wetenhall, J.M. Race, M.J. Downie, Int. J. Green. Gas Cont. **30**, 197–211 (2014)
16. M. Witek, J. Nat. Gas Sci. Eng. **27**, 374–84 (2015)
17. Polish Geological Institute. *Assessment of Formations and Structures Suitable for Safe CO_2 Geological Storage Including the Monitoring Plans. Final report.* (2013) (in Polish)
18. H. Mahgerefteh, G. Denton, Y. Rykov, Institution of Chemical Engineers Symposium Series **154**, 869 (2008)
19. A. Ghasemkhani, A. Anvari-Moghaddam, J.M. Guerrero, B. Bak-Jensen, *Proceedings of IEEE PES Innovative Smart Grid Technologies* (ISGT, 2016)
20. M. Chaczykowski, A.J. Osiadacz, Arch. Min. Sci. **57**, 1, 23–38 (2012)
21. E. Mechleri, S. Brown, P.S. Fennell, N. Mac Dowell, Chem. Eng. Res. Des. **119**, 130–9 (2017)
22. R.J. Stewart, V. Scott, R.S. Haszeldine, D. Ainger, S. Argent, Gren. Gases: Sci. Technol. **4**, 4, 481–94 (2014)

Towards energy neutrality of wastewater treatment plants via deammonification process

Kamil Janiak[1,2,*], *Andrzej* Łojek[1], and *Mateusz* Muszyński-Huhajło[1]

[1]Wroclaw University of Technology, Faculty of Environmental Engineering, Wybrzeże Wyspiańskiego 27, 50-370 Wroclaw, Poland
[2]Municipal Water and Sewage Company, ul. Na Grobli 14/16, 50-421 Wroclaw, Poland

Abstract. Energy neutrality of wastewater treatment plants is possible with constant and consistent optimization and implementation of new technologies. In recent years new process called deammonification has been discovered and implemented in treatment of side streams rich in nitrogen. With its implementation on wastewater treatment plants it is possible to remove nearly all nitrogen from side stream (even 30% of overall nitrogen load) in less energy consuming way. Additionally, thanks to lower nitrogen load to main stream reactors it is possible to optimize them to further lower energy consumption. This article presents simulation studies of deammonification implementation and main stream reactor optimization in case of medium Polish WWTP (115 000 p.e.). With removal of 20% of nitrogen in side stream via deammonification and subsequent main line optimization it is possible to save 5000 euro/year by lowering sludge retention time, oxygen concentration in main stream reactors. When additional COD is precipitated in primary clarifiers with iron coagulants, 55 000 euro/year can be saved in case of energy costs which states for most of the energy costs. However, when coagulant and disposal costs are included savings are on the level of 25 000 euro/year.

1 Introduction

Wastewater sector use 1%–3% [1–3] of total energy used by industry. Energy consumption states for 11–17% of overall costs of wastewater treatment [4]. The exact percentage depends on the size of the plant: lower energy participation is typical for large WWTPs (> 100 000 p.e.), therefore energy neutrality is goal worth achieving. In municipal wastewater treatment, average energy consumption needed to meet strict requirements is at the level of 0.5–0.8 kWh/m^3 [5, 6] while chemical energy stored in wastewater is 5–6 times higher [7]. Additionally, large amounts of energy are also stored in potential, kinetic forms [8]. This implies that wastewater treatment plants can be energy self-sufficient.

To achieve energy neutrality:

a) WWTP must be equipped with sludge anaerobic digestion and efficient method for carbon removal from wastewater to prevent its oxidation in biological part, such as chemically assisted primary sedimentation (CAPS). This is necessary for efficient

* Corresponding author: kamil.janiak@pwr.edu.pl

energy production and low energy use by aeration.

b) Important energy consuming processes must be well optimized for minimal energy use [9].
c) Wastewater must be rich in biodegradable carbon to enable high carbon load to anaerobic digestion without carbon deficiency for nutrient removal.

Latter condition is the biggest obstacle as typical wastewater doesn't contain excess carbon and many WWTPs experience carbon limitations even without elevated carbon removal in primary sedimentation [10]. Removal of nitrogen from digester liquor via deammonification process opens new field for WWTP energy consumption optimization in main line. Firstly, lower N concentration in wastewater directed to activated sludge process allows for lower oxygen concentration and lower sludge age, hence lower energy consumption. Secondly, as higher COD/N ratio is achieved, larger COD load can be removed in primary clarifiers without negative consequences for denitrification but with positive consequences for energy balance.

1.1 Aim and structure of paper

Aim of this paper is to present potential for energy balance improvement in main line thanks to implementation of deammonification process in side-stream. Presented studies are based on simulations of Polish WWTP. In order to quantify effects of deammonification implementation and further optimization of main line, three states of WWTP were simulated:

a. current state – WWTP without side-stream deammonification – this is real state of existing WWTP. For this state mathematical model was calibrated.
b. WWTP with side stream deammonification – this is fictional state and is represented in scenario 1.
c. WWTP with side stream deammonification, CAPS and main line optimization – this is fictional state and is represented in scenarios 2–5.

2 Materials and methods

2.1 WWTP and wastewater characteristics

Information about plant current state and wastewater characteristics were collected during model calibration and verification and are based on real experimental data. N-removal efficiency and effluent composition for partial nitritation-Anammox (PN-Anammox) process were gained during operation of this processes in a pilot-scale installation treating real reject water from sludge dewatering. Effluent composition, oxygen demand and operational costs for mainstream processes were calculated based on simulation results or literature reports if mentioned.

2.1.1 Plant description: current state and assumed Scenarios.

Existing, medium-size municipal Polish WWTP (115 000 p.e.) with calibrated ASM1 model was chosen for this study. Plant is composed of two A2O reactors with two primary clarifiers and two secondary clarifiers as well as anaerobic digestion reactors producing methane which is further used for energy production in CHP (combined heat and power) units. Digested sludge is being dewatered and transported to utilization facility. Mentioned model represents only biological part of the plant, which contains two parallel anoxic-oxic reactors and secondary clarifiers, excluding primary sedimentation and all sludge handling processes as well as deammonification process. Real operational parameters used as background for performed analysis are presented in table 1. Detailed model description is presented in section 2.2. Characteristics of wastewater after mechanical treatment and reject water are presented in table 2.

Total nitrogen (TN) for considered WWTP is set by Polish legislation at 10 g N/m^3 without ammonium nitrogen limitation [11]. SRT (sludge retention time) is adjusted according to temperature, at 10°C or 15°C required SRT is 30 d and in 20°C SRT is 15d. Currently, plant has major problems to meet TN criterion as its average concentration in the effluent is very close to the limit. Dissolved oxygen (DO) level in the aeration zone is kept at 1.5 mg O_2/L which is considered optimal in these conditions.

Nitrogen load in reject water stream from sludge dewatering was assumed 22.7% of daily N-load for whole plant, which was close to real value.

Table 1. Selected WWTP parameters.

Parameter	Unit	Value
SRT	d	15–30
Aeration zone volume	m^3	13 811 (HRT* ≈ 1 d)
Anoxic zone volume	m^3	7944 (HRT ≈ 0,57 d)
Internal recirculation	m^3/d	≈150 000 (≈ 1090% Q_i)
DO concentration	mg O_2/L	1.5
CHP electric efficiency	%	36

* - hydraulic retention time

Five fictional scenarios of possible plant development towards energy neutrality were investigated assuming side-stream deammonification process implementation (20% of N-load removal) and improved COD removal by CAPS resulting in higher gas production. As a background for each Scenario implementation, plant's state after technological optimisation was used. Following scenarios were analysed:

- Scenario 1 – side-stream deammonification implementation, no CAPS
- Scenario 2 – side-stream deammonification implementation, CAPS enhancing COD removal in primary clarifiers by 20%
- Scenario 3 – side-stream deammonification implementation, CAPS enhancing COD removal in primary clarifiers by 30%
- Scenario 4 – side-stream deammonification implementation, CAPS enhancing COD removal in primary clarifiers by 35%
- Scenario 5 – side-stream deammonification implementation, CAPS enhancing COD removal in primary clarifiers by 40%

In scenarios that assumes implementation of side-stream treatment of reject water, two-stage PN-Anammox technology was chosen. Process characteristics and efficiency were collected by authors during over 6-month stable operation of a pilot-scale installation treating real reject water. In this period, average TN removal reached 88%. These data were used in simulation studies (see Table 2).

Scenarios 2–5 assumed addition of inorganic coagulant ($FeCl_3 \cdot 6H_2O$) and anionic polyelectrolyte to enhance primary sedimentation. Such action increases amount of COD directed to anaerobic digestion with primary sludge stream and improve gas production, therefore WWTP energy balance. Coagulant dosage and COD removal efficiency was assumed at the same level as presented by De Feo et al. [12]. For 20%, 30%, 35%, 40% improvement in COD removal, ferric chloride doses were assumed 16.3, 32.5, 51.3 and 98.9 mg/L respectively. Anionic polyelectrolyte dose was the same in all Scenarios and assumed 0.1 mg/L. Coagulant and polyelectrolyte prices in Poland were 71.7 €/ton and 2 500.0 €/ton respectively.

DOI: 10.1051/e3sconf/20172200070

2.1.2 Wastewater characteristics

Characteristics of wastewater after mechanical treatment were real data collected during normal plant operation. Raw and treated reject water were average data collected during operation of mentioned pilot-scale deammonification installation. In scenarios considering deammonification and/or CAPS implementation, influent composition to biological reactor was calculated based on mentioned real data and assumed level of deammonification and/or precipitation efficiency. Selected medium characteristics are summarized in table 2.

Table 2. Characteristics of wastewater after mechanical treatment (Wastewater – real data, Scenario 1 and Scenario 4 - calculated) and raw/treated reject water (real data).

Parameter	Unit	Wastewater	Reject water	PN/Anammox effluent	Scenario 1	Scenario 4
Flow	m^3/d	14 000	272	272	14 000	14 000
TN	$g N/m^3$	64	748	90	51	48
COD	$g O_2/m^3$	531	552	313	526	316
BOD_5	$g O_2/m^3$	305	170	17	302	181

2.2 ASM1 model description and calibration

A calibrated ASM1 model, representing municipal WWTP described in previous sections, was used in presented study to represent only biological part of the plant excluding primary sedimentation. This tool allowed to model autotrophic and heterotrophic reaction, with a facultative consumption of oxygen or nitrate as electron acceptor, without phosphorus removal [13]. Model was calibrated based on collected operational data and intensive measuring campaign (two weeks) under steady-state operating conditions. Model accuracy was satisfactory, despite quite poor results for nitrate concentrations, and could be used in this study. All simulations were performed for three representative temperatures: 10°C, 15°C and 20°C.

2.3 Cost calculations

To evaluate potential savings, a WWTP energy balance was created for each Scenario as for the plant current state. As model considered only processes related to carbon and nitrogen removal, energy balance was limited to energy consumption for mainstream aeration, side-stream deammonification and energy production in CHP from surplus sludge generation thanks to CAPS. In Scenarios including CAPS also coagulant and polyelectrolyte costs were included. In all cases, economical balance included amount of produced sludge and its disposal costs. Assumed energy cost was 0.093 €/kWh, which is typical value for industry in Poland.

Mainstream aeration costs were calculated based on real data of existing aeration system and oxygen demand calculated using available model. Energy use for aeration was calculated based on reactor geometry, αFSOTE, diffuser and blower characteristics (all these values were real data determined experimentally [data not published]). Energy consumption for aeration in deammonification process was assumed 0.8 kWh per kilogram of nitrogen is required in this process, as calculated by Wett [14].

Energy production in CHP units were crucial element of WWTP energy balance. Specific methane production rate was assumed 0.35 L_N per 1 g COD anaerobically degraded in digestion process [15]. Specific energy production from and CHP efficiency were assumed 3 kWh/m^3 CH_4 and 36% respectively. Additional mass of chemical sludge due to precipitation process use were calculated based on coagulant characteristics. Digested sludge disposal costs were assumed 23.3 €/ton, which is a typical value.

3 Results

3.1 Current state

Simulations results confirmed that plant has major issue with meeting the legal requirements considering effluent TN concentration (Figure 1). At cold temperatures 10°C, TN exceed limit value and reach 11.9 mg N/L. As temperature rises, TN level drops slightly below allowed value and is 9.9 and 9.7 mg N/L at 15°C and 20°C respectively. Majority of effluent total nitrogen concentration was N-NO_x fraction (~70%) as the result of insufficient carbon availability for denitrification process. No problems with nitrification process were noticed. DO concentration (1.5 mg O_2/L) was enough to oxidize most of ammonium as its concentration was around 1.0 mg N/L. Current average energy demand for mainstream aeration is 81 kWh/day which generates an annual cost of around 66 500 € (Figure 4).

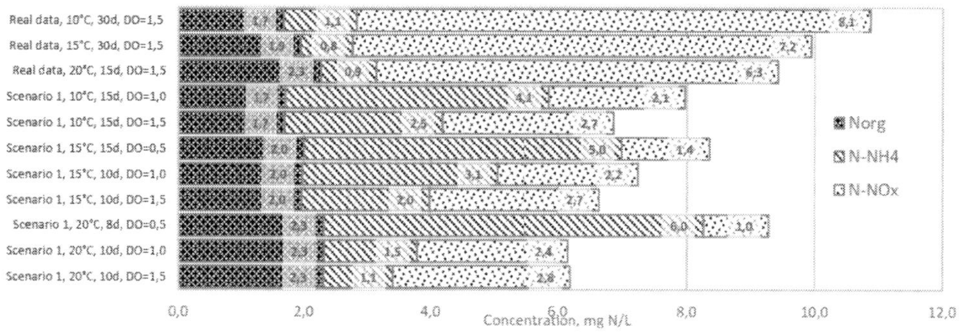

Fig. 1. Nitrogen concentration in WWTP effluent – current state and Scenario 1: implementation of side-stream deammonification.

3.2 Scenario 1 – deammonification implementation and further optimization

In this Scenario, plant was virtually equipped with a side-stream deammonification installation. As the mainstream reactor N-load has decreased by 20%, effluent quality was better and further plant optimization potential was released. In a series of simulations, various SRT and DO values were tested in whole temperature range to pick mainstream reactor setup providing lowest aeration costs and meeting TN limit. In coldest period, SRT equal to 15 days and DO = 1.0 mgO_2/L allowed to decrease TN concentration in the effluent to 7.9 mg N/L. In 15°C (SRT 10 days, DO = 1.0 mgO_2/L) and 20° (SRT 8 days, DO=0.5 mgO_2/L) TN level was 7.3 and 9.3 mg N/L respectively (Figure 1). Average energy demand for aeration (mainstream + side-stream) in optimized setup is 75 kWh/day which yearly cost around 61 100 € (Figure 4). This scenario is not best solution in terms of energy balance and cost balance (figure 3 and 5.).

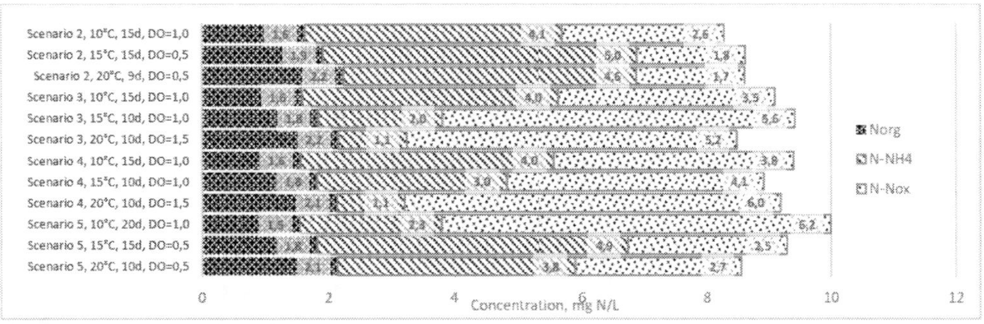

Fig. 2. Nitrogen concentration in WWTP effluent – Scenarios 2–5: side-stream deammonification, main line optimization and CAPS.

3.3 Scenario 2 - 5 – deammonification + CAPS and further optimization

In this Scenarios, beside deammonification implementation, also CAPS was used to increase energy production. Mainstream operational setup was optimized separately for each Scenario as presented in section 3.2. Figure 2 represents nitrogen concentrations in effluent for optimized parameters in each variant.

As additional chemical treatment was used, except energy cost for aeration, also chemicals price was added alongside with chemical sludge disposal costs (figure 4). Savings due to higher energy production were also included. All these values varied depending on coagulant dosage and amount of additional COD removed in primary clarifiers.

Energy calculations showed that Scenario 5 provides biggest savings in case of energy purchase (figure 3), while overall cost calculations, revealed that Scenario 3 allowed to achieve the best economical result, as appropriate balance between SRT, DO and enhanced COD removal was achieved (figure 5).

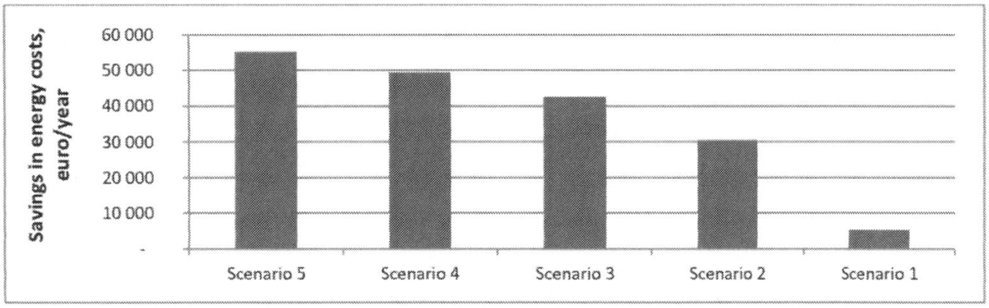

Fig. 3. Savings in energy costs in different scenarios (current state as reference, presented values are mean values from all scenarios variants).

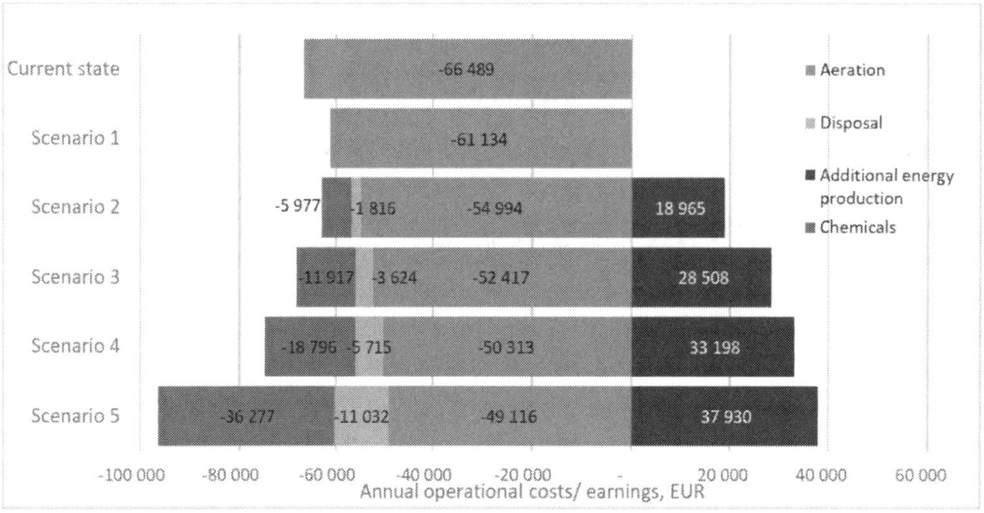

Fig. 4. Overall costs balance (presented values are mean values from all scenarios variants).

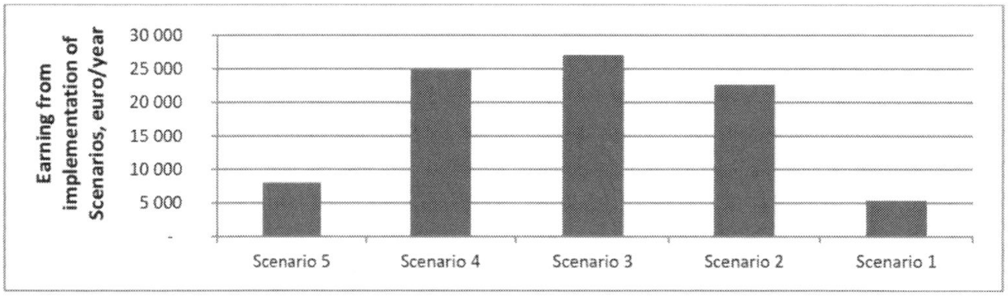

Fig. 5. Overall cost balance (current state as reference, presented values are mean values from all scenarios variants).

4 Conclusions

Following conclusions can be made:

- implementation of side stream deammonification changes wastewater composition through removal of important load of nitrogen and this opens potential for WWTP energy and cost balance optimization.
- with 20% of total nitrogen removed in deammonification, 40% additional COD can be precipitated in primary settlers in addition to COD removal without coagulants (70% overall COD removal efficiency).
- With side stream deammonification and without CAPS (Scenario 1) optimization allows for moderate savings (slightly higher that 5 000 euro/year) in energy and costs. This savings are due to lower SRT and oxygen concentrations.
- Implementation of CASP allows for much higher savings. In case of additional 40% COD removal (Scenario 5) 55 000 euro/year can be saved in energy costs. This Scenario is most energy efficient. However high doses of coagulants are required and overall savings including costs of coagulant and sludge disposal are on the level of 7 500 euro/year.
- Scenario 3 is most profitable in terms of overall costs. Overall savings in this Scenario are on the level 25 000 euro/year.

5 Assumptions and limitations

Research presented in this paper has few limitations and some assumptions had been made:

1. Phosphorus removal was not included in model. It was assumed that phosphorus will be removed biologically in every case despite high removal of organic matter in primary settlers. This assumptions was based on low phosphorus concentration in raw wastewater and steady state calculations [not presented]. Lack of phosphorus modelling has no influence on energy balance.
2. It was assumed that change in composition of wastewater due to deammonification implementation will have no influence on kinetic parameters of biomass.
3. It was assumed that all precipitated COD (scenarios 2–5) will be converted anaerobically.
4. It was assumed that pH won't change significantly due to coagulation because wastewater contains enough buffer capacity.

The authors gratefully acknowledge co-funding from the National Centre for Research and Development (grant no PBS2/B9/25/2014) and Municipal Water and Sewage Company Wroclaw.

References

1. Water Research Fundation and Electric Power Research Institute, *Electricity Use and Management in the Municpial Water Supply and Wastewater Industries* (2013)
2. EPA Office of Water *Wastewater Management Fact Sheet, Energy Conservation p.7* (2006)
3. T.P. Curtis, *Environmental Microbiology* (2nd edition Mitchell, R. Gu, J.D., Eds. Wiley-Blackwell: Hoboken, NJ, 2010)
4. J. Haslinger, S. Lindtner, J. Krampe, Wat. Sci. Tech **74**, 11 (2016)
5. J. Foley, D. de Haas, K. Hartley, P. Lant, Wat. Res. **44**, 1654–1666 (2010)
6. G. Venkatesh, H. Brattebo, Energy **36**, 792–800 (2011)
7. D.J. Batstone, T. Hulsen, C.M. Mehta, J. Keller, Chemosphere **140**, 2–11 (2015)
8. C. Power, A. McNabola, P. Coughlan, Sustaine. Energy Technol. Assess **7**, 166–177 (2014)
9. O. Nowak, S. Keil, C. Fimml, Wat. Sci. Tech. **64**, 1 (2011)
10. Y. Zhang, X.C. Wang, Z. Cheng, Y. Li, J. Tang, Chemosphere **144**, 689–696 (2016)
11. Rozporządzenie Ministra Środowiska z dnia 18 listopada 2014 r. w sprawie warunków, jakie należy spełnić przy wprowadzaniu ścieków do wód lub do ziemi oraz w sprawie substancji szczególnie szkodliwych dla środowiska wodnego (in Polish)
12. G. De Feo, S. De Gisi, M. Galasso, Desalination **230**, 229–238 (2008)
13. M. Henze, W. Gujer, T. Mino, M. Loosdrecht, IWA Scientific and Technical Report No. 9, (IWA, 2000)
14. B. Wett, Wat. Sci. Tech. **53**, 12 (2006)
15. S. Salem, D. Berends, J. Heijnen, M. Van Loosdrecht, Wat. Res. **37**, 1794–1804 (2003)

Study of heat exchange processes in a screw ash cooler installation

Barbara Janowska[1,*], *Paweł* Regucki[1], *Artur* Andruszkiewicz[1], and *Wiesław* Wędrychowicz[1]

[1]Wrocław University of Science and Technology, Faculty of Mechanical and Power Engineering, Wybrzeże Wyspiańskiego 27, 50-370 Wrocław, Poland

Abstract. The paper presents an example of the thermal-flow analysis of a bottom ash cooler cooperating with a circulating fluidized bed (CFB) boiler. The authors present the methodology of the conducted studies and examples of thermal balance calculations basing on measurement results. The comparison of two different ash coolers indicates a significant relationship between the outlet temperatures of the ash and cooling water depending on the construction and arrangement of the analysed ash coolers. The analysis presented in this paper enables determining the optimal rotation speed of the screw in order to achieve the required temperature of the slag depending on the actual working parameters of the CFB boiler. It provides an opportunity to optimize the heat exchange conditions in order to obtain the highest acceptable water temperature at the outlet of the ash cooler while adhering to safety regulations. The hot water could further be used for internal sanitary purposes or to support heating systems operating in the power plant.

1 Introduction

Conventional and thermal power plants are the major producers of electricity and heat in the world power engineering sector. Their period of operation, which is often calculated at 30–40+ years, has a significant impact on the natural environment, as well as on the health and functioning of local communities [1]. The long utilisation period under varying heat-flow conditions signifies that power units require continuous maintenance and improvement of exhaust gas purification technologies [2].

Campbell et al. in [3] have stated that the efficiency of coal-fired power plants decreases with age. Although the continuity of maintenance practices can keep a power plant's efficiency high in the early years of operation, the long-term utilization of the boilers and their auxiliary systems leads to the gradual degradation of their performance as a result of e.g. corrosion or deposition of heat exchange surfaces, leakages in water-steam systems or the deterioration of pump characteristics.

* Corresponding author: barbara.janowska@pwr.edu.pl

On the other hand, the deterioration of the overall efficiency of power units manifests in the increased consumption of fossils, which has a direct influence on the increase of the production of wastewater, waste heat, as well as greenhouse gases and other pollutants.

Henderson in [4] has stated that the improvement of the efficiency of a large number of older coal- and oil-fired power plants operating around the world would provide major savings in CO_2 emission. In order to improve the overall efficiency of old power units, Ryabchikov et al. in [5] have suggested a number of specific actions connected with e.g. the raising of heat exchanger efficiency, the retrofit of hydraulic systems and the improvement or sustenance of their reparability and maintainability. Błaszczyk et al. in [6] have found that the overall efficiency of a power plant encompasses the efficiencies of the various power unit components. Among others, they have pointed out that heat loss minimization is the most important factor affecting the loss of power plant efficiency, and there are many areas of potential heat loss in a power plant.

These observations and comments have led to the Authors focusing their attention on an analysis of the thermal-flow processes taking place in ash coolers. Ash coolers are used for the preliminary cooling down of the slag produced in CFB boilers. As a result of combustion processes, hot slag is produced and then must be removed from the combustion chamber and pre-cooled before final storage. Moreover, the mass flow rate of extracting slag is used to control the pressure in the dense fluidized layer above the primary air nozzle grid in the furnace of the CFB boiler. The mass flow rate of the discharged slag can vary from a few to several kg/s, depending on the types of boilers and their nominal power, as well as on the mass flow rate and technical analysis of the fuel. For technical reasons, the slag transported from the boiler must be cooled down to a temperature of about 100–150°C. The construction, number and location of the ash coolers are closely related to the combustion chamber type and are important elements of any power boiler [7]. Studies and measurements indicate that the ash coolers are a significant source of waste heat [8]. Nichols et al. in [9] have estimated that the potential efficiency improvement for power plants in the USA resulting from the revitalization of the ash removal system would be at a level of 0.1 %.

In the relevant literature, different types of fluidized bed ash coolers [10, 11] have been analysed, together with an estimation of their influence on the thermal economy of the power unit [12]. In the presented paper, two different types of ash coolers cooperating with a CFB boiler are considered. Following a discussion of the study's methodology, examples of thermal balance calculations basing on the measurement results are analysed. This enables determining the optimal heat exchange conditions required to achieve the necessary temperature of the ash depending on the actual working parameters of the boiler.

2 Construction of bottom ash coolers

2.1 Two screw bottom ash cooler

Among the many types of bottom ash coolers utilized in power plants operating with a CFB boiler, the solution based on two screws is very popular. In the considered case, due to security reasons, the circulating fluidized bed boiler is equipped with two independent bottom ash coolers (called SAC-L and SAC-R, respectively, for the left and right bottom ash cooler). Figure 1 provides a schematic image of the most important elements of a SAC-R ash cooler – two screws designated by the acronyms SC-L and SC-R on the left scheme. The right part of Figure 1 presents a side view of the bottom ash cooler with an indication of the hot slag inlet (1), the outlet of the cooled slag (6), the inlet and the outlet of the cooling water flowing through the shaft and blades of the SC-L screw (respectively (2) and (3)), the inlet and the outlet of the cooling water which flows between the walls (casing) of the ash cooler (respectively (4) and (5)).

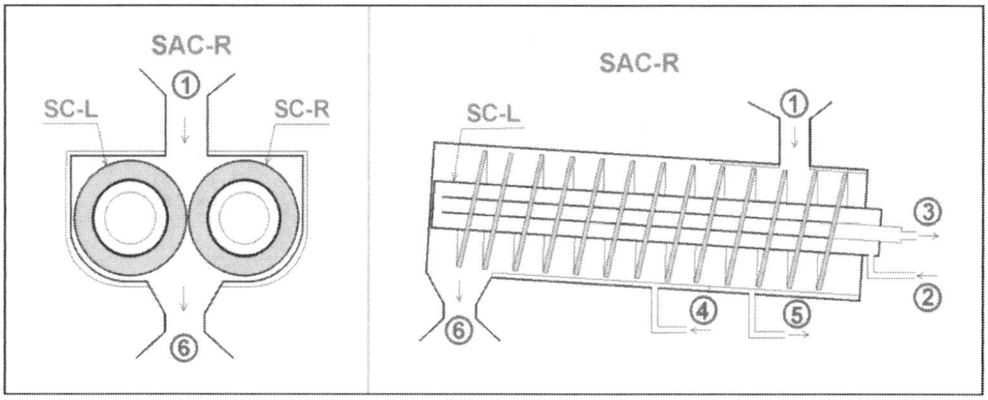

Fig. 1. Scheme of the bottom ash cooler (SAC-R) with internal cooling water distribution: two screws (SC-L and SC-R), the inlet and the outlet of the slag (1 and 6 respectively), the inlet and the outlet of the cooling water (2, 4 and 3, 5 respectively).

The hot slag is delivered directly from the combustion chamber (1), after which it is split between two screw coolers (SC-L and SC-R), and then transported to point (6). The mass flow rate of the removed slag is controlled by the rotation speed of the screws. During its movement along the screw, the slag is gradually cooled by the cooling water circulating in the blades, the shaft and between the walls (the casing). The mass flow rates of the cooling water are constant and independent of the rotation speed of the screws.

2.2 Bottom ash cooler with a single screw

The other construction of the bottom ash cooler is based on a single screw solution. The geometrical parameters of this type of ash cooler are smaller than in the above-presented case of the two screw ash cooler. Thus, single screw ash (SSA) coolers are usually arranged in a set of a few units in order to satisfy the thermal requirements of the CFB boiler. Such a solution provides the additional possibility of removing the hot ash from selected locations in the combustion chamber's bottom. The scheme of a bottom ash cooler with a single screw is presented in Figure 2.

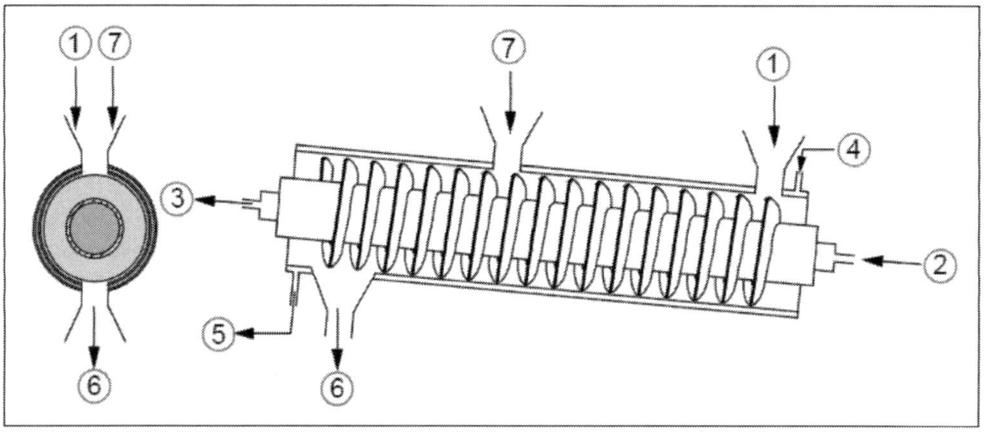

Fig. 2. Scheme of a bottom ash cooler with a single screw: the inlets and the outlet of the slag (respectively 1, 7 and 6), the inlets and the outlets of the cooling water (respectively 2, 4 and 3, 5).

This type of bottom ash cooler has two hot slag inlets: at the beginning and in the middle of the screw. Under normal operational conditions, only the inlet at the beginning of the screw is used, but sometimes the hot ash is delivered also through the middle inlet, connected with Intrex-type superheaters by ash removal channels. The volumetric flow rates of the cooling water circulated in screw ash coolers are constant over time. An example of the arrangement of eight SSA coolers is shown in Figure 3.

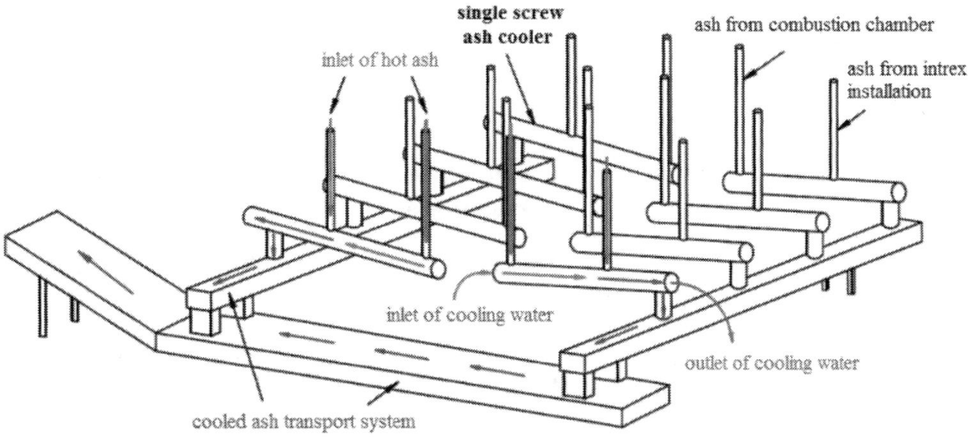

Fig. 3. Scheme of eight single ash coolers linked with a CFB boiler.

3 Measurements and comments

The analysis of the thermal flow processes occurring in the bottom ash coolers presented in this paper is based on measurement data collected on real objects. The volumetric flow rates were measured by an ultrasonic flowmeter (Microsonic Portaflow330), while the thickness of the pipe walls using a thickness gauge (Metrisonic SONO M610). Three different methods were applied to conduct the temperature measurements:

- the PT100 contact thermometer,
- the LaserLiner ThermoSpot-Vision optical pyrometer,
- the TESTO 875-2 infrared camera.

The total heat flux transferred from the hot ash to the cooling water was calculated using the following formula (in kW):

$$Q = c_w \sum_k \Delta t_{w,k} \cdot q_{v,k} \cdot \rho_w = c_a \cdot \Delta t_a \cdot q_{m,a} \qquad (1)$$

where: c_w and c_a is the specific heat of the water and the ash, respectively (in kJ/kg·K); Δt_w and Δt_a – the temperature difference of the cooling water and the ash, respectively, calculated at the inlets and outlets of the ash cooler; $q_{v,k}$ – the volumetric flow rates of the cooling water flowing through the screws and outer walls, respectively (in m^3/s); $q_{m,a}$ – the mass flow rate of the ash (in kg/s); ρ_w – the density of the water calculated for individual temperatures of the cooling water.

In all the calculations, the values of specific heats were used, equal to c_a = 0.958 kJ/kg·K and c_w = 4.190 kJ/kg·K. The value of the water density was calculated for the particular temperature of the cooling water and oscillated between 998.2 and 971.8 kg/m^3 (for t_w equals 20 and 80°C, respectively).

3.1 Two screw bottom ash cooler

Table 1 presents the mean temperatures of the cooling water and ash measured at the outlets of the SAC-R ash cooler versus the ash load of the cooler L_{ash}. The symbols $t_{out,SC-L}$ and $t_{out,SC-R}$ denote the outlet temperatures of the water from the left and right screw coolers, respectively (see Fig. 1), and $t_{out,wall}$ – the outlet temperature of the water flowing between the outer walls (casing). The symbols $t_{in,ash}$ and $t_{out,ash}$ denote the inlet and the outlet temperatures of the ash, respectively. The ash load L_{ash} is expressed in the percentage of the nominal (design) load of ash cooler.

Table 1. Mean temperatures of the cooling water and ash measured at the outlets of the SAC-R ash cooler versus the ash load of the cooler L_{ash}.

ash load L_{ash}, %	rotation speed of screw ω, rpm	outlet temperature of cooling water, °C			temperature of ash, °C	
		$t_{out,SC-L}$	$t_{out,SC-R}$	$t_{out,wall}$	$t_{in,ash}$	$t_{out,ash}$
20	2.04	35.2	35.5	40.6	391.9	69.7
40	4.08	42.5	41.6	47.3	515.3	82.3
60	6.12	47.0	46.0	53.9	581.3	130.7
90	9.18	51.7	52.4	70.8	592.8	209.8

The volumetric flow rates $q_{v,SC-L}$, $q_{v,SC-R}$ and $q_{v,wall}$ were constant over time and equaled 19.56, 18.91 and 13.25 m^3/h, respectively. The inlet temperature of the cooling water fluctuates in a range of between 31.8 and 33.0°C. Comparing the outlet temperatures of the cooling water, one can note that the highest are obtained in the flow between the outer walls of the cooler. Despite this, the mean temperature of the water leaving the ash cooler, calculated over the volumetric flow rates, oscillates at between 38.7 and 51.1°C (for 20 and 90% of the ash load, respectively). Such a range of temperatures is useless for direct utilisation in sanitary or auxiliary installations. This means that either the heat taken by the cooling water in the ash cooler is released in the cooling towers or the warm water is directed to an additional heating system where its temperature will raise up to ~ 90°C. It is worth noting that for 90% of the ash load the outlet temperature of the ash exceeds the technological limit of 150°C, which can be dangerous for the ash transport installation.

Table 2 presents the mean values of the heat flux transferred from the hot ash to the cooling water flowing through the left and right screw coolers and outer walls.

Table 2. Mean values of the heat flux transferred from the hot ash to the cooling water flowing through the left and right screws of the SAC-R cooler and outer walls.

ash load L_{ash}, %	rotation speed of screw ω, rpm	heat flux to cooling water, kW				heat flux from hot ash, kW
		through SC-L	through SC-R	through outer walls	total	
20	2.04	65.5	61.7	123.8	251.0	198.2
40	4.08	215.3	243.1	238.5	697.0	569.0
60	6.12	308.8	340.9	336.0	985.8	903.0
90	9.18	426.1	423.5	580.0	1429.7	1219.9

Comparing the data from Table 2, one can find that the relative error between the heat fluxes calculated from formula (1) for the cooling water and the ash ranges from between 8 and 21%. This discrepancy may be due to the fact that a constant value of the specific heat c_a was applied to the calculations. It seems that in this case the value typically quoted in the

 DOI: 10.1051/e3sconf/20172200071

relevant literature should be 10% higher. It is worth noting that c_a is strongly dependent on the mineral composition of the burned fuel.

3.2 Single screw ash cooler

The thermal-flow measurements presented in this subsection were made on a single screw ash (SSAC) cooler (the left one marked in red and blue in Figure 3). In this construction of the ash cooler, the cooling water flows inside the screw and the outer cover of the ash cooler according to the slag transport direction. Figure 4 shows the relationship between the outlet temperatures of the cooling water from screw $t_{w,s}$ and outer cover $t_{w,c}$ for the SSAC cooler and the L_{ash} ash load. The left diagram refers to a situation in which the additional inlet of hot ash in the middle of the screw is closed (closed Intrex installation), while the right one refers to a case involving an open Intrex installation. The volumetric flow rates of water through the screw and outer cover amounted to 3.90 and 6.87 m^3/h, respectively. The inlet temperature of the cooling water for the left and right diagrams was constant, equaling 29.6 and 28.7°C, respectively.

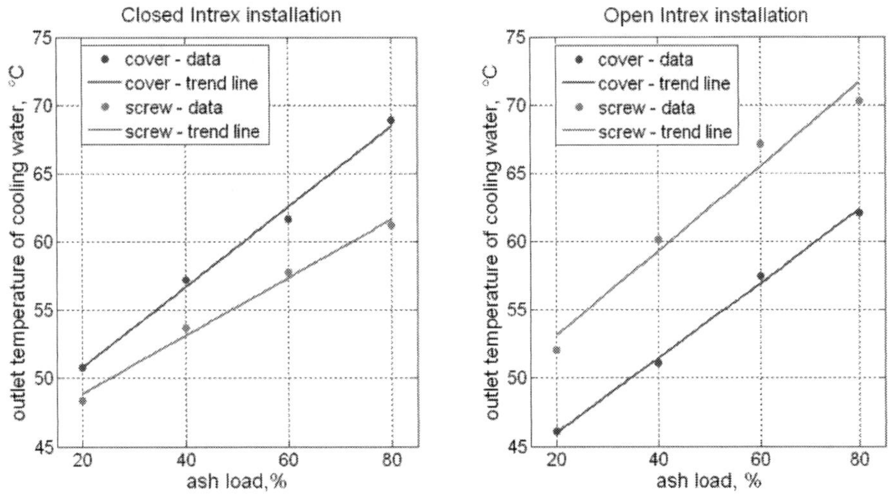

Fig. 4. Outlet temperatures of the cooling water from screw $t_{w,s}$ and outer cover $t_{w,c}$ for the SSAC cooler versus ash load L_{ash}. The left diagram refers to a closed Intrex installation, while the right one refers to an open Intrex installation.

The presented results indicate that higher outlet temperatures of the cooling water are obtained in an open Intrex installation (the right diagram). This is due to the additional portion of hot ash added in the middle of the screw. Despite this, the difference between the mean outlet temperatures in both cases is rather small, amounting to ~ 2°C. An interesting observation involves the fact that for the open Intrex installation higher temperature values are obtained for cooling water flowing inside the screw, in opposition to the case in which the Intrex installation is closed. This effect could be linked to the fact that in a closed Intrex installation the ash is more efficiently cooled down by the water flowing in the cover (due to the larger surface of heat exchange and volumetric flow rate of the cooling water, as well as the lower ambient temperature), which is confirmed by the data presented in Tables 3 and 4.

It is worth noting that the highest mean outlet temperature of water obtained for L_{ash} = 80% is approximately 67°C (for L_{ash} = 60%, it is 60°C). This is a relatively high temperature value, signifying that the hot water can be further used in sanitary installations or in low-temperature heat exchanges.

Tables 3 and 4 present the mean values of the heat flux transferred from the hot ash to the cooling water flowing through the screw and cover of the SSAC cooler in closed and open Intrex installations, respectively.

Table 3. Mean values of the heat flux transferred from the hot ash to the cooling water flowing through the screw and cover of the SSAC cooler in a closed Intrex installation.

ash load L_{ash}, %	**rotation speed of screw** ω, rpm	**heat flux to cooling water**, kW			**heat flux from hot ash**, kW
		through screw	**through cover**	**total**	
20	2	144.7	109.0	253.6	244.2
40	4	179.4	142.3	321.7	309.8
60	6	224.2	170.3	394.6	391.2
80	8	258.4	183.3	441.7	423.2

Table 4. Mean values of the heat flux transferred from the hot ash to the cooling water flowing through the screw and cover of the SSAC cooler in an open Intrex installation.

ash load L_{ash}, %	**rotation speed of screw** ω, rpm	**heat flux to cooling water**, kW			**heat flux from hot ash**, kW
		through screw	**through cover**	**total**	
20	2	166.0	83.5	249.5	262.5
40	4	219.7	108.8	328.5	327.9
60	6	255.9	127.8	383.8	392.4
80	8	293.6	167.9	461.5	462.5

In the case of a closed Intrex installation, the overall heat fluxes are always lower than for an open Intrex installation. This is because extra hot ash is added in the middle of the ash cooler. The relative error between the heat fluxes calculated for the cooling water and the hot ash in both tables varies by less than 5 %. Figure 5 presents the outlet temperatures of the ash in closed and open Intrex installations.

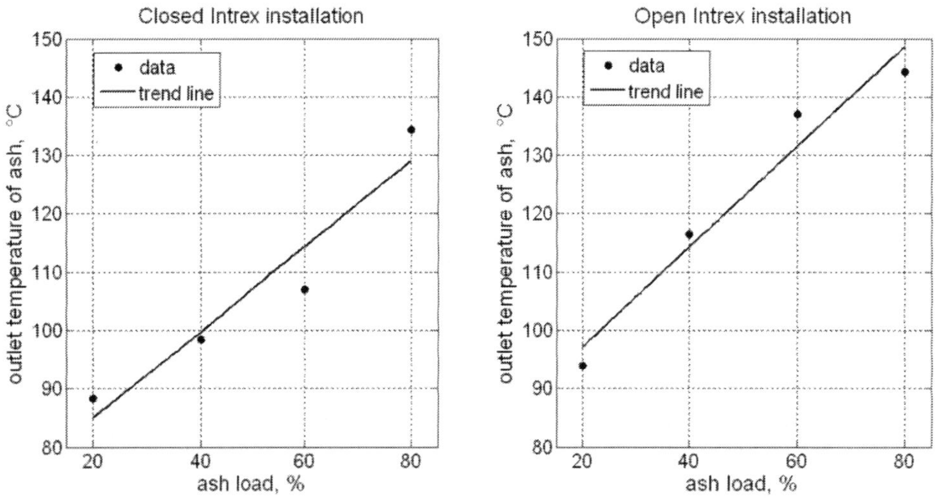

Fig. 5. The outlet temperatures of the ash from the SSAC cooler versus the ash load L_{ash}. The left diagram refers to a closed Intrex installation, the right one – to an open Intrex installation.

4 Closing remarks

The presented analysis of two examples of ash coolers indicates that the amount of heat transferred from the hot slag to the cooling water is relatively high. In the case of the two screw bottom ash cooler, the heat flux varies from 0.25 up to 1.43 MW, which is a significant value in the thermal balance of a boiler. Unfortunately, the outlet temperatures of the cooling water are below 52°C; thus, it is impossible to directly utilize this source of waste heat efficiently in the sanitary or auxiliary installations of a power unit. This means that either the heat taken by the cooling water in the ash cooler is released into the atmosphere in the cooling towers or the warm water is directed to an additional heating system (e.g. in secondary heat exchanger) where its temperature will increase up to useful range (~ 70–90°C).

In the case of the single screw ash cooler, higher values of the outlet temperatures of the cooling water could be achieved. This type of ash cooler can deliver water with an outlet temperature at a level of 60–67°C and, at the same time, preserve the restriction imposed on the outlet temperature of the slag at the required level (below 150°C). The warm water from this heat exchanger could be utilized in the sanitary or auxiliary installations of a power unit. The only disadvantage lies in the insufficient value of the water's volumetric flow rate. In the first case, the volumetric flow rate of the cooling water through the ash cooler amounts to approx. 52 m^3/h, while in the other case – only 11 m^3/h.

References

1. S. Fleischli, https://www.nrdc.org/experts/steve-fleischli/message-epa-time-modernize-americas-power-plants-cooling-systems-included (Last cited 4 January 2016)
2. R.J. Campbell, P. Folger, Ph. Brown, Cong. Res. Serv. Report, **R42950** (2013)
3. R.J. Campbell, Cong. Res. Serv. **7–5700** (2013)
4. C. Henderson, http://www.iea-coal.org.uk/documents/83185/8784/Upgrading-and-efficiency-improvement-in-coal-fired-power-plants,-CCC/221 (2012)
5. A. Yu. Ryabchikov, K.É. Aronson, Yu.M. Brodov, S.I. Khaet, S.N. Blinkov, N.V. Zhelonkin, Power Technol. Eng. **44**, 3 (2010)
6. A. Błaszczyk, J. Głuch, A. Gardzilewicz, Polish Maritime Res. **18**, 3 48–54 (2011)
7. M.A. Malek, *Heating boiler operator's manual: maintenance, operation and repair* (McGraw-Hill, New York, 2007)
8. P. Regucki, A. Andruszkiewicz, W. Wędrychowicz, B. Engler, J. Perspectives in Science **7**, 353–356 (2016)
9. C. Nichols, G. Vaux, C. Zaremsky, J. Murphy, M. Ramezan, DOE/NETL-2008/1329 (2008)
10. Z. Man, B. Rushan, X. Qinggui, Powder Technol. **201**, 114–122 (2010) DOI: 10.1016/j.powtec.2010.03.004
11. B. Zeng, X. Lu, L. Gan, M. Shu, Powder Technol. **212**, 151-160 (2011) DOI: 10.1016/j.powtec.2011.05.005
12. B. Zeng, X. Lu, H. Liu, Energy **35**, 3863–3869 (2010) DOI: 10.1016/j.energy.2010.05.040

DOI: 10.1051/e3sconf/20172200072

Indirect contact freeze water desalination for an ice maker machine – CFD simulation

Harith Jayakody[1, *], *Raya* Al-Dadah[1], and *Saad* Mahmoud[1]

[1]School of Mechanical Engineering, University of Birmingham, Birmingham, B15 2TT, U.K

Abstract. To offer for potable water shortages, sea water desalination is a potential solution for the global rising demand for fresh water. The latent heat of fusion is about one-seventh the latent heat of vaporisation, thus indicating the benefit of lower energy consumption for the freeze desalination process. Limited literature is reported on computational fluid dynamics (CFD) on freeze desalination. Therefore, analysing and investigating thermodynamic processes are easily conducted by the powerful tool of CFD. A single unit of ice formation in an ice maker machine was modelled using ANSYS Fluent software three-dimensionally. Energy, species transport and solidification/melting modules were used in building the CFD model. Parametric analysis was conducted using the established CFD model to predict the effects of freezing temperature and the geometry of the ice maker machine; on ice production and the freezing time. Lower freezing temperatures allowed more ice production and faster freezing. Increasing the diameter and the length of the freezing tube enabled more ice to be produced.

1 Introduction

Water is a fundamental element that contributes imperatively in the growth of human society [1]. The approximate value of freshwater supply for the entire water source on Earth is 3%, indicating the scarcity of water. Water shortage concerns is constantly escalating over major parts around the world, causing an excess rise in water consumption due to the increase in population, variations in socioeconomic conditions, and the rise in demand of water for industrial/agricultural use [2].

The result of increase in processes for improving water quality due to the development of technology has been observed over the past years [3]. Freshwater being produced by the elimination of dissolved minerals from sea water is the process of desalination; as it proves to be an answer to the water shortage issue [4]. Thermal and membrane methods are the two main practises used for desalination [5]. Membrane methods incorporate the reverse osmosis (RO) technology where membrane fouling is sternly noticed. Multi-stage flash desalination (MSF) and multiple-effect distillation (MED) comprise of the thermal methods but high energy consumption is a challenge in these methods [6]. Vapour compression

* Corresponding author: hej246@bham.ac.uk

technology (VCD) is used for small-scale applications [7]. These desalination methods have the disadvantages of high energy consumption and high operating costs [8].

In comparison to other desalination processes, significance for freeze desalination (FD) is perceived owing it has many advantages [9-14]. The salts are rejected in freeze desalination as the formation of ice prevents inclusion of any solutes. This is due to the nature of the ice crystal lattice only formed by pure water without any solutes, which is called the water solidification phenomena [9-10]. The latent heat of fusion is 335 kJ/kg and the latent heat of vaporization is 2256.7kJ/kg, thus indicating the low energy usage by the freeze desalination process [11-12]. A wide range of construction methods and a range of materials are acceptable in freeze desalination due to is low operating temperatures in comparison to other distillation processes that permits corrosion and scale formation [12]. Regasification of liquefied natural gas (LNG) can provide the cold energy that can be used in the freeze desalination process [14].

The salt separation technique and the salinity surge of the remaining solution after the freezing process was not studied by researchers. Therefore, in this paper, a three dimensional CFD model has been created of the formation of ice in an ice maker machine. The brine separation and the ice layer growth was also modelled using CFD where the ice in the form of pure water was generated and the brine in the remaining solution was also examined. The CFD model that has been created was then used to conduct parametric study to investigate the outcomes of freezing temperature and the geometry of the ice maker machine; on production of ice and freezing time.

2 CFD modelling

A single ice unit formation in an ice maker machine has been modelled using ANSYS Fluent in order to carry out computational fluid dynamics study by indirect contact freezing. Therefore as the ice unit is formed with pure water, the brine is then rejected to the rest of the solution. A combination of the modules; solidification/melting, species transport and energy (heat transfer) are used for the ice formation and the separation of brine from the ice to the rejected aqueous solution. In order to simulate the solidification process, ANSYS Fluent uses an enthalpy-porosity method. Thus, in the porous zone, the porosity value is correspondent with the liquid fraction at the liquid-solid mushy zone [15]. The key equations (1)-(4) used to model the freeze desalination process are discussed below.

Firstly, the species transport equation has been used in conjunction with the 'scheil' rule [15, 16] due to the fact that at micro-scale, this enables the separation of species. The diffusion of solute species in the solid is presumed to be zero when using this rule, due to the fact that as the salt water freezes, the ice (solid) formed remains as pure water. Therefore the equation for species transport is:

$$\frac{d}{dt}(\rho Y_{i,liq}) + \nabla\cdot\left(\rho\left[\beta\overrightarrow{v_{liq}}Y_{i,liq} + (1-\beta)Y_{i,sol}\right]\right) = \nabla\cdot\left(\rho\beta D_{i,m,liq}\nabla Y_{i,liq}\right) - K_i Y_{i,liq}\frac{d}{dt}(\rho(1-\beta)) + \frac{d}{dt}(\rho(1-\beta)Y_{i,liq}) \tag{1}$$

Where, the liquid volume fraction is β and ρ is the density of fluid. The liquid velocity is denoted by $\overrightarrow{v_{liq}}$. Solid and liquid mass fractions are $Y_{i,sol}$ and $Y_{i,liq}$ respectively. K_i is the partition coefficient of the solute which is the mass fraction in the solid to liquid ratio at the boundary and the following equation relates these parameters:

DOI: 10.1051/e3sconf/20172200072

$$Y_{i,sol} = K_i Y_{i,liq} \tag{2}$$

When using the 'scheil' rule, the dependant variable is solved using $Y_{i,liq}$. $D_{i,m}$ is the mass diffusion coefficient for species in the mixture.

When solidification problems are incorporated with species transport, the energy equation becomes:

$$\frac{d}{dt}(\rho H) + \nabla \cdot (\rho \vec{v} H) = \nabla \cdot (k \nabla T) + \frac{(1-\beta)^2}{(\beta^2 + \varepsilon)} A_{mus\,h} \vec{v} \tag{3}$$

Where, the enthalpy is *H*, velocity is $\vec{v}$ and the mass fraction coefficient is k. The liquid volume fraction is β, the temperature is denoted by *T* and the mushy zone constant is represented by $A_{mus\,h}$. In order to avoid being divided by zero when $\beta = 0$, a small number of 0.001 is given and it is represented by ε.

Thermal and solutal buoyancies are important factors that need to be considered when modelling freeze desalination. These buoyancies arise when there are more than one species in the mixture due to the density constantly changing with temperature and species composition. Natural convection flows are used to calculate thermal buoyancy and it develops when gravity changes the deviations in density with temperature. Consequently, as the density changes with species composition, solutal buoyancy arise and therefore the solutal buoyancy body forces are obtained from equation (4).

$$\vec{F}_s = \rho_{ref} \vec{g} \sum_{i=0}^{N_s} \beta_{s,i} (Y_{l,i} - Y_{ref,i}) \tag{4}$$

The mass fraction of the liquid phase and the reference mass fraction are represented by $Y_{l,i}$ and $Y_{ref,i}$ respectively. The gravity is denoted by $\vec{g}$, the reference density is represented by ρ_{ref} and $\beta_{s,i}$ is the solutal expansion coefficient. N_s is the total number of solute species. The combination of thermal and solutal buoyancies is the total body force. Hence, modelling thermal and solutal buoyancies allows to accurately calculate the overall solidification behaviour of salt water. Freezing of salt water is an example of such a problem where there are buoyancy induced flows in solidification when multi-components are present. This is called the Boussinesq approach.

3 CFD results

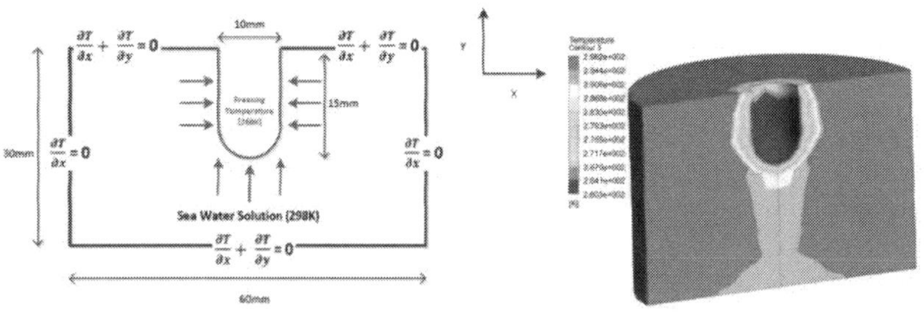

Fig. 1. (a) Schematic diagram of the problem (left) and (b) temperature contours of the developed 3D CFD model's cross sectional view (right).

The configuration used to carry out CFD analysis is shown in figure 1(a) where a single unit of ice formation for a commercially available ice maker machine is modelled. The container holds 35g/L concentration of salt water with an initial temperature of 298 K, where it has been cooled from the freezing tube (Figure 1(a)) at 260 K and the rest of the walls are considered to be adiabatic. The developed 3D CFD model shown in figure 1(b), is set-up with laminar incompressible flow conditions using a pressure based solver. The key modules used to simulate this process were solidification/melting and species transport. The experiment was run for 10 minutes of real time using transient analysis.

The temperature contours are shown in figure 2(a) where the solution has been cooled at a freezing temperature of 260K. In figure 2(b), the distribution of temperature from the freezing location to the rest of the solution is shown after 10 minutes of freezing time. A 5.5mm thickness solid ice layer has been formed at the outer surface of the freezing tube; where the temperature at the bottom of the ice layer is 260 K and at the top is about 268 K. In the ice layer formed; it is seen that the temperature is increasing from the ice layer in contact with the freezing surface, to the ice layer in contact with the rest of the solution and then it remains constant in the rest of the solution.

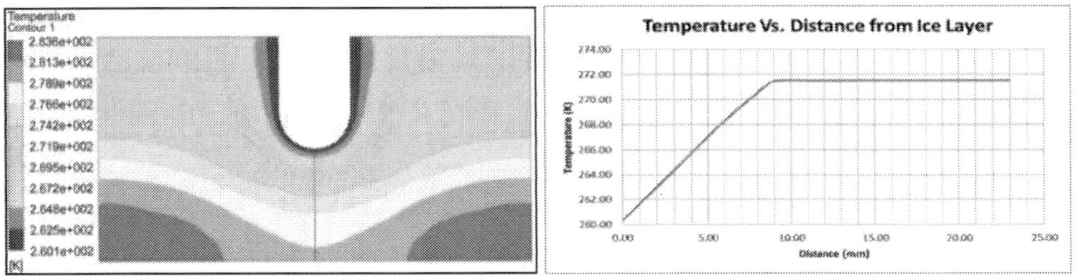

Fig. 2. (a) Temperature contours (left) and (b) temperature vs. distance from the freezing tube (right) after 10 minutes of freezing.

The solidification contours are shown in figure 3(a) where this indicated the liquid phase fraction of the solution. Minimal liquid fraction is indicated by the thick blue layer representing solid ice and the rest is the brine solution shown in red/yellow, indicating high amount of liquid fraction. In the ice layer of 5.5 mm thickness, the liquid fraction is zero as expected and is shown in figure 3(b).

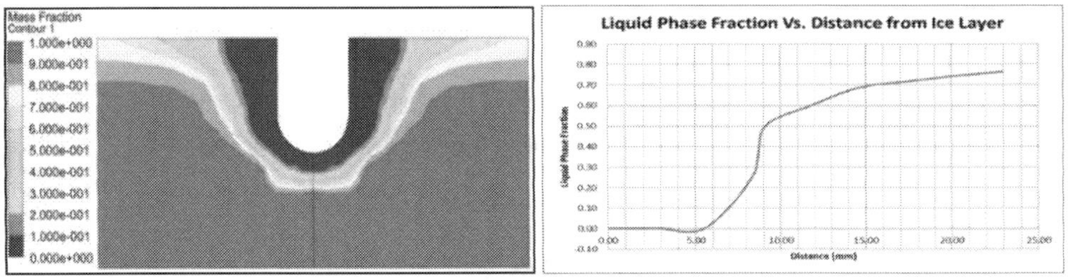

Fig. 3. (a) Liquid phase fraction contours (left) and (b) liquid phase fraction vs. distance from the freezing tube (right) after 10 minutes of freezing.

The salt water (brine) contours are shown in figure 4(a) and the graph of brine concentration versus the distance from the freezing surface is shown in figure 4(b). At the top point of the 5.5 mm ice layer, the brine concentration is 0.03%. The brine concentration then increases away from the ice layer to about 5.9% in the rest of the solution. Therefore it is seen that the brine concentration is negligible in the ice produced and the brine salinity

DOI: 10.1051/e3sconf/20172200072

increases for the rest of the solution during the freezing process. The variations in the brine concentration in the rest of the solution are due to thermal and solutal buoyancies.

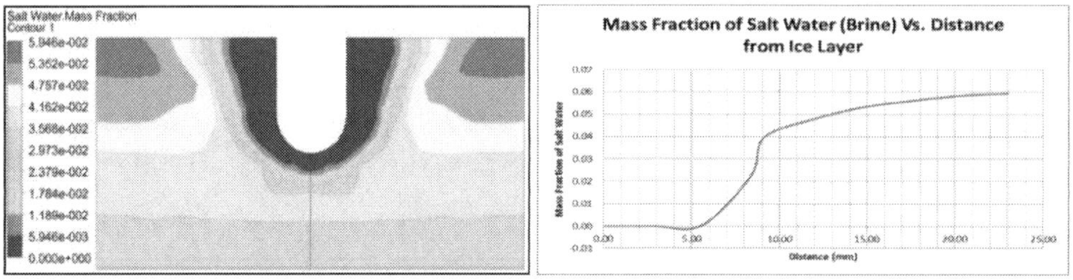

Fig. 4. (a) Salt water (brine) mass fraction contours (left) and (b) salt water (brine) mass fraction vs. distance from the freezing tube (right) after 10 minutes of freezing.

The pure water contours are shown in figure 5(a) and the graph of pure water concentration versus the distance from the freezing surface is shown in figure 5(b). Maximum concentration of pure water is present in the ice layer formed. The variations in the pure water mass fractions in the rest of the solution are due to buoyancies.

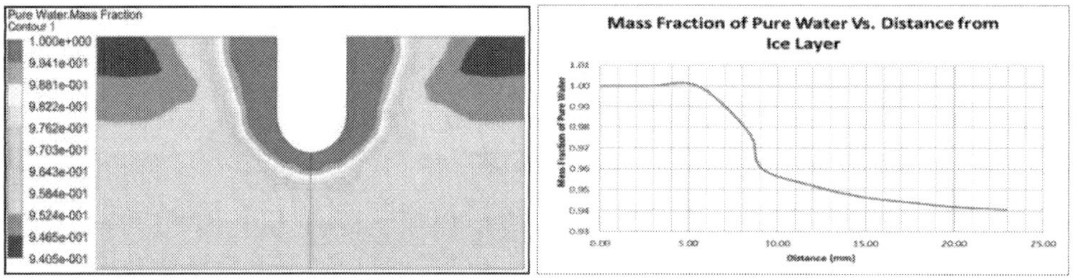

Fig. 5. (a) Pure water mass fraction contours (left) and (b) pure water mass fraction vs. distance from the freezing tube (right) after 10 minutes of freezing.

4 Parametric analyses

Parametric analysis has been carried out using the established CFD model to predict the effects of freezing temperature and the geometry of the ice maker machine; on ice production and freezing time.

4.1 Freezing temperature

Investigation of changing the freezing temperature from 230 K, 260 K and 270 K was carried out and the average solution temperature and solidification (liquid phase fraction) graphs are shown for different freezing temperatures in figure 6. It is quite apparent from figure 6(a) that when the freezing temperature is reduced, the average solution temperature decreases and allows an increased solidification rate as seen from figure 6(b). After 10 minutes of freezing; for 230 K, 260 K and 270 K of freezing temperatures, a liquid phase fraction of 0.72, 0.75 and 1 is reached respectively. This proves that lower temperatures enabled faster solidification and at 270 K freezing temperature, no solidification is seen after 10 minutes; due to the lower freezing point of salt water.

DOI: 10.1051/e3sconf/20172200072

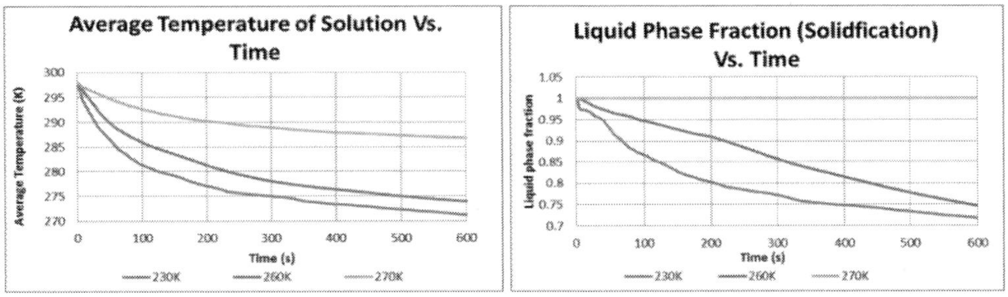

Fig. 6. (a) Average temperature of solution (left) and (b) liquid phase fraction (right) vs. time graphs for different freezing temperatures.

The temperature and the solidification (liquid phase fraction) contours for the solution are shown in figure 7 for freezing temperatures of 230K, 260K and 270K after 10 minutes of real time freezing. The variations in solidification are clearly shown in the figure; where at lower freezing temperatures, more ice production can be observed. The volume of ice produced increases significantly with lower freezing temperatures, as the ice volumes at 230K and 260K are 5690 and 3580mm^3 respectively and no ice formation is seen with a 270K freezing temperature after 10 minutes of freezing.

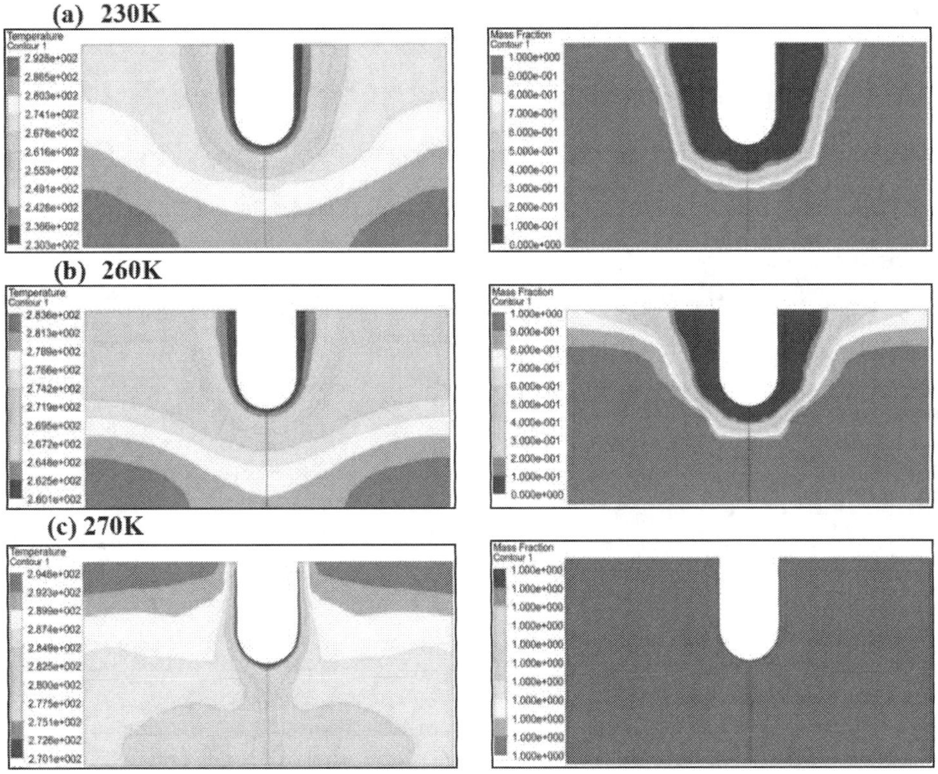

Fig. 7. Contours for temperature (left) and liquid phase fraction (right) for freezing temperatures of (a) 230 K, (b) 260 K and (c) 270 K.

4.2 Geometry of the ice maker machine

The geometry of the freezing surface of a single unit has been changed to investigate the effect on ice production. The diameter and the length of the freezing tube have been varied and the solidification (liquid phase fraction) contours are shown in figure 8. The volumes of ice produced for the freezing tube dimensions of 5 mm diameter is 842 mm^3 and for 20 mm diameter is 3188 mm^3. When the length of the freezing tube was 10mm and 20mm, the ice volumes of ice produced were 1267 and 2770 mm^3 respectively. Therefore it is concluded that as the diameter and the length of the freezing tube is increased, the ice production also increases due to the increase in the freezing surface area.

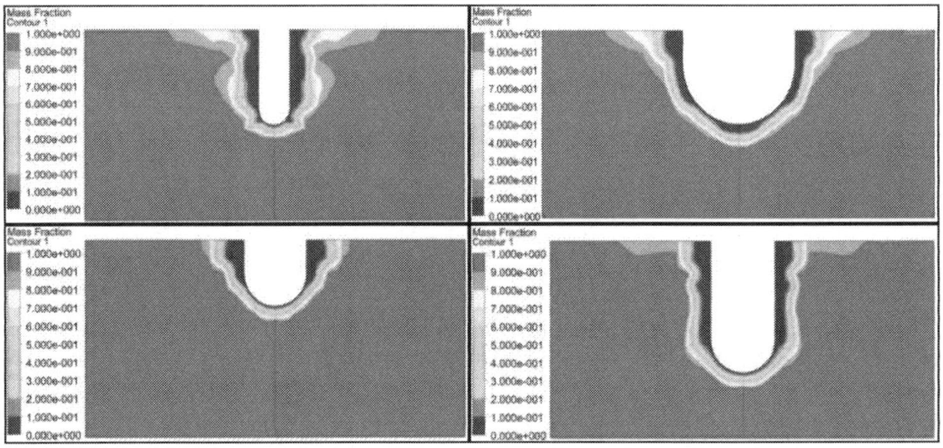

Fig. 8. Liquid phase fraction contours for (a) 5mm diameter (top left), (b) 20mm diameter (top right), (c) 10mm length (bottom left) and 20mm length (bottom right).

5 Conclusions

The freeze desalination process occurs by the formation of ice as pure water; where the ice crystal lattice is formed purely of water and the salts are rejects by the nature of ice. Hence in the freeze desalination process, any solutes are prevented from entering the developing ice crystal lattice. A prodigious potential is displayed for freeze desalination due to its benefits over other desalination methods. Lower energy consumption enables freeze desalination to stand out among other desalination processes; where the latent heat of vaporisation (2256.7 kJ/kg) is much higher than the latent heat of fusion (333.5 kJ/kg). Another major advantage of this process is that scale formation and corrosion are diminished due to its low operating temperatures. Furthermore, regasification of liquefied natural gas (LNG) can be done to exploit the cold energy that can be used in the freeze desalination process.

Limited literature is presented for computational fluid dynamics (CFD) on freeze desalination. Hence, investigating and analysing thermodynamic processes are done by the powerful tool of CFD. A single unit of ice formation in an ice maker machine was simulated using the ANSYS Fluent software three-dimensionally. Solidification/melting, species transport and energy modules were used in constructing the CFD model.

Parametric analysis was done by expanding the established CFD model to predict the effects of freezing temperature and the geometry of the ice maker machine; on ice production and the freezing time. Hence, lower freezing temperatures allowed more ice production and faster freezing. The 230 K freezing temperature allowed 37.1% more ice to be produced compared to the 260 K freezing temperature. Additionally, increasing the diameter and the length of the freezing tube enabled more ice production. Thus, the geometry of 20mm freezing tube diameter permitted 73.6% more ice to be produced compared to the 5mm freezing tube diameter geometry. The geometry of 20 mm freezing tube length gave 2.19 times more ice compared to the 10mm freezing tube length geometry.

The authors would like to thank the School of Mechanical Engineering, University of Birmingham, UK; for offering a PhD scholarship to carry out the research on modelling and investigation of the freeze desalination process.

References

1. M. Pedro-Monzonís, A. Solera, J. Ferrer, T. Estrela, J. Paredes-Arquiola, J. Hydrol. **527**, 482–493 (2015)
2. T.I.E. Veldkamp, Y. Wada, H. de Moel, M. Kummu, S. Eisner, J.C.J.H. Aerts, P.J. Ward, Glob. Environ. Chang. **32**, 18–29 (2015)
3. Y. Ramírez, L.A. Cisternas, A. Kraslawski, J. Clean. Prod. **148**, 223–232 (2017)
4. H. March, WIRE. Water, **2.3**, 231–243 (2015)
5. M. Rahman, M. Ahmed, X.D. Chen, Sep. Purif. Rev. **35**, 2, 59–96, (2006)
6. J. Chang, J. Zuo, K.J. Lu, T.S. Chung, Water Res. **102**, 282–293 (2016)
7. M.C. Fragkou, J. McEvoy, Desalination **397**, 1–8 (2016)
8. A. Chafidz, E.D. Kerme, I. Wazeer, Y. Khalid, A. Ajbar, S.M. Al-Zahrani, J. Clean. Prod. **133**, 631–647 (2016)
9. D.G. Randall, J. Nathoo, J. Water Process Eng. **8**, 186–194 (2015)
10. R. Fujioka, L.P. Wang, G. Dodbiba, T. Fujita, Desalination **319**, 33–37 (2013)
11. A.A.A. Attia, Desalination 254, 1–3, 179–184 (2010)
12. P.M. Williams, M. Ahmad, B.S. Connolly, D.L. Oatley-Radcliffe, Desalination **356**, 314–327 (2015)
13. Z. Lu, L. Xu, DESWARE, **2** (2014)
14. W. Cao, C. Beggs, I.M. Mujtaba, Desalination **355**, 22–32 (2015)
15. ANSYS Fluent Theory Guide. 15th ed. USA: ANSYS, Inc. (2013)
16. V.R. Voller, D. Brent, C. Prakash, Int. J. Heat Mass Transfer **32**, 9, 1719–1731 (1989)

Simulation of thermal environment in a three-layer vinyl greenhouse by natural ventilation control

Tea-Hwan Jin[1], *Ki-Yeol* Shin[1,*], *Si-Won* Yoon[2], *Yong-Hoon* Im[2], and *Ki-Chang* Chang[2]

[1]School of Mechanical Engineering, Yeungnam University, 712-749 Gyeongbuk, Republic of Korea
[2]Convergence Research Center for Smart Farm Solution, Korea Institute of Science and Technology, 25451, Gangwon, Republic of Korea

Abstract. A high energy, efficient, harmonious, ecological greenhouse has been highlighted by advanced future agricultural technology recently. This greenhouse is essential for expanding the production cycle toward growth conditions through combined thermal environmental control. However, it has a negative effect on farming income via huge energy supply expenses. Because not only production income, but operating costs related to thermal load for thermal environment control is important in farming income, it needs studies such as a harmonious ecological greenhouse using natural ventilation control. This study is simulated for energy consumption and thermal environmental conditions in a three-layered greenhouse by natural ventilation using window opening. A virtual 3D model of a three-layered greenhouse was designed based on the real one in the Gangneung area. This 3D model was used to calculate a thermal environment state such as indoor temperature, relative humidity, and thermal load in the case of a window opening rate from 0 to 100%. There was also a heat exchange operated for heating or cooling controlled by various setting temperatures. The results show that the cooling load can be reduced by natural ventilation control in the summer season, and the heat exchange capacity for heating can also be simulated for growth conditions in the winter season.

1 Introduction

As the proportion of farmers using greenhouses has increased recently, studies on the farm household income of domestic and foreign greenhouses have been actively carried out. Through the analysis of the thermal environment of various greenhouse models, diverse greenhouse models are actually applied [1] and attempts to increase productivity through an automation system have been made. In Korea, as in other countries, combined thermal environment control technology for the greenhouse to increase the productivity of cultivated crops through the smart farm is necessary in order to improve the profitability of farm households. There is also a growing need for research and development on the total

* Corresponding author: shinky@ynu.ac.kr

service platform that is associated with the operation of smart farms, such as energy and agriculture management.

Although heating the facilities is necessary in winter and cooling is required to avoid high temperatures and humidity in summer, enormous energy costs are required in order to control the thermal environment according to the growth conditions and it is a factor that hurts the profits of farm households, so it is necessary to establish various energy saving measures for the facilities. Recently, in order to reduce the energy cost, renewable energy resources using solar heat and geothermal heat are being employed [2, 3]. However, a disadvantage of renewable energy sources is that they may not be able to supply energy in a timely manner due to the volatility of the output. This problem may lead to a shortage of supply energy as well as failure to supply necessary energy in a timely manner.

In this study, the thermal environment in a greenhouse through natural ventilation was simulated before controlling the thermal environment through various energy sources. Through the 3D modeling and analysis of the three-layer vinyl greenhouse, the annual energy load is calculated and the energy load reduction and the combined control of temperature and humidity are confirmed through the opening and closing control function of the windows. Using TRNSYS, a three-layer vinyl greenhouse for strawberry cultivation located in Sangcheon-myeon, Gangneung-si, Gangwon Province was simulated. Through this simulation, the indoor temperature and indoor humidity in the enclosed greenhouse were confirmed, and a decrease of the cooling load depending on the change of the indoor temperature, indoor humidity, and natural ventilation through the on/off control of the side windows and skylights was also confirmed. Based on the calculation results, this study aimed to propose a method of operating an environmentally friendly greenhouse through natural ventilation as well as reducing the amount of required energy through thermal and environmental control.

2 3D modeling of the 3-layer greenhouse

2.1 3D modeling of the greenhouse

The 3D modeling of the real greenhouse, which was the analysis target of this study, was simulated through TRNBUILD of TRNSYS and TRNSYS3d of Sketch-up 2016, as shown in Fig. 1. In the case of TRNSYS, it has high accuracy and usability for energy loads of various general buildings as a technology for calculating the heat and energy flow of the building and the energy loads such as heating and cooling for the proper environmental control.

In addition, since it is possible to set the boundary conditions such as the ambient condition and solar radiation condition during the user's desired period, it is possible to supplement limitations of field experiments. Thus, recently, it is frequently used in modeling not only for general buildings but also for agricultural buildings [4].

In this study, the factors related to structures such as each frame and vinyl covering constituting the greenhouse were constructed through the TRNSYS3d of the Sketch-up, and the thermal properties of each material, boundary conditions, and on/off control of windows were constructed through TRNBUILD as shown in Table 1 and Fig. 2.

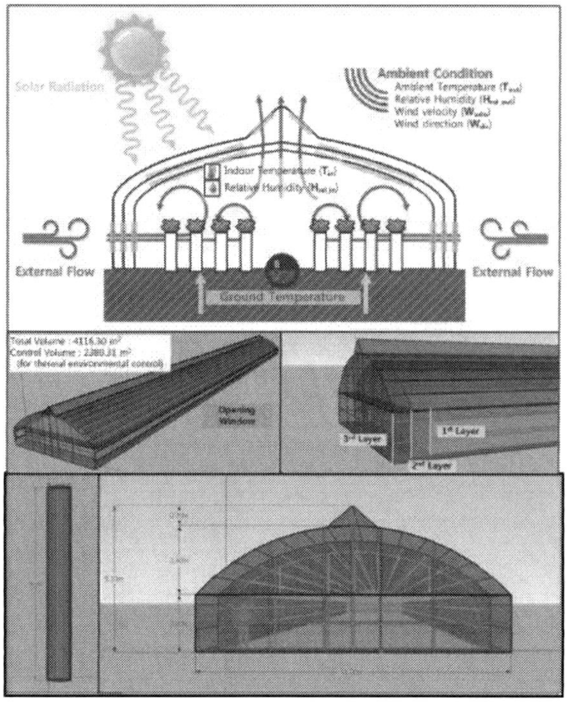

Fig. 1. 3D modeling of 3-layer greenhouse.

Table 1. Input design condition of the 3D modeling.

Weather DB	Calculation		Result
Solar Radiation Temperature Humidity Wind Atmospheric Pressure	Surface Info	Slope	Beam Radiation
		Azimuth	Diffuse Radiation
			Incidence Angle
	Atmospheric Condition	Ambient Temperature	Ambient Temperature
		Dewpoint Temperature	Sky Temperature
		Relative Humidity	
	Window Type	Wind Velocity , Direction	Mass Flow
		Atmoshperic Pressure	Window Open Signal

The ambient conditions of this study were determined using the weather data of the Korea Meteorological Administration for the northern part of Gangneung-si, which is the area nearest to the three-layer vinyl house for strawberry cultivation in Sangcheon-myeon, Gangneung-si. The weather data of the Korea Meteorological Administration were composed of the ambient temperature, indoor humidity, dew point temperature, solar radiation, wind velocity, wind direction and atmospheric pressure. The incoming solar radiation on each side of the greenhouse was calculated from solar radiation, the external ambient temperature and external relative humidity. The internal temperature was calculated using the external ambient temperature, the dew point temperature, calculated direct solar radiation and diffuse solar radiation. The internal thermal environment was simulated through incoming solar radiation on each side and the internal temperature. The wind speed, wind direction, and atmospheric pressure were used to calculate the amount of ventilation through each window.

 DOI: 10.1051/e3sconf/20172200073

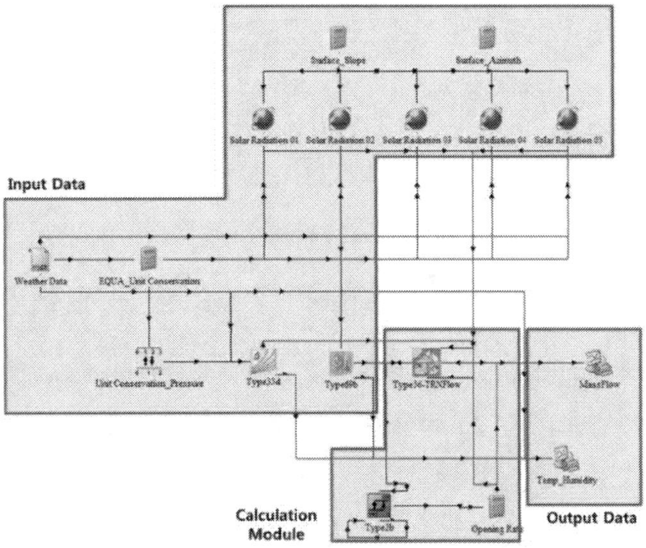

Fig. 2. TRNSYS input schematic diagram of 3-layer greenhouse.

2.2 Modeling of thermal and air nodes

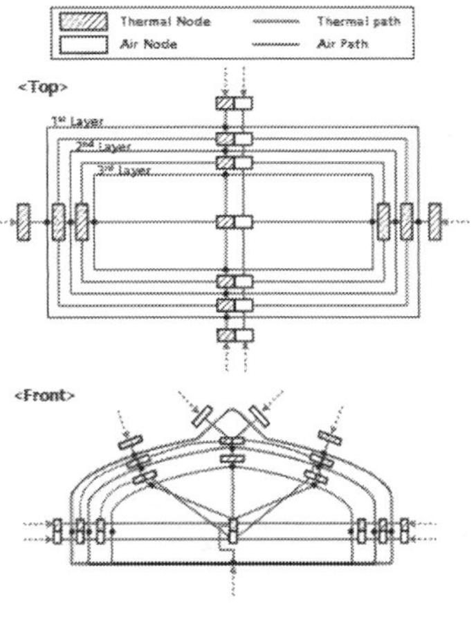

Fig. 3. Thermal and air node configuration.

For the greenhouse analysis for the thermal and environmental control, the analysis model is composed of the thermal-nodes and air-nodes by distinguishing between calculation for thermal heat balance and calculation for mass balance of air ventilation. The thermal-nodes consist of inner and outer walls, which are classified according to whether they are exposed to wind and solar radiation. The outer walls are influenced by the ambient conditions such as solar radiation, the ambient temperature, and the internal temperature. For the inner walls, the condition of the thermal environment is simulated and calculated taking into account the influences of conduction, convection and radiation energy through adjacent surfaces. Fig. 3 shows thermal and air node configuration.

In the study, boundary conditions were constructed by considering both the ground temperature and the internal temperature of the indoor air in contact with the ground in order to simulate the changes of the internal temperature depending on the floor surface, taking into account the effect of the ground temperature.

The air-node is for calculating the air flow rate according to the on/off control of the side windows. The window is opened from the bottom to the top. The width of the window and the height of the window (fully opened) were entered for each window located at each node, and the windows are controlled to be opened or closed according to the change of the indoor temperature.

3 Window opening model of the 3-layer greenhouse

3.1 Window opening model

The TRNFLOW module of TRNSYS was used to calculate the exit flow rate of air depending on the on/off control of the greenhouse windows. In TRNFLOW, the variables in Table 2 are used to calculate the mass flow rate according to the on/off control of the windows [5]. As shown in Fig. 3, the width of the window is constant, and the flow rate changes depending on the window opening rate which is increased by the on/off control of the windows. At this time, when the mass flow rate from Zone 1 to Zone 2 is $\dot{m}_{12}$ and the mass flow rate from Zone 2 to Zone 1 is $\dot{m}_{21}$, each mass flow rate is calculated by the following equation [6, 7].

Table 2. Main variables of window opening model.

Variable	Equation
Pressure	$\mathrm{p} = \frac{\rho}{2} \cdot C_p \cdot {v_o}^2$
Pressure difference	$\Delta\mathrm{p} = \frac{\rho}{2} \cdot \left(\frac{V}{C_d}\right)^2$
Mass flow rate	$\dot{m} = C_s \cdot (\Delta p)^n$
Real wind speed coefficient	$C_w = 0.08 \cdot \left(\frac{Gr}{Re^2}\right)^{-0.38}$

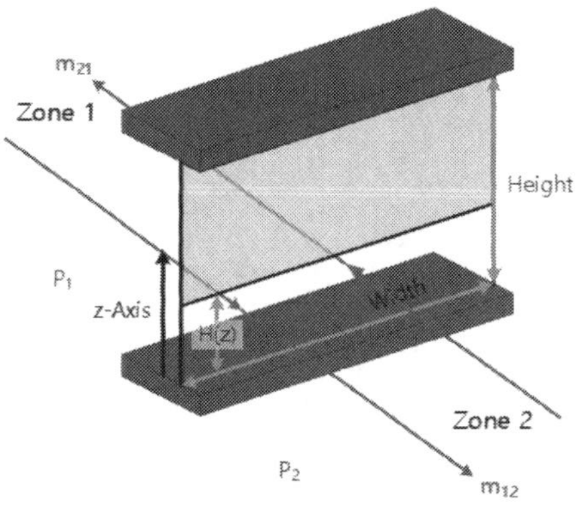

Fig. 4. Control window opening model.

$$\dot{m}_{12} = C_d \cdot \int_0^H \sqrt{2\rho(z) \cdot f_{12}(z)} \cdot H(z) \cdot dz \qquad (1)$$

$$f_{12} = \begin{cases} \Delta p(z) & ,if\ \Delta p(z) > 0 \\ 0 & ,if\ \Delta p(z) < 0 \end{cases} \tag{2}$$

$$\dot{m}_{21} = C_d \cdot \int_0^H \sqrt{2\rho(z) \cdot f_{21}(z)} \cdot H(z) \cdot dz \tag{3}$$

$$f_{21} = \begin{cases} -\Delta p(z) & ,if\ \Delta p(z) < 0 \\ 0 & ,if\ \Delta p(z) > 0 \end{cases} \tag{4}$$

Thus, the air flow rate through ventilation influences the convection heat transfer and the internal temperature of the greenhouse changes. When the quantity of heat is $\dot{Q}_{vent}$, it can be calculated by the equation in (5) below.

$$\dot{Q}_{vent} = \dot{V} \cdot \rho \cdot C_p(T_{vent-air} - T_{indoor-air}) \tag{5}$$

Depending on the on/off control of the windows, $\dot{Q}_{vent}$ flows in or out of each node in Fig. 3, and consequently, the conditions of the thermal environment inside the greenhouse through natural ventilation are newly calculated [8].

4 Results of thermal environment analysis

4.1 Analysis of the thermal environment in the enclosed greenhouse

The calculation results of the thermal environment conditions such as the temperature and humidity inside the greenhouse through the 3D shape modeling are shown in Fig. 5.

For the internal temperatures of the enclosed 3-layer vinyl greenhouse without air-conditioning and heating facilities, the annual mean temperature is 17.7°C, the maximum temperature is 48.8°C, and the minimum temperature is -4.7°C. For seasonal temperatures, in winter (January to February and November to December), the mean temperature is 6.42°C, the maximum temperature is 23.5°C, and the minimum temperature is -4.7°C. In summer (July to August), the mean temperature is 31.5°C, the maximum temperature is 48.8°C, and the minimum temperature is 21.9°C. In spring and autumn (March to June and September to October), the mean temperature is 20.4°C, the maximum temperature is 44.8°C and the minimum temperature is 0.7°C.

In winter, since the temperature inside the greenhouse is low due to the influence of the low ambient temperature and the insignificant changes of the ground temperature, the heating load is expected to be large during winter months. In particular, it was also found that even though the ambient temperature is high, the indoor temperature is not significantly affected by it because of the influence of the ground temperature. In particular, it is predicted that the indoor temperature will be lowered due to the influence of the ground temperature at the time of sunrise or sunset, and as a result, freezing of the crops may occur. In summer, the indoor temperature is observed to be high because solar radiation is abundant, the angle of incidence is close to a vertical angle and the ground temperature is high. It can be seen that the humidity inside the greenhouse is well calculated according to the change of the internal temperature.

 DOI: 10.1051/e3sconf/20172200073

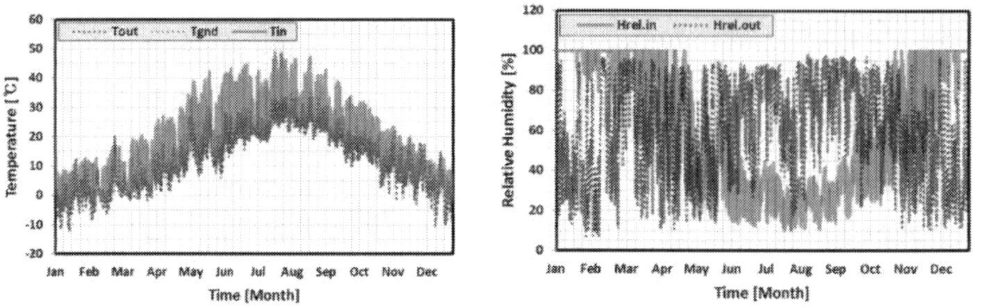

Fig. 5. Annually thermal environment conditions in closed greenhouse.

4.2 Analysis of the thermal environment by natural ventilation control

In order to control the indoor temperature inside the greenhouse in summer, a natural ventilation system that opens and closes side windows was modelled. The annual indoor temperature, humidity and mass flow rate according to the opening rate of the window are shown in Fig. 6. The setting temperature for the window on/off control was 25°C, the opening and closing time was differentiated by the time unit of 15 minutes, and the window opening rates were divided into 25%, 50%, 75%, and 100%. In the calculation results, the window on/off control is conducted from April to October of the year, and the operating time of the on/off control is different according to the opening rate during a total of 8760 hours. The opening rate is 25% for 1522 hours, 50% for 1495.5 hours, 75% for 1487.75 hours, and 100% for 1484.25 hours. The reason why the control time differs according to the opening rate can be confirmed by the mass flow rate graph. The mass flow rate changes according to the opening rate, so the amount of $\dot{Q}_{vent}$ varies accordingly. For this reason, the control operation time changes depending on the opening rate in order to match the setting temperature.

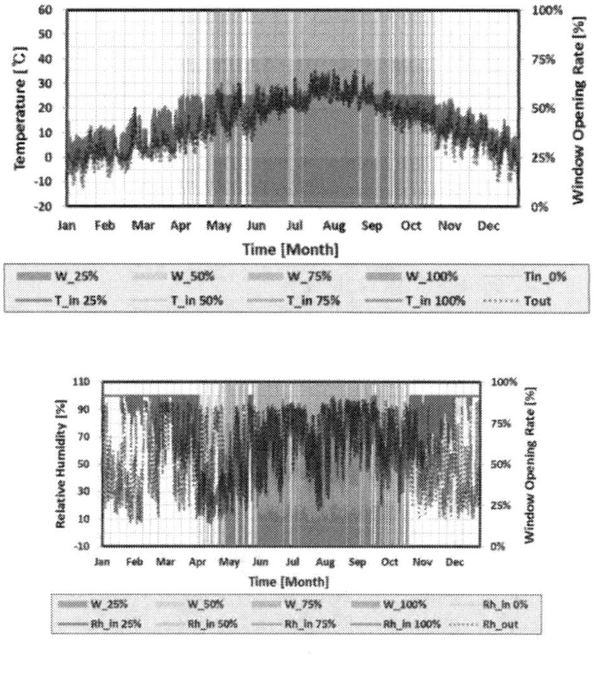

DOI: 10.1051/e3sconf/20172200073

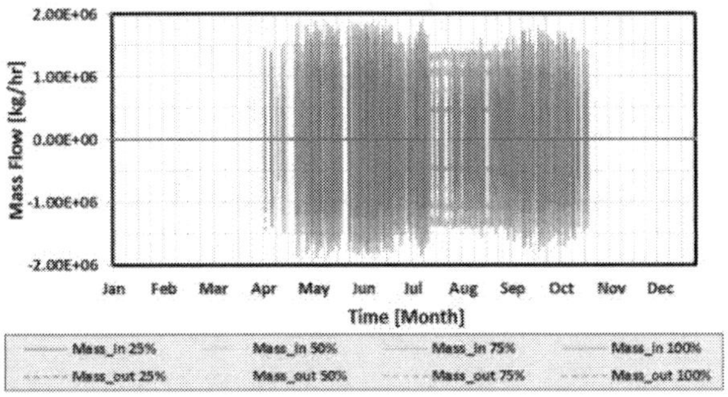

Fig. 6. Annually thermal environment conditions by natural ventilation control.

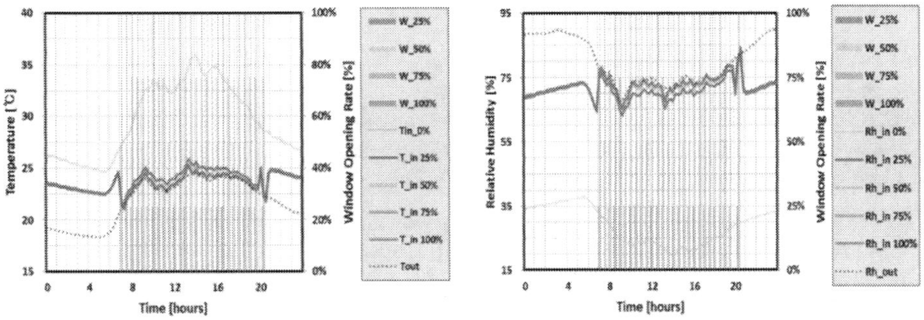

Fig. 7. Daily thermal environment conditions by natural ventilation control.

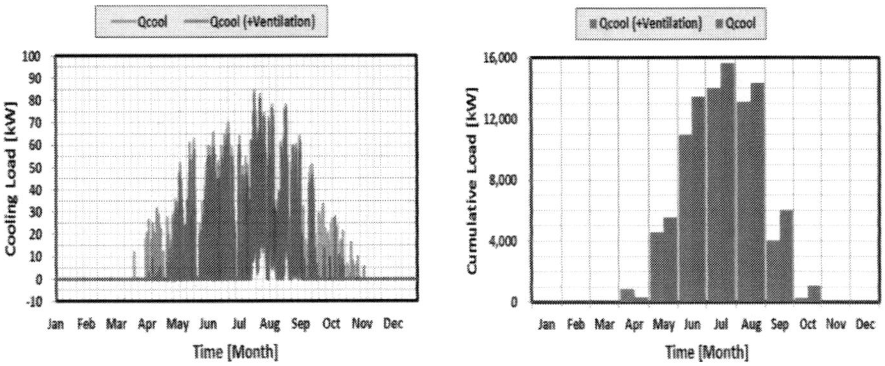

Fig. 8. Energy load comparison by only cooling and cooling with ventilation.

In order to examine daily thermal environment conditions, the calculation results according to the opening rate in July are presented in Fig. 7. Although it is controlled according to the on/off control of the windows, similarly to the ambient conditions, it can be seen that there is a slight difference according to the opening rate. Actually, temperature difference of about 1°C and humidity difference of about 5% were found between the opening rate of 25% and the opening rate of 100%.

4.3 Thermal load analysis for natural ventilation control

As described above, the differences in the thermal environment conditions of the greenhouse according to the opening rate were examined. At the same time, the cooling loads for the case of single cooling and for the case of cooling with natural ventilation control were calculated and compared with the cooling load of the enclosed greenhouse. The results of calculation of the total load are shown in Fig. 8 for the cases where single cooling was performed at 25°C and where cooling load was simulated with natural ventilation control. The cooling load of the enclosed greenhouse and the cooling load by natural ventilation control were found to be 56,386 kW and 47,859 kW, respectively, so it was confirmed that the cooling load can be reduced by approximately 17.82% (8,527 kW) when natural ventilation control is used.

5 Conclusions

In this study, 3D modeling of a 3-layer greenhouse was carried out, and the internal thermal environment conditions of the greenhouse according to the window on/off control were investigated. First, the thermal environment conditions in the enclosed greenhouse were examined, and the thermal environment conditions inside the greenhouse were compared by adjusting the window opening rate to 25%, 50%, 75%, and 100%. In particular, when the windows were controlled by setting the window opening rate to be 50% at the indoor temperature of 18°C and 100% at 23°C, it was confirmed that the cooling load of about 17.82% (8,527 kW) can be reduced in the case of cooling with natural ventilation control through the window on/off control compared to the cooling load of the enclosed greenhouse. These results show that it is possible to maintain not only the thermal environment conditions that meet the growth conditions, but also reduce the cooling load through natural ventilation.

Based on the above results, it is predicted that it will be possible to maintain the energy environment that meets the growth conditions of the crop through the surrounding ecological environment, and simulate a more effective thermal environment when the method is used together with renewable energy resources. In the future, we plan to conduct further research on the ways to reduce the heating load through natural ventilation during heating by adding heat exchanger control for heating during winter months.

This work was supported by the National Research Council of Science & Technology (NST) grant by the Korea government (MSIP) (No. CRC-15-01-KIST).

References

1. H.G. Mobtaker, Y. Ajabshirchi, S.F. Ranjbar, M. Matloobi, Renew. Energy **96**, 509 (2016)
2. M. Asim, J. Dewsbury, S. Kanan, Ener. Proc. **91**, 702 (2016)
3. R.H.E. Hassanien, M. Li, W.D. Lin, Renew. Sus. Ener. Reviews **54**, 989 (2016)
4. S.B. Lee, I.B. Lee, S.W. Homg, I.H. Seo, P.J. Bitog, K.S. Kwon, T.H. Ha, C.P. Han, J. Kor. Soc. Agr. Eng. **54**, 113 (2012)
5. F. Allard, V. Dorer, H.E. Feustel, E.R. Garcia, M. Grosso, M.K. Herrlin, L. Mingsheng, J.C. Phaff, Y. Utsumi, H. Yoshino, AIVC Tech. Note **29**, 115 (1990)
6. E. Dascalaki, M. Santamouris. A. Argiriou. C. Helmis. D.N. Asimakopoulos.

 DOI: 10.1051/e3sconf/20172200073

K. Papadopoulos. A. Soilemes, Sol. Energy **55**, 327 (1995)

7. A. Weber, *TRNFLOW, A module of an air flow network for coupled simulation with TYPE 56* (Solar Energy Laboratory, Univ. of Wisconsin-Madison, 2009)
8. A. Weber, *TRNSYS 17, Multizone Building modeling with Type 56 and TRNBuild* (Solar Energy Laboratory, Univ. of Wisconsin-Madison, 2012)

Temporal and spatial complementarity of wind and solar resources in Lower Silesia (Poland)

Jakub Jurasz[1], *Marcin* Wdowikowski[2], *Bartosz* Kaźmierczak[3,*] *and Paweł* Dąbek[4]

[1] AGH University, Faculty of Management, 30-059 Kraków, 30 Mickiewicza Av., Poland
[2] Institute of Meteorology and Water Management - National Research Institute, 01-673 Warsaw, Podleśna 61, Poland
[3] Wrocław University of Science and Technology, Faculty of Environmental Engineering, PlacGrunwaldzki 13, 50-377 Wrocław, Poland
[4] Wrocław University of Environmental and Life Sciences, Institute of Environmental Development and Protection, Grunwaldzki 24, 50-363 Wrocław, Poland

Abstract. This paper investigates the concept of temporal and spatial complementarity of wind and solar resources in Lower Silesia (south-wester Poland). For the purpose of our research we have used hourly load and energy yield from photovoltaics and wind turbines covering period 2010-2014. In order to assess the spatial complementarity we have divided the considered voivodeship into 74 squared regions with maximal area of 400 km^2. The obtained results indicate an existence of temporal complementarity on a monthly time scale and a positive correlation between load and wind generation patterns (also on a monthly time scale). The temporal complementarity for hourly time series in relatively low but has potential to smooth the energy generation curves.

1 Introduction

The advent of large scale variable renewable energy sources (VRES) mainly in form of photovoltaics and wind turbines is a result of a transition from power system based upon finite fossil fuels to a sustainable one utilizing the abundant and inexhaustible renewable resources. However, the inherently intermittent nature of solar radiation and wind speed makes them a non-dispatchable energy sources. Therefore, their integration to the national power systems dominated by relatively slowly ramping-up and down coal are nuclear power plants is an arduous and complicated task. Considering the increasing installed capacity in photovoltaics (PV) and wind turbines (WT) and the fact that they start to play a significant role in covering the energy demand a need arises to address the problem of integration [1].

So far, various approaches and methods have been presented which aim at facilitating the process of VRES integration. Accordingly to [2] this problem can be solved by applying following solutions:

*Corresponding author: bartosz.kazmierczak@pwr.edu.pl

- interconnecting geographically distributed inherently variable energy sources (e.g. solar, tidal or wind), where a spatial distribution smoothing effect can be observed. Such effect has been presented by *inter alia* [3] in case of interconnected European energy system;
- to couple variable but temporally complementary (solar and wind) energy sources in one hybrid generator, or to connect non-dispatchable power source with a dispatachble one (hydropower) to fill the gaps between demand and generation. The concept of coupling variable PVs with hydropower in form of run-off-river power plant with pondage has been presented in [4];
- to apply so called demand side management (DSM) od demand-response management (DRM) which can shift flexible loads to a time when more electricity will be available from VRES [5];
- to couple VRES generators with energy storage (on a smaller scale batteries [6], fuel cells [7], compressed air [8], pumped storage hydroelectricity [9] to overcome the problem of mismatch between supply and demand;
- to over-size the VRES peak generation in such a way that when the supply is greater than demand the remaining power can be used to produce hydrogen for transportation and heating purposes;
- to store the energy in electric vehicles which are believed to play a significant role in future transport systems [10];
- to apply forecasting methods, for predicting the energy/power generation from selected VRES sources like wind [11] solar [12] to better plan the power system operation.

From the perspective of preparing a strategy for the power system development all above mentioned concepts may play a significant role and heavily impact the decisions which are about to be made. Usually, it is not sufficient to choose only one solution because the intermittent nature of solar and wind generation and added to it variability of energy demand is too. Therefore it is very common to combine the effect of complementary in time domain solar and wind resources with energy storage in form of batteries or pumped storage hydroelectricity depending on the installed capacity and load to be covered.

Unfortunately, the spatial and temporal complementarity of wind and solar resources is not a common problem addressed by meteorologist or developers of national power systems (the advent of VRES in many energy systems is an issue of several last years). Therefore, the aim of this paper is to present the current state of the art in area of studies on the temporal and spatial complementarity between wind and solar energy sources. Both from the worldwide and national (Poland) perspective. To expand the current state of knowledge we will perform an analysis of spatial and temporal complementarity between wind and solar resources on the area of Lower Silesia which is a district laying in the south-western part of Poland. Apart from assessing their complementarity on an hourly, daily and monthly time scale we will address the problem of their relation to the load curve.

1.1 Worldwide studies

Usually in the scientific literature the impact of variable renewable energy sources temporal and spatial complementarity is investigated in order to show that the sudden drop in power generated in one source can be replaced by a simultaneous increase in power provided by another source. A common way to assess the complementarity on a various temporal and spatial scales is to use the coefficient of correlation (CC) which ranges from -1 to 1. Where "-1" denotes a perfect complementarity and "1" means that two considered sources tend to exhibit exactly the same energy generation patterns. In reality the CC values close to "-1" are not observed. The idea laying behind the energy sources temporal complementarity can be explained by means of a two sine functions from which one is out of phase by π (see Fig. 1).

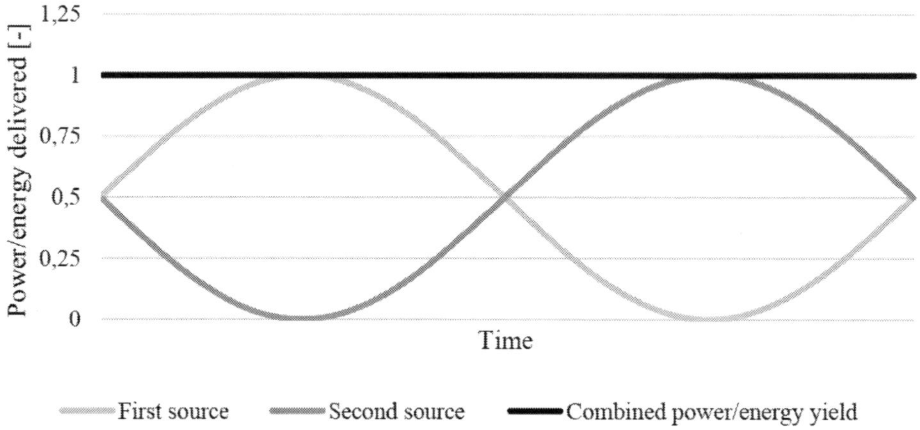

Fig. 1. Ideal temporal complementarity between two energy sources.

The available scientific literature concentrated mainly on individual sites for which measurements on solar radiation and wind speed were available as well as on larger regions (country, district level) where data has been interpolated based on measurements available from several stations.

For an Ontario province of Canada [13] a three years long hourly time series of solar radiation and wind speed were used to assess their spatial and temporal complementarity. The results presented clearly show that spatial dispersion of wind and solar resources smooths their power output. Another research performed for Italy based on a one year long hourly time series [14] based on a Monte Carlo approach proved their strong temporal complementarity on a monthly time scale. Interesting study performed for northeast Brazil [15] has shown that availability of solar and wind resources during the dry period can be used to reinforce hydropower in order to increase the water retention for irrigation purposes. An analysis performed for China's solar and wind resources [16] has shown that their proper spatial distribution can reduce the occurrence of zero and peak power yield hours. A paper [17] analysing solar and wind complementarity with regard to the load demand in Australia has shown that combined resources can effectively deliver energy to the grid when it experiences peaks. [18] applied a simulated annealing method to optimally distribute the solar and wind generators over the Iberian Peninsula and obtained results support the existence of a beneficial spatio-temporal complementarity. Another analysis [19] performed for a Fernando de Noronha Island (Brazil) has shown that both processes (wind speed and solar radiation) demonstrate persistent correlations and cross-correlations. Over the recent months an important paper [20] which summarizes the recent research in area of VRES integration and ways of mitigating their inherently variable nature has been published. This review paper concludes that there exist three climatic regimes which drive the variability of renewable energy sources and the load.

1.2 Related works on solar/wind complementarity in Poland

In Poland the problem of wind and solar resources complementarity has not been widely addressed. This results from the fact that renewable energy sources (RES) in Poland are commonly perceived as a threat to the coal dominated energy sector. In general opinion they not only may disrupt its operation but in consequence decrease the demand on national abundant hard and brown coal extraction. However, the European Union regulations oblige Poland to increase the share of RES in our energy portfolio. This however, cannot be entirely realized based on dispatachble (biomass, geothermal or biogas) resources from several mainly economic reasons. Therefore, the picking up role of PVs and WTs in Polish energy sector in unavoidable.

Hereunder, we summarize the so far conducted research. A study published in the year 2013 [21] for solar and wind conditions similar to Polish concludes that an even mix in terms of installed capacity in PVs a WTs should be optimal for Polish power system considering the temporal complementarity of those two resources. A three years later published paper [22] investigated the temporal complementarity of wind, solar and water resources in Piła (north-western Poland). The obtained results have shown that a beneficial complementarity between wind and solar energy on a monthly scale and between solar and hydropower on an annual level. A research conducted in the area of Energa – Operator SA (a local power system operator) [23] was based on real data from existing wind and solar parks. In general when high power output was observed in case of wind generation than the power generated was significantly smaller and *vice versa.* Another research paper [24] analysed the temporal complementarity of wind, solar and hydrokinetic energy source in two locations (Lębork and LądekZdrój). The observed there values of solar and wind complementarity were not very spectacular however it is important to note that this analysis was conducted form a wind speed measured at 10 meters above sea level which is rather typical for residential not commercial scale wind turbines. The results presented in [22, 24] has been expanded for additional locations in Poland and authors developed a mathematical model which aims at selecting the optimal distribution of wind and solar parks in Poland (considering the spatio-temporal complementarity) aiming at minimizing the variability of their combined energy yield. Presented in the next sections analysis are a follow up of the results presented in [21-25] and investigate the spatial and temporal complementarity of wind and solar resources in Lower Silesia.

2 Data and methods

First of all it is important to underlie that for the purpose of our research we have divided the considered region (Lower Silesia voivodship) into 74 smaller squared regions (20 by 20 kilometres) which centres of gravity fit into its borderlines. However, the data from ground measurements is not available for all selected locations. Therefore the first step of our analysis was to compare satellite and ground measurements in terms of their coherence. Results of conducted analysis are presented on Fig. 2 and 3. Clearly the solar radiation measurements obtained from satellite and ground stations exhibit significantly greater correlation. The relatively low values of coefficient of determination (R^2) in case of wind speeds can be explained by the fact that wind is locally definitely more variable. What is more ground measurements are usually representative for selected location whereas satellite ones have much more coarse resolution (20x20 kilometres). Considering above we have decided to base our research on satellite measurements.

For the purpose of our calculations we have used hourly irradiation and wind speed time series available at [26] covering year 2010-2014. We have also used six time series obtained from [27] which were generated based on the ground measurements. Time series from [26], were then validated against those from [27], (please refer to section 2.2 for deeper analysis). The obtained results justified the application of satellite generated time series from [26] for the purpose of our research. In case of wind speed time series (measured at 10 m.a.s.l.) we have used the power law [28] to calculate the wind speed which might be observed at the hub height (100 meters) of the considered Vestas V90 wind

 DOI: 10.1051/e3sconf/20172200074

turbine. To juxtapose energy yield patterns with energy demand we have used hourly energy demand time series representative for Poland - such data has been downloaded from [29]. Over the year 2014 the cumulative energy consumption in Lower Silesia amounted to 13.8 TWh.

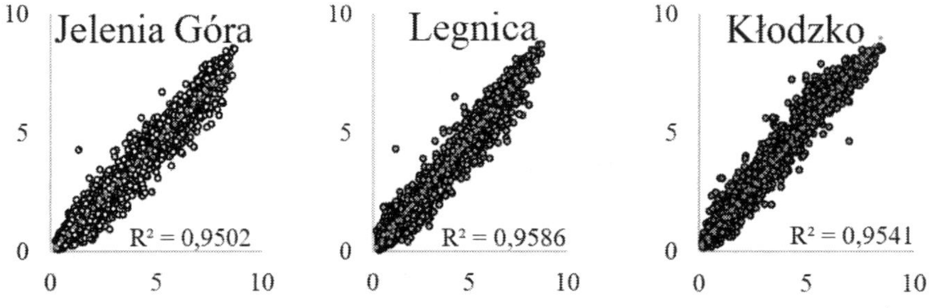

Fig.2. Solar irradiation measurements from satellite (OY) and ground measurements (OX) measurements both in [kWh/day] – daily sums over 2010-2014.

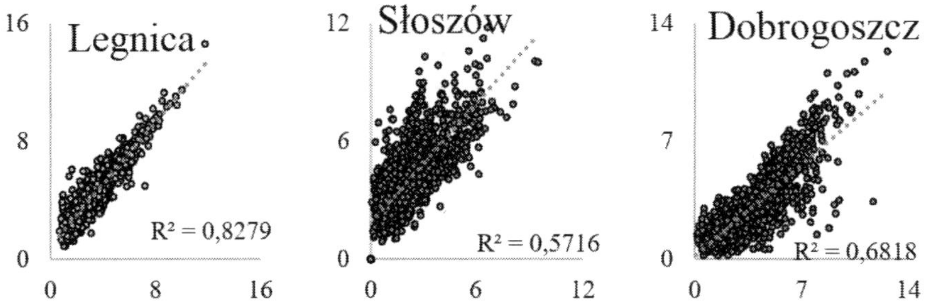

Fig.3. Wind speed measurements from satellite (OY) and ground measurements (OX) measurements both in [m/s] - daily meansover 2010-2014.

2.1 Energy yield estimation

The energy yield from $E^{PV}_{i,j,k}$ (kWh) a photovoltaics generator P^{PV}_{k} (kW) in given location k and time i,j(here time step t = 1hour)depends on global horizontal irradiation $H_{i,j,k}$ (kWh/m^2) and air temperature $T_{i,j,k}$ [30], temperature dependent efficiency reduction factor μ, cell temperature in the standard testing conditions T^{STC} [31] and the overall efficiency η^{PV} of all remaining system components (inverter efficiency, wire losses, shading etc).

$$E^{PV}_{i,j,k} = \frac{H_{i,j,k}}{H^{STC}}\left[1 - \mu\frac{1}{{}^{\circ}C}\left(T_{i,j,k} - T^{STC}\right)\right] P^{PV}_{k}\eta^{PV} t \qquad (1)$$

The energy generated $E^{WT}_{i,j,k}$ (kWh) by a given type of wind turbine over time t = 1 hour (here Vestas V90 with rated power P^{WT}_{k} =2000 kW), depends on the observed wind speed $v_{i,j,k}$(m/s) at the hub height as well as the parameters of the turbine itself especially: power curve $p(v)$ (kW), wind cut in v_{cut-in}, rated v_{rated} and cut off $v_{cut-off}$speeds (m/s).

$$E^{WT}_{i,j,k} = \begin{cases} 0\ for\ v_{i,j,k} \in (0, v_{cut-in}) \cup \left(v_{cut-off}, +\infty\right) \\ P^{WT}_{k} t\ for\ v_{i,j,k} \in \left(v_{rated}, v_{cut-off}\right) \\ P^{WT}_{k} p(v) t\ otherwise \end{cases} \qquad (2)$$

 DOI: 10.1051/e3sconf/20172200074

3 Results and discussion

For the purpose of spatio-temporal analysis of wind and solar energy source complementarity we have neglected the maximal capacity of each region to accommodate given energy source and assumed a uniform distribution of 2 MW wind parks and 2 MW PV installations across the whole voivodship (in each squared region 2 MW in WTs and 2 MW in PVs). After that we have investigated the value of the coefficient of correlation between aggregated (from all 74 regions) wind and solar energy yield as well as the observed load on an hourly, daily and monthly time scale. Obtained results are presented in Table 1 whereas Figure 3 visualizes the annual variability of those three phenomena.

As can be observed in Table 1 the all considered phenomena (load, and energy of solar radiation and wind speed) does not exhibit any significant correlation on an hourly time scale. This implies that from the perspective of the whole voivodeship it is very unlikely that the wind or solar generation will follow the load profile. This will result in a situation in which energy surpluses (supply from WTs and PVs is greater than demand) and deficits (supply from WTs and PVs is smaller than the demand) will occur. In such circumstances it might be very hard to optimize the operation of the whole energy system. Therefore, it will be mandatory to use some form of energy storage which will compensate for the varying energy output and changing energy demand. On a daily time scale (which considered the daily sums of energy generated by PVs and WTs as well as the whole daily energy consumption) we have observed a negative correlation between the load and the energy yield from PVs. This may indicate that usually days with higher observed values of irradiation will be simultaneously days when lower demand is observed. However, this not necessarily must be true because the energy demand patterns are changing and it is being observed that not only low temperatures (winter period) lead to an increase in energy consumption but also high ones [32]. This indicates a strong potential for PVs to cover peak loads resulting from intensive utilization of air-conditioning.

Table 1. Wind and solar energy complementarity and their relation to the load.

	Time scale					
	Hourly		Daily		Monthly	
	Solar	Load	Solar	Load	Solar	Load
Wind	0.0975	0.1631	-0.299	0.1319	-0.6714	0.7264
Solar	X	0.1835	X	-0.4999	X	-0.9210

The results of the energy yield and demand based on monthly means is presented on Fig. 4. As can be observed the load pattern does not exhibit any significant variability (for further reading on its changes we refer the interested readers to [32]).Therefore on a monthly time scale PVs exhibit a significant negative correlation with the energy demand. However, a positive (beneficial) positive CC value between load and wind energy is also observed. This indicated that the annual pattern of wind energy generation tends to follow the load. The analysis on a monthly scale also provides us with information on the complementarity between solar and wind energy sources. Observed here value of CC (-0.67) indicates that this interchangeability is not perfect and there is a need to couple both of those energy sources with a large scale energy storing device. An additional correlation analysis was performed for combined energy yield from PVs and WTs with the energy demand. The observed values of CC ranged from slightly above 0 to 0.2 which shows that energy generation from those two variable energy sources does not exhibit any significant correlation with load. This lead to a conclusions that without other measures/techniques in form of: energy storage, V2G, DMS or energy yield forecasting the integration of those to VRES based purely on their complementary will be a very complicated task.

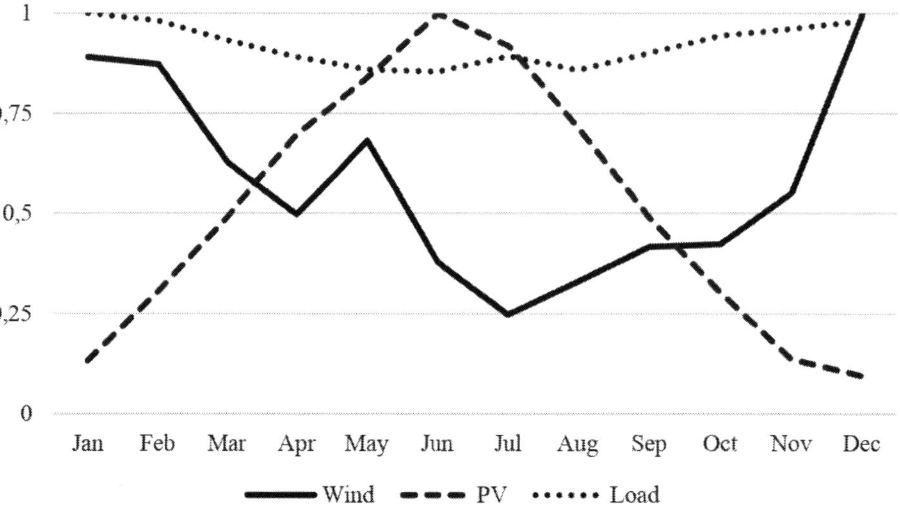

Fig.4. Normalized monthly values of wind and solar energy yield from all 74 regions juxtaposed to the load.

On the Fig. 5 we have presented separately the coefficient of correlation between hourly energy yield from PVs and wind generation in function of distance between individual considered sites. Results indicate that in the investigated area the solar and wind energy sources does not exhibit complementarity on a spatial scale. Values of CC smaller than 1 may only lead to smoothing the energy generation curve. Such phenomena can be observed based on the coefficient of variability (CV) which for hourly wind generation values has been reduced by 6% (from 128% in case of mean CV calculated for each *k* separately to 122% calculated for aggregated wind generation from all sites). For PVs this smoothing effect was slightly smaller and amounted to 3% (a reduction from 88% to 85%).

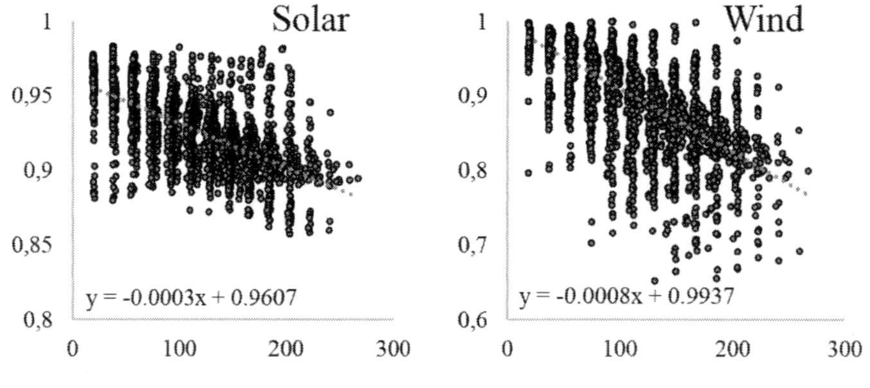

Fig.5. Solar and wind energy yield spatial complementarity, (OY) coefficient of correlation, (OX) distance between sites in kilometres.

4 Conclusions

Conducted here analysis is the first step to assessing the possibility of covering the energy demand of whole Lower Silesia based purely on wind, solar and hydropower (abundant here due to the proximity of mountain ranges) and energy storage in form of pumped-storage hydroelectricity. Conducted preliminary calculations has shown that existing temporal complementarity of wind and solar energy sources combined with energy storage in form of PSH may be a good basis for PV-WT-PSH hybrid development.

References

1. L.E. Jones, *Renewable energy integration: practical management of variability, uncertainty, and flexibility in power grids* (Academic Press, 2014).
2. M.Z. Jacobson, M.A. Delucchi, Energ policy, **39**, 3 (2011).
3. T. Brown, D. Schlachtberger, A. Kies, M. Greiner, *Sector Coupling in a Highly Renewable European Energy System* (15th Wind Integration Workshop, Vienna, 2017).
4. J. Jurasz, B. Ciapała, Appl Energ, **198** (2017).
5. A. Kies, B.U. Schyska, L. von Bremen, Energies, **9**, 11 (2016).
6. J. Jurasz, J. Mikulik, Architecture, Civil Engineering, Environment, **4** (2015).
7. R. Szczerbowsk, B. Ceran, E3S Web Con, **10**, 00090 (2016).
8. A. Cavallo, Ener, **32**, 2 (2007).
9. J. Jurasz, J. Mikulik, M. Krzywda, E3S Web Con, **17**, 00035 (2017).
10. X. Wu, X. Hu, S. Moura, X. Yin, V. Pickert, J. Power Sour, **333** (2016).
11. J. Kleissl, *Solar energy forecasting and resource assessment* (Academic Press, 2013).
12. X. Wang, P. Guo, X. Huang, Energy procedia, **12** (2011).
13. C.E. Hoicka, I.H. Rowlands, Renew Energ, **36**, 1 (2011).
14. F. Monforti, T. Huld, K. Bódis, L. Vitali, M. D'isidoro, R. Lacal-Arántegui, Renew Energ, **63** (2014).
15. P. De Jong, A.S. Sánchez, K. Esquerre, R.D.A. Kalid, E.A. Torres, Renew Sust Energ Rev, **23** (2013).
16. Y. Liu, L. Xiao, H. Wang, S. Dai, Z. Qi, Sci China Technol Sc, **56**, 3 (2013).
17. Y. Li, V.G. Agelidis, Y. Shrivastava, *Wind-solar resource complementarity and its combined correlation with electricity load demand* (4th IEEE Conference on Industrial Electronics and Applications, 2009).
18. S. Jerez, R.M. Trigo, A. Sarsa, R. Lorente-Plazas, D. Pozo-Vázquez, J.P. Montávez, Energy Procedia, **40**, 48-57 (2013).
19. P.S. dos Anjos, A.S.A. da Silva, B. Stošić, T. Stošić, Physica A, **424** (2015).
20. K. Engeland, M. Borga, J.D. Creutin, B. François, M.H. Ramos, J.P. Vidal, Renew Sust Energ Rev, **79** (2017).
21. R. Hyrzyński, M. Karcz, M. Lemański, K. Lewandowski, S. Nojek, Acta Energetica, **4**, 17 (2013).
22. J. Jurasz, A. Piasecki, Acta Energetica, **2**, 27 (2016).
23. M. Bajor, P. Ziołkowski, G. Widelski, Acta Energetica, **3**, 28 (2016).
24. J. Jurasz, A. Piasecki, M. Wdowikowski, E3S Web Con, **10**, 00032 (2016).
25. J. Jurasz, J. Mikulik, Przegląd Elektrotechniczny **93**, 7 (2017).
26. http://www.soda-pro.com
27. http://www.imgw.pl/
28. E.W. Peterson, J.P. Hennessey Jr, J Appl Meteorol, **17**, 3 (1978).
29. http://www.pse.pl/

DOI: 10.1051/e3sconf/20172200074

30. M. Hanif, M. Ramzan, M. Rahman, M. Khan, M. Amin, M. Aamir, ISESCO J Sci Technol, **8**, 14 (2012).
31. J.A. Duffie, W.A. Beckman, *Solar engineering of thermal processes* (New York Wiley Interscience, 1991).
32. J. Jurasz, J. Mikulik, A. Piasecki, Przegląd Elektrotechniczny, **92**, 9 (2016).

Anaerobic co-digestion of sewage sludge and molasses

Katarzyna Kalemba[1,*], and *Krzysztof* Barbusiński[1]

[1]Silesian University of Technology, Faculty of Energy and Environmental Engineering, Poland

Abstract. The efficiency of simultaneous digestion of sewage sludge and by-product of refining sugar beets (molasses) was investigated. The study was conducted for 28 days under mesophilic conditions. 0.5%, 1%, 1.5%, 2% and 3% (m/m) of molasses was added to the mixture of sludge. The result of the study showed that addition of molasses had positive effect the biogas production. The biggest biogas yield was achieved in sample with 0.5% of molasses (95.69 mL/g VS). In this sample biogas production increased by 21% in comparison with reference sample (without molasses). The biggest methane content (73%) was also observed in the sample with 0.5% of molasses. For comparison in reference sample was produced biogas with 70% content of methane. The dose over 0.5% of molasses caused inhibition of fermentation process. The minimal degree (38%) of degradation of organic matter was achieved in reference sample (38.53%) and in sample with 0.5% of molasses (39.71%) but in other samples was in the range of 35.61-36.76 % (from 3% to 1%, respectively). Digestion process have adverse effect on dewatering properties of sludge. Before co-digestion capillary suction time was from 31 s to 55 s, and after process increased from 36 s to 556 s (from 0% to 3% of molasses, respectively).

1 Introduction

Anaerobic co-digestion is process of simultaneous digestion two or more substrates come from different source, occur in four phases: hydrolysis, acidogenesis, acetategenesis and methanogenesis. The addition of co-substrate is the option to overcome the drawback of mono-digestion e.g. low organic load of sewage sludge, low organic load and high nitrogen concentration of animal manures, improper material and relatively high concentration of heavy metals in organic fraction of municipal solid waste, lack of nitrogen in crops and agro-industrial wastes and high concentration of nitrogen and long chain fatty acids in slaughterhouse wastes. Co-digestion is more economical process than mono-digestion because enhance the methane production. Moreover, co-digestion process lead in one fermentation tank could produce more methane than fermentation process conduct in two digesters. Another benefits of co-digestion are acceleration the digestion process and bringing higher degradation rates [1, 2].

* Corresponding author: katarzyna.kalemba@polsl.pl

In the world in 2015/2016 year (from October to September) was produced 165,8 million t of sugar from sugar beet [3]. From each kilogram of sugar beet is obtained 75% water and 25% of dry substance. From dry substance is formed a product and the rest is turned into a by-product. The main by-product of the sugar factory are: pulp, carbonation-lime residue and molasses. First of all the sugar beets are washed, sliced and soaked in hot water to begin the process that separates sugar from the rest of the plant. As a result of extraction is made the hot sugary liquid and pulp. To hot sugary liquid is added lime and CO_2 to purification. The mixture is filtered so the runny syrup is made. As a by-product is obtained carbonation-lime residue. The runny syrup is evaporated and crystallization. At the end, hot water is sprayed over the light brown crystals to remove the remaining molasses, leaving pure naturally white sugar crystals. The sugar crystals are then dried [4]. Beet molasses contain 48% sugar (namely sucrose and invert sugar) and 79.5% dry substances. It is the most valuable by-product of the sugar factory because is used as a carbon source in the yeast fermentation and ethanol fermentation, as animal feed, as fertilizer and for biogas production [5, 6]. There are several reports of co-digestion molasses with other substrates [7–9] but is no report about co-digestion of molasses with sewage sludge.

The main objective of this paper was to examine the process of anaerobic co-digestion of sewage sludge and molasses. The attention was focused on biogas production, the composition of biogas, the VFA/A ratio and dewatering properties. The efficiency of the process in terms of molasses ratio was also examined.

2 Methods

Characteristics of the substrates

Mixed thickened surplus activated sludge from municipal sewage treatment plant and by-product of refining sugar beets (molasses) was used in this investigation. Surplus activated sludge was obtained from sludge treatment system. Then, surplus activated sludge was inoculated by fermenting sludge from the digester. Molasses was obtain from yeast plant. The sample of molasses were dissolved by deionized water because of thick texture. The characteristics of supernatant liquor and molasses are shown in table 1. The sludge mixture (surplus activated sludge with 50 g of fermenting sludge) was mixed with 2, 4, 6, 8 and 12 g of molasses what corresponds to 0.5%, 1.0%, 1.5%, 2.0% and 3.0% ratio of molasses in the samples. The sludge without co-substrate was a reference. The weight of each sample was 400 g.

Table 1. Characteristics of supernatant liquor and molasses.

Parameter	Unit	Supernatant liquor	Molasses
pH	-	7.1	8.5
COD	mgO_2/L	603.2	13300.0
Volatile fatty acid (VFA)	mg CH_3COOH/L	< 28	545
Alkalinity (A)	mg $CaCO_3/L$	475	935

Laboratory installation

The sludge fermentation study was conducted for 28 days under static conditions, in 12 airtight glass flask (reactors) with a capacity of 500 mL each, at the mesophilic conditions (37°C). To keep temperature constant, the flasks were placed in a water bath. The installation was equipped with a device to collect biogas. The flasks were not mixed, therefore the mixture was stirred 2 times a day without disconnecting devices to receive biogas.

Analysis

Before and after the experiment the following analyses were carried out:

- total solids (TS) and volatile solids (VS) according to PN-EN 12879:2004 [10] and PN-EN 12880:2004 [11];
- pH according to PN-EN 12176:2004 [12];
- total alkalinity according to PN-91/C-04540/05 [13];
- volatile fatty acid made based on cuvette test Hach Lange;
- capillary suction time according to PN-EN 14701-1 [14].

The volume of biogas was recorded every day and the composition of the produced biogas was measured by Multitec 540 analyser (Sewerin).

3 Results and discussion

Efficiency of co-digestion

The results of the study showed that addition of molasses have positive influence on the cumulative biogas production. Fig. 1 shows the biogas yields in term of VS of the tested ratio of molasses. After 28 days of co-digestion the biggest biogas yield was reached in a sample with 0.5% of molasses (95.69 mL/g VS). In this sample biogas yield increased by 21% in comparison to the reference sample (79.17 mL/g VS). Additionally, in sample with 0.5% of molasses continued increase of biogas production was observed. However, enhancement the content of molasses over 0.5% resulted the inhibition of biogas production due to organic overloading but in sample with 1% of molasses the biogas production (75.55 mL/g VS) was similar to the reference sample. The lowest biogas yield was noticed in the sample when molasses fraction was 3% (27.20 mL/g VS).

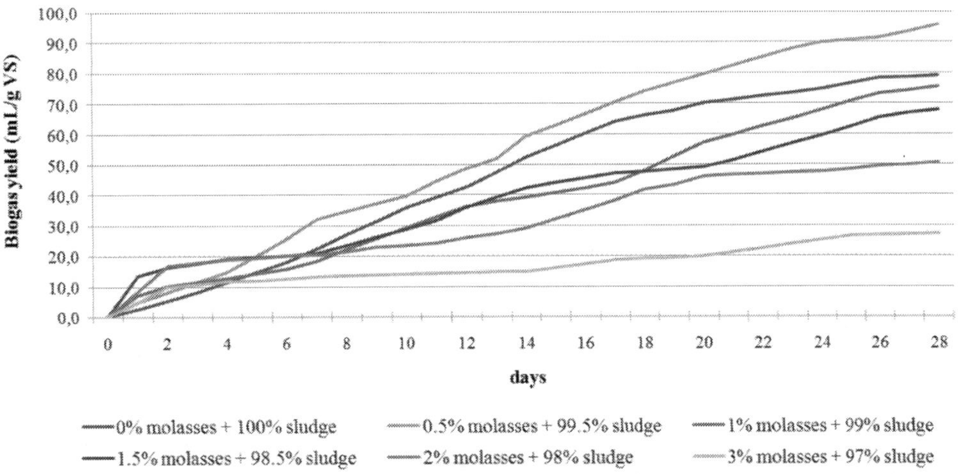

Fig. 1. Biogas yield during anaerobic digestion.

The addition of molasses had impact in the composition of biogas. Biogas produced in the anaerobic digestion of sewage sludge usually contains about 60% methane [15]. The results of this study confirm that addition of substrates had a positive influence on methane content in biogas (Fig. 2). In the reference sample biogas contained up to 70% of methane. The biggest content of methane in biogas (up to 73%) was detected in the sample with 0.5% of molasses. After 14 day of fermentation process in reference sample and in the sample with 0.5% of molasses the same methane content (65%) was reached. This situation continued up to 22 day of fermentation when methane content in both of sample reached 70%. Later methane content in reference sample remained on the constant level. However, in sample with 0.5% of molasses the ratio of methane enhanced to 73%. It could be explained by a long time adaptation of microorganisms to molasses, probably due to the low susceptibility of molasses to biological degradation under anaerobic conditions. The lowest methane content (28%) was observed in the sample with 3% of molasses. As expected, this result was in agreement with the amount of biogas yield.

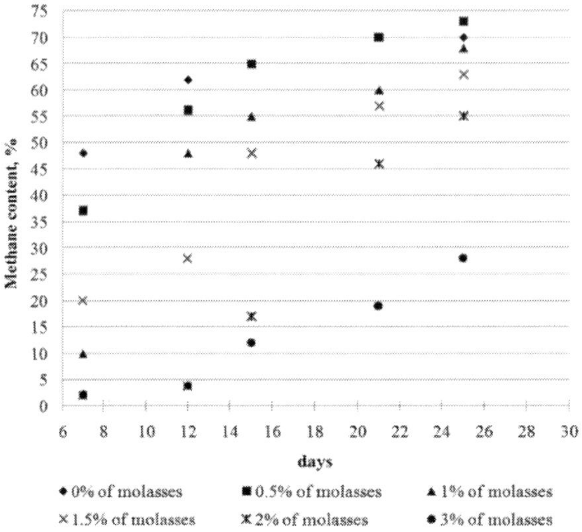

Fig. 2. The methane content in biogas during anaerobic digestion.

The degree of degradation of organic matter was calculated based on the changes in concentration of volatile solids before and after anaerobic stabilization (Tab. 2). The biggest degree of degradation (39.71%) was observed in sample with the biggest production of biogas (0.5% of molasses). According to Environmental Protection Agency the minimal reduction of volatile solids should be 38% [16]. In reference sample and in the sample with 0.5% of molasses the minimal degree of anaerobic stabilization was achieved. In the other sample sufficient degradation of organic matter was not reached but results (35.80%-36.76%) were very close to minimal degree of degradation of organic matter.

Table 2. The degree of degradation of organic matter.

The amount of molasses, %	Volatile solids, g/L		The degree of anaerobic stabilization, %
	Before co-digestion	After co-digestion	
0	31.66	19.46	38.53
0.5	42.66	25.72	39.71
1.0	43.88	27.75	36.76
1.5	44.17	28.40	35.71
2.0	47.71	30.72	35.61
3.0	50.36	32.33	35.80

The change of VFA/A ratio

Concentration of VFA has a significant influence on the efficiency of anaerobic digestion. VFA is an important mid-product in the formation of methane. The result of study showed that the amount of the volatile fatty acids had effects the anaerobic digestion process. The higher ratio of molasses the higher concentration of VFA (Tab. 3). After co-digestion process in sample over 0.5% of molasses accumulation of VFA was observed. As a result, the methanogenesis was inhibited and pH was decline. Therefore, decrease of biogas production in sample over 0.5% was detected.

Examining concentration of VFA and A is recommended strategy to control co-digestion process. The VFA/A ratio up to 0.3 provides a balance between acidogenic and methane bacteria [17]. The results of the study confirmed this literature report. In the samples with 1%, 1.5%, 2% and 3% of molasses where ratio was over 0.3 the biogas yield, the methane content in biogas and the reduction of organic matter was lower than in reference sample.

Table 3. The physical-chemical analysis of supernatant liquid.

The amount of molasses, %		0	0.5	1.0	1.5	2.0	3.0
pH	Before co-digestion	7.1	7.1	7.1	7.3	7.2	7.3
	After co-digestion	7.8	7.9	7.5	7.4	7.0	5.7
A, mg $CaCO_3$/L	Before co-digestion	475	400	375	450	375	475
	After co-digestion	2690	3019	2805	3063	2755	2465
VFA, mg CH_3COOH/L	Before co-digestion	28	66	117	166	210	269
	After co-digestion	73	163	1106	1982	2743	3521
VFA/A	Before co-digestion	0.06	0.17	0.31	0.37	0.56	0.57
	After co-digestion	0.03	0.05	0.39	0.65	1.00	1.43

Dewatering properties

The addition of molasses impacted dewatering properties of sludge. The capillary suction time (CST) test is widely used method to determine the filterability and the easiness of removing moisture from slurry and sludge [18]. Generally, digestion process have adverse effect on dewatering properties of sludge. The capillary suction time enhanced after co-digestion process in each sample (Fig. 3). In the reference sample CST increased from 31 s to 36 s but in other samples expansion was more significantly. After co-digestion, with increasing content of molasses, CST increased from 111 s to 556 s (from 0.5% to 3% of molasses, respectively).

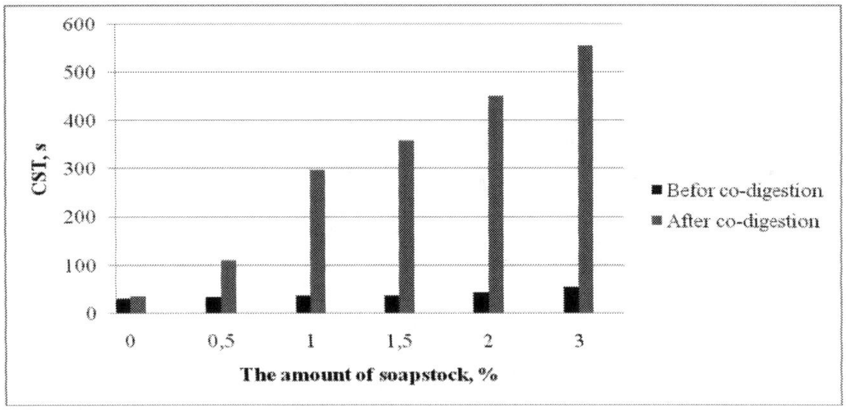

Fig. 3. Changes of CST value for different contents of molasses.

Comparison of biogas production for a different co-substrates at their optimum content

The authors of this publication also done research process of co-digestion with other organic co-substrates. In this investigations were used: waste meat, soapstock, by-product after biodiesel production and waste from fat separator. These compounds have not yet been widely investigated as substrates for fermentation with sewage sludge. However, the results of such applications can be found in a few publications [19–22]. Table 4 shows the results of the maximum biogas production and methane content in biogas under optimum conditions of the fermentation process for different organic co-substrates. Methane content in biogas was the biggest using waste meat (76%). The similar methane content (72–73%) was reached in soapstock, by-product after biodiesel production and molasses investigation. The lowest methane content in biogas was achieved in waste from fat separator investigation. The maximum biogas production using molasses was 95.69 mL/g VS and was only slightly higher than in case of application of waste from fat separator. For other substrates the biogas production was higher. However, compared to other substrates, the maximum biogas production and a relatively high methane content using molasses was obtained with the smallest content of this substrate (0.5%) in a mixture with sewage sludge. Therefore, molasses appears to be a substrate, which should be considered for use in wastewater treatment plants in the process of co-digestion, especially in the case when the digesters have a small supply of additional capacity.

Table 4. The maximum biogas production and methane content in biogas for different organic co-substrates.

Organic co-substrates	**Maximum biogas production (mL/g VS)**	**Methane content in biogas (%)**	**Content of substrate in the sample (%)**
Waste meat	170.30	76	3
Soapstock	117.17	73	10
By-products after biodiesel production	99.91	72	10
Molasses	95.69	73	0.5
Waste from fat separator	88.68	70	3

4 Conclusion

Addition of molasses had positive influence on biogas production. The biggest biogas yield was reached in sample with 0.5% of molasses and was 21% higher than in reference sample. Co-digestion process of sludge and molasses influenced on biogas composition. In sample with 0.5% of molasses methane content increased to 73%. The higher dose of molasses than 0.5% caused inhibition of the digestion process, thus 0.5% of molasses is optimum dose. According to the EPA regulations, the degrees of degradation of organic matter obtained in reference sample and in sample with 0.5% of molasses were significant, because it exceeded 38%. In the other samples sufficient degradation of organic matter was not reached, but obtained results (35.61%-36.76%) were very close to 38%. The result of VFA/A ratio indicated that 0.5% of molasses is the optimum dose because higher content caused inhibition of co-digestion. Generally, digestion process have adverse effect on dewatering properties. After co-digestion, with increasing content of molasses, CST increased from 36 s to 556 s (from 0% to 3.0% of molasses, respectively). This information should be taken into account in the design of sludge dewatering process.

Based on the obtained results it can be stated that molasses is a substrate which can be used in wastewater treatment plants in the process of co-digestion, especially in the case when the digesters have a small supply of additional capacity.

References

1. J. Mata-Alvarez, J. Dosta, M.S. Romero-Güiza, X. Fonoll, M. Peces, S. Astals, Renew. Sust. Energ. Rev. **36**, 412–427 (2014)

2. P. Aichinger, T. Wadhawan, M. Kuprian, M. Higgins, Ch. Ebner, Ch. Fimml, S. Murthy, B. Wett, Water Res. **87**, 416–423 (2015)
3. Sugar production, United State Department of Agriculture (USDA), November 2016
4. M. Asadi, *Beet-Sugar Handbook* (John Wiley & Sons, Inc. 2007)
5. Ch. Fang, K. Boe, I. Angelidaki, Bioresour. Technol. **102**, 1005–1011 (2011)
6. X. Wu, H. Lin, J. Zhu, Bioresour. Technol. **136**, 351–359 (2013)
7. I. Rodriguez-Verde, L. Regueiro, M. Carballa, A. Hospido, J.M. Lema, Sci. Total Environ. **497–498**, 475–483 (2014)
8. P. Kongjan, S. O-Thong, I. Angelidaki, Eng. Life Sci. **13**, 2, 118–125 (2013)
9. Ch. Fang, K. Boe, I. Angelidaki, Bioresour. Technol. **102**, 1005–1011 (2011)
10. PN-EN 12879:2004 Charakterystyka osadów ściekowych. Oznaczanie strat przy prażeniu suchej masy osadu (Characteristics of sludge. Determination of the loss of ignition of dry mass).
11. PN-EN 12880:2004 Charakterystyka osadów ściekowych. Oznaczanie suchej pozostałości i zawartości wody (Characteristics of sludge. Determination of dry residue and water content).
12. PN-EN 12176:2004 Charakterystyka osadów ściekowych. Oznaczanie wartości pH. (Characteristics of sludge. Determination of pH value).
13. PN-91/C-04540/05 Badania pH, kwasowości i zasadowości. Oznaczanie pH, kwasowości i zasadowości mineralnej i ogólnej w osadach ścieków miejskich. (Tests for pH, acidity and alkalinity. Determination of pH, mineral and total acidity and alkalinity in sludge of urban waste waters).
14. PN-EN 14701-1 Charakterystyka osadów ściekowych. Właściwości filtracyjne. Część 1: Czas ssania kapilarnego (CST). (Characteristics of sludge. Filtration properties. Capillary suction time; CST).
15. R. Wickham, B. Galway, H. Bustamante, L.D. Nghiem, Int. Biodeter. Biodegr. **113**, 3–8 (2016)
16. U.S. EPA, Pathogen and vector attraction reduction requirements, EPA/832/R-93/003, Washington, September, 1994.
17. J. Bień, K. Wystalska, *Osady ściekowe. Teoria i praktyka* (Wydawnictwo Politechniki Częstochowskiej, 2011)
18. O. Sawalha, M. Scholz, Environ. Technol. **28**, 12, 1377–1386 (2007)
19. K. Kalemba, K. Barbusiński, Ochr. Sr. **38**, 4, 21–24 (2016)
20. M. Torrijos, P. Sousbie, L. Badey, F. Bosque, Water Sci. Technol. **66**, 10, 2237–2242 (2012)
21. J.A.S. López, M.Á.M. Santos, A.F.C. Pérez, A.M. Martín, Bioresour. Technol. **100**, 5609–5615 (2009)
22. S. Luste, S. Luostarinen, Bioresour. Technol. **101**, 2657–2664 (2010)

Single and competitive adsorption of OMP_s by carbon nanotubes – mechanism and fitting models

Gabriela Kamińska[1,*], *Mariusz* Dudziak[1], *Jolanta* Bohdziewicz[1], and *Edyta* Kudlek[1]

[1]Institute of Water and Wastewater Engineering, Silesian University of Technology, Gliwice, Poland

Abstract. The adsorption of three organic micropollutants (diclofenac – DFN, pentachlorophenol – PCP and octylphenol – OP) on two kinds of carbon nanotubes (single walled carbon nanotubes – SWCNT and single walled carbon nanotubes with amine group – SWCNT-NH_2) was investigated, in single and bicomponent solution at pH 5. SWCNT-NH_2 had three times lower specific surface area than SWCNT. Significant differences were observed in sorption capacity of SWCNT and SWCNT-NH_2 for given chemicals. The sorption uptake changes in the following order: OP > PCP > DFN for SWCNT and DFN > PCP > OP for SWCNT-NH_2. A few times higher adsorption of OP on SWCNT came from low OP solubility in water in comparison to PCP and DFN. While, higher adsorption of DFN and PCP on SWCNT-NH_2 was a result of electrostatic attraction between dissociated form of these chemicals and positively charged SWCNT-NH_2 at pH 5. In adsorption from bicomponent solution, significant competition was observed between PCP and DFN due to similar adsorption mechanism on SWCNT-NH_2. Opposite tendency was observed for SWCNT, DFN did not greatly affect adsorption of PCP and OP since they were very easily absorbable by sigma-sigma interaction.

1 Introduction

Organic micropollutants (OMPs) such as endocrine disrupting compounds, pharmacuticlas, pesticides, polycyclic aromatic hydrocarbons, surfactants usually enter into wastewater after their industrial and domestic application [1–3]. Therefore, they occur in effluents, since the technologies used at conventional wastewater treatment plant does not allow their complete elimination [4, 5]. This is due to many of these contaminates are highly hydrophobic like PAHs, hardly soluble in water and toxic, like pesticides. A consequence of these properties is OMPs' bioaccumulation in environment, for example in animal's tissues and some environmental parts like sediments, soils [6, 7].

Last years many efforts have been made to find the most favourable process for micropollutants removal from different aqueous media. Among many methods to remove these contaminants, membrane filtration, advanced oxidation processes and adsorption have been widely proposed [8–11]. Although advanced oxidation and membrane filtration are very

* Corresponding author: gabriela.kaminska@polsl.pl

 DOI: 10.1051/e3sconf/20172200076

effective for the elimination of OMPs, these methods led to the generation of highly concentrated retentate and formation of undesirable for water oxidation by-products [12, 13].

Therefore, adsorption is considered as superb process, since the high-quality effluent is produced. As an alternative for well-known activated carbon, carbon nanotubes are considered as highly effective adsorbent for OMPs removal. This is due to much faster adsorption of contaminants on carbon nanotubes than on activated carbon [14]. Most of the adsorption studies were conducted for single adsorbate to describe adsorption mechanism or adsorbent potential. This approach does not bring useful information about real adsorption behaviour, for example in environmental water matrix containing a few different contaminates. In that conditions, we can observe cooperative or competitive behaviour, depending on affinity of coadsorbates to the adsorbent surface [15].

The aim of this study was to better understand the competition behaviour between OMPs with different properties (diclofenac, pentachlorophenol, octylphenol) during adsorption on carbon nanotubes. For that purpose, two kinds of experiments were conducted: (1) adsorption of OMPs from single solution, (2) adsorption of OMPs from bicomponent solution.

2 Materials and methods

2.1 Materials

Selected OMPs i.e. 4-octylphenol (OP), diclofenac sodium salt (DFN) and pentachlorophenol (PCP) were purchased from Sigma-Aldrich (Poland) in analytical purity grade. Their physico-chemical properties are presented in Table 1. The stock solution of OMPs was prepared with methanol (1 g/L). The adsorbate feed solutions (single and bicomponent) were prepared by diluting the stock solution with pure water. pH of feed solution was adjusted to 5 using 0.5 M NaOH.

Table 1. Selected properties of micropollutants.

Compound	Chemical structure at pH 5	Molecular mass[1] (g/mol)	log K_{ow}[1] (-)	Solubility in water 20°C [1] (mg/L)	pKa[1] (-)	Equivalent width[2] (nm)	Molar volume (cm^3/g)
4-tert-octylphenol (OP)	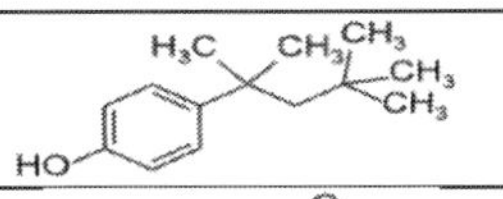	206.32	4.12	7	10.38	0.56	214.96
Diclofenac sodium salt (DFN)	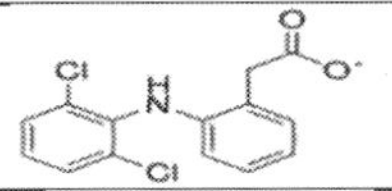	296.15	4.51	50	4.15	0.48	493.58
Pentachloro phenol (PCP)	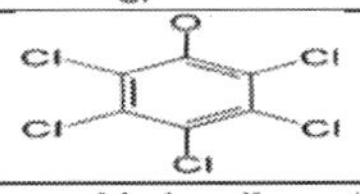	266.34	4.40	10	4.70	0.28	134.34

1 – https://pubchem.ncbi.nlm.nih.gov/compound/2336 2 – calculated with ChemBio3D Ultra 12.0

2.2 Carbon nanotubes and their characterization

Two different types of carbon nanotubes (CNTs): single walled carbon nanotubes (SWCNT) and single walled carbon nanotubes with amine group (SWCNT-NH_2) were used. CNTs were purchased from Chengdu Organic Chemistry Ltd. Characteristic of the nanotubes as provided by the manufacturer is shown in Table 2. Nitrogen adsorption-desorption isotherms of CNTs were estimated at 77 K by means of a volumetric adsorption analyzer ASAP 2010

(Micrometrics USA) and on this basis the BET surface area and porous structure parameters were calculated. Streaming potential of SWCNT was studied using electrokinetic analyser SurPASS™ 3 and then zeta potential was calculated from the Smoluchowski equation.

2.3 Adsorption isotherms experiments

The adsorption isotherms of PCP, DFN and OP were determined at 20°C in single and bicomponent solution system at pH 5. Two concentration levels of OMPs were used i.e. 0.1 and 0.5 mg/L. Experimentally, an adequate amount of a sorbent (from about 10 to 100 mg) was added into stoppered flasks containing 50 mL of single or bicomponent OMPs solution and shaked until equilibrium was reached. Then a sorbent was separated from the solution using PTFE syringe filter with a pore size of 0.45 μm and the resulting filtrates were analyzed for micropollutants concentration. The analytical procedure included solid phase extraction and chromatographic determination with HPLC. The amount of studied OMPs adsorbed on the nanotubes was determined by the following equation:

$$q_e = \frac{(C_0 - C_e) \cdot v}{m} \tag{1}$$

where: q_e (mg/g) is adsorbed amount of the adsorbate, C_0 and C_e (mg/L) are initial and equilibrium concentrations of the adsorbate, respectively, m (g) is weight of the sorbent, v (L) is a volume in which the adsorption was performed.

2.4 Data analysis

In this study, three common adsorption isotherm models including: (1) Langmuir model, (2) Freundlich model, and (3) Dubinin-Radushkevich model were used to fit experimental data. Each of the model is briefly described below.

The Langmuir equation assumes monolayer adsorption, where molecules interact only with the surface of sorbent. The form of the Langmuir isotherm is represented by the following equation:

$$q_e = \frac{a_L \cdot b \cdot C_e}{1 + b \cdot C_e} \tag{2}$$

where: a_L (mg/g) is the maximum adsorption capacity, and b (L/mg) is the Langmuir fitting parameter.

The Freundlich model is empirical and well describes adsorption on heterogeneous surface energy system. The model has significant importance for chemisorption and some cases of physisorption and can be written as:

$$q_e = K_f \cdot C_e^{\frac{1}{n}} \tag{3}$$

where: K_f ((mg/g) (L/mg)^n) is the Freundlich adsorption coefficient, n is the number describing surface heterogeneity and sorption intensity.

The Dubinin-Radushkevich model, which was originally proposed as an empirical adaptation of the Polanyi adsorption potential theory, has been the fundamental equation to quantitatively describe the adsorption of gases and vapours by microporous sorbents. The equation, based on the postulate that the mechanism for adsorption in micropores is that of pore-filling rather than layer-by-layer surface coverage, generally applies well to adsorption systems involving only Van der Waals forces and is especially useful to describe adsorption on microporous adsorbent. The Dubinin-Radushkevich equation is given in the following form:

$$\ln q_e = \ln a_{DR} - E^{-2} \cdot \varepsilon^2 \quad (4)$$

where: a_{DR} (mg/g) is the amount of adsorbate that can be adsorbed in micropores, can be obtained by plotting lnq_e as a function of ε^2, E (kJ/mol) is the adsorption energy, can be read from the slope of the line, ε is adsorption potential, is defined as:

$$\varepsilon = R \cdot T \cdot \ln\frac{C_s}{C_e} \quad (5)$$

where: R (8.314 $J \cdot mol^{-1} \cdot K^{-1}$) is the ideal gas constant, T (K) is temperature, C_s (mg/L) is the solubility in water.

The models were fitted to the experimental data using the Microsoft Excel spreadsheet with the Solver add-in. The least-squares method was used to determine the constants of the equations.

In order to compare competition strength between coadsorbates in bicomponent solutions following equation was used:

$$A = \frac{(K_{dA} - K_{dC})}{K_{dA}} \cdot 100\% \quad (6)$$

where: A (%) is competition strength K_{dA} and K_{dC} (L/g) are adsorption distribution coefficient for studied OMPs without and with coadsorbate, respectively, calculated for selected equilibrium concentration, obtained from equation:

$$K_d = \frac{q_e}{C_e} \quad (7)$$

3 Results

3.1 Morphology and texture of CNT

The pore size distribution (PSD) of the SWCNT and SWCNT-NH_2 are shown in Fig. 1 and their physical structure properties are listed in Table 2. SWCNT have three times larger surface area than SWCNT-NH_2 that suggests better sorption potential of SWCNT. From PSD, SWCNT are dominated with micropores, while the SWCNT-NH_2 structure is bidispersive, in large part consists of micropores and average fraction of mesopores.

The zeta potential versus pH is presented in Fig. 2. The charge of both kinds of nanotubes was positive at pH around 4 and negative at pH around 7. What is more important, the SWCNT-NH_2 have higher electric charge than SWCNT, which is attributed to dissociated form of amine group (NH_3^+) on their surface. The isoelectric point was found at pH 4.5 and 6.3 for SWCNT and SWCNT-NH_2, respectively.

Table 2. Textural characteristic of studied carbon nanotubes.

Name	Outer diameter [1] (nm)	Length [1] (µm)	Purity [1] (wt%)	Specific Surface Area (m^2/g)	Total Pore Volume (cm^3/g)	Total Area in Pores (m^2/g)
SWCNT	< 2	5–30	95	1124.26	0.88	945.56
SWCNT-NH_2	< 2	5–30	90	410.63	0.60	257.52

1 – Data provided by manufacturer

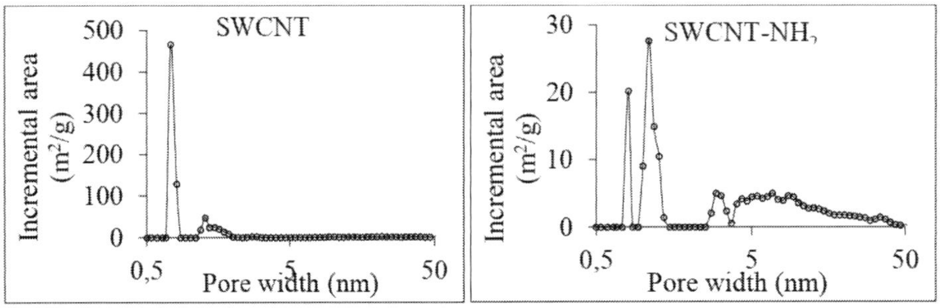

Fig. 1. Pore size distribution of SWCNT and SWCNT-NH_2.

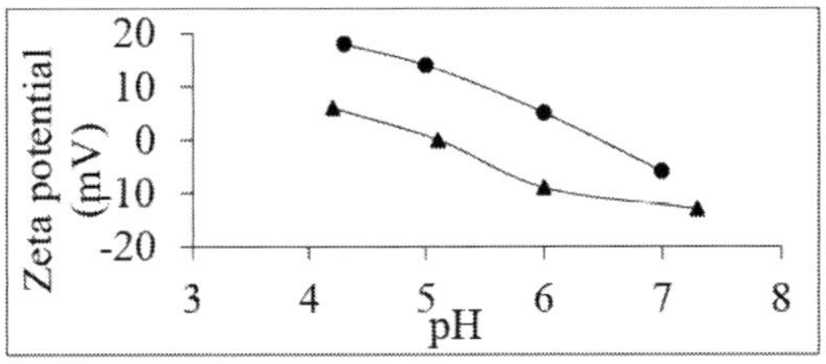

Fig. 2. Zeta potential curves of SWCNT (triangles) and SWCNT-NH_2 (circles).

3.2 Single adsorption of OMPs from water – isotherms and mechanism

The adsorption isotherms yield the most important information about the distribution of the adsorbate molecules between the liquid and the solid phase when the adsorption process reaches equilibrium. Fig. 3 shows the isotherms of DFN, PCP, and OP on SWCNT and SWCNT-NH_2. The sorption of some OMPs on SWCNT occurs in different manners than on SWCNT-NH_2. According to, the Giles classification, OP adsorption isotherms correspond to the L1 and L2 type for SWCNT and SWCNT-NH_2 respectively. This type of isotherm is typical for microporous adsorbents [16], that is consistent with structural properties of SWCNT and SWCNT-NH_2 obtained from nitrogen sorption desorption measurements. The initial curvature in the L type isotherm shows that as more sites in the adsorbent are filled it becomes increasingly difficult for a bombarding adsorbate molecule to find an empty site available. This implies that molecules are adsorbed non-vertically or there is lack of strong competition from the solvent. The shapes of the DFN and PCP isotherm curves resembles the L2 type and the H2 type for SWCNT and SWCNT-NH_2. The H type isotherm is a special case of the L curve, in which the solute has high affinity for the adsorbent surface because of ion-ion attraction [17, 18]. Thus, we can notice that adsorption mechanism of given OMPs was similar for SWCNT (due to L – shape of isotherm), while for SWCNT-NH_2 two different sorption behaviours occurred, first when OP is an adsorbate, the second for PCP and DFN.

DOI: 10.1051/e3sconf/20172200076

Significant differences were observed in sorption capacity of SWCNT and SWCNT-NH_2 for given OMPs. The sorption uptake changes in the following order: OP > PCP > DFN for SWCNT and DFN > PCP > OP for SWCNT-NH_2, respectively. In order to elucidate the factors controlling adsorption mechanism and sorption affinity of OMPs we should consider their hydrophobicity, molecular weight, ionic form and solubility. A few times higher adsorption of OP on SWCNT in comparison to the two other adsorbates can be ascribed to the different solubility in water, which is 7, 10, 50 mg/L for OP, PCP and DFN, respectively. Affinity of the adsorbate to sorption is larger when solubility is low [19], therefore it explains very high sorption uptake of OP and very low uptake of DFN by SWCNT. Effect of molecular weight and log K_{ow} of studied OMPs seems to be negligible because these properties were similar (Table 1). For SWCNT-NH_2, PCP and DFN were found as easily absorbable compounds, while OP as hardly. It can be explained by chemical nature of these compounds at pH 5. When solution pH is around 4.5, both PCP and DFN are dissociated, in contrast to OP (Table 1). When we consider that surface charge of SWCNT- NH_2 is positive at pH 5, we can assume that sorption of PCP and DFN was enhanced by ionic attraction. Mechanism does not occur for SWCNT due to neutral form at pH 5.

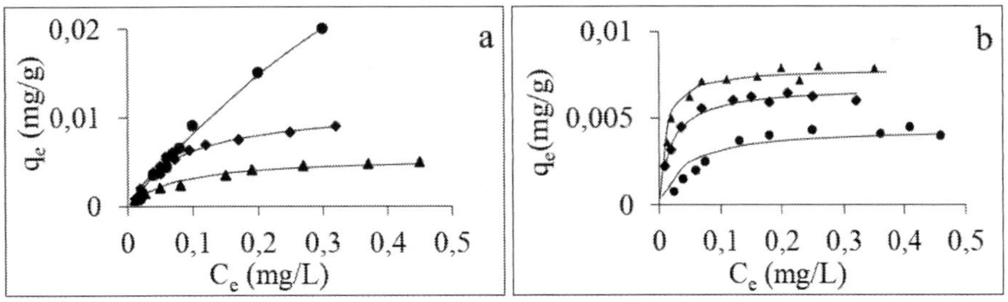

Fig. 3. Adsorption of PCP (diamonds), OP (circles) and DFN (triangles) by SWCNT (a) and SWCNT-NH_2 (b) from single adsorbate solution.

The Langmuir, Freundlich and Dubinin-Radushkevich models were used to describe the experimental data, and the relevant parameters derived from these models are presented in Table 3. The fitting degree (based on correlation factors R^2) decreases in the following order: Langmuir, Dubinin-Raduskevich and Freundlich. Therefore, the subsequent discussion will be mostly concentrated on the parameters obtained from Langmuir and Dubinin-Radushkevich computations. The values of adsorption energy for both SWCNT and SWCNT-NH_2 lie in the range of physical adsorption, apart from OP adsorption on SWCNT. Higher values of energy for SWCNT-NH_2 can suggest that this sorbent mainly induces the adsorption of OMPs coming from strong electrostatic interactions [20].

From the value of maximum adsorption capacity derived from Langmuir and Dubinin-Radushkevich was found that OP and PCP were better adsorbed by SWCNT than SWCNT-NH_2. Maximum adsorption capacity of PCP was 0.011 mg/g and 0.0068 mg/g on SWCNT and SWCNT-NH_2, respectively, and for OP: 0.074 mg/g (SWCNT) and 0.0044 mg/g (SWCNT-NH_2). On the one hand, it can be a consequence of the differences in structural properties of the adsorbents. As many authors suggest higher sorption capacity is found for the adsorbents with high specific surface area [14, 21]. Specific surface area (SSA) of SWCNT is around three times higher than SWCNT-NH_2, thus both OP and PCP were better adsorbed by SWCNT than SWCNT-NH_2. An interesting exception is noticed for DFN adsorption, because higher maximum sorption capacity was observed for SWCNT-NH_2. This suggests that sorption behaviour, when DFN is adsorbate, is different for SWCNT and SWCNT- NH_2. This fact was also revealed by isotherm type, discussed above. In the other words, normal dependence of SSA and sorption capacity does not apply for DFN adsorption by SWCNT-NH_2, suggesting the presence of specific interactions between adsorbate and adsorbent, like for an example ionic interactions and Van der Walls forces.

Table 3. Parameters of Langmuir, Freundlich and Dubinin - Radushkevich equations and correlation coefficients for the adsorption of OMPs on carbon nanotubes from single solution.

Model		Langmuir			Freundlich			Dubinin-Radushkevich		
Adsorbent	Adsorbate	a_L (mg/g)	b (L/mg)	R^2 (-)	K_f ((mg/g) L/mg)^n)	n (-)	R^2 (-)	a_{DR} (mg/g)	E (kJ/mol)	R^2 (-)
SWCNT	DFN	0.0054	14.99	0.974	0.0079	2.15	0.9715	0.0126	12.9	0.9907
	PCP	0.011	12.61	0.993	0.0160	2.23	0.937	0.0237	10.0	0.9445
	OP	0.074	1.25	0.992	0.0579	1.16	0.986	0.0817	7.0	0.9700
SWCNT-NH_2	DFN	0.0082	68.99	0.963	0.0087	8.99	0.867	0.0126	18.3	0.8891
	PCP	0.0068	50.00	0.975	0.089	4.34	0.849	0.0098	14.1	0.9278
	OP	0.0044	25.00	0.965	0.0039	9.99	0.893	0.0084	15.8	0.9063

3.3 Competetive adsorption of OMPs

Fig. 4 presents adsorption isotherms of DFN, PCP and OP from bicomponent adsorbate solution. Shape of these isotherms, both for SWCNT and SWCNT-NH_2 was similar to the single adsorption system. It means that second adsorbate in the solution did not cause significant change in adsorption mechanism of given OMPs. On the contrast, adsorption magnitude changed greatly as an effect of competition between some adsorbates. It was observed that the Langmuir and Dubinin-Radushkevich equations described the adsorption of OMPs with better correlation coefficient (Table 4), meaning that heterogeneous surface or porous structure played an important role in their adsorption and different binding sites with several adsorption energies were involved. Excluding adsorption of OP on SWCNT, adsorption energy of studied OMPs is higher than 8 kJ·mol^{-1}, indicating physical sorption. Comparing the values of adsorption energy for single and bicomponent solutions we can notice that adsorption energy was slightly lower for competitive adsorption. Take, for an example, adsorption of DFN by SWCNT-NH_2, energy is 18.3 kJ·mol^{-1}; 11.2 kJ·mol^{-1}; 14.1 kJ·mol^{-1} for single solution and for bicomponent solutions with PCP and OP, respectively. Similarly, from Langmuir and Freundlich models, the constants b and n, describing the affinity of binding sites and adsorption intensity, are lower for bicomponent system than for single, independently from adsorbent.

DOI: 10.1051/e3sconf/20172200076

Table 4. Parameters of Langmuir, Freundlich and Dubinin - Radushkevich equations and correlation coefficients for the adsorption of DFN, PCP and OP on carbon nanotubes from single and bicomponent solutions.

Adsorbent	**Adsorbate solution system**	**Langmuir**			**Freundlich**			**Dubinin-Raduskevich**		
		a_L (mg/g)	b (L/mg)	R^2 (-)	K_f ((mg/g)(L/mg)^n)	n (-)	R^2 (-)	a_{DR} (mg/g)	E (kJ/mol)	R^2 (-)
SWCNT	DFN	0.0055	14.00	0.978	0.0078	2.15	0.9715	0.0126	12.9	0.9907
	DFN&PCP	0.0045	8.50	0.995	0.0057	2.08	0.946	0.0123	11.2	0.973
	DFN&OP	0.0029	6.98	0.965	0.0034	2.00	0.938	0.0072	10.05	0.9769
SWCNT - NH_2	DFN	0.0079	102.00	0.967	0.0087	8.99	0.867	0.0126	18.3	0.8891
	DFN&PCP	0.0031	20.50	0.951	0.0042	2.50	0.9095	0.0088	11.2	0.9463
	DFN&OP	0.0062	43.00	0.979	0.0083	3.48	0.865	0.0134	14.1	0.942
SWCNT	PCP	0.0112	12.61	0.993	0.016	2.23	0.937	0.0237	10.0	0.9445
	PCP&OP	0.0080	5.00	0.992	0.01	1.419	0.973	0.0125	7.0	0.9832
	PCP&DFN	0.153	5.63	0.997	0.02	1.64	0.9803	0.0237	10.0	0.9897
SWCNT - NH_2	PCP	0.0068	50.00	0.975	0.0087	8.99	0.8831	0.0097	14.1	0.9278
	PCP&OP	0.0055	30.00	0.940	0.007	3.34	0.8255	0.0093	11.2	0.7829
	PCP&DFN	0.0024	18.00	0.984	0.002	3.47	0.936	0.0032	12.0	0.9504
SWCNT	OP	0.0752	1.22	0.992	0.0579	1.16	0.987	0.0817	7.0	0.970
	OP&PCP	0.088	5.00	0.993	0.0108	1.43	0.972	0.0146	7.0	0.984
	OP&DFN	0.0347	2.34	0.988	0.036	1.32	0.975	0.0361	7.0	0.9868
SWCNT - NH_2	OP	0.0047	25.0	0.965	0.0392	9.99	0.893	0.0084	10.0	0.906
	OP&PCP	0.0031	17.0	0.974	0.0044	2.50	0.899	0.0052	10.0	0.955
	OP&DFN	0.0023	18.0	0.983	0.0019	8.99	0.955	0.0031	12.0	0.949

 DOI: 10.1051/e3sconf/20172200076

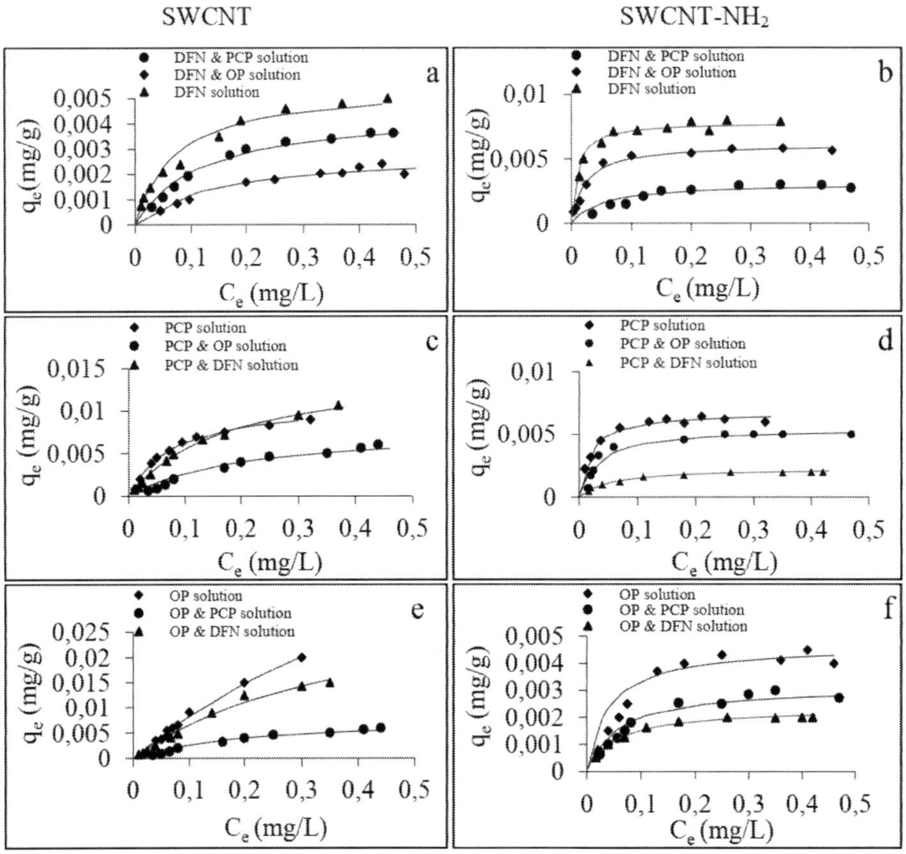

Fig. 4. Adsorption isotherms of DFN (a, b), PCP (c, d) and OP (e, f) by SWCNT and SWCNT-NH_2 from single solution and two bicomponent solutions. The solid lines represent Langmuir model fit.

It was found that between some of OMPs intense competition occurred, expressed as competition strength (Table 5). More specifically, for SWCNT stronger competition was observed between DFN and OP with competition strength of 61.7% than between the DFN and PCP with competition strength of 31%, while for SWCNT- NH_2, opposite tendency was observed i.e. competition strength was 69% and 42% between DFN and PCP and between DFN and OP respectively (Fig. 4a, 4b). Similar results were obtained for PCP adsorption on SWCNT-NH_2 in the presence of DFN and OP (Fig. 4d). In that case, competition strength was at the level of 70% and 23% for DFN and OP as coadsorbates, respectively. Interesting exception was observed for PCP adsorption on SWCNT, in the presence of DFN. Competition strength was 4.4% suggesting negligible impact of DFN on PCP adsorption. These results can be explained when we compare adsorption mechanism of given coadsorbates. Adsorption of DFN, in the presence of OP and PCP, on SWCNT was very low. Lower solubility of OP and its undissociated form (when pH<pKa) resulted in more preferential adsorption of OP than DFN on SWNCT. Similarly, DFN adsorption was inhibited by PCP as coadsorbate. Probably due to a role of ∏-∏ electron-donor-acceptor interaction and hydrogen bonds in PCP adsorption [22]. Strong competition between DFN and PCP on SWCNT-NH_2 was a result of the same mechanism of their adsorption, which was mainly electrostatic attraction. As was documented in the first part of this study, DFN was hardly adsorbable on SWCNT and easily on SWCNT-NH_2, mainly due to its high solubility in water and dissociated form at pH 5. Therefore, dominant adsorption strength of DFN is ionic attraction, that occurs only for SWCNT-NH_2 due to its positive charge at given

 DOI: 10.1051/e3sconf/20172200076

solution pH. The same adsorption mechanism was documented for PCP and therefore strong competition between these adsorbates was observed.

Separate discussion has to be done for OP adsorption with its coadsorbates (Fig. 4e and 4f). As was mentioned, OP was very easily adsorbable by SWCNT from single solution system. Moreover, competition between OP and DFN in bicomponent solution, was relatively low with competition strength at the level of 3%. This effect relates to negligible adsorption of DFN on SWCNT. Higher competition about sorption sites was observed between OP and PCP. Due to both sorbates can be adsorbed by hydrogen bonding and ∏-∏ interactions. Similar adsorption mechanism of PCP and DFN induces mutually competition. On the one hand, OP should be more preferentially adsorbed on SWCNT, on the other hand for two coadsorbates with different size, the smaller one can access the porosity in greater amount [23]. Thus, PCP with smaller size than DFN (expressed as equivalent width and molar volume – Table 1) inhibits DFN uptake. Surprisingly, coadsorbate size did not affect adsorption of OP on SWCNT-NH_2. A careful observation of Fig. 4f shows that competition was higher with coadsorbate molecules with higher size (DFN) than with smaller size (PCP). In other words, considering size of coadsorbates (PCP and DFN) we could have expected that higher competition will occur with PCP due to its lower steric hindrance. In fact, competition effect on SWCNT-NH_2 was clearly controlled by ionic attraction, that was stronger for DFN.

Table 5. Competition strength (calculated for selected equilibrium concentration) between coadsorbates in bicomponent solution.

Bicomponent solution	**SWCNT**	**SWCNT- NH_2**
DFN&PCP	31.0%	69.8%
DFN&OP	61.7%	42.4%
OP&PCP	73.7%	32.8%
OP&DFN	3.0%	51.5%
PCP&DFN	4.4%	70.0%
PCP&OP	56.0%	22.8%

4 Conclusions

The adsorption behavior of OMPs in single and bicomponent solution was evaluated. Adsorption of three OMPs by SWCNT and SWCNT-NH_2 were compared with each other. From the shape of isotherms, similar adsorption mechanism of OMPs was observed for SWCNT and two different behaviors for SWCNT-NH_2 – first for OP, the second for PCP and DFN. In addition, the loadings of OMPs on adsorbents decreased in the following sequence: OP > PCP > DFN for SWCNT and DFN > PCP > OP for SWCNT-NH_2. Favorable adsorption of OP and PCP by SWCNT was an effect of their low solubility in water and sigma-sigma interactions. Easy adsorption of DFN and PCP by SWCNT-NH_2 was assigned to electrostatic attraction between dissociated form of PCP and DFN and positively charged sorbent surface at pH 5.

It was found that the values of adsorption energy for both sorbents indicated physical adsorption, apart from OP adsorption on SWCNT. Higher values of energy for SWCNT-NH_2 can suggest that this sorbent mainly induces the adsorption of OMPs coming from strong electrostatic interactions.

From results obtained in bicomponent solution, different competitive effects were observed. Take, for an example adsorption of DFN with its coadsorbates PCP and DFN by SWCNT, OP inhibited DFN adsorption much stronger than PCP. For comparison, when DFN was adsorbed by SWCNT-NH_2, opposite tendency was observed, stronger inhibition of DFN adsorption was caused by PCP. This is due to control of PCP and DFN adsorption on SWCNT-NH_2 by electrostatic attraction. on the other hand, between some coadsorbate, a lack or negligible competition was observed, like for OP and PCP adsorption on SWCNT with DFN as coadsorbate. Comparing the values of adsorption energy for single and bicomponent solutions we can notice that adsorption energy was slightly lower for competitive adsorption.

The paper has been prepared within the frame of the National Science Centre project based on decision no DEC-2013/11/B/ST8/0439.

References

1. M. Clara, G. Windhofer, P. Weilgony, O. Gans, M. Denner, A. Chovanec, M. Zessner, Chemosphere **87**, 1265 (2012)
2. V. Belgiorno, L. Rio, D. Fatta, C. Rocca, G. Lofrano, A. Nikolaou, V. Naddeo, S. Meric, Desalination **215**, 166 (2007)
3. Y. Luo, W. Guo, H. Ngo, L. Nghiem, F. Hai, Sci. Total Environ. **473–474**, 619 (2014)
4. N. Barco-Bonilla, R. Romero-González, P. Plaza-Bolaños, Sci. Total Environ. **447**, 381 (2013)
5. M. Gardner, S. Comber, M.D. Scrimshaw, E. Cartmell, J. Lester, B. Ellor, Sci. Total Environ. **437**, 363 (2012)
6. M. Carballa, G. Fink, F. Omil, J. Lema, T. Ternes, Water Res. **42**, 287 (2008)
7. M. Gavrilescu, K. Demnerová, J. Aamand, S. Agathos, F. Fava, New Biotechnol. **32**, 147 (2015)
8. M. Dudziak, Pol. J. Environ. Stud. **15**, 35 (2006)
9. G. Kamińska, J. Bohdziewicz, J.I. Calvo, P. Prádanos, L. Palacio, A. Hernández, J. Membrane Sci. **493**, 66 (2015)
10. E. Borowska, E. Felis, J. Kalka, Chem. Eng. J. **304**, 852 (2016)
11. J. Bohdziewicz, G. Kamińska, Water Sci. Technol. **68**, 1306 (2013)
12. A.D. Coelho, C. Sans, A. Aguera, M.J. Gomez, S. Esplugas, M. Dezotti, Sci. Total Environ. **407**, 3572 (2009)
13. L. Juan, F. Acero, J. Benitez, F.J. Real, F. Teva, Chem. Eng. J. **289**, 48 (2016)
14. G. Kamińska, J. Bohdziewicz, Environ. Prot. Eng. **42**, 161 (2016)
15. D. Zhang, B. Pan, M. Wu, H. Zhang, H. Peng, P. Ning, B. Xing, Environ. Pollut. **16**, 178 (2012)
16. P. Strachowski, M. Bystrzejewski, Colloid. Surface A. **467**, 113 (2015)
17. J. L. Sotelo, G. Ovejero, A. Rodrguez, S. Alvarez, J. Galan, J. Garcia, Chem. Eng. J. **240**, 443 (2014)
18. C.H. Giles, T.H. MacEwan, S.N. Nakhwa, D. Smith, J. Chem. Soc. **60**, 3973 (1960)
19. S. Nam, D. Choi, S. Kim, N. Her, K. Zoh, J. Hazard. Mater. **270**, 144 (2014)
20. T. Thiebault, R. Guégan, M. Boussafir, J. Colloid. Interf. Sci. **453**, 1 (2015)
21. B.N. Bhadra, P.W. Seo, S.H. Jhung, Chem. Eng. J. **201**, 27 (2016)
22. P. Devi, A.K. Saroha, Bioresource Technol. **169**, 525 (2014)
23. S. Masson, M. Gineys, S. Delpeux-OUldriane, L. Reinert, S. Giuttonneau, F. Deguin, L. Duclaux, Micropor. Mesopor. Mat. **234**, 24 (2016)

Stability assessment of the chemical composition of the treated mining water used to replenish the cooling circuit in Jaworzno III Power Plant – Power Plant II

Marcin Karpiński[1,*], *Ewa* Kmiecik[1]

[1]AGH University of Science and Technology, Faculty of Geology, Geophysics and Environment Protection, Al. Mickiewicza 30, 30-059 Kraków, Poland

Abstract. In Poland, electricity is still produced mainly in conventional power plants where fuel and water are materials necessary to generate the electricity. Even in modern power plants operating according to the principles of the sustainable development, this involves a high intake of water and considerable production of wastewater. This, in turn, necessitates the application of some technological solutions aimed at limiting the negative impact on the environment. The Jaworzno III Power Plant – Power Plant II is located in Jaworzno, Silesian Province, Poland. In order to minimise the negative impact on the surface water, the plant replenishes the cooling circuit with the mining water obtained from the closed-down Jan Kanty mine. The paper presents a stability assessment of the chemical composition of the treated mining water used to replenish the cooling circuit based on the data from 2007–2017.

1 Introduction

Production of electricity in Poland is carried out mainly in conventional power plants than e.g. in hydropower plants [1], so the necessary resources include fuel and water. Generation of electricity in conventional power plants is carried out in a three-stage energy conversion process. The chemical energy stored in fuel and released in the combustion process in the boiler is converted into steam thermal energy. Subsequently, superheated steam is used to move the blades of the turbine driving the generator in which mechanical energy becomes converted into electricity [2]. Additionally, certain part of energy carried by steam may be used for heating of water in district heating networks.

Electricity generation requires large quantity of water, however in modern plants, it is carried out in accordance with the sustainable development principle in the manner that is least detrimental to the environment. Due to the high purity requirements for water used in process circuits and for the protection of the quality of surface waters that are often a source of drinking water, power plants use their own water treatment plants and wastewater treat-

* Corresponding author: karpinski@agh.edu.pl

ment plants. Some amount of water is reused to supplement the losses in the circuits with lower quality requirements [3, 4].

2 Jaworzno III Power Plant – Power Plant II characterisation

The Jaworzno III – Power Plant II operated by TAURON Wytwarzanie S.A. is located in south-western part of Jaworzno, ca. 2.5 km away from Przemsza river. In the period between 1995 and 2000, due to increased environmental protection requirements, it underwent major upgrade and was repurposed as a municipal combined heat and power plant [5]. The maximum electric output of the Power Plant II is 190 MWe, while the maximum thermal capacity is 318 MWt. The power plant is equipped with three fluidized bed boilers: two fired with hard coal and coal sludge and one fired with biomass. All boilers are coupled with district heating and condensing turbines which are equipped with single-stage DH exchangers [6].

3 Water circuits

There are three main water circuits in the power plant: water/steam circuit (boiler circuit), cooling circuit and district heating circuit (fig. 1) [7].

The power plant operates a closed cooling circulation system with a maximum capacity of 9720 m^3/h. The water taken up from the cooling water channel is pumped to the condenser where it is heated. Next, the water is pumped to the cooling tower and, after cooling, it is recycled to the cooling water channel. Evaporation of water in the natural draft cooling tower results in condensation of the circulating liquid [8]. The entire condensed cooling water must be removed in the form of blowdown from the cooling circuit and replenished with treated water [7].

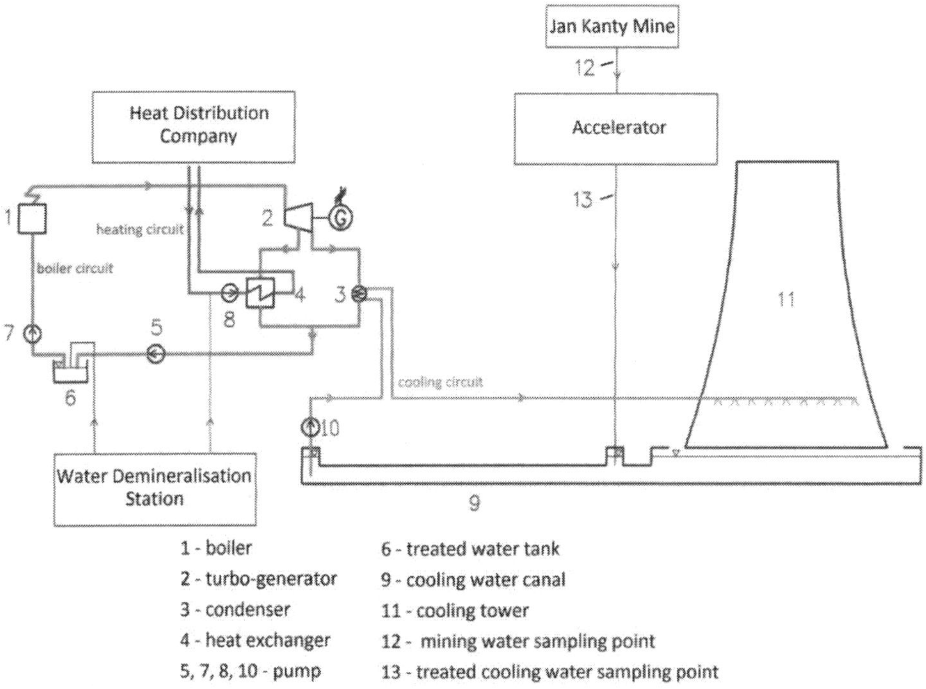

Fig. 1. Water circuits in the power plant Jaworzno II [7].

DOI: 10.1051/e3sconf/20172200077

4 Water sources

Apart from fuel, water is the basic material used in the process of electric and thermal power generation from non-renewable sources. Consumption of water in conventional power plants with open or closed cooling water system is between 1.7 and 8.0 m^3/MWe [9]. An average-sized conventional power plant consumes several million cubic meters of water, the majority of which is used in the cooling processes [2].

The plants utilizing natural water resources are obliged to abide by the water protection laws. Due to increasing prices of water intake and use of the environment, the regulations enforce rational water management. The environment protection is especially one of the biggest issues in the Silesia province. The heavy industry located in this area has an effect on ground and surface water [10]. In order to reduce costs and negative environmental impact, the Power Plant II uses, for water production and cooling circuit replenishment, the mine water from the defunct Jan Kanty Mine held by Spółka Restrukturyzacji Kopalń and surface water from Biała Przemsza river in the total amount of 4,600,000 m^3/year [11]. A stationary drainage system is maintained at the Jan Kanty Mine, while the inflow of mine water amounts to ca. 30 m^3/minute [12]. Drainage of this defunct mine results from the necessary protection against flooding of the TAURON Sobieski Mining Plant [13].

5 Cooling water treatment

The water temperature for the needs of the cooling circuit should be low. Additionally, it must not contain chemicals causing scale to deposit on heating surfaces and should be non-corrosive towards materials from which the components of the circuit are made. The water used to supplement the losses in the cooling circuit is prepared in the coagulation and decarbonisation processes occurring in accelerators [14]. In the process of water preparation for cooling circuit purposes, it is very important to bring water to carbonate calcium equilibrium. This reduces water's aggressive properties and prevents precipitation of calcium carbonate [15].

The raw water originates from the Jan Kanty mine, from a carboniferous aquifer and is classified as fresh and semi-fresh water. It is characterized by low mineral content of approx. 1000–1200 mg/l, concentration of sulphate of approx. 310 mg/l and chloride of approx. 115 mg/l [16]. The presence of these ions adversely affects the operation of power engineering equipment. They promote the corrosive character of water towards metals and concrete. Due to high concentration of sulphates and chlorides which are not removed in the carbonate removal process, the water in the cooling circuit requires continuous treatment. The resulting blowdown is pumped to the power plant's wastewater treatment station and, subsequently, in accordance with the integrated permit, it is discharged to Przemsza river [17, 18].

6 Methodology

The testing of raw and accelerator-treated mine water chemical composition was carried out on a monthly basis in the period between 2007–2017. The samples were collected by the personnel of the Power Plant II in accordance with plant's quality system procedures (PN-EN ISO 5667-13:2004, 2011). Sampling points are shown on the fig. 1.

The water was subsequently analyzed by the laboratory of TAURON Wytwarzanie (PCA accreditation certificate No. AB 688). A maximum of 15 parameters were measured in the samples, while this paper concerns ions whose increased concentration is typical for mine water: sulphate (SO_4^{2-}) and chloride (Cl^-) ions as well as pH and electrical conductivity (EC).

Sulphate concentration was determined by turbidimetry method, chloride concentration was determined by Mohr method (PN-ISO 9297: 1994), the pH and EC values was determined by potentiometric (PN-EN ISO 10523: 2012) and conductometric (PN-EN 27888: 1999) methods respectively [15, 19]. Trend analysis methodology is presented e.g. in [20].

7 Data analysis

In order to evaluate raw and treated water chemical composition stability, the results of 114 analyses from the period between 2007 and 2017 were used. In March and April 2014 as well as August, October, November and December 2016 the power plant did not use mine water and its composition was not analyzed. Table 1 presents basic descriptive statistics for the entire data set.

Table 1. Basic descriptive statistics for the EC, pH, chloride and sulphate ions concentration in mining water (M) and treated water (T).

The value of statistics	EC [µS/cm]		pH [-]		Chlorides [mg/l]		Sulphates [mg/l]	
	M	T	M	T	M	T	M	T
Mean	1147	1351	9.55	7.55	141.3	141.7	374.5	383.6
Std. deviation	85	39	0.47	0.20	16.5	4.0	51.5	41.4
Minimum	663	1173	8.20	7.13	46.1	128.0	243.0	297.0
Maximum	1391	1444	11.00	8.07	184.6	156.2	495.0	464.0
Range	728	271	2.80	0.94	138.5	28.2	252.0	167.0

The analysis of sequence plots for EC (fig. 2), pH (fig. 3), chloride (fig. 4) and sulphate (fig. 5) ions concentration in raw mine water shows that there are no statistically significant monotonous trends. It means that raw mine water is characterized by stable chemical composition in respect of these parameters. The sulphate concentration chart (fig. 5) over time reveals three groups of results related to changes in the drainage system at the Jan Kanty Mine [21] and reduction of pyrite oxidation processes [22, 23].

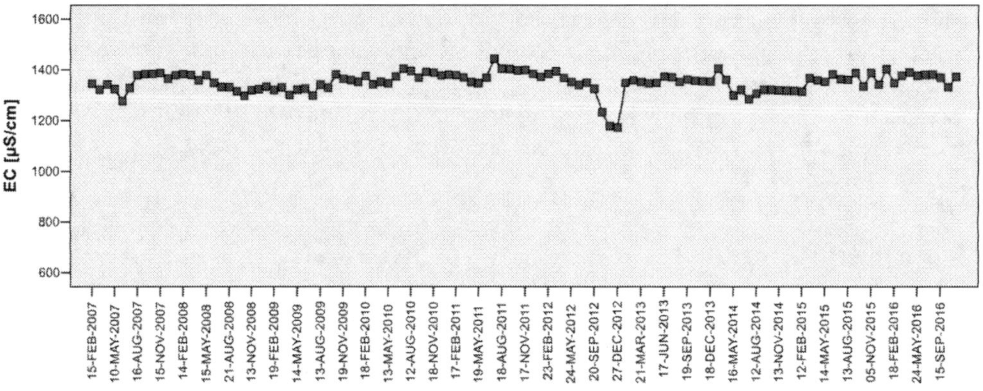

Fig. 2. Conductivity changes in mine (M) water in 2007–2017.

 DOI: 10.1051/e3sconf/20172200077

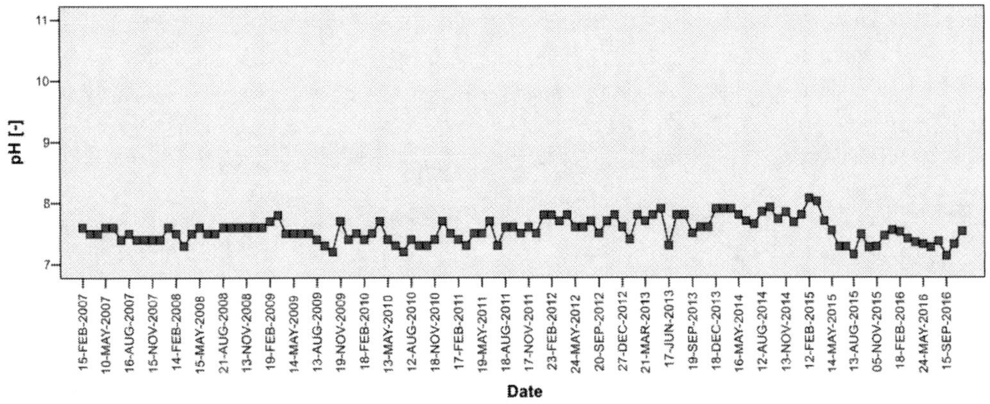

Fig. 3. pH changes in mine (M) water in 2007–2017.

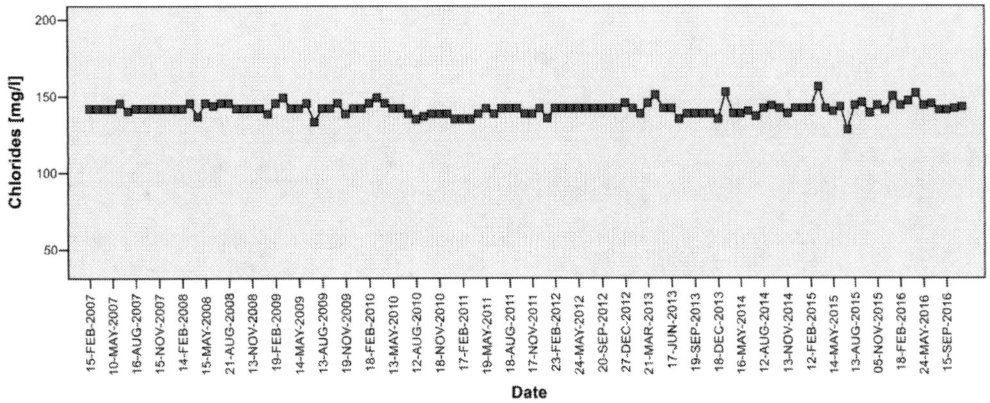

Fig. 4. Fluctuation of chlorides concentration in mine (M) water in 2007–2017.

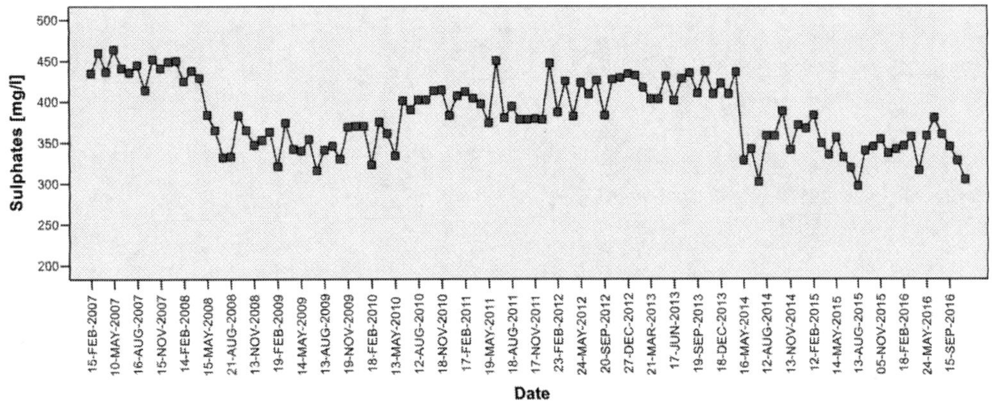

Fig. 5. Fluctuation of sulphate concentration in mine (M) water in 2007–2017.

The analysis of sequence plots for EC (fig. 6), pH (fig. 7), concentration of chloride (fig. 8) and sulphate (fig. 9) ions concentration in treated water (used to make-up losses in the cooling circuit) shows also that there are no statistically significant monotonous trends. It means that treated water is also characterized by relatively stable chemical composition in respect of these parameters. The parameter variability charts over time show only single results with reduced EC, pH and Cl^- and SO_4^{2-} concentrations. During months in which the analyses were carried out, a mix of mine water and water sourced from Biała Przemsza river was used to make-up the losses in the cooling circuit that are characterized by clearly better chemical parameters: lower electrical conductivity and concentration of chloride and sulphate ions.

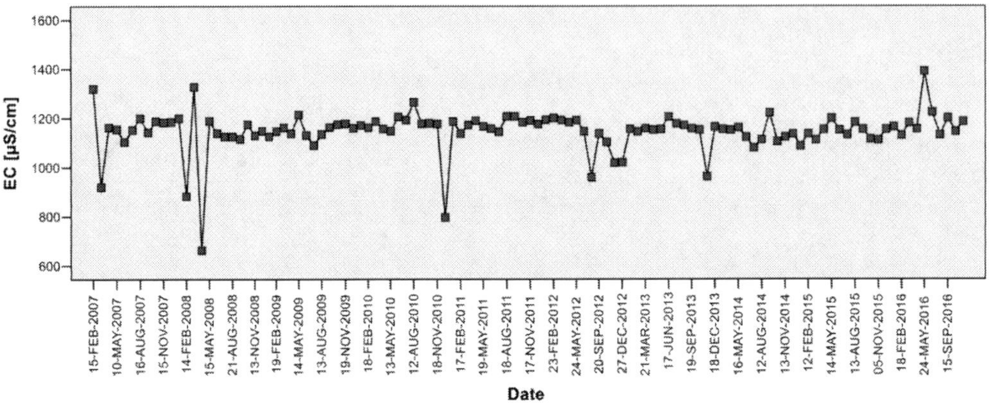

Fig. 6. Conductivity changes in treated (T) water in 2007–2017.

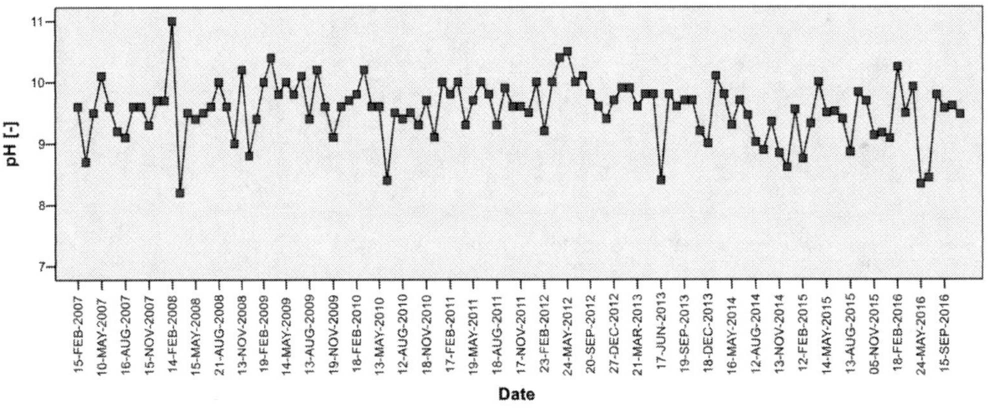

Fig. 7. pH changes in treated (T) water in 2007–2017.

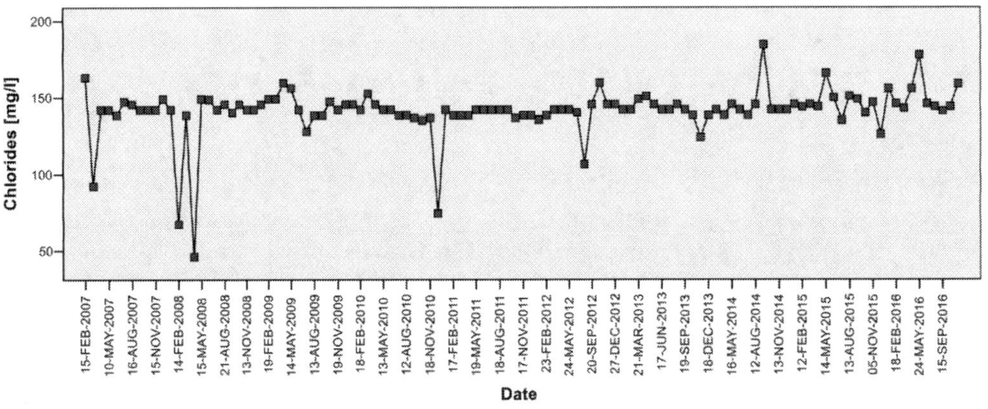

Fig. 8. Fluctuation of chlorides concentration in treated (T) water in 2007–2017.

DOI: 10.1051/e3sconf/20172200077

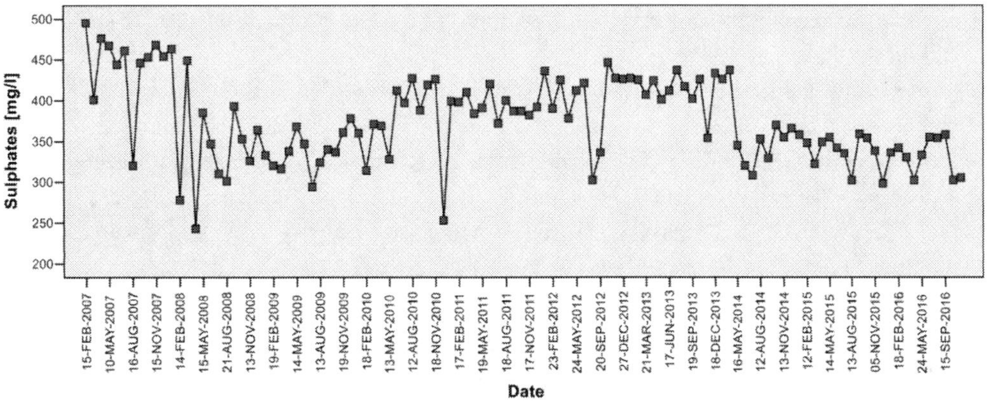

Fig. 9. Fluctuation of sulphate concentration in treated (T) water in 2007–2017.

8 Summary

This paper presents the evaluation of stability of chemical composition of mine water used at the Power Plant II. The analysis was carried out for Cl^- and SO_4^{2-} ions as well as electrical conductivity and pH. The analysis of chemical composition changes in time did not reveal any statistically significant trends. For sulphate ions in raw water, three groups of results were found that may be related to drainage activities at the Jan Kanty mine. For treated water analyses, the results deviating from the average are related to treatment of mine water and water taken up from Przemsza river simultaneously. This confirms that mine water and treated water are characterized by stable chemical composition.

Mine water typically shows increased chloride and sulphate concentrations as compared to surface water. Therefore, they also show considerable corrosive activity. At the same time, they have relatively stable chemical composition depending mainly on the physical and chemical properties of the geological location. Its chemical composition is not affected by a number of factors which are of importance for surface water, e.g. current season, thawing, rainfall or presence of contamination from wastewater [15]. Relative chemical stability of mine water allows for easier determination of chemical dosage during water treatment.

The authors would like to thank branch managers Małgorzata Sztyrak and Mirosław Żbik (TAURON Wytwarzanie S.A.) for providing data for the purposes of this paper.
This work has also been conducted under AGH-UST statutory research grant No. 11.11.140.797.

References

1. A. Operacz, B. Tomaszewska, Environ. Sci. Pollut. **23**, 19, 18953–18959 (2016)
2. D. Sąkol-Sikora, Energet. Ciepl. i Zaw. **9** (2011)
3. J.R. Rak, B. Kucharski, Environ. Prot. Eng. **35**, 2, 15–21 (2009)
4. D. Sąkol-Sikora, Joint Meeting of Water Management Councils of the Water Regions: Little Vistula and Upper Oder and the Presidencies of the Standing Committee for the Participation of the Society Conf Proc, Toszek (2009)
5. http://www.tauron-wytwarzanie.pl/oddzialy/jaworznoiii/Strony/opis.aspx (15.03.2017)

6. Environmental Impact Review – Jaworzno III Power Plant, Sozoprojekt (2009) (in Polish)
7. Operating Instructions of turbogenerators, Nr CKI 2005/320/Er (mat. TAURON) (in Polish)
8. R. Laskowski, A. Smyk, J. Lewandowski, A. Rusowicz, A. Grzebielec, Energy **107**, 725–733 (2016)
9. T. Jamrocha, VIII Discussion Forum Diagnostics and chemistry for power industry Conf. Proc. Szczyrk, 161–171 (2011)
10. T. Suponik, Environ. Prot. Eng. **41**, 1, 15–27 (2015)
11. Jaworzno III Power Plant – Water budget (mat. TAURON) (2015)
12. A. Czapnik, A. Jasińska, E. Janson. Biuletyn PIG. **436**, 55–60 (2009)
13. J. Siemieniec, J. Wodecki, M. Gałęziowski, *Monograph* SRK (2005) (in Polish)
14. A. Puszkarewicz, J. Kaleta, A. Piech, Environ. Prot. Eng. **35**, 4, 41–48 (2009)
15. E. Sierakowski, J. Mrożek, *Control of water and steam in power industry* (Wyd. Energopom. Elektr., 1998) (in Polish)
16. A. Olszówka, Environmental Impact Review of starting production in the area closed down mine Jan Kanty, EC Katowickie Przedsiębiorstwo Geologiczne (2012) (in Polish)
17. Operating instructions for decarbonisation and water coagulation stations, Nr CKI 2005/295/Er (mat. TAURON) (in Polish)
18. Integrated permits for fuel combustion installation Jaworzno III Power Plant in Jaworzno, Nr 360/OS/2012 (in Polish)
19. http://www.pca.gov.pl/akredytowane-podmioty/akredytacje-aktywne/laboratoria-badawcze/AB%20688,plik.html (20.03.2017)
20. E. Kmiecik, S. Stefaniak, J. Szczepanska, I. Twardowska, Proc. Of Spie. **5586**, 126–135 (2004)
21. P. Bukowski, I. Augustyniak, SGEM2013 Conf. Proc. **2**, 25–32 (2013)
22. R.J. Bowell, IMWA Symp. Proc. **2**, 75–91 (2004)
23. K. Labus, S. Skoczynska, Geol. Q **57**, 3, 561–566 (2013)

E3S Web of Conferences **22**, 00078 (2017) DOI: 10.1051/e3sconf/20172200078
ASEE17

Air temperature gradient in large industrial hall

Michał Karpuk[1], *Aleksander* Pełech[1], *Edward* Przydróżny[1], *Juliusz* Walaszczyk[1], and *Sylwia* Szczęśniak[1,*]

[1]Wroclaw University of Science and Technology, Faculty of Environmental Engineering, ul. Norwida 4/6, 50-370 Wroclaw, Poland

Abstract. In the rooms with dominant sensible heat load, volume airflow depends on many factors incl. pre-established temperature difference between exhaust and supply airflow. As the temperature difference is getting higher, airflow volume drops down, consequently, the cost of AHU is reduced. In high industrial halls with air exhaust grids located under the ceiling additional temperature gradient above working zone should be taken into consideration. In this regard, experimental research of the vertical air temperature gradient in high industrial halls were carried out for the case of mixing ventilation system The paper presents the results of air temperature distribution measurements in high technological hall (mechanically ventilated) under significant sensible heat load conditions. The supply airflow was delivered to the hall with the help of the swirl diffusers while exhaust grids were located under the hall ceiling. Basing on the air temperature distribution measurements performed on the seven pre-established levels, air temperature gradient in the area between 2.0 and 7.0 m above the floor was calculated and analysed.

1 Introduction

To calculate of volume airflow in mechanical ventilation systems, it's necessary to establish maximum allowed air temperature increase between air supply and exhaust zone. This issue is quite easy for the rooms with relatively low height ($h \leq 4$ m) and for the rooms where exhaust element is located in users' operating zone. It's much more complicated in the higher rooms ($h > 4$ m), where exhaust elements are located near the ceiling. The reason for this is potential temperature change along with the height of the room. Too small temperature increase assumed for calculations will result in unreasonable air volume and its treatment necessity. On the other hand, when temperature increase over operating zone (without impairing conditions in this zone) will be taken into account, lower air flow rates will be necessary for ventilation (resulting in air treatment costs reduction). This problem is crucial in rooms where due to constructional or technological requirements, air must be blown from the top of the room and exhaust grids are high above occupied zone.

Vertical air temperature gradient in occupied zone is one of the most important factors defining thermal comfort [1]. There are numerous publications related to temperature stratification in the operating zone [1, 2, 5, 7, 8]. Research results of the operating zone are available to provide best possible conditions for the people and technology used in the room.

From energy consumption point of view, it's more important to estimate maximum temperature gain in the entire room than the difference between exhaust and supply air temperature. Obviously we have to provide appropriate conditions in the operating zone. Air temperature increase over the operating zone (θ) is closely related to the issue described as room average temperature gradient (δt). Discussed issue was not described satisfactory for higher rooms (h>4m) with significant heat load.

2 Theory

2.1 Air flow rate calculation

Mechanical HVAC systems serving rooms with dominant sensible heat load and required air flow rate depends on the sensible heat load value and maximum value of temperature increase between exhaust air and supply air zones. Air volume can be calculated using following equation :

$$V = \frac{Q_{shl}^{max}}{\rho \cdot c_p \cdot \Delta t^{max}} = \frac{Q_{shl}^{max}}{\rho \cdot c_p \cdot (t_e - t_s)^{max}} \quad (1)$$

where: V – air flow rate; m^3/s, Q_{shl}^{max} – maximum sensible heat load in the room, kW, ρ – air density; kg/m^3, c_p – air specific heat capacity; kJ/kgK; t_s, t_e – supply and exhaust air flow temperatures ,°C;

Temperature increase Δt^{max} for air flow volume in high rooms with dominant sensible heat load can be expressed in the slightly different form:

$$\Delta t^{max} = \{[t_{oz} + \delta t(h_e - h_{oz})] - t_s\}^{max} \quad (2)$$

where: t_{oz}, t_s - occupied zone air temperature, supply air temperature,°C; h_e, h_{oz} – elevation of exhaust grids and occupied zone, m; δt – average vertical air temperature gradient over operating zone, K/m (°C/m).

2.2 Vertical air temperature gradient over occupied zone

Average air temperature gradient in the room (δt) is the air temperature change along with the elevation, related to the distance from agreed baseline (usually floor level) measured perpendicularly to this baseline [2, 5, 6]. It's expressed in K/m. In this article, we set baseline at the level of 2 m (theoretical height of operating zone). Average air temperature gradient can be calculated using below formula :

$$\delta t = \frac{\theta}{h_e - h_{oz}} = \frac{t_e - t_{oz}}{h_e - h_{oz}};\ \frac{K}{m} \quad (3)$$

where: Θ- air temperature increase above operating zone, K; t_e, t_{oz} – air temperature above occupied zone (h_e) and occupied zone air temperature, °C. Illustration of relationships (1, 2 and 3) are presented in Figure 1.

 DOI: 10.1051/e3sconf/20172200078

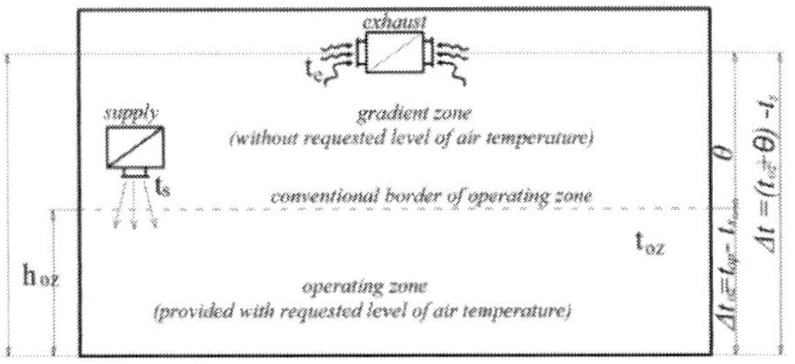

Fig.1. Illustration of symbols and assumptions for formulas 1–3.

There are two zones in rooms with dominant sensible heat load. The first one – operating zone – should be provided with the requested level (or range) of air temperature. This demand comes from an expected thermal comfort or technological expectations. This demand is connected with thermal comfort and/or technological requirements. The second one – let's call it transitional – exists over operating zone where we don't expect to keep thermal requirements. In this zone we may observe thermal stratification and additional temperature increase. Part of the sensible heat load coming from electrical lights, technology, walls and windows can be transferred into this zone through natural convection. Basing on available literature [1–3, 7, 9] additional temperature increase can be estimated.

Temperature increase above operating zone (Θ) is:

- ✓ directly proportional to room specific heat load ($q = \frac{Q_{shl}}{a \cdot b \cdot h}; \frac{W}{m^3}$) and share of the heat generated in the room or transferred to the room over operating zone within total room heat load ($\frac{Q_{shl}^{h_{oz} \div h_e}}{Q_{zbj}}$),
- ✓ inversely proportional to room air change rate ($\psi = \frac{V}{a \cdot b \cdot h}$) and heat sources surface ratio within total room surface ($\frac{\Sigma a}{A}$).

Fig.2. presents changes of average air temperature gradient in the function of distance from the operating zone, calculated according to the guidelines presented in [2, 5]. The conclusion from this chart is : average temperature gradient decreases with the distance from the operating zone.

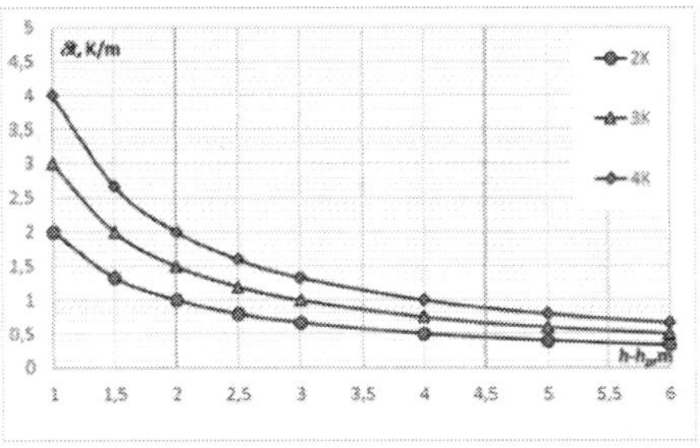

Fig. 2. Average room temperature gradient in function of the distance from operating zone (height difference $h\text{-}h_{oz}$) for different values of temperature increase Θ.

3 Facility and research methods

3.1. Building architecture

Facility chosen for the research was a technological hall located in the northern part of Wroclaw. Examined building has got only two external surfaces: roof and eastern wall. External walls 550mm thick are made from foam concrete bricks without additional thermal insulation. Hall floor is directly based directly on the ground. Steel roof construction was designed in the cylinder shape. There are four bowed sky roof points 3 × 10 m size, made from partially transparent panels. Gate 2.6 × 3.3 m size is situated in the external wall.

At the roof/wall joints there are 26 glow discharge tubes. Additionally, under the ceiling there are 24 halogen lamps situated over transportation paths. There are 29 injection moulding machines 4.3 × 1.2 × 2.0 m each. There are maximum 18 operators working in the hall. The hall is equipped with mixing ventilation system without air cooling capability.

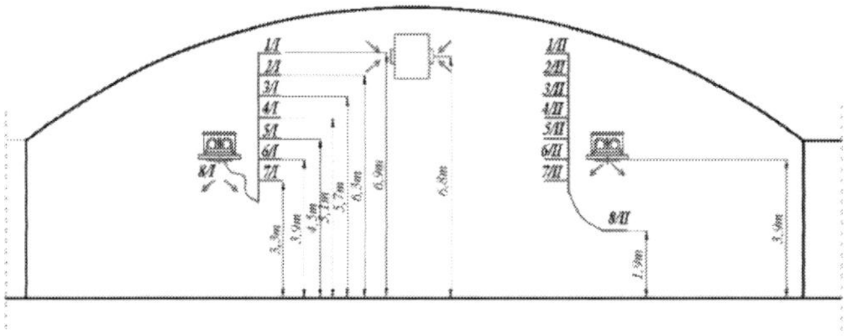

Fig. 3. Location of supply air diffusers and exhaust grids in the examined hall including the layout of air temperature measurement points.

3.2. Thermal characteristics

Dominant pollution emitted in the hall is heat, mentioned in the article as sensible heat. For the purpose of this article the sensible heat balance of the hall during summer and winter season was calculated using the method described in [5]. Sensible heat balance calculation was carried out for July, September and for different ambient temperatures of winter season under different ambient air temperature conditions. Sensible heat balance calculation was used to lead initial vertical air temperature stratification and location of measurement points. Table 2 presents basic hall data.

Table 2. Examined hall data.

Hall space	m^3	12351
Floor surface area	m^2	1774,5
Roof surface area	m^2	1897/120
Ventilation type		No cooling
Maximum ventilation air volume air flow rate	m^3/h	100 000
Maximum air change ratio in the hall.	h^{-1}	5,8
Range of specific sensible heat load value in the hall	W/m^3	7–13
Share of heat load from sources located in the operating zone in total sensible heat balance	%	90
Share of heat load coming from the roof and sky roof panels	%	10

3.3. Description of measurement station

To provide the best possible measurement results the special measurement, with 8 temperature sensors mounted on the each post, was built in industrial hall. One of the sensors on the post had adjustable position, other seven sensors had fixed position with gauge of 60 cm. Figure 3 shows configuration of measurement station and numbering of the sensors including vertical dimension of the position of sensors. Sensor 8/I was used to measure the temperature of supplied air, sensor 8/II was measuring temperature in operating zone. Measurements were done between April 2014 and August 2014. Posts were located in the most representative spots. Initial measurements indicated locations where ventilation system worked inefficiently or was too intensive.

Temperature measurements were done every 10 minutes using sensors PT1000 class A with measurement accuracy of 0.15°C.

4 Results and discussions

4.1. Air temperature in technological hall

Fig.4. and Fig.5. present air temperature changes and average air temperature gradient observed in April 2014 (Fig. 4) and in July and August (Fig. 5) 2014. Charts present air temperature changes in the operating zone (sensor 8/II), temperature near exhaust points (sensor 1/I, 1/II) and temperature nearby air supply points. (sensor 8/I). For clarity of the chart, sensors located in transition zone were not presented.

In April 2014 temperature increase value in the operating zone was approximately at the level of 6°C and it was the highest during the entire measurement session. Lowest temperature of supply air flow was about 18°C. During this period, temperature in the operating zone was at the level of 26°C and temperature of exhaust air was at the level of 29°C. Average air temperature gradient in the hall (δt) was 0.7 Km^{-1}.

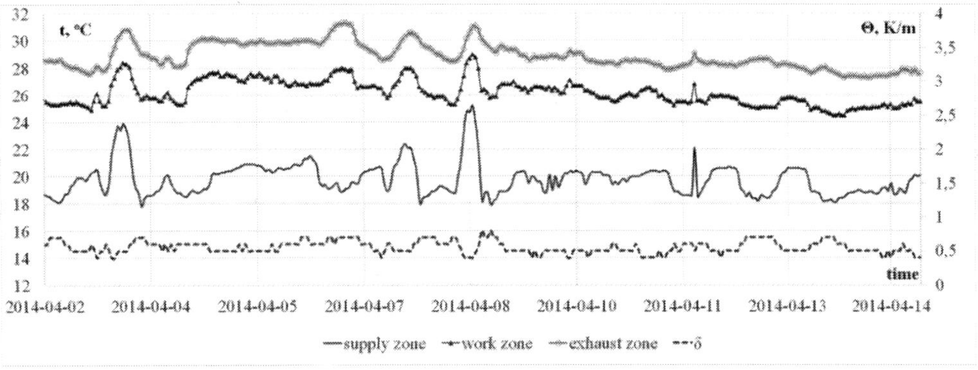

Fig. 4. Air temperature and average temperaure gradient in the hall in April 2014.

Temporary increase of supply air flow temperature (caused by ambient air temperature increase) was resulting in temperature increase in the operating zone and exhaust air. At the same time Θ value was dropping and, as a consequence, average temperature gradient value δt was also decreasing. In April 2014 supplied air temperature resulted from the heating of the mixture of fresh and recirculated air, so constant air temperature in the operating zone was provided. In July and August 2014 average air temperature increase was at the level of 1.8 K and this was the lowest value during entire measurements. The lowest supply air flow temperature was 22.9°C. During this period air temperature in the operating zone was 25°C and exhaust air temperature was 28°C. Average temperature gradient was 0.6 Km^{-1}. The highest difference, in comparison to the April session, was visible in the significant temperature oscillation depending on the time of the day/night. The reason for those variations was no air cooling capability so practically supply air temperature was equal to current ambient temperature.

Fig. 5. shows the temperature variation during July/August 2014. Increase of the supply air temperature was directly resulting in increase of the temperature in the operating zone and exhaust air measurement points. Increasing air temperature in the hall resulted in the smaller temperature growth above operating zone and lower value of average air temperature gradient in the hall.

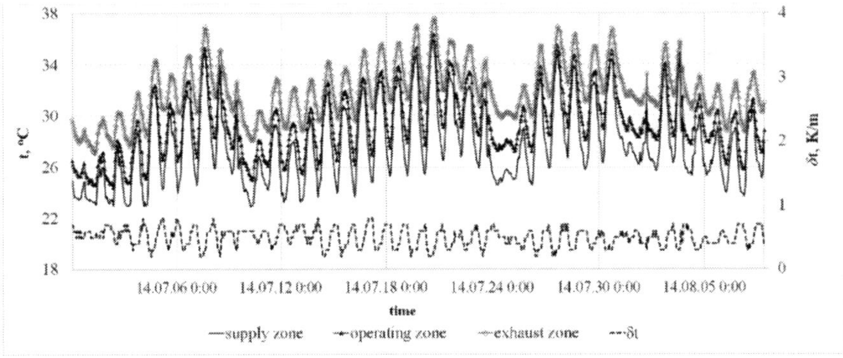

Fig. 5. Air temperature and average air temperature gradient in the hall during July and August 2014.

Fig. 6 and Fig. 7 show air temperature and gradient variations for two selected days in April and two selected days in July.

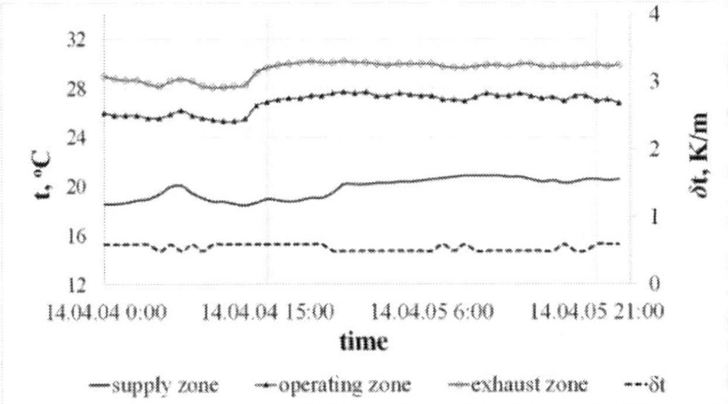

Fig. 6. Air temperature and average air temperature gradient in the hall on 4th and 5th April 2014.

Both charts confirm that in April air temperature in the hall was almost constant. This was caused by the treatment (heating) of fresh and recirculated air flow mixture. During July/August air temperature in the hall was directly depending on the ambient temperature. Variations between day and night are clearly visible on the charts. It can be noticed that along with the growth of supply temperature the average air temperature gradient in the operating zone decreases. This applies also to the transition zone (above operating zone) air temperature.

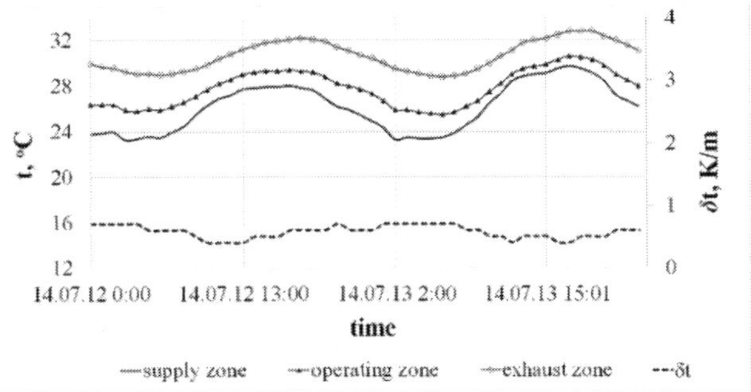

Fig. 7. Air temperature and average temperature gradient in the hall on 14th and 15th July 2014.

Relationship between air temperature increase in the hall (Δt) and temperature increase in operating zone (Δt_{oz}) is presented in Fig.8. The chart includes all values of Δt ($\Delta t = t_e$-t_s) for the different increases of Δt_{oz}. Higher temperature in the operating zone resulted in the higher total temperature increase in the hall.

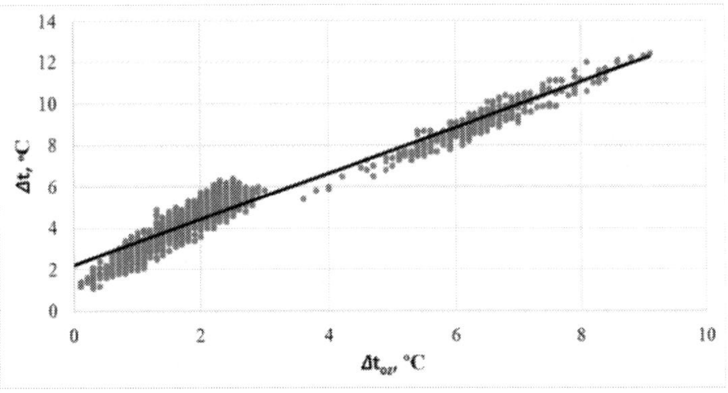

Fig. 8. Temperature increase in operating zone in function of temperature increase in the hall.

Fig. 9 presents relationship between average air temperature gradient in the hall (δt) and supply air temperature (t_s). In this case we've observed inversely proportional relationship which was confirmed in the charts in Fig. 4–Fig. 7. Statistically this relationship is not firm but in our opinion it takes place. Possibly it might have been caused by an uncontrolled variation in air volume made by hall operators. This conclusion needs further research to be fully confirmed.

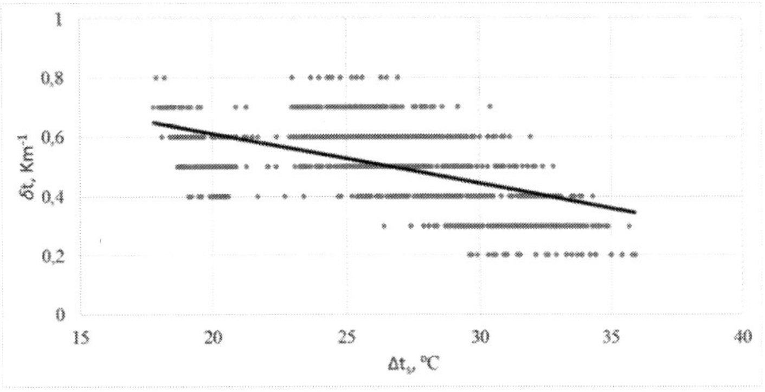

Fig. 9. Average temperature gradient in the hall related to supply air temperature.

DOI: 10.1051/e3sconf/20172200078

In Fig. 10. we present temperature changes at different hall elevation in relation to air temperature and temperature increase in the operating zone. Quick air temperature change in relation to the distance from operating zone takes place within 2 m above this zone. Four metres above floor level the air temperature stabilises and temperature increase is minor. Such dependence was previously described [2, 3, 7–9]. The higher was the temperature in the operating zone, the higher was the temperature in the highest hall point, however technological sensible heat load at that time had the lowest value during the whole year.

Fig. 11. presents variability of average air temperature gradient in relation to the distance from the operating zone. Presented nature of variability of δt stays in the relationship with characteristics presented in Figure 2. Major drop of δt value is observed within 1.5–3.0 m from the operating zone. At more than 4m above the floor, the intensity of value change is dropping.

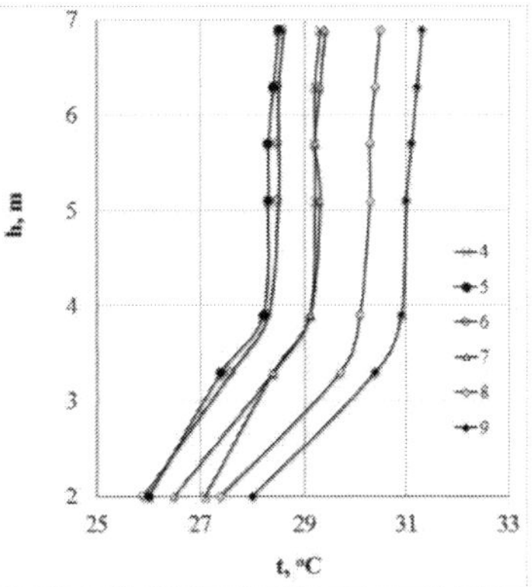

Fig. 10. Air temperature change at different elevation in relation to air temperature and temperature increase in the operating zone.

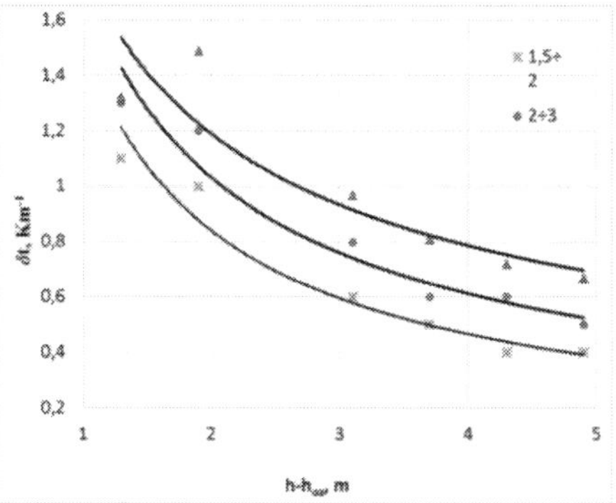

Fig. 11. Average temperature gradient in dependence with distance from the operating zone.

5 Conclusions

1. Air temperature stratification in the hall appears during minimum and maximum heat load in the hall.
2. Relationship between the hall air temperature increase and air temperature increase in the operating zone was determined. This relationship gives an opportunity for easier estimation of the air temperature increase, necessary for air flow rate calculation in the halls with similar technological and thermal features for AHU without air cooling treatment.
3. Average air temperature gradient is inversely proportional to supply air temperature where during summer time ambient air or ambient/recirculated air is blown into the hall.
4. Significant air temperature growth over operating zone appears within 2m distance. Above this line, temperature increase is minor.

This publication presented data obtained under the research project "Air temperature gradient in mechanically ventilated rooms with mixing ventilation" Nr N523 748040 funded by National Science Centre of Poland.

References

1. H. Awbi, *Ventilation of building* (London and New York: Spon Press, 2003)

2. V.V. Baturin, *Osnowy promyschlennoj ventilacii* (Pofizdat, Moskva, 1990)
3. H. Goodfellow, E. Tahti, *Industrial Ventilation* (Academic Press, 2001)
4. T. Hayashi, R.H Howell, M. Shibata, K. Tsuji, *Industrial Ventilation and Air Conditioning* (CRC Press Inc, Boca Raton, Boston, 1985)
5. A. Pełech, *Wentylacja i Klimatyzacja. Podstawy* (Oficyna Wydawnicza Politechniki Wrocławskiej, 2009)
6. S. Przydróżny, *Wentylacja* (Wydawnictwo Politechniki Wrocławskiej, 1991)
7. M. Xu, T. Yamanaka, H. Kotani, Indoor Air **11**, 2 (2001)
8. T. Gil-Lopez, M.A. Galvez-Huerta, P.G O'donohe, J. Castejon-Navas, P.M Dieguez-Elizondo, Energ. Buildings **140**, 371 (2017)
9. B. Iskandriawan, Energ. Buildings **21**, 2 (2010)

The analysis of the possibility of using 10-minute rainfall series to determine the maximum rainfall amount with 5 minutes duration

Bartosz Kaźmierczak[1,*], *Katarzyna* Wartalska[1], *Marcin* Wdowikowski[2], and *Andrzej* Kotowski[1]

[1]Wrocław University of Science and Technology, Faculty of Environmental Engineering, Plac Grunwaldzki 13, 50-377 Wrocław, Poland
[2]Institute of Meteorology and Water Management – National Research Institute, Podleśna Street 61, 01-673 Warsaw, Poland

Abstract. Modern scientific research in the area of heavy rainfall analysis regarding to the sewerage design indicates the need to develop and use probabilistic rain models. One of the issues that remains to be resolved is the length of the shortest amount of rain to be analyzed. It is commonly believed that the best time is 5 minutes, while the least rain duration measured by the national services is often 10 or even 15 minutes. Main aim of this paper is to present the difference between probabilistic rainfall models results given from rainfall time series including and excluding 5 minutes rainfall duration. Analysis were made for long-time period from 1961–2010 on polish meteorological station Legnica. To develop best fitted to measurement rainfall data probabilistic model 4 probabilistic distributions were used. Results clearly indicates that models including 5 minutes rainfall duration remains more appropriate to use.

1 Introduction

In both newly designed and existing sewage systems, according to the European standard PN-EN 752:2008, it is recommended to verify the design assumptions by using hydrodynamic simulations. Designers insufficient access to rainfall databases is the main barrier to model safe drainage systems in Poland. The most commonly required at the entrance to hydrodynamic models are storm and heavy rainfall hyetographs with 5 minutes time resolution. The maximum precipitation models (in the form of IDF or DDF curves) are the grounds for the creation of a model precipitation to use in hydrodynamic models [1–8].

In Poland, an access to the precipitation source data is a matter for the Institute of Meteorology and Water Management (IMWM), the owner of the largest number of meteorological stations in the country. The extremely valuable research material on precipitation, due to the number of the stations and the length of the measurement series (often decades), are paper pluviographic strips, on which precipitation were recorded by 2010. It is possible to read information about rainfall height from the paper strips even with

* Corresponding author: bartosz.kazmierczak@pwr.edu.pl

a resolution time of 5 minutes. Many maximum precipitation models have been developed and applied in engineering practice on the basis of this records [9–12].

Since 2007, the IMWM has been recording precipitation by digital rain gauges. Time series are currently recorded with a 10-minute time resolution, which is a matter of concern because in designing and modeling of sewage systems it is needed information about shorter precipitation, lasting 5 minutes [1].

In the paper [13], an attempt has been made to develop two maximum precipitation models for Legnica, with the use of generalized exponential distribution (GED) – considered to be the best probability distribution for analyzed meteorological station [11]. The first model based on rainfall data with durations ranging from 5 minutes to 6 days, while the second model was developed on the basis of rainfall data of durations from 10 minutes to 6 days and the amounts of precipitation with the duration of 5 minutes were extrapolated. The comparison, using relative residual mean square error (RRMSE), between predicted precipitation amounts and measurement data revealed that the first precipitation model accurately maps the measurement results. For rainfall with a duration of 5 minutes, the differences between the predicted and measured precipitation amounts at the level of RRMSE = 2.9% were noticed. The results of using the second model to determine precipitation heights with a 5 minute duration (parameters extrapolation) were not correlated with the measurement results (RRMSE = 41.5%) and therefore, the GED distribution is inappropriate for estimating precipitation amounts for the duration of $t = 5$ min and it is requested to increase the temporal resolution of recorded rainfall for 5-minute or shorter intervals.

In this paper, the utility of the other, applied in practice, probability distributions was examined, i.e. Fréchet, gamma, log-normal and Weibull distributions [11], in order to analyze the possibility of using 10-minute rainfall series to determine the maximum rainfall amount with 5 minutes duration.The analysis was carried out using archival pluviographs from Legnica from the time span 1961–2010. Distribution parameters were calculated (for rainfall duration from 10 to 8640 minutes) and the model describing the dependence of the amount of rainfall on its duration and specified exceedance probability was subsequently established. Then, precipitation amounts for the 5 minutes duration were computed. The obtained results were compared with the measurement data.

2 Materials and methods

The meteorological station of IMWM in Legnica is a part of national measurement and observation network at hydrological and meteorological service. This synoptic station is located on the south-eastern outskirts of the city, at an elevation of 122 m above the sea level. Fields and wasteland are the main land use in both municipality and rural areas around the station [11]. In this study, the archival pluviographs from this IMWM meteorological station from the time span 1961–2010 were used as a research material.

A selection of data has been done in order to determine the relationship between the amount of rainfall from its duration and probability of exceedance $h(t,p)$. Elaborating archival pluviographs authors limited period of analysis to months from May to October (V–X) [9–11].

From the tested 50-years period top 50 maximum amounts (h, mm) of rainfall were set apart for each of the following rainfall durations: $t = 5$, 10, 20, 30, 40, 50, 60, 90, 120, 180, 360, 720, 1080, 1440, 2160, 2880, 4320, 5760, 7200 and 8640 minutes. The isolation has been done on the basis of the Peak-Over-Threshold (POT) method [14–19]. Selected top rainfall heights were listed in descending order. In compliance with following formula:

$$p(m,N) = \frac{m}{N+1} \tag{1}$$

where m is the sequence number within a decreasing ordered string of the number of N, for each rainfall amount there were assigned the empirical probability of exceedance from $p = 0.020$ (for the highest value) to $p = 0.980$ (for the lowest value) [9, 10]. The rainfall heights recorded and selected values of empirical probability assigned to them are shown in Table 1.

Table 1. The amount of rainfall (h, mm) recorded for selected values of empirical probability (m = 1, 5, 10, 25 and 50).

t, min	p 0.020	0.098	0.196	0.490	0.980
5	16.2	11.2	9.8	7.6	4.6
10	23.5	17.5	14.1	10.9	8.3
20	31.7	25.1	20.1	14.5	11.3
30	32.0	26.9	24.5	16.0	12.5
40	37.4	28.2	25.2	17.6	13.0
50	39.8	31.0	25.2	18.2	13.7
60	40.6	31.6	25.9	19.9	14.2
90	41.0	36.2	29.4	21.6	16.5
120	49.6	39.6	32.2	22.8	18.2
180	57.6	40.7	34.5	26.3	20.0

t, min	p 0.020	0.098	0.196	0.490	0.980
360	57.7	47.2	40.9	30.6	23.9
720	74.9	51.8	44.6	35.8	28.6
1080	77.3	57.7	51.6	41.3	32.4
1440	77.3	66.3	57.7	46.4	35.7
2160	114.8	77.3	61.1	49.4	38.7
2880	129.3	97.8	74.3	53.0	41.3
4320	143.1	97.8	77.6	57.4	41.3
5760	157.0	116.0	83.9	61.9	47.1
7200	158.5	121.1	86.0	69.9	51.7
8640	167.9	132.3	91.9	71.4	53.7

For the description of survey data, the following probability distributions have been used: Fréchet, gamma, log-normal and Weibull distribution. The logarithms of their likelihood functions appear as follows:

$$\mathrm{Ln}L = \alpha^n \beta^{n\alpha} \prod_{i=1}^{n} (x_i - \gamma)^{-(\alpha+1)} \exp\left[-\sum_{i=1}^{n} \left(\frac{\beta}{x_i-\gamma}\right)^{\alpha}\right] \quad (2)$$

$$\ln L = (\alpha - 1)\sum_{i=1}^{n} \ln(x_i - \gamma) - n\ln\Gamma(\alpha) - n\alpha\ln\beta - \frac{1}{\beta}\sum_{i=1}^{n}(x_i - \gamma) \quad (3)$$

$$\ln L = -\sum_{i=1}^{n} \ln(x_i - \gamma) - n\ln\alpha - \frac{n}{2}\ln(2\pi) - \frac{1}{2\alpha^2}\sum_{i=1}^{n}(\ln(x_i - \gamma) - \beta)^2 \quad (4)$$

$$\ln L = n\ln\alpha - n\alpha\ln\beta + (\alpha - 1)\sum_{i=1}^{n} \ln(x_i - \gamma) - \sum_{i=1}^{n}\left(\frac{x_i-\gamma}{\beta}\right)^{\alpha} \quad (5)$$

where α, β and γ are parameters of distributions.

Quantiles of the random variable analyzed distributions takes the form of the following formulas:

$$h_p = \gamma + \beta(-\ln(1-p))^{-\frac{1}{\alpha}} \quad (6)$$

$$h_p = \gamma + \beta t_p \quad (7)$$

$$h_p = \gamma + e^{\alpha + \beta t_p} \quad (8)$$

$$h_p = \gamma + \beta(-\ln p)^{\frac{1}{\alpha}} \quad (9)$$

The Anderson-Darling test for statistics [20, 21] was used for establishing the coincidence of theoretical distributions with measured data:

$$A^2 = -n - \frac{1}{n}\sum_{i=1}^{n}(2i-1)\left[\ln F(X_i) + \ln\left(1 - F(X_{n-i+1})\right)\right] \quad (10)$$

where: x_i – i-th value in the decreasing ordered random sample, $F(x)$ – cumulative distribution function for the theoretical distribution. The null hypothesis H_0 (when measurement data were adequate for tested theoretical distribution), was set at a significance level of 0.05 if the A^2 was less than the critical value A_{kr}^2. The alternative hypothesis was taken otherwise. For Fréchet, gamma, log-normal and Weibull distribution

 DOI: 10.1051/e3sconf/20172200079

and for N = 50 the critical value amounts to respectively: A_{kr}^2 = 0.757, A_{kr}^2 = 0.762, A_{kr}^2 = 0.752 and A_{kr}^2 = 0.757.

For the evaluation of the aptitude of investigated distributions and for describing the survey data, relative residual mean square error (11) was used:

$$RRMSE = \sqrt{\frac{1}{n}\sum_{i=1}^{n}\left(\frac{h_{o,i}-h_{p,i}}{h_{p,i}}\right)^2} \cdot 100\% \qquad (11)$$

where: h_t – the theoretical amount of rainfall (mm), h_m – amount of rainfall from measurements (mm).

3 Results

Calculation results of particular parameters of distributions were presented in Tables 2–5. The parameter estimates were determined by numerical maximization of the log-likelihood functions (2)–(5). The calculations were carried out for each of 20 durations of maximum precipitation amounts analyzed in the paper. All distributions fulfils the compliance criterion A^2 for each of the 20 analyzed rainfall durations (except log-normal distribution for the duration t = 7200 min).

Table 2. Calculation results of parameters of Fréchet distribution.

t, min	α	β	γ	A^2	$RRMSE$	t, min	α	β	γ	A^2	$RRMSE$
5	9.011	15.797	-8.90	0.152	3.0	360	2.297	10.502	17.86	0.436	6.4
10	2.269	3.863	6.05	0.379	3.7	720	2.987	15.784	18.24	0.292	2.3
20	1.551	3.424	9.73	0.582	11.3	1080	2.744	15.314	22.71	0.386	3.3
30	1.640	4.090	10.63	0.514	12.1	1440	4.479	31.384	11.68	0.323	3.0
40	2.007	6.010	9.84	0.403	7.2	2160	2.747	20.546	25.93	0.359	3.0
50	2.444	7.951	9.01	0.383	4.5	2880	1.652	14.144	35.06	0.305	7.1
60	3.675	13.577	4.59	0.261	3.1	4320	2.266	22.305	29.40	0.232	3.3
90	2.385	8.407	11.66	0.325	6.1	5760	1.474	15.110	41.59	0.581	10.2
120	1.795	5.881	15.58	0.467	7.0	7200	1.709	18.962	41.96	0.609	7.5
180	1.977	7.770	15.90	0.489	5.4	8640	1.829	21.802	42.79	0.399	6.3

Table 3. Calculation results of parameters of gamma distribution.

t, min	α	β	γ	A^2	$RRMSE$	t, min	α	β	γ	A^2	$RRMSE$
5	2.113	1.776	4.37	0.176	3.0	360	0.927	10.167	23.89	0.289	3.2
10	0.868	4.065	8.29	0.262	2.6	720	0.945	11.190	28.59	0.408	2.2
20	0.684	7.210	11.29	0.500	3.3	1080	0.969	11.555	32.39	0.249	2.2
30	0.803	6.934	12.49	0.320	4.2	1440	0.938	13.962	35.69	0.699	3.4
40	0.917	6.954	12.99	0.138	2.5	2160	0.932	16.676	38.59	0.474	3.2
50	0.997	6.735	13.69	0.194	2.3	2880	0.831	23.882	41.29	0.163	3.3
60	1.103	6.576	14.18	0.275	2.5	4320	0.975	22.401	41.29	0.423	3.6
90	0.922	8.001	16.49	0.414	3.3	5760	0.866	28.871	47.09	0.366	3.9
120	0.955	7.894	18.19	0.404	2.9	7200	0.666	35.982	51.69	0.410	3.8
180	0.885	9.458	19.99	0.177	2.4	8640	0.802	32.908	53.69	0.266	4.3

Table 4. Calculation results of parameters of log-normal distribution.

t, min	*α*	*β*	*γ*	*A²*	*RRMSE*	*t*, min	*α*	*β*	*γ*	*A²*	*RRMSE*
5	0.508	1.458	3.24	0.156	2.9	360	0.917	1.974	22.89	0.370	5.1
10	0.953	0.951	7.94	0.344	3.1	720	0.800	2.214	26.83	0.262	2.1
20	1.147	1.113	11.00	0.492	7.2	1080	0.859	2.205	30.94	0.340	2.9
30	1.092	1.272	12.17	0.421	8.0	1440	0.694	2.579	32.30	0.303	2.8
40	0.978	1.514	12.47	0.295	5.1	2160	0.817	2.565	36.37	0.377	2.9
50	0.914	1.624	13.05	0.359	3.7	2880	1.075	2.530	40.26	0.198	4.0
60	0.741	1.900	12.79	0.261	2.7	4320	0.870	2.819	39.09	0.207	3.0
90	0.873	1.762	15.63	0.262	4.6	5760	1.085	2.737	46.13	0.392	4.7
120	0.962	1.667	17.66	0.369	4.0	7200	1.229	2.625	50.53	0.776	7.4
180	1.026	1.739	19.42	0.421	4.2	8640	1.076	2.828	52.13	0.375	4.8

Table 5. Calculation results of parameters of Weibull distribution.

t, min	*α*	*β*	*γ*	*A²*	*RRMSE*	*t*, min	*α*	*β*	*γ*	*A²*	*RRMSE*
5	1.442	3.963	4.52	0.204	3.3	360	0.986	9.374	23.89	0.281	3.2
10	0.937	3.431	8.29	0.261	2.6	720	1.002	10.578	28.59	0.350	2.2
20	0.806	4.455	11.29	0.596	3.7	1080	1.014	11.258	32.39	0.217	2.1
30	0.890	5.286	12.49	0.339	4.4	1440	1.028	13.216	35.69	0.565	2.6
40	0.969	6.291	12.99	0.142	2.6	2160	0.983	15.436	38.59	0.479	3.2
50	1.019	6.765	13.69	0.186	2.4	2880	0.896	18.855	41.29	0.150	3.1
60	1.101	7.506	14.18	0.220	2.4	4320	0.996	21.810	41.29	0.405	3.7
90	0.986	7.336	16.49	0.369	3.2	5760	0.916	24.003	47.09	0.351	3.8
120	0.980	7.477	18.19	0.403	3.0	7200	0.787	21.279	51.69	0.527	3.9
180	0.943	8.168	19.99	0.185	2.5	8640	0.881	24.885	53.69	0.261	4.2

Following the equations (6)–(9) and the parameters listed in Tables 2–5, the precipitation amount with any exceedance probability and selected rainfall duration ranging from 5 to 8640 minutes can be calculated. In this respect, it should be noted that the current 10-minute time step of rainfall registration will not allow in the future to estimate parameters for precipitation with the duration of 5 minutes. So these parameters will have to be extrapolated.

The parameters β and γ have a clearly marked trend. Since there is no dependency trend $\alpha(t)$, the mean value of $\bar{\alpha} = 2.312$, $\bar{\alpha} = 0.894$, $\bar{\alpha} = 0.953$ and $\bar{\alpha} = 0.954$ was assumed in the calculations (respectively for Fréchet, gamma, log-normal and Weibull distribution).

Based on the calculated distribution parameters there were prepared plots (Fig. 1) showing their dependence on the rainfall duration (from 10 to 8640 minutes).

Finally, a models describing the dependence of the amount of rainfall on its duration and a specified exceedance probability, based on the quantiles (6)–(9), takes the form of:

$$h_p = 2.763t^{0.300} + 3.578t^{0.204}(-\ln(1-p))^{-0.433} \quad (12)$$

$$h_p = 5.253t^{0.256} + 1.304t^{0.355}t_p(\alpha, p) \quad (13)$$

$$h_p = 4.843t^{0.261} + e^{0.927t^{0.127}+0.953t_p(p)} \quad (14)$$

$$h_p = 5.253t^{0.256} + 1.879t^{0.283}(-\ln p)^{1.049} \quad (15)$$

 DOI: 10.1051/e3sconf/20172200079

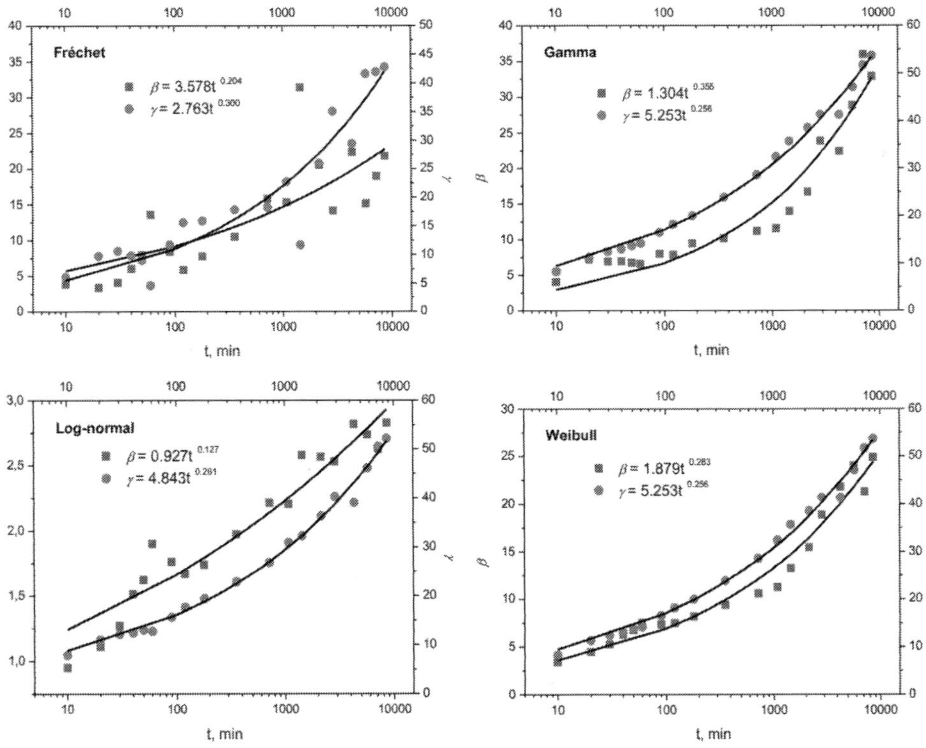

Fig. 1. The dependence of the parameters β and γ of the rainfall duration.

Based on the obtained formula (12)–(15), precipitation amounts for the 5 min duration were computed. The received results were compared with the measurement data and the precipitation amount calculated using formula (6)–(9) with the parameters from Tables (2–5). The fit quality of the equations for rainfall data from Legnica is shown in the *h-h* plot (Fig. 2).

Analysis of Figure 2 shows that precipitation amounts calculated from equations (12)–(15) differ significantly from measurement results and from the results obtained from the models (6)–(9) – already similar to measured rainfall data.

There were also calculated relative residual mean square error statistics, covering the entire range of data (t = 5 min). In this case R_{RMSE} = 3.0%, 3.0%, 2.9% and 3.3% for h calculated by (6–9) and R_{RMSE} = 44.3%, 30.9%, 47.3% and 37.6% for h calculated by (12–15). Deviations from the measurements at the level of several dozen percents disqualify analyzed method of determining precipitation amounts with the duration of 5 minutes.

 DOI: 10.1051/e3sconf/20172200079

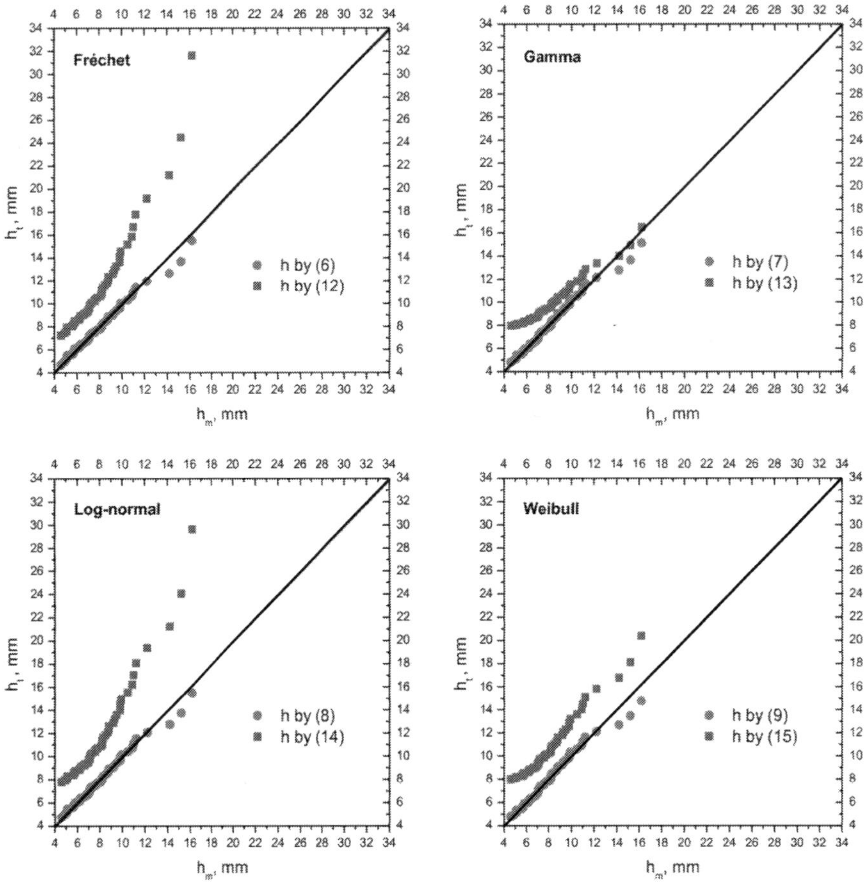

Fig. 2. The *h-h* plots for the distributions.

4 Conclusions

In this paper archival pluviographic records from long term period 1961–2010 were used to develop tow types of probabilistic maximum precipitation models. The area of this research was polish meteorological station in Legnica in Lower Silisia voivodeship. The first group of models were developed on the basis of rainfall data durations from 10 minutes to 6 days. 5 minutes duration precipitation amounts were extrapolated. Models were constructed on four probabilistic distriburions : Fréchet, gamma, log-normal and Weibull. In the second solution only 5 minutes rainfall duration values were computed using all four probability distribution quantiles. Both models were compared with measurement rainfall data using relative residual mean square error. The conducted analysis allowed to draw the following conclusions:

- The maximum precipitation model (6)–(9) with the parameters shown in Table from 2 to 5 thoroughly describes the measurement rainfall data for 5 minutes duration in Table 1. The R_{RMSE} indicator varies from 2.9% for log-normal to 3.3% for Weibull distribution.
- In the case of using models from (12) to (15) to determine precipitation heights with a duration of t = 5 min (parameters extrapolation) obtained results were not correlated with the measurement results - R_{RMSE} from 30.9% to 47.3%. Precipitation amounts with a duration of t = 5 min can be estimated with some approximation based on the

precipitation heights data for the duration of t = 10 min. The analysis of measurement data indicates that during the period considered, the maximum precipitation with a duration of t = 5 min represented on average 68% of maximum rainfall heights with a duration of t = 10 min. However, this estimation should be treated only indicatively, since this variability ranged in the case of the largest 50 precipitations from 55% to 75%. In order to draw up reliable conclusions, further research in this field involving other measurement stations is needed.

- In the light of the obtained results, it is requested to increase the temporal resolution of recorded by IMWM rainfall for 5-minute or shorter intervals.

Including all above described method should be regarded as inappropriate for estimating precipitation amounts for the duration of t = 5 min. Good quality rainfall model should be based on the 5 minutes rainfall durations that had been measured not extrapolated.

The work was realized within the allocation No. 0401/0006/17 awarded for Faculty of Environmental Engineering Wroclaw University of Science and Technology by Ministry of Science and Higher Education in years 2017–2018.

References

1. A. Kotowski, *The principles of safe dimensioning of sewer systems* (Seidel-Przywecki, Warsaw, 2015)
2. E. Burszta-Adamiak, M. Mrowiec, Water Sci Technol **68**, 1 (2013)
3. A. Kotowski, K. Wartalska, M. Nowakowska, Ochrona Środowiska **38**, 1 (2016)
4. B. Kowalska, D. Kowalski, G. Łagód, M. Widomski, *Modelling of Hydraulics and Pollutants Transport in Sewer Systems with Exemplary Calculations in SWMM* (Lublin University of Technology, Lublin, 2013)
5. S. Y. Park, K. W. Lee, I. H. Park, S. R. Ha, Desalination **226**, 1–3 (2008)
6. T. G. Schmitt, *Kommentar zum Arbeitsblatt A 118 Hydraulische Bemessung und Nachweis von Entwässerungssystemen* (DWA, Hennef, 2000)
7. T. G. Schmitt, M. Thomas, KA – Wasserwirtschaft, Abwasser, Abfall **47**, 1 (2000)
8. D. Słyś, A. Stec, Environment Protection Engineering **38**, 4 (2012)
9. A. Kotowski, B. Kaźmierczak, A. Dancewicz, *The modeling of precipitations for the dimensioning of sewer systems* (Polish Academy of Sciences, Warsaw, 2010)
10. E. Bogdanowicz, J. Stachý, *Maximum rainfall in Poland. Design characteristics* (The Publishing House of the Institute of Meteorology and Water Management, Warsaw, 1998)
11. B. Kaźmierczak, M. Wdowikowski, Periodica Polytechnica Civil Engineering **60**, 2 (2016)
12. R. Suligowski, *Maximum credible precipitation at the Kielce Upland* (The Jan Kochanowski University, Kielce, 2013)
13. B. Kaźmierczak, K. Wartalska, M. Wdowikowski, E3S Web of Conferences **17**, 00038 (2017)
14. W. Jakubowski, Meteorology Hydrology and Water Management **3**, 2 (2015)
15. A. Mailhota, S. Lachance-Cloutierb, G. Talbota, A.C. Favrec, J Hydrol **476**, 7 (2013)
16. C. Onyutha, Journal of Urban and Environmental Engineering **6**, 2 (2012)
17. W. L. Shinyie, N. Ismail, A. A. Jemain, Water Resour Manag **28**, 11 (2014)
18. S. Wi, B. J. Valdés, S. Steinschneider, T. W. Kim, Stoch Env Res Risk A **30**, 2 (2016)

DOI: 10.1051/e3sconf/20172200079

19. A. G. Yilmaz, I. Hossain, B. J. C. Perera, Hydrol Earth Syst Sc **18** (2014)
20. R. D'Agostino, M. A. Stephens, *Goodness of Fit Techniques* (Marcel Dekker, New York, 1986)
21. S. Hongjoon, J. Younghun, J. Changsam, H. Jun-Haeng, Stoch Env Res Risk A **26**, 1 (2012)

 DOI: 10.1051/e3sconf/20172200080

Comparing energy payback and simple payback period for solar photovoltaic systems

Will Kessler[1,*]

[1]N Am Board of Cert Energy Practitioners, REDA LLC, 533 Congress St, Portland, ME 04101 USA

Abstract. Installing a solar photovoltaic (PV) array is both an environmental and a financial decision. The financial arguments often take priority over the environmental because installing solar is capital-intensive. The Simple Payback period (SPB) is often assessed prior to the adoption of solar PV at a residence or a business. Although it better describes the value of solar PV electricity in terms of sustainability, the Energy Payback period (EPB) is seldom used to gauge the merits of an installation. Using published estimates of embodied energies, EPB was calculated for four solar PV plants utilizing crystalline-Si technology: three being actual commercial installations located in the northeastern U.S., and a fourth installation based on a simulated 20-kilowatt roof-mounted system, in Wrocław, Poland. Simple Payback was calculated based on initial capital cost, and on the availability of avoided electricity costs based on net-metering tariffs, which at present in the U.S. are 1:1 credit ratio, and in Poland is 1:0.7 credit ratio. For all projects, the EPB time was estimated at between 1.9 and 2.6 years. In contrast, the SPB for installed systems in the northeastern U.S. ranged from 13.3 to 14.6 years, and was estimated at 13.5 years for the example system in Lower Silesia, Poland. The comparison between SPB and EPB shows a disparity between motivational time frames, in which the wait for financial return is considerably longer than the wait for net energy harvest and the start of sustainable power production.

1 Introduction

The adoption of solar photovoltaic (PV) generation at a given site is a blend of feasibility and motivations – both economic and environmental. The technical aspects can include: roof orientation and pitch, solar access, roof composition, land usage, geology, soil character, proximity to the power grid, and many others. On the other hand, financial variables include questions of access to working capital, cost of installation, and the cost of avoided electricity. The Simple Payback Period (SPB) is a commonly cited measure of the

* Corresponding author: tankfullofsun@gmail.com

time in years, between initial outlay of capital on a solar installation and the return on initial investment. The Simple Payback period of a solar PV installation may be expressed:

$$SPB = (c_{\text{system}})\,(Q_{\text{year}}^{-1})\,(PR^{-1})\,(P^{-1})\,(c_{\text{elec}}^{-1}) \quad (1)$$

where c_{system} is the un-subsidized installed system cost, Q_{year} is the total annual irradiation for an optimally-inclined array at the location [1], PR is the system's annual DC to AC performance ratio [2], P is the system DC capacity in Watts, and c_{elec} is the cost of electricity that is avoided by the system's local production. The usefulness of the SPB calculation is that it is an intuitive expression of expenses and returns. However it carries a high degree of variability due the fluctuation of c_{elec}, i.e. the market electricity rate. Historically, SPB for solar PV installations has decreased, as the cost of producing PV modules and c_{system} has decreased [3, author's industry observation].

The merits of a solar PV installation as an energy investment are described with an analogous term, called Energy Payback (EPB), which may be expressed:

$$EPB = (E_{\text{M}} + E_{\text{INV}} + E_{\text{BOS}})\,(Q_{\text{year}}^{-1})\,(PR^{-1})\,(P^{-1}) \quad (2)$$

where E_{M}, E_{INV}, and E_{BOS} represent the embodied energy of the solar modules, the DC-to-AC inverters, and Balance of System (BOS), respectively [4]. Where SPB estimates the wait time for the recoup of initial capital outlay, the EPB expresses the time required for the PV array to generate an amount of energy equivalent to its initial energy inputs, and start net energy production.

This work presents estimated EPB and SPB calculations, for three completed solar photovoltaic installations, and one typical installation based on published estimates of embodied energy and current industry and market parameters. In assessing the differences between EPB and SPB, the question arises of how best to motivate and accelerate adoption of sustainable, local PV harvest, given its comparatively short Energy Payback period.

2 Methods

Simple and Energy Payback periods were calculated for three installed systems in the northeastern U.S. including a ground-mounted array located in Thornton, NH (lat. 43.936740, long. – 71.683367), a pitched-roof array located in Franconia, NH (lat. 44.234389, long. – 71.756667), and a flat-roof array located in Concord, NH (lat. 43.218414, long. – 71.536698). For geographical comparison, payback periods were simulated for a model 20 kW DC array located in Lower Silesia Poland (lat. 51.108752, long. – 17.059876).

All SPB estimates assumed electricity escalation rates of 3% [5], a 0% discount rate, zero cash incentives, and used the electricity tariffs applicable for each customer at the time of system installation, as shown in Table 1. Mean solar production for U.S. locations was modelled using the American National Renewable Energy Laboratory's typical meteorological year data sets (TMY2, TMY3) [6]. Mean solar production for the Polish site was modelled from the Photovoltaic Solar Electricity Potential in European Countries data set [1]. The simulated SPB for the Lower Silesian PV system also assumed solar production was consumed behind the meter 50% of the time for 1:1 cost offset, and exported 50% of the time for 1:0.7 cost offset based on the recently enacted Odnawialne Źródła Energii (Renewable Energy Sources) laws. An installed cost of 7PLN per WDC was used [author's industry observation].

Energy payback period was calculated using primary energy in all instances, assuming a grid efficiency factor of 0.35, and balance of materials records from the three existing projects in the northeastern U.S. Specifically, Balance of System (E_{BOS}) was calculated as

DOI: 10.1051/e3sconf/20172200080

the sum of embodied energies by mass of Aluminum, Steel, Copper, PVC, and Concrete materials used in the solar PV system, based on estimates by Frankl et al. [7]. For all four systems, embodied energy of poly-Si and mono-Si photovoltaic modules (E_M) followed from Peng et al. [4], and embodied energy of the PV inverter by AC wattage (E_{INV}) followed from estimates published by Alsema [8]. Energy payback for the simulated 20 kilowatt system referenced the same estimates of E_M [4] and E_{INV} [8]. However E_{BOS} estimates were based on a linear extrapolation from the roof-attached Franconia array, using an empirically determined E_{BOS} coefficient of 95.7 kWh/m^2 of array.

3 Results

The estimated SPB and EPB for installed and modelled PV systems are presented in Tables 1 and 2, respectively.

Table 1. Estimated Simple Payback periods for three actual and one simulated solar PV system.

Generator	Commissioned	System size kW DC	Solar production	Elec. Tariff $/kWh	Simple Payback
Hubbard Brook Research Foundation Thornton, NH USA	2015	40.56	44,800 [6]	$0.177	13.3
Gale River Motel Franconia, NH USA	2016	13.78	15,540 [6]	$0.170	13.8
Grappone Conf. Center Concord, NH USA	2016	143.6	160,000 [6]	$0.126	14.6
Sample roof mounted Mono-Si PV array Optimal orientation Wrocław, Poland	--	20.0	19,000 [1]	$0.139* *[based on 3.96 PLN = 1 USD]	13.5

Table 2. Estimated Energy Payback along with estimate sources for three actual and one simulated solar PV system.

Generator	Primary energy prod. (kWh)	Estimated primary energy inputs			Total	Energy Payback years
		PV module	BOS	Inverter		
Hubbard Brook Research Foundation	128,000	213,600 [4, 10]	80,000 [7]	13,200 [8]	306,800	2.4
Gale River Motel Franconia, NH USA	44,400	71,200 [4, 10]	8,000 [7]	4,300 [8]	83,500	1.9
Grappone Conf. Center Concord, NH USA	457,100	888,800 [4, 10]	40,400 [7]	38,500 [8]	967,700	2.1
Sample roof mounted Mono-Si PV array Optimal orientation Wrocław, Poland	54,300	123,600 [4, 10]	11,000	7,800 [8]	142,400	2.6

Table 3. Balance of System summary for driven-post, ground-mounted solar system in Thornton, NH USA.

Material	Aluminum		Steel		PVC		Copper		Total
kg	1,418	Racking system	1,256	Posts	85	Conduit	37	Conductors	
	69	Fasteners	73	Channel strut	9	Misc.			
	162	Conductors	92	Misc. electrical					
			7	Fasteners					
subtotal	1,649		1,428		94		37		
kWh embodied [7]	64,900		12,700		1,700		700		80,000

Table 4. Balance of System summary for pitched-roof solar system in Franconia, NH USA.

Material	Aluminum		Steel		PVC		Copper		Total
kg	183	Racking system	12	Fasteners	1	Conduit	8	Conductors	
	12	Fasteners	8	Misc. electrical					
			4	Conduit					
subtotal	195		24		1		8		
kWh embodied [7]	7,660		213		19		152		8,000

Table 5. Balance of System summary for ballasted, flat-roof solar system in Concord, NH USA.

Material	Aluminum		Steel		Concrete		PVC		Copper		Total
kg	31	Racking system	2,524	Racking system	9,973	Ballasts	22	Conduit	149	Conductors	
	59	Conductors	26	Fasteners							
	7	Fasteners	266	Channel strut							
			286	Conduit and wireways							
			123	Misc. electrical							
subtotal	98		3,225		9,973		22		149		
kWh embodied [7]	3,840		28,700		4,520		408		2,900		40,400

Detailed balance of materials lists are shown in Tables 3, 4, and 5. For the systems in operation in the northeastern U.S., simple payback ranged from 13.3 years to 14.6 years. For the simulated system in southwest Poland, simple payback was estimated at 13.5 years. The estimated EPB for existing installations ranged from 1.9 years for a pitched-roof attached-railing system in Franconia, to 2.4 years for a pile-driven ground mount array in Thornton. The existing ballasted flat-roof system in Concord showed an EPB of 2.1 years. The simulated roof mounted PV array in Lower Silesia (Wrocław) had an EPB of 2.6 years.

 DOI: 10.1051/e3sconf/20172200080

4 Discussion

Between the years 1977 and 2013, the average cost of crystalline-Si photovoltaic cells dropped from 76.67 USD/W, to 0.74 USD/W [3], and this trend continues in 2017, as average PV cell prices drop below 0.30 USD/W [author's industry observation]. Nonetheless the capital investment required to purchase and install a distributed solar PV generator is great enough that out-of-pocket costs and simple payback scenarios guide customers' decision-making. Comparisons of EPB and SPB show significantly shorter wait times for the harvest and return of 100% of the energy invested in the solar equipment and the start of net energy harvest, compared to the wait for financial returns.

Fig. 1. Ground-mounted solar photovoltaic arrays at the Hubbard Brook Reseaerch Foundation. The system's estimated energy payback period of 2.4 years was significantly less than the simple payback period, 13.3 years. Note the driven-post system reaches soil depth of 2.4m, and requires no concrete to install.

Energy payback improves alongside advancements in module manufacturing, solar installation techniques, and cell efficiencies. For example, the EPB times calculated in the present study for the northeastern U.S. and Poland are significantly shorter than the EPBs surveyed for systems of similar or greater productivity that were installed between 2003 and 2006 [9]. Furthermore, the E_{BOS} estimated for the ground-mounted array at Hubbard Brook, pictured in Figure 1, was 319 kWh/m^2, small in comparison to previously published estimates of ground mounting balance of system embodied energies [10, 11]. However, it is unlikely that continuing technological advancement will decrease the EPB below certain physical thresholds due to constraints such as resource scarcity, transportation and mobilization costs, and labor costs of installation.

Incentivizing renewable energies by shortened SPB, or low cost loan presents a valid goal based on the physical merits of low-EPB solar generation. In mature solar markets, two policies have been successfully adopted and replicated, and may be worth long term implementation: a) Production-based incentives such as a mandated offer or a feed in tariff guarantee, and b) Low-cost financing programs, which allow the expense of equipment and installation to be spread out over the lifetime of the solar equipment.

The author would like to thank Marta A. Kessler for her patience and swiftness, without which little if any of the solar field work or post-project investigations would be possible.

References

1. M. Šúri, T.A. Huld, E.D. Dunlop, H.A. Ossenbrink, Sol. Energy **81**, 1295–1305 (2007)
2. B. Marion, J. Adelstein, K. Boyle, H. Hayden, B. Hammond, T. Fletcher, B. Canada, D. Narang, D. Shugar, H. Wenger, A. Kimber, L. Mitchell, G. Rich, T. Townsend, IEEE Phot Spec Conf **CP-520-37358** (2005)
3. Bloomberg Finance L.P., *Sustainable Energy in America, 2015 Factbook*, http://www.bcse.org/wp-content/uploads/2015-Sustainable-Energy-in-America-Factbook.pdf (accessed 28 March 2017)
4. J. Peng, L. Lu, H. Yang, Renew. Sust. Energ. Rev. **19**, 255–274 (2013)
5. Energy Information Administration, *Electric Power Annual – State Data Tables,* https://www.eia.gov/electricity/annual/ (accessed 28 March 2017)
6. A. P. Dobos, *PVWatts version 5 manual* (2014)
7. P. Frankl, A. Masini, M. Gamberale, D. Toccaceli, Prog. Photovoltaics **6**, 137–146 (1998)
8. E. Alsema, Energy pay-back time and CO_2 emissions of photovoltaic systems, in Markvart and Castañer (Eds.), *Practical Handbook of Photovoltaics: Fundamentals and Application*, 1097-1117 (Elsevier, Kidlington, UK, 2003)
9. W. Kessler, S. Dornblut, Solar Today **11/12**, 32–35 (2009)
10. N. Jungbluth, M. Stucki, R. Frischknecht, S. Büsser, Photovoltaics, in Dones et al. (Eds.), Sachbilanzen von Energiesystemen: Grundlagen für den ökologischen Vergleich von Energiesystemen und den Einbezug von Energiesystemen in Ökobilanzen für die Schweiz, Ecoinvent Report **6** - XII (ESU-services Ltd, Uster, CH, 2010)
11. J.M. Mason, V.M. Fthenakis, T. Hensen, H.C. Kim, Progress in Photovoltaic Research and Applications **14**, 179–90 (2006)

E3S Web of Conferences **22**, 00081 (2017) DOI: 10.1051/e3sconf/20172200081
ASEE17

Analysis of the impact of simulation model simplifications on the quality of low-energy buildings simulation results

Marcin Klimczak[1,*], *Jacek* Bojarski[2], *Piotr* Ziembicki[3], and *Piotr* Kęskiewicz[1]

[1]Wroclaw University of Science and Technology, Faculty of Environmental Engineering, ul. C.K. Norwida 4/6, 50-373 Wroclaw, Poland
[2]University of Zielona Góra, Faculty of Mathematics, Computer Science and Econometrics, ul. Prof. Z. Szafrana 15, 65-516 Zielona Góra, Poland
[3]University of Zielona Góra, Faculty of Civil Engineering, Architecture and Environmental Engineering, ul. Prof. Z. Szafrana 15, 65-516 Zielona Góra, Poland

Abstract. The requirements concerning energy performance of buildings and their internal installations, particularly HVAC systems, have been growing continuously in Poland and all over the world. The existing, traditional calculation methods following from the static heat exchange model are frequently not sufficient for a reasonable heating design of a building. Both in Poland and elsewhere in the world, methods and software are employed which allow a detailed simulation of the heating and moisture conditions in a building, and also an analysis of the performance of HVAC systems within a building. However, these systems are usually difficult in use and complex. In addition, the development of a simulation model that is sufficiently adequate to the real building requires considerable time involvement of a designer, is time-consuming and laborious. A simplification of the simulation model of a building renders it possible to reduce the costs of computer simulations. The paper analyses in detail the effect of introducing a number of different variants of the simulation model developed in DesignBuilder on the quality of final results obtained. The objective of this analysis is to find simplifications which allow obtaining simulation results which have an acceptable level of deviations from the detailed model, thus facilitating a quick energy performance analysis of a given building.

1 Introduction

A rise in energy consumption in a global scale is unavoidable [1–9]. It is a measure of success of the economy and community of a given country. The highest increase of energy consumption concerns the consumption of electric energy and fossil fuels (mainly in transportation). It is also very likely to reverse the current, decreasing trend concerning heat demand in urban agglomerations due to the more and more popular technologies of using heat for the production of other forms of energy, for example, cooling (most frequently in

* Corresponding author: marcin.klimczak@pwr.edu.pl

the form of the so called "chilled water") for the needs of air-conditioning systems in buildings serving various purposes. This may result in increasing the production of district heat (generated and distributed in the district heating systems).

Rationalisation of heat use can and should be implemented in two different directions. The first is connected with enhancing the efficiency of heat sources and heating networks, whereas the other concerns a reduction of the demand for this type of energy by affecting and convincing designers of buildings and installations, as well as their users, to increase their activity and to take appropriate actions in order to optimize energy efficiency of buildings being designed and also to implement actions that would improve the technical parameters of already existing buildings. It should be underlined that optimisation of energy efficiency of buildings cannot be performed without employing advanced computer technologies, including building energy simulations and modern BIM technologies (Building Information Modelling). Currently, there are many types of programmes available on the market, which are used for energy optimisation of buildings, heating networks and district heating substations. These applications vary in their capabilities, the quality of results obtained and in the level of their complexity. In fact, the latter in many cases limits the utilization of these applications only to specially trained engineers. Moreover, professional software, especially BIM type systems, is very expensive to buy and to operate and this fact hinders or sometimes even makes it impossible for smaller design offices to use it. In this context, it is of crucial importance to develop a methodology for introducing simplifications to the building simulation model that would not affect adversely the quality of simulation results, and at the same time would make it possible to shorten the time and reduce the costs of building energy simulations.

2 Building energy analysis

In order to minimize energy consumption in a building, it is necessary to know its energy characteristics, consumption profile and understand how heat sources and the heating system work. In reality, it is difficult to obtain all the information while analysing the building or making its inventory, since quite frequently the documentation of the building is not up-to-date or incomplete and the performance data are not available. The problem can be solved by running an identification analysis of the system, i.e., the building, by means of developing computer simulation models and their calibration with the use of data obtained when the building is in use. A similar methodology should be used in order to analyse in detail the operation of the heat source, the heating system and domestic hot water supply.

In the literature, the notion of model is defined in many different ways. For the needs of this study, the definition given by Krupa [4] was adopted. According to the definition, a model is a set of variables and relations among them which are represented as an equation or a system of equations. The basic aims for which models of systems are developed include: description and explanation of the system's operation, prediction of its behaviour, impact on the environment and also optimization of its parameters.

According to Malinowski, the level of detail is one of the most important elements taken into consideration during the development and verification of the model [5]. It can be stated that the larger the model, the more realistic it is and simulation results resemble the reality. Practice shows, however, that a reasonable introduction of limitations in the level of data detail to the model is both economically and technically justifiable. Identification of a system considering numerous details requires much effort and is time consuming and this in turn results in higher costs which must be substantiated by the weight of the problem. In addition, once a detailed model is developed and implemented, the costs of the necessary software and the time of its operation are increasing. The costs of testing also rise, the number of data increases as well as the difficulties in manipulating the system's features. It happens frequently that satisfactory results are obtained after the first run of computer simulation and an introduction of further details to the model does not result in their better quality. If such instances are avoided, the time for the development of the model and time of simulations are reduced. According to Krupa, model equations can be solved analytically, numerically or by simulation [4]. In the latter case, the characteristic feature is that independent variables of the model are always the variables which correspond to the input values of the real system. Thanks to this the "experiment" run in the computer system, in structural terms, corresponds exactly to the experiment in the real system. A simulation may "condense" the time to such an extent that it is possible to simulate a couple of years of the system's operation within a very short time. It also allows an "expansion" of the system operation time and in consequence a detailed structure of the changes in the real system, which could not be observed in real time, can be examined. The essential advantages of computer simulations make it possible to experiment, check and compare new systems or to make proposals as to the introduction of changes to the existing systems. This method also enables an examination of hypothetical systems, whose testing in any other way would be dangerous or impossible.

It should be emphasized that a complete identity of model outputs and the real system should not be expected. It follows, among others, from the fact that the model expresses not all, but only the significant features of the system, and what is considered to be significant depends on the modelling objectives. Moreover, exact values of all real system inputs are generally not known. The fact, whether the differences observed between model outputs and system outputs permit the use of the model, is determined by the results of compatibility tests, whose content depends on the intended use of the model.

2.1 Simulation software

The issues of energy simulation of buildings, heat sources, heating systems and domestic hot water supply have been studied for more than 40 years. The studies have resulted in the development of a great number of computer programmes and applications of different functionality and level of complexity, from spreadsheets to specialist simulation tools (e.g. DesignBuilder, EnergyPlus, Polysun, TSOL), which combine various aspect of building design [1, 11].

All simulation calculations were made with the use of EnergyPlus simulation software, which is intended for carrying out building energy analysis. It has an added capability of making a detailed simulation of the heat load and annual energy consumption. By using EnergyPlus, the demand for heating, cooling and electric energy, which must be supplied to the building in order to maintain the required standards of comfort, can be simulated on the basis of a detailed description of the building, its installations and fittings. Moreover, an analysis of primary energy consumption renders it possible to determine the energy efficiency of the building and this, in consequence, may constitute the basis for drawing conclusions as to the introduction of constructional and functional changes in the building that would reduce energy consumption. It is worth emphasizing that at present it is the most modern and most advanced calculation tool designed for that purpose. It should be underlined that EnergyPlus is a software that does not have a graphical user interface. The application comprises only an engine which includes complete calculation (simulation) procedures and defined input and output as simple ASCII text files. This solution enables development of the graphical user interface (GUI) by external companies, thus it allows designers to focus on its intuitive character and functionality and relieves them from programming the computational algorithms (which are programmed by EnergyPlus programmers). Additionally, input and output data defined as simple text files facilitate the exchange of data between the engine and the GUI. This ensures stability of the whole package and improves the reliability of the programme as it allows an analysis of the correctness of the input and output data at each stage of the system operation.

Advanced calculation capabilities of EnergyPlus and its popularity have created interest in developing applications being a graphical user interface, which facilitate data entry and display of simulation results both in a text and graphical form (charts, visualizations of heat flow, etc.) Currently, many applications of this type are being developed, but DesignBuilder software is one of the most popular and offers most capabilities. The software is dedicated to developing 3D models of buildings with the use of a rich database of predefined building materials, construction of walls and installations. This, in combination with the inheritance of the features by successive elements defining the level of detail of the building description, allow a quick and precise modelling of buildings.

3 Simulation models of the building under analysis

Simulation analyses were performed for a multi-family low energy residential building localized in the 2nd climate zone (the city of Legnica), in which the external design temperature is -18°C. The information on the building comprising its architecture, constructional parameters (internal and external walls, floors, flat roofs, glazing, etc.), type of use, localization, etc., were adopted on the basis of the design documentation. The fundamental parameters of the building are as follows:

- number of flats: 30
- dimensions of the building (width x length): 12.5 x 36.5 m
- total floor area: 2 305.10 m^2
- total building volume: 6 856.57 m^3
- heated area: 1 987.28 m^2
- heated building volume: 6 035.76 m^3
- maximum hourly power: 26.1 kW

All variants of the building energy simulation were made on the basis of identical meteorological data (typical meteorological years and statistical weather data for Poland for building energy calculation published by the Ministry of Infrastructure and Development in 2008) for the region where the entity being modelled is situated. The models considered the actual location of the building in relation to the four cardinal points of the compass. The results of analyses presented in the paper comprise only energy demand for heating.

For the purpose of analysing the feasibility of introducing simplifications of the building description to the simulation model, four variants of simulation calculations were developed into which building description simplifications were introduced. A statistical analysis of simulation results was carried out for each variant of the simulation model (hourly data comprising, among others, temporary heating power and heat consumption). Figure 1 illustrates graphically the simulation model variants. The "Vsc" variant was the reference simulation model (base model). The model comprises a complete description of the building including all internal walls (divided into thermal zones), separation of stairways, flats and bathrooms in each of them. Additionally, shading elements on the facade were considered in the reference simulation model. Temperatures in individual, separated thermal zones, the amount of ventilation air and other essential parameters for the heat energy balance calculations were adopted in accordance with applicable regulations and standards.

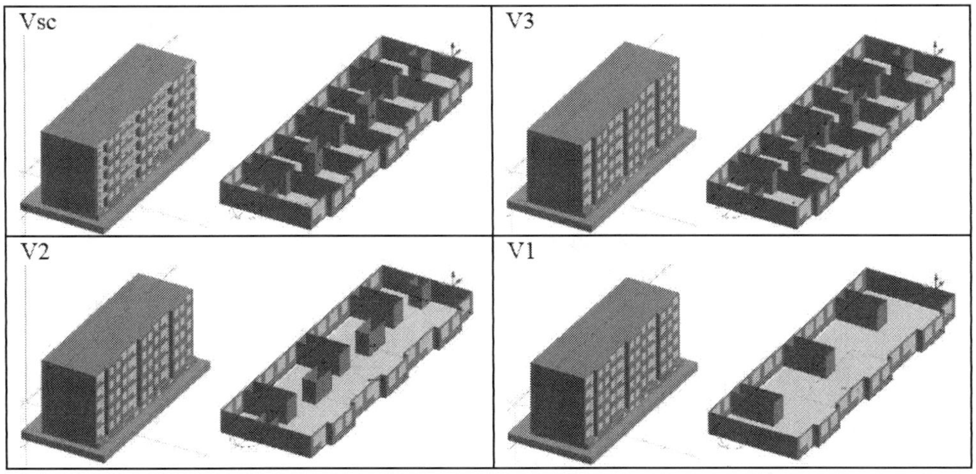

Fig. 1. Simulation model variants.

In the "V3" variant of the simulation model, shading elements (without changing the internal construction) were excluded, and this affected the heat gains caused by solar insolation. However, in the "V2" variant of the simulation model internal walls separating individual flats were additionally included, and thus the living area at each storey (except for the bathrooms) was treated as a whole. The simulation model in the "V1" variant differs from the "V2" variant in that it does not include "bathroom" thermal zones in each of the flats. This forced using an average temperature for the whole heated building volume (except for the stairways). In all the variants of the simulation model, averaged internal heat gains from occupancy, equipment, lighting of premises were adopted. Each zone was assigned the values of internal heat gains on the basis of the applicable regulations and standards.

In the preliminary simulation analyses, directly neighbouring buildings were taken as shading elements. An option of the programme was employed, which allows defining the geometry and precise location of such shading elements. The programme calculates in a ten-day step the annual shading caused by these buildings. However, in the final versions of the simulation model variants, the effect of shading exerted by the neighbouring buildings was excluded, since this parameter had a negligible effect on the final simulation results.

4 Building simulation results

Statistical analysis was performed using R Statistical Software [12] and ggplot package [13]. Table 1 summarizes building simulation results for all individual simulations. The maximum hourly heating power demand for the reference variant was 26.09 kW, whereas specific energy consumption was 18.6 kWh/(m^{2}a). The exclusion of shading (variant V3) caused an increased possibility of using solar heat gains. This in consequence resulted in a slight decrease of the maximum hourly heating power and a reduction in energy consumption. An additional exclusion of internal walls (variant V2) brought about a slightly smaller possibility of obtaining solar gains and reduced the possibility of heat accumulation in internal walls. In consequence, the required maximum hourly heating power (in comparison with Vsc) as well as energy consumption increased. The exclusion of the "bathroom" zone (variant V1) resulted in a slight decrease in the living zone temperature (an average temperature when bathrooms are included should be 20.2°C,

 DOI: 10.1051/e3sconf/20172200081

whereas the temperature of 20.0°C was adopted for calculations). This caused a reduction in energy consumption (compared to the Vsc variant by 8%). However, it was not relevant for the maximum hourly power, which did not change in comparison with the previous variant.

Table 1. Summary of simulation results for individual variants.

Variant	Specific energy consumption, kWh/(m^{2}a)	Maximum hourly power, kW
Vsc	18.6	26.09
V3	17.9	26.03
V2	18.2	26.28
V1	17.1	26.28

Figure 2 illustrates temporary heating demand (in an hourly step) for the simulated Vsc and V1 variants. It should be pointed out that the direction of changes in the model operation cannot be determined explicitly from the graph (for instance, the reduction of heat consumption in comparison with the reference model). Thus, coefficients, which allow correction of the results obtained in the case the simulation models are simplified, cannot be clearly determined.

Figure 3 presents charts which describe the relations between the reference variant and other individual variants. The solid straight line represents the exact convergence of the data of individual models. It should be noted that for high power due to low external temperatures, in all the cases the data are convergent (and this causes such insignificant changes in the maximum hourly heating power). For lower values of power, the differences between individual variants are affected by solar gains and heat accumulation in walls of the building. It is most visible in the chart illustrating the relations between the Vsc and V3 variants. In this case, an increase in heat demand is seen in the Vsc variant compared to the V3 variant, in which the possibility of using solar gains by excluding external shields was increased. In the remaining cases, changes in power occur on both sides (variants), but they do not cause significant changes in the cumulative energy demand.

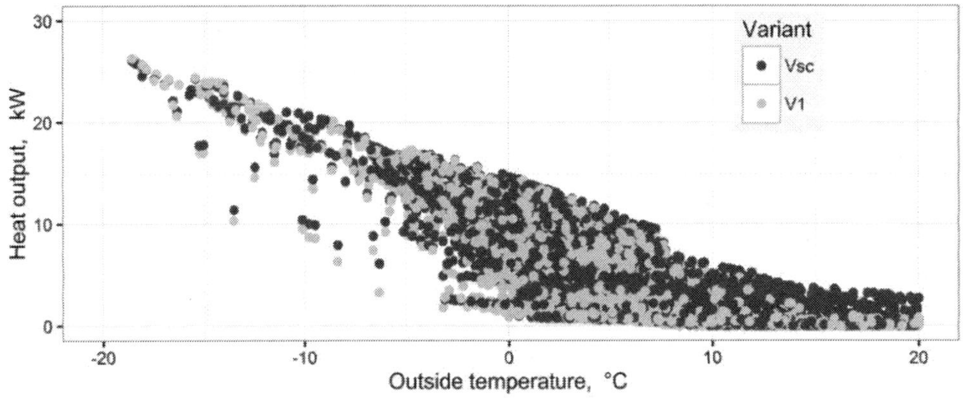

Fig. 2. Power distribution versus external temperature for the Vsc and V1 variants.

Figure 4 shows histograms illustrating the frequency of the results [of total 8760 (365days *24hours) results] with power deviation of a specific value (difference in power at the same hour in a year) in kW between the reference Vsc variant and other variants.

It should be noted that most of the results differ insignificantly (less than 0.25 kW) from the value calculated in the reference model (at the maximum hourly power of 26 kW).

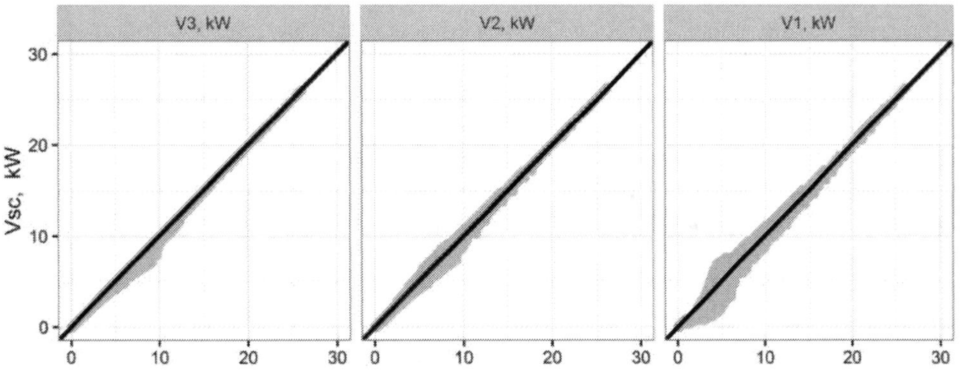

Fig. 3. The relation between the Vsc reference variant and other variants.

Also, in this case a distinct one-sided power reduction is noted when the existing elements which shade the building are excluded. As predicted, the biggest differences occur when extreme simulations are compared. The differences in hourly power demand range from 1.5 to 2.5 kW in less than 10% of observations.

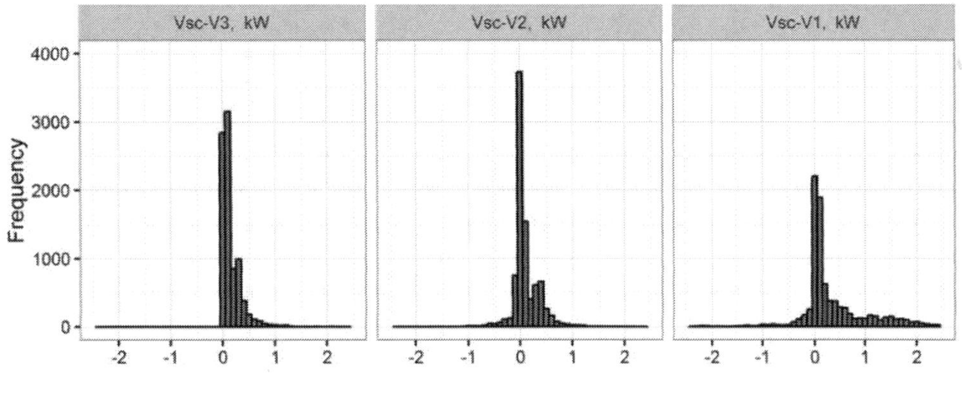

Fig. 4. Histogram illustrating the frequency of the results with power deviation between the Vsc reference variant and other simulation model.

5 Conclusions

While developing a building simulation model, it should always be taken into consideration how accurate and detailed should the model of the real building be and what discrepancies will follow from individual simplifications that are adopted. The adoption of a very detailed model involves a significant increase in the time required for the development and carrying out simulations, which translates into decidedly higher costs. The article presents a selected range of applicable simplifications of the model. Attention was focused exclusively on the construction of the building. The smallest differences in heat demand between the reference

Vsc model and the simplified model were obtained when the effect of shading exerted by the building's own elements (e.g. balconies) was excluded. In the event such shading occurs on the north-facing side, its exclusion will not have a significant impact on heat demand. In the event of shading on the south-facing side, considerable glazing of the building and in the calculations of cooling demand, it must definitely be considered whether or not shading should be included in the model. Such an inclusion will unfortunately increase the time required for entering data into the simulation programme. In the event, not individual spaces of the building, but the building as a whole is analysed, it is worth considering a simplification of the construction by excluding the division into thermal zones. Such a solution reduces significantly the time needed for the development of the model and does not bring about significant changes in the heat demand calculation results. When an analysis of individual spaces (e.g. for the purpose of selecting individual heating and ventilation solutions) is made, such a simplification is inadmissible. However, a possibility of combining several zones of similar internal temperatures should always be considered as it will simplify the model considerably and reduce the time required for its development. An introduction of too many zones may hinder simulation of such a model or, in an extreme case, it may necessitate running the simulation with the division into two or three periods.

References

1. V. Harish, A. Kumar, Renewable and Sustainable Energy Reviews **56**, 1272–1292 (2016)
2. IEA. *World Energy Outlook 2012*, International Energy Agency (2012)
3. Y. Keho, Energy Policy **91**, 233–246 (2016)
4. K. Krupa, *Modelowanie symulacja i prognozowanie* (WNT, 2008)
5. P. Malinowski, *Modelowanie hydrotermiczne i optymalizacja systemów zaopatrzenia w ciepło* (Politechnika Wrocławska, 2009)
6. J. Malko, Energy Policy Journal **18**, 1, 5–14 (2015)
7. L. Perez-Lombard, J. Ortiz, C. Pout, Energy and Buildings **40**, 3, 394–398 (2008)
8. K. H. Solangi, M. R. Islam, R. Saidur, N. A. Rahim, H. Fayaz, Renewable and Sustainable Energy Reviews **15**, 9, 2149–2163 (2011)
9. C. Wolfram, O. Shelef, P. J. Gertler, *How will energy demand develop in the developing world? Working Paper 17747* (National Bureau of Economic Research, 2012)
10. P. Ziembicki, J. Bernasiński, M. Klimczak, i in. *Rozproszone kogeneracyjne źródła energii dla budynków* (Oficyna Wydawnicza Uniwersytetu Zielonogórskiego, Zielona Góra, 2013)
11. M. Trcka, J.L. Hensen, Automation in Construction **19**, 93–99 (2010)
12. R Core Team. R: A Language and Environment for Statistical Computing. R Foundation for Statistical Computing, Vienna, Austria (2016)
13. H. Wickham. ggplot2: Elegant Graphics for Data Analysis. Springer-Verlag New York (2009)

 DOI: 10.1051/e3sconf/20172200082

The Quaternary groundwater as the low temperature energy source for heat pumps in Małopolska Province

Ewa Kmiecik[1,*], *Barbara* Tomaszewska[1], and *Justyna* Mazurkiewicz[1]

[1]AGH University of Science and Technology, Faculty of Geology, Geophysics and Environmental Protection, al. Mickiewicza 30, 30-059 Kraków, Poland

Abstract. The paper presents the assessment of using Quaternary groundwater as the low-temperature energy source for heat pumps in the Małopolska Province. The analysis demonstrates that the Małopolska region has potential to use Quaternary groundwater in low-temperature systems both in terms of potential well flow rates and water temperature. Water as a heat source must also meet the relevant requirements in terms of the physical and chemical parameters indicated by manufacturers of heat pumps as important for the proper operation of installations. The data for this study were obtained from the Polish Hydrogeological Survey (the PHS is carried out by the Polish Geological Institute – National Research Institute). In some cases, physical and chemical properties of water do not meet the requirements indicated by heat pump manufacturers. These requirements may be met by using recycled water or another medium whose parameters do not affect the life of the installation.

1 Introduction

In low-temperature geothermal installations, groundwater can be used in open systems, in which energy is transported by water (double-well and single-well installations) [1, 2]. Groundwater can be used in water/water systems if certain basic conditions are met: sufficient well flow rate (approx. 1.2–1.5 m^3/h per each 5 kW of heating power), water temperature in the range from +4 to +20°C and suitable physical and chemical composition of water [2].

If the mineral content of water is not excessively high and it is not aggressive towards the materials used in the installation, and the values of selected indicators do not exceed the limits set by the manufacturer in question, groundwater can be used as a source of low-temperature heat and pumped directly into the heat pump evaporator [1–5]. Unfortunately, in many cases these requirements cannot be met; in such cases, an intermediate circuit heat exchanger must be used at the low-temperature heat source side, before the heat pump [1–4, 6–8]; this heat exchanger should be individually tailored to each system.

* Corresponding author: ewa.kmiecik@agh.edu.pl

DOI: 10.1051/e3sconf/20172200082

In this paper potential opportunities for using groundwater from Quaternary formations in double-well and single-well low-temperature systems assisted by water/water heat pumps in the Małopolska Province are analysed. Its temperature and well flow rates enable such groundwater to be used as the low-temperature heat source [2, 9, 10].

The paper focuses on the analysis of physical and chemical properties of the waters tested with respect to the requirements indicated by the three leading heat pump manufacturers (Ochsner Wärmepumpen GmbH, Viessmann Ltd. and Dimplex Ltd.). To this purpose, data were obtained from the Central Hydrogeological Database (so-called Bank Hydro). Those were physical and chemical analyses from the years 2000–2013 for 838 wells that were used for extracting water from Quaternary formations; they concerned selected physical and chemical indicators: pH, electrical conductivity (EC) and concentrations of chlorides (Cl^-), sulphates (SO_4^{2-}), nitrates (NO_3^-), bicarbonates (HCO_3^-), iron (Fe), manganese (Mn), ammonium (NH_4), oxygen (O_2), free (aggressive) carbonic acid (CO_2 agg.) and aluminium (Al).

2 Study area characteristics

The Małopolska Province is situated in the south of Poland. It has an area of 15,183 km^2 (1,518,279 hectares), which corresponds to 4.8% of total country area [11]. The area in question includes several units that differ in terms of their geological structures. Its southernmost part are the Inner and Outer Carpathians, which include Paleogene formations, Cretaceous flysch and Oligocene formations. The middle part is the Miocene Carpathian Foredeep, while in the northern and western parts Mesozoic structures are present: the Jurassic and Triassic Silesian-Cracow Monocline, the Cretaceous Miechów Basin and the Paleozoic structures of the Upper Silesian Foredeep [12–14]. Carpathians and their foreland (the Carpathian Foredeep) cover the overwhelming majority of the area (ca. 80%) [12, 15, 16].

The amount of groundwater resources that have been developed and can be extracted in the Małopolska Province has been estimated at 74,697.10 m^3/h (as at 31 December 2015), of which 40,893.79 m^3/h (54.75% of resources) is located in Quaternary formations, 9,220.03 m^3/h (12.34%) in Tertiary formations, 12,974.03 m^3/h (17.37%) in Cretaceous formations and 11,609.25 m^3/h (15.54%) in older formations [17]. As concerns their hydrochemical properties, waters from Quaternary formations usually contain multiple mineral ions. Bicarbonates, sulphates, calcium, magnesium and sodium dominate. Most Quaternary groundwater reservoirs do not have insulating covers in top aquifer layers, which facilitates the penetration of pollutants from the surface [18–19].

3 Research methodology

In order to evaluate the possibility of using groundwater from Quaternary formations in low-temperature geothermal systems in the Małopolska Province, data from 838 wells present in the Central Hydrogeological Database were analysed; the relevant physical and chemical analyses of water had been conducted between 2000 and 2013. Locations of these wells are shown in Fig. 1.

Tests of chemical composition of water in Quaternary wells in the years 2000–2013 were conducted by different laboratories with different methodologies and with different limits of quantification, which may be reflected in a greater variability of the parameters analysed. Unfortunately, no data on the testing methodology used are included in the Central Hydrogeological Database. Table 1 shows minimum and maximum values of the parameters analysed, which are compared to the limits indicated by three leading manufacturers of heat pumps in Poland – Ochsner Wärmepumpen GmbH, Viessmann Ltd. and Dimplex Ltd., which offer also water/water pumps.

Table 1. Characteristics of physical and chemical parameters of groundwater (Quaternary formations, Małopolska Province, based on the Central Hydrogeological Database) (n.d. – not determined, "+" – the material is usually resistant, "0" – may lead to corrosion where the value of most factors is "0", "–" – use is not recommended) [10].

Parameter	Unit	Chemical composition		Limits for physical and chemical parameters according to								
				Ochsner Wärmepumpen GmbH			Viessmann Ltd.			Dimplex Ltd.		
		min	max	*value*	copper	stainless steel	*value*	copper	stainless steel	*value*	copper	stainless steel
pH	–	3.6	13.1	< 6 6–8 > 8	0 + –	0 + 0	< 7.5 7.5–9 > 9	0 + 0	0 + +	< 7.5 7.5–9 > 9	0 + 0	0 + +
Cl^-	mg/dm^3	0.95	2,739.2	< 100 100–200 > 200	+ 0 –	+ + –	< 300 > 300	+ 0	+ 0	< 300 > 300	+ 0	+ 0
NO_3^-	mg/dm^3	<DL	551.7	n.d.	n.d.	n.d.	< 100 > 100	+ 0	+ +	< 100 > 100	+ 0	+ +
SO_4^{2-}	mg/dm^3	<DL	1,967	< 50 50–100 > 100	+ 0 –	+ + 0	< 70 70–300 > 300	+ 0 –	+ + 0	< 70 70–300 > 300	+ 0 –	+ + 0
CO_2 agg.	mg/dm^3	0.2	70	< 5 5–20 > 20	+ 0 –	+ + 0	< 5 5–20 > 20	+ 0 –	+ + 0	< 5 5–20 > 20	+ 0 –	+ + 0
Fe	mg/dm^3	<DL	48.39	> 0.2 (Fe with Mn)	–	–	< 0.2 > 0.2	+ 0	+ 0	< 0.2 > 0.2	+ 0	+ 0
Mn	mg/dm^3	<DL	62.3	> 0.05	–	–	< 0.1 > 0.1	+ 0	+ 0	< 0.1 > 0.1	+ 0	+ 0
O_2	mg/dm^3	0.4	10.2	< 1 1–8 > 8	+ 0 –	+ + +	< 0.2 > 0.2	+ 0	+ +	< 2 > 2	+ 0	+ +
EC	µS/cm	8.4	12,915.5	> 600	–	+	< 10 10–500 > 500	0 + –	0 + 0	< 10 10–500 > 500	0 + –	0 + 0
HCO_3^-	mg/dm^3	68.5	774	n.d.	n.d.	n.d.	< 70 70–300 > 300	0 + 0	+ + 0	< 70 70–300 > 300	0 + 0	+ + 0
NH_4	mg/dm^3	<DL	16.2	< 2 2–20 > 20	+ 0 –	+ + +	< 2 2–20 > 20	+ 0 –	+ + 0	< 2 2–20 > 20	+ 0 –	+ + 0
Al	mg/dm^3	<DL	0.26	n.d.	n.d.	n.d.	< 0.2 > 0.2	+ 0	+ +	< 0.2 > 0.2	n.d.	n.d.

4 Data analysis

The largest number of facilities that extract water from Quaternary formations and that are included in the Central Hydrogeological Database are located in the Outer Carpathians and within the Carpathian Foredeep, especially within large urban areas such as Kraków, Nowy Sącz, Oświęcim and Nowy Targ (Fig. 1). The distribution of these facilities in Małopolska is not uniform, but this does not mean that there are no wells used to extract water from Quaternary formations in other areas. There may be wells with depths less than 30 m, which pursuant to applicable laws do not require hydrogeological documentation to be compiled after geological work has been conducted and thus are not included in the Central Hydrogeological Database.

DOI: 10.1051/e3sconf/20172200082

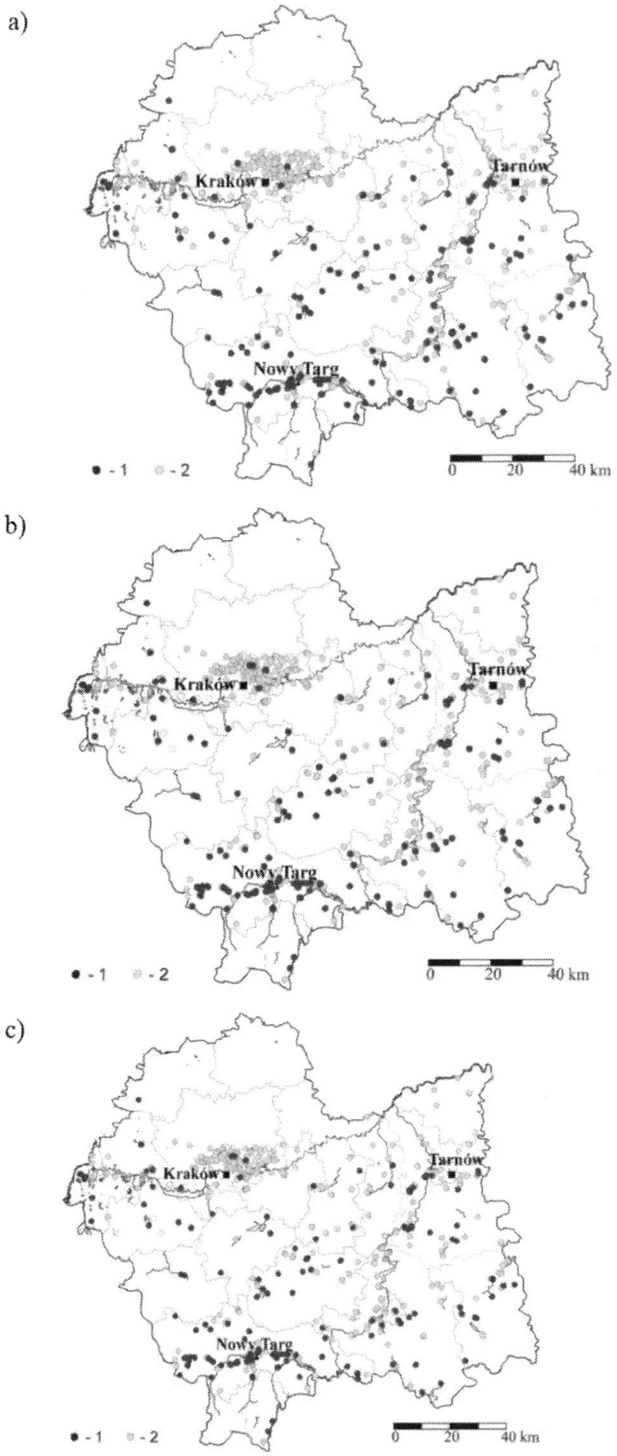

Fig. 1. Quaternary groundwater intakes in Małopolska (based on data from the Central Hydrogeological Database, 2000–2013). Water quality in relation to the requirements indicated by: a) Ochsner Ltd.; b) Viessmann GmbH; c) Dimplex Ltd. (1 – the manufacturer's requirements are met; 2 – limits are exceeded and an intermediate heat exchanger must be used).

The overall depth of wells in Quaternary formations in the database analysed ranged up to a maximum of 71 m. These wells are used to extract water from the depth (sampling depth) of up to 33 m. The aquifer, which is a layer up to approximately 30 m thick and occurs at depths of up to 70 m, consists mainly of sands and gravels with admixtures of pebbles, sands (clayey, with gravel, with pebbles and silty), tills, gravels (with pebbles, sandy) and other formations (rock debris, mud, dust, regolith, etc.).

Waters from Quaternary formations within the Małopolska Province exhibit highly variable physical and chemical parameters (Tab. 1, Fig. 1). In view of the requirements indicated by Viessmann and Dimplex, water from 27.57% of wells can be used directly in their installations assisted by a water/water heat pump; according to Ochsner's requirements, 30.96% of wells are suitable (Fig. 1). In the remaining cases (69.04% of wells according to the requirements indicated by Ochsner and 72.43% of wells according to Viessmann's and Dimplex's requirements), water from Quaternary formations may be used in installations assisted by a water/water heat pump, but an intermediate heat exchanger has to be installed before the heat pump as a result of parameter limits being exceeded. Locations of wells that meet the requirements of individual producers are shown in Fig. 1. Cases where limits are exceeded for the parameters tested can be observed locally, especially in large urban areas, i.e. Kraków, Tarnów and Nowy Targ.

Thus the water analysed can be used in both single-well and double-well installations by any manufacturer, but in many cases an intermediate heat exchanger has to be used before the heat pump in order to ensure the long and safe operation of the entire system.

Surveys conducted annually by the Polish Organisation for the Development of Heat Pump Technology (PORT PC) (2016) indicate that out of the overall number of units sold in 2015, only 0.4% were water/water installations (86 units) [20]. Such a low share of this type of heat pumps may be related to the still fairly high investment expenditure required (it may reach double the price of an installation that uses traditional energy carriers in many cases), complex legal regulations requirement for and the fact that possibilities of using underground water in such systems have not been researched so far [2, 6]. The Geological and Mining Law [21] regulating principles of exploration, recognition and exploitation of groundwater, especially the development of a geological project as well as its submission to the aforementioned administration authority are obligatory. From the other hand, other procedures must comply with regulations of the Water Law [22]. The spent water, that is discharged into waters or to the ground is considered as wastewater, and its quality have to meet the requirements included in documents pertaining to the protection of natural environment. These installations can undoubtedly help to reduce the emission of harmful substances into the atmosphere [23] and thus contribute to improving air quality in the Małopolska region. Additionally, the use of heat pumps for district heating purposes is often mentioned (following the models set in other European countries) as an energy-efficient solution for producing heat [24].

5 Summary

Groundwater used as the low-temperature heat source in systems assisted by heat pumps should, inter alia, meet the requirements related to its physical and chemical parameters indicated by installation manufacturers. This is the main issue related to ensuring the safe and proper operation of a groundwater-based system.

The assessment conducted demonstrates that in the Małopolska Province, groundwater from Quaternary formations could be used in low-temperature systems both in terms of potential water flow rates (several m^3/h) and water temperatures (from +6 to +20°C). Geological and hydrogeological conditions are favourable as well. Physical and chemical

DOI: 10.1051/e3sconf/20172200082

water parameters in some of the wells analysed do not meet the requirements indicated by heat pump manufacturers.

The fact that the manufacturers' requirements concerning physical and chemical indicators are not met does not preclude the use of such water resources in systems used for producing energy. A solution used in many installations are intermediate heat exchangers placed before the heat pump, which transfer the energy stored in groundwater to a medium whose parameters do not affect the service life of the installation.

The selection of a specific heating system assisted by a water/water heat pump should be based on an up-to-date analysis of the chemical composition of water, preferably conducted by an accredited testing laboratory. The analysis should cover all parameters indicated by the manufacturer of the heat pump in question as well as parameters that may affect corrosion or clogging processes; in this manner, a number of problems that may occur during the operation of the entire system can be avoided.

Despite the possibilities of using groundwater from shallow aquifers in low-temperature systems, such installations still have the smallest share in the heat pump market in Poland.

The authors would like to thank OCHSNER Wärmepumpen GmbH, Viessmann Sp. z o.o. and Dimplex Sp. z o.o. for providing technical materials for the purposes of this paper. This work has also been conducted under AGH-UST statutory research grants Nos. 11.11.140.797 and 11.11.140.321.

References

1. B. Sanner, GHC Bull. **22**, 2, 19–25 (2001)
2. J. Mazurkiewicz, E. Kmiecik, B. Tomaszewska, Pol. Geol. Rev. **63**, 10/2, 926–930 (2015)
3. K. Niedbalska, P. Bukowski, A. Haładus, SGEM **5**, 1, 549–556 (2015)
4. K. Rafferty, GHC Bull. **22**, 1, 16–24 (2001)
5. V. Pisarev, *Design of a heat pump systems for heating* (UR Rzeszow, 2013) (in Polish)
6. M. Forsen, *Heat pumps–technology and environmental impact* (SVEP, 2005)
7. M. Rubik, *Heat pumps in low-temperature geothermal systems* (Multicobooks, Warsaw, 2011) (in Polish)
8. Viessmann (http://www.viessmann-us.com/) (access 2017)
9. J. Mazurkiewicz, E. Kmiecik, B. Tomaszewska, *Analysis of the possibility to use the quaternary groundwater in the low-temperature geothermal systems in Małopolska. Part II. Zawoja-3 Case Study*, Pol. Geol. Rev. (to be published)
10. J. Mazurkiewicz, *Evaluation of prospective areas for low–temperature geothermal water utilization in Malopolska Voivodship, Poland* (PhD defense, to be published) (in Polish)
11. SCO, *Area and population in the territorial profile in 2016* (Stat. Gov., Warszawa, 2016) (in Polish)
12. A.P. Barbacki, W. Bujakowski, L. Pajak, *Atlas of geothermal water reservoirs in Malopolska* (MEERI PAS, Krakow, 2006)
13. S. Satora, Infra. Eco **2**, 61–70 (2007)
14. J. Mazurkiewicz, E. Kmiecik, B. Tomaszewska, Geol. Explor. Tech. Geoth. Energ. Sust. Dev. **2**, 45–58 (2013)
15. W. Bujakowski, B. Tomaszewska, M. Miecznik, Renew. Energ. **99**, 420–430 (2016)

16. B. Tomaszewska, M. Rajca, E. Kmiecik, M. Bodzek, W. Bujakowski, K. Wątor, M. Tyszer, Desalin. **406**, 74–82 (2017)
17. S. Pergół, J. Sokołowski, *Balance of exploitation resources of underground waters of Poland, as of December 31, 2016* (PGI–NRI, Warsaw, 2016) (in Polish)
18. R. Bereś, G. Cieśla, L. Czarnecka, B. Dębska, K. Dulemba, M. Dzierko, E. Gondek, A. Główka, K. Gołębiowska, M. Janik, D. Kapustka, R. Listwan, D. Łęczycka, M. Ogar, B. Pająk, T. Reczek, K. Synowiec, E. Żelazowski, D. Żmuda, *State of the Environment Reports in Malopolska 2002* (GIOS, Warsaw, 2003) (in Polish)
19. S. Satora, *Influence of geological environment on design of drilled wells and hydrogeological regime of tapped underground waters* (UR Krakow, Krakow, 2008)
20. POHPTD (www.portpc.pl) (access 2017)
21. Geological and Mining Law (Journal of Laws 2016 item 1131)
22. Water Law (Journal of Laws 2015 item 469)
23. D. Borge–Diez, A. Colmenar–Santos, C. Perez–Molina, A. Lopez–Rey, Energ. **36**, 8, 5027–5036 (2015)
24. R. Lund, U. Persson, Energ. **110**, 1, 129–138 (2016)

Modification of Rhodamine WT tracer tests procedure in activated sludge reactors

Marta Knap[1,*], and *Piotr* Balbierz[1]

[1]Wroclaw University of Science and Technology, Faculty of Environmental Engineering, Wybrzeze Wyspianskiego 27, 50-370 Wroclaw, Poland

Abstract. One of the tracers recommended for use in wastewater treatment plants and natural waters is Rhodamine WT, which is a fluorescent dye, allowing to work at low concentrations, but may be susceptible to sorption to activated sludge flocs and chemical quenching of fluorescence by dissolved water constituents. Additionally raw sewage may contain other natural materials or pollutants exhibiting limited fluorescent properties, which are responsible for background fluorescence interference. This paper presents the proposed modifications to the Rhodamine WT tracer tests procedure in activated sludge reactors, which allow to reduce problems with background fluorescence and tracer loss over time, developed on the basis of conducted laboratory and field experiments.

1 Introduction

Mixing and the intensity of mass exchange processes may have significant impact on the hydraulic efficiency of the activated sludge (AS) reactors and as a result the effectiveness of the wastewater treatment processes [1, 2]. Determination of the hydrodynamic characteristics of AS reactors allows to detect mixing problems and is required during construction of wastewater treatment plant (WWTP) mathematical models. One of the methods used to determine the hydrodynamic models of AS reactors are tracer tests.

Due to the specific properties of the activated sludge, the raw sewage composition, dynamic working conditions and large reactors volumes, tracer experiments in AS reactors are time-consuming, cost- and labor-intensive. Morover, all stages of tracer tests including the choice of proper tracer, design and conduct of the experiment, as well as interpretation of results require more careful consideration. One of the tracers recommended for use in WWTPs, wetlands and natural waters is Rhodamine WT (RWT) [3].

1.1 Rhodamine WT

RWT is a fluorescent xanthene dye which is generally used as water tracer in various hydrodynamic studies (surface and ground water studies, rivers, estuaries, reservoirs and lakes studies, wetlands and wastewater treatment studies) for determining the hydraulic

* Corresponding author: marta.knap@pwr.edu.pl

parameters and contaminant fate and transport. RWT is a useful tracers because it can be relatively easily detected with a field fluorometer at concentrations as low as 0.01 μg/dm^3. Moreover RWT is non-toxic, non-carcinogenic and non-mutagenic even at relatively high concentrations of 2 mg/dm^3 [4]. RWT has low sensitivity to photochemical decay and pH-values, indicated by a constant maximal fluorescence in the pH range between 5 and 8 [5].

RWT is commonly used as a conservative tracer in different hydrodynamic studies, including AS reactors hydrodynamic characterization. However, the literature on RWT reactivity, and in particular susceptibility to sorption to sediments, is somewhat conflicting. Some studies [6] consider the RWT as a conservative tracer, whereas other studies [7, 8] aimed at comparison of RWT and other conservative tracer such as LiCl and Br suitability in determination of hydraulic characteristics of constructed and natural wetlands, showed that RWT is less stable than LiCl and Br.

Thorough laboratory research on the RWT sorption characteristics has shown that the extent of sorption is dependent on different factors such as sediment type, organic matter concentration, initial RWT concentration, pH, ionic strength and the ionic composition of the solution [4, 9–12]. In addition, RWT is made up of two isomers [4, 9, 10] with different sorption properties and different emission spectra, which are equally distributed in RWT solution. Isomer 1 (para) is an almost conservative tracer, while isomer 2 (meta) is a non-conservative tracer due to its sorption to sediments [4, 9, 10].

Studies have shown that another dye tracer Rhodamine B is not recommended for tracer tests due to susceptibility to sorption to sediments, but was used for specific surface area measurement of activated sludge [13].

However, there is no information about the RWT behaviour in the AS reactors. Therefore, the aim of this paper is to evaluate the suitability of RWT as a conservative tracer in hydrodynamic studies of AS reactors, based on its susceptibility to sorption to AS flocs and chemical quenching of fluorescence by dissolved water constituents assessed in laboratory batch tests using several concentrations of RWT and AS from two WWTPs in Poland.

1.2 Tracer tests

Tracer tests are conducted by introduction of a controlled input function (tracer addition) to the stream entering the AS reactor and subsequent measurements of tracer concentration at the outlet of the reactor as a function of time. Results of tracer tests provide approximate information about the mixing regime in reactors, based on the analyses of tracer response curves and determination of the hydraulic parameters, such as the hydraulic retention time (HRT), the number of tanks-in-series or the dispersion number, using numerical methods, e.g. the method of moments or simulation method.

Selection of the proper tracer for a paricular application has a key importance for tracer test success, as it determines further steps including tracer dose selection, samples preparation and detection method. To facilitate the selection of tracer the following criteria, that the ideal tracer should meet, were formulated [14–16]:

- physicochemical properties similar to the properties of wastewater (e.g. density),
- conservative (not absorbed on particles in wastewater and exposed reactor surfaces, non-reactive in any way in the reactor),
- non-toxic, non-carcinogenic and non-mutagenic at used concentrations,
- easy and inexpensive detection at low concentrations,
- low price and availability on the market.

 DOI: 10.1051/e3sconf/20172200083

In practice, selection of the tracer that simultaneously meets all of the above criteria is nearly impossible. Therefore, the choice of tracer is usually a compromise between its positive and negative properties.

Although RWT is susceptible to some extent to sorption to sediments and chemical quenching of fluorescence by dissolved water constituents, it provides easy detection with a field fluorometer at low concentrations, relatively low prices, and has no negative influence on living organisms. The general procedure for performing tracer tests using RWT is presented elsewhere [3].

Taking into account the numerous problems and doubts involved in conducting tracer tests in AS reactors, the authors conducted additional experiments and suggested modification the general tracer test procedure to adapt it to the specificity conditions of AS reactors in WWTPs.

2 Material and methods

A series of laboratory and field tests were conducted to determine background fluorescence interference and the loss of tracer due to sorption to AS flocs in real WWTPs conditions.

Background fluorescence and RWT concentrations were analysed in prefiltered samples using filter fluorometer (Quantech Base) at an excitation wavelength of 550 nm and an emission wavelength of 580 nm.

The 20% RWT aqueous solution (Keystone) was used to prepare RWT calibration curves and in all laboratory batch tests.

2.1 Background fluorescence determination

Background fluorescence in wastewater and its variability over time was determined in field studies conducted at two Leszno and Wroclaw WWTPs (Poland).

The test at WWTP in Leszno was carried out for 50 hours. Samples were taken at one sampling point, the inlet to the anaerobic reactor (AN), every 30 min for the first 2 h and every 2 h for another 48 hours. All samples were immediately filtered through medium filter paper and subsequently analysed for fluorescence (at RWT parameters).

The test at WWTP in Wroclaw was carried out for 7 h (the typical tracer test time). Samples were collected at three sampling points, the inlet to the oxidation reactor (OX), the outlet of the OX and the internal recirculation stream every 1 h for 7 hours. All samples were immediately filtered through medium filter paper and subsequently analysed for fluorescence.

2.2 Laboratory sorption studies

RWT susceptibility to sorption to AS flocs and the tracer loss over time was determined in laboratory batch tests conducted on AS samples collected from Leszno and Wroclaw WWTPs.

Three sorption experiments were conducted on AS samples collected from the bioreactor of Wroclaw WWTP. In the first and second tests, three laboratory reactors equipped with a mechanical stirrer were filled with 3.0 dm^3 of AS, then appropriate quantities of RWT (20% aqueous solution) were added so that the resultant concentrations were 100, 50 and 10 μg RWT/dm^3. Samples were collected every 30 min. for 10 h (first test), and every 30 min. for the first 4 h and every 1 h for another 6 h (second test). All samples were immediately filtered through medium filter papers and subsequently analysed for fluorescence.

 DOI: 10.1051/e3sconf/20172200083

In the third sorption experiment the three laboratory reactors were filled with 3.0 dm^3 of AS at different concentrations. Sludge concentrations in reactors R1 and R3 were respectively approximately two times lower (dilution with tap water) and higher (thickened by sedimentation) compared to reactor R2.

Then appropriate quantities of RWT (20% aqueous solution) were added so that the resultant concentration was 50 μg RWT/dm^3. Samples were collected every 30 min. for the first 2 h and every 1 h for another 6 h. All samples were immediately filtered through medium filter papers and subsequently analysed for fluorescence.

The fourth sorption experiment was conducted on AS sample collected from the bioreactor of Leszno WWTP. Three laboratory reactors were filled in 3.0 dm^3 of AS, then appropriate quantities of RWT (20% aqueous solution) were added so that the resultant concentrations were 100, 50 and 10 μg RWT/dm^3. Samples were collected every 30 min. for 10 h. All samples were immediately filtered through medium filter papers and subsequently analysed for fluorescence.

3 Results and discussion

3.1 Background fluorescence determination results

Background fluorescence in wastewater and its variability over time determined in field studies conducted at Leszno and Wroclaw WWTPs is shown at Figure 1. The emission values read in the samples were converted to the corresponding RWT concentrations (based on the RWT calibration curve).

The average background fluorescence in samples taken at Leszno WWTP corresponds to RWT concentration of 0.18 μg/dm^3 with the standard deviation of 0.014 μg/dm^3. The average background fluorescence in samples taken at Wroclaw WWTP at the inlet and outlet of OX reactor corresponds to RWT concentration of 0.18 μg/dm^3, whereas in samples taken from internal recirculation to RWT concentration of 0.26 μg/dm^3, with the standard deviation of 0.018, 0.012 and 0.010 μg/dm^3, respectively.

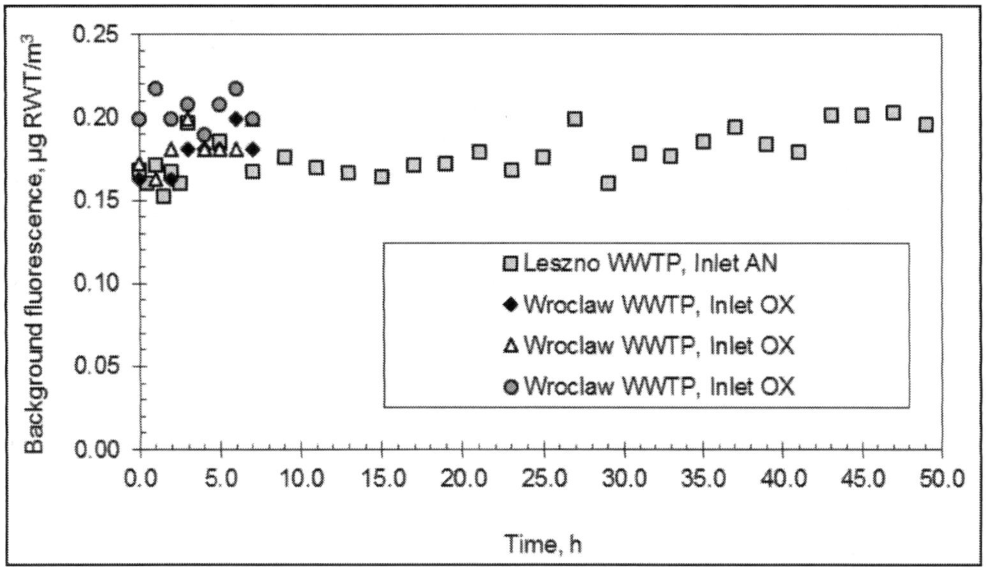

Fig. 1. Background fluorescence interference at Leszno and Wroclaw WWTPs.

 DOI: 10.1051/e3sconf/20172200083

The results suggest that the average background fluorescence should be determined and taken into account during tracer test analysis, whereas its temporal variation usually may be neglected. The RWT fluorescence should be adjusted by manually subtracting the background component or automatically resetting the background fluorescence to zero (if the fluorometer has such a function). Due to the varied composition of municipal wastewater, the authors suggest that the background fluorescence should be determined in field experiments for each WWTP, directly before or if possible simultaneously to the actual tracer tests.

3.2 Sorption studies results

The relative RWT loss over time, due to sorption to sediments and/or chemical quenching of fluorescence by dissolved water constituents (effects which are indistinguishable in this experiment), recorded during the first test was shown in Figure 2. In all the experiments, the effect of the initial RWT concentration on the degree of RWT loss was negligible, therefore, only the results of one experiment at three concentration were presented in details (Fig. 2) and in further considerations the average values of three concentrations were used.

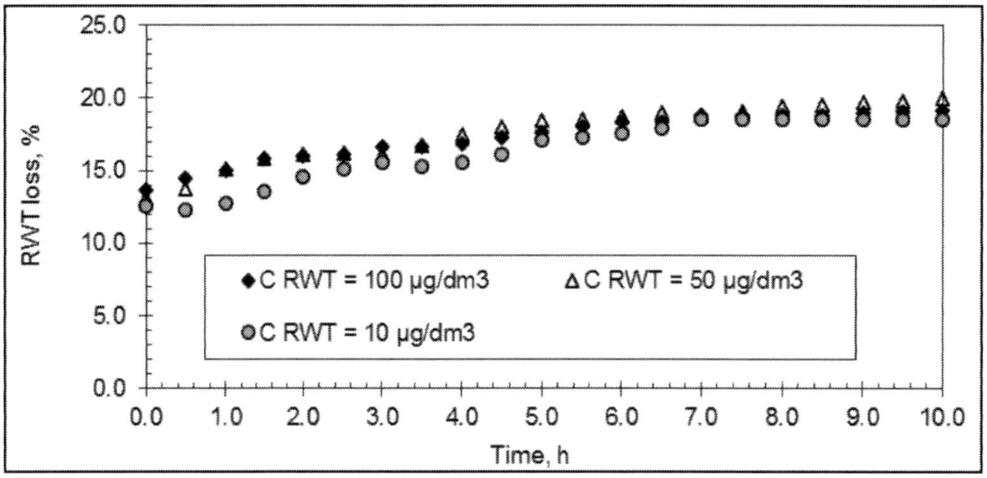

Fig. 2. RWT concentration loss over time in test I (MLSS = 3.20 g TSS/dm^3 and 2.10 g VSS/dm^3).

The average relative RWT losses over time, due to sorption to sediments and/or chemical quenching of fluorescence by dissolved water constituents, recorded during the first, second and fourth test were shown in Figure 3. The degree of RWT loss in these tests was significantly different, despite relatively similar sludge concentrations, and amounted to 17.1%, 21.6%, and 10.6% in the first, second and fourth test, respectively. These results suggest that it is not the sludge concentration, but mainly the composition of activated sludge and wastewater that has a significant effect on RWT loss. Also the initial RWT concentration in the samples that were taken immediately after RWT addition was respectively 12.5%, 17.1% and 9.5% lower than theoretical concentration, which suggest that the initial loss of RWT is instantaneous.

The relative RWT loss over time, due to sorption to sediments and/or chemical quenching of fluorescence by dissolved water constituents, recorded during the third test at various sludge concentrations was shown in Figure 4. The RWT loss over time at relatively low sludge concentrations in reactors R1 (MLSS = 1.75 g TSS/dm^3 and 1.27 g VSS/dm^3) and R2 (MLSS = 3.40 g TSS/dm^3 and 2.47 g VSS/dm^3) amounted to 15.2% and 16.9%, respectively, and did not differ significantly. Whereas, a significantly higher tracer loss of

DOI: 10.1051/e3sconf/20172200083

28.7% was observed in reactor R3 filled with sludge at the high concentration (MLSS = 6.908 g TSS/dm^3 and 4.952 g VSS/dm^3).

The obtained results suggests that the RWT loss is independent of the initial tracer concentration, but is strongly dependent on the activated sludge and wastewater composition, and to a limited extent on sludge concentration (increase in the amount of adsorbed tracer occurs only at sufficiently high organic matter concentration). In order to model the variability of the RWT loss over time, the linear equations were matched to the results of the experiments (Fig. 3 and 4). The RWT loss over time is linear at specified intervals.

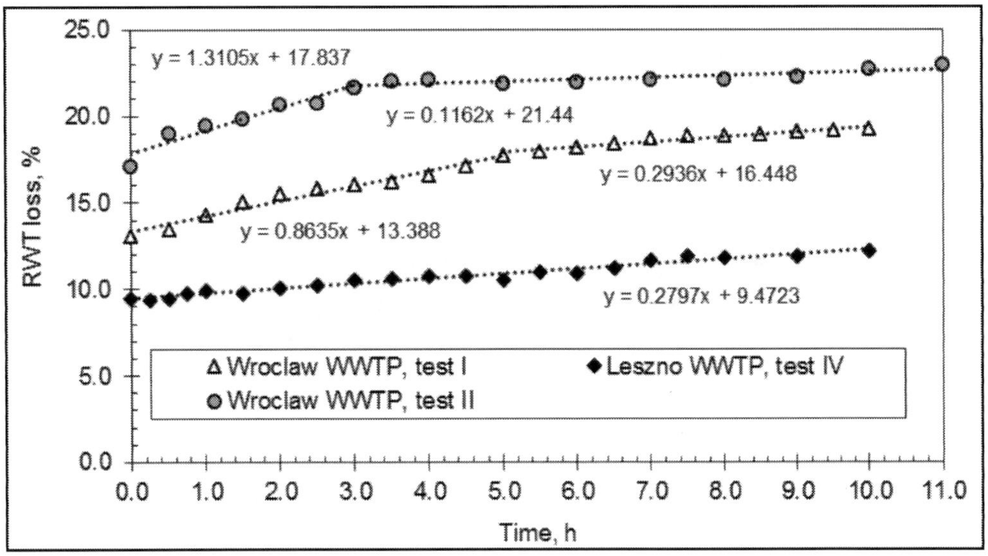

Fig. 3. Average RWT loss over time in test I (MLSS = 3.20 g TSS/dm^3 and 2.10 g VSS/dm^3), test II (MLSS = 3.52 g TSS/dm^3 and 2.20 g VSS/dm^3), test IV (MLSS = 3.8 g TSS/dm^3 and 2.53 g VSS/dm^3).

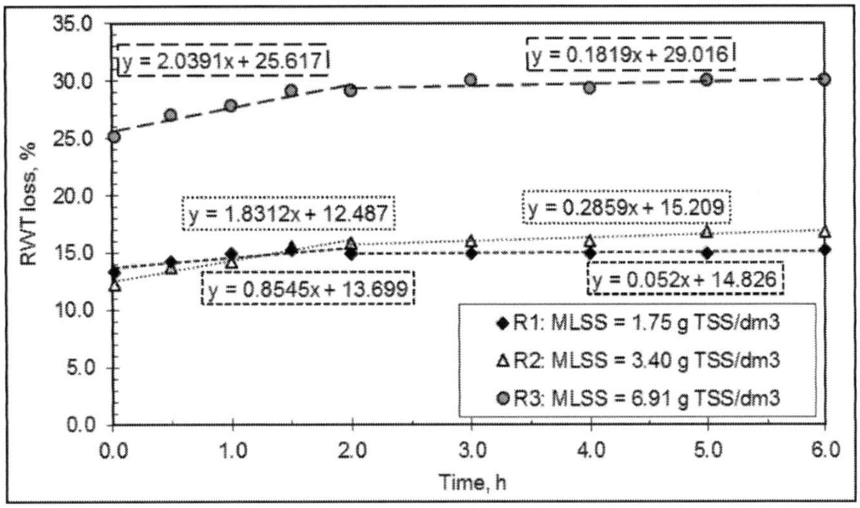

Fig. 4. RWT loss over time in reactor I (MLSS = 1.75 g TSS/dm^3 and 1.27 g VSS/dm^3), reactor II (MLSS = 3.40 g TSS/dm^3 and 2.47 g VSS/dm^3), reactor III (MLSS = 6.91 g TSS/dm^3 and 4.95 g VSS/dm^3).

Due to the varied composition of municipal wastewater and activated sludge, the authors suggest that RWT susceptibility to sorption on sediments over time should be verified in laboratory tests for each WWTP, directly before or if possible simultaneously to the actual tracer tests.

4 Conclusions

Although RWT is a non-conservative tracer, and may be susceptible to some extent to sorption to sediments and chemical quenching of fluorescence by dissolved water constituents, but it provides easy detection with field fluorometers at low concentrations, relatively low costs, and has no negative influence on living organisms, therefore it can be successfully used for tracer tests in AS reactors

The results of conducted studies suggest that RWT is susceptible to sorption to AS flocs and/or to chemical quenching of fluorescence by dissolved water constituents in a mixture of activated sludge and wastewater. The RWT loss is independent of the initial tracer concentration, but is strongly dependent on the activated sludge and wastewater composition, and to a limited extent depends on sludge concentration (increase in the amount of adsorbed tracer occurs only at sufficiently high organic matter concentration). The RWT loss over time is linear at specified intervals.

Consequently, the use of RWT as a conservative tracer may by the reason for poor mass recovery and errors in the tracer tests interpretation. Some researchers claim that the poor tracer mass recovery and the non-conservative character of RWT does not adversely affect the quality of the tracer test results, especially if the method of moments is used for interpretation [6, 8, 17]. However, if the results interpretation is based on a simulation method that involves fitting the model curve directly to the tracer test results these discrepancies may be important.

To facilitate the interpretation of RWT tracer test in AS reactors and to eliminate the interferences caused by background fluorescence as well as adsorption to AS flocs and/or chemical quenching of fluorescence by dissolved water constituents, the following modification of the standard tracer tests procedure are suggested:

- Determination of background fluorescence based on the results of field experiments. The background fluorescence should be properly accounted for when analysing tracer tests samples. Knowledge about background fluorescence is also necessary to correctly estimate the required tracer mass. Ideally, background fluorescence should be an insignificant fraction of the RWT concentration in the tracer test.
- Determination of RWT loss over time due to adsorption to AS flocs and/or chemical quenching of fluorescence by dissolved water constituents based on the results of a batch laboratory experiment. The RWT loss over time should be properly modelled and taken into account in the tracer test interpretation.

References

1. Metcalf & Eddy, *Wastewater Engineering: Treatment and Resource Recovery* (G. Tchobanoglous, F.L. Burton, H.D. Stensel, ed., McGraw-Hill Education, 2014)
2. Y. Le Moullec, O. Potier, C. Gentric, J.P. Leclerc, Chem. Eng. Sci. **63**, 2436–2449 (2008)
3. USGS *Fluorometric procedures for dye tracing,* United States Government Printing Office, Washington (1986)
4. D. Vasudevan, R.L. Fimmen, A.B. Francisco, Environ. Sci. Technol. **35**, 4089–4096 (2001)

5. W. Käss, *Tracing technique in geohydrology* (A.A Balkema, Rotterdam, Brookfield, 1998)
6. L. Pang, M. Close, M. Noonan, Ground Water **36**, 1, 112–122 (1998)
7. A.Y.-C. Lin, J.-F. Debroux, J.A. Cunningham, M. Reinhard, Ecol. Eng. **20**, 75–88 (2003)
8. F.E. Dierberg, T.A. DeBusk, Wetlands **25**, 8–25 (2005)
9. B. J. Shiau, D. A. Sabatini, J. H. Harwell, Ground Water **31**, 913–920 (1993)
10. D.J. Sutton, Z.J. Kabala, A. B. Francisco, D. Vasudevan, Water Resour. Res. **37**, 1641–1656 (2001)
11. T. Kasnavia, D. A. Sabatini, Ground Water **37**, 376–381 (1999)
12. D.A. Sabatini, T.A. Austin, Ground Water **29**, 341–349 (1991)
13. B.L. Sorensen, R.J. Wakeman, Water Res. **30**, 115–121 (1996)
14. J. Makinia, *Mathematical modelling and computer simulation of activated sludge systems* (London, New York, IWA Pub, 2010)
15. E. Iller, *Badania znacznikowe w inżynierii procesowej* (WNT, Warszawa, 1992)
16. Metcalf & Eddy, *Wastewater Engineering: Treatment and Reuse* (G. Tchobanoglous, F.L. Burton, H.D. Stensel, ed., McGraw-Hill Education, 2004)
17. D. Giraldi, M. de'Michieli Vitturi, M. Zaramella, A Marion, R Iannelli, Ecol. Eng. **35**, 265–273 (2009)

Determination of the solar transmittance for the translucent shutter with PCM in liquid and solid state

Anna Komerska[1,*], *Dariusz* Ksionek [1], and *Marian* Rosiński[1]

[1]Warsaw University of Technology, Faculty of Building Services, Hydro and Environmental Engineering, ul. Nowowiejska 20, 00-653 Warsaw, Poland

Abstract. This article presents results of the energy performance of an external translucent shading component integrated with a phase change material. A proposed technology is able to accumulate considerable amounts of energy in the latent heat by absorbing solar energy. Due to selective optical properties, much of the visible light is still transmitted through the facade. Experimental measurements were carried out in a laboratory set-up - testing thermal chamber, located in the Faculty of Building Services, Hydro and Environmental Engineering at Warsaw University of Technology. The main result of the experimental study was the evaluation of the average solar transmittance in the whole measured spectrum, as well as in the infrared and visible light. Since the shift in optical properties was observed when the material was undergoing a phase transition, the average spectral transmittances were measured for different states of matter of the PCM material. The tested shutter showed abilities to reduce and modulate daylight and solar heat gains in the indoor environment, which could contribute to considerable energy savings.

1 Introduction

The fenestration components are still responsible for a significant share in the building energy consumption. Glazed façades, characterised by poor thermal inertia and high heat transfer coefficient, present a weaker energy performance if compared to opaque building elements. During high solar radiation, they play a major role in the internal solar heat gains leading to an increase in the energy need for cooling, overheating problems or thermal discomfort. The most common solutions to decrease solar heat gains applied by engineers focus mainly on the conventional shadings with adequate shading strategy. However, some new solutions are currently under development. New technologies aimed at reducing the solar transmittance include suspended films, aerogel insulation, solar cells or smart materials able to modulate their properties such as thermo- electro- or photochromic materials [1]. The main drawback of most of them is still high investment cost. Another idea based on material's switchable properties and translucent features could be a shading system with a phase change material.

* Corresponding author: anna.komerska@is.pw.edu.pl

Some of the PCMs are able to accumulate considerable amounts of energy in the latent heat. Due to material's selective optical properties, it absorbs considerable amount of infrared light while much of the visible light is still transmitted through the façade [2–5]. Researchers investigated different PCM glazing systems depending on the way the phase change material was integrated into the façade. One of the solution was to fill the window cavity with a PCM [6–10], another idea was to integrate it with a shading and create PCM blinds or shutters [2, 11–16]. The main advantages of this solutions reported by the researchers, was the improvement of the thermal inertia of the façade and the reduction and time shift of the internal solar heat gains.

The main role of shading technologies with a PCM is to decrease the level of the solar heat gains by the absorption of the infrared part of the solar radiation and transmission of the visible light. The investigation of the thermal energy assessment of the proposed shutter proved positive effects on the façade [16]. This paper is however, focused only on the optical properties of the investigated technology. The average spectral transmittances in visible and infrared light were determined through experimental measurements conducted in a thermal chamber in controlled environment.

2 Materials and methods

2.1 System description

The technological concept of the analysed shading is a translucent shutter, consisted of phase change material enclosed in a polycarbonate panel. Due to selective properties of the PCM the shutter is able to absorb considerable amounts of solar radiation in the infrared part of the spectrum and therefore prevent room from high solar heat gains. The PCM also transmits more visible light that IR so the visual comfort inside the building could be still maintained. Moreover, in solid state, high scattering properties of the PCM prevent occupants from a glare risk. The thermo-optical properties of the PCM are however not constant. The shutter is a type of a responsive behaviour shading with optical coefficients dynamically changing along with the temperature and the phase change state.

The tested panel was designed for an external application with dimensions 70 cm x 70 cm, cavity of 1 cm and weight of approximately 9 kg/m^2. A paraffin based PCM was selected with the nominal melting temperature of 31°C, with the heat storage capacity of around 165 kJ/kg. During sunny days, when the panel is exposed to high solar radiation, it absorbs heat and increases its temperature, up to its meting temperature. Since the latent heat represents much higher value than the heat capacity, equal to approx. 2 kJ·kg^{-1}·K^{-1}, the panel maintains its temperature for considerably long time until it completely liquidises. To avoid reducing daylighting during cloudy days, it can be used also as a movable shading.

2.2 Experimental set- up

The experiments were conducted in a testing thermal chamber (
Fig. *1*a), located in the Faculty of Building Services, Hydro and Environmental Engineering at Warsaw University of Technology. The chamber is equipped with a ventilation system ensuring stable thermal conditions inside the test cell. Solar radiation is provided by a solar simulator consisted of four metal halide lamps with the total power of 1.6 kW. Special reflectors ensured acceptable uniformity of the irradiance level over the tested sample.

The tested panel was mounted horizontally in front of the solar simulator. During the measurements, lamps provided an irradiation of approximately 1000 W/m^2 classified as one sun. Solar spectrum of AM = 1.5 according to standards ISO 9050, ASTM G173 is used in

 DOI: 10.1051/e3sconf/20172200084

solar industry for all standardized testing and therefore is acknowledged as a reference in solar simulators. Two pyranometers LP PYRA were used to measure the solar radiation transmitted through the tested panel in two different locations. The transmittance for visible and infrared spectrum was additionally measured by applying on the pyranometers suitable optical filters: UV and VIS filter – Heliopan 62 mm Infra Red 715 nm used for measuring transmittance in the infrared light, UV and IR filter – Heliopan Digital UV/IR used for measuring transmittance in visible light. The temperature of the surface of the panel as well as the temperature in the cell was monitored with the use of resistance thermal sensors Pt100. An aluminium shield was placed over the sensors in order to protect them from direct radiation. The state of the PCM was defined through temperature sensors as well as through visual assessment.

The measurement protocol applied in the characterization of the solar transmittances presents however, some limitations. Main issues that could influence the results are related to the spectral distribution of the solar simulator which does not ideally reflect the solar spectrum or to the uniformity of the generated light.

a) b)

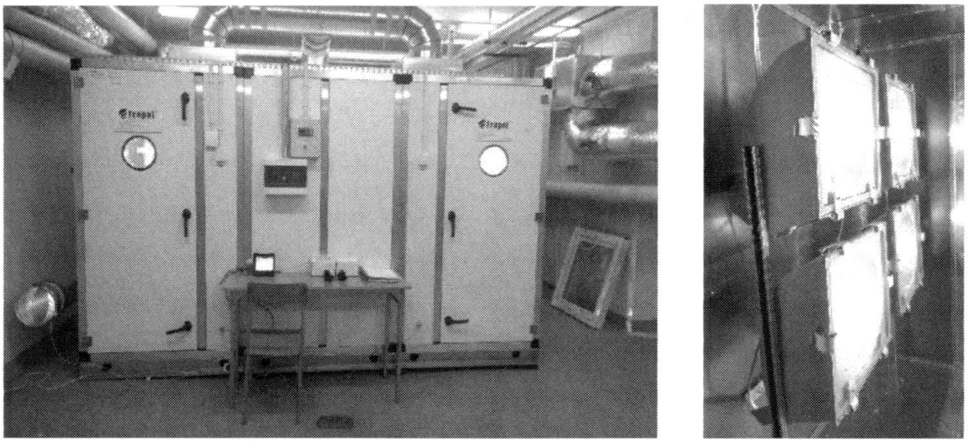

Fig. 1. a) view of the thermal chamber; b) view of the solar simulator.

2.3 Methods – measuring parameters

The experimental procedure was composed of subsequent measurements. Firstly, tests were conducted with the absence of the PCM shutter in order to obtain the level of total and spectral (VIS and NIR) solar radiation generated by solar simulator. The visible spectrum was measured in a spectral range of 380–705 nm, the infrared spectrum: 705–3000 nm. Measured values were afterwards used as a reference to calculate the transmittance coefficients of the panel. Second stage focused on the determination of the total and spectral (VIS and NIR) radiation transmitted through the panel. The full set of measurements was taken for solid and afterwards for liquid state of the PCM. Temperatures and irradiances were collected every 30 seconds. The data used for calculations represents 5 minutes of continuous measurements of each stage. The solar transmission coefficient was calculated as the ratio between solar radiation transmitted and incident on the panel. The average coefficient value represents a mean value of 10 subsequent measurements. Measurement inaccuracy of solar irradiances caused by non-ideal characteristics of optical filters was increased through a correction factor, determined through additional measurements and the technical datasheet. For the sake of brevity, the methodology of this procedure is not included in this paper.

Fig. 2 a) measurements of the solar irradiance transmitted through PCM panel; b) measurements of the spectrum of the solar simulator using optical filters; c) tested shutter with liquid PCM.

3 Results and discussion

The average solar transmittances, were summarised in Table 1. The level of the measured solar irradiances collected from both pyranometers is shown in

Fig. 3. In order to compare transmittances for each studied case, solar irradiances were categorised with the state of aggregation of the PCM and the spectral band.

Table 1. Average transmittance coefficients of the investigated shutter.

State of PCM	t_{tot} [-]		t_{vis} [-]		t_{nir} [-]	
	upper sensor	lower sensor	upper sensor	lower sensor	upper sensor	lower sensor
liquid	0.82	0.82	0.88	0.88	0.74	0.79
solid	0.38	0.36	0.37	0.35	0.31	0.32

Results showed selective optical properties of the PCM shutter. The solar transmittances varied depending on the state of aggregation of the PCM as well as on the analysed spectral band. When PCM was in solid state the transmittance of the panel represented a low value of approx. 0.37 due to high scattering properties and the light absorption of the crystalized PCM. As expected, the optical properties of the investigated panel changed when the PCMs went through a phase transition process. Melting of the PCM resulted in an increase of the average total transmitted irradiance from 373 W/m^2 to 823 W/m^2, visible irradiance from 201 W/m^2 to 497 W/m^2, IR irradiance from 118 W/m^2 to 291 W/m^2. Moreover, the investigated shutter proved to transmit more energy in visible light than in infrared band. The level of transmitted solar radiation through the shutter with solid PCM was found to be approximately 1.71 times higher for visible than for infrared light. In a case when the PCM was completely melted, the ratio represented similar value of 1.70.

As shown in Table 1 small variances in solar transmittances between both sensors can be observed. This issue could be caused by the inhomogeneity of the PCM material. When PCM solidifies, it decreases its volume. If the crystallisation process runs too fast, air cannot escape to the top of the panel and therefore small air bubbles are trapped inside the phase change material.

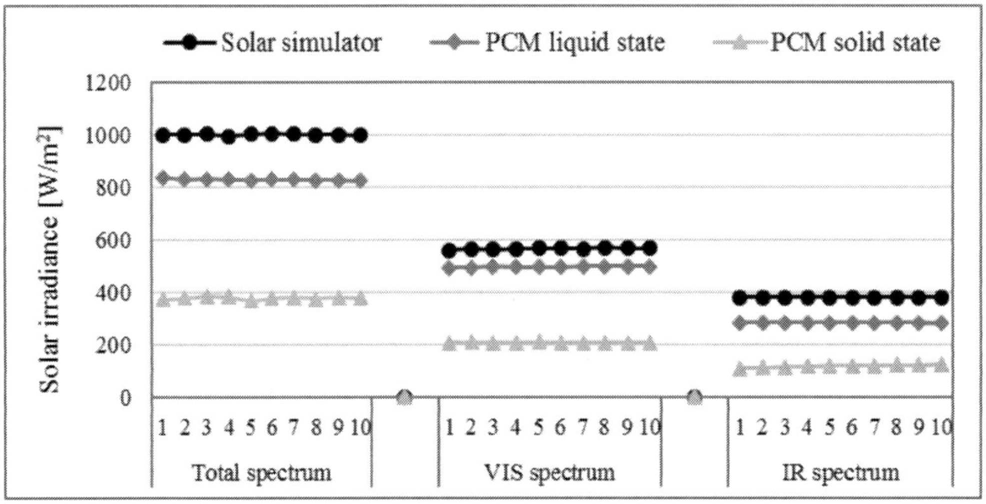

Fig. 3 Total and spectral solar radiation incident and transmitted through the tested shutter in a function of 10 subsequent measurements.

4 Conclusions

The analysed panel is designed to act as a solar shading that improve the thermal performance of the glazed façade. The technological concept is based on selective optical properties of the paraffin wax for different spectral bands. Especially PCM's ability to absorb infrared light and transmit high value of visible light. Such system can decrease the internal solar heat gains, prevent form the glare risk however, still provide daylight and visual comfort inside the room.

The study was focused on the determination of the transmittance coefficients in visible and infrared band. The results showed high potential of the panel in reducing solar heat gains while still providing visible light transmittance. The solar transmittances varied depending on the state of aggregation of the PCM as well as on the analysed spectral band. In solid state the panel transmitted approximately 31% of the IR and 36% of the visible light. When the PCM became liquid, the transmittances increased reaching value of 77% and 88% respectively. Both in solid and liquid state the transmittance coefficients in the infrared spectrum were therefore lower than for the visible band.

This research was performed with the financial support of Warsaw University of Technology, Faculty of Building Services, Hydro and Environmental Engineering, grant No.: 504/02638/1110/42.000100 and grant No.: 504/02639. Authors would like to thank the company Frapol for their kind help in developing and building a thermal chamber. Especially, we acknowledge the contribution and support of Mr Grzegorz Mars.

References

1. P.J. Bjørn, A. Hynd, A. Gustavsen, D. Arasteh, H. Goudey, R. Hart, Sol. Energ. Mat. Sol. Cells **96**, 1–28 (2012)
2. H. Weinlader, A. Beck, J. Fricke, Sol. Energy **78**, 177–186 (2005)
3. F. Goia, M. Zinzi, E. Carnielo, V. Serra, Energy Procedia **30**, 428 – 437 (2012)
4. B.L. Gowreesunker, S. B. Stanković, S.A. Tassou, P.A. Kyriacou, Energy Build. **61**, 239–249 (2013)
5. D. Li, Z. Yumeng, Z. Li, H. Qi, Energy Build. **108**, 381–386 (2015)
6. F. Frontini, J. Pfafferot, S. Herkel, D. Schwrtz, CISBAT (2011)
7. S. Li, Y. Zhou, K. Zhong, X. Zhang, J. Xing, International Journal of Low-Carbon Technologies Advance Access (2013)
8. S. Grynning, F. Goia, E. Rognvik, B. Time, Passivhus norden (2012)
9. F. Goia, M. Perino, V. Serra, Sol. Energy **100**, 217–233 (2014)
10. F. Goia, L. Bianco, Y. Cascone, M. Perino, V. Serra, Energy Procedia **48**, 1272–1281 (2014)
11. H. Weinlaeder, W. Koerner, M. Heidenfelder, Energy Build. **43**, 2468–2475 (2011)
12. N. Soares, J. J. Costa, A. Samagaio, R. Vicente, World Renewable Energy Congress (2011)
13. E.M. Alawadhi, Energy Build. **47**, 421–429 (2012)
14. R. Ryu, S. Janghoo, K. Yongseong, Contemp. Eng. Sci. **7**, 21, 1005–1012 (2014)
15. T. Silva, V. Romeu, F. Rodrigues, A. Samagaio, Energy Build. **88**, 110–121 (2015)
16. A. Komerska, L. Bianco, V. Serra, S. Fantucci, M. Rosiński, Energy Procedia **78**, 3452–3457 (2015)

Analysis of profitability of using a heat recovery system from grey water discharged from the shower (case study of Poland)

Sabina Kordana[1], and *Daniel* Słyś[1,*]

[1]Rzeszow University of Technology, Department of Infrastructure and Sustainable Development, al. Powstańców Warszawy 6, 35-959 Rzeszów, Poland

Abstract. The paper analyses the profitability of the use of Drain Water Heat Recovery units. An original simulation model was used for this purpose, and a detached residential building located in Poland was selected as the test facility. The conducted analysis proved that the type of the hot water heater has decisive influence on the profitability level of such an investment. Application of the abovementioned technology is particularly profitable, when water is heated with the use of an electrical device. When the energy source in the system is a gas water heater, the obtained calculation results are not as favourable, and the period of investment return in many cases exceeds the expected service life of these devices. Moreover, the analysis demonstrated that the potential energy savings, and thus also the financial savings, may be in both cases increased as a result of simultaneous intake of water from various water taps.

1 Introduction

The current living standards and the increasing consumerism result in the excessive exploitation of natural resources, including fossil resources, while the gradual depletion of conventional fuels is identified as one of the challenges that the future generations will be faced with. The problem is additionally exacerbated by the climate changes [1, 2] which have been observed in recent years and which are enforcing the application of innovative technologies in all facets of life [3, 4]. In view of that, it is necessary to strive to reduce energy use at particular stages of energy generation, processing and consumption [5–7].

In the case of residential building the energy use in individual buildings can vary widely [8, 9]. This is because it depends on the size of the building and its type, as well as the demographic characteristics of residents and their income. Not without significance is also the location of the building and its equipment. A significant impact on the energy usage in the household has also the price of individual energy carriers, which also can be of decisive importance in the decision making process for the use of energy-efficient installations.

The above factors affect not only the total energy usage in residential buildings, but also determine the structure of its usage in individual buildings. Recent years have witnessed an increasingly large role of domestic hot water production in the energy consumption

* Corresponding author: daniels@prz.edu.pl

 DOI: 10.1051/e3sconf/20172200085

balance. For developed countries, it represents on average from over a dozen even to several dozen percent of total energy consumption in buildings [10, 11], which renders it a perfect subject for scientific research aiming to reduce household energy demand.

The aim of the research described in this article is to determine the cost-effectiveness of installing a Drain Water Heat Recovery (DWHR) unit below the shower in a single-family residential building. Analyses have been carried out for various types of instantaneous water heaters and different shower head water flow rates. The simultaneity of water intake from two water taps has also been taken into account. This allowed for an unequivocal assessment of the cost-effectiveness of adopting the examined system in various functioning conditions.

2 Theory and calculation

The idea behind the operation of DWHR unit is the transfer of thermal energy contained in grey water to the incoming cold tap water, and their effectiveness depends primarily on the design of the heat exchanger and the rates of water and drain water flow through the unit [12]. Although scientists from all over the world are engaged in developing new, high--performance structures of DWHR units [13–15], vertical counter-flow heat exchangers are still the subject of most interest, as they are characterized by the highest efficiency of heat recovery, in many cases exceeding 50%. A big advantage of DWHR heat exchangers is the opportunity of cooperation with devices designed to acquire energy from other unconventional energy sources. For example, in [16] there is a description of the installation, in which water led into the sanitary facilities is heated by DWHR unit and solar collectors. Wallin and Claesson [17], in turn, examined the possibility of cooperation of such heat exchanger with a heat pump.

DWHR heat exchangers predominantly cooperate with instantaneous water heaters, one of the most widely used devices in housing. However, adopting a system for heat recovery from grey water requires a change in the way of routing particular pipes of the internal water and sewage installation in the building, in relation to the traditional solution of such installations, in which cold water is fed both to the shower mixing valve and hot water heater (Variant 0). DWHR units are usually mounted on the outlet of grey water from the shower, and heated while passing through the heat exchanger water is supplied to the hot water heater and/or shower mixing valve. Additionally, we can distinguish two different methods of preparation of domestic hot water supplied to the other points of use where mixed water is used:

• Variant 1 – heating water by using a separate heat source (Fig. 1),

• Variant 2 – heating water by using the same heater, from which the domestic hot water is led to the shower mixing valve (Fig. 2).

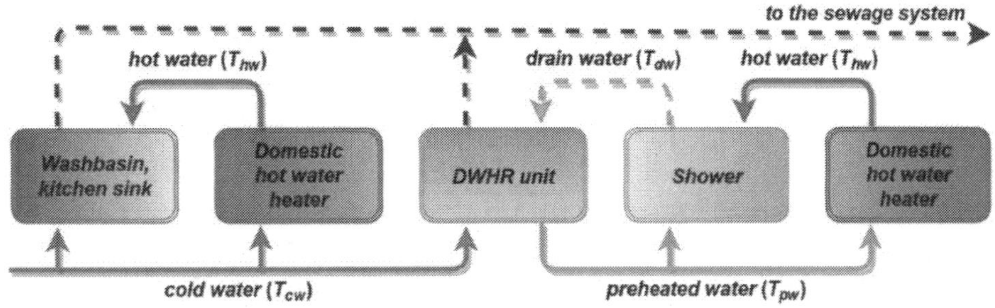

Fig. 1. Diagram of the installation with two hot water heaters (Variant 1).

In order to determine the financial performance of applying a grey water heat recovery system cooperating with the instantaneous electric or gas hot water heater, an original simulation model described by Słyś and Kordana [12] was used. In this study, a simplified assumption was made that during the flow of water through the pipes there are no heat losses, so the adopted temperatures are constant over time. In addition, it was assumed that the efficiency of heat recovery from grey water is not determined by the presence of impurities it may contain.

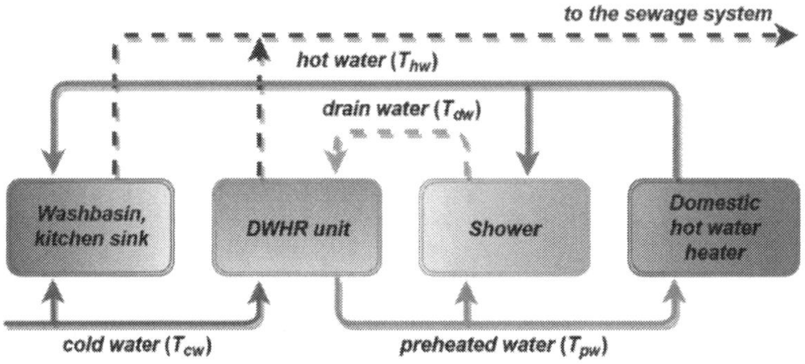

Fig. 2. Diagram of the installation with only one hot water heater (Variant 2).

Numerical model output variables are financial indicators:
• *NPV* – Net Present Value, €,
• *DPP* – Discounted Payback Period, years.

In the case of grey water heat recovery systems the value of *NPV* may be defined for a specific time period on the basis of Eq. (1).

$$NPV = \sum_{t=0}^{n} \left[\frac{S_t - INV_t}{(1+r)^t} \right] \tag{1}$$

where: S_t – savings in the year t, €; INV_t – investment in the year t, €; r – the discount rate; n – number of years of the operation of the system, years.

The value of *DPP* can be, in turn, determined according to Eq. (2).

$$DPP = Y + \frac{|NPV_Y|}{CF_{(Y+1)}} \tag{2}$$

where: Y – the number of full years before the return of investment, years; NPV_Y – unrecovered expenditures set at the beginning of the year (Y + 1), €; $CF_{(Y+1)}$ – discounted cash flow in the year (Y+1), €.

The analysis was performed for the single-family dwelling house located in Rzeszow. The study assumed that each of the residents is using the shower once a day and the consumption of water and energy for water heating is in the particular case the same for each installation user. To determine the effect of specific parameters of installation using on the results of the study, the calculations were carried out for different number of residents (R), the variable duration of showering (l), and also for different values of the flow rate of water from the shower head (q_{mw}). The values of these parameters, as well as other data used in calculations, are the following:

• the number of residents (R): 3, 4, 5 persons,
• shower length (l) (to determine *DPP*): 3.5 min, 6.0 min, 8.5 min,
• the flow of mixed water (q_{mw}): 6.0 L/min or 8.0 L/min,

- investment in the year 0 (INV_0): €900.00,
- the discount rate (r): 5%,
- number of years of the operation of the system (n): 15 years,
- the price of electricity in the year 0 (C_{e0}): 0.13 €/kWh,
- the price of natural gas in the year 0 (C_{g0}): 0.04 €/kWh,
- annual increase in electricity and natural gas prices (i_e): 2%,
- the temperature of hot water (T_{hw}): 55°C,
- the temperature of cold water (T_{cw}): 10°C,
- the temperature of mixed water (T_{mw}): 40°C,
- the efficiency of the system in the case of the use of electric water heater (η_e): 0.99
- the efficiency of the system in the case of the use of gas water heater (η_g): 0.85.

In the case when the source of energy for water heating are the two independent instantaneous water heaters (Variant 1), the amount of heat recovered from grey water does not depend on taking water from washbasin tap or kitchen sink tap. In a situation where there is only one device designed for hot water heating in the installation (Variant 2), it is necessary to take into account the simultaneity of water intake from two or three points of use. In order to determine the impact of such occurrence on financial efficiency of the investment it was assumed that 80% of the time of using the heat recovery system only water used for showering needs is heated. In the remaining period of time there is simultaneity of water intake from the two taps. It was assumed that the temperature of mixed water (T_{mw}) is in both devices the same and the flow rate of water from the mixing valve of the second sanitary facility is equal to 6.0 L/min.

3 Results and discussion

Figure 3 presents the obtained results of calculations of Net Present Value (*NPV*) and Discounted Payback Period (*DPP*). The first column (Fig. 3a, 3c, 3e) shows the findings of an analysis carried out for the event that two separate energy sources in the installation are utilised for water heating. Figures 3b, 3d and 3f, in turn, depict the findings of an analogous analysis conducted with an assumption of simultaneous water intake from two different water taps.

The obtained results clearly indicate that the type of hot water heater used has a large impact on the cost-effectiveness of the undertaking that is installation of a DWHR heat exchanger below the shower. The use of a DWHR unit is particularly advantageous when the energy source for heating water is an instantaneous electric hot water heater. In this case the *NPV* reaches negative value only if the total consumption of water used in the analysed household for the needs of taking a shower is low. In the event the installation has two separate heat sources and daily water consumption for that purpose exceeds 82.0 L at q_{mw} = 8.0 L/min, or 75 L at q_{wm} = 6.0 L/min, the undertaking is in turn cost-effective, since $NPV > 0$.

In a situation where hot water is prepared by a gas heater, the project turned out to be financially advantageous only when the water consumption during taking a shower is high and far exceeds 200.0 L per day. Therefore, such solution can be applied in the case of shower heads characterized by high mixed water flow rate and also in the case where total time of taking a shower is long.

The differences in received *NPV* values for particular types of water heaters are the result of differences in the prices of both analysed energy carriers. The price of electricity is so high in comparison with the costs that have to be born for the delivery of natural gas allowing to produce the same amount of thermal energy that, despite the fact that the efficiency of the system for the hot water heating is in the first case by almost

 DOI: 10.1051/e3sconf/20172200085

15 percentage points higher, the costs for the supply of electricity used for hot water heating are much higher than for the supply of natural gas.

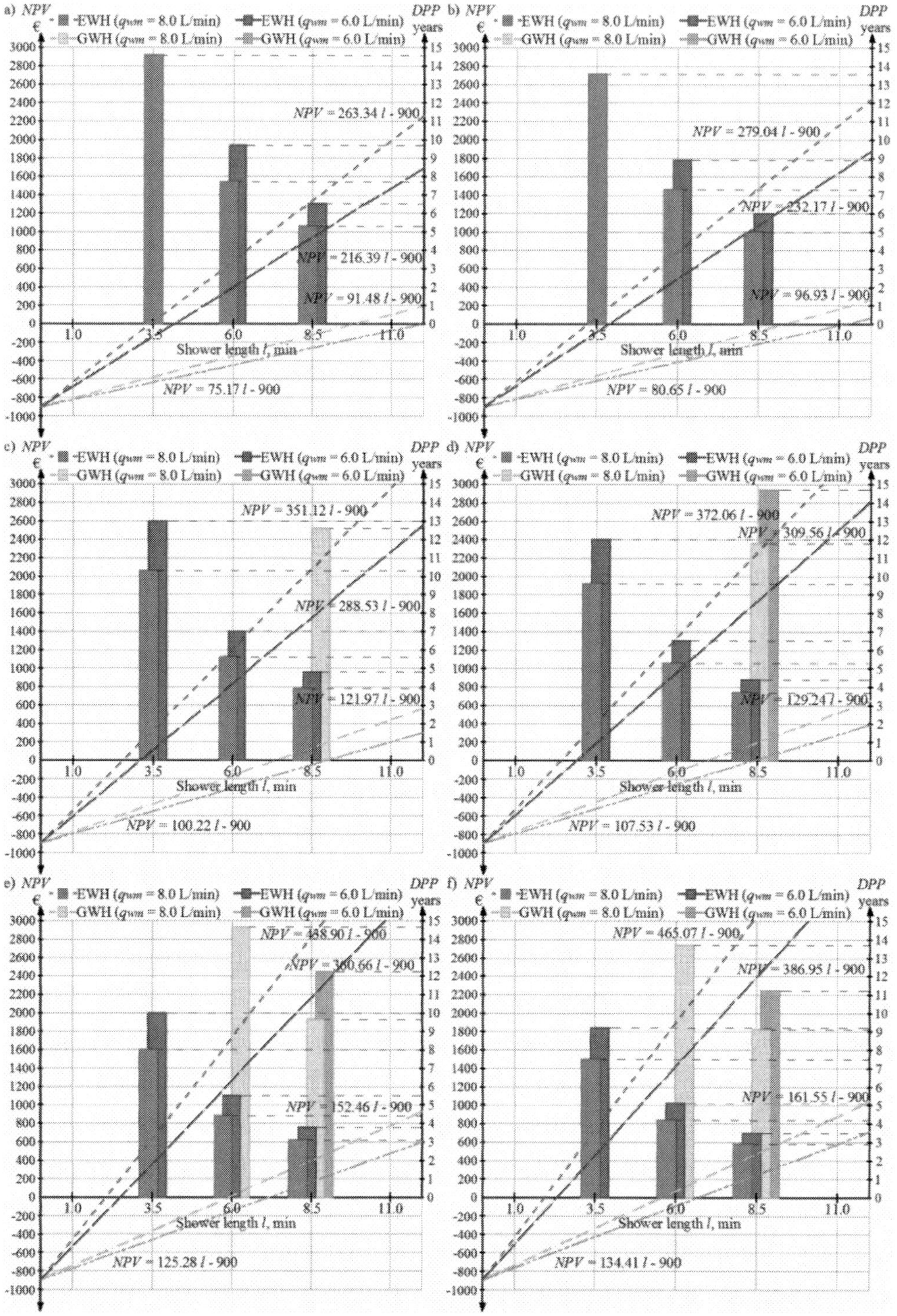

Fig. 3. Summary of results of the calculation of *NPV* and *DPP*: a), b) 3 persons; c), d) 4 persons; e), f) 5 persons (EWH – electric water heater, GWH – gas water heater).

 DOI: 10.1051/e3sconf/20172200085

This is reflected in the amount of potential savings, and consequently in the results of calculation of the Net Present Values, which in the case of an electric water heater significantly exceed the values obtained for gas device. The higher the shower head water flow rate, the bigger the difference. For the analysed time interval, differences exceed at times even €2.000,00.

The amount of recoverable thermal energy was determined on the basis of a graph of the dependence of the energy recovered from grey water on the flow rate of both media (offered by the manufacturer of the analysed heat exchanger). In the situation where in the DWHR unit the water is heated only to be used for the purpose of taking a shower, the flow rates of water and drain water through the heat exchanger are equal, and account for 6.0 L/min or 8.0 L/min, depending on the shower head settings. In the first case taken into consideration (q_{mw} = 6.0 L/min) the amount of heat energy recovered from drain water (E_r) is equal to 7.4 kW. Increasing the water flow through the heat exchanger to 10.0 L/min, resulting from the intake of water from the other sanitary facilities (Variant 2), causes the increase of the amount of heat recovered from drain water by about 1.0 kW. The temperature of preheated water (T_{pw}) will be in such a situation lower by almost 6°C, however, the amount of heated water will increase by nearly 70%, making the final energy demand to be reduced, and the financial performance of the project will be higher than the efficiency obtained from water consumption only from the shower.

The analysis also demonstrated that the potential savings can be increased as a result of simultaneous intake of water from two different water taps. It should be noted, however, that the Net Present Values characterizing Variant 2 are just slightly higher than those obtained with the assumption that only water used for the purpose of taking a shower flows through the heat exchanger. It is clear that in the analysed case, the way the water is heated has a marginal impact on financial efficiency of a project involving the installation of the DWHR unit. For this reason, it should not be the determinant of the decision of using a heat recovery system from grey water.

Also, in the case when mixed water flow (q_{mw}) is equal to 8.0 L/min, the simultaneity of water intake from two sanitary facilities results in the increase of the amount of heat recovery. The temperature of preheated water (T_{pw}) is in this case lower by 4.2°C, which is the result of a minor proportion of water used in the second device in comparison with the previous case, but the predicted increase in the *NPV* values is comparable. In the described situation, simultaneous use of the third device is not advisable, as the water flow rate through the heat exchanger would reach a value which is not included in the graph characterizing the analysed heat exchanger. The lack of available characteristics for such values of water flow through the device can be tantamount to a sharp decrease of its effectiveness. The solution to this problem may be, however, to change the way of water pipes routing, which means supplying water only to the domestic hot water heater. Shower mixing valve would be in this case supplied with cold water, which would allow reducing the amount of water transported through the heat exchanger.

The Discounted Payback Period (*DPP*) was designated for shower length (l) of 3.5 min, 6.0 min and 8.5 min, but only for those investment options for which the *NPV* reaches positive values, as in other cases its length exceeds the technical lifetime of the device (Fig. 3). It is evident that the value of *DPP* is determined by the amount of water used for the needs connected with taking a shower and the type of device used to heat water. It is also easy to note that the *DPP* value decreases with the increase of the Net Present Value of the project.

In order to assess the investment risk associated with the use of a heat recovery system from grey water discharged from the shower in the single-family dwelling house, the investment sensitivity analysis was also performed [18]. The study was conducted for the case where the shower length for one person is 6.0 minutes, and the shower is used by

DOI: 10.1051/e3sconf/20172200085

4 people. It was assumed that in the heat exchanger only water used for the purpose of taking a shower is heated.

The conducted analysis was based on the determination of the investment sensitivity indexes (*ww*), showing how huge impact on the received *NPV* value has a change in the value of individual independent variables by 1%, and the relative safety margins (s_{mr}), defining the permissible deviation of the variable from the base value, where the project remains profitable. It was additionally found how the level of profitability of the analysed project will change in the situation when the given independent variable will change by a specific percentage value.

The values of the sensitivity indexes (*ww*) and the relative safety margins (s_{mr}) were determined for those cases for which the *NPV* reaches positive values, which indicates the profitability of the analysed project. The summary of the results of these calculations is presented in Table 1.

Table 1. The results of sensitivity analysis for the case when water is heated by electricity.

The analysed independent variable	Sensitivity index (*ww*)	Relative safety margin (s_{mr})
Water flow from the shower head q_{mw} *= 6.0 L/min*		
Investment	–1.08	92.33%
Electricity prices	2.08	–48.01%
Water flow from the shower head q_{mw} *= 8.0 L/min*		
Investment	–0.75	134.07%
Electricity prices	1.75	–57.28%

The sensitivity analysis showed that in the case of electric hot water heater the level of profitability of the project is only slightly determined by the values of the analysed independent variables. It is also worth pointing out that the sensitivity of the undertaking to changes of installation operating costs is higher than to changes in investment outlays, as evidenced by the higher absolute value of the sensitivity index and the smaller safety margin. In addition, the difference in sensitivity indexes values calculated for both of the independent variables increases when the shower head water flow rate decreases. It is the consequence of reducing the basic value of the *NPV* with the decrease of water consumption while maintaining the same duration of taking a shower.

The observations above are also verified by the results of a research that assumed changes in values of the analysed independent variables by specific percentages, the results of which are shown in Figure 4. A change in the values of those variables even by 35% will not, in the case of an electric water heater, cause a decrease of the *NPV* value to below zero.

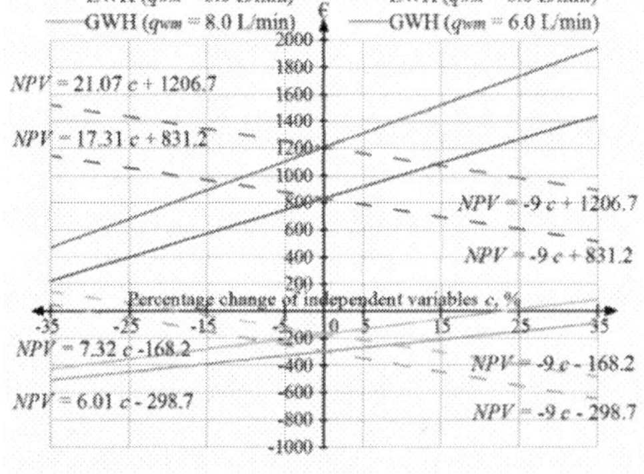

Fig. 4. Graphical form of sensitivity analysis for variables linearly changing return on investment (solid line – electricity/gas prices changes, dashed line – investment changes).

In the case of a gas hot water heater, where the basic value of the *NPV* < 0, the undertaking would become cost-effective only when the investment outlays were decreased to €601.00 and €732.00, respectively for shower heads with flow rates of 6.0 L/min and 8.0 L/min. Energy prices, in turn, would have to increase by 50% and 23% respectively.

4 Conclusions

The conducted financial analysis of the application of a DWHR unit cooperating with various types of instantaneous water heaters in a single-family residential building has demonstrated a large impact of operating parameters of the installation on the undertaking cost-effectiveness. In view of that, there is a need to emphasise the necessity of carrying out a thorough technical and financial analysis for each case of adoption of this solution.

The results of the analysis of financial efficiency of the investment, which were received by using the developed calculation model, can serve as a guide when deciding on the application of the system described in the article, both for individual users of internal installations and for the designers of this type of installation.

References

1. B. Kaźmierczak, A. Kotowski, Theor. Appl. Climatol. **118**, 285–296 (2014)
2. M. Starzec, J. Dziopak, M.I. Alexeev, Water and Ecology **1**, 41–50 (2015)
3. K. Pochwat, D. Słyś, J. Dziopak, *Installation of retention facilities* (EP 15461517.3, 2011)
4. M. Zeleňaková, G. Markovič, D. Kaposztásová, Z. Vranayová, Procedia Eng. **89**, 1515–1521 (2014)
5. M.A. Sayegh, J. Danielewicz, T. Nannou, M. Miniewicz, P. Jadwiszczak, K. Piekarska, H. Jouhara, Renew. Sustain. Energy Rev. **68**, 1183–1192 (2017)
6. A. Stec, S. Kordana, Resour. Conserv. Recycl. **105**, 84–94 (2015)
7. M.H. Wahlström, B. Hårsman, Energy Build. **102**, 58–66 (2015)
8. A.N. Binks, S.J. Kenway, P.A. Lant, B.W. Head, J. Clean. Prod. **135**, 892–906 (2016)
9. J.L. Reyna, M.V. Chester, S.J. Rey, Energy **112**, 742–754 (2016)
10. A. Bertrand, R. Aggoune, F. Maréchal, Appl. Energy **192**, 110–125 (2017)
11. A. Bertrand, A. Mastrucci, N. Schüler, R. Aggoune, F. Maréchal, Appl. Energy **186**, 152–166 (2017)
12. D. Słyś, S. Kordana, Energy Build. **71**, 1–11 (2014)
13. S. Bartkowiak, R. Fisk, A. Funk, J. Hair, S.J. Skerlos, J. Green Build. **5**, 85–94 (2010)
14. S. Morales-Ruiz, J. Rigola, C. Oliet, A. Oliva, Appl. Energy **179**, 1006–1019 (2016)
15. E. Tian, Y.L. He, W.Q. Tao, Appl. Energy **188**, 586–594 (2017)
16. K. Tanha, A.S. Fung, R. Kumar, Appl. Therm. Eng. **90**, 444–459 (2015)
17. J. Wallin, J. Claesson, Energy Build. **80**, 7–16 (2014)
18. W. Rogowski, *Rachunek efektywności inwestycji* (Wolter Kluwer, Cracow, 2008)

 DOI: 10.1051/e3sconf/20172200086

Change-over natural and mechanical ventilation system energy consumption in single-family buildings

Maria Kostka[1,*], and *Małgorzata* Szulgowska-Zgrzywa[1]

[1]Wrocław University of Science and Technology, Faculty of Environmental Engineering, Wybrzeze Wyspiańskiego 27, 50-370 Wrocław, Poland

Abstract. The parameters of the outside air in Poland cause that in winter it is reasonable to use a mechanical ventilation equipped with a heat recovery exchanger. The time of spring, autumn, summer evenings and nights are often characterized by the parameters of the air, which allow for a natural ventilation and reduce the electricity consumption. The article presents the possibilities of energy consumption reduction for three energy standards of buildings located in Poland, ventilated by a change-over hybrid system. The analysis was prepared on the assumption that the air-to-water heat pump is the heat source for the buildings.

1 Introduction

1.1 Residential ventilation solutions in Poland

A typical ventilation in single-family buildings in Poland is implemented in one of two ways – natural/gravity or mechanical. In the natural system the driving force of the air exchange is the pressure difference on the building partition, caused by the difference of internal and external temperature and by the wind. This is the basic system used over many years in the Polish construction industry. Along with the change of regulations related to the quality of buildings, their thermal insulation improvement and energy demand reduction, the currently used mechanical systems are becoming popular. This type of ventilation in detached buildings reduces the energy consumption needed in the natural system for heating the outdoor infiltration air rate. The air exchange is provided by two fans placed in the air handling unit (AHU), equipped also with an exhaust air heat recovery exchanger, air filters and a heater. Usage of the heat recovery is crucial and justifies the choice of the mechanical systems. In case of the gravity ventilation used in low-energy buildings, cost of outdoor air heating become a significant operating cost [1]. Mechanical ventilation allows reducing the cost of ventilation and ensures the constant air exchange efficiency, which is not possible in the case of natural ventilation in the Polish climate. Fluctuations of the outside air temperature in Poland cause that the efficiency of natural ventilation is variable during the year. In winter the volume of the exchanged air is higher

* Corresponding author: maria.kostka@pwr.edu.pl

than during the rest of the year and the reason is the large difference between the internal and external temperatures. It has a significant impact on the energy demand and the building heating costs. In the warmer periods of the year, when the temperature differences decrease, the efficiency of the natural system is also getting lower. It may also happen that the air exchange ceases completely.

The parameters of outside air in Poland indicate that a combination of the mechanical and the natural ventilation in detached buildings can be cost-effective and allow keeping the thermal comfort and good indoor air quality. Low temperatures in winter cause that it is reasonable to use mechanical ventilation equipped with a heat recovery from the exhaust air. Also during summer, in case of a high outside temperature and days of windless weather, the use of fans and a heat recovery system is justified because it maintains a proper air exchange in the building and reduces overheating of the interiors. The time of spring, autumn, summer evenings and nights are often characterized by the parameters of air, which allow for the natural building ventilation. The combination of the natural and the mechanical systems is realized by the hybrid ventilation.

1.2 Hybrid ventilation

The hybrid ventilation is only a general name of system which combines the natural and the mechanical processes, but does not define the forms of control strategy. Their common operation can be implemented in a variety of ways: seasonal switching, switching as a result of a measured parameter (e.g. outside air temperature, indoor temperature, indoor CO_2 concentration) or simultaneous work of systems. In the professional literature, there is lack of information about the design of such systems. There are also very few examples of existing solutions and their energy efficiency. The hybrid ventilation in residential constructions was conducted by M. Kim and J. Hwang [2]. According to their definition, the system with a mechanical air exhaust and a natural air supply is classified as a hybrid system. For an apartment located in Seoul they attempted to determine the airflow which provides comfortable thermal conditions and good air quality, also taking into account the energy savings. A similar system with a mechanical exhaust was studied by D. Jreijiry, A. Husaunndee and C. Inrad [3]. They developed hybrid system control strategies for the housing sector. For buildings located in Greece and France their research conducted that a hybrid ventilation system enables decrease of more than 50% of the exposure of the occupants to high concentrations of CO_2 while reducing at the same time the electrical energy consumed by the fan by 90%. K. Miachou and all [4] conducted a comparative study of natural and hybrid systems for dwellings located in Athens. They showed a slight advantage of the natural system in the air exchange. Their research has been performed in the framework of the research project RESHYVENT [5]. This program defined the design objectives for hybrid ventilation systems and outlined the solutions proposed by the four industrial and scientific consortia for four different European climates zones.

1.3 The justification of the topic

According to the report of Central Statistical Office of Poland, the percentage share of households in the energy consumption structure occupied the first place in 2014 and exceeded 30% [6]. On a national scale, with the construction of approximately 70,000 new single-family buildings each year, solutions that combine natural and mechanical ventilation can provide significant reductions in the energy needs. Due to the differences between the countries that have taken up the research topic earlier, including climatic conditions, typical building structures or even residents' expectations, the results of the research should not be directly applied to Poland. For this reason, there is a need to assess the energy potential of the hybrid systems operating in the temperate climate of our country.

2 Climate in the selected cities

Poland is located in the temperate climate zone, in the transitional area between the temperate oceanic climate on the west and the temperate continental climate on the east. Due to the influence of the continental and oceanic air masses, as well as the direct influence of the Baltic Sea and the presence of mountain areas, the parameters of the outside air in Poland are various in the different regions of the country. The assessment of the energy potential of hybrid ventilation systems should therefore be conducted on a case-by-case basis rather than being generalized to the national results.

Figure 1 show the hourly average outside air parameters for 3 selected Polish cities located in the different regions of the country: Suwałki (influence of the continental air masses), Gdańsk (influence of the Baltic Sea) and Wrocław (temperate region). Figure 2 shows the duration of the outside air temperature for this locations.

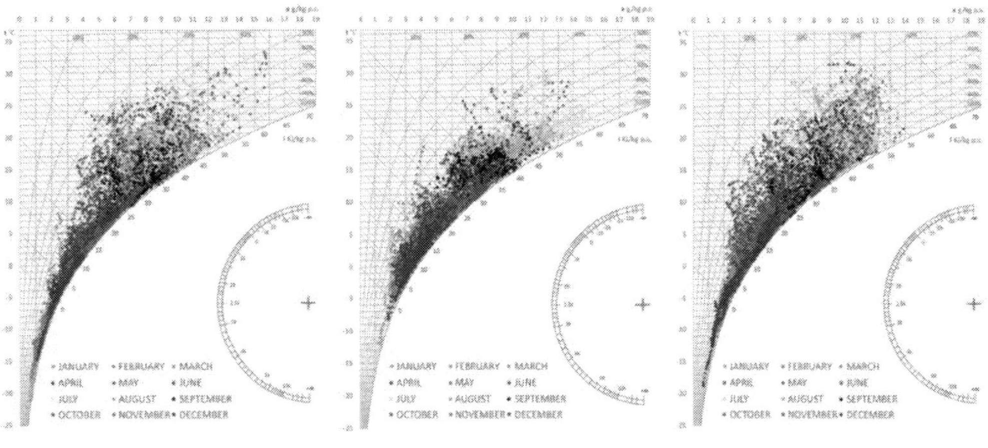

Fig. 1. The parameters of outside air – Suwałki (left), Gdańsk (center), Wrocław (right).

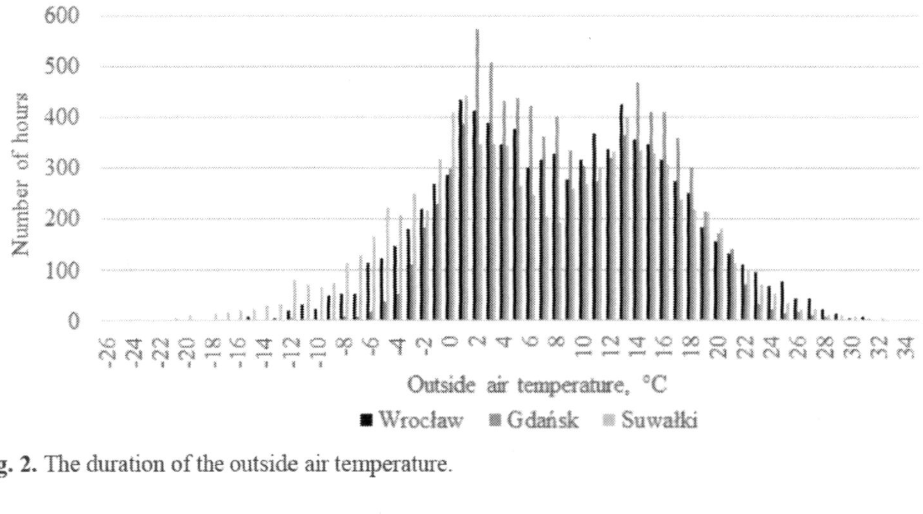

Fig. 2. The duration of the outside air temperature.

According to the presented meteorological data [7], Gdańsk is characterized by the mildest parameters of the outdoor air. Temperatures in the range of 1–20°C occurs there for about 85% of the year, while in Wroclaw about 75%, and in Suwałki only about 66% of the year. Wroclaw has the highest number of hours with temperatures above 20°C – over 7% of the year, while in Suwałki it is about 5% and in Gdańsk less than 4%. The longest duration of the temperature 0°C and less occurs in Suwałki – about 29% of the year. In Wrocław it is about 18%, and in Gdańsk only about 11%. Demonstrated differences indicate a higher climate potential for the hybrid system in the areas characterized by a mild climate. In these locations, the longer the time of the natural ventilation system usage the lower the energy savings associated with the mechanical ventilation work is expected.

3 The research problem

The article presents the energy benefits of using a hybrid ventilation system instead of a full-year mechanical ventilation with the heat recovery. The analysis was performed for a single-family building with a usable area of $160m^2$. The ventilation air flow rate is 250 m^3/h, which corresponds to a building air change rate ACH = $0.5h^{-1}$. The first step of the calculations was to analyse the work of the building in the variant with the mechanical ventilation with the heat recovery (*MW*). The reason for this is that the heat recovery is an essential element that increases the energy efficiency of a building. Then, for the hybrid ventilation (*HW*), the energy savings potential of the final energy demand for the building's heating was determined. In this case the hybrid ventilation was implemented as temporary switching from the mechanical ventilation (*MW*) to the natural system (*NW*). The decision to switch between the ventilation systems in the applied algorithm is based on the total final energy consumption of the building. It is defined further below, in Table 3. The analysis was performed for the climatic data presented in point 2 of this article. The heat source in the analysed case was an air-to-water heat pump.

3.1 Building energy class

As a result of the increase of the energy efficiency requirements, the demand for usable energy for heating the newly designed single-family houses in Poland is decreasing. An analysis of the hybrid ventilation impact on the building's energy efficiency has been carried out on three variants of a detached house, described by the building's heat losses because of the transmission (Q_{tr}). These variants are: a standard building B9 (Q_{tr} = 9 kW), a low energy building B6 (Q_{tr} = 6 kW) and an ultra-low energy building B3 (Q_{tr} = 3 kW). The standard building is the one whose heat transfer coefficients meet the minimum requirements for single-family houses in Poland and the usable energy index for heating (*EU*) is c.a. 100 kWh/(m^2·year). The low-energy building is the one in which the heat transfer coefficients guarantee that the *EU* index is smaller than 40 kWh/(m^2·year). The ultra-low-energy building requires the *EU* index at about 15 kWh/(m^2·year). Table 1 summarizes the buildings' heat losses by transmission (Q_{tr}) and the usable energy indexes (*EU*) for the three variants of the building energy classes in different climate conditions.

 DOI: 10.1051/e3sconf/20172200086

Table 1. The static heat losses and the usable energy index.

	Gdańsk		Wrocław		Suwałki	
Q_{tr} kW	EU_{NW} kWh/(m²year)	EU_{MW} kWh/(m² year)	EU_{NW} kWh/(m² year)	EU_{MW} kWh/(m² year)	EU_{NW} kWh/(m² year)	EU_{MW} kWh/(m² year)
9	88	61	97	67	122	85
6	61	36	68	40	87	51
3	36	13	41	15	53	19

3.2 Mechanical ventilation system

The air volume for the mechanical ventilation system is 250 m³/h. Figure 3 shows the configuration of the air handling unit (AHU). The AHU is assumed in the cross-counter flow heat exchanger with the 80% efficiency. The frost protection of the heat exchanger is an electric preheater, operated when the temperature of the exhaust air (measured by the sensor T_e) drops below 5°C and at the same time outside air temperature is below 0°C. The supply air temperature is controlled by an electric heater to a temperature T_s = 16°C. The heaters work with the fluent regulation of power. The electrical power of the fans is 160W.

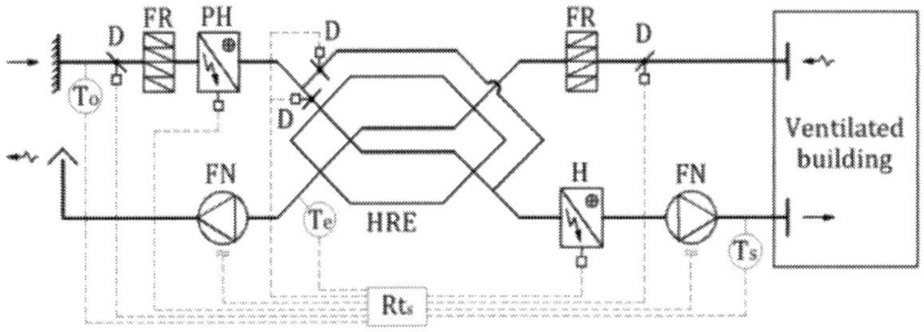

Fig. 3. AHU configuration (D – damper, FR – filter, FN – fan, HRE – heat recovery exchanger, PH – preheater, H – heater, T_o – outside air temperature sensor, T_s – supply air temperature sensor, T_e – exhaust air temperature sensor, Rt_s – supply air temperature regulator).

3.3 Heating system

The calculations were made on the assumption that the building is heated by an air-to-water heat pump. Selection of the heat pump in each of the analyzed variations was based on the thermal characteristics of the building equipped with the mechanical ventilation system with a heat recovery exchanger. In each of the climatic zones and for each building energy class, the power of the pump was chosen so that the bivalent point was -10°C. Table 2 shows the heat pomp COP data for the heating water supply temperature 35°C and 45°C. The analysis assumes that the building is equipped with an underfloor heating with the parameters of heating agent 40/30°C. The minimum outdoor air temperature required to the heat pump is T_a = -15°C.

Table 2. The COP of the heat pump for the heating water supply temperature 35°C and 45°C.

T_a, °C	**-15**	-7	**2**	7	**10**	**12**	**20**
COP for 35°C	1.89	2.62	3.25	4.29	4.40	4.63	5.29
COP for 45°C	1.50	2.10	2.80	3.10	3.40	3.50	4.10

3.4 Methods of calculations

The simulations in this article were made in hourly steps based on the climate data described in point 2. The total final energy demand for the buildings with the natural ventilation as well as with the mechanical ventilation with the heat recovery were calculated. The total final energy demand for the building with the natural ventilation comes from the electricity required to run the heat pump for the building's heating ($Q_{HP,NW}$). The total final energy demand for the building with the mechanical ventilation comes from the electricity required to run the heat pump for the building's heating ($Q_{HP,MW}$) and the electricity required for the central air supply (the power for the fans and the electric preheater - $Q_{EL,MW}$). The electricity demand for the operation of the air-to-water heat pump for heating the building was determined based on the seasonal coefficient of the performance (SCOP). The SCOP of the air-to-water heat pump was calculated taking into account the partial load correction of the equipment and the operation in the stand-by mode. The steps of the calculations were described in the Table 3. For the building with the hybrid ventilation, the decision in which ventilation system (*NW* or *MW*) it should be operating was based on the comparison of the total final energy consumption for both solutions ($Q_{K,NW}$) and ($Q_{K,HW}$) in every hour. Next, the share of the working time of the ventilation system with the heat recovery (τ_{MW}) and the savings results from the use of the hybrid ventilation, both the quantitative (ΔQ_{HW}) and the percentage (β_{HW}), were calculated.

Table 3. The steps of the analysis.

The ventilation system	Building with the mechanical ventilation	Building with the hybrid ventilation	
	MW	*NW*	*MW*
The total final energy consumption of the building calculated for each hour in year (1–8760 h)	$Q_{K,MW} =$ $= Q_{HP,MW} + Q_{EL,M.W.}$	$Q_{HP,NW}$	$Q_{HP,MW} + Q_{EL,MW}$
		$Q_{K,HW} = min(Q_{HP,NW}, Q_{EL,MW}+Q_{HP,MW})$	
The savings resulting from the use of the hybrid ventilation	-	$\Delta Q_{HW}=\Sigma_i(Q_{HP,MW}+Q_{EL,MW}-Q_{HP,NW})$ *where: i – the hours in year when* $Q_{HP,NW}<(Q_{EL,MW}+Q_{HP,MW})$ $\beta_{H.W.}=\Delta Q_{H.W}/(Q_{HP,M.W.}+Q_{EL,M.W})$	

4 Results

The results of the analysis for all the variants are summarized in Table 4. The following values were shown: the total final energy demand during the year for the building with the mechanical ventilation with the heat recovery ($Q_{HP,MW}$ and $Q_{EL,MW}$), the savings results from the work of the hybrid ventilation system ($\Delta Q_{H.W.}$ and $\beta_{H.W.}$) and the time of the work of the mechanical ventilation per year in percentage ($\tau_{M.W.}$).

For the standard energy class (B9) building, the annual savings results due to the hybrid ventilation ($\beta_{H.W.}$) are between 9% and 10%. Quantitatively ($\Delta Q_{H.W.}$), the highest savings were observed for Suwałki, 950 kWh per year. Because in this location the total final energy demand for the heat pump and the ventilation unit is the highest, the share of the energy saving is the smallest. For the low energy class (B6) building, the annual energy savings (β_{HW}) range from 13% to 15%. These savings are quantitatively slightly higher than in the standard building (B9). For the ultra-low energy class (B3) building, the annual energy savings (β_{HW}) are between 18% and 28%. Although the amount of the energy savings is lower than in other energy standards, the savings potential is the highest in these buildings.

Table 4. The results of the analysis.

The climate data	$Q_{HP,MW}$	$Q_{EL,MW}$	ΔQ_{HW}	τ_{MW}	β_{HW}
	kWh/year	kWh/year	kWh/year	%	%
Building B9 – Suwałki	7496	2983	950	38%	9%
Building B9 – Wrocław	5934	2172	677	47%	8%
Building B9 – Gdańsk	5210	1805	727	46%	10%
Building B6 – Suwałki	4591	2983	961	37%	13%
Building B6 – Wrocław	3572	2172	708	46%	12%
Building B6 – Gdańsk	3116	1805	762	45%	15%
Building B3 Suwałki	1773	2983	852	44%	18%
Building B3 - Wrocław	1341	2172	791	45%	23%
Building B3 - Gdańsk	1140	1805	817	42%	28%

In addition to the annual total final energy demand, it is interesting to discuss the monthly statistics of the potential savings in the total final energy demand due to using the hybrid ventilation instead of the full-year mechanical ventilation. These results are summarized in Figure 4.

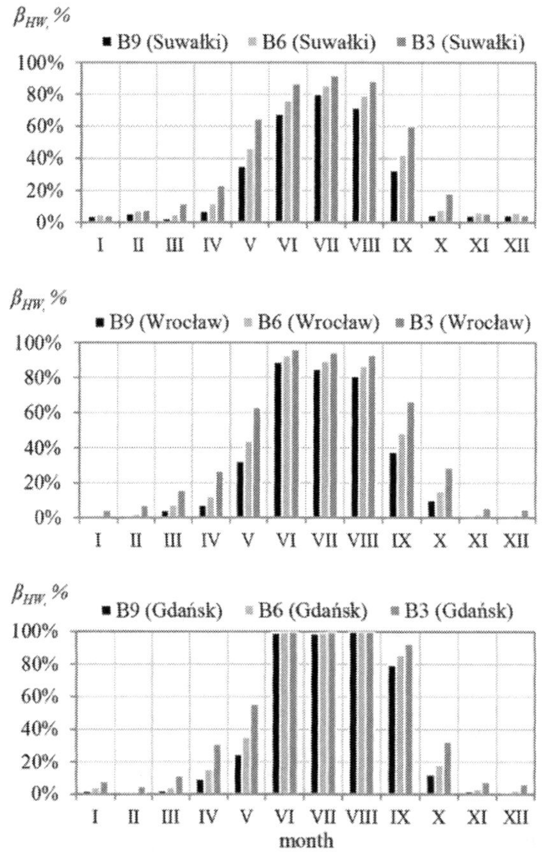

Fig. 4. The reduction in the finally energy demand in the building as a result of the hybrid ventilation.

In the winter months, the savings resulting (β_{HW}) from the periodical shutdown of the mechanical ventilation are low. The percentages of these savings are the highest in Suwałki (up to 4%), which results from the energy consumption in the protection system for the heat recovery exchanger before the frost. It is also clear that these savings are even higher when the building has a better energy class. During the transition periods III-V and IX-X the savings resulting (β_{HW}) from the switching between the mechanical and the natural ventilation are already significant. For the ultra-low-energy building (B3) in the months V and IX, the savings (β_{HW}) up to 60% comparing to the system with the continuous mechanical ventilation were found. In summer, the natural ventilation is a better system. In Gdansk, it would be advisable to completely shut off the mechanical ventilation system. In Wroclaw and in Suwałki the natural ventilation allows for a significant reduction in energy demand (from 65% to 95%), although the ventilation unit work is periodically profitable.

5 Conclusions

Single-family buildings represent a large share of the residential market in Poland and they are impacting significantly the energy consumption in Poland. The climate in Poland varies in the different regions and the systems which are sensitive to the change of the ambient air temperature require the individual analysis. An example of such a system is the mechanical ventilation which periodically can be more costly than the natural ventilation. A detailed analysis of the energy effectiveness of using the hybrid ventilation (i.e. the periodic replacement of the mechanical ventilation by the natural ventilation) can only be carried out together with an analysis of the heating system and for the specific location of the building.

In this study, the analysis of the ventilation system operation was carried out in cooperation with an air-to-water heat pump. Both systems are very sensitive to the ambient air temperature data and in this case the impact of climate data on the system efficiency is particularly important. The significant potential of the energy savings in the total final energy demand (for the building's heating system) has been demonstrated for all the analyzed climates. It should be noted that the tendencies in the construction industry, enforced by the legal regulations, are that the buildings will have better and better energy classes, and the air-to-water heat pumps will become more and more popular in Poland. The presented analysis showed that the lower the demand for the usable energy for heating, the greater the effect of introducing the hybrid ventilation.

From a practical point of view, the analytical and research work are needed to develop both the technical solutions and the control algorithms for the hybrid ventilation systems in the single-family buildings.

Acknowledgments

This work is supported by funds from the Ministry of Higher Education for research and development works of young scientists and doctoral students. Project number: 0402/0098/16.

References

1. M. Kostka, M. Szulgowska-Zgrzywa, Rynek Instalacyjny, **3** (2016)
2. M. Kim, J. Hwang, Energ. Buildings **41**, 579–586 (2009)
3. D. Jreijiry, A. Husaunndee, C. Inard, Sol. Energy **81**, 227–239 (2007)
4. K. Niachou, S. Hassid, M. Santamouris, I. Livanda, Energ. Buildings **37**, 503–513 (2005)

5. RESHYVENT, Cluster Project on Demand Controlled Hybrid Ventilation in Residential Buildings with Specific Emphasis of the Integration of Renewables, European Joule Project, Contract No. ENK6-CT2001-00533 (2002)
6. http://stat.gov.pl/en/topics/environment-energy/energy/energy-efficiency-in-poland-in-years-2004-2014,5,12.html [access: 21.03.2017]
7. http://mib.gov.pl/2-Wskazniki_emisji_wartosci_opalowe_paliwa.htm# [access: 01.2017]

Removal of copper ions from aqueous solutions by means of micellar-enhanced ultrafiltration

Izabela Kowalska[1], *Aleksandra* Klimonda[1,*]

[1]Wroclaw University of Science and Technology, Faculty of Environmental Engineering, Wybrzeże S. Wyspiańskiego 27, 50-370 Wrocław

Abstract. The aim of the study was to assess the usefulness of micellar-enhanced ultrafiltration (MEUF) for removal of copper ions from water solutions in comparison with classic ultrafiltration process. The tests were conducted in a semi-pilot membrane installation with the use of ultrafiltration module KOCH/ROMICON® at a transmembrane pressure of 0.05 MPa. The effect of concentration of copper ions on ultrafiltration process efficiency was investigated. The second part of the tests concerned the removal of copper ions by MEUF under wide range of anionic surfactant concentration (0.25, 1, and 5 CMC (critical micelle concentration)). Concentration of copper ions in model solutions was equal to 5, 20, and 50 mg Cu/L. Furthermore, the effect of surfactant leakage to the permeate side during filtration was evaluated. Conducted experiments confirmed effectiveness of MEUF in copper ions removal. For the highest copper concentration in the feed (i.e. 50 mg/L), the average concentration of copper ions in the permeate ranged from 1.2–4.7 mg Cu/L depending on surfactant concentration. During filtration experiments, UF module exhibited stable transport properties for model solutions containing copper. For the highest concentration of metal, the decrease of permeate flux did not exceed 11% after 60 minutes of filtration. In the presence of the surfactant, a slight deterioration of transport properties was observed.

1 Introduction

Heavy metal compounds are commonly present in industrial wastewater streams. Copper contamination of effluent is typical for metal cleaning and electroplating, paints and pigments, mining, smelting and petroleum industries [1]. Due to the harmful health and environmental effects, removal of copper from industrial wastewater is required.

Treatment of solutions containing copper may be conducted by means of various methods as: chemical precipitation [2, 3], ion-exchange [4, 5], adsorption [5–7] and membrane techniques (reverse osmosis, electrodialysis) [8–10]. However, some of these processes may not be cost effective [10]. The main inconvenience of chemical precipitation is the formation of copper hydroxide precipitate and further, necessity of this compound disposal. Ion-exchange and adsorption are preferred methods for the purification of diluted effluents due to limited adsorption/ion exchange capacity.

* Corresponding author: aleksandra.klimoda@pwr.edu.pl

Micellar enhanced ultrafiltration (MEUF) has been widely tested for treating mixtures containing heavy metals. MEUF is a pressure-driven membrane process combined with an addition of surface active agent in concentration exceeding critical micellar concentration (CMC). An important surfactants feature is ability to forming micelles which size is generally greater than UF membrane pore size. In the MEUF process, the mechanism of solubilisation of organic/mineral compounds in micelle cores as well as bounding ions on micelle surface is being used. As a result of surfactant addition to the mixture containing copper, the metal ions embedded into surfactant aggregate do not pass through the membrane.

MEUF has been widely tested in removal of organic [11], cationic [12] and anionic [13] compounds from water solutions, however the main field of surfactant aided ultrafiltration experiments is removal of heavy metal ions. Anionic surfactant sodium dodecyl sulphate (SDS, CMC ~ 2250 mg/L, molecular weight of 288.37 Da) is commonly used for this purpose. Literature data report high efficiency of MEUF process with SDS [14, 15] in removal of heavy metal ions. In this study, due to the significantly smaller CMC (800 mg/L) and the greater value of molecular weight (348.48 Da), sodium dodecylbenzenesulfonate (SDBS) has been chosen for the experiments.

The experimental investigation was focused on evaluation of suitability of MEUF in copper ions removal comparing to classic UF process. As an important limitation of micellar enhanced ultrafiltration, the surfactant penetration to the permeate side was studied.

2 Experimental

In the first stage of research procedures, membrane permeability was determined via measuring distilled water volume flux at various transmembrane pressure (TMP). Next, copper ions removal process by means of classic UF was performed. In order to enhance rejection of copper ions, sodium dodecylbenzenesulfonate was added to the feed solution in wide range of concentration in the next stage of the tests. In order to verify the proneness of the membrane to fouling, permeate flux was measured during the tests.

Model solutions containing 5, 20 and 50 mg Cu/L were prepared from distilled water and copper sulphate ($CuSO_4$). The concentration of copper in the permeate was determined using bicinchoninate method (DR/2000 spectrophotometer, Hach).

For MEUF experiments anionic surfactant sodium dodecylbenzenesulfonate has been employed. SDBS critical micelle concentration was determined via surface tension measurements using Lauda tensiometr and amounted to 800 mg/L at 22°C. The molecular weight of SDBS monomer was equal to 350 Da. The concentration of SDBS in the feed solutions amounted to 0.25, 1 and 3 CMC. Potentiometric titrator 785 DMP Titrino (Metrohm) has been used for measurement of surfactant concentration.

The ultrafiltration processes were conducted using capillary Koch/Romicon module with polysulfone membranes of the cut-off of 5 kDa (Table 1). Schematic diagram of the membrane installation was presented in previous study [16]. Membrane filtration was performed under the pressure of 0.05 MPa.

Table 1. Characteristics of UF module.

Property	KOCH/ROMICON 1" Hollow Fiber Module
Membrane cut-off, kDa	5
Membrane polymer	polysulfone
Membrane surface area, m^2	0.09
Module diameter, mm	25.4
Module length, mm	457
Number of capillaries	66
Inner diameter of capillaries, mm	1.1

DOI: 10.1051/e3sconf/20172200087

Ultrafiltration was performed in an open loop system where the retentate from the membrane was recycled back to the feed tank. The feed was pumped to the membrane module by the circulation pump.

3 Result and discussion

3.1 Ultrafiltration

Retention coefficient of copper ions and relative flux obtained over the UF experiments are plotted in Fig. 1. As can be seen, during the 60-min filtration cycle, the module exhibited stable permeability – J/J_0 values were in the range of 1.03–1.08, 1.08–1.11 and 0.89–0.92 in experiments with 5, 20 and 50 mg Cu/L, respectively.

Depending on copper ions concentration in the feed, the module exhibited various separation efficiency. The highest retention coefficient (55–74%) was achieved for solution containing 5 mg Cu/L, while the lowest separation of contaminant (from 14 to 23%) was received in process with solution of 20 mg Cu/L.

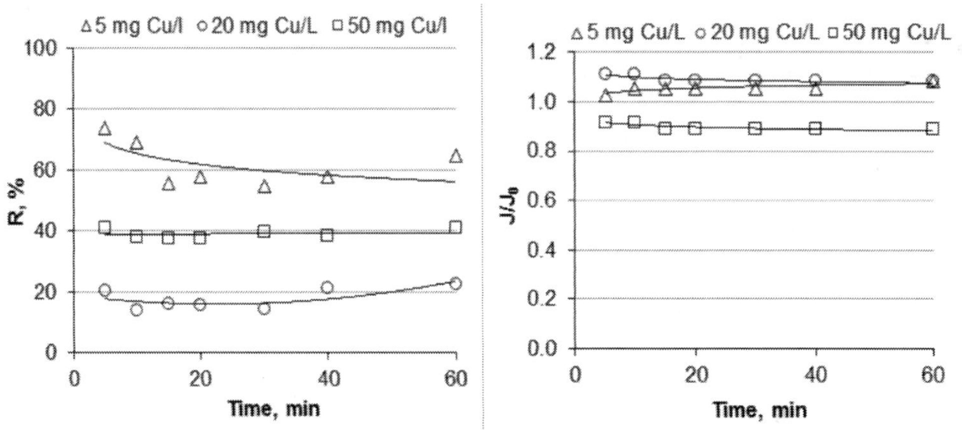

Fig. 1. Retention coefficient (R) and relative flux (J/J_0) during the membrane filtration (transmembrane pressure = 0.05 MPa).

Obtained concentration of copper ions in the permeate was in a range of 1.3–2.3, 15.5–17.1 and 29.5–31.3 mg Cu/L for feed solutions of 5, 20 and 50 mg Cu/L, respectively. According to Bernat et al., [17] separation of copper by ultrafiltration membrane could be attributed to forming concentration polarisation layer near the membrane surface by copper hydroxides. Created active layer may act as an additional membrane, thus part of the copper do not pass through the membrane.

3.2 Surfactant enhanced ultrafiltration

Fig. 2 presents concentration of copper ions in averaged permeate samples collected during the 60-min MEUF process. The presence of anionic surfactant significantly improved membrane selectivity. This phenomenon is ascribed to a mechanism of electrical adsorption of positive charged copper ions on negative charged SDBS micelle surface. Hence, a sieve separation of SDBS aggregates (and bounded copper ions) is possible since micelle size is substantially greater than membrane pore.

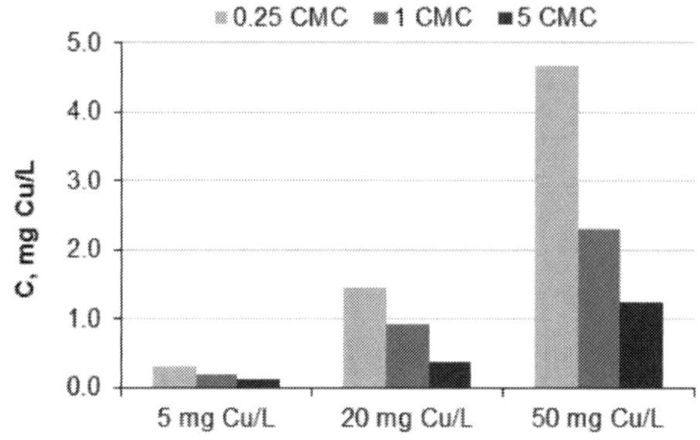

Fig. 2. Concentration of copper in the permeate (Transmembrane pressure = 0.05 MPa).

All conducted experiments allowed to achieve the copper retention coefficient higher than 90%. As can be seen, an addition of small quantity of SDBS (0.25 CMC) resulted in reducing the copper concentration from 5, 20 and 50 mg Cu/L to 0.29, 1.46 and 4.68 mg Cu/L, respectively. An increase of surfactant dosage in the mixture led to further improving of the membrane selectivity – the highest separation of Cu was obtained in test with SDS concentration of 5 CMC (0.11, 0.36 and 1.24 for feed of 5, 10 and 50 mg Cu/L, respectively). It should be also mentioned that the initial concentration of metal ions in the feed affects on the removal efficiency. With an increase of the concentration of copper ions, a decrease of retention coefficient was observed. Treatment of mixture of 50 mg Cu/L comparing to experiments with 5 and 20 mg Cu/L led to significant increase of copper concentration in the permeate. According to Huang et al. [18], at high metal concentration, the micelles became more and more saturated and less micelle surface is available for electrostatic adsorption at higher metal concentrations.

Copper retention coefficients obtained during experiments corresponds to data reported in tests with the application of SDS. Huang et al. [18] achieved copper ions separation on polyethersulfone membrane (feed of 50 mg Cu/L, 1 CMC SDS, pH = 7) amounted to 100%. Li et al. [19] have reported copper removal around 90% in MEUF process with the use of SDS in concentration of 1 CMC and hydrophilic membrane with cut-off 10 kDa; however when the lower surfactant concentration was applied, obtained retention coefficients were not satisfying.

As a serious limitation of the MEUF full-scale application, surfactant monomers penetration to the permeate side as well as membrane permeability deterioration due to the surfactant fouling has been investigated. Fig. 3 shows SDBS retention coefficient and relative flux obtained during the MEUF tests. The analysis of data obtained showed that presence of anionic surfactant in the feed affected on membrane hydraulic properties. Permeate flux decreased about 35% comparing with distilled water flux during the filtration of solution containing SDBS in concentration of 1 CMC. Surfactant application in a dose of 0.25 CMC and 5 CMC resulted in similar J/J_0 values, i.e. about 0.75–0.85, however the causes of flux decease may be different – when surfactant is in the form of monomers, the main reason of membrane properties change is monomer adsorption on membrane pores, while in solutions of 5 CMC, the permeability deterioration may be attributed to polarization concentration layer created by SDBS micelles near the membrane surface. It should be also mentioned that permeate flux was stable during the MEUF process. Corresponding to Mondal [20], the phenomenon of blocking membrane pores by SDBS particles is complete in the initial stages of membrane filtration.

 DOI: 10.1051/e3sconf/20172200087

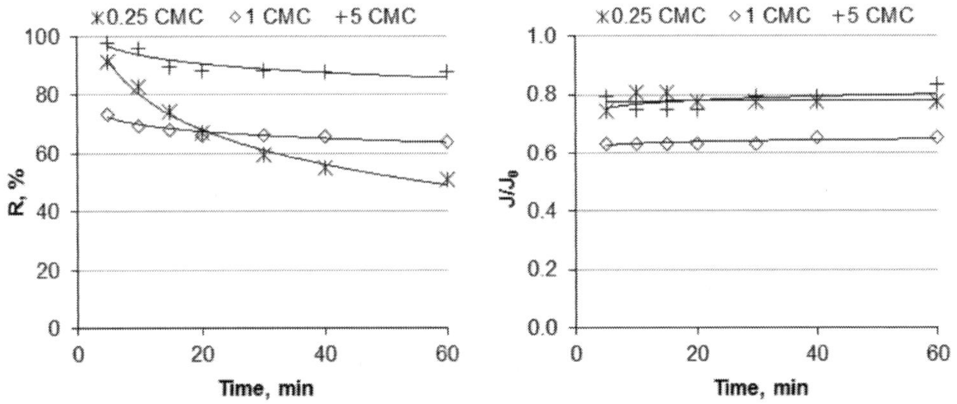

Fig. 3. SDBS retention coefficient (R) and relative flux (J/J_0) during the MEUF (Transmembrane pressure = 0.05 MPa).

Ke et al. [21] reported twofold decrease of permeate flux comparing to distilled water flux at initial stage of cadmium treatment process with SDS (100 mg Cd/l, SDS 1 CMC, polysulfone hollow fiber membrane, cut-off 6 kDa). During the 30-min filtration process, further deterioration of permeability was observed. In the same experimental work, increasing SDS concentration from 1 CMC to 4 CMC value resulted in reduced permeate flux from 3.5 to 2.5 x 10^{-8} $m^3/m^2{\cdot}s{\cdot}Pa$.

It should be noted that the highest SDBS retention (88–98%) was reached when SDBS concentration in the feed amounted to 5 CMC value, i.e. when the surfactant occurs in the form of micelles. Although at the beginning of membrane filtration module exhibited good selectivity of SDBS, during the process decrease of retention coefficient was observed in all test. Surfactant concentration in the permeate achieves a maximum value around the CMC value, which is in accordance with literature data [22, 23]. Particularly pronounced drop, i.e. from 91 to 51% and from 74 to 64% was noticed for solution of 1 and 5 CMC, respectively.

4 Conclusions

- The results obtained proved the surfactant aided ultrafiltration exhibited much better copper ions retention in comparison with classic UF process.
- Micellar-enhanced ultrafiltration with the use of anionic surfactant sodium dodecylbenzenesulfonate is an effective method of copper ions removal from water solutions. Copper rejection in all experiments was very satisfactory and exceeded 90%. The results obtained corresponds to data reported for process enhanced by frequently used anionic surfactant SDS.
- Increase in surfactant concentration affects the rejection improvement of copper ions. In tests with the feed of 50 mg Cu/L, enhancement SDBS dosage from 0.25 CMC to 5 CMC resulted in drop of copper concentration in permeate from 4.68 to 1.24 mg Cu/L.
- During the MEUF experiments surfactant monomer leakage to the permeate side was observed. In the course of membrane filtration this phenomenon has been intensified. The most significant drop of surfactant retention, i.e. from 91% at the beginning of the process to 51% after the 60 min of filtration was observed in experiments with SDBS concentration below the CMC value.

Acknowledgments

This work was supported by a grant (No. 0401/0064/16) from the Department of Environmental Engineering, Wrocław University of Science and Technology.

References

1. M.A. Barakat, Arabian J. Chem. **4** (2011)
2. F. Fu, Q. Wang, J. Environ. Manage. **92**, (2011)
3. L. Ye, L. Chai, Q. Li, X. Yan, Q. Wang, H. Liu, RSC Adv. **6** (2016)
4. M. Ravathi, M. Saravanan, A. Chiya, M. Velan, Clean-Soil Air Water **40**, 1 (2012)
5. M. Bilal, J.A. Shah, T. Ashfaq, S. Mubashar, H. Gardazi, A.A. Tahir, A. Pervez, H. Haroon, Q. Mahmood, J. Hazard. Mater. **263** (2013)
6. N. Ferrah, O. Adberrahim, M.A. Didi, D. Villemin, Desalination **269** (2011)
7. H. Nadaroglu, E. Kalkan, N. Dmir, Desalination **251** (2010)
8. X. Chai, G. Chen, P-L. Yue, Y. Mi, J. Membr. Sci. **123** (1997)
9. M. Arbabi, N. Golshani, Int. J. Epidemiol. **3**, 3 (2016)
10. M. Thomas, B. Bialecka, D. Zdebik, J. Ecol. Eng. **37** (2014)
11. S.V. Jadhav, K.V. Marathe, Can. J. Chem. Eng. **91** (2013)
12. P. Yenphan, A. Chanachai, R. Jiraratananon, Desalination **253** (2010)
13. R. Camarillo, I. Asenico, J. Rincón, Desalination Water Treat. **6** (2009)
14. J.-H. Huang, G.-M. Zeng, Y.Y Fang, Y.-H. Qu, X. Li, J. Membr. Sci. **326**, 2 (2009)
15. E. Samper, M. Rodriguez, M.A. De la Rubia, D. Prats, Sep. Purif. Technol. **65** (2009)
16. I. Kowalska, Sep. Purif. Technol. **124** (2014)
17. X. Bernat, I. Sánchez, F. Stüber, C. Bengoa, A. Fortuny, A. Fabregat, J. Font, Proceedings of European Congress of Chemical Engineering (ECCE-6), Copenhagen, 16–20 September 2007
18. J. Huang, F. Yuan, G. Zeng, X. Li, Y. Gu, L. Shi, W. Liu, Y. Shi, Chemosphere **173** (2017)
19. C-W. Li, C-K. Liu, W-S. Yen, Chemosphere **63** (2006)
20. S.D.S. Mondal, *Micellar Enhanced Ultrafiltration, Fundamentals & Applications* (Taylor & Francis Group, Boca Raton, 2012)
21. X. Ke, Z. Guang-Ming, H. Jin-hiu, W. Jiao-Yi, F. Yao-Yao, H. Guohe, L. Jianbing, X. Beidou, L. Hongliang, Colloids Surf., A, Physicochem. Eng. Aspects **294** (2007)
22. J. Landaburu-Aguirre, *Micellar Enhanced Ultrafiltration for the Removal of Heavy Meatals from Phosphorous-Rich wastewaters,* http://jultika.oulu.fi/files/isbn9789514299117.pdf
23. H. Gecol, E. Ergican, A. Fuchs, J. Membr. Sci. **241** (2004)

 DOI: 10.1051/e3sconf/20172200088

Influence of the single EGR valve usability on development of the charge directed to individual cylinders of an internal combustion engine

Konrad Krakowian[1,*], *Andrzej* Kaźmierczak[1], *Aleksander* Górniak[1], and *Radosław* Wróbel[1]

[1]Wrocław University of Science and Technology, Faculty of Mechanical Engineering, ul. Braci Gierymskich 164, Wrocław 52-640, Poland

Abstract. Exhaust gas recirculation systems (EGR), aside to a catalytic converters, are nowadays widely used in piston internal combustion engines to reduce nitrogen oxides (NO_x) in the exhaust gas. They are characterized in that a portion of exhaust gases from the exhaust manifold is recirculated (via a condenser), and directed to a particular valve. The valve, depending on the current engine load and speed, doses the appropriate amount of exhaust gas into the exhaust manifold. Moreover, its location has a significant impact on the diverse formation of nitrogen oxides and fumes smokiness from the individual cylinders of the engine, which is a result of uneven propagation of exhaust gas into the channels of the intake manifold. This article contains the results of numerical characterized charges formed in symmetrical intake manifold with a centrally-placed EGR valve. Simulations were performed for the original intake system derived from the two-liter, turbocharged VW diesel engine.

1 Introduction

One of the most difficult problems posed to engineers designing internal combustion engines is to meet stringent emission standards set by the environmentalists. This is due to the fact that road transport has a dominant share in toxic emissions, in particular emission of nitrogen oxides, carbon oxides and hydrocarbons. In order to reduce pollution a further restrictions specifying the permissible limits on the level of toxicity of exhaust gases are continuously introduced. In Europe, they are represented by so-called. EURO standards. Fig. 1 shows how a subsequent EURO standards contributed to a reduction in emissions of mainly NO_x and PM. It has necessitated the producers to intervene in engine design, combustion process as well as utilization of an additional equipment for exhaust gas treatment.

* Corresponding author: konrad.krakowian@pwr.edu.pl, aleksander.gorniak@pwr.edu.pl, radoslaw.wrobel@pwr.edu.pl

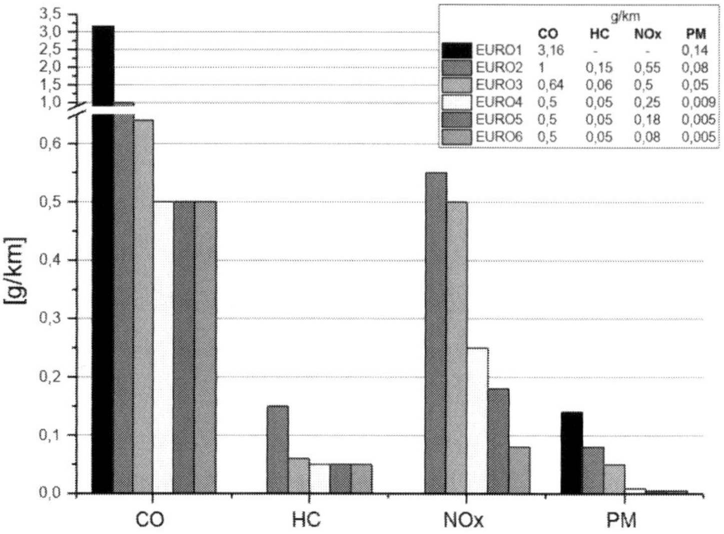

Fig. 1. Changes of allowable exhaust emission limits of subsequent EURO standards.

Designing new engines must involve a reduction of toxic emissions and fuel consumption. Currently in SI engines the reduction of exhaust emissions is achieved by the use of [1]:

- innovative air supply systems,
- innovative shape of the combustion chambers,
- exhaust gas recirculation (EGR)
- high-pressure fuel injection system (Common Rail)
- selective catalytic reduction systems.
- downsizing,
- electronic control systems [3]

One of these actions, which is aimed towards reduction of nitrogen oxides, is the use of a gas recirculation system. Nowadays an external EGR systems are most commonly used. They operation mainly consists on supplying a precisely specify portion of the fumes to the inlet manifold. Such a process is beneficial in terms of chemical reaction taking place in the cylinder due to the fact that the exhaust gas as the inert gas does not participate in the combustion process also have higher specific heat than air thereby significantly reduce the combustion temperature. In consequence less of oxygen reacts with nitrogen which results in limitation of NO_x emission. Additionally the exhaust gases recirculation promotes evaporation of the fuel and the oxidation of unburned hydrocarbons [2]. In the case of the use of EGR in a multi-cylinder internal combustion engine, the most important factor is to supply charge for individual cylinders of the same composition in order to achieve a similar combustion process. The decisive influence on it has the shape of the intake manifold and the position of the EGR valve. This article presents the results of numerical computation of the charges formed in a symmetrical intake manifold with a centrally-placed EGR valve.

2 Characterization of numerical investigations

The investigation was performed on a intake manifold with the EGR which was a member of turbocharged, two-liter, sixteen-valve diesel engine with the designation 2.0 TDI BKD of Volkswagen Group. The default drive unit was equipped with a single EGR valve at the

 DOI: 10.1051/e3sconf/20172200088

inlet of the intake manifold in such a way that the direction of flow of the exhaust valve was perpendicular to the main air supply channel.

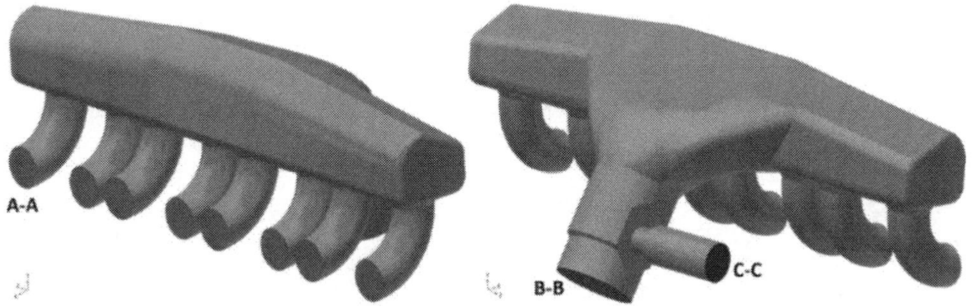

Fig. 2. Model of the intake manifold internal volume from the EGR valve designed for the flow rate numerical analysis.

Figure 2 shows the 3D model of the intake manifolds internal volume and the EGR valve marked cross sections: *A-A* manifolds outlet ducts, *B-B* of the air inlet, the *C-C* the gas inlet to the EGR valve. In order to inspect the creation of the charge and its distribution in the intake manifold a flow simulations were performed for each pair of channels feeding a single cylinder. Simulations were performed for two cases of extreme speeds and two positions of EGR valve ranging 1mm and 4mm, characterizing the degree of the exhaust gas flow. The process of the charge development and its distribution within the inlet manifold was verified by serous of CFD flow simulation, performed for each pair of ducts feeding single cylinder. Simulations were executed for two cases of extreme speeds and two positions of EGR valve – i.e. 1 mm and 4 mm, characterizing the degree of the exhaust gas flow.

The analysis of the charge development in the intake manifold with the involvement of exhaust gas from the EGR valve was performed with aid of the finite volume method using ANSYS FLUENT 15.0. All the analyzed cases were dissolved in 300–350 iterations of the steady state flow with use of the k-epsilon turbulence model. The boundary conditions indicated flow rate of the inlet manifold and the inlet gas of the EGR valve. The individual speed values $v_s$are calculated from equations (1) and (2).

$$Q = \frac{V_s \cdot n \cdot \eta_V}{2} \left[\frac{m^3}{s}\right] \quad (1)$$

where:

$Q \left[\frac{m^3}{s}\right]$ - volumetric flow rate

$V_s\ [m^3]$ - swept volume,

$n \left[\frac{obr}{s}\right]$ - rotation speed

$\eta_v\ [-]$ - coefficient of the cylinder filling

$$v = \frac{4 \cdot Q}{\pi \cdot d^2} \left[\frac{m}{s}\right] \quad (2)$$

where:

$v \left[\frac{m}{s}\right]$ - gas flow rate

$d\ [m]$ - the diameter of the inlet duct

Values for extreme speeds achieved for air (v_{air}) and exhaust gases (v_{exh}) corresponding to the extreme values of revolution speeds (i.e. 800 rpm and 3000 rpm), for the inner diameter of intake manifold of $d = 0.05m$ are $v_{air(min)} = 2$ m/s and $v_{air(min)} = 7.5$ m/s in the case of air flow, whilst the extreme speeds of exhaust gases flow are $v_{exh(min)} = 8.7$ m/s and $v_{exh(min)} = 32$ m/s. Additional parameters determining the virtual flow conditions was the temperature of individual gases. For the air ambient temperature was assumed to $T_p = 300$ K while the exhaust gases temperature was $T_s = 450$ K. The composition of charge formed in the manifold was dependent on the volume fraction of air and exhaust gases. The parameters characterizing the air were imported from Ansys Fluent, while the exhaust gases were defined as the percentage of nitrogen – 72%, carbon dioxide – 14%, and water vapor – 14%. The characteristics of the gas parameters required for the calculations are shown in Table 1.

Table 1. Characterization of gas used to calculate.

Parameter	Value	
	Air	Exhaust gases
Density	$1.225\frac{kg}{m^3}$	$1.366\frac{kg}{m^3}$
Specific heat	$1006.43\frac{J}{kg\cdot K}$	$1146.46\frac{J}{kg\cdot K}$
Thermal conductivity	$0.0242\frac{W}{m\cdot K}$	$0.0213\frac{W}{m\cdot K}$
Viscosity	$1.7894e^{-05}\frac{kg}{m\cdot s}$	$1.4834e^{-05}\frac{kg}{m\cdot s}$
Molar mass	$28.966\frac{kg}{kmol}$	$28.84\frac{kg}{kmol}$

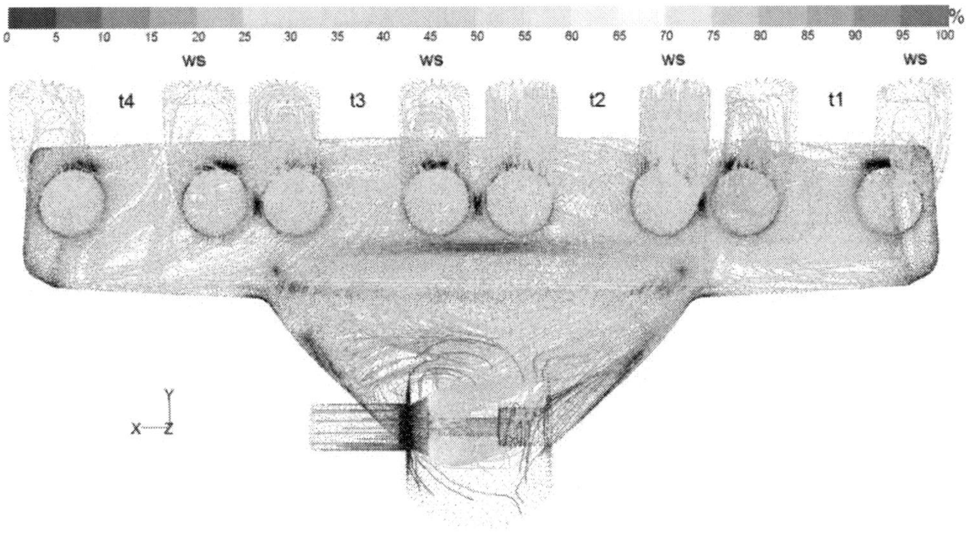

Fig. 3. Charge flow through the intake manifold with streamlines corresponding to percentage share of exhaust gas for channels supplying a second cylinder, for $v_{air(min)} = 7.5$ m/s, $v_{exh(min)} = 32$ m/s, the opening of the EGR valve 1 mm.

Figure 3 and Figure 4 shows an example of the results in the form of flow streamlines representing percentage share of exhaust gases flowing through the intake manifold channels supplying the second cylinder with following boundary conditions:

DOI: 10.1051/e3sconf/20172200088

v_{pmax} = 7.5 m/s, v_{smax} = 32 m/s, and the opening of the EGR valve of 1 mm and 4 mm. Notation *t1, t2, t3* and *t4* corresponds to charge passing into the cylinder through the open duck pair at given time *t* for the sequence of first, second, third and fourth cylinder. Moreover, the denotation *ws* indicates the outlet channels of the manifold associated with swirl vane of the engine head.

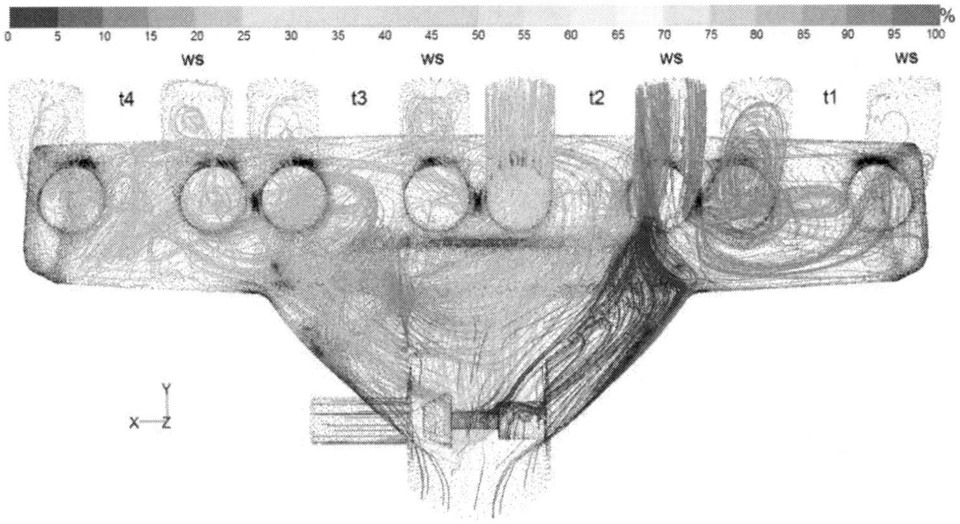

Fig. 4. Charge flow through the intake manifold with streamlines corresponding to percentage share of exhaust gas for channels supplying a second cylinder, for $v_{air(min)}$ = 7.5 m/s, $v_{exh(min)}$ = 32 m/s, the opening of the EGR valve 4 mm.

For each of these cases the charge formed in the intake manifold is not uniform and its degree depends on the permeability of the EGR valve. The results of calculations for the case of Figure 3 illustrate that the lift of the valve disc at a height of 1mm is very good in terms of mixing of the exhaust gases with flowing air. This is due to the increase in the speed of the exhaust gas flowing out of the EGR valve, thereby generating high charge motion above valve plate. The speed increase caused by the reduction of the area of movement, which is the side surface of the truncated cone formed by the valve head and valve-seat. For the case of Figure 4, the lift of the valve equal to 4 mm, there is a significant deterioration in the mixing of the two gases. The reason for the decrease in speed of the exhaust gas flowing out of the increase in the surface area of a side of a truncated cone, which for this case is greater than the surface of gas flow through supply duck. With the decrease in the exhaust gas velocity there is a deviation of the air stream due to perpendicularly directed flow of exhaust gases as well as a small degree of mixing.

Figure 5 shows a cumulative summary of numerical computation of the charge development within a given ducts of the intake manifold under statical condition in terms of percentage distribution of the exhaust gases. Chases from the subpoints *a* and *b* were computed for 1mm lift of the EGR valve, whereas the *c* and *d* for 4 mm lift of the EGR valve.

 DOI: 10.1051/e3sconf/20172200088

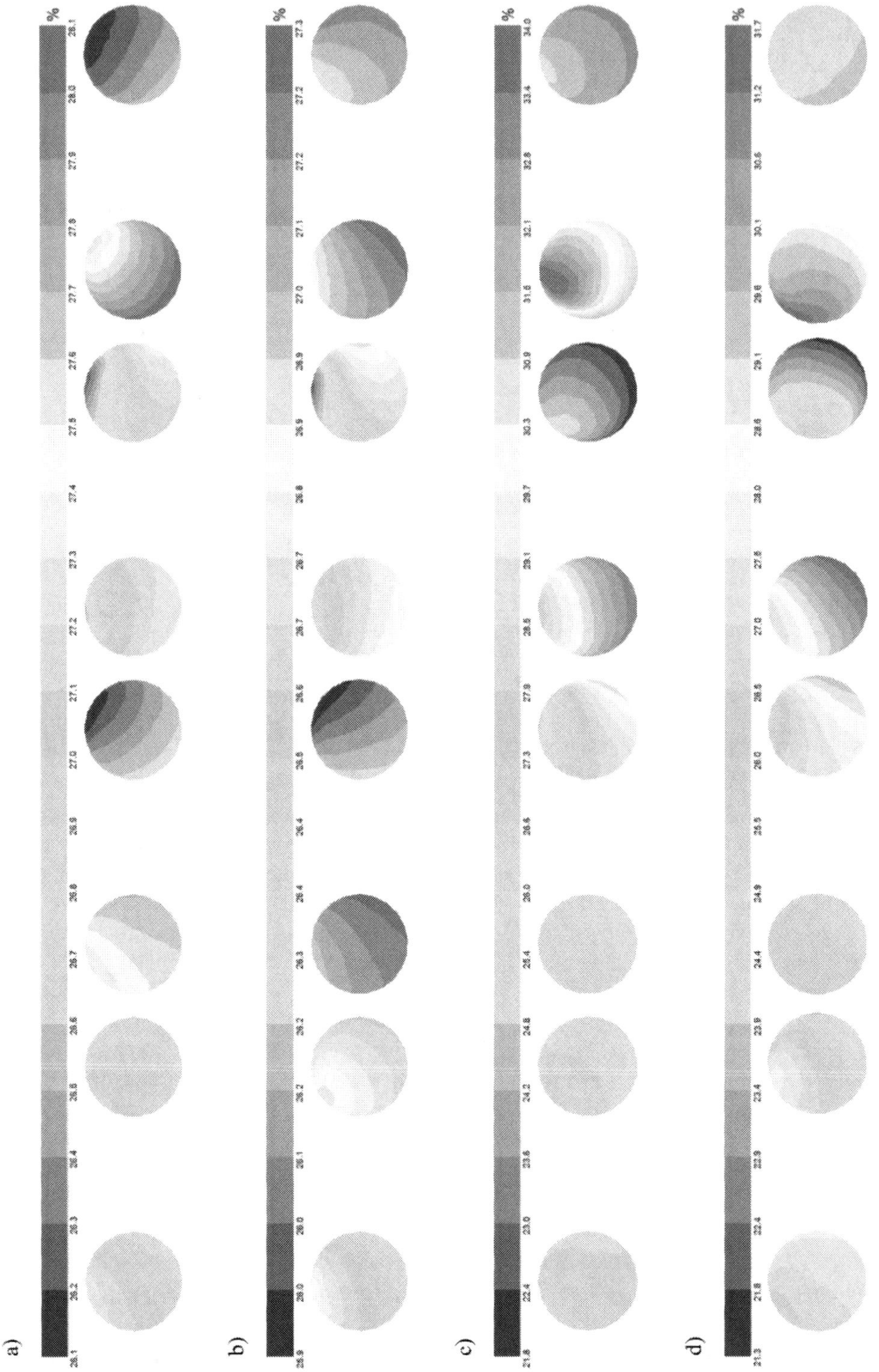

Fig. 5. The percentage of exhaust gas channels in the manifold: a) the EGR valve lift of 1 mm, $v_{air(min)}$ = 2 m/s, $v_{exh(min)}$ = 8.7 m/s, b), the EGR valve lift of 1 mm, $v_{air(max)}$ = 7.5 m/s, $v_{exh(max)}$ = 32 m/s, c) the EGR valve lift 4 mm $v_{air(min)}$ = 2 m/s, $v_{exh(min)}$ = 8.7 m/s, d) the EGR valve lift 4 mm $v_{air(max)}$ = 7.5 m/s, $v_{exh(max)}$ = 32 m/s.

DOI: 10.1051/e3sconf/20172200088

Moreover each case of the valve lift was considered for minimal and maximal velocity of the gases exiting the system. The minimal velocity of air $v_{air(min)}$ and exhaust gases $v_{exh(min)}$ were calculated for the cases *a* and *c* and maximal velocity of air $v_{air(max)}$ and exhaust gases $v_{exh(max)}$ for the cases *b* and *d*. For a detailed analysis, a separate scale for the minimum and maximum values for each visualization was defined.

Simulations showed that the composition of the charge delivered to the individual intake channels of each cylinder is not identical. This phenomenon occurs independently of engine speed and EGR valve opening. mass flow of the gases obtained during the CFD computation allowed to calculate their average percentage share contained in the charge. In both cases of the EGR valve opening at 800 and 3000 rpm the average percentage share of exhaust gas was 27%. From Figure 5 it follows that in the initial phase of charge development there was a stratification of the individual components, which was uniformed during the flow to the cylinder through the intake manifold, the head channels and the intake valve. Therefore, obtaining the same mixtures within the ducts of a single intake manifold is dependent on an identical percentage share of individual gases.

This relationship is not met for any of the pending cases. For comparison, it can be assumed that the calculated average share of exhaust gas is 100% and on that basis determine the average percentage of the uneven distribution in the ducks for individual cases. The read value of the share fumes from Figure 5 are given in Table 2.

Table 2. Percentage uneven distribution of exhaust gases between the outlet duckts of the intake manifold.

Rotation speed [rpm]	EGR valve opening [mm]	t4	t3	t2	t1
800	1	0.37	4.07	0.74	5.37
3000	1	0.93	5.37	0.56	3.52
800	4	2.41	6.85	25.93	26.48
3000	4	2.96	7.41	22.22	17.26

Based on these results it can be concluded that the closed valve provokes the best mixing of the two gases, and the largest divergence occurs for ducks *t3* and *t1* – reaching up to 5.37%. In the case of valve opening unevenness increases having the highest value for ducks t1 and t2 reaching a maximum value of 26.48% for the duck *t1*. This is due to the stratification of the charge. The stream of exhaust gases and air flows almost rectilinearly to the duct – Figure 4. The best mix of gases, regardless of the speed and the opening of the EGR valve, exist for the *t4* pair of ducts. This is due to increased concentration of air flow in the vicinity of the open valve head, which further promotes the formation of a homogeneous mixture.

3 Conclusion

The use of a single exhaust gas recirculation valve in an air supply system to a symmetric inlet manifold results in an uneven percentage share of exhaust gas in the direct duct as well as swirling (ws). In consequence the loads in the individual cylinders has different distribution of exhaust and air in its volume, which results in a different fuel combustion process and therefore, the useful torque obtained from each cylinder. In the case of a higher percentage of exhaust gas in the direct channel than the swirling one, a self-ignition delay and quenching of the flaring flame can occur as a result of lowering the temperature in the central portion of the charge containing a higher concentration of the exhaust gases. As a result, the reduction of nitrogen oxide emissions, with the simultaneous increase of carbon monoxide and Solid particles occurs. An additional factor which enhances uneven concentrations of toxic compounds between cylinders of multi-cylinder engines is the different degree of filling with the fresh load. Confirmation of this fact is carried out on an engines test rig using the intake manifold in which the emission of toxic compounds was measured with engaged and disengaged EGR valve. Figure 6 compares the emission of nitrogen oxides of the tested drive unit for 60Nm load and rotational speeds from 1000 to 2200 rpm

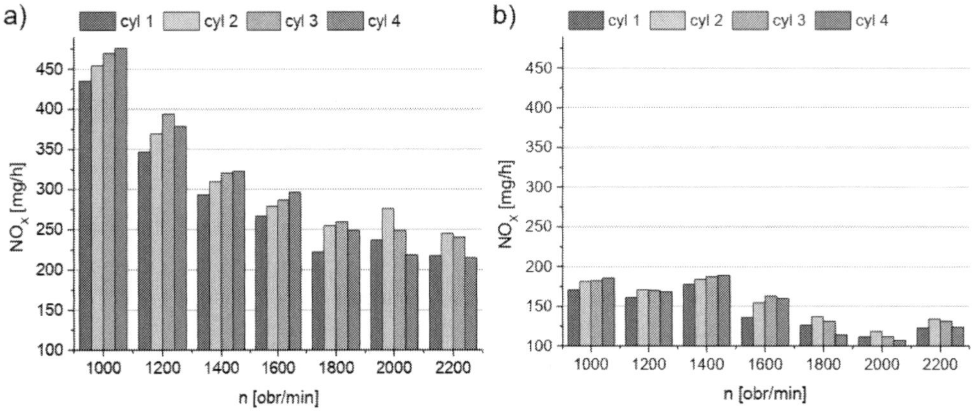

Fig. 6. Nitrogen oxide emissions from individual engine cylinders for a 60 Nm load at: a) EGR valve disabled, b) EGR valve activated.

A solution to compensate the emissions of toxic compounds between cylinders may be the use of selective enrichment of individual loads with exhaust gases by the use of four separate EGR valves whose control could use the electronic crankshaft speed measurement system. The choice of selective percentage of exhaust gas would be based on the measurement of crankshaft acceleration that occurs during successive strokes that determine the correctness of the combustion process.

References

1. A. Bocheńska, C. Bocheński, MOTROL **7**, 15–23 (2005)
2. K. Krakowian, A. Kaźmierczak, A. Górniak, R. Włostowski, KONES **19**, 2 (2012)
3. Z. Sroka, (Oficyna Wydawnicza Politechniki Wrocławskiej, 2013)

Geostatistical methods in the assessment of the spatial variability of the quality of river water

Małgorzata Krasowska[1, *], and *Piotr* Banaszuk[1]

[1]Bialystok Technical University, Faculty of Civil and Environmental Engineering, Department of Environmental Protection and Management, ul. Wiejska 45E, 15-351 Białystok, Poland

Abstract. The research was conducted in the agricultural catchment in north-eastern Poland. The aim of this study was to check how geostatistical analysis can be useful for the detection zones and forms of supply stream by water from different sources. The work was included the implementation of hydrochemical profiles. These profiles were made by measuring the electrical conductivity (EC) values and temperature along the river. On the basis of these results, the authors calculated the coefficient of Moran I and performed semivariogram and found that the EC values are correlated on a stretch of about 140 m. This means that the spatial correlation between samples of water in the stream is readable over a distance of about 140 meters. Therefore it is believed that the degree of water mineralization on this section is shaped by water entering the river channel migration in different ways: through tributaries, leachate drainage and surface runoff. In the case of the analyzed catchment, the potential sources of pollution were drainage systems. Therefore, the spatial analysis allowed the identification pollution sources in a catchment, especially in drained agricultural catchments.

1 Introduction

In recent years many new methods, as well as conceptual and mathematical models were developed to quantify the transfer of non-source pollutions from the catchment to surface waters [1–4]. Among the methods, which allows understanding better phenomena occurring in the geographical space is the geostatistical analysis. The characteristic feature of most natural phenomena is an autocorrelation or spatial dependence a situation when phenomenon being researched within one spatial unit causes an increase or decrease of that phenomenon presence in the adjacent units [5, 6]. This dependence is linked to so-called "first law of geography" by Tobler who in 1970 stated that "everything is related to everything else, but near things are more related than distant things" [7].

Methods based on statistics can be useful to detect the correlation between variables and to identify factors that are not available in direct observation [8]. Therefore, currently, geostatistical methods are used for modeling and forecasting of the pollution of different compartments of the environment and identification of hidden structures of chosen selected measurement series of integrated environmental monitoring [9–11]. Many authors confirm

* Corresponding author: m.krasowska@pb.edu.pl

 DOI: 10.1051/e3sconf/20172200089

that those methods are an efficient tool which lets to define the temporal and spatial dimension of water quality fluctuation, both surface-, and groundwater [12–14]. Yan et al. [15] stated that geostatistical analysis helped to identify the polluted risky regions and could be useful and valuable to pollution control strategies, as well as plan and management on the watershed. Besides, they were also helpful to further research on water quality simulation and validate the simulation accuracy in watershed space. The study by Einax and Soldt [16] and Eneji et al. [17] confirmed the usefulness of geostatistical analyses to assess the water quality and identify sources of pollution. Geostatistical methods can be used to describe the spatial distribution and variability of the quality of riverine ecosystems [18]. Therefore, in this paper, we attempt to use selected geostatistics to verify their usefulness in the detection of the spatial structure of the chemical composition of the stream during the period of high waterlogging. Simultaneously we checked how semivariance analysis and Moran's I could be applied for identifying zones and forms which supply solutes the stream in the agricultural catchment.

2 Study area

The study was performed within a small catchment (187 ha) in the north-east Poland. That stream is a left-bank tributary of Horodnianka River. The catchment is rural with 75% of arable lands, 16% – grasslands, 3.5% – woodlands and 5.5% built-up areas and wastelands. Approximately 60% of the agricultural land is artificially drained, mainly by underground pipes. The analyzed stream has no natural outflow and originates from an underground drainage system. Its length between the origin and the gauging profile was about 1550 m. At this section, the stream has three tributaries (Fig. 1).

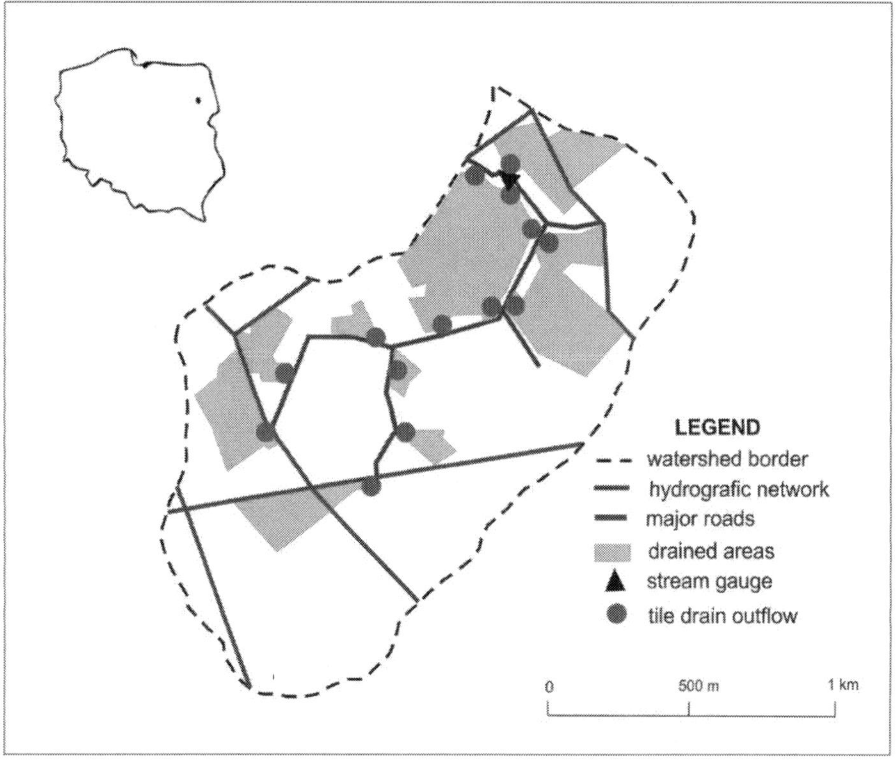

Fig. 1. Location and features of the studied catchment.

3 Methods

For statistical analyses, we used data collected during hydrochemical profiling of stream in March 2010, during a high waterlogging of catchment and high stream flow. They included measurements of electrolytic conductivity, water temperature and the concentration of NO_3^-, Cl^-, $Si_2O_3^{2-}$, performed along the creek course at regular intervals of 10 m. The measurement of electrolytic conductivity and temperature of the water was determined by using Hach Lange HQ40D, while the concentration of ions was determined by spectrophotometry using Slandi LF300.

It was assumed that the conductivity could be employed as a proxy for the content of solutes in water [19]. Furthermore, it was assumed that changes in water chemistry along the stream resulted from a supply of chemicals and were not related to the biogeochemical reactions that occur in the riverbed [20]. This assumption seems to be fulfilled because the measurements were performed in the winter season when biota was dormant phase. To detect the spatial structures underlying the observed variability of the water quality we used Moran's I statistic calculated according to the formula (1):

$$I = \frac{n\sum_i \sum_j w_{ij}(z_i-\bar{z})(z_j-\bar{z})}{\sum_i \sum_j w_{ij} \sum_i (z_i-\bar{z})^2} \quad (1)$$

where:
n – number of observations,
w_{ij} – element of a matrix of spatial weights,
z_i, z_j – variable value at a particular location i or j,
$\bar{z}$ – the mean of the variable [21].

Moran's I statistics [22] belongs to the one of the oldest and most commonly used in studies of spatial autocorrelation. It is interpreted as the coefficient a correlation occurs between the values of the variable in the spatial unit. The value of Moran's I generally varies between 1 and −1. Positive autocorrelation in the data translates into positive values of Moran's I; negative autocorrelation produces negative values. No autocorrelation results in a value close to 0. Moran's coefficient can be used in the assessment of the pattern of dispersion, randomness, and concentration. Moran's I values may range from -1 (dispersed) to +1 (clustered) [6, 23, 24]. The result of the Moran analysis is correlogram. From the intersection of correlogram line with the x-axis, we can evaluate the type of spatial distribution [25]. The SAM 4 software was used to calculate the spatial correlation [26].

The semivariance is defined as half the average squared difference between values (2):

$$\gamma = \frac{1}{2N(h)} \sum_{\alpha=1}^{N(h)} [z(u_\alpha) - z(u_\alpha + h)]^2 \quad (2)$$

where:
$N(h)$ – number of pairs of sampling locations located at distance h from one another
Z – the value of the variable u_α at the sampling location α [27].

Empirical semivariogram was made to characterize the spatial continuity which exists in the analyzed data set. Semivariogram model was characterized by three parameters:
- range, which represents the large-scale correlation,
- nugget effect, which represents uncorrelated small-scale variability,
- sill, which is the upper bound of γ [25, 28].

In practice analysis of semivariograms allows determining the decrease of self-similarity with increasing distance between measurements. Semivariograms models were made by using Surfer 9.

DOI: 10.1051/e3sconf/20172200089

4 The results

4.1 Hydrochemical profiles

During the hydrochemical mapping of the stream measurements ofelectrolytic conductivity (EC) and temperature (T) along its were made (Fig. 2).

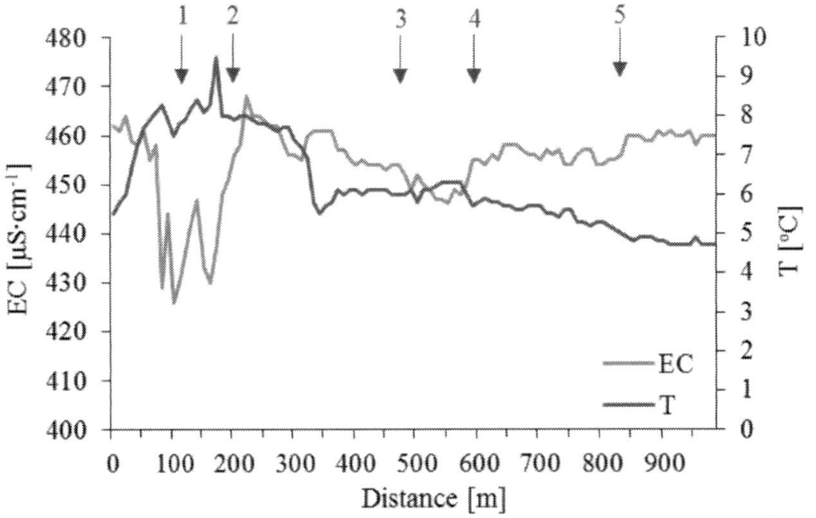

Fig. 2. Variation in the electrolytic conductivity (EC) and temperature (T) of water down the stream (potential sources of pollution: 1, 2, 4 – tributaries; 3, 5 – drain tile).

We observed a sharp decrease in EC value at 100–150 meter stream (from the source), with the fast recovery to initial values tens meters further (with one peak at 250 m), where EC was aligned at about 450 $\mu S \cdot cm^{-1}$ and did not show any clear trend. The water temperature showed the highest value up to 150 m after that there was a marked fall. Besides, along with the water course, an evident variability in physio-chemical properties of water was observed (Tab. 1). NO_3^- changed from 14 $mg \cdot dm^{-3}$ at the 150 meter of the water course to 30 $mg \cdot dm^{-3}$ at 650 meter. Its variability was similar to changes in $Si_2O_3^{2-}$ (Fig. 3). Based on hydrochemical mapping it is possible to detect zones of the stream supply with solutes.

Table 1. Statistics of the chosen physio-chemical properties measured along the water course.

	EC [$\mu S \cdot cm^{-1}$]	Temp. [°C]	NO_3^- [$mg \cdot dm^{-3}$]	$Si_2O_3^{2-}$	Cl^-
M	457	5.8	25.5	1.9	12.7
$\overline{X}$	457 (58)	6.0 (5.4)	24.9 (31.6)	1.9 (1.5)	12.8 (4.4)
SD	10.8	1.3	6.4	0.4	1.1
CV	2.4	21.7	25.9	21.7	8.5

M – median, $\overline{X}$ – arithmetic mean, SD – standard deviation, CV – coefficient of variation, (in brackets the value of the range)

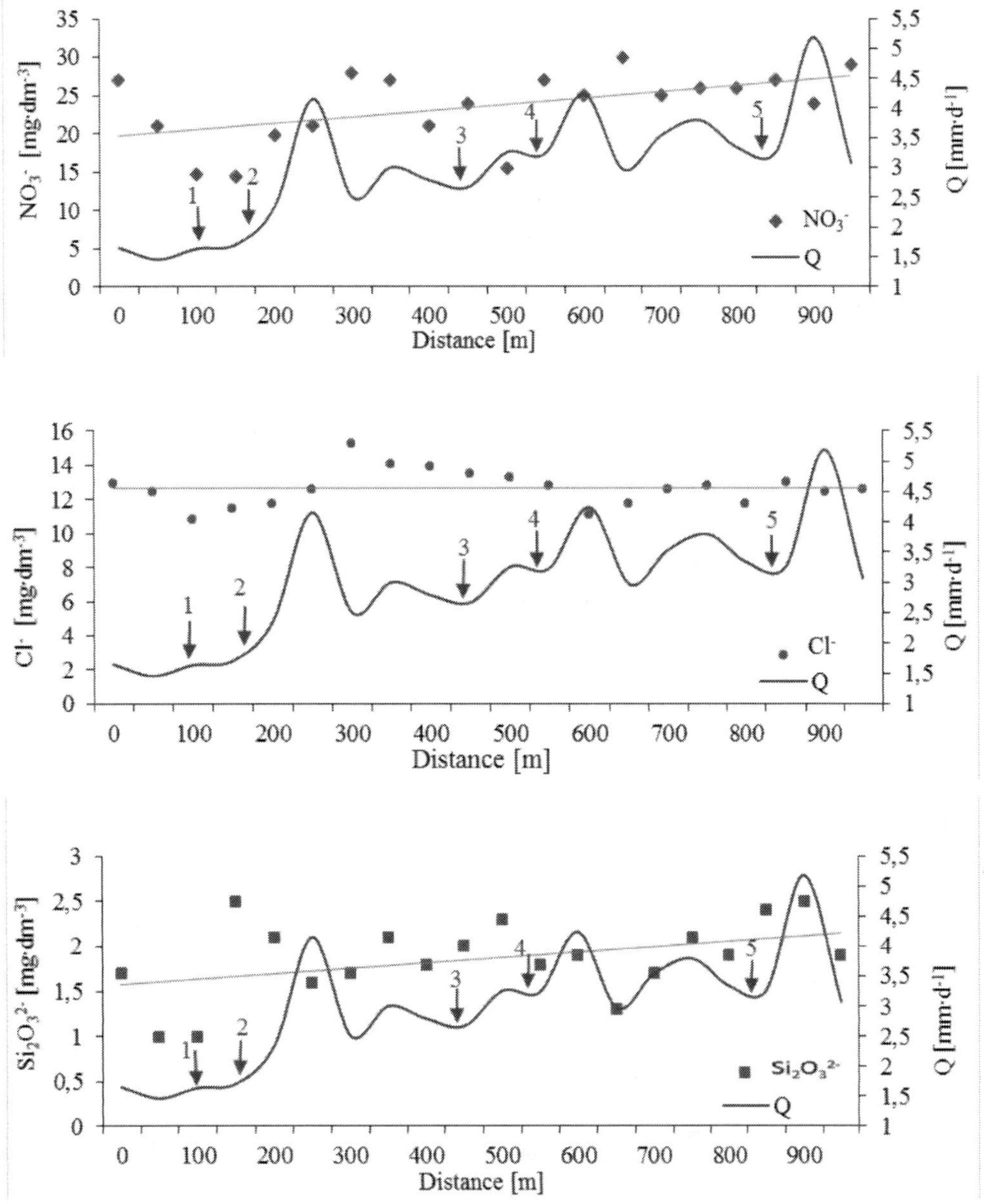

Fig. 3. Dynamics of the stream discharge (Q) and concentration of dissolved components down the stream (potential sources of pollution: 1, 2, 4 – tributaries; 3, 5 – drain tile).

4.2 Semivariance analysis

The semivariogram, which was made for the measurement of electrolytic conductivity showed the presence of autocorrelation at a distance of 100–120 m (Fig. 4). The best-matched model was the spherical model, which the overall formula is (3, 4):

$$\gamma(h) = C_0 + C\left(1{,}5\left(\frac{h}{A_0}\right) - 0{,}5\left(\frac{h}{A_0}\right)^3\right) for\ h \leq A_0 \qquad (3)$$

$$\gamma(h) = C_0 + C \ \ for \ h \geq A_0 \qquad (4)$$

where:
C_0 – nugget effect,
C_0+C – sill
A_0 – range,
h – (distance) between samples.

The value of variance identified the size of the measurement error, which was relatively small and amounted to about 2,5 $\mu S \cdot cm^{-1}$. Whereas sill, which is the upper bound of semivariance was about 82. It was found that the spatial correlation (A0) between the samples of water in the stream is readable on the stretch to about 140 m. At this point, an increase in the value of semivariance has been "detained."

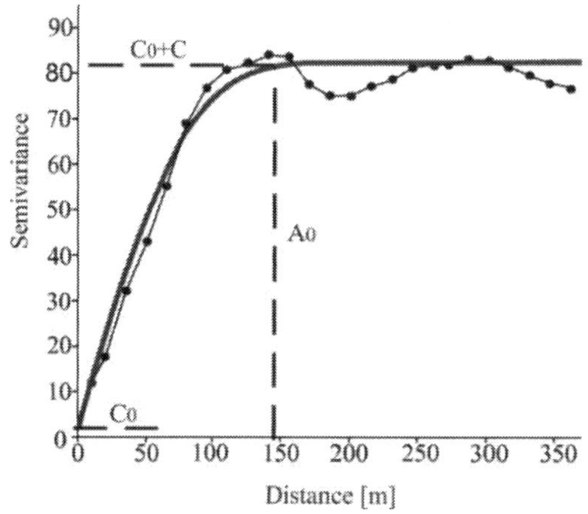

Fig. 4. Semivariogram with their model for analyzed electrolytic conductivity.

4.3 Moran's analysis

There are similar conclusions about Moran's statistics. Correlogram made for EC measurements indicated that the autocorrelation coefficient is positive on the stretch to about 140 m. This is the point where the intersection of a line with the X-axis occurred (Fig. 5). The correlation between the variables decreases from +0.2 to -0.2 at a distance of about 240 m, and then rise again and reaches a low positive value at a distance of 735 m. Such a correlogram results indicate gradual structure creation. Moreover, the cyclical fluctuation was discovered at every 140–200 m.

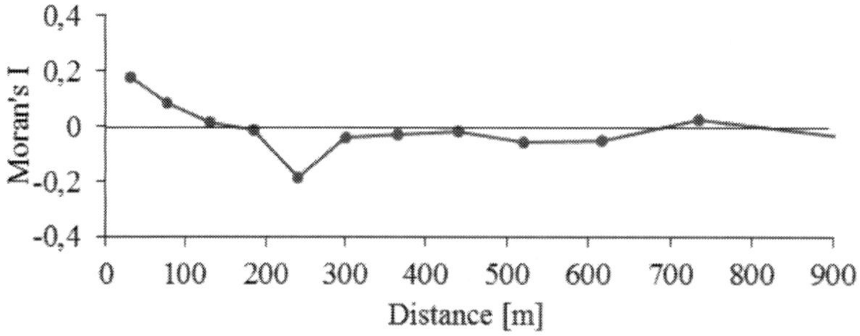

Fig. 5. Correlogram showing the variation of spatial autocorrelation in the time function for the measurement of electrolytic conductivity.

5 Summary

The study was conducted in the winter half-year, under the conditions of high catchment wetness when intensive movement of solutes to the stream occurs [19, 20, 29, 30]. Contaminants can migrate in different ways: with surface runoff, shallow subsurface, and deeper groundwater flow. In agricultural catchments, which in the north-eastern Polish are largely drained, drainage systems play an important role in the movement of solutes from the catchment to stream [30]. The agricultural catchments water from drains are usually are contaminated with nutrients [31].

The research shows that the sources and routes of migration in the basin, which are rich in components dissolved, had an impact on the mineralization of the streamwater. By hydrochemical mapping, it has been found that the variability of electrolytic conductivity in the stream water was driven mainly by water inflow from surface tributaries and tile drains. On the basis of hydrochemical mapping we can only approximately zones which supply waters from the catchment. In this paper was performed geostatistical analysis for the precise identification of these forms. The geostatistical methods used in this study fully confirmed observations made in the field. The analysis of the semivariance and Moran's spatial correlation showed that on the section length of 140 m, there was a spatial correlation, which reflected the real influence of solutes supplied by tributaries and tile drains regularly placed every 140 m. This is the maximum range of the autocorrelation, which is a real similarity between samples [20]. The impact of the sources of dissolved substances was fading over this distance. The geostatistical methods, have shown the occurrence of relatively frequent and regular source of solute, which are situated along the water course. Thus they can be used for quite correct identification of the hidden spatial structures that affect the water quality, as it was stated by Stach [20], Liu et al. [13], Eneji and Onuche [17], Basu et al. [14], Yan et al. [15].

The acknowledgments: research was funded by Ministry of Science and Higher Education as a part of the project S/WBiIŚ/01/17.

References

1. B. Namysłowska-Wilczyńska, A. Wilczyński, Ochr. Sr. **2**, 65 (1997)
2. S.P. Inamdar, M.J. Mitchell, Water Resour. Res. **42** (2006)
3. M.J. Gromiec, (IMGW Warszawa, 2008)
4. H.S. Wheater, N. McIntyre, T. Wagener, (Cambridge University Press, 2008)
5. R. Bivand, PWN Poznań, 23–28 (1980)
6. K. Janc, Dokumentacja Geograficzna **33**, 76–83 (2006)
7. H.J. Miller, Ann. Assoc. Am. Geogr. **94**, 284–289 (2004)
8. J.B. Racine, H. Reymond, (PWN, Warszawa, 1977)
9. P. Goovaerts, (Oxford University Press, 1997)
10. B. Nas, Polish J. of Environ. Stud. **18**, 6, 1073–1082 (2009)
11. A.S. Gharbia, S.S. Gharbia, T. Abushbak, H. Wafi, A. Aish, M. Zelenakova, F. Pilla, J. Geosci. Environ. Protect. **4**, 89–103 (2016)
12. H. Bu, X. Tan, S. Li, O. Zhang, Ecotoxicol. Environ. Saf. **73**, 5 (2010)
13. W.C. Liu, H.L. Yu, C.N. Chung, Int. J. Environ. Res. Public Health **8**, 4, 1126–1140 (2011)

14. S. Basu, K.S. Lokesh, Int. J. Water Res. Environ. Eng. **5**, 10, 591–596 (2013)
15. C.A. Yan, W. Zhang, Z. Zhang, Y. Liu, C. Deng, N. Nie, PLoS ONE **10**, 3 (2015)
16. J.W. Einax, U. Soldt, Elsevier **46**, 1, 79–91 (1999)
17. I.S. Eneji, A.P. Onuche, R.S. Ato, J. of Environ. Protect. **3**, 915–921 (2012)
18. B.G. Kotak, A. Selinger, Manitoba Model Forest Report 05-2-63, (2006)
19. M. Stolarska, Acta Geographica Lodzienisa, **94** (2008)
20. A. Stach, Przegląd Naukowy Inż. i Kszt. Środ. Wyd. SGGW Warszawa **11**, 2 (25), 130–143 (2003)
21. P.A. Longley, M.F. Goodchild, D.J. Maguire, D.W. Rhind, (Wyd. PWN, Warszawa, 2006)
22. P.A.P. Moran, Biometrika **37**, 17–23 (1950)
23. K.P. Overmars, G.H.J. de Koning, A. Veldkamp, Ecol. Model. 164, 2, 257–270 (2003)
24. X.N. Huo, H. Li, D.F. Sun, L.D. Zhou, B.G. Li, Int. J. Environ. Res. Public Health 9, 3, 995–1017 (2012)
25. M.J. Fortin, M.R.T. Dale, (Cambridge University Press, 2005)
26. T.F. Rangel, J.A.F. Diniz-Filho, L.M. Bini, Ecography **33**, 1, 46–50 (2010)
27. E. Gringarten, C.V. Deutsch, Springer Berlin **33**, 34, 507–534 (2001)
28. N.Z. Sun, A.Y. Sun, (Springer-Verlag, New York, 2015)
29. M. Klein, Earth Surf. Process. Landforms **6**, 173–178 (1981)
30. P. Banaszuk, M. Krasowska, A. Kamocki, Springer-Verlag Berlin, 63–77 (2011)
31. C.S. Tan, C.F. Drury, M. Soultani, I.J. van Wesenbeeck, H.Y.F. Ng, J.D. Gaynor, T.W. Welacky, Wat. Sci. Tech. **38**, 4–5, 130–110 (1998)

Influence of sputtering deposition parameters on electrical and optical properties of aluminium-doped zinc oxide thin films for photovoltaic applications

Ewelina Krawczak[1,*], *Agata* Zdyb[1], *Slawomir* Gulkowski[1], *Alain* Fave[2], and *Erwann* Fourmond[2]

[1]Lublin University of Technology, Faculty of Environmental Engineering, Institute of Renewable Energy Engineering, ul. Nadbystrzycka 40B, 20-618 Lublin, Poland
[2]Institut des Nanotechnologies de Lyon –INL, UMR CNRS 5270 CNRS, Université de Lyon, INSA de Lyon, F-69621 Villeurbanne, France

Abstract. Transparent Conductive Oxides (TCOs) characterized by high visible transmittance and low electrical resistivity play an important role in photovoltaic technology. Aluminum doped zinc oxide (AZO) is one of the TCOs that can find its application in thin film solar cells (CIGS or CdTe PV technology) as well as in other microelectronic applications. In this paper some optical and electrical properties of ZnO:Al thin films deposited by RF magnetron sputtering method have been investigated. AZO layers have been deposited on the soda lime glass substrates with use of variable technological parameters such as pressure in the deposition chamber, power applied and temperature during the process. The composition of AZO films has been investigated by EDS method. Thickness and refraction index of the deposited layers in dependence on certain technological parameters of sputtering process have been determined by spectroscopic ellipsometry. The measurements of transmittance and sheet resistance were also performed.

1 Introduction

Transparent Conductive Oxides (TCO) films play a significant role as an transparent electrode in the optoelectronic and photovoltaic devices, especially in thin films solar cells [1–6]. The most widely used TCO, both in research and industry, are Indium Tin Oxide (ITO) films, because they are characterized by a good electrical and optical properties. However, future availability and cost issues require to replace it by another TCO. The main goal of the studies on TCO structures is to obtain high transmittance of the material, above 80% over a large spectral range, with resistivity below 10^{-3} Ω·cm and wide band gap [6, 7]. In order to improve the efficiency in some optoelectronic applications the optical transparency of TCO films should be extended to the ultraviolet and infrared spectral ranges. One of the TCO materials that is widely investigated in recent years is zinc-oxide ZnO, which is non-toxic,

* Corresponding author: e.krawczak@pollub.pl

abundant and relatively cheap material. Modified ZnO material, for example aluminum doped zinc oxide (AZO, ZnO:Al) can be a substitute of ITO in the top electrode or can be a part of double layer back contact in the thin film solar cell. It is also of high importance to obtain ZnO structures doped with the third group element, such as B, Ga, In or Al, since it leads to improvement of electrical conductivity of the layer [8].

Apart from magnetron sputtering belonging to physical vapour deposition methods (PVD), variety of techniques used for ZnO:Al production can be distinguished: sol-gel deposition, spray coating, pulsed laser deposition (PLD) and aerosol-assisted MOCVD technique [9-14]. In order to achieve aluminum doped zinc oxide layers with good electro-optical properties on inflexible or flexible substrate [15], RF and DC magnetron deposition or co-sputtering deposition can be used [16–21]. Numerous of technological parameters have the influence on the quality of the obtained thin films such as: power applied to the cathode, the pressure in the chamber during deposition process, target-substrate distance, type and flow of the process gases, substrate temperature or even positioning of the substrate relative to the target.

In the present study ZnO:Al layers were obtained by RF magnetron sputtering method on clean soda lime glass substrates and their composition was determined by EDS (Energy Dispersive Spectroscopy) method. Ellipsometry measurements provided information on the thickness of the films produced under different conditions of the deposition process. Basing on these data the rate of growth was estimated and the influence of AZO thickness on index of refraction was studied. The refractive index (n) of semiconducting material like AZO is one of the optical constants that is essential in optoelectronic studies and applications. In design of thin film photovoltaic devices determination of n for an outer transparent electrode is of high importance since it influences the amount of light that propagates into the p-n junction and can be potentially utilized. Spectroscopic transmittance measurement confirmed very good transparency of ZnO:Al films. Resistivity of the obtained layers, also determined in this work, plays an important role especially if they are meant to be applied as an electrode in thin film devices.

2 Experimental

The transparent conductive Al doped ZnO films were deposited onto soda lime glass (SLG) substrate using RF (Radio Frequency) magnetron sputtering method. Alliance Concept AC450 magnetron sputtering system was used in the experiment in order to find optimum parameters of the sputtering process. The radio frequency (RF) power was applied to 4 inches in diameter and 6 mm thick ceramic target composed by 98 wt.% ZnO and 2 wt.% Al with purity of 99.99%, purchased from BIMO Tech. Before the process soda lime glass substrates were ultrasonically and chemically cleaned in detergent, isopropyl alcohol, ethanol, deionized water and then dried in a nitrogen gas. The clean substrate was immediately mounted in the magnetron deposition chamber. Substrates were placed under the target directly onto unheated work holder, which is presented in Fig. 1. The static mode of substrate holder was used. The target – substrate distance was equal to 90 mm. In this work AZO films were obtained at room temperature. The substrate holder temperature was kept at constant level during the deposition process due to water cooling system.

 DOI: 10.1051/e3sconf/20172200090

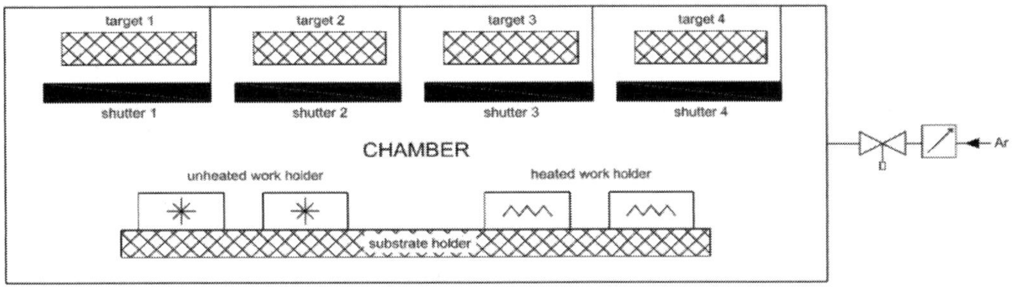

Fig. 1. The inside of the magnetron sputtering system chamber.

The magnetron chamber was evacuated to the base pressure of 1.0 x 10^{-2} Pa and then presputtering process was carried out to clean the target surface and remove encountered impurities. Stability improvement of sputtering conditions was also achieved at the beginning. The RF power during the films deposition process varied in the range of 120–200 W (1.5–2.5 W/cm^2). The working pressure was stabilized at 1.57 Pa by flow of the pure Ar gas, which was set at 20 sccm. EDS detector was applied to achieve EDS profile and analyze energy spectra of the layers. Film thickness and refractive index were measured by spectroscopic ellipsometer Jobin-Yvon Horiba UVISEL model, working between 1.5 and 5 eV (around 830 to 250 nm wavelength) with use of new amorphous model. The sheet resistance of the obtained samples was measured at room temperature using RM3000+ four point probe equipment from Jandel Engineering Limited.

3 Results

Figure 2 shows the example of EDS diagram of the ZnO:Al deposited film. The qualitative analysis was made to obtain the mass and weight fractions of the elements existing in the prepared film. The composition of elements was as follows: Zn w.% – 68.06, O w.% – 29.65, Al w.% – 2.29. It should be noticed that all prepared samples revealed the similar composition and irregularities were not detected.

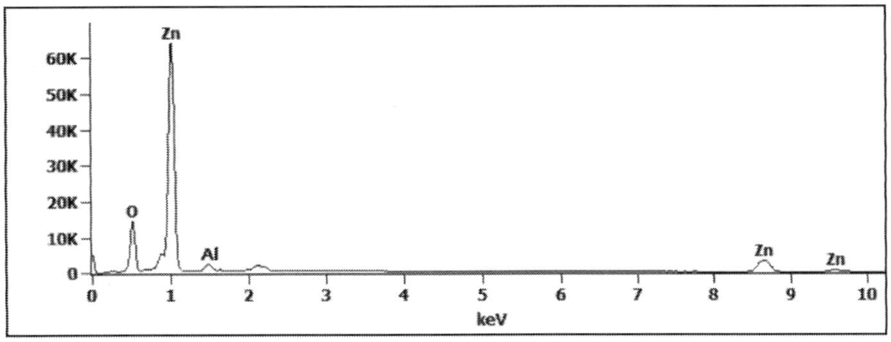

Fig. 2. EDS profile of the Al doped ZnO sample prepared using magnetron sputtering system.

Several series of samples were prepared at three different values of deposition power under the same working pressure at room temperature during different sputtering time. The influence of the sputtering time on the thickness of the AZO films is presented in Fig. 3 for three different deposition powers: 1.5 W/cm^2, 2 W/cm^2, 2.5 W/cm^2. Ellipsometry measurements provided the following thickness values obtained for 30, 40 and 45 min deposition time, respectively: 286 nm, 413 nm, 463 nm (power 1.5 W/ cm^2), 322 nm,

 DOI: 10.1051/e3sconf/20172200090

539 nm, 622 nm (power 2 W/cm^2), 490 nm, 823 nm, 1178 nm (power 2.5 W/cm^2). The dependence shows that the thickness of the deposited films increases for higher power values and longer deposition time. The estimated growth rate values change from average 8.8 $nm \cdot min^{-1}$ for the power of 1.5 W/cm^2, 11 $nm \cdot min^{-1}$ for 2 W/cm^2 to 18 $nm \cdot min^{-1}$ for 2.5 W/cm^2. One can note that linearity is better for high power value. The highest sputtering ratio equal to 19.7 $nm \cdot min^{-1}$ was obtained for the longest deposition time.

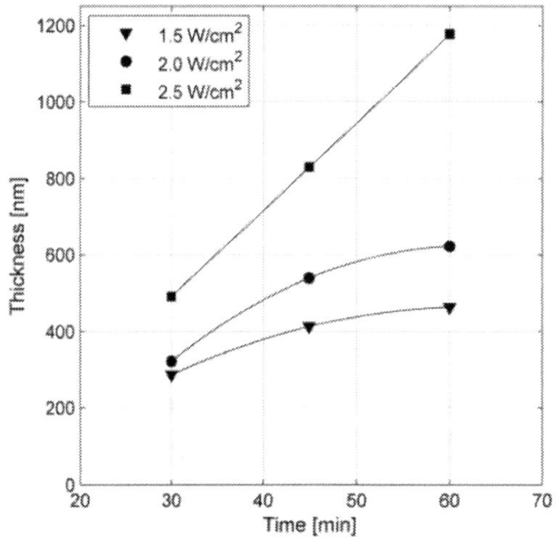

Fig. 3. The thickness of the obtained AZO films as the function of the deposition time for different magnetron power.

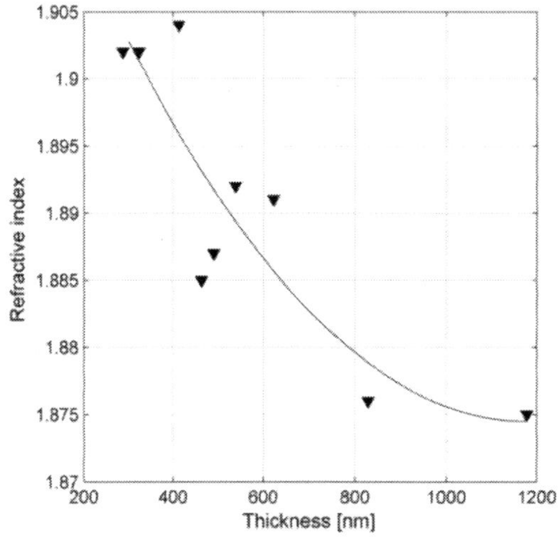

Fig. 4. The refractive index of ZnO:Al films as a function of the thickness, measured at $\lambda = 633$ nm.

The refractive index (n) was obtained from ellipsometry measurements for AZO films of different thickness. Index of refraction values are shown in Fig. 4 as a function of the thickness of the deposited layers at wavelength of 633 nm. The obtained n values decrease from 1.905 to 1.875 with increasing thickness of the films and are lower than those for ZnO i.e. 1.95–2.1 in visible range [22]. The observed relation corresponds to dependency reported in the literature [23]. Basing on the refractive index values transmittance and reflectance at normal incidence can be determined according to Fresnel formulas (Eq. 1, 2):

 DOI: 10.1051/e3sconf/20172200090

$$T = \frac{4n_1n_2}{(n_1+n_2)^2}, \quad (1)$$

$$R = \left(\frac{n_2 - n_1}{n_1 + n_2}\right)^2, \quad (2)$$

where $n_1 = 1$ in the air, n_2 – index of refraction of the layer. Taking into account $n_2 = 1.905$ or $n_2 = 1.875$ the transmittance is equal to 90.3% and 90.7% and reflectance values to 9.7% and 9.3% respectively. Experimentally obtained transmittance spectra presented in Fig. 5 reveal mean values above 90% both for the lowest and highest film thickness. Very good transmittance and homogeneity of the ZnO:Al films were also exhibited before [24].

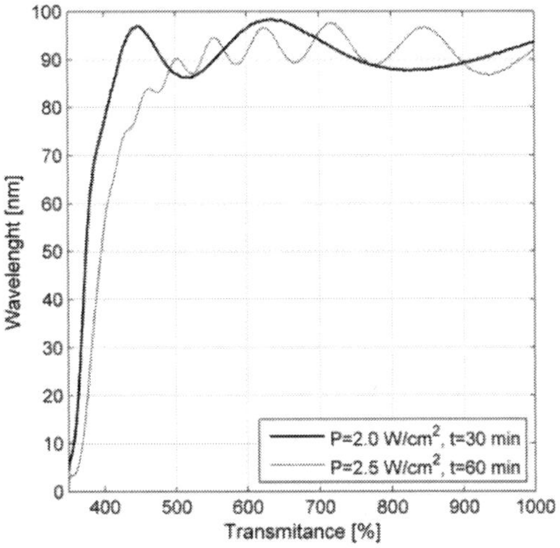

Fig. 5. Transmitance spectra of ZnO:Al layers of thickness 322 nm and 1178 nm (pure glass transmittance is subtracted).

Electrical properties of AZO films were also subject of our investigations. The correlation between sheet resistance and thickness of the deposited AZO layers is shown in Fig. 6. The sheet resistance decreases with increasing thickness of the layers. According to all obtained results these dependences are analogues for all RF powers used under the same pressure of the process. Here we considered the films obtained at the discharged power of 2.5 W/cm^2 since in this case we achieved the lowest range of the resistance value. As expected, the sheet resistance decreases very significantly as the thickness of AZO film increases from the 491 nm to 830 nm. Beyond this range of thickness the resistance drop is not so considerable with further AZO thickness increase. According to all obtained results these dependences are analogues for all RF powers used under the same pressure of the process. Taking into account prospect applications of ZnO:Al layers, the sheet resistance value is expected to lay below 50 Ω/sq. The improvement of the results we obtained is probably possible by lowering the chamber pressure or heating of the samples.

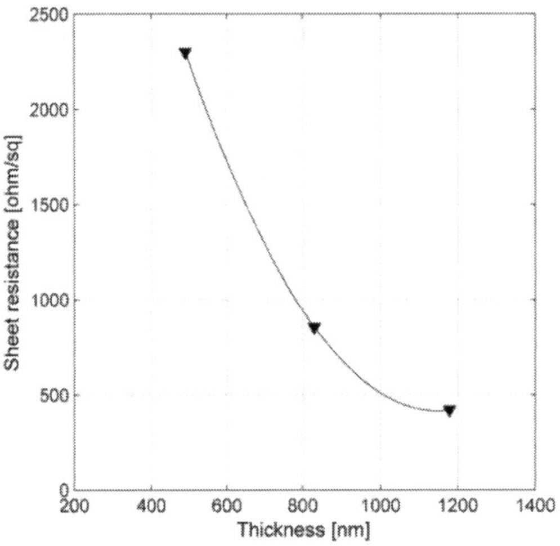

Fig. 6. Sheet resistance of ZnO:Al films as a function of the thickness.

4 Summary

ZnO layers doped with about 2.3 w.% Al were obtained in magnetron sputtering system and their properties were studied. The growth rate during deposition was estimated to be in the range of 8.8-18 nm/min depending on the magnetron power and time of the sputtering process. Refractive index of AZO films is directly related to deposited film thickness and changes from 1.905 to 1.875 for thicker layers. These values were used to calculate transmittance and reflectance according to Fresnel formula, which together with experimental data, allowed confirming good transparency of the layers. It has been also shown that higher power applied to the cathode material results in the lower resistivity of the thin film so the sheet resistance drops below 500 Ω/sq for the power of 2.5 W/cm^2 used during deposition. Further conductivity improvements are necessary to make the produced AZO films useful in photovoltaic applications.

References

1. H.J. Cho, S.U. Lee, B. Hong, Y.D. Shin, J.Y. Ju, H.D. Kim, M.Park, W.S. Choi, Thin Solid Films **518**, 2941–2944 (2010)
2. J. Muller, B. Rech, J. Springer, Mi. Vanecek, Sol. Energy **77**, 917–930 (2004)
3. J.J. Ding, H.X. Chen, S.Y. Ma, Physica E **42**, 1861–1864 (2010)
4. S.J. Tark, M.G. Kang, S. Park, J.H. Jang, J.C. Lee, W. M. Kim, J.S. Lee, D. Kim, Curr. Appl. Phys. **9**, 1318 (2009)
5. W. Beyer, J. Hüpkes, H. Stiebig, Thin Solid Films **516**, 147–154 (2007)
6. J. Yoo, J. Lee, S. Kim, K. Yoon, I.J. Park, S.K. Dhungel, B. Karunagaran, D. Mangalaraj, J. Yi, Thin Solid Films **480–481**, 213–217 (2005)
7. B.G. Choi, I.H. Kim, D.H. Kim, K.S. Lee, T.S. Lee, B. Cheong, Y.-J. Baik, W.M. Kim, J. Eur. Ceram. Soc. **25**, 2161–2165 (2005)
8. P. Nunes, E. Fortunato, P. Tonello, F. Braz Fernandes, P. Vilarinho, R. Martins, Vacuum **64**, 281–285 (2002)

9. S. Kuprenaite, A. Abrutis, V. Kubilius, T. Murauskas, Z. Saltyte, V. Plausinaitiene, Thin Solid Films **559**, 156–164 (2016)
10. T. Schuler, T. Krajewski, I. Grobelsek, M. A. Aegerter, Thin Solid Films **502**, 67–71 (2006)
11. T. Schuler, M.A. Aegerter, Thin Solid Films **351**, 125–131 (1999)
12. S. Kuprenaite, T. Murauskas, A. Abrutis, V. Kubilius, Z. Saltyte, V. Plausinaitiene, Surf. Coat. Tech. **599**, 19–26 (2016)
13. C.G. Granqvist, Sol. Energy Mat. Sol. C **91**, 1529–1598 (2007)
14. Y. Liu, L. Zhao, J. Lian, Vacuum **81**, 18–21 (2006)
15. W. Li, H. Hao, M. He, J. Xing, H. Gao, J. Dong, Surf. Coat. Tech. **258**, 991–995 (2014)
16. Z.A. Wang, J.B. Chu, H.B. Zhu, Z. Sun, Y.W. Chen, S.M. Huang, Solid State Electron. **53**, 1149–1153 (2009)
17. H. Ko, W. Tai, K. Kim, S. Kim, S. Suh, Y. Kim, J Cryst Growth **277**, 352–358 (2005)
18. H. Liu, H. Qiu, X. Chen, M. Yu, M. Wang, Curr Appl Phys **9**, 1217–1222 (2009)
19. C. Guillen, J. Herrero, Vacuum **84**, 924–929 (2010)
20. E. Miorin, F. Montagner, S. Battiston, S. Fiameni, M. Fabrizio, Aip Conf. Proc. **10**, 1–5 (2010)
21. K. Ellmer, R. Cebulla, R. Wendt, Surf. Coat. Technol. **98**, 1251–1256 (1998)
22. W.L. Bond, Jpn. J. Appl. Phys. **36**, 1674–1677 (1965)
23. S. Ilican, M. Caglar, Y. Caglar, Mat. Sc. **25**, 709–718 (2007)
24. A. Zdyb, E. Krawczak, P. Lichograj, Opt. Appl. **46**, 181–185 (2016)

Bioarchitecture – a new vision of energy sustainable cities

Alicja Krzemińska[1,*], *Anna* Zaręba[1], and *Anna* Dzikowska[2]

[1]University of Wrocław, Faculty of Earth Sciences and Environmental Development, pl. Uniwersytecki 1, 50-137 Wrocław, Poland
[2]The International University of Logistics and Transport in Wrocław, ul. Sołtysowicka 19B, 51-168 Wrocław, Poland

Abstract. Transformation of the natural environment will press the humanity to search for the new look at the problems of architecture and urban design. Nowadays passive houses construction is a standard and green roofs are incorporated in the design of contemporary cities. That's why city cluster will be successively transformed into sustainable bionic systems, which allows to protect the nature and stop further degradation and exploitation of public green space. The good examples of contemporary trend of designing in harmony with nature are energy sustainable underground buildings of Malcolm Wells, who in 60s designed his first energy sufficient construction. The underground cities and rock houses were built from the early beginning of architecture, with significant examples of cities: Sanmenxia in China in Henan Province, Matmata (Tunisia), Cappadocia (Turkey), Uplisciche (Georgia) or Brlhovce (Slovakia) etc. The underground buildings and cities, blending in with the background of topography, have a positive influence on the landscape and are energy sustainable. Climate responsive design materials create effective insulation, which allows to maintain the stable temperature inside the buildings. Bioarchitecture improves the microclimate in the neighborhood through increasing oxygen concentration in atmosphere and limiting of CO_2 emission. Bioarchitecture represents new direction in changing the design priorities towards being closer with nature and it's needs.

1 Introduction

People have for centuries adapted their habitat to environmental conditions, guided mainly by the instinct of survival. To build the first primitive homes they used natural materials that they found or hunted. Homes were concentrated in settlements that provided people with a sense of security. They were built both on the ground and underground, for example, in the rocks. Underground housing structures can be found both in hot areas as well as where temperatures are low. The earth is a good thermal insulator, so that the temperatures in the underground homes provided them with adequate thermal comfort.

Contemporary environmental problems, including climate changes, environmental pollution, habitat changes and progressive destruction of the landscape, will certainly contribute to a new look at the issues of house construction as well as urban design, in a different way than before. Conventional perception and design of urban structure in the next decades will get changed, and ecological values and landscape values will be one of the top priorities for new trends in design [1]. First and foremost, the paradigm of development

* Corresponding author: alicja.krzeminska@uwr.edu.pl

and space aesthetics will change, where the bionic system will be a determinant of its quality. However, the mentality of people must change, and the permission to build zero energy bionic houses, including houses built entirely or partly underground, that so far has often been in the zone of taboos, has to change as well.

2 Historical conditioning. Earth sheltered buildings

Recovering the harmony between nature and the world, in which man performs constant transformation is difficult. The cities, in which we live most often do not resemble green enclaves, and buildings – often in spite of their beauty, do not create unity with the natural environment. Negative changes in the surrounding environment force us to change our thinking and to enter another dimension of home and human settlements design (in both cities and villages). Current design trends are changing, as our ecological consciousness becomes a new paradigm of design, and design patterns take into account the coexistence of nature in the city and the creation of bionic construction systems – and therefore there are more and more buildings with green roofs and vertical gardens, which play important ecological functions in the city cluster. The modern and visionary projects inspired by nature were famous for Zaha Hadid, whose vision of futuristic buildings fits in perfectly with the new design challenges. Nowadays, more and more boldly is being talked about houses that are located underground or in hollow rocks. Breaking the taboos and the negative attitude of people to live underground, the natural environment might, in the future, recover the lost areas that have been changed or devastated as a result of human activities. In fact, it is a return to the original solutions, where protection against the changing weather conditions, but also the maintenance of a specific balance between human activity and nature, is an important element.

The idea of constructing underground houses is not new. All around the world a lot of that type of buildings can be found, including settlements and small towns. Examples of this may be houses in China, especially in the Provinces: Hunnan, Henan, Shanxi and Gansu [2]. Nearly 10 million people live there. An example of such a city is Sanmenxia in Henan Province, which, in 2005, was put on the National List of Monuments by Chinese authorities, assessing its cultural and historical value very high [2, 3]. The first settlements in this area were established in the Bronze Age (around 2000–1500 BC), but in fact the city flourished during the reign of the Ming dynasty in the 14th century. The first inhabitants built the settlement almost entirely underground to survive under difficult climatic conditions. As a result, the town of around 10,000 houses was carved into a loess layer. The average house is situated about 6–7 meters under the ground surface, and its length is about 10–12 meters. The house geography is almost identical in all cases – a kitchen room, a bathroom, several bedrooms and chambers, dedicated to domestic animals, i.e. pigs and/or poultry. The whole was designed in the form of a cube, where its center – a kind of yard being a representative part of the house, did not have a roof. Besides, there was also a well in every house, which during the heavy rains also served as a receiver of excess water. The roof of the house was actually an area that could be used in a variety of ways – for example, an additional usable area for the householders, or it was partly used for farming purposes. These houses are very strong and energy efficient – they have quite a constant temperature inside, which gives a great comfort of use, and they are sound proof and earthquake resistant. There was, however, no direct connection between the houses, so than no additional connecting structures, such as tunnels. To get to another home, one had to go to the surface and then go down there, using the stairs. Similar idea of building is represented by sandstone houses in Matmat, located in central Tunisia, on the edge of the Sahara desert in the landscape of the Jabal Matmata Mountains. The hight of the house is approximately 8–10 metres, while the chambers are about 3 metres high [3–5]. The difference in conceptual design of the building is mainly the shape of the courtyard – unlike in Sanmenxia, in Matmata the courtyard is usually round. Also the height of the building and the rooms are different with different parameters – the buildings in Mamat are larger in size. Equally interesting is the village of Kandovan, situated in East Atrpatakan Province, Sahad Mountain, about 50 km from Tabriz in Iran, where the houses were carved into volcanic rocks, and the oldest of them could have been built up to 700 years ago. They have an exceptional structure and shape, which resemble a little the termite nests. Some of them are two- three storey constructions, while the rooms are about two metres high. Due to its unique architecture and its landscape, this place was listed on the National Hertiage Society, Iran [3, 6].

In Lijiashan, near the town of Qikou in Shanxi Province (China), there are examples of well-preserved cave houses that have been hollowed out on the slopes of Jiashan Moutain. These are the houses, where the decorative front wall with windows and large entrance doors is built of bricks, while the rest of the building is hollowed in the rock space [3]. The layout of chambers, however, is different from the one in Sanmenxia – they were arranged in a straight line rather than on the perimeter of the square. Moreover, the courtyards are open to the view of scenic landscape. As individual homes are located on terraces - the whole structure resembles in a sense multi-storey houses. This results in a unique combination of the house and the surrounding space – the usable space of the home becomes optically enlarged, and its direct opening to the landscape suggests that it belongs to the very place. It also gives an impression of the greater sense of security of the people living in these settlements, than in previously discussed cases.

Also very interesting constructions of rock houses made in volcanic tuff can be found in Slovakia, in Brhlovce. The very first information on the community dates back to 1245 [7]. The houses are made, excavated in rock as a whole and have usually one or two small windows and an entrance door. In Georgia, in Uplistsikhe, situated in the neighbourhood of the city of Gori in the region of Kartvelian, on the rocky banks or the river Mtkwari (Kura), there is a unique ancient rock city, whose construction is dated for about the 5th century B.C. The homes have surprisingly large rooms, which are very often located on several levels of interconnected, excavated, corridors and stairs. The Uplisciche complex has very good infrastructure – including narrow roads and stairs leading to other homes [8]. The treasure of Turkey, Cappadocia, is also unique. Its underground, excavated in sediment rocks (tuff), houses, cloisters, churches, granaries, are self-sufficient underground cities, with very well-designed, internal infrastructure, that allow residents to survive with adequate food supplies and to contact each other without going out to the surface. These homes are characterized by excellent thermography [3, 9].

The largest and most interesting cities of Cappadocia, Kaymakli and Derinkuyu, are the towns of monks, connected by a 9 kilometre tunnel. Interestingly, these cities had a more than ten levels of communication with each other. Not only people, but also livestock could stay in specially designated rooms. The city needed to have a proper ventilation system, enabling people to live underground.

The above mentioned examples are characterised by the diversity (structure, division of space, or location, etc.), which, in majority of cases, results from their location and the climatic, physiographic, cultural and political conditions of a given space.

3 Modern realizations of the concept of underground houses

In today's world, there are many examples of buildings, that have been built underground and play either residential or service-utilitarian role – for example offices and shops are located underground. Subway stations, that are located under many cities, are also part of the subterranean model, in the same way like bunkers, that were built mostly in Europe as well as in the United States or China during the Cold War. Nevertheless, it is difficult to talk about their ecological function, as they were created mainly for military purposes, while shelters were built for the civilian population. At present, with the design paradigm shifting to conscious protection of natural environment and landscape values, many designers adapt their projects to new challenges, which gains recognition among ecologists and supporters of sustainable architecture.

Malcolm Wells was a great proponent of the construction of underground houses. He was an architect, who, in the 1960s, as an answer to a construction of a highway in the neighbourhood of his office, designed his office underground, and later designed many similar houses [10]. Wells had many followers, who used the concept of sustainable design in harmony with nature. Among them we can list William Morgan, who built a house on a dune, or who, properly speaking, built this house into a dune, in order to achieve greater stabilization of the building, or Peter Vetsch – the author of a futuristic mini housing estate, in northern Switzerland, in the town of Dietikon, where he designed and built nine houses, sharing a courtyard with a pond, that perfectly fits into the surrounding landscape at the same time increases its attractiveness [11]. A residence, built on the Greek island of Kea near Athens, designed by Marina Stassinopoulos and Konstantios Daskalakis is an interesting investment [12]. The project assumed complete adaptation to the specific environmental and geomorphological conditions of the island and special attention to the landscape. The building, in spite of geometric shapes, harmonises with the landscape due to its minimalist design. Nowadays, not only single underground homes are built, but also underground towns, such as Coober Pedy in Australia, about 680 km south of Alice Springs. Coober Pedy is a mining village that was established after the First World War with the arrival of opals miners to the area [13]. The houses were completely excavated in the rock and connected to a network of corridors resembling narrow roads. Today the city has about 3,500 inhabitants. There are houses, cafes and a theatre. The town functions in a perfect way and at the same time it has become a tourist attraction.

A new trend in ecological construction is, among others, construction of so-called Eartships, whose technology was developed and applied in practice by a famous US architect Michael Reynolds. The construction of these homes was based on recycled materials available to every human being, including mostly used car tires, cans, plastic and glass bottles, sand and earth [14]. At present there are about 3,000 of these types of buildings all over the world, mainly in the United States (most of them in New Mexico) as well as in Canada, Japan, Mexico, Bolivia, Australia, Spain, India, UK, Poland etc. [15]. Earthships are low-budget buildings, that can function in any climate. They are energy efficient, very well insulated from the environment and varied in shapes. These buildings can be built both on the ground and beneath it, where the insulation from the environmental conditions is bigger. The philosophy of building this type of building is based primarily on self-sufficiency, which means not only the reduction of utility bills, but also the on-the-spot production of food, and the use of water in continuous circulation. The height and width of the building depends only on the creativity and imagination of the designer and contractor and on construction possibilities. These houses do not affect the quality of the landscape in a negative way, and with a proper design they can also enrich it.

Underground houses are also a good investment of capital, because they will certainly not lose value. By observing the real estate market, there are more and more luxurious residences, which, apart from the living space above the ground level, have – often a multi-storey – extended underground part of the building. London can serve as an example of a city in which we find many such offers of this kind. Furthermore, it can be predicted, that in several decades building buildings underground cities may become a part of every-day reality.

In 1996, ACUUS (Associated Research Centers for the Urban Underground Space) was established, whose aim was to focus on raising awareness and promoting use of underground space of our cities and changing the paradigm of the sustainable development of urban tissue.

[16]. The organization conducts open discussion in a variety of environments – both at the university level, at the art and decision level, on urban and suburban urban space integration. Perhaps by promoting a new way of thinking about urban development, it is possible to achieve the state of sustainable development in urban space?

4 Advantages and disadvantages of underground houses and cities

Houses, built underground, in spite of their great advantages, have disadvantages and often generate a lot of controversy and emotion. It has been argued that underground housing is associated with humidity, darkness, insects, rodents, etc. It is often a taboos subject. Moreover, the psychological aspect plays a very important role, where the feeling of security, control or ordinary claustrophobia can determine our emotions, related with this type of houses. Also, the materials, from which we make houses, are not indifferent to people. For example, underground houses cannot be made of wood but wood can be put on the walls. When deciding, whether to build a house completely or partially underground, all the pros and cons of such investment should be taken into account. Also a cost account should be done. Advantages and disadvantages of constructing underground houses have been presented in the table 1.

Table 1. Advantages and disadvantages of the houses underground construction.

Advantages	Disadvantages
• lower construction costs compared to conventional one, • lower home maintenance costs, • checked for a multi-years housing system, • better use of the terrain spatial potential • resistance to earthquakes and tornadoes, • safe live environment in extreme weather condiotions • fire resistance, • high thermal efficiency; • energy saving, • large acoustic insulation, • use of less materials than in conventional houses, and in case of Eartship homes use of recycled materials • leaving the area for plants and animals - supporting natural ecological processes and maintaining high quality habitats • Significant impact on the CO_2reduction in the atmosphere by reducing the energy demand and leaving on the earth surface vegetated areas, biotopes support and biological diversity, • minimal impact on the landscape.	• psychical barriers - especially in constructions without windows, • in many cases, too small amount of day light, • poor design of the building may appear problems with ventilation and air quality in individual rooms, • in the case of poor building insulation problems with moisture, rodents and insects may occur, • during the flood there is a possibility of drowning inhabitants in the underground house, • there is a possibility of house collapsing in case of structural defects and misuse of the area above the house (roof, roof with woody vegetation), • The need for a very good roof protection of the house with special anti-corrosion foil, so that the plants do not damage its structure, • In the case of building the housing estates or underground cities will be needed dedicated technical and infrastructural solutions, that can significantly increase costs, • The lack of clearly defined building regulations, or their total absence in many countries in Europe and worldwide.

The construction of underground cities nowadays is an enormous challenge in terms of technic, infrastructure and logistics. The relocation of cities, roads, factories underground will cause many problems, including water supply, sewage, waste disposal, ventilation, lightening or psychosomatic barriers [17]. Part of the technical solutions we already have, because on the one hand there are experiences of our ancestors living in the underground cities or the experiences of Coober Pedy residents in Australia. We also have well-developed subway lines in which people live every day, or shopping malls, where there are parts at levels of at least minus 1. People's psychological barrier to use the underground is slowly breaking. People often do not even realize how much time they really spend underground. On the other hand, these cities may be more exposed to terrorist attacks because of their location.

5 Conclusion

Urban fabric in every corner of the world annually grows at a rapid pace. In the world one can observe two dominant phenomena. The first is related to the movement of people from villages to urban areas, so that the living space in the city starts to shrink. The second, called urban sprawl, means quite an uncontrolled spillover of urban structure, due to the increasing number of population in the city. In spots of green squares there are buildings, greenery is transferred to roofs and vertical gardens are created. But it's still not enough. We eliminate nature from our cities and destroy the natural environment. It is predicted that in the year 2050 the human population may reach over 9 trillion people, and at least 2/3 of them will live in cities [16]. In the emerging megacities and metropolis, space became a scarce commodity. We are already seeing a rise in flats and house prices in big metropolises, such as in Tokyo, Mexico City, London, Paris, New York, etc. There are increasing problems with road infrastructure and communication. This problem will grow year by year due to the systematic increase in the number of cars – in some cities their numbers increase faster than the population [16]. This situation is reflected in the deteriorating air quality, and the construction of high and near situated skyscrapers only aggravates it, rising the problem of urban ventilation system. Modern, bustling day and night cities and towns are extremely energy-consuming, contributing indirectly to the increase of CO_2 in the atmosphere. Landscape values are often severely disturbed, which indirectly influences the comfort of the people. Due to urban overcrowding and environmental problems, it would be better to consider the possibility of a gradual relocation of all or part of the land-based investment, including the entire road infrastructure and housing. Of course, this process will be progressing gradually and for many years, but thanks to its initiation and further realization we will gain not only new surfaces for the lives of people on our planet but will also have a positive impact on the world of plants and animals. Leaving on the surface of green gardens or natural forests and habitats will have a huge impact on the quality of the natural environment in every dimension, including landscape one. So we are left with a change of the cities design paradigm of and meeting the nature.

References

1. K. Ronka, J. Ritola, K. Rauhala, Tunnel. and Underground. Space Technology **13**, 1, 39–49 (1998)
2. A. J. Anselm, Energ. and Build. **40**, 1214–1219 (2008)
3. S.A. Alkaff, S.C. Sim, M.N.E. Efzan, Renew. and Sustain. Energ. Review **60**, 692–713 (2016)

4. A.A. Al-Mumin, Energ. and Build. **33**, 103–111(2001)
5. A.A. Al-Tameemi, D.J. Harris, Energ. and Build. **36**, 251–260 (2004)
6. M. Khodabakhshian, Proced. Engineering **165**, 649–657 (2016)
7. L. Kralovičová, I. Šimková, J. Vlčko, T. Durmeková, M. Brček, V. Boháč, Acta Geol. Slov. **6**, 1, 41–50 (2014)
8. UNESCO: http://whc.unesco.org/en/tentativelists/5234/ on line 17.03.2017
9. A. Erdem, Y. Erdem, *Underground space use. Analysis of the Past and Lessons for the Future*, 35–39 (2005)
10. K. Labs, Underground Space **1**, 135–156 (1976)
11. H. Altan, M. Hajibandeh, K. Anissa, T. Aoul, *A. Deep, Passive Design, ZEMCH: Toward the Delivery of Zero Energy Mass Custom Homes*, 209–236 (2016)
12. A. Benardos, I. Athanasiadis, N. Katsoulakos, Tunnel. and Underground. Tech. **41**, 46–52 (2014)
13. A. De Munari, D.P.S Capão, S.S. Richards, A.I. Schäfer, Desalination **248**, 72–82 (2009)
14. M. Reynolds, *Eartship: How to build your own*, **1** (1990)
15. K. Ip, A. Miller, Renew. Energ. **34**, 2037–2043 (2009)
16. D. Kaliampakos, Proced. Engineer. **165**, 205–213 (2016)
17. D. Kaliampakos, A. Benardos,WIT Transactions on the Built Environment **102**, 1743–3509 (2017)

Influence of environmental pollution with creosote oil or its vapors on biomass and selected physiological groups of microorganisms

Teresa Krzyśko-Łupicka[1,*], *Krystyna* Cybulska[2], *Paweł* Kołosowski[2], *Arkadiusz* Telesiński[2] and *Adam* Sudoł[1]

[1]University of Opole, Faculty of Natural Sciences and Technology, ul. Kominka 6a, 45-092 Opole, Poland
[2]West Pomeranian University of Technology Szczecin, Faculty of Environmental Management and Agriculture, ul. Słowackiego 17, 71-434 Szczecin, Poland

Abstract. Survival of microorganisms in soils from treatment facility and landfill of wooden railway sleepers contaminated with creosote oil as well as in two types of soils with different content of organic carbon, treated with creosote oil vapors, was assessed. Microbiological assays including determination of: the biomass of living microorganisms method and the number of proteolytic, lipolytic and amylolytic microorganisms were carried out under laboratory conditions. Chromatography analysis of the soil extract from railway sleepers treatment facility was performed using GC/MS. The highest biomass and the number of tested microorganisms were determined in soils from wooden railway sleepers landfill, while the lowest in soil from the railway sleepers treatment facility. Vapors of creosote oil, regardless of the soil type, significantly increased only the number of lipolytic bacteria.

1 Introduction

Development of railways in Poland is not only associated with investments in more and more modern fleet, but also with the maintenance of an adequate standard of railway lines. Achieving this goal requires incurring expenditures on production and subsequent utilization of used railway sleepers needed for the construction of railway tracks. A particular environmental problem is caused by treatment and storage of wooden sleepers impregnated with creosote oil [1].

Creosote oil, also known as the impregnation oil, is a mixture of products derived from the distillation of coal tar and boiling at 200–360°C temperatures [2]. It is a mixture of hundreds of compounds, of which only a small fraction (less than 20%) is present in an amount exceeding 1%. One of the most important groups of compounds forming the creosote oil structure are aromatic hydrocarbons, including polycyclic aromatic

* Corresponding author: teresak@uni.opole.pl

hydrocarbons (PAHs). Compounds in the creosote oil are characterized by high durability and ability to bioaccumulation [3, 4], which not only causes changes in the surrounding environment, but also threatens the health of humans and animals leading to increased cancer incidence [5]. Highly toxic effect of these substances on plants was also confirmed [1].

Legislation on the composition of creosote oil is not uniform in different parts of the world. Creosote oil used to impregnate the wood is classified according to the AWPA (American Wood-Preservers' Association) as of P1 and P2 types, while according to WEI (Western European Institute for Wood Preservation) as types: A, B and C. In Europe, only B and C types are approved for use now. They differ with physicochemical properties, which may affect the impregnation process quality and the way of the treatment [6].

In Poland, there are lot of areas contaminated with creosote oil as a result of its penetration into the soil and water from the long operating sleeper treatment facilities, or factories involved in wood impregnation. These post-industrial areas are also abundant in PCBs, benzene, toluene, and mineral oils.

Creosote oil is a substance difficult to recycle due to its low solubility in water and its removal from the environment is usually expensive and often ineffective. From contaminated areas, it is removed by physical, chemical or biological means. One method may be bioremediation, which consists of the controlled and supported activity of microorganisms using the hydrocarbon contamination as a source of carbon and energy [7]. There have been many studies on the degrading abilities of microorganisms for the disposal of toxic components of creosote oil, when searching for the most efficient method of environment cleaning. Robak et al. [8] indicated that high assimilation capacity in relation to the hydrocarbons are shown by yeast of *Yarrowia lipolytica*. However, after the period of bioremediation, very different results of reducing their concentration in the soil from 11% to 70% of the initial content, were achieved. The method of soil bioremediation in situ using eukaryotic microorganisms contributes to a remarkable reduction in the amount of petroleum hydrocarbons contained in the soil and the use of *Yarrowia lipolytica* strains speeds up the cleaning process [9].

Experiments by Sabate et al. [10] have shown the opportunities to biodegrade of creosote oil components by microorganisms present in the soil. They allow to get more effective degradation of compounds in the soil; not only those derived from creosote oil, as well as other contaminants arising from industrial production processes.

Due to the threat posed to the surrounding environment by creosote oil, to understand the biotic relationships within a contaminated environment is a must.

The aim of the study was to evaluate the survival of microorganisms in soils from treatment facility and landfill of wooden railway sleepers contaminated with creosote oil and in two types of soils with different content of organic carbon, treated with creosote oil vapors.

2 Materials and methods

The research material consisted of soil samples from the treatment facility and landfill of railway sleepers as well as soils collected from the topsoil of uncontaminated areas in the West Pomeranian province in:

- Lipnik – granulometric composition of loamy sand and organic carbon content at the level of 0.87%,
- Ostoja – light loam with organic carbon content 1.10%.

Both types of soil were treated with creosote oil vapors from crushed wooden railway sleepers. For this purpose, 500g of the soil was sealed for 21 days in desiccators with

 DOI: 10.1051/e3sconf/20172200092

fragmented railway sleepers. The control sample consisted of untreated soil not subjected to vapor treatment.

Microbiological assays were performed in laboratory including determination of:

- biomass of living microorganisms by means of physiological method by Anderson and Domsch using Ultragas U4S analyzer [11]. The amount of living microorganisms biomass recalculated onto g^{-1} DM soil was achieved from the equation (1):

$$x = 40.4y + 0.37 \tag{1}$$

where: x – amount of C contained in the biomass of living microorganisms [mg C · g^{-1} DM soil]; y – maximum initial CO_2 production [$cm^3 \cdot h^{-1} \cdot g^{-1}$ soil].

- number of microorganisms: proteolytic (LMP) on Frazier's medium, lipolytic (LML) on the medium with tributyrin, and amylolytic (LMA) on Waksman's medium [12] by means of Koch's decimal dilutions culture method. The cultures were incubated at 22–25°C for 3 to 7 days. The number of microorganisms are given as colony forming units (CFU) recalculated onto 1 gram of dry matter (DM) [CFU·g^{-1} DM].

At the same time, chemical analysis of averaged soil sample from railway sleepers treatment facility was performed applying GC/MS technique [13, 14].

The results were processed using one-way variance analysis (ANOVA) in Statistica 10 software applying Duncan's test.

3 Discussion

The number and activity of physiological groups of microorganisms in soils from the treatment facility and landfill of wooden railway sleepers as well as soils treated with creosote oil vapor, was varied (Fig. 1). The highest biomass of live microorganisms amounting to 15.3 mg C g^{-1} was determined in the soil from the landfill of railway sleepers, but it did not significantly differ from the microbial biomass determined in the soil from the Ostoja. In contrast, it was significantly higher than that in soil from Lipnik. Achieved results suggest that creosote oil in soil from the railway sleepers landfill had no significant effect on the amount of biomass.

The lowest biomass of living microorganisms of approximately 7 mgCg^{-1} was recorded in soil from the railway sleepers treatment facility; it was almost by 50% lower as compared to the biomass determined in soil from the railway sleepers landfill and soil from Ostoja.

It is worth noting that the creosote oil vapors, as compared to the control, did not affect the biomass of living microorganisms in the soils from Ostoja and Lipnik.

The tested samples were similar in terms of the number of proteolytic and amylolytic microorganisms, while varied with the number of lipolytic microorganisms.

The highest number of proteolytic microorganisms was determined in the soil from the railway sleepers landfill, which was similar to that in the Ostoja soil exposed to vapors, as well as in control. The lowest number of proteolytic-activity microorganisms was determined in the soil from the sleeper treatment facility ($4.0 \cdot 10^3$ CFU·g^{-1} DM). Creosote oil vapors, as compared to the control, did not affect the number of proteolytic microorganisms in the tested soils from Ostoja and Lipnik (Fig. 2).

 DOI: 10.1051/e3sconf/20172200092

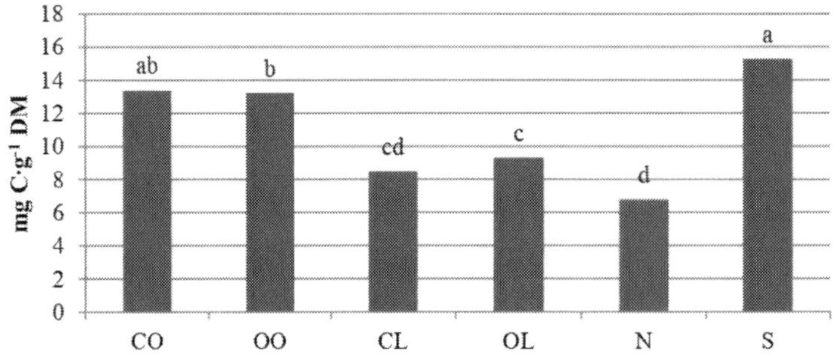

Fig. 1. Biomass of living microorganisms (mg C·g^{-1} DM) in soils from the treatment facility and landfill of railway sleepers as well as soils subject to the creosote oil vapors treatment. CO – soil from Ostoja (control); OO – soil from Ostoja subject to the creosote oil vapor treatment; CL – soil from Lipnik (control); LO – soil from Lipnik subject to the creosote oil vapors treatment; N – soil from the railway sleepers treatment facility; S – soil from the railway sleepers landfill treated with the creosote oil. a, b..d -values denoted with the same letters do not differed statistically (P < 0.05).

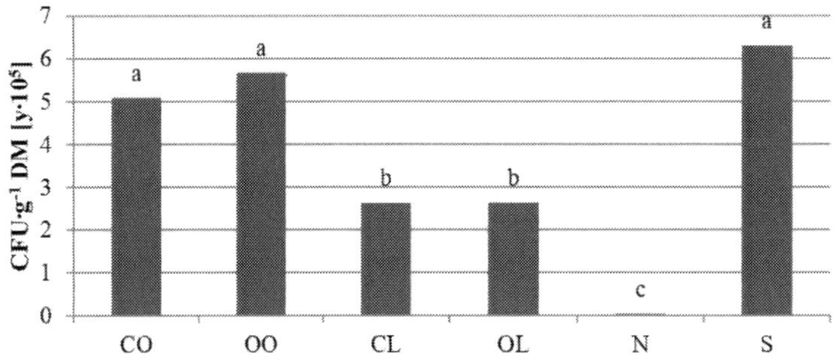

Fig. 2. Number of proteolytic microorganisms (LMP) [CFU·g^{-1} DM of soil] in soils from the railway sleepers treatment facility and landfill as well as soils subjected to creosote oil vapors treatment. CO – soil from Ostoja (control); OO – soil from Ostoja subject to the creosote oil vapor treatment; CL – soil from Lipnik (control); LO – soil from Lipnik subject to the creosote oil vapors treatment; N – soil from the railway sleepers treatment facility; S – soil from the railway sleepers landfill treated with the creosote oil. a, b…d - values denoted with the same letters do not differed statistically (P < 0.05).

A similar trend in the microorganisms number oscillation was recorded for amylolytic activity. Significantly the highest number (7.8·10^5 CFU·g^{-1} DM) was determined in soil from the railway sleepers landfill, whereas the lowest in the soil from railway sleeper treatment facility. Creosote oil vapors, as compared to the control, significantly stimulated (by 32%) development of microorganisms showing amylolytic activity in the loamy soil from Lipnik. In contrast, they had no significant effect on the number of these microorganisms in sandy soil from Ostoja.

 DOI: 10.1051/e3sconf/20172200092

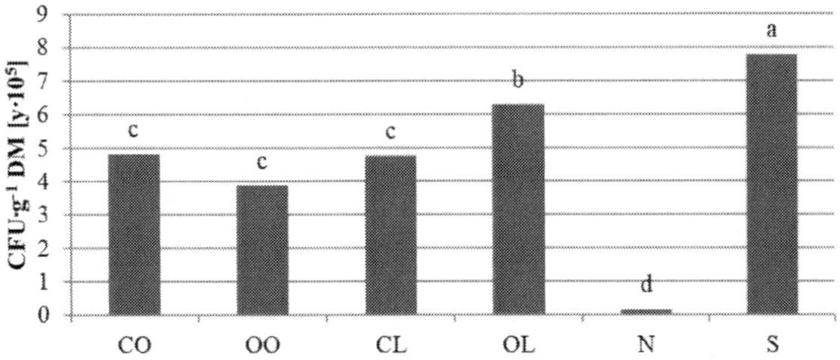

Fig. 3. Number of amylolytic microorganisms (LMA) [CFU·g^{-1} DM of soil] in soils from the railway sleepers treatment facility and landfill as well as soils subjected to creosote oil vapors treatment. CO – soil from Ostoja (control); OO – soil from Ostoja subject to the creosote oil vapor treatment; CL – soil from Lipnik (control); LO – soil from Lipnik subject to the creosote oil vapors treatment; N – soil from the railway sleepers treatment facility; S – soil from the railway sleepers landfill treated with the creosote oil. a, b….d - values denoted with the same letters do not differed statistically ($P < 0.05$).

The lipolytic-activity was shown by the most numerous group of physiological microorganisms in the tested samples. Their averaged number in analyzed samples was three times higher than the amylolytic bacteria and approximately 7 times higher than proteolytic microorganisms.

Significantly the highest number of lipolytic microorganisms ($2.4{\cdot}10^6$ CFU·g^{-1} DM) was determined in the soil from railway sleepers landfill and loamy soil from Lipnik subjected to the creosote oil vapors treatment (2.1 10^6 CFU·g^{-1} DM), while the lowest number of these microorganisms was determined in soil from the treatment facility ($2.5{\cdot}10^4$ CFU·g^{-1} DM), i.e. 55-fold less than average value for all tested samples. There was a significant increase in the number of lipolytic microorganisms in sandy soil from Lipnik treated with creosote oil vapors, as compared to the control. However, their numbers were at a similar level to that in the loamy soil from Ostoja (control).

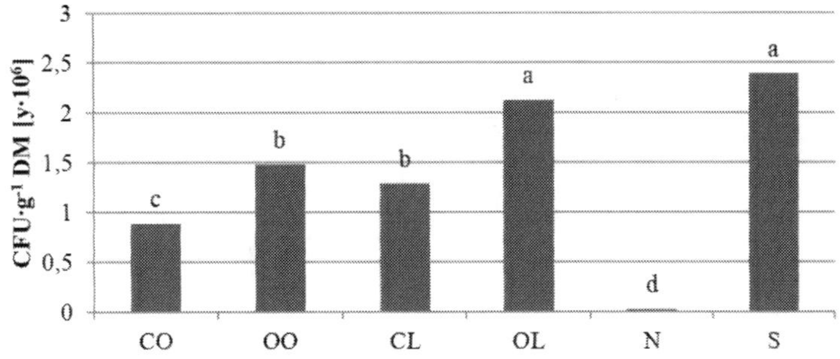

Fig. 4. Number of lipolytic microorganisms (LML) [CFU·g-1 DM of soil] in soils from the railway sleepers treatment facility and landfill as well as soils subjected to creosote oil vapors treatment. CO – soil from Ostoja (control); OO – soil from Ostoja subject to the creosote oil vapor treatment; CL – soil from Lipnik (control); LO – soil from Lipnik subject to the creosote oil vapors treatment; N – soil from the railway sleepers treatment facility; S – soil from the railway sleepers landfill treated with the creosote oil. a…..d - values denoted with the same letters do not differed statistically ($P < 0.05$).

The lowest level of all investigated microbiological indicators was recorded in the soil from the railway sleeper treatment facility, indicating that it is a hostile environment for microorganisms. Chemical analysis showed that the pH of the soil was strongly acidic (pH 4.1), and creosote oil content was 1183 mg/kg DM, including 672 mg/kg DM of PAHs and 511 mg/kg DM of n-aliphatic hydrocarbon with C8-C40 chain length. Presence of aliphatic and polycyclic aromatic hydrocarbons (PAHs), and particularly high concentration of benzo(a)pyrene (62 mg/kg DM) and benzo(k)fluoranthene (175.3 mg/kg DM), leads to a biological imbalance as the result of an increase in the amount of organic carbon with the deficit of available forms of nitrogen and phosphorous necessary for the development of microorganisms (Table 1).

Table 1. Quantitative and qualitative composition of polycyclic aromatic hydrocarbons (PAHs) and C8-C40 n-aliphatic hydrocarbons in soil from the railway sleepers treatment facility [mg · kg^{-1} DM].

WWA [mg · kg^{-1} DM]		**n-alkany (C8–C40) [mg · kg^{-1} DM]**	
Compound	**Amount**	**Compound**	**Amount**
Naphtalene	0.1	C8	0.0
Acenaphtylene	0.0	C10	0.2
Acenaphten	0.5	C12	0.6
Fluorene	28.1	C14	28.8
Fenantrene	46.0	C16	14.1
Anthracene	132	C18	149.7
Fluorantene	84.6	C20	10.0
Pyrene	4.8	C22	80.9
Benzo(a)anthracene	5.7	C24	28.8
Chryzene	88.8	C26	11.4
Benzo(b)fluoranthene	0.0	C28	18.8
Benzo(k)fluoranthene	175.3	C30	114.8
Benzo(e)pyrene	0.0	C32	4.9
Benzo(a)pyrene	61.9	C34	2.8
Indeno(1,2,3-cd)pyrene	43.9	C36	1.7
Dibenzo(a,h)anthracene	0.0	C38	0.6
Benzo(g,h,i)perylene	0.0	C40	43.0
Summary	**671.7**		**511.4**

4 Results

Creosote oil, due to its antiseptic properties, is commonly used as a impregnating substance to protect wooden elements used in industry, such as railway sleepers, fragments of bridges, coastal infrastructure, traction poles, ship hulls, log houses and fences, playground and yards. Components of creosote oil adversely affect the fundamental ecological functions of the soil: lead to the transformation in quantitative and qualitative composition of microorganism populations; inhibit the growth, photosynthesis and respiratory processes at plants; and also have direct and indirect impact on animal and human organisms.

Many studies indicate mutagenic and toxic properties of creosote oil components [15]. Chemically, the creosote oil is a mixture of compounds such as aliphatic hydrocarbons, aromatic hydrocarbons, PAHs, heterocyclic compounds containing sulfur, oxygen, nitrogen, and carboxyl group [16]. PAHs have mutagenic and carcinogenic action to humans and plants, as well as aquatic and land animals [17]. Aromatic amines are also toxic as they are rapidly absorbed through the skin. In turn, nitrogen heterocyclic compounds, e.g. quinoline, due to the high water solubility, can lead to its contamination [18].

Substances with high volatility contained in the creosote oil also pose a threat to the environment, because they are difficult to eliminate. The vapor pressure of the individual oil components is from 12 700 Pa for benzene to $2.0 \cdot 10^{-10}$ Pa for dibenzoanthracene [19].

Populations of microorganisms living within the soil can utilize a dose of PAHs as a carbon or energy source, converting them into biomass, water and CO_2. The formation of biologically inaccessible oxidized metabolites and PAHs fractions was documented by [10]. As expected by other researchers [20, 21], biomass of living organisms in contaminated soil and ground within the railway sleepers landfill, also increased. This fact was confirmed by our findings, since the highest biomass and number of proteolytic, amylolytic and lipolytic microorganisms were also determined in soils from the landfill of timber railway sleepers, despite the presence considerable amounts of benzo(k)fluoranthene (175.3 $mg \cdot kg^{-1}$ DM) and benzo(a)pyrene (62 $mg \cdot kg^{-1}$ DM) in that environment. Permissible total PAH content in soils is 250 $mg \cdot kg^{-1}$ DM, and the amount of individual compounds cannot exceed 50 $mg \cdot kg^{-1}$ DM [22].

Benzo(a)pyrene is the main carcinogenicity indicator of a soil [23], but relative carcinogenicity indices for other PAHs have been also determined. According to these data, dibenzo(a,h)anthracene the presence of which in the tested soil was not confirmed, is more carcinogenic as compared to benzo(a)pyrene [24].

The creosote oil vapors had no significant effect on the biomass of living microorganisms in tested soils, regardless of their structure as compared to the uncontaminated soils, but only stimulated the growth of lipolytic microorganisms, as confirmed by studies of Simarrao et al. (2013) [25]. However, in the soil with a higher content of organic carbon, number of amylolytic microorganisms increased. Due to the serious threat posed to the surrounding environment by creosote oil, the biotic relationship in a contaminated environment have to be recognized.

5 Conclusions

1. The highest biomass and the number of proteolytic, amylolytic and lipolytic microorganisms were determined in soils from wooden railway sleepers landfill, while the lowest in soils from the railway sleeper treatment facility.
2. Creosote oil vapors had no effect on the biomass of living microorganisms in tested soils, regardless of their structure, as compared to the non-contaminated soil.
3. In soils exposed to creosote oil vapors, number of lipolytic microorganisms was significantly higher and that of proteolytic ones did not change as compared to the uncontaminated soils.
4. Creosote oil vapors, compared to the control, significantly stimulated the growth of amylolytic microorganisms in the soil with a higher organic carbon content, but did not affect their number in sandy soil.

References

1. P. Witomski, FFP, B **34**, 37–44 (2003)
2. Karta charakterystyki preparatu: Centrala Obrotu Towarami Masowymi DAW– Bytom Sp. z o.o. Bytom (2010)
3. S. Moret, G. Purcaro, L. S. Conte, Sci. Total Environ. **386**, 1–8 (2007)
4. Dyrektywa Komisji 2011/71/UE z dnia 26 lipca 2011 (2011)
5. S. Karlehagen, A. Andersen, C. G. Olhson, Scand. J. Work. Environ. Health **8**, 26–29 (1992)
6. PN-EN 13991, (2004)
7. M. Vinas, J. Sabate, M.J. Espuny, A.M. Solanas, Appl. Environ. Microbiol. **71**, 7008–7018 (2005)

8. M. Robak, T. Boruczkowski, W. Drożdż, Z. Lazar, M. Baranowska, D. Prządo, M. Steininger, Ochr. Śr. **33**, 27–33 (2011)
9. T. Janeczko, W. Bąkowski, E. Walczak, M. Robak, J. Dmochowska-Gładysz, E. Kostrzewa-Susłow, Ann. Microbiol. **65**, 1097–1107 (2015)
10. J. Sabate, M. Vinas, A. M. Solanas, Chemosphere **63**, 1648–1659 (2006)
11. J.P.E. Anderson, K.H. Domsch, Soil Biol. Biochem. **10**, 195–221 (1978)
12. Manual DIFCO, Difco Laboratories, Detroit, Michigan, 48232 USA, 10th ed. (1984)
13. T. Krzyśko-Łupicka, T. Ciesielczuk, M. Chwałowska, Acta Sci. Pol. Biotechnol. **12**, 5–18 (2013)
14. Cz. Rosik-Dulewska, T. Krzyśko-Łupicka, T. Ciesielczuk, Ł. Kręciło, Pol. J. Chem. Technol. **17**, 17–22 (2015)
15. E. Elovaara, P. Heikkila, P. Pyy Mutanen, V. Riihimaki, Occup. Environ. Med. **52**, 196–203 (1995)
16. C. Melber, J. Kielhorn, I. Mangelsdorf, WHO, CICADS 62 (2004)
17. J.H. McCann, B.M. Greenberg, K.R. Solomon, L. Aquat. Toxicol. **50**, 265–274 (2000)
18. J. Kurzepa (red.), *Chemia organizmów żywych* (Radom. Tow. Nauk., Radom, 2014)
19. R. von Burg, T. Stout, J. Appl. Toxicol. **12**, 153–156 (1992)
20. G. Bengtsson, N. Törneman, J.R. De Liphay, S.J. Sorensen, Environ. Microbiol. **65**, 91–100 (2013)
21. S. Mukherjee, H. Juottonen, P. Siivonen, C.L. Quesada, P. Tuomi, P. Pulkkinen, K. Yrjälä, ISME J. **8**, 2131–2142 (2014)
22. Dz. U. Nr 165, poz. 1359 – Rozporządzenie Ministra Środowiska z dnia 9 września 2002
23. A. Sapota, Podst. i Met. Oceny Środ. Pracy **2**, 179–208 (2002)
24. I.C. Nisbet, P.K. LaGoy, RTP **16**, 290–300 (1992)
25. R. Simarroa, N. Gonzáleza, L.F. Bautistab, M.C. Molina, J. Hazard. Mater. **262**, 158– 167, (2013)

The occurrence of heavy metals and metal-resistant bacteria in water and bottom sediments of the Straszyn reservoir (Poland)

Eliza Kulbat[1,*], *Aleksandra* Sokołowska[1]

[1]Gdańsk University of Technology, Narutowicza 11/12, 80-233 Gdańsk, Poland

Abstract. The aim of this study is to investigate the distribution of selected heavy metals and metal-resistant bacteria in water and bottom sediments of the surface drinking water reservoir for Gdańsk. The following sequence of metals in regard to metal concentration in sediments can be written down: Zn > Pb > Cu > Cd. The evaluation of metals accumulation was performed using the Müller index, to indicate the bottom sediment's contamination and geochemical classification of sediment quality according to Polish standards. The Müller geochemical index was changing in a wide range: < 1–4.1. Although the maximum value of Müller's geochemical index determined for copper indicates that the sediment is 'strongly contaminated', in general the analysed bottom sediments were classified as the I and II category according to Polish geochemical standards. From the microbiological side a significant part of heterotrophic bacteria isolated from the bottom sediment and surface water (raw and treated water) showed a resistance to 0.2 mM and 2 mM concentrations of zinc, copper and lead. The highest percentages of metal-resistant bacteria were recorded in the sediments of the reservoir (60%–88%). The share of metal-resistant strains in the raw water was significantly lower (34%–61%). The results indicate also that water treatment processes may contribute to the selection of resistant strains.

1 Introduction

Heavy metals are one of the most widespread contaminants in the environment. They cause the environmental pollution from sources such as industrial effluents, mine tailings, land application of fertilizers, atmospheric deposition and leaching of metal ions from the soil into lakes and rivers by acid rain [1–3]. The authors of many papers report the impact of heavy metals on aquatic communities including bottom sediments of rivers and lakes [4–9]. Numerous natural and anthropogenic factors affect the chemical composition of surface water sediments [10–12]. In the case of flow-through lakes often the most important material is the suspended and dissolved one, transported by supplying river, especially passing through urban, rural or agricultural areas [9]. The composition of sediments depends on geological structures, geomorphology and weathering process, so the

* Corresponding author: ekul@pg.gda.pl

background values of individual compounds are characteristic for each region. Sediments accumulate contaminants like metals and can serve as secondary sources of pollution to the ecosystems they are connected with. The potential mobility of heavy metals depends on the physio-chemical conditions like pH and Eh level, temperature or presence of organic and mineral ligands [9, 13–15]. In addition, flow interruptions, storm events and other perturbations such as ice scouring can result in heavy metals being released again from the sediments. The desorption and transformation of contaminants into more bioavailable chemical forms may pose ecological risk. Although some heavy metals are required for the normal functioning of bacteria, fungi, plants and animals [16, 17] most can be toxic to all branches of life. The environmental pollution as well as the widespread use of antibiotics promote the formation of the mechanisms of resistance in bacteria. Drinking-water treatment plants use a number of treatment methods to improve water quality: e.g. flocculation, sedimentation, filtration and disinfection. Among the processes, chemical disinfection contributes greatly to the control of microorganisms from treatment plant to point of use [18]. It can promote however the survival of bacteria resistant to chlorine compounds, antibiotics as well as heavy metals.

Bottom sediments are good indicators of the contamination degree of the water environment. The assessing of concentrations of heavy metals and comparing them with background or reference value can be applied to determine the sediment contamination. There have been numerous criteria of sediment quality and some of them have been found to be a relatively good predictors of site contamination [19, 20]. The most commonly used are Muller's index of geoaccumulation (I_{geo}) [9, 15, 21], but in some countries (Poland, Germany) also the regional criteria are employed [22, 24].

The main objectives of this study are: to determine the concentration of heavy metals including Zn, Cu, Pb and Cd in the surface sediments of Straszyn lake, to assess the threshold of heavy metals contamination in sediments using the sediment quality criteria and to determine heavy metal resistant bacteria (Cu, Zn, Pb) in bottom sediments, raw water and drinking water.

2 Study area

The study was conducted on Straszyn Lake located on the Radunia river (the northern Poland) on the eastern edge of the Kashubian Lakes District. This reservoir has been constructed in 1910 mainly to produce energy. Situated here a small hydroelectric power plant with a capacity of 2450 kW is the oldest one on the Radunia river and remains in operation until today. The Radunia's catchment area equals to 837 km^2 and as much as 67.6% of the area is agricultural land. The total volume of Straszyn reservoir is 3 mln m^3, surface area amounts to 0.75 km^2 and the average retention time is 170 hours [25]. Since 1986 the main function of this reservoir is drinking water provision for several districts of Gdańsk.

The Straszyn surface intake is one on the biggest in Gdańsk and supplies the drinking water to more than 30% of the inhabitants [26]. During the research the raw water was pretreated with the ozone (0.5–0.8 gO_3/m^3, 3 min.). After the coagulation (10% solution of $Al_2(SO_4)_3$ in summer or 5% solution of NaOH in winter) and flocculation processes in two-staged labyrinthine chambers, the water was filtered through sand and then intermediate ozone and filtration through activated carbon (up to 6m/h) was applied. Prior to delivery to the network, polyphosphates and lime water were added to the treated water and then it was disinfected with mixture of Cl_2 and ClO_2.

 DOI: 10.1051/e3sconf/20172200093

3 Materials and methods

3.1 Samples collection

Heavy metals concentration (Zn, Cu, Pb, Cd) was determined in sediments and presence of potentially heavy metals resistant bacteria was determined in the bottom sediments, raw water and water in distribution system.

3.1.1 Sample collection of sediments

Sediments sampling sites were located in the mainstream of reservoir. The ordinal numbers of the points increase from the Radunia river inflow to the water intake at the dam. The sampling locations are shown in Fig. 1.

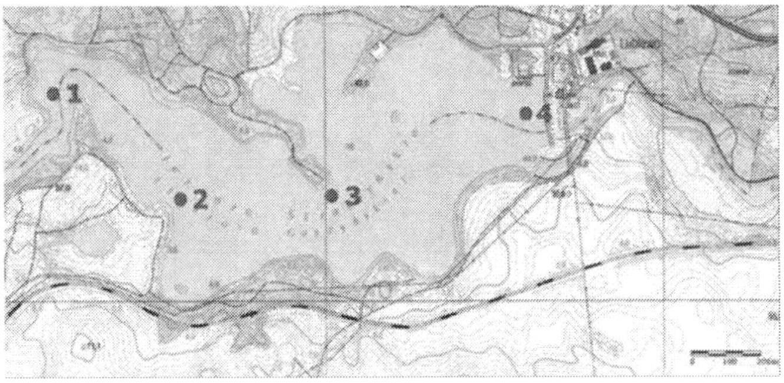

Fig. 1. Positions of sampling sites.

Surface sediment sampling was performed between May and October 2013, 8 times, in approximately 20 days intervals. Sediment samples of 10–15 cm thickness were collected using the KC Kajak sediment core sampler (Denmark Research Equipment) and stored in acid washed plastic containers. After collection samples were air-dried and sieved (2 mm plastic sieve) in order to remove major detritus. A sterilized stainless steel 10 cm diameter sampler was used for collecting sediments for the microbiological analyses.

3.1.2 Sample collection of water

The raw water was collected after the screens, and the treated water was collected at five points chosen on the length of the network – there were downloaded 8 series of samples in total. The water samples were collected directly into sterile 500 ml glass bottles.

3.2 Laboratory analyses

The samples of sediments were analyzed in terms of grain size, organic matter content and selected heavy metals content (Zn, Cu, Pb, Cd).

Grain size distribution was performed using wet sieving method (sand: 1–0.05 mm, clay silty: 0.05–0.002 mm, clay < 0.002 mm) in Analysette 28 Image Sizer. Organic matter content was determined using loss of ignition (LOI) technique at 605°C.

For metal analyses the sediment samples were dried at 105°C (to constant weight) and sieved through a 0,2 mm sieve. In order to determine the content of Zn, Cu, Pb and Cd the sediment samples were extracted with aqua regia ($HCl:HNO_3$ = 3:1) using Büchi digestion system K-438 [27]. All the concentrated acids used for the experiments were supra pure (Merck) quality. Zn, Cu, Pb and Cd standard solutions were prepared from 1000 $mg \cdot L^{-1}$

Merck standard solution (Certipur®). Blank reagent and reference materials analysis were done following the same experimental procedure.

Total metals concentrations in sediments were measured by flame atomic absorption spectrometry (FAAS) using Vario 6 (air/acetylene flame), equipped with a single element hollow cathode lamp and a deuterium lamp for the background correction. All measurements were performed three times and reported as mean values. Replicate measurements agreed within < 5%. Accuracy of the analytical method was assessed using the sediment reference materials (river clay sediment LGC6139). During metals analyses, calibration solutions, blanks and standards were re-run every 12 samples. In studies of bacterial resistance to heavy metals there were used sulfate salts of zinc, copper and lead. The total number of bacteria was isolated on R2A agar [28]. Heterotrophic bacteria in water and sediment samples were enumerated by spread plate count. The plates were incubated at 22°C from 3 to 7 days. The sulfate salts of zinc, copper and lead were used as follows: the appropriate volumes of stock 10 mM $ZnSO_4$, 10 mM CuSO4 and 10 mM $Pb(NO_3)_2$ solutions, sterilized via filtration, were added to a 100 ml sterile R2A liquid medium 15 minutes before inoculation in order to give a final concentration in solution 0.2 mM and 2 mM. The percentages of metal-resistant bacteria were determined by comparing the number of bacteria on the R2A medium with Cu, Pb and Zn with the number of bacteria on the R2A medium without them. 100 g sediment samples were mixed with 100 ml of Ringer solution with 0.28% sodium pyrophosphate additive [29, 30]. The solution was shaken vigorously for 20 minutes with a type 358S laboratory shaker and decanted for 2 minutes.

4 Results and discussion

Grain size analyses revealed that sediments sampled from the site 1 (upper part of reservoir, Fig. 1) were coarser (fraction ‘sand’ 47% and ‘clay silty’ 49%) than sediments taken from the sites 2–4. Sediments from the middle and lower part of reservoir were finer, with the fraction ‘clay silty’ representing about 83%-91% of the sediment mass. The organic matter content (express as LOI) presented variability in the range of 10.6–14.8%. Zn, Cu, Pb and Cd occurred in the investigated sediments in a wide range of concentration. Taking into consideration the criterion of the concentration in sediments, the following sequence of metals can be identified: Zn > Pb > Cu > Cd. The content of Zn in the sediments of the Straszyn reservoir were characterized by considerable variability (15.31–92.47 mg kg^{-1}dry wt). The mean Zn contents for sampling points were lower in the upper part of the reservoir compared to the middle and lower part (Fig. 2). Concentrations of Cu were in the range 2.34–25.65 mg kg^{-1}dry wt; similar on the stations 2–4 (mean values 20.40–22.05 mg kg^{-1}dry wt) and lower on the station 1 (mean value 3.81 mg kg^{-1}dry wt). The highest Pb content (mean value 33.75 mg kg^{-1}dry wt) was detected in the sediments from the middle part. Concentration of Cd were in the range 0.04–1.70 mg kg-1dry wt. Higher contents of Cd were found also in the middle and lower part of reservoir (from stations 2–4). Fig. 2 shows changes in the total metal content in sediments at the sampling sites.

Similar results were found in other investigations conducted in the sediments of lakes in the northern and northeastern Poland. There were observed though significant differences in heavy metals concentrations in bottom sediments in Poland: sediments of lakes and rivers located in southern Poland were more strongly contaminated by various metals in comparison to northern Poland [9, 24, 31].

DOI: 10.1051/e3sconf/20172200093

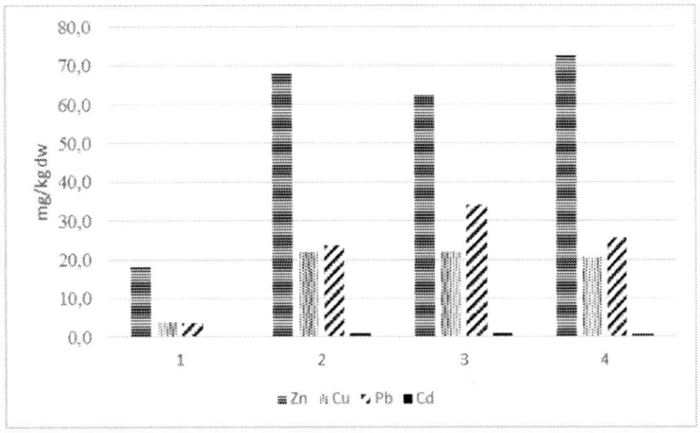

Fig. 2. Changes in the concentrations of zinc, copper, lead and cadmium in sediments at the sampling sites.

The earlier studies (1999–2000) of heavy metals in sediments of Straszyn lake [32] indicated that the concentrations of Zn, Cu, Pb and Cd were higher than the average ones for the Pommeranian Region. In our reserch, (conducted 14 years later) the concentrations of zinc, lead and cadmium were significantly lower and the content of copper was at a comparable level. The improvement of the quality of sediments was probably caused by the modernization and expansion of municipal facilities located in the catchment of the Radunia River: wastewater treatment plants in Przodkowo (2001–2002) and in Somonino (2008–2009). The reports of Gdańsk Inspectorate of the Environmental Protection [33] provided that in 2000 water quality of Radunia river was maintained in I and II class according to the Polish standards, but the concentration of copper was too high. The report on 2013 assesses the condition of the Radunia River's water in accordance with the WFD as good. Kuriata-Potasznik [9] reported that a river flowing through a lake is a key factor responsible for the input of the majority of available fraction of heavy metals.

According to the correlation coefficients, accumulation of heavy metals in the Straszyn reservoir sediments was mainly related to content of fine fraction of sediments: silty clay (r^2 = 0.92–0.99) and clay fractions (r^2 = 0.68–0.87). Similar results were obtained for Cu, Zn, Pb, Cd and Ni by Szarek-Gwiazda and Strzebońska [15, 34]. Additionally a strong negative correlation was found between metals and sand fraction (r^2 = -0.90 – -0.99). The high correlation coefficients between metals were also calculated (r^2 = 0.86–0.98) (Tab. 1).

Table 1. Coefficients of correlation between heavy metals concentration and between heavy metals concentration and parameters affecting metals distribution in the sediments.

Parametr		Zn	Cu	Pb	Cd
		r^2			
Zn			0.97	0.88	0.95
Cu				0.95	0.98
Pb					0.86
Grain size	0.05–1.00	-0.99	-0.96	-0.92	-0.90
	0.002–0.05	0.99	0.97	0.93	0.92
	< 0.002 mm	0.87	0.75	0.71	0.68
LOI		0.23	0.21	0.45	0.00

The assessment of sediment enrichment with metals was carried out using Polish geochemical classification [24] and Müller's geoaccumulation index I_{geo} [35].

The sediment metals concentrations presented by [36] were used as a local geochemical background in all the calculations.

According to the first geochemical classification of river and lake sediments in Poland, proposed by the Polish Geological Institute [24] sediments deposited at the analysed sites in the Straszyn reservoir were classified with respect to:

- Zn and Pb falling into the first class in all the sampling points
- 47% of the samples were classified to the I and 53% to the II category in terms of Cu
- 75% of the samples were classified to the I and 25% to the II category in terms of Cd

Bojakowska [24] also reported, that the bottom sediments in northern Poland (rivers: Narew, Noteć, Łeba, Słupia) were classified in I and II category.

To establish contamination status of the Straszyn reservoir sediments, also the geoaccumulation index was calculated and classification in I_{geo} classes was carried out:

$$I_{geo} = log_2\, C_n/1.5B_n \tag{1}$$

C_n – concentration of the element "n" in the fraction < 0.002 mm (clay)

B_n – background value for this element for this fraction of river sediment

The factor 1.5 is employed to take into account possible variations in the background data.

The index of geoaccumulation consists of seven grades, whereby the highest grade reflects 100-fold enrichment above the background values. The geoaccumulation classes and corresponding contamination intensity listed by Förstner [22] were used.

In this study the values of I_{geo} were calculated from using grain-size fraction of < 0.2 mm. The Müller's geochemical index changed in a wide range: the lowest values for each metal were observed in the sampling point 1 and the highest – near the dam. The maximum value of the Müller's geochemical index was determined for copper: it was in the range 3–4 which means that the sediment was 'strongly polluted'. The values of I_{geo} for Zn and Pb were classified in grades 1-3: 'moderately or strongly polluted' were the samples of sediments in the middle and lower part of reservoir. Sediment of Straszyn reservoir was classified as 'unpolluted to moderately polluted' because of I_{geo}< 1 for Cd. Although the bottom sediments in Poland were classified in grades 3–4 for Cu, 3–4 for Pb, 4–6 for Zn, 5–6 for Cd in river Odra [37], sediments of lake Symsar (northern Poland) were found as 'uncontaminated' and 'moderately contaminated' [9].

The graph 3 shows the mean percentages of heavy metals resistant bacteria. Bacteria isolated from sediments, raw water and drinking water displayed the highest resistance to zinc in the concentration 0.2 mM (respectively 88% , 62% and 83%). In case of lead (0.2 mM) and copper (0.2 mM) the percentages of strains resistant to these metals were lower and reached respectively 81% and 70% of the bacteria isolated from sediment, 54% and 50% of the bacteria isolated from the raw water and 75% and 66% in case of strains isolated from tap water. The resistance to the higher concentration (2 mM) of copper was shown by 60% of the strains isolated from sediments, 35% of the strains isolated from the raw water and 58% from tap water. In case of bacteria resistant to Zn and Pb (2 mM), the figures were respectively 67% and 69% for deposits, 49% and 42% for raw water and 64% for tap water in both categories (Fig. 3).

High lead levels in drinking water are still a concern for households serviced by lead pipes in many parts of North America and Europe [38]. Copper pipes and zinc-coated components are also commonly used in water distribution systems. Many researchers point out that primarily copper and zinc ions inhibit the growth of microorganisms, but bacteria resistant to these metals are often resistant to antibiotics and disinfectants as well [39, 40].

According to Schwartz et al. [40] the relationship between bacteria resistance to heavy metals, antibiotics and disinfectants was observed. The disinfection process, in particular the usage of chlorine compounds may result first of all in the survival of strains resistant to both multiple-antibiotic and heavy metals [41]. One of mechanisms, by which sub-inhibitory levels increase the risk of selection of ARB, is the chemical stress [42].The application of chlorine solution in the disinfection process leads to the abundance of

 DOI: 10.1051/e3sconf/20172200093

antibiotic-resistant bacteria and antibiotic-resistant genes in opportunistic bacteria [43, 44]. Disinfection efficiency does not remain the same throughout the supply system, as the chlorine concentration varies. Responses range from lethality/complete inhibition at high concentrations, selective survivability of resistant populations at sub-inhibiting concentrations, to responses triggering biochemical stress at much lower (sub-inhibitory) concentrations. Some of surviving bacteria may have increased the innate resistance to those impacts [45].

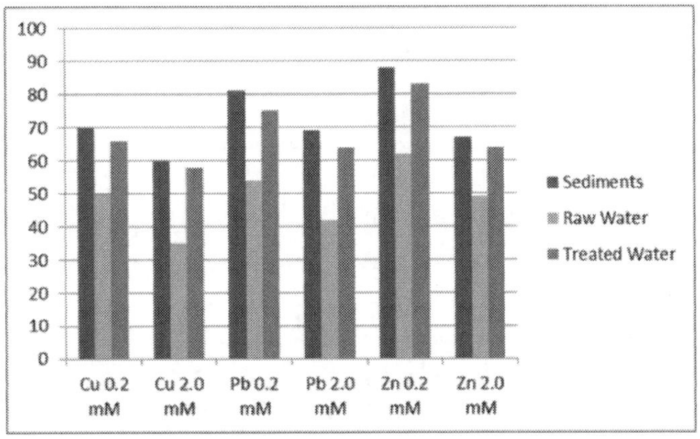

Fig. 3. The percentage of heterotrophic bacteria isolated from bottom sediments, surface raw water and tap water resistant to zinc, lead and copper.

5 Conclusions

The concentrations of metals in the Straszyn Lake measured in this research were lower than the contents reported by Kulbat [32] in studies conducted in 1999-2000, except copper. Accumulation of Zn, Cu, Pb and Cd in sediments was mainly related to content of fine fraction of sediments. The results of this study proved that sediments of Straszyn reservoir not exceed the levels of I and II class according to Polish geochemical standards proposed by Bojakowska [24]. However the maximum value of the Müller's geochemical index was determined for copper: it was in the range 3–4 which means that the sediment was 'strongly polluted'. The sediments were 'moderately or strongly polluted' of zinc and lead and unpolluted to moderately polluted' of cadmium. Bacteria isolated from sediments, raw water and drinking water displayed the highest resistance to zinc, probably because content of this metal was the highest, and the multi-stage water treatment technology has contributed to the selection of bacteria resistant to various impacts, including heavy metals.

Acknowledgments

This study was financially supported by the National Scientific Center, Poland, grant No. 1438/B/P01/2011/40.

References

1. R.A. Wuana, F.E. Okieimen, Ecology, (2011) http://dx.doi.org/10.5402/2011/402647
2. A.S. Mohammed, A. Kapri, R. Goe, Heavy-Metal Pollution: *Sources, Impact and Remedies*. A book chapter, (2011)
3. S. Zheng, X. Zheng, C. Chen, PLoS One **7**, 11 (2012)

4. A. Farkas, C. Erratico, L. Vigano, Chemosphere **68**, 761–768 (2007)
5. P. Kelderman, A.A. Osman, Wat. Res. **41**, 4251–4261 (2007)
6. A. Spain, Rev. U. Res. **2**, 1–6 (2003)
7. J. Bai, B. Cui, B. Chen, K. Zhang, W. Deng, H. Gao, R. Xiao, Ecol. Model. **222**, 301–306 (2011)
8. M. Wojtkowska, J. Bogacki, A. Witeska, Sci. Total Environ. **551–552**, 387–392 (2016)
9. A. Kuriata-Potasznik, S. Szymczyk, A. Skwierawski, I. Cymes, Water **8**, 358 (2016)
10. J. Eggleton, K.V. Thomas, Environ. Int. **30**, 973–980 (2004)
11. W. Guo, X. Liu, Z. Liu, G. Li, Proc. Environ. Sci. **2**, 729–736 (2010)
12. S. Huo, K.M. Yeager, B. Xi, Y. Qin, Z. He, F. Wu, J. Environ. Sci. **31**, 1–11 (2015)
13. M. Świderska-Bróż, *Mikrozanieczyszczenia w środowisku wodnym* (Wyd. P. Wroc. 1993) (in Polish)
14. D.L. Sparks, Elements **1**, 193–197 (2005)
15. E. Szarek-Gwiazda, A. Czaplicka-Kotas, E. Szalinska, Clean – Soil, Air, Water **39**, 4, 368–375 (2011)
16. S. Frassinetti, G. Bronzetti, L. Caltavuturo, M. Cini, C.D. Croce, J. Environ. Pathol. Toxicol. Oncol. **25**, 3, 597–610 (2006)
17. R.A. Festa, D.J. Thiele, Biology Curr. Biol. **21**, 21 (2011)
18. D. Berry, C. Xi, L. Raskin, Curr. Opin. Biotechnol **17** (2006)
19. D.D. MacDonald, C.G. Ingersoll, T.A. Berger, Arch. Environ. Contam. Toxicol. **39**, 20–31 (2000)
20. G.A. Burton, Limnology **3**, 65–75 (2002)
21. Z. Banu, S.A. Chowdhury, D. Hossain, K. Nakagami, J. Wat. Res. Prot. **5**, 239–248 (2013)
22. U. Förstner, W. Ahlf, W. Calmano, Wat. Res. Tech. **28**, 8–9, 307–316 (1993)
23. W. Calmano, J. Hong, U. Förstner, Wat. Sci. Tech., **28**, 8–9, 223–235 (1993)
24. I. Bojakowska, G. Sokołowska, Przegląd Geologiczny **46**, 1 (1998) (in Polish)
25. G.A. Best, T. Bogacka, E. Niemirycz, *International river water quality, Pollution and restoration* (E&FN SPON, London, 1997)
26. www.giwk.pl (2017)
27. J. Sastre, A. Sahuquillo, M. Vidal, G. Rauret, Anal. Chim. Acta **462**, 59–72 (2002)
28. O. M. Zacheus, M.J. Lehtola, L.K. Korhonen, P.J. Martikainen, Water Res. **35** (2001)
29. F.H. Chapelle, *Water microbiology and geochemistry* (J. Wiley&Sons, 1993)
30. K. Jankowska, Olańczuk-Neyman, E. Kulbat, Polish J. of Environ. Stud. **15**, 6 (2006)
31. W. Tylman, K. Łysek, M. Kinder, J. Pempkowiak, Water Air Soil. Pollut. **216** (2011)
32. E. Kulbat, in: *Proc. VII Int. Symp on Water Manag. and Hydraulic Eng.* 369–379 (2001)
33. Raport o stanie środowiska woj. pomorskiego, WIOŚ, (2000 and 2003) (in Polish)
34. M. Strzebońska, A. Kostka, E. Helios-Rybicka, E. Jarosz-Krzemińska, Pol. J. Environ. Stud. **24**, 3 (2015)
35. Müller, Geojournal **2**, 108–18 (1969)
36. J. Lis, A. Pasieczna, *Atlas geochemiczny Pobrzeża Gdańskiego* (1999) (in Polish)

37. L. Boszke, T. Sobczyński, G. Głosińska, A. Kowalski, J. Siepak, Pol. J. Environ. Stud. **13**, 5 (2004)
38. E.J. Kim, J.E. Herrera, D. Huggins, J. Braam, S. Koshowski, Water Res. **45** (2011)
39. R.G. Crounse, W.J. Pories, J.T. Bray, R.L. Mauger, App. Environ. Geochem. (1983)
40. T. Schwartz, W. Kohnen, B. Jansen, U. Obst, FEMS Microbiol. Ecology **43**, 3 (2003)
41. S. Jia, P. Shi, Q. Hu, B. Li, T. Zhang, X.X. Zhang, Environ. Sci. Technol. **49** (2015)
42. J-J. Huang, H-Y. Hu, Y-H. Wu, B. Wei, Y. Lu, Chemoshpere **90** (2013)
43. R. Shrivastava, R.K. Upreti, S.R. Jain, K.N. Prasad, P.K. Seth, U.C. Chaturvedi, Ecotoxicol. Environ. Saf. **58** (2004)
44. P. Shi, S. Jia, X.X Zhang, T. Zhang, S. Cheng, A. Li, Water Res. **47** (2013)
45. S. Khan, T.K. Beattie, C.W. Knapp, Chemosphere **152** (2016)

Determining residential energy consumption-based CO_2 emissions and examining the factors affecting the variation in Ankara, Turkey

Melike Kus[1], *Perihan* Akan[2,*], *Merih Aydinalp* Koksal[2], and *Gulen* Gullu[2]

[1]Nature Conservation Centre, 06460, Cankaya, Ankara, Turkey
[2]Department of Environmental Engineering, Hacettepe University, 06800, Beytepe, Ankara, Turkey

Abstract. Energy demand of Turkey has been showing a remarkable increase in the last two decades due to rapid increase in population and changes in consumption trends. In parallel to the increase in energy demand, the CO_2 emissions in Turkey are also increasing dramatically due to high usage of fossil fuels. CO_2 emissions from the residential sector covers almost one fourth of the total sectoral emissions. In this study, CO_2 emissions from the residential sector are estimated, and the factors affecting the emission levels are determined for the residential sector in Ankara, Turkey. In this study, detailed surveys are conducted to more than 400 households in Ankara. Using the information gathered from the surveys, the CO_2 emissions associated with energy consumption of the households are calculated using the methodology outlined at IPCC. The statistical analyses are carried out using household income, dwelling characteristics, and household economic and demographic data to determine the factors causing the variation in emission levels among the households. The results of the study present that the main factors impacting the amount of total energy consumption and associated CO_2 emissions are household income, dwelling construction year, age, education level of the household, and net footage of the dwelling.

1 Introduction

Since the industrial revolution, the rapid increase in the use of fossil fuel, rapid population growth and the destruction of forests have led to an increase in greenhouse gas (GHG) concentrations in the atmosphere. Climate change stemming from this increase has become an agenda item on a global scale especially in the last two decades. It is now widely accepted that increasing average temperatures with a rapid acceleration are a consequence of GHG emissions from human activities. This increase in temperature causes the climate

* Corresponding author: apakan@hacettepe.edu.tr

characteristics to change in different parts of the world [1]. The main gases that cause climate change are carbon dioxide (CO_2), methane (CH_4), and nitrous oxide (N_2O), which account for 99% of total GHG emissions. Owing to fossil fuel burning and deforestation, concentration of atmospheric CO_2 has increased from 280 ppm at preindustrial revolution era to 406 ppm in February 2017 [2, 3]. Carbon dioxide emission mainly results from the use of fossil fuels and also the sector with the largest share in the formation of carbon dioxide emission is energy production [4]. Historically, there is a link between development and energy consumption, and access to appropriate energy sources is considered as an important sustainable development indicator [5]. Population and income growth are the main factors leading to an increase in energy consumption. Projections show that the world population will rise 8.3 billion in 2030 year. This situation indicates that 1.3 billion more people should be provided with energy supply [6].

Turkey's economy is an energy-intensive economy when compared to similar economies. Looking at the development of energy consumption in Turkey, it is seen that annual demand increase is 8% [7]. While total energy consumption in the country is 87 million tonnes of oil equivalent (Mtoe) in 2011, this value reached 99 Mtoe in 2015 [6]. As a result, Turkey's GHG emissions have increased significantly since 1990. GHG emissions of 187 million tonnes of CO_2 equivalent in 1990 increased to 468 million tonnes of CO_2 equivalent in 2014. As a CO_2 equivalent, total GHG emissions increased in the range of 125% in 2014 compared to 2009 levels. The amount of CO_2 emissions in the residential sector in 2007 was 34 Mtonnes [8]. While the equivalent CO_2 emission per capita in 1990 was calculated as 3.7 tonnes/person, this value was calculated as 6.1 tonnes/person in 2014 [9]. As of 2011 year, total GHG emissions were estimated at 422.4 million tonnes of CO_2 equivalent. Table 1 presents total GHG emissions as million tonnes of CO_2 equivalent for each sector from 1990 to 2011. While the largest share of CO_2 emissions in 2011 is energy-related emissions with 71%, followed by industrial operations with 13%, waste with 9% and agricultural activities with 7% [10].

Table 1. Sectoral GHG emissions, millions tonnes [9].

	1990	1995	2000	2005	2010	2011
Energy	133	162	213	242	285	301
Industrial Processes	15	24	24	29	54	56
Agricultural Operations	30	29	28	26	27	29
Waste	10	24	33	33	36	36
Percent Increase based on 1990	-	27%	58%	77%	113%	124%

The highest CO_2 emissions among the sectors stem from electricity generation. The main reason for this high share resulting from the electricity production is that the majority of electricity in Turkey is generated at fossil fuelled power plants as can be seen from Table 2 [8].

Table 2. Share of fuel type for electricity generation in Turkey from 2011 to 2015 [6].

Primary Energy Source	2011	2012	2013	2014	2015
Thermal	75%	73%	72%	80%	69%
Hydraulic	23%	24%	25%	16%	26%
Geothermal	0.3%	0.4%	0.6%	0.9%	1.3%
Wind	1.7%	2.6%	2.4%	3.1%	3.7%

Energy consumption in Turkey has been showing a dramatic increase in parallel to the increase in population and industrial development. Energy consumption in 2015 (98,811 ttoe) is 14 % higher than 2011 (86,952 ttoe) [6]. According to 2011 data, the residential sector has a significant share of 24% in total electricity consumption (Table 3).

 DOI: 10.1051/e3sconf/20172200094

Table 3. Distribution of net electricity consumption by sectors in Turkey from 2011 to 2015 [11].

Primary Energy Source	2011	2012	2013	2014	2015
Residential	24%	23	23%	22%	22%
Commercial	16%	16%	19%	19%	19%
Government Offices	4%	5%	4%	4%	4%
Industrial	47%	47%	47%	47%	48%
Street Lighting	2%	2%	2%	2%	2%
Others	7%	7%	5%	6%	5%

In order to better understand the environmental impacts of energy consumption, the factors that affect consumption activities need to be identified and better assessed. Households are an important target in this respect, because consumers have been playing an important role in the increase of production structures resulting in the consumption of intensive energy and resource. However, the residential sector is a difficult sector to examine due to various reasons such as the sector consists of units with various sizes, geometries and thermal properties and household behaviour varies greatly. This situation affects energy consumption widely. In addition, the collection of residential energy consumption data is more difficult due to the confidentially issues as well as conducting detailed energy consumption measurements at homes brings an extra cost to the studies.

Households are responsible for a significant portion of the CO_2 emissions due to use of fossil fuels for space and hot water heating, transportation, and electricity [12–14]. Households consume energy in two ways: direct (space heating, lighting, vehicle fuel, etc.) and indirect (consumed energy for production, transport and disposal of consumer goods) [15–17]. Investigations have shown that the household energy consumption is influenced by such factors as income level [18–23], that the location of the residence is the urban or rural area [15] and dwelling type, size and age [24–27], as well as geographical conditions [15, 16, 19], household size and properties [27], and age and education level of the household head. The most influential factor is estimated as the income level of the household in some studies [19, 21, 23]. Increased income levels can lead to the increase in the use and number of electrical appliances, and thus the household electricity consumption and the energy consumption for heating/cooling and hot water increase [28]. Some of the literature findings related to the studies performed to investigate the factors affecting energy consumption in the households are presented below. Aydinalp et al. (2003) found that factors affecting the electricity consumption owing to electrical household appliances, lighting and cooling are income, type of building, housing ownership, and population of living place as well as number of adults and children living in the residence [29]. Druckman and Jackson (2008) stated that energy consumption and carbon dioxide emissions in residential areas have a strong relationship with income level [19]. According to Rosas et al. (2010), in Mexico, inequality in income distribution reflected to the electrical household appliances ownership, household energy consumption and CO_2 emissions [23]. The aim of the study is to assess households on the basis of the factors that cause increased energy consumption and CO_2 emissions in Turkey and to shed light on the complex relationship between household energy consumption, lifestyles, and development.

2 Methodology

The methodology of this study includes multiple steps which can be divided into three main stages; determination of study area and data gathering and survey development as well as data analysis. The following sections describe these three stages in detail.

2.1 Study area

The study was conducted in Ankara, the capital city of Turkey. Ankara is the second biggest city and has a significant proportion in energy consumption in the country. The study was conducted in 2011 when the population of the city was approximately four million eight hundred ninety-one thousand.

2.2 Survey development and administration

This study aimed at calculating the CO_2 emissions from direct energy consumption and determine the factors affecting the level of direct energy consumption. A survey inquiring about the condition of the building, socio-economic condition of the household and aspects determining the energy consumption level was administered by publishing on a webpage designed for the survey and by distributing as printed questionnaires. The printed materials were distributed in primary schools located in the neighbourhoods with different socioeconomic characteristics. Prior to distribution of the surveys, the students were informed about climate change and the role of energy consumption in global warming with a presentation and their questions were answered in order to increase the intergenerational influence in the survey. A total of 451 surveys were obtained, and 396 surveys with electricity consumption data was used in the analysis.

2.3 Data analysis

In order to develop a more systematic approach, the research participant households were clustered according to their structural features. Clustering is a method used in market researches for a long time. Categories such as social class, income level, geographical location, number of adults and children in the household, age and education level of the household head have been used in clustering method [30, 31]. In similar researches on household energy consumption, it is determined that income is the dominant determinant [19, 23, 26]. Thus, the questionnaires are clustered into three income levels (low income, middle income and high income). In the study, CO_2 emissions are calculated by using energy sector guideline of theIntergovernmental Panel on Climate Change (IPCC), published to assist national GHG inventory preparations. Tier 1 method, which is based on calorific value and CO_2 emission factor assumptions for different fuel types, is used as there are no country specific values for these factors. In Tier 1 method, CO_2 emissions are calculated by transforming fuel consumption amount given in mass or volume into Tera joule using the lower calorific value and multiplying the result with the related CO_2 factor [32]:

$$\text{Emission}_{CO2} = \text{Fuel Consumption } (FC)_{fuel} \text{ x } CO_2 \text{ Emission Factor } (EF)_{fuel} \quad (1)$$

where; Emission_{CO2} is the amount of CO_2 emitted from fuel usage (kg), Fuel Consumption $(FC)_{fuel}$ is the amount of fuel consumed (TJ), CO_2 Emission Factor $(EF)_{fuel}$ is the CO_2 emission factor for the specific fuel consumed (kg CO_2/ TJ).

Emissions are calculated by multiplying the amount of fuel used for heating the house and water by the household with the lower calorific value published by Turkish Statistical Institute for the specific type of the fuel used. Electricity consumption is either provided with the exact value of the latest monthly consumption in kWh, or the price of the consumption. The price of consumption is transformed into kWh by estimating via regression analysis with the existing kWh data. In the survey, main fuel used for cooking is asked. If the main fuel is electricity or natural gas, emissions from this consumption are not calculated as they are covered under electricity and natural gas consumptions. If it is bottled gas, it is assumed that a 4-people family consumes a 12 litter bottle of gas per month, and the emissions are calculated estimating the amount of bottle gas consumed in line with the household size. The fuel consumption is asked in terms of the type of the fuel and monthly fuel cost and transformed into litters using the average price of the related fuel.

3 Results and discussion

This section summarizes the obtained results from the stages of the study. In this section, annual energy consumption amounts according to the sources, annual residential CO_2 emission amounts according to the sources, correlation between income and energy consumption from different sources and factors affecting total energy consumption examined.

3.1 Average energy consumption

Results of the study showed that the highest energy consumption in the households in Ankara is 73 GJ per home which is the total main space and water heating energy consumption. Second highest consumption is for secondary heating, which means additional heating sources are used to back up main heating. Electricity consumption and private vehicle energy consumption has an insignificant share in the total energy consumption as seen in Table 4.

Table 4. Annual energy consumption amounts according to the sources.

End-use Energy Consumption	Average Consumption (GJ)	Standard Deviation (GJ)	Lowest Consumption (GJ)	Highest Consumption (GJ)
Main space and water heating	72.724	32.056	16.744	200.000
Secondary heating	1.725	6.793	0	54.428
Electricity	850	402	52	2.578
Private car	652	856	0	6.975
Total	76.015	31.777	18.198	203.148

3.2 CO_2 emissions

When energy consumption based CO_2 emissions are examined, heating and hot water energy consumption has the highest emission amount (4.34 tonnes) per household. Electricity consumption related CO_2 emissions ranks second (1.69 tonnes), followed by private car usage and auxiliary heating energy consumption emissions as seen in Table 5.

Table 5. Annual residential CO_2 emission according to the sources.

End-use Energy Consumption	Average Emission (tonnes)	Standard Deviation (tonnes)	Lowest Emission (tonnes)	Highest Emission (tonnes)
Main space and water heating	4.24	1.81	0.94	11.25
Secondary heating	0.15	0.68	0.0	5.43
Electricity	1.69	0.85	0.11	5.43
Private car	0.60	0.78	0.0	6.21
Total	6.68	2.28	1.97	15.56

3.3 Correlations between energy consumption and income

Income has the highest impact on energy consumption [16, 18–20, 23, 26]. In order to analyse the correlation between the income and energy consumption related CO_2 emissions, Spearman's Rank Correlation Test is used. As the significance probability (p) in the correlation test gets lower than 0.05, the correlation between the factors increases [21]. Results of Spearman Rank Correlation Test showed that there is statistically meaningful correlation between income and energy consumption (for all sources of consumption) and CO_2 emissions as seen in Tables 6 and 7. Main space and water heating energy consumption, electricity consumption and energy consumption from private vehicle usage is positively correlated with the income, while auxiliary heating energy consumption is negatively correlated.

Table 6. Correlation between income and energy consumption from different sources.

Energy Consumption Source	Correlation	Statistical Significance (p)
Main space and water heating	0.4159	0.0000
Secondary heating	-0.2333	0.0000
Electricity	0.2628	0.0000
Private car	0.3848	0.0000
Total	0.3584	0.0000

($p < 0.05$)

Table 7. Correlation between income and CO_2 emissions from different sources.

Energy Consumption Source	Correlation	Statistical Significance (p)
Main space and water heating	0.2411	0.0000
Secondary heating	-0.2380	0.0000
Electricity	0.2641	0.0000
Private car	0.4176	0.0000
Total	0.2894	0.0000

($p < 0.05$)

Total energy consumption and CO_2 emissions decline as the income level increases. The lowest income group has a total of 5.7 tonnes CO_2 emission, while the mid-income group has 6.58 tonnes and the high income group has 7.69 tonnes of CO_2 emissions per year.

3.4 Factor affecting residential energy consumption and related CO_2 emissions

Spearman's rank correlation test is used to determine the factors affecting energy consumption and CO_2 emissions. The results of the analysis showed that the age of the building, the floor the resident is located [19, 26], household size [29], age and education level of the household head [19] and net usage area of the residence [28] are the factors affecting the energy consumption for space and water heating as seen in Table 8. The total energy consumption is affected from the age of the building, age and education of the household head and net usage area of the residence [21]. A significant proportion of the total energy consumption is done for heating the house and water, thus the factors affecting them are similar.

Table 8. Factors affecting total energy consumption.

Factor	Correlation	Statistical Significance (p)
Age of the building	0.1299	0.0098
Age of household head	0.1459	0.0037
Education of household head	0.2153	0.0000
Net area of the residence	0.1696	0.0007

($p < 0.05$)

Similar to the study conducted by Chen et al. [26], the analysis showed that energy consumption increases as the age of the building increases as seen in Table 9. Furthermore, energy usage increases as the net usage area of the residence increases [21, 33] as well as the age of the household head increases. In general, there is a positive correlation between the education level of the household and total energy consumption; however, energy consumption reaches to the highest level in its income cluster in the households with illiterate household heads.

Table 9. Factors affecting CO_2 emissions from total energy consumption.

Factor	Correlation	Statistical Significance (p)
Age of the building	0.1304	0.0098
Age of household head	0.1420	0.0049
Education of household head	0.1344	0.0077
Age of the building	0.1895	0.0002

($p < 0.05$)

It is determined that CO_2 emissions from total energy consumption is affected from the type and the age of the building, age and education level of the household head and net usage area of the residence.

4 Conclusion

In this study, it was aimed to determine the amount of CO_2 emissions due to the energy consumption in the residential sector in Ankara and the factors that affect the energy consumption at homes. According to the results of this study, there is a significant relationship between energy consumption, as well as CO_2 emissions, and income. This relationship is positive with respect to energy consumption and emissions from main space and water heating, electricity consumption, and private vehicle use, while it is negatively correlated with energy consumption and emissions from secondary heating. It has been determined that factors influencing the total energy consumption and CO_2 emissions at the homes are the building construction year, the age and education of the household head, and the floor and net usage area of the dwellings.

References

1. L. Lohmann, Development Dialogue, **48** (2006) [Online]. Available: https://www.cbd.int [Accessed March 2017]
2. J. Peñuelas, J. Sardans, M. Estiarte, R. Ogaya, J. Carnicer, M. Coll, A. Barbeta, A. Rivas-Ubach, J. Llusiá, M. Garbulsky, I. Filella, A.S. Jump, Glob. Chang. Biol. **19**, 2303–2338 (2013)
3. National Oceanic and Atmospheric Administration [Online] Available: https://www.esrl.noaa.gov/gmd/ccgg/trends/. [Accessed March 2017]
4. H. Hens, G. Verbeeck, B. Verdonck, Energy Build. **33**, 275–281 (2001)
5. UN. *Indicators of Sustainable Development: Guidelines and Methodologies* (New York: United Nations Publications, 2007)
6. Ministry of Energy and Natural Resources (MENR) [Online] Available: http://www.eigm.gov.tr [Accessed March 2017]
7. M. Balat, Energ. Convers. Manage. **51**, 1998–2011 (2010)
8. T. Keskin, Turkey's Climate Change National Action Plan Development Project, Ankara http://iklim.cob.gov.tr (2010)
9. National Inventory Report, TURKEY Greenhouse Gas Inventory, 1990 to 2008. Ankara: Turkish Statistical Institute (2014)
10. National Inventory Report, TURKEY Greenhouse Gas Inventory, 1990 to 2008. Ankara: Turkish Statistical Institute (2011)
11. Electricity Distribution and Consumption Statistics of Turkey (TEDC) [Online] Available: http://www.teias.gov.tr/ [Accessed March 2017]
12. S. Tscharaktschiew, G. Hirte, Reg. Sci. Urban Econ. **40**, 498–516 (2010)
13. Z.H. Feng, L.L. Zou, Y.M. Wei, Energy **36**, 656–670 (2011)
14. J. Sánchez-Chóliz, R. Duarte, A. Mainar, Ecol. Econ. **62**, 308–318 (2007)
15. W. Abrahamse, L. Steg, J. Econ. Psychol. **30**, 711–720 (2009)
16. A.C. Kerkhof, R.M. Benders, H.C. Moll, Energy Policy **37**, 1509–1517 (2009)
17. R. Kok, R.M. Benders, H.C. Moll, Energy Policy **34**, 2744–2761 (2006)
18. S.R. Tyler, Atmos. Environ. **5**, 809–816 (1996)
19. A. Druckman, T. Jackson, Energy Policy **36**, 3177-3192 (2008)
20. R. Duarte, A. Mainar, J. Sánchez-Chóliz, Energy Economics **1**, 176–185 (2009)
21. S. Chen, N. Li, H. Yoshino, J. Guan, M.D. Levine, Energy Build. **5**, 1063–1070 (2010)
22. Z. Donglan, Z. Dequn, Z. Peng, Energy Policy **38**, 3377–3383 (2010)
23. J. Rosas, C. Sheinbaum, D. Morillon, Energy Sustain. Dev. **14**, 127–133 (2010)
24. L. Lutzengier, Natl. Acad. Sci. Lett. 77–91 (1997)
25. D.A. Poyer, L. Henderson, A.P. Teotia, Energ. Econ. **19**, 445–463 (1997)
26. S. Chen, H. Yoshino, N. Li, Energy Build. **1**, 136–146 (2010)
27. S. Chen, H. Yoshino, M.D. Levine, Z. Li, Energy Build. **41**, 1063–1070 (2009)
28. M. Aydinalp, V.I. Ugursal, A.S. Fung, Appl. Energy **79**, 159–178 (2004)
29. M. Aydinalp, V.I. Ugursal, A.S. Fung, Int. J. Glob. Energy Issues **3**, 302–315 (2003)
30. F. Reusswig, H. Lotze-Campen, K. Gerlinger, *Changing global lifestyle and consumption patterns: The case of energy and food.* The ICFAI University Press. 197–210 (2005)

31. D. Hogan, Eur. J. Pop. **8**, 109–123 (1992)
32. IPCC, Guidelines for National Greenhouse Gas Inventories (2006)
33. Y.G. Yohanis, J.D. Mondol, A. Wright, B. Norton, Energy Build. **40**, 1053–1059 (2008)

 DOI: 10.1051/e3sconf/20172200095

Experimental determination with EPR-methods of dose loads on local population in inhabited localities adjacent to the tailing pond "Koshkar-ata" (Kazakhstan)

Kairat Kuterbekov[1,*], *Nurllah* Sadykov[2], *Kassym* Zhumadilov[1], *Rano* Nurgaliyeva[1], *Aset* Kabyshev[1], *Erbolat* Abseitov[1], and *Askar* Kurmanzhanov[1]

[1]L.N. Gumilyov Eurasian National University, Astana, Kazakhstan
[2]Sh. Esenov Caspian University of Technology and Engineering, Aktau, Kazakhstan

Abstract. EPR study of TE from teeth donors in KOSHKAR-ATA region. During the first year we received22 tooth samples from local people in KOSHKAR-ATA region; the teeth were removed by dentists in accordance with their medical prescriptions. Obtained data showed that radiation signal in the analyzed samples corresponds to doses not exceeding 0.3 Gy; only one sample carriers the signal of (0.35±0.15)Gy.

1 Introduction

EPR method was chosen as a basic one forexperimental determination of dose loads on local population.It is known that EPR-dosymetry of teeth enamel (TE) has been recommended by IAEA and WHO as a basic method for assessment of various-type radiation impacts on population. High precision and reliability of the method has been proved by inter-laboratory experiments "Intercomparison 1, 2, 3" specially performed under the aegis of IAEA and "Intercomparison 4" performed in 2005–2006 initiated by the University of Hiroshima. Almost all leading laboratories allover the world took part in these experiments.

Considerable preliminary work was performed for metrological and legal provisions: certification of a spectrometer, certification of a laboratory for the right to run EPR-dosimetry of TE in Kazakhstan, participation of the laboratory in international comparison experiments.

Direct measurements of doses are vital since in many cases there are sets of unfavorable (climatic, social, moral) factors that , like radiation, may negatively influence public health and available indirect dose assessments are quite different for many authors. It is also important that information provided by EPR dose reconstruction may influence on conclusions of experts regarding radiological situation in various regions of the country.

It should be mentioned that in the field of EPR-dosimetry there is no conventional internationally accepted standard to set requirements to equipment, instruments, materials,

* Corresponding author: kkuterbekov@gmail.com

 DOI: 10.1051/e3sconf/20172200095

sample preparation methods, spectrum registration and calibration conditions, algorithms for their processing, etc. There are national standards available, for instance, Russian State Standard ГОСТ Р 22.3.04-96 accepted in Kazakhstan, but by now this standard became obsolete to considerable extent.

Calibration of γ-source dose rate. Results considerably depend on calibration of dose rate generated by the γ-source with respect to a standard source of the IAEA Laboratory of Dosimetry. In the beginning of 2004 there was calibrated a ^{137}Cs γ-source of INP laboratory within IAEA project IDAS 2004. In accordance with the procedure, three alanine (from IAEA kit) dosimeters were irradiated with a dose exceeding 100 Gy and the forth one served for estimations of background radiation.

*Adjustment of the method for TE sample preparation to teeth samples of population from inhabited localities near KOSHKAR-ATA.*To gain pure teeth enamel we used IS method with treatment in alkaline solution that is quite widely used with various modifications in many countries since the method intensifies TE extraction. But still possible influence of such treatment on EPR-radiation signal has not been studied in details. We have employed a method that involves additional irradiation. Chosen 4 samples of low-caries teeth were additionally irradiated with quite high dose of about 5 Gy to assure clear registration of radiation signal by EPR, and then split up into halves. One of the halves of each tooth was treated with IS in presence of NaOH or KOH of various concentrations during different times for different samples, and then dentine was removed. In the second halves dentine was removed using only mechanical methods (diamond cut). After that registered EPR dosimetric signals for two methods were compared.

In the experiments [1]it was shown that treatment in 5N alkali solution up to 6-8 hours do not influence practically the magnitude of radiation EPR signal from carbonate radicals $CO_2^{\cdot -}$; at this it facilitates dentine removal. Long-term treatment (for 24–28 hours and more) softens dentine, but significantly reduces EPR signal (to 4 –50%); at this KOH acts more effectively than NaOH.

2 KOSHKAR-ATA region

EPR study of TE from teeth donors in KOSHKAR-ATA region. During the first year we received 22 tooth samples from local people in KOSHKAR-ATA region; the teeth were removed by dentists in accordance with their medical prescriptions.

First set of four teeth was contaminated with non-removable paramagnetic admixture that shows wide intensive line what prevents from determination of informative radiation signal.

Nevertheless, we succeeded to find a technique enabling us to assess doses in those samples; the results are available in Table 1.

EPR study of TE from teeth donors in KOSHKAR-ATA region. During the first year we received 22 tooth samples from local people in KOSHKAR-ATA region; the teeth were removed by dentists in accordance with their medical prescriptions.

First set of four teeth was contaminated with non-removable paramagnetic admixture that shows wide intensive line what prevents from determination of informative radiation signal.

Nevertheless, we succeeded to find a technique enabling us to assess doses in those samples; the results are available in Table 1.

Table 1. Doses determined in the first set of taken teeth.

Sample No.	Name, ID	Age of donor,	No. of tooth	Weig ht mg	Numbe r of registra tions	Dos e mG y	Number of assess ments	Averag e, Relative Error	Necessity of irrad.
1	2	3	4	5	6	7	8	9	10
1	Rysbayev , 63201	1932	5	104.4	5	0.15 mG y	6	±20.3%	recommend ed
2	Reshetnik ov, 63202	1954	5	85.9	4	Not defi ned	4	>100%	necessary
3	Makholov a*, 63003 63004	1960	1 left, 1 right incisor	46.6 54.6	2-for frontal 2-for lingual	0.90 0.04	2	± 4% ± 40%	Not necessary
4	Musayeva 63005	1952	7	70.4	5	0.30		± 34%	necessary

Three tooth were found unsuitable for EPR-analysis due to considerable damage with caries and small amount of TE on them (required to analysis value is 50–10 mg). All other samples were taken in accordance with presented above method.

Obtained data showed that radiation signal in the analyzed samples corresponds to doses not exceeding 0.3 Gy; only one sample carriers the signal of 0.35±0.15 Gy.

Characteristics of samples and qualitative results of measurements for several of them are presented in Table 2 and Table 3.

Table 2. Doses determined for some of samples in the 2nd set.

Sample number	63211	63212	63214	63215	63216	63218
Name of donor	Turekhanov S.	Masimova Zh.	Dudukalov V.	Belyakova S.	Stelmakh S.	Stobva A.A.
Age of donor, years	46	34	53	63	57	37
Dose estimatio n, Gy	< 0.3	< 0.3	< 0.3	0.35±0. 15	0.3±0. 15	< 0.3

One may conditionally compare these data with calculated radiation contamination with plutonium. Let us suppose there is plutonium in soil in amounts 1 kBq/kg (i.e. quite heavy contamination) and all 1 kg has been intaken into a human body. Let us calculate maximal number of defects (in this case – free radicals) during a year due to irradiation from plutonium. Energy of an alpha-particle is approximately $5 \cdot 10^6$eV and, if energy per a bond is 50 eV, one α-particle produces ~ $1 \cdot 10^5$ defects. During a year 1kBq of Pu in a body will generate $1 \cdot 10^5 \cdot 3 \cdot 10^7 \cdot 10^3$ what comprises ~ $3 \cdot 10^{15}$ defects/kg.

From Table 3 one can see that we deal with the values of the same order of magnitude and, keeping in mind that their direct comparison is not scientifically correct, the obtained data indicate the necessity of thorough investigation.

Table 3. Information on teeth donors from Western Kazakhstan dose accumulated according to EPR-dosymetry data.

No.	Sample No.	Donor name	Birth year	Address	Years of current residence	Toot, h	Sample weight, mg	Irradiation dose, mGy	Notes
1	63206	Makhanbetov A.	1952	v. Yeralievo	50	1 low right	60,1	< 170	5 measurements
2	63209	AituarovZh.	1960	v. Priozerniy	20	1 low right.	56,0	< 180	
3	63211	Turekhanov S.	1957	v. Akshukur	46	5 upper left	28,2	< 355	
4	63212	MasimovaZh.	1969	27-10-89	15	5 upper right	56.7	1640±9 %	
5	63213	Turzhakova B.	1937	Kt.Tyube	20	2 low right	7.2		Not evaluated
6	63214	Dudukalov V.	1950	11-6ª-65	30	6 low right	91.5	2000±3 %	
7	63215	Belyakova S.	1940	7-5-105	30	6 upper right	61.4	< 170	
8	63216	Stelmakh S.	1946	15-5-10	25	3 low right	71.1	< 140	6 measurements
9	63217	TuleshovaM.	1949	4-65-2	15	5 low right	6.8		Not evaluated
10	63218	Stobva A.A.	1966	14-20-67	15	7 low right	94.4	< 105	5 measurements
11	63219	Bazarbayev	1963	11-27-118	12	7 upper left	109.2	< 92	
12	63220	Kazakh N.	1963	28-25-19	11	7 upper left	137.3	< 75	
13	63221	Basbayev N.	1964	v. Akshukur	12	2 low right	81.5	480±	
14	63222	Nareshov S.	1958	3ª-5-7	15	6 upper right	125.3	< 80	
15	63223	Smirnov T.	1954	4-27-18	17	6 upper left	200	< 50	

 DOI: 10.1051/e3sconf/20172200095

Measurements were made at room temperature, frequency 9.827 GHz, SHF – power 25 mW, field modulation frequency 100 кHz, field modulation amplitude 4.85 G, field sweep 96.4 G, sweep time 21 s, time constant 41 ms, amplification coefficient 10^5, field resolution 1024 points, resolution of analog-digital conversion – 13 bit. 50 savings were made for each spectrum registration. The spectrum was usually registered three times and every time the sample was shaken (to get better averaging of orientation of crystallographic axes with respect to the constant field).

Acknowledgments

This work was supported by the Ministry of Education and Science of the Republic of Kazakhstan within the program of funding research activities through grants for 2015–2017 «Development of Hydrogen Energy and Technology in the Republic of Kazakhstan».

References

1. K.A. Kuterbekov, et al. 16th *Radiochemical Conference-2010, Marianske Lenzne, Czech Republic, CHLSAC* 104(S), 181 (2010)

Electron thermal EMF for $Na_xCu_{2-x}S$

Kairat Kuterbekov[1,*], *Malik* Balapanov[2], *Rais* Ishembetov [2], *Marzhan* Kubenova [1], *Talgat* Baitasov [1], *Asset* Kabyshev[1], *Aidos* Azhibekov[1], *Bekmyrza* Kenzhebatyr[1], and *Temirulan* Alibay[1]

[1]L.N. Gumilyov Eurasian National University, Astana, Kazakhstan
[2] Bashkir State University, Ufa, Russia

Abstract. In the present study, the temperature dependences of the thermoelectromotive force (thermo-emf) in copper selenide, substituted in a small concentration, were studied. The results of the measurements showed that the thermo-emf coefficient of the samples increases, and the conductivity decreases with increasing silver concentration in its composition. These results allow - with optimal selection of the doping regime and protective coatings - to develop on the basis of nanostructured copper selenide an effective thermoelectric for use at temperatures of 20–500°C as p-type semiconductors suitable for increasing the efficiency of thermoelectric generators.

1 Introduction

Modern thermoelectric converters have a number of advantages over traditional electric generators: simplicity of design, absence of moving parts, noiselessness of operation, high reliability, possibility of miniaturization without loss of efficiency.

In order for thermoelectric generators to become more competitive than conventional power sources, thermoelectric materials should achieve high performance ZT $\geq$ 4 [1]. This indicator is a guide in the search and synthesis of new promising materials that can become the basis for industrially produced thermoelectric devices in the foreseeable future (20 to 25 years).

The dimensionless thermoelectric figure of merit, characterizing the efficiency of materials in thermoelectric devices, is determined by the formula

$$ZT = \alpha^2 \sigma T / \chi \quad (1)$$

where α – is the coefficient of thermal emf (Seebeck coefficient); σ – is electrical conductivity; χ – is the thermal conductivity of the material.

Achieving the optimal combination of all three properties at the same time to obtain a high thermoelectric figure of merit is a complex task.

At present, among the industrially produced thermoelectric materials, the most common is doped bismuth telluride $(Bi_{1-x}Sb_x)_2(Se_{1-y}Te_y)_3$, which has a Q-factor of about 1 (ZT≈1) at

* Corresponding author: kkuterbekov@gmail.com

room temperature [2]. The highest values of Z are strongly doped semiconductors or semimetals with an electron concentration of the order of 10^{19}–10^{20} см$^{-3}$ [2].

2 Experiment

The results of experimental studies of the dependence of the Thermo-electromotive force (emf) of nanocomposite and microcrystalline copper and silver chalcogenides differing in chemical composition, the synthesis method and the particle size, as well as thermo-emf of nanocomposite and microcrystalline copper and silver chalcogenides in the temperature range 20–500°C are presented.

The choice of the objects of research $Na_xCu_{2-x}S$ (x = 0.05, 0.1, 0.15, 0.2) is due to their use as p-type semiconductors suitable for increasing the efficiency of thermoelectric generators [1].

To study the thermoelectric properties of solid alloys of copper selenides, they were prepared by solid-state synthesis in an inert atmosphere and by low-temperature chemical synthesis. To obtain nanocrystalline powders, a technique was used, described in detail in a number of papers [3, 4].

The registration of thermoelectric properties was carried out with the aid of an experimental setup for studying the Seebeck coefficient and electrical resistance (Model ZEM-3) shown in Fig. 1.

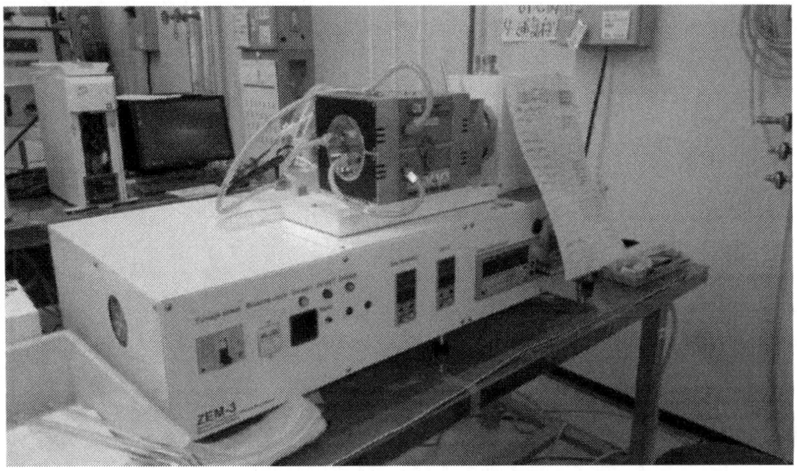

Fig. 1. Experimental setup for studying the Seebeck coefficient and electrical resistance.

The values of the electronic thermal-emf coefficient for the solid solution $Na_xCu_{2-x}S$ (x = 0.05, 0.1, 0.15, 0.2) were measured as a function of temperature. The sign of the coefficient e is positive, which, taking into account the rule of choice of the sign for all samples, corresponds to the motion of electron holes from the hot end of the sample to the cold one.

Figures 2–5 show the dependence of the coefficient α_e of the electronic thermo-emf for the composition $Na_{0,05}Cu_{1,95}S$, $Na_{0,1}Cu_{1,9}S$, $Na_{0,15}Cu_{1,85}S$ and $Na_{0,2}Cu_{1,8}S$ in the range from room temperature to 750 K. For all samples, With an increase in temperature, an increase in the electronic thermal-emf coefficient α_e is observed.

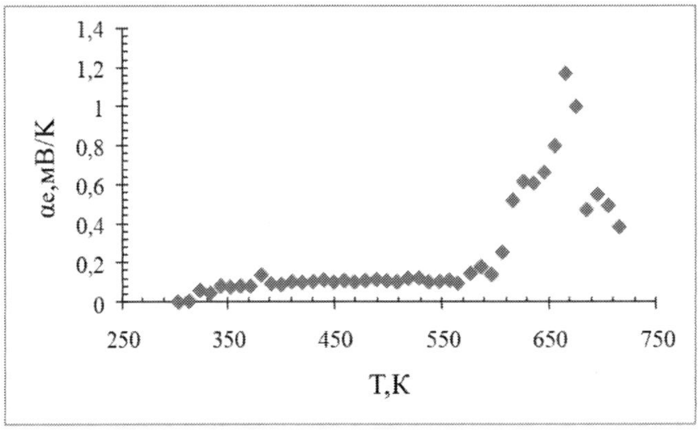

Fig. 2. Temperature dependence of the coefficient of electron thermal emf $Na_{0,05}Cu_{1,95}S$.

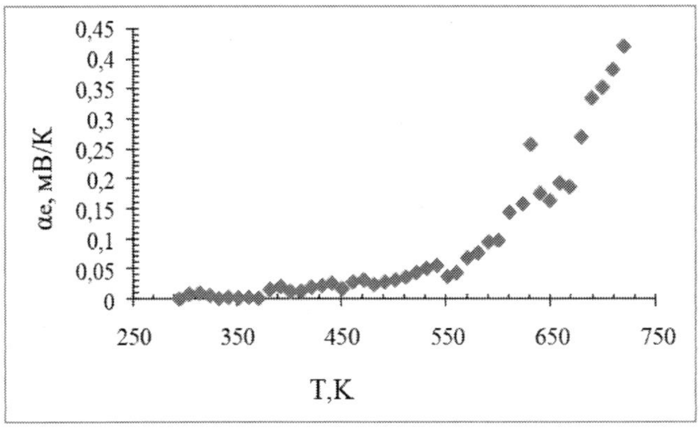

Fig. 3. Temperature dependence of the coefficient of electron thermal emf $Na_{0,1}Cu_{1,9}S$.

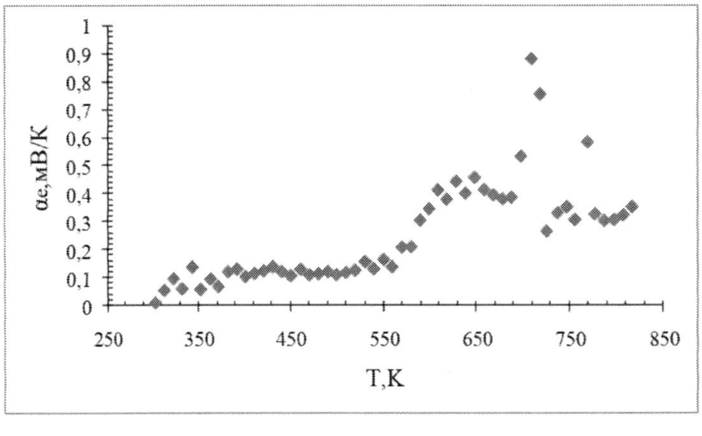

Fig. 4. Temperature dependence of the coefficient of electron thermal emf $Na_{0,15}Cu_{1,85}S$.

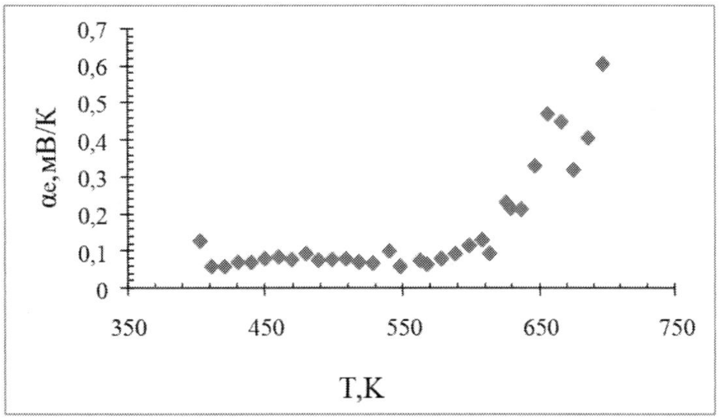

Fig. 5. Temperature dependence of the coefficient of electron thermal emf $Na_{0,2}Cu_{1,8}S$.

Fig. 6 shows the temperature dependence of the emf of the electrochemical cell $Cu/CuBr/Na_{0.2}Cu_{1.8}S/C$. The observed dependence is linear in nature, which indicates the constancy of the entropy of the metal atoms in the compound within the temperature range 630–680 K. From the values of the cell emf, one can judge the degree of non-stoichiometry of the compound, in this case the deviation from the stoichiometric composition is $\delta \approx 0.01$ in the chemical formula of the sample $Na_{0.2}Cu_{1.8-\delta}S$.

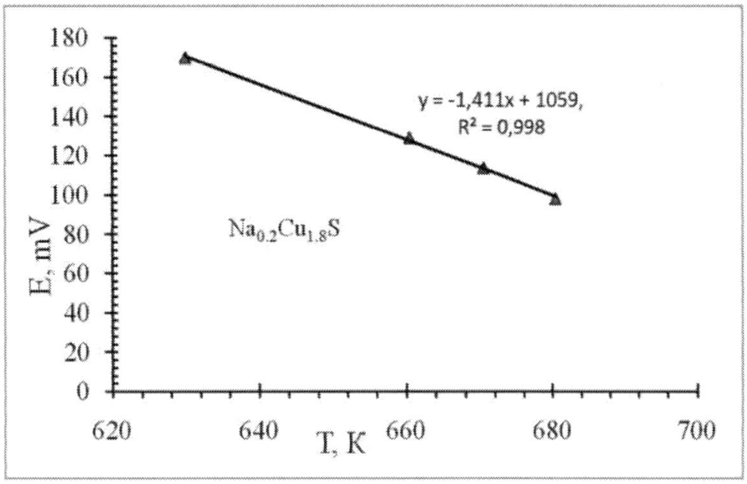

Fig. 6. Temperature dependence of the emf of an electrochemical cell $Cu/CuBr/Na_{0.2}Cu_{1.8}S/C$.

3 Conclusion

Thus, results of measurements have shown that temperature dependences of coefficient of electronic copper of thermo-eds nanocrystal sulfides which cationic sublattice has been alloyed by sodium, with the general chemical formula $Na_xCu_{2-x}S$ (x = 0.05; 0.1; 0.15; 0.2) in the field of temperatures from 20°C to 500°C. The sign of coefficient thermo-eds α_e for all samples was positive that taking into account the rule of the choice of a sign for all samples corresponds to the movement of electronic holes since the hot end of a sample on cold. In general, the coefficient thermo-eds with temperature increase from room temperature to about 650–660 To grows. In the neighborhood of temperature 660 K there is

 DOI: 10.1051/e3sconf/20172200096

a maximum, probably, the nature of which, probably, is connected with the happening phase transition. The highest values of coefficient electronic thermo-eds in 1.2 mV/To are reached at the same time at structure of $Na_{0,05}Cu_{1,95}S$.

The results of the measurements showed that the thermo-emf coefficient of the samples increases, and the conductivity decreases with increasing silver concentration in its composition. These results allow – with optimal selection of the doping regime and protective coatings – to develop an effective thermoelectric on the basis of nanostructured copper selenide for use at temperatures (400–600) K as p-type semiconductors suitable for increasing the efficiency of thermoelectric generators.

This work was supported by the Ministry of Education and Science of the Republic of Kazakhstan within the scientific-technical program for 2015–2017 Development of Hydrogen Energy and Technology in the Republic of Kazakhstan.

References

1. M. Zebarjadi, K. Esfarjani, M.S. Dresselhaus, Z.F Ren, G. Chen, Energy Environ. Sci. **5**, 5147 (2012)
2. C. Han, Z. Li, S. Dou, Chin. Sci. Bull. **59**, 2073 (2014)
3. M.Kh. Balapanov, R.Kh. Ishembetov, K.A. Kuterbekov, et al., Inorganic Materials **50**, 930 (2014)
4. R.H. Ishembetov, JU.H Yulaeva, M.Kh Balapanov, T.I. Sharipov, R.A. Yakishibayev, Perspektivnyie materialy **12**, 55 (2011)

 DOI: 10.1051/e3sconf/20172200097

Application of regression tress for prediction of water conduits failure rate

Małgorzata Kutyłowska[1,*]

[1]Wrocław University of Science and Technology, Faculty of Environmental Engineering, Wybrzeże Wyspiańskiego 27, 50-370 Wroclaw, Poland

Abstract. This paper presents the results of predicting the failure rate of water distribution pipes and house connections in a selected city in Poland by means of regression trees. Several regression tree models were built as part of modelling. Optimal models were selected (separately for each of the water conduit types) via an analysis of the so-called costs. The regression tree structure comprised independent variables, i.e. predictors (length, diameter, year of construction and material). The failure rates of the two types of water conduits were the dependent variable. The optimal models were characterized by the lowest costs and a relatively simple tree architecture. Operational data from the years 2001–2012 were used to determine the experimental (real) values of the failure rate and to build regression tree models. The optimal models included eight divided nodes and nine end nodes. The ranking of the significance of the parameters showed that length was the predictor responsible for division on the successive tree levels. The obtained high consistency (0.99) of the real data with the predicted ones indicates that the regression tree method can be used to analyze and assess the failure rate of water conduits.

1 Introduction

Mathematical modelling has been applied in many spheres of life already for many years [1, 2]. Especially engineering problems are solved using mathematical tools [3, 4]. Environmental engineering is a field which has been developing very rapidly. Among the many branches of broadly understood environmental engineering one should distinguish the sphere of municipal management, i.e. water supply systems, sewage disposal systems and waste management systems. The three areas mentioned can be said to belong to the critical infrastructure since their role is absolutely vital for the security and quality of life of inhabitants [5, 6]. Many of the phenomena connected with operating hydraulics and the design of water distribution systems [7] or sewage disposal systems [8] have already been thoroughly explored and described in the literature on the subject. Therefore today increasing importance is attached to the proper management and operation of municipal systems. Hence it is necessary to use the available mathematical tools to forecast the dynamically changing parameters of water distribution network operation. Reliability indices, e.g. the failure rate, not only should be determined on the basis of operational data,

* Corresponding author: malgorzata.kutylowska@pwr.edu.pl

but also relatively easily and quickly implementable algorithms and models should be employed to predict such random variables. The failure rate has been usually determined on the basis of the available operational data [9, 10] or through model studies [11], statistical models [12, 13] and other less conventional modelling methods [14–16]. So far the regression tree methodology has not been widely used to assess the reliability of municipal system operation or to predict selected indicators. This is why the author decided to take up this subject. The main aim of this paper is to show the possibilities of applying the regression tree algorithm to the analysis of the frequency of failures of water conduits. This method has already found application in many other fields, e.g. in the assessment of the degree of damage to buildings [17], in the modelling of the rate of failure of pumping systems [18] and in broadly understood economy [19]. Thus it seemed worthwhile to find out whether it is suitable for assessing the frequency of failures of water supply networks.

Moreover, this paper is a complement to the research presented in another work by the author [20], in which the failure rate was predicted (for the same water distribution network) by means of artificial intelligence. Thus it will be possible to compare the effectiveness of the regression tree method with that of the artificial neural network algorithm.

2 Materials and methods

Regression trees and classification trees are used to predict respectively quantitative and qualitative variables. This method of data analysis and prediction began to be used in the 1960s, but in 1984 it was popularized by L. Breiman [21]. Generally speaking, a regression tree (RT) or a classification tree (CT) is a directed graph having a root and nodes (leaves), in which the conditions regarding variables are checked, and branches comprising decision rules. As a rule, the regression tree method is easier to implement and makes the analysis of the results easier than the classification tree method [21]. An analysis using a tree building algorithm consists in finding a set of logical division conditions, and relations between the predictors and the dependent variable, which leads to prediction results [21]. The advantage of using RTs and CTs is the relatively easy interpretation of the results and accurate predictions [22]. Moreover, regression tree models are resistant to outliers, which for various reasons may occur in the operational data acquired from water companies. When outliers appear, they are isolated in small nodes. If there are only a few such values, they can be neglected [21].

The failure rate (λ, fail./(km·a)) of water supply pipes was predicted using the regression tree method. Since the failure rate of the distribution pipes (λ_r) and that of the house connections (λ_p) were to be predicted separately it was necessary to build two different tree models. Rates λ were the dependent variables while the predictors (independent variables) were: length, diameter, year of construction and material, separately for the distribution pipes and the house connections.

Operational data for the years 2001–2012, obtained from a water company in one of the medium-sized cities in Poland, were used to determine the actual failure rate λ and to predict it by means of regression trees. When building a regression tree model, one should bear in mind that too complex trees are difficult to interpret. Therefore, as in any other approach so in the case of the RT method, the aim is to create as simple models as possible, but fully rendering the interrelations between the dependent variables and the predictors. The values of the experimental dependent variables and the predictors are presented in table 1.

Table 1. Dependent variables and predictors.

Length, km	Diameter, mm	Year of construction	Material	λ, fail./(km·year)
House connections				
23.4–50.2	20–100	1961–2012	cast iron, steel, galvanized steel, PE, PVC	0.23–1.59
Distribution pipes				
57.3–88.7	80–200	1961–2006	cast iron, steel, PE, PVC	0.10–0.57

The optimal regression tree model was selected on the basis of the resubstitution of costs, where the expected squared error is calculated from the relation [21]:

$$R(d)=\frac{1}{N}\sum_{i=1}^{N}\left(y_i-d(x_i)\right)^2, \qquad (1)$$

where training sample Z consists of points (x_i, y_i), for i = 1, 2,..., N. The calculations are performed for the same data set which was used to build model d [21]. In this case training sample Z consists of such independent variables (x) as length, diameter, material and year of construction of water pipe as well as dependent variable (y) – failure rate (see table 1). A regression tree is created through iterative divisions in the nodes so as to minimize the cost [21]. The notion of cost (in the RT method [22]) is a generalization of the idea that the model with the smallest error is characterized by the best prediction. The measure of cost is a ratio of the incorrectly defined cases to all the cases. Thus the optimal model should be characterized by the lowest cost. An important element of the analysis and the choice of tree size is V-fold cross-validation, consisting in dividing the data into V randomly selected disjoint parts. Using the remaining (V-1) parts of the data as training cases, the dependent variable is predicted and the prediction error is calculated. The tree of a given quantity is calculated V times. Each time cases with one subsample excluded (to serve as a test sample) are used in the calculations. Thus each of the subsamples is used V-1 times in the training sample and only once as the test sample. The cross-validation cost (CV cost) is calculated as the average cost for V test samples. This average is a CV cost estimate [22]. In this study 10-fold cross-validation was used. A given tree must be created many times and a large data set is needed to perform the divisions mentioned above. The tree structure (the number of branches and nodes) depends on the number of divisions responsible for the best prediction. Divisions are performed until the nodes are homogenous or comprise a specified number of cases. The computations presented in this study were carried out using the Statistica 12.0 software.

3 Results and discussion

Ten and twelve regression tree models were built to predict respectively failure rates λ_r and λ_p. The methodology described above was used to select optimal regression tree models. Their structures are shown in figure 1. There were 8 divided nodes and 9 end nodes in each of the models for predicting the failure rate of respectively distribution pipes and house connections. Under close examination the models differ considerably in their architecture. Differences are visible already on the third level of division, i.e. in nodes (leaves) number 4, 5, 10 and 11. One can say that the tree model for distribution pipes is more complicated since most (6 out of 9) of the end nodes are on the last level of division, which may indicate that until the very end of division the decision rules were not explicit. The architectures of the regression trees (fig. 1) do not seem to be too complex, considering the number and type of independent variables which make up the model. It should be mentioned that regression trees can be connected into assemblies of trees to form a random forest. As a rule, an assembly of trees gives better prediction results than a single, even most complicated, tree [22]. In the case of a random forest, it would be necessary to include many other variables in the vector of predictors so that the complicated model architecture reflected the relations between the predicted variable and the independent variables. This is naturally limited by the acquirability of many operational variables. It seems that information on the pressure prevailing in the water supply network and other data on the pipelines (e.g. the number and kind of failures) and even data seemingly not connected with the failure rate problem, such as water production and demand or losses, should be added to the vector of predictors.

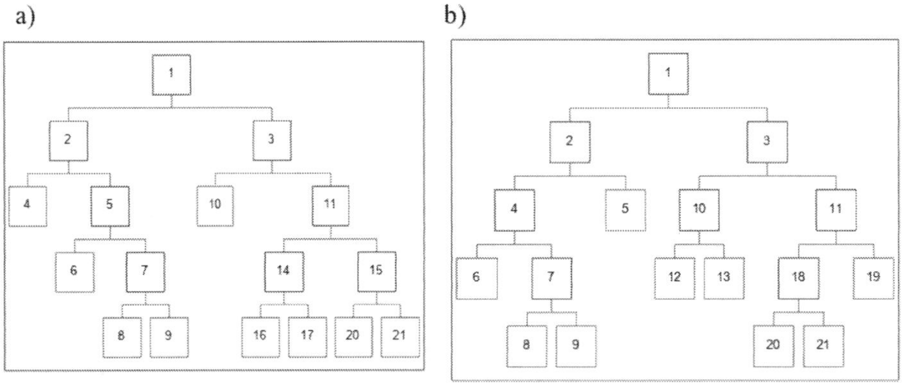

Fig. 1. Optimal structure of regression tree, a) distribution pipes, b) house connections.

Cost sequence diagrams after cross-validation for the RT models are shown in figure 2. Models no. 3 and 4, for which the costs amounted to 0.000333 and 0.001086, were selected as optimal for the prediction of rate λ_r and λ_p, respectively.

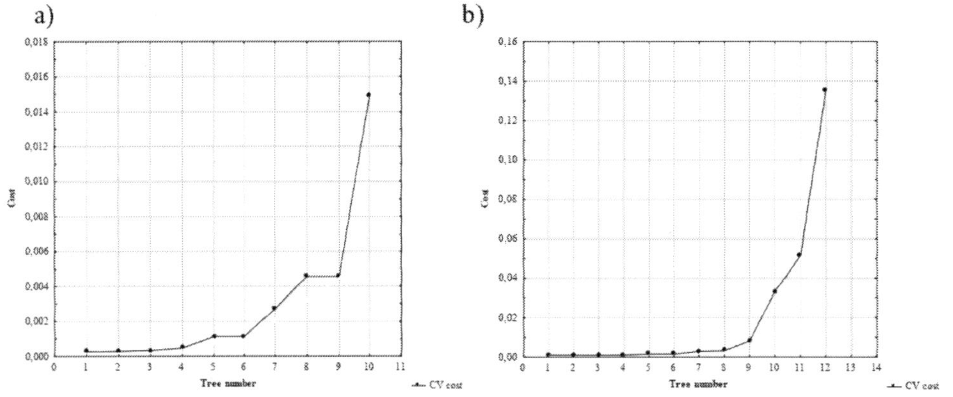

Fig. 2. Cross-validation cost (CV), a) distribution pipes, b) house connections.

It was found that the costs would increase as model complexity decreased, e.g. models 10 and 12 had only one end node. On the other hand, when selecting an optimal model one should also take into account the simplicity of the regression tree architecture. It is not a great achievement when one selects a model characterized by very low resubstitution cost

 DOI: 10.1051/e3sconf/20172200097

values, but whose structure is very complicated (e.g. there are 10–20 end and divided nodes).

For the division into branches and levels it is essential to determine the so-called importance, i.e. rank the predictors according to their significance on the scale of 0–1. This approach is helpful in identifying the variables having a significant prediction power with regard to dependent variables [21, 22]. When predicting the failure rate of house connections (λ_p), the independent variables were ordered as follows: conduit length, material, year of construction and diameter. The importance of the above predictors amounted to respectively: 1.00, 0.07, 0.03 and 0.01. In the case of the regression tree used to model rate λ_r, the significance ranking was a little bit different: pipeline length (1.00), year of construction (0.08), diameter (0.04) and material (0.02). It is apparent that the conduit length predictor dominated in the optimal models, considering that its value was by two orders of magnitude higher than the values of the other independent variables. This fact is quite inexplicable since it is rather such independent variables as conduit material, diameter and laying year should be the predictor having a greater influence on the prediction of the dependent variable (the failure rate). The research so far has shown that not pipeline length, but the other variables are relatively important in the assessment of the failure rate of water conduits [23, 24]. However, an analysis of figure 3, illustrating the real (experimental) and predicted (by the RT model) failure rates of respectively the house connections and the distribution pipes, shows that despite the quite surprising dominance of one dependent variable (length) the modelling results are simply ideal. The regression tree method considerably differs from other regression algorithms, such as support vector machines or K-nearest neighbours, which may affect the quality of modelling and the choice of a particular independent variable as the dominant parameter in the prediction of the failure rate of water conduits.

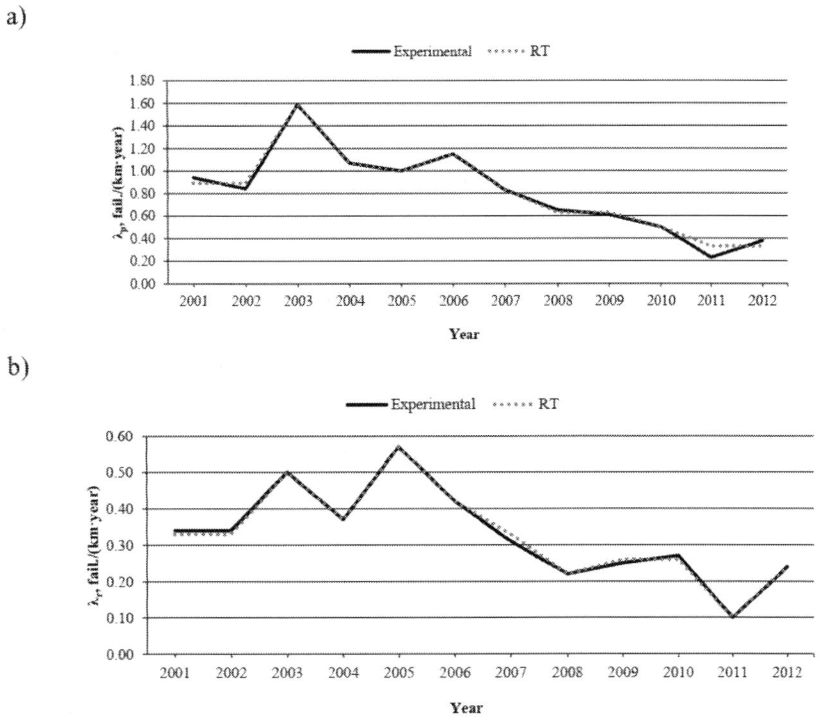

Fig. 3. Experimental and predicted failure rates, a) house connections, b) distribution pipes.

Almost ideal convergence between the real failure rate and the predicted one was observed for the house connections (fig. 3a). Coefficients R and R^2 amounted to respectively 0.994 and 0.988. Only in 2011 the predicted failure rate was slightly higher than the real one. The Spearman rank correlation, which is a nonparametric measure of statistical interdependence between variables, is also high, amounting to 0.995. When the RT modelling results are compared with the results of predicting rate λ_p by means of artificial neural networks (ANN) [20], it becomes apparent that the latter method generates greater divergences between experimental and predicted data. However, one should note that the ANN modelling methodology is somewhat different the RT methodology. In the case of the RT algorithm, a whole data set was used to build the model. Whereas ANN models are created on the basis of a data set which is divided into two subsets: a training subset and a testing subset. The prediction results for the testing set are characterized by larger real-predicted value errors due to the fact that the ANN model previously "saw" only the training set data. Therefore it seems necessary to broaden the present research by carrying out the validation of the RT model on the basis of the operational data from the next years. In this way one can conclusively determine whether the regression method is more advantageous as regards the correct prediction of the failure rate of water supply pipelines.

The prediction of the rate of failure of the distribution pipes (fig. 3b) is very good. The relative prediction error amounted to about max. 6.5%, which is a satisfactory result. The correlation between the experimental results and the ones predicted by the RT method was $R^2 = 0.996$. The Spearman rank correlation is at a similar level as in the analysis of the failure rate of the house connections, amounting to 0.993. The rank correlation values indicate that the variables (the experimental and the predicted failure rates) form an increasing function. The analyses show that the RT algorithm is quite a universal approximator. The other regression methods, e.g. artificial neural networks [20], are not always able to make a correct prediction of data slightly divergent from the other values in the set (e.g. the values of rate λ_r in 2011). As mentioned above, the resistance of the RT method to outliers is undoubtedly its advantage over the artificial intelligence algorithm.

However, when analyzing the suitability of regression methods for predicting failure rates one should take into account not only the agreement between the predicted data and the real ones, but also the type of pipeline and its influence on the reliable operation of the whole water distribution network. Because of the higher costs of repair and the social costs resulting from a drop in network pressure, interruptions in water supply and other failure events, failures of a trunk water main or a distribution pipe have more significant consequences than a failure or 10–20 house connections at the same time. Since the prediction results are promising, it seems that in the future the regression tree method should be considered as an alternative to other regression methods for predicting and analyzing the failure rate of water supply pipelines.

Failure rate prediction using mathematical modelling, e.g. regression trees or other regression methods as artificial intelligence or support vector machines, could be useful for water companies. The modelling results could be used by water utilities to e.g. plan the modernization schedule or to propose the proper renovation method for specific water pipe section. In such case including the street name and exact localization of occurred damage seems to be reasonable and useful especially for exploiters who are interested a lot of in practical and technical aspects. We see that at this stage there many additional facts and data which should be taken into consideration during forecasting of failure frequency. Moreover, this kind of failure rate assessment (by means of RT) allows us to distinguish and choose the most important variables (see significance ranking described above). This knowledge is also very useful for water companies, because in relatively simple way we point out what kind of data and with what accuraccy should be registered.

4 Conclusions

The results of predicting the rate of failures of the distribution pipes and house connections in one of the Polish cities by means of the regression tree method have been presented. The subject seems to be important for the correct and quick estimation of the reliability level of municipal systems. The created RT models can be useful in cases when it is necessary to determine the failure frequency in order to make a decision on planned repairs of water supply pipelines. The modelling methodology presented here is something of a novelty in comparison with the approach used so far to predict the failure rate of water conduits. A survey of literature on this subject revealed that the regression tree method is not widely used to assess the level of operational reliability of water supply networks, which induced the author to take up this subject. The main conclusions emerging from the modelling are as follows:

- ideal convergence (at the level of 0.99) between the experimental data and the values predicted by the regression tree models was obtained for both the distribution pipes and the house connection;
- the optimal RT models, selected on the basis of a minimal cost analysis, comprised 8 divided nodes and 9 end nodes; despite the fact that the number of nodes was the same in both the model for predicting the failure rate of house connections and the one for predicting the failure rate of the distribution pipes, the architecture of the two models was considerably different: different nodes (leaves) would undergo division and the number of nodes on the particular tree levels was different; the tree model for the failure rate of house connections seems to be less complicated since its final nodes are uniformly distributed between the last and the last but one level of the tree, which may indicate that the final divisions were performed uniformly and the decision rules were quite explicit;
- the lowest costs were recorded for the models the structures of which were quite complicated (10–20 divided and end nodes); the criterion for selecting the optimal model cannot be based on costs only since the aim is also to minimize model complexity; as regards regression trees, it is better to replace the very complicated structure of a single tree with a random forest; this seems to be sensible for predicting random variables, which reliability indices undoubtedly are;
- the analysis of the ranking of significance showed that length is an independent variable having the greatest influence on the tree structure; the importance of this parameter amounted to 1.00 and that of the other parameters was lower by as many as two orders of magnitude; it was precisely the length of pipelines which was responsible for divisions in the nodes on the successive levels of the tree;
- the comparison of the results of modelling by means of respectively regression trees and artificial neural networks [20] indicates that the latter method is less resistant to outliers, whereby the convergence between the predicted values and the real ones is lower.

The work was realized within the allocation No. 0401/0006/17 awarded for Faculty of Environmental Engineering Wroclaw University of Science and Technology by Ministry of Science and Higher Education in years 2017–2018.

References

1. Q. Meng, C. J. Cieszewski, M. Madden, Gisci. Remote Sens. **44**, 2 (2007)
2. I. Kasprzyk, A. Grinn-Gofroń, A. Strzelczak, T. Wolski, Sci. Total Environ. **409**, (2011)
3. F. Ünes, M. Demirci, Ö. Kisi, Period. Polytech-Civ. **59**, 3 (2015)

4. B. Tchórzewska-Cieślak, K. Pietrucha-Urbanik, Eksploat. Niezawodn. **18**, 2 (2016)
5. B. Tchórzewska-Cieślak, Global Nest J. **16**, 4 (2014)
6. Y. Omar, A. Parker, J. Smith, S. Pollard, Sci. Total Environ. **576**, (2017)
7. E. Avila-Melgar, M. Cruz-Chávez, B. Martinez-Bahena, Wat. Sci. Technol. **16**, 6 (2016)
8. B. Kaźmierczak, M. Wdowikowski, Period. Polytech-Civ. **60**, 2 (2016)
9. M. Kutyłowska, M. Orłowska-Szostak, Water Practice and Technology **11**, 1 (2016)
10. M. Bakker, E.A. Trietsch, J.H.G. Vreeburg, L.C. Rietveld, Wat. Sci. Technol. **14**, 6 (2014)
11. M. Iwanek, D. Kowalski, M. Kwietniewski, Ochr. Sr. **37**, 4 (2015)
12. H. Osman, K. Bainbridge, J. Perform. Constr. Fac. **25**, 3 (2011)
13. A. Scheidegger, J.P. Leitao, L. Scholten, Water Res. **83**, (2015)
14. M. Aydogdu, M. Firat, Water Resour. Manag. **29**, 5 (2015)
15. K.N. Fleming, B.O.Y. Lydell, Reliab. Eng. Syst. Safe. **86**, 3 (2004)
16. G. Kabir, G. Demissie, R. Sadiq, S. Tesfamariam, Knowledge-Based Systems **85**, (2004)
17. A. Malinowska, Nat. Hazards **73**, 2 (2014)
18. M. Bevilacqua, M. Braglia, R. Montanari, Reliab. Eng. Syst. Safe. **79**, 1 (2003)
19. A. I. Irimia-Dieguez, A. Blanco-Oliver, M. J. Vazquez-Cueto, Procedia Economics and Finance **23**, (2015)
20. M. Kutyłowska, Eng. Fail. Anal. **47**, (2015)
21. L. Breiman, J. H. Friedman, R. A. Olshen, C. J. Stone, Classification and regression tress (Chapman & Hall/CRC, Boca Raton, USA, 1984)
22. Statistica 12.0, Electronic Manual
23. K. Shahata, T. Zayed, Struct. Infrastruct. E. **8**, 11 (2012)
24. G. Pelletier, A. Mailhot, J.P. Villeneuve, J. Water Res. Pl-ASCE. **129**, 2 (2003)

The measures for achieving nZEB standard of retrofitted educational building for specific polish location – case study

Jerzy Kwiatkowski[1,*], *Maciej* Mijakowski[1], and *Adrian* Trząski[1]

[1]Warsaw University of Technology, Faculty of Building Services, Hydro and Environmental Engineering, ul. Nowowiejska 20, 00-653 Warszawa, Poland

Abstract. Most of the EU member states have already set a definition of nZEB for new buildings and some of the countries also done it for existing buildings. As there is no definition of nZEB for existing buildings in Poland, the paper will include various considerations of such a standard. Next, a case study of educational building retrofitting to a proposed nZEB standard will be presented. The aim of the paper is to present what measures can be used in order to decrease energy consumption in existing building. The measures are divided into three parts: architectural and construction, installations and energy sources. Thus a complexity of the solutions are presented. As the nZEB standard is related to available energy sources, also an influence of local condition will be considered. Building chosen for analysis is located in an area under historic protection which makes the work even more difficult. It was proved that used solutions were chosen not only to reduce energy demand or increase energy production from renewable energy sources, but also to increase social and aesthetic features of the building.

1 Introduction

The Recast of the Directive on the Energy Performance of Buildings (the EPB Directive) [1] came into force on 9 June 2010. EU member states should until 9 June 2012, publish the relevant laws and administrative regulations necessary to implement its provisions. European regulatory efforts towards increasing energy efficiency of buildings are focusing on building requirements. All new public building after 31.12.2018 must be nearly zero energy buildings. Such requirement can be very hard to implement in public buildings as the energy needs for heating, cooling and lighting to provide internal comfort are typically very high. It is even harder to retrofit existing public buildings to nearly zero energy standard. In this paper an integrated energy design (IED) process in public building retrofitted to nZEB standard will be presented. In the first step a definition of nZEB standard is given, next an integrated energy design (IED) process is shortly described and in the last part an example of use of IED in educational building retrofitting process is given.

* Corresponding author: jerzy.kwiatkowski@is.pw.edu.pl

2 nZEB definition

Directive 31/2010/EU [1] requires the introduction of nearly zero-energy buildings but does not specify a minimum or maximum harmonized requirements and detailed guidance framework procedure for calculating the energy performance of building. Member states have to define what that means exactly for them. Taking into account the local conditions in the definition does not preclude the adoption of a uniform methodology in all Member States.

The Directive defines a nearly zero-energy building as a high energy performance building and requires the determination of the ratio of primary energy. Very low or almost zero energy demand of the building should be covered from renewable energy sources or renewable sources produced on-site.

In Poland a nZEB standard has been defined by primary energy indicator (EP) and set in Ordinance on technical conditions to be met by buildings and their location [2]. For public buildings, like educational building maximal value of EP has been state as sum of three indicators:

- for heating and domestic hot water – 45 kWh/m^{2}a;
- for cooling – 25*($A_{f,c}$/A_f) kWh/m^{2}a (where $A_{f,c}$ is a cooled area and A_f is a total regulated temperature area);
- for lighting – 25 kWh/m^{2}a for time of light operation < 2500 h/year and 50 kWh/m^{2}a for time of light operation ≥ 2500 h/year.

In total maximum EP may vary from 70 to 120 kWh/m^{2}a.

From the other hand in a Directive 31/2010/EU a requirement of cost optimal parameters is given. Hence in the KODnZEB project [3] a study of the cost optimal EP value for educational building estimation has been done. The calculation has shown that cost optimal EP indicator should have value between 40 and 55 kWh/m^{2}a [4].

Despite existing requirements or analysis for the use of KODnZEB project a value of primary energy indicator equal to 20 kWh/m^{2}a has been set as nZEB standard for two chosen buildings. Such a stringent requirement was defined as a goal was to prove that educational building can be retrofitted with very high energy requirement reduction.

3 Retrofitting measures

An energy needs of the buildings are sum of a heat needs for heating and hot water preparation, cool needs for cooling and electricity for lighting and technical systems (pumps, vents). In complex retrofitting a modernization of each energy needs has to be taken into consideration.

An energy consumption for heating depends on thermal quality of envelopes (external walls, roofs or floors on the ground), windows and ventilation system (natural, mechanical, with or without heat recovery). In order to decrease energy needs for heating an additional layer of insulation material on walls or roofs or replacement of windows can be proposed. It is also required to use controlled ventilation with heat recovery.

The heat needs for hot water preparation are related with consumption. The best solution in reducing such needs is use of water efficient taps.

In the lighting system two main measures can be proposed: replacement non-efficient lighting bulbs to LED installation and use of control system (occupancy sensor, zoning).

For the technical system consumption an energy efficient pumps and vents with inverter can be proposed. It is also recommended to design low pressure ventilation system in case of mechanical ventilation with heat recovery.

4 Case building description

In the KODnZEB project two building were chosen in order to verify possibility of retrofitting them to nZEB standard. One of the buildings is educational building constructed in 1970. Building is located in Warsaw (Poland) in highly urbanized area. The building location with its surroundings is presented in Fig. 1.

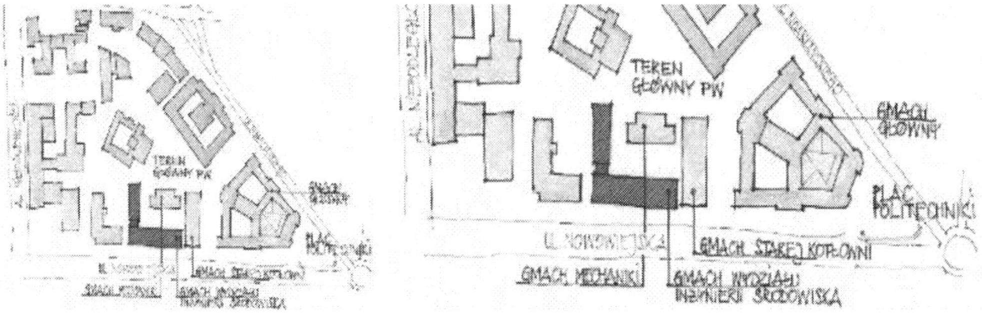

Fig. 1. Location of analysed building on the Warsaw University of Technology campus (marked with red).

The south façade of the building is oriented to the street and city tram line. Thus beside the overheating problem during spring and autumn (July and August are free of classes) the outside noise are reason of uncomfortable environment in the internal spaces. The others façades of the building are oriented to Warsaw University of Technology internal campus so the outside noise is not a problem. Some overheating occurs in the spaces oriented to the west.

The building has two wings: lower with 8 over ground levels and higher with 11 over ground levels. A total heated area of the building is equal to 18 850 m^2. A variety of different function of spaces can be identified within the building: lecture rooms, auditoria, chemical and biological laboratories, mechanical laboratories, computer laboratories, offices and corridors and sanitary units. In the Fig. 2 a layout of the example floor is presented.

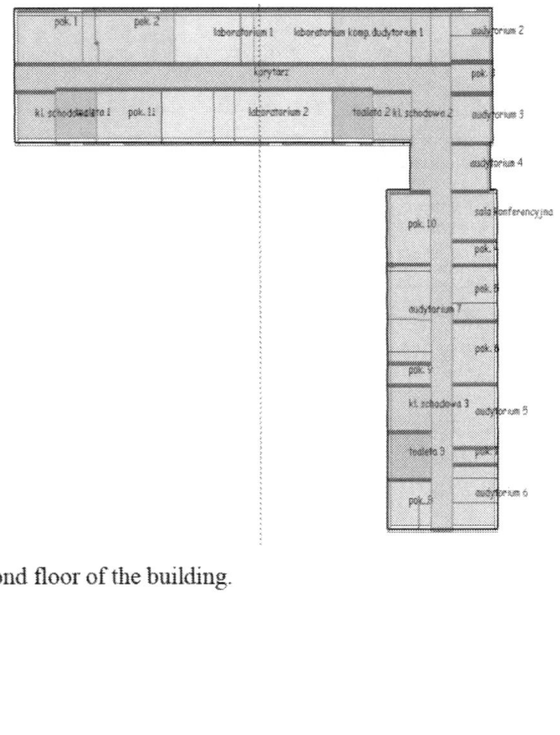

Fig. 2. A layout of second floor of the building.

It can be noticed that different spaces are mix with each other and it is no possible to divide building into functional spaces. A such variety of functions makes the retrofitting of this building very complex, as in each space different occupancy, lighting or internal gain scheduled occurs.

The insolation of the building do not fulfil current building regulation in Poland. Also natural ventilation and low quality lighting system are the reason of low building energy characteristic. Thus some measures are required for decrease energy needs and to achieve nZEB standard.

As it was described earlier definition of nZEB standard is based on primary energy. Thus a primary resource factor (PRF) of energy source will have a huge influence on final results. This building is supplied in electricity from central electricity network with the $PRF_{el} = 3.00$, and from photovoltaic system with $PRF_{PV} = 0.00$. The heat is supplied for district heating network with $PRF_{DH} = 0.68$. It must be added that in Poland a primary resource factor for fossil fuels like coal, oil or gas is equal to 1.00. Thus a use of district heating as heat source is reasonable. Also the PRF value for the electricity shows that use of compressor heat pump for heating can increase primary energy in comparison with district heating.

4.1 Energy standard of existing building

Energy simulations are carried out in the program for the comprehensive thermal analysis of buildings DesignBuilder [5]. The software is compatible with the guidelines CIBSE AM11 "Building energy and environmental modelling". DesignBuilder uses the latest Energy Plus simulation engine for energy calculation. Energy Plus is the U.S. DOE building energy simulation program for modelling building heating, cooling, lighting, ventilating, and other energy flows.

In the program a numerical 3D model of the building can be created. Next each envelope is described by heat transfer coefficient and for windows also total solar transmission and light transmission. For defined zones following data can be set: occupancy, internal loads, light quality, ventilation rates, HVAC systems and usage schedules of each parameter. In the last stage the weather data are set. This detailed building model allows the calculation of the energy consumption for heating, cooling and artificial lighting.

In the table 1 an energy characteristic of the existing building has been showed.

Table 1. The energy characteristic of the existing building.

EP indicator [kWh/m^{2}a]	
Heating and ventilation	72.2
Domestic hot water	8.8
Cooling	0.9
Lighting	60.6
Auxiliary energy	12.8
PV installation	-6.5
Sum	148.9

It can be noticed that highest energy needs are for heating and ventilation and for lighting. Therefore the main goal school be decrease consumption in this area.

 DOI: 10.1051/e3sconf/20172200098

4.2 Retrofitting measures

As it can be noticed the main energy needs are related with space heating. Therefore the first idea was to add extra insulation on the walls and roofs and replace existing windows to ones with lower U value. Such measures will result in decreasing of transmission heat loses.

The other high energy consumption in connected with ventilation system. Thus a mechanical ventilation with heat recovery was considered. However the occupancy of spaces within a building is different and centralized ventilation could work not properly. In addition the building is on constant operation any measure that needs construction work inside a building should not be applied. Thus a decentralized ventilation with heat recovery was proposed. A individual ventilation unit will be placed in each of room. But Polish regulation forbidden and inlet and exhaust of ventilation air to be close to each other. Such situation would occur in this case. It was then proposed to use individual inlet for each ventilation unit and common exhaust duct installed on the outside façade.

The increase of the thermal properties of external envelopes could increase the problem of overheating during mid seasons and in office rooms during summer. Thus an idea of external shadowing devices was considered. But due to the fact that whole campus in under historic protection such modernization has to be agreed. It was stated that the façade of the building should be as simple as possible and any external elements cannot be visible.

The need of individual ventilation with external ducts, need of shading devices and restriction from historical protection office makes to consider double skin façade (at least on the south and west orientation). It occurs that such solution solve a problem with shading and gave possibility to hide external ventilation ducts. A double skin façade decreases outside noise. In order to ventilate space within double skin façade a low-speed fans were placed at the top of the façade.

The energy efficient lighting system with LED light sources and control system like occupancy sensor and zoning were used in order to decrease electricity consumption in the building.

In the Fig. 3 a scheme of double skin façade is presented.

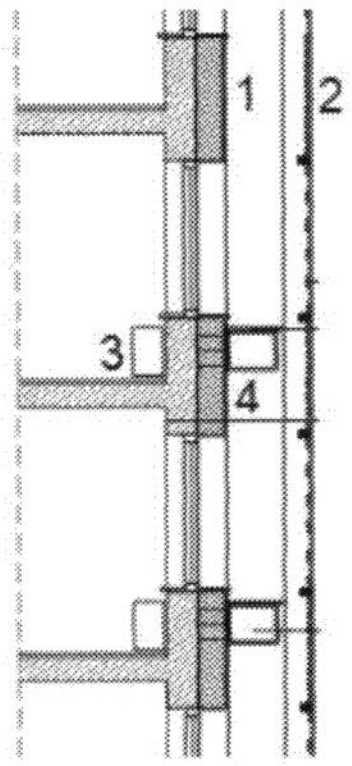

Fig. 3. A scheme of double skin façade (1-extra insulation, 2-PV panels, 3-individual ventilation units, 4-exhoust ventilation ducts).

All proposed systems were designed with the energy efficient auxiliary devices (pumps or fans) in order to not increase of electricity consumption in a significant way.

The analysis of energy sources was also done. It was concluded that due to the low value of the primary resource factor of the district heating any change of the heating source is groundless. Thus in order to decrease primary energy needs related with electricity consumption a photovoltaic installation was proposed. The PV panels were proposed to be installed on the roof and on the south and west façade, but only down to upper part of the surrounding trees.

However the low energy needs of the building do not result in comfortable spaces. Even if thermal comfort is maintained the aesthetics and friendless of the internal spaces is required. Thus some rearrangement of common spaces was proposed with extra social spaces and green plant introduction inside a building.

4.3 Energy standard of retrofitted building

In order to verify energy standard of a building with all proposed modernization an energy modelling has been repeated. All described measures were included in new building model. In the table 2 an energy characteristic of the building after retrofitting has been showed.

Table 2. The energy characteristic of the retrofitted building

EP indicator [kWh/m^{2}a]	
Heating and ventilation	8.9
Domestic hot water	3.9
Cooling	4.4
Lighting	10.4
Auxiliary energy	12.2
PV installation	-27.6
Sum	12.2

It can be noticed that primary energy needs decreased and total EP indicator is equal to 12.2 kWh/m^{2}a, what meets set goal of 20 kWh/m^{2}a.

The highest decrease can be noticed in heating and lighting consumption. The cooling needs increased what was connected with higher thermal resistance of building and an extra cooling system for all office spaces. It can be also noticed that energy production from PV installation was increased by 4.2 times. It must be stated that without electricity production from PV installation EP indicator would be equal to 39.8 kWh/m^{2}a. Thus it should be concluded that without renewable energy sources achieving nZEB standard in not possible.

5 Conclusions

It was shown that the use of IED process in building retrofitting project is possible and gives tangible benefits. Through common work of architects, constructor and HVAC engineers a chosen measures reflect in EP indicator decrease from 148.9 kWh/m^{2}a for existing building to 12.2 kWh/m^{2}a for retrofitted building.

The planned actions include modernization of building elements (external walls, roofs, windows) as well as technical systems (HVAC, hot water, lighting and RES). Moreover, modernization concept includes some improvements in social use of the building e.g. new atrium with spaces for rest and group activities. Façades and planned atrium will be designed using "green architecture" concept.

Acknowledgment

This paper was performed with the financial support of EEA Financial Mechanism 2009-2014 – Operational Program PL04 "Saving energy and promoting renewable energy source" within the framework of the" project KODnZEB: "Thermo-modernization of two chosen Public Buildings according to nZEB Standards".

 DOI: 10.1051/e3sconf/20172200098

Through the funds EEA and Norwegian funds, Iceland, Liechtenstein and Norway contribute to reducing social and economic inequality and strengthening bilateral relations with beneficiary countries in Europe. These three countries closely cooperate with the EC on the basis of the Agreement on the European Economic Area (EEA Agreement).

References

1. Directive 2010/31/EU of the European Parliament and of the Council of 19 May 2010 on the energy performance of buildings (OJ L 153, 18.06.2010)
2. Ordinance of the Minister of Transport, Construction and Maritime Economy of 5 July 2013 amending the regulation on technical conditions to be met by buildings and their location (Dz.U. 2013 pos. 926)
3. KODnZEB project, Thermo-modernization of two chosen Public Buildings according to nZEB Standards, Poland, (2016)
4. J. Kwiatkowski and J. Rucinska, The use of cost optimal methodology to determine nZEB standard of public building, 6th International Symposium on Energy Challenges & Mechanics (ISECM) – towards a big picture, (14-18.08.2016, Inverness, Scotland)
5. http://www.designbuilder.co.uk/documents/DesignBuilder-flyer-v1a.pdf

DOI: 10.1051/e3sconf/20172200099

Research of CO_2 concentration in naturally ventilated lecture room

Marta Laska[1,*], *Edyta* Dudkiewicz[1]

[1]Wroclaw University of Science and Technology, Faculty of Environmental Engineering, Wybrzeże Wyspiańskiego 27, 50-370 Wrocław, Poland

Abstract. Naturally ventilated buildings especially dedicated for educational purposes need to be design to achieve required level of thermal comfort and indoor air quality. It is crucial in terms of both: health and productivity of the room users. Higher requirements of indoor environment are important due to the level of students concentration, their ability to acquire new knowledge and willingness to interact with the lecturer. The article presents the results of experimental study and surveys undertaken in naturally ventilated lecture room. The data is analysed in terms of CO_2 concentration and its possible influence on users. Furthermore the outcome of the research is compared with the CO_2 concentration models available in the literature.

1 Introduction

The one of the main building task is to maintain comfortable and healthy conditions for its users. It can be accomplished on various ways depending whether the building is mechanically or naturally ventilated. The issues of indoor air quality and carbon dioxide as an indicator of ventilation efficiency in mechanically ventilated buildings are described in the literature [1-3]. However there is still huge number of existing naturally ventilated buildings were human is a main source of contaminants and where CO_2 as the sufficient indicator can be used for optimisation of indoor conditions.

As people spend most of their time indoors, IAQ has been identified as a crucial aspect of building utilization in many publication [1, 3–6]. The CO_2 concertation is described by the quantity of carbon dioxide in a million particles of air – ppm (particle per million). Other words the CO_2 concentration is an indicator of the amount of fresh air delivered to the room from the outside. The reference outdoor value is 400 ppm that is 0.04% of Earth's atmosphere [7, 8]. The human body exhales even about 2–4% [7, 9, 10] this is why in enclosed areas dedicated for human being, the concentration of carbon dioxide can be higher. The acceptable level of carbon dioxide is related to building category and is defined in European Standard 13779:2008 [11]. The standard states that the typical ranges of CO_2 concentration above the outdoor level are: for building category I: < 400 ppm, category II: 400-600, category III: 600-1000 and IV: > 1000. The literature also indicates [8], following

* Corresponding author: marta.laska@pwr.edu.pl

Venture Industries Sp.z o.o., that the outdoor level of CO_2 for rural areas is 300 to 350 ppm, while for cities and urban areas the level increases to 400 ppm.

The CO_2 as an indicator of indoor conditions has been described in the literature. The [12] presents the numerical study of CO_2 dispersion from breathing in an auditory room. The study indicate that CO_2 concentration accumulates on a lower levels of the space and decreases with height. It also states that the level of carbon dioxide describes the effectiveness of the ventilation systems which is responsible for diluting also more harmful indoor contaminants the human is exposed to. The [5] presents the research outcome on naturally ventilated school rooms in terms of carbon dioxide level. The literature [5, 13] indicates that CO_2 concentration is an important factor influencing the performance of learning ability and concentration. Publications [4, 7, 10] point out that when the threshold defined by WHO as acceptable of 1000ppm is exceeded, room users can suffer of headaches, tiredness, lack of concentration or throat and nose ailments [1, 4].

In the paper [4] authors present the results of measurements of carbon dioxide concentrations conducted in the school buildings located in two different climates: Białystok (Poland) and Belmez/Córdoba (Spain). They developed a model for CO_2 level estimation that is further applied and validated for the lecture room located in one of the buildings at the Wroclaw University of Science and Technology (Poland). The outcome of this investigation is described in this paper.

2 Description of the building and lecture room

The measurements were conducted in a building situated in Wrocław city, Poland, located at the altitude 110–125 m above sea-level. The climate is mild climate, moderately warm with significant rainfall throughout the year. The average annual temperature is 8.4°C, and the average rainfall is 551 mm. The warmest month of the year is July with an average temperature of 17.7°C, while January has the lowest average temperature throughout the year and it amounts to -2.1°C.

The investigated lecture room is located in D-2 building (fig. 1) – one of the twin buildings that belong to the oldest part of main campus of Wroclaw University of Science and Technology in Wroclaw [14–16]. The layout of both these twin 5-floor buildings is quadrangular with internal courtyard. The buildings were designed according to the doctrine of socialist realism, and therefore are characterized by monumental, some squat proportions and monotony. However their architecture, composition and economy of measures used define them as high level construction works [16]. Both buildings are founded on two-storey rusticated base course, over which a regular rhythm of window openings is shaped. The building body is crowned with a strongly accented cornice and steep ceramic roof. Facade architecture is very modest concerning the details [15, 16]. The building D-2 was constructed in the 1955's [14–16].

The lecture room located on the highest storey and has a form of auditorium (fig. 2). The dimensions of the room plan are 21.2 x 16.9 m. The room height varies from 7.9 m to 5.5 m. A volume of the space is 2400 m^3. The room is dedicated for 300 students. The area per student is 1.19 m^2, what is lesser that national recommended minimum value 1.5 m^2 per person [4]. The room has 4 facing west windows with dimensions of 1.74 x 3.25 m. The windows are equipped in interior vertical blinds to protect the room against excessive sunlight. Additionally, the natural light is delivered into the lecture room via 20 skylights mounted in the flat roof (fig. 2). The lecture room is naturally ventilated and is served with a classic radiator-based water central heating system.

 DOI: 10.1051/e3sconf/20172200099

Fig. 1. View of inter-faculty laboratory-didactic building D-2 from front and from the internal courtyard side [15, 16].

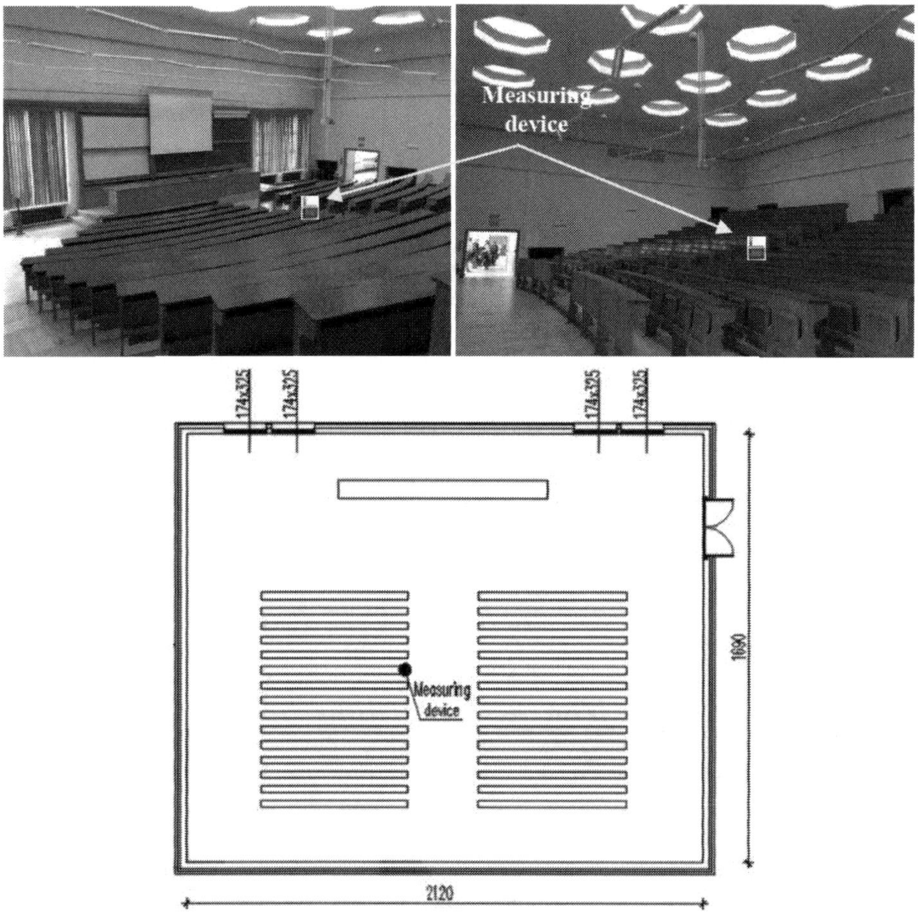

Fig. 2. Lecture room in building D-2.

3 Research methodology

The measurements of CO_2 concentration in the auditorium were conducted in 1-minute interval for 6 days in March and April 2016, during the classes for students. The length of each set of measurements lasts about 90 minutes that corresponds to the length of the lecture at the university. The equipment used for the experiment was Indoor Air Quality Data Logger Rotronic CL11. The accuracy of the data logger is ±0.3°C for the temperature range from 0°C to +50°C, <2.5% RH for the range from 10 to 90% and ±30 ppm ±5% of the measured carbon dioxide value. The measuring and recording device was placed, following [17, 18] at 1.2 m above the floor level, in the middle of the auditorium. Thus the data was collected from the central location of the room, with a special attention of being away from direct influence of room occupants. Additionally the indoor parameters and outdoor atmospheric pressure was measured by precise logger SensoData550. The accuracy of the device regarding the atmospheric pressure in the range of 500 to 1500 hPa is ± 3 hPa.

The measured data was applied to a model describing the carbon dioxide concentration presented by Krawczyk at all. in [4]. The CO_2 concentration is described by the following equation:

$$C_{CO2in}=[T_{in}/(P_{in}(1+ACH\cdot t)]\cdot\{C_{CO2in(t=0)}(P_{in(t=0)}/T_{in(t=0)})+$$
$$+[g\cdot N\cdot R/(\mu_{CO2}\cdot V)+C_{CO2out}\cdot P_{out}\cdot ACH/T_{out}]\cdot t\} \quad (1)$$

where:
- C_{CO2} is carbon dioxide concentration [ppm],
- $T_{in/out}$ is a temperature of indoor/outdoor air [K],
- $P_{in/out}$ is an air pressure of indoor/outdoor [Pa],
- ACH is the air change rate [1/h],
- t is a time measured from the beginning of the experiment [h],
- g is a CO_2 gain from a person [g/h·os],
- N is a number of people in the room [-],
- R is a gas ideal constant [J/mol·K],
- μ_{CO2} is a molar mass of carbon dioxide an air [g/mol],
- V is a volume of the space [m^3].

When the calculations were completed, the next step of the research was to compare the outcome with the measured data. This comparison allowed to validate the applicability of the model to the rooms type of auditorium.

4 Results of measurements and calculations of CO_2 concentration

As presented in the literature [19], in naturally ventilated buildings indoor air quality is in very close correlation with the quality of outdoor air and indoor contaminants. In the auditory room main pollution comes from occupants. For the investigated domain, the main parameters of outdoor air and the number of people occupied the room are presented in Table 1. The results of measurements and calculations of carbon dioxide concentration based on equation (1) are presented on Fig. 3. As mentioned before, the CO_2 level was calculated in 1-minute interval. Each time the former result was taken as the initial value for the next time range. The increment and variability of CO_2 concentration for all tested days are presented on Fig. 4. The percentage of occupation, the temperature of indoor air and the relative humidity at the initial stage, respectively *T0* and *RH0*, and after 90 minutes from the beginning of the experiment, respectively *T90* and *RH90,* are presented in Table 2. This table also presents the comparison between measured and calculated (following equation (1)) carbon dioxide concentration for the room. The variable of CO_2 concentration with the index *0* indicates the initial value (start point) of the measurements, while the variable with the index *90* indicates the final value – measured after 90 minutes from the beginning of the experiment.

As arises from presented in Fig. 3 variations of CO_2 concentration, and the literature mentioned before, in the lecture room the incensement of CO_2 level depends on the number of room occupants.

Table 1. Outdoor air parameters during measurements and number of people occupied a lecture room.

	17.03	18.03	31.03	1.04	7.04	8.04
Average outdoor air temperature [°C]	6	8	10	3	14	10
Average outdoor air atmospheric pressure [hPa]	1030	1016	1013	1016	1010	1012
Wind speed [km/h] and wind direction	7.4 W	31.5 WSW	9.3 WSW	7.4 NNW	13 SSW	6 ENE
Number of people in the room	77	42	113	59	55	44

The proposed by [4] model accurately reflects the increase in concentration of carbon dioxide in the auditory room. The most convergent values were obtained when the room has not been airing during the lecture, namely for three days: 17th, 18th and 31st of March.

The highest room air temperature increment within 90 minutes of lecture was measured on 31st of March and reached 1.56 K. That was also a day when the room was the most occupied. The high increment of indoor temperature of 1.34 K was recorded also on 7th of April when the external temperature was the highest from the investigated period and reached +14°C. For this day also the variations of CO_2 contaminations are the highest. This could be caused by the short-term airings of the room and quite strong wind pressure blowing from the South and South – West, and thus form the widow side of the auditorium. The noise caused by the impact of blinds on the windows and walls due to the strong wind caused the necessity of closing windows and allowed only a brief airing.

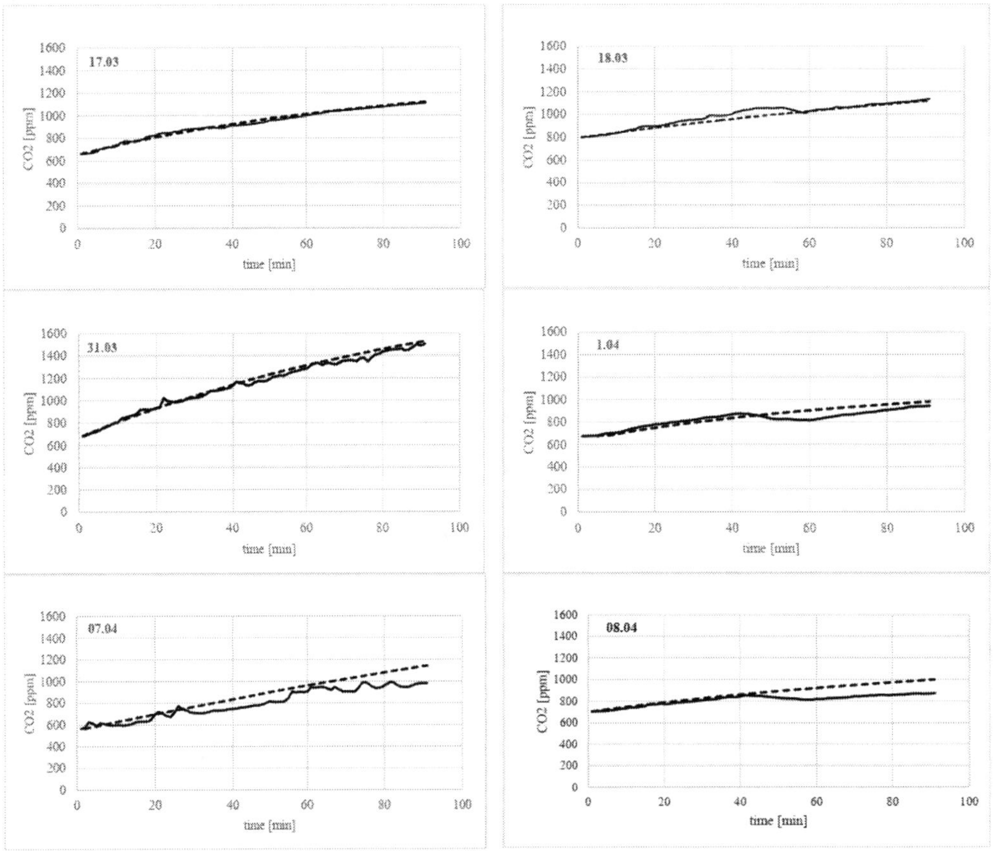

Fig. 3. Measured concentration of CO_2 (constant line) and estimated concentration of CO_2 (dashed line) in lecture room for the investigated 6 days.

The increase in the temperature of indoor air in the room is dependent on both – number of occupants due to heat gains form the human body and also on the intensity of solar radiation. The skylights are not equipped in any solar protection system, therefore during the days when the insolation is high, heat gains from the solar radiation increases the interior temperature.

During two days, namely 1st and 8th of April the low speed of the wind and its direction allowed the longer airing. The windows has been opened after 40 minutes from the beginning of the lecture. It was observed that during the first 15–17 minutes the CO_2 level was decreasing of about 49 and 34 ppm respectively, but after that time the contamination started to increase again.

The variability of the relative humidity also was observed. The highest levels of this parameter were measured on 31st of March, during the day with the highest occupancy. At this date also the CO_2 concentration and the level of indoor temperature was the highest. During the investigated period the initial concentration of CO_2 (*T0*) in the auditorium was measured at the range between 568 and 800 ppm. In days when the widows were closed, after 90 minutes the CO_2 concentration has increased from 335 to 817 ppm up to the maximum recorded value of 1502 ppm on 31st of March. This value is recognized by WHO as maximum permissible for indoors [4].

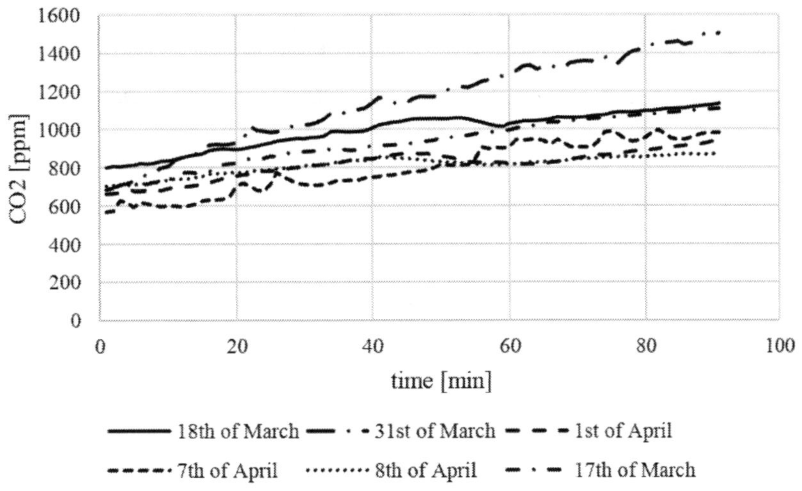

Fig. 4. Calculated concentration of CO_2 in lecture room for 6 investigated days.

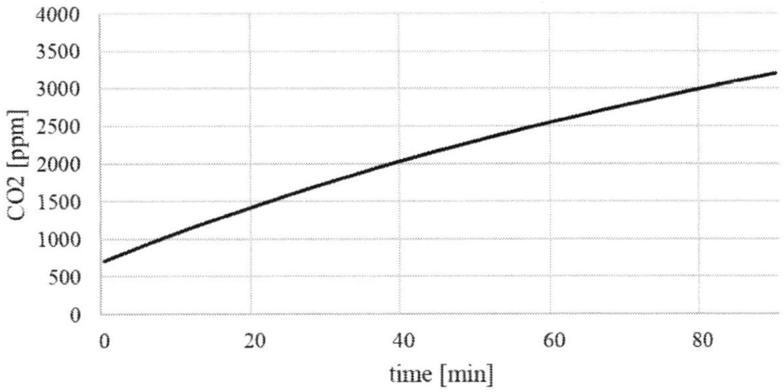

Fig. 5. Calculated concentration of CO_2 in lecture room for fully occupied room.

DOI: 10.1051/e3sconf/20172200099

The presented in [4] and applied in this investigation the formula (1) allows (following the authors [4]) to calculate the CO_2 concentartion levels in lecture rooms with any number of occupants. Therefore the additional calculations for carbon dioxide contamination for fully occupied auditorium were undertaken. The results are presented on Fig. 5. As one can notice, after 90 minutes from the begining of the experiment, the CO_2 level increases about 2492 ppm and reaches 3196 ppm. The acceptable level for rooms is 1400 ppm (1000 ppm above the external contamination [4, 9, 11]), or defined by WHO is 1500 ppm. In this case the acceptable treshold was exceeded after 20–22 minutes from the beginning of the experiment.

The CO_2 level in the room is connected with the occupant's satisfaction of the indoor environment. Therefore the questionnaire survey has been conducted among room users. The outcome showed that in general students were satisfied with the indoor environment and did not suffer strong discomfort conditions like headaches, dizziness, eye, nose or throat irritation, dry or itching skin, difficulty in concentration. However some percentage of respondents experienced sleepiness and general fatigue that may have also different background than CO_2 level. The analysis of the questionnaire survey is not a subject of this article and will be raised and described in another publication.

Table 2. Temperature, relative humidity and concentration of CO_2 in the lecture room.

Date	**17.03**		**18.03**		**31.03**		**1.04**		**7.04**		**8.04**	
Percentage of occupation [%]	26%		14%		38%		20%		18%		15%	
T0 [°C]	20.94		21.56		22.22		21.83		21.83		22.33	
T90 [°C]	22		22.33		23.78		22.44		23.17		23.06	
Air temperature increment [K]	1.06		0.77		1.56		0.61		1.34		0.73	
RH0 [%]	30		29.9		37.3		35.5		38.8		34.9	
RH90 [%]	33.2		32.4		41.5		37.6		40.5		36.8	
Relative humidity increment [%]	3.2		2.5		4.2		2.1		1.7		1.9	
CO_2 0 [ppm]	663	663*	800	800*	685	685*	676	676*	568	568*	704	704*
CO_2 90 [ppm]	1110	1119*	1135	1119*	1502	1528*	941	989*	996	1149*	873	999*
CO_2 increment [ppm]	447	456	335	319	817	843	265	313	428	581	169	295

* the value calculated from the equation (1)

5 Conclusion

The paper presented the results of measurements and calculations of carbon dioxide level in an auditorium. The investigation was conducted during the 6 lecture days in one of the lecture rooms at Wroclaw University of Science and Technology, Poland. The measured data was used to validate the model derived for school rooms by [4]. It showed good agreement and thus proved its applicability also for auditorium room types.

The values of CO_2 concertation were also compared with permissible levels of CO_2 and some conclusions from the questionnaire survey, related to personal discomfort were mentioned. The assessment showed that during the lecture time, the concentration was mostly on proper level and no strong complains arisen from the room users. However the calculations based on the theoretical model indicated that under full occupation the indoor environment would fail to serve healthy conditions. Therefore is important to sensitize the lecturer and students that natural ventilated rooms need to be aired as often as possible, as the indoor conditions strongly influences the comfort and productivity of room users.

References

1. T. Lu, A. Knuutila, M. Viljanen, X. Lu, Build. Environ. **45** (2010)
2. T.N. Quang, C. He, L.D. Knibbs, R. de Dear, L. Morawska, Energ. Buildings, **85** (2014)
3. M.K. Kima, L. Baldini, H. Leibundgut, J.A. Wurzbacher, N. Piatkowski, Energ. Buildings, **87** (2015)
4. D.A. Krawczyk, A. Rodero, K. Gładyszewska-Fiedoruk, A. Gajewski, Energ. Buildings, **129** (2016)
5. M. Griffiths, M. Eftekhari, Energ. Buildings, **40** (2008)
6. A. Ricardo, V.P. de Freitas, Energ. Buildings, **81** (2014)
7. D. Grobel, M. Konieczny, A. Łobzowski LAB-EL Elektronika Laboratoryjna Sp. J. http://www.label.pl/po/co2_1.html
8. K. Kaiser, Rynek Instalacyjny, **9** (2010)
9. A. Pełech, *Wentylacja i klimatyzacja – podstawy* (Oficyna Wydawnicza Politechniki Wrocławskiej, 2011)
10. D. Grobel, A. Łobzowski, W. Szkolnikowski, POMIAR, **3** (2006)
11. European Standard 13779:2008: Wentylacja budynków niemieszkalnych – Wymagania dotyczące właściwości instalacji wentylacji i klimatyzacji
12. K.A. Papakonstantinou, C.T. Kiranoudis, N.C. Markatos, Energ. Buildings, **34** (2002)
13. D.G. Shendell, R.Prill, W.J. Fisk, M.G. Apte, D. Blake, D. Faulkner, Indoor Air, **14** (2004)
14. J. Majczyk, A. Tomaszewicz, Architectures **1**, 41 (2015)
15. M. Burak, K. Dackiewicz, J.M. Pawlikowski, *Budynki i laboratoria Politechniki Wrocławskiej* (Politechnika Wrocławska, Wrocław, 2014)
16. S. Bocheński, J. Piskozub, Journal of Heritage Conservation, **41** (2015)
17. M. Hussin, M.R. Ismail, M.S. Ahmad, J. King, Saud Univ. Eng. Sci. (2014)
18. P.O. Fanger, *Komfort cieplny* (Arkady, 1974).
19. V. Turanjanin, B. Vucicevic, M. Jovanovic, N. Mirkov, I. Lazovic, Energy, **77** (2014)

DOI: 10.1051/e3sconf/20172200100

The importance of monitoring renewable energy plants: three case histories

Renato Lazzarin[1,*]

[1]DTG, University of Padova, 36100 Vicenza, Italy

Abstract. Many renwable energy plants are put into operation without providing a monitoring system to evaluate their performance over time. Then if is often difficult to realise the bad working of the system and the loss of efficiency results in an economic loss. Three quite different examples are reported to illustrate how real performance can be lower than designed due respectively: 1. To bad settings of the parameters; 2. To a hurried commissioning that did not reveal the mistakes in the design of the plant; 3. To a failure of a single component over time.

1 Introduction

Renewable energy plants are seldom simple plants except small installations such as DHW solar plants or small photovoltaic plants. They often include the renewable section and conventional auxiliaries, usually boilers. Hot and/or cold storage is usually present and a suitable control system governs the plant. The plant manager is frequently another person than the designer and he does not always possess a comprehensive understanding of the system control logic.

Many renewable energy plants are put into operation without providing a monitoring system to evaluate their performance over time. It happens that seasonal performance may be well below the values recorded during commissioning or planned in the design. Or wrong setting of parameters or the failure of some components prejudices system performance: the system goes on operating but at a lower efficiency than designed. In the absence of a supervision it is difficult to realise the bad working of the system. In fact the auxiliary provides the service and it is not straightforward to decide whether the energy bill, that often covers a whole month or more different needs than the considered plant, is appropriate. The loss of efficiency results in an economic loss that might go on for years, losing the waited advantages of a more expensive, but potentially also more efficient installation.

Three quite different examples are reported to illustrate how real performance can be lower than designed due respectively:

1. To bad settings of the parameters;

* Corresponding author: renato@gest.unipd.it

 DOI: 10.1051/e3sconf/20172200100

2. To a hurried commissioning that did not reveal the mistakes in the design of the plant;
3. To a failure of a single component over time.

2 The first example: a motor driven heat pump HVAC plant

The "San Nicola" building, in the historic heart of the town of Vicenza, has been hosting since 1999 the Department of Management and Engineering of the University of Padova (DTG). The building has a net floor area of about 4,200 m^2, and a volume of 14,300 m^3.

The building HVAC system consists of:

- Two pipes fan-coil heating and cooling system.
- Ventilation and air conditioning system (21,700 m^3/h AHU for humidity control and supply of outdoor air). The AHU is made of a cross flow heat exchanger, the pre-heating section, the cooling/dehumidification section, the post heating section.
- Extraction system, for exhaust air from the toilets.

The heating and cooling production is obtained through a gas engine driven heat pump and two condensing boilers.

The air to water heat pump can operate both in heating and cooling mode by means of a four-way inversion valve, and it is driven by a gas reciprocating engine. It is possible to modulate the heating/cooling capacity in a continuous way by varying the engine speed from 1800 to 2600 rpm, acting on the throttle butterfly valve. A further modulation down to 40% of the rated heating/cooling power is obtained through the inlet valve unloader on the compressor cylinders. The rated cooling capacity is 276 kW with a gas consumption of 22 Nm^3/h, while the rated heating capacity is 380 kW with a gas consumption of 19 Nm^3/h(respectively in summer for 35°C outdoor air and evaporator inlet/outlet 12/7°C, and in winter for 10 °C outdoor air and condenser inlet/outlet 40/45°C).

Heat recovery from the engine is provided from the lubricant oil, the cooling water and the exhaust gases, as to achieve a rated recovery of 126 kW when working in cooling mode and 109 kW when in heating mode, available at about 70°C. The EER in refrigeration mode is 4.5, with a PER (Primary Energy Ratio, i.e. the ratio of the cooling/heating capacity over the fuel consumption in terms of primary energy) of 1.4; in heating mode the COP is quite high (over 5) with a PER of 1.5.

The whole machine (heat pump and engine) is then equipped with a double hydraulic circuit: the heat pump one, providing chilled water (7/12°C) in summer and warm water (45/40°C) in winter, and the engine heat recovery system, from now on called the high temperature circuit, providing hot water (70/65°C). The control is given by a digital display controller that drives the engine speed control and the inlet valve unloader of the cylinders. The system has two inertial storages, 3000 l on the heat pump circuit and 1500 l on the heat recovery circuit, to let the heat pump working on the smoothest possible load profile. Two condensing boilers of 285 kW are installed to supplement the heat pump.

During wintertime the fan-coils are served by the heat pump circuit (45°C), while the pre and post heating coils of the AHU are served by the high temperature circuit (70°C). The two condensing boilers C1 and C2 are activated as integration of the heat pump (or as backup in case of heat pump failure) on two different thresholds of the distribution return temperature of the heat pump circuit (fan coils); C2 can also be activated to integrate the high temperature circuit, based on a temperature threshold of the heat recovery inertial storage (Fig. 1).

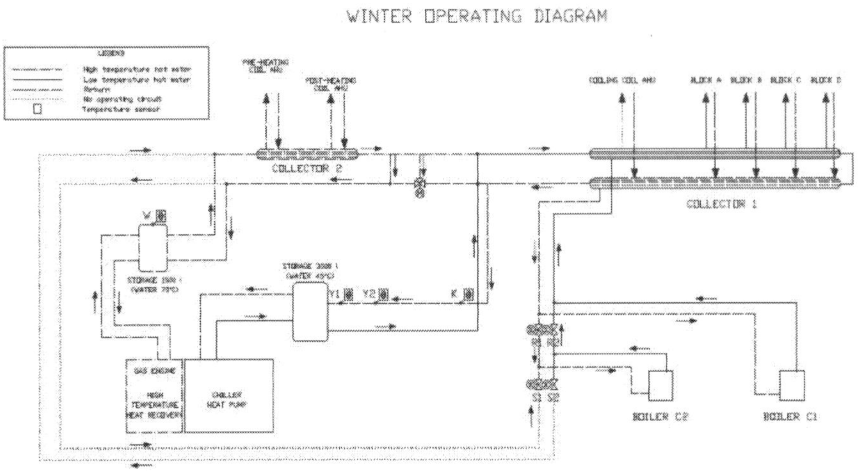

Fig. 1. HVAC plant winter operating diagram.

An analysis of the first year of operation revealed some malfunctions of the plant due to mistakes in the commissioning phase and to lack of a correct maintenance and operation of the system and suggested that the plant had a low efficiency. In fact heating was mainly provided by the boilers, which were set to produce water at 75°C because of a bad control and of a missing temperature sensor, so they were not running in condensing mode. The heat pump was working mainly with on-off cycles instead of part load operation, with a consequent de-rating in terms of COP, increased thermo-mechanical stresses on the engine and a poor use of the heat recovery. Owing to errors in the control strategy presented above, this resulted in a continual use of boiler C2 and incidental use of the heat pump (some minutes per hour). This was due to a mistake in the wiring of the valves' drives and led to relevant energy losses in the first year of operation. Finally, small troubles (arisen from calibration errors on several temperature controllers) did not allow the correct operation according to the design control strategy previously described.

Some test days were recorded by monitoring the natural gas consumption and the outdoor temperature, to obtain the gas consumption in terms of Nm^3 per "degree-hour" (dh) to evaluate the different performance of the heat pump and the condensing boilers in winter heating. Three test days were recorded in the design operating mode (heat pump priority) and three test days were recorded running only the condensing boilers. The results are summarised in Table 1. Condensing boiler mode leads to a consumption exceeding over 50% that of the design operating mode, highlighting the great efficiency of the engine heat pump system. The consumption expressed as Nm^3 per dh was used to evaluate the annual consumption by means of the dh data for the town of Vicenza (around 28,000, from 15th October to 15th April). The annual heating average cost passed from 15,059 € in design operating mode to 21,830 € with condensing boilers.

Table 1. Comparison between the natural gas consumption per dh in design operating mode and in condensing boiler mode (averaged on the three days of each test run).

Mode	Gas consumption in the test run [Nm^3]	Degree hours [dh]	Consumption per degree-hour [Nm^3/dh]
Design operation (HP + condensing boiler)	70.10	91.1	0.77
Condensing boilers	87.40	77.0	1.14

 DOI: 10.1051/e3sconf/20172200100

3 The second example: ground water heat pump for a historical building

The Basilica Palladiana is a historic building, designed by Andrea Palladio in the 16th century, sited in one of the most famous square in the centre of Vicenza. It was deeply retrofitted between 2007 and 2012 to make it a cultural centre for the city. The new HVAC plant is an open loop water source heat pump system using underground water as heat source or as heat sink. Two wells were built (distance about 50 m) to produce and inject the water from the layer (40 m deep).

The use of underground water in heat pump plants is strictly regulated (and till 2015 forbidden) in the territory of Vicenza for environmental reasons. So the realization of the plant here described was allowed by local Authority to have a benchmark for the use of underground water for air conditioning uses. For such reasons the installation of a data logging system was prescribed in order to evaluate the energy and environmental performance of the HVAC plant. The underground water is available at 14°C and used as heat pump source (during heating season to supply hot water at 45–40°C) or as chiller heat sink (during cooling season to supply cold water at 7–12°C). The project provides three main circuits (Fig. 2) :

- underground water layer circuit: it is an open loop with the pumps to circulate the well water (two pumps, one backup, 18 kW nominal power at 70 m water column head each, controlled by inverter). Sand filters are enclosed as well. This circuit exchanges heat with primary circuit by means of two stainless steel heat exchangers (one backup), 700 kW nominal power each. The control logic is set up to operate (by an inverter) the production well pump as a function of a set point of the condensation/evaporation water circuit;

- primary circuit: it allows the heat exchange between the underground water and the water of the evaporator/condenser circuit of the heat pump/chiller avoiding the direct use of the underground water. This is done by two pumps (cooling pump in Fig. 2, one backup, constant flow rate – 90 m^3 h^{-1} –, 5 kW power, 12 m head). The heat pump/chiller is an electrical water/water one, six scroll compressors set up in two parallel circuits, R410A as refrigerant; the compressors operate by on/off and step by step logic.

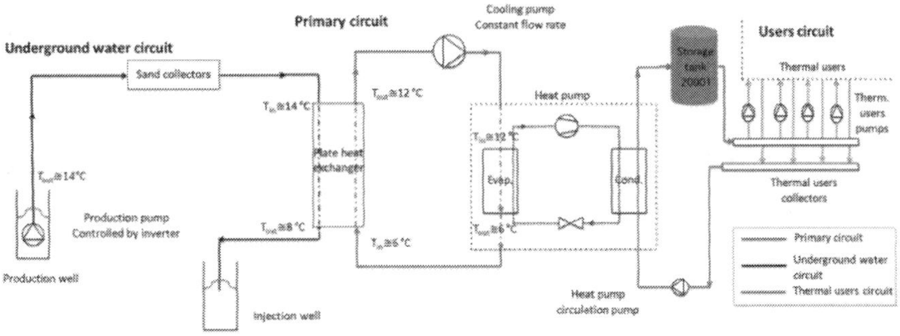

Fig. 2. Scheme of the underground water open loop heat pump plant with the three main circuits.

- users circuit: hot (45°C) or cold (7°C) water exiting the heat pump/chiller feeds a storage tank (2000 l) that is connected by two pumps (one backup, constant flow rate – 85 m^3 h^{-1} –, 3 kW, 8 m head). Finally, pumps, controlled with inverters, bring hot/cold water to each user circuit.

The showroom is set up by radiant floor as water terminal unit and displacement ventilation with very low velocity diffusers fed by two air handling units (15000 m^3 h^{-1} each). A smaller size air handling unit (1500 m^3 h^{-1}) supplies the ventilation of the ticket office. The plant is connected to the local district heating network (heating backup) by a 600 kW (nominal thermal power) plate heat exchanger. The backup service during cooling season is made by a water/water electrical chiller (rated cooling power 330 kW, rated electrical power 66.9 kW, EER=4.95). Such chiller feeds chilled water directly to users collectors and transfer the condensation heat to the primary circuit. Finally, direct expansion air conditioners (CDZ) serve two technical rooms (transformers and general switchboard and Uninterruptible Power Supply (UPS) rooms): they are cooled by the ground water circuit as well.

The first period of the monitoring activity was April-September 2014. Data analysis led to the following remarks:

- electrical energy consumption of the primary circuit pumps was extremely high (3000 kWh_{el} per month, 44% of the total electricity consumption). This was due to the unfavorable control logic of the pumps operating 24 h a day even in period with very low or null cooling load. Such situation was mainly caused by the fact that the CDZ need to be cooled by underground water in every season, so CDZ are responsible for 22% of the total energy rejected into the ground water circuit in the summer period;
- the backup chiller consumed 300 kWh_{el} per month constantly only for stand-by operation;
- electrical consumption of the CDZ was quite constant in the range 1400–1700 kWh_{el} per month. It is clear that the system has an inefficient operation mode when the only cooling load is the CDZ;
- considering that the HVAC plant did not operate during the hottest period (June and July), the electrical consumption in the whole period April-September 2014 is quite high (35500 kWh_{el} excluding the backup chiller), especially due to the continuous operation of the primary circuit pumps even when the HVAC plant was off;
- energy performance indexes (EER_{eq} = energy efficiency ratio of heat pump/chiller equipment; EER_{tot} = energy efficiency ratio of the whole plant that is considering the auxiliary needs) resulted to be very low: EER_{eq} = 3.23 instead of a predicted value of 5; EER_{tot} = 1.21 strongly penalized by low partial load operation of the plant;
- underground water consumption index Q_{wat} (defined as the ratio between the underground water produced and the useful cooling energy) was higher than designed: 223 l kWh_{cool}^{-1} instead of 172 l kWh_{cool}^{-1}. The flow rate control capacity was not sufficient (a consumption by 1000 m^3 $month^{-1}$ during the off operation periods was surveyed). The large ground water flow rate with limited ΔT was due to a ground circuit pumps oversizing (100 kPa was the real head at the maximum flow rate (80 m^3 h^{-1}) against 500 kPa nominal head of the two installed pumps). This caused a very discontinuous operation of the production well pumps with poor modulation capacity and very high on-off frequency (20–30 on-off per day during very low cooling needs periods).

Because of the very poor energy performance and technical problems detected during the first period of operation, in October 2014 the local Authority approved a series of technical interventions (realized between April and August 2015):

- the extension of tube in the injection well in order to avoid excessive oxygenation and bubbling of water that could cause metals precipitation and following well obstruction (frequently occurred during 2014 and 2015);
- inserting two sensors and a balancing valve in the production well to verify the temperature and piezometric gradients and to regulate and optimize the field operation of pumps;
- modifying the operation logic of the production well pumps by increasing the operation range and making the second pump operate by the inverter as well;
- modifying the condensation heat recovery system of the ground water chiller and the backup chiller connecting directly the primary circuit to the post-heating coils;
- installing a separate dry-cooler to dissipate the CDZ condensation heat and modifying the control logic of the primary circuit pumps making them operate strictly connected to the real users cooling needs.

The effects of the above described interventions are summarized in Fig. 3. The EER improvement was higher than 40% in summer operation. The system EER amelioration was really impressive, passing from 1.21 to 3.80. The main contribution was the strong reduction in electrical energy consumption of the primary circuit pumps, also separating the CDZ from the ground water circuit.

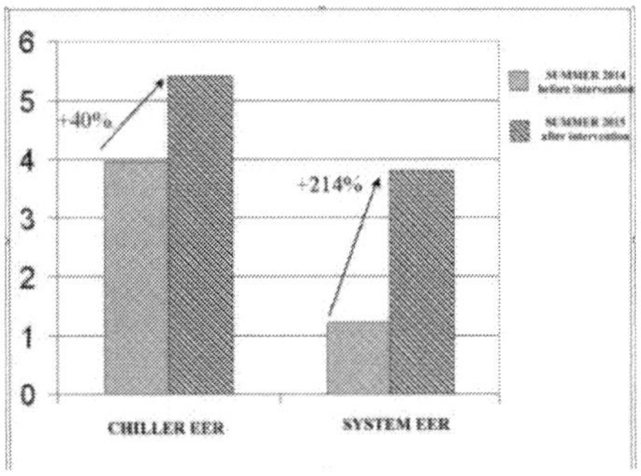

Fig. 3. Comparison of the Energy Efficiency Ratio for the two periods before and after the interventions.

4 The third example: dual source heat pump heating of a school building

By the town of Agordo in the geographical area of the Dolomiti mountains in a valley at 611 m asl, where the climate is severe during wintertime (3376 degree-day a High School Building was built up starting its operations in autumn 2009. The building has a total floor area of 5,680 m^2, an outward surface of 13,608 m^2 and an enclosed gross heated volume of 19,644 m^3; the envelope is well insulated, the outer walls and the roof allowing an average thermal transmittance of 0.16 W/(m^{2}K), the floor having a thermal transmittance to the ground of 0.4 W/(m^{2}K) and the glazing system having a thermal transmittance of 1.38 W/(m^{2}K).

The plant was designed to serve for the sole purpose of heating and ventilation, due to the severe climatic condition and the use of building, closed in summertime; moreover the demand for hot tap water is negligible and is fulfilled by means of electric water heaters. For sake of simplicity a reduced functional diagram of the plant is shown in Fig. 4 reporting only the main hydraulic streams and energy flows within the plant.

The HVAC plant is divided into the space heating section and the ventilation section. The heating section has two ammonia-water absorption heat pumps (HP3 and HP4 in Fig. 4) with geothermal exchangers in parallel (960 m, 6 x 160 in a row, of vertical tube heat exchangers designed with double-U pipes with a outer diameter of 32 mm and a thickness of 2.9 mm), producing thermal energy at 45 °C. Moreover the plant includes the setting up of 50 m^2 of flat type solar heat collectors (4 arrays in parallel, each of those made of 5 modules in series), and can be operated in four different modes, the control acting on a "threshold radiation" calculated as a function of the solar collectors features and the desired outlet temperature. The ventilation section serving the AHU (Air Handling Unit) hot batteries is also made up of two ammonia-water absorption heat pumps (HP1 and HP2 in Fig. 4) with geothermal heat exchangers (750 m, 6 x 125 m in a row, of vertical tube heat exchangers). In order to serve the AHU the heat pumps produce thermal energy at 55–60°C. The heat recuperator downstream of the AHU is a static cross flow type with an efficiency of 50%.

DOI: 10.1051/e3sconf/20172200100

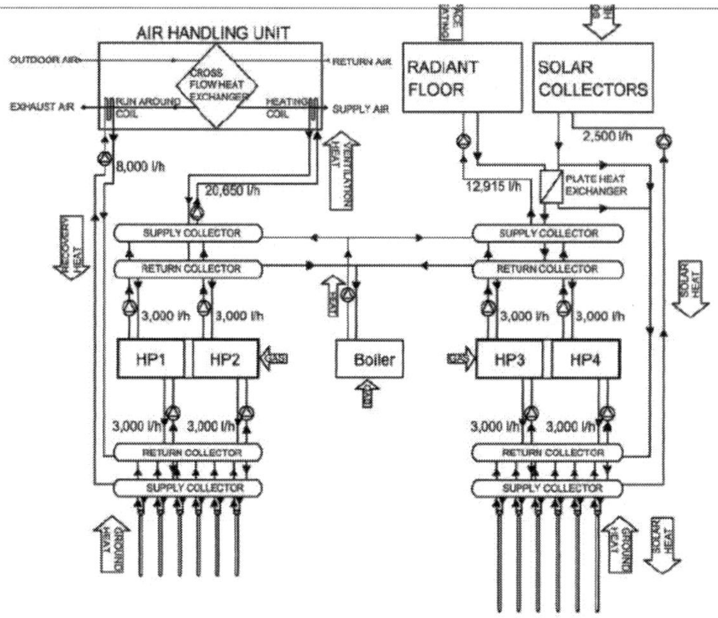

Fig. 4. Schematic of the dual source heat pump for heating a school building.

In order to increase thermal recovery still further when the external temperature exceeded 0°C, run-around coils were installed at the outlet of two out of four, for a global volume flow of 20600 m^3/h out of 25000 m^3/h. The run-around coils can operate on the exhaust flow of a heat exchanger, either sensible or latent; the heat recovered can be sent to the evaporator of the absorption equipment.

The monitoring of the plant was set-up with the cooperation of the building's and the plant's controller designers. The oldest data available on the PC that controls the plant, accessible from remote terminal via VPN after authentication, was from the begin of May 2012. During the period May 2012 – April 2016 the following cumulative energy flows (mass flow times the temperature difference between the inlet and the outlet, via simple thermal energy meters) were logged hourly and the results are presented in the next section on a monthly basis:

- Condenser and evaporator of each heat pump (at the collectors);
- Ground circuits, separately for ventilation and space heating;
- Primary circuit of AHU heating coils and run-around coils;
- Solar circuit;
- Primary circuit of the radiant floor.

Values of Natural Gas (NG) consumption were obtained from the natural gas bills, with all consumption being attributable to the heating/ventilation system.

As the surveys of the first two years of operation were satisfactory from the point of view of performance, no other analysis of the recorded data was operated till recently. Moreover the gas consumption changed not so much from one season to the other, supposedly according the degree-days of the various years.

The analysis of the last 4 years revealed quite a different situation. Whereas the ventilation section behaved in a similar was, this was not the case of the space heating section (fig. 5). Just in the middle of 2013 – 14 winter heat pumps did not more operate with a sudden fall in the Primary Energy Ratio (PER). Most of the energy was suddenly supplied by the boiler unlike during the previous periods. During the left part of the season the direct solar energy contribution raised system performance. However in the following winter, the solar section was completely excluded and the heat pumps occasional contribution appears as an auxiliary integration to a system where the priority is assigned to the boilers.

 DOI: 10.1051/e3sconf/20172200100

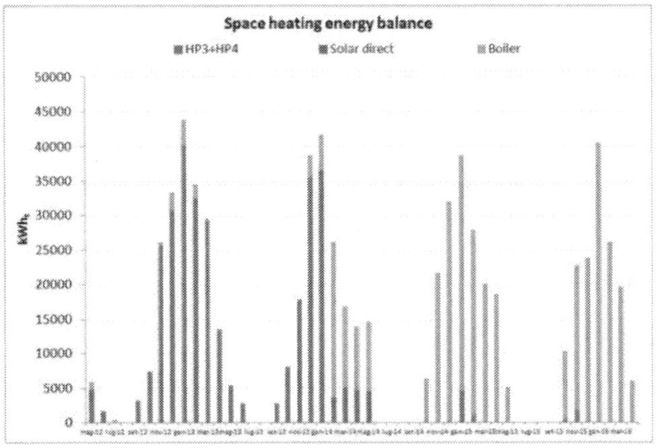

Fig. 5. Space heating energy balance of heat pumps (HP), solar direct and boilers during 4 heating seasons.

Seemingly at first the solar section was excluded from the heat pump circuit due to a failure of a solar collector. Even after fixing the solar collector problem, no circulation was reactivated between solar section and heat pump and some solar direct contribution was provided. The final choice of the management was to rely completely on boilers. The result has been a reduction of the seasonal PER to values lower than 1. The worsening was partially hidden by some energy saving measures such as the inside temperature reduction during Christmas holidays.

5 Conclusions

The examples demonstrate from different points of view the paramount importance of adequate monitoring of a plant particularly for renewable energy, generally less known by technicians. According to the first example a motor driven heat pump was used as an auxiliary to a condensing boiler (just the opposite of correct operation) and due to bad temperature setting even the condensing boiler did not work in condensing mode. The second case revealed how high the electricity requested to drive pumps might be to cool even a small section of the plant. Finally the third example illustrates how the simple control on the energy bill may make acceptable the plant working, when a favourable season or energy saving measures concealed bad operation of the plant even with the exclusion of the solar section and of part of the heat pumps.

References

1. F. Busato, R.M. Lazzarin, M. Noro, Applied Thermal Engineering **31**, 1648–1654 (2011)
2. M. Noro, A. D'Ascanio, R. Lazzarin, Problems and solutions of an underground water source heat pump system for a historical valuable building: an energy analysis. Proc. 12th Rehva World Congress. Aalborg (Denmark), 22–25 May, 2016
3. R, Lazzarin, M. Noro, Energy and economic analysis of an integrated multi-source heat pump system for a school building, 12th IEA Heat Pump Conference, Rotterdam (2017)

Batch orographic interpolation of monthly precipitation based on free-of-charge geostatistical tools

Ondrej Ledvinka[1,*]

[1]Czech Hydrometeorological Institute, Hydrology Database and Water Budget Department, Na Sabatce 2050/17, 143 06 Prague 412, Czechia

Abstract. The effects of possible climate change on water resources in prescribed areas (e.g. river basins) are intensively studied in hydrology. These resources are highly dependent on precipitation totals. When focusing on long-term changes in climate variables, one has to rely on station measurements. However, hydrologists need the information on spatial distribution of precipitation over the areas. For this purpose, the spatial interpolation techniques must be employed. In Czechia, where the addition of elevation co-variables proved to be a good choice, several GIS tools exist that are able to produce time series necessary for climate change analyses. Nevertheless, these tools are exclusively based on commercial software and there is a lack of free-of-charge tools that could be used by everyone. Here, selected free-of-charge geostatistical tools were utilized in order to produce monthly precipitation time series representing six river basins in the Ore Mountains located in NW Bohemia, Czechia and SE Saxony, Germany. The produced series span from January 1961 to December 2012. Rain-gauge data from both Czechia and Germany were used. The universal kriging technique was employed where a multiple linear regression (based on elevation and coordinates) residuals were interpolated. The final time series seem to be homogeneous.

1 Introduction

In hydrological studies dealing with climate change and its effect on water resources, the need of having long precipitation time series is of primary focus because precipitation is generally thought of as a basic source of water in many regions. This particularly applies to the territory of Czechia, for instance, which is known to be a source area of many European rivers, while, in fact, no large river crosses this country that would be, on the other hand, beneficial in cases of prolonged drought occurrences. Furthermore, hydrologists prefer to have the information on spatial distribution of precipitation during rainfall events or, more generally, of their means in different places in prescribed time horizons. However, traditionally, the precipitation totals were (and still are) measured at selected points where rain gauges were established. It is clear that this was for economic reasons because we are

* Corresponding author: ledvinka@chmi.cz

not able to measure everywhere, and it was also found that, in the special case of the mean, it is not necessary to have too many stations spread over an area of interest because there is a limit above which one cannot reach desirable increment of information (e.g. [1]). Radar and satellite precipitation data can be treated only as auxiliary because these measurements started much later (usually in the 1980s) than rain-gauge observations. Therefore, it is necessary to come up with a method capable of estimating the spatial distribution of precipitation using the point records.

For these purposes, spatial interpolation techniques evolved in the past and, especially, the field of geostatistics emerged. Geostatistical applications are not new in hydrology. Rather, it can be said that hydrologists alongside hydrogeologists importantly contributed to the development of geostatistics (e.g. [2–5]). Groundwater hydrologists were probably the first who applied methods of geostatistics but, naturally with the rise of the attention paid to climate change, the shift towards applications to surface water problems and climatology is evidenced by many papers published in the past few decades (e.g. [6,7]). Besides commercial software, the geostatistical techniques were also implemented in freely available tools, which is beneficial to students and other scientists who cannot pay much for them. Thus, the present contribution aims to show how selected tools can be wisely utilized in order to provide valuable information that can be further used in different studies dealing with climate change. The issue of spatial interpolation of precipitation totals in the Ore Mountains located in Central Europe served as an example.

The Ore Mountains stretched along the borders between NW Bohemia, Czechia and SE Saxony, Germany are known to be one of the regions that are sparsely measured regarding precipitation. Also, many gaps occur in time series representing points. This naturally poses many issues to hydrologists studying climate change impacts on small mountainous basins here. Therefore, the task was to estimate reliable time series suitable for climate change studies for selected more or less anthropogenically uninfluenced small river basins. Although so-called technical precipitation series do exist in Czechia that are uninterrupted in the daily time step and should be homogeneous (see [8–10]), similar data could not be gathered from Germany. For this reason, the monthly time step was selected as basic that still allows one to study changes in the annual course of precipitation. The selection of suitable basins was made based on the existence of relatively long discharge series so as to also allow for the comparison of long-term patterns in both precipitation and discharge in some of the future works.

Table 1. Basic morphometric characteristics of six selected small river basins/catchments located in the Ore Mountains and delineated by the water-gauging stations represented by long discharge series. Crosses indicate the east. For the definition of Graveli coefficient see [11].

Station (ID)	River	Area [km^2]	Max. elevation [m a.s.l.]	Outlet elevation [m a.s.l.]	Mean elevation [m a.s.l.]	Mean slope [decimal degrees]	Graveli coefficient
Klingenthal (530020)	Zwota (Svatava)	58.62	930	541	707	9.17	1.52
Svatava (208200)	Svatava	291.64	988	406	673	8.81	1.81
Chaloupky (209100)	Rolava	20.06	950	815	904	2.28	1.50
Stará Role (210100)	Rolava	126.35	1008	392	735	4.43	1.83
Rothenthal$^+$ (568350)	Natzschung (Načetín B.)	75.94	918	540	758	5.56	1.48
Třetí Mlýn$^+$ (218000)	Chomutovka	44.91	914	485	770	7.76	1.77

DOI: 10.1051/e3sconf/20172200101

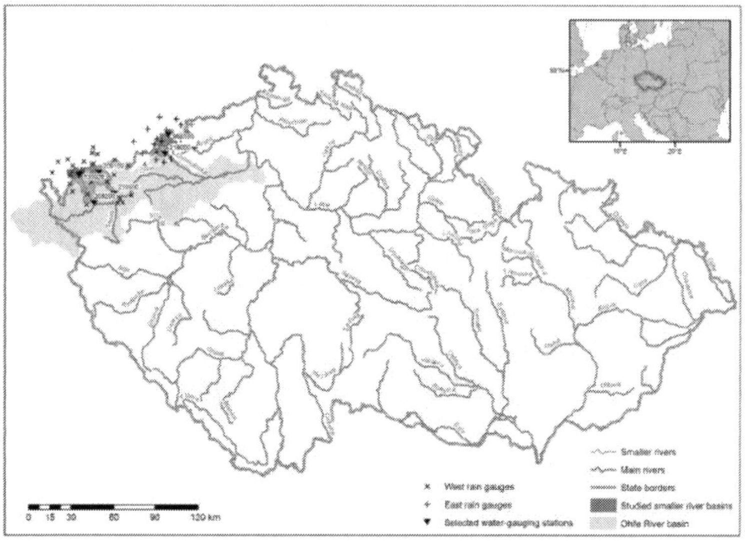

Fig. 1. Location of six studied small mountainous river basins in relation to the borders of Czechia and the larger Ohře (Eger) River basin. Except for the Natzschung, all investigated rivers are left-hand tributaries of the Ohře. All German and Czech rain gauges are depicted from which the data were used, but it does not mean that each of them was available for every month during interpolations.

2 Material and methods

2.1 Area of interest and gathered material

As described above, smaller river basins located in the Ore Mountains were selected. The final selection of six basins was determined by relatively long instrumental observations of discharge that was, at the same time, believed to be not influenced, or only slightly influenced, by human activities in such places. During processing the accessible material, the area of interest was divided into two parts, simply four basins in the western part delineated by water-gauging stations 530020, 208200, 209100 and 210100, and two basins in the eastern part closed by stations 568350 and 218000. The basic information on the morphometric characteristics of these basins is given in Table 1. Note that some of the values provided in the table may slightly differ from those officially published (e.g. at http://hydro.chmi.cz/ismnozstvi/ or in [12]). This is because the values in Table 1 come from the author's own derivation from the GIS material described further below. The location of the basins relative to the borders of Czechia and the larger, and thus better known, Ohře (Eger) River basin is depicted in Fig. 1. Two of the western basins, particularly 530020 and 209100, are in fact the subbasins of the Svatava River basin closed at the profile 208200 and the Rolava River basin closed at the profile 210100, respectively.

Precipitation totals measured at several rain gauges close to or within the basins were gathered from the Czech Hydrometeorological Institute (CHMI) and the German weather service called Deutscher Wetterdienst (DWD). Regarding the Czech data, the so-called technical precipitation series were acquired with the assistance of their main author Petr Štěpánek. These data were in the daily time step, were homogenous and without any missing values. They were and still are derived from the originally recorded precipitation series (see [8,9]). Since the technical series start on 1 January 1961, all the remaining analyses focused on the period 1961–2012. The year 2012 was chosen arbitrarily, but it was also due to the accessibility of the German discharge series from stations 530020 and 568350 (see [12]). However, there is no problem with exteding the length of precipitation series towards recent times in some of the future works. The German precipitation data were somewhat problematic since no technical series (or their equivalent) were available for the purposes of this study. Instead, the data published on the FTP server linked to on the website of the DWD's Climate Data Centre (see http://www.dwd.de/) were used. The daily data were too gappy and therefore monthly data were rather processed in the subsequent analyses. This means that also the Czech data had to be aggregated so that they finally formed monthly data. The total number of rain gauges included in the analyses was 32 (west) + 24 (east) = 56, of which 18 (west) + 11 (east) = 29 rain gauges represented the German territory. However, it is worth noting that the German rain gauges without any data for a given month did not enter the interpolation process.

Because the chosen interpolation technique required elevation data, a digital elevation model (DEM) represented as a grid layer was used to provide the necessary information. Most of DEMs available for the territory of Czechia are unfortunately clipped by its borders, which is undesirable when investigating river basins that cross the borders. Therefore, an alternative DEM had to be used. Here, a somewhat older DEM, downloaded at the end of the 2000s by the author from the website of the Czech company called ARCDATA PRAHA (https://www.arcdata.cz/), served as an elevation input to the geostatistical models. This DEM is based on the well-known SRTM product, but it was further processed by ARCDATA experts so that a number of possible errors was reduced. The approximate spatial resolution of the DEM was 75 m.

The last input to the geostatistical models were the geographical coordinates pertaining to the rain gauges. This information was provided as the metadata, for the German stations available online together with the precipitation data themselves. For the Czech territory, this information was acquired from the CHMI's CLIDATA database or directly from Petr Štěpánek.

2.2 Methods and tools utilized

All computations were carried out within the environment of the R statistical software [13] which is freely available online for all. Besides a number of packages that are intended to handle spatial data such as grids and vector features and are helpful when tranforming various coordinate systems to a common working system (see e.g. [14] for an overview), mainly the functions of packages 'automap' [15], which needs the package 'gstat' [16–18] to automatically perform kriging in general, and 'MASS' [19] were applied to the precipitation and elevation data and coordinates. It is known that the precipitation totals measured at points spread over a mountainous area correlate with elevation, but the correlation is higher if the elevation representative of a wider vicinity of a rain gauge is studied instead of the elevation corresponding to the place where the rain gauge is situated (see e.g. [20]). Therefore, also the 'raster' package (and its function 'focal()'; see [21]) had to be applied in order to compute mean elevation values representative of places up to 6 and 12 km away from the rain gauges. These distances were derived experimentally, while looking at the values of the Akaike Information Criterion (AIC; [22]) after bulding multiple linear regression (MLR) models. In particular, the package 'MASS' was used to determine the values of AIC, which should be the lowest for the right model. The package was also helpful when deciding on the explanatory variables that should enter the MLR models fed to the universal kriging models to which the 'krige()' function of the 'gstat' package switches according to the formula specified.

The procedure itself can be summarized as follows:

- Having four explanatory variables (E_6 – elevation typical of places up to 6 km far from a rain gauge; E_{12} – elevation typical of places up to 12 km far from a rain gauge; X – longitude of a rain gauge; Y – latitude of a rain gauge) try to build a MLR model:

$$P = \alpha + \beta_1 E_6 + \beta_2 E_6^2 + \beta_3 E_6^3 + \beta_4 E_{12} + \beta_5 E_{12}^2 + \beta_6 E_{12}^3 + \beta_7 X + \beta_8 Y + \varepsilon \qquad (1)$$

where α and β indicate nine parameters to be estimated (i.e. intercept and different slopes specified by the subscripts), ε is the residual component and the superscripts indicate powers whose inclusion has been shown beneficial for several mountainous areas (see e.g. [20]). The estimated parameters can in turn be used for the derivation of precipitation estimates $\hat{P}$ for every cell of the DEM where also the coordinates are known. Here, only two rectangular areas for the western and the eastern part of the Ore Mountains were selected (Fig. 2).

- It is not necessary to have all the explanatory variables in the model. Therefore, using the package 'MASS' and its function called 'stepAIC()', decide which of the variables should be included in the MLR model.
- The model with the lowest value of AIC should enter the 'autoKrige()' function, where the (semi)variogram is estimated by the generalized least squares (GLS) method based on the work [23].

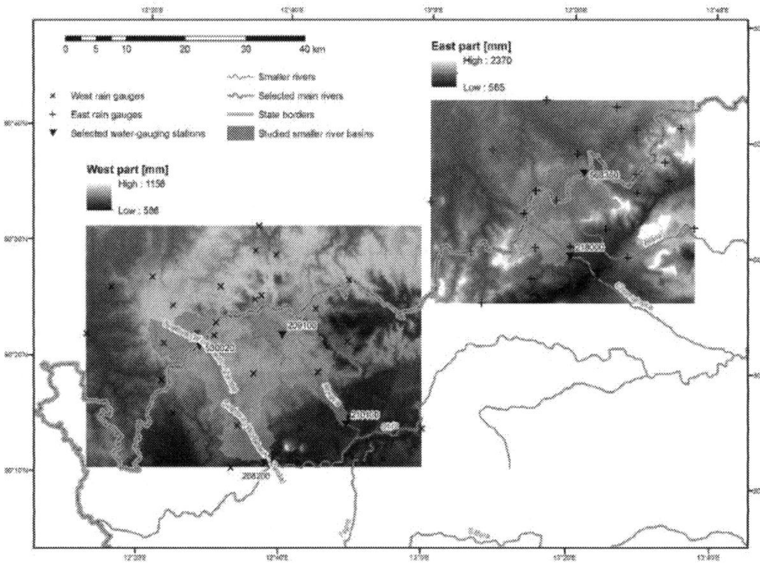

Fig. 2. Closer look at the area of interest with resulting long-term (1961–2012) grids of annual precipitation totals in greyscale separately for the western part and the eastern part of the Ore Mountains. Note that some of the extreme precipitation values may be far from true due to the absence of rain gauges in some places. However, it is believed that within the catchments the totals are much more reliable.

- Because the maximum number of the explanatory variables here was 8, the minimum number of rain gauges, when estimating the value of precipitation for a given DEM cell, was 9. The maximum radius, where the suitable rain gauges had to be searched, was set to 50 km.
- Finally, it is necessary to determine the precipitation totals that should represent the selected river basins. This may be accomplished using the 'zonal.stats()' function of the 'spatialEco' package [24].
- Repeat the procedure for all months. Since there were 624 months in the period 1961–2012 and, in fact, two regions were investigated, the procedure was run 1248 times. Because R is known to be relatively slow, the author recommends to parallelize the computation for larger data sets.

2.3 Homogeneity of produced series

Climatological time series should be homogeneous, which means that the different measuring equipment, the vicinity of a station or changing observers should not have any influences on the population from which the time series is drawn. Therefore, it was desirable to test the homogeneity of the new precipitation series. For this purpose, another R tool, developed in Canada [25], was used. Because this tool works with daily precipitation data, the monthly time series were disaggregated to daily, dividing the monthly totals by the corresponding number of days, which gave the average totals for every day. Then, the tool could be applied to these artificial daily precipitation series.

Table 2. Long-term monthly and annual precipitation totals (in mm) gained after computing spatial means for each month of the period 1961–2012 following universal kriging applied to precipitation totals known for rain gauges depicted in Figs. 1 and 2. The computations run separately for the western basins and for the eastern basins (these indicated by cross).

Station/ Season	530020	208200	209100	210100	568350⁺	218000⁺
01	80.3	73.8	85.5	75.6	69.8	68.5
02	67.8	60.7	67.7	60.6	63.7	60.4
03	73.4	64.1	64.8	62.7	71.8	66.9
04	61.9	53.4	51.0	49.8	66.4	61.1
05	74.9	68.9	61.5	63.6	87.4	81.1
06	93.6	82.5	73.4	75.0	98.5	94.8
07	98.9	88.6	82.6	83.3	111.5	103.4
08	95.5	83.6	80.9	77.2	106.0	99.7
09	76.6	69.3	66.3	66.2	80.6	75.5
10	67.9	61.5	63.2	58.4	64.2	60.5
11	77.4	69.6	72.0	67.7	75.6	71.7
12	93.5	84.3	86.5	80.6	90.8	90.4
Year	961.7	860.4	855.5	820.8	986.3	934.1

3 Results and discussion

According to the main task, complete monthly precipitation time series representing six smaller river basins located in the Ore Mountains were produced, considering both Czech and German rain gauges. The series cover the period 1961–2012, but it does not mean that the analyses could not be conducted also for more recent years. The original rain-gauge data do exist for the years 2013–2016 (see e.g. http://www.dwd.de/). Not all the aspects of the time series can be presented here. Rather, the resulting data were summarized in the form of long-term annual precipitation as the grid layers added to the other geographical features in Fig. 2. Also, long-term monthly precipitation totals for all the six basins are presented in Table 2. The final data are prepared for further analyses.

From Fig. 2, it can be seen that the pattern of the long-term annual precipitation somewhat mimic the distribution of elevation. It is caused by the method of universal kriging itself where the dependence of precipitation on elevation, besides coordinates, was the main source of information when building the models. Moreover, some extremes can be observed in Fig. 2. Whereas the western part looks more realistic, the eastern part includes questionable values (cf. [26]). It may be caused by the fact that there are some places without any rain gauges at all. The resulting grid for the eastern part of the Ore Mountains can then reflect some overestimation of precipitation totals by the geostatistical model utilized. Nevertheless, it is believed that regarding the investigated river basins, the estimated totals are reliable as shown in Table 2.

Table 2 reveals interesting features of the precipitation distribution in the Ore Mountains. It seems that the eastern part is more humid that the western part, which may be supported by the less annual precipitation over the catchment 209100 (also located higher) compared to the situation in the east. Among other aspects, one may notice that the summer months are dominating. However, one exception occurs. It is the catchment 209100 where the winter precipitation is higher from a long-term perspective.

Regarding their quality, all the new series revealed homogeneity according to the Canadian R tool. This feature of the series signify that the next analyses will be valuable also concerning the assessment of the possible anthropogenic influences on discharge.

Last but not least, it is important to note that the selection of the monthly time step actually allowed this type of analysis. Choosing a finer step such as daily, would cause several problems. For instance, the spatial dependence of precipitation totals at different rain gauges would vanish, especially for summer months when predominantly convective rainstorms occur. During these situations, only small areas are usually hit and the kriging methods combined with the regression based on the elevation are no longer recommended (see e.g. [7, 27]).

4 Conclusion and recommendations

For the purposes of future analyses dealing with possible climate change effects in the Ore Mountains on their water resources, six smaller river basins were selected that are believed to be anthropogenically uninfluenced (or at least very slightly influenced). Because the future analyses are planned so that the precipitation totals should be compared with discharge, it was necessary first to pay attention to the issue of the spatial interpolation of data recorded at various rain gauges that existed or still exist in the area of interest (both in Czech and German territories). The method of universal kriging was chosen that incorporated the dependence of monthly precipitation on elevation, longitude and latitude. The monthly time step was selected because of the good accessibility of such data. Free-of-charge geostatistical tools available online in the form of different packages for the R statistical software were utilized. Finally, the monthly precipitation series representing all the six basins were produced that span from January 1961 to December 2012. Furthermore, all the new series were tested for their homogeneity, and it may be concluded that the series are homogeneous, which undoubtedly signify a good start of the next analyses.

On the other hand, some important features of the new series were not studied, but it is highly recommended to do so before their definite use. For instance, the utilized geostatistical technique was not compared to any of its deterministic counterparts. The validation of the results was not performed either. This may be done via a cross-validation approach or through a hydrological model as described in [7].

The author would like to thank Hana Kourková and Pavel Kukla from the CHMI's Surface Water Department. They substantially helped the author when looking for the suitable river basins in the Ore Mountains. Petr Štěpánek's help with gathering the technical precipitation series from Czechia, namely from other than climatological stations, is appreciated as well.

References

1. D.P. Lettenmaier, Water Resour. Res. **12**, 1037 (1976)
2. J.P. Delhomme, Adv. Water Resour. **1**, 251 (1978)
3. P.K. Kitanidis, in *Handb. Hydrol.*, edited by D. R. Maidment (McGraw-Hill, New York, 1993), p. 20.1-20.39
4. P. K. Kitanidis, *Introduction to Geostatistics: Applications to Hydrogeology* (Cambridge University Press, Cambridge; New York, 1997)
5. Y. Rubin, *Applied Stochastic Hydrogeology* (Oxford University Press, Oxford, New York, 2003)
6. G. Hudson, H. Wackernagel, Int. J. Climatol. **14**, 77 (1994)
7. S. Ly, C. Charles, A. Degré, Biotechnol. Agron. Société Environ. **17**, 392 (2013)
8. P. Štěpánek, P. Zahradníček, R. Huth, Idojaras **115**, 87 (2011)
9. P. Štěpánek, P. Zahradníček, A. Farda, Idojaras **117**, 123 (2013)
10. O. Ledvinka, Acta Hydrol. Slovaca **16**, 199 (2015)
11. J. Krešl, in *For. Amelior.*, edited by O. Riedl and D. Zachar (Elsevier, Amsterdam, 1984), pp. 7–90
12. LfULG, *Hydrologisches Handbuch* (Landesamt für Umwelt, Landwirtschaft und Geologie, 2012)
13. R Core Team, *R: A Language and Environment for Statistical Computing* (R Foundation for Statistical Computing, Vienna, Austria, 2017)
14. R. Bivand, E.J. Pebesma, V. Gómez-Rubio, *Applied Spatial Data Analysis with R*, 2nd ed. (Springer, New York, 2013)
15. P.H. Hiemstra, E.J. Pebesma, C.J.W. Twenhöfel, G.B.M. Heuvelink, Comput. Geosci. **35**, 1711 (2009)
16. E.J. Pebesma, C. G. Wesseling, Comput. Geosci. **24**, 17 (1998)
17. E.J. Pebesma, Comput. Geosci. **30**, 683 (2004)
18. B. Gräler, E. Pebesma, G. Heuvelink, R J. **8**, 204 (2016)
19. W.N. Venables, B.D. Ripley, *Modern Applied Statistics with S*, 4th ed. (Springer, New York, 2002)
20. C. Portalés, N. Boronat, J.E. Pardo-Pascual, A. Balaguer-Beser, Int. J. Climatol. **30**, 1547 (2010)
21. R.J. Hijmans, *Raster: Geographic Data Analysis and Modeling* (2016)
22. S. Konishi, G. Kitagawa, *Information Criteria and Statistical Modeling* (Springer, New York, 2008)
23. N. Cressie, J. Int. Assoc. Math. Geol. **17**, 563 (1985)
24. J.S. Evans, *spatialEco* (2016).
25. X.L. Wang, H. Chen, Y. Wu, Y. Feng, Q. Pu, J. Appl. Meteorol. Climatol. **49**, 2416 (2010)
26. R. Tolasz, editor, *Climate Atlas of Czechia* (Czech Hydrometeorological Institute; Palacký University Olomouc, Praha; Olomouc, 2007)
27. S. Ly, C. Charles, A. Degré, Hydrol. Earth Syst. Sci. **15**, 2259 (2011)

 DOI: 10.1051/e3sconf/20172200102

Thermodynamic analysis of a combined-cycle solar thermal power plant with manganese oxide-based thermochemical energy storage

Qi Lei[1], *Roman* Bader[1], *Peter* Kreider[1], *Keith* Lovegrove[1], and *Wojciech* Lipiński[1,*]

[1]The Australian National University, Research School of Engineering, Canberra, ACT, 2601, Australia
[2]IT Power, Canberra, ACT, 2612, Australia

Abstract. We explore the thermodynamic efficiency of a solar-driven combined cycle power system with manganese oxide-based thermochemical energy storage system. Manganese oxide particles are reduced during the day in an oxygen-lean atmosphere obtained with a fluidized-bed reactor at temperatures in the range of 750–1600°C using concentrated solar energy. Reduced hot particles are stored and re-oxidized during night-time to achieve continuous power plant operation. The steady-state mass and energy conservation equations are solved for all system components to calculate the thermodynamic properties and mass flow rates at all state points in the system, taking into account component irreversibilities. The net power block and overall solar-to-electric energy conversion efficiencies, and the required storage volumes for solids and gases in the storage system are predicted. Preliminary results for a system with 100 MW nominal solar power input at a solar concentration ratio of 3000, designed for constant round-the-clock operation with 8 hours of on-sun and 16 hours of off-sun operation and with manganese oxide particles cycled between 750 and 1600°C yield a net power block efficiency of 60.0% and an overall energy conversion efficiency of 41.3%. Required storage tank sizes for the solids are estimated to be approx. 5–6 times smaller than those of state-of-the-art molten salt systems.

1 Introduction

Concentrating solar power (CSP) technologies offer the major benefit of enabling inexpensive thermal energy storage, which allows CSP plants to operate round-the-clock as well as on-demand. Hence, CSP systems are expected to play a central role in future sustainable power systems, particularly in the earth's sunbelt [1]. In this study, we explore the theoretical efficiencies achievable with a solar-driven combined cycle power plant, including a topping Brayton and a bottoming Rankine power cycle, with high-temperature thermochemical energy storage. In combined Brayton-Rankine power cycles, the benefit of high turbine entrance temperatures of gas turbines (> 1000°C [2]) is combined with the

* Corresponding author: wojciech.lipinski@anu.edu.au

high thermal efficiencies of steam Rankine cycles at intermediate operating temperatures, to maximize the operating temperature window and hence the thermal efficiency of the power system. Consequently, combined Brayton-Rankine power cycles are the most efficient commercial power conversion technology to date, with thermal efficiencies approaching 65% [3, 4].

MnO_x is considered as a potential thermochemical energy storage material, due to its suitable reaction temperature range, high storage density, and low cost [5]. In practice, doping of MnO_x with other metals may be required to improve potential issues with pure MnO_x, such as sintering at high temperatures and limited reaction kinetics. However, MnO_x is chosen here as a model material to develop a framework and reference for the assessment of future advanced material compositions. The reversible reduction and oxidation of MnO_x between Mn_2O_3 and MnO occurs via the steps:

$$MnO_2 \leftrightarrow Mn_2O_3 \leftrightarrow \alpha\text{-}Mn_3O_4 \leftrightarrow \beta\text{-}Mn_3O_4 \leftrightarrow MnO \quad (1)$$

The temperatures at which these transitions occur depend on the oxygen partial pressure in the gaseous environment. The Gibbs free energy minimization method has been applied, using the FactSage software [6], to calculate the transition temperatures listed in Table 1. The results show that the required temperatures for the forward (reduction) reaction steps decrease with decreasing oxygen partial pressure in the gas phase, while for the backward (oxidation) reaction steps the opposite applies. Reduction in N_2 gives access to the Mn_3O_4/MnO transition below the slagging temperature. This transition involves a high enthalpy of reaction, and hence strongly increases the mass-specific storage capacity of the solids.

Table 1. Approximate transition temperatures for the redox steps shown in Eq. (1) and associated enthalpy changes.

Substance	$T_{transition}$ (°C)			Δh (kJ kg^{-1})
	N_2	Air	O_2	
Mn_2O_3/α-Mn_3O_4	700	870	963	214
α-Mn_3O_4/β-Mn_3O_4	1173	1173	1174	80
β-Mn_3O_4/MnO	1450	1580	1580	897
Slagging	1625	1580	1580	-

The storage system can be operated with any two reversible MnO_x redox pairs in Eq. (1). For example, with the redox pair Mn_2O_3/MnO, operated over the temperature range 750–1600°C, the thermochemical storage process is described by:

Reduction:

$$Mn_2O_3(750^\circ C) \rightarrow 2MnO(1600^\circ C) + 1/2O_2(1600^\circ C); \quad \Delta\bar{h}^\circ = 304.7 \text{ kJ mol}^{-1} \quad (2)$$

Oxidation:

$$2MnO(1600^\circ C) + 1/2O_2(1600^\circ C) \rightarrow Mn_2O_3(750^\circ C); \quad \Delta\bar{h}^\circ = -304.7 \text{ kJ mol}^{-1} \quad (3)$$

2 Model system

A schematic of the model system is shown in Fig. 1. The system components are grouped into a storage block, a power block, and a heat exchanger block. The main components of the storage block are the reduction and oxidation reactors and the storage tanks for the reduced and oxidized manganese oxide particles. During on-sun operation of the system,

 DOI: 10.1051/e3sconf/20172200102

oxidized particles are heated and thermally reduced in the reduction reactor. The energy source for the endothermic reduction reaction is concentrated solar radiation, provided by a heliostat field and a secondary concentrator (not modeled). Oxidized particles are fed to the reduction reactor in a gas flow. A cyclone separator (separator 1) is used to separate the particles from the gas flow after the reduction reactor. The hot reduced particles are stored in the reduced particle storage tank, while the hot gases provide the heat to run the power cycle. During off-sun operation, reduced particles are cooled and re-oxidized to generate hot gases that provide the heat to run the power cycle. Reduced particles are pre-cooled in heat exchanger 3 before they are fed to the oxidation reactor in a gas flow. A cyclone separator (separator 3) is used to separate the particles from the gas flow after the oxidation reactor.

The reduction reaction (Eq. 2) is conducted in inert gas, modeled as nitrogen, while the oxidation reaction (Eq. 3) is conducted in pure oxygen. Each reactor is operated with a closed gas loop. In the inert gas loop over the reduction reactor, oxygen is accumulated, which needs to be continuously removed from the gas loop with a gas-gas separator (separator 2) [7, 8]. The separated oxygen is pressurized with a gas compressor (compressor 2) and stored in a storage tank. During off-sun operation, the stored oxygen is released and expanded via a throttle, and used in the gas loop over the oxidation reactor to compensate for the oxygen consumed by the oxidation reaction.

The power block comprises a topping solar-driven indirectly-heated air Brayton cycle and a bottoming steam Rankine cycle heated by the exhaust heat from the Brayton cycle. The Rankine cycle includes a single-stage reheating process.

The heat exchanger block couples the storage block with the power block. In heat exchanger 1, the hot gas exiting the reduction reactor heats the compressed air from the Brayton cycle during on-sun operation. In heat exchanger 3, the hot particles exiting the reduced particle storage heat the air from the Brayton cycle. The oxidation reactor serves simultaneously as a reactor and heat exchanger, exchanging heat between the oxidizing particle/gas flow and the air from the Brayton cycle.

The power system is designed to operate round-the-clock at a constant power, with 8 hours of on-sun operation and 16 hours of off-sun operation. The particle and gas flow rates entering the reduction reactor (state points N2 and M5 in Fig. 1) are adjusted such that the system generates a constant power output.

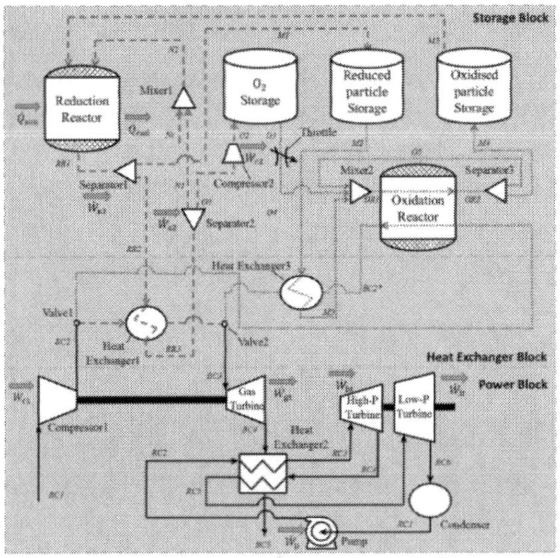

Fig. 1. Schematic of the model system (black solid lines: on- and off-sun operation; red dashed lines: on-sun operation; blue dotted lines: off-sun operation).

DOI: 10.1051/e3sconf/20172200102

3 Thermodynamic model

3.1 Model assumptions

The thermodynamic model of the system shown in Fig. 1 is formulated using the following modeling assumptions. The system is assumed to be at steady state and the air standard assumptions are used. Kinetic and potential energy changes are neglected. The heat loss from the receiver is modeled as black body radiation emission. Heat losses from other system components and connections are neglected. Pressure drop in tubes or heat exchangers are neglected. The thermodynamic effect of oxygen is neglected. System design parameters are listed in Table 2.

Table 2. System design parameters. State points are shown in Fig. 1.

Parameter	Value
Solar power input	100 MW
Solar concentration ratio at the receiver	3000
Direct normal solar irradiation	1000 W/m^{-2}
Brayton cycle compression ratio	10
Efficiency of compressors, turbines, pumps	88%
Efficiency of separators	60% [8, 9]
Pressure at BC1	1 atm
Pressure at BC5	1 atm
Pressure at RC1	9.58 kPa
Pressure at RC2	100 atm [10]
Temperature at BC1	25°C
Temperature at BC3	1580°C
Temperature at RC3 and RC5	565°C
Temperature difference between BC3 and RR2	20°C
Temperature difference between BC5 and RC2	20°C
Temperature of Reduction Reactor	1600°C
Temperature of Oxidation Reactor	750°C
Pressure in solid storage	1 atm
Pressure in gas storage	20 atm

3.2 Model equations

For the reduction and oxidation reactors, the overall steady-state energy balance reduces to:

$$\dot{Q} = \sum_{\mathrm{r}} \dot{n}_{\mathrm{r}} \left(\bar{h}_{\mathrm{f}}^{\circ} + \Delta\bar{h}(T)\right)_{\mathrm{r}} - \sum_{\mathrm{p}} \dot{n}_{\mathrm{p}} \left(\bar{h}_{\mathrm{f}}^{\circ} + \Delta\bar{h}(T)\right)_{\mathrm{p}} \tag{4}$$

where $\dot{Q}$ is the heat rate supplied to the reactor, $\dot{n}$ denotes the molar flow rates into and out of the reactor, $\bar{h}_{\mathrm{f}}^{\circ}$ denotes the enthalpy of formation of a substance at reference temperature $T_{\mathrm{ref}} = 298$ K and standard pressure p° = 1.013 bar, and $\Delta\bar{h}(T)$ denotes the enthalpy change of a substance between temperatures T and T_{ref}.

The reduction reactor is assumed to be a cavity-type radiation receiver that behaves like a black body. The radiative heat loss by the reactor and hence its absorption efficiency depend on the solar concentration ratio, *C*. The absorption efficiency is given by:

$$\eta_{\mathrm{b}} = 1 - \sigma T_{RR}^{4} / CI \tag{5}$$

where σ is the Stefan-Boltzmann constant, T_{RR} is the reduction reactor temperature, and I is the direct normal irradiation ($I = 1$ kWm^{-2}).

For the compressors, the feedwater pump, the gas turbine and the two steam turbine stages, and separator 1, the overall energy balance is:

$$\dot{Q} - \dot{W} = \sum_{\text{OUT}} \dot{n}\Delta\bar{h} - \sum_{\text{IN}} \dot{n}\Delta\bar{h} \tag{6}$$

For heat exchangers, the energy balance simplifies to:

$$\dot{n}_{\text{h}}(\bar{h}_{\text{h,in}} - \bar{h}_{\text{h,out}}) = \dot{n}_{\text{c}}(\bar{h}_{\text{c,out}} - \bar{h}_{\text{c,in}}) \tag{7}$$

The total amount of a medium stored in a storage tank is calculated by integrating the mole balance for each single storage medium:

$$n_{\text{net}} = (\dot{n}_{\text{in}} - \dot{n}_{\text{out}})t \tag{8}$$

where n_{net} is the number of moles of storage medium in the storage.

Separators 1 and 3 are modeled as cyclone gas-solid separators, based on the design by Wang et al., 2006 [9]. Separator 1 operates at 1600°C and 1 atm. The gas-solid mixture enters the separator from the side. The gas exits the separator at the top, while the solids exit at the bottom. The pressure drop over the separator is calculated according to the model by Shepheard and Lapple 1939 [11]:

$$\Delta p = 0.003\rho g v_{\text{in}}^2 H_{\text{v}} \tag{9}$$

$$H_{\text{v}} = K d_7 d_8 / d_2^2 \tag{10}$$

where ρ is the density of the gas, g is the gravitational acceleration, v_{in} is the gas velocity at the separator inlet, H_{v} is the inlet gas velocity head, and K is a cyclone configuration and operating condition parameter. Parameters d_2, d_7 and d_8 are dimensions of the separator. The inlet and outlet enthalpies of the gas flow can then be obtained from their temperature and pressure. The electrical pumping power required to operate separator 1 is calculated from Eq. (6). Separator 3 has negligible influence on the system performance due to the relatively small oxygen gas flow rate.

Separator 2 separates O_2 and N_2 in the temperature range of 330–370°C and at a pressure of 1 atm. Ezbiri et al., 2015 [8] showed the feasibility to separate O_2 from N_2 at temperatures of 327 to 627°C. The minimum required energy to separate two gas components in a mixture corresponds to the Gibbs free energy change involved when the two components are mixed. On a molar basis of the component with the smaller molar fraction, x, in the mixture, this yields [12]:

$$\dot{W}_{\text{sp},x} = -RT[(1-x)\ln(1-x) + x\ln x]/x \tag{11}$$

where R and T are the ideal gas constant and the absolute temperature of the mixture, respectively. The actual electrical power required to separate the gas mixture is obtained from:

$$\dot{W}_{\text{sp2}} = \dot{W}_{\text{sp},x} / \eta_{\text{sp2}} \tag{12}$$

where η_{sp2} is the efficiency of separator 2.

The required size of each storage tank is estimated based on the molar flow rate of the material stored in the tank:

$$\phi V = M\dot{n}t_{\text{on}} / \rho \tag{13}$$

where ϕ is the particle volume fraction [13], ρ and M are the mass density and the molar mass of the stored material, respectively, $\dot{n}$ is the charging molar flow rate, and t_{on} is the charging time.

The off-sun operation molar flow rate $\dot{n}_{off}$ is calculated via conservation of mass:

$$\dot{n}_{on}t_{on} = \dot{n}_{off}t_{off} \tag{14}$$

The combined power cycle operates under the same conditions during on-sun and off-sun operation. The net thermal efficiency of the power cycle is calculated as:

$$\eta_{th} = (\dot{W}_{gt} + \dot{W}_{st,H} + \dot{W}_{st,L} - \dot{W}_{cp1} - \dot{W}_{pm}) / \dot{Q}_{he1} \tag{15}$$

where $\dot{W}_{gt}$, $\dot{W}_{st,H}$, $\dot{W}_{st,L}$, $\dot{W}_{cp1}$, $\dot{W}_{pm}$ are the power of gas turbine, high-pressure steam turbine, low-pressure steam turbine, compressor 1 and pump respectively, while $\dot{Q}_{he1}$ is the heat transfer rate in heat exchanger 1.

The overall solar-to-electric energy conversion efficiency of the system is:

$$\eta_{overall} = [(\dot{W}_{gt} + \dot{W}_{st,H} + \dot{W}_{st,L} - \dot{W}_{cp1} - \dot{W}_{pm})t_{total} - (\dot{W}_{sp1} + \dot{W}_{sp2} + \dot{W}_{cp2})t_{on}] / (\dot{Q}_{sun}t_{total}) \tag{16}$$

where $\dot{W}_{sp1}$, $\dot{W}_{sp2}$, $\dot{W}_{cp2}$ are the power input into separator 1, separator 2, and compressor 2, respectively, while $\dot{Q}_{sun}$ is the solar power input. Separator 3 only separates a small amount of carrier gas which requires negligible energy input.

3.3 Solution methods

The model equations are solved with Matlab [14]. The thermophysical properties of water and air are obtained from RefProp Mini [15]. Thermochemical data for O_2, N_2, Mn_2O_3 and MnO are obtained from Binnewies and Milke [16].

4 Results

4.1 Power cycle analysis

The following results are calculated using the design parameters listed in Table 2. The off-sun operation duration can be set to 16 hours by changing the ratio of N_2 and Mn_2O_3 molar flow rates (n) at the entrance of the reduction reactor. More N_2 means more solar energy is used directly for on-sun operation, while more Mn_2O_3 means more energy is stored for off-sun operation. Numerical modelling shows that with n = 2.96, the off-sun operation duration is 16 hours.

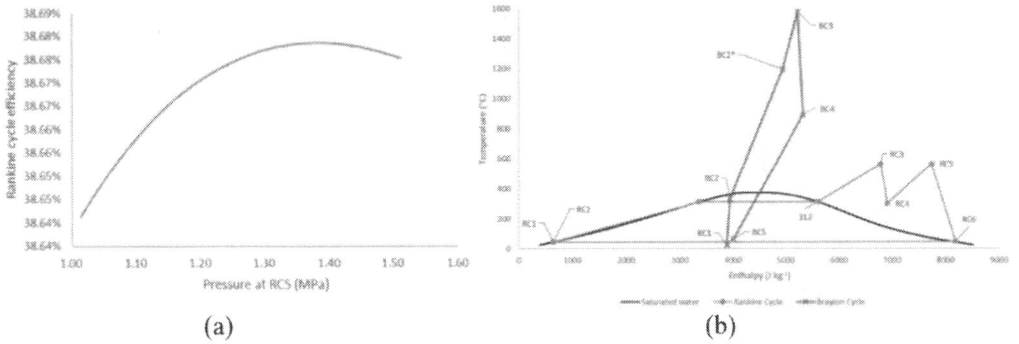

Fig. 3. (a) The effect of p_{RC5} on Rankine cycle efficiency; (b) *T-s* diagram of the combined cycle for the power cycles (BC2* is only for off-sun operation).

The Rankine cycle has a single-stage reheating process. The pressure at the low-pressure steam turbine outlet, P_{RC5}, can be optimized numerically for maximum cycle efficiency (Fig 3a). The resulting maximum Rankine cycle efficiency of 38.68% is obtained when p_{RC5} = 1.38 MPa.

The *T-s* diagram for the combined power cycle is shown in Fig. 3b. During off-sun operation, the Brayton cycle is heated in two steps, which is the only difference from on-sun operation. The flow rates of the combined cycle working fluids are the same for on-sun and off-sun, while the flow rates of the thermal storage materials are different.

4.2 Efficiencies and storage tank sizes

Fig. 4 shows energy balance and work allocation of the system over a 24-hour period. The total daily solar energy input to the system is 800 MWh. The largest energy loss of 186.08 MWh/day is due to blackbody radiation losses from the solar receiver, which operates with an absorption efficiency, η_b, of 76.74%. Additional main energy losses are due to the power input to compressors and separators. Compressors 1 and 2 and separator 2 require the largest amounts of work, amounting to 151.92 MWh/day, 19.92 MWh, and 18.88 MWh, respectively. The daily net work output is 330.32 MWh. The work requirement of the feedwater pump of the Rankine cycle is negligible. The predicted thermodynamic efficiency of the combined cycle η_{th}, is 60.04%. The predicted overall solar-to-electric energy conversion efficiency, $\eta_{overall}$, is 41.30%.

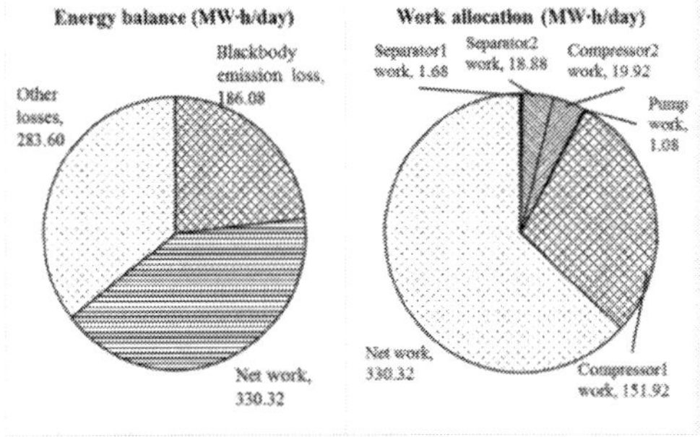

Fig. 4. Energy balance and work allocation of the system.

The required volume of the O_2 storage is 1.55×10^4 m^3 at a storage pressure of 20 atm, while the volume of reduced particle storage (MnO) and oxidised particle storage (Mn_2O_3) are 2.83×10^2 m^3 and 3.81×10^2 m^3, respectively. The MnO and Mn_2O_3 storage volumes are significantly smaller than those of a 120 MW molten salt power plant designed for round-the-clock operation, which are around 1.8×10^3 m^3 each [17]. On the other hand, the required size of the O_2 storage tank is one order of magnitude larger and may prove to be uneconomically large.

5 Summary and conclusions

In this paper, a thermodynamic model has been presented for a solar thermal power plant with a combined power cycle and high-temperature thermochemical energy storage system based on manganese oxide redox cycling. The gas-solid molar flow ratio at the reduction reactor inlet was determined for steady round-the-clock operation with 8 hours of on-sun and 16 hours of off-sun operation. The Rankine cycle reheating pressure was optimized for maximum power cycle efficiency. The overall solar-to-electric energy conversion efficiency and the net power block efficiency were determined to be 41.3% and 60.0%, respectively. Main energy losses in the system are due to blackbody emission by the solar reactor, and power demands by the pumps, compressors, and separators in the system, in addition to the heat rejected by the power cycle. For a system with 100 MW solar energy input, the volumes of reduced and oxidized particle storage tanks are estimated to be approx. 5–6 times smaller than those of a comparable molten salt-based plant [17]. On the other hand, the required O_2 storage volume of 1.51×10^4 m^3 is comparatively large.

This paper establishes the methodology to determine the efficiency of solar combined cycle power plants with thermochemical energy storage and provides an estimation of the theoretical system efficiency for an exemplary plant design. To obtain more accurate predictions of the annual plant efficiency, year-round simulations for a particular location need to be conducted and the component models need to be refined to take into account additional irreversibilities, e.g. due to flow friction, additional heat losses, incomplete chemical reactions, part-load efficiencies, etc. It is also recommendable to study different system parameters, such as the solar concentration ratio and the reduction temperature, and to consider alternative system configurations, including the possibility of operating both chemical reactors with atmospheric air, which would eliminate the potential challenges associated with the storage of gaseous oxygen and gas-gas separation.

Financial support by the Australian Renewable Energy Agency, grant no. 2014/RND005, is gratefully acknowledged.

References

1. International Energy Agency, *Technology Roadmap Solar Thermal Electricity* (2014)
2. A. Kribus, R. Zaibel, D. Carey, A. Segal, J. Karni, Sol. Energy **62**,121 (1998)
3. S. Baykal, P. Rufli, R. Bolliger, F. Fusaro, H. Kujawski, *ASME Turbo Expo* (2016)
4. Siemens, Siemens Gas Turbines, *Siemens AG*, (2017). Available: http://www.energy.siemens.com/hq/en/fossil-power-generation/gas-turbines
5. B. Ehrhart, E. Coker, N. Siegel, A. Weimer, Energy Procedia **49**, 762 (2014)
6. C. Bale, E. Bélisle, P. Chartrand, S. Decterov, G. Eriksson, K. Hack, I. Jung, Y. Kang, J. Melançon, A. Pelton, C. Robelin, S. Petersen, Calphad **33**, 295 (2009)

DOI: 10.1051/e3sconf/20172200102

7. D. Meixner, D. Brengel, B. Henderson, J. Abrardo, M. Wilson, D. Taylor, R. Cutler, J. Electrochem. Soc. **9**, D132 (2002)
8. M. Ezbiri, K. Allen, M. Gàlvez, R. Michalsky, A. Steinfeld, ChemSusChem. **8**, 1966 (2015)
9. B. Wang, D. Xu, K. Chu, A. Yu, Appl. Math. Model. **30**, 1326 (2006)
10. Siemens AG, Industrial Steam Turbines, Duisburg (2013)
11. C. Shepherd, C. Lapple, *Air Pollution Control: A Design Approach. In Cyclones* (1939)
12. M. Hänchen, A. Stiel, Z. Jovanovic, A. Steinfeld, Ind. Eng. Chem. Res. **51**, 7013 (2012)
13. D. Kunii, O. Levenspiel, *Fluidization Engineering* (1991)
14. The MathWorks, Inc., *MATLAB and Statistics Toolbox Release 2015b* (2015)
15. E. Lemmon, M. Huber, M. McLinden, *Reference Fluid Thermodynamic and Transport Properties (REFPROP MINI)* (2015)
16. M. Binnewies, E. Milke, *Thermochemical Data of Elements and Compounds* (2002)
17. J.I. Burgaleta, S. Arias, D. Ramirez, *Proceedings of 2011 SolarPACES, Conference* (2011)

E3S Web of Conferences **22**, 00103 (2017) DOI: 10.1051/e3sconf/20172200103
ASEE17

The effectiveness of the biodegradation of raw and processed polystyrene by mealworms

Karol Leluk[1], *Beata* Hanus-Lorenz[1,*], *Justyna* Rybak[1], and *Magdalena* Bożek[1]

[1]Wrocław University of Science and Technology, Department of Environmental Protection, Wybrzeże Wyspiańskiego 27, 50-370 Wrocław, Poland

Abstract. In our studies biodegradation of four variants of polystyrene was performed. We tested: raw material (PS), processed polystyrene (PSr), building insulation material (EPS) and food packaging boxes (PSp). Materials were characterized by means melt flow ratio (MFR), shore hardness and gloss. The biochemical assessment of macromolecules (proteins, lipids and sugars) in the mealworms organisms fed with tested forms of polystyrene allowed us to set how efficient and beneficial the biodegradation of types of polystyrene is. We also evaluated the variability of bacterial community in larval guts by the use of denaturing gradient gel electrophoresis (DGGE) on the bacterial DNA of 16S rRNA genes amplified in polymerase chain reaction (PCR). The results suggest that EPS and PSp polystyrene are the most digestible for *T. molitor* larvae. The metabolic degradation of polystyrene is probably strictly connected with the changes in biodiversity of gut bacteria.

1 Introduction

With increasing global consumption and their natural resistance to degradation, plastic materials and their accumulation in the environment is of increasing concern. As a rule, widely used polymer plastics do not undergo degradation process in environment [1]. The huge part of waste plastics is disposed of (72%) on a landfill, whereas the remaining part is incinerated or recycled [2]. Since polystyrene is a high molecular weight synthetic hydrophobic polymer, it is not easily biodegradable. In order to simplify and improve its biodegradation rate the addition of starch is usually applied [3]. Although polystyrene is believed to be a non-biodegradable material, it has been proved recently that mealworms, the larvae of *Tenebrio molitor* Linnaeus L. (Coleoptera: Tenebrionidae) are able to chew and digest styrofoam, a common polystyrene product [4]. The yellow mealworm is commercially available in Poland. Yang et al. [4] confirmed polystyrene biodegradation in

* Corresponding author: beata.hanus-lorenz@pwr.edu.pl

the larval gut of this larvae and proved the presence of petroleum based plastic-degrading process in the environment. The studies on biodegradation by mealworms of raw and processed polystyrene have never been conducted.

The aim of our experiment was to compare the effectiveness of biodegradation of four different forms of polystyrene (raw material, processed polystyrene, construction material used for the insulation purposes and food packaging boxes) by mealworms which were fed on all four PS species. In order to better understanding the processes occurring in the investigated materials some analytical techniques were used to fully characterize all four species. In our experiment we based on the mass loss of studied materials as well as lipid, total sugar and carbohydrates contents in larval body. We also evaluated the variability of bacterial community in larval guts by the use of denaturing gradient gel electrophoresis (DGGE) on the bacterial DNA of 16S rRNA gene amplified in polymerase chain reaction (PCR), a commonly used molecular technique for rapid fingerprint analysis of the microbial community. We assessed the changes in biodiversity of gut microorganisms presented in the mealworms fed on polystyrene. The alterations in gut microbial composition are connected to the diet. Serving polystyrene as only food can implicate such changes.

2 Materials and methods

2.1 Materials

Tenebrio molitor belongs to Tenebrionidae family (Coleoptera, Insecta). The life cycle of *T. molitor* is of variable length, from 280 to 630 days depending on the temperature and humidity. Larvae hatch after 10–12 days (at 18–20°C), molt many times, then change into pupa stage and finally after 3–4 months (at ambient temperature) become mature. The length of larvae is about 2.5 cm, whereas adults from 1.25 to 1.8 cm. *T. molitor* is a pest of grain, flour and food stores. They are typically fed on flour (wheat, oats, maize) supplemented with fresh fruits and vegetables (carrots, potatoes, lettuce) for moisture together with protein sources such as soybean flour, skimmed milk powder or yeast [5]. In our studies larvae of *T. molitor* were obtained from the inbreeding farm in Lublin (Poland).

Samples of raw polystyrene (PS) used in the study were ordered from joint-stock company Krakchemia S.A. (Poland) under the brand name SYNTHOS PS GP 137. Basic properties of the material are found on the manufacturer's website [6]. Prior to first processing cycle, material was dried in 70°C for 3 hours. The second group of material investigated was processed polystyrene (PSr). It was obtained by subsequent processing of the PS by means of extrusion, injection moulding and grinding (in order). The processing cycle was repeated ten times.

The third and the fourth group of material's tested were commercially available. The first one was insulation material – expanded polystyrene (EPS), the second one – packaging material for food products (PSp).

All materials were press moulded into the form of cuboidal plate (100x100x0,1 mm) prior to physico-mechanical characterisation. PS and PSr were also press moulded before being applied to consumption test, but EPS and PSp were applied without any further processing except cutting down PSp to smaller parts of suitable dimensions.

DOI: 10.1051/e3sconf/20172200103

2.2 Methods

2.1.1 Experiment design

Preparation of materials

Polystyrene was processed using one-screw extruder (HAAKE Rheomex 252), injection moulding machine (BOY 35A) and grinder (Wanner). Processing temperature was set to the values typical for PS thermoplastic processing; according to the manufacturer's note. For extrusion process temperatures on the extruder's zones were as follows: 205°C, 220°C, 210°C and 200°C (die zone temperature), whereas for injection moulding temperature ranged 200–250°C. Injection pressure profile was set to diminishing from 700° bar down to 550° bar, cycle time interval was set to 2 s. Press moulding was performed on the laboratory press LabTech LP-20B under the following conditions: process temperature 220°C, plasticization time 60 s, actual process time 120 s, pressure applied: 50 bar.

Melt flow ratio was estimated using MeltFlow Junior (CEAST) apparatus. Conditions were as follows: load (2.16 kg), temperature (210°C), cutting time (30 s), preheating time (120 s). Each measurement was repeated at least three times (until the difference between three measured points was no greater than 10%).

Hardness was measured in Shore D (ShD) for PS and PSr and Shore A (ShA) scale for EPS and PSp samples, by applying 1 kg load to the tip end.

Gloss was measured at 60° (semi-gloss surface) using Multi Gloss 268 Plus (Konica Minolta).

For all resulting values an error was estimated as a standard deviation of the mean value and presented on the corresponding figures as error bars.

Mealworms

T. molitor specimens were divided into 5 groups (20 specimens each,) depending on the type of feeding (raw polystyrene PS, processed polystyrene PSr, EPS and PSp and control). We have selected the larvae of the similar mass (from 0.03-0.04g at the beginning of the experiment). The variants used in tests were as follows:

- larvae consuming oatmeal *ad libitum* (control);
- larvae consuming PS plates;
- larvae consuming PSr plates;
- larvae consuming EPS;
- larvae consuming PSp;

Experiment conditions were as follows: constant temperature 28°C +/- 1°C, and humidity 78–80%. All materials used were weighed in the 1st, 21st and 30th day of experiment in order to assess the mass loss. The complete duration of the experiment was 30 days.

Biochemical analyses

At the end of experiment larvae were homogenized using mortar and pestle with ice-cold 0.1M phosphate buffer pH 7.4, 1 mM EDTA (1:10 w/v). For total sugars aliquots of homogenates were taken and analyzed with anthrone reagent [7]. Required aliquots for total lipid assay were treated with chloroform: methanol (1:2 v/v) and analyzed with vanillin reagent [8]. The rest of homogenate were centrifuged at 11 000 x g (5 min) and in order to assess the soluble protein content Bradford method was assayed [9].

Genomic DNA Extraction

Dissected guts from five larvae were immediately immersed in TE (10mM TRIS-HCl, pH 8.3, 0,1mM EDTA) buffer and homogenized together with glass beads. Homogenates were the source of genomic DNA obtained in enzymatic lysis with proteinase K and

 DOI: 10.1051/e3sconf/20172200103

purified on columns GeneMATRIX Bacterial & Yeast Genomic DNA Purification Kit (EURx, Poland) according to manufacturer protocol. These DNA were used as template in touchdown PCR.

PCR amplification and DGGE analysis
The specific V3 region of bacterial 16S rRNA gene was amplified using a set of primers. The sequences of primers are: primer 1: 5′-CGCCCGCCGCGCGCGGCGGGCGG GGC GGGGGCACGGGGGGCCTACGGGAGGCAGCAG-3′ and primer 2: 5′-ATTACCGCG GCTGCTGG-3′ [10]. PCR amplification was performed with Mastercycler Nexus Gradient Thermal Cycler (Eppendorf AG, Germany) as follows: 1 µl of template DNA, 25mM $MgCl_2$, 200 µM of each dNTP's, 50 pmol of each primer and 0.2 U of Bio-X-Act Short DNA Polymerase (Bioline, UK). The touchdown PCR protocol was: initial denaturation 94°C for 2 min, 20 cycles (1min at 94°C, annealing temperature from 65°C to 56°C decreased by 1°C every 10sec; 1 min at 72°C), 10 cycles (1min at 94°C, 1 min at 55°C, 1 min at 72°C), 7 min at 72°C. Amplification product were analyzed first in 2% agarose gel electrophoresis in 1x TBE (10mM TRIS-borate, pH 8.3, 1mM EDTA) and by ethidium bromide staining.

PCR products were separated in DGGE in 8% polyacryloamide gel using DGEEK-2001-220 system (C.B.S. Scientific, USA) for 1h at 50V followed by 10h at 90 V at 60°C with 1x TAE (10mM TRIS-acetate, pH 8.3, 1mM EDTA) buffer. The denaturants (urea and formamide) range was from 40% to 58% [11]. The gel was stained with 0.1% ethidium bromide and documented in BioVision system.

All statistical analyses were performed in Statistica® software package. Differences in total contents of protein, sugars and lipids for all types of materials with the duration of exposure were tested for each factor using analysis of variance (ANOVA).

3 Results

3.1 Characterisation of the material

As the four materials intended for experiment were of different origin some significant differences in their physical properties were supposed to occur. The first test applied was melt flow ratio which could give an insight into differences in molecular structure. As it presented on the Fig. 1 the sample with the lowest value of MFR is PSp whereas EPS and PS reveal higher (2,5 times) and comparable values. As it was expected, recycling process further increased (of about 50%) MFR value (PSr)

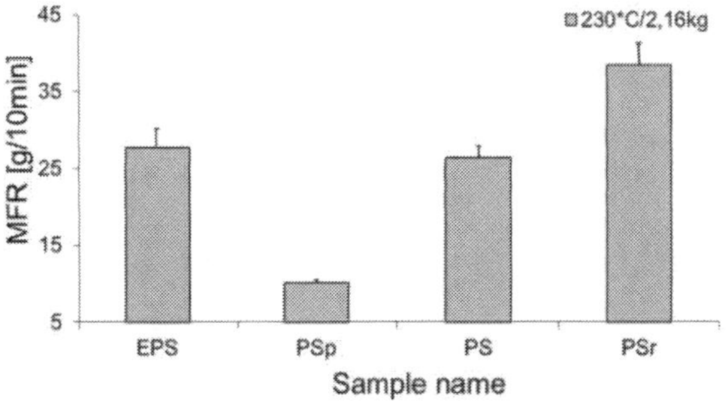

Fig. 1. MFR dependency for all investigated samples.

Also hardness and gloss (Fig. 2) measurements gave predictable results, qualitative score is consistent with the quantitative observation: two samples with the highest value are PS and PSr whereas EPS showed the lowest one.

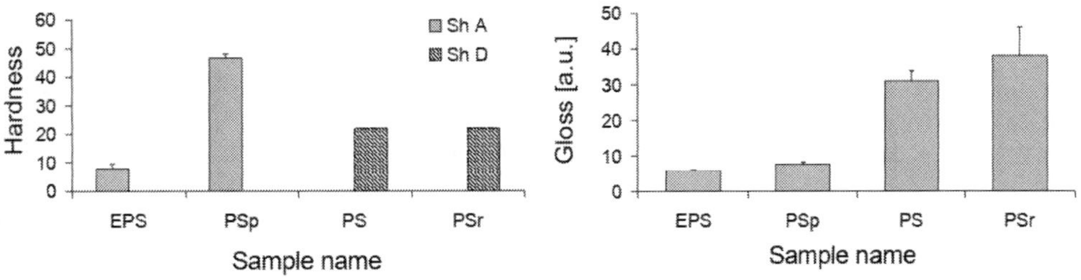

Fig. 2. Shore hardness and gloss results.

3.2 Mass loss measurements

The highest mass loss of studied variants was observed for building waste EPS (19.29% at the end of experiment) and for the food package PSp (16.92% at the end of experiment) (Table 1).

Table 1. Mass loss of materials used in studies (g).

Type of material	Mass in the 1st day of experiment (g)	Mass in the 21st day of experiment (g)	Mass in the 30th day of experiment (final) (g)	Mass loss (%) in 21st day of experiment	Mass loss (%) in 30th day of experiment
PS	1.596	1.578	1.572	1.12	**1.48**
PSr	1.47	1.392	1.385	5.29	**5.75**
PSp	1.12	0.944	0.93	15.65	**16.92**
EPS	0.724	0.608	0.584	15.97	**19.29**

3.4 Biochemical characteristic of mealworms

The concentration of total protein in the body mealworms fed various polystyrene materials at the end of experiment was only slightly lower than the content of protein in control mealworms fed oatmeal (Fig 3). Thus, the concentration of protein remained relatively stable and independent on type of food provided. The total sugars content decreased up to 32.5% of its control value and the sugars level was significantly different ($P < 0.05$) from control values (Fig. 4). The lowest value was obtained for mealworms fed PSr and PS. The total lipids content decreased up to 45% of its control value and it was significantly different ($P < 0.05$) from control values (Fig. 5). The lowest value was obtained for PS, then PSr.

 DOI: 10.1051/e3sconf/20172200103

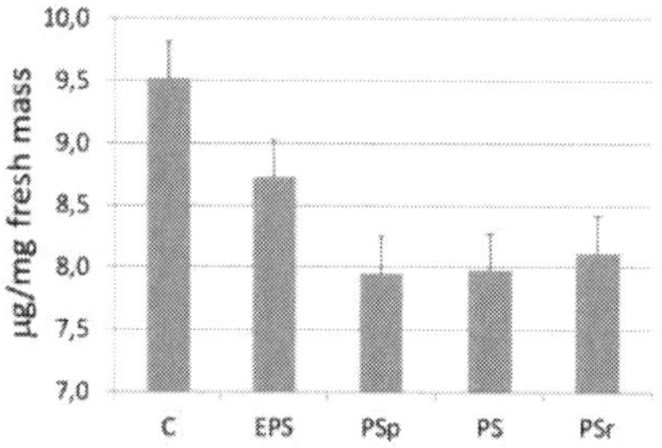

Fig. 3. Protein in larvae fed various polystyrene materials (µg/mg of fresh mass).

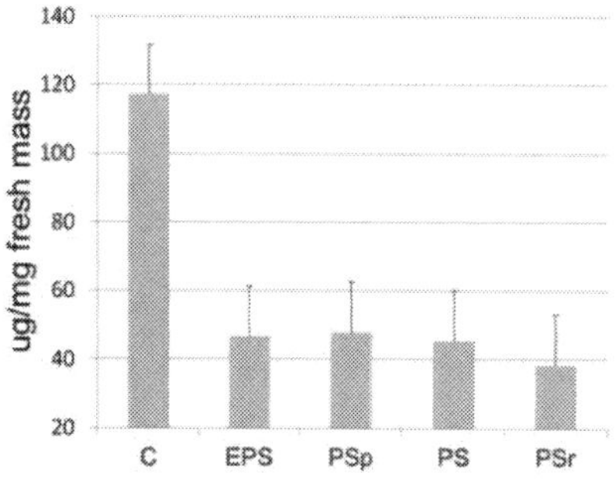

Fig. 4. Sugars in larvae fed various polystyrene materials (µg/mg of fresh mass).

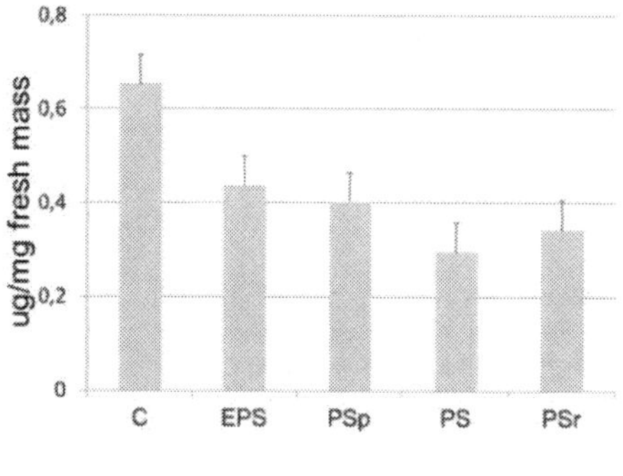

Fig. 5. Total lipids in larvae fed various polystyrene materials (µg/mg of fresh mass).

3.5 Denaturing gradient gel electrophoresis

The DGGE analysis showed the differences in structure of gut microbiome depending on type of food served to mealworms. The presence of four bands 1–4 (Fig 6), which are not identified in the control, corresponds to four species of bacteria which could be probably engaged in polystyrene biodegradation. Widely different pattern on lane C – mealworms fed PS (raw polystyrene) is incomparable to the pattern of other samples. It could be a result of insufficient amount of food for microbiome adaptation as we observed the lowest loss mass of this material, 1.48% only (Table 1).

 DOI: 10.1051/e3sconf/20172200103

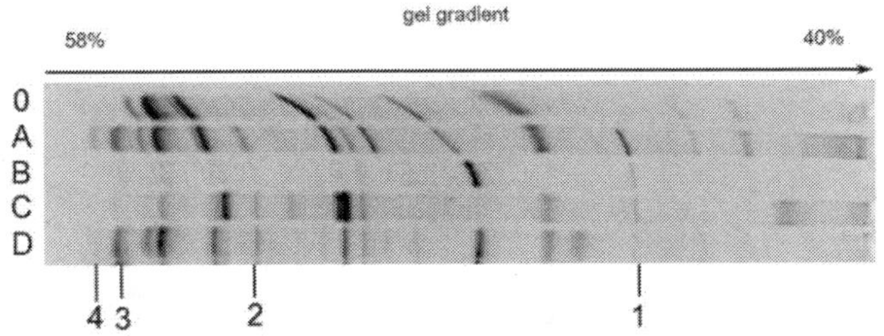

Fig. 6. DGGE gel with denaturant gradient 40-58% amplified in PCR bacterial DNA from gut of larvae fed various polystyrene materials. Lane 0 –control, lane A – EPS, lane B – PSp, lane C – PS, lane D – PSr. 1,2,3 and 4 – bands corresponding to bacterial strains presented in gut of larvae fed polystyrene but not present in guts of control population.

4 Discussion

The characteristic of the material used (PS, PSr, EPS, PSp) was important in term of its availability to mealworms. The most preferable was EPS as the mass loss was the highest during the experiment. This fact was confirmed by hardness measurements in which (refer to Fig. 2) hardness increased in the EPS < PSp < PS < PSr series. This observation is consistent with the mass loss measurement, especially when error bars are taken into consideration. Also, gloss measurement (giving back the information concerning the physical condition of the surface) is consistent with the previous observations; the more glossy surface is the less available for the mealworms for the attack. Contrarily, the raw PS and PSr were the least accessible for mealworms which suggests the form of polystyrene is important as it can be more or less digested by larvae. The results concerning the metabolic response to different food types are in line with the obtained results of mass loss of studied PS species. The most important changes are observed in the levels of lipids, they decreased greatly in mealworms fed PS and PSr suggesting the starvation process might occur as the PS and PSr were nearly not accessible for consumption. The most common source of energy in insects during starvation is the oxidation of fatty acids stored in the form of triglyceride [12] thus, high depletion of lipids suggests its consumption during the starvation period. In this study, levels of total sugars were also affected by the food accessibility. The decrease of sugars was observed again in mealworms fed PS and PSr. Sugars are extremely important for sustaining organs especially for the nervous system [12] although, when the low levels of lipids in the body is achieved the insect has to switch to the use of carbohydrates, which is the case in our experiment. The protein levels are not significantly affected by the food type, although some authors suggest that protein synthesis also decreases during food deprivation [13]. Recent years research evidenced that host diet [14] and some environmental factors change [14] the structure of insect gut bacterial communities. In this work we showed connection between biodiversity of gut microbiota of *T. molitor* larvae and type and formula of consumed food material. Results of studies on biodegradation polystyrene [4] implicate changes in gut microbiota structure. In our research we obtain genomic DNA by homogenization and enzymatic lysis of whole guts to limit oxygenation to provide DNA of anaerobic and aerobic strains. The DGGE analysis showed predicted changes in set of gut bacteria which are adapted to different types of polystyrene as only source of food. MFR investigation was applied to check if there exist a consistency between polymer chain length (which is reflected in MFR value) and availability of the materials to worm larvae. As it is seen on the Fig. 1 recycling increases of about 50% MFR (compare bars respective to PS and PSr samples) what is common for such operation. PS used for the study was intended for extrusion and press moulding purposes. The value stays in good agreement with the manufacturers declaration. On the other hand, the material which was used to made thermoformed packaging revealed the lowest MFR value and EPS about 250% higher and comparative to PS.As the MFR results stay in agreement with the state of the art, they are inconsistent with the other presented. The withdrawn conclusion is that chain length does not affect the mass loss ratio thus has no significant influence on the material consumption.

5 Conclusions

To summarise, although previous studies suggested the possibility of biodegradation of polystyrene, not all kinds of this material are bioavailable to the same extent for mealworms. Some of them are ready to consumption such as EPS (expanded) and PSp (packaging material for food products) and the rate of biodegradation is quite efficient, other like raw PS and PSr (processed) are difficult for chewing which is with line with their characteristic (hardness and gloss). It has been proved that metabolic response to different types of food delivered was adequate to the loss of mass of studies polystyrene forms. What is more, during DGGE analysis we identified 4 additional bands in the gut of *T. molitor* fed polystyrene corresponding to bacteria which could be probably responsible for the biodegradation of this material as they are absent in the gut of control mealworms. The changes in gut biodiversity even in starved mealworms need further studies.

The work was realized within the allocation No. 0401/0068/16 awarded for Faculty of Environmental Engineering Wroclaw University of Science and Technology by Ministry of Science and Higher Education.

References

1. H.K. Webb, J. Arnott, R.J. Crawford, E.P. Ivanova, Polymers **5**, 1 (2013)
2. T. Faravelli, M. Pinciroli, F. Pisano, G. Bozzano, M. Dente, E. Ranzi, J. Anal. Appl. Pyrolysis **60**, 103 (2001)
3. H.A. Pushpadass, R.W. Weber, J.J. Dumais, M.A. Hanna, Bioresource Technol. **101**, 7258 (2010)
4. Y. Yang, J. Yang, W.-M. Wu, J. Zhao, Y. Song, L. Gao, R. Yang, L. Jiang, Environ. Sci. Technol. **49**, 6 (2015)
5. J. Hardouin, G. Mahoux, *Zootechnie d'insectes* (BEDIM, Gembloux, Belgique, 2003)
6. https://synthosgroup.com/files/user/ifs/tds_synthos_ps_gp_137_en.pdf - access date 2017-03-20
7. E. Van Handel, JAMCA **1**, 1 (1985)
8. E. Van Handel, JAMCA **1**, 2 (1985)
9. Bradford, Anal. Biochem. **72**, 6 (1976)
10. G. Muyzer, E.C. De Waal, A.G. Uitterlinden, Appl. Environ. Microbiol. **59**, 3 (1993)
11. Y. Wang, Y. Zhang, Ann. Entomol. Soc. Am. **108**, 941 (2015)
12. D. Renault, F. Hervant, P. Vernon, Phys. Entomol. **27**, 291 (2002)
13. G. Bosquet Comp. Biochem. Phys. **58**, 377 (1977)
14. J. Jung, A. Heo, Y.W. Park, Y.J Kim, H. Koh, W. Park. J. Microbiol. Biotechnol. **24**, 7 (2014)
15. D.R. Colman, E.C. Toolson, C.D. Takacs-Vesbach, Mol. Ecol. **21** (2012)

Modern proposal of methodology for retrieval of characteristic synthetic rainfall hyetographs

Paweł Licznar[1,*], *Ewa* Burszta-Adamiak[2], *Janusz* Łomotowski[2], and *Justyna* Stańczyk[2]

[1]Wroclaw University of Science and Technology, Faculty of Environmental Engineering, Wybrzeze Wyspianskiego 27, 50-370 Wroclaw, Poland
[2]Wroclaw University of Environmental and Life Sciences, Institute of Environmental Engineering, pl. Grunwaldzki 24, 50-363 Wroclaw, Poland

Abstract. Modern engineering workshop of designing and modelling complex drainage systems is based on hydrodynamic modelling and has a probabilistic character. Its practical application requires a change regarding rainfall models accepted at the input. Previously used artificial rainfall models of simplified form, e.g. block precipitation or Euler's type II model rainfall are no longer sufficient. It is noticeable that urgent clarification is needed as regards the methodology of standardized rainfall hyetographs that would take into consideration the specifics of local storm rainfall temporal dynamics. The aim of the paper is to present a proposal for innovative methodology for determining standardized rainfall hyetographs, based on statistical processing of the collection of actual local precipitation characteristics. Proposed methodology is based on the classification of standardized rainfall hyetographs with the use of cluster analysis. Its application is presented on the example of selected rain gauges localized in Poland. Synthetic rainfall hyetographs achieved as a final result may be used for hydrodynamic modelling of sewerage systems, including probabilistic detection of necessary capacity of retention reservoirs.

1 Introduction

Computerized hydrodynamic models enable to conduct complex hydraulic analyses of even very complicated drainage networks. The aim of these analyses is often to recognize not only storm water runoff through the network of underground sewers, but also e.g. rainwater surface runoff or water retention within the system [1–5]. The application of such computerized hydrodynamic models of storm sewer systems places entirely new demands as regards input precipitation data. In practice, when it is not possible to feed the model with uninterrupted precipitation time series at high resolution, standardized rainfall hyetographs are used. They are designed on the basis of empirical formulae of design rainfall intensities (IDF type curves – Intensity-Duration-Frequency) [6, 7]. Today one of

* Corresponding author: pawel.licznar@pwr.edu.pl

DOI: 10.1051/e3sconf/20172200104

the most popular standardized hyetographs is Euler's type II. This model, recommended for simulating dynamic sewage, is also used in Poland due to the lack of reliable equivalents. Euler's type II is relatively easy to design, however its application raises numerous doubts [8, 9]. The alternative to using model (standardized) rainfall may be generating synthetic rainfall time series with the use of random generators. For example, Güntner et.al [10] and Licznar et al. [11] showed multifractal properties of rainfall registered by selected rain gauges and prepared multiplicative random cascade models generating synthetic rainfall time series at high time resolution. Nevertheless, due to advanced scientific sources needed do design random cascade models, practical application of synthetic rainfall time series in hydrodynamic modelling of sewerage systems, e.g. in order to perform probabilistic calculation of retention reservoirs necessary capacity, remains infrequent. [5].

The analyses of precipitation temporal changes towards developing reliable model patterns have attracted numerous researchers from various parts of the world. One of the most recent announcements in this area is a paper by Hassan El-Sayed [12] in which synthetic rainfall hyetographs (WRRI) were presented, and categorized into four groups depending on rainfall total duration (6 hr, 12 hr, 18 hr, 24 hr). They were compared to the standard Soil Conservation Service (SCS) distributions of type I, II, III and IA. Significant differences were shown in runoff computed on the basis of the hyetographs and the abovementioned SCS models. It indicates the need to reject simplified global standardized rainfall models and the necessity to find their local equivalents. Unfortunately, research in this area is still not common in Poland. Only analyses within one city – Warsaw – may be listed, conducted by Barszcz [13] and Licznar and Szeląg [9]. Also not since long there have been reports in literature about research on real urban catchments in Poland, which verify the accuracy of synthetic rainfall time series, based on local rainfall models [5, 14].

Having in mind the depicted state of knowledge and needs as regards feeding hydrodynamic models with rainfall data, it seems necessary to intensify the work on local standardized rainfall hyetographs. Thus, the aim of undertaken research was to develop unified methodology for creating extreme rainfall events, which would enable accurate representation of local precipitation dynamics. The analyses performed were also aiming at answering the question whether precipitation registered in the area of a single city shows some similarity which would be describable with the use of a simplified pattern, or a set of patterns.

2 Methods

2.1 Rainfall data

Pluviographic measurement results registered by gauges installed in two meteorological stations (A and B) in the centre of a city in northern Poland constituted research material used in this paper. The identification of characteristic rainfall scenarios began with selecting storm rainfalls from available local rainfall time series. Station A provided a compilation of rainfall volume from the years 1966–1993. The data was a series of sums of 5-minute rainfalls and was collected from all available pluviographical strips with resolution of 0.1 mm rainfall depth. Rainfall time series usually comprised months from April till October with temperatures above zero, when pluviograph exploitation was possible.

In case of station B, rainfall series from the years 2010–2015 were available. They were registered with the use of electronic tipping bucket rain gauges. The gauge was equipped with a heating system, which enabled precipitation registering also during winter. The

DOI: 10.1051/e3sconf/20172200104

equipment provided annual 1-minute rainfall time series with resolution of 0.2 mm rainfall depth.

Together from both gauges rainfall time series from 34 years were collected. They made a substantial collection of local rainfall register for the implementation of assumed method. Storm rainfalls were separated from available rainfall time series, according to standard criteria included in guideline ATV-A118P [15]. As a minimum limit of total depth of storm rainfalls 10 mm was adopted, and minimum time between single rainfalls was established as at least 4 hours. Moreover, for time series to be established as part of rainfall event as regards time and depth, minimum value of rainfall depth – 0.1 mm in 5 minutes – was assumed to separate rainfall from dry periods. As a result, 236 storm rainfalls were separated from rainfall time series, corresponding to their frequency at almost 7 rainfalls a year. Those rainfalls had depths in the range of 10.0 to 118.3 mm and duration from 35 to 3634 minutes.

Due to observed differences in rainfall time and depth, double normalized cumulative storm rainfall hyetographs were prepared for further analysis of separated storm rainfalls collection and for the identification of typical (meaningful, standardized) hyetographs. The duration and depth of cumulative hyetographs were normalized successively for the range from 0 to 100% for time and from 0 to 1 for rainfall depth.

2.2 Statistical analyses

The lack of apparent correlation between rainfall depth, as well as total time, and the shape of cumulative histogram prompted the authors to analyze the full collection of 236 rainfalls and to implement advanced data mining. The tools used were cluster analysis algorithms. The aim of cluster analysis is to divide the collection to a number of groups, characterized by certain qualities. Moreover, the groups encompass subsets of studied population, fulfilling the conditions of disjointness and completeness [16]. In this paper, cluster analysis was used to divide the set of 236 double normalized cumulative storm rainfall hyetographs into nondetermined and a priori determined number of groups. To this result, hierarchical cluster analysis and non- hierarchical cluster analysis – k-means method – was implemented accordingly. Calculations were made in Matlab.

Agglomerative methods initially assume that each element is a cluster of its own. When the methods are applied, the closest (and thus the most similar) elements are combined into new clusters, joining elements and clusters created at earlier stages, until one master cluster is created. It was necessary to define the distance between elements and created clusters, as well as the rules of linking, determining which clusters are similar enough to be combined. In the analyses the most popular and natural metrics were used: Euclidean metric and squared Euclidean distance. Euclidean metric d(x,y) of elements x and y is given by formula (1) [16]:

$$d(x,y) = \sqrt{\sum_{i=1}^{p}\left(x_i - y_i\right)^2} \tag{1}$$

where: $x = (x_1,\dots, x_p)$ and $y = (y_1, \dots, y_p)$. Formula (1) for $p = 2$ and $p = 3$ is equivalent to distance of two points x and y in plane and space.

Single linkage method was used in analyzing the set of selected rainfalls. The method is also called nearest neighbour clustering, as the distance between two clusters is determined by the distance between two closest neighbours belonging to different clusters. Dendrograms illustrate the structure of a set of elements according to diminishing similarity

 DOI: 10.1051/e3sconf/20172200104

between its elements (and thus increasing linking distance). In conducted research, the number of clusters was established after analyzing a dendrogram which had been made as a result of hierarchical agglomerative method. Thus, eventually not only specific storm rainfalls assigned to given clusters were obtained, but also average cumulative rainfall hyetographs. On the basis of average normalized cumulative storm rainfall hyetographs, standardized rainfall hyetographs were developed for their practical application in modelling and designing drainage systems.

3 Results and discussion

Fig. 1 presents, respectively, histograms of time and total depth of selected rainfalls. Cross-check shows that rainfalls of duration up to approximately 500 min and of depths in the range from 10 to 20 mm were the most common. Specific occurrences, however, differ not only in time and depth, but also in the shape of hyetographs. Therefore an attempt was made to divide the set of all standardized hyetographs with the use of cluster analysis. The result of agglomerative method applied to the set of selected rainfalls is presented in Fig 2. It is a dendrogram showing the neighbourhood (i.e. similarity of shapes) in the group of all 236 standardized hyetographs. On the basis of this hyetograph, for link length set at about 2.5, it can be concluded that there are five different groups of rainfall clusters with similar hyetograph shapes within the cluster. Based on this, in the last stage of research consisting in grouping by k-means method, the number of clusters was assumed at k = 5. The graphs of average cumulative storm rainfall hyetographs for 5 clusters selected by k-means method are shown in Fig. 3. For each of the clusters, its percentage in the total set of 236 storm rainfalls, as a measure of frequency with which storm rainfalls of a similar hyetograph shape take place, was determined. The analyses demonstrate that the most frequent is type 1 cluster, comprising 1/3 of the whole set. Type 3 is also highly frequent (28%). Type 5 and 2 clusters constitute a similar and much lower percentage in the whole set, amounting to only ca. 15–16%. The smallest cluster is type 4 cluster (only 8% of cases).

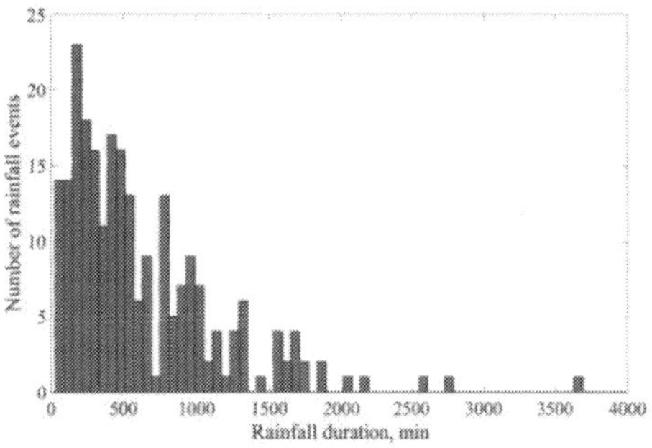

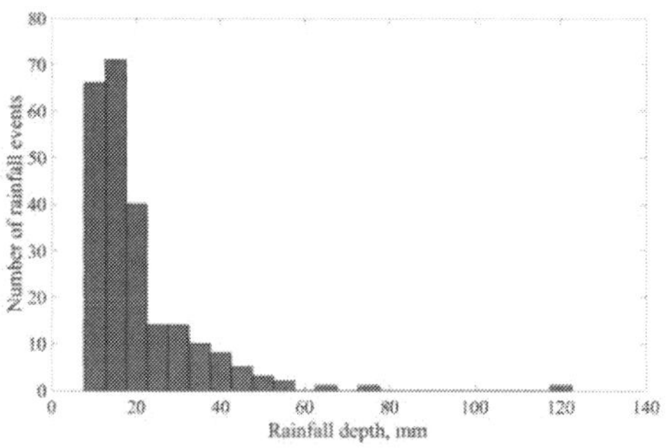

Fig. 1. Histogram of rainfall duration (left side), b) rainfall depth (right side) for a set of 236 storm rainfalls.

DOI: 10.1051/e3sconf/20172200104

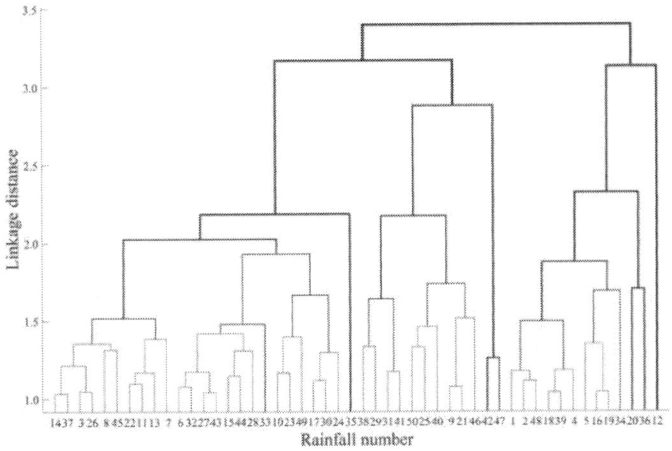

Fig. 2. The result of agglomerative analysis of selected storm rainfalls as a dendrogram (axis y shows linkage distances for individual rainfalls and rainfall groups, for better legibility some branches present hyetograph subsets.

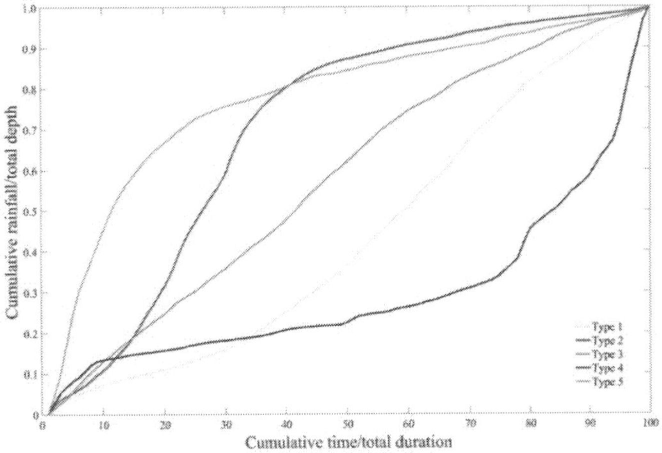

Fig. 3. Cumulative dimensionless hyetographs of storm rainfalls for 5 clusters determined with k-means method.

Presented research results prove that naturally changeable rainfalls registered in a single city area are somewhat similar, which enables to record the shape of hyetographs in time with the use of patterns. It is useful in designing which strives at generalizing (standardizing, normalizing) rainfall distribution curves. In case of the analyzed city, 5 characteristic types were separated from rainfall time series recorded in 10 years. Bearing in mind the probability philosophy in designing drainage networks, as well as limitations resulting from e.g. the costs of proposed engineering solutions, the focus must now concentrate on adjusting drainage system to the most frequent rainfall scenarios. In the analyzed case that would be model rainfall type 1, 2 and 3 (with relatively equal instant rainfall intensities), which correspond to the time series of approximately 76% of extreme rainfall events. Only afterwards the attention of an engineer may focus on modifying or rebuilding drains and system facilities (i.e. retention reservoirs, pumping stations etc.) in order to provide entirely proper functioning of drainage system for model rainfalls according to type 5 and type 4 hyetographs.

It is worth noticing that real rainfall hyetographs were not actually similar to the one commonly used in Poland, namely Euler's type II. As a result, in group 5 of selected model rainfalls only type 2 model was similar to Euler's type 2. To a lesser extent the similarity was noticeable for type 5 model hyetograph. Both types (2 and 5) constituted only 15% of extreme rainfall events in the general rainfall base.

Due to the fact that selected hyetograph types differ from one another not only as regards maximum instant rainfall intensity, but also in the location of peak in hyetograph, it is recommended that future research should aim at showing the influence of hyetograph shape (type of designated pattern) on rainwater runoff simulation. The influence may be significant, as suggested in the study by Mazurkiewicz et al. [17], performed for different standardized hyetographs known from literature.

4 Conclusions

The hydrodynamic model may be a useful tool in hydraulic analysis of drainage systems, provided that it will be fed with rainfall data of the most resemblance to the local (real) rainfall character. A visible increase of interest in rainfall observation and processing its results for urban hydrology (with registering high resolution rainfall data) reduces the access barrier to rainfall observation time series. This means that research material for analyses with the use of the methodology presented in this paper is becoming increasingly richer. The results of complex studies performed on the set of 236 rainfalls registered in urban areas, with depths in the range from 10.0 to 118.3 mm and duration from 35 to 3634 minutes, enabled to formulate the following conclusions and generalizations:

1. Available registered rainfall time series should be treated as a base to build a local rainfall model. The model may later be used to create synthetic rainfall hyetographs, which have the same statistical characteristics as real rainfalls. Such compilation of data may then be used as future potential rainfall scenarios, applicable to computerized simulations.

2. In order to make the modelling of urban drainage networks more closely corresponding to real conditions, it is necessary to include at the input apparently random character of rainfalls, which is reflected in hyetograph. By using a specific method of rainfall event classification, it is possible to group rainfalls into a certain, limited number of standardized hyetograph types.

3. The classification method used in the analyses, namely cluster analysis, proved to be an effective tool to assess varied shapes of normalized hyetographs in dedicated rainfall subsets of different total duration ranges. Dendrograms achieved as a result of the analysis enable to quickly select characteristic hyetographs and identify especially atypical rainfall distributions in time.

4. Euler's type II model rainfall hyetograph was not consistent with the shape of real rainfall hyetographs. In separated types of model rainfalls it resembled type 2 model the most and to a lesser extent type 5. Both types constituted only ca. 15% of storm rainfalls in the rainfall database.

Encouraging initial results of studies in which standardized hyetographs were prepared, substituting real precipitation time series with synthetic pattern hyetographs from the local model, should be verified for other rainfall data. The data should be recorded in other cities and regions, with different local precipitation conditions. According to the authors, presented methodology is worth implementing and verifying on precipitation database from other countries, especially registered with rain gauges operating in radically different climatic conditions.

References

1. S.S. Cipolla, M. Maglionico, I. Stojkov, Ecol. Eng. **95**, 876–887 (2016) doi.org/10.1016/j.ecoleng.2016.07.009
2. P.A. Versini, D. Ramier, E. Berthier, B. de Gouvello, J. Hydrol. **524**, 562–575 (2015) doi.org/10.1016/j.jhydrol.2015.03.020
3. M., Sterrena, A. Rahmana, G. Ryanb, J. Environ. Manage. **141**, 1 (2014) doi.org/10.1016/j.jenvman.2014.02.013
4. E. Burszta-Adamiak, M. Mrowiec, Wat. Sci. Tech. **68**, 1 (2013) doi: 10.2166/wst.2013.219
5. P. Licznar, Ochrona Środowiska, **35**, 2 (2013) (in Polish) http://www.os.not.pl/docs/czasopismo/2013/2-2013.html
6. A. Kotowski, *The principles of safe dimensioning of sewage systems volume I, Sewage systems* (Seidel-Przywecki Publishing House, 2015) (in Polish)
7. B. Kaźmierczak, A. Kotowski, J. Hydrol. **525**, 345–351 (2015) doi.org/10.1016/j.jhydrol.2015.03.063
8. P. Licznar, Gaz, Woda i Technika Sanitarna, **6** (2009) (in Polish)
9. P. Licznar, B. Szeląg, Ochrona Środowiska, **36**, 3 (2014) (in Polish) http://www.os.not.pl/docs/czasopismo/2014/3-2014.html
10. A. Güntner, J. Olsson, A. Calver, B. Gannon, Hydrol. Earth Syst. Sc. **5**, 2 (2001) doi:10.5194/hess-5-145-2001
11. P. Licznar, J. Łomotowski, D.E. Rupp, Atmos. Res. **99** (2011) doi.10.1016/j.atmosres.2010.12.014
12. E.A.H. El- Sayed, Ain Shams Engineering Journal, 2017 (to be published)
13. M. Barszcz, Woda-Środowisko-Obszary Wiejskie **12**, 39 (2012) (in Polish)
14. B. Kaźmierczak, A. Kotowski, *Verification of storm water drainage capacity in hydrodynamic modeling* (Scientific papers of the Institute of Environment Protection Engineering of the Technical University of Wrocław Series of monographs No. 57, Publishing House of Wroclaw University of Technology Wrocław, 2012) (in Polish)
15. T.G. Schmitt, *Kommentar zum Arbeitsblatt A 118 Hydraulische Bemessung und Nachweis von Entwässerungssystemen. DWA*, Hennef (2000)
16. D.T Larose, *Discovering knowledge in data. An Introduction to Data Mining* (John Wiley & Sons, Inc., Hoboken, New Jersey and Canada, 2005)
17. K. Mazurkiewicz, M. Skotnicki, M. Sowiński, *Developing standardized hyetographs in simulating outflow in urban catchments* (Monographs of Water Management Committee of the Polish Academy of Sciences, 39, 2016) (in Polish)

AUTHOR INDEX

Abdul Qyyum Muhammad 531
Abouchabana Nabil 1
Abourida Btissam 282
Abseitov Erbolat 757
Acikgoz Caglayan 49
Adamski Wojciech 1478
Agata Zdyb 717
Aillerie Michel 1217
Akan Perihan 748
Alaktiwi Abdulsalam 945
Al-Dadah Raya 572
Ali Wahid 1099
Alibay Temirulan 762
Al-Mansour Fouad 847
Amanowicz Lukasz 10
Anastasakis Kostas 339
Andruszkiewicz Artur 564
Andrych-Zalewska Monika 1241
Anisimov Sergey 1021, 1036
Ar Irfan 81
Arman Hasan 1539, 1547
Arora Amarpreet Singh 18
Augustyn Grzegorz 25
Axcell Brian 265
Aydemir Ali 33
Azhibekov Aidos 762
Bader Roman 806
Baetens Robin 41
Baitasov Talgat 762
Bakowski Andrzej 89
Balapanov Malik 762
Balbay Senay 49
Balbierz Piotr 57, 664
Banaszkiewicz Kamil 65
Banaszuk Piotr 709
Bandurski Karol 73, 1029
Barbusinski Krzysztof 599, 1368
Barrios-Padura Ángela 419
Bartan Ayfer 81
Bartkiewicz Ewelina 1602
Bartkiewicz Lidia 1368
Basaran Filik Ümmühan 363
Beccarelli Paolo 379
Bekkouche SMA 96
Belarche Lahoucine 282
Belcik Maciej K. 103
Bellahcene Lahcene 96
Bellstädt Daniel 33
Beltran-Galindo Javier 324
Benato Alberto 273, 1209
Bering Slawomira 111, 436
Besler Maciej 120, 208
Bezyk Yaroslav 128
Biczak Robert 1384
Bieda Boguslaw 137
Biedron Izabela 224
Birch Calum 191
Bisak Agnieszka 1639
Blasiak Przemyslaw 1115
Bogacki Marek 1169
Bohdziewicz Jolanta 606
Bojarski Jacek 649
Boruszko Dariusz 159, 257
Bozek Magdalena 815
Brychcy Ewa 1036
Bryjak Marek 1233
Bryszewska-Mazurek Anna 354
Büchele Richard 33
Buda-Ozóg Lidia 1256
Bugaj Andrzej 166
Burszta-Adamiak Ewa 823
Buse Hauke 1305
Cambron Philippe 183
Carvell Aaron 191
Celades Irina 200
Cepinski Wojciech 120, 208
Ceran Bartosz 216
Chacartegui Ricardo 419, 1279
Chaczykowski Maciej 548
Chang Ki-Chang 580
Chauhan Amisha 200
Cheknane Ali 96, 1217
Chen Xin 1509
Chilton John 379
Chlebowska Stys Anna 1297
Chojniak Joanna 224
Chong Min Yee 962
Cichowicz Robert 232
Ciobanu Gabriela 240, 487
Cwiek Barbara 1107
Cybulska Krystyna 731, 1384
Czarny Anna 103
Dabek Lidia 249, 1194
Dabek Pawel 590
Dabrowski Wojciech 159, 257
De Jong Wiebren 339
Debowski Marcin 1594
Delpech Bertrand 200, 265
Demirel Ibrahim Halil 1539, 1547
den Boer Emilia 143, 151
den Boer Jan 143
Dendys Marta 1400
Destro Nicola 273
Doghmi Hicham 282
Dorodnikov Maxim 128
Drabik Anna 1288
Drewnowski Jakub 291
Duangkamol Na-Ranong 1123
Dudkiewicz Edyta 346, 782
Dudziak Mariusz 606

AUTHOR INDEX

Dupont Valerie 191
Duzinkiewicz Kazimierz 513
Dyjakon Arkadiusz 143
Dzierzgowski Mieczysław 299
Dzikowska Anna 724
El Alami Mustapha 449
El Hajjaji Ahmed 1
El-Khozondar Balkess 308
Englart Sebastian 316
Etoua Evina Ghislaine 324
Fave Alain 717
Ferenc Zbigniew 332, 1202
Fernandez-Lopez Maria 339
Fidorów-Kaprawy Natalia 346, 354
Figueiredo Patric 174, 504
Fijewski Michal 1073, 1079
Filik Tansu 363, 411
Filik Ümmühan Basaran 411
Filyarovskaya Viktoriya 370
Fleiter Tobias 33
Flor Jan-Frederik 379
Fourmond Erwann 717
Fröhlich Siegmund 388, 1305
Gabouze Nourddine 1217
Gajewski Andrzej 396
Garcia-Valls Ricard 987
García-Vílchez Encarna 884, 891
Garlicka Agnieszka 1639
Gerek Emine Esra 403
Gerek Ömer Nezih 363, 403, 411
Gil-Baez Maite 419
Glowacka Anna 428, 436
Gnutek Zbigniew 1146
Gogina Elena 861
Golub Adam 442
Gomar Salvador 200
Górniak Aleksander 701
Gounni Ayoub 449
Green Toby 191
Grochowski Michal 513
Grosser Anna 456
Grzebyk Tomasz 1470
Grzegorzek Martyna 465
Grzesik Katarzyna 137
Gulkowski Slawomir 717
Gullu Gulen 748
Gurgul Agnieszka 473
Gwozdziej-Mazur Joanna 480
Haddadi Mourad 1
Hajduk Anna 1594
Hamerla Milosz 73
Hanus-Lorenz Beata 815
Harijan Khanji 191
Harja Maria 240, 487
Harnisz Monika 495
Hertle Ellen 174, 504
Hirsch Piotr 513
Holtra Anna 522
Hoshi Masaharu 1580, 1587
Hrad Marlies 143
Hussain Arif 531, 1099
Ignatowicz Katarzyna 1093, 1264
Ilkan Mustafa 1610
Im Yong-Hoon 580
Ishembetov Rais 762
Ismail Sherif 540
Isoli Niccolo 548
Ivannikov Alexander 1580, 1587
Jaderko Karolina 915
Jadwiszczak Piotr 1656
Janiak Kamil 556
Janowska Barbara 564
Janus Magdalena 111
Jarnuszewski Grzegorz 1384
Jayakody Harith 572
Jedlikowski Andrzej 1021
Jedrczak Andrzej 151
Jimin Hong 531
Jin Tea-Hwan 580
Jouhara Hussam 200, 265
Jurasz Jakub 25, 590
Jurczyk Krzysztof 25
Kabsch-Korbutowicz Malgorzata 1423
Kabyshev Aset 757
Kabyshev Asset 762
Kajdan Tomasz 1384
Kajl Stanislaw 324
Kalemba Katarzyna 599
Kaminska Gabriela 606
Karpinski Marcin 617
Karpuk Michal 625
Kawasaki Hiroyuki 1524
Kazmierczak Andrzej 701
Kazmierczak Bartosz 590, 634
Kazymbet Polat 1580
Kenzhebatyr Bekmyrza 762
Kepa Urszula 1313
Keskiewicz Piotr 649
Kessler Will 643
Khailov Artem 1580
Kheiri Abdelhamid 449
Kiper Justyna 428, 436
Kirszensztejn P. 899
Klimczak Marcin 649
Klimonda Aleksandra 695
Kmiecik Ewa 617, 657
Knap Marta 57, 664
Kneer Reinhold 174, 504
Kobel Przemyslaw 143
Kobus Dominik 1297

AUTHOR INDEX

Koczyk Halina 73
Kogut Janusz 1256
Kokoszka Wanda 1256
Koksal Merih Aydinalp 308, 748
Kolasinski Piotr 1115, 1146
Kolosowski Pawel 731, 1384
Komerska Anna 672
Koparal A. Savas 403
Kopczyk M. 899
Korbiel Tomasz 25
Kordana Sabina 678
Korzeniewska Ewa 495
Kostka Maria 686
Kotowski Andrzej 634
Koukou Maria K. 1408
Kowalska Izabela 695
Krakowian Konrad 701
Krasowska Malgorzata 709, 1005
Krawczak Ewelina 717
Krawczyk Piotr 1639
Kreider Peter 806
Krzeminska Alicja 724, 1563
Krzysko-Lupicka Teresa 731
Ksionek Dariusz 672
Kubenova Marzhan 762
Kucieba Malgorzata 1131
Kucukali Serhat 81
Kudlek Edyta 606
Kulbat Eliza 739
Kurmanzhanov Askar 757
Kus Melike 748
Kuterbekov Kairat 757, 762, 1580
Kutylowska Malgorzata 767
Kwasniowski Stanislaw 1555
Kwiatkowski Jerzy 775
Lach Janusz 1563
Lam Su Shiung 962, 1530
Lamarche Louis 324
Laska Marta 782, 876
Lazzarin Renato 790
Ledvinka Ondrej 798
Lee Moonyong 531, 1099
Lei Qi 806
Leluk Karol 815
Lettner Christian 1271
Lewkowicz Marek 1131
Licznar Pawel 823
Liew Rock Keey 962, 1530
Lipinski Kamil 830
Lipinski Wojciech 806
Litu Loredana 240
Liu Guiqing 839
Liu Jing 1226, 1509
Lojek Andrzej 556
Lomotowski Janusz 823
Lovegrove Keith 806
Lubon Wojciech 1059
Luin Blaž 847
Lv Yang 854
Lymberis Kostas 1408
Ma Nyuk Ling 962
Mabrouk Mohamed Tahar 449
Machi Justyna 1478
Maciejewska Monika 1352, 1360
Mahmoud Saad 572
Maihankuri Sheidi 191
Majewska-Nowak Katarzyna 465
Makisha Nikolay 861
Maletka Marek 876
Malicki Marcin 1457
Malik Daniel 1059
Malinowski Pawel 159, 257
Malmur Robert 867
Maloufi Nabila 1217
Marcinkowski Tadeusz 65
Martínez-García Herminio 884, 891
Martyla Agnieszka 899
Maslon Adam 907
Masson Christian 183
Maziejuk Miroslaw 1360
Mazur Aleksandra 1321
Mazur Jacek 111, 436
Mazurek Wojciech 354
Mazurkiewicz Justyna 657
Mendrek Barbara 224
Midor Katarzyna 915
Mijakowski Maciej 775
Mikulik Jerzy 25
Milani Massimo 922
Minh Le Quang 531
Miodonski Stanislaw 930, 955
Misinski Jacek 1336
Mohammed H.A. 970, 1610
Molczan Marek 1478
Molina-Huelva Marta 419
Montorsi Luca 922
Morawski Antoni Waldemar 111
Mori Hidetoshi 1524
Mozia Sylwia 111
Mrowiec Bozena 938
Mróz Tomasz M. 1107
Mucha Walter 1154
Mularewicz Piotr 1384
Musa Abdullatif 945
Muszynski-Huhajlo Mateusz 556, 955
Nalepa Beata 495
Nam Wai Lun 962
Natschläger Thomas 1271
Neczaj Ewa 456
Nejad Marjan B. 970

AUTHOR INDEX

Niemierka Elzbieta....1656
Niestepski Sebastian....495
Nogaj Kinga....980
Nogalska Adrianna....987
Nowakowska Monika....996
Nurgaliyeva Rano....757
Obersteiner Gudrun....143
Ofman Piotr....1005
Oleniacz Robert....1169
Oleskowicz-Popiel Czeslaw....1463
Orlenko Sergei....1580
Orlowska-Szostak Maria....1013
Orlowski Ryszard....1013
Osinska Adriana....495
Osinska-Broniarz Monika....899
Osman Mohammad Shahril....1530
Osyczka Piotr....1162
Ouzaouit Meriem....282
Pacak Anna....1021
Palaszynska Katarzyna....1029
Pandelidis Demis....1021, 1036
Panesar Angad....1043
Pasiecznik Iwona....1051
Pasonl Lukasz....1313
Pastuszka Józef S.....1154
Pawlik Marcin....25
Pedišius Nerijus....1086
Pelech Aleksander....625
Pelka Grzegorz....1059
Pelletier Francis....183
Petelin Stojan....847
Phang Xue Yee....962
Picheta-Oles Anna....249
Piekarska Katarzyna....103, 1065, 1344
Piekutin Janina....442
Platkowski Maciej....1384
Plaza Grazyna....224
Podedworna Jolanta....1639
Pogorzelec Marta....1065
Polarczyk Iwona....1073, 1079
Popovski Eftim....33
Porowski Mieczyslaw....1029
Poškas Povilas....1086
Poškas Robertas....1086
Price-Allison Andrew....191
Przekop R. E.....899
Przydrózna Aleksandra....1431
Przydrózny Edward....625, 1431
Puchlik Monika....1005, 1093
Qyyum Muhammad Abdul....1099
Rabhi Abdelhamid....1
Radomski Bartosz....1107
Radziszewski Leszek....89
Raji Abdulganiy....191
Rajski Krzysztof....1036
Rak Józef....1115
Rakhypbekov Tolebay....1587
Ratanaporn Yuangsawad....1123
Regucki Pawel....564, 1131
Rodríguez-Artacho Salvador....1279
Rodziewicz Joanna....1329
Rogala Zbigniew....1146
Rogula-Kopiec Patrycja....1154
Rogula-Kozlowska Wioletta....1154
Rohlfs Wilko....174, 504
Rola Kaja....1162
Romero Fernando....200
Rosinski Marian....672
Rucinska Teresa....428
Rumin Rafal....25
Rusanowska Paulina....1594
Rybak Justyna....815
Rybka J. D.....899
Rzeszutek Mateusz....1169
Rzeznik Ilona....1178
Sadeghi O.....970
Sadykov Nurllah....757
Sahel Djemal....96
Salata Aleksandra....1194
Salman Chaudhary Awais....1186
Sambor Aleksandra....332, 1202
Sanchez-Silva Luz....339
Sannad Mohammed....282
Schiro Fabio....273, 1209
Schmied Elisabeth....143
Schwede Sebastian....1186
Sehee Oh....1099
Sekret Robert....980, 1417
Selmane Naceur....1217
Shi Luyang....1226
Shin Ki-Yeol....580
Siekierka Anna....1233
Simson Grzegorz....1005
Sirvydas Arunas....1086
Sitarska Magdalena....370
Skretowicz Maria....1241, 1248
Skrzycki Maciej....120
Skrzypczak Izabela....1256
Skvortsov Valeriy....1580, 1587
Slowik Marta....1256
Slys Daniel....678, 1321
Smyk Joanna....1264
Sobieczky Florian....1271
Sokolowska Aleksandra....739
Soltero Victor Manuel....1279
Sowizdzal Anna....1059, 1288
Sówka Izabela....128, 1297
Spangemacher Lars....1305
Sroka Zbigniew....1248
Stanczyk Justyna....823

AUTHOR INDEX

Stanczyk-Mazanek Ewa 1313
Stathopoulos Vassilis 1408
Stec Agnieszka 1321
Steenackers Gunther 1139
Stefani Matteo 922
Stefanowicz Ewelina 354
Stepanenko Valeriy 1580, 1587
Stoppato Anna 273, 1209
Strek Michal 1384
Striugas Nerijus 1086
Struk-Sokolowska Joanna 1005, 1329
Studzinski Jan 1368
Su Man Huan 962
Sudol Adam 731
Suwala Wojciech 1500
Sytek-Szmeichel Katarzyna 1639
Szalanski Pawel 1336
Szczepaniak Wlodzimierz 473, 1051
Szczesniak Sylwia 625, 1344
Szczurek Andrzej 1352, 1360
Szelag Bartosz 249, 1368
Szerzyna Slawomir 1478
Szlachta Malgorzata 1376, 1493
Szulc Jowita 73
Szulecka Adriana 1169
Szulgowska–Zgrzywa M. 354, 686
Tachos Nikolaos 1408
Tahan Antoine 183
Takao Masuda 1123
Talbi Mosbah 945
Tarnowski Krzysztof 111, 436
Tatsuda Kou 839
Tawfik Ahmed 540
Telesinski Arkadiusz 731, 1384
Teleszewski Tomasz Janusz 1392
Teruoki Tago 1123
Thorin Eva 1186
Tomaszewska Barbara 657, 1288, 1400
Torres David 183
Toyoda Shin 1580
Traczewska Teodora 370, 1470
Traxler Patrick 1271
Trusz-Zdybek Agnieszka 103, 1344
Trzaski Adrian 775
Tsolakoglou Nikolas P. 1408
Tsubaki Noritatsu 839
Turski Michal 980, 1417
Tyszer Magdalena 1400
Uchronski Mariusz 1352
Umiejewska Katarzyna 830
Uqaili Mohammed 191
Urbanowska Agnieszka 1423
Urbanska Klaudia 143
Valincius Vitas 1086
Valverde Jose Luis 339
Van Buren Simon 174, 504
Van Riet Freek 1139
Velázquez Ramón 1279
Venturelli Matteo 922
Verhaert Ivan 41, 1139
Vrachopoulos Michalis Gr. 1408
Walaszczyk Juliusz 625, 1431
Walery Maria Jolanta 1441
Wang Haifeng 854
Wartalska Katarzyna 634
Wdowikowski Marcin 590, 634
Wedrychowicz Wieslaw 564
Wei Feng 1099
Wei Shanshan 854
Wisniewski Jacek 1478
Wisniowska Ewa 1449
Wlodarczyk-Makula Marla 1449
Wojdyga Krzysztof 1457
Wojtkowiak Janusz 10, 1463
Wójtowicz Patryk 1376, 1493
Wolf Mirela 370, 1470
Wołoszyk Czeslaw 436
Wolska Malgorzata 1478
Wolski Pawel 1486
Wong Chee Chung 1530
Wróbel Radoslaw 701, 1241
Wronska Ilona 1384
Wu Yupeng 379
Wyrwa Artur 1500
Xue Puning 1509
Yalcinkaya Fatma 1517
Yamamoto Tatsumi 1524
Yan Jinyue 1186
Yek Peter Nai Yuh 1530
Yilmaz Seval 403
Yoneyama Yoshiharu 839
Yoon Si-Won 580
Yuksel Ibrahim 1539, 1547
Zablocka-Malicka Monika 473
Zabnienska-Góra Alina 1631
Zaczynska Ewa 103
Zajac Pawel 1555
Zajiczek Zaneta 1360
Zamorska-Wojdyla Dorota 522
Zareba Anna 724, 1563
Zathey Maciej 1297
Zawieja Iwona 1572
Zhang Huibo 1226
Zhou Zhigang 1509
Zhumadilov Kassym 757, 1580, 1587
Zielinski Marcin 1594
Ziembicki Piotr 649
Zimoch Izabela 1602
Zmarzly Marcin 291
Zubeer Swar A. 970, 1610

AUTHOR INDEX

Zubrowska-Sudol Monika..1639
Zukowska Adrianna 987
Zukowski Miroslaw 1648
Zwierzchowski Ryszard...................................299, 1621
Zylka Radoslaw...257

International Conference on Advances in Energy Systems and Environmental Engineering (ASEE17)

E3S Web of Conferences Volume 22 (2017)

Wroclaw, Poland
2 - 5 July 2017

Part 2 of 2

Editors:

Bartosz Kazmierczak
Malgorzata Kutylowska
Katarzyna Piekarska
Hussam Jouhara
Jan Danielewicz

ISBN: 978-1-5108-5194-8

Printed from e-media with permission by:

Curran Associates, Inc.
57 Morehouse Lane
Red Hook, NY 12571

Some format issues inherent in the e-media version may also appear in this print version.

Printed by Curran Associates, Inc. (2017)

For additional information, please contact EDP Sciences – Web of Conferences at the address below.

EDP Sciences – Web of Conferences
17, Avenue du Hoggar
Parc d'Activité de Courtabœuf
BP 112
F-91944 Les Ulis Cedex A
France

Phone: +33 (0) 1 69 18 75 75
Fax: +33 (0) 1 69 28 84 91

contact-edps@webofconferences.org

Additional copies of this publication are available from:

Curran Associates, Inc.
57 Morehouse Lane
Red Hook, NY 12571 USA
Phone: 845-758-0400
Fax: 845-758-2633
Email: curran@proceedings.com
Web: www.proceedings.com

TABLE OF CONTENTS

PART 1

USING A COUPLED INDUCTOR CONTROLLED BY FUZZY LOGIC TO IMPROVE THE EFFICIENCY OF A BUCK CONVERTER IN A PV SYSTEM 1
Abouchabana Nabil, Haddadi Mourad, Rabhi Abdelhamid, El Hajjaji Ahmed

EXPERIMENTAL INVESTIGATION AND CFD SIMULATION OF MULTI-PIPE EARTH-TO-AIR HEAT EXCHANGERS (EAHES) FLOW PERFORMANCE 10
Amanowicz Lukasz, Wojtkowiak Janusz

URBAN STORMWATER – GREYWATER MANAGEMENT SYSTEM FOR SUSTAINABLE URBAN WATER MANAGEMENT AT SUB-WATERSHED LEVEL 18
Arora Amarpreet Singh

ESTIMATION OF THE TEMPERATURE SPATIAL VARIABILITY IN CONFINED SPACES BASED ON THERMAL IMAGING 25
Augustyn Grzegorz, Jurasz Jakub, Jurczyk Krzysztof, Korbiel Tomasz, Mikulik Jerzy, Pawlik Marcin, Rumin Rafal

ASSESSING HIGH SHARES OF RENEWABLE ENERGIES IN DISTRICT HEATING SYSTEMS – A CASE STUDY FOR THE CITY OF HERTEN 33
Aydemir Ali, Popovski Eftim, Bellstädt Daniel, Fleiter Tobias, Büchele Richard

A NEW TEST PROCEDURE TO EVALUATE THE PERFORMANCE OF SUBSTATIONS FOR COLLECTIVE HEATING SYSTEMS 41
Baetens Robin, Verhaert Ivan

INFLUENCE OF SILANE ON THE STRUCTURE OF POLYSTYRENE PREPARED BY SOL-GEL COATINGS VIA UV CURING 49
Balbay Senay, Acikgoz Caglayan

COMPARISON OF METHODS FOR SOLIDS RETENTION TIME DETERMINATION AND CONTROL 57
Balbierz Piotr, Knap Marta

USE OF CEMENT-FLY ASH-BASED STABILIZATION TECHNIQUES FOR THE TREATMENT OF WASTE CONTAINING AROMATIC CONTAMINANTS 65
Banaszkiewicz Kamil, Marcinkowski Tadeusz

THE INFLUENCE OF MULTIFAMILY APARTMENT BUILDING OCCUPANTS ON ENERGY AND WATER CONSUMPTION – THE PRELIMINARY RESULTS OF MONITORING AND SURVEY CAMPAIGN 73
Bandurski Karol, Hamerla Milosz, Szulc Jowita, Koczyk Halina

ENVIRONMENTAL IMPACT ASSESSMENT OF COAL POWER PLANTS IN OPERATION 81
Bartan Ayfer, Kucukali Serhat, Ar Irfan

VARIABILITY OF ROAD TRAFFIC NOISE RECORDED BY STATIONARY MONITORING STATIONS 89
Bakowski Andrzej, Radziszewski Leszek

THE EFFECT OF THE THERMAL INERTIA ON THE THERMAL TRANSFER IN BUILDING WALL 96
Bellahcene Lahcene, Cheknane Ali, Bekkouche SMA., Sahel Djemal

CYTOTOXICITY AND GENOTOXICITY PROPERTIES OF PARTICULATE MATTER FRACTION 2.5 µM 103
Belcik Maciej K., Trusz-Zdybek Agnieszka, Zaczynska Ewa, Czarny Anna, Piekarska Katarzyna

THE USE OF MOVING BED BIO-REACTOR TO LAUNDRY WASTEWATER TREATMENT 111
Bering Slawomira, Mazur Jacek, Tarnowski Krzysztof, Janus Magdalena, Mozia Sylwia, Morawski Antoni Waldemar

ENERGY SAVING BY USING NATURAL ENERGY FROM THE SHALLOW GROUND DEPTHS – MANY YEARS OPERATING RESULTS 120
Besler Maciej, Skrzycki Maciej, Cepinski Wojciech

EFFECTS OF CLIMATE FACTORS AND VEGETATION ON THE CO_2 FLUXES AND $\sigma^{13}C$ FROM RE-ESTABLISHED GRASSLAND 128
Bezyk Yaroslav, Dorodnikov Maxim, Sówka Izabela

APPLICATION OF STOCHASTIC APPROACH BASED ON MONTE CARLO (MC) SIMULATION FOR LIFE CYCLE INVENTORY (LCI) OF THE RARE EARTH ELEMENTS (REES) IN BENEFICIATION RARE EARTH WASTE FROM THE GOLD PROCESSING: CASE STUDY 137
Bieda Boguslaw, Grzesik Katarzyna

FOOD WASTE IN CENTRAL EUROPE – CHALLENGES AND SOLUTIONS 143
den Boer Jan, Kobel Przemyslaw, Dyjakon Arkadiusz, Urbanska Klaudia, Obersteiner Gudrun, Hrad Marlies, Schmied Elisabeth, den Boer Emilia

PERFORMANCE OF MECHANICAL BIOLOGICAL TREATMENT OF RESIDUAL MUNICIPAL WASTE IN POLAND 151
den Boer Emilia, Jedrczak Andrzej

ORGANIC MATTER AND HEAVY METALS CONTENT MODELING IN SEWAGE SLUDGE TREATED WITH REED BED SYSTEM 159
Boruszko Dariusz, Dabrowski Wojciech, Malinowski Pawel

ENERGY AND ENVIRONMENTAL EVALUATION OF COMBINED COOLING HEATING AND POWER SYSTEM 166
Bugaj Andrzej

MODELING THE DEFROST PROCESS IN COMPLEX GEOMETRIES – PART 1: DEVELOPMENT OF A ONE-DIMENSIONAL DEFROST MODEL 174
van Buren Simon, Hertle Ellen, Figueiredo Patric, Kneer Reinhold, Rohlfs Wilko

COMPARISON OF POWER CURVE MONITORING METHODS 183
Cambron Philippe, Masson Christian, Tahan Antoine, Torres David, Pelletier Francis

AN INTEGRATED LOW CARBON ENERGY SOLUTION TO COOKING FUEL, TAILORED TO NIGER STATE'S RURAL POPULATION 191
Carvell Aaron, Price-Allison Andrew, Birch Calum, Green Toby, Harijan Khanji, Maihankuri Sheidi, Raji Abdulganiy, Uqaili Mohammed, Dupont Valerie

ACID EMISSIONS MONITORING NEEDS IN CERAMIC TILE INDUSTRY: CHALLENGES DERIVED FROM NEW POLICY TRENDS 200
Celades Irina, Gomar Salvador, Romero Fernando, Chauhan Amisha, Delpech Bertrand, Jouhara Hussam

INDUSTRIAL APPLICATIONS OF THE AIR DIRECT-CONTACT, GRAVEL, GROUND HEAT EXCHANGER 208
Cepinski Wojciech, Besler Maciej

THE USE OF MULTI CRITERIA ANALYSIS TO COMPARE THE OPERATING SCENARIOS OF THE HYBRID GENERATION SYSTEM OF WIND TURBINES, PHOTOVOLTAIC MODULES AND A FUEL CELL 216
Ceran Bartosz

THE EFFECT OF BIOLOGICALLY AND CHEMICALLY SYNTHESIZED SILVER NANOPARTICLES (AGNPS) ON BIOFILM FORMATION 224
Chojniak Joanna, Biedron Izabela, Mendrek Barbara, Plaza Grazyna

USE OF CFD MODELLING FOR ANALYSING AIR PARAMETERS IN AUDITORIUM HALLS 232
Cichowicz Robert

POWER PLANT WASTES CAPITALIZATION AS GEOPOLYMERIC BUILDING MATERIALS 240
Ciobanu Gabriela, Litu Loredana, Harja Maria

ASSESSMENT OF THE POSSIBILITY OF USING DATA MINING METHODS TO PREDICT SORPTION ISOTHERMS OF SELECTED ORGANIC COMPOUNDS ON ACTIVATED CARBON 249
Dabek Lidia, Szelag Bartosz, Picheta-Oles Anna

ENERGY CONSUMPTION MODELING DURING DAIRY SEWAGE PRETREATMENT 257
Dabrowski Wojciech, Zylka Radoslaw, Malinowski Pawel, Boruszko Dariusz

A REVIEW ON WASTE HEAT RECOVERY FROM EXHAUST IN THE CERAMICS INDUSTRY 265
Delpech Bertrand, Axcell Brian, Jouhara Hussam

EU ENERGY POLICIES ACHIEVEMENT BY INDUSTRIES IN DECENTRALIZED AREAS 273
Destro Nicola, Stoppato Anna, Benato Alberto, Schiro Fabio

MIXED CONVECTION IN A THREE-DIMENSIONAL VENTILATED CAVITY WITH TWO ISOTHERMAL HEATING SECTIONS 282
Doghmi Hicham, Abourida Btissam, Belarche Lahoucine, Sannad Mohammed, Ouzaouit Meriem

THE USE OF MATHEMATICAL MODELS FOR DIAGNOSIS OF ACTIVATED SLUDGE SYSTEMS IN WWTP 291
Drewnowski Jakub, Zmarzly Marcin

ANALYSIS OF A COMBINED HEATING AND COOLING SYSTEM MODEL UNDER DIFFERENT OPERATING STRATEGIES 299
Dzierzgowski Mieczyslaw, Zwierzchowski Ryszard

INVESTIGATING THE WATER CONSUMPTION FOR ELECTRICITY GENERATION AT TURKISH POWER PLANTS 308
El-Khozondar Balkess, Koksal Merih Aydinalp
EXPERIMENTAL STUDY OF AIR EVAPORATIVE COOLING PROCESS USING MICROPOROUS MEMBRANES 316
Englart Sebastian
CO_2 CONCENTRATION AND OCCUPANCY DENSITY IN THE CRITICAL ZONES SERVED BY THE VAV SYSTEM 324
Etoua Evina Ghislaine, Kajl Stanislaw, Lamarche Louis, Beltran-Galindo Javier
THE INFLUENCE OF MECHANICAL GEAR ON THE EFFICIENCY OF SMALL HYDROPOWER 332
Ferenc Zbigniew, Sambor Aleksandra
TEMPERATURE INFLUENCE ON THE FAST PYROLYSIS OF MANURE SAMPLES: CHAR, BIO-OIL AND GASES PRODUCTION 339
Fernandez-Lopez Maria, Anastasakis Kostas, De Jong Wiebren, Valverde Jose Luis, Sanchez-Silva Luz
THE IMPACT OF THE HOT TAP WATER LOAD PATTERN IN THE INDUSTRIAL HALL ON THE ENERGY YIELD FROM SOLAR COLLECTORS 346
Fidorów-Kaprawyl Natalia, Dudkiewicz Edyta
THE ANALYSIS OF THE DIFFERENCES BETWEEN THE RESULTS OF THE THERMAL RESPONSE TEST AND THE DATA FROM THE OPERATION OF THE BRINE-TO-WATER HEAT PUMP'S VERTICAL EXCHANGER 354
Fidorów-Kaprawy Natalia, Stefanowicz Ewelina, Mazurek Wojciech, Szulgowska-Zgrzywa Malgorzata, Bryszewska-Mazurek Anna
SOLAR RADIATION – TO – POWER GENERATION MODELS FOR ONE-AXIS TRACKING PV SYSTEM WITH ON-SITE MEASUREMENTS FROM ESKISEHIR, TURKEY 363
Filik Tansu, Basaran Filik Ümmühan, Gerek Ömer Nezih
MICROBIOLOGICAL STIMULATION OF PHYTOREMEDIATION PROCESS USING SALVINIA NATANS TO MERCURY CONTAMINED WATER 370
Filyarovskaya Viktoriya, Sitarska Magdalena, Traczewska Teodora, Wolf Mirela
DYNAMIC ENVIRONMENTAL CONTROL MECHANISMS FOR PNEUMATIC FOIL CONSTRUCTIONS 379
Flor Jan-Frederik, Wu Yupeng, Beccarelli Paolo, Chilton John
CLOSING OF WATER CIRCUITS – A GLOBAL BENCHMARK ON SUSTAINABLE WATER MANAGEMENT 388
Fröhlich Siegmund
SEASONAL COEFFICIENT OF PERFORMANCE FOR GROUND SOURCE HEAT PUMP AND GROUNDWATER ONE IN BIALYSTOK 396
Gajewski Andrzej
CARBON AND ENERGY FOOTPRINT OF ELECTROCHEMICAL VINEGAR WASTEWATER TREATMENT 403
Gerek Emine Esra, Yilmaz Seval, Koparal A. Savas, Gerek Ömer Nezih
SIZING PROCEDURES FOR SUN-TRACKING PV SYSTEM WITH BATTERIES 411
Gerek Ömer Nezih, Filik Ümmühan Basaran, Filik Tansu
NATURAL VENTILATION SYSTEMS TO ENHANCE SUSTAINABILITY IN BUILDINGS: A REVIEW TOWARDS ZERO ENERGY BUILDINGS IN SCHOOLS 419
Gil-Baez Maite, Barrios-Padura Ángela, Molina-Huelva Marta, Chacartegui Ricardo
THE SLAG ORIGINAL FROM THE PROCESS OF SEWAGE SLUDGE INCINERATION SELECTED PROPERTIES CHARACTERISTIC 428
Glowacka Anna, Rucinska Teresa, Kiper Justyna
ENERGY CROPS CULTIVATED ON THE SLAG FROM INCINERATION OF THE SEWAGE SLUDGE ENERGY VALUE ASSESSMENT 436
Glowacka Anna, Tarnowski Krzysztof, Bering Slawomira, Mazur Jacek, Kiper Justyna, Woloszyk Czeslaw
APPLICATION OF POROUS MATERIALS IN OIL SUBSTANCES SEPARATION FROM WATER 442
Golub Adam, Piekutin Janina
INFLUENCE OF PCMS ON THERMAL BEHAVIOR OF BUILDING WALLS: EXPERIMENTAL STUDY USING THE WALLS OF A REDUCED SCALE ROOM 449
Gounni Ayoub, Mabrouk Mohamed Tahar, Kheiri Abdelhamid, El alami Mustapha
PRETREATMENT METHODS AS A MEANS OF BOOSTING METHANE PRODUCTION FROM SEWAGE SLUDGE AND ITS MIXTURES WITH GREASE TRAP SLUDGE 456
Grosser Anna, Neczaj Ewa

ELECTRODIALYSIS POTENTIAL FOR FRACTIONATION OF MULTICOMPONENT AQUEOUS SOLUTIONS 465
Grzegorzek Martyna, Majewska-Nowak Katarzyna
INCINERATION, PYROLYSIS AND GASIFICATION OF ELECTRONIC WASTE 473
Gurgul Agnieszka, Szczepaniak Wlodzimierz, Zablocka-Malicka Monika
MONITORING OF WATER SUPPLY CONNECTIONS AS AN ELEMENT TO REDUCE APPARENT LOSSES OF WATER? 480
Gwozdziej-Mazur Joanna
REMOVAL OF OXYTETRACYCLINE FROM AQUEOUS SOLUTIONS BY HYDROXYAPATITE AS A LOW-COST ADSORBENT 487
Harja Maria, Ciobanu Gabriela
MONITORING OF DRUG RESISTANCE AMPLIFICATION AND ATTENUATION WITH THE USE OF TETRACYCLINE-RESISTANT BACTERIA DURING WASTEWATER TREATMENT 495
Harnisz Monika, Korzeniewska Ewa, Niestepski Sebastian, Osinska Adriana, Nalepa Beata
MODELLING THE DEFROST PROCESS IN COMPLEX GEOMETRIES – PART 2: WALL-FUNCTION BASED COUPLING TO A MULTI-REGION CFD SOLVER 504
Hertle Ellen, van Buren Simon, Figueiredo Patric, Kneer Reinhold, Rohlfs Wilko
MULTICRITERIA OPTIMIZATION APPROACH TO DESIGN AND OPERATION OF DISTRICT HEATING SUPPLY SYSTEM OVER ITS LIFE CYCLE 513
Hirsch Piotr, Duzinkiewicz Kazimierz, Grochowski Michal
POLLUTION OF SOILS (PB, CD, CR, ZN, CU, NI) ALONG THE RING ROAD OF WROCLAW (POLAND) 522
Holtra Anna, Zamorska-Wojdyla Dorota
A NOVEL DESIGN OF REACTIVE DISTILLATION CONFIGURATION FOR 2-METHOXY-2-METHYLHEPTANE PROCESS 531
Hussain Arif, Abdul Qyyum Muhammad, Minh Le Quang, Jimin Hong, Lee Moonyong
COMPREHENSIVE STUDY FOR ANAMMOX PROCESS VIA MULTISTAGE ANAEROBIC BAFFLED REACTORS 540
Ismail Sherif, Tawfik Ahmed
CONCEPTUAL DESIGN OF MULTI-SOURCE CCS PIPELINE TRANSPORTATION NETWORK FOR POLISH ENERGY SECTOR 548
Isoli Niccolo, Chaczykowski Maciej
TOWARDS ENERGY NEUTRALITY OF WASTEWATER TREATMENT PLANTS VIA DEAMMONIFICATION PROCESS 556
Janiak Kamil, Lojek Andrzej, Muszynski-Huhajlo Mateusz
STUDY OF HEAT EXCHANGE PROCESSES IN A SCREW ASH COOLER INSTALLATION 564
Janowska Barbara, Regucki Pawel, Andruszkiewicz Artur, Wedrychowicz Wieslaw
INDIRECT CONTACT FREEZE WATER DESALINATION FOR AN ICE MAKER MACHINE – CFD SIMULATION 572
Jayakody Harith, Al-Dadah Raya, Mahmoud Saad
SIMULATION OF THERMAL ENVIRONMENT IN A THREE-LAYER VINYL GREENHOUSE BY NATURAL VENTILATION CONTROL 580
Jin Tea-Hwan, Shin Ki-Yeol, Yoon Si-Won, Im Yong-Hoon, Chang Ki-Chang
TEMPORAL AND SPATIAL COMPLEMENTARITY OF WIND AND SOLAR RESOURCES IN LOWER SILESIA (POLAND) 590
Jurasz Jakub, Wdowikowski Marcin, Kazmierczak Bartosz, Dabek Pawel
ANAEROBIC CO-DIGESTION OF SEWAGE SLUDGE AND MOLASSES 599
Kalemba Katarzyna, Barbusinski Krzysztof
SINGLE AND COMPETITIVE ADSORPTION OF OMPS BY CARBON NANOTUBES – MECHANISM AND FITTING MODELS 606
Kaminska Gabriela, Dudziak Mariusz, Bohdziewicz Jolanta, Kudlek Edyta
STABILITY ASSESSMENT OF THE CHEMICAL COMPOSITION OF THE TREATED MINING WATER USED TO REPLENISH THE COOLING CIRCUIT IN JAWORZNO III POWER PLANT – POWER PLANT II 617
Karpinski Marcin, Kmiecik Ewa
AIR TEMPERATURE GRADIENT IN LARGE INDUSTRIAL HALL 625
Karpuk Michal, Pelech Aleksander, Przydrózny Edward, Walaszczyk Juliusz, Szczesniak Sylwia
THE ANALYSIS OF THE POSSIBILITY OF USING 10-MINUTE RAINFALL SERIES TO DETERMINE THE MAXIMUM RAINFALL AMOUNT WITH 5 MINUTES DURATION 634
Kazmierczak Bartosz, Wartalska Katarzyna, Wdowikowski Marcin, Kotowski Andrzej

COMPARING ENERGY PAYBACK AND SIMPLE PAYBACK PERIOD FOR SOLAR PHOTOVOLTAIC SYSTEMS 643
Kessler Will
ANALYSIS OF THE IMPACT OF SIMULATION MODEL SIMPLIFICATIONS ON THE QUALITY OF LOW-ENERGY BUILDINGS SIMULATION RESULTS 649
Klimczak Marcin, Bojarski Jacek, Ziembicki Piotr, Keskiewicz Piotr
THE QUATERNARY GROUNDWATER AS THE LOW TEMPERATURE ENERGY SOURCE FOR HEAT PUMPS IN MALOPOLSKA PROVINCE 657
Kmiecik Ewa, Tomaszewska Barbara, Mazurkiewicz Justyna
MODIFICATION OF RHODAMINE WT TRACER TESTS PROCEDURE IN ACTIVATED SLUDGE REACTORS 664
Knap Marta, Balbierz Piotr
DETERMINATION OF THE SOLAR TRANSMITTANCE FOR THE TRANSLUCENT SHUTTER WITH PCM IN LIQUID AND SOLID STATE 672
Komerska Anna, Ksionek Dariusz, Rosinski Marian
ANALYSIS OF PROFITABILITY OF USING A HEAT RECOVERY SYSTEM FROM GREY WATER DISCHARGED FROM THE SHOWER (CASE STUDY OF POLAND) 678
Kordana Sabina, Slys Daniel
CHANGE-OVER NATURAL AND MECHANICAL VENTILATION SYSTEM ENERGY CONSUMPTION IN SINGLE-FAMILY BUILDINGS 686
Kostka Maria, Szulgowska–Zgrzywa Malgorzata
REMOVAL OF COPPER IONS FROM AQUEOUS SOLUTIONS BY MEANS OF MICELLAR-ENHANCED ULTRAFILTRATION 695
Kowalska Izabela, Klimonda Aleksandra
INFLUENCE OF THE SINGLE EGR VALVE USABILITY ON DEVELOPMENT OF THE CHARGE DIRECTED TO INDIVIDUAL CYLINDERS OF AN INTERNAL COMBUSTION ENGINE 701
Krakowian Konrad, Kazmierczak Andrzej, Górniak Aleksander, Wróbel Radoslaw
GEOSTATISTICAL METHODS IN THE ASSESSMENT OF THE SPATIAL VARIABILITY OF THE QUALITY OF RIVER WATER 709
Krasowska Malgorzata, Banaszuk Piotr
INFLUENCE OF SPUTTERING DEPOSITION PARAMETERS ON ELECTRICAL AND OPTICAL PROPERTIES OF ALUMINIUM-DOPED ZINC OXIDE THIN FILMS FOR PHOTOVOLTAIC APPLICATIONS 717
Krawczak Ewelina, Agata Zdyb, Gulkowski Slawomir, Fave Alain, Fourmond Erwann
BIOARCHITECTURE – A NEW VISION OF ENERGY SUSTAINABLE CITIES 724
Krzeminska Alicja, Zareba Anna, Dzikowska Anna
INFLUENCE OF ENVIRONMENTAL POLLUTION WITH CREOSOTE OIL OR ITS VAPORS ON BIOMASS AND SELECTED PHYSIOLOGICAL GROUPS OF MICROORGANISMS 731
Krzysko-Lupicka Teresa, Cybulska Krystyna, Kolosowski Pawel, Telesinski Arkadiusz, Sudol Adam
THE OCCURRENCE OF HEAVY METALS AND METAL-RESISTANT BACTERIA IN WATER AND BOTTOM SEDIMENTS OF THE STRASZYN RESERVOIR (POLAND) 739
Kulbat Eliza, Sokolowska Aleksandra
DETERMINING RESIDENTIAL ENERGY CONSUMPTION-BASED CO_2 EMISSIONS AND EXAMINING THE FACTORS AFFECTING THE VARIATION IN ANKARA, TURKEY 748
Kus Melike, Akan Perihan, Koksal Merih Aydinalp, Gullu Gulen
EXPERIMENTAL DETERMINATION WITH EPR-METHODS OF DOSE LOADS ON LOCAL POPULATION IN INHABITED LOCALITIES ADJACENT TO THE TAILING POND "KOSHKAR-ATA" (KAZAKHSTAN) 757
Kuterbekov Kairat, Sadykov Nurllah, Zhumadilov Kassym, Nurgaliyeva Rano, Kabyshev Aset, Abseitov Erbolat, Kurmanzhanov Askar
ELECTRON THERMAL EMF FOR $NA_XCU_{2-X}S$ 762
Kuterbekov Kairat, Balapanov Malik, Ishembetov Rais, Kubenova Marzhan, Baitasov Talgat, Kabyshev Asset, Azhibekov Aidos, Kenzhebatyr Bekmyrza, Alibay Temirulan
APPLICATION OF REGRESSION TRESS FOR PREDICTION OF WATER CONDUITS FAILURE RATE 767
Kutylowska Malgorzata
THE MEASURES FOR ACHIEVING NZEB STANDARD OF RETROFITTED EDUCATIONAL BUILDING FOR SPECIFIC POLISH LOCATION – CASE STUDY 775
Kwiatkowski Jerzy, Mijakowski Maciej, Trzaski Adrian
RESEARCH OF CO_2 CONCENTRATION IN NATURALLY VENTILATED LECTURE ROOM 782
Laska Marta, Dudkiewicz Edyta

THE IMPORTANCE OF MONITORING RENEWABLE ENERGY PLANTS: THREE CASE HISTORIES 790
Lazzarin Renato
BATCH OROGRAPHIC INTERPOLATION OF MONTHLY PRECIPITATION BASED ON FREE-OF-CHARGE GEOSTATISTICAL TOOLS 798
Ledvinka Ondrej
THERMODYNAMIC ANALYSIS OF A COMBINED-CYCLE SOLAR THERMAL POWER PLANT WITH MANGANESE OXIDE-BASED THERMOCHEMICAL ENERGY STORAGE 806
Lei Qi, Bader Roman, Kreider Peter, Lovegrove Keith, Lipinski Wojciech
THE EFFECTIVENESS OF THE BIODEGRADATION OF RAW AND PROCESSED POLYSTYRENE BY MEALWORMS 815
Leluk Karol, Hanus-Lorenz Beata, Rybak Justyna, Bozek Magdalena
MODERN PROPOSAL OF METHODOLOGY FOR RETRIEVAL OF CHARACTERISTIC SYNTHETIC RAINFALL HYETOGRAPHS 823
Licznar Pawel, Burszta-Adamiak Ewa, Lomotowski Janusz, Stanczyk Justyna

PART 2

APPLICATION OF MULTI-ENZYMATIC HYDROLYSIS FOR IMPROVING THE EFFICIENCY OF THE BIOGAS PRODUCTION IN SOLID WASTE FERMENTATION PROCESS IN OSTRÓDA WWTP 830
Lipinski Kamil, Umiejewska Katarzyna
SYNTHESIS OF MESOPOROUS CERIUM COMPOUND FOR CO_2 CAPTURE 839
Liu Guiqing, Tatsuda Kou, Yoneyama Yoshiharu, Tsubaki Noritatsu
MODELING AN IMPACT OF ROAD GEOMETRIC DESIGN ON VEHICLE ENERGY CONSUMPTION 847
Luin Blaž, Petelin Stojan, Al-Mansour Fouad
THE TRANSMISSION CHARACTERISTICS OF INDOOR PARTICLES UNDER DIFFERENT VENTILATION CONDITIONS 854
Lv Yang, Wang Haifeng, Wei Shanshan
SCIENTIFIC APPROACH AND PRACTICAL EXPERIENCE FOR RECONSTRUCTION OF WASTE WATER TREATMENT PLANTS IN RUSSIA 861
Makisha Nikolay, Gogina Elena
TRANSFER RESERVOIR AS A NEW SOLUTION FOR TRANSFER OF STORMWATER TO WATER RECEIVERS 867
Malmur Robert
VARIOUS METHODS OF HEAT SUPPLY FOR A BUILDING WHICH IS OPERATED PERIODICALLY DURING THE YEAR 876
Maletka Marek, Laska Marta
APPROACH TO THE IMPLEMENTATION OF LINEAR-ASSISTED-BASED BATTERY CHARGERS 884
Martínez-García Herminio, García-Vílchez Encarna
ON THE MODELLING OF LINEAR-ASSISTED DC-DC VOLTAGE REGULATORS FOR PHOTOVOLTAIC SOLAR ENERGY SYSTEMS 891
Martínez-García Herminio, García-Vílchez Encarna
THE INFLUENCE OF THE SMALL PLATINUM CLUSTERS ON HYDROGEN SORPTION PROPERTIES 899
Martyla Agnieszka, Przekop R. E., Osinska-Broniarz Monika, Kopczyk M., Rybka J. D., Kirszensztejn P.
ANALYSIS OF ENERGY CONSUMPTION AT THE RZESZÓW WASTEWATER TREATMENT PLANT 907
Maslon Adam
ANALYSIS OF THE ENERGY POTENTIAL OF MUNICIPAL SOLID WASTE FOR THE THERMAL TREATMENT TECHNOLOGY DEVELOPMENT IN POLAND 915
Midor Katarzyna, Jaderko Karolina
CFD ANALYSIS OF A FULL-SCALE CERAMIC KILN MODULE UNDER ACTUAL OPERATING CONDITIONS 922
Milani Massimo, Montorsi Luca, Stefani Matteo, Venturelli Matteo
CO-DIGESTION OF SEWAGE SLUDGE FROM EXTERNAL SMALL WWTP'S IN A LARGE PLANT 930
Miodonski Stanislaw

NANOMATERIALS IN THE ENVIRONMENT 938
Mrowiec Bozena
CYCLE ANALYSIS OF MCFC/GAS TURBINE SYSTEM 945
Musa Abdullatif, Alaktiwi Abdulsalam, Talbi Mosbah
SHORTCUT NITRIFICATION/PARTIAL NITRITATION START-UP FOR REJECT WATER TREATMENT IN A SBR 955
Muszynski-Huhajlo Mateusz, Miodonski Stanislaw
PRODUCTION OF BIO-FERTILIZER FROM MICROWAVE VACUUM PYROLYSIS OF WASTE PALM SHELL FOR CULTIVATION OF OYSTER MUSHROOM (PLEUROTUS OSTREATUS) 962
Nam Wai Lun, Su Man Huan, Phang Xue Yee, Chong Min Yee, Liew Rock Keey, Ma Nyuk Ling, Lam Su Shiung
INFLUENCE OF NANOFLUIDS ON THE EFFICIENCY OF FLAT-PLATE SOLAR COLLECTORS (FPSC) 970
Nejad Marjan B., Mohammed H.A., Sadeghi O., Zubeer Swar A.
THE INFLUENCE OF USING HEAT STORAGE WITH PCM ON INLET AND OUTLET TEMPERATURES IN SUBSTATION IN DHS 980
Nogaj Kinga, Turski Michal, Sekret Robert
ATMOSPHERIC CO_2 CAPTURE FOR THE ARTIFICIAL PHOTOSYNTHETIC SYSTEM 987
Nogalska Adrianna, Zukowska Adrianna, Garcia-Valls Ricard
THE MATHEMATICAL MODEL OF RAINWATER CATCHMENT IN WROCLAW 996
Nowakowska Monika
IMPACT ASSESSMENT OF TREATED WASTEWATER ON WATER QUALITY OF THE RECEIVER USING THE WILCOXON TEST 1005
Ofman Piotr, Puchlik Monika, Simson Grzegorz, Krasowska Malgorzata, Struk-Sokolowska Joanna
CALIBRATION OF THE COMPUTER MODEL DESCRIBING FLOWS IN THE WATER SUPPLY SYSTEM; EXAMPLE OF THE APPLICATION OF A GENETIC ALGORITHM 1013
Orlowska-Szostak Maria, Orlowski Ryszard
ANALYSIS OF FREEZE PROTECTION METHODS FOR RECUPERATORS USED IN ENERGY RECOVERY FROM EXHAUST AIR 1021
Pacak Anna, Jedlikowski Andrzej, Pandelidis Demis, Anisimov Sergey
ENERGY DEMAND AND THERMAL COMFORT OF HVAC SYSTEMS WITH THERMALLY ACTIVATED BUILDING SYSTEMS AS A FUNCTION OF USER PROFILE 1029
Palaszynska Katarzyna, Bandurski Karol, Porowski Mieczyslaw
PROPOSITIONS OF IMPROVEMENT OF THE CROSS-FLOW M-CYCLE HEAT EXCHANGERS IN DIFFERENT AIR-CONDITIONING APPLICATIONS 1036
Pandelidis Demis, Anisimov Sergey, Rajski Krzysztof, Brychcy Ewa
ORGANIC RANKINE CYCLE – REVIEW AND RESEARCH DIRECTIONS IN ENGINE APPLICATIONS 1043
Panesar Angad
CHEMICAL TREATMENT OF WASTEWATER FROM FLUE GAS DESULPHURISATION 1051
Pasiecznik Iwona, Szczepaniak Wlodzimierz
THE INFLUENCE OF HEAT SINK TEMPERATURE ON THE SEASONAL EFFICIENCY OF SHALLOW GEOTHERMAL HEAT PUMPS 1059
Pelka Grzegorz, Lubon Wojciech, Sowizdzal Anna, Malik Daniel
CONCENTRATION OF POLYCYCLIC AROMATIC HYDROCARBONS IN WATER SAMPLES FROM DIFFERENT STAGES OF TREATMENT 1065
Pogorzelec Marta, Piekarska Katarzyna
IMPACT OF THE CIRCULATION SYSTEM ON THE ENERGY BALANCE OF THE BUILDING 1073
Polarczyk Iwona, Fijewski Michal
IMPACT OF THE CIRCULATION SYSTEM ON DOMESTIC HOT WATER CONSUMPTION 1079
Polarczyk Iwona, Fijewski Michal
INVESTIGATION OF THE BIOFUEL FLUE AND PRODUCER GASES CLEANING EFFICIENCY USING ESP 1086
Poškas Robertas, Sirvydas Arunas, Poškas Povilas, Striugas Nerijus, Pedišius Nerijus, Valincius Vitas
SEASONAL CHANGES IN QUALITY OF WASTEWATER FROM FRUIT AND VEGETABLE INDUSTRY 1093
Puchlik Monika, Ignatowicz Katarzyna
A NOVEL VORTEX TUBE-BASED N_2-EXPANDER LIQUEFACTION PROCESS FOR ENHANCING THE ENERGY EFFICIENCY OF NATURAL GAS LIQUEFACTION 1099
Qyyum Muhammad Abdul, Wei Feng, Hussain Arif, Ali Wahid, Sehee Oh, Lee Moonyong

THE CHOICE OF PRIMARY ENERGY SOURCE INCLUDING PV INSTALLATION FOR PROVIDING ELECTRIC ENERGY TO A PUBLIC UTILITY BUILDING – A CASE STUDY 1107
Radomski Bartosz, Cwiek Barbara, Mróz Tomasz M.

NUMERICAL MODELLING OF MULTI-VANE EXPANDER OPERATING CONDITIONS IN ORC SYSTEM 1115
Rak Józef, Blasiak Przemyslaw, Kolasinski Piotr

RECOVERY OF USEFUL CHEMICALS FROM PALM OIL MILL WASTEWATER 1123
Ratanaporn Yuangsawad, Duangkamol Na-Ranong, Teruoki Tago, Takao Masuda

NUMERICAL MODELLING OF SERIES-PARALLEL COOLING SYSTEMS IN POWER PLANT 1131
Regucki Pawel, Lewkowicz Marek, Kucieba Malgorzata

A SIMPLIFIED MODEL OF A MECHANICAL COOLING TOWER WITH BOTH A FILL PACK AND A COIL 1139
Van Riet Freek, Steenackers Gunther, Verhaert Ivan

EFFECT OF OPERATING CONDITIONS ON PERFORMANCE OF SILICA GEL-WATER AIR-FLUIDISED DESICCANT COOLER 1146
Rogala Zbigniew, Kolasinski Piotr, Gnutek Zbigniew

ORGANIC AND ELEMENTAL CARBON BOUND TO PARTICULATE MATTER IN THE AIR OF PRINTING OFFICE AND BEAUTY SALON 1154
Rogula-Kopiec Patrycja, Pastuszka Józef S., Rogula-Kozlowska Wioletta, Mucha Walter

THE USE OF LICHENS IN POST-SMELTING DUMPS RECLAMATION – PRELIMINARY RESULTS OF EXPERIMENTAL CULTIVATION OF SELECTED SPECIES ON SLAG SUBSTRATE 1162
Rola Kaja, Osyczka Piotr

ASSESSMENT OF THE AERMOD DISPERSION MODEL OVER COMPLEX TERRAIN WITH DIFFERENT TYPES OF METEOROLOGICAL DATA: TRACY POWER PLANT EXPERIMENT 1169
Rzeszutek Mateusz, Szulecka Adriana, Oleniacz Robert, Bogacki Marek

STUDY ON WATER EVAPORATION RATE FROM INDOOR SWIMMING POOLS 1178
Rzeznik Ilona

PROCESS SIMULATION AND COMPARISON OF BIOLOGICAL CONVERSION OF SYNGAS AND HYDROGEN IN BIOGAS PLANTS 1186
Salman Chaudhary Awais, Schwede Sebastian, Thorin Eva, Yan Jinyue

SEDIMENTS FROM STORMWATER DRAINAGE SYSTEM AS SORBENTS OF ORGANIC AND INORGANIC POLLUTANTS 1194
Salata Aleksandra, Dabek Lidia

THE INFLUENCE OF HYDRAULIC CONDITIONS ON COAGULATION PROCESS EFFECTIVENESS 1202
Sambor Aleksandra, Ferenc Zbigniew

GAS FIRED BOILERS: PERSPECTIVE FOR NEAR FUTURE FUEL COMPOSITION AND IMPACT ON BURNER DESIGN PROCESS 1209
Schiro Fabio, Stoppato Anna, Benato Alberto

EXPERIMENTAL STUDY OF OPTICAL AND ELECTRICAL PROPERTIES OF ZNO NANO COMPOSITES ELECTRODEPOSITED ON N-POROUS SILICON SUBSTRATE FOR PHOTOVOLTAIC APPLICATIONS 1217
Selmane Naceur, Cheknane Ali, Gabouze Nourddine, Maloufi Nabila, Aillerie Michel

OPTIMIZATION FOR ENERGY EFFICIENCY OF UNDERGROUND BUILDING ENVELOPE THERMAL PERFORMANCE IN DIFFERENT CLIMATE ZONES OF CHINA 1226
Shi Luyang, Liu Jing, Zhang Huibo

HYBRID CAPACITIVE DEIONIZATION WITH ANION-EXCHANGE MEMBRANES FOR LITHIUM EXTRACTION 1233
Siekierka Anna, Bryjak Marek

EVALUATION OF THE CORRELATION BETWEEN CONCENTRATION OF VOLATILE ORGANIC COMPOUNDS AND TEMPERATURE OF THE EXHAUST GASES IN MOTOR VEHICLES 1241
Skretowicz Maria, Wróbel Radoslaw, Andrych-Zalewska Monika

ANALYSIS OF APPLICATION OF ALTERNATIVE DRIVE SYSTEMS FOR INTERNATIONAL HEAVY-DUTY TRANSPORT ON WROCLAW-DRESDEN-PRAGUE ROUTES 1248
Skretowicz Maria, Sroka Zbigniew

ENVIRONMENTAL ASPECTS AND RENEWABLE ENERGY SOURCES IN THE PRODUCTION OF CONSTRUCTION AGGREGATE 1256
Skrzypczak Izabela, Kokoszka Wanda, Buda-Ożóg Lidia, Kogut Janusz, Slowik Marta

THE INFLUENCE OF GLYCERIN ON NITROGEN REMOVAL IN WASTEWATER TREATMENT WITH ACTIVATED SLUDGE 1264
Smyk Joanna, Ignatowicz Katarzyna

ADAPTIVE HEAT PUMP AND BATTERY STORAGE DEMAND SIDE ENERGY MANAGEMENT 1271
Sobieczky Florian, Lettner Christian, Natschläger Thomas, Traxler Patrick

BIOMASS UNIVERSAL DISTRICT HEATING SYSTEMS 1279
Soltero Victor Manuel, Rodríguez-Artacho Salvador, Velázquez Ramón, Chacartegui Ricardo

ENVIRONMENTAL ASPECTS OF THE GEOTHERMAL ENERGY UTILISATION IN POLAND 1288
Sowizdzal Anna, Tomaszewska Barbara, Drabik Anna

CHARACTERISTICS OF SELECTED ELEMENTS OF THE AIR QUALITY MANAGEMENT SYSTEM IN URBAN AREAS IN POLAND 1297
Sówka Izabela, Kobus Dominik, Chlebowska Stys Anna, Zathey Maciej

HYDRAULIC OPTIMIZATION AND MODELING OF HYDRO-CYCLONE-SYSTEMS FOR TREATMENT AND PURIFICATION OF ANY KIND OF WATERS 1305
Spangemacher Lars, Fröhlich Siegmund, Buse Hauke

CHANGES IN MICROBIOLOGICAL COMPOSITION OF SOILS AND SOIL CONTAMINATION WITH DRUG-RESISTANT BACTERIA CAUSED BY THE USE OF SEWAGE SLUDGE IN NATURE 1313
Stanczyk-Mazanek Ewa, Pasonl Lukasz, Kepa Urszula

EVALUATING THE FINANCIAL EFFICIENCY OF ENERGY AND WATER SAVING INSTALLATIONS IN PASSIVE HOUSE 1321
Stec Agnieszka, Mazur Aleksandra, Slys Daniel

COD FRACTIONS CHANGES IN THE SBR-TYPE REACTOR TREATING MUNICIPAL WASTEWATER WITH CONTROLLED PERCENTAGE OF DAIRY SEWAGE 1329
Struk-Sokolowska Joanna, Rodziewicz Joanna

LABORATORY TESTS OF OVERPRESSURE DIFFERENTIAL SYSTEMS FOR SMOKE PROTECTION OF LOBBIES 1336
Szalanski Pawel, Misinski Jacek

PRELIMINARY SANITARY ANALYSIS OF SUPPLY AND EXHAUST AIR OF VENTILATION UNITS WORKING AT SPECIAL ROOMS 1344
Szczesniak Sylwia, Trusz-Zdybek Agnieszka, Piekarska Katarzyna

DETECTION OF COMFORTABLE TEMPERATURE BASED ON THERMAL EVENTS DETECTION INDOORS 1352
Szczurek Andrzej, Maciejewska Monika, Uchronski Mariusz

MEASUREMENT OF ACETATES IN AIR USING DIFFERENTIAL ION MOBILITY SPECTROMETER 1360
Szczurek Andrzej, Maciejewska Monika, Zajiczek Zaneta, Maziejuk Miroslaw

PREDICTION OF WASTEWATER QUALITY INDICATORS AT THE INFLOW TO THE WASTEWATER TREATMENT PLANT USING DATA MINING METHODS 1368
Szelag Bartosz, Barbusinski Krzysztof, Studzinski Jan, Bartkiewicz Lidia

IMPLEMENTATION OF FERRIC HYDROXIDE-BASED MEDIA FOR REMOVAL OF TOXIC METALLOIDS 1376
Szlachta Malgorzata, Wójtowicz Patryk

INTEGRATED ASSESSMENT OF SOIL QUALITY AFTER APPLICATION OF THE BIOGAS FERMENTATION RESIDUES – A LABORATORY EXPERIMENT 1384
Telesinski Arkadiusz, Cybulska Krystyna, Platkowski Maciej, Strek Michal, Jarnuszewski Grzegorz, Wronska Ilona, Mularewicz Piotr, Kajdan Tomasz, Biczak Robert, Kolosowski Pawel

A NUMERICAL INVESTIGATION OF LAMINAR FORCED CONVECTION IN A SOLAR COLLECTOR WITH NON-CIRCULAR DUCT 1392
Teleszewski Tomasz Janusz

COMPARISON OF THE AVAILABILITY OF GROUNDWATER INFORMATION SOURCES IN POLAND WITH OTHER EUROPEAN COUNTRIES. KNOWLEDGE INVENTORY FOR HYDROGEOLOGY RESEARCH – PROJECT KINDRA 1400
Tomaszewska Barbara, Dendys Marta, Tyszer Magdalena

EXPERIMENTAL PROCESS INVESTIGATION OF A LATENT HEAT ENERGY STORAGE SYSTEM WITH A STAGGERED HEAT EXCHANGER WITH DIFFERENT PHASE CHANGE MATERIALS FOR SOLAR THERMAL ENERGY STORAGE APPLICATIONS 1408
Tsolakoglou Nikolas P., Koukou Maria K., Vrachopoulos Michalis Gr., Tachos Nikolaos, Lymberis Kostas, Stathopoulos Vassilis

A METHOD OF DETERMINING THE THERMAL POWER DEMAND OF BUILDINGS CONNECTED TO THE DISTRICT HEATING SYSTEM WITH USAGE OF HEAT ACCUMULATION 1417
Turski Michal, Sekret Robert

THE EFFICIENCY OF MACROPOROUS POLYSTYRENE ION-EXCHANGE RESINS IN NATURAL ORGANIC MATTER REMOVAL FROM SURFACE WATER 1423
Urbanowska Agnieszka, Kabsch-Korbutowicz Malgorzata

INTAKE POWER MEASUREMENT AS A CRITERION FOR CONTROL OF HVAC SYSTEMS 1431
Walaszczyk Juliusz Jerzy, Przydrózna Aleksandra, Przydrózny Edward

THE EFFECT OF INFLATION RATE ON THE COST OF MEDICAL WASTE MANAGEMENT SYSTEM 1441
Walery Maria Jolanta

SELECTED HEAVY METALS SPECIATION IN CHEMICALLY STABILISED SEWAGE SLUDGE 1449
Wisniowska Ewa, Wlodarczyk-Makula Marla

ABSORPTION COOLING SOURCES ATMOSPHERIC EMISSIONS DECREASE BY IMPLEMENTATION OF SIMPLE ALGORITHM FOR LIMITING TEMPERATURE OF COOLING WATER 1457
Wojdyga Krzysztof, Malicki Marcin

INVESTIGATIONS OF HOT WATER TEMPERATURE CHANGES AT THE PIPE OUTFLOW 1463
Wojtkowiak Janusz, Oleskowicz-Popiel Czeslaw

THE APPLICATION OF IMPEDANCE MEASUREMENT TO ASSESS BIOFILM DEVELOPMENT ON TECHNICAL MATERIALS USED FOR WATER SUPPLY SYSTEM CONSTRUCTION 1470
Wolf Mirela, Traczewska Teodora, Grzebyk Tomasz

THE EFFECTIVENESS OF REMOVING PRECURSORS OF CHLORINATED ORGANIC SUBSTANCES IN PILOT WATER TREATMENT PLANT 1478
Wolska Malgorzata, Szerzyna Slawomir, Machi Justyna, Molczan Marek, Adamski Wojciech, Wisniewski Jacek

RHEOLOGICAL PROPERTIES OF DISINTEGRATED SEWAGE SLUDGE 1486
Wolski Pawel

HOW EFFECTIVE IS AERATION WITH VORTEX FLOW REGULATORS? PILOT SCALE EXPERIMENTS 1493
Wójtowicz Patryk, Szlachta Malgorzata

PROSPECTS FOR THE USE OF SMR AND IGCC TECHNOLOGIES FOR POWER GENERATION IN POLAND 1500
Wyrwa Artur, Suwala Wojciech

AN APPLICATION OF DATA MINING IN DISTRICT HEATING SUBSTATIONS FOR IMPROVING ENERGY PERFORMANCE 1509
Xue Puning, Zhou Zhigang, Chen Xin, Liu Jing

MECHANICALLY ENHANCED ELECTROSPUN NANOFIBERS FOR WASTEWATER TREATMENT 1517
Yalcinkaya Fatma

INFLUENCE OF CROSS-SECTIONAL RATIO OF DOWN COMER TO RISER ON THE EFFICIENCY OF LIQUID CIRCULATION IN LOOP AIR LIFT BUBBLE COLUMN 1524
Yamamoto Tatsumi, Kawasaki Hiroyuki, Mori Hidetoshi

MICROWAVE PYROLYSIS USING SELF-GENERATED PYROLYSIS GAS AS ACTIVATING AGENT: AN INNOVATIVE SINGLE-STEP APPROACH TO CONVERT WASTE PALM SHELL INTO ACTIVATED CARBON 1530
Yek Peter Nai Yuh, Liew Rock Keey, Osman Mohammad Shahril, Wong Chee Chung, Lam Su Shiung

ENERGY PRODUCTION FOR ENVIRONMENTAL ISSUES IN TURKEY 1539
Yuksel Ibrahim, Arman Hasan, Demirel Ibrahim Halil

THE ROLE OF ENERGY SYSTEMS ON HYDROPOWER IN TURKEY 1547
Yuksel Ibrahim, Arman Hasan, Demirel Ibrahim Halil

ZERO ENERGY BUILDINGS IN THE LOGISTICS WAREHOUSE SYSTEMS 1555
Zajac Pawel, Kwasniowski Stanislaw

ENERGY SUSTAINABLE CITIES. FROM ECO VILLAGES, ECO DISTRICTS TOWARDS ZERO CARBON CITIES 1563
Zareba Anna, Krzeminska Alicja, Lach Janusz

ENERGY POTENTIAL OF THE MODIFIED EXCESS SLUDGE 1572
Zawieja Iwona

EVALUATION OF EXTERNAL AND INTERNAL IRRADIATION ON URANIUM MINING ENTERPRISE STAFF BY TOOTH ENAMEL EPR SPECTROSCOPY 1580
Zhumadilov Kassym, Ivannikov Alexander, Khailov Artem, Orlenko Sergei, Skvortsov Valeriy, Stepanenko Valeriy, Kuterbekov Kairat, Toyoda Shin, Kazymbet Polat, Hoshi Masaharu

DOSIMETRY STUDY OF EAST KAZAKHSTAN RESIDENTS BY TOOTH ENAMEL EPR SPECTROSCOPY 1587
Zhumadilov Kassym, Ivannikov Alexander, Skvortsov Valeriy, Stepanenko Valeriy, Rakhypbekov Tolebay, Hoshi Masaharu

ENHANCEMENT OF SEDIMENTATION AND COAGULATION WITH STATIC MAGNETIC FIELD 1594
Zielinski Marcin, Debowski Marcin, Hajduk Anna, Rusanowska Paulina

OPTIMIZATION OF ENERGY COST IN WATER SUPPLY SYSTEM 1602
Zimoch Izabela, Bartkiewicz Ewelina

A REVIEW OF PHOTOVOLTAIC CELLS COOLING TECHNIQUES 1610
Zubeer Swar A., Mohammed H.A., Ilkan Mustafa

CHARACTERISTICS OF LARGE THERMAL ENERGY STORAGE SYSTEMS IN POLAND 1621
Zwierzchowski Ryszard

IMPACT OF EXTERNAL CONDITIONS ON ENERGY CONSUMPTION IN INDUSTRIAL HALLS 1631
Zabnienska-Góra Alina

INTENSIFICATION OF ANAEROBIC DIGESTION EFFICIENCY WITH USE OF MECHANICAL EXCESS SLUDGE DISINTEGRATION IN THE CONTEXT OF INCREASED ENERGY PRODUCTION IN WASTEWATER TREATMENT PLANTS 1639
Zubrowska-Sudol Monika, Podedworna Jolanta, Bisak Agnieszka, Sytek-Szmeichel Katarzyna, Krawczyk Piotr, Garlicka Agnieszka

ENERGY EFFICIENCY OF A SOLAR DOMESTIC HOT WATER SYSTEM 1648
Zukowski Miroslaw

WATER PIPE NETWORK AS A HEAT SOURCE FOR HEAT PUMP INTEGRATED INTO A DISTRICT HEATING 1656
Jadwiszczak Piotr, Niemierka Elzbieta

Author Index

DOI: 10.1051/e3sconf/20172200105

Application of multi-enzymatic hydrolysis for improving the efficiency of the biogas production in solid waste fermentation process in Ostróda WWTP

Kamil Lipiński[1,*], and *Katarzyna* Umiejewska[1]

[1]Warsaw University of Technology, Faculty of Building Services, Hydro and Environmental Engineering, ul. Nowowiejska 20, 00-653 Warsaw, Poland

Abstract. Biomass fermentation is one of the important sources of renewable energy in EU. Application of multi-enzymatic hydrolysis process enables a significant increase in efficiency of biogas production. The main goal of the paper is to present the results of the pilot scale research performed in WWTP in Ostróda. The fixed combination of three enzymes was continiously introduced: amylase, lipase and protease. Research aimed at verifying the impact of enzyme dose on sludge digestion process and on the amount of biogas produced. Statistical analysis of the research results allows to determine the influence of dosing the enzymes in mesophilic digestion on the biogas production.

1 Introduction

Sewage sludge is an inherent product of the operation on WWTP. The most common method of stabilization is anaerobic digestion. The efficient production and use of biogas in a WWTP is a source of heat and power. Multi-enzymatic hydrolysis is a biological technology of primary disintegration of sewage sludge. It accelerates and improves the efficiency of the natural process of hydrolysis [1] occurring in the first stage of sludge digestion, uses a number of naturally produced [2] or additionally dosed forms of [3] enzymes: amylase, protease [4] and lipase [5], which reduce the activation energy of the hydrolysis of selected organic compounds. As a result, the percentage of biodegradable organic matter increases, which provokes the increase in the efficiency of biogas production.

The aim of the study was to determine whether the method of multi-enzymatic hydrolysis improves the efficiency of the sludge digestion process from the treatment plant in Ostróda and if it increases the biogas production during the mesophilic digestion.

* Corresponding author: kamil.lipinski.pl@gmail.com

2 Materials and methods

The research was conducted in a pilot scale in the period 05.10.2011–31.07.2012 in the Ostróda WWTP belonging to PWiK Ostroda sp. z. o.o.

There were three series of tests:

- Series I. Start-up and stabilization of the mesophilic digestion process in the period of 05.10.2011–08.12.2011
- Series II. Sludge mesophilic digestion with the dosage of the enzymes with 20% of dry solids concentration, in the period of 09.12.2011–26.03.2012
- Series III. Sludge mesophilic digestion with the dosage of the enzymes with 10% of dry solids concentration in the period of 27.03.2012–31.07.2012

The model consisted of a feed tank with a capacity of 1130 litres for the continuously supplied primary and excess sludge, mixed in equal proportions. The sludge from the feed tank was directed to a digester with a capacity of 1410 litres. In Series I the temperature in the digester was 38°C, HRT = 35 days. In the Series II and III, the enzymes were dosed to the feed tank in an amount of 20 and 10% dry mass. In the Series III, the temperature was lowered to 35°C and the sludge HRT was reduced to 29 days. Digested sludge was directed to the final tank.

The scope of the research included the measurement of the parameters in raw and digested sludge and the volume of produced biogas. The daily volume of biogas was measured and calculated to the volume of the biogas produced from dry organic mass contents ($Q_{ODM\ INF}$). For each day of the test, the digestion index was calculated. This index demonstrates the effectiveness of decomposition of organic compounds in the sludge.
Samples were collected at the feed tank (mixed sludge) and the final tank (digested sludge). The analysis of sludge parameters were made according to the following standards:

- Determination of dry matter PN-EN 12880:2004
- Determination of the loss of ignition of dry mass % – PN-EN 12879:2004
- Determination of COD - PN-74/C-04578.03

3 Results

3.1 Series I – start-up and stabilization of mesophilic digestion process

After the start-up, the reactor worked in batch process and then in continuous one. The temperature in reactor was 38°C, and the HRT was 35 days. Table 1 shows the sludge quality before and after digestion process during series I and the calculated digestion index.

At the beginning of the research, the dry matter content (DM) in the raw sludge was in the range of 5–6%. The content of the organic dry matter (ODM) remained at a high level oscillating around 78-80%. The COD value was usually over 60 000 mg/L. The dry matter content of digested sludge, which first was low (1.7%), experienced a period of rapid growth to a 4.3%, then declined and finally began to stabilise at around 3.1%. The content of dry organic matter decreased from approximately 73% (the beginning of the study) to about 63%. The actual value of the digestion index, at the start of work is very low 2–6.5% and next grew and ranged between 42.8–56.7. The COD value decreased from 56 000 mg/L (the beginning of the study) and was about 32 000 mg/L.

Table 1. shows the sludge quality before and after digestion process during series I and the calculated digestion index.

Date	Sludge						
	From feeding tank			After digestion			
	DM	ODM	COD	DM	ODM	COD	Digestion index
	%	%DM	mg/L	%	%DM	mg/L	%
2011-10-17	6.1	77.0	60 000	1.7	75.8	-	6.5
2011-10-25	5.6	73.7	32 100	4.3	73.3	56000	2.0
2011-11-15	5.6	76.3	60 000	3.3	63.6	31400	45.7
2011-11-22	5.0	78.7	60 000	3.1	63.9	32800	42.8
2011-11-24	5.1	78.2	59 600	3.1	64.0	31900	44.6
2011-11-29	4.9	79.7	60 000	3.1	63.7	31400	49.7
2011-12-01	5.2	80.0	60 000	3.1	63.4	32600	53.6
2011-12-06	5.2	80.7	60 000	3.1	62.4	32400	56.7

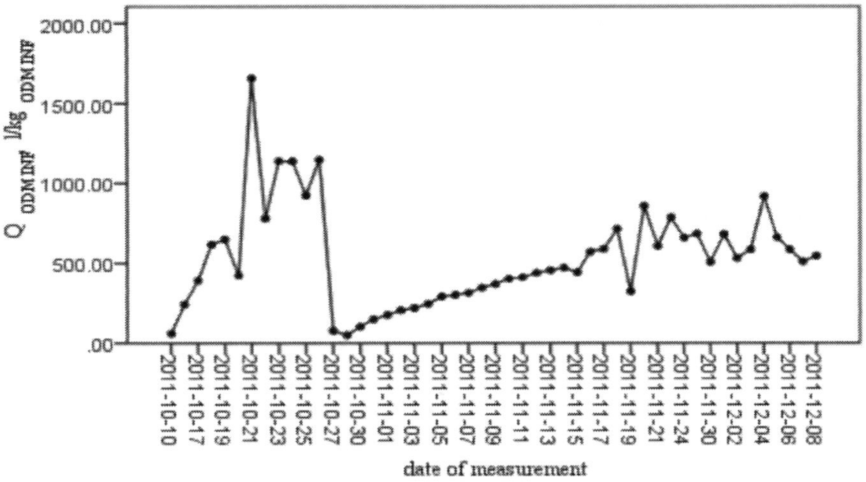

Fig. 1. The volume of biogas produced in the Series I.

In the series I, a far reaching variation of biogas production might be observed (Fig. 1). The stabilisation of the digestion process was preceded by some significant changes in the biogas production expressed by $Q_{ODM\ INF}$ 400–900 l/kg $_{ODM\ INF}$.

3.2 Series II – enzyme dosage of 20% contents dry solid

Before the start of the series II, a stabilization work without the addition of enzymes was achieved. Thus, decisions have been taken about dosage of enzymes. Enzymes were dosed directly into the feed tank in an amount of 20% of dry matter. Table 2 shows the sludge quality before and after the digestion process during series II and the calculated digestion index.

The dry matter content in the raw sludge in series II showed considerable variation, the trend of decreasing the level of 5–6% to 3–4%. The content of organic dry matter remained at a high level oscillating between 78% and 85%, but in this case, at the end of the

 DOI: 10.1051/e3sconf/20172200105

experiment, the occurrence of a significant downward trend could be observed. Sludge COD value was located outside the range of the measurement (60000 mg/L) not as often as in the series I. In February, in a period of intense cold, the sludge COD dropped significantly to the level of 30800 mg/L. Later, it returned to the level of 40000–50000 mg/L.

The decrease in dry matter content in the raw sludge caused the decline in dry matter content in sludge after the mesophilic digestion process to the level of 2.4–3.2%. The content of dry organic matter in the digested sludge was 62–63% before starting the dosage of enzymes and increased to 64–68% as a result of dosing enzymes. The calculated actual digestion index has a high variability within the range of 27.3–65.3% (average 52.1%).

COD value of digested sludge remained relatively stable compared to the changing raw sludge that was supplied. However, it decreased from above 28600 mg /L from the period before the frost to the level of 24100–27400 mg/L.

Table 2. Parameters in the raw and digested sludge obtained in the series II.

Date	Sludge						
	From feeding tank			After digestion			
	DM	ODM	COD	DM	ODM	COD	Digestion index
	%	% DM	mg/L	%	%DM	mg/L	%
2011-12-13	5.0	82.8	60 000	3.0	63.9	32300	58.9
2011-12-20	5.2	78.5	60 000	3.2	62.4	31300	61.7
2011-12-27	4.9	82.4	60 000	3.0	62.8	31400	64.5
2012-01-03	4.7	83.4	58 500	3.1	63.2	31600	65.8
Sludge digestion with the dosage of the enzymes with 20% of dry solids concentration							
2012-01-10	4.6	82.1	54 400	3.0	64.7	37000	37.7
2012-01-17	4.9	82.0	58 400	2.9	64.6	26000	37.7
2012-01-24	5.4	83.9	37 800	2.9	66.4	10600	38.5
31.01.2012-13.02.2012 The period of very low temperatures							
2012-01-31	5.6	84.9	60 000	2.9	66.5	28900	41.3
2012-02-07	4.0	84.3	48 300	2.9	67.1	27800	37.8
2012-02-14	5.1	84.9	60 000	2.8	68.5	26700	35.7
2012-02-21	6.0	84.4	60 000	2.8	68.0	27000	35.5
2012-02-28	4.5	82.2	51 100	2.6	67.8	29300	28.8
2012-03-06	4.0	82.4	52 100	2.8	65.1	26400	44.6
2012-03-08	3.0	76.5	30 800	2.4	67.6	24100	8.7
2012-03-13	2.9	79.9	41 200	2.7	63.6	25700	49.9
2012-03-22	3.7	77.9	46 600	2.6	66.1	27400	51.5

In the Series II there was an increase in the average daily production of biogas during the dosing of enzymes (Fig. 2). The significance of this observation becomes apparent when comparing the value of $Q_{ODM\ INF}$ before and after enzyme dosing. The efficiency of biogas production increased from an average of 600 l/kg of dry organic matter to the level of 900 l/kg of dry organic matter. In addition, it was observed that although initially the decline in the content of the raw sludge dry mass was not included in the enzymes dosage which therefore exceeded 20%, significant differences in the biogas production efficiency were not observed. The dose of enzymes, which was calculated based on previous laboratory tests, could therefore be too high. That is why the dose of enzymes with 20% to 10% dry solid concentration was changed, in order to optimise the dose in the future.

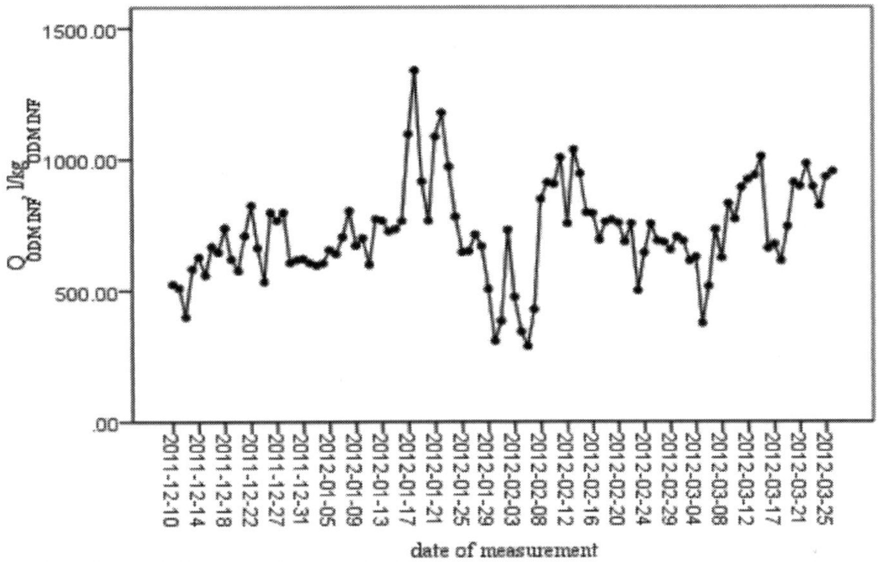

Fig. 2. The volume of biogas produced in the Series II.

3.3 Series III – enzyme dosage of 10% contents of dry solid

The purpose of Series III is to check if a lower dose of enzymes (content 10% dry solid) will have a comparable effect on the efficiency of biogas production. In this series, the sludge HRT was reduced to 29 days, and the temperature was lowered to 35°C. Table 3 shows the sludge quality before and after the digestion process during Series III and the calculated unit of fermentation.

Dry matter content in the raw sludge in the Series III showed a large variation in the range of 2.1–5.9%. In this period, the organic dry matter content in the raw sludge was 80%. A noticeable decrease to 65.3% was recorded in the summer. In July, the dry matter content significantly increased – more than 5% while the content of organic dry matter decreased to less than 70%. Sludge COD value dropped to the level of 27700 mg/L, then rose to the level of the maximum measurable excess of 60000 mg/L.

Digestion conditions during the Series III remained stable, despite the changes in the quality of raw sludge. The dry matter content was in the range of 2.2–3.6%. The dry organic matter was lower than in the Series II – in the range of 54.7–63.3%. The calculated actual digestion index exceeds 50%.The value of COD of the sludge was in the range 19000–30500 mg/L.

DOI: 10.1051/e3sconf/20172200105

Table 3. Parameters in the raw and digested sludge obtained in the Series III.

Date	Sludge						
	From feeding tank			After digestion			
	DM	ODM	COD	DM	ODM	COD	Digestion index
	%	% DM	mg/l	%	% DM	mg/l	%
2012-03-29	4.4	78.8	54200	2.6	62.0	30 500	58.4
2012-04-10	3.3	79.2	41400	2.5	63.3	25 000	52.0
2012-04-17	2.1	72.4	27700	2.4	61.7	25 300	53.3
Sludge digestion with the dosage of the enzymes with 10% of dry solids concentration							
2012-04-24	4.2	80.8	60000	2.4	59.6	29 100	53.4
2012-05-08	2.8	78.9	42300	2.3	62.0	21 900	50.3
2012-05-15	2.7	76.4	40000	2.5	59.2	19 000	55.5
2012-05-22	2.7	77.3	36200	2.5	57.9	21 700	45.7
2012-05-29	3.0	78.3	50100	2.3	59.9	22 600	58.6
2012-06-05	3.8	77.0	53800	2.4	58.4	29 000	58.1
Batch work							
2012-06-12	3.0	77.8	38300	2.2	62.2	23 400	53.0
2012-06-22	4.2	73.8	61200	2.6	54.7	22 800	57.1
2012-06-28	5.9	73.6	60000	2.8	53.9	24 100	58.1
2012-07-05	3.9	69.6	41400	3.0	56.6	27 000	43.0
2012-07-13	5.0	70.9	60000	3.1	55.8	26 000	48.2
2012-07-19	5.4	66.5	60000	3.2	54.7	28 600	39.2
2012-07-26	5.2	65.3	60000	3.6	62.2	28 100	12.6

In the Series III there was a decrease in the average daily production of biogas because the dry matter content in the raw sludge decreased. Organic compounds are the material for methanogenic bacteria. The decrease in the enzyme dose could cause a $Q_{ODM\ INF}$ decrease, which could be associated with the lower efficiency of multi- enzymatic hydrolysis process. Finally, reducing HRT from 35 to 29 days could reduce the efficiency of biogas production. The reasons for the decline in production are therefore complex and require further analysis.

DOI: 10.1051/e3sconf/20172200105

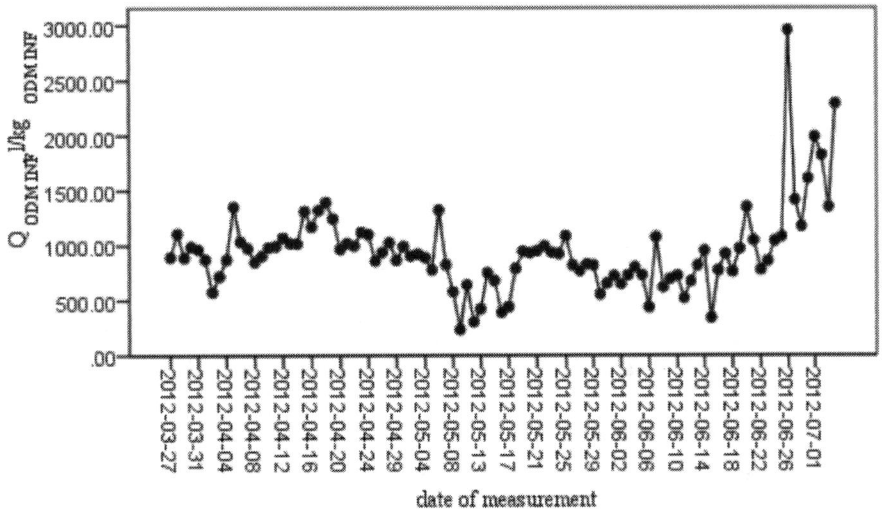

Fig. 3. The volume of biogas produced in the Series III.

During the series III we have managed to achieve an intriguing but inconclusive research. Double dose reduction enzymes with 20% contents of dry matter to 10%, influenced the course of the process, but in a decidedly ambiguous way. The amount of biogas produced $Q_{ODM\ INF}$ in Series II and III was on the same level.

4 Discussion

Verification of hypotheses

To systematize conducted analysis, focused on the influence of multi-enzymatic hydrolysis on the efficiency of biogas production, two following research hypotheses were elaborated.

H I. In the process of mesophilic sludge digestion, enzyme dosing increases the efficiency of biogas production.

In the first stage of hypotheses verification, mean values and standard deviation of $Q_{ODM\ INF}$ were compared. In the series II and III (dosing of enzymes) higher mean value $Q_{ODM\ INF}$. The difference of $Q_{ODM\ INF}$ value in series was initially estimated at 219.7 l/kg $_{ODM\ INF}$ (34.5% increase).

Table 4. Biogas production $Q_{ODM\ INF}$ and enzymes dosing.

Dosage of the enzymes	Mean $Q_{ODM\ INF}$	Standard Deviation
No (Series I)	636.0 l/kg $_{ODM\ INF}$	122.6 l/kg $_{ODM\ INF}$
Yes (Series II and III)	855.7 l/kg $_{ODM\ INF}$	331.2 l/kg $_{ODM\ INF}$

Second stage of the analysis in the process of verification of potential relation between variables was generating a model of linear regression and testing the stastical significance of generated model. Statically important model (Chi^2 = 0.000, R^2 = 0.086, ANOVA significance = 0.000) was obtained. It allowed to estimate following linear regression equation, assessing statically important, positive influence of enzyme dosing on biogas production in mesophylic digestion.

$$Q_{ODM\ INF} = 636.0 + 219.7 \cdot e [l/kg_{ODM\ INF}] \quad (1)$$

e: dosing of enzymes (0 = no; 1 = yes) (dummy variable)

H II. In the process of mesophilic sludge digestion, linear effect of enzyme dosing on biogas production is observed.

To assess the size of optimal enzymes dose, mean values and standard deviations of $Q_{ODM\ INF}$ for different sizes of doses effects were compared. For analysed data, the highest level of $Q_{ODM\ INF}$ can be observed for size of dose 10% of dry solids concentration in fermentation chamber.

Table 5. Biogas production $Q_{ODM\ INF}$ and size of enzymes dose.

Size of enzyme dosage	Mean $Q_{ODM\ INF}$	Standard Deviation
no dosage	636.0 $l/kg_{ODM\ INF}$	122.6 $l/kg_{ODM\ INF}$
10% of dry solids concentration	899.1 $l/kg_{ODM\ INF}$	411.7 $l/kg_{ODM\ INF}$
20% of dry solids concentration	827.1 $l/kg_{ODM\ INF}$	263.2 $l/kg_{ODM\ INF}$

To verify observation, model of linear regression was generated. Preliminary analysis suggested considerably low explanatory potential of the linear model of enzyme dose influence ($Chi^2 = 0.001$; $R^2 = 0.037$; ANOVA significance = 0.001). To analyse the potentially non-linear relation between size of enzymes dose and biogas production, the size of dose data were recoded into two 0–1 "dummy variables" for each size of enzyme dosage: e_{20} (20% of dry solids concentration) and e_{10} (10% of dry solids concentration). The second model that was generated was statistically significant ($Chi^2 = 0.000$, $R^2 = 0.096$, ANOVA significance = 0.000). It allowed to estimate following linear regression equation, assessing statically important, positive influence of size of dose of enzymes on biogas production in mesophylic digestion of sewage sludge.

$$Q_{ODM\ INF} = 636{,}0 + 263{,}1 \cdot e_{10} + 191{,}1 \cdot e_{20} [l/kg_{ODM\ INF}] \quad (2)$$

e_{10}: Enzyme dose 10% of dry solids concentration (0=no) (1=yes) (dummy variable)
e_{20}: Enzyme dose 20% of dry solids concentration (0=no) (1=yes) (dummy variable)

The production of biogas was proven more effective for 10% dry solids enzyme dose. The relation between variables was presented with three equations, representing separate influence of 10% ds and 20% DM enzyme dose. The positive influence of optimal dose of enzymes, 10% of dry solids, on biogas production was estimated on 263.1 $l/kg_{ODM\ INF}$ (41% increase). Last equation, $Q_{ODM\ INF\_e10}$, represents the negative influence of enzymes over the optimal 10% dry solids dose.

$$Q_{ODM\ INF} = 636.0 + 263.1 \cdot e_{10} [l/kg_{ODM\ INF}] \quad (3)$$

$$Q_{ODM\ INF} = 636.0 + 191.1 \cdot e_{20} [l/kg_{ODM\ INF}] \quad (4)$$

$$Q_{ODM\ INF\ E10} = 899.1 - 191.1 \cdot e_{20} [l/kg_{ODM\ INF}] \quad (5)$$

5 Conclusion

Studies show that there is a positive relationship between enzyme dosing and the increased productivity of biogas production in mesophilic fermentation. The results allow us to estimate that the enzyme dosage allows for an increase in the production of biogas from kilogram of organic dry mass of 48%. The analyses indicated that in the case of mesophile fermentation, statistically there is no significant relationship between the increase of enzyme dose and the increase of biogas production. The data obtained suggests that the increase in the dose of more than 10% of dry mass may even cause a decrease in biogas production.

Acknowledgment for dr. R. Wasiak for preparing modern application of multi-enzymatic hydrolysis technology and SEEN Technology for sharing the data.

References

1. W. Parawira Crit. Rev. Biotechnol. **32**, 2, 172–186 (2011)
2. S. Yu, G. Zhang, J. Li, Z. Zhao, X. Kang, Bioresour. Technol. **146**, 758–761 (2013)
3. S. Kavitha, S. Adish Kumar, K.N. Yogalakshmi, S.Kaliappan, J.Rajesh Banu, Bioresour. Technol. **150**, 210–219 (2013)
4. Q. Yang, K. Luo, X.M. Li, D. B Wang, W. Zheng, G.M Zeng, J.J Liu, Bioresour. Technol. **101**, 2924–2930 (2010)
5. A. Donoso-Bravo , M. Fdz-Polanco, Process Biochem. **48**, 936–940 (2013)

DOI: 10.1051/e3sconf/20172200106

Synthesis of mesoporous cerium compound for CO_2 capture

Guiqing Liu [1,*], *Kou* Tatsuda[2], *Yoshiharu* Yoneyama[1], and *Noritatsu* Tsubaki[1]

[1]University of Toyama, Faculty of Engineering, Gofuku 3190, Toyama-shi, Toyama 930-8555, Japan
[2]JGC Plant Innovation Co., Ltd, 1-13-1, Saido, Kounan-ku, Yokohama, Kanagawa Prefecture

Abstract. A mesoporous adsorbent was simply synthesized by adding alkaline substances to cerium(III) nitric hydrate. The surface characteristics of the synthesized cerium compound were determined with BET, XRD and TEM analysis. It was found that although the specific surface areas of the synthesized cerium compounds were among about 120–200m^2 per gram (BET area) which were smaller than the common used zeolite 13X (BET area 743 m^2/g) and activated carbon (BET area 1079 m^2/g), but the cerium compounds had excellent performances for CO_2 adsorption as well as the CO_2 desorption.

1 Introduction

Many investigations have shown that carbon dioxide (CO_2) emissions are responsible for climate change. So many advanced countries have promoted the reduction of CO_2 emitted from large stationary sources including coal-based power plants, steel industry plants, biogas plants, and chemical plants. The development of both economical and environmental benign ways to separate and storage the CO_2 has become a global challenge. Two major approaches have been implemented as CO_2 separation technology: carbon dioxide capture/sequestration (CCS) and sustainable utilization of captured CO_2 as a raw material. The latter will play an important role within the next decade because CO_2 is a non-toxic, abundant, easily available, and renewable carbon resource that can be converted into various useful chemical products. A state-of-the-art process for capturing CO_2 from largely exhausted flue gases has been developed by employing amine-based, regenerative chemical absorption using aqueous amines. This process has a high efficiency for CO_2 capture, but it is energy intensive, complex and requires a high-energy input for regenerating and handling the corrosive solvents that result from the oxidation of amine-based solutions.

Compared with the conventional amine-based absorption, a process for capturing CO_2 using novel solid adsorbents capable of energy-efficient reversibility is far more advantageous. In this regard, various porous materials with fine structural and chemical properties have been widely investigated including zeolites, activated carbon, amine-modified silica materials, and metal organic frameworks. Those materials have been

* Corresponding author: liugq@eng.u-toyama.ac.jp

practical candidates that allow us to perform pressure swing adsorption (PSA) and temperature swing adsorption (TSA) processes for CO_2 capture from a variety of large emission sources.

Many studies have suggested that synthetic zeolites and amine-modified mesoporous materials could be applicable for CO_2 capturing in the post-combustion flue gas with low concentration of CO_2 and low pressure region. Chaikittisilp and Jones have reported that amine-modified mesoporous materials were effective for the CO_2 capturing from dilute CO_2 containing flue gas (< 10%) under ambient condition [1]. Yu and Cheng succeeded in the fabrication of amine-functional porous silica microspheres, which showed good CO_2 adsorption-desorption performance under high temperature flue gas condition (in the range of 348–383 K) with excellent cyclic stability [2]. These novel porous materials may useful under some conditions, though, the conditions of flue gases (CO_2 concentration, pressure, gas components, etc.) may sensitively limit the range of applications, thereby adsorbents must be selected appropriately according to the target emission sources. For instance, the high pressure gasification process may exhaust high concentration of pre-combustion CO_2 gas with an elevated pressure. Capturing the CO_2 from high pressure flue gases would profit in terms of energy-efficiency in the PSA process, thus the adsorbents should be suitable to capture sufficient amount of CO_2 under high pressure region. Current PSA adsorbents such as zeolites and amine-modified mesoporous materials may not be most suitable materials for capturing high concentration of CO_2 under high pressure condition. Activated carbon still shows low durability against water vapour whereas mesoporous carbon materials specially designed for CO_2 capture is also high cost.

In this study, we focused on the use of metal oxides with larger true densities than other porous materials, in particular, we considered cerium compounds such as cerium oxide and cerium hydroxide as an alternative, efficient adsorbent for the CO_2 capture process. Cerium oxide, known as ceria, has been a subject of thorough investigations, and has attracted a lot of attention mainly in the field of catalysts and catalytic supports [3]. Many attempts have been made to improve the catalytic activity of ceria by enhancing its thermal stability and surface area with ordered porous structures. However, the practical application of ceria to CO_2 capture has not been fully investigated even though conventional ceria has the potential to adsorb and desorb the CO_2 under ambient conditions owing to its strong ewis-base sites [4].

In order to develop a high efficient cerium compounds adsorbents for capturing CO_2, the primary interest is to enhance the surface area of cerium compounds. Several approaches have been proposed for increasing the surface area. These approaches include the soft-templating method using poly-block copolymers or organic structure-directing agents, the hard-templating method, which usually replicates the ordered mesoporous structures from SBA-15, KIT-6 or MCM-48, the self-assembly of surface modified ceria nanoparticles, and supercritical drying under CO_2 conditions. The above methods provide ways to enhance the surface area of ceria to around 200 m^2g^{-1}. However, these processes are complex, high cost and high energy consumption. Therefore, a simple synthesis route is desirable for the production of high surface area ceria from both industrial and environmental standpoints.

In this study, we tried a simple preparation of mesoporous cerium compounds with a high surface area of up to 206 m^2g^{-1}. The mesoporous ceria was prepared based on the so-gel method, where the product was simply precipitated within a short reaction period at room temperature under optimized synthesis conditions. Compared with the commercially available non-porous ceria, zeolite 13X, and activated carbon, the prepared mesoporous cerium compounds showed the results with high performance of CO_2 adsorption-desorption.

DOI: 10.1051/e3sconf/20172200106

2 Experiments

2.1 Simple synthesis of mesoporous cerium compounds without template

The following raw materials were used as provided: using sodium hydroxide or ammonia as alkali sources and cerium (III) nitrate 6-hyrate ($Ce(NO_3)_3 \cdot 6H_2O$, Kanto Chemical Co., Ltd) as the cerium source. To compare the CO_2 adsorption capacity of the cerium products, commercially available non-porous ceria, typical activated carbon and zeolite 13X were used as reference materials.

The mesoporous cerium compound samples were synthesized by the following procedure: cerium (III) nitrate 6-hyrate ($Ce(NO_3)_3 \cdot 6H_2O$ was added to ethanol (99.5% C_2H_5OH, Kanto Chemical Co., Ltd) and stirred for 2h to obtain a colloid (A). NaOH was mixed with distilled water with various OH/ H_2O ratios. Then the NaOH(B) solution was added to the A and mixed thoroughly using a magnetic stirrer at room temperature until the suspension became faint yellow. The resulting solution was then separated with a centrifugation. The sediment product was washed with distilled water until it approached a neutral pH and thereafter dried at room temperature in a vacuum desiccator over one night.

2.2 Characterization

The wide-angle powder X-ray diffraction (XRD) patterns of the solid products were collected by a powder X-ray diffractometer (Ultima IV, Rigaku Corp.). The crystalline morphology and crystal size of the products were observed using a transmission electron microscope (TEM, H-7650, Hitachi High-Technologies Corp.) with an accelerating voltage of 100kV. The pore characteristics of the samples were evaluated by performing N_2 adsorption-desorption measurement using an automatic gas adsorption apparatus (3 Flex, SHIMADZU Corp.). To evaluate the surface areas, the Brunauer-Emmett-Teller (BET) method was applied. The pore size distribution curves were calculated with the Dollimore-Heal (DH) method.

2.3 CO_2 capturing experiment

A simplified schematic diagram of the experimental setup is shown as Figure 1.The experiments were carried out under a pressure of 0.1, 0.5, 1.0, and 1.5 MPa respectively, at room temperature. The sample amount was fixed to 3.0g. Before adsorption operates the system was purged with pure N_2 gas for 1h. The concentrations of CO_2 in the input and output were determined by GC /TCD (GC-8A, SHIMADZU Corp.) The capturing capacities for CO_2 were calculated by the following equation.

$$Q\ (\text{mmol/g}) = \frac{\frac{V \times 44}{22.4}\left[\int_0^t (C_0 - C_t)dt\right]}{m} \tag{1}$$

where Q is the CO_2 capturing capacity, mmol/g; V is the flow rate of the gas, L/min; m is the weight of the adsorbent sample, g; t is the adsorption time, min.; C_0 and C_t is the CO_2 concentration in the input and output, respectively.

DOI: 10.1051/e3sconf/20172200106

2.4 Adsorbent regeneration performance evaluation

The best synthesized sample (Sample No. 2) and zeolite 13X were selected to carry out circulating adsorption-desorption experiments to compare their regenerating performance. For each cycle, at first, the adsorption experiment was carried out at 298 K, under 1.0 MPa. When the adsorption reached the saturating point, the pressure in the reactor was released to atmospheric pressure, and then the reactor was purged with pure nitrogen gas for 1 hour to desorb the captured CO_2. The adsorption /desorption circulating experiments were carried out 10 cycles for each sample.

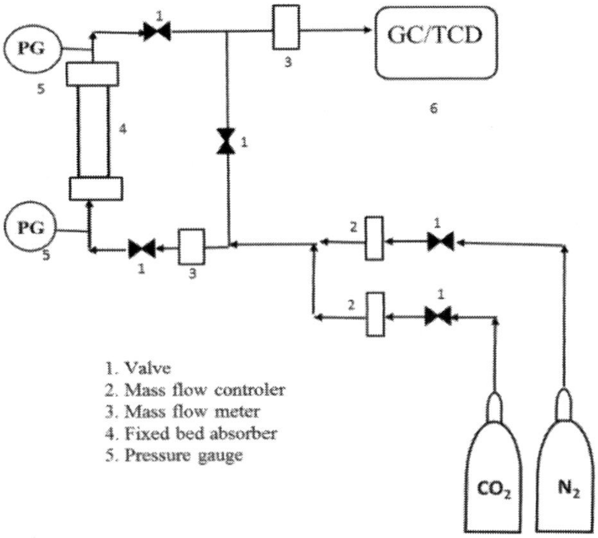

Temperature	K	298
Pressure	MPa	0.1 , 0.5, 1.0

Fig. 1. Fixed-bed adsorption system for CO_2 adsorption-desorption experiments.

3 Results and discussion

3.1 Material synthesis and characterization

3.1.1 Synthesis of mesoporous cerium compound

During the stirring, the colour of the precipitate suspension changed from brown to faint yellow as shown in Figure 2.

 DOI: 10.1051/e3sconf/20172200106

Fig. 2. The color change of the precipitate suspension for the sample with continuous stirring.

The reaction process could be explained as follows: at the initial stage of the reaction, hydrolyzed $Ce(NO_3)_3$ mixed with NaOH formed $Ce(OH)_3$

$$Ce^{3+} + 3OH^- \rightarrow Ce(OH)_3 \qquad (2)$$

here, the color of the aqueous suspension was turned into purple due to the presence of $Ce(OH)_3$ species. When the reaction was further proceeded, the color eventually turned into faint yellow color, and the precipitation of CeO_2 obtained. Thus at the latter stage of the reaction, most of $Ce(OH)_3$ formed during hydrolysis-condensation subsequently by the following oxidation step with the oxygen in the air,

$$4Ce(OH)_3 + O_2 \rightarrow 4CeO_2 + 6H_2O \qquad (3)$$

The mesoporous ceria was synthesized based on sol-gel chemistry at room temperature in highly alkaline solutions with several different NaOH concentrations (see Table 1).

Table 1. BET surface area, total pore volume, average crystallite size of the products obtained.

Sample No.	**Synthesis condition**		S_{BET} **(m² g⁻¹)**	PV **(cm³ g⁻¹)**	d **(nm)**
	Alkali	M (mol L⁻¹)			
1	NaOH	0.11	152	0.29	5.9
2	NaOH	0.22	206	0.21	3.0
3	NaOH	0.28	182	0.34	3.6
4	NaOH	0.42	190	0.42	4.1
5	NaOH	0.84	191	0.41	3.7
6 (calcinated at 873K for 10 h)	NaOH	0.22	20	0.03	10.3

M : mol concentration of aqueous solution ; S_{BET} : surface area calculated using BET method ; PV : pore volume obtained from the volume of N_2 ; d : the average crystallite size evaluated from Scherrer's equation.

Table1 shows that synthesis of the cerium compound by 0.22 M NaOH solution obtained a sample with highest surface area up to 200 m^2g^{-1}, and the average crystallite size evaluated by Scherrer's equation was the smallest. The surface area of this sample is excellent comparing with those reported by literature references.

3.1.2 XRD analysis

Figure 3 shows the results of wide-angle powder XRD measurements of the synthesized solid products and a commercially available ceria (non-porous ceria). The obtained broad XRD patterns of the products were identified as pure crystalline ceria with six noticeable peaks located at 2θ= 28.5°, 32.6°, 47.1°, 55.8°, 68.7°, and 76.5°, which can be indexed to the (111), (200), (220), (311), (400), and (331) reflections [4], respectively. For the product obtained by 0.22M NaOH, after it was calcinated at 873 K, the peak intensity greatly enhanced. The crystallite size of the calcinated ceria greatly increased from 3.0 nm to 10.3 nm owing to the further crystal growth of ceria induced by the calcination at high temperature.

3.1.3 TEM morphological features

The morphological features as well as particle sizes were observed by TEM. The TEM images of three synthesized samples are shown in Fig. 4. It is noticed that synthesized products were clearly agglomerates composited by fine particles. The shapes and sizes of the particles depended upon the concentration of alkaline solution.

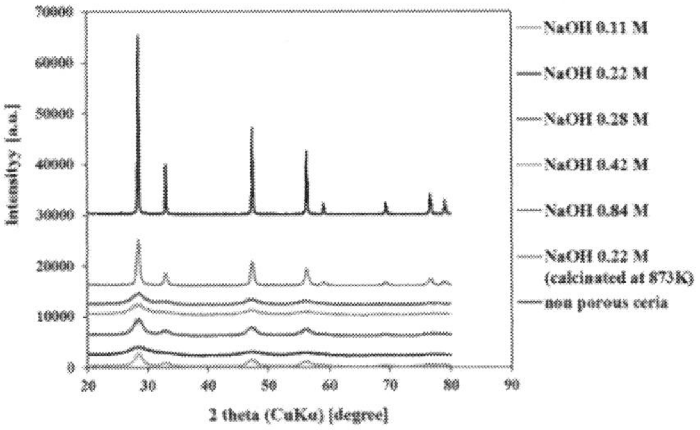

Fig. 3. XRD patterns of ceria synthesized with NaOH.

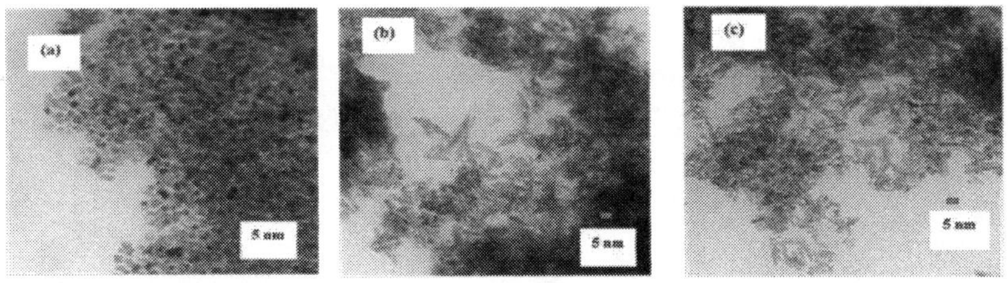

Fig. 4. TEM images of ceria synthesized with NaOH. (a) Sample no.2 (NaOH 0.22 M), (b) Sample no.3 (NaOH 0.42 M), (c) Sample no.4 (NaOH 0.84 M).

3.1.4 Pore size distribution

The pore size distributions of the synthesized products were obtained by nitrogen adsorption/desorption isotherms calculated using the Dollimore Heal (DH) method (see Fig.5). Except the sample obtained by 0.22M NaOH, which gave a sharp distribution for the pore size, all the samples gave broad distributions for pore size which covered the meso-porous region.

 DOI: 10.1051/e3sconf/20172200106

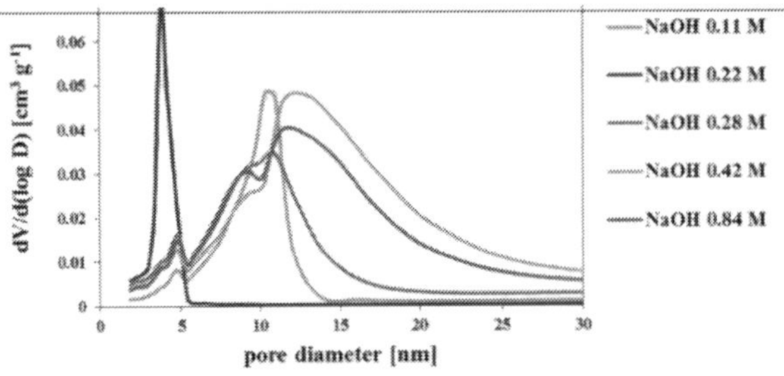

Fig. 5. Pore size distribution of ceria calculated using the Dollimore Heal (DH) method.

3.2 CO_2 capture

3.2.1 CO_2 capturing capacity

The CO_2 capturing capacities of the best synthesized product as well as commercial available adsorbents were compared shown on Fig.6, and their surface areas and pore sizes shown on Table 3. Although the surface area of the synthesized ceria (non-calcinated) is smaller than both the zeolite 13X and activated carbon, it had more excellent CO_2 capturing performance expecially under high pressure.

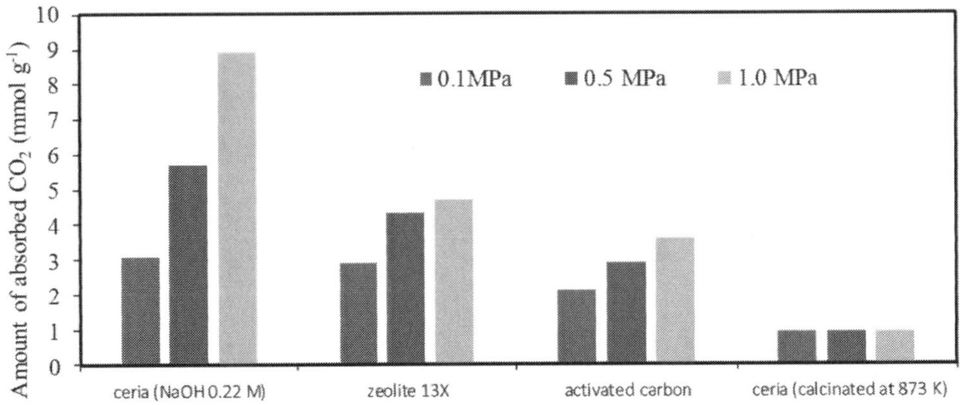

Fig. 6. Capturing capacities of adsorbents for CO_2 (298 K, 0.1MPa, 0.5 MPa, 1.0 MPa).

Table 3. Compare of amount of CO_2 capturing capacities among the different adsorbents.

Sample	S_{BET} (m^2g^{-1})	*PD* (nm)	Amount of adsorbed CO_2 ($mmol\ g^{-1}$)		
			0.1 MPa	0.5 MPa	1.0 MPa
Ceria (NaOH 0.22 M)	206	3.8	3.1	5.7	8.9
zeolite 13X	743	1.0	2.9	4.3	4.7
activated carbon	1079	2.9	2.1	2.9	3.6
Ceria (calcinated at 873 K)	20	4.0	0.9	0.9	0.9

3.2.2 Regeneration

Circulating adsorption-desorption experiments were carried out using the synthesized cerium product (sample No. 2) and comparing with the zeolite 13X. The results of CO_2 capturing capacities after operating up to ten cycles are shown on Fig. 7. From the experimental results, it can be concluded that the synthesized cerium product has not only a high adsorption capacity for CO_2 capture at room temperature under a high pressure condition, but also the good regenerating performance at room temperature under atmospheric pressure. Therefore, mesoporous ceria compound may be available a high efficient and energy-saving adsorbent for CO_2 PSA-capture from flue gas under a high pressure.

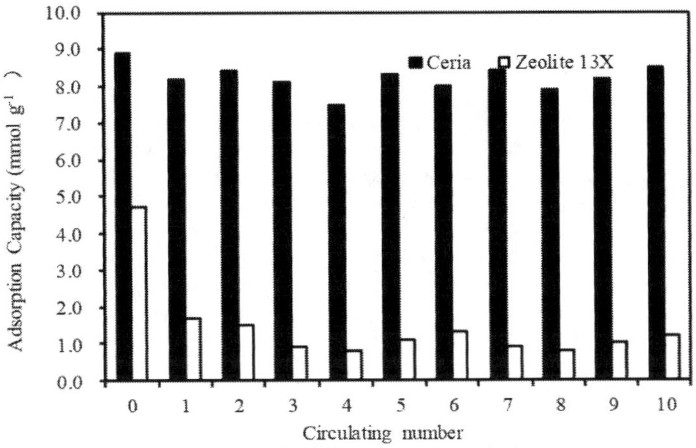

Fig. 7. CO_2 adsorption- desorption circulating tests at 298 K in the pressure range of 0.1 MPa ~1.0 MPa.

4 Conclusion

A high performance mesoporous cerium compound as the adsorbent for CO_2 capture has been successfully synthesized based on sol-gel method without template. The morphologic features and the porous characteristics of the synthesized cerium compound products are significantly affected by the concentration of alkaline solution added to the hydrolyzed $Ce(NO_3)_3$ to obtain $Ce(OH)_3$. Although the surface area of the synthesized cerium compound product (even though the largest one was up to 206 m^2g^{-1}) is smaller than that of the commercially available zeolite 13X and activated carbon, the synthesized products have higher adsorption performance than the commercial adsorbents especially under higher pressure. Because of its excellent desorption performance at room temperature and under atmospheric pressure, the synthesized mesoporous cerium compound is promising for CO_2 capture with the pressure swing adsorption (PSA) process as a high performance adsorbent.

References

1. W. Chaikittisilp, H. Kim, C.W. Jones, Energy Fuels, **25** (2011)
2. J.Yu, Y.Le, B. Cheng, RSC Adv. **2** (2012)
3. R.J. Gorte, AIChe J. **56** (2010)
4. Y. Kamimura, M. Shimomura, A. Endo, J. Colloid Interface Sci. **436** (2014)

Modeling an impact of road geometric design on vehicle energy consumption

Blaž Luin[1,*], *Stojan* Petelin[1], and *Fouad* Al-Mansour[2]

[1]Faculty of Maritime Studies and Transportation, University of Ljubljana, Slovenia
[2]Jozef Stefan Institute, Jamova cesta 39, 1000 Ljubljana, Slovenia

Abstract. Some roads connect traffic origins and destinations directly, some use winding, indirect routes. Indirect connections result in longer distances driven and increased fuel consumption. A similar effect is observed on congested roads and mountain roads with many changes in altitude. Therefore a framework to assess road networks based on energy consumption is proposed. It has been shown that road geometry has significant impact on overall traffic energy consumption and emissions. The methodology presented in the paper analyzes impact of traffic volume, shares of vehicle classes, road network configuration on the energy used by the vehicles. It can be used to optimize energy consumption with efficient traffic management and to choose optimum new road in the design phase. This is especially important as the energy consumed by the vehicles shortly after construction supersedes the energy spent for the road construction.

1 Introduction

The European Union puts a lot of effort on policies for energy efficiency and to decrease emissions in transport [1]. Transport sector is contributing nearly a third of CO_2 emissions and energy consumption within the EU [2]. Most efforts to combat this problem are directed towards traffic mode shift, use of energy-efficient vehicles and alternative fuels [3]. As a result there was a decline of overall emissions by 3% in 2012 and by 2014 average light vehicle emissions were below the targets set for 2015. These values were obtained in part as the EU requires member states to record data on registering motor vehicles [4]. Similar policies are being practiced elsewhere in the world, such as in the USA and China [5, 6]. Such transport policies also focus on limiting road traffic in cities, increasing use of public transport and on shifting from road to other modes of transport. Such approaches work well in densely populated areas, but in sparsely populated areas road traffic is still, necessarily, the transportation mode of choice. An example of a study of reducing emissions in a densely populated area is the article [7], where a case study from London is described. An alternative to such approaches can be better traffic management that could result in less congestion and lower emissions.

When roads are preferred option and shift to other modes is difficult to achieve, more focus should be put on relation between greenhouse emissions and road network

* Corresponding author: blaz.luin@fpp.uni-lj.si

configuration including traffic management. Traffic simulation and emissions models are a tools that should be applied to optimize the traffic network.

Traffic microsimulation models need to be coupled with emission models in order to obtain individual and aggregated vehicle emissions. Vehicle emission models can be divided into microscopic and macroscopic. Macroscopic emission models are used to estimate emissions based on traffic volume and average speed, ignoring instantaneous accelerations. More detailed are the microscopic emission models that rely on inputs such as speed, acceleration, vehicle weight and road grade.

Regardless if the model is microscopic or macroscopic, most models provide no direct physical model of fuel consumption, rather tables or functions that are based on statistical analysis of different variables that have an impact on emissions [8, 9]. One example of a macroscopic model is the ARTEMIS study that involves CO2 and NOX emissions dependant on vehicle speed and driving cycles [10, 11]. A statistical analysis of the dependence of truck fuel consumption on speed and road grade was conducted in an experimental study by [12] conducted on highways in realistic traffic conditions. Fuel volume flow was measured using the vehicles' CAN (Controller Area Network) bus interface. Such results can be used for comparative purposes and for model tuning but not for direct estimation of instantaneous fuel consumption for different types of vehicles that is required for microsimulation. For general energy demand forecasting models with lower level of detail are used. In [13], where a long-term forecasting was simulated, a straight-forward model based on mileage travelled and average fuel economy was applied.

Notable microsimulation models are Versit+, PHEM, CMEM and VT-Micro. Versit+ is a model that is based on a vehicle type, velocity and acceleration [14] that was developed by the TNO from Netherlands. It has interface with AIMSUN traffic microsimulation software. The AIMSUN is based on the Gipps model therefore velocity must be differentiated to obtain acceleration [15]. PHEM is an emissions model that was developed by the University of Graz and has interface for the SUMO microscopic traffic simulator [16–18]. CMEM and VT are models that were development in the USA. The CMEM was developed during the 90s and was afterwards, while the VT (Virginia Tech) is a newer model. Both of them are based on velocity, acceleration and factors that depend on vehicle types.

Since new fuels and electric vehicles are becoming increasingly popular, calculating energy impact of traffic can be difficult. A lot more straight-forward approach would be to rely on the energy required by the vehicle directly. Among all approaches, the VSP (vehicle specific power) model was the only one that was directly based on power required by the vehicle [19, 20]. Vehicle specific power is based on energy conversions, changes of potential and kinetic energy. Power needed for acceleration results in a change of kinetic energy and is therefore affected by congestion; road grade, too, has an impact on potential energy change. Apart from the energies, air and rolling resistances are also modelled. There are two frequently mentioned formulas for VSP in the literature. The first was developed by Jimenez and is based directly on changes of energy, and further derived to include drag and rolling resistance coefficients and accelerations [19]. Another formula was developed by Zhai; this also introduces several coefficients [20]. The main difference between the two is that Jimenez uses wind velocity as an independent variable, which is the reason this formula is preferred; in case of unidirectional tunnels the air velocity is usually in the direction of travel, which results in lower air resistance as opposed to the open road.

The downside of the VSP model is that it does not account for engine idling, which can be significant in case of congested traffic. According to [21] and [22], fuel consumption during engine idling is between 0.5 and 1.5 l/h for light vehicles and a bit over 3 l/h for heavy vehicles. The VSP is also suitable for adaptation to analysis of alternative drivetrains as it is based on energy consumption with a clear physical background. It is not a regressive model of fuel use. It merely estimates power at the wheels. The power at the engine should be higher as transmission and engine efficiencies should be considered as well. A relationship between the vehicle specific power and fuel consumption rate (flow) was conducted by [23, 24].

An exhaustive study involving vehicle efficiency defines separate ratios for powertrain, vehicle and primary energy efficiency [25]. It can be deduced that due to many different fuels that are being used, it is more feasible to have a model that is based on energy and then to calculate primary energy use from the energy at the wheels.

2 Extraurban case study - Karawanks tunnel

An analysis of the Karawanks road tunnel connecting northern Slovenia with southern Austria was carried out. It is a single tube tunnel with bidirectional traffic connecting motorways on both sides of the border. It is located under the Karawanks mountain ridge, which is approximately 2500 m high. This particular route is interesting as all alternative routes are much longer and include many serpentine segments and changes in elevation (as can be seen in figures 3 and 4). In figure 4 the axis represents the geographical longitude and latitude. The road coordinates were obtained from OpenStreetMap, while the elevation data are from GMTED 2010 database [26]. It is evident that the tunnel route (black) is a lot shorter than the alternative route because it crosses the mountain ridge (the white route in figure 4). It is, however, still longer than the hypothetical ideal route (dashed line). Length and elevation comparison of the routes are shown in table 1.

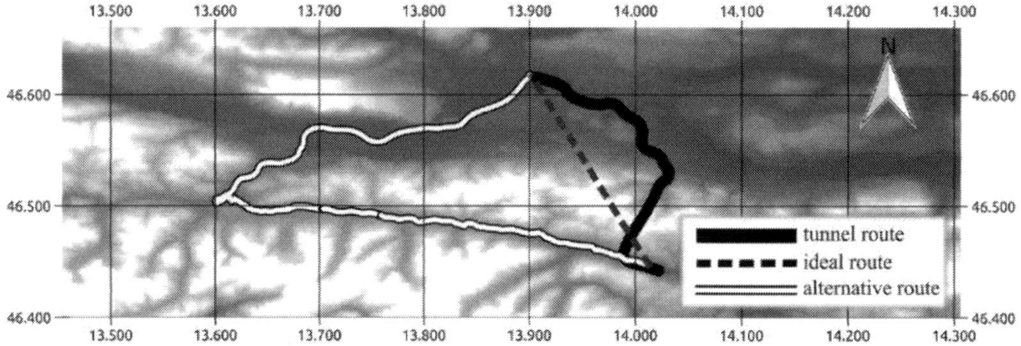

Fig. 3. Tunnel, alternative and ideal routes.

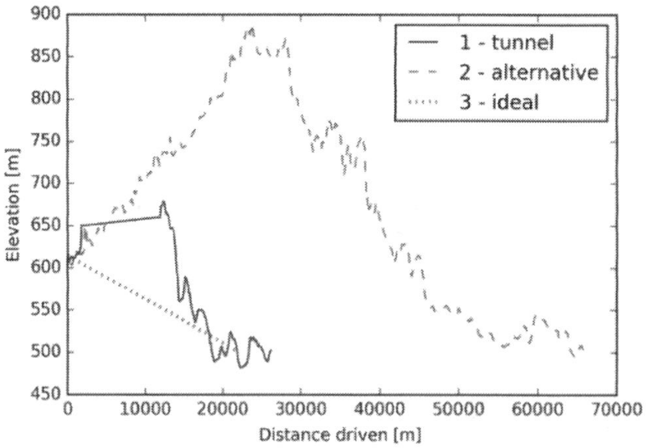

Fig. 4. Elevation profile of the routes.

 DOI: 10.1051/e3sconf/20172200107

Table 1. Basic route characteristics.

	Tunnel Route	Alternative Route	Ideal Route
Minimum Elevation	481 m	495 m	510 m
Maximum Elevation	679 m	890 m	622 m
Route Length	26.2 km	66.2 km	21.7 km
Tunnels	8019 m	-	-

It would be possible to split traffic data into several time periods, but since detailed traffic data were not available analysis using an average number of vehicles per day was conducted: separating them into personal vehicles, busses, light commercial vehicles and heavy goods vehicles. Vehicle weight for each class was approximated. An overview of traffic data can be seen in the table 2. As can be seen, the traffic volume is relatively low with a high share of personal vehicles.

After obtaining the traffic data, we assigned each vehicle class an approximate weight and length. The vehicle length is needed for microsimulation as well.

Table 2. Traffic data.

	Daily traffic	Approximate weight [t]	Vehicle length [m]
Personal vehicles	7144	1.5	5
Busses	96	20	11
Light goods vehicles (up to 7t)	884	4	7
Heavy goods vehicles (over 7t)	1164	40	12

After obtaining vehicle trajectories, VSP calculation along the route was carried out. The VSP is highly dependent on road grade and accelerations. Changes in road grade have many peaks that are not evident in reality because of the resolution of satellite elevation data and the very steep changes in elevation in these narrow Alpine valleys. These peaks will not necessarily yield wrong results as there are significant changes in elevation on the actual road which are spread over longer distances compared to those of the satellite based elevation data. If the analysed road was on a flat area there such irregularities would be minimized.

Calculated average values for energy consumption that were obtained by pre-processing microsimulation vehicle trajectories of each vehicle in a class are shown in table 4, where route 1 is the tunnel route, 2 is the alternative route and 3 is the ideal route.

Table 3. Mean energy consumption per trip for each vehicle class in MJ/t.

Route	1	2	3
Personal vehicles	14.5	39.3	13.3
Busses	13.6	37.4	12.5
Light goods vehicles (up to 7t)	13.6	37.4	12.5
Heavy goods vehicles (over 7t)	13.1	36.3	11.3

Based on data from table 4 the daily amount of energy consumed by all vehicles on all 3 routes can be calculated according to the eq. 7. The results can be seen in table 6, which shows that the tunnel route is about 3 times more efficient than the alternative route (230% more energy is consumed). An interesting overview is the average daily power needed to drive all the vehicles that pass on the route for all three options. The difference between the

DOI: 10.1051/e3sconf/20172200107

tunnel and alternative routes is 18 MW which is an order of magnitude of a smaller electrical power station. This means that the impact of road geometry on fuel consumption is far from negligible. I

Table 4. Total energy consumption per route.

Route	1	2	3
Total Energy[GJ]	893	2431	736
Increase over ideal road [%]	21	230	0
P [MW]	10.3	28.13	8.5

4 Urban case study – Celovska street, Ljubljana

Celovska street is one of major streets in the city of Ljubljana that connects northern bypass ring motorway with city centre. Most of the street has 2 lanes per direction and includes several semaphored intersections. Hourly traffic volume values were obtained from city authorities and are based on inductive loop counters on the street [27]. Average monthly hourly traffic volume from bypass towards the centre can be seen in figures 4 and 5.

A simulation was carried out using predefined flow of moderate 400 vehicles per hour which is normal daily traffic (excluding peaks) plus random trips across the network. 3 different traffic light control programmes were evaluated among which are: 1 – fixed 90 second cycle programme for all traffic lights, 2 – adaptive programme with extensions of priority phases and maximum allowed cycle duration of 125 s and 2 – an adaptive programme with phase extensions combined with forced phase switching and synchronisation among junctions. Goal of this simulation is to assess impact of energy consumption on different traffic light management programmes. It is evident that according to the model, there are dramatic differences between noncongested traffic (programmes 2 and 3) and congested (programme 1).

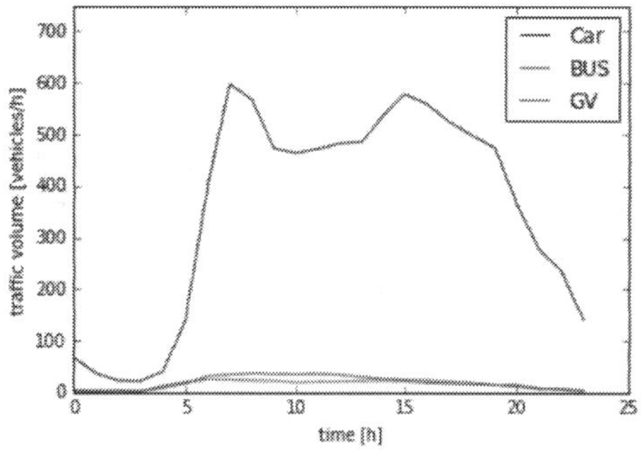

Fig. 4. Hourly traffic, weekdays.

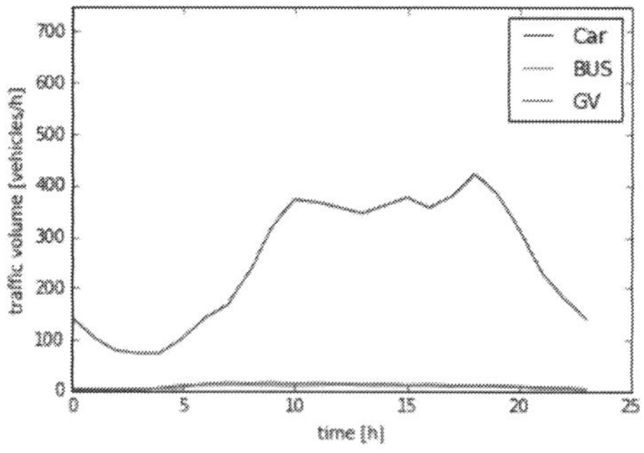

Fig. 5. Hourly traffic, weekends.

Table 5. Total hourly consumption per traffic regime.

Route	1	2	3
Total Energy[GJ]	22.9	9.7	7.4
Average trip duration [s]	2075	665	365
P [MW]	6.4	2.7	2.0

To summarize, avoiding congested stop and go traffic can significantly reduce energy consumption of the motor vehicles and it can be achieved by optimization of traffic light control programmes. These results are still preliminary as more detailed traffic data are needed. It should be noted that modelled traffic control programmes are not currently used programmes but are an evaluation of different possible traffic control programmes on a realistic traffic network with real traffic volume.

5 Conclusions

A methodology to evaluate an impact of road network configuration on energy consumed by the vehicles using the vehicle specific power model presented. It has been shown that it is possible to save enormous amounts of energy by making road networks more energy efficient, by building more direct, shorter routes without significant changes in road grade or by controlling traffic lights more efficiently. First is especially evident in mountainous areas where open roads must pass through significant variations in elevation. This results in longer routes with higher road grades, causing higher energy consumption. In the cities same effect was observed by avoiding congestions with more efficient control of traffic lights.

Vehicle energy consumption was modelled with an assumption that no braking energy is recovered. In order to simulate electric and plug-in hybrid vehicles, simple modifications to the VSP model are needed.

References

1. E. Commissie, *White paper, Roadmap to a single European Transport Area, Towards a competitive and resource efficient transport system*. COM (2011)
2. Statistical Office of the European Communities, *Smarter, greener, more inclusive?: indicators to support the Europe 2020 strategy*. Luxembourg: Publications Office of the European Union, (2015)
3. "ODYSSEE – MURE 2012. Trends and policies for energy savings and emissions in transport." (Jul-2015)
4. P. ten Brink, New Clim. Policies Eur. Union Intern. Legis. Clim. Dipl. **15**, 179 (2010)
5. A. Gambhir, L. K. C. Tse, D. Tong, and R. Martinez-Botas, Appl. Energy **157**, 905–917 (2015)
6. N. Bianco and F. Litz, "Reducing greenhouse gas emissions in the United States using existing federal authorities and state action," The World Resources Institute, 10 G Street, NE Suite 800 Washington, D. C. 20002 USA, (2010)
7. D. Metz, Case Stud. Transp. Policy (2015)
8. K. Ahn, H. Rakha, A. Trani, M. Van Aerde, J. Transp. Eng. **128**, 2, 182–190 (2002)
9. A. Sjödin, M. Jerksjö, *Evaluation of European road transport emission models against on-road emission data as measured by optical remote sensing* (2008)
10. R. Joumard, M. Rapone, M. Andre, and others (2006)
11. M. André, M. Keller, \AAke Sjödin, M. Gadrat, I. M. Crae, *The ARTEMIS European tools for estimating the pollutant emissions from road transport and their application in Sweden and France*. na (2008)
12. O. Franzese, D. Davidson, Oak Ridge Natl. Lab. Tenn. USA (2011)
13. A. Sadri, M. M. Ardehali, K. Amirnekooei, Energy **77**, 831–843 (2014)
14. R. Smit, R. Smokers, E. Rabé, Transp. Res. Part Transp. Environ. **12**, 6, 414–422 (2007)
15. P. Hidas, Road Transp. Res. **14**, 4 (2005)

16. M. Behrisch, L. Bieker, J. Erdmann, and D. Krajzewicz, "Sumo–simulation of urban mobility," *in The Third International Conference on Advances in System Simulation (SIMUL 2011), Barcelona, Spain, Third Int. Conf. Adv. Syst. Simul. SIMUL 2011 Barc. Spain*, (2011)
17. D. Krajzewicz, M. Behrisch, P. Wagner, R. Luz, M. Krumnow, in *Modeling Mobility with Open Data*, Springer, 203–221 (2015)
18. D. Krajzewicz, J. Erdmann, M. Behrisch, L. Bieker, Int. J. Adv. Syst. Meas. **5**, 3&4, (2012)
19. J.L. Jimenez-Palacios, "Understanding and quantifying motor vehicle emissions with vehicle specific power and TILDAS remote sensing," Massachusetts Institute of Technology (1998)
20. H. Zhai, H.C. Frey, N.M. Rouphail, Environ. Sci. Technol. **42**, 21, 7985–7991 (2008)
21. H.C. Frey, A. Unal, N. M. Rouphail, J.D. Colyar, J. Air Waste Manag. Assoc. **53**, 8, 992–1002 (2003)
22. S.M.A. Rahman, H.H. Masjuki, M.A. Kalam, M.J. Abedin, A. Sanjid, S. Imtenan, J. Clean. Prod. **69**, 208–215 (2014)
23. G. Song, L. Yu, Transp. Res. Rec. J. Transp. Res. Board **2139**, 11–20 (2009)
24. G. Song, L. Yu, Transp. Res. Rec. J. Transp. Res. Board **2233**, 90–98 (2011)
25. M. Åhman, Energy **26**, **11**, 973–989 (2001)
26. J.J. Danielson, D.B. Gesch, "Global multi-resolution terrain elevation data 2010 (GMTED2010)," US Geological Survey (2011)
27. "Traffic counting data from Celovska street." JP LPT Ljubljana (2016)

The transmission characteristics of indoor particles under different ventilation conditions

Yang Lv[1,*], *Haifeng* Wang[1], and *Shanshan* Wei[1]

[1]School of Civil Engineering, Dalian University of Technology, 116024 Dalian, China

Abstract: In modern society, ventilation is an important method for removing indoor particles. This study applies the parameter of attenuation index to analyze the effect of the removal of indoor particles in the two typical ventilation strategies called ceiling exhaust and slit exhaust strategy. Experiment was conducted in a chamber and riboflavin particles were used as the indoor particles source, instantaneous microbial detection (IMD) used to measure the particulate concentration. Conclusions can be found that air exchange rate is an important factor affecting the indoor particle concentration distribution. In the process of indoor free settling(air exchange rate is 0 h^{-1}), the deposition rate were 0.086 h^{-1}, 0.122 h^{-1}, 0.173 h^{-1} for the particles of 0.5–1.0 μm, 1.0–3.0μm and 3.0–5.0 μm. When the air exchange rate increased to 2.5 h^{-1}, the differences in the attenuation index is significant. There was also a significant linear relationship between air exchange rate and attenuation index. Furthermore, the effect of the slit exhaust strategy on the removal of coarse particles is more remarkable as the increasing air exchange rate.

1 Introduction

As the worsening of atmosphere particle pollution, indoor particle pollution has become increasingly severe due to the building exterior infiltration and other reasons. In modern society, people spend 80% of the time indoors [1, 2], hence people has begun to realize the importance of indoor particulate pollution on human health. Epidemiological studies showed that short or long-term exposure to particles would lead to widespread adverse effects on human mortality and morbidity, even if the concentration was lower than the national control standards [3–9]. Researches showed that all-cause mortality, cardiovascular mortality and lung cancer mortality significantly increased with the annual average concentration of particulate matter [5].

There are many factors that influence the distribution of indoor particle concentration, such as different forms of air supply, air exchange rate and gravity deposition will cause the diffusion and transmission of indoor particle. Zhang and Chen studied particle transportation and distribution in ventilation chamber that found that different forms of ventilation had a great impact on indoor particle concentration distribution [10]. Munat investigated and analyzed influence that different airflow forms to particle resuspension, the results showed that displacement ventilation system had small resuspension harm [11].

* Corresponding author: lvyang@dlut.edu.cn

DOI: 10.1051/e3sconf/20172200108

Zhao B compared the diffusion characteristics of indoor particulate matter in displacement ventilation and mixing ventilation using numerical simulation method [12].

Although a large number of studies have been conducted on the diffusion and deposition characteristics of indoor particles, there is no good method to evaluate the removal effect of indoor particle under different ventilation strategies. Therefore, the attenuation index is defined to express the decay rate of indoor particles in this paper. It has important scientific significance for solving the problem of indoor particulate pollution to take into account the characteristics of indoor particle transfer and sedimentary and the relationship between air exchange rate and the attenuation index of particles with different particle sizes in the slit exhaust and ceiling exhaust strategy.

2 METHOD

2.1 Materials and equipment

In the experimental study of indoor particulate matters, JIS-11 powder was used as experimental particles in Japan [13]. JIS-11 test powder is made of volcanic ash, soil, etc., belonging to non-biological particles. In this study, however, the riboflavin powder was used as experimental materials instead of indoor particle, considering the concentration of non-biological particles in the background value is higher and the impact on the health of laboratory personnel is not clear [14]. Riboflavin is a special kind of biological particles, emitting fluoresce when exposed to light sources with certain specific wavelength. Previous studies have shown that the riboflavin powder concentration is at a low level in the indoor air that does not affect the results basically [14].

Instantaneous Microbial Detection-Air (IMD-A 200-1, Bio Vigilant Systems) can distinguish between biological and non-biological particles in the air particles based on the optical sensor and fluorescence detection technology, which was used to indoor particle concentration detection in the room. In addition, the instrument can realize continuous detection, the measurement of the particle size range 0.5 μm ~ 1 μm, 1 μm ~ 3 μm, 3 μm ~ 5 μm, 5 μm ~ 7 μm, 7 μm ~ 10 μm, 10 μm ~ 15 μm.

2.2 Ventilation systems and experiment process

This study analyzes the transmission characteristics of indoor particles in the two typical ventilation strategies including ceiling exhaust and ground slit exhaust strategy (as illustrated in Fig. 1), which the ceiling exhaust strategy is a traditional ventilation way. Fig. 2 shows experimental chamber size (L × W × H = 5.37 m × 2.74 m × 2.25 m) and the distribution of measuring points and instruments. In this experimental chamber, the delivery outlet was located in the center of the ceiling; a ceiling exhaust port located in the corner of the ceiling and two slit exhaust port located in the junction of walls and floors. The delivery outlet of two ventilation system was the same that the size was 0.05 m × 0.10 m. The exhaust port size in ceiling exhaust system was 0.05m × 0.10m and the two exhaust port size of the slit exhaust system was 4.2 m × 0.005 m and 4.53 m × 0.005 m, respectively.

DOI: 10.1051/e3sconf/20172200108

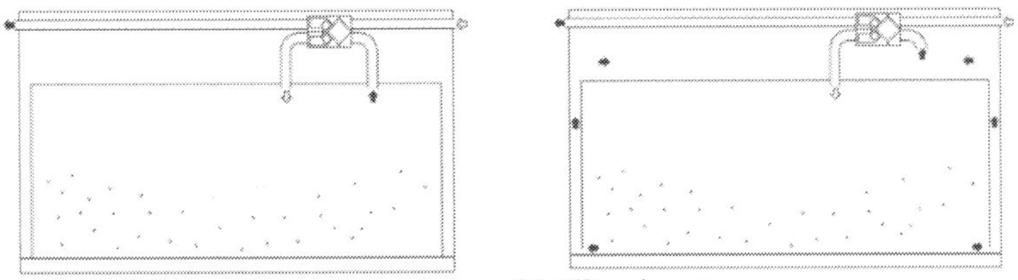

(a) Ceiling exhaust (b) Slit exhaust

Fig. 1. Two typical ventilation strategies.

As shown in Fig. 2, the indoor particles were put in a measuring flask that was fixed at the corner near the ceiling of the room. Nitrogen gas was blown into the measuring flask to disperse particles at a rate of 10 L/min. The airborne particulate matters were mixed by two electric fans in order to ensure the distribution of house dust particles as evenly as possible in the room. Taking into the account of the height of the child, the position of the monitoring point was set at 0.1m and 1.0 m height. During the experiment, the window was sheltered by aluminum foil corrugated board to prevent solar radiation, and the walls were insulated by expanded polystyrene board. The experimental chamber was closed and stable for 12 hours before the release of particles in order to ensure that the indoor background concentration was unchanging. Then the particles were released and the fans were turned on for an hour so that the indoor particles are evenly mixed. Stop particle released and turn off the fan, whereafter, the ventilation system was opened (the air change rate was 0.75 h^{-1} and 2.5 h^{-1}) after the free settlement is maintained for 8 hours until the indoor particle concentration was in a stable state.

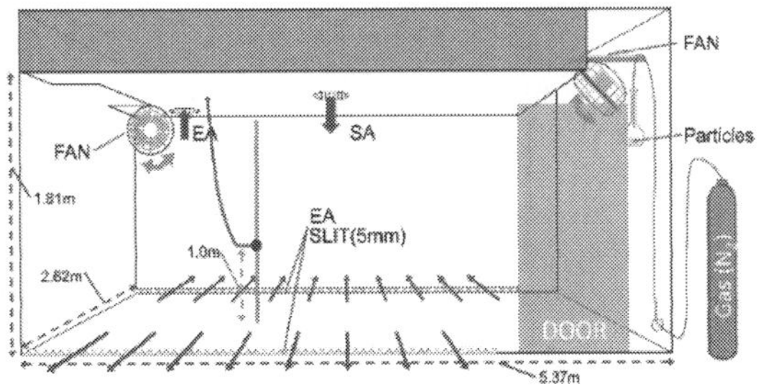

Fig. 2. The experimental chamber size and the distribution of measuring points.

3 Results and analysis

3.1 The effect of air exchange rate on particle distribution

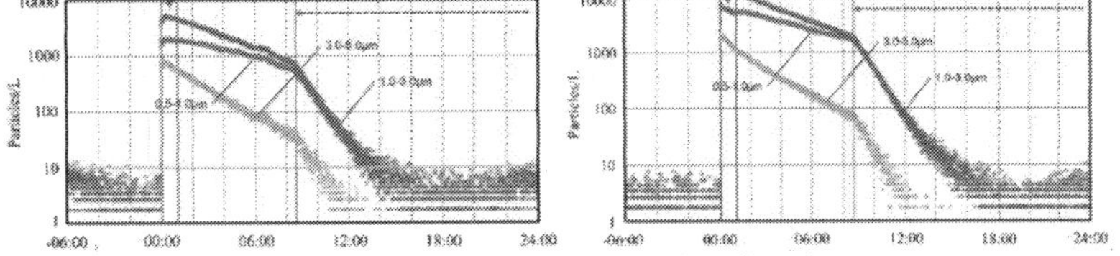

(a) Ceiling exhaust (b) Slit exhaust

Fig. 3. The concentration changes characteristics of two ventilation modes (0.75 h^{-1})

 DOI: 10.1051/e3sconf/20172200108

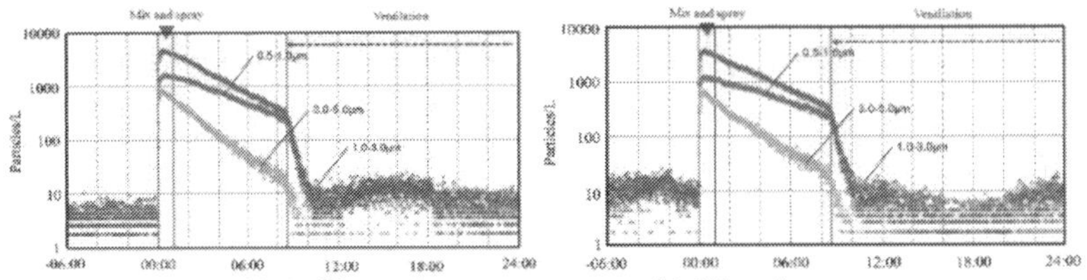

(a) Ceiling exhaust (b) Slit exhaust

Fig. 4. The concentration changes characteristics of two ventilation modes (2.5 h^{-1}).

In order to further study the influence factors of indoor particulate matter distribution, the concentration changes characteristics of two ventilation modes at different air changes rate were analyzed and compared. During the operation of the ventilation system, the attenuation rate of indoor particulate matter concentration is obviously accelerated, but there was no obvious difference in the time that indoor particles concentration of two ventilation modes achieve stability, as shown in Fig. 3 and Fig. 4. At the time of the air exchange rate is 0.75 h^{-1}, the stable time of fine particles is 7h, and the coarse particles is about 4h. When the air exchange rate is 2.5 h^{-1}, the time required for the concentration of indoor particles to stabilize is about 0.67 h. Therefore, air exchange rate is an important factor affecting the distribution of indoor particulate matter concentration. As the increasing of air exchange rate, the elimination rate of indoor pollutants can also greatly accelerate, but ventilation mode has little effect on the elimination time.

3.2 Indoor particle concentration prediction model

The indoor airborne particles are affected by air resistance, Magnus force, Basset force, slip shear lift force, pressure gradient force, electrophoretic force, Brown diffusion force and virtual mass force [15], and temperature and humidity, ventilation frequency and ventilation mode are also crucial factors to determine the power of the force, which have an effect on the diffusion and deposition process of particulate matter. Murakami et al. simulated the particle trajectories in the two regions of ventilation considering only the effects of gravity, buoyancy and flow resistance, who pointed out that indoor airflow pat, ventilation condition and particle properties had a great influence on the distribution and movement of indoor particles [16]. The air velocity affects the distribution and deposition of indoor particles under different ventilation modes, and the smaller the air velocity is, the higher the particle deposition rate is. As the increase of the air velocity, the particles are not only difficult to deposit, but also cause the resuspension of particles in virtue of indoor turbulent flow. When the air velocity is very large, the indoor particle concentration approaches to the outdoor concentration continuously [17]. In this paper, the attenuation index is defined to better represent the removal effect of indoor particulate matter under different conditions, considering the influence of sedimentation, resuspension, air exchange rate and other parameters.

In the course of our experiment, the temperature and humidity changed little, the chemical reaction of the particles was slow, the condensation and phase change could be ignored, and the distribution of aerosol particle size was also quite stable. Therefore, the prediction model of indoor particles concentration can be established according to the mass balance or quantity balance equation:

$$\frac{dC}{dt}=aC_1-aC-kC+rC \tag{1}$$

 DOI: 10.1051/e3sconf/20172200108

where, C – the indoor particle concentration; C_1 – the outdoor particle concentration; r – the particle resuspension and other effects; k – the indoor deposition rate.

In this paper, the biological particles used in the experiment are very low in outdoor air, which can be ignored. The above equation can be simplified as:

$$C(t)=C_0e^{-\lambda t} \tag{2}$$

where, C_0 – the initial concentration of indoor particles, λ – the attenuation index of particles, which indicates the decay rate of indoor particles,

$\lambda=a+k-r$.

In the process of free settlement, the air exchange rate is zero, the indoor air disturbance is little and the resuspension process can also be ignored. The attenuation index is equal to the deposition rate of indoor particles, therefore, the Equation 2 can be expressed as:

$$C(t)=C_0e^{-kt} \tag{3}$$

3.3 The particle attenuation index of different air exchange rate

The indoor particles attenuation index of different air exchange rate (0 h^{-1}, 0.75 h^{-1}, 2.5 h^{-1}) under slit exhaust and ceiling exhaust strategies is indicated in Fig. 5. It can be found that the attenuation index of indoor particles is increasing with air exchange rate, which has been discussed in the previous paper. When the air exchange rate is 0 that is the free settling process of the particles in the room, the attenuation index is equal to the deposition rate of indoor particles. Thus, the deposition rate of three kinds of particle sizes in the free settling process can be obtained, as shown in the Fig. 5. For the particle size of 0.5–1.0 μm, 1.0–3.0 μm and 3.0–5.0 μm, indoor deposition rates are 0.086 h^{-1}, 0.122 h^{-1} and 0.173 h^{-1} respectively and the settlement rate is increasing with particle size. A number of studies have found that particles deposition is the main reason for indoor particles concentration decreased in the process of free settling without the effect of static electricity and high air temperature difference, and particle diffusion and gravity are the most important mechanisms for particle deposition. The attenuation rate of particle concentration in different particle size ranges varies widely. Brown diffusion dominates for the small particle, which results in the distribution of indoor particles uniform, and the deposition rate low. Large size particles are not evenly distributed in the interior due to obviously affected by gravity, hence the distribution in the lower part of the room is more and the deposition rate is higher [15, 17, 18].

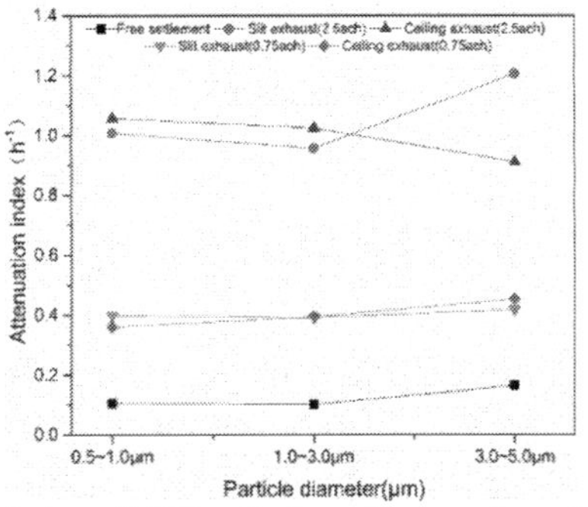

Fig. 5. The particle attenuation index of different air exchange rate under two ventilation modes.

At the time of the air exchange rate is 0.75 h^{-1}, indoor air velocity is also small and the main influence factors is still gravity for ceiling exhaust and slit exhaust strategies, so the difference of indoor three particle size range attenuation index is not obvious. When the air exchange rate increased to 2.5 h^{-1}, there are significant differences in the attenuation index of the two ventilation strategies, especially for coarse particles in the range of 3.0–5.0 μm. Fine particles are mainly controlled by the Brown diffusion effect, less affected by flow organization. The effect of gravity on coarse particles is more significant. As for the slit exhaust, the direction of the air flow is the same as that of gravity, which accelerates the deposition and removal of the particles, resulting in the increase of the coarse particles attenuation index. For the ceiling exhaust, the direction of gravity and airflow is in the opposite direction. The effect of gravity will hinder the particle removal and air lift will also impede the particle deposition process, which leads to the decrease of the coarse particles attenuation index.

3.4 The relationship between air exchange rate and particle attenuation index

The relationship between air exchange rate and particle attenuation index is compared as shown in Fig. 6, which can be found that the linear correlation between them is very significant (the correlation coefficient R^2 is above 0.99). With the increase of ventilation rate increases, therefore, the particle concentration will be lower and the particle propagation distance is shorter resulting in better indoor environment [19]. As for the fine particles with smaller diameters (0.5–1.0μm, 1.0–3.0μm), there was no significant difference in the correlation between air exchange rate and particle attenuation index (the correlation coefficient is about 0.36), the correlation coefficient of the ceiling exhaust is slightly higher than that of the slit exhaust. For the coarse particles of 3.0–5.0 μm, the correlation coefficient of slit exhaust is 0.4232 which significantly higher than 0.2879 that of the ceiling exhaust, indicating that the removal efficiency of coarse particles is better as the increase of ventilation rate under the condition of slit exhaust strategy, and this effect is mainly due to the gravity and particle flow superposition effect.

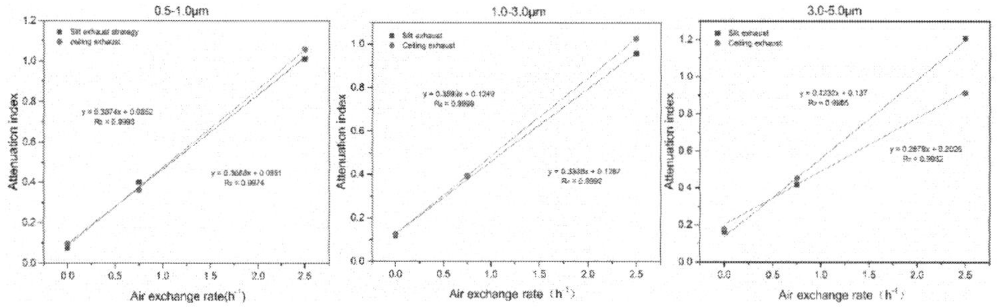

Fig. 6. The relationship between air exchange rate and different particle attenuation index under two ventilation modes.

4 Conclusion

This paper studies the variation characteristics of indoor particulate matter concentration under two ventilation strategies named the slit exhaust and the ceiling exhaust for the sake of eliminating the indoor particulate pollutants quickly. Specific conclusions are as follows:

Air exchange rate is an important factor affecting the indoor particle concentration distribution. In the process of free settlement (air exchange rate is 0 h^{-1}), the deposition rate is 0.086 h^{-1}, 0.122 h^{-1} and 0.173 h^{-1} respectively for 0.5–1.0 μm, 1.0–3.0 μm, 3.0–5.0 μm particles, which is increasing with particle size. When the air exchange rate is 0.75 h^{-1}, the difference of the particle attenuation index between the slit exhaust and the ceiling exhaust strategy is not obvious. However, there a significant differences when the air exchange rate increased to 2.5 h^{-1}, especially for the coarse particles in the range of 3.0–5.0 μm.

There is a significant linear relationship between air exchange rate and the attenuation index of particles. The fine particles of 0.5–1.0 μm and 1.0–3.0 μm has little significant difference under the two ventilation strategies, the correlation index of around 0.36. For the coarse particles of 3.0–5.0 μm, the correlation of slit exhaust (0.4232) is significantly higher than that of the ceiling exhaust (0.2879), indicating that the elimination effect slit exhaust strategy on the coarse particles is better significantly under the condition of increasing air exchange rate.

Acknowledgement

This research is supported by the China Environmental Protection Project (201509063).

References

1. P.L. Jenkins, T.J. Phillips, E.J. Mulberg, S.P. Hui, Atmos. Environ. **26** (1992)
2. J. Robinson, W.C. Nelson, USEPA (1995)
3. L. Yang, *Communication and control of indoor air pollution* (2014)
4. D.W. Dockery, et al. New Engl. J. Med. **329** (1993)
5. C.A. Pope III, et al. Jama, **287** (2002)
6. Q.B. Guo, X.F. Cheng, H. Hou, et al. Enviro. Monit. China, **26** (2010)
7. W.F. Wang, J. Yu, D.D. Xu, et al. Enviro. Monit. China, **29** (2013)
8. Y.X. Wang, J.P. Niu, G.W. Ding, M.M. Noordin, X.Y. Chen, J. Environ. Health, **24** (2007)
9. J.A. Tinker, D. Roberts, In Proceedings of Indoor Air, **4** (1999)
10. Z. Zhang, Q. Chen, Atmos. Environ. **40** (2006)
11. E. Mundt, Build. Environ. **36** (2001)
12. B. Zhao, Y. Zhang, et al. Build. Environ. **39** (2004)
13. Y. Liu, K. Ikeda, et al. J. Archit. Plann. Environ. Eng. AIJ, **483** (1996)
14. G. Kurihara, Yoshino, et al. In Proceedings of AIVC2010 Conference
15. A.C.K. Lai, W.W. Nazaroff, J. Aerosol. Sci. **31** (2000)
16. S. Murakami, S. Kato, S. Nagano, et al. ASHRAE Winter Meeting (1992)
17. Q.S. Zong, Anhui University of Technology (2012)
18. J.G. Crump, J.H. Seinfeld, J. Aerosol. Sci. **12** (1981)
19. B. Zhao, Z. Zhang, X. Li, Build. Environ. **40** (2005)

Scientific approach and practical experience for reconstruction of waste water treatment plants in Russia

Nikolay Makisha[1,*], and *Elena* Gogina[1]

[1]Moscow State University of Civil Engineering (MSUCE), Yaroslavskoye shosse, 26, Moscow, Russia, 129337

Abstract. Protection of water bodies has a strict dependence on reliable operation of engineering systems and facilities for water supply and sewage. The majority of these plants and stations has been constructed in 1970-1980's in accordance with rules and regulations of that time. So now most of them require reconstruction due to serious physical or/and technological wear. The current condition of water supply and sewage systems and facilities frequently means a hidden source of serious danger for normal life support and ecological safety of cities and towns. The article reveals an obtained experience and modern approaches for reconstruction of waste water and sludge treatment plants that proved their efficiency even if applied in limited conditions such as area limits, investments limits. The main directions of reconstruction: overhaul repair and partial modernization of existing facilities on the basis of initial project; - restoration and modernization of existing systems on the basis on the current documents and their current condition; upgrade of waste water treatment plants (WWTPs) performance on the basis of modern technologies and methods; reconstruction of sewage systems and facilities and treatment quality improvement.

1 Introduction

A long-term study of approaches to wastewater treatment held in MSUCE shows that the most common strategy in the field of construction and reconstruction of WWTPs is the maximum increase of capital and operating costs. The reform of the utilities sector in Russia has not led to its improvement, moreover, after the transfer of responsibility for the majority of treatment facilities to the municipalities, their operation and the condition deteriorated significantly due to the following reasons:

- lack of qualified staff that may carry out operation of the sewage treatment facilities;
- lack of appropriate investments for operation and quality improvement of wastewater treatment facilities;
- lack of manufacturing base for production of equipment for treatment facilities;
- lack of clear program of sewage treatment systems modernization.

* Corresponding author: makishana@mgsu.ru

2 Modern requirements

According to the state water account, total discharge of wastewater into surface water bodies consists of the following positions:

- Polluted wastewater – 182.4 million cubic meters;
- Without purification – 22.0 million cubic meters;
- Insufficiently purified – 160.4 million cubic meters;
- Standard clean 3.6 million cubic meters;
- Required treatment – 10.5 million cubic meters.

Table 1. Amount of pollutants in wastewater per person worldwide.

Countries	BOD [kg per person per year]	Suspended solids [kg per person per year]	Total nitrogen [kg per person per year]	Total phosphorus [kg per person per year]
Denmark	20–25	30–35	5–7	1.5–2
Brazil	20–25	20–25	3–5	0.6–1
Egypt	10–15	15–25	3–5	0.4–0.6
Germany	18–22	30–35	4–6	1.2–1.6
India	10–15	-	-	-
Italy	18–22	20–30	3–5	0.6–1
Sweden	25–30	30–35	4–6	0.8–1.2
Turkey	10–15	15–25	3–5	0.4–0.6
USA	30–35	30–35	5–7	1.5–2
Russia	14.6	23.7	2.9	1.2

Table 2. Wastewater features in Russia and Europe.

Pollutants	Russia	Europe
BOD_5, mg/l	135–176	300–350
NH_4, mg/l	26–35	30–50
Total nitrogen, mg/l	30–38	50–80
BOD_5 /N ratio	4–5	6–7
Phosphates, mg/l	2.3–3.4	7–10
Total phosphorus, mg/l	4–9	10
BOD_5 /P ratio	20–34	30

Most of municipal treatment facilities of housing and communal services have a significant degree of technical wear and obsolescence. Thus, if general condition of WWTP is considered only 5% of the discharged wastewater meet the requirements; 20% – qualify for the main indicators, but do not remove nutrients to the necessary values;

75 percent do not meet the requirements for wastewater discharge into water bodies of different values.

There are several reasons why this situation happens. The main factor is the quality of sewage that comes to WWTPs. Table 1 shows data on the number of impurities per person per year in different countries of the world. In Russia, the amount of pollution per capita is several times less than in other countries and this fact determines the quality of waste water, which comes at the sewage treatment plant (table 2) [1–3].

The pollutants ratio needed for normal biological treatment is far from required. This factor has a significant influence on the quality of treated wastewater, as well as the technologies that can and should be applied in the Russian Federation.

We cannot dwell also on the quality requirements of treated wastewater for discharge into the fishery reservoirs (Table 3) [4–6].

Table 3. Requirements for treated waste water in Russia.

Pollutants	Limited concentration for discharge in water bodies
BOD_5 [mg/l]	3.0
$N\text{-}NH_4$ [mg/l]	0.39
$N\text{-}NO_3$ [mg/l]	9.1
$N\text{-}NO_2$ [mg/l]	0.02
$P\text{-}PO_4$ [mg/l]	0.2
Suspended solids [mg/l]	5–6

3 Current conditions

A factor of a top importance is efficiency of technological schemes applied on the wastewater treatment plants, which fail to ensure a quality of treatment stated in the regulation documents especially when removal of nitrogen and phosphorus is concerned. Efficiency of ammonia removal may be, in particular, from 20 to 40%, however if up-to-date technologies implemented that value should be at least 80–90%. Similar solutions exist to reduce the concentration of phosphate in treated water [7–9].

Summarizing of above said the following issues could be highlighted:

- Age of WWTP in Russia – the majority of them were launched before year 1990;
- The quantitative ratio of the major contaminants BOD:N:P does not satisfy the required proportion for the biological purification of wastewater;
- Overall reconstruction of WWTP in most cases requires large costs that leads to slow, partial or insufficient reconstruction;
- The use of modern technology with no consideration of local conditions;
- The poor operation of treatment facilities.

The research carried out in MSUCE showed that if operation of well functioning sewage treatment plants is examined the following factors of influence may be pointed out (fig. 1).

The main goals of the WWTP reconstruction to aim at are the following:

- Improvement of the ecological status of water bodies of the Russian Federation
- Achievement the required quality of treatment
- Capacity increase of WWTP;
- Automation of WWTP;
- Reduction of sewage sludge amount.

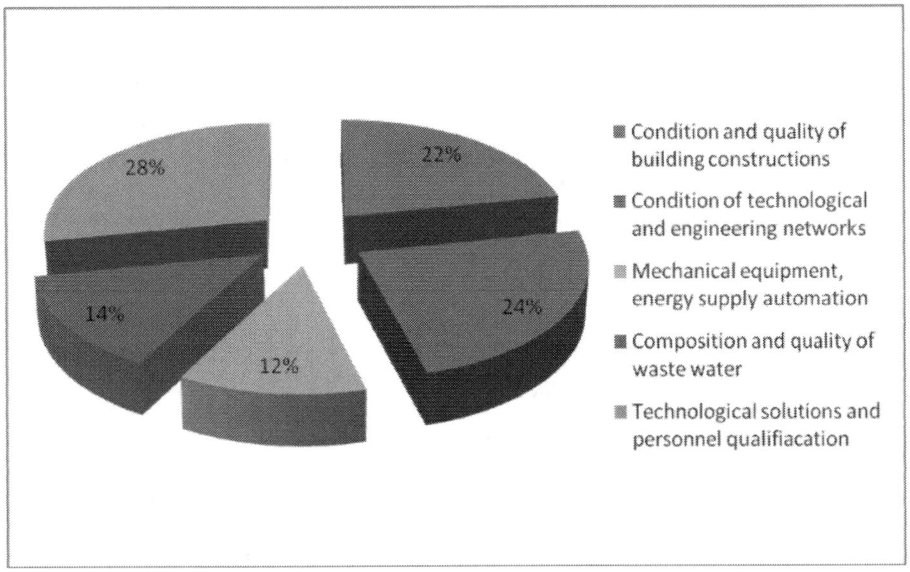

Fig. 1. Factors that influence the waste water treatment plats' operation.

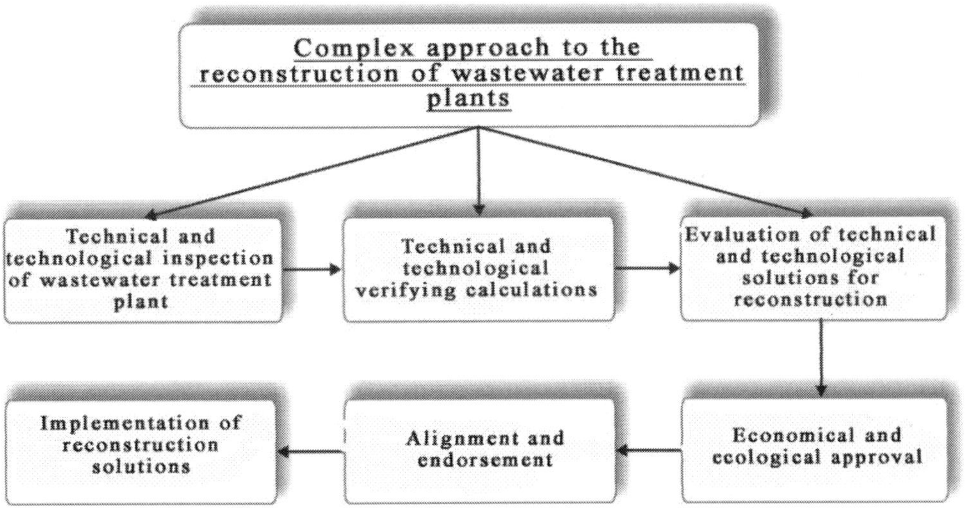

Fig. 2. Stages of reconstruction.

At the same time there is a strong need to focus on the priority reconstruction steps including the following measures [10–14]:

- Flow and concentration control of wastewater delivered to treatment facilities;
- Complex treatment of wastewater providing removal of nutrients;
- Sludge treatment of sewage sludge.

If the goals and measures of modernization are set together solutions to this situation are seen as follows:

- Complex approach – upgrade of the entire treatment facility, not single elements (fig. 2)
- Application of modern technologies considering local (Russia in general or one of its regions in particular) operating conditions of WWTP;
- Design of sewage treatment plants to meet regulatory requirements for wastewater discharges in Russian Federation.

In each case, before design begins, it is necessary to conduct pre-design comprehensive study of the object to estimate its current state, quality and amount of incoming wastewater. Only afterwards, it normally becomes possible to recommend how to reconstruct the WWTP with maximum use of existing facilities and to implement modern technologies of sewage treatment suitable for the particular object.

Methods of treatment facilities reconstruction of medium and small towns largely differ from the techniques of reconstruction of WWTP for large and mega cities mainly because of irregularity of waste water inlet and wastewater quality. For example, large industrial factories have slighter impact on entire volume of wastewater in the big city than in small one.

4 Approaches of reconstruction

First stage of reconstruction of systems and structures of water supply and sanitation has to be aimed at gradual improvement of quality of purified water, without shutting down facilities and their decommissioning.

Within the following reconstruction process, the problem of how to increase the capacity of sewage treatment plants has to be solved first (the quality should comply with so called "full biological treatment of waste water"), quality improvement actions to meet current requirements (BOD, suspended solid, nitrogen and phosphorus compounds etc.) go next [14–16].

It should not be forgotten that existing regulatory document contain almost no instructions on how to run the process of reconstruction and to implement designed solutions for systems and facilities sanitation. Moreover, verifying calculations that based on these documents may not consider current requirements for wastewater treatment.

Numerous researches carried out in MSUCE showed that if a model (typical) projects is applied for several WWTP the sequence and solutions for reconstruction for each of these WWTP may be very different due to specific local conditions. This is the main difference between the design of new treatment facilities and reconstruction of existing structures.

Existing wastewater treatment plants differ from each other by many factors [17–19]:

- Amount of incoming wastewater;
- Quality of wastewater, which depends on type of the industrial enterprises in the city, the capacity and quality of local industrial wastewater treatment plants;
- Degree of physical and moral deterioration of the structures;
- Year of construction and launch;
- Features of WWTP operation;
- Conformity of installations and treatment technology to design decisions;
- Previously conducted reconstruction of individual units or structures.

The complex of works on restoration and reconstruction of the existing sewage treatment plants involves a number of organizational, technological, technical and economic activities, the composition of which depends on the capacity of the treatment facilities and the expected results of the reconstruction. The main activities are [20–22]:

- Examining of the current documents of existing treatment plants and task definition for the reconstruction according to future development of the city.
- Determination of composition and quantity of incoming wastewater;
- Inspection of the facilities, tanks and equipment in order to determine their current state and efficiency of the performance;
- Verifying calculation in order to define current capacity of the plants;
- Evaluation of pre-design solutions for reconstruction on the basis on acquired data and technical task assignment;

- Project evaluation in accordance to given recommendation;
- Construction and assembly works (with the simultaneous technological supervision);
- Commissioning of equipment and wastewater treatment technology with the development of guidance (rules) for operation of sewage treatment facilities.

5 Conclusions

1. Reconstruction of the entire WWTP requires large investments; however there are several ways to reduce costs and to raise overall efficiency of the process.
2. Reconstruction requires reasonable approach in the way to implement the best suitable technology for the specific local conditions.
3. Reconstruction requires a certain procedure to be followed. If this sequence is ignored the aims of the reconstruction may be partially achieved or not achieved in total.

References

1. A.G. Pervov, A.P. Andrianov, T.P. Gorbunova, A.S. Bagdasaryan, Petr. Chem. **55**, 10, 879–886 (2015)
2. I. Gulshin, A. Kuzina, IJAER **10**, 21, 42618–42623 (2015)
3. A. Volkov, V. Chulkov, R. Kazaryan, M. Fachratov, O. Kyzina, R. Gazaryan, AMM **580-583**, 2281–2284, (2014)
4. A.G. Pervov, A.P. Andrianov, E.B. Yurchevskiy, Petr. Chem. **55**, 10, 871–878 (2015)
5. E. Gogina, A. Pelipenko, MATECCONF **73**, 03007 (2016)
6. V.N. Varapaev, S.A. Doroshenko, A.Y. Trotsko, A.V. Doroshenko, IJAER **10**, 21, 42588–42592 (2015)
7. O. Kuzina, E. Pankratov, V. Tkachev, MATECCONF **86**, 05023 (2016)
8. V. Orlov, A. Andrianov, AMM **580–583**, 2398–2402 (2014)
9. V.N. Varapaev, A.V. Doroshenko, I.Y. Lantsova, Pr. Eng. **153**, 816–823 (2016)
10. N. Zaletova, Y. Voronov, N. Makisha, IJAER **10**, 21, 42454–42455 (2015)
11. A.A. Kulakov, Wat. Ecol. **1**, 61, 26–40 (2015)
12. N. Makisha, E. Gogina, AMM **361–363**, 628–631 (2013)
13. E.S. Gogina, O.A. Ruzhitskaya, O.V. Yantsen. AMR **919–921**, 2145–2148 (2014)
14. N. Makisha, E3sconf **6**, 01002 (2016)
15. A.G. Pervov, A.P. Andrianov, DWT **35**, 1–3, 2–9 (2011)
16. E. Gogina, I. Gulshin, AMM **580-583**, 2367–2369 (2014)
17. A. Volkov, O. Kuzina, Pr. Eng. **153**, 838–843 (2016)
18. N. Makisha, Pr. Eng. **165**, 1087–1091 (2016)
19. A.A. Volkov, A.V. Sedov, P.D. Chelyshkov, D.A. Lysenko, A.V. Doroshenko, IJAER **10**, 22, 43269–43272 (2015)
20. N. Makisha, Pr. Eng. **165**, 1092–1097 (2016)
21. E. Gogina, I. Gulshin, Pr. Eng. **153**, 189–194 (2016)
22. A.G. Pervov, A.P. Andrianov, V.A. Chukhin, R.V. Efremov, IJAER **10**, 22, 43517–43525 (2015)

Transfer reservoir as a new solution for transfer of stormwater to water receivers

Robert Malmur[1,*]

[1]Institute of Environmental Engineering, ul. Brzeźnicka 60a, 42-200 Częstochowa, Poland

Abstract. With frequent heavy rainfalls in summer in Poland and fast-melting snow in spring leading to flooding of sewage systems (due to excessive filling levels in water receivers or difficulties with temporary retention of the excess stormwater), a variety of systems are being developed to facilitate transfer of the stormwater to water receivers. Outflow of the excess stormwater is usually ensured by the use of gravitational outflow collectors that connect stormwater drains with waterways. The transfer occurs during intensive precipitation, when the excess wastewater overflows through stormwater drains and is transferred directly to water receivers in order to relieve wastewater treatment plants or to minimize diameters of sewers. These systems are useful wherever the filling levels in waterways are not very high or the sewerage system is located relatively high with respect to the water receiver i.e. outflow collector is located on a steep slope. In such cases, the stormwater that flows through a waterway cannot be returned to the outflow collector. If the gravitational flow is impossible e.g. due to the excessive filling level of water receiver, stormwater can be transferred by means of a variety of modern solutions, such as retention and transfer reservoirs. These reservoirs are supposed to ensure partial retention of the excess stormwater and transfer of this water to water receivers, either gravitationally or forced mechanically, depending on the filling level in the waterway. Furthermore, these reservoirs prevent wastewater from being returned to the system during suddenly rising levels. One of the solutions is offered by the reservoir presented in this paper. The transfer reservoir for the stormwater presented in this paper might be successfully used in modernization of current sewage systems, ensuring the reliability of operation and a more effective wastewater transfer than the systems used to date. All the reservoirs of this type are characterized by similar design and function and guarantee that the wastewater might be transferred regardless of the conditions in the water receiver. An essential feature of these reservoirs is the use of the effective method to control suction and pumping units.

* Corresponding author: rmalmur@is.pcz.czest.pl

1 Introduction

Various countries all over the world use different solutions to the problem of wastewater transfer. There are plethora of solutions in the form of stationary pumping stations and portable pumping stations (e.g. in China). However, the outflow of excess wastewater from precipitation occurs gravitationally through outflow collectors that connect stormwater with waterways. Transfer of the wastewater directly to the water receiver during intensive precipitation occurs through stormwater drains. The aim of the use of stormwater drains is to relieve wastewater treatment plant and to avoid the necessity of the use of high diameters of sewers [1, 2]. These systems are useful wherever the filling levels in waterways are not very high or the sewerage system is located relatively high with respect to the water receiver. In such cases, the stormwater that flows through a waterway cannot be returned to the outflow collector.

A flap return valve is often installed at the outlet from the outflow collectors to prevent from flooding of the protected areas. Flap valves prevent from the backflow in the stormwater system caused by high levels of water in rivers [3]. However, in practice, these valves are not always used or are left without adequate supervision and maintenance. Consequently, poor state of repair usually limits efficient operation of these valves [4].

Unsecured outlets of outflow collectors or inefficiently operated flap valves have led in recent years to numerous instances of flooding in residential areas of many cities in Poland.

Another problem is observed in areas directly adjacent to rivers or artificial dams where raised water levels are much higher than land elevation. Such depressed areas with respect to constant filling on the other side of the embankment are deprived of natural conditions for water outflow from their drainage basin.

In such cases, building adequate stations for transfer of water from water receiver is critical. In practice, the pumping stations are located at the points of wastewater discharge.

In any case, if the surface water (also including stormwater from precipitation) is released directly to the water receiver (river), the protection in the form of flap valves installed on the river side of the embankment should be monitored and maintained at regular intervals. However, they become inaccessible in case of raised river levels [2].

The filling levels that make gravitational outflow impossible lead to the need for building a system of transfer of wastewater from the protected area to the water receiver.

The inefficient and often unreliable solutions of wastewater transfer have become the reason for starting the research on new and more efficient solutions to this problem.

Retention and transfer reservoirs for stormwater took over the function of the effective transfer of wastewater from precipitation to the water receiver. Wastewater transfer occurs if gravitational outflow is inhibited periodically or permanently.

Periodical inhibition of the transfer of wastewater from precipitation can be caused by high filling levels in the river, which are also observed in the period of intensive precipitation. Furthermore, permanent inhibition is observed if the dewatered areas are located in the depression with respect to the filling levels in the receiver due to the accumulation of water [5, 6].

Therefore, retention and transfer reservoirs can be used in the following cases [3]:

- in separated sewerage system at the outlets of main collectors in the stormwater system (Fig. 1a);
- in the combined sewer system at the stormwater channel outlets (Fig. 1b):
- in the combined sewer system at the stormwater channel outlets with simultaneous pumping of sanitary wastewater to wastewater treatment plant (Fig. 1c);
- in the case of collectors used to transfer wastewater from the treatment plant (Fig. 1d).

One of the most recent solutions of stormwater transfer to receivers is the use of the reservoir presented in this paper.

 DOI: 10.1051/e3sconf/20172200110

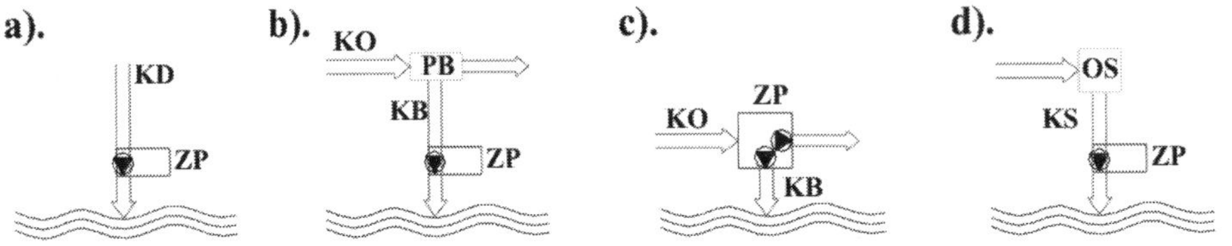

Fig. 1. Design of basic systems of reservoirs operated within different wastewater system arrangements. ZP – transfer reservoir; ZO – combined sewer channel; KB – stormwater channel; KD – stormwater collector; PB – stormwater drain; OS – wastewater treatment plant; KS – collector used to discharge wastewater after treatment.

2 Characteristics of the transfer reservoir

The transfer reservoir for stormwater discussed in this study can be built near channels of water receivers where the wastewater is supposed to be transferred to. The reservoir is supposed to be built near the flood embankment or embedded in the embankment [7, 8]. The reservoirs of this type can be used if the water receiver is located higher than the sewerage system and direct outflow of sewage to the receiver would be impossible.

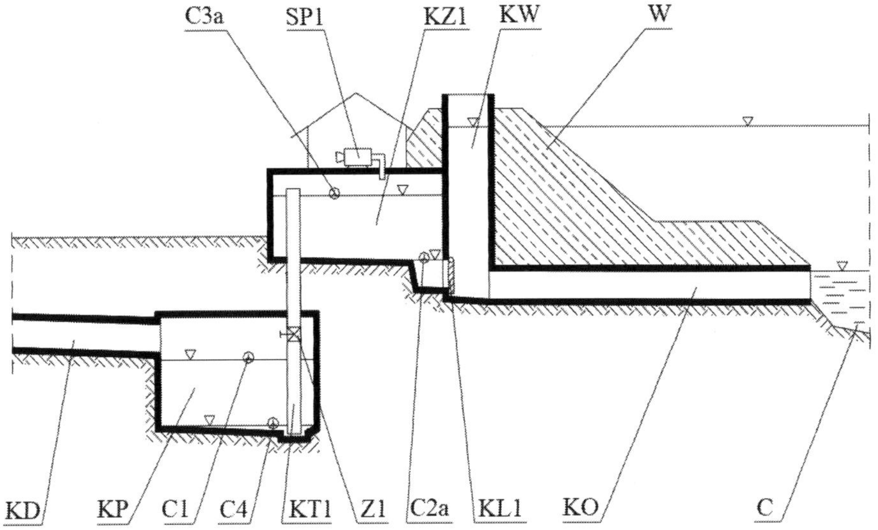

Fig. 2. Transfer reservoir of stormwater.

Principally, the reservoir is comprised of the following components: flow chamber, two pump chambers and tower chamber (Fig. 2 and 3). Wastewater to the reservoirs are transported to the flow chamber through outflow collector and discharged from the tower chamber to a water receiver through an outflow collector. The pump chambers are equipped in one-way flap valves, with the flow oriented towards the tower chamber. The flow chamber is connected with the pump chambers through pumping channels build in lower part of the flow chamber and in front part of pump chambers. Pump chambers contain cut-off valves that prevent from air access in particular cases. The reservoir is also equipped in electrical pump-compressor unit located over the pump chamber and controlled by means of sewage level sensors build in the pump chambers and the flow chamber. The pumping units comprise of air-tight pump chamber and pump-compressor unit connected by means of suction and pumping channel through the roof to the pump chamber. The pump channels are connected to the hydraulic system between one-way flap valves build in the bottom zone with the flow oriented towards the tower chamber, whereas the pumping channels are connected with the flow chamber.

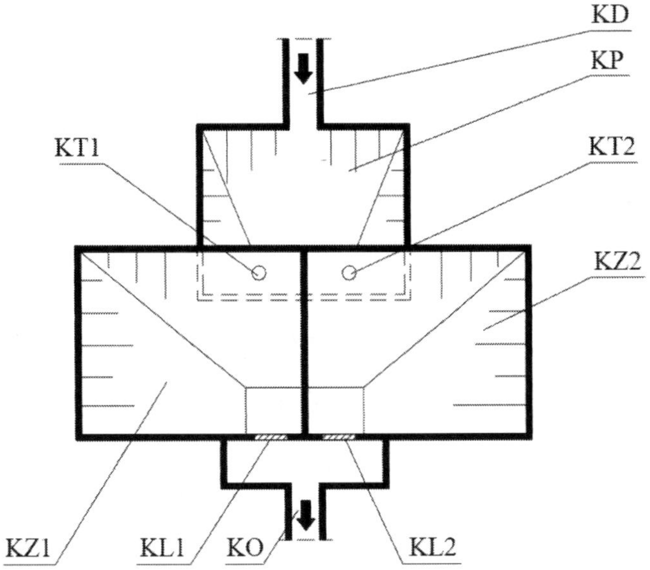

Fig. 3. Top view of the transfer reservoir.

An essential structural feature of the reservoir is that the flow chamber is built below the pump chambers. With this solution, sewage are transported to the flow chamber and then, through the pumping channels, to the pump chambers. Depending on the sewage level in the water receiver, sewage from pump chambers are transferred gravitationally or forced by pressure.

3 The design of transfer reservoir

Sewage flows to the reservoir through the inflow collector (KD) to the flow chamber (KP) (Fig. 2). There is a small cavity at the end of the flow chamber (KP), where pumping channels (KT1 and KT2) are connected. These channels are used to connect the flow chamber (KP) with pump chambers (KZ1, KZ2) (Fig. 3) through the wholes made in the front part of these chambers. The top of the pumping channels (KT1, KT2) reaches almost to the roof of the pump chambers (KZ1, KZ2). Pumping channels (KT1, KT2) have cut-off valves (Z1, Z2), build in the upper part of the flow chamber (KP). Pump chambers (KZ1, KZ2) are connected with the tower chamber (KW) through one-way flap valves (KL1, KL2) which operate automatically, oriented with the flow towards the tower chamber (KW).

The roof of the pump chamber (KZ1, KZ2) contains two pump-compressor units (SP1, SP2), connected through pipes with above-roof space of these chambers. The pump-compressor units (SP1, SP2) are of similar design and their flow rate equals at least the equivalent maximum inflow Q_{dopmax} present in a particular drainage basin. The pump chambers (KZ1, KZ2) are built as air-tight, they have similar volume and are designed next to each other.

The tower chamber (KW) is higher than the other chambers and reaches the height at least equal to the height of the crest of the flood embankment (W) of the waterway (C). Sewage from the tower chamber (KW) flows to the waterway (C) through an outflow collector (KO). The outflow collector (KO) runs along the bottom of the waterway (C) under the flood embankment (W), whereas the outlet of the collector is located under the water surface in the water receiver (C).

Sewage level sensors (C1, C2a, C2b, C3a, C3b, C4) that control operation of pump-compressor units (SP1, SP2) are built in pump chambers (KZ1, KZ2) and flow chamber (KP). The (C1) sensor is installed in the flow chamber (KP) at the height of several centimetres below the bottom of the inflow collector (KD), whereas the (C4) sensor is located at the bottom of the chamber, at the height of ca. 5 cm over the lower part of the pumping channels (KT1, KT2). Sensors (C2a, C2b, C3a and C3b) are built in the pumping chambers (KZ1, KZ2). Lower sensors (C2a, C2b) are located at the upper level of the flap chamber (KL1, KL2), whereas the upper sensors (C3a, C3b) are located at the height of ca. 5 cm below the upper part of the pumping channels (KT1, KT2).

4 The principle of operation of transfer reservoir

The sewage from sewage system flows through inflow collector (KD) to the flow chamber (KP), filling it up to the sensor level (C1) (Fig. 4). As soon as this level is reached, the pump-compressor unit (SP1) is actuated at the suction side. The pump-compressor unit (SP1) sucks the air out of the pump chamber (KZ1) and then, through a pumping channel (KT1), with the (Z1) cut-off valve open, starts sucking the sewage out of the flow chamber (KP). Flap valves (KL1, KL2) are closed as a result of the pressure from the water column in tower chamber (KW). The sewage pumped in the pumping channel (KT1) fills the pump chamber (KZ1) until the level of the (C3a) sensor (Fig. 5) is reached. With the signal from the C3a sensor, the pump-compressor unit (SP1) switches from suction into pumping and the cut-off valve (Z1) is closed in the pumping channel (KT1). The pressure in the below-roof space of the pump chamber (KZ1) rises, causing the flap valve (KL1) to open and sewage to be pumped to the tower chamber (KW). With low filling level in the waterway (C), the sewage opens the flap valve (KL1) with its own pressure force and flows through tower chamber (KW), outflow collector (KO) and finally gets into the water receiver (C). Furthermore, with elevated levels, pressure of the sewage is insufficient to open the flap valve (KL1). Therefore, the use of the pump-compressor unit (SP1) is needed (Fig. 6).

Capacity of the pump-compressor unit (SP1) was chosen so that emptying the pump chamber (KZ1) is possible with the outflow rate slightly higher than the measurable inflow. The pump-compressor unit (SP1) empties the pump chamber (KZ1) until the level of the C2a sensor is reached (Fig. 7).

Signal from the (C3a) sensor also causes the pump-compressor unit (SP2) to switch on at the suction side in the pump chamber (KZ2). When the pump chamber (KZ1) is being emptied, sludge from the flow chamber (KP) is sucked to the pump chamber (KZ2) through the pumping channel (KT2), with the cut-off valve open (Z2). Filling the pump chamber (KZ2) is continued until the time of reaching the level of the sensor (C3b), when the pump-compressor unit (SP2) switches into pumping. The cut-off valve (Z2) in the pumping channel (KT2) closes, preventing from unwanted air loss. At the same time, the pump chamber (KZ1) is emptied until the level reaches the sensor (C2a) and the sewage can be accepted again. The operation of the pump chambers (KZ1, KZ2) occurs alternately, which means that when one of them is filled, the other one is emptied. Pump-compressor units (SP1, SP2) operate until the sewage in the flow chamber (KP) reaches the level of the sensor (C4).

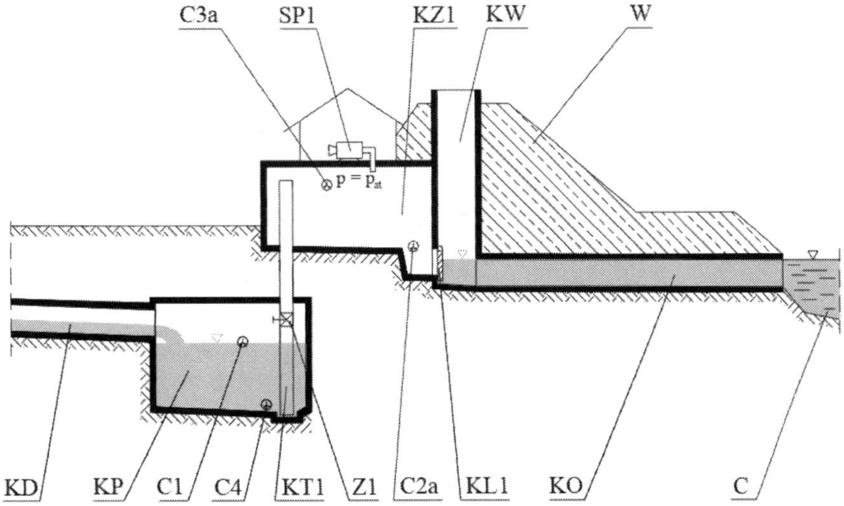

Fig. 4. Filling the flow chamber (KP) up to the level of the sensor (C1) in the transfer reservoir.

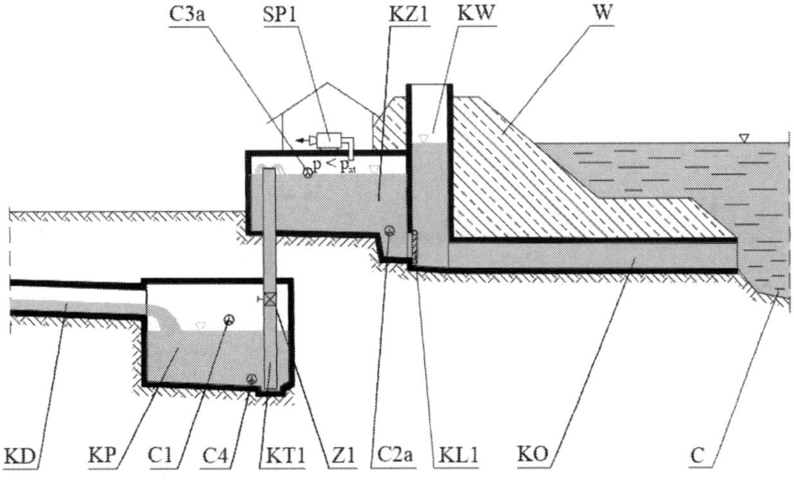

Fig. 5. Filling the pump chamber (KZ1) up to the level of the C3a sensor in the transfer reservoir.

While designing this type of the transfer reservoirs, it should be noted that the sensor (C4) should be located above the lower part of the pumping channel (KT1, KT2), preventing from unnecessary suction of the air out of the flow chamber (KP). Sewage from the flow chamber (KP) after switching on the pump-compressor unit by means of the signal from the sensor (C1) is sucked with a particular delay since the pump-compressor unit sucks the air out of the pump chamber (KZ1) first and then sucks out the sewage. Therefore, the (C1) sensor should be mounted at the height that prevents (during this delay) from filling the flow chamber (KP) to the level of the inflow collector (KD) bottom. The pump-compressor units (SP1, SP2) should be selected so that at the moment of switching on, the time of suction of the air from the pump chambers (KZ1, KZ2) is shortest possible.

 DOI: 10.1051/e3sconf/20172200110

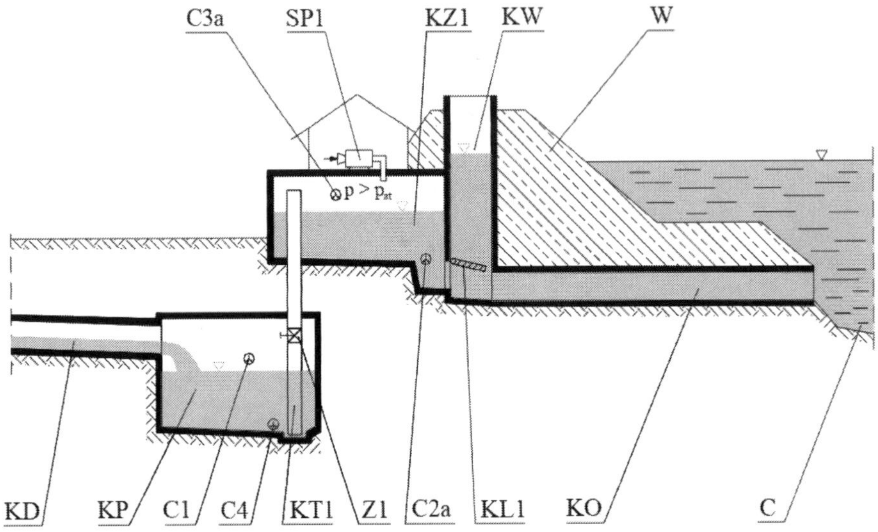

Fig. 6. Pneumatic flow of sewage in the transfer reservoir.

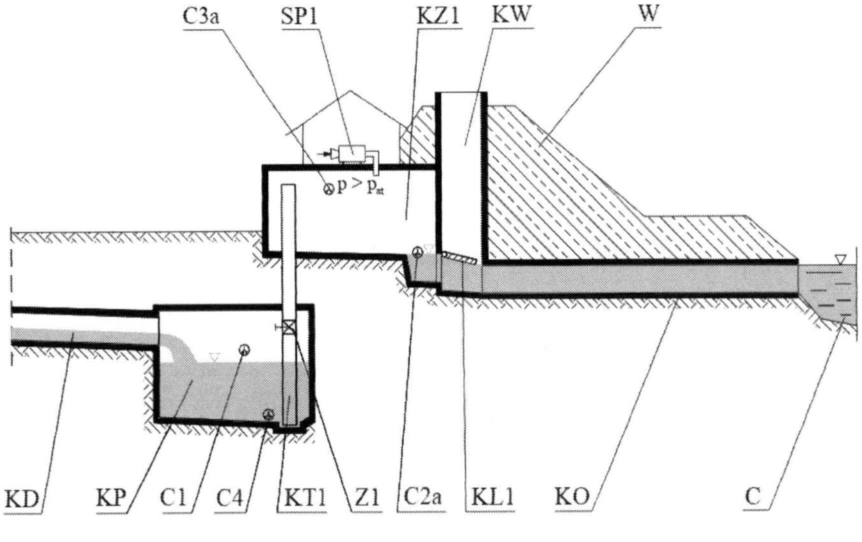

Fig. 7. Emptying the pump chamber (KZ1) to the level of the sensor (C2a) in the transfer reservoir.

5 Conclusions

Present solutions for stormwater transfer to water receivers used in the most of the cities of Poland are obsolete, unreliable and, apart from the extension, need urgent modernization. The unreliability of these systems is manifested primarily with frequent flooding in the sewage system in the case of elevated levels in water receivers.

With respect to the number of problems to be solved in the area of municipal investments connected with sewage disposal, it is very important and urgent that the methods of design and implementation of modern and safe solutions are developed.

Regardless of the size of the drainage basin and the used sewage system, transfer reservoirs should be viewed today as one of the most basic components in contemporary systems of the transfer of the stormwater to water receivers. Building such facilities involves considerable investments while proper operation is possible only if appropriate design and high-quality suction and pumping units are used as they consume considerable amounts of electricity to transfer the wastewater from lower to higher level or to transfer the wastewater up and to the location that ensures its gravitational outflow at the same time.

The transfer reservoir for the stormwater presented in this paper might be successfully used in modernization of current sewage systems, ensuring the reliability of operation and a more effective wastewater transfer than the systems used to date. All the reservoirs of this type are characterized by similar design and function and guarantee that the wastewater might be transferred regardless of the conditions in the water receiver. An essential feature of these reservoirs is the use of the effective method to control suction and pumping units.

Using the reservoir needs a thorough analysis of the data about the drainage basin, sewage system, equipment and facilities involved while taking into consideration minimum costs of implementation of a particular investment [9].

The design of retention and transfer reservoirs in the stormwater systems represents the basic solution in terms of the effective transport of stormwater. Adequate determination of the sufficient retention capacity is of key importance to the process of design with consideration of their effect on both reliability and costs of investment and operation. The analysis of the method of dimensioning of retention and transfer reservoirs demonstrated that the major effect on the required retention capacity is from the precipitation data used in the calculations.

Regardless of the method of dimensioning of retention and transfer reservoirs, their performance should be increased through application of the dynamic control of wastewater outflow intensity. Temporal and spatial irregularity of real precipitation offers many opportunities connected with adjustment of control equipment to the current hydraulic load of the system, both in the reservoir filling and discharge phases.

The retention and transfer reservoir presented in this study is one of the proposals for the solutions to the problem of periodical flooding of urban areas [8]. The system of air suction and pumping used in this reservoir ensures more reliable operation than the system of wastewater pumps installed since it does not go into direct contact with wastewater that contains mechanical pollutants and are often chemically aggressive. The reservoir of this type, after small adaptation, can represent the solution which is alternative to those used in the sanitary waste pumping stations.

Application of this type of the reservoirs requires an in-depth analysis of the data on drainage basin, sewage system, equipment and facilities involved so that minimization of costs of implementation of the investment is achieved.

The initial version of the design of the transfer reservoir assumed almost ideal conditions of its operation. The cycle of filling and emptying of transfer chambers necessitated maintaining balance between these processes. This involved maintaining constant value of flow rate at the inflow of wastewater to the currently filled chamber and ensuring the same or greater outflow from the emptied chamber. This was supposed to ensure that the cyclic operation of the reservoir remains undisturbed. With another approach to this problem, the times of filling and emptying the transfer chambers should be equal. Reaching such conditions in practice is impossible, thus the effect of non-cyclic operation of the chambers had to occur. With varied and lower than nominal values of rates of flow of wastewater to the reservoirs, only partial rather than complete filling of the second chamber occurs during emptying of the first one. Consequently, after decompression of the air contained in the first chamber, it is re-filled, while the other chamber remains filled partially. This is repeated until the second chamber in these stages is filled entirely and switches in (by means of the sensor) the air compressor which empties the chamber. However, the real problem is the risk of locking the accumulation capacity of the transfer chambers. This might occur when the second chamber is filled with an insignificant volume while the process of emptying in the first chamber has already begun. In such cases, both chambers are being emptied and cannot receive wastewater to be retained. Consequently, the natural retention occurs in the reservoir inflow channel. If natural conditions allow for a safe yet short-term channel detention, causing raised filling level at the inflow, this operational mode of the transfer reservoir is acceptable.

The designed transfer reservoir can be also used to transfer the volume of wastewater from the drainage basin located higher to the basin situated at a lower level.

Acknowledgement

This scientific study was financed within the research project BS-PB 401-306/11.

References

1. M. Mrowiec, *Effective dimensioning and dynamic control of sewage reservoirs* (Monograph **171**, Published by University of Czestochowa, 2009) (in Polish)
2. W. Geiger, H. Dreiseitl, *New methods of draining storm water* (Projprzem, Warszawa 1999) (in Polish)
3. J. Łomotowski, *Rain water and extreme events* (Seidel-Przywecki, 2011) (in Polish)
4. W. Błaszczyk, M. Roman, H. Stamatello, *Sewerege* (Arkady, Warszawa 1974) (in Polish)
5. R. Malmur, A. Kisiel, M. Mrowiec, J. Kisiel, CzT. **18**, 105, (2008)
6. M. Mrowiec, T. Kamizela, M. Kowalczyk, Environ. Prot. Eng. **35**, 2 (2009)
7. R. Malmur, Rocz. Ochr. Sr. **13** (2011)
8. R. Malmur, *The energy saving transfer reservoir in a drainage system Geminus-ET* (The Polish Patent Office, Patent No. 209453. Warszawa, 2011) (in Polish)
9. M. Mrowiec, A. Kisiel, R. Malmur, J. Kisiel, Pol. J. Environ. Stud. **16**, 2A, part III (2007)

E3S Web of Conferences **22**, 00111 (2017) DOI: 10.1051/e3sconf/20172200111
ASEE17

Various methods of heat supply for a building which is operated periodically during the year

Marek Małetka[1,*], and *Marta* Laska[2]

[1]Glasgow Caledonian University, School of Engineering and Built Environment, Cowcaddens Rd, Glasgow G4 0BA, United Kingdom
[2]Wroclaw University of Science and Technology, Faculty of Environmental Engineering, Wybrzeże Wyspiańskiego 27, 50-370 Wrocław, Poland

Abstract. Stand-alone buildings operated periodically require heat supply for hot water and heating purposes to be carefully analyzed in terms of the technical capabilities, the energy and financial outlays. The paper presents the analysis of heat supply for hot water purposes and central heating in the stand-alone cloakroom building located in Poland. The analysis is undertaken for different variants of heat delivery for a building from electric heaters, gas boiler and district heating solutions to renewable sources applications, namely solar panels and heat pumps. For each solution, usage of usable, final and primary energy was calculated. Also the financial analysis for investments and energy costs were carried out. This analysis has been done in according to SPBT and NPV method for different levels of building use.

1 Introduction

1.1 Domestic hot water

Because of growth of sophisticated methods of heat obtain a slight inconvenience in system's designing process can appear. One of the fundamental problems is to term hot water demand for specific type of a building like stand-alone cloakroom. In the Polish regulations and other publications there are lacks of clear specified values of hot water demand which should be accepted for this kind of a building. The authors [1] presents the usage of domestic water in Solar City and compare it with former studies in other parts of Canada and North America. The [2] compared usage of DHW (Domestic Hot Water) in Spain, Japan, United States and China. The data for Switzerland, Hungary and Finland can be also found in [3]. However all these publications describe the research for residential buildings. Some data dedicated for stand-alone cloakroom can be found in [4–7]. However it can be seen that each author gives different data, which are set on their own experiences (Table 1).

As can be noticed in Table 1 the highest value of hot water demand is given by [4] which is much bigger than value given in polish standard [5]. Similar data can be found in

* Corresponding author: marek.maletka@gcu.ac.uk

 DOI: 10.1051/e3sconf/20172200111

[5, 6]. The lowest need for hot water in building used for sport activities is defined in [7] and to avoid overestimation for the following analysis this value will be taken.

Table 1. Summary of hot water demand according to various authors.

Literature source	Hot water demand	Hot water demand	Hot water temperature
	l/s	l/person	°C
[4]	0.167	50	45
[5]	0.150	45	60
[6]	0.133	40	60
[7]	0.073	22	55

1.2 Heating

The heating system design for a building which operates constantly during the year, the calculation procedure is determined by European Standard [8] were the external design temperatures for winter period are defied. However when the periodic operation of a building with only summer and transitional period is under considerations there is no information in engineering guidelines about the value of external temperature that should be applied in calculations. Similarly, when building works under non-occupied mode and the indoor temperature is maintained on lower heating set point temperature. Therefore 3 options has been investigated under the study, however the detailed analysis of these options is not a subject of this paper. The chosen method for further heat load calculations is based on stand-by indoor temperature and the external average of the minimum temperatures in the transition period [9, 10].

2 Building characteristic

The object which is considered is a stand-alone cloakroom building located next to the football pitch. The property is located in Lower Silesia District (Poland) in the second climate zone for which design external temperature during the winter is -18°C. The exact values of design parameters used for calculations are described in section 3.

The design building has an heating area of 105.7 m^2. It contains of two players cloakrooms, two showers, two toilets, referees changing room, coach room, disabled WC, store and technical room. The design conditions assume usage of hot water for 32 people at one time (beak between matches). This amount assumes the presence of 15 people in each player's locker room and two referees. The cloakroom is opened from March to October, from Monday to Saturday between 8:00 a.m. and 8:00 p.m., and on Sundays from 10:00 a.m. to 7:00 p.m. Frequency of water usage is 15 minutes for each 1,5 hour.

3 Energy investigation

The heat demand calculation for a building for a central heating purpose was based on the European Standard PN-EN 12831:2006. Because of the periodical utilisation of the building, there is no need to maintain the indoor temperature as a comfort one during the winter season.

First operational scheme is applied for spring and fall (between March to May and from September to October). It assumes the external temperature of -0.3°C [9]; the internal temperature is set on 20°C or 24°C depending on the room type [10]. For these assumption the design heat loss is 9601 W.

Second operational scheme is applied for winter when the building is non-occupied. The external temperature is defined on -18°C. For this condition, to avoid freezing of the building installation, it is needed to provide the stand-by temperature of 5°C. For these conditions in winter time the design heat loss equals 9019 W. Therefore for further considerations the higher value of 9.6 kW was taken into account.

The paper presents different variants of heat delivery for a building, namely: Solution (Sol.) 1, Option (Opt.) 1.1 Electric hot water heaters with the local electric heating; Opt. 1.2 Electric hot water heaters with central heating; Sol. 2 Gas boiler with domestic hot water cylinder; Sol. 3 District heating substation; Sol. 4, Opt. 4.1 Flat plate solar panels, Opt. 4.2 Vacuum solar panels; Sol. 5, Opt. 5.1 Brine-to-water heat pump with horizontal ground collector, Opt. 5.2 Brine-to-water heat pump with vertical the bore holes.

3.1 Usable energy

The total annual demand for usable energy for the building was calculated in a basis of the Regulation of Minister of Infrastructure and Development [11]. The usable energy only for heating system equals 19006 kWh/a.

The usable energy for DHW was calculated in two different ways. The first approach followed the current Regulation of the Minister of Infrastructure and Development [11] that assumes the relation of DHW demand with the heated area of the building. Following this method the value of the usable energy is 252.6 kWh/a.

The second method bases on designed DHW use by the reference unit during the day. It is largely described in the previous Regulation of Minister of Infrastructure [12]. The energy value calculated by this method is 61945 kWh/a.

As one can notice, the value of useable energy calculated by a hot water consumption method based on the number of people in the building [12] is significantly different than that which is contained in the current Regulation [11]. This difference is largely due by water demand and reference method. In the first case the individual requirement is 0.25 dm^3/m^2 per day and the entire surface of the heated space is 105.7 m^2. It gives us 26.4 dm^3 of daily hot water demand. Relative to the second method this value is almost similar to that used by one person during one shower. Due to the type of use of the building which is located at the sports field where usually take place team games it can be concluded that during the day more than one person will take a shower.

The second issue aroused a considerable doubt in the usable energy demand calculation in the case of rigidly specified time of the building usage which is 365 days during the year. The legislature in this case didn't foresee the possibility of changing this value which in case of periodically operated building may appear significant differences in the results. For this reason, it can be concluded that the value calculated using the water consumption related to the number of people in a building is more likely while the value obtained in the first method may greatly underestimate the final results. Therefore the following calculations are based on the second methodology.

3.2 Final energy

The annual demand on the final energy was calculated according to the current Regulation of Minister of Infrastructure and Development [11]. Electric power and working time for specific auxiliary equipment was estimated on the basis data contained in the Regulation. As it is impossible to define the average daily amount of DHW used by the facility users, therefore the final energy consumption has been presented in the level of DHW usage

 DOI: 10.1051/e3sconf/20172200111

where 100% indicates the maximum and 0% the minimum value of DHW use. Summary of final energy demand for each solution is presented in the Fig. 1.

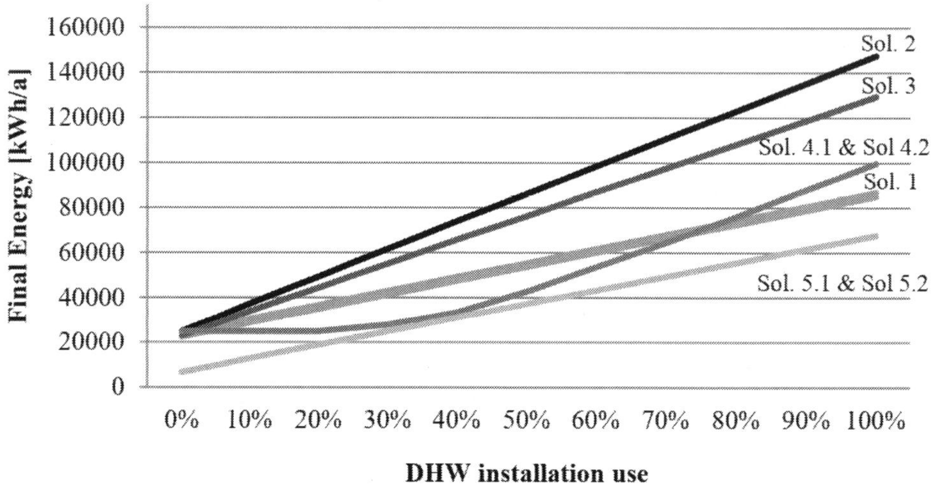

Fig. 1. Final energy demand according to percentage of DHW use.

Fig. 1 shows that solution 2 uses the highest final energy for any partial DHW demand. For the largest use of DHW this value is twice higher than that for solution five – based on brine-to-water heat pump. Slightly smaller demand has the solution with district heating – Sol. 3. The most interesting option in the energy point of view is solar panels (Sol. 4) where the demand for the final energy is changing in nonlinear way, namely: in the range of 0%–30% the final energy demand is higher only than fifth solution assuming heat pump; at 40% the final energy is comparable with Sol. 5; in the range of 50%–100% the final energy grow much faster than the demand for other solutions. What is more it reaches a greater value than electric heater for hot water use (Sol. 1), above 80%.

The difference between solutions results from the efficiency of the various elements of the system like: type of the heat source, its transfer and also method of heat storage.

3.3 Primary energy

Demand on primary energy has been calculated following [11] for particular solution. It takes into account the primary energy factor for electricity that varies depending on the source of energy. The outcome is presented in Fig. 2.

As can be noticed on the chart (Fig. 2) the highest demand on primary energy has Sol. 1 in both options and Sol. 5. It is interesting as the solution 5 assumes usage of renewable, highly efficient energy source. The heat pump converts electrical energy and therefore, according the [11], the value of 'effort coefficient' is the same as for any other electrical device. At the same time the heat pump solution characterizes the lowest demand on the final energy and thus the lowest operating costs.

According to the above chart the least impact on the environment has third solution based on a district heating network. Sol. 3 generates less than half of primary energy comparing the worst one – Sol. 1 and 5.

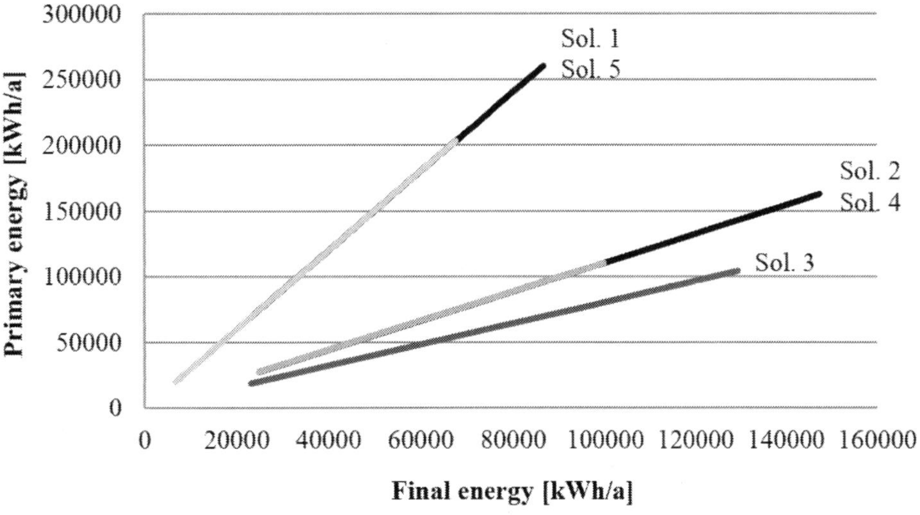

Fig. 2. Annual primary energy demand according to final energy.

4 Financial analysis

4.1 Investment cost

For each solution the investment costs based on prices in Polish Zloty (PLN) valid for 2016 were calculated (The average exchange rate is 1 PLN = $0.259). The summary is presented in Table 2.

Table 2. Investment costs.

Sol. 1		Sol. 2	Sol. 3	Sol. 4		Sol. 5	
Opt. 1.1	Opt. 1.2			Opt. 4.1	Opt. 4.2	Opt. 5.1	Opt. 5.2
PLN	PLN	PLN	PLN	PLN	PLN	PLN	PLN
57157	46043	33824	34161	91970	133286	115208	135640

4.2 Operating cost

Annual operating costs for each solution were calculated for final energy values and energy tariffs for electricity, gas, and district network heat, established by polish companies. All the tariffs are valid for December 2016.The calculation outcome is presented at Fig. 3.

The highest operating cost generate solutions based on electric hot water heaters (Sol. 1) Little less cost has district heating (Sol. 3) and gas boiler (Sol. 2). The most advantageous from the operating costs point of view is utilization of heat pumps.

The Fig. 4 shows price of 1 kWh for each analysed solution according to DHW use. The lowest and almost constant prices generate solutions with gas boiler, solar panel and heat pump (solution 2, 4 and 5). The 1 kWh price varies in a nonlinear way. All the solutions except electric heaters are characterized by minor differences with high DHW use. For solution one with low DHW use 1 kWh price is almost three times higher. It's caused by high fixed fees for this solution. This difference is changing a little with increasing of DHW

 DOI: 10.1051/e3sconf/20172200111

use. Change of 1kWh price for each solution is generally connected with fixed fees which user has to carry on despite the lack of DHW use.

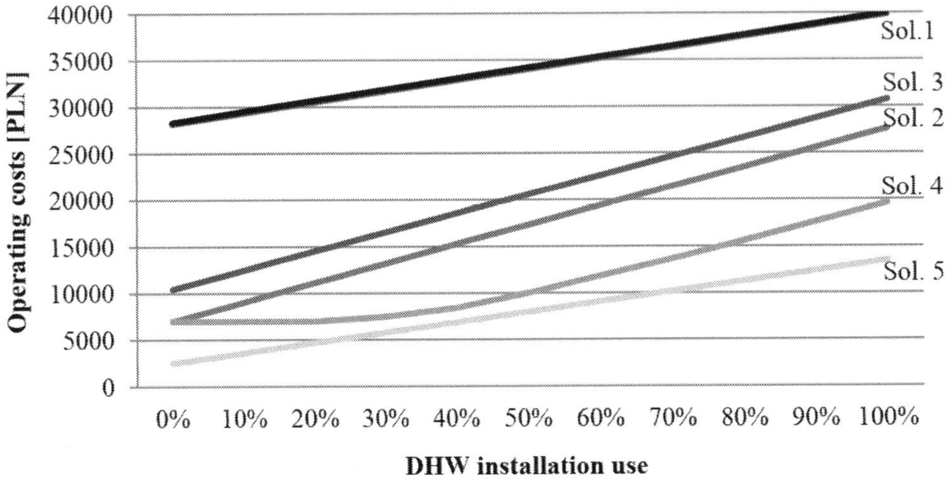

Fig. 3. Operating costs according to DHW use.

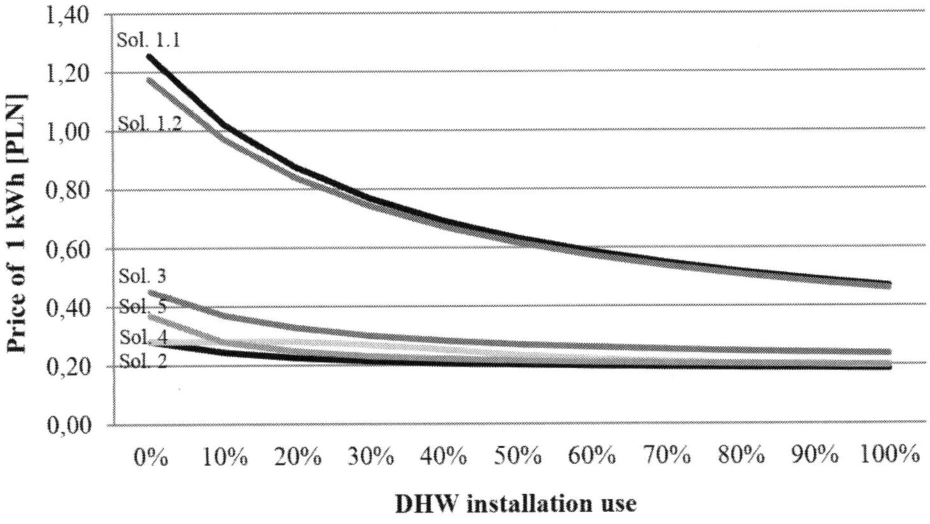

Fig. 4. Price of 1kWh according to DHW use.

4.3 SPBT and NPV calculations

To determine the economic efficiency of particular solutions Simply Pay Back Time (SPBT) and Net Present Value (NPV) were calculated. All the calculations relate to the cheapest solutions from investment point of view, namely Sol. 2. The discount rate was determined based on the average inflation rate since 2001 given by the National Polish Bank.

SPBT value depends generally on the DHW use and it decreases with increasing use. As can be seen on the Table 3 the investment cost of the first and third solutions will never payback compared to Sol. 2. Solutions with solar panels and heat pump have specified payback time. Both options of the fifth solution achieve rational payback time (<15 years) at 30% of DHW use. For Sol.4 this payback time is reached at 40% of DHW use and higher.

Table 3. SPBT value according to the cheapest solution – Sol. 2.

Solution		SPBT										
		0%	10%	20%	30%	40%	50%	60%	70%	80%	90%	100%
		year	year	year	year	year	year	year	year	year	year	year
Sol. 1	Opt.1.1	-	-	-	-	-	-	-	-	-	-	-
	Opt.1.2	-	-	-	-	-	-	-	-	-	-	-
Sol. 3	-	-	-	-	-	-	-	-	-	-	-	-
Sol. 4	Opt.4.1	-	28.2	14.1	10.2	8.5	8.0	7.7	7.5	7.4	7.3	7.2
	Opt.4.2	-	48.3	24.1	17.5	14.5	13.6	13.2	12.8	12.6	12.5	12.4
Sol. 5	Opt.5.1	18.0	14.9	12.6	11.0	9.7	8.7	7.9	7.2	6.7	6.2	5.7
	Opt.5.2	22.6	18.6	15.8	13.7	12.2	10.9	9.9	9.0	8.3	7.7	7.2

The most interesting is comparison between flat solar panels (Sol. 4, Opt. 4.1) and heat pump with horizontal ground collector (Sol. 5, Opt. 5.1). For the DHW use lower than 30% heat pump is more profitable than solar panels. This situation is changing for when DHW use is between 30%–60% then solar panels are more profitable. For 60% the payback value is more or less equal and for more than 70% heat pump is again more profitable.

Summarizing for low and very high DHW demand solution with heat pump is the most reasonable, but when the water use is more centred solar panels is better option.

Table 4. NPV value after15 years of building use.

Solution		NPV										
		0%	10%	20%	30%	40%	50%	60%	70%	80%	90%	100%
		PLN	PLN	PLN	PLN	PLN	PLN	PLN	PLN	PLN	PLN	PLN
Sol. 1	Opt.1.1	-	-	-	-	-	-	-	-	-	-	-
	Opt.1.2	-	-	-	-	-	-	-	-	-	-	-
Sol.3		-	-	-	-	-	-	-	-	-	-	-
Sol.4	Opt.4.1	-	-	-	11225	25295	31065	33817	36821	38315	39056	39796
	Opt.4.2	-	-	-	-	-	-	-	-	-	-	-
Sol.5	Opt.5.1	-	-	-	8991	20759	32526	44294	56061	67828	79596	91363
	Opt.5.2	-	-	-	-	327	12094	23862	35629	47396	59164	70931

According to NPV Sol. 1, 3 and Opt. 4.2 is not profitable compared to gas boiler for period of 15 years in any DHW use level and it will never pay-back. Only solutions with heat pump (Sol. 5) and planar solar panels (Sol. 1, Opt. 1.1) achieve positive values for 2.7% of discount rates after considering working time. However those profits are reachable only when DHW use level is more than 30% or 40% in case of heat pump system with vertical bore holes. It is caused definitely higher investment costs.

5 Conclusions

The paper presented the analysis of heat supply for stand-alone cloak room building. Different option of heat delivery has been investigated in terms of usage, final and primary energy demand together with financial analysis based on SPBT and NPV methods.

As it was mentioned at the beginning of the analysis there are certain difficulties in designing the heat source in these types of buildings. These difficulties are caused by a lack of accurate values of hot water consumption when given available literature values are divergent. Regarding heating, there is lack of a clearly defined procedure and design values for buildings operating only during summer and transitional period.

Also energy calculations are somehow difficult as the current Polish regulations refer the energy usage for DHW purposes to the heated area of the building.

Therefore to reflect the reasonable energy usage and heat demand, and thus to calculate costs and profitability of chosen solution in such buildings the individual analysis as presented in the paper should be carried out. However even then it is not easy to determine the superiority of one solution over another.

References

1. J.C. Evarts, L.G. Swan, Energ. Buildings, **58** (2013)
2. L. Zhang, J. Xia, J.E. Thorsen, O. Gudmundsson, H. Li, S. Svendsen, Energy, **116** (2016)
3. J. de Santiago, O. Rodriguez-Vialon, B. Sicre, Predictions, Energ. Buildings (to be published, http://dx.doi.org/10.1016/j.enbuild.2017.02.045)
4. H. Recknagel, E.-R. Schramek, *Compendium of Heating, Air-conditioning, Domestic Hot Water, Refrigeration* (Omni Scal, Wrocław, 2008) (in Polish)
5. PN-92/B-01706 Domestic water installations – Design requirements
6. Vaillant, Designer's guide (in Polish)
7. S. Mańkowski, *DHW system design* (Arkady, Warszawa, 1981) (in Polish)
8. PN-EN 12831:2006 Heating systems in buildings – Method for calculation of the heat load (in Polish)
9. Polish Ministry of Infrastructure and Civil Engineering - Typical meteorological years and statistical climatic data for building energy calculations (in Polish) (http://mib.gov.pl)
10. Rozporządzenie Ministra Infrastruktury z dnia 12 kwietnia 2002 w sprawie warunków technicznych, jakim powinny odpowiadać budynki i ich usytuowanie (z późn. zmianami) (Dz.U. z 2015 poz. 1422, Warszawa, 2015)
11. Rozporządzenie Ministra Infrastruktury i Rozwoju z dnia 27 lutego 2015 w sprawie metodologii obliczania charakterystyki energetycznej budynku lub części budynku oraz świadectw charakterystyki energetycznej (Dz. U. z 2013 r. poz. 1409, Warszawa, 2015)
12. Rozporządzenie Ministra Infrastruktury z dnia 6 listopada 2008 r. w sprawie metodologii obliczania charakterystyki energetycznej budynku i lokalu mieszkalnego lub części budynku stanowiącej całość techniczno-użytkową oraz sposobu sporządzania wzorów świadectw charakterystyki energetycznej (Dz. U. z 2006 r. poz. 1118,Warszawa, 2008)

E3S Web of Conferences **22**, 00112 (2017) DOI: 10.1051/e3sconf/20172200112
ASEE17

Approach to the implementation of linear-assisted-based battery chargers

Herminio Martínez-García[1,*], and *Encarna* García-Vílchez[1]

[1]Eastern Barcelona School of Engineering (Escola d'Enginyeria de Barcelona Est, EEBE), Department of Electronics Engineering, Tech. University of Catalonia (UPC – BarcelonaTech), Diagonal-Besòs Campus. Eduard Maristany Ave., nº 10 – 14, E-08019 – Barcelona, Spain

Abstract. This article shows the proposal of a linear-assisted regulator or linear-&-switching hybrid regulator with a constant switching frequency. The control loop of the system is based on the current-mode technique. The main disadvantage of a regulator with current-mode control is the inherent instability of the loop when switch duty ratios are greater than 0.5. In order to make stable the proposed linear-assisted regulator, the article shows the technique based on a slope compensation. This kind of voltage regulators is especially devoted to battery chargers in renewable energy systems.

1 Introduction

As it is well known, series *linear regulators* have been structures widely used in power supply systems for decades, providing output voltages with low or moderate consumes and currents [1, 2], especially devoted, among other applications, to battery chargers in renewable energy systems. This kind of voltage regulators has several advantages that lead their use. However, in spite of these advantages, linear regulators suffer from some serious drawbacks. One significant example is that the efficiency of these structures hardly exceeds the 50% and the series–pass transistor has to support all the current demanded by the load. Thus, in high power applications, this component has to be dimensioned (both electrically and thermically) in order to dissipate powers with high values. This point increases the price of the supply system significantly. As a result, they are not recommendable in some supply systems, especially for high power.

The alternative is also widely known: *DC–DC switching converters* [3–5]. The advantages of these structures are also evident opposite to linear regulators: the efficiency of them, in spite of not arriving at the 100% due to the omnipresent circuit losses, is near to this optimal value. However, the design and implementation of this sort of converters is a more complex process than in linear regulators, especially their control loops when both line and load regulations are desired. In addition, the intrinsic switched nature of these converters produces ripples in the output voltage and an increment of the EMIs in neighboring electronic systems.

* Corresponding author: herminio.martinez@upc.edu

 DOI: 10.1051/e3sconf/20172200112

Last years, a third possibility has been presented: *linear–&–switching hybrid regulators* or *linear–assisted regulators* [6, 7]. These regulators are compact structures, which make good use of the advantages from previous alternatives (linear regulators and switching converters). With them, some of the aforementioned drawbacks are minimized as, for instance, the low efficiency and the high power dissipation in linear regulators, or the complexity in the design of the control for switching converters.

In general, linear–assisted regulators presented in literature are *variable–frequency structures* [8–10]. In this case, the switching frequency depends on the parameters of the regulator and is not fixed by an external clock. This drawback affects the design of the regulator and the selection of its components. In addition, this feature is not desirable in some applications because variable switching frequency could insert into the bandwidth of the neighbouring electronic circuits or even of the electronic system that the regulator supplies. As a consequence, with the design of a *constant–frequency linear–assisted regulator*, it is possible to choose the desired switching frequency independently of the component values.

In this article, a proposal of constant–frequency linear–assisted regulator, especially suitable as DC-DC regulator for photovoltaic solar-powered facilities, is presented. Section II presents a review of variable–frequency linear–assisted regulators, Sections III and IV show the proposal of the constant–frequency linear–assisted regulator, revealing the problem of its stability and the solution adopted to avoid it. Finally, Section V concludes with some simulation results that validate the proposed structure.

2 Review of variable-frequency linear-assisted regulators

As we can see in Fig. 1, the proposed configuration makes use of an analog comparator that controls the conduction or cut of the transistor Q_1 and fixes the switching frequency. Note that the objective of the switching converter is to provide the excess of current that the linear regulator does not supply. Consequently:

$$I_{out} = i_{reg}(t) + i_L(t) \tag{1}$$

Just not considering in a first approximation any hysteresis in the comparator CMP_1, the operation of the circuit is just as it is explained as follows: if the load current is below a boundary current value, that we name as *threshold switching current*, I_γ, the output of the comparator CMP_1 is held low. Thus, the switching converter will be disabled and, therefore, the current through the inductance L_1 will be zero. As a result, the linear regulator supplies the load R_L, providing all the output current ($I_{reg} = I_{out}$).

However, in the moment that load current increases and goes slightly beyond this limit current I_γ, the comparator output will pass to high level, increasing the inductance current in a linear form. Taking into account expression (1) and that the load current is constant (equal to V_{out}/R_L), the regulator current $i_{reg}(t)$ will tend to decrease also linearly (Fig. 2), just reaching a value below I_γ. In this moment, the comparator changes its output from high to low level, the transistor Q_1 goes to off and forces the inductance current to decrease. Then, when the current in the inductance decreases to a value in which $i_{reg}(t) > I_\gamma$, the comparator changes from low to high level, and the cycle starts again.

Both reference voltage V_{ref} and shunt resistor R_{lim} determine the value of the threshold switching current I_γ. Fixing I_γ, the value of the R_{lim} can be obtained by expression (2).

$$R_{\lim} = \frac{V_{ref}}{I_\gamma} \tag{2}$$

 DOI: 10.1051/e3sconf/20172200112

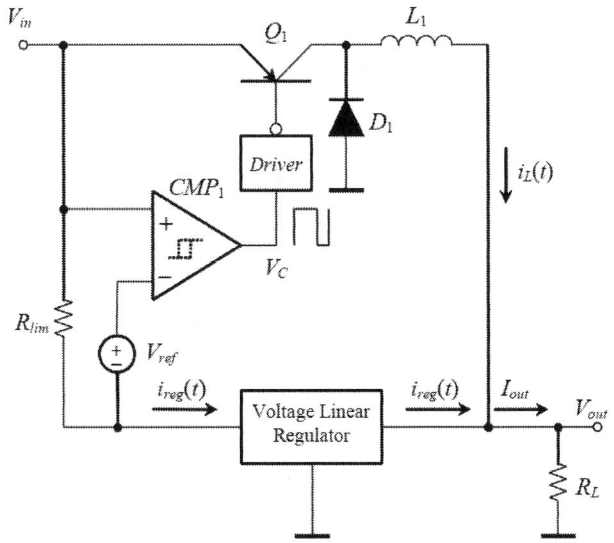

Fig. 1. Basic structure of a self–switched hybrid regulator that supplies a load R_L.

With the objective of not decreasing the efficiency of the regulator, the value of the dissipated power by the internal transistor of the linear regulator must be reduced to the utmost. For this reason, the current I_γ has to be the minimum and necessary value to make the linear regulator work properly, without penalizing its good regulation characteristics.

The delays of the electronic circuits determine a little hysteresis that limits the maximum switching frequency. Nevertheless, with the objective of fixing this maximum frequency to a suitable value (in order not to increase the switching losses significantly), it is convenient to add a hysteresis to the comparator CMP_1.

The experimental implementation of the self-switched hybrid regulator presented in Fig. 1 can be found in [9] and [10]. Fig. 3 shows the waveforms of a variable-frequency linear-assisted regulator with V_{in} = 10 V. In the same figure, it is observed the response of the system to an input voltage step from 10 V to 13 V in the time instant t = 20 μs, and a step change of the load resistor of the 50% (from 10 Ω to 5 Ω) in t = 40 μs.

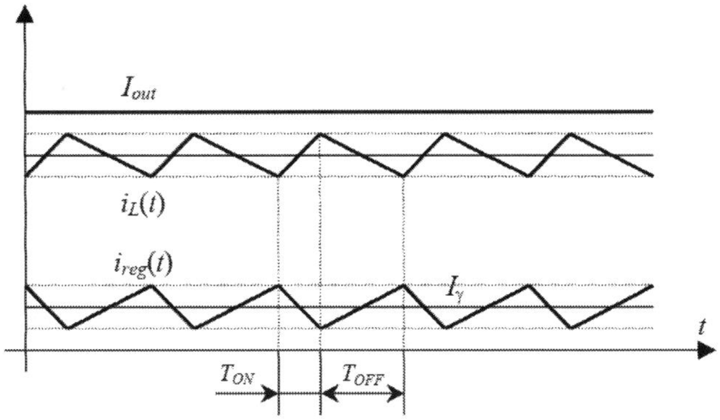

Fig. 2. Currents through the load resistor, inductance L_1 and linear regulator in the steady state.

 DOI: 10.1051/e3sconf/20172200112

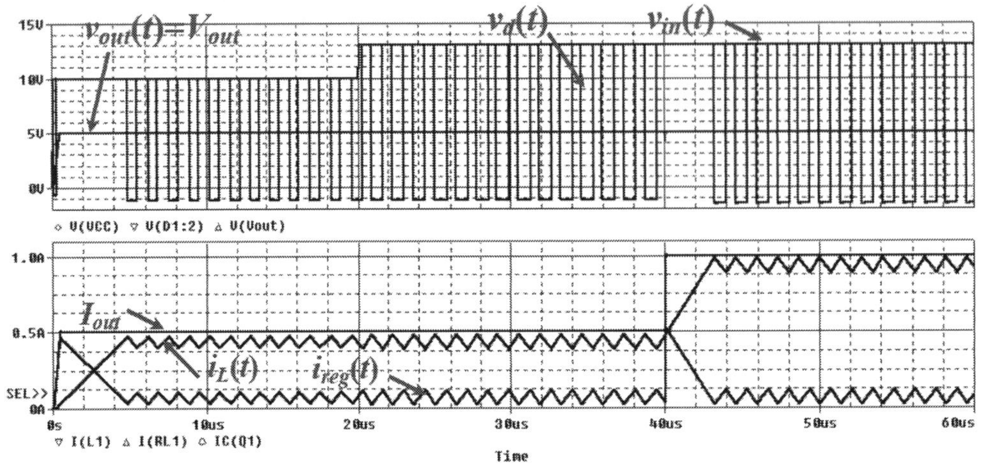

Fig. 3. Transient response of the variable–frequency linear–assisted regulator with V_{in}= 10 V. It can be seen the response of the circuit to an input step from 10 V to 13 V in the time instant t = 20 μs, and when there is a decrement of the load resistance from 10 Ω to 5 Ω in t = 40 μs.

The switching frequency is given by (3), being the upper and lower switching threshold levels of the comparator (Schmitt trigger) V_H and V_L, respectively.

$$f = \frac{R_{\lim}}{L_1} \frac{V_{out}}{V_H - V_L} \left(1 - \frac{V_{out}}{V_{in}} \right) \quad (3)$$

From equation (3) we obtain the switching frequency is a function of the hysteresis of the comparator, input and output voltages and R_{lim} and L_1 values. This dependency from the aforementioned values is an important drawback. Note that the switching frequency depends on the possible disturbances in V_{in} and tolerances and dispersions in passive components of the regulator.

3 Constant-frequency linear-assisted voltage regulator

Fig. 4 shows a constant-frequency linear–assisted regulator. Similar to the variable-frequency regulator shown in Fig. 1, this topology has a switching converter and a linear regulator. However, now the switching frequency is fixed by the external clock signal V_{CLK}. This signal is applied to the set input of a RS flip-flop and the output of the comparator is applied to the reset input of the bistable. The output of the flip-flop is applied to the base or gate of the switch Q_1.

 DOI: 10.1051/e3sconf/20172200112

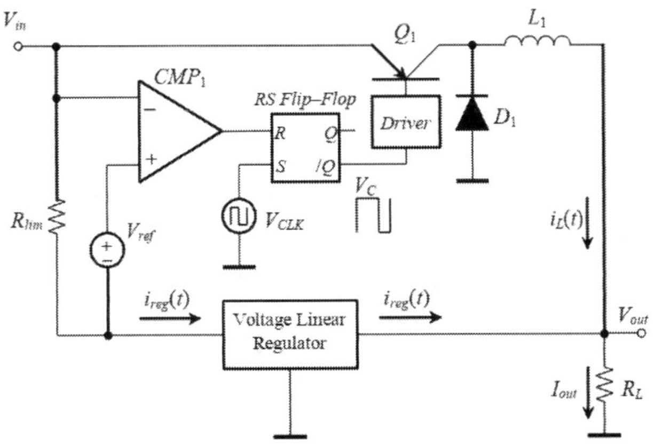

Fig. 4. Schematic of the constant–frequency linear–assisted regulator that has been used to validate the proposed structure.

Fig. 5 shows the waveforms of the constant–frequency linear–assisted regulator with V_{in}=15 V. In the same figure, it is observed the response to an input voltage step from 15 V to 18 V in the time instant t = 400 μs, and a step change of the load resistor of the 50% (also from 10 Ω to 5 Ω) in t = 800 μs. The output voltage is fixed, thanks to the linear regulator, to 7 V and the current I_γ has been adjusted in this case to 100 mA.

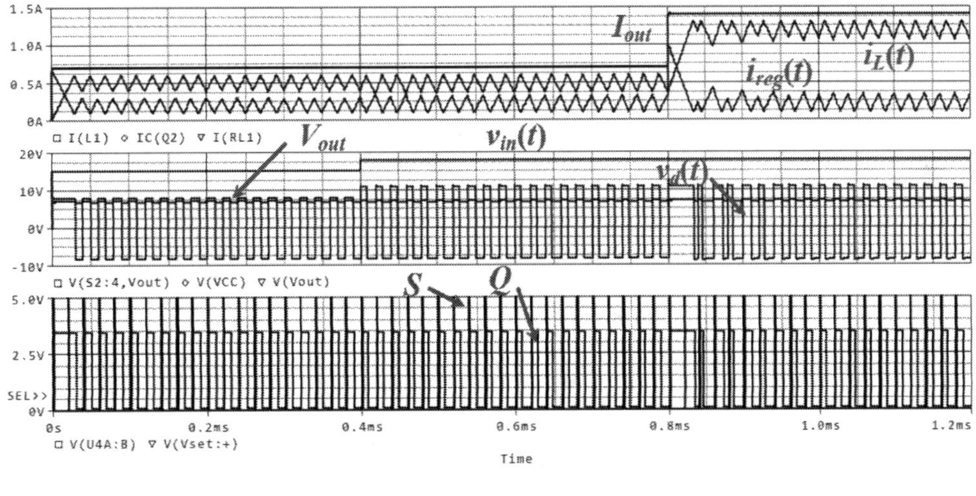

Fig. 5. Transient response of the constant–frequency linear–assisted regulator with V_{in} = 15 V. It can be seen the response of the circuit to an input step from 15 V to 18 V in the time instant t = 400 μs, and when there is a decrement of the load resistance from 10 Ω to 5 Ω in t = 800 μs.

The clock signal determines the start of T_{ON} interval and the output of the comparator determines the beginning of T_{OFF}.

The current sensing of the linear regulator acts as a control loop and establishes the switch duty ratio in a similar way that the well-known PWM modulation. Fig. 6 shows in detail the waveforms of the regulator in the steady state.

DOI: 10.1051/e3sconf/20172200112

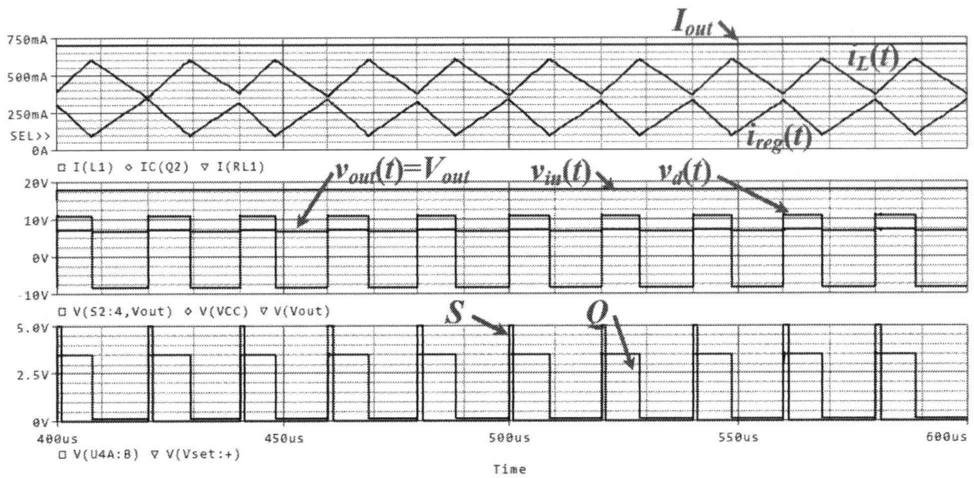

Fig. 6. Detail of the waveforms of the constant-frequency linear-assisted regulator in the steady state.

Actually, note that, in this regulator, the control is a current-mode control. With regard to the classical current–mode control, in this regulator, variations of the inductor current are measured indirectly thanks to the measure of the variations in the linear regulator.

As it is well known, the current–mode control has an important limitation: for switch duty ratios greater than 0.5 the system is unstable [5]. To prevent this limitation in the current-mode control, a slope compensation has to be added to the control loop in order to provide stability, to prevent subharmonic oscillations, and to provide a feedforward property.

Fig. 7 shows the constant-frequency linear-assisted regulator with a system that implements this slope compensation by means of an integrator with reset.

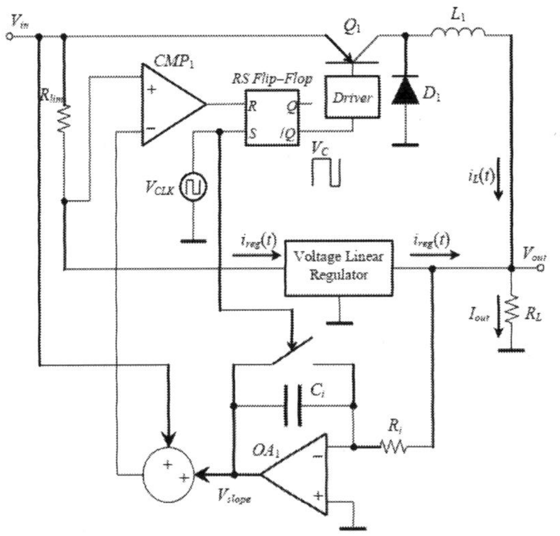

Fig. 7. Schematic of the constant-frequency linear-assisted regulator with the implementation of the slope compensation.

4 Conclusions

This paper has shown the proposal of a linear–assisted regulator or linear-&-switching hybrid regulator with constant switching frequency. With regard to the linear–assisted regulators with variable switching frequency, the presented structure eliminates the disadvantages derived from having variable switching frequency.

The control loop presented in the article is based on the current–mode technique that has as a main drawback, as usual in DC-DC converters, the instability of the loop when switch duty ratios are greater than 0.5. This work has introduced the technique based on slope compensation in order to make stable the linear–assisted regulator. The proposed modifications guarantee the stability and make possible this new mode of work in this kind of regulators.

This work has been partially supported and funded by the Spanish Ministerio de Economía y Competitividad by projects DPI2013-47799-C2-2-R, and DPI2016-78957-R, the LOGIMATIC EU project H2020-Galileo-2015-687534.

References

1. R.K. Dokania, G.A. Rincón-Mora, Electronic Letters **38**, 22, 1300–1302 (2002)
2. V. Grupta, G.A. Rincón-Mora, P. Raha, IEEE International SoC Conference, 311–315 (2004)
3. R.W. Erickson, D. Maksimovic, *Fundamentals of Power Electronics* (2nd Ed., Kluwer Academic Publishers, 2001)
4. J.G. Kassakian, M.F. Schlecht, G.C. Verghese, *Principles of Power Electronics* (Ed. Addison–Wesley, 1991)
5. N. Mohan, T.M. Underland, W.P. Robbins, *Power Electronics: Converters, Applications and Design* (3rd edition. Ed. John Wiley & Sons, 2003)
6. G. Villar, E. Alarcón, F. Guinjoan, A. Poveda, ISCAS 2005 (2005)
7. X. Zhou, P.L. Wong, P. Xu, F.C. Lee, A.Q. Huang, IEEE Transactions on Power Electronics **15**, 6, 1172–1182 (2000)
8. H. Ertl, J.W. Kolar, F.C. Zach, IEEE Transactions on Industrial Electronics **44**, 1, 116–123 (1997)
9. H. Martínez, A. Conesa, ECCTD 2007 (2007)
10. A. Conesa, H. Martínez, J.M. Huerta, 12th EPE 2007 (2007)

On the modelling of linear-assisted DC-DC voltage regulators for photovoltaic solar energy systems

Herminio Martínez-García[1,*], and *Encarna* García-Vílchez[1]

[1]Eastern Barcelona School of Engineering (Escola d'Enginyeria de Barcelona Est, EEBE), Department of Electronics Engineering, Tech. University of Catalonia (UPC – BarcelonaTech), Diagonal-Besòs Campus. Eduard Maristany Ave., nº 10 – 14, E-08019 – Barcelon, Spain

Abstract. This paper shows the modelling of linear–assisted or hybrid (linear & switching) DC/DC voltage regulators. In this kind of regulators, an auxiliary linear regulator is used, which objective is to cancel the ripple at the output voltage and provide fast responses for load variations. On the other hand, a switching DC/DC converter, connected in parallel with the linear regulator, allows to supply almost the whole output current demanded by the load. The objective of this topology is to take advantage of the suitable regulation characteristics that series linear voltage regulators have, but almost achieving the high efficiency that switching DC/DC converters provide. Linear-assisted DC/DC regulators are feedback systems with potential instability. Therefore, their modelling is mandatory in order to obtain design guidelines and assure stability of the implemented power supply system.

1 Introduction

Series linear regulators have been structures widely used in power supply systems for decades, providing supplies with low or moderate currents and consume [1–2]. This kind of voltage regulators has several advantages that lead their use. However, in spite of these advantages, linear regulators suffer from some serious drawbacks. As a result, they are not recommendable in some supply systems, especially for high power.

The alternative, DC/DC switching converters [3–4], has evident advantages opposite to linear regulators. However, the design and implementation of this sort of converters is a more complex process than in linear regulators, especially their control loops when both line and load regulations are desired. Furthermore, the intrinsic switched nature of these converters produces ripples in the output voltage and an increment of the EMIs in neighbouring electronic systems.

The use of linear-assisted hybrid DC/DC voltage regulators takes advantages of previous alternatives [5–7]. Apart from this, some of the aforementioned drawbacks are minimized as, for instance, the low efficiency and the high power dissipation in linear regulators, or the complexity in the design of the control for switching converters. Among

* Corresponding author: herminio.martinez@upc.edu

 DOI: 10.1051/e3sconf/20172200113

other applications, this kind of voltage regulators can be of interest in modern high-efficiency battery chargers.

In this article, a proposal of variable frequency linear–assisted hybrid DC/DC regulator (or linear-switching hybrid DC/DC regulator) and its modelling is presented. The system has a control loop to maintain a constant output voltage that could induce instabilities as a function of the values of the circuit components and the values of the load (in a similar way that linear regulators). Thus a stability study is necessary to obtain design guidelines for this kind of DC/DC voltage regulators and assure the stability of the implemented power supply system.

2 Structure of a linear-assisted DC/DC voltage regulator

The proposed configuration of a linear-assisted DC/DC voltage regulator makes use of an analogue comparator that controls the conduction or cut of the transistor Q_1, just as it can be seen in Fig. 1, and fixes the switching frequency. Note that the objective of the switching converter is to provide the excess of current that the linear regulator does not supply. Consequently:

$$I_{out} = i_{reg}(t) + i_L(t) \tag{1}$$

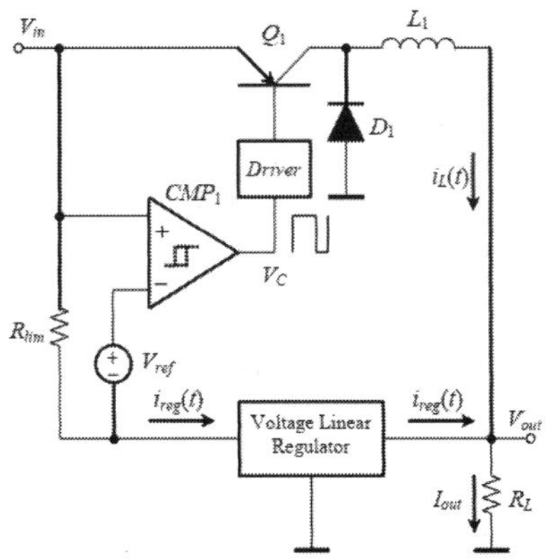

Fig. 1. Basic structure of a linear-assisted DC/DC voltage regulator or self-switched hybrid regulator that supplies a load R_L.

Just not considering in a first approximation any hysteresis in the analogue comparator CMP_1, the operation of the circuit is just as it is explained as follows: if the load current is below a boundary current value, that we name as threshold switching current, I_γ, the output of CMP_1 is held low. Thus the switching converter will be disabled and I_L will be zero. As a result, the linear regulator supplies the load R_L, providing all the output current (that is, $I_{reg} = I_{out}$).

However, in the moment that load current increases and goes slightly beyond this limit current I_γ, the comparator output will pass to high level, increasing I_L in a linear form. Taking into account expression (1) and that the load current is constant (equal to V_{out}/R_L), $i_{reg}(t)$ will tend to decrease also linearly (Fig. 2), just reaching a value below I_γ. In this moment, the comparator changes its output from high to low level, cutting Q_1 and forcing I_L to decrease. Then, when the current in the inductance decreases to a value in which $i_{reg}(t)>I_\gamma$, the comparator changes from low to high level, repeating the cycle again.

V_{ref} and R_{lim} determine the value of the threshold switching current I_γ according to $R_{lim}=V_{ref}/I_\gamma$. With the objective of reducing the value of the dissipated power by the internal transistor of the linear regulator to the utmost, this current I_γ has to be the minimum and necessary in order to make the linear regulator work properly, without penalizing its good regulation characteristics. It will also increase the efficiency of the system, even with significant load currents.

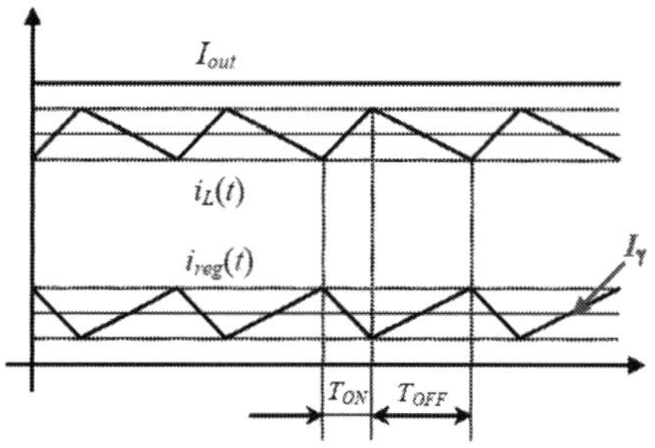

Fig. 2. Currents flowing through R_L, inductance L_1 and linear regulator in the steady state of the linear-assisted DC/DC voltage regulator shown in Fig. 1.

The delays of the electronic circuits determine a little hysteresis that limits the maximum value of the switching frequency. Nevertheless, with the objective of fixing the maximum value of this switching frequency to a suitable value, it is convenient to add a hysteresis to CMP_1.

A practical implementation of the self–switched hybrid regulator presented in Fig. 1 can be found in [8]. Fig. 3 shows the transient of a variable-frequency linear-assisted voltage regulator with V_{in} = 10 V. Having into account that the upper and lower switching threshold levels of the comparator (Schmitt trigger) are V_H and V_L, respectively, the switching frequency is given by:

$$f=\frac{R_{\lim}}{L_1}\frac{V_{out}}{V_H-V_L}\left(1-\frac{V_{out}}{V_{in}}\right) \tag{2}$$

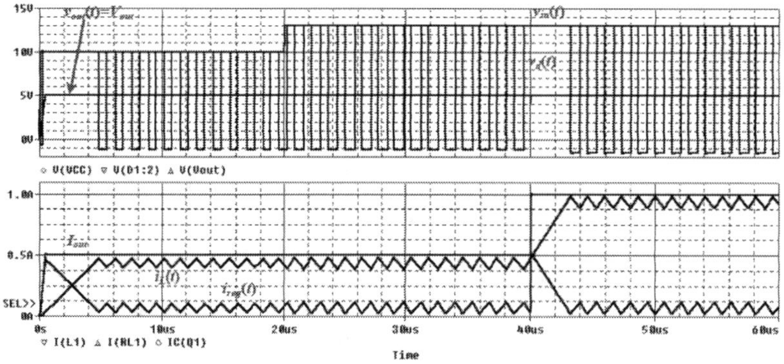

Fig. 3. Transient response of the variable-frequency linear-assisted voltage regulator with V_{in} = 10 V. It can be seen the response of the circuit to an input step from 10 V to 13 V in t = 20 µs, and when there is a decrement of the load resistance from 10 Ω to 5 Ω in t = 40 µs.

DOI: 10.1051/e3sconf/20172200113

3 Modelling of linear-assisted DC/DC voltage regulators

It is important to highlight that in previous discussion the effect of an output capacitance has not been considered. Effectively, these linear–assisted voltage regulators work properly when the load capacitance is null. In spite of this, some capacitance can be suitable in order to eliminate possible transitory peaks in the output voltage when the regulator supplies variable loads (that is, high slew–rate in the output current). This output capacitor improves the response of the converter appreciably. However, the effect of an output capacitor in parallel with the load, a parasitic output capacitance or a capacitance load induces a trend to instability. These instabilities produce and important and considerable increment of the output-ripple around the desired average output voltage that is observed in linear-assisted voltage regulators.

Fig. 4 shows a transient response of the linear-assisted voltage regulator in the same conditions that Fig. 3 but with an output capacitance C_L = 100 nF and an equivalent series resistance (ESR) equal to 10 mΩ. Note that, with these conditions, expression (1) is not satisfied due to the inclusion of the C_L network. As a consequence, the obtained waveforms in the circuit are not as the reflected in Fig. 3. In addition, the current $i_{reg}(t)$ increases and thus the efficiency of the regulator is reduced significantly.

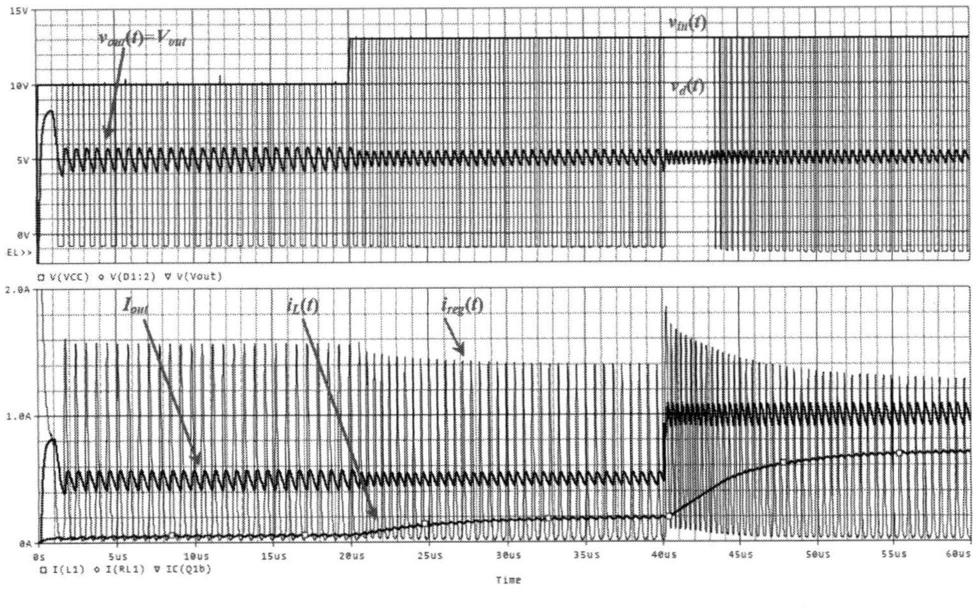

Fig. 4. Transient response of the linear–assisted voltage regulator with an output capacitance C_L = 100 nF. It can be seen the response of the circuit to an input step from 10 V to 13 V in t = 20 µs, and when there is a decrement of the load resistance from 10 Ω to 5 Ω in t = 40 µs.

Therefore, the C_L has to be an agreement value. This value must be high enough but providing stability of the supply system. Even more, the authors have observed that in a linear-assisted voltage regulator, the frequency response of the system (and thus the stability conditions) is highly sensitive to the load conditions and the output capacitor connected (or parasitic capacitance existing) in parallel with the load resistor. Some parameters determine the aforementioned instabilities. However, the most important are the values of the output capacitance C_L, its equivalent series resistance, the load resistor R_L and the inductor L_1.

All these points make necessary to analyze the regulator in order to obtain design approaches that guarantees its stability. The model of the linear-assisted DC/DC voltage regulator is presented in Fig. 5. For the operational amplifier is considered its low-pass response in block H_1, with a differential gain (A_d) and a dominant pole (ω_O), and its output resistance (r_{oa}). For transistor (working in its linear zone) is considered the base-emitter dynamic resistance (r_π) and a current gain (β). Thus the block H_2 contains only the association of resistors r_{oa} and r_π.

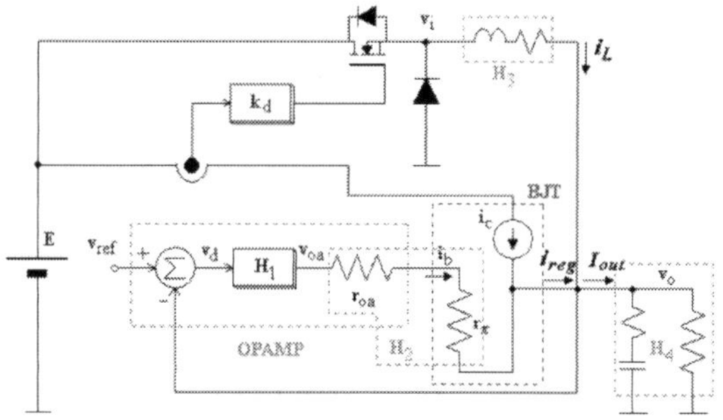

Fig. 5. Proposed model of the linear-assisted DC/DC voltage regulator.

The r_π value is calculated as usual from equation (3):

$$r_\pi = \frac{dv_{be}}{di_b} = \frac{\beta}{I_{Cq}}\frac{v_T}{\lambda} \tag{3}$$

The k_d block makes sense of collector current and obtain the conducting ratio of switching transistor as equation (4).

$$k_d = \frac{\Delta d}{\Delta i_c} = \frac{1}{i_{c,\max}} \tag{4}$$

The inductor small signal voltage is v_i and defined as usual:

$$v_i = (d \cdot E + e \cdot D) - v_o \tag{5}$$

The inductor current (i_L in Fig. 5) is obtained thanks to the voltage v_i and the block H_3, which models the inductor and its series resistance. Finally, with output current (i_{out}) and block H_4, the output voltage (v_o) is obtained. In the block H_4 the load resistance and output capacitor with its ESR is modelled.

In Fig. 6 is presented the flow graph of the modelled linear–assisted voltage regulator. Applying the Mason rule we obtain the transfer function given by:

$$G(s) = \frac{v_o(s)}{v_{ref}(s)} \tag{6}$$

In table 1 the Maple® code is presented in order to obtain the transfer function of equation (6). Subsequent to this, the result is simulated with Matlab® to obtain the poles zeros map of the system (*pzmap* function). The used simulation values are close to the experimental values of the implemented prototype (table 2).

 DOI: 10.1051/e3sconf/20172200113

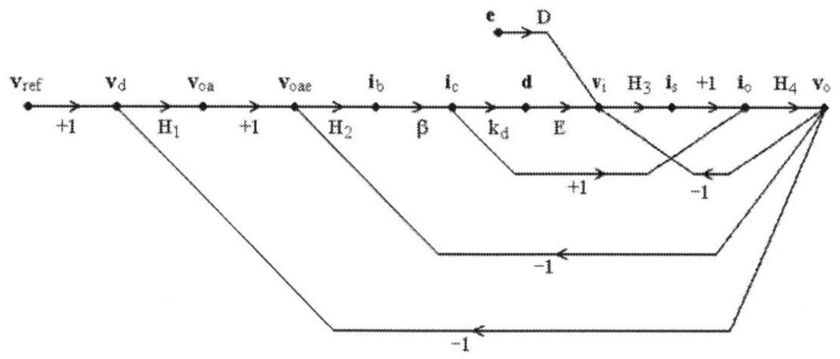

Fig. 6. Proposed model of the linear-assisted DC/DC voltage regulator.

Table 1. Maple® code used to obtain the transfer function given in equation (6) for the linear-assisted DC/DC voltage regulator.

restart;
H1:=Aoa/(1+s/woa);
H2:=1/(roa+rd);
H3:=1/(rl+s*L);
H4:=R*(1/C/s+rc)/(R+1/C/s+rc);
P1:=H1*H2*beta*kd*E*H3*H4;
A1:=1;
P2:=H1*H2*beta*H4;
A2:=1;
	L1:=-H3*H4;
	L2:=-H2*beta*kd*E*H3*H4;
	L3:=-H2*beta*H4;
	L4:=-H1*H2*beta*kd*E*H3*H4;
	L5:=-H1*H2*beta*H4;
	g:=(P1*A1+P2*A2)/(1-(L1+L2+L3+L4+L5));
	g1:=simplify(g);
	g2:=sort(g1,s);

Table 2. Simulation values used in the obtained model.

Parameter	Value	Parameter	Value	Parameter	Value
A_{oa}	$2\cdot10^5$ V/V	R_L	2 Ω	$I_{cq} = I_\gamma$	50 mA
ω_{oa}	$2\cdot\pi\cdot10$ rad/s	L_1	100 μH	β	100
r_{oa}	100 Ω	v_t	25 mV	r_d	$(\beta\cdot v_t)/(\lambda\cdot I_{cq})$
E	12 V	λ	1	k_d	$1/100\cdot10^{-3}$

It is observed that the system can be stable or unstable with different values of the load resistor, the inductance L_1, the load capacitance C_L and its ESR.

4 Simulation and experimental results

Using the model proposed in previous section, some interesting simulations have been carried out in order to determine the stability of the system as a function of different circuit's parameters in order to corroborate the implementation behaviour. In particular, in Fig. 7 the root locus diagram is presented. In particular, we can find three curves for three different values of the capacitor's ESR (1 mΩ, 10 mΩ and 100 mΩ). In the three plots, the capacitor values goes from C_L = 100 nF to C_L = 1000 μF. Note that when the ESR has a low value, the system tends easily to the instability when the capacitance value is higher than 1 μF. On the other hand, when the capacitor's ESR is relatively high the stability of the system is assured, even when output capacitance is high.

In Fig. 8 a root locus diagram is shown for three different values of the inductor L_1 (200 μH, 100 μH and 50 μH). As previous two figures, in the three plots the capacitor values goes from 100 *n*F to 1000 μF. Note that when the load resistor has a high value, the system tends also to the instability when the capacitance value increases.

 DOI: 10.1051/e3sconf/20172200113

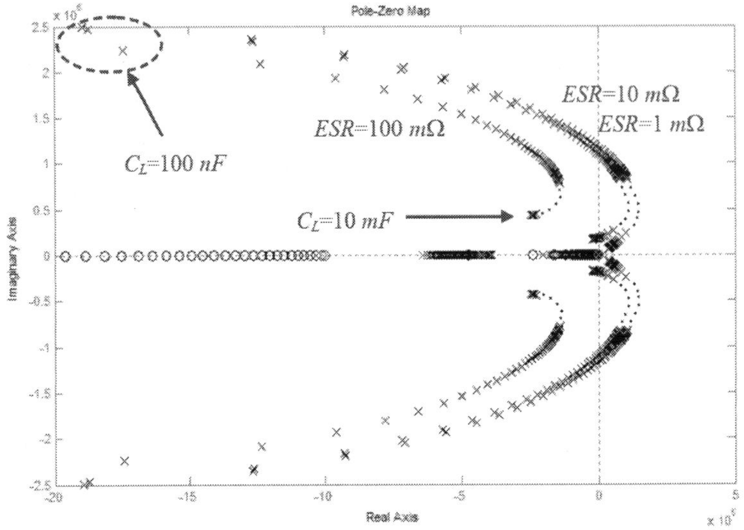

Fig. 7. Root locus diagram for different values of capacitor ESR and output capacitance C_L.

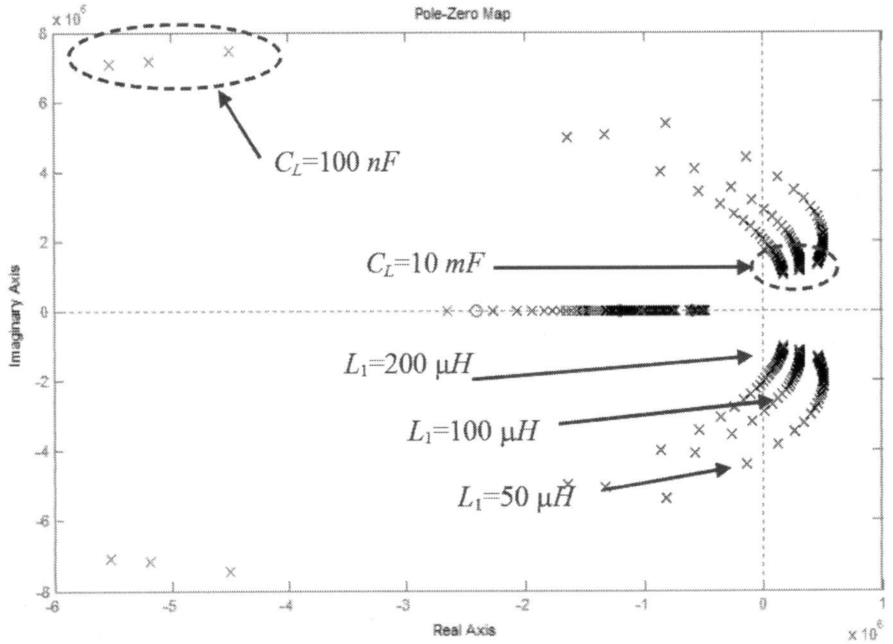

Fig. 8. Root locus diagram for different values of the inductor L_1 and output capacitance C_L.

Varying different circuit parameter, other set of plots can be obtained in the same way that the obtained in previous figures. However, the aforementioned parameters (C_L, capacitor's ESR, L_1 and load resistance) are the four more important that determine the stability of the linear-assisted voltage regulator. Other parameters as, for instance, the inductor's series resistance, the open loop gain of the op-amp, or its output resistance have an effect on the dynamics of the regulator but actually do not lead the converter to the instability.

5 Conclusions

In this article, the modelling of a linear–assisted voltage regulator has been shown. The aforementioned regulator has a control loop to maintain a constant output voltage that could induce instabilities as a function of the values of the circuit components and the load, in a similar way that others regulators.

The analysis done reveals that C_L, capacitor's ESR, inductance L_1 and load resistance are the four more important parameters that determine the stability of the linear-assisted voltage regulator. In addition, from the proposed study we can improve the stability of the regulator and obtain the influence of its different parameters.

This work has been partially supported and funded by the Spanish Ministerio de Economía y Competitividad by projects DPI2013-47799-C2-2-R, and DPI2016-78957-R, the LOGIMATIC EU project H2020-Galileo-2015-687534.

References

1. R.K. Dokania, G.A. Rincón-Mora, Electronic Letters **38**, 22, 1300–1302 (2002)
2. V. Grupta, G.A. Rincón-Mora, P. Raha, IEEE International SoC Conference, 311–315 (2004)
3. R.W. Erickson, D. Maksimovic, *Fundamentals of Power Electronics* (2nd Ed., Kluwer Academic Publishers, 2001)
4. J.G. Kassakian, M.F. Schlecht, G.C. Verghese, *Principles of Power Electronics* (Ed. Addison-Wesley, 1991)
5. G. Villar, E. Alarcón, F. Guinjoan, A. Poveda, ISCAS 2005 (2005)
6. X. Zhou, P.L. Wong, P. Xu, F.C. Lee, A.Q. Huang, IEEE Transactions on Power Electronics **15**, 6, 1172–1182 (2000)
7. H. Ertl, J.W. Kolar, F.C. Zach, IEEE Transactions on Industrial Electronics **44**, 1, 116–123 (1997)
8. A. Conesa, H. Martínez, J.M. Huerta, 12th EPE 2007 (2007)

The influence of the small platinum clusters on hydrogen sorption properties

Agnieszka Martyła[1,*], R. E. Przekop[2,*], *Monika* Osińska-Broniarz[1], M. Kopczyk[1], J. D. Rybka[2], and P. Kirszensztejn[3]

[1]Institute of Non-Ferrous Metals, Division in Poznan, Central Laboratory of Batteries and Cells, ul. Forteczna 12, 61-362 Poznań, Poland
[2]Centre for Advanced Technologies, A. Mickiewicz University, ul. Umultowska 89c, 61-614 Poznań, Poland
[3]Faculty of Chemistry, A. Mickiewicz University, ul. Umultowska 89b, 61-614 Poznań, Poland

Abstract. Hydrogen sorption abilities of Pt-B_2O_3/Al_2O_3 systems with different molar ratio of oxides obtained by sol-gel method were examined. Platinum was introduced by surface impregnation. Main goal of the research was to check an influence of metallic component (platinum) on sorption properties of B_2O_3/Al_2O_3 binary oxides. The oxide systems were characterized using XRD and TEM. Hydrogen adsorption was tested in the volumetric system and TPD measurements were taken. Results show that the amount of adsorbed hydrogen depends not only on the amount of platinum in the system but also on the type of oxide support and mainly on the content of boria.

1 Introduction

The problem of safe and effective storage of hydrogen is one of the major technological barriers currently inhibiting the widespread adoption of hydrogen as an energy carrier. Practical problems with the storage of hydrogen in both gaseous and liquid form appear to make solid state storage the most attractive proposed solution. Due to this fact the search for potential solid state hydrogen storage media is attracting a significant amount of research interest from both academia and industry. Various types of materials have been proposed and are currently being investigated, including microporous materials that can store physisorbed molecular hydrogen at low temperatures [1, 2], MOF [3] and carbons [4, 5]. S.-H. Jhi et al. [6] considering the value of desorption enthalpy, made a suggestion to use B_2O_3, HBO_2 and H_3BO_3 as hydrogen storage materials.

From authors' previous research focused on examination of usability of B_2O_3/Al_2O_3 modified sol-gel systems as hydrogen sorbents [7] it has been concluded that their sorption properties depend on their textural parameters. Next, it was decided to use this binary system as a support for metal possessing hydrogen sorption ability (Pt) and to study its effect in hydrogen sorption. Electrocatalysts consist of a monolayer of Pt on carbon-supported metal or metal-alloy nanoparticles and are primarily envisaged for fuel-cell applications. Fuel cells are expected to become soon one of the major sources of clean energy; yet, despite significant advances in recent years, the efficiency of existing

* Corresponding authors: agnieszka.martyla@claio.poznan.pl, r.przekop@gmail.com

technology is still inadequate, and the available electrocatalysts have high Pt content [8]. Carbon nanotubes (CNTs) and fullerenes functionalized with the metal adatoms have been emerging as an important genre of composite materials for hydrogen storage applications because of the enhanced storage capacity and improved storage kinetics they exhibit [9, 10]. By far the most exclusively studied nanotube–metal composites are those functionalized with transition metals, such as Ti, V, Pd and Pt [11].

The present study concentrates on the effect of boria on the dispersion and hydrogen sorption properties of platinum supported on B_2O_3/Al_2O_3 systems. Former studies of metal catalysts in binary oxide systems obtained by sol-gel method [12] show that in comparison to conventional (non sol-gel) supports, dispersion of the metal phase [13] is higher. On that base it was assumed that satisfying values of hydrogen adsorption on Pt-BAl systems can be achieved. The oxide supports were obtained using sol-gel technique in an aqueous environment. Platinum was introduced by impregnation (1% of platinum in ratio to weight of support). The catalysts with platinum was reduced by thermo-programmed reduction in the hydrogen atmosphere. Samples characterization performed with XRD, TEM, BET and TPD technique was followed by hydrogen sorption.

2 Material and methods

2.1 Preparation

The binary compositions B_2O_3/Al_2O_3 with a molar ratio between 0.25:1 and 1:1 were obtained by the sol-gel method [7, 14].

Corresponding amounts of metallic phase precursor (H_2PtCl_6*H_2O) were deposited on the surface of oxide gel support by total incipient wetness method.

The precursor of the metal phase was a standard water solution of hexachloroplatinic acid, obtained according to the procedure described in [15]. Table 1 presents the compositions and the names of the obtained materials.

Table 1. The compositions and the names of the obtained materials.

Molar ratio (B:Al)	0.25:1	0.5:1	1:1
Sample	B/Al 0.25	B/Al 0.5	B/Al 1.0
Sample with 1% Pt	1% Pt /BAl 0.25	1% Pt/ BAl 0.5	1% Pt/ BAl 1.0

2.2 Characterization

A comparative study of platinum on binary oxide systems obtained by the sol-gel method was performed. The influence of Pt on oxide systems properties was analysed by using the XRD, TEM microscopy, BET and TPD. Finally, the hydrogen adsorption capacity of the materials was studied in low temperatures.

The X-ray powder diffraction patterns of samples were taken on Philips PW1050 diffractometer, with Ni filtered CuKα radiation. Reflections position, half bandwidth as well as intensity were calculated by the APD Philips program. Prior to the XRD investigation, the samples were heated at 773K in air atmosphere for 6 hours.

The structures of the oxides with platinum were characterized by a transmission electron microscopy (JEOL 200 CX *Jeol*).

The porous structure was determined by low temperature (77K) nitrogen adsorption measurements carried out on the Accelerated Surface Area and Porosimetry System model 2010 V2.00 D made by Micromeritics, using 200–300 mg of materials with the grain size fraction between 0.1 and 0.2 mm. Prior to nitrogen adsorption, all samples were outgassed at 623 K at 0.4 Pa until the constant weight. Both adsorptive and desorptive branches of the isotherm were taken in the range of p/p_0 0–1.

The measurements were carried out in a glass flow system [16] equipped with a gradientless microreactor [17]. The composition of the stream leaving the reactor was measured using thermal conductivity detector (TCD). In the study a sample of 250 mg was used. Measurements were preceded by drying the samples in situ in a helium stream (0.5 cm^3/s) with temperature rising at 0.85 K/s, from 290 to 573 K, drying was continued at 573 for 0.5 h. The reduction is carried out in a H_2 stream of 1.0 cm^3/s with temperature rising at 0.17 K/s, from 290 to 623 K. After reaching the maximum temperature, the sample was reduced by another 2.0 h. Afterwards, the sample temperature was decreased to 573 K and the adsorbed hydrogen was next removed from the sample in the reactor by passing a stream of helium (0.5 cm^3/s, 573 K, 0.5 h).

The H_2 adsorption measurements were carried out by the volumetric method. A particular attention was paid to keep up to the standards developed for the measurements of this type to eliminate errors that could lead to wrong interpretation [18]. In this method the amount of adsorption is inferred from pressure measurements that are performed on the gas before and after the adsorption takes place, as we described in [7]. The effects of Pt on the binary systems and methods of their synthesis on the adsorption capabilities were analyzed.

3 Results and discussion

X-Ray diffraction (XRD) is often used in combination with transmission electron microscopy (TEM) in the characterization of supported platinum catalysts. Crystal/particle size information obtained from fresh or aged samples generally correlates with catalytic activity.

On the basis of the poorly developed XRD spectrum (Fig. 1) it is difficult to conduct a far reaching discussion as the solids appear to be poorly crystallized, with rather weak and broad peaks. Visible on X-ray diffraction peaks could be either attributed to mixed oxides (A_9B_2, A_2B) or H_3BO_3 or B_2O_3 or a combination of them in a poorly crystallized state [19]. The peak near 2 θ – 45⁰ is shifted to higher angles when the loading of boron increases. This observation is consistent with the work of J. Grimblot [19], who attributes the reflexes to the plane {113} alumina and then stabilized at 2 θ~50°. With the increase in the participation of component B_2O_3 can be seen an increase in the amorphous structure of the system. This amorphous structure can be accounted for by the type of structure of the polymeric gel and the homogeneous dispersal of B_2O_3 in the gel matrix [7].

The characteristic diffraction peaks of Pt {111}, {200} crystal face are observed at 2 θ values of ca. 39.9, 46.5, respectively. Increasing the amount of boric component also affects the image of the metal phase. For the system with the greatest number of B_2O_3 observation of the presence of platinum is impossible (Fig. 1 C).

Analysis of HR-TEM micrographs excluded presence of ordered structures typical for mesoporous silica systems like SBA or MCM [20, 21] – Fig. 2. BAl gels (in the entire range of quantity of modifier incorporated – B_2O_3) have unordered structure that is characteristic for amorphous material. In other words, mesoporous present in gels are distributed randomly in the bulk. An attempt of metallic phase crystallites size observation proved that platinum dispersion is very high, approximately to a monoatomic level and/or the metallic phase is present as a monolayer on the support surface. It was not possible to detect any crystallites of platinum at micrographs with 1 nm = 4.5 px 15 resolution.

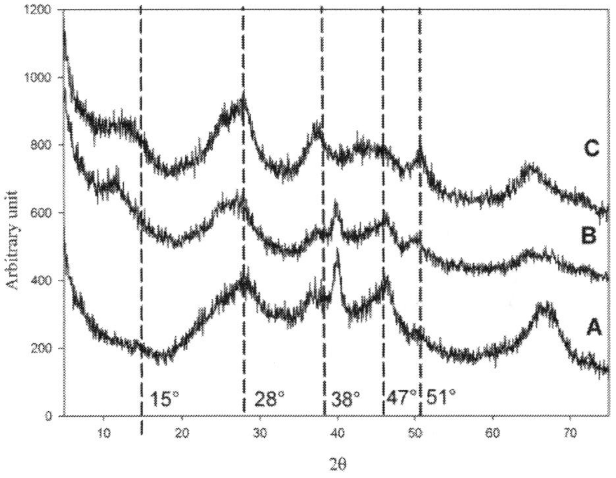

Fig. 1. XRD patterns of Pt doped oxides: A – 1% Pt/BAl 0.25, B – 1% Pt/BAl 0.5, C – 1% Pt/BAl 1.0.

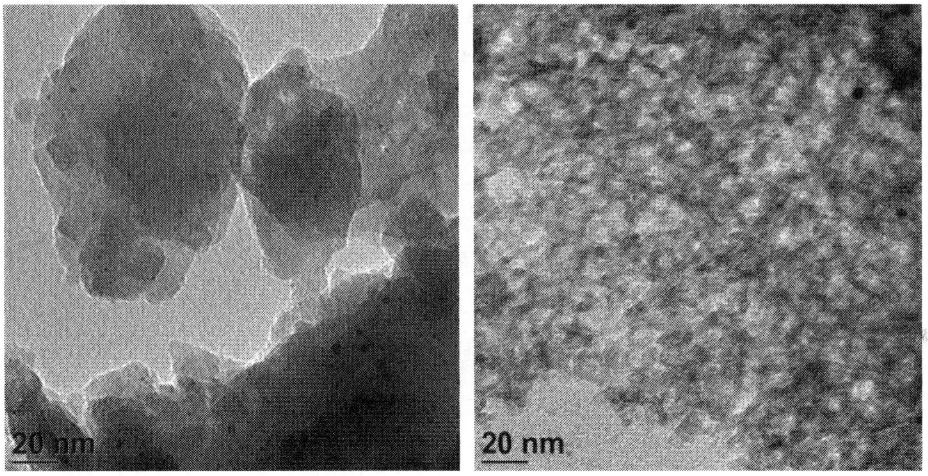

Fig. 2. HR-TEM micrographs of 1% Pt/BAl 0.25 and 1% Pt/BAl 1.0 systems.

Binary oxides gel system B_2O_3-Al_2O_3 clearly differs from the previously studied systems where the modifier oxides were such as SnO_2 or GeO_2. This difference is associated not only with the acidic nature of B_2O_3 but also to the fact that the reaction with water (during impregnation with an aqueous solution of H_2PtCl_6) leads to the formation of boric acid, which acts not as a proton donor, but as a Lewis acid accepting OH^- ions. On the other hand, the transmission IR spectrum of the binary oxide systems (subjected to a dehydration and dehydroxylation at temperatures above 500°C) indicate that the surface of only the hydroxyl groups are bonded to the boron (3685–3693 cm^{-1}). D_2O adsorption on this surface leads to the bandwidth derived from stretching vibration of O-D (2723 cm^{-1}) resulting from the adsorption of D_2O to the Lewis acid centers [15]. On the FTIR spectra, a band of 2723 cm^{-1} is visible even after the process of thermal desorption 700°C – indicating strong binding with boron - which is the center of a Lewis acidity. In the binary system oxide B_2O_3-Al_2O_3 the Lewis acid sites are largely responsible for the strong binding precursor of the metallic phase ($[PtCl_6]^{2-}$ Lewis base). Consequently, the formation of crystallites H_2PtCl_6 on the support surface is to a large extent limited. A reflection of this is the lack of visible crystallites (Pt) on the TEM images after reduction in hydrogen atmosphere.

Surface area of platinum supported materials is substantial with regard to hydrogen sorption measurements that are related to that parameter. The influence of metallic phase on textural properties can be therefore observed this way. The BET surface areas of BAl and Pt/BAl samples with different B:Al ratios are listed in table 2. Results in table 2 show that presence of the boron oxide component caused a decrease in BET surface area accompanied by an increase in content of B_2O_3 [7]. Parallel to reduce the specific surface area, pore diameter decrease occurs. The process of plugging the pores of a smaller diameter may result from a surface diffusion process H_3BO_3/B_2O_3 not bound in the gel network. These results are in agreement with those in the literature [22, 23].

The introduction of the metal component onto the surface of a binary oxide systems leads to evident changes of texture. Incorporation of platinum to binary oxide systems results in noticeable decrease in surface area and change in pore size. In the case of a system with the highest content of boron oxide, specific surface area it is almost three times lower compared to a system with the lowest amount of boron oxide.

Table 2. Experimental details summary of the nitrogen and hydrogen sorption of the samples.

System	Surface area BET [m²/g]	Cumulative Pore Volume [cm³g⁻¹]	Average Pore Diameter [nm]	The dispersion of the Pt on the support, calculated from TPD measurements	H_2 [μmol/g] of adsorbent dynamic 73-600K	H_2 [μmol/g] of adsorbent static 73 K at the temp. liq N_2	H_2 [μmol/g] B_2O_3
BAl 0.25	367	0.54	8.4	-	11.5	251	1188
BAl 0.5	248	0.45	7.3	-	9.6	136	560
BAl 1.0	282	0.34	4.8	-	5.9	187	650
1% Pt BAl 0.25	205	0.44	9.8	1.16	48.5	328	1552
1% Pt BAl 0.5	207	0.37	9.1	0.97	40.6	176	725
1% Pt BAl 1.0	85	0.14	8.6	0.46	22.9	81	282

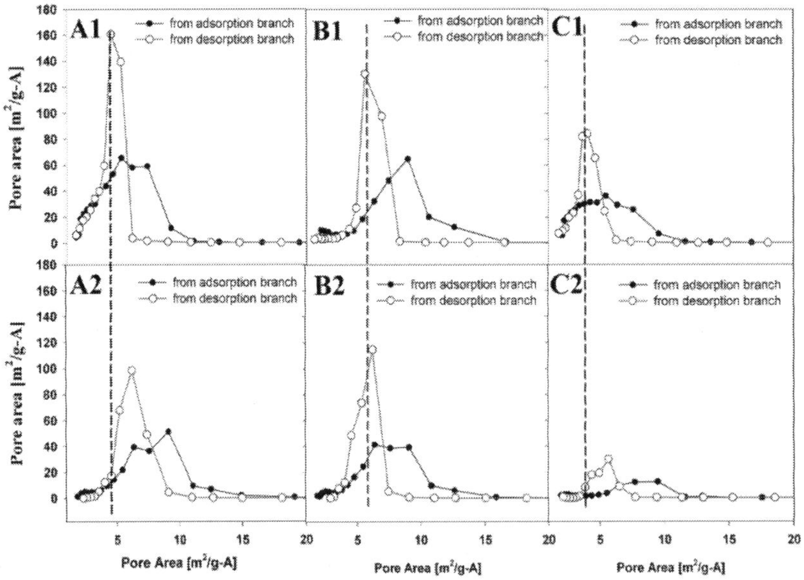

Fig. 3. A comparison of pore volume distributions for the Pt-B/Al samples: A1 – 1% Pt/BAl 0.25, B1 – 1% Pt/BAl 0.5, C1 – 1% Pt/BAl 1.0, A2 – 1% Pt/BAl 0.25, B2 – 1% Pt/BAl 0.5, C2 – 1% Pt/BAl 1.0.

The introduction of the acidic aqueous solution of H_2PtCl_6 leads to interact with B_2O_3 which exists outside the structure of the gel network, eg. the formation of a boric acid, which due to the surface diffusion causes an filling pores [24] (Tamman and Huttig

 DOI: 10.1051/e3sconf/20172200114

temperature are respectively ~ 86 and ~ 57°C). In other words, the drastic decrease in surface area and pore volume is not related to by blocking the pores of by crystallites of the metal phase.

As already mentioned, above system differs from the earlier cases (Al_2O_3/M_yO_x M = Sn, Ge), where there was an matching crystallite size (observed by TEM) to the size of the pores existing on the surface of the carrier [25]. Pores occupation by metal clusters is energetically favorable. Locating of small metal crystallites in porous structure (Fig. 3) can also hinder high-temperature migration, contributing to their stabilization. This can be confirmed by disappearance of micropores and shift in pore distribution towards larger values (Fig. 3) in case of metal-loaded support. Also bimodal distribution disappears. This indicates that micropores of these materials are blocked by fitting small metal crystallites.

The measurements of hydrogen adsorption of Pt/BAl samples are presented in Fig. 4. Measurements of hydrogen sorption properties in Pt-B_2O_3/Al_2O_3 systems with different oxide ratio were carried out using two techniques. Temperature-programmed desorption of preadsorbed hydrogen (TPD-Hads) is the most often used method for characterization of supported metal catalysts [26, 27]. As a rule, the measurements are carried out in flow systems and at a linear temperature rise. Pre-adsorption of hydrogen is usually performed at room temperature, whereupon the reactor with the examined sample is flushed with an inert gas to remove weakly adsorbed hydrogen and the remaining hydrogen is examined by the TPD method. As a result of such a procedure, commonly obtained TPD-Hads spectra characterize merely the hydrogen adsorbed irreversibly at room temperature and leave aside the hydrogen less strongly adsorbed [26]. The pioneer TPD-Hads studies, which consider not only the strongly adsorbed hydrogen, but also the weakly adsorbed hydrogen, were reported by Konvalinka et al. [28] and Stockwell et al. [29]. The second method bases on volumetric measurements.

Maximum of the adsorbed capacity is observed for 1% Pt/BAl 0.25. This support had the highest the surface area (see table 2). It is known that this property assured the homogenic spreading of Pt and, as a result, largest adsorption capacity. Addition of boron oxide leads to a surface area decrease in the gel systems and the adsorption capacity of hydrogen for Pt-deposited samples decreased remarkably with an increase in B_2O_3 content. The change in the surface can be related to differences in porosity of BAl systems. The pore volume is decreasing with the increase in the concentration of boria content and the platinum dispersion increases with the decrease in the support pore – table 2 [30], while the amount of adsorbed hydrogen does not increase. The factor that plays a key role in deciding the final hydrogen capacity of metal doped oxides is the place which platinum occupies. When the metal particles occupy pores they may not effectively adsorb hydrogen. This results in lower storage capacity. The same was observed by R. Zacharia [11] in carbon nanotubes. Increase in boria content causes change in textural properties and consequently, platinum locates in micropores, therefore the hydrogen storage capacity is lower.

In case of Pt, the particles surfaces an active site for hydrogen chemisorption [29]. A single H_2 molecule is chemisorbed to Pt atom on the oxide surface either dissociatively or molecularly. The dissociative adsorption is favorable energetically and is followed by the weakening of the Pt-oxide bond. Out of two adsorbed H_2, one can be adsorbed dissociatively and the other is chemisorbed molecularly. The presented results reveal an important effect of transition of metal atom supported on oxide surface and advance our understanding of the molecular and dissociative adsorption of hydrogen for efficient hydrogen storage. Stoichiometry of hydrogen adsorption on Pt assumes a model in which Pt:H relation is 1:1, however, for Pt crystallites smaller than 1 nm the ratio is different [31, 32]. In the case of small metal species a strong adsorption of stoichiometry 1:2 (Pt:H) may occur.

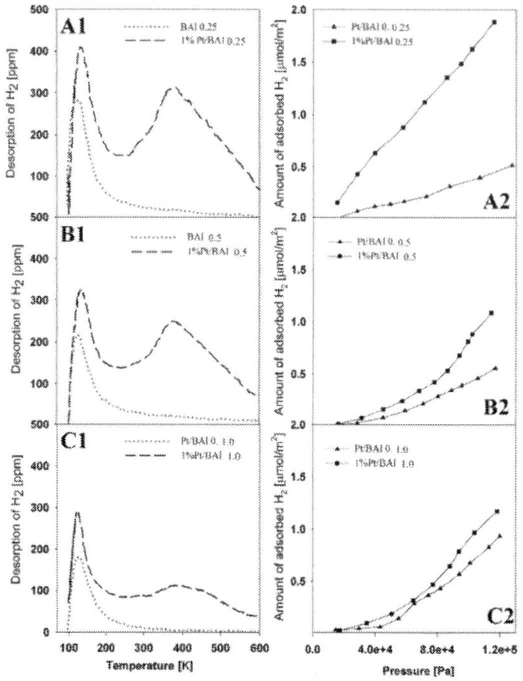

Fig. 4. Hydrogen adsorption isotherms: A. TPD method: A1 – 1% Pt/BAl 0.25, B1 – 1% Pt/BAl 0.5, C1 – 1% Pt/BAl 1.0. B. Volumetric method: A2 – 1% Pt/BAl 0.25, B2 – 1% Pt/BAl 0.5, C2 – 1% Pt/BAl 1.0.

Comparing results of two measurement methods one can conclude that the static (volumetric) technique provides higher values of hydrogen adsorption. In that case the support and its porous structure are of importance. The dynamic (flow) method of measurement relies on a gas flow with a zero hydrogen partial pressure. In that case the stoichiometry is different from 1:1. Also multilayer adsorption can possibly take place.

4 Conclusions

In summary, the study on the hydrogen storage behavior of platinum doped Al_2O_3/B_2O_3 systems at 77 K temperature has been conducted. The storage capacity observed for Pt doped oxides is 3 μmol/m^2. The comparison of the storage capacity of the Pt doped oxides to that of B_2O_3/ Al_2O_3 systems (0.8 μmol/m^2) indicates that presence of metal has enhanced the hydrogen storage capacity of oxides nearly four times.

In the binary system oxide B_2O_3-Al_2O_3 the Lewis acid sites (which are associated with boron component) are largely responsible for the strong binding precursor of the metallic phase ($[PtCl_6]^{2-}$ Lewis base). The compensation of the individual hydrogen storage capacity of metal-particles suggests that the enhancement of storage capacity in metal-doped oxides resulted from the spill-over effect. The results issued from this study suggest that metal-doped oxides can be envisaged as more pragmatic solution for hydrogen storage.

The study was financially supported by statutory activity of the IMN Division in Poznan CLAiO funded by the Ministry of Science and Higher Education and in cooperation with Centre for Advanced Technologies Adam Mickiewicz University.

References

1. K.M. Thomas, Dalton Trans. **9**, 1487 (2009)
2. K.M. Thomas, Catal. Today. **120**, 389 (2007)
3. M.P. Suh, H.J. Park, T.K. Prasad, D.W. Lim, Chem. Rev. **112**, 2, 782 (2012)
4. M. Bastos-Neto et all., Energy Environ. **5**, 8294 (2012)
5. S.H. Barghi, T. Theodore, T.T. Tsotsis, M. Sahimi, Int. J. Hydrogen Energy **39**, 3 (2014)
6. H. Jhi, Y.K. Kwon, Phys. Rev. B. (2005) B71:035408
7. A. Martyla, B. Olejnik, P. Kirszensztejn, R. Przekop, Int. J. Hydrogen Energy **36**, 8358 (2011)
8. R.R. Adzic et all., Top Catal. **46**, 249 (2007)
9. Y. Li, R.T. Yang, J. Phys. Chem. C. **111**, 29, 11086 (2007)
10. S.J. Park, S.Y. Lee, Int. J. Hydrogen Energy **35**, 23, 1348 (2010)
11. R. Zacharia et all., Chem. Phys. Letters **434**, 286 (2007)
12. A. Martyla, Gel system Al_2O_3-SnO_2 in heterogeneous catalysis and fuel cells. PhD Dissertation, Poznan (2004)
13. P. Kirszensztejn, A. Martyla, B. Olejnik, A. Szymkowiak, Przem. Chem. **8–9**, 562 (2003)
14. R. Przekop, P. Kirszensztejn, J. of Non-Crystalline Solids **402**, 10, 128 (2014)
15. R. Przekop, P. Kirszensztejn, React. Kinetics Mech. Catal. (to be published)
16. J. Zielinski, J Catal. **76**, 1, 157 (1982)
17. J. Zielinski, React. Kinet. Catal. Lett. **17**, 69 (1981)
18. D. P. Broom, Hydrogen sorption measurements on potential storage materials: Experimental methods and measurement accuracy. EUR 23242 EN (2008) ISBN 978-92-79-08345-7
19. F. Dumeignil, M. Rigole, M. Guelton, J. Grimblot, Chem. Mater. **17**, 2369 (2005)
20. M. Shanmugam, A. Abilarasu, T. Somanathan, Int. J. Chem Tech Research. **7**(2), 499 (2015).
21. W. Qian, H. Wang, J. Chen, Y. Kong, Materials **8**, 1752 (2015).
22. P. Kirszensztejn et all., Appl. Catal. A Gen. **245**, 1, 159 (2003)
23. P. Kirszensztejn et all., React. Kinet. Catal. Lett. **82**, 2, 287 (2004)
24. D.A. Morris, C.H. Bartholomew, A. Rev. Catal. **5**, 145 (2015)
25. P. Kirszensztejn et all., Pol. J. Environ. Stud. **15**, 74 (2006)
26. L. Znak, J. Zieliński, Polish J. Chem. **78**, 11–12, 2197 (2004)
27. L. Znak, J. Zieliński J. Appl. Catal. A: Gen. **334**, 1–2 (2008)
28. J.J.A. Konvalinka, J. J. F. Scholten, J. Catal. **48**, 374 (1977)
29. D.W. Stockwell, J. Catal. **113**, 317 (1988)
30. R. Przekop, A two-component gel Al_2O_3-B_2O_3 – synthesis and properties compared similar oxide support materials. PhD Dissertation, Poznan (2007)
31. J. Freel, J. Catal. **25**, 1–3, 139 (1972)
32. S.E. Wanke, P.C. Flynn, Catal. Rev.-Sci. Eng. **12**, 93 (1975)

Analysis of energy consumption at the Rzeszów Wastewater Treatment Plant

Adam Masłoń[1,*]

[1]Rzeszów University of Technology, Department of Environmental Engineering and Chemistry, al. Powstańców Warszawy 6, 35-959 Rzeszów, Poland

Abstract. Wastewater treatment plants can be classified as energy-intensive facilities, as they account for up to 35 percent of municipal energy consumption. Pumps and aeration systems consume a significant portion of energy within the wastewater plants in particular. The cost of energy consumption for wastewater treatment processes reaches up to 40% of the total operating cost. In case of the WWTPs with the activated sludge systems, about 50% of energy is used for aeration and mixing purposes. At WWTPs, energy consumption is often correlated with the magnitude and type of pollutant load, which can influence the treatment methods and technologies used in the WWTP. In many cases wastewater treatment plants are operated without optimized measures for process optimization. A detailed study of the energy consumption should be executed in order to determine the optimization potential. This paper presents the energy consumption in municipal wastewater treatment plant in Rzeszów (Poland). In the year 2016, parameters of raw and treated wastewater were tested. The data related to energy consumption in plants allowed us to determine the energy intensity coefficients. Total consumption was measured. Indicators of energy consumption per cubic meter and removed load were calculated.

1 Introduction

Wastewater treatment is an energy-intensive process. Strict standards for discharge often require energy intensive, advanced treatment technologies to be applied. As a result, the number of plants using the advanced treatment processes has been increased. Wastewater treatment plants use for up to 35 percent of the total municipal energy consumption. Electricity demand for wastewater treatment can reach up to 1% of total consumption for the whole country. Pumps and aeration systems consume a significant portion of energy within the wastewater plants in particular. The cost of energy consumption for wastewater treatment processes reaches up to 40% of the total operating cost. At the WWTPs with the activated sludge process used, about 50% of energy is used for aeration and mixing purposes [1–4]. At the wastewater treatment plants, energy consumption is often correlated to the magnitude and type of pollutant load, which can influence the treatment methods and technologies used in the WWTP. In many cases, wastewater treatment plants are operated

* Corresponding author: amaslon@prz.edu.pl

without optimized measures for processes optimization. A detailed study of the energy consumption should be carried out in order to determine the optimization potential. As a result, the most potent measures can be determined for the particular plant [3–5]. The new challenge for the WWTPs is to minimize energy consumption maintaining the quality of the discharged water. Energy optimization could be obtained through reduction of the electric energy consumption of the individual units of the WWTP, but also via recovering energy from sewage sludge or wastewater [6–8].

The anaerobic digestion is a proven technology for sewage sludge treatment, which makes it possible to generate renewable energy from the same process. When anaerobic digestion technology is applied, microorganisms break down the organic matter contained in the sludge and convert it into a biogas, a mixture including mainly methane and carbon dioxide, which can be then used for electricity, heat and biofuel production. At the same time, the sludge is stabilized and its dry matter (mass) content is reduced. The benefits of anaerobic digestion process of sewage sludge are widely recognized and the technology has been well established in many countries now. Different options for use of the biogas from sewage sludge conversion exist, namely conversion to heat or power (electricity). Today, high quantities of biogas produced in anaerobic digestion in municipal wastewater treatment have enormous potential for being used worldwide [5]. Anaerobic digestion optimization is a common practice, often implemented to increase the energy self-sufficiency of WWTPs. [8]. Not only is the anaerobic digestion of sewage sludge important for maximizing of the energy production, as it also bears a great relevance within the scope of minimizing the overall treatment costs at WWTPs. Utilization of biogas for power and fuel as natural gas has many environmental benefits since it can be substituted for fossil fuels to produce electricity and vehicle fuel, reducing the carbon footprint of WWTP operations [9]. Electricity and heat autonomy indicate the ratio of energy generated to energy used in the WWTP. Larger plants achieve higher levels of autonomy due to the fact that processes there are much more efficient (higher production rate, limited losses). Complete heat autonomy is already being achieved by numerous facilities nowadays, while complete energy autonomy today is achieved only by very few, advanced and sophisticated plants [5].

This paper presents the energy consumption at the municipal wastewater treatment plant in Rzeszów (Poland, Podkarpackie Province). In the year 2016, parameters of raw and treated wastewater were tested too. The data related to energy consumption at wastewater treatment plants allowed us to determine the energy intensity coefficients. The total consumption was measured. Indicators of energy consumption per cubic meter and removed load were calculated as well.

2 Characteristics of the Rzeszów WWTP

The Rzeszów WWTP (Poland, Podkarpackie Province) was built between 1974–1988 and has been modernized several times since then. The average daily processing capacity estimated is Q = 62 500 m^3/d and 400 000 PE (Population equivalents). Back in 2015, the plant has been completely modernized and expanded with a new biological wastewater treatment stream created, the effect of which was to change the capacity of the facility up to 54 500 m^3/d and 398 000 PE.

The plant's process lines composed of a step screen, two horizontal grit separators, two radial primary clarifier units, two activated sludge reactors and four radial secondary clarifier units. Biological wastewater treatment involves integrated nutrient removal based on a conventional multistage activated sludge with a pre-emptive denitrification and denitrification in a five stages Bardenpho process. The sewage sludge line includes mechanical excessive sludge thickener, four closed aerobic digestion tanks, belt filter press and installation for storage and incineration of biogas obtained in the sewage sludge fermentation processes. Biogas is burned within a co-generation system and then electricity and heat are generated as a result (CHP system).

3 Methods of energy consumption analysis

Energy profile analysis for the wastewater treatment plant in Rzeszów was carried out on the basis of operating results recorded during the year 2016. Analysis of the removed pollutant load, gathered through wastewater treatment, in relation to energy consumption was defined. Based on the data, consumption of electricity and heat energy were identified, as well as the energy balance for the WWTP was identified. In the present paper, following energy performance indicators of wastewater treatment have been adopted:

- EI_d – daily electricity consumption [MWh/d]
- HEI_d – daily heat energy consumption [MWh/d]
- TEC – total energy consumption indicator [kWh per m^3 wastewater]
- EI – electricity consumption indicator [kWh per m^3 wastewater]
- HEI – heat energy consumption indicator [kWh per m^3 wastewater]
- TER_d – daily total energy recovery [MWh/d]
- ER_d – daily electric energy recovery [MWh/d]
- HER_d – daily heat energy recovery [MWh/d]
- TEIR – total energy recovery indicator [kWh per m^3 wastewater]
- EIR – electric energy recovery indicator [kWh per m^3 wastewater]
- HEIR – heat energy recovery indicator [kWh per m^3 wastewater]

Indicators of energy recovery determine the generation of electricity or heat within the CHP system, taking into account the daily wastewater inflow and production of sewage sludge generated during the wastewater treatment process.

The electricity consumption (electric energy consumption) – EI – refers to the inflow (kWh per m^3 wastewater), often called "specific energy consumption", it is the most commonly used parameter to assess the energy effectiveness of wastewater treatment plants.

Electricity consumption values in relation to the population equivalent (PE) per year, as well with a reference to the amount of pollutants removed from wastewater, were determined. The electricity consumption related to removed BOD_5 (Biological Oxygen Demand) and COD (Chemical Oxygen Demand) loads was exhibited.

4 Evaluation of energy consumption

4.1 Electricity and heat energy consumption

Throughout the year 2016, the total quantity of treated wastewater was higher than 15 600 000 m^3, with an average of 42 631 m^3/d. High flow ratios were recorded during the spring months (snow melt processes) and during the autumn period.

The total energy consumption amounted to 13.39 GWh, while electricity consumption was measured at the level of 7.21 GWh and accounted for 53.8% of total consumption in 2016. The total heat energy consumption of Rzeszów WWTP was equal to 6.18 GWh, shared among the heating demand for the sludge line (anaerobic digesters, drying) and for the facility buildings and offices. The highest energy consumption was recorded in the winter months and stemmed directly from the high heat demand for heating buildings and sewage sludge before the fermentation process in anaerobic digesters. The daily electricity energy consumption was recorded at the level between 17.37 and 22.16 MWh/d, with the average value of 19.7 MWh/d during the research period. The daily heat consumption ranged between 11.75 and 21.16 MWh/d with an average value of 16.92 MWh/d. The sum of electricity and heat was on the level 32.2–40.96 MWh/d. The highest values of the EI_d parameter were observed in June and for HEI_d in January and December 2016. The lowest values of EI_d were recorded in the period from September–December, while HEI_d was the lowest from June to August. The obtained energy usage indicators recorded per 1 cubic meter of wastewater gave results such as 0.367–0.557 kWh/m^3 (EI), 0.314–0.509 kWh/m^3 (HEI) and 0.762–0.981 kWh/m^3 (TEC). The average indicators amounted to 0.468, 0.397 and 0.865 kWh/m^3 for electricity, heat energy and total energy consumption indicator respectively. Energy consumption indicators at the wastewater treatment plant did not reflect the variation of the flow of wastewater (fig. 1). The energy intensity at the wastewater treatment plant in Rzeszów is similar to other facilities of this type. At the Krosno WWTP in Poland (Q = 21 000 m^3/d, 117 000 PE) in the years 2010–2015 the electricity and heat energy consumption indicators reached the level 0.37–0.54 and 0.3–0.6 kWh/m^3 respectively, in turn the TEC indicator was on the level 0.67–1.11 kWh/m^3 [10]. The HEI value was 0.62 kWh/m^3 at the Klimzowiec WWTP (Poland) the size of which is defined by the following values: 200 000 PE and Q = 29 200 m^3/d [11]. At the Sheboygan WWTP (USA), with average treatment capacity of 37 854 m^3/d, the electricity consumption was shaped at the level of 0.43 kWh/m^3 [12]. Moreover, the electric energy consumption indicator recorded is comparable with that of other countries – 0.45 kWh/m^3 in USA, 0.52 kWh/m^3 in Switzerland, 0.53 kWh/m^3 in Spain, 0.56 kWh/m^3 in Singapore, 0.64 kWh/m^3 in United Kingdom or 0.67 kWh/m^3 in Germany [13].

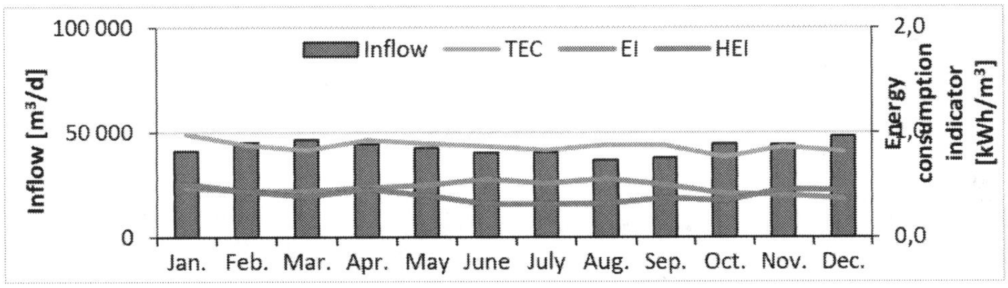

Fig. 1. Change of wastewater inflow and energy consumption indicators at Rzeszów WWTP.

The relationship between energy consumption indicators and the flow wastewater were determined (fig. 2). The electricity consumption decreases with the increasing inflow of wastewater. Very strong statistical significant relationship between EI parameter and wastewater inflow (r = 0.87) made the correlation between HEI and inflow moderate (r = 0.49). For the relationship between total energy consumption and inflow the correlation coefficient value was weak (r = 0.34).

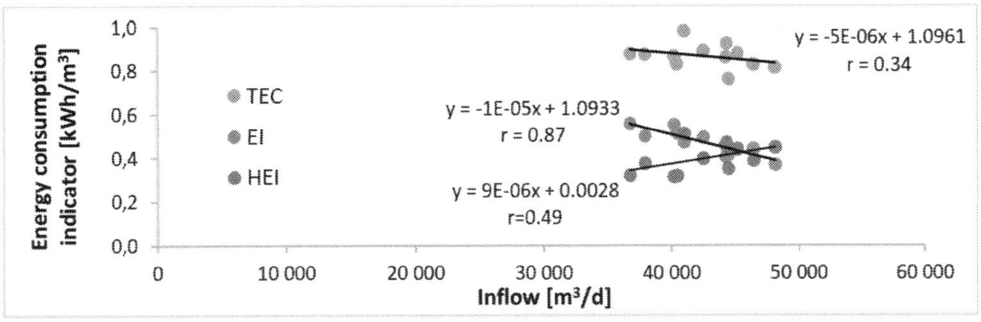

Fig. 2. Relationship between inflow and energy consumption indicators.

Electricity consumption per population equivalent and year was 26.11 kWh/(PE·y), 22.41 kWh/(PE·y) and 48.2 kWh/(PE·y) for electricity, heat energy and total energy consumption respectively. The actual population equivalent of Rzeszów WWTP was shaped at the level of 276 099 PE. For the sake of comparison, electricity, heat energy and total energy consumption in Krosno WWTP was at the level of 30.53 kWh/(PE·y), 28.58 kWh/(PE·y) and 59.11 kWh/(PE·y) [10].

Many others specific parameters can be used to express the energy demand efficiency of a WWTP. The electricity consumption related to removed BOD_5 and COD loads exhibited itself in values ranging from 1.03 up to 1.57 kWh/BOD_5 and from 0.49 up to 0.68 kWh/COD. According to the literature, for large plants with the energy consumption parameter value higher than 100 000 PE, the corresponding value was equal to 0.69 kWh/COD [3]. Electricity consumption increases also with the increasing influent pollution concentration (COD, BOD_5) [1].

4.2 Wastewater energy recovery and energy production

Energy production, with the use of process involving burning of biogas generated from the sewage sludge, has been commonly used at WWTPs now [1, 8]. Daily biogas production reached the level from 5473 m^3 in October to 8041 m^3/d in June, with the average value of 6637.9 m^3/d (fig. 3). During the year 2016 the total biogas generation at the plant was equal to 2427489 m^3. The biogas production rate (BPR) was contained in the range from 0.123 to 0.2 m^3 per m^3 wastewater.

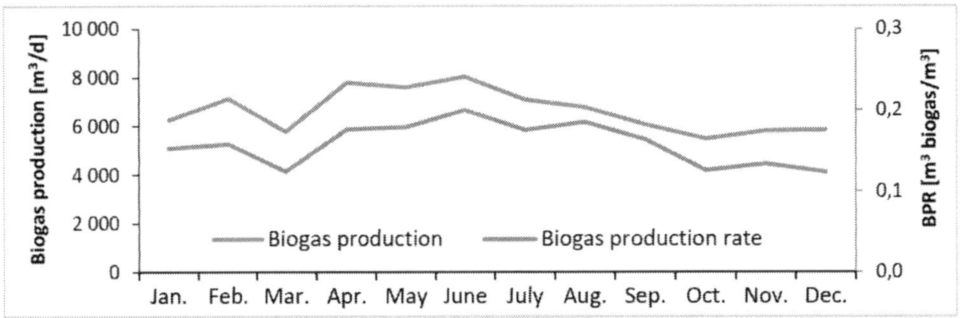

Fig. 3. Volume of biogas production, achieved from the sewage sludge at Rzeszów WWTP.

The installed CHP system produces ca. 1035 kW of electricity and ca. 1530 kW of heat. Furthermore, another two boilers produce 1440 kW of heat. Daily electric energy (ER_d) and daily heat energy recovery (HER_d) were within the range of 11.78–17.44 MWh/d and 11.75–20.34 MWh/d respectively. The results show that daily total energy recovery varies from 26.0 to 37.27 MWh/d with the average value of 30.65 MWh/d.

The total energy generation from biogas was contained between 3.72 and 5.67 kWh/m^3 with the average value of 4.67 kWh/m^3. In relation to the above, the electricity and heat energy production was as high as 2.02–2.48 kWh/m^3 and 1.57–3.25 kWh/ m^3 respectively (fig. 4). At other plants electricity generation from biogas reached the level of 1.9–4.8 kWh/m^3 (Mielec WWTP, Poland) [14], 3.82–4.51 kWh/m^3 (Krosno WWTP) [10]. However at these plants, in anaerobic digesters, the process of co-fermentation of sewage sludge with waste fats takes place quickly, which enhances the production of biogas and methane. Literature also indicates the low electric energy production rates, when biogas is used, e.g. 0.2 kWh/m^3 (Poprad WWTP, Slovakia), 1.21 kWh/m^3 (Trnava WWTP, Slovakia), 2.0 kWh/m^3 (Bratislava Central WWTP, Slovakia) [1].

 DOI: 10.1051/e3sconf/20172200115

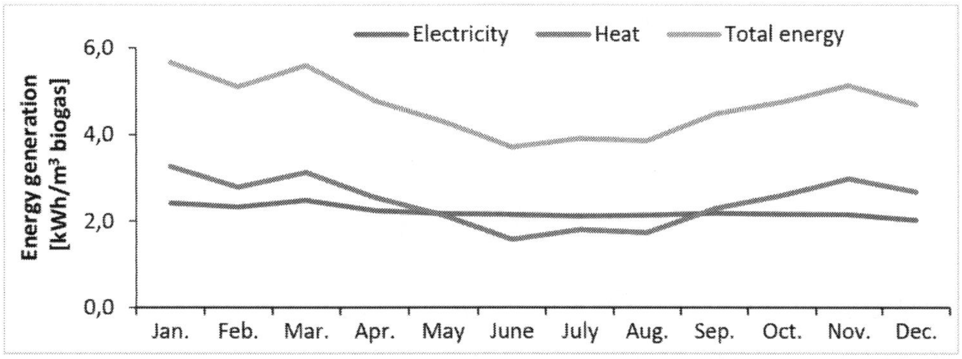

Fig. 4. Electricity, heat and total energy generation from biogas at wastewater treatment plant.

Analysis of energy recovery indicators allowed us to determine the energy generation efficiency of wastewater treatment process. During the research period, the total energy recovery indicator was placed at the level of 0.52–0.77 kWh/m^3 with the average value of 0.52 kWh/m^3. The highest values of TEIR were observed in January and April 2016. Values of EIR and HEIR indicators have allowed us to obtain results such as 0.25–0.43 kWh/m^3 and 0.26–0.41 kWh/m^3. The average of electric energy and heat energy recovery indicators was 0.3 and 0.31 kWh/m^3 respectively (fig. 5).

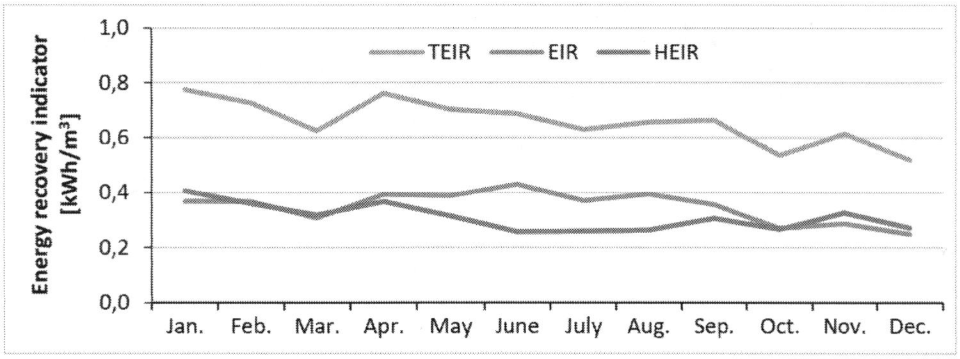

Fig. 5. Energy recovery indicators recorded at wastewater treatment plants in Rzeszów.

At other plants energy recovery in relation to treated wastewater is diversified. In case of the Krosno WWTP, during the years 2010–2015, energy recovery indicators were at the level of 0.25–0.40 kWh/m^3 (EIR), 0.3–0.6 kWh/m^3 (HEIR) and 0.55–0.89 kWh/m^3 (TEIR) [10]. For the sake of comparison, the total energy recovery at the Steinhof WWTP (Germany) with treatment capacity of 61 643 m^3/d amounted to 0.858 kWh/m^3 via Combined-Heat-Power [12].

4.3 Energy balance at wastewater treatment plant

The anaerobic digestion technology plays an important role in energy self-sufficiency of a WWTP [8]. The biogas energy makes it possible to meet 74.3% of the electrical and 95.5% of the needs, in relation to the overall energy consumption of the wastewater treatment plant. The rate between total energy generation and consumption in case of the Rzeszów WWTP was shaped at the level between 71.1–91.6% with the average value of 83.6%. Maximum electrical energy self-sufficiency was noted in February (83.2%) and April (83.9%) and minimum was recorded in October (65.5%). Heat production at plant resulted from demand for heating of the anaerobic digesters and of buildings and offices

 DOI: 10.1051/e3sconf/20172200115

during the winter period. Minimum self-sufficiency referring to heat energy was seen in December (73.8%), whereas maximum values 100% were recorded in February–March and June–September (fig. 6–8).

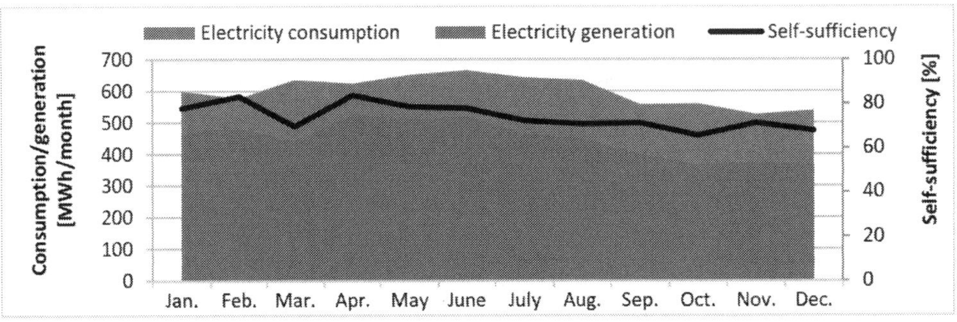

Fig. 6. Electricity balance at the Rzeszów WWTP, during the year 2016.

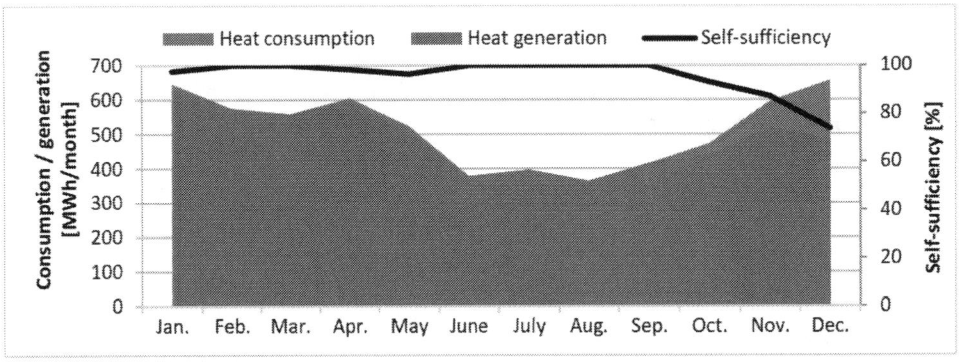

Fig. 7. Heat energy balance at the Rzeszów WWTP, during the year 2016.

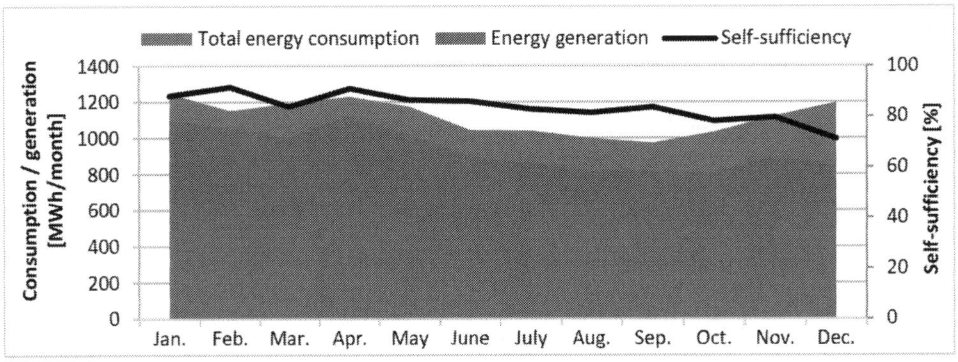

Fig. 8. Total energy balance at the Rzeszów WWTP, during the year 2016.

After comparing the obtained results with the literature data, it can be concluded that Rzeszów WWTP has a very high energy potential. For the sake of comparison, we may state the average degree of cover the electricity consumption at the Krosno WWTP was achieved at the level of 70% [10]. At the WWTPs in Slovakia, energy self-sufficiency was equal to 66% (Bratislava Central WWTP), 64% (Trnava WWTP) and 52% (Martin WWTP) [1].

 DOI: 10.1051/e3sconf/20172200115

5 Conclusions

Wastewater treatment is an energy-intensive process and energy consumption is often the main operational cost component for the WWTPs. Electricity and heat energy is consumed in wastewater treatment systems every year, and there is a large potential of saving the said energy within that process. Furthermore, unit energy cost of wastewater treatment may be significantly affected by wastewater treatment capacity and technology. In this paper, energy consumption of wastewater treatment plant in Rzeszów was explored. The results show that the total energy consumption of Rzeszow WWTP in 2016 year was placed at the level of about 0.865 kWh/m^3 and 48.2 kWh/(PE·y). The average indicators for electricity and heat energy consumption amounted to 0.468 and 0.397 kWh/m^3 respectively. The electric energy consumption was much lower, than the corresponding values present at other plants and in other countries. This result may be tied to the very advanced treatment technologies used and to the low water quality requirements of wastewater treatment in Rzeszów. Anaerobic digestion technology with CHP system used at the Rzeszów WWTP plays an important role in energy self-sufficiency. The biogas energy makes it possible to meet 74.3% of the electrical and 95.5% of the heat needs, in relation to the overall energy consumption of the wastewater treatment plant. The relation between total energy generation and consumption at Rzeszów WWTP had an average value of 83.6%. During the research period total energy recovery indicator was placed at the level 0.52 kWh/m^3. The average of electric energy and heat energy recovery indicators were recorded at the level of 0.3 and 0.31 kWh/m^3. Identification of the WWTP energy balance was the first, but very important step in the process of developing better biogas utilisation scheme. The data describes the real situation seen at the Rzeszow WWTP, but the similar results can be expected at other WWTPs in Poland.

References

1. I. Bodík, M. Kubaská, Environ. Prot. Eng. **39**, 2 (2013)
2. D. Panepinto, S. Fiore, M. Zappone, G. Genon, L. Meucci, Appl. Energ. **161** (2016)
3. S. Longo, B.M. d'Antoni, A. Chaparro, A. Cronrath, F. Fatone, J.M. Lema, M. Mauricio-Iglesias, A. Soares, A. Hospido, App. Energ. **179** (2016)
4. J. Henriques, J. Catarino, J. Clean. Prod. **142** (2017)
5. N. Bachmann, *Sustainable biogas production in municipal wastewater treatment plants* (IEA Bioenergy, 2015)
6. K.J. Chae, J. Kang, Energ. Convers. Manage. **75** (2013)
7. J. Frijns, J. Hofman, M. Nederlof, Energ. Convers. Manage. **65** (2013)
8. G. Silvestre, B. Fernández, A. Bonmatí, Energ. Convers. Manage. **101** (2015)
9. Y. Shen, J.L. Linville, M. Urgun-Demirtas, M.M. Mintz, S.W. Snyder, Renew. Sust. Energ. Rev. **50** (2015)
10. K. Trojanowicz, Ł. Karamus, Forum Eksploat. **4**, 85 (2016)
11. P. Banaszek, Kierunek Wod-Kan. **3** (2014)
12. H. Wang, Y. Yang, A.A. Keller, X. Li, S. Feng, Y. Dong, F. Li, Appl. Energ. **184** (2016)
13. F. Hernández-Sancho, M. Molinos-Senante, R. Sala-Garrido, Sci. Total Environ. **409** (2011)
14. A. Masłoń, S. Pazdro, W. Mroczek, Forum Eksploat. **4**, 79 (2015)

E3S Web of Conferences **22**, 00116 (2017)
ASEE17

DOI: 10.1051/e3sconf/20172200116

Analysis of the energy potential of municipal solid waste for the thermal treatment technology development in Poland

Katarzyna Midor[1,*], and *Karolina* Jąderko[2]
[1]Silesian University of Technology, Faculty of Organisation and Management, Institute of Production Engineering, ul. Roosevelt 26, 41-800 Zabrze, Poland
[2]Ekologia Przedsiębiorczość Innowacje Spółka z o.o., ul. Kuhna 17, 42-256 Olsztyn k/Częstochowy, Poland

Abstract. The problem of overproduction of waste has been a local issue for many years. Since the new environment law came into effect, the current approach to waste management has changed significantly. The accessible technological possibilities of thermal waste treatment with the energy recovery set a new area of research over the process of choosing effective and rational way of calorific waste management. The objective of this article is to provide assessment results of the analysed energy potential in waste management system in the form of calorific waste stream. In includes all the activities and actions required to manage municipal solid waste from its inception to its final disposal i.e. collection, transport, treatment and disposal. The graphical representation of waste flow indicates the lost opportunities of waste energy recovery. Visual research method was supported and founded on value stream mapping. On the basis of the results were presented the directions of further improvement of calorific waste stream mapping for the purposes of implementation the thermal treatment technology in the selected waste management region.

1 Introduction

Due to the growing interest in the use of wastes for energy production, tools for effective management of a stream of high-calorie waste. New trends in waste management such as recycling logistics or ecologicstics cause a dynamic development of tools for assessing and analysing streams of combustible waste.

The ecological concept of logistics in a waste management system allows us to not only integrate processes, but also improve the flow of waste. In light of this approach managing a stream of waste is aimed at finding optimal solutions which take into account a number of conditions that minimise the negative effects on the environment and society, but also affecting its economic efficiency.

Effective and efficient transposition of logistics onto the field of waste management causes the practical use of tools from this area streamlines the flow of waste from the place of their creation to the place where they are ultimately managed. This article directs

* Corresponding author: katarzyna.midor@polsl.pl

attention to the possibility of using value stream mapping in high-calorie waste management. The effect of this work is an attempt at illustrating the flow of caloric waste in a waste management system.

According to the initial concept, value stream mapping is a tool used for developing a complete method of flow management in a production system. Using the right symbols, it allows us not only to visualise the processes, but most importantly notice and eliminate wastage [1–4]. Using the value stream mapping tool we must first and foremost understand what a value is in the system. In the case of waste management, we should look for values in the properties of wastes that allow for their effective reuse. Waste recycling as the most ecological form of their management is based on the wastes' energetic and material properties, which determine the choice of an effective processing method.

2 Methodology

One of the aims of the research was an attempt at transposing a selected production engineering tool onto the area of waste management. Due to the cognitive and analytical nature of the research, the study was carried out in the following stages:

1. Choosing a regional waste management system and identifying the waste removal chain.
2. Analysis of the waste management system, especially the high-calorie waste stream.
3. Drawing up and analysing the current state map, and presenting the proposed changes and improvements

2.1 Choosing a regional waste management system and identifying the waste removal chain

The Silesian Voivodeship has been divided into four regions in terms of waste management. This division is justified by the amount of waste produced as well as the total efficiency of the already existing and planned waste management facilities. Waste logistics within a waste management region encompasses primarily such waste streams as:

- mixed municipal solid waste;
- remains from the sorting of municipal solid waste and mechanical-biological municipal waste processing if they are intended for storage;
- green wastes, which, according to the law, must be processed within the region in designated regional waste processing plants. An exception to this are supra-regional plants which are supposed to be incineration plants that allow for thermal processing of the wastes with energy recovery.

Due to the research goal set, the 2nd Municipal Waste Management Region (MWMR 2) was chosen, characterised by the biggest population and, therefore, the largest amount of municipal solid waste produced among the regions.

The quantitative-qualitative analysis of the municipal waste collected in the MWMR 2 was conducted on the basis of borough reports on the realisation of waste management tasks for 2014. Table 1 presents the amount of municipal waste collected that was subjected to the particular management processes according to the recovery and disposal process list listed in the Waste Management Act (Journal of Laws (Dz.U.) 2013 item 21).

The boroughs' reports show that in 2014 131 426.8 Mg of waste was disposed, which amounted to 16% of all the municipal solid waste collected, and the remaining 84%, i.e. 705 929.9 Mg of waste was recycled. Worth noting are the R12 and R13 recycling processes which accounted for 72% of the waste treated. Taking into account the fact that the R12 process, as defined by the current law [6–8], pertains to the exchange of wastes in

 DOI: 10.1051/e3sconf/20172200116

order to treat them using any of the processes from items R1 - R11, and the R13 process to the storage of wastes prior to any of the processes enumerated in items R1 - R12 (excluding the initial storage prior to collection), we should assume that these processes can cause a decrease in value - both material and energetic.

Table 1. Municipal waste management in MWMR 2.

Method of management of the municipal waste collected	Mass [thous. Mg]
D10 – Waste disposal process – on-land thermal treatment	10.4
D13 – Waste disposal process – blending or mixing prior to submission to any of the operations numbered D 1 to D 12 D 1 to D 12	34 454.79
D5 – Waste disposal process – storage in specially designed landfills (e.g. placement into lined discrete cells, capped and isolated from one another and the environment, etc.)	84 042.98
D8 – Waste disposal process – biological treatment not specified in another item resulting in the formation of final compounds or mixtures which are disposed of using any of the processes enumerated in items D1–D12	12 954.63
R1 – Waste recycling process – use mainly as a fuel or other means of energy production	12.3
R10 – Waste recycling process – on-land treatment resulting in benefit to agriculture or environmental improvement	10 144.1
R12 – Waste recycling process – exchange of wastes in order to treat them using any of the processes enumerated in items R1 - R11	597 311.98
R13 – Waste recycling process – storage of wastes prior to any of the processes enumerated in items R1 - R12 (excluding initial storage prior to collection)	8 077.03
R3 – Waste recycling process – recycling or regeneration of organic substances which are not used as solvents (including composting and other biological processing)	56278.04
R4 – Waste recycling process – recycling or recovery of metals and metal compounds	13.28
R5 – Waste recycling process – recycling or recovery of other non-organic materials	34 093.06
R9 – Waste recycling process – oil re-refining or other reuses of oil	0.07
Total	**837392.663**

Source: own elaboration based on [5]

2.2 Itemising the streams of high-calorie waste

The characteristics of municipal solid wastes, which can potentially become an energetic resource, depends on several factors. These include: the type of development, saturation of the area with non-residential buildings (including business premises), as well as the technical furnishings of the buildings and their heating [9, 10]. Equally important to the composition of the wastes are such elements as: affluence of the inhabitants, season, backyard green waste composters, and selective collection of recyclable waste conducted by the inhabitants. These and other factors affect the morphological composition of municipal solid waste and, consequently, their net caloric value. Therefore, the material composition of waste is connected with its fuel characteristics. Each of the constituents, when studied separately, is characterised by such properties as [10, 11]:

- specific dampness;
- organic substance content;
- gross caloric value.

Separating the properties of wastes enables us to classify them into a group of combustible wastes, which are the input of the R1 thermal treatment process [12].

Quantitative analysis of the waste management system takes into account the total amount of waste collected within the boroughs which belong to MWMR 2 in 2014,

DOI: 10.1051/e3sconf/20172200116

detailing the codes of caloric wastes (Table 2) including paper, plastic, rubber, textile, wood waste, as well as mixed waste 19 12 12, 20 03 01, which are subject to a separate analysis. This division is justified by the need to develop a separate methodology for the analysis of caloric properties of mixed wastes (i.e. 19 12 12, 20 03 01). In practice it is assumed that the ballast from processing 19 12 12 wastes is 50%. This fraction, constituting the leftovers from the sorting process, is a high-calorie, problematic ballast deposited into landfills which simultaneously presents a large, unused energetic potential suitable for thermal processing. The effect of this study are data contained in Table 2.

Table 2. Juxtaposition of wastes with energetic potential in Region 2.

Code	Group, subgroup and types of waste	Mass [Mg] in 2014
15	Waste packaging; absorbents, wiping cloths, filter materials and protective clothing not otherwise specified	
15 01	Packaging (including separately collected municipal packaging waste)	
15 01 01	Paper and cardboard packaging	11326.96
15 01 02	Plastic packaging	4475.73
15 01 03	Wooden packaging	18.8
15 01 09	Textile packaging	0
16	Wastes not otherwise specified in the list	
16 01	End-of-life vehicles from different means of transport (including off-road machinery) and wastes from dismantling of end-of-life vehicles and vehicle maintenance (except 13, 14, 16 06 and 16 08)	
16 01 03	End-of-life tyres	23.4
16 01 19	Plastic	0
17	Construction and demolition wastes (including excavated soil from contaminated sites)	
17 02	Wood, glass and plastic	
17 02 01	Wood	3.4
17 02 03	Plastic	3.1
19	Wastes from waste management facilities, off-site waste water treatment plants and the preparation of water intended for human consumption and water for industrial use	
19 12	Wastes from the mechanical treatment of waste (e.g. sorting, crushing, compacting, pelletising) not otherwise specified	
19 12 01	Paper and cardboard	0
19 12 04	Plastic and rubber	0
19 12 08	Textiles	0
19 12 10	Combustible waste (refuse derived fuel)	0
19 12 12	Other wastes (including mixtures of materials) from mechanical treatment of wastes other than those mentioned in 19 12 11	113321.3
20	Municipal wastes including separately collected fractions	
20 01	Separately collected fractions (except 15 01)	
20 01 01	Paper and cardboard	3655.59
20 01 10	Clothes	36.6
20 01 11	Textiles	13.3
20 01 38	Wood other than that mentioned in 20 01 37	85.5
20 01 39	Plastics	3787.04
20 03	Other municipal wastes	
20 03 01	Mixed municipal waste	491431.51

Source: [5]

 DOI: 10.1051/e3sconf/20172200116

3 Analysis of the energetic potential of municipal wastes

The research results presented indicate that separating caloric waste from the municipal waste stream allows us to present the energetic potential of wastes collected in the waste management region subject to this study.

Transposing the value stream map commonly used for production processes allows for presenting the problem and exposing the value, which, in this case, is the caloric value of the collected wastes. Figure 1 is the result of an attempt at organising the data gathered.

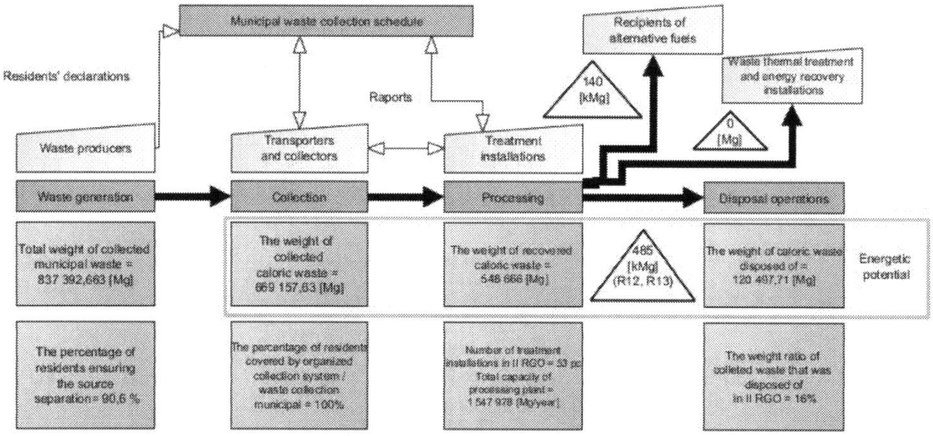

Fig. 1. Value stream map of the waste management system. Source: own elaboration based on [5].

The analysis of caloric stream flow in the waste management system of the 2nd MWMR indicates that the lack of a thermal municipal waste processing facility causes wastage in form of unused energetic potential of the collected waste. The scale of this wastage has been calculated based on the accepted division of energetic potential presented in [5] (Fig. 2):

- theoretical potential including all the caloric municipal wastes collected in a given region;
- available potential, a part of the theoretical potential reduced by the mass of caloric wastes subjected to recycling and recovery (R1–R11) and including the disposal processes - D as well as other recycling processes, i.e. R12 and R13, which pertain to waste exchange and storage;
- technical potential, a part of the available potential that has been reduced by the share of problematic ballast from the sorting process of wastes with the codes 19 12 12 and 20 03 01, which, according to the accepted assumptions [5] amounts to 50%. Therefore, technical potential refers to the part of collected calorie waste that is disposed of through storage and which constitute potential input material for effective energy recovery.

DOI: 10.1051/e3sconf/20172200116

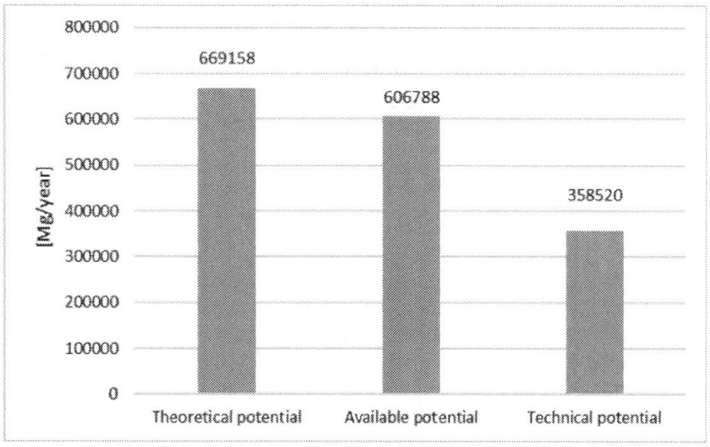

Fig. 2. Energetic potential of wastes in MWMR 2. Source: own elaboration.

4 Conclusion

Analysis of the waste management system with the use of value stream mapping allows us to graphically present the flow between the main parts of the system. The ability to single out a selected stream of waste as well as assign it with a value (energetic or material) allows us to pinpoint and eliminate wastage in terms of wastes' unused potential. A strong point of this method is the reduction of flow time and minimisation of reserves. An attempt at transposing the method onto the area of waste management increases the role of recycling and disposal, which affects the efficiency of the waste management system. Presenting the energetic potential lost by disposing of the wastes at the landfill sets out the possible courses of improvement action and modifications to the system.

Creating maps allows us to concentrate implementation efforts on the areas that require a more detailed analysis and improvement. In the case of waste management we can consider streams of a selected group of wastes in the entire system through quantitative and qualitative presentation of flows from the place of origin to the place of recycling/disposal. Additionally, the cataloguing of the system's infrastructure allows us to present the unused processing power in the system or lack thereof.

After identifying selected aspects connected with the functioning of the 2nd MWMR it is recommended that technology allowing for managing high-calorie wastes be implemented into the system. Introducing new technologies for the retrieval of energy from wastes can significantly contribute to reducing the mass of combustible wastes deposited in landfills. The qualitative analysis of combustible wastes gathered confirmed the unused energetic potential and the validity of further research on the effectiveness of implementing a thermal waste processing facility in the 2nd MWMR. Finding the matching efficiency and determining the optimal location for such a facility will allow for increasing the systems efficiency, taking into account the economic, ecological and social circumstances.

This article was prepared within the statutory research titled "Methods and tools for improving products and services on the selected examples" work symbol 13/030/BK_17/0027 performed at Silesian University of Technology, Institute of Production Engineering.

References

1. M. Rother, J. Shook, *Learning To See – Value Stream Mapping to Create Value and Eliminate Muda* (Lean Enterprise Institute, Brookline, MA, 1999)
2. *Value stream* (http://logistyczny.com/biblioteka/kaizen-lean/item/893-strumien-wartosci - 22.02.2017) (in Polish)
3. M. Zasadzień, Z. Nawrot, Manage. Syst. Prod. Eng. **4** (2016)
4. K. Midor, G. Klimasara, Syst. Wspomag. Inż. Prod. **4** (2016)
5. K. Jąderko, B. Białecka, *Technological-logistic model of energy utilization process of waste* (P.A.NOVA, Gliwice 2016) (in Polish)
6. Act of 14 December 2012 on waste from 2012 (Journal of Laws of the Republic of Poland from 2013 item 21) (in Polish)
7. Regulation of the Minister of Economy of 8 January 2013 on the criteria and procedures for the acceptance of waste for disposal at a landfill of a type (Journal of Laws of the Republic of Poland from 2013 item 38) (in Polish)
8. Waste management plan for Silesian Voivodship 2014 (in Polish) (annex to the resolution of the Silesian Regional Assembly No. IV / 25/1/2012 of 24 August 2012)
9. R. Zarzycki (ed.), *Energy from waste* (Polska Akademia Nauk, Łódź, 2008) (in Polish)
10. K. Skalmowski (ed.), *Waste management guide* (Verlag Dashofer, Warszawa, 2010) (in Polish)
11. J. Wandrasz, A. Wandras, *Formed fuels* (Seidel-Przywecki, Warszawa, 2006) (in Polish)
12. Directive 2010/75/EU of the European Parliament and of the Council of 24 November 2010 on industrial emissions (integrated pollution prevention and control) (in Polish)

CFD analysis of a full-scale ceramic kiln module under actual operating conditions

Massimo Milani[1], *Luca* Montorsi[1,*], *Matteo* Stefani[1], and *Matteo* Venturelli[1]

[1]University of Modena and Reggio Emilia – Department of Sciences and Methods for Engineering – via Amendola 2 – Pad. –Morselli, 42122 Reggio Emilia, Italy

Abstract. The paper focuses on the CFD analysis of a full-scale module of an industrial ceramic kiln under actual operating conditions. The multi-dimensional analysis includes the real geometry of a ceramic kiln module employed in the preheating and firing sections and investigates the heat transfer between the tiles and the burners' flame as well as the many components that comprise the module. Particular attention is devoted to the simulation of the convective flow field in the upper and lower chambers and to the effects of radiation on the different materials is addressed. The assessment of the radiation contribution to the tiles temperature is paramount to the improvement of the performance of the kiln in terms of energy efficiency and fuel consumption. The CFD analysis is combined to a lumped and distributed parameter model of the entire kiln in order to simulate the module behaviour at the boundaries under actual operating conditions. Finally, the CFD simulation is employed to address the effects of the module operating conditions on the tiles' temperature distribution in order to improve the temperature uniformity as well as to enhance the energy efficiency of the system and thus to reduce the fuel consumption.

1 Introduction

The ceramic industry is well known to be characterized by energy intense processes and among them the kiln is one of the main responsible of the primary energy consumption. Recent regulations for the energy consumption limit the fuel or electricity use [1] and therefore, the design of the kiln plays a fundamental role in meeting the regulations' requirements.

Due to the complex physical phenomena that take place in industrial ceramic kiln, simulation tools are an important instrument in the design process of the kiln; thus, different approaches have been adopted for the optimization of the kilns' performance. In [2] a theoretical formulation for the prediction of the kiln operating characteristics has been proposed under regime conditions and many simplifications to the physical phenomena had to be made in order to close the mathematical model. Similar approach is adopted also in [3] in which the main focus was the assessment of the heat transfer between the hot air flow and the kiln walls. The numerical simulation becomes a valuable tool in investigating the performance of complex physical systems including different physical phenomena. Lumped

* Corresponding author: luca.montorsi@unimore.it

and distributed numerical analysis has been extensively adopted for the simulation of complex systems under time dependent operating conditions. In [4] a novel concept for the hydrogen production has been investigated by modelling the entire proposed plant and different operations have been compared in order to define the best energy efficiency configuration. Similarly, in [5] the numerical analysis of the entire cogeneration systems based on aluminium-water combustion was used to determine the best turbine operating point under different working conditions. The modelling approach has also been employed for the validation of integrated waste-to-energy systems with low environmental impact and the advantages compared to traditional technologies have been outlined [6]. Despite the dimensional approximation, the 0D/1D models proved also to be able to account for complex physics involving heat transfer phenomena while including the layout of the full system. Nevertheless, the lumped and distributed parameter model has to assume a large number of quantities that can be difficult to derive from experimental measurements or theoretical approaches. Thus, the combination of this modelling approach to a full multi-dimensional approach of the involved physics is paramount in addressing accurately the behaviour of the analysed systems [7–9].

In this paper, the CFD simulation of the full geometry of a ceramic kiln model is carried out. A module of the preheating section as well as a module in the firing section are analysed. Insulating materials and alumina rollers and burners are included in the simulation and the effect on the tile temperature distribution is investigated. Furthermore, radiation is accounted for in the modelling in order to assess the contribution of the radiative heat transfer in the total heat absorbed by the ceramic tiles during the preheating and firing regions of the kiln. The results are useful not only to better model the heat transfer processes in the lumped and distributed parameter model of the real kiln, but also to better understand the main heat transfer contributions that influence the tiles' temperature profile and thus the product quality.

2 Test case

The ceramic kiln simulated in this paper is real production facility located in the ceramic district in Emilia Romagna – Italy. It is characterized by a production rate of approximately 5,000 kg/h of tiles and it is designed for a continuous working load of about 8,700 hours per year. The entire ceramic kiln includes 43 modules with a length of 2.1 m each and it can be subdivided into the following 5 sections: pre-heating, firing, fast cooling, slow-cooling and outlet section, see Figure 1.

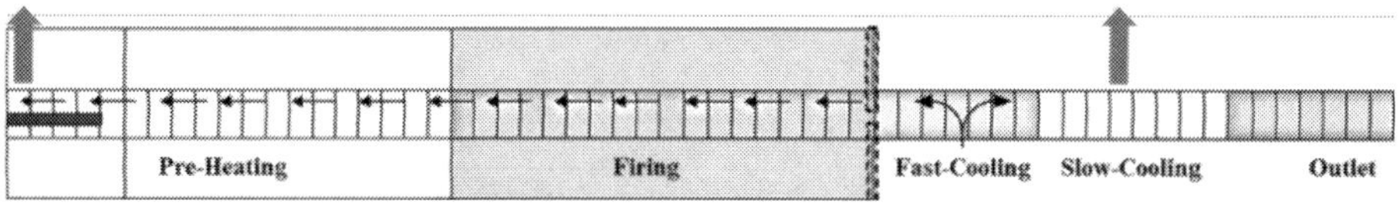

Fig. 1. Schematic of the entire ceramic kiln.

This study focuses on one module in the preheating section and one module in the firing section. These section are critical in order to reach the quality of the tiles and determine the energy consumption of the kiln. In particular, an accurate temperature control is mandatory for achieving the desired product quality and avoiding scraps.

In the pre-heating section, no burners are employed and the heating is due to the hot air flow from the burners' zone. As far as the energy consumption is concerned, the most important zone is the firing where the burners are installed. The burners are fuelled by natural gas and the fuel rate is controlled for each group of 8 burners. One group of burners belongs to two modules and there are two groups of burners for each module that are installed above and below the rollers' plane respectively. The different control of the burners in the chambers above and below the rollers is critical for achieving the correct temperatures in these volumes. At the end of the firing section, a fireproof wall is positioned in order to separate the cooling zone to the firing zone; main task of this wall is to partition the flow between the firing and fast-cooling sections. Its height is adjusted in order to always direct the flow from the cooling to the firing section and regulate the temperature at the interface of these two regions.

An important characteristic of the analysed kiln is the opposite direction of the hot air flow and the tile motion. A mentioned before, the air is inducted from the cooling zone by means of a fan located at the beginning of the kiln before the flue gas stack. Therefore, the pressure in the cooling zone is higher than the one in the firing section, which is usually characterized by values below the atmospheric pressure. Thus, the leakages in the walls of the modules allow the cold ambient air to enter the kiln chambers influencing the temperature of the hot air flow and thus the energy consumption of the entire system. This effect is not taken in the present analysis into account. The remaining portion of air in the fast-cooling section exits the kiln through the slow-cooling section and the final outlet stack. An important feature of the analysed kiln is the counter-current heat transfer mode.

3 Numerical modelling

The CFD analysis is carried out by means of the Star CCM+ code licensed by CD-Adapco. Particular care in the mesh construction as well as in the modeling of the heat transfer is paid. Ideal gas law is assumed to model the air thermal behavior, while the energy equation is solved for the enthalpy and coupled to the Navier-Stokes equation to solve the fluid flow. Dynamic simulations are performed in order to account also for the tiles' motion. Figure 2 shows the real geometry of the analysed modules, while Figures 3a) and 3b) depict the computational domains adopted for the simulation of the preheating module and the firing module respectively. The latter domain includes also the grid for the EKO burners, which are designed to preheat the oxidizing air by recovery the heat from the hot air flow drawn from the kiln chambers.

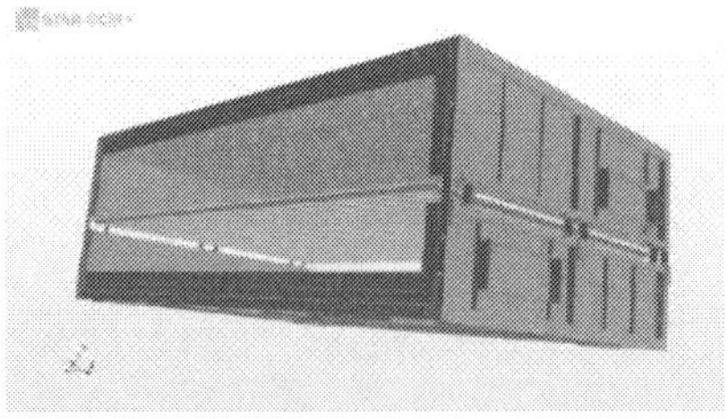

Fig. 2. Geometry of the analysed kiln's module.

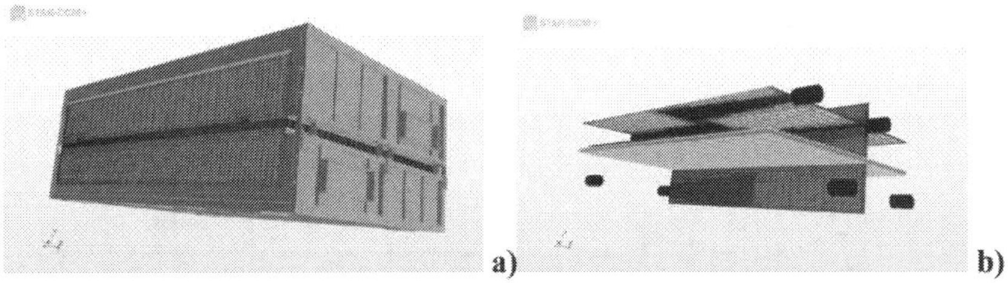

Fig. 3. Computational domain for the module in the a) preheating and b) firing sections.

Both grids are constructed using a trimmed hexahedral mesh with proper refinements particularly in the regions where the burners' flame is supposed to take place, see Figure 3b). The total number of cells used in the CFD domain is approximately 4 million. The mesh used in the simulation is the best trade-off between result's accuracy and computational effort. The fluid domains consist of the upper and lower chambers, the alumina rollers, the tiles' material and eventually the EKO burners. Turbulence is accounted for by means of the two zonal version of the k-ω model, known as the shear stress transport model [9]. This approach demonstrated in previous studies a good accuracy in predicting the flow field for multi-phase flows in hydraulic components [10].

The radiation contribution on the total heat transfer is calculated by using the Surface-to-Surface (S2S) radiation model that model the thermal radiative heat transfer between surfaces of arbitrary complexity that form enclosed spaces. The medium that fills the space between the surfaces is non-participating, thus, it does not absorb, emit, or scatter radiation. The radiation properties and the thermal boundary conditions that are imposed on each surface define uniquely the amount of radiation that a surface receives and emits. The radiation energy transfer to or from each surface is calculated from the radiation transport equation and the boundary conditions [12, 13].

The combustion process for the burners in the firing section module is modelled according to the Eddy Brake Up (EBU) approach. This model characterizes the reacting flow system by a specified number of species and chemical reactions. The individual species in the global reaction are assumed to be transported at different rates according to their own governing equations. The Standard EBU model uses Eqn. (1) as the governing transport equation. The reaction rate is modelled through an expression that takes the turbulent micromixing process into account by means of dimensional arguments. Thus, for a reaction of the form:

$$\nu_F F + \nu_O O \rightarrow \nu_{P1} P_1 + \nu_{P2} P_2 + \ldots\ldots + \nu_{Pj} P_j \tag{1}$$

the rate of fuel depletion is assumed to be:

$$R_F = -\frac{\rho}{M_F}\left(\frac{1}{\tau_R}\right) A_{ebu} \min\left[\bar{Y}_F, \frac{\bar{Y}_O}{s_O}\right] \quad \text{moles} / \left(\text{m}^3\text{s}\right) \tag{2}$$

$$s_O \equiv \nu_O M_O / \nu_F M_F; \; s_{Pi} \equiv \left|\nu_{Pi}\right| M_{Pi} / \nu_F M_F; \; \tau_R = \tau_{mix} = \frac{k}{\varepsilon}$$

where k is the turbulent kinetic energy and ε its dissipation rate. The *min* operator on the right-hand side of Eqn. (2) indicates that the concentration of the limiting reactant should be used to determine a mass fraction scale when calculating the reactant consumption rate. Eqn. (2) essentially states that the integrated micromixing rate is proportional to the mean (macroscopic) concentration of the limiting reactant divided by the time scale of the large eddies.

Figure 4 shows the boundary conditions adopted for the simulations. Since many of the data necessary for properly setting the CFD boundary conditions were not available from experimental measurements, the values are derived from a previous simulation of the overall ceramic kiln under actual operations using a lumped and distributed parameter model described in [13]. In particular, the inlet mass flow rate for the upper and lower chambers as well as the relating outlet pressure values are taken in the 0D/1D at the exact location as in the real module.

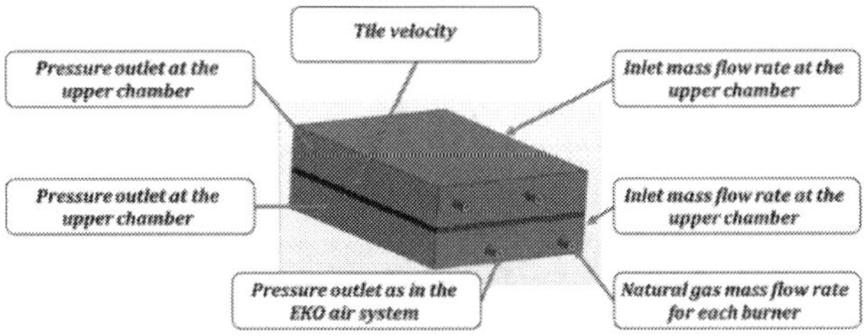

Fig. 4. Boundary conditions for the computational domain.

Furthermore, the lumped approach was used to estimate the natural gas flow rate for each burner for the module in the firing section of the kiln; this value would be very difficult to measure experimentally. The effect of the tile motion is also accounted for in the analysis using the sliding interface technique available in STAR-CCM+. Therefore, the thermal exchange between the tiles and the surrounding environment takes place through the connecting interfaces that updates on a time step basis to address the tile translation on the alumina rollers. The tile motion speed is set as in the real kiln, i.e. 0.35 m/s; therefore, it takes approximately 60 s for a single tile to go through the considered modules. The effects of the insulating material on the global thermal balance of the kiln module are determined by means of a global heat transfer coefficient at the walls. This heat transfer coefficient is calculated by including the convection with the ambient air on the external side of the kiln wall, α_{amb}, and the thermal conductivity through the different layers of insulation, λ_i, along with the corresponding layers' thickness, th_i. The global thermal conductivity of the walls, $C_{Th,wall}$, results as follows:

$$C_{Th,wall} = \frac{1}{\frac{1}{\alpha_{amb}} + \sum_i^{n_{layers}} \frac{th_i}{\lambda_i}} \tag{6}$$

4 Results and discussion

The detailed calculation of the flow field within the kiln module makes it possible to evaluate accurately the temperature distribution of the tiles as well as the air in the upper and lower chambers. In particular, the temperature distribution of the ceramic material should be as uniform as possible in order to enhance the quality of the final product. Figure 5 shows the temperature of the tiles on the top surface after 60 s of simulation time for both analysed modules. Figure 6 depicts the same results on the bottom surface. The tile top and bottom temperature distributions result to be not uniform and they are strongly affected by the hot air flow direction. If this effect could be expected for the preheating module, it is less trivial for the firing section. Furthermore, the lateral surface of the tiles is characterized by the highest temperature values, see Figure 7. This result can be explained by the larger local surface to volume ratio of the tile side and by the effects of the extremes of the alumina rollers. In fact, the alumina rollers are at a higher temperature than the tile's one and close to the value of the hot air flow. Figure 7 shows also the interaction between the tiles and the rollers, since the alumina that is covered by the tiles is characterized by a lower temperature since the tile protect them from the radiation of the upper chamber and absorb heat due conduction through the contacting areas between them.

Figure 8 plots the temperature distribution on a cut plane through the axes of the burners that equips the module in the firing section of the kiln. The flames of the burners result strongly influenced by the hot air flow even though their velocity is relatively small compared to the flame one.

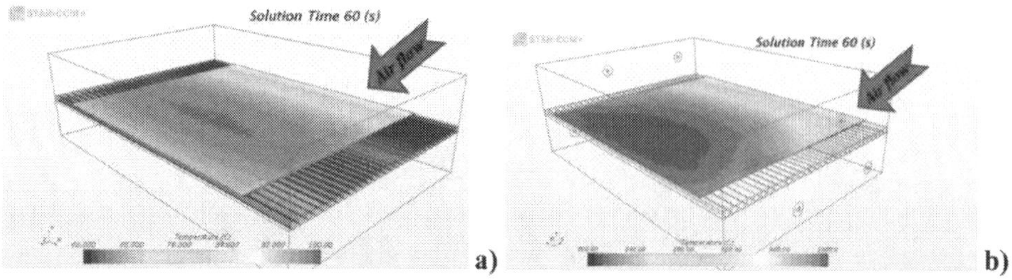

Fig. 5. Temperature distribution on the top surface of the tile for the a) preheating module and b) the eight-burner module.

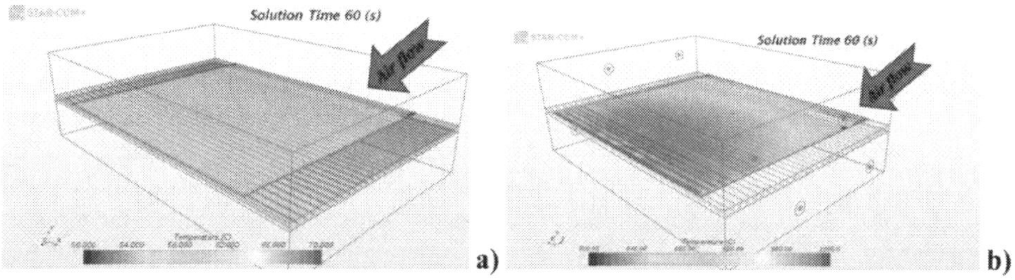

Fig. 6. Temperature distribution on the bottom surface of the tile for the a) preheating module and b) the eight-burner module.

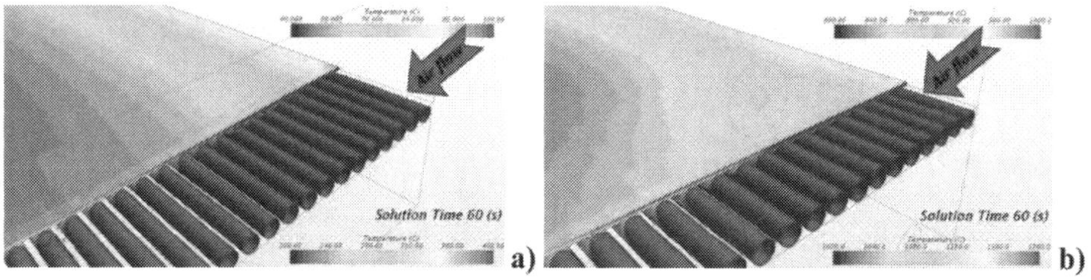

Fig. 7. Temperature distribution on the lateral surface of the tile and on the alumina rollers for the a) preheating module and b) the eight-burner module.

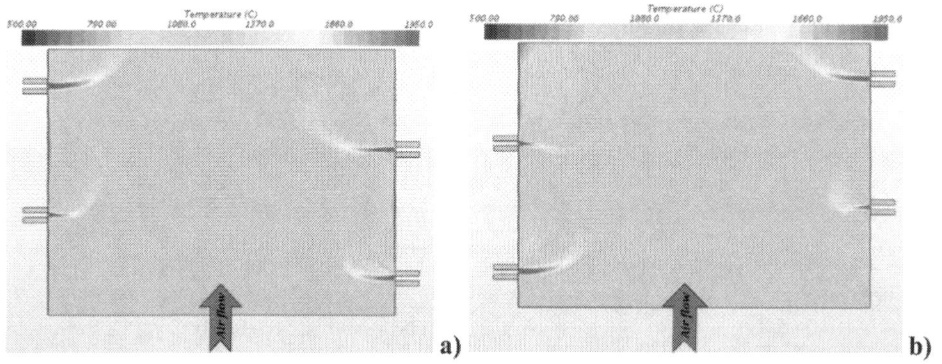

Fig. 8. Temperature distribution on a cut plane through the burners' axis of a) the upper chamber and b) the lower chamber for the eight-burner module.

 DOI: 10.1051/e3sconf/20172200117

Nevertheless, the central regions of the upper and lower chambers demonstrate to have a uniform temperature close to 1300°C. Colder spots can be noticed close to the walls in between the burners' flames that act like a shield for the hot air stream. Therefore, recirculation regions form at low temperature due to the heat dissipation through the walls.

Figure 9 plots the time histories of the characteristic temperatures for the tile. The difference between the average value of the temperature on the top and bottom surfaces of the tile is close to 10°C for the preheating module and reaches 30°C in the firing module. This difference is very important to control, since it may influence significantly the quality of the product and it is very difficult to be measured experimentally. The simulations demonstrate that even though the thickness of the tile is small, i.e. less than 10 mm, the temperature gradient between the top and the bottom of the tile is significant. Furthermore, the highest temperature in the tile surface is reached at the front side, where the tile is more strongly affected by the hot counter-current flow. The temperature at the side results to be approximately 40°C larger than the volume average temperature, whose value demonstrate to be close to the tile bottom temperature.

Finally, Figure 10 shows the time histories of the total heat transfer to the tile and the contribution of the radiative heat transfer. The heat radiated from the surrounding walls to the tiles results to be the main responsible of the total heat transfer, i.e. 81% and 72% for the preheating and firing modules respectively. In fact, the temperature difference between the tile surfaces and the surrounding ones remains very large in both analysed modules. Furthermore, the relative velocity of the flow and the tiles is rather slow radiative heat transfer, thus limiting the convection heat transfer from the hot stream to the tile surfaces. These figures add an insight in the thermal behaviour of the tile firing within the ceramic kiln; moreover, they are used to improve the accuracy of the predictive capabilities of the lumped and distributed parameter model. In fact, the view factors of the radiating surfaces are calculated in the 0D/1D model in order to determine the correct contribution of radiation on the total heat transfer to the tile on the basis of this CFD result.

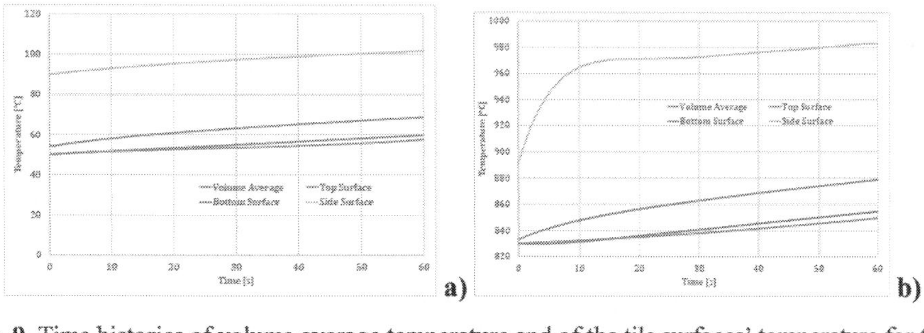

Fig. 9. Time histories of volume average temperature and of the tile surfaces' temperature for the a) preheating module and b) the eight-burner module.

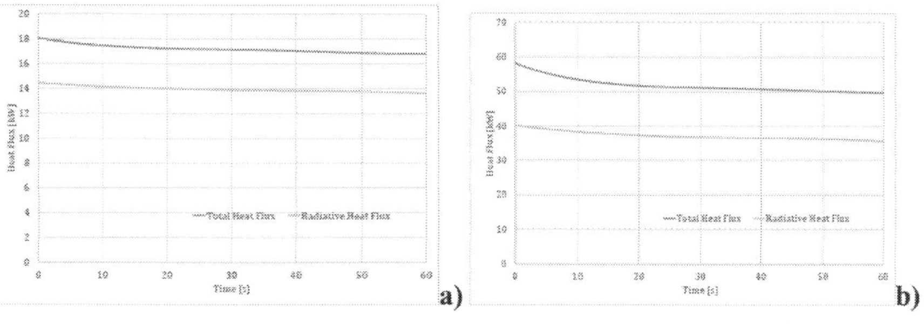

Fig. 10. Time histories of total heat flux absorbed by the tile and the contribution due to radiation the a) preheating module and b) the eight-burner module.

5 Conclusion

In this paper, the CFD analysis of a full-scale module of an industrial ceramic kiln under actual operating conditions has been carried out. A module in the preheating section and one in the firing region have been investigated and the thermal behaviour of the tiles in these parts of the kiln has been assessed.

In the analysis the combustion in the burners of the firing module has been addressed by using the Eddy Brake Up model for the oxidation of natural gas with air; the Surface-to-Surface (S2S) radiation model has been adopted for calculating the thermal radiative heat transfer between surfaces of the ambient surrounding the tiles.

Numerical results demonstrated that the temperature distribution on the top and bottom surfaces of the tile is not uniform despite the small thickness and a difference of approximately 10°C and 30°C is highlighted in the preheating and in the firing modules respectively. The hottest part of the tile proved to be the front side surface with a temperature 40°C higher than the rest of the surfaces.

Finally, the contribution of the radiative heat transfer to the total heat absorbed by the tile was calculated to be close to 81% in the preheating module and to 72% in the firing section. This result is paramount for improving the predictive capabilities of the lumped and distributed parameter model of the entire kiln, as well as to address the thermal behaviour of the tile firing process.

The present work is part of Design for Resource and Energy efficiency in CerAMic Kilns (DREAM) project. The project has received funding from the European Union's Horizon 2020 research and innovation programme under grant agreement No 723641.

References

1. A. Christos, T, Theocharis, Appl. Therm. Eng. **21**, 1231–1249 (2001)
2. S. Kaya, E. Mançuhan, K. Küçükada, Appl. Energ. **86**, 325–332 (2009)
3. A. Mezquita, J. Boix, E. Monfort, G. Mallol, Appl. Therm. Eng. **65**, 102–110 (2014)
4. S. Mercati, M. Milani, L. Montorsi, F. Paltrinieri, Appl. Energ. **97**, 686–694 (2012)
5. F. Franzoni, S. Mercati, M. Milani, L. Montorsi, Int. J. Hydrogen Energ. **36**, 2803–2816 (2011)
6. M. Milani, L. Montorsi, M. Stefani, Waste Manage. Res. **32**, 7, 614–625 (2014)
7. J.M. García-Oliver, X. Margot, M. Chávez, A. Karlsson, Math. Comput. Model. **54**, 7–8, 1732–1737 (2011)
8. Y.R.. Lee, H.S. Choi, H.C. Park, J.E. Lee, Comput. Chem. Eng. **82**, 202–215 (2015)
9. F. Franzoni, M. Milani, L. Montorsi, SAE Tech. Paper 2008-01-2390 (2008)
10. F.R. Menter, AIAA Paper, No93-2906, 1–21 (1993)
11. R. Siegel, J.R. Howell, *Thermal Radiation Heat Transfer* (Third Edition, Hemisphere Publishing Co., 1992)
12. J.P. Holman, *Heat Transfer* (Eighth SI Metric Edition, McGraw Hill, 2001)
13. M. Milani, L. Montorsi, M. Stefani, M. Lizzano, R. Saponelli, J. Environ. Manage. (2017, to be published)

Co-digestion of sewage sludge from external small WWTP's in a large plant

Stanisław Miodoński[1,*]

[1]Faculty of Environmental Engineering, Wroclaw University of Science and Technology, Wybrzeże Wyspiańskiego 27, 50-370 Wroclaw, Poland

Abstract. Improving energy efficiency of WWTPs (Waste Water Treatment Plants) is crucial action of modern wastewater treatment technology. Technological treatment process optimization is important but the main goal will not be achieved without increasing production of renewable energy from sewage sludge in anaerobic digestion process which is most often used as sludge stabilization method on large WWTP's. Usually, anaerobic digestion reactors used for sludge digestion were designed with reserve and most of them is oversized. In many cases that reserve is unused. On the other hand, smaller WWTPs have problem with management of sewage sludge due to lack of adequately developed infrastructure for sludge stabilization. Paper shows an analysis of using a technological reserve of anaerobic digestion reactors at large WWTP (1 million P.E.) for sludge stabilization collected from smaller WWTP in a co-digestion process. Over 30 small WWTPs from the same region as the large WWTP were considered in this study. Furthermore, performed analysis included also evaluation of potential sludge disintegration pre-treatment for co-digestion efficiency improvement.

1 Introduction

Dynamic development of renewable energy is one of the priority of energy Policy In Poland. By 2020, renewable energy will be 15% of the total electrical energy produced in Poland [1]. Between 2010 and 2014 the contribution increased from 10.2% to 11.8% [2]. All activities focused on renewable energy production increase are good way to reduce greenhouse gases emission and are part of sustainable development policy. Anaerobic digestion is one of the most popular methods for renewable energy production. In this process, because of proper conditions and special equipment, activity of naturally occurring bacteria leading to organic carbon mineralization is multiplied. Digestion takes place when organic materials decompose in an oxygen-free environment. During digestion, various microbes use the organic matter such as animal manure, sewage sludge, wasted food and other organics in the absence of oxygen. The process can be controlled and enhanced through chemistry and engineering [3]. The co-digestion process consists on simultaneous digestion

* Corresponding author: stanislaw.miodonski@pwr.edu.pl

a few materials where one is the basis material (e.g. sewage sludge, swine manure) and the others are additional material (e.g. food waste, grease trap sludge) [4,5]. A specific case of co-digestion is situation when basis and additional materials are both sewage sludge, for example basis material form large WWTP and additional material are form transported from medium and small WWTPs located near a large plant. Despite similar characteristics of all used materials, operated process will be still considered as a co-digestion.

Paper presents potential development proposals for use available anaerobic digestion reactors volume for co-digestion process with sewage (excess) sludge form small and medium WWTP, localized nearby a large object. This scenario helps to solve the problem of sewage sludge produced in small and medium WWTP and allows to increase renewable energy production on large WWTP (in next sections called as "Central Plant"). The study and technological analysis were based on real data from a large WWTP in Lower Silesia Region in Poland (1 million P.E.) and medium and small WWTP localized in nearby towns. As a next step of performed analysis, also potential impact of thermal disintegration of excess sludge before an anaerobic digestion process was investigated.

2 Materials and methods

2.1 Central Plant, basis sludge for digestion

Central Plant is a municipal WWTP with capacity of 1000 000 p.e. Plant is equipped with an 6 anaerobic digestion reactors fed with mixed of primary and excess sludge. Limited amount of produced gas can be stored, otherwise it is immediately used for energy production in CHP units.

2.2 Small WWTPs, additional sludge for digestion

For performed analysis, objects form North east region of Lower Silesia located nearby to the Central Plant were selected. Analysis includes WWTPs from:

- Oleśnicki district,
- Oławski district,
- Trzebnicki district,
- Wołowski district,
- Wrocławski district.

In the area taken into consideration, 31 objects were identified. Most of this plants are objects without primary sedimentation. This plants produce only excess sludge which has relatively poor gas production parameters during digestion process.

2.3 Basis of the analysis

In this experiment impact of distance between Central Plant and selected smaller plants was tested along with further advantages from excess sludge disintegration use. For further analysis following variants were adopted:

- Variant 0 – basis scenario. No additional sludge delivered and none excess sludge disintegration method used.
- Variant 0D – No additional sludge delivered. Thermal disintegration of excess sludge used.
- Variant 25 –Sludge was delivered only form smaller WWTP localized within 25 km radius from the Central Plant. No disintegration of excess sludge used.

- Variant 25D –Sludge was delivered form smaller WWTP localized within 25 km radius.. Thermal disintegration of excess sludge used.
- Variant 50 – Sludge was delivered form smaller WWTP localized within 50 km radius from the Central Plant. No disintegration of excess sludge used.
- Variant 50D –Sludge was delivered form small and medium WWTP localized within 50km radius from the Central Plant. Thermal disintegration of excess sludge used.

2.4 Kind of disintegration

Presented assumptions were created based on use of high efficiency thermal or thermo-chemical sludge disintegration. That kind of disintegration allows to increase gas production form excess sludge by 60% [11]. Due to necessary transfer this process form laboratory to full scale object, 50% increase in gas production was assumed.

3 Results

3.1 Technological reserves of Central Plant's digesters.

For analysis, real data from last three years of anaerobic digestion process operation at the Central Plant was used. Based on this data average and percentile 85% was determined. Obtained values were basis for further proceedings presented in following sections. Table 1 summarizes sewage sludge parameters divided into thickened primary sludge and thickened excess sludge.

Table 1. Sewage sludge parameters.

	Flow	Concentration			Load		
		COD	TSS	VSS	COD	TSS	VSS
	m^3/d	gO_2/m^3	$kgTSS/m^3$	$kgVSS/m^3$	kgO_2/d	kgTSS/d	kgVSS/d
Thickened primary sludge							
Avg.	708.9	75 136	57.7	44.5	53 069	40 709	31 337
P85%	742.4	86 434	68.4	49.0	53 776	41 416	29 806
Thickened excess sludge							
Avg.	396.2	63 086	56.1	40.2	25 318	22 146	15 829
P85%	507.1	67 236	59.9	42.4	30 616	28 486	19 681
Mixed sludge							
Avg.	1 105.1	68 546	57.0	42.9	75 573	62 855	47 166
P85%	1 184.7	78 286	62.3	45.1	81 793	67 986	48 454

Table 2 shows digested sludge parameters collected in the same period as previous data. in the same way and from the same. Parameters determination was also the same as in primary and excess sludge. In analysed period average gas production was 20 142 Nm^3/d and gas had methane concentration level at about 58%.

One of the most important parameters describing work regime of anaerobic digestion reactor is Organic Loading Rate (OLR). Safe operating range if from 0.7 kg VSS m^3d^{-1} to 5.7 kg VSS m^3d^{-1} [6]. Average OLR form analysed period was 1.1 kg VSS m^3d^{-1}. In 2016 increase of OLR was noticed (average in 2016 1.41 kg VSS m^3d^{-1}). This was due to the exclusion from the operation one of the anaerobic digestion reactor, due to technical issues.

Table 2. Digested sludge parameters.

	Flow	**Concentration**					**Load**		
		COD	Alk.	VFA	TSS	VSS	COD	TSS	VSS
	m^3/d	gO_2/m^3	$gCaCO_3/m^3$	gCH_3COOH/m^3	kg/m^3	kg/m^3	kgO_2/d	kg/d	kg/d
Avg.	1 044	34 696	3 520	306	35.0	21.3	36 197	36 574	22 243
P85%	1 082	37 943	3 709	287	39.2	23.2	38 685	38 868	23 095

Hydraulic Retention Time (HRT) was next parameter analysed in this study. Usually, HRT value should be about 15–20 d for mesophilic digestion [7]. For further consideration maximal HRT value (20 d) was assumed. Average HRT is 42d but that value since 2013 drops systematically as HRT is being shortened. For that reason, reliable HRT was adopted like its average value in 2015 (38.2 d). In 2016 HRT was shortened from reason explained above.

Parameter describing effect of anaerobic digestion and sludge stabilization is percentage of organic matter reduction. In EPA guidelines inform that parameter for properly stabilized sludge should be over 38%. [8]. Fig. 1 shows monthly average of VSS reduction during analysed period. Only one value was lower than 38% (black square), in other cases values were clearly higher than the minimum value.

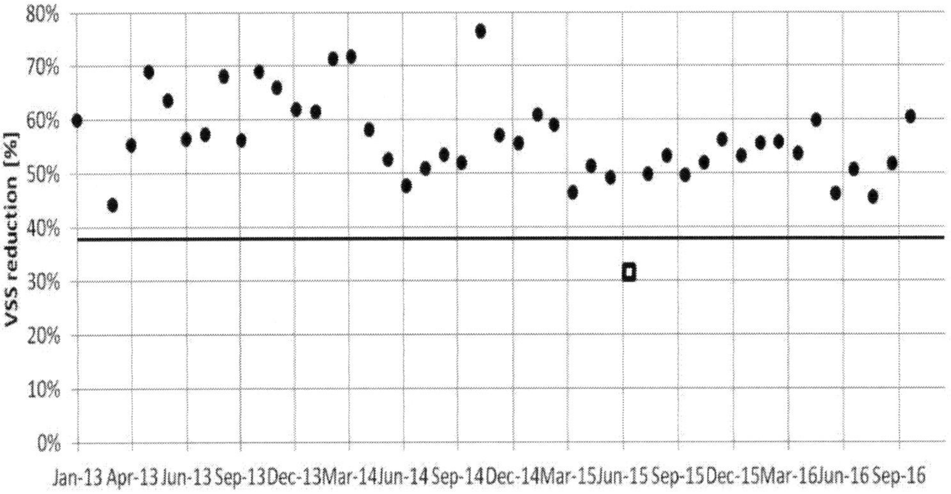

Fig. 1. Monthly average VSS reduction.

Last of the analysed parameters was Volatile Fatty Acids (FVA) to alkalinity ratio in digested sludge. Extreme value of this parameter is 0.3 $gCH_3COOH/gCaCO_3$. Average value for analysed period was 0.087 and that value is almost 3.5 times lower then limit value.

Performed multi-parameter analysis shown that the anaerobic reactors work under stable conditions and whole process have possibility and potential for addition of external stream of sewage sludge.

3.2 Sludge production on analysed area

Initially, for the analysis all WWTPs from 6 districts were used and 31 objects were identified. According to the assumptions of the analysis, two distance range from the Central Plant was chosen: 25 km and 50 km. At this stage 3 objects located in bigger distance than 50 km away and were removed from the list. Among remaining 28 smaller WWTPs, 14 of them was closer to the Central Plant than 25 km. Fig. 2 shows localization of analysed plants with circles marked with 25km and 50km radius and cantered on Central Plant. Based on data from National Program for Municipal Waste Water Treatment (NPMWWT) mass of sewage sludge from each WWTP was determined [9]. Typical VSS and TSS values for considered sludge streams were assumed, 75% and 15% respectively. Transportation cost was assumed equal 3.00 PLN/(km Mg)$^{-1}$ Effect of the calculation was shown in table 3.

Table 3. Daily sludge production form medium and small WWTP.

Distance	TSS	VSS	Sludge volume	Transportation cost
km	kg/m^3 d	kg/m^3 d	m^3/d	PLN/d
25	25 238	18 929	168	16 691
50	30 563	22 922	204	24 068

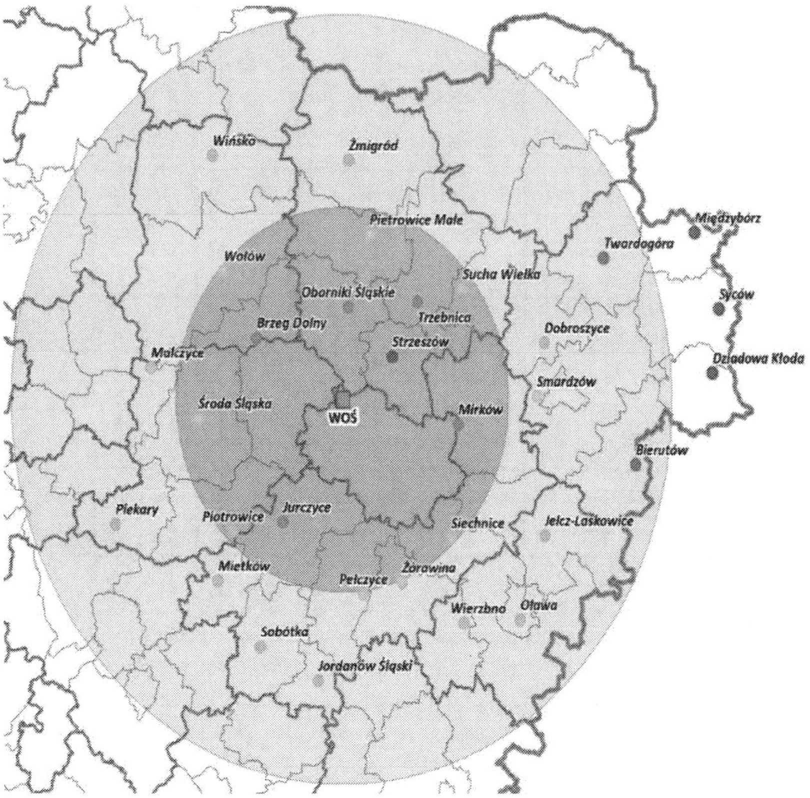

Fig. 2. Localization of analysed plants.

3.3 Increasing of gas production

To estimate amount of gas produced and disintegration process effects it is necessary to divide gas production from primary and excess sludge streams at the Central Plant. It is generally known, that considering anaerobic digestion process there is no difference between excess sludge from WWTP with or without primary sedimentation [10]. For that reason, excess sludge streams from Central and smaller plants will be considered in similar way, regardless of primary sedimentation use in selected plants. Gas production efficiency rates for excess and primary sludge was assumed on 0.2 and 0.54 $Nm^3/kgVSS_{in}$, respectively. In case of disintegration process used for excess sludge pre-treatment, gas production rate was higher (0.3 $Nm^3/kgVSS_{in}$). Based on this information and data form Table 1, gas production was estimated. For excess sludge, it was about 3 165 Nm^3/d and for primary sludge more than 16 976 Nm^3/d. Tab. 4 summarizes gas production analysis for all warrants. As expected, huge influence on amount of produced gas had potential use of disintegration process. Increase of gas production for "50D" scenario reached 42% of the basic value. Amount of sludge from radius greater them 25km and lower than 50km is so small (only more than 5325 kgTSS/d its about 21%) that gas production in "25D" scenario is greater than in "50" scenario.

As the last step of performed analysis, impact of greater amount of excess sludge from smaller plants on digestion process at the Central Plant was investigated. New technological parameters were calculated for each of proposed scenarios.

Table 4. Gas production analysis.

Var.	Primary sludge			Excess sludge			Mixed sludge		
	VSS	Gas production		VSS	Gas production		VSS	Gas production	
	kgVSS/d	m^3/kgVSS	m^3/d	kgVSS/d	m^3/kgVSS	m^3/d	kgVSS/d	m^3/kgVSS	%
0	31 337	0.54	16 976	15 829	0.20	3 166	47 166	20 142	-
0D					0.30	4 749		21 725	8%
25				34 758	0.20	6 952	66 095	23 928	19%
25D					0.30	10 427		27 403	36%
50				38 751	0.20	7 750	70 088	24 726	23%
50D					0.30	11 625		28 602	42%

In Table 5 parameters for anaerobic digestion process for all scenarios were described. None of analysed parameters exceed the limit values mentioned before. HRT in the extreme scenarios ("50" and "50D") were below 24 days. That is safe value but in case of anaerobic digestion of wastewater sludge, long HRT is required considering excess sludge rather than primary sludge. Because transported sludge is dewatered, possibility of increasing sludge concentration in anaerobic digestion reactors above actual value exists. Currently, average sludge concentration in digesters is about 55–60 $kgTSS/m^3$. Increasing the sludge concentration above 70 $kgTSS/m^3$ allows you to prolong HRT up to 30 days.

OLR increased by 50% ("50" and "50D") to 1.8 $kgVSS/m^3d$. Achieved value is not even close to limit value (5.7 $kgVSS/m^3d$) and will not affect whole process. As expected, average gas production efficiency was slightly decreased

 DOI: 10.1051/e3sconf/20172200118

Table 5. Technological parameters of anaerobic digestion process in selected scenarios.

Var.	Mixed sludge			Technological parameters		
	TSS	VSS	Gas prod.	HRT	OLR	Gas prod.
	$kgTSS/m^3$	$kgVSS/m^3$	m^3/d	d	$kgVSS(m^3d)^{-1}$	$m^3/kgVSS$
0	62 855	47 166	20 142	35.4	1.21	0.43
0D	62 855	47 166	21 725	35.4	1.21	0.46
25	88 093	66 095	23 928	25.2	1.69	0.36
25D	88 093	66 095	27 403	25.2	1.69	0.41
50	93 418	70 088	24 726	23.8	1.80	0.35
50D	93 418	70 088	28 602	23.8	1.80	0.41

4 Conclusions

- Wastewater sludge management overview in region revealed significant potential of available additional streams of sewage sludge in medium and small WWTP localized in a short distance from the Central Plant.
- Sewage sludge form analysed region do not fully cover the possibilities of anaerobic digestion process at the Central Plant. Anaerobic reactors on plant are much oversized. To cover the possibility of the process, it will be necessary to transport sewage sludge from a greater distance than assumed range of 50 kilometres. In author's opinion, such solution might not be economically justified but further analysis should be performed.
- Technological analysis confirmed that significant part of gas production comes from primary sludge. Incensement of excess sludge amount by 2.5 times do not change this situation and primary sludge will still have the biggest share in the overall gas production.
- Disintegration might improve gas production from excess sludge by 50%. Significant increase of excess sludge amount will make the disintegration process more profitable, especially considering fact that the biggest cost factor in disintegration process implementation is the cost of buying a patented technology, not size of the specific installation.
- Transport of dewatered excess sludge form smaller WWTPs localized nearby has a lot of advantages. It is possible to use thermal-pressure disintegration, which requires dewatered sludge as influent. That solution eliminated of dewatering sludge on large WWTP before anaerobic digestion. Transport that kind of sludge also allow to increase sludge concentration in anaerobic digestion reactors and prevention of excessive HRT decrease. Dewatering also significantly lowers transport costs.
- When disintegration will be implemented, it is possible to increase gas production by over 40% It should significantly improve the energetic balance of the large WWTP.
- There are no reports in literature about use sludge from several objects in co-digested in one plant.

This research are co-financed by Municipal Support Program for Higher Education, Science and Sector of Economic Activity MOZART" edition 2016/2017.

References

1. Polityka Energetyczna Polski, Ministerstwo Gospodarki (2010)
2. G. Berent-Kowalska, J. Kacprowska, I. Moskal, A. Jurgas, Energia ze źródeł odnawialnych w Polsce, GUS (2015)
3. A. Costa, Ch. Ely, M. Pennington, S. Rock, C. Staniec, J. Turgeon, Anaerobic Digestion and its Applications, (EPA/600/R-15/304 2015)
4. R. Braun, A. Wellinger, *Potential of Co-digestion* (IEA Bioenergy 2007)
5. G. Esposito, L. Frunzo, A. Giordano, F. Liotta, A. Panico, F. Pirozzi, Rev. Environ. Sci. Biotechnol. **11**, 325–341 (2012)
6. G. Tchobanoglous, *Wastewater Engineering: Treatment Reuse* (Metcalf & Eddy, Inc. 2003)
7. J. Bień, *Osady Ściekowe Teoria i Praktyka* (Częstochowa 2002)
8. Environmental Regulations and Technology: Control of Pathogens and Vector Attraction in Sewage Sludge (EPA; Cincinnati 2003)
9. The National Program for Municipal Waste Water Treatment (NPMWWT) from years 2014–2015 (KZGW 2016)
10. K. Iskra, S. Miodoński, Gaz, Woda i Technika Sanitarna **5**, 184–188 (2015)
11. C. Bougriera, C. Albasib, J.P. Delgenèsa, H. Carrèrea, Chem. Eng. Process **45**, 711–718 (2006)

Nanomaterials in the environment

Bozena Mrowiec[1,*]

[1]University of Bielsko-Biala, Institute of Environmental Protection and Engineering, ul. Willowa 2, 43-300 Bielsko-Biala, Poland

Abstract. This paper considers engineered nanomaterials, deliberately engineered and manufactured to have certain properties and have at least one primary dimension of less than 100 nm. Materials produced with the aid of nanotechnologies are used in many areas of everyday life. Researches with nanomaterials have shown that the physiochemical characteristic of particles can influence their effects in biological systems. The field of nanotechnology has created risk for environment and human health. The toxicity of nanoparticles may be affected by different physicochemical properties, including size, shape, chemistry, surface properties, agglomeration, solubility, and charge, as well as effects from attached functional groups and crystalline structure. The greater surface-area-to-mass ratio of nanoparticles makes them generally more reactive than their macro-sized counterparts. Exposure to nanomaterials can occur at different life-cycle stages of the materials and/or products. The knowledge gaps limiting the understanding of the human and environment hazard and risk of nanotechnology should be explained by the scientific investigations for help to protect human and environmental health and to ensure the benefits of the nanotechnology products without excessive risk of this new technology. In this review are presented the proposal measurement methods for NMs characteristic.

1 Introduction

Development directions and an exponential growth of nanotechnology and the wider application of nano-scale materials are referred to as the next industrial revolution. In the 21st century nanotechnology is seen as a transformative technology, which has the potential to stimulate scientific innovation and it being important in our daily lives from human health to electronics [1–4, 31]. It was stated that more than 1600 consumer products are listed in a database to be containing engineered nanomaterials (ENMs) [5]. The International organization for Standardization (ISO) classified nanomaterials into three main groups: nanoparticles (all three dimensions between 1 and 100 nm); nanoplates (one dimension between 1 and 100 nm); and nanofibers (two dimensions between 1 and 100 nm). Manufactures ENMs comprise seven main classes: carbonaceous nanomaterials (e.g. carbon nanotubes); semiconductors (e.g. quantum dots); metal oxides (e.g. zinc oxide); nanopolymers (e.g. dendrimers); nanoclays; emulsions (e.g. acrylic latex); and metals (e.g. silver) [2, 8, 31]. In order from the most to the least often produced they can be presented as

* Corresponding author: bmrowiec@ath.bielsko.pl

follows: carbon black > TiO_2 > SiO_2 > ZnO > Fe and FeO_x > Al_2O_3 > CeO_2 > CNT > Ag [2, 6, 7]. The most common applications of engineered nanopatricles are: personal care products, sunscreens, pigments, polishing agents, photocatalysts, detergents, agrichemicals, window panes and ceramic tiles (TiO_2, ZnO, CeO_2, SiO_2, Al_2O_3,); catalysts, bactericides, groundwater remediation, paints, food packaging, pharmaceutical products, medical implants, textiles (Au, Ag, Cu, Fe); electronics, solar panels, bioimaging (CdSe, CdTe); drug delivery, chemical sensors (multi-funcjonal polimers – dendrimers); super capacitors, hydrogen storage, ultra-high-strength materials, automobile parts (carbon nanotubes, fullerenes) [4, 10, 20, 22, 26]. The extensively using of nanomaterials in diverse consumer products causes their release and subsequently accumulate in the environment. Most assessments suggest very low ENMs mass concentrations (< µg/L) are likely. However, the data on the prevalence of engineered nanoparticles (ENPs) use and the release of ENPs into the environment is limited. This lack of data poses a significant hurdle to accurate risk assessment for NMs. However, there is an ongoing debate about the potential risks of ENM and nanotechnologies. It is because, that the nanomaterials can be released to the environment at any stage of life cycle of products (LCA- Life Cycle Assessment), from the manufacture, use, and disposal or recycling processes, therefore nanotechnology could lead to environmental hazards and adverse health effect [11, 20, 21, 31]. Due to the rapidly increasing production and use of ENM it is self-evident that aspects of danger must be fully recognized, because it may have important long-term consequences for the human health and the environment.

This paper is a review, which focuses on the issues of the impact of nanomaterials on the environment and possibilities of their determination and analyses in environmental samples for risk assessment.

2 Nanomaterials sources

Release of nanomaterials to the environment can be derived from natural or anthropogenic sources. Photochemical reactions, soil erosion by wind or water, forests fires or volcanic eruption and shedding skin and hair by animals, are known as the natural processes which produce and emit very little amount of nanoparticles.

The anthropogenic sources are:

1. Production processes of engineered nanoparticles and nanomaterials. Release during production may occur through leaks into water and air in closed systems or open production processes.

2. Handling and use of nanomaterials and nanoproducts. Handling and use covers several process-related stages e.g. handling of powders, diffuse emission from production plants, mechanical treatment of nanomaterials.

3. Aging of nanoproducts. Aging encompasses all processes taking place in the environment such as selective degradation, wash-out, increased brittleness of the material.

4. Utilization of waste containing nanomaterials (end of life). This group include mainly unwanted or incidental release waste re-use or recycling, waste disposal and incineration, wastewater treatment, and agricultural use of sewage sludge [1, 4, 9, 11, 12, 22].

Currently yet, there are limited data on the nature and magnitude of potential releases from different sources, transport ways and transformations in the environment.

3 Toxicity of nanomaterials

Nanotoxicology represents a new branch of toxicology to address the adverse health effects caused by nanomaterials [12]. But, still is lack a fundamental understanding of how nanomaterials interact with living system and, thus, science is not yet in a position to assess the relevant end-points for nanomaterial toxicity. At the same time, we are faced with an onslaught of new nanomaterials for which testing or screening of toxicity is required. To resolve this situation, reliable methods for assessment of nanomaterial toxicity are needed. [11].

Buzea et al. [12] stated, that the most important parameters which determine the adverse effects of the nanomaterials influence on living organisms are their dose, dimension (size), and durability. Besides, they are also important characteristics, such as: solubility, concentration and agglomeration, shape, surface area, chemical and crystalline structure, surface coating and functionalization. The smaller a particle, the greater its surface area to volume ratio and the higher its chemical reactivity and biological activity. Because of their large surface area, nanoparticles will, on exposure to tissue and fluids, immediately adsorb onto their surface some of the macromolecules they encounter. This may, for instance, affect the regulatory mechanisms of enzymes and other proteins [12, 14, 17–19]. Studies on nanotoxicity are mainly concentrated on empirical evaluation of the toxicity of various nanomaterials, with less regard given to the relationship between nanoparticle properties and toxicity. Therefore scientific articles show different correlations between various physico-chemical properties of nanomaterials and the associated negative effects on living organisms. This approach gives very limited information, and could not be considered adequate for developing assessment of toxicity of seemingly similar nanomaterials.

Nanomaterials released into the environment can react with the components of air, water and soil, which causes, among others, changes in particle charge, surface properties or ability to aggregate [23]. The US National Research Council has indicated that researches with respect to nanomaterials should focus on identifying the so-called. "critical elements of interaction", which are necessary to assess exposure and the risk posed by designed nanomaterials. These critical elements include the physical, chemical and biological changes, which ultimately affect the durability of nanomaterials, their bioavailability/absorption, reactivity and toxicity [24].

Of course, not all nanonaterials are toxic. Sometimes studies performed on the same type of nanomaterials are in disagreement, some studies show their biocompatibility, while others prove their potentially hazardous nature [12–15]. For determine the toxicity of various nanomaterials, it is necessary compare their toxicity to effects caused by known toxic particles. However, the database of studied nanomaterials is still limited. The published results show that the carbon nanotubes (CNT) and fullerenes are very toxic [12, 26–29]. CNT causing significantly more damage to the lungs than carbon black or silica Their aggregates and some carbon blacks can be cytotoxic as asbestos [12]. Also to the group of highly toxic nanomaterials are recommended quantum dots (CdSe, CdTe) and nanogold. To the group of medium toxicity are classified zinc oxide, aluminium oxide, iron oxide, silver and dendrimers, [26–29]. Silver nanoparticles in form of aggregates are more toxic than asbestos [12]. On the other hand, nanomaterials such as titanium oxide, yttrium iron oxide, zirconium oxide, silicon dioxide, silica, silicon nanowires and nanoclay particles are considered to be less toxic [12, 26–29].

Literature data indicate that one mechanism of toxicity of nanoparticles is likely to be induction of reactive oxygen species and the consequential oxidative stress in cells and organs. Response in the form of oxidative stress may appear at different rates depending on the type of nanomaterial. It is thought that the particles of the transition metals greatly accelerate the process. Free radicals - generated in the process of oxidative stress – do cause damage of biological structures - proteins, lipids and DNA molecules. Also the oxidative stress could induce an inflammatory response in cells via activation of transcription factors [15, 22, 25]. Nanomaterials uptake by the body occurs after inhalation exposure (neuronal

 DOI: 10.1051/e3sconf/20172200119

uptake, translocation across lung epithelium), oral (ingestion) and dermal exposure depending on the characteristics of the nanomaterial under investigation [12, 15, 22].

Currently, a significant problem in studies of nanomaterials toxicity is the fact, that in the majority the most of the in vitro studies are short-term tests, while the impact of nanomaterials on human health and the environment takes place in the conditions of long-term exposure, [30].

4 Nanometrology

Possibility of a correct assessment of ENP fate, transport, and toxicity in the environment depends on the ability to analyze ENPs in complex matrices. There are many options for analyzing ENPs in simple matrices. However, validation and application of these tools to analyse of environmental samples, is relatively underdeveloped [20]. The used methods must permit to detect nanomaterials and analyse their physicochemical properties in the media in which humans and ecosystems are exposed to them, such as air, water, soil, consumer products and nanocomposities. Also, the methods must be useful to assess the risk of nanomaterials including toxicological and ecotoxicological studies. It is required, that tools are able to detect nanopatricles in the relevant medium, including cells, fluids and plant tissue [22].

Methods for the nonomaterials analysing can be classified as: detection, quantification, and characterization methods.

- Detection determines ENMs presence or absence. The detection of ENMs alone has limited use, but may be helpful for specific studies such as rapidly screening nano-containing products.
- ENMs quantification to obtain answers to some specific questions is required, for example: What is the degree of ENP uptake by exposed organisms?
- Characterization methods provide additional details on the physical properties (size, shape) and chemical composition of the NMs. These data help in the analyses of ENMs fate and transformation. But now even, measurement of properties such as size, shape, and reactivity are complicated for environmental samples [20]. Table 1 shows the recommended analytical methods for the determination of nanomaterials characteristics.

Essential problems in nanomaterials analysis concern on: 1) mass detection limit, because engineering nanomaterials are expected to enter in to the environment at very low concentrations (ppt); 2) size detection limit, because in most dimensions of ENMs are between 1–100 nm (often smaller than 20 nm); 3) aggregation state, because some NMs are not expected to preserve monodisperse state in the environment, the degree of their dispersion/aggregation is not static and will likely vary in time and is need ability to discern aggregated form single particle material; 4) naturally occurring nanomaterials, because their concentration in the environment are several orders of magnitude above that of engineered nanomaterials (ppm vs. ppt), and some natural nanoparticles have similar elemental composition and moreover morphologies to ENMs and natural nanoparticles tend to be very polydisperse and can interact with ENMs in the environment [20].

Table 1. Parameters for characterization of nanomaterials in the environment and proposal of possible analytical techniques, [11, 20, 22, 29, 32–35].

<table>
<tr><th>Physiochemical properties</th><th>Analytical techniques</th></tr>
<tr><td>Size (distribution)</td><td>AFM, CE, DLS, EM, FI-FFF, FS, HDC, MALS, NTA, SAXS, SEC, SEM, SP-ICP-MS, TEM, UV-vis, XRD</td></tr>
<tr><td colspan="2">Technical and analytical limitations: introduction of artifacts from sample drying (TEM/SEM); no elemental specificity (DLS); inability to differentiate between ENMs and NNPs of similar elemental composition (SP-ICP-MS, TEM, SEM); obstructed by high background of natural particles (SP-ICP-MS, TEM, SEM, DLS, FFF)
Necessity: analysis of samples in situ with minimal sample preparation; elemental specificity to differentiate between dissimilar nanomaterials; requires another measured property to differentiate between particles of similar elemental composition</td></tr>
<tr><td>Surface chemistry/charge/area/speciation</td><td>AES, AFM, BET, EELS, FTIR, NMR, SEM, STM TEM, XPS, Solid-state, Zeta potential by DLS, Fluorescence labelling</td></tr>
<tr><td colspan="2">Technical and analytical limitations: original coating may have been replaced or overcoated in the environment (NMR, FTIR, Zeta potential); ensemble techniques unable to characterize individual particle populations without prior fractionation steps (FTIR, NMR, Zeta potential
Necessity: ability to differentiate between different particle populations in situ; knowledge of how surface groups are attached may help determine if original coating persists</td></tr>
<tr><td>Concentration related to particle number or mass</td><td>FS, GC-MS, MALS, NTA, ICP-MS, SP-ICP-MS, UV-vis</td></tr>
<tr><td colspan="2">Technical and analytical limitations: unable to determine aggregates form single particle without parallel imaging/sizing technique; NTA is nonspecific for particle type
Necessity: require knowledge pertaining to aggregation state of ENMs</td></tr>
<tr><td>Elemental composition</td><td>GC-MS, ICP-MS, ICP-OES, SP-ICP-MS, SEM/EDX, XPS</td></tr>
<tr><td colspan="2">Obstacles to accurate detection/characterization: unable to discern particles of natural or engineered origin; may require acidification, eliminating particle integrity (ICP-MS, ICP-OES); sample preparation may alter sample representativeness
Potential need: determination of elemental composition in situ with additional sample preparation</td></tr>
<tr><td>Agglomeration/aggregation state</td><td>AFM, CLSM, DLS, EM, FFF, FS, NTA, SEM, SP-ISP-MS, TEM, UV-vis (for plasmonic nanoparticles)</td></tr>
<tr><td>Shape</td><td>AFM, CLSM, EM, FFF, SEM, TEM, UV-vis (for plasmonic nanoparticles)</td></tr>
<tr><td>Structure/crystallinity</td><td>ED, HR-TEM, SAED, SAXS, XRD</td></tr>
<tr><td>Stability over time/dissolution</td><td>DLS, UV-vis, ICP-AES, ICP-MS, Colorimetric assays</td></tr>
<tr><td>Dosing metric</td><td>Variable</td></tr>
<tr><td>Uptake</td><td>ICP-AES, ICP-MS, TEM, Fluorescence labelling, Flow cytometry, NAA</td></tr>
</table>

Abbreviations:

AES – Auger electron Spectroscopy; AFM – Atomic Force Microscopy; BET - Brunauer, Emmett, Teller Method; CE – Capillary Electrophoresis; CLSM – Confocal Laser Scaning Microscopy; DLS – Dynamic Light Scattering; ED – Electron Diffraction; EDX – Energy Dispersive X-ray; EELS – Electron Energy Loss Spectrometry; EM – Electron Microscopy; FFF – Field Flow Fractionation; FI-FFF – Flow FFF; FS –Fluorescence Spectroscopy; FTIR – Fourier-TransformInfrared

Spetroscopy; GC-MS – Gas Chromatography Mass Spectrometry; HDC – Hydrodynamic Chromatography; HR-TEM – High-resolution TEM; ICP-AES – Inductively Coupled Plasma Atomic Emission Spetrocsopy; ICP-MS – Inductively Coupled Plasma Mass Spectrometry; ICP-OES – Inductively Coupled Plasma Optical Emission Spectrocsopy; MALS – Multi-AngleLightScattering; NAA – Neutron Activation Spectroscopy; NMR – Nuclear magneticresonanse Spectroscopy; NTA – Nanoparticle-Tracking Analysis; SAXS – Small Angle X-ray Scattering; SEC – Size Exclusion Chromatography; SAED – Selected Aera Electron Diffraction; SEM – Scanning Electron Microscopy; SP-ICP-MS – Single Particle ICP-MS; STM –Scanning Tunnelling Microscopy; TEM – Transmission Electron Microscopy; UV-vis – Ultraviolet-Visible Spectroscopy; XPS – X-ray Photoelectron Spectroscopy; XRD – X-ry Diffraction.

5 Summary

Nanotechnology is a multi-disciplinary field requiring the involvement and collaboration of scientists in many kind of disciplines such as material science, physics, chemistry, engineering, and social science. For this reason, the problem of nanosafety studies also depends on close cooperation between researchers from area of material science, chemistry, biology, toxicology and epidemiology, and risk assessment as well as representatives of industry, and policy makers. Also in this respect, education and upbringing of new generation of specialists in the field of nanosafety in the environment are needed. Additionally, activities to ensure the nanosafety must be integrated with nanotechnology progress. The most significant limitations are difficulties of ENMs separation from their natural counterparts. Solving of these technical problems for the proper risk assessment of nanomaterial and them characterization and monitoring in the environment are required. At the moment practical abilities of analysis and reliable quantification of nanomaterials in environmental samples are far away from robust, applicable and reproducible methods allowing their reliable monitoring.

References

1. A. Maynard, Nanotoday **1**, 2 (2006)
2. G.E. Batley, J.K. Kirby, M.J. Mclaughlin, Accounts Chem. Res. **46**, 3 (2013)
3. S. Singh, H. Singh Nalwa, J. Nanosci. Nanotechnol. **7**, 9 (2007)
4. C. Remedios, F. Rosario, V. Bastos, Journal of Botany (2012)
5. F. Part, G. Zecha, T. Causon, E.K. Sinner, M.Huber-Humer, Waste Manage. **43** (2015)
6. Y. Kim, Environ. Health Toxicol. **29** (2014)
7. A.L. Holder, E.P. Vejerano, X. Zhoub, L.C. Marr, Environ. Sci. Processes Impacts, **15** (2013)
8. S.J. Klaine, P.J. Alvarez, G.E. Batley, T.F. Fernandes, R.D. Handy, D.Y. Lyon, S. Mahendra, M.J. McLaaughlin, J.R. Lead, Environ. Toxicol. Chem. **27**, 9 (2008)
9. S. Smita, S.K. Gupta, A. Bartonova, M. Dusinska, A.C. Gutleb, Q. Rahman, Environ. Health **11**, 1 (2012)
10. A. Keller, W. Vosti, H. Wang, A. Lazareva, J. Nanopart. Res. **16** (2014)
11. K. Savolainen, U. Backman, D. Brouwer, B. Fadeel, T. Fernandes, T. Kuhlbusch, R. Landsiedel, I. Lynch, L. Pylkkanen, Nanosafety in Europe 2015-2025: (Finnish Institute of Occupational Health, Helsinki 2013) (www.ttl.fi/en/publications/electronic_publications/pages/default.aspx)
12. C. Buzea, I.I. Pacheco Blandino, K. Robbie, Biointerphases **2**, 4 (2007)

13. Nanotechnology Risk Governance (International Risk Governance Council, Geneva 2007)
14. S.K. Brar, M. Verma, R.D. Tyagi, R.Y. Surampalli, Waste Manage. **30** (2010)
15. S. Singh, H. Singh Nalwa, J. Nanosci. Nanotechno. **7** (2007)
16. B. Kolodziejczyk, Brief for GSDR 2016 https://sustainabledevelopment.un.org/content/documents/9539GSDR_Nano_brief%204.pdf
17. G. Bystrzejewska-Piotrowska, J. Golimowski, P.L. Urban, Waste Manage. **29** (2009)
18. M. Crane, R.D. Handy, J. Garrod, R. Owen, Ecotoxicology, **17** (2008)
19. D. Ramakrishna, R. Pragna, The Journal Of The International Federation Of Clinical Chemistry And Laboratory Medicine **22**, 4 (2011)
20. U.S. Environmental Protection Agency, Detection and Characterization of Engineered Nanomaterials in the Environment: Current State-of-the-Art and Future Directions (Washington, 2014)
21. G.V. Lowry, E.M. Hotze, E.S. Bernhardt, D.D. Dionysiou, J.A. Pedersen, M.R. Wiesner, B. Xing, J. Environ. Qual. **39** (2010)
22. European Commission Health and Consumer Protection Directorate-General. Directorate C – Public Health and Risk Assessment C7 – Risk assessment (2006)
23. A. Elsaesser, C.H. Howard, Adv. Drug Deliver. Rev. **64** (2012)
24. G.V. Lowry, K.B. Gregory, S.C. Apte, J.E. Lead, Environ. Sci. Technol. **46** (2012)
25. J. Swiatek-Prokop, Prace naukowe Akademii im. Jana Dlugosza w Czestochowie, Seria: Edukacja Techniczna i Informatyczna, **VII** (2012)
26. N. Musee, Hum. Exp. Toxicol. **30**, 8 (2010)
27. N. Musee, Environmental International, **37** (2011)
28. G. Bystrzejewska-Piotrowska, J. Golimowski, P.L. Urban, Waste Manage. **29** (2009)
29. F. Part, G. Zecha, T. Causon, E.K. Sinner, M. Huber-Humer, Waste Manage. **43** (2015)
30. M. Maliszewska-Mazur, Ochr. Środ. i Zasob. Natur. 45 (2010)
31. P. Szewczyk, K. Midor, Zeszyty Naukowe Politechniki Slaskiej, Seria: Organizacja i Zarzadzanie **71,** 1917 (2014)
32. S. Bandyopadhyay, J.R. Peralta-Videa, J. L. Gardea-Torresdey, Environ. Eng. Sci. **30**, 3 (2013)
33. O. Posth, M.F. Hansen, U. Steinhoff, L. Bogart, P. Southern, P. Svedlindh, C. Gruttner, L. Fernandez Barquin, W. Szczerba, F. Ludwig, N. Gehrke, O. Kazakova, C. Johansson, *In proceedings of IMEKO XXI World Congress* (Czech Technical University, Prague, 2015)
34. P.C. Lina, S. Lina, P.C. Wanga, R. Sridharb, Biotechnol. Adv. **32**. 4 (2014)
35. M. Hassello, J.W. Readman, J.F. Ranville, K. Tiede, Ecotoxicology, **17** (2008)
36. J.K. Patra, K.H. Baek, J. Nanomater. 2014 (2014)

 DOI: 10.1051/e3sconf/20172200120

Cycle analysis of MCFC/gas turbine system

Abdullatif Musa[1,*], *Abdulsalam* Alaktiwi[1], and *Mosbah* Talbi[1]

[1]University of Tripoli, Faculty of Engineering, Department of Marine Engineering, Tripoli, Libya

Abstract. High temperature fuel cells such as the solid oxide fuel cell (SOFC) and the molten carbonate fuel cell (MCFC) are considered extremely suitable for electrical power plant application. The molten carbonate fuel cell (MCFC) performances is evaluated using validated model for the internally reformed (IR) fuel cell. This model is integrated in Aspen Plus^{TM}. Therefore, several MCFC/Gas Turbine systems are introduced and investigated. One of this a new cycle is called a heat recovery (HR) cycle. In the HR cycle, a regenerator is used to preheat water by outlet air compressor. So the waste heat of the outlet air compressor and the exhaust gases of turbine are recovered and used to produce steam. This steam is injected in the gas turbine, resulting in a high specific power and a high thermal efficiency. The cycles are simulated in order to evaluate and compare their performances. Moreover, the effects of an important parameters such as the ambient air temperature on the cycle performance are evaluated. The simulation results show that the HR cycle has high efficiency.

1 Introduction

Fuel cells are electrochemical energy conversion devices which typically run on hydrogen or methane or methanol and produce electricity, heat and benign emissions (water and, in the case of methane and methanol, CO_2). The fuel cells used for stationary energy production are typically high temperature fuel cells (HTFCs) such as and molten carbonate fuel cell (MCFC).

MCFC technology was established about 30 years ago and has been developed considerably fast in the USA, Korea, Japan and Europe during the last 10 years. In the literature there have been several studies published on MCFC systems and their analyses [1–7].

In [1] a system modeling was carried out to compare different schemes for the molten carbonate fuel cell (MCFC) technology for an auxiliary power unit (APU) purposes in the marine field. The reference plant had a power of 500 kWe provided by the full cell (FC) stack, plus roughly 100 kWe from a gas turbine. The results were compared to the power needs and raw materials on-board stock capabilities of three reference cases of commercial ships. Considering the size of the selected ships, these could profitably benefit of high efficiency MCFCs for APU purposes, without major ship plant adjustments for feeding the

* Corresponding author: a.musa@na.uot.ly

required diesel oil, while demi-water needs to be recycled within the APU to make the latter less impacting on the ship infrastructure.

In this paper thermodynamic model for internally reformed MCFC is developed. Moreover, newly multistage (MS) and heat recovery (HR) cycles are introduced: the multistage cycle consisting of two stages of compression with intercooling and two stages expansion with reheating. Moreover in the heat recovery (HR) cycle, the air leaving the first stage compressor is at high temperature. This temperature is higher than the inlet intercooler temperature. Therefore, a regenerator is used to preheat water by outlet air compressor. The waste heat of the outlet air compressor and the exhaust gases of turbine are recovered and used to produce steam, this steam is injected in the gas turbine, resulting in a high specific power and a high efficiency of the HR cycle.

South Tripoli Gas Power Plant is one of the network feeder plants in Tripoli, Libya . The power plant was constructed by ABB a German-Swiss company in 1994, and it consists of five gas turbine power units. One of the gas turbine power unit GT 14 selected as a basic cycle for this study. The ambient temperature in Libya may reach 40 C in summer. So it is an important to study effects of ambient temperature changes on the performance of the introduced cycles. The aim of this work is to find the best configuration for a several cycles. Therefore, the basic ,multistage and heat recovery cycles are simulated in order to evaluate and compare their performances.

2 Configuration cycle description

The high temperature inside the cells stack allows for reforming the methane (CH_4) directly inside the cell if the steam is provided at the inlet. The heat necessary for this reforming reaction is delivered by the electrochemical reaction in the cell. The gas turbine unit GT14 of south Tripoli gas turbine power plant is selected as a basic cycle for this study. This power plant was constructed by ABB a German-Swiss company in 1994. The basic cycle consists of compressor , combustion chamber, and gas turbine. The compressed air enters the combustion chamber, where its burned at essentially constant pressure. The gasses is expanded at the gas turbine until they reach the ambient pressure.

2.1 Multistage cycle configuration

The multistage (MS) cycle configuration consisting of two stages of compression with intercooling and two stages expansion with reheating. Moreover, this cycle consisting of the MCFC stack. The compressed air from the compressor (AC1) is split into two parts. Part of the air is sent to the intercooler, the second stage compressor (AC2), the combustor (comb1) and the gasses is expanded at the high pressure gas turbine (GT1). The remaining part of the air is mixed with the exhaust gasses of GT1 and then enters the MCFC stack. By considering that , the MCFC stack exit gasses which contains a major part of air, and CH_4 are sent to the second combustor (Comb2) and the low pressure gas turbine (GT2) (see Fig. 1a).

2.2 Heat recovery cycle configuration

The heat recovery (HR) cycle is similar to the multistage cycle (Fig. 1a), except that, the air leaving the first stage compressor (AC1) is at high temperature, so regenerator (HE1) is used to preheat water by the outlet air compressor (AC1). The waste heat of the major outlet air compressor (AC1) and the exhaust gasses of gas turbine (GT2) are recovered and used to produce steam. This steam is injected in the high pressure gas turbine (GT1) (see Fig. 1b).

3 Water gas shift and methane reforming reactions

The chemical reactions occur in the model are assumed to be in equilibrium. The electrochemical reaction is implemented as the following:

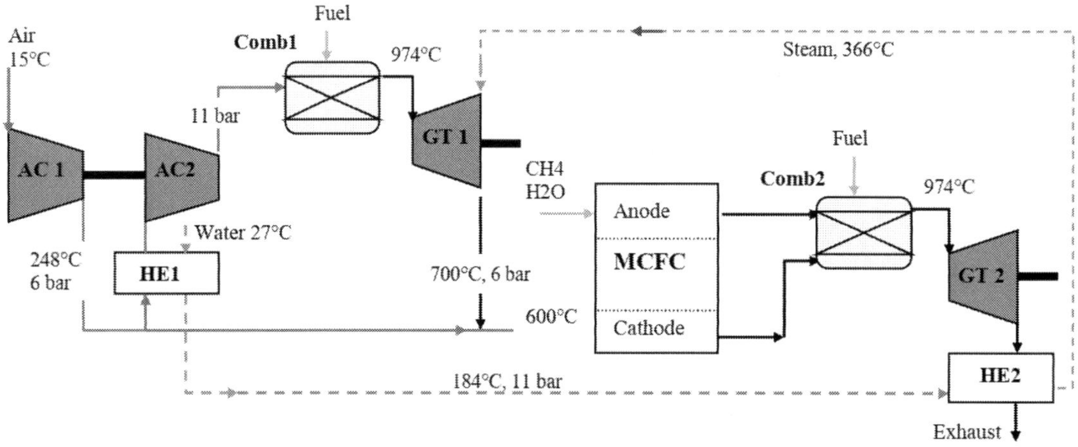

Fig.1. The MCFC/gas turbine cycles: layout A and B (the dotted line is for B).

Table 1. Setting parameters of the cycle.

Setting parameter	Value	Setting parameter	Value
Current density	0.250 Acm^{-2}	Pressure ratio	11
Active cell area	250 m²	Pressure drop in combustor	0.2 bar
Total fuel utilisation rate	85%	Pressure drop in MCFC	0.05 bar
Compressors isentropic efficiency	80%	Pressure drop in heat exchangers	0.02 bar
Gas turbine and pump isentropic efficiencies	85%	Gas turbine and compressor mechanical efficiencies	98%
Fuel flow rate	6.349 kg/s	Air flow rate	394.5 kg/s

$$CO_2 + \tfrac{1}{2}O_2 + 2e^- \rightarrow CO_3^{2-} \; cathode \quad (1)$$

$$H_2 + CO_3^{2-} \rightarrow H_2O + CO_2 + 2e^- \; anode \quad (2)$$

$$H_2 + \tfrac{1}{2}O_2 \rightarrow H_2O \; overall\ reaction \quad (3)$$

The MCFC stack operate at a temperature high enough to reform the steam reforming inside the anode. The electrochemical and water gas shift reactions are exothermic and, on the other hand, the methane reforming reaction is a strongly endothermic reaction. The chemical reactions in the steam reforming and water gas shift are shown below:

$$CH_4 + H_2O \Leftrightarrow CO + 3H_2 \tag{4}$$

$$CO + H_2O \Leftrightarrow CO_2 + H_2 \tag{5}$$

In case of IR-MCFC the equilibrium molar flow rate along the anode and cathode can be calculated using the stoichiometric balance of the steam reforming, water shift and electrochemical reactions. So the mole fractions along the anode and cathode can be calculated as follows:

$$\gamma_{CH_4} = \frac{\dot{n}_{CH_4}}{\dot{n}_{tot,an}} = \frac{\dot{n}_{CH_4,inlet} - x}{\dot{n}_{an} + 2x + z} \tag{6}$$

where: $\dot{n}_{an} = \dot{n}_{CH_4 inlet} + \dot{n}_{H_2O,inlet} + \dot{n}_{CO,inlet} + \dot{n}_{H_2,inlet} + \dot{n}_{CO_2 an,inlet} + \dot{n}_{N_2,an}$ is the inlet molar flow rate of the mixture at the anode input

$$\gamma_{H_2O} = \frac{\dot{n}_{H_2O,inlet} - x - y + z}{\dot{n}_{an} + 2x + z} \tag{7}$$

$$\gamma_{CO} = \frac{\dot{n}_{CO,inlet} + x - y}{\dot{n}_{an} + 2x + z} \tag{8}$$

$$\gamma_{H_2} = \frac{\dot{n}_{H_2,inlet} + 3x + y - z}{\dot{n}_{an} + 2x + z} \tag{9}$$

$$\gamma_{CO_2 an} = \frac{\dot{n}_{CO_2 an,inlet} + y + z}{\dot{n}_{an} + 2x + z} \tag{10}$$

$$\gamma_{N_2,an} = \frac{\dot{n}_{N_2,an}}{\dot{n}_{an} + 2x + z} \tag{11}$$

$$\gamma_{O_2} = \frac{\left(\dot{n}_{O_2,inlet} - z/2\right)}{\left(\dot{n}_{ca} - \left(\frac{3z}{2}\right)\right)} \tag{12}$$

where: $\dot{n}_{ca} = \dot{n}_{O_2,inlet} + \dot{n}_{CO_2 ca,inlet} + \dot{n}_{N_2}$ is the molar flow rate of the mixture at the cathode input

$$\gamma_{CO_2 ca} = \frac{\left(\dot{n}_{CO_2 ca,inlet} - z\right)}{\left(\dot{n}_{ca} - \left(\frac{3z}{2}\right)\right)} \tag{13}$$

$$\gamma_{N_2} = \frac{\dot{n}_{N_2}}{\left(\dot{n}_{ca} - \left(\frac{3z}{2}\right)\right)} \tag{14}$$

The electrochemical, steam reforming and water gas shift reactions rates should be calculated using the definitions of equilibrium constant of the steam reforming (K_f), equilibrium constant of the water gas shift (K_p) ,and fuel utilization factor (z) as follows:

$$K_f = \frac{\left(\dot{n}_{H_2,inlet} + 3x + y - z\right)^3 \left(\dot{n}_{CO,inlet} + x - y\right)}{\left(\dot{n}_{CH_4,inlet} - x\right)\left(\dot{n}_{H_2O,inlet} - x - y + z\right)} \left(\frac{p_{an}}{\dot{n}_{an} + 2x + z}\right)^2 \tag{15}$$

 DOI: 10.1051/e3sconf/20172200120

$$K_p = \frac{(\dot{n}_{H_2,inlet} + 3x + y - z)(\dot{n}_{CO_2 an,inlet} + y + z)}{(\dot{n}_{CO,inlet} + x - y)(\dot{n}_{H_2O,inlet} - x - y + z)} \tag{16}$$

$$z = u_f(\dot{n}_{H_2 in} + \dot{n}_{CO in} + 4\dot{n}_{CH_4 in}) \tag{17}$$

Please note that, the values of (x , y and z) are unknowns, but they can be determined by solving equations (15), (16) and (17).

4 MCFC model description

The fuel cell is treated as a single control volume in which the steady state flow energy equation is applied with the assumption of negligible change of kinetic energy and potential energy. To determine actual cell performance, the overpotential must be deducted from the Nernst potential (E) which represents the ideal performance.

$$V_{cell} = E - V_{loss} \tag{18}$$

The internal resistances of the MCFC stack can be calculated as in [5]. But the ohmic cell (r_{ohm}), anodic reaction ($r_{pol,an}$) and cathodic reaction ($r_{pol,ca}$)resistances are reflected in the functions of the reaction temperatures and partial pressure of gas constituents. Please note that, the Ohmic cell resistance is calculated using an Arrhenius equation as function of the operating temperature.

$$r_{ohm} = 0.5\exp\left[3016 \times \left(\frac{1}{T_{cell}} - \frac{1}{923}\right)\right] \tag{19}$$

$$r_{pol,an} = 2.27 \times 10^{-5}\exp\left(\frac{6435}{T_{cell}}\right) P_{H_2}^{-0.42} P_{CO_2}^{-0.17} P_{H_2O}^{-1} \tag{20}$$

$$r_{pol,ca} = 7.505 \times 10^{-6}\exp\left(\frac{9298}{T_{cell}}\right) P_{O_2}^{-0.34} P_{CO2}^{-0.09} \tag{21}$$

The stack cell voltage can be found by using the following equation

$$V_{cell} = \frac{RT}{2F}\ln K - \frac{RT}{2F}\ln\left(\frac{p_{H_2O}\, p_{CO_2 an}}{p_{H_2}\, p_{CO_2 ca}\sqrt{p_{O_2}}}\right) - i \times (r_{ohm} + r_{pol,an} + r_{pol,ca}) \tag{22}$$

The electrical power produced by the fuel cell is calculated by:

$$W = V_{cell} I \tag{23}$$

The cycle efficiency (η_{Cycle}) is defined as the ratio of the power produced by the fuel cells (W_{MCFC}) and the total gas turbines, minus the total compressor power, to the (LHV) of the total amount of fuel (Q_{tot})

$$\eta_{Cycle} = \frac{W_{MCFC} + W_{turbs} - W_{comps}}{Q_{tot}} \tag{24}$$

 DOI: 10.1051/e3sconf/20172200120

5 Results and discussion

The cycle's layout under investigation is simulated under the standard operating conditions shown in table 1. The gas turbine unit GT14 of south Tripoli gas turbine power plant, Tripoli, Libya is selected as a basic cycle (basic cycle) for this study. The multistage (MS) cycle and the heat recovery (HR) cycle are introduced in this work. The MS cycle consisting of two stages of compression with intercooling and two stages expansion with reheating. In the HR cycle, the air leaving the first stage compressor is at high temperature, and this temperature is higher than the inlet intercooler temperature. Therefore, regenerator is used to preheat water by outlet air compressor. The waste heat of the outlet air compressor and the exhaust gases of turbine are recovered and used to produce steam. This steam is injected in the gas turbine. The aim of this analysis is to investigate and compared the performance of the cycles. As it's known, the temperature and cell voltage are the core of any fuel cell modeling and gas turbine cycle, and referring to the flow sheets shown in Fig.1. The effect of the ambient temperature on the performances of the cycles are investigated.

Fig. 2 shows the changes in the specific work of the basic cycle with the ambient temperature. Increasing the ambient temperature decreases the inlet air density, resulting an increase of the specific work of the compressor (Wcomp1). The increase in the specific work of the compressor causes the decrease in the specific net work of the cycle (Wnet). There is no effect of the ambient temperature on the specific work of the gas turbine (WGT). Fig. 3 shows the changes in the specific net work of the cycles with the ambient temperature. The multistage (MS) and heat recovery (HR) cycles have a higher specific network. This is mainly caused by the reduction of the specific compressor (AC1) work (see Fig.4). In the HR cycle, the waste heat of the major outlet air AC1 and the exhaust gases of gas turbine (GT2) are recovered and used to produce steam and then is injected in GT1. This means the gas turbine power of the HR cycle is higher than of the MS cycle (see Fig. 5). The increase in the power of gas turbine in the HR cycle is caused by the mass flow rate of steam which is injected and expanded in the GT1.

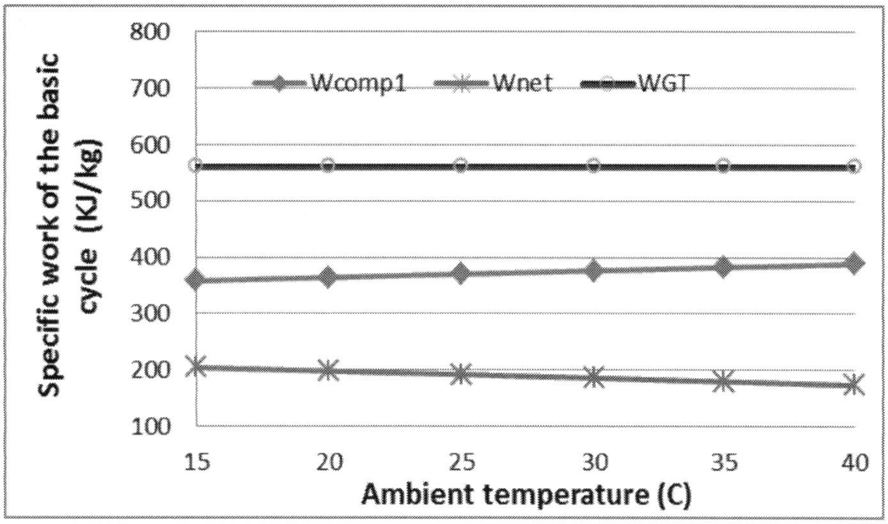

Fig. 2. The changes in the specific work of the basic cycle with the ambient temperature.

DOI: 10.1051/e3sconf/20172200120

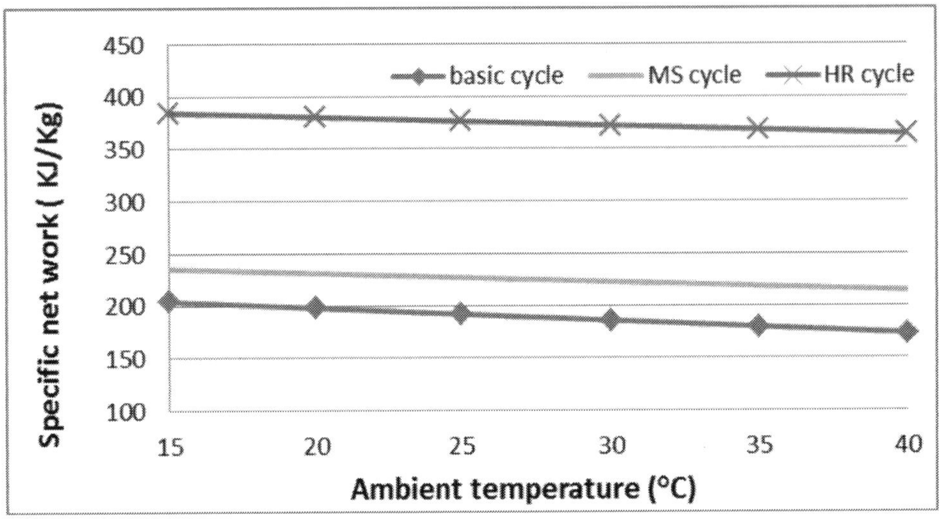

Fig. 3. The changes in the specific net work of the cycles with the ambient temperature.

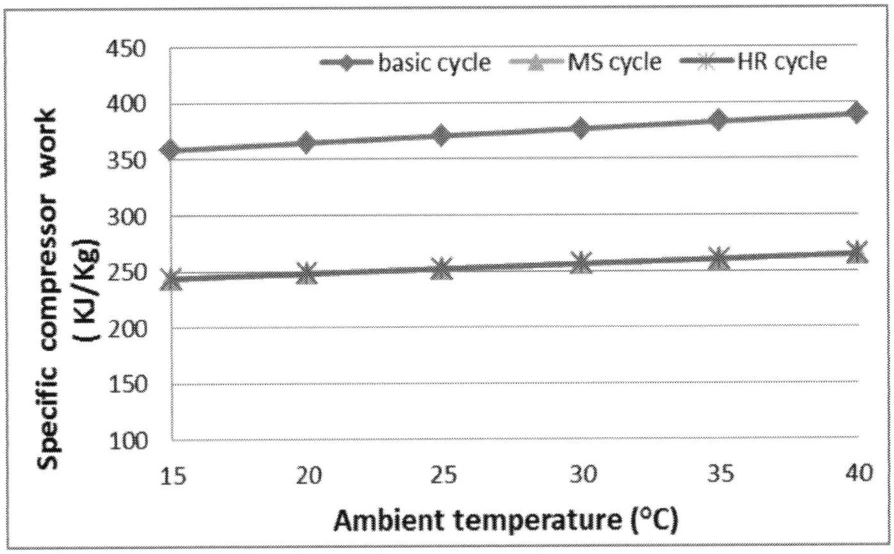

Fig. 4. The changes in the specific compressor (AC1) work of the cycles with the ambient temperature.

 DOI: 10.1051/e3sconf/20172200120

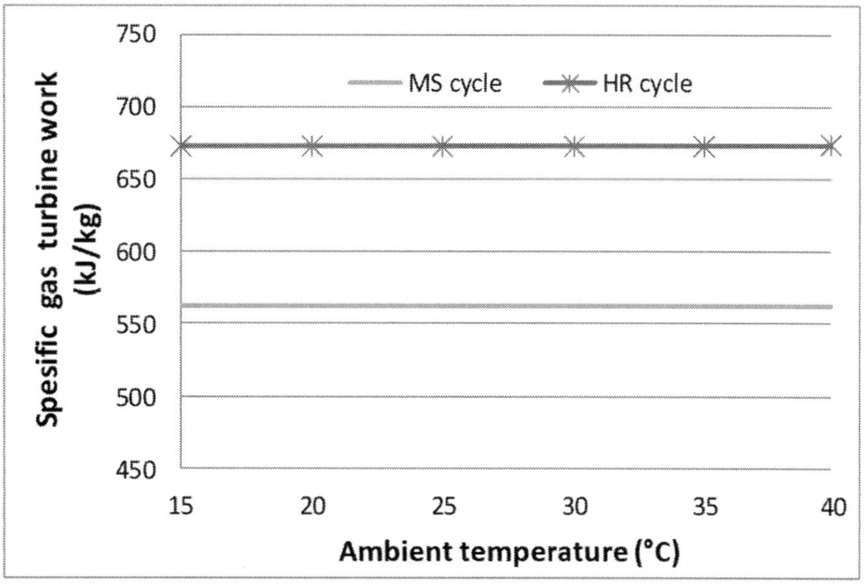

Fig. 5. The changes in the specific gas turbine work of the cycles with the ambient temperature.

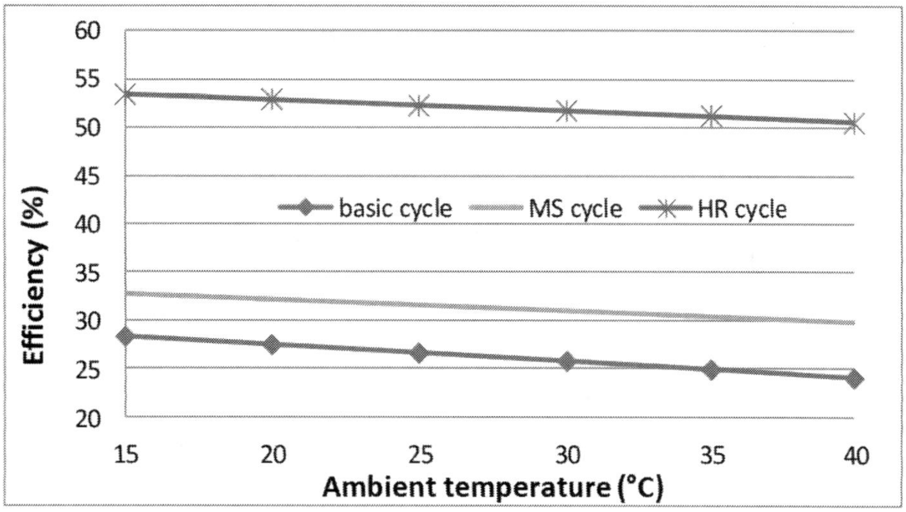

Fig. 6. the changes in the efficiency of the cycles with the ambient temperature.

Fig. 6 shows the changes in the efficiency of the cycles with the ambient temperature. The efficiency of the MS and HR cycles is higher than of the basic cycle. Though the total fuel consumption is higher in the MS and HR cycles, the power of fuel cell stack and gas turbines is also higher, causing the increase in efficiency. The efficiency of the HR cycle is higher than of the MS cycle, because of the recuperation of the waste heat of the major outlet air AC1 and the exhaust gasses of gas turbine (GT2). As more mass flow rate of steam is injected in the high pressure gas turbine, more power can be produced resulting in an increase of the HR cycle efficiency and a decrease of the exhaust gasses temperature (see Fig. 7).

 DOI: 10.1051/e3sconf/20172200120

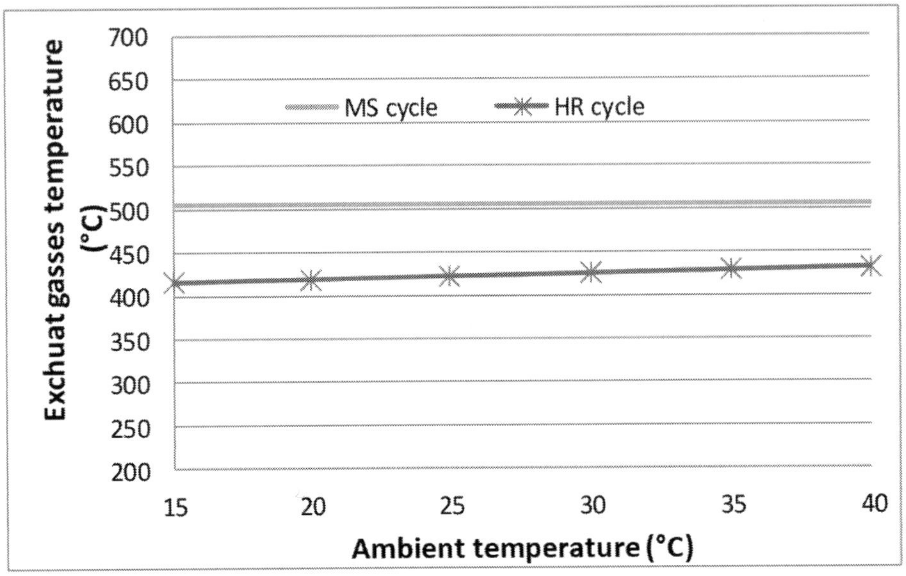

Fig. 7. Exhaust gasses temperature of the cycles as a function of the ambient temperature.

6 Conclusions

High temperature fuel cells such as the solid oxide fuel cell (SOFC) and the molten carbonate fuel cell (MCFC) are considered extremely suitable for electrical power plant application.The molten carbonate fuel cell (MCFC) performances is evaluated using validated models for the internally reformed (IR) fuel cells. This model is integrated in Aspen Plus™. Therefore, several MCFC/Gas Turbine cycles are proposed and investigated. The gas turbine power unit GT14 of south Tripoli gas turbine power plant Tripoli, Libya is selected as a basic cycle for this study. Moreover, a new cycles are introduced: a multistage (MS) a heat recovery (HR) cycles. The MS cycle consisting of two stages of compression with intercooling and two stages expansion with reheating. In the HR cycle, the air leaving the first stage compressor is at high temperature. This temperature is higher than the inlet intercooler temperature. Therefore, a regenerator is used to preheat water by outlet air compressor. The waste heat of the outlet air compressor and the exhaust gases of turbine are recovered and used to produce steam. This steam is injected in the gas turbine, resulting in a high specific power and a high thermal efficiency. The basic ,multistage and heat recovery cycles are simulated in order to evaluate and compare their performances. Moreover, the effects of an important parameters such as the ambient air temperature on the system performance are evaluated. The results indicated that, an increasing the ambient temperature decreases the inlet air density, resulting an increase of the specific work of the compressor. The results indicate that the efficiency is higher in case of the MS and HR cycles compared to the basic cycle for the same ambient temperature. The highest specific work and cycle efficiency are on the contrary obtained for the HR cycle. As more mass flow rate of steam is injected in the high pressure gas turbine, more power can be produced resulting in an increase of the HR cycle efficiency. The recuperation of the waste heat of the major outlet air AC1 and the exhaust gases of gas turbine (GT2) is benefited to the HR cycle.

The authors acknowledge Tripoli University, Tripoli, Libya for the support of this work.

References

1. S. Bensaid, S. Specchia, F. Federici, I. J. Hyd. Ener. **34**, 2026–2042 (2009)
2. T. Araki, T. Ohba, S. Takezawa, K. Onda, Y. Sakaki, Pow. Sour. J. **158**, 52–59 (2006)
3. A. Musa, M. De Paepe, I. J. Hyd. Ener. **33**, 4665–4672 (2008)
4. S. Kimijima, N. Kasagi, JSME. J. **series B**, 48–1 (2005)
5. V. Verda, F. Nicolin, I. J. Hyd. Ener. **35**, 794–806 (2010)
6. E. Bocci, F. Orecchini, A. Di Carlo, Pow. Sour. J. **160**, 835–841 (2006)
7. A. Musa, H. Steeman, M. De Paepe, ASME. J. **4**, 65–71 (2007)
8. M. Baranak, H. Atakül, Pow, Sour. J. **172**, 831–839 (2007)

DOI: 10.1051/e3sconf/20172200121

Shortcut nitrification/partial nitritation start-up for reject water treatment in a SBR

Mateusz Muszyński-Huhajło[1,*], and *Stanisław* Miodoński[1]
[1]Faculty of Environmental Engineering, Wroclaw University of Science and Technology, Wybrzeże Wyspiańskiego 27, 50-370 Wroclaw, Poland

Abstract. For many wastewater treatment plants (WWTPs), side-stream treatment of reject water from digested sludge dewatering is a feasible opportunity to improve N-removal efficiency without costly plant expansion. Biological nitrogen removal over nitrite or combined partial nitritation (PN)-Anammox process has recently become a popular treatment method for such ammonium-rich streams. Shortcut nitrification and PN start-ups were successfully performed in a pilot-scale SBR treating real reject water. In all performed experiments, effective nitrate production inhibition occurred in less than 20 days due to operational conditions selection and without advanced control system. pH adjustment in the PN reactor allowed to achieve NO_2-N /NH_4-N ratio suitable for Anammox process (1.24±0.07).

1 Introduction

High quality standards for wastewater treatment plant (WWTP) effluent induce constant development of new technologies, especially considering biological nutrient removal (BNR). In the last decade, researcher's attention attracted alternative technologies for nitrogen removal, especially connected with treating high-strength ammonium streams (i.e. reject water from sludge dewatering), named as side-stream treatment [1]. Specific characteristics of this medium, such as high ammonium concentration and temperature, create a convenient opportunity to avoid further nitrite oxidation in the second stage of nitrification by selective NOB (*Nitrite Oxidizing Bacteria*) inhibition. Such approach can be described as shortcut nitrification if the main goal is to oxidize whole ammonium load into nitrite. Main advantages of this process are: reduction of specific cost of 1g NH_4-N oxidation due to lower oxygen consumption, less organic carbon required in further denitrification if such occurs later (i.e. SHARON process) [2], production of nitrite-rich medium that can be a useful source of free nitrous acid (FNA) [3] and activated sludge production enriched with AOB (*Ammonia Oxidizing Bacteria*) for mainstream bioaugmentation purposes [4]. Despite only slight difference in the amount of ammonium load oxidized, it seems to be important to distinguish a specific type of shortcut nitrification – partial nitritation (PN) used as a necessary pre-treatment preceding Anammox process. In this case the main goal is to oxidize part of the ammonium load, achieving NO_2-N/NH_4-N ratio in the effluent around 1.32, reported as stoichiometric in Anammox reaction [5].

* Corresponding author: mateusz.muszynski-huhajlo@pwr.edu.pl

DOI: 10.1051/e3sconf/20172200121

Rising popularity of this process can be proven by over 100 PN/Anammox installations operating worldwide by 2014 [6].

Paper contains detailed description of selected factors affecting NOB activity in wastewater treatment systems that may be used to achieve severe inhibition of this nitrifiers group. Based on gathered theoretical information, a simple start-up strategy for shortcut nitrification/PN was developed and successfully tested in a pilot-scale SBR treating real reject water from sludge dewatering at one of Polish WWTPs.

1.1 Nitrification and factors affecting NOB activity.

Nitrification is one of the fundamental biological processes in wastewater treatment and key element of nitrogen removal processes. This two-stage biochemical, autotrophic process of ammonia (NH_3) oxidation to nitrate (NO_2^-) by AOB and further nitrite oxidation to nitrate (NO_3^-) by NOB. According to reaction stoichiometry, complete oxidation of 1 g of NH_3-N to nitrate requires 4.57 g of oxygen, where 3.43 g O_2 is used for nitritation, which also consumes alkalinity due to hydrogen ions production (7.14 g $CaCO_3$ per gram NH_3-N oxidized) [7].

In conventional activated sludge systems, where complete nitrification is key element in the nitrogen removal process, process conditions are usually adjusted to meet the requirements for stable growth and activity of both groups of nitrifiers, however, in some advanced treatment technologies only nitritation is considered as enough for N-removal. AOB and NOB, which has no phylogenetic relation between themselves, reveal slight differences in environmental conditions tolerance range and growth kinetics which might be used to induce selective inhibition of NOB. Selected factors that can be used to promote only AOB activity are discussed in this point.

1.1.1 Temperature

Temperature has significant impact on nitrification process. A common phenomenon is periodical nitrite build-up observed, both in natural waters [8] and activated sludge systems [9] during elevated temperature period in summer. This fact can be explained by different activity characteristics of two nitrifiers groups: AOB show greater activity than NOB at higher temperatures (> 14°C) [10]. Nitrifiers net growth rate at 25°C equals 0.86 and 0.77 d^{-1} for AOB and NOB, respectively. As the temperature rises, this difference increases to about 28% at 30°C and 38% at 35°C in favour of the AOB [11]. As reject water from sludge dewatering usually have relatively high temperature, this feature is commonly used as one of main nitrifiers selection mechanism in side-stream treatment technologies.

1.1.2 Sludge Retention Time (SRT)

In optimal conditions, AOB grow faster than NOB as minimal doubling time for each group, 7–8 h for ammonia oxidizers and 12–13 h for nitrite oxidizers [12]. SRT, as one of fundamental operational parameters characterising activated sludge systems, can be a useful tool to reduce NOB presence in biomass structure. Setting SRT higher than minimal SRT required for keeping AOB in the reactor and lower than minimal required for NOB will wash-out them from the system. Minimal SRT for each nitrifiers groups depends on various parameters so it is very important to maintain properly balanced operating conditions which promote only AOB.

1.1.3 Dissolved oxygen (DO) concentration

Availability of oxygen, the final electron acceptor in nitrification process, is one of its limiting factors. In general, DO concentration around 2 g O_2/m^3 is traditionally considered as optimal for complete nitrification, both for technological and economic reasons. By eliminating nitrite oxidation from the whole process, oxygen consumption can be reduced by 25%. Like in case of growth kinetics, two groups of nitrifiers also has different affinity constants for oxygen. It was shown that AOB has higher oxygen affinity. Oxygen saturation constant are 0.3 and 1.1 g O_2/m^3 for AOB and NOB respectively [13] however, range of values found in the literature for this parameter is quite wide. Possible cause of such discrepancies is the difference between oxygen concentration in the liquid phase and the interior of the flocs. These values are therefore dependent on the density of the flock structure, their size and diffusion efficiency [14]. Summarizing, DO concentrations below 1 g O_2/m^3 result in a gradual elimination of NOB from the system [15].

1.1.3 pH, Free Ammonia (FA) and Free Nitrous Acid (FNA)

pH can be crucial parameter in nitrification process control, as hydrogen ions concentration has impact on nitrifiers enzyme activity [16]. It is generally accepted that optimal pH range for AOB and NOB are 8.2–0.3 and 7.9–0.4 respectively [17]. In solutions containing ammonium and/or nitrite nitrogen also their ion forms are present as FA and/or FNA, depending on their initial concentrations, pH value and temperature [18].

Both substances are known to have inhibitory effect on nitrification process and can be used as a selective force for NOB inhibition due to different inhibition threshold values for each nitrifiers group. For AOB, FA and FNA levels resulting in activity loss are above 10 mg NH_3/dm^3 and 0.42–1.72 mg HNO_2/dm^3. NOB are much more vulnerable as complete inhibition can be observed in concentrations over 1 mg NH_3/dm^3 and 0.026–0.22 mg HNO_2/dm^3. Exact values may vary depending on source due to possible biomass adaptation [19] but general conclusion maintains the same – AOB are more resistant to these factors.

Proper control of FA and FNA concentration was used a powerful tool to obtain shortcut/PN in many different processes, especially in reject water treatment. Different approaches on shortcut nitrification/PN were presented in literature, are mostly connected with FA [20–22] rather than FNA [16, 23] as the selective factor.

2 Materials and methods

2.1 Experimental set-up

2.1.1 Pilot scale SBR and treated medium characteristics

Described experiments were performed in a 150 dm^3 SBRs treating real reject water from digested sludge dewatering. Reactor was operated at predefined time cycles, consisting of filling, reaction (mixing and aeration), sedimentation, decantation phases. Reactor was equipped with number of sensors connected to SCADA system providing parental control of following parameters: temperature, pH (only base solution addition), DO level, filling/decanting volume and mixing intensity. NH_4-N /NO_3-N ion-selective electrode probe was installed in the reactor without connection to SCADA and was used only to control N-compounds concentrations by the operator.

2.1.2 Treated medium characteristics

Pilot installation treated real reject water from a belt press without flushing water (used for belt cleaning) to avoid excessive dilution. Activated sludge after mesophilic digestion (HRT ~40 d) was the belt press influent and origin of reject water. Due to intermittent operation of dewatering unit, reject water was collected periodically in a storage tank to provide supply of influent medium for SBR. After filling, medium composition was determined. During start-up period, typical influent ammonium concentration, and alkalinity were equal 622.6 ± 11.7 g NH_4-N/m^3, 53.8 ± 1.8 val/m^3 respectively during Start-up 1 and 779.3 ± 13.8 g NH_4-N /m^3, 68.6 ± 1.4 val/m^3 during Start-up 2.

2.1.3 Analytical methods

Collected samples were filtered with syringe filters, pore size 1.2 μm and after that ammonium, nitrite and nitrate nitrogen concentrations were measured using Hach photometric cuvette tests and a DR3900 spectrophotometer. Total suspended solids (TSS) and volatile suspended solids (VSS) concentrations were determined using direct weight method according to Standard Methods guideline. Nitrification rate tests were performed in the pilot-scale reactor in process conditions (test procedure and data analysis guides were adopted from [24]).

2.2 Experiment procedure

Experiments were conducted in pilot-scale SBR to test shortcut nitrification/PN start-up strategies in different pH conditions. Shortcut nitrification start-up was performed with a conventional, fully nitrifying activated sludge (CAS) from the same WWTP and second experiment was conducted as an attempt to achieve PN from a completely nitrifying enriched sludge treating the same reject water.

To ensure cultivation of similar nitrifying bacteria species as present in the mainstream reactor of the plant, during all experiments SBRs were seeded daily with activated sludge from the main reactor. Each seed portion was equal 20% of TSS mass in the SBR added once a day.

2.2.1 Start-up 1: From CAS to shortcut nitrification at elevated pH (FA inhibition)

Start-up strategy was based on combination of several factors affecting NOB activity (average values): low DO (0.68 ± 0.04 g O_2/m^3), high temperature (25.14 ± 0.04°C), short SRT (2.92 ± 0.21 days) and finally, elevated pH (7.81 ± 0.01) to maintain high concentration of FA. Process inoculum, CAS from a municipal WWTP, was seeded into pilot-scale reactor and a single nitrification rate test was performed after few hours of sludge adaptation to new conditions. After that, initial nitrogen loading rate (NLR) was calculated based on observed nitrification rate, aeration phase length and medium NH_4-N concentration. In the next days of experiment, NLR was increased based on effluent composition data (daily) and periodical nitrification rate test (performed at least once a week) until it reached 0.5–0.6 kg NH_4-N·(m^3·d)$^{-1}$ when further increasing was stopped and reactor was operated in stable conditions.

DOI: 10.1051/e3sconf/20172200121

2.2.2 Start-up 2: From enriched sludge to PN at low pH (FNA inhibition)

Goal of this experiment was to selectively inhibit NOB activity in a completely nitrifying sludge treating real reject water. Prior to described start-up, nitrification was maintained for over 3 months with average effluent composition: 0.6 ± 0.5, 0.5 ± 0.2 and 636.2 ± 19.9 mg N/dm^3 of NH_4-N, NO_2-N and NO_3-N respectively. At the beginning of PN start-up, NLR was 0.64 kg NH_4-N ·(m^3·d)$^{-1}$ and specific AUR (*Ammonium Uptake Rate*) was equal 0.34 g NH_4-N·(g VSS·d)$^{-1}$. During complete nitrification period, DO, pH, temperature and SRT were 3.29 ± 0.11 g O_2/m^3, 7.02 ± 0.08, 24.98 ± 0.03°C and 6.1 ± 0.5 days respectively.

As alkalinity present in reject water is sufficient for oxidation of about half of NH_4-N load, to restrain ammonium oxidation, solution used or pH control was changed from 8% $NaHCO_3$ to 5% NaOH. Another step on way to achieve PN was change in reactor operating conditions. Influent load was the same as during complete nitrification operation. While reactor was operated in stable conditions after switch to PN, between 45th and 60th day borderline pH set point adjustment was performed to achieve proper NO_2-N/NH_4-N ratio in the effluent, suitable for the Anammox process. During PN period, average DO, pH, temperature and SRT were 0.81 ± 0.05 g O_2/m^3, 7.03 ± 0.01 (6.44 ± 0.03 after pH set point adjustment), 24.99 ± 0.02°C and 3.13 ± 0.28 days respectively.

3 Results

3.1 Start-up 1

Mainstream CAS, used as inoculum, had initial specific AUR: 0.07 g NH_4-N·(g VSS·d)$^{-1}$ determined in first nitrification rate test after sludge acclimatisation. During Start-up 1, target NLR (0.6 kg NH_4-N·(m^3·d)$^{-1}$) was reached in 25 days and period of stable operation started. Nitrate presence in the effluent occurred only at days 1–10. After 30th day nitrate presence significantly dropped, due to higher NH_4-N concentration in the influent resulting in higher FA level supressing NOB activity more effectively. After 30th day average effluent composition was 33.3 ± 10.3, 416.6 ± 20.5 and 11.1±4.1 mg N/dm^3 of NH_4-N, NO_2-N and NO_3-N respectively. Average nitrite accumulation rate (NAR) was 97 ± 1% while specific AUR was 0.41 ± 0.04 g NH_4-N ·(g VSS·d)$^{-1}$. Average FA and FNA were 2.91±0.31 mg NH_3/dm^3 and < 0.015 mg HNO_2/dm^3 respectively. Figure 1 presents effluent composition and NAR during whole experiment.

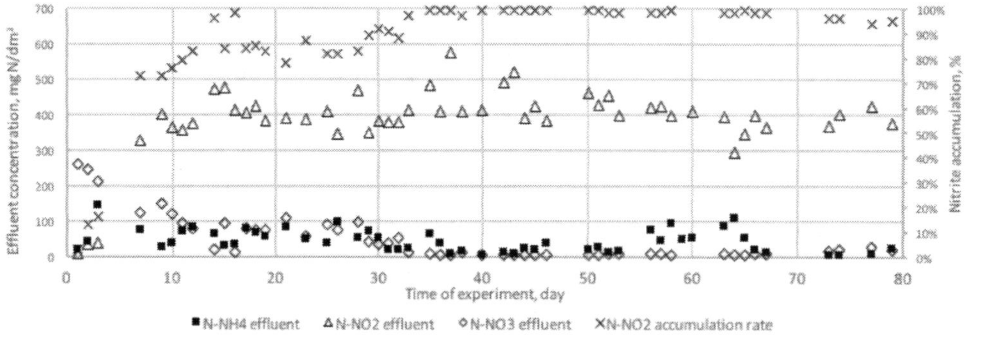

Fig. 1. NAR and effluent composition during Start-up 1.

3.2 Start-up 2

Operating conditions and typical effluent composition during complete nitrification period preceding this start-up were presented in section 2.2.2. At 6th day after start-up experiment begun, solution for pH control was changed, until this time complete nitrification occurred. Due to alkalinity limitation, NH_4-N build-up started alongside with nitrite accumulation. Increased NH_4-N and NO_2-N concentrations resulted in higher FA and FNA level which induced very effective NOB inhibition as NAR grow up to almost 80% in a week. Almost complete suppression of NOB activity can be observed since 30th day of experiment as only residual NO_3-N can be noticed in the effluent. After 30th day, average effluent was 336.6 ± 22.5, 466.3 ± 10.8 and 6.6 ± 0.9 mg N/dm^3 of NH_4-N, NO_2-N and NO_3-N respectively. In this period, average specific AUR was 0.39 ± 0.04 g NH_4-N·(g VSS·d)$^{-1}$ and NAR was 99 ± 0.3%. To achieve NO_2-N/NH_4-N ratio suitable for Anammox process, borderline pH was finally set to 6.4 at 60th day. Since then, average nitrite/ammonium ratio was 1.24 ± 0.07 which can surely be considered as suitable as Anammox process. Until 45th day average FA and FNA levels were 1.51 ± 0.21 mg NH_3/dm^3 and 0.09 ± 0.01 mg HNO_2/dm^3 respectively. In period after pH adjustment, average concentrations of these substances were 0.64 ± 0.03 mg NH_3/dm^3 and 0.38 ± 0.01 mg HNO_2/dm^3.

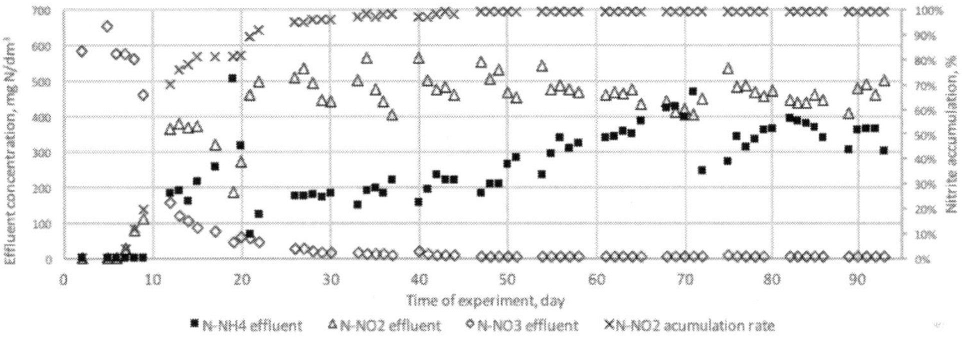

Fig. 2. NAR and effluent composition during Start-up 2.

4 Conclusions

- Shortcut nitrification/PN start-up strategy based on simple control of operational conditions in the reactor has proven to be successful and allows to obtain stable NOB activity suppression within two weeks.
- Combination of low DO level (0.7 gO_2/m^3), high temperature (25°C), short SRT (3 days) and FA inhibition in start-up based on conventional activated sludge was sufficient to achieve over 76% and 98% NAR 10 and 30 days since process start-up, respectively (Start-up 1).
- In performed experiments, elevated FNA concentrations (Start-up 2) induced more effective NOB inhibition than elevated FA concentrations. Nitrate presence in experiment where FNA was favoured as selective factor was lower and less variable than in start-ups based on FA inhibition. No differences in AOB activity were noticed between process conducted under FA or FNA inhibition.
- Suitable for Anammox process influent composition was obtained in PN reactor (average NO_2-N /NH_4-N ratio was 1.24 ± 0.07) by adjusting borderline pH set point, thus controlling the alkalinity added with base solution used for pH control.

References

1. M.C. van Loosdrecht, D. Brdjanovic, Science **344**, 1452–1453 (2014)
2. C. Hellinga, A.A.J.C. Schellen, J.W. Mulder, M.C. van Loosdrecht, J.J. Heijnen, Water Sci. Technol. **37**, 135–142 (1998)
3. J. Guangming, O. Gutierrez, Z. Yuan, Water Res. **45**, 3735–3743 (2011)
4. M. Herrero, D. C. Stuckey, Chemosphere **140**, 119–128 (2015)
5. M. Strous, J. Heijnen, J. Kuenen, M. Jetten, Microbiol. Biot. **50**, 589–596 (1998)
6. S. Lackner, E.M. Gilbert, S.E. Vlaeminck, A. Joss, H. Horn, M.C. Van Loosdrecht, Water Res. **55**, 292–303 (2014)
7. G. Tchobanoglous, *Wastewater Engineering: Treatment Reuse* (Metcalf & Eddy, Inc. 2003)
8. M. von der Wiesche, A. Wetzel, Water Res. **32**, 1653–1661 (1998)
9. Z. Tonkovic, 19th Biennial International Conference, Water Quality International, 165–172 (1998)
10. M.S.M. Jetten, S. Logemann, G. Muyzer, L.A. Robertson, S. de Vries, M.C. van Loosrecht, J.G. Kuenen, A Van Leeuw, J. Microb. **71**, 75–93 (1997)
11. I. Jubany, J. Lafuente, J.A. Baeza, J. Carrera, Water Res. **43**, 2761–2772 (2009)
12. S. Philips, S. Wyffels, R. Sprengers, Appl. Microbiol. Biot. **59**, 557–566 (2002)
13. U. Wiesmann, Adv. Biochem. Eng. Biot. **51**, 113–154 (1994)
14. E.V. Munch, P. Lant, J. Keller, Water Res. **30**, 2, 277–284 (1996)
15. Y. Ma, Y. Penga S. Wanga, Z. Yuanb, X. Wang, Water Res. **43**, 563–572 (2009)
16. J. Claros, E. Jiménez, D. Aguado, Water Sci. Technol. **67**, 2587–2594 (2013)
17. S. Park, W. Bae, J. Chung, S.C. Baek, Process Biochem. **42**, 1671–1676 (2007)
18. A.C. Anthonisen, R.C. Loehr, J. Water Pollut. Con. F. **48**, 835–852 (1976)
19. S. Suthersan, Water Pollution Research Journal of Canada **21**, 257–266 (1986)
20. Y. Jeanningros, S.E. Vlaeminck, Water Sci. Technol. **61**, 1393–1400 (2010)
21. H. Li, S. Zhou, G. Huang, B. Xu, Process Saf. Environ. **92**, 199–205 (2014)
22. G. Ciudad, R. Gonzalez, C. Bornhardt, C. Antileo, Water Res. **41**, 4621–4629 (2007)
23. D.J. Kim, D.W. Seo, S.H. Lee, Bioprocess Biosyst. Eng. **35**, 441–448 (2012)
24. M.C. van Loosdrecht, P.H. Nielsen, C.M. Lopez-Vazquez, D. Brdjanovic, *Experimental Methods in Wastewater Treatment* (IWA Publishing, 2016)

Production of bio-fertilizer from microwave vacuum pyrolysis of waste palm shell for cultivation of oyster mushroom (*Pleurotus ostreatus*)

Wai Lun Nam[1], *Man Huan* Su[1], *Xue Yee* Phang[1], *Min Yee* Chong[1], *Rock Keey* Liew[1], *Nyuk Ling* Ma[2], and *Su Shiung* Lam[1,*]

[1]Pyrolysis Technology Research Group, Eastern Corridor Renewable Energy Group, School of Ocean Engineering, Universiti Malaysia Terengganu, 21030 Kuala Nerus, Terengganu, Malaysia
[2]School of Fundamental Science, Universiti Malaysia Terengganu, 21030 Kuala Nerus, Terengganu, Malaysia

Abstract. Microwave vacuum pyrolysis of waste palm shell (WPS) was performed to produce biochar, which was then tested as bio-fertilizer in growing Oyster mushroom (*Pleurotus* ostreatus). The pyrolysis approach generated a biochar containing a highly porous structure with a high BET surface area (up to 1250 m^2/g) and a low moisture content ($\leq$ 10 wt%), exhibiting desirable adsorption properties to be used as bio-fertilizer since it can act as a housing that provides many sites on which living microorganisms (mycelium or plant-growth promoting bacteria) and organic nutrients can be attached or adsorbed onto. This could in turn stimulate plant growth by increasing the availability and supply of nutrients to the targeted host plant. The results from growing Oyster mushroom using the biochar record an impressive growth rate and a monthly production of up to about 550 g of mushroom. The shorter time for mycelium growth on whole baglog (30 days) and the highest yield of Oyster mushroom (550 g) was obtained from the cultivation medium added with 20 g of biochar. Our results demonstrate that the biochar-based bio-fertilizer produce from microwave vacuum pyrolysis of WPS show exceptional promise as an alternative growing substrate for mushroom cultivation.

1 Introduction

Malaysia is one of the global leading palm oil producer which have palm oil plantation covering 80% of the agricultural area in Malaysia. The production of crude palm oil is approaching 20 million tonnes yearly [1] in order to satisfy the world demand on oil and fats [2]. As a result, there are also huge amount of by-products such as mesocarp fiber (MF), palm kernel shell (PKS), and empty fruit bunch (EFB) being generated as oil palm wastes and they become a disposal problem for palm oil industry.

* Corresponding author: lam@umt.edu.my

Microwave vacuum pyrolysis (MVP) is an innovative pyrolysis approach recently developed to transform the oil palm wastes into potentially useful products such as biochar, bio-oil and biogas. The pyrolysis process is performed using microwave heating in vacuum environment. Biochar is a main product derived from MVP after the evaporation of volatile matter and moisture content from the feedstocks. It is a carbon dense material with high porosity and large surface area. The high carbon content in biochar provides a long term storage solution because of its stable carbon bonds [3]. Biochar gives many environment benefits such as improving water quality by retaining agrochemicals in soils, mitigating climate change by storing the carbon in soil, and serving as a simple way to manage agricultural waste [4]. Its good adsorption property could contribute to soil amendment by enhancing the nutrients and water retention, and consequently boost up the growth rate and product yield of crops [5].

Mushroom possesses an antioxidant property which helps to reduce the oxidative damage in human body and provides protection against disease such as cancer and cardiovascular disease [6]. The high amount of carbohydrate, protein, multivitamins (i.e. vitamin B, C, D) and minerals (i.e. iron, calcium, potassium) present in the mushroom has made it a target food source for health-conscious consumers [7]. In particular, oyster mushroom (*Pleurotus* ostreatus) represents a favourite mushroom that is suitable to be grown in Malaysia. It adapts to moderate temperature and thus it can be cultivated in the lowlands [8]. The low production cost of oyster mushroom allows it to be supplied to consumers all around the year [6]. Due to the nutritional and medical values of mushroom, the demand for the mushroom in Malaysia is about 50 tonnes/day [8]. However, the average production of mushroom (e.g. 50 kg/day) is only enough to supply to the local consumers instead of exporting to other country [8]. In order to increase the product yield and quality of mushroom, it is hypothesized that biochar can act as an enhancer to boost up the growth of mushroom. Thus, this study was performed to produce biochar from microwave vacuum pyrolysis and then apply the biochar produced as bio-fertilizer for mushroom cultivation.

2 Materials and methods

2.1 Collection and pre-treatment of waste palm shell

Waste palm shell (WPS) was collected from TDM Plantation SDN BHD located in Sungai Tong, Setiu, Terengganu. The waste was dried in oven at 110°C for 24 hours in order to remove the moisture content.

2.2 Production of biochar from microwave vacuum pyrolysis of WPS

Microwave vacuum pyrolysis was performed to produce biochar. It has been reported that higher surface area and pore volume was produced from microwave pyrolysis compared to conventional heating [5]. Figure 1 shows the schematic diagram of the microwave vacuum pyrolysis. A 2.45 GHz (model: Samsung) microwave oven was modified to generate microwave radiation as heating source. 300 g of WPS was filled into a 1 Litre quartz reactor with 100 mm length and 123 mm diameter and then placed into a microwave oven to perform microwave vacuum pyrolysis. Air mixtures inside the oven were removed to create a vacuum condition before the start of pyrolysis experiments. The microwave power of oven was then set at 700 W and then the heating and pyrolysis were performed for 1 hour to transform the WPS into biochar.

DOI: 10.1051/e3sconf/20172200122

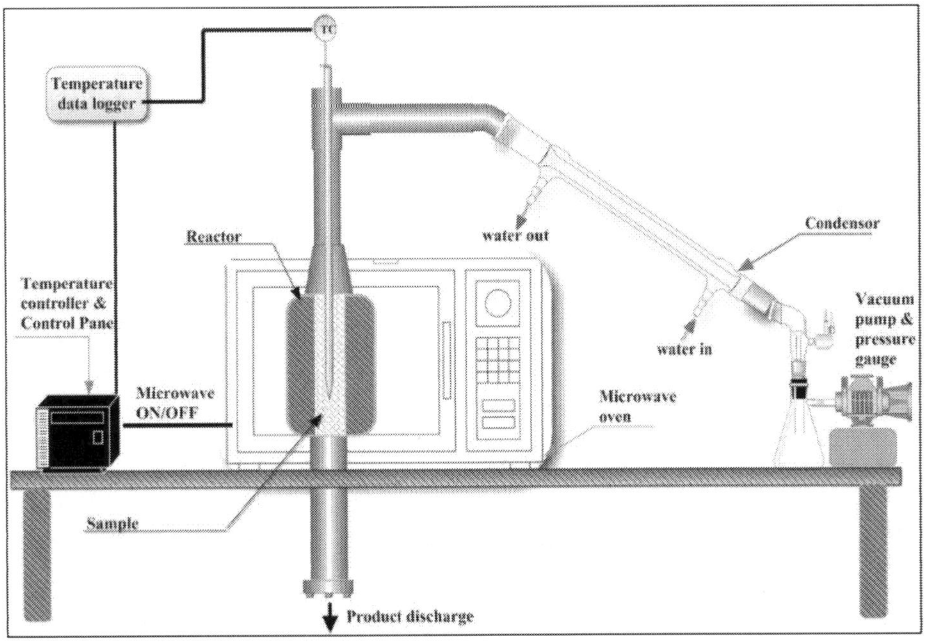

Fig. 1. Schematic diagram of microwave vacuum pyrolysis.

2.3 Characterization of waste palm shell and biochar

Different analysis was conducted to determine the chemical composition and physical surface structure of the WPS and biochar. Thermo-gravimetric Analysis (TGA) and CHNS elemental analysis were performed to quantify the proximate contents (moisture content, volatile matter, fixed carbon and ash) and elemental content (carbon, hydrogen, nitrogen and sulphur) of the WPS and biochar [9]. The surface morphology of biochar was examined by Scanning Electron Microscopy (SEM) and multipoint Brunauer, Emmett and Teller (BET) to determine the porous characteristics (i.e. surface area, pore width, and total pore volume).

2.4 Media formulation for mushroom cultivation

The biochar produced was ground into smaller pieces (2–4 mm) using a grinder and then it was mixed with other ingredients to form the mushroom cultivation medium. Different amounts of biochar (10 g, 20 g, and 30 g) were mixed thoroughly with rice bran, calcium carbonate, and sawdust to produce three different sets of cultivation medium as shown in Table 1. The lignocellulosic content of the rice bran and sawdust is a suitable substrate to provide nutrient for mushroom growth [10] and calcium carbonate is used to balance the pH of the mixture. Water was then added to the mixture to increase moisture content inside the mixture [11]. A polyethylene cylindrical bag was used to make a baglog that was then filled with 1 kg of the cultivation medium, then the baglog was sealed using PVC neck with cover. The bag was then sterilised at 121°C for 1 hour in the autoclave. After sterilisation, the baglogs were cooled to room temperature and then the oyster mushroom spawn was injected into the baglogs. The baglogs were then placed in a storage house for further observation. The experiments were performed in triplicates to improve the accuracy and

 DOI: 10.1051/e3sconf/20172200122

reproducibility of the data collected. Analysis was done with the observation and measurement of the growth of mycelium on a weekly basis and the yield of mushroom over a period of 35 days.

Table 1. Media formulation for mushroom cultivation.

Ingredient	Media Formulation			
	Control (g)	Mixture 1 (g)	Mixture 2 (g)	Mixture 3 (g)
Biochar	-	10	20	30
Rice bran	87	86	85	84
Calcium carbonate	12	12	12	12
Sawdust	901	892	883	874
Mushroom seed	1/10	1/10	1/10	1/10

2.5 Determination of mycelium growth rate and yield of Oyster mushroom

The cultivation period was set in one month for mycelium growth. Observation was performed on the growth of mycelium every week. The analysis on mycelium growth rate was based on the length of the mycelium grown along the baglog. The length of the mycelium growth was measured and expressed as a percentage of total height of mycelium colonized the baglog. At the 35-days harvest period, the mushrooms were harvested. The total weight of mushroom yield (fresh mature mushroom harvested) was determined and average mushroom yield was calculated. The harvesting time could be different for each baglog depending on the mushroom growth [11].

3 Results and discussion

3.1 Characterization of waste palm shell (WPS) and biochar

Table 2 shows the elemental and proximate content of WPS and the biochar produced by microwave vacuum pyrolysis. The results showed that WPS consisted mainly of carbon (50.5%) with low contents of hydrogen (7.0%) and nitrogen (3.1%). Proximate analysis showed that WPS was dominated by volatile matter (66.5%) followed by fixed carbon (27.6%) with a small amount of moisture content (1.9%) and ash (4%).

After the pyrolysis process, the biochar produced was found to have a high content of fixed carbon (83%), indicating the production of a carbon rich material. The biochar contained a low amount of volatile matter (9.4%), possibly due to the evaporation of volatile chemical derived from pyrolysis decomposition occurred at high temperature.

 DOI: 10.1051/e3sconf/20172200122

Table 2. Elemental and proximate content (wt%) of WPS and biochar.

Elemental Content	Weight (wt%)	
	WPS[a]	Biochar
Carbon	50.5	70.3
Hydrogen	7.0	3.6
Nitrogen	3.1	0.6
Oxygen [b]	39.4	25.5
Sulphur[c]	0.0	0.0
Proximate Content		
Moisture	1.9	4.0
Volatile Matter	66.5	9.4
Fixed Carbon [d]	27.6	82.6
Ash[e]	4.0	4.0

[a]WPS – Waste Palm Shell.
[b]Determined by mass difference.
[c]Sulphur content is not detectable by the minimum detection limit of the CHNS analyser (< 0.05 wt%) and thus it is assumed to be 0 wt% in calculation.
[d]Calculated by mass difference (i.e. Fixed carbon = 100 wt% – wt% of volatile matter – ash).
[e]Obtained by combustion at 950°C with 10 min of holding time.

3.2 Surface morphology of waste palm shell and biochar

Figure 2 demonstrates the SEM micrograph of biochar obtained from our microwave vacuum pyrolysis approach. Pores were observed on the surface of the biochar compared to the WPS which was non-porous and had a rough surface. This indicates that the microwave vacuum pyrolysis has caused the formation of a highly porous structure within the biochar. The highly porous structure suggests that the biochar could be utilised as an adsorption media.

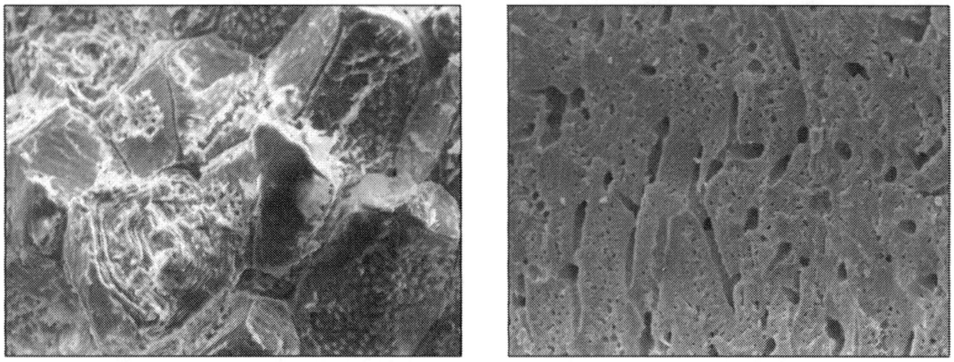

WPS (Before) Biochar (After)

Fig. 2. SEM micrograph of WPS and biochar.

3.3 N_2 adsorption-desorption isotherm and porous characteristics of biochar

Figure 3 shows the N_2 adsorption-desorption isotherm of the biochar produced and the porous characteristics are presented in Table 3. According to IUPAC classification, the isotherm curve of the biochar showed a Type IV isotherm which indicates the presence of mesopores with a pore size from 2 nm to 50 nm. High surface area was detected for the biochar (1225 m^2/g) and this indicates that it has high potential to be used as nutrient adsorbent. Thus, the biochar was tested for its use as a supporting ingredient in mushroom cultivation due to the presence of porous structure and good adsorption capacity

Table 3. Porous characteristics of biochar.

Porous Characterisstics	Biochar
Surface area	1225 m^2/g
Total pore volume	0.75 m^3/g
Pore size	3.2 nm

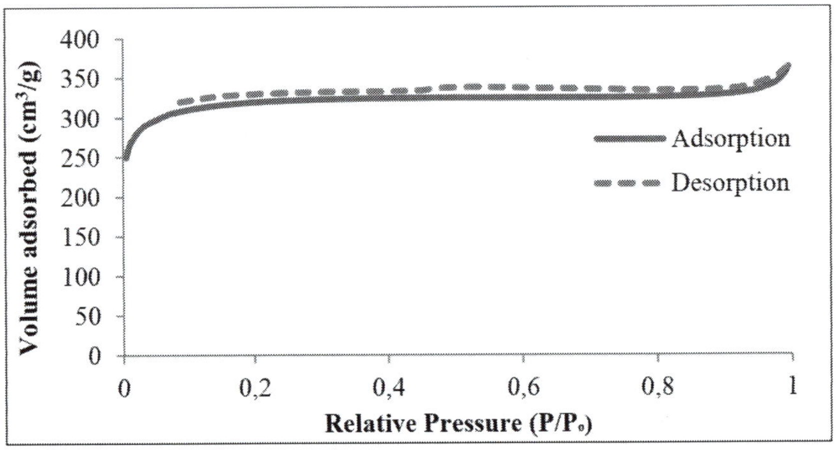

Fig. 3. N_2 Adsorption-Desorption Isotherm of biochar.

3.4 Biochar as bio-fertilizer for mushroom cultivation

The highly porous structure with a high BET surface area (up to 1225 m^2/g) of the biochar produced provided many sites to absorb and retain the desirable nutrient for the mycelium to grow. Mycelium is a vegetative growth in fungi and appears as white microscopic threads filaments (FAO, 2009). The surface of the biochar was fully distributed with pores which support its potential use as a bio-fertilizer. It was observed that mycelium was fully grown in 28–30 days within the baglog added with biochar compared to that observed in the baglog without the biochar (30–35 days).

Table 4 shows the different growth rates of the mycelium of Oyster mushroom observed from the baglog added with different amounts of biochar. It showed that the baglog added with biochar had a faster mycelium growth rate meanwhile the control showed the slowest growth rate. This suggests that the biochar added to the mushroom cultivation medium enhanced the mycelium growth.

The mushroom yields are based on weight of the Oyster mushroom harvested from each of the baglog in one month. As shown in Table 4, mixture 2 records the highest yield, producing 550 g of mushroom. It showed an impressively product yield as the total yield of mushroom is more than 50% compared with the control baglog. The porous structure of the biochar had contributed a large adsorption surface area for water and nutrient retention. The increased level of nutrient availability was likely to have enhanced the growth of mycelium. The strong water-holding capacity of the biochar could also supply enough moisture during fruiting stage of the mushroom, thus reducing the mortality of young mushroom yield due to the water shortage [12]. Therefore, our results showed that biochar can be a potential bio-fertilizer to enhance the mycelium growth and mushroom yield.

Table 4. Mycelium growth rate on weekly basis.

Time (day)	Mycelium Growth (Height of mycelium colonized the baglog)			
	Control	**Mixture 1**	**Mixture 2**	**Mixture 3**
0	Innoculum on the surface	Innoculum on the surface	Innoculum on the surface	Innoculum on the surface
7	Mycelium fills 25% baglog	Mycelium fills 25% baglog	Mycelium fills almost 50% baglog	Mycelium fills 25% baglog
14	Mycelium fills > 25% baglog	Mycelium fills 50% baglog	Mycelium fills 50% baglog	Mycelium fills almost 50% baglog
21	Mycelium fills 50% baglog	Mycelium fills 75% baglog	Mycelium fills whole baglog	Mycelium fills 75% baglog
28	Mycelium fills 75% baglog	Mycelium fills whole baglog	Mushroom start to grow	Mycelium fills > 75% baglog
35	Mycelium fills whole baglog	Mushroom start to grow	Mushroom start to grow	Mycelium fills whole baglog

Table 5. Weight of Oyster mushroom harvested in one month.

Baglog	**Harvest (g/month)**			**Average (g/month)**
	Set 1	**Set 2**	**Set 3**	
Control	199.8	205.4	290.8	232.0
Mixture 1	466.3	461.8	513.1	480.4
Mixture 2	563.5	551.7	534.8	550.0
Mixture 3	449.6	498.3	404.8	450.9

4 Conclusion

The high lignocellulose content and carbon content of WPS indicates its suitability as a precursor for production of biochar. Microwave vacuum pyrolysis converted WPS into a biochar with high surface area and pore volume which contributed a strong adsorption capacity on water and nutrient uptake. The biochar based bio-fertiliser enhanced the mycelium growth and the yield of oyster mushroom. Mixture 2 with 20 g of biochar showed the fastest mycelium growth (21 days to fully filled the baglog) and produced highest yield of Oyster mushroom (550 g/month) for the first harvest after 1 month. This indicates the huge potential of the biochar to be used as a bio-fertilizer for Oyster mushroom cultivation.

References

1. MPOB, Production of Crude Palm Oil for the month of December 2016, (2016)
2. M.F. Awalludin, O. Sulaiman, R. Hashim, W.N.A.W. Nadhari, Renew. Sust. Energ. Rev. **50** (2015)
3. N.D. Van Beilen, Inquiries Jour. **8** (2016)
4. I.B. Initiative, Environmental Benefits of Biochar (2017)

5. J. Li, J. Dai, G. Liu, H. Zhang, Z. Gao, J. Fu, Y. He, Y. Huang, Biomass Bioenerg. **94** (2016)
6. A.N.M. Rashidi, T.A. Yang, IJAST, Engineering and Information Technology, **6** (2016)
7. B.A. Wani, R. Bodha, A. Wani, J. Med. Plants Res. **4** (2010)
8. M.Z.M. Amin, A. Harun, M. Wahab, ETMR, **9b** (2014)
9. S.S. Lam, R.K. Liew, X.Y. Lim, F.N. Ani, A. Jusoh, Int. Biodeter. Biodegr. **113** (2016)
10. Y. Sibel, C.Y. U¨ Mit, D.G. Engin, T. Ali, Process Biochemistr. **38**, 301–306 (2002)
11. L. Marlina, S. Sukotjo, S. Marsudi, Procedia Chem. **16** (2015)
12. W. Yang, F. Guo, Z. Wan, Saudi J. Biol. Sci. **20** (2013)

Influence of nanofluids on the efficiency of Flat-Plate Solar Collectors (FPSC)

Marjan B. Nejad[1], *H. A.* Mohammed[2,*], *O.* Sadeghi[1], *Swar* A. Zubeer[2]

[1]Department of Thermofluids, Faculty of Mechanical Engineering, Universiti Teknologi Malaysia, 81310 UTM Skudai, Johor Bahru, Malaysia
[2]Department of Energy Engineering, Technical College of Engineering, Duhok Polytechnic University (DPU), 61 Zakho Road, 1006 Mazi Qr, Duhok-Kurdistan Region, Iraq

Abstract. A numerical investigation is performed using finite volume method to study the laminar heat transfer in a three-dimensional flat-plate solar collector using different nanofluids as working fluids. Three nanofluids with different types of nanoparticles (Ag, MWCNT and Al_2O_3 dispersed in water) with 1–2 wt% volume fractions are analyzed. A constant heat flux, equivalent to solar radiation absorbed by the collector, is applied at the top surface of the absorber plate. In this study, several parameters including boundary conditions (different volume flow rates, different fluid inlet temperatures and different solar irradiance at Skudai, Malaysia), different types of nanoparticles, and different solar collector tilt angles are investigated to identify their effects on the heat transfer performance of FPSC. The numerical results reveal that the three types of nanofluid enhance the thermal performance of solar collector compared to pure water and FPSC with Ag nanofluid has the best thermal performance enhancement. For all the cases, the collector efficiency increased with the increase of volume flow rate while fluid outlet temperature decreased. It is found that FPSC with tilt angle of 10° and fluid inlet temperature of 301.15 K has the best thermal performance.

1 Introduction

Solar energy is one of the most environmentally friendly and beneficial forms of renewable and sustainable energy. The application of solar heating systems has been developed in the past few decades in order to benefit from sustainability, reasonable initial cost and simple construction. Solar collectors are one of the most significant components of these structures. A solar collector can be considered as one of the specific type of heat exchanger that converts solar radiation or insolation into heat. Among diverse kinds of solar thermal collectors, flat-plate solar collectors have been mostly utilized. This is due to the fact that they have relatively lower cost and simpler mechanical structure [1].

Several numerical, analytical and experimental investigations on the thermal performance of flat-plate solar collector have been carried out by many authors. Dovic and

* Corresponding author: hussein.dash@yahoo.com

Andrassy [2] carried out two-dimensional and three-dimensional numerical simulations of flat-plate solar water collectors with and without tubes and a corrugated plate solar collector. It was seen that the corrugated absorber plates of chevron type and the solar collectors without tubes have better thermal efficiency and performance. Jafarkazemi and Ahmadifard [3] theoretically investigated the effects of fluid flow rate, temperature, type of working fluid and thickness of back insulation on exergy and energy efficiency of flat-plate solar collector. It was concluded that using water with inlet temperature approximately 40°C more than ambient temperature and a lower mass flow rate enhances the overall efficiency of flat-plate solar collector. Eismann and Prasser [4] evaluated the absorber edge effect in 2D numerical simulations of a flat-plate solar collector. It was found that edge effects must be significantly taken into consideration in order to derive more precise collector efficiency and heat loss coefficients as well as material costs. Karanth et al. [5], Manjunath et al. [6] and Selmi et al. [7] carried out a 3D numerical investigation of a flat-plate solar collector using Discrete Transfer Radiation (DTRM) with the help of Computational Fluid Dynamics (CFD) simulation method. It was seen that the water and absorber plate temperatures along flow length decreases with the increase of flow velocity and show similar linear trends for all the chosen velocities.

Dagdougui et al. [8] numerically and theoretically conducted thermal analysis and performance optimization of a flat-plate solar collector (FPSC). They observed that the maximum outlet water temperature decreased with the increase of water flow rate whereas the collector efficiency showed an opposite behavior. Alvarez et al. [9] and Fan et al. [10] presented experimentally, numerically and analytically thermal characteristics of conventional fin-and-tube solar collector and corrugated flow channel type of solar collector. It was observed that the efficiency of flat-plate solar collector with serpentine ducts was higher than that of parallel ducts. Tagliafico et al. [11] and Cruz-Peragon et al. [12] conducted a comprehensive review on the thermal models, CFD analysis and characterization of flat-plate solar collectors respectively. Other researchers [13, 14] studied the heat transfer and thermal characteristics of flat-plate solar collectors.

Inevitably, flat-plate solar water collectors have relatively low thermal efficiency and outlet temperatures. One of the practical methods for increasing solar collector efficiency is substituting the absorbing fluid, water, by high thermal conductivity fluids called nanofluids. Nanofluids are suspension of metallic or nonmetallic nanometer-sized material (nanoparticle, 1–100 nm) in a base fluid such as water [15]. Nanofluids have superior thermophysical properties in comparison to conventional fluids. To illustrate, they have higher viscosity, thermal conductivity and convective heat transfer coefficients [16]. Faizal et al. [17] and Alim et al. [18] theoretically investigated the energy, economic and environmental impacts of using metal oxides nanofluid for a flat-plate solar collector. They show that CuO nanofluid contributes to higher solar collector efficiency and area reduction. They demonstrated that 4.34% reduction of entropy generation, 22.15% enhancement of heat transfer coefficient and 1.58% pumping power penalty when using CuO nanofluid compared to that of water. Yousefi et al. [19-21] carried out three different experimental analyses using MWCNT-H_2O and Al_2O_3-H_2O nanofluids in a flat-plate solar collector. They pointed out that he efficiency of flat-plate solar collector increased by 28.3% using 0.2 wt% Al_2O_3 and 15.63% using 0.4 wt% Al_2O_3 with surfactant compared to water. Experimental analyses on performance of a flat-plate solar collector using different types of nanofluid were carried out by Polvongsri and Kiatsiriroat [22], Said et al. [23] and Chaji et al. [24]. It was concluded that the effects of nanofluids with low volume fractions were insignificant on the pumping power and pressure drop of a solar collector. Nasrin and Alim [25] numerically examined the effects of using Al_2O_3 and Cu nanofluid in a solar collector having the flat-plate cover and sinusoidal corrugated absorber.

It can be noticed from the comprehensive literature review that there are no numerical studies using computational fluid daynamics (CFD) on the thermal efficiency of FPSC using nanofluids as working fluids. Hence, this numerical study will focus on the thermal performance enhancement of flat-plate solar collector when applying three different nanofluids, namely Al_2O_3, MWCNT and Ag nanofluid at different volume fractions of 1–2% and d_{np}= 20 nm by means of CFD analysis and real meteorological data taken in Skudai, Malaysia.

2 Theoretical analysis of FPSC

2.1 Energy analysis

In steady state simulation, the actual useful energy output of a collector gain by the working fluid is calculated using:

$$Q_u = \dot{m}C_p(T_{fo} - T_{fi}) \tag{1}$$

where $\dot{m}$ is the inlet mass flow rate, C_p is the heat capacity of the fluid, T_{fo} is the fluid outlet temperature and T_{fi} is the fluid inlet temperature.

The inlet mass flow rate of the working fluid can be determined by using:

$$\dot{m} = \rho\dot{V} \tag{2}$$

The useful energy gain by the working fluid also can be expressed in terms of heat removal factor, F_R, absorbed solar radiation, S, and collecor overall loss coefficient, U_L, to exhibit the effect of optical properties of the collector and heat losses. Hence, the following equations are used for the useful energy and collector efficiency [26]:

$$Q_u = A_c F_R[S - U_L(T_{fi} - T_a)] \tag{3}$$

$$S = G_T(\tau\alpha) \tag{4}$$

$$F_R = \frac{\dot{m}C_p}{A_c U_L}\left[1 - exp\left(-\frac{A_c U_L F'}{\dot{m}C_p}\right)\right] \tag{5}$$

$$F' = \frac{1/U_L}{W\left[\frac{1}{U_L[D+(W-D)F]} + \frac{1}{C_b} + \frac{1}{\pi D_i h_{fi}}\right]} \tag{6}$$

$$F = \frac{tanh[m(W-D)/2]}{m(W-D)/2} \tag{7}$$

$$m = \sqrt{\frac{U_L}{k\delta_p}} \tag{8}$$

where A_c is the surface area of the solar collector, F' is the collector efficiency factor and F is the standard fin efficiency. $\tau\alpha$, is transmittance-absorptance product (optical properties of the collector), which is assumed to be 0.87. T_a is the ambient temperature of Skudai, which is 301.15 K, on October 15, 2016 and G_T is the total incident solar irradiance on the tilted surface, which is calculated in Skudai (latitude is 1.5333°N and longitude is 103.6667 °E) on October 15 at different times of the day, based on the following correlations [26]:

$$G_T = G_b R_b + G_d\left(\frac{1+\cos\beta}{2}\right) + G\rho_g\left(\frac{1-\cos\beta}{2}\right) \tag{9}$$

where G_b is the beam solar irradiance, G_d is the diffuse solar irradiance, G is the direct normal solar irradiance (total solar irradiance) on a horizontal surface ($G = G_b + G_d$), R_b is

DOI: 10.1051/e3sconf/20172200123

the geometric factor (ratio of beam radiation on the tilted surface to that on a horizontal surface), β is the collector tilt angle and ρ_g is the ground reflectance, which is assumed to be 0.4. The geometric factor, R_b, is calculated from the following equation:

$$R_b = \frac{\cos(\phi-\beta)\cos\delta\cos\omega+\sin(\phi-\beta)\sin\delta}{\cos\phi\cos\delta\cos\omega+\sin\phi\sin\delta} \quad (10)$$

where ϕ is latitude, δ is declination (see [26], Table 1.6.1), calculated from approximate equations [27–29] and ω is hour angle calculated from 15° spin of the earth multiplied by number of hours from/to solar noon. During morning, ω has negative value and afternoon it has positive value.

The instantaneous solar collector efficiency (η_i) is defined as the ratio of the useful energy gain to the total incident solar irradiance on the tilted collector surface:

$$\eta_i = \frac{Q_u}{A_c G_T} = \frac{\dot{m}C_p(T_{fo}-T_{fi})}{A_c G_T} \quad (11)$$

$$\eta_i = F_R(\tau\alpha) - F_R U_L \frac{T_{fi}-T_a}{G_T} \quad (12)$$

According to [19–21], $F_R(\tau\alpha)$ and $F_R U_L$ are called absorbed energy parameter and removed energy parameter respectively.

Convective heat transfer coefficient of the working fluid in the solar collector can be examined by:

$$h = \frac{q''}{(T_w-T_f)} \quad (13)$$

where q'' is actual heat flux derived from useful heat gain, T_w is the tube average wall temperature and T_f is average fluid temperature.

2.2. Pressure drop and pumping power

The constant fluid flow for flat-plate solar collector system is provided with the help of forced convection by electrical pump. The pressure drop of the solar collector is calculated from the following Darcy equation [30, 31]:

$$\Delta P = f\frac{\rho V^2}{2}\frac{\Delta l}{D_i} + K\frac{\rho V^2}{2} \quad (14)$$

where K is the loss coefficienct owing to entrance and exit effects, bends, elbows, valves, etc and ρ is the density of the fluid, f is the friction factor and V is the average flow velocity of the fluid (m/s) which are determined from the following equations [30, 31]:

$$f = \frac{64}{Re},\quad Re = \frac{\rho V D_i}{\mu} = \frac{4\dot{m}}{\pi D_i \mu},\quad V = \frac{\dot{m}}{\rho\pi D_i^2/4} \quad (15)$$

The pumping power is derived from the following equation:

$$pumping\ power = \left(\frac{\dot{m}}{\rho}\right) \times \Delta P \quad (16)$$

3 Numerical modeling

3.1 Physical model

The schematic diagram of a flat-plate solar collector, having 6 horizontally inclined absorber (riser) tubes is shown in Fig. 1. In this study, it is assumed that the flow is distributed uniformly through 6 tubes by headers. Moreover, the header tubes are neglected due to the assumptions that they contain a small area of the solar collector [26]. Hence, the numerical model consists of one of the absorber tubes with its corresponding absorber plate (absorber fin). The cross-section of the numerical model is depicted in Fig. 2. The diameter and thickness of the tube are 0.01 m and 0.0009 m. The width of the absorber plate is taken as 0.14 m. The lengths of absorber plate and tube are considered as 1.5 m.

Mass-flow-inlet boundary is used for inlet flow of the tube and the outlet flow is determined by pressure-outlet boundary. The solid (absorber plate) and fluid domains are bounded with the help of wall boundary conditions. The conjugate heat transfer from the tube to working fluid derives from interface wall between absorber tube and fluid with coupled thermal condition. A constant heat flux is applied at the top surface of the absorber plate. The heat flux is equivalent to solar radiation absorbed by the collector. The walls of the absorber plate (side surfaces) are assumed to be insulated. Bottom surface of the absorber plate is assumed to have free convection with h = 16 W/m^2·K and laminar flow is considered for the CFD model. Copper is selected for both absorber plate and tube material and all the thermal-physical properties are assumed constant.

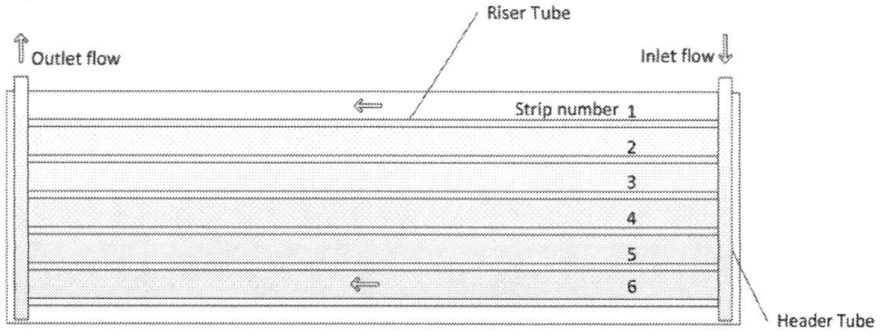

Fig. 1. Schematic diagram of the physical model.

The 3D steady state, incompressible and laminar flow governing equations of continuity, Navier-Stokes (momentum equation) and energy are considered for the case under study and are taken from Patankar [32] and its not written here because of lack of space.

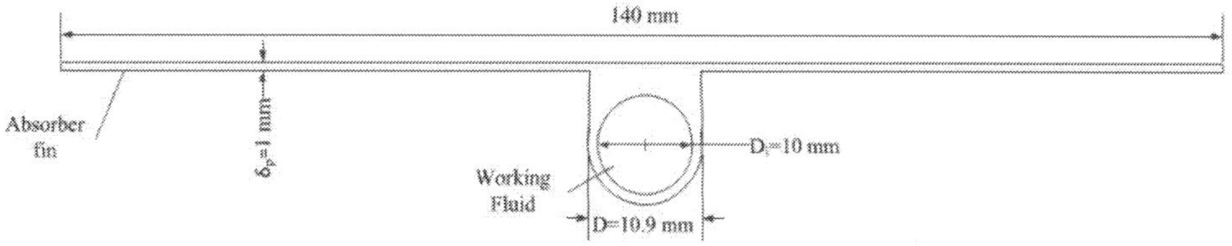

Fig. 2. Cross section of a fin-and-tube collector.

DOI: 10.1051/e3sconf/20172200123

3.2 Grid testing and model validation

The grid independent test was performed for a fin-and-tube collector using water as a working fluid with volume flow of 0.04 L/min to ensure that the solution does not depend on the grid size and quality. As shown in Fig. 3a, if the number of elements for the mesh increases from 2178677 to 4284555, the average water outlet temperature will not change significantly. Thus, the number of 2178677 elements for the non-uniform grid has been adopted for the computational analysis. Grid layout for the present numerical model is depicted in Fig. 3b.

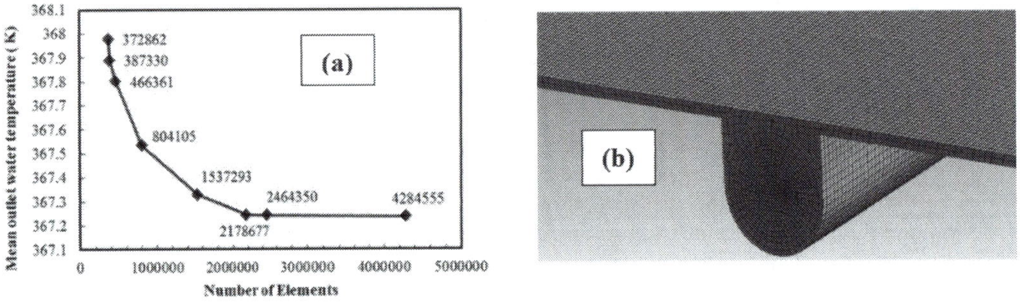

Fig. 3. (a) Grid Test for a fin-and-tube collector, (b) grid layout used in the present work.

The present numerical solution was validated by comparing the temperature distribution of absorber plate and water with the results of Manjunath et al. [6]. The validation results show a very good agreement between the two investigations as shown in Fig. 4a-b. The pressure drop throughout the solar collector obtained from the present numerical study with the correlations of Darcy [31] was also validated as shown in Fig. 4c. It is clearly seen that there is a good agreement between the numerical simulation and the analytical Darcy equation.

The thermophysical properties of nanofluids (density, heat capacity, thermal conductivity and viscosity are calculated based on the correlations given in these References [33–36].

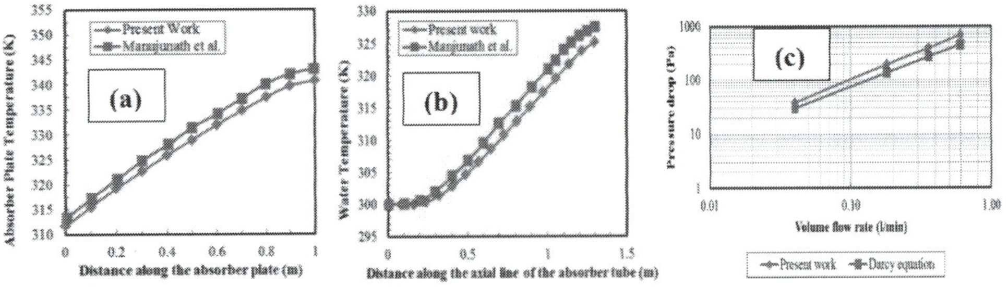

Fig. 4. Comparison of the present results with the results of Manjunath et al. [6], temperature distribution of (a) absorber plate, (b) water and (c) Pressure drop of FPSC.

DOI: 10.1051/e3sconf/20172200123

4 Results and discussion

4.1 The effect of different tilt angles of solar collector

The simulation was done using Al_2O_3 nanofluid with 1 wt% volume fraction. Fig. 5a and Fig. 5b present the outlet fluid and absorber plate average temperatures versus the inlet volume flow rate of the fluid for different tilt angles of solar collector respectively. Clearly, when the tilt angle of solar collector increases from 10° to 70°, the temperatures decrease. This is due to the fact that Malaysia is located near the equator and has a sunny climate; hence, its geographical coordinates lead sun to be perpendicular to the plane of the solar collector with almost small tilt angle. Consequently, the solar collector having tilt angle of 10° appears to gain more useful energy as depicted in Fig. 5c. As shown in Fig. 5d, the solar collector efficiency does not change significantly with respect to tilt angle since the efficiency is defined as the ratio of the useful energy gain to the total incident solar irradiance on the tilted collector surface. Obviously, both the useful heat gain and solar irradiance increase with the decrease of collector tilt angle; thus, the ratio remains unchanged. Furthermore, it is apparent that the average temperatures decrease with the increase of volume flow rate while the useful heat gain and collector efficiency increase.

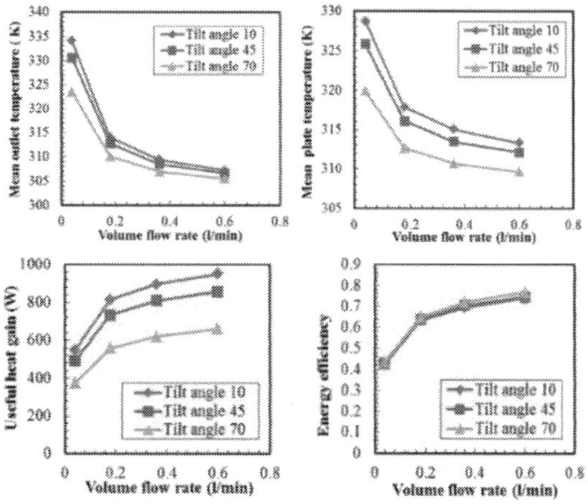

Fig. 5. Variation of (a) mean outlet fluid, (b) plate temperature, (c) useful heat gain and (d) energy efficiency with volume flow rate for different tilt angles of solar collector.

4.2 The effect of different types of nanoparticles

The influences of implementing three different nanoparticles, namely silver (Ag), MWCNT and Al_2O_3 dispersed in water as a base fluid with volume fractions of 1% on the thermal performance of FPSC are discussed. Fig. 6a and Fig. 6b show the outlet fluid and absorber plate average temperatures respectively, using different nanofluids and water as working fluids. The figure clearly shows that when using nanofluids rather than water as working fluids, the outlet and plate temperatures increase. Higher density and thermal conductivity and lower specific heat of nanofluids compared to that of water contribute to higher outlet temperature which in turn leads to an increment in useful heat gain and thermal efficiency as depicted in Fig. 6c and Fig. 6d. Due to the fact that Ag nanofluid has the lowest specific heat and highest density compared to other two nanofluids, it has the highest values for the temperatures and collector efficiency.

 DOI: 10.1051/e3sconf/20172200123

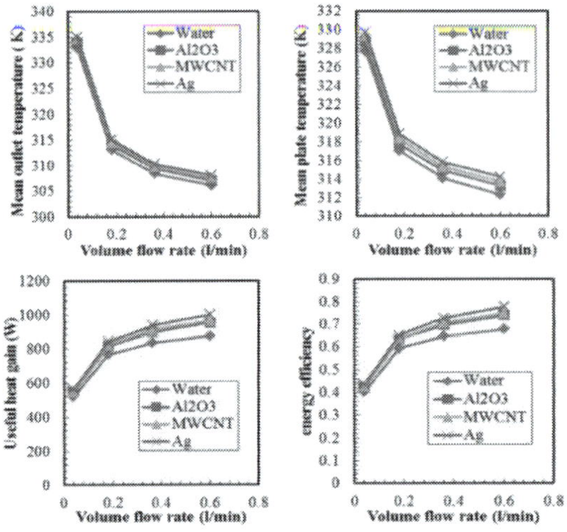

Fig. 6. Variation of (a) mean outlet fluid, (b) plate temperature, (c) useful heat gain and (d) energy efficiency with volume flow rate for different nanofluids and water.

Fig.7a shows the convective heat transfer coefficient (h) of different nanofluids and water. It is noticed that heat transfer coefficient is augmented compared with that of water. Consequently, nanofluids, which have higher thermal conductivity, gain more heat from the absorber plate; results in reduction of thermal boundary layer and hence an increase in heat transfer coefficient. Similarly, Ag nanofluid has the highest value of (h). The figure clearly reveals that (h) is increased with the increase of volume flow rate. As shown in Fig. 7b and Fig. 7c, small increments in pressure drop and pumping power are resulted when using nanofluids compared to that of water. Furthermore, increasing fluid inlet volume flow rate leads to an increase in pressure drop and consequently to an increase in pumping power.

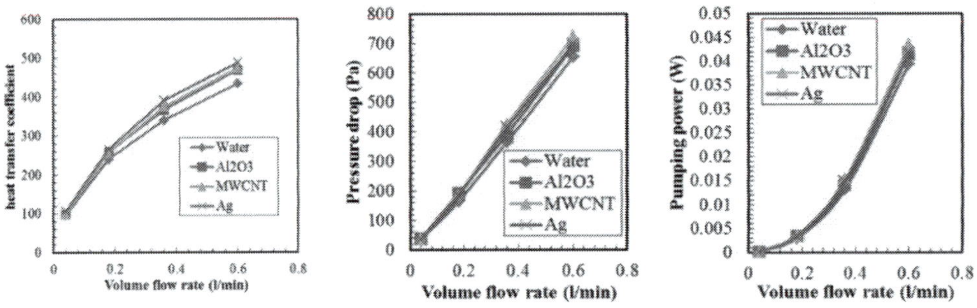

Fig. 7. Variation of (a) heat transfer coefficient, (b) pressure drop and (c) pumping power with volume flow rate for different nanofluids and water.

4.3 The effect of different solar Irradiance at Skudai, on October 15

The simulation was done using Ag nanofluid with 2 wt% volume fraction and collector tilt angle of 10°. Fig. 8a and Fig. 8b show the outlet fluid and absorber plate average temperatures versus the inlet volume flow rate of the fluid for different times of the day and absorbed solar radiations (*S*) respectively. Apparently, at 1 PM with the highest (*S*), the maximum temperatures are generated while the minimum ones are generated at 8 AM with the lowest (*S*). At 9 AM and 5 PM, the values of temperatures are almost equal owing to the same values of (*S*). The similar trend is observed for useful heat gain as shown in Fig. 8c and Fig. 8d. The solar collector efficiency does not change significantly at different times

 DOI: 10.1051/e3sconf/20172200123

of the day. Due to the fact that both useful heat gain and solar irradiance change at different times of the day, the ratio virtually remains unchanged.

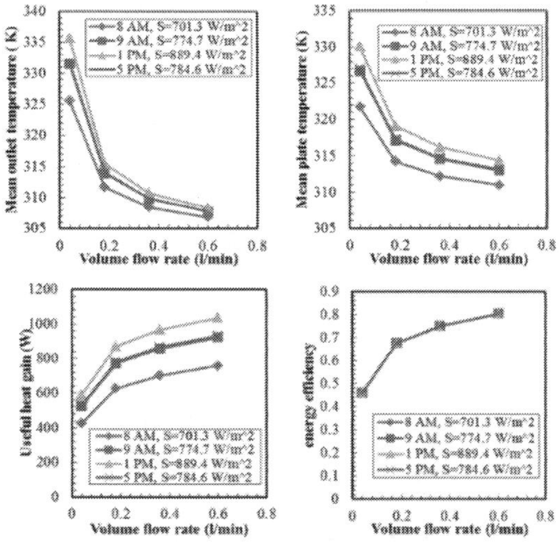

Fig. 8. Variation of (a) mean outlet fluid, (b) plate temperature, (c) useful heat gain and (d) energy efficiency with volume flow rate at different times of October 15, 2016 and absorbed solar radiation.

5 Conclusions

From the present numerical study on the influence of different nanofluids on the efficiency of flat-plate solar collectors, it is observed that the FPSC with tilt angle of 10° shows the best thermal performance. It is also shown that with increasing inlet volume flow rate, absorber plate and fluid outlet temperature decrease while the useful heat gain, solar collector efficiency and heat transfer coefficient increase. Moreover, the pressure drop and hence pumping power increase with the increase of volume flow rate. It is concluded that the three types of nanofluid enhance the thermal performance of solar collector compared to pure water. Apparently, FPSC with Ag nanofluid shows the best thermal performance enhancement. The efficiency of solar collector using Ag nanofluid with inlet volume flow rate of 0.6 L/min increased by 17.65% compared to that of water. A small increment in pressure drop is observed when using nanofluids. The highest solar irradiation and therefore the best thermal performance is observed at 1 PM in Skudai, Malaysia.

References

1. K. Solangi, M. Islam, R. Saidur, N. Rahim, Renew. Sust. Energy Revi. **15**, 2149 (2011)
2. D. Dović, M. Andrassy, Solar Energy **86**, 2416 (2012)
3. F. Jafarkazemi, E. Ahmadifard, Renewable Energy **56**, 55 (2013)
4. R. Eismann, H.M. Prasser, Solar Energy **95**, 181 (2013)
5. K.V. Karanth, M. Manjunath, N.Y. Sharma, Proc of World Congress on Eng. **III**, 6 (2011)
6. M. Manjunath, V.K. Karanth, Y.N. Sharma, Lect. Notes in Eng Comp Sci. **2192**, 2245 (2011)
7. M. Selmi, M.J. Al-Khawaja, A. Marafia, Renewable Energy **33**, 383 (2008)

8. H. Dagdougui, A. Ouammi, M. Robba, R. Sacile, Renew. Sust. Energy Revi. **15**, 630 (2011)
9. A. Alvarez, O. Cabeza, M. Muñiz, L. Varela, Energy **35**, 3707 (2010)
10. J. Fan, L.J. Shah, S. Furbo, Solar Energy **81**, 1501 (2007)
11. L.A. Tagliafico, F. Scarpa, M. De Rosa, Renew. Sust. Energy Revi. **30**, 526 (2014)
12. F. Cruz-Peragon, J. Palomar, P. Casanova, Renew. Sust. Energy Revi. **16**, 1709 (2012)
13. O. Turgut, N. Onur. Int. Commu. Heat Mass Transfer **36**, 274 (2009)
14. N. Akhtar, S. Mullick, Solar Energy **66**, 349 (1999)
15. S. Choi, ASME-Publications-Fed **231**, 99 (1995)
16. K.V. Wong, O. De Leon, Advances in Mechanical Engineering **20**, 240 (2010)
17. M. Faizal, R. Saidur, S. Mekhilef, M. Alim, Energ. Conver. Manag. **76**, 162 (2013)
18. M. Alim, Z. Abdin, R. Saidur, A. Hepbasli, Energy and Buildings **66**, 289 (2013)
19. T. Yousefi, E. Shojaeizadeh, F. Veysi, S. Zinadini, Solar Energy **86**, 771 (2012)
20. T. Yousefi, F. Veisy, E. Shojaeizadeh, S. Zinadini, Exp. Therm. Flui. Sci. **39**, 207 (2012)
21. T. Yousefi, F. Veysi, E. Shojaeizadeh, S. Zinadini, Renewable Energy **39**, 293 (2012)
22. S. Polvongsri, T. Kiatsiriroat, Heat Transf. Eng. **12**, 250 (2013)
23. Z. Said, M. Sajid, R. Saidur, N. Rahim, Int. Commu. Heat Mass Trans. **48**, 99 (2013)
24. H. Chaji, Y. Ajabshirchi, E. Esmaeilzadeh, M. Kahani, Modern Applied Science **7**, 60 (2013)
25. R. Nasrin, M. Alim, Heat Transfer-Asian Research **43**, 61 (2014)
26. J.A. Duffie, W.A. Beckman, Solar Engineering of Thermal Processes (2006)
27. P. Cooper, Solar Energy **12**, 333 (1969)
28. J. Spencer, Search **2**, 172 (1971)
29. M. Iqbal, *An introduction to solar radiation* (Elsevier, 1983)
30. H. Garg, R. Agarwal, Energ. Conver. Manag. **36**, 87 (1995)
31. F. White, *Fluid Mechanics* (5th edt., 2003)
32. S.V. Patankar, D.B. Spalding, Int. J. of Heat Mass Transfer **15**, 1787 (1972)
33. Y. Xuan, W. Roetzel, Int. J. Heat Mass Transfer **43**, 3701 (2000)
34. S.Q. Zhou, R. Ni, Applied Physics Letters **92**, 093123 (2008)
35. S.M. Vanaki, H.A. Mohammed, A. Abdollahi, M. Wahid, J. Mol. Liq. **196**, 32 (2014)
36. H.A. Mohammed, A.A. Al-aswadi, H.I. Abu-Mulaweh, Heat Trans-Asian Res. **40**, 287 (2011)

 DOI: 10.1051/e3sconf/20172200124

The influence of using heat storage with PCM on inlet and outlet temperatures in substation in DHS

Kinga Nogaj[1,*], *Michał* Turski[1], and *Robert* Sekret[1]

[1]Faculty of Infrastructure and Environment, Czestochowa University of Technology, ul. Brzeznicka 60A, 42-200 Czestochowa, Poland

Abstract: The main objective of this article is to indicate the direction of development of new generation heating systems that use phase change materials, and the important criteria needed when choosing a phase change material. The work contains a detailed classification of materials using the latent heat of organic and inorganic PCM. This references the technical possibilities of existing heat storage technologies. A specific objective was adopted to determine the effect of using heat storage with PCM on inlet and outlet temperatures in substation in district heating systems. The scope of the study included determining the parameters of the heat distribution network as a function of an outdoor air temperature within the range of -20°C to + 12°C. The object of analysis was chosen to be the heating system parameters: supply 120°C and return 60°C. It is located on the surface of 160km^2, and supplies heat to 240,000 residents. The total length of the district heating network is 170 km. Based on the study, it was found that the most advantageous material that accumulates heat depends on the return temperature in the heating network. For the above analyzed case, the return temperature was in the range of 46°C to 57°C. The analysis showed that the most preferred materials using heat of phase change, have possible applications in heating networks and received a return temperature including salt hydrates, such as $MgSO_4 \cdot 7H_2O$ and $Na_2S_2O_3 \cdot 5H_2O$. The introduction of stored heat for the district heating system with the phase change material in the form of salt hydrates, allows the return temperature in the district heating to remain at temperatures compatible with the adopted regulatory table for temperatures outside the standard heating season.

1 Introduction

The current development of district heating systems in light of changing regulations, presented in [1], seeks to increase the efficiency of existing systems, through the use of heat

* Corresponding author: k.nogaj@is.pcz.pl

 DOI: 10.1051/e3sconf/20172200124

accumulators. Heat storage in this type of system can be realized through the use of sensible and latent heat, as shown in [2–4]. The use of sensible heat is usually a solution related to the heat storage source and is accomplished in buildings by means of accumulators stratification filled with water. Increasingly, more common technologies are using latent heat.

Storage technologies evolve continuously, and therefore the accumulation of heat through the material using the heat of phase change appears to be particularly interesting.

Phase change materials are organic and inorganic materials. Organic materials include paraffin, non-paraffin (e.g. fatty acids). Inorganic materials include salt hydrates, salt composites and alloys [5]. The classification of phase change materials is shown in Figure 1.

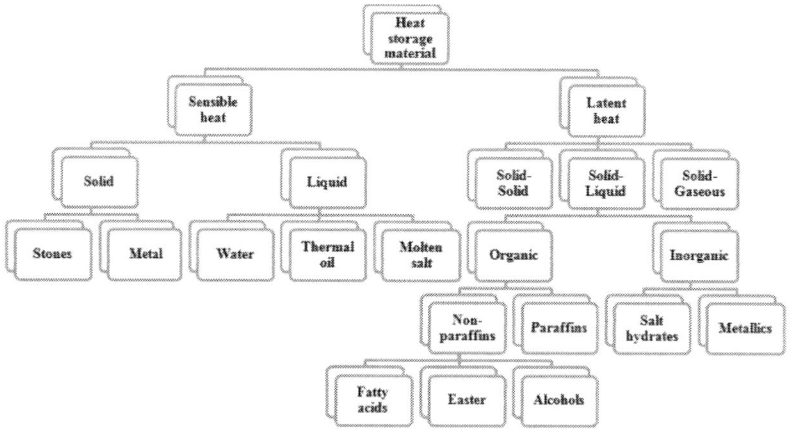

Fig. 1. Classification of phase change materials [2].

The selection of material utilizing the heat of phase change in the heat accumulator is the acceptance of the criteria for evaluating the individual aspects. Phase change materials must have high latent heat and high thermal conductivity. The melting point of materials using latent heat should be in the range of the operating conditions of work [6]. These materials should be subject to melting, minimum supercooling, should be chemically stable, non-toxic and should not cause corrosion. Depending on the application, the use of latent heat should be selected based on the content's melting point [7]. For the purpose of heat accumulation in heating systems, most matching characteristics determine the salt hydrates and paraffins [8].

2 Methodology

The heating system which was adopted for analysis is located in the third climate zone. The adjustment range of the heat distribution network is in the range of outside air temperatures from -20°C to + 12°C. The heating system parameters were chosen as the objects of analysis: supply 120°C and return 60°C. The heating system is located on a surface of 160 km^2, and supplies heat to 240,000 residents. The total length of the district heating network is 170 km. An analysis of the substation was also conducted.

The analyzed substation, is an intermediate node (connected heating). It is single function node, in which hot water is prepared in electric heaters. The analysis was performed in three stages. The first phase was concerned with the designation of the operating parameters in relation to the outside temperature to the choice of material using heat of phase transitions. In the second stage, the possible set of heat to be stored in the heating system was determined. In the third step, the optimum volume of heat storage in order to maintain the return temperature at a constant level corresponding to the regulatory table was determined.

The first stage of analysis determined the return temperature in the district heating network, which allowed for determining the amount of the heat that is not picked up by the end users with respect to the design conditions. The parameters of the heat distribution network has been established in accordance with the regulatory table for temperatures outside in the standard heating season. Then, the obtained data were compared with design values. On the basis of these results the phase change material was chosen for the analysed case.

In the second stage, the calculated computational amount of heat consumption by end users allowed the determination of the amount of heat that was possible to store in a district heating network. Return temperatures in the heating network were the values of the deliverables of heat supplied to end users. The dynamic state of the heating system also made it possible to calculate the heat flux in order to cover the thermal needs of end users, depending on the temperature of the outside air. The heat flux in this analysis, was equal to 1,206.30 kg/s.

In the third step the choice of the optimum volume of the heat storage was made with the estimated amount of heat able to be stored in a substation, which depended on the specific heat of the PCM material in both solid and liquid form, and the enthalpy changes during the phase change.

3 Results and analysis

By joining the selection of phase change material used in heat stores the return temperature of the district heating network should first be determined. Figure 2 shows the return temperature design conditions (blue) for a standard heating season (red). Conditions of the design return temperature in heating networks are based on the outside temperature assuming 100% heat consumption by users. Standard heating season refers to the actual values, which are dependent on consumption of heat related to heat demand in relation to outside temperatures.

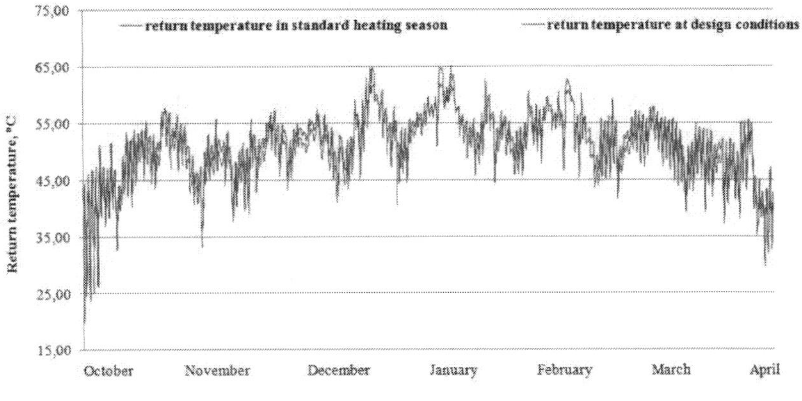

Fig. 2. Summary of return temperature in the district heating network for a standard heating season and the design of the return temperature.

Comparing the values of the two states in the district heating network, it can be seen that the return temperature in both cases (standard heating season – blue, and the design value of the return temperature of the district heating network – red) is in the range of 46°C to 57°C, as shown in figure 3. These temperatures occurred with the greatest frequency in the months between December and March. Other months of the heating season are, characterized by slightly lower return temperatures (below 46°C). The return temperatures for the standard heating season are 3% higher than the design conditions for the system.

Whatever the reason for this estimated discrepancy, it is possible to reduce disparities through the use of distributed heat storage located at the ends of the heating network.

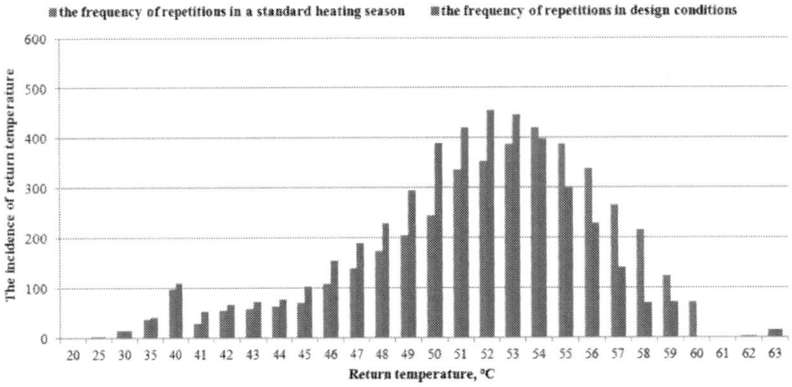

Fig. 3. Figure ordering the incidence of return temperatures in the district heating network.

a. Choosing a phase change material for a representative heating system

Based on the analysis of the frequency of return temperatures in the district heating system, the temperature of phase change was determined and, the material using the heat of phase change requires heat accumulation in the heating system. The analysis shows that the most common return temperature of the district heating network is in the range is of 46°C to 57°C and for that range of temperatures, a material utilizing heat of phase change will be selected.

Table 1 shows the salt hydrates and paraffins that have a carbon number from 23 to 26. All selected materials undergo a phase change within the temperature range of the system's operation and have certain physical parameters chemical and technical. Highlighted are those compounds which, in accordance with the methodology proposed in the paper [8] obtained the highest number of points.

Table 1. The melting point and latent heat of phase change: salt hydrates and paraffin.

Material	Melting point (°C)	Latent heat of fusion (kJ/kg)	Thermal conductivity (W/m*K)	Density (kg/m³)	The resulting points of selected materials (according to the method [8])
$Mg(NO_3)\cdot 4H_2O$	47	142	0,125	552	385,5
$Ca(NO_3)\cdot 4H_2O$	47	153	0,191	540	385,5
$Fe(NO_3)_3\cdot 9H_2O$	47	155	not available	not available	367,5
$Na_2SiO_3\cdot 4H_2O$	48	168	0,265	not available	376,5
$K_2HPO_4\cdot 3H_2O$	48	99	not available	not available	367,5
$Na_2S_2O_3\cdot 5H_2O$	48,5	210	not available	not available	367,5
$MgSO_4\cdot 7H_2O$	48,5	202	0,163	490	385,5
$Ca(NO_3)_2\cdot 3H_2O$	51	104	not available	not available	367,5
$Zn(NO_3)_2\cdot H_2O$	55	68	not available	not available	367,5
$FeCl_3\cdot 2H_2O$	56	90	not available	not available	367,5
$Ni(NO_3)_2\cdot 6H_2O$	57	169	not available	not available	367,5
Paraffin of carbon number of 23	47,5	232	0,21	910	382,5
Paraffin of carbon number of 24	50,6	255	0,21	910	382,5
Paraffin of carbon number of 25	49,4	238	0,21	910	382,5
Paraffin of carbon number of 26	56,3	256	0,21	910	382,5

b. The amount of heat for the storage of the analyzed system

In the standard heating season, the greatest possible amount of heat to be stored (above 20000 GJ/a) was observed in December, January and February, as shown in Figure 4. In the remaining months, the amount of heat able to be stored had various amounts ranging between 3700 GJ/a and 19000 GJ/a. Given the lack of data on full months in October and April (the analysis concerned the heating season, ie. the period from October 10 to April 8) it can be concluded that the highest possible amount of heat to be stored falls during the heating season (December 22.5 thous. GJ/a, January 25.9 thous. GJ/a, February 21.7 thousand. GJ/a), and the lowest during the transitional period (March and November of approximately 19 thous. GJ/a, October 8,8 thous. GJ/a April and a 3.7 thous. GJ/a).

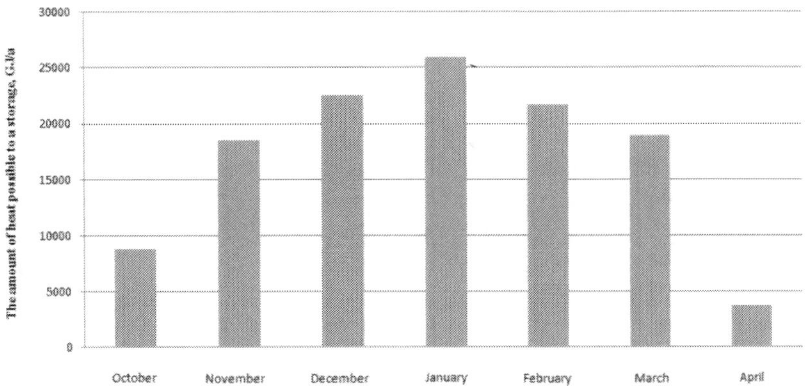

Fig. 4. The amount of heat available for storage in a standard heating season.

c. Selecting the optimum volume of the heat storage accumulator using the phase change material

For the analyzed substation it is estimated that the greatest possible amount of heat to be stored is available when the outside air temperature reaches a value in the range of -2°C to +5°C. There it was observed, that the scope of the return temperature in the system reaches in the range of 46–57°C.

The selection of the optimum volume of the heat storage by using phase change material is shown in Figure 5. The blue color indicates the temperature of the water supply system and the brown color the return water temperature. Also plotted are characteristic amounts of heat available to be stored in the analyzed system, which is shown by a green curve. The course of heat storage with phase change material within the temperature range of the installation in relation to ambient air temperature is marked in red.

When thermal energy storage systems using heat of phase change have the biggest influence on the amount of heat absorbed during the process of charging and discharging there is a temperature change in the system. Until the desired temperature in the system is reached, stored heat uses the heat capacity of water. In a specific temperature range (analyzed case 46°C–57°C, where $\Delta T = 7$ K), the phase change and the proper storage of heat are both taking place. After exceeding the range of the optimum conditions for phase change, heat is again stored using the specific heat of water which is contained in the heat accumulator.

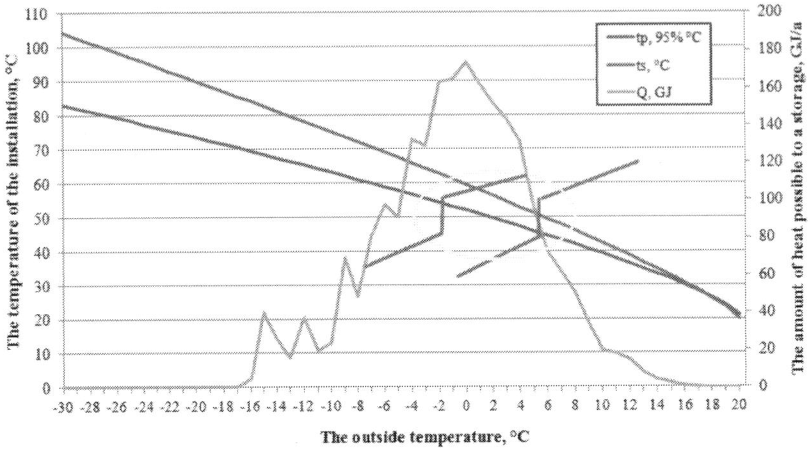

Fig. 5. Selection of the optimum volume of the heat accumulator in the analyzed heat substation.

The advantage of the latent heat above over the sensible heat is that it is easily assessable when comparing the even volume and mass required to store a certain amount of heat. With respect to the unit of weight and volume use of the phase change material stores heat in the heating network can increase the amount of heat possible to be stored by 30%, in relation to aqueous heat storage.

4 Summary

The analysis states, that choosing the most favorable material accumulating heat depends on the return temperature heating network. Therefore, in any case, the choice of material using heat of phase change should be performed individually.

In this analysis, the most preferred materials for the heat storage utilizing latent heat in the heating network are salt hydrates, i.e. $Na_2S_2O_3{\bullet}5H_2O$ and $MgSO_4{\bullet}7H_2O$ and paraffins having a carbon number of 24 and 26.

The analysis also showed that for the analyzed heating system present in significant quantities produced but not received amount of heat (approx. 120 thousand. GJ per heating season), which can be accumulated and used in periods of increased demand for heat. This amount represents approximately 5% of the total heat production.

Heat storage using phase change materials provides a much higher density of storage at a lower temperature difference between the loading and discharge of the heat accumulator.

The use of heat storage with heat of phase change, keeps the return temperature in the system at a required level in accordance with the regulatory table.

References

1. M. Turski, R. Sekret, Ryn. En. **4**, 27–34 (2015)
2. A. Abhat, Sol. En. **30**, 4, 313–332 (1983)
3. P. Feliński, M. Turski, R. Sekret. COW **10**, 416–421 (2013)
4. J. Xu., R.Z. Wang, Y. Li. Sol. En. **103**, 610–6380 (2014)
5. G. Ziskind, Adva. in Ther. En. Stor. Sys. 307–324 (2015)

DOI: 10.1051/e3sconf/20172200124

6. J.K. Basakayi, V.V.C.P. Storm, Internat. Jour. of Adva. in Engin. & Tech. **7**, 3, 692–700 (2014)
7. L.F. Cabeza, I. Martorell, L. Miro, A.I. Fernandez, C. Barreneche, Advan. in Ther. En. Stor. Sys. 1–26 (2015)
8. K. Nogaj, M. Turski, R. Sekret, COW **48**, 3, 91–95 (2017)

DOI: 10.1051/e3sconf/20172200125

Atmospheric CO_2 capture for the artificial photosynthetic system

Adrianna Nogalska[1,*], *Adrianna* Zukowska[2] , and *Ricard* Garcia-Valls[1]

[1]Universitat Rovira i Virgili, Av. dels Països Catalans 26, 43007 Tarragona, Spain
[2]Uniwersytet im. Adama Mickiewicza, Chemistry Faculty, ul. Umultowska 89B, 61-614 Poznań, Poland

Abstract. The scope of these studies is to evaluate the ambient CO_2 capture abilities of the membrane contactor system in the same conditions as leaves works during photosynthesis, such as ambient temperature, pressure and low CO_2 concentration, where the only driving force is the concentration gradient. The polysulfone membrane was made by phase inversion process and characterized by ESEM micrographs which were used to determine the thickness, asymmetry and pore size. Besides, the porosity of the membrane was measured from the membrane and polysulfone density correlation and hydrophobicity was analyzed by contact angle measurements. Moreover, the compatibility of the membrane and absorbent solution was evaluated, in order to exclude wetting issues. The prepared membranes were introduced in a cross flow module and used as contactor between the CO_2 and the potassium hydroxide solution, as absorbing media. The influence of the membrane thickness, absorbent stirring rate and absorption time, on CO_2 capture were evaluated. The results show that the efficiency of our CO_2 capture system is similar to stomatal carbon dioxide assimilation rate.

1 Introduction

Growth of carbon dioxide concentration in Earth's atmosphere is one of the main issues troubling the modern world. Since the Industrial Revolution started, global average level of CO_2 concentration rose from 280 ppm to 404 ppm (May, 2016) and it is the highest recorded in last 650,000 years [1]. Fossil fuel combustion emission, which is the main cause of the carbon dioxide atmospheric growth, is increasing year by year [2]. These factors are forcing us to invent new technologies and develop ones already existing to reduce global warming effects. The best inspiration for scientists is always nature. Our planet was taking care of the balance of CO_2 since billions of years by, among others, photosynthesis.

The approach of the present work is to capture CO_2 from atmosphere, such as photosynthetic organisms do. The system works as a leaf, where CO_2 is captured directly from the air through the membrane pores and is passed to the next compartments to be finally converted to methanol or hydrocarbons and used as fuel in fuel cells.

* Corresponding author: adrianna.nogalska@urv.cat

 DOI: 10.1051/e3sconf/20172200125

The first step, which is CO_2 fixation, determines the speed of the entire process. CO_2 assimilation rate in stomata reaches between 15–40 μmol/m^2*s [3–6]. Stomata are small adjustable pores situated on the leaf surface. The central role of stomata is regulating gas exchange between the inside of the leaf and the external environment, e.g. during photosynthesis it allows CO_2 to diffuse into the leaf. The system designed for carbon dioxide capture process mimics stomata function. It consists on a porous polysulfone membrane contactor which uses a basic solution to absorb carbon dioxide in form of carbonate or bicarbonate. As CO_3^{2-} and HCO_3^- will be further converted to methanol by electro reduction, there is no need for a recovery of the absorbent. The membrane contactor is permeable for the gas and provides high contact surface area. Recently they started to be used for carbon dioxide capture as a hybrid of membrane separator and chemical absorption.

In the membrane contactors, the porous membrane separates gas and liquid phase and the absorption process is driven by pressure or concentration gradient. Mass transfer process includes following steps: convective-diffusive transport of solute from the bulk gas to the membrane surface, transfer through the membrane pores to the liquid membrane surface and a convective-diffusive transfer of a solute to the bulk liquid followed by the physical or chemical absorption. The overall mass transfer coefficient consists of individual resistances. The resistances are: the gaseous phase boundary layer ($1/k_G$), the membrane ($1/k_m$) and the liquid phase boundary layer ($1/k_L$). The overall process is well-known as a "resistance in series" model and overall mass transfer resistance ($1/K_o$) can be obtained by adding the individual resistances [7].

$$\frac{1}{K_0} = \frac{1}{k_G} + \frac{1}{k_m} + \frac{1}{k_l} \qquad (1)$$

A precise knowledge of mass transfer coefficient is essential for simulation, industrial design and use of membrane contactors. Simplified equation for its theoretical value is:

$$k_m = \frac{D * \varepsilon}{\tau * z} \qquad (2)$$

In this equation D stands for effective diffusion coefficient of CO_2, ε is the membrane porosity, z is membrane thickness and τ is a tortuosity coefficient. It is essential to know that this is simplified equation and deviations from reality are highly possible to appear [8].

A great amount of polymers have been studied as materials for membranes in CO_2 capture studies including e.g. polyacetylenes [9], polyaniline [10], poly(ethylene oxide) [11], polyolefin [12], polypropylene [13] and others [14]. Polysufone, chosen in the study, is a thermally and chemically stable polymer, easy to handle. Besides, its use in membrane preparation process results in production of membranes with specific morphology, which is easy to control by changing production conditions [15].

Previous studies on polysulfone membrane contactors for ambient CO_2 capture were performed in the system equipped with peristaltic pump in order to decrease the absorbent mass transfer resistance by increase of the liquid flow rate. In the current studies experiments ware carry out in static conditions, where the liquid flow is zero, to evaluate the impact of the pump. Membranes with the same morphology with different thickness, macrovoids size and the hydrophobicity were used in the research. Besides the influence of the membrane characteristics, absorbent liquid resistance and absorption time were analysed.

2 Materials and methods

2.1 Materials

Polysulfone (SIGMA-ALDRICH) in form of pellets was used for membrane preparation with NMP $\geq$ 99,0% (SIGMA-ALDRICH). Potassium Hydroxide (85%, pellets) (SIGMA-ALDRICH) were used as absorption liquid for CO_2.

2.2 Flat sheet membranes preparation

Polysulfone membranes were made via phase inversion method. The preparation conditions are summarized in Table 1. Polymeric solution used for the fabrication consisted of 20% wt. polysulfone dissolved in 1-Methyl-2-pyrrolidone during 48 h stirring. Resulting solution was cast on a glass support by using a casting knife of particular thickness and immediately immersed into water, where membranes precipitate. Prepared flat sheet membranes were dried at ambient conditions. As the only changed factor in the process is thickness, further membranes names are associated with the size of the casting knife used in the preparation, as fallows: membrane 250, membrane 200 and membrane 100.

Table 1. Membrane preparation process details.

Process	Immersion precipitation
Polymeric solution	20% Polysulfone in NMP
Coagulation bath	Distilled water
Casting knife [µm]	250, 200, 100
Support	Glass

2.3 Environmental scanning Electron Microscopy (ESEM) and morphology analysis

The morphology of fabricated membranes was investigated with Environmental Scanning Electron Microscopy (FEI Quanta 600). Membranes were fractured in liquid nitrogen and fixed to the support suitable for the cross-section analysis. The ESEM micrographs were recorded with 350 magnification.

Obtained micrographs were analyzed with Ifme [16] and ImageJ software, where membrane asymmetry, mean pore size, microvoid size and thickness is calculated based on the cross section structure of the membrane. Moreover, the overall porosity was determined by using densities of the membrane and the polymer, according to the equation:

$$\varepsilon = \left(1 - \frac{\rho_m}{\rho_p}\right) 100\% \quad (3)$$

where ρ_m corresponds to the membrane density and ρ_p to the polysulfone density (1.24 g/cm^3).

As membrane wetting is an unwanted phenomena in gas-liquid membrane contactor operations, membrane hydrophobicity and its affinity for the absorbent solution needs to be evaluated. Therefore, membrane swelling and contact angle measurements of both, water and absorbent solution were done.

 DOI: 10.1051/e3sconf/20172200125

Membrane (2 cm^2) was immersed into the water and absorbent solution for 24 h and swelling was calculated based on the membrane weight before (w_1) and after immersion (w_2) by the following equation:

$$Swelling = \frac{(w_2 - w_1)}{w_1}100\% \quad (4)$$

Hydrophobicity was analyzed by contact angle measurements (Dataphysics OCA 15EC). A 3 µL droplet of milli-q water or potassium hydroxide solution was placed on the bottom surface of the membrane. The measurements were performed only on the bottom surface of the membranes, because it is in contact with the solvent during operation. A contact angle was calculated from a digital image by SCA software included in the apparatus.

2.3 CO_2 absorption studies

Experiments were performed to determine the influence of speed of solution stirring, the membrane thickness and time of exposition of membrane contactor on the CO_2 absorption rate from ambient air. Firstly, membranes of different thickness were examined and placed in a module with capacity of 100 ml. The absorbent was KOH aqueous solution in the concentration of 0.64 M. It was stirred with different RPM ranging in 0-1000. The time of experiment was fixed to be 1 hour. Bottom side of the membrane was in contact with the absorbent solution while top was exposed to the air (Figure 1.) with a contact area of 19 cm^2. Samples were collected and analysed for CO_2 content by using CO_2 ion-selective electrode (Thermo Scientific connected with Thermo Scientific Orion Dual Star pH/ISE Benchtop meter). The absorption flux was calculated based on the following equation:

$$J_{CO2} = \frac{n_{CO2}}{t * A} \quad (5)$$

where n_{CO2} is CO_2 moles in the absorbent solution of 100 ml [µmol] obtained from the measurements, t is the time of experiment [s] and A is the area of the flat sheet membrane [m^2].

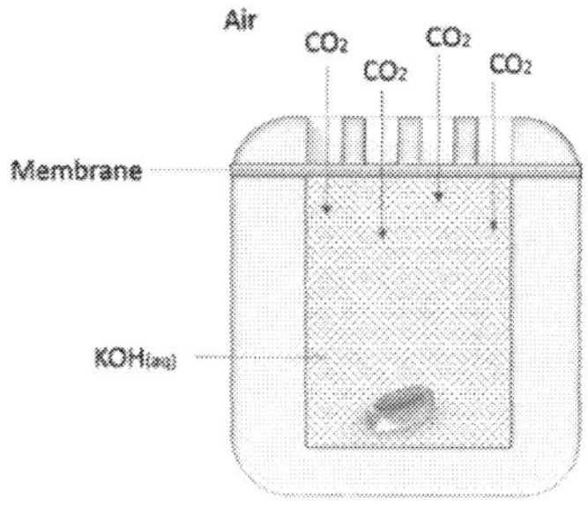

Fig. 1. Module scheme.

Next, membrane with the best absorption properties and the most effective rpm – absorption relation was used to analyse the absorption in function of time. For that purpose, nmp 200 membrane was placed into the system and experiments were conducted without stirring in range of time: 15, 30, 45, 60, 120 min. Collected samples were analysed the same way as previously.

DOI: 10.1051/e3sconf/20172200125

3 Results and discussion

3.1 Morphology analysis

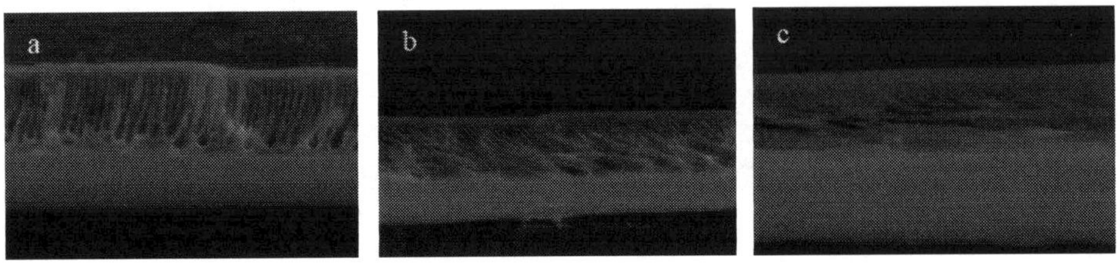

Fig. 2. ESEM micrographs of the membranes cross seccion: a) 250, b) 200, c) 100.

Resulting membranes have fingerlike macrovoids in porous media. Membranes chracteristics obtained from the analysis of ESEM micrographs are listed in Table 2. It shows that membranes vary in thickness and have increasing macrovoids size respectively to the thickness. The porosity and asymmetry do not change significantly.

Table 2. Membranes characteristics.

Membrane	Thickness [μm]	Assymetry [%]	Mean pore size [μm]	Mean macrovoids size [μm^2]	Porosity [%]
100	44.84 [± 1.4]	15	16.89 [± 2.4]	40 [± 26]	66.73
200	88.09 [± 2.4]	14	17.63 [± 1.6]	400 [± 271]	72.53
250	124.19 [± 2.6]	15	11.99 [± 0.3]	610 [± 228]	73.70

3.2 Swelling and contact angle results

The pores wetting is an undesirable phenomena in gas – liquid membrane contactor operations. For that reason we need to exclude the blocking influence of the absorbent solution on the pores. The analysis was done with use of absorbent solution and water as reference. Swelling (Table 3.) and contact angle of the water and KOH solution (Table 4.) were measured.

Table 3. Swelling test measurements results.

Adsorbed liquid	Membrane	Swelling [%]
Water	100	10.96 [±0.55]
	200	55.09 [±2.75]
	250	38.58 [±1.92]
KOH	100	1.21 [±0.06]
	200	n. d.
	250	n. d.

Swelling test results shows that membranes are capable to absorb water but no absorbent solution what is in agreement with contact angle measurements.

Table 4. Water and KOH solution contact angle measurements.

Contact angle liquid	Membrane	Contact angle [°]
Water	100	66.52 [±2.72]
	200	81.85 [±0.82]
	250	67.82 [±1.37]
KOH	100	71.02 [±7.03]
	200	86.72 [±4.74]
	250	75.58 [±5.09]

Despite the prepared membranes are hydrophilic, water contact angle is lower than the one corresponding to absorbent solution. This occurs due to difference in liquid surface tension, thus the interaction with the membrane surface. The tests demonstrate that absorbent solution is adequate for working with polysulfone membranes. The membrane life time is expected to be long as far as, in the tests performed, the polymer does not seems to be affected by the solutions. It might be due to that membrane wetting is very low and, on the other hand, the fouling does not occur in these liquid membrane contactors operations.

3.3 CO_2 absorption results

3.3.1 Influence of the membrane thickness and stirring rate on the contactor absorption capacity

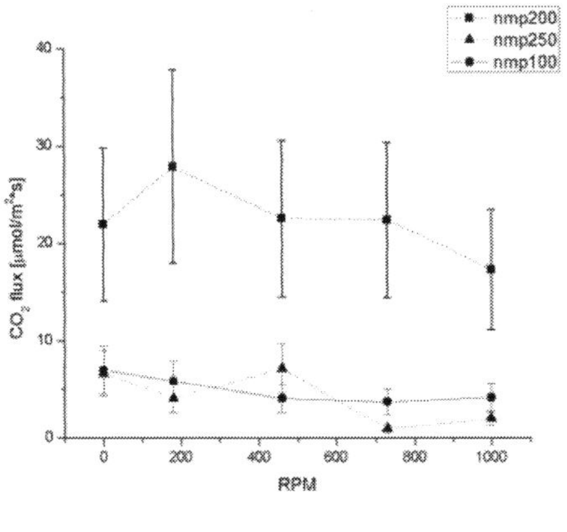

Fig. 3. CO_2 absorption in a function of RPM by different membranes.

It was expected that increasing the mixing of absorbent solution would have a positive influence on the absorption rate, as the fluctuation of the absorbent below the membrane could increase the amount of unreacted potassium hydroxide molecules in the contact with the surface. Nevertheless, studies show that the impact of the mixing is negative or not significant. It looks like the influence of the solutions mass transfer resistance during the static operation of the module is low as the solution remains in the same container. A previously studied case, where the absorbent solution was in a separated flask, and it was circulated through the contactor by the use of pump, revealed the positive influence of the absorbent fluctuations on the absorption rate.

DOI: 10.1051/e3sconf/20172200125

The mass transfer resistance in the membrane can be studied by using of different thickness and hence macrovoids size, as the obtained membranes possess similar internal morphology. The CO_2 passage rate through the membrane phase depends on the distance (membrane thickness), free space (macrovoids size) and the material interaction with absorbent solution. From the study we can conclude that the membrane 200 shows the best results in absorption capacity over the rest of studied membranes. This membrane shows the highest hydrophobicity and macrovoids big enough to the CO_2 pass. Even though the macrovoids are large and open, the distance between air and absorbent is too big when we consider membrane 250. The case of membrane 100 is the opposite, the distance is short but there is not enough free space for CO_2 to pass.

3.3.2 Change of absorption in time

The best conditions from previous experiment were chosen to study the membrane contactor system abilities to absorb CO_2 in passage of time. The experiments were performed with the membrane 200 and without used stirring, as its impact is negligible. In the plot (Figure 4) the change in CO_2 content in the final absorbent solution and its flux through the contactor are displayed together in order to understand the functioning of the system.

The absorbed CO_2 amount is rising within the studied range of time. The system would keep increasing the CO_2 content until the saturation state. The CO_2 flux is decreasing in time and the plateau was achieved after one hour of experiment. The results were expected and seem logic, because according to the equation (5), for the flux, moles are divided by time and area.

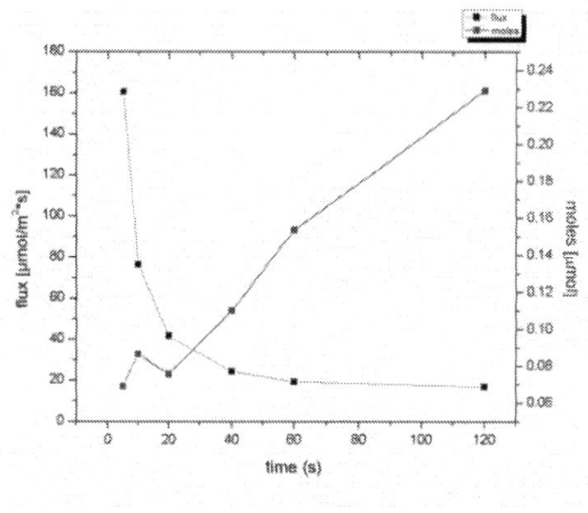

Fig. 4. CO_2 absorption flux and CO_2 content change as a function of time.

3.3.3. Natural stomata vs artificial stomata

As the system is designed based on the natural leaf structure, thus the obtained results will be compare to the natural stomatal CO_2 assimilation rate. For the reference we choose the grape vine leaf assimilation rate value which is 20 μmol/m²·s, as reported by J. Martínez-Lüscher et al. [3]. The range obtained by us is depending on the experimental conditions ranging in 6–160 μmol/m²·s. Nevertheless, in the studies of the influence of time presented above, we see the system achieved stabilization with membrane 200 after 1 h of experiment reaching the value of 20 μmol/m²·s, which should be used as the reference. Results show that the CO_2 assimilation rate in natural stomata is the same range as in our artificial system.

4 Conclusions

Porous polysulfone membranes with different thickness were fabricated and characterized with the purpose to use them as contactors in ambient carbon dioxide capture studies. Compatibility of the membranes and absorbent solutions was evaluated by swelling tests and contact angle measurements. According to the results, we can conclude that polysulfone membranes are compatible to be used with potassium hydroxide solution. Furthermore, the influence of membrane thickness, absorbent solution mixing rate and experimental time on the system absorption capacity were investigated. From the study we can conclude that the membrane nmp 200 shows visible superiority in absorption capacity over the rest of studied membranes, due to the best thickness / macrovoids size ratio and the best hydrophobic properties. Nevertheless, the results showed that the impact of the absorbent solution mixing is not significant. It looks like the influence of the solutions mass transfer resistance during the static operation of the module is low as the solution remains in the same container. As expected, the absorbed CO_2 amount was rising within the studied range of time. The comparison between natural and the artificial stomata revels that the CO_2 assimilation is within the same range. Thus, the designed stomata mimicking unit can be used further in artificial photosynthetic systems.

This work is financially supported by Martin Franquez scholarship by University Rovira y Virgili.

References

1. J.G. Canadell, C. Le Quere, M.R. Raupach, C.B. Field, E.T. Buitenhuis, P. Ciais, T.J. Conway, N.P. Gillett, R.A. Houghton, G. Marland, Proc. Natl. Acad. Sci. U.S.A. **104**, 18866–18870 (2007)
2. A.P. Ballantyne, C.B. Alden, J.B. Miller, P.P. Tans, J.W.C. White, Nat. Geosci. **2**, 831–836, (2009).
3. J. Martínez-Lüscher, F. Moralez, M. Sanchez-Diaz, S. Delrot, J. Aguirreolea, E. Gomès, I. Pascual, Plant Sci. **236**, 168–176 (2015)
4. W. Tezara, V.J. Mitchell, S.D. Driscoll, D.W. Lawlor, Nature **401**, 914–917 (1999)
5. P. Teszlaka, M. Kocsis, K. Gaal, M.P. Nikfardjam, Sci. Hor. **159**, 41–51 (2013)
6. K. Radglou, Ann. Sci. For. **53**, 268–278 (1996)
7. V.Y. Dindore, D.W.F. Brilman, F.H. Geuzebroek, G.F. Versteeg, Sep. Purif. Technol. **40**, 133–145 (2004)

8. H.F. Svendsen, E. Favre, J. Membrane Sci. **407–408**, 1–7 (2012)
9. S.A. Stern, J. Membrane Sci. **94**, 1–65 (1994)
10. K.H.G. Illing, R.J. Wakeman, A. Jungbauer, J. Membrane Sci. **184**, 69–78 (2001)
11. B.D. Freeman, H. Lin, J. Membrane Sci. **239**, 105–117 (2004)
12. P.H.M. Feron, A.E. Jensen, Sep. Purif. Technol. **27**, 231–242 (2002)
13. A. Witek-Krowiak, A. Dawiec, S. Modelski, D. Podstawczyk, Int. J. Chem. Eng. Appl. **3** (2012)
14. Z.X. Hongqun Yang, M. Fan, R. Gupta, R.B. Slimane, A.E. Bland, I. Wright, J. Environmen. Sci. **20**, 14–27 (2008)
15. C. Torras, *Obtenció de membranes polimèriques selectives* (in *Chemical Engineering Dep.* 2005, Universitat Rovira i Virgili, Tarragona)
16. C. Torras, R. Garcia-Valls, J. Membrane Sci. **233**, 119–127 (2004)

The mathematical model of rainwater catchment in Wroclaw

Monika Nowakowska[1,*]

[1]Faculty of Environmental Engineering, Wroclaw University of Science and Technology, Department of Water and Sewage, Wybrzeże Wyspiańskiego 27, 50-370 Wroclaw, Poland

Abstract. On the basis of field studies of canalized storm water catchment of Gaj and Tarnogaj estate in Wrocław the rules of construction of hydrodynamic models in SWMM program were presented. The process of identification of hydrological and hydraulic parameters of the model, in the course of its calibration and validation were presented. To assess the quality of model the rates and statistical criteria were proposed to compare the results of stream simulation and volume of runoff of effective precipitation with the results of the measurements.

1 Introduction

Escalating in recent years extreme weather phenomena such as incidental or prolonged rainfall and associated with them flooding cause significant economic losses. In the existing, modernized or newly designed sewage systems the verification of hydraulic capacity of networks and objects is now recommended, including overflows from channels towards hydrodynamic modeling at different scenarios of precipitation load.

Modeling of systems reliability recommended by standard PN-EN 752:2008 is rarely applied in Poland [1] – with regard to the verification of the frequency of overflow from channels and even required by law according to the regulation of the Minister of the environment from 2014 [2] – with regard to the verification of the frequency of storm water overflows. This is mainly the result of the lack of sufficient basis of modeling methods as well as appropriate output databases with regard to monitoring of precipitation and overflows in sewage systems [3]. For instance, the level of integration of sub-catchment area is as yet intuitively. Usually channels with diameter smaller than 0.5 m are omitted and the width of the hydraulic sub-catchment is designated from several different formulas. In calibration and validation of hydrodynamic models data from the short period of observation are taken into consideration (i.e. few months) and the assessment of the quality of hydrodynamic models is based on various statistical indicators for comparison of the simulation results of sewage with the measurement results.

Based on the example of local data regarding the precipitation and streams of storm water in Wrocław the principles of the construction of mathematical model for drainage the area in SWMM program were presented.

* Corresponding author: monika.nowakowska@pwr.edu.pl

 DOI: 10.1051/e3sconf/20172200126

2 Material and methods of studies

For field research the canalized catchment of Gaj and Tarnogaj estate in Wrocław were selected, with the area F = 104 hectares – giving runoff to indoor drainage collector KD1 (with diameters from 0.3 m to 1.4 m). From the total area of estates, the dimpled areas were excluded, including allotments from which the sewage runoff of precipitation taking place into the ditches. The total length of the inventoried channels was 17731 m and the number of drains 509. This gives an average spacing of drains 34.8 m, which can be considered as representative for the cities.

On the stage of identification, as a minimum diameter of channels taking $d_{min} \geq 0.3$ m, 75 catchment area were discharged. In sub-catchments sealed areas were distinguished – not having retention F_d (roofs), sealed areas-with retention F_a (asphalt road) and $F_{k.b}$ (carriageways with concrete block or cobblestone) and areas not sealed ("green" areas) – with retention F_{nu}. Then degrees of sealed area were calculated in each sub-catchments: total sealing degree: $\%F_u = F_u/F_i$, degree of area sealing without retention: $\%F_{ubr} = F_{ubr}/F_u$ and degree of area sealing with retention: $\%F_{uzr} = F_{uzr}/F_u$. Slope of area (i_p) in the sub-catchments were adopted as weighted average-from the area of the terrain, streets, squares and roofs (tab. 1).

Table 1. Summary of characteristic area and slope of sub-catchments.

Area no.	Mark of catchment area	Node no.	Catch ment area F_i	Sealed areas F_u				Areas not sealed F_{nu}	$\%F_u$	$\%F_{ubr}$	$\%F_{uzr}$	i_p
				F_d	F_a	F_b	ΣF_u					
			ha						%			%
1	P1	18	0.9019	0.1175	0.1926	0.3315	0.6416	0.2603	71.1	18.3	81.7	1.20
4	P4	45	1.3590	0.6212	0.0000	0.7378	1.3590	0.0000	100.0	45.7	54.3	1.18
9	P9	65	0.2985	0.0000	0.1178	0.0407	0.1585	0.1400	53.1	0.0	100.0	1.11
12	P11	73	2.4492	0.6616	1.3501	0.0858	2.0974	0.3517	85.6	31.5	68.5	1.21
16	P 14.1	119	0.2220	0.0000	0.0929	0.0260	0.1189	0.1031	53.6	0.0	100.0	1.12
20	P 16.1	99	3.4074	1.2146	0.2309	1.5695	3.0150	0.3924	88.5	40.3	59.7	1.10
21	P17	121	0.5730	0.0000	0.1896	0.1891	0.3787	0.1943	66.1	0.0	100.0	0.17
31	P21.1	142	1.0254	0.0000	0.8203	0.1230	0.9434	0.0820	92.0	0.0	100.0	0.33
32	P22	170	2.2621	0.1900	0.0414	0.1015	0.3330	1.9291	14.7	57.1	42.9	0.22
33	P23	170	1.8855	0.0000	0.2625	0.3150	0.5775	1.3080	30.6	0.0	100.0	0.68
34	P24	182	3.4543	0.4098	0.0609	0.0895	0.5602	2.8941	16.2	73.2	26.8	0.18
37	P26	242	0.6718	0.0000	0.1671	0.1537	0.3208	0.3510	47.8	0.0	100.0	1.01
43	P 28.2	238o	1.3525	0.3852	0.0000	0.0553	0.4405	0.9120	32.6	87.4	12.6	0.72
56	P38	322	0.7092	0.0263	0.4255	0.2128	0.6646	0.0446	93.7	4.0	96.0	1.81
68	P46	382	5.3379	1.3345	0.4404	0.8195	2.5944	2.7435	48.6	51.4	48.6	0.55
71	P48	454	0.5189	0.0000	0.1920	0.0327	0.2247	0.2942	43.3	0.0	100.0	0.92
75	P51	505	1.0265	0.1721	0.2344	0.1875	0.5940	0.4325	57.9	29.0	71.0	1.20
SUM/AVERAGE			103.98	23.66	15.41	22.44	61.50	42.47	59.2	38.5	61.5	0.93
MIN			0.2220	0.0000	0.0000	0.0260	0.1189	0.0000	14.7	0.0	12.6	0.17
MAX			5.3379	1.3345	1.3501	1.5695	3.0150	2.8941	100	87.4	100	1.81

Individual sub-catchments were organized into 75 network nodes, to which rainwater is drained (fig. 1).

 DOI: 10.1051/e3sconf/20172200126

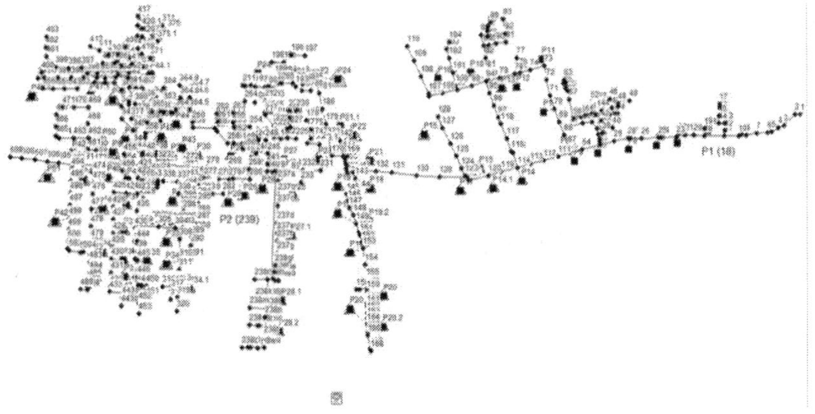

Fig. 1. The scheme of the analyzed storm water drainage in SWMM program.

As the basis to determine hydraulic width of sub-catchment (W), the basic form of the formula was taken:

$$W = x\sqrt{F} \qquad (1)$$

where:

F – area of storm water catchment/sub-catchment, m^2.

In subject literature, the formula (1) is most commonly used with multipliers for values: $x = 1.0$ [6–8] or $x = 1.5$ [9–11]. In Wroclaw's conditions to render the parameter W it turned out that the best value was $x = 1.6$, which will be demonstrated later in this study.

To conduct simulation calculations in SWMM program it was necessary to create databases involving precipitation of rain in catchment and measurement of sewage streams in KD1 collector. For metering sewage streams two ultrasonic flow meters were used. Nivus type flow meter (P1 – fig. 1) was installed in drain no. 18 on the section of KD1 with a diameter of 1.4 m. The second type of flow meter Teledyne (P2) was installed in the drain no. 239, on the section KD1 with a diameter of 1.2 m.

The rain gauge type TRwS was a device used to register rainfall in the catchment located on the southern boundary of the catchment area of Gaj and Tarnogaj estate at a distance of about 620 m from the center of gravity of the tested catchment (fig. 1). Registry of precipitation was conducted for a period of 2 years, from July 29th 2013 to July 19th 2015. Precipitation with duration time $t > 45$ min and height $h > 10$ mm were taken for calibration and validation of the model. Both torrential rains - those of short duration (convective K), as well as those of long-term (frontal F and lowland N), which hyetographs have a continuous character were taken into consideration. The chosen precipitation was assigned later to their frequency of occurrence in Wrocław. As a criterion to determine the frequency of precipitation occurrence was taken the probabilistic model of maximum height in Wrocław (based on the distribution of Fisher-Tippett type III) – for $C \in [1; 100]$ years [12], and for precipitation occurring more frequently than once a year ($C \in [0.1; 1)$) a physical model was used [3]. These precipitations were detailed about episodes of time duration equal to time of flow of sewage in collector KD1 to the intersection of flowmeter P1 ($t \approx 45$ min) and P2 ($t \approx 15$ min). The parameters of 8 precipitation selected for calibration and validation of the model are given in table 2.

DOI: 10.1051/e3sconf/20172200126

Table 2. The parameters of precipitation for calibration (no. 1–5) and validation of model (no. 6–8).

Number, type and date of precipitation	Time		Time t min	Σh mm	C years	h_{45} mm	C_{45} years	h_{15} mm	C_{15} years
	from	to							
1. F 17.05.2014	09:45	17:09	444	15.1	0.55	5.6	0.36	3.0	0.27
2. N 23–24.07.2014	23:34	15:09	935	13.1	0.41	3.4	0.19	1.7	0.13
3. N 1–2.09.2014	15:27	13:47	1340	13.4	0.40	1.7	<0.1	1.0	<0.1
4. N 11.09.2014	8:01	22:54	894	11.7	0.38	2.4	0.12	1.6	0.11
5. N 22–23.10.2014	01:00	01:28	1468	28.4	0.75	2.2	0.11	1.5	0.11
6. K 29.07.2013	20:26	22:12	106	15.7	0.85	14.6	1.1	13.4	2.4
7. K 17.10.2014	18:37	20:12	95	9.7	0.52	8.6	0.59	6.1	0.61
8. K 19.07.2015	19:57	20:39	42	23.8	5.1	23.8	4.7	22.1	18.4

To assess the quality of the model selected statistical measure were used for comparing the results of measurements and calculations of the outflow (Q), such as [3–5, 13–15]:

- special rate of correlation RS:

$$RS = \sqrt{\sum_{i=1}^{n}(2 \cdot Q_{p,i} \cdot Q_{o,i} - Q_{o,i}^2)/\sum_{i=1}^{n} Q_{p,i}^2} \tag{2}$$

- relative residual error WBR:

$$WBR = \sqrt{n \cdot \left(\sum_{i=1}^{n}(Q_{p,i} - Q_{o,i})^2\right)}/\sum Q_{p,i} \tag{3}$$

- average value error SWS:

$$SWS = \bar{Q}_o/\bar{Q}_p \tag{4}$$

- relative error of maximum streams ΔQ_{max}:

$$\Delta Q_{max} = \frac{Q_{max,p} - Q_{max,o}}{Q_{max,p}} \cdot 100\% \tag{5}$$

Indexes "p", and "o" in formulas (2–5) mean respectively measurements and calculations. The value "n" corresponds to the number of extracted compartments-averaged values Q in a registered hydrographic of runoff. Depending on the value of the rate, the model can be qualified for the specified category. Ranges of values of rates RS and WBR together with corresponding categories of models were listed in table 3. For SWS rates and ΔQ_{max} the ranges of values for the appropriate category of model were not specified. The only known limit values appropriate for situation when the model perfectly reproduces reality: SWS = 1.0 i $\Delta Q_{max} = 0$.

Table 3. Categories for the classification of the models.

Category of model	Rates-ranges of values	
	RS [-]	WBR [%]
Excellent	1.00–0.99	0–3
Very good	0.99–0.95	3–6
Good	0.95–0.90	6–10
Average	0.90–0.85	10–25
Dissatisfying	< 0.85	> 25

3 Calibration of model

Calibration of hydrodynamic model was based on determining the values of parameters:

- hydraulic – roughness rate of channels (n - to Manning formula) and roughness rate of sealed (n_{pu}) and unsealed (n_{pnu}) catchment areas,
- hydrological – the height of retention on sealed (h_{pu}) and unsealed (h_{pnu}) catchment areas.

Hydrogeological parameters of model (infiltration) were estimated by model of expert. As a result of the simulation calculations in SWMM program for 5 long-term precipitation calibration (type F and N-table 2) the empirical parameters of the model were determined. In table 4 the literature (output) and resulting values of the parameters of model of the tested drainage system were listed.

Table 4. The value of the calibration parameters of the model of tested drainage system.

Parameter	**The width of the hydraulic sub-catchment**	**Rate of roughness - channels**	**Rate of roughness -sealed area**	**Rate of roughness -unsealed area**	**Height retention -sealed area**	**Height retention -unsealed area**	**The initial intensity infiltration-**	**Final intensity of infiltration-Horton's model**	**Constant recession -Horton's model**	**Lad drying time-Horton**
Symbol	W	n	n_{pu}	n_{pnu}	h_{pu}	h_{pnu}	I_{in0}	I_{ink}	r	T_s
Unit	m	s/m$^{1/3}$	s/m$^{1/3}$	s/m$^{1/3}$	mm	mm	mm/h	mm/h	h^{-1}	d
Literature data for calibration	various formulas	0.013–0.020	0.011–0.050	0.1–0.8	1.3–2.5	2.5–7.5	8–254	0.2–20	2–7	2–14
Results of model's calibration in SWMM program	$1.6\sqrt{F}$	0.020	0.020	0.3	2.0	5.0	90	10	4	7

About the optimum values of parameters of the model proclaim the achieved values of indicators: RS, WBR, SWS and ΔQ_{max}, adopted as the statistical criteria to evaluate the quality of the model. However, the most important was the achievement of the compliance of simulated values with the values measured with respect to the balance of the volume of runoff (V), as shown in Chapter 4. In table 5 the final results of the calculation accuracy of the model for 5 precipitation calibrations and three forms of formula for the parameter W. The best results were achieved for $W3$. On the basis of the value of the indicators RS and WBR the calibrated model was assessed on the border of grades good and very good (according to the criteria from the table 3).

Figure 2 shows, for example, a histogram of precipitation and hydrographs of sewage streams in collector KD1 (intersection P1) for the latest lowland precipitation (N) from October 22nd to 23rd 2014.

 DOI: 10.1051/e3sconf/20172200126

Table 5. The results of the calculations of the calibration model.

Formula	Precipitation data FLOWMETER	17.05.2014		23–24.07.2014		1–2.09.2014		11.09.2014		22–23.10.2014	
		P1	P2	P1	P2	P1	P2	P1	P2	P1	P2
$W1 = \sqrt{F}$	RS [-]	0.977	0.930	0.977	0.970	0.967	0.973	0.958	0.938	0.980	0.976
	WBR [%]	4.0	7.3	5.9	7.2	6.0	5.9	6.8	9.8	4.7	5.2
	SWS [-]	0.940	1.030	0.989	1.113	0.993	1.011	1.057	1.147	1.018	1.100
	ΔQ_{max} [%]	14.9	20.8	0.9	-6.4	9.8	0.9	18.2	13.7	16.5	12.3
$W2 = 1.5\sqrt{F}$	RS [-]	0.985	0.949	0.983	0.970	0.971	0.976	0.968	0.954	0.982	0.978
	WBR [%]	3.2	6.2	5.1	7.2	5.6	5.4	6.0	8.5	4.5	5.0
	SWS [-]	0.943	1.032	0.993	1.117	0.997	1.014	1.061	1.150	1.019	1.099
	ΔQ_{max} [%]	11.5	14.5	-0.4	-10.6	6.9	-6.3	14.5	7.1	16.1	11.8
$W3 = 1.6\sqrt{F}$	RS [-]	0.986	0.951	0.983	0.970	0.972	0.977	0.969	0.956	0.982	0.979
	WBR [%]	3.1	6.1	5.0	7.3	5.6	5.4	5.9	8.3	4.4	5.0
	SWS [-]	0.943	1.033	0.994	1.117	0.997	1.014	1.061	1.151	1.020	1.099
	ΔQ_{max} [%]	11.0	13.6	-0.6	-11.2	6.5	-7.4	14.0	6.3	16.0	11.7

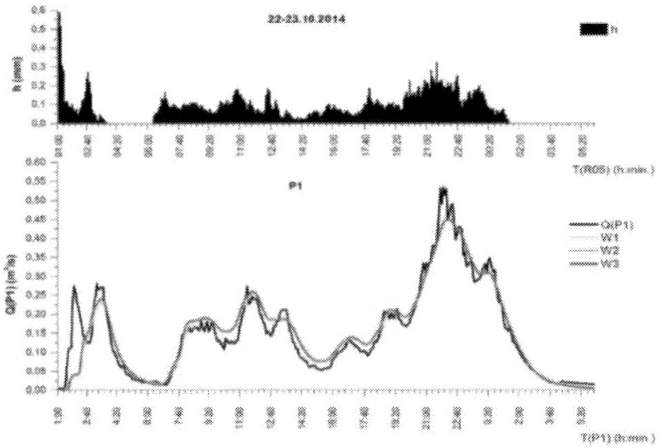

Fig. 2. A histogram of precipitation and hydrographs of flow into the collector KD1 (P1) for the latest lowland precipitation from October 22nd to 23rd 2014.

4 Validation of model

Validation of tested hydrodynamic model of drainage system consists in checking the accuracy of the already calibrated model on the 3 short-lived intense convective precipitation (K-tab. 2) with criterion of compatibility of the balance volume of runoff (*V*). Figure 3 shows an example of histogram of convective precipitation from July 19th 2015 and hydrographs of drainage (measured and simulated) in collector KD1 in the intersection of flowmeter P1.

 DOI: 10.1051/e3sconf/20172200126

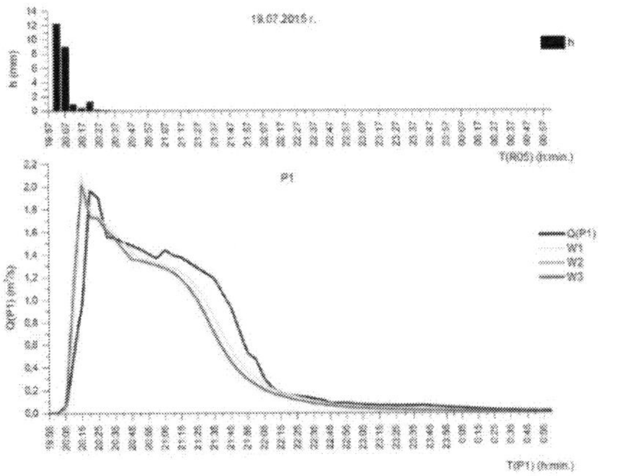

Fig. 3. A histogram of precipitation and hydrographs of flow into the collector KD1 (P1) – for convective precipitation from July 19th 2015.

Table 6 presents the results of a calculation parameter *W* for 3 validated precipitation (K). Accuracy of mapping of tested phenomena precipitation-outflow through the created model of canalized storm water catchment were presented cumulatively for 8 precipitation, i.e. 5 from calibration and 3 from validation. Figure 4 shows the correspondence with the volume balance of runoff (*V*).

Table 6. The results of the validation calculations.

Formula	**Precipitation data /flowmeter**	**29.07.2013**	**17.10.2014**		**19.07.2015**
		P1	**P1**	**P2**	**P1**
$W1 = \sqrt{F}$	RS [-]	0.988	0.994	0.992	0.970
	WBR [%]	2.2	1.1	1.6	3.0
	SWS [-]	1.122	1.070	1.089	0.932
	ΔQ_{max} [%]	-3.0	3.3	5.4	-7.2
$W2 = 1.5\sqrt{F}$	RS [-]	0.978	0.996	0.995	0.962
	WBR [%]	3.0	0.9	1.2	3.3
	SWS [-]	1.098	1.083	1.102	0.883
	ΔQ_{max} [%]	-7.7	-0.3	-0.8	-3.2
$W3 = 1.6\sqrt{F}$	RS [-]	0.978	0.996	0.995	0.961
	WBR [%]	3.0	1.0	1.3	3.4
	SWS [-]	1.090	1.084	1.103	0.876
	ΔQ_{max} [%]	-5.5	-1.0	-1.7	-2.8

 DOI: 10.1051/e3sconf/20172200126

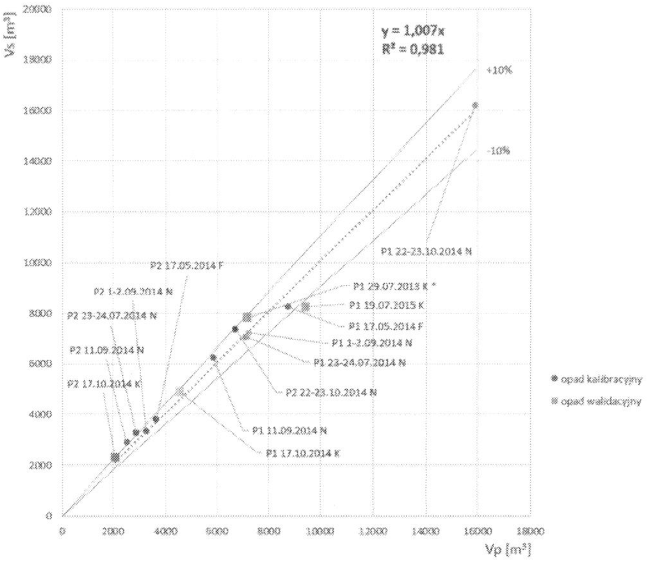

Fig. 4. The volume balance of runoff for precipitation from calibration and validation model.

Figure 4 shows that for 8 precipitations (from calibration and validation model) the volume of runoff describes the simple equation $V_s \approx V_p$ where $R^2 = 0.981$. This means high compliance of the simulated (*S*) and measured (*P*) volumes of precipitations runoff – 14 results of measurements and simulation is in the range accuracy ± 10%.

5 Summary and conclusions

In this paper the rules for the mathematical model of the drainage system in Wrocław area in SWMM program were showed. Selected results of studies about identification, calibration and validation of empirical parameters of the model along with the verification of formulas to determine the width of the hydraulic (W) sub-catchment were presented. Discussed results of this study have in methodical part universal character and are used to build models of other hydrodynamic urban catchment. Identification of the major parameters of empirical models (n, n_{pu}, n_{pnu}, h_{pu} and h_{pnu}) should be carried out in the course of their calibration on long-term precipitation (frontal and lowland) and in the course of the validation of the models on the intense precipitation (convection) with the compatibility criterion of the balance volume of runoff (V). To assess the quality of models the optional statistical indicators can be used (RS, WBR, SWS and ΔQ_{max}) describing the fitting accuracy of simulated and measured values of waste water streams.

The work was realized within the allocation No. 0401/0069/16 awarded for Faculty of Environmental Engineering Wroclaw University of Science and Technology by Ministry of Science and Higher Education in years 2016–2017.

References

1. PN-EN 752:2008, *Drain and sewer systems outside buildings* (PKN, Warszawa, 2008)
2. Rozporządzenie Ministra Środowiska z dnia 18 listopada 2014 r., Dz. U. RP z dnia 16 grudnia 2014 r., poz. 1800

3. A. Kotowski, *Basics of safe dimensioning of drainage areas* (Seidel-Przywecki, Wroclaw, 2015)
4. M. Nowakowska, *Identification of hydrological and hydraulic parameters of urban catchment area in SWMM hydrodynamic modeling* (Oficyna Wydawnicza Politechniki Wrocławskiej, Wroclaw, 2016)
5. M. Nowakowska, A. Kotowski, *Methodology and rules for modelling drainage of urban areas* (Oficyna Wydawnicza Politechniki Wrocławskiej, Wroclaw, 2017)
6. M. Mrowiec, *Effective dimensioning and dynamic regulation of sewerage retention tanks* (PCz., Czestochowa, 2009)
7. M. Skotnicki, M. Sowiński, *Verification of the method for determining the hydraulic width of a sub-catchment on the example of a selected urban catchment* (PW., Warszawa, 2009)
8. D. Słyś, A. Stec, Environ. Prot. Eng. **38**, 99–112 (2012)
9. M. Zawilski, G. Sakson, Gaz, Woda Tech. Sanit. **11**, 32–36 (2010)
10. M. Zawilski, G. Sakson, Gaz, Woda Tech. Sanit. **9**, 321–323 (2011)
11. M. Zawilski, Gaz, Woda Tech. Sanit. **6**, 28–32 (2010)
12. A. Kotowski, B. Kaźmierczak, A. Dancewicz, *Modeling precipitation for sewage dimensioning* (PAN, **68**, Warszawa, 2010)
13. I.R.A. Green, D. Stephenson, Hydrologie Sciences J. **31**, 395–411 (1986)
14. W. James, *Rules for responsible modeling* (CHI Publications, Guelph, 2003)
15. M. Ozga-Zielińska, J. Brzeziński, *Practical hydrology* (PWN, Warszawa, 1997)
16. ASCE, *Design & Construction of Urban Stormwater Management Systems* (New York 1992)
17. U.K. Maheepala, A.K. Takyi, B.J.C. Perera, J. Hydrology **245**, 32–47 (2001)
18. R. Mccuen, P. Johnson, R. Ragan, *Hydrology* (Federal Highway Administration, Washington, 1996)
19. PIT R., *Infiltration Through Disturbed Urban Soils and Compost-Amended Soil Effects on Runoff Quality and Quantity*. United States Environmental Protection Agency, Washington (1999)
20. L.A. Rossman, *Storm Water Management Model. User's Manual.* United States Environmental Protection Agency, Washington (2010)

Impact assessment of treated wastewater on water quality of the receiver using the Wilcoxon test

Piotr Ofman[1], *Monika* Puchlik[1,*], Grzegorz Simson[2], *Małgorzata* Krasowska[3], and *Joanna* Struk-Sokołowska[1]

[1]Bialystok Technical University, Faculty of Civil and Environmental Engineering, Department of Technology in Engineering and Environmental Protection, ul. Wiejska Street 45E, 15-351, Poland
[2]Białystok Waterworks Ltd, ul. Produkcyjna 102, 15-680 Białystok, Poland
[3]Bialystok Technical University, Faculty of Civil and Environmental Engineering, Department of Environmental Protection and Management, ul. Wiejska Street 45E, 15-351, Poland

Abstract. Wastewater treatment is a process which aims to reduce the concentration of pollutants in wastewater to the level allowed by current regulations. This is to protect the receivers which typically are rivers, streams, lakes. Examination of the quality of treated wastewater allows for quick elimination of possible negative effects, and the study of water receiver prevents from excessive contamination. The paper presents the results of selected physical and chemical parameters of treated wastewater from the largest on the region in north-eastern Poland city of Bialystok municipal wastewater treatment and Biała River, the receiver. The samples for research were taken 3–4 a month in 2015 from two points: before and after discharge. The impact of the wastewater treatment plant on the quality of the receiver waters was studied by using non-parametric Wilcoxon test. This test determined whether the analyzed indicators varied significantly depending on different sampling points of the river, above and below place of discharge of treated wastewater. These results prove that the treated wastewater does not affect the water quality in the Biała River.

1 Introduction

According to many researchers, sewage treatment plants have a negative impact on soil and water environment of receivers [1–4]. The most commonly used method of municipal wastewater treatment is activated sludge technology [5, 6]. This process, although gives significant environmental and economic advantages, is not without flaws. One of the main problems of biological wastewater treatment is the inability to guarantee a complete decomposition of mineral and organic substances [7]. Exploitation problems during the operation of sewage treatment plants result in a periodic worsening of the receiver due to the discharge of treated wastewater to an insufficient degree [8]. In Polish law, the quality

* Corresponding author: m.puchlik@pb.edu.pl

requirements for purified wastewater in larger facilities are limited to monitoring of easily and hardly biodegradable compounds, total amounts of nitrogen and phosphorus, as well as suspended solids [9, 10]. In an aqueous environment, biogenic substances are important, including oxygen-derivatives of nitrogen and phosphorus, which are supplied to the environment with treated wastewater [11]. These compounds aggravate the eutrophication of surface waters, which in consequence, causes an aerobic condition [12]. The result is a clear deterioration in water quality, decrease in the utility value of the watercourse, while for a longer time, irreversible changes in the ecosystem and the decline of biodiversity [13]. Environmental protection should consist in discharging sewage into waters with pollution loads permissible from the point of view of the ability to keep their self-cleaning in environment [14, 15].

The aim of the study was to analyze the impact of treated wastewater discharged from the sewage treatment plant in Bialystok on the water quality of Biala river. The study was undertaken due to the fact that this area is one of the Polish Green Lungs. It is characterized by a low level of industrialization. It has unique natural areas and ecological systems that are some of the most valuable in Poland and Europe.

2 Materials and methods

Located in north-eastern Poland, mechanical-biological wastewater treatment plant purifies sewage from more than 300.000-inhabitants agglomeration. Its average capacity is approximately 70 000 m^3/d. The main source of contamination affecting the drains are municipal and industrial sewage from the city of Bialystok, as well as adjacent areas. The work of the plant is based on active sludge technology assuming the load of BOD_5 equal to 30 000 kg O_2/d, suspension load of 55 000 kg/d, total Kjeldahl nitrogen 6 000 kg/d and total phosphorus 850 kg/d. Organic compounds removal characterized by BOD_5 in 99%, while suspension removal in 99%, total nitrogen and phosphorus 90% [16]. After treatment, wastewater from the municipal wastewater treatment plant is discharged into the Biała river (Figure 1). The catchment of Biała river covering over 133 km^2 area is also a receiver of majority of rainfall water and surface runoff, which causes deterioration of its chemical as well as health and hygiene conditions. The purity of Biała river is very important, because it is a left-hand tributary of the Supraśl river, which in turn is a right tributary of the Narew river. Therefore, studies were undertaken upon the impact of the discharge of treated wastewater on the quality of Biała river. Samples of water above and below the discharge point (Figure 1) were taken 3–4 times a month from January to December 2015 (Figure 1) in accordance with Polish standards [17]. Physicochemical analyses included: temperature, acidity, BOD_5, COD, total Kjeldahl nitrogen (TKN), ammonia ($N\text{-}NH_4$), nitrate ($N\text{-}NO_3$), nitrite ($N\text{-}NO_2$), total nitrogen, chloride (Cl^-), total phosphorus, phosphate (PO_4), total suspended solids (TSS), and dissolved oxygen (O_2). All tests were performed in certified laboratory of Bialystok Waterworks Ltd. according to APHA [18]. The paper presents results of the monthly average for the entire study period. The applied statistical analysis takes into account the general measures of listed parameters, including: arithmetic mean, median, minimum, maximum, and standard deviation. In order to verify the impact of the sewage treatment plant on the water quality in the receiver, the nonparametric Wilcoxon test for pairs of dependent variables was performed on the basis of achieved results. The determinant for the test is the lack of normal distribution of variables taken for analysis. The variables was not characterized by normal distribution according by Kolomogorov-Smirnov test.

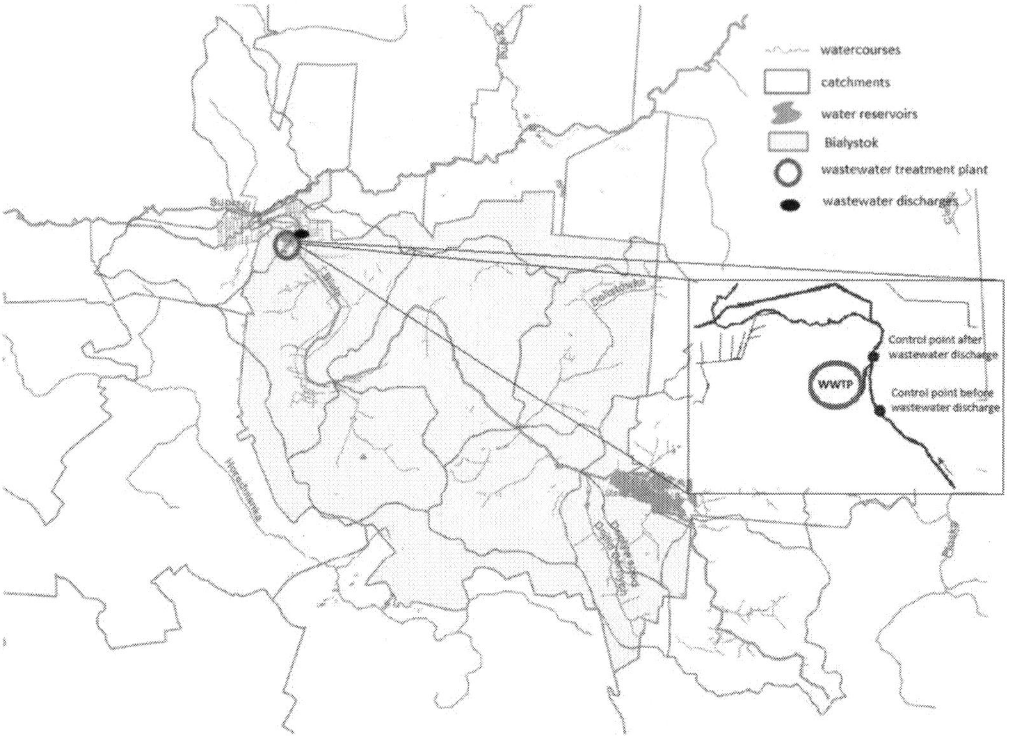

Fig. 1. Location of the research object and the arrangement of measurement and control points against the hydrographic division according to Tyszkewska and Kardel [19].

3 Results

The average temperature of water in the receiver in 2015 before discharge of treated wastewater, changed from 3.4 to 18.0°C during the study. The value of arithmetic mean (9.6°C), median (8.5°C), and standard deviation (5.2°C) indicate seasonal variations of this parameter. The temperature of water after the wastewater discharge was clearly higher as compared to the point before discharge. This parameter varied from 11.0 to 20.8°C. Values of arithmetic mean and median were similar amounting to 14.9 and 13.3°C, respectively, while standard deviation was 3.8°C. The acidity of the receiver's water before and after treated wastewater discharge point did not show remarkable changes. Minimum and maximum values recorded before and after wastewater discharge were equal and amounted to 7.7 and 8.0, respectively. A slight difference was observed only for median that before discharge point took a value equal to 7.9, whereas after discharge – 7.8. The quantity of organic compounds expressed as BOD_5 parameter before discharge of treated wastewater was slightly higher as compared to values recorded after the discharge point. Values of arithmetic means and medians for BOD_5 indicator before and after discharge amounted respectively to 3.4 and 2.8 mg $O_2 \cdot dm^{-3}$ as well as 3.1 and 2.6 mg $O_2 \cdot dm^{-3}$. The range of BOD_5 oscillations was from 1.8 to 8.2 mg $O_2 \cdot dm^{-3}$ before and from 1.6 to 4.7 mg $O_2 \cdot dm^{-3}$ after the wastewater discharge point. In the case of COD, slightly higher values of this parameter were observed after the treated wastewater discharge point. Arithmetic mean values before and after wastewater discharge were close to each other (32.8 and 35.8 mg $O_2 \cdot dm^{-3}$), medians were respectively equal to 24.0 and 33.0 before and after discharge. The COD values before wastewater discharging point varied in the range from

 DOI: 10.1051/e3sconf/20172200127

16.0 up to 116.0 mg $O_2 \cdot dm^{-3}$, whereas after the discharge from 22.0 to 52.0 mg $O_2 \cdot dm^{-3}$. Considering the standard deviations at both points (28.2 before and 10.2 mg $O_2 \cdot dm^{-3}$ after), it can be stated that supply of treated sewage stabilizes the naturally occurring changes in this parameter in surface waters. Water in the receiver was characterized by higher values of determined nitrogen forms after wastewater discharge point. The concentration of total nitrogen before and after discharge of wastewater was largely determined by changes in the quantity of nitrate and total Kjeldahl nitrogen (TKN). Among studied nitrogen forms, nitrate was characterized by the highest concentration and at the same time the highest variability. Arithmetic means for this parameter before and after discharge of sewage were respectively 3.8 and 4.8 mg $N \cdot dm^{-3}$, while the respective variability range was from 2.5 to 4.9 mg $N \cdot dm^{-3}$ before and from 4.9 to 9.4 mg $N \cdot dm^{-3}$ after wastewater discharge point. The second largest, in terms of concentration, form of nitrogen was the total Kjeldahl nitrogen, the mean concentrations of which before and after the discharge of wastewater were as follows 2.2 and 3.9 mg $N \cdot dm^{-3}$. Like for nitrate, range of total Kjeldahl nitrogen (TKN) variations before sewage discharge was lower and amounted from 1.7 to 3.0 mg $N \cdot dm^{-3}$, whereas after the discharge point from 2.5 to 9.7 mg $N \cdot dm^{-3}$. The slightest change concerned concentration of ammonium and nitrite. Arithmetic mean value before and after the discharge of wastewater for ammonia were equal to 1.0 and 1.8 mg $N \cdot dm^{-3}$, while for nitrite was characterized by mean value of 0.09 mg $N \cdot dm^{-3}$ before and 0.15 mg $N \cdot dm^{-3}$ after the wastewater discharging point. Average concentrations of chloride ions before and after the discharge of treated sewage were different and amounted respectively to 114.7 and 131.4 $mg \cdot dm^{-3}$. Much larger range of this parameter variability was recorded before (73.2–147.0 $mg \cdot dm^{-3}$) than after wastewater discharge point (108.0–158.0 $mg \cdot dm^{-3}$). Changes in phosphates and total phosphorus were subject to identical alterations before and after the discharge of treated sewage. Values of these parameters were similar at both points. The arithmetic means for these parameters before and after discharge of wastewater for phosphates were 0.9 and 0.9 mg $P \cdot dm^{-3}$, while for total phosphorus 0.3 and 0.3 mg $P \cdot dm^{-3}$. The value of total suspended solids after discharge of wastewater was significantly reduced in relation to that observed before discharging point. The average value of this parameter before discharge was 22.3 $mg \cdot dm^{-3}$, while after discharge 6.1 $mg \cdot dm^{-3}$. Similar differentiation characterized the variation ranges observed at both points. Before discharge of wastewater, the amount of total suspended solids (TSS) varied in the range from 3.5 to 140.0 $mg \cdot dm^{-3}$, whereas after discharge, the TSS quantity was within the range from 2.0 to 14.3 $mg \cdot dm^{-3}$. The concentration of dissolved oxygen in the waters of the receiver after discharge of wastewater underwent a slight decrease. The average concentration of this parameter before discharge of treated wastewater was 9.2, while after that point 8.8 mg $O_2 \cdot dm^{-3}$. Before discharge of treated wastewater, concentration of dissolved oxygen varied from 5.8 to 7.8 mg $O_2 \cdot dm^{-3}$. The minimum concentration of dissolved oxygen after the discharging point was slightly higher and amounted to 7.8 mg $O_2 \cdot dm^{-3}$, whereas maximum one was lower as compared to the value before discharging point, i.e. 10.3 mg $O_2 \cdot dm^{-3}$. Basic statistical measures for individual tested parameters are summarized in Table 1.

Table 1. Descriptive statistics for water quality before and after the discharge of treated wastewater.

Parameter	Unit	Average		Median		Minimum		Maximum		Standard deviation	
		B	A	B	A	B	A	B	A	B	A
Temp.	°C	9.6	14.9	8.5	13.3	3.4	11.0	18.0	20.8	5.2	3.8
pH	-	-	-	7.9	7.8	7.7	7.7	8.0	8.0	-	-
BOD_5	mg $O_2 \cdot dm^{-3}$	3.4	2.8	3.1	2.6	1.8	1.6	8.2	4.7	1.7	0.8
COD		32.8	35.8	24.0	33.0	16.0	22.0	116.0	52.0	28.2	10.2
TKN	mg $N \cdot dm^{-3}$	2.2	3.9	2.0	3.1	1.7	2.5	3.0	9.7	0.4	2.0
$N\text{-}NH_4$		1.0	1.8	1.0	1.2	1.0	1.0	1.0	6.8	0.0	1.7
$N\text{-}NO_3$		3.8	4.8	4.3	4.7	2.5	1.1	4.9	9.4	0.8	2.8
$N\text{-}NO_2$		0.09	0.15	0.07	0.17	0.05	0.04	0.14	0.27	0.04	0.08
TN		5.9	8.9	6.0	9.2	4.6	5.3	7.3	12.3	0.8	2.4
Cl^-	$mg \cdot dm^{-3}$	114.7	131.4	114.0	126.0	73.2	108.0	147.0	158.0	23.0	16.9
TP	mg $P \cdot dm^{-3}$	0.3	0.3	0.2	0.2	0.2	0.2	0.8	1.0	0.2	0.2
PO_4		0.9	0.9	0.6	0.6	0.6	0.6	2.6	3.0	0.6	0.7
TSS	$mg \cdot dm^{-3}$	22.3	6.1	10.0	4.5	3.5	2.0	140.0	14.3	39.6	4.0
O_2	mg $O_2 \cdot dm^{-3}$	9.2	8.8	9.1	8.6	5.8	7.8	11.7	10.3	1.7	0.9

B - before; A - after

The analysis of results applying Wilcoxon test (Table 2) showed differences between temperature of the wastewater before and after the discharge of treated sewage. Temperature changes were most evident in winter. Water temperature in the receiver during this period is usually significantly lower than the temperature of treated wastewater, which causes it to increase after the point of wastewater discharge. Another changed parameter was BOD_5. Smaller amounts of organic compounds expressed with this indicator were observed after wastewater discharge, which indicates the water dilution due to the supply of treated wastewater. In the case of total Kjeldahl nitrogen, ammonia, and nitrite, higher concentrations were recorded after the point of treated sewage discharge. The results of the test indicate a slight enrichment of water in these forms of nitrogen and, consequently, total nitrogen (TN). It has also been an increase in the concentration of chloride ions in water of the receiver after wastewater discharge point. This is due to the fact that these ions are not removed from wastewater as a result of life processes of microorganisms mostly occurring in the active sludge. The last parameter, for which statistical differences before and after discharge were observed, were total suspended solids (TSS). There was a reduction in the concentration of suspended solids, like BOD_5 value in waters after the wastewater discharging point. This indicates the dilution of water in the receiver with treated sewage. For other tested quality parameters, no significant changes were recorded after discharge of treated wastewater point.

Table 2. Results of the Wilcoxon test for pairs of variables.

Pair of variables	T	p
°C& °C	0.00	0.00
pH & pH	3.50	0.14
BOD & BOD	6.00	0.03
COD & COD	13.00	0.08
TKN & TKN	1.00	0.00
$N-NH_4$ & $N-NH_4$	0.00	0.03
$N-NO_3$ & $N-NO_3$	23.00	0.37
$N-NO_2$ & $N-NO_2$	11.00	0.05
TN & TN	4.00	0.01
Cl^- & Cl^-	1.50	0.01
TP & TP	11.00	0.17
PO_4 & PO_4	11.00	0.17
TSS & TSS	9.00	0.03
O_2 & O_2	14.00	0.17

T – result of the Wilcoxon test; p - probability

High concentrations of biogenic substances in receivers of industrial and municipal wastewater are a worldwide problem. Tests carried out by Chen et al. [1] showed an increased content of ammonia and total nitrogen in the tributaries of the Yellow River in China. However, in Turkey in the Mamasin reservoir, high concentrations of nitrate, nitrite and ammonia were found. The source of these substances, in addition to farming operations, was municipal wastewater [2]. Increasing the concentration of chloride ions and nitrite after discharge of treated sewage was observed by Lewandowska-Robak et al. [4] during research conducted in the area of sewage treatment plants with enhanced removal of biogenic substances in Tuchola Forest. Slightly lower growth in COD value in waters of the San river after the discharge of treated sewage was indicated by Policht-Latawiec et al. [20]. Kanownik and Rajda [3] reported that the treated wastewater does not affect the pH value in Jaszczurówka stream, which is the receiver of wastewater from mechanical-biological treatment plant in Jaszczurowa.

4 Conclusions

The analyzed results related to the relationship sewage treatment plant – receiver, allow to draw the following conclusions:

1. In the analyzed period, municipal wastewater treatment plant in Bialystok provided an efficient removal of impurities in accordance with the requirements of the water permit.
2. Discharge of treated wastewater affects the values of temperature, BOD_5, total Kjeldahl nitrogen, ammonia, nitrite, total nitrogen, chloride ions, and total suspended solids in waters of the receiver, but it does not change other tested water quality parameters.
3. Water below the sewage discharge point revealed an increase in all forms of nitrogen, however not all of the observed changes were statistically significant.
4. Supply of treated wastewater reduces the amount of organic compounds (defined by BOD_5) and the concentration of total suspended solids in water of Biała river due to the dilution of the receiver's water and the water self-purification processes.
5. Taking into account results of the contents of individual indicators in Biała river before and after wastewater discharge points, no adverse effects of the sewage treatment plant

on water quality in the receiver, were recorded, despite the fact that the ratio of the amount of sewage discharged to the average flow in the river is unfavorable for the receiver.

6. The increase in the concentration of nitrogen compounds for discharge is noticeable due to the fact that the water permit is based on standards, not taking into account the size of the receiver nor his opportunities for self-cleaning.
7. Unsatisfactory status of water quality in Biała river due to its central location in 300.000-inhabitants agglomeration of Bialystok and furthermore meandering river through a large area of the city. In addition, the river is a receiver of surface runoff and rainwater from the whole catchment area, which classifies it to only good status. Therefore, to improve the quality of surface waters in urbanized catchments, integrated management and planning activities, involving the development of appropriate action programs dependent on natural conditions and needs of the economy, should be led.

The research has been carried out in the framework of project No. MB/WBiIŚ/13/2015 and project No. MB/WBiIS/6/2016 financed from the funds for science MNiSW and as well as by courtesy and help of Białystok Waterworks Ltd.

References

1. J.S. Chen, D.W. He, N. Zhang, S.B. Cui, Environmental Monitoring and Assessment, **93** (2004)
2. H. Elhatip, Ö. Güllü, Environ. Geol. **48** (2005)
3. W. Kanownik, W. Rajda, Gas, Water and Sanitary Engineering, **9** (2011)
4. M. Lewandowska-Robak, Ł. Górski, T. Kowalkowski, H. Dąbrowska-Naskręt, I. Mieszkowska, Engineering and Environmental Protection **14**, 3 (2011)
5. M. Delkash, B.E. Bakhshayesh, H. Kazemian, Micropor. Mesopor. Mat. **214** (2015)
6. S. De Gisi, G. Lofrano, M. Grassi, M. Notarnicola, Sustainable Materials and Technologies, **9** (2016)
7. I. Evan-Ezra, M. Beliavski, S. Tarre, C. Dosoretz, M. Green, Desalination **272** (2011)
8. J. Długosz, J. Gawdzik, Proceedings of ECOpole **7**, 1 (2013)
9. K. Maj, A. Masłoń, Magazine of Civil Engineering, Environment and Architecture, **62** (2015)
10. Y. Lin, M. Guo, N. Shah, D.C. Stuckey, Bioresource Technol. **215** (2016)
11. K. Pawęska, K. Kuczewski, 3rd Polish General Congress of Environmental Engineering (2009)
12. W. Miernik, D. Młyński, A. Wałęga, K. Chmielowski, R. Karwacki, Infrastructure and Ecology of Rural Areas, **1** (2016)
13. E. Neverova-Dziopak, P. Cierlikowska, Environmental Protection **2**, 36 (2014)
14. B. Kaźmierczak, Annual Set Environment Protection **15** (2013)
15. C. Przybyła, J. Bykowski, Annual Set Environment Protection, **11** (2009)
16. http://www.wobi.pl/
17. PN-ISO 5667-10:1997, Water quality - Sampling - Guidance of sampling plants in Poland
18. E.W. Rice, R.B. Baird, A.D. Eaton, L.S. Clesceri, Standard Methods for the Examination of Water and Wastewater, American Public Health Association (APHA),

DOI: 10.1051/e3sconf/20172200127

American Water Works Association (AWWA), Water Environment Federation (WEF), 22 nd Edition (2012)

19. S. Tyszewska, I. Kardel, Study hydrographic valley Biała river, Białystok City Hall, (2009)
20. A. Policht-Latawiec, W. Kanownik, D. Łukasik, Infrastructure and Ecology of Rural Areas, **1** (2013)

Calibration of the computer model describing flows in the water supply system; example of the application of a genetic algorithm

Maria Orłowska-Szostak[1,*], and *Ryszard* Orłowski[1]

[1]Faculty of Civil and Environmental Engineering, Gdansk University of Technology, ul. G. Narutowicza 11/12, 80-233 Gdansk, Poland

Abstract. The paper discusses some relevant aspects of the calibration of a computer model describing flows in the water supply system. The authors described an exemplary water supply system and used it as a practical illustration of calibration. A range of measures was discussed and applied, which improve the convergence and effective use of calculations in the calibration process and also the effect of such calibration which is the validity of the results obtained. Drawing up results of performed measurements, i.e. estimating pipe roughnesses, the authors performed using the genetic algorithm implementation of which is a software developed by Resan Labs company from Brazil.

1 Introduction

The calibration process is characterized by indeterminacy which results from the fact of rarely distributed measurement points and also from the fact that there are not known neither searched characteristics of the system components (especially the roughness of pipes) nor water partitions occurring during performing these measurements.

The aim of each of the different known calibration method is removing (reducing the magnitude) aforementioned indeterminacy of the task. Each method requires performing a variety of measurements used in the calibration. The efficiency of the calibration depends, among others, on the number and location of measurement points. Moreover, frequently used method of reducing the number of unknown roughnesses and partitions [1] is grouping of pipes of virtually identical roughness coefficients and grouping of junction nodes with virtually identical timing patterns of water partitions.

The efficiency of the calibration also depends on the method of optimal use of the above mentioned measurements (intuitive method, methods/algorithms implemented numerically) when estimating characteristics (parameters appearing in the analytical descriptions of these characteristics) of the system components.

Among the above-mentioned calibration methods, the most popular is the one which can be called "direct"; this method uses measurements of pressure and flow at selected places (nodes) of the system. In small and medium-sized systems the nodes for measuring are selected intuitively (see an expert selection in agreement with the water company's

* Corresponding author: maria.o@pg.gda.pl

 DOI: 10.1051/e3sconf/20172200128

employees, in the way described later in this paper, in the description of the calibration carried out for the exemplary water distribution system). For the larger systems, it is reasonable to optimize the number and location of junction nodes at which we perform the measurements (see e.g. the theoretical basis of *sampling design* [2]).

Drawing up the results of performed measurements, i.e. estimating pipe roughnesses, the author performed using the genetic algorithm [1, 3–6] implementation of which is a software developed by Resan Labs company from Brazil.

2 The modelled exemplary water distribution system

An exemplary system of water transport and distribution supports the following four places in Wejherowo commune: Sopieszyno, Ustarbowo, Gowino and Pętkowice. Water is taken from quaternary forms by water intakes located in Sopieszyno and Pętkowice (Fig. 1).

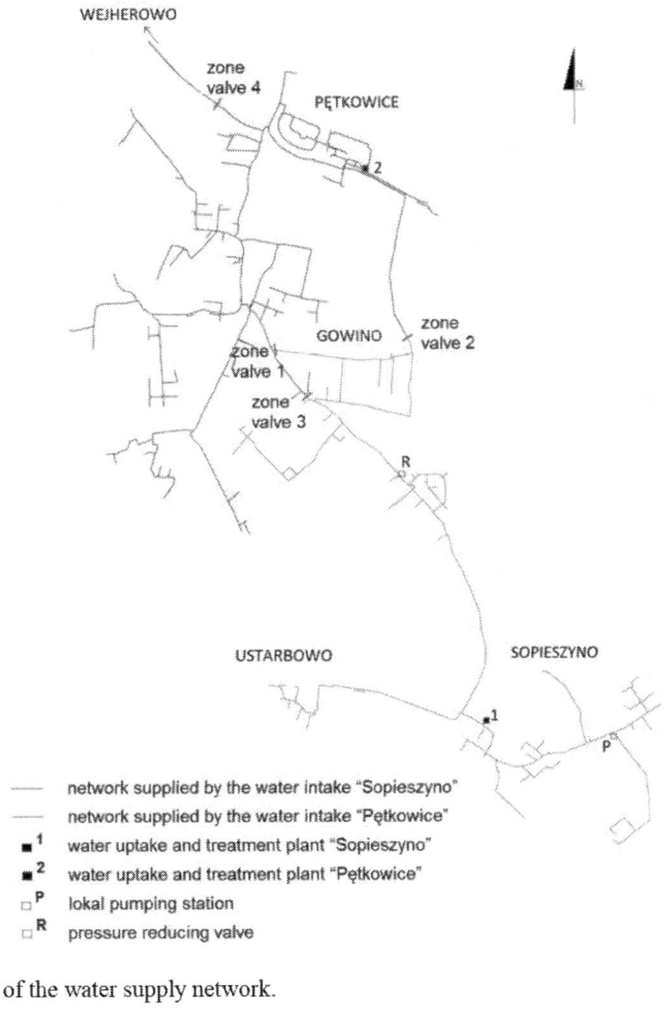

Fig. 1. Scheme of the water supply network.

3 The scope and the manner of key data collection

The computer model of flows in the exemplary water transport and distribution system, presented in the article, was created with the use of the EPANET software. The data describing the current condition of the system include, i.a. system geometry, diameters, material and the age of pipes, technological schemes and methods of the regulation of the operation of objects and fittings which work with the water supply system, current water partitions [7] and current character of water demand in the individual towns supplied by the system as well as parameters characterising the quality of water supplied to consumers. In the process of collecting such data obtained from the water supply company some restrictions appeared concerning these data assortment and accuracy. In such conditions, it was attempted to use the information obtained maximally in terms of the required credibility of the model coming into existence.

The system structure and the pipe materials are typical for municipal systems built in Poland in the 90's of the 20th c. [8]. A substantial part of the network is made of PVC. These are older pipes. On the second place there are pipes the age of which does not exceed ten years. These pipes are made of PE.

The numerous measurements of flow rate and pressure were carried out at the water in-takes in Sopieszyno and Pętkowice and also at selected nodes of the water distribution system (see chap. 4). The measurements were recorded every hour [7, 9]. Due to the impossibility of obtaining exact data on water demands in respective nodes, they were determined on the basis of the average water consumptions by individual households, obtained from the several years' quarterly readings of water meters. Thus, the weight distribution of base consumptions in the water supply system could be determined.

The time patterns of inequality of hourly water consumption for individual nodes were determined with taking into consideration some specific information about water consumption of individual objects, their sizes etc.

4 Calibration of the computer model

Taking into account the assortment of measurements, the applied method of the calibration of model can be classified as the direct method, described in chapter 1, i.e. the method which uses the measurements of pressure and flows at selected nodes of the water distribution system.

The drawing up of the results of above-mentioned measurements, i.e. estimation of roughness of pipes forming a water distribution network as well as emitter coefficients, was carried out using a genetic algorithm. It was applied EPANET Calibrator software, developed by Resan Labs from Brazil [3,6,10].

The authors of the EPANET Calibrator software give the form of the objective function in optimization task of calibration of model of water transport and distribution system. The criterion of the optimization task is to minimize the said objective function [10]

$$MinFO = \sum_{i=1}^{N} \left[\frac{\Delta P_i}{\frac{\sum_{i=1}^{N} P_i^{'}}{N}} \right]^2 + \sum_{j=1}^{M} \left[\frac{\Delta Q_j}{\frac{\sum_{j=1}^{M} Q_j^{'}}{M}} \right]^2 \quad (1)$$

where: $P'i$ – set of pressure values, $Q'i$ – set of flow values, ΔPi – difference of pressure between simulated and observed data, ΔQi – difference of flow between simulated and observed data, N – number of monitored nodes, M – number of monitored pipes.

The coordinates of the decision vector can be roughnesses of pipes and emitter coefficients, which in the objective function occur in the implicit form. In turn, implicit restrictions in this optimization task are conservation equations describing the flows in the

DOI: 10.1051/e3sconf/20172200128

system. Due to formal reasons, mentioned coordinates of the decision vector are under restrictions with explicit character. The optimal solution is the decision vector meeting the above criterion.

The consumption of water in time and space of the supplied area and also the initial roughnesses of pipes have been determined as precisely as possible due to their high importance for the convergence and the stability of the calculations algorithm. Localization of measuring points was made following the selection of sites with the highest pressure fluctuations (primarily terminal nodes of the network branches) and in the areas with the largest water consumptions (around the "centre of gravity of the system").

During calibration, in subsequent simulations, pipe roughnesses are corrected and emitter coefficients are determined which affect the flows calculated in the course of the simulations because the flows are taking into account the differences between initially assumed and estimated partitions involving leakages. In the software used with the EPANET Calibrator plug, emitter coefficients constitute the basis for estimating outflows q from the water supply system. Thus, the author of the said software plug provides the following dependence [11]:

$$q = AC_d\sqrt{2gh} \tag{2}$$

where A is an orifice area, C_d is a discharge coefficient and g is gravitational acceleration.

It must be underlined here that the analogous dependences of a similar nature and form may be found in the various examples of source literature [12–17].

As we recalled in chapter 1, we are creating the groups of pipes of similar roughness in order to reduce the number of unknowns / searched roughnesses.

The rule of the automatic calculations control during calibration consists in it, that in subsequent iterations, the said coordinates of a decisive vector are determined according to the genetic algorithm which provides for searching for a final solution satisfying a criterion in the form (1).

Calibration may be conducted independently of the number of measurement points and generally there is even no obligation of measuring the flow rates of intakes supplying the system with water, because the discharges of intakes and water partitions are balanced during the fulfilment of the estimation algorithm, during the simulations of flows. These are the important assets of the process under discussion because they facilitate model calibration and the calibration accuracy depends on our decision concerning number and locations of the measurement points.

Some more essential findings and remarks about calibration are given in the last chapter devoted to conclusions. Only brief calibration results are provided in this chapter, which ensure that the model adequately describes the operation of the water supply system.

Fig. 2 and fig. 3 show the comparisons of observed 24 h flows and pressures with simulated ones before and after calibration, at two exemplary pipes (comparisons of flows) and nodes (comparisons of pressures). The pipes and nodes are located close to both water intakes (practically at their outlets) Pętkowice (fig. 2) and Sopieszyno (fig. 3).

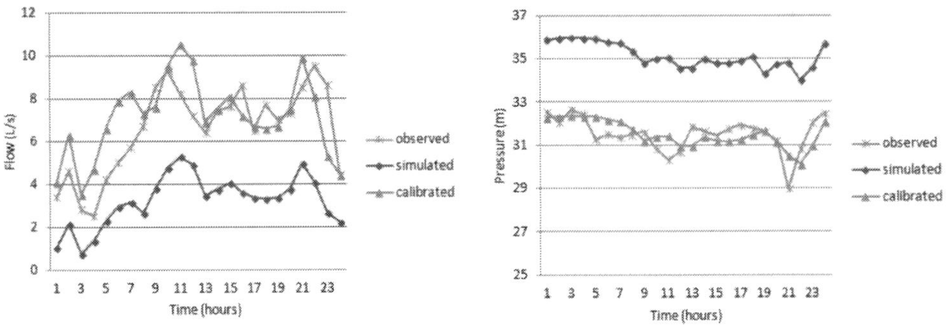

Fig. 2. Comparison of observed flows and pressures with simulated ones before and after calibration, respectively at exemplary pipe and node in Pętkowice village; Initial population: 200, Generations: 400, Mutation probability: 0.01, Crossover probability: 0.95.

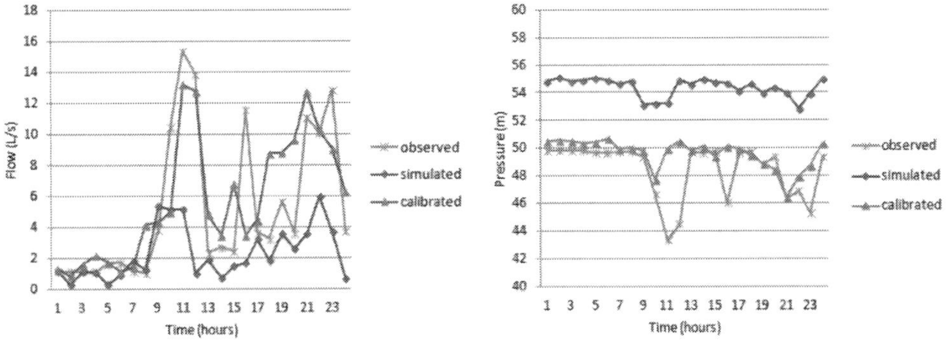

Fig. 3. Comparison of observed flows and pressures with simulated ones before and after calibration, respectively at exemplary pipe and node in Sopieszyno village; Initial population: 200, Generations: 400, Mutation probability: 0.01, Crossover probability: 0.95.

The results of performed calibration, i.e. estimation of the pipes roughnesses and the determination of the emitters coefficients and so the accuracy of the system operation simulation are very good what was not possible to be stated with the reference to not-calibrated model, although starting values of the roughnesses of pipes and the partitions of water were assumed with the great care. Differences between pressures and flows simulated with the help of calibrated model and ones that were measured do not exceed permissible values given at the work [18], and they even are definitely smaller.

DOI: 10.1051/e3sconf/20172200128

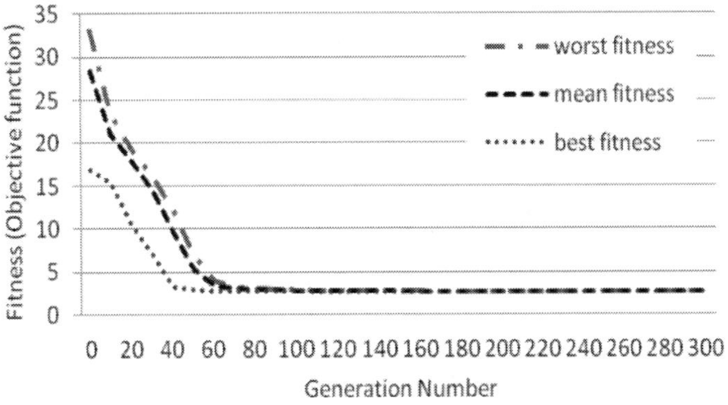

Fig. 4. Characteristics of the convergence of the applied genetic algorithm.

As can be seen (fig. 4), calculations conducted with the EPANET Calibrator software were not only stable, but also were characterized by a very good convergence. In contrast with often applied "manual" roughnesses adjustments while making the calibration, these calculations are proceeding automatically and quickly, not requiring the interference of the engineer drawing the computer model up. It also helps to avoid the tendency of setting the local minimum of the objective function.

5 Conclusions

Calculations conducted with the genetic algorithms are not only stable, but also are characterized by a very good convergence of the optimization process of calibration. The graph presented in the fig. 4 is demonstrating a very good convergence which is even better than assumed while starting the calibration.

Significant elements affecting the convergence of the calibration process of the model are: applying presently standard methods of reducing the number of unknown roughnesses, carefully prepared starting data recalled in chapter 3 as well as the advantages of the genetic algorithm.

The calibration of the model, using measurements of pressure and flow in selected locations of the system, turned out to be completely sufficient.

The method of locating measuring points described in the article was sufficient. It was based on the expert selection in agreement with the water company's employees. Namely, the localization of measuring points was made following the selection of sites with the highest pressure fluctuations (primarily terminal nodes of the network branches) and in the areas with the largest water consumptions (around the “centre of gravity of the system”). More sophisticated methods of sampling design are desirable only for very large and complex water supply systems.

In the course of the calibration there are calculated the emitter coefficients which affect calculated flows; the flows are taking into account the differences between initially assumed and estimated partitions involving leakages.

The leakages determined by means of the software with the EPANET Calibrator plug depend on pressures in the respective pipelines of water supply network and their diameters, as well as other characteristic data included by means of relevant coefficients in the formulas describing emitter coefficients. There are very similar approaches to this question in the source literature.

References

1. S. Lingireddy, L. E. Ormsbee, Civ. Eng. Environ. Syst. **19**, 1, 13–39 (2002) http://dx.doi.org/10.1080/10286600212161
2. C.A. Bush, J.G. Uber, J. Water Res. Pl-Asce **1249**, 6, 334–344 (1998) http://dx.doi.org/10.1061/(ASCE)0733-9496(1998)124:6(334)
3. N. Abe, P.B. Cheung, *EPANET Calibrator. User Guide* (2008)
4. E. Keedwell, S.T. Khu, Eng. Appl. Artif. Intel. **18**, 461–472 (2005) http://dx.doi.org/10.1016/j.engappai.2004.10.001
5. D. Nardo, M. Di Natale, C. Gisonni, M. Iervolino, J. Water Supply Res. T. **64**, 1, 36–46 (2015) http://dx.doi.org/10.2166/aqua.2014.004
6. W. Stanisławski, M. Zmarzły, Pomiary Automatyka Robotyka **12**, 61–64 (2010)
7. The materials used and the source documentation: geodesic map of the area with the specified water supply infrastructure acquired from the Municipal Department of Geodesy; details of the water supply network, data on water consumption in selected time periods and pressure measurements data acquired from the water company (2016)
8. M. Kwietniewski, M. Tłoczek, L. Wysocki et al., *Rules for selection of materials and design solutions for the construction of water pipes* (Publishing House Izba Gospodarcza Wodociągi Polskie, Bydgoszcz, Poland, 2011) (in Polish)
9. E. Zamościńska, A. Koncewicz, *The computer modeling of flows occurring in the water supply system of the Wejherowo Commune; suggestions in relation to the modernization of the system from a point of view of hydraulics* (Master's thesis done under the direction of M. Szostak-Orłowska, Department of Sanitary Engineering, Faculty of Civil and Environmental Engineering, Gdańsk University of Technology, Poland (2016) (in Polish)
10. N. Abe, P.B. Cheung, *Epanet Calibrator – An integrated computational tool to calibrate hydraulic models* (Integrating Water Systems. Boxall & Maksimovic (eds) Taylor & Francis Group, London, ISBN 978-0-415-54851-9, 2010)
11. N. Abe, *Avaliacao e deteccao de vazamentos Em tempo Real atraves de metodo inverso* (Ministerio da Educacao, Universidade Federal de Mato Grosso do Sul Centro de Ciencias Exatas e Tecnologia. Campo Grande, Brasil, 2010) (in Portuguese)
12. B. Greyvenstein , J.E. van Zyl, J. Water Supply Res. T. **56**, 2, 117–124 (2007) https://doi.org/10.2166/aqua.2007.065
13. Bakogiannis, A. Tzamtzis, *Modeling Of District Metered Areas With Relatively High Leakage Rate. The Case Study Of Kalipoli's DMA* (11th International Conference on Hydroinformatics, HIC 2014, New York City, USA, 2014) http://academicworks.cuny.edu/cc_conf_hic/384
14. G. Germanopoulos, Civil Eng. Syst. 171–179 (1985) http://dx.doi.org/10.1080/02630258508970401
15. A. Vela, R. Perez, V. Espert, *Incorporation of leakage in the mathematical model of a water distribution network* (Proceedings of the 2nd international conference on computing methods in water resources, Computational Mechanics Publication, 245–257, 1991)
16. R. Burrows, G. Mulreid, M. Hayuti, *Introduction of a fully dynamic representation of leakage into network modeling studies using Epanet* (Advances in Water Supply Management, C. Maksimovic, D. Butler, A. Memon (eds) Taylor & Francis Group,

DOI: 10.1051/e3sconf/20172200128

London, ISBN 978-90-5809-608-1, 2003)
http://dx.doi.org/10.1201/NOE9058096081.ch12

17. M. Tabesh, A.H. Asadiyani Yekta, R. Burrows, Int. Ser. Prog. Wat. Res. **23**, 477–492 (2009) http://dx.doi.org/10.1007/s11269-008-9284-2

18. T.M. Walski, D.V. Chase, D.A. Savic et al., *Advanced water distribution modeling and management* (Haestad Press, Waterbury, CT USA, 2003)

DOI: 10.1051/e3sconf/20172200129

Analysis of freeze protection methods for recuperators used in energy recovery from exhaust air

Anna Pacak[1,*], *Andrzej* Jedlikowski[1], *Demis* Pandelidis[1], and *Sergey* Anisimov[1]

[1]Wrocław University of Science and Technology, Faculty of Environmental Engineering, Department of Air Conditioning, Heating, Gas Engineering and Air Protection, ul. Norwida 4/6, 50-373 Wrocław, Poland

Abstract. In this study theoretical analysis of the heat and mass transfer in counter-flow recuperators used for energy recovery in air handling units (AHU) under sub-zero outdoor air temperature operating conditions is presented. The most probable variants of year-round heat exchanger operation performance, which characterized by existence of three active heat and mass transfer zones (*"dry"*, *"wet"*, *"frost"*), and effect of the latent heat of water vapour condensation on the realization of these variants was determined. It was established, that the frost tends to take place with increasing temperature effectiveness of the heat exchanger. Two main techniques of the frost prevention (preheating and bypassing the outdoor airflow) were described and analysed. The values of critical outdoor temperatures and outdoor-to-return airflow rate ratio were determined on the base of parametric frosting limits analysis conducted under different inlet return airflow conditions. The comparison of the heat recovery efficiency and additional energy consumption for the air treatment in the AHU is presented.

1 Introduction

Nowadays the increased global energy consumption, many legal requirements and environmental protection, resulted in the rational management of energy resources. For this reasons, some actions are used to improve the effectiveness of existing and designed air conditioning and ventilation systems, well-known as the major consumers of electricity and heat [1]. For this reason, running costs of air conditioning system can be significantly reduced by implementing heat recovery. Heat recovery can be simply defined as the collection and re-use of waste heat arising from any technical or physical process that would otherwise be lost. This process can contribute to reduce the overall energy consumption and provide energy for other purposes [2]. The counter-flow heat exchanger can be used to perform this function in heat recovery technology.

Unfortunately, counter-flow heat exchangers are very often exposed to freezing when a heat recovery system is used in cold climates. Additionally, the presence of a frost layer in the heat exchanger core decreases significantly the heat recovery system performance. For this reason, frosting limits, freezing processes and different frost-preventing techniques

[*] Corresponding author: anna.pacak07@gmail.com

have been extensively studied in recent years. Nasr et al. [3] analysed two heat exchangers: with a permeable membrane and impermeable membrane under frosting conditions. Wang et al. [4] described a "partial bypass" technique. The authors concluded that this action facilitates the heat transfer augmentation. Freund et al. [5] considered a preheated air strategy to change the air temperature and prevent frost formation on the coil. Besides, the authors compared the performance of five frost control strategies used for improve energy recovery in ventilation systems. Nasr et al. [6] studied the performance of two cross-flow heat exchangers under different operating conditions. They presented the values of frosting limit and defrosting time ratio. Moreover the authors evaluated the effects of two defrosting methods on energy consumption of ventilation in three cold cities. They concluded, that the air preheating method performs better than the outdoor air bypassing method for frost prevention.

Nevertheless, the problem of frost formation occurrence inside the counter-flow heat exchangers is still not solved and the most popular frost control techniques based on the outdoor air preheating or adjusting the outdoor-to-return airflow rate ratio are still facing challenges under extreme outdoor air temperature conditions.

2 Theoretical analysis

In order to analyse the counter-flow heat exchanger performance it is necessary to establish the most probable variants of its year-round operating conditions. The most often analysed variant is the "*dry*" variant of heat exchange performance. In this case, the surfaces of both heat exchanger channels are dry. Such conditions occur usually in the summer. Although the "*dry*" heat transfer conditions always occur in the outdoor air channel, the processes in the exhaust air channel may run differently [1]. For instance, there is a possibility of combined sensible and latent heat transfer on the surface of the return air channel, especially under the cold climate conditions. It means that a part of the plate surface is "*wet*" due to water vapour condensation in the form of water film. In this case the latent heat of the water vapour condensation can be additionally recovered. That is why heat recovery effectiveness is higher comparing to the "*dry*" mode. On the other hand, water vapour condensation in the form of the frost layer, which can be observed under sub-zero ambient air temperature conditions, reduces the heat exchanger efficiency and increases the pressure drop on the return airflow side, because the passage may be blocked by frost.

In this regard, a detailed analysis of coupled heat and mass transfer inside the counter-flow plate heat exchanger under frosting operating conditions was performed on the base of the original ε-NTU model [7]. Parametric numerical analyses were conducted to investigate heat and mass transfer in the counter-flow plate heat exchanger under sub-zero outdoor air temperature operating conditions. Three active heat and mass transfer areas were established in a counter-flow plate heat exchanger (see Fig. 1).

The detail analysis of these particular heat and mass transfer zones creation revealed the most probable variants of year-round operating conditions of the counter-flow heat exchanger. The implementation of the particular variant of heat and mass transfer inside the return airflow channel of the counter-flow plate heat exchanger depends upon the relations of the temperatures in two decisive zones on the return airflow channel surface (in the "*cold zone*" and in the "*hot zone*") and the value of the inlet return airflow dew point temperature (see Fig. 1). In order to illustrate and explain clearly the processes occurring inside the heat exchanger channels, the numerical simulations were carried out under conditions with constant inlet outdoor and return airflows temperatures (t_{1i}= -20°C, t_{2i} = 20°C) and variable inlet return airflow relative humidity (RH_{2i} = (10–90)%). The results of these numerical investigations are presented in Figs. 2(a) and (b).

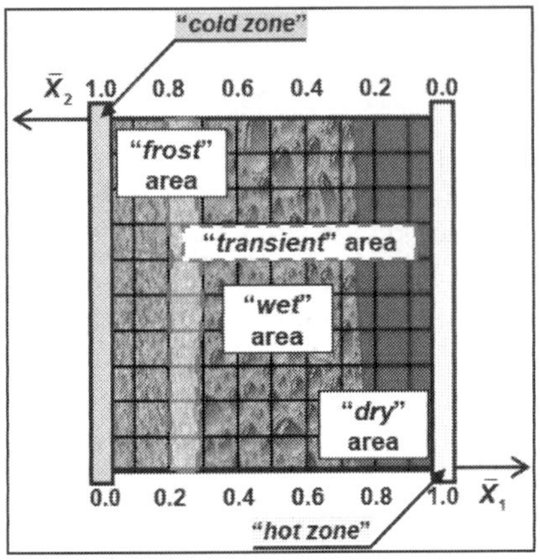

Fig. 1. Variants of heat and mass transfer zones.

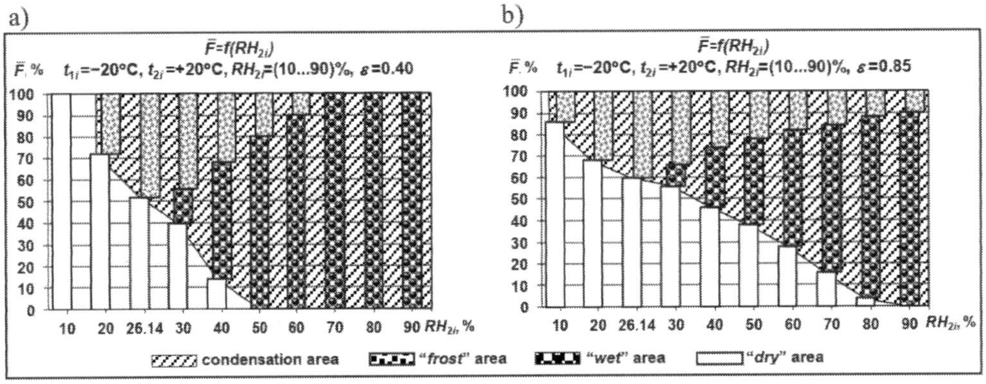

Fig. 2. The percentage distribution of the heat and mass transfer areas inside the counter-flow plate heat exchanger channels determined for the inlet outdoor airflow temperature t_{1i} = -20°C expressed as a function of the return air relative humidity (at the temperature effectiveness: a) η = 0.40, b) η = 0.85).

The data presented in Figs. 2(a) and (b) confirm clearly the obvious fact, that a higher value of the inlet return airflow relative humidity results in a larger area of the plate surface on which water vapour condensation takes place.

A "*dry*" zone at the entrance region dominates at low values of the return airflow relative humidity, but increasing the relative humidity RH_{2i} reduces the dimension of this "*dry*" zone and at the values of RH_{2i} > (53–90)% this zone disappears.

It can be also seen, that in the case of the sub-zero inlet return airflow dew point temperature $t_{2i}^{DP} < 0°\text{C}$, which corresponds to the relative humidity $RH_{2i} \leq 26.15\%$, the water vapour condensation may occur *only* in the form of the frost layer and increasing the inlet relative humidity RH_{2i} under such conditions reduces "*dry*" zone at the expense of increasing the "*frost*" zone. In the case of the above-zero dew point temperature of the return air $\left(t_{2i}^{DP} > 0°\text{C}\right)$ the water vapour condensation may be observed both in the form of the water film and in the form of the frost layer. It should be noted, that the "*frost*" zone reaches maximum at the value of inlet return airflow dew point temperature equalled to $t_{2i}^{DP} = 0°\text{C}$ (RH_{2i} = 26.14%). At first sight, it could be surprising, that the dimension of the "*frost*" zone decreases with the increasing the inlet relative humidity of the return air higher than 26.14%. The reason for this phenomenon can be explained by the analysis of the interfacial heat and mass transfer in the return airflow channel. The latent heat of water vapour condensation in the form of liquid film increases the local plate temperature in the return air channel. In turn, a higher value of the local plate temperature leads to a smaller air side temperature gradient at the gas-liquid interface and thus a lower value of the local sensible heat flux and, as a consequence, a smaller temperature drop of the return air in its flow direction. That is why the local plate temperature at the exit region of return airflow channel increases with increasing RH_{2i} and thus a frost formation occurs on the more remote exit region of the plate surface in the return air channel [7].

So, an increase of inlet relative humidity of the return air in the range of value from 10% to 90% leads to an increase of the *total* water vapour condensation zone due to natural reduction of the *"dry"* zone. Increasing the inlet relative humidity of the return air under such conditions causes a decrease of the *"frost"* zone at the exit region due to preliminary expansion of the *"wet"* zone, although the total area of condensation is increasing. Moreover, it should be noted, that increasing the inlet relative humidity of the return in the range of RH_{2i} = 10–90% causes an increase of the temperature effectiveness ε due to additional latent heat recovery at the expanse of the condensation zone expansion [7].

It can be seen in Fig. 2(a), that in the case of the minimum temperature effectiveness (ε = 0.40) frost-free operation is observed only at the values of the return air relative humidity in the following ranges $RH_{2i} \leq 10\%$ and $RH_{2i} \geq 70\%$ but at maximum temperature effectiveness (ε = 0.85, see Fig. 2(b)) frost-free operation is completely impossible under analysed range of the return air relative humidity variations.

The reason of such heat exchanger performance behaviour is that a higher temperature effectiveness results in a higher risk of frost formation at the expense of the temperature decrease of the outlet return airflow and plate surface in the "*cold zone*" (below water freezing point temperature). It should be noted, that the largest "*frost*" area occurs under conditions corresponding to the range of the return air relative humidity RH_{2i} = 20–50%. Unfortunately, this range corresponds to the normal (recommended) indoor air conditions in winter season.

3 Frost prevention methods

There are two typical methods of frosting protection for counter-flow plate heat exchangers: preheating of intake (outdoor) air and bypassing the outdoor airflow.

Preheating of ambient air is usually realized in electrical air heater to the safe values of the inlet outdoor air temperature (called as threshold temperature) defined as minimum intake air temperature at which the local temperature of the plate in the "*cold zone*" is above 0°C. In this case there is no any frost area on the heat exchanger's surface. The values of these temperatures are presented in Table 1.

In the case of bypassing, the ratio of the outdoor airflow to exhausted airflow is reduced to the level at which frost layer is not observed on the plate surface of the return airflow channel. Threshold values of the outdoor airflow to returned ratio are presented in the Tab. 2 for different values of outdoor air temperature.

Table 1. The threshold values of the outdoor air temperature for return air temperature equal to t_{2i} = 20°C and different temperature effectiveness of the counter-flow plate heat exchanger used for energy recovery from exhaust air (without bypassing G_1/G_2 = 1.0) [7].

Return air parameters		**Outdoor air parameters**			
relative humidity, RH_{2i}, %	dew point temperature, t_{2i}^{DP}, °C	the threshold values of outdoor air temperature t_{trshld}, °C, for different temperature effectiveness of counter-flow plate heat exchanger			
		ε = 0.50	ε = 0.60	ε = 0.70	ε = 0.80
10	-12.5	-23.3	-20.6	-18.2	-16.1
20	-3.6	-11.5	-9.5	-7.7	-6.2
26.14	0.0	-6.6	-4.9	-3.5	-2.2
30	+1.9	-7.6	-5.8	-4.3	-3.0
40	+6.0	-10.0	-8.1	-6.5	-5.0
50	+9.3	-12.5	-10.5	-8.7	-7.1
60	+12.0	-15.0	-12.8	-10.9	-9.2
70	+14.4	-17.5	-15.1	-13.1	-11.2
80	+16.4	-20.0	-17.5	-15.3	-13.3
90	+18.3	-22.5	-19.9	-17.5	-15.4

Table 2. The threshold values of the outdoor airflow to returned ratio (G_1/G_2) at different values of outdoor air temperature for return air temperature equal to t_{2i} = 20°C and different heat recovery efficiency η of the counter-flow plate heat exchanger [7].

Return air parameters	**Threshold values of the outdoor airflow to returned ratio (G_1/G_2)**	**Outdoor air temperature, °C**							
relative humidity, RH_{2i}, %		t_{1i}	η	t_{1i}	η	t_{1i}	η	t_{1i}	η
40	1.00	-10.0	0.50	-8.1	0.60	-6.5	0.70	-5.0	0.80
	0.80	-11.2	0.47	-9.3	0.55	-7.7	0.63	-6.4	0.71
	0.60	-12.8	0.42	-11.0	0.48	-9.4	0.53	-8.1	0.58
	0.40	-16.6	0.34	-13.9	0.37	-11.6	0.40	-9.6	0.43
	0.20	-22.2	0.20	-18.0	0.21	-14.3	0.22	-10.7	0.23
	0.05	-28.5	0.06	-23.1	0.06	-18.2	0.06	-13.3	0.06
50	1.00	-12.5	0.50	-10.5	0.60	-8.7	0.70	-7.1	0.80
	0.80	-13.8	0.46	-11.8	0.55	-10.1	0.63	-8.7	0.70
	0.60	-15.7	0.41	-13.8	0.47	-12.0	0.53	-10.7	0.57
	0.40	-19.8	0.33	-17.0	0.36	-14.5	0.39	-12.3	0.42
	0.20	-26.5	0.19	-21.9	0.20	-17.7	0.21	-13.7	0.22
	0.05	-36.9	0.05	-29.7	0.06	-23.2	0.06	-16.6	0.06
60	1.00	-15.0	0.50	-12.8	0.60	-10.9	0.70	-9.2	0.80
	0.80	-16.4	0.46	-14.3	0.54	-12.4	0.62	-10.9	0.70
	0.60	-18.5	0.41	-16.4	0.47	-14.6	0.52	-13.1	0.56
	0.40	-23.0	0.33	-20.0	0.36	-17.4	0.39	-15.1	0.42
	0.20	-32.2	0.19	-26.7	0.20	-21.6	0.21	-16.7	0.22
	0.05	-45.2	0.05	-36.3	0.05	-28.2	0.06	-19.8	0.06

Temperature effectiveness and heat recovery efficiency for the purpose of frost protection analysis were defined as follows:

DOI: 10.1051/e3sconf/20172200129

Temperature effectiveness of heat exchanger:

$$\varepsilon = \frac{t_{1o} - t_{1i}}{t_{2i} - t_{1i}} \tag{1}$$

Heat recovery efficiency:

$$\eta = \frac{t_{1o} - t_{1i}}{t_{2i} - t_{1i}} \left(\frac{G_1 \cdot c_{p1}}{G_2 \cdot c_{p2}} \right) \tag{2}$$

From the above results, one can see that the values of heat recovery efficiency η and threshold ratio of the outdoor airflow to exhausted drop with decreasing of the ambient air temperature.

4 Heat recovery efficiency analysis

The comparison of the heat recovery efficiency and energy demand for the air treatment in the AHU with counter-flow plate heat exchanger under two examined frosting control strategies is presented in Fig. 3.

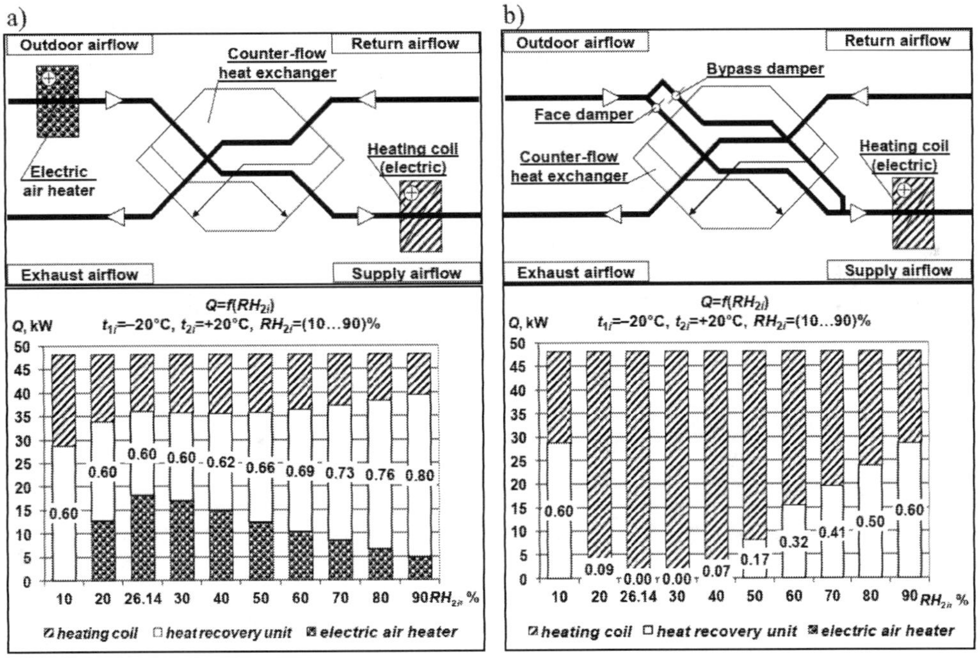

Fig. 3. The comparison of the heat recovery and additional energy consumption for the air treatment in AHU with counter-flow plate heat exchanger expressed as a function of the return air relative humidity under different frosting control methods: (a) preheating (b) bypass damper system.

The energy demand for the air treatment in the AHU was calculated for the value of the volumetric airflow rate equalled to 1 m/s. As one can see in Fig. 3, the performance of the heat exchanger equipped with the *bypass dampers* is characterized by very low heat recovery efficiency. In this regard, the AHU with the heat exchanger equipped with the *bypass dampers* needs to provide significantly more total power for the air treatment than in the case of the preheating.

To sum it up, it can be said that the performance of preheating technique is more efficient than and bypass dampers method of frosting protection.

5 Conclusions

In this study a theoretical analysis of the heat and mass transfer in the counter-flow plate heat exchanger used for energy recovery in AHU under sub-zero outdoor air temperature operating conditions was presented. Two main techniques of the frost prevention (preheating and bypassing the outdoor airflow) were described and analysed on the base of the modified ε-NTU model. Unique aspects of the most probable variants of year-round heat exchanger operation performance, characterized by existence of three active heat and mass transfer areas ("*dry*", "*wet*", "*frost*"), were revealed. The results of the conducted investigations are summarized as follows:

- The formation and variability of different active heat and mass transfer zones strongly depends on the inlet return air relative humidity and inlet outdoor airflow temperature.
- The values of critical outdoor air temperatures and outdoor-to-return airflow rate ratio for frost-free operating conditions were determined at different inlet return airflow parameters (see Tab. 1 and 2).
- Inlet return airflow dew point temperature plays crucial role in the frosting limit. The most unfavourable operating conditions at sub-zero outdoor air temperature occur at the value of inlet return air dew point temperature $t_{2i}^{DP} = 0°C$. In the case of inlet return air temperature $t_{2i} = +20°C$ such conditions correspond to the relative humidity $RH_{2i} = 26.14\%$. Unfortunately, this air relative humidity corresponds to recommended indoor air conditions in winter season.
- The frost tends to take place with increasing temperature effectiveness ε (heat recovery efficiency η) of the heat exchanger.

The obtained results may help to reduce the costs of the supply air treatment in AHU and increase the period of time when heat recovery units are effectively operated. However, more research is needed to find the best frost protection method, which allows for maximum heat recovery from return air.

The conference participation was financed from the statutory funds of Faculty of Environmental Engineering, Wroclaw University of Science and Technology from the project 0401/0066/16.

Nomenclature

c_p – specific heat capacity, J/(kg·K); F – heat or heat and mass transfer surface area, m^2; F_o – total surface area, m^2; $\bar{F} = (F/F_o)\cdot 100\%$ – heat or heat and mass transfer surface area relates to total surface area, %; G – moist air mass flow rate, kg/s; t – temperature, °C; L_{X_1} – channel of heat exchanger in X_1 direction, m; L_{X_2} – channel length of heat exchanger in X_2 direction, m; Q – heating capacity, kW; X_1 – Cartesian coordinate in X_1 direction (along outdoor airflow direction), m; $\bar{X}_1 = X_1/L_{X_1}$ – relative X_1 coordinate, dimensionless; X_2 – Cartesian coordinate in opposite to X_1 direction, m (along return airflow direction); $\bar{X}_2 = X_2/L_{X_2}$ – relative X_2 coordinate, dimensionless; α – convective heat transfer coefficient, W/(m^2 K); ε – temperature effectiveness of heat exchanger, dimensionless; η – heat recovery efficiency, dimensionless.

Subscripts: 1 – outdoor airflow; 2 – return airflow; *DP* – referenced to dew point temperature; *i* – inlet; *o* – outlet; *trshld* – threshold value of outdoor air temperature.

Acronyms: AHU – Air Handling Unit; $\mathrm{NTU} = \alpha F / (G c_p)$ – Number of Transfer Units; *RH* – Relative Humidity, %.

References

1. S. Anisimov, A. Jedlikowski, M. Skrzycki, Proceedings of the Xth International Scientific Conference „Quality of Indoor Air and Environment", 302–308, Volgograd-Budapest May 13-20.2012,
2. CARBON TRUST, Published in the United Kingdom, August 2011, accessed August 21, 2013: http://www.carbontrust.com/resources/guides/energy efficiency/heat-recovery, http://www.carbontrust.com/media/31715/ctg057_heat_recovery.pdf
3. M.R. Nasr, F. Fathieh, D. Kadylak, R. Huizing, R.W. Besant, C.J. Simonson, Exp. Thermal Fluid Sci. **77**, 100–115 (2016)
4. C.-C. Wang, K.-Y. Chen, J.-S. Liaw, C.-Y. Tseng, Int. J. Heat Mass Transf. **55**, 5367–5372 (2012)
5. S. Freund, S.A. Klein, D.T. Reindl, HVAC&R Res. J. **9**, 493–508 (2003)
6. M.R. Nasr, M. Kassai, G. Ge, C.J. Simonson, Appl. Energy **151**, 32–40 (2015)
7. A. Jedlikowski, S. Anisimov, J. Danielewicz, M. Karpuk, D. Pandelidis, Int. J. Heat Mass Transf. **108**, 585–613 (2017)

Energy demand and thermal comfort of HVAC systems with thermally activated building systems as a function of user profile

Katarzyna Pałaszyńska[1,*], *Karol* Bandurski[1], and *Mieczysław* Porowski[1]

[1]Poznan University of Technology, Faculty of Civil and Environmental Engineering, Institute of Environmental Engineering, ul. Berdychowo 4, 60-965 Poznań, Poland

Abstract. Thermally Activated Building Systems (TABS) are a way to use building structure as a thermal energy storage. As a result, renewable energy sources may be used more efficiently. The paper presents numerical analysis of a HVAC system with TABS energy demand and indoor thermal comfort of a representative room in a non-residential building (governmental, commercial, educational). The purpose of analysis is to investigate the influence of a user profile on system performance. The time span of the analysis is one year – a typical meteorological year. The model was prepared using a generally accepted simulation tool – TRNSYS 17. The results help to better understand the interaction of a user profile with TABS. Therefore they are important for the development of optimal control algorithms for energy efficient buildings equipped with such systems.

1 Introduction

Thermally activated ceilings (a part of TABS – Thermally Activated Building Systems) is a solution making it possible to shift in time thermal loads (concerning both heating and cooling) by means of using the accumulation potential of construction elements. It also makes it possible to change thermal comfort by directly impacting the average temperature of internal surfaces.

Lines with water or air are embedded in the ceiling and act as heat exchangers transferring heat to and from occupied zone and storing thermal energy TABS do not only transfer or receive heat (mainly by radiation [1]) but also affect thermal characteristics of a building. The most common TABS solution is ceiling activation.

The development of simulation models is necessary to create design guidelines and system controls. Adjusting a system with TABS is difficult due to the high thermal inertia resulting from a significant heat capacity.

* Corresponding author: katarzyna.palaszynska@put.poznan.pl

DOI: 10.1051/e3sconf/20172200130

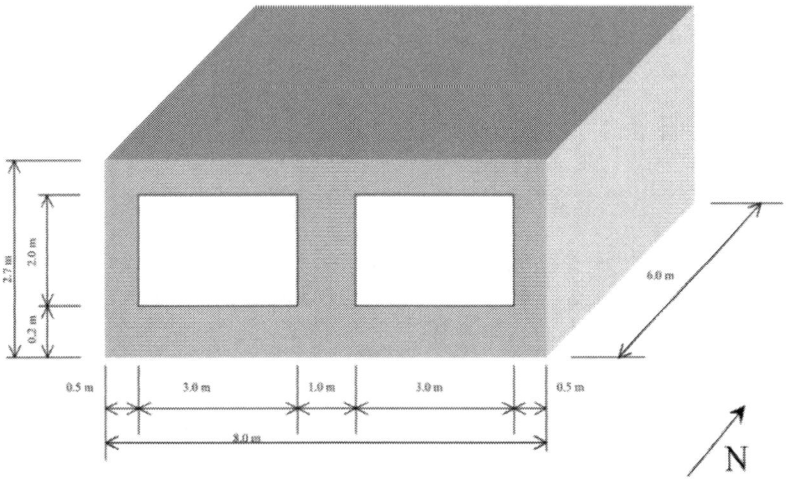

Fig. 1. The modeled room [9].

Research on systems with thermally activated ceilings have been conducted for about 20 years. Olsen et al. [2–4] conducted research on modeling TABS and proposed a dimensioning method for thermally activated ceilings. This research lead to the development of the ISO 11855-4:2012 standard [6]. On the other hand Rijksen et al. [5] conducted simulation and empirical research on TABS, providing general guidelines for determining the required cooling capacity for an entire office building with such a system, located in a temperate climate. Lehmann et al. [7] demonstrated a significant impact of the adopted hydraulic system and the applied control strategy on the efficiency of using the energy from a system with thermally activated ceilings. Saelens et al. [8] analyzed the occupant behaviour impact on the energy efficiency of TABS during the cold season for the climate conditions of Belgium. The assessment parameter for the heat comfort of the simulation model was temperature in a given room.

The presented paper solves two optimization problems of a HVAC system with TABS. Solution is based on one-zone office model. The optimization criterion for the first problem was thermal comfort and for the second problem – it was the minimum primary energy demand with constraints concerning thermal comfort. The decision variable of both optimization problems was the supply water temperature of a thermally activated ceiling.

2 Model and assumptions

The analyzed object is a room (Fig.1) (in an office building) with the following parameters:

- **dimensions**: area = 48.0 [m^2], height = 2.7 [m];
- **exterior wall**: facing South (S), area = 2.6 [m^2], heat transfer coefficient U = 0.35 [W/(m^2K)];
- **exterior wall windows**: two pieces = 2.0 x 3.0 [m] (12 [m^2]), heat transfer coefficient U = 2.89 [W/(m^2K)];
- **infiltration**: 0.1 [1/h];
- **ventilation (hygienic needs)**: 2.0 [1/h], supply temperature 22°C.

The modeled room is based on a case study from RADTEST [9]. The simulation was performed with TRNSYS 17 [10]. The used building model was TRNSYS, type 56 using the mode of the most detailed calculation for heat exchange by radiation:

DOI: 10.1051/e3sconf/20172200130

- direct radiation through an external partition – geometric distribution;
- diffuse and long wave radiation – based on view factors.

Other assumptions (as fixed parameters) in simulation models of the performed analysis:

- weather data: Typical Meteorological Year for Poznań, MeteoNorm Reference Year;
- comfort condition: category B acc. to EN 15251, indoor air temperature: from 20°C to 24°C;
- thermal comfort model: Fanger acc. to ISO 7730, additional assumption: constant air velocity = 0.1 [m/s], clothing = 1 [clo], metabolism = 1 [met];
- heat transfer with other zones of the building was neglected (adiabatic partition wall);
- zone heat capacity: air heat capacity added to room equipment heat capacity – estimated as five times the heat capacity of the air alone acc. to the rule of thumb in building performance simulations;
- type of building –internal heat gains patterns: governmental, commercial, educational (Fig. 2).

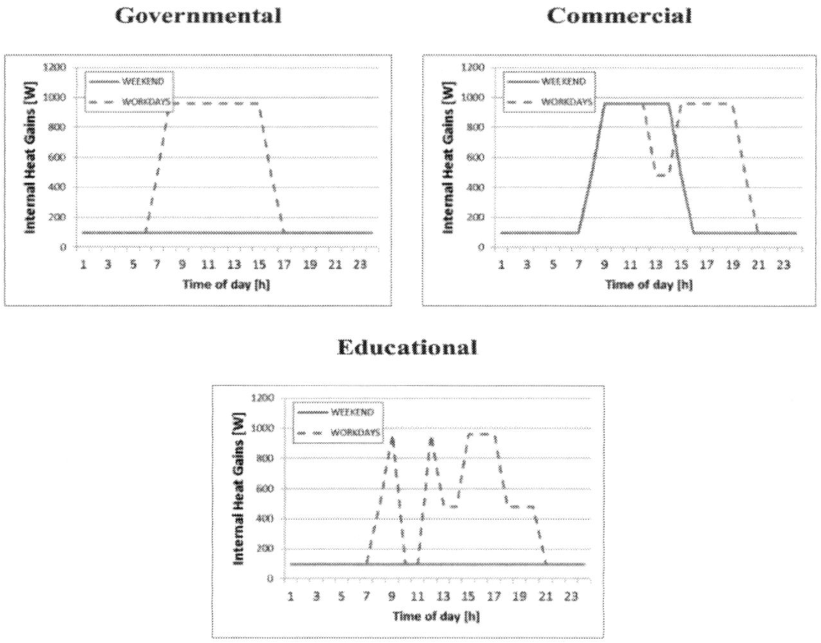

Fig. 2. Internal heat gains patterns.

Values of internal heat gains are based on ISO 13790, annex G. The gains are divided into radiation (30%) and convection (70%) heat gains acc. to human body's thermal balance and [11].

3 Optimization criteria

There are two optimization problems:

- First optimization criterion:

  - decision variables: TABS supply water temperature $t_{in} \in$ <20, 24> °C;
  - optimization criterion: minimum number of hours of thermal discomfort per year = thermal comfort of PMV $\in$ <-0.5, 0.5> (minimum number of hours of thermal discomfort).
- Second optimization criterion:
  - decision variables: TABS supply water temperature, $t_{in} \in$ <20, 24> °C;
  - constraints: PMV $\in$ <-0.5, 0.5>;
  - optimization criterion: minimum annual primary energy demand E_p = min.

Decision variables for optimization models are presented in Table 1.

Table 1. Decision variables for optimization models.

Designation	**TABS supply water temperature t_{in}**	**Type of building/user profile**
G_20	20	office in a governmental building
G_21	21	
G_22	22	
G_23	23	
G_24	24	
C_20	20	office in a commercial building
C_21	21	
C_22	22	
C_23	23	
C_24	24	
E_20	20	office in an educational building
E_21	21	
E_22	22	
E_23	23	
E_24	24	

The usable energy demand (usable heat and usable cooling) and the primary energy demand were calculated based on the zone heat balance and the following equations:

$$0 = \frac{dC_{zone}}{dt} + Q_{inf}(t) + Q_{vent}(t) + Q_{gain\_conv}(t) + Q_{sun\_air}(t) + Q_{surf\_conv}(t) + Q_{H-AC\_air}(t) \tag{1}$$

$$Q_{system}(t) = Q_{water\_TABS}(t) + Q_{H-AC\_air}(t) \tag{2}$$

$$E_p = \sum_i \frac{w_i}{\eta_{i,t}} \cdot Q_{system}(t) \tag{3}$$

where:

t – time;
C_{zone} – zone heat capacity (air, furniture, equipment);
Q_{inf} – heat supplied by infiltration;
Q_{vent} – heat supplied by ventilation (in this case $T_{vent\text{-}air} = 22°C$);

DOI: 10.1051/e3sconf/20172200130

Q_{gain_conv} – heat supplied by convection through internal heat sources;
Q_{sun_air} – heat gains of the solar energy which, directly after passing through a window, are immediately transferred as convective heat gains into air;
Q_{surf_conv} – heat coming from zone partition surfaces to air by convection;
$Q_{H-AC\_air}(t)$ – heat supplied to the zone by the air line of a HVAC system (in TRNSYS corresponding to = $-Q_{SENS}$ which is negative in case of the demand for heating to a desired air temperature, and positive in case of the demand for cooling to a desired air temperature);
$Q_{water\_TABS}(t)$– heat supplied to TABS by water (in TRNSYS corresponding to $-Q_{ALFL}$ which is negative in case of water heating the ceiling, i.e. $T_{inlet} > T_{outlet}$, and positive in case of water cooling the ceiling i.e. $T_{inlet} < T_{outlet}$;
$Q_{system}(t)$ – heat energy demand of the whole comfort system (air conditioning and HVAC excluding ventilation)
E_p – annual primary energy demand for a HVAC system
w_i – input coefficient of the non-renewable primary energy for the production and supply of an energy carrier (or the final energy) (i→ H relates to heat, i→ C relates to cold); w_H = 1.1 and w_C = 3 [12]
$\eta_{i,t}$ – seasonal average total efficiency of production (conversion): i→ H of heat for heater blowers, i→ C of cold for intercoolers); $\eta_{H,t}$ = 0.81 and $\eta_{C,t}$ = 1.9 [12]

4 Calculation results and interpretation

For the first optimization problem with the optimization criterion being the thermal comfort, results of the simulation are presented in Table 2 (dark fields: 0 hours of discomfort; the brighter the field, the more hours of discomfort). The optimization criterion is the minimum number of hours of discomfort (decreasing preference criterion). The optimal variants for different types of buildings (internal heat gains patterns) are:

- Governmental: minimum number of hours of discomfort n = 102 h for t_{in} = +22°C – constant throughout the year (max n = 558 h, t_{in} = 24°C)
- Commercial: minimum number of hours of discomfort n = 119 h for t_{in} = +21°C – constant throughout the year (max n = 770 h, t_{in} = 24°C)
- Educational: minimum number of hours of discomfort n = 79 h, t_{in} = +22°C – constant throughout the year (max n = 363 h, t_{in} = 24°C)

It should be noted that for all three types of buildings – optimal variants are in the temperature range of t_{in} = +21 to 22°C.

For the second optimization problem with the optimization criterion being the minimum annual primary energy demand (E_p = min.), results of the simulation are presented in Table 3. The optimal variants for different types of buildings (internal heat gains patterns) are:

- Governmental: E_{p_min} = 1720 kWh/a for tin = +20°C, (E_{p_max} = 4364 [kWh/a] for tin = +24°C)
- Commercial: E_{p_min} = 2093 kWh/a for tin = +20°C, (E_{p_max} = 5370 [kWh/a] for t_{in} = +24°C)
- Educational: E_{p_min} = 1496 kWh/a for tin = +20°C, (E_{p_max} = 4039 [kWh/a] for t_{in} = +24°C)

Table 2. Number of hours of discomfort for different types of buildings.

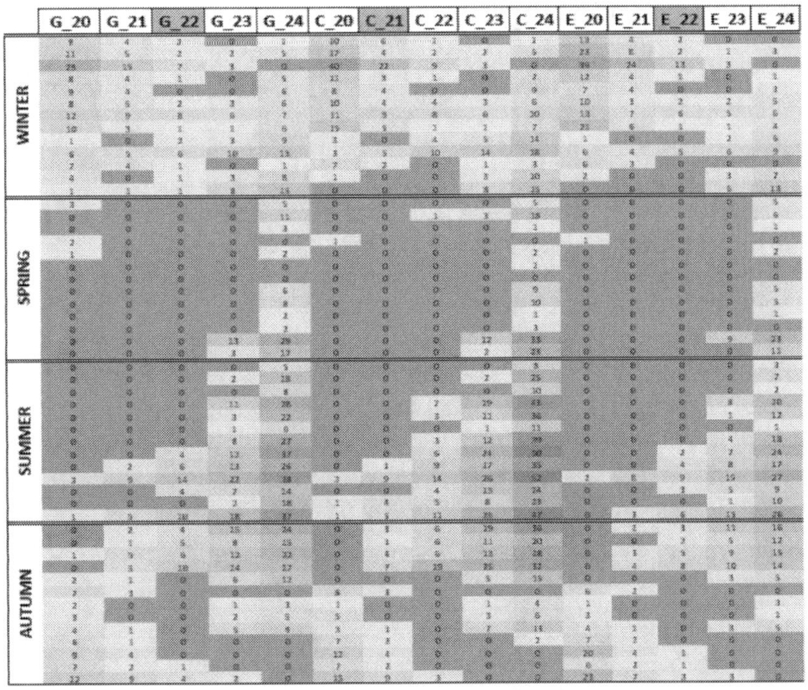

Table 3. Calculation results of the annual energy demand for a HVAC system with TABS: a) useful heat, b) useful cold, c) primary energy, d) percentage difference between optimal variant and others.

a)

T_inlet	Q_heat [kWh/a]		
[°C]	G	C	E
20	199	130	209
21	72	41	76
22	18	9	19
23	4	2	4
24	0	0	0

b)

T_inlet	Q_cool [kWh/a]		
[°C]	G	C	E
20	923	1214	768
21	1296	1688	1124
22	1743	2228	1559
23	2239	2806	2045
24	2764	3401	2558

c)

T_inlet	E [kWh/a]		
[°C]	G	C	E
20	1728	2093	1496
21	2144	2721	1878
22	2777	3530	2487
23	3541	4433	3234
24	4364	5370	4039

d)

T_inlet	E [%]		
[°C]	G	C	E
20	-38%	-23%	-40%
21	-23%	0%	-25%
22	0%	30%	0%
23	28%	63%	30%
24	57%	97%	62%

5 Conclusion

Based on the one-zone office model the simulation analysis and optimization task have been performed. The following conclusions can be defined:

- type of building (internal heat gains patterns) significantly affects the optimal comfort temperature of the TABS supply water. For the analyzed types of buildings, the optimum comfortable temperature equals t_{in} = +21 to 22°C
- the comfortably optimal TABS supply water temperature is a function of parameters of the external climate (seasons). The analysed temperature range of t_{in} +20 to 24°C is the most suitable for spring and summer
- type of building (internal heat gains patterns) significantly affects the annual primary energy demand. For the energetically optimal variants, the annual primary energy demand for a Commercial building (2093 kWh/a) is 40% higher than the same demand for an Educational building (1496 kWh/a)
- energetically optimal TABS supply water temperature equals t_{in} = +20°C for all of the analyzed type of buildings.
- having analyzed the two optimization criteria (concerning comfort and energy), the operationally optimal TABS supply water temperature equals t_{in} = +21°C.
- changing temperature of TABS supply water, relative to comfortably optimal TABS supply water temperature, has a significant effect on energy consumption in the range of -38% to +57% for governmental, in the range of -23% to +97% for commercional and -40% to 62% for educational (Table 3d)

The undertaken research does not cover the whole subject. Further studies will extend the assumptions of this simulation model. This particularly applies to:

- taking into account a model with more zones (not only one-zone) ;
- changing thermal insulation and heat capacity of external wall;
- introducing a (adjustable/fixed) shading;
- introducing ventilation with a variable air flow (depending on requirements) – DCV (Demand Controlled Ventilation) ;
- taking into account changes in clothing (clothing insulation) throughout the year;

References

1. J. Romaní, A. De Gracia, L.F. Cabeza, Energy Build. **127** (2016)
2. B.W. Olesen, K. Sommer, B. Düchting, ASHRAE Trans. **108** (2000)
3. P. Weitzmann, E. Pittarello, B.W. Olesen, *Proceedings of the 8th symposium on Building Physics in the Nordic Countries* (2008)
4. B. Olesen, ASHRAE J. **50** (2008)
5. D.O. Rijksen, C.J. Wisse, A.W.M. van Schijndel, Energy Build. **42**, 3 (2010)
6. ISO 11855-4 *Building Environment Design – Design, Dimensioning, Installation and Control of Embedded Radiant Heating and Cooling Systems - Part 4: Dimensioning and calculation of the dynamic heating and cooling capacity of Thermo Active Building Systems* (2012)
7. B. Lehmann, V. Dorer, M. Gwerder, F. Renggli, J. Tödtli, Appl. Energy **88**, 1 (2011)
8. D. Saelens, W. Parys, R. Baetens, Build. Environ. **46** (2011)
9. M. Arechmann, G. Zweifel, *RADTEST - Radiant Heating and Cooling Test Cases* (2003)
10. S.A. Klein et al., *TRNSYS 17: A Transient System Simulation Program* (2009)
11. M.H. Hosni, B.W. Jones, H. Xu, ASHRAE Trans. **105** (1999)
12. M. Porowski, Ciepłownictwo, Ogrzewnictwo, Wentylacja **42**, 3 (2011)

 DOI: 10.1051/e3sconf/20172200131

Propositions of improvement of the cross-flow M-Cycle heat exchangers in different air-conditioning applications

Demis Pandelidis[1,*], *Sergey* Anisimov[1], *Krzysztof* Rajski[1], and *Ewa* Brychcy[2]

[1]Wroclaw University of Science and Technology, Department of Environmental Engineering, Wybrzeże Wyspianskiego 27, 50-370, Wroclaw, Poland
[2]Consultant, Researcher of Aurae Cooling Technologies Limited, Hong Kong

Abstract. This paper presents results of mathematical simulation of the heat and mass transfer in the two different Maisotsenko Cycle (M-Cycle) heat and mass exchangers used for the indirect evaporative cooling in different air-conditioning systems. A two-dimensional heat and mass transfer model is developed to perform the thermal calculations of the indirect evaporative cooling process, thus quantifying the overall heat exchangers' performance. The mathematical model was validated against the experimental data. Numerical simulations reveal many unique features of the considered units, enabling an accurate prediction of their performance. Results of the model allow for comparison of the two types of heat exchangers in different applications for air conditioning systems in order to obtain optimal efficiency.

1 Introduction

In recent years, the increase in summer temperatures, improved insulation of the buildings, and a growth of indoor facilities have led to an increased requirement for air conditioning in buildings. Conventional mechanical vapor-compression air-conditioning systems consume a large amount of the electrical energy that is largely dependent upon a fossil fuel. This mode of air conditioning is, therefore, neither sustainable nor environmentally-friendly. Due to the increasing need for air conditioning and the growing interest in energy savings, seeking ways to reduce fossil fuel consumption and to increase usage of the renewable energy during air-conditioning process in building sector is a matter of great importance. Due to this fact, this paper focuses on analysis of two types of Maisotsenko cycle heat and mass exchangers (HMXs), which are able to generate significant savings in the air conditioning systems and therefore reduce the energy consumption. First unit is a currently produced device, while the second one is the proposition of an improvement proposed by authors.

* Corresponding author: demis.pandelidis@pwr.edu.pl

 DOI: 10.1051/e3sconf/20172200131

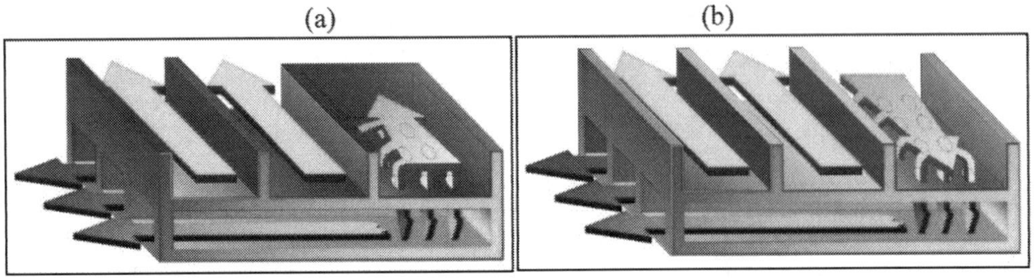

Fig. 1. Analyzed HMXs: (a) original cross-flow HMX (HMX1); (b) modified cross-flow HMX (HMX2).

The HMXs considered are shown schematically in Figure 1. The first device (HMX1 – Fig. 1(a)) is currently produced by Coolerado Corporation, where the working and the primary air flows in the dry channels are flowing parallel to each other. The second unit (HMX2- Fig. 1(b)) is the modification of the first cross-flow HMX, where entrance to the dry channel for the working and the product air flow are placed on the opposite sides of the HMX, which allows for implementation in different configurations in air-conditioning systems. Both units can be used as an individual cooling coils, which are the only source of cooling power in the system. However, in many air conditioning systems, especially in the public buildings and offices, the individual rooms typically use fan coil units to provide individual comfort for the occupants (Fig. 2(a), (b), (d)). The exhaust air for the systems with the original HMX is removed to from the outside or it is partly recirculated to the primary air flow (Fig. 2(a)). The modified HMX allows for the system to be implemented in the supply-exhaust flow stream in the air-handling unit (Fig. 2(b)). Therefore it can operate on the exhaust air, which is colder than the ambient air. The system presented in Figure 2(b) can be replaced with a system having a heat pump, presented in Figure 2(c). The operation base in this case is the same as in supply-exhaust system with the fan coil units. In more humid climates, there is a possibility to use the original HMX with the desiccant wheel (Fig. 2(d)), which dries the air flow and increases its temperature [1]. The arrangements presented require different operational conditions for the presented HMXs, with different temperatures and relative humidities of the primary and the working air-flow entering each exchanger. It is essential to establish the conditions for which it is more reasonable to use one HMX instead of another.

2 Methods

All the analysis is based on the numerical model based on the modified ε-NTU method. All the details of the model are presented in [2], along with the validation, transformation of the heat balance equations and detail analysis of the algorithm. Due to the fact that the model was presented in the past, it will be omitted in this paper.

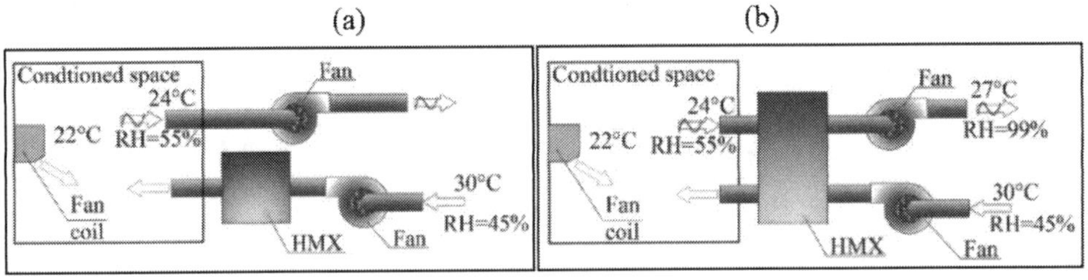

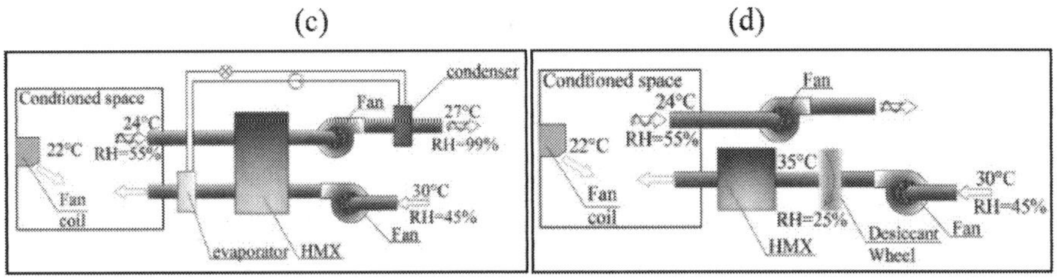

Fig. 2. Analyzed heat exchangers in different applications for air conditioning systems: (a) HMX1 in air conditioning system with fan coils; (b) HMX2 in air conditioning system with fan coils; (c) HMX2 in air conditioning system with heat pump; (d) HMX1 in air conditioning system with desiccant wheel.

3 Results and discussion

The reference operating conditions for the analyzed exchangers are presented in Table 1. The considered exchangers will be compared for the three hypothetical inlet conditions, representing different operational possibilities for the air conditioning system:

1. Both exchangers have the same inlet parameters for the working and the product air channels: $t_{1i}= t_{3i}$; $RH_{1i}=RH_{3i}$; (the purpose of this study is the general comparison between the two exchangers for the same operational conditions),

2. For HMX1 the working and the product air streams' inlet temperature and humidity are the same, while HMX2 have different values of the primary and the working air inlet temperature and humidity (comparison between the HMX1 working in the supply air system presented in Fig. 2(a) and HMX2 working in the supply-exhaust air system shown in Fig. 2(b) and (c)). In this case, the inlet temperature of the working air flow is lower than the inlet temperature of the product air flow for HMX2,

3. HMX1 has hot and dry inlet air conditions (the same for the product and the working air flow), while HMX2 has different values for the primary and the working air inlet temperature and humidity (comparison between HMX1 working in the desiccant system presented in Fig. 2(d) and HMX2 working in the supply-exhaust air system visible in Fig. 2(b) and (c)).

Table 1. The reference operating conditions for the analyzed exchangers.

Length, m	Width, m	Channel height, mm	Working to primary air ratio, -	Primary air stream velocity, m/s
0.5	0.5	3.0	1.0	3.0

Three main parameters (indices) have been selected to study the operational performance of the investigated HMX configurations:
• temperature of outlet primary air flow;
• the wet bulb thermal effectiveness, which is defined as the ratio of the difference between intake and outlet process air temperature to the difference between intake process air temperature and its wet bulb temperature [3, 4];

$$\varepsilon_{WB} = \frac{t_{1i} - \overline{t}_{1o}}{t_{1i} - t_{1i}^{WB}} \tag{1}$$

• the specific cooling capacity per cubic meter of the HMX structure

$$\hat{Q}=Q_1/V_{HMX} \tag{2}$$

where $Q_1 = G_1 c_{p1}\left(t_{1i} - \bar{t}_{1o}\right)$ is the cooling capacity of the HMX and V_{HMX} is the volume of the HMX structure, m^3.

3.1 Analysis at different operational conditions

A set of simulations was performed in order to compare the effectiveness of the considered exchangers in the different operational conditions in the air conditioning systems. Figure 3(a) presents comparison of systems visible in Figure 2(a)–(c). For the purpose of this analysis, HMX1 is using only on ambient air, while in HMX2 the ambient air enters only the product air channels and the working air channels are operating on the exhaust air from the conditioned space. Depending on the type of conditioned space, the exhaust air may have a different temperature and relative humidity; however the temperature inside the conditioned space is colder than the ambient temperature, while its relative humidity is usually a little higher. Simulations were conducted for an ambient air temperature equal 30°C, and the relative humidity of the ambient air was varied from 25 to 50%. The exhaust air temperature was changed from 21 to 26°C, with 1°C increments, while its relative humidity was changed from 40 to 60%. The charts (Fig. 3(a), (c) and (d)) are presented for constant levels of inlet temperatures of the primary air flow (t_{1i}), which are listed in frames in the upper part of the charts. For HMX1 the inlet temperatures represent different operation in air conditioning system (operation only on ambient air or in a system with a desiccant wheel. HMX2 operates on the ambient air flow in the primary air channels and exhaust air from the conditioned spaces in the working air channels. The X axis in Figs. 3(a), (c) and (d) represents the temperature of the exhaust air, which enters the channels of the HMX2. It's important to mention, that the parameters of the exhaust air don't affect the outlet air temperatures obtained by HMX1 (for this unit $t_{1i}= t_{2i}$), therefore values of t_{1o} for this unit are presented as a constant lines. Blue lines represent the outlet air temperatures obtained by the HMX1 for constant inlet air temperature ($t_{1i}= t_{2i}$) and different inlet relative humidities of the working air flow (listed in the frame on the right side of each chart). The black lines represent the outlet temperatures obtained by the HMX2 for constant level of inlet primary air temperature (t_{1i}), variable inlet working air temperature (presented on the X axis) and variable inlet working air relative humidity (listed in frame on the right side of each chart). It can be seen, that the inlet parameters have significant impact on the efficiency of considered systems. For moderate climate conditions (RH_i = 45 to 50%) outlet temperatures obtained by HMX1 were equal 21.8°C and 20.8°C. At the same time, for the typical indoor conditions (t = 24 to 26°C; RH_i = 50 to 60%) HMX2 achieved outlet temperatures varying from 19.0°C (t_{3i} = 24°C and RH_{3i} = 50%) to 21.75°C (t_{3i} = 26°C and RH_{3i} = 60%). This shows that HMX2 operating on the exhaust air stream at the entrance to the wet channel has more effective performance than the HMX1, which operates only on the ambient air. This phenomenon can be explained by analyzing the heat flux profiles (Fig. 3(b)). Although the heat flux profile for the HMX2 operating on the exhaust air is similar to the profiles characteristic for the parallel-flow exchangers, the heat transfer rate is higher than the heat transfer rate of the same unit using ambient air. The higher value of the heat transfer rate, caused by the lower temperature of the working air, results in more efficient cooling of the primary air. As a result, HMX2 can achieve higher effectiveness than the HMX1 operating on outside air in moderate climates (Fig. 3(a)). However, in more dry climatic conditions (RH_i = 35 to 40%) HMX1 achieves higher efficiency than HMX2 using exhaust air (for most of the typical indoor conditions). It can be seen that the detail analysis of the ambient and indoor air parameters is essential to achieve the highest efficiency of the designed air-conditioning system. The next set of simulations was performed in order to compare the HMX1 operating in system with the desiccant wheel (see Fig. 2(d)) with HMX2 operating on the exhaust air (see Fig. 2(b) and (c)). The desiccant wheel dehumidifies the ambient air and it additionally increases its temperature. Depending on the material used for the wheel structure and the range of the dehumidification process, the air temperature may increase or decrease depending on the conditions. For HMX2, the inlet temperature of the primary stream was assumed to be 30°C, while for HMX1 two types of the inlet conditions were selected for this study: dehumidification with a small increase in inlet temperature ($t_{1i} = t_{3i}$ = 3 2.5°C; RH_i = varied) and dehumidification with a high increase in inlet temperature t_{1i} = t_{3i} = 35.0°C; RH_i = varied). The results of the comparison are shown in Figure 3(c) and (d), respectively. It can be observed, that for the inlet temperature equal to 32.5°C and $RH_i \leq 25\%$, HMX1 achieves lower outlet temperatures than HMX2 operating on the exhaust air with the typical indoor conditions (Fig. 3(c)).

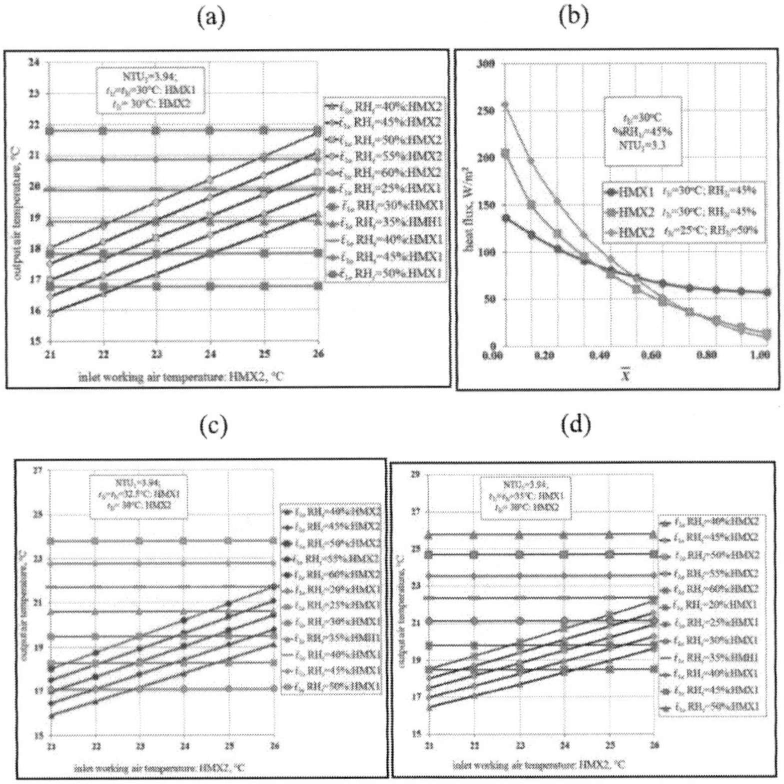

Fig. 3. Simulation results for HMXes operating in different inlet conditions: (a) outlet temperatures HMX1: t_{1i} = 30°C; RH_i = 25 to 50%; HMX2: t_{1i} = 30°C; t_{3i} = 21 to 26°C; RH_i = 40 to 60%; (b) heat flux distribution inside the primary air channels; (c) outlet temperatures: HMX1: t_{1i} = 32.5°C; RH_i = 20 to 50%; HMX2: t_{1i} = 30°C; t_{3i} = 21 to 26°C; RH_i = 40 to 60%; (d) outlet temperatures: HMX1: t_{1i} = 35°C; RH_i = 20 to 50%; HMX2: t_{1i}=30°C; t_{3i} = 21 to 26°C; RH_i = 40 to 60%.

When the temperature of the inlet air is raised to the higher level after dehumidification, the HMX1 obtains higher outlet temperatures (Fig. 3(d)). In this case, for the inlet temperature equal to 35.0°C and RH_i = 25%, HMX1 achieves higher outlet temperatures than HMX2 operating using exhaust air with typical indoor conditions. However, for relative humidities equal to 20%, HMX1 shows better performance than HMX2 for the most of the typical indoor conditions (Fig. 3(d)). This shows that the inlet air stream for HMX1 has to be dehumidified to very low values of the relative humidity to overcome the performance of HMX2 operating on the exhaust air. Table 2 presents the outlet temperature, specific cooling capacity and the wet bulb effectiveness obtained by HMX1 and HMX2 for the selected inlet parameters. The analysis of Table 2 illustrates another interesting observation: HMX1 for the inlet conditions of t_{1i} = 32.5°C; RH_i = 20% obtained lower outlet temperature than HMX2 (for t_{1i} = 30°C; t_{3i} = 24°C and RH_{3i} = 40%), however its wet bulb effectiveness is 32.5% lower. HMX2 (for t_{1i} = 30°C; t_{3i} = 24°C and RH_{3i} = 40%) obtained a lower outlet temperature than HMX1 (for t_{1i} = 35.0°C and RH_i = 20%), however its specific cooling capacity is 4.4 kW/m^3 lower. This shows that the wet bulb effectiveness and the specific cooling capacity aren't adequate to describe the performance of the HMXs operating in different arrangements in the air-conditioning system. The higher specific cooling capacity for the air with higher inlet temperature is caused by the higher difference between inlet and outlet product air temperatures. Greater temperature difference, for the higher temperature of the inlet air flow causes a very intensive evaporation process in the wet channels, which improves the efficiency of cooling process and results in greater cooling capacity. However, the additional cooling capacity can be used only for the reduction of the primary air flow temperature to level before the air stream is heated in dehumidification process and thus it doesn't give any energy benefit. An explanation of the trends in the wet bulb effectiveness presented in Table 2 is that the wet bulb effectiveness refers to the wet bulb temperature of the product air. In case of HMX2, the air stream entering the wet channel has a different temperature, relative humidity and therefore different wet bulb temperature. In this case, the efficiency factor based on the primary air's wet bulb temperature is not informative and can lead to an incorrect conclusions. Therefore it shouldn't be used for the description of indirect evaporative exchangers with different primary and working air inlet parameters.

Table 2. Outlet temperature, specific cooling capacity and wet bulb effectiveness of HMX1 and HMX2.

	HMX1: t_{1i} **= 32.5°C;** RH_i **= 20%;**	**HMX1:** t_{1i} **= 35.0°C;** RH_i **= 20%;**	**HMX2:** t_{1i} **= 30°C;** t_{3i} **= 24°C;** RH_{3i} **= 40%;**	**HMX2:** t_{1i} **= 30°C;** t_{3i} **= 24°C;** RH_i **= 50%;**
$\bar{t}_{1o}$, °C	17.1	18.5	18.17	19.35
$\hat{Q}$, kW/m^3	33.4	35.5	31.1	28.0
ε_{WB}, %	114.5	115.0	147.0	132.6

It can be seen on the basis of Figure 3, that in many cases the HMX2 can show a better performance than HMX1, especially for the colder indoor parameters.

4 Summary and conclusions

This study presents the theoretical analyses of the M-Cycle cross-flow heat exchangers used for the indirect evaporative cooling in air-conditioning systems. The units were compared in different operational conditions, representing potential different applications in air-conditioning systems. Modified ε–NTU-model based on the accurate assumptions was proposed for the mathematical analysis of the heat and mass transfer processes inside the exchangers. The basic operating parameters were given in Table 1. The discrepancies in

DOI: 10.1051/e3sconf/20172200131

effectiveness characterizing the considered HMXes were explained on the basis of the unique features of heat and mass transfer phenomenon occurring in each of the units. It was established that: The energy efficiency of the considered HMXs strongly depends on the inlet air parameters. HMX2 operating on the exhaust air with the typical indoor parameters in the working air channels shows higher effectiveness than HMX1 operating on ambient air in moderate climate conditions; HMX1 operating in the system with the desiccant wheel shows higher efficiency than HMX2 operating on indoor air, when air stream is highly dehumidified (RH*i* = 25% and lower). The material used for the desiccant wheel construction should allow for intense dehumidification of the air stream and a relatively small temperature increase in order to obtain the lowest outlet temperatures; In systems, where the air stream is additionally cooled by a mechanical compression system, it is reasonable to partly recirculate the exhaust air stream in systems with HMX1. The wet bulb effectiveness and the specific cooling capacity are not informative when comparing the indirect evaporative exchangers operating on the different primary and working air inlet parameters.

Nomenclature

c_p – specific heat capacity of moist air, J/(kg·K); G – moist air mass flow rate, kg/s; NTU – number of transfer units, NTU=$\alpha F/(Gc_p)$; $\hat{Q}$ – specific cooling capacity respected to 1 m^3 of the exchanger structure, kW/m^3, W; RH - relative humidity, %; t – temperature, $\bar{t}$ – average temperature, °C; 1 – main (primary) air flow; 2 – working air flow in the wet channels (product part of exchanger); 3 - working air flow in the dry channels (pre-cooling part of exchanger); 4 – working air flow in the wet channels (pre-cooling part of exchanger); i – inlet; o – output; *WB* - wet bulb temperature; $\overline{X}$ – relative X coordinate, dimensionless, $\overline{X} = X/l$;

Acknowledgement

Dr. Demis Pandelidis received financial support for his research from resources for scientific work for years 2016–2019 from Polish Ministry of Science and High Education (program "Iuventus Plus"), project number IP2015 058274.

References

1. A. Hasan, Appl. Energ. **89** (2012)
2. D. Pandelidis, S. Anisimov, Energy Conv. Manag. **90** (2015)
3. B. Riangvilaikul, S. Kumar, Energy Build. **42** (2010)
4. W.M. Worek, M. Khinkis, D. Kalensky, V. Maisotsenko, HT2012, Proc. of the ASME 2012 Summer Heat Transfer Conf., HT2012-58039

 DOI: 10.1051/e3sconf/20172200132

Organic Rankine cycle – review and research directions in engine applications

Angad Panesar[1,*]

[1]School of Computing, Engineering and Mathematics, University of Brighton, BN2 4GJ, UK

Abstract. Waste heat to power conversion using Organic Rankine Cycles (ORC) is expected to play an important role in CO_2 reductions from diesel engines. Firstly, a review of automotive ORCs is presented focusing on the pure working fluids, thermal architectures and expanders. The discussion includes, but is not limited to: R245fa, ethanol and water as fluids; series, parallel and cascade as architectures; dry saturated, superheated and supercritical as expansion conditions; and scroll, radial turbine and piston as expansion machines. Secondly, research direction in versatile expander and holistic architecture (NO_x + CO_2) are proposed. Benefits of using the proposed unconventional approaches are quantified using Ricardo Wave and Aspen HYSYS for diesel engine and ORC modelling. Results indicate that, the implementation of versatile piston expander tolerant to two-phase and using cyclopentane can potentially increase the highway drive cycle power by 8%. Furthermore, holistic architecture offering complete utilisation of charge air and exhaust recirculation heat increased the performance noticeably to 5% of engine power at the design point condition.

1 Introduction

Diesel engines are projected to remain one of the primary power source in trucking, marine and rail sectors, hence, future engine developments will emphasise on CO_2 reductions, alongside regulated emissions standards (e.g. Euro 6). Theoretical studies have shown indirect methods to be the more efficient technique for waste heat to power conversion for exhaust gases [1]. Organic Rankine Cycles (ORC) is stated as the preferred system when comparing the overall conversion efficiency and technology readiness level for highway trucking [2]. Exhaust gas recirculation (EGR) and post-turbine exhaust are being targeted due to higher quality and/or quantity. However, coolant and Charge Air Cooler (CAC) heat require innovative approaches (e.g. direct engine block heat recovery) [3].

ORC demonstrations of multiple heat recovery in series, parallel and cascade arrangements are taking advantage of the broader choice of working fluids and advancements in expanders. For relatively low, medium and high temperature heat sources, R245fa by Domingues et al. [4], ethanol by Larsen et al. [5] and MM by Fernandez et al. [6] has been suggested. Additionally, with emphasis on environment, R1234yf was indicated optimal from a thermo-economic approach by Yang et al. [7]. Experimental

* Corresponding author: a.s.panesar@brighton.ac.uk

results with efficiencies approaching 80% with scroll expander by Wang et al. [8], 70% with screw expander by Zhang et al. [9] and 65% with piston expander by Seher et al. [10] have been presented. Furthermore, the trucking sector is integrating prototypes to address issues relating to durability, packaging and controls [11]. ORCs were the most demonstrated technology in the 5–15 kW range, offering thermal efficiencies around 8–12% and a potential of 4–6% reduction in fuel consumption. Specific system weight and cost following standardisation are projected to be 10 kg/kW and 300 £/kW [12–14].

This paper consists of two parts. Firstly, a conceptual and technical review of automotive ORCs is presented focusing on the pure working fluids, thermal architectures, expanders and application specific challenges. The review is divided by the common classifications of working fluids. Consideration is given to the associated expander inlet conditions (illustrated using Fig. 1) and heat recovery arrangements (illustrated using Fig. 2). Secondly, research direction in versatile expander and holistic architecture (NO_x + CO_2) are proposed to partially address the challenges hindering ORC implementation. This extends the earlier works which linked process requirements to the fundamental fluid properties and process integration guidelines to optimised energy conversion [3]. Engine and ORC modelling was performed in Ricardo Wave and Aspen HYSYS (using Peng-Robinson equation of state) [15, 16]. The relative simulation results, against the published ORC conclusions, quantified the benefits of using methods less common within the automotive sector and unconventional approaches.

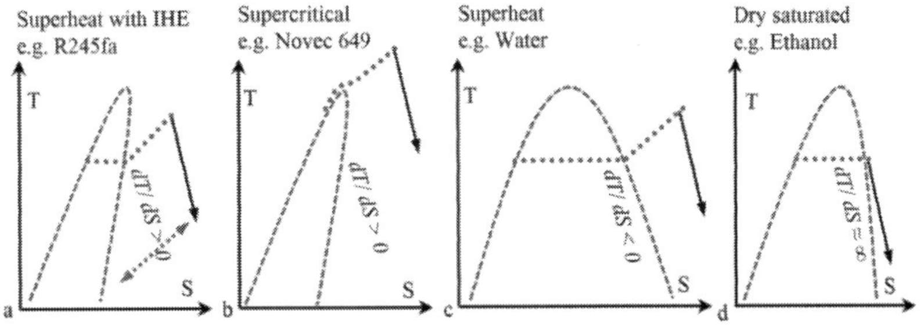

Fig. 1. Illustrative literature ORC operation (a) Recuperated (b) Supercritical (c) Superheated (d) Dry vapour.

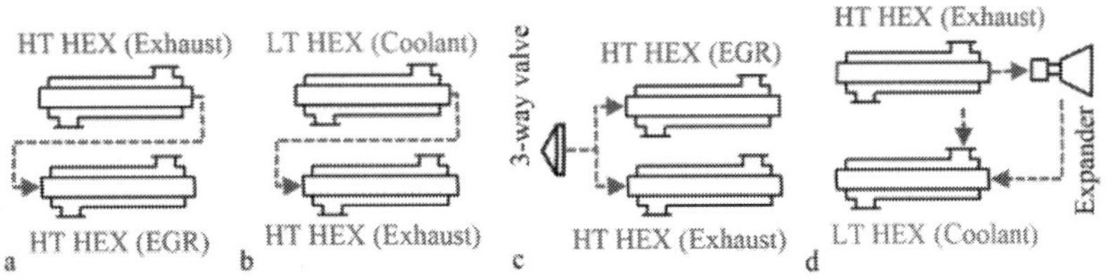

Fig. 2. Illustrative literature heat recovery architecture (a, b) Series (c) Parallel (d) Cascade.

2.1 Review – refrigerants

Using nearly dry saturated vapour expansion (Fig. 1d), Wang et al. simulated a cascade architecture (Fig. 2d), where the interconnected High Temperature (HT) cycle and Low Temperature (LT) cycles recovered exhaust heat using R245fa and coolant heat using R134a [17]. Despite a low 7% ORC thermal efficiency due to the LT coolant, a 3% point increase in gasoline engine thermal efficiency was calculated. Cascade architectures allowed higher heat exploitation, however, the challenges relating to system size, control and complexity may be prohibitive for automotive applications. Hussain and Brigham also simulated coolant and exhaust heat using R245fa, however, in a series architecture (Fig. 2b) [18]. The residual exhaust heat coming out of the evaporator was also used for preheating. The power generated during city drive cycle could only provide 1/3rd of the accessory load requirement due to the on-off conditions of the considered hybrid gasoline engine.

Cummins demonstrated EGR and exhaust parallel architecture (Fig. 2c), initially using a superheated recuperated R245fa cycle (Fig. 1a). In the continued work, the ORC system was packaged on a truck employing low rates of EGR and high efficiency Selective Catalytic Reduction (SCR) system [11]. The system delivered 4.5-6% fuel efficiency improvements, the turbine output power was mechanical to the crankshaft through a gearbox and the condenser was a direct air cooled condenser at the front of the engine cooling module. R245fa has a high Global Warming Potential (GWP) of 1030. This coupled with the lack of non-flammable alternatives has led to intense development efforts by the chemical companies to design improved molecular makeup refrigerants [19]. Cogswell et al. has examined a 5 kW supercritical recuperated ORC using Novec 649, a 4th generation low GWP refrigerant (Fig. 1b) [12]. Furthermore, Cummins have replaced 245fa with R1233zd(e) [11].

In summary, for refrigerant as the working fluid, the majority of publications have selected R245fa, a drying fluid due to higher molecular complexity. R245fa played a key role in the initial resurgence of automotive ORC due to a positive stationary ORC experience and the possibility to modify existing sector hardware. R245fa was better suited for heat recovery < 300°C and condensing temperatures < 50°C.

Finally, scroll expanders, screw expanders and radial turbines, all have been demonstrated using refrigerants. For scroll expanders under 10 kW output, Barccoa et al. have shown efficiencies up to 70% with Pressure Ratio (PR) of 5.5:1 and maximum cycle temperature of 190°C [20]. With improvements in rotor profile of screw expanders, Leibowitz et al. presented efficiencies up to 70% at engine RPMs for under 25 kW output [21]. Subcritical turbine envelope with maximum inlet conditions, pressure ratio and maximum speed of 27 bar, 230°C, 10.8:1 and 70,000 rpm was demonstrated by Halliwell [22]. Furthermore, supercritical radial inflow turbine with 15 primary and splitter vanes was modelled by Cogswell et al. [12]. Inlet conditions, peak efficiency, pressure ratio, power output and speed was 28 bar, 260°C, 80%, 12:1, 10 kW and 78,000 rpm.

2.2 Review – inorganics

Using water as the working fluid, Katsanos et al. simulated exhaust only and series exhaust then EGR architectures (Fig. 2a) for a truck engine, offering average fuel consumption improvements of 4.3% and 6.8% [23]. The majority of the heat needed for evaporation came from exhaust and the superheating was covered by the high EGR rate (Fig. 1c) at the mid-speed mid-load engine condition (B50). Cooper et al. additionally showed a low sensitivity to engine speed and load [13]. At high EGR rate and high-speed high-load engine condition, Arunachalam et al. theoretically evaluated Rankine cycle architectures, even combining CAC [24]. They showed that the power produced from EGR only heat recovery was 72% of the cycle which recovered all the three gaseous sources.

Simulations investigating the controllability of the Rankine cycle for exhaust only architecture in highway drive cycle with practical operating pressures was presented by Weerasinghe et al [14]. The higher fuel consumption improvements reported were due to a steam storage reservoir between the evaporator and the expander, providing an energy buffer through periods of varying engine operation. With HT heat favoring water and LT heat favoring 4th generation refrigerants (e.g. R1234yf), the combination of these two different classes of fluids were also proposed in water-organic cascade architecture by Bae et al. [25]. Additional inorganic cycles in literature included, supercritical CO_2, ammonia and Kalina cycle, however, due to the requirement of a low condensing temperature, excessive system pressures for equal performance to refrigerants and complex packaging, respectively, these may be inappropriate options for truck applications [26, 27].

In summary, Rankine cycle was stated to be effective only for HT heat (e.g. series exhaust then EGR architecture) usually > 400°C and paired with condensing temperature > 100°C. Water gave the highest thermal efficiency for a given ΔT of source, however, required expander PR was higher for a given evaporator and condenser temperature compared to refrigerants [3]. Screw expanders were also modelled for two-phase water expansion in the HT cascade section offering noticeable improvements [28, 29].

2.3 Review – hydrocarbons

AVL initially performed fluid comparisons study between R245fa and ethanol, highlighting a favorable case for ethanol. In the continued work, they demonstrated a superheated ethanol cycle with IHE for trucks using exhaust only architecture due to low EGR rates [30]. For fixed evaporation (120°C) and condensation (70°C) temperatures, average fuel consumption improvement of 3.8% was shown. Seher et al. compared working fluids at a condensation temperature of 100°C using a piston expander and turbine in the EGR and exhaust parallel architecture [10]. Ethanol was considered as a suitable fluid as it provided high performance using either piston expander or turbine. Edwards et al. investigated EGR only, exhaust only and EGR and exhaust parallel architectures using ethanol [31]. The controlled use of EGR and exhaust showed the greatest simulated fuel consumption improvement of 4.3%. The ORC condenser was integrated with the HT engine radiator and all architectures utilised the same piston expander map.

For series coolant and exhaust architecture, compared to refrigerants, contrary conclusions can be drawn for hydrocarbons from the simulations of Vaja and Gambarotta [32]. Thermodynamic attributes using benzene showed that the series coolant and exhaust architecture gave an improvement only $1/10^{th}$ higher than exhaust only architecture. Hence, the marginal drop in efficiency for a reduction in cost and complexity favoured the simpler architecture.

In summary, ethanol in EGR and exhaust parallel architectures was the preferred hydrocarbon at 300–400°C heat (i.e. typical exhaust gas temperatures). Hydrocarbons generally decompose quickly, thus reducing their overall toxicity risk, they also have GWP < 20 [19]. Hydrocarbons do present the challenge of flammability, however, improved technologies to reduce risk and use in the context of a closed system has led to a revived interest. For exhaust only architecture, alkanes have also been recommended due to excellent thermo-physical and environmental characteristics by Shu et al. [33]. In high ΔT applications, studies demonstrated efficient piston expanders with engine RPMs and high inlet temperature/pressure (380°C/32 bar) as suitable alternative to turbines for water and ethanol [10].

2.4 Research direction – vapour and two-phase expanders

In this section, the fundamental results for two-phase expansion are presented and compared against superheated expansion. Furthermore, included in the discussion are the challenges and recommendations for two-phase expansion. Heat recovery was considered from a turbocharged, 12.8 L truck engine model (Euro 6 with high efficiency SCR) at the low-speed mid-load condition (A50). The ORC inputs, assumptions and sizes presented in Table 1, which correspond to realistic values, were utilised. Simulations were conducted

DOI: 10.1051/e3sconf/20172200132

using cyclopentane, the proposed fluid in this study, and ethanol, a common literature fluid. Heat transfer areas (approximated as *UA*, product of the heat transfer coefficient and area) and expander size (approximated as PR) were fixed. Expansion and pumping efficiencies were kept constant at relatively lower values (55%).

Table 1. Boundary conditions, assumptions and sizes considered during modelling.

Common in all simulations		**Specific to A50 1160 rpm, 131 kW**		**Specific to B50 1440 rpm, 158 kW**		**Specific to B50**	
Parameters	Values	Parameters	Values	Parameters	Values	Parameters	Values
$T_{cool\ air\ in}$	30°C	T_{exh}	380°C	T_{exh}	375°C	$UA_{exh\ HEX}$	1000 W/°C
$T_{cool\ air\ out}$	50°C	m_{exh}	0.176 kg/s	m_{exh}	0.223 kg/s	$UA_{CAC+EGR}$	1680 W/°C
η_{pump}	55%	$c_{p\ exh}$	1.14 kJ/kg°C	T_{CAC}	90°C	UA_{cond}	2000 W/°C
η_{exp}	55%	$UA_{exh\ HEX}$	600 W/°C	m_{CAC}	0.223 kg/s	PR_{exp}	16:1
		UA_{cond}	1100 W/°C	T_{EGR}	430°C	T_{cond}	50°C
		$VFR_{exp\ X=1}$	19:1	m_{EGR}	0.033 kg/s	$\Delta P_{HEX\ fluid}$	0.6 bar
				η_{trans}	94%	$\Delta P_{cond\ fluid}$	0.2 bar

Fig. 3a presents the expansion Volume Flow Ratio (VFR), net power and thermal efficiency results for cyclopentane and ethanol with expansion inlet condition of liquid ($X = 0$) to dry saturated vapour ($X = 1$) expansion. With reduction in vapour fraction, the thermal efficiency decreased (11 to 8.7%), due to the reduction in the expansion enthalpy change. Nonetheless, the reduction in the net power of cyclopentane was limited to below 15% (4.7 vs. 5.6 kW) when a vapour fraction ≥ 0.4 was maintained. For approximately 1/6th of the duration in highway drive cycle, power is not generated due to the insufficient temperature and pressure conditions at the conventional expander inlet. Hence, a working fluid and expander combination, which can tolerate two-phase expansion then potentially offers a partial means to recover more frequent power.

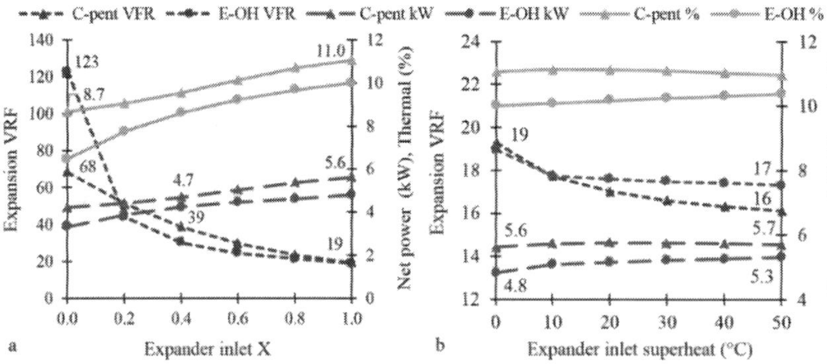

Fig. 3. Simulation results for (a) Two-phase expansion (b) Superheated expansion.

The two-phase expansion was compared against the conventional approach of superheated expansion. Fig. 3b presents the results for cyclopentane and ethanol of dry saturated vapour to 50°C superheated expansion. At the dry saturated vapour condition, cyclopentane offered 15% additional net power (5.6 vs. 4.8 kW) for equal *UA* and VFR values to ethanol. With increasing superheat, the net power results converged (Δ15 to Δ7%) due to the opposing shape of the T-S curve for the two fluids. As a result, for organic fluids in high ΔT applications, the design point corresponded close to the dry saturated vapour condition.

The VFR with cyclopentane for the two-phase expansion (X = 0.4) was noticeably higher (2 times) than the dry saturated vapour expansion (39:1 vs 19:1). This may not necessarily correspond to the requirement of a linear increase in the design point VFR. This can be attributed to the density of the two-phase being much higher (3.7 times) compared to the dry saturated vapour. As a result, operational VFR of the two-phase expansion is expected to be greater than the design point VFR of the dry saturated vapour expansion. Cyclopentane offered an equal VFR to ethanol at X = 1 with lower PR (16:1 vs. 19:1) and a relatively lower VFR variation between X = 0–1 (3.6 vs. 6.5 times), providing a more practical solution.

The fluid side of the exhaust heat exchanger (HEX) typically includes two temperature measurements, one at the inlet (T_{in}) and the other at the exit (T_{out}). With the two-phase expansion, the HEX only requires an additional in-build intermediate temperature measurement (T_{inter}). When $T_{inter} \approx T_{out}$, it can be assumed the HEX exit is in a two-phase state (i.e. $X > 0.4$). Due to the limited impact on net power and VFR for $X > 0.4$, the above approach is expected to suffice, limiting the challenge relating to the ORC control.

The two-phase (X = 0.4) expansion process is expected to be slightly inefficient when compared to the dry saturated vapour expansion. If the two-phase expansion efficiency was assumed to be 50%, the net power generated was still 75% of the design point. This indicates that despite the decreasing net power by reducing vapour fraction, power can be produced more frequently in a transient drive cycle using expanders that can expand two-phase as well as vapour. Finally, assuming power generation potential for 2/3rd of the untapped transient highway drive cycle, the implementation of expander tolerant to two-phase increased the drive cycle power by 8%.

2.5 Research direction – holistic NO_x and CO_2 reducing thermal architectures

Modern emission compliant truck engines with high efficiency SCR systems offer three gaseous sources for potential heat recovery. CAC for cooler intake temperatures and engine downsizing, EGR for NO_x reduction, and post-turbine exhaust gases. As such, emission standards and CO_2 reduction cannot be uncoupled and has to be addressed on the same engine platform, since techniques for NO_x reduction have a high impact on engine thermal efficiency and the available waste heat. There is a potential challenge in integrating exhaust gases with CAC, due to lower quality (75–150°C) and transient flows (boost pressures 1.5–3 bar), and EGR, due to lower quantity (≈ 15% of exhaust) and higher quality (50–100°C > exhaust). In this section, holistic revision of the engine and ORC thermal architecture is presented with the objective of proposing an integrated heat recovery, engine downsizing and NO_x limiting platform. Simulations were conducted using cyclopentane at B50, an operating condition which corresponds closely to the expected ORC design point in highway trucking.

Since CAC and EGR thermal conditions differ amongst the engine applications, Fig. 4a presents an ORC utilising a combined medium temperature HEX. This is used for cooling the blended charge air and exhaust recirculation, and is located post-compressor in the engine platform as shown in Fig. 4b (SCR excluded for illustration). Due to the available cooling margin in the engine cooling module, the LT engine radiator was modified into the ORC condenser and the air temperature rise was limited to 20°C. Compared to the exhaust HEX (UA = 1000 W/°C), the combined HEX (UA = 1680 W/°C) was larger in size due to a cooling requirement to 60°C and relatively medium quality heat. Nonetheless, the larger size can be partially compensated since these streams are already a part of the engine cooling module. Additionally, this architecture reduces the number of thermal components, offers suitable medium quality heat for series heat recovery, and utilises only higher quality exhaust heat.

 DOI: 10.1051/e3sconf/20172200132

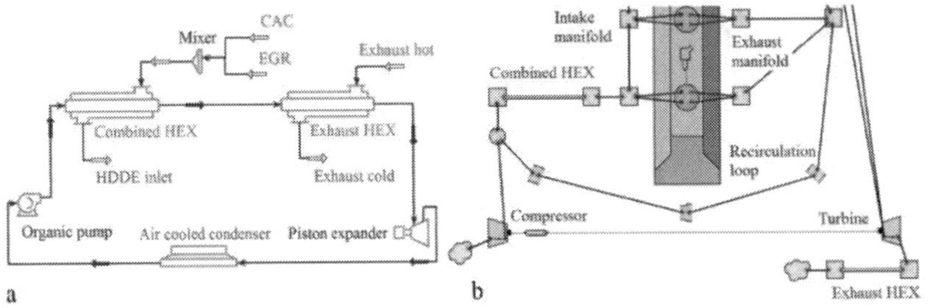

Fig. 4. (a) Holistic architecture modelled in Aspen HYSYS (b) Modified truck engine modelled in Ricardo Wave.

In the simulated case, heat recovery was 20.2 and 57.9 kW in the combined and the exhaust HEX, resulting in the complete utilisation of CAC and EGR heat (Table 1). The minimum temperature difference was 10 and 20°C for combined and the exhaust HEX. Such minimum temperature difference values offer a suitable trade-off between the heat transfer area and the ORC power [3]. The maximum and minimum cycle temperature/pressure combination were 180°C/20.5 bar and 50°C/1 bar during vapour expansion condition. The ORC system was 10.1% efficient and generated a net power of 7.9 kW, which corresponded to 5% of engine power. These results were compared against ethanol for exhaust heat recovery architecture only, a total improvement of 36% (Δ 2.1 kW) was recorded.

3 Conclusion

Despite an increased risk, ORCs have demonstrated higher overall conversion efficiency and technology readiness level for CO_2 reductions in highway trucking. The 8–12% thermally efficient systems for 5–15 kW output have shown 4–6% reduction in fuel consumption. The specific system weight and cost following standardisation are projected to be 10 kg/kW and 300 £/kW. Firstly discussed was, the use of exhaust, EGR and coolant as heat sources; R245fa, ethanol and water as working fluids; series, parallel and cascade as thermal architectures; dry saturated, superheated and supercritical as expansion inlet conditions; and scroll, radial turbine and piston as expansion machines. Furthermore, emerging ORC approaches included: use of 4th generation refrigerants, innovative methods for LT heat recovery, and two-phase water expansion in HT cascade section. Secondly, research direction in versatile expander and holistic architecture (NO_x + CO_2) were proposed. By linking process requirements to the fundamental fluid properties and process integration guidelines to optimised energy conversion, ORC modelling in Aspen HYSYS and engine modelling in Ricardo Wave was conducted. Results indicated that, the implementation of versatile piston expander tolerant to two-phase and using cyclopentane can potentially increase the highway drive cycle power by 8%. Implementation challenges relating to: high expansion VFR, two-phase ORC control and expansion inefficiency were also considered. Finally, holistic revision of the engine and ORC thermal architecture was presented with the objective of proposing an integrated heat recovery, engine downsizing and NO_x limiting platform. This reduced the number of thermal components, offered suitable medium quality heat for series architecture, resulted in complete utilisation of charge air and recirculation heat, utilised only high quality exhaust heat and increased the ORC performance to 5% of engine power at the design point.

References

1. Z. Xu, J. Liu, J. Fu, C. Ren, SAE Int. 10.4271/2013-01-1648 (2013)

2. C. Sprouse, C. Depcik, Appl. Therm. Eng. **51**, 711–722 (2013)
3. A. Panesar, Doctoral thesis, 10.13140/RG.2.1.4559.0248 (2015)
4. A. Domingues, H. Santos, M. Costa, Energy **49**, 71–85 (2013)
5. U. Larsen, L. Pierobon, F. Haglind, C. Gabrielii, Energy **55**, 803–812 (2013)
6. F.J. Fernández, M.M. Prieto, I. Suárez, Energy **36**, 5239–5249 (2011)
7. M.H. Yang, R.H. Yeh, Applied Energy **149**, 1–12 (2015)
8. H. Wang, R. Peterson, K. Harada, E. Miller, Energy **36**, 447–458 (2011)
9. Y.Q. Zhang, Y.T. Wu, G.D. Xia, C.F. Ma, Energy **77**, 499–508 (2014)
10. D. Seher, T. Lengenfelder, J. Gerhardt, N. Eisenmenger, Aachen Colloquium, October 8–12, Aachen, Germany (2012)
11. D.W. Stanton, SAE Int. J. Engines **6**, 1395–1480 (2013)
12. F.J. Cogswell, D.W. Gerlach, T.C. Wagner, J. Mulugeta, IMECE, November 11–17, Denver, USA (2011)
13. C. Cooper, F. Kamakate, T. Reinhart, M. Kromer, Reducing Heavy-Duty Long Haul Combination Truck Fuel Consumption and CO_2 Emissions (2009)
14. R. Weerasinghe, R. Stobart, S. Hounsham, App. Therm. Eng. **30**, 2253–2256 (2010)
15. Ricardo Software, WAVE version 8.1 (2015)
16. Aspen Technology Software, HYSYS version 8 (2015)
17. E.H. Wang, H.G. Zhang, Y. Zhao, B.Y. Fan, Energy **43**, 385–395 (2012)
18. Q.E. Hussain, D.R. Brigham, IDETC/CIE, August 15–18, Montreal, Canada (2010)
19. J.M. Calm, G.C. Hourahan, ICR, August 21–26, Prague, Czech Republic (2011)
20. R. Braccoa, S. Clementeb, D. Michelib, M. Reini, ECOS, June 26–29, Perugia, Italy (2012)
21. H. Leibowitz, N. Stosic, I.K. Smith, ASME IMECE, November 5–10, Chicago, USA (2006)
22. Halliwell, International Conference on Turbochargers and Turbocharging, May 15–16, London, UK (2012)
23. C. Katsanos, D. Hountalas, T. Zannis, E. Yfantis, SAE Int. 10.4271/2010-01-0315 (2010)
24. P.N. Arunachalam, M. Shen, M. Tuner, P. Tunestal, SAE Int. 10.4271/2012-01-1602 (2012)
25. S. Bae, H. Heo, J. Park, H.Y. Lee, SAE Int. 10.4271/2013-01-0046 (2013)
26. H. Tian, G.Q. Shu, H. Wei, X. Liang, SAE Int. J. Engines **6**, 766–776 (2013)
27. M. He, X. Zhang, K. Zeng, K. Gao, Energy **36**, 6821–6829 (2011)
28. I.K. Smith, N. Stosic, E. Mujic, A. Kovacevic, Journal of Process Mechanical Engineering **225**, 117–125 (2011)
29. J. Song, C.W. Gu, Applied Energy **156**, 280–289 (2015)
30. H. Teng, J. Klaver, T. Park, G. Hunter, SAE Tech. 10.4271/2011-01-0311 (2011)
31. S. Edwards, J. Eitel, E. Pantow, P. Geskes, Int. J. Commer. Veh. **5**, 395–406 (2012)
32. I. Vaja, A. Gambarotta, Energy **35**, 1084–1093 (2010)
33. G. Shu, X. Li, H. Tian, X. Liang, Applied Energy **119**, 204–217 (2014)

Chemical treatment of wastewater from flue gas desulphurisation

Iwona Pasiecznik[1,*], *Włodzimierz* Szczepaniak[1]

[1]Wroclaw University of Science and Technology, Faculty of Environmental Engineering, Wybrzeze Wyspianskiego 27, 30-370 Wroclaw, Poland

Abstract. The article presents results of laboratory tests of removing boron and arsenium from non-ideal solutions using double-layered magnesium/aluminium hydroxides (Mg/Al Double-Layered Hydroxide – DLH) produced with nitrate-chloride method. In research, wastewater from an installation for flue gas desulfurization was examined. Double-layered hydroxides are perfect absorbents for anionic compounds. The research proved high effectiveness of preparation with reference to arsenium, as well as confirmed the effect of presence of sulfatic and arsenate ions on the effectiveness of boron removal. On the basis of research on absorption kinetics a theoretical dose of DLH/NO_3-Cl/M preparation was calculated and compared with a dose that ensures emimination of boron below the limit standarized by the national regulations. Application of double-layered magnesium/aluminium hydroxides for boron elimination from industrial wastewater requires significantly higher doses of preparation than those calculated in model investigations. It is due to the priority of removal of multivalent ions, such as sulfatic, arsenate or phosphate ions, by DLH/NO_3-Cl/M.

1 Introduction

The composition of wastewaters emerging during production of non-ferrous metals, including copper, is significantly diverse. They consist almost all substances, both in soluble form or as suspensions. They are mostly neutral inorganic materials, soluble components of raw materials, dusts, additives, auxiliary agents, etc. [1].

Copper production in Poland is performed using technologies of shaft furnace and suspension furnace. In the shaft process copper concentrate is remelted to copper stone, then subjected to convertoring. Gases from shaft furnace (so-called furnace top gases) are subjected to wet dust cleaning and directed to heat and power generating plant, for energetic use. Gases from heat and power generating plant are desulfurized in flue-gas desulfurizing installation. Wastewaters from this process contain significant amounts of heavy metals and boron. Sludges generated in the process of wet dust cleaning are lead-bearing raw material temporarily deposited in landfills in the area of metal-works. It was found that eluates from such landfills are characterized by high concentrations of boron

* Corresponding author: iwona.pasiecznik@pwr.edu.pl

and arsenium. Scheme of technological process of refined copper production is presented in Figure 1 [1].

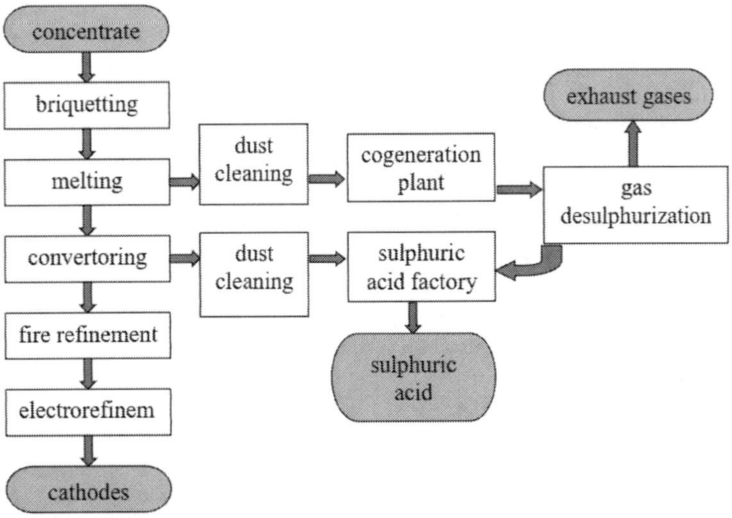

Fig. 1. Scheme of technological process of refined copper production [1].

In case of the wastewater plant in one of Polish copper mills, an additional source of boron in influent wastewater is a temporary landfill of lead-bearing concentrate and a slagheap. Drainage and rain waters from these objects flow into a detention-sedimentation tank, and then to the wastewater plant. Boron content in draining and rain water from the lead-bearing concentrates landfill and from the slagheap varies from a dozen or so to a few hundred mg/L. Limiting amounts of water flowing into the tank and wastewater plant is difficult to achieve from the technological point of view, therefore it is impossible to lower the boron concentration in processed wastewater leaving the wastewater plant by lowering boron load in raw wastewater [1].

The problem of eliminating boron from wastewater is connected, among others, with processing wastewater from the copper industry. It should be considered as an important issue, especially in regions having a main source of boron penetration into the environment. Lack of effective method of processing waste having high boron concentration is one of the main causes of high concentration of this element in nearby surface waters. Wastewater disposed to the environment causes contamination of potable water and water used for irrigation of crops in which boron concentration should be limited for health reasons [2]. Due to this, reduction of boron concentration in industrial wastewater below the values defined by regulations is necessary.

Boron can be eliminated from water or wastewater using various technologies, but most of the methods proposed have limitations and can be used only conditionally. Chemical precipitation may be used for elimination of boron but it requires relatively high levels of pH (pH > 9), that causes high saltness of solution and generates significant amounts of sludge [3–7]. Besides, high costs of chemical substances, as well as the costs of sludge utilization and disposal make this process of boron elimination inadvisable in many cases [8, 9]. Activated carbon, impregnated activated carbon and silty materials, due to their high sorption capacity, proved to be ineffective and unprofitable when used in a large scale [10, 11]. A liquid-liquid extraction technology offers a possibility of retrieving boron for market purposes from industrial wastewater, but it is only profitable when the initial concentration of boron in solution is at least 3 g/L [12, 13]. Ion-exchange resin Amberlite IRA 743 shows excellent effectiveness of borate ion removal, but high costs of technological system, as well as costs of regeneration, considerably limit its use [14, 15]. Reverse osmosis and electrodialysis allow for effective boron removal but these processes require increasing pH of influent solution above 9 pH, that causes high potential with colloidal scaling and fouling, therefore limiting application of these technologies for boron removal from wastewater [16]. Obtaining the boron concentration value in potable water at the level normalized by WHO (0.5 mg/L) requires application of multistage RO system [17]. Numerous difficulties and economical disadvantages of processes mentioned above were the impulse for new research works oriented on establishing effective and economically reasonable methods of boron removal, since according to the World Health Organization [18], boron is an element that cannot be removed using conventional methods of water and wastewater processing. One of recently tested technologies of boron removal are double-layered hydroxides [19].

Double-layered hydroxides (DLH) are a kind of synthetic silty materials with a structure described by the formula $M(II)_{1-x}M(III)_x(OH)_2(A^{n-})_{x/n}\cdot mH_2O$, where both bivalent M(II) and trivalent M(III) cations form positively charged layers. These layers remain in equilibrium with A^{n-} anions in hydrated interlayer areas [19]. Removal of boron/borates or arsenate ions may be done by exchange with ions from the hydrated layer or by complexation of hydroxyl groups on the surface of DLH layers. Double-layered hydroxides are considered excellent adsorbents of anionic compounds [20]. Numerous studies showed that DLH are effective in processes of remediation of land-water environment, especially if anionic contaminants of water are concerned [21–23].

Double-layered Mg/Al hydroxides have high boron removal capacity and hold promise of high effectiveness in boron removal from waters and wastewaters. Application of DLH allows to remove this element effectively from solutions having its concentration below 50 mg/L, and down to the limits recommended by the WHO in potable water (< 0.5 mg/L) [18].

The aim of laboratory studies was to determine the effectiveness of removing borate and arsenate ions from wastewater generated in flue-gas desulfurizing installation, using magnesium/aluminium hydroxide obtained with nitrate-chloride salts.

2 Materials and methods

Preparatics of original DLH substance, prepared using magnesium nitrate(V), aluminium chloride and sodium hydroxide, is described in detail in literature [19]. Detailed research performed with the preparation mentioned above led to some appropriate modifications of original DLH preparation, and in consequence obtaining stable preparation, efficiently removing boron and avoiding significant saltness of solution, denoted as DLH/NO_3-Cl/M in subsequent text [24]. The chemical formula of prepared double-layered magnesium/aluminium hydroxide (DLH/NO_3-Cl/M), obtained using chloride-nitrate method, differs from the DLH formula given in literature: $Mg_2AlNa_{1,4}(OH)_{7,57}Cl_{0,03}(NO_3)_{0,8}\cdot xH_2O$ [19], mostly by the number of sodium and chloride ions.

The research was conducted on wastewater from the flue-gas desulfurizing installation in one of Polish copper mills. The composition of raw wastewater is presented in Table 1. Wastewaters from flue-gas desulfurizing installations are strongly acidic (pH about 1.7) and contain high concentrations of heavy metals.

DLH/NO_3-Cl/M was added to wastewater from the flue-gas desulfurization installation in amounts of 1.5–50 g/L. The first series of tests was performed without modification of the wastewater pH, whereas in the second series for the same doses of double-layered magnesium/aluminium hydroxide the initial pH of wastewater was adjusted to 6.0–7.0 with sodium hydroxide (20% NaOH). A mixture of wastewater and DLH/NO_3-Cl/M was mixed for 1 hour applying the rotary speed of 300 – 400 rpm, then concentrations of boron and arsenium were determined in supernatant separated by percolation. Time of contact of the preparation with solution was assigned on the basis of earlier research [24] that showed that effectiveness of boron removal increases with time of contact of solution with DLH/NO_3-Cl/M, and at the same time it is worth stressing that straight majority of boron was sorbed during the first 40–60 minutes of the process.

Table 1. Average composition of wastewater from flue-gas desulfurizing installation.

Kind of contaminant	**Average concentration [mg/L]**	**Kind of contaminant**	**Average concentration [mg/L]**
Boron	3.69	Zinc	477
Arsenium	143	Iron	2.27
Cadmium	124	Mercury	1.556
Copper	0.767	Sulphates	2493
Lead	3.16	Nitrates	3.5
Nickel	0.096	Dissolved substances	7426
Chromium	0.101	Dry residue	7356
Cobalt	0.038	Suspensions	70

3 Results and discussion

Double-layered hydroxides introduced into model solutions prepared with distilled water, due to their chemical properties, affect the reaction of solutions. They are hydroxide preparations, and that causes that solutions gain an equilibrium about a pH value of 10. Therefore, the pH value might have a significant effect on boron ions removal. Due to this, tests with non-ideal solutions were performed in two variants: without initial pH adjustment in wastewater, and with its adjustment to the range of 6.0 - 7.0 (the range was determined for model solutions showing the most efficient boron ions removal [24]). The results obtained are presented in Tables 2 and 3.

Table 2. Results of boron and arsenium removal from wastewater generated in the installation for flue-gas desulfurization, for various doses of DLH/NO_3-Cl/M, without initial pH adjustment in wastewater.

Dose of DLH/NO_3-Cl/M [g/L]	5	10	20	30	50	70
Wastewater pH	1.7	1.7	1.7	1.7	1.7	1.7
pH after addition of DLH/NO_3-Cl/M	5.4	5.51	6.71	7.35	9.14	9.53
Wastewater pH after the process	6.97	7.11	8.88	9.24	10.99	11.2
Boron concentration after the process [mg/L]	2.37	2.17	1.94	1.63	1.57	0.89
Arsenium concentration after the process [mg/L]	138	126	45.8	34	29.6	11.1

Results of the research are presented graphically in Figs. 2 and 3. On the Figure 3, showing the effect of DLH/NO_3-Cl/M dose on the final arsenium concentration in wastewater (with and without pH adjustment) it is clearly visible that pH of wastewater had a significant effect on arsenium ions removal. The preparation sorbed arsenate ions far better from wastewater with initial pH adjusted. Differences in effectiveness of arsenium removal were even as great as 80%, for the same doses of double-layered hydroxides. It should also be noted that the preparation studied, despite of high concentration of contaminants in wastewater from flue-gas desulfurization, removed arsenate ions from solutions with high effectiveness. Arsenium was sorbed by DLH/NO_3-Cl/M even in 99%.

Table 3. Results of boron and arsenium removal from wastewater generated in the installation for flue-gas desulfurization, for various doses of DLH/NO_3-Cl/M, with initial pH adjustment in wastewater.

Dose of DLH/NO_3-Cl/M [g/L]	5	10	20	30	50	70
Wastewater pH	1.7	1.7	1.7	1.7	1.7	1.7
pH adjusted with 20% NaOH	6.05	5.9	6.86	6.41	6.23	6.25
pH after addition of DLH/NO_3-Cl/M	7.3	7.71	9.26	9.68	10.0	10.45
Wastewater pH after the process	9.2	9.48	11.19	11.32	11.50	11.64
Boron concentration after the process [mg/L]	2.48	1.95	1.86	1.83	1.56	0.93
Arsenium concentration after the process [mg/L]	67	18.3	10.7	8.3	5.2	1.1

Arsenium is a very toxic element and its concentrations even so small as a few mg/L may be very dangerous. The acceptable concentration of arsenium in potable water in Poland, according to directives of the World Health Organization [18] and a directive of Health Minister [2], is 0.01 mg/L. Acceptable arsenium concentration in wastewater dumped to waters or soil [25] is 0.1 mg/L for all kinds of industrial wastewater, except wastewater from the glass industry, where the acceptable limit was established at the level of 0.3 mg/L. Tests on model solutions proved that even small amounts of arsenium (concentrations ratio $As^{+5}/B = 1:5$) lower the effectiveness of boron removal by about 30% [24]. The effectiveness of boron removal with nitrate-chloride preparation was changing exponentially with the increase of As concentration. Arsenate ions significantly limited borate ions removal [24] and initial adjustment of pH in solution did not make any difference. Borate ions at the initial pH of 1.7, as well as after its adjustment to 6.0–7.0 were removed with the same effectiveness. These results confirm that the presence of other ions in solution made a competition in removing borate ions by DLH preparation [24]. Multivalent ions were the first exchanged with anions from hydrated layer of double-layered hydroxides. Therefore the initial stages of wastewater processing should be planned (knowing their elemental composition) in a manner that would allow to eliminate other contaminants first.

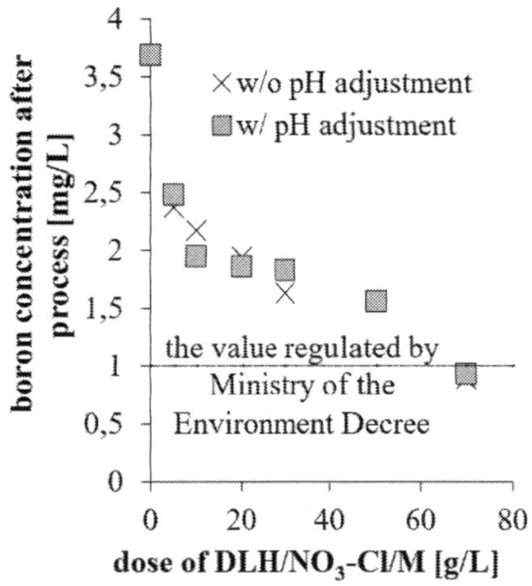

Fig. 2. Relation between final boron concentration in wastewater from Solinox and the dose of DLH/NO_3-Cl/M; wastewater pH adjusted.

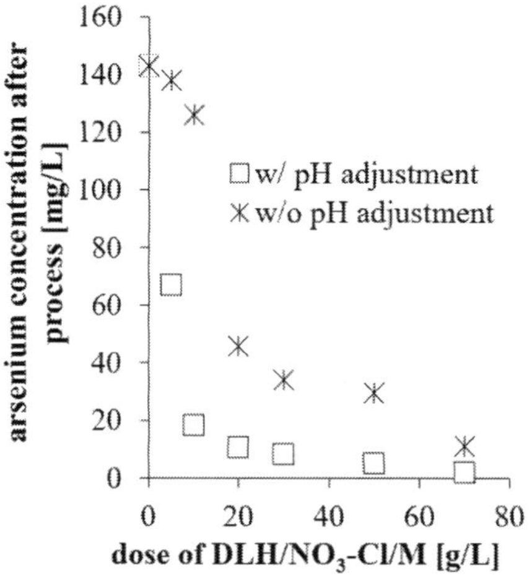

Fig. 3. Relation between final arsenium concentration in wastewater from Solinox and the dose of DLH/NO_3-Cl/M; wastewater pH adjusted.

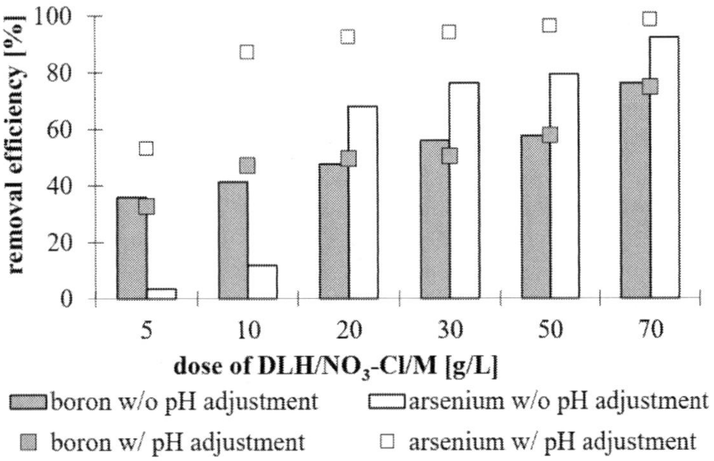

Fig. 4. Comparison of effectiveness of boron and arsenium removal from wastewater generated in flue-gas desulfurization installation depending on the dose of DLH/NO_3-Cl/M, with and without initial pH adjustment.

On the basis of adsorption kinetics [24] a theoretical dose of DLH/NO_3-Cl/M was determined, using the formula [26]:

$$x = \frac{x_m bC}{1 + bC} \tag{1}$$

where: x - amount of substance adsorbed, per unit of adsorbent [mg/g]

$$x = \frac{(C_o - C)}{m} \tag{2}$$

m – dose of adsorbent [g/L]; C – adsorbate concentration after the process [mg/L]; C_0 – initial adsorbate concentration [mg/L]; x_m – capacity of monolayer (maximum number of ions that can be retained by the unit mass of adsorbent [mg/g]; b – parameter of Langmuir's isotherm [L/mg].

The basis for calculation of the dose of double-layered magnesium/aluminium hydroxide was the boron adsorption isotherm. The process of boron adsorption on the surface of DLH/NO_3-Cl/M was properly described by Langmuir's model. It was confirmed by a high value of determination coefficient $R^2 = 0.9996$, as well as the mass of substance adsorbed per unit mass of adsorbent, almost identical to the calculated value [24]. For boron concentration in wastewater of 3.69 mg/L and required concentration after the process (1.0 mg/L), a theoretical dose of DLH/NO_3-Cl/M was 27.0 g/L.

A dose of DLH/NO_3-Cl/M that allowed to remove boron below 1.0 mg/L was determined as 70 g/L. On the ground of obtained results it was concluded that the calculated dose proved to be too small to remove boron below the level of 1.0 mg/L. It might be caused by the fact that a theoretical dose was calculated for model solutions, whereas the composition of wastewater tested (extremely high concentrations of heavy metals) might deteriorate boron removal on the surface of double-layered magnesium/aluminium hydroxide, due to presence of other ions hindering the process of boron removal (especially arsenate [24] and sulphate ions [27, 28]).

Technology of processing wastewater generated in described copper mill, based on neutralization and chemical precipitation of contaminants, enables processing wastewater with the effectiveness of 90% and boron removal in 50%, that is not always sufficient to achieve a concentration of boron in processed waste at the level below 1.0 mg/L [1]. Processed wastewater is directed from the wastewater plant to nearby stream, where boron concentrations often exceed values normalized by the national regulations. Boron concentration downstream from the wastewater outlet increases by about 29%–63% compared with water upstream the wastewater plant [1]. Application of double-layered magnesium/aluminium hydroxides as a technology of wastewater post-processing ensures effective removal of borate ions after previous processes of chemical precipitation.

4 Conclusions

Removal of boron from non-ideal solutions and achieving its concentrations below the limits established in regulations requires far greater doses of double-layered magnesium/aluminium hydroxides than a theoretical dose calculated for model solutions.

High concentrations of contaminants in water or wastewater make the process of boron removal difficult. This is due to the priority of removing multivalent ions before borate ions by DLH/NO_3-Cl/M.

DLH/NO_3-Cl/M preparation removes arsenate ions from solutions very effectively and it could be applied for co-removal of arsenium from water solutions.

In non-ideal solutions, characterized by high concentrations of contaminants, double-layered magnesium/aluminium hydroxides could be used in post-processing of waters and wastewaters, both for boron and arsenium removal.

References

1. M. Mańczak, M. Balbierz, A. Jaksik, *Appendix to legal agreement delimiting the special ways of water usage in the area of dumping wastewater to the Białynia stream by the 'Energetyka' Co. in Lubin,* Report 12/2005 – Institute of the Environmental Engineering, Wrocław Technical University, Wrocław (2005) (in Polish)
2. Health Minister Decree of 13.11.2015 r., Dz.U. 2015, poz. 1989 (2015) (in Polish)
3. I. Pasiecznik, M. Mańczak, Gaz, Woda i Technika Sanitarna **83**, 5, 12–13 (2009) (in Polish)
4. Y. Xu, J-Q. Jiang, Ind. Eng. Chem. Res. **47**, 16–24 (2008)
5. A.G. Pervov, A.P. Andrianov, R.V. Efremov, A.V. Desyatov, A.E. Baranov, Desalination **157**, 377–384 (2003)
6. A. Withers, Desalination **179**, 11–24 (2005)
7. I. Pasiecznik, M. Mańczak, Forum Eksploatatora **6**, 40–41 (2007) (in Polish)
8. J.R. Sims, F.T. Bingham, Soil Sci. Soc. Am. Proc. **32**, 364–369 (1968)
9. R. Kern, R.G. Gast, Soil Sci. Soc. Am. J. **47**, 1116–1121 (1983)
10. L.V. Rajakovic, M.D. Ristic, Carbon **34**, 6, 769 (1996)
11. J. Kluczka, J. Ciba, J. Trojanowska, M. Zolotajkin, M. Turek, P. Dydo, En. Viron. Prog. **26**, 1, 71 (2007)
12. D.E. Garrett, F.J. Weck, A.J. Marsh, H.R. Foste, U.S. Patent 3,111,383, November 19, (1963)
13. R.R. Grinstead, Ind. Eng. Chem. Prod. Res. De. Vel. **11**, 454 (1972)
14. M.O. Simonnot, C. Castel, M. Nicolai, C. Rosin, M. Sardin, H. Jauffret, Water Res. **34**, 1, 109 (2000)

15. Y. Xiao, B.Y. Liao, W.G. Liu, Y. Xiao, G.H. Swihart, Chin. J. Chem. **21**, 1073 (2003)
16. M.R. Pastor, A.F. Ruiz, M.F. Chillo´n, D.P. Rico, Desalination **140**, 145–152 (2001)
17. K. Fukunaga, M. Matsukata, K. Ueyama, S. Kimura, Membrane **22**, 4, 211 (1997)
18. World Health Organization, Guidelines for Drinking Water Quality, 4th ed., Geneva (2011)
19. J.Q. Jiang, Y. Xu, J. Simon, K. Quill, K. Shettle, Ind. Eng. Chem. Res. **46**, 4577 (2007)
20. Y.T. Liu, T.Y. Chen, M.K. Wang, P.M. Huang, P.N. Chiang, J.F. Lee, Applied Clay Science **48**, 485–491 (2010)
21. F. Konvanda, E. Kovacsova, D. Kolousek, Collect. Czech. Chem. Commum, **64**, 1517–1528 (1999)
22. Y. Seida, Y. Nakano, Water Res. **36**, 1306–1312 (2002)
23. R. Nomura, T. Mori, E. Kanezaki, T. Yaburani, Int. J. Mod. Phys. B. **17**, 1458–1463 (2003)
24. I. Pasiecznik, Doctoral Thesis, *Application of double-layered magnesium/aluminium hydroxides for removal of boron from solutions* (Institute of the Environmental Engineering, Wrocław Technical University, Wrocław, 2010) (in Polish)
25. Minister of Environmetal Protection Decree of 18.11.2014 r., Dz. U. 2014, poz. 1800 (2014) (in Polish)
26. W. Adamski, *Modeling of water purification systems* (Polish Scientific Publishers PWN, Warszawa, 2002) (in Polish)
27. I. Pasiecznik, M. Mańczak, W. Szczepaniak, *The effect of sulphate ions on boron removal from water solutions* (Polish environmental engineering five year after the access to the European Union, **T. 1**, 177–182, Lublin, 2009) (in Polish)
28. I. Pasiecznik, *Boron removal from aqueous solutions in presence of sulphate ions* (Interdisciplinary Problems in Environmental and Protection Enginnering, Publishing House of the Wrocław University of Technology, 325–332, Wrocław, 2011) (in Polish)

DOI: 10.1051/e3sconf/20172200134

The influence of heat sink temperature on the seasonal efficiency of shallow geothermal heat pumps

Grzegorz Pełka[1,*], *Wojciech* Luboń[1], *Anna* Sowiżdżał[1], and *Daniel* Malik[1]

[1]AGH-UST, Faculty of Geology, Geophysics and Environmental Protection, al. Mickiewicza 30, 30-059 Cracow, Poland

Abstract. Geothermal heat pumps, also known as ground source heat pumps (GSHP), are the most efficient heating and cooling technology utilized nowadays. In the AGH-UST Educational and Research Laboratory of Renewable Energy Sources and Energy Saving in Miękinia, shallow geothermal heat is utilized for heating. In the article, the seasonal efficiency of two geothermal heat pump systems are described during the 2014/2015 heating season, defined as the period between 1st October 2014 and 30th April 2015. The first system has 10.9 kW heating capacity (according to European Standard EN 14511 B0W35) and extracts heat from three vertical geothermal loops at a depth of 80m each. During the heating season, tests warmed up the buffer to 40°C. The second system has a 17.03 kW heating capacity and extracts heat from three vertical geothermal loops at a depth of 100 m each, and the temperature of the buffer was 50°C. During the entire heating season, the water temperatures of the buffers was constant. Seasonal performance factors were calculated, defined as the quotient of heat delivered by a heat pump to the system and the sum of electricity consumed by the compressor, source pump, sink pump and controller of heat pumps. The measurements and calculations give the following results: – The first system was supplied with 13 857 kWh/a of heat and consumed 3 388 kWh/a electricity. The SPF was 4.09 and the average temperature of outlet water from heat pump was 40.8°C, and the average temperature of brine flows into the evaporator was 3.7 °C; – The second system was supplied with 12 545 kWh/a of heat and consumed 3 874 kWh/a electricity. The SPF was 3.24 and the average temperature of outlet water from heat pump was 51.6°C, and the average temperature of brine flows into the evaporator was 5.3°C. To summarize, the data shown above presents the real SPF of the two systems. It will be significant in helping to predict the SPF of objects which will be equipped with ground source heat pumps.

1 Introduction

In recent years in Poland the constant progress of the heat pump sector has been observed in contrast to many previous years. In 2015, it was expressed by an average 14% growth of

* Corresponding author: gpelka@geol.agh.edu.pl

 DOI: 10.1051/e3sconf/20172200134

sales compared with 2014. The most spectacular growth (70%) was in air heat pumps in 2015 and this trend is expected to be continued [1, 2].

The Educational and Research Laboratory of Renewable Energy Sources and Energy Saving in Miękinia is a unit of the Faculty of Geology, Geophysics and Environment Protection AGH-UST. At the Faculty, research connected with renewable energy sources has been conducted for many years [3–8]. In 2012, a research laboratory that would improve the quality of the research work was established.

The Laboratory is located in the former administrative building of the porphyry mine in Miękinia (35 km north-west of Kraków) The building has been completely redesigned, achieving energy-efficient building for the RES laboratory. Currently the building is heated by four heat pumps extracting heat from the ground. Heat from the ground is taken by vertical heat exchangers and a horizontal ground exchanger.

The heating system in Miękinia is measured by heat production, electricity consumption and energy taken from the heat source. Inlet and outlet temperatures for each of the heating circuits are metered. PoWoGaz flow meters are used with rotor transducers, PT100 resistance temperature probes and SchellCount energy counters. Everything is archived in the memory of the DigiEnergy monitoring system and it is possible to export data and process them. This system has also an implemented algorithm for heat calculation using flow rate and a temperature difference of the fluid being measured. The collected data were evaluated.

2 Heat pumps parameters

Heat pumps are heating devices which have high work efficiency. In order to specify the work of heat pumps, we use the following parameters:

• COP (Coefficient of Performance) – calculated according to PN-EN 14511 [9]. It is defined as a ratio of heating output divided by the electrical power input. It is measured in determined conditions of work. To calculate COP, the electrical power required from all components such as the compressor, controller, circulation pumps or fans of heat sources and sink are taken into account. For the circulation pump on the side of the heat sink, the electrical power demand needed to overcome the hydraulic resistance of the condenser is taken into account. This coefficient characterizes heat pump's work in the strictly defined conditions of temperature of heat source and heat sink.

• SCOP (Seasonal Coefficient of Performance) is calculated according to PN-EN-14825 [10]. This coefficient describes the annual average efficiency factor of heat pumps. SCOP shows how efficient a specific heat pump will be in a designated heating season and for designated heating demand profile. Seasonal Coefficient of Performance can be estimated at the design stage of building and this calculation can be used to predict future exploitation cost. SCOP is use to optimize a heating installation with heat pumps.

• SPF (Seasonal Performance Factor) is defined as ratio of heat production and electricity consumption during specified period. There are several types of SPF, which account for some additional elements of installation such as the pump of a central heating system, additional electric heaters, etc. [11, 12].

The most useful coefficient for ending user is Seasonal Performance Factor as it represents a real test of the efficiency of an installation. The bigger ratio responds to the lower costs of exploitation related with preparing heat by the user. On paper the SPF is defined as the ratio of heat generated by the heat pump and electricity consumption to work of all elements, such as compressor, controller, pumps of heat source and heat sink, electric heater installed on heat pump (this device was shut off during the time of test). Calculated SPF is equal to SPF 3 according to the methodology presented by Miara et. al [13].

DOI: 10.1051/e3sconf/20172200134

3 Heating system in the laboratory - description

The heat pumps installed in the AGH-UST Educational and Research Laboratory of Renewable Energy Sources and Energy Saving in Miękinia work in three different heating systems. Two heat pumps prepare heat at a medium temperature (35°C–40°C) and another two prepare heat with a higher temperature (55°C). Low temperature's heat pumps reach 10.9 kW heating power with B0W35 conditions and COP is 4.4. The heat pumps producing high temperature heat have 17 kW power in B0W35 conditions and their COP is also 4.4. Each of the heat pumps has an independent heat source system. Three heat pumps have borehole heat exchangers and one of them has a horizontal ground source. The first one has three borehole heat exchangers (each of them at a depth 83 m) and heat or chill is distributed on the building using the ventilation ducts (system A – Fig 1). The second heat pump has three borehole heat exchangers (each of them at a depth 87 m) and heat or chill is distributed by underfloor heating (system B). The third heat pump has three borehole heat exchangers (each of them at a depth 100 m depth) as a source and the fourth has the closed-loop located 1.5 meters underground (600 m^2 of area). Two high temperature heat pumps share together one buffer tank and heat pumps produce high temperature heat for domestic hot water and to distribute heat to the building by means of radiators. The low temperature heat pumps work for independent buffer tanks (system C).

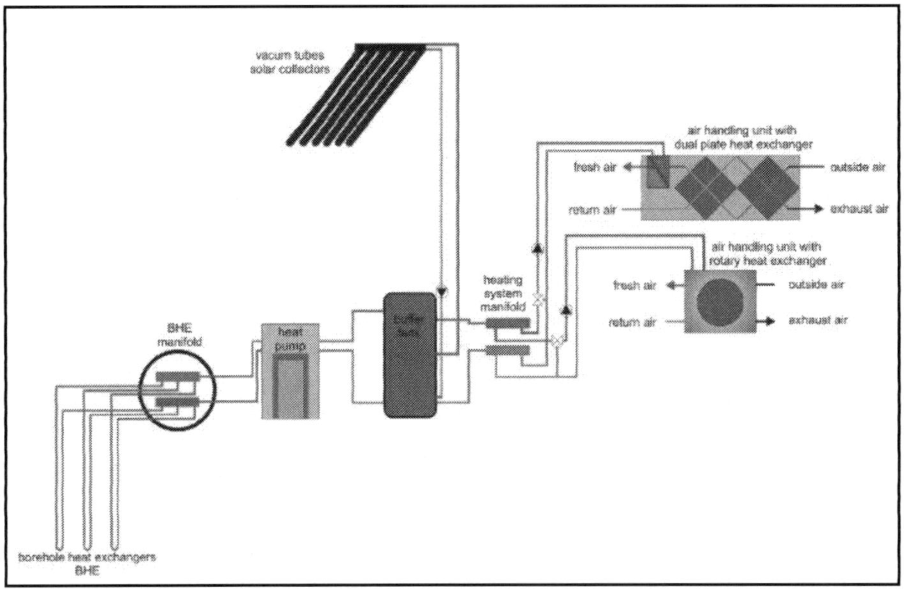

Fig. 1. A diagram of the system A in Laboratory in Miękinia.

DOI: 10.1051/e3sconf/20172200134

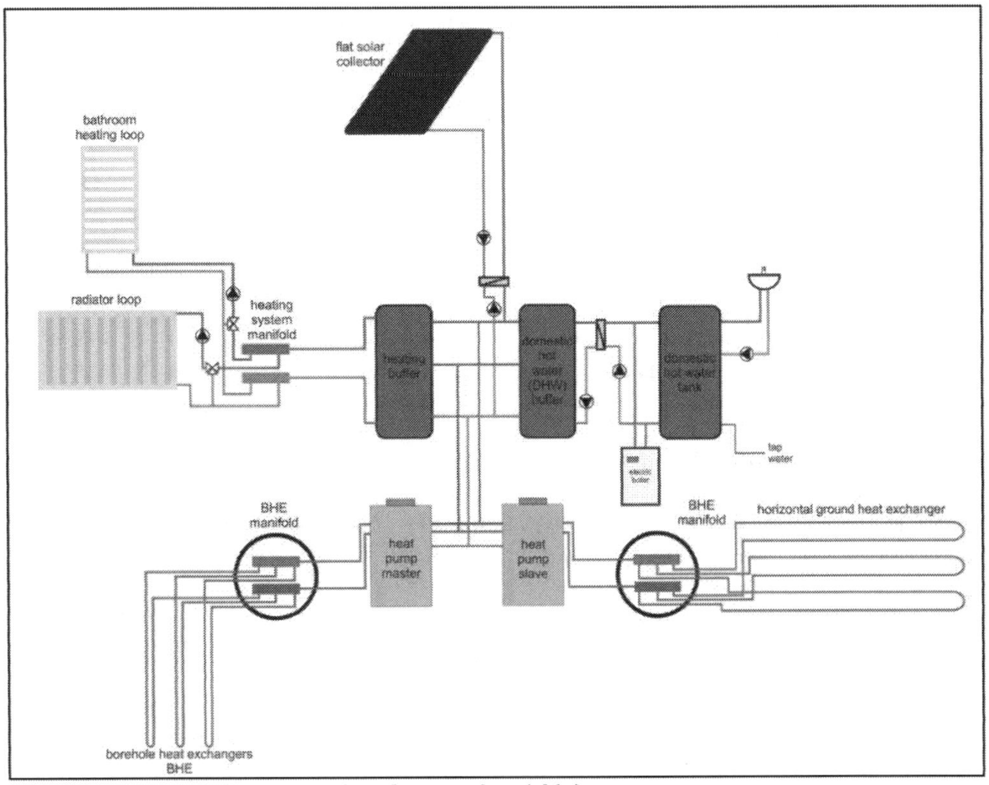

Fig. 2. A diagram of the system C in Laboratory in Miękinia.

4 Heating system operation – results of research

The heating system in the Laboratory in Miękinia is equipped with a control system and is able to archive data such as the heating power of each heating system, power demand of particular devices, supply and return temperature, temperature inside a buffer etc. Heating power is measured by impeller flow sensor and resistance temperature sensor and is calculated in DigiEnergy program. Electricity consumption is measured by one-phase and three phase meter. Data of the two heat pumps working during one heating season was analysed to determine the influence of heat sink temperature on the Seasonal Performance Factor. The heating season was defined as the time interval from 1th October 2014 30th April 2015.

Results of the research for system A show the real values of SPF which are attainable in heating systems with heat pumps producing low temperature heat. In this system, heat is used to supply the air handling units. During the test, temperature was set at constant value. The heat pump maintained a temperature of water in a buffer tank at 40°C. In the case of system A, there existed a real possibility of increasing the SPF factor by connecting a set water temperature dependent on outdoor temperature and create the heating curve. System C in comparison to system A during the heating season provided relatively less heat. System A delivered 13 857 kWh of heat, consumed 3387 kWh of electricity, while system C (with the vertical exchanger and a higher heating power device) supplied only 12 545 kWh of heat, consumed 3874 kWh of electricity. The reason for this was less demand for heat generated in the system C. During the analysed period, radiators were closed most of the time and demand for hot domestic water wasn't high. The water temperature in system C was set at 50°C, which corresponds to the convection radiator installation. The SPF of System A was 4.09, and the system C was 3.24. These results confirm that a floor-heating system is more appropriate for heat pumps than radiators [14]. The other energy parameters were presented in table 1.

Similar studies were conducted by Nouvel et. al. [15]. In this work the authors created a map of potential values of SPF for the operation parameters similar to the operation parameters in system C. This paper shows that heat pumps with borehole heat exchangers as a heat source and heat sink temperature of 50°C, located around Miekinia can be on the level 3.4. It is very similar to the results presented in this paper. The results of research published by the Coimbra University in Portugal [16] coincided with the results obtained by the heat pump in system A. A heat pump with seven borehole heat exchanger, each one with 125 meters of depth, heated water to 40°C. Both the research in Portugal and that conducted by the Fraunhofer Institute for Solar Energy Systems ISE [13] show that heating systems with ground sources heat pumps achieve similar results.

The Seasonal Performance Factor of heat pumps at 4.09 and 3.24 receive during research in Laboratory in Miękinia may be used to conduct estimates for potential future users. The efficiency of heating systems with heat pumps depends on the heat sink. In the future, heating seasons need to be researched to increase the performance factors and it is intended that this study will be ongoing.

Table 1. Energy parameters of heating systems during 1th October 2014 to 30th April 2015 heating season.

Parameter	System A	System C
Average outside temperature [°C]	5.3	
Cooling power [W/m]	33.8	48.3
Average temperature of evaporator supply [°C]	3.7	5.3
Average temperature of return from evaporator [°C]	0.8	1.1
Average temperature of brine [°C]	2.3	3.2
Average temperature of a sink supply [°C]	40.8	51.6
Average temperature of return from a sink [°C]	33.9	44.6
An average temperature of a sink[°C]	37.3	48.1
Average of difference between sink and energy source [°C]	35.1	44.9
The electricity demand [kWh/a]	3387.7	3874.0
The generated energy [kWh/a]	13857.1	12545.0
The energy from ground [kWh/a]	10399.1	11495.3
Seasonal Performance Factor	**4.09**	**3.24**

5 Summary and conclusions

The AGH-UST Educational and Research Laboratory of Renewable Energy Sources and Energy Saving in Miękinia utilizes shallow geothermal heat by means of heat pumps. It is a good example of a practical use of heat pumps.

The Seasonal Performance Factor of heat pumps receive during research in Laboratory in Miękinia may be used to perform estimates for potential future users. The efficiency of heating systems with heat pumps is depending on the heat source. Research of heat pumps operation to increase the performance factors will be undergoing in the future.

Acknowledgments

The paper was prepared under AGH-UST statutory research grant no. 11.11.140.321 and grant no. 11.11.140.704.

References

1. B. Kepinska, *Geothermal Energy Use – Country Update for Poland, 2013–2015* (EGC Proceeding, Strasbourg, France, 2016)
2. W. Górecki, E. Hałaj, J. Kotyza, A. Sowiżdżał, W. Luboń, G. Pełka, D.Woś, M. Kaczmarczyk, M. Hajto, M. Kaczmarczyk, P.Lachman, *Effective use of renewable energy sources* (2015)
3. W. Górecki, A. Sowiżdżał, M. Hajto, A. Wachowicz-Pyzik, Environ. Earth Sci. **74**, 7487–7495 (2015)
4. A. Sowiżdżał, W. Luboń, G. Pełka, *Prospective areas for geothermal energy resources utilization In the Polish Lowlands* (Conf. mat. from the IGD, Slovakia, 2009)
5. A. Sowiżdżał, B. Papiernik, G. Machowski, M. Hajto, Geol. Quart. **57**, 729–744 (2013)
6. A. Sowiżdżał, Carpath. J. Earth Env. **10**, 237–246 (2015)
7. A. Sowiżdżał, Appl. Ecol.Environ. Res. **14**, 2, 555–574 (2016)
8. A. Sowiżdżał, M. Kaczmarczyk, Geol. J. **51**, 65–76 (2016)
9. PN-EN 14511-3:2013-12, *Air conditioners, liquid chilling packages and heat pumps with electrically driven compressors for space heat ing and cooling - Part 3: Test methods*
10. PN-EN 14825:2016-08, *Air conditioners, liquid chilling packages and heat pumps, with electrically driven compressors, for space heating and cooling. Testing and rating at part load conditions and calculation of seasonal performance*
11. P. Lachman, InstalReporter **3**, 42–47 (2013)
12. R. Nordman, A Zottl, *SEPEMO-Build - a European project on seasonal performance factor and monitoring for heat pump systems in the building sector* (2011)
13. M. Miara, D. Gunther, T. Kramer, T. Oltersdorf, J. Wapler, *Heat pump efficiency. Analysis and Evaluation of Heat Pump Efficieny in real-life conditions* (2011)
14. I. Sarbu, C. Sebarchievici, Energies **9**, 4, 228 (2016)
15. R. Nouvel, M. Cotrado, D. Pietruschka, *European Mapping of Seasonal Performances of Air-Source and Geothermal Heat Pumps for Residential Application*, CISBAT (2015)
16. A.D. Carvalho, *High Efficiency Ground Source Heat Pump Systems for Sustainable Building Space Conditioning*, PhD Thesis (2015)

 DOI: 10.1051/e3sconf/20172200135

Concentration of polycyclic aromatic hydrocarbons in water samples from different stages of treatment

Marta Pogorzelec[1,*], and *Katarzyna* Piekarska[1]

[1]Wroclaw University of Science and Technology, Faculty of Environmental Engineering, Wybrzeże Wyspiańskiego 27, 50-370 Wrocław, Poland

Abstract. The aim of this study was to analyze the presence and concentration of selected polycyclic aromatic hydrocarbons in water samples from different stages of treatment and to verify the usefulness of semipermeable membrane devices for analysis of drinking water. For this purpose, study was conducted for a period of 5 months. Semipermeable membrane devices were deployed in a surface water treatment plant located in Lower Silesia (Poland). To determine the effect of water treatment on concentration of PAHs, three sampling places were chosen: raw water input, stream of water just before disinfection and treated water output. After each month of sampling SPMDs were changed for fresh ones and prepared for further analysis. Concentrations of fifteen polycyclic aromatic hydrocarbons were determined by high performance liquid chromatography (HPLC). Presented study indicates that the use of semipermeable membrane devices can be an effective tool for the analysis of aquatic environment, including monitoring of drinking water, where organic micropollutants are present at very low concentrations.

1 Introduction

Many of organic micropollutants present in potential sources of drinking water e.g. polycyclic aromatic hydrocarbons (PAHs), pesticides, polychlorinated biphenyls (PCBs) and dioxins are considered to be hazardous to human, because of their toxic, genotoxic and carcinogenic potential [1-5].

Polycyclic aromatic hydrocarbons are a class of organic chemical consisting of two or more fused aromatic rings and do not contain heteroatom or carry substituents. PAHs belong to the group of persistent organic pollutants (POPs). These are organic pollutant contaminants that are resistant to degradation, can remain in environment for a long period and have the potential to cause adverse environmental effects [6]. PAHs present in water intended for consumption generally come from coatings, tar, bitumen or water distribution pipelines [7]. Once ingested, polycyclic aromatic hydrocarbons are rapidly

* Corresponding author: marta.pogorzelec@pwr.edu.pl

DOI: 10.1051/e3sconf/20172200135

absorbed through the digestive tract and distributed in various tissues, and also the fetus. The mechanism by which these compounds act as carcinogens is still unclear and several theories have been formulated [7–9]. The International Agency for Research on Cancer (IARC) categorize benzo[a]pyrene and other PAHs in Group 1 (carcinogenic to humans), 2A and 2B (possible or probable carcinogens to humans) [7, 8].

In order to ensure health safety, it is necessary to carry out treatment processes of water intended for water supply purposes. The direct and immediate risk to health and life caused by the presence in water of pathogenic microorganisms makes it unthinkable to abandon the process of disinfection. As a result, the parameters proposed for the by-produscts of disinfection should not be so restrictive as to impair its use. And there comes another threat, although not all PAHs are highly harmful, in the processes of water treatment, they can form other dangerous compounds defined as disinfection by-products, which have high mutagenic and carcinogenic potential [10–12].

Organic micropollutants are present in water at very low concentrations, often below the detection limit, therefore samples should be concentrated prior to analysis. Semipermeable Membrane Devices (SPMDs) are designed for concentrating the hydrophobic organic compound present in water or air samples. SPMDs contain neutral lipid, the most often triolein, sealed within a thin wall of low density polyethylene (LDPE) tube. These membranes mimicking the biological membranes, thus allow selective diffusion of dissolved low molecular weight organic compounds, and their condensation in triolein. The technique is important for estimating risks associated with exposure of living organisms to contaminants, because allows to easily determine the time-weighted average concentration of the analytes dissolved and absorbed biologically for long periods of time [13–20]. Many studies have shown that this technique is appropriate for sampling water from various areas like the marine environment [21, 22], rivers [23–27] and lakes [15]. Long running water monitoring using SPMDs allows the assessment of changes in them over the years [23]. However, there is only a few publications describing the use of SPMDs for samples from a water treatment plant [28]. Extracts obtained through SPMDs are suitable for carrying out a variety of chemical and biological analysis [29–32]. The combination of semipermeable membrane devices and bioassays allows to determine the toxic and genotoxic potential of environmental samples. The effectiveness of the method was demonstrated for short-term bioassays like Microtox^{TM} [24, 28] and Daphnia pulex test for assessing toxicity [24], as well as Vibrio harveyi test and MutatoxTM [24] for examination of water samples mutagenicity.

2 Materials and methods

2.1 SPMDs deployment

Semipermeable membrane devices of the standard configuration (91.4 cm long, 2.5 cm wide, membrane thickness 70–95 µm, containing 1 ml of 99% purity triolein) from ExposMeter AB (Tavelsjő, Sweden) were mounted on a metal spiders (Fig. 1.A.), then placed in perforated stainless steel containers (Fig. 1.B.), deployed in a surface water treatment plant called "Mokry Dwór", located in Lower Silesia, nearby Wrocław (Poland). Three places were chosen for sampling: A- pumping station "Czechnica" as a raw water input, B- pipeline with a stream of water just before ozone disinfection and C- clean water tank as a treated water output. Technological system of water treatment in "Mokry Dwór" is shown in Fig. 2. Characteristics of the sampling points is described below in Table 1.

In the study, samples were taken monthly over a period of about five months (II-VI: February 2016–July 2016), which means that after each month membranes were exchanged. Exceptions were membranes labeled "III-IV" (March 2016–April 2016), which, due to technological problems at the water treatment plant, were remained in water for about eight weeks. There were also field blank control membranes for each month (K), which were exposed to the same conditions like regular ones during instalation. Each membrane represents the state of water in a given month.

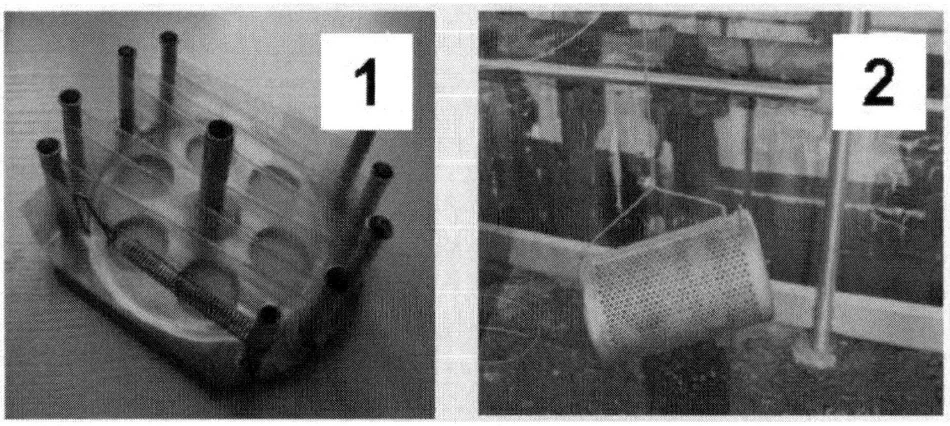

Fig. 1. Installation of semipermeable membrane devices (SPMDs): A- on metal spider, B- in stainless steel container.

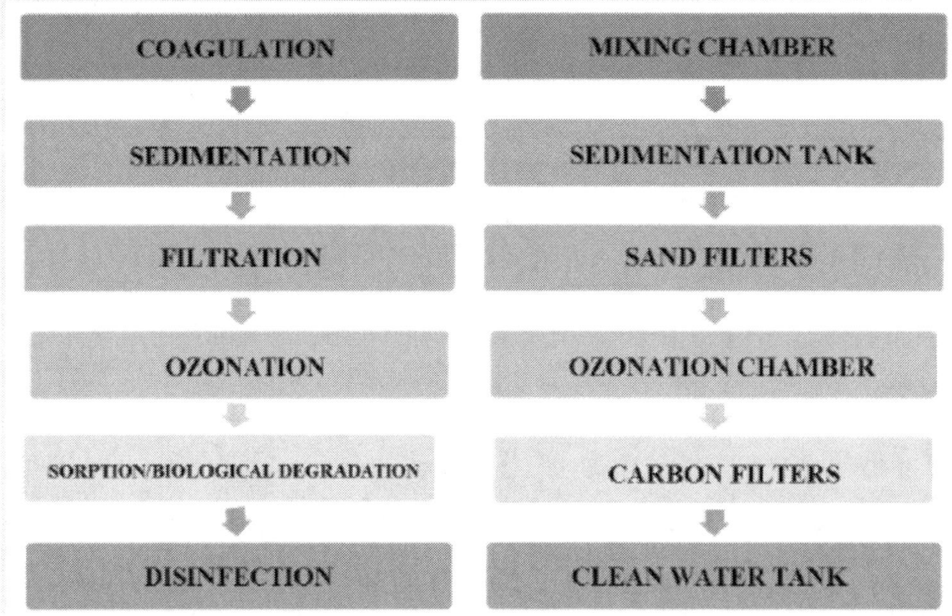

Fig. 2. Technological system used in "Mokry Dwór" Water Treatment Plant.

DOI: 10.1051/e3sconf/20172200135

Table 1. Characteristics of sampling places.

Sampling place	Symbol used in 'Results' tables	Characteristic
Pumping station "Czechnica": **raw water input**	**A**	The water that flows into the pumping station "Czechnica" is surface water without any treatment from the Oława River. The river has a length of 91.7 km and a catchment area of 1167.4 square kilometers. The source of Oława is in the Sudeten Foreland at an altitude of about 315 meters above sea level. Its composition is heavily dependent on the seasons. This water is characterized by high content of suspensions, colloids and dissolved organic compounds. In comparison with infiltrated or groundwater is characterized by a low content of minerals such as manganese and iron.
Pipeline: **water just before ozonation**	**B**	The water stream leaving the sand filter is free of suspensions, both those present in the raw water input, as well as those formed during the coagulation process. Organic compounds affecting the taste and odor are still present. To remove them effectively on a carbon filters, ozone is used to breaking them into smaller molecules, which increases the biodegradability of organic compounds. Disinfection by-products (generated during ozonation) are adsorbed on a bed of coal or decomposed by a biomass in the bed.
Clean water tank **water after treatment**	**C**	In the clean water tank there is a water after complete treatment and disinfection with chlorine dioxide. The water from this tank is collected and pumped directly to the water supply network. This water meets the requirements of water intended for human consumption in accordance with the Regulation of the Minister of Health in Poland. The dose of disinfectant dispensed before the clean water tank must be large enough to prevent secondary bacterial contamination of the water supply system.

2.3 Samples preparation

After each one month of sampling SPMDs were transported to the laboratory where mechanical and chemical cleaning was carried out according to the good SPMDs practice [14, 26, 32]. Then membranes were dialyzed separately in 130 cm^3 of n-hexane to release compounds bounded in membrane. After 18 hours of dialysis in 18°C the solvent was replaced with fresh one (130 cm^3) and dialysis was continued for another 8 hours. Then, both dialysates were mixed and n-hexane has been completely removed using the rotary evaporation. The same procedures were carried out for field blank membranes and for one unused, clean membrane (sample "K_0"). In order to assess the effect of the solvent, n-hexane in amount corresponding with the amount used per one membrane (260 cm^3) was poured into round bottom flasks and also evaporated (sample "hex").

2.4 Quantitative analysis of polycyclic aromatic hydrocarbons

Dry extracts were drenched with 4 cm^3 of cyclohexane, then evaporated to dryness under nitrogen and drenched with 1cm^3 of acetonitrile. Concentrations of PAHs were determined by high performance liquid chromatography (HPLC) with fluorescence detection on liquid chromatograph AT 1200 Agilent Technologies with fluorescent detector and spectrophotometric detector DAD.

Operation conditions of liquid chromatograph

- separation on the chromatographic column:
  - *Zorbax Eclipse PAH 5,0 μm; 150 x 4,6 mm from Agilent Technologies*
- volume of injection: 10 μl
- gradient elution with acetonitrile/water
  - 50:50 1 min
  - 100:0 16 min
  - 100:0 25 min
  - 50:50 30 min
- column temperature: 30°C
- duration of the analysis: 30 min
- mobile phase flow: 1 cm^3/min
- fluorescence detector parameters: wavelength: Ex 229 nm; Em. 400 nm

Table 2. Limits of detection and limit of quantification for polycyclic aromatic hydrocarbons determined on AT 1200 liquid chromatograph from Agilent Technologies with fluorescence detector.

Compound	Abbreviation used in 'Results' tables	Limit of detection (LOD) [ug/ml]	Limit of quantification [ug/ml]
Naphthalene	**N**	0.033	0.100
Acenaftylen	**Ace**	0.067	0.200
Acenaften	**Ac**	0.033	0.100
Fluoren	**Fl**	0.007	0.020
Fenantren	**Fen**	0.003	0.010
Anthracene	**A**	0.003	0.010
Fluoranthene	**Flu**	0.007	0.020
Pyrene	**Pyr**	0.003	0.010
Benzo [a] anthracene	**B[a]A**	0.003	0.010
Chryzen	**Chr**	0.003	0.010
Benzo [b] fluoranthene	**B[b]F**	0.007	0.020
Benzo [k] fluoranthene	**B[k]F**	0.003	0.010
Benzo [a] pyrene	**B[a]P**	0.003	0.010
Dibenzo [a, h] anthracene	**D[a,h]A**	0.007	0.020
Benzo [g, h, i] perylene	**B[g,h,i]P**	0.007	0.020
Indeno [1,2,3-c, d] pyrene	**I[1,2,3-c,d]P**	0.003	0.010

3 Results and discussion

The results of quantitative analysis of fifteen polycyclic aromatic hydrocarbons in water samples from various stages of potabilization are shown in Table 3 and Table 4. Polycyclic aromatic hydrocarbons are removed in a very efficient way during the water treatment process that is used in the "Mokry Dwór" Water Treatment Plant. The number, concentrations and complexity of PAHs are varied with each month. In all analyzed month, the amount of each detected PAH has decreased, in most cases even below the detection limit of the analytical method, with the progress of the treatment process. The results obtained for the solvent (n-hexane) used for samples preparation for analysis (cleaning, dialysis, dissolving material) indicates that it did not have any additional contaminants and its impact on the results was negligible. Similar

 DOI: 10.1051/e3sconf/20172200135

conclusions can be drawn from the analysis of the results obtained for control membranes that have undergone exactly the same procedures as the regular membranes. This shows that the test material was not contaminated during the preparation of the samples for analysis.

Table 3. Concentrations of fifteen polycyclic aromatic hydrocarbons [μg/sample] in water samples from various stages of potabilization (A: pumping station- raw water input; B: pipeline-stream of water just before ozonation; C: clean water tank) and in control membranes (K), in particular months (II, III–IV).

	K0	II A	II B	II C	II K	III-V A	III-IV B	III-IV C	III-IV K
N	n.d[a].	**0.148**	<LOD	<LOD	<LOD	<LOD	<LOD	n.d.	n.w.
Ace	n.d.	n.d.	n.d.	n.d.	n.d.	n.d.	n.d.	n.d.	n.w.
Ac+Fl	<LOD[a]	<LOD	<LOD	<LOD	<LOD	<LOD	<LOD	n.d.	p.o.
Fen	**0.096**	**0.980**	**0.412**	**0.080**	**<LOD**	**0.276**	**0.071**	**0.060**	**0.041**
A	<LOD	**0.018**	<LOD	n.d.	<LOD	<LOD	<LOD	n.d.	p.o.
Flu	<LOD	**2.162**	**2.065**	**0.108**	**0.048**	**0.900**	**0.606**	**<LOD**	p.o.
Pyr	**0.024**	**0.818**	**1.044**	<LOD	<LOD	**0.552**	**0.321**	**0.016**	p.o.
B[a]A	<LOD	**0.145**	**0.283**	<LOD	<LOD	**0.130**	**0.160**	<LOD	p.o.
Chr	<LOD	**0.178**	**0.163**	<LOD	<LOD	**0.171**	**0.173**	<LOD	n.w.
B[b]F	<LOD	**0.054**	**0.035**	**0.028**	<LOD	**0.050**	**0.044**	<LOD	p.o.
B[k]F	<LOD	**0.024**	**0.010**	<LOD	<LOD	**0.038**	**0.036**	<LOD	p.o.
B[a]P	<LOD	<LOD	<LOD	<LOD	<LOD	**0.039**	**0.010**	<LOD	n.w.
D[a,h]A	n.d.	<LOD	<LOD	<LOD	n.d.	n.d.	<LOD	n.d.	n.w.
B[g,h,i]P	n.d.	<LOD	<LOD	n.d.	n.d.	<LOD	n.d.	n.d.	n.w.
I[1,2,3-cd]P	n.d.	**0.020**	**0.010**	n.d.	n.d.	**0.027**	**0.015**	n.d.	n.w.

[a]n.d – not detected
[b]LOD: limit of detection

Table 4. Concentrations of fifteen polycyclic aromatic hydrocarbons [μg/sample] in water samples from various stages of potabilization (A: pumping station- raw water input; B: pipeline-stream of water just before ozonation; C: clean water tank), and in control membranes (hex, K) in particular months (V, VI).

	hex	V A	V B	V C	V K	VI A	VI B	VI C	VI K
N	n.d.[a]	n.d.	n.d.	n.d.	n.d.	n.d.	<LOD	**0.393**	n.d.
Ace	n.d.	n.d.	n.d.	n.d.	n.d.	n.d.	n.d.	n.d.	n.d.
Ac+Fl	<LOD[b]	<LOD	<LOD	<LOD	<LOD	<LOD	<LOD	<LOD	<LOD
Fen	**0.024**	**0.265**	**0.137**	**0.146**	**0.063**	**0.210**	**0.138**	**0.111**	**0.052**
A	<LOD	<LOD	<LOD	<LOD	<LOD	<LOD	n.d.	<LOD	<LOD
Flu	n.d.	**1.150**	**1.012**	<LOD	<LOD	**1.173**	**0.210**	<LOD	<LOD
Pyr	**0.026**	**0.431**	**0.228**	**0.029**	<LOD	**0.544**	**0.213**	**0.025**	<LOD
B[a]A	<LOD	**0.122**	**0.077**	<LOD	<LOD	**0.140**	**0.044**	<LOD	<LOD
Chr	<LOD	**0.121**	**0.140**	**0.013**	n.d.	**0.142**	**0.122**	<LOD	n.d.
B[b]F	n.d.	**0.086**	**0.144**	**0.028**	**0.017**	**0.100**	**0.179**	<LOD	<LOD
B[k]F	<LOD	**0.045**	**0.058**	<LOD	<LOD	**0.059**	**0.047**	<LOD	<LOD
B[a]P	n.d.	**0.020**	<LOD	<LOD	<LOD	**0.029**	<LOD	<LOD	n.d.
D[a,h]A	n.d.	<LOD	n.d.	n.d.	n.d.	n.d.	<LOD	<LOD	n.d.
B[g,h,i]P	n.d.	n.d.	n.d.	n.d.	n.d.	<LOD	<LOD	n.d.	n.d.
I[1,2,3-cd]P	n.d.	**0.035**	**0.022**	n.d.	n.d.	**0.047**	**0.032**	n.d.	n.d.

4 Conclusions

Due to the low concentration of micropollutants in water, it is necessary to fit appropriate technique for concentration of them, to be detectable during the chemical and biological analysis. This study indicates that the use of semipermeable membrane devices can be an effective tool for the analysis of drinking water, in which organic micropollutants are found at very low concentrations. The proposed method uses SPMDs, which are placed directly at the place of sampling, what eliminates logistical problems.

Presented study shows, that water treatment processes carried out in "Mokry Dwór" Water Treatment Plant effectively reduce the concentration of polycyclic aromatic hydrocarbons, which may be a precursor of many dangerous substances.

Nevertheless, to accurately assess the changes in water with the progress of the treatment process, chemical analysis for the detection of other potential compounds in the water, with emphasis on disinfection by-products, including trihalomethanes, should be extended. Furthermore, to assess the effects of substances contained in test samples on living organisms, biological studies including toxicity and genotoxicity tests should be carried out.

What is more, to be able to draw accurate conclusions about the impact of weather conditions on the water, analysis should be conducted for an extended period, at least throughout the year.

The research financed by the funds of the Ministry of Science and Higher Education for young scientist and doctoral students order number: B50557.

References

1. J. Fawell, M.J. Nieuwenhuijsen, Brit. Med. Bull. **68**, 199–208 (2003)
2. L. Kowal, M. Świderska-Bróż, *Water treatment* (PWN, Warszawa, 2007) (in Polish)
3. M. Świderska-Bróż, *Micropollutants in aquatic environment* (Wydawnictwo Politechniki Wrocławskiej, Wrocław, 1993) (in Polish)
4. T.M. Traczewska, *Biomonitoring of the mutagenicity of microcontaminants in drinking water* (Oficyna Wydawnicza Politechniki Wrocławskiej, Wrocław 2002) (in Polish)
5. T. Ohe, T. Watanabe, K. Wakabayashi, Mutat. Res. **567**, 109–149 (2004)
6. J.C. Igwe, P.O. Ukaogo, Journal of Natural Sciences Research **5**, 7, 117–132 (2015)
7. S. Scicca, G.O. Conti, Mediterr. J. Nutr. Metab. **2**, 157–162 (2009)
8. I. Abdel-Shafy, M.S.M. Mansour, Egypt J. Petrol. **25**, 107–123 (2016)
9. H.T. Rengarajan, P. Rejendran, N. Nanakumar, B. Lokeshkumar, P. Rajendran, I. Nishigaki, Asian Pac. J. Trop. Biomed. **5**, 3, 182–190 (2015)
10. D.B. McGregor, Methods for Assessing the Effects of Mixtures of Chemicals, 523–542 (1987)
11. S.D. Richardoson, Trend Anal. Chem. **22**, 666–684 (2003)
12. C. Zani, D. Feretti, A. Buschini, P. Poli, C. Rossi, L. Guzella, F. Di Caterino, S. Monarca, Mutat. Res. **587**, 26–37 (2005)
13. J.N. Huckins, J.A. Lebo, M.W. Tubergen, G.K. Manuweera, V.L. Gibson, J.D. Perry, United States Patent, Patent Number **5**, 098573 (1992)
14. J.N. Huckins, *Fundamentals of SMPDs. Monitors of organic chemicals in the environment*, Springer 29–43 (2006)

15. Y. Lu, Z. Wang, J. Huckins, Aquat. Toxicol. **60**, 139–153 (2002)
16. B. Vrana, G. Schüürmann, Environ. Sci. Technol. **36**, 290-–296 (2002)
17. L. Šetková, J. Hajšlová, P.-A. Bergqvist, V. Kocourek, R. Kazda, P. Suchan, J. Chromatogr. A **1092**, 170–181 (2005)
18. K. Booij, H.M. Sleiderink, F. Smedes, Environ. Toxicol. Chem. **17**, 1236–1245 (1998)
19. B. Vrana, A. Paschke, P. Popp, G. Schüürmann, Environ. Sci. Pollut. R **8**, 27–34 (2001)
20. J.N. Huckins, G.K. Manuweera, J.D. Petty, D. Mackay, J.A. Lebo, Environ. Sci. Technol. **27**, 2489–2496 (1993)
21. A. Marrucci, B. Marras, S.S. Campisi, M. Schintu, Mar. Pollut. Bull. **75**, 69–75 (2013)
22. K. Pazdro, Rocz. Ochr. Sr. **4**, 189–205 (2002)
23. J.F. Mueller, M.R. Mortimer, J. O'Brein, T. Komarova, S. Carter, Mar. Pollut. Bull. **63**, 73–76 (2011)
24. D. Sabaliūnas, J.R. Lazutka, I. Sabaliuniene, Environ. Pollut. **109**, 251–265 (2000)
25. D. Sabaliūnas, A. Södergren, Environ. Pollut. **96**, 195–205 (1997)
26. M. Schaefer, *Pharmaceutical and Personal Care Products in the Sacramento River-Report, Standard Operating Procedure: Preparation of SPMDs for Dialysis*, SOP E-14 (2009)
27. W.L. Cranor, D.A. Alvarez, S.D. Perkins, G.A. Tegerdine, R.C. Clark, J.N. Huckins, *Use of semipermeable membrane device (SPMD)technology for a probabilistic assessment of hydrophobic organic contaminants in selected reaches of Virginia rivers* (U. S. Geological Survey, 2005)
28. G. Gilli, T. Schiliro, Pignata, D. Traversi, E. Carraro, C. Baiocchi, R. Aigotti, D. Giacosa, E. Fea, Chemosphere **61**, 1691–1699 (2005)
29. B. Karacik, O.S. Okay, B. Henkelmann, G. Pfister, K.-W. Schramm, Mar. Pollut. Bull. **70**, 258–265 (2013)
30. J.D. Petty, J.N. Huckins, S.B. Jones, T.P. Boyle, Chemosphere **41**, 311–321 (2000)
31. E.S. Emelogu, P. Pollard, C.D. Robinson, F. Smedes, L. Webster, I.W. Pliver, C. McKenzie, T.B. Seiler, H. Holler, C. F. Moffat, Chemosphere **90**, 210–219 (2013)
32. D. Alvarez, *Guidelines for the use of the semipermeable membrane device (SPMD) and the polar organic chemical integrative sampler (POCIS) in environmental monitoring studies*, (U.S. Geological Survey, Techniques and Methods 1–D4 2010)

Impact of the circulation system on the energy balance of the building

Iwona Polarczyk[1,*], and *Michał* Fijewski[1]

[1]Wroclaw University of Science and Technology, Faculty of Environmental Engineering, Wybrzeze Wyspianskiego 27, 50-370 Wroclaw

Abstract. The efficiency of the hot water system is one of the factors necessary to determine the overall efficiency of the building. From the calculative point of view, it is easy to make. The article presents how working of the circulation system has an influence on the efficiency of domestic hot water system. The differences in the results was presented and based on calculations of various methods, the measurements results was also taken into account. The attention was especially paid to the possibility of using ultrasonic flowmeter for measuring the flow and energy.

1 Introduction

Referring to hot water installations, according to the Regulation of the Minister of Infrastructure with regard to technical conditions which buildings and their location should conform to, should ensure that the water temperature is not lower than 55°C and not higher than 60°C at the water points, which should allow for the periodic thermal disinfection at temperatures in the range 70°C–80°C. A correctly designed and insulated hot water system allows the recipient to obtain water at the correct temperature at the outlet point. Otherwise, not only will the warming-up time increase, but also the water consumption will increase, which in turn will result in increased energy consumption charges borne by the user/recipient.

2 Measurements

The selection of facilities was not accidental. Two buildings of collective residence – student houses T-15 and T-16 were selected for the measurements, characterized by "twin" construction and equipped with heating hubs with the same power and operating parameters. The circulation system did not work in one of the buildings (T-16). In the building with efficient circulating installation, its work was "nonstop".

The parameter characterizing the system of domestic hot water preparation can be the utility efficiency factor. The efficiency factor η is defined as the ratio of heat energy

* Corresponding author: iwona.polarczyk@pwr.edu.pl

DOI: 10.1051/e3sconf/20172200136

consumed in hot water by the recipients to the heat needed to prepare it and maintain a suitable temperature.

$$\eta = \frac{Q_{CWU} + Q_{CYRK}}{Q_{DOST}} \qquad (1)$$

where:
Q_{CWU} – The heat flux required to heat cold water to the required temperature
Q_{CYRK} – The heat flux needed to maintain the correct hot water temperature
Q_{DOST} – The heat flux supplied to hot tap water

Photo 1. T15 Heating station.

Photo 2. T16 Heating station.

 DOI: 10.1051/e3sconf/20172200136

Study work of the hot water installation was undertaken using three non-invasive measuring devices mounted in the heating node of the buildings. This was:

- Flexim's FLUXUS F601 portable ultrasonic flowmeter,
- Fuji's Portaflow C portable ultrasonic flowmeter,
- Wisensys wireless measurement platform.

3 Analysis of measurement results

In order to analyze the operation of hot water installation and circulation in the T-15 and T-16 dormitories, the results of measurements of corresponding days from both buildings were compiled, i.e. from 30.04.2013 to 24.07.2013. Due to the lack of reading from the measurement equipment, several data from the T-15 building are missing from 14.06 to 19.06 and 5.07 to 10.07.

3.1 The efficiency of the installation

3.1.1 Comparing daily efficiency

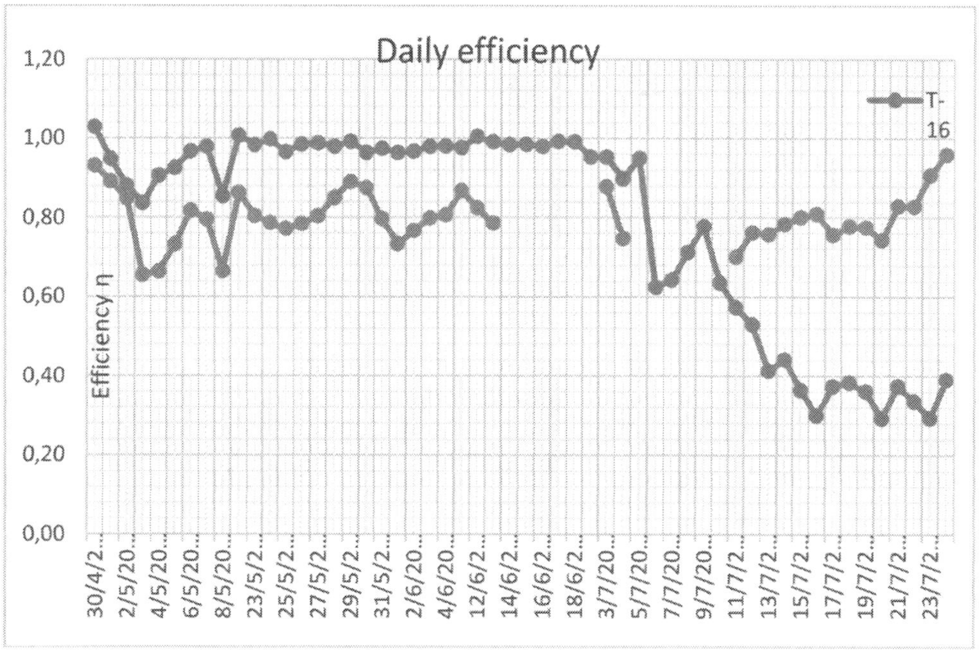

Fig. 1. Comparing T-15 and T-16 daily efficiency.

 DOI: 10.1051/e3sconf/20172200136

3.1.2 Comparing weekly efficiency

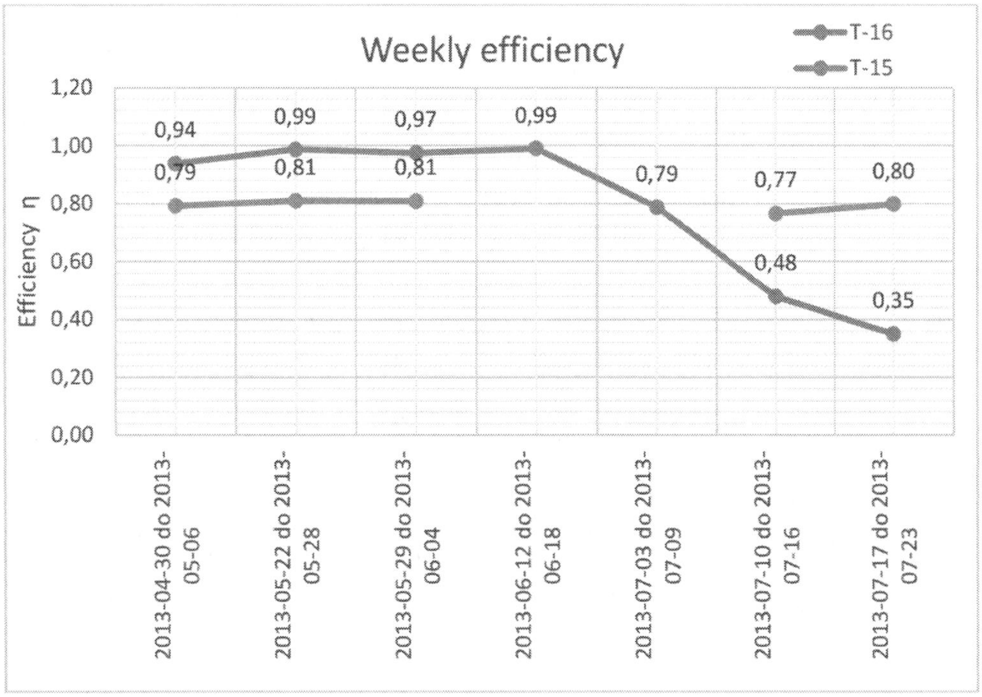

Fig. 2. Comparing T-15 and T-16 weekly efficiency.

3.1.3 Comparing monthly efficiency

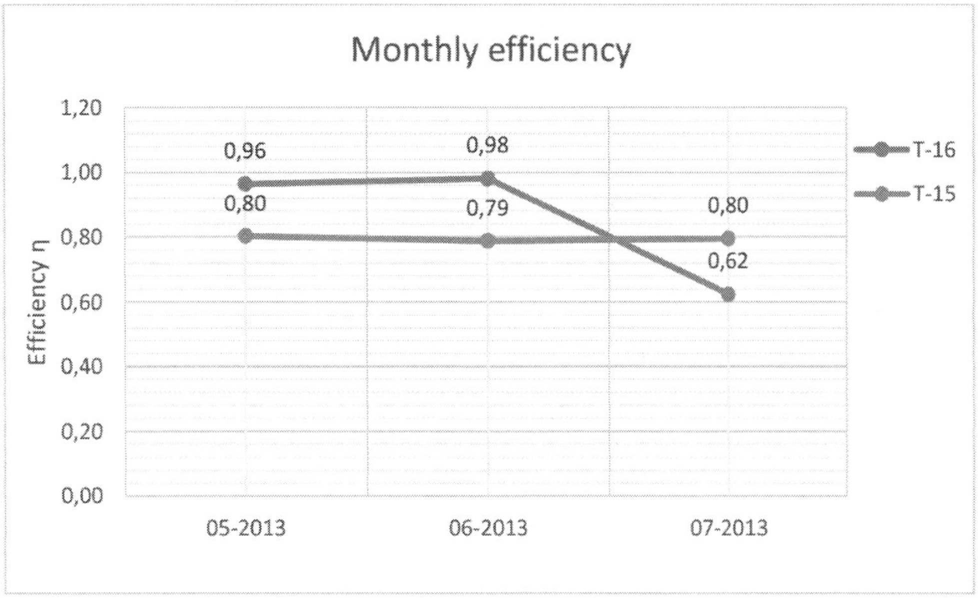

Fig. 3. Comparing T-15 and T-16 monthly efficiency.

3.2 Conclusions from the analysis of the efficiency of the installation

Two distinctly different result areas can be noted for the T-16 building on the diagrams showing the efficiency of the hot water system. This is due to the drastic change in the number of dormitory users – the holidays start in July. Consequently interpretation of the results should be carried out for two periods – full and partial residence. In addition, the nature of the water intake for a more periodic change is also changing in a facility such as a student dormitory with the change in the number of hot water recipients.

In times of full occupancy, the T-16 building paradoxically achieves a higher efficiency where the circulation system does not work, while there is less efficiency in the T-15 building:

- T-15 average total efficiency of the hot water installation. – 0.80
- T-16 average total efficiency of the hot water installation – 0.93

This situation should be explained by the presence of continuous water consumption and bad control of the circulation pump operation (nonstop work). In long periods of hot water consumption, the work of the circulating system is unnecessary, resulting in energy losses – lower efficiency of the hot water installation.

During the period of partial residence, in July the efficiency is kept constant, similar to the period of full residence in the T-15 building, while the T-16 building is clearly falling. During this period there are significantly fewer users using hot water in both buildings. The heat output taken at the T-15 is mainly used to maintain circulating water temperature (low domestic water consumption). On the other hand, for the T-16 building, the heat consumption depends on the instantaneous demand for hot water and the need to prepare it for the correct temperature parameters. Clearly this is the lack of circulating installation confirmed by the decline in efficiency of the hot water system.

4 Summary

The circulating system used in the building affects the efficiency of hot water installation.

The efficiency of the domestic hot water installation depends to a large extent on the operation of the circulation system. The analysis showed that in the building with an operating efficiency of circulation, the hot water installation is lower than in a building with circulation off. Of course this is the case if there are a large number of users in the building and they use hot water. When hot water is collected sporadically, the situation is different, and then the efficiency of such installation is about 80%. All the collected power is then intended to maintain the temperature of the circulating water.

Buildings that do not have a circulation system at the time of limited domestic hot water consumption show efficiency of about 60%, and with increased use of water, efficiency is at the level of 90–95%. For large domestic hot water partitions the efficiency factor reaches the highest value.

According to the Ordinance of the Minister of Infrastructure and Development the circulation installation is required in every building, except for single-family houses, farm buildings and individual recreation. According to the analysis, it is important because it maintains the level of energy consumption at a reasonably low level. However, efforts should be made to reduce energy and heat losses, for this purpose use of night and summer circulation limits (for buildings such as dormitories and other buildings used periodically). This will reduce the energy costs of heating circulating water during these periods and will result in lower tap water consumption.

DOI: 10.1051/e3sconf/20172200136

References

1. I. Polarczyk, M. Fijewski, Research project: Evaluation of the condition and efficiency of the plant of domestic hot water by method of direct measurements, Reg. No. N N523 617839 (2016)
2. Ordinance of the Minister of Infrastructure of 12 April 2002 on the conditions to be met by buildings and their location, Journal of Laws No. 75, item 690 (2002–2015)
3. Ordinance of the Minister of Infrastructure and Development of 3 June 2014 on the methodology for calculating the energy performance of a building and a dwelling or a part of a building constituting a whole independent technical and utility as well as the method of drawing up and specifying certificates of their energy performance, Journal of Laws, item 888 (2014)
4. Ordinance of the Minister of Infrastructure of 6 November 2008 on the methodology for calculating the energy performance of a building and a dwelling or part of a building constituting a whole independent technical and utility as well as the manner of drawing up and specifying certificates of their energy performance, Journal of Laws, item 1240 (2008)

Impact of the circulation system on domestic hot water consumption

Iwona Polarczyk[1,*], and *Michał* Fijewski[1]

[1]Wroclaw University of Science and Technology, Faculty of Environmental Engineering, Wybrzeze Wyspianskiego 27, 50-370 Wroclaw

Abstract. The article presents the results of researches for domestic hot water in two, the same buildings – students dormitories. There was not working circulation system in one of the objects. Characteristics of hot water intake and the size of the design flows were determined. The influence of circulation loop existence on the size of hot water consumption and characteristic values which are describing the size of hot water abstraction in analyzed buildings was pointed out.

1 Introduction

Energy conservation is one of the main topics discussed nowadays. It results not only from concern for the environment, but most of all from the desire to decrease the amount of energy fees, which are gradually increasing. Hence, it is important to strive for maintenance of energy consumption at a rationally low level. This has to be considered as early as during the stage of designing water supply installations. In case of warm water supply installation, one of the key reasons for energy loss is defective operation of the circulating system. Appropriately designed circulating system should prevent water temperature decrease in DHW pipes. What is more, it should also ensure appropriate water temperature at the outlet in the place of the point-of-use. Yet when the design of the circulating system is faulty or when the system itself does not operate properly, the effect is decrease of the temperature of water at the point-of-use below the required parameters. What follows is not only increase of energy loss connected with the necessity of heating water up to the appropriate temperature, but also tap water losses and higher warm water bills for water, which is not necessarily warm.

2 Measurements

The research analysis covered hot water supply installations in two "twin" buildings – student dormitories T-15 and T-16 with a similar construction. They were equipped with two-stage series-parallel heating substation and the same level of central heating and warm water power. The circulating system was not working in one of the buildings (T-16).

* Corresponding author: iwona.polarczyk@pwr.edu.pl

The measurements were taken with the use of three non-invasive tools installed in the heating substations:

- portable ultrasonic flow meter FLUXUS F601 provided by Flexim,
- portable ultrasonic flow meter Portaflow C provided by Fuji,
- wireless temperature measurement platform provided by Wisensys.

For the needs of the article the presented results encompass only two months in which the buildings were occupied by their users -May and June. The obtained measurements allowed for preparation of charts juxtaposing average hourly water consumption levels during periods such as: 24 hours, a week and a month. Colour blue represents water consumption in the building in which the circulating system was not working (T-16). Failures and blackouts took place during measurements, what is reflected by lack of measurements. Figure 1 presents the measurement diagram.

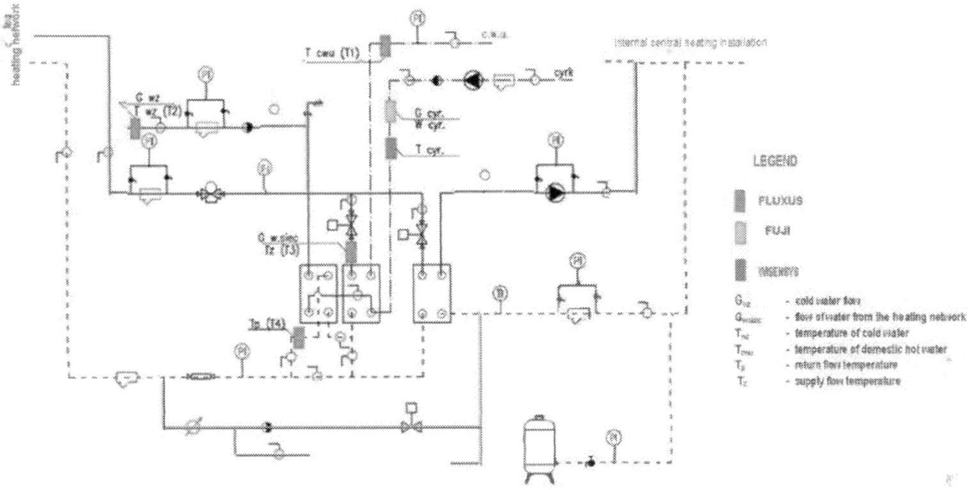

Fig. 1. Diagram displaying measurement system in the heating station of T-15 and T-16 buildings. T-16 building is not equipped with working circulating system.

3 Consumption of domestic hot water

The conducted research allowed to collect a series of measurement data on the basis of which it was possible to draw up a series of analyses devoted to operation of domestic hot water installation. One of the main and the most basic indicators of a given domestic hot water installation's efficiency is its hot water consumption. A comparative analysis regarding water consumption in various time periods was conducted on the basis of measurements related to two identical buildings as far as their construction, number of inhabitants and nature of water consumption are concerned and differing only in regard to functioning of the circulating system.

 DOI: 10.1051/e3sconf/20172200137

3.1 Comparison of daily water consumption

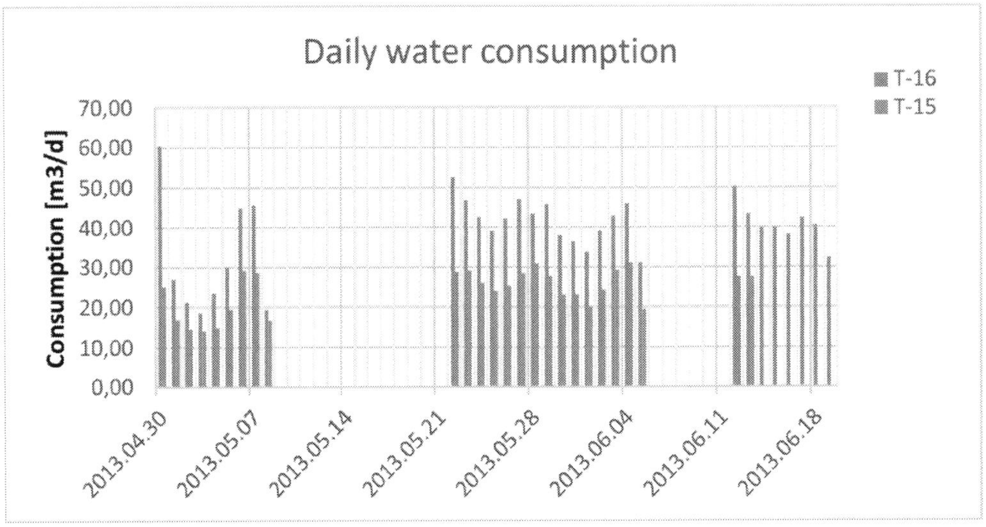

Fig. 2. Comparison of daily water consumption.

3.2 Comparison of weekly water consumption

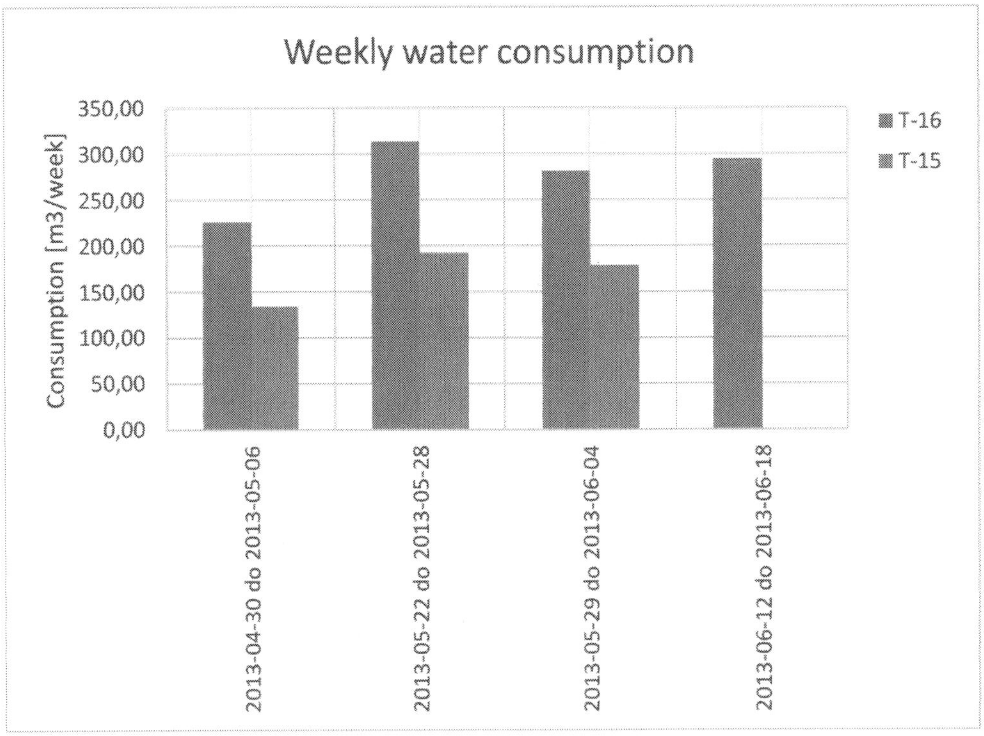

Fig. 3. Comparison of weekly water consumption.

DOI: 10.1051/e3sconf/20172200137

3.3 Comparison of monthly water consumption

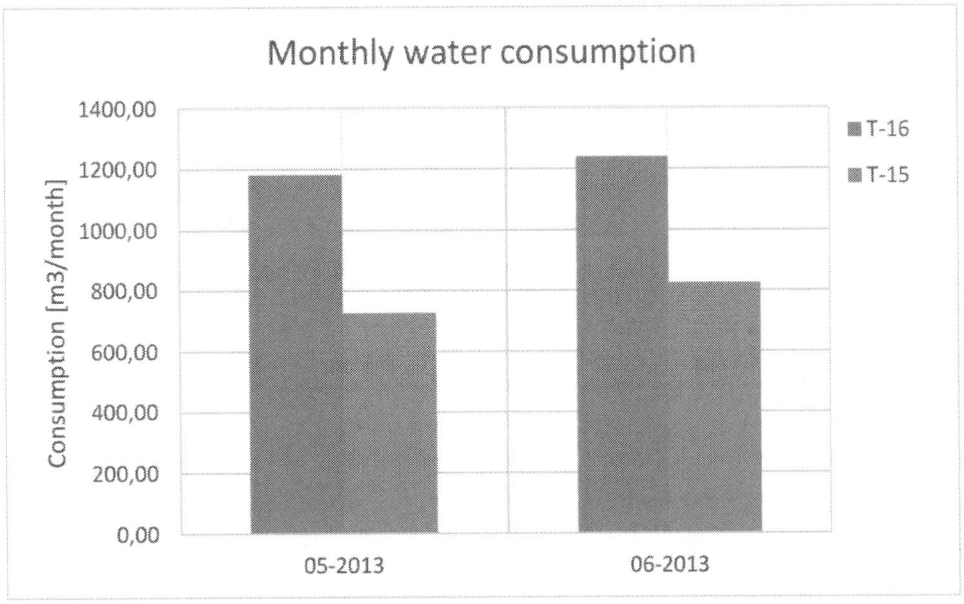

Fig. 4. Comparison of monthly water consumption.

4 Energy consumption generated by DHW preparation

The level of hot water consumption is not the only indicator of energy-saving character of a given installation's operation. Another important factor is level of thermal energy consumption generated by the need to heat the utilized water and to maintain its appropriate temperature (circulation).

A comparison of thermal power uptake from the heating network concerning the analyzed objects during specified periods (a week, a month) was based on the collected measurement data.

 DOI: 10.1051/e3sconf/20172200137

4.1 Comparison of daily energy consumption

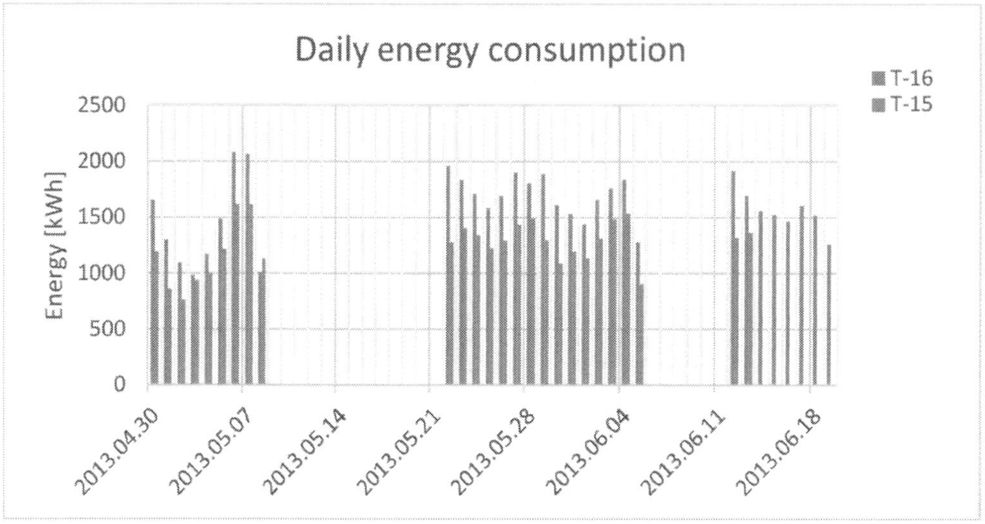

Fig. 5. Comparison of daily energy consumption.

4.2 Comparison of weekly energy consumption

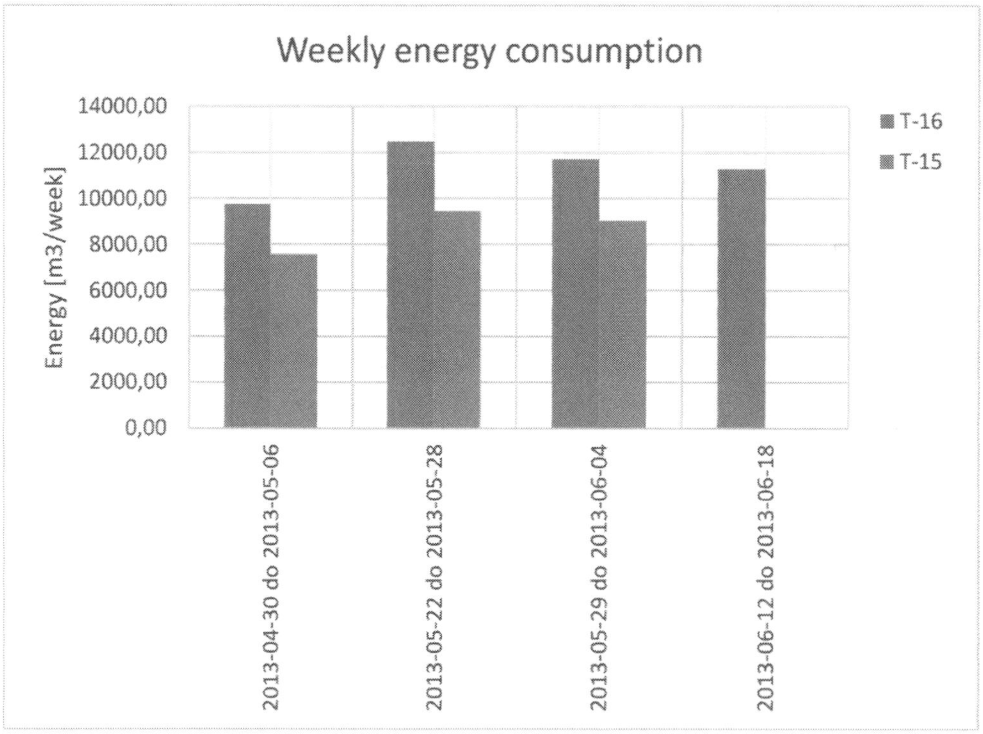

Fig. 6. Comparison of weekly energy consumption.

DOI: 10.1051/e3sconf/20172200137

4.3 Comparison of monthly energy consumption

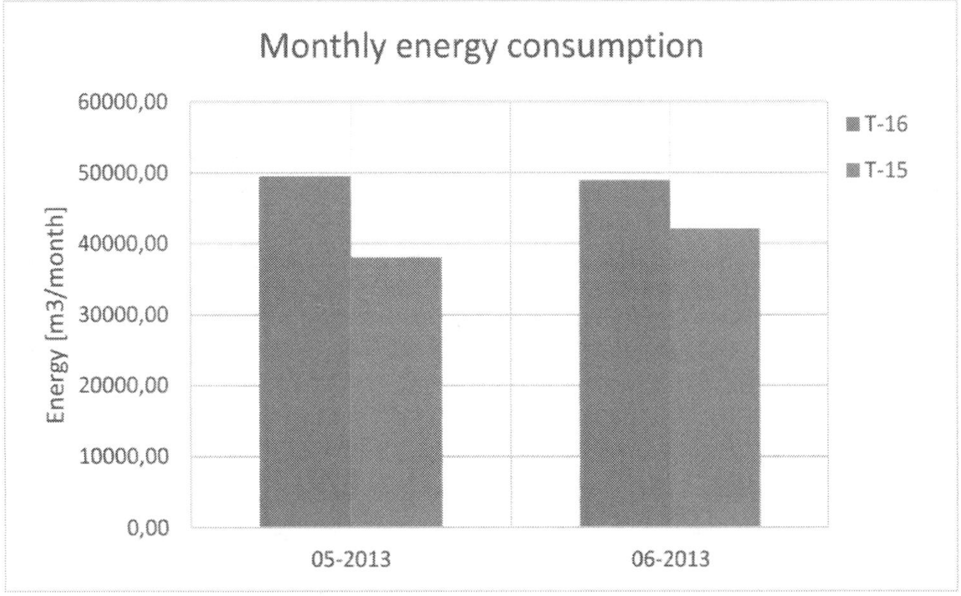

Fig. 7. Comparison of monthly energy consumption.

5 Interpretation of the presented results of measurements

The presented analysis of domestic hot water consumption in two identical buildings T-15 and T-16, one of which was not equipped with a working circulating system, revealed concurrence with the expected results. Hot water consumption in student dormitory T-16 (without circulation) is greater than in the building with well-functioning circulating system. However, the very value of the revealed difference in water consumption was exceptionally surprising.

The analysis of daily water consumption showed significant fluctuations in consumption levels depending on the day of the week. However, the observed fluctuations were convergent for both buildings, what confirms the impact of various factors on water consumption in the objects – water consumption is thus subject to sociological correlations.

Considerable divergence in water consumption levels, reaching even up to 2.5 times, can be observed in daily periods. It is related to the dynamics of water consumption and multiple factors influencing it. Hence, 24-hours period does not seem to be a very meaningful research period.

A much more representative period for comparing water consumption levels is a week or a month. Extending the research period eliminates the daily divergences in DHW consumption caused by various factors influencing its dynamics and value. The differences in hot water consumption levels in all considered time periods revealed by the presented charts fluctuate at around 60%.

Operating circulating system increasing comfort of using hot water and (what was proven above) leading to a significant decrease of its consumption is connected with the need to supply thermal energy to DHW installation. The necessity of heating water, especially during periods when it is not necessary for the user, decreases the energy efficiency of DHW installation. However, the presented charts displaying energy consumption in representative periods reveal that despite the worst variant of the circulating system's operation (nonstop) the energy consumption in T-15 building is smaller by approx.

20%. It is an evidence of the user's expectations, even a young one being still a student, concerning the quality of hot water. Appropriate water temperature is mandatory for water consumer, even if in order to obtain the required operating temperature of DHW installation he or she has to waste more time as well as water.

6 Summary and conclusions

According to the Regulation of the Minister for Infrastructure and Development, circulating installation is required in every building with the exception of family residential buildings, farm buildings and individual recreation buildings. Effective circulating system has significant impact not only on water consumption but also on energy consumption related to the process of hot water preparation. The presented results are also interesting due to the possibility of direct comparison of water consumption in two very similar installations with regard to water use, amount and character of its uptake.

The analysis indicates that the differences in water consumption levels amount to even 80% in daily periods. It is related to a longer time of waiting for hot water with appropriate temperature. It also translates into direct increase of consumption of energy necessary for heating that water and higher operating costs.

References

1. I. Polarczyk, M. Fijewski, Research project: Evaluation of the condition and efficiency of the plant of domestic hot water by method of direct measurements, Reg. No. N N523 617839 (2016)

Investigation of the biofuel flue and producer gases cleaning efficiency using ESP

Robertas Poškas[1,*], *Arūnas* Sirvydas[1], *Povilas* Poškas[1], *Nerijus* Striūgas[2], *Nerijus* Pedišius[3], and *Vitas* Valinčius[4]

[1]Lithuanian Energy Institute, Nuclear Engineering Laboratory, Breslaujos 3, Kaunas LT44403, Lithuania
[2]Lithuanian Energy Institute, Laboratory of Combustion Processes, Breslaujos 3, Kaunas LT44403, Lithuania
[3]Lithuanian Energy Institute, Laboratory of Heat Equipment Research And Testing, Breslaujos 3, Kaunas LT44403, Lithuania
[4]Lithuanian Energy Institute, Plasma Processing Laboratory, Breslaujos 3, Kaunas LT44403, Lithuania

Abstract. The use of biofuel has been increasing in Europe over the last years, and the reason for that is acceptable cost and the least negative impact on the environment. However, NO_x and emissions of fine particulates are important, and biofuel is still a disadvantage compared to oil and natural gas fired systems. Usually, flue gas is filtered in multicyclones or fibre filters before discharge into the atmosphere. Yet, in the case of fine particulates, the filters of such type do not show high effectiveness, thus electrostatic precipitators are used. In this comparative study on biofuel (wood pellets), the collection efficiency of solid particles from a class 3 boiler (50 kW) and from a gasification unit (100 kW) was investigated. Although releases of solid particles from modern boilers are low, a combination of such a boiler with an electrostatic precipitator may reduce the releases of particles to the minimum, and the collection efficiency of the electrostatic precipitator obtained during the investigation was ~98–99%. There is a big difference in particle concentrations comparing the systems with flue gas and producer gas. As the working conditions in the test section with producer gas were harder, it led to lower efficiency of the electrostatic precipitator (~75%).

1 Introduction

During the last decades, biofuel has been started to be widely used in many countries for various types of heating appliances. Biofuel is also a renewable energy source that makes the least negative impact on the environment in comparison with some other types of fuel [1]. During combustion of any liquid or solid fuel, particles are formed from the ash contained in the fuel. This is also valid for biofuels where the ash content varies from a fraction of a percent to more than ten percent by weight in the dry substance. Some of the

* Corresponding author: Robertas.Poskas@lei.lt

 DOI: 10.1051/e3sconf/20172200138

ash usually deposits in the furnace, whilst another part is carried out as cinder with the flue gas into the atmosphere [2]. Therefore, one of the main disadvantages using biofuel in comparison with some other types of fuels (gaseous, liquid) are the emissions of solid particles, which especially increase in the ambient air during the heating season. Solid particles can penetrate into the internal organs of human and can cause serious illnesses [3].

Since environmental protection problems are becoming more relevant, the requirements for production of small combustion appliances (with a nominal heat output up to 500 kW) are being coordinated by EN303-5:2012 standard. In general, this standard aims to increase the efficiency of boilers and to reduce the emissions. It also indicates the limiting emission values of SO_2, NO_x, CO, organic gaseous carbon and solid particles depending on the boiler class.

Usually, before discharge into the atmosphere, flue gas is filtered (cleaned) in multicyclones or fibre filters. However, such types of filters are not very effective in the case of fine or ultrafine particulates; therefore, for that purpose electrostatic precipitators (ESPs) are used [4]. The advantages of electrostatic precipitators are low aerodynamic resistance and high efficiency [5]. Therefore, electrostatic precipitators are one of the most commonly employed particulate control devices for collecting particulate emissions from boilers, incinerators and from many other industrial installations.

In this comparative study on biofuel (wood pellets), the collection efficiency of solid particles from a European standard class 3 boiler (50 kW nominal capacity) and from a gasification unit (100 kW nominal capacity) using an electrostatic precipitator is investigated.

2 Experimental setups

2.1 Experimental setup with boiler

Flue gas (solid particles) was generated by incinerating wood pellets in the boiler. The commercial boiler of class 3 was used. It is an automatic device with a nominal power of 50 kW and can be used for incineration of pellets, wood, pea-coal, and cereals (Fig. 1).

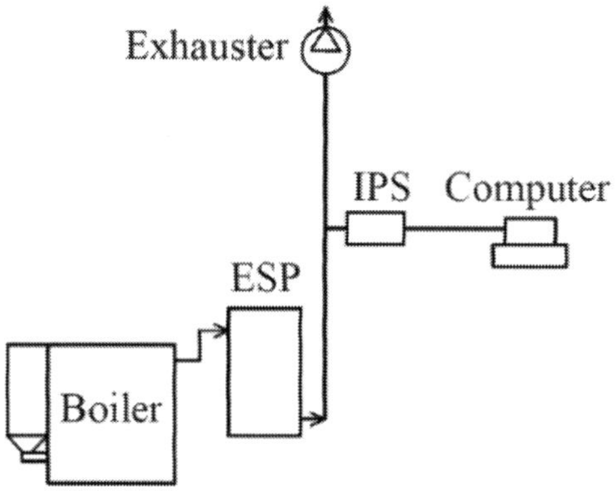

Fig. 1. Schematic view of the experimental setup with the boiler.

The fuel and air supply for combustion are adjusted automatically basing on the set power of the boiler. During the experiments, only wood pellets made from different kinds of wood were incinerated. Flue gas temperature exhausted from the boiler was about 150°C, and the chimney draft ~ 25 Pa (automatically kept by the flue gas exhauster). Such draft gives a flue gas velocity of approx. 1.9 m/s in the flue gas pipe of 180 mm diameter (flow rate ~180 m^3/h).

DOI: 10.1051/e3sconf/20172200138

From the boiler, flue gas enters the ESP, where the solid particles are captured and then cleaned flue gas is exhausted into the atmosphere. The total height of the insulated flue gas pipe was about 15 m. The discharges of flue gas into the atmosphere are always at some distance from the boiler, and this section is the most important for measurement. Therefore, for particle measurements using an infrared particle sizer (IPS), the distance of ~8 m from the ESP was selected. The measurement results were further processed using a computer.

2.2 Experimental setup with gasification unit

The main components of this experimental setup were the gasification unit (nominal power 100 kW), the plasma unit, the gas cooler, the electrostatic precipitator and the boiler (Fig. 2).

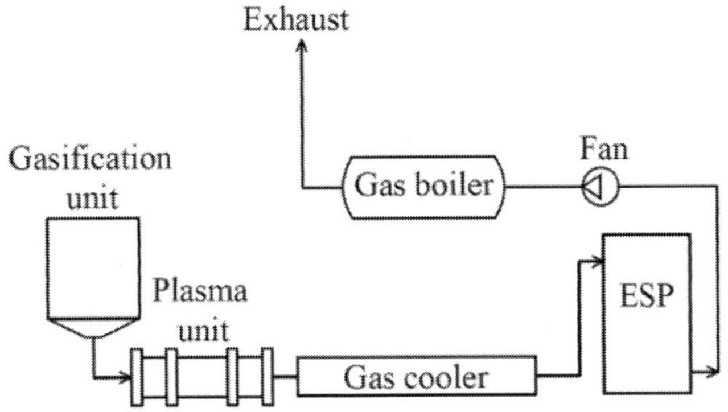

Fig. 2. Schematic view of the experimental setup with the gasification unit.

The process of wood pellets gasification (as well as generation of solid particles) took place in the gasification unit. The producer gas (flow rate ~ 62 m^3/h) with a temperature of about 550°C entered the air plasma unit, which is needed for the neutralization of hazardous chemicals (dioxins, furans, etc. [6]), and existed as generated producer gas. The processed hot producer gas of about 1100°C from the plasma unit was supplied to the water-cooled gas cooler, where the temperature of the gas was decreased, and the gas with a temperature of ~160°C entered the electrostatic precipitator. The cooled gas further flows via the electrostatic precipitator where the solid particles were captured from the gas flow. After passing the ESP, cleaned gas was transported by a fan to the gas boiler, and prepared flammable gas-air mixture was combusted. A more detailed description of the experimental setup is presented in [6].

The concentration of solid particles during these experiments was measured at the same time in the points before and after the ESP by applying the gravimetric method only.

2.3 Electrostatic precipitator

In the experiments with the boiler and the gasification unit was the same precipitator (see Fig. 3) used, and it was designed at the Lithuanian Energy Institute. Its frame is made from carbon steel bars. The inlet to the ESP is made in the upper part, and the outlet is positioned in the lower part. Inside the frame, two parallel stainless steel pipes (i.e. collection electrodes) with a diameter of 120 mm (each) and a length of ~ 1000 mm are installed between two holding planes. A single nichrome wire (i.e. discharge electrode) of 0.2 mm is stretched between the special holders inside the center of each stainless steel pipe. The distance between the discharge electrodes is about 160 mm. The tension of the electrodes is ensured by stainless steel springs.

DOI: 10.1051/e3sconf/20172200138

In order to avoid flue gas leakages between the holding planes and the collecting electrodes, the contact perimeters between them are sealed.

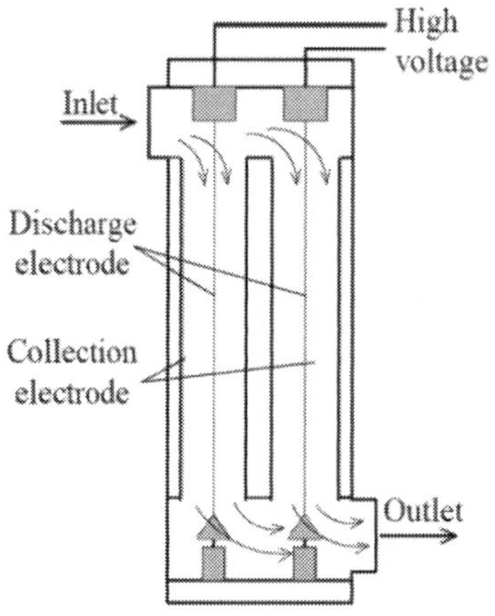

Fig. 3. Schematic cross-section view of the ESP used in the experiments.

During the experiments each of the discharge electrodes was connected to separate high-voltage power supply units (Glassman Series FR) with adjustable output voltage in the range between 0–30 kV. So, in total two high-voltage supply units were used. A more detailed description of the ESP is presented in [7]. The experiments were carried out using the negative potential of the high-voltage power supply units.

To reveal the peculiarities of the collecting efficiency of the ESP, different number of collecting electrodes (1 or 2) were used. This allowed having different flow velocities. When one collecting electrode was used, the other pipe with the electrode was closed using special plates, and flue gas could flow only via the open pipe.

The current-voltage characteristic (high-voltage supply units used in the experiments had a feature enabling them to measure the current in an electrode depending on the supplied voltage) of one ESP electrode is shown in Fig. 4.

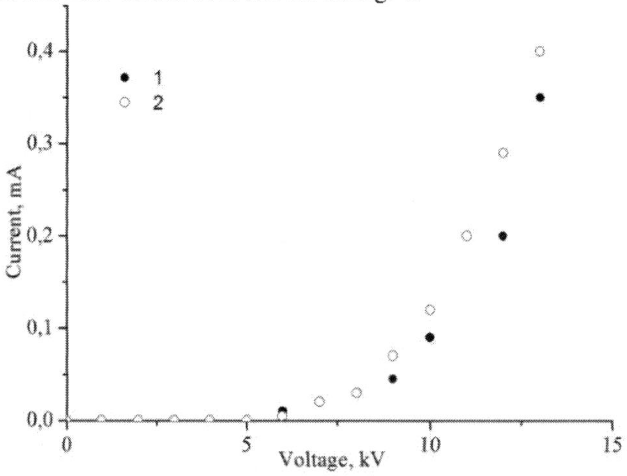

Fig. 4. Current-voltage characteristic of one ESP electrode: 1 – experimental data with the flue gas from the boiler; 2 – experimental data with the producer gas.

From the Figure it is evident that for flue gas from the boiler and for producer gas, the current is registered starting from the voltage of ~6 kV. This means that from this point the electrostatic discharge, which ionizes the flue gas around the discharge electrode (i.e.

creates corona), occurs. Although the character of the experimental data is rather similar in both cases, the layout of the experimental data is steeper for producer gas compared to flue gas from the boiler.

2.4 Measurement techniques

The concentrations of particles during the experiments with flue gas from the boiler were measured using the infrared particle sizer (IPS, made by "Kamika instruments") and applying the gravimetric method. The IPS is an instrument for direct measurements of total particle concentration and concentration of particle fractions in flue gas channels. It is composed of a measuring head and an electronic computer-controlled measurement unit. An IPS sizer ensures only 1D measurements and can measure particles in the range of 0.4–300 μm (this range is divided into four subranges). The measuring head of the device is also equipped with a special holder for the filter, which can be used for the determination of concentration of solid particles by the gravimetric method. The IPS also measures flue gas temperature, concentration of solid particles, flue gas velocity and flow rate.

Distribution of particles by fraction in producer gas was not measured using the IPS because the gas was rather hot, humid and "dirty", and could foul the optical system of the device. Therefore, for producer gas only, the total concentrations of particles were measured before the inlet and after the outlet of the ESP using automatic isokinetic samplers Isostack Basic.

3 Results and analysis

Relative quantity of solid particle distribution in flue gas without the ESP is presented in Fig. 5.

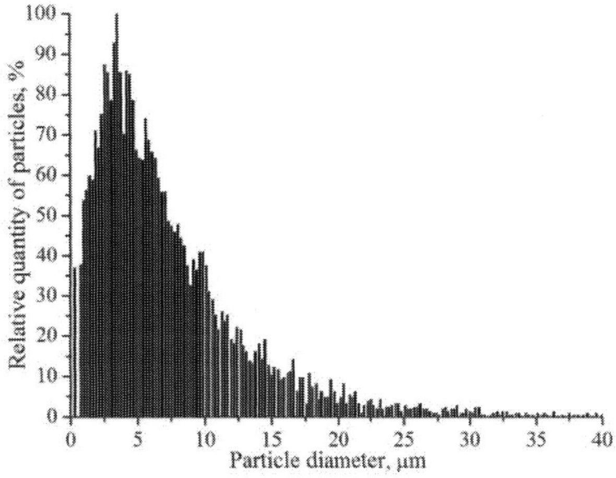

Fig. 5. Relative distribution of particles in flue gas without the ESP.

The Figure shows that flue gas mainly contains solid particles with diameters from 0.3 to ~ 20 μm. The highest quantity of the particles in the distribution is at the diameter of about 2–6 μm (similar as presented in [8]).

Variation of the concentrations of solid particles during the steady-state experimental regime in producer gas and flue gas for the measurements during the experiments are presented in Fig. 6. Results obtained using the gravimetric method show that the initial concentration of solid particles in producer gas is rather high and is scattered in the range of ~140–220 mg/m^3, while the average value is about 180 mg/m^3.

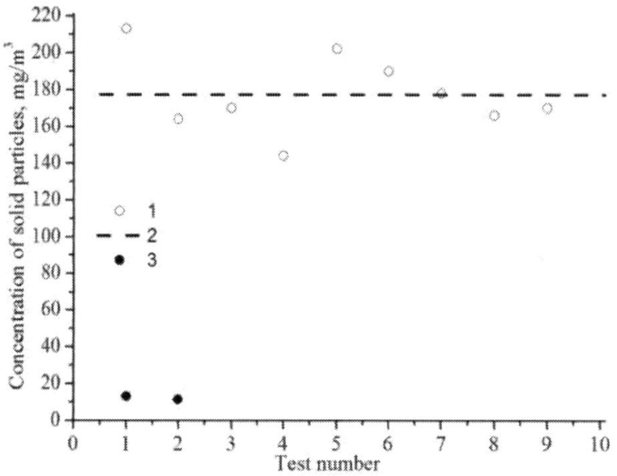

Fig. 6. Concentration of solid particles (for reference O_2 value of 10%) at the inlet to the ESP: 1 – in producer gas; 2 – average value in producer gas; 3 – in flue gas.

The concentration obtained for solid particles in flue gas is rather low – about 13 mg/m³. Similar results were obtained in [8] where particle concentrations from the modern 50 kW boilers fired by wood pellets vary in the range of ~11–40 mg/m³. So, particle concentrations in flue gas are about fourteen times less in comparison with producer gas. The limiting values for such class of the boiler are much higher (cf. EN303-5:2012).

Total collection efficiency E_{Ti} presented in Fig. 7 was calculated using the equation:

$$E_{Ti} = (1 - C_i / C_o) \cdot 100 \quad (1)$$

where C_o is concentration of the particles obtained without the ESP; C_i means concentration of the particles obtained with the ESP at a certain voltage.

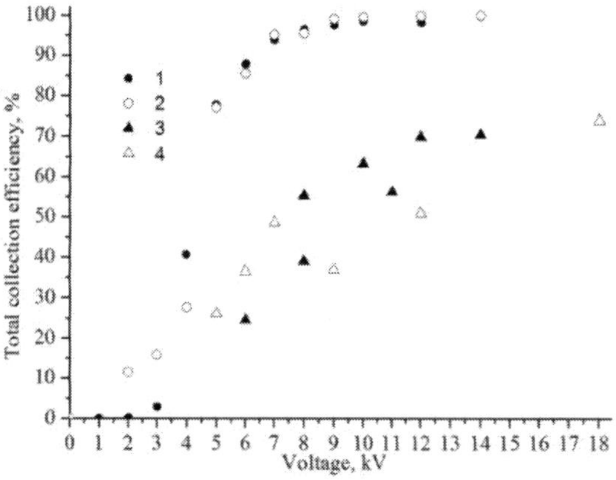

Fig. 7. Total collection efficiency of particles with the applied voltage to the ESP and in the case of a different number of collection electrodes: 1 – one collection electrode (flue gas); 2 – two collection electrodes (flue gas); 3 – one collection electrode (producer gas); 4 – two collection electrodes (producer gas).

The results with the flue gas show that in the region of voltages till 6 kV (corona discharge registered), the collection efficiency just slightly depends on the number of used collection electrodes.

From the voltage equal to or higher than 6 kV, the collection efficiency sharply increases and does not depend on the number of used collecting electrodes. The collection efficiency increases from about 75% at 6 kV to about 98–99% at 14 kV. A further increase of the voltage does not give any tangible increase of the total collection efficiency.

For the case with producer gas, the dependency on used collection electrodes is not evidently expressed in the whole range of the applied voltages (5–18 kV) either. Due to variations of the initial concentration of solid particles (Fig. 6), the collection efficiency in the case of one or two collecting electrodes is slightly scattered. With the increase of voltage, the collection efficiency is gradually increasing, and there is no sudden increase in the efficiency as it was observed for flue gas. For producer gas, the highest reached efficiency is only about 75% at a voltage of 18 kV. Further increase of the voltage does not increase the collection efficiency. This is due to the fact that the working conditions in the test section with producer gas is harder and less stable in comparison with fuel combustion in the boiler and therefore leads to lower collection efficiency of the electrostatic precipitator.

4 Conclusions

After the investigation of the biofuel flue from a class 3 boiler and producer gas cleaning efficiency using an ESP, the following conclusions have been drawn:

- The initial concentration of solid particles in flue gas was about 13 mg/m^3 and about 180 mg/m^3 in producer gas.
- The collection efficiency of the ESP used in the experiments practically did not depend on the number of used collection electrodes.
- At the voltage of the corona discharge, a sharp increase of the collection efficiency for solid particles in flue gas is observed, and further increase of the voltage gave only a slight increase of the collection efficiency of the ESP.
- At the voltage of the corona discharge, no sharp increase of the collection efficiency for solid particles in producer gas was observed; however, further increase of the voltage gave a gradual increase of the collection efficiency.
- The maximal total collection efficiency of the ESP was about 98–99% for flue gas and about 75% for producer gas.

References

1. AEBIOM Stat. Rep. (2015)
2. J. Pettersson, M. Strand, L. Lin, *Charging- and removal efficiency of an ESP in a 250 kW biomass boiler* (ICESP, Nürnberg, 2011)
3. J.T. Zelikoff, L.C. Chen, M. Cohen, K. Fang, T. Gordon, Y. Li, C. Nadziejko, R.B. Schlesinger, Inh. Toxic **15**, 131–150 (2003)
4. K.L. Parker, *Applied electrostatic precipitation* (Springer, Netherlands, 1997)
5. A.A. Varonos, J.S. Anagnostopoulos, G.C. Bergeles, J. Electr. **55**, 111–133 (2002)
6. Z. Kavaliauskas, V. Valincius, G. Stravinskas, M. Milieska, N. Striugas, J. Air & Wast. Manag. **65**, 1292–1296 (2015)
7. R. Poskas, A. Sirvydas, P. Poskas, V. Simonis, J. Jankauskas, Mechan. (to be published)
8. N. Pedisius, M. Praspaliauskas, M. Valantinavicius, T. Vonzodas, V. Zaleskas, E. Lemanas, D. Cepauskienė, Pow. Eng. **62**, 268–281 (2016)

Seasonal changes in quality of wastewater from fruit and vegetable industry

Monika Puchlik[1,*], and *Katarzyna* Ignatowicz[1]

[1]Bialystok Technical University, Faculty of Civil and Environmental Engineering, Department of Technology in Engineering and Environmental Protection, ul. Wiejska 45E, 15-351, Poland

Abstract. The paper aimed at evaluating the seasonal changes in quality of wastewater from facilities producing fruit and vegetable juices, processed and frozen products, and vegetable concentrates. The study revealed that wastewater from fruit and vegetable industry contain large amounts of organic substances expressed as BOD_5 (minimum – 500 mgO_2/dm^3, maximum – 6 100 mgO_2/dm^3) and COD (minimum – 806 mg O_2/dm^3, maximum – 7 732 mg O_2/dm^3), while is deficient in nitrogen and phosphorus. Considerable seasonal oscillations in sewage load disposed by industry to sewerage, were observed. An increase of 50%–60% wastewater concentrations was found between June and October in 2013–2016 as compared to the remaining months.

1 Introduction

High-quality water becomes or will become a rare product in many branches. Food production and processing require large amounts of water of different qualities. The water re-use during food production and processing is common practice with probable increase in near future [1, 2]. Facilities of the food industry having the specificity of particular branch can be a source of many threats for all elements of natural environment – soil, water, air, plants, animals, and human. The foodstuff industry is a division involved in acquiring and processing of natural resources and is one of the links of food economy composed of many elements that are closely associated one to another [3]. The food industry wastewater, like household sewage, is characterized by organic substances expressed as COD and BOD_5 [4, 5]. Organic compounds contained, however, are decomposed to a lesser extent. They are mainly natural biologically-decomposable high-molecular proteins, carbohydrates, and fats, along with depending on raw material, inorganic salts, admixtures from the soil, as well as disinfection agents. Quantity of wastewater oscillated within very broad limits in reference to the unit amounts of raw material processed of a final product. The wastewater disposal is often uneven and is characterized by sinusoidal shape. In some facilities, it can be reduced

[*] Corresponding author: m.puchlik@pb.edu.pl

to particular seasons (seasonal work, campaign operation). There are about 2400 facilities in Poland; over 90% are small, while 6% are large facilities of all processing plants registered in REGON [6]. The processed 1015–1020 thousand tons of fruits and 1070 thousand tons of vegetables in 2013–2014 [7]. The largest amount of processed products in Poland are concentrated juices, frozen products, jams, and puree, fruit and vegetable processing industry is very common, whereas the identification of the food industry facilities interaction to the environment is extremely difficult. It results, among others, from this branch structure in Poland that is characterized by great degree of fragmentation. Therefore, the activities upon the determination of seasonal changes in quality of wastewater from fruit and vegetable industry, were undertaken based on the environment protection issues. It should consist in disposal of wastewater with contaminants loads permissible from a point of view of self-clearing ability of the environment maintenance [8, 9].

2 Characteristics of raw wastewater from fruit and vegetable industry

Fruit and vegetable processing is characterized by seasonal activity that falls mainly between June and October [10]. The activity associated with processing and conserving fruits and vegetables involved 88% entities, while remaining 12% were involved in juices and beverages production [11]. Considering the specificity and diversity of the processed materials, wastewater from fruit and vegetable industry is difficult to characterize [12, 13]. The largest amounts of wastewater are generated in the first stage of production [13]. Wastewater arising from washing the fruits and vegetables are usually used once. In larger facilities, water used in the first stage is pre-purified in sieves and disposed into the sewerage after washing all fruit and vegetable lot, and then replaced with new water batch. In opinion of Nawirska (2007), wastewater generated in fruit and vegetable industry facilities are characterized by high amounts of carbohydrates and minerals as well as variable composition depending on the raw material and season [3, 12, 14]. Its composition is also affected by washing and line disinfection processes [15]. Washing introduces solid, colloid, and dissolved contaminants into the wastewater, which is dependent on the type of raw material processed, but also on technological process applied. At the same time, difficult to determine quantities of washing and disinfection agents penetrate to wastewater. Depending on the site of wastewater generation, following stages can be distinguished: raw material washing, purification, peeling, blanching, cooling (post-cooling medium), or filling. Composition and amounts of wastewater also depend on: type, quality, setup and size of a washer, type and origin of raw material [15, 16]. Wastewater generated during fruit concentrates production is characterized by reaction from 5.8 to 9.4 along with COD value from 1 030 to 5 630 mgO_2/dm^3 [17]. In facilities producing diverse assortment, e.g. salads, puree, marinades, the acidity range can be from 4.9 to 7.7, while COD value from 5 260 to 270 mgO_2/dm^3. These levels depend on production technology and currently processed raw material [17]. Wastewater from fruit and vegetable processing is deficient in nitrogen and its main contaminant is organic substance [10,15). According to literature, the BOD_5 value can amount from 500 to 5 000 mgO_2/ dm^3, whereas during the production season, it even exceeds 5 000 mgO_2/dm^3 [10, 18–20]. These concentrations are much higher than in household-farm wastewater [10, 21, 22], while comparable to those reported by Demirel et al. [23], Puchlik et al. [3] and Neczaj et al. [24] from dairy industry. Seasonal oscillations in quality of wastewater from fruit and vegetable industry became problem for many conventional treatment plants, to which they are disposed [25–32].

 DOI: 10.1051/e3sconf/20172200139

3 Materials and methods

Experiments were carried out in Department of Technology in Engineering and Environmental Protection and certified laboratory of Białystok Waterworks Ltd. in Białystok according to APHA [33]. Samples of raw wastewater were collected according to Polish norm [34] 4 times a month since January till December in 2013–2016 from the facility in north-eastern Poland producing fruit and vegetable juices, processed and frozen products, and vegetable concentrates. Physicochemical analyses included: acidity, BOD_5, COD, total nitrogen, total phosphorus, total solid suspension (TSS). Achieved average monthly results from physicochemical determinations of raw wastewater were subjected to statistical analysis using Statistica 13.1 software in order to confirm the linear dependencies for the production season (since June till October) and out of this period (since January till May and since November till December). Pearson correlation at the significance level of α=0.05 was applied assuming values from [-1, 1] interval. The calculations take into account general measures of the above parameters including arithmetic mean, median, minimum, maximum, and standard deviation.

4 Results

Studies upon the composition of raw wastewater in selected fruit and vegetable processing plants were carried out in 2013-2016. Wastewater from facilities producing a wide range of assortment was analyzed. These facilities produce fruit and vegetable juices, processed and frozen products, and vegetable concentrates on a national scale, and the production is performed the whole year. Results from physicochemical determinations of raw wastewater made possible to analyze in details the changes in wastewater quality within 2013–2016. Tables 1–2 present study results for raw wastewater, which were divided into the campaign and out of this period. It was found that intensified production of juices and processed products in summer causes an increase in organic compounds concentrations expressed as COD and BOD_5, as well as increase in the total solid suspension in wastewater. Tables 1 and 2 present significant differences for both periods at intensified juice and processed vegetables production and out of the production campaign, during which frozen and concentrated vegetables dominated. Wastewater from fruit and vegetable industry collected since June till October were characterized by much higher values of COD and BOD_5 as compared to that collected since January till May and December till January. The highest average levels of BOD_5 and COD were recorded during the production campaign (Table 1). The BOD_5 levels varied from 500 up to 6100 $mgO_2 \cdot dm^{-3}$, while COD from 1256 to 7732 $mgO_2 \cdot dm^{-3}$ during the intensified juice production since June till October. Meanwhile out of the campaign, values of BOD_5 were from 535 to 4000 $mgO_2 \cdot dm^{-3}$ and COD from 806 to 6436 $mgO_2 \cdot dm^{-3}$. Sewage disposed from the fruit and vegetable processing facility were characterized by low concentrations of total nitrogen. The mean content for the production campaign period was 62.1 $mgN \cdot dm^{-3}$, while for the remaining period 58.9 $mgN \cdot dm^{-3}$. Similar statistical distribution of total nitrogen concentrations within the whole study, was observed. Maximum total phosphorus content amounting to 178.4 $mgP \cdot dm^{-3}$ was observed during the campaign, whereas minimum equal to 7.6 $mgP \cdot dm^{-3}$ – out of the production campaign. When analyzing the total solid suspensions concentration, a considerable increase was found since June till October (825.4 $mg \cdot dm^{-3}$). This increase in total solid suspension in summer was associated with production intensification. Enhancing production since June till October also resulted from introduction of fresh fruits and vegetables (strawberry, carrot, beet, apple, cherry, sweet cherry, currant, gooseberry), which in consequence caused the increase in contaminants concentrations in wastewater. Decrease in the contamination levels was recorded in the remaining period, which was caused by a stable production of frozen and concentrated vegetables. Acidity of wastewater from fruit and vegetable industry oscillated from 3.7 up to 10.2 during the campaign, while from 3.2 to 9.8 out of that period. Analysis of the total solid suspensions revealed their increase during the production intensification, whereas decrease in winter and spring. A strong linear dependence between total solid suspension vs. values of COD and BOD_5 was found (Table 3).

Table 1. Quality of raw wastewater from fruit and vegetable industry during the production campaign in 2013–2016.

Parameter	**Unit**	**Average**	**Median**	**Standard deviation**	**Minimum**	**Maximum**
COD	$mgO_2 \cdot dm^{-3}$	2820.0	2604.0	1254.3	1256.0	7732.0
BOD_5	$mgO_2 \cdot dm^{-3}$	2361.3	2319.0	1020.6	500.0	6100.0
TSS	$mg \cdot dm^{-3}$	825.4	698.6	502.0	239.0	2367.0
Total Phosphorus	$mgP \cdot dm^{-3}$	17.5	12.8	26.3	7.9	178.4
Total Nitrogen	$mgN \cdot dm^{-3}$	62.1	65.3	16.8	29.0	88.3
pH	-	-	-	-	3.7	10.2

Table 2. Quality of raw wastewater from fruit and vegetable industry out of the production campaign in 2013–2016.

Parameter	**Unit**	**Average**	**Median**	**Standard deviation**	**Minimum**	**Maximum**
COD	$mgO_2 \cdot dm^{-3}$	2113.9	1819.0	1162.6	806.0	6436.0
BOD_5	$mgO_2 \cdot dm^{-3}$	1660.3	1208.0	1012.1	535.0	4000.0
TSS	$mg \cdot dm^{-3}$	649.9	377.0	660.9	165.0	3268.0
Total Phosphorus	$mgP \cdot dm^{-3}$	13.6	11.6	5.0	7.6	32.9
Total Nitrogen	$mgN \cdot dm^{-3}$	58.9	60.0	13.8	32.7	106.0
pH	-	-	-	-	3.2	9.8

Table 3. Correlation of variables for raw wastewater from fruit and vegetable industry in 2013–2016.

Variable	**COD**	**BOD_5**	**TSS**	**Total Phosphorus**	**Total Nitrogen**
COD	1.00	0.86	0.45	0.13	0.10
BOD_5	0.86	1.00	0.43	0.09	0.21
TSS	0.45	0.43	1.00	0.15	0.19
Total Phosphorus	0.13	0.09	0.15	1.00	0.09
Total Nitrogen	0.10	0.21	0.19	0.09	1.00

5 Conclusions

The study results subjected to analyses allow for drawing the following conclusions:

1. In the case of tested parameters, it is evident their remarkable variability due to the type of production process in fruit and vegetable processing plant. It is characterized by the presence of intensified production period, when fruit and vegetable juices are produced, processed products during the campaign, as well as frozen and concentrated vegetables (out of the campaign) along with the

downtimes of technological lines, when conservation works and washing of devices is performed.

2. Remarkable seasonal oscillations in the concentrations of contaminants disposed with the industrial wastewater into the sewerage. Since June till October, they increased by about 50%–60% as compared to the remaining periods.
3. Organic substance expressed as COD and BOD_5 as well as total soil suspensions were the general contaminants present in wastewater from fruit and vegetable processing facility.
4. Wastewater from fruit and vegetable industry, as compared to the typical sewage from sewerage, was characterized by much higher values of BOD_5 and COD, as well as much lower levels of total nitrogen and phosphorus.
5. Calculations revealed that COD to BOD_5 ratio for raw wastewater from analyzed facility was lower than 2 indicating that the sewage is readily bio-degradable.
6. Achieved results from studies upon raw wastewater from the facility producing fruit and vegetable juices, processed products, and frozen and concentrated vegetables, confirm the necessity to carry out the qualitative and quantitative studies of disposed wastewater.

The research has been carried out in the framework of project No. MB/WBiIŚ/13/2015 and financed from the funds for science MNiSW and as well as by courtesy and help of Białystok Waterworks Ltd.

References

1. R.M. Kirby, J. Bartram, R. Carr, Food Control **14**, 5 (2003)
2. A. Kotowski, B. Kaźmierczak, M. Wdowikowski, Environ. Prot. Eng. **35** (2013)
3. M. Puchlik, J. Struk-Sokołowska, E3S Web Of Conferences (to be published)
4. W. Dąbrowski, M. Puchlik, Annual Set Environment Protection, **12** (2010)
5. K. Ignatowicz, M. Puchlik, Annual Set Environment Protection, **13** (2011)
6. GUS Central Statistical Office, Entities of the national economy registered in the REGON register, declaring their activity according to particular legal forms and PKD200, Quarterly information on entities of the national economy in REGON register declaring the economic activity (2015)
7. GUS Central Statistical Office, Environmental Protection, Information and statistics. Warszawa 2013–2015 (2015)
8. R. Al-Isawi, M. Scholz ,Y. Wang, 15th IWA International Conference on Wetland Systems for Water Pollution Control, Conference Proceedings, **1** (2016)
9. J.A. Alvarez., D. De la Varga, A. Pascual, D. V. Oirschot, R. Kilian, P.N. Carvalho, H. Brix, 15th IWA International Conference on Wetland Systems for Water Pollution Control, Conference Proceedings, **1** (2016)
10. M. Soroko, Water, Environment, Rural Areas, **3** (2003)
11. S. Kierczyńska, Yearbooks X, **4** (2008)
12. A. Nawirska, Agro Industry **3** (2007)
13. M. Strzelczyk, A. Steinhoff-Wrześniewska, A. Rajmund Pol. J. Chem. Tech. **12**, 4 (2010)

14. G. Murtaza, A. Ghafoor, M. Qadir, G. Owens, M.A. Aziz, M. H. Zia, Saifullah Pedosphere **20**, 1 (2010)
15. M. Puchlik, Journal of Ecological Engineering **17**, 1 (2016)
16. A. Nawirska, L. Szymański, Publishing House of Agricultural University in Wroclaw, (2002)
17. D. Kopiec, Communal Review, **2** (2007)
18. B. Koziorowski, *Treatment of industrial wastewater* (WTN, Warsaw, 1980)
19. F. Meinck, H. Stoof, Kohlschutter, *Industrial wastewater* (1975)
20. B. Talik, B. Chudzik, IMUZ **20**, 3 (2000)
21. M. Puchlik, K. Ignatowicz, W. Dąbrowski, Journal of Ecological Engineering **16**, 1 (2015)
22. B. Bartkiewicz, K. Umiejewska, *Treatment of industrial wastewater* (PWN, 2010)
23. B. Demirel, O. Yenigun, T.T. Onay, Proc. Biochem. **40** (2005)
24. E. Neczaj, M. Kacprzak, T. Kamizela, J. Lach, E. Okoniewska, Desalination, **222** (2008)
25. K. Ignatowicz, J. Piekarski, D. Kozłowski, Annual Set Environment Protection, **17** (2015)
26. P. Kogut, J. Piekarski, K. Ignatowicz, Annual Set Environment Protection, **16** (2014)
27. C.A. Arias, 15th IWA International Conference on Wetland Systems for Water Pollution Control, Conference Proceedings, **1** (2016)
28. R.G. Andries, A.T. de Matos, W. Freitas, 15th IWA International Conference on Wetland Systems for Water Pollution Control, Conference Proceedings, **1** (2016)
29. J.S. Bays,15th IWA International Conference on Wetland Systems for Water Pollution Control, Conference Proceedings, **1** (2016)
30. K. Haarstad, J. Bavor,. 15th IWA International Conference on Wetland Systems for Water Pollution Control, Conference Proceedings, **1** (2016)
31. P. Haase, A. Carter, E. Corwin, P. Myer, 15th IWA International Conference on Wetland Systems for Water Pollution Control, Conference Proceedings, **1** (2016)
32. E. Jiao, Ch. Gao, Q. Zhang, Y. Tian, , 15th IWA International Conference on Wetland Systems for Water Pollution Control, Conference Proceedings, **1** (2016)
33. E.W. Rice, R.B. Baird, A.D. Eaton, L.S. Clesceri, Standard Methods for the Examination of Water and Wastewater, American Public Health Association (APHA), American Water Works Association (AWWA), Water Environment Federation (WEF), 22nd Edition (2012)
34. PN-ISO 5667-10:1997, Water quality – Sampling – Guidelines for the sampling of wastewater in Poland

 DOI: 10.1051/e3sconf/20172200140

A novel vortex tube-based N_2-expander liquefaction process for enhancing the energy efficiency of natural gas liquefaction

Muhammad Abdul Qyyum[a, 1], *Feng* Wei[a, 1], *Arif* Hussain[1], *Wahid* Ali[1], *Oh Sehee*[1], and, *Moonyong* Lee[1,*]

[1] Yeungnam University, School of Chemical Engineering, Gyeongsan 712-749, Republic of Korea

Abstract. This research work unfolds a simple, safe, and environment-friendly energy efficient novel vortex tube-based natural gas liquefaction process (LNG). A vortex tube was introduced to the popular N_2-expander liquefaction process to enhance the liquefaction efficiency. The process structure and condition were modified and optimized to take a potential advantage of the vortex tube on the natural gas liquefaction cycle. Two commercial simulators ANSYS® and Aspen HYSYS® were used to investigate the application of vortex tube in the refrigeration cycle of LNG process. The Computational fluid dynamics (CFD) model was used to simulate the vortex tube with nitrogen (N_2) as a working fluid. Subsequently, the results of the CFD model were embedded in the Aspen HYSYS® to validate the proposed LNG liquefaction process. The proposed natural gas liquefaction process was optimized using the knowledge-based optimization (KBO) approach. The overall energy consumption was chosen as an objective function for optimization. The performance of the proposed liquefaction process was compared with the conventional N_2-expander liquefaction process. The vortex tube-based LNG process showed a significant improvement of energy efficiency by 20% in comparison with the conventional N_2-expander liquefaction process. This high energy efficiency was mainly due to the isentropic expansion of the vortex tube. It turned out that the high energy efficiency of vortex tube-based process is totally dependent on the refrigerant cold fraction, operating conditions as well as refrigerant cycle configurations.

1 Introduction

The refrigeration and liquefaction in LNG plant demands the high capital investment and consumes a tremendous amount of energy. It normally occupies about 35% of the capital cost, and up to 50% of the operating costs [1]. In terms of energy consumption, liquefaction of 1 kg natural gas consumes 1,188 kJ of energy [2], which is equivalent to the 30–35% of the total required energy for the LNG production. This energy consumption varies according to cite conditions and the type of available liquefaction processes. Now

* Corresponding author: mynlee@yu.ac.kr

[a] These two authors contributed equally to this work.

DOI: 10.1051/e3sconf/20172200140

a days, several processes are available for baseload natural gas liquefaction with different energy efficiency, capacity, complexity, environmental and safety impact. The N_2-expander process as one of representative natural gas liquefaction processes has a high environmental and safety impact but the major issue is the low energy efficiency.

The optimization of design and operational parameters is one of the most popular approach for improving the energy efficiency of N_2 refrigeration processes. Various optimization techniques have been reported to enhance the energy efficiency of the N_2 refrigeration process [3-6]. Another alternative to improve the energy efficiency of the LNG plant is to enhance the refrigeration cycle units like compressors, cryogenic heat exchangers, and expansion devices.

By improving the refrigeration cycle units like compressors, cryogenic heat exchangers, and expansion devices is another main alternative approach to improve the energy efficiency of LNG plant [7]. A vortex tube with its great potential for liquefaction applications was first introduced by Georges J. Ranque (1933) and experimentally investigated by German physicist Rudolf Hilsch (1947) [8] hence it is known as Ranque-Hilsch vortex tube. The significant merits of the vortex tube are, compactness with no moving parts, low in cost, maintenance free and adjustable cold and hot streams [9, 10]. The construction of the vortex tube consists of inlet nozzle(s), diaphragm, vortex generator, chamber, cylindrical tube, conical valve, hot outlet and cold outlet. The simple schematic diagram of vortex tube is shown in Figure 1.

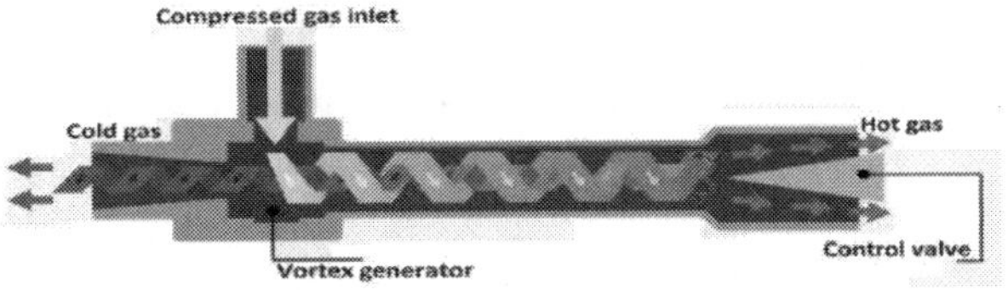

Figure 1. The schematic diagram of vortex tube [11].

The working principle of vortex tube depends upon the pressure gradient that causes the energy separation (in terms of hot and cold) from compressed gas. The compressed gas is introduced tangentially into the tube chamber through the one or more nozzles. The swirls more like a typhoon is formed through the interchangeable vortex generator. The compressed gas leaves the tube from the two outlets: one is from the cold side and other from the hot side. A small conical control valve at the hot gas side is installed to control the temperatures, cold and hot fraction corresponding to the specified application of vortex tube. It has been demonstrated that the refrigeration cycle associated with isentropic expansion has higher COP (coefficient of performance), which is defined as the ratio of useful cooling effect provided to required compression energy, in comparison with isenthalpic expansion-based refrigeration cycles [12], as proven in Maxwell thermodynamic relation (i.e., $dH = TdS + VdP$). Mohiuddin and Elbel [13] has investigated that the overall expansion process in the vortex tube is an isentropic.

This study addressed the potential benefits of vortex tube coupled with turbo expander to enhance the energy efficiency of N_2 expansion LNG process. The structure and design parameters of the proposed process were optimized to achieve the maximum benefits of proposed vortex tube-based configuration corresponding to the minimum required compression power. The modified knowledge-based optimization method was proposed and successfully applied to optimize the proposed LNG process.

 DOI: 10.1051/e3sconf/20172200140

2 Vortex tube-based LNG process: simulation and process description

The energy separation behavior was rigorously modeled using the CFD software ANSYS FLUENT®. The proposed LNG process has been then modeled in a commercial simulator ASPEN HYSYS® by embedding the CFD model results.

2.1 CFD model of vortex tube

In the present study, a CFD model was used to investigate the energy separation in the vortex tube with Nitrogen at cryogenic temperature as the working fluid (refrigerant) using the CFD software ANSYS FLUENT®. The standard k-ε turbulence model was used to investigate the flow behavior in the vortex tube. For meshing, ICEM-CFD application was used to generate the 5mm hexahedral structural mesh. The CFD model was based on the experimental investigation for an Exair™ 708 slpm vortex tube by Skye et al. [14]. The total temperature distribution for working fluid nitrogen at inlet temperature 139 K and 7 bar pressure is shown in Fig. 2. Inlet conditions for vortex tube were chosen according to the available properties data of nitrogen at cryogenic temperatures [15]. The static pressure at the cold exit boundary was fixed at 2 bar and the static pressure of hot exit boundary was adjusted in the way to vary the cold mass fraction. The CFD results which were embedded in the ASPEN HYSYS®, are summarized in Table 1.

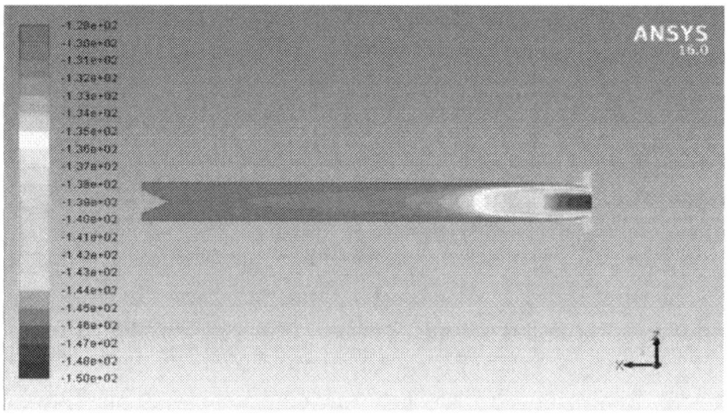

Figure 2. Total temperature contours of nitrogen at 139K and 7 bar.

Table 1. CFD results for working fluid Nitrogen.

Property	Value
Inlet temperature (K)	139
Inlet Pressure (bar)	7
Cold side temperature (K)	123
Cold side Pressure (bar)	2
Cold mass fraction	0.25
Hot side temperature (K)	145
Hot side Pressure (bar)	3

2.2 Process simulation and description

The process simulation basis and feed conditions are summarized in Table 2. The well-known Peng-Robinson [16] was used to calculate the thermodynamic properties and the Lee-Kesler [17] EOS was used to calculate the enthalpies and entropies. The proposed LNG process is demonstrated in Fig. 3. Note that in Fig. 3, different streams with the name as 'stream-x' (x = 1, 2, 3, 4......) are used for the process description. In this process, nitrogen as a refrigerant stream-1 was compressed to high pressure 77.5 bar (stream-2) through five compressor stages each equipped with after-coolers. To avoid the high compression power and reduce the irreversibility of the process, the compression ratio was chosen in the practical range 1:3. A hybrid cooling system combined by air/water cooler (AC-1) and a cryogenic plate fin exchanger (CHE-01) was used to cool down the high pressure stream-2 before entering the expander. After expansion, the stream-5 at 139 K and 7 bar was introduced into the vortex tube. The cold side stream of vortex tube (stream-6) was used to cool down stream-3 (3–4) through CHE-01 and stream-8 was again introduced into the expander K-3. Stream-9 was used to liquefy the compressed natural gas through CHE-02. The feed natural gas was compressed to 95 bar through a booster compressor K-NG. Stream-10 from CHE-02 as a superheated vapor and stream-7 from CHE-01 were mixed in the mixer and recycled to achieve complete cycle. LNG product was obtained with 8% boil-off gas at the pressure of slightly higher than atmospheric, i.e., 1.209 bar.

Table 2. Process simulation basis and feed conditions.

Feed natural gas [18]	Value
Temperature (K)	303
Pressure (bar)	50
Flow rate (kg/h)	1
Composition	Mole %
Methane	91.30
Ethane	5.40
Propane	2.10
i-Butane	0.50
n-Butane	0.50
i-Pentane	0.01
n-Pentane	0.01
Nitrogen	0.20
After-coolers outlet temperature (K)	303
LNG Tank pressure (bar)	1.209

DOI: 10.1051/e3sconf/20172200140

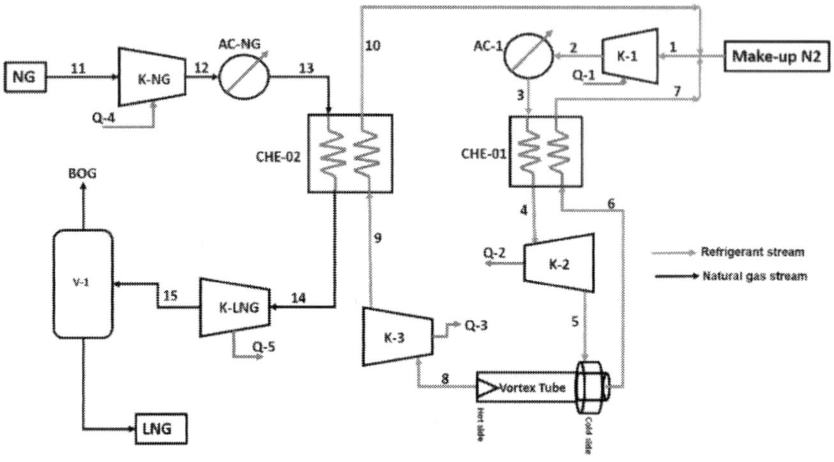

Figure 3. The process flow diagram of vortex tube-based LNG process.

3 Process Optimization

The knowledge-based optimization method [18, 19] was modified to employ for the optimization of the proposed LNG process. Figure 4 illustrates the modified knowledge-based optimization algorithm used in this study.

The minimization of specific compression energy in the proposed LNG process was chosen as an objective function for optimization. Since the refrigerant flow rate and operating pressures have pronounced impact on the overall required compression power and process irreversibility, these were chosen as the key decision variables in the optimization of the proposed LNG process. Table 3 shows these decision variables with lower and upper bounds.

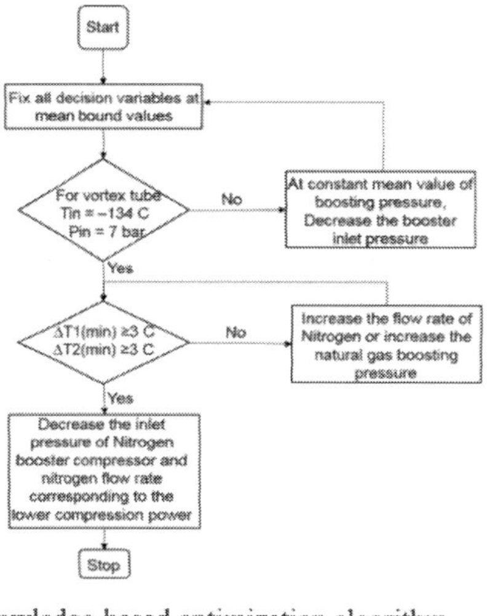

Figure 4. Proposed modified knowledge-based optimization algorithm.

The refrigerant pressure before and after the turbo-expander booster compressor is being designated as 'P_1' and 'P_2', respectively. The minimum internal temperature approach (MITA) value as a major constraint was chosen as 3 °C in both cryogenic exchangers considering the LNG cryogenic exchangers transfer the heat with MITA value as small as 1–3°C [20].

DOI: 10.1051/e3sconf/20172200140

Table 3. Decision variable bounds.

Decision variables	Lower bound	Upper bound
Boosting pressure of natural gas, P_{NG}, bar	70	120
Pressure of stream-1, P_1, bar	25	45
Pressure of stream-2, P_2, bar	55	110
Flow rate of nitrogen, m_{N2}, kg/hr	4.5	9.5

Objective function was formulated as:

$$Minimize\ f(X) = Min.\left(\sum_{i=1}^{4} wi/m_{LNG} \right)$$

Where 'X' is the vector of decision variables, $X = (P_{NG}, P_1, P_2, m_{N2})$

4 Optimization Results

Table 4 compares the optimization results of the proposed and conventional N_2-expander processes. The results showed the specific compression energy can be saved significantly up to 20% in comparison with the commercial N_2-expander LNG process.

Table 4. Optimization results.

Parameters	Proposed vortex tube - based LNG process	Conventional N_2-Expander process [18]
Boosting pressure of natural gas, P_{NG}, bar	95	50
Pressure of stream-1, P_1, bar	32	–
Pressure of stream-2, P_2, bar	77.50	100
Flow rate of nitrogen, m_{N2}, kg/hr	5.618	8.257
Specific compression energy (kJ/kg-LNG)	2143.80	2681.64
Specific energy savings	20%	–

4.1 Optimization results in terms of composite curves

The composite curve matching technique is widely used to measure the efficiency of any process where cooling and heating are dominantly involved. For an energy efficient liquefaction process with low specific compression energy, each hot (natural gas) and cold (refrigerant) composite curve should be located as closely as possible. The composite curves of proposed LNG process (Figures 5a and 5b) also illustrate the clear difference with the composite curves of the commercial N_2-expander LNG process (Figures 5c and 5d).

DOI: 10.1051/e3sconf/20172200140

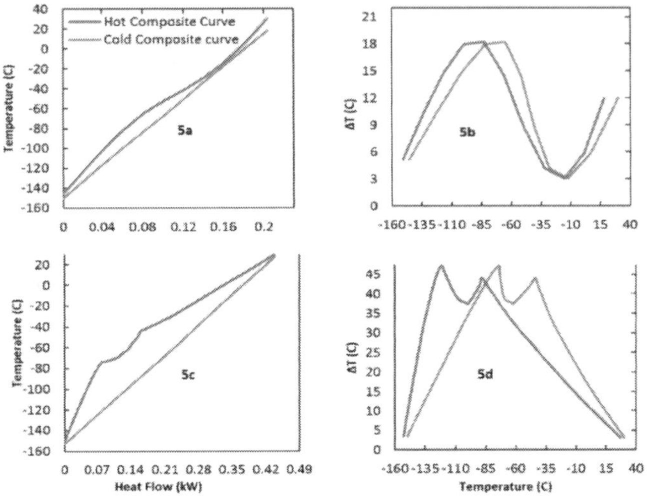

Figure 5. Comparison of composite curves.

5 Conclusions

An enhancement in the refrigeration cycle of LNG process, it improves the energy efficiency in terms of required specific compression power. The proposed vortex tube-based LNG process showed superior performance to liquefy natural gas with significantly less energy in an eco-friendly manner in comparison with the existing well-established N_2-expander LNG process. The modified knowledge-based optimization algorithm proposed was successfully applied to achieve the maximum benefit of newly added expansion device (vortex tube) as well as to make this liquefaction process feasible on the commercial scale. Based on the optimized results, it was found that the specific compression power could be reduced by improvement in an expansion step of LNG process. By using other heuristic evolutionary algorithm, further minimization in specific compression power of proposed LNG process might also be possible. It was also found that the isentropic expansion efficiency of vortex tube could be further improved by optimizing the geometric and operational parameters with respect to LNG process. It turned out the vortex tube has a promising potential to enhance the refrigeration effect as well as energy efficiency of the industrial gases liquefaction processes.

This research was supported by Basic Science research Program Foundation of Korea (NRF) funded by the Ministry of Education (2015R1D1A3A01015621) and also by Priority Research Centers Program through the National Research Foundation of Korea (NRF) funded by the Ministry of Education.

References

1. X. Wang and M. Economides, *Liquefied Natural Gas (LNG)*, in *Advanced Natural Gas Engineering* (Gulf Publishing Company, 2009).
2. A.J. Finn, G.L. Johnson, H.L. Tomlison, J. Hydrocar Process, **78**, 47 (1999).
3. C.W. Remeljej, A.F.A. Hoadley, Energy, **31**, 2005 (2006).
4. P. Venkataraman, *Applied Optimization with MATLAB Programming* (Wiley Publishing, 2009).

5. A. Aspelund, T. Gundersen, T. J. Myklebust, M.P. Nowak, A.J. Tomasgard, Comput. Chem, **34**, 606 (2010).
6. M.S. Khan, M. Lee, Energy, **49**, 146 (2013).
7. A. Mortazavi, C. Somers, A. Alabdulkarem, Y. Hwang, R. Radermacher, Energy, **35**, 3877(2010).
8. N. Bej, K.P. Sinhamahapatra, Int. J. Refrig, **45**, 13 (2014).
9. Y. Xue, M. Arjomandi, and R. Kelso, Int. J. Refrig, **36**, 1730 (2013).
10. Eiamsa-ard, S. and P. Promvonge, Renew. Sust. Energ. Rev, **12**, 1822 (2008).
11. *AiRTX Vortex Tubes*. [cited 2017 2/13]; Available from: http://www.airtx.net/airtx-vortex-tubes-review.
12. A. Mortazavi, C. Somers, Y. Hwang, R. Radermacher, P. Rodgers, S. Al-Hashimi, J. Appl. Energ, **93**, 125(2012).
13. M. Mohiuddin, S. Elbel, 2014.
14. H. M. Skye, G.F. Nellis, S.A. Klein, Int. J. Refrig, **29**, 71 (2006).
15. T. Dutta., K. Sinhamahapatra, and S. Bandyopadhyay, J. Fluid, **2013**.
16. D. Y. Peng, D.B. Robinson, Ind & Eng. Chem. Fund, **15**, 59 (1976).
17. B.I. Lee, M.G. Kesler, AIChE. J, **21**, 510 (1975).
18. M. S. Khan, S. Lee, M. Getu, M. Lee, J. Nat. Gas. Sci. Eng, **23**, 324 (2015).
19. M. S. Khan, S. Lee, G.P. Rangaiah, M. Lee, J. Appl. Energ, **111**, 1018 (2013).
20. M. M. F. Hasan, I.A. Karimi, H.E. Alfadala, H. Grootjans, AIChE. J, **55**, 150 (2009).

The choice of primary energy source including PV installation for providing electric energy to a public utility building - a case study

Bartosz Radomski[1,*], *Barbara* Ćwiek[1], *Tomasz M.* Mróz[1]

[1] Poznań University of Technology, Institute of Environmental Engineering, Piotrowo 3a, 60-965 Poznań, Poland

Abstract. The paper presents multicriteria decision aid analysis of the choice of PV installation providing electric energy to a public utility building. From the energy management point of view electricity obtained by solar radiation has become crucial renewable energy source. Application of PV installations may occur a profitable solution from energy, economic and ecologic point of view for both existing and newly erected buildings. Featured variants of PV installations have been assessed by multicriteria analysis based on ANP (Analytic Network Process) method. Technical, economical, energy and environmental criteria have been identified as main decision criteria. Defined set of decision criteria has an open character and can be modified in the dialog process between the decision-maker and the expert - in the present case, an expert in planning of development of energy supply systems. The proposed approach has been used to evaluate three variants of PV installation acceptable for existing educational building located in Poznań, Poland – the building of Faculty of Chemical Technology, Poznań University of Technology. Multi-criteria analysis based on ANP method and the calculation software Super Decisions © has proven to be an effective tool for energy planning, leading to the indication of the recommended variant of PV installation in existing and newly erected public buildings. Achieved results show prospects and possibilities of rational renewable energy usage as complex solution to public utility buildings.

1 Selection of a photovoltaic system - a case study

Due to their simplicity and broad application range, PV systems have become one of the most desirable options to assist the energy balance of the building - green energy. The design of the PV system should cover a number of technical, economic, energy and environmental aspects. Fig. 1. presents the proposed overall algorithm of planning photovoltaic systems, which includes various stages of implementation of the investment.

* Corresponding author: bartosz.an.radomski@doctorate.put.poznan.pl

1.1 Selection and rate of location for future investment

This article shows the result of multi-criteria analysis of the choice of the photovoltaic system for existing building - Didactic Centre of the Faculty of Chemical Technology (CDWTCh), located on the Warta Campus at Poznan University of Technology, Poland. The main building entrance is located on the south side, while the north side is the underground car park. Over the east and west side glass façades, 15-meter high racks were installed: one talks of Double Skin Façade (Fig. 2). Their task is to reduce the excessive direct sunlight premises, resulting in increased demand for cooling energy. During the usage of the facility, it turned out that the mounted glass hindered the cleaning process of the windows and the two building facades. What is more, as a result of insufficient transparency of the glass, the interior is not receiving enough daylight. As a result of the complications presented, the idea of replacing the second facade with a useful photovoltaic installation, structurally supported on the existing rack has been considered. Bearing in mind the building location, the area which may take a photovoltaic system, its slope (vertical wall) and the azimuth, there are two rectangle-shaped areas measuring 46 x 15 m, that can be used for PV panel installation. That would give a total area of nearly 1 400 m^2.

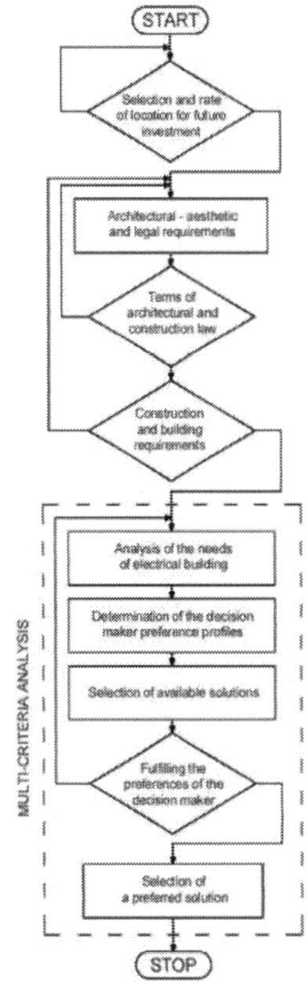

Fig. 1. The general algorithm of planning PV systems.

a) North Direction b) South Direction

Fig. 2. Building CDWTCh, satellite photos indicating "Double skin facade".

Google Maps - access 2016-10-08

1.2 Terms of architectural and construction law

The fulfilment of the architectural, aesthetic and legal conditions for existing buildings during the construction of the new PV installation may be the biggest challenge. In this paper there are two solutions taken into consideration. The first is based on the body of traditional PV panels, leaving space for windows and the second constitutes a full façade, without voids using PV panels with daylight transparency of approx. 70%. In the case of public buildings, aesthetics can become a key factor and influences the choice of recommended solution. The idea of replacing the existing unfavourable second façade and increasing the profitability of the investment determines the search for solutions that meet the conditions of building construction (usage of existing frames). Hence the choice of specific models of PV panels and their arrangement can vary significantly.

1.3 Analysis of the electricity demand of educational building

The total electricity consumption in the building CDWTCh in 2016 was derived using the BMS monitoring data. An annual electricity consumption profile was created on the basis of collected measurements (Fig. 3.).

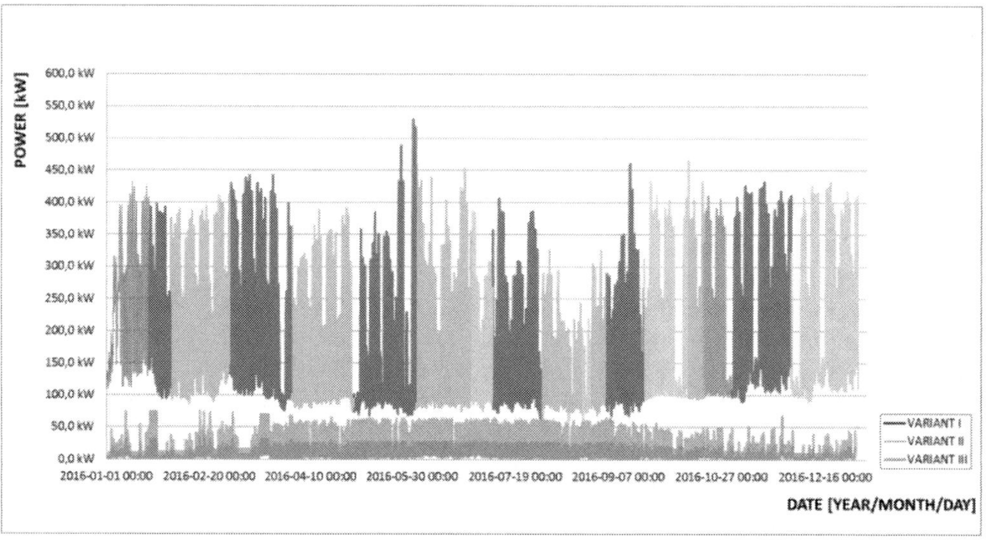

Fig. 3. Annual electricity consumption profile of the building CDWTCh - data from the BMS *

* Periods marked in gray represent an estimate of the value of electricity consumption in the facility, due to the lack of measurements within the prescribed time

The analysis of the electricity consumption in the facility CDWTCh leads to following conclusions:

• daily electrical power needed to supply the building varies according to the season in the range of 70-450 kW for days and within the 70-300 kW during the weekend, while the night-time demand of power is stable throughout the year and varies between 70 -100 kW,

• during days off, the mean instantaneous electric power consumed by the building varies from 60 to 100 kW,

• during the year, there are several peaks of power where electric power drawn exceeds the the value of 450 kW,

• the maximum instantaneous electric power reported in the building was 525 kW (30th May), and the minimum 60 kW (13th August),

• annual electricity consumption in the building CDWTCh over the year considered was 1 720.3.

1.4 Selection of available solutions

After fulfilling all the necessary conditions to meet the demands of a potential photovoltaic installation and after determining the power demand of the object, an analysis of available market solutions was carried out. On the basis of market research, three variants of photovoltaic panels have been selected.

Variant I – SOLARWORLD SW 260 poly

The first solution stands out thanks to its lowest price. The maximum power of one module with dimensions of 1675x1001x33 mm and a weight of 18 kg is Pmax = 260 Wp (STC

 DOI: 10.1051/e3sconf/20172200141

conditions). Each module has three zones with its own bypass diode, with shady approx. 40% they are excluded from use. This increases the efficiency of the panels and extends their service life, while minimizing the risk of damage. Within each module there are 60 polycrystalline cells. The set is based on high-quality components, including 8 inverters Fronius SYMO 15,0-3-M.

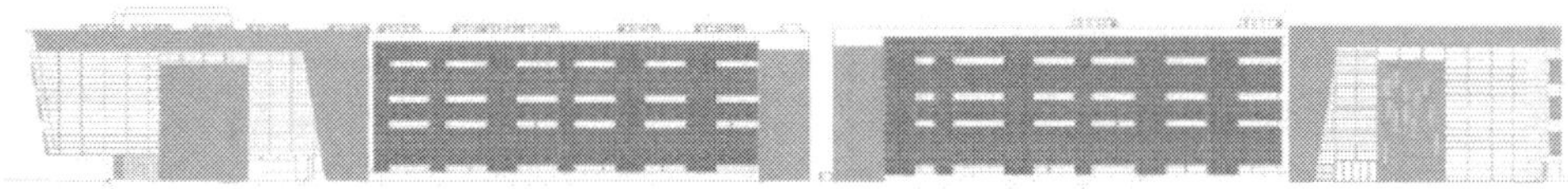

Fig. 4. Design of S-E and N-W facades using SW 260 poly panels.

Variant II – SOLARWORLD SW XL mono

The second variant is characterized by a high solar to electricity energy conversion efficiency, which can go as high as $\eta = 16.3\%$. A high maximum output of one module Pmax = 325 Wp (STC) can be achieved thanks to its large surface area of monocrystalline cells of 1.77m2, with dimensions of 1993x1001x33 mm and weight 21.6 kg. Each module includes 72 high-performance monocrystalline cells. Just as in Variant I, the product also has 3 bypass diodes. The set is based on high-quality components, including 8 inverters Fronius SYMO 15,0-3-M.

Fig. 5. Design of S-E and N-W facades using SW XL mono panels.

Variant III – MLsystem C60 mono

The third variant is characterized by a high architectural and practical quality. The structure of photovoltaic panels is designed to achieve the expected transparency, which in this case results in the penetration of sufficient daylight to the rooms overshadowed by second elevation (assuming 74% transparency). The modules with dimensions of 2400x1000 mm made of tempered glass have 40 monocrystalline cells (125x125 mm) with a high conversion efficiency of $\eta = 23.7\%$. Despite the high efficiency due to the incomplete coverage of cell modules, a maximum power of a selected module is only Pmax = 138 Wp (STC).

Fig. 6. Design of S-E and N-W-panels facades using MLsystem C60 mono panels.

All three variants considered have their specific advantages: variant I – the lowest investment cost, variant II the largest nominal power and variant III the best aesthetics. Hence, the choice of the recommended variant starts to be the decision problem, which can be solved using multi-criteria decision aid (MCDA) approach.

1.5 The set of decision criteria

The choice of the electricity supply variant for a public building depends on a number of technical, economic, energy and environmental factors. For such a defined decision problem, a set of evaluation criteria (features) has been proposed (Tab. 1). A detailed description of the selected evaluation criteria can be found inter alia in [1, 2, 5, 6].

Table 1. A set of criteria and sub-criteria decision-making for the selection of the power supply of electricity using PV installations with an indication of the preferences and numerical values

A set of criteria and sub-criteria decision-making						Values		
Lp.	Criteria	Sub-criteria	Symbol	Unit	Preferences direction	Variant I	Variant II	Variant III
I.	Technical	Realization difficulties	K^{T}_{UR}	pt.	Decreasing	5	5	3
		Degree of existing construction load	K^{T}_{SOIK}	pt.	Decreasing	3	3	5
		Architectural and construction conditions	K^{T}_{AB}	pt.	Increasing	3	2	5
II.	Economic	Investment cost	K^{EK}_{NI}	zł	Decreasing	714 290	781 230	2 433 212
		Discounted payback period (DPP)	K^{EK}_{DPP}	years	Decreasing	22.0	25.0	none
		Net present value (NPV)	K^{EK}_{NPV}	zł	Increasing	91 244	3 558	-3 512 070
III.	Energy	Amount of "produced" electricity	K^{EN}_{SRIPEE}	kWh/year	Increasing	107 300	110 278	50 156
		PV peak electric power for STC	K^{EN}_{SME}	kWp	Increasing	174.2	179.1	78.7
		Efficiency of conversion of the module	K^{SS}_{WSKM}	%	Increasing	15.51	16.29	23.7
IV.	Environmental	Rate of return of embedded energy (EPBT)	K^{SS}_{EPBT}	year	Decreasing	8.30	13.70	11.00
		Reduction of CO_2 emission	K^{SS}_{REZ}	tonnes	Increasing	1 321	874	520
		Reduction of EP consumption	K^{EN}_{SRRZEP}	kWh/year	Increasing	246 791	253 640	115 360

2 Multi-criteria analysis

As a multi-criteria decision support tool the analytical network process (ANP) has been employed. The detailed description of ANP method can be found in [3, 4]. ANP method is related to the AHP method, one of the best known and most widely used in the world [5, 6]. The set of evaluation criteria (features) has an open character and is created in the dialog mode between the decision-maker and the expert - in the present case, an expert in planning the development of energy supply systems.

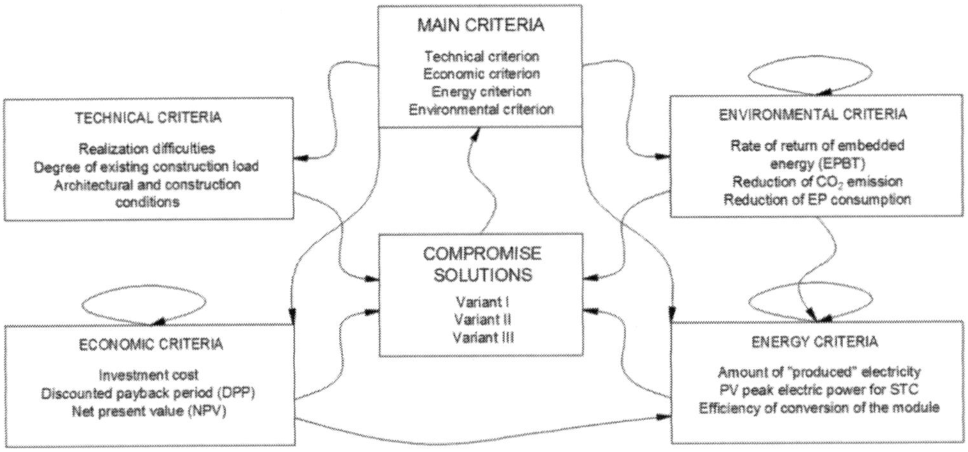

Fig. 7. The structure of the analytic network process (ANP) of defined decision problem.

2.1 Determination of the decision-makers preference profiles

The analysis of ANP for the evaluation of pairs of individual criteria, sub-criteria and variants in all specific relationships related to the decision-making network (Fig.7) includes the preference model of the decision maker - Investor. Pairwise comparisons are made with respect to all combinations of interconnections between the elements and their groups. The ANP method is based on a 9-level scale of comparison, analogous to the AHP method [3].

 DOI: 10.1051/e3sconf/20172200141

2.2 Calculations

The practical software that uses the ANP method is the Super Decisions © program [7]. The latter has been used in the calculations.

In accordance with the principles of the ANP method, after creating the decision-making network (Fig. 7), one has to determine the concentrations of elements and establish connections between them (Tab. 2). Super-matrixes are the fundamental tools used to structure the work in the ANP method. They represent preferences indicate an advantage of the elements on the left side of the matrix over elements located in the upper part. Preference vector "a" calculated from pairwise comparisons of the data elements defines the influence of another component element of the model. When the element has no effect on the other, the preference is not defined and shall be replaced by zero (Tab. 2).

Table 2. Super matrix defining the relationship between the components of the network structure

Node label	**Variants**	**Main criteria**	**Technical criteria**	**Economical criteria**	**Energy criteria**	**Environmental criteria**
Variants	0	1	0	0	0	0
Main criteria	1	0	0	0	0	0
Technical criteria	0	0.121509	0	0	0	0
Economical criteria	0	0.389051	0	0.142857	0	0
Energy criteria	0	0.172403	0	0.142857	0.166667	0.280833
Environmental criteria	0	0.317036	0	0	0	0.13501

The preferences vectors are calculated successively and then introduced in the columns of unweighted network decision-making super-matrix. The essence of the ANP method is a synthesis and presentation of the final results in the form of priorities for all possible dependencies, and coupling cycles between network elements. The result is the creation of super-matrix' limits, comprising the final preference of each element of the model over the other elements.

2.3 Calculation results

The final result of conducted calculations is a ranking of variants (Tab. 4). The "raw" column presents the results derived from super-matrix limits (Tab. 3), the "normalized" column contains the normalized values of super-matrix limits, whereas the "target" column accounts for the values relative for all the variants considered.

Table 4. The results of the analysis of the ANP for the defined decision problem

	Target	Normalized	Raw
Variant I	**1.000000**	0.444702	0.138158
Variant II	**0.796717**	0.354302	0.110073
Variant III	**0.451980**	0.200996	0.062444

The results of the calculations indicate that the most compromise way of energy supply to the analyzed building with the use of a photovoltaic system is the realization of Variant I.

The highest priority (Tab. 3) among the main criteria has the economic criterion (0.12087), followed by: environmental criterion (0.09849), the energy criterion (0.05356) and technical criterion (0.03775). The most important sub-criteria are capital expenditures (0.07051), the amount of "produced" electricity (0.06025) and the rate of return of embedded energy - EPBT (0.0486). Sub-criteria with the least significance are: degree of existing construction load (0.00593), realization difficulties (0.00941) and the efficiency of conversion of the module (0.01501).

Table 3. Date from the super matrix limits defined decision problem

Compromise solutions	Priority
Variant I	**0.13816**
Variant II	0.11007
Variant III	0.06244

Main criteria	Priority
Economical criterion	**0.12087**
Energetic criterion	0.05356
Environmental criterion	0.09849
Technical criterion	0.03775

Sub-criteria	Priority
Realization difficulties	0.00941
Degree of existing construction load	0.00593
Architectural and construction conditions	0.02241
Investment cost	**0.07051**
Discounted payback period (DPP)	0.03022
Net present value (NPV)	0.03022
Amount of "produced" electricity	**0.06025**
PV peak electric power for STC	0.03207
Efficiency of conversion of the module	0.01501
Rate of return of embedded energy (EPBT)	**0.04860**
Reduction of CO_2 emission	0.02342
Reduction of EP consumption	0.03061

2.4 Discussion of calculations results

The results of multi-criteria analysis leads to the following conclusions:

• variants I and II demonstrate the best economic performances over the long time period taken into consideration (25 years),

• the investment cost and the amount of electricity produced in the PV panels are determinants of a favourable outcome for those variants,

• the sub-criterion related to the amount of solar energy converted to electricity directly affects all environmental sub-criteria: EPBT, reduction of CO2 emission and reduction of EP consumption,

• the Sub-criterion for PV peak electric power affect the investment cost and the amount of converted energy,

• variant III based on transparent PV panels has the lowest rank and should be excluded from further analysis.

3 Summary

The choice of the electricity supply system for the public buildings with the use of PV installations should be treated as a decision problem due to its complexity. It can be solved using multi-criteria aid methods and this paper presents a new approach to the conceptual design phase of such systems. The proposed multi-criteria analysis performed using the method of analytical network process (ANP) and the calculation program Super Decisions © has proven to be an effective tool for energy planning, leading to indications for the recommended variant of PV panels based electricity supply system for public buildings. The proposed approach has been employed in the context of a case study analysis – the choice of PV panels based electricity supply system for an educational buildings located in the city of Poznań, Poland. The technical, economic, energy and environmental criterion are the four main decision criteria that have been identified as most relevant for the defined decision problem. The proposed set of main criteria and sub-criteria has an open character and can be subject to changes in the process of modelling the preferences of the decision maker.

References

1. J. Bernal-Agustin, R. Dufo-Lopez, Renew Energ, **6** (2006)
2. K. P. Bhandari, J.M. Collier, R.J. Ellingson, D.S. Apul, Renew Sust Energ Rev, **47**, 133 (2015)
3. T.L. Saaty, *Encyclopedia of Operations Research and Management Science* (Springer US, 2001)
4. T.L. Saaty, *Decision Making with Dependence and Feedback The Analytic Network Process*, (RWS Publications, 1994)
5. B. Radomski, T.M. Mróz, W. Grządzielski, COW, **47/2**, 47 (2016)
6. W. Grządzielski, T.M. Mróz, B. Radomski, Trans. IFFM (to be published)
7. http://www.superdecisions.com/

Numerical modelling of multi-vane expander operating conditions in ORC system

Józef Rak[1], *Przemysław* Błasiak[1], and *Piotr* Kolasiński[1,*]

[1]Wrocław University of Science and Technology, Department of Thermodynamics, Theory of Machines and Thermal Systems, Wybrzeże Wyspiańskiego 27, 50-370 Wrocław, Poland

Abstract. Multi-vane expanders are positive displacement volumetric machines which are nowadays considered for application in micro-power domestic ORC systems as promising alternative to micro turbines and other volumetric expanders. The multi-vane expander features very simple design, low gas flow capacity, low expansion ratios, an advantageous ratio of the power output to the external dimensions and are insensitive to the negative influence of the gas-liquid mixture expansion. Moreover, the multi-vane expander can be easily hermetically sealed, which is one of the key issues in the ORC system design. A literature review indicates that issues concerning the application of multi-vane expanders in such systems, especially related to operating of multi-vane expander with different low-boiling working fluids, are innovative, not fully scientifically described and have the potential for practical implementation. In this paper the results of numerical investigations on multi-vane expander operating conditions are presented. The analyses were performed on three-dimensional numerical model of the expander in ANSYS CFX software. The numerical model of the expander was validated using the data obtained from the experiment carried out on a lab test-stand. Then a series of computational analysis were performed using expanders' numerical model in order to determine its operating conditions under various flow conditions of different working fluids.

1 Introduction

Organic Rankine Cycle systems (ORCs) are the Clausius-Rankine (CR) vapour power plants adopting low boiling working fluids. Its operation principle is the same as classic CR systems, however the system design is different. ORCs utilize smaller heat exchangers instead of steam boiler and condenser. Also, a gas turbine or volumetric expander is applied instead of steam turbine. Compared to CR systems, ORCs feature smaller overall dimensions and system power. ORC systems provide the opportunity for heat harvesting from renewable and waste sources featuring low (40-250 °C) and medium temperature (250-500 °C) of the carrier (such as e.g. solar, geothermal or industrial waste heat) and convert this energy into electricity, useful heat and cool. Research and development works

[*] Corresponding author: piotr.kolasinski@pwr.edu.pl

on ORC systems are currently proceeding in a number of R&D and scientific institutions around the world. As the diversification of the energy supply is nowadays of great interest especially the design of small- and micro-power ORC systems is one of the most actively developed research topics. Despite numerous studies, reliable and safe small- and micro-power ORC systems, possible for application in e.g. domestic applications, are still not fully developed.

A literature review [1, 2] shows that the ORC systems can be classified according to different criteria. By temperature of the heat source, they can be classified into systems powered by high-, medium-, and low temperature sources. By power, ORCs can be classified into systems featuring high power (500 kW or more), medium power (100-500 kW), small power (10-100 kW) and micro power (0.5-10 kW). By the type of the applied expander, ORCs can be classified into systems adopting turbines or systems adopting volumetric expanders.

Small- and micro-power ORC systems dedicated to domestic use by the prosumers utilize heat sources featuring low thermal power and low carrier temperature. Temperature and output characteristics of these sources are often floating, which translates into adverse and changing conditions of the heat supply, and negatively influences the ORC system operation. In small- and micro-power ORCs mainly volumetric expanders are applied (e.g. scroll expanders). However, there are also attempts to adopt micro turbines into these systems. Domestic ORC systems are still under research and development, there are no commercially viable solutions currently available and the majority of these systems are at the level of lab-prototypes. By the technical configuration, these systems can be classified into power plants and CHPs (Combined Heat and Power). Ongoing research and development works on small- and micro-power ORCs are focused on the selection of working fluid, heat exchangers, pumps and expanders design. Small- and micro-power ORC systems designed for domestic use by prosumers should be cheap and reliable. The main part of the ORC system cost is the price of heat exchangers and the expander. The micro turbine design requires high precision workmanship that directly translates into the high price of such machines. In order to provide optimum operating conditions and high efficiency of the micro turbine, the working medium flow through the machine should be maintained on a high and stable level. Such flow conditions can be obtained only if special construction high-output pumps are applied. In small and micro-power domestic ORC systems it is necessary to minimize the power and external dimensions of the expander. Small- and micro-power turbines feature very high rotational speeds, which leads to complications in the design of the clutch connecting the expander and electrical generator.

Compared to micro turbines, volumetric expanders feature a lower range of operating pressure and lower gas flow capacity. In the prototypes of small and micro-power ORC systems, scroll, piston and other types of volumetric expanders are adopted. Scroll expanders are complex in design, their manufacturing requires advanced engineering facilities in order to maintain the high quality of scrolls workmanship, which directly translates into the high price of such expanders. Prototypes of scroll expanders adopted in ORC systems are manufactured in oil-free versions. Compared to the scroll expander, piston expander has a much simpler design, but requires lubrication, valve timing, and as a result of its cyclical operation, vibrations are generated.

In turn, the multi-vane expander design is very simple, which translates into low production costs. This machine features an advantageous ratio of the power output to the external dimensions. The application of special construction materials gives the opportunity for eliminating the need for lubrication. Compared to the other types of volumetric machines and micro turbines, multi-vane machines features a lower gas flow capacity and lower expansion ratios. Moreover, it can be easily hermetically sealed, which is one of the key issues in the ORC system design. Multi-vane machines are insensitive to the

 DOI: 10.1051/e3sconf/20172200142

negative influence of the gas-liquid mixture expansion which, in the case of powering ORC systems by heat sources featuring floating thermal characteristics, is a great advantage. Multi-vane compressors and expanders are applied in many industrial applications, including mining, refrigeration and pneumatic systems. Multi-vane expanders feature power outputs of several hundred W to approximately 5 kW. Maximum gas pressure on the inlet to multi-vane expander reaches approximately 10 bar. These machines feature rotational speeds of 3000 rev/min.

The above described features of multi-vane expanders indicate that these machines may be a promising alternative to micro turbines and other volumetric expanders applied in small- and micro-power CHP ORC prototypes designed for domestic generation of electricity and useful heat by prosumers. The literature review shows issues that are insufficiently recognized and scientifically described related to: operating of multi-vane expander with different low-boiling working fluids, internal leakages of working fluid in expanders' working chambers, design of the inlet and outlet ports, optimization of the expander geometry, lowering the expanders' internal friction and a gas-liquid mixture expansion. The solution of the raised problems requires comprehensive study and experimental analysis. Thus, the authors decided to carry out the experimental and numerical study on these issues. In the following part of this paper the description of the experimental test-stand, the multi-vane expander and the experimental results together with the description and results of numerical analysis are presented. The numerical investigations showed the insight in the flow phenomena in the expanders' working chambers and indicated the differences in these phenomena for analysed working fluids. Moreover, obtained results showed the expanders' design imperfections, i.e. sharp edges of the inlet and outlet ports causing working medium turbulence and internal leakages, reducing expanders' output power. The numerical model was validated with experimental data what proved its suitability and allowed to use the model in testing modifications to the machines design or finding the optimum working parameters and working fluid. In the paper a numerically obtained results of the multi-vane expander operation for various working media is presented. All refrigerants were modelled for a range of pressure and temperature parameters to find the optimal working point for the Organic Rankine Cycle.

2 Description of the experimental test-stand, the multi-vane expander and the experiment results

The experimental investigations were carried out using a prototype of domestic combined heat and power (CHP) ORC system featuring the thermal power of 18 kW and electric power of 300 W. A general view of the test-stand is presented in Fig. 1.

The main test-stand components are: a plate evaporator (1), a plate condenser (2), a working fluid pump (3), a reservoir of a working fluid (4), and a micro-vane expander (5). The working fluid is R123. The heat source for the test-stand is hot water (heated in a gas boiler) which temperature can be regulated in the range of 40—95 °C. The condenser is cooled by the cold water in the open cycle.

The working principle of the test-stand can be described as follows. At first, the hot water is pumped from the gas boiler to the evaporator and circulates between the plates. Then, the liquid R123 is pumped from the reservoir to the evaporator where it is evaporated. Obtained pressurized vapor drives the multi-vane expander. After the expansion, the vapor is cooled and liquefied in the condenser by the cold water circulating between the plates. At last, liquefied working fluid flows to the reservoir and the cycle is therefore complete.

The applied expander is an air motor which was adapted for low-boiling working fluid by application of stainless steel bearings, PTFE seals and graphite vanes. The expander

DOI: 10.1051/e3sconf/20172200142

features the maximal power of 300 W. The main expander components are: the cylinder, the rotor, vanes and two end covers. The gas is supplied to the expanders' working chambers through the inlet port and flows out of the expander through the outlet port. Fig. 2 shows a general view of the expander. Detailed description of the expander is presented in [3].

Fig. 1. A general view of the test-stand.
1 – plate evaporator, 2 – plate condenser, 3 – working fluid pump, 4 – working fluid reservoir, 5 – multi-vane expander.

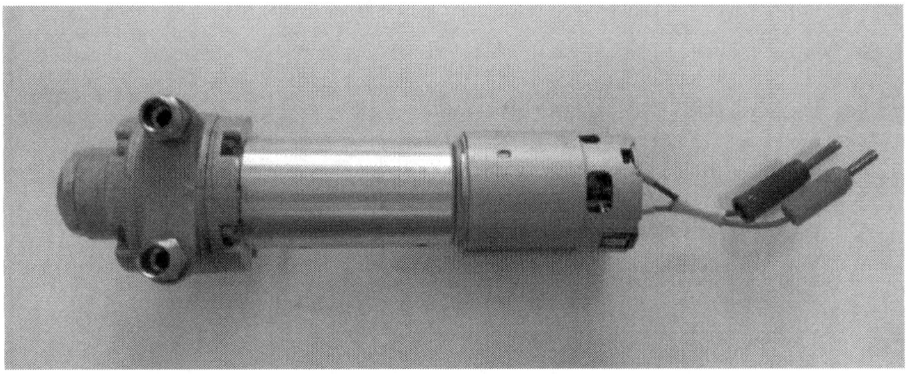

Fig. 2. A general view of the expander.

In order to obtain the experimental data needed for validation of expanders' numerical model, a series of experiments with varied measurement conditions were carried out using the test-stand.

The experimental series were performed for the values of the heat source temperature ranging between 55 °C and 85 °C and for the working fluid pressure ranging between 2.0 and 5.2 bar. The experimental results showed that the expander indicated work varies in the range of 0.96—4.18 kJ/kg while its internal efficiency varies in the range of 17.2—58.3 %

 DOI: 10.1051/e3sconf/20172200142

depending on the experimental conditions. The more detailed description of the experiment and experimental results is presented in [3]. The results from the experiment were in a good agreement with the numerical predictions so the model was validated for a given range of temperature and pressure parameters.

3 Numerical modelling

3.1 Description of the numerical model

A transient operation of the multi-vane expander was modeled numerically using transient boundary conditions. The model complies the rotational motion of the rotor and the vanes as well as the vanes radial position change during the process. This allows for an in-depth calculation of the working medium inside a working chamber including the internal leakages. The inlet and the outlet pipes remain stationary and they relate to the rotors domain through the interface (see Figure 3).

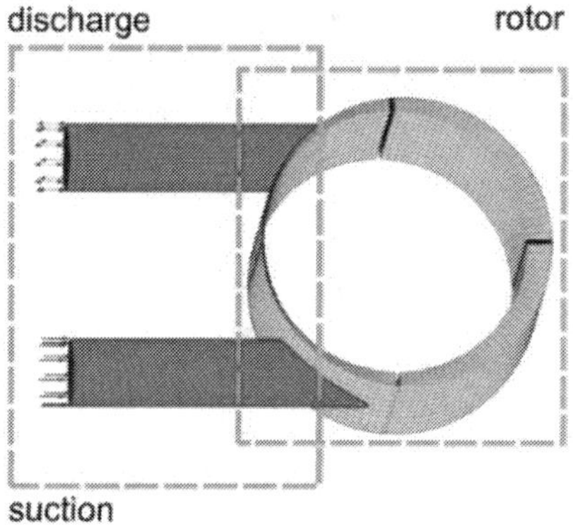

Fig. 3. A numerical mesh of the numerical model.

The working fluid flow was modelled as a compressible, turbulent and single-phase ideal gas. The specific heat capacity of each tested medium was set as a function of temperature. The wall boundary was assumed adiabatic due to the short time spent by a portion of the fluid inside the machine (see Table 1).

Tab. 1. Boundary conditions and model assumptions

Parameter	Value	Remarks
Inlet temperature	40-90°C	with step of 10 degrees
Inlet pressure	Saturation pressure at the inlet temperature	-
Condenser Temperature	20°C	-
Condenser Pressure	Saturation pressure at 20°C	depending on the refrigerant
Time step	5.5×10^{-5} s	1 degree of revolution
Total time	0.1 s	5 full revolutions
Turbulence model	*k*-epsilon	Standard model
Wall heat transfer	Adiabatic	-
Vane-to-wall clearance	40 µm	constant

DOI: 10.1051/e3sconf/20172200142

Parameters at the inlet to the expander were selected to be temperature and pressure. For each working fluids the inlet temperature was varied in a range of 40-90°C with step of 10 degrees. The inlet pressure was taken as a saturation pressure of the working fluid at the inlet temperature.

The rotor rotational speed was assumed constant at a nominal value of 3000 rpm for all considered cases. The temperature at the outlet of the discharge socket was assumed 20°C for all the cases and the pressure was set to the saturation pressure given by the medium properties.

4 Results and discussion

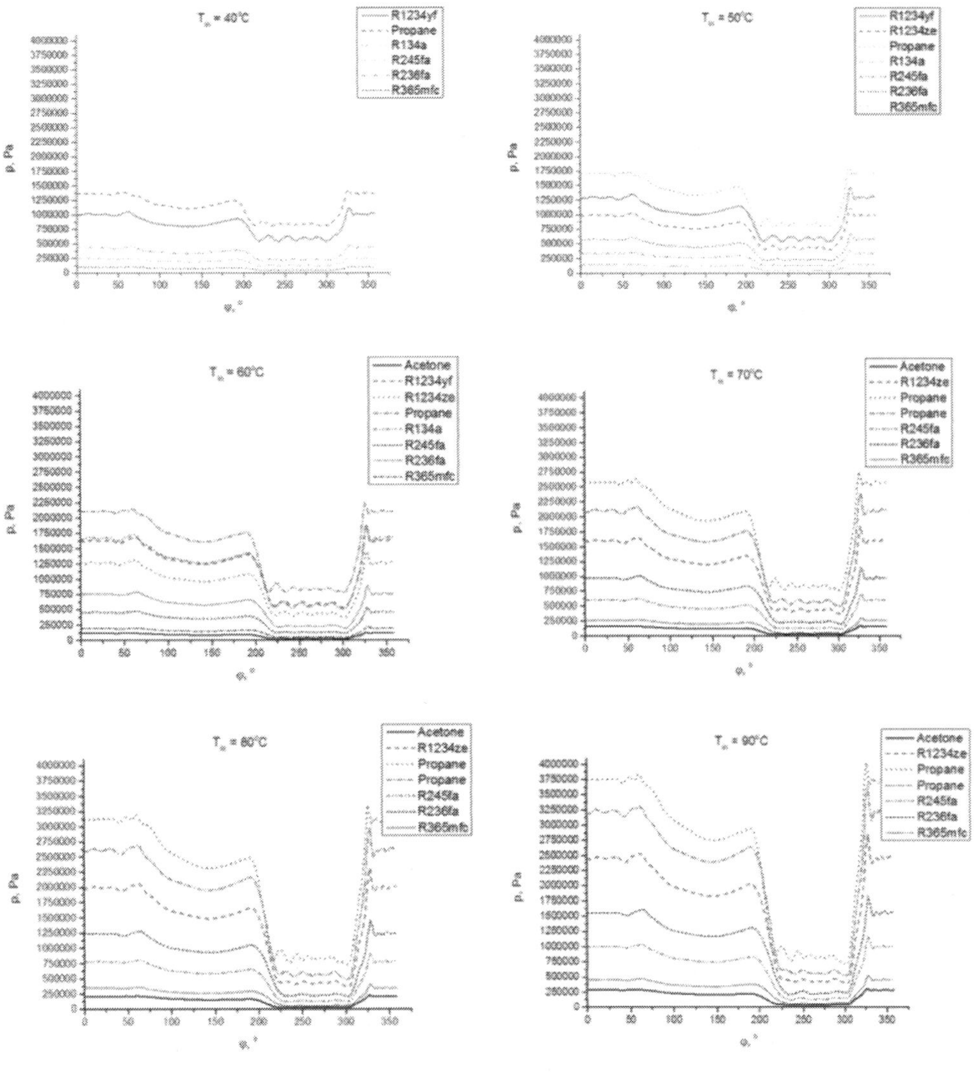

Fig. 4. Pressure distribution against rotational angle for different working fluids and inlet temperatures.

In Fig. 4 pressure distribution in a working chamber versus rotational angle, for different working fluids and different inlet temperatures, is presented. With increasing the

DOI: 10.1051/e3sconf/20172200142

inlet temperature the inlet pressure (i.e. saturation pressure) increases and the area under the curve in the p-φ diagram increases. Moreover the saturation pressure differs for each working fluid. The highest inlet pressure occurred for propane and the lowest for acetone. Presented curves show a standard variation except the strange recompression effect which is visible in the vicinity of φ = 200°. This behaviour stems from the improper construction of the expander and it was described in detail in [3]. This effect is more visible for working fluids with a higher saturation pressure and it results in decrease of efficiency of the expander.

In order to better understand the influence of a working fluid on the expander's performance, indicated work was calculated as a difference between the inlet and the outlet enthalpy. Fig. 5 shows the results of indicated work versus pressure ratio for working fluids used. Almost all curves (except the one for acetone) characterize of a similar shape, i.e. they first steeply increase achieving the maximum in the range of σ = 3-4 (only propane for σ ≈ 2) and then gradually decrease with increasing σ. Different behaviour presents the curve for acetone where the minimum is visible for σ ≈ 6,5. From these results one can infer that optimal pressure ratio for the rotary vane expander is located in the range of σ = 3-4 and the best performance can be accomplished with propane for σ ≈ 2. However, more detailed investigation of the results is shown in Fig. 6 where the efficiency of the expander versus pressure ratio is plotted. It is obvious now that using propane as a working fluid in the expander analysed is very ineffective and features very low efficiency of the order of 5%. However, Fig. 6 also confirms that the optimal pressure ratio is located in the range of σ = 3-4 for which the efficiency is in the range of 12-35%. The highest efficiency can be achieved for the acetone, however the pressure ratio has to be increased to about σ ≈ 5.

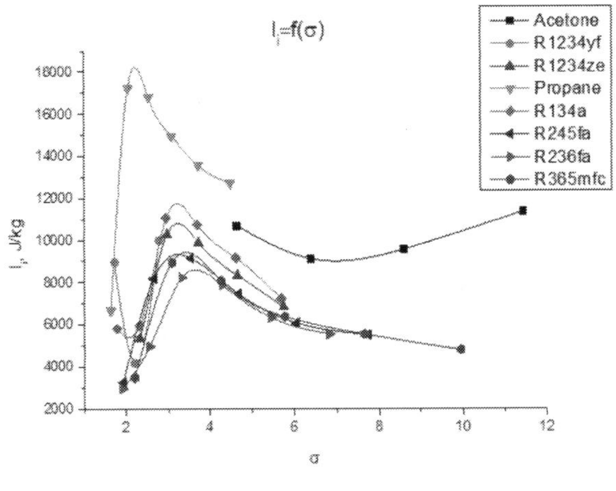

Fig. 5. Indicated work against pressure ratio for different working fluids.

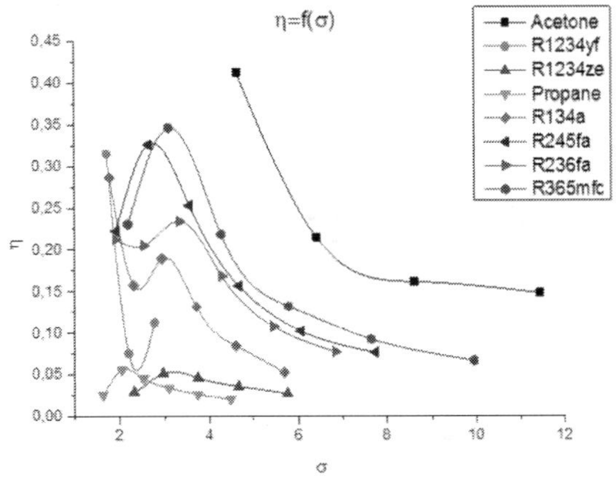

Fig. 6. Efficiency against pressure ratio for different working fluids.

5 Summary and conclusions

In the paper the numerical investigation of rotary vane expander was presented. The experimental survey was carried out using a prototype of domestic ORC system equipped with small multi-vane expander featuring a power of 300 W and utilizing R123 as a working fluid. The main aim of experimental investigations was gathering the input data needed for creating the numerical model of the expander and analysis of its operating conditions with a variety of working fluids. Additionally, during the experiment the operation of the expander was kept under observation. It was found that the expander was operating continuously and without serious problems. The experimental results confirmed that multi-vane expanders are suitable to small power domestic ORC systems and proved the usability of the numerical model.

The numerical survey on the expander operation was carried out for different working fluids i.e. R1234yf, R1234ze, R134a, R245fa, R236fa, R365mfc, Acetone and Propane. During the numerical simulations the value of temperature on the expanders' inlet was varied similarly as during the experiments. The results of numerical analysis showed that for the expander analysed and working fluids used, the optimal pressure ratio is in the range of $\sigma = 3$-4. Corresponding efficiency is in the range of 12-35% depending on working fluid. This results confirms the findings of survey presented in [4], i.e. the multi-vane expanders feature the highest power and efficiency within this certain range of pressure ratios. Thus, increasing the pressure ratio is not profitable and results in decreasing the power and efficiency of the expander. Additionally it can be concluded that the domestic ORC systems adopting multi-vane expanders should be filled with working fluids featuring moderate pressures at the considered level of the heat source temperatures (40—90 °C).

References

1. T.C. Hung, T.Y. Shai, S.K. Wang, Energy **22**, 661-667 (1997)
2. P. Kolasiński, Energies **8**, 3351-3369 (2015)
3. P. Kolasiński, P. Błasiak, J. Rak, Energies **9**, 606 (2016)
4. Z. Gnutek, P. Kolasiński, J Eng Gas Turb Power Journal **135**, 061901 (2013)

Recovery of useful chemicals from palm oil mill wastewater

Yuangsawad Ratanaporn[1], *Na-Ranong* Duangkamol[2,*], *Tago* Teruoki[3] and *Masuda* Takao[4]

[1]Rajamangala University of Technology Krungthep, Faculty of Engineering, 2, Nanglinchee Rd., Thunmahamek, Satorn, Bangkok 10120, Thailand
[2]King Mongkut's Institute of Technology Ladkrabang, Faculty of Engineering, 1, Chalongkrung 1, Ladkrabang, Bangkok 10520, Thailand
[3]Tokyo Institute of Technology, Graduate School of Science and Engineering, 2-12-1, Ookayama, Meguro-ku, Tokyo 152-8552, Japan
[4]Hokkaido University, Faculty of Engineering, N13W8, Kita-ku, Sapporo, Hokkaido 060-8628, Japan

Abstract. A two-step process consisting of pyrolysis of dried sludge and catalytic upgrading of pyrolysed liquid was proposed. Wastewater from a palm oil mill was separated to solid cake and liquid by filtration. The solid cake was dried and pyrolysed at 773 K. Liquid product obtained from the pyrolysis had two immiscible aqueous and oil phases (PL-A and PL-O). Identification of chemicals in PL-A and PL-O indicated that both phases contained various chemicals with unsaturated bonds, such as carboxylic acids and alcohols, however, most of the chemicals could not be identified. Catalytic upgrading of PL-A and PL-O over $ZrO_2{\cdot}FeO_x$ were separately performed using a fixed bed reactor at various conditions, T = 513-723 K and mass of catalyst to feed rate = 0.25-10 h. The main components in the liquid products of PL-A upgrading were methanol and acetone whereas they were acetone and phenol in the case of PL-O upgrading. More than 15% of carbon in raw material was deposited on the catalyst. To reduce the carbon deposition, the used catalyst was treated with air at 823 K. This simple treatment could reasonably regenerate the catalyst only for the case of PL-A catalytic upgrading.

1 Introduction

In South East Asia countries, palm oil and by-products from palm oil refinery process are mainly used as a feedstock of biodiesel production [1]. Due to high demand of biodiesel as alternative energy, scale of palm oil production has been remarkably increased in the recent years. To produce 1 ton of crude palm oil, about 5-7.50 tons of water is consumed and more than 2.5 tons of wastewater is generated and must be treated before discharging to the environment [2]. Since this wastewater is a brown liquid containing large amount of suspended solids and has high COD and BOD values, a large system is usually installed for biological treatment.

[*] Corresponding author: dnaranong@hotmail.com, duangkamol.na@kmitl.ac.th

Wu T. Y. et al. [3] reported that the suspended solid in wastewater generated from palm oil mills mainly consisted of macromolecules of several organic compounds, such as protein (12.75%), carbohydrate (29.55) and lipid (10.21%) [3]. Pyrolysis is usually applied to decompose macromolecular compounds to obtain mixtures of small molecules in liquid and gas. The pyrolysed liquid may be used, without further treatment, as fuel in an incinerator. On the other hand, it may be catalytically upgraded to the more stable liquid containing less groups of chemicals and used as an alternative chemical feedstock [4-5]. It should be possible to recover useful chemicals from the palm oil mill wastewater by application of a series of pyrolysis and catalytic upgrading steps.

In this study, a two-step process consisting of the pyrolysis of sludge separated from palm oil mill wastewater and catalytic upgrading of the pyrolysed liquid was proposed and its performance, mainly focused at the catalytic upgrading of the pyrolysed liquid, was investigated. Since iron based catalysts were successfully applied to recover ketones and phenol from palm shell-derived oil and domestic sludge derived black water and the addition of zirconia could improve the stability of this catalyst [6-8], zirconia-iron oxide catalyst ($ZrO_2 \cdot FeO_x$) was selected as a candidate for the proposed process. Effects of the operating parameters of the catalytic upgrading, reaction temperature and the ratio of catalyst weight to feed flow rate (W/F), on the product distribution were investigated. In addition, the reusability of the spent catalyst was also examined.

2 Experimental

2.1 Preparation and characterization of pyrolysed liquid

Figure 1 shows schematic diagram of overall process for recovering useful chemicals from palm oil mill wastewater. Wastewater was obtained from a palm oil mill located in Chonburi Province, Thailand. It was filtrated through a filter paper in order to obtain solid cake and liquid filtrate. The obtained solid cake was dried at 373 K overnight and then pyrolysed in a quartz reactor at 773 K in a flow-through nitrogen atmosphere. Condensable liquid was collected by using a series of three traps immersed in an ice bath. After mixing the liquid trapped in each ice trap, the mixture was settled until it was clearly separated to two layers, aqueous phase at the lower layer and oil phase at upper layer. For further discussion, the pyrolysed liquids (PL) in the aqueous phase and in the oil phase were named as "PL-A" and "PL-O", respectively. The liquid filtrate obtained from the filtration, name as F-water, was used instead of pure water in the reaction test in section 2.3.

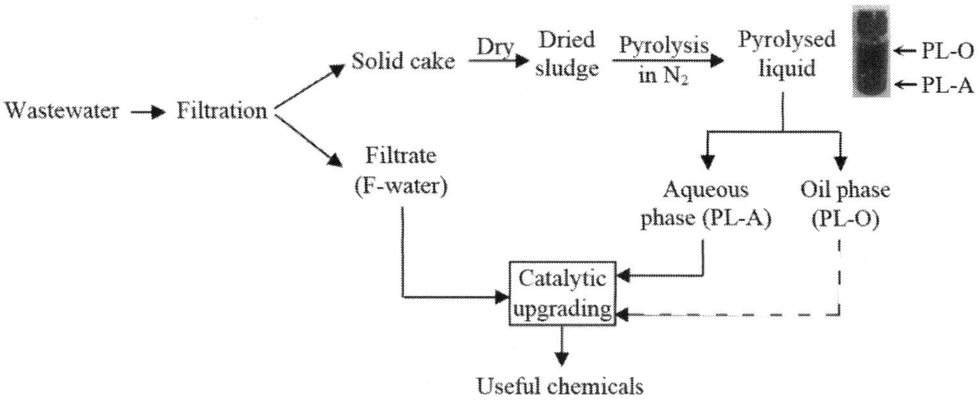

Fig. 1. FTIR functional group of (a) wastewater (b) PL-A and (c) PL-O

Functional groups of the chemicals in the wastewater, PL-A and PL-O was observed using a Fourier transform infrared spectrometer (FT-IR). Identification of the chemicals in both pyrolysed oil, PL-A and PL-O, was performed using a gas chromatograph equipped with mass spectrometer (GC-MS). Quantifications of the components in PL-A and PL-O were performed using a gas chromatograph equipped with a flame ionization detector (GC-FID).

2.2 Preparation and characterization of zirconia-iron oxide catalyst

Zirconia-iron oxide catalyst was prepared by co-precipitation technique described in our previous work [8]. $ZrO(NO_3)_2{\cdot}8H_2O$ and $Fe(NO_3)_3{\cdot}8H_2O$ purchased from Kanto Chemicals (Japan) were used as precursors. NH_3 solution (Merck) used as precipitating reagent and was slowly dropped into the continuously stirred solution of the precursors until pH of the solution reached at 7. The precipitate was separated from the mixture using vacuum filtration. The obtained solid was dried at 383 K overnight and then calcined at 773 K for 4 h. The obtained catalyst, $ZrO_2{\cdot}FeO_x$, was ground and sieved to obtain the particle sizes in the range of 300 – 850 μm. Crystalline structure of the catalyst was observed using X-ray diffractometer (XRD; Bruker AXS).

2.3 Catalytic upgrading and reusability test

The reaction test of each pyrolysed liquid was performed using a fixed bed reactor operated at 1 atm for 1 h. The pyrolysed liquid and F-water were separately fed to the reactor using a syringe pump. Nitrogen was used as a carrier gas. The outlet of the reactor was connected to a series of two ice traps to collect condensable liquid product. Incondensable product was collected using a gas bag. As for PL-A catalytic upgrading test, PL-A and F-water were fed to the reactor with the mass ratio of PL-A to F-water = 1:8. On the other hand, in the case of PL-O catalytic upgrading test, PL-O was mixed with benzene (PL-O to benzene = 1:3 by wt) before being fed to the reactor, the mass ratio of PL-O to F-water was set at 1:1. The temperature of the catalytic upgrading was varied from 513-723 K and the ratio of catalyst weight to feed flow rate (W/F) was varied from 0.25 to 10 h. Quantitative analysis of hydrocarbons in the liquid and gas obtained from the catalytic upgrading was performed using GC-FID. Quantification of inorganic gases, CO and CO_2, was performed using a gas chromatograph equipped with a thermal conductivity detector (GC-TCD).

For the test of catalyst reusability, after the reaction test at T = 593 K, W/F = 2 h using a fresh catalyst was performed for 1 h, air was fed to the reactor at 823 K for 2 h in order to regenerate the catalyst. The products obtained from the reaction test using at the fresh and the regenerated catalysts were analysed using GC-FID and GC-TCD.

3 Results and Discussions

3.1 Characterization of wastewater and product from pyrolysis

Wastewater was separated to solid cake and liquid by filtration. Gravimetrical analysis showed that solid cake (18.75 %wt) and liquid filtrate (81.25 %wt) was obtained from the filtration. Pyrolysis of dried solid cake at 773 K produced char (34.5 %wt), pyrolised liquid (44.8 %wt) and gas (20.7 %wt). The pyrolysed oil was then separated to water-soluble fraction, PL-A (66.74 %wt) and water-insoluble fraction, PL-O (33.26 %wt). Quantification of carbon content using GC-FID revealed that PL-O had carbon content of 72.78 %C while PL-A had only 16.04 % of C in the solution.

 DOI: 10.1051/e3sconf/20172200143

Figure 2 shows FTIR spectra of wastewater, PL-A and PL-O. In spectrum of the wastewater, two peaks were observed. Broad peak at 3300-3600 cm^{-1} was assigned to O-H stretching vibration, indicating water, alcohols and carboxylic acids [9]. Another strong peak around 1500 cm^{-1} was assigned to C=C stretching in aromatics [10]. In addition, the identification using GC-MS showed the presence of pyridine, phenol and derivatives of phenol. As for PL-A, the spectrum also shows broad peak at 3300-3600 cm^{-1} and it was assigned to the O-H stretching vibration of water, alcohols and carboxylic acids. The small peaks in the range of 2800-3000 cm^{-1} was assigned to C-H stretching in saturated hydrocarbons [10]. The peak at 1611 cm^{-1} was assigned to C=O stretching of carboxylic acids, ketones and aldehydes [11]. This result agreed well with GC-MS analysis, where acetic acid, propanoic acid, butanoic acid, hydroxyl acetone and pyrrolidinone were detected. Spectrum of PL-O shows peaks at 2800-3000, 1611, 1400 and 650 cm^{-1}. They were assigned to C-H stretching in saturated hydrocarbons, C=O stretching of carboxylic acids, ketones and aldehydes, vibrations in CH_2 and CH_3 groups [11] and C-H bending in aromatics, respectively. These chemicals were also detected by GC-MS. Compositions of PL-A and PL-O quantified by GC-FID were summarised in Table 1. In PL-A, only methanol, acetic acid and acetone showed large peak which were calculated easily. Since various peaks with very small area appeared in the chromatogram of PL-A, these were sum as "others-I". Similarly, there are various small peaks appeared in the chromatogram of PL-O and these peaks were sum as "others-II". It should be noted that the components presented in others-I and others-II were different. Due to the difference of chemical and physical properties of PL-A and PL-O, catalytic upgrading tests were separately performed.

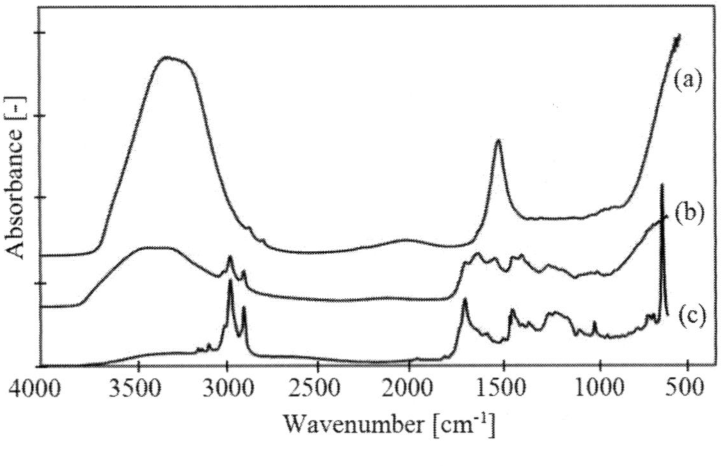

Fig. 2. FTIR spectra of (a) wastewater, (b) PL-A and (c) PL-O

Table 1. Main components in PL-A and PL-O.

PL-A		PL-O	
Component	Composition (%C)	Component	Composition (%C)
methanol	6.32	acetone	0.39
acetic acid	27.24	phenol	6.60
acetone	2.40	m-cresol	1.85
others-I	64.04	others-II	91.16

3.2 Catalytic upgrading

3.2.1 Catalytic upgrading of PL-A

Figure 3 shows effect of W/F on composition of the product obtained from the catalytic upgrading of PL-A at various temperatures. The catalytic upgrading in this temperature range could reduce the amount of others in PL-A and others were completely removed from PL-A at high W/F, for example at W/F = 10 h, T = 583 K. Effect of W/F on the amount of acetic acid was similar to the effect of W/F on the amount of others, the amount of acetic acid monotonically decreased when W/F was increased and acetic acid was completely removed at high enough W/F and T. On the contrary, the amount of acetone increased until it reached the maximum and then it decreased with further increase of W/F. Only the amount of methanol was insignificantly changed due to the catalytic upgrading. At all temperatures, solid residue and gas were the main products obtained from the reaction of PL-A. It should be noted that the quantification analysis revealed that CO_2 was the main composition in the gas product. The dependences of acetic acid, acetone and gas on W/F implied that acetone is produced from the decomposition of acetic acid and further goes to complete oxidation at high W/F and T. This behaviour was supported by the mechanism reported in the previous work, acetic acid can be directly converted to acetone and CO_2 over ZrO_2-FeO_x [13]. The highest yield of acetone was obtained at W/F = 2 h and T = 593 K. To obtain the highest purity of acetone in the liquid product (> 95 %), the catalytic upgrading at W/F = 6 h and T = 593 K is suggested. Since large amount of carbon deposited on the catalyst, regeneration of the used catalyst was further investigated in section 3.3.

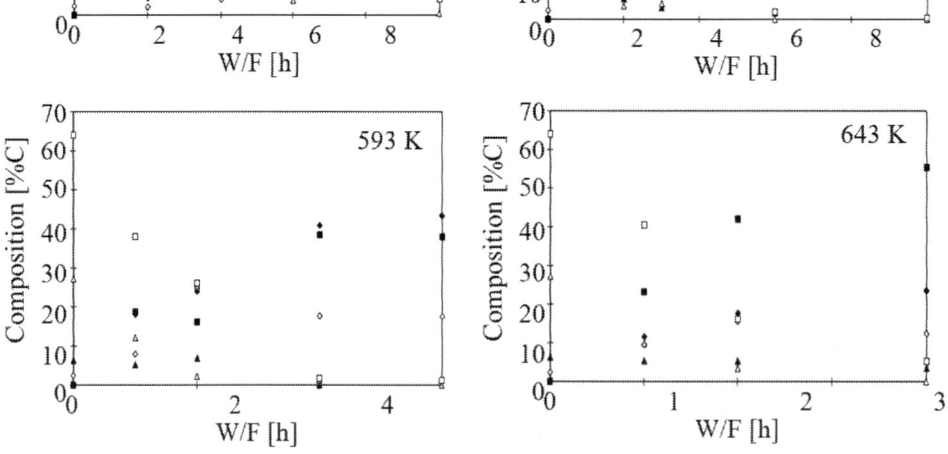

Fig. 3. Effect of W/F on product composition obtained from catalytic upgrading of PL-A (▲: methanol, ◇: acetone, Δ: acetic acid, □: others in liquid, ◆: gas, ■: solid residue)

 DOI: 10.1051/e3sconf/20172200143

3.2.2 Catalytic upgrading of PL-O

The solid cake obtained from the filtration of palm oil mill wastewater should consist of small debris of fibre and shell. As reported in the previous work [6], acetic acid, methanol, acetone, phenol and cresols were successfully recovered from palm shell derived pyrolysed oil. Although almost all of chemicals in PL-O were not able to identify and the amount of these unidentified compounds was as high as 91.16 %, the chemicals should be originated from lignin, hemicellulose and cellulose, which are similar to the case of palm shell derived pyrolysed oil. Therefore, the performance of $ZrO_2{\cdot}FeO_x$ on PL-O upgrading was focused on the formation of acetic acid, methanol, acetone, phenol and cresols.

Figure 4 shows that the amounts of acetone, phenol and m-cresol were slightly affected by the catalytic upgrading. The amounts of acetone and phenol reached their maximum and then decreased with W/F. As temperature was increased, the maximum yields of acetone and phenol shifted to the lower W/F. Although the highest yield of phenol (19 %C) was obtained at W/F = 0.25 h, T = 723 K, the amount of others in liquid product was high. As for acetic acid and methanol, they were not detected in the products at all conditions. The amounts of others, solid residue and gas were significantly affected by the catalytic upgrading. The amount of solid residue increased with the increase of W/F. At the same W/F, the amount of solid residue at the lower temperature was less than the amount of solid residue obtained at the higher temperature. For example, at W/F = 6 h, the amount of solid residue reduced from 45 %C to 38 %C when temperature was increased from 593 to 693 K. This effect was considered as a result of rapid rate of CO_2 formation from the oxidation of solid residue at high temperature.

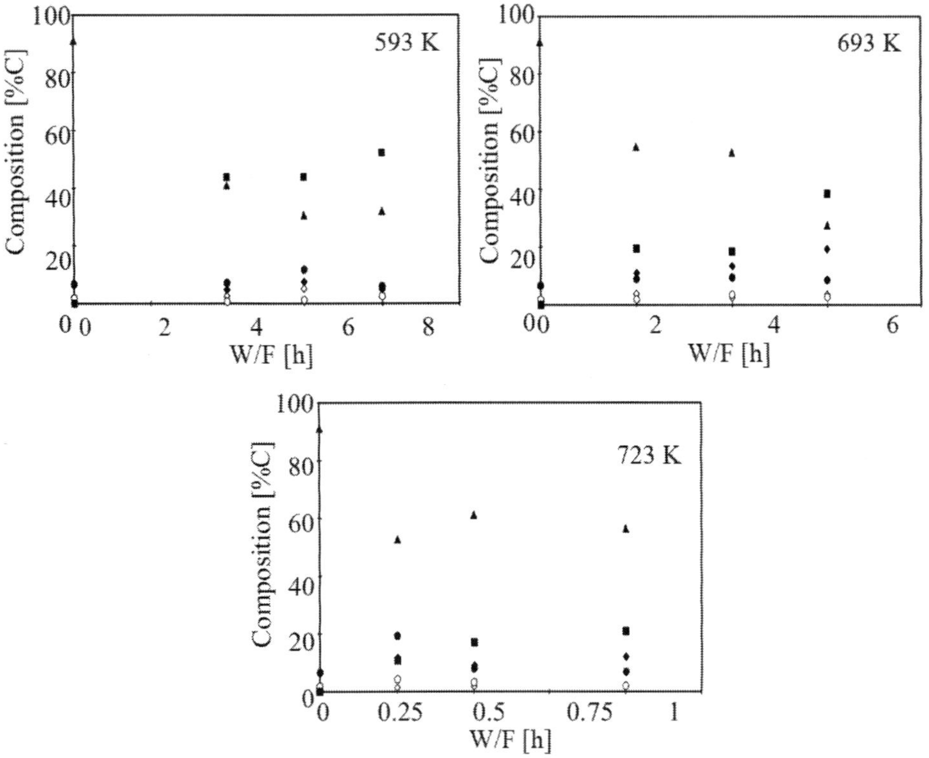

Fig. 4. Effect of W/F on product composition obtained from catalytic upgrading of PL-O (○: m-cresol, ●: phenol, ◇: acetone, ▲: others in liquid, ◆: gas, ■: solid residue)

3.2.3 Conditions for upgrading of PL-A and PL-O

Based on the experimental results in sections 3.2.1 and 3.2.2, effects of T and W/F on product distribution of PL-A and PL-O upgrading were analyzed. The conditions those provided the highest total amount of the target compounds were T = 593 K, W/F = 2h for PL-A and T = 593 K, W/F = 5h for PL-O. The predicted compositions of the products obtained from PL-A and PL-O upgrading were summarized in Table 2.

Table 2. The main product obtained from the optimum condition of the reaction.

Component	Composition (%C)	
	PL-A	PL-O
Methanol	6.88	-
Acetone	24.71	5.15
Acetic acid	2.27	-
Phenol	-	11.74
m-cresol	-	1.23
others in liquid	26.11	30.56
gas	23.93	7.44
residue	16.10	43.88

3.3 Catalyst regeneration

Figure 5 shows XRD patterns of the fresh $ZrO_2{\cdot}FeO_x$ and the $ZrO_2{\cdot}FeO_x$ after using for catalytic upgrading of PL-A. In both spectra, pattern of hematite structure was observed. These results implied that FeO_x in the catalyst had hematite structure and this crystalline structure was stable under the reaction condition. In order to study the reusability of catalyst, the used $ZrO_2{\cdot}FeO_x$ was regenerated by treating with air 823 K for 2 h. Products obtained from the catalytic upgrading of PL-A using the fresh and the first cycle regenerated catalysts were compared in Figure 6. The yield of acetone obtained from the reaction over the regenerated catalyst was slightly lower than that of the fresh catalyst. The yield of acetone decreased from 25%C to 20 %C whereas the amount of methanol remains constant at about 10%C. The reaction over regenerated $ZrO_2{\cdot}FeO_x$ produce more solid residue, comparing the reaction over the fresh catalyst.

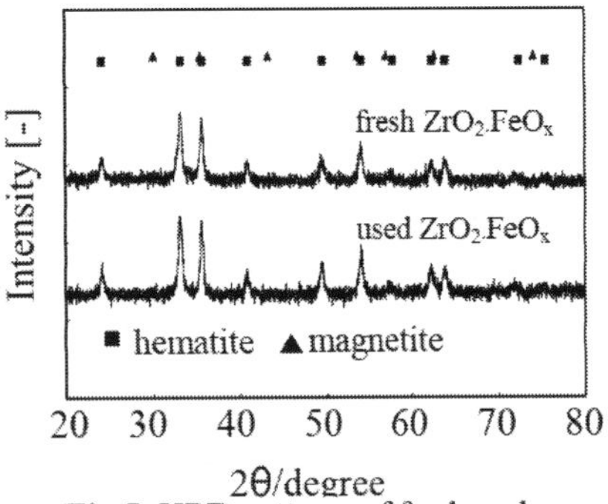

Fig.5. XRD patterns of fresh and used catalysts

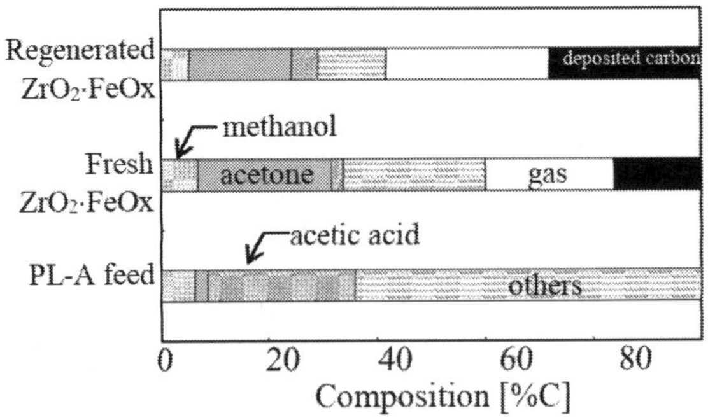

Fig.6. Product distributions obtained from catalytic upgrading of PL-A with fresh and regenerated catalysts (T = 593 K, W/F = 2 h)

4 Conclusions

A two-step process, consisting of pyrolysis of dried sludge and catalytic upgrading of pyrolysed liquid (PL), was applied to recover methanol, acetone, acetic acid, phenol and m-cresol from palm oil mill wastewater. Since PL obtained from the pyrolysis was easily separated to aqueous and oil phases, the catalytic upgrading of each phase was performed using $ZrO_2 \cdot FeO_x$, separately. Investigation of effect of operating condition (temperature (T) and ratio of catalyst weight to feed flow rate (W/F)) on product distribution revealed that the highest yields of target compounds were obtained at T = 593 K, W/F = 2h for PL-A and at T = 593 K, W/F = 5h for PL-O. At the optimum condition of PL-A upgrading, the yields of methanol, acetone and acetic acid were 6.88, 24.71 and 2.27 %C, respectively. On the other hand, acetone (5.15 %C), phenol (11.74 %C) and m-cresol (1.23 %C) were recovered from PL-O at its optimum condition. Although carbon deposition on the catalyst was severe, the activity of the catalyst was reasonably recovered by oxidation with air at 823 K.

Nomenclatures

F mass flow rate of PL-A or PL-O (g/h)
W weight of catalyst (g)

Acknowledgement

Financial supports by Thailand Research Fund and King Mongkut's Institute of Technology Ladkrabang are gratefully acknowledged.

References

1. D. Sheil, A. Casson, E. Meijaard, M. V. Noordwijk, J. Gaskell, J.Sunderland-Groves, K. Wertz, M. Kanninen, *The impacts and opportunities of oil palm in Southeast Asia* (CIFOR, Bogor Barat, 2009)
2. S. A. Muyibi, T. Tajari, M. S. Jami, M. K. Amosa, Adv. Environ. Biol., **8**(3), 590 (2014)
3. T. Y. Wu, A. W. Mohammad, J. Md. Jahim, N. Anuar, Biotechnol. Adv., **27**, 40 (2009)
4. Y. Kim, W. Parker, Bioresour. Technol., **99** , 1409 (2008)
5. S. Yaman, Energ. Convers. Manage., **45**, 651 (2004)
6. D. Na-Ranong, R. Yuangsawad, T. Tago, T. Masuda, Korean J. Chem. Eng., **25**, 426 (2008)
7. T.Masuda, Y. Kondo, M. Miwa, T. Shimotori, S.R. Mukai, K. Hashimoto, M. Takano, S. Kawasaki, S. Yoshida, Chem. Eng. Sci., **56**, 897 (2001)
8. E. Fumoto, Y. Mitzutani, T. Tago, T. Masuda, Appl. Catal. B: Env., 68, 154 (2006)
9. R. O. Arazo, D. A. D. Genuino, M. D. G. de Luna, S. C. Capareda, Sustain. Environ. Res., **27**, 7 (2017)
10. B. B. Krishna, B. Biswas, P. Ohri, J. Kumar, R. Singh, Renew. Energ., **98**, 238 (2016)
11. J. Alvarez, G. Lopez, M. Amutio, M. Artexe, I. Barbarias, A. Arregi, J. Bilbao, M. Olazar, Fuel Process Technol., **149**, 169 (2016)
12. M. B. Valenzuela, C.W. Jones, P. K. Agrawal, Energy Fuels, **20**, 1744 (2006)
13. S. Funai, T. Tago, T. Masuda, Catal. Today, **164**, 443 (2010)

Numerical modelling of series-parallel cooling systems in power plant

Paweł Regucki[1,*], *Marek* Lewkowicz[1] , and *Małgorzata* Kucięba[2]

[1]Faculty of Mechanical and Power Engineering, Wrocław University of Science and Technology, Wybrzeże Wyspiańskiego 27, 50-370 Wrocław, Poland
[2]PGE GiEK, Turów Power Plant Section, Młodych Energetyków 12, 59-916 Bogatynia, Poland

Abstract. The paper presents a mathematical model allowing one to study series-parallel hydraulic systems like, e.g., the cooling system of a power boiler's auxiliary devices or a closed cooling system including condensers and cooling towers. The analytical approach is based on a set of non-linear algebraic equations solved using numerical techniques. As a result of the iterative process, a set of volumetric flow rates of water through all the branches of the investigated hydraulic system is obtained. The calculations indicate the influence of changes in the pipeline's geometrical parameters on the total cooling water flow rate in the analysed installation. Such an approach makes it possible to analyse different variants of the modernization of the studied systems, as well as allowing for the indication of its critical elements. Basing on these results, an investor can choose the optimal variant of the reconstruction of the installation from the economic point of view. As examples of such a calculation, two hydraulic installations are described. One is a boiler auxiliary cooling installation including two screw ash coolers. The other is a closed cooling system consisting of cooling towers and condensers.

1 Introduction

In the current economic situation, the flexible work of a power unit requires a quick reaction to the power demand of the electrical grid. The market situation often forces the boiler to run with maximum thermal output, which also entails the work of hydraulic systems with maximum efficiency. Despite the fact that the daily consumption of fresh water by a professional power plant can reach the level of several thousand cubic meters, there are situations when the amount of water circulated in cooling systems is insufficient. This is usually observed in transient stages or periods when, e.g., the power unit works with variable thermal output or the ambient temperature is very high, making efficient heat exchange in cooling towers difficult. Such situations also often arise from the fact that the efficiency of coal-fired power plants gradually decreases with age [1].

In response, the Environmental Protection Agency (EPA) alerts that power plants are one of the largest consumers of natural water reservoirs in the USA, recommending, in September 2013, new standards for control of carbon dioxide (CO_2) emissions from new

* Corresponding author: pawel.regucki@pwr.edu.pl

power units burning fossil fuels and regulations concerning the protection of water resources. It is also worth mentioning that the deterioration of the overall power unit efficiency results in the greater consumption of fossil fuels and consequently higher CO_2 emission (as well as that of other pollutants) [2]. Keeping in mind that the improvement of heat exchange processes could have a direct influence on the overall efficiency of a power unit, Ryabchikov *et al.* in [3] suggested a number of specific actions, among which the retrofit of cooling water installations is one of the most important.

An investigation of hydraulic installations indicates that most of them are designed to work under nominal parameters which do not take into account the cooling performance reserves [4, 5]. Later, during the long period of their utilization, the gradual deterioration of their operating parameters is observed, resulting from, e.g., changes in the operating characteristics of force pumps or the contamination of pipelines and heat exchanger surfaces by mineral deposits. Other important factors affecting the operating parameters of hydraulic systems are connected with various minor losses caused by installed valves and other fittings. In this situation, either the power unit will not reach its maximum nominal thermal output under the standard operational conditions or the hydraulic system must be modernized. Nichols *et al.* in [6] estimated that the modernization of cooling system performance in American power plants could lead to improving their potential efficiency by about 0.2-1.0 %. Similar estimations made for APEC countries by Boncimino *et al.* in [7] indicated that the improvement of feed water heaters and condensers could update the overall power unit efficiency by a value of approximately 0.8 %.

The modernization of an existing cooling installation is usually an expensive solution. Therefore, such a decision should be preceded by a careful analysis of various variants of the prospective reconstruction. Because hydraulic installations are usually complicated systems consisting of many series-parallel branches, their proper investigation should be based both on the measurements of the actual operating parameters and the development of a mathematical model of flow rates in individual pipelines. The analytical approach establishes a set of non-linear algebraic equations which can be solved with numerical techniques. As a result of the iterative process, a set of volumetric flow rates of water through all branches of the investigated hydraulic system is obtained. Calculations indicate the influence of changes in the pipeline geometrical parameters on the total cooling water flow rate in the analysed installation. Such an approach makes it possible to analyse different variants of the modernization of the studied system as well as enabling the indication of its critical elements, together with the directions of the flow in all branches.

2 Mathematical model of a series-parallel cooling system

The fundamental issue in modelling hydraulic installations is connected with the calculation of volumetric flow rates in each branch of the system. The scheme of the simplest series-parallel hydraulic system is presented in Fig. 1.

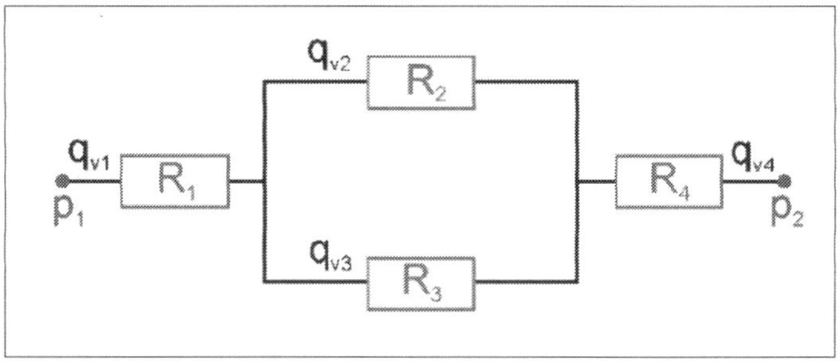

Fig. 1. Scheme of the simplest series-parallel hydraulic system (q_v – volumetric flow rate, R - flow resistance, p – absolute pressure)

Due to the fact that water installations are usually complicated systems consisting of many series-parallel branches, the mathematical modelling of a hydraulic system leads to a set of non-linear algebraic equations. In the model, the following three types of equations can appear (the examples of equations given below conform to the scheme presented in Fig. 1):

• an equation which describes volumetric flow rate conservation law in an individual joint occurring in the system (e.g., $q_{v1} = q_{v2} + q_{v3}$ and $q_{v4} = q_{v2} + q_{v3}$),

• an equation which describes pressure losses in an individual closed loop of the hydraulic system (e.g. $R_2 \cdot q_{v2}^2 - R_3 \cdot q_{v3}^2 = 0$). In the mesh, a "+" or "-" sign is taken when the flow in the considered branch is clockwise or counter-clockwise, respectively),

• an equation which describes a pressure drop Δp along a selected path connecting the inlet and the outlet of the system (e.g. $\Delta p_{12} = (p_1 - p_2) = R_1 \cdot q_{v1}^2 + R_3 \cdot q_{v3}^2 + R_4 \cdot q_{v4}^2$). In this case, knowledge of the inlet and outlet pressure (p_1 and p_2, respectively) is required.

For example, the scheme presented in Fig. 1 requires four algebraic equations (listed above), which allow one to calculate 4 unknown volumetric flow rates: q_{v1}, q_{v2}, q_{v3} and q_{v4}.

The mathematical model also requires information about flow resistances R_i, which have to be assigned to each branch of the system (schematically marked by green boxes in Fig. 1). The flow resistance of an individual branch of the installation (denoted by index "*i*") is calculated as the sum of local minor and friction losses:

$$R_i = \frac{8\rho}{\pi^2 d_i^4}\left(\sum_k \xi_{i,k} + \lambda_i \frac{l_i}{d_i}\right) \tag{1}$$

where: $\xi_{i,k}$ – coefficients of minor losses and λ_i – coefficient of friction losses which occur on "*i*" branch; d_i and l_i – diameter and length of the "*i*" branch, respectively [8]. The values of ξ coefficients are usually selected basing on literature data or measurements. The values of λ coefficients are calculated from Blasius' formula in the range of turbulent flow [8, 9]. It is worth mentioning that the formula for λ includes information about q_v, which is, at the beginning, an unknown value. Therefore, in the process of solving, the values of λ_i (and hence flow resistances R_i) must be updated before each iterative step. For this reason, the set of non-linear algebraic equations has to be solved using iterative techniques, e.g. Newton's method [10], which enables determining the values of volumetric flow rates q_{vi} in the process of successive approximations with any predetermined precision ε.

3 Hydraulic installation cooperating with bottom ash coolers

3.1 Description of the installation and mathematical model

A cooling water installation cooperating with two bottom ash coolers of a circulating fluidized bed (CFB) boiler (called SAC-L and SAC-R, respectively for left and right bottom ash cooler) is discussed as the first example of a series-parallel hydraulic system. Figure 2 provides a schematic image of the most important elements of a SAC-R ash cooler – two screw coolers designated by the acronyms SC-L and SC-R on the left scheme. The right part of Figure 2 presents a schematic side view of the bottom ash cooler with an indication of the hot slag inlet (1), the outlet of the cooled slag (6), the inlet and outlet of the cooling water flowing through the shaft and blades of the SC-L screw (respectively (2) and (3)), the inlet and outlet of the cooling water which flows between the walls (casing) of the ash cooler (respectively (4) and (5)). The mass flow rate of the hot slag is directly delivered from the combustion chamber (1), after which it is split between two screw coolers (SC-L and SC-R), and then transported to point (6). During its motion along the screw, the slag is gradually cooled by the cooling water circulating in the blades, the shaft and between the walls (casing).

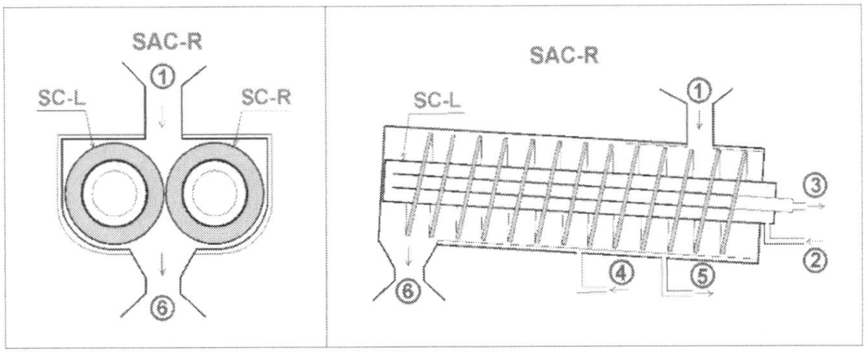

Fig. 2. Scheme of the bottom ash cooler (SAC-R) with internal cooling water distribution: two screw coolers (SC-L and SC-R), the inlet and outlet of the slag (1 and 6 respectively), the inlet and outlet of the cooling water (2,4 and 3, 5 respectively).

The series-parallel system of cooling water propagation to two bottom ash coolers of a CFB boiler is shown in Fig. 3.

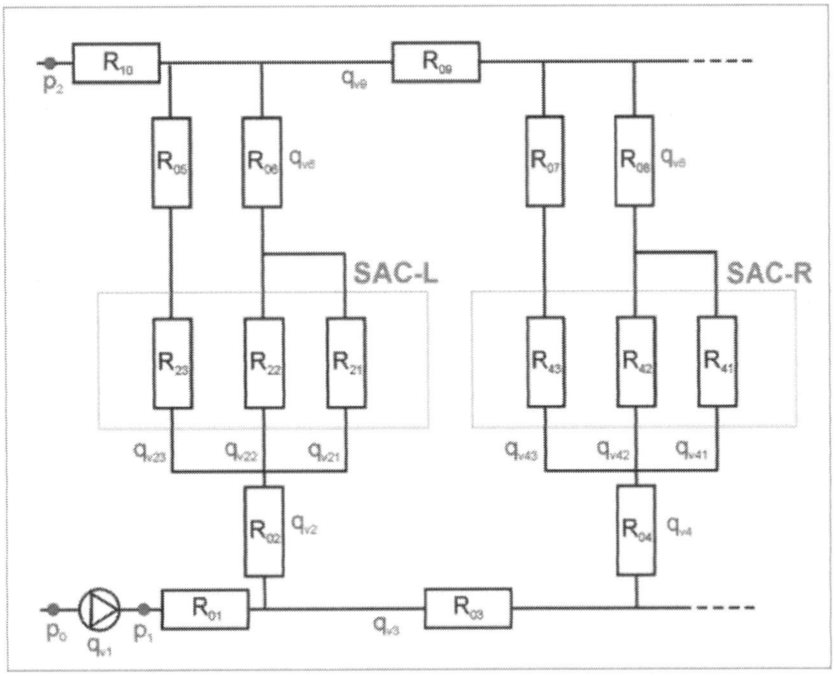

Fig. 3. An example of a series-parallel system of cooling water propagation to the two bottom ash coolers of a CFB boiler.

Each bottom ash cooler (SAC-L and SAC-R) consists of two screw coolers (SC-L and SC-R), supplied by cooling water volumetric flow rates denoted as q_{v21} and q_{v22}, q_{v41} and q_{v42}, respectively. The volumetric flow rates which cool down the walls of coolers are q_{v23} and q_{v43}, respectively. Cooling water is taken from the common collector and pumped into the system through a pump, which has a static pressure p_1 at the inlet to the installation. The static pressure at the exit from the system is denoted by p_2. The pressure difference $\Delta p_{12} = (p_1 - p_2)$ forces the cooling water to circulate in all the branches of the system. The pump discharge is calculated based on the pressure difference $\Delta p_p = (p_1 - p_0)$. Cooling water is delivered to (and removed from) each ash cooler by pipes with a diameter of ϕ160 (parts R_{01}, R_{03}, R_{09} and R_{10}), and next by pipes ϕ120 in diameter (parts R_{02} and R_{04} - R_{08}). The pipes which directly supply cooling water to the screws and walls of the ash coolers have a diameter of ϕ60 (parts R_{21}-R_{23} and R_{41} -R_{43}).

Basing on the scheme presented in Fig. 3, it is possible to note eleven non-linear algebraic equations, including:

- 5 equations describing the water distribution in individual joints,
- 5 equations describing the pressure losses in the closed hydraulic meshes,
- 1 equation describing the pressure drop in the selected branch between the inlet and the outlet of the system (p_1 and p_2, respectively).

The system of the abovementioned algebraic equations was solved iteratively by applying Newton's method [10] with precision $\varepsilon = 10^{-3}$.

3.2 Numerical results and discussion

The aim of the numerical modelling was to indicate those elements of the analysed hydraulic installation whose improvement could significantly increase the overall volumetric flow rates of cooling water flowing through the bottom ash coolers (q_{v2} and q_{v4}, respectively). It is worth mentioning that for long-pipe installation, the major source of pressure drop are friction losses connected with the pipe parameters, such as their diameter, length and absolute roughness. Due to the fact that traces of pipelines are already fixed at the level of the boiler's construction, the only parameters which could be changed in the existing installation, basing on the results of the mathematical model, are connected with the tube diameters. The starting point for numerical modelling was the existing hydraulic installation with the initial volumetric flow rates through the bottom ash coolers equalling q_{v2} = 38.26 m^3/h and q_{v4} = 42.52 m^3/h. The pump discharge, forcing the flow in the cooling water system, amounted to Δp_p = 0.325 MPa. The numerical simulation focused on the influence of the changes in the pipe diameters in the selected branches on the values of the volumetric flow rates q_{v2} and q_{v4}.

In the first variant, the diameters of the pipes in branches R_{21}-R_{23} and R_{41} -R_{43} were enlarged from the initial value of ϕ60 to ϕ80, ϕ100 and ϕ120, respectively. The results of the modelling are presented in Table 1.

Table 1. Results of the numerical modelling basing on the increase of the pipe diameters in branches R_{21}-R_{23} and R_{41} -R_{43}.

pipe diameter ϕ, mm	**volumetric flow rate** q_{v2}, m^3/h	**volumetric flow rate** q_{v4}, m^3/h	**pump discharge** Δp_p, MPa
60	38.26	42.52	0.325
80	39.76	44.08	0.301
100	40.78	45.41	0.272
120	41.02	45.65	0.267

The increase of the diameters results in the increase of the volumetric flow rates q_{v2} and q_{v4}. It is worth noting that the reduction of the flow resistances in the system also results in the decrease of pump discharge Δp_p required to force the flow. An interesting observation is connected with the fact that the linear increase of the diameter does not reflect in the linear increase of q_{v2} and q_{v4}. Enlargement to ϕ80 results in the increase of the volumetric flow rate by Δq_{v2} = 1.50 m^3/h and Δq_{v4} = 1.56 m^3/h when further changes generate the increase at the level of:

- $\phi 80 \rightarrow \phi 100$: $\Delta q_{v2} = 1.02$ m^3/h and $\Delta q_{v4} = 1.33$ m^3/h,
- $\phi 100 \rightarrow \phi 120$: $\Delta q_{v2} = 0.24$ m^3/h and $\Delta q_{v4} = 0.24$ m^3/h.

The last change of diameter from ϕ100 to ϕ120 caused the increase of q_v by only 0.24 m^3/h through both bottom ash coolers. This means that further enlargement of the diameter of these pipes does not have a significant influence on the volumetric flow rates q_{v2} and q_{v4}.

In the second variant, the diameters of the pipes in branches R_{01}, R_{03}, R_{09} and R_{10} were enlarged from the initial value ϕ160 to ϕ200 and ϕ250. The results of the modelling are presented in Table 2.

Table 2. Results of the numerical modelling basing on the increase of the pipe diameters in branches R_{01}, R_{03}, R_{09} and R_{10}.

pipe diameter ϕ, mm	**volumetric flow rate** q_{v2}, m^3/h	**volumetric flow rate** q_{v4}, m^3/h	**pump discharge** Δp_p, MPa
160	38.26	42.52	0.325
200	43.00	48.05	0.202
250	43.60	48.71	0.162

In this variant, the enlargement of the pipe diameters results in the increase of the volumetric flow rates:

- $\phi 160 \rightarrow \phi 200$: $\Delta q_{v2} = 4.74$ m^3/h and $\Delta q_{v4} = 5.53$ m^3/h,
- $\phi 200 \rightarrow \phi 250$: $\Delta q_{v2} = 0.60$ m^3/h and $\Delta q_{v4} = 0.66$ m^3/h.

This is a much larger increase than for the first variant, which means that the reduction of the flow resistances on these branches could increase the flow rates of cooling water through the bottom ash coolers more efficiently (by up to 14.5 %).

4 Hydraulic installation with condensers and cooling towers

4.1 Description of the installation and mathematical model

Another example of a series-parallel hydraulic system would be a closed cooling water system consisting of condensers and cooling towers. This kind of hydraulic installation usually appears in professional power plants located far from large natural sources of water (like seas, rivers or lakes). The daily water demand of power units often reaches a level of several thousand cubic meters, which signifies that knowledge about water distribution in hydraulic systems could help not only in better water management but also in improving heat exchange processes occurring in condensers and cooling towers.

The scheme of a closed cooling installation is presented in Fig. 4. Warm water from condensers is pumped to the cooling system by twelve pipes denoted as external volumetric flow rates (q_{vz1} - q_{vz12}). Through the pipelines, water is delivered to five cooling towers marked with black circles. Inside each cooling tower, water is additionally divided between a core and ring sections. Each branch of the system is described by the individual flow resistance R_i and volumetric flow rate q_{vi} (not marked in the scheme). The values of R_i were calculated basing on the technical documentation and literature data. Due to the fact that the pipelines are made from concrete, the value of the friction losses coefficient λ_i in (1) are calculated from the Altšul formula using the value of absolute roughness k = 10^{-4} m [8].

The aim of the mathematical model was to determine the values of internal volumetric flow rates q_{vi}. To do this, a set of 41 non-linear algebraic equations was created:

- 28 equations describe water distribution in individual joints,
- 4 equations describe pressure losses in the closed hydraulic meshes,
- 9 equations describe the pressure drop along traces connected to two outlets of the system.

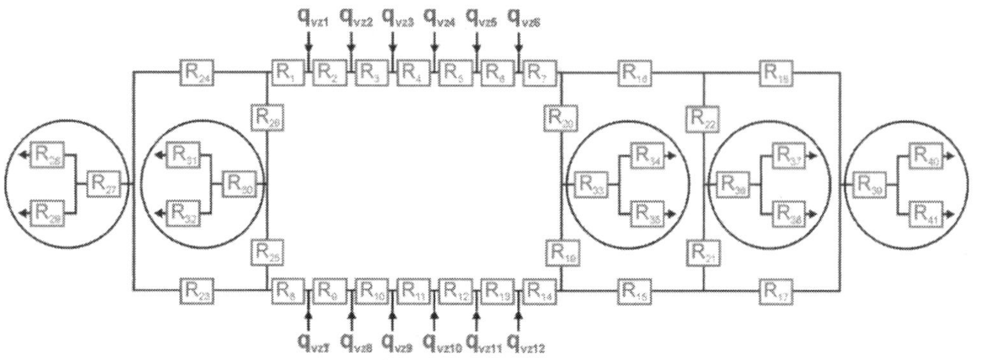

Fig. 4. Scheme of a closed cooling water system with external volumetric flow rates (from condensers) denoted by q_{vz} and flow resistances R_i (and connected internal volumetric flow rates q_{vi}). The five cooling towers are marked with black circles.

It is worth mentioning that for an open hydraulic installation it is not necessary to know the pressure at the inlet of the external volumetric flow rates q_{vz} because equations describing the pressure drop can be drawn between the outlets of the system (which have atmospheric pressure). A novelty with respect to the mathematical model described in subsection 3.1 is the need to determine the direction of the flow in all branches of the network. This is due to the fact that there are several inlets and outlets for cooling water. Therefore, in equations describing the pressure drop, the terms q_{vi}^2 were replaced by the expressions ($q_{vi} \cdot |q_{vi}|$), which not only enable determining the value of q_{vi} but also the direction of the flow (through the positive or negative sign of q_{vi} values). The initial values of the volumetric flow rates and its flow directions are, in the further iterative process, adjusted to satisfy the equations of the mathematical model. The system of the abovementioned algebraic equations was solved by applying Newton's method with precision $\varepsilon = 10^{-3}$. As a result of the calculations, volumetric flow rates to individual cooling towers were obtained: q_{v27}, q_{v30}, q_{v33}, q_{v36} and q_{v39}.

4.2 Numerical results and discussion

An example of measurements and numerical results from the mathematical model of the closed cooling water system is presented in Table 3.

Table 3. Comparison of the measurements and numerical results from the mathematical model of the closed cooling water system. The total external volumetric flow rate q_{vz} = 151,100 m^3/h.

cooling tower	**volumetric flow rate** measurement, m^3/h	**volumetric flow rate** numerical result, m^3/h	**relative error, %**
1 (q_{v27})	27107	26986	0.45
2 (q_{v30})	31338	29087	7.74
3 (q_{v33})	25219	24599	2.52
4 (q_{v36})	35977	36612	-1.73
5 (q_{v39})	31459	33816	-6.97

The presented data indicates that the largest discrepancies between the measurements and numerical simulation are for the second and fifth cooling tower. The maximum value of relative error is at the level of 7.74 %. This is very high agreement taking into account the complexity of the series-parallel system reflected in the number of non-linear algebraic equations. The numerical solution shows that the external volumetric flow rates q_{vz4}-q_{vz6} and q_{vz10}-q_{vz12} were directed only to cooling towers nos. 3-5 (q_{v33}, q_{v36}, q_{v39}) in contrast to the volumetric flow rates q_{vz1}, q_{vz2} and q_{vz7}, q_{vz8}, which only cooperate with cooling towers nos. 1 and 2 (q_{v27} and q_{v30}). The volumetric flow rates q_{vz3} and q_{vz9} are divided between the left and right group of cooling towers (1-2 and 3-5 respectively). Knowledge about flow resistances R_i in each branch enables numerically analysing the changes in the structure of the flows through the hydraulic system when e.g. one of the power units is turned off (which is simulated by the individual value of $q_{zi} = 0$) .

5 Closing remarks

A mathematical model of a series-parallel hydraulic system enables the numerical study of the influence of individual components of the installation on the resulting volumetric flow rates. A set of non-linear equations is always individually established for the analysed hydraulic system. Such an approach makes it possible to study the impact of its various modifications on the generated volumetric flow rates. This information provides the investor with an opportunity to decide which of the analysed options gives the best economic effects as the cost of the improvements is often compared to the expected return in increased efficiency as a primary determinant of whether to go forward with a modernization.

References

1. R.J. Campbell et al., *Prospects for coal in electric power and industry*, Tech. rep., Congressional Research Service Report, R42950 (2013)
2. C. Henderson, P. Baruya, *CO_2 reductions from CCTs and CO_2 capture*, CCC/200, London, UK, IEA Clean Coal Centre (July 2012)
3. A. Ryabchikov et al., Power Technology and Engineering, **44 3** (2012)
4. M.A. Malek, *Heating boiler operator's manual: maintenance, operation and repair*, McGraw-Hill companies, New York (2007)
5. K.E. Heselton, *Boiler operator's handbook*, Lilburn, GA: Fairmont Press, New York, (2005)
6. C. Nichols et al., *Reducing CO_2 emissions by improving the efficiency of the existing coal-fired power plant fleet*, US Department of Energy National Energy Technology Laboratory, DOE/NETL-2008/1329 (2008)
7. G. Boncimino et al., *Costs and effectiveness of upgrading and refurbishing older coal-fired power plants in developing APEC economies*, Asia-Pacific Economic Cooperation, Energy Working Group Project EWG 04/2003T, (June 2005), available: http://www.egcfe.ewg.apec.org/projects/UpgradePP_Report_2005.pdf
8. F.M. White, *Fluid mechanics*, McGraw-Hill companies, New York (2010)
9. Y. Nakayama, R.F. Boucher, *Introduction to fluid mechanics*, Butterworth-Heinemann (2000)
10. J.H. Mathews, K.D. Fink, *Numerical methods using Matlab*, Prentice Hall (1999)
11. R.B. Baker, H.A. Shryock, *A comprehensive approach to the analysis of cooling tower performance*, SPX Marley, Thermal Science (2013)

A simplified model of a mechanical cooling tower with both a fill pack and a coil

Freek Van Riet[1,*], *Gunther* Steenackers[2], and *Ivan* Verhaert[1]

[1]University of Antwerp, Faculty of Applied Engineering, Energy & Materials in Infrastructure and Buildings, Campus Groenenborger, Groenenborgerlaan 171, 2020 Antwerp, Belgium
[2]University of Antwerp, Faculty of Applied Engineering, Op3Mech, Campus Groenenborger, Groenenborgerlaan 171, 2020 Antwerp, Belgium

Abstract. Cooling accounts for a large amount of the global primary energy consumption in buildings and industrial processes. A substantial part of this cooling demand is produced by mechanical cooling towers. Simulations benefit the sizing and integration of cooling towers in overall cooling networks. However, for these simulations fast-to-calculate and easy-to-parametrize models are required. In this paper, a new model is developed for a mechanical draught cooling tower with both a cooling coil and a fill pack. The model needs manufacturers' performance data at only three operational states (at varying air and water flow rates) to be parametrized. The model predicts the cooled, outgoing water temperature. These predictions were compared with experimental data for a wide range of operational states. The model was able to predict the temperature with a maximum absolute error of 0.59°C. The relative error of cooling capacity was mostly between ±5%.

1 Introduction

Simulations benefit the design process of cooling networks. Especially the cooling source, such as mechanical draught cooling towers, should be selected well to maximize the overall performance of the network. The models used in the simulations, should meet the following requirements: be fast to calculate and be easy to parametrize using manufacturers' data only.

Great efforts have been made to develop such models for mechanical draught cooling towers. Pioneers in these developments were, among others, Walker et al. [1], Merkel [2] and Poppe [3] . More recent and further simplified models can also be found in literature for both direct and indirect contact, and both counter-flow and crossflow cooling towers (see [4] for more information about the differences between these types of cooling towers). These models are mostly based on the *e*-NTU (effectiveness - Number of Transfer Units) approach. Indeed, Lebrun et al. [5] made a universal model for both direct and indirect contact cooling towers, and evaporative condensers, based on the *e*-NTU approach. They compared the model with experimental data and found relative errors of the cooling capacity smaller than (except some outliers) ±8%. This model is characterized by four

* Corresponding author: freek.vanriet@uantwerpen.be

parameters, that can be determined using manufacturers' data. Stabat et al. [6] modelled an indirect contact cooling tower in a more physical way and found a comparable model structure with two parameters after simplification. They included a temperature correction on the dynamic viscosity of the process water. The relative errors of the predicted cooling capacity that they found, were of the same order of magnitude (mostly between ±5%). Lu et al. [7] considered the *e*-NTU relation as a black-box with six parameters. Their model was intended for online control, not for simulations.

However, to the authors' best of knowledge, no simplified model exists of a cooling tower with both a cooling coil and a fill package. As in the left side of Figure 1 can be seen, this type of cooling tower is a combination of a crossflow direct (fill pack) and parallel-flow indirect (coil) contact cooling tower. This type of cooling tower provides a better heat transfer and was invented by Carter [8] and patented by Baltimore Aircoil Company, Inc. The structural differences in design do not allow to use existing models. This article describes a new model for this type of cooling tower and its validation. The model predicts the temperature of the cooled process water.

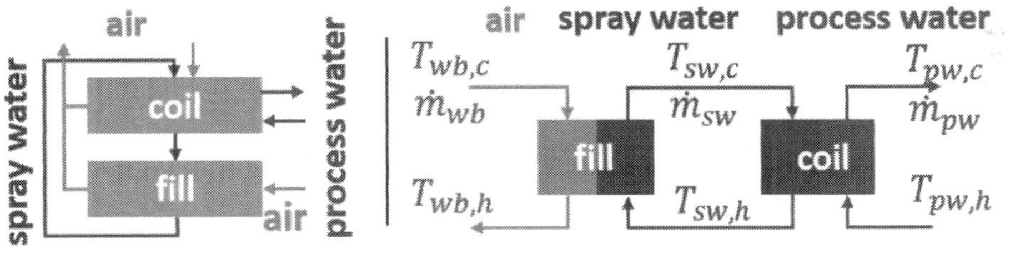

Fig. 1. Schematic representation of the cooling tower. Left: mass flow in the real installation (based on [8]). Right: Resulting flows when air flow through coil is neglected.

2 Model development

In this section, the first part handles the general derivation of the model and how to use it for simulations. This means that the parameters characterizing this model, are assumed to be known. These parameters should, however, be determined using manufactures' data. This so-called *parametrization* will be discussed in the second part of this section.

2.1 Model description

The air flow in the cooling coil (see right side of Fig. 1) is neglected to simplify further derivations. Two arguments are given. The first is, that the fill is designed to cool the spray water and hence most evaporative cooling will occur here. The second is that the coil will be fully covered with water, as the air and spray water are in parallel flow. This allows to treat the coil as a spray water/process water counterflow heat exchanger. The fill component is treated as an evaporative air/spray water crossflow heat exchanger. In the following derivations, the subscripts '*wb*', '*sp*', '*pc*', '*h*', '*c*' refer to ('wet bulb') air, spray water, process water, hot and cold, resp. The variable T is used as temperature (°C), $\dot{m}$ as mass flow rate (kg/s), h as specific enthalpy (J/kg), $\dot{C}$ ($=c\,\dot{m}$, with c the thermal capacity) is the capacitive flow rate (W/K) and $\dot{Q}$ the heat flux (W).

For the coil component, a conventional counterflow *e*-NTU model is used. The corresponding relation between the effectiveness (ε_{coil}) and NTU_{coil} is (see e.g. [9]):

$$\varepsilon_{coil} = \frac{1-\exp(-NTU_{coil}\cdot(1-\hat{C}_{coil}))}{1-\hat{C}_{coil}\cdot\exp(-NTU_{coil}\cdot(1-\hat{C}_{coil})} \qquad (1)$$

With:

$$\varepsilon_{coil} = \frac{\dot{Q}}{\dot{Q}_{coil,\max}} \qquad (2)$$

$$\dot{Q}_{coil,\max} = \dot{C}_{coil,\min}\cdot(T_{pw,h}-T_{sw,c}) \qquad (3)$$

$$NTU_{coil} = \frac{UA_{coil}}{\dot{C}_{coil,\min}} \qquad (4)$$

$$\hat{C}_{coil} = \frac{\dot{C}_{coil,\min}}{\dot{C}_{coil,\max}} \qquad (5)$$

$\dot{C}_{coil,min}$ and $\dot{C}_{coil,max}$ are respectively the smallest and largest capacitive flow rates of $\dot{C}_{sw}$ and $\dot{C}_{pw}$. $\dot{Q}_{coil,max}$ is the theoretical maximum of the heat flux. The overall heat-transfer coefficient, UA_{coil}, is assumed to vary with varying process water flow rate, $\dot{m}_{pw}$. By doing so, the influence of the Reynolds number and Prandtl number on the heat-transfer coefficient are simplified, as discussed by [6] (see equations 24 – 29). As a further simplification, the dynamic viscosity is considered to be constant. When doing so, the equation can be rewritten as:

$$UA_{coil} = p_1\cdot\left(\frac{\dot{m}_{pw}}{\dot{m}_{pw,ref}}\right)^{p_2} \qquad (6)$$

in which p_1 and p_2 are parameters that should be estimated for a particular installation (see section 2.2). $\dot{m}_{pw,ref}$ can be any $\dot{m}_{pw}$, it is only used for scaling. Note that the effect of $\dot{m}_{sw}$ is not taken into account as such, as it is fixed during operation.

Finally, the energy balance of the process water is:

$$\dot{Q} = \dot{C}_{pw}(T_{pw,h}-T_{pw,c}) \qquad (7)$$

The fill package is modelled as a crossflow direct contact cooling tower. Again, it is based on an *e*-NTU model. However, an adjustment is made to include the sensible heat flux because of the spray water evaporation. An often used approach, is to treat the air as a fictitious fluid, having the temperature of the wet bulb of the air when entering ($T_{wb,c}$) and having a *fictitious* specific thermal capacity (c_{wb}) that takes into account the latent heat. This latter is defined as (see e.g. [5] or [6] for more information):

$$c_{wb} = \frac{h_{wb,h} - h_{wb,c}}{T_{wb,h} - T_{wb,c}} \tag{8}$$

An *e*-NTU relation that is appropriate for crossflow heat exchangers, should be used. Which relation to use depends on, among other things, whether the fluids are mixed or unmixed. Note that these definitions say something about the mixing *within* a particular fluid, not *between* the two fluids. From a technical point of view, neither of this two theoretical situations will occur, as discussed by Digiovanni and Webb [10]. However, for simplicity, the fill component is treated here as an unmixed-unmixed heat exchanger. This means that the lateral heat flux (because of diffusion, convection and bulk flow mixing) is neglected. Therefore, the relation of Triboix [11] (which is an improvement compared to the widespread approximate *e*-NTU relation for unmixed-unmixed heat exchangers in e.g. [9]) can be used:

$$\begin{cases} \underline{\text{if}}\ \hat{C}_{fill} > 0.3 \text{ and } NTU_{fill} > 1: \\ \varepsilon_{fill} = (1 + 0.44 \cdot (1 - \hat{C})) \cdot \left(1 - \left(\dfrac{1}{0.92 + (\pi \cdot \hat{C}_{fill}^{0.15} \cdot NTU_{fill})^{1.25}}\right)^{0.4}\right) \\ \underline{\text{else}}: \\ \varepsilon_{fill} = 1 - \exp\left(\dfrac{\exp(-\hat{C}_{fill}^{1.15} \cdot NTU_{fill})}{\hat{C}_{fill}^{1.15}}\right) \end{cases} \tag{9}$$

The next equations that describe the fill are analogue to (2) – (7). The overall heat-transfer coefficient is assumed to be dependent on $\dot{m}_{wb}$:

$$UA_{fill} = p_3 \cdot \left(\frac{\dot{m}_{wb}}{\dot{m}_{wb,ref}}\right)^{p_4} \tag{10}$$

This equation was also used in [5] for a direct contact cooling tower. Note that the fill component can be seen as a direct contact cooling tower. p_3 and p_4 should be known from the parametrization (see below). Again, an arbitrary mass flow rate $\dot{m}_{wb,ref}$ is used for scaling. No cold loss is assumed, i.e. the heat flux in (7) is also used in the is the energy balance for the fill:

$$\dot{Q} = C_{wb} \cdot (T_{wb,c} - T_{wb,h}) \tag{11}$$

 DOI: 10.1051/e3sconf/20172200145

To apply above equations for the simulation of $T_{pw,c}$ the steps of [5] can be followed. The inputs are $T_{wb,c}$, $T_{pw,h}$, $\dot{m}_{wb}$ (can be varied during operation to control e.g. $T_{pw,c}$) and $\dot{m}_{pw}$. An iteration is required to find c_{wb}, as it is not known because $T_{wb,h}$ and hence also $h_{wb,h}$ are no inputs of the model when simulating (see (8)). Contrary to the model in [5], the energy balance of the spray water ($\dot{Q} = \dot{C}_{sw}\,(T_{sw,h} - T_{sw,c})$) is necessary to eliminate $T_{sw,c}$ in (3) and $T_{sw,h}$ (appears in analogue equation as (3) but for fill, not shown) for the calculation of the heat flux. By combining this energy balance with (2) and (3), and the analogue equations for the fill, the heat flux can be calculated without $T_{sw,c}$ and $T_{sw,h}$, as:

$$\dot{Q} = \frac{\varepsilon_{coil}\varepsilon_{fill}\dot{C}_{coil,\min}\dot{C}_{fill,\min}\dot{C}_{sw}\left(T_{pw,h} - T_{wb,c}\right)}{\dot{C}_{sw}\left(\varepsilon_{coil}\dot{C}_{coil,\min} + \varepsilon_{fill}\dot{C}_{fill,\min}\right) - \varepsilon_{coil}\varepsilon_{fill}\dot{C}_{coil,\min}\dot{C}_{fill,\min}} \tag{12}$$

Finally, $T_{pw,h}$ can be obtained by using (7).

2.2 Model parametrization

To find the parameters p_1, p_2, p_3 and p_4 of a particular installation, manufacturers' data should be used. In this data, the following variables at different operational states can be found: $T_{wb,c}$, $T_{pw,h}$, $\dot{m}_{wb}$ and $\dot{m}_{pw}$. Manufacturers do not provide any information of the spray water temperatures. Therefore, when parametrizing, $T_{sw,c}$ is approximated as the mean between its maximum possible temperature $T_{sw,c,max}$ and its minimum possible temperature $T_{sw,c,min}$. These can be calculated as:

$$\begin{aligned} T_{sw,c,\max} &= \min\left(T_{wb,c}, T_{wb,h} - \dot{Q}/\dot{C}_{sw}\right) \\ T_{sw,c,\min} &= \max\left(T_{pw,c}, T_{pw,h} - \dot{Q}/\dot{C}_{sw}\right) \end{aligned} \tag{13}$$

From here, the conventional calculations for heat exchanger parametrization can continue. For the coil, this means calculating $\dot{Q}_{coil,max}$ (3), ε_{coil} (2) and NTU_{coil} (from the inverted form of (1), see [9]). Finally, (4) is used to determine UA_{coil}. To be able to calculate p_1 and p_2, this should be done two times: each time with data from the installation at different $\dot{m}_{pw}$. The biggest $\dot{m}_{pw}$ can be chosen to be $\dot{m}_{pw,ref}$. Filling this two times in in (6) results in a set of two equations, from which p_1 and p_2 can be calculated. Note that p_1 is the overall heat-transfer coefficient of the coil at the reference process water flow rate. p_2 characterizes the effect of the process water flow rate on the overall heat-transfer coefficient, as described earlier. Analogue steps can be taken to calculate the parameters p_3 and p_4 that characterize the fill. See [11] for the inverted form of (9). The data at two different air flow rates should be used. The interpretation of p_3 and p_4 is analogue to p_1 and p_2. In total, the data at only three operational states is required, as one state can be used in common for both the coil and the fill.

DOI: 10.1051/e3sconf/20172200145

3 Model validation

3.1 Data description

The data of an FXVS 1212C-30-T-*K* cooling tower of Baltimore Aircoil Company was used to validate the model. All the used data is publically available at [12]. This data complies with the CTI-201RS standard [13]. As a result, the cooling capability error of the data is less than 5%. In this dataset, only data at nominal air flow rates are provided. However, the data of other installations are used (ending with *K, L, M, O* and *P*) to emulate the data at different air flow rates. Indeed, these different installations are identical, with the exception of their nominal air flow rate (they are equipped with different fans). By combining the data of all installations, a dataset of 245 operational states was obtained. The data at three of these states were selected (most extreme variations of $\dot{m}_{pw}$ and $\dot{m}_{wb}$ in the dataset, see Table 1) for the parametrization. The rest of the dataset was used to validate the model.

Table 1. Selected operational states for the parametrization of the model. For all states is $\dot{m}_{sw}$=*54.6* kg/s. The data in the first two rows were used to determine p_1 and p_2, and the data in the first and last row, were used to determine p_3 and p_4.

$\dot{m}_{pw}$ (l/s)	$T_{pw,h}$ (°C)	$T_{pw,c}$ (°C)	$T_{wb,c}$ (°C)	$\dot{V}_{wb}$ (m^3/s)
96.7	47.2	40.2	32.2	53.9
28.2	20	14	10	53.9
71.4	47.2	40.2	32.2	33.4

3.2 Results and discussion

All calculations were performed with Matlab R2015a. The parameters p_1, p_2, p_3 and p_4 were calculated as 8.6e+05 W/K, 0.918, 1.2e+06 W/K and 0.852, resp. The comparison between data and simulations are shown in Fig 2. It can be seen that most data lies between ±5 % relative error of the predicted heat flux. All the errors lay between 5.5% and -8.5%. This is similar to the accuracy of the existing models for the conventional cooling towers [5],[6].

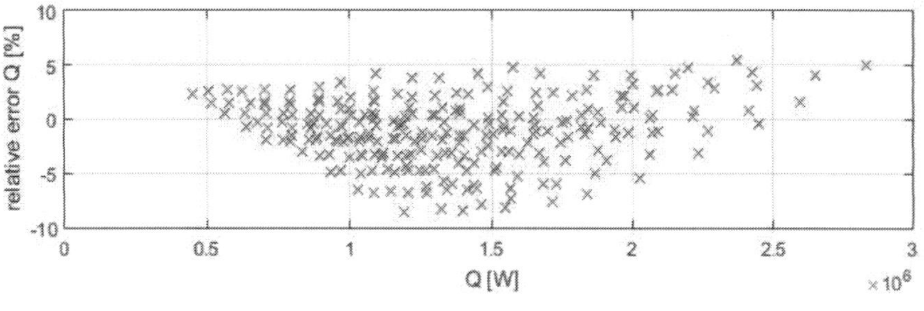

Fig. 1. . The results of the simulation. The relative error is defined as the heat flux of the data minus the heat flux predicted by the model, relative to the heat flux of the data.

A small asymmetry in the errors should be noted. For the smallest and largest heat fluxes, the model tends to underestimate the heat flux. For medium heat fluxes, on the contrary, the model tends to overestimate the heat flux slightly more than it underestimates it. This latter observation is also true for the complete dataset, i.e. the mean error of the dataset is -1%.

Two suggestions are given to bring this mean error closer to zero, or in general, to calculate the parameters $p_1 - p_4$ more precisely. First of all, the temperatures of the spray water can be calculated more exactly. In order to do so, an iterative procedure should be performed on the parametrization of the model. (13) can be used as an initial guess and after the calculation of the parameters, (3) can be used to calculate the simulated spray water temperature. Secondly, the parameters can be estimated by a nonlinear equation solver, when the performance of the cooling tower is known at more than three operational states. However, upper results show acceptable accuracy when using only three extreme operational states.

4 Conclusions

A new model was developed for a mechanical draft cooling tower with both a cooling coil and a fill pack. The model needs the data of the cooling towers' performance at only three operational states (at different process water and air flow rates) for its parametrization. The model is validated for a wide range of operation. The outgoing process water temperature is predicted with a maximum absolute error of 0.59 °C. Some suggestions are given to further improve the accuracy of the model.

References

1. W. H. Walker, W. K. Lewis, and W. H. McAdams, *Principles of Chemical Engineering* (1923)
2. F. Merkel, *VDI-Zeitschrift*, **70**, 123–128 (1925)
3. M. Poppe and H. Rögener, *VDI-Wärmteatlas*, 1–5 (1991)
4. A. K. . Mohiuddin and K. Kant, *Int. J. Refrig.*, **19**, 52–60 (1996)
5. J. Lebrun, C. Aparecida Silva, F. Trebilcock, and E. Winandy, **1**, 25–31 (2002)
6. P. Stabat and D. Marchio, *Appl. Energy*, **78**, 433–451 (2004)
7. L. Lu, *Int. Compress. Eng. Conf. Refrig. Air Cond. Conf. Purdue Univ. USA*, no. 65, 9 (2002)
8. T. P. Carter, patent US5435382 A, Baltimore Aircoil Company, Inc. (1995)
9. J. Mitchel, *Principles of Heating, Ventilation and Air Conditioning in Buildings*. John Wiley Sons Inc. (2013)
10. M. A. Digiovanni and R. L. Webb, *Heat Transf. Eng.*, **10**, 61-70 (1989)
11. A. Triboix, *Int. Commun. Heat Mass Transf.*, **36**, 121–124 (2009)
12. https://www.baltimoreaircoil.eu/products/FXVS-1212-1218
13. http://www.cti.org/certification.php

Effect of operating conditions on performance of silica gel-water air-fluidised desiccant cooler

Zbigniew Rogala[1*], *Piotr* Kolasiński[1], and *Zbigniew* Gnutek[1]

[1]Department of Thermodynamics, Theory of Machines and Thermal Systems, Wrocław University of Science and Technology, Wybrzeże Wyspiańskiego 27, 50 - 370 Wrocław, Poland

Abstract. Fluidised desiccant cooling is reported in the literature as an efficient way to provide cooling for air-conditioning purposes. The performance of this technology can be described by electric and thermal Coefficients of Performance (COP) and Specific Cooling Power (SCP). In this paper comprehensive theoretical study was carried out in order to assess the effect of operating conditions such as: superficial air velocity, desiccant particle diameter, bed switching time and desiccant filling height on the performance of fluidised desiccant cooler (FDC). It was concluded that FDC should be filled with as small as possible desiccant particles featuring diameters and should not be operated with shorter switching times than optimum. Moreover in order to efficiently run such systems superficial air velocities during adsorption and desorption should be similar. At last substantial effect of desiccant filling height on performance of FDC was presented.

1 Introduction

Fluidised desiccant cooling is reported as an efficient way to provide cooling for air-conditioning purposes [1]. This technology allows to convert low-temperature heat [2] (even lower than 50 °C [3]) into cooling effect. Scheme of simple fluidised desiccant cooler (FDC) is presented in the Fig. 1. The system consists of two fluidised beds filled with desiccant (in this case silica gel) which are periodically supplied with hot or cold air. The fan (4) blows the air through the air cooler (5). Then the air enters fluidised bed (3). Due to adsorption the air is dehumidified and warmed up. The dehumidified air is cooled again (2) and at last it is humidified (1) what results in air temperature drop. Process of direct evaporative cooling can be replaced by indirect process so-called Maisotsenko cycle [4] which may provide less humid air for better thermal comfort. Simultaneously in the second fluidised bed the desiccant undergoes regeneration. The fan (8) blows the air through recovery heat exchanger (7) and the air heater (6). Then the hot air enters the fluidised bed, where due to high air temperature, desorption of water from the desiccant takes place. At last outlet regeneration air flows again through recovery heat exchanger (7). For better understanding described processes are illustrated on psychrometric chart which is presented in Fig. 2. FDC requires electric power to run fans and thermal power to regenerate

* Corresponding author: zbigniew.rogala@pwr.edu.pl

desiccant after adsorption. Therefore its performance can be described by thermal and electric COPs which are the ratios of cooling output to thermal or electric input. Another useful parameter describing the performance of FDC is Specific Cooling

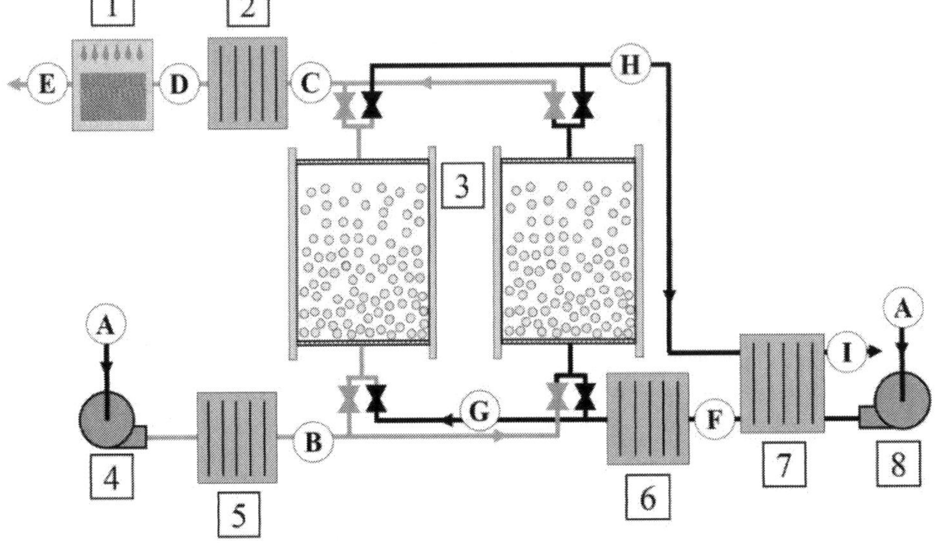

Fig. 1. Scheme of simple fluidised desiccant cooler (FDC). 1 – Humidifier, 2 – Air cooler, 3 – Fluidised beds, 4 – Electric fan, 5 – Air cooler, 6 – Air heater, 7 – Heat recovery, 8 – Electric fan. Letters in circles correspond to processes presented in Fig.2.

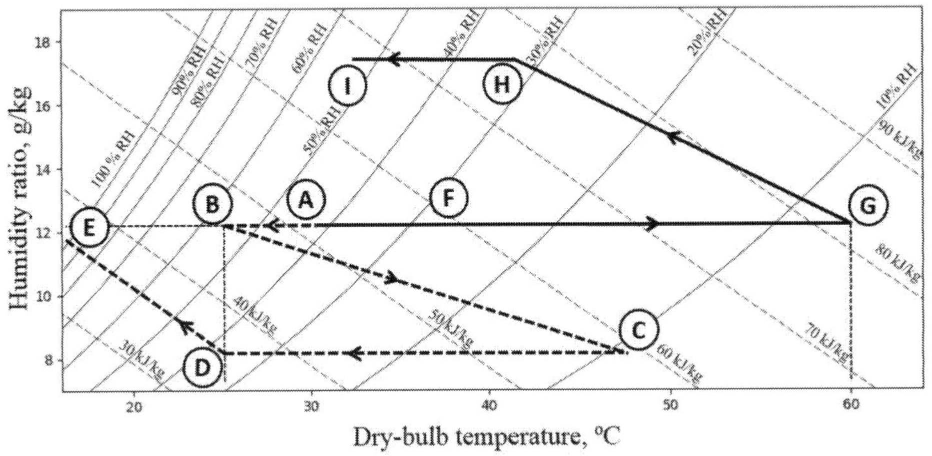

Fig. 2 Air processing in FDC. Letters in circles correspond to Fig. 1.

Power (SCP) which is cooling power obtained per unit mass of desiccant. Assuming that entire change of desiccant water uptake generates cooling effect and pressure drop on fluidised bed is described by [1]

$$\Delta P = H_d (1-\varepsilon)\,(\rho_p - \rho_g) g \tag{1}$$

thermal and electric COPs and SCP can be expressed as follows

$$COP_{el} = \Delta w\ H_{ev}\ \eta_{fan}\ /\ [(U_{ads} + U_{des})\ g\ t_{sw}] \tag{2}$$

$$COP_{th} = \rho_d\ \Delta w\ H_{ev}\ H_d\ /\ (\rho_g\ \Delta T_{mean}\ U_{des}\ c_g\ t_{sw}) \tag{3}$$

$$SCP = \Delta w\ H_{ev}\ /\ t_{sw} \tag{4}$$

Where ε is void fraction of the fluidised bed, g is gravitational acceleration, Δw is change of desiccant water uptake, H_{ev} is heat of evaporation, ΔT_{mean} is mean air temperature difference between outlet and inlet during desorption and c_g is specific heat of air. The performance of FDC mainly depends on adsorption/desorption processes taking place in fluidised beds. On the other hand these processes depend on operating conditions such as bed switching time t_{sw}, desiccant particles diameter d_p, superficial velocity U_{des} and U_{ads} and mass of desiccant M_d in the fluidised bed. Influence of these parameters on FDC COPs and SCP were not yet investigated. Few prototypes of FDC were already presented in the literature [3, 5] and their electric COP and SCP did not exceed 0.37 and 285 W/kg, respectively. It was probably caused by improper selection of operating conditions. Analysis of the effect of abovementioned operating conditions will provide recommendations and guides on efficient FDC operating. Moreover, it contributes to better understanding of FDC operation and its limitations.

2 Results and discussion

The theoretical analysis of adsorption was carried out basing on modelling method presented in [1] while desorption was calculated basing on [6-8]. Basing on abovementioned references computational code was implemented in Python and run explicitly. The modelling was performed for regeneration temperature of 60 °C, cooling temperature of 25 °C and humidity ratio of 12 g/kg. Parameters which were kept constant during the modelling are presented in Table 1.

Table 1. Constant parameters used in modelling

Parameter	Symbol and unit	Value	Parameter	Symbol and unit	Value
Air density	ρ_g, kgm^{-3}	1.2	Fluidised bed diameter	D, m	0.28
Desiccant density	ρ_d, kgm^{-3}	800	Fan efficiency	η_{fan}, -	0.8

Moreover some of the parameters were varied during modelling (see Table 2).

Table 2 Parameters which were varied during the modelling

Operating condition		d_d, mm	t_{sw}, s	U_{ads}, ms^{-1}	U_{des}, ms^{-1}	H_d, m
Desiccant particle diameter Fig. 2	d_d	1 – 5	optimum	2	2	0.05
Switching time Fig 3.	t_{sw}	1, 3, 5	200 - 1500	2	2	0.04
Superficial air velocities during adsorption and desorption Fig 4.	U_{ads}, U_{des}	1	optimum	1 - 6	1 - 6	0.04
Superficial air velocity during adsorption and desiccant filling height Fig.5	U_{ads}, H_d	1	optimum	2-6	2	0.01 – 0.1

Firstly, the effect of desiccant particle diameter on performance of FDC was investigated. In FDC desiccant is simply poured into fluidised bed. During fluidisation each single particle has a direct contact with the air and therefore some of the limitations [9, 10] of other adsorption technologies are not present. On the other hand, desiccant particle diameter affect minimum fluidisation velocity [1] and termination velocity [11]. In order to maintain the proper FDC operation the fluidisation velocity must be kept between these velocities, otherwise either fluidisation will not occur (too low superficial air velocity) or desiccant may clog the upper sieve of fluidised bed and increase the pressure drop (too high superficial air velocity). Modelling was carried out for particles diameters ranging from 1 to 5 mm and its results are presented in Fig. 2. The parameters of modelling were set accordingly to Table 1. Superficial velocities during adsorption and desorption were set to 2 ms^{-1} and desiccant filling height H_d in each bed was set to 0.05 m. Switching time is initially unknown and therefore range of switching times is analyzed at first and then the best result is selected. Optimum switching time is also presented in Fig. 2. Thermal COP (see Fig. 2a) is nearly constant, regardless of desiccant particle diameter and equals to ca. 0.5. On the other hand optimum electric COP and SCP strongly depend on desiccant particle diameter (see Figs 2b and 2c). For desiccant particle of 5 mm electric COP and

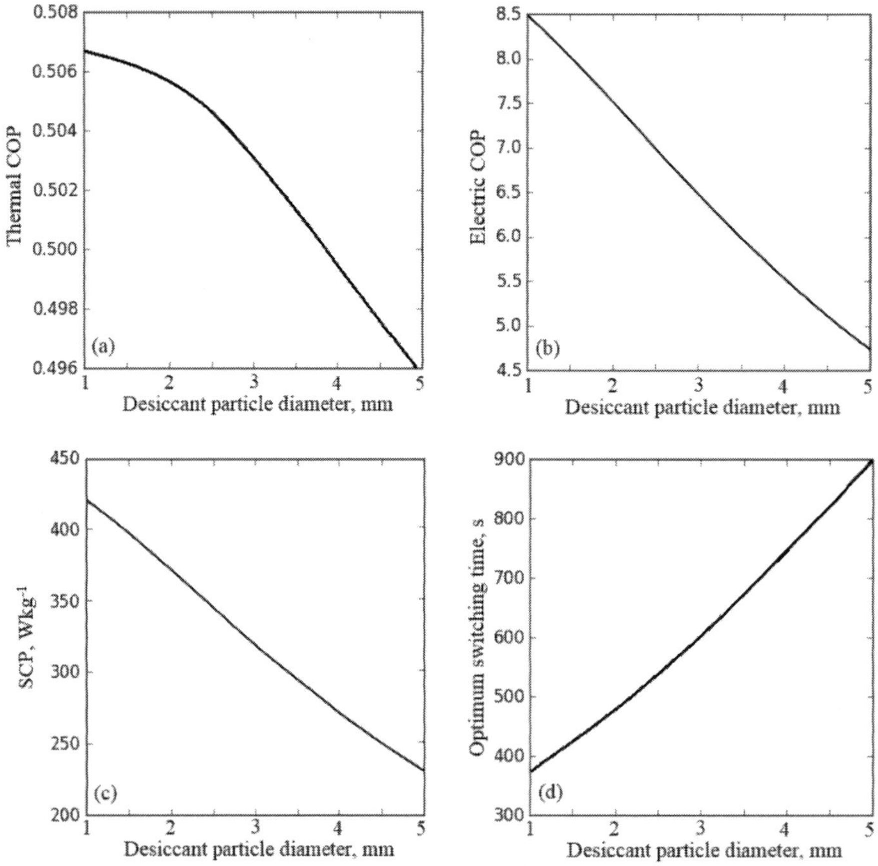

Fig. 1. Performance of fluidised desiccant cooler with respect to desiccant particle operated at optimum switching time: (a) Thermal COP, (b) Electric COP, (c) SCP, (d) Optimum switching time.

 DOI: 10.1051/e3sconf/20172200146

SCP are equal to 4.7 and 240 W/kg while for desiccant particle of 1 mm they are equal to 8.5 and 420 W/kg, respectively. The increase of electric COP and SCP for smaller desiccant particle diameters is caused, among other factors, by decrease of optimum switching time (Fig. 2d) which drops from 900 s for 5 mm particle to less than 400 s for 1 mm particle. Analysis of the effect of desiccant particle diameter on performance of FDC leads to conclusion that the performance increases with decreasing particles diameter. On the other hand, as was already mentioned, desiccant particle diameter affect the minimum fluidisation and termination velocity of particles. Limitations related to this phenomena should be investigated in practice. Another important factor affecting performance of FDC is switching time. It is already known that for certain switching time SCP and electric COP achieve the highest possible values. Nevertheless the effect of switching time on FDC performance is still unknown. Therefore the analysis in this matter was carried out and its results are presented in Fig. 3. The modelling was carried out for superficial air velocities equal to 2 ms^{-1} (during adsorption and desorption) and desiccant filling height of 0.04 m. Desiccant particles featuring diameters of 1, 3 and 5 mm were analyzed. Thermal COP (see Fig. 3a) increases with the increase of switching time for each analyzed desiccant particle

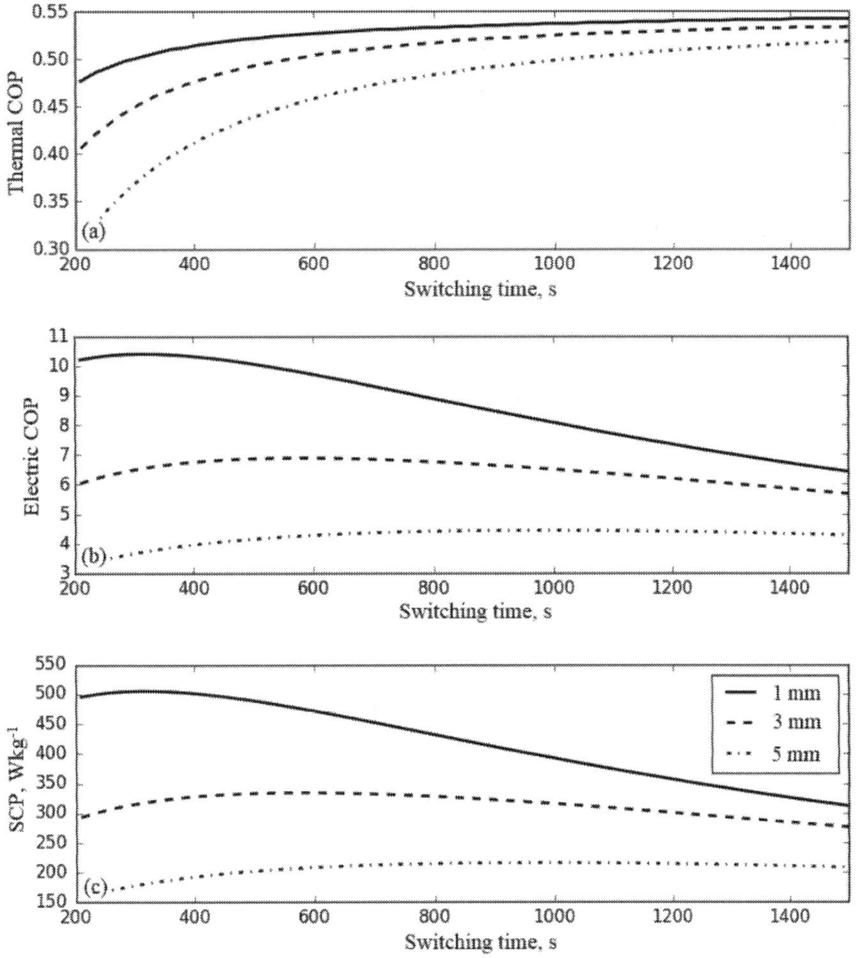

Fig. 2. Performance of fluidised desiccant cooler with respect to switching time for desiccant particle diameters 1 mm, 3 mm and 5 mm: (a) Thermal COP, (b) Electric COP, (c) SCP.

 DOI: 10.1051/e3sconf/20172200146

diameter. Moreover, thermal COP is greater for smaller desiccant particles. For desiccant particle diameter of 1 mm thermal COP remains high in the investigated range of switching times and its maximum value reaches ca. 0.54. Electric COP (see Fig. 3b) and SCP (see Fig. 3c) for certain switching time reach their maximum value. The larger desiccant particle is the lower maximum electric COP and SCP are and the longer optimum switching time is. Such characteristics results from kinetics of adsorption. Initially adsorption is rapid but it decreases with time as desiccant water uptake increases towards equilibrium. In case of operating on switching times shorter than optimum performance of FDC is generally deteriorated and such operating conditions are undesirable. On the other hand operating on switching time longer than optimum causes the increase of thermal COP but at the cost of electric COP and SCP. In short, switching time should be not shorter than optimum switching time which for desiccant particle of 1 mm in considered case is about 350 s.

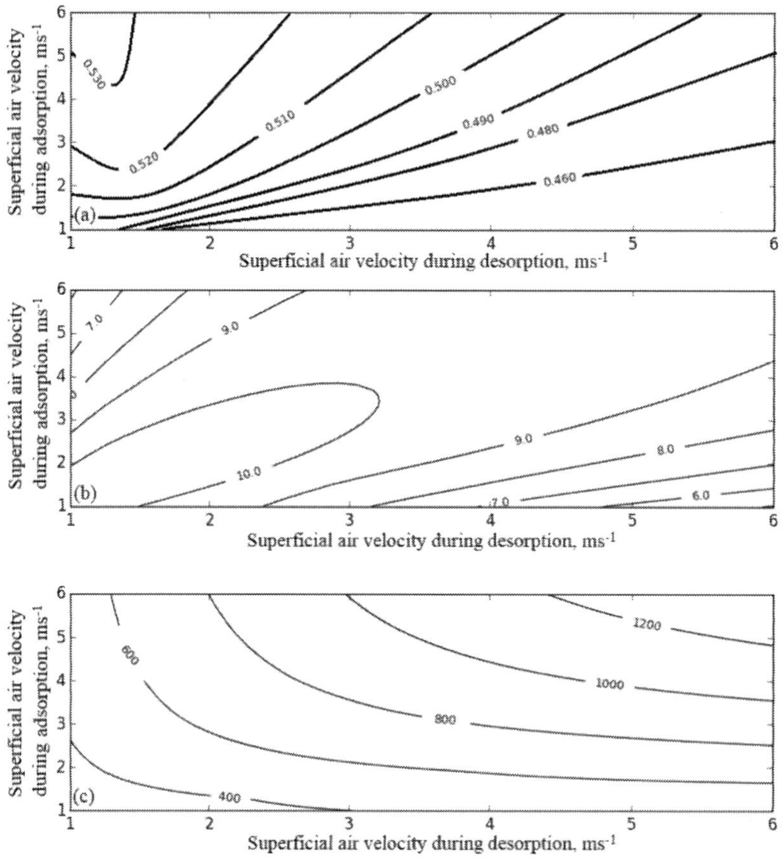

Fig. 3. Performance of fluidised desiccant cooler with respect to superficial air velocities during adsorption and desorption, operated at optimum switching time: (a) Thermal COP, (b) Electric COP, (c) SCP.

Longer switching times result in increased thermal COP while electric COP and SCP are decreased. FDC can be operated with range of superficial air velocities. Lower limit is minimum fluidisation velocity and upper limit is termination velocity. Superficial air velocity affects electric demand of fans and kinetics of adsorption. Moreover kinetics during adsorption and desorption is different. Therefore superficial air velocity may differ during these processes. Analysis of the effect of superficial air velocity was carried out for

 DOI: 10.1051/e3sconf/20172200146

superficial air velocities ranging from 1 to 6 ms^{-1} and its results are presented in Fig. 4. The modelling was performed for desiccant particle diameter of 1 mm and desiccant filling height of 0.04 m. Switching time was set to optimum in each case. Thermal COP of FDC (see Fig. 4a) is highest for high superficial air velocity during adsorption and low during desorption. When both superficial air velocities are the same, thermal COP remains nearly constant irrespective of superficial air velocities. Electric COP of FDC (see Fig. 4b) achieves the highest values for both superficial air velocities of ca. 2 ms^{-1}. Electric COP also remains high when superficial air velocities during adsorption and desorption are similar. At last SCP (see Fig. 4c) is the highest when both superficial velocities are ca. 6 ms^{-1}. In order to keep electric COP high FDC should be operated with similar superficial air velocities during adsorption and desorption. These velocities should be low in order to

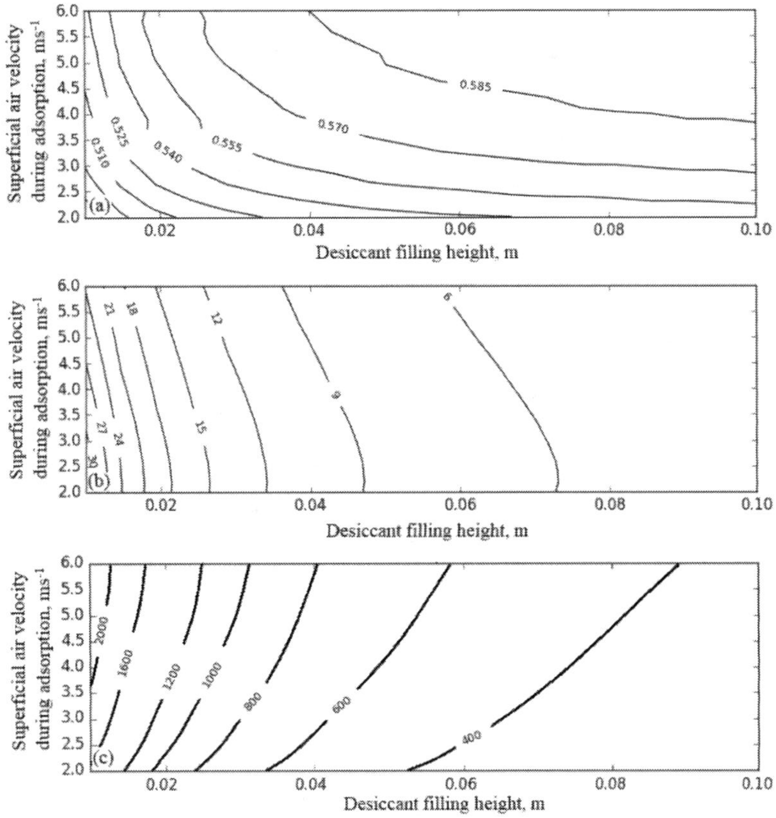

Fig. 4. Performance of FDC with respect to superficial air velocity during adsorption and desiccant filling height, operated at optimum switching time: (a) Thermal COP, (b) Electric COP, (c) SCP.

obtain higher thermal COP or high in order to obtain high SCP. At last, the effect of desiccant filling height on FDC performance was investigated. Desiccant filling height represents the mass of desiccant poured into the fluidised bed. This portion of desiccant will be either cooled or heated by the air stream flowing through the fluidised bed. Due to the limitations of superficial air velocity the amount of heating or cooling provided by the air stream is also limited. Therefore, the effect of desiccant filling height is analyzed along with superficial air velocity during adsorption. The modelling was carried out for desiccant particle diameter of 1 mm and superficial air velocity during desorption of 2 ms^{-1}.

Switching time was set to optimum. The modelling results are presented in Fig. 5. As it can be observed, the electric COP and SCP depend strongly on desiccant filling height. For higher fillings dependence on superficial air velocity becomes substantial too. Electric COP is generally higher than 9 for desiccant filling height lower than 0.04 m. On the other hand, SCP decreases with the increase of filling height and superficial air velocity during adsorption. Desiccant filling height has opposite influence on thermal and electric COP. Thermal COP increases for higher fillings while electric COP drops. In the authors opinion it is of great importance to keep electric COP high which limits the increase of desiccant filling height.

3 Conclusions

The effect on performance of FDC of desiccant particle diameter, switching time, superficial air velocity and desiccant filling height was modelled and analyzed. Thermal and electric COP and SCP were investigated. The modelling results provided useful information on FDC efficient operating. Conclusions and recommendations are as follows:

• Desiccant particle diameter should be as small as it is possible. Technical aspects related to minimum fluidisation and termination velocities should be taken into account to avoid clogging.
• Switching time should not be shorter than optimum. Optimum switching time allows to obtain maximum electric COP and SCP. Switching time longer than optimum allows to increase thermal COP at the cost of SCP and electric COP. Optimum of bed switching time is not constant and depends on operating conditions. The effect of the individual parameters should be further investigated.
• In order to keep high electric COP FDC should be run with similar superficial air velocities during adsorption and desorption. In order to obtain higher thermal COP superficial velocities should be lower, on the other hand to obtain higher SCP the superficial air velocities should be higher.
• Desiccant filling height has opposite influence on electric and thermal COP. Increase of filling height leads to increase of thermal COP and SCP but on the other hand it leads to deterioration of electric COP.

1. C-H. Chen, G. Schmid, C-T. Chan, Y-C. Chiang, S-L. Chen, Appl. Therm. Eng. **89**, 229 (2015)
2. A.M. Hamed, W.R.A. Rahman, S.H. El-Emam, Energ. **35**, 2468 (2010)
3. C-H. Chen,S-S. Ma, P-H. Wu, Y-C. Chiang, S-L. Chen, Appl. Energ. **155**, 708 (2015)
4. D. Pandelidis, S. Anisimov, W.M. Worek, P. Drąg, Energ. Buildings **140**, 154 (2017)
5. Y-C, Chiang, C-H. Chen, Y-C. Chiang, S-L. Chen, Appl. Energ. **175**, 199 (2016)
6. Y.C. Lee, L.C. Weng, P.C. Tseng, C.C. Wang, Heat Mass Transf. **51**, 441(2014)
7. C.C. Ni, J. Y. San, Int. J. Heat Mass Tran. **45**, 1839 (2002)
8. A.A. Pesaran, A.F. Mills, Int. J. Heat Mass Tran. (1986)
9. S. Salvatore, A. Sapienza, A. Frazzica, A. Freni, I.S. Girnik, Appl. Energ. **134**, 11 (2014)
10. V. Brancato, L. Gordeeva, A.Sapienza, A. Freni, A. Frazzica, Appl. Therm. Eng. **105**, 28 (2016)
11. Y. Wen-Ching, *Handbook of fluidization and fluid-particle systems* (Marcel Dekker, New York, 2003)

DOI: 10.1051/e3sconf/20172200147

Organic and elemental carbon bound to particulate matter in the air of printing office and beauty salon

Patrycja Rogula-Kopiec[1,1], *Józef S.* Pastuszka[2], *Wioletta* Rogula-Kozłowska[1,3], and *Walter* Mucha[2]

[1]Institute of Environmental Engineering, Polish Academy of Sciences, 34 M. Skłodowska-Curie St., 41-819 Zabrze, Poland
[2] Department of Air Protection, Silesian University of Technology, 22B Konarskiego St., 44-100 Gliwice, Poland
[3]The Main School of Fire Service, Faculty of Fire Safety Engineering, 52/54 Słowackiego St. 01-629 Warsaw, Poland

Abstract. The aim of this study was to determine the role of internal sources of emissions on the concentrations of total suspended particulate matter (TSP) and its sub-fraction, so-called respirable PM (PM_4; fraction of particles with particle size ≤ 4 μm) and to estimate to which extent those emissions participate in the formation of PM-bound elemental (EC) and organic (OC) carbon in two facilities - one beauty salon and one printing office located in Bytom (Upper Silesia, Poland). The average concentration of PM in the printing office and beauty salon during the 10-day measurement period was 10 and 4 (PM_4) and 8 and 3 (TSP) times greater than the average concentration of PM fractions recorded in the same period in the atmospheric air; it was on average: 204 μg/m^3 (PM_4) and 319 μg/m^3 (TSP) and 93 μg/m^3 (PM_4) and 136 μg/m^3 (TSP), respectively. OC concentrations determined in the printing office were 38 μg/m^3 (PM_4) and 56 μg/m^3 (TSP), and those referring to EC: 1.8 μg/m^3 (PM_4) and 3.5 μg/m^3 (TSP). In the beauty salon the average concentration of OC for PM_4 and TSP were 58 and 75 μg/m^3, respectively and in case of EC - 3.1 and 4.7 μg/m^3, respectively. The concentrations of OC and EC within the those facilities were approximately 1.7 (TSP-bound EC, beauty salon) to 4.7 (TSP-bound OC, printing office) times higher than the average atmospheric concentrations of those compounds measured in both PM fractions at the same time. In both facilities the main source of TSP- and PM_4-bound OC in the indoor air were the chemicals - solvents, varnishes, paints, etc.

[1] Corresponding author: patrycja.rogula-kopiec@ipis.zabrze.pl

1 Introduction

The quality of indoor air in different types of rooms depends among others on the physiological activity of people staying there, the type and intensity of the functions served within such rooms and the type and effectiveness of the ventilation system. In service facilities, where food is prepared (cooking, frying), or coal furnaces or fireplaces are being used for heating purposes, and most of all where people smoke tobacco, the quality of indoor air is much worse compared to outdoor one. These concern, among other, air pollution by particulate matter, PM, and especially its fine particles [1-5]. Fine PM particles freely migrate into rooms interiors, which means that their concentration in the indoor air should be similar (or greater) compared to its ambient concentrations (outdoor air). This tendency can occur depending on the presence and effectiveness of the internal sources of PM and other pollutants emission (eg. some gases reacting with each other and with PM components), room airing (preventing or promoting accumulation of PM inside the building) or the efficiency of room ventilation system [6-10].

Due to the harmfulness associated with the emissions of different types of air pollutants in different factories and other workplaces, the concentrations of specific substances, including PM, are being routinely measured. However, there are many facilities, where, despite heavily polluted air, there is no standards or legal requirement warranting the assessment of this pollution. This anticipated, significant air pollution in the room interiors can be caused by both: the fact that such objects are located in highly polluted region, (where at certain, described above conditions the accumulation of indoor pollution occurs) and by the presence of internal sources of emissions. The need to recognize the quality of indoor air in such areas is objectively justified when both mentioned sources of emissions (the outdoor and the indoor one) are constantly present.

Silesia Agglomeration is one of Europeans regions with the most polluted air. Concentrations of PM and PM components in its urban locations in some periods are several times higher compared to those meet in the big cities of Western, Northern and Southern Europe [11-13]. Compared to other Polish and European regions Silesia region is also heavily industrialized and urbanized. Within the Upper Silesia Agglomeration, which consists of 14 cities, the total number of its residents exceeds 2 million people, more than 60% of the working population employed in the so-called services sector. Among service facilities, where significant internal sources of air pollution are normally found printing office and beauty salons are definitely the leading one. In such places, due to the nature of the performed services, a very high concentrations of various types of organic pollutants can be suspected. Given the fact that Silesia region is characterized by a very high concentrations of PM and relatively high concentrations of PM-bound carbon [13,14], and interesting topic was to recognize to which extent the activities carried out in such facilities enhance the formation and structure of PM-bound carbon compounds in the indoor air.

The PM-related Total Carbon (TC) consists of elemental carbon (EC) and organic carbon (OC) [15,16]. PM-bound EC is mostly of primary origin, emitted mainly as a result of incomplete combustion of solid fuels, or the pyrolysis of biological material [17], while the PM-bound OC in a one part consist of primary matter (primary organic carbon, POC), and partly is the product of the transformation of gaseous precursors in the air (secondary organic carbon, SOC) [16,18,19]. The sources of POC include combustion processess, including the combustion of: gasoline (in internal combustion engines), coal from biomass and also vegetation, including pollen, spores, crushed plant material, as well as bacteria and viruses [20]. The intensity of SOC creation depends on the direction and dynamics of the chemical reactions occurring in the air, in particular the transformation of volatile organic compounds (VOCs) [21]. The expected, significant emission of VOCs, but also PM-bound OC will probably occur in rooms where the printing is being carried out, using various types of paints, solvents and but also laser printing, welding, etc. but also there, where different types of treatments and beauty services are being carried out using the chemicals rich in organic acids, phenols, hydrocarbons, alcohols, but also those requiring the usage of laser, milling machines, and so on. Such activities will influence not only the changes in the contribution of PM-bound OC and EC in the indoor air in relation to atmospheric air but also the contributions of POC and SOC in PM-bound OC.

The aim of the study was to determine the role of internal sources of OC and EC on the concentrations of PM-bound EC and OC in two facilities - one beauty salon and one printing office located in a city of Upper Silesia (Poland).

2 Materials and methods

To conduct this study, a single printing office and beauty salon were typed. Both were located in Bytom city (one of the 14 cities located in Silesia Agglomeration with an area of 69 km^2 and 170 thousand of inhabitants). Both facilities were located within the same housing estate, surrounded by the typical settlement, without any large source of emission in the closest proximity, like industrial/manufacturing facility or traffic artery. Selected printing office providing printing services like large format printing on a variety types of paper and films, but also other services like thermal treatment (welding) the edges of the selected products (banners, posters). The measurements campaign was conducted in a hall with an area of about 50 m^2, which run printing, at a distance of about 3 m in a straight line from the printing machine. The hall is located in a new building, which remaining part (about 15 m^2) is a customer service, social room (10 m^2) and a toilet. The building has a natural ventilation system and two large (PVC) windows on each wall. In the summer, when the measurements were carried out the door between the hall and the office together with the exit door (emergency exits from the hall) and windows were constantly opened. All printouts from the machine are hanging to dry in the open air inside the hall, at the time when the machine is kept on a standby, for example when cyclical printing areas are sprayed with the appropriate solvent. The beauty salon is located in a residential building, adopted for service activities. Measurements were carried out in one among five rooms belonging to the salon (in each a different kind of treatments is being run) – the office, where pedicure and manicure are performed. The room has an area of about 30 m^2. The room is equipped with a seat for customers and a beauty table. Measurements were conducted at a distance of approximately 3 m from the table. The building is equipped with a gravity ventilation and PVC windows, most of which are closed, and the room is ventilated only after the working day is end (after 18.00). During work a beautician uses miling machine, nailfiles, and a number of cosmetics containing hydrocarbons, alcohols, esters, phenols, and the like. Nail coatings and gels are cured under UV lamp.

PM samples were collected in the period of 18-31 July 2015 in the salon and in the period of 18-29 August 2015 in the printing office in a 8-hour cycles, usually between 10.00-18.00, (normal working day). All measurements were performed during summer period, because in winter the concentrations of PM and PM-bound OC and EC are overshadowed by the emissions from the combustion of coal and biomass for heating purposes [14,16]; this emission is very high in this region and masks the influence of other sources of PM and PM-bound carbon components. In each measurement site a set of two identical, calibrated PM samplers were used, one collecting (PM4) and the second one collecting (TSP). Such set was located inside the room as well as the outside. In this arrangement, at each measurement point within a single room, ten measurements were performed using Gilian sampling pumps (Sensidyne, St. Petersburg, FL).

Samples were collected on Whatman quartz filters (QMA, ø25 mm) (GE Medical Systems Polska, Life Sciences, Warsaw, Poland). The mass of a PM sample was determined by filter weighing on a microscale (resolution of 1 μg) (Radwag; Radom, Poland) before and after the exposure. Before weighing, chemical composition on the filter was conditioned in the weighing room for 48 h under constant conditions (humidity of 45±5 % and air temperature of (20±2°C). After the collection of PM samples and gravimetric analysis, a 1.5 cm^2 fraction of each collected sample was taken for tests on organic carbon (OC) and elemental carbon (EC) contents using a thermal-optical method. The analytical methods including precision verification procedures, has been described in detail elsewhere [16].

3 Results and discussion

Inside the beauty salon and printing office the 8-h concentration of PM4 and TSP as well as their carbon components (TSP- and PM4-bound OC and EC) were generally significantly higher than in atmospheric air. In both locations the differences in the 8-h concentrations between the indoor and outdoor were statistically significant ($p < 0.05$ at $p_{assump.} = 0.05$; U Mann-Whitney test). Those concern both the PM fractions and their carbonaceous species. In the printing office the average concentration of PM4 and TSP in the indoor air was 10 times and 8 times higher compared to outdoor concentrations, respectively; analogous relationship concern beauty salon were this differences were 4 (PM4) and 3 (TSP) times, respectively - Table 1. In case of TSP- and PM4-bound OC the ratios of the average concentrations of those fractions between the printing office's and a beauty salon's interiors and ambient air do not exceed 5; and for EC it was in range 1.66-2.83 (Table 1).

In both facilities the impact of internal emission sources of PM and PM-bound OC and EC on their indoor concentrations was clearly marked. The 8-h concentration of PM and PM-bound OC and EC in both facilities were not correlated with the 8-h concentrations of PM and PM-bound OC and EC measured at the same time outside those buildings, which excludes the substantial influence of atmospheric air on indoor one (Table 2). In the printing office the effect of the internal emissions on PM and PM-bound OC/EC concentrations was much more pronounced than in the beauty salon and much clearer in case of PM than PM-bound OC/EC.

In the printing office the relative (compared to atmospheric/outdoor air) increase in the concentrations of indoor PM and PM-bound OC wasn't comparable. So, the increase in the concentrations of PM in the indoor air may be related to the workers and clients activities within the printing office, the erosion of different mechanical elements of the printing machine, and the emission associated with the processing of finished prints, etc. Thus, within such area the concentration of other than EC and OC components (like non determined in this work minerals, metals and metalloids, etc.), increase probably even greater (compared to the outdoor air concentrations). It is also very likely that in the specific conditions of high concentrations of ozone and ammonia typical for printing office interiors [22] and the free flow of sunlight from the outside at the same time (open doors and windows) the transformations of other than VOCs (the precursors of OC) compounds emitted during the printing and thermal processing of prints, like the precursors of secondary (inorganic) aerosol, ie. nitrogen oxides. The 8-h concentrations of TSP- and PM4-bound OC in indoor air in the printing office were not correlated with the 8-h concentrations of TSP and PM4, which confirms that the increase in the 8-h concentration of PM in the air did not proceed in same way as in case of 8-h OC concentrations (Table 2).

In the beauty salon, this situation was different. A strong and positive correlations between TSP and TSP-bound OC as well as in case of PM4 and PM4-bound OC testify that their 8-h levels grew at a similar (proportional) rate. In addition, an increase in the concentrations of PM-bound OC in indoor air vs. outdoor air was more pronounced than the increase in the concentrations of PM (Table 1). In case of beauty salon a strong and direct impact of emissions of various kinds of organic compounds that are components of the cosmetics were reflected by an increase in the concentration of OC bounded to both: fine and coarse PM (the contribution of PM4 in TSP in indoor air was about 65% - Table 1 and R2 for the correlation between the 8-h concentrations of PM4-bound OC and TSP-bound OC in the printing office came to 0.9 - Table 2).

In both facilities the effect of the internal activity on the concentration of indoor PM-bound EC was strictly observed. In case of printing office it was probably the result of EC emission during the thermal processing of prints (welding) done on different kinds of foils (the correlation between the 8-h concentrations of PM4 and PM4-bound EC and the lack of correlation between the 8-h concentrations of TSP and TSP-bound EC). In the beauty salon interiors, the probably explanation for almost three times higher concentrations of indoor EC concentrations compared to outdoor ones for both fine and coarse PM is almost continuous use the incense and scented candles [23,24].

Table 1. The concentrations and ratios of PM_4, TSP and related carbon compounds (elemental, EC and organic, OC) in the printing office and in the beauty salon.

[a] PRINTING OFFICE	PM_4	OC	EC	TSP	OC	EC	PM_4/ TSP	[b] OC/OC	[c] EC/EC
Ave±st.dev. Indoor	203.87± 320.55	38.05± 9.10	1.83± 0.84	318.95± 456.46	56.21± 8.79	3.46± 1.26	0.65± 0.19	0.68± 0.14	0.53± 0.16
Ave±st.dev. Outdoor	19.36± 7.46	10.89± 2.38	0.86± 0.54	33.48± 10.46	13.26± 3.74	1.40± 0.64	0.58± 0.10	0.84± 0.12	0.59± 0.16
Ave±st.dev. Indoor/Outdoor	10.29± 15.53	3.60± 0.97	2.83± 1.90	8.21± 10.03	4.67± 1.84	2.81± 1.32	-	-	-
BEAUTY SALON	PM_4	OC	EC	TSP	OC	EC	PM_4/ TSP	[b] OC/OC	[c] EC/EC
Ave±st.dev. Indoor	92.84± 50.44	58.00± 24.82	3.12± 2.45	135.99 ± 59.95	75.38 ± 32.24	4.66 ± 3.01	0.68± 0.15	0.79± 0.11	0.67± 0.22
Ave±st.dev. Outdoor	24.81± 5.80	14.76± 5.83	1.68± 0.34	43.66 ± 10.12	20.27 ± 3.85	2.92 ± 0.66	0.59± 0.16	0.72± 0.23	0.58± 0.10
Ave±st.dev. Indoor/Outdoor	3.88± 2.19	4.66± 2.74	1.81± 1.15	3.37± 2.06	3.82± 1.76	1.66± 1.08	-	-	-

[a] in each facility, both inside and outside, the number of measurement data for one fraction n=10; Total number of samples in each facility n=40;
[b] The mass contribution of PM_4-bound OC in the TSP-bound OC
[c] the contribution of PM4-bound EC in the TSP-bound EC

Table 2. The coefficient of determination (R^2) indicating the strength of the linear correlation between the 8-h concentrations of PM4, TSP and their carbon components (elemental, EC and organic, OC carbon) inside and outside the printing office and the beauty salon.

Location; PM fraction	Relation	R^2	
		Printing office	Beauty salon
Indoor; PM_4	OC *vs* EC	0.469	0.524
	OC *vs* PM4	0.299	0.884
	EC *vs* PM4	0.626	0.532
Indoor; TSP	OC *vs* EC	0.195	0.761
	OC *vs* TSP	0.000	0.842
	EC *vs* TSP	0.021	0.561

DOI: 10.1051/e3sconf/20172200147

Indoor PM_4 and TSP	PM4 *vs* TSP	0.716	0.884
	OC *vs* OC	0.238	0.876
	EC *vs* EC	0.468	0.798
Indoor and Outdoor; PM_4	PM4 *vs* PM4	0.015	0.030
	OC *vs* OC	0.132	0.123
	EC *vs* EC	0.042	0.216
Indoor and Outdoor; TSP	TSP *vs* TSP	0.282	0.018
	OC *vs* OC	0.082	0.020
	EC *vs* EC	0.306	0.000

Beside the influence of internal emission sources on the concentrations of PM and PM-bound OC and EC in the printing office and beauty salon, we also examined their impact on the qualitative composition of PM-bound OC, specifically the contribution of SOC and POC in the PM-bound OC in the indoor vs. outdoor. Because there is no simple method for separation of SOC and POC in PM-bound OC, 8-h concentrations of TSP- and PM4-bound SOC and POC in indoor and in outdoor air, for both sites, were computed from the formulas **[25]**:

$$POC=(OC/EC)_{min} \times EC \quad (1)$$
$$SOC=OC-POC \quad (2)$$

where: OC and EC are 8-h OC and EC concentrations for the same fraction in each location; $(OC/EC)_{min}$ is the lowest OC/EC for the same fraction in each location.

Both in case of printing office as well as in case of beauty salon the contribution of SOC in the OC was higher in the outdoor than in indoor air; this concerns OC associated with both PM fractions (Fig. 1 and 2). Considerably higher contribution of POC in OC in the indoor air compared to outdoor one, in both analyzed cases, shows the direct influence of PM-bound OC and EC emissions on POC concentrations. The high concentrations of PM-bound EC in both areas contributed, among others, to the fact that the more volatile and semi-volatile compounds from the group of OC was adsorbed onto PM particles (rich in EC concentrations, and having large surface area). In the atmospheric air, during the conditions of strong solar radiation and high air temperature in the summer, these compounds undergo photochemical reactions, and forms the SOC.

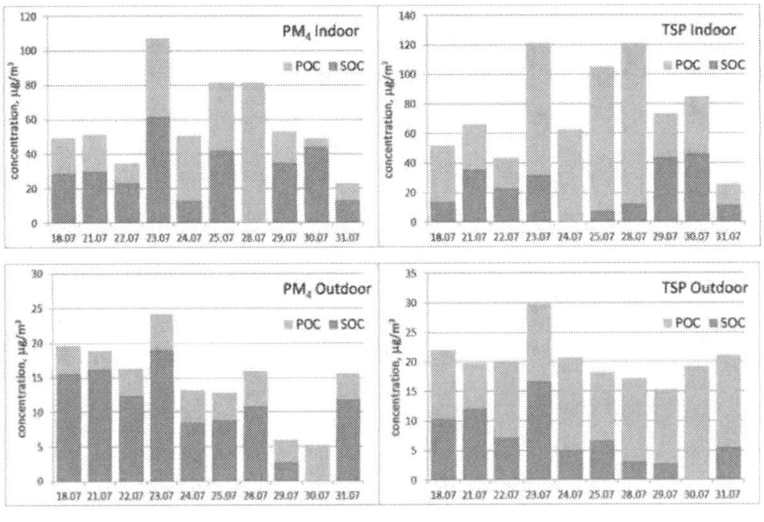

Fig. 1. The 8-h concentrations of SOC and POC in the beauty salon.

DOI: 10.1051/e3sconf/20172200147

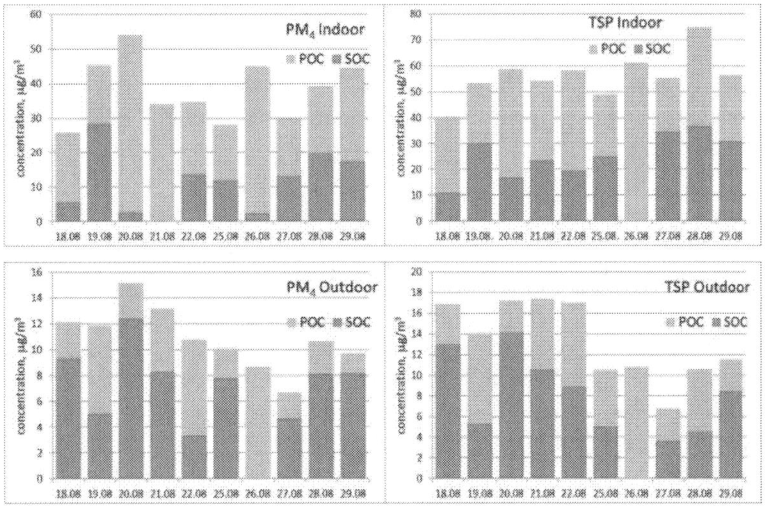

Fig. 2. The 8-h concentrations of SOC and POC in the printing office.

4 Conclusions

It has been shown that the indoor concentrations of total particulate matter TSP and the inhalable PM_4 fraction in the printing offices and beauty salons, can be very high, and even up to 8 times higher compared to the outdoor ones. In the printing office this could be attributed not only to the internal emissions of PM-bound organic compounds but also due to the mechanical processes of the printing machines working inside such buildings. An increase in the concentrations of PM in the printing office could be also connected with the production of a secondary inorganic aerosol generated by the conversion of nitrogen oxides, ozone and ammonia which are constantly present in such service facilities in very high concentrations. The mentioned phenomena may be although just a seasonal effect (measurements were conducted in a very warm season of the year). In the beauty salons, where human activity is restricted to the hand movement of the beautician an increase in the PM concentrations depends on the emissions of a various substances used during the beauty treatments.

The emissions of organic compounds related to the indoor activities in both service facilities significantly increase the contribution of secondary and primary matter, like SOC and POC in the OC mass. In both facilities, the contribution of POC in OC mass was much greater than in the atmospheric air. This is indirectly due to the relatively high concentrations of EC in the indoor air of both facilities and the adsorption of volatile and semi-volatile organic compounds on the surface of the EC particles.

Acknowledgements

Work was supported by the DEC-2013/09/N/ST10/04224 project, which was financed by the National Science Center (NCN).

References

1. J. S. Pastuszka, U. Kyaw Tha Paw, DO. Lis, A. Wlazło, Atmos Environ **34** (2001)

2. A. Zwoździak, I. Sówka, B. Krupińska, J. Zwoździak, A. Nych, Build Environ **66** (2013)
3. A. Mainka, E. Zajusz-Zubek, K. Kaczmarek, Int J Environ Res Pub He **12** (2015)
4. H. R. Chao, D. E. Que, Y. Y. Gou, C. Y. Chuang, T. Y. Chang, Y. C. Hsu, Aerosol Air Qual Res **16** (2016)
5. M. Oliveira, K. Slezakova, C. Delerue-Matos, M. do Carmo Pereira, S. Morais, Environ Pollut **208** (2016)
6. L. Morawska, T. Salthammer, *Indoor environment: airborne particles and settled dust* (John Wiley & Sons, 2006)
7. F. R. Spellman, *The Science of Air. Concepts and Applications* (Taylor&Francis Group, 2009)
8. M. Lippman, *Environmental toxicants: human, exposures and their health effects* (Wiley, 2007)
9. E. Błaszczyk, W. Rogula-Kozłowska, K. Klejnowski, I. Fulara, D. Mielżyńska-Švach, Air Qual Atmos Hlth (2016)
10. G. Majewski, K. Kociszewska, W. Rogula-Kozłowska, H. Pyta, P. Rogula-Kopiec, W. Mucha, J. S. Pastuszka, Atmosphere **7** (2016)
11. W. Rogula-Kozłowska, K. Barbara, K. Krzysztof, S. Szopa, Arch Environ Prot **39** (2013)
12. W. Rogula-Kozłowska, G. Majewski, P.O. Czechowski, Environ Monit Assess **187** (2015)
13. J.S. Pastuszka, W. Rogula-Kozłowska, E. Zajusz-Zubek, Environ Monit Assess **168** (2010)
14. J.S. Pastuszka, W. Rogula-Kozłowska, K. Klejnowski, P. Rogula-Kopiec, Atmosphere **6** (2015)
15. A. Satsangi, T. Pachauri, V. Singla, A. Lakhani, K. M. Kumari, Atmos Res, **113** (2012)
16. K. Klejnowski, B. Błaszczak, A. Krasa, B. Mathews, P. Rogula- Kopiec, W. Rogula-Kozłowska, [In] Ochrona Powietrza w Teorii i Praktyce (Ed. J. Konieczyński, IPIŚ PAN, 2014)
17. A. M. Jones, R. M. Harrison, Atmos Environ **39** (2005)
18. R. M. Harrison, J. Yin, Sci Total Environ **249** (2000)
19. M. Sillanpää, A. Frey, R. Hillamo, A. S. Pennanen, R. O. Salonen, Atmos Chem Phys **5** (2005)
20. W. F. Rogge, L. M. Hildemann, M. A. Mazurek, G. R. Cass, B. R. Simoneit, Environ Sci Technol **27** (1993)
21. J.H. Seinfeld, S.N. Pandis, *Atmospheric chemistry and physics: from air pollution to climate change* (John Wiley & Sons, 2006)
22. *Encyclopaedia of Occupational Health and Safety* (Ed. J. M. Stellman, International Labour Office,1998)
23. A. Manoukian, E. Quivet, B. Temime-Roussel, M. Nicolas, F. Maupetit, H. Wortham, Environ Sci Pollut Res **20** (2013)
24. EPA, *Candles and Incense as Potential Sources of Indoor Air Pollution: Market Analysis and Literature Review* (2001)
25. L.M. Castro, C.A. Pio, R.M. Harrison, DJ.T. Smith, Atmos. Environ. (1999)

DOI: 10.1051/e3sconf/20172200148

The use of lichens in post-smelting dumps reclamation – preliminary results of experimental cultivation of selected species on slag substrate

Kaja Rola[1] and *Piotr* Osyczka[1,*]

[1] Jagiellonian University, Institute of Botany, Kopernika 27, 31-501 Kraków, Poland

Abstract. Conventional reclamation interventions of post-smelting slag dumps being undertaken so far either failed or produced poor results. Certain lichens, especially of the genus *Cladonia*, are known as effective colonisers of bare ground in anthropogenic habitats. The paper presents preliminary results of the experiment aimed at the evaluation of lichen usefulness in reclamation interventions. The cultivation in vivo involving transplantation of lichens directly on slag substrate was established in 2015. Five species, i.e. *Cladonia rei*, *C. cariosa*, *C. pyxidata*, *C. subulata*, *C. macilenta*, were transplanted into 32 cuvettes filled with sterilised slag substrate. The sample weight of 2 and 6 g were used and half of cuvettes were regularly supplied with 2% malt solution. The first important symptoms at the present stage of the experiment are as follow: the growth of thalli has appeared only in the case of first three species; *C. rei* shows the most effective development; cuvettes with 6 g sample weight are characterized by higher coverage of fresh lichen thalli; lichen biomass are visually higher in cuvettes treated with malt solution. The results give us reason to believe that lichens could be successfully used as an alternative element during planning of slag dumps reclamation in the future.

1 Introduction

1.1 Environmental problems of post-industrial dumps

Mining and heavy industry have substantially transformed the landscape and contaminated the environment. Large quantities of wastes have been deposited in post-industrial areas as dumps. According to the origin and chemical properties, the dumps associated with the extraction of zinc and lead ores can be generally divided into three main types [1, 2]: post-mining dumps which constitute a mixture of dolomites, limestone, clays, sands and gravels, post-flotation dumps with a flotation sludge and finely ground dolomite, and post-smelting dumps consisting of slag from blast and muffle furnaces composed of dolomite crumbs, sinter slag, coke residue, Pb-rich and Zn-rich wastes. The latter dumps represent completely

* Corresponding author: piotr.osyczka@uj.edu.pl

DOI: 10.1051/e3sconf/20172200148

artificial environment characterised by extreme concentrations of toxic elements and highly adverse habitat conditions [1, 3]. Due to high toxicity, they have negative impact on the surrounding environment and human health and constitute a serious problem requiring solution in many countries, particularly those with intensive industrial development and using primitive metal-ores processing technologies. Although in recent years metal industry has been reduced and/or modernised in Poland, many dumps remain in the landscape of Upper Silesia and are still partly exploited. The necessity of reducing the negative impacts on the local environment and indigenous populations cause the need for reclamation interventions. Regrettably, previous efforts toward natural compensation either failed or produced poor results in case of dangerously toxic slag dumps.

1.2 Lichens as pioneer organisms

Lichens (lichenized fungi) constitute a symbiotic association composed of two units, a lichen-forming fungus (mycobiont) and an alga and/or a cyanobacterium (photobiont), forming a combined organism. Compared with autonomic fungi, lichens have many peculiar features, of which the most important are distinct and highly variable morphology and anatomy, specific metabolic pathways, and exclusive methods of reproduction. Lichens occur at all latitudes and grow on almost any type of substrate. The structure and physiology of lichens decide their success in the colonisation of environments extremely hostile to life [4]. Lichens have been defined as stress tolerators [5] and they are effective and rapid colonisers of bare ground [6] Their pioneer nature is associated not only with natural sites characterised by harsh climate conditions, but also with anthropogenic and strongly affected habitats. There are various adaptation mechanisms facilitating successful colonisation of such areas (see e.g. [7]). Some epigeic lichens with modest ecological requirements are known to be the first and most important colonisers of contaminated substrates [8-11] (Fig. 1).

Fig. 1. Post-industrial wastes with associated lichens. **A** – solid sinter slag overgrown by the most resistant lichens; **B** – initial stage of succession of post-smelting dump with lichens as the first colonisers; **C** – dense sward of lichen assemblage covering slag substrate.

1.3 Ecological role of lichens in spontaneous succession of post-smelting dumps

The natural regeneration of post-smelting dumps is extremely sluggish. The process begins from artificial and sterile habitat and progresses towards the poorly developed xerothermic grasslands. The initial stages of succession are completely devoid of vascular plants, whereas cryptogamic organisms, mainly lichens and mosses, constitute the main visual component of vegetation [8, 11]. The evidences have been provided that certain lichens are indifferent to extreme contamination and thus they are well adapted to slag substrate [9].

The spontaneous and rapid appearance of lichens in the initial successional stages of affected sites should be regarded as a positive phenomenon crucial for gradual and finally stable plant colonisation. They initiate physical weathering of solid slags and accelerate the natural process of soil formation, consolidation of the upper substrate layer and stabilization of the slopes (Fig. 1). Because of immense hygroscopic properties of thalli, lichens retain water droplets and atmospheric vapour, which improves local moisture conditions. Moreover, lichens act as providers of organic matter and fixed nitrogen, and therefore, they contribute to the fertility of the substrate [4].

1.4 The purpose of the experiment

It has been already postulated that the natural lichen cover should be taken into account when reclamation activities at post-smelting dumps are planned [9]. In relation to this problem, the purpose of our experiment is to evaluate the possibility of practical use of lichens in reclamation interventions aimed at accelerate the process of spontaneous regeneration of dumps. To accomplish this, we transplanted lichen thalli of selected species onto sterilized slag substrate and culture them in vivo.

2 Materials and methods

2.1 Lichen species selected for experimental cultivation

Lichens from the genus *Cladonia* are involved in the experiment. Certain representatives of the genus occur frequently on post-smelting dumps, achieve high coverage and frequently grow lushly forming dense swards [12]. Generally, these lichens as a rule consist of two kinds of thalli: a primary thallus in the form of horizontal squamules and vertical secondary thalli, called podetia. The occurrence of *Cladonia* lichens within metal-enriched areas has been frequently documented [6, 9, 12, 13]. Fruticose form of growth and relatively high biomass of *Cladonia* cause that some of them appear to be the most effective colonisers of post-smelting dumps.

Specifically, we selected five species: *Cladonia rei* Schaer., *C. cariosa* (Ach.) Spreng., *C. pyxidata* (L.) Hoffm., *C. subulata* (L.) F. H. Wigg., *C. macilenta* Hoffm. All of them are common in Poland and have a high ability to vegetative reproduction enabling efficient propagation. The first three species have special ability to grow in highly contaminated habitats [14-16] and thus are considered as successful and rapid colonisers of highly disturbed habitats [8, 11]. They constitute the major component of specific cryptogamic assemblages on post-smelting dumps in Poland [12] (Fig. 1). Lichen thalli intendent for transplantation were collected both from slag dump and adjacent semi-natural habitat. The latter two species, i.e. *C. subulata* and *C. macilenta*, are confined to open and sunny habitats. These lichens do not appear on slag dumps; however, they are associated with psammophilous grasslands in anthropogenic areas [10].

2.2 Experiment design and procedure

The cultivation in vivo experiment involving transplantation of lichens directly on slag substrate gives us the opportunity to recognize natural potential for growth of lichen thalli under stress conditions. The cultivation has been established in the Botanical Garden of the Jagiellonian University (Kraków) in 2015. The slag substrate was obtained from various parts of post-smelting dump in Ruda Śląska town. The chemical properties of the substrate are provided in Table 1.

DOI: 10.1051/e3sconf/20172200148

Table 1. Selected chemical properties of slag substrate used in the experiment (based on own earlier studies: [9, 11, 14]); min and max determined values are given.

	Concentration of heavy metals (μg/g; total and exchangeable forms)						
	Zn_{tot}	Pb_{tot}	Cd_{tot}	As_{tot}	Zn_{ex}	Pb_{ex}	Cd_{ex}
Min	60766	20050	21.1	4280	785	3932	1.93
Max	99721	22845	90.3	4665	817	4340	2.14
	Acidity and nutritional elements content						
	$pH_{(KCl)}$	C_{org} (%)	N_{tot} (%)	K (μg/g)	Mg (μg/g)	Ca (μg/g)	
Min	6.39	3.75	0.076	931	14068	7975	
Max	6.48	4.24	0.080	960	14519	8281	

The material was sterilised in 180°C for 24 hours to exclude the presence of any live propagules. Subsequently, the substrate was placed into 32 cuvettes of 38×24×7 cm. Lichen thalli of all selected species were powdered into the form of propagules, which were evenly scattered on the substrate surface. The sample weight of 2 and 6 g were used. Half of the cuvettes have been regularly, i.e. once a week, supplied with 2% malt solution for potential support of fungal hyphae growth. The arrangement of cuvettes with lichen material are demonstrated on Fig. 2.

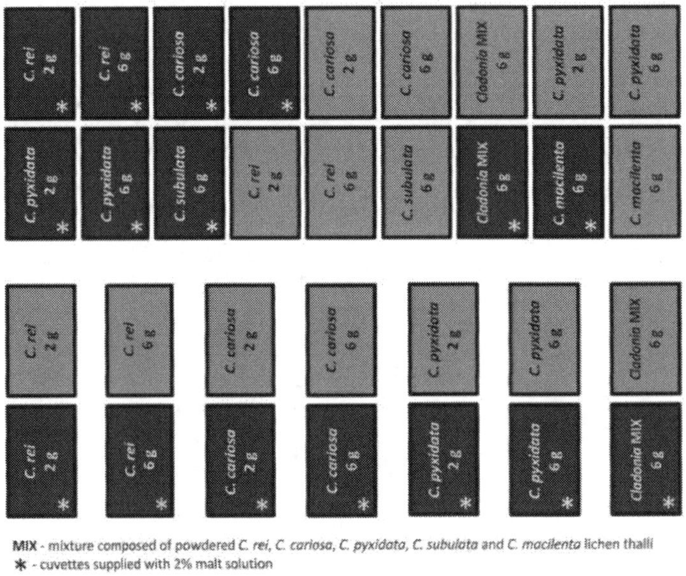

Fig. 2. Schematic layout of cuvettes in the cultivation experiment. The information on lichen species, sample weights and malt treatment are provided.

3 Preliminary results

The first symptoms of lichen thalli re-growth were observed half year after propagules transplantation (Fig. 3). At the present stage, one and a half year after the beginning of the experiment, regenerated thalli occur in the cuvettes with *C. rei*, *C. cariosa*, *C. pyxidata* and mix of *Cladonia*. Generally, cuvettes with 6 g sample weight are characterized by higher coverage of fresh lichen thalli compared to the cuvettes with 2 g; however, in the case of *C. rei* these differences become less apparent over time.

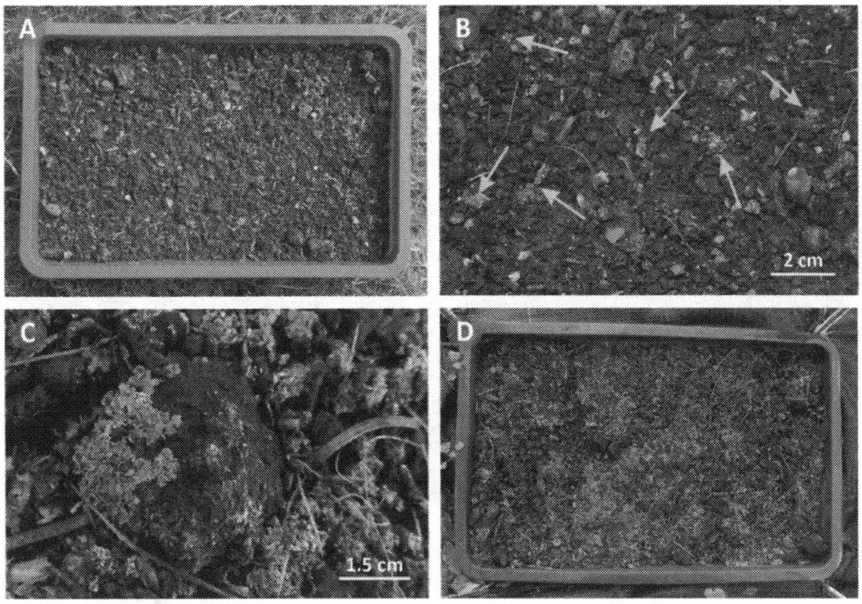

Fig. 3. The experiment of *Cladonia* lichens cultivation on slag substrate. **A** – the cuvette with scattered lichen propagules at the initial stage of cultivation; **B** – development of *Cladonia rei* primary squamules observed after six months from the beginning of experiment; **C** – colonization of solid sinter slag by *Cladonia* primary squamules in the cuvette; **D** – compact cover of *C. rei* in the cuvette after one and a half year.

The application of malt solution considerably improved lichen re-growth. Considering the samples with the same species and sample weight, the coverages of fresh lichen thalli are always higher in the cuvettes treated with this solution compared to those untreated (Table 2).

Table 2. Maximum coverage of lichen thalli in the cuvettes after 1.5 year of the experiment.

	sample weight 2 g		sample weight 6 g	
	without malt solution	with malt solution	without malt solution	with malt solution
Cladonia rei	28.31%	43.28%	17.87%	45.03%
Cladonia cariosa	5.12%	12.42%	6.15%	34.73%
Cladonia pyxidata	18.63%	29.18%	20.55%	41.47%
Cladonia macilenta	-	-	0%	0%
Cladonia subulata	-	-	0%	0%
Mix of *Cladonia*	-	-	8.52%	24.34%

Cladonia rei demonstrates the most effective development since regardless of sample weight and treatment its coverage in cuvettes are the highest among all tested species. Moreover, only in case of this species secondary thallus development has been already observed after one year and at the moment numerous podetia are present in the cuvettes. In contrast to these three aforementioned lichens, the thalli of *C. macilenta* and *C. subulata* showed no signs of regeneration. These results are additional argument that *C. rei*, *C. cariosa* and *C. pyxidata* are well adapted to extremely contaminated slag substrate and their occurrence on the dumps is not accidental. In case of cuvettes with assemblage of *Cladonia* only squamules of primary thallus have been developed till now. Therefore unambiguous determination of the material is not possible at this moment.

4 Discussion

Environmental rehabilitation of disturbed sites is a challenge facing the heavy-metal industry worldwide. Low levels of nutrients and organic matter, pH extremes, contamination and often difficult microclimatic conditions all hinder revegetation of post-smelting dumps. The most alarming are enormous contents of heavy metals which often can be detected on a percentage scale rather than at a trace level (see Table 1). Additionally, dark colour of the substrate contributes to accumulation of heat and high temperature amplitudes, which makes slag dumps extremely unfavourable for vegetation [3]. Therefore, the colonisation and natural regeneration of the considered dumps is extremely sluggish. Moreover, steep and unstable slopes of dumps additionally hinder reclamation treatments performed using conventional techniques. Our several previous studies showed that lichens play a fundamental role as effective pioneers in the initial stage of succession [9-12]. Consequently, lichen swards should be considered as an essential stage of succession, without which the development of stable plant cover would not be possible. We believe that the pioneering nature and unique adaptability to the contaminated substrate of certain *Cladonia* species could find practical application in reclamation interventions of slag dumps.

The results of earlier reclamation attempts confirm the need to search unconventional working methods. The support of plant cover development on post-smelting dumps only by means of 'rapid greening' that relied on sowing and planting mixtures of grasses, legumes and trees taxa either failed or resulted in a poor effect. Initially, such procedure seemed to be effective but after several years the intended effect of phytostabilization was not achieved. The patches of upper soil substrate slid of the slope under the influence of rainfall and erosion and this consequently led to uncovering the contaminated slag again [3, 17]. More radical methods of remediation, theoretically possible to adapt, are practically sporadically used, because of the large size of the dumps and enormous costs related to implementation. Considering the problem in the long term, it can be observed that the most stable and diverse plant cover has developed on these fragments of dumps without any human influence.

Recent studies show that the spontaneous succession of insensitive or resistant to contamination species is in fact the most effective form of restoring the functionality of degraded habitats [18]. Natural regeneration often prevails over sophisticated and costly technical reclamation as species that are naturally well adapted to local habitat conditions do not require subsequent treatments to maintain their population. Compact cover of pioneer cryptogams is a very important element during restoration process [11, 19]. Initiation and acceleration of succession processes at its earliest stages may prove to be an ecologically appropriate way of ecosystem restoration of strongly disturbed habitats.

 DOI: 10.1051/e3sconf/20172200148

5 Conclusions

Lichens have not been considered as an useful biological tool in reclamation of areas degraded by heavy metal industry so far. Since certain species spontaneously colonize such areas and frequently constitute compact cover in the initial stage of succession, it seems to be reasonable take them into account when reclamation intervention are planned. The recognition of the potential for spontaneous regeneration of lichen thalli on slag substrate is the first step necessary for future unconventional remediation practices. This can be of particular importance in case of steep slopes of slag dumps on which the colonisation of vascular plants are nearly impossible even after several decades after end of wastes deposition. Our preliminary observations (Fig. 3) give us reason to believe that lichens could be successfully used as an alternative element during planning of slag dump reclamation in the future.

References

1. F. Maciak, *Ochrona i rekultywacja środowiska* (Szkoła Główna Gospodarstwa Wiejskiego, Warsaw, 1996)
2. World Bank Group, *Pollution Prevention and Abatement Handbook 1998. Toward Cleaner Production* (The International Bank for Reconstruction and Development/The World Bank, Washington, 1999)
3. K. Skubała, Arch. Environ. Prot. **37**, 55–74 (2011)
4. T. H. Nash III, *Lichen biology* (Cambridge University Press, Cambridge, 2008)
5. J. P. Grime, *Plant Strategies and Vegetation Processes* (Wiley, Chichester, 1979)
6. O. Gilbert, Lichenologist **22**, 87–101 (1990)
7. M. Bačkor, S. Loppi, Biol. Plantarum **53**, 214–222 (2009)
8. D. Cuny, F. O. Denayer, B. de Foucault, R. Schumacker, P. Colein, C. van Haluwyn, Environ. Pollut. **129**, 289–297 (2004)
9. P. Osyczka, K. Rola, Cent. Eur. J. Biol. **8**, 876–887 (2013a)
10. K. Rola, P. Osyczka, Environ. Monit. Assess. **186**, 5897–5910 (2014)
11. K. Rola, P. Osyczka, M. Nobis, P. Drozd, Ecol. Eng. **75**, 332–342 (2015)
12. K. Rola, P. Osyczka, M. Nobis, Herzogia **27**, 121–135 (2014)
13. O. W. Purvis, C. Halls, Lichenologist **28**, 571–601 (1996)
14. Osyczka, K. Rola, Environ. Sci. and Pollut. R. **20**, 5076–5084 (2013b)
15. P. Osyczka, K. Rola, A. Lenart-Boroń, P. Boroń, Cent. Eur. J. Biol. **9**, 579–591 (2014)
16. P. Osyczka, K. Rola, K. Jankowska, Ecol. Indic. **61**, 766–776 (2016)
17. B. Tokarska-Guzik, A. Rostański, Natura Silesiae Superioris, Suppl. 2001, 5–18 (2001)
18. K. Prach, R. J. Hobbs, Restor. Ecol. **16**, 363–366 (2008)
19. K. Prach, Appl. Veg. Sci. **6**, 125–129 (2003)

Assessment of the AERMOD dispersion model over complex terrain with different types of meteorological data: Tracy Power Plant experiment

Mateusz Rzeszutek[1,*], *Adriana* Szulecka[1], *Robert* Oleniacz[1], and *Marek* Bogacki[1]

[1] AGH University of Science and Technology, Department of Environmental Management and Protection, Al. Mickiewicza 30, 30-059 Krakow, Poland

Abstract. The accuracy of air pollutants dispersion modelling results depends on the quality of the input data, including the representativeness of the meteorological data. The paper presents the results of the AERMOD model validation using data from Tracy Power Plant experiment (Nevada, USA) with various meteorological data sources, including WRF modelling system outputs. The highest efficiency of the AERMOD model performance was found using site-specific meteorological data and the results from the WRF model. In general, the AERMOD modelling system inadequately represents the concentration levels of tracer gas (SF_6) at receptors situated below the emitter height in areas of complex topography.

1 Introduction

AERMOD is an air pollutant dispersion model developed by the American Meteorological Society (AMS) / Environmental Protection Agency (U.S. EPA) for regulatory purposes in the near field studies (up to 50 km) and complex terrain [1]. It is a steady-state model, in which the plume of emitted pollutants spreads both horizontally and vertically in accordance with the Gaussian distribution [2]. The model is adapted for air pollutant dispersion modelling in complex terrain with the variability of vertical wind profile, temperature and turbulences in the planetary boundary layer (PBL) taken into account [3]. It belongs to the second generation of dispersion models in which Gaussian formulations are used for stable conditions, and for unstable situations a semi-empirical approach is adopted [4].

Fundamental concern during modelling in complex terrain is involved with limited availability of representative meteorological data that can be used in the model. Most conducted field experiments that aimed at the AERMOD validation, were carried out using site-specific meteorological data, including temperature and wind field information in the vertical profile [3], [5-7]. More often, however, access to such representative meteorological data is limited. Therefore, the results of calculations using prognostic meteorological models are increasingly used as input data to the AERMOD model [8-11].

* Corresponding author: rzeszut@agh.edu.pl

The guidelines on air quality models [1] allow the use of prognostic meteorological data if their quality is acceptable and both site-specific and National Weather Service (NWS) data are unavailable or impossible to collect.

Previous studies concerning the integration of AERMOD and WRF models [9] have shown, that prognostic meteorological data perform well compared to the site-specific measurements. However, in these studies and in the case of the work [4-5], the analysis of the AERMOD performance was limited to the high-end concentrations that are important from the perspective of legal regulations. Such approach may be insufficient, since many researches use coupled AERMOD dispersion model with WRF to achieve different goals [8, 10, 11]. Furthermore, three of the experiments conducted in the study [9] referred to a small number of receptors (2 to 7). Instead of stable and non-reactive tracer gas, the SO_2 and Pb were used, for which both the determination of the emission rates and background concentrations may be a source of additional uncertainty, as highlighted by Dresser and Huizer [6].

Therefore, in the study, a more detailed evaluation of the AERMOD modelling system was conducted in the area of complex topography using Tracy Power Plant experiment (based on 106 receptors and SF_6 as a tracer gas) considering different sources of meteorological data, including those derived from the WRF model.

2 Materials and methods

2.1 Description of Tracy Power Plant experiment

The Tracy Power Plant experiment is one of four campaigns commissioned by U.S. EPA for the development of dispersion models in irregular mountainous areas [12, 13]. Tracy power plant is located east of Reno (Nevada) in the Truckee River valley, surrounded by mountain peaks reaching relative height of 950 m (Fig. 1).

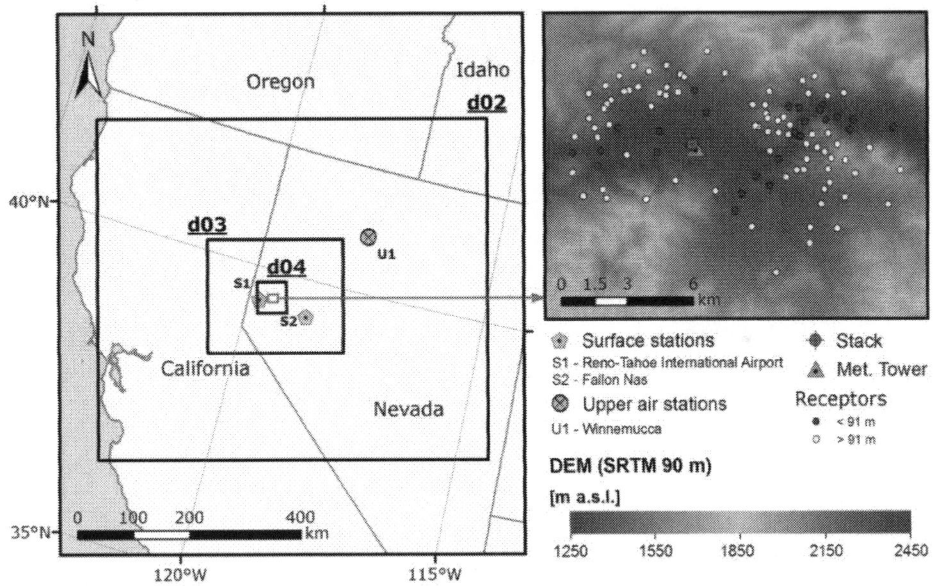

Fig. 1. Location of the emitter and measurement points used in the Tracy Power Plant experiment with the boundaries of three inner domain (d02-d04) in the WRF model

Field trials in this experiment were conducted in August 1984 using SF_6 tracer gas. This marker was released into the atmosphere through a 91-m power plant stack. The experiment was carried out in 14 single campaigns, during which a series of 128 hours of data was collected at 106 receptors located both below and above the emitter height (26 and 80 receptors, respectively). Meteorological measurements were collected from an instrumented 150-m tower and extended above this height using a Doppler acoustic sounder and a tethersonde [12, 14].

2.2 Configuration of the WRF model and simulation design

Simulation of meteorological parameters in this study was performed using the WRF v3.7 modelling system described in [15]. Initial and boundary meteorological conditions were provided using NCEP North American Regional Reanalysis dataset (ds608.0) [16] produced every 3 hours. This dataset was chosen due to its fine spatial resolution (32x32 km) and the availability for the period of Tracy Power Plant experiment. The simulation was performed in the coarse domain (d01) comprising the territory of the western USA with a spatial resolution of 36x36 km. Three one-way nested domains with the following resolutions: 12x12 km (d02), 4x4 km (d03) and 1x1 km (d04) were set to include the area of research (Fig. 1). The calculations were conducted at 44 assigned vertical levels with the lowest layer of 20 m above ground to ensure comparability with observations from meteorological tower.

2.3 Settings and computational variants in AERMOD

In the study a number of model simulations were carried out with respect to various meteorological data. Table 1 summarizes the computational variants with a short description of used meteorological data.

Table 1. Description of calculation variants with regard to the input data

Variant	Description of meteorological data	Tools
FW	Fallon Nas (S2) and Winnemucca (U1)	AERMET
RW	Reno - Tahoe International Airport (S1) and Winnemucca (U1)	AERMET
RWO-L	Reno - Tahoe International Airport (S1), Winnemucca (U2) and only 10 m layer from meteorological tower	AERMET
RWO-R	Reno - Tahoe International Airport (S1), Winnemucca (U2), all layers from meteorological tower and Doppler acoustic sounder	AERMET
WRF1	From WRF, domain with the resolution of 1x1 km (d04)	MMIF/AERMET
WRF4	From WRF, domain with the resolution of 4x4 km (d03)	MMIF/AERMET

Meteorological data from surface stations (S1, S2), upper air (U1) and meteorological tower were processed using AERMET v15181 preprocessor to determine planetary boundary layer parameters, such as the Monin-Obukhov length, convective velocity scale, temperature scale, mixing height, and surface heat flux. Subsequently, they were preprocessed using AERMET in accordance with the U.S. EPA guidelines [1, 17]. The only exception was the height of vertical layers, which were assigned in accordance to the height of the meteorological tower measurements. Calculations of SF_6 atmospheric dispersion were conducted using AERMOD v15181 with Complex Terrain Algorithms [3].

2.4 Comparison of meteorological data and model evaluation methods

Validation of meteorological conditions was limited to 128 hours of valid meteorological tower observations. Calculation period was synchronized with the measurement time at the tower. The comparative analysis was based on the wind field representation from meteorological stations (at 10 m a.g.l.), meteorological tower (at 10 and 100 m a.g.l.) and the WRF modelling results at the location of meteorological tower (at 100 m a.g.l.). Final verification of WRF modelling outputs compliance with meteorological tower observations was based on the analysis of wind field displacement (WD) calculated as reported in [9].

Evaluation of the AERMOD model with respect to different meteorological data sources was conducted to determine the ability to predict the upper tail end concentration distributions and to represent the distribution of observed 1-hour concentrations. Model capability to predict the high-end distribution of the highest 1-hour concentrations was analysed using Robust Highest Concentration (RHC) values [18]. The evaluation of model ability to predict the 1-hour concentration distribution was conducted on the subset of 200 of the highest 1-hour concentrations using statistical model evaluation parameters: FB (Fractional Bias), NMSE (Normalized Mean Square Error), NAD (Normalized Absolute Difference) and FAC2 (Factor of Two of Observations). For calculation of these parameters the quantitative limitation of the sample was applied because of the impossibility to determine the appropriate lower concentration threshold. Attempts to establish the lower limit of compared concentrations resulted in significant differences of sample sizes [19].

3 Results and discussion

3.1 Comparison of meteorological data

Near surface wind field varies significantly in complex terrain, as it highly depends on the terrain features and local airflows (e.g. slope and valley winds) induced by cyclic changes in the amount of energy reaching the surface of the earth [20, 21]. High variability of surface wind speed and direction during the experiment in the area of study is shown in Fig. 2. Considered surface weather stations S1 and S2 are located 22.3 and 68.9 km from the emission source, respectively, and the dominant wind direction varies from WS direction at the meteorological tower location to SE and WN at the location of S2 surface station. Mean wind speed is also lower for the meteorological tower observations (1.67 m s^{-1}) during experimental campaigns compared to the measurements at S1 and S2 stations (2.48 and 2.15 m s^{-1}, respectively).

Wind roses obtained at the height of 100 m a.g.l. (Fig. 3) show, that the WRF model inaccurately represents dominant directions of wind flow at the location of meteorological tower, but in general mean wind speed values are comparable with observations. Overall, the WRF model performance was considered to be acceptable, as the values of wind field displacement calculated at 10 m a.g.l. do not exceed the spatial resolution of modelling domains (reaching 0.889 km for WRF1 and 0.877 km for WRF4). Nested grid with a higher spatial resolution does not improve the accuracy of wind field characteristics, which, to a certain degree, may result from model configuration [22].

DOI: 10.1051/e3sconf/20172200149

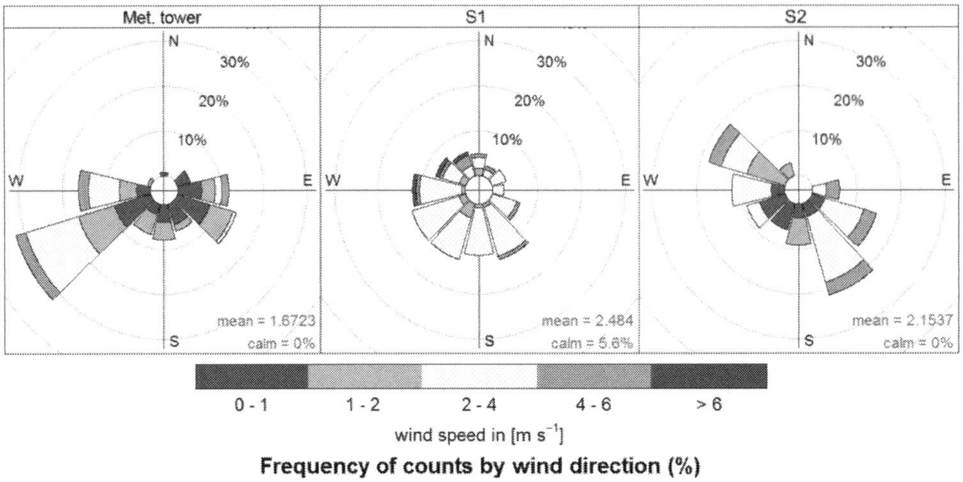

Fig. 2. Wind roses at the height of 10 m a.g.l. for meteorological tower, S1 and S2 meteorological stations

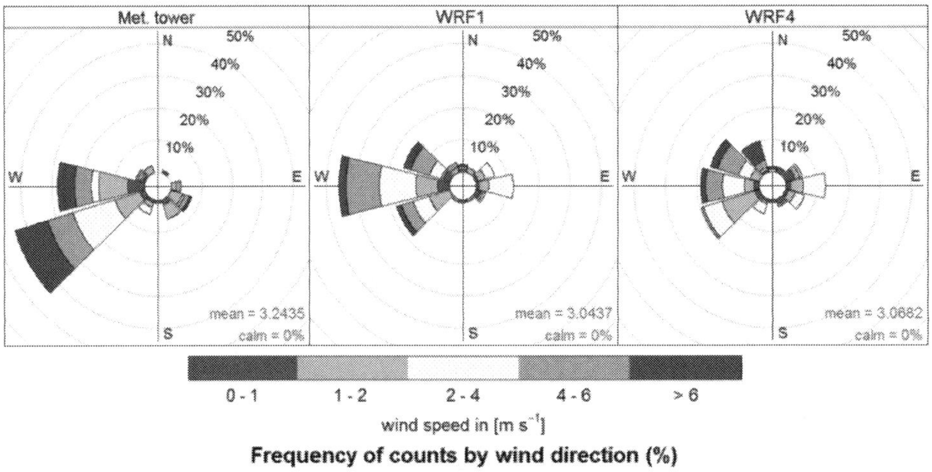

Fig. 3. Wind roses at the height of 100 m a.g.l. for meteorological tower observations and the results of WRF modelling at the tower location

3.2 Evaluation of the AERMOD model with regard to the maximum values

Determined statistical parameters of the model evaluation (Tab. 2) indicate that for the most of concerned calculation variants, the AERMOD model accurately represents the high-end distribution of the highest 1-hour concentrations both for the whole sample (all receptors) and for receptors located only above the emitter height (relatively high comparability of modelled and observed RHC values). Similar conclusions, but with no receptor grouping performed and using only site-specific data are presented in [5-7], [14] as well. The RWO-L and RW variants are the only exceptions for which the value of |FB| is greater than 0.3 and therefore falls outside the limits of good air quality models for rural areas as specified in [19]. The assessment results for the RWO-L variant indicate a strong, even 3-fold overestimation of modelled values, which results directly from the large difference in average wind speed between 10 and 100 m (Fig. 2 and 3). It should be noted that the use of meteorological data from surface stations (at 10 m a.g.l.) representative in terms of the location, will not provide adequate results for air pollutants dispersion modelling for

DOI: 10.1051/e3sconf/20172200149

regulatory purposes. Therefore, the meteorological representativeness of input data for the AERMOD model according to the guidelines [1] should refer to the vertical wind field data as well.

Table 2. One-hour RHC_m (modelled) compared with RHC_o (observed) in μg m^{-3} for analysed variants and groups denoting receptor location

Variant	All receptors			Receptors < 91 m			Receptors > 91 m		
	RHC_m	Ratio*	FB	RHC_m	Ratio*	FB	RHC_m	Ratio*	FB
FW	13.92	0.93	0.07	0.91	0.10	1.65	13.92	1.03	-0.03
RW	8.80	0.59	0.52	1.48	0.16	1.45	8.67	0.64	0.43
RWO-L	42.24	2.82	-0.95	12.98	1.38	-0.32	42.91	3.18	-1.04
RWO-R	13.07	0.87	0.14	4.98	0.53	0.61	12.56	0.93	0.07
WRF4	17.57	1.17	-0.16	2.84	0.30	1.07	17.57	1.30	-0.26
WRF1	13.70	0.91	0.09	2.14	0.23	1.26	13.70	1.02	-0.02
RHC_o	14.98			9.38			13.49		

* RHC_m / RHC_o

The analysis show, that the prognostic meteorological input data can be used for regulatory purposes, as most concentrations calculated in the AERMOD model for the WRF variants are characterized by slight overestimation. It is indicated by negative value of FB that stays within the limits of good air quality models. However, studies described in [9] show, that the use of WRF data does not always lead to the overestimation of the highest 1-hour observations, and the site-specific data generally yield better results. On the other hand, the use of data derived only from the NWS stations located at a great distance from the emission source (for which, as indicated in Section 3.1, the wind field differs significantly compared to the meteorological tower observations), can yield both good (FW) and poor (RW) results. Acceptable quality of the FW variant results is probably related to the several occurrences of similar meteorological events responsible for the high-end distribution of the highest 1-hour concentrations. Therefore, the RHC parameter is not sufficient to clearly determine which of the analysed solutions is appropriate.

The outcomes obtained for receptors below the emitter height indicate that AERMOD poorly represents the actual concentration levels at these points, generally with significant underestimation of the results. This behaviour of the model is disturbing, especially considering the use of the AERMOD model for regulatory purposes over complex terrain.

3.3 Similarity assessment of the 1-hour concentration distributions

Distributions of 200 of the highest observed concentrations with respect to the modelled values for analysed calculation variants (Fig. 4) generally confirm the conclusions drawn in Section 3.2. However, they show that the use of data from NWS (the RW and FW variants) generally results in a large number of significantly underestimated values. Despite the fact that the upper part of 1-hour concentrations distribution is similar to the observations for the FW, the remaining results are strongly underestimated (at least 2-fold).

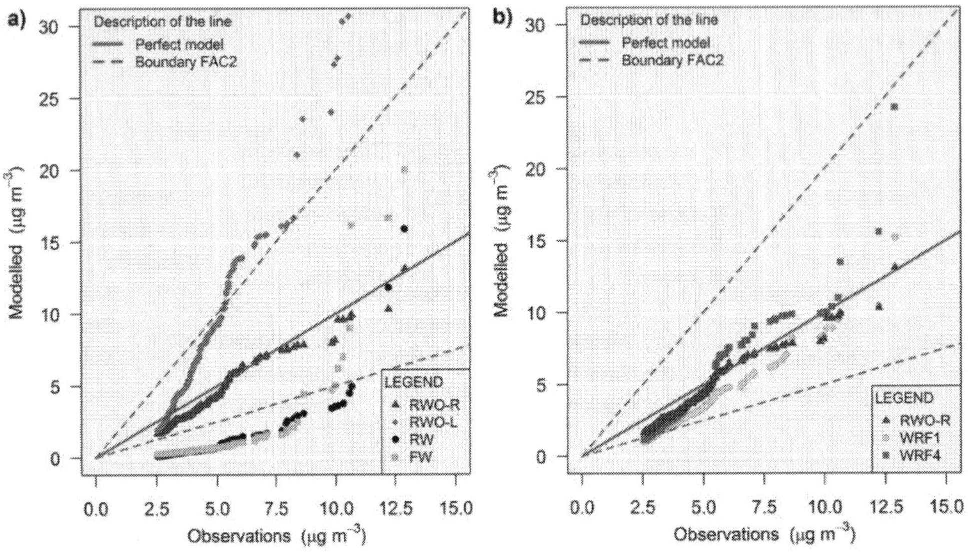

Fig. 4. Q–Q plots for the AERMOD results (200 of the highest 1-hour concentrations) corresponding to different meteorological input data for Tracy Power Plant experiment: a) site specific (all layers and one layer) and surface meteorological data, b) site-specific (all layers) and WRF meteorological data

With regulatory purposes of the AERMOD model kept in mind, it should be noted that the calculation results of concentrations ought to be rather slightly overestimated than underestimated [7]. Such results were obtained using the WRF meteorological data at the spatial resolution of 4x4 km (Fig. 4). The outcomes indicate the use of coarser WRF input data from 4 km grid instead of 1 km grid may be a better solution, since the WRF model does not always produce better results for finer resolution grids in the areas of complex orography and without additional data assimilation [21]. This is confirmed by the values of statistical model evaluation parameters (Tab. 3) for receptors located above the emitter height, as they are noticeably worse for the WRF1 compared to the WRF4 variant.

Table 3. Selected parameters for statistical evaluation of dispersion model evaluation for analysed variants and groups denoting receptor location

Variant	Receptors < 91 m				Receptors > 91 m			
	FAC2	FB	NMSE	NAD	FAC2	FB	NMSE	NAD
FW	0.00	1.84	30.21	0.92	0.05	1.23	3.03	0.65
RW	0.00	1.79	20.92	0.90	0.01	1.41	4.38	0.71
RWO-L	0.08	0.89	1.46	0.51	0.82	-0.56	1.32	0.29
RWO-R	0.01	0.99	1.54	0.50	1.00	0.12	0.02	0.07
WRF4	0.00	1.59	8.38	0.79	0.95	0.10	0.14	0.14
WRF1	0.00	1.55	7.67	0.77	0.72	0.36	0.15	0.19

The results shown in Table 3 indicate, that the most accurate results were obtained for receptors located at the height exceeding 91 meters using site-specific meteorological data and for 4 km grid from the WRF model. For both of these variants, the statistical evaluation parameters meet the acceptance criteria for dispersion models in rural areas [19]. Therefore, the application of WRF model as the substitution in the absence of representative NWS and site-specific measurements in complex terrain is appropriate, as indicated in recent update of the guidelines on air quality models [1].

However, the above statements refer only to receptors located above the height of the emission source, similarly as in the work [9], and the model assessment indicate that the AERMOD model underestimates concentrations for receptors below the emitter height for all of the considered variants. Findings in the corresponding studies conducted for simple terrain experiments [5, 14] provide opposite conclusions. However, it should be emphasized that in the Tracy Power Plant experiment used in the study, many receptors below the emitter height were located in the valley behind the mountain ridges (Fig. 1), and the formulations used in AERMOD for complex terrain applications are not suitable for these situations [3]. This fact may constitute a significant problem, since the highest concentrations are not always associated with the receptors located above the emitter height, and the calculation results for those below the emitter will be underestimated, even with the use of measurements from the near surface atmospheric layer.

4 Summary

The results of calculations and analysis presented in the study allow to conclude, that the use of site-specific and the WRF model meteorological data is appropriate for air dispersion modelling using the AERMOD model for complex terrain applications. The use of the data from 4 km grid is suggested, as there is no improvement in the quality of modelling results and the computational costs increase significantly for finer resolutions. According to the U.S. EPA guidelines [1], the application of the WRF model in air dispersion modelling for regulatory purposes is recommended. However, this approach is appropriate in the absence of spatially representative meteorological data and with lack of information about the wind field characteristics in the vertical profile of the atmosphere.

For the Tracy Power Plant experiment results, a strong underestimation of concentrations at receptors below the emitter height was found. This is probably associated with the large number of receptors belonging to this group and located in the valleys between the mountain ranges, but a proper explanation of the cause of this issue requires further and more detailed research in the areas of complex topography.

Additionally, the analysis carried out in the study indicate, that the use of RHC parameter alone provide insufficient information to determine which set of meteorological data contributes to better performance of the dispersion model. Furthermore, the assessment focused solely on the analysis of the high-end distribution of the highest 1-hour concentrations may be conditioned by the accidental occurrences of several similar meteorological events responsible for these concentrations.

The paper has been prepared within the scope of the AGH UST statutory research no. 11.11.150.008.

References

1. U.S. EPA, *Revisions to the Guideline on Air Quality Models: Enhancements to the AERMOD Dispersion Modeling System and Incorporation of Approaches to Address Ozone and Fine Particulate Matter* (Federal Register, 40 CFR Part 51, Appendix W, 2017)

2. A.J. Cimorelli, S.G. Perry, A. Venkatram, J.C. Weil, R.J. Paine, R.B. Wilson, R.F. Lee, W.D. Peters, R.W. Brode, J. Appl. Meteorol., **44**, pp. 682–693 (2005)
3. A. Venkatram, R.W. Brode, A.J. Cimorelli, R.F. Lee, R.J. Paine, S.G. Perry, W.D. Peters, J.C. Weil, R.B. Wilson, Atmos. Environ., **35**, pp. 4211–4221 (2001)
4. K. Juda-Rezler, Arch. Environ. Prot., **36**, 1, pp. 3–28 (2010)
5. S.G. Perry, A.J. Cimorelli, R.J. Paine, R.W. Brode, J.C. Weil, A. Venkatram, R.B. Wilson, R.F. Lee, W.D. Peters, J. Appl. Meteorol., **44**, 5, pp. 694–708 (2005)
6. A.L. Dresser, R.D. Huizer, J. Air Waste Manage. Assoc., **61**, 6, pp. 647–659 (2011)
7. A.S. Rood, Atmos. Environ., **89**, pp. 707–720 (2014)
8. G.A. Abril, S.C. Diez, M.L. Pignata, J. Britch, Atmos. Pollut. Res., **7**, 1, pp. 180–189 (2016)
9. U.S. EPA, *Evaluation of Prognostic Meteorological Data in AERMOD Applications* (Report EPA-454/R-16-004, Research Triangle Park, NC 27711, 2016)
10. A. Kumar, R.S. Patil, A.K. Dikshit, S. Islam, R. Kumar, J. Clean. Prod., **116**, pp. 1–8, (2015)
11. L. Turtos Carbonell, G.C. Mastrapa, Y.F. Rodriguez, L. Alvarez Escudero, M. Sanchez Gacita, A B. Morlot, I.B. Montejo, E.M. Ruiz, S.P. Rivas, L.T. Carbonell, Atmos. Pollut. Res., **4**, 1, pp. 64–74 (2013)
12. L.E. Truppi, *EPA complex terrain model development: Description of a computer data base from the full scale plume study, Tracy Power Plant, Nevada* (Report EPA/600/S3-86/068, Research Triangle Park, NC 27711, 1987)
13. U.S. EPA, *Support Center for Regulatory Atmospheric Modeling - Preferred and Recommended Models* (available at: https://www.epa.gov/scram/air-quality-dispersion-modeling-preferred-and-recommended-models, accessed: 25-mar-2017)
14. U.S. EPA, *AERMOD: Latest features and evaluation results* (Report EPA-454/B-03-001, Research Triangle Park, NC 27711, 2003)
15. W.C. Skamarock, J.B. Klemp, J. Dudhi, D.O. Gill, D.M. Barker, M.G. Duda, X.-Y. Huang, W. Wang, J.G. Powers, *A Description of the Advanced Research WRF Version 3* (Report NCAR/TN–475+STR, Boulder, CO, 2008)
16. F. Mesinger, G. DiMego, E. Kalnay, K. Mitchell, P.C. Shafran, W. Ebisuzaki, D. Jovic, J. Woollen, E. Rogers, E.H. Berbery, M.B. Ek, Y. Fan, R. Grumbine, W. Higgins, H. Li, Y. Lin, G. Manikin, D. Parrish, W. Shi, Bull. Am. Meteorol. Soc., **87**, 3, pp. 343–360 (2006)
17. U.S. EPA, *Guidance on the Use of the Mesoscale Model Interface Program (MMIF) for AERMOD Applications* (Report EPA-454/B-15-001, Research Triangle Park, NC 27711, 2016)
18. W.M. Cox, J.A. Tikvart, Atmos. Environ., **24A**, 9, pp. 2387–2395 (1990)
19. S. Hanna, J. Chang, Atmos. Phys., **116**, 3–4, pp. 133–146 (2012)
20. L. Giovannini, D. Zardi, M. de Franceschi, F. Chen, Int. J. Climatol., **34**, 4, pp. 1111–1131 (2014)
21. L. Giovannini, G. Antonacci, D. Zardi, L. Laiti, L. Panziera, Energy Procedia, **59**, pp. 323–329 (2014)
22. Y. Zhang, X. Zhang, L. Wang, Q. Zhang, F. Duan, K. He, Atmos. Environ., **124**, pp. 285–300 (2016)

 DOI: 10.1051/e3sconf/20172200150

Study on water evaporation rate from indoor swimming pools

Ilona Rzeźnik[1,*]

[1]Poznan University of Technology, Institute of Environmental Engineering, ul. Berdychowo 4 61-131 Poznań, Poland

Abstract. The air relative humidity in closed spaces of indoor swimming pools influences significantly on users thermal comfort and the stability of the building structure, so its preservation on suitable level is very important. For this purpose, buildings are equipped with HVAC systems which provide adequate level of humidity. The selection of devices and their technical parameters is made using the mathematical models of water evaporation rate in the unoccupied and occupied indoor swimming pool. In the literature, there are many papers describing this phenomena but the results differ from each other. The aim of the study was the experimental verification of published models of evaporation rate in the pool. The tests carried out on a laboratory scale, using model of indoor swimming pool, measuring 99cm/68cm/22cm. The model was equipped with water spray installation with six nozzles to simulate conditions during the use of the swimming pool. The measurements were made for conditions of sports pools (water temperature 24°C) and recreational swimming pool (water temperature 34°C). According to the recommendations the air temperature was about 2°C higher than water temperature, and the relative humidity ranged from 40% to 55%. Models Shah and Biasin & Krumm were characterized by the best fit to the results of measurements on a laboratory scale.

1 Introduction

For years, there has been increase in the number of pools, in Poland. The new objects are usually indoor pools and they are built as aqua parks, recreational pools or swimming pools with additional attractions, so their offer is directed to wider community. In 2015 year there were 736 indoor swimming pools in Poland: 12 olympic pools, 66 sports pools (25×16 m), 489 smaller pools (dimensions 25×12,5m) and 169 training and recreational pools (min. 16m) [1]. The most important phenomenon, in swimming pools is the water evaporation process. The indoor humidity directly impacts on user's comfort and durability of buildings construction. Wherefore, the most important is calculation of vapor production, what is necessary to proper design of HVAC system. Too high air humidity causes a feeling of breathlessness among users and it intensifies corrosion of the building. The oversized ventilation system increases the investment and operational costs and may cause a sense of

* Corresponding author: ilona.rzeznik@put.poznan.pl

coolness among users. To dimension of HVAC systems, the mathematical models describing the evaporation of water in the occupied and unoccupied indoor swimming pools are used.

There are many formulas for calculation the water evaporation rate in occupied and unoccupied swimming pools, but their results differ from each other [1-18]. Additionally the problem is estimation the influence of using intensity on vapor production [2]. The aim of the study was experimental verification of published formulas for calculation the water evaporation rate in indoor swimming pools.

2 Mathematical modeling

The water evaporation rate was calculated according to Dalton's law:

$$\dot{g}_w = \varepsilon \cdot (p_w^{"} - p_w) \tag{1}$$

or

$$\dot{g}_w = \sigma \cdot (x" - x) \tag{2}$$

where:

$\dot{g}_w$ – water evaporation rate ($kg \cdot m^{-2} \cdot h^{-1}$),
ε, σ – evaporation coefficients ($kg \cdot m^{-2} \cdot h^{-1} \cdot Pa^{-1}$),
$p_w^{"}$ – water vapor partial pressure at water temperature in boundary layer at saturation state (Pa),
p_w – water vapor partial pressure at the air temperature (Pa),
$x"$ – humidity content in boundary layer at water temperature ($kg \cdot kg^{-1}$),
x – humidity content for saturated air at pool temperature ($kg \cdot kg^{-1}$).

Mass flow rate is given by:

$$\dot{m}_w = \dot{g}_w \cdot F \tag{3}$$

where:

$\dot{m}_w$ – mass flow rate ($kg \cdot h^{-1}$)
F – evaporation area (m^2).

The water evaporation phenomenon in swimming pools were studied in laboratory [3, 4, 5] and in real conditions [6, 7]. Based on the literature study, Shah [8] reported the area of examined swimming pools, which ranged from $0{,}073m^2$ to $425m^2$. Those pool differed not only in the area but also in their shape. Asdrubali [3], Moghiman & Jodat [4] and Tang & Etzion [5] conducted measurement in perpendicular pools, Sartori [9] in cylindrical.

The equations of mass flow rate in the swimming pool halls, which were experimental verified in this study are presented in table 1 and 2. Differences in results are mainly due to the different way of calculating the evaporation factors ε, σ, which are a function of Sherwood and Schmidt number and turbulence level.

 DOI: 10.1051/e3sconf/20172200150

Table 1. Equations of mass flow rate in unoccupied swimming pools

Author	Equation
Malicki [10]	$\dot{m}_w = \frac{(a+0{,}0174\cdot w)\cdot 760}{p_b}\cdot(p_w^{"} - p_w)\cdot F$ (4) a = 0,022 for t_{pw} < 30°C a = 0,028 for t_{pw} < 40°C t_{pw} - temperature of water surface (°C) w – air velocity (m·s^{-1}) p_b – atmosphere pressure $p_w^{"}$, p_w , p_b (mmHg)
Ferencowicz [11]	$\dot{m}_w = \frac{(22{,}9+17{,}4\cdot w)\cdot 760}{(1000\cdot p_b)}\cdot(p_w^{"} - p_w)\cdot F$ (5) $p_w^{"}$, p_w , p_b (mmHg)
Sprenger [12]	$\dot{m}_w = (25 + 19\cdot w)\cdot(x" - x)\cdot F$ (6)
Dienelt [2]	$\dot{m}_w = (30 + 14\cdot w - 0{,}26\cdot t_s)\cdot(x" - x)\cdot F$ (7) t_s – average air temperature (°C) $t_s = \frac{(t_{wg}+t_{ph})}{2}$, t_{wg} - temperature of boundary layer (°C) t_{ph} – temperature of air in hall of pool (°C) w < 1 m·s^{-1}
Biasin & Krumm [13]	$\dot{m}_w = \left[-0{,}059 + \left(0{,}0105\cdot\frac{p_w^{"}-p_w}{1{,}333}\right)\right]\cdot F$ (8) $p_w^{"}$, p_w (hPa)
Shah [14]	$\dot{m}_w = K\cdot\rho_w\cdot(\rho_p - \rho_w)^{0{,}333}\cdot(x" - x)\cdot F$ (9) $(\rho_p - \rho_w) < 0 \rightarrow \rho_w - \rho_p$ $(\rho_p - \rho_w) > 0{,}02 \rightarrow K = 35$ $(\rho_p - \rho_w) < 0{,}02 \rightarrow K = 40$
Smith [7]	$\dot{m}_w = \frac{70+0{,}35\cdot w}{r}\cdot(p_w^{"} - p_w)\cdot F$ (10) $p_w^{"}$, p_w (hPa) r – latent heat of vaporization of water (kJ·kg^{-1})
Carrier [15]	$\dot{m}_w = \frac{42{,}6+37{,}6\cdot w}{r}\cdot(p_w^{"} - p_w)\cdot F$ (11) $p_w^{"}$, p_w w (hPa)
VDI 2089 [13]	$\dot{m}_w = \frac{\varepsilon\cdot(p_w^{"}-p_w)\cdot F}{1000}$ (12) $\varepsilon = 5$ – for peaceful surface of water $p_w^{"}$, p_w (hPa)

DOI: 10.1051/e3sconf/20172200150

Table 2. Equations of mass flow rate in occupied swimming pools

Author	Equation
Besler [16]	$\dot{m}_w = 2{,}8 \cdot \frac{(a+0{,}0174 \cdot w) \cdot 760}{p_b} \cdot (p_w^{"} - p_w) \cdot F$ (13) $a = 0{,}022$ for $t_{pw} < 30$ °C $a = 0{,}028$ for $t_{pw} < 40$ °C $p_w^{"}$, p_w, p_b (mmHg)
Labhom [13]	$\dot{m}_w = L_b \cdot (25 + 19 \cdot w) \cdot (x" - x) \cdot F$ (14) $L_b = (3 \div 3{,}5)$
Kappler [2]	$\dot{m}_w = k_p \cdot (25 + 19 \cdot w) \cdot (x" - x) \cdot F$ (15) $k_p = 1{,}10$ – small private indoor swimming pools $k_p = (1{,}10 \div 1{,}15)$– general purpose pools $k_p = 1{,}20$ – school and hotel pools
VDI 2089 [13]	$\dot{m}_w = \frac{\varepsilon \cdot (p_w^{"} - p_w) \cdot F}{1000}$ (16) $\varepsilon = 20$ – public pools $\varepsilon = 28$ – recreational pools $p_w^{"}$, p_w (hPa)
Recknagel [17]	$\dot{m}_w = \sigma \cdot (x" - x) \cdot F$ (17) $\sigma = 20$ – general purpose pools $\sigma = 30$ – pools with artificial wave
Biasin & Krumm [13]	$\dot{m}_w = \left[0{,}119 + \left(0{,}01995 \cdot 6 \cdot \frac{N}{F}\right) \cdot \frac{p_w^{"} - p_w}{1{,}333}\right] \cdot F$ (18) $p_w^{"}$, p_w (hPa) N – number of users
Shah [18]	$\dot{m}_w = \left(5{,}95 \cdot \frac{N}{F} + 1{,}2\right) \cdot K \cdot \rho_w \cdot \left(\rho_p - \rho_w\right)^{0{,}333} \cdot (x" - x) \cdot F$ (19) For $\frac{F}{N} > 45$ $\dot{m}_w = \left(14{,}85 \cdot \frac{N}{F} + 1\right) \cdot K \cdot \rho_w \cdot \left(\rho_p - \rho_w\right)^{0{,}333} \cdot (x" - x) \cdot F$ (20) For $\frac{F}{N} > 4{,}5 \div 4$ $\dot{m}_w = 2{,}5 \cdot K \cdot \rho_w \cdot \left(\rho_p - \rho_w\right)^{0{,}333} \cdot (x" - x) \cdot F$ (21) Dla $\frac{F}{N} < 4{,}5$ $\left(\rho_p - \rho_w\right) < 0 \rightarrow \rho_w - \rho_p$ $\left(\rho_p - \rho_w\right) > 0{,}02 \rightarrow K = 35$ $\left(\rho_p - \rho_w\right) < 0{,}02 \rightarrow K = 40$ N – number of users

DOI: 10.1051/e3sconf/20172200150

3 Experimental procedure

The tests were carried out in laboratory conditions, using model of indoor swimming pool, measuring 0,99m/0,68m/0,22m. The area of the water surface was 0,6732 m^2 (fig. 1) and is in the range of pools area given by Shah[8]. Studied model was equipped with water spray installation with six nozzles to simulate conditions during the use of the swimming pool. The air provided to the pool chamber was pre-dried and heated. Temperature and relative humidity of air in the pool chamber during measurement were set at constant level and maintained by ventilation unit and duct heater. To heat the water in model of pool was used the flow heater controlled by thermostat.

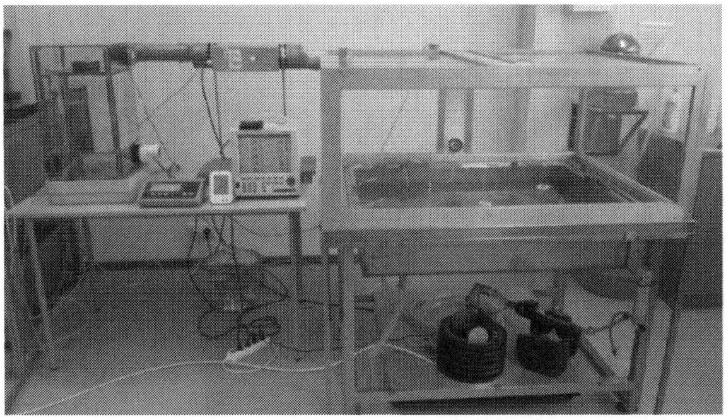

Fig. 1. Experimental stand.

Studied pool model was equipped in 12 thermocouples for air and water temperature measurements (accuracy ± 0,1°C). The temperatures readings were collected and recorded by Omega RD-MV100 recorder. The LOG 32 TH recorder (measurement range 0-99%; accuracy ± 3%) was used to measure the relative humidity of air. Measurements of air velocity were made at point located 0,3m above the water's surface as recommended [9], using the PL-135HAN anemometer, produced by Voltcraft (measurement range of 0,1-25,0 m·s^{-1}; accuracy ± 5%)
The pan of swimming pool model was put on electronic laboratory scale ICS425, produced by METTLER TOLEDO (accuracy ±1 g). This allowed for the determining the mass of evaporated water. The measurements were made for conditions of sports pools (water temperature 24°C) and recreational swimming pools (water temperature 34°C). According to the recommendations the air temperature was about 2°C higher than water temperature [3], and the relative humidity ranged from 40% to 55%, as recommended [3, 9]. Air velocity above surface of water was at constant level 0,1 m·s^{-1} and typical for these experiments [3].

4 Results and discussion

Results of the measurements were compared with values of water evaporation rate calculated according to equations (4-21). The calculated water evaporation rate were expressed per 1 m^2 water's surface. The value for unoccupied sports swimming pool are shown in Fig. 2 and for recreational swimming pool are shown in Fig. 3.

 DOI: 10.1051/e3sconf/20172200150

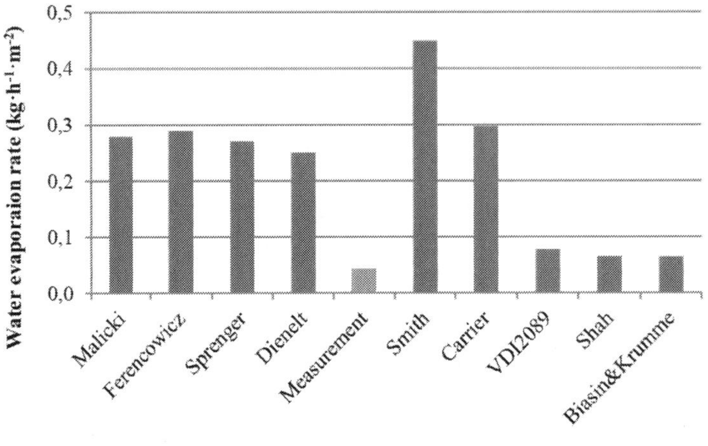

Fig. 2. Water evaporation rate in unoccupied sports swimming pools, with conditions described above.

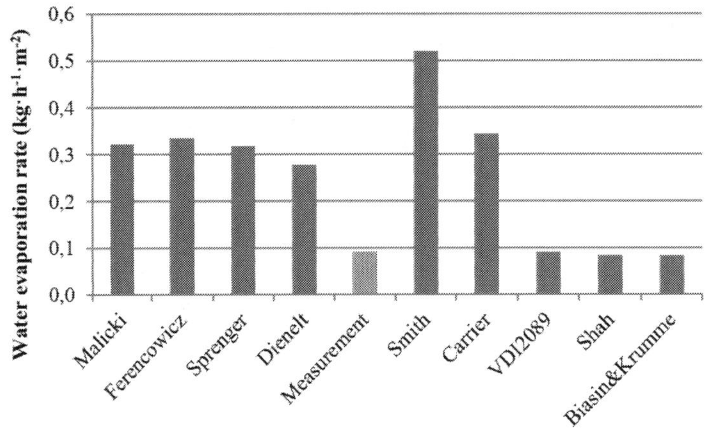

Fig. 3. Water evaporation rate in unoccupied recreational swimming pools, with conditions described above.

During study for occupied swimming pools was simulated the intensity of use at six level. The number of level was equal to the number of nozzles used during experiment. The results of comparisons between experimental data and the values calculated using published equation are shown in fig. 4 for sports swimming pools and in fig. 5 for recreational.

 DOI: 10.1051/e3sconf/20172200150

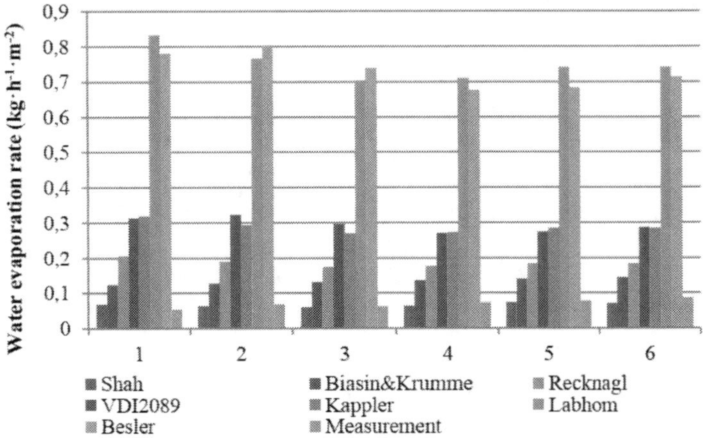

Fig. 4. Water evaporation rate in occupied sports swimming pools, depending on intensity of use, with conditions described above.

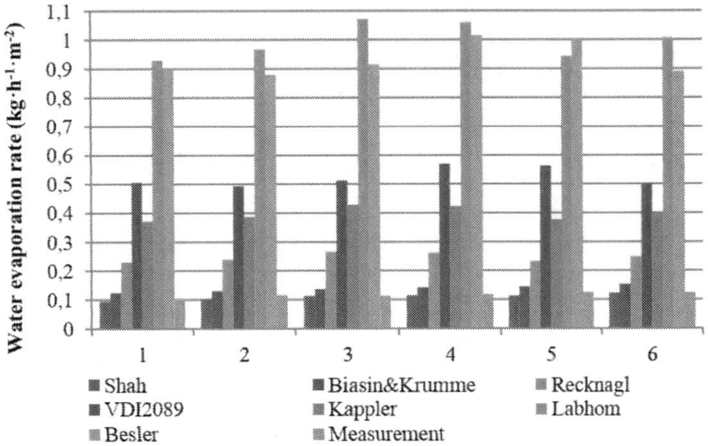

Fig. 5. Water evaporation rate in unoccupied recreational swimming pools, depending on intensity of use, with conditions described above.

5 Conclusions

Based on the results obtained during study, it was found that Shah and Biasin & Krumm models were the best fitted to results of measurements in laboratory conditions, both in unoccupied and occupied swimming pools. At high air and water temperatures the indicated equations show better fit to experimental data. The underestimate was up to 12% for Shah model and to 18% for Biasin & Krumm equations. In case of sports swimming pools conditions the differences was higher. The overestimate for unoccupied pools was similar for both equations and it was equal to 32%. For occupied pools during low intensity of use the calculated water evaporation rates was higher about 22% for Shah equation and even 57% for Biasin & Krumm model. When the intensity of use increased the overestimation of indicated both equations decreased. For Biasina & Krumm, the overestimation was 40% at maximum measured intensity, and for Shah's equations at maximum measured intensity overestimation turned into underestimation about 22%).

On the basis of conducted measurements for further analysis and calculations of vapor production are recommended equations published by Shah (9) and (19-21).

5 References

1. Swimming pools in Poland, Prepared by the Ministry of Sport and Tourism, Warsaw, 2015, p. 165. (in Polish)
2. K. Ratajczak, E. Szczechowiak, Heating Engineering Heating Ventilation, **41**, 4 (2010) 141-148 (in Polish)
3. F. Asdrubali, Energy and Buildings, **41** (2009) 311-319
4. M. Moghiman, A. Jodat, Iranian Journal of Mechanical Engineering, **8**, 1 (2007) 19-30
5. R. Tang, Y. Etzion, Building and Environment, **39** (2004) 77-86
6. K. Ratajczak, E. Szczechowiak, CLIMA 2016 - proceedings of the 12th REHVA World Congress: volume 5. Aalborg: Aalborg University, Department of Civil Engineering.
7. C. C. Smith, C. O. G. Lof, R. W. Jones, ASHRAE Transactions , **99**, 2 (1993) 864-874
8. M. M. Shah, Energy and Buildings, **49** (2012) 306-309
9. E. Sartori, Solar Energy, **68**, 1 (2000) 77-89
10. M. Malicki, Ventilation and air conditioning, PWN, Warsaw 1980 (in Polish)
11. J. Ferencowicz, Wentylacja i Klimatyzacja, PWN 1974
12. H. Recknagel, E. Sprenger, Compendium of Heating and Air Conditioning, Omni Scala Wroclaw 2008 (in Polish)
13. M. Jaskólski, Z. Micewicz, IPPU MASTA, Gdańsk 2000 (in Polish)
14. M. M. Shah, Energy and Buildings, **35** (2003)
15. C. C. Smith, C. O. G. Lof, R. W. Jones, Solar Energy, **53** (1994)
16. G. Besler, Heating Engineering Heating Ventilation,, **5** (1972)
17. H. Recknagel, E. Sprenger, Heating and air conditioning. Guide, Arkady, Warsaw 1994
18. M. M. Shah, Heating Piping Air Conditioning Engineering, **76** (2004)

Process simulation and comparison of biological conversion of syngas and hydrogen in biogas plants

Chaudhary Awais Salman*[,1], *Sebastian* Schwede[1], *Eva* Thorin[1] *Jinyue* Yan[1,2]
[1]School of Business, Society, and Engineering, Mälardalen University, 883, SE-72 123 Sweden
[2]School of Chemical Science and Engineering, Royal Institute of Technology, SE 1 0044, Sweden

Abstract. Organic waste is a good source of clean energy. However, different fractions of waste have to be utilized efficiently. One way is to find pathways to convert waste into useful products via various available processes (gasification, pyrolysis anaerobic digestion, etc.) and integrate them to increase the combined efficiency of the process. The syngas and hydrogen produced from the thermal conversion of biomass can be upgraded to biomethane via biological methanation. The current study presents the simulation model to predict the amount of biomethane produced by injecting the hydrogen and syngas. Hydrogen injection is modelled both in-situ and ex-situ while for syngas solely the ex-situ case has been studied. The results showed that 85% of the hydrogen conversion was achieved for the ex-situ reactor while 81% conversion rate was achieved for the in-situ reactor. The syngas could be converted completely in the bio-reactor. However, the addition of syngas resulted in an increase of carbon dioxide. Simulation of biomethanation of gas addition showed a biomethane concentration of 87% while for hydrogen addition an increase of 74% and 80% for in-situ and ex-situ addition respectively.

1 Introduction

Waste biomass is an exclusive source of renewable energy and can be utilized in various ways like incineration, gasification, anaerobic or aerobic digestion. Anaerobic co-digestion of biodegradable municipal solid waste to produce biomethane is a proven technically feasible option[1]. However, the major challenge with source separated waste fractions is the significant amount of substrate with high ligno-cellulose content, which is not suitable for biodegradation under anaerobic conditions. Therefore, finding the alternate solutions for efficient utilization of different waste fractions is of importance. One of the options is to separate non-biodegradable waste, which is further converted by thermochemical processes such as pyrolysis, gasification, etc. for syngas production. The two processes i.e. anaerobic digestion and gasification/pyrolysis can be integrated to increase the efficiency of biogas plants, efficient utilization of various fractions of waste within one facility and decrease the complexity of carbon dioxide separation in the downstream processes [2]. The syngas/H_2

*corresponding author: chaudhary.awais.salman@mdh.se

produced from these thermochemical processes can subsequently be converted to biomethane via biological methanation also known as methanogenesis process [3].

Biomethanation is the production of biomethane through biologically catalyzed reactions. Biological methanation is carried out by methanogenic archaea, which utilizes acetate or; carbon dioxide and hydrogen as a substrate [4]. Methanogens usually favor the conversion path of CO_2 and H_2 to methane and only methanogens are able to convert acetate to methane. The biological conversion of hydrogen is following the Sabatier reaction: $CO_2 + H_2 \rightarrow CH_4 + H_2O$ with the Gibbs free energy of -131 kJ/mol [5]. The conversion of hydrogen by biological Sabatier reaction can either be in-situ by introducing the hydrogen into an anaerobic digester or ex-situ i.e. in an external reactor with methanogens to favor the production of biomethane [4], [3].

The production of pure hydrogen is an energy intensive process either by hydrolysis of water or from thermochemical conversion, such as gasification or pyrolysis, of organic matter for example lingo-cellulosic biomass. The hydrogen can then be separated from the syngas containing CO, H_2, CO_2 and CH_4 by membrane separation of. The syngas itself can be used as a substrate for the biological conversion to biomethane [6]. The biological conversion of syngas can be carried out by a range of microorganisms which can simultaneously support the production of methane, hydrogen, and acetate [7]. Westman et al. [8] report that for the biological conversion of syngas through microorganisms can be carried out via two pathways with following six reactions with reaction 1 & 2 follow the H_2/CO_2 pathway:

$$CO + H_2O \rightarrow CO_2 + H_2 \ (\Delta G = -20 \text{ kJ/mol}) \quad (1)$$

$$CO_2 + 4H_2 \rightarrow CH_4 + H_2O \ (\Delta G = -131 \text{ kJ/mol}) \quad (2)$$

The second pathway follows the conversion of CO to acetate followed by conversion of acetate to methane via following reactions [9].

$$4H_2 + 2CO_2 \rightarrow CH_3COOH + 2H_2O \ (\Delta G = -104 \text{ kJ/mol}) \quad (3)$$

$$CH_3COOH + 2H_2O \rightarrow 4H_2 + 2CO_2 \ (\Delta G = 95 \text{ kJ/mol}) \quad (4)$$

$$4CO + 2H_2O \rightarrow CH_3COOH \rightarrow CH_4 + CO_2 \ (\Delta G = -176 \text{ kJ/mol}) \quad (5)$$

$$CH_3COOH \rightarrow CH_4 + CO_2 \ (\Delta G = -31 \text{ kJ/mol}) \quad (6)$$

2 Model Development

A number of authors have studied mathematical modeling of the biological conversion of syngas or hydrogen [10]–[13]. However, most of the modeling work is based on the development of a mathematical model for the biological methanation with the well-known ADM1 as the basis for these models [14]. Simulation of chemical processes gives detailed insights into the process streams and will be of great use for the further processing of information regarding the process. By considering these issues, this work is performed to develop the simulation of the process based on biochemical reactions instead of developing a mathematical model. Furthermore, a simulation model is established to predict the amount of biomethane from the anaerobic digestion and biological methanation process. The model is developed in Aspen Plus ®, which is a leading software for the development simulation models for various chemical processes [15]. The developed model can be incorporated with other thermochemical processes such as pyrolysis and gasification to measure the benefits and impact of integrating different processes.

Three cases have been considered for comparison. (a) In-situ hydrogen addition to the digester, (b) Ex-situ hydrogen addition in external bioreactor, and (c) Ex-situ syngas addition in the external bioreactor.

Process simulation for the anaerobic digestion is adopted from [16] and [17]. The model is selected because it can be incorporated in flowsheet based software such as Aspen Plus® with fewer complexities. The model can then be able to extend and integrate with different thermochemical processes such as pyrolysis and gasification. The model is originally developed to predict the biogas production from biodegradable substrates. The model is based on the biochemical reactions with first-order reaction kinetics and also includes the pH-value and take ammonia inhibition into account. All the reaction kinetics and inhibitions are implemented into the Aspen Plus model through Fortran calculator blocks. The detail about all the chemical reactions used in the model can be found in [16], and the rate equations and their kinetic parameters are taken from [17]. The model predicts the biogas from the organic substrate by their carbohydrates, proteins, lipids and water content.

The model is then modified to introduce the hydrogen into the process. The design scheme for both in-situ hydrogen addition and ex-situ hydrogen addition is displayed in Figure 1. For in-situ upgrading of biogas, hydrogen is added in stoichiometric ratio i.e. 4 moles of hydrogen per each mole of CO_2 present. The Sabatier reaction is added to the previous set of reactions. For the ex-situ processes, the digestate is separated into liquid and solid. The solid digestate with unconverted nutrients is utilized in the external reactor along with biogas from the digester and stoichiometric amounts of hydrogen. The reactions with their first order rate equations are added to the model to convert the hydrogen to biomethane biologically.

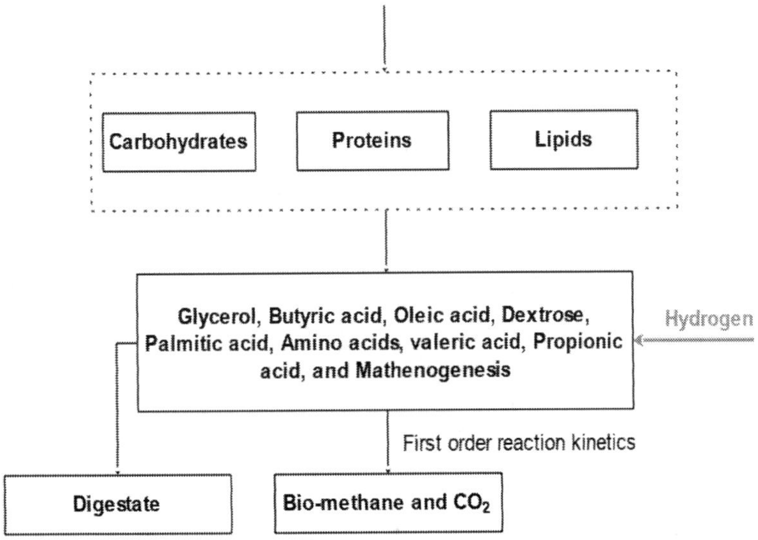

(a)

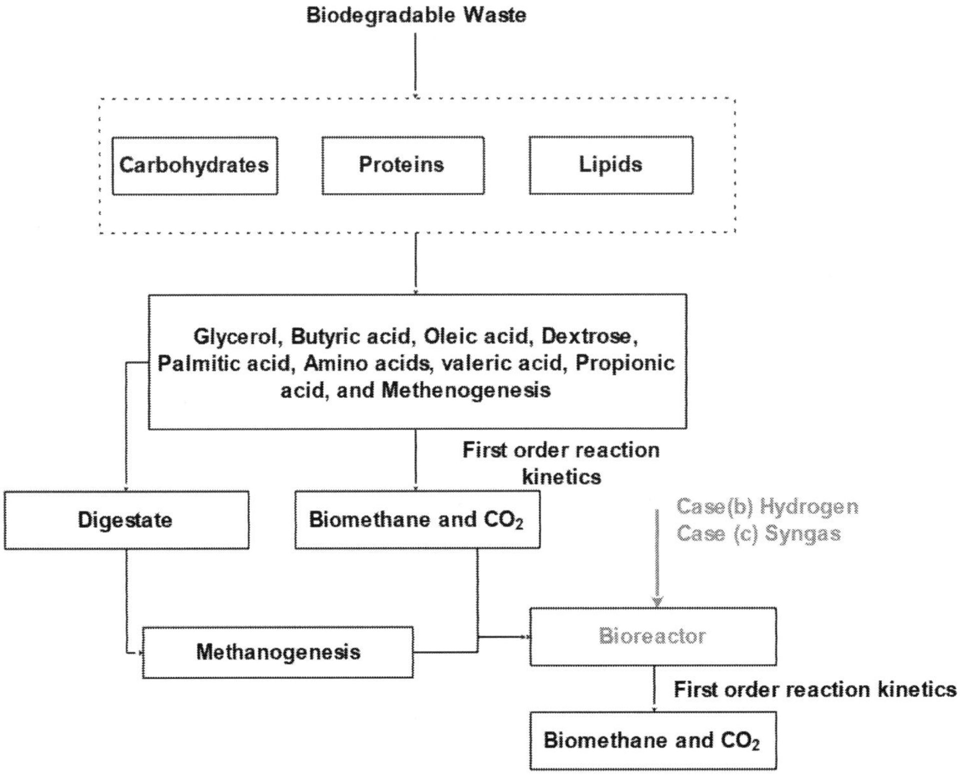

(b), and (c)

Figure 1: Scheme for designing the model and process simulation of anaerobic digestion with methanation of hydrogen (a) in-situ hydrogen addition (b) ex-situ hydrogen addition (c) Ex-situ syngas addition

For syngas addition to the bioreactor, the same modeling approach is used as described for ex-situ hydrogen addition in Figure 1 (b). Carbon monoxide is added in a stoichiometric amount of water present in the external bioreactor at a molar ratio of 1:1 and hydrogen to carbon dioxide mole ratio is 4:1 respectively. The organic composition of biodegradable waste used for the simulation of hydrogen and syngas methanation is presented in Table 1.

Table 1: Organic composition of biodegradable waste used for simulation

Organic composition of waste [18]			
Carbohydrates, %	50	Moisture,%	80
Proteins, %	35	Dry matter TS,%	20
Lipids, %	15	VS (% TS)	66

3 Results

The model for anaerobic digestion is validated with data taken from two plants and has been displayed in Table 2. Both plants are located in Sweden and run under mesophilic conditions. The primary substrate used in the Växtraft biogas plant is a source-separated organic municipal waste, and for Eskilstuna, the biogas plant is operated with waste sludge. The results obtained from the model are in agreement with the plant data. It should be noted that the organic composition of feed for the both biogas plants is assumed from the

 DOI: 10.1051/e3sconf/20172200151

literature shown in Table 1. The actual composition may give different results for the process simulation.

Table 2: Difference between simulation results for anaerobic digestion and biogas plant data, taken from [19], [20].

	Volume of digesters, m^3	Feed, t/y	HRT, days	Methane production, MWh	Simulation results, MWh	Error, %
Växtraft biogas plant	4000	23000	20	15000	14641	-2.38
Eskilstuna biogas plant	3600	72560	25	8603	9380	5.18

Figure 2 shows the results obtained for the simulation of methanation of hydrogen. A stoichiometric amount of hydrogen i.e. 4 moles per each mole of carbon dioxide is added to both the in-situ and external reactor. The increase in biomethane concentration over the course of 30 days has been simulated. The results show that the concentration of biomethane will increase by the addition of hydrogen in both cases. However, the conversion of biomethane is faster in the case of the external methanation reactor. The final concentration of up to 80% biomethane is achieved for ex-situ conversion of hydrogen.

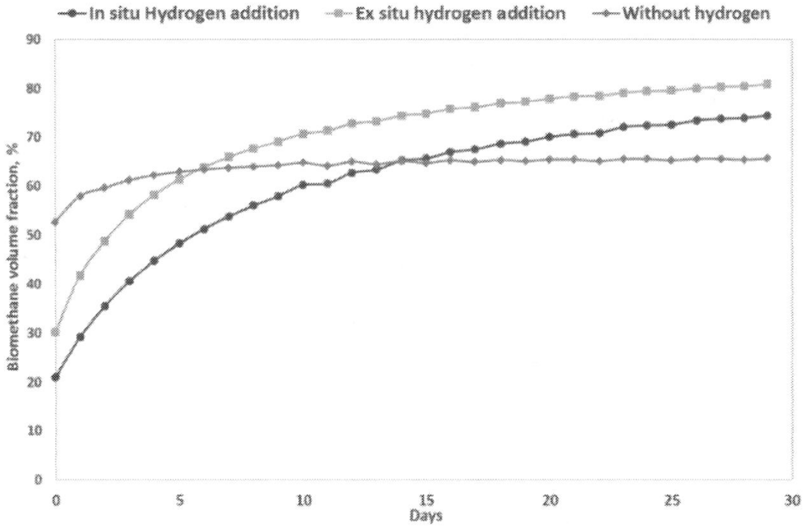

Figure 2: Change in biomethane concentration in biogas for all the cases (without hydrogen addition, hydrogen addition in the digester and external reactor)

Figure 3 displays the conversion of hydrogen for both cases. It can be seen that the model indicates that the hydrogen is not fully convertible via Sabatier reaction biologically. However, the conversion of hydrogen is higher in the case of the ex-situ biological reactor (85%) after 30 days as compared to in-situ, where 80% of the supplied hydrogen was utilized.

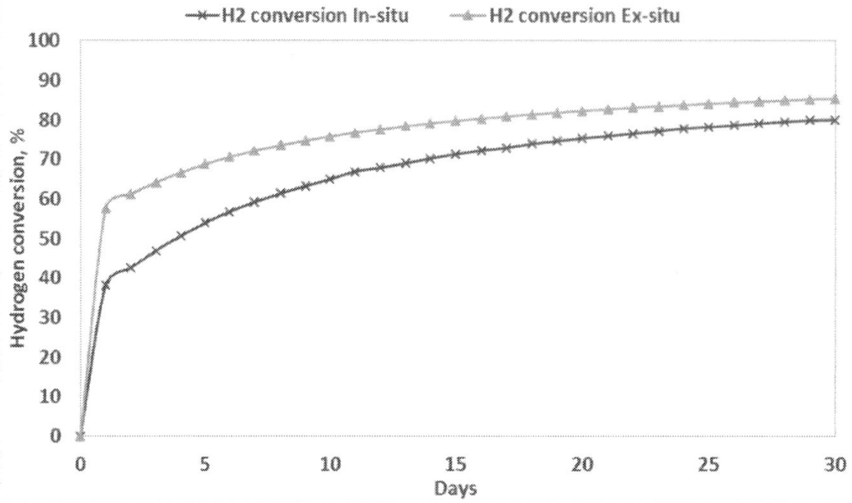

Figure 3: Conversion of hydrogen in the biological reactor

Figure 4 presents the change in volume fraction of biomethane with syngas addition. Initially, the simulation shows that a higher concentration of biomethane can be achieved by the addition of syngas as compared to the case with the addition of hydrogen. The model also indicates that the injection of syngas increases the concentration of biomethane as high as 95%, but the concentration decreases and stabilizes around 87%. This decrease is a result of the CO_2 increase through reaction 5 and 6, and also the incoming carbon dioxide with syngas also makes the final biogas slightly rich in CO_2. The model also predicts that ex-situ syngas addition yields 13 % more biomethane than ex-situ hydrogen addition and 23% more biomethane than the in-situ addition of hydrogen.

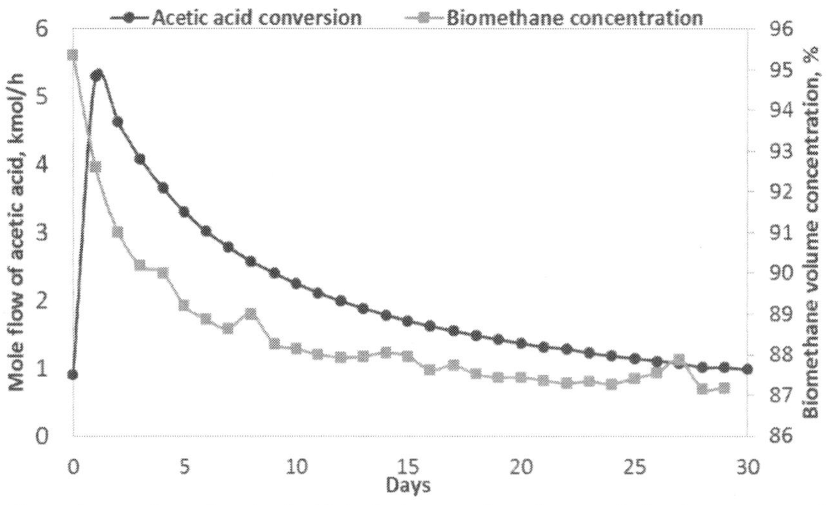

Figure 4: Biomethane volume fraction by biological conversion of syngas in Ex-situ reactor and Molar flow (kmol/h) of acetic acid by the addition of syngas in ex-situ reactor (Waste input of 3600 kg/h)

The simulation of the biological conversion of syngas also shows the full conversion of hydrogen and carbon monoxide in the very beginning which indicates that the syngas utilization adopted to the acetogenic pathway for the biological conversion as described by reactions 3-6. The increase and conversion of acetic acid by the result of syngas addition is

 DOI: 10.1051/e3sconf/20172200151

displayed in Figure 4. It can be seen that almost all the generated acetic acid from the syngas addition is converted to biogas.

4 Discussion

The in-situ upgrading of biogas offers a less complex option for biological methanation, but the process conditions have some limitations regarding the size of the digester, the production rate of CO_2, and also requires thorough process design to inject the stoichiometric amount of hydrogen. Simulations for the above cases have been run for the Växtraft biogas plant. The biogas plant has an annual input of 23000 tons per year of organic waste with biomethane production capacity of 15000 MWh annually [20]. The stoichiometric amount of hydrogen and syngas is injected into the process, and results are displayed in Figure 5. The results show that the biological methanation process can achieve 46-85% increase in biomethane.

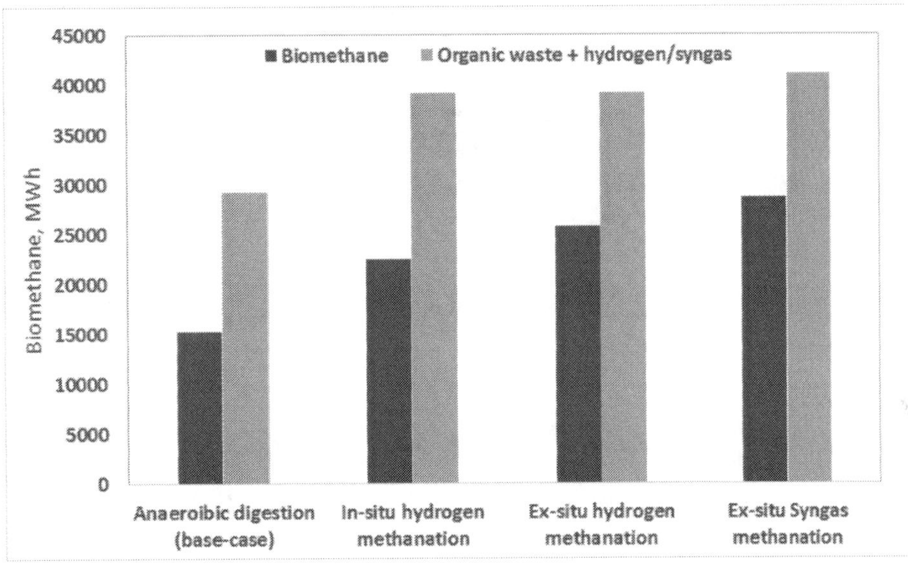

Figure 5: Increase in biomethane by biological methanation for a biogas plant

The increase is highly dependent on the amount of substrate i.e. hydrogen and syngas available for the process. Also, the production of hydrogen and syngas from thermochemical conversion processes are an energy intensive process and requires a significant amount of heat and power. Addition of hydrogen or syngas also affects the pH and buffer capacity due to the removal of CO_2 from the liquid. Furthermore, the operation of an external biological reactor also requires work to setup and maintain a stable-running process. Hence, comprehensive process integration must have been carried out to evaluate the technical feasibility of thermochemical and biochemical process integration.

5 Conclusion

A simulation study of biological methanation of syngas and hydrogen was performed in this work for both ex-situ and in-situ bioreactor configurations. Highest conversion of hydrogen and syngas was achieved in ex-situ or separate methanation reactors as compared with the in-situ process.

References

1. G. M. Guebitz, A. Bauer, G. Bochmann, A. Gronauer and S. Weiss, Biogas sci & Tech, (2015)
2. C. Awais, S. Schwede, E. Thorin and J. Yan, E.Procedia (2017), (To be published)
3. G. Luo and I. Angelidaki, Biotechnol. Bioeng. **109**, 2729–2736 (2012).
4. A. J. Guneratnam, E. Ahern, J. A. FitzGerald, S. A. Jackson, A. Xia, A. D. W. Dobson and J. D. Murphy, Bioresour. Technol. **225**, 308–315 (2017)
5. G. Leonzio, Chem. Eng. J. **290**, 490–498 (2016)
6. K. T. Klasson, M. D. Ackerson, E. C. Clausen and J. L. Gaddy, Enzyme Microb. Technol. **14**, 602–608 (1992)
7. K. T. Klasson, B. B. Elmore, J. L. Vega, M. D. Ackerson, E. C. Clausen and J. L. Gaddy, Appl. Biochem. Biotechnol. 24–25 (1990)
8. S. Westman, K. Chandolias and M. Taherzadeh, Fermentation **2** (2016)
9. S. S. Navarro, R. Cimpoia, G. Bruant and S. R. Guiot, Front. Microbiol. **7** (2016)
10. A. Bensmann, R. Hanke-Rauschenbach, R. Heyer, F. Kohrs, D. Benndorf, U. Reichl and K. Sundmacher, Appl. Energy **134**, 413–425 (2014)
11. L. Jürgensen, E. A. Ehimen, J. Born and J. B. Holm-Nielsen, Bioresour. Technol. **178**, 323–329 (2015)
12. M. K. Nikoo and N. A. S. Amin, Fuel Process. Technol. **92**, 678–691 (2011)
13. I. Tilla and E. Dace, Energy Procedia **95**, 475–481 (2016)
14. D. J. Batstone, J. Keller, I. Angelidaki, S. V Kalyuzhnyi, S. G. Pavlostathis, A. Rozzi, W. T. M. Sanders, H. Siegrist and V. A. Vavilin, Water Sci. Technol.**45**, 65–73 (2002)
15. "Aspentech." [Online]. Available: http://www.aspentech.com/. Accessed: (2016)
16. K. Rajendran, H. R. Kankanala, M. Lundin and M. J. Taherzadeh, Bioresour. Technol. **168**, 7–13 (2002)
17 I. Angelidaki, L. Ellegaard and B. K. Ahring, Biotechnol. Bioeng. **63**, 363–372 (1999)
18. U. Sonesson and H. Jönsson, Urban biodegradable waste amount and compostiion - Case study Uppsala (1996).
19. E. Nordlander, J. Holgersson, E. Thorin, M. Thomassen and J. Yan, Int. Conf. Appl. Energy 1661–1675 (2011)
20. S. R. Esteves, Västerås (Växtkraft) Biogas Plant CASE STUDY – Source Segregated Biowastes, (2007).

Sediments from stormwater drainage system as sorbents of organic and inorganic pollutants

Aleksandra Sałata[1,*] and *Lidia* Dąbek[1]

[1]Kielce University of Technology, Department of Environmental, Geomatics and Energy Engineering, aleja Tysiąclecia Państwa Polskiego 7, 25-314 Kielce, Poland

Abstract. The study presents the results of tests aimed at determining variations in concentrations of polycyclic aromatic hydrocarbons (naphthalene, acenaphthylene, acenaphthene, fluorene, phenanthrene, anthracene, fluoranthene, pyrene, benzo[a]anthracene, chrysene, benzo[b]fluoranthene, benzo[k]fluoranthene, benzo[a]pyrene, dibenzo[a,h]anthracene, benzo[g,h,i]perylene and indeno[1,2,3-c,d]pyrene) and heavy metals (copper, cadmium, chromium, nickel, lead and zinc) in the sediments from the stormwater sewer system. Results of this study were then compared to Polish regulations and established ecological screening values to determine their potential risk to environment. The results suggest that stormwater sediments are moderately contaminated with PAHs and heavy metals according to domestic regulations and highly contaminated according to ecological benchmarks.

1 Introduction

Water quality issues remain invariably at the centre of research and environmental protection institutions because of their importance for the functioning of ecosystems. Sediments accumulated in the aquatic environment (lakes, rivers, seas, stormwater and stormwater reservoirs) are a valuable source of information on the ecological state of the environment. The fact is that sediments are widely recognized as a place of deposition of chemical pollutants entering the aquatic environment and thus as an essential component of their circulation in nature. Despite this, the assessment of environmental threats using this tool still seems insufficient and still neglected [1,2]. Therefore, the assessment of the quality of sediment should not be overlooked in the improvement of the aquatic environment, especially since the recommendation of the Water Framework Directive (EU 2000/60/EC) for 2015 was the need to achieve good ecological and chemical water status.

One of the components of the water environment functioning within the urban agglomerations is the sediment accumulating in various elements of the rain drainage system: in sewers, wells, street inlets and settlers located in the rainwater treatment plant assigned to each urban catchment. These sediments have a major influence on the processes running through the sewage system causing significant fluctuations of the charge in the inflow into the stormwater treatment plant in short periods of time. This transforms directly

* Corresponding author: asalata@tu.kielce.pl

 DOI: 10.1051/e3sconf/20172200152

into the load of pollutants entering the treatment plant or overflow, and thus the receiver. Research conducted to determine the qualitative composition of sludge is used to determine the level of release of chemical impurities [3].

Sediment-forming matter consists of organic and mineral compositions and anthropogenic compounds. Such varied compositions are determined by the type of particles and their origin, transport processes and chemical and biochemical changes occurring in the water, sediment, and the boundary of these phases [4]. The composition of sediments, including the presence of organic matter, has a significant impact on their sorption capacity. These capacities and the influence of pollutants present in rain waters with slurry of grains, sand, gravel, plant remains and other sedimentation components of the solid phase forming the precipitate are possible due to the crystalline structure of the components, the presence of active functional groups predominantly in amorphous phases [5]. The composition of stormawter and of sediments accumulated in the system depends on many factors. Those include the type of the catchment management programme, the roadway system, the vehicular traffic volume, the occurrence of failures and environmental disasters, the frequency and manner of cleaning of the catchment (roads, squares) surface, the application of de-icing agents during winter months, parameters of atmospheric precipitation, season of the year, the sewer system hydraulics and structure and its facilities (separators, sedimentation tanks, street gullies, and sumps), the way of the sewer system operation, and also the fraction green belts constitute in the drained area. Sediments may hamper the flow of stormwater and they also can increase the pollutant load transported to the stormwater treatment plant, stormwater overflow structures, or directly to the receivers, where they create a major hazard to the biosphere and the adjacent land environments. Due to the chemical and biochemical processes, a part of pollutants may migrate to the waters again [6]. The most commonly occurring impurities in water sediments are heavy metals and polycyclic aromatic hydrocarbons (PAHs). In the aquatic environment, heavy metals can occur in two basic forms, they can either be solved, or bound to the solid phase particles. Metal occurrence in either of those forms depends on many factors including: the chemical form of metals delivered to the aquatic ecosystem, conditions prevailing in the liquid phase (pH, redox potential, temperature, biological activity), surface reactivity, and grain composition of the solid phase. Tests on aquatic sediments are very often conducted to assess the environmental pollution with heavy metals, because after they get into the stormwater, a majority of them migrate to the sediments. In polluted aquatic sediments, elevated concentrations of cadmium, chromium, copper, nickel, zinc or lead are found most often. Their presence in the aquatic environment may result both from natural processes occurring in the atmosphere, which provides a dispersed source of pollutants, and from anthropogenic processes (transportation, industries, mining, municipal and industrial sewage, and also agriculture) which constitute point emission sources [7-9]. Polycyclic aromatic hydrocarbons are on the list of the most dangerous substances, which was compiled by the American Environmental Protection Agency (EPA). The list is updated and adjusted to the environmental conditions in many countries [10]. Polycyclic aromatic hydrocarbons are commonly classified into one of two groups based on their molecular structure. Low molecular weight (LMW) PAHs have three or fewer aromatic rings, while high molecular weight (HMW) PAHs have four or more rings. Differences in the structure and size of individual PAHs result in substantial variability in the physical and chemical properties of these substances. The toxicity of PAHs depends on a number of factors, including the species, route of exposure, and molecular structure of the PAH. In general, LMW-PAHs are considered to be acutely toxic and noncarcinogenic to aquatic organisms, whereas, HMW-PAHs are generally not acutely toxic to aquatic organisms, but a number of them are carcinogenic [11].

In the literature on bottom sediments and stormwater sediments, the geochemical criterion is used to assess the degree of pollution. The criterion is one of the chemical methods of evaluating the sediment quality. Chemical methods utilise the content of trace elements and chemical properties of sediments, which determine the potential capacity of sediments to bind toxic components, or the mobilisation of the latter from sediments. The criterion involves the comparison of the pollutant content in the sediment with that found in natural or low-polluted sediments, which provides the so-called geochemical background. Recent years have witnessed developments in ecotoxicological investigations methods for the aquatic environment, and also in assessing the hazards related to pollutant occurrence in water bodies receiving wastewater and surface runoff. In many countries, ecotoxicological criteria have entered legislation on sewage and hazardous wastes. Numerous methods have been developed to determine quality standards for bottom sediments. Sediment Quality Guidelines (SQGs) are based on both chemical and biological aspects. Poland still lacks legal regulations on the assessment of the harmful effect of chemical compounds in water sediments for the organisms that inhabit this environment. Many approaches have been made to establish SQGs. They rely on different theoretical or empirical basis used to assess the hazard posed by the presence of organic compounds and heavy metals in bottom sediments. The most widely used are TEL and PEL parameters, which are components of the empirical method applied to benthic organisms. The TEL, Threshold Effects Level, specifies the upper limit of pollutant concentration range, below which the harmful effect on aquatic organisms is found relatively seldom. The PEL, Probable Effects Level, is defined as the lower limit of pollutant concentration, which may significantly adversely affect the living organisms [12-15].

Sediment classification proposed by Bojakowska [16] incorporates ecotoxicological parameters TEL and PEL used in the USA and Canada. Sediments are categorised in a four-level scale, where: I. Unpolluted sediments (none of the harmful component content exceeds the TEL parameter), II. Low-polluted sediments (the content of at least one harmful component is higher than the TEL parameter, the content of all components is lower than the PEL parameter) III. Polluted sediments (the content of at least one harmful component exceeds the limit of Class II). IV. Strongly-polluted sediments (the content of at least one harmful component exceeds the limit of Class III). Sediments categorised as Class I and Class II can be freely applied to both aquatic and land environments. They are most often used as soil fertilisers, or to restore damaged banks of water reservoirs and rivers. Class III sediments can be relocated to designated sites and managed to a limited extent. Due to the elevated concentrations of hazardous compounds, those can be applied to industrial crops, or may be excluded from agricultural use. Strongly-polluted Class IV sediments should be treated before they are placed in the environment, or they should be stored at protected refuse sites under controlled conditions.

2 Materials and methods

2.1 Study site and sampling

The sampling stations including of four sedimentation tanks located in the urban area of Kielce city were studied. The facilities have different construction, operating time and land use. The largest and the oldest is the Piekoszowska catchment S1 (open tank) representing a wide range of land uses including both low – and high – density residential areas, industrial areas and agricultural, partly forested ones. The three other catchments are classified as typically industrial - the Jesionowa facility S2 (closed tank), and typically residential - the Witosa facility S3 (open tank) and the Kaczmarka facility S4 (closed tank) (Table 1).

Table 1. Main technical parameters of catchments selected to study.

Parameter	S1	S2	S3	S4
Construction year	1992	2001	2003	2002
Tank type	open	closed	open	closed
Total catchment area (ha)	804.6	383.9	132.0	224.0
Residential area (%)	63	33	89	78
Industrial area (%)	9	65	0	14
Meadows, agricultural area (%)	27	0	0	0
Forests (%)	1	1	0	0
Green areas (%)	0	1	11	8

The sediment samples were collected during winter season 2015, in accordance with the PN-EN ISO 5667-15:2009 standard method [17]. Samples were collected using a stainless steel standard bottom grab sampler or Eijkelkamp equipment for sites with difficult access, and placed in acid-washed plastic containers (~1L). Between the series of sample collecting, all equipment was rinsed with water from the reservoir. Then, the samples were transported to the laboratory and homogenized using a solvent-cleansed stainless steel bucket and a spoon. The physical and chemical parameters were measured immediately afterwards.

2.2 Analytical techniques

The pH and temperature were measured promptly after samples homogenization in accordance with the PN-EN 12176:2004 method using SevenMulti™ meter (Mettler Toledo) [18]. Organic content, measured as a loss of ignition (LOI), was determined in accordance with the PN-78/C-04541 method [19], which involved drying the sediment at 105°C to a constant weight, and then heating it at 550°C for 2h. Sediment samples of 0.2g were mineralized in 7ml nitric acid using microwave oven (Multiwave 3000, Anton Paar). Concentrations of heavy metals: cadmium, copper, chromium, nickel, lead and zinc were determined by atomic emission spectrometry with inductively coupled plasma ICP Optima 8000 (Perkin Elmer) with certified multi element standards. Sediment samples with a mass of 1g were extracted in dichloromethane to determine the total sum of PAHs. These compounds are grouped in two categories based on their molecular weights and these are low and high. Low Molecular Weight PAHs (LMW) containing two- and three-rings with molecular weight from 152 $g{\cdot}mol^{-1}$ to 202 $g{\cdot}mol^{-1}$ include: naphthalene (NAPT), acenaphthylene (ACNY), acenaphthene (ACEN), fluorene (FLUR), phenanthrene (PHEN) and anthracene (ANTH). The High Molecular Weight (HMW) PAHs have four-rings to seven-rings with molecular mass ranging from 228 g·mol-1 to 278 g·mol-1 include: fluoranthene (FLTH), pyrene (PYR), benzo[a]anthracene (B(a)A), chrysene (CHRY), benzo[b]fluoranthene (B(b)F), benzo[k]fluoranthene (B(k)F), benzo[a]pyrene (B(a)P), dibenzo[a,h]anthracene (D(ah)A), benzo[g,h,i]perylene (B(ghi)P), and indeno[1,2,3-c,d]pyrene (INPY). Concentrations of PAHs were identified in samples in accordance with the PN-EN 15527:2008 method [20]. To determine the reliability of the PAH data, including the efficiency of extraction methods and interferences that may be introduced during sample preparation, deuterated internal standards were used (d-8 NAPHT and d-12 B(a)A). These standards ensure that the accuracy of the results obtained is independent of the recovery degree. The next step included wetting the sediment sample with acetone, adding acetone with deuterated standards, intensive stirring of the sample and undisturbed evaporation of the acetone. This procedure makes it possible to achieve a similar binding form between standards added and the sediment matter. All PAH concentrations were reported on a dry weight basis.

3 Results and discussion

The mean of pH, content of mineral (SM) and organic matter described as a loss of ignition (LOI) together with moisture content were measured in the sediment samples. The pH values ranged 7.14÷7.30 and were comparable with those of silts from inland water bodies, in which about neutral pH conditions (6.5 ÷ 7.5) are maintained. Sediments were characterised by a high moisture content of 78% ÷ 86%, mineral matter content up to 97%, and organic matter content ranging from 3% to 5%.
The content of heavy metals (zinc, cadmium, chromium, nickel, lead and copper) in the sediments from the each site geochemical classification are presented in Table 2. The data indicate that the sediments of concern showed varied metal content. The highest average concentrations were recorded for zinc. The zinc content ranged from 127.4 mg·kg^{-1} for S4 site, to 375.6 mg·kg^{-1} for S1 site. The lead content comparable in every location at level ~70 mg·kg^{-1}. The concentrations of other metals were markedly lower. In general, the highest concentrations were recorded for the tank S1, which is the most contaminated. Comparing the results of tests on metal concentrations in the analysed sediments with the quality standards for soil and land stipulated in the Regulation of the Minister of Environment [21], it can be stated that concentrations of all examined heavy metals are much lower than the permissible values for urbanised areas categorised as Group B. In accordance with the regulation, the concentrations range as follows, for zinc: 300 – 720 mg·kg^{-1}, cadmium: 4 – 10 mg·kg^{-1}, lead: 100 – 200 mg·kg^{-1}, copper: 100 – 200 mg·kg^{-1}, nickel: 50 – 210 mg·kg^{-1} and chromium: 150 – 380 mg·kg^{-1}.

Table 2. Heavy metal concentrations in sediments compared with local background values. Geochemical assessment of stormwater sediments wg Bojakowska [16]

Metal	S1	S2	S3	S4	Class I	Class II	Class III	Class IV
mg·kg^{-1}					mg·kg^{-1}			
Zinc	**375.6**	**221.2**	**199.5**	**127.4**	125	300	1000	>1000
Cadmium	0.1	0.2	0	0	0.7	3.5	6	>6
Chromium	**56.8**	40.3	31.1	24.3	50	100	400	>400
Nickel	13.85	14.8	11.55	9.2	16	40	50	>50
Lead	**85.05**	**99.15**	**61.65**	**88.7**	30	100	200	>200
Copper	**39.8**	**47.2**	**38.15**	**45.8**	20	100	300	>300

As regards cadmium, chromium and nickel for sites S2, S3 and S4 respectively, the sediments were classified as unpolluted belonging to Class I. With respect to the content of zinc, however, the sediments were categorised as Class II in every case, except location S1 – where sediments were assigned to class number III . Taking into account geochemical parameters, the whole tested material can be considered medium-polluted, i.e. the sediments may pose an immediate hazard to the aquatic environment and the organisms living there. Elevated concentrations of heavy metals may result from many processes that occur in the aquatic environment, to a large extent, they depend on sediment composition, thus on the character of the catchment.

The total content of PAHs with geochemical assessment in the examined sediments was presented in Table 3. The highest concentrations of polycyclic aromatic hydrocarbons were observed in sediment samples from S4 site and the lowest for S1 site – 83.40 mg·kg^{-1}

and 39.30 mg·kg^{-1}, respectively. Comparing the results with the standards on soil and land quality [21], the values reported were below the permissible ones, specified as 200 mg·kg^{-1}. Consequently, the sediments do not pose a hazard to the natural environment. However, in addition to the total content, also the content of individual PAH compounds was taken into account. The material of concern was categorised as Class IV – for all samples. As regards Class IV, i.e. strongly polluted sediments, it is recommended that they should be treated before they are disposed of into the environment. Otherwise, they should be stored at protected refuse sites under controlled conditions.

Table 3. PAHs concentrations in sediments compared with local background values. Geochemical assessment of stormwater sediments wg Bojakowska [16]

PAH	S1	S2	S3	S4	Class I	Class II	Class III	Class IV
	mg·kg^{-1}				mg·kg^{-1}			
ACEN	0.22	**0.55**	0.22	0.44	0.01	0.1	0.5	>0.5
ACNY	**2.20**	**2.97**	**3.19**	**13.31**	0.01	0.15	0.5	>0.5
PHEN	**7.15**	**10.89**	**6.60**	**8.36**	0.05	0.5	1	>1
FLTH	**14.63**	**15.62**	**12.10**	**26.62**	0.1	1.5	3	>3
FLUR	**0.81**	**2.42**	**0.92**	**1.96**	0.02	0.15	0.5	>0.5
ANTH	**1.73**	**2.07**	0.81	**5.29**	0.05	0.25	1	>1
PYR	**12.77**	**14.03**	**10.01**	**26.80**	0.1	1.5	2.5	>2.5
CHRY	0.92	1.15	**4.60**	1.84	0.1	0.75	2.8	>2.8
B(a)A	0.58	1.15	0.69	0.58	0.1	0.7	1.6	>1.6
B(b)F	0.92	1.50	**6.33**	2.42	0.1	0.75	3	>3
B(k)F	0.12	0.69	2.19	1.15	0.1	0.75	3	>3
B(a)P	0.92	1.15	**5.87**	**1.73**	0.1	0.75	1.6	>1.6
D(ah)A	0.00	**0.81**	**2.30**	**0.46**	0.01	0.15	0.25	>0.25
B(ghi)P	**0.58**	**1.04**	**2.42**	**1.38**	0.05	0.2	0.4	>0.4
INPY	**0.58**	**0.81**	**4.26**	**1.38**	0.05	0.2	0.4	>0.4
∑PAHs	**39.30**	**50.60**	**55.20**	**83.40**	1	7.5	15	>15

4 Conclusion

The sediments accumulated in the stormwater sewer system cause problems that are related not only to the system operation, but also to the hazard they may pose when delivered to the receiver. Because of their properties, the sediments are characterised by a high affinity for accumulating and storing toxic pollutants in their structure, such as heavy metals and also organic compounds like polycyclic aromatic hydrocarbons. The results of tests show significant differences in the physico-chemical composition of the sediments of concern. As expected, the analysis of the pollutant content revealed the occurrence of heavy metals, namely zinc, cadmium, chromium, nickel, lead and copper, and also of polycyclic aromatic hydrocarbons. The results describing metal concentrations in the sediments were referred to the soil and land quality standards stipulated in the Regulation of the Minister of Environment. The comparison shows that the concentration values of all analysed heavy metals are much lower than the permissible values for urbanised areas. In accordance with the geochemical classification, however, the sediments of concern were categorised as Class I or in some cases class II, that is low-polluted. Consequently, they may pose an immediate hazard to the aquatic system and the organisms inhabiting it. The analysis of the content sum of PAH organic compounds shows high concentrations of those substances from 39.30 mg·kg^{-1} to 83.40 mg·kg^{-1}, which indicates their accumulation in the examined sediments. Comparing the test results with the soil and land quality standards, it can be concluded that, as it was the case for heavy metals, the values are lower than the permissible ones, i.e. 200 mg·kg^{-1}. As a result, the sediments can be considered as not causing any environmental hazard. Detailed geochemical assessment of the stormwater sediments, however, indicates a different classification. Sediments are categorised as Class IV, i.e. strongly polluted, which means it is recommended to treat them before disposal into the environment or to store them at protected refuse sites under controlled conditions.

The work reported herein was undertaken as part of the statutory project No. 05.0.10.00/2.01.01.01.0026 (MNSP.IKTW.15.001).

References

1. E. Daly, P.M. Bach, A. Deletic. *Adv Water Resour.*, **74**, 148-155, (2014)
2. J.V. Loperfido, G. B. Noe, S. T. Jarnagin, D. M. Hogan. *J Hydrol.*, **519**, 2584-2585 (2014)
3. A. Jlilati, K. Jaromin, M. Widomski, G. Łagód. *Proceedings of ECOpole*, **3(1)**, 147-152 (2009)
4. E. Szalińska. Role of bottom sediments in the assessment of the quality of continental waters environment *Publishing house of Cracow University of Technology*, Monograph **396**, (2011) (in Polish)
5. M. Gdaniec-Pietryka. Physical speciation and the mobility of analytes from polycyclic aromatic hydrocarbons and polychlorinated biphenyls on the bottom sediment-water phase boundary (in Polish). *Doctoral thesis. Faculty of Chemistry, Gdansk University of Technology, Department of Analytical Chemistry*, (2008) (in Polish)
6. J. Królikowska, A. Królikowski. Rainwater. Disposal, management, pre-treatment and use. *Seidel – Przywecki Publishing house*, (2012) (in Polish)
7. D. Foster, S. M. Charlesworth. *Hydrol Process.*, **10**, 227-261, (1996)
8. A. El-Mufleh, B. Béchet, I. Basile-Doelsch, C. Geffroy-Rodier, A. Gaudin, V. Ruban. *Envirol Sci Pollut Res*, **21**, 9764-9776 (2014)
9. P. Bigus, M. Tobiszewski, J. Namieśnik. *Mar Pollut Bull.*, **78**, 26-42, (2014)
10. X. Yuan, L. Zhang, L. Li, C. Wang, J. Ji. *Catena,* **119**, 52 – 60, (2014)
11. J.L. Bertrand-Krajewski, C. Becouze, A. Dembélé, M. Coquery, C. Cren-Olivé, B. Barillon, P. Dauthuille, J. Chapgier, M. Grenier-Loustalot, P. Marin. *Proceedings of the 11th International Conference on Urban Drainage*, (2008)
12. POLYCYCLIC AROMATIC HYDROCARBONS (PAHs), Canadian Sediment Quality Guidelines for the Protection of Aquatic Life, *Canadian Environmental Quality Guidelines Canadian Council of Ministers of the Environment*, (1999)
13. Bojakowska I., Gliwicz T., Małecka K. Results of geochemical tests on Poland's aquatic sediments in the years 2003-200. *Inspection of Environmental Protection.* Warsaw (2006) (in Polish)
14. Bojakowska I., Sokołowska G. Geochemical classes of aquatic sediment purity. *Polish Geological Review.* **46 (1)**, 49-54, (1998) (in Polish)
15. D.D. McDonald, C.G. Ingersoll, T.A. Berger. *Arch Environl Con Tox.*, **39**, (2000)
16. I. Bojakowska. Criteria of evaluation of pollution of aquatic sediments. *Polish Geological Review,* **49**, nr 3, (2001) (in Polish)

17. PN-EN ISO 5667-15:2009. Water quality. Sampling. Part 15: Guidance on the preservation and handling of sludge and sediment samples (in Polish)
18. PN-EN 12176:2004. Characterisation of Sludge. Determination of pH value (in Polish)
19. PN-78/C-04541. Water and Wastewater. Determination of Dry Residue, Residue on Ignition, Loss on Ignition and Dissolved Substances, Dissolved Mineral Substances and Dissolved Volatile Substances (in Polish)
20. PN-EN 15527:2008. Characterisation of Waste. Determination of Polycyclic Aromatic Hydrocarbons (PAH) in Waste Using Gas Chromatography Mass Spectrometry (GC/MS) (in Polish)
21. Regulation of the Minister of Environment of 9 September 2002 on soil and land quality standards, *Journal of Laws*, No. **165**, item 1359, (2002) (in Polish)

The influence of hydraulic conditions on coagulation process effectiveness

Aleksandra Sambor[1,*], and *Zbigniew* Ferenc[1]

[1]Wroclaw University of Science and Technology, Faculty of Environmental Engineering, Wybrzeze Wyspianskiego 27, 50-370 Wroclaw, Poland

Abstract. This paper presents the impact that small changes in the hydraulic installation between the flocculation chamber and the sedimentation tanks have on coagulation process effectiveness. This study has shown significant improvements in the parameters of the treated water. The research was conducted in two treatment systems: reference and test, in order to compare the changes that were introduced in the time period between January and May 2016. The hydraulic conditions between the flocculation chamber and the sedimentation tank were changed in the test system, leaving the reference system unchanged for comparative purposes. The height-wise positioning of the sedimentation tank relative to the flocculation chamber resulted in a formation of a cascade at the flocculation chamber drain at a height of 0.60m. Air was therefore introduced into the water, forming an air-water mixture, which disturbed the flow between the devices. It was found that floc transported by the pipeline was broken down, which hampered sedimentation in the sedimentation tank. This was confirmed by the analysis of chosen parameters from treated water. After changes in the hydraulic system, changes in water turbidity were noticed, indicating an increase in post-coagulation suspension separation effectiveness. Consequently, an increase in organic carbon removal was found relative to the reference system. This change influenced changes in UV_{254} absorbance to a much lesser extent.

1 Introduction

Coagulation is a process that has been commonly used for many years in water treatment technology, especially for surface water. This process aims to effectively remove organic substances [1,2], especially oxidation or disinfection by-product precursors (OBP/DPB) [3], whose formation potential is related to UV absorbance at 254 nm (UV_{254}) [4]. Due to the very high variability in surface water content during the year, and the varying levels of contamination dependent on source location, it is important to ensure maximum coagulation effectiveness, which influences subsequent processes in the water treatment system and their effectiveness. Optimization of the coagulation process is even more difficult due to the presence of many types of coagulants on the market and the influence on many factors on this process [5-7].

* Corresponding author: aleksandra.sambor@pwr.edu.pl

Despite many studies worldwide concerning methods of conducting this process, its effectiveness, and factors influencing organic substance removal, little attention is focused on the influence of hydraulic conditions during water transport between the devices used for coagulation and sedimentation processes.

In analyses, the impact of improper water transport and flow conditions from one treatment point to another is not taken into account, and the possibility of breaking flow continuity is not considered at all.

This work presents the effect of a design error in the hydraulic installation between the flocculation chamber and the sedimentation tank on coagulation process effectiveness.

The hydraulic installation details are described in the article by Szerzyna et al [8].

Problem description

The existing water treatment system is of a conventional configuration:

coagulation→sedimentation→ sand filters→ ozonation → carbon filters → disinfection.

During operation a low flocculation process effectiveness was found. Two possible causes were identified:

- suboptimal operating settings of the slow mixing
- error in the water transportation system from floculation chamber → sedimentation tank.

Changes in coagulant type and mixer operating parameters did not improve process efficiency, and therefore a decision was made to change the hydraulic installation.

2 Hydraulic conditions of the water treatment system

Prior to change, water from the slow mixing chamber drained by a PVC pipe of a DN110 diameter, then straight down via a 0.7 m PVC DN110 pipe and a 0.7 PCV DN63 pipe, and then via a PVC DN63 7.5m long pipeline into the sedimentation tank. The height-wise positioning of the sedimentation tank relative to the flocculation chamber resulted in a formation of a cascade at the flocculation chamber drain at a height of 0.60m (fig. 1).

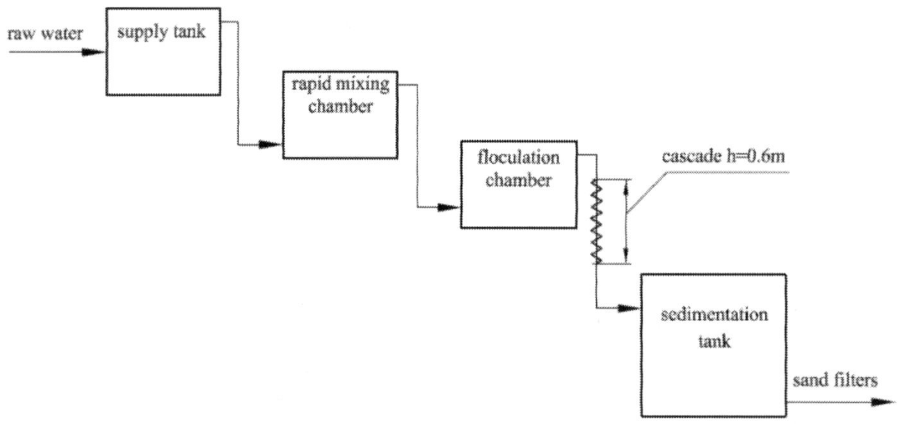

Fig. 1. Schematic of test water treatment system before changes.

Air was therefore introduced into the water, forming an air-water mixture, which disturbed the flow between the devices. It was found that floc transported by the pipeline was broken down, which hampered sedimentation in the sedimentation tank. In order to increase flocculation process effectiveness, it was decided to eliminate the cascade between the floculation chamber and sedimentation tank. After analyzing the technical possibilities and rejecting variants requiring changes in water levels in the floculation chamber, it was decided to generate linear or local hydraulic loss head of 0.60 mH_2O.

2.1 Calculation assumptions - case studies

Calculation were conducted according to PN-76/M-34034.
Hydraulic losses are:

$$\Delta h = \Delta h_l + \Delta h_m \tag{1}$$

where: Δh_l – a linear hydraulic loss head, m,
Δh_m – a local hydraulic loss head, m.

$$\Delta h_l = \lambda \cdot \frac{L}{d_w} + \frac{v^2}{2g} \tag{2}$$

$$\Delta h_m = \xi \frac{v^2}{2g} \tag{3}$$

where: λ – a nondimensional coefficient of linear resistance,
ξ – a nondimensional coefficient of local resistance,
L – the length of a pipeline, m,
v – the average flow rate of liquid, m/s,
g – the acceleration of gravity, m/s^2.

For this calculation the following assumptions were made:

- the average volume flow $Q_{śr} = 3.0\ m^3/h$,
- the average flow rate of liquid $v = 0.4 \div 0.6$ m/s.
- generation of hydraulic loss head of 0.6 mH_2O,
- avoidance of any sudden changes in cross-section, so that the transported liquid does not undergo significant accelerations.

2.2 Generation of local hydraulic loss head

The potential local hydraulic loss generators were taken to be a DN50 plug valve or a DN50 butterfly valve. The generator location was taken to be in the PVC DN63x2.4 pipeline between the floculation chamber and the sedimentation tank, after the existing cascade (fig. 2).

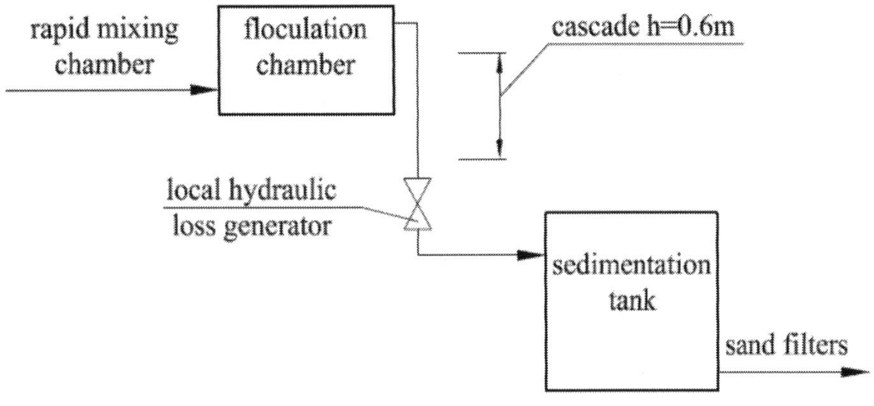

Fig. 2. Schematic with the local hydraulic loss generator placed in the system.

Based on equation (3), assuming a constant flow velocity in the installation, the required local resistance coefficient value for a one-point generator was calculated to be $\xi = 122{,}5$. This value of ξ could be obtained by restricting the plug valve by 55^0 or restricting the butterfly valve by 60^0. However, such a large restriction would result in sudden local fluid acceleration, which would have an adverse impact on the floc-water mixture transported by this pipeline. Furthermore there would be no guarantee of ensuring a constant volume flow.

2.3 Generation of linear hydraulic loss head

The linear hydraulic loss generator was taken to be an flexible PE hose coiled helically around a vertical axis. The generator location was taken to be in the PVC DN63x2.4 pipeline between the floculation chamber and the sedimentation tank, after the existing cascade (fig. 3). The hose diameter was calculated from the assumed range of velocities and average volume flow rate.

$$d_w = \sqrt{\frac{4Q_{śr}}{\pi v}} \tag{4}$$

The range of acceptable inside diameters d_w of the hose was from 42.10 mm to 51.52 mm. For further considerations it was assumed that d_w = 50 mm (DN50).

According to studies by Weissbach, for flow in a circular arc, the local resistance coefficient ξ is:

$$\xi = \left[(0.131 + 0.163\left(\frac{d}{R}\right)^{3.5}\right] \cdot \frac{\Theta}{90^0} \tag{5}$$

where: d – the inside diameter of a pipe, m,
R – radius of curvature, m,
Θ – arc angle, ^{0}.

Equation (5) does not take into account linear losses, which should be calculated according to (2). Due to the use of a flexible hose, which when coiled was deformed due to its own mass and the mass of water, an equivalent diameter was calculated according to (6):

$$d_z = \frac{4A}{U} \tag{6}$$

where: d_z – the equivalent diameter of a pipe, m,
A – the cross-section area of pipe, m^2,
U – wetted diameter, m.

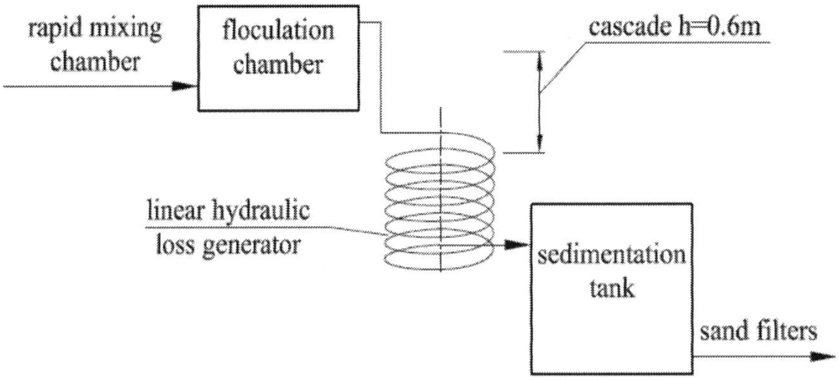

Fig. 3. Schematic with the linear hydraulic loss generator placed in the system .

Based on equation (6) the equivalent diameter of the pipeline was d_Z = 48.1mm. The bending pitch diameter of the coiled hose was assumed to be 2R = 0.9m. After calculating per (5) the local resistance coefficient caused by bending of pipeline, the local hydraulic loss head for one coil was:

$$\Delta h_{m\ (for\ 360)} = 5.63 \cdot 10^{-3}\ m.$$

Value λ – of a nondimensional coefficient of linear resistance calculated according to the Colebrook – White formula (7):

$$\lambda = \left[-2log\left(\frac{2.51}{Re\sqrt{\lambda}} + \frac{k}{3.71d}\right)\right]^{-2} \quad (7)$$

where: k – the absolute roughness of a pipeline, m,
Re – the Reynolds number.

Inserting values into the equation (7) gave the linear resistance coefficient value of $\lambda = 0.064$. The linear hydraulic loss head for one coil taking into account the equivalent diameter, the coefficient λ and a unit length of L=2.83m, amounted to:

$$\Delta h_{l\,(for\,L=2.83)} = 40.43 \cdot 10^{-3}\ m.$$

Therefore the total hydraulic losses for a unit length Δh_1 according to (1) amounted to:

$$\Delta h_1 = \Delta h_{l\,(for\,L=2.83)} + \Delta h_{m\,(for\,360)} = 40.43 \cdot 10^{-3} + 5.63 \cdot 10^{-3} = 46.06 \cdot 10^{-3}\ m.$$

The required correction was obtained by the use of 13 coils with a coil diameter of D=1m. The total length of DN50 flexible PE hose amounted to 36.9 m.

3 Verification of improvements during flocculation

The change in water flow conditions from the flocculation chamber into the sedimentation tank caused a decrease in turbulence, and therefore yielded better conditions for flocculation and transport of flocs. Consequently, there was an almost 10% increase in organic carbon removal effectiveness (fig. 4). This was likely connected with laminar floc flow and limiting of floc break-up that occurred in the installation before the change.

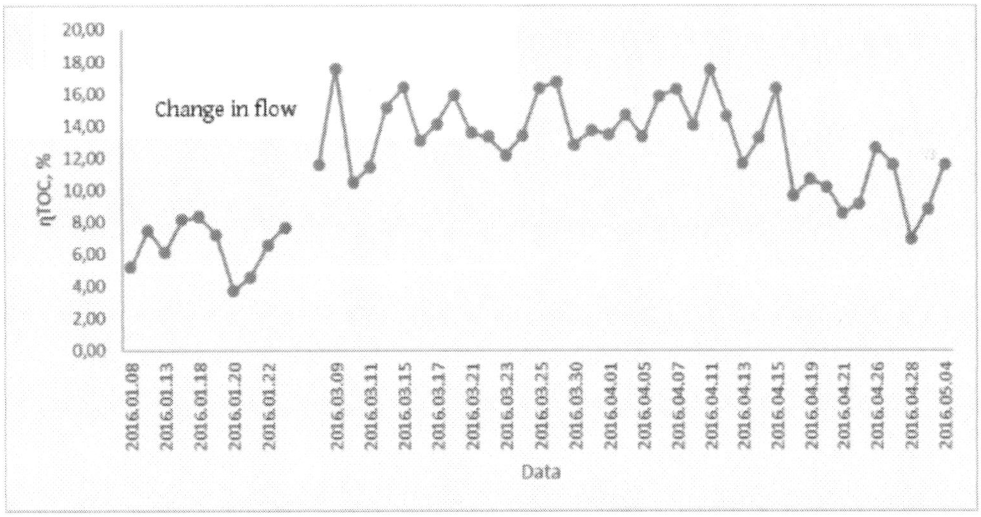

Fig. 4. TOC removal effectiveness during coagulation and sedimentation processes (test system - laboratory results).

The increase in effectiveness that was found was not accompanied by a decrease in source water quality, which might have improved the effectiveness of the coagulation process. However, no effect of installation change was found on UV_{254} absorbance and water colour. This is indicates removal of smaller organic substances which do not affect water colour.

The improvement in flocculation conditions contributed however to changes in turbidity reduction effectiveness for water undergoing sedimentation and coagulation (fig. 5). In the initial phase after changes in the hydraulic flow, the turbidity reduction effectiveness was comparable to that found in the second (reference) water treatment system. After about 3

 DOI: 10.1051/e3sconf/20172200153

weeks, the inner surfaces of the pipeline became covered with sludge/biofilm, which contributed to increases in removing suspended particles after coagulation and sedimentation processes. This condition continued for several months of operation, which may indicate a permanent improvement in sedimentation conditions, and therefore an improvement in the turbidity reduction effectiveness.

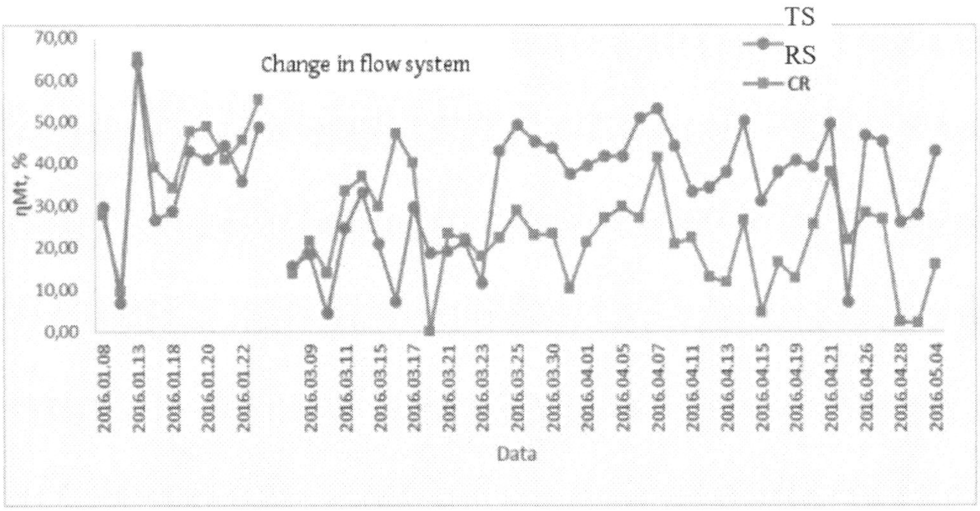

Fig. 5. Comparison of reduction in turbidity during coagulation and sedimentation processes (TS-test system, RS-reference system).

The resulting improvement in water quality after coagulation and sedimentation is connected with an increase in flocculation, which took place in the pipes connecting the flocculation chamber with the sedimentation tank.

4 Conclusions

The incorporation of the linear hydraulic loss generator significantly improved water quality after coagulation and sedimentation. Consequently, an increase in removing organic carbon and a decrease in water turbidity was observed, reducing disinfectant usage.

However, this change had no impact on UV_{254} absorbance and water color.

This increase in organic substance removal effectiveness from water clearly shows that the water treatment process is not only dependent on the performance of the individual processes, but also the conditions in which the medium is transported between the individual devices.

Acknowledgments

This publication was made possible by the National Center for Research and Development grant (PBS3/B9/44/2015) "Research on effectiveness of new water treatment technology as a step towards a shift in thinking about water utility sector" (WODTECH), and thanks to involvement of both project consortium members (Wroclaw Municipal Waterworks and Drainage Company and Wroclaw University of Science and Technology).

References

1. E. E. Chang, P. C. Chiang, H. J. Hsing, S. Y. Yeh, Practice Periodical of Hazardous, Toxic and Radioactive Waste Management, (2007)
2. B. Bolto, D. Dixon, R. Eldridge, S. King, K. Linge, Water Res., **36** (2002)
3. L. Qi, H. Wang, X. Zheng, G. Li, Desalination and Water Treatment, **50** (2012)
4. P.C. Chiang, E.E. Chang, C.H. Liang, Chemosphere, **46** (2002)
5. M. Świderska-Bróż, M. Rak, Arch.Ochr. Środ., **30**, 2 (2004)
6. S. Lee, J. Cho, H. Shin, B. Son, S. Chae, J. Water Supply Res. T., (2003)
7. Y-L. Lin, P-Ch. Chiang, E-E Chang, J. Hazard Mater, **146** (2006)
8. S. Szerzyna, M. Mołczan, M. Wolska, W. Adamski, J. Wiśniewski, International Water Association (2016).

DOI: 10.1051/e3sconf/20172200154

Gas fired boilers: Perspective for near future fuel composition and impact on burner design process

Fabio Schiro[1,*], *Anna* Stoppato[1], and *Alberto* Benato[1]

[1]University of Padova, Department of Industrial Engineering, Via Venezia n.1, Padova, Italy

Abstract. The advancements on gas boiler technology run in parallel with the growth of renewable energy production. The renewable production will impact on the fuel gas quality, since the gas grid will face an increasing injection of alternative fuels (biogas, biomethane, hydrogen). Biogas allows producing energy with a lower CO_2 impact; hydrogen production by electrolysis can mitigate the issues related to the mismatch between energy production by renewable and energy request. These technologies will contribute to achieve the renewable production targets, but the impact on whole fuel gas production-to-consumption chain must be evaluated. In the first part of this study, the Authors present the future scenario of the grid gas composition and the implications on gas fed appliances. Given that the widely used premixed burners are currently designed mainly by trial and error, a broader fuel gas quality range means an additional hitch on this design process. A better understanding and structuring of this process is helpful for future appliance-oriented developments. The Authors present an experimental activity on a premixed condensing boiler setup. A test protocol highlighting the burners' flexibility in terms of mixture composition is adopted and the system fuel flexibility is characterized around multiple reference conditions.

1. Introduction

Heating is a fundamental necessity especially in countries with cold climates, and gas fired boilers are currently satisfying large part of this heat demand in the domestic sector. Gaseous fuels are forecasted to be widely used to feed gas boilers in the next years [1]. Among them, premixed boilers allow to obtain higher energy efficiency and lower pollutant emission with respect to non-premixed boilers [2]. For this reason, premixed boilers are becoming mandatory in the EU except for special circumstances [3]. Significant efforts have been done in order to develop and industrially apply the technology of these devices. At the same time, a significant commitment was placed on increasing renewable energy production. The renewables production will have a direct impact also on the quality of the fuel gas feeding heating appliances in the near future, since the gas grid will face an

* Corresponding author: fabio.schiro@dii.unipd.it

increasing injection of alternative fuels (mainly biogas, biomethane, hydrogen) [4-7]. In particular, hydrogen production from electrolysis can mitigate, on the one hand, the issues related to the fluctuations in renewable energy production (particularly massive in the wind farms) and, on the other hand, can help to cover the mismatch between energy offer and request. This path will reduce the overall CO_2 emission and contribute to achieve the renewable production targets, but the impact on every element of the production-to-consumption chain need to be evaluated. In the first part of this study the Authors discuss the future scenario of the grid fuel gas compositions. A qualitative evaluation of the implications on premixed burners design will be conducted basing on existing studies. Premixed boilers will be the leading technology for gas heating in the next future and burner design is currently driven by trial and error methodology [8-9]. A broader fuel gas quality range means an additional hitch on the burner design process: a better understanding and structuring of the today's burner design process would be helpful in the future appliance-oriented research and developments. In the second part of the paper, the Authors present an experimental performance assessment of cylindrical premixed burners coupled to a very common commercial premixed boiler setup. A test protocol highlighting the burners' flexibility in terms of mixture composition is adopted and the system fuel flexibility is characterized around multiple reference conditions.

2. Perspective for future grid fuel composition

2.1 General context

Gaseous fuels composition is expected to vary in the next years, and the existing gas quality standards need to evolve in parallel. Guiding factors for this process are the security of energy supply and the reduction of overall CO_2 emission.

2.2 Hydrogen

The recent developments in renewables, especially in wind power farms, make the electricity storage issue more and more important for the future energy system efficiency. Among the most trusted and consolidated storage systems are the pumped-hydro energy storage and the CAES, whose potential is anyway limited due to geographical and technical reasons. An option being intensively studied is the storage of the overabundant electricity into hydrogen, produced by electrolysis, and the subsequent injection through the natural gas network. This technology is often reported as P2G (Power To Gas). It allows to use an already existing huge and geographically branched energy storage system, and potentially avoids to build new electricity transmission lines. Hydrogen volume percentages up to 30% have been taken into consideration according to certain studies [10]. In fact, today, it is necessary to improve the energy and economic efficiency of renewable production, since not all the electricity grids are able to transport the energy produced during the peaks of wind and sun, with the consequent need of forcing the production shut down to avoid electricity grid overloading. Obviously, the hydrogen injected into the gas network could also contribute to produce heat and power in distributed or centralized plants. [4]

2.3 Biogas

The European Union encourages the use of biomass as renewable energy source, in the perspective of greenhouse gases reduction. Biogas is a fuel produced by anaerobic digestion of organic matters, such as animal slurries and energy crops, and it is mainly composed by methane and carbon dioxide. The composition depends on numerous factors such as the process design and the nature of the digested substrate: table 1 reports data from a literature datasheet [11] on fuel properties. Often, biogas fuel is used to produce heat and/or power in boilers or cogeneration units near the biogas production plant. However, an alternative is the injection of biogas into the natural gas grid [5-7] after a treatment process, allowing to use the grid itself for long distance transport and to balance fuel demand and supply by storage.

Table 1. Relevant compositional data of biogas and natural gas *[11]*

	Biogas	**Natural gas**
Methane [Vol%]	60-70	85-92
Heavy hydrocarbons [Vol%]	0	0-9
Nitrogen [Vol%]	0.2	0.3-1.0
Carbon dioxide [Vol%]	30-40	0.2-1.5
Hydrogen sulphide [ppm]	0-4000	1.1-5.9

2.4 Security of supply and strategic aspects

The security of fuel gas supply is a fundamental need especially in Europe after the reduction of fuel reserves and the growth of domestic fuel demand (which is recovering from the economic crisis and forecasted to further increase [1]). In this context, the fuel gas demand can be satisfied by the already mentioned renewable gas production (biogas, biomethane, hydrogen) or by increasing the dependency from foreign energy imports (LNG or pipeline imports). This situation involves both environmental and strategic evaluations, and will add variables to the fuel gas composition, that will be made up by a combination among indigenous gases, renewable gases, imported pipeline or LNG gases [12].

3 Impact on gas devices and on burner design process

Gas boilers and other gas fed appliances are tested, basing on standards, with a reference gas and extreme limit gases [13] (e.g. incomplete combustion and soot limit gas, light back limit gas, flame lift limit gas, overheating limit gas). These limit gases are designed to guarantee a safe operation under variations in gas compositions within the appliance lifetime. The limit gases, anyway, do not represent the allowed distribution gases, and the standard does not define which distributed gas composition and Wobbe index range is covered by the limit gases testing. Distribution gas ranges are usually defined as Wobbe index range, but compositional effects (non-methane hydrocarbons, hydrogen, CO_2, ...) play also a key role on safe operation. Regarding the distribution range, each European country made decisions based on own practice and experience [14]. Due to the complex scenario, any variation on the gas quality standards must be carefully studied and evaluated. Several studies [12,15,16] are focused on understanding the impact of gas quality variations on gas appliances: pollutant emissions (CO, NOx) are related to the fuel quality and potential safety risks have been identified. The specific theme of hydrogen addition has also been explored in literature with reference to domestic appliances [10,17,18]. Hydrogen significantly impacts on the flame dynamics: flame lift, noisiness and combustion control issues were detected in laboratory testing of existing devices. The varied combustion chemistry and consequent different flame speed influenced also pollutant emissions, burner temperature and consequent burner lifetime. Therefore, basing on the above-mentioned works, it is possible to underline that each application has its own characteristic response to the different gas quality specification. The possible future modifications of device design in order to increase the tolerance on gas composition and heating value will require an even more tailor-made approach than that current applied [17]. Active combustion control systems could mitigate the required modifications to application, but these systems were also found to be not completely mature to deal with all fuel components such as hydrogen [10,18]. In conclusion, it is highly probable that to widen the gas specification range an increased fuel flexibility would be required to the gas burner itself.

4 Experimental activities on premixed boiler fuel flexibility

4.1 Introduction

Based on the scenario presented in Section 2 and 3, a consistent research and optimization activity on gas fed appliances is necessary to grant safe operation and full functionality under future boundary conditions. The Authors focused on domestic premixed boilers, which will be widely used for emission and efficiency reasons and whose burner design is particularly complex. The burner is a key component for these devices, because it significantly influences the combustion performance; however trial and error approach still drives the premixed burner design [8-9]. The combustion dynamics governing these burners are not fully understood. Therefore, the design process is founded on manufacturer's know-how and on empirical observations. Additional difficulties are added to the burner development phase by budget, time and production cost factors [19]. These difficulties would be even worsened by the forecasted wider grid gas quality range, discussed in detail in the previous sections. In fact, if the grid fuel gas composition will be more variable in the future, the premixed burners need to correctly operate under the consequent more variable air/gas ratio without problems. In this context, it is useful to generate knowledge on the premixed burners behaviour under variable fuel quality, with the aim of assisting future appliance-oriented burner design processes. The Authors focused on a domestic premixed boiler system and implemented a test protocol, with the aim to overcome the mere pollutant emission testing with nominal gas and with limit gases. An in-depht analysis has been conducted with the aim of gathering more information on combustion behaviour at gradually varying air/gas mixture composition. Emission and flame stability characteristics have been identified.

4.2 Experimental setup

The test rig adopted in the experiments replicates the fundamental structure of a domestic premixed boiler: pneumatic gas valve, modulating blower, Venturi effect air/gas mixer, steel coil heat exchanger, cylindrical burner. A 25 kW nominal thermal capacity has been chosen because it is very common among domestic boilers [20]. A front view of the test rig used is visible in Figure 1.

 DOI: 10.1051/e3sconf/20172200154

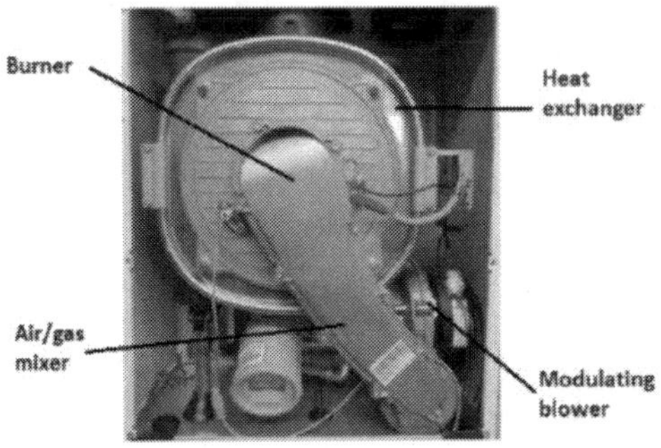

Fig. 1. Front view of the test rig used for the experiments

The emission values reported in this paper are normalized to the air free reference condition. Gaseous emissions are measured by infrared-based analysers (SIEMENS Ultramat 6 for CO and CO_2, repeatability <1% of respective measuring range, linearity deviation <0.5% of full-scale values) and chemiluminescence based analysers (Eco Physics CLD 82Sh for NOx, repeatability <0.5% of reference point, lack-of-fit<0.8% of full-scale value). Gas flows are measured by thermo-mass meters (Brooks SLAMF, repeatability < 0.2% of reference point, flow accuracy ±0.9% of full scale value). For the aim of this study an industrially available premixed steel burner (see figure 2) equipping an on-the-market 25 kW domestic boiler was chosen as reference combustor. Such a burner is designed to fulfil several target requirements, including overall dimensions, pressure drop value, thermo-acoustic behaviour, flashback resistance, and lifetime.

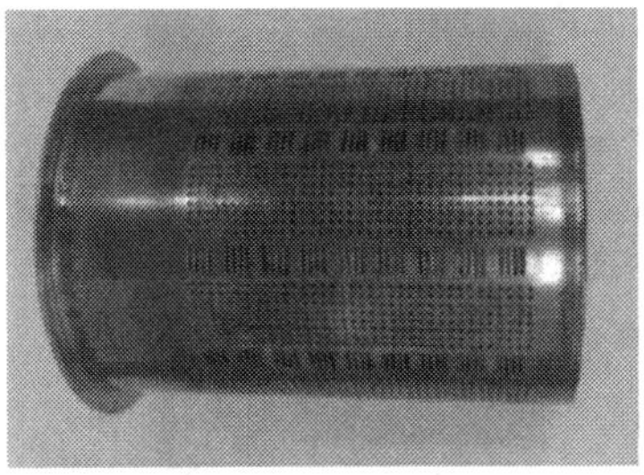

Fig. 2. The reference steel premix burner

4.3 Test protocol

In order to evaluate the response of the system to a fuel gas quality variation and to characterize its flexibility, a test protocol has been developed. Three reference thermal load conditions (maximum load, average load, minimum load) have been selected in a thermal load modulation range of 10:1, and the air/fuel ratio is gradually varied in correspondence to each of these conditions. The CO emission is taken as marker of flame stability and combustion hygiene while the CO_2 emission is directly linked to the fuel/air ratio. Operatively, the boiler fan speed is fixed for each thermal load reference condition to elaborate a fixed air flow while the gas valve setting is gradually varied to regulate the fuel flow. In this way, the response of the boiler to a variable fuel/air ratio mixture is simulated. Experimental results are acquired when the system is in steady-state condition. The fuel gas used is pure methane, but the same methodology can be applied also with other fuel components. The nominal reference conditions are listed in Table 2. The nominal air/gas ratio has been tuned in order to achieve a 9% CO_2 level in the exhaust: this is a typical value for this kind of boilers.

Table 2. Reference testing conditions

Reference condition	Thermal load [kW]	CO_2 setting [%]	Fan speed [rpm]
Maximum load	25.0	9.0	5100
Average load	12.7	9.0	3000
Minimum load	2.5	9.0	1100

4.4 Experimental results

The CO values versus the CO_2 levels for each condition are depicted in Figure 3.

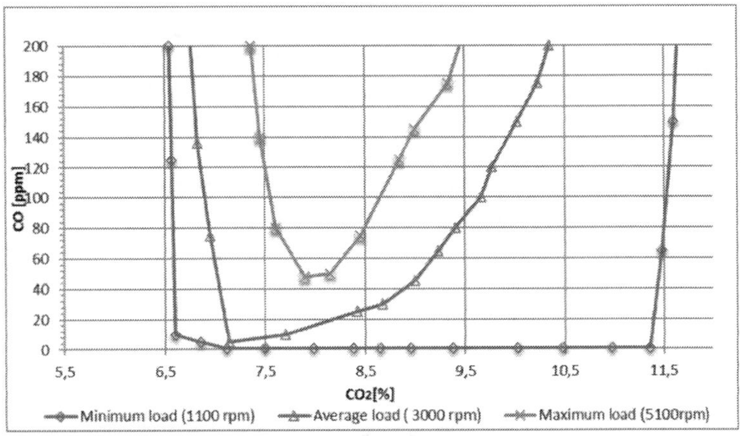

Fig. 3. CO versus CO_2 curves at reference fan speeds

The CO versus CO_2 assumes a macroscopic common shape for all reference conditions: we will refer to this shape in the following with the name "U curve". This curve presents two quickly ascending branches and a region in which the CO emission is low and more or less constant. These results can be referred to some basic premixed combustion mechanisms. The CO increases for low CO_2 values (lean mixtures) because at low gas/air ratio the flame is long, prone to stretching and lifting. Thus, it is affected by edge effects and potential quenching through the heat exchanger walls. The CO increases for high CO_2 values (rich mixtures), on the other side, because increasing gas/air ratio the oxidizer concentration decreases, thus incomplete combustion is favoured. The CO increment for rich mixture conditions is smoother than for lean ones: this is linked to the above-mentioned phenomena, since flame lift condition can be sudden in response to a small ratio variation and quenching through the walls could play a significant role. The results confirm that the gas/air ratio has a fundamental impact on the combustion quality of a premixed boiler. Thus fuel gas composition variations shifting this ratio can significantly increase the CO production. In addition, the U curves get narrowed passing from minimum thermal load through maximum load, suggesting that the decrease of combustion products residence time and the increase of flame length generally disadvantage combustion completion. The fuel flexibility result to be wider al lower thermal load: this suggests that appliance developers should take care about fuel flexibility along whole thermal load modulation range, since emission level and behaviour can significantly differ. The proposed U-curve tracing is a fundamental first step to characterize the system fuel flexibility: comparing the mixture ratio tolerance required for the appliance and the obtained U curve an evaluation on the burner flexibility can be drawn. This method allows also to detect the preferable air/gas ratio with the aim to minimize CO production, and allows a developer to eventually advise different ratio settings for the reference conditions. Tracing the U-curves, obviously, could be helpful not only in characterizing a fixed system but also to compare the performances of different burners in the development phase. Finally, since combustion behaviour is strongly system related, the impact of other components on combustion can also be spotted: the heat exchanger size and volumes, for example, would be interesting parameters to be evaluated in future appliance-oriented research.

5 Conclusions

The transition through a more environmentally friendly energy system will have an impact on grid gas quality. In fact, gaseous fuels composition is expected to significantly vary in the next years due to several factors, as the injection of biogases and hydrogen on the gas grid. This will help mitigating the problems related to renewable energy fluctuation and reducing the overall CO_2 emission, but the impact of gas quality variation on each element of the production-to-consumption chain has to be evaluated. Premixed burners need particular attention in this context, because its design is very complex and currently based mainly on trial and error approach and empirical observations. Each gas device has its own characteristic response to varied inlet gas quality, so any future requirement for increased tolerance on gas composition and heating value will require a tailored activity to optimize the device design. Active combustion control systems could facilitate the transition process, but these systems are not completely mature to deal with all fuel components such as hydrogen. As consequence, a widening of the gas specification range would require an increased fuel flexibility to the gas burner itself. In this context, it is very useful to generate knowledge on the premixed burners behaviour under variable fuel quality, aiming to assist future appliance-oriented developments. Laboratory testing was conducted on a domestic premixed boiler system with the aim to gather more information on combustion behaviour at gradually varying air/gas mixture composition: a test protocol oriented to overcome the mere testing with limit gases was implemented. The tests were run at varying fuel/air ratio at fixed fuel composition, but can be taken as a basis for further testing on different fuels. The results allowed to detect a CO versus CO_2 figure assuming a common macroscopic shape ("U-curve") for all reference conditions. The gas/air ratio is confirmed to play a fundamental role on the combustion quality, thus fuel gas composition variations shifting this ratio can significantly increase the CO production. The tested system resulted to be more fuel flexible at lower thermal loads: appliance developers should take care about fuel flexibility along whole modulation range, since the response can be strongly variable. The proposed U-curve tracing is a fundamental first step in characterizing the system fuel flexibility and detecting the preferable air/gas ratio for minimization of CO production. Furthermore, the method is applicable not only to the characterization of a fixed system but also to detection of the impact of different components on the combustion performance. In conclusion, the future burner design process will necessarily take into consideration an increased fuel quality range, and new test protocols and regulations need to be developed accordingly.

References

[1] European Commission, *EU Reference Scenario 2016 – Energy, transport and GHG emissions - Trends to 2050*(2016).

[2] VHK for European Commision, *Eco-design of Boilers - task 4 - Technical Analysis* (2007).

[3] European Parliament, *Directive 2009/125/EC* (2009).

[4] R. Judd and D. Pinchbeck, *Compendium of Hydrogen Energy,* **vol. 4**, p.165-192 (2016).

[5] E. Hengevelda, . W. van Gemert, J. Bekkering and A. Broekhuis, Biomass Bioenerg, **vol. 67**, p. 363–371(2014).

[6] Department of Energy and Climate Change UK, *Biomethane into the Gas Network: A Guide for Producers* (2009).

[7] A. Singhal, *Biomethane to Natural Gas Grid Injection: A Technological Innovation System Analysis*, KTH Stockholm, MSc Thesis (2012).

[8] H.M. Altay, S.Park, D. Wu, D. Wee, A.M. Annaswamy and A.F. Ghoneim, Proc Combust Inst, **vol. 32**, p. 1359-1366(2009).

[9] Maarten Hoeijmakers, Viktor Kornilov, Ines Lopez Arteaga, Philip de Goey and Henk Nijmeijer, Proc Comb Inst, **vol. 35**, p.1073-1078(2015).

[10] P. Nitschke-Kowsky, Impact of hydrogen admixture on installed gas appliances, *Kuala Lumpur 2012 World Gasconference* (2012).

[11] SGC - Swedish Gas Technology Centre, *Basic data on biogas* (2012).

[12] Bord Gáis Networks, *Report on Gas Quality arrangements in Republic of Ireland*(2008).

[13] European legislation EN 437 - Test gases - Test pressures - Appliance categories.

[14] DNV GL commissioned by Ministry of Economic Affairs Netherlands, *Requirements for gas quality and gas appliances* (2015).

[15] Ernest Orlando Lawrence Berkeley National Lab, *Natural gas variaility in California: environmental impacts and device performance* (2009).

[16] UK Govenment, *Future arrangements for Great Britain's gas quality specifications; Government Response (including Summary of Responses) to consultation on future arrangements for Great Britain's gas quality specifications*, UK Government (2007).

[17] B. Slim, Should we add hydrogen to the natural gas grid to reduce CO_2 emissions? (consequences for gas utilization equipment), *23rd World Gas Conference Amsterdam* (2006).

[18] Health and Safety Laboratory - HSE UK, *Injecting hydrogen into the gas network - a literature research* (2015).

[19] José M. N. Rodrigues and Edgar C. Fernandes, Stability Analysis and Flow Characterization of Multi-Perforated Plate Premixed Burners, *17th International Symposium on Applications of Laser Techniques to Fluid Mechanics* (2014).

[20] VHK for European Commision, *Eco-design of Boilers - task 2 - Market Analysis*(2007).

Experimental study of optical and electrical properties of ZnO nano composites electrodeposited on n-porous silicon substrate for photovoltaic applications

Naceur Selmane[1], *Ali* Cheknane[1,*], *Nourddine* Gabouze[2] ,*Nabila* Maloufi[3], and *Michel* Aillerie[4, 5]

[1]Laboratoire des Semi-conducteurs et des Matériaux Fonctionnels. Université Amar Telidji de Laghouat, Bd des Martyrs. BP37G Laghouat (03000), Algérie
[2]Centre de recherche en technologie des semiconducteurs pour l'énergétique, 02, boulevard Dr Frantz Fanon.Sept Merveilles.Bp 140 Alger, Algérie
[3]Université de Lorraine, LEM3, UMR CNRS 7239, F-57050 Metz, France
[4]Université de Lorraine, LMOPS, EA 4423, F-57070 Metz, France.
[5]Centrale Supelec, LMOPS, F-57070 Metz, France.

Abstract. ZnO films deposited on silicon porous substrates (PS) were prepared by electro-deposition anodization on n type (100) silicon wafer. This ZnO/PS structure combines substrates having specific structural and optical properties (IR emission), with nano-composites of ZnO potentially interesting due to their functional properties (UV emission) to be integrated as constitutive elements of devices in various optoelectronic applications mainly in blue light emitters.With this combined structure, the blue shift in the PL peak is possible and easy to obtain (467nm). The vibration modes of PS and ZnO films on PS substrates (ZnO /PS) were investigated by infrared (FTIR) measurements and their behaviors were analyzed and discussed by considering the structural properties characterized by X-ray diffraction (DRX) and scanning electronic microscopy (MEB).

1. Introduction

ZnO is an important binary compound semiconductor of the II-VI group. It currently grows in the hexagonal wurtzite crystal structure. It has and large free excitation binding energy of 60 meV, at room temperature and a direct wide band gap ranging between 3 and 3.7 eV depending of the elaboration process. Zinc oxide has received considerable attention for a long time due to its many attracting applications such as piezoelectric [1], photovoltaic devices [2-4],gas sensors [5,6], optical catalysis[8],and solar cells [9]. Due to its large direct band gap. This compound is also interesting for the production of blue and UV light emitters used in the optoelectronic devices, such as UV lasers [10], blue to UV light-emitting diodes[10-11] and UV sensors [10]. Several methods are used for preparing ZnO films, such as magnetron sputtering [12-13], pulsed laser deposition(PLD)[14], spray pyrolysis [15,16], sol–gel process [17,18] chemical bath deposition (CBD) [19,20] and

DOI: 10.1051/e3sconf/20172200155

chemical vapor deposition (CVD) [21,22],..etc. However, these methods are not well suited for large area coating, low temperature processing, and low cost. The electro-deposition technique has been proved to be one of the simplest and most effective ways to prepare nicely crystallized ZnO at relatively low temperatures. This technique is widely used in order to growth and functionalize oxide materials with specific chemical and physical properties. The advantages offer by the electro-deposition technique for oxide films [23-25] are numerous as the easy preparation, rapid and simple, low cost, possibility of large-scale deposition, low temperature processing and direct control of film thickness, in comparison with other deposition techniques [12-18]. Generally, in literature [23], the aqueous deposition was performed using either zinc chloride or zinc nitrate on a variety of seeded or seedless substrates. In this work, the nano composites of ZnO have been deposited on n type porous silicon (PS) substrates. This later has been intensively studied since the discovery of its efficient photoluminescence at room temperature in the visible light region with the possibility to adjust its emitting wavelength from red to blue by managing the erosion process [26].
In the present work, the electro-deposition technique was successfully used for the preparation of ZnO on nano-porous silicon substrates with high porosity. Within this device topology, we expect to combine the particular properties of PS with the potential applications of ZnO. For this reason, structural, optical and electrical properties of the structure ZnO /PS were investigated using X-ray diffraction (DRX), Scanning Electro Microscopy (SEM) spectroscopy and FTIR measurements.

2. Experimental set up

2.1 Preparation of the porous silicon substrate

The silicon samples used in this study were cut into 5 x 5 mm^2 squares, from single crystal silicon wafer. The porous silicon (PS) layers are formed on n type (100) c-wafers with resistivity of 1^{-10} Ω.cm and thickness of 500-550μm using the electrochemical anodization. For that, silicon wafers are initially prepared with a bath in Trichloro heated at 50 ^{0}C for 10 min followed by bath in acetone. The cleaning was performed in a bath of ethanol followed by a final rinsed with deionised water. After a drying phase, samples are placed in the N_2 nozzle, and the ohmic contacts are deposited by application of InGa on the non polished face of the substrates mounted on copper support.
The electrochemical anodization is carried out at room temperature with HF(50%)/ C_2H_5OH (99%) for 2 min at a constant current density of 100 mA/cm^2. Because the samples are n type, a back illumination from a 50W lamp mounted at a distance of 10 cm was used during etching process. Fig.1 shows the anodization cell with three electrodes.i.e. The working electrode on the silicon substrate (cathode), a platinium wire as a counter electrode and a reference electrode.

2.2 Deposition of ZnO layer

The ZnO layers were deposited on PS substrates using electrochemical anodization. This technique has been chosen because it gives a non blocking electrode for cathodic reactions in the silicon pores compared with others techniques as spray pyrolysis, vacuum evaporation. The electrolytic solutions were prepared using aqueous solution of 5mM $ZnCl_2$ with a pH about 6.To ensure a good conductivity of the solution allowing optimal conditions for the deposition of ZnO, the samples were introduced at 65^{0}C during time between10 and15 min 0.1 M KCl, controlled by a cyclic voltammetry.

 DOI: 10.1051/e3sconf/20172200155

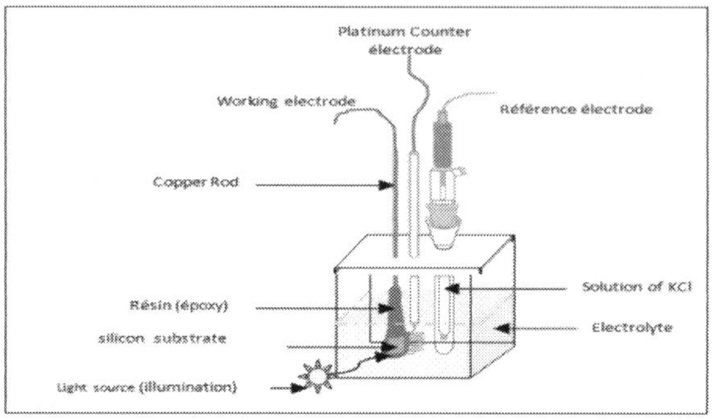

Fig.1. Descriptive schematic of anodization cell.

The important aim of the cyclic voltammetry of our experiment approach is to give the optimal condition to achieve ZnO deposition on silicon substrate using this technique. It is thus possible to know the potential that may be applied at the potentionstat to obtain good ZnO composites quality. ZnO nanocompossites were electrodeposited from an aqueous solution of $ZnCl_2$ and KCl, used as supporting electrolyte, and flowing air as precursors. The electrodeposition conditions were optimized in order to obtain uniform and stoichiometric thin films. The deposition was carried out by adjusting the bath temperature at 65°C, and the pH of electrolyte at 6.5 and the applied potentials from -0.3 to -1.6V /SCE. Fig.2 shows the voltamogram on copper substrate obtained from acidic solution (pH 6.5) containing 5mM $ZnCl_2$ and 0.1M KCl at temperature 65°C. Good quality films of ZnO were obtained at the cathodic potential of -1.3 V/SCE. This value corresponds to the good reduction of the Zn^{+2} ions. The reduction of Zn^{+2} ions leads to a metallic Zn that reacts chemically with ZnO. Therefore, it is observed that there is no current at E= -1.3 V/SCE corresponding to the reduction of Zn^{+2} into metallic Zn, while the value E= -1.3 V/SCE found for the potential is consistent with the literatures [29,36]. All samples reported in this work were prepared at potential -1.3V/SC. All electrochemical experiments were carried using a potensiostat/galvanometer.

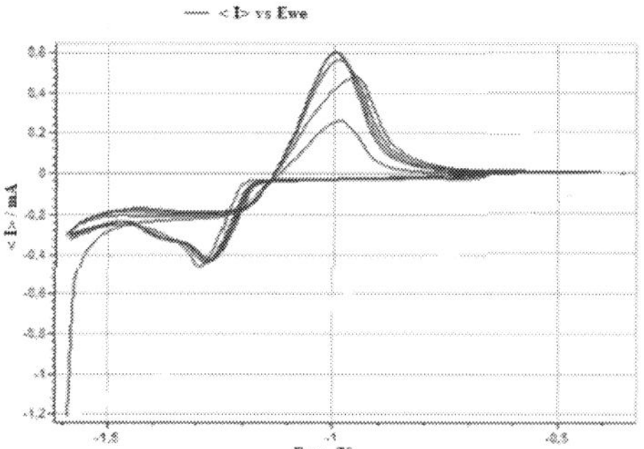

Fig.2. Cyclic voltamograms mesured on ZnO/PS at 0.005 M $ZnCl_2$+0.1 M kCl, T=65^0 C, Scan rate=20mV/S

3. Results and discussion

3.1 Structure and morphology of the ZnO samples

Fig.3 (a,b) shows SEM images of the PS samples at different magnifications. SEM micrographies of deposited ZnO layers on PS are presented in Fig. 4 (a,b). It is clear that crystalline ZnO is deposited along the outer walls of the pores of PS either partially by

 DOI: 10.1051/e3sconf/20172200155

filling or completely covering them Fig .4 (a). Few pores of PS were not covered completely by the ZnO particles. Fig (4-b) shows the particles of ZnO in the pores of PS. This technique of deposition is better than others physical techniques for many reasons The first reason is that these methods of deposition as the vacuum evaporation and sputtering show a blocking of the particles at the entered pores of PS. And the second reason is due to the difficulty to obtain the porous silicon n-type. Because, this type of silicon needs an illumination for increasing the probability of the ionised impurities. We assume the diameter of the pores between 10-40 nm, and the size of ZnO between 4 – 13 nm.

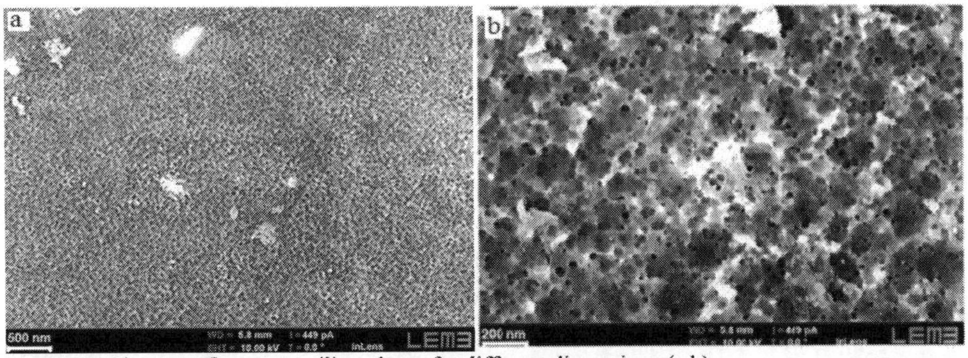

Fig.3. SEM images of n porous silicon layer for different dimensions (a,b)

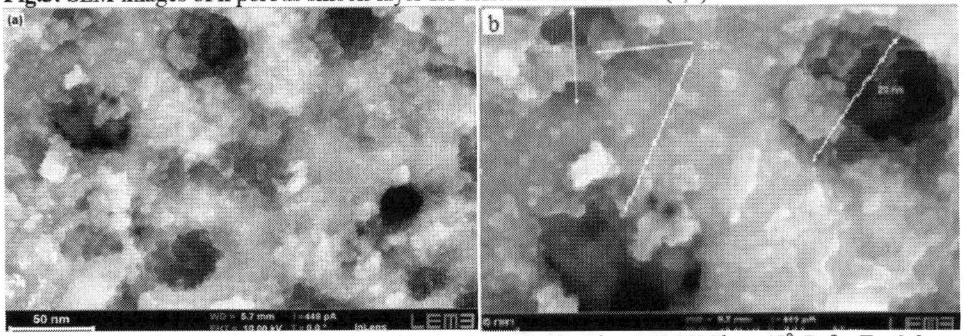

Fig.4. SEM images of Zno deposited on PS from 0.005M $ZnCl_2$ +0.1M KCl at 65^0 C for E=-1.3 V/SCE (a,b)

The DRX of ZnO films prepared on PS substrate is displayed in Fig.5. The most intensive peak diffraction observed at 63◦, can be indexed to n-Si (1 0 0) diffraction. The strong peak intensity of the nano-composites ZnO is located at 34.22◦. Also, others peaks displayed for the ZnO peaks having intensity weaker than the first, located at 31.21, 38.17◦ and 45.32◦ corresponding to the (100),(101) and (102) planes of ZnO, respectively. The XRD analysis indicates that the nanocomposites of ZnO presents a preferential orientation along the n-Si (1 0 0) substrate. It is similar with a strong (0 0 2) peak which located at 34.12◦ and a number of weaker peaks with (1 0 0), (1 0 2), (1 0 1), which indicates that the nanoparticles on Si wafer are of perfect c-axis orientation.

 DOI: 10.1051/e3sconf/20172200155

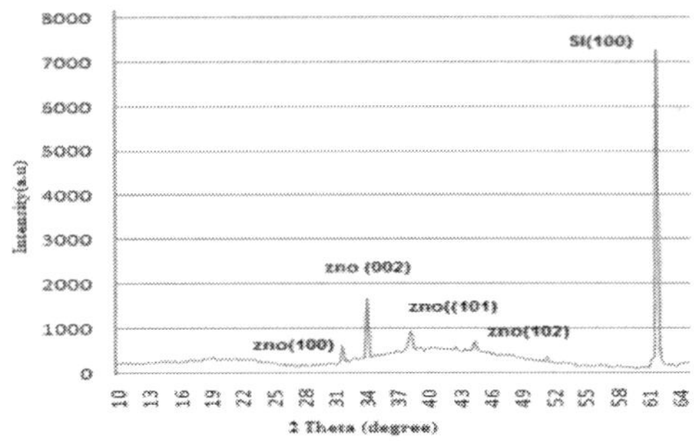

Fig.5. X-ray diffraction patterns of ZnO films prepared on PS substrates at a potential E=-1.3V/SCE for 10min

3.2 Electrical properties:

The solution using to preparing the ZnO, either zinc chloride $ZnCl_2$ or zinc nitrate ($Zn(No_3)$) and their concentrations are important parameters because they affect the electrical properties and all films properties that have been influenced by some parameters as by the proposed technique as: the time of deposition, substrate type, concentration of the solutions, current density or potential, bath temperature. We are interested in this present work by the electrical resistivity of these films because they give us an idea for the application of our thin films.
The carriers are electrons in our samples of ZnO /PS. The electrical Resistivit y ρ (Ω.cm) is a function of the concentration N and mobility μ,

$$\rho = 1/Nq\mu \tag{1}$$

With q: electron charge ($1.6x10^{-19}$ Coulomb).

The relationship between resistance by unit area (square resistance) and resistivity of a thin layer of a thickness d is by the relation (2)

$$P=R_s d \tag{2}$$

We found that the value of square resistance changed vs. the time of deposition of the ZnO films deposited on porous silicon substrate .For the films deposited at 15 min, the square resistance was 385kΩ/∗ ; and it is 442 kΩ/∗ at 10min. Another effect was observed concerning of the films prepared using zinc nitrate $Zn(NO_3)$ deposited at 10min where the resistance was 849 kΩ/∗ , and the films prepared using zinc chloride $ZnCl_2$ the resistance became 384 kΩ/∗ at the same time. We can conclude that the time of deposition is an important parameter when using the electrodeposition technique. We know that there is a net dependence between the time of deposition, the concentration and the resistivity. The resistivity of a semiconductor is inversely proportional to the concentration of impurities. If we increase the deposition time, the concentration also increases; because the probability of the ionized impurities will increase.
This electrical resistance was calculated by the technique of four points. It uses two pairs of electrodes, a pair injecting the current (I) and the second pair measuring the voltage (V). This method reduces the error due to resistances of contact between the electrodes

 DOI: 10.1051/e3sconf/20172200155

3.3 Optical properties

Fig.6 shows the IR spectrum of the silicon after anodization (silicon porous). The porous silicon is evidenced by different vibrations modes. The stretching vibrations bonds of the Si -H; SiH_2; SiH_3; at 2083, 2118, 2140cm^{-1}, respectively. The peaks at 2150-2400 cm^{-1} can be attributed to the composite O_xSiH_y compounds (O_2SiH at 2305 cm^{-1}, O_3SiH at 2366cm^{-1}), these vibrations result firstly by the oxidation of Si-H bonds in the solution, secondly, the contact with the atmosphere, Others peaks were appeared at the region 900-1150 cm^{-1}, attributed of the stretching vibrations of Si-O-Si and Si-O_x[27], which confirming the formation an oxide on the surface.

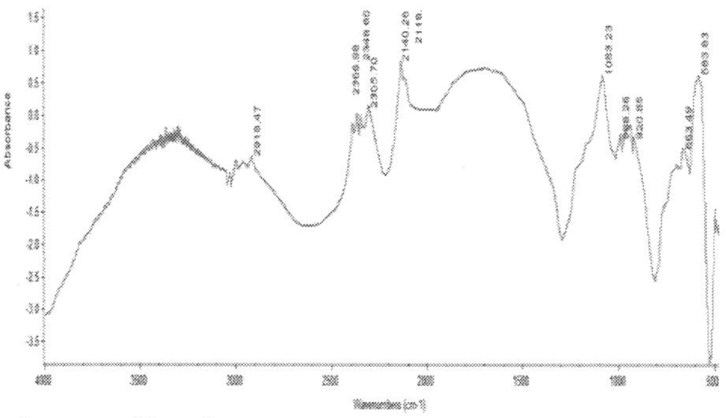

Fig.6. FTIR spectra of porous silicon layer

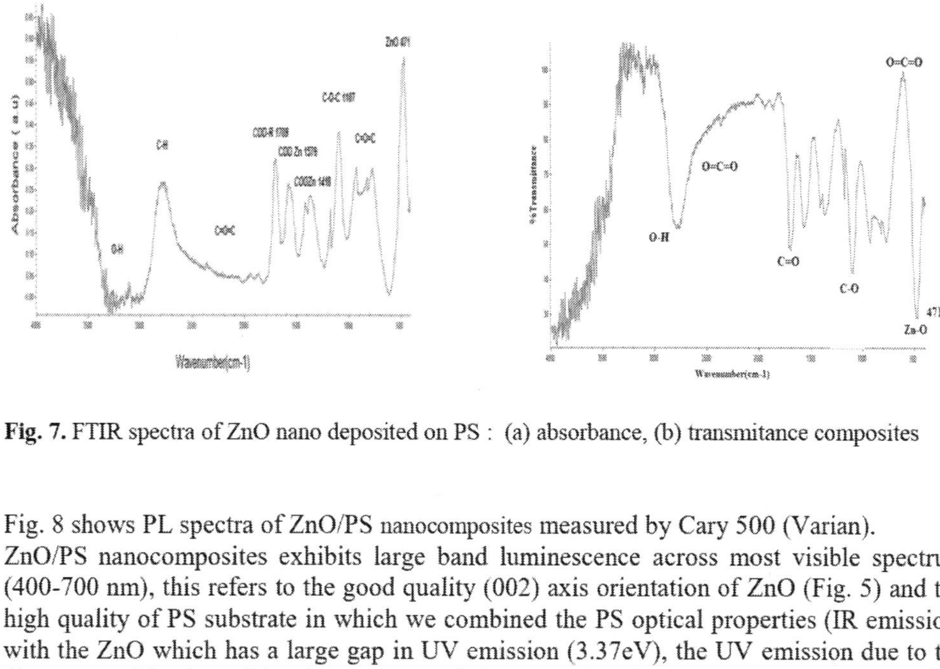

Fig. 7. FTIR spectra of ZnO nano deposited on PS : (a) absorbance, (b) transmitance composites

Fig. 8 shows PL spectra of ZnO/PS nanocomposites measured by Cary 500 (Varian).
ZnO/PS nanocomposites exhibits large band luminescence across most visible spectrum (400-700 nm), this refers to the good quality (002) axis orientation of ZnO (Fig. 5) and the high quality of PS substrate in which we combined the PS optical properties (IR emission) with the ZnO which has a large gap in UV emission (3.37eV), the UV emission due to the direct recombination of the photo generated charge carriers [29, 31],

In PL spectra of ZnO/PS, we can observe an intense PL peak located at (374 nm) caused by wide emission of ZnO (3.37 eV), and the peak located at 623 nm from PS revealing the high porosity of PS.

Red shift in PL peak of ZnO from the original PL emission 367.5nm [32] toward the longer wavelength of 374 nm was attributed to Zn vacancies in ZnO energy band, the peak located at 462 nm correspond to the blue emission caused by the combine by the high quality of the porosity of PS and ZnO optical properties and its UV emission [31].

According to the previous studies the photoluminescence of PS is commonly attributed to the oxidation of PS surface and the quantum confinement of electrons in nanosized particles of C-Si presents in PS. This can be observed clearly in Fig.7 revealing to the different stretching mode of PS, we know that the surface of PS influenced easy by ambience and the PL properties are very sensibly by the surface structure, the entire PS surface is covered by hydrogen, such as SiH_x species Fig.7 which unstable and easy to break. The Si-O and Si-H bonds break and form some defects, which may decrease the Photoluminescence intensity. Fig .4.b shows that a few pores of PS didn't covered completely by the ZnO particles, this refers to many deposition parameters of the electro deposition technique as: the deposition time, type and nature of PS substrate (orientation and resistivity), concentration of $ZnCl_2$, bath temperature etc... . The blue shift in the PL peak is possible and easy to obtain [30, 33, 34] Fig .8. The low dimension of the PS structure may be increase the probability of the recombination of electrons and holes leading to the increasing of photoluminescence of ZnO/PS films leading to used in much optoelectronics devices as the blue and red emitters.

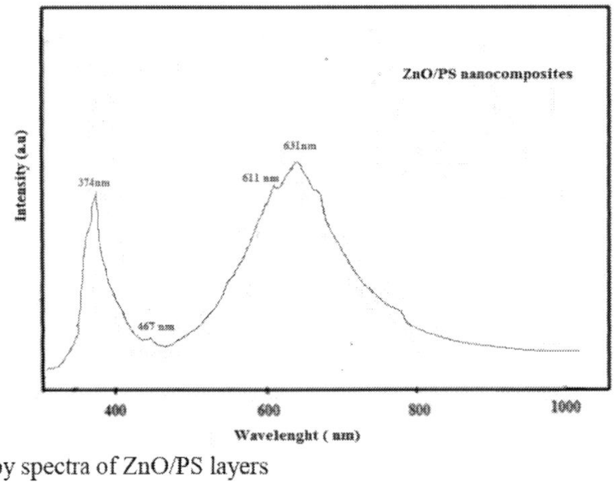

Fig. 8. Spectroscopy spectra of ZnO/PS layers

4. Conclusions

To combine the optical properties of porous silicon with those of zinc oxide which has wide applications for example UV and blue emitters. ZnO nanocomposites were successfully deposited on PS substrates. These nanostructures were prepared by electrodeposition technique using $ZnCl_2$ and the KCl under optimized conditions. This technique has some advantages compared to the physical deposition techniques. By this combine the photoluminescence (PL) of ZnO/PS increased and covered the most of the visible spectra. MEB images shows clearly that the ZnO particles entered in the pores of porous silicon despite the fact that few pores of PS were not covered completely by the ZnO particles. In other words, with the high porosity and the low dimension of the PS layer structure may be increase the photoluminescence (PL), the blue shift in the PL peak is

 DOI: 10.1051/e3sconf/20172200155

possible and easy to obtain. The electrical properties had been investigated, and the FTIR and RDX measurements had been interpreted.

Authors are grateful to the Amar Telidji university, for providing financial support through the laboratory of semiconductors and functional materials. Also, they would like to thank the Centre de Recherche en Technologie des Semi-conducteurs pour l'Energétique. CRTSE, ALGÉRIE for allowing to realize the experimental setup.

References

1. M.H. Zhao, Z.L. Wang, S.X. Mao, Nano Lett.. 587–590,4 (2004).
2. L. Stolt, J. Hedström, J. Kessler, et al., J.Appl. Phys. Lett. 62 597,3 (1993).
3. K. Keis, E. Magnusson, H. Lindström, SE. Lindquist, A.Hagfeldt, Sol. Energ. Mat. Sol. Cells . 73:51–58 ,1 (2002)
4. L. Stolt, J. Hedström, J. Kessler, M. Ruckh, K.-O. Velthaus, and H.-W.J. Appl Phys Lett, vol. 62, no. 6, pp. 597–599, (1993).
5. Changhui Ye, Xiaosheng Fang, Guanghai Li and Lide Zhang, , Appl. Phys. Lett. 85 (15) 3035–3037 (2004).
6. Jian Feng Yan, Hong Wei Liang, You, J. Infrared Millim.Waves 23 (2) 103–106(2004)
7. H. Jacobs, W. Mokawa, D. Kohl, G. Heiland, Surf. Sci.160, 217(1985).
8. S. Yano, R. Schroeder, H. Sakai, Appl. Phys. Lett. 82 (13) 2026–2028(2003).
9. Xiu-hua Fu, Yu-lin Sun, Cai-mei He, Infrared Laser Eng. 38 (5) 803–806(2009).
10. Guo Gang Qin, J. Infrared Millim.Waves 24 165–173 ,3 (2005)
11. R.T. Girard, O. Tjernberg, G. Chiaia, S. Sôderholm, U.O. Karlsson, C. Wigren, H. Nylen, I. Lindau, Surf. Sci. 373 (1997) 409.
12. Paraguay FD, Estrada WL, Acasta DRN, Andrade A, Miki-Yoshida M. Thin Solid Films;350:192(1999).
13. P. Nunes, E. Fortunato, R. Martins, Thin Solid Films 383 277(2001).
14. X.W. Sun, R.F. Xaio, H.S. Kwok, , J. Appl Phys 5776.84 (1998).
15. P. Nunes, E. Fortunato, R. Martins, Int. J. Inorg. Mater 1125, 3 (2001).
16. A. Bedia, F.Z. Bedia, M. Aillerie, N. Maloufi, B. Benyoucef, Energy Procedia, 603 - 609,50(2014)
17. Y. Natsume, H. Sakata, Thin Solid Films 30.372 (2000).
18. W. Tang, D.C. Cameron, , Thin Solid Films 238 .83(1994)
19. Q. Zhang, C.S. Dandeneau, X. Zhou, G. Cao, Adv. Mater. 4087.21 (2009).
20. K .Kakiuchi, Saito M, Fujihara S. Thin Solid Films 516:2026–30(2008).
21. Y. Kashiwaba, F. Katahira, K. Haga, T. Sekiguchi, H. Watanabe, J. Cryst. Growth 221 431(2000).
22. V. Sallet, C. Thaindoume, J.F. Rommeluere, A. Kusson, A. Riviere, J.P. Riviere, O.Gorochov, Materials Letters 126 .53 (2002)
23. N.Ait Ahmed, G.Fortas, H.Hammache, S.Sam, A.Keffous, A.Manseri, L.Guerbous, N.Gabouze, Appl.Surf. Sci 7442.256(2010)
24. Mane RS, Nguyen H-M, Ganesh T, Kim N, Ambade SB, Han S-H. ElectrochemCommun;11:752–5(2009).
25. Li G, Dawa C, Bu Q, Lu X, Ke Z, Hong H . J Phys.Chem C;111:1919–23(2007).
26. Yoshida T, Zhang J, Komatsu D, Sawatani S, Minoura H, Pauporté T, . Adv Funct Mater, 19, 17–43(2009).
27. Y. Zhao et al. / Current Applied Physics 10 (2010) 930–933.
28. Kyeong-Won Park, Jong Hwa Jung, J. Power Sources 379– 385. 199 (2012)

29. Y. He, B. Yang, G. Cheng, Catal. Today 595.98 (2004).
30. Khaldun A, Salman, Khalid Omar,Z.Hassan, Sol. energy 541-547.86(2012).
31. Prabakaran ,R peres,M, Monteiro,T,Fortunato,E.Martin,R, Ferreira,I, . U.Non – Cryst.Solids 354,2181-2185(2008).
32. T.Steinner.*Artech House* . Inc,Norwood (2004).
33. Liu ,Y.L Liu.Y.C,yang .HW.BMa.JG, 2003, J.Phys.D Appl.Phys.36,2705-2708.
34. Umar ,A.karunagaran,B,BSuh, E.K,Hahn.Y.B, *Nanotechnology* 17,4072-4077. (2006)

 DOI: 10.1051/e3sconf/20172200156

Optimization for energy efficiency of underground building envelope thermal performance in different climate zones of China

Luyang Shi[1], *Jing* Liu[1,*], *and Huibo* Zhang[2]

[1]School of Municipal and Environmental Engineering, Harbin Institute of Tech, 202 Haihe Street, Nangang District, Harbin, China

[2]School of Naval Architecture, Ocean and Civil Engineering, Shanghai Jiao Tong University, 800 Dongchuan Road, Minhang District, Shanghai, China

Abstract. The object of this article is to investigate the influence of thermal performance of envelopes in shallow-buried buildings on energy consumption for different climate zones of China. For the purpose of this study, an effective building energy simulation tool (DeST) developed by Tsinghua University was chosen to model the heat transfer in underground buildings. Based on the simulative results, energy consumption for heating and cooling for the whole year was obtained. The results showed that the relationship between energy consumption and U-value of envelopes for underground buildings is different compared with above-ground buildings: improving thermal performance of exterior walls cannot reduce energy consumption, on the contrary, may result in more energy cost. Besides, it is can be derived that optimized U-values of underground building envelopes vary with climate zones of China in this study. For severe cold climate zone, the optimized U-value of underground building envelopes is 0.8W/(m^2·K); for cold climate zone, the optimized U-value is 1.5W/(m^2·K); for warm climate zone, the U-value is 2.0W/(m^2·K).

1 Introduction

The global contribution from buildings towards energy consumption has steadily increased with urbanization ([1]). And underground building is one of big energy consumer and has significant impacts on energy consumption ([2]). The optimization design of energy saving for buildings below ground is also extremely important.

Generally, building envelope features considerably affect the energy efficiency of buildings ([3], [4]). For above-ground buildings, the reduction of energy consumption could be generated by improving thermal performance of the envelopes. However, underground buildings are more advantageous for their better capacity of heat storage, heat stability and smaller variation of temperature ([5]). The heat transfer through underground building envelopes maybe different compared with above-ground buildings.

* Corresponding author: liujinghit0@163.com

The aim of this study is to investigate the influence of thermal performance of envelopes on energy consumption for underground buildings. The heat transfer within underground buildings was simulated using energy simulation tool DeST and the energy cost was also calculated based on the standards of energy efficiency evaluation ([6]). In addition, the optimal U-values of envelopes in underground for different climate zones of China were determined and intercompared for the purpose of building energy-efficiency

2 Simulation model for underground buildings

2.1 DeST

In this study, an effective building energy simulation tool (DeST) developed by Tsinghua University was selected. DeST simulation model is a dynamic building thermal process model which can be used to calculate the heat transfer through the building components and energy consumption yearly.

The modeling process of DeST is to perform a discrete solution for the equations of heat energy conservation. Since the nodes on the exterior surface of envelopes are in direct contact with soil, these nodes must do some special handing. In DeST, the heat transfer between exterior surface of envelopes and surrounding soil can be solved by the method of frequency domain ([7]). The heat government equation for these nodes can be written in Eq.(1).

$$c_p\rho\frac{dt_i}{d\tau}\frac{\Delta x_i}{2}=Q_i(\tau)+\frac{K_i}{\Delta x_i}(t_i-t_{i+1}) \quad (1)$$

Where, c_p is the specific heat capacity of indoor air, J /(kg·°C); ρ is the density of indoor air, kg/m^3; t_a is the temperature of indoor air, °C; K_i is the heart transfer coefficient between soil and the surface of wall *i*, W/(m^2·K); t_i, t_{i+1} represents the temperature of the exterior surface of the wall *i* and the soil, respectively, °C; Δx_i is the thickness of discrete layer, m; τ is the calculated time, s.

In Eq.(1), $Q_i(\tau)$ represents heat flux transferred into buildings from surrounding soil, which can be calculated in Eq.(2) using state space method.

$$Q_i(\tau)=\text{his\_g\_disturb}+t_i(\tau)t_i(\tau)\sum_j g\varphi_{ij,0} \quad (2)$$

Where, his_g_disturb is related to thermal disturbance and thermo-physical property of underground building. And his_g_disturb is a determined quantity for calculated time τ. $\sum_j g\varphi_{ij,0}$ is constant and determined by thermo-physical property of the system.

Therefore, the energy conversation equation for these special nodes on the exterior surface of envelopes can be written in Eq.(3).

$$c_p\rho\frac{dt_i}{d\tau}\frac{\Delta x_i}{2}=\text{his\_g\_disturb}+t_i(\tau)\sum_j g\varphi_{ij,0}+\frac{K_i}{\Delta x_i}(t_i-t_{i+1}) \quad (3)$$

2.2 Simplified model of underground building

In this simulation, the complex underground building is assumed as a simple building composed of three rooms of same size (4.0×4.0×3.3m). Fig.1 shows the layout of underground building with 3.0m depth below the ground. The calculated building is surrounded by rammed clay considered a sort of special component of exterior walls. The thermal conductivity coefficient of rammed clay is 1.16 W/(m·K).

 DOI: 10.1051/e3sconf/20172200156

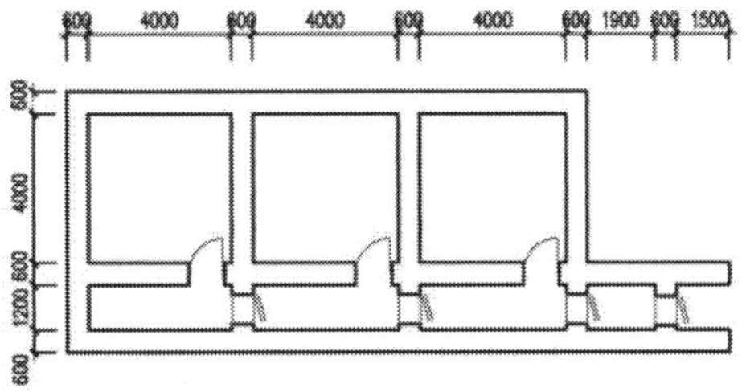

Fig. 1. The layout of simplified underground building

The U-value of underground building envelopes is set based on design standard for energy efficiency of public buildings ([6]). Table 1 shows design standard of thermal performance of envelops for different climate zones of China. Since the heat transfer in underground building and above-ground building is different, design standard of thermal performance of envelops may have different values. The U-value of exterior wall were set in the range of 0.2~3.0W/(m·K) for the purpose of investigating the effect of U-value of exterior wall on the energy consumption.

Table 1. Design standard of U-values of envelops for different climate zones

Climate zone	**U-value** (W/(m^2·K))
Severe cold zone	≤ 0.35
Cold zone	≤ 0.50
Hot summer and cold winter zone	≤ 0.60
Hot summer and warm winter zone	≤ 0.80
Temperate climates zone	≤ 0.8

2.3 Description of modelling parameter and weather condition

Compared with ground buildings, underground buildings have excellent properties such as large capacity of heat storage, and good airtightness. These properties help to keep underground space warm in the winter and cool in the summer. And the indoor temperature in underground is lower than 20 °C in sometimes during summer, and then the room need be heated. Thus, the indoor climate conditions for underground buildings may be set a wide comfortable range. According to the references about ground buildings, indoor temperature for underground buildings in summer and winter was set in a wide range of 20~28°C, 18~22°C, respectively. The air conditioning was used to keep indoor temperature at the set range. The sources of heat and humidity, including people, equipment and lightings, were set based on references of typical office buildings ([7]). And the ventilation was not taken into account in this simulation in order to remove the impacts of fresh air on building loads.

The weather data is used as representative of typical weather variations. Twelve typical cities selected from different climate zones of Chin were studied in this simulation (Fig. 2).

 DOI: 10.1051/e3sconf/20172200156

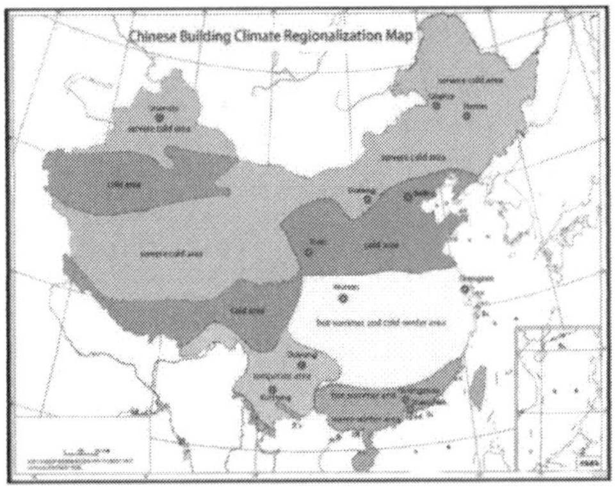

Fig. 2. The map of climate zones in China and location of twelve typical cities

3 Results & discussion

Generally, enhancing thermal performance of envelopes is beneficial for above-ground building because of reduction of heat loss from building envelopes. However, for underground buildings, raising the thermal performance of the envelopes cannot reduce energy consumption, on the contrary, may result in more energy cost. These differences are supposed to be due to large capacity of heat storage of underground buildings which helps to keep the room warm in the winter and cool in the summer.

In summer, decreasing U-value of envelopes can transfer more heat into the surrounding soil effectively, which is helpful to drop indoor temperature significantly. In winter, thermal resistance of exterior walls can effectively prevent heat from transferring into the surrounding soil. But, if U-value of envelopes is too small, the heat generated in the room cannot transfer into soil effectively, even leading to cooling load in winter. In terms of energy conservation, it is not reasonable and available. Therefore, there is a trade–off when the thermal performance of exterior walls of underground buildings is designed for the purpose of energy conversation.

Relationship between energy consumption and heat transfer coefficients of envelopes for different climate zones is plotted in Fig.3, respectively. It is observed that the shapes of the curves which both have single valley in Fig.3 are similar.

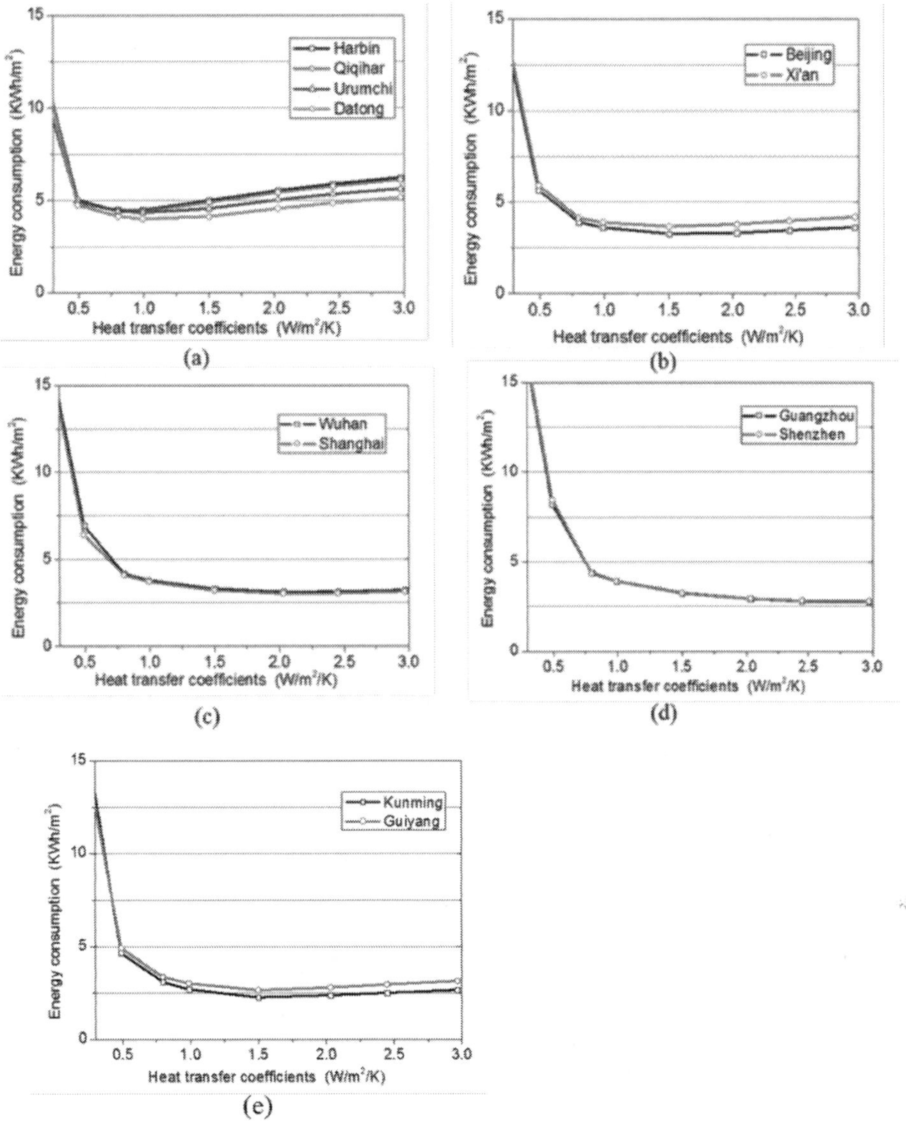

Fig. 3. Relationships between energy consumption and U-value for different climate zones of China

Fig.3 (a) shows calculated results in severe cold climate zone. It is noted that increasing the heat transfer coefficient of envelopes from 0.22 W/(m^2·K) to 1.0 W/(m^2·K) tends to reduce the sum of energy used for heating and cooling, whereas the energy consumption increases when the heat transfer coefficient increases continuously from 1.0 W/(m^2·K). Therefore, energy-efficient optimized value of underground envelopes for bitter cold climate zone is 1.0W/(m^2·K). Similarly, it is can be derives that energy-efficient optimized value of underground envelopes for cold climate zone is 1.5 W/(m^2·K) from Fig.3 (b). From Fig.3 (c)~Fig.3 (e), we observed that for other climate zones (including hot summer and cold winter zone, hot summer and warm winter zone, and temperate climates zone), optimized value of heat transfer coefficients are similar, thus the optimized value can be regard as the same value in the error-allowed range, 2.0W/(m^2·K).

The optimized values of underground envelopes in different climate zones are different. The major cause for this is the difference of the soil temperature around buildings.Fig.4

DOI: 10.1051/e3sconf/20172200156

presents measured data of the soil temperature with 3.2m depth in underground in typical cities of different climate zones in China.

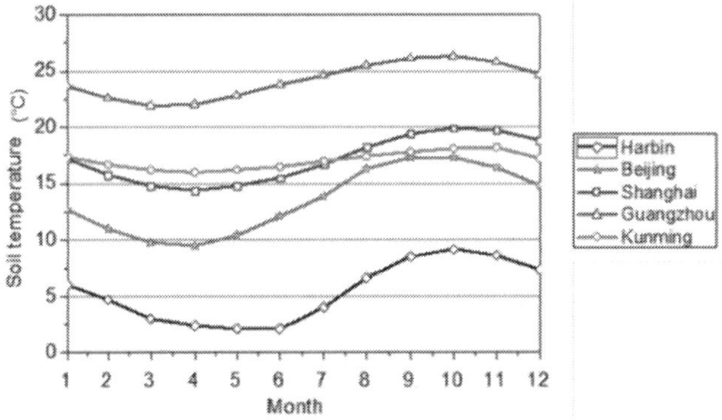

Fig.4. Measured data of the soil temperature of 3.2m depth below the ground

From Fig.4, we observed that the fluctuation of soil temperature in Harbin and Beijing is bigger than other three cities. For Harbin, the average of soil temperature (3.2m depth below ground) is lowest among in all studied cities, about 5°C, which account for basic requirements of good thermal insulation of envelopes below ground. For warm climate, the average soil temperature (3.2m depth below the ground) is higher than 15°C, especially in Guangzhou (the average soil temperature of 3.2m depth below the ground reaches 24°C). For theses climate zones, cooling the interior space is main factor to consider when the thermal performance of underground building envelopes is designed for the purpose of energy saving.

4 Conclusions

The effect of envelope thermal performance on energy consumption for shallow-buried buildings in different climate zones of China was investigated in this paper. DeST was used to simulate the heat transfer within building below the ground. And the energy consumption was also calculated according to the standards of building envelope thermal performance trade-off.

Because of large capacity of heat storage of underground buildings, effect of thermal performance of envelopes on energy consumption for ground building and underground building varies considerably. Raising thermal performance of envelopes below ground cannot reduce energy consumption, on the contrary, may result in more energy cost. When the thermal performance of envelopes in underground is designed for the purpose of building energy conversation, we should make a trade- off synthetically. Besides, it is can be derived that the optimized U-value for underground building envelopes is different in different climate zones of China. For severe cold climate zone, the optimized U-value is $0.8W/(m^2{\cdot}K)$; for cold climate zone, the optimized value is $1.5W/(m^2{\cdot}K)$; for warm climate zone, the optimized value is $2.0W/(m^2{\cdot}K)$.

These conclusions are drawn for the buildings which are buried less than 5.0m depth below the ground. A further work could be conducted to consider the buildings located more than 5.0m depth below the ground in different climate zones of China.

This research was supported by Technology and Research Project of China State Construction Engineering Corporation (CSCEC-2014-Z-1-2-01).

References

1. Pérez-Lombard, L., Ortiz, J., & Pout, C. (2008). A review on buildings energy consumption information. Energy & Buildings, 40(3), 394-398.
2. Casals, M., Gangolells, M., Forcada, N., Macarulla, M., & Giretti, A. (2014). A breakdown of energy consumption in an underground station. Energy & Buildings, 78(4), 89-97.
3. Nielsen, T. R. (2005). Simple tool to evaluate energy demand and indoor environment in the early stages of building design. Solar Energy, 78(1), 73-83.
4. Lin, Y. H., Tsai, K. T., Lin, M. D., & Yang, M. D. (2016). Design optimization of office building envelope configurations for energy conservation. Applied Energy, 171, 336-346.
5. Shao-Hua, L. I., Shang, Z. H., Gong, M. Z., Wang, H. L., & Kun, Y. E. (2012). Calculation of heat shortage capacity of soil temperature field around the pipeline in the preheating process of underground oil pipelines. Contemporary Chemical Industry.
6. Xie Xiaona, Jiang Yi. (2004).Study on heat transfer within underground for thermo physical model in DeST. The national HVAC academic annual meeting.
7. GB 50189-2015. (2015).Design standard for energy efficiency of public buildings.

Hybrid capacitive deionization with anion-exchange membranes for lithium extraction

Anna Siekierka[1,*], *Marek* Bryjak[1]

[1]Wroclaw University of Science and Technology, Faculty of Chemistry, Wyb.Wyspiańskiego 27, Wroclaw, Poland

Abstract. Lithium is considered to be a critical material for various industrial fields. We present our studies on extraction lithium from diluted aqueous solution by novel hybrid system based on a membrane capacitive deionization and batteries desalination. Hybrid CDI is comprised by a lithium selective adsorbent, activated carbon electrode and anion-exchange membranes. Here, we demonstrated implication of various type of anion-exchange membranes and influence their properties on effective capacity and energy requirements in charge/discharge steps. We described a configuration with anion-exchange membrane characterized by adsorption capacity of 35 mg/g of Li^+ with 0.08Wh/g and removal efficiency of 60 % of lithium ions, using novel selective desalination technique.

1 Introduction

The global market for lithium salts is a fast growing and is expected to reach 30$ billion value by 2020. Increasing rapidly demand of Li ions is caused by their application for portable electronics, for electric and hybrid cars and for electrical intelligent grids [1]. Information on individual lithium resources has been collected and divided into two main categories: extraction from natural as minerals, salt lakes, underground brines or seawater, and the recycling of lithium from electronic waste and lithium ions batteries [2]. However, the most promising lithium ions depositions are estimated at groundwater's and brines where its content is evaluated to be $2.55 \cdot 10^{14}$ kg [3].

Seawater desalination process require electric or thermal technology to separate saline water into two streams, fresh water with a low concentration of dissolved salts and a concentrated brine stream. A variety desalination technology have been developed over year. Reverse osmosis and multistage flash distillation require a large electrical energy input which accounts for an approximately 50% of its costs [4]. Forward osmosis is a promising method that utilizes lower temperatures (60°C) than MFD but still requires energy. A search for less energy consuming technique was finalized with launching capacitive deionization based on the use of supercapacitors [7]. Finally, a new concept of a "desalination battery" which has configuration similar to CDI, but with different electrode materials, was investigated. In this case, system is consist of cationic sodium insertion electrode and a chloride-capturing anionic electrode, made of Ag [4]. Batteries electrode

* Corresponding author: anna.siekierka@pwr.edu.pl

pose higher specific capacity and lower energy requirement than capacitive deionization [4].

In this paper, we were investigated a novel combination of membrane capacitive deionization and battery desalination, called hybrid capacitive deionization or ion-pumping process with electrical potential assembly. The goal of the paper is to develop a search for lithium selective system suitable for construction of electrodes and for extraction of lithium ions with higher efficiency. Preliminary application of CDI to extraction lithium from diluted aqueous solutions was confirmed in our previously paper[6]. In this investigation, we evaluated a system built with lithium selective adsorbent as a anode, activated carbonaceous electrode for chloride ions capturing coated with anion-exchange membrane that prevented adsorption released lithium on the surface of counter electrode (co-ions effect [7]). We compare systems with our PVC-aminated anion-exchange membrane with industrial AMX-408 to find the best combination with the highest capacity and lowest energy requirements for selective lithium extraction.

2 Materials and Methods

2.1 Electrodes

2.1.1 Activated Carbon

YP-50F Kuraray Chemical Co. LTD. Osaka, Japan, steam activated coconut carbon (AC) was used throughout this study. Its characteristic can be found in Tab.1.

2.1.2 Lithium adsorbent

Spinel lithium-manganese-titanium oxide (LMTO) with 5% wt. of Ti was prepared according to procedure described in ref. [5]. Its characteristic can be found in Tab.1.

Table 1. Specific surface area and pore volume of LMTO and AC

Sample	Specific surface area S_{BET} (m^2g^{-1})	Pore volume (cm^3g^{-1})	Pore diameter (nm)	Ref.
LMTO	54.9	0.113	1.03	[5]
AC	1672	0.757	1.65	[5]

2.1.3 Preparation of electrode

Both electrodes were prepared by mixing 90% wt. of lithium adsorbent or activated carbon, with 10% wt. of PVC dissolved in DMF (3.5% wt. solution of polymer). Paste was stirred by ultrasonication supply for 30 minutes at room temperature. Finally, slurry was cast on the graphite foil and electrode of 70-80 μm was formed by casting knife. The evaporation of solvent was carried out by 2 h in vacuum dryer. The prepared electrodes were kept in DI water. Before measurement electrode were been activated 3 times in charge-discharge process to remove any contaminants from electrodes.

2.2 Anion-exchange membranes

Amine modified poly(vinyl chloride) membranes were used in the experiments. Selected membranes were obtained by chemical modification of poly(vinyl chloride) by ethylene di-amine (EDA) or diethylene tetra-amine (DETA). Details of modification and preparation can found in ref. [8]. Fundamental properties of PVC-amines membranes are presented at Tab.2

Table 2. Properties of PVC-EDA and PVC-DETA

Sample	[1] Z_{IEC} [mmole/g]	[2] Z_{Cl} [mmole/g]	[3] Z_N [mmole/g]	[4] W_{H2O} [g/g]	[5] MD [%]	[6] P [%]	Ref.
PVC-EDA	3.5	11.8	3.3	0.4	23.9	17.5	[3]
PVC-DETA	2.6	12.0	3.0	0.9	22.6	17.8	[3]

[1] *Ion-exchange capacity was estimated according to acid-base method*
[2] *Chloride content was determined according Schöniger's method*
[3] *Nitrogen content was determined by Kjeldahl's method*
[4] *Water uptake*
[5] *Modification degree was calculated as the relative content of chloride*
[6] *Polarity was taken as a partition of polar component to total surface energy*

Before measurements membranes were kept in ethanol:water solution (40:60 v:v) to activated functional groups. After 24 h were washed by DI water. To compare anion-exchange properties of obtained membranes, Neosepta AMX-408 membrane was used. Its characteristic can be found on the manufacturer web page [9].

2.3 Layout of hybrid CDI system

2.3.1 Hybrid CDI system configuration

To study desalination performance of chloride salts, a laboratory electrodialyzer FT-ED-100-4 was used. The stack composed of two electrode and anion-exchange membrane while electrodes were divided by spacer of 200μm thickness. As a anode (negatively polarized) spinel lithium-manganese-titanium oxide (LMTO) with 5 % wt. of Ti was mounted [6]. As cathode activated carbon YP-50F was used. LMTO electrode was used for selective capturing Li^+ ions and a AC electrode was used to capture Cl^- ions. To prevent the effect of co-ions adsorption [7] in desorption step of the hybrid CDI testing, anion-exchange membrane was applied on AC electrode.
Desalination performance was evaluated under constant voltage mode (CV). The electrodialyzer was biased by DC power from Konrad KD3005D Digital-Control DC Power Supply with 2V. The fluid properties were monitored with U3401A Digit Dual Display Multimeter and with CPC-501 meter. To investigate selectivity of lithium adsorbent, solutions of LiCl, NaCl and KCl, of 10 mM concentration were used. The adsorption and desorption steps were carried out for 10 min at 25°C.

2.3.2. Calculations section

In all experiments, such metrics as salt adsorption capacity (SAC) and average salt adsorption rate (ASAR) were calculated:

$$SAC\ (\mathrm{mg/g}) = [(C_0 - C_f)\cdot \mathrm{V}]/\mathrm{m} \quad (1)$$
$$ASAR\ (\mathrm{mg/g{\cdot}s}) = SAC/\mathrm{t_{charge}} \quad (2)$$

where, C_o (mg/dm^3) and C_f (mg/dm^3) are initial and final salt concentration, V (dm^3) is the solution volume, and m (g) is the mass of both electrodes, t_{charge} (s) is charging time.

The important parameter for evaluation of the stack working in the CV mode, is charge efficiency Λ that is given by:

DOI: 10.1051/e3sconf/20172200157

$$A=[(C_0-C_f)\cdot V\cdot F]/Q \tag{3}$$

where F is the Faraday's number, F=96485 ($C\cdot mol^{-1}$), Q is a charge stored for charging steps and was calculated by integration of the current and time of charging.

To analyze energy consumption and calculate system capacity, the simple RC circuit with measurement of current change was used. Total energy consumption (E_{total}) was the sum of energy for adsorption (E_{ads}) and for desorption (E_{des}) processes, and were computed from numerical integration of current versus time relationship [10,11]. Next metrics is energy normalized adsorbed salt (ENAS) in units of grams per Joules of energy total and is defined as

$$ENAS=[\int (C_0-C_f)\,dt]/E_{total} \tag{4}$$

Which quantifies the energetic efficiency of the desalination process. Last metrics described hybrid capacitive deionization process is electrical work in Wh per gram of salt and is defined as

$$W=Q\cdot V/[\int (C_0-C_f)\,dt] \tag{5}$$

It expresses consumed energy that has use the system to adsorb or desorb one gram of salt. Q is a charge flow by system during adsorption/desorption process and V is electrical potential between electrodes.

2.3.3 Analytical section

Membranes were analysed using a scanning electron microscope (SEM, Tescan Vega3 SB). All samples were gold-coated for 60 s and 7 nm thickness before SEM imaging to reduce the effect of charging.

3 Results and Discussion

3.1 SEM morphology examination

The surface structure of anion-exchange membranes were determined by using SEM images. In Fig.1 one sees SEM images surface of PVC-DETA and PVC-EDA membranes.

It can be noted the morphology change happened during amination step. It is possible that new functional group were introduced which added some kind of aggregates on the surface and increased their roughness. When comparing the SEM images, the PVC-DETA was characterized by more homogenous diameter of aggregates than the second membrane. The average value was located in the range 28-35 nm.

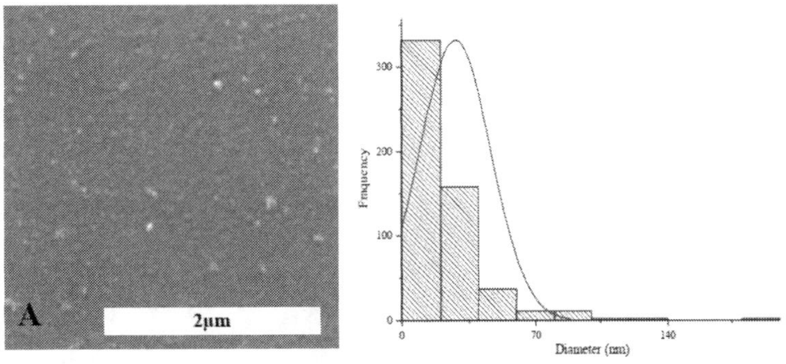

 DOI: 10.1051/e3sconf/20172200157

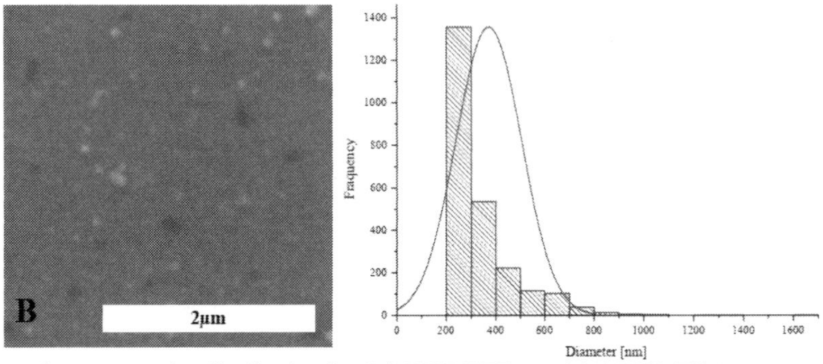

Fig.1. SEM images and aggregate size distribution for (A) PVC-DETA and (B) PVC-EDA.

3.2 Hybrid Capacitive Deionization

To presented the most prominent aspects of the hybrid CDI system the CDI Ragone plot was used. To see implications of the CDI Ragone plot, we compared different anion-exchange membranes at the same system parameters. Additionally, we evaluated the plot for desorption process.

All collected data are showed ideal capacitive deionization performances, according to reference [12]. When compared the results for configuration with and without anion-exchange membranes and noted higher capacity and rate value for system equipped with membrane. Application of anion-exchange membranes provided shift of the plot up (Fig.2A) and caused an opportunity to gain salt adsorption capacity at 45 mg/g for PVC-EDA and for AMX-408 membranes in 10 minutes. At Fig.2B one can observe different behaviour. The stack with PVC-DETA membrane approached zero and thereby released adsorbed lithium to concentrated solution.

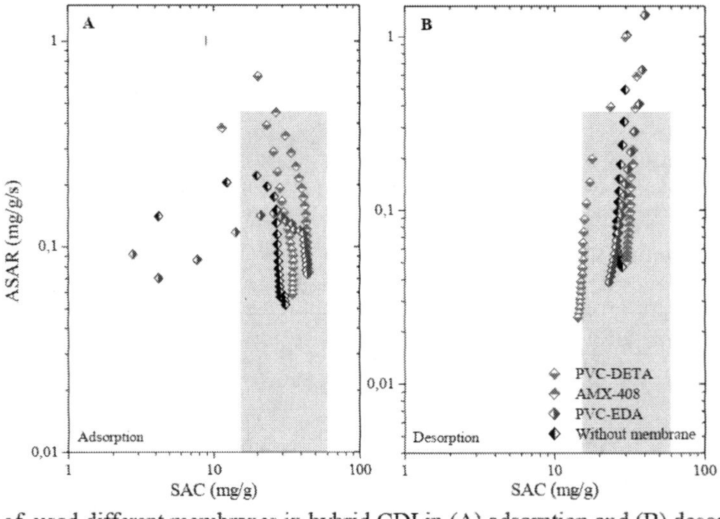

Fig.2. Effect of used different membranes in hybrid CDI in (A) adsorption and (B) desorption processes. The grey region marked at the plot is related with ideal performance described at [12].

Effect of desorption against concentration gradient is also seen in Fig.3. This graph shows removal efficiency of lithium chloride and repeatability of salt extraction. Based on these results it can be deduced that system with anion-exchange membrane directly influenced on removal efficiency of lithium chloride during desorption process. Moreover, it can be noted that layout of hybrid CDI system with PVC-DETA exhibited the higher of 60% removal efficiency allowed to obtain two times higher capacity than configuration with industrial membrane AMX-408. It was also six times more efficient than system without membrane.

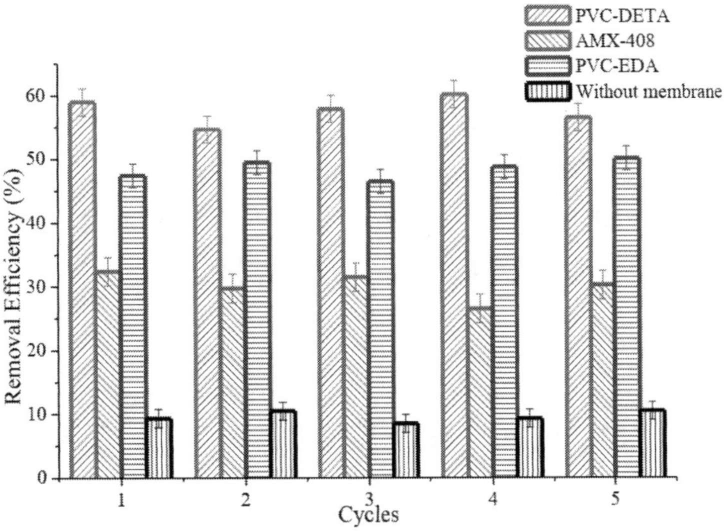

Fig.3. Removal Efficiency of LiCl for system. Time of removal 10 min. Used voltage 2 V. Three times of repeatability.

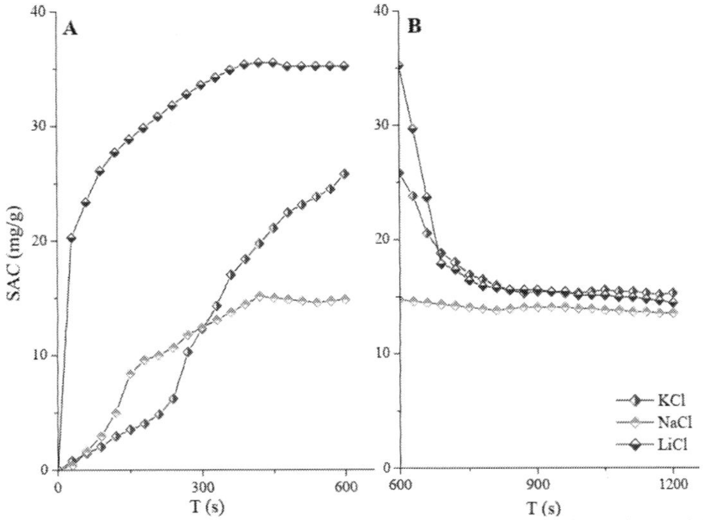

Fig.4. Changed NaCl, KCl and LiCl salt adsorption capacity during (A) adsorption and (B) desorption for configuration with PVC-DETA membrane. 10 minute of charge and disachrge steps, bias 2V.

The most critical parameter fo selective extraction of lithium is its comparision to other monovalents salts. In this case 10 mM of sodium and potassium chlorides were emloyed. The relationships are shown in Fig.4. According to colleted data, it is possible to extract Na and K from aqueous solution but with significantly lower adsorption capacity than for lithium chloride. It can be noted that the removal efficiency is maintained at 40 and 8 % level for KCl and NaCl, respectively. It could be concluded that LMTO had an unique ability for selective extraction of lithium ions from low salinity aqueous solutions.

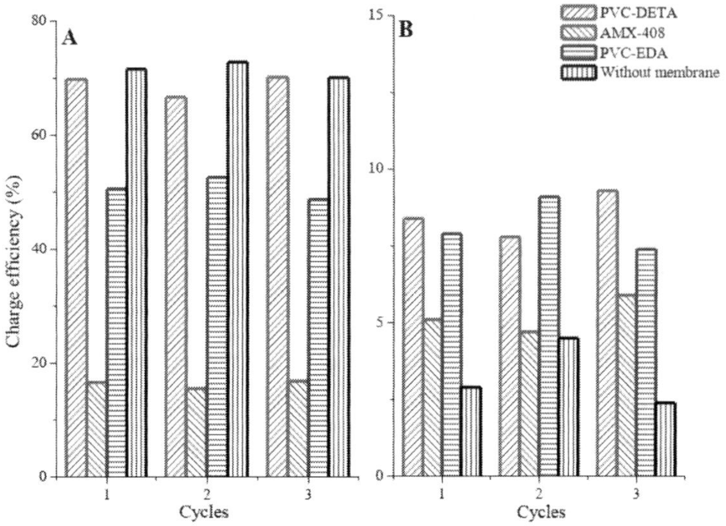

Fig.5. Date of charge efficiency vs. time of (A) adsorption and (B) desorption. Initial salt concentration was 10mM, time of adsorption/desorption 10 min, external voltage at 2V.

The next key metrics describing performances of any CDI system is its charge efficiency, which refers to ration between the amount of salt ions adsorption from bulk solution and the amount of electronic charge transferred between the electrodes. Fig.5. depicts the variation of the charge efficiency for evaluated configuration for charge and discharge steps. Observed differences with characteristic of charge efficiency are resulted from various kinetics of adsorption and desorption. The most stable course charge efficiency have PVC-DETA and without membranes systems. During the discharge step, two important observations can be made after careful inspection of Fig. 5B. Firstly, charge efficiency was not more decreasing for system with higher salt removal degree. This effect could be explained by higher energetic requirements to get greater efficiency of desorption process. It indicated the work against concentration gradient. The second observation was that the charge efficiency of the stack with AMX-408 or without membrane slightly decreased and approached linear shape during the discharge step. We think that phenomenon was caused by minor degree of ions release from electrodes.

More representative metrics determined hybrid CDI system for lithium extraction are work that system has to devote for adsorb or desorb 1 gram of salt. The second metrics is ENAS that shows how much salt can be adsorb/desorb when 1 joules of energy passes the system. The energy requirements data for investigated stacks are collected in Tab.3.

Table 3. Work and ENAS for all membranes during adsorption and desorption process

	Adsorption		**Desorption**	
	Work	ENAS	Work	ENAS
	Wh/g	*mg/J*	*Wh/g*	*mg/J*
PVC-DETA	0.08	3.62	0.64	0.44
AMX-408	0.24	1.15	1.06	0.26
PVC-EDA	0.10	2.61	0.68	0.41
Without membrane	0.07	3.72	1.82	0.15

Based on the above results, stack with AMX-408 is characterized by the higher electrical work during charging step. It can be explained by membrane thickness (AMX-408 has an approximately 200 μm thickness, two time larger than PVC membranes). However, during the discharge step, directly disadvantage of membrane-less system was observed. It could be the co-adsorption effect on counter electrode [7] that caused an increase of electrical work value. Moreover, systems with PVC-DETA was characterized by the most efficiency electrical work ratio, achieved 8 time lower value between charge/discharge steps. The ENAS metrics reach optimum values for that system.

4 Conclusion

The paper presents comparison different type of anion-exchange membranes used for selective extraction lithium ions from diluted aqueous solution in hybrid capacitive deionization or ion-pumping system. The studies pointed that combined system of LMTO selective adsorbent and PVC-DETA anion-exchange membrane allowed to obtain high adsorption capacity that value was 60 % better for lithium chloride than for other alkaline chlorides. Moreover, the described hybrid system showed high energetic efficiency for lithium ions sorption/desorption processes. It reached the values of 0.08 and 0.64 Wh/g, respectively. The developed system had higher capacity and lower energy consumption that configuration with industrial AMX-408 membrane.

References

1. H. Vikström, S. Davidsson, and M.Höök, App. Energy **110** 252 (2013)
2. P. Meshram, B.D. Pandey, and T.R. Mankhand, Hydrometallurgy **150** 192 (2014)
3. T. Ryu, D.-H. Lee, J.C. Ryu, J. Shin, K.-S. Chung, and K. H. Young, Hydrometallurgy **151** 78 (2015)
4. M. Pasta, C.D. Wessells, Y. Cui, and F. La Mantia, Nano Lett., **12** 839 (2012)
5. A. Siekierka, M. Bryjak, and J. Wolska, Desal. Wat. Treat. **64** 251 (2017)
6. M .Bryjak, A. Siekierka, J. Kujawski, K. Smolińska-Kempisty, and W. Kujawski, J. Memb. Separ. Tech **4** 110 (2015)
7. S. Porada, R. Zhao, A. Van der Wal, V. Presser, and P.M. Biesheuvel, Prog. Mater. Sci.**58** 1388 (2013)
8. A. Siekierka, J. Wolska, W. Kujawski, and M. Bryjak, Sep. Sci. Tech. (2017) [to be published]
9. Web page: http://www.astom-corp.jp/en/product/images/astom_hyo.pdf [available: 12.03.2017]
10. M.E. Suss, S. Porada, X. Sun, P.M. Biesheuvel, J. Yoon, and V. Presser, Energy Environ. Sci. **8** 2296 (2015)
11. A. Hemmatifar, J.W. Palko, M. Stadermann, and J.G. Santiago, Water Res. **104** 303 (2016)
12. T. Kim, and J. Yoon, RSC Adv. **5** 1456 (2015)

Evaluation of the correlation between concentration of volatile organic compounds and temperature of the exhaust gases in motor vehicles

Maria Skrętowicz[1], *Radosław* Wróbel[1,*], and *Monika* Andrych-Zalewska[1]

[1]Wroclaw University of Science and Technology, Faculty of Mechanical Engineering, Braci Gierymskich 164, 51-640 Wrocław

Abstract. Volatile organic compounds (VOCs) are the group of organic compounds which are one of the most important air pollutants. One of the main sources of VOCs are combustion processes including fuel combustion is internal combustion engines. Volatile organic compounds are very dangerous pollution, because even in very low concentrations they have significant harmful effect on human health. A lot of that compounds are mutagenic and carcinogenic, in addition they could cause asthma, intoxication or allergy. The measurements of VOCs are quite problematic, because it is required using the specialist analytical apparatus, ex. chromatograph. However, not always it is need to measure the content of that compounds in engine exhaust with high precision and sometimes it is enough only to estimate the level of the concentration. Emission of the VOCs mainly depends on the combustion process in the engine and this determines the temperature of the exhaust gases. In this paper authors tried to determine if the correlation between temperature of exhaust gases and VOCs' concentration exist and is able to determine.

1 Introduction

Motor vehicles, especially in big cities, are a very relevant source of the air pollution. Although producers are forced by law acts to produce low emission engines, the share of the internal combustion engines in the air pollution is still significant [1]. The most important vehicles category, in terms of the numerical amount, is passenger cars which are mostly driven by spark ignition engines [2]. The most harmful exhaust components are: carbon monoxide (CO), carbon dioxide (CO_2), hydrocarbons (HC), nitrogen oxides (NO_x) and in trace amounts sulphur and lead compounds [2]. The main cause of the formation of hydrocarbons is the incomplete combustion which interrupts the chain of elementary reactions [4]. This situation is due to the local or global deficit of the oxygen and disturbances of the combustion process. Deficiency of the air in the combustion chamber could be caused not only by the combustion of rich air-fuel mixture, but also by

* Corresponding author: radoslaw.wrobel@pwr.edu.pl

contamination of inlet air. Disruptions of the burning process are especially in cases of the ignition missing or wall and slotted effects [5]. In addition, the causes of the hydrocarbons emission are oil and fuel evaporation and oil burning in the combustion chamber. Emission of hydrocarbons is characteristic primary of spark ignition engines. Hydrocarbons are very hazardous and toxic for human health. The Euro 5 norm limits emission of hydrocarbons from passenger cars only for spark ignition engines and it is set to 100 mg/km [5].

One of the groups of hydrocarbons is volatile organic compounds (VOCs). Most of these compounds exhibit toxic effects on the human health, especially low boiling-point aromatic hydrocarbons or some of aldehydes. The most vulnerable are the cells of respiratory system, which is the main way of absorption of these compounds. Almost all of VOCs have a mutagenic and carcinogenic effect [7]. The most substantial air pollutants of this group are benzene, toluene, ethylbenzene and xylenes in all isomers. The group of these compounds is called briefly BTEX. This batch has also the most negative impact on the human health.

Nowadays, the measurement of the total hydrocarbons in the fumes is pretty easy and possible to take at most diagnostic stations (with FID analysers). However, in order to evaluate some specific components, such as BTEX, specialist and advanced equipment (e.g. gas chromatography) is required. It is obvious that the more the engine load increases, the more the temperature of the combustion process raises. As the temperature in the combustion chamber increases, the temperature of exhaust fumes at the cylinder outlet continues to rise. Simultaneously, when the temperature in the combustion chamber is higher, the fuel combustion process runs better (i.e. complete combustion). Thus, the higher the temperature is, the fewer hydrocarbons in the exhaust gases are. The aim of this study is to determine if it is possible to estimate the concentration of selected aromatic hydrocarbons (BTEX) in the exhaust based on the exhaust temperature measurement. Thanks to this analysis, it would be easy and quick to estimate the amount of emitted VOCs, excluding complicated analyses that require expensive equipment. The research focuses, among other things, on the measurement of VOCs in the exhaust fumes (determining emission factors), the influence of temperature on the VOCs inside the car cabin and the impact of ambient temperature on the VOCs content in exhaust [7-10].

2 Materials and Methods

The investigations were conducted on the spark-ignition engine Fiat 1.2 on the engine test bench in the Department of Automotive Engineering, Wroclaw University of Science and Technology. The tested engine had 4 cylinders and 12 valves and was equipped with three-way catalyst reactor (TWC). In the Figure 1 the test bench was shown. The test bench was also equipped with a drive shaft and the water-cooled electromagnetic brake EMX 100/10000, which allowed applying the engine torque.

Fig 1. The engine test bench, Fiat 1.2

The engine was powered by the commercial gasoline with 95 octane number. Investigations began when the engine has reached its operating temperature. As it was already mentioned, the engine was cooperating with TWC, but the samples were taken only before the reactor (the investigation focused on the raw exhaust).

As it was written below the aim of this research was to evaluate the correlation between the temperature of exhaust fumes and the concentration of volatile organic compounds (VOCs) in them. The knowledge of the relationship between these two parameters could be helpful while estimating the level of the VOCs content in exhaust fumes without complicated chemical analysis of VOCs. These tests could answer the question if it is possible to get the mathematical equation of these parameters.

In the tests load characteristic of engine with fixed rotational speed (2500 rpm) was performed. The engine was loaded with the electromagnetic break which was coupled to the engine with the torque values from 0 Nm to 80 Nm, every 10 Nm. Changing the engine's torque with constant rotational speed in spark ignition engine, the temperature of the combustion process and thus the temperature of exhaust fumes changes as well. Moreover, the additional fuel consumption (in g/s) was measured.

The high temperature, pressure and other harsh environmental conditions distort the measurement method. In this case thermocouples appear which the best cluster of sensors is. It consists of a number of factors: the lack of need for the placements of contacts soldered or bonded in place of the measurement and a method for mounting the diagnostic element. In case of measurement with the use of semiconductor devices, the installation is more difficult (and hence expensive). Still, it is possible.

During the research, K-type thermocouples (NiCr-NiAl) were used [11]. Parameters of these elements could be used in the temperature's range between -40 to 1000 °C. In this case, LimaTherm thermocouples were used (product no. PTTK-TKB-45-1-SO-150-4m-Si). This kind of sensors is equipped with compensatory wire which facilitates measurement. Thermocouples were placed on, in and out of catalyst (Figure 2).

Fig. 2. The catalyst with thermocouples and gas uptake points

The acquisition system was based on the classic ADC card National Instrument (9215 type). Although this card is not directly designed for temperature measurements, the input voltage range allows performing tests. In a program written in LabView the object language corrective functions was applied. It permitted reading the temperature directly [12]. The measurements were performed at the same time, in order to synchronize the moment of sampling for toxic compounds and temperature.

During the tests, infrared camera was also selectively used so as to confirm the order of magnitude indicated by the thermocouple.

The samples of the fumes for chemical analysis of VOCs were taken from special connectors to Teflon bags by the vacuum pump. Then, the content of the bags was given by semi-automatic aspirator ASP-3II into special tubes filled with active carbon (Figure 3). The aspirator was equipped with the pomp and flow controller. From each bag 10 dm^3 of exhausts were transferred with a fixed airflow rate set at 30 dm^3/h. All samples were moved to the Laboratory of Emission Researches in the Department of Automotive Engineering, Wroclaw University of Science and Technology, where were analysed with a gas chromatography (GC) method. The step preceding the chemical analysis was to prepare the samples by desorption VOCs from active carbon using the carbon disulfide (CS_2) as a solvent. Chromatographic analysis was carried out using a gas chromatograph Varian 450GC equipped with a flame ionization detector (FID), a capillary column and a system of automatic feeding of samples (autosampler).

 DOI: 10.1051/e3sconf/20172200158

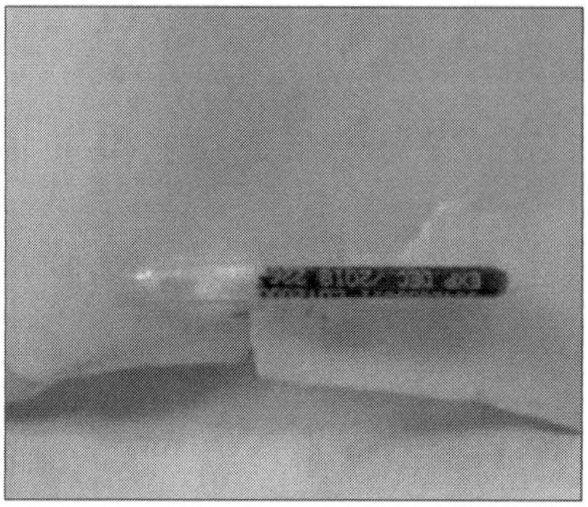

Fig. 3. The tube with active carbon

3 Results

As the results of measurements, VOCs' concentration and exhausts' temperature values before catalyst reactor was received for chosen points of engine's work (as load characteristic). The values received directly from the gas chromatography were obtained as concentrations of every identified VOCs in ppm (parts per million) in the used solvent (CS_2). It was necessary to calculate VOCs' concentrations in sampled fumes in mg/dm^3. Due to the fact that the most important (have been identified in the highest concentrations) and harmful for the human health compounds of VOCs group are benzene, toluene, ethylbenzene and xylenes (BTEX), these compounds were taken for the further consideration. In Table 1 all measurements and calculated results are presented.

Table 1. The concentration of analysed VOCs (in $\mu g/m^3$) and temperature values

Torque, Nm	T, ^{0}C	Benzene	Toluene	Ethylbenzen e	p-,m-,o-xylene	Sum of BTEX
0	374	7.25	22.18	4.82	10.53	44.78
10	399	5.21	15.35	3.66	8.01	32.22
20	453	5.99	18.96	4.70	10.59	40.23
30	482	1.14	4.18	0.69	3.01	9.01
40	507	1.17	4.19	0.63	2.78	8.77
50	533	1.21	4.36	0.73	3.02	9.32
60	561	0.31	1.03	0.12	0.55	2.01
70	577	1.17	3.95	0.55	2.37	8.03
80	621	0.50	1.35	0.15	0.65	2.64

As shown in Table 1, when the engine's load changes, the combustion temperature increases at constant speed (in this test 2500 rpm). Thus, exhaust fumes temperature raises as well. It is known that the combustion is closer to the complete one. Therefore, the hydrocarbon concentration decreases. The linear regression equations were used so as to evaluate the dependency of the VOCs and exhaust temperature. Statistical analyses were performed in Statistica StatSoft software. Every compound (benzene, toluene, ethylbenzene and p-, m-, o-xylene) was examined separately and the sum of BTEX was analyzed as well. While conducting the statistical analysis, the following parameters were taken into consideration:

- absolute value of the correlation coefficient, R,
- coefficient of determination, R^2, which shows how the independent variable (here: the temperature) explains the dependent variable (here: the concentration of VOCs),
- Beta, β, which is equal to the correlation value and shows the direction of the straight line,
- constant term, a.

Based on the above parameters the linear regression equations have been determined. They show the quantitative relationship between analysed variables, i.e. the temperature and VOCs' concentration. The results of all calculations are presented in Table 2. As it is visible in the table, the coefficient of determination in every case exceeds 0.73. It indicates a very good correlation between the analysed variables [13]. This means that on the basis of the exhaust's temperature, using the determined equations, the volatile organic compounds can be estimated with more than 70% of accuracy.

Table 2. Results of statistical analysis of linear regression

Type of the compound	**R^2**	**β**	**a**	**Equation**
Benzene	0,76	-0,8728	16,9543	C_B = 16,9543 – 0,0285 · Temperature
Toluene	0,77	-0,8753	51,6493	C_T = 51,6493 – 0,0864 · Temperature
Ethylbenzene	0,74	-0,8583	12,2160	C_E = 12,216 – 0,0208 · Temperature
p-,m-,o-xylene	0,77	-0,8768	26,0035	C_X = 26,0035 – 0,0427 · Temperature
Sum of BTEX	0,77	-0,8752	106,8230	C_{BTEX} = 106,823 – 0,1784 · Temperature

4 Summary and discussion

The research of the engine operation has shown that it is possible to estimate the concentration levels of the analysed volatile system. However, it should be borne in mind that three-way catalysts are standard equipment in modern vehicles powered by spark-ignition engines.

The analysis was carried out on one engine with fixed engine parameters (e.g. engine capacity, compression ratio) and performance parameters (e.g. torque, rotation speed, inlet airflow, etc.). The outcome of dependencies is true and can be only applied to this engine. In order to generalize the results of the research and to derive the universal (in terms of different types of spark ignition engines) dependencies, several additional aspects need to be considered in subsequent studies. One of the most important points in terms of the engine performance is the application of VOCs emission, despite their concentrations. It is very crucial with regard to the engine control and preparation of the air-fuel mixture. In spark-ignition engines without supercharging, when the engine load changes, the air flow in the intake manifold and the exhaust flow change as well.

In addition, the tests were conducted only in a certain operating range of the engine, at a constant rotational speed of 2500 rpm. Therefore, the obtained dependencies can be only applied for this speed. So as to estimate the VOC content at the full engine range, it is important to prepare and implement an experiment plan where various engine operating points (both load and rotation speed) are analysed. This study goes to show if it is possible to estimate VOCs content based only on exhaust temperatures, or whether an additional engine performance parameter such as rotation speed is required. However, it should be borne in mind that both of these parameters are much simpler to measure comparing to the measurement of volatile organic compounds.

Summing up, it is possible to obtain equations that describe the relationship between the exhaust temperature and the content of volatile organic compounds. The continuation of the research in this realm should focus on the following aspects:

- examination of the dependencies over the engine's full operating range (evaluate whether additional parameters are required to obtain the analysed relationship),
- analysis of VOCs emissions instead of their concentrations (additional measurement of air flow in the intake manifold) – the universality of dependencies,
- testing more engines so as to generalize dependencies,
- consideration of a three-way catalyst which oxidizes volatile organic compounds (hydrocarbons) and thereby changes their content at the tail end of the exhaust pipe,
- changing the place of experiment from the engine test stand (static testing) to performance testing of actual vehicles (e.g. on a chassis dynamometer).

References

1. Central Statistical Office, *Transport. Activity results in 2015* (Warsaw 2016)
2. J. Merkisz, J. Pielecha, *Emission of the particle matter from motorization* (Wydawnictwo Politechniki Poznańskiej, Poznań 2014)
3. J. Merkisz, *Ecological problems of internal combustion engines* (Wydawnictwo Politechniki Poznańskiej, Poznań 1998)
4. Z. Chłopek, *Environmental protection. Vehicles* (Wydawnictwa Komunikacji i Łączności, Warsaw 2002)
5. J. Merkisz, *The impact of motorization on environmental pollution* (Wydawnictwo Politechniki Poznańskiej, Poznań, 1993)
6. Regulation (EC) No. 715/2007 of the European Parliament and of the Council, *Type approval of motor vehicles with respect to emissions from light passenger and commercial vehicles (Euro 5 and Euro 6) and on access to vehicle repair and maintenance information* (Brussels 2007)
7. A. Janicka, *Toxicity assessment of the microatmosphere of the interior environment of a motor vehicle* (Oficyna Wydawnicza Politechniki Wrocławskiej, Wrocław 2013)
8. J. Wang, L. Jin, J. Gao, J. Shi, Y. Zhao, Sh. Liu, T. Jin, Zh. Bai, Ch. Wu, *Sci Total Environ*, **110-1,** 445–446 (2013)
9. K. Kim, J. Szulejko, H.J. Jo, M.H. Lee, Y.H. Kim, E. Kwon, Ch.J. Ma, P. Kumar, *Environ Pollut*, **215, 340-346** (2016)
10. A. Roy, Y. Choi, *Atmos Environ*, **152,** 240-245 (2017)
11. L. Michalski, K. Eckersdorf, J. Kucharski, J. McGhee, *Temperature Measurement* (John Wiley & Sons Ltd, 2002)
12. LabView website, *http://poland.ni.com/labview*
13. M. Korzyński, *Methodology of experiment* (Wydawnictwo WNT, Warsaw 2013)

Analysis of application of alternative drive systems for international heavy-duty transport on Wroclaw-Dresden-Prague routes

Maria Skrętowicz[1,*], *Zbigniew* Sroka[1]

[1]Wroclaw Unviesrity of Science and Technology, Faculty of Mechanical Engineering, Braci Gierymskich 164, 51-640 Wrocław

Abstract. The depletion of the fossil fuels resources, significant increase of the air pollution caused by the use of internal combustion engines, and emission of carbon dioxide which is responsible for the greenhouse effect escalates the development of vehicle's alternative drive systems. Generally, the emphasis is given to the alternative fuels (natural gas CNG, mixture of propane-butane gases LPG, hydrogen, alcohol fuels, biofuels) and hybrid or electric vehicles. Roads between large industrial and commercial centres, i.e. Wroclaw – Dresden – Prague, are used mainly by heavy-duty vehicles. Consequently, the contribution of the road transport to the ecological threat in this realm is significant. The objectives of this research were the assessment of the traffic volume and emission rate of exhaust gases caused by heavy-duty vehicles on the analysed roads and evaluation of the possibility of using existing and alternative drive systems in vehicles driving on the roads in the analysed region.

1 Introduction

Transport of goods is one of the most important aspects of both national and international economy. It enables the exchange of goods between economic centres. According to data from the Central Statistical Office of Poland, there are more than 3,000,000 trucks and almost 330,000 road tractors [1] registered in Poland. In the European Union the largest amount of goods in 2016 was transported in the countries such as Germany (especially the Nordrhein-Westfalen region) with 151trilion tkm, and subsequently the United Kingdom, Spain, France, Italy, Poland, Belgium, the Czech Republic, the Netherlands, Portugal, Finland and Sweden [2].

Apart from the economic and social benefits, the transport (especially the road one) entails serious risks to the environment such as: accidents (threat to the health and life of humans and animals, spills and evaporation of transported substances, fuel and liquid consumables), noise and vibration of surface, not to mention the exhaust fumes emissions, including greenhouse gases, into the atmosphere [3]. According to the report published by The National Centre for Emissions Management (KOBiZE), in 2014 the transport was responsible for 14.06% of the nitrogen dioxide emission into the atmosphere in Poland [4].

* Corresponding author: maria.skretowicz@pwr.edu.pl

Figure 1 shows the share of particular groups of motor vehicles in CO2 emission in Poland for 2014.

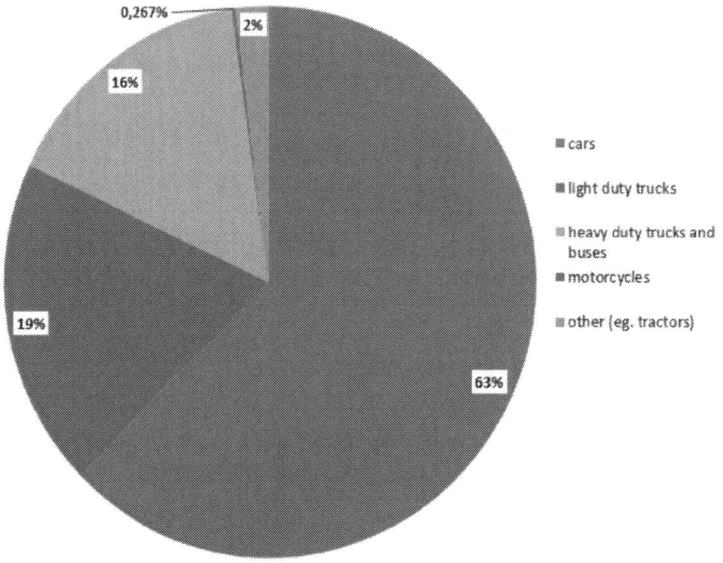

Figure 1 - The share of particular groups of motor vehicles in CO_2 emission in Poland in 2014. Source: The National Centre for Emissions Management (KOBiZE)

Nowadays, the aim is to ensure that the vehicles are equipped with engines that comply with EURO standards which were introduced by the European Union. They define the permitted amount of pollutants emitted into the atmosphere. However, the range of trucks and commercial vehicles in terms of the year of production is so large that one cannot clearly determine the concentrations of toxic compounds in the air at the designated area. The reduction of toxic exhaust emissions and carbon dioxide is carried out in various ways, starting from constructional design solutions (e.g. Variable Valve Timing) through the use of aftertreatment systems (e.g. three-way catalyst, selective catalytic reduction) and ending with alternative fuels and drives. Seeing that, some researches on the possibility of using fuel cell electric trucks across the commercial medium and heavy duty vehicles have been conducted [5]. The studies have shown that aforementioned implementations are possible for many types of vehicles without a loss of engine's performance (which is very important for long-distance transport). Moreover, in many cases these parameters are improved while introducing the analyzed alternative propulsion systems.

One of the solutions that allow the reduction of negative impact of motorization on the environment in a quick and effective way is the power supply of the vehicle with compressed natural gas (CNG). CNG fuel is easily accessible and can be used as the alternative fuel for both diesel and petrol [6,7]. In order to supply the combustion engine with natural gas, one only needs to equip the engine with a special installation. Many truck manufacturers already have such equipped vehicles in their offer (e.g. Fiat, Iveco, MAN, Scania and Mercedes). Needless to say that trucks and vans equipped with natural gas installation would contribute to the significant reduction of emission of harmful substances into the atmosphere. Additionally,
it would decrease the noise level, since CNG engines operate quieter than conventional ones.

The aim of this study is to analyse the possibility of using natural gas as the fuel for vehicles hauling goods on selected routes connecting the three major economic centres: Wroclaw, Dresden and Prague.

 DOI: 10.1051/e3sconf/20172200159

2 Characteristic of the transport and infrastructure in the analysed region

So as to conduct the research, the authors of this work chose routes that connect three large European cities that are located in the neighbourhood: Wroclaw (Poland), Dresden (Germany) and Prague (the Czech Republic).

The goods produced in these centres are received and delivered to the other city with the use of road transport. The market for services is also more developed than in smaller towns. As a result, the greatest concentration of vehicles (passenger cars, light duty vehicles, LCVs, as well as heavy duty vehicles and HDVs) is observed in the vicinity of large cities. It contributes to the higher concentration of air pollution in these areas.

As far as the analysed routes are concerned, many trucks transporting goods between different centres of trade can be observed. It should be noted that traffic intensity on these routes is various, depending on many factors such as: the time of day, day of week and direction of motion. Moreover, the working time requirements of professional drivers have an impact on the amount of moving vehicles.

The vehicles were counted in three measuring points on every selected road. They were toted up based on the following classes:

- Class I – delivery trucks with technically permissible maximum laden mass not greater than 3.5 tones,
- Class II – vehicles with technically permissible maximum laden mass greater than 3.5 tones but not exceeding 12 tones,
- Class III – trucks with technically permissible maximum laden mass exceeding 12 tones.

Within each class of vehicles, trucks were divided in terms of the producer and the type of fuel. It enabled the estimation of toxic emissions into the atmosphere. Thus, the level of air pollution in the area caused by vehicles could be assessed. During the study, petrol stations on the analysed routes and types of fuel which they offer have been listed. Figure 2 shows a map with marked routes that have been analysed. The A4 highway was a route from Wroclaw to Dresden. When it comes to the Dresden-Prague connection, A17 motorway on the German side and the road E55 in the Czech Republic were evaluated. Last but not least, the road section from Prague to Wroclaw was analysed based on the E65 road (in the direction of Jelenia Góra) and then a short stretch of the A4 motorway from "Kostomłoty" exit.

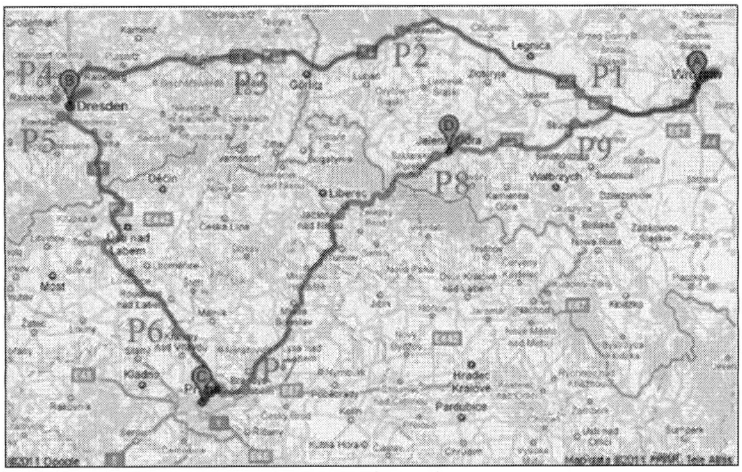

Figure 2 - The map of analysed routes with measuring points.

 DOI: 10.1051/e3sconf/20172200159

Measurements were carried out in two days (28th and 29th April 2010) between 9am and 2pm on the Wroclaw – Dresden road. In case of Dresden – Prague route analysis was conducted between 3:50pm-7:30pm and on Prague – Wroclaw stretch between 6:30 am and midnight. Figure 3 shows a quantitative summary of vehicles in all measuring points. At each point the vehicles were counted for 30 minutes in both directions of travelling.

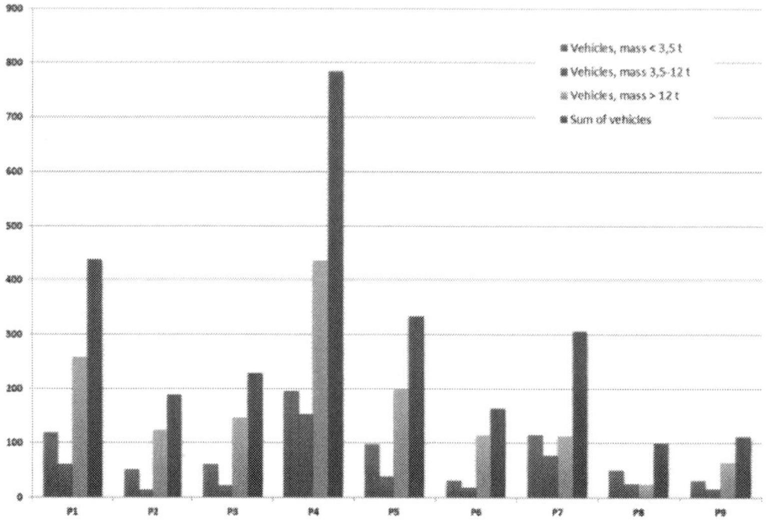

Figure 3 - The number of vehicles measured at the measurement points

As it is shown in Figure 3, the largest number of vehicles was measured in the vicinity of large cities (P1, P4, P5, P7 points). Moreover, the largest share of vehicles belongs to the heavy goods vehicles with a total gross vehicle mass (GVM) exceeding 12 tonnes and then to light duty vehicles designed to carry small loads.

Wroclaw and Dresden are directly connected with the A4 motorway on the Polish side and the BAB 4 motorway on the German side. This route is 270 km long, including 165km length on the Polish side. The whole stretch of the route is free of charge. Measurements on this distance were carried out on 3 points: 2 points on the Polish side (P1 and P2) and 1 point on the German side (P3). Conducting research at two points on the Polish side enabled the verification of the number of vehicles that cross Polish-German border on the route. Additionally, it determined the amount of automobiles that head to A18 motorway (point P1 was located in front of the exit of the A18 motorway and the point P2 just behind the exit - Figure 2).

As far as the point P1 is concerned, it was located at the Pietrzykowice. It is the first exit of the motorway on the Wroclaw - Jędrzychowice section, 43km from Wroclaw. The research was conducted between 9:00-9:30am. At that point, the heaviest traffic on that route has been recorded. There may be two reasons for this fact. Firstly, the test was performed in a relatively close distance from Wroclaw, which is a large economic centre. As it was already mentioned, heavy traffics are in the vicinity of important economic centres. Secondly, the time when the research was conducted is known as the peak traffic hours. When it comes to the point P2, it was located in the meadow, about 45 km from the state border, behind the junction of A18 motorway. In addition, the measurement at point P2 was performed between 11:20am and 11:50 am. The level of traffic at P2 was less than half at the point P1. It could be due to the fact that the part of the vehicles was heading to A18 motorway. Moreover, P2 is remote from major economic centre. On the other hand, point P3 was located in Germany, 15km from Bautzen. The measurement was performed between 1:00 – 1:30pm. At that point, the number of vehicles remained at a similar (slightly higher) level than the point P2, which is still significantly lower than at point P1. It is primary due to the remote distance from the large industrial centre, as it was in the P2 case.

On the basis of the Figure 4, one can conclude that the vast majority (more than 60%) of all types of vehicles on this route were trucks with GVW exceeding 12 tons. Almost 30% of automobiles were delivery trucks and the rest of the vehicles were those with GVW ranging from 3.5 till 12 tones. The researchers identified a total of 18 petrol stations on the road. The location of some stations requires getting off the freeway. The majority of petrol stations offer fuels such as diesel oil (ON), 95-octane and 98-octane gasoline. Some of them sale LPG and bio-additives fuel.

Another stretch of analysis was the Dresden-Prague route which is 148 km long. The first measurement point on that road (point P4) was still on the BAB 4, at the exit from Dresden and in the direction of Prague following the signs. At that point the measurement was performed between 4:00 – 4:30pm. Because of the rush hour and the fact that it was a point located in the nearby of the exit road form the big economic centre, which is Dresden, the measured traffic volume at that point was the greatest of all measuring points. Another point (P5) was located on BAB17. Similarly to the previous route, the greater the distance from the big economic centre, the lower the number of vehicles is around. It decreased by almost half comparing to the point P4. However, it was slightly lower than at the point P1 (not far from Wroclaw). Due to the inconveniences on the road (the reconstruction of road infrastructure and consequently traffic congestions) in the Czech Republic, the last measurement was taken at a point P6. It is located just before the drive-in entry to the capital of the Czech Republic, i.e. Prague. This measurement was conducted in the evening, between 7:00pm and 9:30pm. It could explain the small number of vehicles in that area, comparing to the previous points, despite the short distance from Prague.

On the basis of the results shown in Figure 4, one can detect that Dresden is a thriving economic centre which is burdened with a very high flow of vehicles that transport goods.

During the measurement conducted at the point P4 (in close proximity to Dresden), researchers counted more vehicles than at the points P1 (near Wroclaw) and P6 (near Prague) cumulatively. However, it should be emphasized that a significant impact on such a low score at the point P6 may be caused due to the evening time of measurement. The share of every group of vehicles on that stretch in question is likewise to the Wroclaw-Dresden route. The majority of counted vehicles (60%) accounts for those with total GVM exceeding 12 tonnes. On the Dresden - Prague way, 8 petrol stations have been identified. Surprising as it may seem, they offer diesel oil (ON), gasoline with different octane number, fuel with bioadditives. On the four petrol stations LPG was available.

The third stretch was the route connecting Prague with Wroclaw, with the length of 280 km. So as to leave Prague, one needs to go by the highway and then get off to the national road. This way on the Czech side leads through the hilly terrain with many bends. Due to this fact it is difficult to drive vehicles there, especially those with a weight exceeding 12 tonnes. As a consequence, there are often traffic congestions. The first measure at that section was taken just behind Prague (point P7) between 6:3am and 7:00am. A point P8 was located at the drive-in entry to Jelenia Gora. Vehicles at that point were being counted between 10:00am and 10:30am. The traffic there was small comparing to other points. The increase in traffic was recorded only at the point P9 in the vicinity of Strzegom (measured 11:30am-12:00 am). The increase in the number of vehicles could be explained by the fact that this area is abundant of granite mines. This type of raw materials is transported by trucks.

The last route is characterized by the lowest number of vehicles that are used for transport of goods (Figure 3). However, one should take into account that this route is unsuited for heavy duty vehicles with gross vehicle mass exceeding 3.5 tonnes. Divers of such vehicles usually chose other routes between Wroclaw and Prague, e.g. through Kłodzko or Liberec.

DOI: 10.1051/e3sconf/20172200159

As it is shown on Figure 4, the number of vehicles with GVM exceeding 12 tonnes is almost the same as delivery trucks. Contrary to the other measurement points, at the point P8 vehicles weighing up to 3.5 tonnes prevailed. There were 14 petrol stations located on this stretch with the possibility of fuelling diesel, gasoline, fuels with bioadditives and LPG.

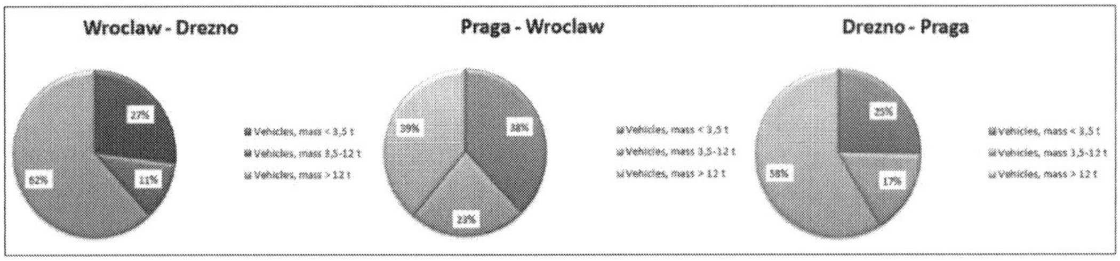

Figure 4 - The percentage share of groups of vehicles on the analysed routes

3 The technical infrastructure of CNG stations for heavy-duty vehicles

The natural gas as a fuel for motor vehicles is becoming more and more popular in the countries of the study. Germany is the country where CNG is the most popular and is willingly used by drivers. There are 913 petrol stations that offer this type of fuel (data for July 2016). In Poland there are only 26 publicly accessible stations, while in the Czech Republic there are 108 [8,9].

Natural gas, which is dedicated to fuel vehicles, is compressed to 200-230 bar. The efficiency of the station plays an important role in fuelling since it has a direct effect on the fuelling time. The basic CNG station is equipped with:

- compressor,
- gas dehumidifier,
- cascade storage,
- steering and control devices,
- distributors.

Self-service petrol stations provide the opportunity to fill the fuel tank at the time comparable to the time of diesel fuelling. Cascade storages, where the compressed CNG is stored, ensure fast fuelling [10]. Another alternative is slow fuel filling stations where there are no cascade storages. The gas is compressed directly by the compressor. This is a much cheaper solution. Nevertheless, the fuelling is much slower and may take up to 8 hours.

While selecting a petrol station, one should specify some basic parameters that are suitable to a specific user group. In case of delivery truck fuelling the best solution is the combined station. It enables fast fuel filling and the opportunity to refuel at night, for instance during a driver's downtime. The station location is also very important factor. It should be selected in shortest possible way so as not to lose any extra time. The location of the CNG station is conditional upon the access to the gas network.

There were several CNG stations near the analysed routes. They were as follows:

- Wroclaw – Dresden route: Wroclaw, Legnica, Zgorzelec, Bautzen, Dresden
- Dresden – Prague route: Heidenau, Usti nad Labem, Prague (5 petrol stations)
- Prague – Wroclaw route: Prague, Milovice, Mlada Boleslav, Semily.

All aforementioned stations are located in the city centres. It means that in order to fuel the vehicle, the driver has to get off the planned route. Moreover, it forces to drive in the city center.

4 Summary and conclusions

The study analysed the possibility of using natural gas as an alternative fuel for vehicles which moves on the Wroclaw - Dresden - Prague route. On every route 3 measurements points were specified. The traffic intensity of the trucks has been determined, as well as the number of petrol stations with the range of available fuels. The advantages of reduction of CO_2 emissions into the atmosphere, as a result of the use of CNG in combustion engines, have been also evaluated. Moreover, the stations with CNG fuelling facilities have been designated near the routes of investigation.

Based on the gasoline, diesel and natural gas properties, the amount of carbon dioxide produced by the combustion of the particular fuel has been calculated. Table 1 shows the main characteristics of fuels and the amount of CO_2 which was produced as a result of the combustion of 1 kg or 1 dm^3 of the single fuel.

Table 1 - Properties of selected fuels and the amount of CO_2 produced by their combustion [7]

Properties	Gasoline	Diesel	Natural gas
Carbon content in 1 kg of fuel [kg]	0.86	0.86	0.76000
Calorific value [MJ/kg]	43.00	43.00	47.70000
Density [kg/dm^3]	0.75	0.83	0.00071
Amount of CO_2 from the combustion of 1 kg of fuel [kg]	3.16	3.16	2.79000
Amount of CO_2 from the combustion of 1 dm^3 of fuel [kg]	2.37	2.62	0.00180

In order to estimate the CO_2 emissions on the analysed routes, length and estimated average fuel consumption for each group of tested vehicles have been assumed and it was: for vehicles < 3.5 t – 10.1 dm^3/100km (diesel) and 11,7 (gasoline); 3.5-12 t – 21.3 dm^3/100km (diesel); > 12 t – 37.8 dm^3/100km (diesel). In case of delivery vehicles with the GVM not greater 3.5t, it was assumed that 25% of this group of vehicles are equipped with spark-ignition engines, while the rest is supplied with the diesel. The research took into account the brands and types of vehicles. Based on the above data and counted vehicles, the fuel consumption on particular route and then CO_2 emission (Equation 1) have been estimated. Assuming that all of the counted vehicles would be powered by the natural gas, CO_2 emissions on each route would be cut by at least 20%

$$CO_2 = N_r \cdot G_r \cdot CO_{2,fuel} \tag{2}$$

where: CO_2 – the amount of CO_2 produced along the entire route
N_r – the number of vehicles on the road [pc.]
G_r – the amount of fuel consumed by one vehicle on the route [dm^3]
$CO_{2,fuel}$ – the amount of CO_2 produced by the combustion of 1dm^3 of fuel

Assuming that all of the counted vehicles would be powered by the natural gas, CO_2 emissions on each route would be cut by at least 20%. Figure 5 shows a comparison of the CO_2 amount produced during the combustion of conventional fuels and combustion of CNG for each route.

DOI: 10.1051/e3sconf/20172200159

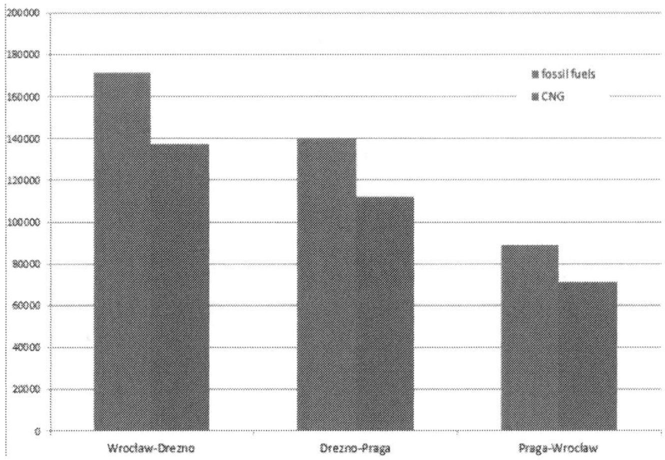

Figure 5 - The comparison of the CO2 amount produced during the combustion of petrol, diesel and natural gas

The analyses have shown that delivery vehicles have a significant environmental impact in the vicinity of the subjected routes. Supplying vehicles with CNG instead of conventional fuels would reduce CO_2 emissions and consequently other pollutants. However, it will be possible only if the appropriate infrastructure is provided. The requirements include quick-fill tank capabilities and close distance from transit routes. It is also crucial to ensure that a sufficient number of CNG stations are set so as to enable vehicles to cover the whole distance only with the natural gas (e.g. station location in every 100 km).

References

1. Central Statistical Office, *Transport. Activity results in 2015*, Warsaw, Poland, (2016); *Transport. Wyniki działalności w 2015 r.*
2. Eurostat regional yearbook – 2016 edition, Luxembourg (2016)
3. Z. Chłopek, *Protection of the natural environment*, Wydawnictwo Komunikacji i Łączności, Warsaw, Poland (2002); *Ochrona środowiska naturalnego*
4. The National Centre for Emissions Management (KOBiZE), *National Inventory Report 2016. Inventory of greenhouse gases in Poland for the years 1988-2014*, Warsaw, Poland (2016)
5. J. Kast, R. Vijayagopal Jr. J.J. Gangloff, J. Marcinkoski, INT J HYDROGEN ENERG, **42**, 4508-4517 (2017)
6. M. Kułażyński, Publication within the Project "The Development of the Potential and Academic Programmes of Wroclaw University of Technology", *Automotive Engineering. Green fuels*, Wroclaw University of Technology, Wroclaw, Poland (2011)
7. M.I. Khan, T. Yasmin, A. Shakoor, Ren. and Sus. En. Rev., **51**, pp. 785–797 (2015)
8. NGVA Europe, the Natural & bio Gas Vehicle Association, *Vehicle catalogue*, Brussel, Belgium (2016); http://www.ngva.eu/
9. https://cng.auto.pl/stacje-cng-w-polsce/ [access: 15.03.2017]
10. J. Sas, K. Kwaśniewski, *Natural gas for vehicles*, Uczelniane Wydawnictwa Naukowo-Dydaktyczne AGH, Cracow, Poland (2004)
11. K. Baczewski, T. Kałdoński, *The fuels for diesel engines*, Wydawnictwo Komunikacji i Łączności, Warsaw, Poland (2004); Paliwa do silników o zapłonie samoczynnym.

DOI: 10.1051/e3sconf/20172200160

Environmental aspects and renewable energy sources in the production of construction aggregate

Izabela Skrzypczak[1,*], *Wanda* Kokoszka[1] , *Lidia* Buda-Ożóg[1], *Janusz* Kogut[2], *Marta* Słowik[3]

[1]Rzeszow University of Technology, Faculty of Civil, Environmental Engineering and Architecture, Rzeszow, Poland
[2]Cracow University of Technology, Faculty of Civil Engineering, Krakow, Poland
[3]Lublin University of Technology, Faculty of Civil Engineering and Architecture, Lublin, Poland

Abstract. The main activity of open pit mining of aggregates are aggregates' exploitation of natural mineral deposits and its modification in order to obtain high-quality aggregates. The development of aggregate production is conditioned by a number of factors. The most important are: documented material resources, mining and manufacturing capabilities, the need of environmental protection (environmental aspects), the subordination of the plan of spatial development, formal and legal issues, as well as economic and financial aspects. While identifying and assessing the environmental impacts of manufacturing aggregates one may distinguish those environmental aspects that have or may have the greatest magnitude of the impact on the environment as a result of industrial activities. Manufacturers producing aggregates located in the areas covered by the special environmental protection require extra diligence in the conduct of mining activities for preservation of natural resources. The article discusses some main environmental aspects of the production of construction aggregates on the example of one of the largest producers of this material in Subcarpathian province of Poland. Environmental protection in production of aggregates may refer to four aspects: the use of natural resources, having excluded land from agriculture and forestry, land reclamation after exploitation, and use of energy from renewable energy sources. The economic and environmental impact of production volume of aggregates is evaluated by the index information capacity method and the method of graphs.

1 Introduction

As a result of the open pit mining activities, exploitation natural aggregates in building and road construction, several components the natural environment such as soil, terrain, water relations are violated. During the service and production processes a lot of energy is used, as well as the waste and pollutant emission to the environment is generated.

* Corresponding author: izas@prz.edu.pl

DOI: 10.1051/e3sconf/20172200160

Companies which produce aggregates distinguish those environmental aspects that may have the greatest magnitude of the impact on the environment as a result of their activities [1–5]. Manufacturers having produced aggregates located in the areas covered by the special environmental protection require extra diligence in the conduct of mining activities for the preservation of natural resources [4,5]. The article discusses some main environmental aspects of the production of construction aggregates on the example of one of the largest producers of this material in Subcarpathian province of Poland. The economic and environmental impact of production volume of aggregates is evaluated by the index information capacity method and the method of graphs.

2 Environmental issues in the production of construction aggregate

The most important restriction in access to resources of natural aggregates is the protection of living nature and a landscape. The strictest form of protection are national parks and wildlife sanctuaries. Protecting wildlife habitats of wild animals and birds is a priority, hence the exploitation of minerals in these areas is impossible, while in landscape parks there are limitations of mining activities. One of the most restrictive forms related to the investment expansion, is an European network of protected areas Natura 2000, which aims to preserve biodiversity by protecting ecosystems. The largest wildlife refuge – the Bieszczady mountains cover about 111,519.5 ha, and appears here as both the area of protection of birds and habitats.

Some restrictions to conducting mining activities, connecting with Natura 2000, may prevent access to a significant amount of mineral resources, and in the case of exploited deposits a need of abandoning exploitation. Every area of the Natura 2000 regardless of whether it has been approved or is only planned is protected by law. The rules prohibit, until approved of the area by the European Commission from carrying out any activity that may aggravate the condition of natural habitats included in such area [7].

Development of the industry of natural aggregates is limited by increasing the size of protected areas. This fact prevents and/or restricts access to the production of aggregates and the environmental factor rises to the primary factor. Natura 2000 law has covered large area, which is a natural resource base of raw materials suitable for the production of aggregates, so the environmental issues significantly reduce the amount of deposits for possible production of aggregates.

2.1 Environmental objectives and targets on the example of the Subcarpathian manufacturer of aggregates

The main environmental issues are discussed on the example of the company, which introduced of classified-aggregates in south-eastern Poland based on natural aggregate. Currently, the basic activity of the company is the mining of natural raw materials, later modification and manufacturing of construction aggregates, according to the European standards. In 2015, this company produced 5.1 million tons of aggregates, which have been used in construction and road building, which is about 5% of national production in Poland [7]. The primary activity of the company is determined by the environmental issues, which have or may have the greatest magnitude of impact on the environment as a result of production. The influence of particular parameter on the environment is performed by the analysis of environmental performance factors. Table 1 presents each factor in relation to the production volume for the years 2013–15.

Table 1. Environmental performance factors [8]

Factor	Type	2013	2014	2015
Waste	Hazardous [t]	13.191	12.864	16.793
	Non-hazardous [t]	370.983	116.657	81.087
Energy efficiency	Electric energy consumption [MWh]	11776	11308	12704
Material efficiency	Fuel consumption [m^3]	2035.5	2153.6	2244
Biodiversity	Industrial areas [ha]	220.9	216.8	205.1
Air pollution emissions	(SO_2, NO_x, PM) [t]	129.38	161.60	143.82

It should be emphasized that the volume of waste generated in all organizational units of the company will not exceed the amount of permitted by decisions of the organs of state administration. Referring to Table 1, one may concluded that the change of production technology contributed to a reduction of electricity and fuel consumption. Introduction of high-performance excavator and crushers, which have replaced reduction in consumption of electricity in 2014. The rise in unit consumption of electricity per ton of production in 2015, is the result of increased demand for grits, which is associated with work of energy-intensive crushers.

Until 2013, reduction in unit fuel consumption is recorded. This is due to the use in the extraction of aggregates excavators with increased efficiency. In addition, changing of technology works is continued through the use of open-pit dump truck, replacing the bulldozers. As a result of these actions the fuel consumption per 1 ton of aggregates production is reduced in relation to 2013 in all company units. It reached (calculated in thousands. liters of fuel per ton of aggregate) 0.45 in 2010, 0.43 in 2011 and 0.42 in 2012 [3]. The increase in fuel consumption per ton of production in 2014-15 is related to the more difficult deposits (deeper located) and the extension of transport routes between mines and processing units. The consequence of higher fuel consumption is an increase in the unit rate of emissions of pollution to the air.

It should be stressed, that the company using natural resources, pays maintenance fees to the National Fund for Environmental Protection and Water Management. Also, local authorities get the fees for using the environment (including emissions, dust and gas, and water consumption fees). In 2015, the total funds granted for the environmental protection amounted to 2.88 million PLN (including 97.1% service charge due to the minerals extraction and 2.9% fee for emissions of pollutants and water consumption). In 2015 unit fee for 1 ton of aggregate production decreased due to the reduction of operational charges.

In 1997-2001, in the frame of reclamation of gravel pits, the company constructed the water reservoir having dammed the Wisłoka river with the weir. The capacity of the reservoir is about 6 million cubic meters of water. This reservoir is currently in use also for leisure purposes, and enables the production of renewable energy in the hydroelectric power plant. However, despite this investment the company did not apply to the aggregate production other renewable energy sources.

2.2 Economic and environmental factors and the production volume

The economic and environmental impact of production volume of aggregates is evaluated by the index information capacity method and the method of graphs.

Index information capacity method is based on the basic mathematical formulas with combinatorics and correlation coefficients of the analysed features. In contrast, the method of graphs is based on the correlation matrix, shows the relationship of selected parameters. The following factors are used for the analysis (see Table 2):

Y	– aggregates production	(10^6 t),
X1	– hazardous waste	(Mg),
X2	– other waste	(Mg),
X3	– electric energy consumption	(MWh),
X4	– environmental fees	(PLN/t),
X5	– index of production efficiency	(m^2/t).

Table 2. Production of construction aggregates factors in subsequent years [8]

Year	Y	X1	X2	X3	X4	X5
2006	4.696	13885	151.100	10485.5	0.550	0.460
2007	4.800	14456	160.300	10635.2	0.560	0.450
2008	5.349	15680	190.500	12335.8	0.530	0.450
2009	5.168	16776	193.700	11607.5	0.490	0.440
2010	6.972	13359	193.700	11968.8	0.520	0.400
2011	5.443	16270	235.240	13177.4	0.489	0.388
2012	5.170	13992	179.983	11264.7	0.518	0.454
2013	4.915	13191	170.983	11776.0	0.523	0.449
2014	5.031	12864	116.657	11308.0	0.578	0.431
2015	5.099	16793	81.087	12704.0	0.556	0.402

The method of graphs analysis [9] – is a graphical method of expressing the relationship between the explanatory variables. It involves searching for a group of variables correlated with each other and to find the variables that are not correlated with other explanatory variables. The next step is to select from this group, variables which are strongly correlated with the response variable and then use them in the model. The model also includes all uncorrelated variables but related to the dependent variable. The correlation matrix for the data from the Table 2 is shown in Table 3.

Table 3. The correlation matrix

	Y	X1	X2	X3	X4	X5
Y	1	-0.099	0.368	**0.385**	-0.339	**-0.589**
X1	-0.099	1	0.104	**0.566**	**-0.428**	-0.340
X2	0.368	0.104	1	0.223	**-0.822**	-0.079
X3	0.385	**0.566**	0.223	1	**-0.457**	**-0.765**
X4	-0.339	**-0.428**	**-0.822**	**-0.457**	1	0.244
X5	-0.589	-0.340	-0.079	**-0.765**	0.244	1

Subsequent variable that is hazardous waste (X1), non-hazardous waste (X2), consumption of electric energy. (X3) and environmental fee (X4) show a weak correlation with the variable Y (the production of aggregates). Based on the correlation matrix it may be concluded that the production of aggregates is influenced primarily by the index of production efficiency (significant correlative relationship) and energy consumption (moderate correlation).

In order to determine which of the variables has a decisive influence on the production volume of aggregates inference based on the method of graphs has been conducted (Figure 1). Schematic relationship between variables is developed. The relationship is based on the correlation matrix taking into account these variables, for which the value of the correlation coefficient is greater than 0.40 (i.e. the average correlation). Determinants are shown in red in Figure 1.

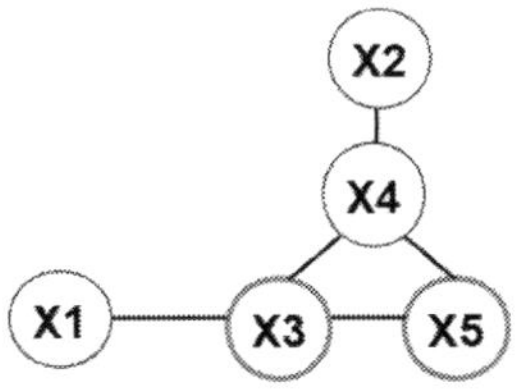

Fig.1. Graphs method

Based on the analysis of graphs it may be observed that the revised explanatory variables can be merged into two groups. With the first group, the main characteristics that influence the volume of aggregates production is a factor of efficiency of the production (variable X5). It is strongly correlated with the endogenous variable (the volume of aggregate production) than the variable X3 (energy consumption). While the variable X1 (hazardous waste), as virtually insulated top, also affects the parameter determining the volume of aggregates production.

Index information capacity method [10-13] consists in determining the indicators of individual and integral to the individual carriers. The basic problem of the method application is to calculate all the possible 1,2, ..., k-1, k - elements combinations of potential explanatory variables of the so-called information capacity of individual indicators. The total number of combinations that may enter the model for the analyzed case is equal to:

$$K = 2^m - 1 \quad (1)$$

where m is the number of exogenous variables, m = 5, so the number of combinations is equal to 31.

Subsequently, the capacity of each individual media presented in the specified combinations according to the formula (2):

$$h_{ji} = \frac{r_i^2}{1 + \sum_{i=1}^{n} |r_{ij}|} \quad (2)$$

where: r_i – the correlation coefficient parameter i with respect to the characteristics of the explanatory y

r_{ij} – the correlation coefficient of explanatory variables i and j

n – the number of correlation coefficients

The next step is the determination of the maximum integral capacity according to the following formula (3):

$$H_j = \sum_{j,i=1}^{n} h_{ji} \quad (3)$$

which corresponds to a combination of C ={**X1, X5**} = 0.798.

This means that the variables X1 (hazardous waste), and X5 (production efficiency), should enter into the assessment model of economic and environmental factors of the aggregates production.

Another value obtained is integral capacity determined for the variables **X1, X3, X5**, C ={**X1, X3, X5**} = 0.764, i.e. the value obtained for hazardous waste energy consumption and production efficiency.

Inference carried out, while capacity indicators information have been used confirmed the results obtained by the graphs analysis. A significant correlation of mineral production with the mining efficiency ratio is generally understood. However, the environmental

 DOI: 10.1051/e3sconf/20172200160

restrictions limit access to the deposits. One of the most restrictive forms of nature protection is European network of protected areas Natura 2000, which aims to preserve biodiversity by protecting entire ecosystems. Nearly 50% of area of the Subcarpathian province is environmentaly protected, and it is planned to increase the number of potential sites for habitat protection [3,6]. Theoretically, Natura 2000 does not oblige to stop the running production on restricted area, but in the case of exploited deposits, the need to refrain from exploitation can appear [1].

The analysis shows that not only the effectiveness of extraction, and the proper management (balancing area of land incorporated into the operation of aggregate with the surface of reclaimed land), but also the energy consumption and the volume of waste are significant factors conditioning the development of the open-pit mining industry.

In such circumstances it would be advisable to use in the production of construction aggregates energy from renewable sources.

3 Recommendations by standards and specifications in the use of aggregates

There are many documents and standards which define requirements of the use of aggregates in Poland. The most important are the technical requirements (WT4) published by the General Director for National Roads and Motorways (GDDKiA) [14], relating to the requirements of the materials used in road construction. These requirements systematise methods of production and use aggregates and asphalt mixtures (MMA), but some of the provisions included therein are ambiguous and raises serious concerns. For example, there is unclear procedure determining the parameter of sensitivity of MMA to water according to the standard PN-EN 1097-6 [15]. Unclear are also some provisions contained in WT-4 [14]. This is an important document ordering issues of mechanically stabilized substructures. However, the specifications written on this basis impose, for example, applying a mineral particle with sand equivalent index at least 40% or 45%, even for the roads having the lowest traffic category. Polish Association of Aggregate Producers (PZPK) repeatedly requested to change this condition, having stressed that overestimated requirement restricted business competition [16]. Another issue is unnecessary limitation of the maximum particle size of the aggregates. Commonly used MMA allows the use of aggregate with particle size up to 16 mm. In this way, one is not able to design a thicker layer, in which aggregate with particle size to 22 mm could be applied. This would rationalize the use of mineral resources, but the lack of adequate equipment for MMA testing laboratories.

In this situation, meeting the manufacturers production and expectations of customers deriving from the technical requirements and specifications become impossible.

As the result, there is a significant increase in production costs and the burden on the environment due to the need for additional storage of materials. It also affects the irrational use of non-renewable resources [16].

Referring to the requirements aggregates used to the concrete production, the specifications should be developed on the basis of currently recommended standard PN-EN 206-1 [17]. It should be noted, that PN-EN 206-1 standard does not explain the water absorption requirements neither for aggregates nor for concrete. Using aggregate with high water absorption, it is necessary to calculate the quantity of effective water in the concrete mix, which means that one should be careful in the use of aggregates with water absorption greater than 3%. The code does not specify the requirements for the mechanical properties of aggregates, as well.

In the case of specifications relating to non-cohesive soils, recommended to the construction of railway embankments and bridgeheads, the specifications define the basic requirements and related tests. This includes a strength parameters (internal friction angle), soil permeability, coefficient of uniformity, density index, organic content, pH index, and water content. These specifications recommend the implementation of backfill aggregate (e.g. gravely sands) with a specific grain size distribution, and some of them also provide the re-incorporation of soil from the excavation, conditioned by fulfillment of the above specification requirements and approval of Contract Engineer. As a rule, indigenous soil is not appropriate, but by addition of the particles of aggregate in the right proportion is able to meet the specifications. This is confirmed by laboratory tests carried out by the Department of Geodesy and Geotechnics by Rzeszow University of Technology in frame of the development of mixtures of indigenous soils with gravely sands and applications as backfill. This solution (mixing indigenous soil with aggregate) is considered effective in terms of both economic and environmental reasons. Both the contractor, saving on the cost of the purchase and transport of aggregate and masses of soil from the excavation, and the environment, gain a lot.

Specifications for aggregates used in bridge and railway construction are often too complex and required parameters overestimated due to the safety reasons. In developing of the requirements for aggregates used both for building construction and linear objects one should be guided primarily by rational use of non-renewable resources, rather than creating additional inconsistent requirements, which do not affect the safety and durability of engineering structures. Such an approach, however, requires full and continuous cooperation of the investor, designer and contractor.

4 Conclusions

One of the most restrictive forms of nature protection is the European network of protected areas by Natura 2000, which objective is to preserve biodiversity by protecting entire ecosystems. This plays a special role in the Subcarpathian region where, as it has been mentioned in the work, the areas covered by such protection constitute almost half of the province. When analyzing the environmental issues discussed on the example of the largest aggregate manufacturer in the Subcarpathian province, one may formulate the following conclusions:

•minimizing the influence of the environmental impact of the production of construction aggregates may be achieved primarily by implementation of modern technology and modernization of the existing mining equipment, which provide for more rational use of natural resources, while reducing significantly the emissions of pollutants into the environment.

• increase of the protected areas prevents or even restricts access to natural resources (by reducing production of aggregates), so the environment protection rises to the primary factor limiting the development of the natural aggregates production.

• in developing the requirements for aggregates used both for building of various types of construction and linear objects one should be guided primarily by rational use of non-renewable resources, rather than creation of additional requirements, which do not affect the safety and durability of engineering structures.

• based upon the analysis which had been carried out by the index information capacity method it may be concluded that the production volume depends on hazardous waste, the efficiency ratio of production and electric energy consumption.

• it is desirable then to use in the production of construction aggregate energy from renewable sources.

• noticeable decreasing trend of hazardous waste is only apparent, because by outsourcing of services outside the company, the waste associated with contract execution, by law, belong to the contractor.

DOI: 10.1051/e3sconf/20172200160

• rational requirements and recommendations specified by standards and specifications for the use of aggregates, would guarantee the reasonable use of non-renewable resources.

On the basis of the analysis performed in the work using the Helwig and Bartosiewicz method, it can be concluded that renewable energy sources can be successfully applied at the production of aggregates. For example, small hydroelectric power stations can be effectively exploited at the manufacturing process of aggregates to decrease the costs of energy.

References

1. G. Bajorek, M. Kiernia-Hnat, I. Skrzypczak, Works of Institute of Ceramics and Building Materials, 5, 9 (2012)
2. B. Berge, The Ecology of Building, Materials, Oxford Architectural Press (2001)
3. West T.R., Kyuho Cho, IAEG, 692 (2006)
4. S. W. Danielsen, E. Kuznetsova, Engineering Geology for Society and Territory, 44, 5 (2014)
5. R. Poulin, R. C. Pakalnis, K. Sinding, Environmental Geology, 23, 3, (1994)
6. http://www.stat.gov.pl/cps/rde/xbcr/gus/PUBL_rs_rocznik_stat_wojew.pdf (2015)
7. First Forum of Aggregates Producers. Warsaw (in Polish) (2009)
8. Kruszgeo: Environmental Declaration, Rzeszow (in Polish) (2006-2015)
9. J. Dziechciarz, *Econometrics*, Publisher University of Economics in Wroclaw (in Polish) (2002)
10. Z. Hellwig, in Z. Gostkowski (ed.), *Toward a system of quantitative indicators of components of human resources development*, UNESCO (1968)
11. P. Kowalik, 2, Publisher of University of Business, Kielce (in Polish) (2011)
12. P. Kowalik, Monographs of Lodz University of Technology, Publisher of Lodz University of Technology (in Polish) (2014)
13. F.S. Hillier, G.J. Lieberman, *Introduction to Operations Research*, 4th Edition, McGraw Hill, New York (1986)
14. Technical Guidelines 4 WT-4 (in Polish)
15. PN-EN 1097-6:2002/AC:2004 Research of mechanical and physical properties of aggregates. Part 6: Determination of grain density and absorption
16. http://inframedia.pl/article_advances/category/kruszywa/1/kruszywa_podaz_a_popyt/1
17. PN-EN 206-1: 2014 Concrete. Part 1: Requirements, properties, production and conformity

The influence of glycerin on nitrogen removal in wastewater treatment with activated sludge

Joanna Smyk[1], *Katarzyna* Ignatowicz[1]

[1] Bialystok University of Technology, Faculty of Civil and Environmental Engineering, Department of Technology in Engineering and Environmental Protection, Wiejska 45E, 15-351 Białystok

Abstract. The paper presents the efficiency of nitrogen removal from wastewater using glycerin as an external carbon source. The study was conducted during the municipal wastewater treatment process in two independent activated sludge chambers of SBR type. One of the chambers contained glycerol as the source of easily available organic compounds. The study showed that the use of glycerol as the external carbon source resulted in a higher efficiency of nitrogen removal in compare with the reactor without external carbon source.

1. Introduction

These sources are divided into internal (present in wastewater), endogenous (produced in the activated sludge chambers due to biomass decomposition), and external (not present in the wastewater)[1]. Internal sources refer to the organic carbon substrates obtained both in the inlet wastewater (as the organic load of entering wastewater) or generated and stored in cells, also known as an endogenous carbon source[2].

Delivery of external carbon sources to treated wastewater often becomes necessary to achieve high efficiency of sewage treatment, which must meet very stringent requirements for reducing the concentration of nitrogen. One of the initial stages of the activities related to the possibility of using the alternative sources carbon for the denitrification process is a review of available waste products and semi-finished products from different branches of the food industry in terms of the high COD/N ratio and high content of easily degradable organic compounds. Mainly the post-production wastewater from distillery, brewery, fishing industry, as well as waste and semi-finished products such as corn syrup, glucose, molasses, beet pulp, raw spirit or fusel oil, are taken into account [1, 3, 4, 5,6].

Another waste product, which has a great potential as an external carbon source is the glycerol fraction, which is a byproduct of biodiesel production. There are numerous reports of the glycerol use as a carbon source suitable for the wastewater treatment process[7, 8, 9, 10],as well as on the use of glycerol to reduce the biomass production [11]. Nowadays, in many countries, diesel oil is supplemented with biodiesel causing a gradual increase of its production, and thereby the amount of waste glycerol fraction [12, 13].At about 10 liters of produced biodiesel, about 1 liter glycerol is generated [14, 15].Glycerol phase, besides glycerol (propane-1,2,3-triol) - 50-60%, contains other substances in its composition such as methanol, mono-, diacylglycerols, free fatty acids and soaps [16].Due to the fact that glycerol originated from biodiesel production has many impurities, therefore it has led to a concomitant decrease in the price of glycerol along with an increase in its production. Consequently, glycerol has become a waste, which should be neutralized, and price for this product is more attractive as compared to other substrates used as external carbon sources in the denitrification and phosphorus removal processes[17, 18, 19,20,21].

The study aimed at demonstrating that the use of glycerol as an external carbon source in wastewater treatment process has a positive impact on the effectiveness of the removal of nitrogen forms from wastewater and can replace other alternative carbon sources.

2. Methods

The study was conducted during the municipal wastewater treatment process in two independent SBR-activated sludge reactors. The active capacity was 10 dm^3, including 6.5$dm^3$was the activated sludge provided by the sewage treatment plant in Białystok, while the remaining quantity originated from the raw mechanically treated wastewater (3.5 dm^3) that were also obtained from the sewage treatment plant in Białystok.

The single cycle of the reactor lasted for 6 hours and included following phases: sewage supply (2 min), mixing (anaerobic) (60 min), aeration (3.5hrs), sedimentation (1h), and decantation (0.5h). During the aeration phase, the compressed air was fed through the diffuser placed at the bottom of the rector; depending on the operating phase, the amount of air was from 0.1 to 3.0 mg O_2 /dm^3, concentration of activated sludge 3.5 kg/m^3, sludge index oscillated within 120-150 cm^3/g, hydraulic load of the chamber was 1.4 m^3/m^3d, whereas the pollution load 0.2 kg COD/m^3d. Glycerin, as a source of easily available organic compounds, in an amount of 100 mg/dm^3 wastewater was added into one of the chambers in each cycle, twenty minutes after the sewage pouring.

The collected sewage samples were filtered immediately after the filtration. Each filtrate was subject to determination of the following items in accordance with applicable methodology [2,4,5,3,22]:

- COD_{Cr} - dichromate PN-74/C-04578.03,
- BOD_5–manometric applyingOxiTop Standard system,
- N-NH_4–spectrophotometryaccording to PN-ISO 7150-1:2002,
- N-NO_3–spectrophotometryaccording to PN-82/C-04576/08,
- Ntot. –spectrophotometryaccording to PN-EN ISO 6878:2006,
- Ptot. –spectrophotometryaccording to PN-EN ISO 6878:2006.

3. Results and Discussion

The results of the R1 reactor wastewater tests without the addition of an external carbon source were included in Table 1, while Table 2 shows the results of wastewater tests from R2 reactor with addition of glycerin as external carbon source. Figure 1 shows the concentration of studied nitrogen forms after individual stages of wastewater treatment.

DOI: 10.1051/e3sconf/20172200161

Table 1. Data and test results of wastewater without addition of glycerin as an external carbon source

REACTOR R1 –without carbon source							
	1	2	3	4	5	6	7
	Raw wastewater	Poured wastewater(20 min) **after 20 min**	Denitrification (20 min) **after 40 min**	Denitrification(20 min) **after 60 min**	Nitrification(1.5 h) **after2,5 h**	Nitrification (2 h) **after4,5 h**	Decantation (0.5h) **after6 h**
COD[mgO_2/dm^3]	530	268	262	259	254	243	35
BOD_5[mgO_2/dm^3]	230	25	20	20	15	8	6
Total nitrogen [mgN/dm^3]	118.0	84.0	68.7	55.2	48.1	24.3	14.4
Nitrate NO_3^-[mgN/dm^3]	2.6	3.9	2.4	3.6	4.6	8.6	8.6
Ammonia NH_4^+[mgN/dm^3]	51.1	58.2	55.3	41.6	33.5	9.7	1.8
Phosphate PO_4 [mgP/dm^3]	18.5	15.2	18	12.8	1.5	1.1	0.8

Table 2. Data and test results of wastewater with addition of glycerin as an external carbon source

REACTOR R2–glycerin							
			Dosage of carbon source				
	1	2	3	4	5	6	7
	Raw wastewater	Poured wastewater (20 min) **after 20 min**	Denitrification (20 min) **after 40 min**	Denitrification (20 min) **after 60 min**	Nitrification (1.5 h) **after2,5 h**	Nitrification (2 h) **after4,5 h**	Decantation (0.5h) **after6 h**
COD[mgO_2/dm^3]	530	262	322	331	279	234	28
BOD_5[mgO_2/dm^3]	230	25	50	45	20	8	2
Total nitrogen [mgN/dm^3]	118.0	84.1	72	69	61.2	16.5	7.6
Nitrate NO_3^-[mgN/dm^3]	2.6	4.8	2.2	2.6	6.6	9.6	5.6
Ammonia NH_4^+[mgN/dm^3]	51.1	42.2	49	45.2	24.4	4.3	0.6
Phosphate PO_4 [mgP/dm^3]	18.5	14.9	16.9	14.9	1.2	0.8	0.5

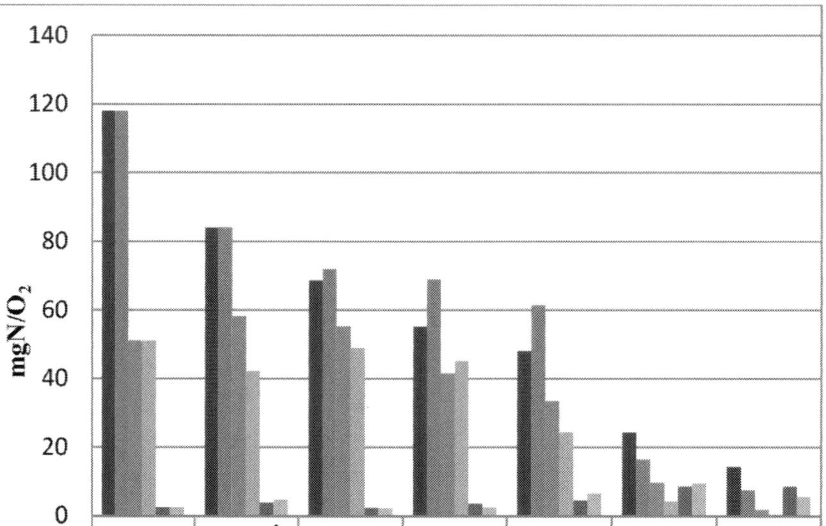

	raw wastewater	poured wastewater (20 min)	denitrification (20 min)	denitrification (20 min)	nitrification (1,5 h)	nitrification (2h)	decantaction (0,5h)
Total nitrogen - R1	118	84	68,7	55,2	48,1	24,3	14,4
Total nitrogen - R2	118	84,1	72	69	61,2	16,5	7,6
Ammonia - R1	51,1	58,2	55,3	41,6	33,5	9,7	1,8
Ammonia - R2	51,1	42,2	49	45,2	24,4	4,3	0,6
Nitrate - R1	2,6	3,9	2,4	3,6	4,6	8,6	8,6
Nitrate - R2	2,6	4,8	2,2	2,6	6,6	9,6	5,6

Fig. 1. Concentrations of tested nitrogen forms after individual phases of wastewater treatment

Studies performed byBernatet al. [12] upon the use of glycerin as an external carbon source in the process of nitrogen removal in SBR-type chambers revealed that the optimum COD/N ratio should amount to 3.0, at which both nitrates removal and biomass production from SBR chambersis the lowest. The COD/N ratio in tested raw wastewater was 4.49, while just before glycerin addition, this COD/N decreased to 3.11, thus close to recommended by above authors.

Value of COD in raw wastewater amounted to 530 mgO_2/dm^3, while BOD_5 230 mgO_2/dm^3. Concentration of the total nitrogen in raw sewage was118mgN/dm^3, ammonia 51.5 $mgN/dm^3$and nitrate2.6mgN/dm^3.

Twenty minutes after the reactors filling, some decrease in COD and BOD_5values were observed –available carbon source were taken by denitrification bacteria. Value of COD in reactor R1 was 268.0mgO_2/dm^3,while value of BOD_5 25.0mgO_2/dm^3. In the reactor R2, similar levels were recorded: COD262.0mgO_2/dm^3, BOD_5 25 mgO_2/dm^3. In both reactors, ammonia concentrations were similar–in reactorR1 84.0mgN/dm^3,while in reactor R284.1mgN/dm^3. Content of nitrates in both reactors increased up to 3.9 mgN/dm^3in R1 and to4.8mgN/dm^3in R2. The increase in the ammonia concentration was also recorded in R1 to 58.2 mgN/dm^3, whereas decrease in R2 amounted to42.2mgN/dm^3.

Then glycerin as the external carbon source was added into the reactor R2 and after 20 minutes of denitrification process, subsequent samples were collected. In R1 reactor, where the wastewater treatment process proceeded without support of the external carbon source, a further slight decrease in COD (262mgO_2/dm^3) and BOD_5 (20 mgO_2/dm^3) was noted. Introducing the carbon source in a form of glycerin caused the increase in COD value in reactor R2 by60.0mgO$_2$/dm^{3}up to322.0mgO$_2$/dm^3. The BOD$_5$value in reactor R2 increased to 50 mgO$_2$/dm^3. Some decrease in the total nitrogen concentration in both reactors was recorded: in R1 68.7 mgN/dm^3, while in R2 72.0mgN/dm^3. In the case of nitrate, there were also lower concentrations as compared to the previous control point. In reactor R1, nitrate content was 2.4mgN/dm^3, in reactorR2 2.4 mgN/dm^3.Concentration of ammonia slightly decreased in R1 to 55.3mgN/dm^3, whereas in reactor R2 the adverse situation was recorded–there was an increase in ammonia up to49.0 mgN/dm^3, however the concentration was still lower than in reactor R1.

After subsequent 20 minutes of anaerobic process of wastewater treatment, further slight decrease in COD in reactor R1 was found (259 mgO$_2$/dm^3);Value of BOD$_5$remained intact (20 mgO$_2$/dm^3). In the reactor with glycerin addition, an opposite situation was observed–there was further COD increase by 9 mgO$_2$/dm^{3}to the level of322mgO$_2$/dm^3.The BOD$_5$value decreased to 45 mgO$_2$/dm^3, but these results were higher than in reactor without carbon source addition. Concentration of the total nitrogen still decreased reaching55.2 mgN/dm^{3}in reactor R1. For R2 reactor with addition of glycerin, the total nitrogen concentration was higher than in reactor R1 and amounted to69.0mgN/dm^{3}giving the difference of 13.8mgN/dm^{3}total nitrogen between both reactors. In both reactors, an increase in nitrates contents to3.6 mgN/dm^{3}was recorded in R1 and2.6mgN/dm^{3}in R2. The ammonia level decreased in R1 to 41.6 mgN/dm^3, while in R2 to 45,2 mgN/dm^3. Difference of the ammonia concentrations at that control point between reactor without carbon source R1 and with glycerin R2 amounted to3.6mgN/dm^3.

Another samples were collected after 1.5hours of wastewater aeration. Value of COD in reactor R1 was 254 mgO$_2$/dm^3, while BOD$_5$ 15 mgO$_2$/dm^3. In reactor R2, despite of decreased COD value by 52 mgO$_2$/dm^3 to the level of 279mgO$_2$/dm^{3}and value of BOD$_5$by 15 mgO$_2$/dm^3 to 20mgO$_2$/dm^3, these numbers were higher than in the reactor without glycerin addition. Also total nitrogen concentration was lower in reactor without carbon source amounting in reactor R1 to 48.1 mgN/dm^3. In the case of reactor R2, concentration of the total nitrogen was61.2mgN/dm^3.Difference between total nitrogen levels between reactors R1 and R2 was 13.1mgN/dm^3. Concentration of nitrates in R1 increased up to 4.6 mgN/dm^3, which resulted from the inhibition of denitrification process due to the supply of oxygen into reactors. In reactor R2,also increase in nitrates concentration was recorded, but these were higher values–even as high as6.6mgN/dm^3. Content of ammonia decreased in reactor R1 to 33.5 mgN/dm^3, whereas in R2 to 24.4 mgN/dm^3. Difference in the ammonia concentrations at that control point between reactor without carbon source (R1) and that with glycerin (R2) amounted to 9.1mgN/dm^3.

Subsequent samples were collected after another 2hoursof wastewater aeration. Value of COD in reactor R1 was 243.0mgO$_2$/dm^3. In reactor with glycerin, there was further intake of available organic compounds, which resulted in lower COD values (234.0 mgO$_2$/dm^3)than in reactor without carbon source addition. In the case of BOD$_5$in both reactors, the same values were recorded– 8.0mgO$_2$/dm^3. Concentration of the total nitrogen in reactor R1 was determined as 24.3mgN/dm^3. For reactor R2, concentration of total nitrogen was16.5mgN/dm^3.Difference in the total nitrogen between both reactors amounted to7.8mgN/dm^3. Nitrate concentration in both reactors increased -in R1 to8.6mgN/dm^3, while in R2 to 9.6 mgN/dm^3. A long duration of wastewater aeration resulted in a significant lowering the ammonia concentration in both reactors as compared to the previous control point. In R1, concentration of this nitrogen form was 9.7 mgN/dm^3, while in R24.3 mgN/dm^3. Difference in the ammonia content between reactor without carbon source R1 and that with glycerin R2 was5.4 mgN/dm^3.

Value of CODin treated wastewater from reactor R1 after decantation process amounted to 35.0mgO$_2$/dm^3, while value of BOD$_5$was 6.0mgO$_2$/dm^3. Despite of adding the external carbon source, the COD value in treated sewage with glycerin addition amounted to $28.0mgO_2/dm^3$, and $BOD_5 2.0mgO_2/dm^3$. These are lower numbers than in reactor R1, where no additional carbon source was used. Concentration of the total nitrogen in treated wastewater from reactor R1 amounted to14.4mgN/dm^3, ammonia 1.8 mgN/dm^3, and nitrates 8.6 mgN/dm^3.Treatment of wastewater in reactor R1 caused the removal of total nitrogen in 87.8% and ammonia in 96.5%. In the case of reactor with glycerin addition, concentration of the total nitrogen in treated wastewater was7.6 mgN/dm^3, ammonia 0.6mgN/dm^3 and nitrate5.6mgN/dm^3.Treating the sewage in reactor R2 resulted in the removal of total nitrogen in 93.6%, while ammonia in98.8%. The use of glycerin in R2 reactor has resulted in a higher efficiency of the wastewater treatment process than in R1 reactor, where no external carbon source was applied. Despite of the increase in the final nitrate concentration in reactors R1 and R2, the reactor with external carbon source addition contained lower nitrate concentration by3.4mgN/dm^3.

Like in the research by Bodiket al. [8], in which results from laboratory and technical scale tests upon the use of glycerin fraction in denitrification process in SBR-type chambers, were presented, No elevated COD concentrations were recorded in the outflow indicating a properly adjusted glycerin dose and its utilization during denitrification process.

4. Conclusions

1. The use of glycerin as an external source of carbon during wastewater treatment has resulted in a higher removal efficiency of nitrogen forms than in reactor where no external carbon source has been applied along with low COD content in the purified wastewater.
2. Despite of the increase in nitrate concentration in sewage treated in both reactors, the use of glycerin has resulted in a decrease in the concentration of nitrates in purified wastewater by3.4mgN/dm^{3}as compared to reactor without supply the external carbon source.
3. Glycerin as a waste product can provide better alternative to other expensive sources of carbon.

References

1. J. Mąkinia, K. Czerwionka, J. Drewnowski, M. Swinarski, M. Chrzanowska, W. Fordoński, For. Ekspl., **2**, 15-20 (2008).
2. W. Styka, P. Beńko, G., W. i Tech. San., **9**, 21-24 (2007).
3. E. Kalinowska, For. Ekspl., **5**, 16-17 (2006).
4. E. Kalinowska,For. Ekspl., **3**, 41-42 (2006).
5. M. Kaszubowska, J. Majtacz, J. Mąkinia, K. Czerwionka, E. Kulbat, Zesz. Nauk.. Inż. Śr., **141**, 78-86 (2011).
6. J. Mąkinia, K. Czerwonka, Guidance for the assessment of alternative sources of coal. Innovative source Coal Support for denitrification in municipal wastewater treatment plants. The project financed by the European Regional Development Fund under the Operational Programme Innovative Economy(in Polish, 2013).
7. Bodik, A. Mlstakova, S. Sedlacek, M., Biores. Technol., **100**, 2452-2456(2009).
8. Grabinska-Loniewska, T. Slomczynski, Z. Kanska, Wat. Res., **19**, 1471–1477(1985).
9. J.C. Akunna, C. Bizeaua, R. Moletta, Wat. Res., **27**, 1303–1312 (1993).
10. J.A. Torà, J.A. Baeza, J. Carrera, J.A. Oleszkiewicz, Chem. Eng. J., **172**, 994–998 (2011).

11. D. Kulikowska, K. Bernat, Biores.Tech., **142**, 297-303 (2013).
12. K. Bernat, D. Kulikowska, K.Żuchniewski, Chem. Eng. J., **267**,324–331 (2015).
13. G.P. da Silva, M. Mack, J. Contiero, Biotech. Adv., **27**, 30–39(2009).
14. D.T. Johnson, K.A. Taconi, Env. Progr., **26**, 338–348 (2007).
15. J. Guerrero , C. Tayà, , A. Guisasola, J. A. Baeza, Wat. Res., **46**,2983–2991 (2012).
16. Melcer, E. Klugmann-Radziemska, A. Ciunel, Arch. Gosp. Odpad. i Ochr. Śr.,] **13**, 1-20(2011).
17. S.S. Yazdani, R. Gonzales,Curr. Op. in Biotech., **18**,213–219 (2007).
18. Escapa, M.F. Manuel, A. Morant, X. Gomez, S.R. Guiot, B. Tartakovsky, Ener. and Fuels, **23**,4612–4618 (2009).
19. W. Janczukowicz, J. Rodziewicz,*Źródła węgla w procesach biologicznego usuwania związków azotu i fosforu, (*Monografie Komitetu Inżynierii Środowiska Polskiej Akademii Nauk, 114, Lublin 2013).
20. K. Ignatowicz, J. Piekarski, D. Kozłowski,Rocz. Ochr. Śr., **17**, 1178-1195 (2015).
21. P. Kogut, J. Piekarski, K. Ignatowicz, Rocz. Ochr. Śr., **16**, 534-545 (2014).
22. W. Dąbrowski, R. Żyłka, P. Malinowski, Envir.Res., 153 (2017)

The project was financed by the Ministry of Science and Higher Education (S/WBiIS/3/2014 and W/WBiIS/9/2016).

Adaptive heat pump and battery storage demand side energy management

Florian Sobieczky[1], *Christian* Lettner[1], *Thomas* Natschläger[1], and *Patrick* Traxler[1]

[1]Software Competence Center Hagenberg, Data Analysis Systems, Hagenberg, Austria

Abstract. An adaptive linear model predictive control strategy is introduced for the problem of demand side energy management, involving a photovoltaic device, a battery, and a heat pump. Moreover, the heating influence of solar radiation via the glass house effect is considered. Global sunlight radiation intensity and the outside temperature are updated by weather forecast data. The identification is carried out after adapting to a time frame witch sufficiently homogeneous weather. In this way, in spite of the linearity an increase in precision and cost reduction of up to 46% is achieved. It is validated for an open and closed loop version of the MPC problem using real data of the ambient temperature and the global radiation.

1 Introduction

Recently, the inclusion of continuously changing electricity market price variability in predictive modelling of efficient demand side energy management including heat pump operation has been studied under the assumption of the customer's ability to shift energy load [4]. In energy management using weather forecasts, model predictive control (*MPC*) has previously been identified as superior over non-predictive methods [3, 4, 12]. So far, however, there has not been any work including the photovoltaic devices, the heat-pump, as well as the solar global radiation altogether in the optimization. We show that this is possible within a purely linear, adaptive *MPC* approach.

1.1 Previous Work related to adaptive MPC

A quickly growing body in the literature on model predictive control (*MPC*) has established this method as a state of the art technique to solve control problems for deterministic systems as well as systems with uncertainty showing its superiority over conventional control loop feedback mechanisms (such as PID) [10]. Closed loop stability analysis has now since long been studied with *MPC* for non-linear systems (e.g. [13, 16]). For domestic energy management of heat energy this method has been successfully applied in both in open and closed loop instances [4, 6, 11]. Adaptive *MPC* techniques, however, in which the recursive estimation procedure systematically adapts the choice of model according to some recursive strategy have only recently gained a similar amount of attention ([7]; [10] 3.4).

DOI: 10.1051/e3sconf/20172200162

1.2 Our strategy

By adopting a two-phase view of the control problem ([2], 6.1), we introduce a method solving a purely linear closed-loop optimal control problem with an adaptive strategy for the parameter estimation of the underlying dynamic constraints. More precisely, we use a linear program with cost function taking into account the return temperature together with a 'comfort-parameter' λ, and corresponding required power p with a (time-changing) cost parameter ξ. Furthermore, photovoltaic and stored electric power is included in the cost function while the battery state of charge is modelled by another linear dynamic constraint. At every instant t when an optimization is carried out, the data for the SI is used from a period starting at some suitably chosen time s in the past. Since these parameters determine the resulting optimal control (the applied power p and therefore also the temperature T), the proposed scheme solves an *adaptive* control problem.

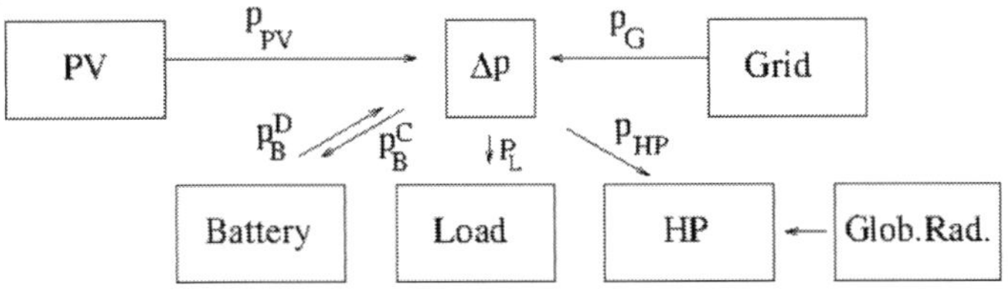

Figure 1: Energy flow diagram for PV, battery, heat-pump, and global solar radiation.

2 Experimental setup

Our model is based on the assumption of there being a linear relationship between the return temperature and the desired room temperature. The reason is the availability of measured values of the return temperature (due to there not being any temperature sensors). Also, the room temperature would have to involve either a more sophisticated building model or more sophisticated assumptions about its role in the model.

2.1 Available Data and Consequence for Model

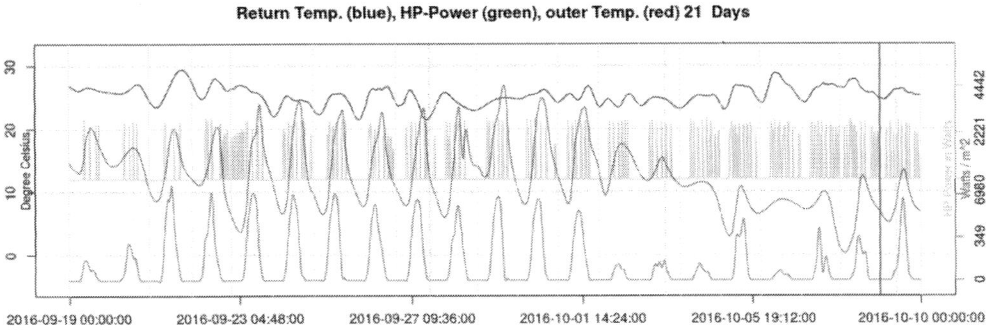

Figure 2: Return temperature data (black) (regularization from the temperature sensor of a heat pump) and its electrical power (green). The ambient temperature (red) on the same scale as the global radiation (magenta) [1], shown as used here. Note drastic cool-off at beginning of October. The `current' moment t is the dark vertical line: 09 Oct 2016, at 00:00 a.m., moment r is 24 hours later.

We are using data collected by the control unit of an end user heat pump in Austria for the return temperature, the electrical power used to drive the heat pump, and the outside temperature: The solar global radiation data was provided by [1] from Linz (143km). While it would be more desirable to have a direct description of the dynamics of the *room temperature*, it is usually not available in domestic heating systems as generic temperature sensors are not standard equipment. We therefore propose a theory using the return temperature of the heater representing the global heat of the whole building: by assuming a linear relationship between the room temperature and the return temperature.

Table 1. Parameters and Variables: *T, SOC* and `decapitalized' powers are subject to optimization.

Variable	Description	Respective Units
p_{HP}, T	State of Heat Pump	Watt, degree Celsius
p_B^C, p_B^D, SOC	State of Battery	Watt (2x), percentage of maximum load
P_L, P_{PV}	Domestic Load, Photovoltaic Power	Watt (2x)
T_a, I	Ambient Temp., Glob. Radiation	degree Celsius, Watt per sq. meters
a, b, c	Parameters of Linear Building Model	Cents per degree Celsius, Cents/KWh (2x)

2.2 The linear model representing the building

Instead of using criteria for the thermal storage of the given building architecture [4] we choose a simple model of the amount of heat delivered to the building which only uses the heating system's return temperature as a general, collective measure for all the whole heat transfer into the building. Using the assumption of a linear system equation of the underlying building model, the exponential cool-off rate a will be assumed to be piece-wise constant in time. Another parameter, b, representing the linear relationship between the thermal power $\epsilon(T_o)p_{HP}$ and the increase in return temperature T (where ϵ is the heat pump's coefficient of performance COP, which depends only on the outside temperature) is characterising the building. A third parameter (c) describes the influence of the global solar radiation (glass house effect) on the change of the room temperature ($T'_{room} = T'_{room,0} + cI$) and, in turn, on the return temperature T (under the assumption of a linear relationship between these two). This is modelled by the linear ODE with parameters a, b, c

$$T'(t) \quad = \quad a\ (T_a(t) - T(t)) \ + \ b\ \epsilon(T_a(t))\ p_{HP}(t) \ + \ c\ I(t). \qquad (1)$$

We will use a discretized (difference-) equation of (1) over a suitably chosen time-interval $[t_0, t]$. The block diagram (see Fig.6) shows the two different used modes (i. open-loop, and ii. closed-loop, [11]) distinguished in terms of whether the output prediction of the optimizer is used in the subsequent system identification step. In other words, the optimization results *from the future* is used to perform a comparison of the resulting identification of the parameters with simpler estimations. The decision variables of the optimiser (Opt) are the set of powers (battery charge/discharge and electric heat-pump power) together with the resulting state of charge (SOC) and return temperature of the heating system. Their history is also used as predictor variables in the system identification (SI) stage via (1). The prediction $\hat{p}$ of the optimal powers from the optimiser yields a direct control directive for the use of battery and heat pump. Furthermore, the two modes for the usage of these two stages' output variables are: *i.) Open-loop:* This mode simply uses the results of the system identification estimation process to determine the optimal length of the time interval of the considered data of p, SOC and T, from the past. Statistics

DOI: 10.1051/e3sconf/20172200162

determining the quality of the parameter estimation, *only*, are used when making the decision of which past to consider. *ii.) Closed-loop:* In this mode, the result of the optimizer is fed back into the system identification process. The data which is used to carry out the system identification is augmented with the optimal solution, giving a new estimate of the system-parameters *a, b* and *c*. Given convergence, the iteration of this process leads to a self-consistent solution for parameters and optimal control of the heat pump and battery. We determine the advantage of using closed-loop mode by comparing the cost of the resulting strategies (the 'energy bill').

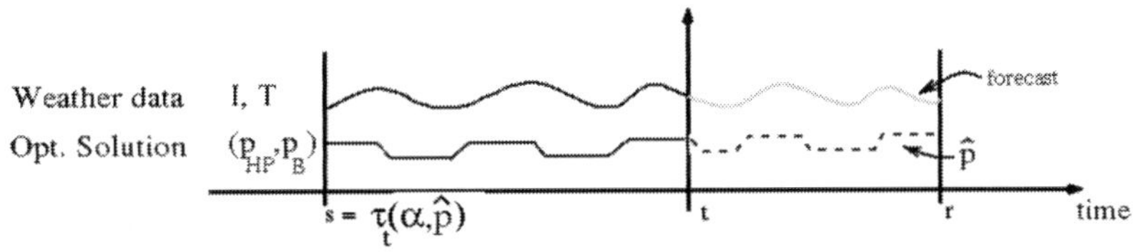

Figure 3: Mutually dependent weather data and optimal solution: *s* depends on opt. solution and v/v.

2.3 Adaption: Criteria for model selection

Adaptive *MPC* allows for switching adaptively to a different model during operation. In the present case, we define for each choice of a measuring procedure in the system identification there to be a corresponding (building) model. The set of considered models only differs in the domain of data on which the identification process is carried out. Data is considered only with timestamps from the interval [*s, t*], where *t* is the present moment (a fixed parameter representing the end of the domain). Therefore the model class(=: $\mathcal{M}_t$) considered for predictions of optimal heat pump and battery usage *at time t* is parametrized by the variable *s* ranging over times up to *t*. We now adopt the view (Fig. 3) of there being an interval of the past [*s, t*] (up to the present *t*), and an interval of the future [*t, r*]. Now, in order to determine the best criteria for selecting the right model at time t among the many in $\mathcal{M}_t$ (with different lengths of the interval [*t-s*]), we use, for a given significance level $\alpha>0$, the hypothesis $H_0 : \widehat{\theta}_{s,t} = \widehat{\theta}_{t,r}$ and define the time $\tau_t(\alpha)$ (in the past, i.e. before *t*)

$$\tau_t(\alpha) = \sup\{ s < t : \text{H}_0 \text{ is rejected under significance level } \alpha \}. \tag{2}$$

$\widehat{\theta}_{s,t}$ is the LS-estimator [14] of the parameters a, b, c on the interval $[s, t]$. The fact that $\widehat{\theta}_{t,r}$ are future predictions make $\tau_t(\alpha)$ depend on the optimal solution $\widehat{p}$ determined by the optimization (see Fig. 3). On the other hand, this optimal solution depends on the parameters delivered by the least square estimation carried out on $[\tau_t(\alpha), t]$. This circular dependence is the basis for the closed loop mode of the *MPC* problem, where an initial value for τ_t is picked by hand, when no optimal solution has yet been calculated. We are using the multivariate two-sample *Hotelling* test. Even though it proves useful to apply univariate two-sample tests (such as the *Welch* test or the t-test) for each parameter (a, b, c), separately (enabling to set separate thresholds), it is then not clear how to combine the resulting distinct time-ranges into a single choice of *s*. So, after choosing a level of significance α for whether each multivariate parameter mean $\widehat{\theta}$ on $[s, t]$ deviates $\alpha-$significantly from that on [*t, r*] (1 day), we measure the p-values from our data for every (discrete) *s* in a properly chosen maximal interval (three weeks) before *t*. A p-value smaller

 DOI: 10.1051/e3sconf/20172200162

than the given threshold indicates significant deviance of the future's mean from the mean of the past for the specific value s.

3 Optimisation

When including the photovoltaic facility in the domestic energy model together with a battery, the relationship between the SOC and the charging power plays a role similar to that of equation (1), namely between the return temperature (reflecting the accumulated heat content) and the electric power used to operate the heat pump:

$$SOC' = \eta_1 \, p_B^C - \eta_2 p_B^D \tag{3}$$

where η_1 and η_2 (set to 0.9 in the experiment) are the (constant in time) degrees of efficiency of the charging and discharging module, respectively.

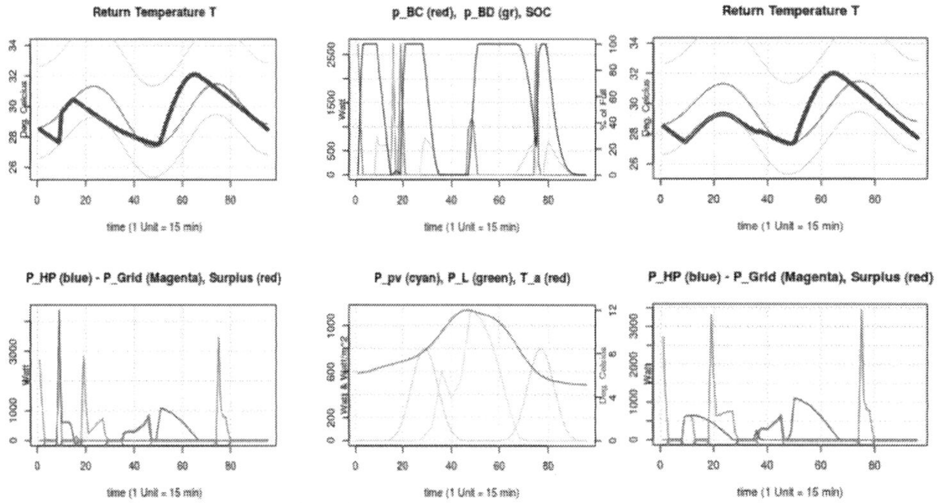

Figure 4: End user heat pump data of return temperature, electrical power used to drive the heat pump, and ambient temperature: The solar global radiation data was provided by [1] from Linz (distance to Stockerau 143km). Left & Middle: Initial return temperatur (blue, thick line) of 28.5 deg. Celsius is increased to over 30 degrees Celsius in early morning to cope with condition of first (high) minimal heat load shortly after 20 time units. The positive surplus power (red) as a result of high global radiation (cyan) indicates ability to feed energy into the grid. Right Column: Parameter a larger by a factor of 1.2 yields increase of cost by 5.2%, as morning high minimum heat demand isn't 'overcome' by first heating up and then cooling off (while p_{HP}=0: see blue, thin line in lower left diagram).

We formulate a linear program with an objective function with two terms, one of which realises the modelling of the heat pump and one which belongs to the battery/PV device. The decision variable into which both of these terms enter is the power $p_B = p_B^C - p_B^D$, related to charging (C) and discharging (D) the battery. We start with the balance of energy flows derived from the schema shown in Fig. 1. The power required from the grid during the i-th elementary time interval is given by

$$\Delta p(i) = P_L(i) + p_{HP}(i) + p_B^C(i) - p_B^D(i) - P_{PB}(i). \tag{4}$$

 DOI: 10.1051/e3sconf/20172200162

The de-capitalized letters refer to decision (optimization) variables, the capitalized ones are fixed, given functions obtained from measurements. The objective function is

$$\tilde{F}(\Delta p, T, SOC) = \sum_i^n \left(\Delta p(i)\xi_P(i)\chi_{\Delta p(i)>0} + \Delta p(i)\xi_S\chi_{\Delta p(i)<0} + \lambda|T(i) - T_{exp}(i)|\right). \quad (5)$$

The second term bears a new parameter, λ, which measures the amount of money the user is willing to pay for an increase of the temperature by one degree Celsius during a unit time interval (cmp.[]). In this form the problem is a mixed integer program. Considering the large number (n=96 for so many 15 min intervals per day) of degrees of freedom and the absolute value sign of the temperature term it is advisable to switch to the following linear form with two additional degrees of freedom (u and v): $F(u,v) = \sum_i^n(u(i) + v(i))$ subject to the following constraints $u(i) \geq \xi_P(i)\Delta p(i)$ and $u(i) \geq \xi_S(i)\Delta p(i)$ with $\xi_P(i)$ the purchase prices, and $\xi_S(i)$ the selling price at time $i \in \{1, \ldots, n\}$. Also, the constraints $v(i) \geq \lambda\,(T(i) - T_{exp}(i))$ and $v(i) \geq \lambda\,(T_{exp}(i) - T(i))$ hold to realise the positivity of the absolute value $|T_{exp} - T|$. Other constraints include the linear system dynamics (1) and (3), as well as hard boundaries of the involved decision variable intervals. As an example for the sensitive nature of the optimal solution on the parameters, consider the pair of solutions in Figure 4, and 5: The return temperature T is given the freedom to range between the margins of a 'rule-of-thumb' temperature (red) from a custom-set 'heat curve' H_0 (pink, Fig.4): $T_{exp} = H_0(T_a) + 1.5(T_{\text{Room}} - 20^oC)$, with margins of 2 degrees below and 4 degrees above.

4 System identification

4.1 An MPC problem in Open-loop and Closed-loop mode

After it has been established that there is high sensitivity of the optimal solution on the parameters (a,b,c), we show how (5) may become a closed-loop identification problem.

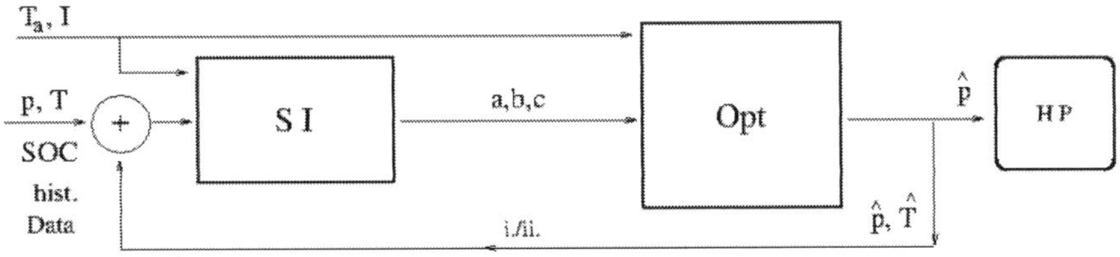

Figure 5: Block diagram showing the control loop with open-loop (i.) and closed-loop (ii.) mode.

The 'weather data' of the ambient temperature T_a and global radiation I is needed both in the system identification (historical data), as well as in the optimization (forecast). Once the forecast for battery p_B^C, p_B^D and return temperature T is available, they can either be passed on to the devices as the final result (i. open-loop), or they are first reused in another SI step (ii. closed-loop). Namely, the fitting of the parameters may now occur on the extended time-interval of the historical (p,T) data together with the future interval holding the optimizer's optimal prediction $(\hat{p},\hat{T})$. We now present the results for the two different modes (Fig..5; compare with [14], Chap. 13.4): **i.)** ***Open-Loop:*** In this case, a separate system identification process is first carried out, before the resulting parameter estimates are used for the optimal choice of power used in the heat pump and for/by the battery (the corresponding temperature and state of charge variables follow from integration of the system equations (1) and (3)). **ii.)** ***Closed-Loop:*** For each instance t for which a prediction of an optimal power and battery use over a future interval is demanded, an iterated sequence (6) of system identification and optimization stages is carried out (once with and once without I) using a `mixing parameter' q. Using the value of the objective function we break-off at step 4 of the iteration-procedure if it doesn't lead to further reduction of the cost:

1. $i = 1, \alpha > 0, \widehat{p} : [t, r] \to \mathbb{R}_+$, set to $\widehat{p} \equiv 0$
2. SI: $\theta_i \ := \ \widehat{\theta}_t(\tau_t(\alpha))$
3. Opt: Obtain $\widehat{p}_{i+1}$, the optimal solution of the linear program on $[t, r]$
4. Increment i; Use $(1-q) \cdot \widehat{p}_{i-1} \ + \ q \cdot \widehat{p}_i$ in next step: GOTO 2. (6)

4.2 Results and Conclusion: Several approaches to reduce cost

For some given significance levels α, we report the interval length $t - \tau_t(\alpha)$ obtained from available data for a specific example. In other words we calculate the maximum number of days possible to reach back in time to perform the proposed system identification and optimization procedure without significant differences in the distribution of estimated parameters on the two intervals $[s, t]$ and $[t, r]$. We use our Stockerau-data for the first 3 weeks in October 2016 using the Benjamini-Hochberg [5] correction for multiple tests for (2). Our interval in the future is $|r - t|$ = 24 hours, the forecast is made at midnight. The selling price of electrical energy into the grid is 3 Cent/Kwh. The purchasing price has been

Table 2. Results of interval size $t - \tau_t(\alpha)$ (=Days, left cell) and cost reduction (right cell) for i.) open loop and ii.) closed loop mode, with global radiation (I) and without (here c=0). The percentages refer to **cost reduction** in comparison with the case where s is categorically chosen to be 21 (=∞, corresp. to α=0). All cost is given in Cents where λ = 0.1 Cent / (15min deg Celsius) has been used.

	i.) Without I		**i.) With I**		**ii.) Without I**		**ii.) With I**	
α	**Days | Cost/ %**		**Days | Cost/ %**		**Days | Cost/ %**		**Days | Cost/ %**	
0.00	21	605/ 0	21	600/ 0	21(0)	605/0	21	600/0.0
0.001	21	605/ 0	21	600/ 0	18(2)	347/43	18(2)	344/**42**
0.002	20	443/27	21	600/ 0	14(4)	374/38	12(2)	378/**37**
0.01	16	420/31	17	378/37	13(3)	377/38	12(2)	370/38
0.02	15	427/29	15	424/29	11(3)	326/46	11(3)	324/**46**
0.05	13	377/38	13	379/37	11(2)	326/46	11(2)	324/**46**
0.1	12	349/42	12	346/42	12(1)	349/42	12(1)	345/42
0.2	11	326/**46**	11	324/**46**	11(1)	326/46	11(1)	324/46
0.3	10	372/39	10	369/39	8(2)	372/39	8(2)	344/**43**

chosen to be 20 Cent/KWh (which is the approximate effective cost at a formal advertised price of about 6 Cent/KWh). As the weather forecasts on the future interval $[t, r]$ we use the exact values ('perfect weather forecast'). As the break-off criterion of the closed-loop procedure we check if the objective function (cost) continuous to decrease. For the period of Fig. 2, triples $(\alpha, \tau_t(\alpha), \text{Cost}(\alpha))$ with $\tau_t(\alpha)$ in days, and cost-reduction in % of the $\alpha = 0$ case) are given in Table 2 for q=0.1. As seen in Figure 2, this aligns with a strong temperature drop and decrease in global radiation around 4 October 2016. The numbers in parentheses in the table refer to the number of closed loop cycles, of which the maximum has been chosen to be 20.

From these results it becomes clear that the adaptability of the linear model representing the building has strong saving potential (> 40%, in this example). The inclusion of the global radiation term involving the intensity I of the global radiation in the model increases the saving potential for the given example by 2 to 5%. Furthermore, applying the hypothesis testing systematically (with the BH-correction, which adds some uncertainty in the localization of the weather change), it shows that there is an optimal significance level α in terms of cost minimization. We note that the actual cost reductions seem to occur in the vicinity of strong input weather data changes (here, about 10 days before t). Finally, the application of a closed loop mode (9) with a cost sensitive break-off criterion may give further cost reduction or reduction in the necessary length of the past $[s, t]$ used for the System Identification. Our conclusion is that the use of an *adaptive MPC* approach for domestic energy management is suitable for reacting to drastic weather changes when remaining within a purely linear framework of optimisation and building model is necessary (as in the case of large scale problems).

Acknowledgements

The research in this paper has been supported by the Austrian Ministry for Transport, Innovation and Technology, the Federal Ministry of Science, Research and Economy, and the Province of Upper Austria in the frame of the COMET center SCCH. Special thanks go to W. Traunmüller and M. Backmann from Blue Sky Wetteranalysen. Thanks also to Ramin Nikzad-Langerodi (FLLL) for a helpful discussion about multiple hypothesis testing.

References

1. Blue Sky Wetteranalysen , Attnang-Puchheim, Austria
2. D. P. Bertsekas: *Dynamic Programming and Optimal Control, Volume I,* 2nd Ed., Athena Scientific, Belmont, Massachusetts Chap. 6.1.2, (2000)
3. G. C. Chasparis, T. Natschlaeger *J. Dyn. Sys., Meas., Control* **139** (2), 2016
4. J. Cigler, Z. Vana, T. Muzik, J. Sulc, L. Ferkl. *2016 Europ. Cont.l Conf.* EUCA (2016) Aalborg, Denmark, 978-1-5090-2590-9
5. Y. Benjamini, D. Yekutieli Ann. Stat. **29** (4), (2001), 1165-1188
6. G. P. Henze, J. Pfafferott, S. Herkel, C. Felsmann Energ. Blds **39**, pp. 221-235 (2007)
7. X. Luo, Y. Song Abstr. Appl. Anal. **2014** (2014), Article ID 869879
8. D.Q. Mayne, J. B. Rawlings, C. V. Rao, P. O. M. Scokaert Autom. **36**, 789—814 (2000)
9. F. De Lillo, F. Cecconi, G. Lacorata, A. Vulpiani EPL **84,** (2008)
10. D. Q. Mayne Automatica **50**, (2014)
11. F. Oldewurtel, C. N. Jones, M. Morari *47th IEEE Conference on Decision and Control* Cancun, Mexico, (2008)
12. F. Oldewurtel, A. Parisio, C. N. Jones, D. Gyalistras, M. Gwerder, V. Stauch, B. Lehmann, M. Morari Energy and Buildings **45**, 15-27, (2012)
13. P. B. Sistu, B. W. Bequette AIChE Journal **42** (12), 3388-3402 (1996)
14. L. Ljung *System Identification, Theory for the User,* Prentice Hall, 2nd Ed. (1999)

DOI: 10.1051/e3sconf/20172200163

Biomass universal district heating systems

Victor Manuel Soltero[1], *Salvador* Rodríguez-Artacho[2], *Ramón* Velázquez[3], *Ricardo* Chacartegui[3,*]

[1]Department of Design Engineering, University of Seville, Seville, 41011, Spain
[2]Department of Constitutional Law, National University of Distance Learning, Madrid, Spain
[3]Department of Energy Engineering, University of Seville, Seville, 41092, Spain

Abstract. In mild climate regions Directive 27/2012 EU application for developing sustainable district heating networks in consolidated urban nucleus is a challenge. In Spain most of the municipalities above 5,000 inhabitants have a reliable natural gas network and individual heating systems at homes. In this work a new heating network paradigm is proposed, the biomass universal heating network in rural areas. This model involves all the economic, legal and technical aspects and interactions between the different agents of the systems: provider company, individual and collective end-users and local and regional administration. The continental region in Spain has 588 municipalities with a population above 1,500 inhabitants close to forest biomass with renewable use. In many of these cases the regulation identifies the ownership of the forest resources use. The universal heating networks are a great opportunity for energy saving of 2,000 GWh, avoiding 2.7 million tons of CO2 emissions and with a global annual savings for end users of 61.8 million of euros. The presented model is easily extrapolated to other small municipalities in Europe. The real application of the model is presented for three municipalities in different locations of Spain where Universal Heating Networks are under development. The analysis show the interest of the integrated model for the three cases with different structural agents and relationships between them. The use of sustainable forest resources, extracted and managed by local companies, strengths circular economy in the region with a potential global economic impact above 200 M€.

1. Introduction

The Energy Efficiency Directive (EED – 2012/27/EU) (*1*) establishes a common framework for the promotion of energy efficiency within the European Union, to ensure the achievement of the European Union headline target of 20% on energy efficiency by 2020.
A barrier for district heating (DH) development is the lack of a sectorial regulation for allowing public or private domain use for thermal infrastructures, similar to the existing sectorial regulation for water and electricity (*2*). For massive implementation of heating networks in consolidated urban nucleus is required a development model integrated for all participating agents. Although optimized networks can result in high environmental, economic, social benefits (*3, 4*) they must be adequately evaluated and identified by all participating agents in order to assure optimized development and maximizing these

* Corresponding author: ricardoch@us.es

benefits and the sustainability of the system. In 2015 there were more than 5,000 district network systems in Europe (*5*). The level of development of district heating networks in Europe varies strongly from an intense presence in Norther and Central Europe to scarce introduction in Southern regions, even when there is interesting potential for their application. This is the case of Spain, where there is an incipient interest in District Heating and Cooling (DHC). Currently there are 240 urban heating and/or cooling networks, of which 202 are officially registered. Most of these of DH/DHC networks correspond with demonstration systems, with a mean length below 1.5 km, of which almost half (47%) are public sector networks, and with very little commercial technology development (*6*).
The first option for implanting DH/DHC systems is the use of surplus heat from nearby power plants and industries for residential heating. In Spain however, medium and large size industries -which can provide a cost-effective surplus energy - are in most cases located too far from urban areas, and it is not feasible in most cases to take advantage of this surplus heat for the residential sector. For covering this demand two different and complementary ways can be followed for new DH/DHC installations in Spain: Biomass District heating networks(*7*) and Natural gas District heating networks (*4*). Potential for biomass heating networks in Spain is high in the continental region, however it has not been developed. This is due to the risk and uncertainties due to legal and regulatory gap for their development combined with adequate models for their development. In this paper are analysed all the subsystems and agents participating in the promotion and development biomass district heating and they are integrated in a holistic model for optimizing the benefits adapted to the local situation. The model is applied to the continental region in Spain

2. Legal sectorial framework for services infrastructures

Within the EU, it is already pointed out that DH network need to be seen and considered as Services of [Economic] General Interest –SEIG-, entering a specific category of services already envisaged for other traditionally monopolistic services such as gas, electricity, water or telecommunications. When referring to the economic difficulties linked to financing infrastructure building and deployment, the gap between sectors is substantial and relevant. In this sense, a comparison between an already existing legal framework for those traditional sectors and the lack of such a framework specifically for DH is an obstacle which better sooner than later, should be removed. Otherwise, competitiveness among sectors that should share the same market rules will never occur in fair and just proportion. So has it been pointed out by European Commission in (*8*) when highlighting that governments could –and should- do much more to leverage the inequality among infrastructure deployments rules and legal framework. It concludes that public intervention is reasonable and justified since it has been a fact in the past. Accordingly, it states that "this has been routinely done in the past for network providers by effectively the granting or auctioning of licenses and this could readily be applied to district heating networks, or indeed any other form of energy infrastructure". More precisely, some competitive disadvantages should be remove. Not by means of eliminating those advantages that traditionally were put in place in order to help out infrastructure financing in order to meet deployment goals and plans. To the contrary, by means of compensating DH projects with the same tools from which others sector have been benefiting from the past. As it is stressed in the report, "incumbent utility suppliers such as gas and electricity pipe and cable networks have a huge competitive advantage over would-be competitors piped heat suppliers by virtue of the various rights such as pre-granted planning permission, ground and road breaking rights, not enjoyed by DH operators". Consequently, if a more fair and well-balance scenario is to be built for all infrastructure players –irrespective of the type of infrastructure- then those same rights traditionally recognized to former incumbents " must be created and assigned in some way to CHP-DH developers if private capital is to create these networks" . It is a matter of jus balancing the unfair treatment. Otherwise, private sector might not have the financial muscle for engaging significant investment. A final remark made by the report states that "unless this is recognized by the EU and remedied, CHP-DH networks will not be built on the scale envisaged as part of the programme to combat climate change and reduce energy dependence". This leads us to the exercise of identifying what elements of those sectorial laws and regulations are common to different sectors so that we can guess a similar content when and if a DH Services Law/Directive was ever foreseen.

3. Universal biomass heating network

In previous section has been presented a context of challenging/non fully developed regulation for district heating systems as for other energy services. In this work is defined a new holistic approach for identifying and optimize the interaction between the multiple agents participating within the district heating. This concept is the Universal biomass heating network. It optimizes the global performance of the system defining the requirements/ interactions and participating agents. It can be defined as a thermal energy generation, distribution and selling system characterized by:

- Universal access for all end-users. Any end-user must have access to the network assuming the payment of the connection and individual preparation costs.
- Investors and promoters assume commercial risks and in counterpart are the owners of the system and exploitation profits.
- System is based on local biomass as fuel. It assures sustainability in the whole process of extraction, transport and operation and promotes circular economy.
- Local administration regulates and supports the global system with regulation, licences and as consumer.
- End user access in free market conditions and quality of supply is warranted by contract with the operator/promoter
- Specific regulation/rules/contracts warranty sustainability of the system and the circular economy in the region

A universal biomass district heating can be modelled as a System where there are four main Subsystems: i) forestry resource, ii) techno-economic, iii) commercial and iv) regulator.
Interaction and articulation between them is presented in Fig.1.

i) Forestry resource subsystem. Fuel is defined by the availability, homogeneity, recurrence, ownership, price and quality. It is the main pillar around which is developed the biomass heating system. The ownership of the resource and the manager of mountain will be the agents who develop functional relationships with other agents.

ii) Regulatory subsystem. It supports and provides legal stability and safety to the global system. While it is desirable a sectorial normative for DH, in many cases it is not developed and systems are regulated by local authorities with environmental and urbanistic competences. For the case under analysis, biomass forestry, the legal subsystem should interact with the mountain manager for assuring supply and maintenance. However this interaction is not always possible because forestry use and biomass recollection/exploitation are adjudicated by economic auction without linking both actors.

iii) Commercial subsystem. It includes the structure of relationships between the Universal District Heating and the end-user. The agents are the end-users and the investor/network developer. It includes economic issues but also agreements in supply quality, service, availability and after sales service including maintenance aspects.

DOI: 10.1051/e3sconf/20172200163

iv) Techno-economic subsystem. Includes design process, detailed engineering, economic and environmental viability analyses and financing. It links the participating agents, the investor/promoted of the Biomass Universal District Heating and the municipalities.

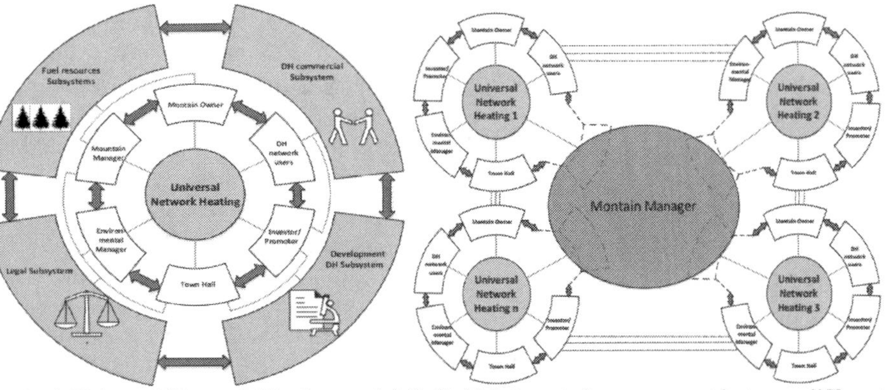

Fig. 1. a) Universal Network Heating model (left). b) Integrated management between different universal heating networks of the forestry resource (right).

4. Biomass universal district heating agents

The agents participating in the Universal Biomass District Heating networks, BioUnivDH, are related in a hierarchal dependence of the four subsystems and they interact dynamically between them. Depending on the country, region or area under of study they can be a set of independent organisms or involved competences can be grouped in one single organism, as is the case of Spain, with Mountain manager and Environmental manager. The following agents are identified in each Biomass Universal Heating Network:

-Forestry resource manager. It has assigned the management of the mountain (independently of the public or private ownership). In Spain is done via the law on Mountains/Forest Act. They have the role of mountain planning and exploitation approval. It launches the processes for treatment and exploitation of the forestry resource. They are regional entities with the vision about recurrent and sustainable production capacity of the forestry resource. Integral management requires the redistribution of resource from areas with excess in production to those with unsatisfied demand, Fig.1 b. In the cases based only on the direct public auction there are difficulties for the development of structured redistribution strategies based on general interest criteria.

-Owner of the forestry resource. Ownership can be public or private. In the case of public ownership with a volume enough for fuel supplying, the system eliminates development and operation risks because the warranty of supply and price. In addition a circular economy is generated from forestry exploitation for heating self-consumption as it combines mountain maintenance and exploitation for covering heating demand.

-Environmental Manager. He supervises and emits the environmental authorization for the generation central. It constrains the urbanistic authorization of the Municipality to evaluation of emissions derived from combustion (NOx, soots) and noise levels.

-Municipality. It has the competences on urbanistic licenses and local regulation. Sometimes if it is also the owner of the soil. It eases the development of the project collaborating/participating with promoters. It also can have competences on taxes, which can be used as support of the network construction. It has an economic counterpart from the installation exploitation because of the uses of subsoil for network spreading. Municipality participation is required for an adequate development and exploitation of the system. Besides, in occasions, Municipality is the owner of the forestry resource. Municipalities are

 DOI: 10.1051/e3sconf/20172200163

the main channel supporting the social interests on the project and they can use it as a tool for social energy policy (i.e. for energy poverty fighting policy measures).

-**Universal district heating users**. In rural environments three main categories of customers can be identified as function of energy requirements and use: permanent users, second residence users (weekends and holidays) and singular buildings. The first group is highly affected by energy costs and depending on the area even by energy poverty, as happens in different rural areas in Spain, meanwhile, in the second group, priority aspects are comfort and availability. Singular public buildings (schools, ambulatory, sport centres, administrative buildings) and singular private buildings (hotels, commercial buildings…) have an energy demand estimated in the range of 8-15%. Generally their incorporation to the heating network is easier due to the professional management of the energy bill. All these users are expecting a fixed price without oscillation which will result in more than 10% of savings after their integration in a biomass heating network. Their main investments are the connection costs and the equipment for heat exchange and measure. One of the main social challenges for new networks development is the social acceptance of the heating central location.

-**Promoter/investor.** They can be public or private promoters. The promoter assumes the investment risk and all the risks of project: regulatory, commercial, resource availability. In some cases, public promoters have difficulties for assuming commercial risks and the end users management.

5. Potential for application in continental region of Spain

For the application of the concept of BioUnivDH the continental region of Spain is considered, Fig.2, including all those villages with a population in the range of 1,000 - 10,000 inhabitants, and disregarding the villages with existing natural gas networks. It implies a set of 501 municipalities. The development of biomass universal district heating systems to cover these municipalities is estimated in 3,744 M€ and it would affect to 1.6 million people, resulting in a global installed power of 8.3 GW and an annual energy of 7,825 GWh, with an estimated CO2 emissions savings of 2,747 t. Main data of applying the model are given in table 1.

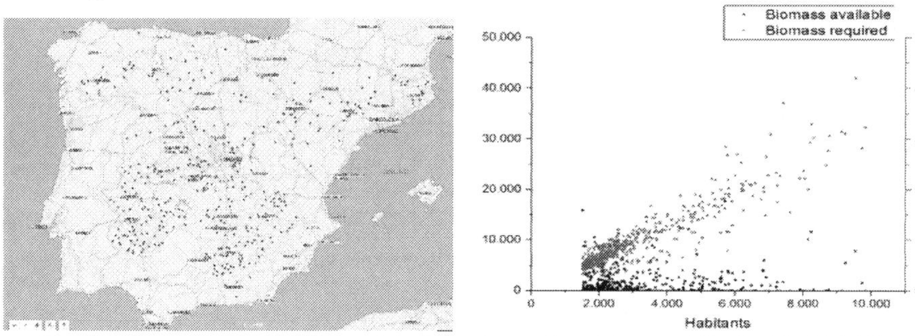

Fig 2. a) Case study area: 501municipalities in continental area of Spain (right). b) Biomass required and available by village (left).

The required biomass for this set of universal biomass district heating is 5.2 million tons, meanwhile these municipalities only have within their territories 787,831 t (Fig. 2). However in the study region where these villages are located, the forestry biomass potential is 6.9 Mtons. This area considers the nearby municipalities disregarded within this study because population size or natural gas network installed make not sustainable the BioUnivDH installation. This circumstance between neighbour territories highlights the relevance of the role of the agent manager of the forestry resource in the redistribution of resource between the production areas and demand areas.

DOI: 10.1051/e3sconf/20172200163

Table 1. Municipalities under study and the expected impact of BioUnivDH

Habitants	Number of villages	Users	Invest M€	Required Biomas (t)	Available Biomass	Jobs created	Energy Poverty index (%)	Incomes M€	CO2 savings (t)
< 2k	131	228 k	558	737 k	197 k	1,229	10.67 – 50.28	9.0	401 k
2k-5k	294	858 k	2,085	2,718 k	452 k	4,530	15.19 – 53.19	33.3	1.497 k
> 5k	76	498 k	1,100	1,592 k	138 k	2,653	12.81- 46.44	19.5	886 k
	501	**1,584 k**	**3,744**	**5,047 k**	**787 k**	**8,412**	**10.67 – 53.19**	**61.8**	**2.747 k**

A relevant result of the analysis is that in the rural areas under study a high percentage of end users are vulnerable to the risk of energy poverty, Fig. 3. It means that a population comprised between 10.67% and 53.19%, dedicates more than 10% of their incomes to energy costs, meanwhile the mean value in Spain is 15%. This situation is justified by the high price of fuel in these not well communicated rural areas, in most cases the use gasoil as fuel for heating, and the minor per capita incomes compared with cities and other regions. In this sense the development of a network of BioUnivDH can be a tool for a social policies, introducing bonuses in places with this combination of costs and incomes. Meanwhile currently there are no social bonuses for gasoil heating; the biomass universal district system could allow to introduce bonuses for low incomes families in a rational and controlled way. Different levels of support depending on the incomes and integrated within the municipal taxes or in the energy bill can be applied to families. Expected savings by end-users for these systems are 10% of energy costs. Considering the savings compared to individual heaters (amortization and maintenance), and discounting the funds dedicated to energy poverty fighting measures, the massive implementation of BioUnivDH in the region under study would allow annual incomes for the municipalities of 61.8 M€, Fig. 4. Biomass consumption would be 311.6 M€ in biomass, and estimated new jobs would be 8,412, in activities of extraction, manipulating and drying of biomass, Fig. 3. It goes to local handwork, which implies that local economy would receive an economic activity above 200 M€. The extended implementation of BioUnivDH in the region would contribute to 2,747 t CO2 savings. Besides it has additional environmental benefits. Currently biomass management is done only where there is direct demand and extraction costs are more competitive. In cases these management is done by public authorities. Intensity of these forestry maintenance treatments is limited by budget availability. The new market created by these systems would be a catalyser for spreading of viable forestry maintenance plans. Another environmental global benefit is associated to the local use of biomass, considering supply in a mean distance below 32 km, and avoiding associated transport emissions (*9*).

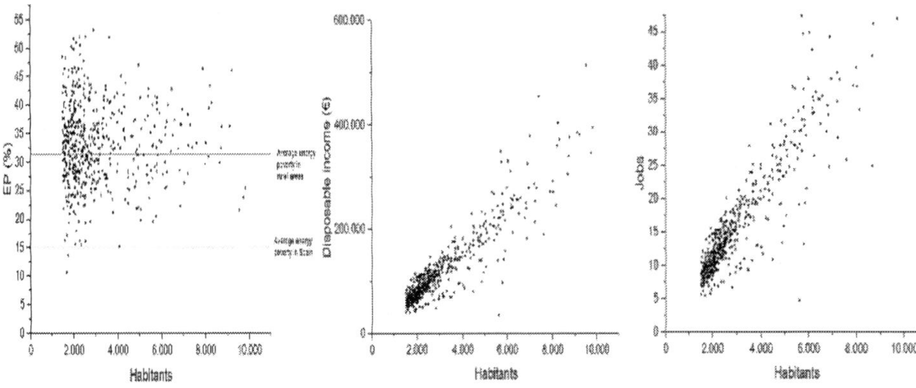

Fig. 3. a) Energy poverty index in the villages (left); b) annual incomes per muncipality (center); c) jobs (right) generated by each BioUnivDH system

6. Study cases

In this section, the application of BioUnivDH is presented disaggregated in three rural municipalities with high forestry resource availability and in where currently universal biomass district heating BioUnivDH are under development, Table 2. They have high heating demand, in some cases with demands similar to central Europe cities. Currently they are using gasoil as main fuel for heating in individual boilers. They have been selected in three different regions of Spain: Municipality A is placed at the South of Spain, municipality B in the North meanwhile municipality C is locate in the central region.

Table 2. Main characteristics of the three municipalities under study BioUnivDH

	Town A	Town B	Town C
Habitants [ud]	1,502	1,919	1,998
Fuel Poverty Index [%]	48.58	50.28	21.29
First houses [ud]	530	745	770
Second houses [ud]	294	305	775
Thermal Power [MW]	7.91	10.11	10.53
Degree Days 20 based [ºC/y]	3,439	4,133	3,359
Energy [MWh/y]	10,090	15,063	13,281
Biomass required [t/y]	6,312	9,306	8,316
Biomass avaiable [t/y]	15,969	8,830	8,799
Network leght [m]	4,792	8,456	11,905
Linear heat density [MWh/y·m]	230.52	192.58	122.25
Invest [M€]	2,9	4.2	5.0
Biomass Price 20% RH [€/t]	73.79	57.74	67.00
Biomass cost [k€/y]	373	430	446

Among these three cases there are great differences in the energy poverty index. Village C, close to Madrid, has a high impact of second residence and tourism, which improves its economic activity and characteristics of energy demand. However in village B, a high rate of unemployment combined with a high heating demand causes a very high level of energy poverty 50.28%. For case A, there is biomass in excess, however, the difficult extraction of forestry in nearby mountains, compared with the other cases, implies that it has the highest biomass price, 73.79 €/t. By other side, this village has the highest building density, with a linear heat density of 230.52 MWh/y·m, almost the double that the case C. The required investments for implementing BioUnivDH in the municipalities under analysis have a wide dispersion, 2.9, 4.2 y 5.0 M€, due to differences in building densities. The participating agents in the BioUnivDH in these municipalities are presented in Table 3, showing the different participation and role of each one. Due to current regulatory framework in Spain, promoter is in charge of obtaining authorizations and establishing alliances with the rest of agents of the BIoUnivDH system in order to give sustainability to the global system. From the villages under analysis, only in case B where the forestry resource owners are the end users, price supply and circular economy is fully assured. Viability is promoted and supported by the municipalities by the integration of municipal buildings in the network.

Table 3. BioUnivDH agents and system structure in the three municipalities under study

	Town A	Town B	Town C
Resource owner	Municipality 79% - Private 21%	Citizens born in Municipality 100%	Municipality 86% - Private 14%
Mountain Manager	Natural park. Assignment of exploitation because of management.	Regional Government. There is no assignment of exploitation. Management taxes	Regional Government. Assignment of exploitation because of management.
Municipality participation	No land contribution for heating generation central. 5 municipal buildings with 6% of whole heating demand.	No land contribution for heating generation central. 8 municipal buildings with 7% of whole heating demand.	It contributes with land for heating generation central. 4 municipal buildings with 5% of whole demand.
Environmental manager	Municipality Regional government	Municipality Regional government.	Municipality Regional government.
User	Saving in energy costs of 20%.	Saving in energy costs of 18%. Assignment of exploitation with additional 5% discount on energy price. Circular economy.	Saving in energy costs of 15%.
Investor/developers	Private with bank financing. It manages licenses/authorizations for network development, heat generation central, environmental impact. It negotiates biomass supply by providers. It manages heating sales contracts with end users.	Private with bank financing. It manages licenses/authorizations for network development, heat generation central, environmental impact. It negotiates biomass supply by end users. It manages heating sales contracts with end users.	Private with bank financing. It manages licenses/authorizations for network development, heat generation central, environmental impact. It negotiates biomass supply by providers. It manages heating sales contracts with end users.

The role of each agent can vary, depending on the specific characteristics of the location, existing regulation and even social perception. To identify the role and benefit of each participant is a key activity for the successful development of a Bio UniveDH. Social, environmental and economic benefits of the implementation of these BioUnivDH in the selected rural locations are presented in table 4. The substitution of the current gasoil fuel boilers for heating by these biomass universal district heating network converts the heating global system in a global null CO_2 emissions system. From the social point of view they allow social energy policy from the municipalities via bonuses on the energy bill depending on the family incomes. At the same it generates circular economy creating jobs and promoting an industry based on forestry exploitation and mountain maintenance. This synergy is increased in case B where the users are the owners and providers of the forestry resource.

Table 4. BioUnivDH impact of BioUnivDH in the three locations under study

	Town A	Town B	Town C
CO_2 emissions save [t/y]	3,435	5,064	4,526
Disposable income [k€/y]	77	124	102
Economic benefits [k€/y]	324	560	408
Jobs [ud]	6	9	8

7. Conclusions

This paper introduces the holistic approach for Universal Biomass District Heating development. It includes all the subsystems and agents involved in the installation and operation of a biomass district heating. It identifies the different roles and benefits for each participating agent and maximizes the whole system benefits, eliminating inefficiencies and barriers, as those related to absence of adequate sectorial regulations for heating networks or inadequate location of resources. The concept has been applied to the central region of Spain, where there are 501 villages with a population in the range 1,000 -10,000 inhabitants and without a natural gas network supply. The analyses show that there is enough capacity for local biomass supply with an adequate interaction with neighbourhood territories where forestry biomass resources are currently under exploited because a reduced local demand. It shows the relevance of the mountain management role. The massive application of the concept to this region has a high potential for energy and CO_2 emissions savings, creation of new direct incomes for municipalities and new jobs. The model estimates a global economic impact above 200 M€ in the region under a concept of circular economy. Three municipalities with the first UnivBioDH systems under construction in this region have been studied and compared, showing the role of the different agents depending of the specific characteristics of the villages and the sustainability of the integrated approach.

References

1. European Parliament, *Off. J. Eur. Union Dir.*, 1–56 (2012).
2. C. Henry, M. Matheu, A. Jeunemaître, *Regulation of network utilities : the European experience* (Oxford University Press, 2001).
3. J. Lizana, C. Ortiz, V. M. Soltero, R. Chacartegui, *Energy* (2016),
4. V. M. Soltero, R. Chacartegui, C. Ortiz, R. Velázquez, *Energy*. 115, 1513–1532 (2016).
5. C. Europe, COGEN Europe- European association for the promotion of cogeneration.
6. Spanish Heat and Cooling Companies Association Report ADHAC, 1–35 (2014).
7. V. M. Soltero, R. Velázquez, R. Chacartegui, M. Carvalho, J. A. Becerra, in *Ecos2015 : 28th International Conference on Efficiency, Cost, Optimization , Simulation and Enviromental Impact of Energy Systems* (2015).
8. D. Andrews *et al.*, Background Report on EU-27 District Heating and Cooling Potentials, Barriers, Best Practice and Measures of Promotion. Report EUR 25289 EN (2012).
9. M. Manzone, A. Calvo, *Renew. Energy*. 108, 250–259 (2017).

Environmental aspects of the geothermal energy utilisation in Poland

Anna Sowiżdżał[1], *Barbara* Tomaszewska[1], and *Anna* Drabik[1,*]

[1]AGH University of Science and Technology, Faculty of Geology, Geophysics and Environmental Protection, al. A. Mickiewicza 30, 30-059 Kraków, Poland

Abstract. Geothermal energy is considered as a strategic and sustainable source of renewable energy that can be effectively managed in several economic sectors. In Poland, despite the abundant potential of such resources, its share in the energy mix of renewable energy sources remains insubstantial. The utilisation of geothermal resources in Poland is related to the hydrogeothermal resources, however, numerous researches related to petrogeothermal energy resources are being performed. The utilisation of each type of energy, including geothermal, has an impact on the natural environment. In case of the effective development of geothermal energy resources, many environmental benefits are pointed out. The primary one is the extraction of clean, green energy that is characterised by the zero-emission rate of pollutants into the atmosphere, what considering the current environmental pollution in many Polish cities remains the extremely important issue. On the other hand, the utilisation of geothermal energy might influence the natural environment negatively. Beginning from the phase of drilling, which strongly interferes with the local landscape or acoustic climate, to the stage of energy exploitation. It should be noted that the efficient and sustainable use of geothermal energy resources is closely linked with the current law regulations at national and European level.

1 Introduction

The utilisation of geothermal energy resources, besides many other positive aspects, carries a number of environmental benefits. Currently, this constitutes the particularly important issue due to the constantly increasing problem of a global pollution in numerous regions of Poland, that is mainly related to the use of fossil fuels. The geothermal energy is environment-friendly and its utilisation does not entail the emission of pollutants [1].

The subject area referring to the environmental impact of geothermal energy utilisation has been discussed so far by many Polish and foreign authors. Predominantly, the positive impact on the environment was highlighted [2-4], nevertheless, the potential risks that might occur during the exploitation of geothermal resources were presented [5-8]. Also, the subject of broadly understood legal frameworks that determine the process of the exploitation of geothermal energy resources was introduced [9, 10]. Environmental perils

* Corresponding author: anna7drabik@gmail.com

may be related to the process of the exploration of geothermal waters (the stage of drilling an exploitive hole) as well as the phase of its exploitation. The paper analyses crucial environmental aspects connected with extraction and utilisation of geothermal energy resources in Polish formal and legal regulations.

2 Geothermal energy in Poland

Geothermal energy is distributed unequally around the world. Especially favourable conditions for the formation of geothermal systems containing vapour deposits – potentially best suited for the industrial application – exist within the edge of lithosphere plates: in the rift and subduction zones. In these areas, the most prospective high-temperature geothermal regions are located. Poland is placed outside the aforementioned zones, and the use of geothermal energy is associated with the occurrence of low-temperature geothermal resources [1, 11].

In our country, there are natural sedimentary-structural pools filled with geothermal waters that temperatures vary from 20°C to even 90°C, and in the extreme case exceed 100°C. Geothermal energy resources in Poland are related to the underground waters of different stratigraphic layers occurring at different depths, within geological units in the Polish Lowland, the Sudetes and the Carpathians. The information about their size is widely known thorough results of numerous research projects [12-19]. The recognition of geological structures for direct or indirect applications of the geothermal energy is an important direction of carried geological research. The most prospective regions for geothermal energy development in Poland are connected with Polish Lowlands and Podhale area (Western Carpathians). The principal resources of geothermal waters in the Polish Lowlands are accumulated in the Mesozoic groundwater horizons. Geothermal waters are accumulated first of all in the Lower Jurassic and Lower Cretaceous formations but significant resources of geothermal energy are assembled also in the Upper Jurassic, Middle Jurassic, Upper Triassic and Lower Triassic formations [13].

Since Poland is endowed with low-temperature geothermal resources, primary applications involve the space heating, bathing and recreation as well as balneotherapy. In 2015, six geothermal district heating plants were operating: in Podhale region and municipalities of Pyrzyce, Mszczonów, Poddębice, Uniejów and Stargard Szczeciński. Moreover, 13 recreation centres are located within the country, where geothermal water is used both for swimming pools and other facilities as well as for preparation of hot domestic water. Geothermal water is also utilised in 10 health resorts for healing and rehabilitation purposes. It is worth noting that numerous projects concerning various geothermal aspects have been under development [20, 21].

3 Law regulations

The text of the internal law in the context of geothermal energy utilisation, has been determined by the conditioning derived from the European Union. Such regulations allow the natural environment to be preserved against the possible negative impact of the geothermal installation.

The notion of geothermal waters is a term commonly used in Polish and world literature. Nevertheless, the Polish law introduces the term of "thermal water", defining it as groundwater that temperature at the outflow of the water intake is not less than 20°C [22 (Art. 5)]. The most important legal act regulating principles of the exploration, recognition and exploitation of geothermal waters remains the Geological and Mining Law [22]. Under the terms of Art. 5 and Art. 10 of the present act, the geothermal waters are treated as minerals and the mining property right is appertained to the State Treasury. Simultaneously, the act constitutes that any mining activities require to obtain the concession, which according to Art. 21.1 [22], is conceded by the proper province marshal. In case of the exploration and recognition of "thermal waters", the obtainment of concession is unrequired, however, the development of a geological project as well as its submission to the aforementioned administration authority are obligatory.

Analysing the process of the acquiring a concession for the exploitation of geothermal waters, the first phase is to develop the project of geological works. The requirements to be met by such a document are prescribed by the relevant Regulation of the Minister of Environment [23]. At this stage, it may be necessary to determine the possible environmental impact of the projected enterprise. As specified in the Regulation of the Council of Ministers [24], drillings with depths exceeding 1,000 m can have a significant effect on the natural environment, therefore it may be required to obtain a decision on the environmental conditions of the project. In such a case, the entrepreneur is obliged to submit to the competent authorities the information card and/or Environmental Impact Report, which specifies the potential impact of investment on the surface and groundwater quality, emission of pollutant into the atmosphere, waste management, water and sewerage management, landscape, noise emissions and electromagnetic fields. At the phase of geological works aimed at the exploration and recognition of geothermal water resources, the regulations delineated in the section VI, chapter 2 of the Geological and Mining Law [22], concerning the operation of mining works are applied. The results of the conducted geological work including the amount of the exploitable geothermal waters resources should be presented in the hydrogeological documentation prepared in the accordance with guidelines stated by the Art. 88 of the Geological and Mining Law [22] as well as by the Regulation of the Minister of Environment on hydrogeological documentation [25]. The hydrogeological documentation of the reservoir of "thermal water" is a strategic document that constitutes the basis for obtaining the consent for the quantity of extracted water and the method of exploitation. In the case of a positive recognition of hydrogeothermal resources, the document has to specify the conditions of spent water re-injection into the rock, with the indication of methods to be applied for the particular water intake to prevent it against colmatation.

The stage of the acquiring a concession for the mining of "thermal waters" is regulated by the section II, chapter 1 of the Geological and Mining Law [22]. While considering the environmental aspect, the application for a such concession, besides the environmental protection requirements [26], includes also a description of how to eliminate or to constrain the negative impact of a potential activity on the natural environment. In addition, it defines several important issues, like information - on the amounts of geothermal resources and planned extraction, geological and hydrogeological conditions of water extraction, as well as for spent water re-injection back into the rock formation. It should be emphasised that the obtained concession is an important document on the sustainable exploitation of the deposit, moreover it remains the basic act determining the treatment of used geothermal waters. What is substantial, according to the Regulation of the Minister of Environment [27], it is required to prepare a technical documentation (for each injection well separately), which is of the responsibility of the mining plant executive. Other procedures must comply with regulations of the Water Law [28]. According to the Art. 9 of the aforementioned act [28], the spent geothermal water, that is discharged into waters or to the ground is considered as wastewater, and its quality have to meet the requirements included in documents pertaining to the protection of natural environment [29-31].

 DOI: 10.1051/e3sconf/20172200164

4 Impact on the environment

Taking into account the legal regulations referring to the sequentially phases of the exploration, recognition and exploitation of geothermal energy resources, it can be noted that environmental care is important during the each stage, and any potential risks are eliminated at the beginning of investment planning. The evaluation of the operating geothermal installation impact on the natural environment can be made dependable on the perspective – on a global or local scale. Namely, considering the aspect of a reduction on the emission of harmful substances into the atmosphere and thus the diminution of fossil fuels utilisation on a global scale, or in the context of the impact of the particular geothermal installation on the local water and land environment. The extraction of geothermal energy may cause an influence on the state of natural environment, therefore potential hazards should be identified and subsequently eliminated or limited in the accordance with the national legislation.

4.1 Impact on the air quality

Due to the use of geothermal energy for space heating purposes in Poland, there is a noticeable improvement of the atmospheric air quality. This is confirmed by the significant ecological effects attained by geothermal installations that started to operate in the 90s of the last century. Also, at the present time when the problem of low-emission is increasing constantly and on the other hand, the preservation of natural environment has a crucial role, the possibility of geothermal energy utilisation remains a particularly important issue. The limitation of emissions of gases such as carbon dioxide (CO_2), carbon monoxide (CO), sulphur dioxide ($S0_2$), and particulate matter (PM10) connected with the incineration of fossil fuels, is of particular importance in Podhale region, which is one of the most frequently visited tourist regions in Poland. According to the Geothermal District Heating Plant - Geotermia Podhalańska [32], the average annual concentration of SO_2 before the implementation of the geothermal project (1994-1998) equalled to 32,6 $\mu g/m^3$. Currently, as the result of growing number of geothermal heat recipients, the concentration of SO_2 amounts to 8-13 $\mu g/m^3$ [32].The matter of PM10 reduction is resembling, the values were 54,3 $\mu g/m^3$ and 43,1 $\mu g/m^3$ before and after the implementation of the geothermal project, respectively [32]. The annual decline of CO_2 emission in 2013 totalled 37,35x10^6 kg [32]. The reduction of CO_2 emissions associated with the development of geothermal district heating in Podhale region in presented on the Fig. 1.

 DOI: 10.1051/e3sconf/20172200164

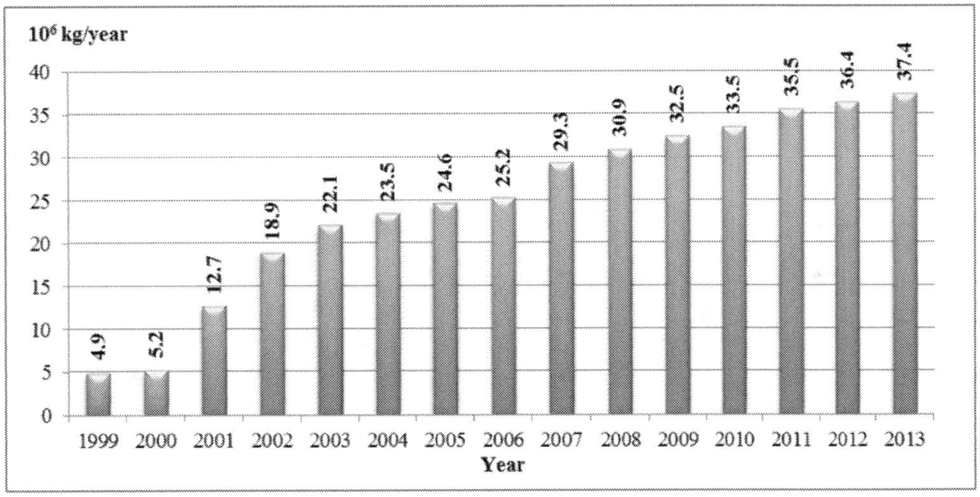

Fig. 1. Reduction of CO_2 emissions in years 1999-2013; based on [32]

Also, the remaining geothermal district heating plants have achieved significant ecological effects, contributing to the improvement of a local air quality. The amount of avoided emission into the atmosphere in 2014, depending on the geothermal plant and type of pollutants, is presented in the Table 1.

Table 1. Delimited emission of pollutants in tonnes/year; based on [33-36]

	SO_2	NO_x	CO	CO_2	dust
Geothermal DH Plant Mszczonów	60	14	*no data*	7,278	145
Geothermal DH Plant Uniejów	5,320	38.06	2.5	299.05	134.72
Geothermal DH Plant Pyrzyce	1,158	257.32	2,759.68	83,608.8	240.99
Geothermal DH Plant Poddębice	6.2	4.4	54.7	4.3	6.3

4.2 Impact on the water and ground environments

The most significant impact on the water and ground environments takes place during the drilling phase (mainly as results of a failure) and exploitation of the geothermal waters (the effect of technological errors). At both stages, the possibility of a contamination of groundwater, rendering them to be unsuitable for use, is considered possible to occur. As a result of conducting the unsustainable exploitation of geothermal water deposits, the fluid can be reduced or even completely exhausted in the definite reservoir. At the same time, the inherent peril associated with the issue of over-exploitation of geothermal water deposits is the problem of possible land deformations (e.g. subsidence, horizontal displacement of rock masses, etc.). Also, as the result of the exploitation of geothermal waters, the rock formation can be cooled down. What is more, the potential risk may be the so-called cold front breakthrough, i.e. the temperature of the reservoir come down as the effect of re-injection of waste waters at much lower temperature (the effect of an incorrectly designed installation). Considering the subject of injecting or discharging spent water into surface water, it is impossible to ignore the potential risk of influence on the mineralisation, pH and temperature of formation waters.

The re-injection of used (cooled down) geothermal waters into the rock formation is the best and the most optimal solution for its disposal. This aspect is particularly required in cases of the extraction of saline waters (usually with mineralisation above 3 g/L). On practical side, the boundary mineralisation of the geothermal water that requires the implementation of injection wells is determined by the proper Regulation of the Minister of Environment [30]. Only the fulfilment of terms contained in the aforementioned regulation, provides an opportunity to obtain the water-legal permit for discharging waste waters into the surface water. Besides the total dissolved solids (TDS), chloride and sulphur content, geothermal waters may contain elevated concentrations of heavy metals and toxic microelements such as chromium, arsenic and boron. Also, the temperature of discharged waters has an significant importance – too high can cause damages to the environment of inland fisheries [31], which is often a factor precluding the entrepreneur from acquiring the permission for dropping waste water into the river.

4.3 Impact on the landscape and acoustic climate

Impact on the local landscape should be considered in the context of the temporary and long-term changes. The short-term impact on the local landscape is principally relevant to the drilling phase - the appearance of drilling rigs, equipment and machinery necessary for running work, as well as the infrastructure for the staff. However, this element is of short duration, usually does not exceed half a year. As for long-term changes, an interference in the landscape is slightly larger. The magnitude of changes in the local landscape largely depends on the current status and use of the area, i.e. mainly whether there is a need for drastic changes in the land use (e.g. deforestation) or the necessity to build transport infrastructure, etc. However, in contrast to exploratory drilling for fossil fuels purposes (gas, petroleum), geothermal wells geared towards obtaining heat energy, are usually located in the close vicinity of recipients - in areas of existing settlements, often with the high-density housing. This is an important factor associated with the limitation of energy losses on the way from the water intake, through the geothermal heat plant to the heat and hot water users.

Another important issue of the analysis proceeded at the exploratory stage is the impact of drilling works on the acoustic climate. Deep hole drilling requires active and uninterrupted 24-hours work. Drilling facilities like compressors, power generators, pumping equipment and, above all, drilling rigs always generate noise that can be disruptive to the local environment. The impact range of the drilling rig in terms of acoustic impact as well as vibration and electromagnetic fields constitutes the important part of the detailed Environmental Impact Assessment.

5 Environmental impacts of Enhanced Geothermal Systems (EGS)

The petrogeothermal energy accumulated within hot dry (or hardly dry) rocks, is exploited on the surface by Enhanced Geothermal Systems (EGS). An Enhanced Geothermal System is an underground reservoir that has been created or improved artificially. Such systems are not currently used in Poland, thus there is no Polish experience in the environmental impact of EGS installation. Nevertheless, environmental aspects concerning the functioning of this type of installation are known worldwide. Most of them are not significantly different from the conventional geothermal installation. However, there are several disparate environmental aspects related only to the EGS installation, that should be identify in order to mitigate their negative impact on the natural environment and make the most of this ecological source of geothermal energy.

Most of potentially important environmental impacts of EGS are associated with ground water use and contamination, and with related concerns about land subsidence and induced seismicity as a result of water injection and production into and out of a fractured reservoir formation. Issues of noise, safety, visual impacts and land use associated with drilling and production operations are also important but fully manageable [37]. To create an EGS, stimulation techniques are needed. The stimulation of EGS is to increase the permeability and to create a loop. Moderated quantity of water is used and it is re-injected in any case so the EGS do not produce wastewater by products [38].

Stimulation techniques may cause induced seismicity that may be sufficiently intense to be felt on the surface. Although induced seismicity can be used as a monitoring tool to understand the effectiveness of EGS operations and shed light on the mechanics of the reservoir, there is some risk that, particularly in seismically quiet areas, operation of an EGS reservoir under pressure for sustained periods may trigger a felt earthquake [37]. EGS projects require access to water during several stages of development and operation. In most advanced EGS applications, surface water will be needed to both stimulate and operate the reservoir and produce the needed circulation patterns. The water may be taken from a nearby high flow stream or river, if available, or collected in a temporary surface reservoir during the rainy season, thusly EGS systems may be approximated as "closed-loop". However, the circulation system is not exactly closed because water is lost to the formation; this lost water must be made up from surface water supplies [37].

The present issue does not concern Polish geothermal installations, which use hydrogeothermal energy, however, together with the growth of geothermal energy sector, the development of EGS technology is expected. Then, the knowledge of these specific and possible impacts on the natural environment will be crucial.

6 Conclusion

Geothermal energy is considered ecological and renewable source of energy that can become an alternative for the Polish energy sector founded on fossil fuels, mainly for the purposes of space heating and hot water supply. The compliance with the formal and legal regulations together with applying the good practice during the various phases of exploration, recognition and extraction of geothermal resources significantly reduces the negative impact on the natural environment of the considered geothermal investments. However, at this stage, it is impossible to omit the issue of the positive impact on the environment of the use of geothermal energy. The avoided emissions of pollutants into the atmosphere (SO_2, NO_x, CO, CO_2, dust) in all geothermal areas in Poland are not only numbers, but also better living conditions and, above all, the possibility to breathe clean air.

The paper has been prepared under the AGH-UST statutory research grant No. 11.11.140.321.

References

[1] B. Tomaszewska, A. Szczepański, Renew. Energ. **21**: 11409-1147 (2014)

[2] R. Shortall, B. Davidsdóttir, G. Axelsson, Renew. Sust. Energ. Rev. **44**: 391-406 (2015)

[3] B. Towler, The Future of Energ.: 237-256 (2014)

[4] A. de Jesus, Geotherm. Power Gener.: 477-498 (2016)

[5] R. DiPippo, Geotherm. Power Pl.: 657-684 (2016)

[6] E. Garrido Schneider, A. Garcia-Gil, E. Vázquez-Suñè, J. Sánchez-Navarro, Sci. Total. Environ. **544**: 354-368 (2016)

[7] R. Allis, C. Bromley, S. Currie, Geothermics **38**, 1: 169-180 (2009)

[8] K. Brown, Sci. Total. Environ. **22**, 1: 61-77 (1981)

[9] E.R. Arias, Geotherm. Pow. Gen.: 763-786 (2016)

[10] S. Haehnlein, P. Bayer, P. Blum, Renew. Sust. Energ. Rev. **14**, 9: 2611-2625 (2010)

[11] A. Drabik, A Sowiżdżał, Geology, Geophysics&Environ. **4** (2017)

[12] W. Bujakowski, B. Tomaszewska, M. Miecznik, Renew. Energ. **99**: 420-430 (2016)

[13] W. Górecki, A. Sowiżdżał, M. Hajto, A. Wachowicz-Pyzik, Environ. Earth Sci. **74**: 7487-7495 (2015)

[14] M. Miecznik, A. Sowiżdżał, B. Tomaszewska, L. Pająk, Geologos **21**, 3 (2015)

[15] A. Sowiżdżał, Carpath. J. Earth Env. **10**: 237-246 (2015)

[16] A. Sowiżdżał, M. Kaczmarczyk, Geol. J. **51**: 65-76 (2016)

[17] A. Sowiżdżał, Appl. Ecol.Environ. Res. **14**, 2: 555-574 (2016)

[18] W. Bujakowski, A. Barbacki, M. Miecznik, L. Pająk, Renew. Energ. **80** (2015)

[19] A. Sowiżdżał, B. Papiernik, G. Machowski, M. Hajto, Geol. Quart. **57**: 729-744 (2013)

[20] B. Kępińska, *Geothermal Energy Use – Country Update for Poland, 2013-2015,* (EGC Proceedings, Strasbourg, France, 2016)

[21] B. Kępińska, *Geothermal Energy Country Update Report from Poland, 2010-2014,* (WGC Proceedings, Melbourne, Australia, 2015)

[22] Ustawa Prawo Geologiczne i Górnicze (Dz. U. 2016 poz. 1131)

[23] Rozporządzenie Ministra Środowiska w sprawie szczegółowych wymagań dotyczących projektów robót geologicznych, w tym robót, których wykonywanie wymaga pozyskania koncesji (Dz. U. 2011 nr 288 poz. 1696)

[24] Rozporządzenie Rady Ministrów w sprawie przedsięwzięć mogących znacząco oddziaływać na środowisko (Dz. U. 2016 poz. 71)

[25] Rozporządzenie Ministra Środowiska w sprawie dokumentacji hydrogeologicznej i dokumentacji geologiczno-inżynierskiej (Dz. U. 2016 poz. 2033)

[26] Ustawa o udostępnianiu informacji o środowisku i jego ochronie, udziale społeczeństwa w ochronie środowiska oraz o ocenach oddziaływania na środowisko (Dz.U. 2016poz. 353)

[27] Rozporządzeniem Ministra Gospodarki w sprawie szczegółowych wymagań dotyczących prowadzenia ruchu zakładów górniczych wydobywających kopaliny otworami wiertniczymi, zatłaczanie wody do górotworu (Dz. U. 2014 poz. 812)

[28] Ustawa Prawo Wodne (Dz. U. 2015 poz. 469)

[29] Ustawa Prawo ochrony środowiska (Dz. U. 2017 poz. 519)

[30] Rozporządzenie Ministra Środowiska w sprawie warunków jakie należy spełniać przy wprowadzaniu ścieków do wód lub do ziemi, oraz w sprawie substancji szczególnie szkodliwych dla środowiska wodnego (Dz. U. 2014 poz. 1800)

[31] Rozporządzenie Ministra Środowiska w sprawie wymagań jakim powinny podlegać wody śródlądowe będące środowiskiem życia ryb w warunkach naturalnych (Dz. U. 2002 Nr 176 poz. 1455)

[32] PEC Geotermia Podhalańska, (11/03/2017: *http://geotermia.pl/ochrona-srodowiska/*)

[33] ZG Mszczonów (11/03/2017: *http://mineralne.pgi.gov.pl/geotermia.html?id=2405*)

[34] Geotermia Uniejów (11/03/2017: *http://www.pnec.org.pl/smart/pdf/uniejow.pdf*)

[35] Geotermia Pyrzyce (11/03/2017: *http://geotermia.inet.pl/ochrona_srodowiska.php*)

[36] Geotermia Poddębice (11/03/2017: *http://www.chronmyklimat.pl/projekty/klimapolka/wiadomosci/geotermia-ogrzewa-poddebice*)

[37] J.W. Tester, B.J. Anderson, A.S. Batchelor, D.D. Blackwell, R. Dipippo, E.M. Drake, J. Garnish, B. Livesay, M.C. Moore, K. Nichols, S. Petty, M.N. Toksoz, R.W. Veatch, *The Future of Geothermal Energy. Impact of Enhanced Geothermal Systems (EGS) on the United States in the 21st Century* (Massachusetts Institute of Technology, 2006)

[38] Geoelec, *Environmental study on geothermal power. Geoelec report* (2013)

 DOI: 10.1051/e3sconf/20172200165

Characteristics of selected elements of the air quality management system in urban areas in Poland

Izabela Sówka[1,*], *Dominik* Kobus[2], *Anna* Chlebowska Styś[1,3] and *Maciej* Zathey[4,5]

[1] Wroclaw University of Science and Technology, Faculty of Environmental Engineering, Unit of Ecologistics and Environmental Risk Management, 50-370 Wroclaw, Wyb. Wyspianskiego 27, Poland
[2] Institute of Environmental Protection – National Research Institute, Air Quality Monitoring Division, 00-548 Warsaw, Krucza 5/11d, Poland
[3] Voivodship Inspectorate of Environmental Protection in Poznan, 61-625 Poznan, Czarna Rola 4, Poland
[4] Wroclaw University of Science and Technology, Faculty of Architecture, Department of spatial planning and settlements processes, 50-317 Wroclaw, B. Prusa 53/55, Poland,
[5] Institute for Territorial Development, 50-527 Wroclaw, J. Wl. Dawida 1a, Poland

Abstract. Most of Europeans living in cities are exposed to concentrations of air pollutants in excess of the thresholds given in the WHO guidelines and EU legislation. Due to this fact, for the urban air quality systems, the mechanisms of proper information and warning of the inhabitants as well as legal, economic and spatial planning instruments should be improved. The analysis of Polish air quality management system and its' selected components (exemplary measures, information-spreading methods, spatial planning instruments) in four selected Polish cities (Wroclaw, Warsaw, Poznan and Cracow) indicated the need to develop effective solutions, among others, in terms of: emission requirements for combustion of fuels of power of up to 1 MW; admission of high emission fuels on the market; legal and coordination issues at the level of implementation of the area development policy and coordination of activities covering issues within the scope of the structure of planning documents including mainly: ambient air protection programs, spatial developments plans in communes and voivodeships, low emission economy plans, plans of sustainable development of public transport, plans of providing heat, electric power and gas fuels to communes, acts of regional parliaments, introducing limitations based on the Environmental Protection Act and strategies of voivodeship.

1 Introduction

Information provided by the World Health Organization (WHO) describes the negative impact of air pollution on human health [1,2] and indicates that most of Europeans living in

* Corresponding author: izabela.sowka@pwr.edu.pl

cities are exposed to concentrations of air pollutants in excess of the thresholds given in the EU legislation and WHO guidelines [3]. Therefore, managing and caring for air quality, especially in urban areas, are one of the current challenges for EU countries.

Urban air quality management (UAQM) is defined as a system that can establish a robust and integrated environmental management life cycle for urban areas of interest; a system that focuses on local actions to deal with urban hot spots of air pollution; activities that identify specific pollution hot spots on the basis of advanced and complex air-quality dispersion modelling and monitoring techniques; and as activities carried out by the government, business entities, industry, NGOs and the society that helps in attaining air quality goals in a specified geographical area [4]. Those activities, which have been carried out for many years, cover conceptional studies on, among others: Air Quality Monitoring Programs (AQMP) as programs aimed at ensuring control of emissions in order to protect the public health and welfare and as systems for designing and implementing monitoring and management activities as well as policies within an urban area, an integrated methodological scheme for evaluation of air pollution control measures schemes, Local Air Quality Management Plans (LAQMP) focusing on the distribution of power at the community level and strong communication and cooperation between entities participating in air quality management, temporary Urban Air Quality Management Plans (e-UAQMP), and software tools. The basic elements of an air quality management system are: air objectives/quality standards (NAAQS), air quality monitoring standards (NAAQM), source apportionment (SA), emission inventory (EI), UAQM strategies, assessment of air pollution exposure and its influence on health and activities carried out by the public. Implementation of UAQM results in creation and introduction of legal provisions to international legal systems, which concern, among others, local and regional areas (or a city) and creation and implementation of programs as well as development of strategies related to, among others, road transport, including: stringent emissions norms, alternative fuels in vehicles, improvement in fuel quality, fitting catalytic converters, use of low and zero emission vehicles, support for cycling, bus lanes, low emission zones, intelligent transport systems, subsidies for registration tax applicable to environment-friendly vehicles, maintenance of road infrastructure [4,5]. The final stage of their application aims at the reduction of pollution emissions and brings benefits to urban residents, such as improvement of the air quality in urban areas, a lower number of hospital patients suffering from bronchitis and asthma and an increase in the average life expectancy [4,6] that are related to it. Thus, it is important that local and national actions are synergic and integrated and that they aim at construction and optimization of an air quality management system, whose synchronized and multi-level operation could eventually result in improved air quality and improved well-being for residents of Polish cities.

The paper shows the characteristics of air quality systems and their exemplary components (exemplary measures, information-spreading methods, spatial planning instruments) applied in selected four Polish cities: Wroclaw, Warsaw, Poznan and Cracow. Shortcomings and proposed measures for improvement of the existing national system of air quality management are also described.

2 Examples and elements of urban air quality management system in Poland

2.1 Air pollution standards and monitoring

According to the Environmental Protection Act (POŚ), [7] agglomerations and big cities (with more than 100,000 residents) constitute zones in which the air quality assessment is carried out in relation to the health of their residents. The remaining zones consist of voivodeships, with the exclusion of those with more than 100,000 residents and agglomerations, for whom the air quality assessment is done in order to protect health and plant life. The assessment in terms of human health takes into consideration the content of SO_2, NO_2, CO, C_6H_6, O_3, PM_{10} and $PM_{2,5}$ particulate matters, heavy metals (As, Cd, Ni, Pb) and benzo(a)pyrene in PM_{10}. The aforementioned types of pollution were assigned limit, target and long-term (in case of ozone) levels of concentration in the air [8]. Due to the significant harmfulness to health of $PM_{2,5}$ particulate matter, standards concerning exposure of residents of big cities and agglomerations to $PM_{2.5}$ particulate matter have been conducted [9]. Short time exposure to very high levels of air pollution can cause serious damage to the human body, and that is why Polish law establishes appropriate alert levels. Information levels have been also established for PM_{10} particulate matter and O_3. An information level is a level of concentration of a given substance in the air, which if exceeded, can cause a threat to human lives resulting from a short exposure of sensitive groups of people to pollution, and that is why it has to be immediately and appropriately announced to the public. An alarm level means that there is such a concentration of a pollutant that if exceeded even for a short time, it can be life threatening. Those levels have been established for O_3, PM_{10}, SO_2 and NO_2 [8].

Air quality monitoring is carried out in Poland by Voivodship Inspectorates for Environmental Protection (WIOS) in accordance with Voivodeship Environmental Monitoring Programs. Supervision over the State Environmental Monitoring (PMS), as part of which measurements referred to in the Voivodeship Programmes are carried out, is maintained by the Chief Inspectorate for Environmental Protection (GIOS). Measurements of pollutant concentrations are done according to referential methodologies indicated in [10]. The results of the measurements at the Voivodeship level are then submitted to the database of GIOS (JPOAT2.0 database). After measurements are accepted at the Voivodeship level, the data is subjected to analysis and selection in terms of compliance with established criteria and possibilities of their use for annual air quality assessment.

Measurement of air pollution belong to the responsibility of WIOS appropriate for the given territory and are supervised by the National Reference and Calibration Laboratory (KLRiW) of GIOS, which belongs to the Air Quality Reference Laboratories (AQUILA). The main tasks of KLRiW are: maintaining reference standards, calibration of analysers, carrying out audits in monitoring stations, carrying out national intercalibration of particulate and gas meters, organisation of training courses and implementation of new measuring and research methods.

The quality of air in cities is established on the basis of measurements carried out by WIOŚ, complemented by mathematical modelling and estimation methods. There are two types of measuring stations on urban territories – background and traffic stations localised in the direct vicinity of roads with high traffic. Their aim is to study the impact of road transport.

During the winter season, Polish cities are faced with problems such as, among others, increased concentration of particulate matters and benzo(a)pyrene and that is why, for the needs of the paper, their annual concentration values between 2010-2016 (Fig.1) were averaged, and daily average of PM_{10} concentrations were compiled (Fig. 2). The annual and daily average values were obtained thanks to measurements done by urban background and traffic stations, located in the following cities: Poznan, Wroclaw, Cracow and Warsaw. The analysis of results of concentration measurements indicates (Fig. 1) that since 2010, PM_{10} particulate matter concentrations are decreasing, even though they are at a level exceeding allowed standards (Cracow traffic and background and Warsaw traffic stations) or are close to the allowable value (Warsaw background, Poznan background and Wroclaw background stations). In some stations, the standard for annual average concentration was exceeded

 DOI: 10.1051/e3sconf/20172200165

(Fig.1), however, this happened mainly in case of average daily concentrations (Fig. 2). The highest concentration levels were found to be present in Cracow and Warsaw. In case of Cracow, this results mainly from the geographical localisation of the city, which makes ventilation of premises impossible, and combustion of fossil fuels like coal (often with low quality) in the residential sector. In case of Warsaw, especially in it central area, increased car traffic plays an important role. High concentration of PM_{10} in the air results in high concentration of one of its components: benzo(a)pyrene. Every year, in all cities studied, the concentration of this substance was above the standard, and the highest concentration was found in Cracow, which was slightly lower in Wroclaw and Poznan, while the lowest in Warsaw (Fig. 1). Due to the above, it can be concluded that what influences the air quality in Cracow, Wroclaw and Poznan is combustion of fuels in the residential sector, while in case of Warsaw it is the road traffic. The measurements of PM10 concentration in 2015 carried out in urban background stations (Fig. 2) show clearly that there is a seasonal variety of PM_{10} concentration in the air. Higher concentrations occur in autumn and winter, and this clearly shows the influence of the residential sector on the quality of air in the analysed cities.

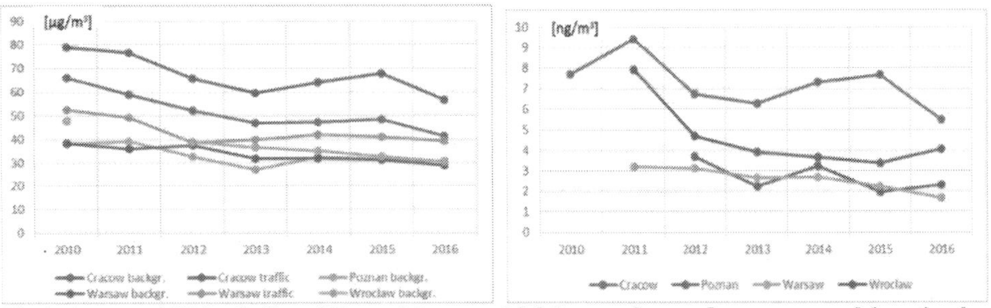

Fig.1. Averaged PM_{10} (left) and Benzo(a)pyrene (right) annual mean [source: own elaboration based on PMS/ GIOS data]

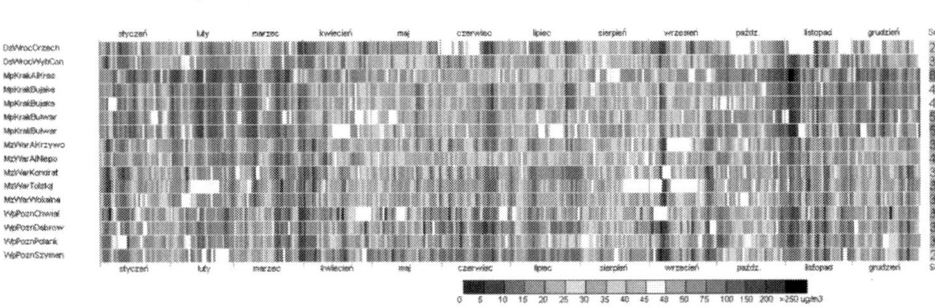

Fig. 2. Daily averages of PM_{10} concentration in 2015 [source: as above]

2.2. Information-spreading methods

Information about the air pollution level is an important element of an urban air quality management system. It is related to fulfilment of legal obligations towards selected public institutions. Those obligations are included in both international acts [11, 12], and national documents [13]. Another factor that contributes to it is the social need for reliable and current information on the quality of chosen components of the environment, including ambient air. An important aspect is also the educational significance of information and its possibility of shaping chosen behaviours of its recipients, such as the choice of a mean of transport [14].

Information systems involve, among others, publication of reports on measurement results and assessment of air quality, presentation of the current state and forecasts concerning air quality for the next few days. The last two elements have an informative and alerting function, which, apart from the decisive and management or the forecasting and planning function are the main purpose of operation of public information systems [15].

Information about the quality of air is provided at different levels and with varied amount of details. At the national level, GIOS operates an information system based on data coming from monitoring stations, which belong to the PMS network, and forecasts of future ozone concentration [16]. WIOS, using their websites, present information concerning their voivodeship area [17-20]. Data presented by units which are officially involved in PMS are subject to strict quality assurance and quality control (QA/QC) requirements related to, for example, the use of reference measurements or equal methods, calibrating and intercalibrating systems, supervision and multi-level assessment of the results as well as assessment of the station's localisation in terms of its representativeness and compliance with the rules.

Information systems are needed the most in situations where the content of a given substance is the highest and thus, the social interest is also the highest. That is when warning messages are published with information on potential health consequences and recommendations, especially for those belonging to sensitive groups. As part of short-term activities plans adopted for Warsaw, Cracow, Wroclaw and Poznan, procedures concerning information activities taken in case of the need to make announcements were established for every level. The level of the alert (from I to IV) depends on the observed (exceeded) or forecasted (risk) pollution concentration in the air. For example, the action plan for Warsaw concerning exceeded ozone concentrations establishes the following procedure of passing information: the Mazovian Voivodeship Inspector for Environmental Protection informs the Voivodeship Crisis Management Team (WZZK) and the Mazovian Voivodeship Management. WZZK then informs the President of the capital city of Warsaw about the incident, and the President submits the information to organisational units of the capital city of Warsaw, including the Crisis Management Team of the capital city of Warsaw. The latter informs local media, local rescue services, hospitals, educational establishments, entities and institutions obliged to act and the society. Procedures adopted for other cities and applicable to other types of pollution are similar.

Selected cities create their own local information systems that are a part of an urban air quality management. One of such cities is Warsaw, which is working on its Warsaw Air Index (WIP) supposed to consist of an algorithm for processing and presenting measuring data from PMS stations and a local network of several non-reference sensors independent from them [21]. The WIP is also supposed to include a prediction system generating information for the city residents as well as supporting the process of making management decisions in the city.

2.3. Spatial planning instruments

Pursuant to the Environmental Protection Act [7], the basic instruments intended got air protection in Poland are regional programs of air protection (POP), which establish their aims, but do not include any financing or control measures. It is also possible to adopt a resolution introducing limitations concerning fuels and fuel combustion systems.

On the basis of the act on spatial planning and management of voivodeships, it is possible to create functional zones and define the rules of spatial development for them [22]. This can be done on the territory of, among others, a capital of a voivodeship and its surroundings – the city's functional zone. Those regulations can also concern another type of a functional zone, for example mountain areas – that are territories which are distinguished by their morphology, transport access and, very frequently, loss of their function. A local spatial development plan can also include arrangements concerning thermal energy by way of indicating sources of heat energy. It can also cover emissions of equipment in relation to heating capacity. In addition, the plan can establish protection zones and introduce limitations there. However, such provisions will only apply to new investments and actions taken after the plan is adopted. Such provisions are not corrective actions applicable to already existing forms of spatial development and heating sources already used.

One of the significant aspects of air protection is the spatial planning process and the need to improve the level of energy efficiency of urban layouts. Even though there is a methodology allowing for calculation of energy efficiency of buildings, it does not apply to cities or wider residential layouts, including suburban zones and the core of functional areas. However, the residential estate's localisation, transport systems and their equipment with technical and social infrastructure have a significant influence on energy consumption of the society living there and business entities operating there. Dispersion of buildings, lack of correlation between settlement development and technical infrastructure, including heating and gas networks, is favourable for dispersion of pollution emission sources and increases the surface emission. If energy consumption is multiplied within several years after the investment is implemented, it will result in such an amount of energy and a size of pollution that will have influence of exceeded allowable substances levels in the air. In addition, costs of such ineffective wat of power supply to a settlement system will be multiplied as well. Current attempts to estimate those costs and already established scientific methods have not been yet implemented in the planning system [23-25]. Another conclusion from the considerations above is that the level of energy efficiency of buildings and also of settlement systems, is responsible for high concentration of pollutions in the air. As it applies to the residential sector, it is also a justification for additional increase of emissions of, especially, benzo(a)pyrene and PM_{10} particulate matter.

As part of local activities, non-obligatory (from the legal point of view) local plans, programs of regeneration and low-carbon economy plans to be adopted in communes can become instruments supporting increased energy efficiency of buildings and functioning of entire housing estates. Those documents can be drawn by communes optionally; however, they are required in the process of applying for funds from the European Regional Development Fund as part of long-term operational programs. Thus, no financial support can be granted if there is no previous concept. Such a plan requires the commune to systematize its activities concerning low-carbon economy and can result in both new investments, changes to the current ways of functioning of the society and the economy in the commune.

Strategies for elimination of air pollution are based on three main fields of public intervention: activities targeted at limitation of emissions by production plants (point emission), elimination of pollution from the residential sector (low-stack emission) and development of public transportation together with minimization of car traffic (linear emission). According to the reports submitted by WIOS, the biggest challenge of all is the size of emissions generated in the residential sector. Point emissions elimination can be done via legal regulations applicable business entities; however, taking into account the experience of Germany, it should be predicted that the significance of the issue with linear emissions in Poland would increase in the future.

According to the statutory regulations, what is currently supposed to be Poland's most effective measures for the limitation of surface pollution is acts adopted by regional parliaments. Their aim is to introduce limitations and prohibitions concerning fuels and systems for their combustion within the territory of selected voivodeships (table 1).

Table 1. The process of preparing the anti-smog resolution regarding the cities under study

City	Voivodeship	Resolution status	Development stage and the territorial reference
Cracow	Lesser Poland Voivodeship	Accepted	The resolution adopted in two stages, separately for the city of Cracow and the rest of the province.
Poznan	Greater Poland Voivodeship	In the development, horizon – the end of 2017	Consultation with local governments, the resolution will cover the entire voivodeship.
Warsaw	Masovian Voivodeship	In the development, horizon – the end of 2017	Consultation with local governments, the resolution will cover the entire voivodeship.
Wroclaw	Lower Silesian Voivodeship	In the development, horizon – the end of 2017	Consultation with local governments. Territorial differences for city of Wroclaw, health resorts and rest zones of the voivodeship.

3 Summary

The basic elements functioning in the urban air quality management system in Poland are constantly developed in compliance with (or aiming at compliance with) requirements referred to in applicable legal regulations. However, legal and coordination issues at the level of implementation of the area development policy should be improved. Environmental protection, including protection of ambient air, is seen as an activity limiting investment processes. Such an approach should be changed mainly by presenting the development policy as a policy improving the well-being of residents and ensuring good environmental situation.

Currently, the biggest difficulties in activities aiming at protection of ambient air are delegation of tasks to self-governments without and legal support at the national level, especially in case of emission requirements for combustion of fuels of power of up to 1 MW and admission of high emission fuels on the market. The current legal solutions also limit the possibilities of controlling natural persons. Voivodeship self-governments, which are responsible for preparation of ambient air protection programs, do not have any coercive means or financial support for communes. Works on coordination of those activities should cover issues within the scope of the structure of planning documents including mainly: ambient air protection programs, spatial developments plans in communes and voivodeships, low emission economy plans, plans of sustainable development of public transport, plans of providing heat, electric power and gas fuels to communes, acts of regional parliaments, introducing limitations based on the Environmental Protection Law and strategies of voivodeship development. The structure of coordination of the development policy should be made simpler and the voivodeship spatial development plan should be integrated with its development strategy. Then, strategic aims concerning air protection would gain a territorial dimension at the stage of their definition.

A key issue within the scope of air quality management in Poland is the need to change the basis of the power production paradigm in Poland from fossil fuels to other energy sources, including renewable ones, and to limit pollution emissions from road transport.

References

1. Review of evidence on health aspects of air pollution (REVIHAAP). Technical report. Copenhagen: WHO Regional Office for Europe 2013

2. Health risk assessment of air pollution – general principles. Copenhagen: WHO Regional Office for Europe; 2016)A. Mecke, I. Lee, J.R. Baker jr., M.M. Banaszak Holl, B.G. Orr, Eur. Phys. J. E **14**, 7 (2004)
3. Air quality in Europe — 2016 report, EEA Report No 28/2016
4. S. Gulia, S.M. Nagendra, M. Khare, I.Khanna, Atm. Poll. Res., **6**, 286 (2015)
5. L. Duque, H. Relvas, C. Silveira, J. Ferreira, A. Monteiro, C. Gama, S. Rafael, S. Freitas, C. Borrego, A.I. Miranda, Atm. Environ. **127**, 196 (2016)
6. A. Miranda, C. Silveira, J. Ferreira, A. Monteiro, D. Lopes, H. Relvas, C. Borrego, P. Roebeling,, Atm. Poll. Res. **6,** 434 (2015)
7. *Environmental Protection Act*, Dz.U. (Polish Journal of Laws) No. 62 pos. 627 with subsequent amendments (2001)
8. Regulation of the Polish Minister of the Environment *on the levels of certain substances in the air*, Dz.U. pos. 1031 (2012)
9. B. Toczko in: *Fine dust in the atmosphere. Compendium of Knowledge of Air Pollution from particulate matter in Poland* (GIOŚ, Warsaw 2016, in Polish)
10. Regulation of the Polish Minister of the Environment *on the assessment of levels of substances in the air*, Dz.U. pos. 1032 (2012)
11. Directive of the European Parliament and of the Council *on establishing a scheme for greenhouse gas emission allowance trading within the Community and amending Council Directive 96/61/EC*, No. 2003/4/EC
12. Directive of the European Parliament and of the Council *on ambient air quality and cleaner air for Europe* (CAFE), No. 2008/50/UE
13. Regulation of the Polish Minister of the Environment *on the way and frequency of updating environmental information*, Dz.U. No. 227 pos. 1485 (2010)
14. D. Kobus, ENVIRON **34**, 69 (2008)
15. J. Oleński, *Information infrastructure of the state in the global economy* (Department of Economics of Warsaw University, Warsaw 2006, in Polish)
16. http://powietrze.gios.gov.pl
17. http://sojp.wios.warszawa.pl,
18. http://air.wroclaw.pios.gov.pl
19. http://powietrze.poznan.wios.gov.pl
20. http://monitoring.Cracow.pios.gov.pl
21. http://www.um.warszawa.pl/aktualnosci/r-ne-aspekty-ochrony-powietrza-w-warszawie
22. *Planning and Spatial Development Act*, Dz.U. No. 80 pos. 717 (2003)
23. S. Czaja S., B. Fiedor in: *Environmental aspects of the urban economy and new instruments in the urban management* (University of Opole , Opole 2000, in Polish)
24. A. Kowalewski, J. Mordasewicz Konfederacja J. Osiatyński, J. Regulski, J. Stępień P., *Report on economic loss and social costs of uncontrolled urbanization in Poland* (Institute of Geography and Spatial Sciences (GaSSI) of PAS, Warsaw 2013, in Polish)
25. P. Śleszyński P., M Andrzejewska, D. Cerić, A. Deręgowska, T. Komornicki, M. Rusztecka, J. Solon, P. Sudra, B. Zielińska, *Analysis of the status and conditions of planning work in municipalities in 2014.* (GaSSI of PAS, Warsaw 2016, in Polish)

Hydraulic optimization and modeling of hydro-cyclone-systems for treatment and purification of any kind of waters

Lars Spangemacher[1,*], *Siegmund* Fröhlich[1], and Hauke Buse

[1]Hochschule Emden/Leer, Constantiplatz 4, 26723 Emden

Abstract. Water is an indispensable resource for many purposes and good drinking water quality is essential for mankind. This article is supposed to show the need for mobile water treatment systems and therefore to give an overview of different mobile drinking water systems and the technologies available for obtaining good water quality. The aim is to develop a simple to operate water treatment system with few processing stages such as multi-cyclone-cartridge and reverse osmosis with energy recuperation, while the focus is set on modeling and optimizing of hydrocyclone systems as the first treatment stage.

1 Introduction

Good water quality is essential for mankind and an indispensable resource for many purposes. Corresponding to the growing number of people on earth, the worldwide water demand also rises continuously. Today, already 20 % of the world's population experiences a substantial lack of accessible drinking water resources – yet the demand is expected to keep rising dramatically over the next years [1].

Statistic figures confirm the urgency for change. Millions of people have died due to insufficient water quantity and drinking contaminated water. According to the WHO, approximately 88 % of worldwide diseases result from polluted water [2] And, following UNO, there are roughly 263 "international water resources" – corresponding to approx. 60 % of global water occurrence – where legal claims to water usage remain unsettled [3]. That leads to serious conflicts and triggers the migration of thousands towards clean water resources.

In the tragic event of natural disasters, political conflicts or extreme climate situations the existing water shortage becomes particularly urgent. As such, tsunamis, hurricanes or earthquakes like, for example, in Nepal in April 2015 have shown that rural places are especially vulnerable to malfunctioning water supply. Although Kathmandu's water supply was re-established within a few days, people living in isolated mountain regions had to wait weeks or even months to receive good drinking water after the reconstruction of water supply systems.

*Corresponding author: lars.spangemacher@hs-emden-leer.de

The aim of this research is to develop a safe and reliable technology, which is capable of supplying drinking water in any of the above mentioned tragic events.

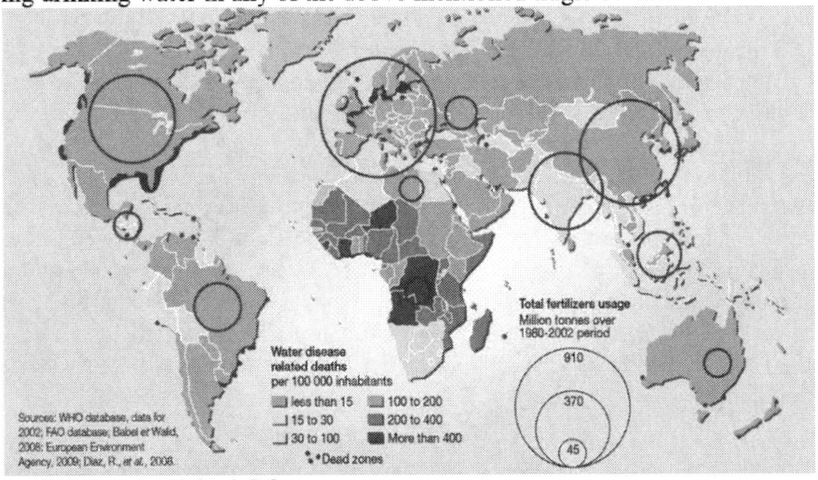

Fig. 1. Water diseases related death [2]

The rising water shortage in many world regions has different causes. They include the rising world population, which boosts the high agricultural water consumption and the increase of water pollution. However, the problem of water shortage affects different populations more or less severely. Due to unbalanced water distribution, unpolluted drinking water is taken for granted by some, whereas other populations regard it as precious commodity. Common strategies to solve that supply problem range from sustainable water usage, multi usage of water up to long distance transport of water. According to many population development prognoses the drinking water demand will rise drastically, because the world population will grow intensively in the next decades.

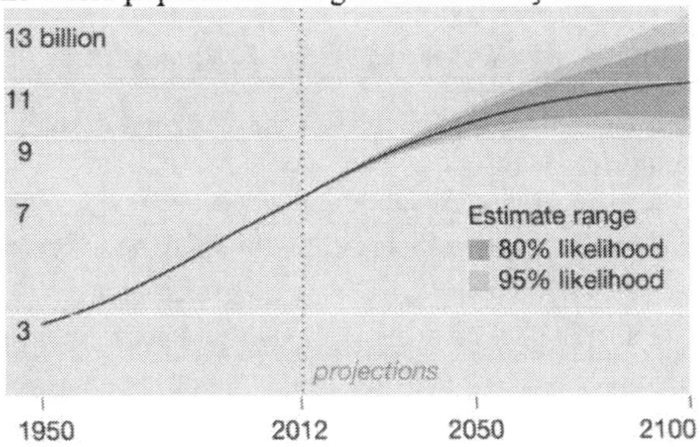

Fig. 2. World population prognosis 2100 [4]

In fact, countries such as Kenya, Egypt, Tunisia, Israel, Saudi Arabia, Jordan Kuwait, Ruanda or Bahamas are especially vulnerable to water scarcity. In the forthcoming years huge countries like China, India, Mexico, Nigeria and others will also be threatened by increasing water stress.

As consequences of those scenarios for the governments of those countries and for the international foreign aid organizations there is huge advantages coming soon

1.1 Catastrophes and other causes of water stress and water shortage

Particular importance results in terms of natural catastrophes, military conflicts and extreme climatic situations. In the past, natural catastrophes caused by earthquakes, tsunamis, hurricanes and others have led again and again to insufficient supply of drinking water. In case of disastrous earthquakes as, for example, in Nepal 2015 the drinking water supply of rural mountain regions was not restored for weeks or months. Acc. UNICEF approximations only 20% of inhabitants had access to safe water sources.

Those catastrophic incidences of insufficient water access are likely to be followed by malnutrition and devitalization among first and foremost children. These conditions can be live-threatening. In 2015, nearly 1,7 million Nepali children were affected. Although in the urban region of Kathmandu the drinking water distribution could be restored within few days [5] far regions in the surrounding mountains were disconnected for weeks or longer from good drinking sources. The restoration of plenty kilometers of the water grid took such a long time.

1.2 Mobile Drinking Water Unit Demand for Difficult Accessible Rural Regions

Nowadays there are several transportable drinking water plants available on the market. The biggest disadvantage of them is their few cubic meters production output per hour. For example, the German Public Social Aid Organization (Technisches Hilfswerk - THW) disposes plants based on a modular system, which contains up to seven process steps at a water production rate of few cubic meters per hour. Beside a rather big space requirement and adequate logistics, the effort of qualified personnel and necessary chemicals and auxiliary materials also quite high. Such a support of far rural regions as in Nepal is realizable only at high expenses.

1.3 State of Art of Mobile Drinking Water Processing

Current mobile plant technologies available on the market for such disaster management are based on several process engineering unit operations [6, 7] as:

- mechanical separation of solids and waste by sieves, filters, etc. [6]
- physical processes as sedimentation, flotation, aeration or adsorption [7]
- aerobic or anaerobic biological water purification [8]
- membrane processes as reverse osmosis or electro dialysis [9]
- vacuum processes as mechanical vapor compression [10]
- thermic processes as distillation or multi-stage-flash-processes [11] or
- water chlorination [12] and
- UV-disinfection [13]

However, the biggest weakness of these state of art systems is their poor water production rate. Additionally, those multistage systems are massive in size and need specially trained personnel to oversee the difficult procedure and ensure good results. That implies, that the produced drinking water with such state of art plants is for the most people in crisis situations as well not technologically as well economically realizable.

An essential process to create a compact, smart and mobile drinking water production unit is reverse osmosis. Reverse osmosis is a unique technology, which can eliminate as well as dissolve inorganic and organic compounds. After a pre-treatment of solid- and oil-elimination this process is able to assure adequate drinking water quality and quantity.

Since more than 40 years diverse processes of water desalination are established to stop the water shortages worldwide. But all of them come with quite high cost investments and huge energy demands to produce drinking water. Therefore, water desalination plants were initially installed mostly in quite rich counties. Most of those seawater desalination plants are located in Middle East, North America and Mediterranean Countries. At this moment there are more than 14.000 stationary large scale seawater desalination plants operating worldwide, which produce a total of over 95 mln. cubic meter per day.

Technological progress took place in those plants, so the lowering of invest costs and energy demand made sure, that the reverse osmosis process is the most successful technique to deliver best quality drinking water. Particularly the implementation of energy recovery modules in the high pressure hydraulic part could lower substantially the energy demand of desalination.

2 A New technological Solution

At the University of Applied Sciences Emden-Leer, we are designing a system for mobile water treatment which draws on approved concepts, yet improving them in several aspects. Drinking water treatment plants must purify raw water, originating from wells or surface waters, in several process steps. There are already various methods implemented, such as filtration (table 1), followed by chemical (chlorination / iodination) or physical (irradiation by ultraviolet light) disinfection process steps.

Table 1. Filtration methods (Drews, A.; TU Berlin - Principles of membrane processes)

Pore size	Process	Filtration pressure	Separation of
> 10 µm	Filter		
> 0,1 µm	Microfiltration	< 2 bar	Bacteria, yeast, particles
100–2 nm	Ultrafiltration	1–10 bar	Macro molecules, protein
2–1 nm	Nano filtration	3–20 bar	Viruses, 2-valent Ions
< 1 nm	Reverse Osmosis	10-80 bar	Salt, small organic molecules

As all established purification processes [6, 14] have a significantly high energy demand, especially in disaster areas without functioning energy infrastructure, usage does not come without problems. The proposed plant concept for mobile drinking water treatment consists of a 2-step treatment process.

Step 1: Reject Separation by means of a new hydrocylone cartridge

Step 2: Reverse Osmosis (RO) with energy recuperation

As one design parameter for the drinking water treatment plant, this process aimed at producing of 20,000 l/h, which is sufficient to supply up to 25,000 people [15]. The system concept is scalable to almost any size due to its modular design – allowing higher production rates of drinking water.

2.1 Step 1 – Hydrocyclone cartridge

A newly developed hydrocyclone cartridge forms the first stage of the water treatment concept. This cartridge will combine many individual hydrocyclones with common hydraulic connections requiring very small space. The separation limit d50 is designed to 5 microns; for efficient separation of both waterborne inert particles and (micro-) organisms.

Hydrocyclones have long tradition in the field of particle separation, especially in mining and basic materials industry, while they are characterized by a maintenance free operation.

Within the scope of a research project different types of hydrocyclones have been tested regarding their separation performance as a first stage for mobile drinking water treatment. From these results a very compact Version was designed. It combines several single hydrocyclones in a small chamber with common hydraulic connections while aiming at a separation limit of d50: 5µm; sufficient to separate inert particles as well as microorganisms from water. Due to patenting this hydrocyclone cartridge cannot be revealed at this time.

2.2 Step 2 – Reverse osmosis

A second stage reverse osmosis (RO) filtration ensures a germ-free drinking water due to the small membranes pore size. Common RO-systems have the disadvantage of consuming high amounts of energy in order to reach the high pressure needed at the primary membrane side. Newly developed large scale seawater desalination plants solve this dilemma by means of mechanical energy-recovery.

By using energy recovery processes the disadvantage of high energy consumption will be compensated to a large extend. Thus, this technology can also operate on temporary power generators (e.g. mobile photovoltaic generators). A modern high pressure pump requires due to its energy recovery approx. 40% less electrical energy [16] compared to traditional pumps without any energy recovery option. This is done by recirculating unused feed water to the pump unit.

The main advantage of using a reverse osmosis – after solid separation by means of hydrocylones – is the possibility to process any sweet water as well as seawater loaded with any kind of organic and inorganic matter to reach a water quality according drinking water ordinance. Depending on the pore size of the RO membrane, microelements can be removed almost completely to ensure good drinking water quality without any hazards regarding human health.

3 Modeling of a hydrocyclone prototype

There are numerous approaches for modeling of hydrocyclones which can be broken down to 7 different categories. Their validity is mostly restricted as most of them were developed by use of empirical data. The following description is limited to the so called thin stream classification with volume concentrations $c_v < \sim 10\%$ [17]

The contemplation by Trawinski provides the following statement with regard to separation size. Particles will be separated in case their sedimentation velocity w_{fz}

$$w_{fz} = \frac{\Delta\rho}{18\eta} d_{St,t}^2 \cdot r_i \cdot w^2 \quad (1)$$

in the centrifugal field is larger than the median introversive radial velocity v_{ri}.

$$v_{ri} = \frac{v_O}{2 \cdot \pi \cdot r_i \cdot h_i} \quad (2)$$

In terms of separation size x_t , here as stoke diameter $d_{St,t}$, the equality of velocities applies:

$$w_{fz} = v_{ri} \quad (3)$$

Formally this correlates to the equality between surface loading and settling velocity. Therefore we get

$$d_{St,t}^2 = \frac{18\eta}{\Delta\rho} \cdot \frac{Vo}{2\pi\, r_i^2 \varpi^2 h_i} \tag{4}$$

The upstream volume flow V_O is proportional to the total volume flow V_A where Trawinski makes the following assumption:

$$V_A \sim d_e \cdot d_o \cdot \sqrt{\Delta\rho/\rho} \sim V_o \tag{5}$$

The idea behind this assumption is that hydrocyclones, with regard to pressure drop, can be seen as a pipe with a median diameter $\sqrt{d_e \cdot d_O}$. Assuming the proportionality $d_e \cdot d_o \sim D^2$ and $h_i \sim L$ we get:

$$d_{St,t} \sim \sqrt{\frac{18\eta}{\Delta\rho}} \cdot \sqrt{\frac{D}{L}} \cdot \frac{\sqrt{D}}{\sqrt[4]{\Delta\rho/\rho}} \tag{6}$$

Here we can see, that $d_{St} \cdot t$ is getting smaller (and separation is becoming better) the smaller the hydrocyclone diameter and the slighter the cyclone is. The difference in pressure which is applied to the hydrocyclone only has a small impact The first factor, where changes in given material systems are not possible, is the characteristic material system combination in all sedimentation equations, if only the stokes condition of viscous particle flow is fulfilled. A similar correlation for the median separation size d_T , which considers the volume flow ration $\frac{V_O}{V_U}$ and the solid volume concentration C_v , has been developed by Neeße and Schubert with the help of their model of turbulent cross-flow classification:

$$d_T = K \sqrt{\frac{\eta m}{\Delta\rho}} \cdot \frac{\sqrt{D}}{\sqrt[4]{\frac{\Delta\rho}{\rho m}}} \cdot \sqrt{\frac{ln\frac{V_o}{V_u}}{(I-c_v)^{4,65}}} \tag{7}$$

K is a specific adaptation constant with the value K=0,12 for the version used by Neeße and Schubert. For the material values ρ_m and η_m density and viscosity of the suspension have to be used. The median suspension density is calculated depending on the volume concentration C_V as follows:

$$\rho_m = \rho_s \cdot \rho_v \cdot + \rho_f \cdot (1 - c_v) \tag{8}$$

whereas ρ_s and ρ_f shall be the solid density and liquid density. The median suspension viscosity for concentrations up to about 30% is calculated by:

$$\eta_m = \eta \cdot \left[1 + \frac{1{,}25 \cdot c_v}{1 - c_v/c_{max}}\right]^2 \tag{9}$$

The solid volume portion $C_{V\,max}$ for the sedimented suspension the value is in the range between 0,63 and 0,84.

Apart from cut size also throughput and pressure drop are important parameters. They are linked together by the following formula:

$$V = K_V \cdot d_e \cdot d_o \cdot \sqrt{\Delta\rho/\rho} \tag{10}$$

with Kv Rj 0,33 ... 0,39 for long construction types *(L/D* > approx. 2)
and Kv Rj 0,22 ... 0,28 for short construction types

The distribution of upstream and downstream is defined by the diameters d_o and d_u. There also exist empirically adapted formulas, e.g. from Tarjan

$$\frac{V_O}{V_U} \approx 0{,}91 \cdot (d_o/d_u)^3 \tag{11}$$

Rietema established the following optimal proportions for dimensioning of hydrocyclones with regard to cut zise, throughput and pressure drop

Table 2. Optimal hydrocyclone proportions according Rietema

d_o/D [in mm]	d_u/D [in mm]	d_e/D [in mm]	L/D [in mm]	$(L-h_i)/D$ [in mm]	Cone angle [in °]
0.25	0.8	0.25	9.63	0.94	6°

4 Experimental results

Tests with model media have been conducted. Therefore the new hydrocyclone prototype has been benchmark tested with a commercially available cyclone. Tests were run at different conditions while changing the parameters input pressure (and respectively input volume flow) as well as downstream nozzle diameter. These tests have shown, that the hydrocyclone prototype is capable in achieving similar or even better results with regard to solid concentration reduction.

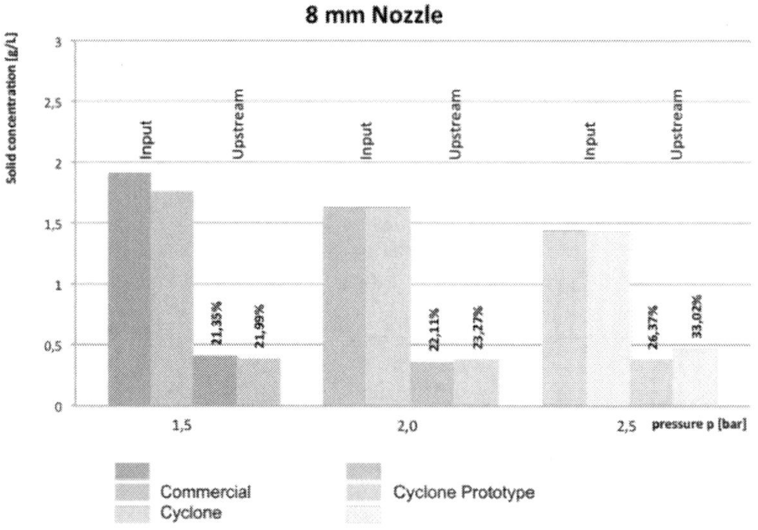

Fig. 3. Results comparing commercially available hydrocyclone with prototype development (here with 8mm downstream nozzle)

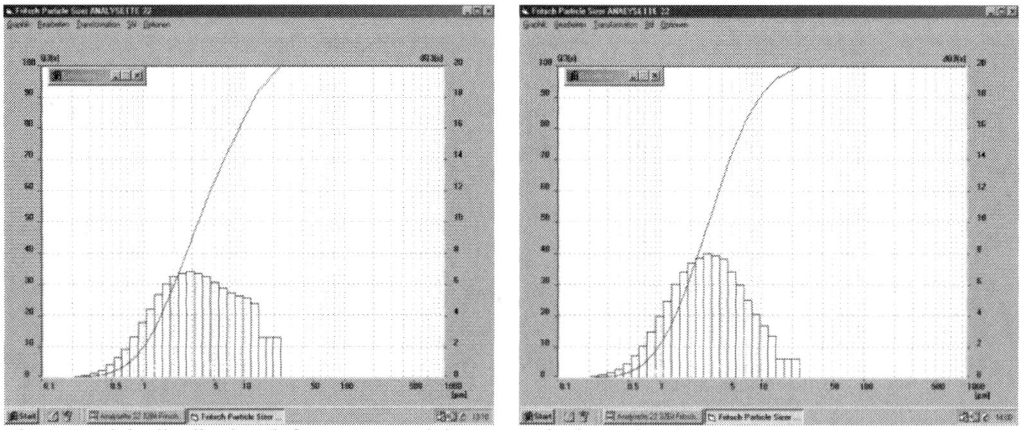

Fig. 4. Particle distribution (left: prototype; right: comm. hydrocyclone)

5 Conclusion

In general we can conclude, that hydrocyclone systems are capable of removing solids and other large particles from water as a first treatment stage for mobile drinking water purification systems, whereas parameters such as downstream nozzle diameter as well as input pressure and respectively the volume flow have a strong impact on the degree of purity of the treated water. Further investigations have to be conducted using water from natural sources to analyze the purification potential

6 Scientific Literature

1. G. Rüschelmeyer "Das Parlament" des Deutschen Bundestags und Bundeszentrale für politsche Bildung (2006) https://www.bundestag.de/dasparlament/2006/25/Thema/017.html
2. http://www.who.int/water_sanitation_health/diseases (2016)
3. M. Kalinin, UN Water Transb. waters: Sharing Benefits, Sharing Respons. http://www.unwater.org/downloads/UNW_TRANSBOUNDARY.pdf (2008)
4. R. Kunzig, Nat. Geogr. http://news.nationalgeographic.com/news/2014/09/140918-population-global-united-nations-2100-boom-africa/ (2014)
5. THW, unknown author, https://www.thw.de/SharedDocs/Meldungen/DE/Einsaetze/international/2015/05/meldung_003_nepal_trinkwasserversorgung_beendet.html?nn=922620 (2015)
6. K. Hancke, *Wasseraufbereitung*, Springer (2003), ISBN:978-3540068488
7. P. A. Schweitzer, *Handbook of Separation Techniques for Chemical Engineers*, 4-161ff, ISBN:0-07-057061 (1996)
8. A. C. van Haandel, J. G. M. v.d. Lubbe, *Handbook of Biological Wastewater Treatment*, 575ff, ISBN:9781780400006 (2012)
9. A. Basile, *Handbook of Membrane Reactors*, Woodhead Publishing, 478ff, ISBN:9780857094148 (2013)
10. F. Lokiec, A. Ophir, The Mechanical Vapor Compression: 38 Years of Experience, IDA World Congress Maspalomas, Gran Canaria – Spain Oct. 21.-26. (2007)
11. Abraha Woldai, *Multi-Stage Flash Desalination, CRC Press - Engineering Systems and Sustainability*, 27ff, ISBN:978-1-4987-21170 (2016)
12. D. J. Gates, *Chlorine Dioxide Handbook - Water Disinfection Series*, ISBN: 978-0898679427 (1998)
13. J. R. Bolton, C. A. Cotton, The Ultraviolet Disinfection Handbook, American Water Works Association, ISBN:9781583215845 (2008)
14. W. Lorch, *Handbook of Water Purification*, ISBN978-0133728712 (1998)
15. P. Gleick, *Basic water requirements for human activities: Meeting basic needs* Pac. Inst. f. stud. in dev., env. and sec. in wat. int., **21**, 83-92 (1996)
16. M. Elimelech, W. –A. Phillip, *The Future of Seawater Desalination: Energy, Technology, and the Environment*, Science, **333** (2011)
17. M. Stieß, *Mechanische Verfahrenstechnik 2*, Springer, 167ff (1997)

 DOI: 10.1051/e3sconf/20172200167

Changes in microbiological composition of soils and soil contamination with drug-resistant bacteria caused by the use of sewage sludge in nature

Ewa Stanczyk-Mazanek[1,*], *Lukasz* Pasonl[1], and *Urszula* Kepa[1]
[1]The Faculty of Infrastructure and Environment, Czestochowa University of Technology, ul. Dabrowskiego 69,42-200, Czestochowa

Abstract. This study evaluated the effect of the use of sewage sludge in nature on biological soil parameters. The study was conducted is field experiment environment (small beds). The sandy soil was fertilized with sewage sludge dried naturally (in heaps) and in solar drying facilities. The fertilization was based on the doses of sewage sludge and manure with the amounts of 10, 20, 30 and 40 Mg/ha. The experiment duration was 3 years. The sanitary status of the soils fertilized with the sludge and manure was evaluated (coliform index, Clostridium perfrinens). Furthermore, the content of pathogenic bacteria was evaluated, with determination of its resistance to first-line antibiotics.

1 Introduction

Sewage sludge may contain a variety of chemical contaminants [1, 2] and very dangerous pathogens that represent a threat to the health of both humans and animals [3]. Sewage sludge is also characterized by substantial amounts of bacteria, fungi, viruses and helminth eggs.

Examinations of the microbiological flora in soil and plants that were fertilized by sewage sludge have unequivocally revealed the presence of pathogenic organisms such as *Escherichia*, *Klebsiella*, *Salmonella*, *Pseudomonas*, *Proteus* and *Shigella*, or fungi species of *Aspergillus* and *Candida* [3, 4].

Several studies have reported presence of drug-resistant bacteria in sewage sludge [5, 6, 7, 8, 9]. Similar to manure, numerous pathogenic forms have been documented to be present in sewage sludge and to acquire and develop resistance patterns with respect to a number of antibiotics used in health care [5, 10].

2 Material and research methodology

Sewage sludge from the wastewater treatment plant in Myszków, Poland, was used as a fertilizer to perform a field experiment. After dehydration (with addition of

* Corresponding author: stanczyk@is.pcz.czest.pl

polyelectrolytes) in presses, part of the samples were subjected to liming and stored in a storage area for 6 months. Other samples were dried after liming in a solar dryer. After this process, sludge was also stored for the period of six months.

One type of soil was fertilized in small beds (sandy soil). Properties of sewage sludge and fertilized soil were consistent with the recommendations for using sewage sludge [11]. For comparison purposes, straw-based cattle manure was also stored for 6 months before the evaluation to fertilize sandy soils.

Biological conditions of research materials used in the field experiments are presented in Table 1. The sanitary status of sewage sludge used for fertilization evaluated before the field experiments allowed for its use in agriculture. *Salmonella* bacteria and helminths eggs were not found in this type of sludge.

The experimental small beds with surface area of 10 m^2 were organized on the sandy soil with reaction of 6.55. Sewage sludge after natural drying (Mn) sewage sludge after drying in a solar dryer (MS), and manure (O) were added (once) to the arable layer (with thickness of 20 cm) using the doses of 10,20,30 and 40 Mg/ha. The control sample was non-fertilized sandy soil (P). The experiment was continued for 3 vegetation season. *Dactylis glomerata* was sown on individual mixtures. Biological sanitary parameters were determined in the soil mixtures (coliform index, *Clostridium perfringens*, *Proteus vulgaris*, presence of *Salmonella* bacteria and helminth eggs (ATT). Presence of drug-resistant bacteria in soils fertilized with different fertilizers was also analysed. This study presents the results of examinations after 3 years from fertilizing.

Table 1. Biological properties of soil, sewage sludge and manure used in the field experiment.

Parameter	Unit	Soil	Sewage Sludge		Manure
			Myszków (after natural drying)	Myszków (after solar drying)	
Escherichia coli	Coliform index	10^{-1}	10^{-5}	10^{-5}	10^{-5}
Proteus vulgaris		10^{-2}	10^{-5}	10^{-4}	10^{-5}
Clostridium perfringens		n.f.	10^{-5}	10^{-4}	10^{-4}
Salmonella sp. bacteria	[CFU/g d.m.]	n.f.	n.f.	n.f.	n.f.
Total bacterial count		$5.2*10^5$	$21*10^9$	$31*10^8$	$29*10^{10}$
Total fungi count		$2.5*10^4$	$25*10^6$	$31*10^3$	$51*10^6$
Total actinobacteria count		$1.4*10^3$	$43*10^6$	$49*10^6$	$55*10^6$
ATT		n.f.	n.f.	n.f.	n.f.

n.f. - no presence was found

3 Research methodology

Sanitary status of the sewage sludge, manure and soil was evaluated according to the commonly accepted standards [12, 13]. Determination of drug resistance of microorganisms in sandy soils fertilized with manure and sewage sludge was based on blood agar plates and MacConkey agar plates. Biochemical tests were used to identify individual species. *Enterobacteriaceae* bacteria were identified using Enterotest and non-fermenting gram-negative bacilli test (Nefermtest, Lachema). Simultaneous tests of drug

 DOI: 10.1051/e3sconf/20172200167

resistance were also carried out. Drug resistance was tested on Mueller-Hinton plates. These plates have been used in bacteriological laboratories for determination of resistance to antibiotics.

Plates with antibiotics were added on Petri dishes with breeding of isolated pathogenic microorganisms: ampicillin (10 µg), gentamicin (10 µg), amikacin (10 µg), ceftazidime (30 µg), amoxicillin with clavulanic acid (20/10 µg), cefotaxime (30 µg). After 15-minute preincubation, plates were placed in a laboratory thermostat with temperature of 37°C per 18 hours. After this time, the results were read. Antibiotics were chosen so that they were consistent with the basic antibiogram (first-line antibiotics) according to the recommendations for test selection of the National Reference Centre for Microbial Drug Sensitivity.

4 Results

The results of determination of the sanitary status of the fertilized soils during the field experiment were compared in Table 2. Furthermore, the results of evaluation of drug resistance of isolated pathogenic bacteria are presented in Table 3-5.

Table 2. Results of evaluation of sanitary indices in fertilized soils.

Object No.	Combination	Coliform index		
		Escherichia coli	*Clostridium perfringens*	*Proteus vulgaris*
1	Control (P)	n.d.	n.d.	10^{-2}
2	PMn10	10^{-3}	10^{-2}	10^{-4}
3	PMn20	10^{-4}	10^{-2}	10^{-5}
4	PMn30	10^{-5}	10^{-3}	10^{-5}
5	PMn40	10^{-5}	10^{-3}	10^{-6}
6	PMs10	10^{-2}	10^{-3}	10^{-3}
7	PMs20	10^{-2}	10^{-3}	10^{-3}
8	PMs30	10^{-3}	10^{-4}	10^{-4}
9	PMs40	10^{-3}	10^{-4}	10^{-4}
10	PO10	10^{-1}	10^{-1}	10^{-2}
11	PO20	10^{-1}	10^{-1}	10^{-3}
12	PO30	10^{-2}	10^{-2}	10^{-3}
13	PO40	10^{-2}	10^{-2}	10^{-4}

n.d.- not determined; geohelminths eggs were not found

Table 3. Results of evaluation of drug resistance of conditionally pathogenic microorganisms isolated from sandy soil after 3 years of fertilization with naturally dried sewage sludge.

Fertilizer dose [Mg/ha]	Type of isolated microorganisms	Susceptibility of the microorganisms to the antibiotic used					
		Amikacin	Amoxicillin with clavulanic acid	Ceftazidime	Cefotaxime	Gentamicin	Ampicillin
10	*Enterobacter sp*	s	r	s	s	s	r
	Photorhabdus luminescens	s	s	s	s	s	r
	Escherichia coli	s	s	s	s	s	r
	Proteus vulgaris	s	s	s	s	s	ms
20	*Enterobacter sp*	s	r	s	s	s	r
	Photorhabdus luminescens	s	s	s	s	s	r
	Escherichia coli	s	s	s	s	s	r
	Proteus vulgaris	s	s	s	s	s	ms
	Klebsiella oxytoca	r	r	s	r	s	r
30	*Enterobacter sp*	s	r	s	s	s	r
	Photorhabdus luminescens	s	s	s	s	s	r
	Escherichia coli	s	ms	s	s	s	r
	Proteus vulgaris	s	s	s	s	s	ms
	Klebsiella oxytoca	ms	r	s	r	s	r
40	*Enterobacter sp*	r	r	s	s	s	r
	Photorhabdus luminescens	s	s	s	s	s	r
	Escherichia coli	s	r	s	s	s	r
	Proteus vulgaris	s	s	s	s	s	ms
	Klebsiella oxytoca	r	r	s	r	s	r

s - susceptible, ms - medium susceptible, r - resistant

DOI: 10.1051/e3sconf/20172200167

Table 4. Results of evaluation of drug resistance of conditionally pathogenic microorganisms isolated from sandy soil after 3 years of fertilization with solar dried sewage sludge

Fertilizer dose [Mg/ha]	Type of isolated microorganisms	Susceptibility of the microorganisms to antibiotics					
		Amikacin	Amoxicillin with clavulanic acid	Ceftazidime	Cefotaxime	Gentamicin	Ampicillin
10	*Alcaligenes faecalis*	s	s	s	s	s	s
20	*Alcaligenes faecalis*	s	s	s	s	s	s
	Achromobacer piechandii	s	s	s	s	s	s
	Pseudomonas alcaligenes	s	s	s	s	s	ms
30	*Alcaligenes faecalis*	s	s	s	s	s	s
	Achromobacer piechandii	s	s	s	s	s	s
	Pseudomonas alcaligenes	s	s	s	s	s	ms
40	*Alcaligenes faecalis*	s	s	s	s	s	s
	Achromobacer piechandii	s	s	s	s	s	s
	Pseudomonas alcaligenes	s	s	s	s	s	ms

s - susceptible, ms - medium susceptible, r - resistant

Table 5. Results of evaluation of drug resistance of conditionally pathogenic microorganisms isolated from sandy soil after 3 years of fertilization with manure

Fertilizer dose [Mg/ha]	Type of isolated microorganisms	Susceptibility of the microorganisms to antibiotics					
		Amikacin	Amoxicillin with clavulanic acid	Ceftazidime	Cefotaxime	Gentamicin	Ampicillin
10	*Enterobacter sp*	s	s	s	s	s	r
	Serratia fonticola	s	ms	s	ms	s	r
	Yersinia rodhei	s	r	s	ms	s	r
20	*Enterobacter sp*	s	s	s	s	s	r
	Serratia fonticola	s	ms	s	ms	s	r
	Yersinia rodhei	s	r	s	ms	s	r
	Photorhabdus luminescens	s	ms	s	r	s	r
	Enterobacter kobei	s	s	s	s	s	r
30	*Enterobacter sp*	s	s	s	s	s	r
	Serratia fonticola	s	ms	s	ms	s	r
	Yersinia rodhei	s	r	s	ms	s	r
	Photorhabdus luminescens	s	r	s	r	s	r
	Enterobacter kobei	s	ms	s	s	s	r
	Citrobacter brakii	s	r	s	ms	s	r
40	*Enterobacter sp*	s	r	s	s	s	r
	Serratia fonticola	s	r	s	r	s	r
	Yersinia rodhei	s	r	s	ms	s	r
	Photorhabdus luminescens	s	r	s	r	s	r
	Enterobacter kobei	s	ms	s	r	s	r
	Citrobacter brakii	s	r	s	ms	s	r
	Edwardsiella ictalurii	s	s	s	r	s	r

s - susceptible, ms - medium susceptible, r – resistant

5 Discussion

The sanitary status of the control soil in the field experiment (Tab. 1) was evaluated based on the recommended standards [12]. The soil can be considered as non-contaminated (no *Escherichia coli* bacteria were observed). Presence of *Clostridium perfringens* was also not found. The coliform index for *Proteus vulgaris* 10^{-2} pointed to insignificant tendencies of the soil for putrefaction. Fertilization with solar dried sludge and manure with the dose of 10 $Mg \cdot ha^{-1}$ led to small soil contamination with *Escherichia coli* compared to the control soil. The use of greater doses of organic fertilizers caused deterioration in the sanitary status

 DOI: 10.1051/e3sconf/20172200167

of soil. Mixtures fertilized with sewage sludge and manure with the doses greater than 10 Mg·ha^{-1} were regarded as contaminated based on the recommended coliform indices [12]. The sludge after natural drying caused less substantial soil contamination with *Escherichia coli* than sludge after solar drying. Based on the *Clostridium perfringens* bacteria count, fertilized soils were classified as non-contaminated only for the use of manure with the dose of 10 Mg·ha^{-1}. Sludge after solar drying led to higher contamination of soil mixtures with these microorganisms compared to sewage sludge dried naturally. These sporulated forms are likely to have been resistant to the processes of solar drying. Introduction of all organic fertilizers to soils resulted in an increase in the tendencies of mixtures for putrefaction compared to control soils without fertilization. The coliform index of *Proteus vulgaris* in the soils from the field experiment demonstrated the lowest contamination with spoilage bacteria after fertilization with sewage sludge after solar drying. The sanitary status of soils was affected by the methodology of the sewage sludge drying used for fertilization. Lower contamination of e.g. *Escherichia coli* in the field experiment was caused for sewage sludge after solar drying. The method of solar drying is one of the most effective methods to process sewage sludge, leading to e.g. reduction in the number of pathogens [14].

In the sandy soils fertilized with sewage sludge, presence of pathogenic drug-resistant bacteria was found after three years. A particular threat during the natural use of sewage sludge and manure is connected with presence of conditionally pathogenic microorganisms in these organic materials, which are dangerous for people and animals with acquired drug-resistance. Similar results of the studies have been documented by other researchers [5, 9]. Statistical analysis of the results obtained for the field experiment presented in Tables 3 to 5 demonstrated varied level of drug resistance of pathogenic bacteria isolated from soil mixtures to the antibiotics used in the study.

Amikacin-resistant bacteria were found in the case of fertilization with the sludge after natural drying. They were not present after fertilization with manure or with the sludge after solar drying. With regard to amoxicillin with clavulanic acid in the soil after the use of the solar-dried sludge, all the bacteria exhibited susceptibility. After fertilization with manure and the sludge after natural drying, the cases of medium susceptibility and resistance to the drug were observed. Susceptibility to cefotaxime was found in all the potentially pathogenic bacteria isolated from sandy soil fertilized with the sludge after solar drying. In the mixtures fertilized with manure and the sludge after natural drying, the cases of medium susceptibility and resistance to the drug were observed. The most of the bacteria in the soils fertilized with the sludge after solar drying were susceptible to ampicillin. Furthermore, fertilization of the soil with manure and the sludge after natural drying caused, in the most of the cases, the development of pathogenic forms of bacteria which were resistant to this antibiotic. Similar findings concerning the resistant pathogenic forms were obtained in studies Krzysko-Lupickiej (15).

In the case of ceftazidime and gentamicin, all the analyses revealed the sensibility of the isolated conditionally pathogen forms to these antibiotics. The field experiment showed that the increase in the dose of fertilizes, especially sewage sludge, led to the increase in resistance of pathogenic forms. Similar findings concerning the resistant pathogenic forms in the soils fertilized with sewage sludge were obtained in previous studies [6, 9].

6 Conclusions

1. The study showed that sanitary contamination of soil fertilized with the municipal sewage sludge and cattle manure increased with the increase in their dose.

2. Sewage sludge and manure used with the doses of 10 $Mg \cdot ha^{-1}$ did not cause a significant sanitary contamination in the soils. Manure led to lower contamination of the fertilized soils compared to sewage sludge.
3. The use of municipal sewage sludge and manure for fertilization of sandy soils led to the presence of conditionally pathogenic drug-resistant microorganisms.
4. It was found that many of the isolated microorganisms (especially in sewage sludge compared to manure) were characterized by considerable resistance to the commonly used first-line antibiotics.
5. The highest resistance in bacteria isolated from soils fertilized with sewage sludge and manure was found for ampicillin (ca. 100% drug-resistant forms). The highest sensitivity of the most of conditional pathogens was found for ceftazidime.
6. The three-year field experiment showed that the survival rate of the potentially pathogenic drug-resistant forms in soils may be relatively extended and the contact with polluted ground represents a real threat to the environment, humans and animals.

References

1. P. Wolski, I. Zawieja, Desalin Water.Treat., Volume **52**, Issue 19-21, 3725-3731 (2014)
2. Zawieja, P. Wolski, Rocz. Ochr. Sr, Volume **15**, Issue 1, 2054-2070, (2013)
3. R.P. Singh, M. Agrawal, Waste. Manage., **28**, 347–358 (2008)
4. H.G. Schlegel, General microbiology (in Polish), PWN, Warsaw (2005)
5. C. S. Holzel, K. Schwaiger, K. Harms, H. Kuchenhoff, A. Kunz, Meyer K., C. Muller, J. Bauer, Environ. Res., **110**, 318–326 (2010)
6. T. Nalewajek, Analysis of the content drug-resistant microorganisms in sewage sludge and soils fertilized with sewage sludge. Master's thesis (in Polish), Czestochowa University of Technology (2010)
7. F.F. Reinthaler, G. Feierl, H. Galler, D. Haas, E. Leitner, F. Mascher, A. Melkes, J. Posch, I. Winter, G. Zarfel, E. Marth, Water. Res., **44**, 1981 – 1985 (2010)
8. F.F. Reinthaler, J. Posch, G. Feierl, G. Wust, Haas D., G. Ruckenbauer, F. Mascher, E. Marth, Water. Res., **37**, 1685–1690 (2003)
9. E. Stańczyk-Mazanek, T. Nalewajek, M. Zabochnicka, Arch. Environ. Prot., **Vol.38** nr 1, 97-102 (2012)
10. J. Venglovsky, N. Sasakova, I. Placha, Bioresource. Technol., **100**, 5386–5391 (2009)
11. The Ordinance of the Minister of Environment as of 6 February 2015 on municipal sewage sludge. Journal of Laws 2015 No. 0, pos. 257 (in Polish)
12. Z. Kańska, A. Grabińska-Łoniewska, M. Łebkowska, E. Rzechowska, Laboratory exercises in sanitary biology. Part 1. (in Polish), Warsaw University of Technology (2006)
13. J. Mrozowska, R. Galimska-Stypa, A. Małachowska-Jutsz, E. Zabłocka-Godlewska, Laboratory in general and environmental biology (in Polish), Silesian University of Technology (1999)
14. N.K. Salihoglu, V. Pinarli, G. Renew. Energ., **32**, 1661–1675 (2007)
15. T. Krzysko-Lupicka, M. Myslek, K. Blaszczyk, Proceedings of ECOpole, **9** (**2**), 633-639 (2015)

Acknowledgments

The research have been conducted at the Institute of Environmental Engineering,

Częstochowa University of Technology BS – PB – 401/301/13.

 DOI: 10.1051/e3sconf/20172200168

Evaluating the financial efficiency of energy and water saving installations in passive house

Agnieszka Stec[1,*], *Aleksandra* Mazur[1] and *Daniel* Słyś[1]

[1]Department of Infrastructure and Sustainable Development, Rzeszow University of Technology, al. Powstańców Warszawy 6, 35-959 Rzeszów, Poland

Abstract. The article contains the outcomes of the Life Cycle Cost analysis for alternative energy and water sources utilized in passive buildings. The solutions taken into account included: heat pumps, solar collectors, photovoltaic panels, Drain Water Heat Recovery units, Rain Water Harvesting Systems and Greywater Recycling Systems. In addition, air pollution emission reduction was also calculated for all the installation variants analyzed. The analysis have shown that the systems under consideration could serve as alternatives for traditional installations. Their use has resulted in reductions in the consumption of fossil fuels and natural water resources, thus contributing to environmental improvements.

1 Introduction

The observable economy growth in several countries over the last few years have resulted in improved life quality, reduced poverty as well as increased aspirations and expectations. These factors have together with increased populations led to the over exploitation of natural resources and growing emissions [1]. The forecasts are that such trends will persist into the future if no suitable actions are taken [2]. The issue of climate change, as an outcome of both natural and anthropogenic factors, cannot be overlooked [3, 4]. The observed enormity of the damaging impacts of the changes emphasizes the significance of steps, aimed at minimizing the exploitation of non-renewable resources [5]. Since one of the largest global consumer of energy is the civil engineering sector [6, 7], the quest for efficient and cheap technologies that enable reductions in water and energy consumption by the sector has become significant [8]. Developments in energy-saving constructions in recent years such as the passive or zero energy have effected changes in the share of various components of energy consumption in buildings and consequently contributed to significant increases in demand for energy for water heating purposes [9, 10]. Reductions in the amount of energy derivable from fossil fuels has become possible due to the application of solutions that make use of renewable sources of energy, especially solar and waste [11-13] .

A key aspect of sustainable construction, besides all measures aimed at reducing the amount of energy needed for general as well as water heating, is the limitation of tap water use [14]. This is an accomplishable step through the use of alternative sources of water such

* Corresponding author: stec_aga@prz.edu.pl

 DOI: 10.1051/e3sconf/20172200168

as the use of rainwater and graywater sources [15-17]. Such models for managing water resources of varied quality parameters is compatible with contemporary requirements for water and sewage economy that is being implemented in accordance with the principles of sustainable development. With the awareness of the magnitude of the issue at stake, studies aimed at ascertaining the profitability of the application of systems that enable reductions in demand for tap water as well as energy both for general and water heating in passive buildings were undertaken. As a result of the prevailing climatic conditions in Poland, heating systems, example gas boilers, for passive house are most often designed for use of conventional sources of energy. Passive house besides being designed to ensure a comfortable microclimate indoors, ought to be economical both in terms construction and in exploitation since this guarantees possibilities of popularising the conceptual standards for passive house construction while attracting the attention of potential investors. Consequently, an effective financing was determined for standard passive house in Poland, including for similar buildings fitted with devices for minimizing energy and water consumption through the use of alternative solutions such as solar panels, photovoltaic panels, heat pumps, graywater recycling system, rainwater harvesting system and Drain Water Heat Recovery unit. Reductions in emissions of pollutants accruing to the use of renewable sources of Energy were also calculated for each of the installation variants covered by the study.

2 Methodology

The Life Cycle Costs methodology was chosen as the tool for the financial analysis. *LCC* costs for the variants of installation analyzed are based on the formula (1). This methodology allowed to take into account all the costs connected with investment (*INV*) and use of a system (OC_t) and the residual value (*RV*). However, according to the guidelines in the work [18] in justified cases when the life of the system exceeds the length of the period of analysis, the residual value can be omitted. Taking this into account and the assumed lengths of the *LCC* analysis, the *RV* value was not considered.

$$LCC = INV + \sum_{t=1}^{T} (1+r)^{-t} \cdot OC_t + RV \qquad (1)$$

where:
INV - investments, €;
OC_t - operating costs in the year *t*, €;
RV- residual value, €;
T - duration of the *LCC* analysis, years;
r - constant discount rate;
t - another year of the system use.

Exploitation costs that result from Energy uses for purpose of heating water and living space as well as the associated polluting emissions were, for all the investment variants covered by the study, determined using the ArCADia-TERMO 6.6 software. While the simulation model described in [19] was applied for calculations in all variants using rainwater harvesting systems, the model described in the publication [20] used waste water heat recovery system in all the variants.

3 Case Study

The object of the study is a detached, family residential building located in Rzeszów, south-east Poland. The building has two floors above ground level with a total usable surface area of 138 m^2. The building in question meets all the requirements stipulated by the German Passivhaus Institut in Darmstadt (PHI) for all constructions erected in passive standards. The building had a system of mechanical inflow-exhaust ventilation with heat recuperation installed. The building also envisaged the installation of underfloor heating system due to the prevailing climatic conditions in Poland. The heating capacity requirement of the building, amounting to 4.4 kW was determined based on the PN-EN12831standards [21]. The heat capacity needed for heating utility hot water, which ranged from 3.2 to 9.2 kW depending on the adopted number of residents and on the daily demand for hot water per person was also determined.

In order to obtain maximum efficiency of the installed non-conventional systems in the analysed building, studies were conducted for several investment options that involves varied combinations of the solutions. Installation arrangements for a few selected options are illustrated in the charts.

- Variant 0 – traditional installation with a dual-functional gas boiler supplied with water from public networks and its sewerage and rainwater discharged through the drains.
- Variant 1 – installation with underground heat pump and solar collectors supplied with water from public networks and its sewerage and rainwater discharged through the drains.
- Variant 2 – installation with underground heat pump and solar collectors supplied with water from public networks, fitted with a system for use of rainwater and its sewerage discharged through the drains (Fig. 1).

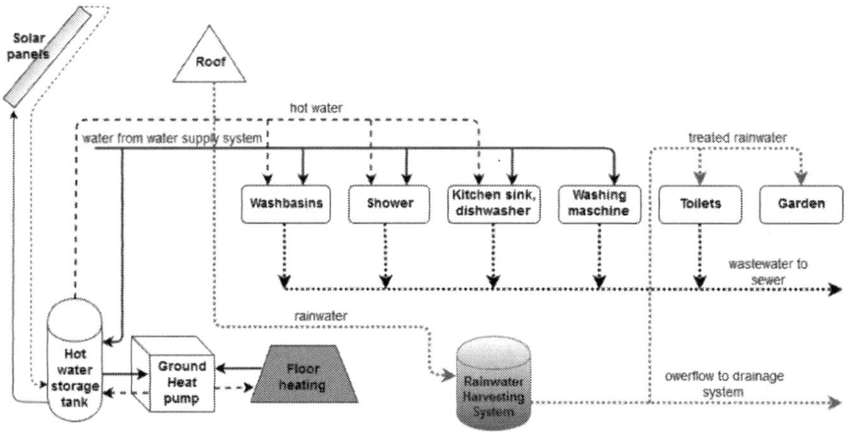

Fig. 1. The solution of installation in Variant 2.

- Variant 3 – installation with underground heat pump and solar collectors supplied with water from public networks, equipped with a system for recycling greywater and its sewerage and rainwater discharged through the drains.
- Variant 4 – installation with air heat pump and photovoltaic panels supplied with water from public networks and its sewewrage and rainwater discharged through the drains.
- Variant 5 – installation with air heat pump and photovoltaic panels supplied with water from public networks, equipped with a system for using rainwater and its sewerage discharged through the drains.

DOI: 10.1051/e3sconf/20172200168

- Variant 6 – installation with air heat pump and photovoltaic panels supplied with water from public networks, equipped with a system for recycling greywater and its sewerage and rainwater discharged through the drains (Fig. 2).
- Variant 7 – installation with dual-functional gas boiler supplied with water from public networks, equipped a system for heat recovery from waste water and a system for recycling greywater and its sewerage and rainwater discharged through the drains (Fig. 3).
- Variant 8 - installation with dual-functional gas boiler supplied with water from public networks, equipped with a system for recovering heat from waste water and its sewerage and rainwater discharged through the drains.

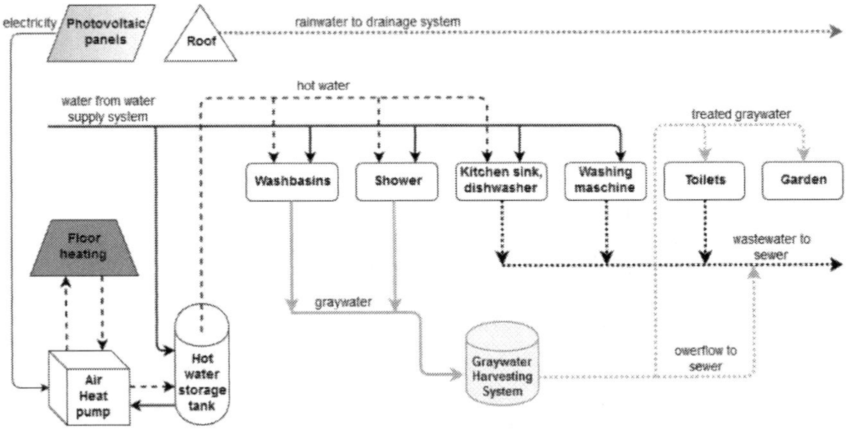

Fig. 2. The solution of installation in Variant 6.

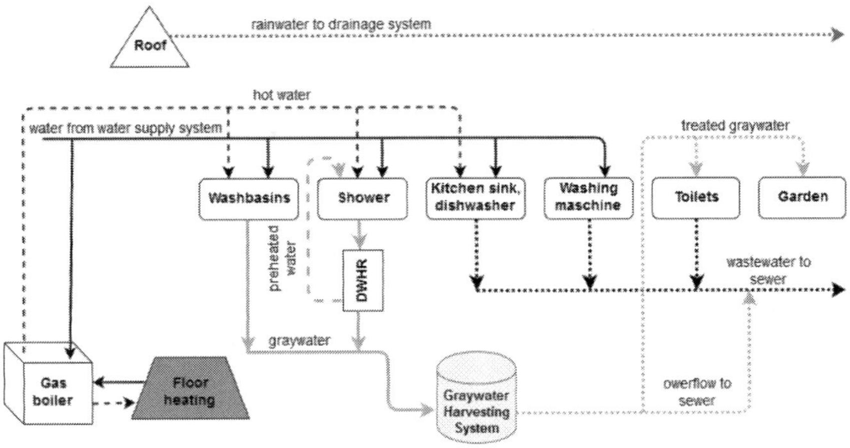

Fig. 3. The solution of installation in Variant 7.

The study uses parameter values presented in table 1 to enable it determine the annual water and energy needs, resulting from the exploitation of analyzed installation options. In order to reflect their impacts on the eficiency of the financial investments involved, calculations concerning varied combinations of duration of using the shower, and varied number of users, which in turn affect the volume of water consumption and amount of discharged waste water in each variant were done.

The profitability analysis for each of the installation variants exploited in the building understudied was carried out and present in table 2. The unit cost price for water, electricity, gas, including the cost of discharging sewage, and rainwater through the drains was established based on current prices set by the relevant Municipal Companies in Rzeszów. Investment costs were, on the other hand, calculated relying on prices provided by producers of various systems and components of the analysed installations.

Table 1. Key data adopted for the calculations.

Parameter	Parameter value
Daily water consumption for green areas watering	1.25 L/m²/day
Daily water consumption for toilet flushing	35 L/person/day
Daily water consumption in washbasins	12 L/person/day
Daily water consumption in washing machine	16 L/person/day
Hot water temperature	55°C
Mixed water flow from the showerhead	7 dm³/min
Roof area	125 m²
Shower length l_{sh}	5, 8, 12 min/person/day
Surface of green areas	300 m²
The number of occupants O_c	3, 4, 5 persons

Table 2. Data used in the calculation of *LCC* costs.

Parameter	Parameter value
Analysis period T	20 years
The annual increase in electricity prices	4%
The annual increase in gas prices	5%
The annual increase in the prices of purchase of water from the water-pipe network	6%
The annual increase in the prices of rainwater discharge to the sewage network	4%
The annual increase in the prices of sanitary sewage discharge to the sewage system	6%
The cost of purchasing electricity in the year 0	0.133 €/kWh
The cost of purchasing water from the water-pipe network in the year 0	1.04 €/m³
The cost of sanitary sewage discharge to sewage network in the year 0	0.894 €/m³
The cost of discharge of rainwater to the sewage network in the year 0	0.719 €/m³
The cost of purchasing gas in the year 0	0.471 €/m³
The discount rate r	5%

3 Results and discussion

The results of the study, presented in table 3 have made it possible to specify the most financially benefitting installation variant, whose installation in the analyzed building would enable the achievement of the highest energy and water savings. The *LCC* values obtained indicate that the variant is, from all the instances considered, the option in which air heat pump powered with electric energy from photovoltaic panels as well as rainwater harvesting system (Varant 5). The exception was in two cases in which the installation was exploited by 3 people, while the showering time lasted 5 or 8 min/person/day. The most financially benefitting option is, in such circumstances, the variant 6 in which alternative

 DOI: 10.1051/e3sconf/20172200168

source of water namely, recycling of greywater was applied. Despite the fact that the variant was associated with more than 2.5-fold higher financial inputs *INV* than variant 0, the savings attained during its exploitation made the exploitation costs OC_t to be almost 3-fold lower when compared to variants in which traditional system solutions were applied. The use of air heat pumps powered with energy from photovoltaic panels resulted in reductions in annual energy needs for heating water and living spaces by as much as 90% than from external networks. It was observed that the difference between the *LCC* values for both variants 5 and 6 was insignificant which offers room for choice depending on the investor's preferences regarding the use of alternative sources namely, greywater or rainwater.

Table 3. Results of calculations.

	Variant	l_{sh} = 5 min/person/day			l_{sh} = 8 min/person/day			l_{sh} = 12 min/person/day		
		INV €	OC_t €	*LCC* €	*INV* €	OC_t €	*LCC* €	*INV* €	OC_t €	*LCC* €
3 persons	0	5080	16730	21810	5115	19366	24481	5115	22918	28033
	1	14004	13719	27723	16008	14827	30835	16328	17164	33491
	2	15498	11488	26986	17502	12596	30099	17983	14933	32916
	3	17107	9774	26881	19111	10940	30052	19776	13301	33077
	4	11465	8876	20341	11488	9838	21326	12186	11158	23344
	5	12960	6645	19604	12983	7607	20589	13841	8927	22768
	6	14569	4930	19499	14592	5951	20543	15634	7318	22952
	7	8885	12278	21163	8919	14237	23156	9264	17239	26503
	8	5781	16224	22005	5816	18124	23939	5816	21102	26917
4 persons	0	5115	19950	25065	5115	23485	28600	5225	28205	33430
	1	16008	15483	31491	16328	18328	34655	17381	20240	37621
	2	17502	13182	30684	17822	16027	33849	19167	17939	37106
	3	19111	11578	30690	19431	14468	33899	20960	16419	37379
	4	11716	10740	22457	12414	12043	24457	13029	13786	26815
	5	13210	8440	21650	13908	9743	23651	14684	11486	26169
	6	14820	6836	21656	15517	8183	23700	16477	9965	26442
	7	8919	14994	23914	8919	18000	26919	9374	21993	31367
	8	5816	18899	24714	5816	21860	27676	5926	25814	31739
5 persons	0	5115	23170	28285	5225	27592	32817	5225	33489	38714
	1	16328	17655	33983	17381	20637	38019	18226	24185	42411
	2	17822	15059	32881	18875	18041	36917	19881	21589	41470
	3	19431	13793	33224	20485	16817	37301	21674	20406	42080
	4	12414	12605	25019	13256	14237	27493	13781	16413	30193
	5	13908	10009	23917	14750	11641	26391	15436	13816	29252
	6	15517	8743	24260	16360	10416	26776	17229	12633	29862
	7	8919	18017	26936	9029	21763	30793	9374	26744	36119
	8	5816	21880	27695	5926	25584	31510	5926	30524	36450

The highest *LCC* index value was obtained in variants where underground heat pump worked in association with slar collectors, Variant 1, Variant 2, Variant 3. This is mainly due to the initial high investment costs, which are not compensated for by energy saves during the 20-year period of exploitation. The exploitation of options could limit annual energy concumption by as much as 34 – 49%, depending on the number of persons and level of hot water consumption. The little reduction in energy needs in comparison with variants with photovoltaic panels is primarily due to the large variability of sunshine prevailent in Polish climatic conditions during the year, which calls for necessity to use substantial amount of energy from external public power supply networks during the autumn and winter periods. However, if the three variants were compared based on their limited demand for water from public supplies the more favourable, assuming the system is used by three individuals with the q_{sh} = 5 min/person/day or q_{sh} = 8 min/person/day, turned out to be Variant 3 which uses recycled greywater, while Variant 2 which uses rainwater harvesting system proved to be better in all other cases.

Results of calculations for variants 7 and 8, which use waste water heat recovery through the installation of Drain Water Heat Recovery unit at the shower's outlet have shown that they are more profitable than variant 0 if the shower is used by 3 persons and does not last longer than 5 minutes. Increased hot water uses due to extensions of bathing duration and increased number of residents results in greater amount of energy being recovered from waste water and in consequence increased financial efficiency of investments. Variant 7, which uses greywater recycling system besides the DWHR unit is more benefitting financially. The application of waste water heat recovery system in the analyzed building limits annual energy needs for hot water heating, which in the case being considered was only 9-17%.

The assumptions for a passive building is that it is environmentally friendly, which is why ecological, besides economical criteria should be taken care of while deciding the installation system. It was observed, in course of analyzing pollutant emissions, that Variants 4,5 and 6 had the least environmental impact. This being due to the fact that the energy needs of such buildings are met without resorting to fossil fuels as the electricity needed to power the heat pumps and ancillary equipment is supplied by photovoltaic panels. The DWHR heat exchanger applied in Variants 7 and 8 has enabled the reduction of annual carbon dioxide emissions by about 8 – 16% compared to Variant 0, depending on the of residents and amount of hot water consumption. Much better results were obtained in Variants 1, 2 and 3, where CO_2 emissions were reduced to 14 – 29% compared to Variant 0 thanks to the use of solar collectors and underground heat pumps.

4 Conclusion

The analyses have indicated the appropriateness of use of Life Cycle Cost (LCC) methodology for assessing investment projects, and could the choice of the most financially viable option. Reliance on the level of initiation investment costs, as it is currently practiced in majority of cases in Poland can lead to the choice of variants that would generate high costs of operation in the long-run. This was corroborated by studies conducted in family homes, where alternative water and energy sources were applied. Although the variant with traditional installations (Variant 0) was characterized by the least financial engagement, its operating costs calculated for all instances significantly exceded similar costs for unconventional installation options. The lowest level of LCC index was, in majority of the cases considered, obtained in the variant in which the energy needed of central and water heating was sourced from air heat pumps driven by electricity from photovoltaic panels, while alternative source of water for non-potable uses was rainwater (Variant 5). The use of this source of energy in the passive building being analyzed contributed to reductions in annual CO_2 emissions by over 95% compared to Variant 0.

The studies have indicated that the use of heat pumps, solar collectors and panels as well as rain water and greywater can have significant impacts on reducing energy and drinkable water consumption in family homes. Their application would therefore contribute to limiting the exploitation of fossil sources for energy purposes, including natural water resources, thus leading to improved state of the natural environment.

References

1. S. Manju, N. Sagar. Renewable and Sustainable Energy Reviews **73**, 594-609 (2017)
2. Ş. Kılkış. J. Clean. Prod. **130**, 222-234 (2016)
3. D.I. Stern, R.K. Kaufmann. Climatic Change **122**, 257-269 (2014)
4. B. Kaźmierczak, A. Kotowski. Theor. Appl. Climatol. **118**, 285-296 (2014)
5. A. Figueiredo, J. Figueira, R. Vicente, R. Maio. Build. Environ. **103**, 276-288 (2016)
6. L. Pérez-Lombard, J. Ortiz, C. Pout. Energ. Buildings **40**, 394-398 (2008)
7. K. Sperling, B. Möller. Appl Energy **92**, 831-842 (2012)
8. A. Figueiredo, J. Kämpf, R. Vicente. Energ. Buildings **118**, 181-196 (2016)
9. J. Schnieders, W. Feist, L. Rongen. Energ. Buildings **105**, 71-87 (2015)
10. A.J. Marszal, P. Heiselberg, J.S. Bourrelle, E. Musall, K. Voss, I. Sartori, A. Napolitano. Energ. Buildings **43**, 971-979 (2011)
11. J. Rekstad, M. Meir, E. Murtnes, A. Dursun. Energ. Buildings **96**, 149-161 (2015)
12. Directive 2010/31/EU of the European Parliament and of the council of 19 May 2010 on the energy performance of buildings (recast). J. Eur. Union, 13–35 (2010)
13. Ramadan, M., Lemenand, T., Khaled, M., 2016. Energ. Buildings **128**, 575-582.
14. K. Urbaniec, H. Mikulčić, N. Duić, R. Lozano. J. Clean. Prod. **130**, 1-11 (2016)
15. A. Stec, S. Kordana. Resour. Conserv. Recy. **105**, 84-94 (2015)
16. G. Markovič, D. Káposztásová, Z. Vranayová. Transactions on Environment and Development **10**, 242-249 (2014)
17. D. Słyś, A. Stec. Ecol. Chem. Eng. S **21**, 623-635 (2014)
18. Life Cycle Cost Handbook. Guidance for Life Cycle Cost Estimation and Analysis. Office of Acquisition and Project Management, U.S. Department of Energy, Washington. Available: www.energy.gov/sites/prod/files/2014/10/f18/LCC%20Handbook%20Final%20Version%209-30-14.pdf (accessed February 2017).
19. D. Słyś. Water Environ. J. **23**, 318-325 (2009)
20. S. Fidewicz. Biblioteka Źródłowa Energetyki Prosumenckiej. Available: www.klaster3x20.pl
21. PN-EN 12831. Heating systems in buildings. Method for calculation of the design heat load (2006)

COD fractions changes in the SBR-type reactor treating municipal wastewater with controlled percentage of dairy sewage

Joanna Struk-Sokołowska[1*], and *Joanna* Rodziewicz[2]

[1]Bialystok University of Technology, Department of Technology in Engineering and Environmental Protection, 15-351 Wiejska St. 45a, Bialystok, Poland
[2]University of Warmia and Mazury in Olsztyn, Department of Environment Engineering, 10-719 Warszawska 117a, Olsztyn, Poland

Abstract. The aim of study was to investigate the influence of percentage of dairy wastewater in the municipal wastewater on the changes of COD fractions during the cycle of SBR-type reactor. The scope of the research included physicochemical analyses of municipal wastewater without dairy wastewater, dairy wastewater, mixture of municipal and dairy wastewater as well as treated sewage. Both the concentrations and the proportions between COD fractions changed in the SBR cycle. In raw municipal and dairy wastewater - X_S, insoluble hardly bio-degradable fraction of COD dominated (49.6 and 64.5% respectively). In treated wastewater S_I, COD for dissolved compounds that are not biologically decomposed (inert) (from 62.1 to 74.6%) dominated, while X_S fraction was from 19.1 to 24.4%. The consumption rate of organic compounds depended on the type of COD fraction, SBR cycle phase and the percentage of dairy wastewater. The highest rates of organic compounds consumption were noted in the phase of mixing. In the case of fraction S_I, no differences in concentration in the SBR cycle time, were found. Concentration of COD in treated wastewater was from 34.8 to 58.9 $mgO_2 \cdot L^{-1}$ (efficiency wastewater treatment from 96.0 to 98.6%).

1 Introduction

The dairy industry is one of the biggest sources of industrial wastewater in Europe [1-2]. It generates wastewater characterized by a high content of organic compounds expressed as biochemical oxygen demand (BOD) or chemical oxygen demand (COD). The biodegradability of wastewater depends, among other things, on the availability of organic compounds (in the forms of dissolved compounds or suspensions) for microorganisms, substrate type, adaptation of the system, presence of toxic compounds, and temperature [3-4]. The most frequently used parameters characterizing the biodegradability of wastewater include the COD/BOD_5 ratio. Its value above 2.5 indicates a high share of hardly biodegradable or non-biodegradable impurities, while a value below 1.8 indicates the

* Corresponding author: j.struk@pb.edu.pl

susceptibility of wastewater to biological decomposition [5]. However, in many cases this information is insufficient.

Dairy wastewater treatment uses membranes, coagulation, flotation and electrochemical, methods. In the case of biological treatment, anaerobic and aerobic methods are commonly used, as well as co-treatment of dairy and municipal wastewater, which most frequently operate using activated sludge technology. This results from the fact that dairy wastewater is characterized by a C:N:P ratio suitable for supporting denitrification and biological dephosphatation, and is source of so-called volatile fatty acids (VFA) in the process line [1, 6-9].

For understanding the processes of decomposition of organic compounds and purification of wastewater, COD fractioning is helpful [10]. It allows for a detailed characterization of the composition of wastewater in regard of their biodegradability, as well as changes in the individual fractions during the specific technological processes. This information is necessary for modelling the operation of a wastewater treatment plant.

Four COD fractions are distinguished in municipal wastewater: a fraction of dissolved easily biodegradable compounds – S_S; a fraction of dissolved compounds which are not biologically decomposed (inert) – S_I; a fraction of insoluble hardly biodegradable compounds – X_S, and a fraction of undissolved compounds which are non-biodegradable (inert) – X_I.

The goal of the research was to determine changes in the values and relationships between the individual COD fractions during the treatment of municipal wastewater with a controlled share of dairy wastewater in laboratory reactors of the SBR type.

2 Methods

The tests were carried out using 8 reactors of the SBR type (R1-R8), to which the municipal wastewater (MW) and the dairy wastewater (DW) were supplied. The activated sludge used in the studies was collected from an SBR type reactor in Hajnówka (Poland), treating municipal wastewater which contain a 10% share of dairy wastewater. The average volume of milk processed in the plant, from which the DW was collected, amounted to 160 $m^3 \cdot day^{-1}$.The water consumption index per the amount of the milk processed was 3.36 $m^3 \cdot m^{-3}$. In dairy plant, wastewater connected with the production of ripening cheeses, creamed cheeses of Swiss type, butter, cottage cheeses, powdered milk, and cream, is generated.

The wastewater supplied to the individual laboratory reactors of the SBR type differed in the MW and DW percentages (R1 – 100% MW, 0% DW; R2 – 95% MW, 5% DW; R3 – 90% MW, 10% DW; R4 – 85% MW, 15% DW; R5 – 80% MW, 20% DW; R6 – 70% MW, 30% DW; R7 – 50% MW, 50% DW; R8 – 0% MW, 100% DW). The volume of the MW-DW mixture was 5 L, whereas the volume of the activated sludge and the treated wastewater remaining in the reactor after a previous cycle - 8 L. The activated sludge concentration was maintained at the level of 3.7 $kg \cdot m^{-3}$.

For 72 hours (nine 8-hour cycles), the contents of reactors were subjected to technological regime, in which the sequence and duration of the purification process phases was the same as in the sewage treatment plant, from where sludge and municipal wastewater was collected (prior to its merging with dairy sewage stream). The SBR phases included: filling and mixing - 20 min., aeration and mixing - 300 min., sedimentation - 100 min., decanting - 60 min. Then relevant studies were conducted, which lasted another 3 days (9 cycles). Samples for analysis of physicochemical properties were collected in cycle 1, 4, and 9. The concentration of the dissolved oxygen, pH, and wastewater temperature were measured directly in the reactors (using a Hach HQd Meter and IntelliCAL Smartprobes, Germany) (Table 1). Determinations of COD, BOD_5 and total suspended solids were carried out according to the APHA [11]. The COD fractions were calculated according to methods described by Struk-Sokołowska [12].

3 Results and discussion

The classification of COD into easily biodegradable substance and hardly biodegradable or non-biodegradable substance in the dissolved form and in suspensions is commonly used in the modelling of wastewater treatment processes. According to the literature data, the share of the dissolved fraction in the total COD for domestic wastewater amounts to 21-35% on average, while that of the suspended fraction – 65-79%. Biodegradable organic compounds constitute on average 60-80% of all organic substances. The S_S and X_S fractions have the highest effect for biological wastewater treatment processes. Moreover, the S_S fraction may be used by the microorganisms directly, while the X_S requires a previous hydrolysis under the influence of extracellular enzymes, thanks to which these substances are transferred into the dissolved fraction. The COD value of the S_I and X_I fractions undergoes the smallest changes as a result of the activity of microorganisms. A part of the X_I fraction is removed together with the excess biomass [13-16].

Table 1. The values of pH, temperature and dissolved oxygen in raw wastewater and after the various phases of SBR.

Parameter	Sample	R1	R2	R3	R4	R5	R6	R7	R8
pH	A	7.0	6.9	6.9	6.9	6.8	6.8	6.7	6.5
	B	7.0	7.0	7.0	6.9	6.9	6.9	6.9	6.9
	C	7.0	7.0	7.0	7.0	7.0	6.9	6.9	7.0
	D	7.2	7.2	7.2	7.1	7.1	7.1	7.1	7.1
	E	7.2	7.1	7.1	7.1	7.1	7.1	7.1	7.1
	F	7.1	7.1	7.1	7.1	7.1	7.1	7.1	7.1
	G	7.1	7.1	7.1	7.1	7.1	7.1	7.1	7.1
Temperature [^{0}C]	A	22.8	22.8	22.8	22.8	22.8	22.8	22.8	22.8
	B	22.8	22.8	22.8	22.8	22.8	22.8	22.8	22.8
	C	22.8	22.8	22.8	22.8	22.8	22.8	22.8	22.8
	D	22.8	22.8	22.8	22.8	22.8	22.8	22.8	22.8
	E	22.8	22.8	22.8	22.8	22.8	22.8	22.8	22.8
	F	22.9	22.8	22.8	22.9	22.8	22.9	22.9	23.0
	G	23.1	23.0	23.1	23.1	23.0	22.9	23.0	23.1
Oxygen concentration [$mgO_2 \cdot dm^{-3}$]	A	3.1	3.6	2.3	2.9	3.2	4.1	5.1	7.3
	B	0.1	0.1	0.1	0.1	0.1	0.1	0.1	0.1
	C	0.1	0.0	0.1	0.0	0.1	0.1	0.1	0.1
	D	1.0	1.0	0.9	1.1	0.9	0.6	1.0	1.0
	E	3.1	3.0	3.2	3.2	3.0	3.2	3.1	3.3
	F	0.2	0.1	0.2	0.1	0.1	0.2	0.2	0.1
	G	0.0	0.1	0.1	0.0	0.1	0.1	0.1	0.0

A – raw wastewater. **B** – wastewater after filling SBR. **C** – wastewater after 20 min. mixing. **D** – wastewater after 150 min. aeration and mixing. **E** – wastewater after 300 min. aeration and mixing. **F** – wastewater after sedimentation. **G** – wastewater after decantation

In the tests carried out, the increase in the share of dairy wastewater was accompanied by an increase in the COD value in the raw wastewater; also, the ratios between the individual fractions changed. This resulted from the diversification of municipal and dairy wastewater (Table 2). In the municipal wastewater (R1, 100% MW), fractions undergoing biodegradation both in their dissolved form and as suspensions prevailed. The X_S fraction constituted 49.6% (449.2 $mgO_2 \cdot dm^{-3}$), while the S_s fraction amounted to 31.3%

(284.1 $mgO_2 \cdot dm^{-3}$). In the dairy wastewater (R8, 100% DW), undissolved fractions prevailed, both the biodegradable fraction and the inert fraction. The X_S fraction constituted 64.5%, and its value was six times higher than that of the municipal wastewater (2698.6 $mgO_2 \cdot dm^{-3}$). The X_I fraction constituted 21.5% (899.5 $mgO_2 \cdot dm^{-3}$) compared to 16.5% in the municipal wastewater, and its concentration was approximately six times lower (149.7 $mgO_2 \cdot dm^{-3}$). The S_S fraction constituted 13.2% in the dairy wastewater. However, its value was two times higher than that of the municipal wastewater (551.4 $mgO_2 \cdot dm^{-3}$). Both in the municipal wastewater and in the dairy wastewater, the lowest percentage was characterised by the S_I fraction, 2.6 and 0.9% (23.9 and 36.6 $mgO_2 \cdot dm^{-3}$), respectively.

The largest drop of the COD value in the treated wastewater occurred as a result of the dilution with sludge and the treated wastewater remaining in the reactor after a previous cycle, with the mixture. It amounted to 81.6% (R1) to 93.2% (R8). It corresponded to the values of the total COD from 166.5 $mgO_2 \cdot dm^{-3}$ (R1) to 285.8 $mgO_2 \cdot dm^{-3}$ (R8). The biodegradable fractions S_S and X_S prevailed. However, with a lower share of dairy wastewater, higher values were found for the dissolved fraction (R1-R3), while with a higher amount of dairy wastewater, higher values were found for the undissolved fraction (R4-R8).

The percentages of the individual COD fractions of municipal wastewater with various shares of dairy wastewater varied in comparison to other wastewater from the food industry. The studies performed by Chiavol *et al.* [17], the share of biodegradable fractions in wastewater from an olive oil press, subjected to treatment process in SBR reactors, amounted to 66.6% (S_S 29.2%; X_S 37.4%), while that of non-biodegradable fractions – 33.4% (S_I 9.9%; X_I 23.5%). Municipal wastewater with dairy wastewater had a higher share of X_S and lower shares of S_I and X_I. On the other hand, Rodriguez *et al.* [18] while analysing the composition of wastewater from the food industry, found that the shares of non-biodegradable fractions (inert) S_I and X_I in wastewater from the processing of tomatoes, sugar beet, potatoes, and from winemaking amounted to 18.4, 18.9, 28.8 and 20.9%, respectively.

In the phase of 20-minute mixing, organic compounds were consumed in denitrification and biological dephosphatation processes (Table 1-2) [19]. Depending on their availability (dissolved or suspended compounds), they were consumed at various rates. In this phase, a decrease in the value of the S_S fraction in reactors R1, R2, R3 and R5 was observed. The highest value (56.3 $mgO_2 \cdot dm^{-3}$) was found in R3, and the lowest one (41.6 $mgO_2 \cdot dm^{-3}$) – in R5. In R4 and R6, the S_S fraction value decreased slightly, but simultaneously its percentage in the total COD increased. In the reactors with significant shares of dairy wastewater (R7 – 50% of dairy wastewater; R8 – dairy wastewater only), an increase in the S_S fraction value was found, accompanied by an increase in its percentages. This was probably caused by the hydrolysis of organic compounds present in the composition of the X_S fraction, which prevailed in these reactors [14-15]. The compounds which were not consumed by microorganisms directly passed to the S_S fraction. As a consequence, the highest share of the S_S fraction (60.6%) in the wastewater after the 20-minute mixing phase was found in R8. Values and percentages of the X_S and X_I fractions in R3, R4, R6, R7 and R8 decreased, while in R2 their increase was found. In R1 and R5, the values of the X_S and X_I fractions decreased, while their percentage in the total COD increased. Values of S_I did not change in wastewater collected from each SBR. and their percentages increased.

In the results of 150-minute aeration and mixing of the SBR contents (the duration of the whole phase amounted to 300 minutes), a decrease in the value and percentage of the S_S fraction was found. R1 and R5 were exceptions; the value of the S_S fraction decreased there, although its percentage increased. Simultaneously, the largest decrease in the percentage of the X_I fraction was found in these reactors. The highest values of the S_S

DOI: 10.1051/e3sconf/20172200169

fraction (42.7 $mgO_2 \cdot dm^{-3}$) were found in R4, while the lowest one (14.6 $mgO_2 \cdot dm^{-3}$) – in R2. In the case of the X_S and X_I fractions, decreases in the value and the percentage was found in six reactors. In R3 and R4, their values decreased, while their percentages increased. The S_I values did not change, and its percentages increased, which resulted from the use of organic compounds from other fractions.

Table 2. The mean values of COD fractions in raw wastewater and after the various phases of SBR.

SBR	COD fractions	A	B	C	D	E	F	G
		$mgO_2 \cdot dm^{-3}$						
R1	S_S	284.1	72.1	48.1	32.1	12.0	0.9	<0.1
	S_I	23.9	23.9	23.9	23.9	23.9	23.9	23.9
	X_S	449.2	52.9	43.6	16.2	4.7	9.1	8.3
	X_I	149.7	17.6	14.5	5.4	1.6	3.0	2.8
	Total COD	906.9	166.5	130.1	77.6	42.2	36.9	35.9
R2	S_S	327.6	124.6	43.6	14.6	7.4	1.8	<0.1
	S_I	24.4	24.4	24.4	24.4	24.4	24.4	24.4
	X_S	505.7	17.1	48.1	27.1	5.9	8.2	10.0
	X_I	168.6	5.7	16.0	9.0	2.0	2.7	3.3
	Total COD	1026.3	171.8	132.1	75.1	39.7	37.1	37.7
R3	S_S	300.3	76.3	56.3	16.3	7.2	5.2	<0.1
	S_I	23.7	23.7	23.7	23.7	23.7	23.7	23.7
	X_S	566.4	65.4	43.7	25.4	9.5	3.1	8.3
	X_I	188.8	21.8	14.6	8.5	3.2	1.0	2.8
	Total COD	1079.2	187.2	138.3	73.9	43.6	33.0	34.8
R4	S_S	284.7	60.7	58.7	42.7	2.7	0.9	<0.1
	S_I	26.3	26.3	26.3	26.3	26.3	26.3	26.3
	X_S	748.6	89.3	24.6	20.6	10.6	9.1	10.0
	X_I	249.5	29.8	8.2	6.9	3.5	3.0	3.3
	Total COD	1309.1	206.0	117.8	96.5	43.1	39.3	39.6
R5	S_S	289.6	84.6	41.6	25.6	6.4	0.9	<0.1
	S_I	24.4	24.4	24.4	24.4	24.4	24.4	24.4
	X_S	843.7	82.1	75.1	7.7	6.9	10.8	8.3
	X_I	281.2	27.4	25.0	2.6	2.3	3.6	2.8
	Total COD	1438.9	218.5	166.1	60.3	40.0	39.7	35.5
R6	S_S	278.9	52.9	51.9	13.9	2.5	0.2	<0.1
	S_I	26.1	26.1	26.1	26.1	26.1	26.1	26.1
	X_S	987.8	80.4	14.8	6.1	10.8	9.8	6.7
	X_I	329.3	26.8	4.9	2.0	3.6	3.3	2.2
	Total COD	1622.1	186.2	97.7	48.1	43.0	39.4	35.0
R7	S_S	385.9	60.9	76.9	28.9	3.1	2.2	<0.1
	S_I	30.1	30.1	30.1	30.1	30.1	30.1	30.1
	X_S	1180.8	105.8	39.8	4.4	20.2	6.1	8.3
	X_I	393.6	35.3	13.3	1.5	6.7	2.0	2.8
	Total COD	1990.4	232.1	160.1	64.9	60.1	40.4	41.2
R8	S_S	551.4	52.4	102.4	24.4	1.5	1.4	<0.1
	S_I	36.6	36.6	36.6	36.6	36.6	36.6	36.6
	X_S	2698.6	147.6	22.6	8.9	28.5	15.3	16.7
	X_I	899.5	49.2	7.5	3.0	9.5	5.1	5.6
	Total COD	4186.1	285.8	169.1	72.9	76.1	58.4	58.9

A – raw wastewater. **B** – wastewater after filling SBR. **C** – wastewater after 20 min. mixing. **D** – wastewater after 150 min. aeration and mixing. **E** – wastewater after 300 min. aeration and mixing. **F** – wastewater after sedimentation. **G** – wastewater after decantation

Further aeration and mixing (150 min) of the SBR content in all reactors led to a decrease in the values and percentages of the S_S fraction. The highest value (12.0 $mgO_2 \cdot dm^{-3}$) was found in R1 (100% MW), while the lowest one (1.5 $mgO_2 \cdot dm^{-3}$) –

in R8 (100% DW). In these reactors, the S_S fraction constituted 28.4 and 2.0%. respectively. The value of the S_I fraction did not change, similarly as in the earlier phases, while simultaneously its percentage in the total COD increased. The values of the X_S and X_I fractions in R1-R5 decreased, while in R6-R8, they increased. This probably resulted from an increase in the availability of organic substances in the form of suspensions (R1-R5), while in the reactors R6-R8 the total amount of organic compounds could be so high in relation to the demand of the microorganisms that a part of the energy was stored in synthesised extracellular polymeric compounds, thereby increasing the amount of the undissolved COD fractions [1, 14-15]. Depending on the changes in the other fractions, the percentages of X_S and X_I decreased (R1-R3) or increased (R4-R8).

In the sedimentation phase only a slight decrease in the value of the total COD in the treated wastewater occurred (below 1% in relation to the value after the aeration phase - except R7 and R8). Despite this, a further decrease in the percentages and the values of the S_S fraction in all reactors was found. The lowest value (0.2 $mgO_2 \cdot dm^{-3}$) was found in R6, the highest one (5.2 $mgO_2 \cdot dm^{-3}$) – in R3. The percentages of the S_S fraction constituted 0.5 and 15.8% of the total COD, respectively. Similarly as in the earlier phases, the S_I fraction did not change its value, and its percentage increased. In the case of the X_S and X_I fractions, an increase in their values and percentages in R1. R2 and R5 was found (in R1: X_S from 9.1 to 24.7%. X_I from 3.0 to 8.1%; in R2: X_S from 8.2 to 22.1%. X_I from 2.7 to 7.3%; in R5: X_S from 10.8 to 27.2%. X_I from 3.6 to 9.1%). In other reactors a decrease in the values and the percentages was found. The highest X_S value (15.3 $mgO_2 \cdot dm^{-3}$) was confirmed in R8, while the lowest one (3.1 $mgO_2 \cdot dm^{-3}$) – in R3. In the same reactors, the highest and the lowest value of the X_I fraction was ascertained (5.1 and 1.0 $mgO_2 \cdot dm^{-3}$, respectively). Similarly as the sedimentation phase, the decantation phase did not cause significant changes in the value of the total COD. Total efficiency of the removal of organic compounds was from 96.0% (R1) to 98.6% (R8). No presence of the S_S fraction was found in the treated wastewater, proving the use of easily biodegradable organic compounds irrespective of the dairy wastewater share. The S_I fractions prevailed (from 62.14% – R8 to 73.06% – R7). The X_S fraction constituted from 9.14% (R6) to 28.35% (R8), while the X_I fraction – from 6.80% (R7) to 9.51% (R8) (Table 2).

4 Conclusions

1. Fractions of insoluble hardly biodegradable compounds and undissolved compounds which are non-biodegradable were the prevailing COD fractions of dairy wastewater originating directly from the plant.
2. In raw wastewater the fraction of dissolved easily biodegradable compounds and the fraction of insoluble hardly biodegradable compounds prevailed, while in treated wastewater – the fraction of dissolved compounds which are not biologically decomposed, irrespective of the dairy wastewater share.
3. It was found that an increase in the percentage of dairy wastewater is accompanied by an increase in the value of the total COD and the fraction of insoluble hardly biodegradable compounds share.
4. Organic compounds belonging to the fraction of dissolved compounds which are not biologically decomposed were consumed by microorganisms directly, while those from the fraction of insoluble hardly biodegradable compounds had to be hydrolysed first.
5. Irrespective of the dairy wastewater share, a high efficiency of the use of organic compounds by microorganisms was ascertained (above 96.0%).

 DOI: 10.1051/e3sconf/20172200169

The research has been carried out in the framework of project No. S/WBiIS/3/2014 of the Technical University of Bialystok, Poland and Project No. 18.610.008-300 of the University of Warmia and Mazury in Olsztyn, Poland and financed from the funds for science MNiSW.

References

1. A. A. L. Zinatizadeh, Y. Mansouri, A. Akhbari, S. Pashaei, Chem. Ind. Chem. Eng., **17**, no. 4, 485-495 (2011)
2. B. Demirel, O. Yenigun, T. T. Onay, Process Biochem., **40**, no. 8, 2583–2595 (2005)
3. A. Mielcarek, W. Janczukowicz, K. Ostrowska, T. Jóźwiak, I. Kłodowska, J. Rodziewicz, M. Zieliński, J. Inst. Brew., **119**. no. 4, 242–250 (2013)
4. S. Mathieu, P. Etienne, Water Res., **34**, no. 4, 1233-1246 (2000)
5. Z. Sadecka, E. Płuciennik-Koropczuk, Rocz. Ochr. Środowiska, **13**, 1157–1172 (2011)
6. J. M. Garrido, F. Omil, B. Arrojo, R. Méndez, J. M. Lema, Water Sci. Technol., **43**, no. 3, 249-256 (2001)
7. W. Janczukowicz, J. Rodziewicz, K. Czaplicka, I. Kłodowska, A. Mielcarek, J. Environ. Sci. Health. A. Tox. Hazard. Subst. Environ. Eng., **48**, no. 7, 809–816 (2013)
8. K. Ostrowska, W. Janczukowicz, J. Rodziewicz, A. Mielcarek, Rocz. Ochr. Środowiska, **15**, 1411–1425 (2013)
9. D. Karadag, O. E. Köroğlu, B. Ozkaya, M. Cakmakci, Process Biochem., **50**, no. 2, 262–271 (2015)
10. I. Pasztor, P. Thury, J. Pulai, Int. J. Environ. Sci. Technol., **6**, no. 1, 51–56 (2009)
11. APHA-AWWA-WEF, *Standard Methods for the Examination of Water and Wastewater* (18th ed. Washington. USA. 1992)
12. J. Struk-Sokołowska, Rocz. Ochr. Środowiska, **13**, 2015–2032 (2011)
13. W. Dąbrowski, M. Puchlik, Rocz. Ochr. Środowiska, **12**, 735–746 (2010)
14. P. Ginestet, A. Maisonnier, M. Spérandio, Water Sci. Technol., **45**, no. 6, 89–97 (2002)
15. E. Morgenroth, R. Kommedal, P. Harremoës, Water Sci. Technol., **45**, no. 6, 25-40 (2002)
16. G. Koch, M. Kühni, W. Gujer, H. Siegrist, Water Res., **34**, no. 14, 3580–3590 (2000)
17. A. Chiavola, G. Farabegoli, F. Antonetti, Biochem. Eng. J., **85**, 71–78 (2014)
18. L. Rodríguez, J. Villaseñor, F. J. Fernández, Water Sci. Technol., **55**, no. 10, 63–70 (2007)
19. A. Mielcarek, J. Rodziewicz, W. Janczukowicz, A. J. Thornton, T. Jóźwiak, P. Szymczyk, J. Environ. Sci., **38**, 119–125 (2015)

 DOI: 10.1051/e3sconf/20172200170

Laboratory tests of overpressure differential systems for smoke protection of lobbies

Paweł Szałański[1,*], *Jacek* Misiński[1]

[1]Department of Air Conditioning, Heating, Gas Engineering and Air Protection, Faculty of Environmental Engineering, Wrocław University of Science and Technology, 4/6 C.K. Norwida st. 50-373 Wrocław, Poland

Abstract. Paper presents the methodology of laboratory tests for ventilation overpressure differential systems for smoke protection of lobbies. Research area consists of two spaces representing the lobby and the area under fire equipped with proper ventilation installation. This allows testing of overpressure differential systems for smoke protection of lobbies. Moreover, piece of laboratory tests results for two selected smoke protection systems for lobbies are presented. First one is standard system with constantly opened transfer-damper mounted between lobby and area under fire. Second one - system with so called "electronic transfer" based on two dampers (supplying air to a lobby and to unprotected area alternatively). Opening and closing both dampers is electronically controlled. Changes of pressure difference between lobby and fire affected area during closing and opening doors between those spaces is presented. Conclusions, concerning the possibility of meeting the time period criteria of pressure difference stabilization required by standards, are presented and discussed for both systems.

1 Introduction

Pressure differential systems (PDS) are the method implemented to maintain safe conditions (lack of smoke) in protected areas, especially in escape routes. The task of PDS is to provide proper pressure gradient to obtain proper flow direction of clean air (free of smoke). In pressurization system pressure should decrease: starting from maximum value in the protected space to lower values in next spaces. Full examination of the system requires not only laboratory tests, but also final adjustment and tests in existing building. Results of the laboratory tests should present that the PDS is able to maintain proper pressure gradient and air flows for full range of operating conditions and for particular air tightness range of protected area.

Operation of two types of PDS's are analyzed in the present paper. First one is standard system with constantly opened transfer-damper mounted between lobby and fire affected area (System 1). Second system with so called "electronic transfer" [1] based on two dampers (supplying air to a lobby and to unprotected area alternately) which opening

* Corresponding author: pawel.szalanski@pwr.edu.pl

sequence is electronically controlled (System 2). In recent years such a systems appeared on Polish market [1].

2 Laboratory test procedures

Following papers [2], [3], [4], [5] are showing basic aspects of pressure differential systems laboratory testing procedures designed to keep smoke behind a leaky physical barrier in a building, such as doors or other openings. Meanwhile, in recent years the group of specialists (the Technical Committee CEN/TC 191/SC1) is working on changes in standard [6] to provide uniform test procedure for pressure differential systems. The considerations contained in above mentioned sources and information about the Technical Committee actual work results on EN12101-6 standard are referring mainly to the case of stairwell protection. However, in this paper the investigation guidelines of pressure differential systems for lobbies are described. It was decided that laboratory testing procedures should be similar to those described in above mentioned sources. In such a case the following tests should be performed:

- first functionality test, **Fu** - opening and closing the door cycle repeated 20 times in the time sequence: $t_1 = 1$ s – door opening time, $t_2 = 6$ s – door opened, $t_3 = 3$ s – closing time, $t_4 = 6$ s – door closed,
- durability test, **Du** –opening and closing the door cycle repeated 10,000 times in the time sequence: $t_1 = 1$ s – door opening time, $t_2 = 0$ s – door opened, $t_3 = 3$ s – closing time, $t_4 = 0$ s – door closed,
- second functionality test, **Fu** - opening and closing the door cycle repeated 20 times in the time sequence: $t_1 = 1$ s – door opening time, $t_2 = 6$ s – door opened, $t_3 = 3$ s – closing time, $t_4 = 6$ s – door closed,
- oscillating test, **Osc** - opening and closing the door cycle repeated 20 times for various values corresponding to door opening angles without waiting time ($t_2 = 0$ s and $t_4 = 0$ s).

These tests should be performed in the order given above, using the same PDS kit and the same test rig. Replacement of system components is not allowed. According to above mentioned papers, criteria for evaluation of PDS test results are proposed:

1- during each cycle sequence of the functionality test and after last cycle of oscillating test the time period to reach the lower limit of the nominal flow rate $V=0,9 \cdot V_n$ (flow rate through the door between test room 2 and 3) should be not longer than $\Delta t_v = 3$s after the door is fully opened (Fig. 1),

2 - during each cycle sequence of the oscillating test and after last cycle of oscillating test the upper limit for the pressure criterion equals to $\Delta p_{max}=1,2 \cdot \Delta p_n$ should not be exceeded during the time not longer than $\Delta t_p=3$s Where: Δp_n - nominal pressure differential to be maintained by the PDS kit with closed door between room 2 and 3 (Fig.1),

3 – in steady state, the measured pressure difference between the test rooms 2 and 3, with the door closed, should not be less than 80% of the nominal pressure and not less than 35 Pa as the absolute limit for the system where $\Delta p_n=50$Pa.

3 Description of the laboratory

Test rig schematics are shown in Fig. 2 and 3. Below described layout of room, ducts systems and measuring system is the part of laboratory built for testing different PDS's

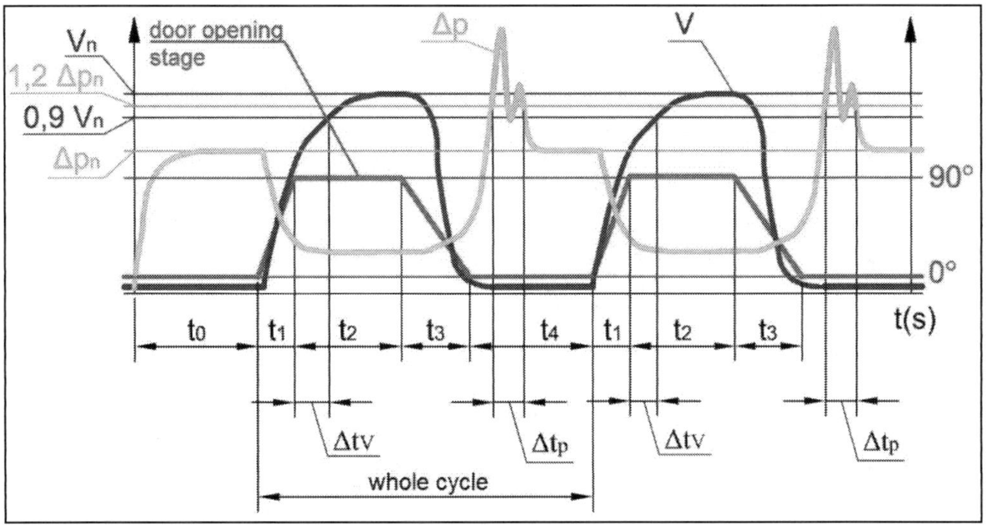

Fig. 1. Pressure and volume flow rate criteria based on [2,3,6]

intended for protection of escape routes. The case described in this paper concerns over-pressurisation of lobby in escape route. Test room 2 corresponds to protected area with increased pressure – lobby. Test room 3 corresponds to fire affected area. The volume of each room is about 124m^3 and 138m^3 respectively. Height of the rooms is 3,8m.

System 1 has a duct supplying air only to the room 2. The idea of the system 2 configuration is based on supplying air to both rooms (2 and 3) by the fan placed in room 4 through branched duct system (Fig.3). Each duct branch has its own motorised volume control damper and flow meter. The fan supplies outside air to both rooms alternately.

Supplied air flows from test room 2 (protected area) to the room 3 (fire affected area). In case of System 1 air flows from room 2 to 3 through door and through transfer damper (adjustable pre-set opening in the wall) - Fig.2. In case of System 2 air flows through the door - Fig. 3.

There are two methods of air release from room 3:

- natural air release through openings in external wall,
- mechanical air release – fan exhaust system.

The first case is described in this paper.

Supply fan inlet duct is equipped with conical inlet and four wall pressure tappings to measure the fan capacity. Before system investigation there was room 2 air tightness determined using testing tool P.A.N.D.A., Airflow. Leakage flow rate V_l for overpressure of 50Pa was measured with this tool.

Different values were measured, especially:

- pressure drop along inlet cone,
- dynamic pressures on measuring crosses in VAV dampers (rooms 2 and 3),
- pressure difference:
  - o between rooms 2 i 3,
  - o test room 2 - reference pressure,
- opening angle of damper blades (opening angle of escape door).

Air flows through the ducts were calculated basing on measured pressure drop on inlet cone (for known characteristic) and pressure difference on measuring crosses. All measurements and calculations were based on pressure transducers readings. Measuring equipment of presented pressure difference (Fig.4 and 5) gives results with accuracy of ±1Pa.

There is PC station implemented for controlling the damper simulating door and collecting data in each test.

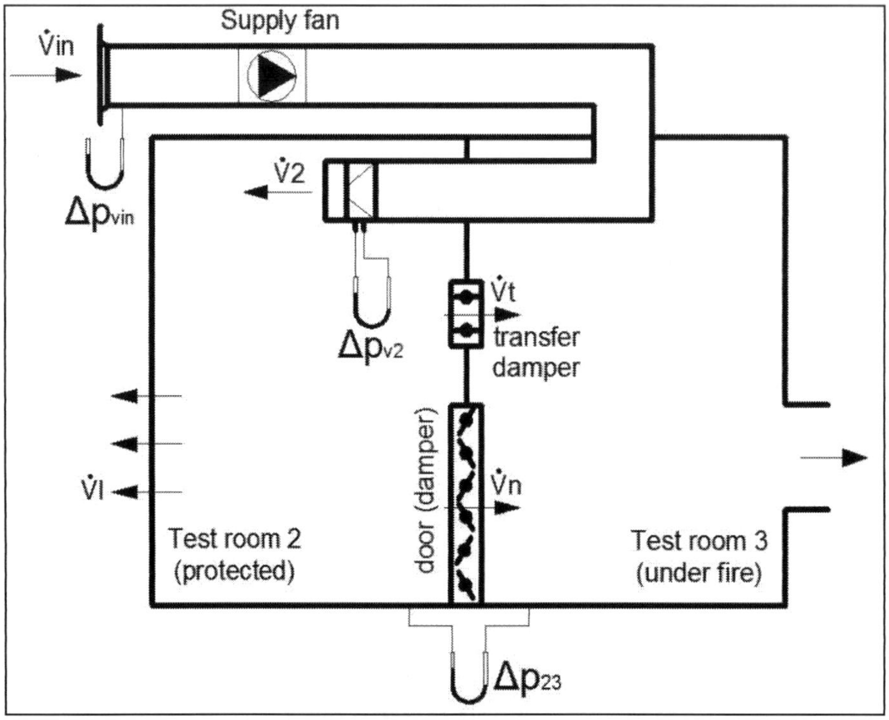

Fig. 2. Diagram of a test rig for standard system (system 1)

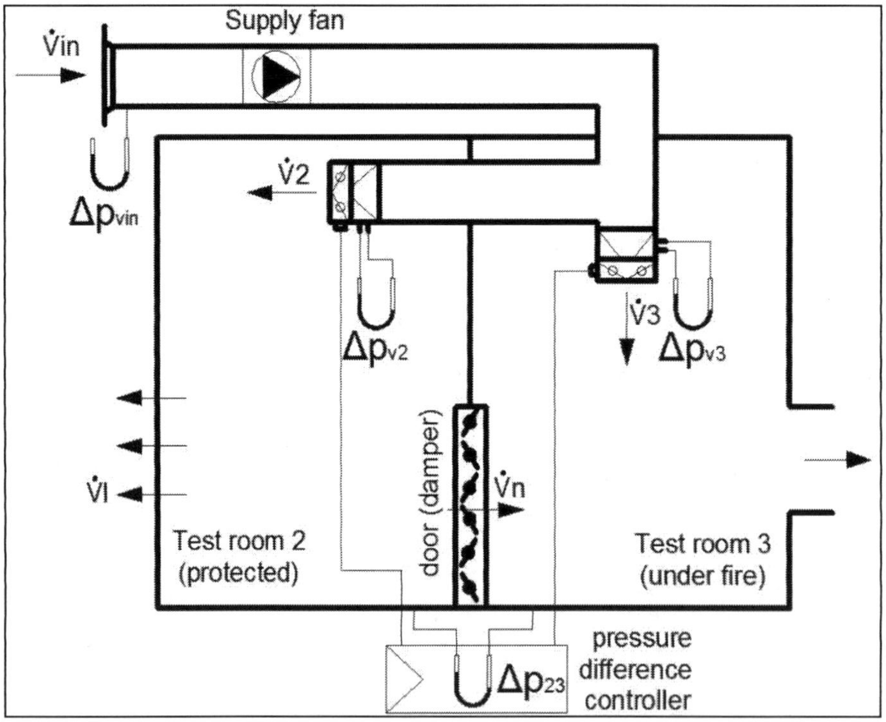

Fig. 3. Diagram of a test rig for system with "electronic transfer" (system 2)

 DOI: 10.1051/e3sconf/20172200170

4 Description of examined systems

The following section presents selected results of the measurements carried out on the above described test rig. The results are presented for two tested PDS's for smoke protection of lobbies.

System 1 is a standard system with constantly opened transfer-damper mounted between lobby and area under fire. The system contains:

- air supply fan operating at constant speed,
- fixed (constantly opened) transfer vent.

The system does not require electronic controllers. With closed doors between the zones, the air flow should cause the pressure drop at the transfer vent equal to the required pressure difference between the zones. While, when the door is opened, the speed in the door opening should reach the required value (Eq 2m/s).

System 2 with so called "electronic transfer" is based on two dampers supplying air to the lobby and to the area under fire alternately. Their position is electronically controlled. Adjustment of overpressure values in the protected area by changing damper's positions is performed by the SMVEST6 controller by Control System Company [7]. The system contains:

- air supply fan operating at constant speed,
- pressure transducer for measuring the pressure difference between the protected and fire affected zone,
- two motorized multileaf dampers on air supply ducts, one for the protected zone and the other for the fire affected zone,
- controller of pressure difference between zones.

Pressure difference controller is set to keep constant value of pressure difference between zones by proper positioning of the air dampers. Software dedicated to contol the system has self-learning function. It is suitable for controlling different types of dampers with different characteristics of fire protection areas. The supply air flow to both zones varies depending on the current state of the door opening between the zones. In the case where the air release path from the protected area is closed, the air damper to the zone is in the minimum opening position, providing overpressure and covering air leakage. In the same time, the second damper is in a position of considerable opening and directs most of the supplied air to unprotected area. When the door is opened, the pressure in the lobby rapidly decreases. This causes opening the damper supplying air to the lobby and closing the damper to the unprotected area. Consequently, the amount of air supplied to the protected zone increases to achieve the required air speed in the door between the lobby and the fire affected area. Investigated system was set to maintain average air velocity of 2m/s in 2m^2 cross-section of opened door. It means that the necessary air flow was equal at least V_n = 14 400m2. The other air flows V_{in}, V_2, V_3, V_t and pressure differences Δp_{v2}, Δp_{v3}, Δp_{in} were result of system operation. Above mentioned results were obtained with room air leakage up to V_l = 600m^3/h at 50Pa overpressure in the room.

5 Sample test results

Below you will find selected test results for both systems. Measurements for System 2 are preliminary and have been carried out during the development works on the SMVEST6 by Control System Company [7]. Both systems have been set up to provide:

- flow velocity of at least 2m/s in the open doorway,
- overpressure Δp_n = 45Pa ± 20% in the protected area against the unprotected area.

 DOI: 10.1051/e3sconf/20172200170

Figures 4 and 5 show the measured variations in the pressure difference between protected and unprotected zones. Presented diagrams show part of functionality test results. One can see that the investigated systems fulfil above mentioned requirements of criteria 1 in presented cycles of functionality test. Preliminary analysis of test results carried out by authors shown that in 3s (Δt_v) after fully door opening system reaches at least 90% of capacity enough for obtaining 2m/s in the doorway and in this state of work system maintains required capacity V_n. However, results are preliminary and need confirmation.

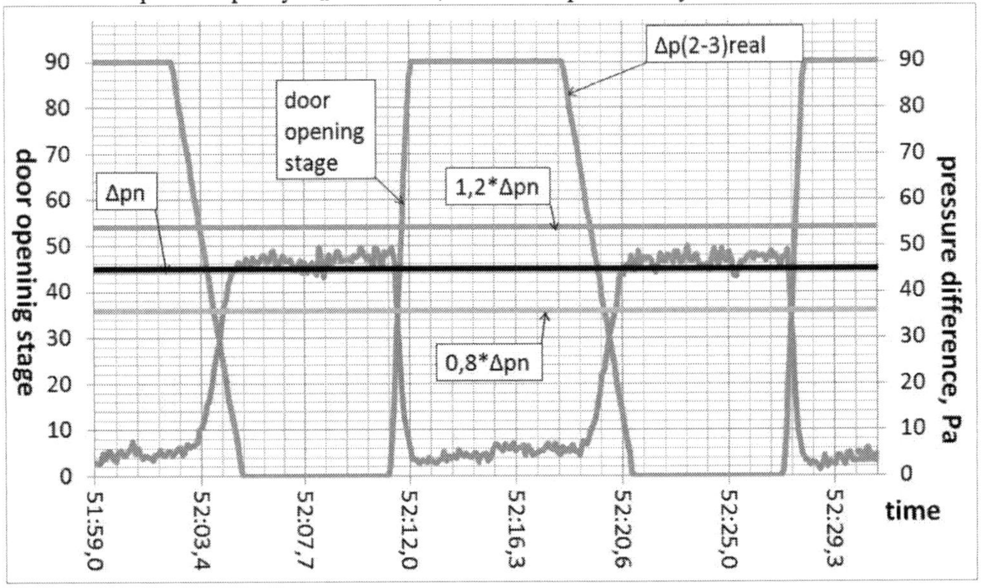

Fig. 4. System 1 - differential pressure - functionality test (2 selected cycles), abbreviations: $\Delta p_{(2-3)real}$ - measured difference pressure between lobby and unprotected zone, Δt_p - pressure regulation time period

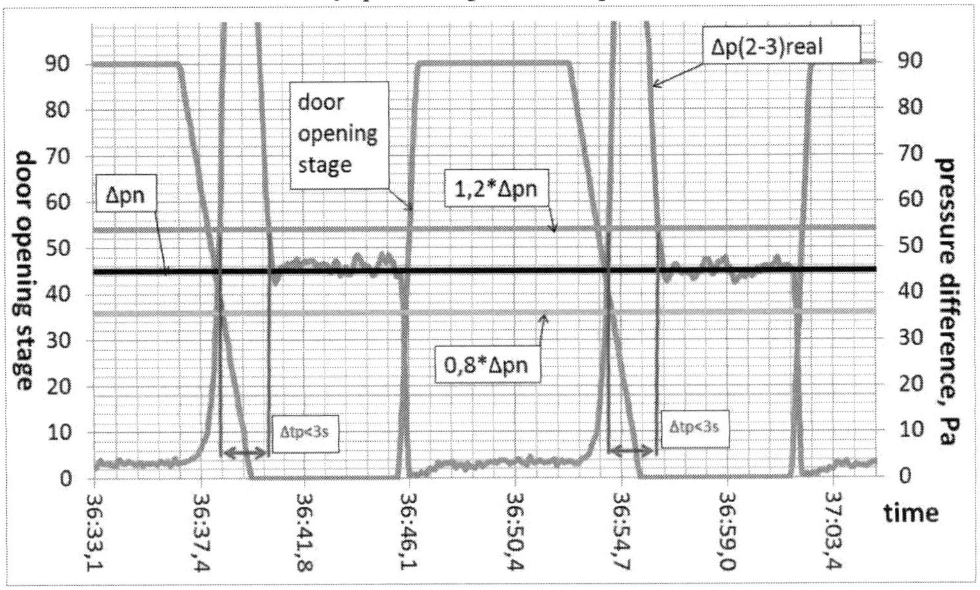

Fig. 5. System 2 - differential pressure - functionality test (2 selected cycles), abbreviations: $\Delta p_{(2-3)real}$ - measured difference pressure between lobby and unprotected zone, Δt_p - pressure regulation time period

After closing the door (closure time: 3s), each system responds in different way. For system 1, when the door was closed, there is no increase of the value of pressure difference between the analyzed zones. On the other hand, in the system 2 the door closure causes a sudden increase of this pressure difference and few seconds period of exceeding the value of $\Delta p_{max} = 1{,}2 \cdot \Delta p_n$. In the analyzed cycles it lasts no longer than $\Delta t_p = 3s$. Similar effects were observed during examination of System 2 with mechanical (fan) air release from unprotected area.

6 Summary

Based on the results of the tests, the comparison of two analyzed PDS's for smoke protected lobbies was presented. Major discrepancy between them is the variation of the pressure difference between the protected and unprotected zones while closing the door. Manufacturers of systems with electronically controlled transfer, e.g. [1], are pointing out advantages:

- limiting the dimensions of the air transfer units,
- proper controlling of the overpressure in protected escape routes,
- constant monitoring of the operation of equipment performance, thereby reducing the duration of the commissioning and performance tests on site.

The above mentioned features should be considered as advantages, especially because often the dimensions of lobbies do not allow for the installation of transfer vents. However, one should consider that the system without constantly open transfer-vent causes temporary increase of the pressure difference between the protected and unprotected zones. Too high pressure difference between these spaces is unfavorable. This may make escape from fire affected area impossible during few seconds. Therefore these systems must be built so that the time of exceeding the permissible pressure differences is no longer than required by the regulations. These should be subject of appropriate testing. The authors of such systems suggest that these are suitable for small lobbies. In such a case we are planning to repeat tests for different volumes of lobbies to confirm that lobby volume does not significantly influence on the range of pressure difference variations between zones. The research carried out at the laboratory does not only serve to study existing pressure differential systems, but also provides opportunities for their development. The presented principles of anti-smoke inspection systems do not cover all work scenarios and installation variants. The actual operation of the lobby during evacuation may differ significantly from the scenario set. Already at the design stage, the designer should have access to the information about building leakage range for which planned system has been tested. During PDS testing, the range of stable operation of the system should be specified, ie the maximum and minimum leakages of the protected zone. Test should be repeated for adjustable leakages of test room (protected zone).

Above described test rig is the part of laboratory built by Control System company in Błonie near Wrocław in cooperation with Wroclaw University of Technology and FlaktBovent Company. Authors would like to thank these Companies for sharing the laboratory to give us possibility to carry out research of different systems based on our own ideas and also to take part in research of a new products supplied by manufacturers.

DOI: 10.1051/e3sconf/20172200170

References

1. Website:http://smay-products.1000i.pl/wp-content/uploads/2017/02/Smay_Roznicowanie_Cisnienia_Przewodnik__2013-a__v523-PL_.pdf, 28.03.2017r.
2. J. Misiński, M. Sompoliński, P. Szałański, Air, Heat and Energy in Buildings (2014), p. 135-140
3. J. Misiński, M. Sompoliński, P. Szałański, INSTAL, **356**, 11 (2014), p. 42-44
4. P. Głąbski, SITP Seminar (The Association of Fire Engineers and Technicians Poland), „Fire protection - spring 2010", 18-21 march 2010, Zakopane, s.1-5
5. P. Głąbski, Fire Protection published by SITP (The Association of Fire Engineers and Technicians Poland), 2 (2010), s.17-19
6. EN 12101-6:2007 Smoke and heat control systems – Part 6: Specifications for pressure differential systems - Kits
7. SMVEST6 Data Sheet, Control System

Preliminary sanitary analysis of supply and exhaust air of ventilation units working at special rooms

Sylwia Szczęśniak[1,*], *Agnieszka* Trusz-Zdybek[1], and *Katarzyna* Piekarska[1]

[1]Wydział Inżynierii Środowiska, Politechnika Wrocławska, Wyb. Wyspiańskiego 27, 50-370 Wrocław, Poland

Abstract. This article is an attempt to estimate concentration of bacteria and fungi in characteristic places in two air handling units. The research was conducted with the units equipped with heat recovery exchanger and recirculation of exhaust air. Additionally, microbiological analysis was done for indoor and fresh air. We performed measurements regarding supply and exhaust air volume. This means that the air velocity inside ventilator chamber was known. The mean concentration of bacteria and fungi was specified for air filters on the supply and exhaust side. The efficiency of air filters was verified by the concertation of bacteria and fungi in fan chamber. This article confirms that bacteria content in supply air is always lower than in exhaust air despite air recirculation. In this work we proved that the position of Petrie plate is essential for bacteria and fungi measurement.

1 Introduction

Actually in Poland, there aren't any of guidelines of concentration limits of bacteria, fungi etc. for public utility buildings. The terms are set only for clean rooms. It is obvious that all ventilation and air-conditioning systems are the base of growing and hibernation of microorganisms. Source of primary pollution in HVAC systems is ambient air. Ambient air is vital for keeping proper hygienic conditions and is always processed in Air Handling Units before supply to the ventilated room. Secondary source of pollution as comes to bacteria and fungi present in ventilation systems are people and animals. Furthermore humidity condensing from ambient and internal air can help in growth of microorganisms. Main factors having influence on and types of microbiota are: emission source type, microbes emission size, distance from emission source, microbes survivability, weather conditions (temperature, wind speed, humidity, precipitation, natural light, UV level) and season [1–3].

* Corresponding author: sylwia.szczesniak@pwr.edu.pl, agnieszka.trusz-zdybek@pwr.edu.pl

 DOI: 10.1051/e3sconf/20172200171

2 Research facility

To establish the quantities of germs and fungi we've investigated two ventilation systems serving two non typical room in public buildings. Pollution sources were people and animals. There are water containers in the rooms with possibility of evaporation (depending on current heat/humidity balance). First investigated system (called System A) is a typical ventilation system without cooling capability, equipped with air recirculation and heat recovery. Second investigated system (called System B) is equipped with cooling module, air recirculation and heat recovery.

2.1. Ventilation system „A"

Figure 1 presents scheme of system A. Dashed line shows air flow direction on supply and exhaust side. Air velocity in supply air fan module was w^A_s = 1,42 m/s and in exhaust fan module w^A_e = 2,16 m/s.

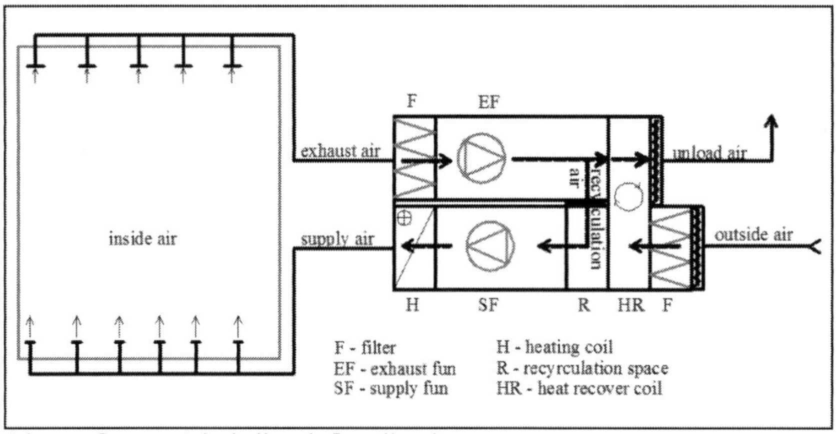

Fig. 1. Scheme of system A including air flow directions.

System A is used during autumn, winter and spring. System is not operating during summertime. During this period, room is ventilated naturally using opening roof. System uses air recirculation and heat recovery. Ambient air share in entire air volume equals 38% and is constant during work of the system. Air recirculation results in cost savings due to reduced heating requirements.

2.2. Ventilation system „B"

Figure 2. presents system B. Dashed line shows air flow directions on supply and exhaust side. Air velocity in supply fan module was w^B_s = 3,59 m/s, and in exhaust fan module w^B_e = 3,18 m/s.

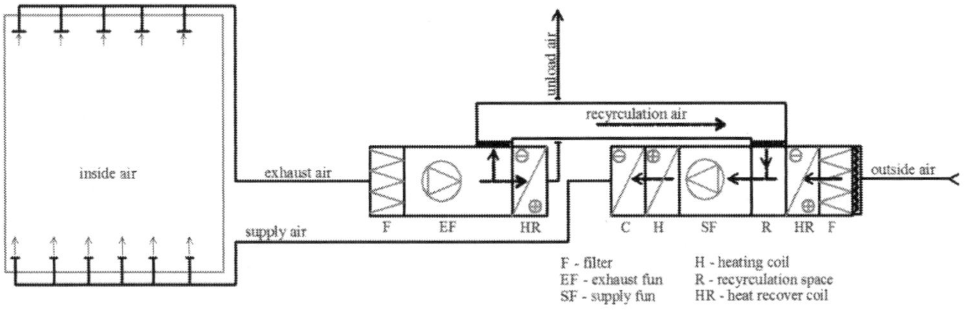

Fig. 2. Scheme of system A including air flow directions.

Ventilation system B is all year operating. This system is also equipped with heat recovery system and air recirculation. Ambient air share in total air volume equals 39% and is constant across the whole year.

3 Description of the methods used in research

Sanitary analysis was made twice - during operation and shutdown of the system in autumn and early spring. Differences were based on air filters contamination and weather conditions. To establish general germs and fungi quantities, measurements were done in supply and exhaust fan modules. Due to the difference between ambient air conditions and air pushed through fan sections, Petri plates were located in three positions: 1 – plate based horizontally on the level, 2 – plate based vertically with counter flow ground location, 3 – plate based vertically with the ground along the airflow. Plates configuration is presented in Fig. 3. Exhaust fan module was chosen for detailed measurements of Petri plates configuration and it's influence to the final results. Exhaust fan module removes room air containing primary and secondary pollutions.

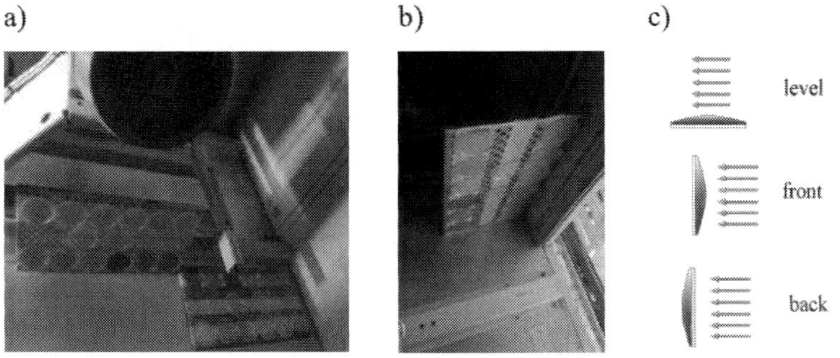

Fig. 3. a, b – Pictures presenting measurement plates in exhaust fan module.

Fig. 3. c – Petri plates configuration against air flow.

For assessment of air microbiological contamination we've chosen general quantity of mesophilic germs cultured on Trypticasein Soy LAB-AGAR™ substrate provided by Biocorp. Air tests to provide assessment of microbiological pollution due to many dangers for health or even life were done using three different substrates dedicated for fungi cultivation provided by Biocorp Czapek-Dox, Sabouraud Dextrose LAB-AGAR™, Malt Extract LAB-AGAR™. Bacteriological contamination of air filters and heat exchangers

DOI: 10.1051/e3sconf/20172200171

surface was done by measurement of general quantity of psychrophilic and mesophilic germs on substrate Trypticasein Soy LAB-AGAR™ by Biocorp.

3.1. Pollution measurements – Assessment of microbiological purity of the air, air filters surface and heat exchangers

Microbiological purity was carried out using sedimentation method (Koch method). This method is based on gravity sedimentation of microorganisms on specially prepared plates. Sanitary analysis of the air filters was done using swabs.

Open plates were left for sedimentation for the period of 15 minutes. After this time, plates were transported to the lab for incubation. Swabs were done on the surface of 25 cm^2 using sterile swab stick. After that operation, stick was placed into solution of 0,9% NaCl (4.5 cm^3). Vortex mixer was used for surface cultures.

Apart from drawing method, samples were incubated in optimal conditions for microorganisms grow: mesophilic bacteria in 37 °C for 48 hours, psychrophilic bacteria in 22 °C for 72 hours, fungi in 26 °C for 5 days.

For sedimentation method Omelinski formula was used to calculate general quantity of microorganisms in 1 m^3 of air (assuming that number of sedimenting microorganisms on 100 cm^2 plate within 5 minutes equals the quantity of microorganisms in 10 dm^3 of air).

In swab method used for air filters examination, we've counted adult colonies and calculated CFU number for the surface of 25 cm^2 , taking into account the volume of solution used in vortex mixing operation (4.5 cm^3) and culture volume on the surface (4.5 cm^3).

4 Results and Discussion

4.1. Pollution measurements using sedimentation method in supply and exhaust fans modules.

Figures 4÷7 present measurements results of general quantity of mesophilic bacteria and fungi in supply and exhaust fan modules during their operation and shutdown.

Highest TNB (Total Number of Bacteria) value for system A (Fig. 3) was measured during exhaust fan shutdown in October 2016 and it equaled 12575 CFU/m^3, lowest value (393 CFU/m^3) was noticed in exhaust fan module in March 2017. During fan operation, highest TNB value (19020 CFU/m^3) was noticed in October 2016 in „back" configuration of the measurement plates. Lowest TNB value (2778CFU/m^3) was noticed with standard „level" configuration of the plates „level".

For system B (Fig. 5) lowest, fan shutdown TNB value was measured for supply fan in March 2017 and it was 236CFU/m^3, highest value (1729CFU/m^3) was measured in exhaust fan module at the same time. During fan operation, lowest and highest values were measured at exhaust fan. Highest value (14574CFU/m^3) was measured with „front" plate configuration in October 2016. Lowest value (629CFU/m^3) was observed in March 2017 r. for standard plate configuration.

In all cases, general quantity of mesophilic bacteria, calculated on a basis of highest indicated value (independent from plates configuration) is higher for measurements taken during operation of the fans (Fig.4 and 5) and is between 629 CFU/m^3 (for exhaust fan and standard plates configuration) and 14574 CFU/m^3 (for exhaust fan and „front" plate configuration. The only discrepancy in measurement results was noticed in October 2016. During fan shutdown results were close to highest TNB value for operating fan with „back" plates configuration. (Fig. 4). Highest values for operating fan were not noticed for standard

DOI: 10.1051/e3sconf/20172200171

(level) Petri plates configuration. In system A (Fig. 4) highest values (19020 CFU/m^3) were noticed in „back" configuration of the plates. In system B (Fig. 5) highest quantities (14574 CFU/m^3) were observed in front configuration of the plates.

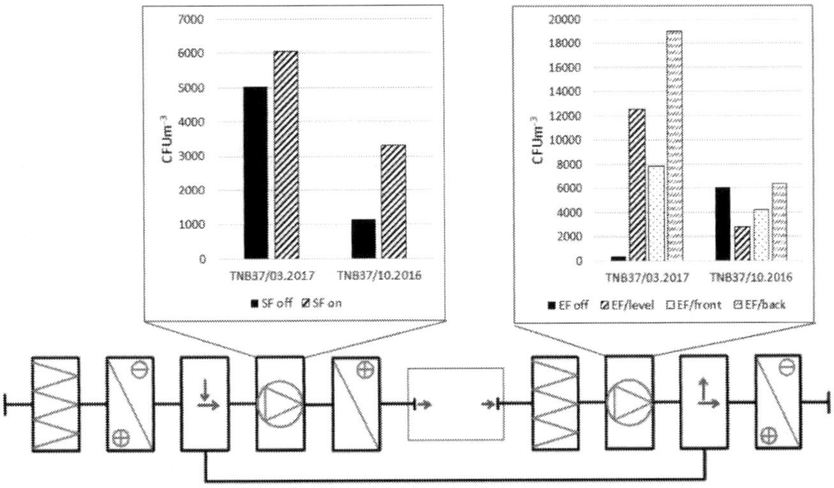

Fig. 4. Mesophilic bacteria measured in supply and exhaust fan modules for system A.

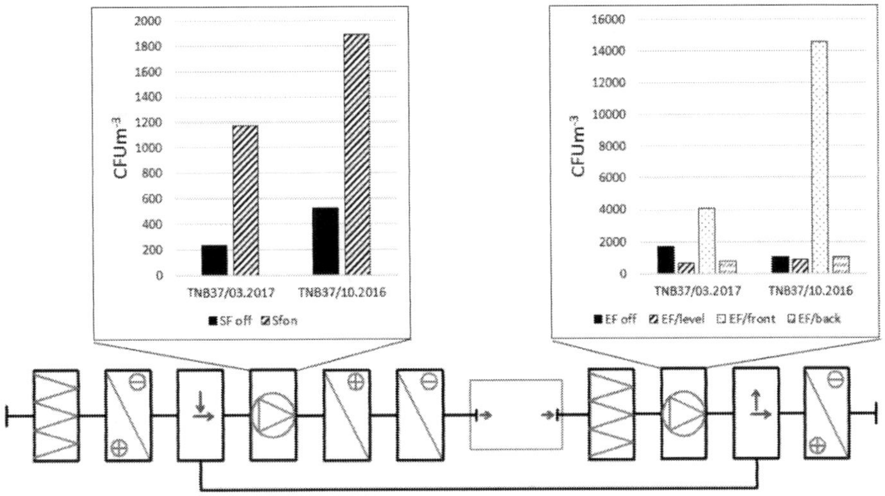

Fig. 5. Mesophilic bacteria measured in supply and exhaust fan modules for system B.

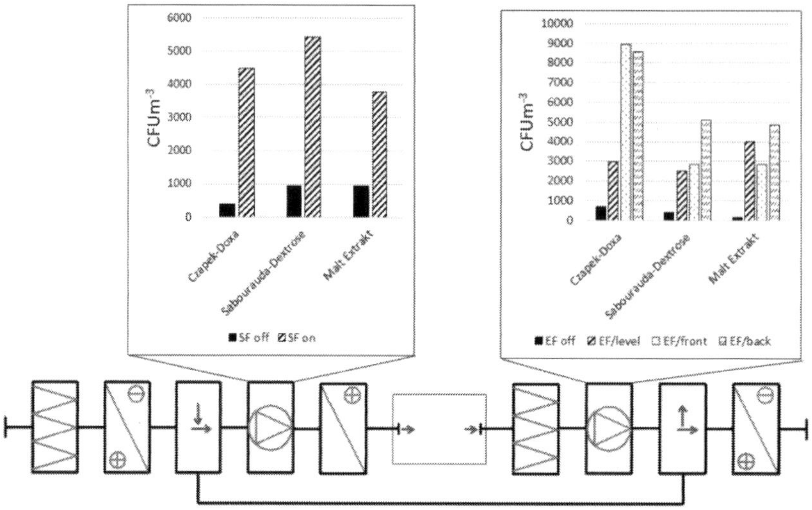

Fig. 6. General quantity of fungi measured in supply and exhaust fan modules in system A.

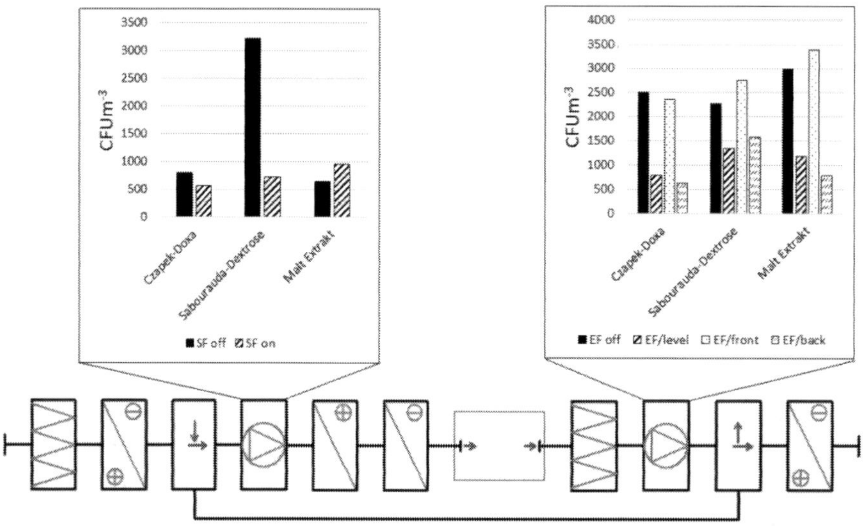

Fig. 7. General quantity of fungi measured in supply and exhaust fan modules in system B.

Data presented above (Fig. 6 and 7) suggest that general fungi quantity may not depend on operation/shutdown of the fan. Only in supply fan module TNF (Total Number of Fungi) values were lower during shutdown of the fan (independently from used substrate). In system A (the same like for mesophilic bacteria) higher values were noticed for operating fan. Highest value (8960 CFU/m^3) was measured for „front" plates configuration , lowest value (2515 CFU/m^3) -for standard plates configuration. Opposite results were measured for system B. Highest quantities in this case were observed for operating exhaust fan (3380 CFU/m^3) with substrate Malt-Extrakt an „front" plates configuration. Lowest results (550 CFU/m^3) were observed with operating supply fan and Czapek-Doxa substrate. What's important, that supply fans were examined only with standard plates configuration. We have to take into account that standard („level") configuration of Petri plates didn't bring highest results of mesophilic bacteria. In this case we can conclude that other plates configuration could result in higher measurements than values taken during shutdown of the fan. In system A highest values (8960 CFU/m^3) were achieved in „front" plate configuration and Czapek-Doxa substrate. For two other substrates higher values were measured for „back" plate configuration. In system B highest values were measured for „front" plates configuration 380 CFU/m^3. At this stage of research we couldn't find any correlation between used substrate and maximum/minimum TNF value. In supply parts (for both A and B systems) (Fig. 6 and 7) highest TNF values were observed for substrate Sabourauda Dextrose. On the other hand, for exhaust fan in system A, highest TNF values were observed on Czapek-Doxa substrate (8960 CFU/m^3) but only in „front" and „back" configuration. In system B highest TNF values were noticed with substrate Malt Extrakt but only in „front" plates configuration.

3.2. Pollution measurements carried out on air filters and heat exchangers.

Fig.8. and Fig.9. present measurements results of general quantities of bacteria taken from air filters and heat recovery heat exchangers in system A and B. Below the results we present the scheme of the systems and location of measurement points. Axis Y is logarithmic.

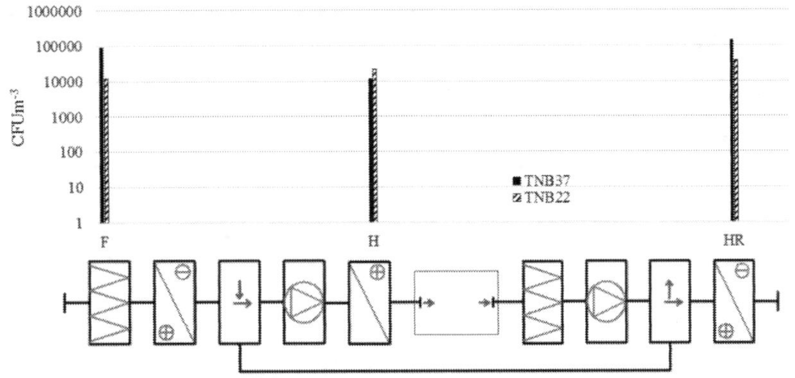

Fig. 8. Spores quantity measured in supply and exhaust fan modules in system A.

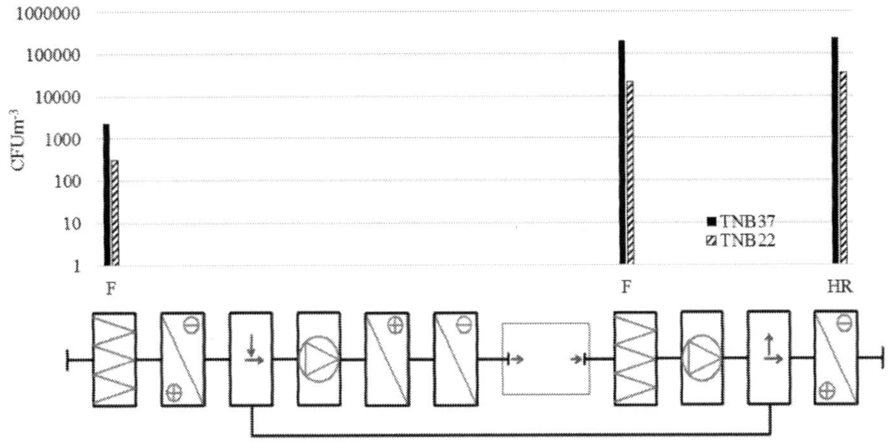

Fig. 9. Spores quantity measured in supply and exhaust fan modules in system B.

TNB value of mesophilic bacteria was higher (independently from examined option) except one case (air heater) than TNB of psychrophilic bacteria. In system A (Fig.8) highest

TNB value of mesophylic bacteria (142000 CFU/m^3) was observed on heat exchanger based in exhaust air unit. Lowest value was measured for ambient air 118000 CFU/m^3. In system B (Fig.9) highest value (233000 CFU/m^3) was notice on heat recovery coil , lowest 2300 CFU/m^3) for ambient air filter. Measurements made on a basis swab samples from air filters and heat exchangers (Fig. 8 i 9) indicate that general quantity of mesophilic bacteria in both systems is higher on exhaust (dirty) side of the systems. This conclusion agrees with general rule of sanitary role of ventilation systems. We have to bear in mind that both systems use air recirculation so they partly re-use „dirty" air. Important thing is that in system A (where it was possible to examine air heater) general quantity of mesophilic and psychrophilic bacteria was higher on the air filter than on air heater. In some sense this proves efficiency of the initial air filter which role is to clean the air from bigger dirt and dust particles. Heater wall temperature may also influence the quantity of bacteria.

5 Conclusions

1. General quantity of bacteria and fungi is always lower in air supply part, despite air recirculation.
2. Standard plates configuration (level) was never resulting in highest bacteria quantities.
3. For air velocity of w=2,16 m/s maximum general quantity was measured for „back" plates configuration. For air velocity of w=3,18m/s –„front" configuration was resulting in highest values. This leads to a conclusion that air velocity and plates configuration may influence the sanitary assessment of ventilating air but this conclusion should be confirmed in further research activities that will be carries on specially prepared research station.

In most cases, maximum pollution values were observed during operation of the fans. During shutdowns, pollution quantities were lower than maximum. This relationship is not obvious for air supply fan, where only standard plates configuration (level) was possible to apply. In this case we have to remember that higher values were achieved for different plates configuration.

Initially we have to confirm that pollution values were always higher during operation of the fans.

References

1. A. Flannigan, Brian, Samson., A. Robert , Miller, J.David, *Microorganisms in Home and Indoor Work Environments* (CRCnetBASE,2011).
2. A.A. Abdel Hameed, M.I Khoder, Y.H. Ibrahim, Y.Saeed, M.E Osman, S.Ghanem, , SCI TOTAL ENVIRON, **414,** 696 (2012).
3. A. Kiliszczyk, B.Podlaska, K.Sadowiec, B. Zielińska-Polit, M. Rychtel, S. Russel, *Water-environment-rural areas,* **T13**, 79 (IT-P w Falentach, 2013).
4. K. Bródka, M. Sowiak, A. Kozajda, M. Cyprowski, I. Szadkowska-Stańczyk, MED PR, **63(3)**, 303 (2012).
5. H. Awbi, *Ventilation of building*, (London and New York: Spon Press, 2003).
6. A. Pełech, *Wentylacja i Klimatyzacja. Podstawy*, (OWPW , 2009).
7. S. Przydróżny, *Wentylacja*, (WPW, 1991).

Detection of comfortable temperature based on thermal events detection indoors

Andrzej Szczurek[1], *Monika* Maciejewska[1,*], and *Mariusz* Uchroński[2]

[1] Wroclaw University of Science and Technology, Faculty of Environmental Engineering, Wybrzeże Wyspiańskiego 27, 50-370 Wrocław, Poland
[2] Wroclaw Centre for Networking and Supercomputing, Wybrzeże Wyspiańskiego 27, 50-370 Wrocław, Poland

Abstract. This work focussed on thermal comfort as the basis to control indoor conditions. Its objective is a method to determine thermal preferences of office occupants. The method is based on detection of thermal events. They occur when indoor conditions are under control of occupants. Thermal events are associated with the use of local heating/cooling sources which have user-adjustable settings. The detection is based on Fourier analysis of indoor temperature time series. The relevant data is collected by temperature sensor. We achieved thermal events recognition rate of 86 %. Conditions when indoor conditions were beyond control were detected with 95.6 % success rate. Using experimental data it was demonstrated that the method allows to reproduce key elements of temperature statistics associated with conditions when occupants are in control of thermal comfort.

1 Introduction

Thermal comfort is defined as "conditions of mind which expresses satisfaction with the thermal environment" [1]. It is a key component of the indoor environment quality and the major criterion to evaluate the performance of building systems from the occupants perspective. Thermal discomfort is a serious problem. It can create dissatisfaction, reduces the performance as well as productivity of workers and affects their health. Hence, ensuring appropriate thermal conditions is an important goal in the operation of modern buildings.

The service sector is now one of the largest and fastest-growing branches of the economy. Its rapid expansion is visible in both employment levels and the dynamic development of the office buildings. Despite a great effort and high rate of energy use for the indoor environment conditioning, a significant portion of the users are unsatisfied with thermal conditions inside commercial buildings [2-4].

Thermal environment within the modern office is determined at two levels – centrally, for the whole building and locally, at room level, by the occupants. Centralized heating systems are typically designed for the peak load conditions and their performance is not sufficiently analyzed during mid-season. The design typically obeys standards that specify

* Corresponding author: monika.maciejewska@pwr.edu.pl

conditions in which majority of the occupants find their environment thermally acceptable. However, the assumption that all occupants have the same comfort preferences is unrealistic [1]. At the local level, it is expected that occupants will moderate their own comfort by dressing and using air temperature regulators or thermal comfort regulators [5].

Air temperature regulators are often available in modern offices. Some of them also take humidity into account. Radiators, convectors or room thermostats allow to adjust the temperature [6, 7]. However, the temperature control seems to poorly fulfil the occupant expectations. Regulators allow to efficiently maintain the indoor temperature/humidity within the demanded range. But sometimes, it is impossible to create comfort conditions in this way. Hence, there are proposed methods using thermal comfort regulators, which operate in accordance with the individual preferences of occupants.

The differences in the assessment of thermal comfort cause that offices are mostly controlled automatically according to the commonly used thermal comfort indices, e.g. the predicted mean vote (PMV) [8,9,10,11]. The PMV value is defined by four environmental variables (indoor air temperature, mean radiant temperature, relative air velocity, and humidity) and two person-dependent variables (activity level and the value of clothing worn). Approaches based on PMV exhibit are difficult in practical applications. For example, they intend to determine the comfort needs of average occupants rather than those of the specific users sharing a room. They require a large amount of environmental data whose retrieval is very costly. It is often difficult to obtain the precise values for the required person-dependent data. PMV often differs markedly and systematically from the actual mean vote. Recently, the development of automatic systems for the control and assessment of the thermal comfort is focused on the application of artificial intelligence (neural networks, fuzzy logic, genetic algorithms, reinforcement learning etc.) [5, 12, 13].

The solution of thermal comfort control problem is important and it has high application potential [14,15]. However, several questions need to be addressed. (1) Is the control of thermal comfort indeed necessary? Occupants may have low expectations regarding thermal environment in the office and the cost of control system may not pay off. (2) What are the heating/cooling preferences of occupants? The most acceptable temperatures need to be specified. (3) What is the level of acceptance of temperature variations? The maintenance of temperature in a narrow range requires significant energy inputs, while it was found that static environments do not necessarily result in appreciably higher levels of occupants' satisfaction. (4) Shall the control of thermal environments be based on constant conditions (steady-state) or adaptive comfort principles. Open character of these questions encouraged us to perform this study. The aim of the presented work is to create a method to determine thermal preferences of office occupants.

2 Method

2.1 Assumptions

Thermal comfort is evaluated individually and subjectively, since it depends on occupants' age, gender, physiological and psychological factors. The group as a whole can express its collective opinion about thermal environment e.g. when individual occupants use a facility for thermal conditions adjustment. Since devices for temperature control are easily accessible (e.g. by pilot) each user of the room can use them to push indoor conditions towards what he recognizes as thermal comfort. The air temperature behaviour inside room in such circumstances reflects the group's opinion about thermal comfort. In other words, by examination of some period of active control there may be identified thermal conditions that are acceptable for the all group members. In our study, correction of indoor conditions towards comfortable temperature is attained using local cooling unit. Each occupant can switch the unit on/off and change its settings upon wish. It was taken for granted, that occupants do not provide false feedback votes.

We assumed that time series of temperature should be a source of information about thermal and cooling preferences of occupants. Sensor technology was used to obtain the relevant data. In order to extract the needed information, we made use of the fact that temperature control caused periodic changes of this parameter. The operation of majority of AC units dedicated for local use is controlled by an electronic thermostat. The thermostat is an on/off switch that commands the compressor. When the desired temperature (user-defined set-point plus or minus a bit) is achieved the compressor is switched off. But, when the indoor temperature goes beyond the set point range, compressor is switched on again. This results in compressor running in on-off cycle. Due to this kind of control, when AC unit operates, room temperature oscillates around the set-point [16].

The key element of this work is a method of thermal event detection. Thermal event is defined by the period of time when the temperature in the room exhibits the characteristic periodic variations. Thermal events allow to determine the time when thermal conditions were under or beyond control of occupants. The number and length of these periods informs whether occupants have low or high expectations regarding the thermal environment in the office, whether control of thermal comfort is indeed necessary. Information about thermal preferences are included in the probability density function of temperature recorded during thermal events. It provides exhaustive statistical description of the temperature which is comfortable for the room users. This function informs also about most preferable temperature as well as the range of temperature variation which is accepted by the office users.

2.2 Pattern of thermal event

Thermal events were distinguished on the basis of their patterns. Fourier analysis was applied in order to determine the pattern of temperature variation during thermal events in the room. In general, Fourier analysis transforms a signal from time domain to the frequency domain (called a spectrum or transform). After Fourier analysis the signal is expressed as weighted sum of periodic components with different frequencies. When signal and its Fourier transform are replaced with discretized counterparts, it is called the discrete Fourier transform (DFT). The DFT is widely used in signal processing because of very fast algorithm for computing it which is called Fast Fourier Transform (FFT) [17, 18].

DFTs were computed for temperature time series divided into segments. We considered segments of various size, namely consisting of n_s=32, 64, 128, 256 and 512 data points. For time series division, the sliding window technique was utilized. The window, with the size of segment, was moved along the time series with a step of one data point.

Two kinds of temperature data segments were distinguished. In the first group, there were segments associated with thermal events, when thermal conditions were under control of the room users. They were labelled "under control". The temperature data segment belonged to this group, when at least 50 % of segment plus 1 element fell on the time of thermal event. The second group consisted of segments associated with conditions other than thermal events, i.e. when temperature was beyond control of occupants. It was labelled "beyond control". The level of thermal acceptance was unknown in these circumstances. Similar, as temperature data segments, discrete Fourier transforms of these segments belong to two groups: "under control" and "beyond control".

It was proposed that the pattern of particular thermal conditions is the mean discrete Fourier transform of temperature data associated with these conditions. The pattern is obtained by averaging spectra computed for all segments of temperature time series recorded when the particular thermal conditions occurred. Using this principle, there were constructed the pattern of thermal events i.e. „under control" conditions and the pattern of „beyond control" thermal conditions.

2.2 Criterion for thermal event detection

The amplitude of discrete Fourier transform was identified as the feature with high discriminatory potential with respect to the patterns of thermal conditions indoors. In case of "under control" pattern the amplitude increase was expected in the part of DFT which corresponds to the frequency of the AC unit compressor switching. Important to note, this frequency was not fixed, but belonged to a certain range The magnitude and dynamics of room heat gains/losses may vary in time, causing change of the period of temperature variations. In "beyond control" pattern, the amplitude increase was not expected.

Let (f3, f4) be the frequency range where the amplitude of "under control" pattern is elevated. Other frequency range, selected as reference is (f_1, f_2). Let the amplitude of discrete Fourier transform of the test data set of temperature be A(f). The following criterion was proposed in order to evaluate the similarity of DFT of the test data with the pattern of "under control" and " beyond control" conditions.

If $max(A(f_1, f_2)) < max(A(f_3,f_4))$, then the test spectrum is belongs to group "under control" otherwise it belongs to group "beyond control".

2.3 Recognition of thermal event

We propose to apply thermal event detection approach for determining comfort temperature in a room. Although the thermal event detection is based on data collected over a period of time, it may be applied for determining the kind of thermal conditions with the temporal resolution of temperature data collection. The procedure of determining thermal conditions in a particular time point consists of the following steps:

1. Picking a segment of temperature time series, such that the time point in question is one of middle data points in this segment.
2. Computing Discrete Fourier transform of temperature data segment.
3. Assigning the obtained DFT to one of two categories "comfort" or "unspecified", based on the amplitude criterion introduced in Sec. 2.2.
4. Labelling the time moment in question with the label assigned to the spectrum in pt. 3.

2.4 Performance assessment

The performance of the method is assessed by comparing true and recognized category of thermal conditions in time point. For each data point, there are two possibilities, namely "under control" and "beyond control". We used two kinds of error as performance measures. They are defined for the period of time which consists of n time moments. In this period, temperature was under control during n_1 time moments and beyond control during n_2 time moments. The number of correctly recognized data points in the first group was n_{UC} and in the second group it was n_{BC}. The error e_{UC} refers to the fact of not recognizing 'under control' conditions, while they existed

$$e_{UC} = 1-(n_{UC}/n_1) \tag{1}$$

and the error e_{BC} refers to the fact of not recognizing "beyond control" conditions while they existed

$$e_{BC} = 1\text{-}(n_{BC}/n_2) \quad (2)$$

Two errors were applied due to imbalanced size of two categories "under control" and "beyond control", the second one being considerably greater.

The method of thermal events detection was optimized. The optimisation consisted in search for the best combination of parameters $\{n_s, f_1, f_2, f_3, f_4\}$. The minimization of e_{UC} and e_{BC} jointly was chosen as the criterion of finding the best combination of parameters.

All scripts were written in Python.

3 Experimental part

Measurement study was carried out. It consisted in indoor air monitoring in an air-conditioned room and recording human interactions with the air conditioning unit.

The observation comprised a selected room of IT engineers in Wroclaw Centre for Networking and Supercomputing. The room is occupied by 10 people, whose working hours are 9:00-17:00. The layout of the room and its dimensions are shown in Fig. 1.

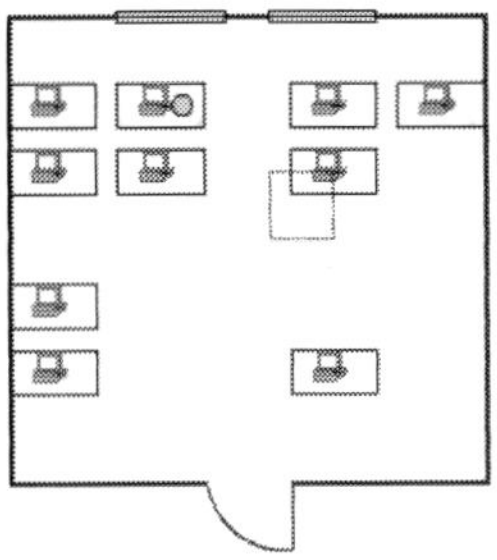

Fig. 1. The schema of the room where the study was performed. Green dot indicates the measurement point and dotted line shows the location of AC unit.

The room is fitted with the AC unit, which is the ceiling cassette ACK30A+A4LC30C, the cooling only model. The unit is equipped with rotary compressor controlled by the electronic thermostat. Two modes of operation are available: COOL, DRY & FAN. Temperature set point may be changed and the fun speed can be regulated by the user. The cassette is mounted in the ceiling, in central part of the room, as shown in Fig.1. In course of the study, there were made notes about the switch on/off moments of the AC unit. Its working parameters have not been recorded. We assumed that various temperature set points, modes of operation and fun speeds have been tried to achieve thermal comfort.

Temperature was the indoor air parameter considered in this study. It was measures by air quality monitoring instrument, HD 37B17D, equipped with NTC 10 kΩ temperature sensor. The instrument offers measurement accuracy of ±0.2°C or ±0.15% of the measure, and measurement resolution ±0.2°C. The sensor device was mounted on the table in the central part of the room, as shown in Fig. 1. Measurements were performed continuously. The data were recorded with temporal resolution of 30 s. The results of the measurement study were the basis for developing the method of comfort temperature detection based on thermal events.

The measurement study was carried out in the summer season of 2016, namely from the 1st June to 30th September. In this period AC unit was frequently used due to hot temperature outdoors and considerable heat gains resulting from solar radiation. For that reason, our attention was focused on the cooling process. It should be noticed, that methodology proposed in this work may be also applied in case of heating.

 DOI: 10.1051/e3sconf/20172200172

4 Results

In Fig. 2 there is shown the daily record of temperature in the room. The period of time when occupants were in control of thermal conditions via AC unit is characterised by the variation of temperature around the set-point.

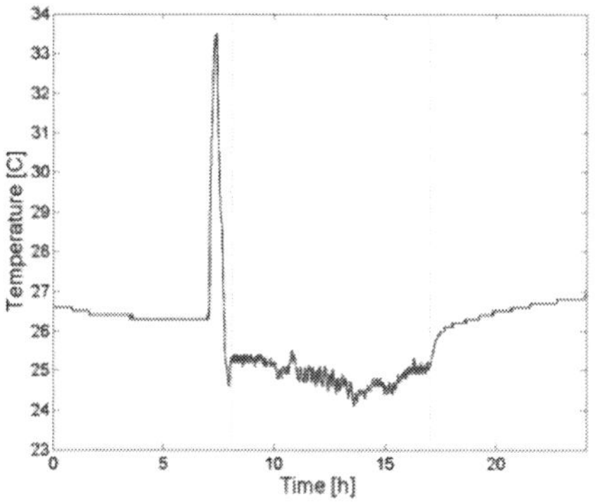

Fig. 2. Time series of indoor air temperature. Thermal event is displayed between dotted lines.

The pattern of thermal events when thermal environment was under control is shown in Fig. 3, left and the pattern when thermal environment was beyond control is shown in Fig. 3, right. Individual plots refer to spectra computed for temperature data segments of various lengths. The amplitudes of individual spectra were scaled with reference to the maximum amplitude.

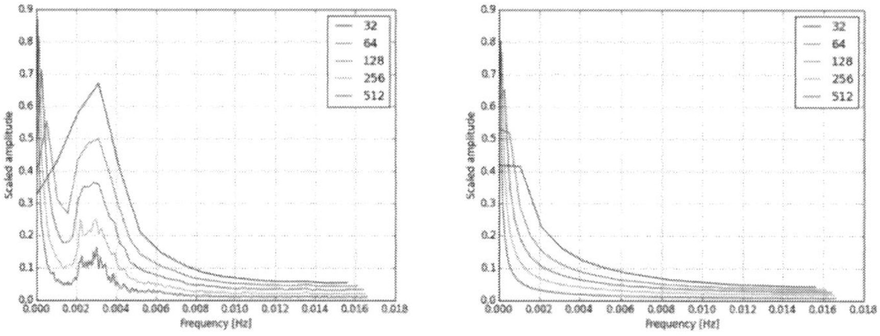

Fig. 3. Patterns of "under control" and "beyond control" conditions. The mean DFT spectra of temperature data refer to time series segments composed of 32, 64, 128, 256 and 512 elements.

As shown in Fig. 3 the patterns of "under control" and "beyond control" conditions are different. In certain frequency range, approximately $f \in (0.0015, 0.005)$ the amplitude is elevated in case of DFT representing "under control" conditions, as compared with the DFT representing "beyond control" ones. This observation indicates that the cycle of AC unit compressor operation varied in quite wide range, from 3 to 17 min. This observation also confirms that the amplitude criterion, presented in sec.2.2 has a considerable discriminatory potential. Its optimization in the studied case involved search in the following ranges of values: $n_s \in \{32, 64, 128, 256, 512\}$; $f_1 \in \{0{:}0.0001{:}0.0009\}$; $f_2=f_3$ and $f_2 \in \{0.001{:}0.0001{:}0.0024\}$; $f_4=0.005$. The exhaustive search of parameters' space was done.

The error analysis showed that the detection of thermal events was most successful when using temperature data segments composed of $n_s=32$ elements. The best values of f_1 and f_2, were $f_1=0$ and $f_2 \in \langle 0.001, 0.002 \rangle$. For any (f_1, f_2) combination in this range of values, the e_{UC} and e_{BC} errors were constant. The error values were $e_{UC} = 14\%$ and $e_{BC} = 4.4\%$.

 DOI: 10.1051/e3sconf/20172200172

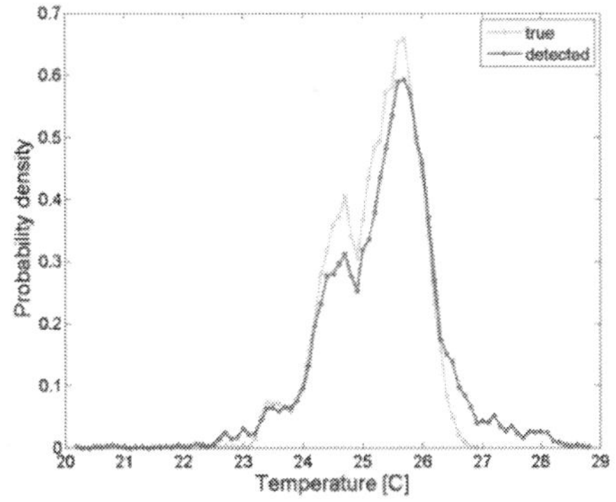

Fig. 4. Empirical probability density function for temperature in "under control" conditions, true (green) and recognized with thermal events detection method (red).

Based on our study, there were identified 47 thermal events. Except for 2 cases, there was recorded one thermal event per day. The event lasted from 33 min to 10h 8 min, with the average length of 6 h. The overall duration of conditions when thermal comfort was under control was 286 h. This constituted about 42.5 % of the work-time hours in the study time. Based on thermal events detection the respective numbers were 265 h and 39.4 %. The method underestimated by 3 %, which is negligible. From this result we may infer that in the considered room, control of thermal comfort is highly needed. Occupants have high expectations towards the quality of thermal environment.

In Fig. 4 we show, the empirical probability density function for temperature, when thermal environment is under control. They are based on raw data and results of recognition by the thermal events detection method presented in this work. As shown by the green plot, the essential feature of probability density function in "under control" conditions is the bimodal character. One observes two maxima, one at temperature 24.7 °C and the other at 25.7 °C. This fact may be interpreted as resulting from the existence of two kinds of dominating thermal preferences amongst people who occupy the room. Namely, part of the group is satisfied when air temperature is maintained around 24.7 °C and the other prefers 1 degree warmer conditions. As shown in Fig. 4, the red plot displays the same bimodal character. Namely, the estimation of comfort temperature based on thermal events detection allows to retrieve this key information about comfortable temperature in an indoor space.

Clearly the studied office is the case where the temperature is maintained in quite a narrow range of 24°C to 26°C, when under control. The spread is 2 degrees. The occupants are not too willing to accept high range of temperatures. Partially it may be due to low speared of individual expectations towards thermal comfort. As compared to optimum temperature detection performance we were not so much successful regarding, the retrieval of the range of temperature which is accepted by the office users. The event detection method qualified too many high temperature occurrences as being examples of "under control" conditions. Although in the lower temperature end the match was much better, there is still room for the detection performance improvement.

5 Conclusions

The work presents a method which allows to determine comfortable temperature in offices, which are fitted with local heating/cooling units controlled by an electronic thermostat. The method consists in detection of temperature events, associated with their use. The detection is based on temperature monitoring and Fourier analysis of the recorded data.

The method was developed and optimised using measurement data collected in the summer season 2016 in an air conditioned office occupied by IT engineers.

We achieved thermal events recognition rate of 86 %. Conditions other than thermal events were recognized with the efficiency of 95.6 %.

It was demonstrated that the method allows to correctly retrieve the key characteristics of the probability density function for temperature in comfort conditions. In this study we detected that room occupants favour two comfort temperatures, which indicates inhomogeneous thermal preferences in this group. We also found that, they prefer temperatures within small range of about 2 °C.

The method is computationally efficient. It may be directly applied to detect comfort temperature in any space where thermal environment may be controlled by occupants.

Further work is planned, aimed at improving the performance of thermal events detection. Alternative recognition approaches and additional parameters of indoor air will be taken into consideration.

Acknowledgments

This research was supported in part by E-SCIENCE.PL Infrastructure.

References

1. P.O. Fanger, *Thermal Comfort* (Danish Technical Press,1970).
2. W. Guo, M. Zhou, Proc., 2009 IEEE Int. Conf. Systems, Man and Cybernetics, SMC 2009, 3883–3888 (2009)
3. O. Seppanen, W. J. Fisk, D. Faulkner, Proc., American Soc. Heating, Refrig. Air-Condition. Eng., 680–686, (2005)
4. F. Jazizadeh, G. Kavulya, L. Klein, B. Becerik-Gerber, B. Int. Workshop on Computing in Civil Engineering, 161–168 (2011)
5. J. Liang, F. DU, Energ. Convers. Manage., 49(4), 517-528 (2008)
6. D.P. Wyon, In: D. Clements-Croome (ed.) *Creating the productive workplace.* (London and New York: E & FN SPON, 2000)
7. S. Karjalainen, O. Koistinen, Build. Environ. **42**, 2880–2887 (2007)
8. P.O. Fanger, , ASHRE Trans. 73, II, (1967)
9. ANSI/ASHRAE STANDARD 55-2004
10. ISO/DIS 7730-2003
11. M.A. Humpreys, J.F. Nicol, Energ. Buildings, **34**(6), 667-684 (2002)
12. Z.L. Baus, S.N. Nikolovski, P.Z. Marcic, J. Electr. Eng., **59**(1), 34-39 (2008)
13. D. Kolokotsa, A. Pouliezos, G. Stavrakakis, C. Lazos, Build. Environ., **44**(9), 1850-1863 (2009)
14. S. Karjalainen, O. Koistinen, Build. Environ. 42(8), 2880-2887 (2007)
15. E. Arens, M.A. Humphreys, R. de Dear, H. Zhang, Build. Environ. **45**(1), 4-10 (2009)
16. H. Zhou, L. Qiao, Y Jiang, H. Sun, Q Chen, Energ. Buildings, **111**, 233–241 (2016).
17. J. W. Cooley, J. W. Tukey, Math. Comput. **19**, 297-301 (1965)
18. W. Press, S. Teukolsky, W.T. Vetterline, B.P. Flannery, *Numerical Recipes: The Art of Scientific Computing*, (Cambridge Univ. Press, Cambridge, 2007)

Measurement of acetates in air using differential ion mobility spectrometer

Andrzej Szczurek[1], *Monika* Maciejewska[1,*], *Żaneta* Zajiczek[1] and *Mirosław* Maziejuk[2]

[1] Wroclaw University of Science and Technology, Faculty of Environmental Engineering, Wybrzeże Wyspiańskiego 27, 50-370 Wrocław, Poland
[2] Military Institute of Chemistry and Radiometry, al. Chruściela 105, 00-910Warsaw, Poland

Abstract. Volatile organic compounds are one of the most important group of air pollutants. Potential health and environmental problems resulting from their emission prompted the requirement for monitoring these species. It motivates development of new measurement techniques which are fast, cost effective, reliable and field deployable. One of novel approaches is ion mobility spectrometry. It dwells on ion separation in electric field, based on differences in ion mobility. Many variants of this method are developed. In this wok, differential ion mobility spectrometry (DMS) was considered in respect of acetate measurements in air. It was demonstrated that DMS offers linear response to methyl, ethyl, propyl and butyl acetate in concentration range from 0.3 ppm to 7 ppm. Positive ions spectrum has to be utilised for this purpose. We showed that fragments of DMS spectrum which secure linearity are compound-specific. The obtained results are promising from the application point of view.

1 Introduction

Volatile organic compounds (VOCs) are believed to have short-term and long-term adverse effects on environment and human health. Therefore they are classified as a one of the most important group of air pollutants. Although natural (biological) emissions tend to be larger anthropogenic sources are a greater concern in many environments. VOCs are emitted to the atmosphere from a number of outdoor and indoor sources [1, 2]. Typical outdoor sources are industrial processes (especially those involving solvents e.g. printing, spray painting, coil coating, wood treatment etc.), vehicle emissions, evaporative loss from petrol storage, and evaporation from landfill sites. The key indoor sources include paints, cleaning supplies, furnishings, glues, permanent markers and printing equipment. They are particularly important when ventilation is limited.

VOCs are important environmental contaminants because of their mobile, persistent and toxic properties. Some of them are harmful to human health, with several classified as carcinogenic (e.g. benzene) and toxic (e.g. methyl alcohol). Prolonged human exposure to VOCs can lead to throat and eyes irritation, respiratory problems, headaches and liver, kidneys, or central nervous system damage. These substances can also cause serious

* Corresponding author: author@email.org

environmental damage [3]. They play important roles in stratospheric ozone depletion, contribute to local or transboundary formation of photochemical oxidants in the boundary layer of the troposphere, leading to environmental degradation, enhancing the global greenhouse effect. VOCs have a potentially negative impact on industrial safety. A lot of them are fire hazards or can form an explosive mixture in air.

Potential health and environmental problems resulting from the emission and use of VOCs has prompted requirement to monitor these pollutants, particularly in relation to compliance with limit values, determination of the emission control effectiveness, risk assessment or assessing trends in time or space. A requirement to monitor implies the availability of well establish analytical methods and techniques. Today, there are a number of instruments capable of identifying and quantitating specific chemical species. If qualitative and quantitative analysis is required then infrared analyzers, Fourier-transform infrared spectrometers (FT-IR spectrometers) or gas chromatographs with appropriate detector (usually FID or tuned mass spectrometer MS or other mass-selective detector) can be employed [4, 5]. These instruments offer very good accuracy, sensitivity, selectivity and repeatability. Unfortunately, they are expensive, time-consuming and require trained, experienced personnel. An alternative method is the electronic nose (e-nose) system that consists of an array of chemical gas sensors [6]. However, different VOCs may produce similar patterns, thus the discrimination capability of this approach is limited. The problems with targeted VOC measurements cause that determination of total VOCs is acceptable to comply with legislation. The two most widely used technologies to measure total VOCs are Flame Ionization Detection (FID) and Photo Ionization Detection (PID).

Large demand for sensitive and selective detection of VOCs in the gas phase for routine laboratory analysis, emission determination, industrial processes examinations, safety monitoring, surveillance and homeland security is so important, that novel methods and techniques are required. They should be objective, fast, cost effective, reliable and field deployable. It is worth mentioning that less accurate, but simpler, quicker and more convenient methods are entirely satisfactory and desirable for many applications.

One of novel approaches is based on a Ion Mobility Spectrometry (IMS). It stands for a set of laws, methods and instrumentation which may be used to characterize analytes based on the movement of ensembles of gaseous ions derived from these analytes in an electric field, through a supporting gas atmosphere [7]. This technology combines high sensitivity and relatively low cost of a single analysis with a fast response to tested gas. Unfortunately, drawbacks such as low resolving power (not very high selectivity), high false positive rates, weaknesses in the robustness regarding a poor linear range, “memory effects” from contamination and interference from matrices reduce the applicability of the IMS system for detection of harmful chemicals. Recently, differential mobility spectrometry (DMS) has proven to be a valuable addition to mobility spectrometry (MS) analyses.

Differential mobility spectrometry (DMS) is a method of separating ions based on the difference between ion mobility in high and low electric fields in gases at or near atmospheric pressure. DMS separates ions based upon subtle differences in their chemical structures [7-10]. For the development of DMS, it is important to know what role in the mechanism of analytical signal generation play functional groups. In organic chemistry, functional group is defined as the defined group of atoms within molecule, arranged in a specific manner that gives the compound its characteristic properties. It is the centre of chemical reactivity. The same functional group will undergo the same or similar chemical reaction(s) regardless of the size of the molecule it is a part of. It is important to recognize the influence of various functional groups on differential mobility spectra [11,12].

The aim of this paper is to demonstrate that DMS allows to distinguish organic compounds that contain various functional groups and determine their concentration. Our study was focused on acetates. These compounds have found broad application in textile industry. They are widely used as solvents, precursors of important chemicals, essential ingredients in many products and materials e.g. in coatings, inks and fragrances in food production. Acetates are popular because they have inoffensive, often sweet odors, they are inexpensive, and they usually have low toxicity. Various reasons cause that information about these compound is needed. Acetates examined in this work are described by the following general formula: $CH_3COO–R$. In this study, we focused on the following alkyl groups R, methyl (CH_3–), ethyl (C_2H_5–), propyl (C_3H_7–) and butyl (C_4H_9–). The functional groups included in the molecules of acetates can be precursors of different ions which are responsible for differential mobility spectra. On this basis, there is possible to perform quantitative and qualitative analysis of acetates in air.

2 Experimental part

2.1 Differential ion mobility spectrometer

Differential ion mobility spectrometry (DMS) is one of approaches developed in the domain of IMS. It is founded on the observation that in strong electric fields (exceeding 12 kV/cm) ion mobility displays nonlinear increase with the field [8]. This property is applied for ions separation and characterization.

Essential issues in DMS development are: 1) ionization of gas sample 2) evoking oscilating movement of ions in high amplitude, high frequency, asymmetric electric field caused by the potential difference, called separation voltage; 3) ions drift compensation by applying constant component, called compensation voltage, which prevents their neutralization on chamber electrodes, 4) collection of ions which passed the drift region on collecting electrodes. Ions which have different mobility can be passed through the drift area at different compensation voltages.

DMS technique may be applied for the quantitative characterization of gas samples. If the residence time of sample molecules in reaction region is fixed, the peak intensity of product ions in the mobility spectrum is proportional to the concentration of sample molecules. However, the linear range of the relationship is limited by a number of factors. A very important one is spectrum variation upon the analyte concentration change. Upon sample concentration increase, there are formed proton bound dimers, trimers and even more complex ions in addition to protonated monomers.

In this work, we focussed on demonstrating that there are fragments of DMS spectrum where the instrument response is proportional to the concentration of selected acetates. Moreover, these fragments are compound-specific.

Measurements were performed with the DMS spectrometer, construction of the Military Institute of Chemistry and Radiometry, Poland. [13]. In the device, gas ionization is achieved by the β-emitter, nickel electrode covered with radioactive nickel ^{63}Ni. The DMS spectrometer itself has planar geometry. The drift region is formed by two electrode-plates (5 x 25 mm) separated by a 0.5 mm gap. A high amplitude, high frequency, asymmetric voltage waveform, generated by HSV generator (2 MHz) is applied to electrodes, crosswise to the direction of the carrier gas flow. The separation voltage amplitude is from 100 to 1600 V (peak to peak), which corresponds to from 6,5 to 104 Td. Additionally, constant weak compensation voltage is applied to electrodes. The compensation voltage is from -30 to 8 V. At the end of the drift zone there are located two collecting electrodes (5 x 5 mm), for collecting positive and negative ions respectively. Two types of ions are neutralized in parallel. The internal gas flow through the detector is 2.0 L/min. Chamber entrance temperature is 318 K, chamber temperature is 323 K.

 DOI: 10.1051/e3sconf/20172200173

2.2 Measurements

Volatile organic compounds considered in this study were methyl, ethyl, propyl and butyl acetate. They were examined in the following concertation ranges: methyl acetate (from 0.21 to 6.15 ppm), ethyl acetate (from 0.25 to 6.08 ppm), propyl acetate (from 0.22 to 4.28 ppm) and butyl acetate (from 0.39 to 5.44 ppm).

Standard gas mixtures containing individual acetates were prepared using an evaporation method. The setup dedicated to gas mixtures preparation was a part of the measurement setup, shown in Fig. 1. In the setup, purified air was used as the carrier gas and the cleaning agent. Purified air was attained by blowing room air through a set of filters, with a pump.

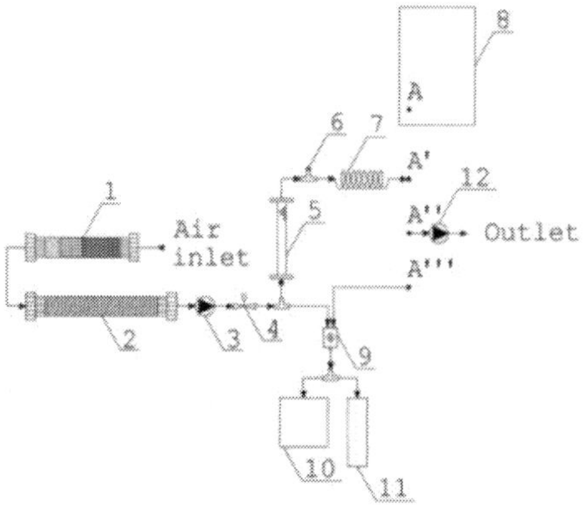

Fig. 1. Measurement set-up. It consists of the following elements: 1. filter (active carbon, molecular sieve, soda lime), 2. filter (silica gel), 3. membrane pump, 4. regulation valve, 5. rotameter, 6. T-connector, 7. Teflon coil, 8- Tedlar bag (50 L), 9- two-way valve, 10-DMS spectrometer, 11-PID meter, 12- membrane pump.

For preparing standard gas mixture, the adequate amount of liquid analyte was injected, with the syringe, to the injection port mounted in Teflon tubing. The analyte evaporated while purified air passed through Teflon tubing and the coil. The resulting gas mixture was collected in the Tedlar bag (position A'). The purified air flow was adjusted by the valve and controlled by the rotameter. The concentration of analyte in gas mixture was estimated based on the amount of injected analyte, gas flow and its duration.

During measurement, gas mixture from Tedlar bag (position A'') was continuously delivered to the spectrometer (constant rate (0.5 l/min)) and photoionization detector (PID), in parallel. The time required to measure an individual gas sample with the spectrometer was 30 s. In this period of time the separation voltage was changed (in steps equal 25 V) and at each discrete value of separation voltage full range of compensation voltage was scanned (in steps equal 74 mV). The response, ionic current was measured on both collecting electrodes for each combination of separation and compensation voltage. Subsequent gas samples were measured with a time interval of 2 min. On average, standard gas mixture was sampled 15 times, although in individual cases there were made from 6 to 20 repetitions. PID meter was employed as reference, to provide the true concertation of analyte in gas mixture. The device operated in a continuous manner and the measurement data was recorded with time resolution of 1 s.

Prior to measurements and after measuring each standard gas mixture, Tedlar bag used for its preparation was cleaned by filling the bag with pure air and emptying it (position A''). The operation was repeated three times. Measuring instruments were cleaned by blowing purified air through them. This operation was performed prior, between and after measurements of standards gas mixtures.

 DOI: 10.1051/e3sconf/20172200173

3 Approach to measurement data analysis

DMS spectrum consists of values ionic current recorded on the collecting electrode of DMS spectrometer in the predefined range of separation voltage (SV) and compensation voltage (CV). For data recording purposes these voltages are discretized, which results in a finite and countable set of combinations (SV, CV). The aim of measurement data analysis performed in this work was to evaluate all (SV, CV) combinations regarding linearity of the relationship between the concentration of analyte in gas sample and the ionic current on the collecting electrode. For this purpose, we applied univariate linear regression represented by the following formula

$$c_A = a_0 + a_1 I_{(SV,CV)} + \varepsilon \quad (1)$$

where c_A is the concentration of particular analyte, $I_{(SV,CV)}$ is the ionic current recorded in response to analyte concentration c_A at voltage combination (SV, CV), a_0 and a_1 are model coefficients and ε is random error.

The quality of the linear fit between analyte concentration and ionic current was evaluated using coefficient of determination (R^2) and root mean square error of concentration assessment (RMSE). Both performance measures were determined for the complete set of measurements data applied for model parameterization.

3 Results and discussion

The results presented in this section are based on positive ions DMS spectra. In case of negative ions, there was not observed satisfactory degree of linearity between analyte concentration and the instrument response. Promising results were achieved exclusively for positive ions.

In Fig. 2 there are presented DMS spectra of purified air (Fig. 2a) and standard gas mixture, which contained methyl acetate at the concentration of 1.33 ppm (Fig. 2b). Based on the difference between plots, it may be observed that DMS spectrometer responded to methyl acetate. As a result of positive reactant ions interaction with the analyte molecules, M, at least the protonated molecules MH^+, also called monomer ion were formed and recorded.

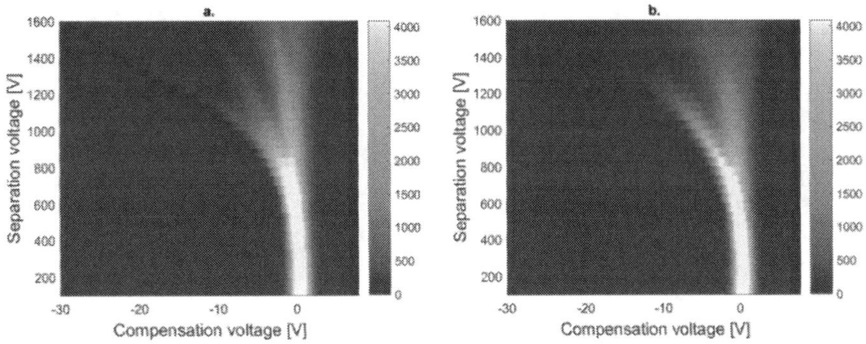

Fig. 2. DMS spectrum of positive ions recorded in response to **a.** purified air, **b.** gas mixture containing methyl acetate at the concentration of 1.33 ppm.

 DOI: 10.1051/e3sconf/20172200173

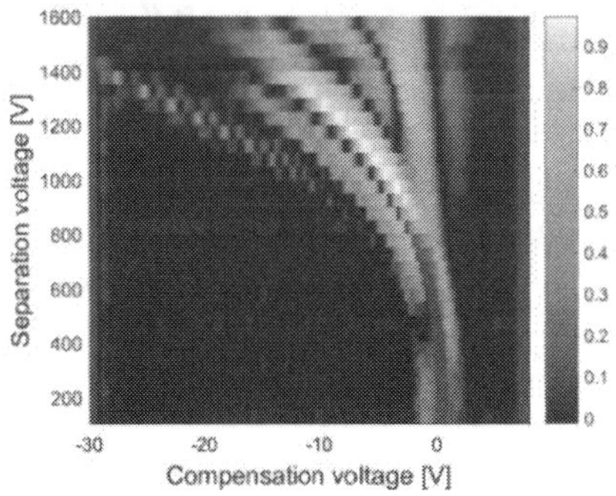

Fig. 3..Mapping of R^2 coefficient in (SV,CV) domain, for ethyl acetate measured with DMS spectrometer, using positive ions.

In Fig. 3. there is shown mapping of R^2 coefficient in (SV,CV) domain, for ethyl acetate. It is representative for the results obtained for other acetates. Based on Fig. 2b. there are zones in the spectrum, where the DMS response may be considered as linearly proportional to the concentration of analyte. They are indicated by high values of R^2 coefficient. As revealed by the comparison of DMS spectrum (Fig.2b) and R^2 map, the location of zones is not directly deducible from the spectrum. The identification requires additional measurement data processing in addition to the visual investigation of spectrum.

In Fig. 4 there are presented fragment of positive ions DMS spectrum which are best in respect of linearity between the analyte concentration and the ionic current. Fragments include combinations of separation voltage and compensation voltage (SV,CV), where the liner fit was characterized by R^2>0.85. This cut off threshold was chosen arbitrarily. Linearity zones in DMS spectrum were individually distinguished for methyl acetate, ethyl acetate, propyl acetate and butyl acetate.

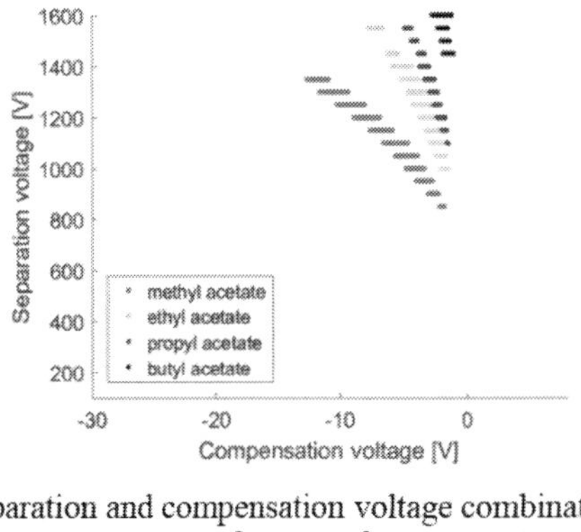

Fig. 4. Selection of best separation and compensation voltage combinations (SV,CV) for acetates. The selection was based on the criterion of R^2>0.85. R^2 indicates the quality of linear fit between analyte concentration and ionic current measured on the positive ions collecting electrode for the particular combination (SV,CV).

As presented in Fig. 4, the linearity of DMS response and analyte concentration was observed at different combinations (SV, CV) for different acetates. Based on the comparison of DMS spectrum fragments distinguished for methyl, ethyl, propyl and butyl acetate, it could be noted that for acetate with methyl group, linearity was achieved at relatively small separation voltages and relatively high compensation voltages. With the increase of the size of functional group in acetate molecule, from methyl to butyl group, linearity zones moved towards higher separation voltages and lower compensation voltages. In particular, the results obtained for butyl acetate suggest that more combinations (SV, CV) would be available for the compound determination when extending the separation voltage range. The shift of peaks towards lower absolute values of compensation voltage indicates decreasing difference of ion mobility between low-field and high-field.

In Fig. 5.a there is presented the quality measure for linear fits, coefficient of determination (R^2)' and in Fig. 4.b we show the analyte concentration prediction error (RMSE), which is made while applying linear models. Box plots were used to display the spread of these measures for models associated with best (SV,CV) combinations, as shown Fig. 4. The performance measures were compared for methyl, ethyl, propyl and butyl acetate.

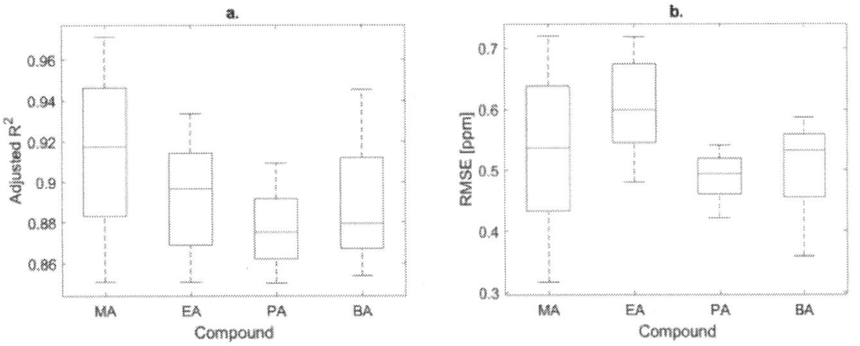

Fig. 5. a. Quality measure of the fit (R^2) for the pool of liner models associated with best combinations (SV,CV), see Fig. 4. **b.** Error of analyte concentration prediction using models associated with best combinations (SV,CV), see Fig. 4. Names of analytes were abbreviated: MA- methyl acetate, EA- ethyl acetate, PA- propyl acetate and BA- methyl acetate.

As shown in Fig. 5a, the response of DMS to acetates was highly linear, if the appropriate part of spectrum was exploited. Best linearity was achieved for methyl acetate. In this case, the average R^2 for the group of best combinations (SV,CV) was R^2=0.92. Second best results were attained for ethyl acetate, average R^2=0.90. Butyl acetate was third (R^2=0.88) and for propyl acetate the linearity was weakest (R^2=0.87). When using these linear models acetates concentration prediction errors were in a range from 0.3 to 0.7 ppm, (see Fig.5 b.).

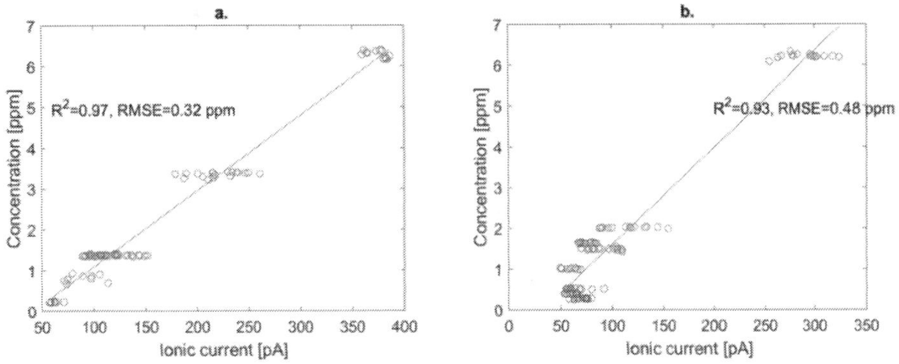

Fig. 6 Best linear fit between analyte concentration and the ionic current on positive ions collecting electrode. The fit was obtained for the best combination of (SV,CV). **a.** methyl acetate (-3.4297, 950), **b.** ethyl acetate (-2.5, 1100).

In Fig. 6 there are displayed best linear fits achieved for two acetates, namely methyl and ethyl acetate. They illustrate the fact that high linearity was attained in spite of considerable spread of DMS response, while exposing spectrometer to an individual standard gas mixture in a continuous manner. It can be noticed that in case of ethyl acetate (Fig.6 b.) measurements done at low concentrations (less than 1 ppm) did not fully match the linear relationship. The same problem was observed in case of propyl acetate measurement and it

 DOI: 10.1051/e3sconf/20172200173

has to be further examined. In case of methyl acetate (shown in Fig. 6a) as well as butyl acetate this kind of deviation was not observed.

4 Conclusions

VOCs are an important group of pollutants. In order to control their environmental impact there are needed novel methods of measurement. In this wok, differential ion mobility spectrometry was considered in respect of measuring acetates in air. These compounds are released from a variety of sources, in particular indoors.

The performed study demonstrated that DMS offers linear response to methyl, ethyl, propyl and butyl acetate in the concentration range from 0.3 ppm to 7 ppm. Positive ions spectrum has to be utilised for this purpose.

Achieving linearity requires that the appropriate part of spectrum is used. We showed that fragments of DMS spectrum, where proportionality to analyte concentration is achieved, are compound-specific. Upon increase of the size of functional group in acetate molecule, the useful part of spectrum moves towards higher separation voltage and lower compensation voltage.

The best linear dependencies were featured by the following coefficient of determination and root mean square errors of analyte concentration prediction: methyl acetate (R^2=0.97, RMSE=0.32 ppm), ethyl acetate (R^2=0.93, RMSE=0.48 ppm), propyl acetate (R^2=0.91, RMSE=0.42 ppm), butyl acetate (R^2=0.95, RMSE=0.36 ppm).

References

1. Y. Pang, M. Fuentes, P. Rieger, Atmos. Environ 122, 686-695 (2015)
2. S.J. Maisey, S.M. Saunders, N. West, P.J. Franklin, Atmos. Environ. 81, 546-553 (2013).
3. S.Chatani, S.N. Matsunaga, S. Nakatsuka, Atmos. Environ. 120, 38-50 (2015).
4. M. Kheirmand, A. Barkhordari, M.H. Mosaddegh, Z. Farajzadegan, Int. J Env. Health Eng. 3, 27 (2014).
5. Chien-Hou Wu, Chien-Tai Feng, Yu-Shiu Lo, Tsai-Yin Lin, Jiunn-Guang Lo, Chemosphere, 56, 71–80 (2004).
6. A. Szczurek, M. Maciejewska, Gas Sensor array with broad applicability, in Sensor Array,(InTech, 2012)
7. G.A. Eiceman, Z. Karpas, H.H. Hill Jr, Ion Mobility Spectrometry, 3rd edition (CRC Press/Taylor & Francis Group, Boca Raton, 2014).
8. A.A. Shvartsburg, Differential ion mobility spectrometry: Nonlinear ion transport and fundamentals of FAIMS, (CRC Press, Boca Raton, 2009)
9. E.V. Krylov, E.G. Nazarov, R.A. Miller, Int. J. Mass Spectrom. 226, 76–85 (2007).
10. B.B. Schneider, E.G. Nazarov, F. Londry, P. Vouros, T.R. Covey, Mass Spectrom Rev. 35(6), 687-737 (2016).
11. J.L. Willems, M.M. Khamis, W.M. Saeid, R.W. Purves, G. Katselis, N.H. Low, A. El-Aneed, Anal. Chim Acta 933 164-174 (2016).
12. A.Szczurek, M. Maziejuk, M. Maciejewska, T. Pietrucha, T. Sikora, Sens. Actuators B 152, 137–146 (2016)
13. M. Maziejuk J. Puton M. Szyposzyńska, Z. Witkiewicz, Talanta 144, 1201–1206, (2015).

Prediction of wastewater quality indicators at the inflow to the wastewater treatment plant using data mining methods

Bartosz Szeląg[1,*], *Krzysztof* Barbusiński[2], *Jan* Studziński[3], *Lidia* Bartkiewicz[1]

[1]Kielce University of Technology, Faculty of Environmental, Geomatic and Energy Engineering, Tysiąclecia Państwa Polskiego Av. 7, Kielce, Poland
[2]Silesian University of Technology, Institute of Water and Wastewater Engineering, Konarskiego Street 18, Gliwice, Poland
[3]Systems Research Institute Polish Academy of Science, Newelska Street 6, Warszawa, Poland

Abstract. In the study, models developed using data mining methods are proposed for predicting wastewater quality indicators: biochemical and chemical oxygen demand, total suspended solids, total nitrogen and total phosphorus at the inflow to wastewater treatment plant (WWTP). The models are based on values measured in previous time steps and daily wastewater inflows. Also, independent prediction systems that can be used in case of monitoring devices malfunction are provided. Models of wastewater quality indicators were developed using MARS (multivariate adaptive regression spline) method, artificial neural networks (ANN) of the multilayer perceptron type combined with the classification model (SOM) and cascade neural networks (CNN). The lowest values of absolute and relative errors were obtained using ANN+SOM, whereas the MARS method produced the highest error values. It was shown that for the analysed WWTP it is possible to obtain continuous prediction of selected wastewater quality indicators using the two developed independent prediction systems. Such models can ensure reliable WWTP work when wastewater quality monitoring systems become inoperable, or are under maintenance.

1 Introduction

The operation of the wastewater treatment plant is a complex process which is meant to ensure effective and relatively stable purification of wastewater. That can prove problematic due to a high variation in the quantity and quality of raw wastewater conveyed by the sewer system to the treatment facility. To maintain adequate level of pollutant reduction in wastewater, it is necessary to develop tools for predicting both the quantity and quality of influent wastewater. That will make it possible to take steps to correct, in advance, the technological parameters in the treatment facilities, mainly in bioreactors.

[*] Corresponding author: bszelag@tu.kielce.pl

In modern wastewater treatment plants (WWTPs), the values of indicators that describe wastewater quality are measured with on-line systems according to a pre-set time step, or determined in laboratory tests. In both cases, however, problems may arise with obtaining time series of measurement data at a constant resolution. In practice, that leads to a situation in which the data providing an input to a mathematical model for the optimisation of the operation of WWTP individual components are not available. In order to ensure effective performance of WWTPs, it is advisable to develop autonomous systems for the prediction of influent wastewater quantity and quality.

Key indicators of wastewater quality, which at the same time provide a basis of WWTP efficiency assessment, include the following: biochemical oxygen demand (BOD_5), chemical oxygen demand (COD), total suspended solids (TSS), total nitrogen (TN) and total phosphorus (TP). To predict wastewater quantity and quality within separate and combined sewage systems, physical mathematical models, based on the differential equation systems were developed [1]. However, due to the fact that it is necessary to collect detailed information on the catchment and meteorological conditions, the problems with the model calibration may arise. Therefore, to model the quantity and quality of wastewater flowing from the catchment to the WWTP, data mining methods are used. On the basis of the literature data [2,3], it can be stated that to simulate total suspended solids (TSS), total nitrogen (TN) and total phosphorus (TP) in influent wastewater ARIMA-type models and Artificial Neural Networks (ANN) can be employed. To predict TSS, Verma et al. [4] used a number of models, such as Support Vector Machines (SVM), Random Forests (RF), Multivariate Adaptive Regression Splines (MARS) etc. The differences between simulated and measured values turned out to be small. Additionally, the analyses performed by Minsoo et al. [5] confirmed the applicability of the k-NN method for the prediction of quality indicators of influent wastewater. In addition to the above-mentioned methods for WWTP operation modelling hybrid models [6] are used. The originate from the combination of two statistical models (the classification and the regression ones, or a few regression-type models).

This study presents a proposal of the modelling of the values of selected quality indicators (BOD_5, COD, TSS, TN, and TP) of influent wastewater with the use of three data mining methods. The modelling is based on the results of last measurements that preceded the predicted quantity, and also on the measured values of daily inflow to the facility. The model could be applied when the systems measuring pollutant indicators become inoperable. In the analyses, 3-year daily time series of flow rate and wastewater quality indicators measured at the inflow to the wastewater treatment plant located in the city of Rzeszów were used.

2 The object of investigation

Rzeszów, an urban-industrial agglomeration, lies in the south-eastern Poland. The city has mainly the separate sewerage system. Municipal wastewater from the city and adjacent localities is conveyed to the mechanical-biological wastewater treatment plant with the design capacity of 62,500 m^3/d. Continuous monitoring conducted since 2013 provides measurements of parameters describing quantity and quality of influent wastewater and also of effluent at a daily resolution.

3 Methodology

In the study, simulations were conducted for two instances. In the first case, the possibility of predicting wastewater quality indicators (BOD_5, COD, TSS, TN and TP) was considered, based on the results of last measurements of those indicators. In the other case, the possibility of predicting the values of the above-mentioned indicators on the basis of the flow rate measurements at the inflow to the plant was analysed. The advantage offered by this solution is that when the on-line measurement system fails, an alternative option makes it possible to predict selected quality indicators of influent wastewater on the basis of measured flows. Under service conditions of the plant, that allows maintaining continuity in predicting influent wastewater quality indicators, and thus ensures a stable operation of bioreactors by adequate selection of the technological parameters (e.g. oxygen concentration, TSS, the amount of PIX dosage, recirculation ratio, and others). The concept of the system described above is shown in Fig. 1.

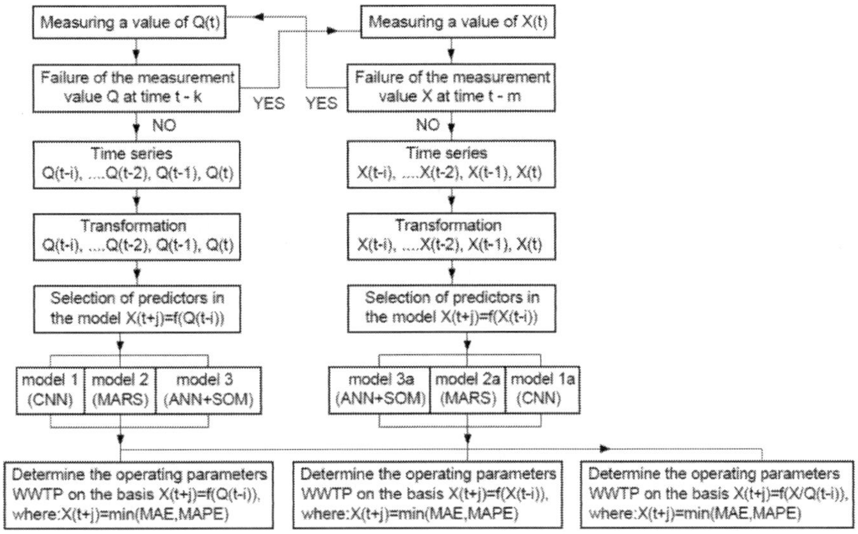

Fig. 1. Diagram of the autonomous measurement and prediction system for influent wastewater quality indicators intended to selected the settings of technological facilities

where: i – time shift between the predicted value and the previous values that describe the predicted value in the model for the prediction of a selected wastewater quality indicator; j – forward time step (in the study it was assumed j = 1) with which a quality indicator of concern is predicted; k, m – failure period in which the results of daily measurements of Q, X are not available.

Based on the concept shown above, statistical models were developed to predict wastewater quality indicators using the MARS method and cascade neural networks (CNN). Additionally, a hybrid model was designed, which is combination of neural network of the multilayer perceptron type and Kohonen self-organizing network. To ensure a correct training process and assessment of the mathematical model performance, 5-fold cross-validation was used. That was preceded by the dataset partitioning into the training set (75%), and the testing and validating set (25%). Prior to the development of statistical models, input and output data were standardised by means of the min-max transformation [7].

The MARS method is one of numerous tools used for solving regression problems [8]. In the classical approach, independent variables are treated uniformly, whereas in the MARS method, variation ranges of the predictors of concern are divided into subranges in which independent variables can have different impacts on the phenomenon investigated. The boundaries of subranges are determined on the basis of threshold values (t). The values of the predictors of concern are segregated into those lower and higher than threshold values t_i. That is performed by means of the basis function which has the following form:

$$h(\boldsymbol{X}) = \alpha_i \cdot (max(0, X - t)) \tag{2}$$

where: h(**X**) – vector of the basis functions for individual variables (x_i) for which the following condition is satisfied:

$$x_i - t_i = \begin{cases} x_i - t_i; for\ x_i > t_i \\ 0; for\ x_i \leq t_i \end{cases} \tag{3}$$

To calculate the values (α_i) of the model parameters, Friedman [8] developed a special algorithm that allows the search of observations to establish threshold values.

Multilayer unidirectional network known as multilayer perceptron [2,9,10] is one of the most widely used structures. It has found multiple applications to the modelling of the quantity and quality of influent wastewater and the operation of technological facilities. One of the modifications of multilayer perceptron–type ANN is a cascade network (CNN). It contains additional weighted connections from each layer input, and from each layer to a successive one. It should be noted that so far CNNs have solely been used to model the performance of wastewater treatment plants. They have never been applied to predict influent wastewater quality. In addition to classic neural networks with adequately modified architecture [6,11], hybrid models have been developed that result from the combination of the classification and regression models. Multilayer perceptron–type ANN provides an example of hybrid models. The most popular type of the network termed as self-organizing map (SOM) is Kohonen network [11]. Due to the algorithm used in a majority of SOMs, neurons representing similar classes are close in proximity to each other, thus producing an organised map. Consequently, it is possible to identify relations between the classes obtained. At the SOM training stage, the number of neurons making the topological layer and the number of training epochs are defined.

In this study, cascade neural network models were developed. In the models, three connections were assumed to join inputs to successive layers of the CNN [12,13]. In individual hidden layers, various activation functions were considered, namely hyperbolic tangent, linear, sine, exponential, logistic, and sigmoidal ones. For the output layer, the linear activation function was adopted. In addition, it was assumed in computations that the number neurons in the successive hidden layers is the same. The optimal number of neurons and values of weights were established for CNN by generating, every time, 500 different models for the adopted independent variables. Out of the trained networks, the one showing the lowest values of the absolute and relative errors in the prediction of wastewater quality indicators was selected. To obtain required generalizing abilities of CNN models, Bayesian regularisation algorithm was employed at the training stage [14]. In the case under consideration, for the hybrid model, Kohonen self-organizing neural networks with the topological layer 2x2 were used as a classifier, whereas the number of epochs during the training stage was 1000.

In the above analyses, to assess the predictive powers of the models predicting BOD_5, COD, TSS, TN, and TP values, the following were used: mean absolute error (MAE), mean relative error (MAPE) and coefficient of correlation (R).

4 Results

Based on the results of measurements of influent wastewater quantity and quality, the ranges of variation were established (Table 1). The data in Table 1 indicate substantial variation in BOD_5, COD, TSS, TN and TP values. Because the indicators of wastewater quality constitute input data for the models describing the kinetics of changes in carbon, nitrogen and phosphorus compounds in the bioreactors, and their values vary to a large extent, it is necessary to predict those indicators. Out of the methods investigated in this study, only the parameter estimation algorithm developed for the MARS model offers the possibility of discarding those predictors the effect of which on the dependent variable is negligible. Therefore, the simulation, with the use of this method, of selected quality indicators of influent wastewater was performed first. On the basis of independent variables specified using the MARS method, predictions of BOD_5, COD, TSS, TN and TP values were made with the other methods.

Table 1. Range of variation in the values of parameters describing influent wastewater in the years 2013-2015

Parameter	Minimum	Maximum	Average
COD, mg/dm^3	159.0	2510.0	927.1
BOD_5, mg/dm^3	38.1	788.0	374.0
TSS, mg/dm^3	80.0	1140.0	430.0
TN, mg/dm^3	21.3	99.0	69.7
TP, mg/dm^3	3.4	37.5	12.45
Q, m^3/d	26973	66773	38658

The values of wastewater quality parameters obtained from the last measurements were used for that purpose. That was followed by simulations of influent wastewater quality indicators based on flow rate measurements.

The analyses demonstrate that in order to compute COD, TN and TP values, it is advisable to take into account two last measurements (i=1,2) of those indicators, which precede the predicted value. For TSS prediction, it is necessary to account for the values from TSS(t-1) to TSS(t-4), whereas for BOD_5 prediction, BOD_5(t-1) and BOD_5(t-3) values are sufficient. The analyses indicate that to predict COD, TSS, TN and TP, the values Q(t-1) and Q(t-2) could be used. To compute BOD_5, it is necessary to account for the values Q(t-1), Q(t-2) and Q(t-3). The results obtained with the use of the MARS method are confirmed by the investigations conducted by Szeląg et. al.[7] with the classification-tree method. Tables 2-4 show the values of parameters of fit (MAE and MAPE) of computations with MARS, CNN and ANN+SOM methods of selected wastewater quality indicators to the measurement results for the testing set. The number of neurons in individual hidden layers in the models ranged from 3 to 6.

As regards the models based on X(t-i), the best computational results were observed for the hyperbolic tangent function in the hidden layer, whereas in the modelling of the concentration of organic compounds, and of nitrogen and phosphorus, the logistic function produced the lowest values of MAE and MAPE.

In hybrid neural networks, the number of neurons in the hidden layer of the designed statistical models ranged 3-8 in separate classes, and the hyperbolic tangent function was most frequently used as the activation function. The data presented in Tables 2-4 show that better predictive powers of independent variables (lower MAE and MAPE values) are shown by the models developed on the basis of wastewater quality indicators, which is true for both MARS and CNN methods. The statistical models developed using ANN+SOM for computations of BOD_5, TSS and TN, based on Q(t-i), have better predictive abilities with respect to wastewater quality indicators.

Taking into account the time necessary for BOD_5 determination and the simulation results of wastewater quality indicators, it can be stated that autoregression models that take into account BOD_5(t-i) values cannot be used in practical applications. Additionally, error values for models relying on BOD_5(t-i) and Q(t-i) values range as follows: MAE = 30.16-41.12 mg/dm^3, MAPE = 8.75-12.39%, and MAE = 27.93-48.10 mg/dm^3, MAPE = 7.62-14.80%, respectively. In models for COD prediction, based on COD(t-i), and solely on Q(t-i), values of errors in the indicator prediction varied as follows: MAE = 93.29-117,54 mg/dm^3, MAPE = 10.26-14.06% and MAE = 92.50-133.02mg/dm^3, MAPE = 10.93-15.89%,respectively. In the models for TSS prediction that account for TSS(t-i), and only Q(t-i), the magnitude of simulation errors were MAE = 49.05-59.01 mg/dm^3, MAPE = 13.09-15.90% and MAE = 50.89-71.88 mg/dm^3, MAPE = 14.62-20.07%, respectively.

Table 2. Values of parameters of fit (MAE and MAPE) of computational results of wastewater quality indicators obtained with MARS method, determined on the basis of X(t-i) and Q(t-i), to measurement results

Indicators	X(t-i)			Q(t-i)		
	MAE	MAPE	R	MAE	MAPE	R
BOD_5	41.12	12.39	0.59	48.1	14.80	0.55
COD	117.54	14.06	0.56	133.02	15.89	0.52
TSS	59.01	15.9	0.55	71.88	20.07	0.38
TN	6.07	9.64	0.67	7.04	11.43	0.66
TP	1.61	13.97	0.65	2.16	19.03	0.45

Table 3. Values of parameters of fit (MAE and MAPE) of computational results of wastewater quality indicators obtained with CNN method, determined on the basis of X(t-i) and Q(t-i), to measurement results

Indicators	X(t-i)			Q(t-i)		
	MAE	MAPE	R	MAE	MAPE	R
BOD_5	33.07	9.67	0.79	37.29	11.29	0.72
COD	99.56	10.26	0.68	112.5	13.14	0.63
TSS	52	13.89	0.7	60.9	16.61	0.63
TN	5.45	8.53	0.74	5.76	8.96	0.72
TP	1.4	12.19	0.66	1.76	15.54	0.56

Table 4. Values of fit (MAE, MAPE) of computational results of selected wastewater quality indicators obtained with hybrid model (ANN+SOM) to measurement results

Indicators	X(t-i)			Q(t-i)		
	MAE	MAPE	R	MAE	MAPE	R
BOD_5	30.16	8.75	0.74	27.93	7.62	0.82
COD	93.29	11.34	0.73	92.5	10.93	0.72
TSS	49.05	13.09	0.75	50.89	14.62	0.73
TN	5.00	7.83	0.78	4.38	6.42	0.84
TP	1.72	15.11	0.62	1.79	15.75	0.55

As regards to TN prediction models based on TN(t-i), and solely on Q(t-i), the error values were similar, namely MAE = 5.00-6.07 mg/dm^3, MAPE = 7.83-9.64% and MAE = 4.38-7.04 mg/dm^3; MAPE = 6.42-11.42%. In addition, the data in Tables 2-4 show that for TP prediction models that involve TP(t-i) and solely Q(t-i), prediction errors range as follows: MAE = 1.40-1.72 mg/dm^3, MAPE = 12.19-15.11% and MAE = 1.76-2.16 mg/dm^3, MAPE =15.54-19.03%, respectively. Predictions of wastewater quality indicators (BOD_5, COD, TSS, TN and TP) listed in Tables 2-4 show that the lowest prediction errors were observed for the MARS method. The greatest congruence between computational results and measurements of BOD_5, COD, TSS, and TN was achieved using the ANN+SOM method. Only for TP, the lowest values of prediction errors were found when the CNN method was employed. Computations performed using the ANN+SOM method demonstrate that the predicted values of COD, TSS, TN and TP based on Q(t-i) do not differ more than 5% from the results produced with the models in which independent variables were wastewater quality indicators. A similar relation was found for BOD_5 values. Those findings can prove useful in case of the measurement system failure.

The value of the coefficient of correlation for BOD_5 prediction model, based on Q(t-i), developed using ANN+SOM method (Table 4) is lower than that obtained by Abyaneh [3] (R = 0.83) with ANN. To predict BOD_5, Abyaneh [3] relied on temperature, pH of influent wastewater, and TSS. Dogan et al. [15] received a better fit (R = 0.92) of BOD_5 simulation results to measurements than Abyaneh [3]. Additionally, the values of MAPE computed for COD prediction model are greater than those (MAPE = 7.35%) obtained by Minsoo et al [5] with the k-NN method. The values of R computed for TSS prediction model turned out to be lower than the results received by Verma and Kusiak [4] (R = 0.93), who used the ANN method and relied on $CBOD_5$ and flow measurements.

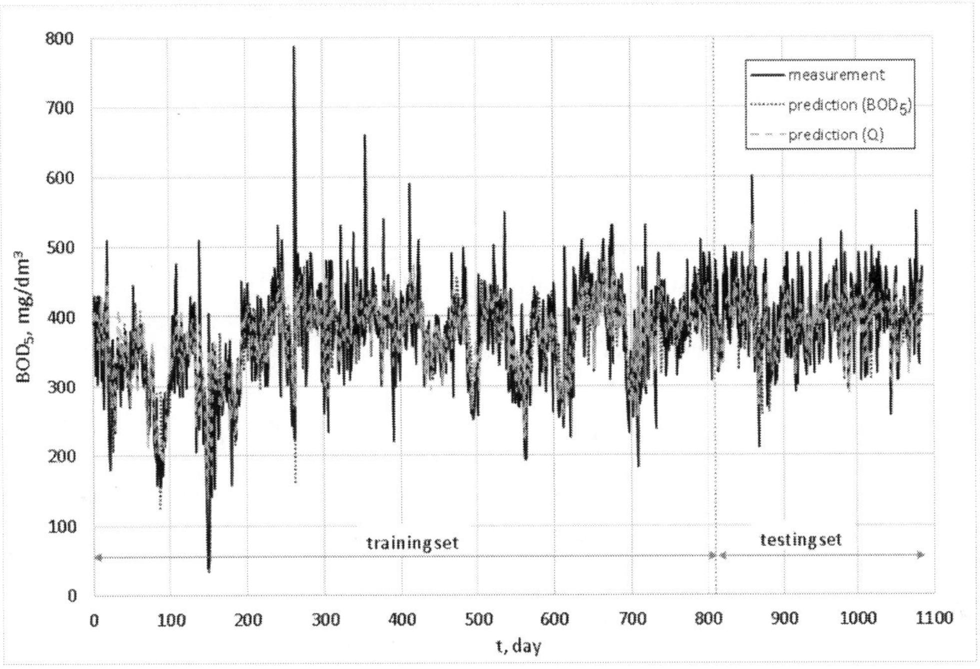

Fig. 1. Comparison of BOD_5 computations with hybrid models and measurements in the period of concern

Values of MAPE in models for the prediction of TN and TP with the ANN+SOM method (Table 4) are greater than those produced with the k-NN method (MAPE = 5.79% and MAPE = 8.87%), obtained by Minsoo et al [5]. The analyses presented above indicate that the parameter which considerably affects the variation in quality indicators is wastewater inflow to WWTP. That is illustrated by the parameters of fit of computational results to measurements obtained in this study, and also by other researchers. At the same time the analyses conducted for this study confirm that it is possible to predict wastewater quality indicators solely on the basis of flow data. The accuracy of predictions is satisfactory, which is illustrated by the values of errors obtained (Tables 2-4). Although the parameters of computations fit to measurements (R, MAE, MAPE) received in this study were worse than those reported by Minsoo et al [5], Dogana et al [15], Verma and Kusiak [4] the models designed in this study can find practical applications. When the number of independent variables in the statistical models is limited to only wastewater inflow data, the models could be employed under failure conditions of the measurement system, or during the system overhaul. Such a scenario was not considered in other studies. As remaining researchers [3,4,5,7] took into account other independent variables in addition wastewater inflow, it is necessary to conduct further research aimed at finding the impact of those variables on the predictive abilities of the models.

5 Conclusions

The study discusses the possibility of modelling the values of selected indicators (BOD_5, COD, TSS, TN, and TP) of wastewater quality at the inflow to the treatment plant on the basis of the data from the last measurements, and also solely on the basis of flow rate values. The computations demonstrated that lower values of errors in the prediction of biochemical and chemical oxygen demand, concentration of nitrogen, phosphorus, and total suspended solids were generated in the statistical model developed using cascade neural networks than with the MARS method. In the majority of cases considered, the best predictive abilities as regards wastewater quality indicators were shown by hybrid models that combine Kohonen neural networks and multilayer perceptron.

Slightly better results of simulations of wastewater quality indicators were obtained when independent variables were the values of wastewater quality indicators rather than the flow values. That refers to models based both on MARS and CNN. For hybrid models, the prediction error values were lower than for MARS and CNN-based models. It was only for TP computations with the use of Q(t-i) that MAE and MAPE values were lower when CNN-based model, rather than ANN+SOM–based one, was used.

It was shown that for the wastewater treatment plant of concern, it is possible to ensure continuity in predictions of the values of selected pollutant indicators using the two independent systems developed in the study. That could ensure reliable operation of the plant facilities when the wastewater quality monitoring system fails or is being upgraded.

References

1. K. V. Gernaey, M.C.M. van Loosdrecht, M. Henze, M. Lind, S.B. Jørgensen, Environ Modell Softw **19**, 9, 763 (2004)
2. A. Dellana, D. West, Environ Modell Softw **24**, 1, 96 (2009)
3. H. Z. Abyaneh, J Environ Health Sci **12**, 40, 2 (2014)
4. A.Verma, X.Wei, A. Kusiak, Eng Appl Artif Intel **26**, 1366 (2013)
5. K.Minsoo, K. Yejin, K. Hyosoo, P. Wenhua, K. Changwon, Front Env Sci Eng **10**, 2, 299 (2016)
6. S. Grieu, F. Thiery, A. Traoré, T. P. Nguyen, M. Barreau, M. Polit, Chemical Engineering Journal **1**, 116, 1 (2006)
7. B. Szeląg, L. Bartkiewicz, J. Studzinski, Ochrona Srodowiska **38**, 4, 39 (2016)
8. W . Zhang, A. T. C. Goh, Geoscience Frontiers **7**, 1, 45 (2016)
9. D Güçlü., Ş. Dursun, Bioproc and Biosyst Eng **33**, 9, 1051 (2010)
10. B. Szeląg, J. Gawdzik, Pol J Environ Stud **25**, 4, 1709 (2016)
11. R. Rustum, A. J. Adeloye, M. Scholz, Water Environ Res **80**, 1, 32 (2008)
12. G.Capizzi, G. L. Sciutto, P. Monforte, C. Napoli, International Journal of Electronics and Telecommunications **61**, 327 (2015)
13. S. B.Al – Batah, M.S Alkhasawneh., L.T Tay., U.K. Ngah, H. H. Lateh, M.T.A Isa, Math Probl Eng **2015**, 2015, 1 (2015)
14. K. Hirschen, M. Schäfer, Computer Methods in Applied Mechanics and Engineering **195**, 7-8, 481 (2006)
15. E. Dogan, A.Ates, E.C. Yilmaz, B. ErenEnvironmental Progress **27**, 4, 439 (2008)

 DOI: 10.1051/e3sconf/20172200175

Implementation of ferric hydroxide-based media for removal of toxic metalloids

Małgorzata Szlachta* and *Patryk* Wójtowicz

Wrocław University of Science and Technology, Faculty of Environmental Engineering, Wybrzeże S. Wyspiańskiego 27, 50-370 Wrocław, Poland

Abstract. Effective removal of inorganic arsenic species is possible by application of the sorption technique with the use of iron-based sorbents. This study investigates the removal of arsenic(III) and arsenic(V) from an aqueous solution by application of a granular ferric hydroxide-based sorbent. The performance of tested media was evaluated based on the batch and fixed-bed adsorption studies. The efficiency of the process was determined with various treatment times, adsorbent doses, initial concentrations of arsenic and various solution temperatures. The obtained adsorption data were fitted with pseudo-first and second-order kinetic models and Langmuir and Freundlich isotherm equations. It was observed that the overall arsenite removal was lower when compared to the arsenate, and all tested operating parameters influenced the process efficiency. The experiments under dynamic conditions showed high treatment capacity and stability of tested adsorbent over a long period of time.

1 Introduction

The pollution of the aquatic ecosystem is a global long-term environmental problem due to the rapid industrial development and urbanisation. Toxic heavy metals or metalloids are widespread and their discharging through air and industrial effluents or leaching from the soil by acid rain into natural waters often is a reason of their severe degradation. Many unwanted contaminants in an excessive concentration in the aquatic environment cause a detrimental effect on human and animals. The accumulation of metal(loid)s species over time in living organisms can be damaging to them and cause serious consequences [1-2].

One of the common metalloid widely known for its high toxicity is arsenic. Cases of severe arsenic poisoning in humans have been reported in many places around the world, including Bangladesh, India, Vietnam, China, Chile, Mexico, Argentina and USA [3-4]. This element occurs as a natural constituent of rock, soil, water and atmosphere as well as may originate from the anthropogenic sources. In natural waters, arsenic is usually present in inorganic forms, as arsenite or arsenate oxyanions. However, trivalent arsenite As(III) is dominant in more reducing conditions, whereas pentavalent arsenate As(V) is mostly present in an oxidizing environment [5-6].

* Corresponding author: malgorzata.szlachta@pwr.edu.pl

 DOI: 10.1051/e3sconf/20172200175

The removal of various arsenic species from contaminated water and wastewater is possible due to the application of oxidation and precipitation techniques, coagulation, ion exchange and sorption methods or membrane separation [7-11]. Each of these treatment techniques has advantages and may offer a good solution for the treatment of arsenic-rich water/wastewater. However, other factors like objectives of the treatment, capital and operating costs, disposal of generated waste are the key elements in the selection and design of the proper treatment process.

Obviously, effective removal of arsenic is possible by the application of the sorption technique with use of iron-based sorbents (IBS). Many studies have confirmed that arsenic has a natural affinity for iron compounds under various operating conditions [12-15]. Therefore, considering adsorption on iron-containing surfaces for treatment drinking water or industrial wastewater with elevated concentrations of arsenic may be the best choice in many cases.

This study investigated the removal of arsenic(III) and arsenic(V) from an aqueous solution by using iron-based sorbent – a granular ferric hydroxide-based material. The performance of the tested media was evaluated based on the batch and fixed-bed adsorption studies carried out on the laboratory scale. The efficiency of the process was determined with various treatment times, adsorbent doses, concentrations of arsenic and various solution temperatures. The obtained adsorption data were fitted with kinetic and equilibrium models.

2 Materials and methods

The ferric hydroxide-based material (CFH12) is a commercial adsorbent (Kemira Oyj, Finland) and was selected for the adsorption laboratory experiments. It is a porous material with a specific surface area of 158 m^2/g. A comprehensive characteristic of the tested adsorbent is provided in our previous work [16-17].

To evaluate the performance of CFH12 in arsenic(III) and arsenic(V) removal from model solutions, batch and fix-bed experiments were conducted.

For the batch tests, synthetic solutions were freshly prepared using Milli-Q water and spiked with As(III) and/or As(V). The adsorption equilibrium experiments were performed using various initial concentrations of As(III) or As(V). The samples were agitated in a shaker incubator until equilibrium was reached. The experiments with various material doses (1–10 g/L) and solution temperatures (5–35°C) were conducted in a similar way, with an initial adsorbate concentration of 20 mg/L. In the kinetics tests, the pre-weighted material was added to 500 mL of a water sample spiked with As(III) or As(V). The tests were performed for five hours using a magnetic stirrer.

The fixed-bed column set-up consisted of custom-made columns loaded with pre-washed granular iron hydroxides. As a feed solution, dechlorinated tap water spiked with arsenic solution was used and passed through the columns downward with a continuous flow rate of 3 mL/min using a peristaltic pump. The samples were collected periodically using an automated samples collector.

The initial pH of the solution was adjusted by the addition of either 0.1M HCl or 0.1 M NaOH. The initial and residual arsenic concentrations were determined using the ICP-OES or ICP-MS technique.

3 Results and discussion

3.1 Kinetics studies

A kinetics study is particularly important in terms of proper assessment of the effect of the treatment time on the performance of the tested material. The mechanism of the process involved as well as the rate of adsorption were analysed based on kinetic parameters determined with the use of the non-linear regression method. Two kinetic models, pseudo-first (Eq. 1) and pseudo-second-order equation (Eq. 2), were used to describe the kinetic data.

$$q = q_e(1 - e^{-k_1 t}) \quad (1)$$

$$q = \frac{k_2 q_e^2 t}{1+k_2 q_e t} \quad (2)$$

where q and q_e is the amount of arsenic adsorbed at time t and equilibrium state(mg/g), respectively; t is the contact time (min); k_1 is the rate constant of the pseudo-first-order adsorption (min^{-1}); k_2 is the rate constant of the pseudo-second-order adsorption (g/mg min).

The adsorption of arsenic species onto CFH12 is shown in Fig. 1, while the calculated parameters of kinetic models for 1 mg/L and 20 mg/L of arsenic(III) and (V) are presented in Table 1.

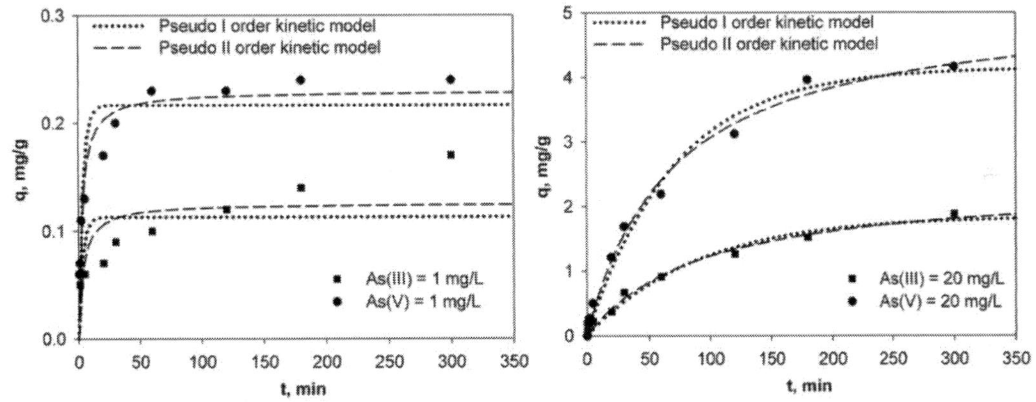

Fig. 1. Adsorption kinetics of As(III) and As(V) on the CFH12; experimental data fitted with pseudo-first and second-order kinetic models

Table 1. Parameters of adsorption kinetic models for As(III) and As(V)

Model	Parameter	As(III) C_0 = 1 mg/L	As(III) C_0 = 20 mg/L	As(V) C_0 = 1 mg/L	As(V) C_0 = 20 mg/L
	q_{exp} (mg/g)	0.17	1.88	0.24	4.16
Pseudo I order kinetic model	k_1 (1/min)	0.313	0.011	0.283	0.014
	q_1 (mg/g)	0.11	1.85	0.22	4.15
	R^2	0.480	0.969	0.856	0.986
Pseudo II order kinetic model	k_2 (g/mg min)	2.188	0.005	1.605	0.003
	q_2 (mg/g)	0.13	2.37	0.23	5.16
	R^2	0.625	0.978	0.938	0.991

The experimental results reveal that the adsorption capacity along with removal efficiency increase with contact time regardless of the initial concentration of arsenic ions. However, at the beginning of the process, the rate of uptake of As(V) was faster comparing to As(III). The percentage removal of 1 mg/L of As(V) after one hour of contact was approximately 97.2%, whereas the removal for As(III) was close to 45%. The extension of the contact time to five hours allows for a slight improvement (1.3%) of As(V) uptake, but in the case of As(III), the increment was considerably higher, and the recorded removal efficiency reached 78.2%. The performance of CFH12 in the treatment solution contained high concentration of arsenic (20 mg/L) resulted in maximum removal of As(III) and As(V) approximately 50% and 85%, respectively, which corresponds to the uptake of 1.88 mg/g and 4.16 mg/g.

The evaluation of the best kinetic models for fitting the experimental data was made based on the correlation coefficient values. Higher values of R^2 were observed for the second-order kinetic equation for both treated solutions (Table 1). Since this model was more representative for simulating the kinetic data than the pseudo first-order kinetic equation, it can be assumed that the adsorption mechanism of As(III) and As(V) follows pseudo-second-order kinetic reaction, and that the chemisorption may be a rate controlling step in the investigated adsorption systems. However, under applied experimental conditions, differences between solutions with lower (1 mg/L) and higher (20 mg/L) concentration of arsenic(III) and (V) were noticeable in the agreement between adsorption capacity predicted by the model and capacity determined experimentally.

3.2 Equilibrium studies

To evaluate the equilibrium adsorption of arsenic(III) and arsenic(V) on the investigated material CFH12, non-linear Langmuir (Eq. 3) and Freundlich (Eq. 4) models were used to fit experimental data.

$$q_e = \frac{q_m b C_e}{1+bC_e} \quad (3)$$

$$q_e = KC_e^n \quad (4)$$

where q_e and q_m are the equilibrium and maximum adsorption capacities (mg/g), respectively; C_e is the equilibrium concentration of arsenic in the solution (mg/L), b is the Langmuir adsorption equilibrium constant (L/mg); K is the Freundlich constant ((mg/g)·(L/mg)), and n is the adsorption intensity parameter.

The Langmuir isotherm equation assumes that there is no interaction between the particular molecules of adsorbate, which leads to monolayer formation on the adsorbent surface. Thus, as saturation is reached, no further sorption can take place. The Freundlich model generally suits better for the multilayer adsorption process. The experimental data fitted with both models are shown in Fig. 2.

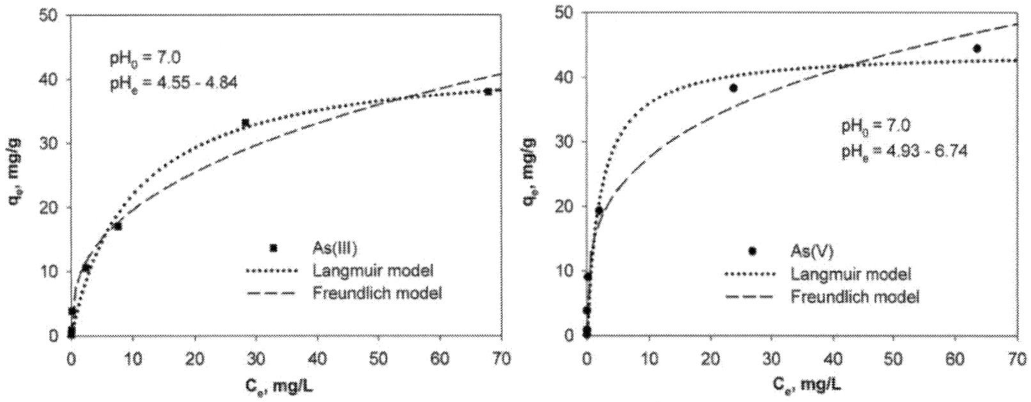

Fig. 2. Adsorption isotherm of As(III) and As(V) on the CFH12; experimental data fitted with Langmuir and Freundlich models

The applicability of the models was assessed based on the calculated values of correlation coefficients. The experimental results were well correlated to the Langmuir equation ($R^2 = 0.986$ for As(III); $R^2 = 0.974$ for As(V)), however in case of As(V) a slightly better fit of the Freundlich model was observed ($R^2 = 0.980$ for As(III); $R^2 = 0.981$ for As(V)). In the Langmuir isotherm calculated adsorption capacities for arsenite and arsenate are 43.75 mg/g and 44.04 mg/g, respectively. In the Freundlich isotherm, adsorption intensity parameters are found to be 2.655 and 3.467 for As(III) and As(V).

3.3 Effect of material dose and solution temperature studies

The removal of arsenic from an aqueous solution was performed with various adsorbent doses and the obtained results are shown in Fig. 3. As expected, the efficiency of the process increases with the increase of the amount of CFH12 applied. The increase of the removal efficiency could be attributed to the availability of a specific surface area and a higher number of adsorption sites at the solid phase. However, under applied experimental conditions a substantial difference was observed between the removal of arsenic(III) and arsenic(V). As the adsorbent dose was higher than 5 g/L, the As(V) removal extent remained almost constant, indicating the saturation of the adsorption sites. The percentage removal of As(III) gradually increases from 52.9% to 93.8% as the material dose increased from 1 g/L to 10 g/L.

In this study, it was observed that the process temperature has an influence on CFH12 performance in arsenic(III) removal, comparing to arsenic(V) for which no significant change in the percent removal was observed (Fig. 3). Raising the solution temperature from 5°C to 35°C increased As(III) removal from approximately 80% to 92%, suggesting an endothermic nature of the process. The results show that the amount of As(V) removed by the CFH12 remained on the same level (99.0–99.7%) within the range of the tested solution temperatures.

DOI: 10.1051/e3sconf/20172200175

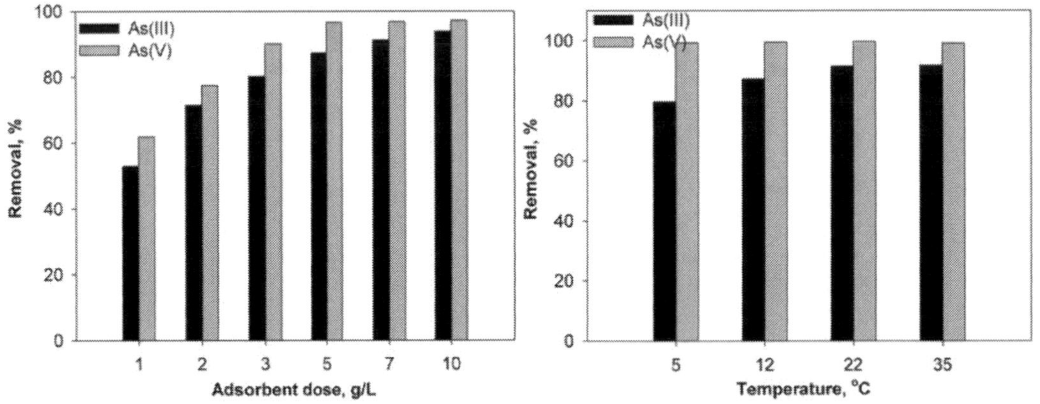

Fig. 3. Effect of the dose of CFH12 (left) and the solution temperature (right) on the removal efficiency of As(III) and As(V)

3.4 Fixed-bed adsorption studies

Granular iron hydroxide CFH12 is a material developed to work under dynamic conditions. Therefore, a series of fixed-bed adsorption tests was run with an empty bed contact time (EBCT) of 23 minutes. This contact time was sufficient enough to maintain a high removal rate of As(III) and As(V) for a very long period of time. The results from the fixed-bed adsorption tests are illustrated in Fig. 4. The curves represent the residual concentration of As in the treated solution ($C_{e\text{-}As}$) as a function of the treated bed volumes (BV) along with the final pH (pH_e). As can be seen, the bed was not exhausted by arsenic during the experiment. The adsorbent CFH12 performed very well and the arsenic concentration in the treated solution remained below 10 μg/L (the maximal permissible level in drinking water) up to about 15,000 bed volumes. After 15,000 BV, the level of As(III) and (V) gradually increased and reached 15 μg/L at approximately 20,000 BV.

Initial pH values of the model solutions were between 7.0 and 7.5. It was observed that at the beginning of the process, in case of both As(III) and As(V), pH decreased rapidly to about 2.5. After 700 BV, the pH volume stabilized at 7.3–7.9 and did not change significantly during the experiment.

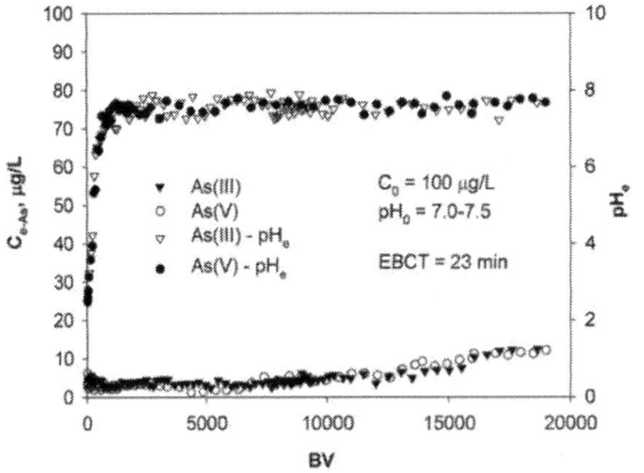

Fig. 4. Changes of solution pH and breakthrough curves for sorption of arsenic(III) and (V) onto CFH12

 DOI: 10.1051/e3sconf/20172200175

4 Conclusions

Tested granular iron hydroxide is a commercial sorbent and its great potential for adsorptive removal of toxic arsenic species from aqueous solutions was confirmed.

The Langmuir and Freundlich isotherm equations were applied for modeling adsorption data and the maximal adsorption capacity calculated by the Langmuir model was comparable for arsenic(III) and arsenic(V). The effect of the contact time, adsorbent dose and solution temperature on the removal of arsenic ions was discussed. It was demonstrated that the percentage removal of arsenic(II) and (V) increased with longer treatment time and significantly higher efficiency of the process was observed for a lower concentration of arsenic present in the solution. The study also reveal that the pseudo-second-order adsorption mechanism is predominant.

The adsorption test results under dynamic conditions demonstrated high adsorbent efficiency and stability over a long period of time. The arsenic concentration in the treated solution remained below 10 μg/dm^3 up to about 15,000 bed volumes.

Hence, the application of CFH12 to adsorptive treatment of arsenic-contaminated water and wastewater should be regarded as a robust and effective solution.

This research was supported by the National Centre for Research and Development grant (2014–2017), "Tools for Sustainable Gold Mining in EU"-SUSMIN, within the ERA-NET ERA-MIN programme.

References

1. X. Hu, Y. Zhang, Z. Ding, T. Wang, H. Lian, Y. Sun, J. Wu, Atmos. Environ. **57**, 146 (2012)
2. P.B. Tchounwou, C.G. Yedjou, A.K. Patlolla, D.J. Sutton, *Molecular, Clinical and Environmental Toxicology, Chapter 6* (Springer 2012)
3. M.M. Rahman, U.K. Chowdhury, S.C. Mukherjee, B.K. Mondal, K. Paul, D. Lodh, B.K. Biswas, C.R. Chanda, G.K. Basu, K.C. Saha, S. Roy, R. Das, S.K. Palit, Q. Quamruzzaman, D. Chakraborti, J. Toxicol. - Clin. Toxic. **39**, 7, 683 (2001)
4. L.M. Camacho, W. Gutierrez, M.T. Alarcon-Herrera, M.D. Villalba, S.G. Deng, Chemosphere, **83**, 211(2011)
5. M. Bissen, F.H. Frimmel, Acta Hydroch. Hydrob. **31**, 97 (2003)
6. P.L. Smedley, D.G. Kinniburgh, Appl. Geochem. **17**, 5, 517 (2002)
7. R.J. Gohari, W.J. Lau, T. Matsuura, A.F. Ismail, *Remediation of Heavy Metals in the Environment, Chapter 12* (CRC Press 2016)
8. A.L.S. Duarte, S.J.A. Cardoso, A.J. Alçada, Sustainabillity **1**, 4, 1288 (2009)
9. S. Fazi, S. Amalfitano, B. Casentini, D. Davolos, B. Pietrangeli, S. Crognale, F. Lotti, S. Rossetti, Rend. Lincei - Sci. Fis. **27**, 1, 51 (2016)
10. D. Mohan, C.U. Pittman Jr., J. Hazard. Mater. **142**, 1 (2007)
11. R. Singh, S. Singh, P. Parihar, V.P. Singh, S.M. Prasad, Ecotox. Environ. Safe. 112, 247 (2015)
12. M. Szlachta, V. Gerda, N. Chubar, J. Colloid Interface Sci. **365**, 213 (2012)
13. T. Mishra, D.K. Mahato, J. Environ. Chem. Eng. **4**, 1, 1224 (2016)
14. J. Qi, G. Zhang, H. Li, Bioresour. Technol. **193**, 243 (2015)

15. A.B. Cundy, L. Hopkinson, R.L.D. Whitby, Sci. Total Environ. **400**, 1-2, 42 (2008)
16. M. Szlachta, P. Wójtowicz, Desalin. Water Treat. **57,** 54, 26376 (2016)
17. M. Szlachta, P. Wójtowicz, Ochr. Środ. **38**, 4, 47 (2016)

 DOI: 10.1051/e3sconf/20172200176

Integrated assessment of soil quality after application of the biogas fermentation residues – a laboratory experiment

Arkadiusz Telesiński[1,*], *Krystyna* Cybulska[2], *Maciej* Płatkowski[1], *Michał* Stręk[1], *Grzegorz* Jarnuszewski[3], *Ilona* Wrońska[2], *Piotr* Mularewicz[4], *Tomasz* Kajdan[4], *Robert* Biczak[5] and *Paweł* Kołosowski[6]

[1]West Pomeranian University of Technology, Faculty of Environmental Management and Agriculture, Department of Plant Physiology and Biochemistry, ul. Słowackiego 17, 71-434 Szczecin, Poland
[2]West Pomeranian University of Technology, Faculty of Environmental Management and Agriculture, Department of Chemistry, Microbiology and Environmental Biotechnology, ul. Słowackiego 17, 71-434 Szczecin, Poland
[3]West Pomeranian University of Technology, Faculty of Environmental Management and Agriculture, Department of Soil Science, Grassland Management and Environmental Chemistry, ul. Słowackiego 17, 71-434 Szczecin, Poland
[4]Biowatt S.A., ul. Blacharska 2, 61-006 Poznań, Poland
[5]Jan Długosz University in Częstochowa, Faculty of Mathematics and Natural Sciences, Department of Biochemistry and Ecotoxicology, al. Armii Krajowej 13/15, 42-200 Częstochowa, Poland
[6]West Pomeranian University of Technology, Faculty of Environmental Management and Agriculture, Department of Construction and Use of Technical Devices, ul. Papieża Pawła VI 3, 71-459 Szczecin, Poland

Abstract. The aim of study was to determine the impact of three different biogas fermentation residues on some chemical and biochemical characteristics in sandy soil. The laboratory experiment was carried out on loamy sand. Residues were added to soil samples in the forms of pulp, drought, and granulate at dosages of 10, 50, and 100 $g \cdot kg^{-1}$. The reference was the soil sample without residues. On day 28, the content of macroelements and heavy metals was determined. In addition, on days 1, 7, 14, 28, and 56, the content of biomass and the activities of some hydrolases and oxidoreductases were assayed. Results showed that the application of all fermentation residues caused an increase in most of the chemical parameters. The highest impact on pH and the content of C_{tot}, N_{tot}, S_{tot}, K, and P was observed in the soil treated with granulate, whereas the increase in the content of heavy metals was the highest after the drought application. The effect of biogas fermentation residues on all hydrolases and *o*-diphenol oxidase activities was mostly significant, but depended on the kind of residues and the day of experiment. Biomass content and the activity of dehydrogenase were increased in the whole experiment.

[*] Corresponding author: arkadiusz.telesinski@zut.edu.pl

1 Introduction

A by-product in the form of fermentation residue is formed during biogas production. The processing of fermentation residue, for example, by thermal drying, pelletization, and drying with lime, enables the use of the resulting product for fertilization, energy, or others [1].

Fertilizer produced in this way reduces energy consumption and limits greenhouse gas emissions, which contribute to climate protection. The chemical composition of the fermentation residue depends on the substrates introduced into the reactor of a biogas plant [2]. The hydrated fermentation residue is quite strongly mineralized compared with the input substrates, which allows rapid absorption of nutrients by the plants [3]. The fermentation residues can be used to fertilize the soil on which the energy crops are grown. Fermentation residue use for the above purposes allows partial closure of elements flux and may reduce the costs of biomass production [4].

There are many positive effects resulting from the agricultural use of the residue; however, there are also many opponents of that method. They argue their reasons with the fear of bacteria and pathogens present in the fermentation residue. Another argument against the use of residue for fertilization is the odor arising from its dispersion on the field [2].

Currently, an increasing number of studies are being conducted in this field, and technological processes aimed at the use of fermentation residues in agriculture or their processing into high-quality organic fertilizers are being tested [5–7].

The aim of this study was to determine the effect of various forms of fermentation residues from biogas production, raw and processed by Biowatt S.A., Poznań, Poland, on selected chemical and biological properties of sandy soil.

2 Material and methods

2.1 Scheme of the experiment

The experiment was carried out in triplicate under laboratory conditions. The testing was performed on a soil material collected from the topsoil of Brunic Arenosol in Agricultural Experimental Station in Lipnik (53°24'N, 14°28'E), located in the West Pomeranian District, Poland. According to the classification by the U.S. Department of Agriculture, the sample was the soil with a granulometric composition corresponding to loamy sand. The soil was air dried and sieved with a mesh size of 2 mm.

Various forms of fermentation residues from biogas production, provided by the technological company Biowatt S.A., specializing in the processing of organic waste into energy and high-quality organic fertilizers, were used in the experiment. The substrate for methane fermentation consisted of corn silage (51%), pig slurry (20%), chicken manure (12%), cattle manure (7.5%), green rye (7.5%), and grains (2%). After the fermentation process, the collected pulp was dried and subjected to granulation according to the technology developed by Biowatt S.A.

Fermentation residues were introduced into the soil samples in three forms: pulp, drought, and granulate, in quantities of 10, 50, and 100 $g \cdot kg^{-1}$. The reference sample was the soil without an addition of fermentation residues. All soil samples were adjusted to 60% maximum water-holding capacity, and they were incubated in tightly closed glass containers at a temperature of 20°C.

 DOI: 10.1051/e3sconf/20172200176

2.2 Determination of soil chemical properties

Soil chemical properties were assayed on day 28. Soil pH (pH in 1 M KCl, and in distilled water) was determined in the samples using the potentiometric method with a pH-meter. The content of organic carbon (C_{org}) was determined using the Tiurin's method; the content of total potassium and heavy metals (Cu, Zn, Mn, Ni, Fe, Cd, Pb, and Co) was determined after mineralization in a mixture of HNO_3+$HClO_4$, using an atomic absorption spectrometer UNICAM-SOLAAR 929, whereas in case of phosphorus, the content was determined using the colorimetric method with MERCK VEGA 400 spectrophotometer. The content of the total sulfur, and total nitrogen (S_{tot}, and N_{tot}) was determined with an elemental analyzer COSTECH ECS 4010 using gas chromatographic techniques [8].

2.3 Determination of soil biological properites

The activity of the selected hydrolytic enzymes, namely, acid phosphatase (EC 3.1.3.2) and alkaline phosphatase (EC 3.1.3.1) [9], and beta-glucosidase (EC 3.2.1.21) [10], as well as oxidoreductive enzymes, namely, dehydrogenase (EC 1.1.1.x) [11] and *o*-diphenol oxidase (EC 1.10.3.1) [12], was determined in the soil samples during the experiment on days 1, 7, 14, 28, and 56. The analyzes were performed using UV-1800 spectrophotometer from Shimadzu Corporation. Furthermore, the content of living microorganism biomass was determined on the same days as per the physiological method of Andersen and Domsch [13].

2.4 Statistical analysis

The results concerning the effect of fermentation residues on biological properties of the soil were analyzed statistically using two-way analysis of variance and were complementary compared with post hoc Tukey's HSD test. The experimental factors were the dose and the type of fermentation residues. The analyzes were performed separately for each measurement term. The level of significance was $P < 0.05$. Statistical analyzes were performed using Statistica 12.5 software (StatSoft).

3 Results and discussion

3.1 Soil chemical properties

The chemical properties of the soil were determined on day 28, which allowed to evaluate the possibility of using the fermetation residues for fertilization purposes. The application of all forms of biogas fermentation residues to the soil resulted in an increase in the soil pH, which grew with an increase in the applied dose. It was the most apparent after the application of granulate (Table 1). A similar relationship was found for the contents of N_{tot}, S_{tot}, and P. However, an increase in the C_{org} content was found to be similar after the application of drought and granulate. Among the parameters presented in Table 1, the highest increase compared with the control soil was noted for K after the application of drought at a dose of 100 $g \cdot kg^{-1}$ dry matter (DM) (714.4%). Zheng et al. [14] reported that the digestates may contain large fractions of free K ions, which, in the case of light, poorly buffered soils may overload the sorption complex and destroy the aggregate. The use of digestate as a fertilizer is considered beneficial because it provides nutrients (N, P, K) and improves the structure of the soil with the addition of organic matter [15]. De Neve et al. [16] showed that the C_{org} in biogas residues is more stable compared with other organic wastes. Moreover, Bardgett and Wardle [17] found that the application of biogas fermentation residues to soil should accurately follow precise agricultural practice bearing in mind the ready availability of N and the crop N demand so as to avoid N loss as NO_3^-, which could be drained into surface waters, leached into ground waters, or denitrified into gaseous forms. An application of fermentation residues results in the introduction of nutrients into the soil in the forms readily available to plants [18]. This may limit the consequential effects of these fertilizers' activity.

Table 1. Chosen chemical parameters in soil treated with biogas fermentation residues.

Form of residues	Dosage ($g \cdot kg^{-1}$ DM)	pH		C_{org} ($g \cdot kg^{-1}$)	N_{tot} ($g \cdot kg^{-1}$)	K ($g \cdot kg^{-1}$)	S_{tot} ($g \cdot kg^{-1}$)	P ($g \cdot kg^{-1}$)
		H_2O	1 M KCl					
Control		7.02	6.45	8.71	0.88	1.55	0.14	3.45
Pulp	10	7.21	6.63	10.20	1.56	10.30	0.14	3.98
	50	7.27	6.70	11.48	0.73	10.44	0.17	4.64
	100	7.67	7.00	13.04	1.98	10.96	0.23	4.83
Drought	10	7.40	6.77	11.26	1.72	11.14	0.21	4.07
	50	7.95	7.09	14.41	2.55	11.53	0.43	4.64
	100	8.19	7.64	24.81	3.88	12.22	0.61	5.11
Granulate	10	7.41	6.80	12.31	2.06	10.89	0.21	4.26
	50	7.97	7.30	14.62	2.89	10.67	0.50	4.78
	100	8.29	7.83	23.96	4.10	11.19	0.75	6.31

The content of the total heavy metals in the experimental soil was also found to have increased compared with that in the control soil after the application of fermentation residues. The observed effect increased with increasing doses of the fermentation residues. Similar results were obtained by Pilarska et al. [18]. Even after the application of the highest dose, the content of the total of all examined heavy metals was less than the acceptable content of the soil, both in Poland and in the European Union. Moreover, Cd was not detected either in the control soil or in the soil containing fermentation residues. In the case of Cu, Zn, and Co, the highest increase was observed after drought application (Table 2), whereas an increase noted for Pb and Ni was at a similar level in the soil with the addition of pulp and drought. It is difficult to conclude unequivocally as to which of the applied forms of fermentation residues affected the increase in the Mn content to the highest degree. According to Insam et al. [2], just the content of heavy metals and pathogens is one of the major factors limiting the use of fermentation residues as organic fertilizers.

3.2. Soil biological properties

The changes to the biological properties of the soil after the application of fermentation residues were not as clear as those of chemical parameters. This is mainly because they were observed over a period of time. Compared with the control soil, the activity of acid phosphatase was subject to a statistically significant increase only on the first day of the experiment after the application of fermentation pulp at a dose of 10 $g \cdot kg^{-1}$ DM and drought at a doses of 10 and 50 $g \cdot kg^{-1}$ DM (by 18.7% and 12.3%, respectively). An inhibition of this enzyme activity, or no significant effect of fermentation residues, was noted in other cases (Fig. 1A). The highest decrease in acid phosphatase activity compared with the control was observed on the day 14 in the soil containing pulp at a dose of 100 $g \cdot kg^{-1}$ DM (64.8%). Alkaline phosphatase activity during the experiment was subject

DOI: 10.1051/e3sconf/20172200176

to numerous statistically significant fluctuations (Fig. 1B). Attention should be paid to a significant increase in this enzyme activity in the last two measurements in the soil containing drought and granulate at the doses of 10 and 50 g·kg^{-1} DM, respectively. Beta-glucosidase activity in the soil containing pulp and granulate was usually significantly lower than that in the control soil (Fig. 1C). This effect was most pronounced for a dose of 100 g·kg^{-1} DM. Clear stimulation of this enzyme activity during the whole experiment after drought application at a dose of 100 g·kg^{-1} DM should be emphasized (from a level of 36.2% to 88.1%).

Table 2. Content of chosen heavy metals in soil treated with biogas fermentation residues.

Form of residues	Dosage (g·kg^{-1} DM)	Cu (mg·kg^{-1})	Zn (mg·kg^{-1})	Mn (mg·kg^{-1})	Pb (mg·kg^{-1})	Co (mg·kg^{-1})	Ni (mg·kg^{-1})
Control		4.35	22.93	210.61	9.91	1.58	3.27
Pulp	10	4.37	30.67	222.82	9.37	1.58	3.86
	50	4.64	31.27	239.78	11.13	1.99	4.39
	100	5.73	34.75	253.18	19.86	2.28	4.64
Drought	10	4.35	31.17	231.00	9.88	2.07	4.17
	50	6.31	43.17	255.54	11.56	2.22	4.33
	100	8.44	46.39	257.72	19.15	2.37	4.64
Granulate	10	4.35	29.56	237.71	9.94	1.70	4.17
	50	4.56	32.10	252.80	10.04	2.04	4.36
	100	5.23	35.56	256.38	12.01	2.12	4.37

The application of pulp at all doses to the soil usually did not affect significantly the dehydrogenase activity (except on day 56, when a dose of 10 g·kg^{-1} DM resulted in a statistically significant increase by 42.4% compared with the control). In turn, drought application at all doses and granulate at doses of 50 and 100 g·kg^{-1} DM caused a statistically significant stimulation of dehydrogenase activity, even reaching 557.9%, during the whole experiment (Fig. 2A). An activity of *o*-diphenol oxidase during the experiment was subject to numerous statistically significant variations (Fig. 2B). However, in the last two terms of measurements, the application of pulp at all doses reduced, whereas the application of drought and granulate increased the activity of this enzyme (Fig. 3B). However, the biomass content after pulp application at a dose of 10 g·kg^{-1} during the whole experiment did not change significantly, whereas an application of higher doses usually initially caused an increase, and then a decrease in this parameter value. The most significant increase in biomass content was noted in the soil containing drought at all applied doses (Fig 3C).

Available literature data indicate a differentiated effect of fermentation residues on biological properties of the soil. In their 3-year field experiment, Bachmann et al. [19] demonstrated the negative effect of fermentation residues on the activity of dehydrogenase and alkaline phosphatase. However, most reports indicate an increase in the content of biomass and the activity of various enzymes involved in C and N fluxes, such as dehydrogenase, alkaline phosphatase, acid phosphatase, and beta-glucosidase after fermentation residue application as organic fertilizer [20–23]. According to Cybulska et al. [24], then effect of fermentation residue from biogas plant on the biological properties of the soil depends on the timing of components released from the bio-waste. Moreover, the level of enzymatic activity occurring in fermentation residues is determined by the type of components and processes which they undergo [18].

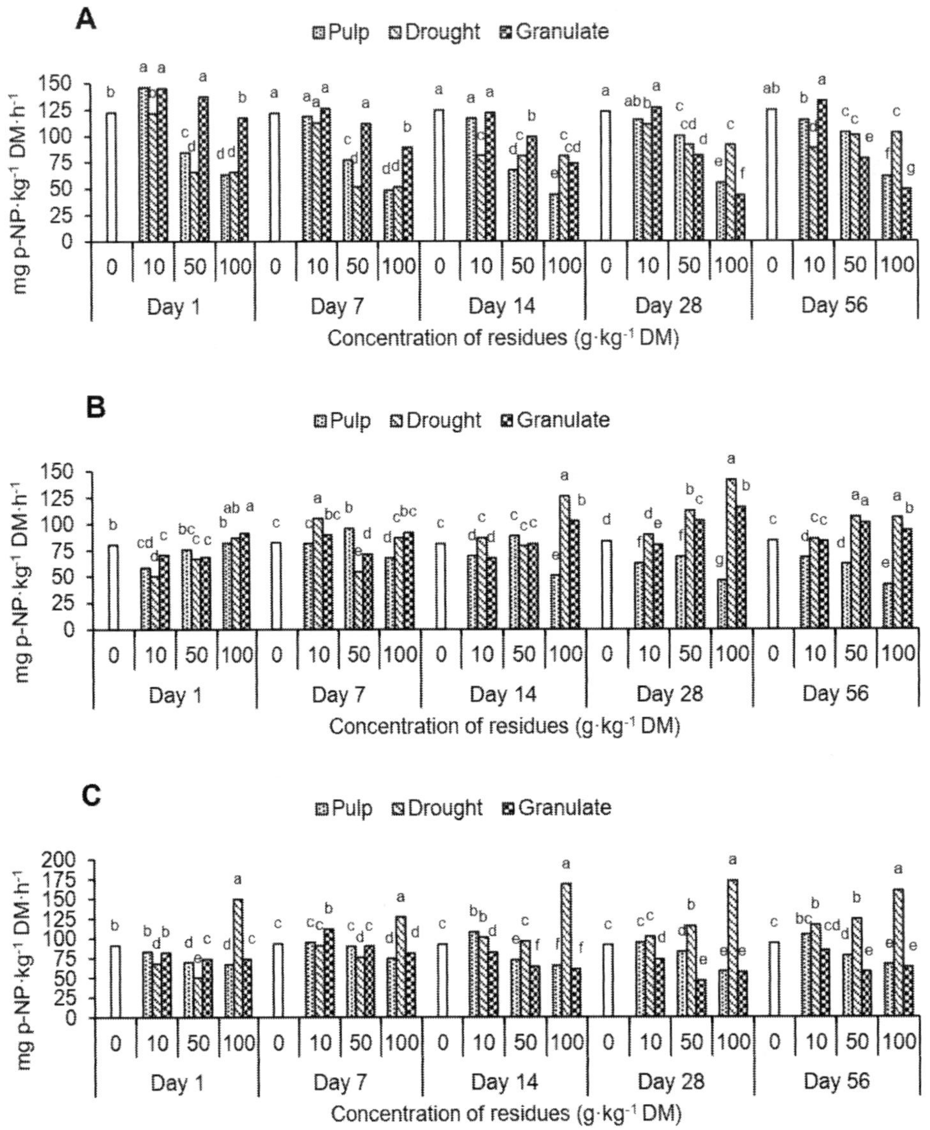

Fig. 1. Activity of hydrolytic enzymes: acid phosphatase (A), alkaline phosphatase (B) and beta-glucosidase (C) in soil treated with biogas fermentation residues; Values denoted with the same letters within the day of experiment do not differ statistically ($P < 0.05$).

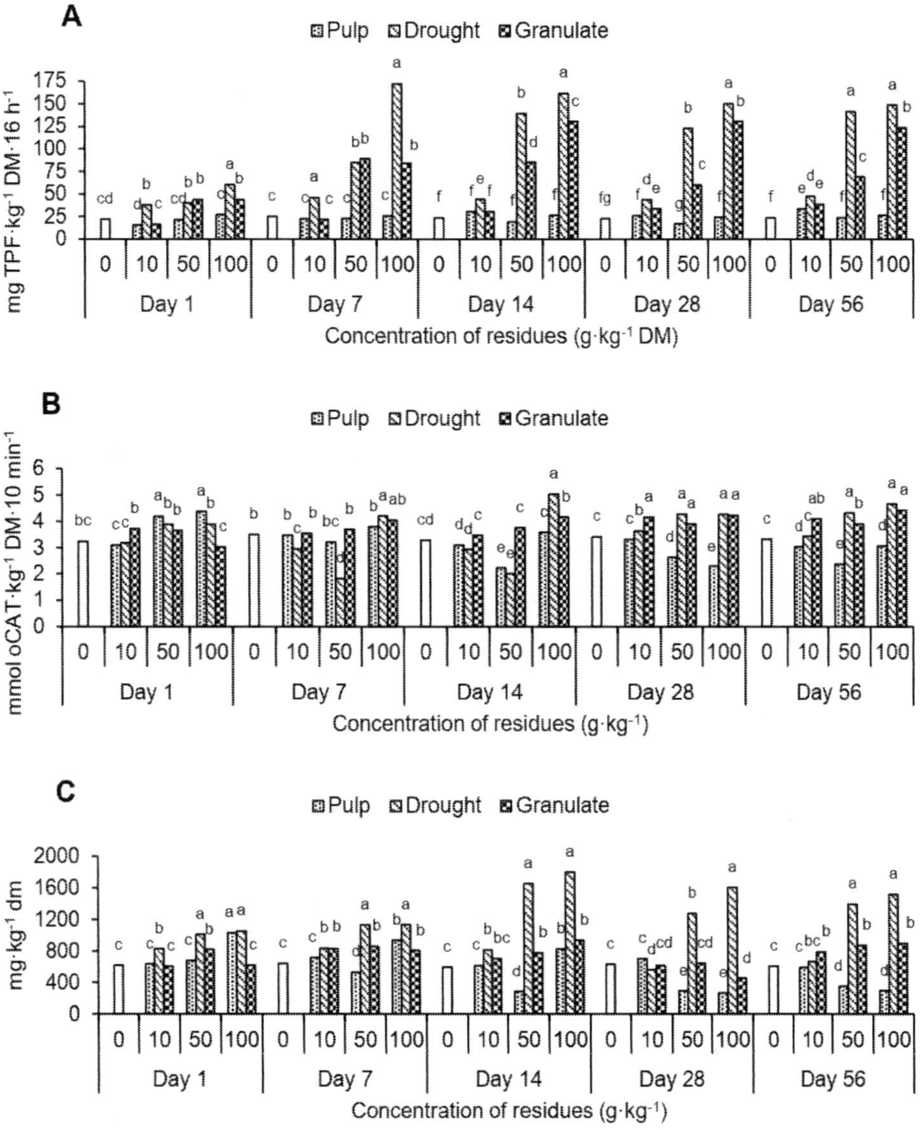

Fig. 2. Activity of oxidoreductive enzymes: dehydrogenase (A), *o*-diphenol oxidase (B) and content of microorganisms biomass (C) in soil treated with biogas fermentation residues; Values denoted with the same letters within the day of experiment do not differ statistically ($P < 0.05$).

4 Conclusions

Application of all biogas fermentation residues caused an increase in most of the chemical parameters. The highest impact on pH and the content of C_{org}, N_{tot}, S_{tot}, P, and K was observed in the soil treated with granulate, whereas the increase in the content of heavy metals was the highest after the drought application. The effect of biogas fermentation residues on all hydrolase and *o*-diphenol oxidase activities were mostly significant, but depended on the type of residues and the day of experiment. Biomass content and the activity of dehydrogenase were found to have increased during the whole experiment, and the highest impact was observed in the soil treated with drought.

References

1. E. Liu, S. Liu, Renev. Sust. Energ. Rev. **72**, 439–444 (2017)
2. H. Insam, M. Gómez-Brandón, J. Ascher, Soil Biol. Biochem. **84**, 1–14 (2015)
3. S.V. Borges de Oliveira, A.B. Leoneti, G.M. Magrini Caldo, M.M. Borges de Oliveira, Biomass Bioenerg. **35**, 2608–2618 (2011)
4. M. Stefaniuk, P. Bartmiński, K. Różyło, R. Dębicki, P. Oleszczuk, J. Hazard. Mat. **298**, 195–202 (2015)
5. J. Abubaker, K. Risberg, M. Pell, Appl. Energy **99**, 126–134 (2012)
6. H. Coban, A. Miltner, F.J. Elling, K.-U. Hinrichs, M. Kästner, Soil Biol. Biochem. **86**, 108–115 (2015)
7. Q. Jin, Y. Yang, A. Li, F. Liu, A.J. Shan, Clean. Prod. **148**, 245–253 (2017)
8. G. Jarnuszewski, Soil Sci. Ann. **67**, 1, 24–31 (2016)
9. M.A. Tabatabai, J.M. Bremner, Soil Biol. Biochem. **1**, 4, 307–310 (1969)
10. F. Eivazi, M.A. Tabatabai, Soil Biol. Biochem. **20**, 601–606 (1988)
11. A. Thalmann, Landwirt. Forsch. **21**, 249–258 (1968)
12. R. Perucci, C. Casucci, S. Dumonet, Soil Biol. Biochem. **32**, 8–9, 1927–1933 (2000)
13. J.P.E. Anderson, K.H. Domsch, Soil Biol. Biochem. **10**, 3, 215–221 (1978)
14. X. Zheng, J. Fan, L. Xu, J. Zhou, Effects of combined application of biogas slurry and chemical fertilizer on soil aggregation and C/N distribution in an Ultisol. PLoS ONE **12**, 1, e0170491 (2017)
15. R. Nkoa, Agron. Sustain. Develop. **34**, 473–492 (2014)
16. S. De Neve, S. Sleutel, G. Hofman, Nutr, Cycl. Agroecosys. **67**, 13–20 (2003)
17. R.D. Bardgett, D.A. Wardle D, *Aboveground-belowground linkages: Biotic interactions, ecosystems processes, and global change* (Oxford University Press, 2010)
18. A. Pilarska, A. Wolna-Maruwka, T. Piechota, K. Pilarski, M. Szymańska, D. Wolicka, Nauka Przyr. Technol. **9**, 2, 19 (2015)
19. S. Bachmann, M. Gropp, B. Eichler-Löbermann, Biomass Bioenerg. **70**, 429–439 (2014)
20. Y.C. Liang, Y.F. Yang, C.G. Yang, Q.Q. Shen, J.M. Zhou, L.Z. Zang, Geoderma **115**, 149–160 (2003)
21. N. Jedidi, A. Hassen, O. van Cleemput, A. MHiri, Waste Manag. Res. **22**, 93–99 (2004)
22. J.A. Alburquerque, C. de la Fuente, M. Campoy, L. Carrasco, I. Najera, C. Baixauli, F. Caravaca, A. Roldan, J. Cegarra, M.P. Bernal, Europ. J. Agron. **43**, 119–128 (2012)
23. M. Gomez-Brandon, M. Fernandez-DelgadoJuarez, M. Zangerle, H. Insam, J. Hazard. Mat. **302**, 267–274 (2016)
24. K. Cybulska, I. Wrońska, T. Kitczak, J. Dłużewska, S. Mahdi Oraibi, H. Czyż, Woda Środ. Obsz. Wiej. **15**, 1, 29–36 (2015)

DOI: 10.1051/e3sconf/20172200177

A numerical investigation of laminar forced convection in a solar collector with non-circular duct

Tomasz Janusz Teleszewski[1,*]

[1]Bialystok University of Technology, Faculty of Civil and Environmental Engineering, Department of HVAC Engineering, ul. Wiejska 45A, 15-351 Bialystok, Poland

Abstract. This paper presents a two-dimensional numerical study to investigate laminar flow in a flat plate solar collector with non-circular duct (regular polygonal, elliptical, and Cassini oval shape) featuring forced convection with constant axial wall heat flux and constant peripheral wall temperature (H1 condition). Applying the velocity profile obtained for the duct laminar flow, the energy equation was solved exactly for the constant wall heat flux using the Boundary Element Method (BEM). Poiseuille and Nusselt numbers were obtained for flows having a different number of geometrical factors. The results are presented and discussed in the form of tables and graphs. The area goodness factor and volume goodness factor are calculated. The predicted correlations for Poiseuille and Nusselt numbers may be a very useful resource for the design and optimization of solar collectors with non-circular ducts.

1 Introduction

A flat plate solar collector absorbs incoming solar radiation, converting it into thermal energy at the surface of a receiver and heat is transferred to the working fluid through the collector. In fact, the collection of solar radiation and its conversion to thermal energy depend on the collector design, including factors such as riser shape, absorber material, and transmissivity of the glass cover [1]. Simulation of the forced convection in the riser is the most essential part of the process of designing a flat plate collector. The problem of determining the Nusselt number for fully developed laminar flow with triangular cross-section has been solved by Schneider and Ledain [2] using the finite element method to solve the momentum and energy conservation equations. Taherian and Yazdanshenas [3] made an experimental investigation of forced convection in a finned rhombic tube of a flat plate solar collector. Theoretical and experimental investigations of an elliptical heat pipe flat plate solar collector were performed by Sivakumar et al [4]. Forced convection heat transfer in smooth and roughened rectangular ducts has been studied extensively in references [5–9]. Ekramian [1] studied numerically the effect of riser shape on the thermal efficiency, finding a collector with circular tubes, had 38.4%, 11.2%, and 6.6% greater thermal efficiency than collectors with triangular, square, and hexagonal ducts respectively.

* Corresponding author: t.teleszewski@pb.edu.pl

Fan et al. [10] analysed numerically and experimentally the flow and temperature distribution in a solar collector with a quadrangular absorber tube. Nasrin and Alim [11] numerically simulated forced convection in a cross-section of a flat plate solar collector with nanofluids, using finite elements with Galerkin's weighted residual technique. Dehghan et al. [12] investigated the effects of radiation heat transfer on the forced convective heat transfer mechanism in solar heat exchangers filled with a porous medium. Helvaci and Khan [13] analysed multiphase flow in a flat plate solar energy collector.

The aim of this work is to study the effect of riser shape on the Nusselt number, Poiseuille number, area goodness factor, and volume goodness factor for laminar forced convection, using boundary element method (BEM). These dimensionless quantities may provide the function of screening the selection of a riser shape before other design criteria are applied. There exist varieties of flat plate collector with different non-circular cross-section such as regular polygonal, elliptical, and Cassini oval. The results obtained may lead to the design of more efficient flat plate solar collectors with non-circular ducts.

2 Mathematical model

The problem being considered concerns laminar forced convection in solar risers. A simplified numerical model based on the boundary element method for fluid flow and heat transfer in solar risers is developed under the following assumptions: two-dimensional, steady, laminar and constant thermal conductivity k and a constant dynamic viscosity μ with incompressible fully developed flow, negligible body forces and viscous dissipation, and negligible radiation heat transfer. Newtonian laminar fluid flow is governed by the usual continuity (1), momentum (2) and energy (3) equations for two-dimensional simulations as follows:

$$\frac{\partial u_z}{\partial z} = 0 \tag{1}$$

$$\mu\left(\frac{\partial^2 u_z}{\partial x^2} + \frac{\partial^2 u_z}{\partial y^2}\right) = \frac{\partial p}{\partial z}, \qquad \frac{\partial p}{\partial x} = \frac{\partial p}{\partial y} = 0 \tag{2}$$

$$\lambda\left(\frac{\partial^2 T}{\partial x^2} + \frac{\partial^2 T}{\partial y^2}\right) = \rho c_p \frac{\partial T}{\partial z} u_z \tag{3}$$

where u_z is the axial velocity, T is temperature, p is pressure, c_p is specific heat capacity and ρ is density. The coordinate system and computational domain are shown in Figure 1.

In the first step the Poisson equation (2) is solved using the boundary element method. The velocity is required to satisfy the no-slip condition on the duct wall ($u_z = 0$). Secondly, the energy equation (3) is solved for temperature distributions in a cross-section of the duct by BEM. The axial derivative $\partial T/\partial z$ is independent of the x and y coordinates. The energy equation (3) requires the Dirichlet or H1 condition (temperature) and the Neumann or H1 condition (heat flux density), and neglects the axial heat conditions in the fluid. In this paper the H1 condition is considered: T = const, at L_h (constant peripheral wall temperature). The H1 condition has found many applications with negligible normal wall thermal resistance [14], such as in tube heat exchangers and in solar energy collection [4, 13]. Details of the solution of Poisson's equations (2–3) using the boundary element method can be found in the literature [15].

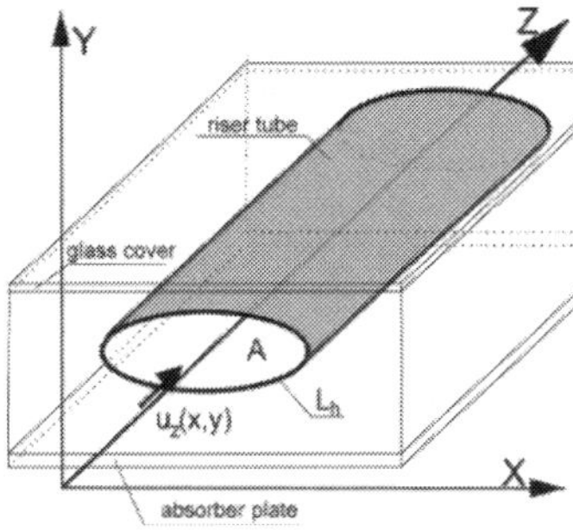

Fig. 1. Schematic diagram of a riser duct in a solar collector.

3 Analysis and modelling

The governing equations (2–3) were solved numerically using the boundary element method, by means of a computer program written by the author in Fortran, using 4000 constant boundary elements and 40602 triangular cells in the duct cross-section. To determine relative differences between numerical and analytical values, the numerical results obtained by BEM were compared with the analytical values of the Poiseuille number Po and the Nusselt number Nu obtained by Shah and London [14] for a circular tube. For 4000 constant boundary elements and 40602 triangular cells in the duct cross-section, the maximum relative differences in the Poiseuille number and Nusselt number were found as ΔPo = 0.001% and ΔNu = 0.005%.

The Poiseuille number defined as the product of the Fanning friction factor f for fully developed flow and the Reynolds number Re, can be expressed as

$$Po = f\frac{\mathrm{Re}}{4}, \qquad f = \frac{8\tau_w}{\rho u_m^2}, \qquad \mathrm{u}_m = \frac{1}{A}\iint_A u dA \tag{4}$$

where τ_w is the peripherally averaged wall shear stress, u_m is the mean velocity, and A is the cross-sectional area. The Reynolds number is equal to

$$\mathrm{Re} = \frac{\rho \mathrm{u}_m D_h}{\mu}, \qquad D_h = \frac{4A}{L_h} \tag{5}$$

where D_h is the hydraulic diameter, and L_h is the wetted perimeter of the duct cross-section. The Nusselt number is defined as

$$Nu = \frac{q_w D_h}{k(T_w - T_b)}, \qquad \mathrm{q}_w = \frac{1}{L_h}\int_{L_h} q dL_h \tag{6}$$

where T_w is the wall temperature, T_b is the bulk temperature, and q_w is the wall heat flux density. In fact, the fully developed laminar Nusselt number is found to be independent of the length of the duct, the Prandtl number Pr and the Reynolds number Re, but dependent on the duct geometry and boundary conditions [13, 14].

The considered non-circular geometries include several common shapes, such as regular polygons, ellipses and Cassini ovals. Shah and London [14] provided an analytical relationship for the Poiseuille number and Nusselt number versus the ratio of the minor axis A to the major axis B. This analytical relationship contains an elliptic integral of the first kind, which can be computed using numerical integration. Shahsavari et al [16] and

Teleszewski and Sorko [17] provided numerical relationships for the Poiseuille number and Nusselt number respectively, depending on the number *n* of sides. No compact relationship has previously been found for the Poiseuille number and the Nusselt number for Cassini ovals. In this paper, a compact relationship for Po and Nu for ellipses, regular polygons and Cassini ovals versus a geometrical factor σ, with correlation coefficient $R^2 \approx 1$, is proposed:

$$Po(\sigma), Nu(\sigma) = \frac{a + b\sigma^{0.5m} + c\sigma^{m} + d\sigma^{1.5m} + e\sigma^{2m}}{1 + f\sigma^{0.5m} + g\sigma^{m} + h\sigma^{1.5m} + i\sigma^{2m}}, \qquad \sigma = A/B \vee n \vee C/D \tag{7}$$

where the coefficients *a*, b, *c*, *d*, *e*, *f*, *g*, *h*, *i*, *n* are given in Table 1. The factor σ represents the ratio of the minor to the major axis for an ellipse ($\sigma = A/B$), the number n of sides for a regular polygon ($\sigma = n$), and the ratio of the coefficients *C* and *D* in the Cassini oval equation ($\sigma = C/D$). Equation (7) is more compact and can predict Po and Nu values for regular polygonal, elliptical and Cassini oval ducts with an uncertainty less than 0.2%. A Cassini oval is defined as

$$\begin{aligned} x &= \cos(\alpha)\sqrt{C^2\cos(2\alpha) + \sqrt{C^4(\cos(2\alpha))^2 - C^4 + D^4}} \\ y &= \sin(\alpha)\sqrt{C^2\cos(2\alpha) + \sqrt{C^4(\cos(2\alpha))^2 - C^4 + D^4}} \end{aligned} \tag{8}$$

where α is the angular coordinate.

Table 1. Coefficients of equation (7) to determine the Poiseuille and Nusselt numbers.

Coefficients of equation (7) to determine the Poiseuille number					
shape	a	b	c	d	e
polygonal	15.45392	-27.24564	17.31518	-4.11344	0
ellipse	19.739	-31.3853	74.6744	-76.4	52.576
Cas. oval	16.012	-33.43074	12.0508	12.5435	-7.17415
	f	g	h	i	m
polygonal	-1.73499	1.0846	-0.257136	0	1
ellipse	-1.5874	3.7212	-3.0711	2.3876	1
Cas. oval	-2.08368	0.61023	1.05387	-0.58033	4
Coefficients of equation (7) to determine the Nusselt number					
	a	b	c	d	e
polygonal	-11.1816	0	2.2561	0	4.8055
ellipse	5.2251	-11.1214	31.7343	-36.338	22.0829
Cas. oval	4.3636	-14.7687	18.6452	-10.3976	2.1576
	f	g	h	i	m
polygonal	0	0.52438	0	1.1011	1
ellipse	-2.1272	6.0374	-6.3329	4.0771	1
Cas. oval	-3.39323	4.23613	-2.2834	0.440526	4

Plots of Po and Nu for various values of σ are shown in Figures 2a and 2b respectively. Similar trends are observed in both cases. The highest values for Nu are observed for the Cassini oval with geometrical factor $\sigma = 0.945$. In Table 2, the Poiseuille and Nusselt numbers are presented for a Cassini oval duct for some values of the geometrical factor σ = C/D.

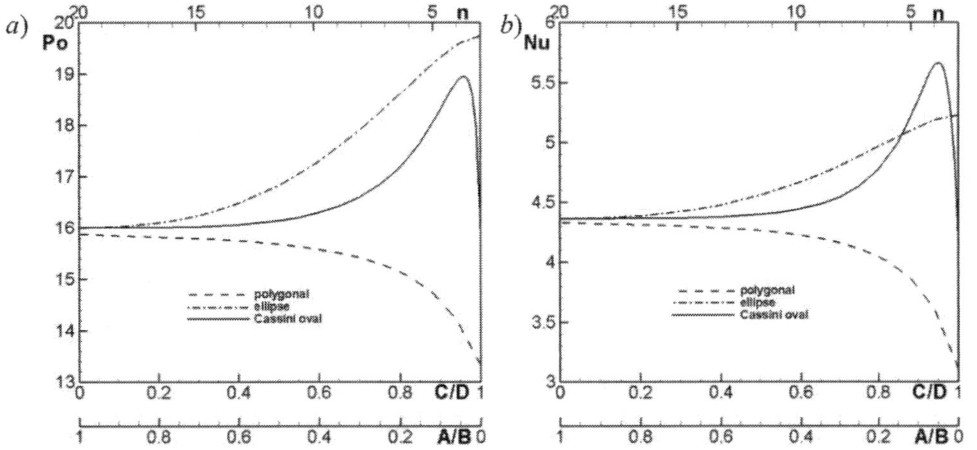

Fig. 2. Plot of Nu (a) and Po (b) vs *σ*.

The 2D dimensionless velocity u/u_m and the dimensionless temperature distribution $T/|T_b|$ with the heatline through the Cassini oval are shown in Figures 3a-b and 4a-b for different values of the geometrical factors: $\sigma = 0.85$ and $\sigma = 0.945$ respectively. As Figures 3a and 4a show, the maximum velocity and minimum fluid temperature occur at the centre of the Cassini oval with $\sigma = 0.85$. In Figures 3b and 4b, for a Cassini oval with geometrical factor $\sigma = 0.945$, it is interesting to observe that two points of maximum fluid velocity and two points of minimum fluid temperature lie on the axis of symmetry $y = 0$.

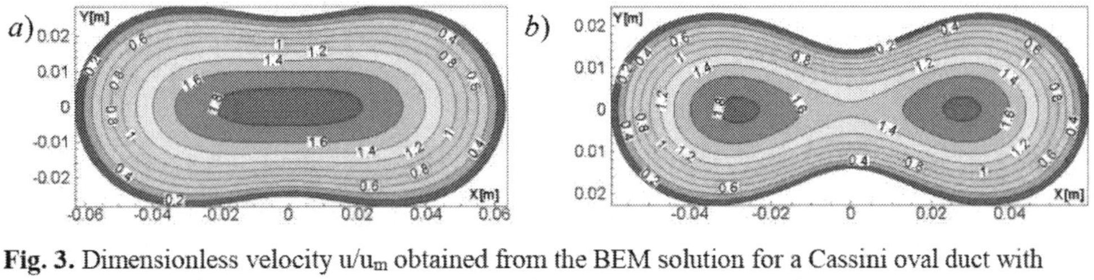

Fig. 3. Dimensionless velocity u/u_m obtained from the BEM solution for a Cassini oval duct with different values of σ: (a) σ = 0.85 (Po = 17.64); (b) σ = 0.945 (Po = 18.88).

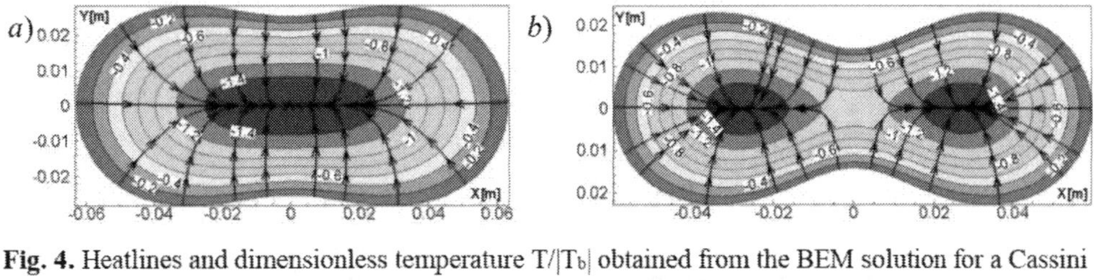

Fig. 4. Heatlines and dimensionless temperature $T/|T_b|$ obtained from the BEM solution for a Cassini oval duct with different values of σ: (a) σ = 0.85 (Nu = 5.00); (b) σ = 0.945 (Nu = 5.66).

Table 2. Poiseuille number and Nusselt number for some values of the factor σ = C/D for a Cassini oval.

Poiseuille number and Nusselt number vs. σ = C/D for Cassini oval						
σ = C/D	0.2	0.4	0.6	0.8	0.945	0.98
Po	16.01	16.06	16.30	17.18	18.88	18.60
Nu	4.37	4.38	4.44	4.77	5.66	5.28

4 Results and discussion

Two classical types of comparison methods for shaped risers have been considered: area goodness factor [14] and volume goodness factor [14, 18, 19].

In order to compare the thermal performances of riser shapes, it is possible to use the area goodness factor [14], which is the ratio of the Colburn factor *j* and the Fanning friction factor *f*, defined as:

$$G=\frac{j}{f}=\frac{Nu}{\mathrm{Pr}^{1/3}\,Po}, \qquad \mathrm{Pr}=\frac{c_p \mu}{k} \tag{9}$$

where Pr is the Prandtl number. However, for fully developed forced convection through a straight duct, both the Poiseuille number Po and the Nusselt number are independent of the Prandtl number Pr. Therefore, for this parameter, one may consider only the variations $\mathrm{Pr}^{1/3}G = \mathrm{Pr}^{1/3}j/f = Nu/Po$ as functions of the shape factor. This form of the area goodness factor was introduced by Ray and Misra [20]. The dimensionless area goodness factor is independent of the hydraulic diameter D_h. The area goodness factors of collectors with regular polygonal, elliptical and Cassini oval ducts as functions of the geometrical factor σ are presented in Figure 5a. A higher area goodness factor is considered desirable because it means that a lower flow area is required for the riser [14, 21]. Figure 5a shows that the Cassini oval risers with $\sigma = 0.945$ have an area goodness factor 28.5%, 18.2%, 13% and 9.9% greater than those of the collectors with triangular, square, elliptical $\sigma = 0.1$, and circular riser tubes, respectively. In fact, for the duct geometries having strong corner effects, the area goodness factor is even lower than for channels without sharp corners [14].

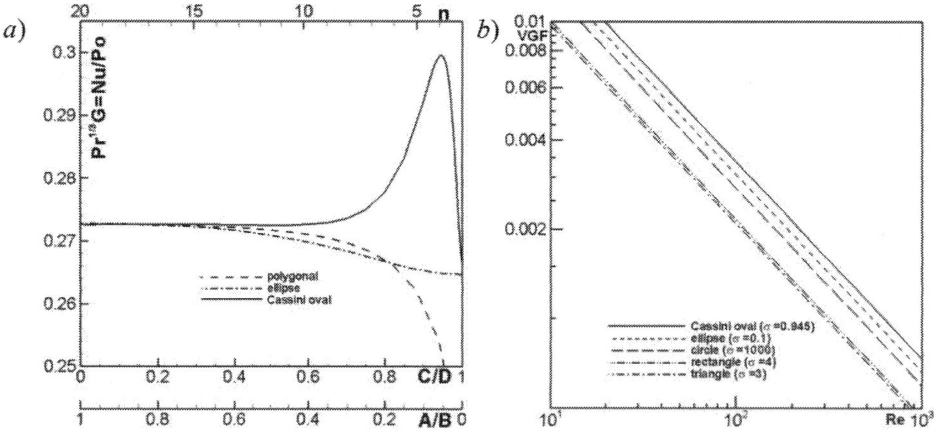

Fig. 5. Comparison methods for riser shape selection: (a) variation of area goodness factor the geometrical parameter factor σ, (b) volume goodness factor vs. Reynolds number for several riser shapes.

 DOI: 10.1051/e3sconf/20172200177

Another comparison method was suggested by Shah and London [14, 18, 19], using the volume goodness factor *VGF* as defined by the following equations:

$$VGF = \frac{\text{St}}{f^{1/3}}, \qquad \text{St}=\frac{\text{Nu}}{\text{RePr}} \tag{10}$$

where St is the Stanton number. This factor is used for comparing different heat exchangers [18, 19, 21]. The higher *VGF* is required the smaller heat transfer. The volume goodness factor is plotted as a function of the Reynolds number in Figure 5b for triangular (σ= 3), square (σ= 4), elliptical (σ= 0.1), circular (σ= 1000) and Cassini oval (σ= 0.945) risers. The simulations were performed for 50% propylene glycol mixtures with constant physical quantities (T_w = 50°C, μ= 0.002 Pas, k = 0.4 W/(mK), ρ = 1025 kg/m^3 and c_p= 3480 J/(kgK)). In Figure 5b the Cassini oval shape with σ= 0.945 demonstrates higher volume goodness factor than other geometries. Figure 5b shows that the Cassini oval risers with σ= 0.945 have an volume goodness factor 62%, 57%, 10.3% and 22.7% greater than those of the collectors with triangular, square, elliptical σ= 0.1, and circular riser tubes, respectively. Plots of the area goodness factor (Figure 5a) and the volume goodness factor (Figure 5b) show that the best performance is obtained for a Cassini oval with geometrical factor σ = 0.945.

Figure 6 is a graphical representation of Nu, Nu/Po and efficiency [1] for regular polygonal ducts versus the number n of sides. In Figure 6, a similar increasing trend is observed for efficiency versus n [1], as for Nu, and Nu/Po.

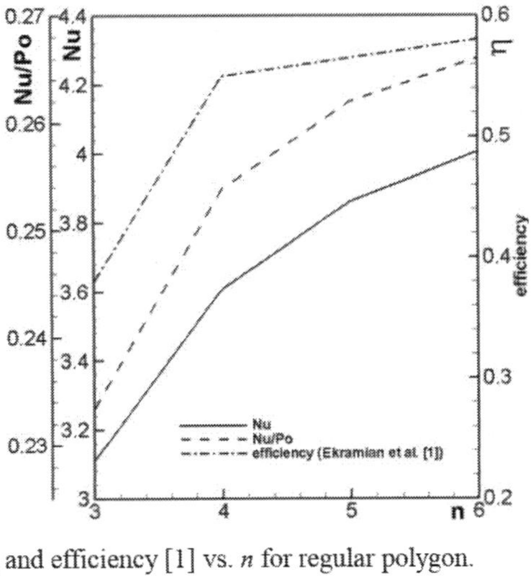

Fig. 6. Plot of Nu, Nu/Po, and efficiency [1] vs. n for regular polygon.

5 Conclusions

This study has focused on fully developed laminar forced convection in regular polygonal, elliptical and Cassini oval riser ducts in solar collectors. A simple compact correlation between the Poiseuille number and Nusselt number versus a geometrical factor σ has been proposed for these riser shapes. The flow is modelled by two-dimensional momentum and energy equations with H1 boundary conditions. The steady state model has been solved numerically using the boundary element method. This is the first time that a relationship for the Poiseuille number and Nusselt number vs. geometrical shape (σ= C/D) has been computed for the Cassini oval. The influence of the analysed non-circular riser shapes on thermal performance was investigated. The results show that the optimum solution for the considered non-circular geometries is a Cassini oval with $\sigma = 0.945$. The results obtained may be utilized in the optimization of riser geometries.

This scientific project was financed by S/WBiIS scientific research funds at Bialystok University of Technology.

References

1. E. Ekramian, S.G. Etemad, M. Haghshenasfard, J. of Fluid Flow, Heat and Mass Transfer **1**, 1 (2014)
2. G.E. Schneider, B.L. Ledain, J. Energy **5**, 1 (1981)
3. H. Taherian, E. Yazdanshenas, In Proceedings of the Second International Green Energy Conference, IGEC2-126 (2006)
4. K. Sivakumar, N.K. Mohan, B. Sivaraman, International J. of Engineering & Technology **4**, 1 (2012)
5. A.M.E. Momin, J.S. Saini, S.C. Solanki, International J. Heat Mass Transfer, **45** (2002)
6. V.S. Hans, R.P. Saini, J.S. Saini, Solar Energy, **84** (2010)
7. S.V. Karmare, A.N. Tikekar, Solar Energy, **84** (2010)
8. A. Kumar, R.P. Saini, J.S. Saini, Renewable Energy, **58** (2013)
9. A. Kumar, Energy Procedia, **57** (2014)
10. J. Fan, L. J. Shah, S. Furbo, Solar Energy, **81** (2007)
11. R. Nasrin, M.A. Alim, J. of Applied Fluid Mechanics **7**, 3 (2014)
12. M. Dehghan, Y. Rahmani, D.D. Ganji, S. Saedodin, M. S. Valipour, S. Rashidi, Renewable Energy, **4** (2015)
13. H.U. Helvaci, Z.A. Khan, Energy Conversion and Management, **106** (2015)
14. R.K. Shah, A.L. London, *Laminar Flow Forced Convection in Ducts, Advanced in Heat Transfer* (Academic Press, London, UK, 1978)
15. C.A. Brebbia, J.C.F. Telles, L.C. Wrobel, *Boundary Element Techniques-Theory and Applications in Engineering* (Springer-Verlag, Berlin, Heidelberg, New York, Tokyo, 1984)
16. S. Shahsavari, A. Tamayol, E. Kjeang, M. Bahrami, ASME J. of Heat Transfer **134**, 9 (2012)
17. T.J. Teleszewski, S.A. Sorko, Procedia Engineering, **157** (2016)
18. J.-H. Doo, H.S. Yoon, M.Y. Ha, Transactions of the Korean Society of Mechanical Engineers B **33**, 6 (2009)
19. S.W. Hwang, D.H. Kim, J.K. Min, J.H. Jeong, Journal of Mechanical Science and Technology **26**, 9 (2012)
20. S. Ray, D. Misra, Int. J. Therm. Sciences, **49** (2010)
21. K.M. Stone, ACRC TR-105 (1996)

DOI: 10.1051/e3sconf/20172200178

Comparison of the availability of groundwater information sources in Poland with other European countries. Knowledge inventory for hydrogeology research – project KINDRA

Barbara Tomaszewska[1], *Marta* Dendys[2], and *Magdalena* Tyszer[1,*]

[1]AGH – University of Science and Technology, Faculty of Geology, Geophysics and Environmental Protection, al. Mickiewicza 30, 30-059 Kraków, Poland
[2] Mineral and Energy Economy Research Institute, Polish Academy of Science, ul. Wybickiego 7, 31-261 Kraków, Poland

Abstract. Regulations of the Water Framework Directive 200/60/EC (WFD) had been applied by European Union countries into their legislation system. However, it does not guarantee that the groundwater research has the same standard and quality in EU countries. KINDRA international research project was launched to assessment of existing groundwater-related practical and scientific knowledge based on a new Hydrogeological Research Classification System (HRC-SYS). This classification is supported by a web-service – the European Inventory of Groundwater Research (EIGR). The main goal of the project is implementation policy of optimization in groundwater research in EU. The preliminary result of survey about groundwater management shows that in Poland is a good state of implementation WFD. Good level of implementation is especially related with groundwater monitoring. It is because a lot of institutions and municipal entities carry out their tasks referring to quality or quantity assessment. Results of their works are published as reports, newsletters, maps, bulletins etc. These materials are potential source of information which can be a valuable contribution to EIGR. However, a lot of information are published only in polish language, so it is impossible to spread this knowledge in Europe.

1 Introduction – about KINDRA

Water is essential „raw material". Access to groundwater resources determines development of a settlement and cities, water is necessary for biological existence of ecosystems and efficient functioning of the environment. Nowadays water is a key-element of interdisciplinary research, especially in connections: water-food-energy-climate. That is why international projects are very important – growth of cooperation and awareness between researchers can protect in effective way this valuable „raw material".

* Corresponding author: mtyszer@agh.edu.pl

KINDRA project was launched in February 2015 and is coordinating by consortium of six partners from EU. One of the objective of this project is to summarize the current state of scientific and practical knowledge in the groundwater research field in European countries. The aim will be achieved by gathering information of the results of research, projects, programs and hydrogeological databases. It will help identify the challenges and priorities for the future research [1]. The result of KINDRA project will be also Knowledge Inventory for Hydrogeology Research (EIGR) [2]. Project is coordinated by the Sapienza University of Rome, Earth Sciences Department (Italy). The main partners are: EFG – European Federation of Geologists, Redi – Environment and Water Agency of Andalusia (Spain), LPRC – La Palma Research Centre for Future Studies S.L. (Spain), UM – University of Miskolc, Faculty of Earth Science and Engineering (Hungary) and GEUS – Geological Survey of Denmark and Greenland (Denmark). European Federation of Geologists (EFG) invited to participate representatives from 20 European countries: Belgium, Croatia, Czech Republic, Denmark, Finland, Greece, Spain, Holland, Ireland, Germany, Polish, Portugal, Serbia, Slovenia, Switzerland, Ukraine, Hungary, Great Britain and Italy (Fig. 1). Countries are represented by experts - specialists in hydrogeology field. They had to select domestic sources of knowledge and information about groundwater for the EIGR database requirement.

Results of KINDRA project and the EIGR database will be analysed and used to support UE policy and implementation of water directives. Currently the most important Framework Water Directive 2000/60/EC was published 23th October 2000. WFD established a framework of action in groundwater management policy [3]. Each of EU countries were obliged to implement FWD regulations into national legislation system. Directive instructs all EU members to the rational using of water resources and protecting this resources with the principles of sustainable development. The strategic objective of EU water policy is to achieve and maintenance good quality and quantity status of the water resources which will ensure: 1) good condition of ecosystems, 2) safe access to water resources for people and environment, 3) precise definition of social and economical request of water supplies, 4) reduce the negative effects of floods and drought, 5) minimize the risk of emergency situations and 6) implementation of integrated water resources management.

Topics related with rules of rational water management and implementation of the WFD are a subject of numerous scientific works. They concern the implementation of the requirements of WFD into national legislation e. g. in the Netherlands [4] or Sweden [5]. Sometimes papers include summaries of the experiences and problems common for European countries [6]. Conclusions from these summaries allow to present guidelines and recommendations for the effective operation in water management [7]. Some of topics are linked with maintaining of good chemical status [8, 9] and good ecosystems conditions [10]. Also important is the awareness of risks and appropriate definition of challenges for the future in individual countries of Europe, e. g. Great Britain [11]. The WFD also helps to regulate activities related with specific areas of water management, e. g. in coastal areas and ecosystems of the seas [12–14] or of flood risk assessment or rivers network management [15]. The WFD regulations correspond with other branches of economy, such as agriculture, especially in context of irrigation in arid regions of southern Europe [16]. Sometimes the WFD regulations determine the subject of researches and allow to define challenges of key-topic for the future development of European economy related with climate changes [17, 18]. Growth of water deficit was the subject of discussion in Poland [19]. The deficit areas occur because of: 1) dangers for sustainable utilization of surface water and groundwater, 2) growth consumption of the groundwater resources, especially in intensively exploited areas, 3) lack of possibilities of recharge municipal water supply system. Witczak and others [20, 21] shows in their papers the role of monitoring in regard to the presence of heavy metals in the water. Tomaszewska and Szczepanski [22] indicate the possibility of using recycled water treatment as a potential resource of drinking water.

Poland joined to the UE nations on 1st May 2004 and has its own experiences with implementation the WFD regulations into legislation system. Maciejewski and Walczykiewicz [23] showed main risks related with transfer the WFD regulations into polish law. The most important of them was lack of information or poor access to existing data, insufficient investment and risk of extending of deadlines. Participation of polish expert in KINDRA project was the opportunity to evaluate the functioning of the WFD regulations in country groundwater management.

2 Materials and Methods

KINDRA project include three stages, divided into six tasks (Fig. 2a). The first stage was to create a framework of methodology which allows to classify the projects, works and actions related to groundwater (Harmonised methodology on Groundwater Research – HRC-SYS). During this stage has been created a web-service which is the kind of metadata repository. It is the European Inventory of Groundwater Research (EIGR). In the second stage was gathered information about groundwater research and their results. All data had been input into EIGR by national experts. This step had allowed to assess and summarize scientific and practical knowledge in the hydrogeological studies in Europe. In the third stage of the project gaps in the knowledge related with groundwater will be identified. After that will be prepared recommendations for: 1) the future policy, 2) better understanding relations between surface water and groundwater and 3) acts to help prevent climate changes. The third stage was started on February 2017.

Methodology of research topics classification (HRC-SYS) had been created because there was need of using the same criteria for different kind of projects. In the first step „key words" from projects had been identified. Sources of "key words" were the most important EU regulations, documents and scientific publications related to groundwater and hydrogeology topics. Then had been identified links between subjects of research. The next step was to create ranking list based on importance of particular phrases, according to score of searching in databases like: Web of Science, Scopus and Google Scholar Web. List of about 200 key words was connected in tree hierarchy (Fig. 2b). Each of words had been linked to the three main category: SC – Societal Challenges, OA – Operational Action and RT – Research Topic. Every of this main category has been divided into five high level group (level 1). The HRC-SYS classification system was based on the assumption of interactions between the three main groups and the five high level groups.

Methodology HRC-SYS had been created for the better classification data and information in the EIGR. This database will allow to gather and management of information about groundwater in Europe. National experts had to input in the EIGR different kind of data e. g.: reports of research projects, institution reports, magazines, bulletins or technical publications linked with groundwater topics. The EIGR will be a database which complement the scientific articles and publications databases: Web of Science, Scopus, and Google Scholar.

DOI: 10.1051/e3sconf/20172200178

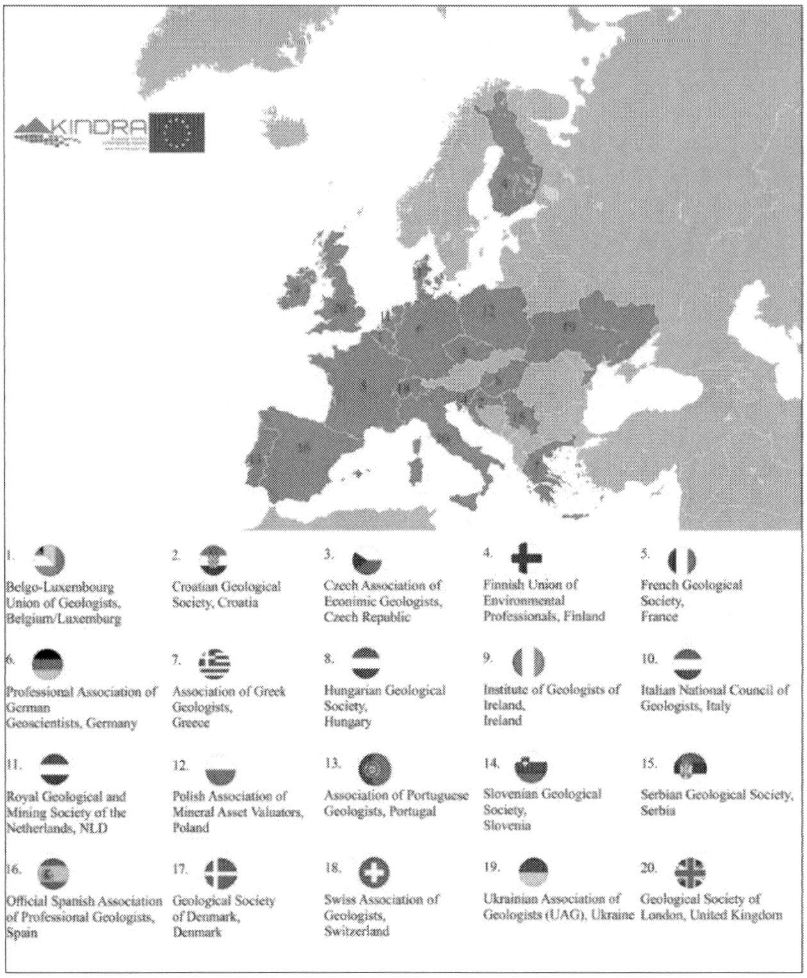

Fig. 1. Partners in the KINDRA project (based on [2]).

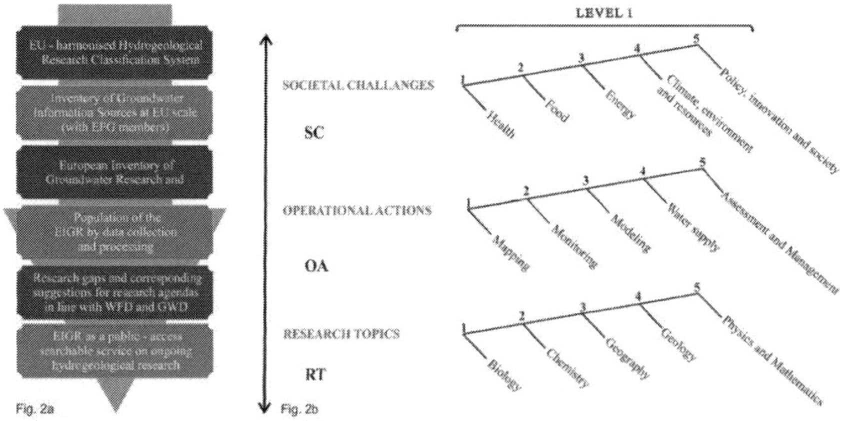

Fig. 2. a) KINDRA project stages, b) Hierarchy tree in HRC-SYS classification (based on [2]).

Stage of the data input to the EIGR has been preceded by the survey on the availability of information about groundwater in European countries [1]. Author of this survey, prof. Eva Hartai from the European Federation of Geologists, had prepared 12 questions divided into four groups (Tab. 1). The answers had been given by national experts. The survey helped them to systematize information for EIGR and make a preliminary assessment of the implementation the WFD regulations into national legislation systems.

Table 1. Groundwater data sources – questionnaire (autor: prof. Eva Hartai, European Federation of Geologists). (Answer by Polish national expert – in brackets).

Institutions dealing with groundwater research/survey	**Journals/archives focused on hydrogeology**
*– How many institutions deal with groundwater research/survey in your country? **(More than 20)***	*– Are there any journals/archives focused on hydrogeology in your country? **(Yes)***
*– Please fill the table for the institutions related to groundwater/survey in your country, indicating the level, the type and the data accessibility at the institution. **(53 institutions, full list available in: [1])***	*– Please list the name of these journals/archives, indicating if they are on-line/printed (O/P) and English/national language (E/N). **(10 issues, full list available in: [1])***
Groundwater withdrawal, availability of data	**Groundwater monitoring, availability of data**
*– What percentage of the drinking water derives from groundwater in your country? **(More than 70%)***	*– How much of your country is covered by groundwater monitoring network/s? **(Between 50-100%)***
*– Are there official data about anthropogenic groundwater withdrawals? **(Yes)***	*– What type of data are collected by the groundwater monitoring network? **(Quantitative and qualitative)***
*– If yes, please indicate those withdrawal types where data are accessible. **(Mining)***	*– Are you aware of any parameters that are presently not monitored, but should be (in addition to those which are recommended by the European Commission)? **(No)***
*– How many observation sites are in the strategic groundwater monitoring network in your country? **(More than 100)***	*– Are the monitored data available on-line? **(Yes)***

3 Results

Methodology HRC-SYS had allowed to indicate the priority category related with groundwater research. There are "Societal Challenges" and all main implementation directives and strategic EU projects which should be linked with them. HRC-SYS "Societal Challenges" include five sub-categories: 1) Health, 2) Food, 3) Energy, 4) Climate, environment and resources and 5) Policy, innovation and society. Every record which had been input to EIGR mandatory had to be classified according to Societal Challenges. For all records was also specified "Operational Action" and "Research Topic", because it helps to identify and assess in the future, trends or gaps in UE legislation and water management policy. The examples of input data and way of its classification had been shown in Tab. 2.

Table 2. An example of the classification of Polish publications according to the HRC-SYS categories.

Publication	Category HRC-SYS		
	Societal challenges	**Operational actions**	**Research topics**
Atlas of the possible use of geothermal waters for combined production of electricity and heat using binary systems in Poland [24]	*Energy*	*Mapping*	*Geology*
Atlas of geothermal resources of Paleozoic formations in the Polish Lowlands [25]	*Climate, Environment and Resources*	*Mapping*	*Geology*
Groundwater monitoring in the border area of the Sudetes in 2005-2012) [26]	*Climate, Environment and Resources*	*Monitoring*	*Geology*
Hydrogeological annual report Polish Hydrogeological Survey (hydrogeological year 2015) [27]	*Climate, Environment and Resources*	*Water Supply*	*Geology*

National experts from Poland inputs in EIGR 50 records:
- National and local reports containing facts and data – 16 inputs
- Hydrogeological maps – 6 input
- Technical reports, Guidelines, Manuals etc. – 9 inputs
- Book and book chapters – 17 inputs
- Atlases – 2 inputs.

A lot of this information is really valuable for groundwater management according to requires of KINDRA project. However, majority of them are not useful in the European scale, because the language of this issues is often only polish. The sources of this information are databases or web-services of institutions or entities whose activity and tasks are concentrated on research of groundwater or water management (Tab. 1).

The survey allowed to systematize knowledge of sources of information about groundwater and hydrogeology in Poland. There was also possibility to compare Poland with other European countries. Number of 53 institutions which operating profile is related with groundwater in Poland seems to be good achievement. It was the first place among surveying countries. Institutions listed in the survey are the entities which control the quality and condition of the surface water and groundwater. There are e.g.: National Inspector of Environmental Protection and Provincial IEP, National Sanitary and Epidemiological Agencies and Provintional SEA, Polish Geological Institute – National Research Institute with Polish Hydrogeological Survey, numerous accredited laboratories and others, including universities carrying out research work in the field of hydrology and hydrogeology.

Other successful area for Poland had been linked with questions about journals focused on hydrogeology. From all countries was 91 magazines including 10 journals from Poland. It was 11% among all of 20 countries which take part in KINDRA project. 3 of these 10 journals are published in polish language and the rest of them are published in English or polish and English. It is nearly 18% among the 40 magazines reported as available in English or English and national language in all countries.

In other areas Poland was on the standard position. It is essential for country from Europe to provide their groundwater management in way which was summarized by the Author of the survey:
- the drinking water supply is groundwater-dominated

- the strategic groundwater monitoring network are managed according to the Water Framework Directive and Groundwater Directive
- most of country are completely covered by groundwater monitoring network
- the monitored data are generally online available.

4 Conclusions

Sustainable water management is a societal challenge and is one of the priorities of European Union policy. All EU countries are obliged to implement the Water Framework Directive regulations. However nowadays there is no system which allows European countries the exchange data and information in area of implementation of EU regulations. There are only specific databases (Web of Science, Scopus, Google Scholar) with information useful for scientific workers. This situation was a reason to launch project KINDRA and repository of practical knowledge database – EIGR. Classification data and information according to HSC-SYS allows to gather data and also conduct qualitative analysis of records. Before EIGR started there was carried out a survey. The results of the survey have shown that there is a high level of implementation of the WFD regulation in EU countries, especially in the field of monitoring of the quality of groundwater.

Poland is one of the countries which implement the WFD regulations with success. There is a lot of points in monitoring network and results of research which are commonly accessible. Poland is one of the countries in which the main drinking water source are groundwater resources. It is optimistic for the future because this group of countries can propose solutions and regulations for policy related to this part of water management.

Currently in Poland there are a lot of possibilities of publishing scientific papers international spread, because of large number of journals published in English. At once there is a large number of institutions and entities which activity concern on the groundwater monitoring and water quality control. Their achievements and practical knowledge can be popularized widespread in Europe. But it is untapped potential because many of reports, bulletins, publications are available only in polish language. An opportunity to improve the situation would be widespread use of English title and abstract together with the national language.

Work funded by the European Union's Horizon 2020 Research and Innovation Programme (grant agreement No 642047).

References

1. B. Tomaszewska, M. Dendys, TPG Geotermia, Zrównoważony Rozwój **2**, 179–187 (2016)
2. Website of KINDRA project: http://kindraproject.eu/, version from 10.03.2017
3. Directive 2000/60/EC of the European Parliament and of the Council establishing a framework for the Community action in the field of water policy
4. J.P.A. Lijzen, P. Otte, M. van Dreumel, Sci. Total Environ. **485–486**, 804–809 (2014)
5. C. Söderberg, J. Environ. Manage. **183**, 90–97 (2016)
6. N. Voulvoulis, K.D. Arpon, T. Giakoumis, Sci. Total Environ. **575**, 358–366 (2017)
7. D. Hering, A. Borja, J. Carstensen, L. Carvalho, M. Elliott, C.K. Feld, A.-S. Heiskanen, R.K. Johnson, J. Moe, D. Pont, A.L. Solheim, W. van de Bund, Sci. Total Environ. **408**, 4007–4019 (2010)

8. W. Brack, V. Dulio, M. Ågerstrand, I. Allan, R. Altenburger, M. Brinkmann, D. Bunke, R.M. Burgess, I. Cousins, B.I. Escher, F.J. Hernández, L.M. Hewitt, K. Hilscherová, J. Hollender, H. Hollert, R. Kase, B. Klauer, C. Lindim, D.L. Herráez, C. Miège, J. Munthe, S. O'Toole, L. Posthuma, H. Rüdel, R. B. Schäfer, M. Sengl, F. Smedes, D. van de Meent, P.J. van den Brink, J. van Gils, A.P. van Wezel, A.D. Vethaak, E. Vermeirssen, P.C. von der Ohe, B. Vrana, Sci. Total Environ. **576**, 720–737 (2017)
9. Y. Reyjol, C. Argillier, W. Bonne, A. Borja, A.D. Buijse, A.C. Cardoso, M. Daufresne, M. Kernan, M.T. Ferreira, S. Poikane, N. Prat, A.-L. Solheim, S. Stroffek, P. Usseglio-Polatera, B. Villeneuve, W. van de Bund, Sci. Total Environ. **497–498**, 332–344 (2014)
10. M. Vlachopoulou, D. Coughlin, D. Forrow, S. Kirk, P. Logan, N. Voulvoulis, Sci. Total Environ. **470–471**, 684–694 (2014)
11. C. Henriques, K. Garnett, E.K. Weatherhead, F.A. Lickorish, D. Forrow, J. Delgado, Sci. Total Environ. **512–513**, 381–396 (2015)
12. Y. Henocque, B. Andral, Mar. Pollut. Bull. **47**, 155–161 (2003)
13. A. Borja, J. Franco, V. Valencia, J. Bald, I. Muxika, M.J. Belzunce, O. Solaun, Mar. Pollut. Bull. **48**, 209–218 (2004)
14. I. Muxika, A. Borja, J. Bald, Mar. Pollut. Bull. **55**, 16–29 (2007)
15. J. Albrecht, Environ. Sci. Policy **55**, 368–375 (2016)
16. M.T. Kahil, J.D. Connor, J. Albiac, Ecol. Econ. **120**, 226–233 (2015)
17. P. Quevauviller, J. Hydrol. **518**, 180–185 (2014)
18. G. Escribano Frances, P. Quevauviller, E. San Martín Gonzalez, E. Vargas Amelin, Environ. Sci. Policy **69**, 1–12 (2017)
19. B. Paczyński, A. Sadurski, *Hydrogeologia regionalna Polski; Wody słodkie. t. I* (Wyd. Państw. Inst. Geol., Warszawa, 2007)
20. S. Witczak, T. Szklarczyk, E. Kmiecik, J. Szczepańska, A. Zuber, K. Różański, M. Duliński, Geol. Q. **51**, 1, 1–16 (2007)
21. S. Witczak, A. Zuber, E. Kmiecik, J. Kania, J. Szczepanska, K. Rozanski, *Tracer Based Study of the Badenian Bogucice Sands Aquifer, Poland In: Natural Groundwater Quality* (W.M. Edmunds, P. Shand (Eds.). Willey-Blackwell, 2008, 335–352)
22. B. Tomaszewska, A. Szczepański, Environ. Sci. Pollut. R. **21**, 11409–11417 (2014)
23. M. Maciejewski, T. Walczykiewicz, Infrastruktu. i Ekol. Terenów Wiejskich **4**, 1, 63–76 (2006)
24. W. Bujakowski, B. Tomaszewska, *Atlas wykorzystania wód termalnych do skojarzonej produkcji energii elektrycznej i cieplnej przy zastosowaniu układów binarnych w Polsce* (Inst. Gosp. Surow. Min. i Ener. PAN. Wyd. Jak, Kraków, 2014)
25. W. Górecki, red. *Atlas of geothermal resources of Paleozoic formations in the Polish Lowlands* (2006)
26. Polish Geological Institute – National Research Institute, Polish Hydrogeological Survey. Groundwater monitoring in the border area of the Sudetes in 2005–2012
27. Polish Geological Institute – National Research Institute, Polish Hydrogeological Survey. Hydrogeological annual report Polish Hydrogeological Survey (hydrogeological year 2015)

 DOI: 10.1051/e3sconf/20172200179

Experimental process investigation of a latent heat energy storage system with a staggered heat exchanger with different phase change materials for solar thermal energy storage applications

Nikolas P. Tsolakoglou[1], *Maria K.* Koukou[1*], *Michalis Gr.* Vrachopoulos[1], *Nikolaos* Tachos[1], *Kostas* Lymberis[2], and *Vassilis* Stathopoulos[3]

[1]Technological Educational Institute of Sterea Ellada, Mechanical Engineering Department, Energy and Environmental Research Laboratory, 344 00 Psachna, Evia, Greece
[2]Z&X Mechanical Installations Ltd, 12 Agapinoros Street, 8049 Paphos Cyprus
[3]Technological Educational Institute of Sterea Ellada, Department of Electrical Engineering, Laboratory of Chemistry and Materials Technology, 344 00 Psachna, Evia, Greece

Abstract. This work investigates melting and solidification processes of four different Phase Change Materials (PCM) used as latent heat thermal storage system. The experimental rig was consisted of an insulated tank, filled with the under investigation PCM, a staggered heat exchanger to supply or extract heat from the PCM cavity and a water pump to circulate Heat Transfer Fluid (HTF). Both charging (melting) and discharging (solidification) processes were conducted for two different HTF flow rates. The main scope of this work was to develop a first approach and to investigate the behaviour of PCM under various load conditions (different HTF flow rates). Results show that different HTF flow rates affect melting and solidification time periods; in both processes time was reduced while HTF flow rate was increased but in differentways due to the transition from conduction to convection heat transfer mechanisms.

1 Introduction

Solar energy has been extensively utilized in residential cooling/heating and DHW production systems. One of the most efficient methods to store thermal energy is Latent Heat Thermal Energy Storage (LHTES) via the integration of PCM, exploiting high values of latent heat [1]. LHTES can store 5–14 times more heat per unit volume than sensible storage materials, such as water [2]. However, these materials have low thermal conductivities requesting large power density. The energy stored can be recovered in a constant temperature which may easily be utilized by Heat Pumps. LHTES materials should exhibit certain properties as thermal, physical, kinetic and chemical. Paraffin waxes

*Corresponding author: mkoukou@mail.teiste.gr, m_koukou@otenet.gr

are promising materials because they exploit efficient energy storage in an optimal temperature range for integration into residential heating/cooling systems. Regarding the design of commercial thermal energy storage units, further investigation is required to assess the physical phenomena and successfully implement PCM. Their proper design must take into account material characteristics, desired storage rate, transient PCM heat exchange and required energy storage capacity. In order to overcome low conductivity, various methods have been proposed including the use of nano-particles or the use of fins to maximize heat transfer [3, 4].

There are many phase change materials in the market and the selection of the appropriate PCM to be used in a certain application must take into consideration various technical and material properties issues. Some of these issues are, melting temperature which should fulfill the application area requirements, latent heat, chemical stability and thermal conductivity. Various fin configurations have been proposed in literature [5–9] as they influence heat transfer mechanisms during phase change process. Dimension, type and number of fins are selected based on a compromise between possible heat transfer increase and energy storage requirement. Numerous authors have investigated both numerically and experimentally heat exchanger configurations with PCM, as latent heat thermal storage systems. Medrano et al. [7] compared different types of heat exchangers with RT35 as PCM and water as heat transfer fluid (HTF). Performance evaluation was based on the required time to fully melt a certain cavity with different types of heat exchangers. Results state that's taggered heat exchanger made of aluminum-fins-and copper-tubes widely used as evaporator or condenser in small air conditioning units, is by large the one with the highest average thermal power (above 1 kW), showing the highest ratio of heat transfer area to external volume.

In this work, a study on melting and solidification time is performed for different PCM and different HTF flow rates using a staggered heat exchanger (HE). The experimental rig comprises an insulated heat storage tank, an HE, an air to water heat pump for cold water supply, a buffer tank with an electric heater for hot water supply, a circulation pump, a three-way temperature control valve, a flowmeter and thermocouples along with a data logger. The HE is made of copper tubes with aluminum fins. Experiments involved melting and solidification of the PCM, were conducted for different HTF flow rates, showing the feasibility of using the PCM as a thermal storage media. Higher efficiency could be achieved by optimizing the performance of the equipment and selecting PCM with higher thermal conductivity.

2 Experimental methodology

2.1 Experimental setup description

An experimental rig was designed and developed for the investigation of melting and solidification processes of four different PCM for different HTF flow rates. Fig. 1 shows the experimental set-up. The experimental rig comprised a well-insulated heat storage transparent glass tank allowing viewing of melting and solidification procedures and also constant monitoring with the use of a camera. The dimensions of the tank were: 600 x 120 x 80 mm (Length x Width x Depth). A staggered heat exchanger was immersed in the PCM cavity (HE technical data are given in Table 1 and a 3D drawing in Fig. 2). Hot water was supplied by an 80 lt boiler with a 4 Kw electric heater and cold water was supplied by a 200 lt water buffer tank connected to an air to water heat pump.

DOI: 10.1051/e3sconf/20172200179

Table 1. Staggered heat exchanger characteristics.

Number of tubes	12
Inner tube diameter (mm)	7.75
Outer tube diameter (mm)	9.525
Tube material	Copper
Fins material	Aluminium
Fin thickness (mm)	0,3
Fin length (mm)	68
Fin spacing (mm)	5.0

The experimental rig also included a circulation pump (Grundfos® ALPHA 2 32-60 180 inverter pump), a three-way temperature control valve (Belimo® LR24A-SR valve and Vector® TCI-W11 controller), a flow meter (rotameter with measuring range of 6–60 lt/h), thermocouples (T type), data logger (National instrument® cDAq – 9174 base, 2 x 9213 16 ch cards, LabVIEW® data logger) and, finally, the necessary piping and valves to regulate water flow. The rig allows replacing easily phase change material and the heat exchanger as well. Thermal insulation has been applied on the outer side of the tank. HTF used is water. The volume between the tubes and the fins and the tank is filled with PCM. The hot HTF supplied from hot water buffer exchanges its heat and melts the PCM during the charging process. The heat is stored in the molten PCM. During discharging process, the cold HTF supplied by the cold water buffer tank flows through the tube and solidifies the molten PCM. The stored energy is released from the PCM as it solidifies.

To analyse HE thermal performance along with charging and discharging process, temperature measurements are performed on HTF inlet and outlet temperature and also within the PCM volume (positioning of the thermocouples are indicated in Fig. 3).

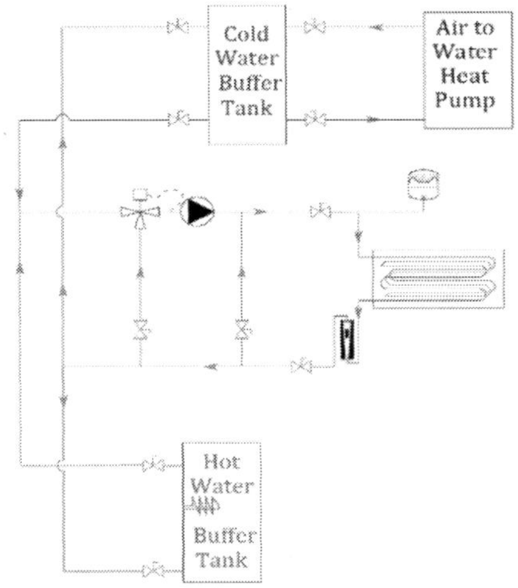

Fig. 1. Experimental rig layout.

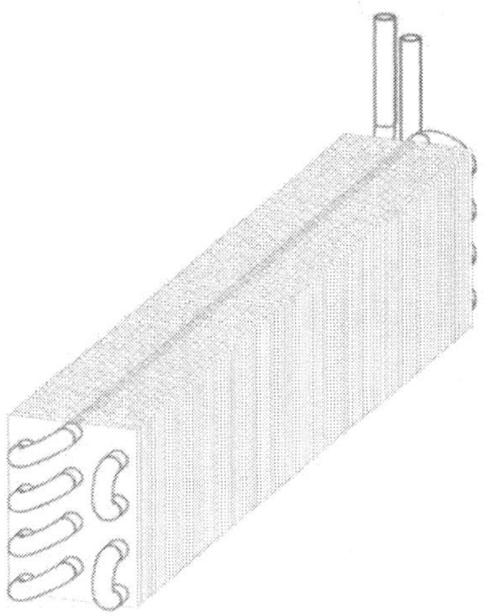

Fig. 2. Staggered heat exchanger 3D drawing.

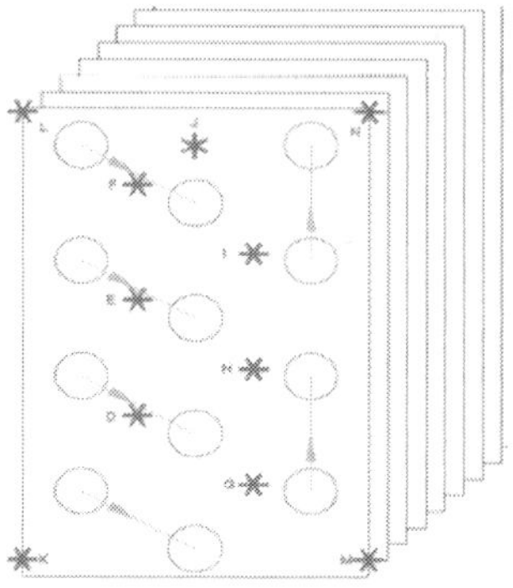

Fig. 3. Positioning of thermocouples within the PCM cavity (among heat exchanger fins).

2.2 PCMs studied

The selection of the PCM used, is an important parameter regarding the design of aLHTES as the amount of energy that may be recovered is directly linked to the nature of the material. The selection of PCM in this work is based on the consideration of low temperature solar thermal applications. The materials tested are commercial products of PCM PRODUCTS® [10] and their detailed thermal and physical properties provided by the company are summarised in Table 2.

Table 2. Thermal and physical properties of paraffin waxes used in the experiments [10].

	A53	A46	A44	A40
Melting point (K)	324,2	318	317	312
Density liquid (kg/m^3)	775	775	775	775
Thermal conductivity solid (W/mK)	0.22	0.24	0.24	0.24
Thermal conductivity liquid (W/mK)	0.22	0.24	0,24	0.24
Specific heat liquid(J/kgK)	1.8	1.8	1.8	1.8
Specific heat solid(J/kgK)	2.4	2.4	2.4	2.4
Latent heat (kJ/kg)	160–180	185–215	250	153

2.3Experimental procedure and conditions tested

The first thing to position within the glass tank was the staggered heat exchanger. The second step was the positioning of the thermocouples in a certain place throughout all experiments (Fig. 3). Hot water was then supplied to the HE and portions of solid PCM were placed onto its surface. In any case, before PCM was inserted the rig was tested for leakages. The rise in temperature of the HE walls caused PCM melting and more quantity was added to a certain volume. This was the fastest way to fill the tank with PCM and to ensure that no air was trapped within its volume. The PCM was then cooled to room temperature and the experimental procedure was performed, always following the same sequence:

(i) Charging (melting): Hot water within the buffer was circulated within the piping at a temperature rise of 8°C more than the melting temperature of the PCM (if A40 was examined the HTF temperature was 48°C). Temperature inlet was controlled by the three-way mixing valve. Melting process was ended when the temperature of the thermocouples at the edges of the exchanger (thermocouples K, L, M, N) exceeded melting temperature more than 3°C (in the case of A40 the temperature should exceed 43°C). Then solidification process was performed

(ii) Discharging (solidification): Subsequently, a discharging experiment starts just after the completion of the charging experiment. Cold water from the buffer tank was circulated with a temperature of 8°C less than the melting temperature of the PCM (e.g. if A40 was examined the HTF temperature was 32°C). Temperature inlet was controlled by the three-way mixing valve. Solidification ended when the temperature of the thermocouples at the edges of the exchanger (thermocouples K, L, M, N in Fig. 3) was less than 3°C than melting temperature (in the case of A40 the temperature should exceed 37°C).

To justify repeatability selected experiments were carried out three times. Fig. 4 shows thermocouples mean temperature variation for three repetitions of a charging and discharging process of A44 using aflow rate of 60 lt/h and an inlet temperature of 52°C. A sufficient repeatability is observed.

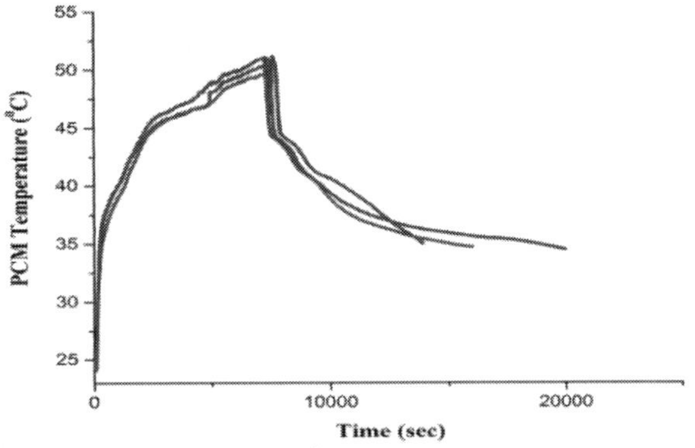

Fig. 4. Repeatability test of A44 using a flow rate of 60 lt/h and inlet water at 52°C.

3 Results and discussion

Melting and solidification process of a material, in both sensible or latent heat storage procedures is related to the energy provided (charging) or extracted (discharging) by the HTF. The amount of energy is dependent on the HTF inlet temperature and flow rate. Both factors are related via the equation:

$$q = \dot{m}_{HTF} Cp_{HTF} (T_{in} - T_{out}) \quad (1)$$

$\dot{m}_{HTF}$ = HTF mass flow rate (kg/s)
Cp_{HTF} = HTF specific heat capacity (kJ/KgK)
T_{in} = HTF inlet temperature (K)
T_{out} = HTF outlet temperature (K)
q = Heat Flux (kJ/sec) (q=Q/t)

As mentioned above, HTF is water hence specific heat capacity is constant. All experiments were performed for two HTF flow rates: (i) 30 lt/h (0.0083 kg/s), and (ii) 60 lt/h (0.0166 kg/s). By Eq. 1 it is then concluded that the larger the value of HTF flow rate (for a same constant inlet temperature) the larger the heat flux by the pipes to the PCM hence the shortest the period to complete the process of charging (melting). Discharging process period is also decreased, respectively. Experimental plots confirmed what expected but not in the same rate. Melting time is affected more than solidification time when doubling HTF flow rate and this is mainly due to the alternation of heat transfer mechanism.

When charging, or discharging a PCM three different thermal and conditional states take place (shown in Fig. 5).The initial condition of the PCM when melting begins is solid. Whence HTF is starting to circulate within the HE pipes, due to conduction (between pipes and PCM) the material starts to melt. Liquid fraction increases as time progresses. Due to buoyancy within the liquid state PCM, convection heat transfer is occurring. This phenomenon increases heat transfer rate. Actually, conduction is the mechanism to transfer heat from liquid to solid PCM, but convection speeds up the process. In between altering state (solid to liquid), latent heat region is present where heat is stored to the PCM without altering its temperature. Finally, above phase change region the PCM acts as a sensible heat storage medium by storing heat and increasing its temperature. During solidification the same procedure is performed respectively.

By Eq.1, the second factor to analyse is temperature difference between inlet and outlet of the HE. As inlet temperature approaches outlet conditions, heat flux is reduced. It is then concluded that at the beginning of each experiment, heat flux takes its maximum value as temperature difference is at its maximum difference. Besides flow rate a key role regarding melting and solidification processes of a PCM is temperature variation between HTF inlet temperature and PCM temperature. These conclusions are based on the assumption that the tank containing the material has adiabatic walls, otherwise thermal losses also contribute to an extent.

During solidification processes all PCM had the same behaviour. The first region to solidify was the volume of the PCM is touch with the HE walls (as seen in Fig. 5). The formation of the solidified PCM in an axial direction round HE pipes though had a significant factor on total discharging process. The formation of this layer eliminated convection heat transfer from HE walls to the PCM. Conduction heat transfer is now the only possible method to complete solidification. As all PCM have low thermal conductivity solidification time decreases rapidly.

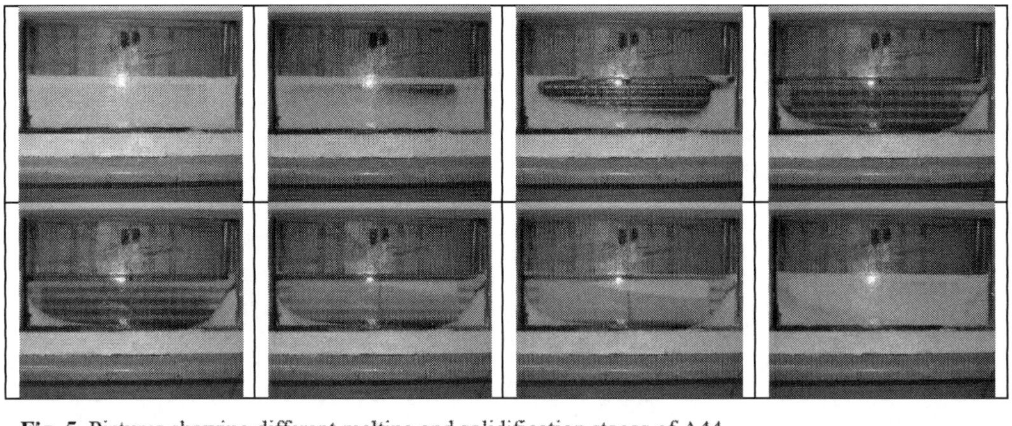

Fig. 5. Pictures showing different melting and solidification stages of A44.

There are also some other factors that affect charging and discharging processes regarding PCM thermal behaviour and these need further discussion and may be analysed in future work. HTF flow rate has to be studied regarding Reynolds number. Heat is transferred from the HTF to the walls of the HE via conduction when flow is laminar and convection, when flow becomes turbulent. HTF flow rate then has to be analysed regarding Reynolds number in a pipe and not simply by a value of volumetric flow rate. Another factor to be further analysed is insulation and especially that of the air above the PCM. If the amount of air is large and convective heat transfer may appear then the phenomenon is altered. Finally, the geometry of the HE or even its materials may be utilised to optimise heat flux and speed up charging and solidification processes. Plots below, show time variation for both charging and discharging processes of A40, A44, A46 and A53. Y axis presents the mean temperature value of the thermocouples placed into the PCM cavity (positioning in Fig. 3).

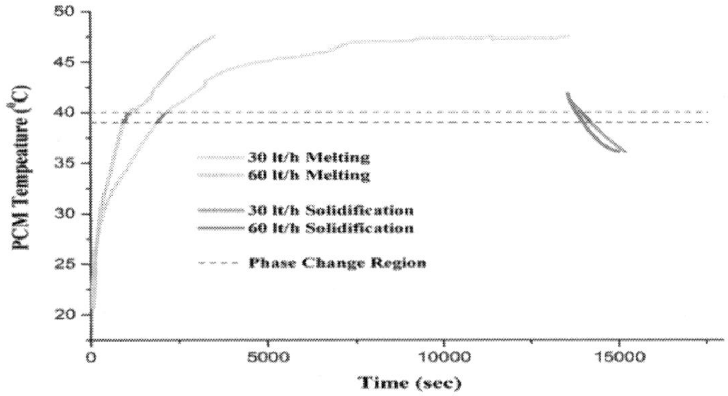

Fig. 6. Temperature profiles inside A40 for charging and discharging and various inlet feed rates.

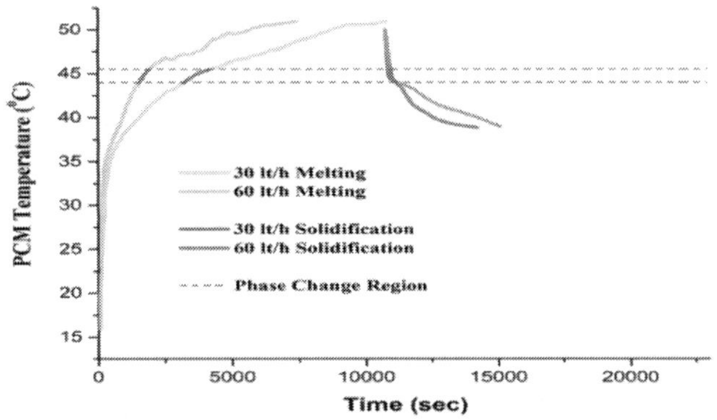

Fig. 7. Temperature profiles inside A44 for charging and discharging and various inlet feed rates.

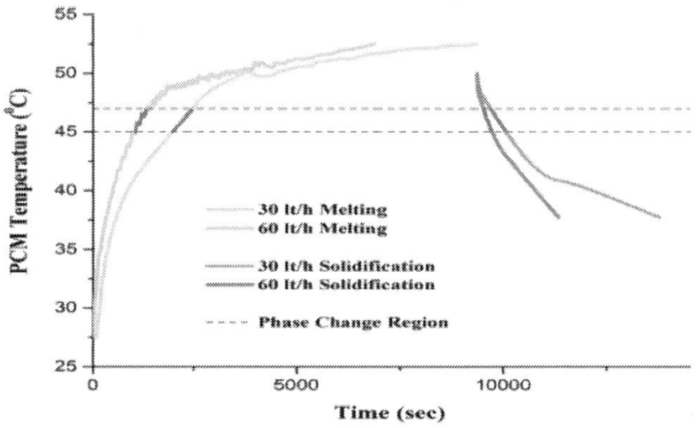

Fig. 8. Temperature profiles inside A46 for charging and discharging and various inlet feed rates.

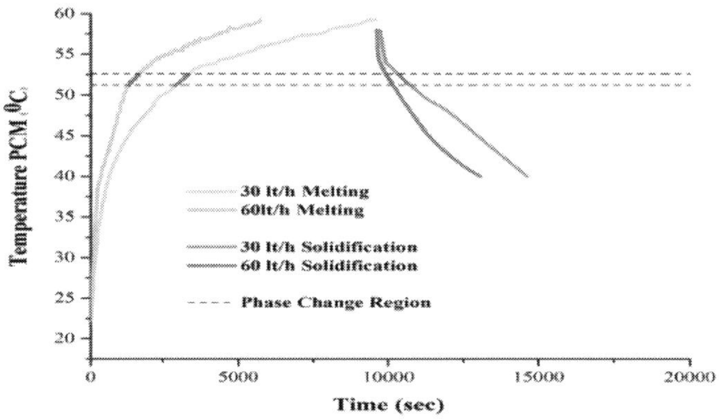

Fig. 9. Temperature profiles inside A53 for charging and discharging and various inlet feed rates.

4 Conclusions

This work focuses on the effect of HTF flow rate regarding charging and discharging processes of a heat storage system for four different PCM with a staggered heat exchanger. All experimental data showed that process time (in both melting and solidification) is reducing when HTF flow rate increases. The phenomenon is not analogous and it is dependent on the variation of the heat transfer mechanism occurring. During melting convection speeds up the procedure whence during solidification the formation of a solid layer around HE pipes, eliminates convective effects and it is only affected by conduction among the PCM.

Acknowledgement

The work is supported by TESSe2b project which has received funding from the European Union's Horizon 2020 research and innovation programme under grant agreement No 680555.

References

1. A. Abhat, Sol. Energy **30**, 4, 313 (1983)
2. A.A. Robaidi, Int. J. Mater. Sci. Appl. **2**, 6, 168 (2013)
3. L. Liu, D. Su, Y. Tang, G. Fang, Renewable and Sustainable EnergyReviews **62**, 305 (2016)
4. M. Lacroix, Int. J. Heat Mass Transfer **36**, 2083 (1993)
5. F. Agyenim, P. Eames, M. Smyth, Sol. Energy **83**, 9, 1509 (2009)
6. A. Castell, M. Belusko, F. Bruno, L.F. Cabeza, Appl. Energy **88**, 11, 4120 (2011)
7. M. Medrano, M.O. Yilmaz, M. Nogués, I. Martorell, J. Roca, L.F. Cabeza, Appl. Energy **86**, 10, 2047 (2009)
8. N.H.S. Tay, M. Belusko, A. Castell, L.F. Cabeza, F. Bruno, Appl. Energy **131**, 377 (2014)
9. S. Paria, S. Baradaran, A. Amiri, A.A.D. Sarhan, S.N. Kazi, J. Therm. Anal. Calorim. **123**, 1371 (2016)
10. http://www.pcmproducts.net

A method of determining the thermal power demand of buildings connected to the district heating system with usage of heat accumulation

Michał Turski[1,*], and *Robert* Sekret[1]

[1]Faculty of Infrastructure and Environment, Czestochowa University of Technology, ul. Brzeznicka 60A, 42-200 Czestochowa, Poland

Abstract. The paper presents a new method of determining the thermal power demand of buildings connected to the district heating system, which included the actual heat demand and the possibility of balancing the thermal power using the thermal storage capacity of district heating network and internal heat capacity of buildings. Moreover, the analysis of the effect of incidence of external air temperature and duration of episodes with the lowest outdoor temperatures on the thermal power demand of district heating system was conducted.

1 Introduction

One way to reduce pollution from dispersed sources of low power (i.e. low emission) can be a significant development of district heating systems. This is caused by the high efficiency of these systems and modern solutions for reducing pollution in sources in comparison to dispersed sources [1].

One of the barriers to the development of district heating systems is that the buildings connected to the district heating network are highly volatile of daily demand for thermal power, what causes problems when is designing power of heat sources in the district heating system. This occurs regardless of the generation of heat networks, or projected scenarios of their development [2–4]. In addition, pay special attention to the very rare cases of a value equal to or below the design temperature of the outside air during the heating season [5, 6]. It can be concluded that the maximum design parameters of the heat source in heating networks are not being met [7]. The effect of oversized power of heat sources is a decline in the profitability of district heating systems [8]. Currently used methods of design power of heat sources of the district heating system based on the design temperature of the outside air. In addition, sources of heat are designed with more power due to the anticipated development of the buildings infrastructure and future demand for thermal power. However, this trend does not necessarily have to be maintained because of the intense thermal modernization of existing buildings and use new building technologies with extremely low heat consumption [9–11]. In the currently used calculation is not taken

* Corresponding author: m.turski@is.pcz.pl

into account the thermal storage capacity of district heating network and internal heat capacity of buildings [12, 13].

In this material it is important to present a new calculation method and a new perspective on problem of current oversizing of heat sources capacities of district heating systems. New method includes the actual heat demand and the possibility of balancing the thermal power using the thermal storage capacity of district heating network and internal heat capacity of buildings.

2 Method of determining the thermal power with usage of heat accumulation

The proposed method consists of seven stages: analysis of the hourly frequency of outside air temperatures, analysis of average daily outside air temperature, analysis of frequency and duration of episodes with the lowest values of outside air temperature, determine the actual thermal power demand of buildings, analyse of the possibility of balancing the thermal power using the thermal storage capacity of district heating network, analyse of the possibility of balancing the thermal power using the internal heat capacity of buildings and determine the reduction in demand for thermal power of district heating system.

On the basis of climatic data for the considered area, designated as data for the standard meteorological year, should be determine the hourly frequency of the temperature in regulation range of the district heating network (e.g. from -20°C to 12°C) during the heating season (e.g. October–April).

Then should be analyzed the average daily temperature of the outside air. Such an analysis should be carried out on the basis of measurement data from the nearest meteorological station or data from weather monitoring of the system. Using the method of decomposition of the time series can be estimated distribution of the forecast external temperature for the coming years. Then, based on the obtained distribution of the value of the coefficient of determination R^2 resulting from the number of seasons to be analyzed, select the forecast based on time series with the best fit.

It is likely that the analysis of the average daily temperature of the outside air using the method of decomposition of the time series will not be precise (coefficient of determination $R^2 < 0.8$) [14–16]. For such a case it is necessary to carry out further analysis of frequency and duration of episodes with the lowest values of outside air temperature. This method involves the determination of a set of the average daily outside air temperature, from the period of possibly the greatest number of years (min. 30), episodes lasting from one to up to five days. The criterion for the designation of the episodes of time should be a temperature below the design temperature (e.g. in the range from -15°C to -20°C). Temperature ranges should be divided as 1K.

If the value of the coefficient R^2 are satisfactory, but the forecasted temperatures are still doubtful (e.g. are too high), it is also recommended to carry out this part of the analysis.

To determine the planned heating demand are required characteristics of district heating network. To their designation may be used of heating curves depending on outside air temperature. In that case, use heating curves of the network side (high performance), as well as heating curves on the installation side (low performance). Return temperatures of both the network and the installation are taken as the result of heat supply to end users. On this basis shall be designated predicted heat fluxes to meet the needs of end users, depending on the temperature of outside air. This method involved the use of the incidence of outside air temperature based on standard meteorological year. This determines the computing standard value of the planned thermal power demand by end users.

In addition to above can be practically used heat capacity of the heating network for the purpose of heat accumulation using increasing the temperature of the water network in

 DOI: 10.1051/e3sconf/20172200180

supply, return or together. In the case of increasing the temperature of supply water the parameter limiting is the maximum operating temperature due to the recommended operating conditions of pre-insulated district heating pipes. It can be assumed that this temperature is up to 130–135°C. The other two methods are not taken into account due to the increased cost of relief tubing or development of each substation. The amount of heat possible to accumulate in the district heating network, by increasing supply temperature can be calculated from equation (1):

$$Q_{A,n} = V \rho\, c_p\, \Delta T\, 10^{-3},\ MJ \qquad (1)$$

where:
$Q_{A,n}$ – amount of heat possible to accumulate in the district heating network, by increasing supply temperature, MJ,
V – water volume of heating network at supply, m³,
ρ – density of water, kg/m³,
c_p – specific heat capacity of water, kJ/(kg K),
ΔT – temperature difference, K.

Compensating time of heating source power reduction using a heat accumulation on the district heating supply can be calculated from the equation (2):

$$\tau_{K,n} = Q_{A,n}/(3600\ \Delta Q_{sr}),\ h \qquad (2)$$

where:
$\tau_{K,n}$ – compensating time of heating source power reduction using a heat accumulation on the district heating supply, h,
$Q_{A,n}$ – amount of heat accumulated in the heating network, MJ,
ΔQ_{sr} – assumption in reduction of heat source power, MW.

Another possibility of using the heat accumulation give buildings connected to the district heating network. The accumulated amount of heat by using the internal heat capacity of buildings connected to the district heating network can be calculated from the equation (3):

$$Q_{A,b} = C_m\, \Delta T_b\, 10^{-6},\ MJ \qquad (3)$$

where:
$Q_{A,b}$ – accumulated amount of heat by using internal heat capacity of buildings connected to district heating network, MJ,
C_m – value of internal heat capacity of building, J/K,
ΔT_b – building internal temperature drop, K.

The value of the internal heat capacity of buildings should be calculated for the total area of heated buildings considered using the equation (4):

$$C_m = K_j\, A_f\, 10^{-6},\ MJ/K \qquad (4)$$

where:
C_m – value of internal heat capacity of building, MJ/K,
A_f – heated area, m²,
K_j – internal heat capacity relative to the surface area of the building element, J/(m² K).

Summary of data to be used in the calculation is presented in Table 1. Example values of the internal heat capacity of the building C_m are calculated for building space available in the GUS [17] for medium-sized heating system occurring in Poland. It should be noted that for each of heating system should be used current and real data of buildings connected to the district heating network.

Table 1. Data used to calculate the internal heat capacity of buildings.

Building structure	**Internal heat capacity relative to the surface area of the building element**	**Heated surface**	**Internal heat capacity of building**
	K_j , J/(m² K)	A_f , m²	C_m , MJ/K
Very light	80000	–	–
Light	110000	–	–
Average	165000	26737	4411.6
Heavy	260000	85151	22139.3
Very heavy	370000	389481	144107.0

Compensation time of source thermal power reduction using the heat accumulated in the buildings can be calculated from equation (5):

$$\tau_{K,b} = Q_{A,b}/(3600\ \Delta Q_{sr}),\ h \tag{5}$$

where:
$Q_{A,b}$ – amount of heat accumulated in buildings, MJ,
ΔQ_{sr} – assumption in reduction of heat source power, MW.

For such presented method of determining the thermal power demand of buildings connected to the district heating system with usage of heat accumulation can be represented graphically compensation times as a function of assumed reducing thermal power of heat source, both for the case of using the thermal storage capacity of district heating network and internal heat capacity of buildings. Example of graphical presentation constructed on the basis of equations 2 and 5 is shown in Fig. 1.

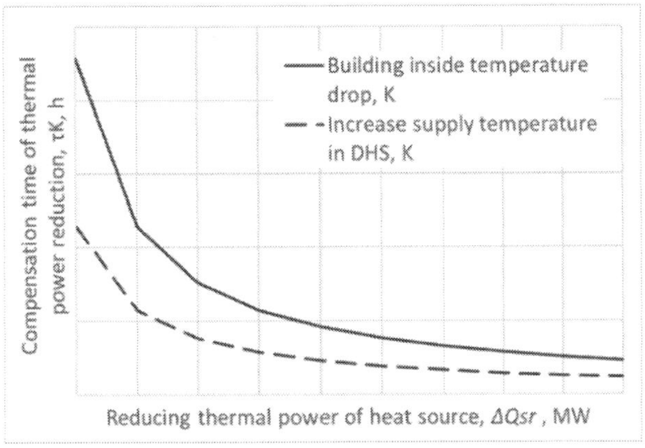

Fig. 1. Example of graphical presentation compensation times as a function of assumed reducing thermal power of heat source.

3 Discussion

The forecast of outside temperature distribution shows a strong correlation in determining the minimum outdoor temperature on the assumed length of the time series. Taking into consideration the value of coefficient of determination R^2 below 0.8 obtained results may not be sufficient to provide more accurately calculation the outdoor temperature for the forecast period.

In that case, an analysis of the outside temperature distribution should be conducted, from as long as possible measurement period.

After determining episodes of temporary occurrence of the minimum values of the outside air temperature usually it turns out that one-day episodes statistically one can assume at 1–2 times per season for average daily outside air temperatures below -15°C, and the average daily outside air temperature below -20°C there are only once every four heating seasons. Considering the occurrence of episodes of multi-days occurs that they appear very rarely, and among them may be relevant episodes two- and three-days.

4 Conclusions

On the basis of presented determining the thermal power demand of buildings connected to the district heating system with usage of heat accumulation can draw following conclusions:

Having regard to the incidence of outside air temperatures, and the frequency and duration of episodes with the lowest values of outside air temperature the most possible source thermal power reduction for the purpose of central heating is 5–10%.

In addition, taking into account the utilization of the thermal storage capacity of district heating network and internal heat capacity of buildings it is possible to balance the heat power reduction for another 5–10% in a maximum period of 72 hours. These values are dependent on the permissible temperature increase in the supply of district heating and building inside temperature drop.

It is known that the heating power of district heating system sources in Poland is oversized. It is also evident that the adaptation of the power of heating sources to the building's energy needs will be beneficial for improving the energy efficiency of the district heating system, reducing transmission losses and reducing pollutant emissions. All this leads to maintaining the profitability of heating companies. The presented method does not exhaust the possibilities of usage of heating accumulation of the system, especially with additional heat accumulators. Therefore, it should be carried out further research into the use of alternative methods of determining the thermal power demand of buildings connected to the district heating system with usage of heat accumulation of the system.

The material has been prepared in terms of Czestochowa University of Technology statutory research BS/PB-407-302/11 'Energy-efficient and environmentally sustainable supply systems of buildings in heat, coldness and electricity'.

References

1. K. Wojdyga, M. Chorzelski, E. Rozycka-Wronska, J. of Clean Prod. **75**, 157–165 (2014)
2. H. Lund, S. Werner, R. Wiltshire, S. Svendsen, J. Thorsen, F. Hvelplund, B. Mathiesen, Energy **68**, 1–11 (2014)
3. M. Turski, R. Sekret, Rynek Energii **119**, 27–34 (2015)

4. M. Turski, R. Sekret, Rynek Energii **122**, 66–74 (2016)
5. K. Wojdyga, Energy and Build. **40**, 2009–2014 (2008)
6. M. Santamouris, C. Cartalis, A. Synnefa, D. Kolokotsa, Energy and Build. **98**, 119–124 (2015)
7. M. Turski, R. Sekret, J. of Power Technol. **96**, 444–448 (2016)
8. A. Colmenar-Santos, E. Rosales-Asensio, D. Borge-Diez, F. Mur-Pérez, Energy **85**, 403–414 (2015)
9. P. Späth, H. Rohracher, Energy Policy **78**, 273–280 (2015)
10. K. Wojdyga, Renew. Energy **34**, 2935–2939 (2009)
11. J. Heier, C. Bales, V. Martin, Renew. and Sustainable Energy Rev. **42**, 1305–1325 (2015)
12. K. Wojdyga, Int. J. of Energy and Power Eng. **5**, 237–244 (2014)
13. K. Klajic, D. Gvozdenac, S. Vukmirović, Energy **45**, 304–311 (2012)
14. P. Piotrowski, Przegląd Elektrotechniczny **90**, 1–4 (2014)
15. E. Musiał, E. Gąsiorek, M. Rojek, Acta Agrophysic **3**, 333–342 (2004)
16. J. Kajewska, M. Rojek, Water-Environment-Rural Areas **36**, 149–160 (2011)
17. I. Adamczyk, K. Andrusiuk, A. Dec, E. Knyszewska, M. Przybylska, B. Różańska, M. Sobczyk, INHABITED BUILDINGS National Census of Population and Housing 2011. Główny Urząd Statystyczny, Departament Handlu i Usług, Warszawa (2013)

The efficiency of macroporous polystyrene ion-exchange resins in natural organic matter removal from surface water

Agnieszka Urbanowska[1,*], and *Małgorzata* Kabsch-Korbutowicz[1]

[1]Wroclaw University of Science and Technology, Faculty of Environmental Engineering, Chair in Water and Wastewater Treatment Technology, Wybrzeze Wyspianskiego 27, 50-370 Wroclaw, Poland

Abstract. Natural water sources used for water treatment contains various organic and inorganic compounds. Surface waters are commonly contaminated with natural organic matter (NOM). NOM removal from water is important e.g. due to lowering the risk of disinfection by-product formation during chlorination. Ion exchange with the use of synthetic ion-exchange resins is an alternative process to typical NOM removal approach (e.g. coagulation, adsorption or oxidation) as most NOM compounds have anionic character. Moreover, neutral fraction could be removed from water due to its adsorption on resin surface. In this study, applicability of two macroporous, polystyrene ion exchange resins (BD400FD and A100) in NOM removal from water was assessed including comparison of treatment efficiency in various process set-ups and conditions. Moreover, resin regeneration effectivity was determined. Obtained results shown that examined resins could be applied in NOM removal and it should be noticed that column set-up yielded better results (contrary to batch set-up). Among the examined resins A100 one possessed better properties. It was determined that increase of solution pH resulted in a slight decrease in treatment efficiency while higher temperature improved it. It was also observed that regeneration efficiency was comparable in both tested methods but batch set-up required less reagents.

1 Introduction

Natural organic matter (NOM) is one of the main group of contaminants present in surface and ground water sources. NOM is defined as a complex mixture of relatively low molecular weight compounds (e.g. amino acids, proteins) with macromolecular humic substances (humic and fulvic acids, humins) originating from natural decomposition of vegetable and animal remains [1–4].

Organic compounds are responsible for water colour, taste, smell and, most importantly, act as disinfection by-product precursors. Moreover, during water treatment they increase coagulant and disinfectant demand as well as create complexes with various anthropogenic

* Corresponding author: agnieszka.urbanowska@pwr.edu.pl

 DOI: 10.1051/e3sconf/20172200181

contaminants. Furthermore, organic contaminants block surface of ion-exchange resins, foul membranes and support microbial growth in water supply network [5]. For these reasons, NOM should be removed from water sources intended for drinking water production [6].

Ion exchange with the use of ion-exchange resins is one of the available methods to remove ionic NOM [7]. NOM properties analysis indicates possibility of two removal mechanisms using ion-exchange resins: ion exchange due to electrostatic interaction between resin ionic groups and functional groups of organic macromolecules or physical adsorption of resulting from van der Waals forces between hydrophobic organic macromolecule centres and polymer matrix [8]. NOM removal efficiency from water using ion-exchange process ranges from 30 to 90% and depends on physico-chemical water properties, resin properties and process set-up [9].

Polystyrene macroporous ion exchange resins are one of the most commonly used in water treatment. Thanks to their structure allowing easy organic anion penetration, they are very useful in organic matter removal from water [10]. Water retention is an important property due to the fact that high moisture anion exchange resins have less compact structure thus allowing “slower” large particles (e.g. organic acids) binding [11, 12]. Water content to ion-exchange capacity ratio strongly affects resin regeneration effectivity – from this point of view lower capacity is better because it leads to higher ratio value and thus easier regeneration.

In this study, two polystyrene macroporous ion exchange resins (BD400FD and Purolite A100) were used to remove organic matter from water and to compare removal process efficiency in various conditions and feed water properties. Moreover, the resin regeneration methods were examined.

2 Materials and methods

Properties of anion exchange resins used in these study can be found in Table 1.

Table 1. Ion-exchange resins characteristics [13, 14].

Resin	Type	Polymer structure	Particle size, mm	Functional group	Total capacity, eq/dm^3	Specific gravity, kg/m^3
A100	weak base	macroporous polystyrene crosslinked with divinylbenzene	0.6–0.85 (mean)	R-N(CH$_3$)$_2$	1.13	1.04
BF400FD	strong base		0.4–0.9 ($\geq$ 95%) 0.59–0.69 (mean)	R-N$^+$(CH$_3$)$_3$	$\geq$ 1.1	1.05–1.09

Two model solutions of different origin but similar properties were subject of this study. Model solution 1 was taken from humic acid rich stream flowing from a peat-bog in the Table Mountains (Poland, sampling point location: 50°27’29.97” N; 16°23’16.87” E) while model solution 2 was prepared by mixing humic acids (Aldrich) with dechlorinated tap water (0.02 g/dm^3 solution).

Organic matter content in examined solutions was determined by colour intensity (absorbance of 350 nm wavelength), UV 254 nm absorbance as well as total (TOC) and dissolved organic carbon (DOC) measurements. TOC (DOC) concentration is the most reliable method for determining the total amount of NOM, UV adsorption at 254 nm monitors the amount of NOM fractions containing aromatic structures in their molecules, while the colour of water is related to the presence of large fractions of NOM. Model solutions properties can be found in Table 2.

DOI: 10.1051/e3sconf/20172200181

Table 2. The properties of model solutions.

Solution	Colour	UV 254 nm absorbance	DOC	Conductivity	pH
	g Pt · m^{-3}	cm^{-1}	g C · m^{-3}	μS · cm^{-1}	-
Model solution 1	99.54	0.693	11.98	337	4.84
Model solution 2	104.58	0.526	11.86	679	7.25

Anion exchange resin preparation for the use in the ion exchange process included converting it into chloride form by soaking it in 12% NaCl solution and then washing it out using distilled water.

Water treatment in batch ion exchange set-up was conducted in 2 dm^3 jar containing 1 dm^3 of examined solution and given resin dose (1, 2.5, 5, 10, 20, 30 or 40 cm^3). Jar content was mixed using magnetic stirrer (135 rpm) to assure equal resin bead distribution and keeping it in suspended state. Stirring for 5 to 60 minutes was followed by 30 min sedimentation. The temperature of solutions amounted to 20°C.

Ion exchange in flow set-up was performed in a 20 cm glass column, 13 mm of internal diameter, containing 15 cm^3 of resin. Examined solution flux velocity through resin bed was 15 cm^3·min^{-1}. Process was carried on for 180 to 300 min. pH and feed temperature influence on NOM removal efficiency was examined in 2–10 and 20–40°C ranges respectively.

Resin regeneration in batch set-up was carried similarly to resin preparation: exhausted resin was soaked in 200 cm^3 of 12% NaCl solution and stirred using magnetic stirrer for 30 min with the intensity keeping resin in suspended state. Afterwards, spent regenerant was decanted and resin was washed using distilled water until no chloride ion could be detected. Resin regeneration in flow mode was preceded by its backwashing with distilled water in order to loosen resin beads and remove gas bubbles. Than regenerant was passed through the resin bed for 120 min with flow velocity 6 cm^3·min^{-1}. Finally resin was washed-out with distilled water till no chlorides were detected in the effluent.

Regeneration efficiency was determined by comparing NOM removal for regenerated and fresh resin.

3 Results

Experiments conducted in order to assess polystyrene ion exchange resins usefulness in NOM removal from water allowed determination of resin dose, its type, contact time and process set-up influence on water treatment efficiency. Results presented in Figure 1 obtained for batch set-up shows that NOM removal efficiency from water improves together with prolonged contact time of resin with contaminants and with increasing resin dose. It was observed that ion exchange intensity (amount of NOM removed in an unit of time) was highest at the beginning of a process with further gradual slowdown. It was observed that optimal stirring time was 40–60 min. Prolonged stirring only slightly improved NOM removal efficiency after 60 min. Kinetic curves show (Figure 1) that UV 254 nm absorbance decreased with increasing resin dose (analogous results for colour intensity and DOC removal were recorded). Together with increased resin amount, more organic anions might be exchanged for chloride ions on resin active sites. Lower resin dose provides less active sites on resin beads of which only part is available for weak organic ions. For example, after stirring model solution 1 for 30 min with A100 resin dosed at 5 cm^3/dm^3 UV 254 nm absorbance decreased to 0.595 cm^{-1} while twice higher resin dose (10 cm^3/dm^3) resulted in absorbance removal to 0.482 cm^{-1}.

 DOI: 10.1051/e3sconf/20172200181

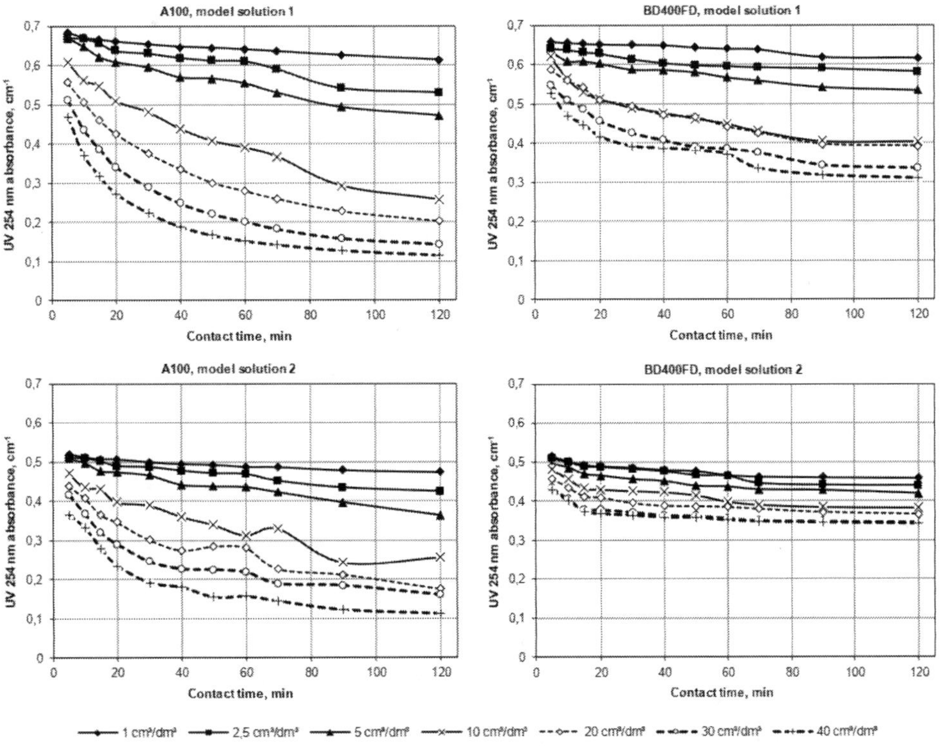

Fig. 1. NOM removal in batch set-up (UV 254 nm absorbance vs contact time).

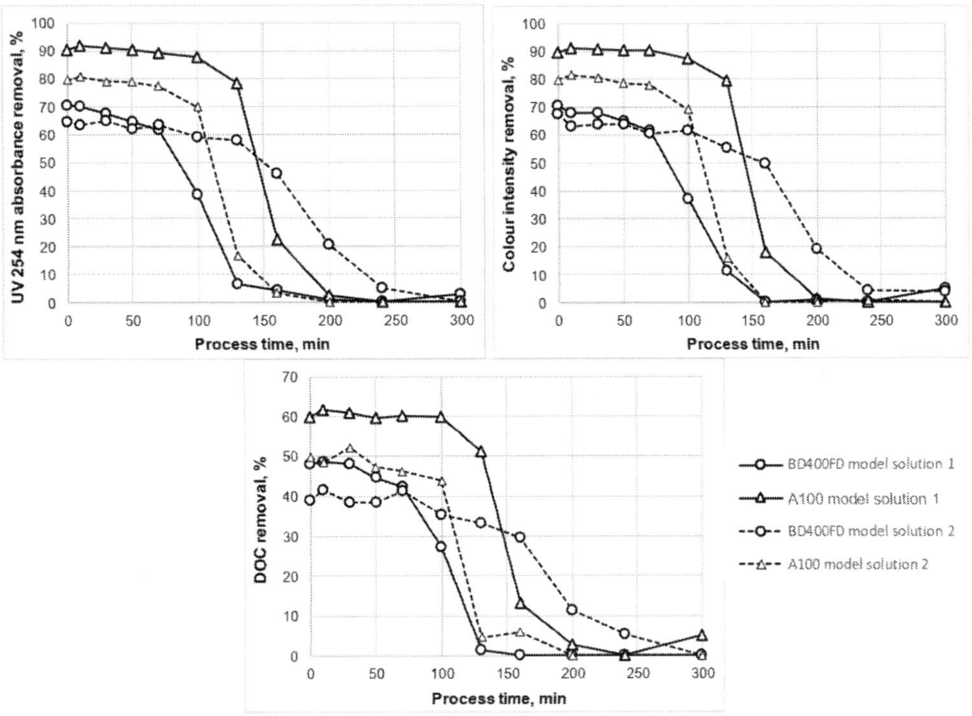

Fig. 2. NOM removal in flow set-up.

Results obtained for flow set-up experiments are shown in Figure 2 and are consistent in terms of nearly constant, low, colour intensity, UV 254 nm absorbance and DOC concentration levels in the first minutes of running process followed by increasing levels up to values equal to those measured for raw solutions. It is important to observe that none of the resins removed NOM in 100%. Moreover, NOM fractions responsible for UV 254 nm absorbance and colour were retained to the same extent, but DOC reduction was noticeably lower. For example for model solution 2, in the first 50 minutes of process organic substances responsible for colour intensity were removed in 79% and 64% of, ones measured as UV 254 nm absorbance in 79% and 62% of, while DOC was eliminated only in 47% and 38% with the use of the A100 and BD400FD resins, respectively.

Comparing the effects obtained in both investigated process set-ups it was observed that flow set-up performed better in terms of water treatment efficiency. It was due to the fact that in the batch mode feed water and resin mixing occurred until reaching equilibrium followed by decantation. Equilibrium state, in which NOM concentration is lower than in initial solution, limited resin ion exchange capacity utilization. In the flow set-up on the other hand at any time "fresh" feed solution contacted resin bed which allowed much better resin capacity utilization.

Resin type was another NOM removal decisive factor. Obtained results shown that A100 was the most effective resin which allowed to remove DOC, colour intensity and UV 254 nm removal up to 62%, 92% and 91% respectively in flow set-up while BDF400FD reached only 49%, 71% and 71%. Both examined resins are macroporous which facilitates penetration by micropollutants and retaining them inside resin structure. In such a situation sorption is decisive when it comes to NOM removal efficiency. Tested resins differ by basicity: A100 is weak base while BD400FD is strong base one. Obtained results are contrary to the literature by which weak base anion exchange resins should perform noticeably worse than strong base when it comes to NOM removal [15]. It was probably caused by BD400FD ion exchange capacity being lower than declared in product sheet.

NOM removal efficiency was also affected by composition of treated water. Obtained results shown that for both resins better treatment effect was achieved for model solution 1. Aldrich NOM (solution 2) which has larger NOM molecules [16], were removed in lower extent as compared to low molecular weight organic compounds typical for natural waters (solution 1). Moreover higher concentration of inorganic anions in solution 2 deteriorated organic anions uptake by resins.

Experiments on solution pH influence on NOM removal efficiency (Figure 3.) demonstrated that the increase of pH up to 10 deteriorated organic macro-anion removal. For example, treating model solution 1 in flow set-up with the use of BD400FD resin at pH 10 after 30 minutes of process resulted in 46% of colour intensity, 43% of UV 254 nm absorbance and 29% of DOC decrease, respectively, which is significantly less than for original solution (pH 4.84: 68% 67%, 48%). It can be assumed that alkalinisation of the feed solution increased concentration of inorganic anions competing with organic ions for resin active sites and furthermore progressing organic compound functional groups dissociation increased particles sizes thus impeding ion exchange.

 DOI: 10.1051/e3sconf/20172200181

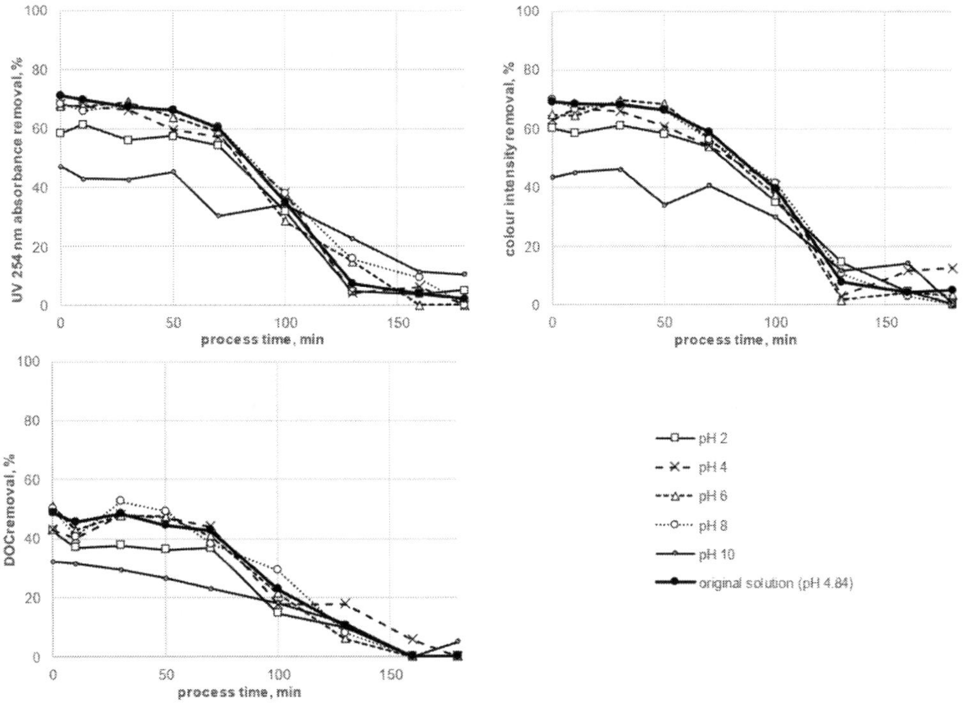

Fig. 3. Solution pH influence on NOM removal from model solution 1 by BD400FD resin in flow set-up.

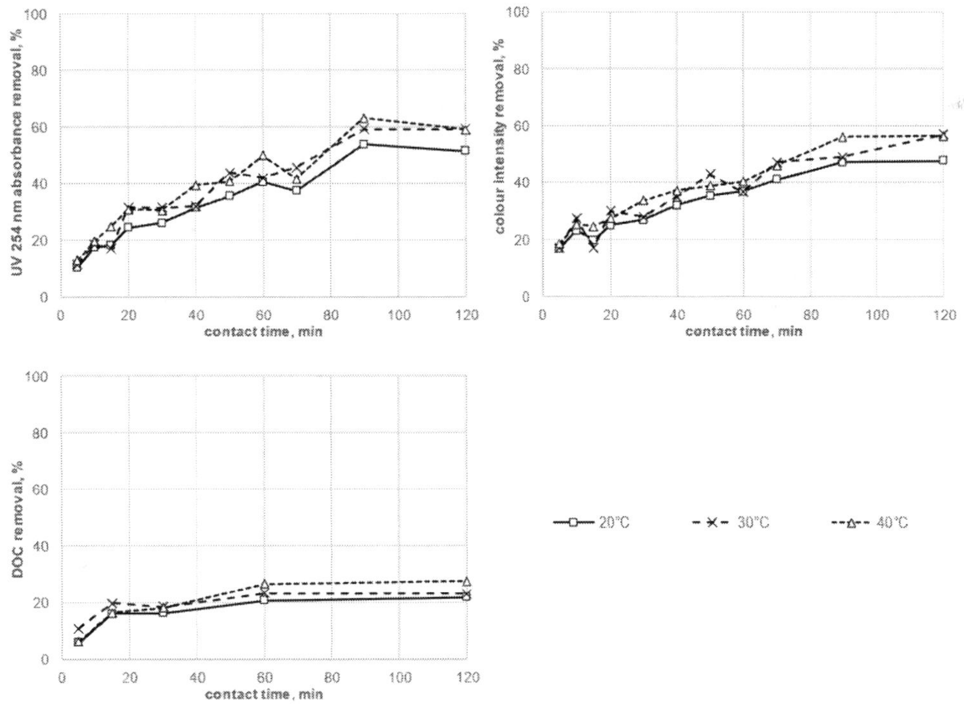

Fig. 4. Temperature influence on NOM removal from model solution 2 by A100 resin in flow set-up.

Results presented in Figure 4. shows that feed solution temperature also influenced ion exchange performance. The increase of temperature from 20°C to 40°C improved ion exchange efficiency which might be due to the fact that higher temperature accelerated exchanging ions movement, loosened resin structure (thus facilitated large ions diffusion into resin), increased functional groups dissociation and decreased pressure loss in resin bed.

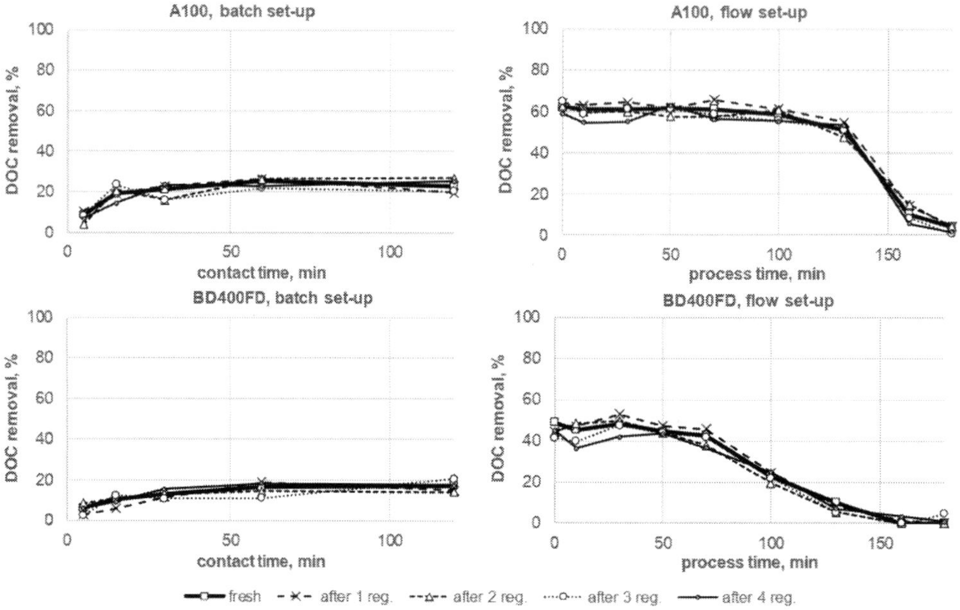

Fig. 5. DOC removal from model solution 1 by fresh and regenerated resins (set-up type indicates both regeneration and NOM removal efficiency tests).

Ion exchange resin regeneration experiments focused on restoring sorption capacity enabled comparison of two regeneration methods. It can be seen (Figure 5) that both flow and batch set-ups gave similar results (no significant difference in fresh and regenerated resin DOC removal efficiency), however the latter one required more regenerating agent, thus making it much less economically effective. In the case of batch tests, desorbed contaminants kept circulating in jar, competing with regenerating agent inorganic ions for resin active sites. On the contrary, in flow set-up desorbed NOM was immediately removed from bed while resin beds constantly contacted fresh regenerating agent.

4 Conclusions

Conducted experiments showed that ion exchange process with the use of synthetic polystyrene resins can be effectively used in NOM removal from water. Resin type and its dosed amount as well as contact time were found to be one of the decisive factors of process effectiveness. Longer contact time and higher resin dose resulted in better treated water quality. A100 was found to be more efficient. Comparison of process set-up influence on NOM removal proven flow set-up to be noticeably more effective. It was shown that physico-chemical properties of treated water were not without significance: the increase of solution pH slightly worsened process performance while the temperature rise improved water treatment efficiency. Moreover, resin regeneration studies revealed that both examined set-ups allowed for similar effect, but with different regenerating agent consumption.

 DOI: 10.1051/e3sconf/20172200181

References

1. A.M. Sillanpää, *Natural Organic Matter in Water: Characterization and Treatment Methods* (Butterworth-Heinemann, London, 2014)
2. A. Matilainen, M. Vepsäläinen, M. Sillanpää, Adv. Colloid Interface Sci. **159** (2010)
3. A. Matilainen, E.T. Gjessing, T. Lahtinen, L. Hed, A. Bhatnagar, M. Sillanpää, Chemosphere, **83** (2011)
4. J.A. Leenheer, J.-P. Croue, Environ. Sci. Technol. **37** (2003)
5. A. Grefte, M. Dignum, E.R. Cornelissen, L.C. Rietveld, Drink. Water Eng. Sci. **6** (2013)
6. J.-Q. Jiang, H.-Y. Wang, Sep. Sci. Technol. **44** (2009)
7. J.P. Croue, G.V. Korshin, M.M. Benjamin, *Characterization of Natural Organic Matter in Drinking Water* (American Water Works Association, 2000)
8. P.S. Bäuerlein, T.L. ter Laak, R.C.H.M. Hofman-Caris, P. de Voogt, S.T.J. Droge, Wat. Res. **46** (2012)
9. N. Ates, F.B. Incetan, Ind. Eng. Chem. Res. **52** (2013)
10. T.H. Boyer, P.C. Singer, Environ. Sci. Technol. **42** (2008)
11. B. Bolto, D. Dixon, R. Eldridge, React. Funct. Polym. **60** (2004)
12. A.A. Zagorodni, *Ion Exchange Materials: Properties and Applications* (Elsevier, 2007)
13. Purolite® A100 Product Data Sheet, http://www.lenntech.com/Data-sheets/Purolite-A100-L.pdf
14. BD400FD Anion exchange resin data sheet
15. E.R. Cornelissen, N. Moreau, W.G. Siegers, A.J. Abrahamse, L.C. Rietveld, A. Grefte, M. Dignum, G. Amy, L.P. Wessels, Wat. Res. **42** (2008)
16. S.A. Parsons, B. Jefferson, P. Jarvis, E. Sharp, D. Dixon, B. Bolto, P. Scales, *Treatment of Waters with Elevated Organic Content* (AWWA Pub., 2007)

DOI: 10.1051/e3sconf/20172200182

Intake power measurement as a criterion for control of HVAC systems

Juliusz Jerzy Walaszczyk[1,*], *Aleksandra* Przydróżna[2], and *Edward* Przydróżny[1]

[1]Wrocław University of Science and Technology, Department of Air Conditioning, Heating, Gas Engineering and Air Protection, ul. Norwida 4/6, Building C-6, 50-373 Wrocław, Poland
[2]Max Fordham, St Andrew's House, 59 St Andrew's St, CB2 3BZ, Cambridge, UK

Abstract. Under certain conditions, a HVAC system may consume less energy, provided that the initial response to the increasing heat gains is the increase in the airflow, while cooling is delayed. This ensures that the energy for cooling and reheat is reduced, while the fan power increases. Note that it is difficult to develop the following approach in the standard sequential control algorithms. On the other hand, the use of more complex algorithms than standard sequential algorithms is generally not encouraged because of the increase in the complexity of the control system and its resistance to respond to the varying parameters of a HVAC system. In this work, therefore, to avoid the following complications, the Intake Power Optimization algorithm is proposed. The Intake Power optimization algorithm is compared to the two sequential algorithms: Air conditioning and Airflow priority. To create the reference point enabling comparisons of the three strategies, the optimal control was established using the Nelder-Mead method. It is shown that the Intake power optimization algorithm provides a better control than the sequential algorithms and almost exclusively performs the optimal control actions. As an aside, the results indicate that the same heat gains, however, evenly distributed between rooms may contribute to the significant reduction of the energy demand.

1 Introduction

Supply Air Temperature (SAT) reset ensures the reduction of the energy demand for cooling and may decrease the energy demand for heating in reheat units. This, however, leads to the increase in the energy demand for air transport [1, 3]. The primary research problem is, therefore, the pursuit of the optimal air flow rates and the optimal supply air temperatures, ensuring that the benefits of reducing the energy demand for air treatment are higher than the losses caused by the increase in the energy demand for air transport.

Ke & Mumma (1997) present four methods of controlling the setpoint of the supply air temperature [4]: (i) as a function of the external air temperature; (ii) depending on the heat gains in rooms; (iii) as a function of the air humidity (as a supporting criterion); (iv) the optimal method in context of the energy demand. The majority of publications devoted to

* Corresponding author: juliusz.walaszczyk@pwr.edu.pl

the SAT reset control strategy describe setting the supply air temperature as a function of the external air temperature [3–6].

An optimal method of setting the supply air temperature is not obvious [5]. Endhal & Johansson (2004) compare different control strategies for setting the supply air temperatures to the optimal control. They show a relationship between the total energy consumption (for ventilation) and the supply air temperature. It appears that too low supply air temperature, relative to the optimal value, corresponds to a low risk of a substantial increase in the energy consumption. Simultaneously, too high supply air temperature, relative to the optimal value, may result in a substantial increase in the energy consumption [7]. Besides Ke & Mumma (1997) state that, the optimization of the primary flow rate and the supply air temperature provides better results in the temperate and dry climate [8].

Diverse SAT reset strategies may be compared, but the optimal control is ultimately of interest. A key research question is: how to select the optimal supply air temperature in the real system? In order to answer this research question, two fundamental issues need to be addressed. First of all, optimization of the supply air temperature often requires working with the model of installation, for instance in methods based on MPC (Model Predictive Control). Note that using a model of HVAC system and performing optimization during standard operation, significantly complicates control system. Second of all, the optimal control is set under defined conditions and usually with constant coefficients of performance. Under real conditions, however, many parameters, which were assumed constant during optimization, are variable, subject to conditions of operation.

1.1 Research idea (problem)

The use of more complex algorithms than standard sequential algorithms is generally not encouraged because of the increase in the complexity of the control system and its resistance to respond to the varying parameters of a HVAC system.

In this work, to avoid the following complications, the Intake Power Optimization algorithm is proposed. The algorithm, based on the instantaneous measurement of power intake, generates control actions based on the actual power intake. The approach presented herein avoids the complexity of standard approaches by omitting the necessity to determine the optimal airflow and the optimal supply air temperature. A precise measurement of the airflow, which remains difficult in the existing VAVs [1], is not required in the Intake Power optimization algorithm.

The Intake Power optimization algorithm is compared to the following two sequential algorithms: Air conditioning priority and Airflow priority. Moreover, in order to create a reference point to compare the three strategies to, the optimal control was established using Nelder-Mead method. It allowed estimating the limit of energy savings and establishing under which conditions algorithms provide the optimal control.

2 Methodology

All the control strategies were compared, based on the results of simulations performed. Each simulation lasted an hour and was repeated for different external air temperatures in the range 13 to 16.5°C with 0.5°C step change. Simulations were performed using the computational model of the VAV system, marked as NW3 (Fig. 1). The computational model was calibrated to the nominal parameters (extracted from the documentation for construction). Parameters of the model, which affect the operation in the steady state, were calibrated. Time constants were selected using the method of trial and error. Simulations do not account for air heating in fans.

 DOI: 10.1051/e3sconf/20172200182

2.1 Heat gains

The heat gains were related to the number of people within rooms. It was assumed that heat gains are constant in time and linearly dependent on the external air temperature. Two variants of heat gains were considered. The variant A comprised of heat gains that prevented use of reheat units and VAV boxes could be fully open. In other words, in variant A, the air treatment could be limited to the central unit only. In variant B, the sum of heat gains was equivalent to the heat gains in variant A. However, the breakdown of the heat gains within rooms varies, i.e. the room 1.27 has significantly higher heat loads than the two remaining rooms (Fig.1).

Note that the variant A is a special case. Wang et al. (2012) propose the optimal control for an air handling unit with a variable airflow, which services one room [1]. They specify zone 1, when there is a complete free-cooling, zone 2, when the airflow is equivalent to the critical value and zone 3, when the critical airflow is below the minimum flow rate. Using variant A allows investigating the presence of zones 1 to 3 within the modelled NW3 system, even though NW3 services more than one room (three rooms).

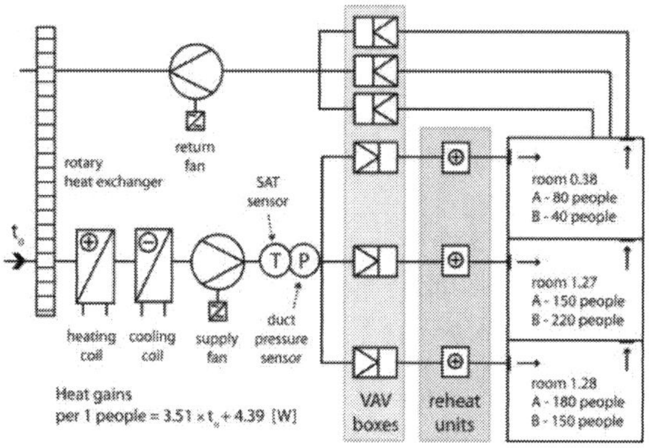

Fig. 1. Schematic of the VAV system NW3 modelled (computationally) in this work.

2.1 Total energy demand

All the control algorithms presented in this paper, were compared based on the total energy demand during one hour of operation. Air conditioning priority, Airflow priority, and optimal control actions were investigated under the system's steady state. Once the steady state is established, the total energy consumption may be calculated directly from the power demand (Eq. (1)).

$$E = P \cdot t \tag{1}$$

where: E – total energy demand [Wh], P – power demand at steady state [W], t – time [h].

The total power demand is a sum of the power demands of all the components of the VAV system. The fan power P_F is calculated using Eq. (2).

$$P_F = \Delta p \cdot V / \eta \tag{2}$$

where: Δp – fan pressure [Pa], V – volume flow rate [m^3/s], η – fan efficiency.

 DOI: 10.1051/e3sconf/20172200182

The thermal power of each air treatment process in the air handling unit is calculated using Eq. (3).

$$P_{term} = V \cdot \rho \cdot c_p \cdot \Delta T / \eta_{term} \quad (3)$$

where: ρ – air density [kg/m^3], c_p – specific heat [J/(kg·K)], ΔT – temperature difference at the inlet and outlet of the device [K], η_{term} – normalized thermal power coefficient.

The value of coefficient η_{term} is estimated from the thermal energy cost and the cost of equivalent electric energy (Eq. (4)). Normalization was performed in order to compare the electric energy and thermal energy.

$$\eta_{term} = \textit{electric energy cost / thermal energy cost} \quad (4)$$

3 Control algorithms

The control of fan speed, as a function of constant pressure, is not energy efficient. The setpoint is usually selected, based on the peak-load, even though the majority of operation is under lower loads [2]. For this reason, all algorithms (Air conditioning priority, Airflow priority and Intake power optimization) have the duct static pressure reset. Simulations were carried out with the assumption that at least one of the VAV dampers was almost fully open.

3.1 Air conditioning priority

The sequential control with the priority set on the air treatment, initially controls the efficiency of a cooling coil, in order to maintain the design conditions within rooms. When the supply air temperature is greater than or equal to the minimum supply air temperature, the control system regulates the operation of the cooling coil, while maintaining the minimum airflow. A further increase in the heat gains leads to maintaining the minimum supply temperature and controlling the volume flow rate. Fig. 2 shows examples of distribution of the control sequence. Each room has an individual PID regulator. The maximum signal from all the regulators controls the efficiency of devices in the air handling unit. The following approach allows cooling the air to the requirements of the room with the highest heat gains. The supply air temperature in the other rooms with lower heat loads may be adjusted in the local reheat units.

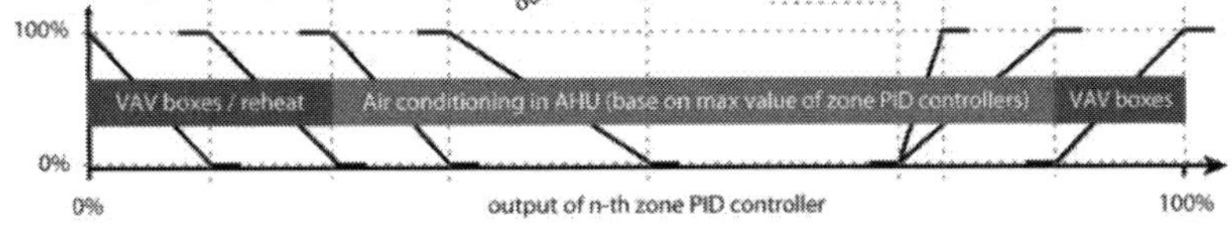

Fig. 2. Control sequences for Air conditioning priority strategy.

 DOI: 10.1051/e3sconf/20172200182

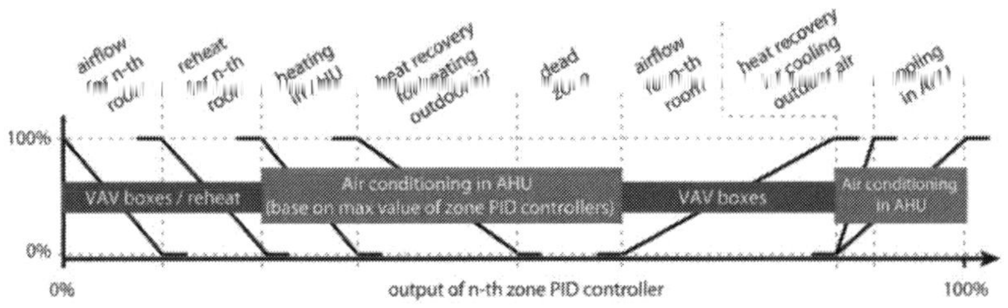

Fig. 3. Control sequences for Airflow priority strategy.

3.2 Airflow priority

The sequential control with priority set on the airflow, initially controls the volume flow rate in order to maintain the design parameters in rooms. When the maximum airflow is supplied to the room, the control system begins cooling in the air handling unit. The sequences of control in the Airflow priority algorithm are shown in Fig. 3.

3.3 Intake power optimization

The air temperature outside the air handling unit is set within limits of the acceptable supply air temperature, based on the instantaneous measurements of power intake in the ventilation system. The algorithm actions may be presented in four steps. Each step has its individual, constant time of operation. Step 1 (modify): the algorithm slightly modifies the setpoint temperature outside the central unit, for example by +0.1°C. Step 2 (wait): the algorithm waits for the system to operate in steady state (corresponding to the set values of the supply air temperature and the airflow). In this version of the algorithm reaching the steady state is not verified and step 2 is treated as a safety margin only. Step 3 (measuring power): the control system measures the instantaneous power intake and averages the value over set time. Step 4 (make decision): the algorithm makes a decision regarding the direction of changes. If the initial modification (+0.1°C) results in the decrease in power intake, the control system will try to introduce the same type of modification. If the initial modification results in the increase in power intake, the control system will reverse the direction of changes to -0.1°C.

The Intake power optimization strategy is characterized by the variable supply air temperature setpoint. The chiller power and the fan power vary as well. In case of the Intake power optimization strategy, the instantaneous power intake must be integrated with respect to the time.

DOI: 10.1051/e3sconf/20172200182

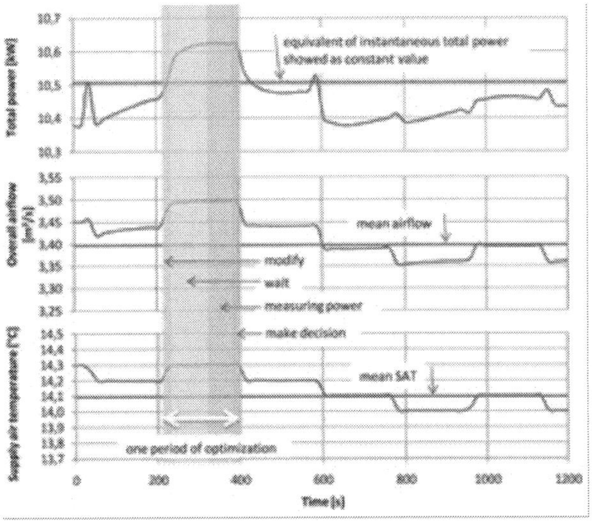

Fig. 4. Total power, overall airflow and SAT charts for Intake power optimization strategy.

3.4 Optimal control actions

The set of optimal control actions was used as a benchmark for the Intake power optimization and the sequential control strategies. Nelder-Mead simplex method, implemented in the Matlab function *fminsearch*, was used on the function P_{HVAC} (Eq. (5)).

$$P_{HVAC}\ (\Delta p_{SD};\ \Delta p_{RD};\ t_o;\ V_1;\ V_2;\ V_3) = P_{termCC} + P_{termR1} + P_{termR2} + P_{termR3} + P_{SF} + P_{RF} \qquad (5)$$

where: P_{HVAC} – total system power [W], Δp_{SD} (Δp_{RD}) – supply (return) duct static pressure [Pa], t_o – external air temperature [°C], V_1 (V_2, V_3) – supply airflow to the room 0.38 (1.27, 1.28) [m^3/s], P_{termCC} – cooling power [W], P_{termR1} (P_{termR2}, P_{termR3}) – reheat power for room 0.38 (1.27, 1.28) [W], P_{SF} (P_{RF}) – supply (return) fan power [W].

4 Results

4.1 Intake power optimization works in non-steady state

With constant heat gains, both the sequential algorithms (Air conditioning priority, Airflow priority) and the optimal control actions, which were determined during optimization, operate in the steady state. The proposed algorithm (Intake power optimization), however, does not operate in the steady state. This is caused by continuous changes in time of the supply air temperature and the airflow. As a result, the control signals fluctuate about the optimal control setpoints. Fig. 4. shows example results of simulation, where the external air temperature was set at 14.5°C. The duration of step 2 (wait) was selected with a safety margin, but at some instances, a HVAC system fails to reach a steady state. The results of simulation indicate that this should not be an issue. Similarly, the algorithm oscillates between the optimal control actions but with a greater standard deviation than the mean supply air temperature.

DOI: 10.1051/e3sconf/20172200182

4.2 Compare control strategy

The air conditioning priority ensures HVAC operation with the minimum airflow that allows assimilation of heat gains within rooms (upper graph in Fig. 5 - 6). As a result, this algorithm contributes to the maximum reduction of energy consumption by fans (Fig. 8 and 11). Based on the results of simulation, it appears that the Air conditioning priority algorithm has a greater total energy demand for the optimal control, particularly at lower external air temperatures. For instance, in variant B, at 14°C, optimal control actions are responsible for the energy consumption 20% lower than the Air conditioning priority (Fig. 12).

The Airflow priority algorithm ensures HVAC operation with relatively low energy consumption for air treatment (Fig. 7 and 10). The reduced energy demand for air treatment, however, corresponds to a very high energy demand for fans, compared to the other algorithms (Fig. 8 and 11). For instance, in variant A, at 16°C, the optimal control actions account for 80% less fan power compared to the Airflow priority (Fig. 8).

The Intake power optimization algorithm has a similar energy demand to the optimal control. Based on the following, the optimal control algorithm must have both a static pressure reset and SAT reset.

4.2 Different profiles of heat gains

There are two aspects worth discussing when considering variants of distribution of heat gains. First of all, there is an aspect of energy conservation. Note that the same heat gains, however, evenly distributed between rooms may contribute to the significant reduction of energy demand. For instance, in variant B, at 13°C, an hour of optimal control uses 8.8 kWh, whereas, in variant A, 2.7 kWh (31%). Similar trends are observed at different setpoint temperatures. For instance, at 16.5°C, an hour of optimal control uses 17.7 kWh in variant B and 12.2 kWh (69%) in variant A.

The second aspect corresponds to the comparability of the control algorithms. The more unevenly distributed heat gains between rooms are, the less apparent are the differences between the different algorithms. In variant A, the differences between the minimum and maximum total energy demand were in the range 0 to 11.6 kWh, whereas in variant B these were in the range 0.6 to 2.6 kWh.

DOI: 10.1051/e3sconf/20172200182

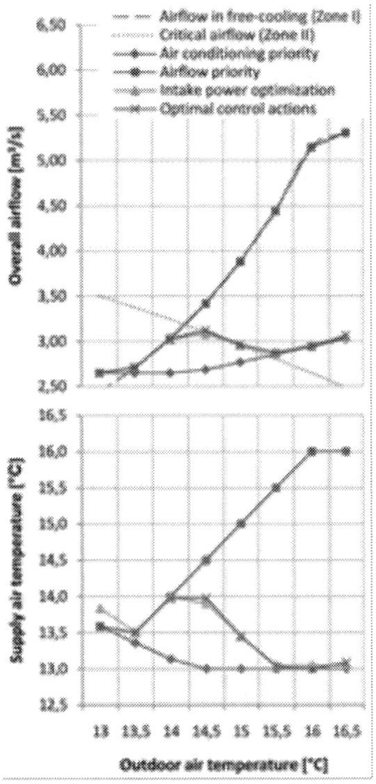

Fig. 5. Overall airflow and SAT by heat gains profile in variant A.

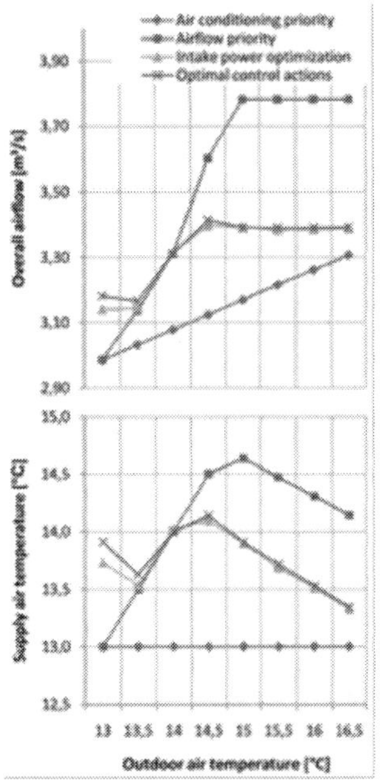

Fig. 6. Overall airflow and SAT by heat gains profile in variant B.

 DOI: 10.1051/e3sconf/20172200182

4.3 One room effect

In [1] Wang et al. (2012) used analytical methods to determine the optimal control for a system servicing one room. They showed relationships that allowed performing the optimal control actions during the economizer strategies. The analytical determination of the optimal control in the remaining modes of operation of ventilation allows to prepare a complete 'control table', which displays the optimal control under any conditions. It is shown that, in variant A, VAV system with three rooms is simplified to one room. As a result, the same control table may be used to determine the optimal control for a more complex system, consisting of more than one room. This, however, applies to the heat gains in variant A. Fig. 5 shows the total supply airflow and the supply air temperature in variant A. Additionally, the optimal airflows, determined analytically for one room (Zone I, Zone II) are marked in Fig. 5. The Intake power optimization algorithm ensures the optimal control because it `goes correctly' through Zone I and Zone II.

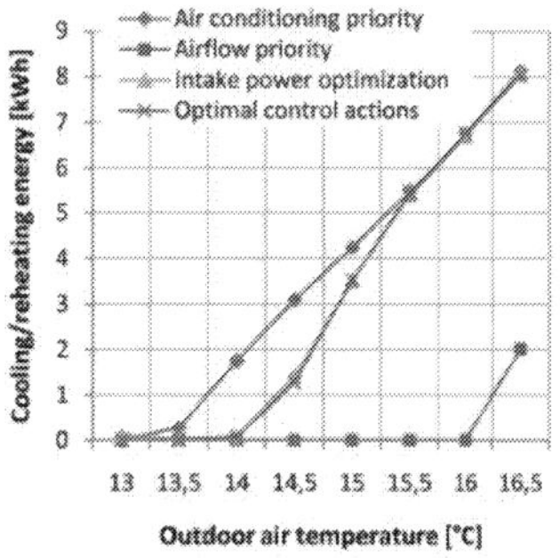

Fig. 7. Cooling/reheating energy in variant A.

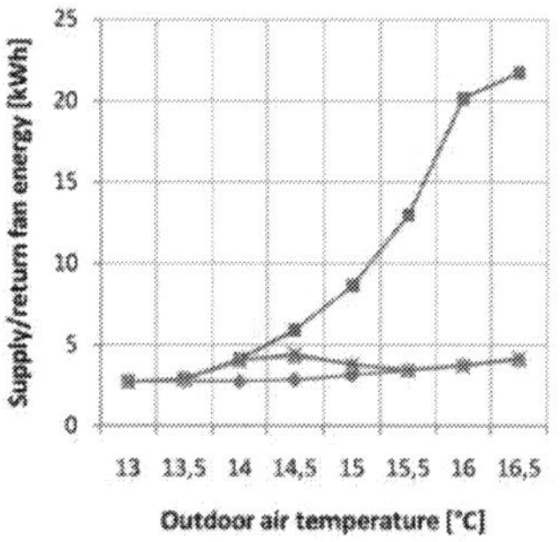

Fig. 8. Fans energy in variant A.

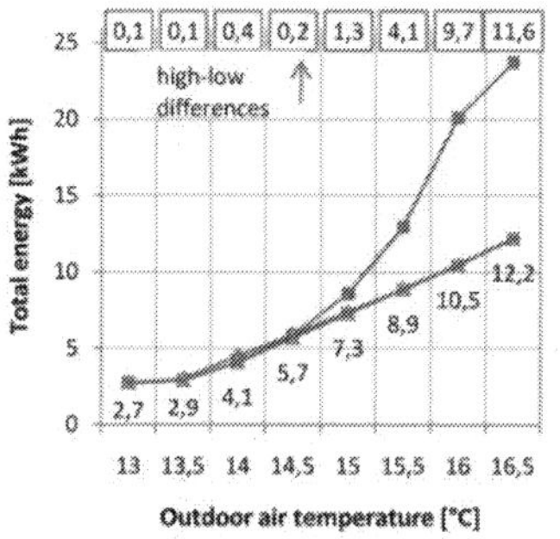

Fig. 9. Total energy in variant A.

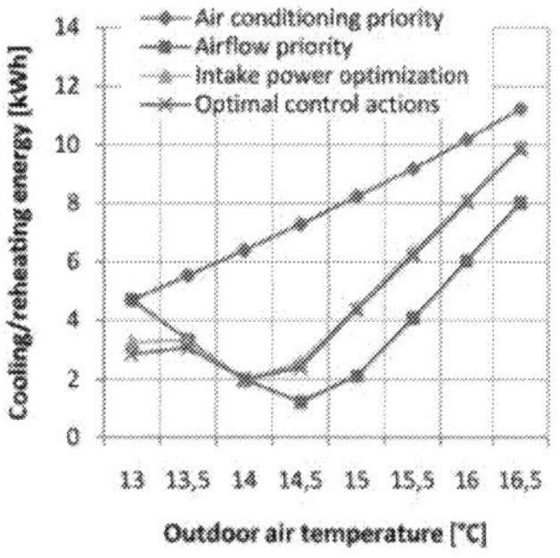

Fig. 10. Cooling/reheating energy in variant B.

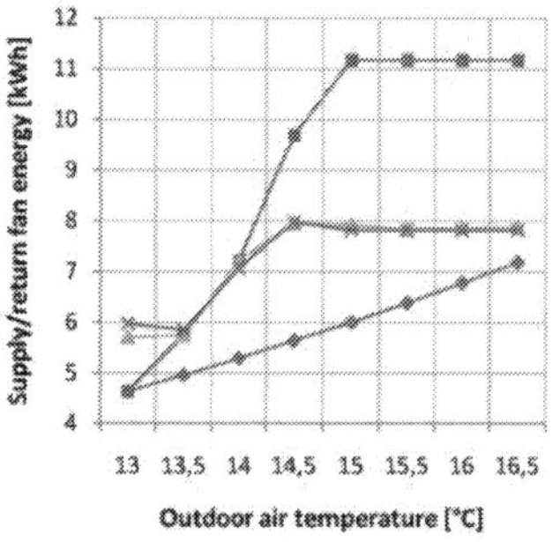

Fig. 11. Fans energy in variant B.

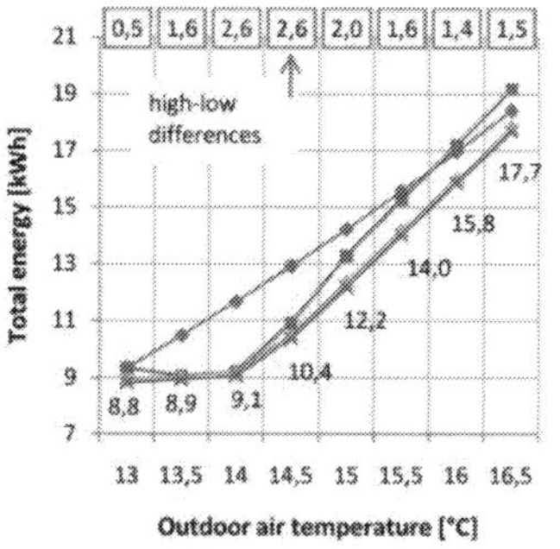

Fig. 12. Total energy in variant B.

It appears that, by calculating the optimal control for a system that services one room only, the lower limit of the energy demand of a more complex system may be established. This lower limit could be used as a criterion to evaluate the energy consumption of ventilation system in practice. If the control algorithm implemented controls the system closely to the optimum, then based on the calculated limit, the entire system can be evaluated. For instance, if the ventilation system operates consuming significantly more energy than the lower limit value, it may indicate that the system was designed for different profiles of heat gains. To avoid too high energy consumption, the following two actions could be undertaken, either the system could be modernised or the use of rooms could be modified to ensure more energy efficient use of the system.

DOI: 10.1051/e3sconf/20172200182

5 Conclusions

In the presented examples, the Intake power optimization algorithm provides a better control than the sequential algorithms and almost entirely performs optimal control actions. It may be deduced that implementing the duct static pressure reset and the SAT reset in the control system strategy enables the optimal control.

Implementing a SAT reset in a real ventilation system is not easy. Control strategies using the model are quite complex and sensitive to the assumed values of coefficients (e.g. efficiency). Therefore, the Intake power optimization strategy is based on measuring the actual power intake of the system. In this way, the complex issue of choosing the optimum control action was modified to the question of proper measurement.

Based on the simulations performed, the performance of Airflow priority algorithm was poor, particularly in variant A. Note that the increase in the supply air temperature to the maximum level is equivalent to the application of the Airflow priority algorithm. This effect would, thus, be negative. As a result, while implementing the SAT reset strategy, the margin of error should be set towards the lower temperatures. In other words, it is better to set the low setpoint supply air temperature than too high. Similar conclusions are drawn in [5,8].

Moreover, the performed simulations indicate that calculating the optimal control for the system servicing one room only may be used to determine the lower limit of energy demand for a more complex system. This limit may be used as a criterion to evaluate the energy consumption of ventilation system in practice. The lower limit allows assessing if the system is designed for the profiles of heat gains that occur most often and if the system is controlled optimally.

The research to follow should be focused on the implementation of the Intake power optimization algorithm in real buildings. First of all, the measurement of power intake could be based on the electric meter, provided that the devices installed are charged with electric energy only. The measurement of the total power intake by chillers operating cyclically and then translating it into the constant value in time could be challenging.

References

1. G. Wang, L. Song, Energy and Building **49**, 310–316 (2012)
2. G. Liu, M. Liu, ASHRAE Transactions **NY-08-055**, 451–457 (2008)
3. G.S. Okochi, Y. Yao, Renewable and Sustainable Energy Reviews **59**, 784–817 (2016)
4. Y. Ke, S.A. Mumma, Energy **22**, 601–614 (1997)
5. Energy Design Resources, *The advanced variable air volume (VAV) system design guide* (2009)
6. J. Murphy, Ashrae Journal, **October**, 18–28 (2011)
7. F. Engdahl, D. Johansson, Energy and Buildings **36**, 205–218 (2004)
8. Y. Ke, S.A. Mumma, International Journal of Energy Research **23**, 371–387 (1999)

DOI: 10.1051/e3sconf/20172200183

The effect of inflation rate on the cost of medical waste management system

Maria Jolanta Walery[1,*]

[1]Bialystok University of Technology, Faculty of Civil and Environmental Engineering, ul. Wiejska 45E, 15-351 Bialystok, Poland

Abstract. This paper describes the optimization study aimed to analyse the impact of the parameter describing the inflation rate on the cost of the system and its structure. The study was conducted on the example of the analysis of medical waste management system in north-eastern Poland, in the Podlaskie Province. The scope of operational research carried out under the optimization study was divided into two stages of optimization calculations with assumed technical and economic parameters of the system. In the first stage, the lowest cost of functioning of the analysed system was generated, whereas in the second one the influence of the input parameter of the system, i.e. the inflation rate on the economic efficiency index (E) and the spatial structure of the system was determined. With the assumed inflation rate in the range of 1.00 to 1.12, the highest cost of the system was achieved at the level of PLN 2022.20/t (increase of economic efficiency index E by ca. 27% in comparison with run 1, with inflation rate = 1.12).

1 Introduction

Optimization models can assist in ensuring that investment strategies will be economically feasible in the optimal decision-making in deciding their choice. Good decision-making is done based on experience, trial and error or comparison between different options of the integrated model of waste management [1–3].

New investments in waste management system are needed due to a number of factors including continuously increasing waste amounts, legal, political and economical demands for efficient utilization of the waste or energy production, and decommissioning of existing waste treatment facilities due to age and stricter environmental regulation [4, 5].

Various economic optimization models for waste treatment have been developed which focus on different parameters. Models focusing on transport, on energy production, have also been developed as well as models which take into account the economies of scale environmental impact, material recovery and social costs. Finally, models combining different criteria for selection of waste treatment methods in multi criteria analysis have been developed [6–8].

* Corresponding author: m.walery@pb.edu.pl

A thorough updated review of the existing models is presented and the main and the crucial parameters to take into account when assessing the economic performance of waste management system.

Yu et al. [9] conducted a study on optimization of the long-term performance of municipal solid waste management system in the form of a bi-targeted mathematical model. In this paper, a linear and dynamic programming bi-targeted model is proposed for decision-making on supporting long-term operations of municipal solid waste management system. The proposed mathematical model simultaneously deals with calculation of the economic productivity and environmental pollution from the municipal solid waste management system within several time periods. The proposed model has been calculated and solved by Lingo Software. The proposed model provided an effective solution for the long-term operational planning of the municipal solid waste management system [10].

Ravindra et al. [11] focused on the analysis of municipal solid waste management system and the methods to minimizing it in Chandigarh in India. The information was collected from the involvement of various public-private stakeholders through interviews and information of the registry related to transportation and solid waste disposal. This study has emphasized that the solid waste recycling system in this city is poor, where negligence of this important section has followed negative consequences. Thus, for environmentally-friendly solid waste management systems, a serious decision-making process and adjustment of operational activities are required. Further, based on system analysis approach the study proposes options for the improvement in current waste management practices in Chandigarh, which could also be adopted by other cities in developing world to reduce the adverse impact of solid waste management on environment and human health.

Kollikkathara et al. [12] presented a system dynamics approach designed for planning for sustainable municipal solid waste management. It attempts to address some of these issues by fitting a model framework for Newark urban region in the US, and running a forecast simulation. The dynamic system developed in this study incorporates the complexity of the waste generation and management process to some extent which is achieved through a combination of simpler sub-processes that are linked together to form a whole. The impact of decision options on the generation of waste in the city, on the remaining landfill capacity of the state, and on the economic cost or benefit actualized by different waste processing options are explored through this approach.

Shmelev [13] investigated three different approaches to the spatial-temporal analysis of the municipal solid waste management systems are used together, namely a life cycle inventory analysis, which helps to identify emission patterns within the municipal solid waste management systems, a multi-criteria optimisation approach, which helps to find compromise solutions among environmentally and economically preferred options, and a geographic information systems approach, which provides a tool for identifying waste management facilities, transportation environmental and social impacts; it also gives an analysis of environmental impacts on valuable ecosystems. The approach provides a new perspective for the analysis of municipal solid waste management systems on the regional scale. The principal novelty of the proposed complex the municipal solid waste strategic management model is an integration of the different types of data: geographical, environmental and economic, using relational database technology.

Soltani et al. [14] focused on the analysis of decision-making in Municipal Solid Waste Management (MSWM) problems such as finding proper waste treatment locations or strategies often requires multiple stakeholders such as government, municipalities, industries, experts, and/or general public to get involved. Multi-criteria Decision Analysis (MCDA) is the most popular framework employed in previous studies on MSWM; MCDA methods help multiple stakeholders evaluate the often conflicting criteria, communicate their different preferences, and rank or prioritize MSWM strategies to finally agree on some elements of these strategies and make an applicable decision. Results of this study show that AHP is the most common approach in consideration of multiple stakeholders and experts and governments/municipalities are the most common participants in these studies.

Shi et al. [15] investigated the a Mixed Integer Linear Programming model with minimizing costs for medical waste reverse logistics networks. The total costs for reverse logistics include transportation cost, fixed cost of opening the collecting centers and processing centers and operation cost at these facilities over finite planning horizons. An improved genetic algorithm method with a hybrid encoding rule is used to solve the proposed model. The efficiency and practicability of the proposed model is validated by an application to an illustrative example dealing with medical waste returned from some hospitals to a given manufacture.

Alshraideh, Qdais [16] focused on the analysis of medical system in Jordan. In this paper, a route scheduling model is proposed to minimize the total travel distance which in turn minimizes transportation cost and reduces emissions. The proposed model takes into account the capacity of trucks, number of visits per week, timing between visits along with the service level required by the hospitals. An optimal routing schedule was found and verified.

2 Research methodology

This paper uses optimization model of the export and disposal of municipal waste [3], as well as a computer program MRGO (Model of Regional Waste Management), which is its implementation. It has been verified by the author and adapted to the needs of the proposed model to optimize the removal and disposal of medical waste.

Medical waste due to the very diverse morphological composition and their potential infectious and toxic properties require a specific action depending on the degree of risk to humans and the environment. Medical waste management system is very different from municipal waste management system in the principles of collection of medical waste at source, storage, means and conditions of transport with regard to the provisions of the European Agreement ADR or processing systems and waste disposal. With this in mind the author has reviewed the structure of the mathematical model MRGO+[3] with a four-element system, and between themselves correlation to the system three-piece, in which the following system objects: source objects (hospitals), intermediate objects (medical waste incinerators) and final facilities located in the area of medical waste incinerator, i.e. the place for the temporary detention of post-process waste from waste incineration process. Such a solution in functioning of medical waste management enabled the achievement of the overall optimization of the system, relating to the effects and costs of servicing the region, not only in terms of local optimization, e.g. in a single *gmina* (PL administrative district). Therefore, it is extremely important to raise funds for the development of a cost-efficient and modern waste management system, in terms of technical and technological solutions as well as environmental protection requirements.

The structure of functioning of medical waste management system adopted by the author, was further verified in terms of the system functional division of transport operational tasks. The infectious medical waste management system does not include waste dumping mode, which normally is capable of a 2–3-fold compaction of waste in lorries of large capacity, minimum several m^3. As stated in the Act on Waste Management, infectious waste is not subject to segregation and composting due to its infectious properties, hence the medical waste management system is not composed of two-stage transportation and redistribution stations (waste processing and distribution to various sites of treatment). For this reason, the author verified multistage transport to the advantage of a single-stage, direct system and in this perspective the elements of the system (stages of activity) are as

follows:
- source forming regions (collection) of medical waste,
- route export of waste from the source areas of collection to an indirect facility in which there are waste treatment processes, independent or combined into a pre-technological and secondary process line,
- routes for post-processing waste disposal from the intermediate object to the end-object, where waste undergoes final treatment processes.

In addition, waste landfill storage - one of the methods of waste disposal commonly used in Poland is not applied in the infectious medical waste management system. Only post-process waste may be disposed of in landfills, and only marked as hazardous type. Therefore, the author introduced the concept of a "post-process thermal treatment waste temporary storage", i.e. adequately protected and separated area by the medical incineration plant, where post-treatment waste is temporarily stored for up to 3 years in accordance with Polish laws, and then exported to a hazardous waste landfill. Therefore, the costs of transport and storage of waste will be incurred once every three years, which was included in this stage of implementation of transport activities proposed by the author of the cost optimization model.

2.1 Case study

In this paper, the optimization model of the export and disposal of medical waste is presented in the dynamic version, taking into account expected changes of input and output parameters of the waste management system as well as its status in given periods of time.

A model region in north-eastern Poland, i.e. the Podlaskie Province, which can be considered representative for other regions, was chosen for the optimization research. The input data, necessary to describe the proposed variations in medical waste management system, has been collected and compiled in the course of analytical and factual study. Optimization study which was based on real data for both technical parameters and economic quantities, allows generalise the results and their implication for other similar regions.

The aim of the study was to analyse the impact of the parameter describing the degree of reduction in the amount of medical waste in the combustion process (wwp) and the unit cost of transport of medical waste (K_{ij}) on the value of the economic efficiency index (E). With the assumed technical and economical parameters of the system the operating range of tests, performed as part of the optimization study was divided into two stages. In the first stage, the lowest cost of operation of the system was calculated, while the second stage resulted in describing the impact of input parameters of the system, i.e. the degree of reduction in the amount of medical waste in the incineration of waste and the unit cost of transport of waste on the economic efficiency index and spatial structure of the system (set of system facilities and related transportation network).

18 sources of waste generation and accumulation (hospitals) within the studied area of the Podlaskie Province were selected for the analysis after taking into account the above mentioned assumptions and environmental conditions. The study also included: four intermediate objects (medical waste incinerators), respectively: IF1 (Suwalki), IF2 (Lomza), IF3 (Bialystok) and IF4 (Hajnowka), where pyrolytic decomposition process of waste will take place, and four end objects (respectively FF1, FF2, FF3, FF4) – areas for temporary storage of post-process waste from the incineration process located in the area of waste incineration facility. The model did not include restrictions on the capacity of intermediate and end objects.

The scope of operational research carried out in the framework of the optimization study was divided into successive stages in order to present options of the proposed model:

Stage I – includes optimization calculations, assuming fixed technical and economic parameters. Sequence 1, made in this stage, was also a comparative course – a benchmark for other solutions and obtained results to compare.

Stage II – included a number of additional runs aimed at determining the impact of the model input parameters of the system on the indicator of expenses of economic efficiency (E) and the spatial structure of the system (system location of objects and their associated waste disposal routes).

The following input data were taken into account:

- economic parameters describing the system (waste transport unitary costs, inflation and discount rate),
- economic parameters describing the objects of the system (capital and operating costs),
- the size reduction of medical waste in the system of indirect objects expressed in the form of the output factor of the process – wwp [%],
- the planned time horizon [t], (duration of model process).

The data relating to the costs of transport, investments and operation of the system objects, necessary for optimisation calculations, derived from existing plants, located in the model region. The calculation was performed by the unit cost of the work presented in Biedugnis et al. [3] taking into account the current prices and fees. The cost of medical waste removal from the source unit to the disposal site, with the adopted technical and operational conditions is K_{ij} = PLN 9.57, and when expressed in unit cost of 1 ton of transport per 1 minute (k) = PLN 1.33/t/min.

The economic efficiency calculations of the method were presented in the work by Biedugnis et al. [3] whose dynamic model related to inflation and discounting of the annual capital and operating costs in each model period. Transport costs are also discounted and adjusted for inflation.

Finding the best solution from the point of view of economic efficiency indicator was set as a priority. The above mentioned criterion takes into account both the selection of waste treatment technologies, as well as search for the best relationship between the location of objects used and the associated waste transport route network, depending on the amount of waste transported in each model periods.

3 Results and discussion

Calculations were carried out in the following runs:

Stage I - run 1 - the run like in the solution with the following parameters: duration of model period, respectively t1 = 5 and t2 = 15 years, the unitary cost of transportation of medical waste in the first and second model period, respectively, 1.33 and 0.44 PLN/t/min, the level of reduction of medical waste in the intermediate facilities expressed as a coefficient of the process output, wwp=10%.

Stage II – in runs 2–5 – the impact of changes in the inflation rate on the cost of the system and its spatial structure was examined. The inflation rate has a significant impact on the cost of operating the system and, consequently, the change in the economic efficiency index (E).

As a result of optimization calculations for the course 1 (Stage I) of the pre-established model system of the 26 facilities (18 – the source of the medical waste, 4 – incineration, 4 – storage of hazardous waste, 55 – possible routes for waste transport), the was a number of facilities selected in model periods I and II: 3/3 incinerators, 3/3 of the landfills and 21/21 waste transport routes, in consequence minimizing the cost of the system. Process levels in intermediate and final facilities in each model period for Stage I are presented in Table 1.

Table 1. The level of processing activities of intermediate and final objects for the 1st run [ton/year].

System facilities	Process	Processing activity level [t/year]	Duration of model studies I=5 years, II=15 years
IF1	incineration	140.400	I
IF1	incineration	148.800	II
IF2	incineration	210.400	I
IF2	incineration	222.400	II
IF3	incineration	434.400	I
IF3	incineration	450.900	II
FF1	storage	14.040	I
FF1	storage	14.880	II
FF2	storage	21.040	I
FF2	storage	22.240	II
FF3	storage	43.440	I
FF3	storage	45.090	II

For runs 2, 3 and 4 (Table 2) with the assumed inflation rate (1.00–1.08) – a scheme of location of system objects, quantity of transported waste and associated transport routes are the same as in the case of run 1.

The increase in the inflation rate in the range from 1.00 to 1.08 results in an increase in the economic efficiency index (E) by ca. 23% from PLN 1,390.60/t to PLN 1,707.20/t, with the assumed duration of the first and second model periods t1 = 5 years and t2 = 15 years and wwp = 10%.

In run 5, with the assumed inflation rate = 1.12, the transport system was changed in particular model periods. There were not only changes in the amount of waste transported to intermediate plants (incineration plants), but also the time (model period) in which these changes took place. Hence, there were two additional transport activities - transport routes of waste from the source of waste to the intermediate site, i.e. in the first model period the vast majority of waste from Augustow (16,300 t/year) was transported to the incinerator in Bialystok. The remaining amount of waste produced (6,200 t/year) was transferred to an incineration plant in Suwalki. In the second model period, the waste in total (23,600 t/year) was transported to the incinerator in Bialystok instead of to the incineration plant in Suwalki. Another additional transport activity occurred in the case of transportation of waste from Dabrowa Bialostocka, in the first and second model year, to the incineration plant in Bialystok and not to Suwalki, where the amount of transported waste did not change.

Table 2 presents the levels of processing activity in intermediate and final objects in the first and second model periods for run 5.

In run 5 (inflation rate = 1.12) the highest economic efficiency index was obtained at the level of 2022.20 PLN/t (increase of E by about 27% in relation to run 1).

Figure 1 shows the values of economic efficiency index E for runs 2–5 depending on the inflation rate. The increase in inflation leads to an exponential increase in E.

Table 2. The level of processing activities of intermediate and final objects for the 5nd run [ton/year] within the model area.

System facilities	Process	Processing activity level [t/year]	Duration of model studies I=5 years, II=15 years
IF1	incineration	116.900	I
IF1	incineration	116.900	II
IF2	incineration	210.400	I
IF2	incineration	222.400	II
IF3	incineration	457.900	I
IF3	incineration	482.800	II
FF1	storage	11.690	I
FF1	storage	11.690	II
FF2	storage	21.040	I
FF2	storage	22.240	II
FF3	storage	45.790	I
FF3	storage	48.280	II

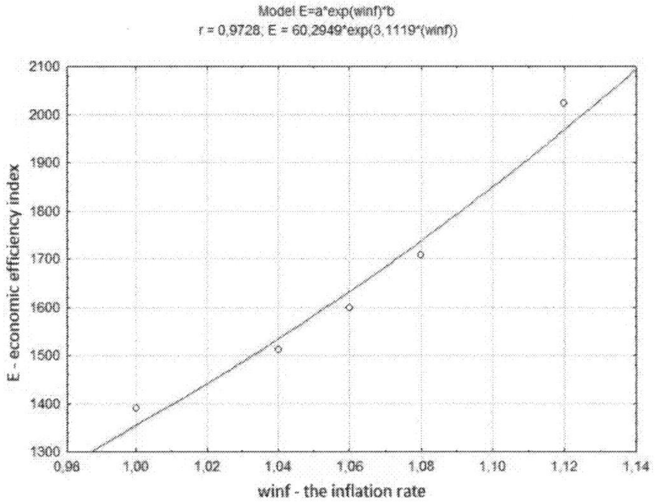

Fig. 1. Correlation of economic efficiency index (E) and inflation rate.

4 Conclusions

Taking into account the inflation rate causes a significant change in the system's operating costs and consequently a change in the economic efficiency index (E). It does not, however change the spatial structure of the system, i.e. the location scheme of the system objects, the quantity of waste transported and the associated waste transport routes. Only for the case with the assumed inflation index = 1.12, the pattern of transport activity changed in particular model periods. There were not only changes in the amount of waste transported to intermediate plants (incineration plants), but also the time (model period) in which these changes took place. With the assumed inflation rate in the range of 1.00 to 1.12, the highest cost of the system was achieved at the level of 2022.20 PLN / t (increase of economic efficiency index E by ca. 27% in comparison with run 1, and inflation rate = 1.12).

The increase in the inflation rate is accompanied by an increase in the economic efficiency ratio (E), which can be described by the following correlation:

$$E(winf) = 60.2949 \cdot e^{3.1119 \cdot (winf)} \text{ [PLN/t]} \quad (1)$$

The research was carried out within the scope of work No. S/WBIIŚ /2/14 and financed from the resources for science of Ministry of Science and Higher Education.

References

1. M. Chaerul, M. Tanaka, A.V. Shekdar, Waste Manage. **28** (2008)
2. N.O. Eriksson, M. Bisaillon, Waste Manage. **31** (2011)
3. S. Biedugnis, P. Podwójci, M. Smolarkiewicz, *Optimization of municipal waste management in micro and macro-region scale* (PAN, Warsaw, 2003)
4. A. Moreo, F.J. DeMicco, L. Xiong, J Qual. Assurance Hosp. Tour. **10**, 1 (2009)
5. A. Skordilis, Resour. Conserv. Recy. **41** (2004)
6. G. Ghiani, D. Laganà, E. Manni, R. Musmanno, D. Vigo, Comput. Oper. Res. **44** (2014)
7. N. Juul, M. Münster, H. Ravn, M.L. Söderman, Wastes: Solut., Treat. and Opp. (2011)
8. S.E. Shmelev, J.R. Powell, Ecol. Econ. **59**, 1 (2006)
9. H. Yu, W.D. Solvang, S. Li, J. Eng. Environ. **6**, 2 (2015)
10. M.A. Shirazi, R. Samieifard, A.M. Abduli, B. Omidvar, J. Environ. Health Sci. Eng. **14**, 8 (2016)
11. K. Ravindra, K. Kaur, S. Mor, J. Clean. Prod. **89** (2015)
12. N. Kollikkathara, H. Feng, D. Yu, Waste Manage. **30**, 11 (2010)
13. S.E. Shmelev, Ecol. Econ. 195–224 (2012)
14. A. Soltani, K. Hewage, B. Reza, R. Sadiq, Waste Manage. **35** (2015)
15. L. Shi, H. Fan, P. Gao, H. Zhang, Adv. Comput. Intelligence, 40–52 (2009)
16. H. Alshraideh, H.A. Qdais, J. Mater. Cycles Waste Manag. 1–11 (2016)

Selected heavy metals speciation in chemically stabilised sewage sludge

Ewa Wiśniowska[1], and *Maria* Włodarczyk-Makuła[1,*]

[1]Częstochowa University of Technology, Department of Chemistry, Water and Wastewater Technology, ul. Dąbrowskiego 69, 42-200 Częstochowa, Poland

Abstract. Selected heavy metals (Pb, Ni, Cd) were analysed in soil, digested sewage sludge as well as in the sludge stabilised with CaO or Fenton's reagent. The dose of Fenton's reagent was as follows: Fe^{2+} = 1g·L^{-1}, Fe^{2+}/H_2O_2=1:100; stabilisation lasted for 2 h. Dose of CaO was equal to 1 g CaO·g d.m.$^{-1}$ Total concentration of all metals in the digested sewage sludge was higher than in the soil. Chemical stabilisation of sludge with Fenton's reagent increased total metal content in the sludge as a result of total solids removal. Opposite effect was stated when the sludge was mixed with CaO. Also chemical fractions of heavy metals were identified (exchangeable, carbonate bound, iron oxides bound, organic and residual). The results indicate that stabilisation of the sludge with Fenton's reagent increased mobility of heavy metals compared to the digested sludge. Amendment of CaO increased percent share of examined metals in residual fraction, thus immobilised them and decreased their bioavailability.

1 Introduction

Digested sewage sludge contains nutrients such as nitrogen or phosphorus compounds which are essential for plant growth. Because of this it can be used in agriculture or land reclamation. The main problem with agricultural use of sewage sludge is contamination of this waste material with heavy metals [1]. This problem may be aggravated if the toxic metals are mobilized in the soil [2]. Total concentration of these contaminants in the sludge may vary between several mg·kg d.m.$^{-1}$ and several thousand mg·kg d.m.$^{-1}$ [1, 3, 4]. Before land application data on heavy metals concentration in the sludge are necessary to evaluate total concentration of these pollutants according to the limit concentrations (Table 1). Permissible heavy metals concentrations in sewage sludge are, however, not suitable for assessment of metals' mobility and potential toxicity. The most suitable method for these purposes is quantification of different chemical fractions of heavy metals in the sludge. Heavy metals occur in sewage sludge and soil in various forms, e.g. soluble, adsorbed, exchangeable, precipitated, complexed [5]. The most frequently used method of heavy metals fractionation – Tessier method – distinguish five fractions of heavy metals: exchangeable, carbonate bound, iron oxides bound, organic and residual [5]. Information on the extractability and fractionation of heavy metals in sewage sludge, especially the sludge chemically stabilized are insufficient. In Polish legislation, as well as in the EU one only

* Corresponding author: mwm@is.pcz.czest.pl

total concentration of heavy metals is limited (Table 1). No limits on individual fractions of heavy metals are set. However the knowledge on individual fractions of metals is important because they affect the bioavailability of individual compounds. It is extremely important in the case of highly toxic compounds such as Pb, Ni or Cd.

Table 1. Permissible total heavy metals concentration in sewage sludge according to the Polish and EU legislation (including proposals of changes in it).

Legal act	Limit concentration [mg·kg d.m.$^{-1}$]						
	Zn	**Cu**	**Ni**	**Cd**	**Pb**	**Hg**	**Cr**
Polish law [6]	2500[1)] 3500[2)] 5000[3)]	1000[1)] 1200[2)] 2000[3)]	300[1)] 400[2)] 500[3)]	20[1)] 25[2)] 50[3)]	750[1)] 1000[2)] 1500[3)]	16[1)] 20[2)] 25[3)]	500[1)] 1000[2)] 2500[3)]
European Union law [7]	2500–4000	1000–1750	300–400	20–40	750–1200	16–25	-
New proposal for a directive, 2000 [8]	2500	1000	300	10	750	10	1000
New proposal for a directive, 2010 [9]	2500	1000	300	10	500	10	1000

1) in agriculture

2) for not agricultural purposes

3) for other purposes permitted by law acts

Concentration of Pb in sewage sludge can reach even 370 mg·kg d.m.$^{-1}$. It is mainly present in the sludge in the residual fraction and adsorbed by Fe oxides and organic matter [3]. Thus it is not susceptible for release [5]. Intake of Pb in soil and sludge is also affected by carbonates. It's exchangeability in these materials is rather low. Nickel concentration in sewage sludge can reach even 4000 mg·kg d.m.$^{-1}$ [3]. The content of this metal in sewage sludge rarely is higher than set by law standards [5, 6], however it is more bioavailable for plants and even at allowable concentrations can be toxic for plants and living organisms. Ni is mainly bound by Fe oxides [3]. Cd is also strongly toxic. Its concentration in the sludge is usually not higher than 100 mg·kg d.m.$^{-1}$ [3]. This compound is however well soluble and it can be present practically in all fractions. It indicates that not only total concentration of heavy metals but also individual fractions are important during assessment of environmental risk of heavy metals [3, 5]. Chemical stabilisation is an interesting alternative to conventional stabilisation methods of sewage sludge. Usually advanced oxidation methods (e.g. Fenton's reagent, peracetic acid etc.) are used for this purpose. They can be used both for improvement of sewage sludge degradability (as pre-treatment method) [10] and for stabilisation of sewage sludge [11]. Because use of strong oxidants usually needs low pH removal of heavy metals from solid phase to reject water may occur [12]. As it was stated by Azhdarpoor et al. [13] leaching of heavy metals from sewage sludge is the most effective under pH of 2–3, at dose of hydrogen peroxide 3 g·L^{-1} and Fe^{2+} 2 g·L^{-1}. During 15 minutes about 80–100% of heavy metals can be leached from the sludge. Cd and Pb leachability was under these conditions comparable. In the studies done by Ito et al. [14] effective leaching of heavy metals from sewage sludge was observed at dose of H_2O_2 equal to 0.2 g·L^{-1} when Fe^{3+} was present in the reaction environment (at pH = 2.5). Use of H_2O_2 alone needed dose of this oxidant equal to 1–5 g·L^{-1}. Use of strong oxidants for sewage sludge is considered mainly as a tool for heavy metals leaching and also for improvement of sludge dewatering [12, 15]. However heavy metals leaching by oxidants is well documented, no detailed study on the changes of the metal fractions in the sewage sludge has been done.

The aim of the study was examination of soil, digested sludge and chemically stabilized (with Fenton's reagent and CaO) sewage sludge for total concentrations of Pb, Cd and Ni and chemical fractions (I-exchangeable, II-carbonates, III-iron oxides bound, IV-organic, V-residual) of the metals.

2 Materials and methods

Soil. Soil samples were collected from the plough layer (0–20 cm) from the farm placed near Częstochowa. They were air-dried and sieved through a 3 mm sieve. In the soil samples grain size analysis, pH in water, pH in KCl (potentiometrically), sorption capacity (by Kappen method), humidity, total suspended solids (TSS) (by owen dry-method, at 105°C), humus concentration (by Tiurin method) were primary analyzed. The measurements were done according to standard methods [16].

Sewage sludge. **Chemical stabilization.** Raw sewage sludge samples were taken from primary sedimentation tank of municipal wastewater treatment plant. Sewage sludge samples were passed through a 3 mm sieve. Sewage sludge was chemically stabilized with Fenton's reagent (Fe^{2+}/H_2O_2=1:100; dose of Fe^{2+} = 1g·L^{-1}) for 2 h, at 20±1°C. CaO was introduced into the sludge at the dose of 1 gCaO·g d.m.$^{-1}$.

Analysis of chemical properties. pH was analysed potentiometrically according to PN EN 12176. Total suspended solids in the sludge (TSS) was analysed according to the PN-EN 12880 :2004.

Heavy metals speciation. In order to study the forms of Pb, Cd and Ni in soil and sewage sludge the modified Tessier extraction procedure was used [5]. It consisted of five steps which give rise to five different fractions – Table 2. The extraction was carried out in glass centrifuge tubes of 50 mL capacity with an initial mass of 1 g oven dried (105°C) fine fraction (< 1mm) of the samples. Supernatants were analysed by atomic absorption spectroscopy. Analyses were performed in triplicate.

Table 2. Method of metal speciation in sewage sludge and soil [5].

Fraction	Form of metal	Parameters of fractionation	Time of extraction, h
I	Exchangeable	1.0 M CH_3COONH_4, pH = 7.0, T = 20°C	2.0
II	Carbonate bound	1.0 M CH_3COONa, pH = 5.0, T = 20°C	4.0
III	Iron oxides bound	0.04 M $NH_2OH{\cdot}HCl$ in 25% CH_3COOH (v/v), T = 95°C	4.0
IV	Organic	0.02 M HNO_3 + 30% H_2O_2 + 3.2 M CH_3COONH_2 in 20% HNO_3 (v/v), pH = 2.0, T = 85°C	5.5
V	Residual	10 M HNO_3 + 30% H_2O_2, T = 100°C	1.5

 DOI: 10.1051/e3sconf/20172200184

3 Results and discussion

Total concentrations of selected heavy metals in soil, digested sludge and chemically stabilized sewage sludge are listed in Table 3.

Table 3. Total heavy metals concentrations [mg·kg d.m.$^{-1}$] in soil, digested sewage sludge and chemically stabilized sewage sludge.

Metal	Digested sludge	Sludge stabilised with CaO	Sludge stabilised with Fenton's reagent	Soil
Pb	103.5 ± 11.2	63.0 ± 9.7	190.7 ± 15.1	31.8 ± 3.2
Cd	14.4 ± 1.3	13.0 ± 2.3	39.1 ± 19.1	4.3 ± 0.7
Ni	113.2 ± 18.2	64.1 ± 14.1	173.8 ± 22.9	19.3 ± 1.8

Total concentration of all examined heavy metals in digested sludge was higher than in soil. Concentrations of Pb and Cd were about three times higher, whereas Ni content in the sewage sludge was about 6 times higher than in the soil. The studied soil was qualified as sandy (sand light-clay). There was a light soil with the low humus concentration (1.5%), acid (pH KCl = 5,2), with the low sorption capacity (13.6 cmol(+)·kg^{-1}) and low humidity (24%). Stabilization of sludge with CaO decreased total content of heavy metals in sewage sludge. It was probably associated with the fact that amendment of CaO increased total solids concentration in the sample, whereas concentration of metals did not change. As a result apparent decrease of heavy metals concentration in sewage sludge occurred. Contrary to stabilization with CaO use of Fenton's reagent increased total concentration of examined heavy metals in sewage sludge. Pb concentration increased from 103.5 to 190.7 mg·kg d.m^{-1}. Concentration of Cd from 14.4 to 39.1 mg·kg d.m.$^{-1}$, whereas Ni from 113.2 to 173.8 mg·kg d.m^{-1}. During stabilisation of sewage sludge with Fenton's reagent decrease of total solids occurred. Because use of Fenton's regent did not affect concentration of heavy metals, but decreased TSS content, concentration of these compounds in the sludge after stabilisation apparently increased (Table 4). Thus it can be stated that fates of heavy metals in sewage sludge during chemical stabilisation are connected with the changes of total solids concentration. In all samples the less abundant contaminant was cadmium.

Table 4. Selected chemical properties of sewage sludge used in the study.

Parameter	Unit	Digested sewage sludge	Sludge stabilised with CaO	Sludge stabilised with Fenton's reagent
pH	-	6.9	10.8	4.5
TSS	gL^{-1}	215.1	389.2	189.8
VSS	%	48.6	32.1	42.5

Chemical stabilisation of the sludge also affected chemical fractions of individual heavy metals. Fractions of Pb, Ni and Cd in sewage sludge, chemically stabilised (with CaO and Fenton's reagent) sewage sludge compared to fractions in soil are presented in Figures 1–3. In relation to Pb, the series followed for soil, digested sewage sludge, sewage sludge stabilised with Fenton's reagent, sludge stabilized with CaO were:
Carbonate bound > organic > iron oxides bound > residual > exchangeable,
Carbonate bound > organic > iron oxides bound > residual > exchangeable,

Carbonate bound > iron oxides bound > exchangeable > organic > residual
Residual > organic > carbonate bound > iron oxides bound > exchangeable.

According to the data presented by Wilk and Gworek [3] Pb in sewage sludge is mainly adsorbed by Fe oxides and organic compounds. It is also present in residual fraction. During the study we have found that Pb was not present in exchangeable fraction of sewage sludge, and the dominant fraction of Pb in digested sewage sludge was the carbonate bound one.

For the Cd the series were followed:
Residual > iron oxides bound > carbonate bound > organic = exchangeable,
Residual > iron oxides bound > exchangeable > organic > carbonate bound,
Residual > iron oxides bound > exchangeable > carbonate bound > organic,
Residual > iron oxides bound > exchangeable > organic > carbonate bound.
Compared to the results obtained by other authors [3] it can be stated that high content of Cd in residual fraction is typical. By the mentioned authors this metal compound percent share is usually the less in exchangeable fraction, whereas in carbonate bound one is usually relatively high. In this study percent share of exchangeable fraction of Cd in the sludge was higher than carbonate bound one. This was probably associated with the diversity of sewage sludge between various wastewater treatment plants.

In the case of Ni the series were:
Carbonate bound > residual = exchangeable > organic > iron oxides bound,
Iron oxides bound > organic > exchangeable > residual > carbonate bound,
Iron oxides bound > exchangeable > carbonate bound > organic > residual,
Residual > iron oxides bound, exchangeable> organic> carbonate bound.
The results obtained in the study for sewage sludge are typical for sewage sludge. As it was stated by Wilk and Gworek [3] Ni in sewage sludge is mainly present in connections with Fe and Mn oxides and organic fraction, as well as in the residual one.

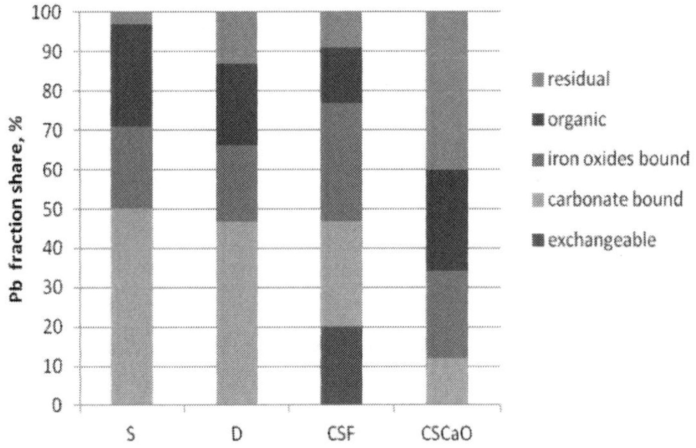

Fig. 1. Speciation of Pb in examined samples: S – soil, D – digested sludge, CSF – sludge stabilized with Fenton's reagent, CSCaO – sludge stabilised with CaO.

DOI: 10.1051/e3sconf/20172200184

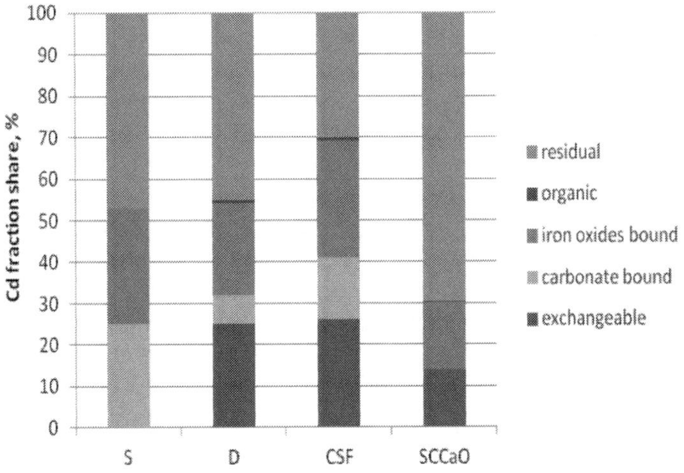

Fig. 2. Speciation of Cd in examined samples: S – soil, D – digested sludge, CSF – sludge stabilized with Fenton's reagent, CSCaO – sludge stabilised with CaO.

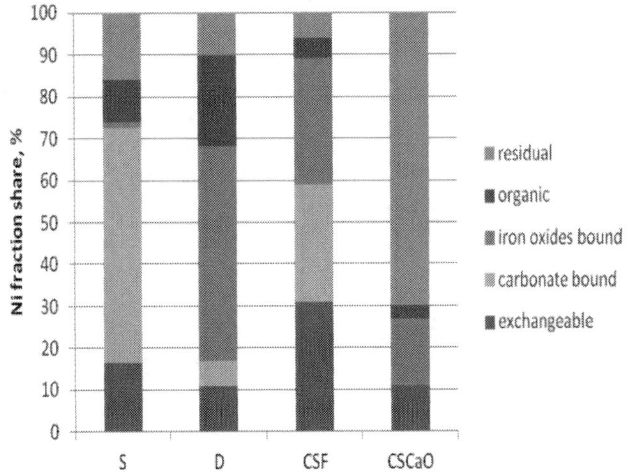

Fig. 3. Speciation of Ni in examined samples: S – soil, D – digested sludge, CSF – sludge stabilized with Fenton's reagent, CSCaO – sludge stabilised with CaO.

Chemical fractions of heavy metals in sewage sludge can be affected by origin of sewage sludge, pH, organic matter content [3]. It is the reason for the differences observed between the results obtained in this study compared to the ones observed by other authors [3, 5].

As can be seen from Fig. 1–3 amendment of CaO increased residual fraction of examined heavy metals in the sludge both compared to the soil and digested sewage sludge. The increase was especially visible in the case of Pb and Ni. Amendment of Ca simultaneously decreased exchangeable fraction of the sludge. Whereas Fenton's reagent increased this fraction percent share in the case of all metals. Percent share of exchangeable fraction was the lowest in the soil (only Ni was present in this fraction in soil). Ni in soil was present mainly as carbonate bound compounds (> 50%). Fraction of iron oxides bound Ni was negligible. Over 10% were residual and exchangeable fractions. Less than 10% were the organic compounds of Ni (Fig. 3). Percent shares of Ni in the digested sewage sludge differed from the ones in soil. The dominant fraction were iron oxides bound and organic. It can be connected with the fact that during sewage sludge treatment ferrous salts are used as coagulants, and the flocks which are formed during the process catch the Ni ions on its surface. Reach organic fraction of the sludge also can adsorb heavy metals present in the sludge. This fraction is significantly less in the soil. As it was mentioned above chemical stabilisation of sewage sludge with Fenton's reagent increased exchangeable fractions of Ni in the sludge. Simultaneously decrease of Ni content in residual and organic fraction occurred. It could be connected with the destruction of solids during chemical stabilisation of the sludge. During 2 h chemical stabilisation of sewage sludge with Fenton's reagent decrease of total suspended solids by about 12% occurred. It doubtless moved persistent compounds of Ni bounded to suspended solids. As a result increase of exchangeable fraction occurred simultaneously with the decrease of Ni content in residual and organic fractions. Amendment of CaO immobilised Ni in the sludge which was with no doubt connected with pH increase (Table 4). Simultaneously percent share of exchangeable fraction of Ni decreased from more than 20% to less than 15%. Iron oxides bound fraction of Ni decreased slightly compared to digested sewage sludge. The residual fraction of Ni stayed the dominant one in the sludge amended with CaO.

Similar phenomenon was observed in the case of Cd (Figure 2). Amendment of CaO also caused the residual fraction of this metal as dominant. Simultaneously percent shares of exchangeable and iron oxides bound fractions of Cd decreased compared to the digested sludge. Carbonate bound fraction of Cd practically did not exist in sludge stabilised with CaO. Contrary to the stabilisation with CaO treatment of the sludge with Fenton's reagent slightly decreased exchangeable fraction of Cd in the sludge. However it must be emphasised that Cd in sewage sludge was more exchangeable in all samples of the sludge than in soil. Percent share of residual fraction of Cd was however comparable both in soil and sewage sludge, in the last sample content of carbonate bound Cd was lower.

Also in the case of Pb (Fig. 1) treatment of sludge with Fenton's reagent increased the mobility of the examined heavy metals. Chemical stabilisation of the sludge with Fenton's reagent also decreased carbonate bound fraction of Pb. Simultaneously percent share of iron oxides bound fraction of Pb increased. It is connected with the changes of sludge properties during stabilisation with strong oxidants under low pH values (pH of chemically stabilised sludge was equal to 4.5). Introduction of Fe (as catalyst in Fenton's reaction) could enrich iron oxides bound fraction compared to the digested sludge. Destruction of residual and organic fractions caused release of not bioavailable Pb to the supernatant and to other fractions of metals, particularly to the exchangeable one.

As in the case of Cd and Ni also Pb percent share in residual fraction increased as a result of CaO stabilisation. Organic fraction was comparable to the one observed in the sludge before chemical stabilisation.

Based on the results obtained during the study it can be stated that use of strong oxidant mobilises heavy metals in sewage sludge, probably as a result of suspended solids destruction as well as a result of pH drop. Contrary to this treatment method amendment of CaO decreases the mobility of heavy metals in sewage sludge, simultaneously decrease apparent decrease of total heavy metals concentration occurs.

4 Conclusions

1. Heavy metals concentrations in chemically stabilized sewage sludge was higher than permissible limits for this kind of waste material established by Polish law standards. Concentration of heavy metals in the sludge was higher than in soil.

2. Amendment of sewage sludge with CaO decreased total concentration of heavy metals in the sludge, as well as decreased their mobility. The dominant chemical fraction of metals in the sludge amended with CaO was the residual one.
3. Chemical stabilization of sewage sludge by Fenton's reagent increased total concentration of heavy metals in sewage sludge. It also mobilised them by increase of percent share of heavy metals in exchangeable fraction.

Funding for this project was provided by BS-PB-402-301/11

References

1. J. Bernacka, L. Pawłowska, A. Krobski, Publishing House of Institute of Environmental Protection, Warsaw (2002) (in Polish)
2. A. Fuentez, M. Loréns, J. Sáez, M. Aguilar, J. Ortuño, V. Mesequer, Bioresour. Technol. **99** (2008)
3. M. Wilk, B. Gworek, Environ. **39** (2009) (in Polish)
4. E. Wisniowska, M. Włodarczyk-Makuła, R. Nowak, Rocz. Ochr. Sr. **18**, 2 (2016)
5. L. Dąbrowska, Chem. Pap. **66**, 3 (2012)
6. Ordinance of the Minister of the Environment on sewage sludge, 06. 02. 2015
7. Council Directive 86/278/EEC of 12 June 1986 on the protection of the environment, and in particular of the soil, when sewage sludge is used in agriculture
8. Working document on sludge 3rd draft, Brussels 27 April 2000
9. Working document on sludge and biowaste, Brussels 21. 09. 2010
10. I. Zawieja, L. Wolny, M. Próba M., Environ. Res. **156** (2017)
11. K. Barbusiński, K. Filipek, Pol. J. Environ. Stud. **9**, 3 (2000)
12. M. Janosz-Rajczyk, E. Wiśniowska, Chem. Pap. **59**, 6b (2005)
13. A. Azhdarpoor, R. Hoseini, M. Dehghani, JHSSS **3**, 4 (2015)
14. A. Ito, K. Takahashi, J. Suzuki, T. Umita, J. Water Environ. Technol. **11**, 4 (2013)
15. R. Devil, J. Baeyens, E. Neyens, Environ. Eng. Sci. **23**, 6 (2006)
16. A. Mocek, S. Drzymała, Publishing House of University of Life Science, Poznań (2010)

Absorption cooling sources atmospheric emissions decrease by implementation of simple algorithm for limiting temperature of cooling water

Krzysztof Wojdyga[1,*], and *Marcin* Malicki[2]

[1]Faculty of Building Services, Hydro and Environmental Engineering of Warsaw University of Technology, ul. Nowowiejska 20, 00-653 Warsaw, Poland
[2]New Energy Transfer, ul. Domowa 6, 02-913 Warsaw, Poland

Abstract. Constant strive to improve the energy efficiency forces carrying out activities aimed at reduction of energy consumption hence decreasing amount of contamination emissions to atmosphere. Cooling demand, both for air-conditioning and process cooling, plays an increasingly important role in the balance of Polish electricity generation and distribution system in summer. During recent years' demand for electricity during summer months has been steadily and significantly increasing leading to deficits of energy availability during particularly hot periods. This causes growing importance and interest in trigeneration power generation sources and heat recovery systems producing chilled water. Key component of such system is thermally driven chiller, mostly absorption, based on lithium-bromide and water mixture. Absorption cooling systems also exist in Poland as stand-alone systems, supplied with heating from various sources, generated solely for them or recovered as waste or useless energy. The publication presents a simple algorithm, designed to reduce the amount of heat for the supply of absorption chillers producing chilled water for the purposes of air conditioning by reducing the temperature of the cooling water, and its impact on decreasing emissions of harmful substances into the atmosphere. Scale of environmental advantages has been rated for specific sources what enabled evaluation and estimation of simple algorithm implementation to sources existing nationally.

1 Introduction

The strategic objective of Poland's energy and environmental policy is to improve energy efficiency in the area of energy production, supply and use [1]. The energy efficiency target is to reduce energy consumption by 20% compared to the 2020 projections [2]. Energy

* Corresponding author: Krzysztof.Wojdyga@is.pw.edu.pl

efficiency issues are at the heart of the European Union, considered to be the most effective method of reducing primary energy consumption, leading directly to reducing emissions to the atmosphere and, consequently, improving air quality. An essential element of energy efficiency improvement is the maximizing usage of heat that has so far been discharged into the atmosphere and convert it into useful energy. For example waste heat can be used for the production of cold. In the world, cooling systems are responsible for from 16% to 50% of electricity consumption [3], with more than 40% share of energy consumption in the residential and service sector [4]. This leads directly to reaching peak demand for electricity in the summer, when the power of the national electricity system is the smallest. This situation can cause local power deficits, examples of which have appeared more than once in Poland. In spite of its efforts in the area of energy efficiency, Poland still uses 3.4 times more energy to produce goods of the same value than Denmark's most energy-efficient economy and 2.2 times more than the EU average [5]. Dominant, in improving energy efficiency, actions focus is on increasing the share of high-efficiency combined heat and power (cogeneration) units that work both for central and local systems. Their part in total energy production should grow in the perspective of a planned 29% increase in energy consumption by 2030 compared to the 2006 level [6], as a technology capable of achieving efficiency of more than 90% [7], which is significantly higher than in the case of the split generation [8] and contributing to the reduction of CO_2 emissions by supplied with, for example, gas [9] and energy production in the place where it is needed, further reducing transport losses [10]. High-efficiency cogeneration is promoted both in terms of investment and exploitation by means of sectoral, regional and national programs. The technology of combined heat, power and cooling (trigeneration) is particularly important here, enabling the production of all necessary media with the highest efficiency – with the lowest consumption of primary energy.

Over the past decade, due to rising energy carriers prices, environmental awareness, introduced bans on the use of hydrofluorocarbon and chlorofluorocarbon refrigerants combined with increased emphasis on energy efficiency and reduction of emissions, the absorption refrigeration technology returned to the dynamic growth retained in previous years by low prices of electricity. Unlike conventional compressor chillers, absorption chillers can be supplied with low quality energy, which can also be obtained from renewable sources, such as solar radiation, geothermal or industrial waste heat. Additionally, refrigerants used in the absorption chillers were found to be environmentally friendly, as opposed to the refrigerants used in conventional electric compressor chillers [11]. The efficiency of the absorption chillers, understood as the ratio of the useful cooling effect to the energy supplying the unit, is lower than the conventional compressor equipment, but the possibility of supplying them with unused or waste heat results in the primary energy consumption being less than conventional solutions [12]. This has also led to speculation about the design of cooling sources that include not only compressor or absorption aggregates, but also hybrid systems that combine compressor units with absorption units in a single installation [13].

2 Model

In absorption chillers, the energy is supplied in the form of heat from the combustion of fuel (e.g. gas, biogas, oil [14]), hot water (e.g. from the district heating network, useless/waste heat from a cogeneration or technological process, solar collectors), steam (e.g. from technological processes, or specially made for powering the appliance) or electrical energy (e.g. by means of a heater). The chilled medium most often is water, its anti-freeze mixture, air or other heat transfer medium. The absorption chiller operates on the principle of using the absorption effect (refrigerant absorption) and the desorption (separation of the refrigerant from the solution). Boiling refrigerant absorbs heat to provide a useful cooling effect [15]. The absorber and desorber system in the absorption chiller is called a chemical compressor and corresponds to the functionality of an electrically powered compressor in conventional chillers.

The efficiency of the absorption chiller (ζ) is defined as the ratio of cooling capacity (Qc) to desorption heat (Qd) at constant temperature of desorption (Td) and evaporation (Tp). The numerical value of ζ is easy to calculate because it depends on the temperature. Based on steady evaporation and desorption temperatures, the efficiency of the absorption chiller circuit is increased when the condensation temperature is lowered. This temperature is mainly dependent on the cooling water temperature (in most commercially available units cooling water flows through the condenser and the absorber in parallel to provide the same cooling water temperature for both exchangers).

In order to assess the effect of decreasing the cooling water temperature, a model of the influence of the cooling water temperature on the consumption of hot water of the lithium-bromide absorption chiller [16] was developed. Hot water consumption for all working points have been gathered from chiller manufacturer (for constant and decreased water temperature), validated by experiment, and reflected in Microsoft Excel spreadsheet. Afterword's water consumption data has been combined with the data collected, at hourly intervals, from the actual operation of the air-conditioning of the office building. The result of the complete model operation is the hourly hot water requirement for the lithium-bromide absorption chiller working for the office building's air conditioning needs in the annual cycle with standard and decreased water temperature. Under Polish climate conditions, the peak demand for air-conditioning occurs from June to August. This is due to the occurrence of the highest values of annual temperature and the highest values of solar radiation in this period. In addition to monthly changes, the demand for cooling also changes hourly or daily, putting the cooling source in constantly variable operational conditions - most of the time significantly lower than nominal. In practice, the demand for air-conditioning cooling depends directly on the outdoor temperature and solar radiation. This is also a key factor affecting the temperature of the outlet water from the cooling system (e.g. cooling towers or drycoolers). When the cooling demand for air conditioning (i.e. the absorption chiller) is reduced, the cooling systems can provide a lower than the nominal cooling water temperature. This leads to an increase in cooling efficiency at partial loads, and consequently to a reduction in heat demand when the cooling water temperature is reduced. This relation is directly connected with lowering the cooling water temperature [17]. In order to assess the full environmental effect, CO_2, SO_x/SO_2, NO_x/NO_2 and total suspended particle emissions were calculated for a absorption chiller supplied with heat from a water boiler of less than 5 MW capacity using coal as a fuel. For this, procedure and adequate emission factors for the combustion of fuels made available by the National Center for Balancing and Emissions Management were carried out for each of the compounds. The effect of the model is also the comparison of the heat consumption values for the absorption chiller. Figure 1 presents hot water usage for chiller working with and without algorithm decreasing cooling water temperature throughout model year and Figure 2 during one summer day.

 DOI: 10.1051/e3sconf/20172200185

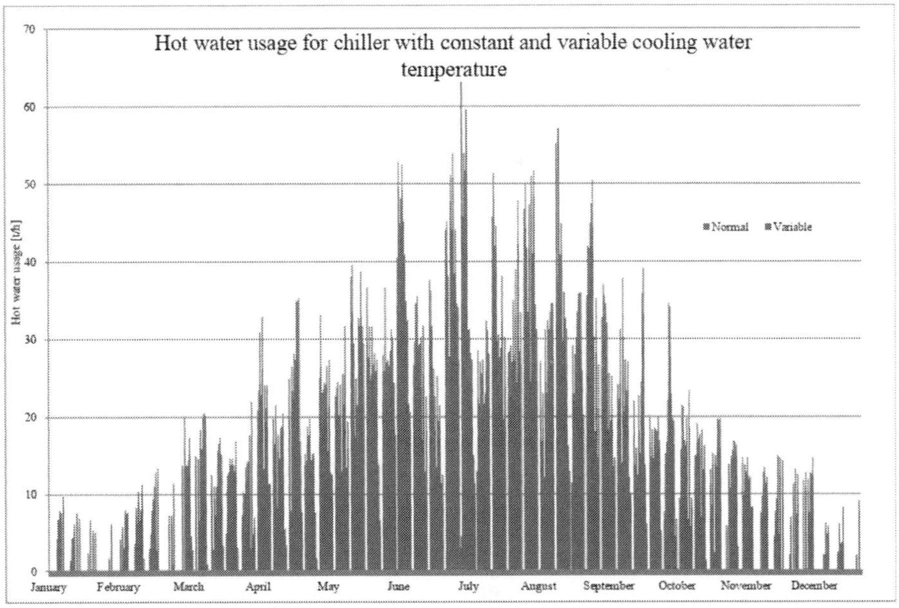

Fig. 1. Absorption chiller hot water usage with normal and variable cooling water temperature during model year.

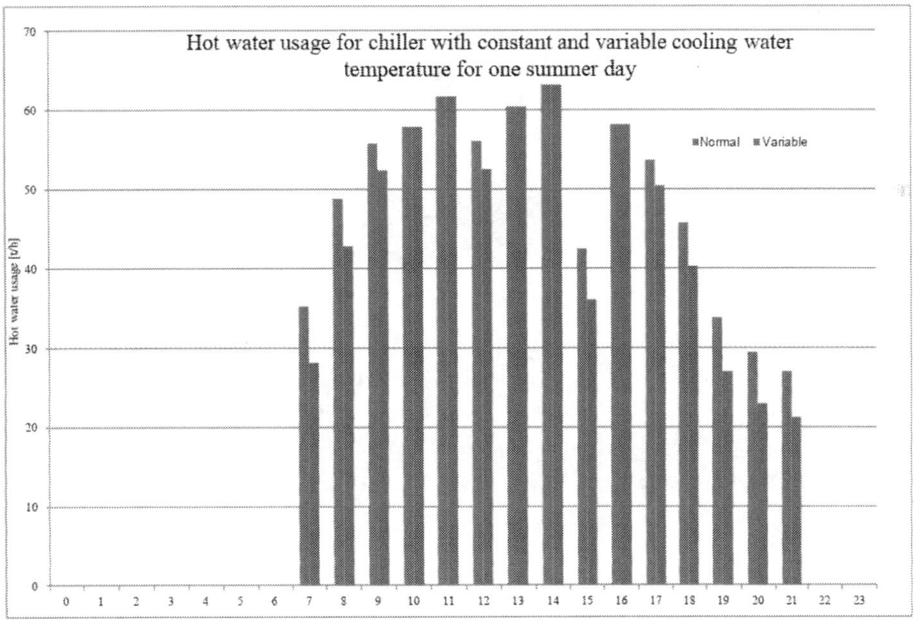

Fig. 2. Absorption chiller hot water usage with normal and variable cooling water temperature during one summer day.

In the analyzed case for 1100 kW chiller supplied by gas CHP unit, the annual savings are equal to 6800.49 t of hot water at 90/70°C (570.56 GJ of heat saved, representing 11.79% of total heat consumption for cooling purposes), which means CO_2 emissions decrease of 35.39 t. For the abovementioned heat source consisting of below 5 MW coal boiler, CO_2 emissions would be reduced by 57.04 t, SO_x/SO_2 by 465.45 kg, NO_x/NO_2 by 85.56 kg, and dust by 128.34 kg per year.

3 Impact

To assess scale of environmental advantages of algorithm for cooling water decrease in absorption chillers implementation, cooling sources, including lithium-bromide chillers, have been quantified and separated by medium (eg. hot water, steam) and by primary energy source (gas, coal, biomas). For each energy source primary energy conversion efficiency has been calculated (whenever data was available from the user) or assessed basing on installed equipment. Data about existing installations has been gathered basing on direct contact with exploitation party, publications and commercial materials study. Totally more than 30 installations were analysed from which 23% were steam chillers and 77% hot water chillers. More than 44% of units were supplied with heat from gas fired CHP unit (mainly based on gas engine) and more than 23% from waste heat or biomass boilers. Rest of the chillers were supplied with heat generated in conventional installations (mainly coal boilers).

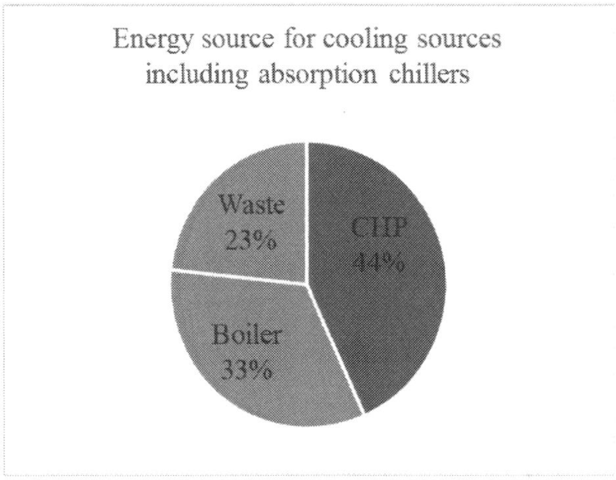

Fig. 3. Energy sources used for energy supply to absorption chillers.

Total heat, for supplying absorption chillers, decrease potential has been calculated for 10 200 GJ yearly what corresponds to reduction of CO_2 emissions by 603.61 t, SO_x/SO_2 by 3.08 t, NO_x/NO_2 by 566.64 kg and dust by 849.96 kg per year. In atmospheric emissions decrease waste heat and biomass installations had been neglected.

Taking into consideration steady absorption chiller market growth and need for energy efficiency improvement yearly new 1 to 3 installations are predicted (market growth by ~5% is considered) from which one will be supplied with CHP unit, one with waste/biomass and one with heat generated from conventional boiler. This allows to predict that within a decade algorithm implementation could lead to atmospheric emissions decrease by 7016 t of CO_2, SO_x/SO_2 by 35.8 t, NO_x/NO_2 by 6.6 t, and dust by 9.9 t.

DOI: 10.1051/e3sconf/20172200185

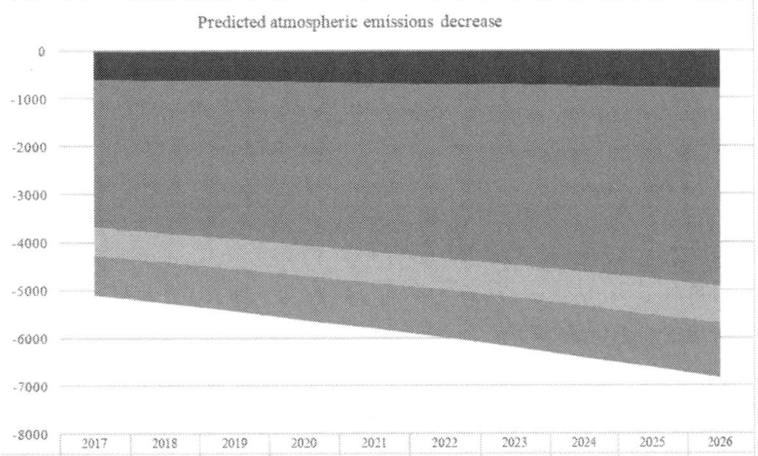

	2017	2018	2019	2020	2021	2022	2023	2024	2025	2026
Dust [kg]	-849,96	-878,01	-906,99	-936,92	-967,83	-999,77	-1032,77	-1066,85	-1102,05	-1138,42
NOx/NO2 [kg]	-566,64	-585,34	-604,66	-624,61	-645,22	-666,52	-688,51	-711,23	-734,70	-758,95
SOx/SO2 [kg]	-3082,56	-3184,28	-3289,36	-3397,91	-3510,04	-3625,87	-3745,53	-3869,13	-3996,81	-4128,71
CO2 [t]	-603,61	-623,53	-644,11	-665,36	-687,32	-710,00	-733,43	-757,64	-782,64	-808,46

Fig. 4. Predicted atmospheric emissions decrease.

References

1. National Development Strategy until 2020 as an attachment to Resolution No. 157 of the Council of Ministers of 25 September 2012
2. Announcement of the Minister of Economy of 21 December 2009 on the state energy policy until 2030
3. R. Saidur, H. Masiuki, M. Jamaluddin, Ener. Pol. **35**, 1050–1063 (2007)
4. Directive 2002/91/EC "Energy Performance of Buildings", Official Journal of the European Commission, 65–71 (2003)
5. W. Dołęga, Ener. Mark. **2**, 24–28 (2014) (in Polish)
6. T. Popławski, Ener. Mark. **1**, 13–18 (2014) (in Polish)
7. L. Fu, XL. Zhai, S.G. Shang, Y. Jiang, H. Li, W.W. Yang, Ener. Conv. And Man. **50**, 977–82 (2009)
8. A. Marantans, Appl. Therm. Eng. **18**, 940–946 (1998)
9. N. Fumo, P. Mago, L. Chamra, Appl. Ener. **86**, 2344–50 (2009)
10 J. Ortiga, J.C. Bruno, A. Coronas, Appl. Therm. Eng. **50**, 1536–42 (2013)
11. R. Saravanan, M.P. Maiya, Appl. Therm. Eng. **18**, 553–68 (1998).
12. F.C. McQuiston, J.D. Parker, *Heating, Ventilation and Air conditioning: analysis and design* (4th edition, John Wiley and Sons Inc., 1994)
13. M. Meckler, Ener. Eng. **95**, 3, 8–13 (1998)
14. W. Kotowski, B. Doliński, Ener. Giga. **6–7** (2013) (in Polish)
15. ASHRE Handbook, *Absorption Cooling, Heating, and Refrigeration Equipment* (1988)
16. M. Malicki, *The influence of chiller parameters change on the trigeneration source efficiency*, Ph. D. Thesis (Warsaw University of Technology 2016) (in Polish)
17. H. Yin, *An absorption chiller in micro BHCP application: model based design and performance analysis*, Ph. D. thesis (Carnegie Mellon University, 2006)

Investigations of hot water temperature changes at the pipe outflow

Janusz Wojtkowiak[1,*], and *Czesław* Oleśkowicz-Popiel[1]

[1]Poznan University of Technology, Faculty of Civil and Environmental Engineering, ul. Berdychowo 4, 60-965 Poznan, Poland

Abstract. In this paper a process of cold water withdrawing from hot water supply pipe systems without recirculation is considered. System of partial differential equations was used to describe the pipe and water temperature changes. An exact solution of a simplified form of the equations was obtained and validated experimentally. The exact solution was applied to calculate the hot water temperature changes at the pipe outflow. Calculations were done for typical pipe materials (PP, PE, Cu), different pipe diameters and lengths as well as for various water flow rates. It was shown that in order to obtain the required hot water temperature in the tap, there is necessary to withdrawn much more (even two times) water from the pipe in comparison to the pipe volume. The reason of such significant water wastes is a heat exchange between hot water flowing inside the pipe and the colder pipe walls. The results can be useful for optimal selection of hot water supply pipes as well as for making decision about applying of hot water recirculating systems.

1 Introduction

Domestic hot water in supply pipes will cool due to heat loss when it will be stagnant for quite some time. In large domestic hot water systems, where the distribution piping is extensive, it will take a relatively long time to remove this cold water from the pipes before the hot water can be available at the tap. This wastes both water and energy are often not acceptable. To overcome this problem, a water recirculation system can be used. Unfortunately some of these systems are very wasteful. According to the regulation [1] the maximum internal volume of the hot water supply pipe for which recirculation system is not required is 3 dm^3. The aim of this paper is to investigate the effect of pipe type as well as an influence of flow rate on hot water temperature changes at the pipe outflow. In particular, we are interested in time needed to obtain a required outflow water temperature and mass of water withdrawn from the pipe before the accepted temperature is reached.

* Corresponding author: janusz.wojtkowiak@put.poznan.pl

DOI: 10.1051/e3sconf/20172200186

2 Theoretical model

Temperature changes of water (T_c) and pipe (T_r) are described in simplified form as [2]:

$$\tau_c \frac{\partial T_c}{\partial t} + \frac{1}{N_c}\frac{\partial T_c}{\partial z^*} = T_r - T_c \quad (1)$$

$$\tau_r \frac{\partial T_r}{\partial t} = T_c - T_r \quad (2)$$

with boundary conditions:

$$T_c\big|_{z=0} = T_0 + \Delta T \quad (3)$$

$$T_c\big|_{t=0} = T_0 \quad (4)$$

$$T_r\big|_{t=0} = T_0 \quad (5)$$

where:
T_c [°C] – water temperature
T_r [°C] – pipe temperature
t [s] – time
N_c [-] – number of heat transfer units
z [m] – axial coordinate ($z = 0$ at pipe inflow)
$z^* = z/L_r$ [-] – dimensionless axial coordinate
L_r [m] – pipe length
τ_c [s] – time constant of water
τ_r [s] – time constant of pipe
T_0 [°C] – initial water and pipe temperature (cold water temperature)
$\Delta T = (T_1 - T_0)$ [°C] – hot and cold water temperature difference
T_1 [°C] – hot water temperature

In Eqs. (1) and (2), a lumped thermal capacity model [3] for pipe and assumption of uniform radial temperature distribution in water are used. The thermal insulation of the pipe is perfect. Functions N_c, τ_c i τ_r were calculated as follows:

$$N_c = \frac{h_c A_r}{m c_c}, \qquad \tau_c = \frac{m_c c_c}{h_c A_r}, \qquad \tau_r = \frac{m_r c_r}{h_c A_r}$$

where:
h_c [W/(m^{2}K)] – heat transfer coefficient at the pipe surface
A_r [m^2] – pipe internal surface
m [kg/s] – water flow rate
c_c [J/(kg K)] – water specific heat
m_r [kg] – pipe mass
c_r [J/(kg K)] – pipe specific heat
m_c [kg] – water mass in pipe

Heat transfer coefficient h_c was calculated using modified Nusselt equation [4] ($Nu=0.024\ Re^{0.786}\ Pr^{0.45}\ [1+2.4254/(L_r/d)^{0.676}]$). Transient temperature responses of water at the pipe outflow $T_c = f(t, z = L_r)$ were obtained from exact analytical solution of Eqs. (1)–(5):

$$T_c = T_0 + \Delta T\, U(N_c, \eta)\exp[-(N_c + \eta)] \qquad (6)$$

where:

$$U(N_c, \eta) = \sum_{n=0}^{\infty}\sum_{k=0}^{n}\frac{\eta^n N_c^k}{n!\,k!}, \qquad \eta = \frac{t - \tau_c / N_c}{\tau_r}$$

3 Model validation

An experimental validation of the theoretical calculation model was accomplished using the experimental set-up shown in Fig. 1. Details of the inlet and outlet of the investigated pipe are shown in Fig. 2. Length of the investigated copper pipe was $L_r = 2.49$ m. Inner and outer diameters of the pipe were $d_1 = 8.183$ mm and $d_2 = 10.0$ mm. The flow rate was measured with the scales and stop watch. The inlet and outlet water temperatures were measured with the transient recorder. Initially in the insulated experimental pipe was cold water, i.e. the water taken from cold water supply pipe. The experimental run was started closing the recirculation hot water loop and opening the supply pipe hot water flow. During the hot water flow the outlet temperature was rising asymptotic until reaching the inlet hot water temperature. The whole temperature history was recorded with the 2 sec step.

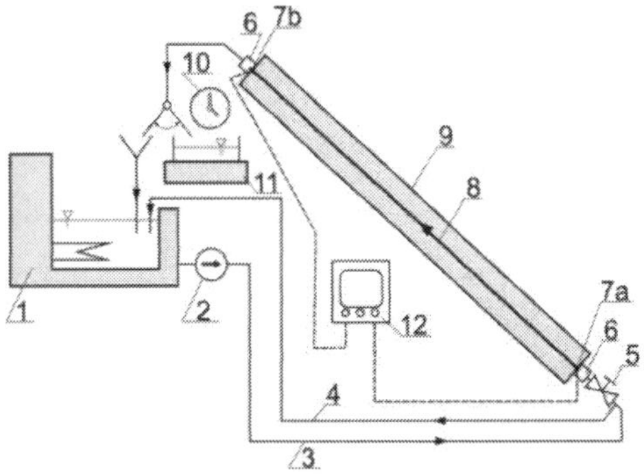

Fig. 1. Experimental set-up, 1 – thermostatic bath, 2 – circulation pump, 3 – supply pipe, 4 – recirculation pipe, 5 – stop and control valve, 6 – connector (thermal insulation), 7a – inflow temperature sensor, 7b – outflow temperature sensor, 8 – investigated pipe, 9 – thermal insulation, 10 – stop watch, 11 – scales, 12 – temperature transient recorder.

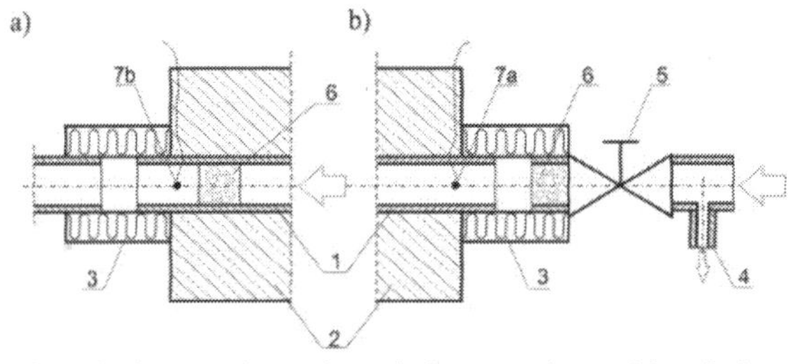

Fig. 2. Details of investigated pipe, 1 – investigated pipe, 2 – thermal insulation, 3 – connector (thermal insulation), 4 – recirculation tap, 5 – stop and control valve, 6 - mixer, 7a – inflow temperature sensor, 7b – outflow temperature sensor.

DOI: 10.1051/e3sconf/20172200186

Comparison between theoretical and experimental results are shown in Figs. 3–5. As is seen, results obtained by means of Eqs. (1)–(5) are about Δt = 10±2 s shifted in time (are retarded) in comparison with our experimental data.

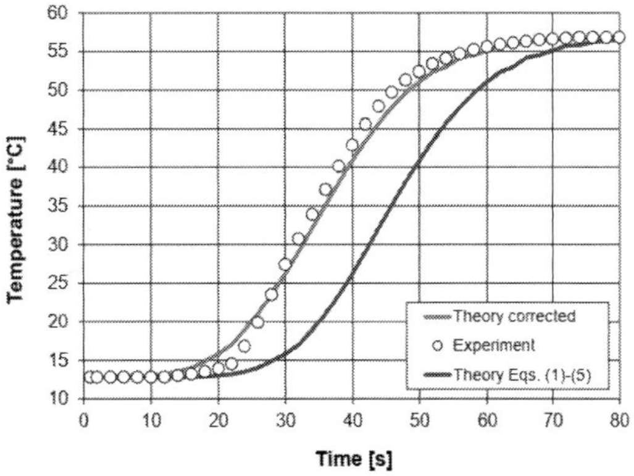

Fig. 3. Comparison of experimental and theoretical results (copper pipe: d_1 = 8.183 mm d_2 = 10 mm, L_r = 2.49 m, w = 0.166 m/s, T_0 = 12.8°C, T_1 = 56.8°C, h_c = 1460 W/(m^{2}K).

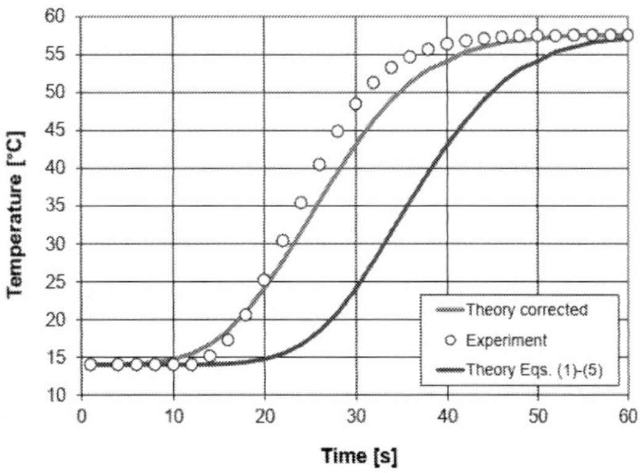

Fig. 4. Comparison of experimental and theoretical results (copper pipe: d_1 = 8.183 mm d_2 = 10 mm, L_r = 2.49 m, w = 0.242 m/s, T_0 = 14°C, T_1 = 57.6°C, h_c = 1960 W/(m^{2}K).

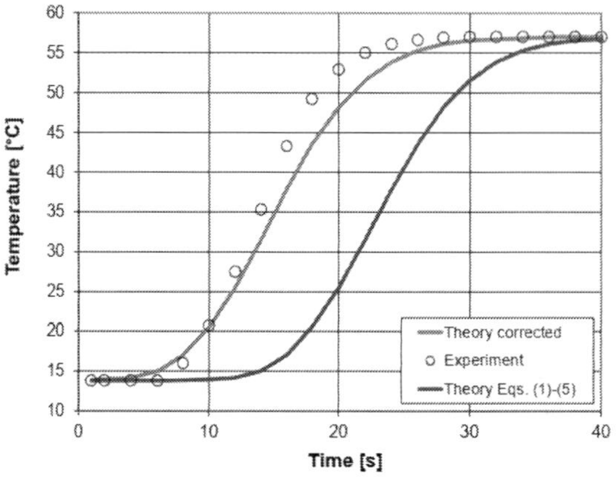

Fig. 5. Comparison of experimental and theoretical results (copper pipe: $d_1 = 8.183$ mm $d_2 = 10$ mm, $L_r = 2.49$ m, $w = 0.435$ m/s, $T_0 = 13.8$°C, $T_1 = 57.0$°C, $h_c = 3110$ W/(m^{2}K).

The differences are slightly higher at larger water velocities and their explanation can be find in simplified form of the theoretical model. Applying simple time correction in form of $(t + \Delta t)$ to the theoretical results it can be obtain a quite good agreement between experimental and corrected theoretical results. Finally, i.e. after "shifting of time axis", the discrepancies between dots and red lines shown in Figs. 3–5 were found as acceptable.

4 Calculation results

4.1 Effect of pipe type

A corrected theoretical model was applied for calculation of water outflow temperature changes for three standard commercial hot water supply pipes (Table 1).

Table 1. Considered pipes: Cu (copper), PE (polyethylene), PP (polypropylene).

Pipe	d_1 [mm]	d_2 [mm]	L_r [m]	A_r [m^2]	ρ_r [kg/m^3]	m_r [kg]	c_r [J/(kgK)]	Q_r [J]	w [m/s]	h_c [W/(m^{2}K)]
Cu	13.0	15	22.6	0.9225	8800	8.743	380	132894	1.23	6150
PE	12.0	16	26.5	0.9797	1290	2.949	1620	190665	1.44	7180
PP	14.4	20	18.4	0.8139	905	2.463	2000	197040	1.00	5100

The pipe lengths L_r shown in Table 1 were obtained assuming that each pipe contains 3 dm^3 of water. According to the regulation [1], this is the maximum value. Symbols d_1 and d_2 mean inner and outer pipe diameters, respectively. The water velocities w were calculated for the flow rate 163 g/s in each pipe. The temperatures of cold and hot water were $T_0 = 20$°C and $T_1 = 60$°C. Pipe thermal capacity Q_r was calculated as $Q_r = m_r\, c_r\, (T_1 - T_0)$. Calculation results in form of outflow water temperature changes are presented in Fig. 6.

DOI: 10.1051/e3sconf/20172200186

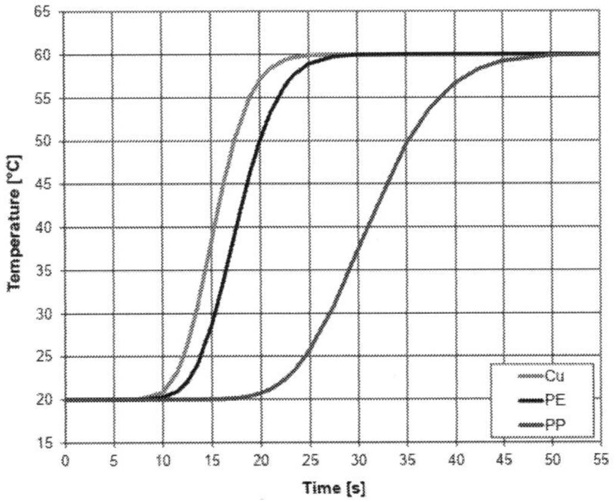

Fig. 6. Water temperature changes at the pipe outflow for 3 different pipes (Table 1) of identical internal volume (3 dm^3), the same water flow rate (163 g/s), cold and hot temperatures 20°C and 60°C, respectively.

In Table 2 times t_{55} needed to obtain outflow water temperature of 55°C, and masses m_{55} of water withdrawing from the pipe until the water outflow temperature reaches 55°C are given.

Table 2. Times t_{55} and masses m_{55} for different pipes (pipes data: see Table 1).

Pipe	t_{55} [s]	m_{55} [kg]
Cu	19.0	3.10
PE	22.4	3.64
PP	38.0	6.21

As one can see in Table 1, the PP pipe has the smallest internal surface A_r, the lowest heat transfer coefficient h_c, and the largest heat capacity Q_r. Therefore the intensity of heat exchange between water and PE pipe is smaller in comparison with the Cu and PE pipes. This is the most important cause of the discrepancies in temperature changes shown in Fig. 6. In Table 2 it is seen that for Cu pipe the smallest values of time t_{55} and mass m_{55} were obtained. For this pipe the mass m_{55} = 3,1 kg is only a little bit larger than the initial mass of cold water in the pipe. Also for Cu pipe, the time t_{55} = 19 s is only 1 s larger in comparison with time necessary to withdrawn the whole cold water from the pipe. For PP pipe, the m_{55} and t_{55} values are two time larger than for Cu pipe. For PE pipe, m_{55} and t_{55} reach intermediate values.

4.2 Effect of water velocity

PE pipe (pipe data: see Table 1) was selected to present the effect of water velocity on outflow water temperature changes. The results are shown in Fig. 7 and Table 3.

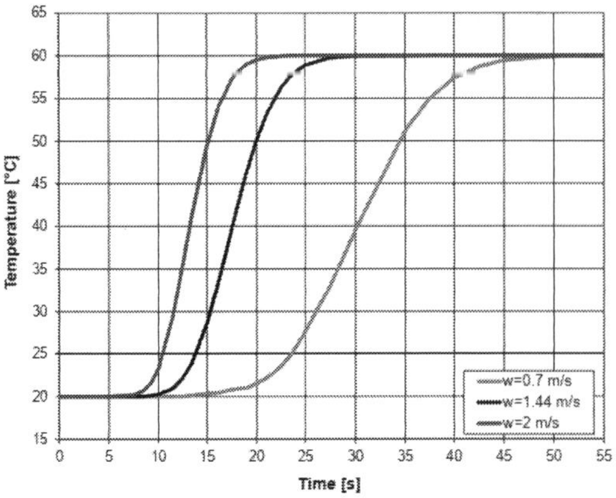

Fig. 7. Water temperature changes at the PE pipe outflow (pipe data: see Table 1) for 3 different velocities: $w = 0.7$ m/s and $h_c = 4160$ W/(m^2 K), $w = 1.44$ m/s and $h_c = 7180$ W/(m^2 K), $w = 2.0$ m/s and $h_c = 9500$ W/(m^2 K).

Table 3. Times t_{55} and masses m_{55} for PE pipe (pipe data: see Table 1) and different water velocities.

w [m/s]	m [g/s]	h_c [W/(m^2 K)]	t_{55} [s]	m_{55} [kg]
0.70	79	4160	37.5	2.97
1.44	163	7180	22.4	3.64
2.00	226	9500	17.3	3.85

As is seen, if water velocity increases, time necessary to reach an acceptable outflow temperature decreases and simultaneously mass of waste cold water increases.

5 Conclusions

The simply theoretical model was experimentally verified and introducing a time correction it was found useful for investigations of hot water temperature changes at the pipe outflow. The best hot water supply pipe should have small diameter and low thermal capacity value. Thermal insulation of the pipe ought to be as good as possible. An effect of the hot water flow rate on time needed to obtain a required outflow water temperature and the mass of cold water withdrawn from the pipe were investigated. It was found that with increasing of water velocity a decrease of time necessary to reach an acceptable outflow temperature and simultaneously an increase of losses of cold water are noticeable. The results ca be applied for appropriate selection of hot water supply pipes, i.e. length, diameter and material. They can be also useful for making decision about applying of hot water recirculating systems.

References

1. Dz. U. z dn. 17.07.2015, poz. 1422, § 120.1, (2015)
2. E.P. Sierov, B.P. Korolkov, *Dinamika parogienieratorov*. Energija, Moskva, **359–362** (1972)
3. J. Wojtkowiak, *Lumped Thermal Capacity Model* (In: *Encyclopedia of Thermal Stresses* Ed. R.B. Hetnarski, Springer, 2808–2817, 2014)
4. S. Kakac, R.K. Shah, W. Aung, *Handbook of Single-Phase Convective Heat Transfer* (Wiley, 4–33 & 4–58, 1987)

The application of impedance measurement to assess biofilm development on technical materials used for water supply system construction

Mirela Wolf[1,*], *Teodora* Traczewska[1], and *Tomasz* Grzebyk[2]

[1]Wroclaw University of Science and Technology, Faculty of Environmental Engineering, Wybrzeze Wyspiańskiego 27, 50-370 Wroclaw, Poland
[2]Faculty of Microsystem Electronics and Photonics, Wrocław University of Technology, Wybrzeze Wyspianskiego 27, 50-370 Wrocław, Poland

Abstract. The lack of biological stability of water which is introduced into the network, leads primarily to its secondary contamination during transport to the consumer. The water that is biologically unstable creates ideal conditions for colonization of the inner surface of pipelines by microorganisms and adhesion of their products (biocorrosion). The studies was conducted using the identified microorganisms isolated from the water supply network which accounted inocula in continuous culture of biofilm in CDC reactor. As a result of studies it was revealed the presence of biofilm formed on different materials polyethylene, polypropylene, polyvinyl chloride, polybutylene. Microbiological biodiversity of organisms inhabiting a biofilm of the diversity of nucleic acids was used. It was observed the amount of the psychrophilic bacteria oscillation in the effluent from the reactor. It was also determined the affinity of various bacteria to the plastic through adhesion measurement using impedance spectroscopy. For impedance measurements apparatus SIGNAL RECOVERY 7280 DSP LOCK-IN AMPLIFIER was used, recording impedance components (real and imaginary). The results will allow for the creation of biosensor systems that can be used in predicting health risks in connection with drinking water and taking corrective actions.

1 Introduction

The main task of the drinking water treatment plants is to maintain its proper purity and stability during transport to the consumer. Ensuring both chemical and biological stability of transported water is one of the key problems related to water quality management. Water quality may be reduced mainly by corrosion of pipes and water supply equipment. Corrosion is an extremely complex process, depending on the quality of the distributed water, hydraulic conditions, the material forming the network, and the presence of

*Corresponding author: mirela.wolf@pwr.edu.pl

 DOI: 10.1051/e3sconf/20172200187

microorganisms. Biological corrosion is associated with degradation of material due to the activity of microorganisms, mainly of those remaining in close contact with the surface of the pipeline, forming the biofilm [1–3].

Mature biofilm is a complex structure comparable to the ecosystem. Microorganisms are suspended in the extracellular polymers, and the exchange of matter and genes occurs through an extensive system of canals. Beside the possibility of detachement and transfer of portions of a microbial biofilm to other parts of the network, further problem is the possibility of long-term survival or proliferation of pathogens (e.g. *Legionella sp.*) in the structures of the biofilm, among others, because of the limited penetration of disinfectants [4–6].

Traditional methods of assessing the microbiological quality of water are usually just the first step of identifying microorganisms, which directs further analyses using molecular biology methods. Standard methods do not reflect the actual species composition, because they are targeted at the bacteria that can be isolated and cultured in the laboratory. It is believed that cultivation techniques are able to obtain from 1–10% of the total bacteria present in the environment [7]. These methods are also tedious and labor intensive. They also require the presence of qualified staff and special care during the sample preparation, whereas the results of those tests are available after a few days [8]. Therefore, an attempt was made to develop methods for a quick analysis that would yield results in real time without having to prepare special additional samples. The proposed process for the continuous measurement will allow in the future for a constant monitoring of stages of the formation of biofilms, which will allow for control over the current risks to the consumer and to take appropriate preventive measures [9].

The article is aimed at explaining the problem that is the formation of biological growth on the surface of water pipes, despite introducing new technical materials. For this reason, the impedance sensor was used, which was colonized by the microorganisms during water flow. The measured impedance value provide a real-time information on the biofilm development.

2 Materials and methods

For the culture of the biofilm we have used a CDC reactor from Biosurface Technologies company (Fig 1A).The contents of the reactor was a synthetic tap water inoculated with a mixture of 11 strains from our own collection [10]. Three segments of pipes made of plastic (polyethylene, polypropylene, polyvinyl chloride, polybutylene) (Figure 1B) and a measuring sensor (Figure 1C) were placed in the reactor. After 14 days of pre-incubation in static conditions, the culture was carried out in flow conditions with constant stirring (100 rpm stirring) at room temperature. The flow rate of water in the system was similar to the actual value and was set at 0.3 m/s. In the work has been used a specially designed conductometric cell created with the procedure shown on Figure 2B.

DOI: 10.1051/e3sconf/20172200187

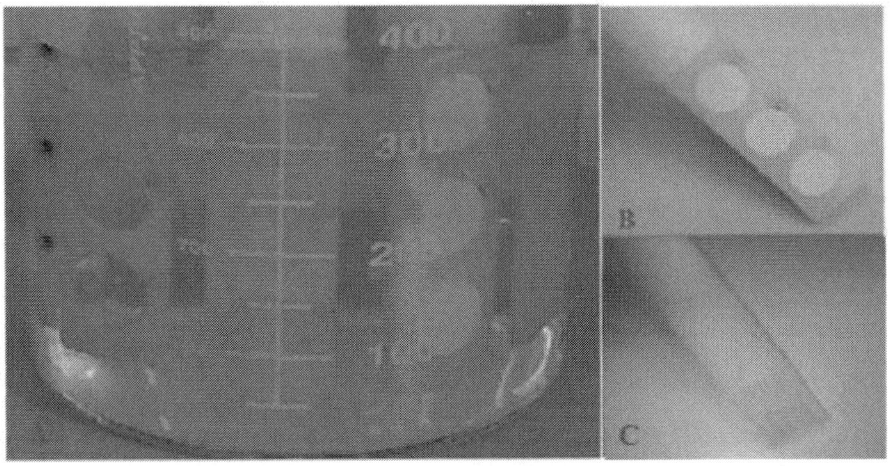

Fig. 1. Reactor with working fluid (A), handle to the technical materials (B-polyethylene), sensor attached to a connector (C).

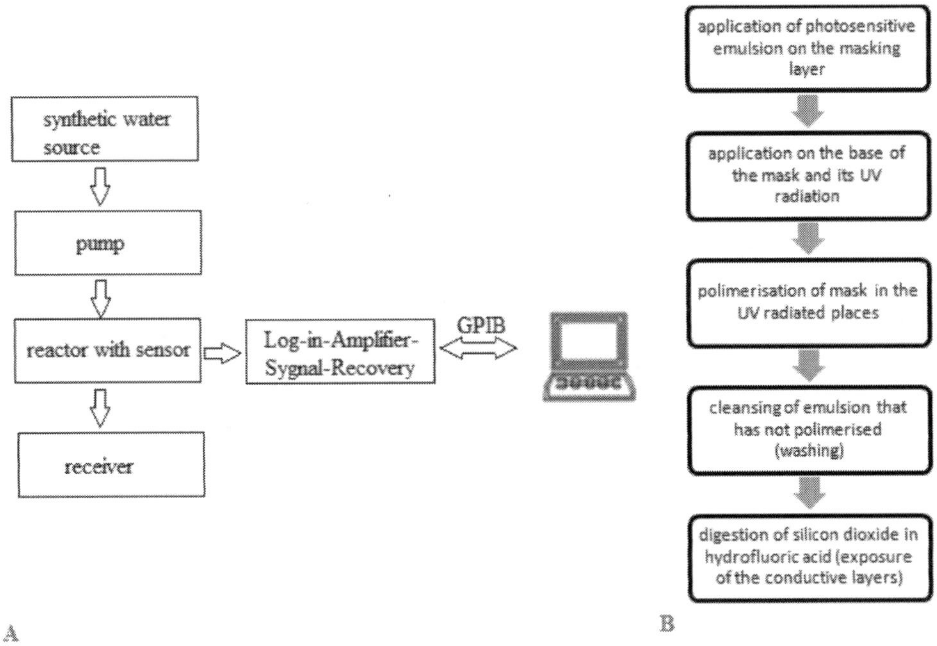

Fig. 2. Diagram of the measuring system-A and the process of fabrication of an impedance sensor using photolithography on ITO medium-B [11, 12].

The adhesion of bacterial cells to the sensor disturbed the current flow between electrodes, what resulted in the change of impedance value. The measurement was carried out using measuring equipment of Signal Recovery 7280 DSP Lock-in-amplifier. At the same time, along the measurement, the bacteria in the outflow of the reactor were quantified daily for 30 days. Using deep inoculation, 1 cm^3 from dilution was applied to the medium with R2A agar. After 72 h of incubation at 20 ± 2°C grown colonies were counted.

Biological growth on test materials and the sensor was monitored using a fluorescent dye DAPI (aqueous solution of 4,6-diamidino-2-phenyl-indol) and using a fluorescence microscope Olympus CX41. DAPI by binding to double-stranded DNA (dsDNA) present in the bacteria forms under the UV light λ = 365 nm is providing a highly fluorescent blue complex. For this purpose, 1 µl of dye was applied onto the test material and then it was incubated in the dark for 20 minutes, then the samples were viewed under a microscope. Live bacterial cells dyed with DAPI are fluorescent blue [13].

An important element was to correlate the changes in impedance with the build-up of the biofilm created by the selected strains, therefore, in the first stage it was necessary to determine the frequency of measurements.

3 Results and discussion

The first attempt was made to determine the most appropriate frequency of the conductometric measurements. For this purpose, a sensor was placed in a homogeneous mixture of 11 strains in synthetic water. Figure 3 presents, as an example, the results of relative impedance for the frequencies from 1 to 100000 Hz. The measurements showed that the frequency has no significant impact on the received signal, as the changes reached about 1%, and the frequencies for which minimum and maximum values occurred did not overlap. The inoculum in a reactor was a mixture of 11 strains, each of which may be received in different ranges. It was impossible to explicitly select, either during the weekly or the daily measurement, for which frequency the changes were the most significant. Therefore, for further measurements the frequency was set at 100Hz, selected based on the literature [14].

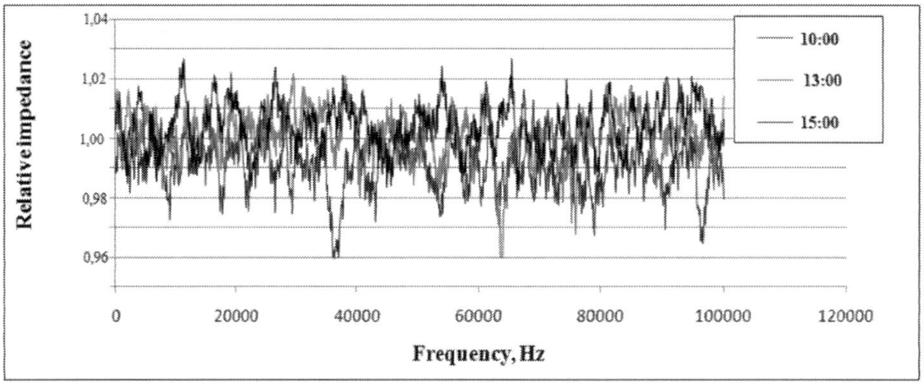

Fig. 3. Relative impedance measurement according to changing frequency.

Subsequently, the measurement of biofilm was carried out continuously in flow conditions for 30 days [15]. The measurements were carried out at intervals of 10 seconds. After the minimal fluctuations in the range from 0.650 to 0.725 in the first 5 days of the experiment, the value of the measured impedance was maintained for 20 days at approx. 0.670. After that time, a systematic increase in the observed impedance value up to 0.850 for the thirtieth day was observed (Fig. 4).The conditions of the experiment reflected the actual system of the water supply network, where the number of bacteria is small and is a mixture of several strains, mainly of psychrophilic bacteria.

DOI: 10.1051/e3sconf/20172200187

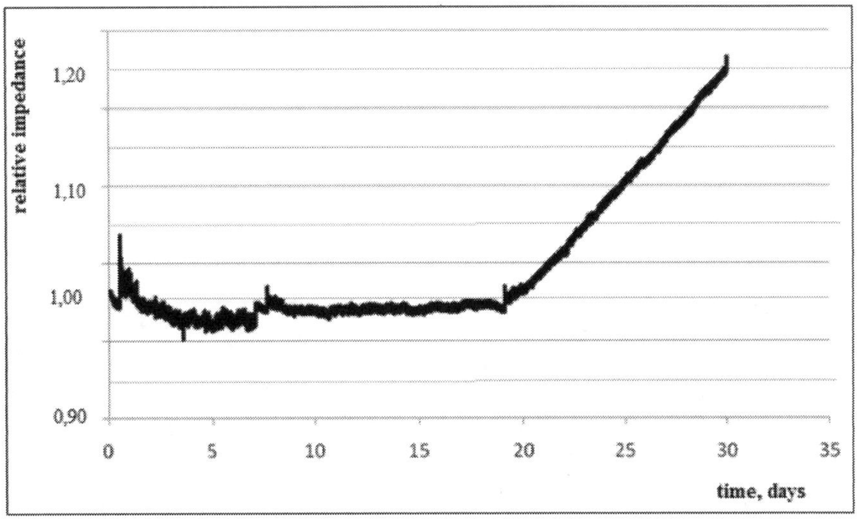

Fig. 4. Relative impedance during strains growth in synthetic water media.

Hadar Ben-Yoav, while measuring the impedance value in the range of 400 kHz to 100 mH, demonstrated that the results depend on the growth phase of biofilm formed by *E. coli* [16]. Similar observations were achieved by [17] in study which involved measurements of impedance before and after the adhesion of cells of bacteria *Pseudomonas stutzeri* on indium-tin-oxide (ITO) plate, for variable range of frequencies [17]. In both cases, a culture of a single strain has been used for study, and the duration of the experiment was a few days. Hadar Ben-Yoav have demonstrated that better effects are obtained for lower frequencies, hence the obtained results differ from those obtained by [17].

Due to potential partial leaching of the bacteria from the system, a quantitative analysis of the water flowing out of the reactor was carried out. This study indicated that the greatest amount of bacteria in the outflow was observed after the change of conditions as a result of the introduction of water flow (Figure 5).

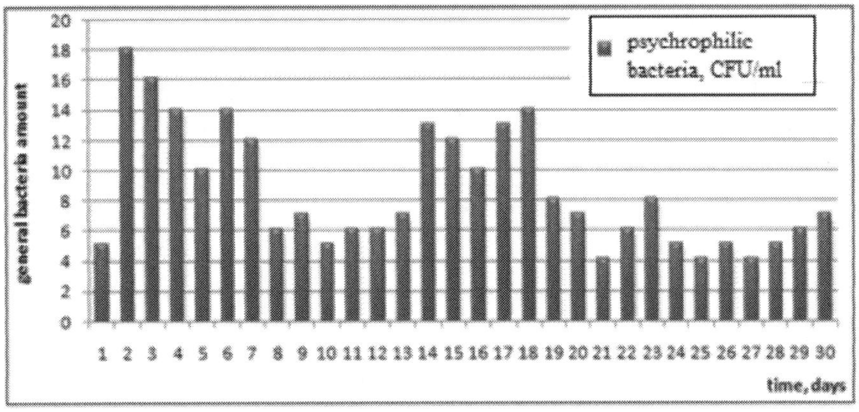

Fig. 5. Quantitative inoculation onto reasoner's 2A agar (2A) of samples from the outflow from the biofilm reactor.

 DOI: 10.1051/e3sconf/20172200187

Bacterial count declined steadily from 18 CFU/ml on the second day to about 6 CFU/ml from 8th to 13th day of the duration of the experiment. Between 14th and 18th day an increase to about 14 CFU/ml was observed, and another decrease of the number of microorganisms. From day 19th to the end of the experiment there was a stabilization of the system and the flushed cells ranged from 4–6 CFU/ml. This suggests stabilization of the system as a result of the process completion of biofilm maturation.

To visualize all the bacteria colonizing the technical materials and the sensor, staining with a fluorescent dye was used (Fig. 6).

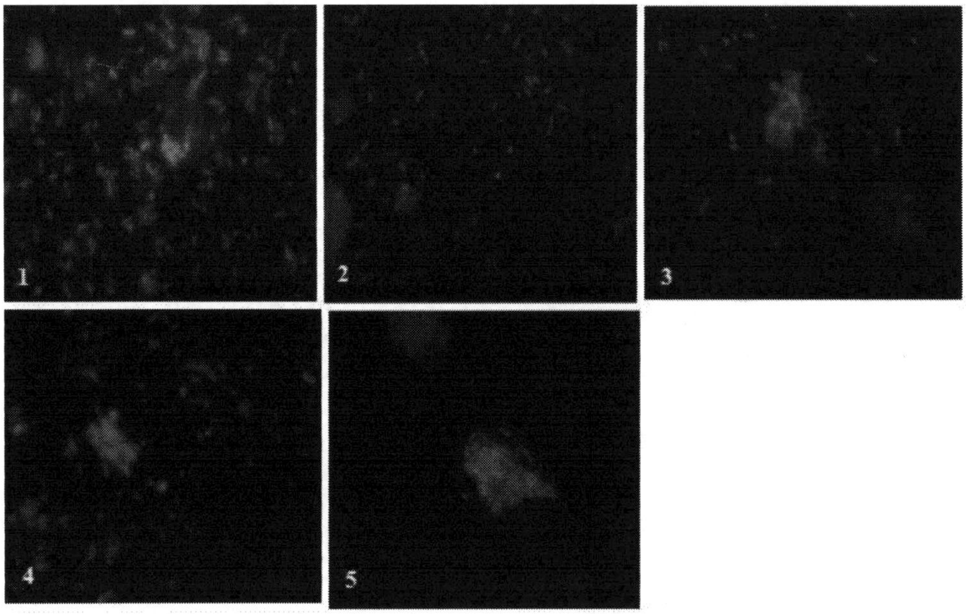

Fig. 6. The total amount of bacteria accumulated on parts of pipeline and sensor; 1 – polypropylene, 2 – polyethylene, 3 – polyvinyl chloride, 4 – polybutylene, 5 – sensor, seen in a fluorescent microscope under 1000x immersion magnification.

The most numerously colonized plastic was the polypropylene (6.1). In turn, the images obtained for polyvinyl chloride (6.3) and polybutylene (6.4) were very similar to those of the sensor (6.5).The least colonized was the area of polyethylene (6.2).

The study confirmed the results of Wang demonstrating that the material from which the water supply network is made of can have a significant effect on the amount, frequency and type of biofilm formation, the presence of pathogens as well as the effectiveness of used disinfectants [18].

The results of studies on the susceptibility to biological growth of technical materials used to build water supply networks show significant discrepancies. Niquettei demonstrated that the materials from synthetic materials (PE, PVC) were less colonized by biofilm than pipes made of iron, copper and steel, however, on the other hand, Zacheusi demonstrated a lack of significant differences in colonization of synthetic materials (PE, PVC) and steel [19, 20]. Schwartz have shown in their work that more bacterial cells are gathered on the surface of polyethylene, what is favorable for a faster growth of biofilm than on the surface of PVC [21]. Not with standing, according to Ginige the most frequently selected material for the water supply network is high density polyethylene, due to its low cost of production and durability [22].

4 Conclusion

It is necessary to select the appropriate material for construction of the water supply network to prevent bacterial growth in distribution systems for drinking water. Corrosion, deteriorating water quality, taste and smell as well as increased numbers of pathogens are just some of the negative effects caused by the presence of biofilm. The total elimination of biofilm from the network seems unlikely, even if the amount of absorbable organic carbon will be reduced to 0.2 mgC/m^3 and it will be manageable to maintain an adequate level of disinfectant in the network. The basic cause is the material used to build the network and its properties. All technical materials have micro cracks on their surface, a result of the manufacturing process, which are susceptible to the adhesion of bacterial cells. Thus, sensors should be built on substrates with physical characteristics similar to the material used for the construction of water supply systems. The measurement of impedance can be correlated with the cell adhesion of bacteria to the surface of the water pipes and, after detailed studies, it could serve as a method for monitoring the dynamics of the biofilm development , providing a tool for microbiological quality control of tap water.

Acknowledgments

The research was finance by the funds of the Polish Ministry of Science and Higher Education for young scientist and doctoral students; order numer: 0402/0082/16.

References

1. M. Ginige, J. Wylie, J. Plumb, Biofouling **27**, 2 (2011)
2. J. Rubulis, T. Juhna, Water Sci.Technol. **55**, 8–9 (2007)
3. D. Berry, Xi Chuanwu, L. Raskin, Curr. Opin. Biotech. **17**, 3 (2006)
4. H. Wang, C. Hu, X. Hu, M. Yang, J. Qu, Water Res. **46**, 4 (2012)
5. J. Wingender, F. Hans-Curt, Int. J. Hyg. and Envir. Healh. **214**, 6 (2011)
6. T. Battin, K. Besemer, M. Bengtsson, A. Romani, A. Packmann, Nat. Rev. Micrbiol. **14** (2016)
7. S. Nevel, S. Koetzsch, W. Hans-Ulrich, N. Boom, F. Hammes, J. Biol. Met. **94**, 2 (2013)
8. Ch. Peng, G. Korshin, R. Valentine, A. Hill, M. Friedman, S. Reiber, Water Res. **44**, 15 (2010)
9. M. Schock, R. Hyland, M. Welch, Envir. Sci. Tech. **42**, 12 (2008)
10. I. Biedroń, M. Traczewska, G. Płaza, Ann. Set Envi. Pro. **18** (2016)
11. Z. Zimniak, M. Marciniak, I. Zubel, Scientific Works of Warsaw University of Technology, **267** (2015) (in Polish)
12. S. Mannsfeld, B. Tee, R. Stoltenberg, Ch. Chen, S. Barman, B. Muir, A. Sokolov, C. Reese, Z. Bao, Nat. Mater. **9** (2010)
13. A. Ferro, T. Mestre, P. Corneiro, I. Sahumbaiev, R. Seruca, J. Sanches, Lab. Invest. **6** (2017)
14. J. Parades, S. Becerro, S. Arana, Sen. Acu. Chem. **195** (2015)
15. S. Nevel, K. Roy, N. Boon, FEMS Microbiol. Ecol. **85** (2013)

16. H. Ben-Yoav, A. Freeman, M. Sternheim, Y. Shacham-Diamand, Electrochim. Acta **56**, 23 (2011)
17. S. Bayoudh, A. Othmane, L. Ponsonnet, H. Ben Ouada, Physicochem. Eng. Asp. **318** (2008)
18. H. Wang, S. Masters, M. Edwards, J. Falkinham, A. Pruden, Env. Sci. Tech. **48**, 3 (2014)
19. P. Niquette, P. Servais P, R. Savoir, Water Res. **24**, 6 (2000)
20. O. Zacheus, E. Iivanainen, T. Nissinen, M. Lehtola, P. Martikainen, Water Res. **34**, 1 (2000)
21. T. Schwartz, S. Hoffmann, U. Obst, Water Res. **32**, 9 (1998)
22. M. Gringe, S. Garbin, J. Wylie, C. Krishna, Plos One, **6** (2017)

The effectiveness of removing precursors of chlorinated organic substances in pilot water treatment plant

Małgorzata Wolska[1,2,*], *Sławomir* Szerzyna[1], *Justyna* Machi[1], *Marek* Mołczan[1,2], *Wojciech* Adamski[1], and *Jacek* Wiśniewski[1]

[1]Faculty of Environmental Engineering, Wrocław University of Science and Technology, Wyb. Wyspiańskiego 27, 50-370 Wrocław, Poland
[2]Wroclaw Municipal Water and Sewage Company, New Technologies Center, ul. Na Grobli 14/16, 50-421 Wrocław, Poland

Abstract. The presence of organic substances in the water intaken for consumption could be hazardous to human health due to the potential formation of disinfection by-products (TOX). The study were carried out in the pilot surface water treatment system consisting of coagulation, sedimentation, filtration, ozonation, adsorption and disinfection. Due to continuous operation of the system and interference with the parameters of the processes it was possible not only assess the effectiveness of individual water treatment processes in removing TOX, but also on factors participating on the course of unit processes.

1 Introduction

The creation of oxidation by-products during water treatment is connected with a very large risk to consumer health [1, 2]. Therefore, studies have been carried out for many years worldwide aiming to identify these oxidation by-products and to determine the factors contributing to the type and amounts of harmful substances created.

Chlorine/sodium hypochlorite is still the most commonly used disinfectant, and therefore it is important to determine the factors contributing to the formation of chlorination by-products, of which the most dangerous are chlorinated organic compounds [3, 4]. The types of Total Organic Halogens (TOX) created depend above all on the type and amount of organic matter contained in the water undergoing disinfection [5, 6]. Therefore, effective removal of organic substances, in particular TOX precursors, in processes preceding disinfection is necessary for limiting health hazards.

The amount of chlorinated oxidation products created is decided by the amount of refractive substances in water, i.e. compounds of a complicated molecular structure. As shown by Chu et al. [7] the amount of TOC precursors is proportional to dissolved organic carbon (DOC) in water undergoing treatment, but above all to UV absorbance at wavelengths of 254 nm and 272 nm. It is also known that organic substances absorbing UV

* Corresponding author: malgorzata.wolska@pwr.edu.pl

light are removed during adsorption and coagulation processes [8-11]. However, rarely, and practically never, do these process ensure a complete elimination of the hazards connected with TOX formation. Limiting the amounts of oxidation by-products can be achieved by increasing the effectiveness of complete mineralization of organic substances, which can be obtained by increasing chemical oxidant dosages and increasing the oxidation time, especially when using ClO_2 [12]. Results of studies [13, 14] have shown that, in fact, free chlorine reacts slowly with organic substances. In operating water treatment systems, increasing the oxidation time is most often impossible, which is compensated by increasing chemical oxidant dosages. In the case of using free chlorine, and to a lesser degree when using ClO_2, the above solution unfortunately results in increasing chloroorganic compound concentrations in water [15–17].

Therefore it is important to maximize organic substance removal in processes preceding disinfection. The aim of this study was determining the effectiveness in removing organic contaminants as potential precursors of chlorinated organic substances in a pilot surface water treatment system and determining the factors contributing to the effectiveness of their removal in unit processes.

2 Subject and methods of study

The study was conducted from January to August in two surface water treatment systems (TS – Test System, RS – Reference system), made up of coagulation, sedimentation, filtration, adsorption and disinfection with pH correction. In the time period preceding study termination, an ozonation process was introduced prior to adsorption. Each of the system operated with a throughput of 3 m^3/h (fig. 1).

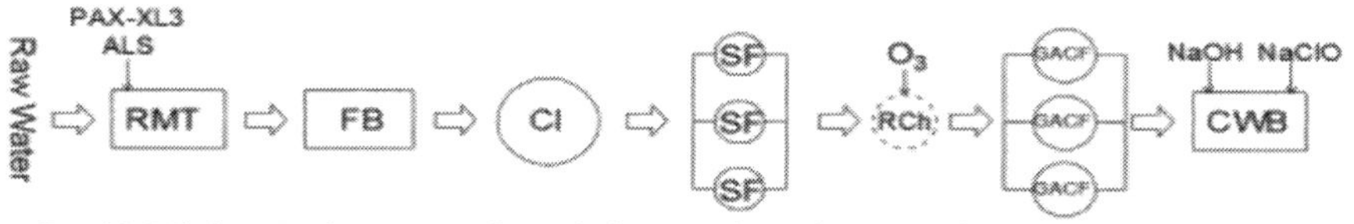

RMT – Rapid Mixing Tank, FB – Flocculation Basin, Cl – Clarifier, SF – Sand Filter, RCh – Reaction Chamber, GACF – Granular Activated Carbon Filter, CWB – Clear Water Basin

Fig. 1. Water treatment system schematic.

The coagulation process used PAX-XL3 (from January to April 13) and ALS (from April 13 to August). The coagulant dosage for both treatment systems was in the range of 1.70–3.78 gAl/m^3. The sedimentation time for post-coagulation sediments was 2.0 h and was constant due to the constant water inflow into the system. The water flow speed through the sand bed at the nominal system throughput was 5 m/h. However, the water – adsorption bed contact time was in the range of 13–18 minutes.

The ozonation process was incorporated into the treatment system only during the final days of the study, and therefore it was not possible to determine its effect on the organic substance removal effectiveness. The object of this study were samples of raw water (supplying both systems) and after each unit process from each system. Samples were taken five times per week (Monday to Friday). The water treatment systems were turned off from February to March. For all water samples the following were determined: temperature, pH, conductivity, basicity, turbidity, color in visible light at wavelengths of 340 nm and 410 nm, UV absorbance at wavelengths of 254 nm and 272 nm, Total Organic Carbon (TOC) and Dissolved Organic Carbon (DOC) concentrations and dissolved oxygen concentrations. For raw water samples and samples leaving the sedimentation tank the zeta potential was

DOI: 10.1051/e3sconf/20172200188

determined. Based on the DOC content and UV_{254} absorbance value the specific ultraviolet absorbance (SUVA) was calculated.

3 Results and discussion

During the entire study period the source water contained dissolved oxygen, with its concentration reducing from April so that during the final period of study it amounted to only 33% of the concentration at the beginning of study, which was directly connected with temperature changes of the source water.

Due to the long study period, the source surface water was characterized by large changes of not only temperature but also content, (tab.1), especially with respect to organic substance content. Among the organic substances in the raw water, the dissolved fraction dominated, which made up 93.7% to 100% of total organic carbon. It is precisely these compounds that are precursors to chlorinated disinfection by-products [18], and an evaluation of their removal effectiveness was the essence of this study. Liming of the risk connected with the formation of disinfection by-products (DBPs) is therefore proportional to the reduction in UV_{254} and UV_{272}, absorbance, which during the study period varied significantly for raw water (tab. 1). As shown by Lee and Westerhoff [19], values of SUVA in source surface water (tab. 1) indicate a low susceptibility of organic substances to removal during the coagulation process.

In spite of this, this process ensured a reduction in total and dissolved organic carbon in the ranges of 2.3–26.4% and 1.8–26.9% and also 3.7–27.1% and 9.2–27.5% respectively for the reference and test systems.

In general, the coagulation process effectiveness in removing organic substances increased with increases in raw water temperature (fig. 2). The deviation from this relationship in the first study period (From January to April) may have been caused by the use a pre-hydrolized coagulant during this time, whose effectiveness is dependent on water temperature to a much smaller degree. These relationships were similar in both systems.

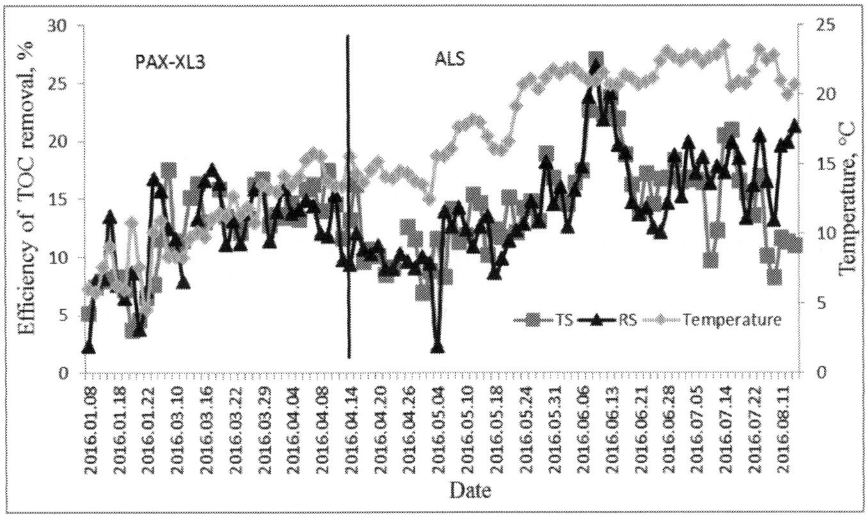

Fig. 2. Variation in Total Organic Carbon removal during coagulation and water temperature.

No unambiguous relationship was found between the organic substance removal effectiveness and the coagulant dosages, regardless of the type of used coagulant. On the other hand, the coagulation effectiveness depended on the degree of raw water contamination and increased with increases in this contamination.

Table 1. Ranges of studied water quality indicators.

		Test System (TS)			Reference System (RS)		
Parameter	**RW***	**CO***	**SF***	**GAC***	**CO***	**SF***	**GAC***
Temperature, °C	4.6–23.5	5.2–23.5	5.9–23.6	7.1–23.7	7.9–23.5	8.3–23.5	9.5–23.6
pH	6.78–8.08	6.77–8.06	6.65–8.03	6.64–7.93	6.63–8.02	6.43–7.91	6.32–7.92
Conductivity, µS/cm	393–735	396–746	396–795	400–747	397–745	396–749	398–750
Basicity, val/m^3	2.16–3.19	2.02–2.96	2.01–2.97	2.01–2.93	1.97–3.00	1.91–2.95	1.93–2.95
Turbidity, NTU	3.7–28.5	0.3–14.3	0.0–3.4	0.0–0.6	0.4–8.4	0.1–0.5	0.0–0.3
UV 254, m^{-1}	5.08–14.76	2.01–13.55	1.21–11.98	2.10–8.04	1.96–10.26	1.24–9.76	2.47–7.32
UV 272, m^{-1}	5.04–11.95	2.51–10.95	1.83–9.68	0.28–6.39	2.46–8.26	1.85–7.83	0.30–5.76
Color 340, g/m^3	5.74–14.46	3.20–12.75	2.89–11.08	0.57–6.20	5.25–9.74	4.55–8.06	0.43–5.21
Color 410, g/m^3	8.05–24.11	2.52–20.24	2.74–18.31	0.15–8.18	2.46–14.61	1.12–10.96	0.95–6.24
Oxygen, gO_2/m^3	4.91–12.21	7.91–12.23	7.17–12.1	5.37–12.7	7.89–11.55	7.14–11.19	5.58–14.28
TOC, gC/m^3	3.23–5.90	2.71–5.69	2.69–5.22	2.01–4.05	3.34–5.04	2.83–4.56	2.35–3.85
DOC, gC/m^3	3.34–5.62	2.42–5.45	2.54–5.03	1.92–4.01	3.14–4.68	2.78–4.46	2.31–3.79
Zeta potential, mV	-15.10– -7.85	-13.50– -6.27	–	–	-12.80– -6.63	-	-
SUVA, m^3/gC·m	1.27–2.69	0.63–2.49	0.43–2.38	0.89–2.07	0.62–2.29	0.44–2.28	0.92–1.95

* RW – raw water, CO – outflow from the settling tank, SF – outflow from sand filters, GAC – outflow from carbon filter

Among the dissolved organic substances removed during the coagulation process, chlorinated organic substance precursors dominated, which is shown by effectivenesses in decreasing UV_{254} and UV_{272} absorbance in the ranges of 3.6–61.8% and 3.9–51.2% for RS and 0.9–60.3% and 0.8–50.2% for TS. The effectiveness in reducing the values of these parameters were similar to each other, which is connected with similar properties of substances absorbing UV radiation at the analyzed wavelengths. On the other hand, differences in obtained effectivenesses between the two treatment systems should be explained by differences in hydraulic flow conditions between subsequent devices for coagulation and sedimentation. The effectiveness of reducing UV absorbance during coagulation depended on the values of these parameters in source water. The effectiveness in removing DOC was proportional to decreases in UV absorbance (fig. 3).

Due to reductions in refractive substances as evaluated by UV_{254}, a 14.2% and 14.7% reduction in specific absorption occurred respectively for the reference and test systems.

A consequence of removing organic substances was a decrease in the coagulated water color intensity. The study evaluated the reduction in color intensity which was determined for absorbance at two wavelengths of 340 nm and 410 nm. The coagulation process provided a very effective reduction of color intensity at both wavelengths. The greater effectiveness found for 410 nm is due to the properties of substances present in water (SUVA values).

DOI: 10.1051/e3sconf/20172200188

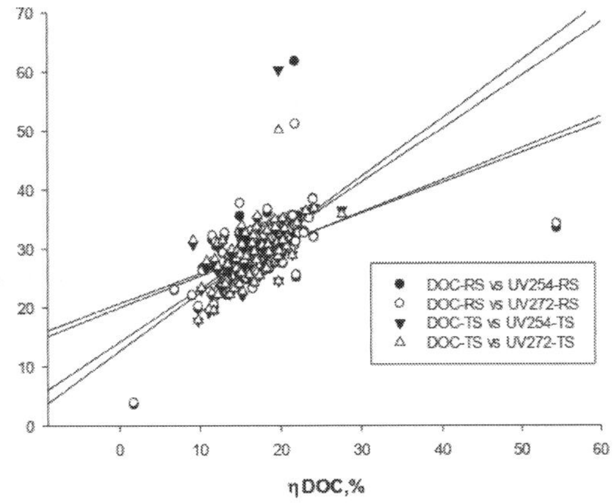

Fig. 3. Reduction in UV absorbance as a function of DOC removal.

The zeta potential for post-coagulation water samples was lower than that found for source water, however no transition of this potential through the isoelectric point was detected. This may point to the possibility of increasing coagulation effectiveness by a significant increase in coagulant dosage, but this is not economically justifiable.

Water coming into GAC beds was characterized by low SUVA values (tab. 1). The effectiveness in removing dissolved organic carbon was in the ranges of 7.3–28.0% and 4.0–35.4% for the reference and test systems respectively. No clear impact of the temperature of treated water on DOC removal effectiveness was found, with the highest effectiveness found mid-January (fig. 4).

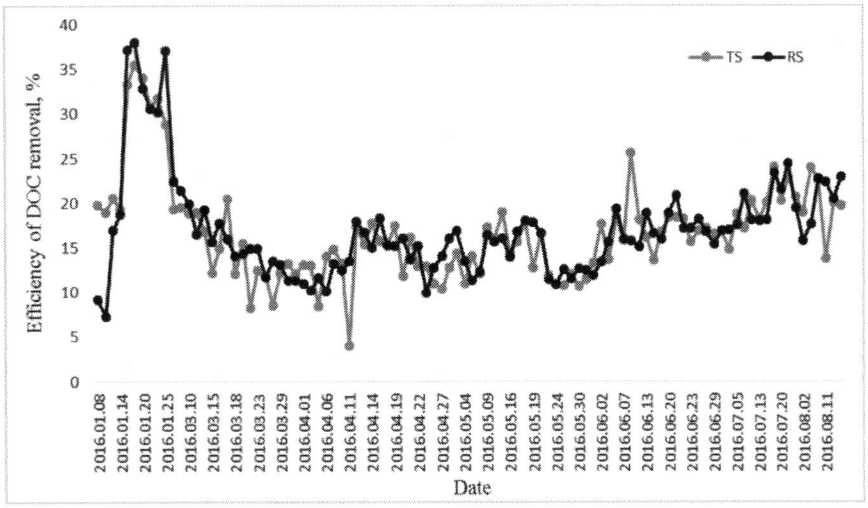

Fig. 4. DOC removal effectiveness during the absorption process.

The reduction in DOC was accompanied by a decrease in ultraviolet absorbance (UV_{254}), and therefore by a decrease in chlorinated organic substance precursors in the range of 18.3–54.3% in the test system and 19.5–54.2% in the reference system. The resulting effectiveness is comparable to that found during the coagulation process, which may point to the fact that large molecular mass compounds were removed during coagulation, while medium molecular mass compounds were removed during adsorption [21], with compounds absorbing UV light dominating the fraction removed in both processes.

Adsorption also provided for a reduction in UV_{272}, whose effectiveness was proportional to the reduction in color intensity at 340 nm, and to a lesser degree at 410 nm (fig. 5). An analogous relationship was found UV_{254} absorbance.

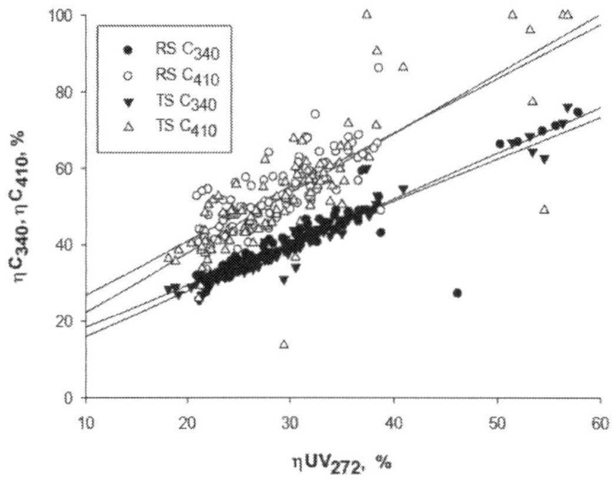

Fig. 5. Reduction in color intensity (340 nm and 410 nm) as a function of effectiveness in removing organic substances exhibiting absorbance at 272 nm.

No unambiguous impact of activated carbon-water contact time on the effectiveness of removing chlorinated organic compound precursors was found. This must be explained by very small differences in the time of water flow through the bed, because, as Esa et al [22] found, the greatest degree in removing organic substances takes place at contact times 20 minutes, with significant changes being observed in the first 20 minutes.

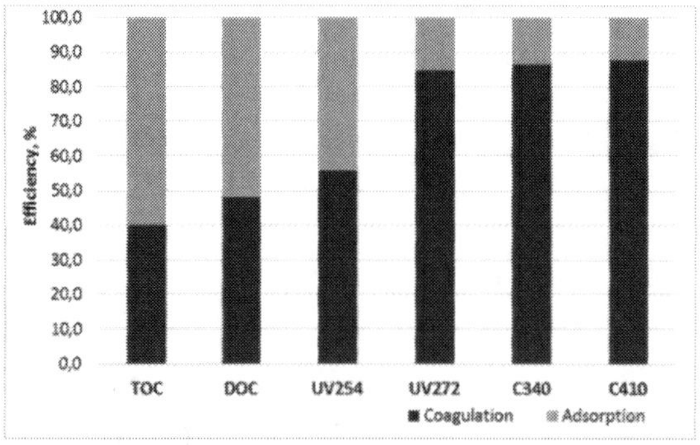

Fig. 6. Average effectiveness in reducing water organic contamination indicators (reference system).

A consequence of reduction in absorbance were further decreases in UV specific absorbance, which in water after adsorption amount to 0.92–1.95 $m^3/gC \cdot m$ and 0.89–2.07 $m^3/gC \cdot m$ respectively for reference and test systems. The reduction in UV absorbance reduces hazards to human health connected with the formation of disinfection by-products. The effectiveness in removing organic substances, and in particular chlorinated organic substance precursors, was influenced by these two described processes. However, no significant effect of filtration through a sand bed on organic substance content was found. Coagulation contributed mainly to removal of colored organic substances and reduction in UV_{272}, and absorption to reductions in UV_{254} absorbance. Ultimately, the adsorption process has a greater effect on reducing total and dissolved organic carbon (fig. 6).Water undergoing disinfection still contained precursors of chlorinated organic substances, however their concentrations were significantly lower than those in source water, and exhibited a lower variability with time. The chlorinated organic substance precursor content in water undergoing disinfection dependent above all on the source water quality (fig. 7). However, no unambiguous effect was found of source water temperature on UV_{254} and UV_{272} absorbance.

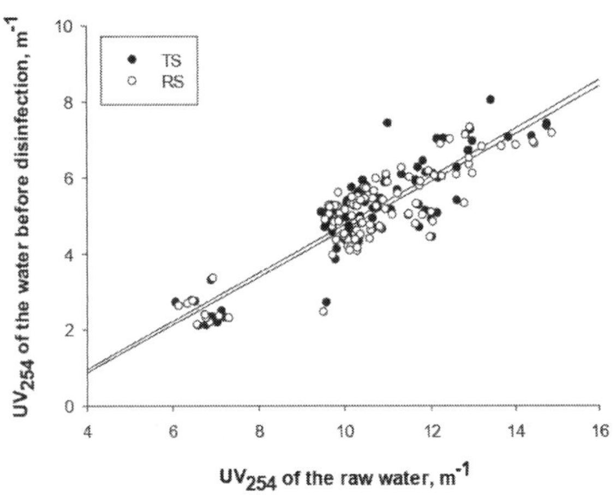

Fig. 7. Relationship between chlorinated organic substance precursor content in raw water and water before disinfection.

4 Summary

The study showed that:

- Chlorinated organic substance precursors, evaluated as content of substances exhibiting ultraviolet absorbance, are most effectively removed in coagulation and adsorption processes, and their effectiveness depends on the source water quality.
- Reducing the effect of source water temperature on the effectiveness of removing chlorinated organic substance precursors during the coagulation process was achieved by the use of a pre-hydrolyzed coagulant.
- Coagulation influenced the removal of colored organic substances to a larger degree than adsorption, which ensured a greater effectiveness in removing total organic carbon.
- The water-adsorption bed contact time had no effect on TOX precursor removal effectiveness.

- The effectiveness in removing chlorinated organic substance precursors in the water treatment system was decided by the source water quality.

This publication was made possible by the National Center for Research and Development grant (PBS3/B9/44/2015) "Research on effectiveness of new water treatment technology as a step towards a shift in thinking about water utility sector" (WODTECH), and thanks to involvement of both project consortium members (Wrocław Municipal Waterworks and Drainage Company and Wrocław University of Science and Technology).

References

1. Y-L. Lin, P-Ch. Chiang, E.E. Chang, J. Hazard. Mater. **146**, 20–29 (2007)
2. B. Kalajdžić, M. Habuda-Stanić, Ž. Romić, M. Kuleš, Global Nest J. **15**, 13–20 (2013)
3. S.D. Richardson, D.M. DeMarini, M. Kogevinas, P. Fernandez, E. Marco, C. Lourencetti, R. Marcos, Environ. Health Persp. **118**, 11, 1523 (2010)
4. J. Fang, J. Ma, X. Yang, C. Shang, Water Res. **44**, 6, 1934–1940 (2010)
5. M.J. Plewa, J.E. Simmons, S.D. Richardson, E.D. Wagner, Environ. Mol. Mutagen. **51**, 8–9, 871–878 (2010)
6. T. Bond, J. Huang, M.R. Templeton, N. Graham, Water Res. **45**, 15, 4341–4354 (2011)
7. W. Chu, N. Gao, D. Yin, Y. Deng, M.R. Templeton, Chemosphere **86**, 11, 1087–1091 (2001)
8. A. Matilainen, M. Vepsäläinen, M. Sillanpää, Adv. Colloid Interfac. **159**, 2, 189–197 (2010)
9. W. Xu, B. Gao, Y. Wang, Z. Yang, X. Bo, Chem. Eng. J. **171**, 3, 926–934 (2011)
10. J. Altmann, A.S. Ruhl, F. Zietzschmann, M. Jekel, Water Res. **55**, 185–193 (2001)
11. K. Yapsakli, F. Çeçen, Process Biochem. **45**, 3, 355–362 (2010)
12. A. Dąbrowska J. Świetlik J. Nawrocki, Water Res. **37**, 5, 1161–1169 (2003)
13. S. Fass, J.C. Block, M. Boualam, V. Gauthier, D. Gatel, J. Cavard S. Benabdallah, V. Lahoussine, Water Res. **37**, 3, 493–500 (2003)
14. G.V. Korshin, W.W. Wu, M.M. Benjamin, O. Hemingway, Water Res. **36**, 13, 3273–3282 (2002)
15. D.A. Reckhow, P.C. Singer, R.L. Malcolm, Environ. Sci. Technol. **24**, 11, 655–1664 (1990)
16. C.M.M. Bougeard, E.H. Goslan, B. Jefferson, S.A. Parsons, Water Res. **44**, 4, 729–740 (2010)
17. H. Gallard, U. Von Gunten, Water Res. **36**, 1, 65–74 (2002)
18. G.V. Korshin, W.W. Wu, M. M. Benjamin, O, Hemingway, Water Res. **36**, 13, 3273–3282 (2002)
19. W. Lee, P. Westerhoff, Water Res. **40**, 20, 3767–3774 (2006)
20. S.K. Bekaroglu, N. O. Yigit, T. Karanfil, M. Kitis, J. Hazard. Mater. **183**, 1, 389–394 (2010)
21. C. Guigui, J.C. Rouch, L. Durand-Bourlier, V. Bonnelye, P. Aptel, Desalination **147**, 1–3, 95–100 (2002)
22. E.S. Melin, H. Ødegaard, Water Res. **34**, 18, 2001 (2000)

 DOI: 10.1051/e3sconf/20172200189

Rheological properties of disintegrated sewage sludge

Paweł Wolski[1,*]

[1]Czestochowa University of Technology, Faculty of Infrastructure and Environment, Institute of Environmental Engineering, ul. Brzeznicka 60a, 42-200 Czestochowa, Poland

Abstract. The rheology of the sludge provides information about the capacity and the flow, which in the case of project tasks for the hydraulic conveying installation is an important control parameter. Accurate knowledge of the rheological properties of sludge requires the designation of rheological models. Models single and multiparameter (Ostwald, Bingham, Herschel-Bulkley'a, and others) allow an approximation of flow curves, and the determination of the boundaries of the flow of modified sludge allows you to control the process compaction or are dewatered sludge undergoing flow. The aim of the study was to determine the rheological parameters and rheological models of sludge conditioned by physical methods before and after the process of anaerobic digestion. So far, studies have shown that the application of conditioning in the preparation of sewage sludge increases shear stress, viscosity as well as the limits of flow in relation to the untreated sludge. Offset yield point by the application of a conditioning agent is associated with decreased flowability tested sludge, which has also been observed by analyzing the structure of the prepared samples. Lowering the yield point, and thus the shear stress was recorded as a result of the fermentation test of disintegrated sludge.

1 Introduction

Rheology is the study of flow, dealing with the phenomena of deformation and flow caused by external mechanical stresses, occurring over a finite time [1]. Familiarizing with properties of bodies and the movement of some elements of a body with regards to the other is of essential importance to hydraulic transport. Determination of the models used for the description of bodies exposed to the effect of force can be used for its characterization e.g. after the process of conditioning or fermentation [2].

Rheological analysis of sewage sludge considered as multi-phase systems with flocculent structure reveals changes in their rheological parameters through the increase in dry matter content. Intervention in the structure of sludge through application of conditioning factors and exposing them to stabilization contributes to changes in the value of stress, viscosity and yield limits. Neglecting of these parameters may generate substantial design errors, thus leading to functional problems with the whole system of sludge treatment [3–5].

* Corresponding author: pwolski@is.pcz.czest.pl

 DOI: 10.1051/e3sconf/20172200189

In the research on rheological properties, apart from determination of flow curves and viscosity, one should also develop rheological models which allow for approximation of flow curves [6, 7]. Among common rheological models used for the description of flow curves is the Ostwald-de Waele power model:

$$\tau = k \cdot (\gamma)^n \tag{1}$$

where the constant k [Pa*s], termed consistency coefficient, and the exponent n, termed the yield exponent, are the rheological parameters determined empirically at a specific temperature. The coefficient k is the measure of viscosity, which means that the higher the k value the more viscous the fluid. Yield exponent n adopts values of $n < 1$ (shear thinning), $n > 1$ (shear thickening) and $n = 1$ (Newtonian fluid).

In the Bingham model, for lower fluid stresses, fluids behave as plastic solid bodies, whereas flow is observed only after application of the shear stress τ_o:

$$\tau = \tau_o + \eta_{pl} \cdot \gamma \tag{2}$$

value of yield stress τ_o [Pa] and plastic viscosity η_{pl} [Pa•s] are parameters that are also determined experimentally. In the case of the three-parameter model, e.g. the Herschel-Bulkley model, apart from the yield stress τ_o, the constant K is also used, similar to n:

$$\tau = \tau_o + K \cdot (\gamma)^n \tag{3}$$

Determination of the curves of flow and viscosity and the analysis of rheological models contributes to the development of precise knowledge about the fluids, which is directly connected with security of sending the fluids [8–10].

The aim of the study was to determine rheological parameters of sewage sludge after physical conditioning, followed by the stabilization process.

2 Methodology

The examinations concerned sewage sludge from the cellulose treatment plant with initial hydration of 98.4%, whereas total solids (TS) was 1.6 g/dm^3. The sonication process was conducted under static conditions using the ultrasound field with intensity of 2.7 W/cm^2; 3.2 W/cm^2 and 3.8 W/cm^2. The initial power of the ultrasonic processor was 1500 W, with vibration frequency of 20 kHz, and maximal wave length of 39.42 μm. Sonotrod emitting an ultrasonic field was placed under the surface of the modified sludge, 2 cm from the bottom of the beaker in which the sludge were conditioned. The volume of the sonicated sludge samples was 500 cm^3, whereas the exposure time was 600 s.

Pre-sonicated sludge were subjected to a fermentation process in parallel in the glass flasks with capacity of 0.5 dm^3 and in the lab-bioreactor with capacity of 7 dm^3 (in both cases the laboratory models of the fermentation chamber). The analysis concerned non-conditioned and non-fermented sludge, sludge after conditioning and fermentation for 5 and 10 days in glass flasks, and sludge after conditioning and fermentation for 25 days in the lab-bioreactor. The fermentation process, both in flasks and bioreactor, was carried out under mesophilic conditions (37°C). In order to mix and maintain the set temperature, the flasks were placed on magnetic stirrers and inserted into a laboratory thermometer. In the bioreactor, the set temperature was maintained with the aid of heaters, while stirring was continued using a stirrer. Parameters and rheological models were determined for sewage sludge at the shear rate of 0-200s^{-1} for the period of 120s, using the cylindrical rheometer RC20.

3 Results

Analysis of the results of the study revealed the effect of the conditioning factor (ultrasound field) and the process of stabilization conducted in the anaerobic environment on changes in shear stresses, viscosity and yield limit. Analysis of the values of the Ostwald model showed a reduction in viscosity, expressed by the consistency coefficient, with elongation of the fermentation time (see Fig. 1). This value was reduced from 0.143 Pas (non-conditioned and non-fermented sludge) to 0.028 Pas on the 10th day of fermentation. In the case of the bioreactor, consistency was 0.053 Pas. A five-time increase in viscosity was observed in the case of sludge exposed to the effect of the ultrasound field. In the model, the value of non-fermented sludge for the lowest ultrasound field intensity (2.7 W/cm^2) was 0.541 Pas and it continued to increase to 0.585 Pas for the highest field intensity (3.8 W/cm^2). The reduction in viscosity for all three values of the wave length was found after fermentation of the samples. The lowest consistency (0.076 Pas) was observed on the 10th day of fermentation for the sludge sonicated with ultrasound field with intensity of 3.8 W/cm^2. The yield exponent n was increasing on each day of the fermentation process. For each tests, the exponent was below unity ($n < 1$), ranging from 0.38 to 0.63.

The values obtained correlated with rheological parameters determined empirically (Fig. 2). The values of viscosity were also rising with elongation of the ultrasound field effect, whereas fermentation caused a reduction in the parameter studied.

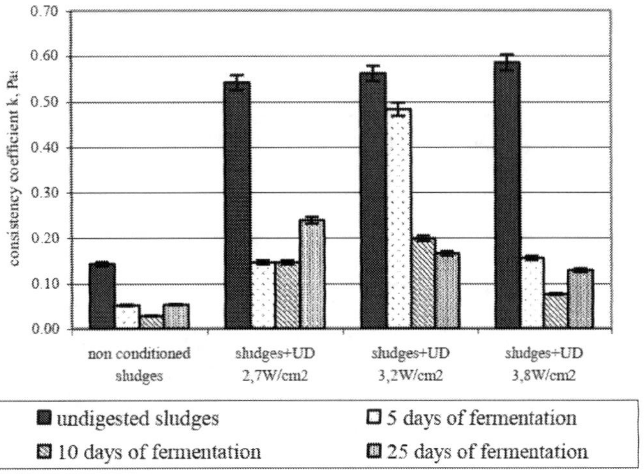

Fig. 1. The dependency of the consistency coefficient determined in the Ostwald model on ultrasound field intensity and fermentation time.

 DOI: 10.1051/e3sconf/20172200189

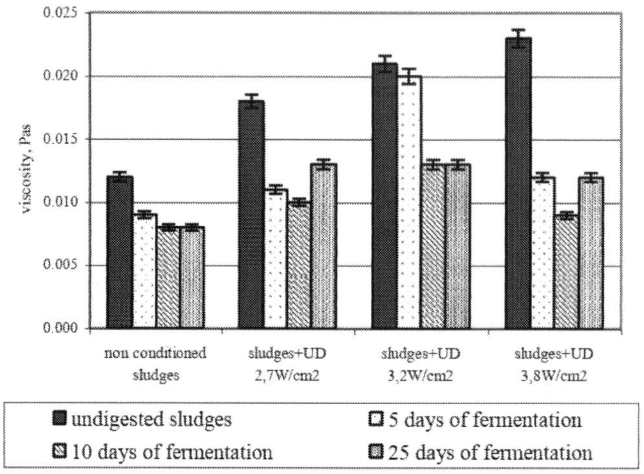

Fig. 2. The dependency of the empirically determined viscosity on ultrasound field intensity and fermentation time.

The analysis of the Bingham model revealed the effect changes in the conditioning factor and fermentation on rheological parameters (Fig. 3). The flow limit in the case of the non-conditioned sludge after fermentation was reduced on consecutive days of the process. This parameter was reduced for three times, from 0.466 Pa (non-fermented sludge) to 0.154 Pa (sludge on 10th day of stabilization). Similar to viscosity, elongation of the yield stress was affected by the use of the energy of the ultrasound field. The most substantial effect of the shift in this rheological parameter was observed in the case of non-fermented sludge after exposure to the ultrasound field with intensity of 3.8 W/cm^2. A substantial reduction in the yield stress was found after exposure of the sewage sludge to fermentation. Yield stress was reducing for each ultrasound field wave length, with the lowest value (0.276 Pa) observed on the 10th day of fermentation after sonication with field intensity of 3.8 W/cm^2. The analogous pattern was found for the Herschel-Bulkley model (Fig. 4). The yield stress values were insignificantly lower compared to yield stress in the Bingham model.

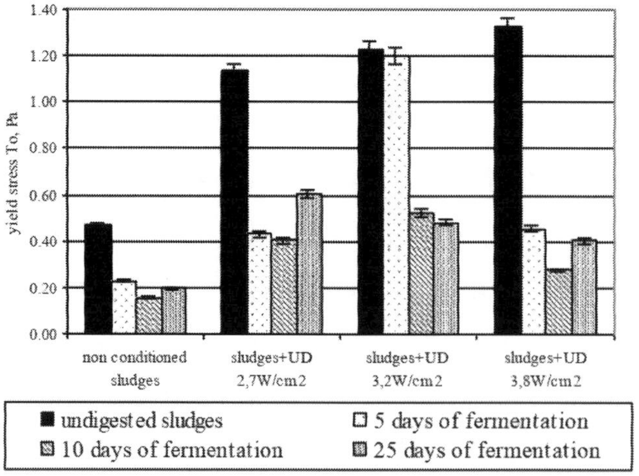

Fig. 3. The dependency of the yield stress determined in the Bingham model on ultrasound field intensity and fermentation time.

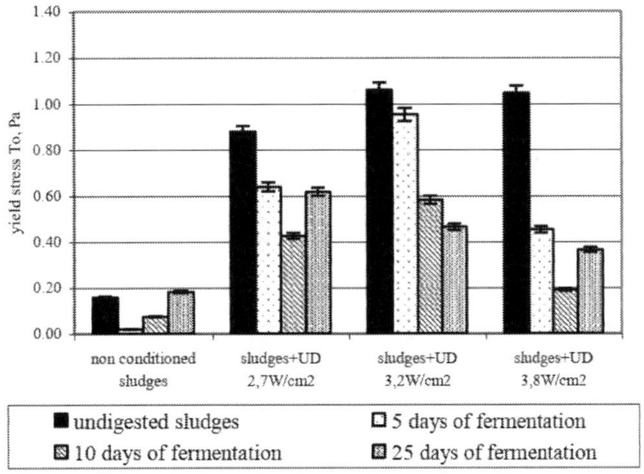

Fig. 4. The dependency of the yield stress determined in the Herschel-Bulkley model on ultrasound field intensity and fermentation time.

The yield stress values determined in the models were correlated with the values of yield stress determined empirically. Values of stresses in sewage sludge increased directly proportionally to the ultrasound field intensity, reaching the highest value (4.616 Pa) at the intensity of 3.8 W/cm^2 (Fig. 5). Similar to viscosity and yield stress, the use of fermentation led to a reduction in shear stress in sludge samples.

It is remarkable that TS of sewage sludge after conditioning with the ultrasound field and fermentation was changing. The increase in daily reduction in TS was observed on consecutive days of the stabilization process (see Fig. 6). However, the most noticeable reduction effect was found for sonicated sludge. On the 25th day of fermentation, the decrease in TS of the sludge conditioned with the ultrasound field with intensity of 3.2 W/cm^2 was 2.26 g/dm^3, whereas this value for the exposure to the field intensity of 3.8 W/cm^2 was 4.69 g/dm^3.

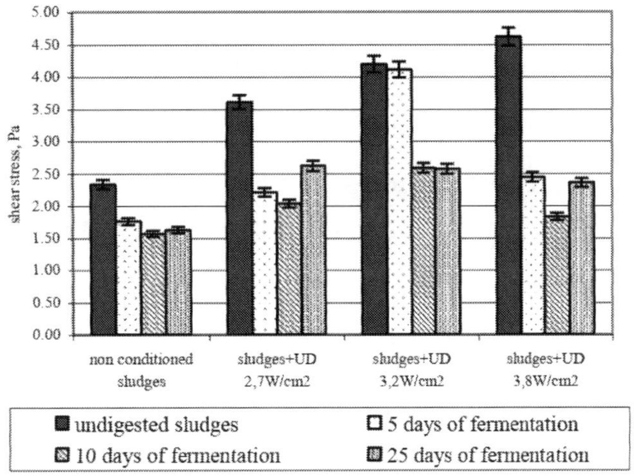

Fig. 5. The dependency of the empirically determined shear stresses on ultrasound field intensity and fermentation time.

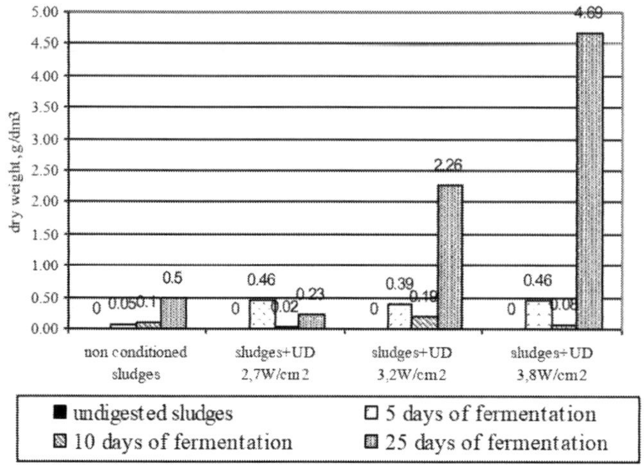

Fig. 6. The dependency of daily decrease in dry matter on ultrasound field intensity and fermentation time.

4 Conclusions

Knowledge of the properties and rheological parameters of fluids is essential to minimize calculation errors connected with their processing and hydraulic transport. It is also important to support dewatering and, more and more often, fermentation, through exposure of sewage sludge to physical conditioning. As observed based on the examinations, conditioning through sonication has an effect on changes in rheological parameters (viscosity, shear stress, yield stress). Fermentation also has a significant effect on hydraulic properties of sewage sludge.
The analysis of the results obtained in this study leads to the following conclusions:

- Consistency coefficient (which represents the measure of viscosity in the Ostwald model) and the empirically determined viscosity rose as ultrasound field intensity increased. The stabilization process led to the reduction in the value of the parameters discussed.
- Values of yield limit and stresses in the sewage sludge were increased after sonication. After the fermentation process, yield limit was reducing as stabilization time was elongating.
- The ultrasound field intensified the process of fermentation, which is reflected by the more substantial daily decrease in TS.

The research was funded by the project No. BS – PB – 401/303/12

References

1. M.M. Sozański, E.S. Kempa, K. Grocholski, J. Bień, Water Sci. Technol. **36**, 69–78 (1997)
2. R. Aranowski, J. Hupka, Ch. Jungnickel, Physicochem. Probl. Miner. Process **44**, 13–22 (2010)
3. B. Fryźlewicz-Kozak, O. Długosz, Inż. Ap. Chem. **54**, 31–32 (2015) (in Polish)

DOI: 10.1051/e3sconf/20172200189

4. B. Fryźlewicz-Kozak, J. Jamróz, M. Pachołek, Inż. Ap. Chem. **54**, 33–35 (2015) (in Polish)
5. P. Travnicek, T. Vitez, P. Junga, E. Krcalova, J. Sevcikova, Pol. J. Environ. Stud. **1209–1212** (2013)
6. M. Ruiz-Hernando, G. Martinez-Elorza, J. Labanda, J. Lorens, Chem. Eng. J. **230**, 102–110 (2013)
7. Y. Dong, Y. Wang, Acta Scientiae Circumstantiae, **678–682** (2012)
8. I. Zawieja, L. Wolny, Annual Set Environ. Protect. **13**, 1719–1730 (2011)
9. F. Markis, J. Baudez, R. Parthasarathy, P. Slatter, N. Eshtiaghi, Chem. Eng. J. **253**, 526–537 (2014)
10. L. Wolny, P. Wolski, I. Zawieja, Desalination **222**, 1, 382–387 (2008)

How effective is aeration with vortex flow regulators? Pilot scale experiments

Patryk Wójtowicz[1,*], and *Małgorzata* Szlachta

[1]Wrocław University of Science and Technology, Faculty of Environmental Engineering, Wybrzeże Wyspiańskiego 27, 50-350 Wrocław, Poland

Abstract. Vortex flow regulators (VFR) are used in urban drainage systems as a replacement for traditional flow throttling devices. Vortex regulators are not only very efficient energy dissipators but also atomizers which are beneficial for sewer aeration. A deficit of dissolved oxygen can be a problem in both natural waters and sewerage. Hydrodynamic flow regulators can boost oxygen concentration preventing putrefaction and improving treatment of stormwater and wastewater. We were first to investigate the aeration efficiency of semi-commercial scale cylindrical vortex flow regulators to determine the potential of their application in environmental engineering and to propose modification to enhance the aeration capacity of basic designs. Different device geometries and arrangements of active outlets for both single and double discharge vortex regulators were tested in a recirculating system. In this study, we present a concise review of the current state of our extensive research on the aeration efficiency of vortex flow regulators and their application in sewerage systems.

1 Introduction

Vortex flow regulators (VFR) (known as hydrodynamic flow regulators) are used in urban drainage systems as a replacement for traditional flow throttling devices. Vortex regulators are very versatile and can be used in gravitational as well as pressurised systems with different configurations. Their application is not limited to storm or wastewater systems. They can be used in other fields too, e.g. in water distribution – for water hammer protection or in hydraulic structures – for limiting outflow from a dam [1].

From a practical point of view, vortex regulators are compact, efficient in throttling, reliable and free from common disadvantages of the traditional regulation devices. To throttle the flow efficiently, they do not need to reduce the pipe cross-section, do not require power supply or moving parts. Vortex devices are fascinating with their simple design and straightforward operational principle, as opposed to a highly complex vortex motion employed, which defies a complete solution for over 100 years now [2–5].

Our research indicated that the atomisation of a liquid by vortex flow regulators accelerates oxygenation and prevents the putrefaction process in wastewater and storm water collection systems [6]. This feature is a positive side effect – intrinsic to the

* Corresponding author: patryk.wojtowicz@pwr.edu.pl

operational principle. A deficit of dissolved oxygen is a known problem in both natural waters and sewerage systems [7, 8]. Therefore, there is an important need to utilise effective and inexpensive methods and devices to restore the oxygen balance [9].

To date, there has been no systematic research available in the literature regarding the aeration efficiency of vortex flow regulators and this was a main motivation for this long-term study. We were first to investigate the aeration efficiency of semi-commercial scale cylindrical vortex flow regulators to determine their application potential in environmental engineering and propose modification to enhance the aeration capacity of basic designs. To obtain research goals both experimental and computational methods were used at laboratory and semi-commercial scale [10]. A wide range of device geometries and arrangements of active outlets for both single and double discharge vortex regulators (patented) were tested. The main assumption was that modifications should not compromise any of the advantages and especially simple structure, reliability and extreme throttling efficiency.

In this paper, we aim to present the current state-of-the-art cylindrical vortex regulator (CVR) designs and a concise summary of several years of research on aeration efficiency of vortex flow regulators.

2 Materials and methods

Semi-commercial scale models of hydrodynamic flow regulators were tested in a recirculating hydraulic system. The experimental setup consisted of a feed tank equipped with a circulation pump, an upper tank and a supply system where the models were tested. The flow rate and pressure drop were continuously recorded while the air core diameter and spray cone angle were determined based on photographs.

Different device geometries and arrangements of active outlets for both single and double discharge vortex regulators were tested. The experimental programme included tests of three types of vortex regulators shown in Figure 1a–c. The first type (CVR – Fig. 1a) is well-known, basic designs originally developed to maximise throttling performance. They discharge only through the bottom outlet. Using the CVR type as a baseline, we proposed two new types with enhanced aeration capability (both have been patented). The first modification was the addition of a second outlet in the top plate (CVR-2 – Fig. 1b). In the next modification (CVR-U – Fig. 1c), only the top outlet is used, although the bottom outlet was open and used to drain the device during low flows.

In the flow regulator, the liquid flows into the device through the tangential inlet, imposing the vortical motion, which is sustained until the liquid reaches the outlet (Fig. 1). An air core is formed in the centre, obstructing the effective cross-section of the outlet. Discharged liquid leaves the device in the form of hollow spray cone.

 DOI: 10.1051/e3sconf/20172200190

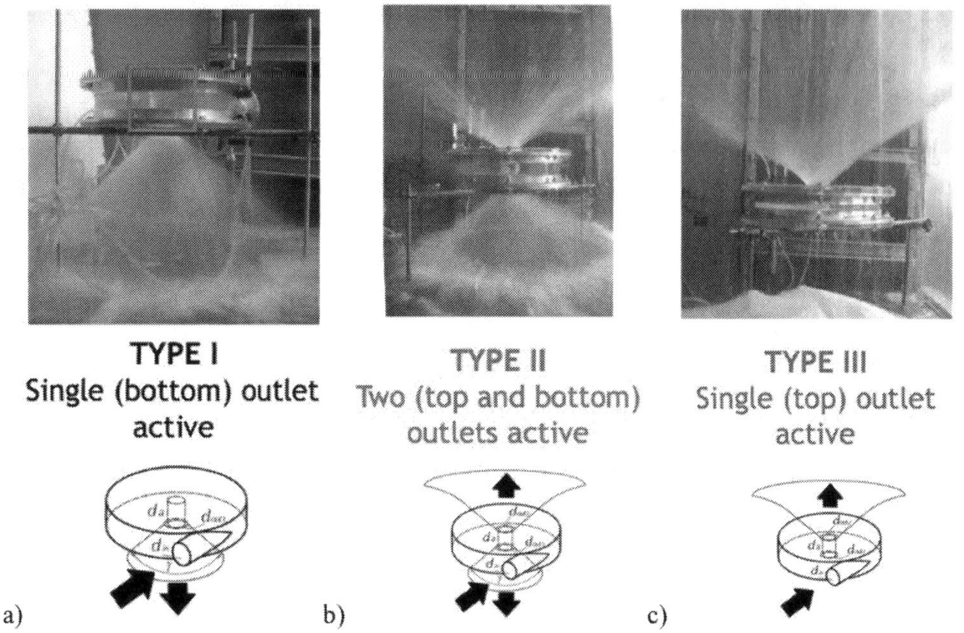

Fig. 1. Overview of three types of cylindrical vortex regulators tested (aircore is indicated by red square).

When the flow rate is low enough, the resistance is minimal and liquid flows freely, resembling "elbow flow". But when the pressure at the inlet increases, the induced vortex motion becomes stronger, intensifying the throttling effect. An interesting fact is that the flow field inside the vortex chamber is like a tornado (Rankine vortex [2]).

The aeration theory applied for vortex flow regulators is based on Henry's law, in particular the oxygen transfer in water or sewage is regulated by the molecular diffusion theory, governed by Fick's law. This general law predicts that the mass transfer rate, in the steady state, is proportional to the gas concentration gradient across the gas-liquid interface [11]. By applying this law to a water-oxygen system together with the film theory, it is possible to obtain the interphase mass transfer rate equation. This relationship allows to extract the values of the oxygen transfer coefficient based on the experimental data.

The absorption tests were performed in compliance with the unsteady state clean water test procedure [12–14] in a recirculating system. The aeration efficiency was measured by means of the overall oxygen transfer coefficient (k_La_{20}, h^{-1}) adjusted to the standard conditions (20°C, 101.325 kPa), standard oxygen transfer rate (SOTR, kgO_2/h) and standard aeration efficiency (SAE, kgO_2/kWh).

Tap water in the system was deoxygenated with the use of sodium sulphite catalysed by cobalt chloride. The dissolved oxygen concentration in the system was monitored with Hach LDO probes at several upstream and downstream locations. The total dissolved solids concentration was monitored by a WTW TetraCon 325 electrical conductivity probe. Barometric pressure as well as water and ambient air temperature were constantly recorded during the experiments. A detailed description of experimental and chemical methods used in this study is included in our recent paper [10].

 DOI: 10.1051/e3sconf/20172200190

3 Results

Each vortex flow regulator type was rigorously tested for several different flow regimes to qualitatively and quantitatively describe its aeration and hydraulic efficiency.

3.1. Flow rate vs. oxygen transfer coefficient and SOTR

We found that the oxygen transfer coefficient (k_La_{20}) and the oxygen transfer rate (SOTR) are in direct proportion to the flow rate and that all regulators types follow the same trend (*cf* Fig. 2 for CVR-2 type).

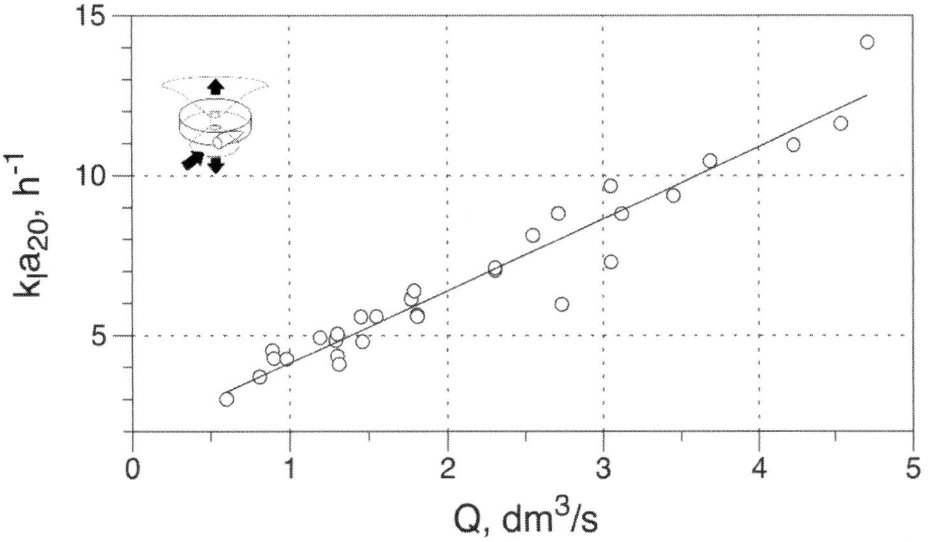

Fig. 2. Flow rate vs oxygen transfer coefficient for CVR-2 type of vortex regulator.

This result can be explained by increased turbulence and renewal rate of the liquid film available for absorption with an increasing flow rate [11]. Oxygen absorption in vortex regulators depends on several competing phenomena and the result is always a superposition of these governing parameters.

3.2. Pressure head vs. standard aeration efficiency

We found that the pressure head before the regulator describes the relationship with standard aeration efficiency (SAE) better than the flow rate. This parameter was decreasing with increasing pressure – approaching a limiting value above approximately 2 meters of the head (Fig. 3). The relationship between these two parameters is described with an exponential function, contrary to correlation between k_La_{20} and SOTR.

DOI: 10.1051/e3sconf/20172200190

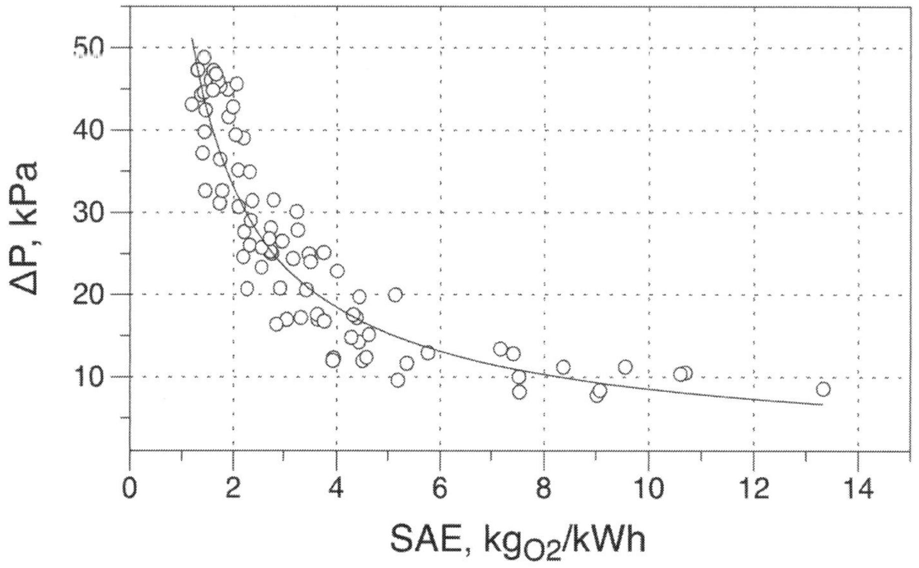

Fig. 3. Standard aeration efficiency vs. pressure head for all cylindrical vortex regulators tested.

3.3. Effect of active outlets: one bottom (CVR) vs. two active outlets (CVR-2)

In the type CVR-2 we were expecting to double the aeration efficiency by adding a second outlet. Indeed, the experiments proved that there was an increase, but not proportional to the number of active outlets. The oxygen transfer coefficient and the standard oxygen transfer rate were improved up to 20%. Slightly better results were obtained for standard aeration efficiency. The addition of the second outlet divided the outflow between the top and the bottom one – this reduced the effective flow per area and the quality of atomisation. Also, the splashing and formation of air bubbles in the downstream basin were decreased. On the other hand, the second spray cone was formed and the contact time was longer. Of course, this effect was limited by the local saturation of droplets as the turbulence and consequently the renewal rate of the liquid film contacting with air was relatively low. We think that the major effect that limited the expected gain in aeration efficiency, was the quality of atomisation.

3.4. Effect of active outlets: one top (CVR-U) vs. two active outlets (CVR-2)

We observe similar competing effects for modified regulators operating with two outlets (CVR-2) and one top outlet (CVR-U). Our experiments indicated that devices operating only with the top outlet had a slightly better aeration efficiency comparing to two active outlets.

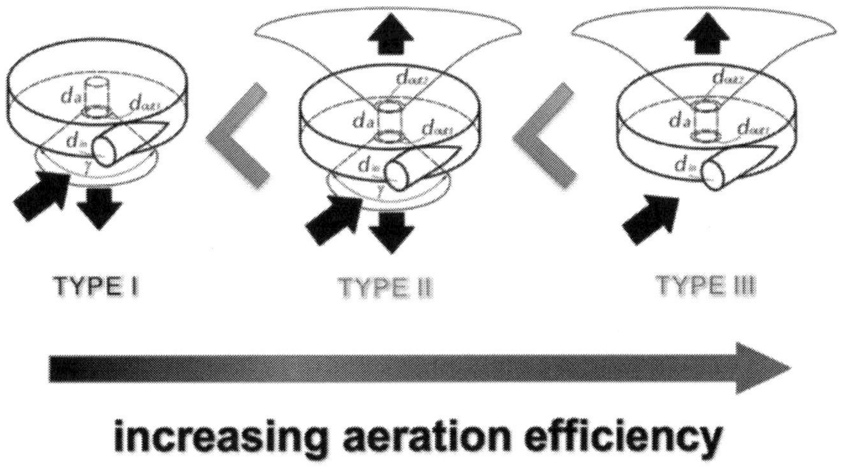

Fig. 4. Comparison of VFR aeration efficiency.

This may prove an argument that the quality of atomisation had a major effect on the aeration efficiency of vortex regulators [15]. In this case, the increase of specific interfacial area in the second active outlet was compensated by the better atomisation and longer contact time for the regulator operating only with the top outlet.

3.5. Effect of active outlets: one bottom (CVR) vs one top outlet (CVR-U)

Comparing two modes of operation with outlets located on the opposite sides (CVR vs. CVR-U) – the oxygen transfer coefficient and standard oxygen transfer rate were improved up to 20% in favour of the regulators discharged with the top outlet (CVR-U). The hydraulics was similar but with the CVR-U type of regulators, the fountain-like discharge combined with inevitable gravity improved the disintegration of the liquid film and increased the contact time of droplets. On the other hand, splashing and air entrainment in the downstream basin were reduced, but the effect was not substantial.

Figure 4 summarises and compares all three types of regulators tested according to aeration efficiency.

4 Conclusions

The oxygen absorption in vortex regulators depends on several competing phenomena – the end result is always a superposition of these effects. We found that the best aeration efficiency can be obtained for a regulator discharged through the top outlet only (CVR-U). In general, the aeration capability of basic devices (CVR) can be increased up to 20% with a negligible effect on throttling performance. The aeration efficiency can be increased even further by combining two chambers and adding more active outlets. The aeration performance is constant and on a par with mechanical aerators and diffuser systems. Future work will concentrate on testing modified vortex regulators (CVR-2 and CVR-U) in full-scale application.

The major part of this research was made possible by a grant from the Polish Ministry of Science and Higher Education N N523 450036.

References

1. P. Wójtowicz, A. Kotowski, J. Hydraul. Res. **47**, 5 (2009)
2. A.K. Gupta, D.G. Lilley, N. Syred, *Swirl flows* (Abacus Press, Tunbridge Wells, Kent, 1984)
3. R.R. Long, J. Meteorol. **15**, 1 (1958)
4. M. Escudier, Annu. Rev. Fluid Mech. **19**, (1987)
5. H.A. Einstein, La Houille Blanche **10**, 4 (1955)
6. P. Wójtowicz, M. Szlachta, Water Sci. Technol. **67**, 12 (2013)
7. A.H. Nielsen, J. Vollertsen, H.S. Jensen, T. Wium-Andersen, T. Hvitued-Jacobsen, Water Res. **42**, 15 (2008)
8. N.A. Jensen, Water Environ. Res. **67**, 6 (1995)
9. J.L. Huisman, N. Weber, W. Gujer, Water Res. **38**, 5 (2004)
10. P. Wójtowicz, M. Szlachta, Chem. Eng. J. **287**, (2016)
11. W.K. Lewis, W.G. Whitman, Ind. Eng. Chem. **16**, 12 (1924)
12. ASCE, Measurement of oxygen transfer in clean water, American Society of Civil Engineers, ANSI/ASCE 2-91, Reston, Va. (2007)
13. DWA, Measurement of the Oxygen Transfer in Activated Sludge Aeration Tanks with Clean Water and in Mixed Liquor, German Association for the Water Environment, (1996)
14. CEN, Wastewater treatment plants W Part 15: Measurement of the oxygen transfer in clean water in aeration tanks of activated sludge plants (EN 12255:15), European Comittee for Standarization, (2003)
15. A. Lefebvre, *Atomization and sprays* (Taylor&Francis, New York, 1989)

Prospects for the use of SMR and IGCC technologies for power generation in Poland

Artur Wyrwa[1,*], *Wojciech* Suwała[1]

[1]AGH University of Science and Technology, Faculty of Energy and Fuels, Al. A. Mickiewicza 30, 30-059 Krakow, Poland

Abstract. This study is a preliminary assessment of prospects for new power generation technologies that are of particular interest in Poland. We analysed the economic competitiveness of small size integrated gasification combined cycle units (IGCC) and small modular reactors (SMR). For comparison we used one of the most widely applied and universal metric i.e. Levelized Cost of Electricity (LCOE). The LCOE results were complemented with the results of energy-economic model TIMES-PL in order to analyse the economic viability of these technologies under operation regime of the entire power system. The results show that with techno-economic assumptions presented in the paper SMRs are more competitive option as compared to small IGCC units.

1 Introduction

The European Union aspires to be a world leader in sustainable energy development. It has notably committed to decrease its CO_2 emissions by 20% and 40% in 2020 and 2030, respectively in relation to 1990 level. In long term the 80–95% CO_2 reduction is foreseen by 2050. For countries which energy systems are heavily dependent on fossil fuels such as Poland this implies ambitious efforts to transform the energy system that may include: fuels and technologies switch to less-carbon intensive ones e.g. renewable energy sources (RES) or nuclear, improving energy efficiency, use of emission reduction technologies, including capture and storage (CCS – Carbon Capture and Storage) [1]. The energy strategy of Poland assumes that in transition period that will last approximately for the next three decades and during which a new EU energy system will be developed (i.e. RES-based and more flexible due to energy storage) coal will be still a dominant fuel for power generation.

Being a country with abundant coal resources the use of domestic fuel guarantees energy security. New ways of utilisation of coal are under consideration which include i.a. coal gasification in CCS-equipped IGCC units. CCS technology involves carbon dioxide capture from flue gases and then its storage in tight underground geological structures. All stages of the process, namely capture, transport, and storage, are mastered and used commercially, but not in the power industry. Fundamental problems are the high cost of carbon capture in power plants, a significant reduction in the power plants efficiency and a social acceptance. It cannot be prejudged at present whether the CCS technology will be

* Corresponding author: awyrwa@agh.edu.pl

commercially available for power sector in the near future since the limited progress in its development has been achieved for the last ten years. However CCS still seems to be the possible way to use fossil fuels under the stringed EU climate regime [2].

The share of coal in the Polish energy mix will gradually decrease. This will give a space for new power generation technologies such as nuclear, which has been considered in the Polish energy policy. The plan of constructing a classical nuclear power plant seems to be delayed due to problems in finding appropriate financial schemes for these highly capital intensive investments. Classical nuclear technologies include reactors with the electric capacity of 1000 MW or more. The current investment costs estimate is ca. 5500 USD/ kW (built AP1000 reactors in the USA) and 7000 USD/ kW (Hinkley Point C in the UK) of installed power while the costs of classic coal technologies are aporox. 2500 USD/kW. Such large investments, coupled with construction delays raise significantly the capital costs. The other interesting alternatives are small modular reactors (SMRs) [3, 4]. SMRs can be developed based on two mature technologies i.e. light water reactors (LWR) or a high temperature reactors (HTR). In December 2012, concrete was poured into the construction of a set of two HTR reactors for a 210 MW electric power plant to be commissioned in 2017. The efforts have been made in the US to implement small-capacity pressurized water reactors. NuScale company plans to launch the first Integral Pressurized Water Reactor in the US in 2024, and the 600 MW power plant built from 12 SMRs in 2025. In Europe, the biggest interest in SMRs lies in the UK.

In this paper we analysed the perspectives of the commercial application of small size IGCC and SMR power plants in Poland. For comparing the costs of different technologies we used one of the most widely applied and universal metric i.e. Levelized Cost of Electricity (LCOE). The LCOE method has been complemented by the system analysis approach in which an energy-economic model TIMES-PL was used to in order to analyse the economic viability of these technologies under operation regime of the entire power system.

2 Description of the method

In this chapter at first main methods used for comparison of economic competitiveness of perspective power generation technologies are briefly described. Secondly, the input data used in calculations are presented. The Chapter ends with presentation of other most critical parameters that have an impact on the results.

2.1 Levelized Cost of Electricity

The Levelized Cost of Electricity (LCOE) is an universal metric widely applied to compare economic competitiveness of energy technologies . It is a ratio of the sum of discounted cost components and discounted electricity generation as presented below:

$$LCOE = \frac{\sum_{t=TP}^{TO}(N_t) * (1+r_t)^t + \sum_{t=TO}^{TK}(OM_t + F_t + CO2_t + L_t) * (1+r_t)^{-t}}{\sum_{t=TO}^{TK} P_t * (1+r_t)^{-t}}$$

$$- \frac{SV_t * (1+r_{TK})^{-TK}}{\sum_{t=TO}^{TK} P_t * (1+r_t)^{-t}} \quad (1)$$

The nomenclatures, superscripts and subscripts are listed in Appendix I and II.

Determining the exact time of analysis and year "0" can be a subject of the discussion. The investment does not start with the start of the construction of the plant but often is preceded by long-term activities related to obtaining the required permits, licenses or even persuading the local community to accept the investment (Fig. 1). Although the costs of individual activities in this period may seem to be insignificant, their cumulative and discounted value may affect the overall capital costs.

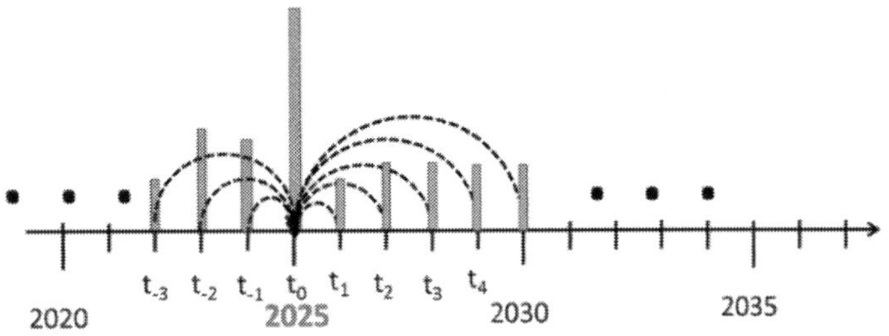

Fig. 1. Discount scheme for the year of production start (2025).

In this study we propose that year "0" is 2025 and this is the time by when the plants are already built and their electricity generation starts.

2.2 Energy-economic model TIMES-PL

TIMES is an economic model generator for energy systems. TIMES-PL is the name of the model of the Polish power system that belongs to the families of national models developed with the use of TIMES generator. It belongs to a class of bottom-up models providing a technology-rich basis for analysing energy system development over a long-term period. Its objective function represents the total costs of the supply of energy services, which are minimized by the model. In this study the mixed integer programming method was used to solve the model as for some technologies only investments in desecrate capacities were allowed. The detailed description of decision variables and equations of TIMES can be found in [5]. TIMES-PL includes all existing power plants as well as combined heat and power plants (the major plants are implemented in the model as separated units whereas renewables and small capacity units are aggregated by fuel and technology types). All technologies included in the model are characterized by a set of technological and economic parameters described in [6]. On top of these technologies two power plant types were added i.e. small size IGCC units and SMRs.

2.3 Input data

The main input data including costs of: power generation technologies, fuels, carbon credits are described in next paragraphs. The influence of discount rates is also discussed. For all calculations the decommissioning costs were set equal to 15% of the total investment cost for nuclear power plants and 5% for all other technologies. It was assumed that the salvage value of technologies after reaching their lifetime was zero.

2.3.1 Techno-economic data describing power generation technologies

Two techno-economic datasets were prepared for calculation of LCOE and for energy system analysis with the use of TIMES-PL, respectively. In the first case all data used were based on [7]. The referenced paper provide an overview of region-specific costs of power technologies for many countries. As also different data sources are indicated in this paper, in our study we used the data provided by the Eurelectric. Two values of investment costs were used for SMR technology (Table 1). The lower one i.e. 3000 USD/kW is very optimistic and served as the benchmark for showing the costs reduction potential due to learning effects with aggressive investments in this technology. The other one represent the forecasted value expected in 2025 when SMRs are planned to be available commercially. Additionally, it was assumed that four SMRs will be integrated to reach the electric capacity of 600 MW.

Table 1. Techno-economic data describing technologies used for LCOE calculation.

Technology	**Net electric capacity**	**Investment OVN**	**O&M variable**	**Operation hours**	**Lifetime**	**CO_2 emission**
	MW	[USD/ kW]	[USD/ kWh]	[hrs/ year]	years	[tCO_2/ MWh]
SMR6005k	600	5 000	16.0	8 000	60	0
SMR6003k	600	3 000	16.0	8 000	60	0
NUC1000	1 000	5 000	12.0	8 000	60	0
Coal1000	1 000	1 900	5.1	7 000	40	0.75
Lig1000	1 000	2 100	5.5	7 000	40	0.8
CCGT300	300	1 500	3.9	3 000	25	0.5
Coal1000_CCS	1 000	3 460	8.7	7 000	40	0.1

In the second case (Table 2) the data had to be adjusted to the requirements of TIMES-PL as new technologies considered in this study were added on the top of those that already are implemented in the model. Also the currency differs as in case of LCOE costs were expressed in USD whereas TIME-PL uses Polish zloty (PLN).

Table 2. Techno-economic data describing new technologies analysed in this study in TIMES-PL.

Fuel/technology	**Net electric capacity**	**OVN**	**O&M fixed**	**O&M variable**	**Efficiency 2015/30/50**	**Lifetime**	**CO_2 emission**
	MW	[PLN /kW]	[PLN /kW]	[PLN /MW]	%	years	kg/GJ
HC/IGCC	205	14 650	206	10.74	42.8/47/52	40	95.9
HC/IGCC+CCS	150	25 237	302	16.21	31.8/37/44	40	9.96
BC/IGCC	205	16 810	225	11.2	42.8/47/51	40	99.87
BC/IGCC+CCS	150	29 460	348	17.8	30.6/35/42	40	10.19
HC/IGCC	600	8000	210	12.0	44/51/52	40	94.19
HC/IGCC+CCS	600	12 360	294	33,5	43/45	40	11.30
BC/IGCC	600	8000	210	12.0	43/49/51	40	109.08
BC/IGCC+CCS	600	12 640	294	36	41/44	40	13.09
Nuclear/SMR	600	18500	315	9,5	36/37/37	50	95.9

DOI: 10.1051/e3sconf/20172200191

2.3.2 Fuel prices

Another parameter that has a significant impact on the competitiveness of IGCC and SMRs technologies is the fuel price. Fuel prices used for LCOE calculation are given for 2025 (year in which operation of power plants starts). TIMES-PL considers the modelling time horizon from 2011 till 2050. The prices presented in Table 3 are given for the base year i.e. 2011. In years coming after 2011 they follow the forecast presented in [8].

Table 3. Fuel prices.

Fuel type/method	LCOE (2025)		TIMES-PL (2011)	
Hard coal	3.60	[USD/GJ]	12.8	[PLN/GJ]
Brown coal	2.12	[USD/GJ]	7.1	[PLN/GJ]
Natural gas	9.76	[USD/GJ]	24.0	[PLN/GJ]
Nuclear fuel	9.30	[USD/MWh]	2.0	[PLN/GJ]

2.3.3 Prices of EU CO_2 emission allowances

Since 2020 electricity generators will have to fully pay for CO_2 emission (in fact they will need first to purchase the EU CO_2 emission allowances for emissions – EUAs on auctions). There are many studies devoted to the analysis of the evolution of EUAs prices in the short, medium and long term. As EUAs prices have a strong impact on results, two scenarios were adopted for carbon pricing in TIMES_PL (Table 4). First one, our "REF", refers to the high emission price scenario presented in [9]. The second, a "HIGH" scenario, refers to the CO_2 emissions allowance prices presented in the scenario "Current Policy Initiatives" of the EU Energy Roadmap 2050 [10].

Table 3. EUAs price scenarios considered in TIMES-PL [PLN/tCO_2].

Scenario	2015	2020	2025	2030	2035	2040	2045	2050
REF	41	62	62	70	74	78	82	87
HIGH	41	62	95	132	165	202	206	210

In the case of LCOE calculation the carbon price was set according to the value forecasted in the HIGH scenario for 2025 i.e. 25 USD/t CO_2.

2.3.4 Discount rates

A discount rate r is a real rate of return on investment. It is also used for calculation of the cost of capital. The most commonly used values are: 5, 8, 10 or 12%, depending on the degree of investment risk. A higher rate is assumed for investments with higher risk [11]. For LCOE calculation we proposed the approach which differentiate discount rates depending on the phase of the investment. During construction, the risk is higher because there is an uncertainty about the completion date and construction costs, so the discount rate should be higher (Table 4). Investment risk is the highest for nuclear power plants because of the possibility of prolonging construction time, slightly lower for coal-fired power plants and the lowest for gas fired power plants with short construction times. When power plants start to operate the risk is much smaller and is mainly related to the variations in electricity and fuel prices as well as operational accidents.

Table 4. Discount rates assumed for power generation technologies (%) for construction period.

	2017	...	2025	2026 onwards
SMR6005k	10	...	10	5
SMR6003k	10	...	10	5
Nuc1000	10	...	10	5
Coal1000	8	...	8	8
CCGT300	5	...	5	10
Lig1000	8	...	8	5
Coal1000 CCS	8	...	8	8

Another approach was used in TIMES-PL where the universal discount rates were used equal to 5% and 8% for all technologies in two different model runs, respectively. At first the investment costs were split into construction years based on the following formulae:

$$Fr_t = \int_{t+1}^{t} \frac{sin\theta_t}{2} \, d\,\theta_t, \qquad \theta_t = \frac{t \cdot \pi}{Tc} \tag{2}$$

Secondly, to costs of capital incurred during power plants construction were calculated.

3 Results

Calculations of LCOE metric with the use of Equation 1 gave the results presented in Table 5. As one can see, the lowest LCOE is for small nuclear reactors with a capital investment of 3k USD/kWe. However, as mentioned above, SMR6003k served as the benchmark and achieving such a low investment cost in the near future might be unrealistic. Another technology is a small reactor with a capital investment of 5k USD/kWe. The lignite-fired power plant has comparable costs to a large nuclear unit. The LCOE for hard coal fired power plant is about 20% higher than for a nuclear power plant (carbon capture and storage system raises costs further by ca. 15%).

Table 5. LCOE calculated for different power generation technologies.

Technology	SMR 6005k	SMR 6003k	NUC 1000	Coal 1000	Lignite 1000	CCGT 300	Coal 1000_CCS
LCOE [USD/MWh]	62.80	47.80	65.42	77.60	66.48	127.67	89.83

Since the same overnight costs for small and large nuclear technology were adopted (SMR6005k and NUC1000) [3], the difference in favour of the former is only due to the shorter construction time and hence to the lower capital costs.

As capital expenditures for power technologies cited in the literature vary a sensitivity analysis of LCOE was performed in which the unit investment costs (OVN) were differentiated for all technologies assuming normal distribution with standard deviations estimated in [7]. The results of 250 simulation runs are presented in Fig. 2.

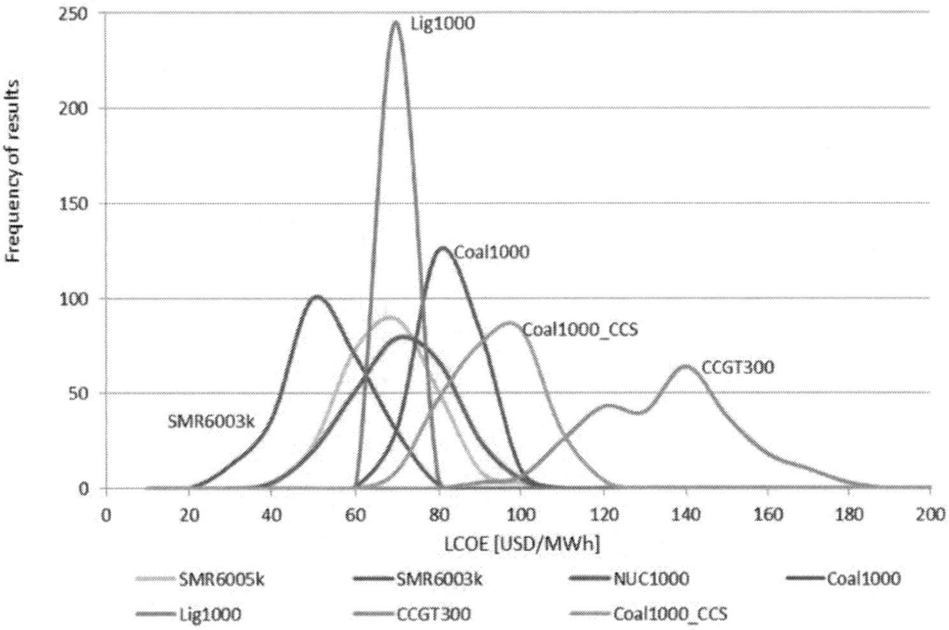

Fig. 2. Frequency distributions of LCOE caused by the change of investment costs (OVN).

The results of sensitivity analysis confirm that nuclear and lignite power plants are the most competitive. The results of TIMES-PL model runs in terms of new capacities constructed in the analysed time horizon are presented in Table 6.

Table 6. New electric capacities [GW] added by the model.

Scenario	Technology	r[%]	2025	2030	2035	2040	2045	2050
REF	Nuclear/SMR	5	-	1.8	2.4	0.6	-	-
HIGH	Nuclear/SMR	5	0.6	1.2	1.8	2.4	3.0	-
HIGH	Nuclear/SMR	8	0.6	1.2	1.8	2.4	3.0	-
HIGH	BC/IGCC+CCS	8				0.6	5.6	-

One can see that results depend on the CO_2 price scenario and discount rates considered. With the lower discount rate SMRs are constructed in both CO_2 price scenarios. In HIGH CO_2 price scenarios SMRs are constructed for both values of the discount rate and in the total maximum electric capacity allowed (the new capacity additions were constrained to reflect e.g. limitation in availability of resources, qualified constructing staff, etc.). Different results, however, were obtained for small IGCC units equipped with CCS, which have not been "constructed" by the model at all. Simulations with larger IGCC/CCS units (with lower investment costs per kW due to the scale effect) were done (only for one discount rate equal to 8%) and for different CO_2 price scenario. In HIGH CO_2 price scenario new capacities in brown-coal fired IGCC units were added by the model. The electricity generation mix for HIGH CO_2 price scenario is depicted in Fig. 3.

DOI: 10.1051/e3sconf/20172200191

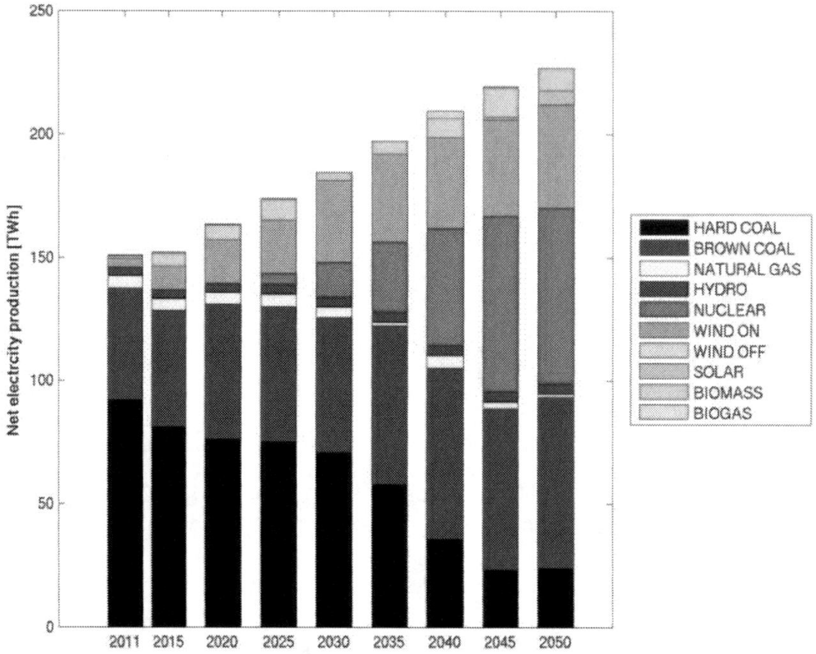

Fig. 3. Net electricity generation split into fuel types for HIGH CO_2 price scenario. For nuclear plants only SMR were considered.

4 Conclusions

Coal-fired units will still dominate power generation in Poland in the next 20–30 years. However, the relative share of coal in the Polish energy mix will gradually decrease, giving a space for new power generation technologies. We analysed the economic competitiveness of two perspective small size power generation technologies i.e. SMRs and IGCC/CCS with electric capacity of about 200 MW. These small units are based on known technologies that have been constructed with the larger capacities. On one hand downsizing to lower capacities increases investment costs per kW of installed power, on the other there are several economic factors such as design simplifications, accelerated learning, reduced construction time and associated risks that compensate the loss due to limited impact of economy of scale. Consequently, the costs of capital incurred during construction period are lower. Both technologies are not commercially available for power generation at present and it is forecasted that the first units can be offered in the market around 2025–2030. One should bear in mind, that both technologies can be used to deliver also other products e.g. heat [4], syngas that may influence their economic performance. However, in this study we focused on power generation. The results show that with techno-economic assumptions presented in this study and taking into account the present and perspective EU climate policy SMRs are more competitive options as compared to small IGCC/CCS units. In fact, the LCOE metric for SMRs is the lowest from all technologies considered in this study. This result is supported by results obtained from the TIMES-PL energy-economic model, which confirmed that SMRs will be able to successfully compete with other technologies once becoming a part of the power system.

This work received financial support from the statutory funding of AGH (no. 11.11.210.217).

Appendix I - Nomenclature

r – discount rate [%],
N – total investment costs [USD],
OM – annual operation and maintenance costs [USD],
F – annual fuel costs incurred [USD],
CO_2 – annual costs of purchasing CO_2 emission allowances [USD],
L – decommissioning costs incurred [USD],
SV – salvage value of the power plant [USD],
P – annual electricity generation [MWh],
Fr – coefficient of allocation of investment expenditures into particular years [%].

Appendix II – Superscript and Subscript

t – time index,
TP – starting year of the construction process,
T0 – year in which construction ends and operation of the plant begins,
TK – year in which operation of the plant ends,
TC – overall construction time (in years).

References

1. L. Gawlik, A. Szurlej, A. Wyrwa, Energy **92**, 172 (2015)
2. S. Selosse, O. Ricci, Applied Energy **188**, 32 (2017)
3. D.T. Ingersoll, Progress in Nuclear Energy **51**, 589 (2009)
4. L. Pieńkowski, *International nuclear energy congress: Warsaw*, **178** (2012)
5. R. Loulou, Computational Management Science **5**, 41 (2008)
6. A. Wyrwa, M. Pluta, S. Skoneczny, T. Mirowski, Lecture Notes in Computer Science **8500**, 489 (2014)
7. IEA OECD, *Projected Costs of Generating Electricity*, Paris (2010)
8. L. Gawlik, Z. Grudzinski, J. Kaminski, P. Kaszynski, D. Kryzia, U. Lorenz, T. Mirowski, E. Mokrzycki, T. Olkuski, U. Ozga-Blaschke, M. Pluta, A. Sikora, K. Stala-Szlugaj, W. Suwała, A. Szurlej, A. Wyrwa, J. Zysk, *Coal for the Polish Power Sector in the Time Perspective up to 2050 – Scenario Analysis* (In Polish). Wydawnictwo Instytutu Gospodarki Surowcami Mineralnymi i Energią PAN, Katowice (2013)
9. DAS, Chancellery of the Prime Minister of Poland, *Optimal Polish energy mix by 2060*, Warsaw (2013) (in Polish)
10. European Commission, Energy Roadmap 2050 – Impact assessment and scenario analysis (2011)
11. H. Khatib, Energy Policy **38**, 5403 (2010)

An application of data mining in district heating substations for improving energy performance

Puning Xue[1], *Zhigang* Zhou[1], *Xin* Chen[1], and *Jing* Liu[1,2,*]

[1]Shool of Municipal and Environmental Engineering, Harbin Institute of Technology, Harbin 150000, China
[2]State Key Laboratory of Urban Water Resource and Environment, Harbin Institute of Technology, Harbin 150000, China

Abstract. Automatic meter reading system is capable of collecting and storing a huge number of district heating (DH) data. However, the data obtained are rarely fully utilized. Data mining is a promising technology to discover potential interesting knowledge from vast data. This paper applies data mining methods to analyse the massive data for improving energy performance of DH substation. The technical approach contains three steps: data selection, cluster analysis and association rule mining (ARM). Two-heating-season data of a substation are used for case study. Cluster analysis identifies six distinct heating patterns based on the primary heat of the substation. ARM reveals that secondary pressure difference and secondary flow rate have a strong correlation. Using the discovered rules, a fault occurring in remote flow meter installed at secondary network is detected accurately. The application demonstrates that data mining techniques can effectively extrapolate potential useful knowledge to better understand substation operation strategies and improve substation energy performance.

1 Introduction

China is the second largest building energy user in the world [1]. District heating (DH) system, an efficient method of supplying heat to consumers, account for 24.0% of total building energy consumption in China [2]. With the acceleration of urbanization, the DH sector presents a tendency of sharp increasing for the past few years. As a result, improving DH energy performance is vital for building energy savings.

Recently, automatic meter reading system has been installed in DH system resulting in large amount of DH-related data. However, the data obtained is rarely fully analysed due to the poor quality of meter readings and the lack of effective data analysis techniques. Data mining is an effective technique to extract new knowledge from big data, while few studies have been performed concerning the application of data mining in DH system.

This paper aims to use data mining method for improving energy performance of DH substations. The method includes two popular data mining techniques, namely, cluster

* Corresponding author: liujinghit0@163.com

 DOI: 10.1051/e3sconf/20172200192

analysis and association rules mining. A case study of analysing the dataset collected from a DH substation in Changchun, China, is conducted to demonstrate its applicability.

2 Methodology

2.1 Technical approach

Hand et al. [3] define data mining as: "The analysis of large observation datasets to find unsuspected relationships and to summarize the data in novel ways so that owners can fully understand and make use of the data". In the past several decades, it has been successfully applied in economics, retails, telecommunication, and financial services [4]. Recently, efforts have also been made to investigate the application of data mining in HVAC field, including building energy consumption prediction [5, 6], building energy management [7, 8], fault detection and diagnosis [9, 10], and occupant behaviour [11, 12].

From a broad view, data mining is the process of the Knowledge Discovery in Database (KDD) and involves the following seven steps [4]: (1) Data cleaning; (2) Data integration; (3) Data selection; (4) Data transformation; (5) Data mining; (6) Pattern evaluation; (7) Knowledge presentation. Steps 1−4 are different forms of data preprocessing and aim to improve data quality and transform the data into suitable format for mining. Step 5 is the essential process for data analysis. And steps 6−7 are the process of post-mining, where the truly interesting knowledge are selected and presented to users.

With the consideration of data characteristics of DH substations, a three-step method is employed to extrapolate unsuspected patterns and associations from DH-related data. In step 1, two-heating-season substation data with a resolution of 10-min are selected from the automatic meter reading system. And one-heating-season data are considered as a dataset. That is, two datasets named dataset 1 and 2 are prepared for mining. It should be mentioned that both step 2 and 3 are performed separately in each dataset. In step 2, cluster analysis is applied to identify distinct heating patterns based on the primary heat supply of the substation. Due to the heat load variation in DH system are both seasonal and daily [13]. Thus, cluster analysis consists of two parts: month clustering and hour clustering. Specifically, data within one month are regarded as a whole data object, and clustering is performed to identify the seasonal heating patterns among all objects. Then, data in an hour are considered as a whole data object, and hour clustering is applied to study the daily operation patterns based on each seasonal heating pattern. In step 3, association rule mining (ARM) is implemented to discover the unsuspected knowledge in the format of association rules based on each heating pattern identified by cluster analysis.

2.2 Cluster analysis

Cluster analysis is the process of grouping data objects into clusters so that objects in the same cluster have high similarity, while objects in different clusters have low similarity [4]. Agglomerative hierarchical clustering is one of commonly used clustering algorithm and is adopted in this study. Using a bottom-up strategy, it begins with every object representing a singleton cluster and recursively merge the closest clusters into a single cluster until certain termination conditions are satisfied [4]. It should be mentioned that users can specify the desired clusters number k as the termination condition. The performance of clustering result is evaluated by Dunn index (DI). The index is defined as the ratio of minimum distance between clusters to maximum distance inside clusters [14]. The larger value of DI means higher inter-cluster distances and lower intra-cluster distances, and then indicates better performance of clustering.

2.3 Association rule mining

ARM aims to identify frequent correlations hidden in dataset. Let $I = \{I_1, I_2, \dots, I_n\}$ be a set of n attributes called itemset. Let $D = \{D_1, D_2, \dots, D_m\}$ be a set of transactions called the database. Each transaction in D contains a subset of the items in I. The association rule can be described by the formula $X \rightarrow Y$, where $X \subset I$, $Y \subset I$, $X \neq \varnothing$, $Y \neq \varnothing$, and $X \cap Y \neq \varnothing$. The itemset X is called the antecedent of the rule, while the itemset Y is called the consequent of the rule [4]. The validity and certainty of the association rule can be measured in terms of its support, confidence and lift. Support is the joint probability of the antecedent and consequent. Confidence is the conditional probability of the consequent, given the antecedent. And lift can be considered as a simple dependence and correlation measurement between the consequent and the antecedent.

In this paper, Apriori algorithm is used to discover association rules. To perform ARM, minimum support threshold and minimum confidence threshold should be pre-specified. Moreover, most of the original dataset is numeric while Apriori algorithm require the data to be categorical. Therefore, data discretization is conducted to transform numeric attributes to categorical values before ARM. Specifically, all numeric attributes are classified into three categories using equal frequency method. And the three categories can be defined as "low", "medium", and "high" according to original numerical magnitude.

3 Results and discussions

3.1 Dataset

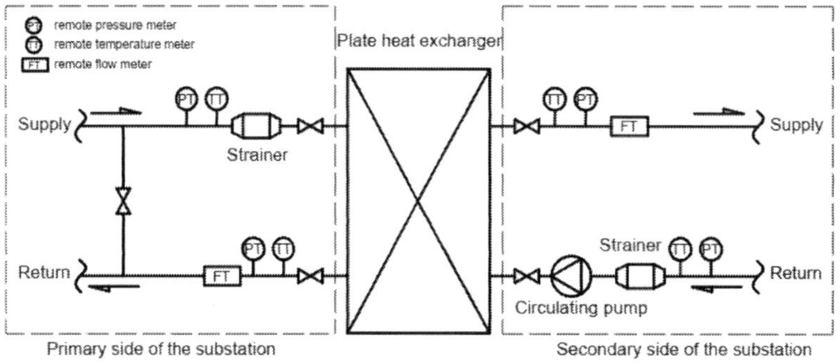

Fig. 1. Schematic view of the study substation.

The study DH substation is located in Changchun, a large provincial capital city in northeast China. Fig. 1 shows the schematic view of the substation. Automated and remote meters are mounted for monitoring and recording real-time operation parameters. Two-heating-season historical data from October 19, 2014 to April 11, 2015 and from October 19, 2015 to April 11, 2016 are collected for case study. The data of 14−15 heating season are called dataset 1 and the data of 15−16 heating season are called dataset 2. As shown in Table 1, a total of 14 attributes are monitored, and data of each attribute are collected with a measuring resolution of 10 min. It should be mentioned that there is not any remote heat meter mounted in the substation, so primary heat and secondary heat are calculated based on corresponding measured values of supply and return temperatures and flow rate separately. In order to make the association rules easier to understand, selecting the attributes under consideration is necessary. The pressure difference ΔP_{1st} and ΔP_{2nd} are calculated to express the pressure characteristics of primary network and secondary network, respectively. Similarly, the temperature difference ΔT_{1st} and ΔT_{2nd} are calculated

 DOI: 10.1051/e3sconf/20172200192

to represent the temperature characteristics of primary network and secondary network separately. As a result, the number of attributes is reduced to 10.

Table 1. The monitored attributes of the dataset.

No.	Nomenclature	Attribute	Unit
1	t	Date	1
2	τ	Time	1
3	$P_{s,1st}$	Primary supply pressure	MPa
4	$P_{r,1st}$	Primary return pressure	MPa
5	$T_{s,1st}$	Primary supply temperature	°C
6	$T_{r,1st}$	Primary return temperature	°C
7	G_{1st}	Primary mass flow rate	t/h
8	Q_{1st}	Primary heat power	MJ/h
9	$P_{s,2nd}$	Secondary supply pressure	MPa
10	$P_{r,2nd}$	Secondary return pressure	MPa
11	$T_{s,2nd}$	Secondary supply temperature	°C
12	$T_{r,2nd}$	Secondary return temperature	°C
13	G_{2nd}	Secondary mass flow rate	t/h
14	Q_{2nd}	Secondary heat power	MJ/h

3.2 Substation heating patterns

Due to data are gathered at a 10-min interval, 4,320 observations of Q_{1st} are collected in each month. If all 4,320 observations are used, a 4,320-dimension sparse matrix is generated. Month clustering on this sparse matrix can take a long time, making such analysis impractical or infeasible. Therefore, a reduced representation of this sparse matrix is necessary. The mean and standard deviations of Q_{1st} are calculated for each month, resulting in a 2-dimension feature dataset. Month clustering on this feature dataset is more efficient yet produce the same analytical results. k, the number of clusters, is set from 2 to 3. Correspondingly, DI values are 1.08 and 1.25, respectively. Therefore, $k = 3$ is selected for clustering. Table 2 shows the clustering results of dataset 1. Three clusters are identified, indicating three distinct seasonal heating patterns. Same clustering results can be obtained from dataset 2.

Table 2. Clustering results of dataset 1.

Cluster ID	Month
A	10, 4
B	11, 3
C	12, 1, 2

Based on the three seasonal heating patterns, hour clustering is then performed. Note that, the original objects in the months partitioned into the same cluster should be analysed together. k is set between 2 and 10. Fig. 2 illustrated the evaluation results of cluster *C* of dataset 1. The DI is maximum when $k = 2$. Therefore, such parameter setting is selected. Fig. 3 shows the clustering memberships. It seems that a majority of objects in cluster *C1* distribute between 11:00–14:00 and 23:00–2:00. The rest of data are basically gathered from 2:00–11:00 and 14:00–23:00, and are collected in cluster *C2*. The box plot in Fig. 4 demonstrates the clustering results. Cluster *C1* and cluster *C2* indicate non-heating period and heating period, respectively. It can be seen that the substation is running under intermittent operation in each day. Same hour clustering results can be obtained from other five seasonal operational patterns of the two datasets. It can be concluded that the intermittent operation is a strategy commonly used during the two heating seasons. Table 3 summarizes the overall results of cluster analysis. By a combination of seasonal and daily operation patterns, each original dataset is disaggregated into 6 clusters. Each cluster indicates a particular operation pattern.

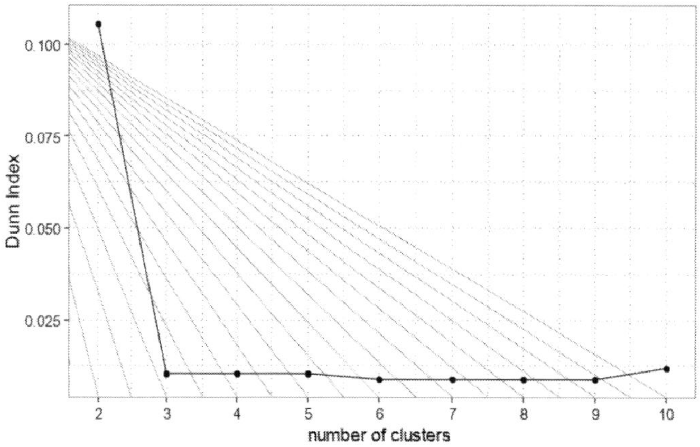

Fig. 2. Clustering evaluation of seasonal heating pattern *C* of dataset 1.

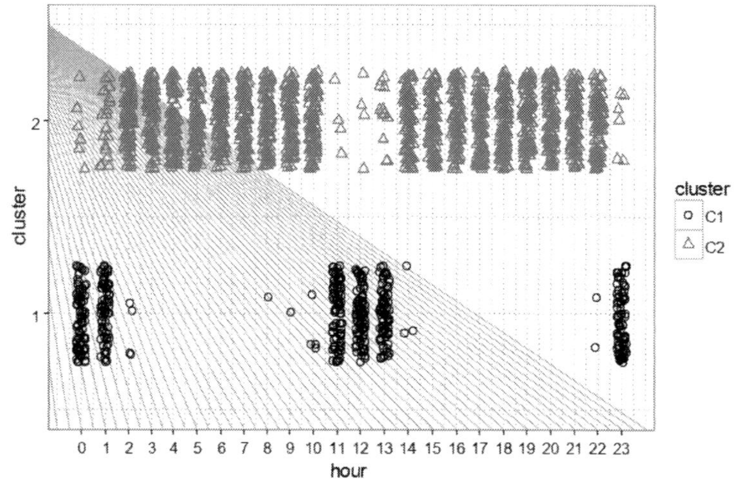

Fig. 3. Clustering results of seasonal heating pattern *C* of dataset 1.

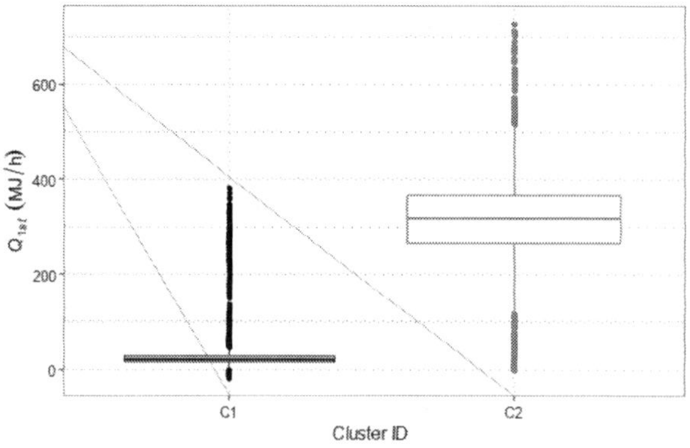

Fig. 4. Cluster features of daily heating pattern *C1* and *C2* of dataset 1.

Table 3. Summary of the clustering results of dataset 1 and 2.

Data set	Heating season	Seasonal operation pattern	Cluster membership	Daily operation pattern	Cluster membership		Cluster ID
1	14–15: 19/10/2014–11/04/2015	*A*	10, 4	Non-heating	7:00-16:00	20:00-4:00	*A1*
				Heating	4:00-7:00	16:00-20:00	*A2*
		B	11, 3	Non-heating	9:00-16:00	21:00-4:00	*B1*
				Heating	4:00-9:00	16:00-21:00	*B2*
		C	12, 1, 2	Non-heating	11:00-14:00	23:00-2:00	*C1*
				Heating	2:00-11:00	14:00-23:00	*C2*
2	15–16: 19/10/2015–11/04/2016	*A*	10, 4	Non-heating	8:00-16:00	21:00-4:00	*A1*
				Heating	4:00-8:00	16:00-21:00	*A2*
		B	11, 3	Non-heating	9:00-15:00	21:00-3:00	*B1*
				Heating	3:00-9:00	15:00-21:00	*B2*
		C	12, 1, 2	Non-heating	11:00-13:00	23:00-2:00	*C1*
				Heating	2:00-11:00	13:00-23:00	*C2*

3.3 Application of association rules

Apriori algorithm is then performed for discovering rules hidden in the 12 patterns respectively. For each pattern, a support of 0.1 and a confidence of 0.6 are set as the minimum thresholds. The support is relatively low to generate the association rules that maybe infrequent but interesting, and the confidence is relatively high to ensure the rules obtained are strong. Besides, the minimum threshold of lift is set as 1 to find positive correlations between the antecedent and consequent. In view of easy interpretation for rules extracted, the number of items in a rule is set to be 2, namely, both the antecedent and consequent of the rule have only one item. Many rules are obvious and can be easily interpreted by domain knowledge. Truly interesting rules need to be further selected.

The rules mined from dataset 1 and 2 with pattern $\Delta P_{2nd} \rightarrow G_{2nd}$ are potential interesting and are given in Table 4. From rules 1–6, it can be observed that the secondary pressure difference and secondary flow rate have a strong association in the substation. In order to provide an insight into these rules, the variation of secondary pressure difference and flow rate in one typical day of heating pattern *C* of 15–16 heating season is plotted in Fig. 5 (a). It can be seen that the variation of secondary pressure difference and flow rate follows the same trend in 15–16 heating season. Moreover, this rule pattern can help to understand the hydraulic condition of substation and be used as a guide of fault detection on the remote meters. This indicates that the secondary pressure difference and flow rate are always strongly associated. Accordingly, if the variation tendency of pressure difference and flow rate is dramatically different, then it can be deduced that the automated and remote meters may have faults.

Table 4. Summary of the rule pattern $\Delta P_{2nd} \rightarrow G_{2nd}$ of dataset 1 and 2.

No.	Antecedent ΔP_{2nd}	Consequent G_{2nd}	Support	Confidence	Lift	Cluster ID	Dataset
1	*low*	*low*	0.499	0.752	1.137	A1	2
2	*high*	*high*	0.818	0.895	1.095	A2	
3	*low*	*low*	0.440	0.893	1.085	B1	
4	*high*	*high*	0.820	0.852	1.040	B2	
5	*low*	*low*	0.658	0.990	1.041	C1	
6	*high*	*high*	0.953	0.989	1.038	C2	
7	*high*	*low*	0.851	1.000	1.000	C2	1

For example, rule 7 is mined from dataset 1. When substation supplies heat to consumers in heating pattern *C* of 14−15 heating season, the secondary pressure difference is high while the secondary flow rate is low. Moreover, the lift value is equal to 1. It means that there is no correlation between the secondary pressure difference and flow rate. The rule disobeys the knowledge extracted from rules 1−6, so it is reasonable to deduce that there is an anomaly in either remote flow meter or remote pressure meter. As illustrated in Fig. 5 (a) and (b), the variation of secondary pressure difference of two heating seasons is identical in heating pattern *C*. It can be inferred that the remote pressure meters work well and data transmission is accurate. Correspondingly, the variation trend of secondary flow rate should be identical to that of secondary pressure difference. However, Fig. 5 (b) shows that there is a continuous irregular oscillation in secondary flow rate. Hence, it can be deduced that a fault occurred in remote flow meter installed at secondary network in heating pattern *C* of 14−15 heating season.

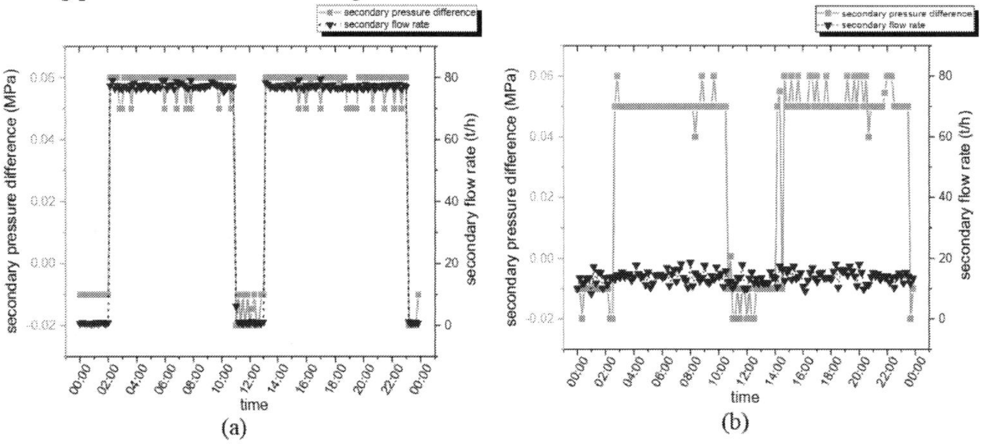

Fig. 5. The variation of secondary pressure difference and flow rate in one typical day of heating pattern *C*: (a) 15−16 heating season; (b) 14−15 heating season.

4 Conclusions

Using data mining techniques, a novel method was proposed to extrapolate useful knowledge from substation operation data. A three-step technical approach is applied to fully analyse the massive data. Two popular data mining techniques, agglomerative hierarchal clustering and Apriori algorithm, are integrated in the method. The main conclusions are: (1) Cluster analysis identifies six distinct heating patterns based on the primary heat of the substation. The heating patterns can help users to further understand the operation strategies of the substation. During the whole heating season, the substation supplies heat to users with intermittent operation in each day. (2) Association rule mining discovers the interesting rules between secondary pressure difference and secondary flow rate, which can provide an insight to fault detection on the remote meters. According to the rules mined, a fault occurring in remote flow meter installed at secondary network is detected accurately. (3) An important aspect of this paper demonstrates the successful use of data mining techniques to discover useful knowledge about substation operation, which can provide essential guidance for improving energy performance of DH substation.

One limitation of this method is that data mining techniques cannot tell the value of the rules discovered, and domain knowledge is still needed to select and interpret the rules for practical applications. Therefore, further study is still necessary and the main focus should be placed on applying more advanced data mining techniques for extracting rules more efficiently. In addition, DH systems have important roles to play in future sustainable energy systems. To fulfil such roles, the current DH systems must evolve to be smart DH systems, in which more information exchanges between interfaces are required. This study demonstrates that data mining can bring significant benefits in data processing and failure detection for DH systems. To implement future smart DH systems, it is highly recommended that future research pay more attention to develop new methodologies based on data mining techniques for energy management, failure detection, operation optimization, and decision-making.

This study was sponsored by Key Projects of Natural Science Foundation of Heilongjiang Province of China (Grant No. ZD2016010). The authors would like to thank Changchun City Heating (Group) Co. Ltd. and Heilongjiang LONG DIAN Electric Co. Ltd. for providing measured data.

References

1. IEA, *Energy Balances of OECD/non-OECD Countries 2014* (2014)
2. Building Energy Research Center of Tsinghua University, *Annual report on China building energy efficiency* (China Architecture & Building Press, Beijing, China, 2015)
3. D. Hand, H. Mannila, P. Smyth, *Principles of data mining* (MIT Press, 2001)
4. J. Han, M. Kamber, J. Pei, *Data mining concepts and techniques, 3rd edition* (Morgan Kaufmann Publishers Inc., 2012)
5. Z. Yu, F. Haghighat, B.C.M. Fung, H. Yoshino, Energ. Buildings **42**, 1637–1646 (2010)
6. C. Fan, F. Xiao, S. Wang, Appl. Energ. **127**, 1–10 (2014)
7. Z. Yu, F. Haghighat, B.C.M. Fung, L. Zhou, Energ. Buildings **47**, 430–440 (2012)
8. F. Xiao, C. Fan, Energ. Buildings **75**, 109–118 (2014)
9. C. Fan, F. Xiao, C. Yan, Automat. Constr. **50**, 81–90 (2015)
10. Z. Du, B. Fan, X. Jin, J. Chi, Build. Environ. **73**, 1–11 (2014)
11. Z. Yu, B.C.M. Fung, F. Haghighat, H. Yoshino, E. Morofsky, Energ. Buildings **43**, 1409–1417 (2011)
12. S. D'Oca, T. Hong, Build. Environ. **82**, 726–739 (2014)
13. H. Gadd, S. Werner, Appl. Energ. **106**, 47–55 (2013)
14. Z. Zhou, *Machine learning* (Tsinghua University Press, Beijing, China, 2016)

Mechanically enhanced electrospun nanofibers for wastewater treatment

Fatma Yalcinkaya[1,*]

[1]Department of Nanotechnology and Informatics, Institute for Nanomaterials, Advanced Technology and Innovation, Technical University of Liberec, Studentska 1402/2, 46117 Liberec, Czech Republic

Abstract. A novel high-performance polyamide 6, polyacrylonitrile and polyvinylidene fluoride nanofibers were fabricated using industrial production Nanospider equipment for liquid filtration as microfilters. The application of nanofibers has been hindered by their poor mechanical strength. This work developed a feasible approach to preparing mechanically strong nanofiber webs. The mechanical strength of the nanofibers was enhanced using special lamination technique on a supporting layer. Experimental results show that the mechanical strength of the nanofibers enhanced more than 5 times while high porosity and liquid permeability retain. The separation results indicate that nanofibers have a potential to be used in liquid filters.

1 Introduction

Membrane-based technologies for wastewater filtration are constantly being designed due to the increase in water pollution, insufficient water resources and imbalance between water supply and consumption. Membrane separation technology has been important over the last few decades due to high separation performance, ease of operation and low cost. There are various techniques to fabricate membranes such as melt extrusion, expanded film, solution casting, track etching, phase inversion, template leaching and thermally induced phase separation. Electrospinning could be considered to use in membrane fabrication. Unlike the melt extrusion, electrospinning is a technique uses solution or melt to produce webs by applying high electrical field. This technique is a simple and inexpensive method to produce functionalized membranes.

Due to unique properties of high porosity, small pore size, large specific surface area, and high selectivity, nanofibers are excellent candidate for application of many areas, such as air filtration, ion-battery separation, biomedical, liquid filtration, so on [1–4]. The mechanical strength of the nanofibers restricts their application in real life. Herein, a special lamination technique was used to improve mechanical properties of the nanofibers. Polyamide 6 (PA6), polyacrylonitrile (PAN) and polyvinylidene fluoride (PVDF) were chosen as nanofiber layer. PVDF is one of the most used polymers in membrane technology due to its outstanding chemical, thermal and oxidation resistance properties [5, 6]. PAN has good characteristics including thermal stability, tolerance to most solvents,

* Corresponding author: fatma.yalcinkaya@tul.cz

and commercial availability. PA6 is good in mechanical properties, water insoluble and easy to process.

The purpose of this article is to improve mechanical properties of the nanofibers and represent the results of research investigations on the application of polymeric microfiltration membranes of various nanofibrous membrane structures for the removal of impurities from industrial wastewater using the cross-flow system. The novelty of this paper is, that nanofiber layers were produced by using the Nanospider industrial equipment and that the layers strongly adhered on the supporting layer without any damage using a special lamination technology to improve application in wastewater remediation.

2 Materials and methods

2.1 Preparation of the nanofibers

8 g of PA6 (BASF B27, Germany) was dissolved in 30 g formic acid and 62 g acetic acid. 14 g of PVDF (Solef, Solvay, Germany) and 8 g of PAN (150 kDa, Elmarco, Czech Republic) were dissolved in 84 g dimethylacetamide and 92 g dimethyl formamide, respectively. To make a viscous mixture, all the solutions were stirred under constant stirring over the night.

Nanofiber layers were prepared from the polymer solutions using Nanospider NS-Lab 1WS500U device (Elmarco, Czech Republic). The experimental conditions are given in Table 1. The fibers were collected on a silicon paper.

Table 1. Experimental conditions of electrospinning.

Samples	Effective applied voltage (kV/cm)	Temperature (C)/ Relative humidity (%)	Width of sample (mm)	Substrate speed (mm/min)
PA6	4.25	22/38	500	25
PAN	4.33	20/22	500	15
PVDF	4.33	20/22	500	15

2.2 Characterization of membranes

The surface of nanofibers was characterized using a scanning electron microscope (SEM, Tescan Vega 3, Czech Republic). Image-j program was used to determine fiber diameter, diameter distribution, and surface porosity. The water contact angle was measured at different places of the samples using a Kruss Drop Shape Analyzer DS4. Thickness of the membranes was measured using a micrometer.

Tensile tests of the nanofibers and composite membranes were performed using a universal testing machine (Labor-Tech s.r.o., CR) with the extension rate of 10 mm/min, 100 mm length and 25 mm width at room temperature with a between the two clamps was 50 mm. Both machine (MD) and cross (CD) directions were tested. The composite membranes were kept under wet conditions for 4 days before testing, and the nanofiber layers were measured under dry condition due to their low mechanical properties after wetting.

 DOI: 10.1051/e3sconf/20172200193

2.3 Lamination of the membranes

The nanofibers were laminated on a supporting layer 20 g/m^2 polypropylene/polyethylene bi-component (Pegas, Czech Republic) using a Meyer RPS-L Mini fusing machine under heat (125°C) and pressure (15 N/cm) at a specific contact time (1.7 m/min) as shown in Figure 1. This process was carried out gently to avoid damaging the structure of the nanofibers such as the fiber diameter and pore size.

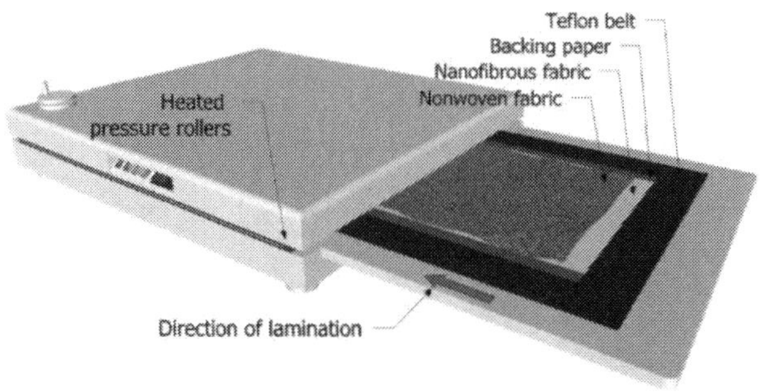

Fig.1. Lamination equipment [7].

2.4 Liquid filtration test

A custom-made cross-flow laboratory module equipped with pressure control and a constant water flow velocity through the cell (70 L/min) was used to filtrate the wastewater using the nanofibrous membranes [8]. The membranes were tested for 15 hours, and the flux was occasionally measured. All of the tests were performed at room temperature. Industrial sludge was used for the filtration test.

The permeate flux (F) and the permeability (k) of the membrane were calculated (Eq. (1) and (2)):

$$F = 1/A\ (dV/dt) \tag{1}$$

$$k = F/P \tag{2}$$

where A is the effective membrane area (m^2), V is the total volume of permeate, P is transmembrane pressure (bar), and t is the filtration time (h) [8].

3 Results and discussion

PA6, PVDF, and PAN nanofibrous composite membranes were prepared by using a special lamination technique. The lamination process carried out gently to avoid the damages of the fiber. As shown in the SEM images, Figure 2, uniform nanofibrous composite membranes were formed. Fiber diameter slightly changed after lamination process without any damage on the surface. PVDF nanofiber has a fiber diameter almost 2 times higher compared to

DOI: 10.1051/e3sconf/20172200193

PAN and PA6. Fiber diameter has importance on the pore size of the material. Lower diameter means smaller pore size.

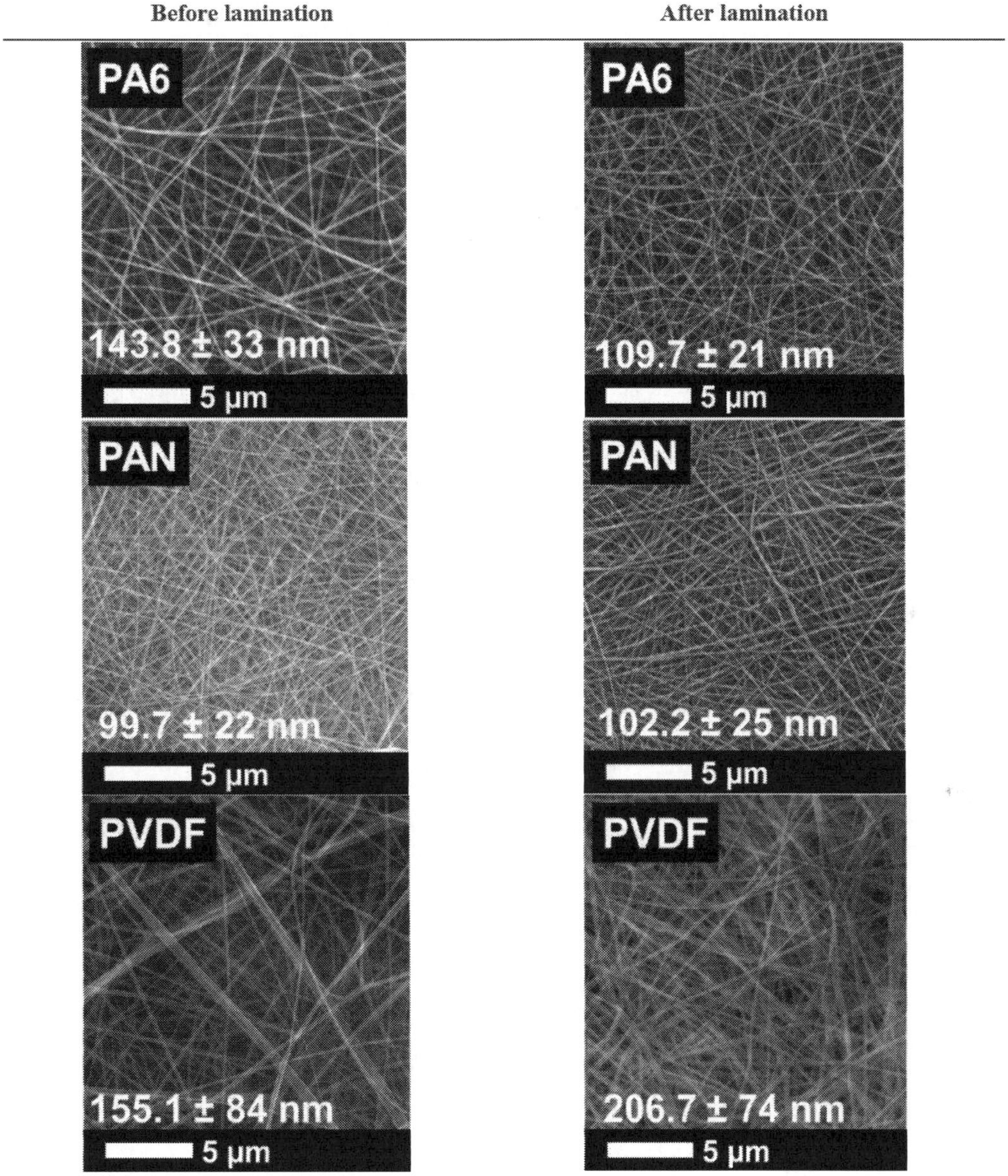

Fig. 2. SEM images of nanofibers before and after the lamination process.

The water contact angle and the porosity of the materials are shown in Table 2. The thickness of the membranes mainly depends on the thickness of the supporting layer.

Table 2. Water contact angle, surface porosity and the thickness of the membranes.

Sample	Water contact angle (°)	Surface porosity (%)	Thickness of the membranes (μm)
PA6	125.8	88.56	259
PAN	79.45	85.13	296
PVDF	133.8	85.37	275

Membrane surface morphology has a significant effect on the permeability of the membranes. It was found that to have an idealized membrane should have a highly porous structure over than 70% of the pore size of the order of 2 μm for membrane distillation [9]. The small surface pore size could eliminate the internal fouling of the membranes. The surface porosity plays an important role to determine liquid flow across the membrane in microfiltration. All the membranes have very high surface porosity over than 85% which is advantages compared to commercial micro filters.

Hydrophobic membranes are resistant to wetting. Kim et al. found that [10] membrane fouling potential lowered by a hydrophilic membrane. On the other side, Yalcinkaya et al. [8] found that super hydrophilic membranes tend to foul easily compared to hydrophobic membranes. Among all membranes, only PAN membranes show the hydrophilic property. Figure 3 demonstrates the permeability of the membranes for the separation of industrial wastewater during 15 h separation. The permeability decline in time due to fouling of the membranes in first 3 h. The hydrophilic PAN membrane shows the lowest permeability compared to hydrophobic PVDF and PA6. The reason could be due to the high affinity of the PAN membrane for solid particles. The permeability results of PA6 and PVDF showed that nanofiber membranes could be a good candidate for wastewater separation.

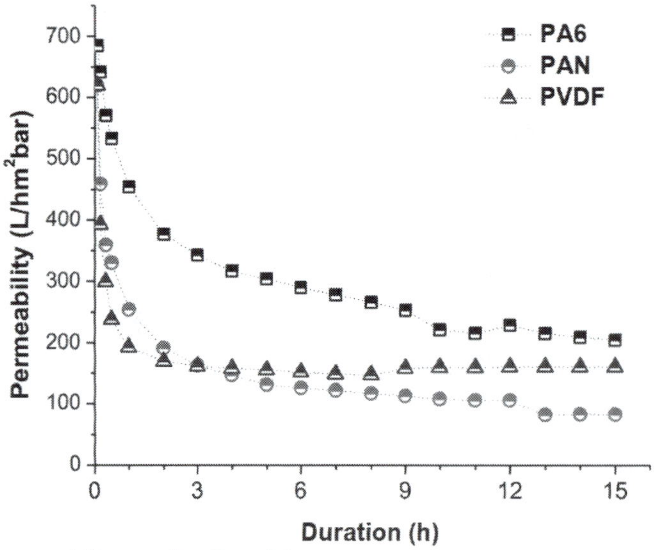

Fig. 3. Membrane permeability as a function of time.

The mechanical properties are one of the key parameters for microfiltration membranes. To determine the mechanical strength of the prepared nanofibrous composite membranes,

 DOI: 10.1051/e3sconf/20172200193

tensile testing was conducted for wet membranes, and presented in Figure 4. Single nanofiber layers were examined under dry condition due to their poor wet strength.

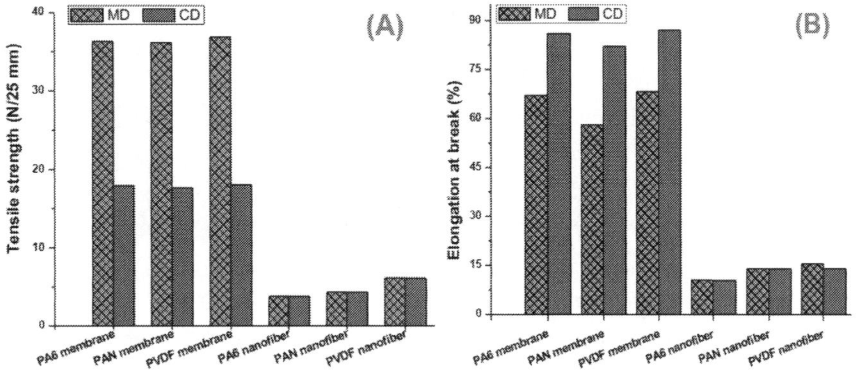

Fig. 4. Mechanical properties of the nanofibers and composite membranes (A) tensile strength, (B) elongation at break.

The tensile test showed that the mechanical strength of the nanofibers had been enhanced more than 5 times. The tensile strength of the membranes mainly depends on the property of the supporting material. The nanofibers have a slight effect on the strength of the composite membrane. It can be concluded that in wet condition of the membrane, it is stronger in the machine direction. When we compared the single-layer nanofibers, PVDF nanofibers are stronger than compared to PAN and PA6. In dry condition, nanofibers have the same strength in both MD and CD due to their anisotropic structure.

4 Conclusion

Novel high flux nanofibrous composite membranes were produced by using electrospinning system. Special lamination technique was applied to improve mechanical properties of the nanofiber web for microfiltration process. The results showed that laminated samples could strength the membrane without any damage on the nanofiber. It could be observed that for the single-layer nanofiber, the tensile strength and the elongation at break increase dramatically after lamination with a supporting material. The filtration of industrial wastewater indicates that the nanofibrous membranes can be suitable for the wastewater treatment with high permeability and good mechanical strength.

Acknowledgement

The results of this work was supported by project LO1201 obtained through the financial support of the Ministry of Education, Youth and Sports in the framework of the targeted support of the "National Program for Sustainability I", the OPR&DI project Centre for Nanomaterials, Advanced Technologies and Innovation CZ.1.05/2.1.00/01.0005 and the Mobility Fond FOM (Fondu Mobilit TUL) 2017.

References

1. T. H. Cho, T. Sakai, S. Tanase, K. Kimura, Y. Kondo, T. Tarao, M. Tanaka, Electrochem. Solid-State Lett. **10**, A 159 (2007)
2. M. Maryskova, I. Ardao, C. A. Garcia-Gonzalez, L. Martinova, J. Rotkova, A. Sevcu, Enzyme Microb. Tech. **89**, 31 (2016)
3. J. Matulevicius, L. Kliucininkas, T. Prasauskas, D. Buivydiene, D. Martuzevicius, J. Aerosol. Sci. **92**, 27 (2016)
4. J. Venugopal, L.L. Ma, T. Yong, S. Ramakrishna, Cell Biol. Int. **29**, 861 (2005)
5. Y. Wang, S. W. Foo, T. S. Chung, Ind. Eng. Chem. Res. **48**, 4474 (2009)
6. W. J. Chen, Y. L. Su, J. M. Peng, X. T. Zhao, Z. Y. Jiang, Y. N. Dong, Y. Zhang, Y. G. Liang, J. Z. Liu, Environ. Sci. Technol. **45**, 6545 (2011)
7. B. Yalcinkaya, F. Yalcinkaya, J. Chaloupek, J. Nanomater. **2016**, Article ID 2694373 (2016)
8. F. Yalcinkaya, B. Yalcinkaya, J. Hruza, P. Hrabak, Sci. Adv. Mater. **9**, 747 (2017)
9. A. G. Fane, R. W. Schofield, C. J .D. Fell, Desalination **64**, 231 (1987)
10. D. S. Kim, J. S. Kang, Y. M. Lee, Sep. Sci. Technol. **39**, 833 (2004)

Influence of cross-sectional ratio of down comer to riser on the efficiency of liquid circulation in loop air lift bubble column

Tatsumi Yamamoto[1,*], *Hiroyuki* Kawasaki[1], and *Hidetoshi* Mori[1]

[1]Dep. Life Sciences and Bioengineering, Univ. Toyama, 3190 Gofuku, Toyama, 930-8555, Japan

Abstract. Loop type bubble columns have good performance of liquid circulation and mass transfer by airlift effect, where the liquid circulation time is an important measurable characteristic parameter. This parameter is affected by the column construction, the aspect ratio of the column, the cross-sectional area ratio of down comer to riser (R), and the superficial gas velocity in the riser (U_{GR}). In this work, the mean gas holdup and the liquid circulation time (T_C) have been measured in four types of loop airlift type bubble column: concentric tube internal loop airlift type, rectangular internal loop airlift type, external loop airlift type, external loop airlift with separator. Air and tap water were used as gas and liquid phase, respectively. The results have demonstrated that the mean gas holdup in riser increases in proportion to U_{GR}, and that it in downcomer changes according to the geometric parameters of each bubble column. T_C has been found to conform to an empirical equation which depends on U_{GR} and the length of draft tube or division plate in the region of $0.33 < R < 1$.

1 Introduction

Various styles of bubble column are used as gas-liquid two phase or gas-liquid-solid three phase reactor in the plants for energy supply and environment conservation. The loop type bubble column is particularly important because of good performance of liquid circulation and mass transfer by airlift effect [1, 2]. In this type bubble column, the liquid circulation time and the gas holdup are the important and characteristic parameters [3, 4], including the column construction, the aspect ratio of the column, the cross-sectional area ratio of down comer to riser (R) and the superficial gas velocity in the riser (U_{GR}) [5]. For a well-understandings of the behavior of the loop type bubble column, we have to collect the data on various styles of the loop type bubble column.

We have already presented the behavior of the mean gas holdups (ε_{GR}, ε_{GD}) in riser and downcomer and T_C measured in rectangular internal loop airlift type bubble column in a previous congress [6]. In this paper, we report these behaviors observed in four types of loop airlift bubble column: concentric tube internal loop airlift type, rectangular internal loop airlift type, external loop airlift type, external loop airlift type with separator.

* Corresponding author: yamamoto@eng.u-toyama.ac.jp

DOI: 10.1051/e3sconf/20172200194

2 Experimental

Four types of airlift bubble column used in this work are illustrated in Figure 1. The details of rectangular internal-loop type were presented previously [6]. The dimensions of bubble column are tabulated in Table 1. Air and tap water at 303K were used as the gas and liquid phase, respectively. A perforated plate with many holes was used as a gas disperser. The non-aerated liquid level (H_D) was kept at 0.12m from the upper end of the draft tube or the division plate. The mean gas holdups (ε_{GR}, ε_{GD}) in riser and downcomer sections were calculated by measuring the liquid height with a static pressure manometer. A tracer technique was used to measure the liquid circulation time (T_C) with a conductance meter and 20wt% sodium chloride aqueous solution.

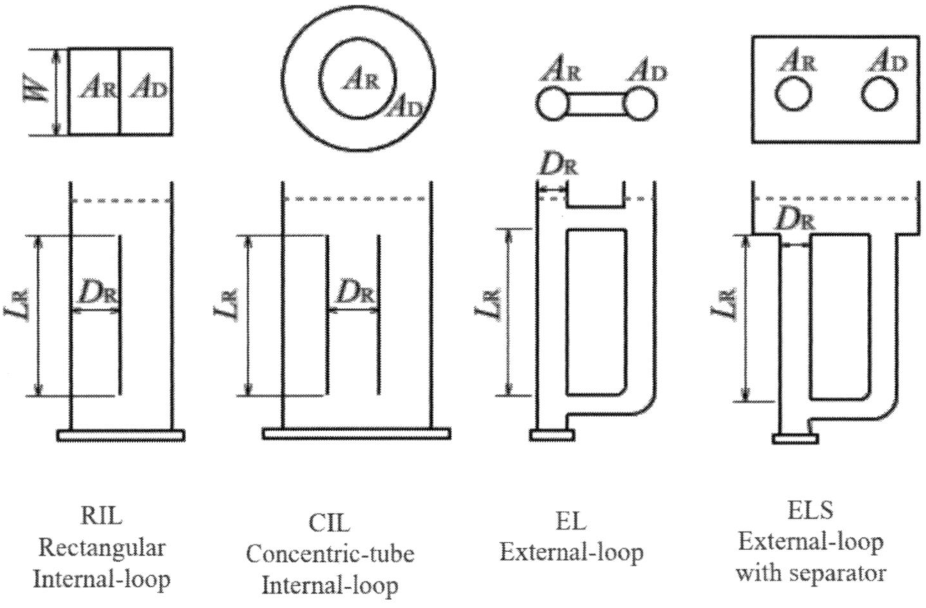

Fig. 1. Four types of loop airlift bubble column employed.

Table 1. Dimensions of loop airlift bubble column employed.

Column type	R (=A_R/A_D) [–]	L_R [m]	D_R [m]	W [m]
Rectangular Internal-loop (RIL)	0.33–3.00	0.77–2.50	0.08–0.24	0.12
Concentric tube Internal-loop (CIL)	1.12–6.20	0.50–4.06	0.06–0.22	-
External-loop (EL)	0.44–3.06	1.03–4.12	0.02–0.15	-
External-loop with Separator (ELS)	0.38–14.06	1.19–4.46	0.04–0.15	-

DOI: 10.1051/e3sconf/20172200194

3 Results and Discussion

3.1 Bubble Distribution Pattern

Shapes of bubble generated through the perforated plate were predominantly spherical. The bubbles rose with a constant velocity without colliding or coalescing one another inside the riser in low U_{GR} region. When U_{GR} was increased, some bubbles entrained into the main liquid flow from the riser to the downcomer. The amount of entrained bubbles decreased when the clearance between the lower end of inner tube or division plate and the bottom of column was small.

3.2 Mean Gas Holdup

In the case of rectangular internal-loop bubble column (RIL) for example, the mean gas holdup in riser ε_{GR} increased in proportion to U_{GR} in the higher range of U_{GR} as demonstrated in Figure 2. The same behaviour was observed in the case of external-loop reactor (EL and ELS). The solid line in Figure 2 shows an Akita-Yoshida equation [7] obtained empirically in the loop bubble column with a concentric-tube using a single hole gas sparger. ε_{GR} was usually larger than ε_{GD} in the observed region of U_{GR}. A relation of ε_{GD} to ε_{GR} depends generally on R and H_C which is the clearance between the bottom of the column and the lower end of the division plate. In the case of $H_C = 0.24$ m, ε_{GD} increased in proportion to ε_{GR} from a threshold value of ε_{GR} as Figure 3, which also suggests that the decrease of R tends to intensify the dependency of ε_{GD} to ε_{GR}. The decrease of cross-sectional area of down downcomer was empirically found to encourage the formation of liquid down flow containing bubbles. The bubble entrainments have been also observed in the external-loop type airlift bubble column, resulting in another threshold value of ε_{GR}. On the other hand, in the case of $H_C = 0.015$ m, ε_{GD} kept to be approximately zero with increasing ε_{GR}.

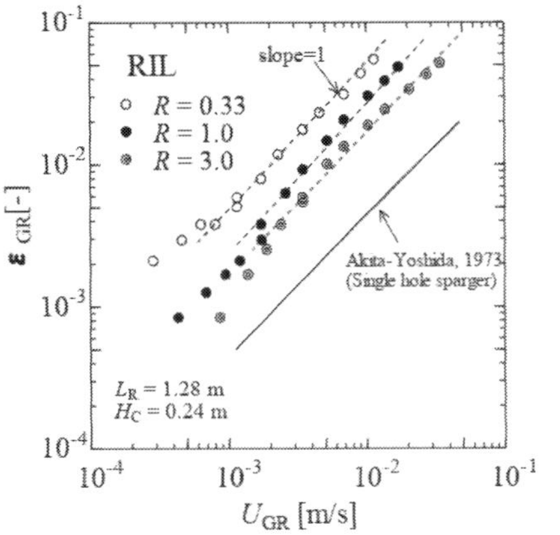

Fig. 2. Changes of ε_{GR} with U_{GR} and R in case of $H_C = 0.24$m.

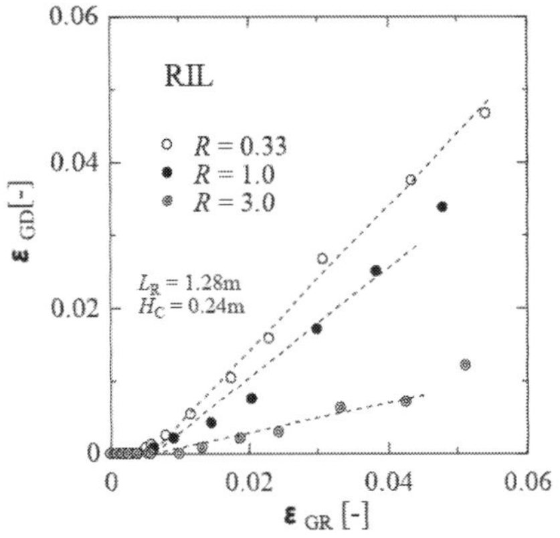

Fig. 3. Relations between ε_{GD} and ε_{GR} in case of $H_C = 0.24$m.

3.3 Liquid Circulation Time

In the internal-loop bubble columns (RIL and CIL), T_C have been found to increase in proportion to the -1/3 power of U_{GR} in the lower U_{GR} region, and T_C was constant with increasing U_{GR} in the higher U_{GR} region larger than 3×10^{-3} m/s. These results stem from the characteristics of liquid flow and entrained bubbles in the column, where bubbles on the riser rise straight without coalescence or entrainment to the downcomer. T_C has been also found to depend on H_C and R. When H_C = 0.24m, the time required to maintain a constant T_C increases with an increase in R as shown in Figure 4. These belabours of T_C were approximately same in the external-loop type columns. All the data of $T_C U_{GR}^{1/3}$ observed in four types of airlift bubble column, are empirically related to L_R as shown in Figure 5.

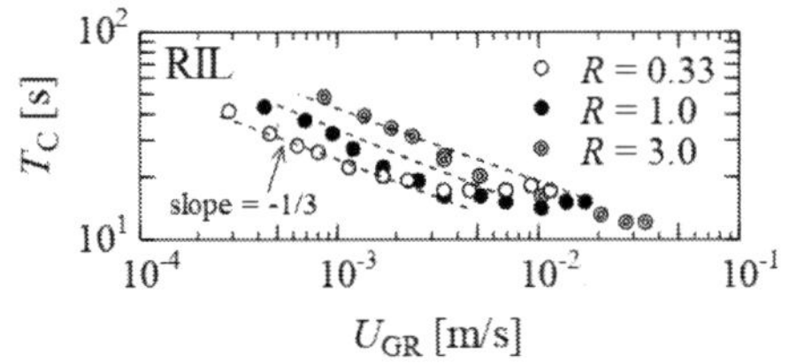

Fig. 4. Dependency of T_C on U_{GR} and R in case of H_C = 0.24m.

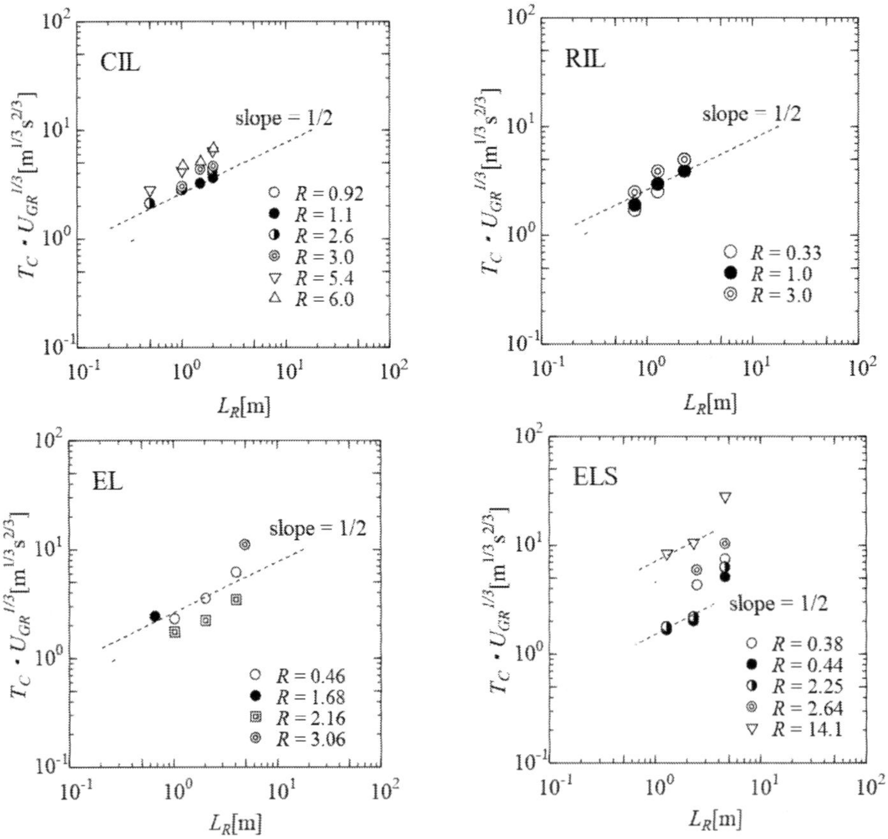

Fig. 5. Relationship between $T_C U_{GR}^{1/2}$ and L_R in the four types of loop airlift bubble column.

The results indicate that the values of $T_C U_{GR}^{1/3}$ are nearly proportional to the 1/2 power of L_R, suggesting that the values of $T_C U_{GR}^{1/3} L_R^{-1/2}$ should be constant. Kawasaki et al. [3] have proposed an empirical equation as

$$T_C = 2.5\ U_{GR}^{-1/3}\ L_R^{1/2} \tag{1}$$

which has been derived from the data using the CIL type bubble column in the region $0.7 \leq R \leq 2.5$. Figure 6 demonstrates that the observed results in the present paper are reasonably explained by Equation (1), suggesting that Equation (1) can be applied to various types of loop airlift bubble column.

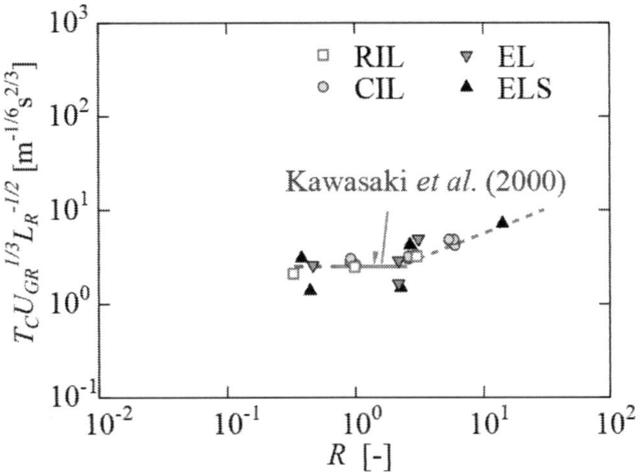

Fig. 6. Relationship between $T_C U_{GR}^{1/3} L_R^{-1/2}$ and R obtained for four types of loop airlift bubble column.

4 Conclusions

Four types of airlift type bubble column have been employed to observe the effects of the geometric parameters on the bubble distribution patterns. In the experiments the superficial gas velocity in the riser (U_{GR}), the gas holdup (ε_{GR}, ε_{GR}), the liquid circulation time (T_C) have been measured and correlated with various geometric parameters. The empirical equation observed in concentric-tube internal-loop airlift type bubble column can be reasonably applied to other three types of loop airlift bubble column.

The authors would like to thank Mr. Takumi Uesugi for his experimental assistance.

References

1. M.Y. Chisti, M. Moo-Young, Chem. Eng. Comm. **60**, 195–242 (1987)
2. M.Y. Chisti, *Airlift Bioreactors* (Elsevier, 1989)
3. H. Kawasaki, T. Yamamoto, H. Kumazawa, Seibutsu-kogaku Kaishi **78**, 115–119 (2000) (in Japanese)
4. L. Meng, Y. Bando, S. Hamano, Y. Oura, M. Nakamura, J. Chem. Eng. Jpn, **38**, 1015–1019 (2005)

DOI: 10.1051/e3sconf/20172200194

5. R.A. Bello, C.W. Robinson, M. Moo-young, Can. J. Chem. Eng. **62**, 573–577 (1984)
6. T. Yamamoto, T. Uesugi, H. Kawasaki, *Proc. CHISA 2014*, P3.199 (Praha, 2014)
7. K. Akita, F. Yoshida, Ind. Eng. Chem. Process Des. Develop. **12**, 76–80 (1973)

 DOI: 10.1051/e3sconf/20172200195

Microwave pyrolysis using self-generated pyrolysis gas as activating agent: An innovative single-step approach to convert waste palm shell into activated carbon

Peter Nai Yuh Yek[1,2], *Rock Keey* Liew[2], *Mohammad Shahril* Osman[1], *Chee Chung* Wong[1], and *Su Shiung* Lam[2,*]

[1]University College of Technology Sarawak, Department of Engineering, 96000, Sibu, Sarawak, Malaysia
[2]Pyrolysis Technology Research Group, Eastern Corridor Renewable Energy Group (ECRE), School of Ocean Engineering, Universiti Malaysia Terengganu, 21030 Kuala Terengganu, Terengganu, Malaysia

Abstract. Waste palm shell (WPS) is a biomass residue largely available from palm oil industries. An innovative microwave pyrolysis method was developed to produce biochar from WPS while the pyrolysis gas generated as another product is simultaneously used as activating agent to transform the biochar into waste palm shell activated carbon (WPSAC), thus allowing carbonization and activation to be performed simultaneously in a single-step approach. The pyrolysis method was investigated over a range of process temperature and feedstock amount with emphasis on the yield and composition of the WPSAC obtained. The WPSAC was tested as dye adsorbent in removing methylene blue. This pyrolysis approach provided a fast heating rate (37.5°C/min) and short process time (20 min) in transforming WPS into WPSAC, recording a product yield of 40 wt%. The WPSAC was detected with high BET surface area (≥ 1200 m^2/g), low ash content (< 5 wt%), and high pore volume (≥ 0.54 cm^3/g), thus recording high adsorption efficiency of 440 mg of dye/g. The desirable process features (fast heating rate, short process time) and the recovery of WPSAC suggest the exceptional promise of the single-step microwave pyrolysis approach to produce high-grade WPSAC from WPS.

1 Introduction

Palm oil represents one of the main agriculture commodity in Malaysia, recording approximately 17 million tons of production in 2016 [1]. Consequently, nearly 53 million tons of wastes palm have been generated every year from the palm oil industry [2]. Each ton of oil palm wastes consists mainly of 22 wt% of empty fruit bunch (EFB), 6 wt% of waste palm shell (WPS) and 14 wt% of palm mesocarp fibre (PMF). These wastes are

* Corresponding author: lam@umt.edu.my

currently under-utilized and simply burned as solid fuel in boiler to generate steam for sterilization. The current practice emits undesirable fly ash and greenhouse gas to the atmosphere which in turn creates environmental pollution. Therefore, oil palm wastes should be converted into useful products such as activated carbon in order to maximize its economic value rather than using it as boiler fuels.

Waste palm shell activated carbon (WPSAC) is a carbonaceous material produced from controlled thermal process in order to increase the level of porosity by high internal surface area and pore volume. The performance of WPSAC is determined by pore size (e.g. microporous, mesoporous, or macroporous) and surface composition (e.g. the presence of oxygen or acidic functional group) influences the adsorption efficiency [3]. Due to these properties, WPSAC has widely been applied in removal or adsorption of carbon dioxide from the exhaust gas streams in power plant and the raw natural gas in agriculture industry [4], removal of phenol [5] and tetracycline [6] in chemical industry, or used as adsorbent to remove heavy metals such as lead (Pb) and cadmium (Cd) [7]. Recently, WPSAC was utilized in technological applications such as supercapacitor and protection from electromagnetic radiation [8, 9].

The present study attempts to utilize an innovative single step approach to produce WPSAC from waste palm shell by microwave pyrolysis using self generated CO_2 as activating agent. Microwave radiation as a heat source provide rapid and targeted heating to the inside waste palm shell by dipole rotation and ionic conduction [10]. The study also examined the influence from varying the amount of waste palm shell as feedstock and the pyrolysis temperature on the process features, the yield and composition of the WPSAC obtained, and its adsorption of methylene blue. These features are important to assess the feasibility of developing a single step method that is energy and cost efficient in producing WPSAC from oil palm waste. There have been limited studies on single step physical activation of WPSAC preparation, and no studies have been reported on the use of self-generated pyrolysis gas for WPSAC preparation via microwave pyrolysis technique.

2 Material and methods

2.1 Source and preparation of waste palm shell

Waste palm shell (WPS) was collected from a palm oil mill located in Sibu, Sarawak. The waste was washed with distilled water to remove dust and impurities. It was then dried in the oven at 110°C for 2 hours to remove moisture and ground to a desired size of 0.15 mm for further analysis.

2.2 Characterization of WPS

WPS was analysed using proximate and ultimate analyses. The proximate analysis was conducted according to the ASTM standards D5142 to determine the contents of moisture, fixed carbon, volatile matter, and ash. Elemental analysis was conducted using a FlashEA 1112 CHNS elemental analyser to determine the carbon, hydrogen, nitrogen and sulphur contents in the WPS, whereas the oxygen content was determined by mass difference.

2.3 Microwave pyrolysis of WPS

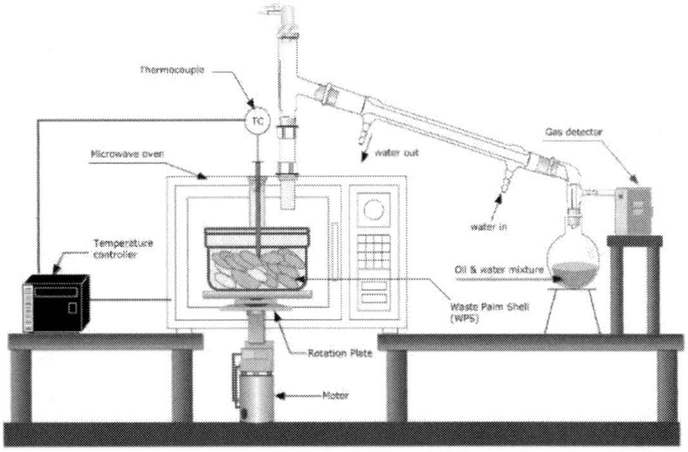

Fig. 1. Schematic diagram of microwave pyrolysis system used to produce activated carbon.

Fig. 1 shows the schematic diagram of the microwave pyrolysis system. Pyrolysis of WPS was performed in a modified microwave oven (Sharp R213CST) with a frequency of 2450 MHz. The microwave oven housed a porcelain reactor that can prevent combustion in the reactor. The pyrolysis experiments were performed in batch and closed operation, and a rotation plate was used to ensure even distribution of microwave irradiation. A Type-K thermocouple was used to measure the temperature and the readings were recorded by a data logger every 5 seconds throughout the pyrolysis process. A gas detector was placed near to the exhaust point to determine the presence of hydrocarbon, CO and CO_2 gases in the pyrolysis gas produced after the pyrolysis operation.

The experiment started by inserting different amounts of WPS into the reactor. Then, the microwave oven was turned on to heat and pyrolyze the WPS for 35 min at 900 W with a rotation speed of 10 rpm. The waste palm activated carbon (WPSAC) was produced as the main product of interest and it was cooled down inside the ceramics reactor. The product yield was calculated based on the equation 1, where Y is the mass yield, m_i and m_f refer to the initial mass and final mass respectively. WPSAC was washed using distilled water and dried in an oven at 105°C. The dried WPSAC was stored for analyses and future use.

$$Y = \frac{m_i}{mf} \times 100\% \quad (1)$$

2.4 Single step activation

$$C + H_2O \rightarrow H_2 + CO \quad (2)$$
$$CO + H_2O \rightarrow H_2 + CO_2 \quad (3)$$
$$C + 2H_2 \rightarrow CH_4 \quad (4)$$
$$C + CO_2 \rightarrow 2CO \quad (5)$$

During carbonization stage, WPS was heated and pyrolyzed and this resulted in the release of volatile matter in the form of pyrolysis gas and the generation of pores on the surface of the solid char formed during the microwave pyrolysis process. The pyrolysis gas was remained in the reactor and this in turn contributed to activation of the solid char to produce

 DOI: 10.1051/e3sconf/20172200195

more mores on its surface, thus producing a highly porous solid char termed "activated carbon". The WPS was converted into WPSAC via one step activation by self-generated pyrolysis gas in which the carbonization and activation occurred simultaneously. Heating in the closed system of ceramics reactor, the moisture (H_2O) content of WPS was likely to evaporate as hot gas and WPS has carbonized by microwave pyrolysis to become hot char. Reaction between hot char of WPS and hot gas generated hydrogen gas and carbon monoxide (Equation 2). Carbon monoxide as one of the product from Eqn. 2 was further reacted with H_2O to produce more hydrogen and carbon dioxide (Equation 3). Both products (i.e. CO_2 and H_2) obtained from Eqn. 3 were reacted with hot char to produce methane gas and carbon monoxide as shown in Equation 4 and 5, respectively. Activation process occurred during reaction between carbon dioxide with hot char as shown in Eqn 4.

2.5 Characterization of waste palm activated carbon

The WPSAC was analysed for their chemical functional group using a Perkin Elmer Spectrum 100 Fourier Transform Infrared (FTIR) spectrometer. A thin KBr (potassium bromide) disc method was used in the FTIR analysis. The analysis was then performed at wavenumber ranging from 400 to 4000 cm^{-1} for each sample [11]. The infrared spectrum obtained was presented in a form of transmittance (%) against wavenumber (cm^{-1}).

Scanning electron microscopy (SEM) analysis of WPSAC was conducted using JEOL JCM6000 to study the surface morphology and porosity. The WPSAC was also characterized using an accelerated surface area and porosimetry system (ASAP 2010, Micromeritics) to obtain N_2 adsorption-desorption isotherm for calculation of BET surface area, pore volume and pore size.

2.6 Adsorption test of methylene blue using WPSAC

The adsorption test of methylene blue (MB) was conducted using the WPSAC obtained. The adsorption test was performed in a beaker containing 0.5 g of WPSAC and 300 mL of methylene blue solution with an initial concentration of 10mg/L. The beaker was agitated at room temperature with a mixer speed of 200 rpm until the equilibrium was reached. The methylene blue solution was centrifuged and the residual concentration of MB in the supernatant solution was analysed using a UV–Vis spectrophotometer at 665 nm. The amount of MB adsorbed per unit mass of adsorbent at equilibrium, q_e (mg/g), was calculated by:

$$q_e = \frac{(C_o - C_e)V}{W} \quad (6)$$

where C_0 and C_e (mg/L) are the liquid-phase concentrations of dye at initial and equilibrium, respectively. V is the volume of the solution in litre, and W is the mass of adsorbent used in gram.

 DOI: 10.1051/e3sconf/20172200195

3 Results and discussions

3.1 Proximate and elemental analyses of waste palm shell

Table 1. Proximate analysis (wt%) and Elemental analysis (wt%)

Description	Waste palm shell (WPS)	Waste palm shell activated carbon (WPSAC)
Proximate analysis (wt%)		
Moisture[a]	4	2
Volatile matter[b]	58	10
Fixed carbon[c]	34	85
Ash[d]	4	3
Elemental analysis (wt%)		
Carbon	50.5	82.8
Hydrogen	6.9	4.1
Nitrogen	3.1	2.8
Sulfur	0.1	0.1
Oyxgen[c]	39.4	10.2

[a] Dry basis
[b] Volatile matter produced from pyrolytic decomposition of hemicellulose, cellulose, and lignin
[c] Calculated by difference
[d] Obtained by combustion at 1000°C

Table 1 shows the proximate and elemental composition of WPS and WPSAC. The WPS was dominated by volatile matter (58 wt%) and fixed carbon (34 wt%) with small amounts of moisture (4 wt%) and ash (4 wt%). The high volatile content suggests that WPS is more likely to be reactive in creating more pores on the carbonaceous material formed from pyrolytic carbonization process [12]. The WPS can be pyrolyzed to produce carbon rich material such as biochar or activated carbon due to its considerable amount of fixed carbon. As shown in Table 1, the fixed carbon, volatiles, moisture and ash content for the resulting WPSAC was 85 wt%, 10 wt%, 2 wt% and 3 wt%, respectively. This indicates that the carbonization of WPS during microwave pyrolysis had reduced the volatile matter of WPSAC to 10% and increased the fixed carbon content to 85%. Relatively high fixed carbon and low ash contents make WPS more profitable if compare to other palm biomass to produce higher yield of WPSAC. It was also found from Table 1 that the carbon content of WPSAC was increased to 82.8 wt%, and the content of oxygen reduced to 10.2 wt% with small amount of hydrogen (6.9 wt%), nitrogen (3.1 wt%) and sulfur (0.1 wt%). Overall, the results from proximate and elemental analysis indicate that the microwave pyrolysis transformed WPS into a carbon rich material with properties comparable to biochar.

 DOI: 10.1051/e3sconf/20172200195

3.2 Chemical functional groups of waste palm shell activated carbon

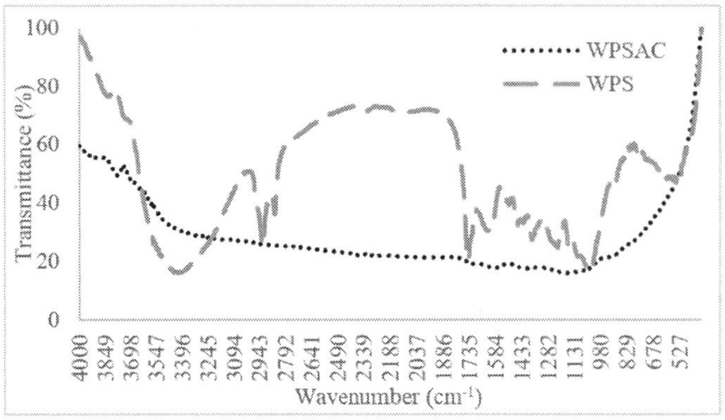

Fig. 2. FTIR spectrum of raw waste palm shell (WPS) and waste palm shell activated carbon (WPSAC).

The FTIR spectra of raw WPS and WPSAC are shown in Fig 2. The spectra of the WPS displayed the following bands 3200–3600 which correspond to O-H stretch in alcohols; 1400–1600 which is assigned to C=C stretch aromatics; the bands at 1000–1300 could be related to C-O groups. Raw WPS showing a wide range of transmittance peak, indicated that a large amount of organic and inorganic matters from cellulose, hemicellulose and lignin. Compare to the almost flat FTIR spectra of WPSAC, indicated that almost all the lignocellulose materials were completely decomposed during carbonization and activation process [13]. Spectra showed great difference at bands 3200–3600 which directly indicates that hydrogen was broadly removed to form graphitic structure; a typical activated carbon structure. However, WPSAC also showing transmittance peak at 1590 cm^{-1} related to functional group of aromatic with C=C stretch bending that cause higher reactions in the methylene blue absorption later.

3.3 Temperature profile of different feedstock amount

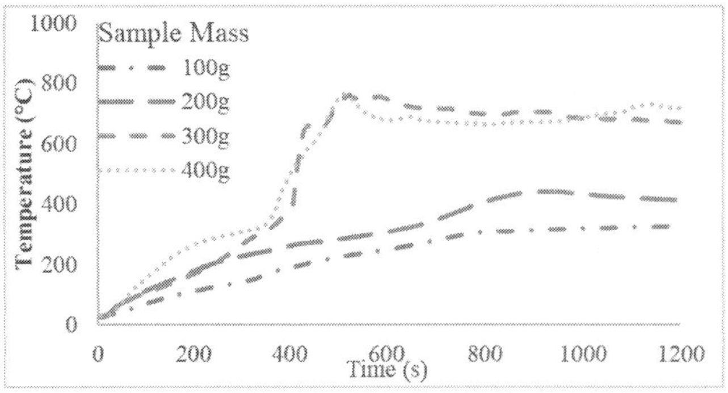

Fig. 3. Temperature profiles for pyrolysis of WPSAC obtained from different feedstock amount.

Fig. 3 shows temperature profiles recorded during pyrolysis of the waste palm shell with different sample mass. It can be observed that comparatively lower temperatures were achieved for 100 and 200 g of sample mass with maximum temperature recorded at 330°C and 450°C respectively. However, when sample mass was increased to 300 g and 400 g, the maximum temperature achieved was increased significantly to 750°C at around 500 s. This was probably because more carbon monoxide and hydrogen generated from the exothermic reaction in Eq 3 and 4. Also, exothermic reaction in more feedstock (300 g and 400 g) is much larger than endothermic reaction (Eq. 2 and 5). Then, more heat energy released to cause rapid increasing of temperature from 300 to 750°C in 300 and 400 g feedstock amount.

Generally, pyrolysis of the biomass was observed to produce biochar, bio-oil vapours, steam, and incondensable gases as by-products. The incondensable gases could consist of high amounts CO and small amounts of H_2 and CO_2, which form pyrolysis gas [14]. In this research, it was thought that the pyrolysis gas produced within the reactor was trapped inside the ceramics reactor and forced to react with hot char and caused exothermic reaction. Besides that, hot char also carried out activation by carbon dioxide shown in Equation 5 as majority carbon monoxide and with minor gases of hydrogen and carbon dioxide were detected by gas alert at gas emission channel. When the pyrolysis temperature was further increased above 450°C, softening and decomposition of the high-molecular-weight compounds occurred that resulted from the depolymerisation of lignocellulosic components such as lignin [15].

3.4 Yields, surface area and adsorption performance of waste palm shell activated carbon

Table 2. Yields, surface area and adsorption performance of waste palm shell activated carbon.

Sample mass (g)	Maximum Temperature (°C)	Yields (wt %)	S_{BET} (m^2/g)	MB adsorption (mg/g)
100	330	43.0	750	160
200	450	35.5	890	250
300	750	31.5	1200	425
400	750	31.0	1252	440

Table 2 shows the yield of WPSAC obtained at different maximum temperature and feedstock amount. The reaction rate in pyrolysis can be expressed in terms of conversion from the raw waste palm shell to WPSAC The conversion (C) of pyrolysis is defined as the ratio of weight loss to the initial raw material weight shown in the Eq. 7,

$$C = \frac{m_i - m_t}{m_i} \tag{7}$$

where m_i is the weight of the initial raw material (WPS) and m_t is the weight of char after pyrolysis. In this respect, the weight loss of the sample or the conversion is an indication of the thermal degradation occurring during pyrolysis. Higher pyrolysis temperature contributed immensely to the volatilization of cellulose, hemicelluloses and lignin which may be responsible for the low WPSAC yield [15].

The release of the volatiles from the carbon structures of the waste palm shell is dependent on the prevailing pyrolysis temperature. At about 450°C, low molecular - weight volatiles are released, thereby WPSAC yields with around 35.5% to 43.0 wt% (Table 1). Further increases in the pyrolysis temperature to 750°C, the small gradual decrease of yields to 31.0 wt% caused by the releasing of high-molecular-weight volatiles. Beyond that, the release of volatiles is almost complete and hence the conversion remains constant. The produced WPSAC had tested as absorbent for Methylene blue adsorption test. The removal of Methylene blue increased with prolonging the contact time to 24 hours and reached the equilibrium stage. The amount of MB adsorbed at the equilibrium time reflects the maximum adsorption uptake of WPSAC under operating conditions. In the present study,

 DOI: 10.1051/e3sconf/20172200195

adsorption equilibrium, q_e reached 440 mg/g with an initial concentration of 10 mg/L.

3.5 Textural and surface morphology

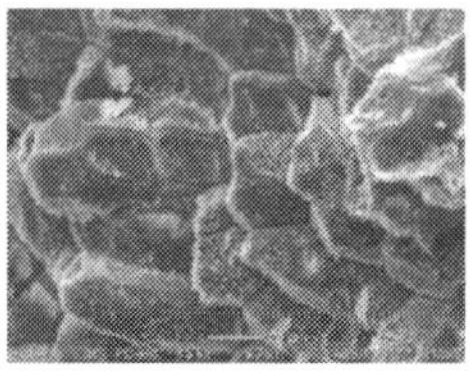

Fig. 4. Raw waste palm shell.

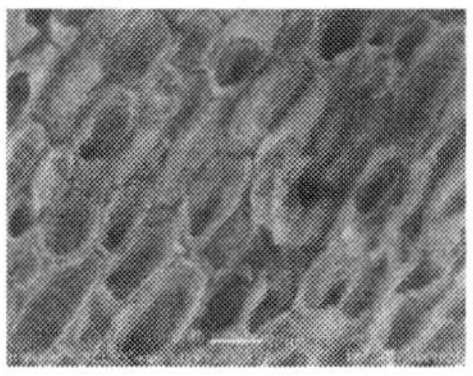

Fig. 5. Uniform pore in 100 g.

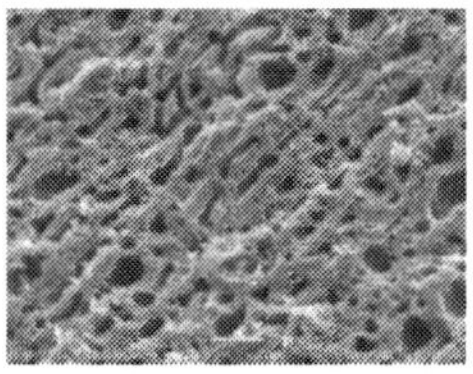

Fig. 6. Irregular pore sizes in 200 g.

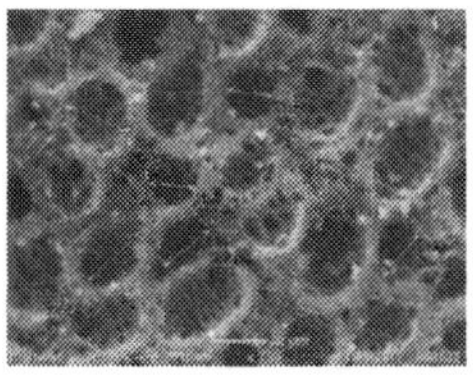

Fig. 7. Uniform macro pore in 300 g and 400 g.

The scanning electron micrograph shows the pores development on the surface of the WPS to WPSAC. WPS is covered with many globular silica bodies that contain sharp, conical agglomerations shown in Fig. 4 [16]. During microwave pyrolysis, silica body was removed from the carbon surface and create a macro pore shown in Fig. 5. With 100 gram of WPS loading and maximum temperature of 330°C, pores with less than 1 µm have been observed on waste palm biochar due to releasing globular silica body from the WPS surface. However, Fig. 6 observed some porous structure with non-regular pore sizes developed on the surface. This is probably due to decomposition of hemicellulose happen when maximum temperature reached 450°C in 200g WPS sample. Fig. 7 shown the effect of pyrolysis temperature appears on the larger pores generated and within it some smaller pores generated by pyrolysis gas activation. The pores on the surface of the WPSAC are well-developed with mainly 2 types of macrospore. Larger pores size range from 15–25 µm and within it some smaller pores ranging from 1–5µm are generated with activation process. Due to this, larger surface area generated and better for the adsorption.

4 Conclusion

In this study, waste palm shell-based activated carbons were produced successfully by self-generated pyrolysis gas as activating agent. From the experiment, quantity of WPS as one of the factor influence the generation of pyrolysis gas to perform the activation. Single-step approach reduced the production cost and time to convert waste palm shell into activated carbon. The results of the present investigation show that WPSAC prepared is potential useful adsorbent for the MB adsorption.

The authors gratefully acknowledge the financial support provided by University College of Technology Sarawak under Grant scheme (Project No. UCTS/RESEARCH/<1/2016/03>(01)).

References

1. C.Y. May, Overview of the Malaysian Oil Palm Industry, Malaysian Palm Oil Board, Economics & Industry Development Division, 2016
2. J.P. Tang, H.L. Lam, M.K. Abdul Aziz, N.A. Morad. Energ.
3. S. Sedghi, S.H. Madani, C. Hu, A. Silvestre-Albero, W. Skinner, P. Kwong, P. Pendleton, R.J. Smernik, F. Rodríguez-Reinoso, M.J. Biggs. Carbon **95**, 144–149 (2015)
4. M.K. Aroua, W.M.A.W. Daud, C.Y. Yin, D. Adinata. Sep. Purif. Technol. **62**, 609–613 (2008)
5. S. Mubarik, A. Saeed, M.M. Athar, M. Iqbal. J. Ind. Eng. Chem. **33**, 115–121 (2016)
6. P. Liu, W.-J. Liu, H. Jiang, J.-J. Chen, W.-W. Li, H.-Q. Yu. Bioresour. Technol. **121**, 235–240 (2012)
7. P. Xu, C.-X. Sun, X.-Z. Ye, W.-D. Xiao, Q. Zhang, Q. Wang. Ecotoxicol. Environ. Saf. **132**, 94–100 (2016)
8. R.K. Gupta, M. Dubey, P. Kharel, Z. Gu, Q.H. Fan. J. Power Sources **274**, 1300–1305 (2015)
9. M. Tripathi, J.N. Sahu, P. Ganesan, J. Jewaratnam. Appl. Thermal Eng. **105**, 605–612 (2016)
10. S.S. Lam, R.K. Liew, C.K. Cheng, H.A. Chase. Appl. Catal. B-Environ. **176–177**, 601–617 (2015)
11. S.S. Lam, R.K. Liew, X.Y. Lim, F.N. Ani, A. Jusoh. Int. Biodeter. Biodeg. **113**, 325–333 (2016)
12. S.S. Idris, N.A. Rahman, K. Ismail, A.B. Alias, Z.A. Rashid, M.J. Aris. Bioresour. Technol. **101**, 4584–4592 (2010)
13. A.F. Rugayah, A.A. Astimar, N. Norzita. J. Oil Palm Res. **26**, 251–264 (2014)
14. H. Guo, F. Peng, H. Zhang, L. Xiong, S. Li, C. Wang, B. Wang, X. Chen, Y. Chen. Int. J. Hydrogen Energ. **39**, 9200–9211 (2014)
15. Q. Jia, A.C. Lua. J. Anal. Appl. Pyrol. **83**, 175–179 (2008)
16. A.M. Abioye, F.N. Ani. Appl. Mech. Mater. **695**, 12–15 (2014)

Energy production for environmental issues in Turkey

Ibrahim Yuksel[1,*], *Hasan* Arman[2], and *Ibrahim Halil* Demirel[3]

[1]Yildiz Technical University, Faculty of Civil Engineering, Department of Civil Engineering, Hydraulics Division, 34220 Davutpasa, Istanbul, Turkey
[2]United Arab Emirates University, College of Science, Geology Dep., P.O. 15551, Al-Ain, UAE
[3]Yildiz Technical University, Civil Engineering Ph.D. Student, 34220, Istanbul, Turkey

Abstract. Due to the diversification efforts of energy sources, use of natural gas that was newly introduced into Turkish economy, has been growing rapidly. Turkey has large reserves of coal, particularly of lignite. The proven lignite reserves are 8.0 billion tons. The estimated total possible reserves are 30 billion tons. Turkey, with its young population and growing energy demand per person, its fast growing urbanization, and its economic development, has been one of the fast growing power markets of the world for the last two decades. It is expected that the demand for electric energy in Turkey will be 580 billion kWh by the year 2020. Turkey's electric energy demand is growing about 6–8% yearly due to fast economic growing. This paper deals with energy demand and consumption for environmental issues in Turkey.

1 Introduction

Turkey has dynamic economic development and rapid population growth. Turkey's population of about 80 million is growing. Along with the economic growth and population increase, significant increases were observed both in primary energy (primary energy sources include lignite, hard coal, oil, natural gas, hydroelectricity, geothermal, wood, animal and plant wastes, solar and wind) and electricity consumption during the 8th Plan period.

Over the last few decades, the numbers of Turkish citizens live in the eastern part of country have migrated to the western part of the country. This caused some problems in the energy utilization and environmental pollution in the western regions in Turkey. In this case, in the coming decades, global environmental issues could significantly affect patterns of energy use in Turkey. As in the world, hydropower has a great role in the global environmental issues such as environmental and air pollutions in Turkey.

* Corresponding author: yukseli2000@yahoo.com or iyuksel@yildiz.edu.tr

 DOI: 10.1051/e3sconf/20172200196

Affordable energy services are among the essential ingredients of economic development, including eradication of extreme poverty as called for in the United Nations Millennium Development Goal (DGs). Modern energy services-mainly provided by liquid and gaseous fuels, as well as electricity-are essential. Convenient, affordable energy is also important for improving health and education, and for reducing the human labour required to cook and meet other basic needs [1–3].

Renewable energy projects developed in scores of developing countries-many with international donor assistance-have demonstrated that renewable energy can contribute to poverty alleviation directly by providing the energy needed for creating business and jobs, turning locally available resources into productive economic assets [2].

However, developing the remaining hydropower potential offers many challenges and pressures from some environmental action groups over its impact has tended to increase over time. Hydropower throughout the world provides 17% of electricity from installed capacity in Turkey. The contribution of hydropower, especially small hydropower (SHP) to the worldwide electrical capacity is more of a similar scale to the other renewable energy sources (1–2% of total capacity), amounting to about 47 GW (53%) of this capacity is in developing countries [2, 3].

In the coming decades, global environmental issues could significantly affect patterns of energy use in Turkey. As in the world, hydropower has a great role in the global environmental issues such as environmental and air pollutions in Turkey. This paper deals with the role of hydropower in energy utilization and environmental pollution in Turkey [3].

A suitable environment will be established, with legislative regulations if necessary, in order for the private sector to fill the gap that will arise as a result of the withdrawal of the state from the sector, in a timely manner and to expedite the start of the new production investments in line with supply and demand projections. Thus, emphasis will be given to privatization of the existing facilities in order to prevent the burden of new investments on the state. The state will be adequately equipped in a way to closely monitor the supply security within the framework of its regulatory and supervisory role and to take measures [4].

2 Hydropower for energy policies and environmental issues

Hydropower is available in a broad range of Project scales and types. Projects can be designed to suit particular needs and specific site conditions. As hydropower does not consume or pollute the water it uses to generate power, it leaves this vital resource available for other uses. At the same time, the revenues generated through electricity sales can finance other infrastructure essential for human welfare. This can include drinking water supply systems, irrigation schemes for food production, infrastructures enhancing navigation, recreational facilities and ecotourism.

Hydropower has very few greenhouse gas emissions compared with other large-scale energy options. Table 1 shows the comparison of energy amortization time and emissions of various energy technologies [1]. In addition, by storing water during rainy seasons and releasing it during dry ones, dams and reservoirs can help control water during floods and droughts. These essential functions protect human lives and other assets. This will be increasingly important in the context of global warming, which implies an expected rising variability in precipitation frequency and intensity. On the other hand, hydropower projects do not export impacts such as acid rain or atmospheric pollution. Environmental impacts are limited to changes in the watershed in which the dam is located. When well-managed, these changes can sometimes result in enhancements, and other impacts can be avoided, mitigated. Hydropower can contribute to mitigating the widespread potential human impacts of climate change [5–7]. There is no single solution to the world's quest for more, cleaner energy and effective water management. Energy and water for sustainable development depend not only on supply choices, but also on how these choices are implemented. It requires the creation of a level playing field among available energy options and global water governance involving all stakeholders in a participatory decision-making process. In adopting their own sustainability guidelines, the members of the

International Hydropower Association are committed to developing and operating their projects, in collaboration with all stakeholders, in a way that is environmentally friendly, socially responsible and economically efficient so that hydropower projects can make a major contribution to achieving sustainable energy and resource development [6, 7]. Hydropower energy is a renewable, sustainable and clean energy in the other alternative energy sources. Moreover, it does not deprive future generations in terms of raw materials, or burdening them with pollutants or waste.

Hydroelectric power plants utilize the basic national and renewable resource of the country. Although the initial investment cost of hydropower seems relatively high, the projects have the lowest production costs and do not depend on foreign capital and support, when considering environmental pollution and long-term economic evaluation [3, 6, 8].

Main aim of the energy policies is to meet the energy needs of increasing population and growing economy in a continuous, qualified and secure manner through primarily private sector investments in a competitive and transparent free market environment. In this context, it is the main target to supply the required energy timely, uninterrupted and at minimum costs while making energy supply planning [9].

A suitable environment will be established, with legislative regulations if necessary, in order for the private sector to fill the gap that will arise as a result of the withdrawal of the state from the sector, in a timely manner and to expedite the start of the new production investments in line with supply and demand projections [10]. Thus, emphasis will be given to privatization of the existing facilities in order to prevent the burden of new investments on the state. The state will be adequately equipped in a way to closely monitor the supply security within the framework of its regulatory and supervisory role and to take measures [11].

It is fundamental to complete the projects, in particular the hydroelectricity power plant projects, covered by the public investment program with the lowest costs and in the fastest manner. For this reason, emphasis will be given to ensure that investment costs reflect the true cost, cross financing among sectors are avoided and cost increases that could arise as a result of delays in projects are prevented.

On the other hand, in order to create a healthy diversification in electricity supply, nuclear energy will also be included among electricity production resources. Before the construction of a nuclear power plant, detailed plans and programs on storage, elimination of wastes and informing the public will be prepared by considering maximum harmony with the free market [3, 6].

Table 1. The comparison of energy amortization time and emissions of various energy technologies.

Technology	Energy pay back time in months	SO_2 emission in kg/GWh	NO_2 emission in kg/GWh	CO_2 in Ton/GWh
Coal fired	1.0–1.1	630–1370	630–1560	830–920
Gas (CCGT)	0.4	45–140	650–810	370–420
Large-hydro	5–6	18–21	34–40	7–8
Micro hydro	9–11	38–46	71–86	16–20
Small hydro	8–9	24–29	46–56	10–12
Wind turbine				
4.5 m/s	6–20	18–32	26–43	19–34
5.5 m/s	4–13	13–20	18–27	13–22
6.5 m/s	2–8	10–16	14–22	10–17
Photovoltaic				
Mono-crystalline	72–93	230–295	270–340	200–260
Multi-crystalline	58–74	260–330	250–310	190–250
Amorphous	51–66	135–175	160–200	170–220

Source: [1].

3 Hydropower for renewable, sustainable and clean energy

As a contributor of air pollution and deforestation, the share of biomass in the renewable energy share is expected to decrease with the expansion of other renewable energy sources [12]. Table 2 shows renewable energy supply and projections for future in Turkey, respectively [13, 14].

The aim of the Renewable Energy Program is to increase privately-owned and operated power generation from renewable sources within a market-based framework, which is being implemented in accordance with the Electricity Market Law and the Electricity Sector Reform Strategy [15]. This program will assist the Directorate of the Ministry of Energy and Natural Resources (MENR) in the preparation of a renewable energy law, as well as to define the required changes and modifications related to legislation such as the Electricity Market Law to better accommodate greater private sector involvement [12, 15].

Turkey has aimed at increasing domestic production by public, private and foreign utilities, at increasing efficiency by rehabilitation of existing plants and acceleration of existing construction programs. On the other hand, there are 436 sites available for hydroelectric plant construction, distributed on 26 main river zones.

At present, the average hydroelectric capacity in Turkey is 45 TWh/yr which corresponds to only 36% of the total economically feasible potential of the country. Up to 2020, it is expected that about 502 new hydropower plants will be constructed to make use of the full available potential [6,16-19].

The distribution of hydroelectric power potential in Turkey is as shown in Table 3. By 2020, it is expected that the total hydro installed capacity will be almost 36,000 MW. Thus, within this framework, it is also expected that the GAP (Southeastern Anatolian Project) will comprise 22 dams and 19 hydroelectric plants and a total generation of about
27 TWh/yr, and the irrigation of an area of 1,785,050 ha will be realized. The GAP is an integrated, multi-sectorial development project, and covers many sectors such as agriculture, industry, transportation, water sports, etc. [20, 21].

Table 2. Renewable energy supply in Turkey.

Renewable energy sources	2000	2005	2010	2015	2020
Primary energy supply					
Hydropower (ktoe)	2 656	4 067	4 903	7 060	9 419
Geothermal, solar and wind (ktoe)	978	1 683	2 896	4 242	6 397
Biomass and waste (ktoe)	6 457	5 325	4 416	4 001	3 925
Renewable energy production (ktoe)	10 091	11 074	12 215	15 303	19 741
Share of total domestic production (%)	38	48	33	29	30
Share of TPES (%)	12	12	10	9	9
Generation					
Hydropower (GWh)	30 879	47 287	57 009	82 095	109 524
Geothermal, solar and wind (GWh)	109	490	5 274	7 020	8 766
Renewable energy generation (GWh)	30 988	47 777	62 283	89 115	118 290
Share of total generation (%)	25	29	26	25	25
Total final consumption					
Geothermal, solar and wind (ktoe)	910	1 385	2 145	3 341	5 346
Biomass and waste (ktoe)	6 457	5 325	4 416	4 001	3 925
Renewable total consumption (ktoe)	7 367	6 710	6 561	7 342	9 271
Share of total final consumption (%)	12	10	7	6	6

Source: [13, 14].

Table 3. Present and potential of hydroelectric power in Turkey.

Present hydropower plants	Number of Power Stations	Total installed Capacity (MW)	Proven Production (GWh/year)	Total Annual Production (GWh/year)
In production > 10 MW	74	193	287	722
In production < 10 MW	68	12,595	33,273	45,208
Under construction > 10 MW	8	45	151	228
Under construction < 10 MW	32	3,152	6,207	10,290
Present total	182	15,985	39,918	56,448
Future Possible Potential				
> 5 MW	164	366	571	1,848
5–10 MW	82	610	897	2,587
10–50 MW	187	4,727	9,234	18,959
50–100 MW	51	3,692	7,734	13,001
100–250 MW	37	5,815	11,824	19,308
240–500 MW	10	3,250	5,620	10,688
500–1000 MW	2	1,053	2,054	3,173
1000 < MW	1	1,200	2,459	3,833
Future Total	534	20,713	40,393	73,398
TOTAL	716	36,698	80,311	129,846

Source: [16].

4 Air pollution, climate change and CO_2 factors for renewable energy in Turkey

Since the Kyoto Protocol in December 1997, at the Third Conference of the Parties to the United Framework Convention on Climate Change (UNFCC), international attention is growing toward a consensus in favor of reducing carbon emission to mitigate climate change [10].

The leading options for achieving cost-efficiency in carbon emission reductions address the design issue of how to implement an emission cap-and-trade system in Turkey. We therefore examine the design of alternatives permits trading programs to address carbon emission related energy-consumption. In this context, tradable emission permits would entitle holders to emit up to a specified level of carbon emissions. By issuing a fixed number of allowances less than business-as-usual current emissions, Turkey could reduce its national CO_2 emissions to meet internationally targets [4, 11, 22].

Emissions of CO_2, SO_2 and NO_X have increased over the 1980s (Table 4) both in absolute terms and relative to GDP. Turkey is the only OECD country were the intensity of emissions of NO_2 rose in this period, while it was one of only two countries that experienced an increase in SO_2 emission intensity.

By the late 1990s, emissions of SO_2 in relation to GDP were double the OECD average, reflecting the heavy share of high-sulfur lignite in power generation and poor quality liquid fuels. Since the end of the 1990s, natural gas usage has been rising rapidly and this has helped concentrations of SO_2 to decline markedly in some of the major urban areas, such as Ankara, Izmir and Istanbul, but they still remain at levels that are double those found on the average in OECD metropolitan areas.

Concentrations in the rest of the country, as measured by a simple average of all monitoring stations, are some 75% higher than in the three metropolitan areas. In addition there has been little downward movement in the estimated concentrations of lead in the air, but levels in Turkey are low compared to those found in a number of European countries [10].

Table 4. Total greenhouse gas emissions in Turkey.

Year	CO_2 (Mt/year)	NOx (kt/year)	SO_2 (kt/year)	CH_4 (kt/year)
1980	73.3	372	17.5	128
1990	133	656	31	150
1995	162	785	33	140
2000	254	1166	45.7	190
2010	460	2046	79.4	216

Mt: Million tons; kt: kilo tons. Source: [10].

5 Conclusions

Energy projects, in particular the hydroelectricity power plant projects, covered by the public investment program with the lowest costs and in the fastest manner. For this reason, emphasis will be given to ensure that investment costs reflect the true cost, cross financing among sectors are avoided and cost increases that could arise as a result of delays in projects are prevented.

On the other hand, in order to create a healthy diversification in electricity supply, nuclear energy will also be included among electricity production resources. Before the construction of a nuclear power plant, detailed plans and programs on storage, elimination of wastes and informing the public will be prepared by considering maximum harmony with the free market

As a large hydro energy project the Southeastern Anatolian Project (GAP) will also integrate, multi-sectorial development project, and covers many sectors such as agriculture, industry, transportation and water sports.

Renewable energy supply in Turkey is dominated by hydropower and biomass, but environmental and scarcity-of-supply concerns have led to a decline in biomass use, mainly for residential heating. As a contributor of air pollution and deforestation, the share of biomass in the renewable energy share is expected to decrease with the expansion of other renewable energy sources such as solar and wind.

In view of the low operational costs of hydropower, related to not having any fuel expense, hydro plants can easily fit load demand, while having a long economic life and low environmental impacts compared with alternative energy resources.

Acknowledgments

This study was supported by Yildiz Technical University, Scientific Research Coordinator, Istanbul, Turkey. Project ID: 1897, Project Number: 2016-05-01-KAP01.

References

1. UNDP, United Nations Development program, *World Energy Assessment Report*, New York: United Nations, 2000
2. I. Yuksel, J. En. So., Part B: Eco., Plan., and Pol. **2**, 2, 113–121 (2007)
3. I. Yuksel, A. Dorum, J. En. So., Part A: Rec. Utiliz. and Env. Eff. **33**, 13, 1221–1229, (2011).
4. I. Yuksel, J. of E. S., Part A: R., Ut., and En. Ef. **31**, 20, 1915–1925 (2009)
5. WEC, World Energy Council, *Survey of Energy Resources*, www.worldenergy.org (2000)
6. I. Yuksel, J. of R. and Sust. En. Re. **12**, 6, 1622 – 1640 (2008)
7. IHA, International Hydropower Association, *The Role of Hydropower in Sustainable Development*, IHA White Paper, 1–140 (2003)
8. O. Paish, Re. and Sust. En. Rev. **6**, 537 – 556 (2002)
9. DPT, State Planning Organization. *Ninth Development plan 2007–2013*, (2006)
10. MEF, Ministry of Environment and Forestry. *First national Communication of Turkey on Climate Change*, 60–150 (2007)
11. MENR, Ministry of Energy and Natural Resources. *Energy Statistics in Turkey*, www.enerji.gov.tr (2007)
12. K. Kaygusuz, A. Sarı, En. So., Part B **1**, 23–35 (2006)
13. IEA, International Energy Agency. *Energy policies of IEA countries: Turkey 2005 review*, Paris: OECD/IEA, (2005)
14. MENR, Ministry of Energy and Natural Resources. *Energy Statistics in Turkey*. http://www.enerji.gov.tr (2009)
15. TEIAS, Turkish Electricity Transmission Corporation. *Annual Report* (2006)
16. DSI, State Hydraulic Works, *Hydropower potential in Turkey* (2007)

17. R. Ozturk, O. Kincay. En. So. **26**, 1141–1156 (2004)
18. O. Yuksek, M.I. Kömürcü, I. Yüksel, K. Kaygusuz, En. Po. **34**, 3093–3103 (2006)
19. K. Kaygusuz, O. Yüksek, A. Sarı, En. So., Part B **2**, 19–29 (2007)
20. GAP, Southeastern Anatolia Project. *Energy production in GAP region*, www.gap.gov.tr, (2006)
21. O. Yuksek, K. Kaygusuz, En. So., Part B **1**, 279–290 (2006)
22. K. Kaygusuz, Cl. **35**, 539–547 (2007)

DOI: 10.1051/e3sconf/20172200197

The role of energy systems on hydropower in Turkey

Ibrahim Yuksel[1,*], *Hasan* Arman[2], and *Ibrahim Halil* Demirel[3]

[1]Yildiz Technical University, Faculty of Civil Engineering, Department of Civil Engineering, Hydraulics Division, 34220 Davutpasa, Istanbul, Turkey
[2]United Arab Emirates University, College of Science, Geology Dep., P.O. 15551, Al-Ain, UAE
[3]Yildiz Technical University, Civil Engineering Ph.D. Student, 34220, Istanbul, Turkey

Abstract. Over the last two decades, global electricity production has more than doubled and electricity demand is rising rapidly around the world as economic development spreads to emerging economies. Not only has electricity demand increased significantly, it is the fastest growing end-use of energy. Therefore, technical, economic and environmental benefits of hydroelectric power make it an important contributor to the future world energy mix, particularly in the developing countries. On the other hand, the hydropower industry is closely linked to both water management and renewable energy production, and so has a unique role to play in contributing to sustainable development in a world where billions of people lack access to safe drinking water and adequate energy supplies. In addition to, approximately 1.6 billion people have no access to electricity and about 1.1 billion are without adequate water supply. However, resources for hydropower development are widely spread around the world. Potential exists in about 150 countries, and about 70% of the economically feasible potential remains to be developed-mostly in developing countries where the needs are most urgent. This paper deals with renewable energy systems and the role of hydropower in Turkey.

1 Introduction

Hydropower contributes one-fifth of the world's power generation. In fact, it provides the majority of supply in 55 countries. For several countries, hydropower is the only domestic energy resource. Its present role in electricity generation is therefore substantially greater than any other renewable energy technology, and the remaining potential, especially in the less developed countries such as Turkey, is vast. While it is not a panacea, in that it is restricted to sites with available water and appropriate geomorphology, hydropower's flexibility and proven technology sets it apart from other.

With an estimated 95% of population growth in the coming decades likely to be in and around cities, the need for sustainable solutions with compact, interconnected power plants will be ever more pressing. Decentralized schemes, however, will remain important for rural electrification programs.

* Corresponding author: yukseli2000@yahoo.com or iyuksel@yildiz.edu.tr

Hydropower can be adaptive and flexible. Depending on the storage capacity involved, a major advantage of hydropower is that generation can be scheduled. Run-of-river schemes can be implemented to provide continuous 'base-load' generation. The operation of a cascade of several run-of-river power plants can be optimized to provide generation when it is needed. This is especially true if there is a reservoir scheme at the head of the cascade. Schemes that include a reservoir are able to store potential energy for production when the demand is highest.

When water resources are not available to replenish reservoirs by natural inflow, pumped-storage schemes have been developed to assist in the storage of energy from other generation sources. Therefore, hydropower can substantially improve efficiency in a mixed power system, reducing emissions from fossil-fuel power plants, and backing up intermittent sources such as wind power [1–5].

Hydropower is available in a broad range of project scales and types. Projects can be designed to suit particular needs and specific site conditions. As hydropower does not consume or pollute the water it uses to generate power, it leaves this vital resource available for other uses. At the same time, the revenues generated through electricity sales can finance other infrastructure essential for human welfare. This can include drinking water supply systems, irrigation schemes for food production, infrastructures enhancing navigation, recreational facilities and ecotourism.

Water is a vital resource that supports all forms of life on earth. Unfortunately, it is not evenly distributed by season or geographical region. Some parts of the world are prone to drought, making water a particularly scarce and precious commodity. In other parts of the world, floods that cause loss of life and property are major problems. Throughout history, dams and reservoirs have been used successfully in collecting, storing and managing water needed to sustain civilization. Hydropower often supports other essential water services such as irrigation, flood control and drinking water supplies. It facilitates the equitable sharing of a common vital resource [4–7].

2 Hydropower and renewable energy for sustainable focus on Turkey

In view of its geographical location, the mainland of Anatolia has throughout historical times been the birthplace of many great civilizations. It has also been prominent as a centre of commerce because of its connections to three continents and the seas surrounding it on three sides. As a result of the varied geographical conditions, Turkey shows great diversity in geological structure, topography, climate and vegetation. The country is subdivided into seven regions: four of them (Black Sea, Marmara, Aegean, Mediterranean) are coastal units, and the remaining three (Central Anatolia, East Anatolia, Southeast Anatolia) are mountainous and subject to harsh climatic conditions [8].

According to the latest population survey (in 2000), the population of Turkey is 67.80 million of which 44.00 million live in urban settlements and 23.79 million live in rural settlements. The annual population growth rate fell below 2% for the first time in history after 1945. The population in urban settlements increases at a rate greater than that in rural settlements. In the last decade, the population in urban settlements has shown a considerable increase, rising from 56% in 1990 to 64.9 % in 2000. On the other hand, the average life expectancy in the country is 71.24 years. The infant mortality rate is 4.73%; and the literacy rate is 85% [8].

Water resources

The amount of precipitation in any particular region usually varies from year to year but, over a long period, the average remains relatively constant. Turkey averages about 643 mm of precipitation annually, but the distribution is quite uneven. The range is from less than 250 mm in the inland areas of central Anatolia to more than 3000 mm in the northeastern Black Sea coastal region. Autumn marks the start of the rainy season, which continues until late spring on the western and southeastern coasts; whereas the Black Sea coast receives rain throughout the year.

This average annual precipitation corresponds to an average of 501 km^3 (501 billion m^3) of water per year. While 274 km^3 of this quantity returns to the atmosphere through evaporation-transpiration, 69 km^3 feeds the aquifers through infiltration from the surface. Thus the average annual surface water potential is 186 km^3, of which 158 km^3 comes from surface runoff and 28 km^3 of groundwater feeds the rivers. With a surface run-off of 7 km^3 volume coming from the neighboring countries, the total surface run-off within the country reaches 193 km^3. However, from the economic and technical points of view, the average exploitable water potential of the country is 110 km^3 per year [8, 9].

In view of the considerable variation in runoff in terms of seasons, years and regions, it is absolutely necessary for the major rivers in Turkey to have water storage facilities, to allow for the use of the water when it is necessary. Consequently, priority has always been given to the construction of water-storage facilities. Significant progress has taken place in the construction of dams throughout the 48 years that have elapsed since the establishment of the State Hydraulic Works (DSI) [9]. With the projects developed primarily by DSI and other institutions engaged in water resources development, water consumption in Turkey reached 39.3 billion m^3 by 2000, corresponding to only 36% of the economically exploitable water resources. During water consumption estimates on a sectoral basis, it is accepted that all of the economically irrigable land will be irrigated with irrigation schemes constructed by the year 2030 and water consumption for irrigation will be 71.5 billion m^3. Hence, while its share in the total consumption was 75% in 1999, the share of irrigation water in the total water consumption will be decreased to 65% by the year 2030, through the utilization of modern irrigation techniques [10].

It has been accepted that the total population of Turkey will reach 110 million by 2030, with an annual increase rate of 2%. Additionally, it is assumed that the per capita water consumption of 280 l/day (in 2003) will reach 540 l/day by 2030. By taking into consideration that about 5.2 billion m^3 water is needed in the tourism sector, the total water consumption for domestic purposes will reach 26.1 billion m^3 by 2030. With the assumption of 4% annual growth rate in the industrial sector, it is expected that industrial water consumption will increase from 4.8 billion m^3 in 2003 to 13.2 billion m^3 in 2030. Thus, considering all of these issues that 100% of the total economically exploitable water resources will be under use by the year 2030 [7, 11].

The water potential of countries is usually evaluated based on water potential per capita. According to international criteria, countries with a water potential greater than 10,000 m^3 per capita per year are accepted as water-rich; countries with a potential of 3,000 to 10,000 m^3 are accepted as self-sufficient; countries with a potential of 1,000 to 3,000 m^3 are accepted as having a water-deficit; and, those with a water potential of less than 1,000 m^3 per capita per year are regarded as water-poor countries [9, 10, 12, 13].

Energy utilization

The basic target of Turkey's national policy on energy is the provision of cheap electrical energy in sufficient amounts and on time, under qualified, reliable and competing conditions of the energy market. The energy policy, determined by 5-year development plans, is as follows [10]:

- Provision of qualified, reliable and cheap energy for sustainability in socio-economic development;
- Provision of safety in energy supply;
- Encouragement of private sector investments and expedition of privatization activities in the power sector; and,
- Addition of new and renewable sources as soon as possible to the energy supply cycle.

Studies to maintain power sector competition, increase productivity and provide transparency are going to be carried out to restructure the power sector and establish organizations through the Electricity Market Law [10, 12–14].

Energy resources

Turkey has a wide variety of conventional and renewable energy resources including lignite, hard coal, asphalted, bituminous shale, oil, natural gas, hydro, biomass, geothermal, wind and solar [15, 16]. Table 1 shows both current situation and future aspect on renewable energy supply, generation and consumption in Turkey [10, 12]. However, most of these are of inadequate quality and quantity. Each of its coal, geothermal and hydro reserves is estimated to be around 1% of world reserves, and its oil and natural gas reserves are negligible compared with the world total [10, 12–14, 17]. Turkey's national energy resources consist mainly of hydraulic (126 TWh/year), lignite (105 TWh/year) and hard coal (16 billion kWh/year) resources, with a total annual average of 248 TWh. Approximately 126 TWh of hydropower potential corresponds to 35,000 MW of generation capacity. Although 35–46% of electricity generation was provided by hydropower before 1993, the share of hydropower decreased to 23% in 2001 [10, 12].

Table 1. Primary energy supply, generetion and final consuption in Turkey.

Renewable energy sources	2000	2005	2010	2015	2020
Primary energy supply					
Hydropower (ktoe)	2 656	4 067	4 903	7 060	9 419
Geothermal, solar and wind (ktoe)	978	1 683	2 896	4 242	6 397
Biomass and waste (ktoe)	6 457	5 325	4 416	4 001	3 925
Renewable energy production (ktoe)	10 091	11 074	12 215	15 303	19 741
Share of total domestic production (%)	38	48	33	29	30
Share of TPES (%)	12	12	10	9	9
Generation					
Hydropower (GWh)	30 879	47 287	57 009	82 095	109 524
Geothermal, solar and wind (GWh)	109	490	5 274	7 020	8 766
Renewable energy generation (GWh)	30 988	47 777	62 283	89 115	118 290
Share of total generation (%)	25	29	26	25	25
Total final consumption					
Geothermal, solar and wind (ktoe)	910	1 385	2 145	3 341	5 346
Biomass and waste (ktoe)	6 457	5 325	4 416	4 001	3 925
Renewable total consumption (ktoe)	7 367	6 710	6 561	7 342	9 271
Share of total final consumption (%)	12	10	7	6	6

DOI: 10.1051/e3sconf/20172200197

Electrical energy consumption is one of the most important indicators of economic development and social welfare. Considering electricity consumption development by sector in Turkey, while the consumption by industry has increased, its share has decreased. During the same period, both consumption and the share increased for both the residential and services sectors; there was not a noticeable change in the transportation sector.

On the other hand, it is important to point out the restructuring policies that are directing the development of new and renewable energy, and giving special emphasis to socio-economical bodies, laws and legal regulations. In this respect, particular attention and priority should be given to the development of the hydroelectric potential in Turkey, since it is the most important natural renewable resource and only 35% of the technically and economically utilizable hydro potential has been developed so far [10, 12–14, 17, 18].

3 Hydropower development and review of Turkey

3.1. Historical review and situation of the last decades

Hydroelectric power plants go back some 90 years and they supply about 34% of the electricity produced in Turkey. The first production began with a 60 kW hydro plant in Tarsus, which was used only for providing lights during the initial years of the Republic of Turkey. The installed capacity taken over by the Republic was 29,664 kW, and the electricity was available only in Istanbul, Izmir, Tarsus and Adapazarı. With the development of industry, usage of electrical energy other than for lighting started in 1930 and large industrial establishments began trying to produce their own electricity [19–23].

Water projects were initiated by the Ministry of Public Works under the leadership of Atatürk in 1932. The Electrical Power Resources Planning and Survey Administration (EIE) was then established in 1935 to define Turkey's energy demand, carrying out surveys and investigations to develop the hydroelectric potential of water resources and other energy resources. The important projects of that period were: Seyhan, Sarıyer, Hirfanlı, Kesikköprü, Demirköprü and Kemer dams and hydro plants. There were altogether 28 hydro plants, sharing 3.2% of the total energy production, by 1940. Etibank and the Bank of the Provinces were involved in the construction of small hydropower plants and the electrification of villages and towns [21–24].

On the other hand, 85% of the total hydro capacity in operation has been developed by DSI , corresponding to 9,931 MW (49 hydro plants) and 35,795 GWh/year respectively. The largest and most comprehensive regional development project ever implemented by DSI in Turkey is "The Southeast Anatolian (GAP) Project", which is located in the region of Southeast Anatolia on the Euprates and Tigris Rivers and their tributaries, which originate in Turkey [25–30]. By 2004, 203 small hydro projects (with installed capacities of less than 10 MW) were at various stages of development. When fully developed, these projects will have a total installed capacity of 849 MW and generate 3623 GWh/year. So far, 70 SHP projects have been put into operation, 6 are under construction and 127 projects are at various planning stages. In other words, 2.4% of the total economic installed capacity and 3% of the energy potential correspond to SHP projects (plants of less than 10 MW) [9, 10, 12, 30, 31].

3.2 Hydropower potential and important of hydropower in Turkey

It is estimated that there is considerable small hydropower (SHP) potential in Turkey. DSI has started a pre-investigation study on "The Place of Small HEPPs Within Estimated Hydroelectric Potential". These studies conclude that an additional technical hydroelectric energy potential of 57 TWh/year could be utilizable. 38 TWh/year of hydroelectric energy potential, corresponding to two-thirds of this additional potential, has been estimated to be economically utilizable, so the total economically utilizable hydroelectric potential of Turkey will reach 164 TWh/year [10, 25, 32-35].

Approximately 50% of the additional potential of 38 TWh (that is, 19 TWh) could be realized as small HEPPs (hydroelectric power plants), with installed capacities of less than 10 MW. The share of SHP potential in the total, which is 3% at present, would be 14%. On the other hand, in accordance with the results obtained from the pre-evaluation study, about 15% of the increase in 126 TWh/year exploitable energy potential might be achieved by developing additional SHP potential. However, this study gives only rough results about the additional SHP potential of the country and the potential must be evaluated more precisely, with comprehensive master plan studies for each hydrological basin [10, 22, 33].

The generation of hydropower provides an alternative to burning fossil fuels or nuclear power, which allows for the power demand to be met without producing heated water, air emissions, ash, or radioactive waste. Of the two alternatives to hydropower, in the last decade, much attention has been given to thermal power production because of the adverse effect of CO_2 emissions. With the increasing threat of greenhouse gases originating from such anthropogenic activities on the climate, it was decided to take action. Thus the Framework Convention on Climate Change was enacted on 21 March 1994 and has been signed by 174 countries to date.

Dams that produce electricity by this most productive renewable clean energy source in the world provide an important contribution to the reduction of air pollution. The result of an investigation held in the USA suggests that the productivity of hydroelectric power-plants is higher than 90% of thermal plants and this figure is twice that of thermal plants. In case of Turkey, the public has been wrongly informed. Some people have claimed that hydro plants do not produce as much energy as planned because of irregular hydrological conditions and rapid sedimentation of reservoirs. It is also claimed that the cost of the removal of dams entirely filled by sediment at the end of their physical lives is not considered in the total project cost, and that there are major problems in recovering the cost of investment and environmental issues [11].

3.3 Current situation and future aspect in Turkey

The Ministry of Energy and Natural Resources carries out the general energy planning studies, using an 'MAED' demand model, and TEIAS, (Turkish Electricity Transmission Company) carries out energy generation expansion planning studies, using the DECADES model. The MAED model [10], which was developed by the International Atomic Energy Agency (IAEA), makes projections of the medium and long-term general electricity demand. It takes into consideration a detailed analysis of social, economic and technical systems. The model is based on low, medium and high case scenarios.

The TEIAS, has prepared the Long-Term Energy Generation Plan, taking into consideration the MAED model demand outcome. According to the Plan, the installed capacity will increase to 57,551 MW in 2010 and to 117,240 MW in 2020. The installed hydropower capacity is anticipated to increase to 18,943 MW in 2010 and to 34,092 MW in 2020. Thus, an additional 1000 MW of hydro capacity should be added to the system annually over the next 20 years. Turkey is thus seeking support for the development of all its economic potential by 2023, which is the 100th anniversary of the foundation of the Turkish Republic [13, 17, 33].

4 Conclusions

Turkey's demand for energy is increasing rapidly and heavily dependent on expensive imported energy resources. As would be expected, the rapid expansion of energy production and consumption has brought with it a wide range of environmental issues at the local, regional and global levels. With respect to global environmental issues, Turkey's GHG emissions have grown along with its primary energy consumption. In this regard, renewable energy resources appear to be the one of the most efficient and effective solutions for clean and sustainable energy development in Turkey.

Turkey's geographical location has several advantages for extensive use of most of these renewable energy sources. Turkey has substantial reserves of renewable energy sources, including approximately 1% of the total world hydropower potential. There is also significant potential for wind power development. Turkey's geothermal potential ranks seventh worldwide, but only a small portion is considered to be economically feasible. This article presents the role of renewables in meeting energy demand of Turkey as a sustainable manner.

Turkey, with its young population and growing energy demand per person, its fast growing urbanization, and its economic development, has been one of the fast growing power markets of the world for the last two decades. Turkey is heavily dependent on expensive imported energy resources that place a big burden on the economy and air pollution is becoming a great environmental concern in the country. In this regard, renewable energy resources appear to be the one of the most efficient and effective solutions for clean and sustainable energy development in Turkey.

On the other hand, renewable energy supply in Turkey is dominated by hydropower and biomass, but environmental and scarcity-of-supply concerns have led to a decline in biomass use, mainly for residential heating. As a contributor of air pollution and deforestation, the share of biomass in the renewable energy share is expected to decrease with the expansion of other renewable energy sources such as solar and wind. There is also significant potential for wind power development. Turkey's geothermal potential ranks seventh worldwide, but only a small portion is considered to be economically feasible. Turkey should be use the renewable energy sources as much as possible to meet the energy demand as a clean and sustainable manner. Of course, the climate change is a reality and country's imported expensive energy demand is increasing rapidly.

References

1. I. Yuksel, J. of En. So., Part A: Rec., Util., and Env., Eff. **31**, 20, 1915–35 (2009)
2. Renew. and Sus. E. Reviw. **12**, 1622–40 (2008)
3. K. Kaygusuz, E. Con. and Man. **43**, 1099–1120 (2002)
4. K. Kaygusuz, E. S. **26**, 215–224 (2004)
5. IEA, International Energy Agency. *World Energy Outlook 2002*, (2002)
6. IHA. *The role of hydropower in sustainable development*, www.hydropower.org (2003)
7. WEC. *Survey of Energy Resources*, www.worldenergy.org (2001)
8. UNDP. *World Energy Assessment Report*, New York: United Nations (2000)
9. DIE. *Statistic yearbook of Turkey in 2003* (Ankara, 2004)

10. DSI. *Statistics on hydropower*, www.dsi.gov.tr (2004)
11. WCD. *Dams and Development*, www.wcd.org (2000)
12. MENR., Ministry of Energy and Natural Resources. 2009. Energy Statistics in Turkey. Available from http://www.enerji.gov.tr.
13. IEA. Energy Policies of IEA Countries: Turkey 2005 *Review* (2005)
14. IHA. *Hydropower information and country report for Turkey*, www.hydropower.org (2006)
15. D. Altınbilek, Wa. Res. Dev. **18**, 9–24 (2002)
16. O. Yuksek, M.I. Kömürcü, I. Yüksel, K. Kaygusuz, En. Pol. **34**, 3093–3103 (2006)
17. TEDAS. www.tedas.gov.tr (2003)
18. TEIAS. *Short history of electrical energy development in Turkey*. http://www.teias.gov.tr (2005)
19. UNDP/W. B. E. S. Mana. Assi. Prog. (ESMAP). *Turkey-Energy and the Environment, Issues and Options Paper*. Report 229, (2000)
20. H. Özgöbek, S. Timuçin, DSI, *HYDRO conf., Dev., Man., Perf.* (2002)
21. F. Adıgüzel, A. Tutus, DSI, *HYDRO conf., Dev., Man., Perf.* (2002)
22. D. Altınbilek, C. Çakmak, DSI, *3rd Inte. Sym., Os.* (2001)
23. C. Çakmak, DSI, *Report on the Development of Hydroelectric Energy* (1998)
24. H.D. Altınbilek, M. Bayram, T. Hazar, *Benefits and Concerns about Dams and Case Studies, 203–212, 67th Annual Meeting of ICOLD* (1999)
25. K. Kaygusuz, En. So. **23**, 775–799 (2001)
26. K. Kaygusuz, En. So. **24**, 803–815 (2002)
27. R. Ozturk, O. Kıncay, En. So. **26**, 1141–56 (2004)
28. I. Yuksel, J. of En. So., Part B: E., P., and Po. **2**, 2,113–21 (2007)
29. I. Yuksel, J. of En. E. & E. **24**, 4–5, 361–70 (2006)
30. H. Ceylan, H.K. Oztürk, Int. J. of Gr. En. **1**, 393–406 (2004)
31. D. Altinbilek, *Hydroelectric development plans in Turkey*, www.dsi.gov.tr/doganyazi.htm (2000)
32. C. Akkaya, *Benefits and Concerns about Dams and Case Studies, 113–122, 67th Annual Meeting of ICOLD* (1999)
33. K. Kaygusuz, Ren. and Sust. En. Revi. **6**, 339–366 (2002)
34. EIE. *Hydroelectric Power Activities of the EIE*, www.eie.gov.tr (2004)
35. K. Kaygusuz, En. So. **21**, 581–588 (1999)

Zero energy buildings in the logistics warehouse systems

Pawel Zajac[1,*], *Stanislaw* Kwasniowski[1]

[1]Wroclaw University of Science and Technology, Faculty of Mechanical Engineering, Wybrzeze Wyspianskiego 27, 50-370 Wroclaw, Poland

Abstract. The article discusses the evaluation of possible reductions of energy consumption in warehouse buildings and the analysis of construction and functioning of modern storage warehouses. For a warehouse the following are presented: operation evaluation measures and indexes, an energy balance, ideas for improvements in terms of energy conservation in particular functioning zones.

1 Introduction

The functioning of warehouses is clearly related to energy consumption in its various forms. Among various forms and varieties of warehouses, the cold and heated warehouses are characterized by relatively high energy demand. Modern warehouse buildings often cover tens of thousands square meters. The energy demand in such types of buildings is enormous. Not only is the economic aspect essential, but also ecological one. The renewable energy sources support natural environment protection. Investments in such types of sources are often funded. Therefore if we assume that a zero energy building is a building that does not consume energy in an annual balance, then logistic warehouse systems seek just such solutions [1].

2 Warehouse energy balance

Many goods require controlled storage conditions. Therefore, warehouse buildings are equipped with devices that enable creating a suitable microclimate inside. Thus these will be cold, air conditioned and heated warehouses [2].

Depending on intended use, cold warehouses may require an internal temperature of even -40°C. The air-conditioned warehouses are the warehouses in which normal conditions are maintained. The heated warehouses are buildings in which the temperature should be above zero $T_w > 0$°C. Obviously there are special warehouses that require maintaining higher temperatures [3].

The present analysis considers typical warehouse buildings in which various goods are stored, equipment is operated and people work. In such a situation we treat warehouse buildings as facilities in which interior temperature is practically constant and its little

* Corresponding author: pawel.zajac@pwr.edu.pl

DOI: 10.1051/e3sconf/20172200198

fluctuations resulting from the working control systems are ignored. External and internal factors determine the condition of the warehouse interior and the goods stored within. Two situations can be distinguished. Case I – a warehouse requires energy supply. Case II – a warehouse requires draining energy. These situations are described by the heat balance equation.

$$Q_{dop}/_{odp} = Q_{przen} + Q_{wew} + Q_{urz} + Q_{o\acute{s}w} + Q_{wew.wen} + Q_{os} - \Delta E_p \quad (1)$$

Symbols in the equation (1) represent: Q_{dop}- heat supplied by warehouse heating system; Q_{odp} – heat removed from interior by interior cooling equipment; $\boldsymbol{Q_p}$ – **heat penetrating through insulated warehouse walls** with different thermal insulation of walls, windows, roof, floor particular areas taking also into account thermal bridges. The flux presents the equation (2) [4].

$$Q_{przen} = Q_{pnn} + Q_{pos} + Q_{pn} + Q_{ok} + Q_{okn} + Q_{pbp} + Q_m \quad (2)$$

Symbols in the equation (2) represent: Q_{przen} – heat of penetration; Q_{pnn} – heat of penetration through sunless walls; Q_{pos} – heat of penetration through floor; Q_{pn} – heat of penetration through sunlit walls; Q_{ok} – heat of penetration through windows; Q_{okn} – heat of penetration through windows or skylights in radiation path; Q_{pbp} – heat of penetration through walls from buildings adjoining to warehouse, e.g.: common-office rooms; Q_m – heat of penetration through thermal bridges. The situation is presented in Fig. 1.

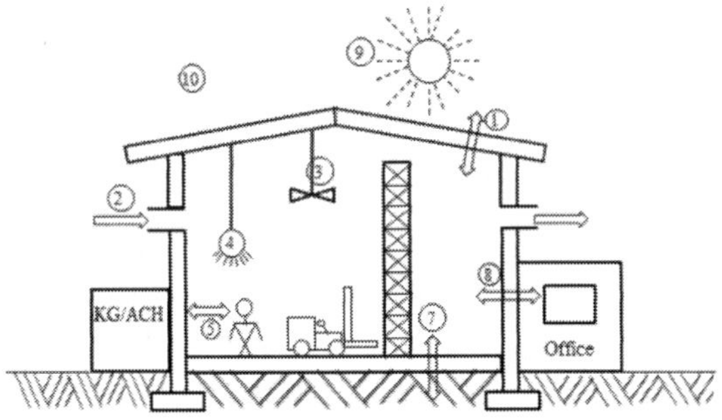

Fig. 1. Diagram of warehouse facility with internal and external energy sources: 1 – heat transfer through walls, 2 – facility ventilation, 3 – warehouse internal air supply devices, 4 – lighting, 5 – energy emitted by persons, 6 – energy emitted by devices, 7 – energy exchangeable with substrate (ground), 8 – energy exchangeable with adjoining buildings of different temperature, 9 – solar radiation energy, 10 – environment, 11 – heating boiler or refrigerating unit, 12 – office.

$$Q_{pnn} = (T_{ot} - T_w)\sum k_i \cdot F_i \quad (3)$$

Symbols in the equation (3) represent: T_{ot} – ambient temperature; T_w – interior temperature; k_i – penetrability of given partition; F_i – area of given partition. The value of heat losses through the so-called thermal bridges, depending on the warehouse structure, is given as a percentage part in relation to the total penetration heat losses. The heat penetration losses Q_{pnn} are attempted to be reduced by using of suitable heat-insulating materials. It is worth to pay attention to the parameter of the heat penetration coefficient k_i that expresses the Peclet dependence (4).

$$k_i = \frac{1}{\frac{1}{\alpha_z} + \sum\frac{s_i}{\lambda_i} + \frac{1}{\alpha_w}} \quad (4)$$

Symbols in the equation (8) represent: α_z – convective heat-transfer coefficient on external wall surface dependent on air flow around speed and surface condition (wet-dry); α_w – convective heat-transfer coefficient on internal wall surface depends on air flow around speed relative to wall; s_i – thickness of distinguished partition; λ_i – thermal conductivity of wall making layer. In case of differentiated walls construction, the heat balance takes into account the area and penetrability of each wall fragment of the same transverse structure. The analysis of convective heat-transfer coefficients $\frac{\alpha_z}{\alpha_w}$ impact on the "k" heat-transfer coefficient k is limited (< 1%) – here decides the wall thermal resistance $\frac{s_i}{\lambda_i}$ that is considerably higher than the thermal resistances on the building's external and internal surfaces.

Q_m- heat penetrating through thermal bridges is estimated by the coefficients Q_p – in vehicles it is assumed as equal to 10% of the penetration heat. In warehouse buildings it is proposed to assume $5\%Q_p$. In the structure the so-called thermal bridges across the wall section are avoided. The thermal bridges can be: ventilating pipes, reinforcement bars, metal connections across wall, various types of connectors, metal door frames, levelling ramp metal platforms.

In many cases it is difficult to avoid such elements. Approx. 1000 times more heat runs away through a steel thermal bridge than through a heat-insulating wall of similar section. In practice it is easy to detect thermal bridges on the basis of thermovision tests.

Designers of warehouse buildings developed such solutions of constructional nodes that reduce the thermal bridge effects to a minimum.

Q_{pos} heat exchange through a floor of cold warehouse. The heat exchange through the floor can lead to ground freeze-up on which it is laid, which can result, in some circumstances, cracking of the floor and foundation (defrosting of frozen ground). Therefore the floor should be resistant to loads and, on the other hand, it should be well insulated so that not to lead to ground frost under it. In such a case it can be necessary to heat up the whole area of the floor. The waste heat of the condenser of the units cooling the warehouse room can be used to heat the floor. The heat flux supplied by the floor heating system, in accordance with the heat exchange theory, will be divided into two fluxes inversely proportional to the thermal resistance $\frac{s_i}{\lambda_i}$ of the insulation over the heating layer and under the heating layer. These layers are separated by a heated pipe screen (underfloor heating). The power of the heating system Q_{pos} will result from the thermal resistance of the insulation between the heating layer and the warehouse interior k_p as well as the assumed temperature difference [5–7]:

$$Q_{pos} = F_p \cdot k_p(T_{wy} - T_w) \tag{5}$$

Symbols in the equation (5) represent: F_p – floor area; k_p – heat penetration of floor with insulation; T_{wy} – heated layer temperature, screen with heating pipes; T_w – warehouse interior temperature.

Q_{pn} – heat penetrating sunlit walls. The flux can be determined on the basis of heat radiation intensity, the area, the absorption coefficient and the walls heat-insulating features or from the equation of heat exchange through penetration, determining earlier the average calculation temperature of heat exchange.

$$T_{obl} = T_{ot} + \frac{A \cdot I}{\alpha_z} \tag{6}$$

Symbols in the equation (6) represent: A – absorption coefficient depending on wall colours: for dark walls A = 0.9, for grey and green ones A = 0.7, for light walls A = 0.5; I – intensity of solar radiation depending on time of day, latitude, season; α_z – convective heat-transfer coefficient on external walls surface 20 W/(m^2 K) .

The nominal solar radiation intensity, called the solar constant, varies between 1.35 and 1.44 kW/m^2. This radiation during crossing the Earth's atmosphere is weakened. The measure of the weakening is the atmosphere transparency coefficient B. Its values depend on the season and the local atmosphere cleanness. Characteristic values [8, 11].

The sunlit area F_{SN} is the area of the walls on which the sunrays fall directly and the roof one decreased by the area of windows on these walls.

$$Q_{pn} = \sum_{\mathrm{i}} F_{SNi} \cdot k_{N_i}(T_{obl} - T_W) \tag{7}$$

Where the I = 1, 2, 3 index means the wall through which heat is exchanged, taking the radiation into account. The sun is usually on the three walls of the warehouse.

Heat penetrating through windows Q_{ok}.

$$Q_{ok} = F_{SZ} \cdot k_{SZ}(T_{ot} - T_W) \tag{8}$$

Symbols in the equation (8) represent: F_{SZ} – area of all window panels; k_{sz} – heat transfer coefficient through window panels that depends on window types, window panels are single or double, glued multilayer. Detailed data about window panels give their manufacturers. The last component of the dependence is the heat penetrating into interior Q_{okp} in the path of radiation through window panels.

$$Q_{okn} = [F_{SZN} \cdot I \cdot a \cdot (F_{SZ} - F_{SZN})I_R] \cdot b \tag{9}$$

Symbols in the equation (9) represent: F_{SZN} – area of sunlit window panels; F_{SZ} – total area of window panels; I – solar radiation intensity; I_R – scattered solar radiation intensity.

Q_{pbp} – heat penetrating through walls from buildings adjoining to warehouse (10):

$$Q_{pbp} = \sum_i F_{pbp} \cdot k_i(T_{wm} - T_{wp}) \tag{10}$$

Symbols in the equation (10) represent: F_{pbp} – wall area; k_i – wall penetrability; T_{wm} – warehouse interior temperature; T_{wp} – adjoining building's interior temperature. Saving of the energy consumption is sought in the roof suitable structure. Windows and skylights are located in the roof; these, light the interior in a natural way (thus reducing energy consumption for lighting). Pipe optical fibres are also used. The warehouse energy consumption is also affected by its location and alignment relative to the compass.

Q_{wen} – heat lost or outflowing due to intended or unintended ventilation of interior. The temperature difference between the interior and environment as well as the pressure difference will decide unintended air exchange. Unintended air exchange is determined by the temperature difference between the interior and the environment as well as the pressure difference. The heat losses in this path are attempted to be decreased by reducing leaks and the use of various airlocks and air curtains.

The warehouse building leaktightness is understood as the ability to prevent air penetration through internal partitions. It regards, in particular, the air movement through joints of external partition elements. It is estimated that the air penetration is responsible for more than 30% of the costs borne for heating and ventilation LSM.

Q_{wew} – heat of object ventilation is given by the dependence (11).

$$Q_{wen} = W \cdot c_p(T_{ot} - T_W)(1 - \eta_c) \tag{11}$$

Symbols in the equation (11) represent: c_p – air specific heat; T_{ot} – ambient temperature; T_W – temperature inside warehouse; W – quantity of air exchanged in a time unit between interior and environment.

The value of the exchange is influenced by the air streams enforced by the ventilation system due to hygienic or technical reasons, as well as the uncontrolled stream through technical holes, doors, gates, airlocks. The forced ventilation can be expressed by: the change rate that is the ratio of the exchanged air quantity to the interior volume $\left[{}^{1}/_{h}\right]$ or $\left[{}^{m^3}/_{h}\right]$.

The warehouse ventilation system should be consistent with the standard PN-83/B-03430/Az3 that defines minimal values of ventilation air supplied to the defined rooms. The fresh air supply to rooms can significantly impact heat losses and that is why the EU regulations provide that the supply-exhaust ventilation of capacity higher than 500 m^3/h must be equipped with an energy recovery system of efficiency exceeding 50%. Since 1st January 2016, in accordance with EU Commission regulation No. 1253/2014 of 7th July 2014 on execution of European Parliament and Council Directive 2009/125/UC in relation to the eco-project requirements for ventilation systems (Journal of Laws EU L 337/8), the minimal thermal efficiency for supply-exhaust ventilation systems must be $\eta_c = 67\%$ and from January 2018 – $\eta_c = 73\%$.

<u>Q_{urz} – heat emitted by devices operating in warehouse</u>.

It regards forklift trucks and various types conveyors. They perform mechanical work the part of which impacts increasing potential energy ΔE_p of the goods stored on the warehouse racks. This part of the energy in the energy balance should be deducted from the work used by internal transport devices.

$$\Delta E_p = \sum_i \dot{m}_i gh \tag{12}$$

Symbols in the equation (12) represent: m_i – total mass of goods placed on warehouse racks in considered time unit; h – storage height from floor level. From the point of view of a heat balance and energy consumption, the electric forklift trucks with electric drive are definitely more advantageous than the trucks with combustion drive and are at least 70% more efficient.

<u>$Q_{ośw}$ – energy consumed by light sources</u>

$$Q_{osw.} = \sum Q_{osw.jedn.} \tag{13}$$

The correct luminous flux density is approx. 150 lx/m^2 – zone lighting control should be used if possible.

<u>$Q_{wew.wen}$ – energy consumed for internal ventilation forced by fans</u>.

Considering the temperature homogenization in the interior and the correct heat dissipation airflows and fans are used to equalize the temperature throughout the warehouse. The energy power consumed by the devices puts the air in motion, but finally, as a result of viscous friction of air particles, it is converted into the heat.

<u>Q_{os} – heat emitted by warehouse employees</u>

$$Q_{os} = \sum Q_{jedn.} \tag{14}$$

Every man working in a cold store emits heat. The heat is emitted through skin and the exhaled air; it is the so-called latent heat.

In the warehouse energy balance (dependence (1)) there are components strictly related to the construction technology of the warehouse wall, roof and floor. While testing the LSM energy consumption it can be noticed that the LSM energy consumption is significantly impacted by technical-organizational solutions. It is not justified to compare the energy consumption by warehouses of various intended use. The work character and primary goals may overrule energy consumption [9, 10].

3 LSM energy consumption evaluation by coefficients

The energy consumption by a warehouse is influenced by: goods turnover, assortment quantities, storage technology, warehouse equipment, work organization in a warehouse, degree of mechanization and automatization, advancement of automatic identification technics, microclimatic requirements, external factors: temperature and humidity, requirements inside warehouse, warehouse size.

Therefore, it seems reasonable to compare the energy consumption with warehouse units rotated in the considered time interval e.g. $\frac{kWh}{EP}$ in kWh per one palette or 24 h. It is also possible to compare the global energy consumption with the warehouse floor area, its cubic volume or total quantity of warehouse nests on the rack storage. In the table 1 we present a comparison of exemplary evaluation indexes of LSM energy consumption that can be a measure of the warehouse energy consumption.

Table 1. Comparison of LSM energy consumption evaluation indexes. Own analysis.

Item	Index	Formula
1	Global energy consumption per warehouse unit	$W_{kjm} = \frac{\sum E_i}{k_{jm}}$
2	Index of energy consumption per warehouse area unit	$W_{Fe} = \frac{\sum F_i}{F}$
3	Index of energy consumption per warehouse cubature unit	$W_V = \frac{\sum F_i}{V}$
4	Index of energy consumption per rack nest in time	$W_g = \frac{\sum F_i}{m}$
where: k_{jm} – number of warehouse units going through the warehouse in specific interval; F – warehouse area; V – volume (storage capacity) $[m^3]$; m – number of rack nests;		

4 Directions of solutions minimizing energy consumption by unit warehouse

While discussing minimization of energy consumption, we keep in mind the minimization of the indexes listed in table 1. It does not concern the absolute values of the energy used by an individual warehouse, but it does concern the consumption comparison with the warehouse unit handled. The consumption will be the highest at low turnover in a big warehouse e.g. state reserves in cold stores. Low values will be in warehouses of quick turnover of a high throughput working in the cross-docking system.

The energy consumption minimization in case of cold stores can be obtained through: 1. minimization of external impacts; 2. minimization of internal heat sources; 3. suitable solutions of warehouse internal infrastructure; 4. suitable warehouse and supporting system work organization; 5. using green energy from own sources and energy recovery.

Ad. 1. The external impact means heat losses through walls and air exchange between the interior and external environment. The state of the situation can be improved through:

- the use of suitable heat insulating materials of economically justified quality and thickness, and avoiding the so-called thermal bridges;

- the ventilation losses can be decreased by reducing controlled air flows, i.e. ensuring the tightness of buildings, traffic openings (doors, gates) surrounded by airlocks with insulating elements (sleeves, pneumatic shafts), openings protected with strip curtains or quick-rotating curtains – activated by proximity sensors. Finally, air curtains are being used – also automatically controlled. The air exchange should take place with as much control as possible through recuperation systems.

Ad. 2. The minimization of the energy supplied to the interior from internal sources regards the energy emitted by personnel, devices supporting warehouse work, lighting, and internal ventilation.

Ad. 3. Energy consumption in unit warehouses also depends on infrastructural solutions: flow racks, chutes, gravitational movement of goods. The transport devices should be equipped with energy recuperation systems, for example forklift trucks that operate intensively in an accelerating and braking cycle should have energy recovery systems. Possible benefits in this regard are presented in Fig. 2.

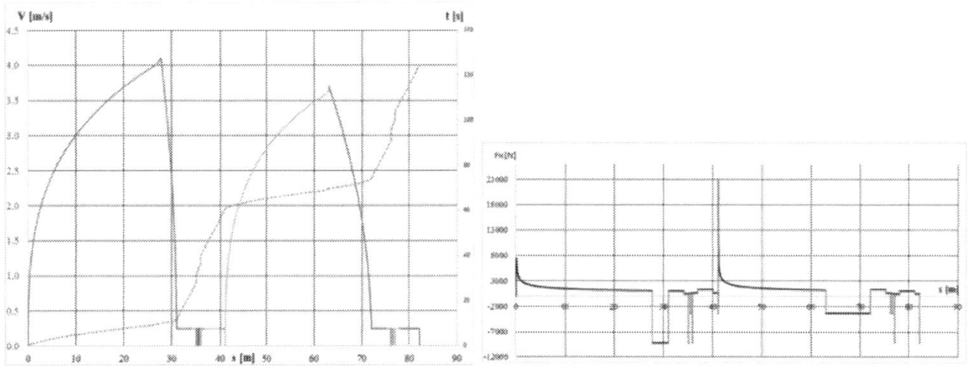

Fig. 2. The run of traffic changes in the function of distance and time of covering of particular movement phases. Own analysis.

Fig. 2 shows the speed profile of a forklift truck travel consistent with standard VDI2198 for Jungheinrich EFG-220 forklift truck so-called minimal time travel and the time of distance covering. Fig. 2b shows run of driving forces and braking forces of the truck during operation in VDI2198 cycle. The fields on the graph between the force graph and the y-axis illustrate the energy consumed by the truck for drive and braking. It is possible, in this way, to calculate precisely the energy consumed by the truck and estimate potential benefits from the energy recovery during braking.

A warehouse full of goods is a big energy accumulator in a form of potential energy contained in the stored goods. That energy should be used. From this point of view, it is advantageous to situate the acceptance zone higher than the picking and release zones.

Ad. 4. Appropriate work organization, warehouse operation management and the arrangement of the goods has a huge impact on the energy consumption. For these purposes ABC and XYZ (analysis of the distribution of goods ABC: according to cumulated turnover, XYZ - according to the accuracy of the forecast) analyses are used. Picking and automatic identification enables considerable savings regarding work and time. The systems of picking processes support: pick-by-light, pick-by-voice. Depending on the warehouse operation character, savings can bring a dispersed system of goods storage in various zones making picking possible for many employees at the same time. Undoubtedly the automated picking systems can be also more advantageous than hand picking.

Ad. 5. The energy supplying the warehouse, ensuring its functioning, should be used as efficiently as possible e.g. by using heat pumps in the cooling and heating systems. Large areas of the roofs (flat areas) of the warehouses are used for the arrangement of solar cells. Only in such a situation can the warehouse, under certain circumstances, achieve zero energy condition in an annual balance.

5 Summary

The goal of this analysis was to identify all essential factors affecting energy consumption in a warehouse. Formally speaking, it is expressed by the energy consumption equation (1), and quantitative trends can be shown by the indexes proposed in Table 1. In many cases, it is possible to shape the warehouse and its equipment in such a way, that the energy consumption, equals zero e.g. in an annual scale. It must be noted that energy consumption in a warehouse is not always the most important goal. The superior goal is always the efficient handling of the flow of goods. In many cases we may agree to extensive automation, often energy-consuming, instead of short handling time and correct picking. In various situations, however, it is worth taking into account the energy consumption of a warehouse facility.

References

1. P. Zajac, *Evaluation method of energy consumption in logistic warehouse systems* (Springer, DOI 10.1007/978-3-319-22044-4, 2015)
2. P. Zajac, *Springer Proceedings in Energy*, (Springer, 195–200, DOI: 10.1007/978-3-319-16901-9)
3. P. Zajac, *The energy consumption in refrigerated warehouses* (Springer, DOI 10.1007/978-3-319-40898-9, 2016)
4. P. Zajac, *Advances in Intelligent Systems and Computing (AISC)* **350**, 345–355, DOI: 10.1007/978-3-319-15796-2_35 (2015)
5. P. Keskiewicz, B. Sniechowska, J. Danielewicz, Instal **2**, 27–30 (2012)
6. J. Danielewicz, K. Golecki, *A guide for the designer of the boiler house* (Publishing House of the Wrocław University of Technology, Wroclaw, 2004)
7. J. Danielewicz, et al., Energy, **108** (2016)
8. P. Zajac, Archives of Civil and Mechanical Engineering **11**, 2, 479–492 (2011)
9. J. Grajnert, S. Kwasniowski, Heat transfer dynamics in refrigerated vehicles, Refrigeration - Polish Engineering Association (NOT) **33**, 37–39 (1998)
10. S. Jebaraj, S. Iniyan, Renewable and Sustainable Energy Reviews **10**, 4, 281–311 (2006)
11. E.-R. Schramek, *Taschenbuch fur Heinzung und Klimatechnik*, (07/08, Herman Recknagel, Industrieverlag-Munchen, 2010)

Energy sustainable cities. From eco villages, eco districts towards zero carbon cities

Anna Zaręba[1,*], *Alicja Krzemińska*[1], and *Janusz Łach*[1]

[1]University of Wrocław, pl. Uniwersytecki 1, 50-137 Wrocław, Poland

Abstract. Minimizing energy consumption is the effect of sustainable design technics as among many others: designing buildings with solar access and natural ventilation, using climate responsive design materials and effective insulation. Contemporary examples of zero-carbon cities: Masdar City, United Arab Emirates and Dongtan, China, confirm technical feasibility of renewable energy by implementation of solar PV and wind technologies. The ecological city - medium or high density urban settlement separated by greenspace causes the smallest possible ecological footprint on the surrounding countryside through efficient use of land and its resources, recycling used materials and converting waste to energy. This paper investigates the concept of energy sustainable cities, examines, how urban settlements might affect building energy design in eco-villages, eco-districts (e.g. Vauban, Freiburg in Germany, Bo01 Malmo in Sweden), and discuss the strategies for achieving Zero Emission Cities principles in densely populated areas. It is focused on low energy architectural design solutions which could be incorporated into urban settlements to create ecological villages, districts and cities, designed with consideration of environmental impact, required minimal inputs of energy, water, food, waste and pollution.

1 Introduction

According to scientific research quoted in the U.S. National Science and Technology Council report [1], 60 to 70% of energy savings in buildings can be achieved through different improvements (such as: passive heating and cooling, heat pumps, water conservation and recycling) and in addition about 30 to 40% of energy can be produced by renewable sources [2]. To achieve the significant energy savings, a simple system of urban metabolism, expressed as a linear combination of inputs (materials, energy, food and water), converted into outputs (waste: solid, liquid and gases) has to be expended to incorporate recycling technics (three R– Reclaim, Reuse and Recycle) [2, 3]. According the European Commission by 2050 it is estimated 80% cut in the EU greenhouse emission (intermediate GHG cuts of 25 % by 2020, 40% by 2030 and 60% by 2040 [4].

* Corresponding author: anna.zareba@uwr.edu.pl

2 Objectives of the research

The paper focuses on the special characteristics of low energy consumption residential areas, introduces energy sustainable technologies, incorporated in architecture and urban scales (eco-villages, eco-district. ZEC), expressing the feasibility of existing design technics. This paper investigates the concept of energy sustainable cities, examines, how urban settlements might affect building energy design in eco-villages, eco-districts (e.g. Vauban, Freiburg in Germany, Bo01 Malmo in Sweden), and discusses the strategies for achieving Zero Carbon Cities principles in densely populated areas. The aim of the studies was to determine the scale and extent of the energy-sustainable architecture phenomenon (ZEB) and highlight the diversity of projects, introduced in different climatic conditions. The research was based on identifying contemporary eco-village, eco-district and eco-city initiatives, through the analysis of the relevant literature, reports, government and policy documents, conference proceedings and websites of international networks relevant to energy-sustainable communities initiatives.

3 ZEB – Zero Energy or Zero Emission Building

The term "ZEB" is related to Zero Energy Building and Zero Emission Building. The Zero Energy Building refers to energy consumed by a single construction (residential or commercial building), the Zero Emission Building refers to the carbon emissions produced by a building, which is reduced to minimum [5]. ZEB types can be divided into Autonomous ZEB (energy self-sufficient free standing structure) and NetZEB (energy neutral building, delivering the same amount of energy it returns) connected to the community network [5, 6]. According to Energy Performance of Building Directive [7] ZEB :"*has a very high energy performance with a low amount of energy required covered to a very significant extent by energy from renewable sources ,including energy from renewable sources produced on-site or nearby*". In the ZEB or nearly ZEB, according to European Commission: "*we will live and work in low-energy, low-emission buildings with intelligent heating and cooling systems. We will drive electric and hybrid cars and live in cleaner cities with less air pollution and better public transport*" [4]. According to the European Union Directive on the energy performance of buildings (Directive 2010/31/EU) all new building ought to be designed as nearly zero-energy houses from 2020 and the European Strategic Energy Technology Plan (SET-Plan) assumed, that at least half of the existing buildings in 25 demonstration cities should be transformed into nearly zero energy buildings by 2020 [8]. A nearly zero-energy building is defined in Article 2 of the EU Directive as "*a building that has a very high energy performance. The nearly zero or very low amount of energy required should be covered to a very significant extent by energy from renewable sources, including energy from renewable sources produced on-site or nearby*" [7].

Energy savings can be achieved through various practices: adaptation the design to local climate condition and topography, use of environmental building design technics, such as: passive solar systems, natural ventilation, proper thermal insulation of the buildings (solid and vaccum insulation, vaccum windows and cool roofs) [8]. Most of energy is produced on-site or off-site, on the basis on Renewable Energy Sources (REN), including: solar (solar photovoltaics, solar-thermal power systems incorporated into buildings), biomass-, wind-, hydro- and geothermal energy systems. Nevertheless off-site technologies, exploiting the wind, hydro or geothermal energy cannot be easily introduced in densely populated area.

Heating, Ventilation and Air-Conditioning (HVAC) systems (heating, hot water, air-conditioning and ventilation systems) are supported by intelligent, computer based, Building energy Management Systems (BMS) [8].

On the basis on the way a given building is supplied with energy, Lund at al. [9] distinguished four types of ZEB: APV-ZEB – building equipped with photo-voltaic system (PV), Wind-ZEB with on-site wind turbines, APV-Solar thermal-heat pump ZEB with combination of PV installation, APV-solar thermal collectors and a heat pump and heat storage system, and A wind—solar thermal-heat pump ZEB.

4 Eco-Village

The term eco-village was coined by Robert Gilman (1991) in the paper "The Ecovillage Challenge" and defined as: a combination of a human habitat and human activities which does not cause environmental damage and support human health [10]. Average eco-villages range from a population of 50 to 150 inhabitants. Ecovillages combine social, ecological and economic aspects. Eco-villages cause minimal ecological impact using alternative systems of: energy, water, transportation and waste-treatment. Dawnson [11] highlights five features, which describe eco-villages principles: non-governmental projects, strong communities, provision and distribution of the natural sources of water, energy and food and strong sense of shared values. In the context of energy distribution the basic assumption is to maximize the use of village's own supplies (combined in micro-grids supplied by renewable energy sources) and reduce the necessity of use of national network. According to data quoted by Hawasly, Corne, Roaf [2010] who investigated the optimizing of energy consumption in the experimental Riccarton Ecovillage (Scotland) consisting of 20 homes, savings in energy consumption can achieve the value 35–40% [12].The Global Ecovillage Network (GEN), established in 1996, defines eco-villages as: "*an intentional or traditional community using local participatory processes to holistically integrate ecological, economic, social, and cultural dimensions of sustainability in order to regenerate social and natural environments*"[13].UN-Habitat, the agency responsible for promoting sustainable communities, has included eco-villages in its database of the top 100 listing of Best Practices since 1998.

4.1 Cloughjordan Eco-village, Ireland

The development began in 2007 in a rural area of the midlands region in Ireland, approximately 1.5 hours from the capital city of Dublin. The eco-village is accessible by train from various urban centres. In 2002 Sustainable Projects Ireland Ltd (SPIL) – a nonprofit organization, identified a site for location of the new Eco-village, to minimize the cost of investment - adjacent to existing town-Cloughjordan. The Village Ecological Charter stipulates the energy requirements for new dwellings (30 percent below the input allowed by the 2006 Building Regulations) and establishes total energy usage for single family house of 84 $kWh/m^2 \cdot yr$ [14]. To achieve this goals the technologies of high level of insulation had to be incorporated in the building's construction. An Cloughjordan Eco-village heating system is based on wood chips and a solar panel field. The eco-village is connected to a sustainable drainage system (SuDS). The entire village, located on the area of 67 acre, when completed, will include mixed used environment, economic and socially friendly sustainable developments. According to the Master Plan, the Cloughjordan Eco-village will incorporate 130 energy efficient residential dwellings, green and educational centre, eco-hostel, allotments and community gardens and leisure facilities.

Located along the village stream the pedestrian greenway will join two green centres with playgrounds [14].

5 Eco-district

Eco-districts are characterized by positive environmental impacts and increase wellbeing of local communities. EcoDistricts™, a nonprofit organization in the US, defines them as "*a whole systems green neighborhood investment strategy that integrates resource efficiency, ecological diversity, mobility, and community wellbeing*" [15]. In eco-districts energy costs are reduced by constructing buildings with low energy consumption in sense of: materials, orientation, insulation and ventilation. The eco-district consists of positive energy buildings, which produce more energy than they consume. Technologies and strategies of sustainability development incorporate: renewable energy, water and zero-waste management, green streets, smart green, rainwater harvesting, greenways (pedestrian paths and bike lanes), TOD (Transport Oriented Development), multimodal transit, urban Eco-farms and new green areas planning. Strong community vitality and active and diverse participation are distinguished features of an eco-district's society. The EcoDistricts™ Framework [15] identifies eight goals of the eco-district such as: Equitable Development (understood as ensuring "fair distribution" of community benefits: job and investment opportunities), Health and Wellbeing (access to recreation and natural areas, healthy food, improving air quality), Community Identity, Access and Mobility (mix-uses, accessible services, promoting active transportation), Energy (efficient dispersed energy system based on renewable energy resources – solar, wind, biomass, geothermal), Water (reuse and recycle water resources, water stormwater management and water discharge system), Habitat and Ecosystem Function (increase biodiversity, connectivity within and beyond the system – ex. through green infrastructure, enhance watersheds) and Materials Management (reduce of waste, use of recycled materials, organic waste compositing).

Eco-districts are created as an alternative for urban sprawl with the principles of reducing energy costs and low energy consumption, through smart uses of building materials, smart orientation of residential blocks and natural ventilation and insulation incorporated in the architecture. Fitzgerald and Lenhart describe eco-districts as "*a missing link between building scale and city or regional scale planning and implementation.*" [16]. Over the last two decades model eco-districts have been born in EU, with significant examples of: Vauban in Freiburg, Western Harbour in Malmö, Hammerby Sjöstad in Stockholm. Freiburg was awarded the Academy of Urbanism's European City of the Year Award in 2010, Malmö gained, among many others awards: a 2010 United Nations World Habitat Award for the revitalization of Augustenborg and in 2012 Malmö was a finalist for the European Green Capital designation [16]. Three of the most significant Northern European eco-districts (mentioned above), were built on post-industrial wasteland.

5.1 Riesfield and Vauban Eco-districts, Freiburg. Transport innovative solutions

Construction of Vauban (south-west Germany), established on the former military area, comprising 2,000 energy sustainable houses began in 1998, Riesfield in 1992. The 70 ha brownfield redevelopment investment was completed in 2010. Both habitats were designed as mix-use eco-suburbs with 3–5 story buildings, high density (in Vauban 122 persons/ha) pedestrian and bicycle friendly transport, new tram line and parking ratio less than 0.5 per house unit [17]. Car usage was limited to minimum through reduce of parking space and marking out parking-free residential streets and car-free zones. Most of the parking spaces are located in garages on the edge of the residential area. Construction of new buildings corresponds to high standards of thermal insulation, as the heating system is based on solar energy and woodchips. The innovative transport solutions include: increase of frequencies of local train, extension of city's tram system, a new transport interchange with 1000-space bike parking, "bike and ride" parking, u-shape, 4 m in width play streets, used as common community space, giving priority to pedestrians. Most of the district is marked a car-free residential area with access to tram stops within a 400- maximum distance from every single household [17].

5.2 Bo01 Eco-district, Malmö, Sweden. Innovation in ecology

Bo01 was launched on a 160 ha former industrial estate and port area with a principle of creating of sustainable, energy self-sufficient model residential unit for over 30,000 inhabitants. Bo01 is a high density urban unit, supplied by 100% renewable energy (sun and wind). Most of the heat for the district is delivered by community heat pump installation from natural underground reservoir. Bo01 Eco-district is equipped with 10 cold and warm wells in an aquifer with temperature of 10–11°C located in the ground at a depth of 40–70 meters[16]. The heat production is supported by 1400 m^2 of solar collectors, connected to residential area heating system. The district comprehensive heating system produces 5,800 MWh of heath, 5,000 MWh cooling and 6,300 MWh electricity, which means it supplies 85,000 m^2 of living space. 2 MW wind turbine, placed in the northern part of the harbor and 120 m^2 photovoltaic solar panels on the buildings supply the Bo01 Eco-district in electricity [16,18]. Rainwater drainage system, treated as a public space, includes waterfalls, gutters, artificial ponds for storage and to purify rainwater. The district topography has been changed to design natural rundown to the sea or to the canal, green roofs were used to reduce the amount of rainwater. Apart from about mentioned ecological technologies, the recycled water, raw materials and waste are reused. The area is characterized by relatively high biodiversity, diverse green areas, public spaces and green roofs [15].

6 Sustainable city principles

The search for the answer about the urban form, which the best suits the principles of sustainable city, has led to two dominant and contradictory theories: the compact and the dispersed city. The compact city theory is connected with a form of high density, mixed urban environment (housing, workplaces, services) based on semi-detached and multifamily housing. The compact city idea enables to reduce energy costs inside urbanized areas through low energy use, centralized heating system and reducing the demands for everyday travelling. The negative aspects of compact city and the belief that it: "*rejects suburban and semi-rural living, neglects rural communities, affords less green and open space, increases congestion and segregation, reduces environmental quality and lessens the power for making local decisions*" have contributed to the development of urban planning concepts which combine attitudes of compact city with dispersed city, such as: urban village, new urbanism, TOD transit-oriented development and smart growth [19, 20].

The development of energy storage network opens the opportunities for local societies to undertake community initiatives, such as decentralized energy supply production. In some countries power providers are obliged to incorporate decentralized energy supply in the comprehensive energy production and distribution system. According to nowadays market requirements, energy production is partially based on municipality governed facilities and individual combined heat and power installations, what creates favourable conditions for the development of social network related to energy [21].

According Eco2 Cities, an initiative launched by the World Bank, as part of the World Bank Urban and Local Government Strategy,: "*Ecological cities enhance the well-being of citizens and society through integrated urban planning and management that fully harnesses the benefit of ecological systems, and protects and nurtures these assets for future generations.*" [22]. To fulfill the requirements of minimal ecological footprint contemporary eco-city ought to be designed with principles of: 1) creating of compact, green and safe mix-used community 2) public transport priority with emphasis of "access to proximity" 3) revitalization of degraded land - postindustrial of other uses 4) supporting local production and agriculture, urban greening, community allotments and gardening 5) creating convenient, safe and economically mix housing 6) creating opportunities for the underprivileged people 7) recycling and resource conservation 8) supporting ecological activities 9) promotion of simple lifestyle 10) ecological education [23].

6.1 Masdar Zero Carbon City

Masdar, covering an area of seven square kilometres, was designed by architects Foster and Partners, as entirely carbon neutral, powered by renewable energy and supplied with electric transportation system settlement for about 40, 000 inhabitants. The Eco-city incorporates five separate units: Masdar City, Masdar Carbon, Masdar Capital, Masdar Power, and Masdar Institute of Science and Technology [24]. The Masdar investment was launched in 2006 with the aim to create low-energy environment, habitat equipped with the newest renewable energy and sustainable technologies and as the demonstration project of the practical usage of renewable energy technologies on a utility scale [25, 26]. The urban grid layout is based on narrow streets shaded by buildings, to reduce the need for air conditioning, covered with solar panels. Among technical solutions to achieve sustainability goals within the city are: electrically powered public transport system, intelligent monitoring system applied in buildings and utility supplied system, recycling of waste and water, composing of biodegradable materials, energy system based on photovoltaic plants, a solar thermal power plant, evacuated thermal tube collectors, a waste-to-energy plant and a solar cooling system [26]. The photovoltaic panels (PV) on the roofs, constructed as first parts of the structures enable to generate energy needed to construct the rest of the building. Masdar is connected to Abu Dhabi national power grid to balance the electricity demands (overproduction during the day and shortfall during the night) [26]. These technical innovations are supplemented by architecture design, such as: location of air wells above courtyards, shading the streets and facades of buildings. The government of Abu Dhabi announced that Masdar would aim "*to be zero-carbon, powered entirely by renewable energy, car-free, and produce net-zero waste*". The sustainability principles include: reduce of energy demand and embodied and operational carbon produced by the city, water by 30% and landfill waste by 50% [24].

6.2 Dongtan Zero Carbon City

Dongtan City on Chongming Island – the world's largest alluvial island is situated about nine miles from Shanghai's financial district. It is estimated, that Dongtan will house 25,000 residents by the 2010 and half of million inhabitants by 2050 [24, 27]. Dongtan is adjacent to the Chongming Dongtan National Nature Reserve registered on the List of Wetlands of International Importance (the Ramsar List). The Eco-city will consist of a city center (formed of three villages) and pedestrian-friendly small mix-use towns, including residential, working and services infrastructure, connected by bicycle routes and public transport. Dongtan is seen as the demonstration project which outlines direction for future development of new Eco-cities in China - Langfang, Wuxi, Tangshan, and Tianjin [27]. According to the Dongtan Master Plan and estimated measures of the impact of the investment on the environment: the Eco-city will have 60% smaller ecological footprint in comparison with other Chinese cities, energy demand will be reduced by 66%, waste to landfills by 83%, 40% of energy will come from bio-energy and no CO2 emissions [27, 28]. Moreover: "*This project is expected to become a showcase for technologies and urban design that help to protect the environment while maintaining economic growth and social development. Dongtan eco-city was designed following an "integrated urbanism" approach that considers the environment, social, and economic aspects to create sustainable communities.*" [27]. Shanghai Institute of Urban Planning and Design for Chongming, Dongtan [2001] established functional zones planned at Dongtan : residential zone, artificial lake, wetland park, bird protection and education zone, education and research park, research and development park, land recreation area, water recreation area, transit centre, organic farm and experimental zone, eco-industry zone, aquatic farm and hatchery, landscaping and urban forestry, modern agriculture area, reserved land and others [27].

Design principles combine:

1. location and building design (compact, medium density residential area consisting of four to eight story buildings built using local and recycled materials),
2. energy system (energy self-sufficient city based on renewable sources: solar, wind, and biomass, using passive design: natural ventilation, natural insulation, green roofs, on-site energy system with photovoltaic cells and micro wind turbines applied in buildings, out-site energy system with wind turbines plant, waste to energy system, combined heat and power plant located on the edge of the city),
3. water management system (comprehensive system of potable water management and reclaimed water management for irrigation and toilets, rainwater storage system, sewage biological treatment system, flood control system)
4. waste management system (recycling of solid waste, composing of organic and human waste used as biogas to generate energy and fertilizer for farming)
5. transport system ("park and way" system, prioritize of public transport with public transport stops within distance of 500 meters from every single house, tram, bus, bicycle, and pedestrian networks, exclusion of Gas and diesel vehicles from the city, hydrogen power stations located in various points in the city)
6. food delivery and production system (local organic farming combined with waste and sewage recycling systems, usage of hydroponic techniques) ecosystem (3.5 km buffer zone between the city and wetlands, 60% of the Dogtan Eco-city reserved for to farms, parks, and wetlands, principle of 27 square meters of green areas per-capita, greening the city program – increasing the biodiversity along planned canals, streams, streets and green roofs)

7. social and economic development strategy (eco-industry, tourism, safe and diversified community goals) [27, 28]

7 Conclusion

Sustainable residential areas are urban units designed with consideration of minimal environmental impact, they cause the smallest possible ecological footprint through efficient use of energy, water and food resources and require minimal amount of waste and pollution. To achieve these principles, a sustainable residential area is basing on local materials and energy, using geographic location to incorporate sunlight, wind and rain into urban technologies and takes advantages of natural ecosystems and vegetation to improve urban microclimates and reduce pollution. Savings in energy consumption are used to enhance the role of urban public spaces and public transport and as results which enables to create new vibrant communities and improve welfare of inhabitants. Medium density sustainable urban settlements offer the best adapted energy self-sufficient solutions incorporated in architecture design. It is estimated, that by 2050, 70% of the world's population will live in cities and cities will accommodate more than 2.5 billion of additional residents. Nowadays, urban settlements are responsible for one-third of energy use worldwide, and over 20% of man-made CO2 emissions. Buildings, using less energy to heat, cool, ventilate, lighting are technically and financially possible. Existing technologies represent spectrum of possibilities for cities to operate at low levels of energy consumption, using renewable energy sources, local materials, air and water flows and incorporating natural ecosystems into urban areas

References

1. National Science & Technology Council, *Federal Research and Development Agenda for Net-Zero Energy, High Performance Green Buildings*, Committee on Technology, , finalreport.http://www.bfrl.nist.gov/buildingtechnology/documents/FederalRDAgendaforNetZeroEnergyHighPerformanceGreenBuildings.pdf (2008)
2. V. Novotny, Water/Energy Interactions of Water Reuse, 37–57 (2012)
3. V. Novotny, Water Sci. Technol. **63**, 1, 184–90 (2011)
4. Communication From the Commission to the European Parliament, COM (2011), 112 final (2011)
5. D. D'Agostino, J. of Build. Eng. **1**, 20–32 (2015)
6. J. Laustsen, *Energy Efficiency Requirements in Building Codes, Energy Efficiency Policies for New Buildings* (2008)
7. Directive 2010/31/EU Off. J. Eur. Union, 13–35 (2010)
8. A. Kylili, P.A. Fokaides, Sustain. Cities and Soc. **15**, 86–95 (2015)
9. H. Lund, A. Marszal, P. Heiselbergb, Energ.Build. **43**, 7, 1646–1654 (2011)
10. R. Gilman, *In Context 1* (1991)
11. J. Dawnson, *Ecovillages. New Frontiers for Sustainability* (2006)
12. M. Havasly, D. Corne, S. Roaf, Arch. Sc. Rev. **53**, 1, 126–140 (2010)
13. Global Ecovillage Network What is an Ecovillage? Retrieved from

http://gen.ecovillage.org (accessed 15.03.2017) (2011)

14. N. Winston, Adv.in Ecopolitics **9**, 85–103 (2012)
15. Ecodistrict Institute, retrieved from https://ecodistricts.org/wp-content/uploads/2013/07/resourceguide_web2.pdf (accessed 15.03.2017) (2012)
16. J. Fitzgerald, J. Lenhart, Env. and Plan. C: Government and Policy **34**, 2, 1–17 (2015)
17. S. Field, *Low Car(Bon) Communities. Inspiring car-free and car-lite urban futures*, 23–39 (2016)
18. G. Austin, Journ. of Green Build. **8**, 34–50 (2009)
19. E. Holden, I.T. Norland, Urban Studies **42**, 12, 2145–2166 (2005)
20. H. Frey, *Designing the City: Towards a More Sustainable Urban* Form (1999)
21. T. Schoor, B. Scholtens, Renew. and Sust. Ener. Rev. **43**, 666–675 (2015)
22. H. Suzuki et al. *Eco2 Cities Ecological and Economic Cities* (2010)
23. M. Roseland, Cities **14**, 197–202 (1997)
24. A.Alusi , R.G.Eccles, A.C.Edmondson, T.Zuzul, *Sustainable Cities: Oxymoron or the Shape of the Future? HBS Working Paper 11-062*, (2011)
25. S. Nader, Energ. Procedia **1**, 3951–3958 (2009)
26. D. Reiche, Energ. Policy **38**, 378–382 (2010)
27. H. Cheng, J.Hu, J. Environ. Monit. **12**, 119–126 (2010)
28. I. Chang, E. Sheppard, Journ. of Urb. Tech. **20**, 1, 57–75 (2013)

Energy potential of the modified excess sludge

Iwona Zawieja[1,*]

[1]Czestochowa University of Technology, Faculty of Infrastructure and Environment, Institute of Environmental Engineering, ul. Brzeznicka 60a, 42-200 Czestochowa, Poland

Abstract. On the basis of the SCOD value of excess sludge it is possible to estimate an amount of energy potentially obtained during the methane fermentation process. Based on a literature review, it has been estimated that from 1 kg of SCOD it is possible to obtain 3.48 kWh of energy. Taking into account the above methane and energy ratio (i.e. 10 kWh/1Nm^3 CH_4), it is possible to determine the volume of methane obtained from the tested sludge. Determination of potential energy of sludge is necessary for the use of biogas as a source of power generators as cogeneration and ensure the stability of this type of system. Therefore, the aim of the study was to determine the energy potential of excess sludge subjected to the thermal and chemical disintegration. In the case of thermal disintegration, test was conducted in the low temperature 80°C. The reagent used for the chemical modification was a peracetic acid, which in an aqueous medium having strong oxidizing properties. The time of chemical modification was 6 hours. Applied dose of the reagent was 1.0 ml CH_3COOOH/L of sludge. By subjecting the sludge disintegration by the test methods achieved an increase in the SCOD value of modified sludge, indicating the improvement of biodegradability along with a concomitant increase in their energy potential. The obtained experimental production of biogas from disintegrated sludge confirmed that it is possible to estimate potential intensity of its production. The SCOD value of 2576 mg O_2/L, in the case of chemical disintegration, was obtained for a dose of 1.0 ml CH_3COOH/L. For this dose the pH value was equal 6.85. In the case of thermal disintegration maximum SCOD value was 2246 mg O_2/L obtained at 80 °C and the time of preparation 6 h. It was estimated that in case of thermal disintegration as well as for the chemical disintegration for selected parameters, the potential energy for model digester of active volume of 5L was, respectively, 0.193 and 0,118 kWh.

1 Introduction

The costs of construction and operation of wastewater treatment plants are very high, therefore economic conditions force the facility operators to treat sludge as a raw material with a certain amount of fertilizer, especially energy, which may give rise to a partial return of financial expenditures. Sewage sludge, as far as possible, can be used for: agricultural purposes, as fertilizer for soils, recultivation of degraded land and landless or thermally

* Corresponding author: izawieja@poczta.onet.pl

 DOI: 10.1051/e3sconf/20172200200

degraded land, i.e. by incineration or co-incineration of sludge with other fuels. Selection of the disintegration method should be preceded by a technical and economic analysis that takes into account the desired disintegration effect, the efficiency of the installation, the operating costs of the disintegration plant, the balance of costs associated with the operation and processing of the sludge [1, 2].

In the case of the disposal of sludge, it is technically and economically justified to use an anaerobic sludge stabilization process, where the energy input is mainly related to the need for optimum process temperature achievement. The demand for heat energy is proportional to the amount of sludge fed. Based on research conducted by Muhammad [3], it was found that as a result of disintegration of sludge, an increase in biogas production was achieved by 36% compared to no use of initial disintegration. There was also a decrease in the volume of sludge, which directly affects the reduction in the cost of purchasing the polymer used for the dewatering process. Pre-treatment through disintegration involves additional costs for the installation and energy consumption of the process. It is difficult to compare all disintegration methods in the economic terms, taking into account the costs and benefits of the lack of data, as most pre-treatment techniques are carried out only on a laboratory scale. It is also difficult to decide which method is more economical and effective, because each has advantages and disadvantages. The main disadvantage of mechanical methods is high energy consumption, which increases the cost of sewage sludge disintegration, while the disadvantage of chemical disintegration methods is the secondary sludge contamination by the addition of chemical reagents and the supplementary cost of their purchase.

It should be noted, however, that pre-treatment increases the amount of biogas that can be used in the cogeneration process for heat and power generation, which depreciates some of the costs incurred for the disintegration of sludge [4].

According to Dhar et al. [5] all thermal methods in the range of 50–90°C, net savings of $ 44–66 per tonne of total solids (TS). The use of disintegration of sewage sludge allows to reduce the cost of wastewater treatment by as much as 50%. [6].

The aim of the study was to determine the energy potential of excess sludge subjected to the thermal and chemical disintegration using an oxidizing agent i.e.: peracetic acid.

2 Methodology

According to the intended purpose of the study, excess sludge was subjected anaerobic stabilization assisted by disintegration. In order to initiate the anaerobic stabilization process, a fermented sludge was used. The sludge samples were collected from a sewage treatment plant with a capacity of 90,000 m^3/d, 314 835 RLM. It is a mechanical and biological treatment plant with increased nutrients removal. Table 1 shows the physico-chemical characteristics of the sludge used in the studies.

Table 1. Selected physico-chemical indicators of excess sludge, which is the basic substrate for the study.

Indicators	Excess sludge
Volatile Suspened Solids (VSS)	8.8 ± 0.15 g/L
Soluble Chemical Oxygen Demand (SCOD)	157 ± 12 mg O_2/L
Volatile Fatty Acids (VFAs)	72 ± 3 mg CH_3COOH/L
Alkalinity	860 ±20 mg $CaCO_3$/L
pH	7.2 ±0.2

An anaerobic stabilization process of excess sludge was carried out in a fermentation chamber made in the form of a glass cylinder with an active volume of 5 L. The test plant consists of a biogas installation, a device providing the optimum mixing speed and keeps the constant temperature of process. The process was carried out at mesophilic temperature of 37°C for 25 days. Anaerobic stabilization was applied to the following sludge:
- Sludge 1 – unconditioned excess sludge;
- Sludge 2 – excess sludge thermally disintegrated at 80°C for 6 h;
- Sludge 3 – excess sludge chemically disintegrated using an acidic reagent 1.0 ml of STERIDIAL/L of sludge.

Excessive sediments have been subjected to the following characteristics by performing physicochemical determinations, i.e.:
• ODM (Organic Dry Matter) according to PN-EN-12879,
• pH using a pH meter from Cole Palmer 59002-00 according to PN-9/C-04540/05,
• Soluble chemical oxygen demand (SCOD) by dichromate method using HACH 2I00N IS spectrophotometer tests according to ISO 7027 in the supernatant of sludge.
• volatile fatty acids (VFAs) by steam distillation in accordance with PN-75 / C-04616/04.
• alkalinity according to PN-91 / C-04540/05.

The process of thermal disintegration of excess sludge was carried out in the low temperature i.e. 80°C for 6 h. In order to perform a thermal modification excess sludge was placed in laboratory flasks with an active volume of 0.5 L, protected from air by a glass stopper with a manometric tube. Flasks were placed in a shaking water bath. Steridial W-15, which is the aqueous solution of peracetic acid (15%), acetic acid (> 36%) and hydrogen peroxide (18–22%), was used during chemical disintegration. The following dose of reagent was used: 1.0 ml of STERIDIAL/ 1 L of sludge. The time of the excess sludge chemical exposition was equal 6h. In the case of chemical disintegration, excess sludge was placed in laboratory flasks with an active volume of 0.5L, protected from air by a glass stopper with a manometric tube and placed in a shaker. The recorded pH of the sludge was corrected chemically to the proper pH of the methane fermentation process. It has been assumed that peracetic acid in aqueous medium is completely decomposed to acetic acid. The value of SCOD of chemically pretreatment sludge by peracetic acid should be estimated according to the following relationship 1.07kgO_2/kgCH_3COOH [7]. The percentage of peracetic acid and acetic acid in the Steridial W-15 solution and their density were taken into account.

3 Results

3.1 Effect of the disintegration process conditions on the sludge physicochemical characteristics

The studies have shown the positive effect of thermal and chemical disintegration on the increase of dissolved organic matter, which was directly reflected in the increase of SCOD value. The selection of the most favorable disintegration conditions was made on the basis of preliminary results. The increase of the volatile fatty acids concentration determined in the thermally modified excess sludge was correlated with the obtained SCOD value. In samples of modified sludge (80°C and 6 h) a value of O_2/L demand was equal 2246 mg/L.A significant growth of the VFAs concentration was obtained in comparison to the initial value of this indicator. The VFAs concentration after the thermal disintegration process was 552 mg CH_3COOH /L. Supporting effects on the degree of hydrolysis of sludge were also reported in the case of chemical disintegrated by peracetic acid at the dose of reagent 1.0 ml CH_3COOOH/L of sludge. SCOD values and VFAs concentration of 2576 mgO_2/L and 387 mgCH_3COOH/L respectively were obtained. In addition, the pH value equal 6.85 of chemically sludge did not need to be adjusted to an optimum value for the further stage of the methane fermentation process.

3.2 Generation of biogas from excess sludge subjected to 25-day methane fermentation

In the methane fermentation process of untreated excess sludge, about 39% digested degree was obtained. As a result of anaerobic stabilization of untreated excess sludge, SCOD and VFAs concentration of 287 mg O_2/L and 164 mg CH_3COOH/L were obtained at 25 day of treatment. At the time of initiation of the process, the values of the tested indicators were 157 mg O_2/L and 103 mg CH_3COOH/L, respectively. The pH on the day of initiating the process (day 0) was 7.62, while in the final day 7.43. Initial and final values of alkalinity were 880 mg $CaCO_3$/L and 1620 $CaCO_3$/L, respectively. During the methane fermentation of non-disintegrated sludge, the highest daily biogas production value of 0.61 L was observed in the 13th day of the process, recording approximately 82% of the methane fraction. Total production and biogas yield of 4.3L and 0.158 L/g VSS was obtained. The biogas energy potential of an average volume content of about 80% methane generated from a 5L of active reactor was estimated to be about 0.0344 kWh. Table 2 presents selected physicochemical indicators of untreated excess sludge during methane fermentation (25 days), while Figure 1 shows the daily production of biogas with volume of methane.

Table 2. Selected physico-chemical indicators obtained for Sludge1 subjected to 25-day methane fermentation.

Indicator/Unit	The methane fermentation time	
	0d	**25d**
Volatile Suspened Solids (VSS), g/L	**8.8** ± 0.15	**5.4** ± 0.14
Alkalinity, mg $CaCO_3$/L	**880** ± 40	**1620** ± 20
Chemical Oxygen Demand (SCOD), mg O_2/L	**157** ± 8	**287** ± 12
Volatile Fatty Acids, mgCH_3COOH/L	**103** ± 12	**164** ± 15
pH	**7.56** ± 0.21	**7.41** ± 0.16

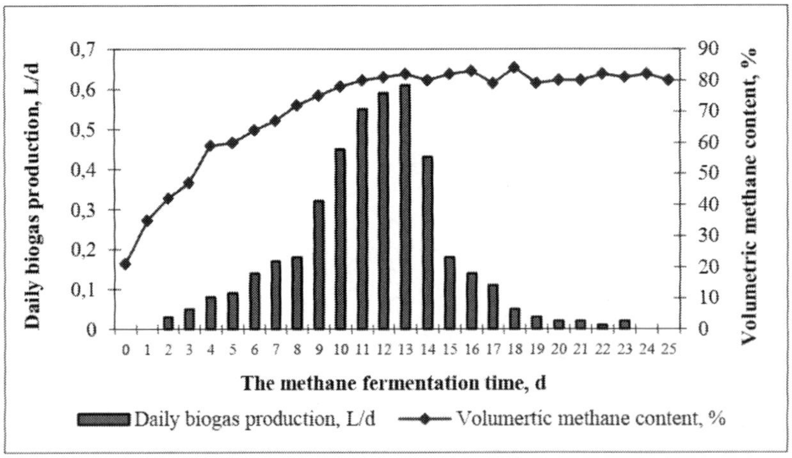

Fig. 1. Daily biogas production with respect to volume of methane during anaerobic stabilization of untreated excess sludge (Sludge 1).

3.3 Generation of biogas from thermally disintegrated excess sludge subjected to 25-day methane fermentation

Methane fermentation of thermally modified excess sludge at temperature 80°C and time 6h (Sludge 2), resulted with 71% degree of sludge digestion. At 25th day the SCOD and VFAs concentrations were 357 mg O_2/L and 241 mg CH_3COOH/L, respectively, while at the first day of the process parameters were equal 2246 mg O_2/L and 552 mg CH_3COOH/L, respectively. The pH during the initiation day was 7.18, the final day was 7.38, and the initial and final alkalinity values were 660 mg $CaCO_3$/L and 2860 $CaCO_3$/L, respectively. The highest daily biogas production value of 1.87L was observed in the 8th day of the process, recording about 83% methane content. A total biogas production of 15.46L and a biogas yield of 0.493 L/g VSS were obtained. The biogas energy potential of an average content of about 80% methane generated from a 5L of active reactor was estimated to be about 0.193 kWh. Table 3 shows the physicochemical indicators of thermally pretreated sludge, during the 25-day methane fermentation process, while Figure 2 shows the daily production of biogas with the volume of methane.

Table 3. Selected physico-chemical indicators obtained for Sludge 2 subjected to 25-day methane fermentation.

Indicator/Unit	**The methane fermentation time**	
	0d	**25d**
Volatile Suspened Solids (VSS), g/L	**8.8** ± 0.15	**2.53** ± 0.18
Alkalinity, mg $CaCO_3$/L	**660** ± 20	**2860** ± 20
Soluble Chemical Oxygen Demand (SCOD), mg O_2/L	**2246** ± 32	**357** ± 16
Volatile Fatty Acids, mgCH_3COOH/L	**552** ± 11	**241** ± 6
pH	**7.18** ± 0.22	**7.38** ± 0.17

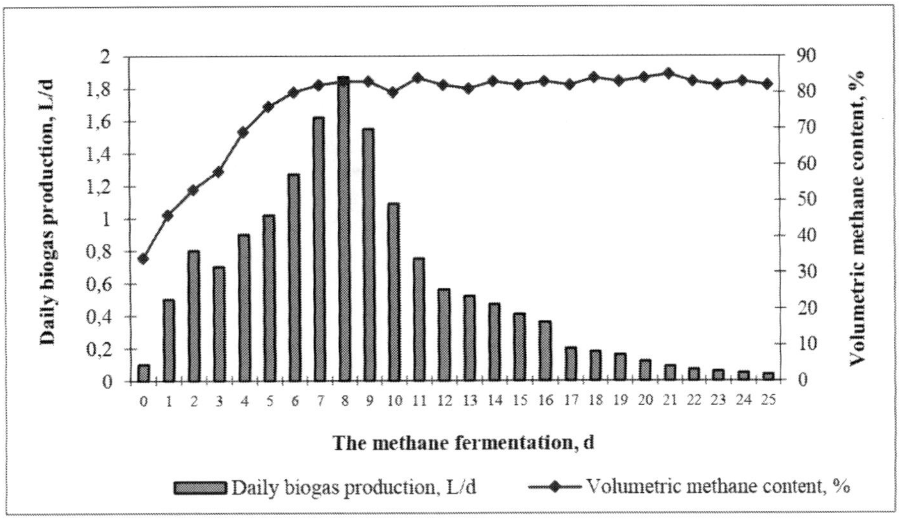

Fig. 2. Daily biogas production with respect to volume of methane during anaerobic stabilization of thermally modified excess sludge (Sludge 2).

DOI: 10.1051/e3sconf/20172200200

3.4 Generation of biogas from chemically disintegrated excess sludge subjected to 25-day methane fermentation

In the next study cycle, excess sludge was chemically modified using the selected dose of reagent, i.e.: 1.0 ml of STERIDIAL/L of sludge (Sludge 3). The 53% digested degree was obtained in the stabilization process of the chemically modified sludge. The SCOD value of 746 mg O_2/L and the concentration of volatile fatty acids (VSAs) of 543 mg CH_3COOH/L were obtained during the 25 day conducting of the process. However, during the initiation of the process (0d) the SCOD and VFAs concentration were 886 mg O_2/L and 487 mg CH_3COOH/L. The pH during the day of initiation of the process (day 0) was 7.38, in the final day pH value was 7.42, while the initial and final alkalinity value were equal respectively 650 and 2560 $CaCO_3$/L. At the seventh day of methane fermentation of chemically disintegrated excess sludge, the highest daily biogas production was observed: 1.63 L with 85% volumetric methane content. Total biogas production of 14.75L and a biogas yield of 0.625 L/g VSS was observed. It was estimated that the biogas energy potential of an average of about 80% methane generated from a 5L of active reactor was about 0.118 kWh. Table 4 presents selected physicochemical indicators of chemically modified excess sludge during the 25-day methane fermentation, while Figure 1 shows the daily production of biogas with volume of methane.

Table 4. Selected physico-chemical indicators obtained for Sludge 3 subjected to 25-day methane fermentation.

Indicator/Unit	The methane fermentation time	
	0d	**25d**
Volatile Suspened Solids (VSS)	**8.8** ± 0.15	**4.08** ± 0.14
Alkalinity, mg $CaCO_3$/L	**760** ± 20	**2680** ± 40
Soluble Chemical Oxygen Demand (SCOD), mg O_2/L	**2575** ± 46	**851** ± 23
Volatile Fatty Acids, mgCH_3COOH/L	**387** ± 9	**478** ± 12
pH	**6.85** ± 0.25	**7.37** ± 0.18

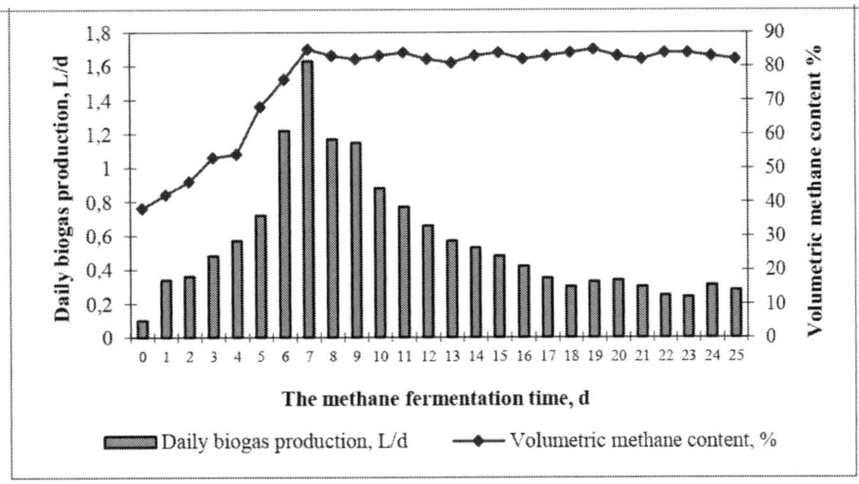

Fig. 3. Daily biogas production with respect to volume of methane during anaerobic stabilization of chemically modified excess sludge (Sludge 3).

Yield of biogas production, and degree of sludge digestion, obtained during 25-day anaerobic stabilization of chemically modified excess sludge, shows a significant increase of susceptibility of modified excess sludge to biodegradation, in comparison to untreated sludge. A significant increase of the energy potential of disintegrated excess sludge, with respect to the untreated sludge was also observed.

4 Conclusions

Since, the stabilization process is spontaneous and the limiting phase is the hydrolysis phase, it is desirable to subject the excess sludge to disintegration processes that affects the increase of energy efficiency of the process.

Based on the results of the research, the following conclusions were made:
1. The application of selected methods of disintegration of excess sludge prior to the anaerobic stabilization process resulted in an increase in process efficiency expressed as an increase in the unit value of biogas production.
2. The application of the studied disintegration methods, in the technological process of sludge treatment, entails both the ecological and economic benefits of the reduced environmental burden through the use of sludge modification methods that do not cause additional pollution of the environment.
3. As a result of methane fermentation of excess sludge modified by thermal disintegration at 80°C and chemical disintegration using the reagent dose of 1.0 ml of STERIDIAL/L of sludge, the biogas yield value was 0.493 and 0.625 L/gVSS, and digested degree about 71% and 53% respectively. It was found that the biogas energy potential about 80% of the volume methane content, which was generated from a 5L reactor volume, was 0.193 and 0.118 kWh respectively, for the thermal and chemically disintegrated sludge.

The research was funded under the project No. BS – PB – 401/303/12.

References

1. G. Zhang, P. Zhang, J. Yang, H. Liu, Bioresource Technology **99**, 9029–9031 (2008)
2. Internet resources, www.huber.com.pl, 07.03.2012.

DOI: 10.1051/e3sconf/20172200200

3. M.S. Shehu, Z.A. Manan, S.R. Wan Alwi, Bioresource Technology **114**, 69–74 (2012)
4. V.K. Tyagi, S-L. Lo, Rev. Environ. Sci. Biotechnol. **10**, 215–242 (2011)
5. B.R. Dhar, G. Nakhla, M.B. Ray, Waste Management **32**, 54 –549 (2012)
6. J. Abelleira, S.I. Pérez-Elvira, J. Sánchez-Oneto, J.R. Portela, E. Nebot, Conservation and Recycling **59**, 52–57 (2012)
7. L. Appels., A. Van Assche, K. Willems., J. Degreve, J. Van Impe, R. Dewil, Bioresource Technology **102**, 4124–4130 (2011)

Evaluation of external and internal irradiation on uranium mining enterprise staff by tooth enamel EPR spectroscopy

Kassym Zhumadilov[1,*], *Alexander* Ivannikov[2], *Artem* Khailov[2], *Sergei* Orlenko[2], *Valeriy* Skvortsov[2], *Valeriy* Stepanenko[2], *Kairat* Kuterbekov[1], *Shin* Toyoda[3], *Polat* Kazymbet[4], *and Masaharu* Hoshi[5]

[1]L.N. Gumilyov Eurasian National University, Munaitpasova 13, Astana 010008, Kazakhstan
[2]A.F. Tsyb Medical Radiological Research Center, 4, Korolev str., Obninsk, 249036, Russia
[3]Department of Applied Physics Faculty of Science Okayama University of Science, 700-0005, Japan
[4]Institute of Radiobiological Research, Astana 010008, Kazakhstan
[5]Hiroshima University, 734-8553, Japan

Abstract. In order to estimate radiation effects on uranium enterprise staff and population teeth samples were collected for EPR tooth enamel dosimetry from population of Stepnogorsk city and staff of uranium mining enterprise in Shantobe settlment (Akmola region, North of Kazakhstan). By measurements of tooth enamel EPR spectra, the total absorbed dose in the enamel samples and added doses after subtraction of the contribution of natural background radiation are determined. For the population of Stepnogorsk city average added dose value of 4 +/- 11 mGy with variation of 51 mGy was obtained. For the staff of uranium mining enterprise in Shantobe settlment average value of added dose 95 +/- 20 mGy, with 85 mGy variation was obtained. Higher doses and the average value and a large variation for the staff, probably is due to the contribution of occupational exposure.

1 Introduction

At the mining of uranium ore in the world a large number of people is employed. Most of the uranium ore is mined in mines and open pits. At that personnel of mines and quarries are at an increased risk of irradiation from the isotopes of uranium and its decay products. Only for some ore fields, the mining with using of the safer method of underground leaching is available. Radiation threat is also represented by piles of waste of uranium deposits.

Kazakhstan is the largest producer of uranium. It is on the first place in world uranium production (39.3% of world production in 2016) and the second place on its reserves [1]. A significant portion of Kazakhstan's uranium reserves (about 18%) is concentrated in the Northern Kazakhstan uranium ore province [2, 3].The bulk of the production of uranium ore from the mid 50-ies of the last century, there were carried out here in the mines and open pits.

In the places of uranium mining radiation-ecological monitoring, as well the

* Corresponding author: zhumadilovk@gmail.com

individual control of doses is typically carried out. However, this is not enough to properly assess the radiation doses. It is therefore important to estimate the doses by independent methods. One of such methods is the EPR (electron paramagnetic resonance) dosimetry using tooth enamel [4–7]. This method can estimate individual accumulated absorbed dose through more than 50 years after exposure to radiation. Dose determination is based on the measured by EPR spectroscopy amount of stable radicals formed in the enamel after exposure to ionizing radiation. This method uses enamel samples prepared from teeth extracted according to medical indications. The error of determination of absorbed dose by this method is about 30-50 mGy [8–12].

Particularly important is the assessment of the radiation effects of uranium mining enterprises on their personnel and on the population of adjacent territories at the North of Kazakhstan. For this aim, a survey was conducted by EPR dosimetry for the personnel of uranium mine placed near Shantobe settlement and for population of this settlement placed in this region. Also such survey was performed for population of Stepnogorsk city in this region, where the uranium ore processing enterprise is placed and waste dumps of this enterprise are located in the vicinity from this city [2].

2 Materials and Methods

2.1 Sample collection

Teeth samples collection according to medical indication were carried from population of Stepnogorsk city and staff of uranium mining enterprise (mine) located near Shantobe urban-type settlement (Akmola region of Kazakhstan) located 180 km to the North from Astana city, the capital of Kazakhstan. Totally, more than a hundred teeth samples were collected with purpose to separate tooth enamel from dentin for EPR dosimetry.

The two groups of enamel samples were prepared from teeth collected as extracted according to medical indications in local dental clinics. A group of samples indicated with number codes 97 St–119 St (23 teeth) was collected from population of Stepnogorsk city. Another group indicated with codes 67 Sh–96 Sh (30 teeth) was collected in Shantobe settlement, part of population of which is engaged as a stuff of uranium mine. Part of these samples 78 Sh–96Sh (19 teeth) was collected from the uranium mine personnel, another part, 67 Sh–77 Sh (11 teeth) – from population of another occupation not directly connected with work in uranium mine. For each sample information was supplied about teeth type (teeth position) and individual information about teeth donors (name, birth date, address, occupation, x-ray diagnostic procedures).

Stepnogorsk city is the center, where the processing of uranium ore from Northern Kazakhstan ore uranium province is performed. This city is located 200 km to north of Astana, capital of Kazakhstan. The ore is delivered from mines of Shantobe, which is located 400 km to the west from Stepnogorsk. Also, it is processing uranium ore concentrates produced in other ore fields of Kazakhstan. Population of Stepnogorsk is about 69 000 (in 2016). Population of Shantobe urban-type settlement is about 3700 (in 2016). Part of them (about 700) is engaged as personal of uranium ore mine placed in 7 km near this settlement [13, 14].

2.2 Sample preparation

Enamel samples were prepared separately from outer (buccal) and inner (lingual) sides of teeth [10, 11]. The assignment of the samples to the inner and outer sides of the teeth carried by the appearance of samples based on the analysis of the shape of the tooth in consultation with the dentists. Enamel was separated from dentin using dental drill. In order to avoid mechanical induced signal, in final stage of enamel separation hard allow burs were used at lowed rotation speed. After separation, enamel was cut to chips 0.5–1.0 mm in size with hard alloy nippers. Sample mass for measurement was in the range of 40–120 mg. The lower mass limit is defined by availability of enamel and by EPR sensitivity limitation. For molars and premolars (teeth with positions 4–8), enamel from both sides of tooth was used for analysis. For incisors and canines (positions 1–3) only enamel from inner side was used, since in enamel from outer side the signal is induced because of the effect of the solar light UV component [15–19].

2.3 EPR spectra measurement and dose determination

EPR spectra measurements were performed with the use of the EPR spectrometer ESP-300 E (Bruker) in X-band (9.8 GHz) equipped by the standard cavity with the following registration parameters: center field 390 mT, sweep width 10 mT, modulation 0.3 mT, microwave power 10 mW, total time of spectra measurement with accumulations 40 min. The third line of the MnO marker sample build in the cavity was together with enamel spectra. Spectra processing was performed with the use the basic software of the spectrometer's build-in computer under manual control similar as it is described in our previous publications [20–23]. The radiation-induced signal was obtained after subtraction of the simulated native signal fitted under the operator control. For such simulation, a spectrum of not irradiated enamel was used.

Dose in enamel was determined from the amplitude of the radiation-induced signal measured as its maximum and normalized by sample mass, MnO marker signal amplitude and by calibration coefficient. Calibration of the enamel dose response was performed using the set of samples prepared from pulled enamel irradiated by a cobalt-60 gamma source in different doses controlled by a tissue-equivalent dosimeter. Therefore, dose in enamel was obtained in terms of tissue equivalent dose according to cobalt-60 gamma calibration.

Additional dose in enamel caused by technogenic radiation was determined by subtraction from the total dose of the contribution caused by natural background radiation since enamel formation with dose rate 0.8 mGy per year [24, 25].

3 Results and discussion

To estimate the dose determination error, the correlation between the doses determined for the enamel from the inside and outside of the teeth has been analyzed (Figure 1). For some samples the increased difference between doses for internal and external enamel was observed (samples #70, 91, 96 with tooth positions 7, 6 and 6 respectively). Such data were excluded from further analysis as outliers. Perhaps this increased difference is due to increased signal in outer enamel because of possible irradiation at X-ray diagnostics of teeth. Since the assignment of the samples to the inner and outer sides of the teeth was carried by the appearance of samples, such assignment may be mistaken, especially for molar teeth at high damages by caries.

Based on the analysis of correlation between the doses determined for the enamel from the inner and outer sides of the teeth, it is possible to get an idea of experimental error in determining the dose. The root mean-square (RMS) of the difference between doses gives the resulting error of both doses for which the difference was estimated. For each value contributing to this difference the error will be $2^{1/2}$ times less. From the estimated value of RMS within 60–76 mGy it follows that error in determining of dose for each enamel sample is about 40–50 mGy.

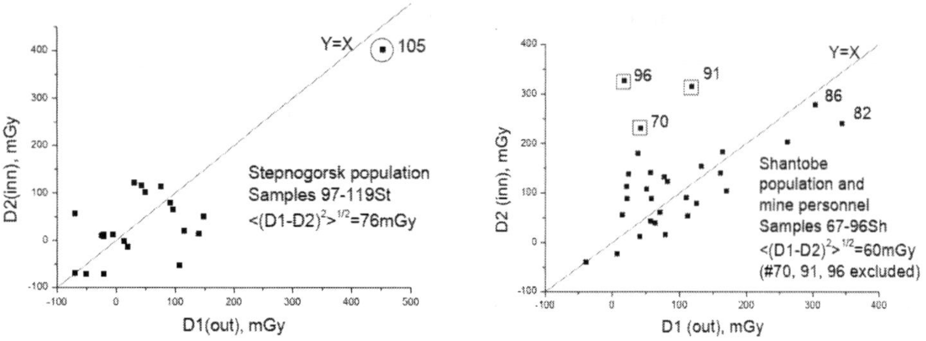

a

b

Fig. 1. Relation between total doses in enamel measured for outer (D1) and inner (D2) sides of teeth: a – for Stepnogorsk population; b – for Shantobe population and mine personnel. Line with unit slope (Y = X) is shown for comparison. The squares mark results, which were rejected from further processing as outliers because of anomalous high difference between inner and outer doses.

Dependencies of the added dose on the age of the enamel are shown in Figure 2. Histograms of the distribution of added doses are shown in Figure 3. The values of average doses, uncertainty of the average and their variation (standard deviation from the mean) are presented in the fields of figures.

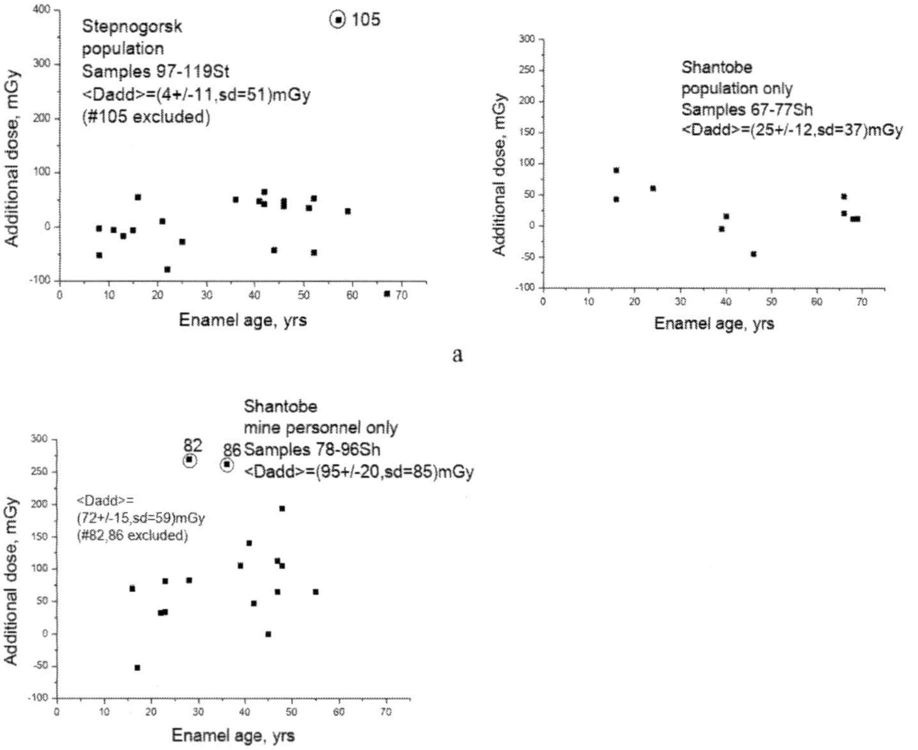

a b

c

Fig. 2. Dependencies of additional doses on the enamel age: a – for Stepnogorsk population; b – for Shantobe population; c – for Shantobe mine personnel. The circles mark increased doses.

 DOI: 10.1051/e3sconf/20172200201

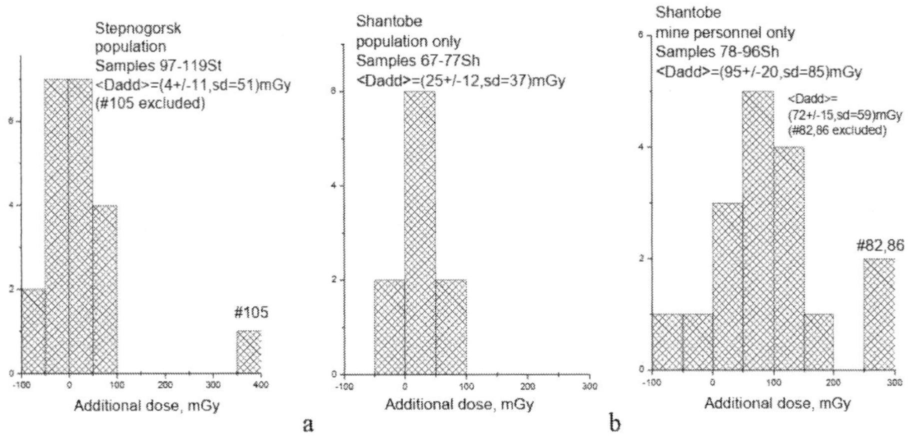

a b c

Fig. 3. Distribution histograms of additional doses: a – for Stepnogorsk population; b – for Shantobe population; c – for Shantobe mine personnel.

On the basis of measurements a statistical analysis of the added dose values was carried out separately for groups of population and for personnel. For the population of Stepnogorsk (97–119 St samples) the average value of the added dose is obtained 4 +/- 11 mGy, with 51 mGy variation. These values were obtained after excluding from the calculation of high-dose value for the sample #115. Probably, the donor of this sample was in places with high radiation. The magnitude of the variation of doses roughly equivalent to the dose estimation error, and this error determines the width of the distribution of doses. The average dose is close to zero. This indicates the absence of man-made radiation for this group of the population.

For the population of Shantobe settlement average added dose is 25 +/- 12 mGy. This value is significantly higher than for the population Stepnogorsk. This is probably due to the proximity of uranium mines. Variation value of 37 mGy is close to the error of the method.

For the personnel of the Shantobe mine average added dose is 95 +/- 20 mGy, which is significantly higher than for the rest of the population Shantobe and for Stepnogorsk population. The variation of the dose values of 85 mGy significantly exceeds the error of the method, and probably is due to the different radiation exposure to staff according to their working conditions. Noteworthy higher dose values for two people, samples #82 and #86, which is probably caused by the special conditions of their work.

According to published data for the results of a conventional personal dosimetry, when working in uranium mines, the level of external exposure of an average of about 10 mGy per year [26]. This value is consistent with the average value obtained by EPR dosimetry, if to assume the work in the mine for about 10 years.

4 Conclusion

On the basis of measurements a statistical analysis of the values added doses separately for groups and staff were carried out. For the population of the Stepnogorsk city dose value from 4 +/- 11 mGy to 51 mGy. For staff of uranium mining enterprise in the Shantobe village average value of dose 95+/-20 mGy, 85 mGy variation. Higher doses and the average value of a large variation of the staff, probably due to the contribution of occupational exposure. Possibly, part of the contribution of radiation due to internal alpha source in the soft tissues. In the next phase of the study it is planned to allocate this contribution by examining EPR enamel samples after etching of the surface layer.

Acknowledgements

This work was supported partly by the Ministry of Education and Science of Kazakhstan (5284/GF4 agreement No. 47) and by Russian Foundation of Basic Research (15-58-45050 IND_a and 16-04-01276), which is greatly appreciated.

References

1. Uranium 2016: Resources, Production and Demand. Joint report by the OECD Nuclear Energy Agency and the International Atomic Energy Agency. http://www.oecd-nea.org/ndd/pubs/2016/7301-uranium-2016.pdf
2. K. Zhumadilov, A. Ivannikov, V. Stepanenko, Sh. Abralina, L. Sadvokasova, A. Akilbekov, A. Morzabayev, T. Rakhypbekov, M. Hoshi, Radiat. Environ. Biophys. **54**, 4 (2015)
3. Mining of uranium (*in Russian*)
http://www.kazatomprom.kz/ru/content/kompaniya/deyatelnost/dobycha-prirodnogo-urana
4. IAEA-TECDOC-1331, Vienna (2002)
http://www-pub.iaea.org/MTCD/publications/PDF/te_1331_web.pdf
5. E. Tielewuhan, A. Ivannikov, K. Zhumadilov, M. Nalapko, D. Tikunov, V. Skvortsov, V. Stepanenko, Sh. Toyoda, K. Tanaka, S. Endo, M. Hoshi, Radiat. Meas. **41**, 7 (2006)
6. K. Tanaka, S. Endo, A. Ivannikov, S. Toyoda, E. Tieliewuhan, K. Zhumadilov, C. Miyazawa, S. Suga, K. Kitagawa, M. Hoshi, J. Radiat. Res. **46**, 4 (2006)
7. K. Zhumadilov, V. Stepanenko, A. Ivannikov, Z. Zhumadilov, S. Toyoda, K. Tanaka, S. Endo, M. Hoshi, Radiat. Environ. Biophys. **47**, 4 (2008)
8. V.G. Skvortsov, A.I. Ivannikov, V.F. Stepanenko, A.F. Tsyb, L.H. Khamidova, A.E. Kondrashov, D.D. Tikunov, Appl. Radiat. Isot. **52**, 7 (2000)
9. P. Fattibene, A. Wieser, E. Adolfsson, L.A. Benevides, M. Brai, F. Callens, V. Chumak, B. Ciesielski, S. Della Monaca, K. Emerich, H. Gustafsson, Y. Hirai, M. Hoshi, A. Israelsson, A. Ivannikov, D. Ivanov, J. Kaminska, Wu Ke, E. Lund, M. Marrale, L. Martens, C. Miyazawa, N. Nakamura, W. Panzer, S. Pivovarov, R.A. Reyes, M. Rodzi, A.A. Romanyukha, A. Rukhin, S. Sholom, V. Skvortsov, V. Stepanenko, M.A. Tarpan, H. Thierens, S. Toyoda, F. Trompier, E. Verdi, K. Zhumadilov, Radiat. Meas. **46**, 6 (2011)
10. K.S. Zhumadilov, A.I. Ivannikov, V.G. Skvortsov, Zh. Zhumadilov, S. Endo, K. Tanaka, M. Hoshi, J. Radiat. Res. **46**, 7 (2005)
11. K. Zhumadilov, A. Ivannikov, K. Apsalikov, Zh. Zhumadilov, V. Stepanenko, V. Skvortsov, G. Berekenova, S. Toyoda, S. Endo, K. Tanaka, C. Miyazawa, M. Hoshi, Radiat. Meas. **42**, 4 (2007)
12. A.I. Ivannikov, D. Sanin, M. Nalapko, V.G. Skvortsov, V.F. Stepanenko, A.F. Tsyb, F. Trompier, K. Zhumadilov, M. Hoshi, Health Phys. **98**, 6 (2010)
13. Stepnogorsk *(in Russian)* – https://ru.wikipedia.org/wiki/Степногорск
14. Shantobe (*in Russian*) - https://ru.wikipedia.org/wiki/Шантобе
15. A.I. Ivannikov, V.G. Skvortsov, V.F. Stepanenko, D.D. Tikunov, A.A. Romanyukha, A. Wieser, Radiat. Prot. Dosim. **71**, 5 (1997)

16. A. Ivannikov, K. Zhumadilov, E. Tieliewuhan, L. Jiao, K.N. Apsalikov, G. Berekenova, Zh. Zhumadilov, Sh. Toyoda, C. Miyazawa, V. Skvortsov, V. Stepanenko, S. Endo, K. Tanaka, M. Hoshi, J. Radiat. Res. **47**, 7 (2006)
17. A.I. Ivannikov, V.G. Skvortsov, V.F. Stepanenko, K.Sh. Zhumadilov, Radiat. Prot. Dosim. **159**, 4 (2014)
18. K. Zhumadilov, A. Ivannikov, K.N. Apsalikov, Zh. Zhumadilov, Sh. Toyoda, E. Tieliewuhan, S. Endo, K. Tanaka, C. Miyazawa, T. Okamoto, M. Hoshi, J. Radiat. Res. **47**, 6 (2006)
19. K. Zhumadilov, A. Ivannikov, Z. Zhumadilov, V. Stepanenko, K. Apsalikov, M. Rodzi, A. Zhumadilova, S. Toyoda, S. Endo, K. Tanaka, T. Okamoto, M. Hoshi, Radiat. Environ. Biophys. **48**, 6 (2009)
20. K. Zhumadilov, A. Ivannikov, V. Stepanenko, S. Toyoda, Z. Zhumadilov, M. Hoshi, J. Radiat. Res. **54**, 4 (2013)
21. A. Wieser, E. Vasilenko, P. Fattibene, S. Bayankin, N. El-Faramawy, D. Ivanov, P. Jacob, V. Knyazev, S. Onori, M.C. Pressello, A. Romanyukha, M. Smetanin, A. Ulanovsky, Radiat. Environ. Biophys. **44**, 9 (2006)
22. M. Hoshi, S. Toyoda, A. Ivannikov, K. Zhumadilov, A. Fukumura, K. Apsalikov, Z.S. Zhumadilov, S. Bayankin, V. Chumak, B. Ciesielski, V. De Coste, S. Endo, P. Fattibene, D. Ivanov, C.A. Mitchell, S. Onori, M. Penkowski, S.P. Pivovarov, A. Romanyukha, A.B. Rukhin, K. Schultka, T.A. Seredavina, S. Sholom, V. Skvortsov, V. Stepanenko, K. Tanaka, F. Trompier, A. Wieser, G. Wolakiewicz, Radiat. Meas. **42**, 9 (2007)
23. A. Ivannikov, S. Toyoda, M. Hoshi, K. Zhumadilov, A. Fukumura, K. Apsalikov, Z.S. Zhumadilov, S. Bayankin, V. Chumak, B. Ciesielski, V. De Coste, S. Endo, P. Fattibene, D. Ivanov, C.A. Mitchell, M Nalapko, S. Onori, M. Penkowski, S.P. Pivovarov, A. Romanyukha, A.B. Rukhin, D. Sanin, K. Schultka, T. Seredavina, S. Sholom, V. Skvortsov, V. Stepanenko, K. Tanaka, F. Trompier, A. Wieser, G. Wolakiewicz, Radiat. Meas. **42**, 5 (2007)
24. V.F. Stepanenko, M. Hoshi, Yu.V. Dubasov, A. Sakaguchi, M. Yamamoto, M. Orlov, I.K. Bailiff, A.I. Ivannikov, V.G. Skvortsov, I.G. Kryukova, K.S. Zhumadilov, K.N. Apsalikov, B.I. Gusev, J. Radiat. Res. **47**, 9 (2006)
25. V.F. Stepanenko, M. Hoshi, A.I. Ivannikov, I.K. Bailiff, K. Zhumadilov, V.G. Skvortsov, R. Argembaeva, A.F. Tsyb, Radiat. Meas. **42**, 7 (2007)
26. Exposure at uranium mining *(in Russian)* http://chornobyl.ru/radiacia/46-uran-dobucha-obluchenie.html

 DOI: 10.1051/e3sconf/20172200202

Dosimetry study of East Kazakhstan residents by tooth enamel EPR spectroscopy

Kassym Zhumadilov[1,*], *Alexander* Ivannikov[2], *Valeriy* Skvortsov[2], *Valeriy* Stepanenko[2], *Tolebay* Rakhypbekov[3], *and Masaharu* Hoshi[4]

[1]L.N. Gumilyov Eurasian National University, Munaitpasova 13, Astana 010008, Kazakhstan
[2] A.F. Tsyb Medical Radiological Research Center, 4, Korolev str., Obninsk, 249036, Russia
[3]Semey State Medical University, Abaya Str., 103, Semey 071400, Kazakhstan
[4]Hiroshima University, 734-8553, Japan

Abstract. The tooth enamel electron paramagnetic resonance (EPR) dosimetry method was used to determine accidental doses of population of settlements in the vicinity of the Semipalatinsk Nuclear Test Site (SNTS), Kazakhstan. The influence of four explosions to the populations was included into this report. The distances between investigated settlements and Ground Zero (SNTS) are in the range of 70–200 km from SNTS. Most of settlements (Dolon, Mostik, Bodene, Cheremushki, Kanonerka) are located near the central axis of radioactive fallout trace from the most contaminating surface nuclear test, which was conducted in 29, August 1949. The other settlements located close to radioactive fallout trace result in a surface nuclear tests in 24, August 1956 (Ust-Kamenogorsk, Znamenka, Shemonaikha, Glubokoe, Tavriya, Gagarino), in 12 august 1953 (Sarzhal) and in 7, August 1962 (Akzhar, Kurchatov, Begen, Semenovka, Buras, Grachi). Tooth samples were extracted according to medical recommendations in a course of ordinary dental treatment.

1 Introduction

In the period from 1949 to 1962 125 nuclear tests (including 25 near-surface nuclear tests) were conducted at the Ground Zero technical site in the territory of Semipalatinsk Nuclear Test Site (SNTS). 11 of surface explosions heavily contaminated a huge territory of East Kazakhstan. Only the data of influence from four explosions to the population was included into this report [1–4].

The date of 29 August 1949 represents the first nuclear explosion, which contaminated a huge territory northeast from the epicenter by radioactive fallout (Dolon, Mostik, Bodene, Cheremushki, Kanonerka villages). The radioactive dust cloud was transferred by the wind and its gradual precipitation formed several radioactive fallout traces [5–8]. There is no data describing the dynamics of the precipitation of the dust from the radioactive cloud on the region adjacent to the SNTS. It is a very complicated procedure to reconstruct the individual and collective radiation doses obtained by the local population. This is

* Corresponding author: zhumadilovk@gmail.com

a sophisticated procedure combining data of the radioactive fallout, individual doses measured by EPR, doses reconstructed from retrospective area dosimetry and individual behavior of inhabitants.

On 24 August 1956, the nuclear explosion with 27 kiloton total yield took place, which contaminated a territory to the east from the hypocenter and near Ust-Kamenogorsk city, Znamenka, Glubokoe, Tavriya, Gagarino and Shemonaikha settlements with radioactive fallout [9, 10]. On 12 August 1953, the first thermonuclear explosion with 400 kiloton total yield took place, which contaminated the territory to the south from the hypocenter and near Sarzhal village with radioactive fallout [4, 11, 12]. On 7 August 1962, the nuclear explosion with 9.9 kiloton total yield took place, which contaminated a huge territory to the east from the hypocenter and near Kurchatov city, Akzhar, Begen, Semenovka, Grachi, Mayskoe and Buras villages with radioactive fallout [3, 13].

EPR dosimetry is one of the useful tools for such dose reconstruction [14, 15]. This method can determine the radiation doses retrospectively even more than 40 years after the exposure event. EPR measures the amount of the stable radicals created by radiation exposure in tooth enamel.

2 Materials and Methods

More than 200 teeth samples were collected from the population residing around SNTS. 105 teeth samples were extracted on the basis of medical indications from the adult residents of Dolon, Mostik, Bodene, Cheremushki, Kanonerka villages, located near the radioactive fallout trace formed as result of the most hazardous nuclear test of 1949 [16] and Semipalatinsk (Semey) city. 57 teeth samples were extracted from residents of Kurchatov City, Akzhar, Begen, Semenovka, Grachi, Mayskoe and Buras villages which are located from 70 to 150 km from SNTS [13]. In 2008 and 2009, 88 teeth samples were extracted according to medical reasons from adult residents of Znamenka, Glubokoe, Tavriya, Gagarino, Shemonaikha settlements and Ust-Kamenogorsk city [9, 10]. 6 teeth samples were collected from Sarzhal village and 8 teeth were collected as controls from the population of Kokpekty village (400 km east of the test site), which was not subjected to any radioactive contamination.

All samples were treated by hard alloy dental drills and diamond saws by separating the enamel from the dentine after the cutting of each tooth sample to buccal and lingual parts [15, 17]. Dentins were removed carefully in order to prevent the sample from heating which can induce an additional EPR signal and significantly change shape of the signal [18–20]. Tooth enamel was crushed by cutting pliers to chips 0.5–1.5 mm in diameter.

The measurements were carried out in the X-band on the EPR spectrometer JEOL JES-FA100 at stabilized room temperature of 21°C. The spectrometer was equipped with a high Q-factor cylindrical TE_{011} cavity model ES-UCX2. The spectrum recording parameters were the same as previously published [21–24]. Specially designed computer software was used for spectra processing and dose estimation [25–27].

3 Results and Discussion

The doses for the residents of Kurchatov City, Akzhar, Begen, Semenovka, Grachi, Buras villages (table 1) were included in this report from the published data [13].

Table 1. Results of dose determination for population as a result of surface nuclear tests in 7, August 1962 in the Semipalatinsk Nuclear Test Site. Dose values and its errors are reflected in mGy.

Settlements	**Number of teeth**	**Archival dose (mSv) (Logachev)**	**EPR average excess dose with enamel formation (mGy)**		**EPR maximal excess dose (mGy)**
			Before 7.08.62	**After 7.08.62**	
Akzhar	8	5–10	31 ± 27	-13 ± 29	202
Begen	10	5–10	82 ± 30	-	153
Buras	4	10	101 ± 34	14 ± 27	187
Grachi	5	10	-17 ± 29	52 ± 26	40
Kurchatov	16	10	41± 35	1 ± 30	180
Mayskoe	4	< 5	50 ± 31	70 ± 27	134
Semenovka	10	10	60 ± 29	-	148
Kokpekty	8	< 0.1	0.01 ± 1.92	0.01 ± 1.92	24

For Kurchatov average excess dose before 1962 is 41 ± 35 mGy and after 1962 is 1 ± 30 mGy. Low doses were found for the group with enamel formed after 1962, the end of atmospheric nuclear tests. The dose values for the group having enamel formed before 1962 are consistent with estimations based on the official registered data indicating high levels of the fallout in the period of 1949-1962. The experimentally determined dose was considered to consist of two contributions: dose from natural radiation background accumulated during a tooth enamel lifetime and dose obtained as a result of nuclear tests (excess dose). The last contribution is subject of the interest for present dose reconstruction. At first, the intensity of the RIS was converted into a dose absorbed by enamel D_{en} (expressed in mGy) calibrated using calibration by a ^{60}Co γ-source. Second, excess dose in enamel was determined by subtraction of contribution of the natural background radiation during the enamel existence after its formation from the absorbed dose in enamel.

For the residents of Dolon absorbed doses were found to be in the range from –24 ± 37 to 496 ± 55 mGy. Negative doses were obtained for some values. It is natural that some of the values become negative according to their statistical distribution determined due to experimental errors. The negative doses probably are the result of the underestimation of uncertainty of the dose assessment. This is because the measurements were performed near the threshold of sensitivity of the EPR method.

For all the samples, excess doses were calculated by equation:

$$D_{ex} = D_{en} - TA * D_b,$$

Where: D_{en} – dose calculated by automatic program, in mGy
TA – teeth enamel age, years
D_b – background dose, 0.8 mGy/year [28, 29].

Uncertainty of dose determination (*Er*) was determined based on formula used in the previous publication [6, 14, 16].

The average excess dose for enamel formed before 1949 for Dolon is 153 ± 54 mGy (table 2), for enamel formed after 1949 the average dose is 25 ± 11 mGy. For Mostik excess dose before 1949 is 19 ± 15 and after 1949 is 44 ± 14. For Bodene excess dose before 1949 is 74 ± 40 and after 1949 is 17 ± 10. For Cheremushki average excess dose before 1949 is 104 ± 34 and after 1949 is 4 ± 27. For Kanonerka average excess dose before 1949 is 48 ± 29 and after 1949 is 23 ± 27. For Semipalatinsk City average excess dose before 1949 is 145 ± 68 and after 1949 is 74 ± 35. The value of average dose for Semipalatinsk City is close to the average results of Dolon village [5, 12]. One of the explanations is the person who worked in a place located close to SNTS and another explanation that they were born in the village that was affected by the fallout from Test Site. The bulk of the excess doses are near the sensitivity threshold of the method. For control samples, excess doses are from –66 ± 39 up to 24 ± 39 mGy, for Dolon from -74 ± 38 up to 440 ± 106 mGy, for Mostik from –64±32 up to 119± 51, for Bodene from –50 ± 38 up to 356 ± 58 mGy, for Semipalatinsk City from 0±46 up to 268±79 (table 2). The experimentally measured individual doses can be compared with the dose reconstruction data, which was demonstrated in previous publications [7, 8] and its amount was about 0.5 Gy for Dolon.

Table 2. Information about samples from vicinity to radioactive fallout traces as a result of surface nuclear tests in 29, August 1949 in the Semipalatinsk Nuclear Test Site.

Settlement	Number of teeth samples	Enamel formed before 1949	ESR average excess dose with enamel formation before 1949 (mGy)	ESR average excess dose with enamel formation after 1949 (mGy)
Dolon	38	17	141 ± 37	25 ± 11
Mostik	23	12	19 ± 15	44 ± 14
Bodene	20	9	74 ± 40	17 ± 10
Cheremushki	5	3	104 ± 34	4 ± 27
Semipalatinsk	9	4	145 ± 68	74 ± 35
Kanonerka	3	1	48 ± 29	23 ± 27
Total	97	47	-	-

Table 3. Information about tooth samples from the settlements located close to radioactive fallout trace as a result of surface nuclear tests in 24, August 1956 in the Semipalatinsk Nuclear Test Site.

Settlements	Archival dose (mSv)	ESR average excess dose (mGy)	ESR maximal excess dose (mGy)
Znamenka	25	41±46	268
Glubokoe	10-15	36±31	83
Tavriya	10-15	17±36	54
Ust-Kamenogorsk	80	20±25	120
Gagarino	10-15	-24±33	47
Shemonaikha	0.1	17±37	110

For Tavriya and Gagarino villages all studied samples have been formed before the date of the nuclear test. For other settlements teeth samples were divided into two parts: before and after nuclear explosion. Low mean excess doses have been found for Gagarino residents, while the highest mean excess dose has been determined for Shemonaikha and Ust-Kamenogorsk residents (Table 3). Given the small number of tooth samples investigated, the deduced doses should not be seen as representative for the whole population in the selected villages. The maximum dose obtained for samples from Shemonaikha was not expected, due to its large distance from the radioactive trace, but may be due to radioactivity released by some of the uranium enterprises located there. The maximum dose obtained for Znamenka village [9, 10] can confirm that this village located close to the radioactive trace. The other settlements included in the study also have uranium enterprises, except for Tavriya and Gagarino, where agriculture is prevalent. Average excess doses in the latter two settlements are consistent with estimations based on the official registered data indicating high levels of fallout in the period of 1949–1962.

 DOI: 10.1051/e3sconf/20172200202

Also data of absorbed doses from residents of Sarzhal village, which is located close to fallout trace after test in 12, August 1953. According to archival data [4] the dose was about 500 mSv and this is much higher than reported in table 4. This can be due to evacuation of residents from Sarzhal village to Ayaguz settlements which is located 150 km from Sarzhal.

Table 4. Information about tooth samples from the settlements located close to radioactive fallout trace as a result of surface nuclear tests in 12, August 1953 in the Semipalatinsk Nuclear Test Site.

Settlements	Number of teeth samples	Archival dose (mSv) (Logachev)	ESR average excess dose (mGy)	ESR maximal excess dose (mGy)
Sarzhal	6	500	79 ± 31	138.2 ± 35

4 Conclusion

Higher average excess doses were determined in Dolon and Semipalatinsk city for residents whose tooth enamel was formed before 1949. Results of dose estimation from Dolon samples are in agreement with the fact that this village is located closer to the axis of the radioactive trace. A result from Semipalatinsk needs special investigation.

Compared to samples from the distant Kokpekty village which was chosen as controls, higher average excess doses have been obtained for Znamenka, Ust-Kamenogorsk city and Shemonaikha. This is in agreement with the fact that Znamenka village and Ust-Kamenogorsk city are located close to the axis of the radioactive trace, but Shemonaikha is located on the distance about 70 km from it. It is necessary to note, that the investigated area is well known for its active uranium processing plant. This may explain that higher values have also been found for samples from Shemonaikha, which is located about 70 km from the center line of the radioactive trace. At Tavriya and Gagarino no uranium enterprises exist and, accordingly, the measured doses are consistent with independent estimates of external doses from the fallout that can be found in the literature.

Acknowledgements

This work partly was supported by Grant-in-Aid for Young Scientists (B) from Japan Society for the Promotion of Science #21710060.) And by the Grant of Ministry of Education and Science of Kazakhstan (5284/GF4 agreement No 47).

References

1. O.I. Artemev, M.A. Akhmetov, V.N. Larin, L.D. Ptitskaya, K.S. Zhumadilov, G.P. Silkina, S.B. Subbotin, NATO advanced science institute series, sub-series: Disarmament technologies **31**, 16 (2000)
2. A.I. Ivannikov, Zh. Zhumadilov, B.I. Gusev, Ch. Miyazawa, L. Jiao, V.G. Skvortsov, V.F. Stepanenko, J. Takada and M. Hoshi, Health Phys. **83**, 13 (2002)
3. K. Gordeev, S. Shinkarev, L. Ilyin, A. Bouville, M. Hoshi, N. Luckyanov, S.L. Simon, J. Radiat. Res. **47**, 7 (2006)
4. V.A. Logachev, M., IzdAT, **470** (2008)
5. K. Zhumadilov, A. Ivannikov, K.N. Apsalikov, Zh. Zhumadilov, Sh. Toyoda, E. Tieliewuhan, S. Endo, K. Tanaka, C. Miyazawa, T. Okamoto, M. Hoshi, J. Radiat. Res. **47**, 6 (2006)

6. A. Ivannikov, K. Zhumadilov, E. Tieliewuhan, L. Jiao, K.N. Apsalikov, G. Berekenova, Zh. Zhumadilov, Sh. Toyoda, C. Miyazawa, V. Skvortsov, V. Stepanenko, S. Endo, K. Tanaka, M. Hoshi, J. Radiat. Res. **47**, 7 (2006)
7. V.F. Stepanenko, M. Hoshi, Yu.V. Dubasov, A. Sakaguchi, M. Yamamoto, M. Orlov, I.K. Bailiff, A.I. Ivannikov, V.G. Skvortsov, I.G. Kryukova, K.S. Zhumadilov, K.N. Apsalikov, B.I. Gusev, J. Radiat. Res. **47**, 9 (2006)
8. V.F. Stepanenko, M. Hoshi, A.I. Ivannikov, I.K. Bailiff, K. Zhumadilov, V.G. Skvortsov, R. Argembaeva, A.F. Tsyb, Radiat. Meas. **42**, 7 (2007)
9. K. Zhumadilov, A. Ivannikov, Z. Zhumadilov, V. Stepanenko, K. Apsalikov, M. Rodzi, A. Zhumadilova, S. Toyoda, S. Endo, K. Tanaka, T. Okamoto, M. Hoshi, Radiat. Environ. Biophys. **48**, 6 (2009)
10. K. Zhumadilov, A. Ivannikov, Z. Zhumadilov, V. Stepanenko, Sh. Abralina, L. Sadvokasova, A. Zhumadilova, S. Toyoda, S. Endo, T. Okamoto, M. Hoshi, Radiat. Meas. **46**, 4 (2011)
11. S. Sholom, M. Desrosiers, A. Bouville, N. Luckyanov, V. Chumak, S.L. Simon, Radiat. Meas. **42**, 4 (2007)
12. K. Zhumadilov, A. Ivannikov, V. Stepanenko, S. Toyoda, Z. Zhumadilov, M. Hoshi, J. Radiat. Res. **54**, 4 (2013)
13. K. Zhumadilov, A. Ivannikov, V. Stepanenko, S. Toyoda, V. Skvortsov, M. Hoshi, Radiat. Prot. Dosim. **172**, 4 (2016)
14. IAEA-TECDOC-1331, Vienna (2002). http://www-pub.iaea.org/MTCD/publications/PDF/te_1331_web.pdf
15. P. Fattibene, A. Wieser, E. Adolfsson, L.A. Benevides, M. Brai, F. Callens, V. Chumak, B. Ciesielski, S. Della Monaca, K. Emerich, H. Gustafsson, Y. Hirai, M. Hoshi, A. Israelsson, A. Ivannikov, D. Ivanov, J. Kaminska, Wu Ke, E. Lund, M. Marrale, L. Martens, C. Miyazawa, N. Nakamura, W. Panzer, S. Pivovarov, R.A. Reyes, M. Rodzi, A.A. Romanyukha, A. Rukhin, S. Sholom, V. Skvortsov, V. Stepanenko, M.A. Tarpan, H. Thierens, S. Toyoda, F. Trompier, E. Verdi, K. Zhumadilov, Radiat. Meas. **46**, 6 (2011)
16. K. Zhumadilov, A. Ivannikov, K. Apsalikov, Zh. Zhumadilov, V. Stepanenko, V. Skvortsov, G. Berekenova, S. Toyoda, S. Endo, K. Tanaka, C. Miyazawa, M. Hoshi, Radiat. Meas. **42**, 4 (2007)
17. A. Wieser, E. Vasilenko, P. Fattibene, S. Bayankin, N. El-Faramawy, D. Ivanov, P. Jacob, V. Knyazev, S. Onori, M.C. Pressello, A. Romanyukha, M. Smetanin, A. Ulanovsky, Radiat. Environ. Biophys. **44**, 9 (2006)
18. A. Romanyukha, V. Nagy, O. Sleptchonok, M.F. Desrosiers, J.J. Jiang, A. Heiss, Health Phys. **80**, 3 (2001)
19. A.I. Ivannikov, D. Sanin, M. Nalapko, V.G. Skvortsov, V.F. Stepanenko, A.F. Tsyb, F. Trompier, K. Zhumadilov, M. Hoshi, Health Phys. **98**, 6 (2010)
20. A.I. Ivannikov, V.G. Skvortsov, V.F. Stepanenko, K.Sh. Zhumadilov, Radiat. Prot. Dosim. **159**, 4 (2014)
21. K.S. Zhumadilov, A.I. Ivannikov, V.G. Skvortsov, Zh. Zhumadilov, S. Endo, K. Tanaka, M. Hoshi, J. Radiat. Res. **46**, 7 (2005)
22. K. Zhumadilov, V. Stepanenko, A. Ivannikov, Z. Zhumadilov, S. Toyoda, K. Tanaka, S. Endo, M. Hoshi, Radiat. Environ. Biophys. **47**, 4 (2008)
23. K. Tanaka, S. Endo, A. Ivannikov, S. Toyoda, E. Tieliewuhan, K. Zhumadilov, C. Miyazawa, S. Suga, K. Kitagawa, M. Hoshi, J. Radiat. Res. **46**, 4 (2006)

DOI: 10.1051/e3sconf/20172200202

24. S. Toyoda, A. Kondo, K. Zhumadilov, M. Hoshi, C. Miyazawa, A. Ivannikov, Radiat. Meas. **46**, 4 (2011)
25. A.I. Ivannikov, V.G. Skvortsov, V.F. Stepanenko, D.D. Tikunov, A.A. Romanyukha, A. Wieser, Radiat. Prot. Dosim. **71**, 5 (1997)
26. V.G. Skvortsov, A.I. Ivannikov, V.F. Stepanenko, A.F. Tsyb, L.H. Khamidova, A.E. Kondrashov, D.D. Tikunov, Appl. Radiat. Isot. **52**, 7 (2000)
27. E. Tielewuhan, A. Ivannikov, K. Zhumadilov, M. Nalapko, D. Tikunov, V. Skvortsov, V. Stepanenko, Sh. Toyoda, K. Tanaka, S. Endo, M. Hoshi, Radiat. Meas. **41**, 7 (2006)
28. M. Hoshi, S. Toyoda, A. Ivannikov, K. Zhumadilov, A. Fukumura, K. Apsalikov, Z.S. Zhumadilov, S. Bayankin, V. Chumak, B. Ciesielski, V. De Coste, S. Endo, P. Fattibene, D. Ivanov, C.A. Mitchell, S. Onori, M. Penkowski, S.P. Pivovarov, A. Romanyukha, A.B. Rukhin, K. Schultka, T.A. Seredavina, S. Sholom, V. Skvortsov, V. Stepanenko, K. Tanaka, F. Trompier, A. Wieser, G. Wolakiewicz, Radiat. Meas. **42**, 9 (2007)
29. A. Ivannikov, S. Toyoda, M. Hoshi, K. Zhumadilov, A. Fukumura, K. Apsalikov, Z.S. Zhumadilov, S. Bayankin, V. Chumak, B. Ciesielski, V. De Coste, S. Endo, P. Fattibene, D. Ivanov, C.A. Mitchell, M Nalapko, S. Onori, M. Penkowski, S.P. Pivovarov, A. Romanyukha, A.B. Rukhin, D. Sanin, K. Schultka, T. Seredavina, S. Sholom, V. Skvortsov, V. Stepanenko, K. Tanaka, F. Trompier, A. Wieser, G. Wolakiewicz, Radiat. Meas. **42**, 5 (2007)

Enhancement of sedimentation and coagulation with static magnetic field

Marcin Zieliński[1], *Marcin* Dębowski[1,*], *Anna* Hajduk[1], and *Paulina* Rusanowska[1]

[1]University of Warmia and Mazury in Olsztyn, Faculty of Environmental Sciences, Department of Environment Engineering

Abstract. The static magnetic field can be an alternative method for wastewater treatment. It has been proved that this physical factor, accelerates the biochemical processes, catalyzes advanced oxidation, intensifies anaerobic and aerobic processes or reduces swelling of activated sludge. There are also reports proving the positive impact of the static magnetic field on the coagulation and sedimentation, as well as the conditioning and dewatering of sludge. In order to be applied in larger scale the published results should be verified and confirmed. In the studies, the enhancement of sedimentation by the static magnetic field was observed. The best sedimentation was noted in the experiment, where magnetizers were placed on activated sludge bioreactor and secondary settling tank. No effect of the static magnetic field on coagulation with the utilization of PIX 113 was observed. However, the static magnetic field enhanced coagulation with the utilization of PAX-XL9. The results suggest that increased sedimentation of colloids and activated sludge, can in practice mean a reduction in the size of the necessary equipment for sedimentation with an unchanged efficiency of the process.

1 Introduction

Due to constantly increasing volume of produced wastewater, changes of their composition and properties, a steady evolution and development in the methods used for wastewater treatment can be observed, nevertheless effective and economically viable methods are still sought [1]. An alternative to the methods used so far, can be an innovative technique based on the use of physical factor, that is a static magnetic field (SMF) [2]. The SMF causes selective ionization, changes in electrical conductance of solutions. Due to this the molecules are polarized and receive and appropriate charge.

It has been proved that this physical factor, accelerates the biochemical processes, catalyzes advanced oxidation, intensifies anaerobic and aerobic processes or reduces swelling of activated sludge [1, 3]. There are also reports proving the positive impact of the SMF on the coagulation and sedimentation [4], as well as the conditioning and dewatering of sludge [1, 5]. The results, however, require verification and confirmation.

These examples confirm that the use of the SMF might become an alternative and prospective method which allows to improve wastewater and sludge treatment. Economic

* Corresponding author: marcin.debowski@uwm.edu.pl

 DOI: 10.1051/e3sconf/20172200203

considerations prompt to use magnetizers as assistive technology systems in water purification, wastewater and sludge treatment [1, 6]. Magnetic fluids activators (MFA) are characterized by simple design. The greatest advantage is the lack of need to supply any form of energy and that the intensity of the magnetic field induction is not weakened over time [7]. For these reason such devices can be operated efficiently over many years. Moreover, magnetizers can be used in system without the need of adaptation period [1].

The aim of the study was to evaluate sedimentation and coagulation efficiency in the bioreactors and in the secondary settling tanks.

2 Materials and methods

The studies were carried out on a laboratory scale. The wastewater used in experiments was a model of dairy wastewater obtained by dissolving skimmed milk powder in tap water. The inoculum was activated sludge from municipal wastewater treatment plant in Olsztyn. The wastewater was characterized by chemical oxygen demand (COD) of 4181 ± 890 mg O_2/dm^3, pH of 4.8 ± 0.2 and temperature 20.2 ± 1.2°C.

The studies were divided into two stages. In stage 1, the influence of the SMF on the sludge sedimentation in the secondary settling tank was investigated. The experiments were carried out in two activated sludge bioreactors, one with magnetizers placed on tank (series 1) and second one without magnetizers placed on tank (series 2) (Fig. 1). Raw wastewater were pumped to bioreactors from retention tank. The mixture of the activated sludge and the effluent was directed to the two secondary radial settling tanks, one with magnetizers placed (variant 1) and second one without magnetizers placed (variant 2).

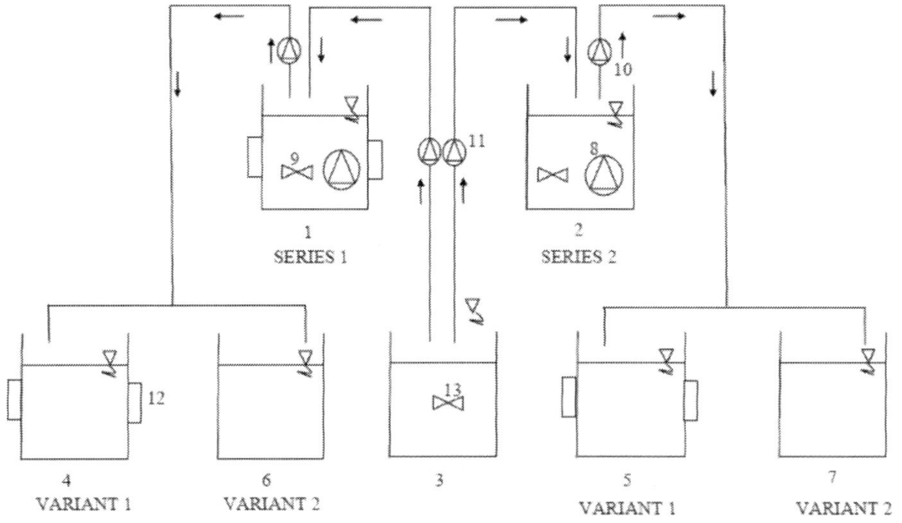

Fig. 1. Scheme of the experiments in stage 1 (1 – activated sludge bioreactor with MFA, 2 – activated sludge bioreactor without MFA, 3 – retention tank, 4, 5 – secondary settling tanks with MFA, 6, 7 – secondary settling tanks without MFA, 8 – aeration pump, 9 – stirrer, 10 – pump discarding effluent, 11 – pumps introducing raw wastewater, 12 – MFA, 13 – stirrer).

The technological parameters of the bioreactors and secondary settling tanks are shown in Table 1. The sedimentation was describe by the volume of activated sludge after 30 minutes in 1 dm^3 of wastewater (cm^3/dm^3) and the sludge volume index (SVI_{30}) (cm^3/g TSS).

Table 1. Technological parameters of the bioreactors and the secondary settling tanks.

Parameter	Unit	Value
Bioreactor		
Volume	dm^3	4.5
Concentration of activated sludge	g TSS/dm^3	3
Organic loading	g COD/g TSS·d	0.15
Hydraulic retention time	day	22.5
Concentration of oxygen	mg O_2/dm^3	3.5
Secondary radial settling tank		
Volume	dm^3	0.04
Surface	dm^2	0.02
Operating height	dm	2
Hydraulic loading	$dm^3/dm^2 \cdot h$	10
Settling time	h	0.5

The magnetizers were placed directly on the circumferences of the bioreactor and secondary settling tank. The magnetizers were built of two parts forming a ring. Permanently magnetized ceramic frits in the MFA emitted the SMF. In order to increase the direct impact of SMF, a set consisting of two rings were placed on bioreactors. The parameters of the MFA were width of the ring of 65 mm, height of the individual ceramic magnet of 45 mm, weight of the individual ring of 1.25 kg, and nominal diameter range of 90–110 mm. The maximal induction of SMF was 6 000 Gauss (0.6 T).

In the stage 2, the effect of the SMF on coagulation was observed. The coagulation was carried out in the chambers. To the chamber the inorganic coagulants were dosed. After dosing of coagulant the rapid stirring, slow stirring (flocculation) and sedimentation take place. Basic technological parameters of the chambers are presented in Table 2.

Table 2. Technological parameters of coagulation chambers.

Parameter	Unit	Value
Volume	dm^3	1
Surface	dm^2	0.7
Height	dm	1.4
Time of rapid stirring	min.	5
Intensity of rapid stirring	rpm	200
Time of slow stirring	min.	20
Intensity of slow stirring	rpm	30
Time of sedimentation	min.	120

The two series of experiments were done. In the series 1, the source of SMF was placed on coagulation chamber. In the series 2, the coagulation chamber was operated without the magnets (control) (Fig 2.).The chambers were equipped with stirrer with adjustable speed. Each series was divided into two variants to test different type of the coagulant. Moreover, four different doses of the coagulants were investigated.

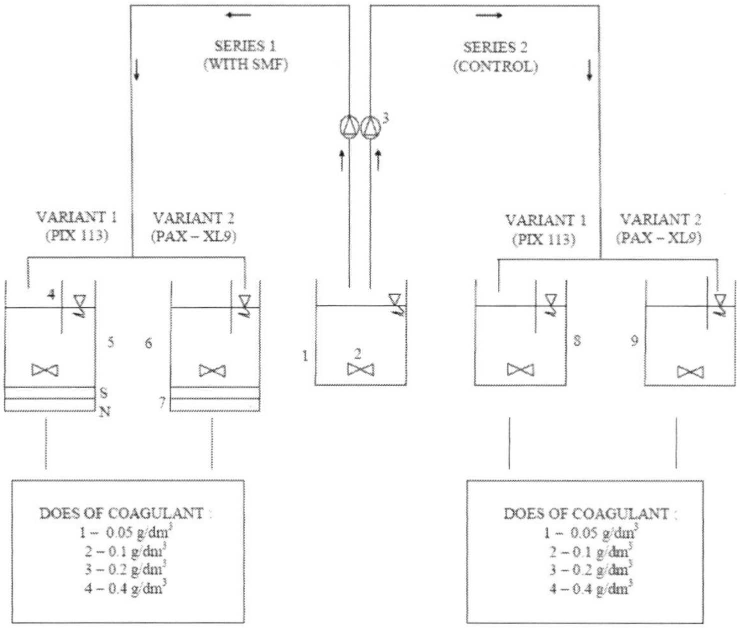

Fig 2. Scheme of the experiment in stage 2 (1 – retention tank, 2 – stirrer, 3 – pumps introducing raw wastewater, 4 – outlet for sample collecting, 5,6 – chambers with magnets, 7 – neodymium magnets, 8,9 – chambers without magnets (controls).

The criterion for separating variant was type of used inorganic coagulant. In the variant 1, the ferric coagulant name PIX 113 was tested. In the variant 2, the aluminum coagulant name PAX - XL9 was tested. For both coagulants four doses were applied: 0.05 g/dm^3, 0.1 g/dm^3, 0.2 g/dm^3, 0.4 g/dm^3. The characteristics of the used coagulants is presented in Table 3.

Table 3. Chemical compositions of the coagulants.

PIX 113	**PAX-XL9**
Total Fe 11.8 ± 0.4%	Al_2O_3 8.5 ± 0.4%
Fe^{2+} 0.4 ± 0.3%	Al_2 4.5 ±0.2
pH < 1	Chlorides 10.0 ± 1.0%
	pH 3.0 ± 0.6 %

In the 2 stage, the effect of the SMF on coagulation was provided by the use of permanent neodymium magnets. Magnets were directed up with the positive side, opposite to the charges of the most natural colloids (Fig. 2). The characteristics of the used neodymium magnets is shown in Table 4.

In order to assess the effectiveness of coagulation supported by the SMF, changes in concentration of the COD in 15 minute intervals were recorded. The standard dichromate method was used to determine the COD. During this method high temperature mineralizer HT200S and spectrophotometer UV—VIS DR 5000 were used.

DOI: 10.1051/e3sconf/20172200203

Table 4. Characteristics of neodymium magnets.

Parameter	Unit	Value
Diameter	dm	0.7
Height	dm	0.02
Energy density of magnetic field	kJ/dm^3	0.286–0.302
Induction of remanence	T	1.21–1.25
Coercion H_{cB} min	kA/m	899
Coercion H_{cJ} min	kA/m	955

Statistical analysis of the results was done with STATISTICA 10.0 PL. The hypothesis on the distribution of each analyzed variable was verified with the Shapiro-Wilk W test. The significance of differences between variables was stated with the use of one-way analysis of variance (ANOVA). The homogeneity of variance in groups was checked with the Levene's test. The significance of differences between analyzed variables was determined with the RIR Tukey test. Differences were found significant at $\alpha = 0.05$.

3 Results and discussion

In the stage 1 the concentration of activated sludge in both series was about 3000 mg TSS/dm^3. Similar values were observed by Hattori et al. (2001). The authors suggest that the magnetic field with induction of 0.08 T seems to not affect the growth of activated sludge [7]. However, Xu and Sun (2008) observed that magnetic field with induction of 0.024 T and 0.06 T improved the quantity of mixed bacteria in medium. The authors also noted that propagation rate of mixed bacteria was the biggest at the SMF induction of 0.06 T. From the presented results it appears that the enhancement of bacteria growth can be observed with magnetic induction lower then 0.1 T. Whereas higher densities of magnetic field does not affect the growth of bacteria [5].

In the studies, the enhancement of sedimentation by the SMF was observed. The best sedimentation was noted in variant 1 of series 1, where magnetizers were used on activated sludge bioreactor and secondary settling tank. The volume of the sludge after 30 minutes of sedimentation was 440 ± 15 cm^3/dm^3 and SVI_{30} was 145.25 ± 4.95 cm^3/g (Fig. 3, 4). The application of magnetizers only on the activated sludge bioreactor decreased the sedimentation properties of the sludge. The volume of the sludge after 30 minutes of sedimentation and the SVI_{30} increased of 200 cm^3/dm^3 and 70 cm^3/g, respectively. The sedimentation of the activated sludge was decreased in the series 2, when the activated sludge bioreactor was operated without influence of the SMF. The volume of the sludge after 30 minutes of sedimentation was 920 ± 12.5 cm^3/dm^3 and 960 ± 9.5 cm^3/dm^3 in the variant 3 and 4, respectively. Similar tendency was observed in case of the SVI_{30}. Hattori et al. (2001) also used SMF to enhance to sludge sedimentation [7]. The authors applied the SMF with induction of 0.08 T. In the experiment, their used the wastewater circulation system in which the activated sludge flow through magnetic field. The SVI_{30} observed by Hattori et al. (2001) was lower then in presented experiments and it was about 90 cm^3/g. The enhancement of sedimentation was also observed by Gokon et al. (2002). The authors have shown, that despite the higher concentration of the activated sludge, the smaller volume was occupied by the sludge after sedimentation in reactor with SMF in comparison to the control reactor [8]. Similar result were also observed for aerobic granular sludge. The magnetic field with induction of 48 mT, significantly decreased SVI from 77.32 cm^3/g to 20.93 cm^3/g [9].

 DOI: 10.1051/e3sconf/20172200203

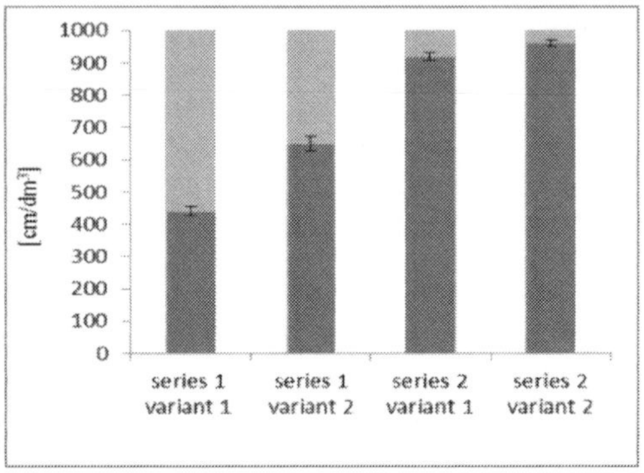

Fig. 3. The volume of the sludge after 30 minutes of sedimentation.

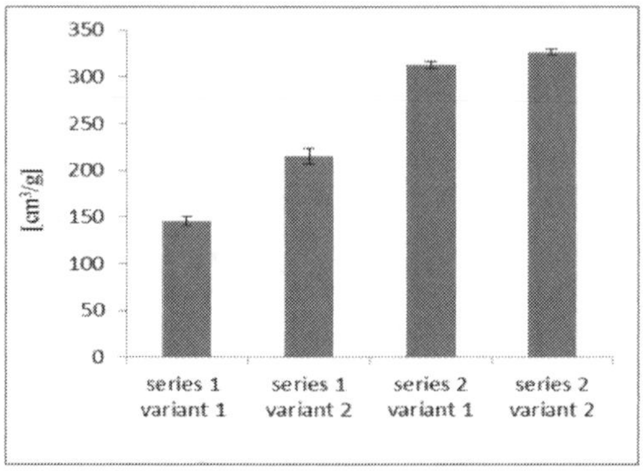

Fig. 4. The SVI30.

In stage 2, no effect of the SMF on coagulation with the utilization of PIX 113 was observed, only the increased dose of coagulant (from 0.05 gFe^{3+}/dm^3 to 0.2 gFe^{3+}/dm^3) enhanced coagulation. The efficiency of coagulation with SMF increased from 28 ± 2.3% to 60 ± 1.5%, when the doses of coagulant were 0.05 gFe^{3+}/dm^3 and 0.2 gFe^{3+}/dm^3, respectively (Fig. 5). In case of the control chambers, the efficiency of coagulation increased from 26 ± 2% to 57 ± 1.2%, when the doses of coagulant were 0.05 gFe^{3+}/dm^3 and 0.2 gFe^{3+}/dm^3, respectively. However, the SMF enhanced coagulation with the utilization of PAX-XL9. The significant differences were observed in the chambers, in which the dose of coagulant was from 0.1 gAl^{3+}/dm^3 to 0.4 gAl^{3+}/dm^3 (Fig. 6). The efficiency of coagulation enhanced by SMF was 52 ± 1.3%, 61 ± 0.8%, 64 ± 0.6% for the doses of 0.1 gAl^{3+}/dm^3, 0.2 gAl^{3+}/dm^3, 0.4 gAl^{3+}/dm^3, respectively. In control, the efficiency of coagulation was 45 ± 2% , 54 ± 1.7% and 58 ± 1.5 for the doses of 0.1 gAl^{3+}/dm^3, 0.2 gAl^{3+}/dm^3, 0.4 gAl^{3+}/dm^3, respectively.

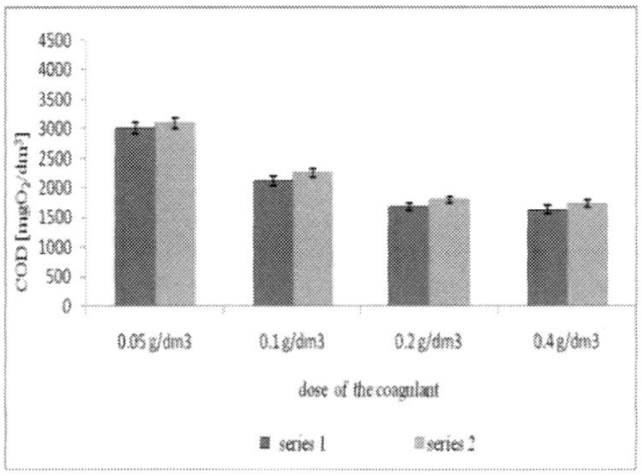

Fig. 5. The COD values depending on the dose of the coagulant PIX 113.

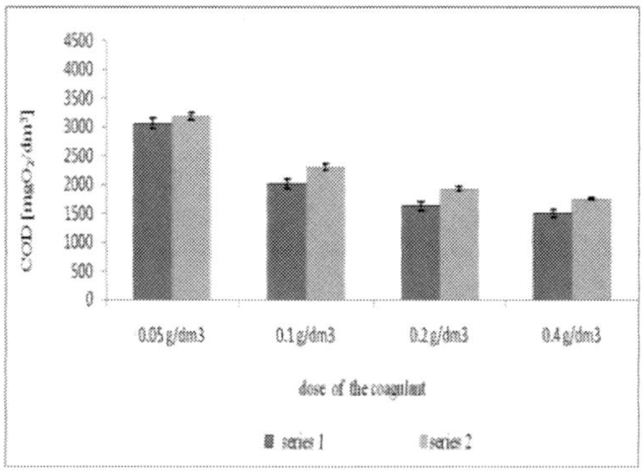

Fig. 6. The COD values depending on the dose of the coagulant PAX-XL9.

The impact of SMF on the rate of COD removal was only found in the case of the highest dose of coagulant PAX-XL9. The rate of the COD removal in the control was 73 ± 9.26 $mgO_2/dm^3 \cdot min$, while in the chamber with magnets it increased to 87 ± 7.23 $mgO_2/dm^3 \cdot min$. (Fig. 8) In the case of the coagulant PIX 113, the rate of COD removal was 74±7.85 $mgO_2/dm^3 \cdot min$ in the control, and 79 ± 11.77 $mgO_2/dm^3 \cdot min$ in the chamber with magnets (Fig. 7) .

DOI: 10.1051/e3sconf/20172200203

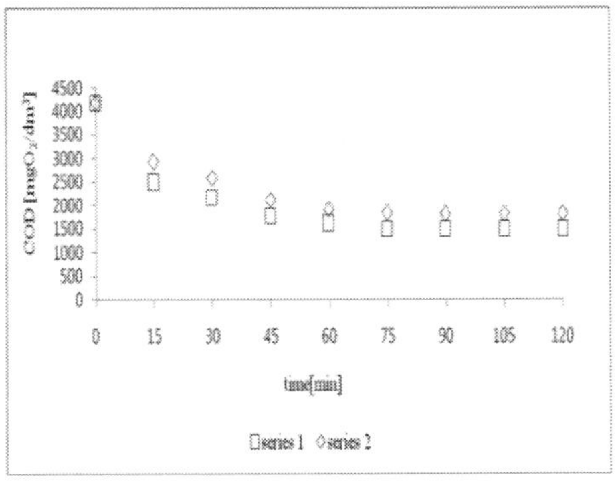

Fig. 7. The changes of COD after dosing 0.4 gFe^{3+}/dm^3 of PIX 113.

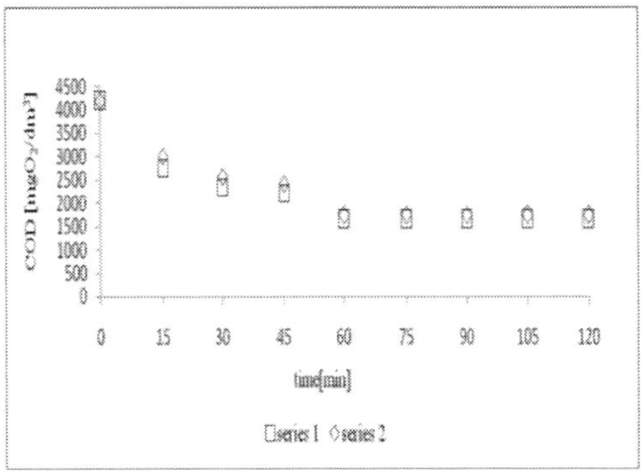

Fig. 8. The changes of COD after dosing $0.4gAl^{3+}/dm^3$ of PAX-XL9.

Hattori et al. (2001) observed similar results with the utilization of coagulant PIX 113. The addition of 0.1% $FeCl_3$ to the reactors with SMF did not increased sedimentation properties. The authors explained that the sludge they used in the experiment had already 0.05–0.10% Fe/dry weight. Therefore, the addition of $FeCl_3$ might not be so influential on sedimentation enhancement with SMF [5]. The same observations was noted by Wang et al. (2012) [9].

4 Conclusions

The application of SMF led to significant enhancement of capability to sedimentation of activated sludge. It was found, that the best sedimentation properties had the activated sludge from bioreactor with SMF, which was concentrated in secondary settling tank with SMF. The volume of activated sludge increased of 560 cm^3/dm^3 in secondary settling tank without magnetizers and bioreactor without magnetizers.

The SMF improved the efficiency and rate of COD removal from wastewater after coagulation with coagulant PAX-XL9. The efficiency of coagulation increased of about 5% by using magnets. The COD removal rate was dependent on the dose of coagulant and was the highest after the dose of 0.4 gAl^{3+}/dm^3. However, no significant impact of the SMF on the technology results of coagulation with PIX 113 was observed. The results suggest that increased sedimentation of colloids and activated sludge, can in practice mean a reduction in the size of the necessary equipment for sedimentation with an unchanged efficiency of the process.

References

1. M. Krzemieniewski, M. Dębowski, M. Zieliński, *Utilization of eletromagnetic microwave radiation nad static magnetic field in wastewater and sludge treatment* (2002) (in Polish)
2. M. Zieliński, A. Cydzik-Kwiatkowska, M. Zielińska, M. Dębowski, P. Rusanowska, J. Kopańska, Water Air Soil. Pollut. **228**, 126 (2017)
3. A. Yadollahpour, S. Rashidi, Z. Ghotbeddin, M. Jalilifar, Z. Rezaee, Jour. of Pure and Appiled Microbiol. **8**, 5, 3711–3719 (2014)
4. H. Yavuz, S.S. Celebi, Enzyme and Microb. Techn. **26**, 1, 22–27 (2000)
5. Y.B. Xu, S.Y. Sun, Biodegradation **19**, 3, 455–462 (2008)

6. M. Zieliński, M. Dębowski,M. Krzemieniewski, A. Brudniak, Environ. Prog. and Susta. Energy **34**, 2, 427–41 (2014)
7. S. Hattori, M. Watanabe, T. Endo, H. Togii, K. Sasaki, World Jour. of Microb. and Biotech. **17**, 279–285 (2001)
8. N. Gokon, A. Shimada A, H. Kaneko, Y. Tamaura, K. Ito, T. Ohara, J. Magn. and Magn. Mat. **238**, 47–55 (2002)
9. X.H. Wang, M.H. Diao, Y. Yang , Y.J. Shi, M.M. Gao, S.G. Wang, Bioresour. Technol. **110**, 105–10 (2012)

Optimization of energy cost in water supply system

Izabela Zimoch[1,*] , and *Ewelina* Bartkiewicz[1]

[1]Silesian University of Technology, Faculty of Energy and Environmental Engineering
Institute of Water and Wastewater Engineering, ul. Konarskiego 18, 44-100 Gliwice, Poland

Abstract. The decreasing amount of fossil fuels and deteriorating air quality forces the governments to introduce a rational energy management in all sectors of economy. At the beginning of the twenty-first century many water supply systems (WSSs) were oversized because of the reduction of water consumption, especially in industry. This resulted in high energy consumption in the pumping stations. Improving pumps operation will decrease energy consumption and also the water prices. The purpose of this paper is to present a method of energy optimization in WSSs. This paper presents an analysis of energy consumption in a selected water supply system. In this study pumps located in water treatment plants and pumping station cooperating with the tanks are analyzed. The study used hydraulic model of the WSS created in MOSKAN-W, which defines pumps' parameters such as flow, head, and efficiency. Using optimization options of calculation software several scenarios of energy costs were prepared.

1 Introduction

Nowadays the high demand for energy is observed, thus means of efficient electricity management become more important subject-matter than ever before. Responsible use of energy is important not only from the economical point of view but increasingly from environmental as well. According to the European Union Directive 2012/27/ EU plan, by 2020 CO_2 emissions should be reduced by 20%, share of energy produced from renewable sources increased to 20%, and the energy efficiency of all production and maintenance processes improved by 20%. European Union countries are forced to reduce energy consumption in all sectors of their economies, thereby water supply companies. Water supply systems (WSS) are required to provide population with water of satisfactory quality and sufficient quantity. For this purpose, water companies use variety of water treatment and transport processes, which are highly electricity consuming. It is estimated that means of water distribution absorb from 3% to 7% [1, 2] of electricity, worldwide. According to Copeland and Carter [3] energy consumption in WSS is distributed as follows: 67% for tap water pumping, 14% for water treatment, 11% for raw water pumping and 8% for in-plant water pumping (e.g. backwash water of filters). While Vilanova and Balestieri [1] state that

* Corresponding author: izabela.zimoch@polsl.pl

 DOI: 10.1051/e3sconf/20172200204

80-90% of electricity consumption, due to water distribution, comes from running of pumping stations. These and many other researchers point the pumping stations to be a significant contributor to electricity consumption. Therefore any efficiency improvements in involved processes would be highly desirable.

There are several methods that have proven to have positive impact on reduction of energy consumption in WSS [2–4]. They can be divided into three groups: (a) implementation of control and monitoring systems SCADA (Supervisory Control And Data Acquisition), (b) installing higher efficiency pumps motors, and drives, (c) generating energy in WSS using alternative energy sources.

SCADA is used in water supply system to, through use of field devices, gather data such as: water flows and pressures, valve settings, tank storage levels, and pump operation states. SCADA's data can be used to determine the optimal parameters of pumps' working states among the others, to minimize energy consumption. For example through adjusting pump settings according to varying water demands. Dispatcher can analyze these data in a real time and control operating parameters of pumps and valves thus maintain the entire water supply system at the highest efficiency level. Trough above mentioned actions water facilities can achieve energy savings of 5% to 10% [4].

Nowadays, many WSS are oversized and rarely work at their full capacity, which often makes pumps to work at their non-optimum yield. In such situations usually throttling valves or a bypass lines are used. However, their use is not the best solution as throttling valves reduce pump efficiency by shifting their operating point away from its optimum efficiency level. Bypass lines redirect part of the flow thus they do not have much influence on pumps' operating point. An alternative to these solutions is Variable Speed Drives (VSD) which can control pump's pressure or flow by varying the pump's rotational speed. Additionally use of VSDs allows reduction of pipe breaks (which often happen when switching pumps) thus increasing likelyhood of secondary water contamination due to breaking a bioflim. These drives can provide energy savings of 30–50% according to some sources [5], or of 10–20% according to Coelho and Andrade-Capos [3]. Yet, the most effective way to reduce electricity consumption would be to replace pumps with new ones with capacity matching water demands. The new pumps should be chosen respectively to system's capacity and operating conditions. If pumps are selected based on an actual flow and pressure a better efficiency improvements than from using VSD, alongside further cost reduction can be expected.

Using renewable energy sources in WSS is becoming more and more common. Excluding costs of implementation, these systems have many environmental and economic advantages. The renewable energy solution used in WSS can be divided into three kinds: (a) solar, (b) wind and (c) hydropower generation. For WSS the more effective and less expensive to implement solution is use of pumps as turbines (PAT). Water turbine is simple system for generating electricity, which is based on using pressure energy and speed energy of transported water. There are two kinds of turbines: (a) impulse turbine which changes the direction of water flow and (b) reaction turbines which converts water pressure and kinetic energy into energy. Many pumps available on the market can be used as turbines (in micro-hydroelectric plants with capacity of up to 100 kW) [5, 11]. However, PAT is very sensitive to changing parameters, especially head, and characterized by low efficiency, between 30% and 60%.

As it can be seen, there are many ways to reduce energy consumption in WSS. Choice of the best economic and environmental solution of energy optimization in WSS depends on reliable analysis of data and working regime of water distribution network. For this propose a hydraulic model of WSS can be used. Nowadays, modern IT techniques enable the integration of SCADA, GIS and CIS systems, resulting in dynamic models of WSS. This is a very difficult process, requiring knowledge about: all systems SCADA, GIS, CIS, the working regime of WSS, and random events (e.g. pipe failures). The software used in modeling beyond the hydraulic calculations has also optimization functions of energy consumption. All the optimization goals of WSS are in majority multi-criteria optimization tasks. There are several types of optimizations which allow to find global extreme of multi-criteria function (Integrated Objective Function-IOF), which is based on stochastic algorithms that use probability theory to define iterative method. Group of these algorithms includes, among others: genetic algorithm, clustering algorithm, evolution algorithm, and simulating annealing, accelerated random search, ant colony optimization artificial bee colony (ABC) algorithm [6–10]. In stochastic optimization algorithm the objective is to find points in which the given integrated objective function (*IOF*) reaches its global extreme. The aim of the optimization in *n*-dimensional space is to find optimum value (decision variable x^{opt}) in a feasible set X_d fulfilling relationships mentioned below:

- finding global minimum: $$\exists_{\mathbf{x}^{opt}} \forall_{\mathbf{x}\in X_d} \; IOF(\mathbf{x}^{opt}) \leq IOF(\mathbf{x}) \tag{1}$$

- finding global maximum: $$\exists_{\mathbf{x}^{opt}} \forall_{\mathbf{x}\in X_d} \; IOF(\mathbf{x}^{opt}) \geq IOF(\mathbf{x}) \tag{2}$$

Therefore, there is a set $X \subset R^n$, which is a compact set, and a function $IOF : X \rightarrow R$ which is a continuous function. When looking for global minimum is for feasible set there is such X that:

$$X_d = \arg\min(x) = \{IOF(x *) = \min IOF(x))\} \tag{3}$$

For this task it should be used formula:

$$\mathbf{X}_t = T_t(X_{t-1}, \mathbf{Y}_{t-1}), \text{ for } t \rightarrow \infty \tag{4}$$

where T_t is the algorithm's mechanism, and Y_t is a random variable. To solve this problem various algorithms would be used, some of which are more or less complex e.g. Pure Random Search (PRS), algorithms autonomous over time, evolutionary algorithm and algorithms not autonomous. Stochastic optimization algorithm funds its use in solving of integrated objective function for variety cases of dynamic exploitation systems, which include water systems.

2 Research object

The research object is a large WSS which supplies water to over 3 million customers. The area covered by the network is hilly with elevation differences equal to 120 m. The system is over 880 km long and it is divided into eight major divisions of networks region. Total average water production is about 330 000 cubic meter per day. Due to the closure of large factories and reduction of water consumption by productions companies, the water production of this WSS has dropped by about 40% in last 10 years. The WSS is composed of 11 water treatment plants (WTP), 15 pumping stations and 9 complexes of storage tanks with total capacity of over 374 000 cubic meters.

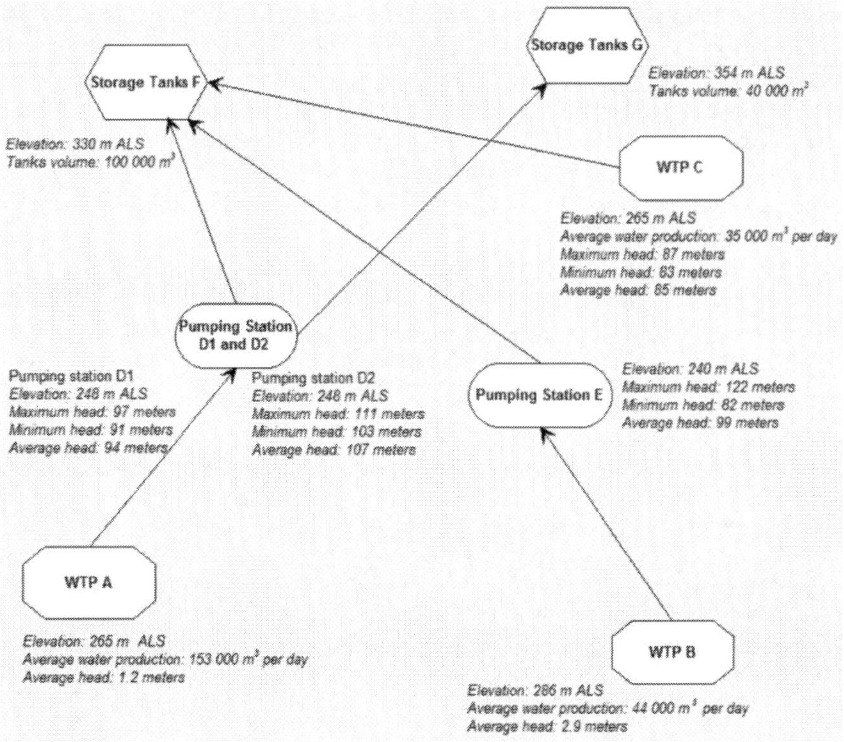

Fig. 1. Schematic diagram of a separate area of the water supply system.

This research looks in particular at the southern area of this WSS (Fig.1), including three main water treatment plants (WTP A, WTP B and WTP C) and four pumping stations (Pumping Station D1, D2 and Pumping Station E) and two reservoirs (Storage Tanks F and Storage Tanks G). Water produced by these three WTPs amounts to about 232 000 cubic meters per day, which is 70 % of total water production of this WSS. Studied WSS uses gravity and pumping to convey water to bulk buyers, delivering in this way about 56 000 cubic meter per day. Storage tanks F situated in central area of whole supply region are a strategic element of WSS, their capacity is 100 000 cubic meters. They are filled by three water treatment plants and distribute water to the northern and south-western part of the water supply area. Elevation differences between storage tanks F and WTPs are measured to be around 70 m, that's why water has to be pumped to the tanks by pumping stations (D1,D2, E) situated in the network and WTP C. Second storage tanks are supplied only by: WTP A (pumping station D2) and WTP C (drinking water pumping station). Storage tanks F transport water to the northern part of the analyzed region.

2.1 Pumping stations

The southern area network includes the following six pumps stations (Table 1), which went through the energy consumption analysis:

a) WTP A contains eight pumps, but water is transported by gravity to pumping stations D1 and D2, this is more economical solution (elevation differences equal to 17 meters);

b) WTP B contains five pumps which are inactive for most of the time as the relatively small amounts of water demanded can be delivered by gravity due to elevation differences with pumping station E;

c) Pumping station D1 contains four pumps. During normal conditions of its operation, there is active only one pump with nominal flow 1400 m^3/h. When power plant, situated between pumping station D1 and storage tank F, connects to the network, a dispatcher switches a pump to a one with nominal flow 2400 m^3/h;
d) Pumping station D2 contains nine pumps and can work at different capacities. Usually only one small pump with nominal flow 1800 m^3/h is working, but in case of increased water demand this pump is switched to one with nominal flow 3600 m^3/h;
e) Pumping station E contains seven pumps, during normal operation there are two active pumps with nominal flow 1400 m^3/h or one pump with nominal flow 2400 m^3/h;
f) WTP C contains nine pumps, during the normal work of the pumping station of drinking water, there is one active pump with nominal capacity 1800 m^3/h, but when the water demand decreases, dispatcher can switch to pump with nominal flow 1167 m^3/h.

3 Research methodology

In this study hydraulic model of water supply system was used to perform a variety of simulations. Hydraulic model used in research is created and computed for average demand values (for period 1-31 October 2016) and checked for maximum water demand and minimum water demand. A water network graph is generated from data, collected using GIS and SCADA, and transferred to modeling software MOSKAN-W. MOSKAN-W created by ISB PAN uses calculation algorithms contained in Epanet 2.0. It also has optimization options developed on the basis of stochastic algorithms. This research uses Random Search Algorithms (RSA)[13] to calculate energy cost in pumping stations, using equation (5):

$$IOF_c = \min_{x \in S} f(\mathbf{x}) \tag{5}$$

where **x** is a pump speed (vector of n decision variables), S is pressure in pumping stations (feasible region) and f is a function defined over S, IOF_c is energy cost. The region S is defined by fuzzy sets [14], where criteria and limitations can be approximated by the trapezoid function shown in Fig. 2.

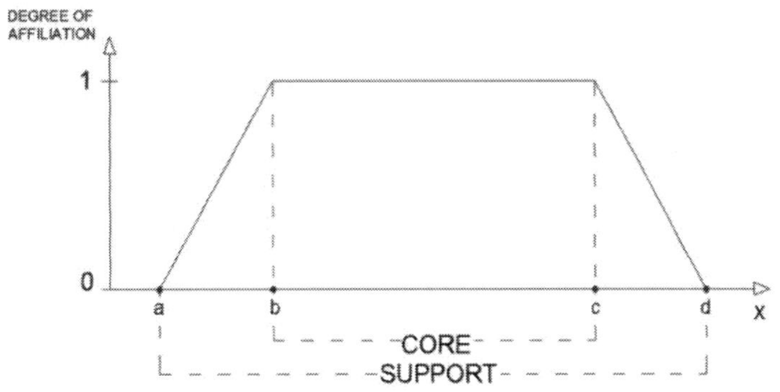

Fig. 2. Trapezoid-Shaped Membership Function.

Core is a simple non-fuzzy set, so-called an universe X and Support is a fuzzy set A. Trapezoid function is described by the following equations (6):

$$\mu_A = \begin{cases} 0; & x \leq a \\ \frac{x-a}{b-a}; & a < x \leq b \\ 1; & b < x \leq c \\ \frac{d-x}{d-c}; & c < x \leq d \\ 0; & x > d \end{cases} \quad (6)$$

where x is variable, a, b, c, d are parameters of affiliation function and μ_A is the affiliation function, which determines to what degree element x belongs to set A. The equations (6) means that:

– if $\mu_A=0$ – variable x does not belongs to the set A
– if $\mu_A=1$ – variable x fully belongs to the set A
– if $0<\mu_A<1$ – variable x partially belongs to the set A

In conducted research the following data was defined for pumping stations: minimum and maximum pressure and range of pump speed, and for the whole area: minimum and maximum energy cost. The simulation was ran for a period of 72 hours for four scenarios:
a) normal network operation when all network objects are working. Optimization for this scenario is carried out for three cases: maximum water demand, minimum water demand, and average water demand;
b) using backup pumps, in this scenario we used pumps before the modernization (in 2015 took place modernization of PS);
c) WTP B shutdown, this situation occurs when the water turbidity rapidly rises to very high values, e.g. in September 2007 water turbidity increased to 285 NTU;
d) WTP A shutdown, this situation happened once in 2003 during power failure.
In scenarios "c" and "d" WTP C increased water production and opened second pump to complement water shortage in storage tanks F and G. In this scenario storage tanks delivered water to pumping station D direction by gravity.

For each object energy cost was put to calculation software (average 0.23 PLN per KW-hr) and run simulations. The results of simulations are shown in Table 1. After that, optimization calculations were started, which results are shown in Fig. 3 and 4. Value above the graph is the result of optimization, while the value below graph is an initial value for calculations, for example in scenario A, initial value for pressure of PS "D1" equals 89.86 meters and value after optimization equals 93.47 meters.

Table 1. Energy cost before optimization for each scenario.

Scenario A		Scenario C	
Daily energy cost	16 981.85 PLN	Daily energy cost	16 568.47 PLN
Annual energy cost	6 198 375.25 PLN	Annual energy cost	6 047 491.55 PLN
Scenario B		Scenario D	
Daily energy cost	18 443.76 PLN	Daily energy cost	15 265.32 PLN
Annual energy cost	6 731 972.40 PLN	Annual energy cost	5 571 841.80 PLN

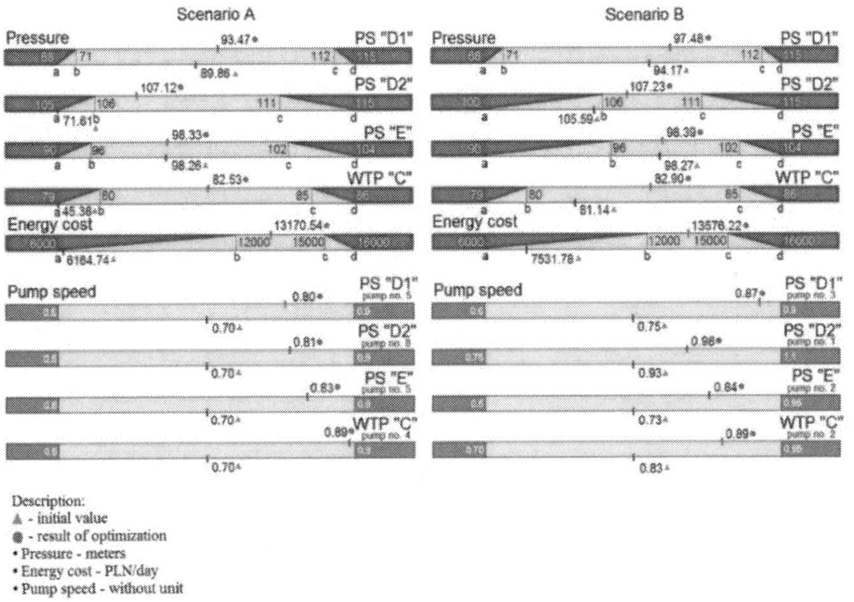

Fig. 3. Optimization results for Scenario A and B.

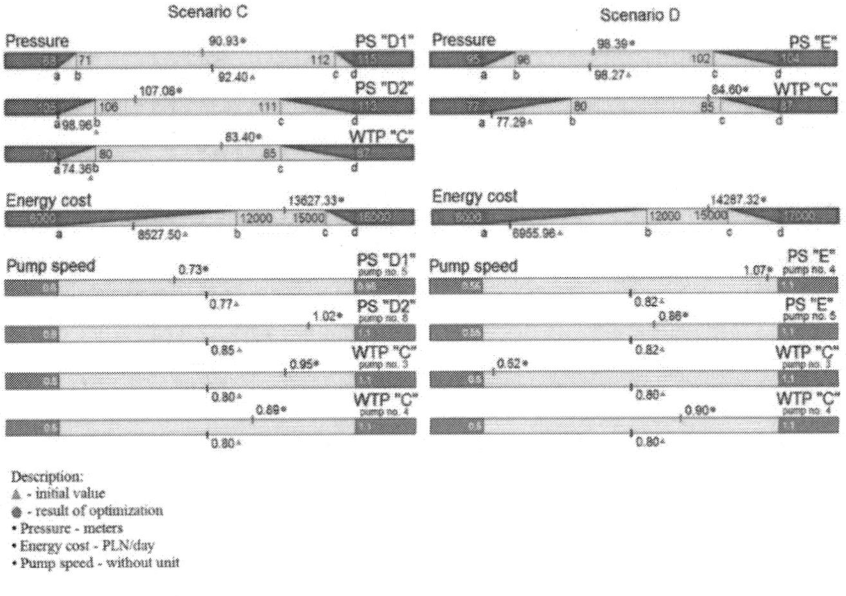

Fig. 4. Optimization results for Scenario C and D.

4 Conclusions

The paper presents optimization of energy consumption in the pumping stations using hydraulic model of WSS. Optimization options used in calculation software are based on stochastic methods with elements of fuzzy sets. The purpose of the study was to find the lowest pump speed for the desired range of pressure and energy cost. The methodology has been tested on a real water distribution network. In this study there were four scenarios conducted to find the most advantageous network operation. All simulations were run for a

DOI: 10.1051/e3sconf/20172200204

given time horizon (72 hours). For each scenario, the following energy costs and costs reduction were achieved:
– scenario A: 13 170 PLN/day – 22.0% reduction
– scenario B: 13 576 PLN/day – 26.4% reduction
– scenario C: 13 627 PLN/day – 17.8% reduction
–scenario D: 14 287 PLN/day – 6.4% reduction

The study was focused on the pump speed control, which has a strong influence on the energy consumption of the pump systems and it is easy to determine with the model. The average reduction in energy cost is about 18%, the highest cost reduction (but not the lowest energy cost) was achieved in scenario B, which presents operation of old pumps, indicating that earlier pump systems had lower efficiency and were oversized. The energy cost difference between scenario A and B, before optimization, is about 8%, which shows that replacing pumps with new ones gives good results, but not the best. The similar effects can be achieved by using other elements of system improvements, such as the use of pumps as turbines to generate energy.

The paper presented has been realized in frame of the research project no. BK-228/RIE-4/2016 co-financed by Ministry of Science and Higher Education.

References

1. M.R.N. Vilanova, J.A.P. Balestieri, Renewable and Sustainable Energy Reviews **30**, 701–714 (2014)
2. B. Coelhon, A. Andrade-Campos, Renewable and Sustainable Energy Reviews **30**, 59–84 (2014)
3. C. Copeland, N.T. Carter, Energy, Congressional Research Service, **2–10** (2017)
4. S. Pabi, A. Amarnath, R. Goldstein, L. Reekie, Electric Power Research Institute, **1–194** (2013)
5. H.M. Ramos, F. Vieira, D.I.C. Covas, Water Science and Engineering **3**, 331–340 (2010)
6. Y. Shen, S. Kiatsupaibul, Z.B. Zabinsky, R.L. Smith, Journal of Global Optimization **38**, 333–365 (2007)
7. A. Ahrari, A.A. Atai, Applied Soft Computing **10**, 1132–1140 (2010)
8. D. Karaboga, B. Basturk, Journal of Global Optimization **39**, 459–471 (2007)
9. R.L. Yang, Journal of Optimization Theory and Applications **104**, 691–716 (2000)
10. D. Bulger, W.P. Baritompa, G.R. Wood, Journal of Optimization Theory and Applications **116**, 517–529 (2003)
11. H.M. Ramos, A. Borga, M. Simao, Water Science and Engineering **2**, 69–84 (2009)
12. J.A. Elias-Maxil, J.P. van der Hoek, J. Hofman, L. Rietveld, Renewable and Sustainable Energy Reviews **30**, 808–820 (2014)
13. Z.B. Zabinsky, Wiley Encyclopedia of Operations Research and Management Science, **1–16** (2009)
14. H.J. Zimmermann, John Wiley & Sons, Inc. **2**, 317–332 (2010)

 DOI: 10.1051/e3sconf/20172200205

A review of photovoltaic cells cooling techniques

Swar A. Zubeer[1,*], *H.A.* Mohammed[1], and *Mustafa* Ilkan[2]

[1] Department of Energy Engineering, Technical College of Engineering, Duhok Polytechnic University (DPU), 61 Zakho Road, 1006 Mazi Qr, Duhok-Kurdistan Region, Iraq
[2] School of Computing and Technology, Eastern Mediterranean University, Famagusta North Cyprus

Abstract. This paper highlights different cooling techniques to reduce the operating temperature of the PV cells. This review paper focuses on the improvement of the performance of the small domestic use PV systems by keeping the temperature of the cells as low as possible and uniform. Different cooling techniques have been investigated experimentally and numerically the impact of the operating temperature of the cells on the electrical and thermal performance of the PV systems. The advantages and disadvantages of ribbed wall heat sink cooling, array air duct cooling installed beneath the PV panel, water spray cooling technique and back surface water cooling are examined in this paper to identify their effective impact on the PV panel performance. It was identified that the water spray cooling system has a proper impact on the PV panel performance. So the water cooling is one way to enhance the electrical efficiency of the PV panel.

1 Introduction

The renewable energy use becomes more popular during the increase of human population and the environmental issues. The solar energy is one of the important type of the renewable energy sources that has attracted many researchers around the world to work on. There are two types of energy that can be produced from the solar energy: electrical energy and thermal energy. The electrical energy can be produced by using photovoltaic (PV) cells. The PV cell directly converts the incident solar irradiance to electricity. The most efficient, sustainable, and eco-friendly systems are the PV modules which convert small part of the solar irradiance to electricity. The remaining part of the solar irradiation then converts into heat, which increases the temperature of the cells and reduces the performance of the PV module [1]. The maximum expected PV cells temperature with solar irradiance of 1000 W/m^2, 70% of absorption rate and no-winds is 60°C while for the winds speed of higher than 4 m/s the PV cells temperature is lower than 40°C [2]. The maximum output power, open circuit voltage, and short circuit current are the main parameters which are influenced by the temperature variation of the cells temperature. Thus, the open circuit voltage and the maximum output power reduce with the temperature increase whereas the short circuit current rises [3]. Chander et al. [4] investigated experimentally the impact of the temperature of the cell on the performance of mono-crystalline silicon photovoltaic. A remarkable effect on the performance of the photovoltaic was presented by the cell

* Corresponding author: szubeer@yahoo.com

DOI: 10.1051/e3sconf/20172200205

temperature. They noticed that the temperature coefficient was negative for the fill factor, open circuit voltage and maximum output power whilst was positive for the short circuit current. Ike [5] reported that the output power of the photovoltaic and the ambient temperature of the site are indirectly proportional. Therefore, the output power produced by the photovoltaic is higher for the low ambient temperature interval than the high ambient temperature interval. Ahmad et al. [6] studied experimentally the impact of the temperature on a poly-crystalline solar panel at hot climate of Malaysia. It was noticed that the temperature increase led to decrease the efficiency and the maximum power. Zaoui et al. [7] investigated experimentally and numerically the impact of the cell temperature on the performance of the photovoltaic panel with constant irradiance value. Results indicated that the efficiency and output power were reduced with the raised temperature of the cell. The decrease in efficiency is caused by the temperature difference and therefore the location of the PV panel is an important factor to be considered. It is shown from the above literature review that the cell temperature has a significant impact on the performance of the small domestic use photovoltaic cells. Therefore, the present study focuses on reviewing the two types of cooling techniques (passive cooling and active cooling) to remove the heat transfer and enhance the performance of the PV cells.

2 Air cooling

2.1 Heat sink

Heat sink is one of the cooling ways which uses a high thermal conductivity metal to remove the heat from the photovoltaic cell. Popovici et al. [8] investigated numerically the temperature reduction of the PV panels during a clear day of summer by using different arrangements of ribbed wall heat sink of air and passive cooling. It was found that the maximum temperature of the panel for the angle 45° was less than that for the angle 135°. The study found that the maximum power produced by PV panel in case of using heat sink was increased by 6.97% and 7.55% compared to the reference case, for angles of the ribs from 90° and 45°, respectively. Farhana et al. [9] studied experimentally the operating temperature variation for the PV module with and without active cooling system to realize the electrical performance of the PV module.

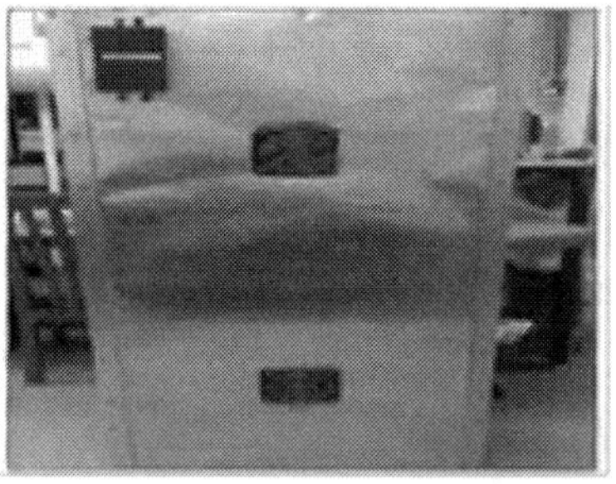

Fig. 1. Back side of PV module.

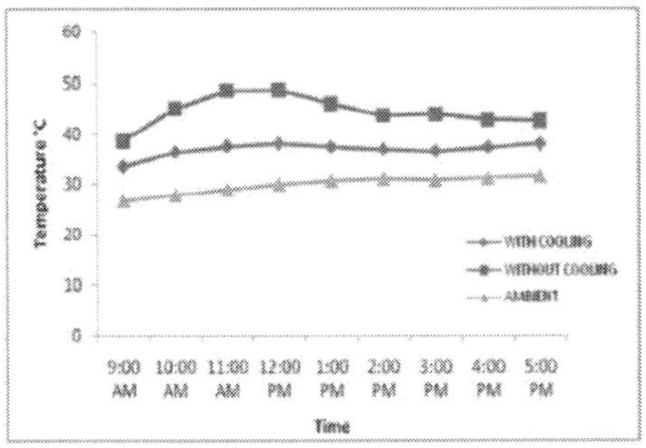

Fig. 2. Comparison between ambient temperature and average module temperature under cooling and without cooling condition.

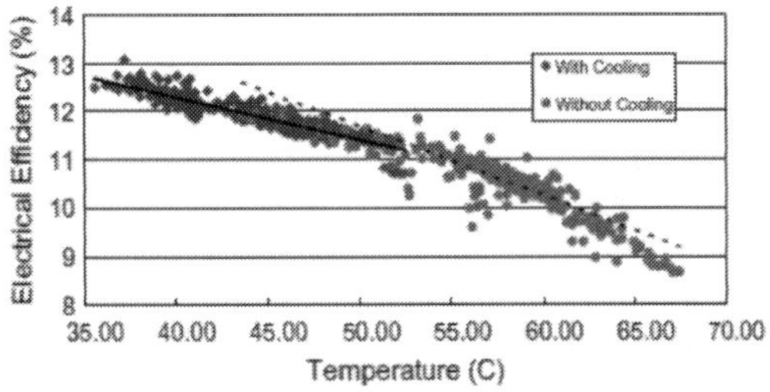

Fig. 3. Electrical efficiency as a function of PV temperature.

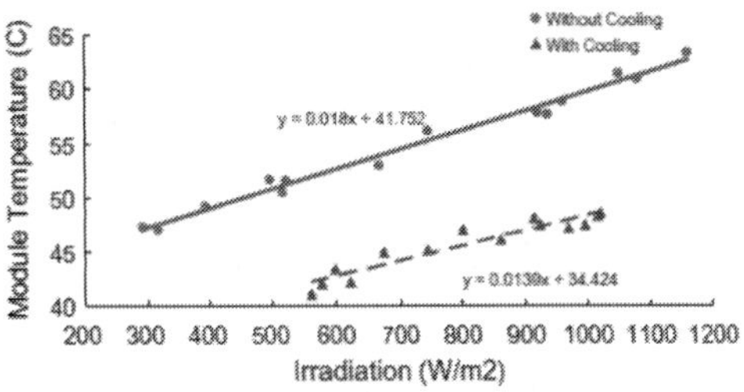

Fig. 4. Module temperature as a function of solar irradiation.

Two multi-crystal silicon solar modules with 13% of peak efficiency at standard condition (25°C, 1000 W/m^2) were used in the test. One of the modules was used as a reference and the other an aluminium was installed at the bottom of the PV panel as a heat sink and a DC brushless fan was installed on the heat sink as shown in Fig. 1. The temperature for the PV module with and without cooling system was higher than ambient temperature by 30% and 70%, respectively as shown in Fig. 2. Hence, the open circuit voltage (V_{oc}) of the PV module with cooling system was little higher than PV module without cooling system.

2.2 Air channels

Several studies investigated the performance of the PV cells with active cooling by using air channels connected to the back of the PV panel. Teo et al. [10] concentrated on the comparison of the PV module electrical efficiency with and without active cooling. The study investigated experimentally and numerically the influence of the operating temperature on the efficiency of the hybrid photovoltaic/thermal solar system. It was noticed that the electrical efficiency was decreased when the cells operating temperature increased for both cooling and non-cooling cases, but for the cooling case the electrical efficiency was higher as shown in Fig. 3. In addition, the experiments found that there was a linear proportional relation between the PV panel temperature and the irradiation as displaced in Fig. 4. Tonui et al. [11] investigated experimentally the performance of PV/T solar collector using forced or natural air circulation to extract heat. The air channel was modified by two different ways to boost the heat transfer from the channel walls to airflow. The first one was inserting a thin flat metallic sheet at the middle of the channel (TMS system), and the second one was joining rectangular fins at the back of the channel (FIN system) as shown in Fig. 5. The study achieved that the modified PVT/Air systems would participate considerably in enhancing the performance of larger applications of PV systems. Ameri et al. [12] investigated experimentally the performance of photovoltaic/thermal air collector. The panels were installed on the channel of air and on the top of a thin metal (aluminium) sheet (TMS) as shown in Fig. 6. The study indicated that the electrical efficiency of the system had a direct relationship with the solar radiation intensity, PV cells temperature, and the power rate consumed by fans. Therefore, the electrical efficiency in case of forced convection was not always increased with number of fans, but there were an optimum number of fans for high electrical efficiency as presented in Fig. 7.

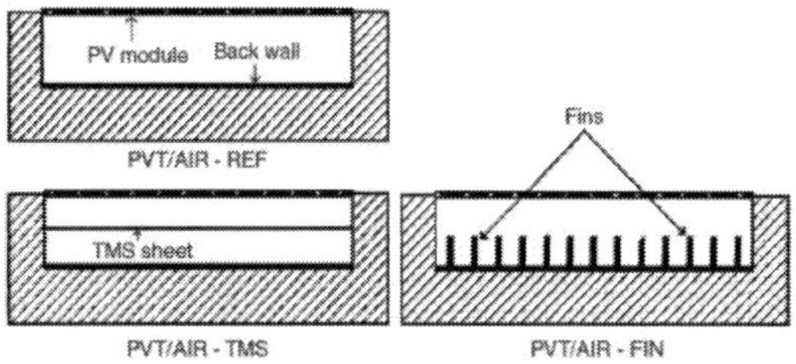

Fig. 5. Cross-sectional view of PVT/AIR collector models. Flow direction is perpendicular to the page.

DOI: 10.1051/e3sconf/20172200205

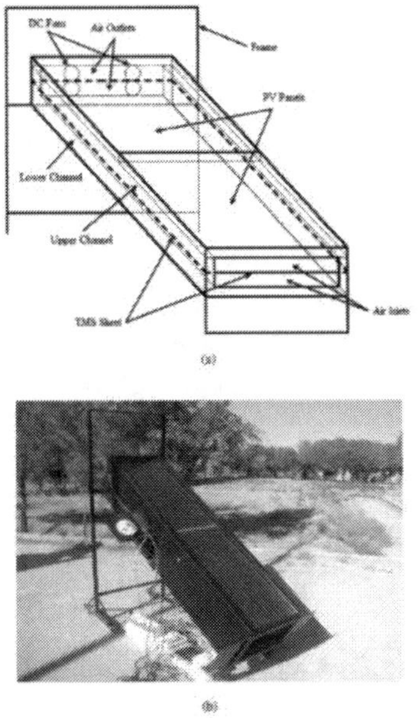

Fig. 6. (a) Schematic diagram of studied PV/T air system. (b) Photograph of experimental setup of the studied PV/T air system.

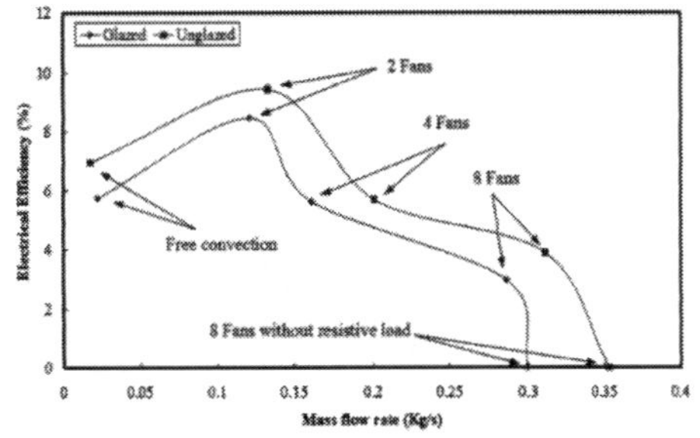

Fig. 7. Effect of air mass flow rate on electrical efficiency of the studied system in similar environmental conditions.

3 Water cooling

3.1 Water spray

Several studies have investigated experimentally the performance of the PV cells with active cooling water. Nizetic et al. [13] investigated experimentally the impact of water spray cooling on the performance of the PV panel in highest solar irradiation level environment. Both sides of the PV panel were cooled at the same time by utilizing twenty nozzles, ten on each side as shown in Fig. 8. The results were measured for three different cases of cooling: front side cooling, rear side cooling and both sides together and compared with non-cooling case. The research indicated that the water spray cooling has achieved a suitable effect on the PV panel performance and the best case was the simultaneous front and back sides cooling PV panel. Lastly, depending on the experimental results, as presented in Table 1, the water spray cooling system had a proper impact on the PV panel performance. Abdolzadeh et al. [14] studied experimentally the impact of the water spray

DOI: 10.1051/e3sconf/20172200205

cooling on the performance of photovoltaic water pumping. The configuration with two modules and 25 lit/h/module water spray was called case 'A' and the configuration with three modules and 5 lit/h/module and 25 lit/h/module water spray were called case 'B_1' and case 'B_2', respectively were used in the test. In case A and B_1 the module temperature was decreased, and the reduction in case A was higher than case B_1 as shown in Fig. 9. The experimental results indicated that the system performance was significantly improved by spraying water on the PV module. Irwan et al. [15] studied experimentally the performance of the PV panel by using water cooling method. Indoor test was carried out by a solar simulator consisted of twenty 500 W halogen lamps. Two units of 50W Monocrystalline PV panel were used in the test. A DC water pump was used to spray water was connected to the front surface of one of the panel and the other panel was used as a base panel. It was observed from the experimental results that the operating temperature of the PV panel with water cooling system was reduced by 5–23°C and the power output was increased by 9–22%. So the water cooling is one way to enhance the electrical efficiency of the PV panel as shown in Figs. 10–11.

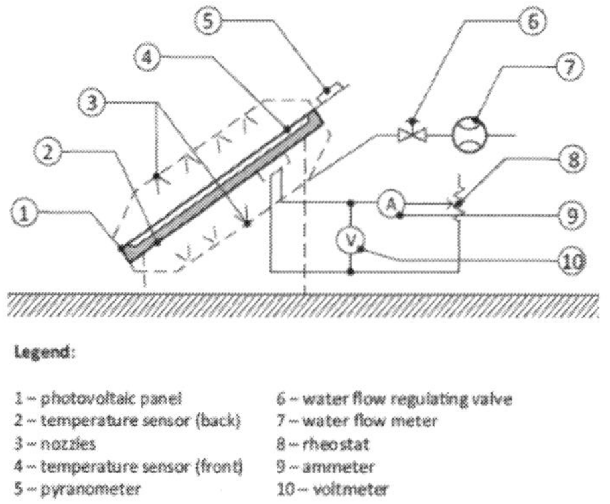

Fig. 8. Schematic layout of the specific experimental setup.

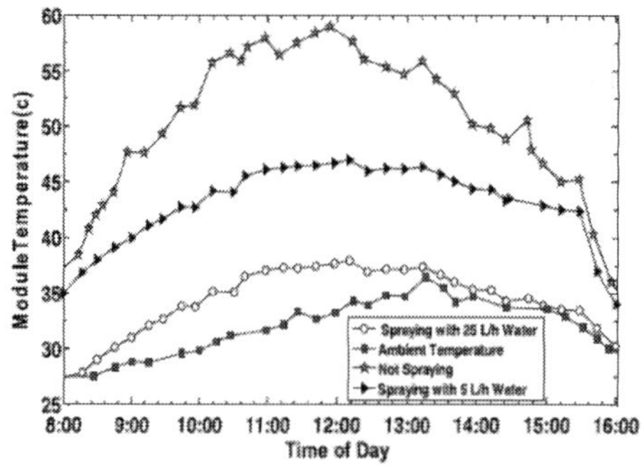

Fig. 9. Effect of water spray on the module temperature.

 DOI: 10.1051/e3sconf/20172200205

Table 1. PV panel mean performance parameters for different examined cooling circumstances.

Applied cooling options	Maximal power output (W)	Relative increase in power output (%)	Effective increase in power output (%)	Average panel temperature (°C)	Electrical efficiency (%)	Effective increase in el. efficiency (%)
Without cooling	35	–	–	56	13.92	–
Back surface cooling	39.9	14.0	5.4↑	33.7	15.59	3.6↑
Front surface cooling	40.1	14.6	6.0↑	29.6	15.42	2.5↑
Simultaneous cooling	40.7	16.3	7.7↑	24.1	15.92	5.9↑

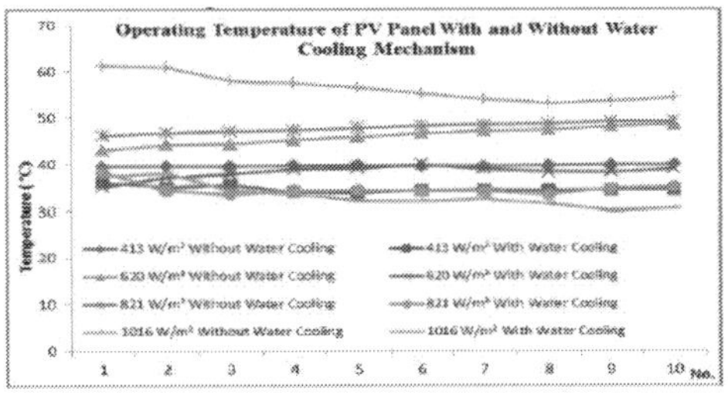

Fig. 10. Operating temperature of PV panel with and without water cooling mechanism.

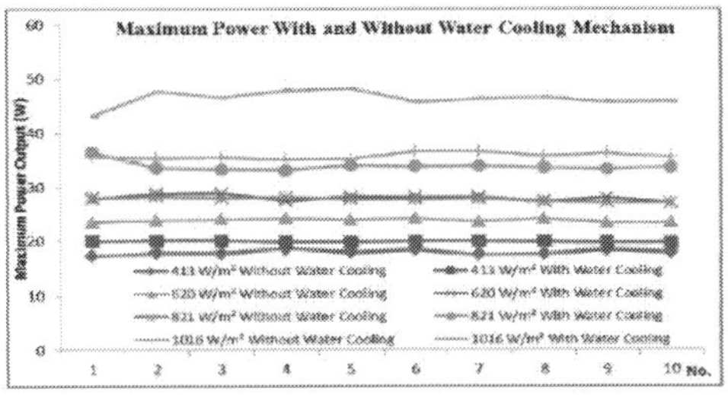

Fig. 11. Maximum power output of PV panel with and without water cooling mechanism.

3.2 Heat exchanger

Several studies have investigated numerically and experimentally the performance of the PV cells using active cooling water with the aid of heat exchanger. Hussien et al. [16] implemented an experimental investigation of improving the electrical efficiency of photovoltaic /thermal system by using water cooling technique. The cooling mechanism was contained of heat exchanger and seven pipes of water attached to the back of the PV panel. The electrical efficiency of the PV panel was improved more in the case of 0.3 L/s water flow rate compared to the other cases of water flow rate as shown in Fig. 12. Bahaidarah. [17] compared the performance of the PV panel with rectangular heat exchanger cooling (RHX) attached to the back with the PV panel without cooling. The maximum cell efficiency for the rectangular heat exchanger cooling was 13.07%, while for uncooled PV cells was 7.82% as shown in Fig. 13. Bahaidarah et al. [18] studied experimentally and numerically the performance of a hybrid PV water cooled system. A heat exchanger was connected to the back of the PV cells to enhance the performance of the PV panel for the climate of Dhahran, Saudi Arabia. The hybrid PV-water cooled system was consisted of a 230 W mono-crystalline type, cooling panel (heat exchanger) was

 DOI: 10.1051/e3sconf/20172200205

connected to the back side of the PV module, and insulated tank to store the cooling water. It was found that the operating temperature of the water cooled PV panel was reduced remarkably to about 20% as shown in Fig. 14 and the electrical efficiency increase was 9%. In addition, the energy collection of the water cooled PV system was almost 4 times more than the PV only system.

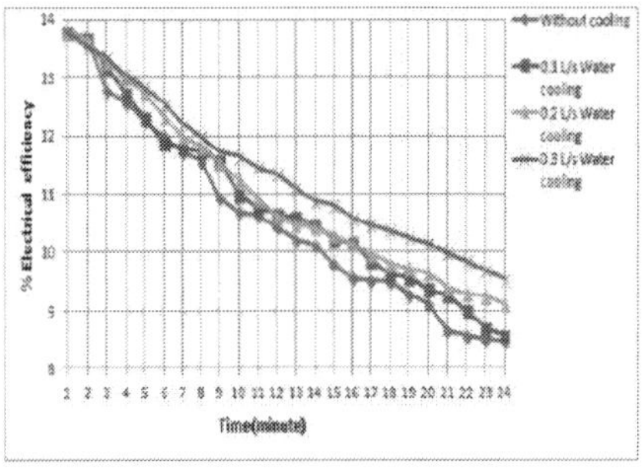

Fig. 12. Effects of water mass flow rate on electrical efficiency of the PV panel.

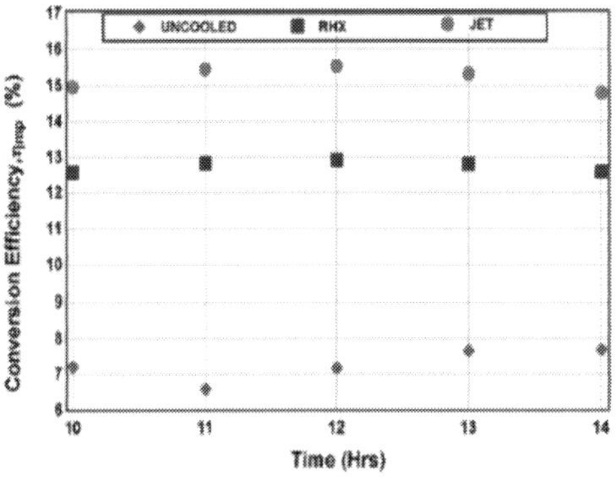

Fig. 13. Conversion Efficiency variation for PV with jet cooling, RHX cooling and an uncooled panel.

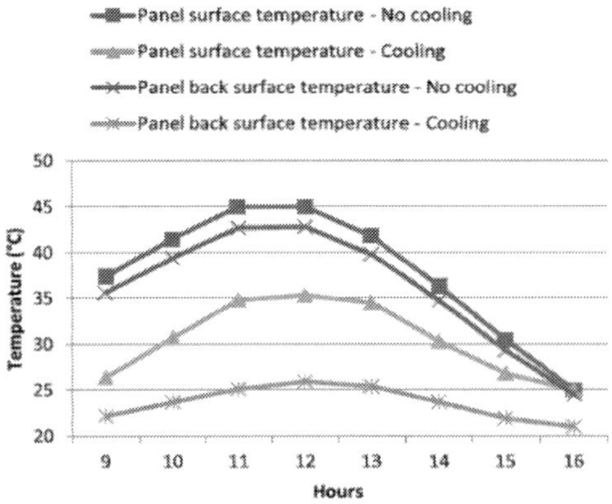

Fig. 14. Conversion Efficiency variation for PV with jet cooling, RHX cooling and an uncooled panel.

 DOI: 10.1051/e3sconf/20172200205

4 Fins cooling

Several studies have investigated numerically and experimentally the performance of the PV cells using different types and shapes of fins. Chandrasekar et al. [19] used aluminium fins combined with cotton wick as a passive cooling system to maintain the temperature of the PV panel. The cooling system was consisted of three aluminium fins (630 × 100 × 60 mm) with cotton wick attached to the back side of the crystalline silicon PV cells as shown in Fig. 15.

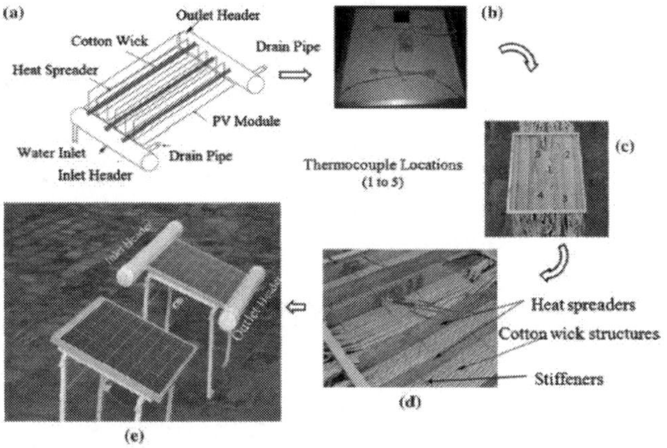

Fig. 15. Photographic view of the experimental PV module with stages of fabrication. (a) Desired rear side of PV module (b) location of thermocouples (c) fins in conjunction with wick structures (d) details of the stiffeners and (e) final fabricated experimental setup with headers.

It was observed from the experimental results that the maximum temperature of the PV panel was decreased 12% by using the cooling system and the output power was increased by 14%. Nehari et al. [20] numerically investigated the proper length of the fins to identify the enhancement of the PV panel power during the passive cooling by phase change material (PCM). It was noted that the fins was remarkably decreased the temperature of the PV cells compared to the case without fins. In addition, the fins length of 25, 30, and 35 mm provide a preferable PV cell cooling. Gotmare et al [21] studied experimentally the performance enhancement of PV cell by using fins cooling under natural convection. Two 37 W PV panel were used in the test and 9 fins of aluminium were attached to the back of the one panel. The experimental results indicated that the cell temperature for the PV panel with fins cooling was decreased by 4.2% compared with the panel without fins and the average output power was enhanced by 5.5% in case of PV panel with fins as shown in Figs. 16–17.

 DOI: 10.1051/e3sconf/20172200205

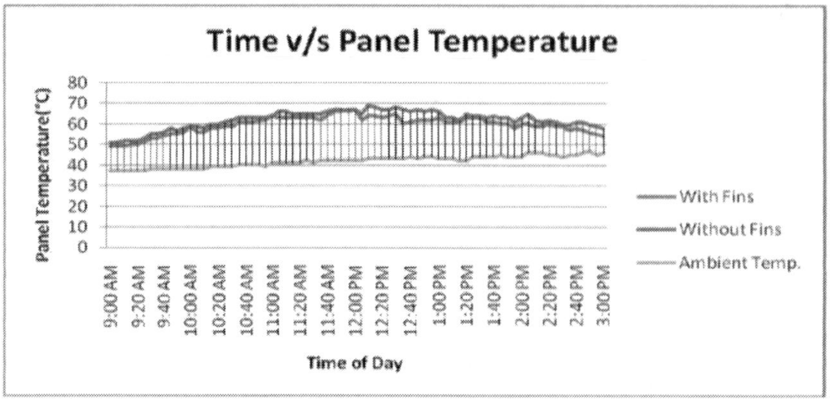

Fig. 16. Comparison of PV panel temperatures with & without fin cooling.

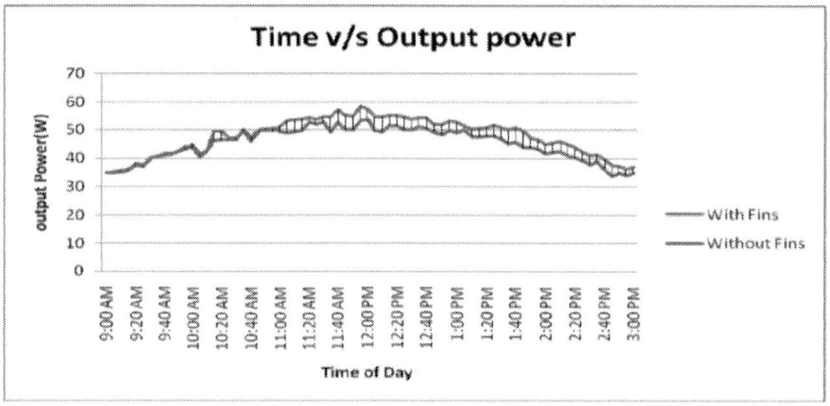

Fig. 17. Comparison of output power with & without fin cooling.

5 Conclusions

The present article has highlighted different techniques of cooling for small domestic use photovoltaic panel. It is concluded that:

- The temperature of the PV cell decreases about 12° C by using heat sink with air cooling.
- In case of air cooling the electrical efficiency of the system does not always increase with increasing the mass flow rate of the air but there is an optimum value of mass flow rate.
- Water spray cooling has a considerable effect on the performance of the PV cell, even for the low flow rate of the water spray the performance of the system enhances remarkably.

- Water cooling has the most impact on the reduction of the operating temperature of the PV cell and improve the electrical performance of the PV panel.
- Cooling system with fins is efficient to reduce the temperature of the PV panel and enhance the electrical efficiency of the PV panel.

References

1. M.M. Rahman, M. Hasanuzzaman, N.A Rahim, Energ. Convers. Manage. **103** (2015)
2. Y. Du, C.J. Fell, B. Duck, D. Chen, K. Liffman, Y. Zhang, M. Gu, Y. Zhu, Energ. Convers Manage. **108** (2016)
3. N.H. Zaini, M.Z. Ab Kadir, M. Izadi, N.I. Ahmad, M.A. M Radzi, N. Aziz, IEEE, (2015)
4. S. Chander, A. Purohit, A. Sharma, S.P. Arvind, M.S. Nehra, M.S. Dhaka, Energ. Reports, **1** (2015)
5. C.U. Ike, Int. J. Eng. Sci, **3** (2013)
6. N.H. Zaini, M.Z. Ab Kadir, M. Izadi, N.I. Ahmad, M.A.M Radzi, N. Aziz, IEEE, (2015)
7. F. Zaoui, A. Titaouine, M. Becherif, M. Emziane, A. Aboubou, Energ. Procedia, **75** (2015)
8. C.G. Popovici, S.V. Hudişteanu, T.D. Mateescu, N.C. Chereche*ş*, Energ. Procedia, **85** (2016)
9. Z. Farhana, Y.M. Irwan, R.M.N. Azimmi, A.R.N. Razliana, N. Gomesh, IEEE, (2012)
10. H. G. Teo, P.S. Lee, M.N.A. Hawlader, Appl. Energ. **90** (2012)
11. J. K. Tonui, Y. Tripanagnostopoulos, Renew. Energ, **32** (2007)
12. M. Ameri, M.M. Mahmoudabadi, A. Shahsavar, Energ. Source, (2012)
13. S. Nizetic, D. Coko, A. Yadav, F. Grubiši-Cabo, Energ. Convers. Manage, **108** (2016)
14. M. Abdolzadeh, M. Ameri, M.A. Mehrabian, Energ. Source, (2011)
15. Y.M. Irwan, W.Z. Leow, M. Irwanto, M. Fareq, A.R. Ameliaa, N. Gomesha, I. Safwati, Energ. Procedia, **79** (2015)
16. H.A. Hussien, A.H. Numan, A.R. Abdulmunem, Mater. Sci. Eng. **78** (2015)
17. H.M.S. Bahaidarah, IEEE, (2015)
18. H. Bahaidarah, A. Subhan, P. Gandhidasan, S. Rehman, Energ. **59** (2013)
19. M. Chandrasekar, T. Senthilkumar, Heat Mass Transfer, (2016)
20. T. Nehari, M. Benlakam, D. Nehari, **60** (2016)
21. J.A. Gotmare, D.S. Borkar, P.R. Hatwar, Int. J. Adv. Manuf. Tech, (2015)

Characteristics of large thermal energy storage systems in Poland

Ryszard Zwierzchowski[1,*]

[1]Department of Heating and Gas Engineering, Faculty of Building Services, Hydro and Environmental Engineering, Warsaw University of Technology, ul. Nowowiejska 20, 00-653 Warsaw, Poland

Abstract. In District Heating Systems (DHS) there are significant fluctuations in demand for heat by consumers during both the heating and the summer seasons. These variations are considered primarily in the 24-hour time horizon. These problems are aggravated further if the DHS is supplied by a CHP plant, because fluctuations in heat demand adversely affect to a significant degree the stable production of electricity at high overall efficiency. Therefore, introducing Thermal Energy Storage (TES) would be highly recommended on these grounds alone. The characteristics of Large (i.e. over 10 000 m^3) TES in operation in Poland are presented. Information is given regarding new projects (currently in design or construction) that apply TES technology in DHS in Poland. The paper looks at the methodology used in Poland to select the TES system for a particular DHS, i.e., procedure for calculating capacity of the TES tank and the system to prevent water stored in the tank from absorbing oxygen from atmospheric air. Implementation of TES in DHS is treated as a recommended technology in the Polish District Heating sector. This technology offers great opportunities to improve the operating conditions of DHS, cutting energy production costs and emissions of pollutants to the atmosphere.

1 Introduction

In District Heating Systems (DHS) there are significant fluctuations in consumer demand for heat during both the heating and the summer seasons. These variations are considered primarily in the 24-hour time horizon. During the heating season the fluctuations are associated with rapid changes in meteorological conditions, such as outside air temperature, solar radiation and wind speed. During the summer season the fluctuations are associated mainly with variable demand for domestic hot water. These variations in the consumer demand for heat cause significant problems in operational terms, forcing frequent changes to boilers' heat power in heat sources at both Heat Only Boilers (HOB) and Combined Heat and Power (CHP) plants, which in turn lead to a reduction in the efficiency of heat and power generation and in the security of heat delivery to consumers.

* Corresponding author: ryszard.zwierzchowski@is.pw.edu.pl

These problems are aggravated further if the DHS is supplied by a CHP plant, because fluctuations in heat demand adversely affect to a significant degree the stable production of electricity at high overall efficiency. Introducing additional thermal capacity to the DHS in the form of Thermal Energy Storage (TES) would be highly recommended on these grounds alone. The EU shares this approach, as the Energy Efficiency Directive (2012/27/EU) supports the broad application of TES in DHS.

Use of TES is essential both in the process of upgrading existing CHP or HOB plants as well as in the construction of new plants. When planning the construction of new plants, the use of TES can significantly reduce their size (heating capacity) and structure, i.e., decrease the number of or eliminate peak boilers. These activities will contribute to a reduction in the investment costs required for heat generating units. However, the greatest economic and environmental effects relate to the current operation of energy sources supplying DHS, and with properly selected and designed TES systems.

2 Accumulation of heat in water, thermally stratified TES

Heat storage in water, thermally stratified TES is one of the easiest ways to collect thermal energy [1–4]. The TES is a tank that stores thermal energy in the form of hot water. From a thermodynamic point of view, water is one of the best substances to store energy in the form of sensible heat. The heat capacity of water (specific heat) is approximately four times higher than air and most inorganic materials, as well as from four to ten times higher than most metals.

2.1 The principle of operation of thermally stratified TES

In TES hot water collects in the top part of the tank and is separated by a relatively narrow thermocline from cold water at the bottom of the TES tank, Fig. 1. The height of this type of tank can be as much as tens of meters, and their capacity reaches tens of thousands of cubic meters. When charging the TES, hot water from boilers (or heat exchangers) pushes the cold water towards the bottom of the tank, which then flows into the boilers. When discharging the TES, return water from DHS called "cold water", displaces the hot water towards the top of the tank and then to the supply pipeline of the DHS. The water level in the TES tank remains practically unchanged, only the amount of hot and cold water in the tank changes, as does the location of the thermocline. In order to avoid oxygen penetrating from the ambient air to the hot water stored in the TES tank, a steam bed is applied above the water in the tank.

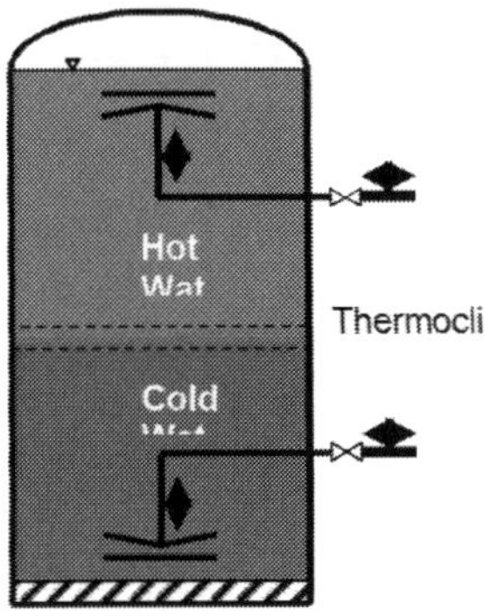

Fig. 1. Principle of Operation of the water, thermally stratified TES.

 DOI: 10.1051/e3sconf/20172200206

2.2 Breakdown of the water, thermally stratified TES

The water is divided in thermally stratified TES due to various criteria, as presented in [15]. So, due to operation pressure, TES systems can be divided into:

- Pressurized TES,
- Non-pressurized TES,
  - TES included in thermal system directly,
  - TES included in thermal system indirectly (through heat exchangers).

In terms of size, TES tanks can be divided into:

- TES – large, with a capacity of more than 10,000 m^3,
- TES – medium, with a capacity of 1,000 to 10,000 m^3,
- TES – small, with a capacity of less than 1,000 m^3.

2.3 Basic tasks of TES

The use of TES in DHS increases operational efficiency, production and delivery of heat to the end user, and thus its cost [1–4]. Hence, programs to modernize DHS should take into account the possibility of accumulating heat in the system through the use of TES. The basic task of the TES system is heat source load leveling when heat demand by consumers is variable [5]. Other main roles are listed below:

- to increase production in cogeneration mode in the CHP plant,
- to produce more electricity in the CHP plant during periods of peak demand,
- to cover peak heat demand,
- to increase the operational efficiency of the energy generating units through stable performance with balanced heat input,
- to reduce fuel consumption through stable operation with a balanced load,
- to reduce pollutant emissions through lower fuel consumption and stable operation with a balanced load,
- to keep maintenance costs low through a low failure rate achieved by stable and continuous operation of the generating units,
- to avoid temporary run of peak boilers to meet changing consumer heat demand,
- to reduce pollutant emissions through avoiding running peak boilers,
- to reduce water losses, as TES plays the role of expansion vessel for DHS.

3 Selection of TES for the District Heating System

To select the best TES system for a particular DHS, it is necessary to collect and analyze operational data for the heat source (CHP or HOB) and for the entire DHS (temperature, pressure, water flows, weather conditions, etc.). The analysis should cover at least the last 2–3 years, to establish the trends and to confirm the existing forecasts regarding future operation of the DHS

The required capacity of the TES tank for full heat accumulation in a 24-hour scale was estimated. The estimation was performed on the basis of the hourly variation in heat demand by the DHS over the period of analysis [1, 5]. A TES system with full heat accumulation is one that is able to compensate for fluctuations in the heat demand of the DHS, while simultaneously providing the operation of energy generating units with constant, stable power.

On the other hand, the calculated volume of the TES tank with full heat accumulation ($V_{TES(cal)}$) should take into account the share of the non-useful space of the tank and its efficiency.

A variant of the TES tank with full accumulation is the basic operational variant. Other variants that should be considered depend on local and national circumstances. For example, for CHP Siekierki beyond the basic variant, the considered variants have taken into account the consequences of participation in the balancing electricity market and sales of electricity with a variable price per day. In turn, for example, in the Nordic countries a variant is often considered that is associated with shutdown of the CHP plant for the weekend. Sometimes, in addition, location-connection variants are considered. They depend on the location of the TES and how it hooked up to the thermal system of the CHP plant. These variants differ mainly in investment and operational costs. The final size of the TES tank verifies the classic economic analysis carried out for each operational and possibly location-connection variants e.g. for three sizes of TES tank [6]:

1) $V_{TES1} = V_{TES(cal)}$,
2) $V_{TES2} = 0{,}8\ V_{TES(cal)}$,
3) $V_{TES3} = 0{,}6\ V_{TES(cal)}$.

Economic analysis includes investment costs of constructing TES tanks of different sizes, taking into account the economic effects that can be obtained in individual, analyzed variants.

4 Characteristics of large TES systems operated in Poland

Large TES systems in DHS are a recent development in Poland. The first large TES system was built in Poland in the CHP Siekierki plant in Warsaw and has been operational since 2009, another four went operational in the period 2011–2015: CHP Cracow, CHP Bialystok, CHP Bielsko Biala [6], CHP Ostroleka, and a sixth is currently in the start-up phase: CHP Torun.

The largest TES in Poland is at CHP Siekierki. It has a water capacity of 30,400 m^3 and a thermal capacity of approx. 1,600 MWh. It is currently in its seventh full heating season. The highest TES tank was built for CHP Bielsko Biala, and its total height including the top of the roof is approximately 64.5 m. The big tank height of the TES is due to the significant elevation variation of the District Heating Network (DHN) in the city.

The first TES system with a capacity of approx. 750 m^3 (thermal capacity approx. 26 MWh), has been in operation since 2002 in PEC in Siedlce [5].

The technical characteristics of large TES systems built in Poland are presented in brief below.

4.1 TES in Siekierki CHP plant in Warsaw

Siekierki CHP plant in Warsaw is a second largest CHP plant in Europe and supply with heat the biggest DHS in EU with pipelines length over 1,800 km and over 19,000 facilities (heat consumers). The DHS covers around 80% of the city heat demand.

Below, short technical characteristic of the TES in Siekierki CHP plant is presented:

1. The design of the TES was completed in 2007 [6]; the TES after a start-up period has been operational since March 2009.
2. Technical parameters of the TES:

• type	full of water, non-pressurized
• volume	30,400 m^3
• diameter	30.0 m
• height of the tank cylindrical part	43.0 m
• radius of the roof sphere	45.0 m

• thickness of insulation	500 mm
• flowrate of charging/ discharging	4,500 Mg/h
• temperature of the network water	98/40°C
• protection against oxygen absorption to water	steam bed
• weight of the tank (without insulation)	c. 53,000 kg

Fig. 2. TES in Siekierki CHP plant.

Fig. 3. Upper orifice in the tank of TES.

Due to application of the TES in Siekierki CHP plant, only in 2009 year i.e. in the first year of the TES operation, overall efficiency of energy generation (heat and electricity) by the plant has increased from 86.6% to 87.3% [6]. Also, temporary run of peak water boilers was reduced by nearly 70%, production in cogeneration mode in the CHP plant was increased significantly and reduction of pollutant emissions i.e. dust, SO_2, NO_x to the atmosphere was observed [6].

From 2009 up to now, the TES system in CHP Siekierki is in continuous operation, both in heating and summer seasons and compensate variable heat demand by consumers in 24-hours mode.

4.2 TES in Cracow CHP plant

The CHP plant in Cracow supply with heat the city with population over 700,000. The DHS consists of pipelines with length over 830 km and 8,500 heat consumers (facilities).

Below, short technical characteristic of the TES in Cracow CHP plant is presented:

1. The TES was designed in 2009 [6]; has been operational since 2011.
2. Technical parameters of the TES:

• type	full of water, non-pressurized
• volume	20,015 m^3
• diameter	23.0 m
• height of the tank cylindrical part	48.2 m
• radius of the roof sphere	34.5 m

 DOI: 10.1051/e3sconf/20172200206

- thickness of insulation 300 mm
- flowrate of charging/ discharging — 2,600 Mg/h
- temperature of the network water — 98/40°C
- protection against oxygen absorption to water — steam bed
- weight of the tank — c. 38,000 kg

Fig. 6. TES in Cracow CHP plant.

4.3 TES in Bialystok CHP plant

The DHS in Bialystok consists of pipelines with length over 270 km and 2,200 heat distribution substations and supply with heat the city with population app. 300,000.

Below, short technical characteristic of the TES in Bialystok CHP plant is presented:

1. The TES was designed in 2009/10 [6]; has been operational since 2011.
2. Technical parameters of the TES:

- type — full of water, non-pressurized
- volume — 12,800 m^3
- diameter — 21.0 m
- height of the tank cylindrical part — 37.0 m
- radius of the roof sphere — 31.5 m
- thickness of insulation — 500 mm
- flowrate of charging/ discharging — 1,500 Mg/h
- temperature of the network water — 98/40°C
- protection against oxygen — steam

absorption to water bed
- weight of the tank (without of insulation): c. 30,000 kg

Fig. 7. TES in Bialystok CHP plant.

4.4 TES in Bielsko Biała CHP plant

The DHS in Bialystok consists of pipelines with length over 170 km and 900 heat distribution substations and supply with heat the city with population over 170,000.

1. The design of the TES was completed in 2011/12 [6]; the TES after a start-up period has been operational since autumn 2013.
2. Technical parameters of the TES:

- type: full of water, non-pressurized
- volume: 21,450 m^3
- diameter: 21.0 m
- height of the tank cylindrical part: 62.0 m
- radius of the roof sphere: 31.5 m
- thickness of insulation: 500 mm
- flowrate of charging/ discharging: 2,500 Mg/h
- temperature of the network water: 98/40°C
- protection against oxygen absorption to water: steam bed
- weight of the tank (without of insulation): c. 44,000 kg

Fig. 8. TES in Bielsko Biała CHP plant.

4.5 TES in Ostroleka power plant

1. The design of the TES was completed in 2013; the TES after a start-up period has been operational since 2014.
2. Technical parameters of the TES:

• type	full of water, non-pressurized
• volume	c.13,000 m^3
• diameter	22.0 m
• height of the tank cylindrical part	37.25 m
• thickness of insulation	500 mm
• flowrate of charging/ discharging	c. 1,500 Mg/h
• temperature of the network water	98/40°C
• protection against oxygen absorption to water	steam bed
• weight of the tank (without of insulation)	c. 35,000 kg

Fig. 9. TES in Ostroleka Power plant.

4.6 TES in Torun CHP plant

1. The design of the TES was completed in 2015/16; it is scheduled to go operational in Fall 2017.
2. Technical parameters of the TES:

• type	full of water, non-pressurized
• volume	13,200 m^3
• diameter	20.0 m
• height of the tank cylindrical part	42.0 m
• thickness of insulation	405 mm
• flowrate of charging/ discharging	c. 1,500 Mg/h
• temperature of the network water	98/45°C
• protection against oxygen absorption to water	steam bed
• weight of the tank (without of insulation)	c. 37,000 kg

Fig. 10. TES in Torun CHP plant.

5 Summary and conclusions

Technologies associated with the use of TES systems are still something of a novelty in the Polish DHS sector. They offer great opportunities to improve the operating conditions of DHS and the economics of the production of heat and electricity, reducing emissions of pollutants into the atmosphere and increasing security of energy supply to consumers. The application of TES in the DHS facilitates the efficient running of energy sources, i.e. it lets

 DOI: 10.1051/e3sconf/20172200206

energy generating units (boilers, turbomachinery) operate at high efficiency and reduces to a minimum the number of start-ups of peak boilers [5].

In the past four years the European Union has adopted two fundamental and harmonized Directives of the EU Parliament and the Council, which create a real basis for achieving the objectives of the EU's energy strategy and protection of the environment by the 2020, i.e.:

1. Directive of the European Parliament and of the Council of 2012/27/EC of 25.10.2012 on Energy Efficiency.
2. Directive of the European Parliament and of the Council 2010/75/EU of 24/11/2010 on Industrial Emissions (integrated pollution prevention and control).

Both Directives are extremely important for the Polish district heating sector, influence its development and modernization, efficiency and competitiveness, and meet environmental concerns. The obligations arising from them will be extremely difficult to fulfil in the Polish heating sector, particularly as regards emissions of gaseous pollutants, which pose a potent threat to the very existence of the sector. Nevertheless, there are relevant support programs in place and widespread information about potential threats [6].

Directive 2012/27/EU, which creates a common policy framework to reduce about 20% of primary energy consumption in the EU by 2020, cites measures to yield significant improvements in energy efficiency in district heating systems. This Directive has absorbed the existing Directive 2004/8/EC on the promotion of cogeneration based on useful heat demand in the internal energy market. Maximization of energy production in cogeneration will require the use of TES systems to improve the overall efficiency of energy production.

Current status of large TES systems i.e. with capacity over 10,000 m^3:

- 5 TES systems are in operation,
- 1 TES system is under commissioning and will go operational in Fall 2017,
- 2 TES systems are at the design stage,
- 2 TES systems are at the pre-design stage.

In addition to large TES systems, there are also a few medium (e.g. CHP Mielec) and small (e.g. CHP Siedlce) TES systems operating in Poland.

References

1. J.A. Dahl, L.R. Hermansson, *Proceedings of the 6th International Conference on Thermal Energy Storage – Calorstock'9., Espoo, Finland*, 403–410 (1994)
2. I. Dincer, M.A. Rosen, *Thermal Energy Storage. Systems and Applications, Second Edition* (John Wiley & Sons, Ltd., Chichester, England, 2011)
3. F. Dinter, M. Ger, R. Tamme, *Thermal Energy Storage for Commercial Applications* (Springer-Verlag, Berlin, 1994)
4. S. Gnuchtel, T. Sander, *7Th International Symposium on District Heating and Cooling. Lund, Sweden* (1999)
5. R. Zwierzchowski, *The International Conference on Energy Efficiency and Control of Air Pollutants from Utilization of Fossil Fuels* (Wroclaw, Poland, 181, 2009)
6. R. Zwierzchowski, Politechnika Warszawska, Prace Naukowe. Inzynieria Srodowiska **64**, 1–136 (2013)

DOI: 10.1051/e3sconf/20172200207

Impact of external conditions on energy consumption in industrial halls

Alina Żabnieńska-Góra[1,*]

[1]Wroclaw University of Science and Technology, Faculty of Environmental Engineering, ul. Norwida 4/6, building C-6, 50-373 Wrocław, Poland

Abstract. The energy demand for heating the halls buildings is high. The impact on this may have the technology of production, building construction and technology requirements (HVAC systems). The isolation of the external partitions, the location of the object in relation to the surrounding buildings and the degree of the interior insolation (windows and skylights) are important in the context of energy consumption. The article discusses the impact of external conditions, wind and sunlight on energy demand in the industrial hall. The building model was prepared in IDA ICE 4.0 simulation software. Model validation was done based on measurements taken in the analyzed building.

1 Introduction

Understanding heat transfer and the temperature distribution through building materials is important for assessing energy use. The heat transfer coefficient is the first and the most considered factor for determining the energy consumption in a building. The value of the heat transfer coefficient must not be greater than the maximum values in Regulation [1]. The thermal insulation requirements apply to the buildings being designed or renovated.

Wind is one of those parameters that characterize the meteorological conditions and which can influence the heat demand variation in the room. This parameter generates a difference in pressure inside and outside the building and thus infiltrates the cold air. The influence of wind depends on the tightness of the partitions (especially windows size), the size of the building and its location (detached, serial, etc). Solar radiation is an additional, noticeable but very variable source of heat and is increasingly important, especially in low-energy construction. Knowledge of the solar energy on a building envelop is crucial for good architecture and energy concept of the building [2]

The analyzed building was built in 2005 and met the requirements of thermal insulation for buildings during this period. Production hall is located in Wroclaw in built-up areas. Due to its production of electrical and electronic components it is an example of a modern light industry hall. It is a one-storey building without cellar, with a height of 7.7 m and a production area of 790 m^2 with additional office and social space. The analyzed hall is a continuation of the existing facility, both in terms of dimensions (height and width) and

* Corresponding author: alina.zabnienska@pwr.edu.pl

 DOI: 10.1051/e3sconf/20172200207

external finishing standards. It has two exterior walls (exposed on north and west) and 12 skylights on the roof.

The article below analyzes the effect of wind and solar gain on an existing building.

2 Building model

As part of the research, the hall's energy model was developed in IDA ICE 4.0. It is an application not only for energy consumption analysis in the building but also for thermal comfort analysis.

IDA ICE 4.0 can be applied to most types of buildings for calculations [3]:

- heat balance (dynamic heat balance in many zones); taking into account heat gains from: insolation, people, appliances, lighting, ventilation, heating and cooling, leaks, thermal bridges and indoor objects such as furniture, office equipment – computers, printers;
- sunlight through transparent seams, with full regard for local shadows, as well as surrounding buildings and other objects;

• air and surface temperatures;

• temperature range, near hot or cold surfaces, taking into account the radiation between surfaces;

• definition of comfort factors: PPD and PMV;

• daylight intensity in rooms;

• moisture and CO_2 concentrations which can be used to control the VAV system;

• vertical air temperature gradient;

• leaks through holes through a fully integrated model air flow (allows examination, e.g. temporary open window or door between rooms);

• airflow, temperature, humidity, pressure and CO_2;

• power and temperature levels of primary and secondary system components;

• Total cost of energy consumption as well as detailed energy bills;

• reports.

Equation based modelling, using the Modelica-like Neutral Model Format (NMF), makes it straight-forward to quickly expand the software with new modelling capabilities. The newly created component models can also easily be shared with other IDA ICE users. Below is an example equation used to determine local wind speed based on wind speed:

$$WindVel = a0\_coeffWindVel_{Ref}\left(\frac{Height}{Height_{Ref}}\right)^{a\_exp} \quad (1)$$

where:
Height – height of the building [m]
Heightref – the reference height (usually 10 m), [m]
a0_coeff – the wind profile coefficient, [-]
a_exp – an exponent of the wind profile, [-]

The wind direction and speed are transmitted through the "climate model" and "Face model".

Computer simulations give the opportunity to model and analyze energy consumption, estimate energy savings resulting from design changes and applied energy saving algorithms. Energy consumption modeling can detect irregularities in the operation of individual installations. It allows estimate performance factors and HVAC systems and assess the reduction in energy consumption in a building. The following analysis uses IDA

 DOI: 10.1051/e3sconf/20172200207

ICE 4.0 software to analyze the impact of building location on energy consumption. All information necessary to determine boundary conditions for the model has been collected and verified. Technical documents as well as their own measurements were used for this purpose.

Construction and parameterization of the model took place in several stages:
1. Reflecting the geometry of the building, including the design of walls, ceilings, windows, shutters, doors and skylights.
2. Parameterization of building components and their thermal properties.
3. Division of rooms into zones and definition of comfort conditions.
4. Define for each zone the heat gains from lighting, appliances and people and work schedules.
5. Identify external conditions such as climate, location, location to the world, and elements shading the building.
6. Build a model of indoor installation, including ventilation unit and define the operating conditions of the equipment.
7. Model validation based on measured data.

Validation of the constructed model was performed by comparing the results of the simulation with the measured data. In order to match the energy model of the production hall to the real object 38 validation tests were performed to correct the following parameters:
- internal gains,
- times the air exchange,
- work schedules of people,
- conditions for the end of the heating season.

As the criterion of correctness, the compatibility of variation was adopted:
- annual electricity consumption [kWh],
- annual heat consumption [kWh],
- air temperature in the hall in a weekly cycle,
- CO_2 concentration in the hall in a weekly cycle.

The results of simulation with measurements in a real object are presented below. The difference between the electricity consumption for the ventilation unit is only 993 kWh, which is 2.21%. The difference between the heat demand is 5985 kWh, which is only 5.1%.

Table 1. Comparison of simulation results with measurements.

Energy	IDA ICE	Real building
Electrical power of the ventilation unit [kWh]	44913	43920
Heat demand [kWh]	117348	123333

3 Energy simulations

3.1 Wind effect

A simulation of the location impact on annual energy consumption was performed for the production hall. For this purpose, equation number 1 was used to calculate the height of the building (7m), the reference height (10m) and the wind profile coefficient (a_0coeffi) and the wind profile exponent (a_exp). After conversion, the values shown in Table 2 were obtained.

Table 2. Heat demand for different locations of the object.

Locations	$a\_0coeffi$	a_exp	$\left(\frac{Height}{Height_{Ref}}\right)^{a\_exp}$	$a\_0coeffi \left(\frac{Height}{Height_{Ref}}\right)^{a\_exp}$	**Heat demand [kWh]**
Ocean	1.3	0.1	0.9650	1.25	118173
Airport	1	0.15	0.9479	0.95	117348
Open area	0.85	0.2	0.9311	0.79	116829
Suburban area	0.67	0.25	0.9147	0.61	115855
City center	0.47	0.35	0.8826	0.41	115047

The heat demand for building heating obtained from the simulations taking into account the location of the building and the wind speed defined in the program are shown in Figure 1. It is evident that there is an increase in heat demand for heating as the wind speed increases due to the change in the location of the constructed model. By comparing the two most extreme locations: the city center (the total wind speed factor A = 0.41) and the ocean shore (factor A = 1.25) the difference in the heat demand for the hall heating is only 2.6%.

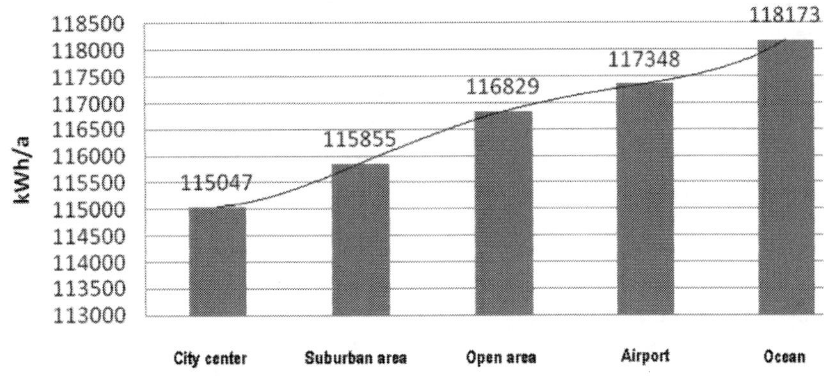

Fig. 1. Energy for heating depending on the wind speed.

In the next step, hall location (suburban zone)was analyzed. An analysis was conducted examining how the lack of wind throughout the year would affect on the hall heat consumption [4]. The results from the simulations showed that in the absence of wind, the heat consumption would change only by 1.45%. Analysis of the results showed that the influence of wind on the location of the production hall depending on the world`s side and urban development is low.

3.2 Sunlight

Solar radiation is an additional, noticeable but very variable source of heat. This parameter is increasingly important, especially in low-energy construction. This applies to direct sunlight, incident on any plane and to radiation scattered on air molecules. In the winter, the dispersed component is usually larger than direct and can account for 60–70% of total radiation [5]. However, the direct sunlight is characterized by a direct radiation component.

As part of the simulation research, the impact of solar radiation on the hall energy consumption was analyzed. The simulations show that due to the specific distribution of the windows, only on the northern wall and only 12 roof skylights, the effect of solar radiation in the winter on the lighting of the production hall is small. During the winter there is practically no direct radiation to the room through the transparent partitions. This is very clearly visible in Figure 2. It is clear that in the southern hours in January direct radiation entering the hall through roof skylights passes through the north-east walls without ever falling to the center of the hall. Hence, it was proposed that reducing the amount of electricity used to light the hall during autumn and winter, by using the building automation controlling the lighting intensity in the hall depending on the hall zone is practically impossible.

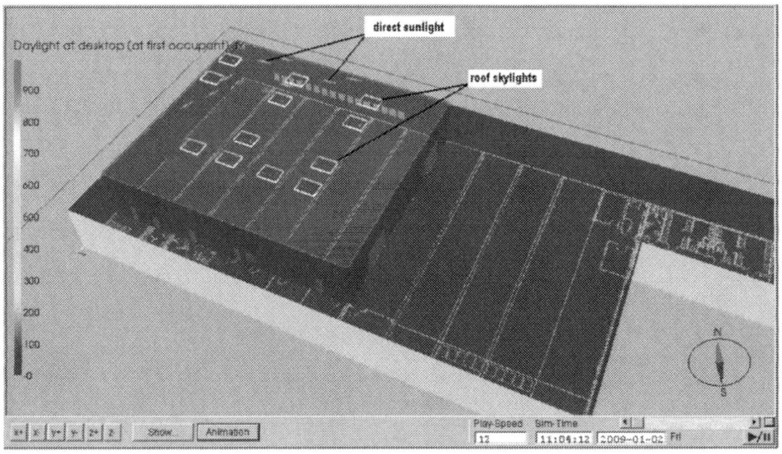

Fig. 2. Influence of solar radiation on the lighting during the winter.

Below in Figure 3 we can see how the sun's position is related to the hall changed with the change of season. At 11:30 in the first days of April and the southern hours of July, direct radiation reaches the hall directly through 12 skylights.

The variability of heat gain from sunlight considering direct radiation and absorbed on the surface of the windows was considered. The monthly variability of heat gains from solar radiation is shown in Table 3 and Figure 4.

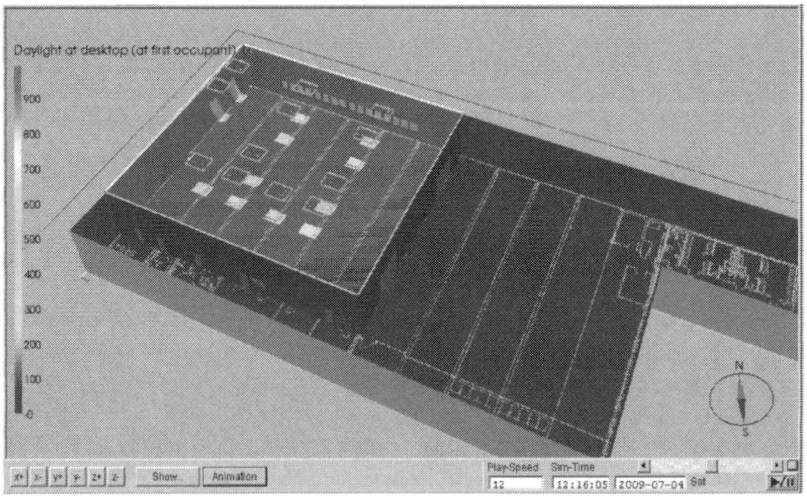

Fig. 3 Influence of solar radiation on the lighting during the summer.

DOI: 10.1051/e3sconf/20172200207

Table 3. Heat loss through structural components.

	Without sunlight			With sunlight		
	walls and floors, [W]	**direct radiation, [W]**	**windows + radiation absorbed, [W]**	**walls and floors, [W]**	**direct radiation, [W]**	**windows + radiation absorbed, [W]**
January	-16246.3	-11.8	-4547.3	-16031.1	685.5	-4224.2
February	-16219.6	-11.7	-4599.6	-15978.8	977.1	-4158.8
March	-15593.2	-11.5	-3963.3	-15412.2	1788.4	-3187.1
April	-14445.4	-11.9	-2928.8	-14078	2680.1	-1805.7
May	-13014.6	-11.3	-2028.5	-12478	3636.8	-514.1
June	-12671.9	-11.9	-1226.4	-12286.5	3780.3	274.4
July	-12834	-12	-1464.9	-12216.1	3752.1	-6.5
August	-12082.2	-11.3	-1124.8	-11640.7	3239.9	184.1
September	-13146.2	-11.9	-1942.7	-12789.6	2210.4	-982.5
October	-14084.9	-11.8	-2804.9	-13820.9	1354.3	-2204.7
November	-15220.5	-11.4	-3776.5	-15193.8	740.7	-3454.9
December	-16341	-12	-4694.9	-16188	630.2	-4406.7
Sum	**-171900**	**-140.5**	**-35102.6**	**-168114**	**25475.8**	**-24486.7**
Annual average value	-14314.2	-11.7	-2916.3	-13997.7	2129.8	-2029.0

 DOI: 10.1051/e3sconf/20172200207

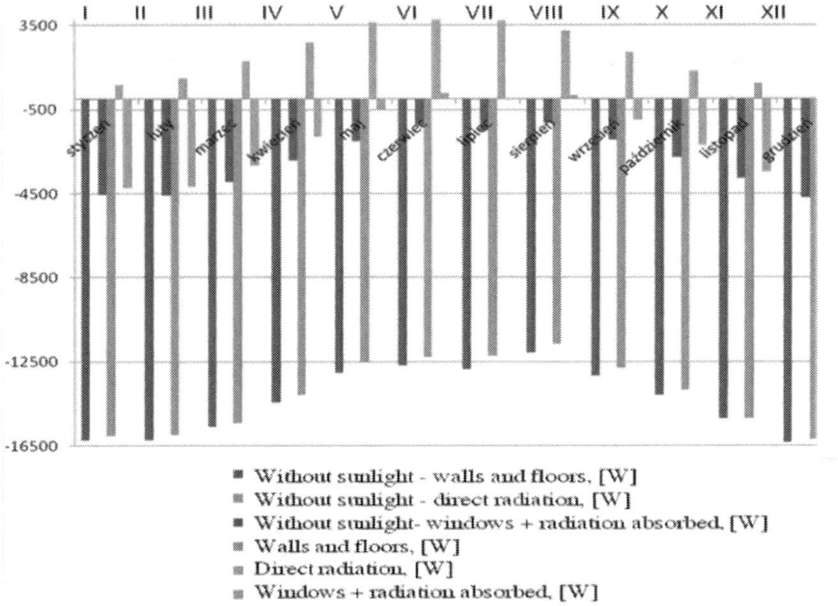

Fig. 4. Monthly variation of heat gains by buildings structure.

Looking at presented variation of heat loss by the structural components it can be considered that the effect of sunlight on the energy consumption is significant. However, the analysis of total heat consumption in the hall showed that when we take into account the heat gains from sunlight in the total annual heat demand for amount of 117616 kWh and in the absence of solar radiation of 113223 kWh in the case of the production hall we will observe only 3.7% increase in heat consumption to heat it.

4 Conclusions

The most popular source of energy is sunlight. This is due to the widest access to energy in such form and the large potential for its use in the form of direct benefits to heating. Unfortunately, high levels of solar radiation do not coincide with the heating period of the building. This requires the designer to refer to solar energy for three distinct strategies: heating, cooling and daylight. The building's enclosure should have high thermal insulation to separate the interior of the building from changing outdoor conditions and consequently to significantly reduce the losses and heat gains of these walls. Additionally, the orientation of the building and its shape, (A/V), can significantly reduce its heat demand.

Looking at the building as a whole the main façade of the production hall is on the south side. In this part offices are located. The production part is on the north side and have limited number of window openings (no direct solar radiation).

Energy simulations have shown that this is the most optimal shape of a building and locations that could be designed for production. The analysis of total heat consumption in the hall showed that when we take into account the heat gains from sunlight in the total annual heat demand for amount of 117616 kWh and in the absence of solar radiation of 113223 kWh in the case of the production hall we will observe only 3.7% increase in heat consumption to heat it. In the absence of wind, the heat consumption would only change by 1.45%. Analysis of the results showed that the influence of wind on the location of the production hall depending on the world`s side and urban development is low.

 DOI: 10.1051/e3sconf/20172200207

References

1. Ordinance of the Minister of Infrastructure regarding the technical conditions to be met by buildings and their placement from the 12 of April 2002, Journal of Laws Iss. **75**, pos. 690 with later changes.
2. D. Chwieduk, *Wytyczne tworzenia koncepcji energetycznej budynku*. Cz T **2-B** (2012)
3. A. Żabnieńska-Góra, *IDA ICE w praktyce. Efektywność układów grzewczych i klimatyzacyjnych*, pod redakcja naukową S. Anisimowa (Ofi. Wyd. PWr. 2009)
4. A. Żabnieńska, *Zarządzanie energia w halach przemysłowych*, PhD thesis (2011)
5. A. Chochowski, F. Krawiec, *Zarządzanie w energetyce. Koncepcje, zasoby, strategie, struktury, procesy i technologie energetyki odnawialnej* (Centrum Doradztwa i Informacji Difin, 2014)

 DOI: 10.1051/e3sconf/20172200208

Intensification of anaerobic digestion efficiency with use of mechanical excess sludge disintegration in the context of increased energy production in wastewater treatment plants

Monika Żubrowska-Sudoł[1], *Jolanta* Podedworna[1], *Agnieszka* Bisak[1], *Katarzyna* Sytek-Szmeichel[1,*], *Piotr* Krawczyk[2], and *Agnieszka* Garlicka[1]

[1]Warsaw University Of Technology, Faculty of Building Services, Hydro and Environmental Engineering, ul. Nowowiejska 20, 00-653 Warsaw, Poland
[2]Warsaw University Of Technology, Faculty of Power and Aeronautical Engineering, ul. Nowowiejska 20, 00-655 Warsaw, Poland

Abstract. The main goal of the study was to evaluate the effects of mechanical sludge disintegration for enhancing full scale anaerobic digestion of municipal sludge. Batch disintegration tests and lab dewatering tests were also performed aiming at determining the release of organic compounds and assessing the impact of disintegration of excess sludge before the fermentation process of mixed sludge on the dewaterability of post-fermented sludge, respectively. In the study a disc disintegrator driven by a motor with a power of 30 kW, revolutions n = 2950 rpm has been used. It was shown that with increase of energy consumed in the disintegration, the increased amounts of organic compounds were released from the sludge. It was also documented that the introduction of the excess sludge disintegration prior to fermentation tank, resulted in a significant increase in biogas production (by an average of 33.9%) and in increase in volatile total solids reduction in the fermented sludge (by an average of 22.7%). Moreover, the obtained results indicate the possibility of obtaining a higher degree of sludge dewatering, which was subjected to anaerobic stabilization with using disintegrated excess sludge.

1 Introduction

The legal requirements, more stringent in recent times in terms of quality of effluent discharged into the receivers (need for highly effective removal of nitrogen and phosphorus from wastewater), and in terms of how to deal with waste sludge appearing in the process of wastewater treatment (retreat from landfilling of sludge and the need to the use of other advanced processes that require complex technology including composting, drying or

* Corresponding author: katarzyna.szmeichel@pw.edu.pl

 DOI: 10.1051/e3sconf/20172200208

incineration) have contributed to a significant increase in energy consumption in the processes of wastewater and sludge treatment, and thus increase the expenses of the operators. Published by Chamber of Commerce "Polish Waterworks" data [1] showed that in 2015 the specific energy consumption for technological processes in the Polish wastewater treatment plants (WWTP) were in the range of 0.12–2.60 kWh/m^3 of the treated wastewater and it was significantly higher than specified for the year 2014 (0.05–1.75 kWh/m^3). Thus, operators began searching for solutions that will allow for the gradual reduction of energy consumption and reducing expenses for its purchase. To decrease the costs of electricity, so-called renewable energy sources (RES) are increasingly used, including mainly biogas, but also wind energy, photovoltaic panels and heat pumps. The first of the above-mentioned RES is closely linked to the effectiveness of the process of anaerobic stabilization of waste sludge. Ensuring the appropriate parameters of the fermentation process and quality of the substrate introduced into anaerobic digester affects the amount of produced biogas as well as its calorific value [2]. Taking under consideration the value of index energy self-sufficiency for Polish WWTP (the ratio of electricity produced at the plant in cogeneration units to the annual electricity demand in the plant) set for 2015 which are equal to 56.7% and 36.6% respectively for medium-sized and large enterprises [1], it becomes important to increase biogas production to increase their share in energy used for the needs of WWTP. It may be done, among others, by co-digestion process, which results in increase of calorific value of substrates entering the anaerobic digester [3]. Another solution is to precede an anaerobic sludge stabilization by a process, which will increase the availability of substrate in the sludge for bacteria carrying out acidogenesis and methanogenesis. The disintegration of excess sludge is such a process. With the pretreatment, an acceleration of hydrolysis by the increase of dissolved components, improvement of biodegradability and sludge dewatering can be achieved [4]. Many different types of sludge disintegration like ultrasonic treatment [5], high pressure homogenization [6], hydrodynamic cavitation [7], Fenton process [8], acidic thermal treatment [9], microwave treatment [10] have been investigated in half-scale and lab scale for the purpose of anaerobic digestion intensification. All these studies have documented the ability to achieve higher production of biogas. For example, Rani et al. [10] using the microwave pretreatment observed 57% increase in biogas production. Boni et al. [5] as a result of ultrasonic disintegration noted 30% increase in biogas production, and Patil et al. [7] using hydrodynamic cavitation got 56% increase in this parameter.

So far, few work describes the results obtained in full scale, and existing indicate that it is not always possible to achieve the expected increase in biogas production. Neis et al.[11] and Rittmann et al. [12] reported 30% and 60% increase in biogas production. On the other hand, Zielewicz [13] reports that only in 3 out of 9 analyzed WWTP achieved a substantial effect of the ultrasonic disintegration on intensification of the fermentation process, and only in one case a significant, amounting to 40%, increase of biogas production has been noted. Therefore, the research in full scale has been undertaken in order to evaluate the effects of mechanical sludge disintegration for enhancing anaerobic digestion of municipal wastewater sludge. Research in technical scale have been preceded by a preliminary stage, the aim of which was to evaluate the release of organic compounds due to mechanical disintegration of excess sludge.

In this study also lab tests have been performed on dewatering of post-fermented sludge subjected to anaerobic stabilization, respectively without and with using of disintegrated excess sludge. Their objective was to assess the potential impact of disintegration of excess sludge before the fermentation process of mixed sludge on the dewaterability of post-fermented sludge, changing the optimal dose of polyelectrolyte for its conditioning prior to dewatering and obtained dewatering efficiency.

2 Methods

2.1 Characteristics of the disintegration apparatus

A full scale disc disintegrator was used in this study. It consists of three discs: two fixed (guide discs) and one rotating connected with a shaft driven by an electric motor (patent No. 211672). Working surfaces of discs have been shaped by the cutting on them constant-width radial channels and five circumferential channels. This configuration creates on discs' working surfaces systems of sectoral protrusions. After placing the rotating disc between guide discs, its sectoral protrusions are located in the circumferential channels of fixed discs. Between the side surfaces of protrusions of rotating disc and the walls of the channels of the fixed disc a gap of constant width is formed, acting as a hydraulic seal. A bipolar electric motor with a power of 30 kW and a nominal rotational speed of 2950 rpm was used to drive the disintegrator. The motor is powered by a frequency converter capable of different speeds in the range 0–4000 rpm. Flow through the device is regulated in the range of 0 to 10 m^3/h. The flow system is equipped with a bypass.

2.2 Disintegration batch tests

The disintegration batch tests were carried out by repeated feeding the disc disintegrator with a 60-litre sample of thickened excess activated sludge (total solids (TS): 5.35 ± 0.67%). The device was operating at a sludge throughput of about 3.0 m^3/h. The process was carried out in a closed circle. After introducing a selected amount of energy, expressed as energy density (ε_L: 70, 140, 210 and 280 kJ/L), to the sludge under the disintegration process, a 0.5 L sample of disintegrated sludge was collected from a continuously stirred buffer tank. Immediately after this the liquid (subsequently called filtrate) was separated from the sludge in each sample. For this purpose the sludge was centrifuged for 30 min at 19621 g (centrifuge MPW-350) and then filtered with 0.45 µm filters. The scope of the analysis included: 1) TS in raw sludge (samples not subjected to the disintegration process), 2) soluble chemical oxygen demand (SCOD) and volatile fatty acids (VFA) concentrations in filtrate before and after disintegration and 3) sludge disintegration degree (DD) according to Nickle and Neis [14]. In the experiment, seven repetition of the disintegration batch test were conducted in 3-weeks intervals. For each of them a new portion of sludge was delivered from a local WWTP.

2.3 Evaluation of impact of thickened excess sludge disintegration on the efficiency of methane fermentation process

The experiment aiming at evaluating the impact of mechanical disintegration of the thickened excess sludge on the efficiency of the process of anaerobic digestion was carried out for 8 months in a full scale at the WWTP of population equivalent equal to 107 700. It was assumed that the whole flow of the thickened excess sludge directed to the digester would be subjected to the process of disintegration, and energy density at which the process will be performed would be selected based on the results of batch disintegration tests (carried out in the initial stage of research). Excess sludge were fed to the disintegrator right from the mechanical thickener, and after disintegration process were directed back to the thickener, and from it to a mix sludge tank, wherein they were mixed with thickened primary sludge. Disintegrator worked for 16 hours a day (at what time excess sludge thickener was working), disintegrating 89 m^3 thickened excess sludge. The mixed sludge in an amount of 118 m^3/d was directed into the digester of working volume of 5000 m^3. The process of anaerobic digestion was carried out at 34°C, retention time 42.4 d. During the experiment the TS of thickened excess sludge directed to the disintegration was 4.08 + 0.61%, and the percentage share of VTS was 80.7 + 2.28%. The values of these indicators characterizing the mixed sludge fed to a digester were respectively 4.95 + 0.31% and 84.6 + 3.11%.

To evaluate the effect of mechanical disintegration of the thickened excess sludge on the efficiency of the process of anaerobic digestion, the concentration of TS and VTS was monitored in sludge introduced to and discharged from the digester. On this basis the reduction of VTS (expressed in kg VTS/d) was determined. Also the observations of the amount of the produced biogas were carried out. The results were compared with the effects of anaerobic digestion obtained before installing the disintegrator (average values of reduction of VTS and the amount of produced biogas from a period of 12 months prior to starting the disintegration have been obtained from the operator of the WWTP).

2.4 Sludge dewatering tests

Tests of dewatering were carried out by filtration on a Büchner funnel and by lab centrifuge, always with use of the pre-flocculated sludge by optimal dose of cationic polyelectrolyte Zetag 8180 (the same polyelectrolyte was used to support the dewatering of post-fermented sludge in industrial scale). The optimal dose of the polyelectrolyte was determined based on the smallest value of capillary suction time (CST). Dewatering on the Büchner funnel was conducted using a sludge sample of 100 mL, a standard filter to determine the specific resistance of filtration (Whatman® quantitative filter paper, ashless, Grade 41) under a pressure of 0.05 MPa and the filtration time 10 minutes. For dewatering using centrifugal force, a laboratory centrifuge type MPW-350 was used working with centrifugal acceleration 19 621 g, at constant spin time of 10 minutes. The dewatering efficiency was measured by determining in each case the TS concentration of the dewatered sludge.

NOTE: It should be emphasized that the concentration of TS in dewatered sludge in the above-specified laboratory conditions can never be compared directly with the values of TS in sludge dewatered using devices on a technical scale. However, they may serve as relative comparisons of changes in the ability of the sludge to release water; in the present case, depending on the fact of the use (or not) the process of disintegration of excess sludge before the fermentation process.

2.5 Analytics

All chemical analyses were performed in duplicates in accordance with APHA Standard Methods [15].

3 Results and discussion

3.1 Effects of mechanical disintegration of excess activated sludge on organic compound release

Mechanical disintegration was used to improve the bioavailability of sludge particulate material. As a result of disintegration, soluble organic compounds are released from the activated sludge flocs that can be readily used to produce methane during anaerobic digestion [16]. From the Table 1 it is evident that with the increase of energy consumed in the disintegration, the increased amounts of organic compounds (SCOD) were released from the sludge. Part of them was VFA. The phenomenon of obtaining VFA was called acidification. It can be seen that although the increase in energy density allowed getting the increase of SCOD concentration, the acidification was observed only to a certain level of energy density. The "inhibition" of acidification was observed at 210 kJ/L.

Table 1. The effect of energy density on SCOD and VFA release from sludge.

Parameter	Unit	Before disintegration	After disintegration			
			70 kJ/L	140 kJ/L	210 kJ/L	280 kJ/L
SCOD	mgO_2/L	92-411	1104–2375	2694–6111	6152–9153	7410–11451
		160 ± 114*	1784±506*	4660±1309*	7659±1046*	9520±1398*
VFA	mg/L	3.0–28	77–239	361–868	590–1000	590–1000
		11 ± 8.1*	123±55*	623±196*	872±133*	886±139*

*) average ± standard deviations

Analyzing the relationship between VFA and SCOD depending on the energy density (Fig.1) it was found that the highest percentage of VFA in the organic compounds released form activated sludge flocs was achieved with a process of disintegration performed at energy density equal to 140 kJ/L. Analyzing relationships between ratio SCOD:VFA and the disintegration degree it could be noted that the minimum values of this ratio occurred when DD remained in the range of 9–17%. Higher values SCOD:VFA achieved for DD < 9% can be attributed to the fact that they were obtained in the initial stage of the disintegration (ε_L-70 kJ/L). Most likely then the breakup of macrostructures of activated excess sludge occurred, resulting mainly in the release of long-chain organic compounds. The increase in ratio SCOD:VFA when DD exceeded the level of 17%, resulted from a significant reduction in the efficiency of acidification, with continuous release of organic compounds other than VFA. It seems it might be affected by two factors: 1) at higher levels of disintegration observed mainly for ε_L = 210 kJ/L and 280 kJ/L the amount of slowly biodegradable organic compounds originated from the lysis of bacterial cells increased, 2) with an increase in disintegration degree the structure of sludge floc changed significantly which no longer shaped dense clusters, but they formed "dispersed" mass of microorganisms and it could limit the propagation of the forces responsible for generating the VFA.

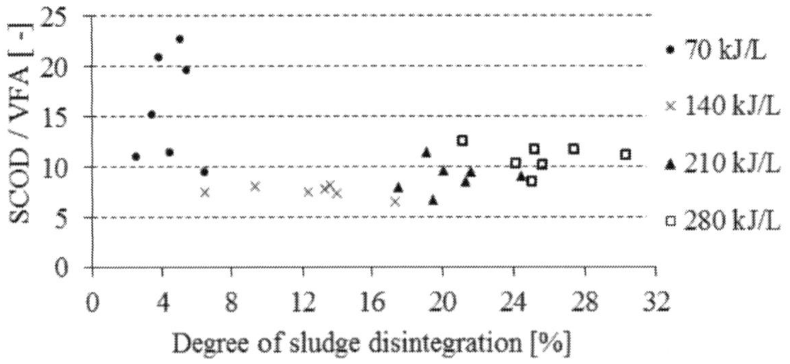

Fig. 1. SCOD:VFA as a function of the degree of sludge disintegration with consideration of energy density level.

3.2 Effects of mechanical disintegration of excess activated sludge on anaerobic digestion of municipal sludge

Taking into account the results of disintegration batch tests, it was assumed that during the experiment the disintegrator will work with an energy density of 140 kJ/L. During installation at WWTP it turned out, however, that the subjecting the entire flow of excess sludge directed to the digester to disintegration with this energy density is not possible using the existing infrastructure in the plant. Therefore it was decided to carry out the experiment by disintegrating the entire flow of excess sludge, but at energy density resulting from the technical conditions at the plant, i.e. 20 kJ/L.

Before starting the experiment, the average values of VTS reduction and the biogas production were 2183 kg VTS/d and 2400 m^3/d respectively (Tab. 2) and therefore a unit biogas production was at a level of 1.1 m^3/kg VTS. The introduction of the disintegrated excess sludge prior to fermentation tank resulted in a significant increase in biogas production by an average of 33.9% (Tab.2). A similar percentage increase in the amount of biogas produced was observed by Neis et al. [11] (full-scale) and Boni et al. [5] (lab-scale) using ultrasonic pre-treatment. From the Figure 2 it is evident that in the initial phase of the experiment abrupt changes in the amount of produced biogas occurred, and after approx. 3.5 months since the start of disintegration the biogas production was relatively stable, remaining in the range of 3040–3421 m^3/d. The effect of application of excess sludge disintegration was also increase in VTS reduction in the fermented sludge by an average of 22.7%. Unit biogas production for the analyzed period was 1.2 m^3/VTS kg.

Table 2. Comparison of VTS reduction and biogas production for the periods before and after the mechanical disintegration of the thickened excess sludge.

Parameter	Unit	Average values	
		Period before disintegration	Period after disintegration
VTS reduction	kg VTS/d	2183	2678
Biogas production	m^3/d	2400	3213
Unit biogas production	m^3/kg VTS	1.10	1.20

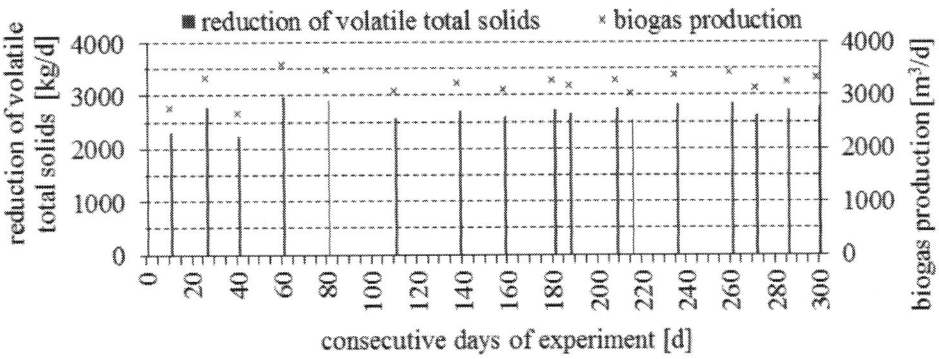

Fig. 2. Reduction of volatile total solids and biogas production on consecutive days of experiment.

Recorded significant increase in biogas production has to be understood in relation to the energy costs necessary to its inducing. Only the analysis of the size of the net increase in energy can give a full picture of rightfulness of the use of disintegration from the point of view of energy balance. We here are dealing with two types of energy. To drive the disintegrator, an energy is converted in the form of electricity, while biogas is a carrier of chemical energy. It is therefore impossible to make a direct comparison between them. Such a comparison is legitimate only if we make the conversion of one form of energy into another. Biogas as a carrier of chemical energy is a fuel of good quality and can be converted into electricity. A typical way to use biogas to produce electricity is its use in a piston engine or gas turbine. In WWTP much more popular form of biogas chemical energy conversion into electricity and heat is a piston engine. The process is carried out, obviously, with a certain efficiency far different from 100%. For medium-sized gas engines with a nominal power of several hundred kilowatts an average electrical efficiency is about 40%. In addition, the engine can produce functional heat in the form of hot water. It can be assumed that the amount of generated heat is equal to the amount of produced electricity. Assuming biogas calorific value is 23 MJ/m^3, the surplus of produced biogas corresponds to the chemical energy of about 5200 kWh/d. This amount of chemical energy allows producing approximately 2080 kWh/d of electricity and the same amount of heat. Electricity consumption by the disintegrator in the analyzed WWTP was significantly lower and equal to approximately 480 kWh. In case of installing in the present plant a piston engine powered by biogas, total energy benefits of using the disintegrator would amount to about 1600 kWh of electricity and 7.5 GJ (2080 kWh) of heat. The heat can be used to maintain the desired temperature in digester or to enhance the process of dewatered sludge drying.

During the experiment it was also observed that disintegration preceded the fermentation process had a positive effect on the rheological properties of fermented sludge which largely affected the work of the pumps feeding the sludge from digester to fermented sludge tank. A significant drawback of the technology, however, was the presence of foam in the digester. The cause of observed foaming was an increase in biogas production and problems with its discharge (too low flow capacity of an installation capturing biogas from digesters). In order to reduce the phenomenon described above there was necessary to increase the diameter of the biogas pipe. Modifications improved the working conditions of a digester.

3.3 Effects of mechanical disintegration of excess activated sludge on sludge dewatering

Table 3. The results of laboratory tests of dewatering post-fermented mixed sludge stabilized anaerobically without and with disintegrated excess sludge.

Sample No.		1	2
Type of sludge fed to the digester		mixed (without disintegration of excess sludge)	mixed (with disintegration 100% of excess sludge)
TS [%] (in the post-fermented sample of sludge collected from WWTP)		2.38	3.06
CST of no-conditioned sludge [s]		249	724
The optimal dose of polyelectrolyte Zetag 8180 [% VTS]		0.86	0.44
CST of sludge conditioned with optimal dose of Zetag 8180 [s]		43.1	20.3
TS in sludge after dewatering [%] (sludge pre-conditioned with optimal dose of polyelectrolyte)	Büchner funnel	8.03	11.5
	Centrifuge	7.62	10.0

DOI: 10.1051/e3sconf/20172200208

Laboratory dewatering tests were performed on 2 samples of fermented sludge delivered from WWTP. The results are shown in Table 3. The sample indicated by No. 1, of the concentration of TS: 2.38%, is the sludge formed by the fermentation of the mixture of primary and the excess sludge which were not subjected to disintegration. The sample No. 2 (TS: 3.06%) was the sludge formed in the fermentation of a mixture of primary and 100% of pre-disintegrated excess sludge, collected after replacing the entire volume of the digester (to ensure that in the entire volume of the digester was sludge together with sludge disintegrated involving excess sludge). The data shown in Table 3 indicate a significant reduction in the optimal dose of the polyelectrolyte for treated post-fermented sludge after anaerobic stabilization using entirely disintegrated excess sludge in relation to the fermented sludge without disintegration of the biological sludge even, that CST of the sludge prior to its conditioning was almost 3 times higher than the one of conventionally fermented sludge. The best indicator of growth of sludge dewaterability during mechanical dewatering is an increase in TS of dewatered sludge. In the analyzed samples of post-fermented sludge a significant increase in this parameter for the sludge pre-flocculated with the optimal dose of polyelectrolyte was obtained for fermented sludge with use of disintegrated excess sludge (in sample 2, for sludge after filtration TS was 11.50%, in relation to the sample 1–8.03%), and after centrifuging 10% respectively, in relation to samples 1–7.62%. The observed changes confirm the increase in the degree of dewatering of sludge fermented with disintegrated excess sludge and document the slightly larger effect of the process performed by pressure filtration.

4 Conclusions

The organic matter degradation and biogas production in the anaerobic sludge digestion were greatly improved by the mechanical disintegration of thickened excess sludge. The performed analyses regarding filtration properties of post-fermented sludge suggested the possibility of obtaining a higher degree of sludge dewatering, which was subjected to anaerobic stabilization with using disintegrated excess sludge (even at the lower dose of used polyelectrolyte) in relation to the sludge fermented without disintegration of excess sludge.

References

1. Benchmarking in Polish Water and Wastewater Sector in 2015 (in Polish)
2. G. Tchobanoglous, F.L. Burton, H.D. Stensel *Wastewater Engineering: Treatment and Reuse* (Metcalf & Eddy Inc., McGraw-Hill, New York, USA, 2004)
3. K. Koch, M. Plabst, A. Schmidt, B. Helmreich, J.E. Drewes, Waste. Manage. **47**, 28–33 (2016)
4. A.V. Ebenezer, S. Kaliappan, S.A. Kumar, I.T. Yeom, J.R. Banu, Bioresour. Technol. **185**, 194–201 (2015)
5. M.R. Boni, E. D'Amato, A. Polettini, R. Pomi, A. Rossi, Waste. Manage. **48**, 209–217 (2016)
6. W. Fang, P. Zhang, J. Ye, Y. Wu, H. Zhang, J. Liu, G. Zeng, Int. Biodeterior. Biodegradation **102**, 126–130 (2015)
7. P.N. Patil, P.R. Gogate, L. Csoka, A. Dregelyi-Kiss, M. Horvath, Ultrason. Sonochem. **30**, 79–86 (2016)
8. S. Kavitha, J.R. Banu, C.D. IvinShaju, S. Kaliappan, I.T. Yeom, Bioresour. Technol. **221**, 1–8 (2016)

9. M. Takashima, Y. Tanaka, J. Environ. Chem. Eng. **2**, 2, 773–779 (2014)
10. R.U. Rani, S.A. Kumar, S. Kaliappan, I. Yeom, J.R. Banu, Waste. Manage. **33**, 5, 1119–1127 (2013)
11. U. Neis, K. Nickle, A. Lunden, J. Environ. Sci. Health A **43**, 13, 1541–1545 (2008)
12. B.E. Rittmann, H.S. Lee, H. Zhang, J. Alder, J.E. Banaszak, R. Lopez, Water Sci. Technol. **58**, 10, 1895–1901 (2008)
13. E. Zielewicz, Gas, Water and Sanitary Engineering **90**, 2, 69–76 (2016) (in Polish)
14. K. Nickle, U. Neis, Ultrason. Sonochem. **14**, 450–455 (2007)
15. *Standard Methods for the Examination of Water and Wastewater* (20th edn. APHA, AWWA & WEF, Washington, DC, USA, 1998)
16. F. Wang, Y. Wang, M. Ji, J. Hazard. Mater. **123**, 145–150 (2005)

Energy efficiency of a solar domestic hot water system

Miroslaw Zukowski[1,*]

[1]Bialystok University of Technology, Department of HVAC Engineering, ul. Wiejska 45E, 15-351 Bialystok, Poland

Abstract. The solar domestic hot water (SDHW) system located on the campus of Bialystok University of Technology is the object of the research described in the current paper. The solar thermal system is composed of 35 flat plate collectors, 21 evacuated tube collectors and eight hot water tanks with the capacity of 1 m^3 of each. Solar facility is equipped with hardware for automatic data collection. Additionally, the weather station located on the roof of the building provides measurements of basic parameters of ambient air and solar radiation. The main objective of Regional Operational Program was the assessment of the effectiveness of this solar energy technology in the climatic conditions of the north-eastern Poland. Energy efficiency of SDHW system was defined in this research as the ratio between the useful heat energy supplied to the domestic hot water system and solar energy incident on the surface of solar panels. Heat loss from water storage tanks, and from the pipe network to the surrounding air, as well as the electrical energy consumed by the pumps have been included in the calculations. The paper presents the detailed results and conclusions obtained from this energy analysis.

1 Introduction

Economic assessment of energy-conversion systems is quite complex. The long-term energy efficiency is the key factor of this analysis. Experimental studies carried out on the existing and operating facilities will allow to develop the energy balance of the whole solar system in a very accurate way. Unfortunately, this procedure is quite expensive. As it turned out, the additional cost of the monitoring and automatic data collection (ADC) of operating parameters of the solar domestic hot water (SDHW) system tested by the author, was equal about 2/3 of the cost of materials and installation of the solar system. For this reason, experimental studies to evaluate the energy efficiency of medium and large solar installations are not often performed.

A large number of actual data from field-testing and monitoring of domestic solar hot water systems in China have been analysed by Yao et al. [1]. They discovered that there are many problems resulting from the relatively large amount of measurement error and use of different methodologies for assessing and monitoring. The authors proposed the methods and indicators for precise estimation of solar hot water systems.

* Corresponding author: m.zukowski@pb.edu.pl

The results of evaluation of water heating system supported by 640 m^2 of solar collectors installed on high-rise housing complex in South Korea have been presented by Yoo in [2]. Unfortunately, as it turned out, the total energy gains from three years of SDHW operation is lower by a half of the design assumptions. A wide gap between theory and practice can be observed in this case. Analysis of measurement results have shown that the collector efficiency ranges from 42% in August to 53% in April. Unfortunately, the author has not performed an assessment of the energy efficiency of this large solar facility.

A solar domestic hot water system installed on a one-family house in Dublin (Ireland) has been tested by Gill et al. [3]. Based on the analysis of the annual results of experimental research the overall efficiency of the solar system has been estimated at about 63%. It should be noted that the key components of heat loss have been taken into account by the authors of this study.

Several studies have investigated the energy and economic efficiency of the whole SDHW system using some simplifications. For example, solar collectors are treated as the only part of the installation that generates heat loss. Some selected studies of this kind are presented below.

Stevanovic and Pucar [4] have estimated the performance of solar water heating system with the f-Chart method. The results of theoretical analysis have been used to determine the level of the subsidy should be offered by the Serbian government to increase the popularity of renewable energy sources (RES) in this country.

The energetic, economic and environmental impacts of different types of water heating systems for the USA typical residential buildings located in Los Angeles, Atlanta, and Chicago have been investigated by Hang et al. [5]. Unfortunately, the performance of solar water heating systems has been determined only based on the thermal characteristics of the flat-plate (efficiency – 55%), and evacuated-tube (efficiency – 59%) solar collectors provided by the manufacturer.

Lin et al. [6] have developed a procedure for estimating the payback period of residential SDHW system. Daily solar radiation and hot water consumption pattern per person have been taken into account in the calculation of annual energy gains for case study in southern Taiwan. The analysis has assumed that the thermal efficiency of the collectors (do not exceeding 40%) is equivalent to the performance of the whole system.

Fifteen different energy supply systems for single family houses in the northern European climate conditions have been analysed by Dziugaite-Tumeniene et al. [7]. The theoretical analysis made by the authors has been based on the integrated assessment model of an energy supply system for an energy-efficient building (ESSINTEGRA). Using this method has allowed to obtain energy efficiency (equal to 31%) of the system consisting of the wood boiler and solar collectors.

Based on the literature review, it can be concluded that there are not much information on the research projects related to the experimental evaluation of the energy efficiency of medium solar systems used for domestic water heating. Thus, the author hopes that this paper will contribute to increase the knowledge on this important issue and provide the necessary data to design and economic analysis of SDHW systems.

2 Description of the research object

Regional Operational Program of Podlaskie Voivodeship on the practical application of renewable energy technologies were being implemented at the Bialystok University of Technology (BUT) in the years 2013–2015. One of the practical results of this project was installation of the solar domestic hot water system (SDHW). Solar facility is located on the roof of the Hotel for Research Assistants on BUT campus in Bialystok (Fig. 1). The solar thermal system is composed of two parallel-connected subsystems: the first includes of

35 flat plate collectors (FPCs), connected in series of 5 units in one section, with a total area A_{FPC} of about 72 m^2 (65.5 m^2 – aperture area), and the second consists 21 evacuated tube collectors (ETCs), connected in series of 3 units in one section, with a total area A_{ETC} of more than 74 m^2 (67m^2 – active area). Eight tanks with the total capacity of about 8 m^3 are used to store the hot water heated by the solar thermal panels.

Fig. 1. Flat plate solar collectors connected in series of 5 units in one section (left side of the fig.), and evacuated tube solar collectors connected in series of 3 units in one section (right side of the fig.).

Solar facility is equipped with hardware for automatic data collection. The following elements are included in the data capture system: 17 heat meters, 4 electricity meters used for recording electric energy consumption of pumps, and 42 PT 500 platinum resistance temperature sensors for monitoring of the temperature distribution in the whole system. Additionally, the weather station located on the roof of the hotel provides measurements of the basic parameters of ambient air and solar radiation. All the data collected by the ADC system are readily available via the Internet. The simplified scheme of the hydraulic circuits of the solar system are shown in Fig. 2.

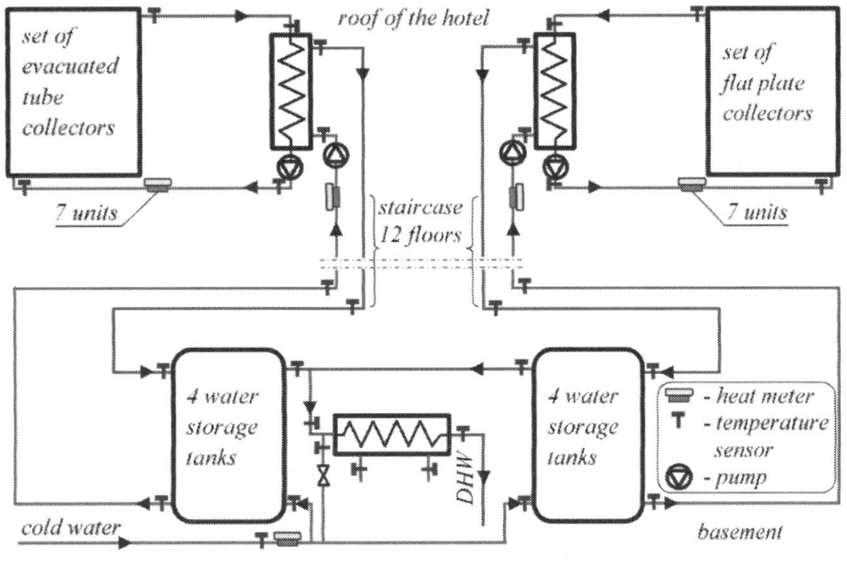

Fig. 2. Simplified diagram of the SDHW system with the location of measuring instruments and sensors.

3 Energy efficiency of a solar domestic hot water system

The main objective of the previously mentioned Regional Operational Program was the assessment of the effectiveness of solar water heating technology in the climatic conditions of the north-eastern Poland. Energy efficiency of SDHW system η_{SYS} has been defined in this research based on Eq. (1) as the difference between the useful heat energy supplied to the tap water system in the hotel E_{DHW} and electrical energy consumed by the pumps E_P divided by solar energy incident on the solar panels E_{SOL}.

As mentioned earlier, ultrasonic heat meters have been utilised to obtain the value of thermal energy in 17 points at the system. Relative measurement error of these devices is small, and according to the manufacturer's information, is only 2%. The data recorded by heat meters have been used to prepare the energy balance. The study covered the period from 1 May 2015 to 30 November 2016.

$$\eta_{SYS} = \frac{E_{DHW} - E_P}{E_{SOL}} \times 100 \tag{1}$$

$$E_{DHW} = \sum_{1}^{nm} \left[m_{DHW} \cdot nd \cdot 24 \cdot 3600 \cdot \left(c_p \cdot \left(T_{FPC} - T_{CW} \right) + c_p \cdot \left(T_{ETC} - T_{CW} \right) \right) \right] \tag{2}$$

$$E_{SOL} = \sum_{1}^{nm} q_{SOL} \cdot nd \cdot 24 \cdot 3600 \cdot \left(A_{FPC} + A_{ETC} \right) \tag{3}$$

where: nm – number of months, m_{DHW} – DHW mass flow rate [kg/s], nd – number of days in a month, c_p – heat capacity of water [J/kg/K], T_{FPC} – temperature of the water flowing from the tanks supplied by FPCs [K], T_{CW} – temperature of the cold water [K], T_{ETC} – temperature of the water flowing from the tanks supplied by ETCs [K], q_{SOL} – solar radiation flux that is incident on the collector surface [W/m^2].

The weather station is located very close to the collectors and total solar radiation sensor is set at the same angle as the panels i.e. 40°. The sampling frequency of the weather parameters is equal to 5 seconds, and other parameters of the system are collected in one-minute time steps. Therefore, it was necessary to average all the results of measurement on a one hour interval. Then, the results have been summed to provide energy gain, energy consumption in the case of pump operation, and hot water consumption within one month period.

Solar radiation energy E_{SOL} that is incident on the two sets of solar collector surfaces of the total area 146.29 m^2 has been calculated using Eq. (3). The total amount of E_{SOL} from 1 May 2015 to 30 November 2016 was equal 1029.72 GJ.

Next, the amount of useful energy gain coming from both arrays of the solar collectors E_{SC} has been determined based on the data recorded by heat meters placed in the substation on the roof. During the reporting period of 19 months the solar collectors provided 397.98 GJ of thermal energy, of which 204.09 GJ has been produced by flat plate collectors, and 193.89 GJ by evacuated tube collectors. The bar chart in Fig. 3 shows that the ETCs (much more technologically advanced) have been characterized by higher power only in the cold season from October to March. Unfortunately, at this period of the year duration and intensity of solar radiation is very low. Heat losses to the environment from flat plate collectors are usually higher compared to from evacuated tubes. However, the overall balance has shown that FPCs produced 5% more energy during the whole analysed time period. Part of this energy has been lost. The vast majority of sources of energy loss are outlined on the block diagram shown in Fig. 4.

DOI: 10.1051/e3sconf/20172200209

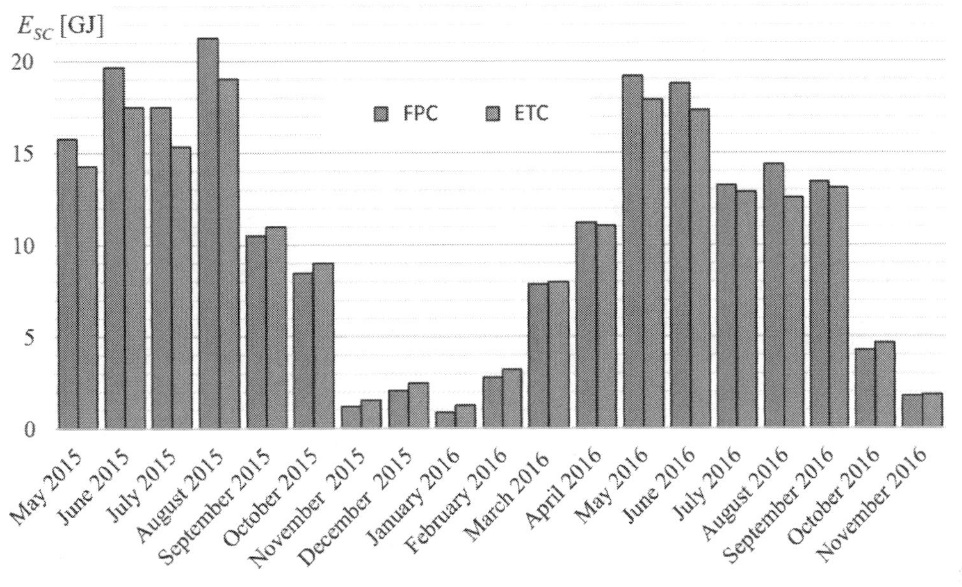

Fig. 3. Solar energy converted into thermal energy by solar panels.

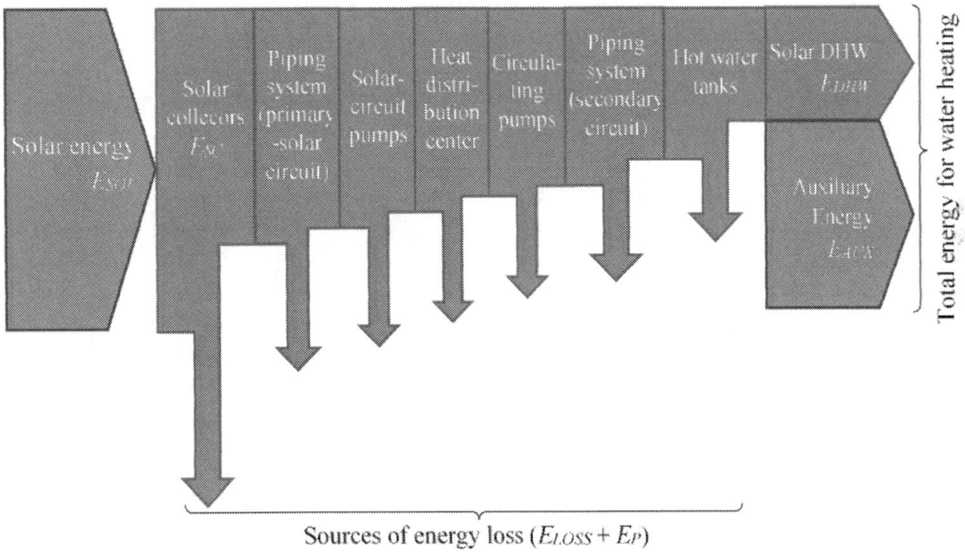

Fig. 4. Energy balance of the SDHW system.

The closed loop solar system under the analysis is divided into primary and secondary circuits, that are separated by heat exchangers. Hence, such a system must be fitted with four pumps, two for each of the collector arrays. All pumps consumed 8.69 GJ of electricity E_P in the period under consideration. This value has reduced energy gain from the solar system. The pumps located in the ETCs closed loop system consumed about 5.5% more of electric energy compared to FPCs system. As can be observed in Fig. 5, this trend intensified during the winter.

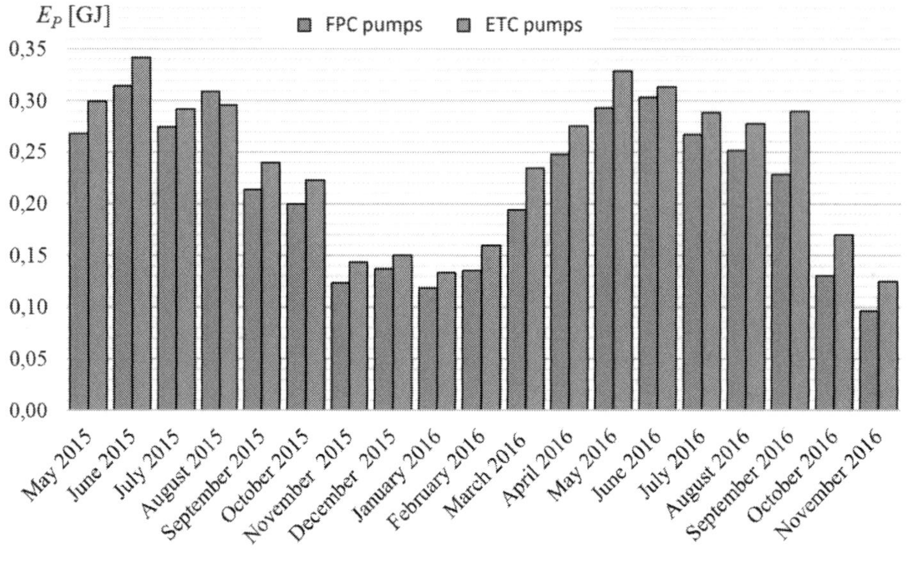

Fig. 5. Consumption of electric energy by the pumps by months.

The useful heat energy supplied to the domestic water system E_{DHW} has been calculated separately for each type of solar panel sub-systems based on Eq. (2) and was equal 344.52 GJ. Then, seasonal energy efficiency of the whole SDHW system has been estimated using the relation given in Eq. (1). Fig. 6 shows the η_{SYS} taken separately for each month during the entire study period. As can be observed, in the warmer months from April to August, the efficiency was the highest and most often exceeded 35%. In the winter η_{SYS} was significantly decreasing and even it dropped up to 10% in January 2016.

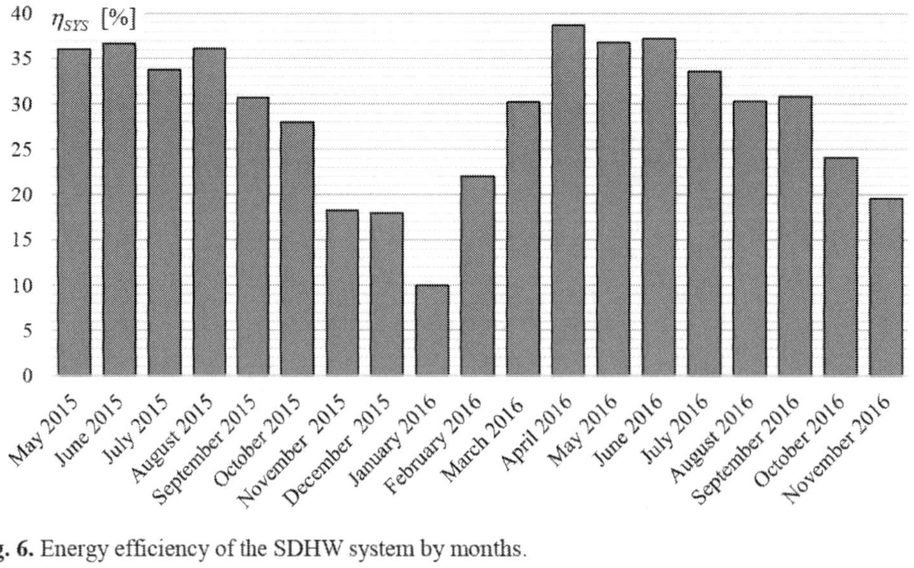

Fig. 6. Energy efficiency of the SDHW system by months.

Eight hot water storage tanks, the hot water pipelines connecting substation on the roof with the second substation in the basement (average length of each of the four DN40 pipes is 86 m), flow control fittings and shut-off valves turned out to be the major sources of energy loss E_{LOSS}. Detailed analysis of the different sources of heat loss was not the main subject of this study. They have been taken as a total. The value of E_{LOSS} was equal 53.46 GJ and has been obtained as the difference between E_{SC} and E_{DHW}. This was only 13.4% of energy gain coming from the solar collectors.

The parameter which indicates what part of energy is provided by the solar system and what must be supplied by an additional power source is a solar fraction *SF*. This value is determined by the following formula [8]:

$$SF = \frac{E_{DHW}}{E_{DHW} + E_{AUX}} \tag{4}$$

The auxiliary energy demand was calculated according to Eq. (5), assuming the design value for hot water temperature T_{DHW} equal to 60°C.

$$E_{AUX} = \sum_{1}^{nm} \left[m_{DHW} \cdot nd \cdot 24 \cdot 3600 \cdot \left(c_p \cdot \left(T_{DHW} - T_{FPC} \right) + c_p \cdot \left(T_{DHW} - T_{ETC} \right) \right) \right] \tag{5}$$

As shown in Fig. 7, SDHW system was able to satisfy around 35% of the hot water demand during the warmer months. However, in winter time, this factor decreased to only about 5%. The monthly average *SF* in the period under consideration was 23.4%. Usually it is assumed that a well-designed solar system should be able to provide around 40% of total energy load for water heating [9]. It proves too small area of solar collectors used in this system. Unfortunately, in medium-rise buildings (5 to 15 storeys) often there is a lack of space on the roof to allocate additional solar panels.

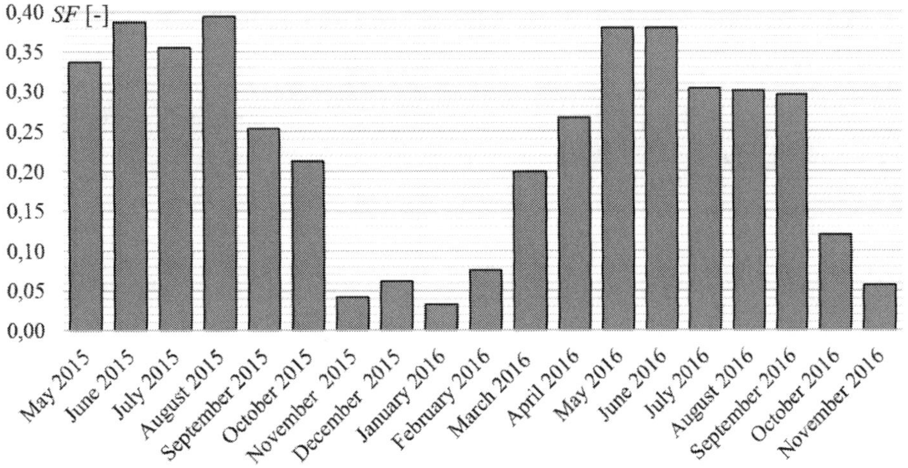

Fig. 7. Changing the monthly average value of the solar fraction.

4 Conclusions and the future work

The main goal of the present research has been to determine the seasonal energy efficiency of the solar domestic hot water system located on the campus of Bialystok University of Technology. This study involving analysis of measurement data that have been collected from 1 May 2015 to 30 November 2016.

It has been found that the energy efficiency of SDHW system was strongly related to the intensity of solar radiation. As indicated by the calculation results η_{SYS} was equal 32.6%

for the time period of this analysis, with 34.7% in the warmer months and 24.2% during the colder months (period from October 2015 to March 2016).

To summarize, it can be concluded that only about 1/3 of the solar energy that falls on the solar panels has been converted into useful heat supplied to the domestic hot water system. So, this level of energy efficiency can be applied to any type of economic evaluations associated with the use of solar collectors for water heating in similar types of buildings in areas with a similar climate to north-eastern Poland.

It is planned that the study on the effectiveness of SDHW system will be continued for the next three years. Besides, the thermal performance of both types of solar collectors according to the standard ISO 9806:2013 will be determined.

This work has been performed within the framework of a Grant of Bialystok University of Technology (Grant No. S/WBIIS/4/2014).

References

1. C. Yao, B. Hao, S. Liu, X. Chen, Energy Procedia, **70** (2015)
2. J.-H. Yoo, Energy and Buildings, **101** (2015)
3. L. Gill, J.M. Mahon, K. Ryan, Solar Energy, **137** (2016)
4. S. Stevanovic, M. Pucar, Energy and Buildings, **54** (2012)
5. Y. Hang, M. Qu, F. Zhao, Energy and Buildings, **45** (2012)
6. W.M. Lin, K.C. Chang, K.M. Chung, Renewable and Sustainable Energy Reviews, **41** (2015)
7. R. Dziugaite-Tumeniene, V. Motuziene, G. Siupsinskas, K. Ciuprinskas, A. Rogza, Energy and Buildings, **138** (2017)
8. *The German Solar Energy Society, Planning and installing solar thermal systems: A guide for installers, architects and engineers*, James and James, UK (2007)
9. Y. Evans, T. Dwyer, CIBSE Journal, **02** (2009)

Water pipe network as a heat source for heat pump integrated into a district heating

Piotr Jadwiszczak[1], and *Elżbieta* Niemierka[1,*]

[1]Wroclaw University of Science and Technology, Faculty of Environmental Engineering, Wybrzeże Wyspiańskiego 27, 50-370 Wrocław, Poland

Abstract. The paper will present a technical analysis of the performance of the Heat Pumps (HP) installed in the domestic water pipe network for a big city scale. The HP integration scheme predicts the domestic water flow as a heat source and the district heating as a heat sink. The technical factors which influence on the estimated thermal power and performance of HP unit will be identified. Additionally, the pros and cons of HP operation in water intake will be determined. The analysis will be based on long-term measurement data from Głogów city.

1 Introduction

The Heat Pump (HP) technology is one of the most dynamically developing branches of Renewable Energy Sources (RES) in heating sector. The HP units, based on thermodynamic refrigeration cycle, use the low grade heat from air, water, geothermal, waste, etc. to generate useful thermal energy and consuming only a small amount of driving energy. The water-to-water HP technology (water source HP) allows to use the water as a low temperature heat source and offers high performance of HP units. The thermal power and performance of water-to-water HP unit depends on two main factors: the flow rate and the temperature of heat source water. Additionally, the heat source water should be available with sufficient quantity and quality, without contaminants, best with constant annual temperature.

2 The idea

The city water network seems to be a profitable heat source for water-to-water HP thermal station located on main pipeline near to the drinking water treatment plant (fig. 1). The favorable conditions for HP technology are: large and stable water flow for whole city and fluid without any contaminants. The temperature of heat source water dependents on type of water intake. The water intakes can be divided into surface and groundwater. In surface intakes the raw water temperature varies strongly with the year seasons, which changes the working conditions and reduces HP performance. The groundwater intakes are more advisable for HP due to the stable temperature during the year. To generate useful thermal energy HP unit need to decrease the domestic water temperature only by 1-3 K [1]. After

* Corresponding author: elzbieta.niemierka@pwr.edu.pl

that the cooled water will be re-heated in non-insulated underground domestic water distribution network, so the cooling phenomenon will not be noticeable by domestic water users. Water-to-water HP at seasonal efficiency SCOP = 5,2 by cooling all domestic water in Poland [2] only by 1 K can generate about 54 PJ of useful renewable heat per year, representing almost 10% of total heat demand in polish residential heating sector [3]. This is a huge available energy potential in heating sector, which can help lower the building's utility bills.

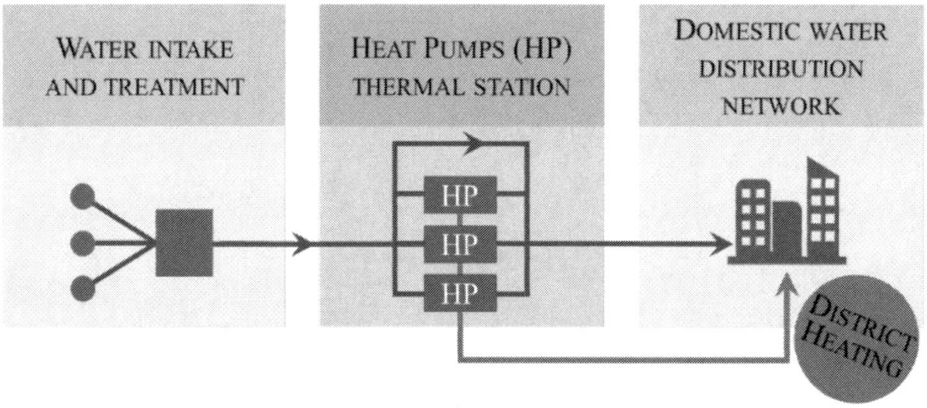

Fig. 1. Water source HP thermal station in domestic water distribution network.

3 Pros & cons list

The domestic water network can be used as a heat source both for small and large scale HP units [4]. The assessing the pros and cons of HP integration in water network must be taken into account (Tab. 1). Pros & cons lists are a quick and easy way to a decision-making and a risk analyse. The pros are related to renewable energy, economics and ecology. The cons are connected with ecological risk, investment and technical issues. It should be noted that economic and ecological factors are on both sides of the list. Low operational cost vs large investment and renewable heat vs risk of water contamination. The proper choice of technical solution and good design allow minimizing issues from the cons list.

Table 1. Pros & cons list.

Pros	Cons
▪ HP units generate the renewable heat from domestic water network ▪ water-to-water HP technology offers higher performance factors	▪ risk of water contamination ▪ large investment cost ▪ risk of emergency shutdowns of HP units due to low water flow
Pros	Cons
▪ water networks are large, widely available and favorable heat source for HP technology ▪ the cooled domestic water will be re-heated in underground network ▪ the water temperature drop will not be noticeable by users'	▪ availability of heat sink ▪ incompatibility parameters of heat sink

4 The city Głogów case study

To illustrate the potential of HP integration into domestic water network the technical and the energy analysis was made on the case of the real water network in city Głogów. The goal is the properly selection of HP units and their parameters to the water network conditions and to supply the maximum amount of renewable heat from the HP thermal station.

The mathematical model was developed to calculate the Coefficient Of Performance (COP), the Seasonal Coefficient Of Performance (SCOP), supplied heat and driving energy consumption of HP units. The model includes: the measured intake water flow and temperature, the domestic water temperature drop (water cooling), the COP characteristics of HP units, the permitted domestic water flows ranges for HP units, the risk of HP units emergency shutdowns caused by low water flow and the weather data.

The study was based on measurement data of intake water flow in water network in city Głogów. The Głogłów city with a population of over 68 000 people has the city scale water network and the district heating systems. The database contains hourly flows and temperatures of intake water for years 2014, 2015 and 2016. The source of water is the groundwater intakes (drilled wheels) with a depth of 40 to 60 m [5] and steady temperature in the year around 10,8°C. The water demand in the city determines the rate and variability of intake water flow. Figures 3, 4 and 5 and table 2 show the statistics of intake water flows for the city of Głogów based on measurement data.

On this basis the number and size of water-to-water, electrically driven HP units were chosen. It should be noted that the large scale one-stage compression HP units can work only in well-defined water flows range, eg. from 50 to 120% of its nominal flow. An exceeding this range causes an emergency shutdown of HP unit. The shutdowns and startups of large scale HP units are the time- and energy-consuming processes. Restarting involves and causes energy losses due to reduced heat production or its total absence during this procedure. For these reasons the HP thermal station design is based on the maximum use of water flow and minimum the risk of HP emergency shutdown. The next factors that should be considered during the selection of HP units are their quantity, thermal power and performance. Consequently, the proper design of HP thermal station requires detailed intake water flows analysis based on measured data.

Table 2. The statistics of water intake flows for the city of Głogów.

	Year 2014	Year 2015	Year 2016
Water intake flow range, m3/h	219-492	206-500	251-499
Median, m3/h	363	345	350
Standard deviation, m3/h	38	35	28

DOI: 10.1051/e3sconf/20172200210

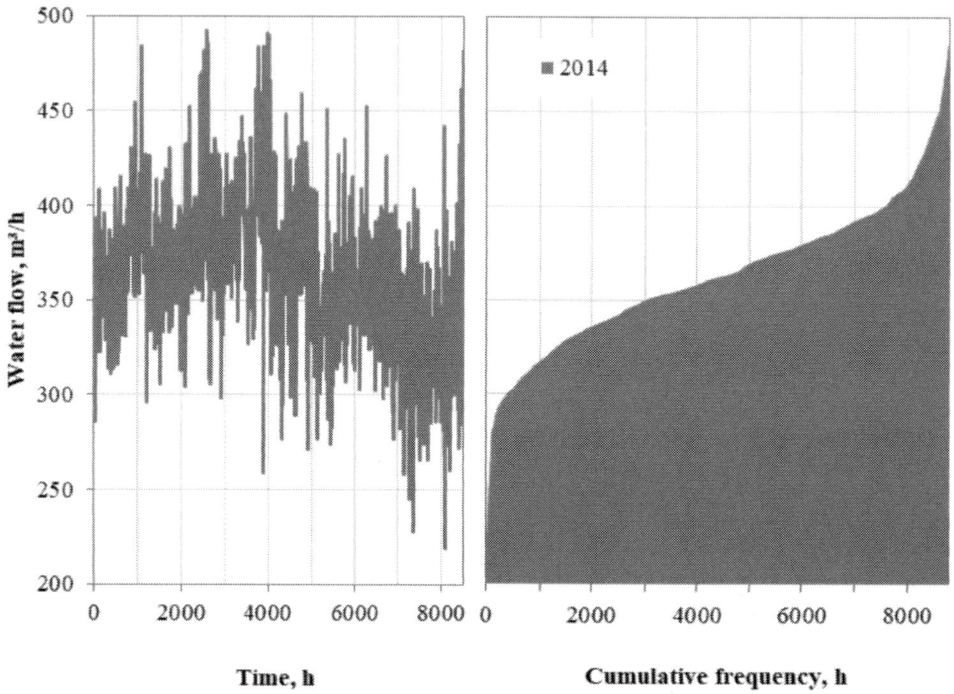

Fig. 2. Intake water flows ratio and its cumulative frequency in year 2014.

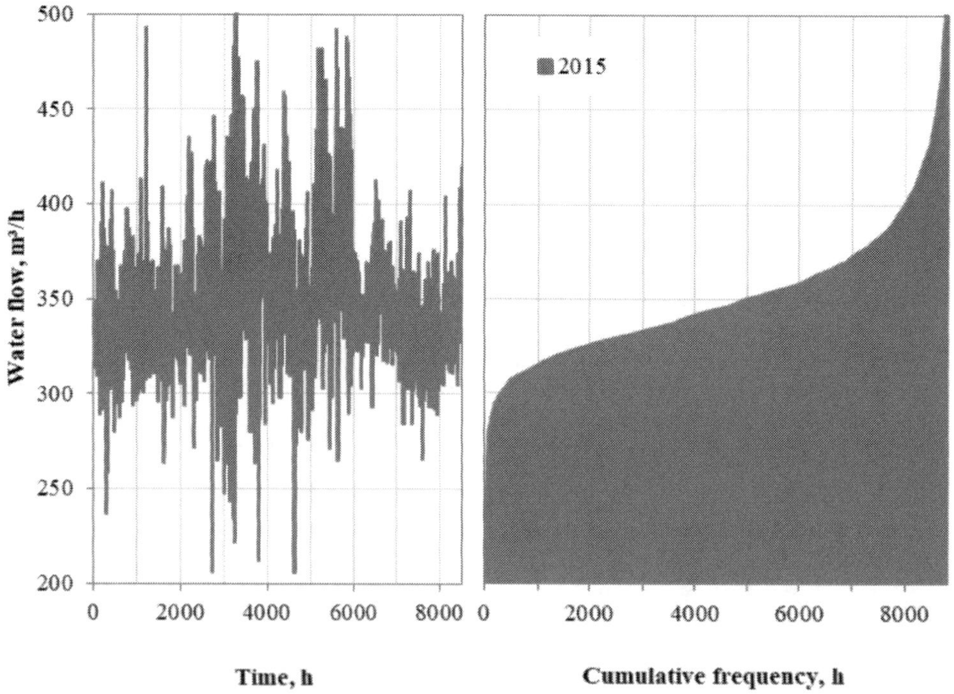

Fig. 3. Intake water flows ratio and its cumulative frequency in year 2015.

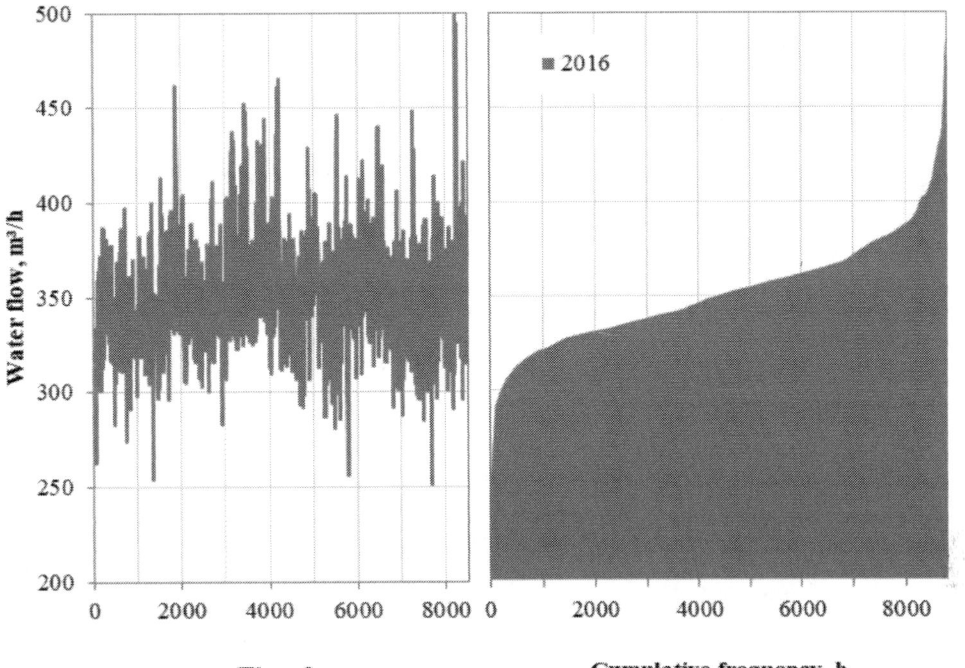

Fig. 4. Intake water flows ratio and its cumulative frequency in year 2016.

In case of Głogów city the 2014-2016 data was taken (Fig. 2, 3 and 4). The full analysis covered 4 variants of HP nominal water flow and 3 variants of domestic water cooling level (temperature drop) for each year in the database (Fig. 5). In total, 36 mathematical model variants were considered.

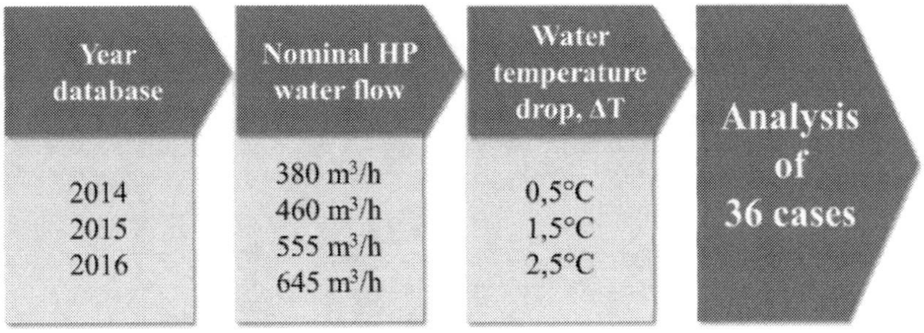

Fig. 5. Flow chart diagram of HP integration analysis cases.

Based on the flow statistics and HP technical data of one the leading large scale water-to-water HP manufacturer, five HP units thermal station was design in Głogów water intake. The analyzed cases require on the one hand applying by-pass as a protection against overflow and on the other hand using HP with minimal flow, which does not cause excessive quantities of emergency shutdown (Fig. 6).

DOI: 10.1051/e3sconf/20172200210

The following theses have been tested:

- the larger HP water flow, the more renewable heat supplied
- the greater domestic water cooling level, more renewable heat supplied
- the larger HP water flow, the greater risk of emergency shutdown.

The analysis included four nominal flows through 5 parallel connected HP by three different water temperature drop levels (0,5; 1,5 and 2,5°C) for years 2014, 2015 and 2016. Using the statistical analysis of flow database, basic parameters of the domestic water network were determined, such as minimum and maximum flow in recent years (flow range), mean flow and standard deviation (which included the range of the majority of observations of the examined feature). Figure 6 shows results and general statistics for analyzed cases. Based on these results, four cases of HP nominal flow were analyzed: 380, 460, 555 and 645 m^3/h. The minimal, safe flows for 5 parallel connected HP units are: 206, 250, 300 and 350 m^3/h.

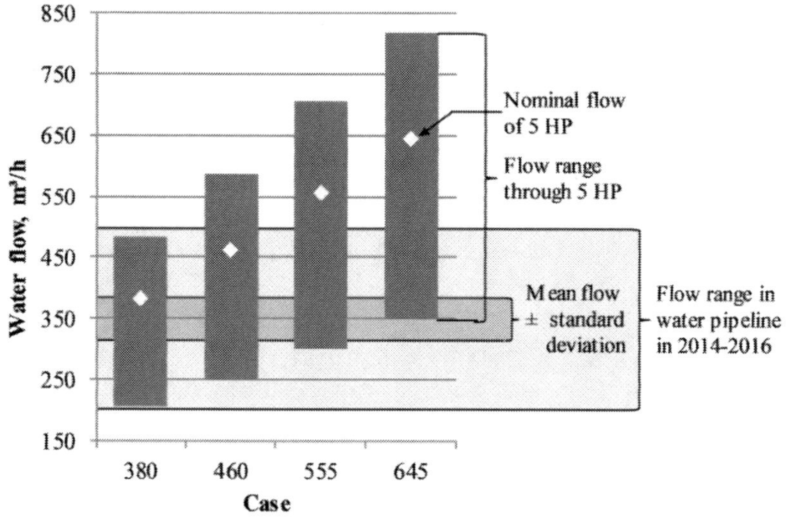

Fig. 6. Graphical analysis of the nominal flow of the HP based on the intakes statistical data.

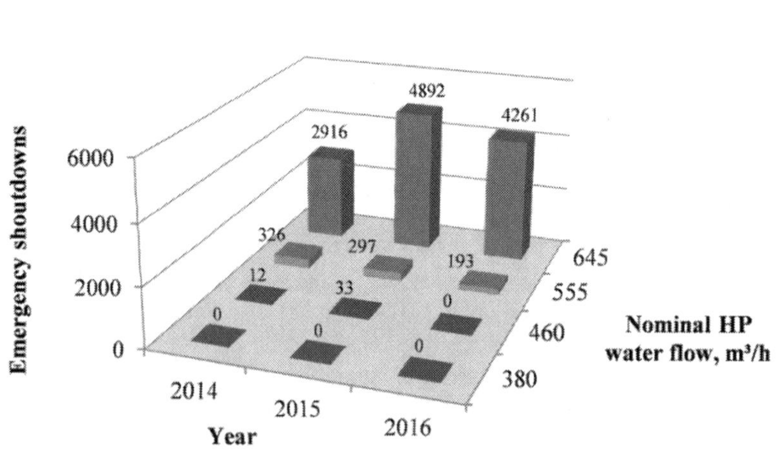

Fig. 7. The number of HP units emergency shutdowns in analyzed years and nominal HP water flows.

 DOI: 10.1051/e3sconf/20172200210

The number of HP emergency shutdowns was calculated in a mathematical model, for the measured variable intake water flows, in each year and for 4 variants of the HP nominal water flows (Fig. 7). The 645 m^3/h case has a huge, unacceptably high number of HP shutdowns and therefore was excluded from the next analysis. Reducing the HP nominal flow to 460 or 380 almost completely eliminates this phenomenon.

The seasonal performance of HP units is described by the SCOP factor. The SCOP for HP thermal station in Głogów was calculated for sink design temperatures 65/45°C, suitable for supplying the own heating system and for supporting the city district heating (Fig. 8). Increasing the HP nominal water flow does not result in a proportional rise of SCOP, but causes the rapid enhance the risk of emergency shutdowns. A change in the HP nominal flow from 380 to 555 m^3/h (by 46%) results in a SCOP increase by only up to 3.9% and the emergency shutdowns also occurs.

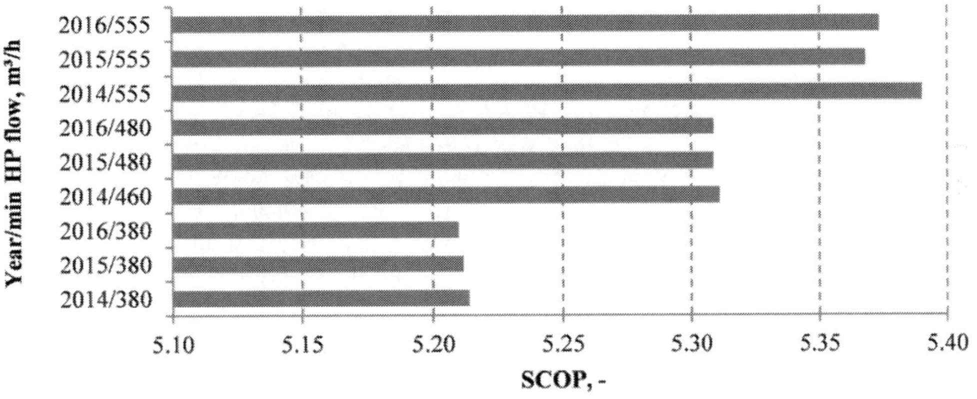

Fig. 8. The HP units SCOP factors of analyzed years and HP nominal water flows.

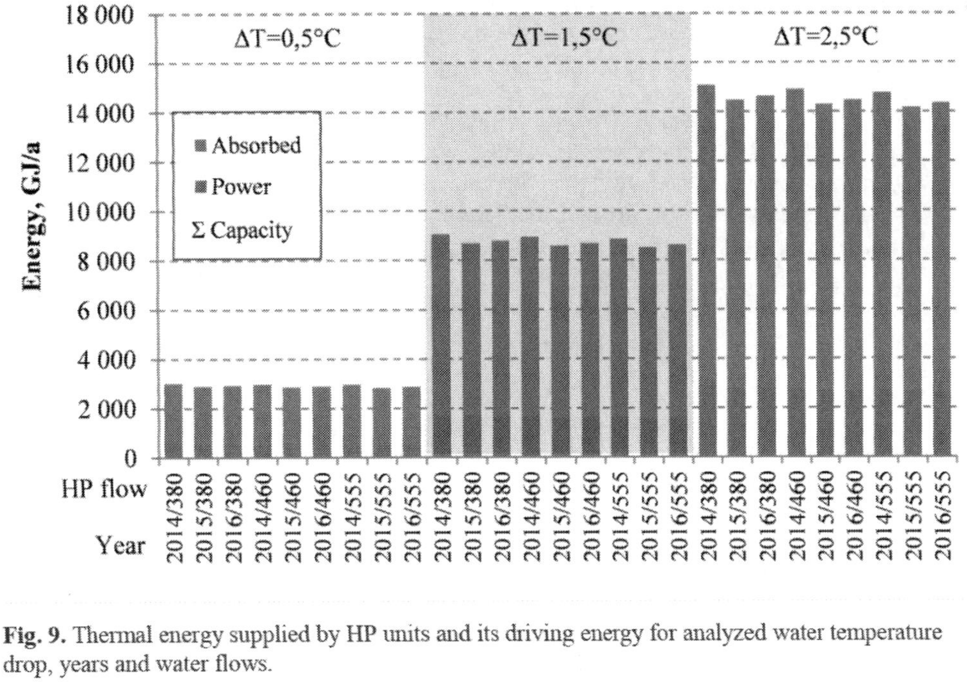

Fig. 9. Thermal energy supplied by HP units and its driving energy for analyzed water temperature drop, years and water flows.

The SCOP of HP unit describes the amount of supplied heat and the quantity of consumed driving energy in working conditions. Fig. 9 presents the thermal energy supplied by HP units and its driving energy (electricity) for analyzed water temperature drop, years and water flows. It is evident, that the domestic water cooling level has a bigger impact on amount of thermal energy supplied by HP than an intake water flow increasing. By change the water cooling level from 0,5 to 2,5°C increases energy supplied up to 400,3%.

5 Conclusions

The water-to-water, electrically driven HP technology enables to use the energy potential of water intakes and domestic water networks for generate of renewable heat. The variability of water intake flow causes the risk of HP unit emergency shutdown. For this reason the sizing of HP units and the selection of HP nominal water flow should be made individually and on base of long-term water flow measurement data. The increasing of HP nominal flow rapidly rise the HP emergency shutdown risk and slightly raises the SCOP. Much better effects of amount of thermal energy supplied by HP can be achieved by increasing the intake water cooling, of course with the risk of freezing.

It should be noted that the analysis of the HP selection on the basis of flow and temperature database should be a common procedure. The database for most cities is available for minimum several years due to the need to monitor the water treatment process. Observing extensive water supply networks also allows for big cities with surface intake to use significantly more water cooling and increase the energy potential due to re-heating in the extensive network.

References

1. A.M. De Pasquale, A. Giostri, M.C. Romano, P. Chiesa, T. Demeco, S. Tani, District heating by drinking water heat pump: Modelling and energy analysis of a case study in the city of Milan, Energy, Volume 118, 2017, Pages 246-263
2. Bank danych lokalnych, https://bdl.stat.gov.pl
3. Central Statistical Office, Energy consumption in households 2012 (2015)
4. J.A. Elías-Maxil, Jan Peter van der Hoek, Jan Hofman, Luuk Rietveld, Energy in the urban water cycle: Actions to reduce the total expenditure of fossil fuels with emphasis on heat reclamation from urban water, Renewable and Sustainable Energy Reviews, Volume 30, 2014, Pages 808-820
5. https://www.pwik.glogow.pl/

AUTHOR INDEX

Abdul Qyyum Muhammad 531
Abouchabana Nabil 1
Abourida Btissam 282
Abseitov Erbolat 757
Acikgoz Caglayan 49
Adamski Wojciech 1478
Agata Zdyb 717
Aillerie Michel 1217
Akan Perihan 748
Alaktiwi Abdulsalam 945
Al-Dadah Raya 572
Ali Wahid 1099
Alibay Temirulan 762
Al-Mansour Fouad 847
Amanowicz Lukasz 10
Anastasakis Kostas 339
Andruszkiewicz Artur 564
Andrych-Zalewska Monika 1241
Anisimov Sergey 1021, 1036
Ar Irfan 81
Arman Hasan 1539, 1547
Arora Amarpreet Singh 18
Augustyn Grzegorz 25
Axcell Brian 265
Aydemir Ali 33
Azhibekov Aidos 762
Bader Roman 806
Baetens Robin 41
Baitasov Talgat 762
Bakowski Andrzej 89
Balapanov Malik 762
Balbay Senay 49
Balbierz Piotr 57, 664
Banaszkiewicz Kamil 65
Banaszuk Piotr 709
Bandurski Karol 73, 1029
Barbusinski Krzysztof 599, 1368
Barrios-Padura Ángela 419
Bartan Ayfer 81
Bartkiewicz Ewelina 1602
Bartkiewicz Lidia 1368
Basaran Filik Ümmühan 363
Beccarelli Paolo 379
Bekkouche SMA 96
Belarche Lahoucine 282
Belcik Maciej K. 103
Bellahcene Lahcene 96
Bellstädt Daniel 33
Beltran-Galindo Javier 324
Benato Alberto 273, 1209
Bering Slawomira 111, 436
Besler Maciej 120, 208
Bezyk Yaroslav 128
Biczak Robert 1384
Bieda Boguslaw 137
Biedron Izabela 224
Birch Calum 191
Bisak Agnieszka 1639
Blasiak Przemysław 1115
Bogacki Marek 1169
Bohdziewicz Jolanta 606
Bojarski Jacek 649
Boruszko Dariusz 159, 257
Bozek Magdalena 815
Brychcy Ewa 1036
Bryjak Marek 1233
Bryszewska-Mazurek Anna 354
Büchele Richard 33
Buda-Ozóg Lidia 1256
Bugaj Andrzej 166
Burszta-Adamiak Ewa 823
Buse Hauke 1305
Cambron Philippe 183
Carvell Aaron 191
Celades Irina 200
Cepinski Wojciech 120, 208
Ceran Bartosz 216
Chacartegui Ricardo 419, 1279
Chaczykowski Maciej 548
Chang Ki-Chang 580
Chauhan Amisha 200
Cheknane Ali 96, 1217
Chen Xin 1509
Chilton John 379
Chlebowska Stys Anna 1297
Chojniak Joanna 224
Chong Min Yee 962
Cichowicz Robert 232
Ciobanu Gabriela 240, 487
Cwiek Barbara 1107
Cybulska Krystyna 731, 1384
Czarny Anna 103
Dabek Lidia 249, 1194
Dabek Pawel 590
Dabrowski Wojciech 159, 257
De Jong Wiebren 339
Debowski Marcin 1594
Delpech Bertrand 200, 265
Demirel Ibrahim Halil 1539, 1547
den Boer Emilia 143, 151
den Boer Jan 143
Dendys Marta 1400
Destro Nicola 273
Doghmi Hicham 282
Dorodnikov Maxim 128
Drabik Anna 1288
Drewnowski Jakub 291
Duangkamol Na-Ranong 1123
Dudkiewicz Edyta 346, 782
Dudziak Mariusz 606

AUTHOR INDEX

Dupont Valerie 191
Duzinkiewicz Kazimierz 513
Dyjakon Arkadiusz 143
Dzierzgowski Mieczysław 299
Dzikowska Anna 724
El Alami Mustapha 449
El Hajjaji Ahmed 1
El-Khozondar Balkess 308
Englart Sebastian 316
Etoua Evina Ghislaine 324
Fave Alain 717
Ferenc Zbigniew 332, 1202
Fernandez-Lopez Maria 339
Fidorów-Kaprawy Natalia 346, 354
Figueiredo Patric 174, 504
Fijewski Michal 1073, 1079
Filik Tansu 363, 411
Filik Ümmühan Basaran 411
Filyarovskaya Viktoriya 370
Fleiter Tobias 33
Flor Jan-Frederik 379
Fourmond Erwann 717
Fröhlich Siegmund 388, 1305
Gabouze Nourddine 1217
Gajewski Andrzej 396
Garcia-Valls Ricard 987
García-Vílchez Encarna 884, 891
Garlicka Agnieszka 1639
Gerek Emine Esra 403
Gerek Ömer Nezih 363, 403, 411
Gil-Baez Maite 419
Glowacka Anna 428, 436
Gnutek Zbigniew 1146
Gogina Elena 861
Golub Adam 442
Gomar Salvador 200
Górniak Aleksander 701
Gounni Ayoub 449
Green Toby 191
Grochowski Michal 513
Grosser Anna 456
Grzebyk Tomasz 1470
Grzegorzek Martyna 465
Grzesik Katarzyna 137
Gulkowski Slawomir 717
Gullu Gulen 748
Gurgul Agnieszka 473
Gwozdziej-Mazur Joanna 480
Haddadi Mourad 1
Hajduk Anna 1594
Hamerla Miłosz 73
Hanus-Lorenz Beata 815
Harijan Khanji 191
Harja Maria 240, 487
Harnisz Monika 495
Hertle Ellen 174, 504
Hirsch Piotr 513
Holtra Anna 522
Hoshi Masaharu 1580, 1587
Hrad Marlies 143
Hussain Arif 531, 1099
Ignatowicz Katarzyna 1093, 1264
Ilkan Mustafa 1610
Im Yong-Hoon 580
Ishembetov Rais 762
Ismail Sherif 540
Isoli Niccolo 548
Ivannikov Alexander 1580, 1587
Jaderko Karolina 915
Jadwiszczak Piotr 1656
Janiak Kamil 556
Janowska Barbara 564
Janus Magdalena 111
Jarnuszewski Grzegorz 1384
Jayakody Harith 572
Jedlikowski Andrzej 1021
Jedrczak Andrzej 151
Jimin Hong 531
Jin Tea-Hwan 580
Jouhara Hussam 200, 265
Jurasz Jakub 25, 590
Jurczyk Krzysztof 25
Kabsch-Korbutowicz Malgorzata 1423
Kabyshev Aset 757
Kabyshev Asset 762
Kajdan Tomasz 1384
Kajl Stanislaw 324
Kalemba Katarzyna 599
Kaminska Gabriela 606
Karpinski Marcin 617
Karpuk Michal 625
Kawasaki Hiroyuki 1524
Kazmierczak Andrzej 701
Kazmierczak Bartosz 590, 634
Kazymbet Polat 1580
Kenzhebatyr Bekmyrza 762
Kepa Urszula 1313
Keskiewicz Piotr 649
Kessler Will 643
Khailov Artem 1580
Kheiri Abdelhamid 449
Kiper Justyna 428, 436
Kirszensztejn P. 899
Klimczak Marcin 649
Klimonda Aleksandra 695
Kmiecik Ewa 617, 657
Knap Marta 57, 664
Kneer Reinhold 174, 504
Kobel Przemyslaw 143
Kobus Dominik 1297

AUTHOR INDEX

Koczyk Halina 73
Kogut Janusz 1256
Kokoszka Wanda 1256
Koksal Merih Aydinalp 308, 748
Kolasinski Piotr 1115, 1146
Kolosowski Pawel 731, 1384
Komerska Anna 672
Koparal A. Savas 403
Kopczyk M. 899
Korbiel Tomasz 25
Kordana Sabina 678
Korzeniewska Ewa 495
Kostka Maria 686
Kotowski Andrzej 634
Koukou Maria K. 1408
Kowalska Izabela 695
Krakowian Konrad 701
Krasowska Malgorzata 709, 1005
Krawczak Ewelina 717
Krawczyk Piotr 1639
Kreider Peter 806
Krzeminska Alicja 724, 1563
Krzysko-Lupicka Teresa 731
Ksionek Dariusz 672
Kubenova Marzhan 762
Kucieba Malgorzata 1131
Kucukali Serhat 81
Kudlek Edyta 606
Kulbat Eliza 739
Kurmanzhanov Askar 757
Kus Melike 748
Kuterbekov Kairat 757, 762, 1580
Kutylowska Malgorzata 767
Kwasniowski Stanislaw 1555
Kwiatkowski Jerzy 775
Lach Janusz 1563
Lam Su Shiung 962, 1530
Lamarche Louis 324
Laska Marta 782, 876
Lazzarin Renato 790
Ledvinka Ondrej 798
Lee Moonyong 531, 1099
Lei Qi 806
Leluk Karol 815
Lettner Christian 1271
Lewkowicz Marek 1131
Licznar Pawel 823
Liew Rock Keey 962, 1530
Lipinski Kamil 830
Lipinski Wojciech 806
Litu Loredana 240
Liu Guiqing 839
Liu Jing 1226, 1509
Lojek Andrzej 556
Lomotowski Janusz 823
Lovegrove Keith 806
Lubon Wojciech 1059
Luin Blaž 847
Lv Yang 854
Lymberis Kostas 1408
Ma Nyuk Ling 962
Mabrouk Mohamed Tahar 449
Machi Justyna 1478
Maciejewska Monika 1352, 1360
Mahmoud Saad 572
Maihankuri Sheidi 191
Majewska-Nowak Katarzyna 465
Makisha Nikolay 861
Maletka Marek 876
Malicki Marcin 1457
Malik Daniel 1059
Malinowski Pawel 159, 257
Malmur Robert 867
Maloufi Nabila 1217
Marcinkowski Tadeusz 65
Martínez-García Herminio 884, 891
Martyla Agnieszka 899
Maslon Adam 907
Masson Christian 183
Maziejuk Miroslaw 1360
Mazur Aleksandra 1321
Mazur Jacek 111, 436
Mazurek Wojciech 354
Mazurkiewicz Justyna 657
Mendrek Barbara 224
Midor Katarzyna 915
Mijakowski Maciej 775
Mikulik Jerzy 25
Milani Massimo 922
Minh Le Quang 531
Miodonski Stanislaw 930, 955
Misinski Jacek 1336
Mohammed H.A. 970, 1610
Molczan Marek 1478
Molina-Huelva Marta 419
Montorsi Luca 922
Morawski Antoni Waldemar 111
Mori Hidetoshi 1524
Mozia Sylwia 111
Mrowiec Bozena 938
Mróz Tomasz M. 1107
Mucha Walter 1154
Mularewicz Piotr 1384
Musa Abdullatif 945
Muszynski-Huhajlo Mateusz 556, 955
Nalepa Beata 495
Nam Wai Lun 962
Natschläger Thomas 1271
Neczaj Ewa 456
Nejad Marjan B. 970

AUTHOR INDEX

Niemierka Elzbieta....1656
Niestepski Sebastian....495
Nogaj Kinga....980
Nogalska Adrianna....987
Nowakowska Monika....996
Nurgaliyeva Rano....757
Obersteiner Gudrun....143
Ofman Piotr....1005
Oleniacz Robert....1169
Oleskowicz-Popiel Czeslaw....1463
Orlenko Sergei....1580
Orlowska-Szostak Maria....1013
Orlowski Ryszard....1013
Osinska Adriana....495
Osinska-Broniarz Monika....899
Osman Mohammad Shahril....1530
Osyczka Piotr....1162
Ouzaouit Meriem....282
Pacak Anna....1021
Palaszynska Katarzyna....1029
Pandelidis Demis....1021, 1036
Panesar Angad....1043
Pasiecznik Iwona....1051
Pasonl Lukasz....1313
Pastuszka Józef S.....1154
Pawlik Marcin....25
Pedišius Nerijus....1086
Pelech Aleksander....625
Pelka Grzegorz....1059
Pelletier Francis....183
Petelin Stojan....847
Phang Xue Yee....962
Picheta-Oles Anna....249
Piekarska Katarzyna....103, 1065, 1344
Piekutin Janina....442
Platkowski Maciej....1384
Plaza Grazyna....224
Podedworna Jolanta....1639
Pogorzelec Marta....1065
Polarczyk Iwona....1073, 1079
Popovski Eftim....33
Porowski Mieczyslaw....1029
Poškas Povilas....1086
Poškas Robertas....1086
Price-Allison Andrew....191
Przekop R. E.....899
Przydrózna Aleksandra....1431
Przydrózny Edward....625, 1431
Puchlik Monika....1005, 1093
Qyyum Muhammad Abdul....1099
Rabhi Abdelhamid....1
Radomski Bartosz....1107
Radziszewski Leszek....89
Raji Abdulganiy....191
Rajski Krzysztof....1036
Rak Józef....1115
Rakhypbekov Tolebay....1587
Ratanaporn Yuangsawad....1123
Regucki Pawel....564, 1131
Rodríguez-Artacho Salvador....1279
Rodziewicz Joanna....1329
Rogala Zbigniew....1146
Rogula-Kopiec Patrycja....1154
Rogula-Kozlowska Wioletta....1154
Rohlfs Wilko....174, 504
Rola Kaja....1162
Romero Fernando....200
Rosinski Marian....672
Rucinska Teresa....428
Rumin Rafal....25
Rusanowska Paulina....1594
Rybak Justyna....815
Rybka J. D.....899
Rzeszutek Mateusz....1169
Rzeznik Ilona....1178
Sadeghi O.....970
Sadykov Nurllah....757
Sahel Djemal....96
Salata Aleksandra....1194
Salman Chaudhary Awais....1186
Sambor Aleksandra....332, 1202
Sanchez-Silva Luz....339
Sannad Mohammed....282
Schiro Fabio....273, 1209
Schmied Elisabeth....143
Schwede Sebastian....1186
Sehee Oh....1099
Sekret Robert....980, 1417
Selmane Naceur....1217
Shi Luyang....1226
Shin Ki-Yeol....580
Siekierka Anna....1233
Simson Grzegorz....1005
Sirvydas Arunas....1086
Sitarska Magdalena....370
Skretowicz Maria....1241, 1248
Skrzycki Maciej....120
Skrzypczak Izabela....1256
Skvortsov Valeriy....1580, 1587
Slowik Marta....1256
Slys Daniel....678, 1321
Smyk Joanna....1264
Sobieczky Florian....1271
Sokolowska Aleksandra....739
Soltero Victor Manuel....1279
Sowizdzal Anna....1059, 1288
Sówka Izabela....128, 1297
Spangemacher Lars....1305
Sroka Zbigniew....1248
Stanczyk Justyna....823

AUTHOR INDEX

Stanczyk-Mazanek Ewa 1313
Stathopoulos Vassilis 1408
Stec Agnieszka 1321
Steenackers Gunther 1139
Stefani Matteo 922
Stefanowicz Ewelina 354
Stepanenko Valeriy 1580, 1587
Stoppato Anna 273, 1209
Strek Michal 1384
Striugas Nerijus 1086
Struk-Sokolowska Joanna 1005, 1329
Studzinski Jan 1368
Su Man Huan 962
Sudol Adam 731
Suwala Wojciech 1500
Sytek-Szmeichel Katarzyna 1639
Szalanski Pawel 1336
Szczepaniak Wlodzimierz 473, 1051
Szczesniak Sylwia 625, 1344
Szczurek Andrzej 1352, 1360
Szelag Bartosz 249, 1368
Szerzyna Slawomir 1478
Szlachta Malgorzata 1376, 1493
Szulc Jowita 73
Szulecka Adriana 1169
Szulgowska–Zgrzywa M. 354, 686
Tachos Nikolaos 1408
Tahan Antoine 183
Takao Masuda 1123
Talbi Mosbah 945
Tarnowski Krzysztof 111, 436
Tatsuda Kou 839
Tawfik Ahmed 540
Telesinski Arkadiusz 731, 1384
Teleszewski Tomasz Janusz 1392
Teruoki Tago 1123
Thorin Eva 1186
Tomaszewska Barbara 657, 1288, 1400
Torres David 183
Toyoda Shin 1580
Traczewska Teodora 370, 1470
Traxler Patrick 1271
Trusz-Zdybek Agnieszka 103, 1344
Trzaski Adrian 775
Tsolakoglou Nikolas P. 1408
Tsubaki Noritatsu 839
Turski Michal 980, 1417
Tyszer Magdalena 1400
Uchronski Mariusz 1352
Umiejewska Katarzyna 830
Uqaili Mohammed 191
Urbanowska Agnieszka 1423
Urbanska Klaudia 143
Valincius Vitas 1086
Valverde Jose Luis 339
Van Buren Simon 174, 504
Van Riet Freek 1139
Velázquez Ramón 1279
Venturelli Matteo 922
Verhaert Ivan 41, 1139
Vrachopoulos Michalis Gr. 1408
Walaszczyk Juliusz 625, 1431
Walery Maria Jolanta 1441
Wang Haifeng 854
Wartalska Katarzyna 634
Wdowikowski Marcin 590, 634
Wedrychowicz Wieslaw 564
Wei Feng 1099
Wei Shanshan 854
Wisniewski Jacek 1478
Wisniowska Ewa 1449
Wlodarczyk-Makula Marla 1449
Wojdyga Krzysztof 1457
Wojtkowiak Janusz 10, 1463
Wójtowicz Patryk 1376, 1493
Wolf Mirela 370, 1470
Woloszyk Czeslaw 436
Wolska Malgorzata 1478
Wolski Pawel 1486
Wong Chee Chung 1530
Wróbel Radoslaw 701, 1241
Wronska Ilona 1384
Wu Yupeng 379
Wyrwa Artur 1500
Xue Puning 1509
Yalcinkaya Fatma 1517
Yamamoto Tatsumi 1524
Yan Jinyue 1186
Yek Peter Nai Yuh 1530
Yilmaz Seval 403
Yoneyama Yoshiharu 839
Yoon Si-Won 580
Yuksel Ibrahim 1539, 1547
Zablocka-Malicka Monika 473
Zabnienska-Góra Alina 1631
Zaczynska Ewa 103
Zajac Pawel 1555
Zajiczek Zaneta 1360
Zamorska-Wojdyla Dorota 522
Zareba Anna 724, 1563
Zathey Maciej 1297
Zawieja Iwona 1572
Zhang Huibo 1226
Zhou Zhigang 1509
Zhumadilov Kassym 757, 1580, 1587
Zielinski Marcin 1594
Ziembicki Piotr 649
Zimoch Izabela 1602
Zmarzly Marcin 291
Zubeer Swar A. 970, 1610

AUTHOR INDEX

Zubrowska-Sudol Monika....................................1639
Zukowska Adrianna 987
Zukowski Miroslaw 1648
Zwierzchowski Ryszard................................299, 1621
Zylka Radoslaw..257